D1153931

LOW CYCLE FATIGUE

A symposium
sponsored by
ASTM Committee E-9
on Fatigue
Bolton Landing (on Lake George), New York
30 September - 4 October 1985

ASTM SPECIAL TECHNICAL PUBLICATION 942
H. D. Solomon, General Electric Company,
G. R. Halford, NASA-Lewis Research Center,
L. R. Kaisand, General Electric Company, and
B. N. Leis, Battelle Columbus Laboratories,
editors

ASTM Publication Code Number (PCN)
04-942000-30

ASTM 1916 Race Street, Philadelphia, Pa. 19103

Library of Congress Cataloging-in-Publication Data

Low cycle fatigue: a symposium / sponsored by ASTM Committee E-9 on
Fatigue, Bolton Landing (on Lake George), New York, 30 September–4
October 1985; H. D. Solomon . . . [et al.], editors.
(Special technical publication; STP 942)
Includes bibliographies and index.
"ASTM publication code number (PCN) 04-942000-30."
ISBN 0-8031-0944-X
1. Metals—Fatigue—Congresses. I. Solomon, H. D. (Harvey D.)
II. American Society for Testing and Materials. Committee E-9 on
Fatigue. III. Series: ASTM special technical publication; 942.
TA460.L678 1987 87-23714
620.1'63—dc19 CIP

Library of Congress Catalog Card Number: 87-23714

NOTE

The Society is not responsible, as a body,
for the statements and opinions
advanced in this publication.

Printed in Baltimore, Md.
January 1988

Foreword

The symposium on Low Cycle Fatigue: Directions for the Future was held in Bolton Landing (on Lake George), New York, 30 September to 4 October 1985. It was sponsored by ASTM Committee E-9 on Fatigue, with support from the Hudson-Mohawk Chapter of AIME and the Eastern New York Chapter of ASM. The symposium was organized by Harvey D. Solomon (chairman), Brian Leis, Gary Halford, and Leonard Kaisand.

The organizing committee thank the international advisory committee for their support and suggestions:

Argentina:	Dr. J. C. Crespi
Australia:	Dr. J. M. Finney
Belgium:	Dr. J. M. Drapier
Canada:	Prof. T. Bui-Quoc
	Prof. A. Plumtree
Federal Republic of Germany:	Prof. W. Bunk
	Prof. Dr. D. Munz
	Prof. Dr. P. Neumann
	Dr. H. Nowack
	Dr.-Ing. W. Schutz
France:	Dr. J. L. Chaboche
	Dr. C. Amzallag
	Prof. J. Lemaitre
Israel:	Dr. A. Berkovits
Japan:	Dr. Y. Asada
	Prof. K. Iida
	Prof. R. Ohtani
	Prof. T. Tanaka
	Prof. T. Yokobori
The Netherlands:	Dr. H. P. Van Leeuiven
	Dr. J. Schijve
People's Republic of China:	Prof. M. Yan
	Prof. X-L Yu
Spain:	Dr. R. R. Jordana
Sweden:	Dr. R. Lagneborg
Switzerland:	Dr. M. Nazmy
United Kingdom:	Dr. C. J. Beevers
	Dr. E. G. Ellison
	Dr. R. H. Jeal
	Prof. J. F. Knott
	Dr. I. L. Mogford
	Prof. K. J. Miller
	Dr. B. Tomkins
	Dr. P. Watson
	Prof. G. A. Webster

The organizing committee also thanks the session chairmen for their assistance: D. F. Mowbray, B. Leis, G. R. Halford, L. F. Coffin, R. C. Bill, K. Wright, R. V. Miner, S. D. Antolovich, J. B. Conway, S. S. Manson, F. Lawrence, H. D. Solomon, R. P. Skelton, J. Wareing, R. Williams, and L. Kaisand.

Related ASTM Publications

Fatigue in Mechanically Fastened Composite and Metallic Joints, STP 927 (1986), 04-927000-30

Fatigue at Low Temperatures, STP 857 (1985), 04-857000-30

Multiaxial Fatigue, STP 853 (1985), 04-853000-30

Rolling Contact Fatigue Testing of Bearing Steels, STP 771 (1982), 04-771000-02

Low-Cycle Fatigue and Life Prediction, STP 770 (1982), 04-770000-30

Methods and Models for Predicting Fatigue Crack Growth under Random Loading, STP 748 (1981), 04-748000-30

Fatigue Crack Growth Measurement and Data Analysis, STP 738 (1981), 04-738000-30

Hold-Time Effects in High-Temperature Low-Cycling Fatigue, STP 489 (1971), 04-489000-30

A Note of Appreciation to Reviewers

The quality of the papers that appear in this publication reflects not only the obvious efforts of the authors but also the unheralded, though essential, work of the reviewers. On behalf of ASTM we acknowledge with appreciation their dedication to high professional standards and their sacrifice of time and effort.

ASTM Committee on Publications

ASTM Editorial Staff

Contents

Microstructural Effects

Multiaxial and Variable Amplitude Loading

NOTCHES

Life Prediction

Introduction

It has been over 30 years since the development of the low cycle fatigue (LCF) law. This symposium was organized to commemorate this technological advance and to honor the pioneers in this very important field. The symposium was not, however, organized only to look backward at past accomplishments but also to look forward. We were fortunate in having Dr. L. F. Coffin, Jr., and Professor S. S. Manson give us their perspectives on low cycle fatigue and their opinions on the future directions of its study.

Indeed, their presence at the symposium was most fitting, since they are two of the most important pioneers in the field, having independently developed the LCF law which bears their names. Their papers were presented at a banquet held in their honor and are given at the beginning of this volume.

Following Coffin's and Manson's papers are those in the order in which they were presented. The first sessions were on cyclic deformation and LCF damage. These were followed by papers on crack propagation, led off with a review of this field by R. P. Skelton. Next came sessions on high temperature LCF and thermal and thermomechanical fatigue. (It is clear from the large number of papers in these two sessions that the bulk of LCF work is for elevated temperature applications.) Microstructural effects, with a review by J. Wareing, were considered next, followed by a session on multiaxial and variable amplitude loading. The penultimate session was on notches. The final session on life prediction tried to tie together LCF and fatigue life prediction.

It is clear from this enumeration of session topics that low cycle fatigue is a complex field with numerous aspects. This has led to a large symposium and a formidably sized publication (72 papers out of over 140 submitted abstracts). It is hoped that these papers will aid in the understanding of low cycle fatigue and will have a role in charting its future directions.

H. D. Solomon

General Electric Corporate Research and Development Center, Schenectady, New York; symposium chairman and co-editor

G. R. Halford

NASA-Lewis Research Center, Cleveland, Ohio; co-editor

L. R. Kaisand

General Electric Corporate Research and Development Center, Schenectady, New York; co-editor

B. N. Leis

Battelle Columbus Laboratories, Columbus, Ohio; co-editor

Future Directions

L. F. Coffin[1]

Some Perspectives on Future Directions in Low Cycle Fatigue

REFERENCE: Coffin, L. F., "**Some Perspectives on Future Directions in Low Cycle Fatigue,**" *Low Cycle Fatigue, ASTM STP 942,* H. D. Solomon, G. R. Halford, L. R. Kaisand, and B. N. Leis, Eds., American Society for Testing and Materials, Philadelphia, 1988, pp. 5–14.

ABSTRACT: This paper on low cycle fatigue (LCF) is divided into three parts. The first provides a brief historical perspective, the second deals with today's scene, and the third speculates on future LCF directions.

KEY WORDS: fatigue, low-cycle fatigue, fatigue perspectives

I am most pleased and honored to participate in this Symposium on Low Cycle Fatigue—Directions for the Future and to present this talk on the future course of fatigue. My involvement with this subject over the last 35 years has been greatly aided by the encouragement and support by many people, but in particular by my wife, Mary, who has provided a balanced perspective for my frustrations and elations derived from my work and who has maintained the home front during my extended trips to technical meetings and conferences. Also the General Electric Corporate Research and Development management, specifically Harvey W. Schadler and Mark G. Benz, with whom I have been associated for a part of my years with GE, has provided the support and environment for performing the work that has led to this recognition today. Finally, we should all acknowledge the tremendous effort that Dr. Harvey Solomon has exerted in almost single handedly conceiving, planning, organizing, and operating this symposium. It has been an excellent conference, and Harvey should be quite proud of the job he has done.

My talk is divided into three parts. The first provides a brief historical perspective of low-cycle fatigue with some specific observations on why the work that Stan Manson and I reported on 31 years ago has been so well received. The second part deals with the scene today and provides a few observations largely brought on by the present meeting. Finally, I would like to speculate on future directions in fatigue, say, from 10 to 30 years from now.

Historical Perspective

The Coffin-Manson or Manson-Coffin relationship was proposed some 31 years ago quite independently by Stan Manson and myself [*1,2*]. This was only one of several relationships that Stan and I have enjoyed over the years as each of us developed our separate viewpoints on extensions of this original concept to account for the complicating effects of temperature, environment, frequencies, waveshapes, thermal cycling, and other design-related concerns. Although our differences in views on many of these topics were a matter of public record, nevertheless, I

[1]Mechanical Engineer, Materials Laboratory, Corporate Research and Development, General Electric Company, Schenectady, NY 12301. Dr. Coffin is now retired.

feel that this competition has provided a stimulus for advancing our knowledge more rapidly than otherwise in this difficult and complex subject.

The world-wide interest generated in the concept of the plastic strain-life relationship has been a very humbling experience. Why did this happen? Since that time I have had other ideas that I felt were fully as useful as this, but none have had the same response. Why was this work so well received? I feel there were several factors that contributed to its wide reception:

1. *Timing*—Our publications appeared at a time when design procedures were undergoing reexamination to provide a more realistic service-related basis. At that time nuclear power plants were beginning to evolve, commercial jet aircraft engines were on the drawing boards, steam turbines were experiencing problems with shell cracking, and the designs of a host of other components were being carefully examined.
2. *It filled a need*—The problem that concerned people at the time was how to account for the cyclic effects of severe thermal stresses or of loads which could result in fatigue failures in finite lives. Some background for these concerns is provided elsewhere [*3*].
3. *The approach was different*—Here strain was used in contrast to previous stress-determined approaches. Strain-controlled testing provided a stimulus for new technique development, materials exploration, and a wide variety of phenomena studies such as ratchetting, cyclic strain aging, softening and hardening, hysteresis loop stabilization, and cyclic stress-strain studies. The use of plastic strain was consistent with more basic views of the then-emerging dislocation theory.
4. *It was simple*—The relationship employed a single, simple engineering property, the reduction in area. Previous approaches to fatigue life prediction were highly empirical and complex; this approach was very straightforward.
5. *It was fun*—The use of strain and the hysteresis loop became a gold mine of interesting things for the experimentalist to do. With the advent of servo-mechanical control systems, the effect of complex hysteresis loops could be studied. This included computer-controlled testing, thermo-mechanical strain cycling, etc. [*4*]. The interface between strain-controlled fatigue testing and the computer has significantly broadened the range of interest in this form of mechanical testing.
6. *The approach provided a common denominator for many disciplines*—The interface between strain-controlled testing and the computer has already been mentioned. Other disciplines which interfaced with this approach included those of design, materials metallurgy, solid-state theory, metallography, and corrosion chemistry.

Present Perspective

An interesting view of the present state of affairs in low-cycle fatigue can be drawn from the symposium at hand. One of the most exciting indicators of the health of the subject is the large number of young people who are in attendance here and actively participating and reporting on their work. This is a healthy sign that interest in the subject is continuing with youthful enthusiasm. I find this most encouraging.

Secondly, it is gratifying to see so many participants here from other countries. This indicates a world-wide interest in the subject. According to my count, some 16 countries are represented. A symposium such as this is a useful forum for becoming acquainted with each other, exchanging views, and stimulating new ideas.

Thirdly, it is a pleasure to see increasing attention being given to the role of environment on low-cycle fatigue, especially at elevated temperature. Although there are still too many papers that concern themselves with creep-fatigue interactions without recognition that environment may dominate the behavior, appreciation for the role of the environment is spreading.

Fourthly, and most importantly, a gradual unification appears to be occurring in viewing the fatigue process between low cycle fatigue and crack growth. For too long a period, these two important areas of the fatigue process have been separated by artificial distinctions such as low cycle fatigue versus fracture mechanics and fatigue versus crack growth, or by organizational differences such as between ASTM Committees E-9 on Fatigue and E-24 on Fracture Testing. These distinctions have been divisive and non-constructive. Fortunately, the view is changing. We are really dealing with one problem by two views which are quite closely related, as Kaisand [*5*] points out. The regular joint meetings of Committees E-9 and E-24 are also helping to bring things back together.

Future Directions

In attempting to identify future directions in low cycle fatigue, I will not distinguish between low cycle fatigue and fatigue more generally, since the distinction may become cloudy as we think far ahead. Also, I will view the subject more from industrial needs and applications rather than from directions for future research. The presentation will be in the form of predictions for the future, say, 10 to 30 years hence.

Prediction 1:

> *Fatigue computer modelling will supplant fatigue testing with the possible exception of cases where very high temperatures and severe environments exist and where time dependency may be a complication. Specific microstructure and defect details will be accounted for in the computational process. Fatigue testing of standardized specimen shapes will be less emphasized. Designs will be based on crack growth analysis, and experimentation will be directed towards life verification for specific component shapes.*

The computer has an overpowering presence and must be recognized as playing an increasingly important role in modelling of the fatigue process. We will be entering the age of computer-aided life prediction. To meet the requirements of this development, the fatigue process must be described quantitatively. This includes crack initiation and early growth (i.e., short cracks). While long cracks are now reasonably well understood and their growth parameters quantified, the short crack problem needs quantification before reliable modelling is possible. Some work along these lines has been reported at this symposium [*6*].

The short crack growth problem has been limited by the lack of experimental tools for their study. In our own laboratory we have been developing a technique for crack growth measurement that is directly applicable to this problem. We call it the *reversing d-c electrical potential method* [*7–10*].

Some details of the reversing d-c electrical potential technique are of interest here. All electrical potential techniques are based on the knowledge that when a curent is caused to flow through a specimen perpendicular to a crack, the potential difference between two fixed probes located on opposite sides of the crack will increase as the size of the crack increases. The reversing d-c potential technique has several features that offer distinct advantages over other potential measurement techniques:

1. Thermoelectric effects are eliminated by reversal of current polarity at half-second intervals.
2. Utilization of a reference probe against which all active probes are normalized compensates for changes in current, temperature, or material properties.
3. The signal-to-noise ratio is enhanced by multiple (16) potential readings before and after each current reversal.

4. Potential ranges from multiple reversals are averaged to obtain a single potential measurement for a given time period, thus enhancing the signal-to-noise ratio.

To determine the shape of a crack it is necessary to obtain a solution for the potential field in a body containing a crack when a current is passed through the body. The potential field has been derived for an ellipsoidal cavity in an infinite solid [*8*]. It is possible to define a semi-ellipse that approximates the shape of any crack if three or more active probe parts are monitored during testing. Software has been developed to (*a*) log data from multiple active probes, (*b*) normalize the readings from the active probes to the reference probe, and (*c*) calculate the depth of the crack from these readings assuming no change in the shape of the crack. This approximate depth is used to calculate the load required to yield the desired stress intensity, *K*, for the next loading cycle. A more accurate determination of crack dimensions is done "off-line" using a fitting program, which allows a change in shape of the best fitting semi-ellipse.

Figure 1 can be used to illustrate the capability of the system designed to monitor crack growth in specimens. A defect 0.64 mm (0.025 in.) deep by 2.54 mm (0.100 in.) long was introduced at the midpoint of a specimen of SA333 Gr6 carbon steel by electrical discharge machining (EDM). Six pairs of active potential probes were attached to the specimen surface as shown in the upper-left portion of Fig. 1. The specimen was subjected to uniaxial cyclic loading for over 10 000 cycles while enclosed in an autoclave containing 288°C (550°F) water with 200 ppb oxygen. The output from each of the probe pairs is shown in the upper-right portion of Fig. 1. Data from the six probe pairs, normalized with respect to similar readings taken from a reference probe pair, are used to calculate, cycle-by-cycle, the dimensions of the ellipse, which describes the shape of the crack at all points during the test. These results are plotted in the lower portion of Fig. 1. In this plot, the upper and lower curves are the calculated length and depth of

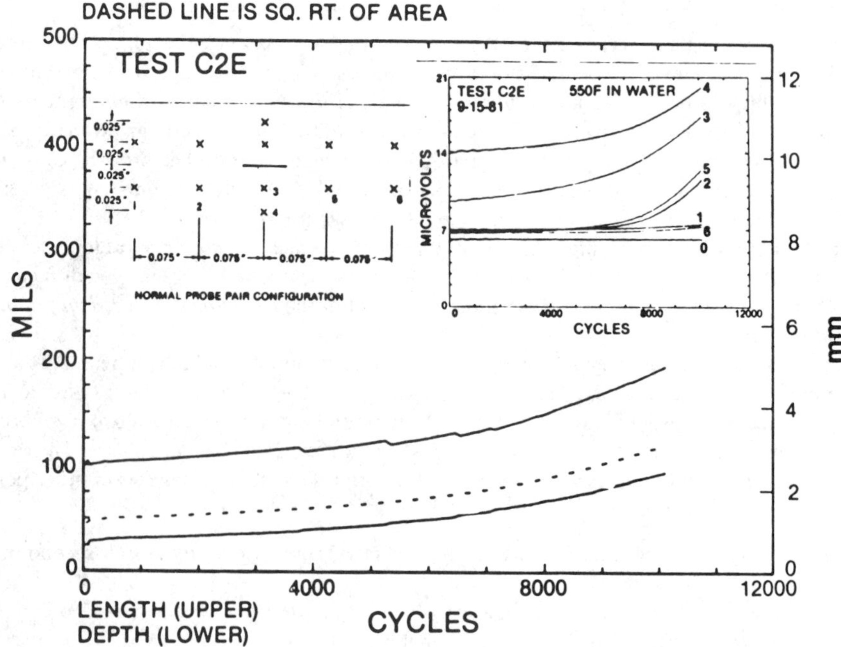

FIG. 1—*Probe positions, measured potential readings, and calculated crack dimensions attained in testing of surface defected Specimen C2E* [8].

FIG. 2—*Fractured surface of Specimen C2E showing agreement of "best fitting ellipse" with crack front* [8].

crack, respectively. The intermediate, dotted curve is the square root of the area of the cracked surface. It will be noted that the aspect ratio, defined as the crack depth divided by the width, changed from 0.25 at the outset of growth to 0.49 at the termination of cycling. The fractured surface of this specimen is shown in Fig. 2. The calculated "best fitting ellipse" is shown in white. The excellence of the fit is obvious [8].

Prediction 2:

> *Better understanding of corrosion fatigue and stress corrosion cracking will exist based on experimental information derived from real-time crack growth rather than from accelerated testing.*

Most stress corrosion and corrosion fatigue studies aimed at solving practical industrial problems are performed under accelerated testing conditions. These tests do not permit full development of the cracking phenomenon, which involves several concurrent processes associated with the crack tip chemistry, local metallurgical considerations, and mechanical factors such as crack tip strain rate and stress intensity factor. Using techniques such as these, it is possible to perform experiments in real time. Assume, for example, that a crack growth of 1 mm in 40 years can be tolerated in a nuclear pressure vessel. This is equivalent to a crack velocity of 7.9×10^{-10} mm/s (1.12×10^{-7} in./h). Crack velocities of this level are possible using techniques such as described above. Figure 3, for example, shows the crack velocity versus period of cycling results for a surface crack in SA533 Class B pressure vessel steel tested at a K_{max} of 22 MPa m$^{1/2}$ (20 ksi in.$^{1/2}$) with $R = 0.1$ using a trapezoidal waveform in high temperature water (288°C

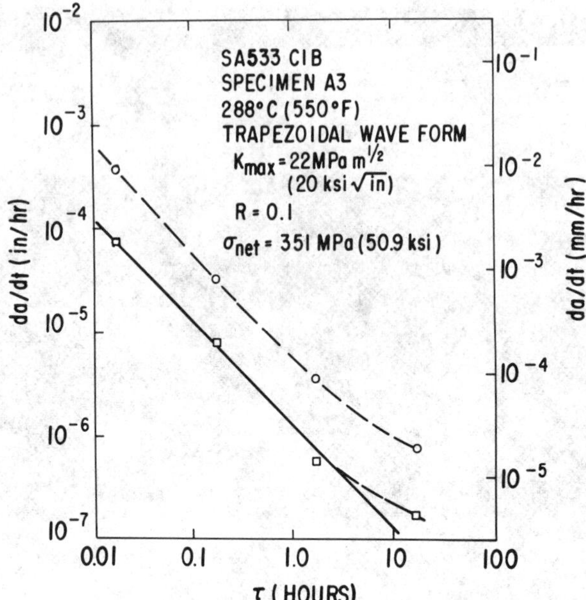

FIG. 3—*Crack growth rate in a SA533 C1B alloy steel specimen tested in 200 ppb oxygen* (dashed line) *and 150 ppb hydrogen* (solid line) *containing water* [11].

[550°F] and 10.3 MPa [1500 psi]) containing either 200 ppb oxygen or 150 ppb hydrogen [*11*]. Note that the lowest reported crack velocity is equivalent to the example selected. To get meaningful information such experiments must be run for up to one month's duration, equivalent to 2×10^{-3} mm (8×10^{-5} in.). Some concern can be raised as to whether information gathered over such short distances is meaningful, and this point must be sorted out.

Prediction 3:

> *High performance components or those subjected to severe environmental conditions will be monitored continuously for possible fatigue damage.*

The design and life prediction assessment of structural components is an "up front" process. That means that the designer must use his best judgement of the design life, using all the tools at his disposal for this purpose. Nevertheless, the process is an extrapolative one where the uncertainties in loading, materials performance, manufacturing processes, and environment may be sufficient to cause eventual problems. In high performance components or those involving public safety, measurement and interpretation of progressive damage should be made during service life using appropriate damage monitors [*12*]. From damage information gathered in the course of component life, estimates can be made of optimum timing for planned outages for repairs, remaining life, and other information valuable for plant operation.

One area where this approach is currently of interest is in the study of damage in nuclear piping. Here damage is defined as cracking in weld heat-affected zones of Type 304 stainless steel due to intergranular stress corrosion cracking. Monitoring for this type of damage is of considerable interest.

One of the monitor forms under development at the GE Research and Development Center involves the use of probes applied to the outside surface of a pipe to follow the growth of a crack on the internal surface. Analysis has shown that, depending on crack shape, internal cracks of 10% or greater of the wall thickness can be readily monitored from the outside. A recent report describes our experience with these procedures [*13*]. Here a laboratory test has been performed on a 102 mm (4 in.) diameter, 5.59 mm (0.22 in.) wall thickness, furnace-sensitized Type 304 stainless steel pipe with an internal defect of 1.50 mm (0.059 in.) depth by 25.4 mm (1.0 in.) length using external probes. Coolant water at 288°C (550°F) and 10.3 MPa (1500 psi) with 200 ppb oxygen was circulated through the pipe while it was axially and cyclically loaded with a 10-min hold period. Using multiple probe pairs and appropriately located current sources and sinks, crack growth rates and crack dimensions of reasonable accuracy could be made. Table 1 gives the comparative crack growth rates in pipe and surface crack specimen tests in the environments indicated, confirming the ability to monitor cracked pipes, at least for the conditions for these tests. The equipment developed for this work is shown in Fig. 4. Programs using these techniques to monitor precracked pipe subjected to weld overlay treatments are about to be undertaken at General Electric's Pipe Test Laboratory in San Jose, California, under EPRI sponsorship.

Prediction 4:

Alloys will be designed primarily to emphasize fatigue resistance rather than as a secondary property.

Generally materials are selected for service for their static strength, ductility, fracture toughness, and creep resistance. Fatigue resistance is most often the dependent variable. Since most service failures are fatigue related, it seems reasonable that fatigue resistance be given much greater attention in the materials selection process.

There are a few examples where this approach has been applied. One is the use of single crystals in manufacturing aircraft engine turbine brackets. Here the elimination of grain boundaries eliminates the problem of time-dependent grain boundary cracking, a common damage process in high temperature, long hold time loadings experienced by these components.

A good example of work aimed at improving fatigue crack growth resistance based on insights regarding crack closure has been reported by Suresh [*14*]. He has undertaken studies aimed at providing a simple quantitative model of the changes in fatigue crack growth characteristics due to the combined influence of deflections in the crack path and relative sliding between the fracture surfaces causing premature contact. This model predicts that significant shifts to the left (to higher values of ΔK in the near threshold da/dN versus ΔK crack growth response) can be brought about by introducing the above fracture surface mismatch. He supports these predictions by observations on an under-aged 7475 aluminum alloy tested in vac-

TABLE 1—*Comparison of crack growth rates in pipe and specimen testing.*

Environment	Test	K_{max} (MPa m$^{1/2}$)	$K_{eff} - K_{max}$ $(1 - R)^{0.5}$	da/dN (mm/cycle)
200 ppb oxygen at 288°C	Specimen S2	16.8	15.9	4.7×10^{-4}
	pipe	19.0[a]	16.5[a]	6.9×10^{-4}
150 ppb hydrogen water at 288°C	Specimen S2	16.8	15.9	2.9×10^{-4}
	pipe	19-20.4[a]	16.5-17.7[a]	4.8×10^{-4}

[a]These values are probably low for reasons discussed in text.

FIG. 4—*Pipe test stand and related pipe crack growth monitoring equipment* [13].

uum. Here the crack path develops a high degree of tortuosity compared with the overaged alloy where the deflections are extremely small. Figure 5 shows the crack path for the underaged alloy, while Fig. 6 shows the comparative crack growth curves for the over- and under-aged alloy together with the model predictions. Both the shift in threshold and the agreement between analyses and measurement are impressive.

Prediction 5:

> *Emphasis in fatigue studies will be given to composite materials tailored for high performance applications combining strength, high temperature and environmental resistance, and fatigue damage tolerance.*

Individual inventiveness, combined with challenges of the business, will lead to new materials capable of combining the best of several properties. These will be composites of special forms not conceived of at present. The demands of the aircraft engine will provide this challenge. The opportunities are great.

Summary

Making the foregoing predictions come to pass will require considerable ingenuity, skill, patience, and dedication. It will also require commitment and resources from research and devel-

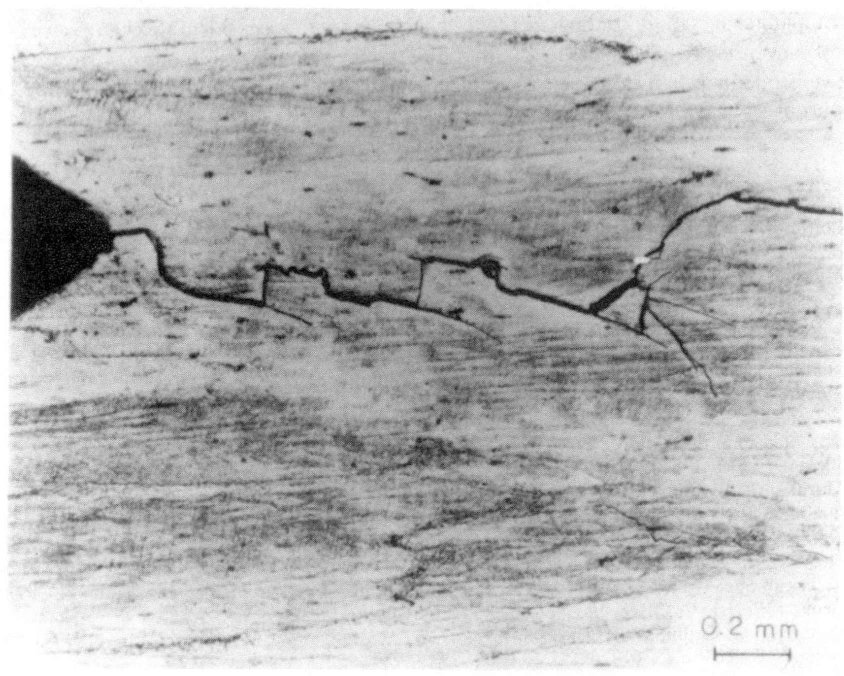

FIG. 5—*Crack path in an "underaged" 7475 aluminum alloy (of yield strength comparable to the -T7351 temper) tested in vacuum (10^{-6} torr) at near-threshold growth rate at* R = *0.1 and a frequency of 30 Hz. Crack growth direction is from left to right* [14].

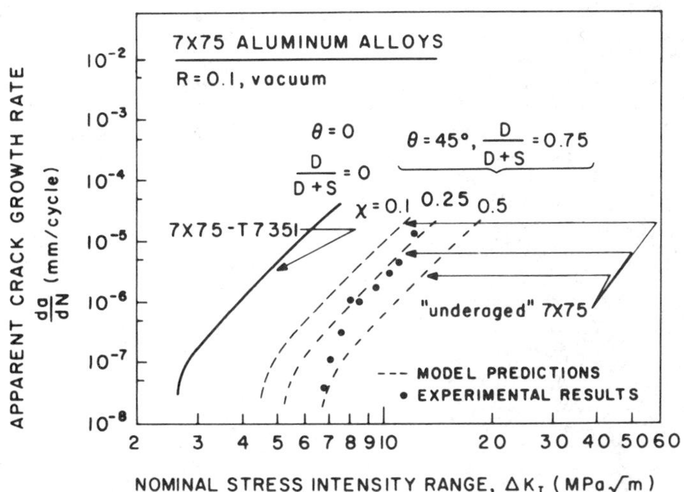

FIG. 6—*Predicted variation of the fatigue crack growth rates for the underaged 7X75 alloys* (dotted lines) *from a knowledge of the upper bound growth rate curve for the 7X75-T7351 alloy* (solid line) *tested* in vacuo. *The points refer to the typical upper bound experimental data for the underaged 7X75 alloys (of yield strength comparable to -T7351 temper) tested* in vacuo [14].

opment management and funding agents. It will be a period of considerable change away from the now conventional approaches in fatigue studies where the emphasis is on sophisticated computer-controlled mechanical testing equipment. It will be an exciting period, and I hope I can be a part of it for a long time.

References

[*1*] Coffin, L. F., "A Study of the Effects of Cyclic Thermal Stresses on a Ductile Metal," *Transactions of ASME,* Vol. 76, 1954, pp. 931-950.

[*2*] Manson, S. S., "Behavior of Materials under Conditions of Thermal Stress," NACA Report 1170, Lewis Flight Propulsion Laboratory, Cleveland, 1954.

[*3*] Coffin, L. F., "Some Reflections on the Early Days of Low-Cycle Fatigue," Report 84CRD218, General Electric Co., Sept. 1984.

[*4*] *Manual on Low Cycle Fatigue Testing, ASTM STP 465,* American Society for Testing and Materials, Philadelphia, 1969.

[*5*] Kaisand, L. R., "A Call for Unification of Low Cycle Fatigue and Fatigue Crack Growth," paper presented to Symposium on Low Cycle Fatigue—Directions for the Future, Bolton Landing, N.Y., 30 Sept.-4 Oct. 1985.

[*6*] Murakami, Y., "Correlation Between Strain Singularity at Crack Tip under Overall Plastic Deformation and the Exponent of the Coffin-Manson Law," this publication, pp. 1048-1065.

[*7*] Catlin, W. R., Lord, D. C., Prater, T. A. and Coffin, L. F., "The Reversing D-C Electrical Potential Method," in *Automated Test Methods for Fracture and Fatigue Crack Growth, ASTM STP 877,* American Society for Testing and Materials, Philadelphia, 1985, pp. 67-85; also Report 83CRD293, General Electric Co., Dec. 1983.

[*8*] Prater, T. A., Catlin, W. R., and Coffin, L. F., "Environmental Crack Growth Measurement Techniques," EPRI Project RP 2006-3, Final Report, Nov. 1982.

[*9*] Prater, T. A. and Coffin, L. F., "Experimental Cycle Analysis of Fatigue Crack Growth in Compact Type Specimen Geometries," TIS Report 82CRD039, General Electric Co., Corporate Research and Development. Feb. 1982.

[*10*] Prater, T. A., Catlin, W. R., and Coffin, L. F., "Surface Crack Growth in High Temperature Water," paper presented to International Symposium on Environmental Degradation of Materials in Nuclear Power Systems - Water Reactors, Myrtle Beach, S.C., 22-25 Aug. 1983.

[*11*] Prater, T. A., Catlin, W. R., and Coffin, L. F., "Effect of Hydrogen Additions to Water on the Corrosion Fatigue Behavior of Nuclear Structural Materials," paper presented to Second International Symposium on Environmental Degradation of Materials in Nuclear Power Systems - Water Reactors, Monterey, Calif., Sept. 1985, forthcoming.

[*12*] Coffin, L. F., "Damage Evaluation and Life Extension of Structural Components," in *Failure Prevention and Reliability—1985,* H-332, P. E. Doepker, Ed., ASME 1985; also Report 85CRD148, General Electric Co., June 1985.

[*13*] Prater, T. A., Catlin, W. R., and Coffin, L. F. "Application of the Reversing dc Electrical Potential Technique to Monitoring Crack Growth in Pipes," paper presented to Specialist Meeting on Subcritical Crack Growth, International Atomic Energy Agency, 15-17 May 1985; also Report 85CRD095, General Electric Co., June 1985.

[*14*] Suresh, S., "Fatigue Crack Deflection in Fracture Surface Contact: Micromechanical Models," *Metallurgical Transactions,* Vol. 16A, 1985, p. 249.

S. S. Manson[1]

Future Directions for Low Cycle Fatigue

REFERENCE: Manson, S. S., **"Future Directions for Low Cycle Fatigue,"** *Low Cycle Fatigue, ASTM STP 942,* H. D. Solomon, G. R. Halford, L. R. Kaisand, and B. N. Leis, Eds., American Society for Testing and Materials, Philadelphia, 1988, pp. 15-39.

ABSTRACT: The thesis is maintained that the next 30 years will witness a proliferation of publications in the area of fatigue, following as a continuum of today's high productivity rate. The driving forces for such high production are discussed, and a "wish list" is provided suggesting priority areas requiring development. Some current projects at Case Western Reserve University are outlined.

KEY WORDS: fatigue, fatigue literature, Manson-Coffin equation, Coffin-Manson equation, Strainrange Partitioning (SRP), cumulative fatigue damage, double-linear damage rule, damage curve approach, mean stress effects, Neuber correction, time-dependency of fatigue, thermal fatigue

Let me start by confirming Lou Coffin's enthusiasm for this meeting. It is easily one of the best meetings I have attended in many years. Special credit must go to Harvey Solomon, its organizer and chairman, who has left no detail unattended and who has solved all the attendant problems.

Although I agreed to provide some after-dinner remarks, I did not choose the title of the talk. While Lou Coffin chose his title, and limited it to *perspectives* on future directions in LCF, Harvey Solomon chose mine, and I am to tell you the *true* future directions for LCF. It is a dangerous ambition to predict the future in this unpredictable world, and I am reminded of the folly of some other would-be Nostradamuses, a few oft-cited of which are shown in Fig. 1. If Lord Kelvin could predict in 1895 that flying machines are impossible, as did Thomas Edison in the same year; if Wilbur Wright thought in 1901 that man would not fly for 50 years; if these renowned and respected people could make such predictions, perhaps a small amount of venturesomeness might not be totally out of order.

Lack of imagination, even in their own fields of expertise, has characterized many otherwise very knowledgeable people. A case close to my experience relates to Dr. George Lewis, for many years director of NACA and for whom the Cleveland NASA Center is named. Figure 2 shows a remark of Lewis's from a 1964 book. Dr. Lewis was concerned about the future of NACA, since it seemed to him that most everything worth doing had already been accomplished for the airplane. Luckily I joined NACA in 1942, and the field was opened to new conquests, among them winding my own wire-resistance strain gages and adapting them to high temperature use—not to forget jet engines, supersonic engines, quiet engines, and a host of other miraculous developments. Obviously the future of NACA was its destiny as the nucleus for NASA, but even as the aeronautics agency, which is the first A in NASA, it had, and has now, a brilliant future of service in many fields, among them low cycle fatigue.

So, with plenty of negative examples to draw on, I shall move cautiously to the remainder of my remarks.

[1]Professor of Mechanical and Aerospace Engineering, Case Western Reserve University, Cleveland, OH 44106.

"HEAVIER-THAN-AIR FLYING MACHINES ARE IMPOSSIBLE."

- LORD KELVIN

1865

"IT IS APPARENT TO ME THAT THE POSSIBILITIES OF THE AEROPLANE, WHICH TWO OR THREE YEARS AGO WAS THOUGHT TO HOLD THE SOLUTION TO THE (FLYING MACHINE) PROBLEM, HAVE BEEN EXHAUSTED, AND THAT WE MUST TURN ELSEWHERE."

- THOMAS ALVA EDISON

1895

"MAN WILL NOT FLY FOR 50 YEARS."

- WILBUR WRIGHT

1901

FIG. 1—*Some not-so accurate predictions.*

The Next Twenty-Five Years

What can I say about the future of LCF that will not embarrass me as the true history unfolds?

I examined a number of reports outlining progress in research on materials behavior and was struck with an illustration in one of David McLean's papers relating to progress in mechanical behavior of materials (Fig. 3). One point that impressed me was the period of time that it took from when Hooke first announced his law that deflection is proportional to load to when Young announced that this constant of proportionality was a material property relating elastic stress to strain. It took approximately 130 years to gain the insight and to publish the results. I also noticed the flattering entry of the Manson-Coffin Law (although McLean used the Arabic-Hebrew convention for writing the designation, so that it has to be read from right to left), and I wondered how long it might have taken Coffin to conclude that the 0.5 exponent he found for the material he tested was a universal constant applied to all materials. If not instantaneous, it certainly was not 130 years!

This thought led me back to a subject I had previously contemplated in the 1964 SESA Murray Lecture, when I expressed some concern over the proliferation of reports on fatigue. Figure 4*a* shows how many reports on fatigue one would have to read if he chose to read all these published between 1950 and 1970. For the period between 1950 and 1960, actual numbers were available on publications as collected by ASTM, but between 1960 and 1970 it was necessary to estimate the numbers based on the approximation that publications were increasing at a rate of about 15% per year, since that seemed to be the trend. An inverted time scale was used to cause the curve to resemble a Wohler fatigue plot, and it did indeed show that reports proliferated

"THIS AIRPLANE (THE DOUGLAS DC-1) INCORPORATES ALL OF OUR LATEST RESEARCH INFORMATION. IT IS OF SEMI-MONOCOQUE CONSTRUCTION AND IS MADE OF HIGH-STRENGTH ALUMINUM ALLOY. THE LANDING GEAR IS RETRACTABLE AND THE ENGINES ARE ENCLOSED IN NACA COWLINGS. THE WING IS DESIGNED FOR BEST AERODYNAMIC AND STRUCTURAL EFFICIENCY, AND IT EMPLOYS FLUSH RIVETS ON ITS FORWARD PORTION TO REDUCE THE DRAG.

WHAT IS LEFT FOR US TO DO? WHAT IS THE FUTURE OF THE NACA?"

-DR. GEORGE W. LEWIS
FIRST DIRECTOR, NACA

FIG. 2—*A quotation from* Wings into Space *by Martin Caiden (Holt, Rinehart & Winston, New York, 1964, p. 22).*

with time just as cycles to failure did with reduced stress. Then, with the help of my student Ramesh Kalluri, we determined what the report proliferation might look like in the future, the results of which are shown in Figs. 4*b* and 4*c*. These figures retain the assumption of a 15% increase per year until the report rate became 2000 per year; from this point we retained the 2000 rate since, after all, there is a limit as to the number of meetings where papers can be presented. Even with this limitation, by the year 2010 there should be 70 000 reports in the fatigue literature. This figure led to one prognosis that I feel comfortable with, one that should lie easy on my bones in the distant future.

The message is shown in Fig. 5, namely that reports will proliferate but that there are several corollaries. One is that if we are uncomfortable about so many reports to read, something could be done about it. We could be more selective about which reports are published or change the way papers are presented (it is surprising we have not yet taken advantage of computer technology to disseminate research results). But, as indicated in Fig. 5, we probably will not change the rules for a while, primarily because too many people find it advantageous to keep generating papers, as will be discussed. Maybe it is just as well. Setting up a hierarchy to limit report publications could do more harm than good. I remember my own early efforts in fatigue research. This field was not traditionally one of the specialities of the Lewis Research Laboratory and there was a hierarchy within NACA that "recommended" leaving this type of research to others. Luckily I was in a position to ignore the "recommendations."

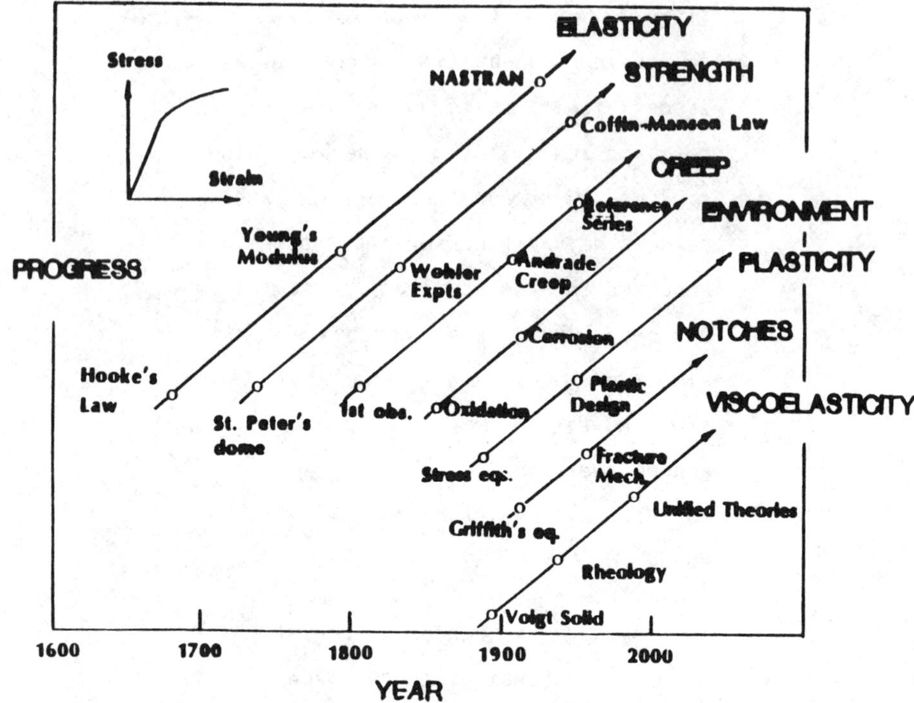

FIG. 3—*Some milestones in mechanical behavior of materials (after D. McLean,* Materials Science and Engineering, *Vol. 26, 1976, pp. 141–152).*

The Driving Forces

Importance of the Problem

Where will these 70 000 reports come from? What will drive them to be published? There are numerous driving forces, of which I can describe but a few. Figure 6 shows some of the more obvious factors.

Consider first the fact that in "fatigue" (low cycle as well as other types) we are dealing with one of the most important problems that govern the viability of modern high-performance power-generating equipment such as turbines, jet engines, nuclear reactors, and rocket engines. High static and dynamic loads, together with high temperatures, force working available materials to the limit of their capacity. All involve load alternations or vibrations, and the most common failure mechanism is fatigue. Reports describing specific failures will abound, because understanding past failures is the surest way to delay or avoid future failures.

Consider, too, that much of the equipment is aging. Many stationary power-generating turbine plants are reaching their design lives—100 000 h, 200 000 h, and even more. Many nuclear power plants are reaching the life limits envisioned by their designers, too. Decisions to extend service, replace parts, reduce severity of loading, retirement, and so forth will take much research. Opinions will differ; seminars will be held; reports will be written.

Then, of course, advanced new thermal equipment is under contemplation, in design, or in early stages of use. As noted in Fig. 6, Item C, the Electric Power Research Institute is now

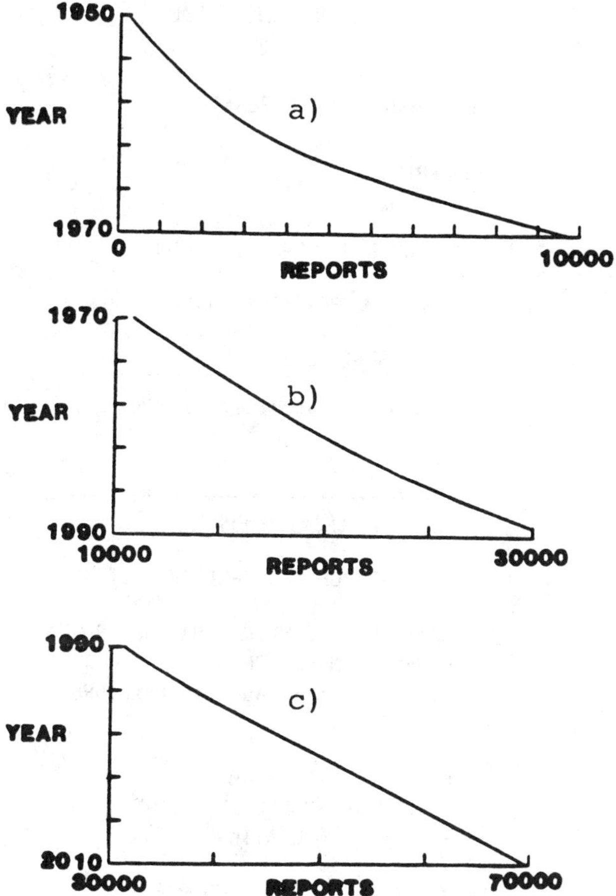

FIG. 4—*Potential report proliferation in fatigue behavior over the next 25 years.*

seriously considering a very-high temperature stationary power-generating turbine to take advantage of new materials, mass flow, and heat-transfer technology. We note also as isolated examples of technology development the Lavie airplane being developed in Israel, the hypersonic airplane under contemplation by the U.S. Air Force, the Liquid Metal Fast Breeder Reactor (both the American model which is temporarily suspended and the European version which is being studied actively), and solar power engines which obviously will require high and fluctuating temperatures. All the equipment involved in these advanced projects will be limited by fatigue problems, and it is easy to see how they will beget families of reports.

Now consider Item D of Fig. 6. Here I am trying to recognize that there is a desperate need for new design codes to replace the codes in use today and to provide guidance for design of future projects. The ASME Nuclear Pressure Vessel Code was adopted many years ago when research on high temperature fatigue in the creep range was based more on speculation than on a solid experimental foundation. Certainly for future projects we are entitled to a design code that not only incorporates the wealth of research that has been generated over the past 20 years, but also

MANSON'S FIRST LAW

"REPORTS WILL PROILIFERATE."

CORROLARIES:

1) COULD DO SOMETHING ABOUT IT.

2) BUT WE PROBABLY WON'T

3) MAYBE IT'S BETTER

FIG. 5—*The one prognosis I can confidently make.*

THE DRIVING FORCES

1) IMPORTANCE OF THE PROBLEM

A) FATIGUE OF CONVENTIONAL POWER GENERATION EQUIPMENT
- JET ENGINES, NUCLEAR REACTORS AND ROCKETS

B) EQUIPMENT IS AGING
- THERMAL AND NUCLEAR POWER PLANTS IN OPERATION

C) ADVANCED THERMAL EQUIPMENT
- EPRI, LAVIE, HYPERSONIC, LMFBR AND SOLAR POWER

D) THE NEED FOR NEW STANDARDS
- REPLACE CURRENT ASME NUCLEAR PRESSURE VESSEL CODE
- FOR NEW EQUIPMENT, E.G. COAL CONVERSION

FIG. 6—*Dominance of fatigue as the driving force for report publications in the future.*

demands additional research to answer questions not yet resolved. Such research will generate more and more reports—especially when new high-performance equipment, such as will be required for efficient coal conversion, require new materials because of their specialized environments and operating conditions. Perhaps it is even best to approach an advanced code by envisioning it for a new application. Then, if it proves satisfactory, it can be used for other applications, such as the Nuclear Pressure Vessel Code, without reflecting too strongly on possible deficiencies of equipment already designed according to earlier codes.

Proliferation of Publication Outlets

Another reason why I expect so many reports to be written in the period ahead is that there has arisen an abundance of vehicles for publication, and once in existence, they are forced, willy-nilly, to generate papers. ASTM, ASME, and similar parent engineering societies have long recognized the importance of fatigue to structural integrity and have strongly fostered publications in this field. In addition, there have arisen, over the past decade or so, a number of specialized journals that in some way touch on fatigue. I list several in Fig. 7, Item *2*. These journals are but a few of the crop of the past two decades. A trip to the library shelves can prove an overwhelming experience to a novice who wants to become familiar with the current fatigue technology.

Utilization of Advanced Equipment

It is mind-boggling to contemplate how much new equipment has been installed in this country, as well as in other countries, concerned with the fatigue problem. In Fig. 7 I list only two entries: (1) the new laboratory recently unveiled by the NASA-Lewis Center and (2) all others (lumped together under the major suppliers of such equipment such as MTS and Instron). I include NASA's new equipment because I am personally familiar with it and because of the cooperative arrangement which permits my students and me to use this equipment. Thirty years ago, together with the help of Bob Smith and Marv Hirschberg, I was responsible for arranging for a modest fatigue laboratory at Lewis. We designed the machines and had them built to our specifications. The new laboratory pales its earlier prototype (upgraded several times over the years). There are more machines of greater sophistication and precision. They were commercially built to exacting specifications, and they are computerized to perform functions virtually impossible to control manually. Figure 8 shows a photograph of just the control panel for eight machines, but one photograph cannot do justice to or reflect the research capabilities not before possible.

THE DRIVING FORCES

2) PROLIFERATION OF PUBLICATION OUTLETS

A) ASTM AND ASME

B) OTHER JOURNALS
EXAMPLES:

FATIGUE & FRACTURE OF ENGINEERING MATERIALS AND STRUCTURES

JOURNAL OF STRAIN ANALYSIS

JOURNAL OF THERMAL STRESS

ETC.

3) ACQUISITION OF ADVANCED EQUIPMENT

A) NASA-LEWIS NEW LABORATORY

B) MTS AND INSTRON

FIG. 7—*Encouragement and equipment availability as driving forces for publications.*

FIG. 8—*NASA-Lewis control panel for some of its computer-controlled, high-temperature fatigue testing machines.*

The Lewis Research Center is not the only laboratory where fatigue testing machines of advanced capability can be found. Everywhere one goes—government laboratories, universities, research institutes, the laboratories of large and small companies, and commercial laboratories that perform services for others—the picture is the same: equipment sophistication and extended capability. To reflect the growth of such equipment without an item-by-item count, it is interesting to examine the corporate growth of the two main producers of this type of equipment in the United States: MTS and Instron. I asked my stockbroker to send me market performance charts for these two companies (Fig. 9). Such charts are quite complicated, and show much information that is of interest to an investor or specialist. But what is of interest to us here are the heavy jagged lines, which together with the logarithmic scale at the right, show how the price of the stocks have increased over the past dozen years. If one wants to study the charts in detail, he can also glean the growth of earnings, using the left logarithmic scale. One conclusion is patently clear: from their lows ten years ago, both companies have grown by more than a factor of 10 in earnings and in stock price in the following decade. It is clear that this growth resulted because of the large amount of equipment they supplied to interested users.

Now it is an obvious corollary that once the high price is paid for the equipment, and once it is installed and operating, the owners will strive for the pay-off. Some of the testing done will be proprietary, of course, but much of it will be published. These publications will contribute to the 70 000 reports I am talking about.

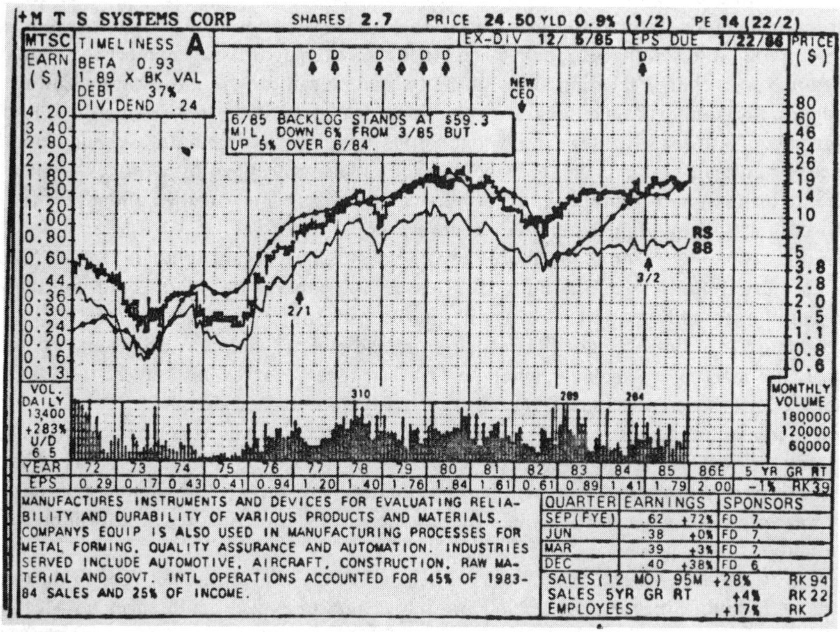

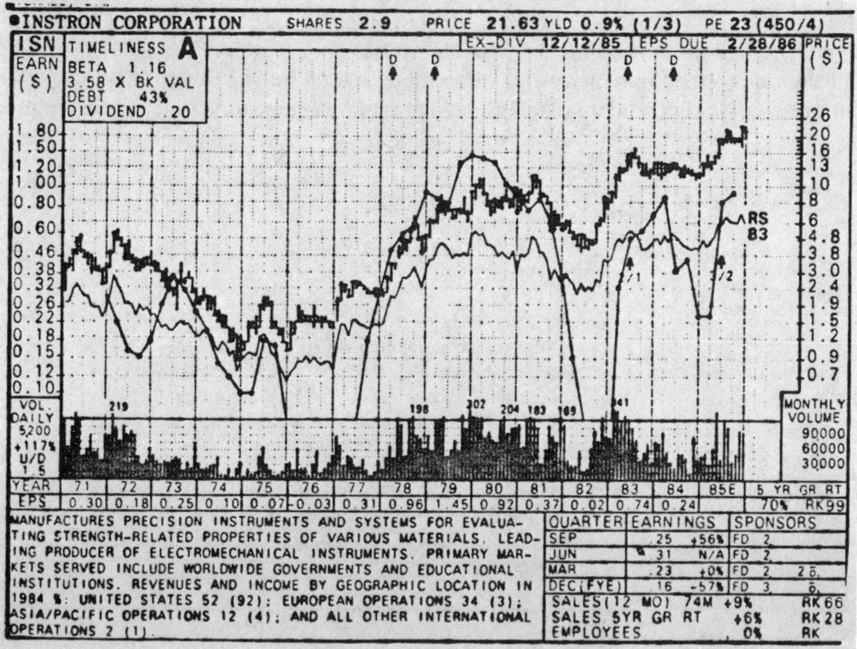

FIG. 9—*Financial growth of two fatigue testing machine suppliers as indicator of equipment proliferation.*

Drive to Organize Meetings and Symposia

An important factor that will contribute to report proliferation is the enormous drive that exists among technical committee members of the various societies to be matchmakers for meetings, symposia, and seminars in their special fields. Give an aggressive engineer the assignment to organize or to chair a meeting in his specialty, and he will direct himself straightaway to his colleagues, his counterparts in other organizations, or experts with whose names he is familiar, citing enticing reasons why they should contribute. A chance to contribute, a chance to learn, a chance to put it all in one place, to travel to exotic places and pleasant climates—whatever is needed to bring in a "voluntary" contribution. Soon a meeting is corralled and all for the good because some papers might otherwise not have been written.

Consider this very meeting. A more active and effective organizer than Harvey Solomon would be hard to find. As shown in Fig. 10, Item A, 140 papers were volunteered, but there was only room for about 80 papers. Few would quarrel with the notion that this meeting has been one of the best we have attended. Every promise the meeting held has been and is being fulfilled. It is a meeting we shall all remember. In the process Harvey has laid the foundation for future meetings where the remaining 60 papers are bound to be offered.

Few of us can forget, for reasons good or not so good, the Symposium on Fundamental Questions and Critical Experiments in Fatigue held in October 1983 in Dallas, Texas. The large number of papers presented there will be published in *ASTM STP 924*. Remember, also, that these were proposed experiments; when they are performed and analyzed, they will be a source of numerous future papers.

In Fig. 10, Item C, we recognize what has become a dependable source of good international papers on fatigue. Since the early 1960s we have come to expect conferences on Mechanical Behavior and on Fracture to alternate, each on a four-year cycle (with a few exceptions). These conferences started in Japan and are sponsored by competing factions in the Japanese technical community, but both have strong international sponsorship as well. Each conference that has been held has provided us with a rich heritage of papers on fatigue from outstanding international experts. The Fifth International Conference on Mechanical Behavior of Materials will,

THE DRIVING FORCES

4) DRIVE TO ORGANIZE MEETINGS AND SYMPOSIA

A) CURRENT MEETING
(140) REPORTS VOLUNTEERED,
(80) ACCEPTED.

B) SYMPOSIUM ON FUNDAMENTAL QUESTIONS AND CRITICAL EXPERIMENTS.

C) FOUR-YEAR CYCLE FOR INTERNATIONAL CONFERENCE ON MECHANICAL BEHAVIOR AND INTERNATIONAL CONFERENCE ON FRACTURE

D) EUROPEAN COMMON MARKET SYMPOSIA

E) AGARD SYMPOSIA

F) INDIVIDUAL VESTED INTERESTS

FIG. 10—*Meetings and symposia as driving forces for report publication.*

for example, be held in Beijing on 3–6 June 1987. I expect it will be a resounding success and will provide substantial increments to our back-file of reading wealth on fatigue.

As noted in Item D of Fig. 10, many papers originate in symposia among the European Common Market Countries. Some papers have distributions restricted to members of the Common Market, at least for some specified minimum period, but eventually they should become available to the world community at large. I expect that several of the papers being presented at this Conference had already circulated for some time among the European members of the Common Market.

Then we have the AGARD publications, the technical and scientific arm of the NATO countries, as noted in Item E of Fig. 10. Many of the joint or collaborative efforts of the technical experts within these allies frequently meet and reveal new findings to each other. The reports are widely distributed. I have myself often participated in these symposia. I remember with special fondness one that was held in Aalborg, Denmark, in 1978. Its sole subject was Strainrange Partitioning and the international verdict was that the method was valuable for fatigue analysis. With somewhat less fondness I also remember that the students of Aalborg University, being somehow informed that the method might have application to military aircraft engines and to nuclear reactors (which they detest), formed human walls to prevent us from entering our conference room. We had to rent space in a country club to continue our conference. But in a sense it was an indirect compliment: Why go to the trouble if the method were not good?

Finally, listed in Fig. 10, Item F, are "individual vested interests," meaning individuals who have special interests in some topic within the global field of fatigue and who will pursue this interest by all possible means, pronouncing their results to anyone who will listen. I also refer to collective individuals who meet regularly to exchange thoughts and data on subjects of mutual interest. An example of this type of source is the International Union of Theoretical and Applied Mechanics. I recently had the pleasure of being invited to present an engineering viewpoint of cumulative fatigue damage analysis at a meeting held in Israel. Many of the participants were much more oriented toward theory than experiment, and it was extremely interesting to interact with them. The proceedings of this symposium will be published as a special issue of the *International Journal of Fracture,* and the paper by Gary Halford and me will be included in that issue.

Emergence of China and Other Asian Technologies

The real sleeper in publication sources may lie, as suggested in Fig. 11, in the emergence of China and other Asian countries as contributors to the Western technological storehouse and

THE DRIVING FORCES

5) EMERGENCE OF CHINA AND OTHER ASIAN TECHNOLOGIES

A) ENTRY OF JAPAN INTO THE TECHNOLOGICAL WORLD 30 YEARS AGO

B) FORTHCOMING CONTRIBUTIONS BY CHINESE ANALYSES AND MEETINGS IN CHINA

C) RUSSIA?

FIG. 11—*Untapped sources of future publications from new players.*

literature. We have already recognized how enriched our backgrounds have become over the past 30 years within which Japan has shared its literature with us. How much more enriched we will become as the vast Chinese and Korean technologies become available to us and as we stimulate them to contribute to our technology. Already an auspicious beginning is being made at this meeting. There are more contributions from mainland China that we commonly find at our meetings, and, from what I have been able to judge, the quality of the contributions is excellent. I also note that they are anxious to host meetings and that the 1987 International Conference on Material Behavior is to be held in Beijing, as mentioned previously. It is even possible that with this stimulus, together with the incipient thawing of political relationships, Russia may become more involved in technical interchange. Already we have benefited strongly from ideas and analyses originating in Russia and other countries within its sphere.

New Materials

Finally, in Fig. 12, I have expressed the recognition that the need for new materials for special applications will provide a rich source of new reports to feed our mill. Not only will their development details require dissemination, but their properties and applications will attract the services of many investigators.

Directionally-solidified alloys and monocrystals have become quite common over the last two decades, but much needs yet to be studied about their individual performance characteristics and failure mechanisms. Similarly, composites have been with us for more than three decades, and have been effectively used in many specialized applications, but a vast arena lies ahead involving new compositions, temperature ranges, and environmental interactions. Sources for reports aplenty!

While ceramics have been a hope for structural applications because of their potential low cost, light weight, and temperature resistance, the hope has been just one arm's-length ahead of widespread use. I remember working on ceramic turbine blades at NACA in the early 1950s. There were fastening problems, foreign object damage problems, and quality control problems—to mention a few. More recently, new compositions, an understanding of how to avoid stress-concentration problems, and greatly improved quality control have brought these materials again to serious attention. Will they have fatigue problems? Many reports will precede a jury's verdict.

THE DRIVING FORCES

6) NEW MATERIALS

A) DIRECTIONALLY SOLIDIFIED ALLOYS, MONOCRYSTALS

B) COMPOSITES

C) CERAMICS

D) SPECIAL Ni-BASE ALLOYS

E) ULTRA-HIGH TEMP. ALLOYS

F) IMPLANT MATERIALS

G) COMPUTER MATERIALS

FIG. 12—*New materials will be studied in fatigue; these will be a source of future reports.*

Nickel-base superalloys have been the backbone of high temperature service components for many years, and as new applications develop new compositions may be needed for specialized service uses. For example, EPRI is studying the possibilities of a long-life, high-temperature turbine for stationary power generation. Current alloys are not quite adequate, and new compositions are being considered. In addition to nickel-base alloys, there are the alloys involving matrixes of the more refractory metals (molybdenum, zirconium, tungsten, etc.). Oxidation problems dominate, but there is hope for practical solutions.

Items F and G of Fig. 12 list more prosaic materials, not involving high temperature. But important applications they are, and there is much to be gained from technological study. Implant materials must function compatibly with human tissue, and they must be both strong and highly ductile to avoid repeat surgery required by fatigue failures. Likewise, computers have their own materials problems—for example, as expressed by Harvey Solomon in his paper at this conference. Dr. Solomon has attempted to apply current high-temperature fatigue approaches to solders used in computers. Such solders creep at relatively low temperatures and have caused failures in relatively short times. While the manner in which he has applied the methods have not led him to be optimistic, he points to measures of current theory that can be modified to make them applicable. It is hoped that future reports will pave the successful way.

A "Wish List"

Having suggested that there will be no shortage of fatigue reports in the next two to three decades, I suppose it would only be fair for me to follow with a list of subjects that I would like to see covered. Such a "wish-list" could be very long, and the time required to discuss the entries would far exceed the time available to me. Perhaps, at least, I can provide a limited list with just a few words of clarification (Fig. 13), emphasizing those aspects that have touched my direct experience over the years.

Emphasize Fundamentals—Practically

It seems axiomatic to urge emphasis on fundamentals. For over 50 years scientifically oriented researchers have sought to understand the exact mechanism of fatigue. We have made considerable progress, but more is possible. The development of scanning electron microscopy (SEM) has given us a tool for studying fatigue surfaces with precision at ultra-high magnifications, and the more recent scanning transmission electron microscope (STEM) has extended this tool to penetrate the material and include some depth. Compositional effects in the immediate vicinity of a fatigue element can give us much information for clarification of the details we lack. The use of radiation equipment, such as Auger and positron emission, have further extended our capabilities. Certainly there should be intense effort to gain understanding through studies of fundamentals.

To carry the aspect of fundamentals to at least two more steps, Items 2 and 3 are also listed in Fig. 13. Item 2 essentially echoes Lou Coffin's call for tools to monitor service equipment with tools that are based in fundamentals. Sample tools listed are electric potential fields, and how they are influenced by initiating and propagating cracks, and acoustic emission—the types of sounds generated as material deforms and cracks. Item 3 of Fig. 13 carries the use of fundamentals at least one step further, namely to unite the technology of crack initiation into the technology of fracture mechanics. In a sense we are already doing this when we assume that there is always present a pre-existing crack or hypothesize the start of such a crack by early fracture of brittle particles. However, small cracks do not necessarily follow the laws of fracture mechanics of large cracks which have been so intensively studied in recent years. Small-crack technology is now receiving considerable attention, but obviously much more must be done.

1) MORE FUNDAMENTAL STUDIES USING NEW TOOLS - SEM, STEM, AUGER AND POSITRON EMISSION, ETC.

2) DEVELOP BETTER TOOLS FOR DETECTING FATIGUE DAMAGE. EXAMPLES:
- ELECTRIC POTENTIAL FIELDS, ACOUSTIC EMISSION

3) MERGE CRACK INITIATION TECHNOLOGY WITH FRACTURE MECHANICS TECHNOLOGY

4) DEVELOP SIMPLE CONSTITUTIVE MODELLING FOR LOW CYCLE FATIGUE AT HIGH TEMPERATURES.
   A) CONTINUATION OF CURRENT EFFORTS TO DETERMINE VALIDITY OF BASIC CONCEPTS AND PROCEDURES
   B) APPLICATION TO SPECIAL CASES TO DEVELOP SIMPLE RULES FOR COMMONLY ENCOUNTERED SITUATIONS.
   C) APPLICATION TO ASYMPTOTIC CONDITIONS

5) DEVELOP FATIGUE MODEL FOR HIGH TEMPERATURE
   A) BASED ON MECHANISTIC UNDERSTANDING
   B) COMBINING THE BEST FEATURES OF CURRENT MODELS
   C) RECOGNIZING SPECIAL CHARACTERISTICS OF INDIVIDUAL MATERIALS OR CLASSES OF MATERIALS

6) APPLICATION OF HIGH TEMPERATURE THEORY TO THERMAL FATIGUE
   A) COMBINING CONSTITUTIVE MODELLING WITH FATIGUE THEORY
   B) RECOGNIZING METALLURGICAL CHANGES

7) APPLICATION TO LOW STRAINS AND LONG TIME CYCLES
   A) CLARIFICATION OF THE ROLE OF ENVIRONMENT AND THE ROLE OF METALLURGICAL INSTABILITIES

8) DEVELOPMENT OF IMPROVED MODEL FOR CUMULATIVE FATIGUE ANALYSIS
   A) BELOW AND IN THE CREEP RANGE.
   B) RECOGNIZING MECHANISTIC GOVERNANCE
   C) PRACTICAL FROM THE POINT OF APPLICATION

9) ROLE OF MEAN STRESS
   A) CLARIFICATION OF MECHANISM
   B) EXPLANATION OF DIFFERENT BEHAVIOR AMONG MATERIALS
   C) DISTINCTION WHEN ACCOMPANIED BY LARGE INELASTIC STRAINS AT HIGH TEMPERATURE, E.G., LARGE CP AND PC STRAINS

10) ROLE OF MULTIAXIALITY
   A) STRESS-STRAIN PATH
   B) METALLURGICAL DEGRADATION
   C) LIFE RELATIONSHIPS

FIG. 13—*A "wish list" of future research.*

Emphasis on Constitutive Behavior

Items 4 to 7 of Fig. 13 list aspects of constitutive behavior that are bound to be useful in the years ahead. While understanding the relation of the rheology of a material to its fracture characteristics is important at temperatures below the creep range, it is at high temperature, where creep may occur and where metallurgical transformations are common, that the need is greatest. Under the stimulus of the Air Force and of the NASA-Lewis Research Center there has recently occurred a flurry of activity to develop high temperature constitutive models. Many such models are available, but the emphasis seems to be settling on a detailed study of two.

While the implementation of these methods is complicated and requires many material constants (much materials characterization as a data base), the progress has been very good. It is important to continue these studies.

Items 4B and 4C of Fig. 13 are identified to urge two specific aspects of this continued study. Even though the generalized application of these constitutive models is very complex, they do serve to enable us to study special cases as a means of developing simple engineering rules for common application, as listed in Item 4B. Specifically one is referred to such uses as are described by Gary Halford and his NASA colleagues at this conference. By applying the more generalized theory, Dr. Halford was able to derive simple formulae for separating the various strainrange components in Strainrange Partitioning theory. These formulae work very well for the material tested, and, once identified, could also be studied for other materials. The concept involved here—deriving the specific from the more general for a few special cases, and extension to general but simple—would seem to have more widespread use.

Item 4C of Fig. 13 urges supplementary use of these complex constitutive models to study asymptotic behavior of repeatedly loaded materials. As they are presently applied, the methods contain many terms that relate mostly to the transient conditions associated with early loadings. These transients do, however, disappear as more and more loadings are applied. Since in most applications we are concerned with fatigue after numerous repetitions of loadings, it would seem appropriate to apply the methodology for these cases; much simplification may result. It is analogous to our current practice in studying materials behavior at room temperature. If we try to incorporate the details of the early loadings when cyclic hardening and softening takes place, we soon bog down in a mass of complexities which usually do not influence the final behavior of the part. Since, by far, the largest fraction of the total life is sustained after the material behavior has stabilized, we make the computations using the stabilized cyclic stress-strain curve, ignoring the early cycles. Special analogous studies should be made at high temperature.

Item 5 of Fig. 13 calls for the development of a high-temperature fatigue theory that is based on the knowledge of constitutive behavior. When we first developed Strainrange Partitioning (SRP), we thought that was what we were doing. At least we tried to recognize that there could be two types of strain: that associated with slip-plane sliding and that associated with grain-boundary sliding. Later we found that the relative importance of the two types of strain varies considerably among materials: some materials do not suffer grain boundary sliding at all, but may still fracture in the grain boundaries because they are brittle and fracture easily when intersected by sliding slip planes. These observations did not invalidate SRP, but showed us that we must be careful to recognize the individual material's characteristics. Many other technologies of high-temperature analysis have grown up over the past decade or more. Some are claimed to predict behavior better than does SRP. As would be expected, I am not convinced, but I am certainly open to objective study of the relative merits of various methods. I think it is very important to narrow down the number of treatments used so that the few chosen remaining methods can be evaluated more generally by the community at large rather than only by their proponents. We should utilize the best features of the current models and should allow in the method the individual behavior of several classes of materials in common usage.

The development of a viable procedure for treating thermal stress problems should benefit tangibly from an informed understanding of constitutive behavior. It is obviously important to know how the material responds to the stresses and strains introduced by mechanical and thermal loadings involved in such problems. Item 6B of Fig. 13 emphasizes that at high temperatures we frequently observe metallurgical transformations which affect both the rheological behavior and fracture characteristics. Each material of interest must be individually studied in the temperature and time range of interest to ferret out its peculiarities. Perhaps the common trends of many materials will eventually permit us to generalize.

Finally, Item 7 of Fig. 13 identifies the importance of studies in the range of low strains and long cycle times. Most of the tests that we conduct involve relatively short failure times. Equipment use is costly, most investigators are anxious to publish in a hurry, and if students are involved the length of the test must be compatible with their thesis time limitations. But service times in the field are long—we hope! Much equipment is intended for 30 and even 40 years of service. In such periods the phenomena are not necessarily the same as in the tests we conduct which are limited to hundreds of hours. Thus I urge that at least some tests be undertaken in the (near-) real-time frame and that theoretical calculations recognize the real-time aspects of the problem. I would be very happy if a large fraction of our future literature would involve this feature.

Improved Cumulative Fatigue Theory

The final topic on my abridged "wish-list" is identified by Items 8, 9, and 10 of Fig. 13, and it relates to the development of a viable cumulative damage theory for use both below and within the creep range. Most fatigue studies limit their focus on material behavior under a single well-defined set of loading conditions. Let us assume we have characterized a material under a large variety of such loadings. How the material behaves under a sequence of different loadings has come to be known as *cumulative fatigue*, even though a single loading also involves the cumulative effects of several fatigue processes. Over the years many theories have been proposed, but none has succeeded in general satisfaction. I have devoted considerable effort to the pursuit of an engineering approach that is based to some extent on mechanistic understanding and yet is simple and practical enough for a designer to use. Item 8 of Fig. 13 expresses the hope of seeing much more widespread effort in this direction. Obviously, unless we know how to put it all together, having the individual parts is of limited use.

Item 9 of Fig. 13 identifies one of the factors involved in putting it all together. Many cyclic loadings involved in a service complex involve conditions characterized by a hysteresis loop which has a mean stress. We now accept that a hysteresis loop, the width of which implies plastic flow, should result in a finite life. That is what the Coffin-Manson (or is it Manson-Coffin) equation is all about. But what does the mean stress do to reduce life, as it is usually found to do? Is it related to opening and closing of microscopic cracks, to added local plasticity in the vicinity of grain boundaries, or to some other phenomenon? And why do various materials react to mean stress in different ways? Some materials follow a linear Goodman-type diagram: for some it is concave and in others even convex. And why, for some materials, is the Goodman-type of diagram independent of life level, while for others it is distinctly life dependent, the geometry of the diagram even changing shape with life level? Mechanistic studies should be very valuable.

An interesting aspect of mean stress effect is listed as Item 9C of Fig. 13. Investigators of high temperature fatigue will recognize that rather large mean stresses can be introduced by combining creep in the tensile half of the loop and plasticity in the compression half (or vice versa). We call the types of strains so induced CP (or PC if the sense is reversed). Do mean stresses under such conditions have the same effect as mean stresses in sub-creep applications? Halford has done some preliminary work in this field and has concluded that some effect is present but limited. More work is needed. Not only will it relate to the practical analysis of such loadings, but it may give us insight as well as into the mechanism that causes mean stress to influence life.

Finally, Item 10 of Fig. 13 lists the role of multiaxiality as a subject that begs for better understanding. Recently there has been considerable emphasis on this subject and, with particular thanks to Keith Miller at Sheffield University in England, several conferences have been held. The next two or more decades will see much written on this important subject—and it has my encouragement. Experiments in this field are more difficult because costly special equipment is often involved, but much of this equipment is in place (one reason for the commercial success of companies like MTS and Instron) and it should be used to advantage. As listed in the figure, I would like to read future reports on the multiaxiality effects on plasticity and creep, on special aspects which cause metallurgical degradation, and on the fatigue life relationships, particularly when creep and plasticity are both involved, as when SRP becomes applicable.

Some Research for Future Reporting

Having given you a list of some research I would like to see done by the technical community at large, let me finish by telling you a little about research that some of my colleagues and I have been doing over the past few years, but which has not as yet been published. In a sense I am listing some of the 70 000 reports I am expecting in the next 25 years (if I can meet this time schedule). I will keep details to a minimum, just mentioning the subject, giving a typical result, and acknowledging the contributor, who in most cases is a student that has recently completed either a Master's or Ph.D. thesis. These items are listed in the Appendix.

Conclusion

The next 25 to 30 years promise exciting developments in the fatigue field, for low cycle and other classifications as well. The field is important, there is much interest, and a large number of experts are anxiously ready to explore generalities and details. I am happy that I have had a role in bringing this field to its present position, and I look forward to continuing participation. Good luck to all of us!

APPENDIX

The following projects are either in progress or have been completed but not yet published.

Modified Universal Slopes Equation

The original Universal Slopes Equation, developed by Marvin Hirschberg (NASA Lewis RC) and me and presented in the 1964 SESA Murray Lecture, has been very useful over the years to estimate fatigue life when only tensile properties of the material were known. It was based on a detailed study of 29 materials which were then tested by us. The vast amount of data generated in the interim period by us and by numerous other investigators throughout the country has led us to re-examine this equation to determine if its accuracy can be improved through the use of all materials information now available. This task was undertaken by U. Muralidharan as part of his Ph.D. dissertation at Case Western Reserve University; some of the results are shown in Fig. 14. In addition to showing the improved equation (which incidentally contains a term including ultimate tensile strength in the plastic line), it also shows a comparison between the predictions of the equation and the experimental data for 50 materials. It is seen that the new equation not only predicts life with a lower spread of error but that such error as is found is better distributed about a zero value. By comparison, the original universal slopes equation seems to produce predominantly positive errors in some life range and predominantly negative errors in other ranges.

Predicting Bending Fatigue from Axial Fatigue Data

Figure 15 shows another project with Muralidharan as collaborator. The problem here is also an outgrowth of a calculation contained in the 1964 Murray Lecture. In treating the bending problem of a rotating circular shaft we found that the integrals generated could not be determined in closed form. By combining the form of the solution for a shape which does lend itself to closed-form solution, we started with a hypothesized solution that could be used for the circular shape. The constants in the equation were then determined from numerical calculations for the

$$\Delta\epsilon = \Delta\epsilon_p + \Delta\epsilon_e$$

$$= 0.0266 \quad D^{0.155}\left(\frac{\sigma_u}{E}\right)^{-0.53}\left(N_f\right)^{-0.56}$$

$$+ 1.17\left(\frac{\sigma_u}{E}\right)^{0.832}\left(N_f\right)^{-0.09}$$

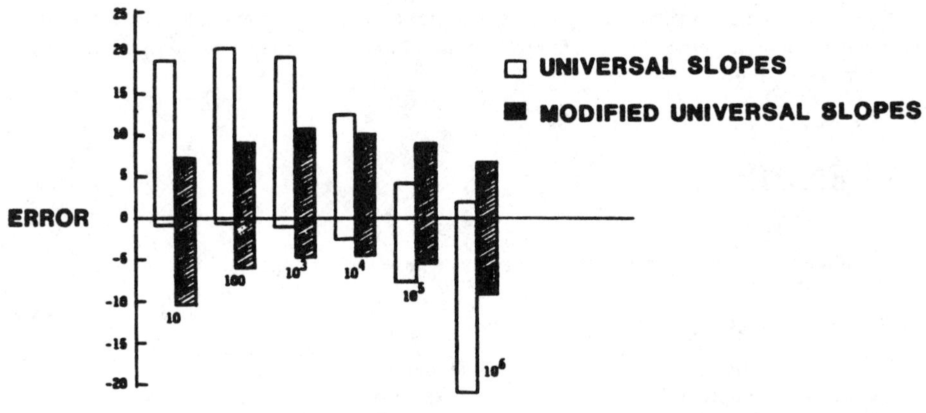

FIG. 14—*A modified universal slopes equation for estimating fatigue properties (collaborator: U. Muralidharan).*

circular shape. In the end we obtained a closed-form solution for the circular shaft, although it was only approximate. The approximation was so close, however, that it was rarely possible to detect a numerical difference between the results obtained by the closed-form formula and very accurate numerical integrations.

Generalized Neuber Equation for Notch Analysis

For many years I have been concerned with the problem of improving the Neuber notch equation for the plastic range, as shown in the first equation of Fig. 16. In 1978, working with Walcher and Gray at Teledyne CAE, I suggested a correction in the form shown by the second equation in Fig. 16. Determining the exponent α in this equation has provided thesis subjects for several students in the intervening years. We used strain gages in some cases, finite-element calculations in others, and finally examined the available literature for other specific studies which were of help to us in generalizing the results. My most recent student on this project was

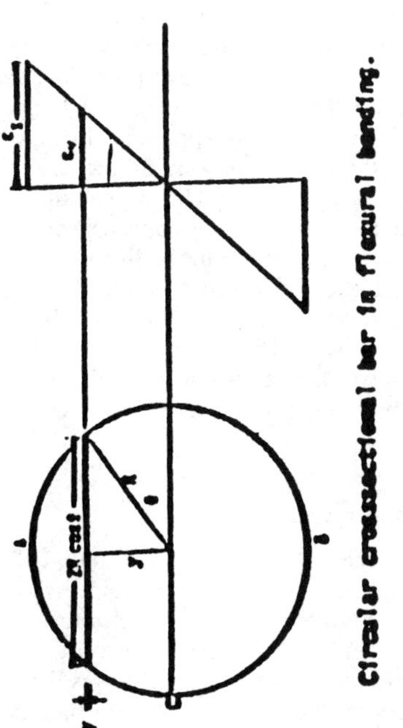

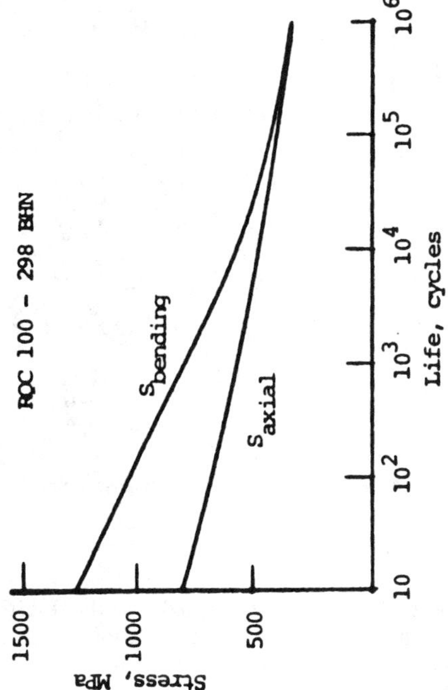

FIG. 15—*Closed-form conversion of axial fatigue to bending fatigue (collaborator: U. Muralidharan).*

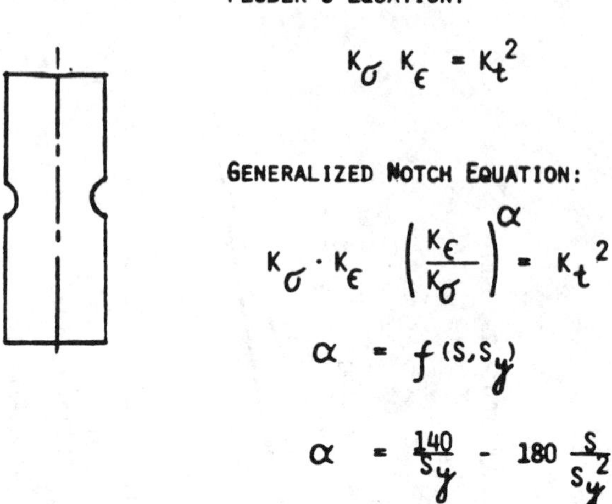

NEUBER'S EQUATION:
$K_\sigma K_\epsilon = K_t^2$
GENERALIZED NOTCH EQUATION:
$K_\sigma \cdot K_\epsilon \left(\frac{K_\epsilon}{K_\sigma}\right)^\alpha = K_t^2$
$\alpha = f(S, S_y)$
$\alpha = \frac{140}{S_y} - 180 \frac{S}{S_y^2}$

FIG. 16—*Correction to Neuber's equation for notch analysis in the plastic range (collaborator: B. Bora).*

Bipin Bora. While I am not yet completely satisfied with the result, a tentative expression for α is given in Fig. 16 which indicates that it depends on both the nominal stress S and the yield stress of the material S_y.

New Damage Equation for Cumulative Fatigue Analysis

Figure 17 shows another area of research that has recently been of special interest to us. Starting in 1963, at the First International Conference on Fracture, Sendai, Japan, we have been concerned with the development of a cumulative damage procedure based on a double-linear damage rule (DLDR). Later, in 1980, Dr. Gary Halford and I modified the original DLDR formulation and also formulated a damage curve approach. Both methods treated the cumulative damage in independent ways but still gave answers that were similar to each other. This year, on the invitation of the International Union of Theoretical and Applied Mechanics (IUTAM) to present to them a status report of engineering approaches to cumulative fatigue analysis, we took the opportunity to unify our past efforts in the independent DCA and DLDR methods. With the collaboration of Halford, the results shown in Fig. 17 were obtained. Figure 17*a* shows how the DCA relations were modified to make them nearly coincident with the DLDR equations. The modified damage curves, which are expressed analytically by two terms, we called DDCA (double-damage curve analysis). The greater flexibility of the DDCA approach did, however, permit us to fit certain experimental data better than we had been able to do previously (Fig. 17*b*).

Improved Formulation for Treating Mean Stresses

In working with cast iron for his Ph.D. thesis, Kurt Heidmann found that, because of the immense difference between the flow properties in tension and compression, all the hysteresis loops generated contained mean stresses. We therefore concerned ourselves with the effect of such mean stresses on fatigue life. The framework was some prior work by Halford and me in our consideration of cumulative fatigue damage, but the approach was limited to a Goodman-type diagram proposed by Morrow in which a straight line results in figures such as Fig. 18*b*. His results did not always show a straight line, so we set about to generalize the formulation.

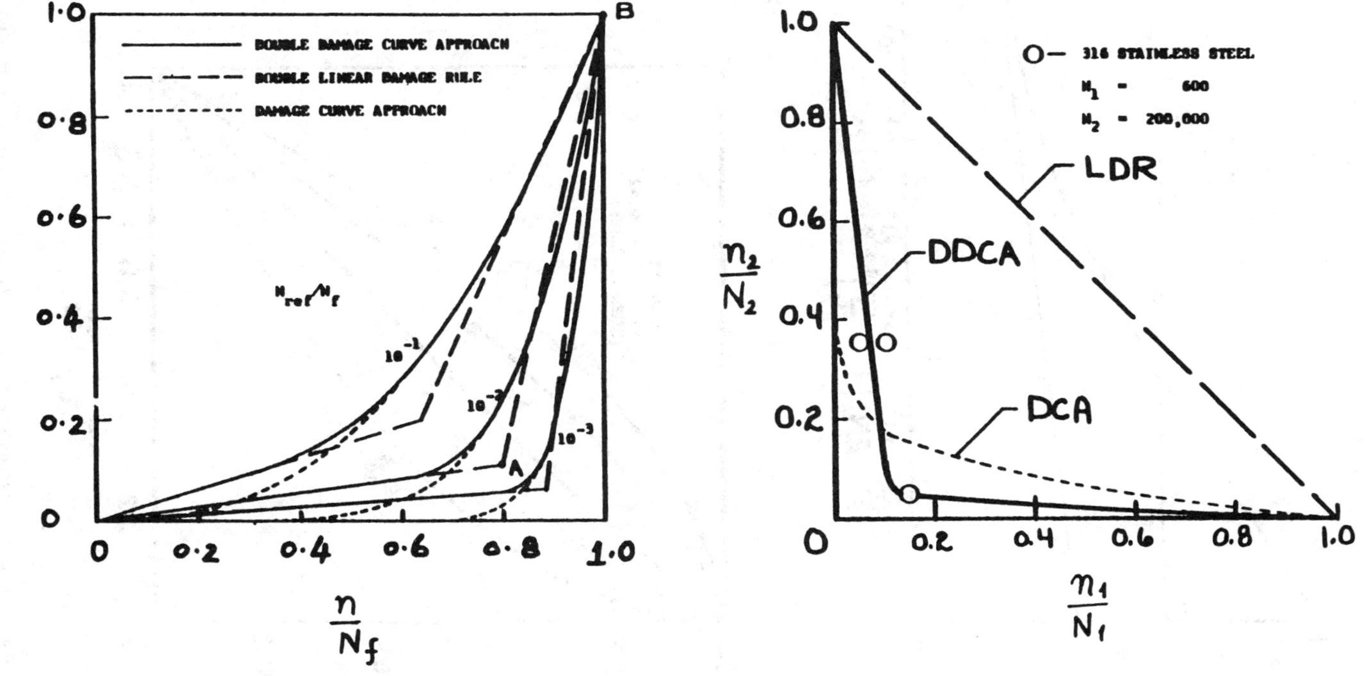

FIG. 17—*New damage equation for cumulative fatigue analysis (collaborator: G. R. Halford).*

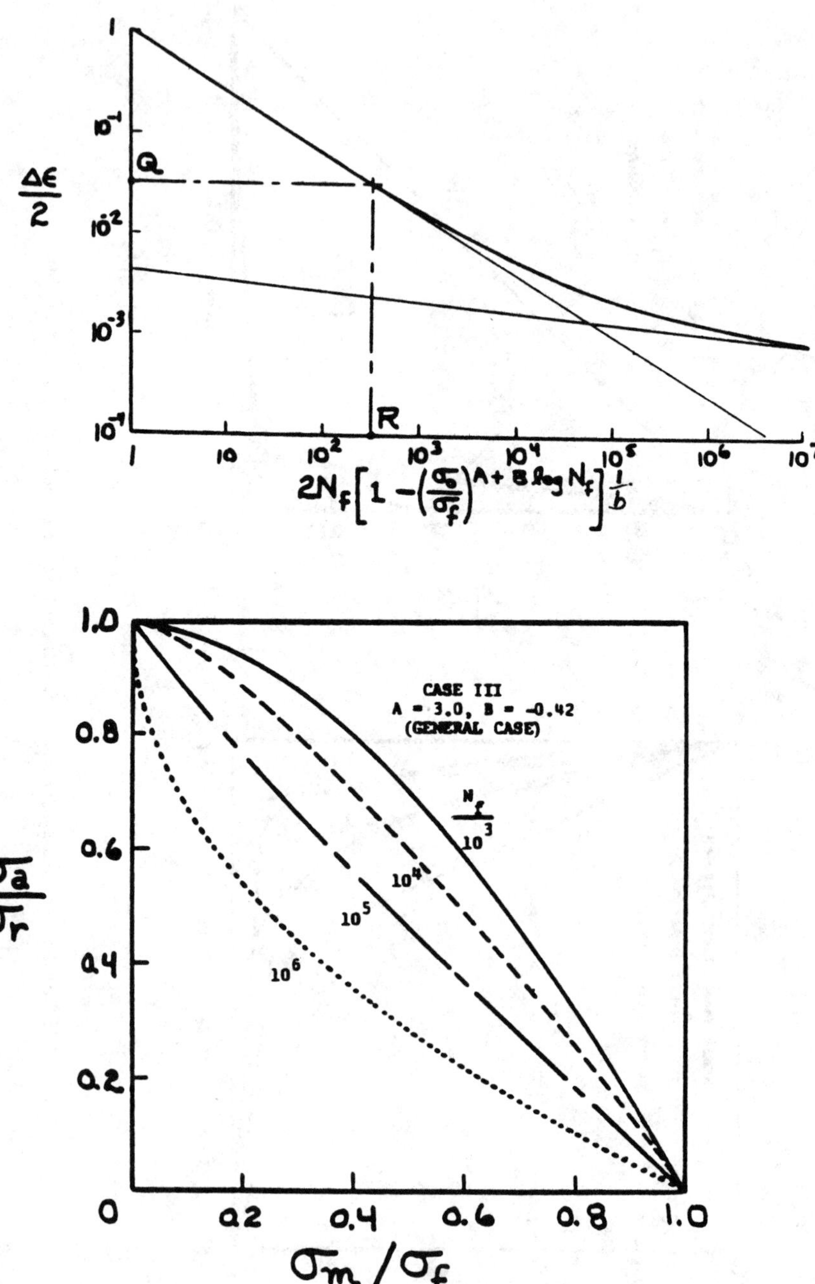

FIG. 18—*New formulation for mean stress effects in fatigue (collaborator: K. Heidmann).*

The result is indicated in Fig. 18*a*. The material characterization is identified for completely reversed loading, consisting of the standard elastic and plastic lines. However, the horizontal life scale is revised to contain a multiplier that includes mean stress. There are two material constants *A* and *B* which can be determined by relatively few tests. Once these constants are known, mean stress effects can be determined at any life level. Different characteristics can result in different life levels (Fig. 18*b*). The Goodman-type diagram can be convex, straight, or concave as life changes from 10^3 to 10^6.

Thermo-Mechanical Fatigue

Thermal fatigue has been a subject of interest for me for many years. Presently we have been using the SRP framework. By defining CP and PC strains as only those containing steady-state creep components, and using the simplified formulas shown in Fig. 19*a*, we have been able to

TENSION: $\dot{\epsilon}_{ss,t} = A_t \, \sigma^{n_t} e^{-\left(\frac{K_t}{T}\right)}$

COMPRESSION: $\dot{\epsilon}_{ss,c} = A_c \, \sigma^{n_c} e^{-\left(\frac{K_c}{T}\right)}$

$\epsilon_t = \int \dot{\epsilon}_{ss,t} \cdot dt$ (CREEP IN TENSION)

$\epsilon_c = \int \dot{\epsilon}_{ss,c} \, dt$ (CREEP IN COMPRESSION)

ϵ_t AND ϵ_c ARE USED TO OBTAIN

$\Delta\epsilon_{CP}$, $\Delta\epsilon_{PC}$, AND $\Delta\epsilon_{CC}$

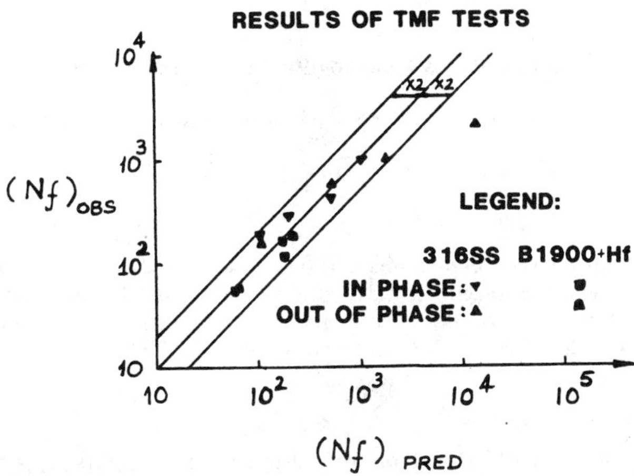

FIG. 19—*Simplified formulation for thermomechanical fatigue analyses in the creep range (collaborators: A. Seren and B. Bora).*

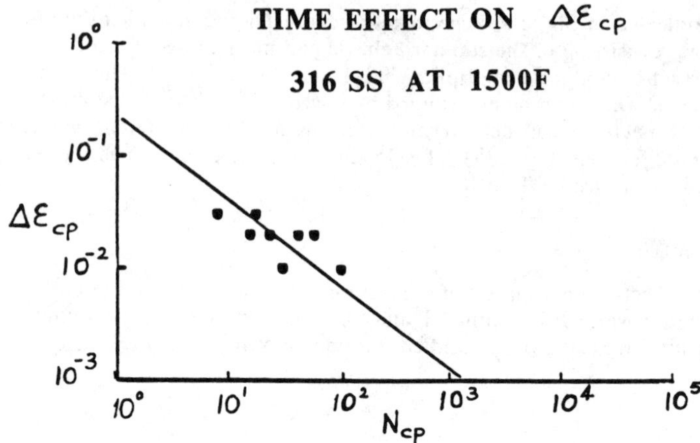

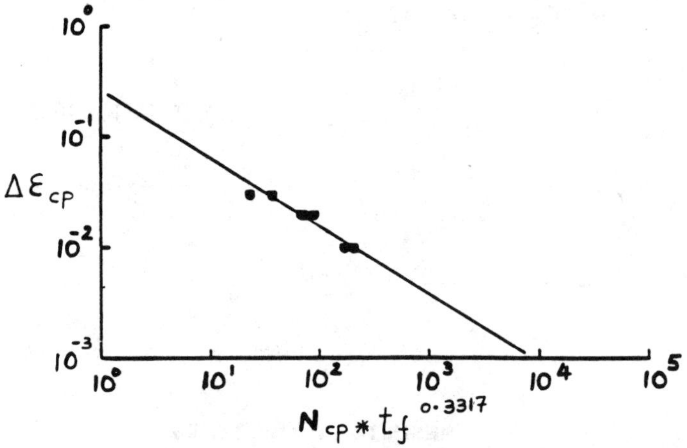

FIG. 20—*Time effects in strainrange components in SRP analyses for long-time application (collaborator: S. Kalluri).*

obtain good correlations between predictions and test data, an example of which is shown in Fig. 19*b*. Two materials are involved here: 316 stainless steel and the nickel-base superalloy B1900 + hafnium. Over the years several students have conducted experiments in this field, but lately it has been Ayden Seren and Bipin Bora, as noted in the figure.

Time-Effects on SRP Components

When we first developed the SRP method we considered that as a first approximation the time factor was already included in the concept that creep requires time. More recently we have been examining the time factor independently, since some of the practical applications would involve very long-time applications when environmental interactions and metallurgical transformations might take place. This subject became the M.S. and Ph.D. focus of Ramesh Kalluri. One of the results is shown in Fig. 20, which relates to 316 SS. In Fig. 20*a*, the Manson-Coffin plot contains considerable scatter when the time factor is not included, but Fig. 20*b* shows a much tighter line when time is included. This figure refers only to the CP strainrange component, but the other components have been studied as well.

Cyclic Deformation

J. Polák,[1] *J. Helešic,*[1] *and M. Klesnil*[1]

Effect of Elevated Temperatures on the Low Cycle Fatigue of 2.25Cr-1Mo Steel—Part I: Constant Amplitude Straining

REFERENCE: Polák, J., Helešic, J., and Klesnil, M., **"Effect of Elevated Temperatures on the Low Cycle Fatigue of 2.25Cr-1Mo Steel—Part I: Constant Amplitude Straining,"** *Low Cycle Fatigue, ASTM STP 942,* H. D. Solomon, G. R. Halford, L. R. Kaisand, and B. N. Leis, Eds., American Society for Testing and Materials, Philadelphia, 1988, pp. 43–57.

ABSTRACT: A 2.25Cr-1Mo steel was subjected to constant amplitude straining in the temperature interval 22 to 550°C. The conventional low-cycle fatigue properties at 22, 350, 450, and 550°C at constant diametral strain rate $\dot{\varepsilon}_d = 5 \times 10^{-3}\ s^{-1}$ were measured. The fatigue hardening-softening curves, cyclic stress-strain curves, and Manson-Coffin curves were obtained at each temperature. A detailed study of the cyclic stress-strain response was performed on a computer-controlled electrohydraulic machine. The temperature dependence of the cyclic yield stress was measured using a multiple step test procedure. The stress-dip procedure in cyclic straining was adopted in order to separate the effective and internal component of the cyclic stress. The dependence of both components on temperature and during fatigue life was measured.

The effective stress component was found to be relatively high in this material and drops only mildly with temperature. The typical temperature dependence of the cyclic stress is due to the internal stress that drops quickly above 450°C and at 550°C is considerably lower than the effective stress. With continued cycling, the decrease of both components contributes to cyclic softening.

Both the cyclic stress-strain response and fatigue life were affected by the strain rate and hold periods within the cycle. The time-dependent effects were studied in detail at 550°C. The frequency dependence of fatigue life at different strain amplitudes was measured. In addition to continuous straining, three basic types of hold times within the cycle with strain or stress relaxations were applied. The experimental data were analyzed, and a set of parameters to correlate the low-cycle fatigue life of the material with various straining conditions was evaluated.

KEY WORDS: low-cycle fatigue, cyclic stress-strain response, fatigue softening, fatigue life, hold time, 2.25Cr-1Mo steel

Two important factors determine the low-cycle fatigue resistance of metallic materials: the level of the cyclic plastic stress-strain response and the resistance to crack initiation and propagation in strain cycling. Both factors are temperature and time dependent.

In constant amplitude straining the cyclic plastic stress-strain response is usually characterized by the fatigue hardening-softening curves for different total or plastic strain amplitudes. Where the long-term mechanical behavior exhibits or is close to saturation the concept of the cyclic stress-strain curve (CSSC) can be used. Although these characteristics were measured for a number of metallic materials, information on the sources of the fatigue strength and its temperature dependence is not satisfactory.

The crack initiation and propagation leading to fatigue fracture is a complex process including fatigue and creep damage in interaction with the environment. A very useful integral char-

[1]Institute of Physical Metallurgy, Czechoslovak Academy of Sciences, 616 62 Brno, Czechoslovakia.

acteristic of fatigue life is the Manson-Coffin curve [*1,2*]. However, in high-temperature cyclic straining a single Manson-Coffin curve did not prove to be satisfactory in predicting fatigue life. The frequency compensated concept [*3*] or the strainrange partitioning method [*4,5*], introducing complementary parameters that characterize the fatigue life of materials under various high-temperature fatigue conditions, have been proposed.

In order to investigate systematically the effect of elevated temperatures on both the cyclic plastic stress-strain response and the fatigue life of metallic materials, low alloy 2.25Cr-1Mo steel was chosen because it is used for numerous components working at temperatures up to 550°C. It is widely applied in the power generation industry; therefore several investigations have already been undertaken [*5–12*] aiming to acquire its high temperature fatigue resistance at different loading conditions.

This contribution reports on the conventional low-cycle fatigue properties of this steel at elevated temperatures, the detailed investigations of the sources of the cyclic stress of this material, and the effect of the time factor on the fatigue life in constant amplitude straining. The study on the behavior of the same material under variable amplitude straining at elevated temperatures is reported in a companion paper [*13*].

Experimental Procedure

The batch of material from which the specimens were made was a plate of 15313.5 steel (2.5Cr-1Mo) having dimensions 1800 by 4200 by 40 mm^3. It was in the normalized (950°C) and tempered condition (tempering at 680°C for 1 h followed by air cooling). Thermal treatment resulted in about 95% tempered bainite and the rest proeutectoid ferrite. Tension tests were performed on cylindrical specimens 8 mm in diameter and 45 mm in gage length. The specimens for cyclic straining had a diameter of 10 mm and a gage length of 5 mm. The gage section was fine ground.

The tension and cyclic tests were conducted in an electrohydraulic closed-loop testing system. The tension tests were run in the stroke control, the cyclic tests with the control of the diametral strain using a diametral extensometer. The constant total amplitude tests were run with a constant diametral strain rate $\dot{\epsilon}_d = 5 \times 10^{-3}\ s^{-1}$. The constant plastic strain amplitude tests, including stress-dip tests and strain relaxation within the loop, were accomplished in an MTS computer-controlled machine under programs written in Basic. In the majority of these tests the diametral strain rate was held constant at $\dot{\epsilon}_d = 2.5 \times 10^{-3}\ s^{-1}$.

The specimens were heated in air in a split resistance furnace. The temperature was controlled using two thermocouples attached symmetrically at a distance of 15 mm from the middle of the specimen. Using this arrangement the temperature of the gage section of the specimen could be held within $\pm$ 2°C even in long-term tests.

The stress-dip procedure for determination of the effective stress in cyclic straining has been applied in room temperature cyclic straining and is described in detail elsewhere [*14*]. It can also be applied in elevated temperature cyclic straining with the control of diametral strain. In a set of subsequent constant plastic strain amplitude cycles at one preselected value of the relative plastic diametral strain, ϵ_{rpd}, within the loop the stress was reduced by $\Delta\sigma$ (different in successive loops), the strain relaxation for a time period of 8 s was followed, and the mean strain relaxation rate $\bar{\dot{\epsilon}}_d$ was calculated; all this was done both in tension and compression. The mean relaxation rate was plotted versus the stress drop, $\Delta\sigma$, and the critical stress drop, $\Delta\sigma_0$, corresponding to the actual zero strain rate was evaluated. The critical stress drop is approximately equal to the effective stress component of the cyclic stress.

Two types of experiments with hold times within the cycle were performed. In the first experiment, apparent longitudinal strain was controlled using a digital waveform generator. Hold times of a desired length could thus be introduced in strain control resulting in stress relaxations at maximum strain or at minimum strain or both. In the second experiment, the diametral plastic strain was controlled and hold times in load control in tension or in compression or both were introduced. This was accomplished in a computer-controlled testing machine under a program written in Basic. The plastic diametral strain amplitude was held constant during testing. Except for hold periods the diametral strain rate was held constant. In the case of hold-in-tension tests, the control mode was switched from strain to load at the point of the same relative plastic strain, ϵ_{rp}, in a loop. The ϵ_{rp} was selected during the first few cycles by the control program itself, according to the desired hold time. The same ϵ_{rp} was used for compression holds and tension and compression holds. During each experiment the maximum stress (both in tension and compression), hold time, and plastic strain amplitude were measured and stored.

Results

Standard tension tests and low-cycle fatigue tests with constant total diametral strain amplitudes were performed at 22, 350, 450, and 550°C with the diametral strain rate $\dot{\epsilon}_d = 5 \times 10^{-3}$ s^{-1}. The mechanical properties derived from the tension tests are summarized in Table 1. In Fig. 1 the fatigue hardening-softening curves at 450°C are shown. The fatigue hardening-softening curves for other temperatures were similar. Short-term initial hardening was always followed by long-term softening. For higher temperatures the softening took place earlier and was more pronounced.

In spite of the pronounced softening, the CSSCs at individual temperatures can be plotted using the values of stress amplitude σ_a and plastic strain amplitude at half-life. For the CSSC and the Manson-Coffin curve, the longitudinal plastic strain amplitude, ϵ_{ap}, equal to twice the diametral plastic strain amplitude, ϵ_{apd}, was used.

Figure 2 shows the CSSCs at four different temperatures. Full lines correspond to the power law approximation

$$\sigma_a = K_b \epsilon_{ap}^{n_b} \tag{1}$$

fitting reasonably well the experimental points at each temperature. The parameters K_b and n_b of the "basic" stress-strain curve were evaluated using the least-squares method (Table 1).

Figure 3 shows the Manson-Coffin plots at four temperatures. For all temperatures investigated, the experimental data fit reasonably well the generalized Manson-Coffin law

$$\epsilon_{ap} = \epsilon_f' (2N_f)^c \tag{2}$$

The parameters ϵ_f' and c, evaluated by the least-squares method, are also shown in Table 1.

To study the temperature dependence of the cyclic yield stress $\sigma_{0.2}'$ in more detail the multiple step test procedure for CSSC determination was used. The diametral plastic strain amplitude was controlled, and at each level the cumulative diametral plastic strain $\epsilon_{cd} = 1$ was reached. Figure 4 shows the stress amplitude plotted at 20 and 550°C versus the cumulative plastic strain. Except for the two smallest plastic strain amplitudes, reasonable saturation behavior was observed and cyclic stress-strain curves using the short-cut procedure could be plotted. From these plots, $\sigma_{0.2}'$ was evaluated. Figure 5 shows $\sigma_{0.2}'$ versus temperature simultaneously with the values derived from the standard tests. The cyclic yield stress, at first, decreases with temperatures above room temperature and exhibits a minimum at 200°C and a maximum at 350°C; above that temperature, it decreases progressively. In order to separate the total cyclic stress into the effective and internal stress, the stress-dip procedure in cyclic straining was applied. For diametral plastic strain amplitude $\epsilon_{ad} = 2.5 \times 10^{-3}$, the experiments were performed at different temperatures. At the highest temperature, the dependence on the number of cycles was followed. Figure 6 plots the mean diametral strain rate, $\bar{\dot{\epsilon}}_d$, after the stress drop versus the stress drop, $\Delta\sigma$, in both tension and compression at 450°C. The stress drop was

TABLE 1—*Monotonic mechanical properties and low-cycle fatigue parameters of the 2.25Cr-1Mo steel at different temperatures.*

Temperature (T), °C	$\sigma_{y0.002}$, MPa	Tensile Strength, MPa	Reduction of Area, %	Fracture Stress, MPa	True Fracture Strain	Total Elongation, %	K_b, MPa	n_b	ϵ_f'	c
22	405	543	78	1322	0.66	28	922	0.158	0.47	−0.53
350	340	476	71	979	0.53	20	946	0.165	1.65	−0.67
450	320	470	73	976	0.56	22	912	0.170	1.86	−0.72
550	295	376	85	872	0.83	24	613	0.138	4.18	−0.86

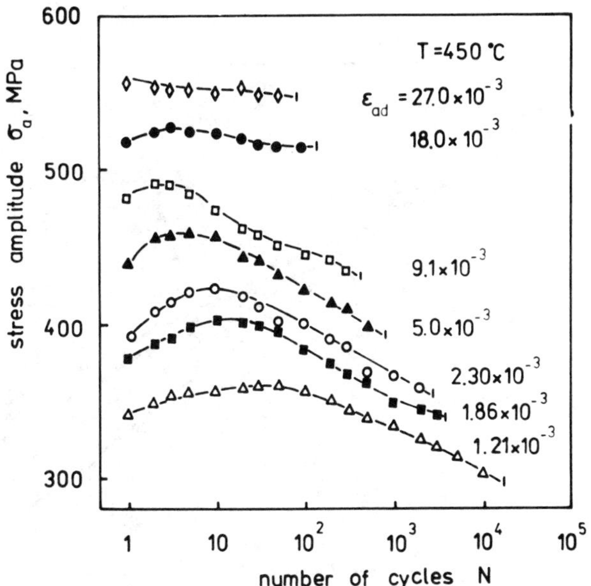

FIG. 1—*Fatigue hardening-softening curves in cycling with constant diametral strain rate* $\dot{\varepsilon}_d = 5 \times 10^{-3}$ s^{-1} *at 450°C.*

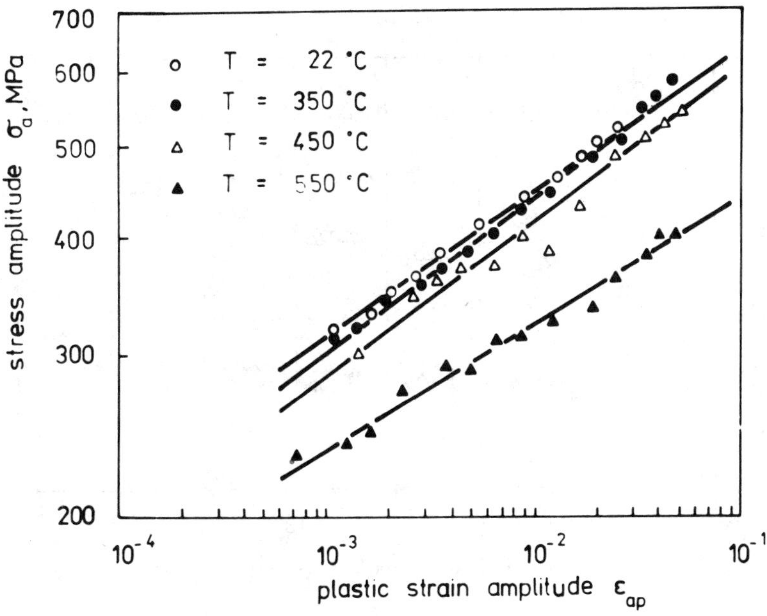

FIG. 2—*CSSCs at different temperatures.*

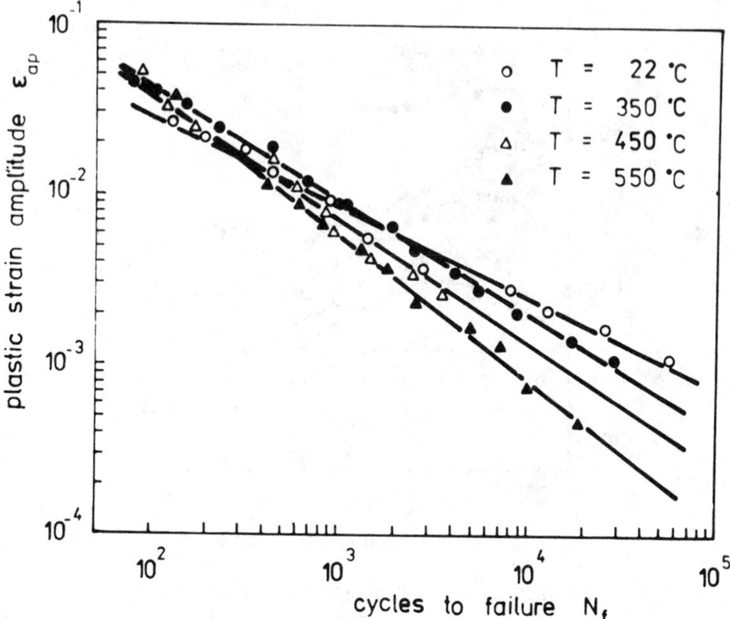

FIG. 3—*Manson-Coffin curves at different temperatures.*

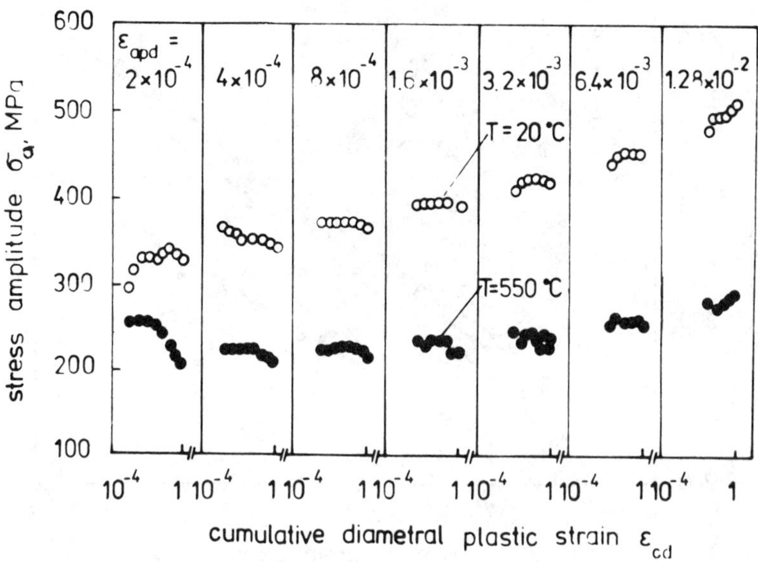

FIG. 4—*Stress amplitude versus cumulative diametral plastic strain ϵ_{cd} in multiple step tests.*

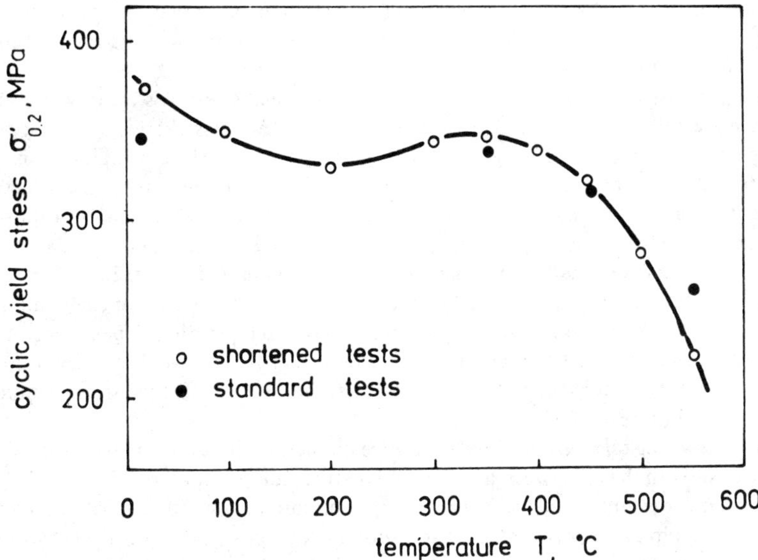

FIG. 5—*Temperature dependence of the cyclic yield stress* $\sigma'_{0.2}$.

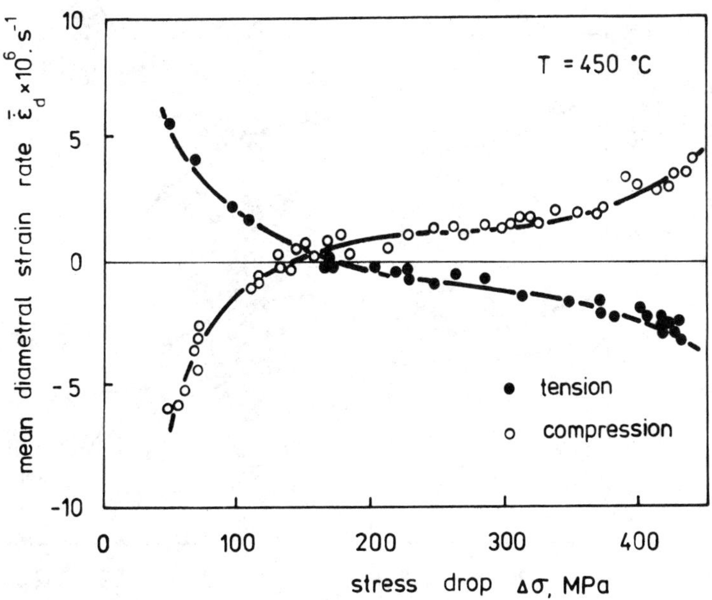

FIG. 6—*Mean diametral strain relaxation rate* $\bar{\dot{\varepsilon}}_d$ *after stress drop plotted versus the stress drop* $\Delta\sigma$ *in tension and compression.*

applied at relative diametral plastic strain in a loop $\epsilon_{rpd} = 1.8\ \epsilon_{apd}$. The point of intersection of the curves measured in tension and compression corresponds to the true zero average strain relaxation rate. The slight shift to a positive $\bar{\dot{\epsilon}}_d$ corresponds to minor cooling of the specimen during relaxation due to the smaller rate of heat generation by the slower straining; this agrees with previous results [*14*].

The temperature dependence of the effective and internal stress was measured at diametral cumulative plastic strain $\epsilon_{cd} = 2$. The critical stress drop $\Delta\sigma_0$ was obtained with highest precision at temperature 550°C, because the strain relaxation rate increases with increasing temperature. From the critical stress drop, $\Delta\sigma_0$, evaluated for each temperature, the effective stress component, σ_e, at relative diametral plastic strain, ϵ_{rp}, was calculated. From this value the effective component of the stress amplitude, σ_{ae}, was found. In Fig. 7 the total stress amplitude, σ_a, its effective σ_{ae} and internal σ_{ai} component in experiments with $\dot{\epsilon}_d = 2.5 \times 10^{-3}\ s^{-1}$ and $\epsilon_{ad} = 2.5 \times 10^{-3}$ are plotted versus temperature. The effective component is high and decreases only mildly with temperature. The characteristic temperature dependence of the total stress amplitude is due to the internal stress component σ_{ai}. Above 450°C the σ_{ai} decreases strongly with temperature; at 550°C it was less than half of the effective stress component for the applied strain rate.

Figure 8 shows the total stress σ for the relative diametral plastic strain in a loop $\epsilon_{pr} = 1.6\ \epsilon_{ap}$ simultaneously with the effective σ_e and internal σ_i stress component versus the number of cycles for cycling with constant diametral plastic strain amplitude 2.5×10^{-3} at 550°C. To make the complete stress-dip experiment, at least ten subsequent loops with stress drop had to be run. Therefore the points in the initial part of the curves in Fig. 8 represent the average values of over ten cycles. That is why the initial hardening is not apparent in this plot. Both components of the stress are decreasing simultaneously during the fatigue life.

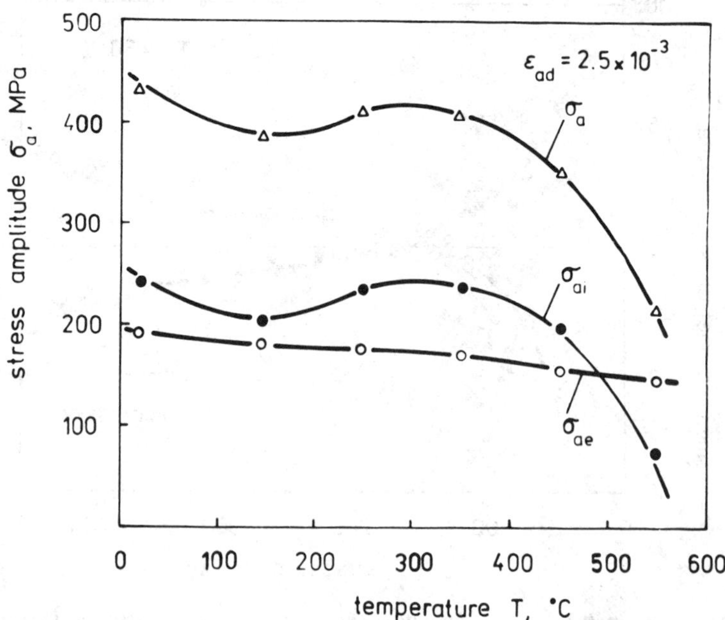

FIG. 7—*Total stress amplitude σ_a, internal stress amplitude σ_{ai}, and effective stress amplitude σ_{ae} plotted versus temperature.*

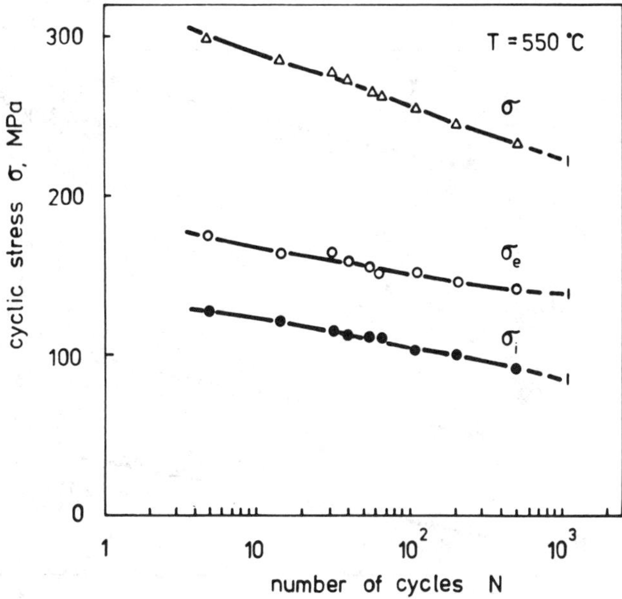

FIG. 8—*Total stress σ, internal, and effective stress components plotted versus number of cycles.*

The time dependence of low-cycle fatigue life was investigated, at all four temperatures studied, for constant diametral strain amplitude by changing the frequency. The systematic dependence of fatigue life on frequency was found for temperatures above 400°C. Therefore further measurements were done at 550°C only where the time dependence was most pronounced.

Figure 9 shows the dependence of fatigue life, N_f, on the frequency of cycling, ν, for three diametral strain amplitudes. The experimental dependence in the range of frequencies under study can be described by a simple power law

$$N_f = N_{f0}\left(\frac{\nu}{\nu_0}\right)^{\gamma} \tag{3}$$

where N_{f0} is the number of cycles to failure at a reference frequency ν_0. The exponents of the power law dependence were 0.19, 0.18, and 0.22; they increase slightly with decreasing strain amplitude.

The effect of strain hold times within the cycle was measured for four basic types of holds during cycling, as schematically depicted in Fig. 10. The apparent longitudinal strain was controlled, and the strain rate was $\dot{\epsilon} = 5 \times 10^{-3}\ s^{-1}$. Tension hold was 60 s, compression hold 60 s. When the hold time was applied both in tension and compression it was for 30 s each so that the total time to run one cycle was equal for all three types of cycles with hold times. The hold at maximum or minimum strain resulted in stress relaxation, and the plastic strain amplitude was slightly higher relative to that measured in continuous cycling with the same total strain amplitude. In Fig. 10 the fatigue lives are plotted versus the plastic strain amplitude evaluated for half-life. The full line is the Manson-Coffin curve obtained earlier (Fig. 3 and Table 1) from the standard test. The experimental points lie in a narrow scatter band, and the Manson-Coffin curve represents a reasonable description of the fatigue life. The only exception was for experi-

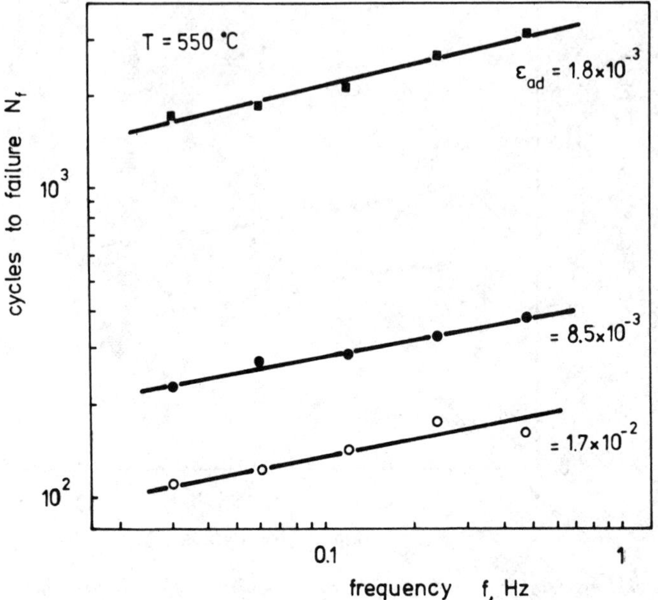

FIG. 9—*Fatigue life versus cycling frequency.*

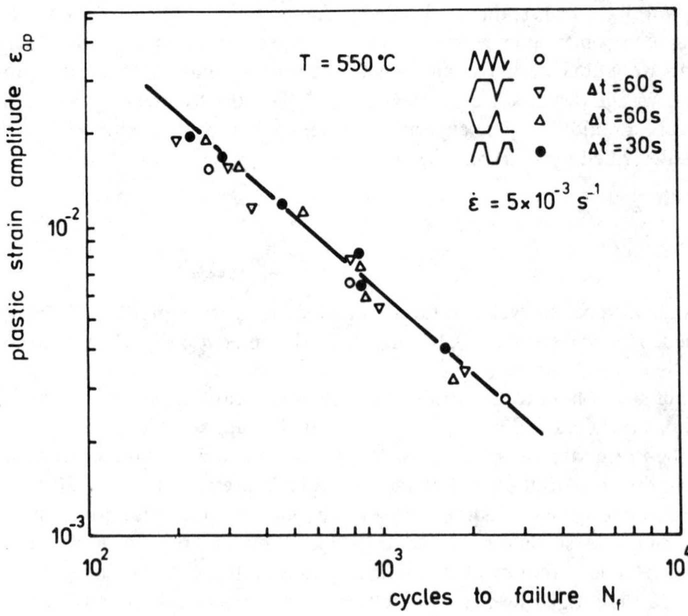

FIG. 10—*Manson-Coffin plot for cycling with strain hold times.*

ments with the hold times in tension that tend systematically to the left from the straight line. This type of hold reduces the fatigue life beyond the scatter of experimental data.

The effect of stress hold periods within the cycle on the fatigue life was measured for all four basic types of hold using the procedure described above. For plastic strain amplitude, ϵ_{ap}, and one value of relative plastic strain, ϵ_{rp}, the continuous cycling experiment and three different hold experiments were performed. As the mode switch to stress control was performed for constant ϵ_{rp}, the relaxation times during the whole fatigue life were approximately constant in spite of a strong fatigue softening.

The Manson-Coffin plots for all four types of cycling are shown in Fig. 11. The scattering in fatigue lives is surprisingly small, and the power law dependencies of the type of the generalized Manson-Coffin law could fit the individual plots. Instead of doing so, however, the strainrange partitioning method (SRP) was applied to obtain the general parameters of the fatigue life curves. The computer least-squares method as described by Hoffelner et al. [*15*] was followed. The equations and parameters are defined in Table 2. The so-called interaction damage rule was used for the calculation of the resulting fatigue life. The simpliest Monte Carlo method was used to fit the data presented in Fig. 11; the best parameter estimation is included in Table 2. The fit was very good as the final dispersion

$$s = \sqrt{\frac{\sum_{i=1}^{n} (\ln N_i(\epsilon_{ap}) - \ln N_i)^2}{n}} \tag{4}$$

had the value $s = 0.05$, which indicates that on the average the particular experimental fatigue life differs only 12% from the fitted one.

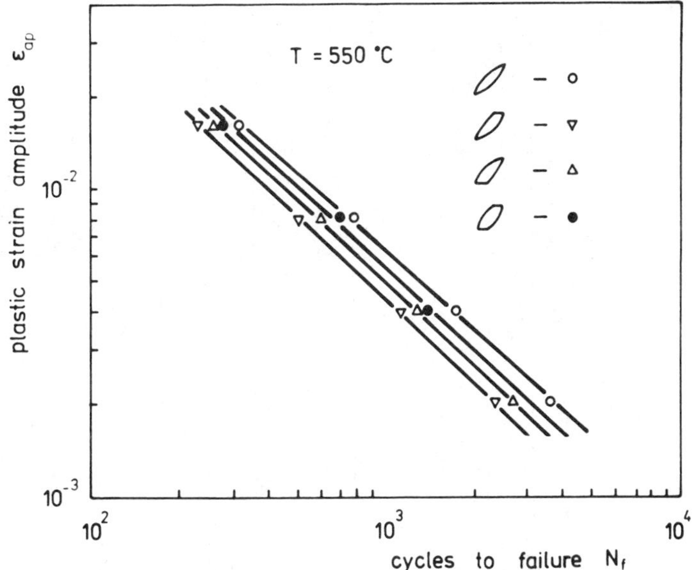

FIG. 11—*Manson-Coffin plot for cycling with stress hold times.*

TABLE 2—*Failure laws and life evaluation rule used in SRP method and parameter values obtained by least squares fit.*

Type of Cycling	pp (plastic-plastic)	cp (creep-plastic)	pc (plastic-creep)	cc (creep-creep)
Failure law	$N_{pp} = A_{pp}\epsilon_{ap}^{\alpha_{pp}}$	$N_{cp} = A_{cp}\epsilon_{ap}^{\alpha_{cp}}$	$N_{pc} = A_{pc}\epsilon_{ap}^{\alpha_{pp}}$	$N_{cc} = A_{cc}\epsilon_{ap}^{\alpha_{cc}}$
F_{ij}	$\frac{\epsilon_{rp,pp}}{2\epsilon_{ap}}$	$\frac{\epsilon_{rp,cp}}{2\epsilon_{ap}}$	$\frac{\epsilon_{rp,pc}}{2\epsilon_{ap}}$	$\frac{\epsilon_{rp,cc}}{2\epsilon_{ap}}$
Damage rule	$\frac{1}{N_f} = \frac{F_{pp}}{N_{pp}} + \frac{F_{cp}}{N_{cp}} + \frac{F_{pc}}{N_{pc}} + \frac{F_{cc}}{N_{cc}}$			
A_{ij}	2.68	2.32	2.67	2.89
α_{ij}	−1.096	−1.106	−1.102	−1.166

Discussion

Cyclic Plastic Stress-Strain Response

The elevated temperature cyclic plastic stress-strain response of the 2.25Cr-1Mo steel in standard constant amplitude testing can be represented by fatigue hardening-softening curves and cyclic stress-strain curves obtained at different temperatures. The detailed investigation of the cyclic stress and its temperature dependence revealed a high effective stress component that decreases only slightly with increasing temperatures up to 550°C. The characteristic temperature dependence of the total cyclic stress is determined by the internal stress component exhibiting a local maximum at 350°C and a sharp decrease above 450°C. This decrease of the internal stress at high temperatures is due to dynamic recovery and formation of the cellular dislocation structure with larger average cell size in comparison with that at room temperature [*16*]. The high relative contribution of the effective stress to the total cyclic stress at high temperatures results in a low exponent for the CSSC (Table 1).

In order to contribute to the explanation of the fatigue softening of this steel the stress-dip procedure was applied during fatigue life. It was found that in the course of cycling both the internal and effective stress decrease. The softening mechanism that is compatible with this finding is the precipitation of carbides rich in the alloying elements from the solid solution. This will result in a smaller effective stress due to the removal of substitution atoms from the solid solution and in simplification of the internal dislocation structure. Both of these processes are promoted by a considerable point defect production during cyclic straining and their annealing out to the sinks. The nonequilibrium point-defects accelerate diffusion and contribute to higher recovery rates [*10,17*].

Fatigue Life

By means of measurement at different temperatures the original concept of Manson and Coffin [*1,2*] was shown to correlate the fatigue life at constant strain rate with the plastic strain amplitude for 2.25Cr-1Mo steel at high temperatures. The Manson-Coffin curve is useful up to 550°C and it changes with temperature much less than, for example, the level of the stress-strain response. The time effects on fatigue life were investigated at 550°C. The quantitative evaluation of the effect of cycling frequency on fatigue life is in agreement with the frequency-modified Manson-Coffin law [*3*]. At constant plastic strain amplitude the fatigue life depends on frequency, ν, according to Eq 3. The average value of the exponent was $\gamma = 0.20$. It increases only mildly with decreasing plastic strain amplitude (Fig. 9).

The weak dependence of fatigue life on the hold periods at extreme values of strain in the Manson-Coffin plot (Fig. 10) agrees with the fact that relative plastic strain achieved during stress relaxation amounts to only several percent of the plastic strain amplitude. Therefore the ratio of creep damage in all types of the strain holds is very small. As a result only hold periods in tension reduced the fatigue life beyond the scatter of the data.

The hold periods of constant stress, however, resulted in considerable reduction of fatigue lives at all plastic strain amplitudes. In the modified procedure used, the ratios F_{ij}, defined in Table 2, were held constant during the whole fatigue life. The strain relaxation times were then approximately constant and the small scatter in fatigue lives was achieved. In the SRP method only the interaction damage rule was applied. The set of parameters derived applying the SRP method allows the prediction of fatigue life at temperature $T = 550°C$. Because of the modified experimental procedure and presumably also due to the slightly different material, the parameter values found in this study differ from those obtained by Brinkman et al. [*8*] for the same type of steel.

Brinkman et al. [7] found that for longer hold periods compression holds were more deleterious to fatigue lifetime than tensile holds; however, the set of parameters given for 538°C [8] exhibits a similar effect of compression and tensile holds on fatigue life as found here (Fig. 11 and Table 2).

Although the SRP method is considered mostly as a phenomenological description allowing one to make a prediction of fatigue life in high temperature straining with the presence of hold times, it would be important to find out the physical reason for shortened fatigue lives in cycling with different types of holds in a particular material. This task could be accomplished if the mechanism of the fatigue initiation and propagation at high temperatures were understood. Some recent results obtained at room temperature suggest that the degree of the slip inhomogeneity under various types of loading can play an important role in initiating cracks capable of further growth. The measurement of fatigue crack growth rates of short cracks in air at high temperatures [12] proved a considerable effect of environment and hold time on fatigue crack growth. In spite of the quantitative evaluations, however, the separation of the crack initiation and crack propagation stage has not yet been done. Therefore the failure criteria in the form of a generalized Manson-Coffin law proved to be a valuable tool for the evaluation of the fatigue resistance of materials in high temperature low-cycle straining.

Conclusions

The experimental study of the elevated temperature low-cycle fatigue behavior of 2.25Cr-1Mo alloy steel led us to the following conclusions:

1. The temperature dependence of the cyclic stress-strain response of the steel is given by the characteristic temperature dependence of the internal stress, decreasing steeply at temperatures above 450°C, and by the high effective stress that decreases mildly and linearly with temperature.

2. The cyclic softening in high temperature cycling is the result of the decrease of both the internal and the effective stress component. Possible mechanisms leading to both internal and effective stress drop during cycling are the removal of alloying elements from solid solution due to carbide precipitation and the enlargement of the dislocation cell structure.

3. The high temperature fatigue life is strongly affected by the strain rate and stress hold times resulting in creep deformation. The strain hold times have only a minor effect. The SRP method can be successfully applied to assess the effect of hold times, and relevant parameters were evaluated allowing fatigue life prediction of the steel under various hold time conditions at high temperature.

References

[1] Manson, S. S. in *Heat Transfer Symposium*, University of Michigan, Engineering Research Institute, 1953, pp. 9-75.

[2] Coffin, L. F., *Transactions of ASME*, Vol. 76, 1954, pp. 931-950.

[3] Coffin, L. F., *Proceedings of the Institute of Mechanical Engineering*, Vol. 188, 1974, pp. 109-119.

[4] Manson, S. S. in *Fatigue at Elevated Temperatures, ASTM STP 520*, A. E. Carden, A. J. McEvily, and C. H. Wells, Eds., American Society for Testing and Materials, Philadelphia, 1973, pp. 744-775.

[5] Halford, G. R., Hirschberg, M. H., and Manson, S. S. in *Fatigue at Elevated Temperatures, ASTM STP 520*, A. E. Carden, A. J. McEvily, and C. H. Wells, Eds., American Society for Testing and Materials, Philadelphia, 1973, pp. 658-667.

[6] Seeley, R. R. and Zeisloft, R. H. in *Fatigue at Elevated Temperatures, ASTM STP 520*, A. E. Carden, A. J. McEvily, and C. H. Wells, Eds., American Society for Testing and Materials, Philadelphia, 1973, pp. 332-342.

[7] Brinkman, C. R., Strizak, J. P., Booker, M. K., and Jaske, C. E., *Journal of Nuclear Materials*, Vol. 62, 1976, pp. 181-204.

[8] Brinkman, C. R., Strizak, J. P., and Booker, M. K., "Characterization of Low-Cycle High Temperature Fatigue by the Strainrange Partitioning Method," AGARD Conference Proceedings No. 234, 1978, Paper No. 150.
[9] Challenger, K. D., Miller, A. K., and Brinkman, C. R., *Journal Engineering Materials and Technology*, Vol. 103. 1981, pp. 7-14.
[10] Challenger, K. D. and Vinning, P. G., *Materials Science and Engineering*, Vol. 58, 1983, pp. 257-267.
[11] Challenger, K. D. and Vinning, P. G., *Journal of Engineering Materials and Technology*, Vol. 105, 1983, pp. 280-285.
[12] Skelton, R. P. and Challenger, K. D., *Materials Science and Engineering*, Vol. 65, 1984, pp. 271-288.
[13] Polák, J., Vašek, A., and Klesnil, M., "Effect of Elevated Temperatures on the Low Cycle Fatigue of 2.25Cr-1Mo Steel—Part II: Variable Amplitude Straining," this publication, pp. 922-937.
[14] Polák, J., Helešic, J., and Klesnil, M., *Fatigue of Engineering Materials and Structures*, Vol. 5, 1982, pp. 45-46.
[15] Hoffelner, W., Melton, K. N., and Wütrich, C., *Fatigue of Engineering Materials and Structures*, Vol. 6, 1983, pp. 77-87.
[16] Polák, J., Klesnil, M., and Helešic, J., *Fatigue and Fracture of Engineering Materials and Structures*, Vol. 9, 1986, pp. 185-196.
[17] Jones, W. B. and Van den Ayle, J. A., *Metallurgical Transactions*, Vol. A11, 1980, pp. 1275-1286.

A. C. Pickard[1] *and J. F. Knott*[2]

Effects of Testing Method on Cyclic Hardening Behavior in Face-Centered-Cubic Alloys

REFERENCE: Pickard, A. C. and Knott, J. F., **"Effects of Testing Method on Cyclic Hardening Behavior in Face-Centered-Cubic Alloys,"** *Low Cycle Fatigue, ASTM STP 942,* H. D. Solomon, G. R. Halford, L. R. Kaisand, and B. N. Leis, Eds., American Society for Testing and Materials, Philadelphia, 1988, pp. 58–76.

ABSTRACT: Several types of test methods are employed to evaluate the cyclic hardening behavior of engineering alloys. Three such tests are the plastic-strain limit (PSL) test, the multiple-step test (MST), and the incremental-step test (IST). In each type of test, the material experiences a different plastic-strain history prior to measurement of the saturation stress. The present paper examines effects of these different histories on cyclic hardening behavior in a number of face-centered-cubic (fcc) alloys, using all three types of test. For 316 stainless steel and a weldable Al-4.5Zn-2.5Mg alloy, behavior in the IST is similar to that in the MST. In a series of experiments carried out on Cu and Cu-Al alloys, significant differences between the different tests are observed; these are explained in terms of the dislocation substructures generated by the strain histories. Differences in behavior are also observed for Al-4.5Zn-2.5Mg in under-aged, peak-aged, and over-aged conditions; these are most pronounced for the under-aged alloy, where slip is planar.

KEY WORDS: fatigue, cyclic-hardening, copper alloys, aluminum alloys, stainless steel

The fatigue properties of engineering materials are frequently assessed by performing tests on smooth specimens under conditions of load control. The application of this information to the assessment of the fatigue life of an engineering component is relatively simple if the applied stresses in the component are below the yield stress of the material. If the stresses locally exceed the yield stress, for instance at stress concentrations, an assessment of the fatigue cycle occurring after yield must be performed before the relevant fatigue information can be selected. A number of methods are available for determining the local stress-strain conditions after yield, elastic-plastic finite-element stress analysis being probably the most versatile technique available. All these methods, however, require that a stress-strain curve for the material should be available. For approximate analysis, a monotonic stress-strain curve may be sufficient, but if significant cyclic hardening or softening occurs the analysis will be incorrect and a cyclic stress-strain curve must be obtained.

Several methods have been proposed for the determination of the cyclic hardening behavior of materials. The present paper investigates the differences between cyclic stress-strain curves generated using three of these methods for two commercial alloys, pure copper, and a series of copper-aluminum alloys of progressively decreasing stacking fault energy.

[1]Rolls-Royce plc., Aero-Engine Division, Derby, England.
[2]Department of Metallurgy and Materials Science, University of Cambridge, England.

Determination of the Cyclic Stress-Strain Curve

Table 1 lists three of the main methods (and one variant) available for determining cyclic stress-strain curves, together with the abbreviations commonly used to denote them and schematic representations of the strain waveforms applied during testing. Strictly, all the methods should be applicable only to fully reversed strain cycling. If total strain limits are used, the methods are in principle valid for zero-to-maximum strain cycling, provided that significant stress-"ratchetting" does not occur.

Plastic-Strain Limit (PSL) Test

During a plastic-strain limit (PSL) (or companion specimen) test, a specimen is cycled between *constant plastic strain limits* to failure. The stress-strain hysteresis loop is monitored during the test, and the extrema for a loop recorded at half of the life to failure of the specimen are taken as a pair of points (one tensile, one compressive) on the cyclic stress-strain curve. The full curve is obtained by testing a series of specimens, each with a different applied plastic strain amplitude.

When a smooth specimen is subjected to cyclic deformation between plastic strain limits until fracture occurs, the stress-strain hysteresis loop for many materials is observed to be constant over the life of the specimen [*1,2*]. A "shakedown period" is usually observed at the start of the test, when the stress amplitude increases or decreases (depending on whether the material cyclically hardens or softens) to the saturation value. During the "shakedown period" the cyclic dislocation structure changes until a stable structure is attained, which persists throughout the saturation region. Near the end of the test, crack propagation produces a significant reduction in area of the specimen and results in a fall in the peak tensile load required to obtain the applied strain. At low alternating strains, some modification of the dislocation structure continues to occur in the saturation period, and at high alternating strains early microcracking may confuse the situation.

The PSL method of determining the cyclic stress-strain curve is valid so long as the half-life stress-strain hysteresis loops are fully saturated; if this is not so, the curve does not describe stable behavior. Single crystals of high-purity alloys occasionally require a significant proportion of the cyclic life to saturate, but most commercial polycrystalline materials saturate quickly [*3*]. Some Cr-Mo steels when tested at high temperature also fail to saturate.

Occasionally, this type of test is performed with total strain limits applied. For materials which saturate quickly, the differences between total and plastic strain limit tests are usually small.

TABLE 1—*Methods for determining cyclic stress-strain curves.*

Type of Test	Abbreviation	Strain Waveform	Strain Limits
Constant Plastic Strain Limit	PSL, Constant PSL	To Failure	Plastic
Multiple Step, Increasing Strain	MST		Plastic
Multiple Step, Decreasing Strain	MST(D)		Plastic
Incremental Step	IST		Total

One major problem with the PSL method is that a large number of specimen tests, and hence considerable testing time, is required to determine a single cyclic stress-strain curve. For this reason several other testing methods which require fewer specimens have been suggested. Details of two of these methods are given below.

Multiple-Step Test (MST)

The multiple-step test (MST) is a development of the PSL test, but a single specimen is tested at a number of different alternating plastic strains. Firstly, a specimen is cycled between *constant plastic strain limits* until a fully saturated loop is obtained. Then the plastic strain limits are *increased,* and cycling continues until another fully saturated loop is obtained. This procedure is repeated until the specimen develops a significant crack. If the material saturates within a very few cycles, a large amount of information can be obtained from one specimen. If, however, the pre-saturation period is prolonged, the advantages over the PSL test are small. For most commercial alloys, saturation occurs in a small number of cycles and the advantages over the PSL test are significant; frequently a complete cyclic stress-strain curve (up to $\pm 1\%$ plastic strain) can be obtained from one specimen. Although the MST is usually performed with increasing plastic strain limits, the same test with *decreasing limits* is occasionally performed and is then referred to as the MST(D). Again, constant total strain limits are occasionally used in MST and MST(D), with little difference observed for materials which saturate rapidly.

Incremental-Step Test (IST)

The incremental-step test (IST) is unlike the previous two tests in that the specimen is cycled between *total strain limits.* It is popular mainly because it has the advantages of requiring few specimens (usually only one), very little testing time, and is semiautomated.

During an IST, a "pattern" of cyclic strains is repeatedly applied to the test specimen until a stable stress "pattern" is generated. The strain pattern consists of a set of strain cycles with linearly increasing and decreasing amplitude, from zero to a predefined maximum strain. The form of the strain pattern is shown in Table 1 and Fig. 1, the latter also showing a typical resultant stress pattern. The tips of the stress-strain hysteresis loops give the locus of the cyclic stress-strain curve.

The main theoretical objection to the IST is that although the strain pattern applied may generate a stable stress pattern, the resulting stress-strain curve may not correspond to that obtained by PSL tests or MST, because the dislocation structure in the specimen may be a hybrid, showing aspects of the structures corresponding to both large and small plastic strains. It has, however, been reported [*1*] that for some commercial alloys, the MST and IST give identical results within experimental scatter. The work described in the present paper was undertaken to observe cyclic hardening behavior in two commercial alloys (Type 316 stainless steel and an Al-Zn-Mg alloy in under-aged, peak-aged, and over-aged conditions) and to examine whether similar behavior was observed for copper and some high purity copper-aluminum alloys.

Experimental Methods

Cyclic tests were performed on smooth specimens of Type 316 stainless steel, a weldable aluminum alloy (Al-4.5wt%Zn-2.5wt%Mg), and copper and copper-aluminum alloys of nominal composition Cu-2wt%Al, Cu-4wt%Al, and Cu-8wt%Al. The aluminum alloy was solution-treated at 450°C, water-quenched, pre-aged for 8 h at 90°C, and then aged at 150°C for 100 min (under-aged), 700 min (peak-aged), and 5000 min (over-aged), such that the under-aged and over-aged conditions possessed approximately the same tensile yield stresses. Analysis of

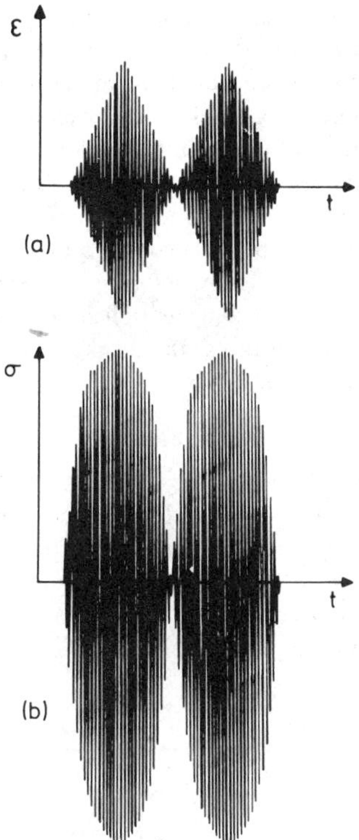

FIG. 1—*IST applied strain and stress response.*

the copper-aluminum alloys showed their true aluminum contents to be 1.98%, 4.10%, and 7.85% by weight, respectively. The average grain sizes of the pure copper and copper-aluminum alloys, measured by linear intercept methods, were 37 μm, 8 μm, 5.4 μm, and 13 μm, respectively.

In the case of the aluminum alloy, the precipitation hardening mechanisms are well known [*14*]. The copper and copper-aluminum alloys are all single-phase materials; hence solution hardening is the only hardening mechanism present. The 316 stainless steel is also single phase, and solution hardening would again be anticipated to be the only mechanism present.

The specimens were held in through-zero tension-compression grips and tested under strain control conditions using a 60-kN MAND servohydraulic testing machine operating under closed-loop control. The tests were carried out in air, at room temperature, and with strain rates on the order of 10^{-2}/s. Plastic-strain-limit and multiple-step tests were performed using a small analog computer to calculate the plastic component of the strain cycle and maintain this component at a constant value during each test or increment [*3*]. The ranges of plastic strain amplitude ($\Delta\epsilon_p/2$) investigated lie between 0.3 and 0.9% for the PSL test, with MST tests running from 0 to 0.9%. The same analog computer was used, for incremental step testing, to combine a slow zero-to-maximum envelope with a fast through-zero cycle to produce the strain waveform blocks shown in Fig. 1. In this case, *total* strain limits were used, no attempt being made to compensate for the elastic strain component. For all the materials tested it was observed that, at most, 20 blocks had to be applied before a stable stress response was obtained.

Optical micrographs were prepared for the aluminum alloy by mechanical polishing followed by light etching in 20% NaOH (aqueous) solution, and for the stainless steel by electropolishing followed by electroetching in 10% oxalic acid (aqueous) solution. Thin foils were prepared for examination in a Philips EM300 microscope from copper and Cu-8Al specimens that had been tested to failure. Disks were cut from cyclically deformed material by spark machining. The disks were thinned by light grinding, followed by jet polishing with 30% HNO_3 in methanol at −40°C until perforation occurred. In all cases, the disks were cut with the plane of the disk normal to the tensile axis of the specimen.

Results

Figures 2, 3, and 4 show the results of MST, IST, and monotonic tensile tests on the aluminum alloy in under-, peak-, and over-aged conditions, respectively. Figure 5 shows similar results for the Type 316 stainless steel. In all cases, cyclic work hardening occurs and cyclic saturation is obtained in the first few cycles. The strain plotted in Figs. 2 to 5 for the cyclic results is half of the total strain in a fully reversed strain cycle.

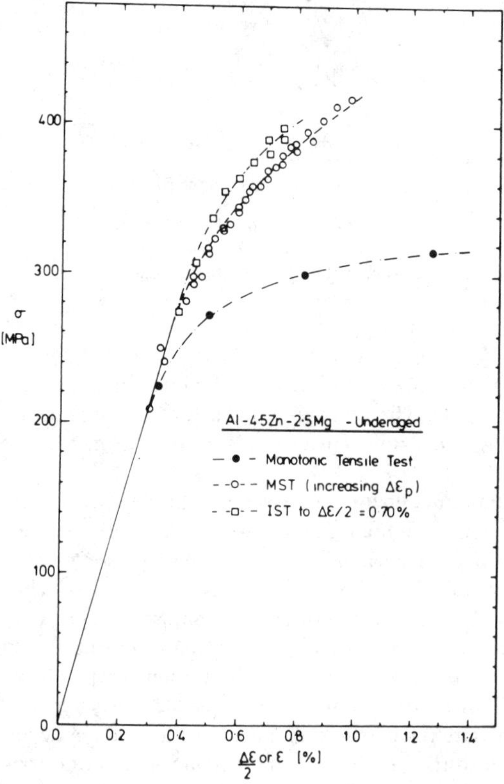

FIG. 2—*Stress-strain curves for under-aged Al-4.5Zn-2.5Mg.*

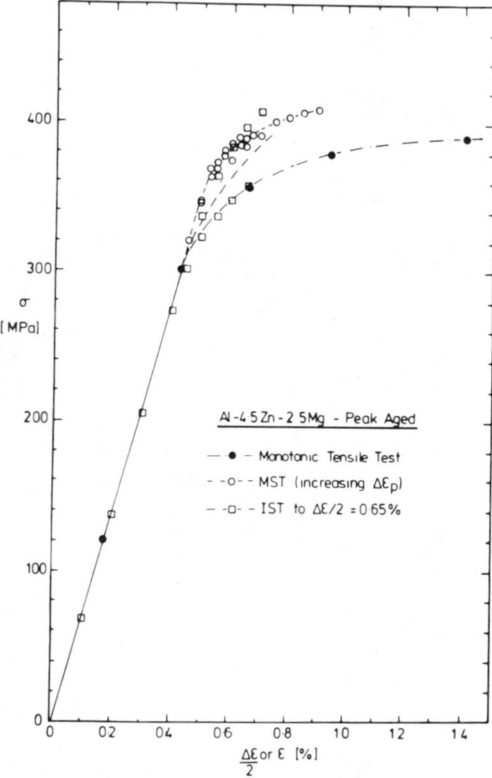

FIG. 3—*Stress-strain curves for peak-aged Al-4.5Zn-2.5Mg.*

The under-aged aluminum alloy shows considerable cyclic work hardening, and the MST curve falls below the IST results, although the differences between the two are small. During IST of this material, approximately 20 blocks had to be applied before a stable stress response was obtained. The peak-aged alloy shows less cyclic work hardening, but again the MST and IST results are similar. In this case, however, the IST curve lies below the MST results and the stress response in the IST stabilized within three blocks, although a mean stress effect is observed, such that the tensile and compressive stress-strain extrema do not fall on the same curve. Comparison of results for the under-aged and peak-aged treatments show that the IST curves for the two materials are almost coincident. The over-aged alloy shows very little cyclic hardening, the MST and IST curves are coincident throughout the cyclic stress-strain envelope explored, a stable stress response being obtained after several IST blocks. Metallographic sections of the alloy in the under-aged, peak-aged, and over-aged conditions revealed coarse and fine distributions of insoluble precipitates; these altered little with aging time. The quantity of fine precipitate appeared to increase in the over-aged material; this phenomenon is due to some of the soluble precipitates coarsening sufficiently to become resolvable optically.

The results for the stainless steel (Fig. 5) show that considerable cyclic work hardening occurs, but there is very little difference between the MST and IST results. A polished and etched section of the steel revealed both coarse and fine distributions of inclusions, the larger particles being elongated in the rolling direction. The smaller inclusions were in general spherical or cuboidal and tended to lie on the grain boundaries.

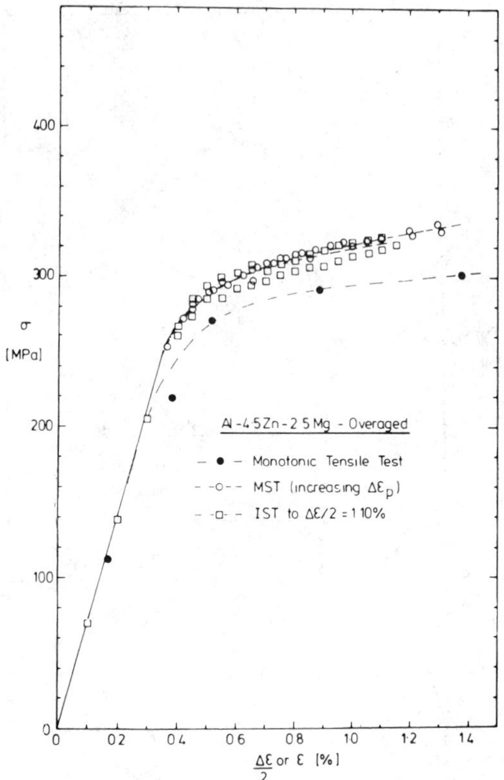

FIG. 4—*Stress-strain curves for over-aged Al-4.5Zn-2.5Mg.*

Figures 6 to 9 show the results of MST, IST, PSL, and monotonic tensile tests on pure copper and the series of copper-aluminum alloys. In all cases, cyclic hardening occurs, and saturation is observed within the first few cycles; in the case of the PSL tests, the half-life loops are all stable.

There is little difference between the MST and PSL results for pure copper. The curve generated by the IST method shows greater hardening at low strains, but coincides with the MST/PSL results at a high strain corresponding to the peak strain in the IST block. Similar results are observed for Cu-2Al, although in this case the difference between the IST and MST/PSL results is much smaller and within experimental scatter. A change in behavior is observed with Cu-4Al; for this material, the MST curve falls well below the PSL results, comparing more closely with the monotonic results at low strains. At high strains, the MST and PSL curves are observed to merge. The IST curve agrees with the PSL results at all strains. Similar results are observed for Cu-8Al, although the IST curve falls below the PSL results at strains below the IST maximum.

Figure 10 shows the dislocation structure in a copper specimen which has been cycled to failure in a PSL test; Figs. 11 and 12 show similar structures in copper specimens deformed in the MST and IST, respectively. In all cases, a dislocation cell structure is observed, with relatively little misorientation between adjacent cells. Figures 13, 14, and 15 show typical dislocation structures for Cu-8Al material that has been tested to failure in PSL, MST, and IST, re-

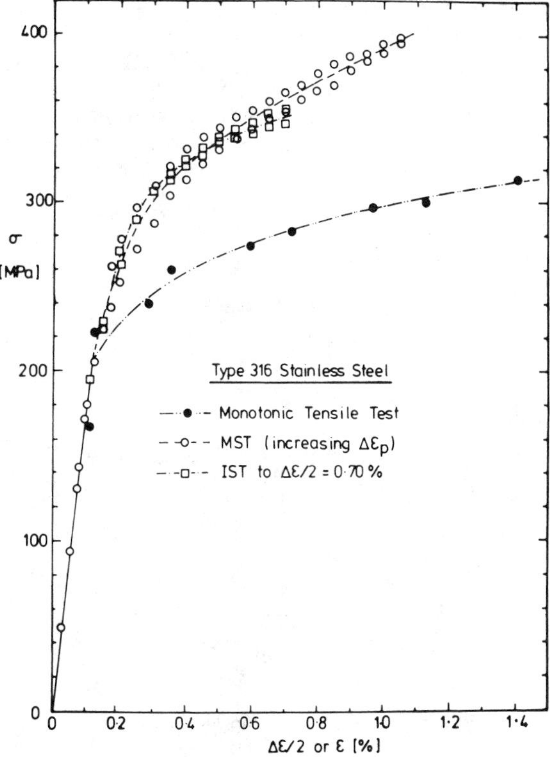

FIG. 5—*Stress-strain curves for Type 316 stainless steel.*

spectively. In all cases, a slip band structure is observed, with stacking faults clearly visible in the slip bands.

Discussion

The results of the cyclic stress-strain curve determinations for Al-4.5Zn-2.5Mg and Type 316 stainless steel show that there is very little dependence of observed cyclic hardening behavior on cyclic strain history for these commercial alloys. Thus the equality of IST and MST, which is generally assumed in determining the cyclic hardening behavior of materials [*1*], is confirmed.

Comparison of the results of MST, IST, and PSL test cyclic stress-strain curve determinations for copper and the copper-aluminum alloys reveal that the above observation is not generally true. For these materials, significant dependence of the observed cyclic hardening behavior on cyclic strain history is observed; these differences may be explained in terms of the dislocation substructures produced by cyclic deformation.

A number of investigations have been conducted to determine the dislocation structures occurring during the low-cycle fatigue of copper [*4–12*] and copper-aluminum alloys [*4–6*]. In the case of copper and Cu-2Al, the general consensus is that cross slip is relatively easy and, for polycrystalline material, cell structures are formed relatively early in the fatigue life at plastic strain amplitudes ($\Delta\epsilon_p/2$) greater than 0.1%. The cells observed tend to be approximately equiaxed, occasionally being elongated in one direction, the cell size decreasing with increasing

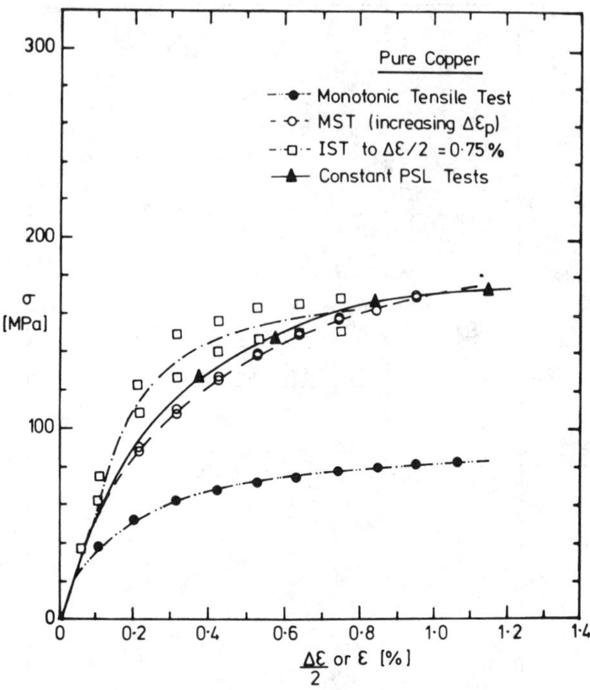

FIG. 6—*Stress-strain curves for pure copper.*

alternating plastic strain. The cells themselves contain very few dislocations, but the cell walls contain large numbers of tangled dislocations and defect structures, dipoles and multipoles being frequently observed. At very low alternating strains, there is a tendency for the cell structure not to be formed, a residual dipole structure usually being found. At high strain amplitudes, the cell structure is pronounced, with tight networks of dislocations and defects in the cell walls. The dislocation structures observed here in copper correspond to this pattern, with well-defined cell structures seen throughout (Figs. 10 to 12). Figure 16 shows the experimental variation of dislocation cell size with alternating plastic strain observed for copper, together with observations from the literature [*4,5,7*]. Also marked are the cell sizes observed for MST and IST material, plotted at the maximum alternating plastic strain applied in each test; within experimental scatter, these points lie on the basic curve for material deformed between constant plastic strain limits. This suggests that, upon cycling a polycrystalline specimen of copper, with $\Delta\epsilon_p > 0.2\%$, a dislocation cell structure is established which is related to the maximum alternating plastic strain generated during the test. Thus, during MST, the dislocation structure established at any particular strain level is that which would have been established in a specimen subjected to that strain level from the commencement of testing. In the case of IST material, however, the dislocation structure which is formed is that which occurs at the *maximum* alternating strain applied during the IST, and all flow stresses observed at lower strain levels are relative to this dislocation structure.

The observations concerning dislocation cell structures provide an explanation for the results shown in Fig. 6. Since the MST involves cycling a specimen between constant plastic strain limits until saturation occurs and then incrementing the alternating plastic strain, the cell struc-

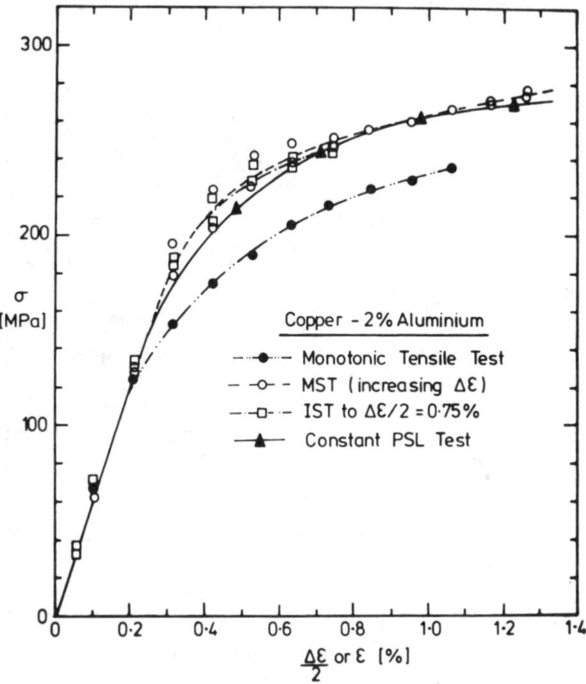

FIG. 7—*Stress-strain curves for Cu-2Al.*

ture which is present at each particular strain level is identical (in terms of cell size, at least) with that which would have been present in a PSL test at that strain. Hence the flow stress observed is also identical. During the IST, the cell structure setup is that of the *maximum* alternating strain; hence the flow stress observed at maximum strain is that which would be predicted from the MST or PSL results at that strain. At lower alternating strains, however, the cell structure does not revert to that which would occur in PSL or MST at those strains, and the flow stresses observed during IST are only slightly lower than the peak value. This finding is supported by the observation by Pratt [*8*] that when a copper specimen is cycled at one strain level for some fraction of its life and is subsequently tested at a lower strain level (in this case, $\Delta\epsilon/2 = 1.0$ to 0.5%), approximately 200 cycles are required for that saturation stress to fall from the high strain level value to within 5% of the expected low strain value. Since the alternating strain level changes continuously during IST, the cell structure will not have time to adjust to the lower strain configuration, but will retain that generated at peak strain.

The test results for Cu-2Al (Fig. 7) shows a general similarity to those for pure copper, although the MST, IST, and PSL tests all give identical results within experimental scatter. It is interesting to note that Saxena and Antolovich [*4*] observe that, in Cu-2.2Al, the rate of change of cell size with alternating plastic strain is considerably smaller than for pure copper. Thus the cell structure formed during incremental step testing will be less dissimilar to the PSL and MST structures than in the case of copper.

The dislocation structures observed during the low-cycle fatigue of Cu-4Al and Cu-8Al are less well documented than those for copper, but it is generally agreed that because of the low stacking fault energies of these two alloys, cross-slip is not easily accomplished and a planar

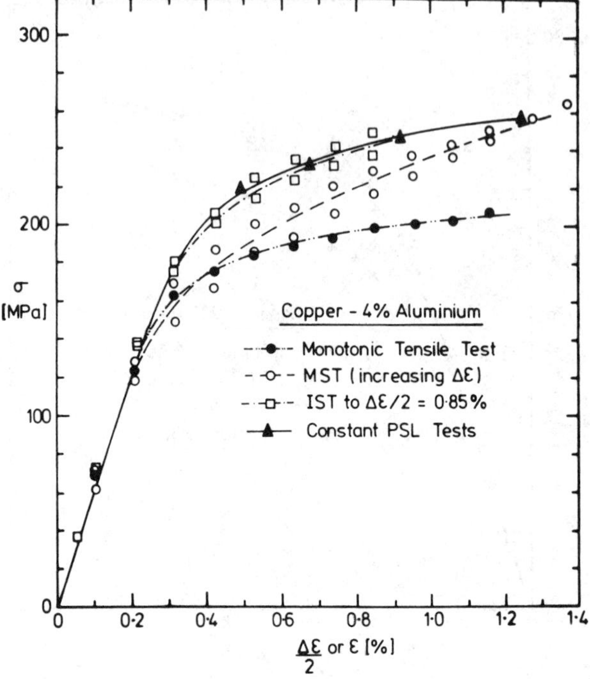

FIG. 8—*Stress-strain curves for Cu-4Al.*

dislocation structure results. At high alternating strains, there is some evidence that cell structures may be formed in these alloys, but generally the dislocations are confined to bands and tangles parallel to the trace of the primary slip planes. The structures observed in Figs. 13 to 15 for specimens of Cu-8Al tested to failure by PSL, MST, and IST are all of the planar array type, with tangled dislocation networks between the slip bands.

Figure 17 shows the variation of slip band spacing with alternating plastic strain observed in the current investigation for Cu-8Al, together with information for Cu-4.2Al [*4*] and Nimonic PE16 [*13*]. The results for MST and IST on Cu-8Al are plotted in terms of the maximum alternating plastic strain applied in each test; these points lie very close to the curve through the PSL values. These observations of slip band spacing provide some understanding of the cyclic hardening behavior of Cu-8Al shown in Fig. 9; since the slip band structures for the MST and IST specimens are similar to those for PSL tests at the maximum MST or IST alternating strain, the flow stresses for the MST and IST tests match those observed in PSL tests at this strain. However, both MST and IST show saturation stresses lower than would be expected from PSL test results at strains below the test maxima.

A possible explanation of this discrepancy can be given using an observation by Feltner and Laird [*2*] that an annealed planar-slip material (Cu-7.5Al) shows cyclic hardening, but the same material after cold working shows cyclic softening to a saturation stress higher than that observed for annealed material. The effect was not observed for pure copper, the saturation stresses being identical in both cases. Thus in the present investigation the saturation stresses observed for Cu-4Al and Cu-8Al should be strain-history dependent. In the case of a specimen deformed between constant plastic strain limits, the initial quarter cycle applied to the specimen may be considered as a tensile prestrain; the subsequent deformation behavior will depend

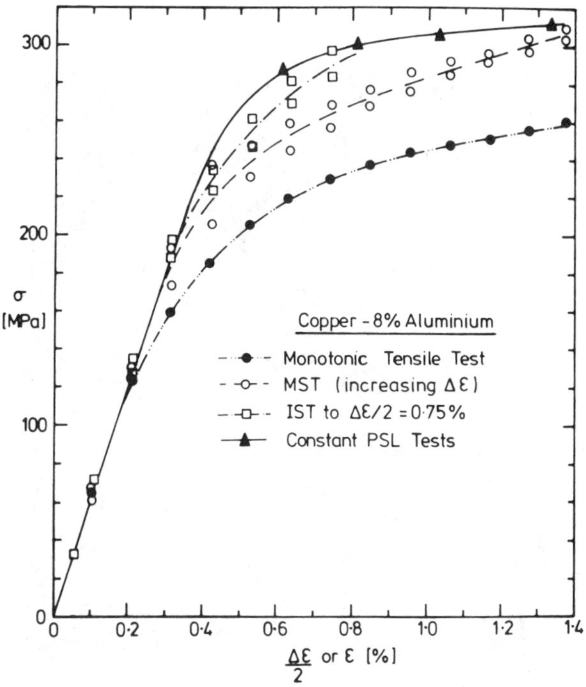

FIG. 9—*Stress-strain curves for Cu-8Al.*

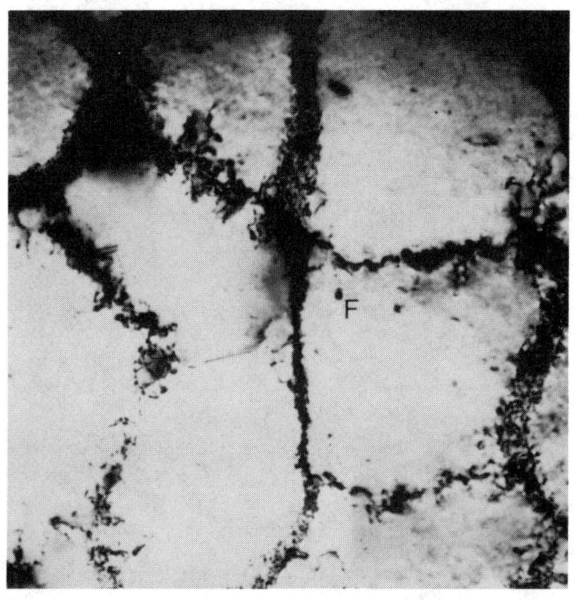

FIG. 10—*Pure copper, PSL test,* $\Delta\epsilon_p = \pm 0.9\%$, $\times 48\,900$.

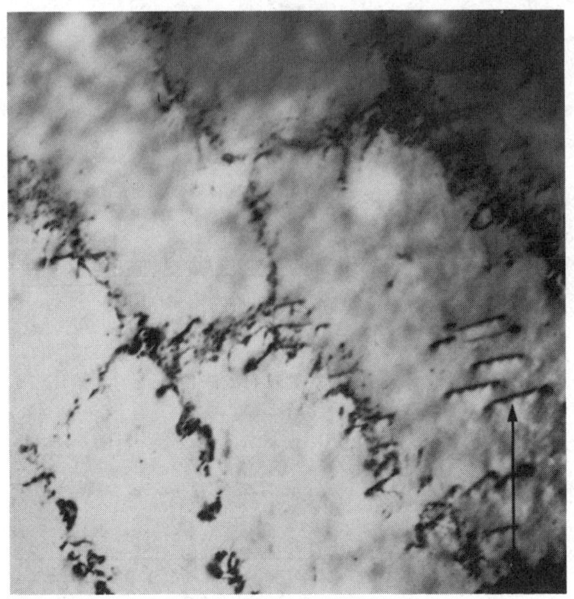

FIG. 11—*Pure copper, MST,* $\Delta\epsilon_p^{max} = \pm 0.65\%$, *×47 300.*

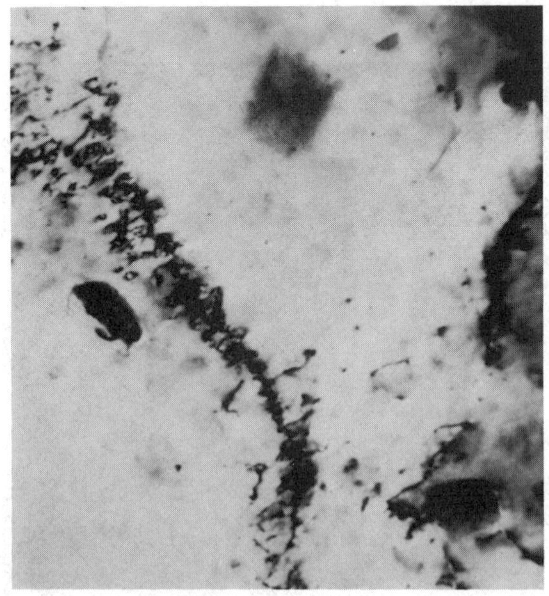

FIG. 12—*Pure copper, IST,* $\Delta\epsilon_p^{max} = \pm 0.45\%$, *×68 900.*

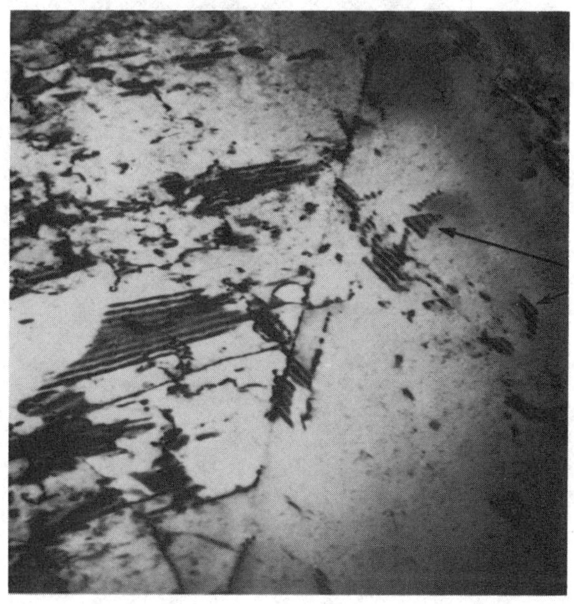

FIG. 13—*Cu-8Al, PSL test,* $\Delta\epsilon_p = \pm 0.5\%$, *×39 000.*

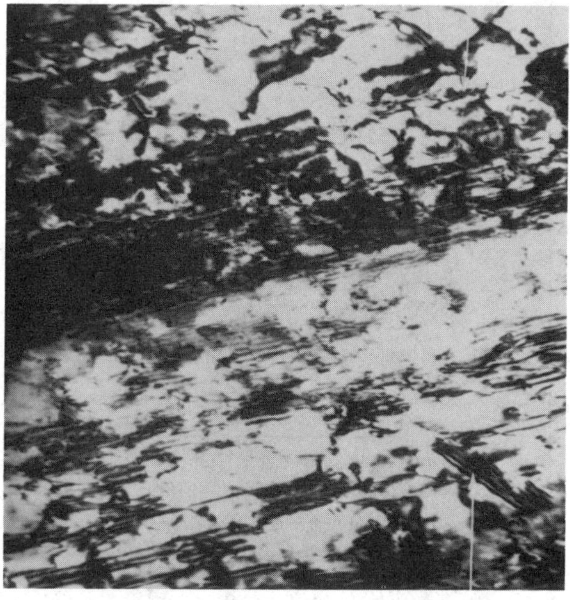

FIG. 14—*Cu-8Al, MST,* $\Delta\epsilon_p^{max} = \pm 0.9\%$, ×62 700.

FIG. 15—*Cu-8Al, IST,* $\Delta\epsilon_p^{max} = \pm 0.25\%$, *×61 500.*

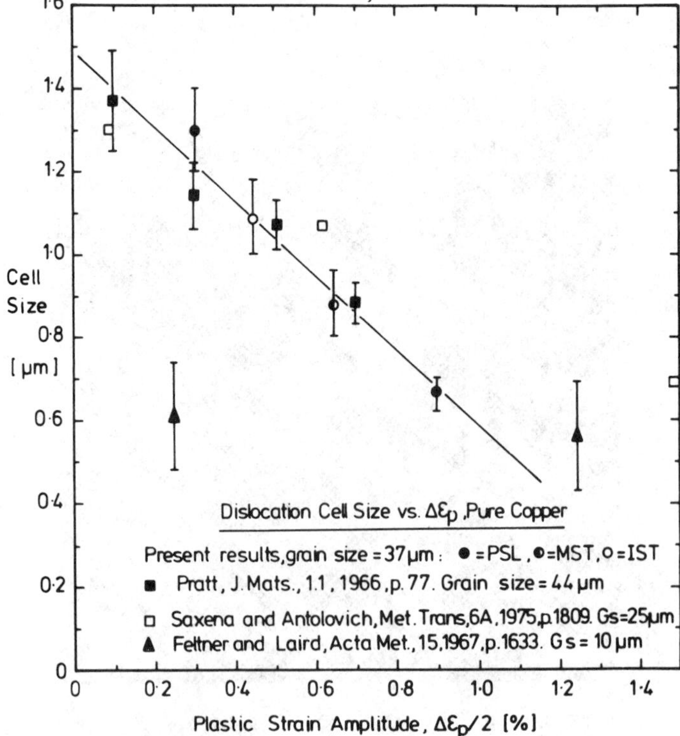

FIG. 16—*Variation of dislocation cell size with alternating plastic strain for pure copper.*

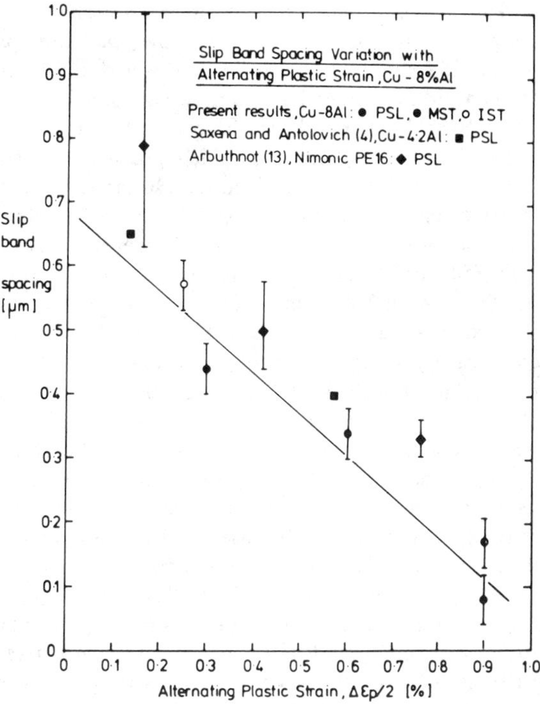

FIG. 17—*Variation of slip band spacing with alternating plastic strain for Cu-8Al.*

on the dislocation structure generated in that quarter cycle. This tensile prestrain will activate dislocation sources on certain favored slip planes, and a slip band structure of dislocations and defects will result [5]. Upon cycling, this slip band structure will adjust, by increasing the dislocation density within, and decreasing the density outside the slip bands.

During MST, the "prestrain" applied at the beginning of the test will be small; the dislocation structure formed on cycling will contain relatively few dislocation tangles, since very little dislocation motion will be necessary to accommodate the applied strains. As the plastic strain amplitude is increased the "prestrain" associated with each increase acts on material that already contains a slip band structure, which is not the situation for PSL tests. This "gentle" formation of a slip band structure should result in fewer dislocation tangles and in consequence a lower stress being required at attain the applied strain. However, as the strain is further increased in the MST, progressive dislocation tangling will occur and the flow stress will rise. Thus it may be coincidental that at the maximum MST plastic strain amplitude of ±1.4%, dislocation tangling has generated a slip band structure with flow stress equal to that occurring in PSL tests at the same strain level. Saxena and Antolovich [*4*] observed that for Cu-6.3Al, a transition from planar to cross slip dislocation structures occurs at an alternating plastic strain of ±1.7%, which is close to that at which the MST and PSL results coincide for both Cu-4Al and Cu-8Al. Arbuthnot [*13*] has made slip-band spacing measurements on Nimonic PE16 and has noted that a "cellular" type of slip band structure is formed at plastic strain amplitudes greater than ±1.1%, coincident with a change in work-hardening behavior of the material. The variation of slip band spacing with alternating plastic strain observed in this material is also very close to that observed for Cu-8Al (Fig. 17) for alternating plastic strains less than ±1%.

For an IST made on Cu-8Al, a dependence of saturation stress on strain history would again be anticipated. During the first incremental "pattern," the "prestrain" applied corresponds effectively to the maximum IST strain, since very few cycles within the pattern are performed before this strain is attained and the dislocation structure will not have saturated. Following the application of a number of IST "patterns," the dislocation substructure corresponds to that which would occur in a PSL test with strain amplitude equal to the maximum IST value. At lower strain amplitudes, more slip bands are present than in the equivalent PSL material, and the amount of slip necessary in each slip band to satisfy the applied strain will be less. Two mechanisms are thus possible: *either,* slip occurs uniformly in each slip band, but because there are more slip bands in the IST case, the stress necessary to operate each slip band is reduced; *or,* each slip band operates to its limiting slip when a threshold stress peculiar to that slip band is exceeded. As more slip bands are available in the IST material, a "weakest link" approach suggests that the stress necessary to obtain the required strain would be lower for the IST case.

The cyclic stress-strain behavior of Cu-4Al (Fig. 8) is generally similar to that of Cu-8Al. The differences between the MST, IST, and PSL results are less marked, however, suggesting that strain history effects may be less significant in this alloy. Since the stacking fault energy of Cu-4Al is significantly higher than that of Cu-8Al, this is not unreasonable.

The above observations of the relationships between cyclic work hardening behavior and dislocation substructure may be used to explain the insensitivity to testing technique of the cyclic stress-strain curves of the commercial alloys. Firstly, materials with stacking fault energy similar to that of Cu-2Al would be expected to show a similar insensitivity to testing technique; this is the case with Type 316 stainless steel. Secondly, the presence of particles which impede dislocation motion and which are not easily cut through will result in the generation of a dislocation cell structure whose characteristic size will depend on the (fixed) particle spacing rather than the applied cyclic plastic strain; this mechanism is suggested for the over-aged aluminum alloy, where both inclusions and precipitates would be expected to impede dislocation motion.

The behavior of the peak-aged aluminum alloy shown in Fig. 3 can be explained by a mechanism similar to that operating in the over-aged material. In this case, however, the precipitates will be cut through during cyclic deformation, the number of particles sheared depending on the alternating plastic strain applied. During incremental step testing the precipitate structure will correspond to that occurring at the peak plastic strain amplitude; at lower alternating plastic strains, the flow stress observed is lower than the MST flow stress, since more particles will have been sheared than in the equivalent MST case. No clear explanation can be offered, though, for the mean stress effect observed during IST of the peak-aged material.

The results for the under-aged aluminum alloy shown in Fig. 2 are directly comparable to those for the low stacking fault energy copper-aluminum alloys; in this case, cut-through of the precipitate structure leads to a planar slip deformation mechanism and considerable cyclic work hardening occurs. The differences between the IST and MST results are smaller than those observed for the copper-aluminum alloys, though, and may effectively be ignored.

Conclusions

Significant differences may occur between cyclic stress-strain curves generated using the PSL, MST, and IST techniques. The nature of these discrepancies has been shown to depend upon the slip characteristics of the material in question. For single-phase materials of low inclusion content which form cellular dislocation structures on cycling, strain history effects are minor and the differences between results generated using the three testing techniques can be explained simply in terms of the stabilized dislocation cell size occurring during the test. For planar slip materials, strain history effects cannot be ignored and the relationship between the results generated by the three testing techniques is more complex.

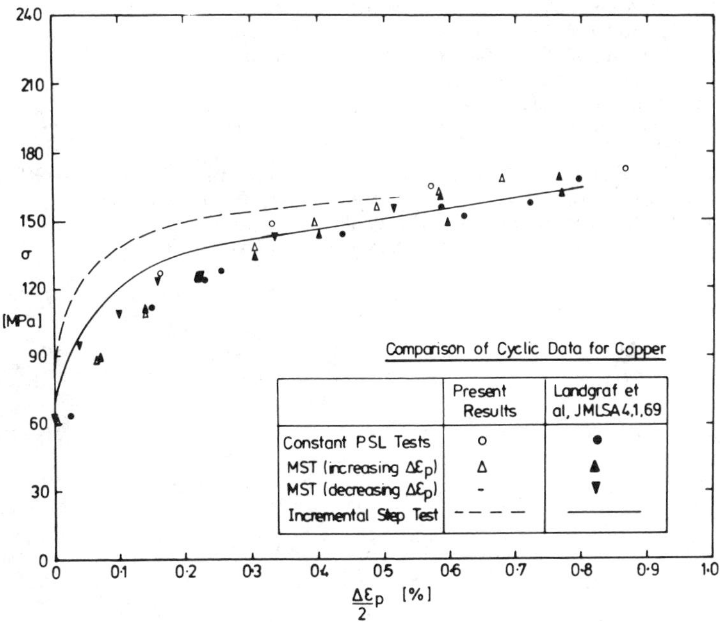

FIG. 18—*Comparison of cyclic data for copper.*

In the case of the commercial alloys investigated, the differences in cyclic hardening behavior observed between MST and IST results follow the same trends as those observed for copper and the copper-aluminum alloys, but the differences are considerably smaller and may effectively be ignored. Thus the equality of the IST and MST which is generally assumed in engineering practice is confirmed for the commercial alloys investigated. It is pertinent to note in this context that Landgraf et al. [*1*] have published the results of PSL, MST, MST (D), and IST on OFHC copper, which would be expected to contain more inclusions than the pure material used in the present work. Figure 18 compares the results for OFHC copper [*1*] and pure copper (this investigation). It is seen that the form of the difference between the IST results and the PSL/MST curve is the same, but the overall difference is considerably less for OFHC copper. The difference between the stress-strain curves for the two materials is presumably due to a difference in grain size. The results for OFHC copper support the conclusion that the discrepancies between cyclic stress-strain curves generated by the different testing techniques are small, but the results in the present paper show that this conclusion does not hold generally.

Acknowledgments

The authors wish to thank Professor R. W. K. Honeycombe, FRS, for provision of research facilities and the Science and Engineering Research Council for financial support.

References

[*1*] Landgraf, R. W., Morrow, J., and Endo, T., *Journal of Materials,* Vol. 4.1, 1969, p. 176.
[*2*] Feltner, C. E. and Laird, C., *Acta Metallurgica,* Vol. 15, 1967, p. 1621.
[*3*] Pickard, A. C., Ph.D. thesis, University of Cambridge, England, 1977.

[4] Saxena, A. and Antolovich, S. D., *Metallurgical Transactions*, Vol. 6A, 1975, p. 1809.
[5] Feltner, C. E. and Laird, C., *Acta Metallurgica*, Vol. 15, 1967, p. 1633.
[6] Feltner, C. E. and Laird, C., *TMS-AIME*, Vol. 242, 1968, p. 1253.
[7] Pratt, J. E., *Journal of Materials*, Vol. 1.1, 1966, p. 77.
[8] Pratt, J. E., *Acta Metallurgica*, Vol. 15, 1967, p. 319.
[9] Shinozaki, D. and Embury, J. D., *Metal Science*, Vol. 3, 1969, p. 147.
[10] Gostelow, C. R., *Metal Science*, Vol. 5, 1971, p. 177.
[11] Basinski, S. J., Basinski, Z. S., and Howie, A., *Philosophical Magazine*, Vol. 19, 1969, p. 899.
[12] Hancock, J. R. and Grosskreutz, J. C., *Acta Metallurgica*, Vol. 17, 1969, p. 77.
[13] Arbuthnot, C. H. D., *Proceedings of ECF4*, p. 407.
[14] Martin, J. W., *Precipitation Hardening*, Pergamon Press, Oxford, 1968.

M. Schwartz[1] *and J. C. Crespi*[1]

Fracture of Pearlite under Conditions of High Deformation Fatigue

REFERENCE: Schwartz, M. and Crespi, J. C., "**Fracture of Pearlite under Conditions of High Deformation Fatigue,**" *Low Cycle Fatigue, ASTM STP 942,* H. D. Solomon, G. R. Halford, L. R. Kaisand, and B. N. Leis, Eds., American Society for Testing and Materials, Philadelphia, 1988, pp. 77–93.

ABSTRACT: The theory of Kettunen and Kocks allows us to obtain a description of the relationship between the saturation stress amplitude, the saturation plastic strain range, and the number of cycles to fracture that fit the experimental data for fully annealed SAE 1038 steel with at least as comparable agreement as that obtained by the Coffin-Manson law. The parameters C and p of the Kettunen and Kocks relationship can be predicted fairly well from two types of tests: the monotonic stress-strain test and the incremental step test. The method of analysis through the linear multiple regression proposed by Bucher and Grozier for ferritic-pearlitic steels satisfactorily predicts the monotonic properties of the material, and the resistance to total strain cycling can be reliably predicted by the Feltner and Landgraf prediction criterion. When controlled strain amplitudes are greater than $\pm 0.37\%$ instantaneous strain hardening is observed during cyclic straining. For smaller strain amplitudes, cyclic softening occurs.

KEY WORDS: fatigue (materials), hardening (materials), softening, stress-strain, fatigue life, ferritic-pearlitic steels

Nomenclature

a_o	Mean area per obstacle
a_s	Mean slip area at saturation
B	Burger's vector
c	Fatigue ductility exponent
d	Average linear ferrite intersection
E	Young's modulus
K, K'	Monotonic strength coefficient, cyclic strength coefficient
n, n'	Monotonic strain-hardening exponent, cyclic strain-hardening exponent
$n_p(\sigma_a)$	Mobile dislocation number, function of the applied stress amplitude
N_f, $2N_f$	Number of cycles to fracture, number of reversals to fracture
$2N_t$	Number of reversals to transition point
% RA	Percent reduction in area
S_u	Ultimate tensile strength
S_{ys}	0.2% monotonic offset yield strength
S'_{ys}	0.2% cyclic offset yield strength
S''_{ys}	0.2% monotonic offset yield strength after a cyclic test

[1]Research Assistant and Head, respectively, Fatigue and Fracture Division, Gerencia Desarrollo, Departamento Materials, Comisión Nacional de Energía Atómica, 1429 Buenos Aires, Argentina.

$\epsilon_f \sigma_f/(n+1) = U_p$	True toughness
ϵ	True monotonic strain response
ϵ_p, ϵ_t, ϵ_f	True monotonic plastic strain, true monotonic total strain, true monotonic fracture ductility
$\Delta\epsilon_e$, $\Delta\epsilon_p$, $\Delta\epsilon_t$, $\Delta\epsilon_{ps}$	Elastic strain range, plastic strain range, total strain range, saturation plastic strain range
$\Delta\epsilon_e/2$, $\Delta\epsilon_p/2$, $\Delta\epsilon_t/2$, $\Delta\epsilon_{ps}/2$	Elastic strain amplitude, plastic strain amplitude, total strain amplitude, saturation plastic strain amplitude
σ	True monotonic stress response
σ_f	True monotonic fracture strength
σ_f'	Fatigue strength coefficient
$\Delta\sigma/2 = \sigma_a$	Applied stress amplitude
$\Delta\sigma_s/2 = \sigma_s$	Saturation stress amplitude
$\dot{\phi}$	Strain rate, monotonic and cyclic ($\dot{\phi} = 2 \cdot \Delta\epsilon_t \cdot f; f =$ frequency)

Introduction

The investigation of the monotonic properties of ferritic-pearlitic structures has shown that with the increment of volume of pearlite, the ultimate tensile strength increases more than the yield stress due to a higher strain-hardening of pearlite [*1*]. It has been observed in tensile tests that pearlite fracture originates with small deformations [*2*]. Cracks begin in colonies where the plates are oriented or nearly oriented in the direction of the applied stress and grow in a direction approximately 45° in relation to the tensile stress [*3*]. On the other hand, it has been shown that fatigue of high purity iron at constant temperature and cyclic strain rate [*4,5*] may produce its softening or hardening depending on the initial metallurgical condition of the metal and the deformation amplitude.

Grozier and Bucher [*6*] have studied the relationship between the fatigue limit obtained under controlled stress conditions and the microstructure and composition of ferritic-pearlitic steels. However, the cyclic behavior of these steels under controlled strain conditions and its relation with the microstructure is not yet satisfactorily described.

Kettunen and Kocks [*7–10*], working on Cu single crystals, have explained some characteristics of the fatigue phenomena, such as cyclic hardening, instantaneous work-hardening, a fatigue limit, and the relation between the saturation stress amplitude, the saturation plastic strain range, and the number of cycles to fracture, based on the statistical work-hardening concept. They propose an equation for the fatigue life which adequately describes their experimental results, considering the cumulative damage concept.

The aim of the present work was to study the relationship between the stress and strain amplitudes under controlled strain conditions, the hardening and softening caused by fatigue, and the fatigue life resistance properties of fully annealed SAE 1038 steel, and to attempt to apply the Kettunen and Kocks approach to polycrystalline materials where a second phase is present.

Material and Experimental Procedures

The chemical composition of the steel used is listed in Table 1. The SAE 1038 steel was heat treated as follows: 1 h at the austenitizing temperature of 870°C (1598°F), cooled down from 870 to 400°C (752°F) in steps of 50°C (90°F) per hour, and then air cooled. Figure 1 shows the optical micrography of 1038 steel, which is a typical microstructure of a ferritic-pearlitic steel in the fully annealed condition.

The average thickness of the cementite plates was 0.25 μm, the spacing between them 0.41 μm, and the average grain size of ferrite 22 μm. Quantitative microscopy was done in scanning electron microscopes in order to perform these observations. The volume fraction of pearlite

TABLE 1—*Chemical composition (weight percent) of fully annealed SAE 1038 steel.*

C	Si	Mn	P	S	Cr	Ni	Cu	Sn	Pb
0.39	0.35	0.75	0.019	0.009	0.19	0.13	0.21	0.03	0.02

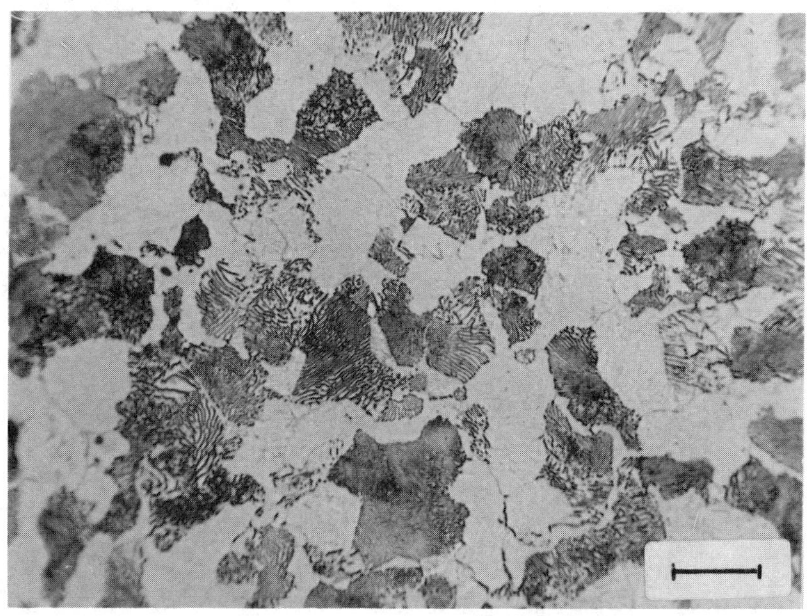

FIG. 1—*Ferritic pearlitic microstructure of fully annealed SAE 1038 steel (scale mark indicates 20 μm).*

was 0.43. The inclusions occupied 1.9% of the total area observed, and their composition was found to be mainly manganese sulfur and calcium aluminate silicate. The average HRA was 48.5.

The tests were conducted in accordance with ASTM Recommended Practice for Constant-Amplitude Low-Cycle Fatigue Testing (E 606). The diameter of the specimens for the tensile and uniaxial push-pull fatigue test was 7.5 mm, and the effective gage lengths were 35 mm and 15 mm, respectively. The fatigue specimens were mechanically polished to an average arithmetic roughness lower than 0.2 μm.

An MTS testing machine (100 kN capacity) was used to perform the monotonic and cyclic strain tests, the last of which was done under controlled total strain amplitude. The tensile tests were done with a strain rate of $\dot{\phi} = 8 \times 10^{-2}\ s^{-1}$.

The cyclic stress-strain curve was determined from decreasing and increasing multiple step testing and from incremental step testing [*11*]. In the first method one specimen was cycled at several levels of decreasing strain amplitude and increasing strain amplitude by a certain number of cycles in order to achieve stability. The cyclic stress-strain curve was then obtained by drawing a smooth curve through the tips of the superimposed stable hysteresis loops. Figure 2 shows the results of the decreasing multiple step test with five levels of strain amplitude begin-

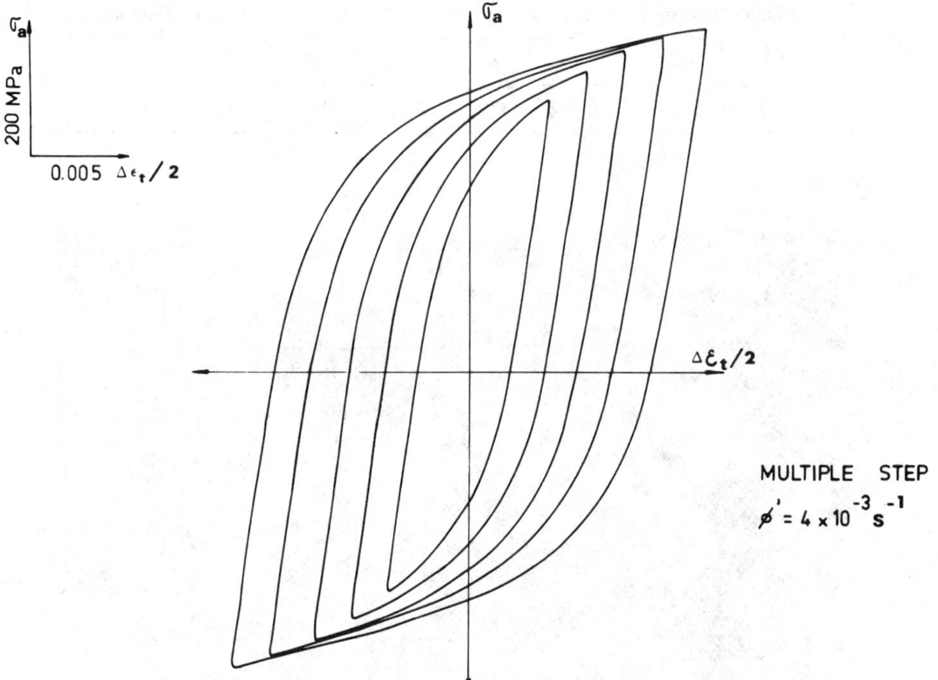

FIG. 2—*Stable hysteresis loops in decreasing multiple step test.*

ning with ±1.2%. In the second method, one specimen was subjected to a series of four to six blocks of gradually decreasing and then increasing strain amplitudes. A maximum strain amplitude of ±1.5% was usually sufficient to stabilize the material cyclically in a quick way. The cyclic stress-strain curve was then determined by the locus of the superimposed hysteresis loop tips. Figure 3 shows the initial prestrain and the first three blocks for the incremental step test; it also shows the cyclic stress amplitude response for each block of this particular cyclic strain program. Figure 4 shows the cyclic stress-strain curve generated by the locus of the 17 superimposed hysteresis loop tips in 1038 steel. These cyclic tests were done with a strain rate of $\dot{\phi} = 4 \times 10^{-3}\ s^{-1}$.

The cyclic strain amplitude-life curve was done by performing three individual tests for each of the following cyclic strain amplitudes: ±0.2, ±0.4, ±0.8, ±1.0, ±1.2, ±1.4, ±1.8, and ±2.0%. The fatigue life test under strain controlled conditions were considered to be over when the cyclic stress amplitude decreased 20% from the saturation value. In all tests minimum-to-maximum cyclic strain amplitude was equal to −1.

Experimental Results

Table 2 shows the mechanical properties of full annealed SAE 1038 steel. After averaging the results of three tensile tests, the monotonic stress-strain curve can be expressed by the relationship

$$\sigma = 994(\epsilon_p)^{0.23} \tag{1}$$

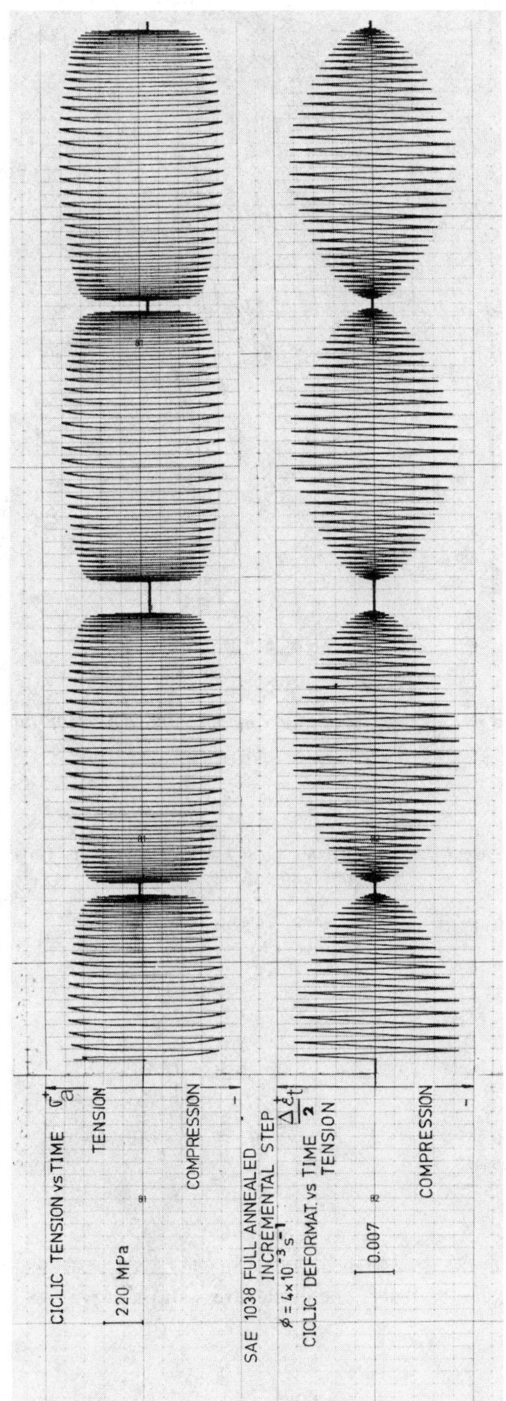

FIG. 3—*Initial prestrain and first three blocks of the incremental step test.*

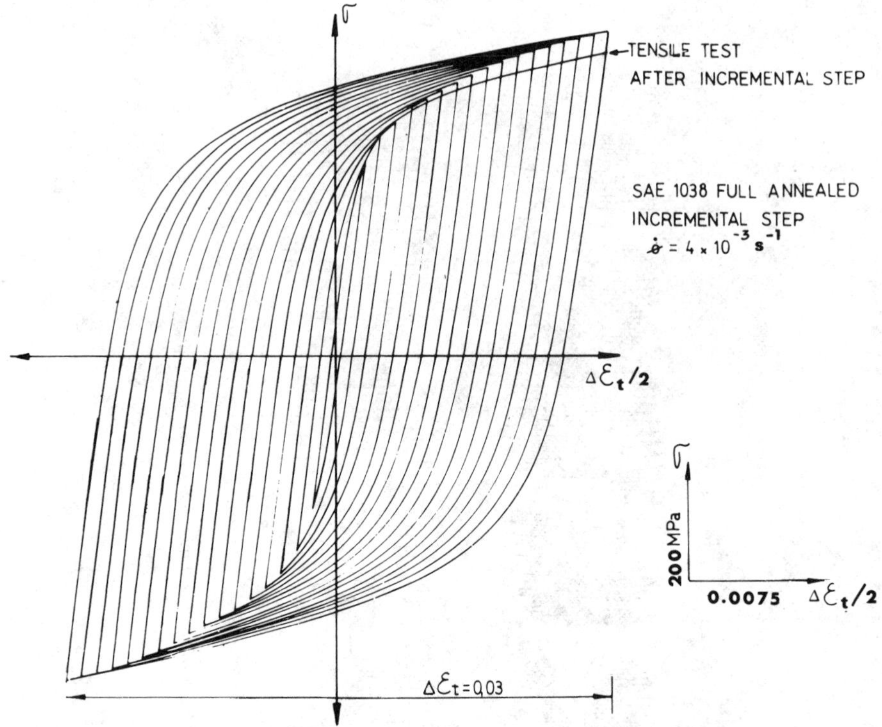

FIG. 4—*Incremental step test and tensile test after incremental step test in fully annealed SAE 1038 steel.*

TABLE 2—*Mechanical properties of fully annealed SAE 1038 steel at room temperature.*

Monotonic Properties	Cyclic Properties
S_{ys} = 347 MPa	S'_{ys} = 354 MPa
n = 0.23	n' = 0.19
K = 994 MPa	K' = 998 MPa
E = 207 GPa	b = −0.53 MPa
S_u = 610 MPa	c = −0.14
% RA = 55.5	$2N_t$ = 23 000
ϵ_f = 0.59	ϵ'_f = 0.449
σ_f = 956 MPa	σ'_f = 1094 MPa
U_p = 458 MN/m^3	

Table 2 also shows the cyclic properties of fully annealed SAE 1038 steel after averaging the results of three incremental step tests. The cyclic stress-strain curve can be expressed by the relationship

$$\sigma_a = 998\left(\frac{\Delta\epsilon_p}{2}\right)^{0.19} \tag{2}$$

Figure 5 compares the monotonic and cyclic stress-strain curves obtained by different types of tests: the incremental step test, the increasing multiple step test, the decreasing multiple step test, and the companion test. These results show that there is no appreciable difference between the various procedures to determine the cyclic stress-strain curve. Figure 5 also shows that the cyclic stress-strain curve is below the monotonic stress-strain curve for strain amplitude values lower than ±0.3% and it is above this curve for strain amplitude values higher than ±0.3%.

The hardening and softening curves were obtained from the tensile peaks of the hysteresis loops of the companion specimens tested at constant strain amplitudes till fracture (Fig. 6). Two different cyclic strain rates were used: $\dot{\phi} = 4 \times 10^{-3}\ s^{-1}$ and $\dot{\phi} = 2 \times 10^{-3}\ s^{-1}$. It was shown that this difference in the cyclic strain rates, selected to diminish the available testing time, do not introduce variation into the final results [*12*].

Rapid cyclic softening was observed for the ±0.2% strain amplitude controlled test, up to the first 100 reversals, then a slow softening followed up to 1000 reversals and finally, a very slow cyclic hardening to final rupture.

For a strain amplitude controlled test higher than ±0.4%, it was observed that the cyclic hardening rate increases as the strain amplitude increases. Furthermore, a definite saturation stress value was not observed. In spite of the absence of a definite saturation stress level for strain amplitude controlled tests higher than ±0.4%, tensile peak stress for 50% life was taken in order to obtain the cyclic stress-strain curve.

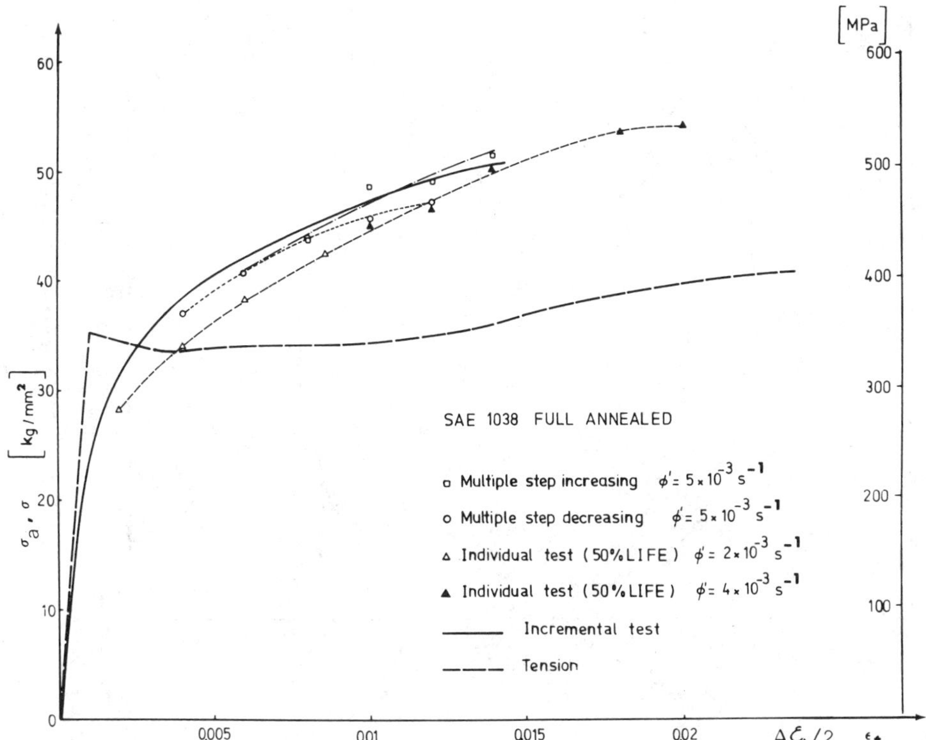

FIG. 5—*Monotonic and cyclic stress-strain curves of fully annealed SAE 1038 steel.*

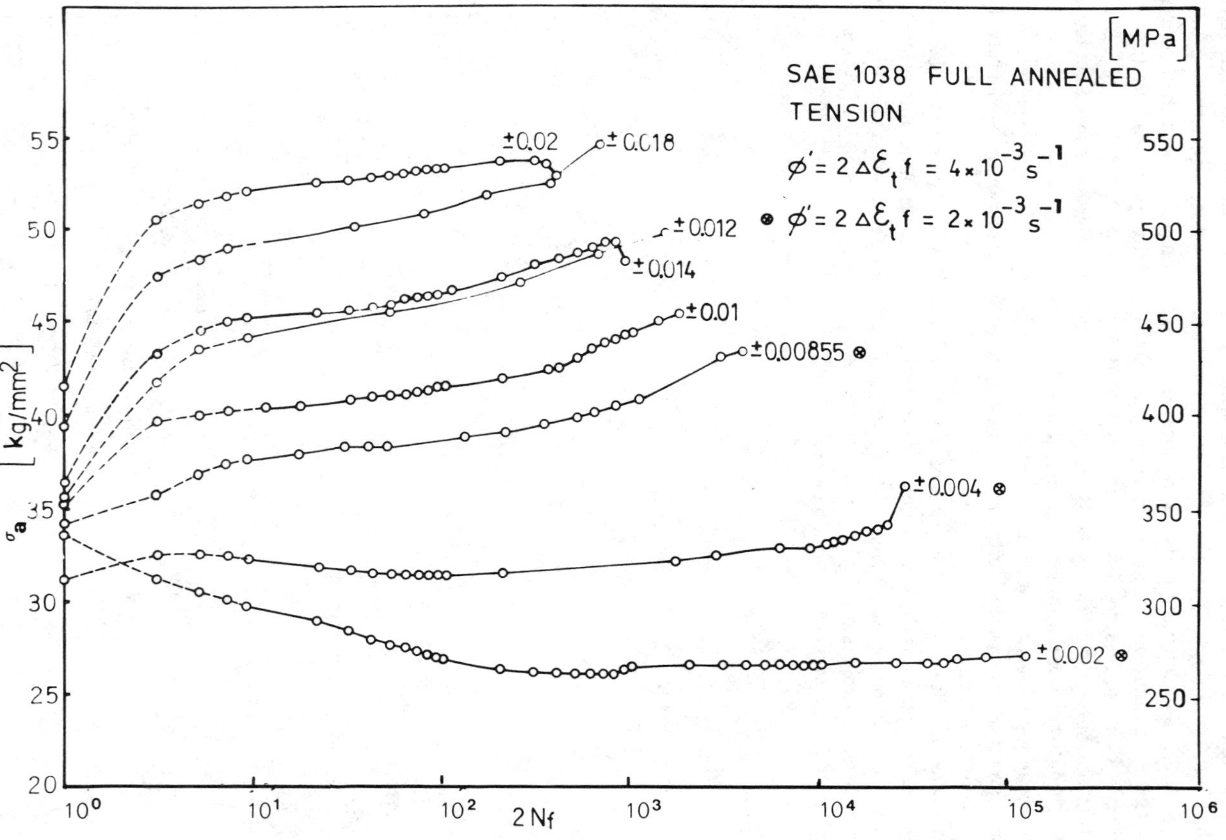

FIG. 6—*Cyclic stress behavior as a function of the number of reversals for fully annealed SAE 1038 steel.*

Bucher et al. [6,13] have found the following relationships between the monotonic properties S_{ys}, S_u, n, and K and the microstructure and chemical composition of ferritic-pearlitic steels, applying the multiple linear regression analysis to their results:

$$S_{ys}\ (\text{ksi}) = 13.29 + 5.90\ (\%\ \text{Mn}) + 10.21\ (\%\ \text{Si}) + 0.220\ (\%\ \text{volume, pearlite}) + 0.476\ (d^{-1/2}) \quad (3)$$

$$S_u\ (\text{ksi}) = 32.26 + 8.23\ (\%\ \text{Mn}) + 14.79\ (\%\ \text{Si}) + 0.627\ (\%\ \text{volume, pearlite}) + 0.34\ (d^{-1/2}) \quad (4)$$

$$n\ (\text{ksi}) = 0.32 - 0.034\ (\%\ \text{Mn}) - 0.025\ (\%\ \text{Si}) - 0.0016\ (d^{-1/2}) \quad (5)$$

$$K\ (\text{ksi}) = 85.77 + 9.93\ (\%\ \text{Mn}) + 17.65\ (\%\ \text{Si}) + 1.259\ (\%\ \text{volume, pearlite}) \quad (6)$$

Table 3 compares the predicted monotonic properties, making use of Eqs 3, 4, 5, and 6 and those obtained experimentally.

From the comparison of the calculated and experimental values, the equations proposed by Bucher et al., in order to predict the monotonic properties of the fully annealed SAE, give satisfactory results.

The fatigue-life curve can be expressed by the relationships

$$\frac{\Delta\epsilon_t}{2} = \frac{\Delta\epsilon_e}{2} + \frac{\Delta\epsilon_p}{2} = \frac{\sigma_f'}{E}(2N_f)^b + \epsilon_f'(2N_f)^c \quad (7)$$

where the elastic behavior can be transformed to $\sigma_a = \sigma_f'(2N_f)^b$ which is Basquin's equation proposed to 1910 [14] and the relation between plastic strain amplitude and life $\Delta\epsilon_p/2 = \epsilon_f'(2N_f)^c$, the Manson-Coffin relationship [15,16].

In order to predict the fatigue-life curve for full annealed SAE 1038 steel the Feltner and Landgraf strain resistance prediction criterion [17] has been adopted. Feltner and Landgraf have shown that the values of the constants ϵ_f', c, σ_f', and b can be determined from the monotonic and cyclic stress-strain curves obtained from two tests: a tension test and an incremental step test (or multiple step test), respectively. Their approach is supported on the experimentally proved relationship developed by Morrow [18] and Feltner and Landgraf. Morrow has shown through energetic considerations that c and b can be given by

$$c = \frac{-1}{1 + 5n'} \quad (8)$$

TABLE 3—*Prediction of monotonic properties of fully annealed SAE 1038 steel.*

Monotonic Property	Calculated Values	Experimental Values	95% Confidence Limits
n	0.23	0.23	...
S_{ys}, MPa(ksi)	324 (47)	347 (50)	±25.5 (3.7)
S_u, MPa(ksi)	566 (82)	610 (88)	±52 (7.5)
K, MPa(ksi)	1058 (153.5)	994 (144)	±124.5 (18)

$$b = \frac{-n'}{1 + 5n'} \tag{9}$$

Landgraf [19] has shown that ϵ_f' should be determined from the cyclic stress-strain curve:

$$\sigma_a = K'\left(\frac{\Delta\epsilon_p}{2}\right)n' = \sigma_f'\left(\frac{\Delta\epsilon_p}{2\epsilon_f'}\right)n' \tag{10}$$

If it is assumed in Eq 10 that $\sigma_f' = \sigma_f$ [18] (i.e., the true fracture stress obtained from a monotonic test) and if a 0.2% cyclic offset yield strength is introduced so that $\sigma_a = S_{ys}'$ when $\Delta\epsilon_p/2 =$ 0.2%, the result is

$$\epsilon_f' = 0.002\left(\frac{\sigma_f}{S_{ys}'}\right)^{1/n'} \tag{11}$$

The accuracy of this equation has been tested by comparing measured and predicted values for several structural steels [17]. Thus all the constants for predicting the low-cycle fatigue resistance of a variety of structural steels can be obtained from a monotonic test and an incremental step test.

The relationships for the experimental and predicted cyclic strain amplitude versus life curve (ϵ-N curve) were

$$\frac{\Delta\epsilon_t}{2} = 0.0071(2N_f)^{-0.14} + 0.45(2N_f)^{-0.53} \tag{12}$$

$$\frac{\Delta\epsilon_t}{2} = 0.0046(2N_f)^{-0.97} + 0.4135(2N_f)^{-0.52} \tag{13}$$

Both curves are shown in Fig. 7. The ϵ-N curve predicted in Fig. 7 following the Feltner and Landgraf strain resistance prediction criterion is very similar to the experimental one for fully annealed SAE 1038 steel. On the other hand, the experimental results indicate that a better adjustment can be obtained if the expression $\sigma_f' = 1.15\sigma_f'$ is selected instead of that suggested by Morrow [18].

Discussion

Kettunen and Kocks [7] explained several characteristics of the fatigue phenomena based on the statistical work-hardening theory [8]. As work-hardening is a consequence of the strength of the obstacle structure that opposes to mobile dislocation movement, the measurement of such hardening gives direct information of the changes occurring in the obstacle structure. These obstacles may be dislocations, reactions between dislocations, solute atoms, precipitates, or inclusions.

This theory gives an interpretation of the cyclic hardening in annealed metals, the fatigue limit, and the S-N curve when the tests are conducted under stress or strain control.

In Cu single crystals, in which the strain amplitude is controlled, the applied stress increases in each semicycle until the 0.2% monotonic offset yield strength is reached. Then σ_a and S_{ys} increase up to the stress level corresponding to the controlling strain amplitude. Values of σ_a increase in each semicycle until the saturation value σ_s is obtained. This hardening, sometimes

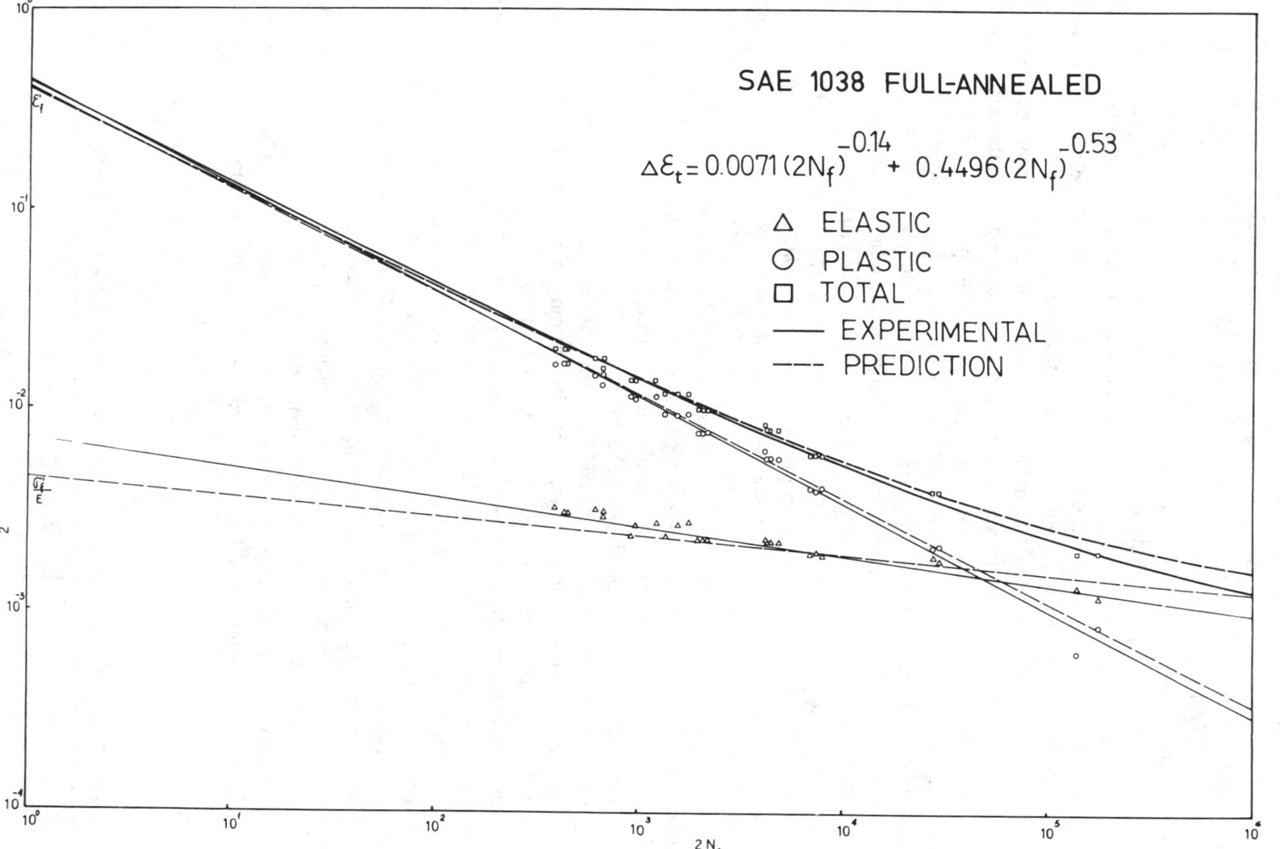

FIG. 7—*Experimental and predicted ε-N curves of fully annealed SAE 1038 steel.*

called "fatigue hardening", can be compared to an "instantaneous work-hardening" which is present in each semicycle. Values of σ_s and S_{ys} are the same [7].

In order to determine the relationship between σ_s and S_{ys} for the fully annealed SAE 1038 steel, the S''_{ys} value, obtained from a tensile test done after the incremental step test, was considered to be a first approximation, instead of the value S'_{ys} (Fig. 4). Table 4 shows the results, considering $S''_{ys} = 372.4$ MPa.

Following Kettunen and Kocks [7] the results shown in Table 4 would indicate that from $\pm 0.37\%$ controlled strain amplitude fully annealed SAE 1038 steel should behave presenting instantaneous work-hardening in each semicycle. In fact this happens with the structural steel tested when $\sigma_s/S''_{ys} \gg 1$, but it was not possible to obtain a definite saturation value for the stress amplitude for all levels of controlled strain amplitude tested higher than $\pm 0.37\%$ (Fig. 6). Burns and Pickering [1] have studied the monotonic behavior of ferritic-pearlitic steels with different pearlite volume fractions. They found that S_u increases more than S_{ys} due to a higher strain hardening of pearlite. Based on these experimental results, it is possible to attempt to explain the behavior observed in 1038 steel when strain amplitude levels are higher than $\pm 0.37\%$, suggesting that not only instantaneous work-hardening is present but progressive cyclic hardening also, due to a higher strain hardening of pearlite as compared with ferrite during fatigue straining.

Kettunen and Kocks [7] have also observed that in Cu single crystals subjected to small strain amplitudes where $\sigma_s/S''_{ys} \leq 1$, "cyclic hardening" will be present; that is, from the very beginning of the testing it controls the plastic straining during saturation. If the strain amplitude chosen is small enough ($\sigma_s/S''_{ys} \ll 1$), the microstructure does not harden at all; this was the main result of Wood and Segal [19] working on annealed metals under alternating plastic strains. In fact, "cyclic softening" was observed on tests done with full annealed SAE 1038 steel subjected to small plastic strain amplitudes on the order of $\pm 0.065\%$, where $\sigma_s/S''_{ys} = 0.4$ (Fig. 6), which is a characteristic behavior of annealed ferritic-pearlitic steels. Keshavan [20] has suggested an explanation for this particular behavior.

In these tests, the controlled strain amplitude lies within the Lüders region of the monotonic stress strain test (Fig. 5). By the end of the initial tensile straining in the first cycle, some of the dislocations will be released from their anchoring Cotrell atmospheres. During the next few cycles, the plastic strain is produced by to-and-fro movements of these free dislocations. As they are not too numerous at the beginning, they do not lead to any perceptible hardening. However, motion of these dislocations within the lattice appears to bring about a gradual release of other interstitial anchored dislocations. Such an action has a two-fold effect on the deformation resistance: (1) there is an increase in the number of mobile dislocations available for plastic deformation, and (2) there is a reduction in the resistance to motion of mobile dislocations offered by the anchored non-mobile dislocations and their stress fields. Both effects can be thought of as contributing to the softening evidence as cycling proceeds.

Kettunen and Kocks [7] have also proposed a relationship between the controlled strain amplitude and fatigue life assuming that the fracture occurs when a certain amount of structural

TABLE 4—σ_s/S''_{ys} *ratio for fully annealed SAE 1038 steel* ($S''_{ys} = 372.4$ *MPa*).

$\Delta\epsilon_t/2$, %	$\Delta\epsilon_{ps}/2$, %	σ_s, MPa	σ_s/S''_{ys}
± 0.37	± 0.2	376.3	1.01
± 0.62	± 0.4	417.5	1.12
± 1.04	± 0.8	469.4	1.26
± 1.25	± 1.0	486.1	1.31
± 1.47	± 1.2	502.7	1.35

damage has accumulated, this damage being dependent on the applied amplitude stress. The total structural change is directly proportional to the number of obstacle intersections with mobile dislocations. The total number of obstacle intersections is proportional to the number of cycles to fracture, N_f and a_s/a_o where a_s is the average free slip area saturation value and a_o the average obstacle area. Then

$$N_f \frac{a_s}{a_o} = C(\sigma_a) \tag{14}$$

where $C(\sigma_a)$ is the relationship between the critical amount of structural change and the probability of generating relevant structural change per dislocation-obstacle intersection.

The ratio a_s/a_o is a function of the ratio σ_a/S_{ys} given by the statistical theory of flow stress and work-hardening [8]. The amount of cumulative damage, $C(\sigma_a)$, depends on the applied stress; if this dependence is known, Eq 14 is the equation of the *S-N* curve.

When a_s/a_o is not known, it can be estimated in the following way: a_s depends upon $\Delta\epsilon_{ps}$, the saturation plastic strain range, through the relationship

$$\Delta\epsilon_{ps} = \frac{a_s}{n_p(\sigma_a)\cdot B} \tag{15}$$

where $n_p(\sigma_a)$ is the mobile dislocation number, B is the amount of Burger's vector, and a_o is inversely proportional to the square of yield stress, S_{ys}, so that Eq 14 can be modified to

$$\Delta\epsilon_{ps}\cdot S_{ys}^2\cdot N_f = C(\sigma_a) \tag{16}$$

When the difference between S_{ys} and σ_s is small, as in cyclic straining at amplitudes capable of leading to fracture, S_{ys} can be replaced by σ_s, then

$$\Delta\epsilon_{ps}\cdot \sigma_s^2\cdot N_f = C(\sigma_a) \tag{17}$$

Equation 17 can be verified by the experimental results obtained from the material tested, fully annealed SAE 1038 steel. Table 5 shows data used to obtain the curve shown in Fig. 8. The mathematical relationship of this curve, when $\sigma_a = \sigma_s$ under controlled plastic strain amplitude, is

$$\Delta\epsilon_{ps}\cdot \sigma_s^2\cdot N_f = C(\sigma_s^p) \tag{18}$$

TABLE 5—*Kettunen and Kocks relationship data (Eq 18).*

$\Delta\epsilon_t/2$, %	σ_s, MPa	$\Delta\epsilon_{ps}$, %	N_f
±2.0	537	±3.38	220
±1.8	520.1	±2.98	340
±1.4	491.2	±2.26	460
±1.2	452.5	±1.93	675
±1.0	438.3	±1.55	981
±0.86	414.0	±1.26	2 096
±0.6	373.8	±0.8	3 676
±0.4	333.2	±0.42	13 869
±0.2	276.8	±0.13	68 539

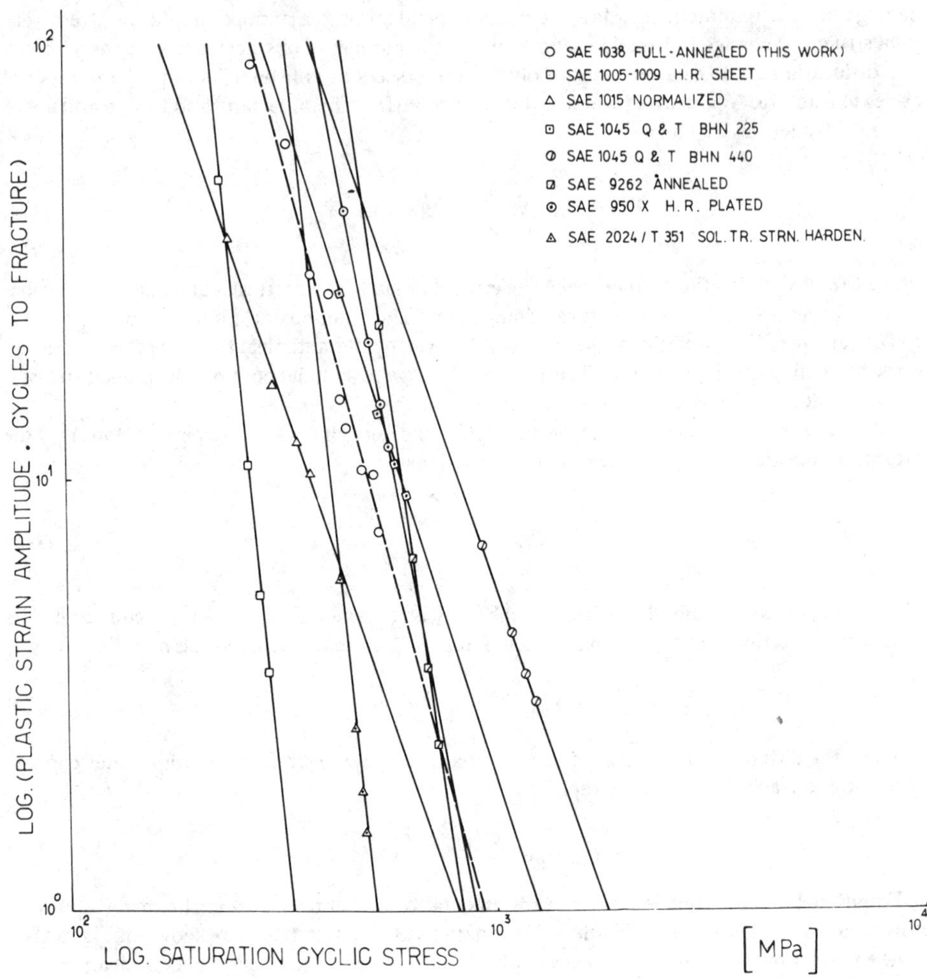

FIG. 8—*Kettunen-Kocks relationship* ($\Delta\epsilon_p$, N_f) *versus* (σ_s) *curves for several engineering alloys.*

Figure 8 also compares the results obtained with several engineering alloys [*21*]. Table 6 indicates the corresponding C and p values of Eq 18 for each alloy considered, where $m = p - 2$ is the slope of the curves shown in Fig. 8.

The Kettunen-Kocks relationship, Eq 18 for fully annealed SAE 1038 steel is

$$\Delta\epsilon_{ps} \cdot \sigma_s^2 \cdot N_f = 5.25 \times 10^3(\sigma_s^{-1.85}) \tag{19}$$

Kettunen and Kocks [*7*], based on data obtained by Coffin on Cu [*22*], have shown that the critical amount of cumulative damage is inversely proportional to the applied stress, σ_s; that is, $p = -1$. Nevertheless, in Table 6, for several engineering alloys it is shown that p may range from -1 to -10; $p = -0.92$ for 1015 steel normalized, and $p = -1.85$ for fully annealed SAE 1038 steel. It is suggested that this different behavior is due to the presence of the second phase, pearlite. The presence of second phases other than pearlite and the different probability of

TABLE 6—C *and* p *values for the Kettunen and Kocks relationship (Eq 18) for several engineering alloys.*

Material	C	$p = m + 2$
SAE 1005-1009 hot rolled sheet	1.85×10^6	-10.31
SAE 1015 steel, normalized	4.55×10^2	-0.92
SAE 1038 steel, full annealed (this work)	5.25×10^3	-1.85
SAE 1045 steel, quenched and tempered, HB 225	3.05×10^2	-1.19
SAE 1045 steel, quenched and tempered, HB 440	4.95×10^3	-0.96
SAE 9262 steel, annealed	1.6×10^7	-5.94
SAE 950X steel	2.15×10^5	-3.68
SAE 2024-T 351 aluminum, solution treated, strain hardened	1.3×10^8	-9.57

generating relevant structural changes per dislocation-obstacle may explain the wide range of p values obtained. On the other hand, the agreement found by Eq 17 to describe the relationship between the saturation plastic strain range, the saturation stress amplitude, and the number of cycles to fracture, as is shown in Fig. 8, is at least comparable to those generally reported for the Coffin-Manson relationship.

It is possible to rearrange Eq 18 and find an expression for C and $m = p - 2$:

$$\Delta\epsilon_{ps} \cdot N_f = C(\sigma_s^m) \tag{20}$$

When N_f and $\Delta\epsilon_{ps}$ are replaced by the following relationships obtained from Eqs 7 and 10, respectively:

$$\Delta\epsilon_{ps} = 2\left(\frac{\sigma_s}{K'}\right)^{1/n} \tag{21}$$

$$N_f = \frac{1}{2}\left(\frac{\Delta\epsilon_{ps}}{2\epsilon_f'}\right)^{1/c} \tag{22}$$

The following relationships are obtained for C and m:

$$C = \left(\frac{1}{K'}\right)^m \cdot \left(\frac{1}{\epsilon_f'}\right)^{1/c} \tag{23}$$

$$m = p - 2 = \frac{1}{n'} + \frac{1}{b} \tag{24}$$

Thus it is possible to predict the values of C and p of the Kettunen-Kocks relationship (Eq 18) if the values of K', ϵ_f', n', c, and b are known. These values can also be obtained from two tests only, the monotonic test and the incremental step test.

From Table 3, the predicted values of C and m for SAE full annealed 1038 steel will be: $C = 9.6 \times 10^4$ and $m = -1.88$, which can be compared with the experimental values: $C = 5.25 \times 10^5$ and $m = -3.85$. It was found that, under the controlled strain amplitude levels tested, a saturation value was not clearly defined. This may be the reason for the differences between the predicted and experimental m values, where $m = p - 2$.

Conclusions

1. The multiple linear regression analysis proposed by Grozier et al. for ferrite-pearlite structures gave a satisfactory result for the prediction of fully annealed SAE 1038 steel monotonic properties.
2. Resistance to total strain cycling can be reliably predicted by the Feltner and Landgraf prediction criterion by considering two parameters, the fatigue ductility coefficient, ϵ_f', and the true fracture strength, σ_f. These parameters can be determined from two types of test, the incremental step test and the monotonic stress-strain test, respectively.
3. When controlled strain amplitudes are greater than $\pm 0.37\%$ instantaneous strain hardening is observed during cyclic straining and a saturation value for the stress amplitude is not clearly defined. For smaller strain amplitudes, cyclic softening occurs.
4. The Kettunen-Kocks theory allows us to obtain a description of the relationship between the saturation stress amplitude, the saturation plastic strain range, and the number of cycles to fracture that fit the experimental data for fully annealed SAE 1038 steel in at least as comparable agreement as with that obtained by the Coffin-Manson law.
5. The parameters C and p of the Kettunen-Kocks relationship can be fairly well predicted from only two types of tests: the incremental step test and the monotonic stress-strain test.

Acknowledgments

The authors wish to thank Establecimientos Santa Rosa, Argentina, for supplying testing material; Ing D. A. Di Bella†, CNEA, Argentina, for testing assistance; and Professor F. B. Pickering, Reader in Metallurgy, Sheffield Polytechnic, England, Professor A. Plumtree, Waterloo University, Ontario, Canada, and Dr. U. F. Kocks, Los Alamos Scientific Laboratory, New Mexico, for valuable discussions. This research has also been supported by the Proyecto Multinacional de Tecnología de Materiales OEA-CNEA.

References

[1] Burns, K. W. and Pickering, F. B., "Deformation and Fracture of Ferrite-Pearlite Structures," *Journal of Iron and Steel Institute,* Vol. 202, 1964, pp. 899-906.
[2] Barnby, J. T. and Johnson, M. R., "Fracture in Pearlitic Steels," *Metal Science Journal,* Vol. 3, 1969, pp. 155-160.
[3] Bruckner, W. H., *Welding Journal Research Supplement,* Vol. 29, 1950, p. 467-s.
[4] Abdel-Raouf, H., Benham, P. P., and Plumtree, A., "Mechanical Behavior and Substructures of Strain Cycled Iron," *Canadian Metallurgical Quarterly,* Vol. 10, 1971, pp. 87-95.
[5] Feltner, C. E. and Lairdm, C., "The Role of Slip Character in Steady Cyclic Stress-Strain Behavior," *Transactions of the American Institute of Mechanical Engineers,* Vol. 245, 1969, p. 1372.
[6] Grozier, J. D. and Bucher, J. H., "Correlation of Fatigue Limit with Microstructure and Composition of Ferrite-Pearlite Steels," *Journal of Materials,* Vol. 2, 1967, pp. 393-407.
[7] Kettunen, P. O. and Kocks, U. F., "Fatigue Studied as a Statistical Theory of Flow Stress and Work-Hardening Phenomenon," *Acta Polytechnica Scandinavica,* Vol. 104, 1971, pp. 1-23.
[8] Kocks, U. F., "A Statistical Theory of Flow Stress and Work-Hardening," *Philosophical Magazine,* Vol. 13, 1966, pp. 541-566.
[9] Kettunen, P. O., "Work-Hardening During Cycling Deformation," in *Proceedings,* Asilomar Conference on Strength of Metals and Alloys, Vol. 1, 1970, pp. 214-218.
[10] Kettunen, P. O. and Kocks, U. F., "On Possible Relation Between Work-Hardening and Fatigue Failure," *Scripta Metallurgica,* Vol. 1, 1967, pp. 13-17.
[11] Raske, D. T. and Morrow, J. D., "Mechanics of Materials in Low-Cycle Fatigue," *Manual on Low-Cycle Fatigue Testing, ASTM STP 465,* American Society for Testing Materials, Philadelphia, 1969, pp. 1-26.

[*12*] Benson, D. K. and Hancock, J. R. "The Effect of Strain Rate on the Cyclic Response of Metals," *Metallurgical Transactions,* Vol. 5, 1974, pp. 1711-1715.

[*13*] Bucher, J. H., Grozier, J. D., and Enrietto, J. F., "Strength and Toughness of Hot-Rolled Ferrite-Pearlite Steels," in *Fracture,* H. Liebowitz, Ed., Academic Press, New York, Vol. 6, Chapter 5, 1969, p. 247.

[*14*] Basquin, O. H., "The Exponential Law of Endurance Test," in *Proceedings,* American Society for Testing Materials, Vol. 10, Part II, 1910, p. 685.

[*15*] Tavernelly, J. F. and Coffin, L. F., Jr., "Experimental Support for Generalized Equation Predicting Low Cycle Fatigue," *Journal of Basic Engineering, Transactions of ASME,* Vol. 84, No. 4, Dec. 1962, p. 533.

[*16*] Manson, S. S., discussion of Ref *15, Journal of Basic Engineering, Transactions of ASME,* Vol. 84, No. 4, Dec. 1962, p. 537.

[*17*] Feltner, C. E. and Landgraf, R. W., "Selecting Materials to Resist Low-Cycle Fatigue," Technical Report 69-DE-59, American Society of Mechanical Engineers Conference, New York, 5-8 May 1968.

[*18*] Morrow, J. D., "Cyclic Plastic Strain Energy and Fatigue of Metals," in *Internal Damping and Cyclic Plasticity, ASTM STP 378,* American Society for Testing and Materials, Philadelphia, 1965, pp. 45-87.

[*19*] Wood, W. A. and Segal, R. L., "Annealed Metals Under Alternating Plastic Strain," *Proceedings of the Royal Society of London,* Vol. A242, 1957, p. 180.

[*20*] Keshavan, S., "Some Studies on the Deformation and Fracture of Normalized Mild Steel Under Cycling Conditions," Ph.D. dissertation, University of Waterloo, Canada, Dec. 1967.

[*21*] Tucker, L. E. and Landgraf, R. W., "Proposed Technical Report on Fatigue Properties for the SAE Handbook," Technical Report 740279, pp. 1-16, presented at the Automotive Engineering Congress, Detroit, 25 Feb.-1 March 1974.

[*22*] Coffin, L. F., Jr., "Cyclic Straining and Fatigue," *Internal Stresses and Fatigue of Metals,* G. M. Rassweiler and W. L. Grube, Eds., Elsevier, Amsterdam, 1959, p. 363.

Y. S. Chung[1] and A. Abel[1]

Low Cycle Fatigue of Some Aluminum Alloys

REFERENCE: Chung, Y. S. and Abel, A., "**Low Cycle Fatigue of Some Aluminum Alloys,**" *Low Cycle Fatigue, ASTM STP 942,* H. D. Solomon, G. R. Halford, L. R. Kaisand, and B. N. Leis, Eds., American Society for Testing and Materials, Philadelphia, 1988, pp. 94–106.

ABSTRACT: Axial strain-controlled tests were performed on five aluminum alloys to provide information on their cyclic behavior. The alloys chosen for this investigation represent a complete survey of the strengthening mechanisms associated with commercial aluminum products. The strengthening microstructures ranged from subgrain and dislocation strengthening in the 1200 alloy, solid solution strengthening by magnesium in the 5083, and the high density of GP zones and finely dispersed precipitates in the case of the 7005, 6351, and 6061 alloys. The five alloys were plant-fabricated with commercial tempers.

One important aspect of the tests was the detailed measurement of the hysteresis loop shape change during cycling and its possible correlation with the fatigue performance. The fatigue results show the effect of microstructure and confirm the presence of a break in the Coffin-Manson plot.

KEY WORDS: low cycle fatigue, aluminum alloys, microstructure, hysteresis loop

Various semi-empirical expressions [*1–10*] are used to describe low cycle fatigue endurance. The Coffin-Manson plot is the most commonly used approach, which relates the plastic strain range, $\Delta\epsilon_p$, to the fatigue life, N_f, and is of the form:

$$\Delta\epsilon_p/2 = \epsilon_f'(N_f)^c \tag{1}$$

where ϵ_f' and c are the fatigue ductility coefficient and exponent respectively.

A break has been observed in the Coffin-Manson plots of some aluminum alloys [*11–15*]; this has been associated with mechanisms affecting the various phases of crack propagation [*12–19*]. The present project was designed to examine this phenomenon by monitoring a number of variables during cycling including the hysteresis loop shape variation.

The change in hysteresis loop shape has been monitored through the measurements of the Bauschinger energy parameter β_E [*22,23*]. For a closed hysteresis loop this is calculated as $(2\sigma_p\Delta\epsilon_p - \int\sigma d\epsilon)/\int\sigma d\epsilon$, where σ_p is the cyclic peak stress and $\Delta\epsilon_p$ is the plastic strain amplitude. Accordingly, when the hysteresis loop shape approaches that of a parallelogram the value of β_E approaches zero. Conversely, the more pointed the hysteresis loop shape, the higher the value of the energy parameter, reaching 1.0 when the loop area equals half the area given by $(2\sigma_p\Delta\epsilon_p)$. It has been shown [*31*] for single crystals of Cu and Cu-Al alloys that rectilinear hysteresis loops are associated with poor, and more pointed loops are associated with improved, fatigue performance.

It seems that β_E gives information about the nature of the deformation processes involved. Very low β_E values would indicate deformation processes corresponding in a rheological model

[1]The University of Sydney, Civil and Mining Engineering, Sydney 2006, Australia.

to a rigid-plastic material, where the energy input of a deformation cycle would be used up in kinetic friction. More pointed hysteresis loop shapes, larger β_E values, would correspond with deformation processes where a larger fraction of the energy input is stored in a recoverable manner. During unloading and reverse loading, the initial departure from a response defined by the elastic modulus of the material is a manifestation of this recoverably stored energy.

The ratio of recoverably stored energy to the energy requirement for the irreversible processes will develop a characteristic value during cycling which must be a structure-sensitive material property. It is quite plausible that the two kinds of deformation processes will produce fatigue damage at a different rate. Thus the interest in the existence of a break in a $\beta_E - N_f$ plot.

Experimental Procedures

The alloys chosen for this investigation represent a complete survey of the strengthening mechanisms associated with commercial aluminum products [*14,20,21*]. The strengthening microstructures ranged from subgrain and dislocation strengthening in the 1200 alloy, solid solution strengthening by magnesium in the 5083, and the high density of GP zones and finely dispersed precipitates in the case of the 7005, 6351, and 6061 alloys. The five alloys were plant-fabricated with commercial tempers. Table 1 lists the details relating to these alloys.

The fatigue tests were performed in a 250 kN Instron, TT-K model, using the automatic transfer points with fully reverse control of total strain amplitude. The load was monitored by a load cell in series with the specimen and the elongation by a 25 mm gage length Instron clip-on type extensometer. The nominal diameter of the specimens varied between 6 and 12 mm with a 12 mm gage section. The tests were started in the tensile direction, and the applied strain rate was approximately 8×10^{-4}/s. The load was continuously recorded on the Instron autograph chart; the load-strain hysteresis loops were monitored on an *X-Y* plotter.

Failure was defined by a 25% reduction in the load-carrying capacity of the specimen. The characteristic fatigue response was plotted on log-log scales by using a reiterated linear "least squares" regression method [*20*]. The positions of breaks in these plots were determined from the extrapolation of the lines derived from the low and the high cycle life regimes.

Experimental Results

The results are presented in two sections relating to general cyclic deformation behavior and fatigue fracture performance respectively.

The cyclic hardening response of the five alloys investigated is shown in Fig. 1, where the cyclic stress-strain curves are plotted together with the corresponding monotonic curves. These results show that the subgrain strengthened 1200 alloy hardens the least amount, the precipitation alloys taking the intermediate position, while the solution strengthened system leads to the greatest degree of cyclic hardening.

More detailed cyclic response is shown in Figs. 2 to 4 where the behavior of the 6351, 5083, and 1200 alloys is illustrated, each alloy taken as a representative of the three strengthening cases. The plots represent the changing values of: peak stress, σ_a, plastic strain range, $\Delta\epsilon_p$, energy absorption per cycle, ΔW, average flow stress, $\bar{\sigma}$, and the energy parameter, β_E, when cycling takes place with a constant total-strain amplitude.

The fatigue performance is shown in Figs. 5 to 9. The values of all the variables which are plotted against the number of cycles to failure, N_f, were taken at 50% of N_f. This arbitrary procedure was adopted in order to compare values taken at relatively stable cyclic states.

Breaks are indicated on a log-log scale in all the investigated variables when they are plotted against the number of cycles to failure. This is illustrated by Fig. 5, where the results obtained on the 6351 alloy are shown. The breaks are sufficiently close to give added strength to the observed break in $\Delta\epsilon_p$ versus N_f (that is, in the Coffin-Manson plot) at around 1000 cycles to

TABLE 1—*Alloy description.*

Alloy	True Fracture Ductility (ϵ_f)	True Fracture Stress, MPa	Fracture Toughness, kJ/m³	Composition (Nominal[a] and Analyzed Matrix[b])								
				Cu	Fe	Mg	Mn	Si	Cr	Zn	Ti	Zr
7005 T5 Rod[c]	0.494	570	35 700	0.10 (0.05)	0.40 (0.01)	1.0–1.8 (1.96)	0.2–0.7 (0.24)	0.35 (0.02)	0.06–0.2 . . .	4.0–5.0 (2.05)	0.01–0.06 . . .	0.08 . . .
6061 T6 Rod[d]	1.335	559	28 500	0.15–0.4 (0.15)	0.70 . . .	0.8–1.2 (1.45)	0.15 (0.04)	0.4–0.8 (0.59)	0.04–0.35 . . .	0.25 (0.03)	0.15 . . .	. . . . . .
6351 T6 Rod[d]	0.905	534	29 200	0.10 (0.05)	0.50 (0.03)	0.4–0.8 (0.51)	0.4–0.8 (0.15)	0.7–1.3 (0.85)	. . . . . .	0.10 (0.05)	0.20 . . .	. . . . . .
5083 H112 Plate[e]	0.150	404	27 970	0.01 (0.01)	0.35 (0.03)	4.86 (6.05)	0.76 (0.14)	0.16 (0.03)	0.08 . . .	. . . (0.03)	. . . . . .	. . . . . .
1200 O Plate[f]	1.5	180	29 370	. . . (0.02)	0.52 (0.02)	. . . (0.02)	. . . . . .	0.15 (0.14)	. . . . . .	. . . . . .	. . . . . .	. . . . . .

[a]Composition and temper information from aluminum association via "Standards for Australian Aluminum Mill Products," Metric Edition, 1973, The Aluminum Development Council.

[b]Composition from microprobe analysis of the matrix.

[c]T5 indicates cooling from an elevated temperature shaping process and then artificially aged.

[d]T6 indicates solution heat treated and then artificially aged.

[e]Strain hardened by cold working.

[f]Full annealed to lowest strength condition.

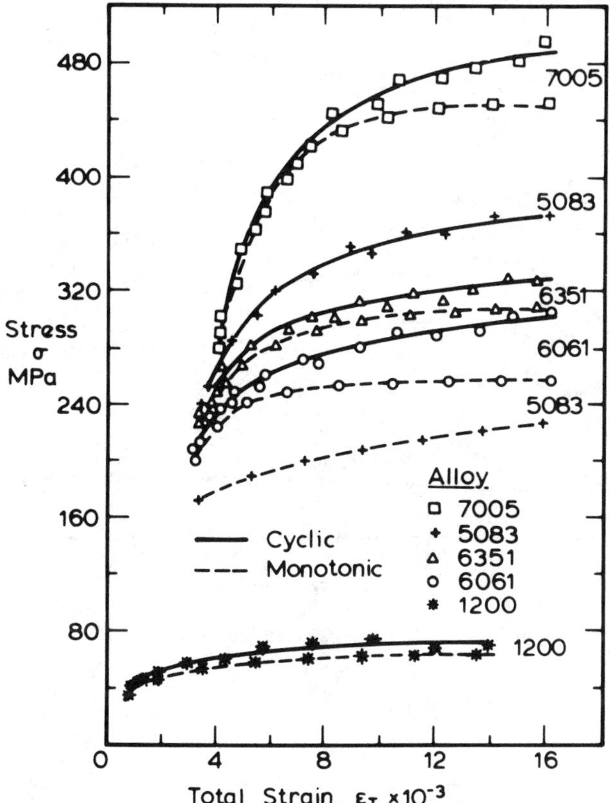

FIG. 1—*Cyclic and monotonic stress-strain curves.*

failure (Fig. 6). At fatigue lives lasting longer than approximately 1000 cycles the deformation processes must change significantly as reflected by the hysteresis loop shape changes indicated in Fig. 7. Similar tendencies are reflected when the total energy absorption to failure, W_f, values are examined (Fig. 8). Finally, the results plotted in the form of an *S-N* curve are illustrated in Fig. 9.

The obtained cyclic parameters together with some calculated values are presented in Tables 2 and 3. The following expressions are used in calculating the values in these tables:

$$\sigma_a = \sigma_f' N_f^b \tag{2}$$

where σ_f' and b are the fatigue strength coefficient and exponent respectively and

$$W_f = W_f' N_f^{d'} \tag{3}$$

where W_f' is the fatigue toughness coefficient and d is the fatigue toughness exponent (d' is simply equal to $d + 1$).

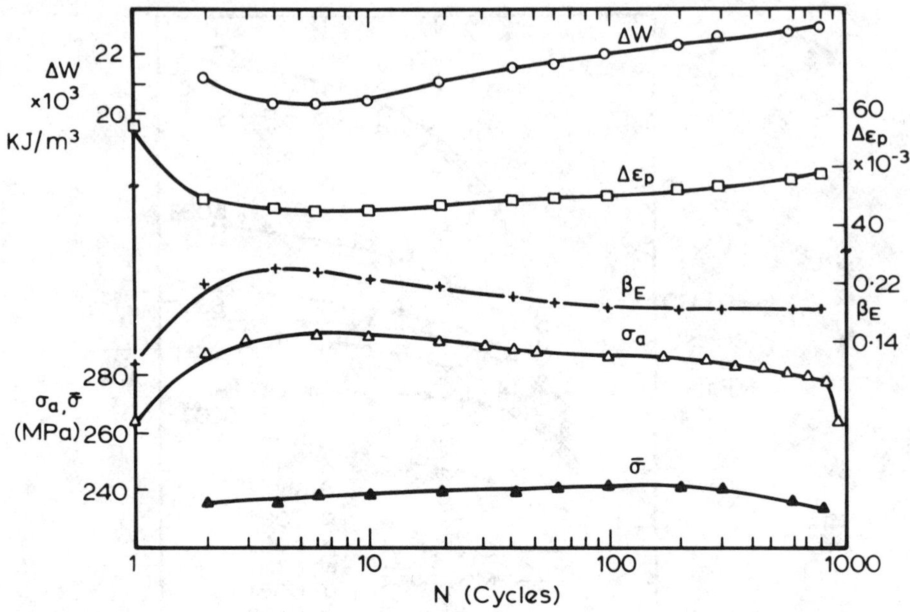

FIG. 2—*Cyclic response of Alloy 6351;* $\epsilon_T = 0.9\%$.

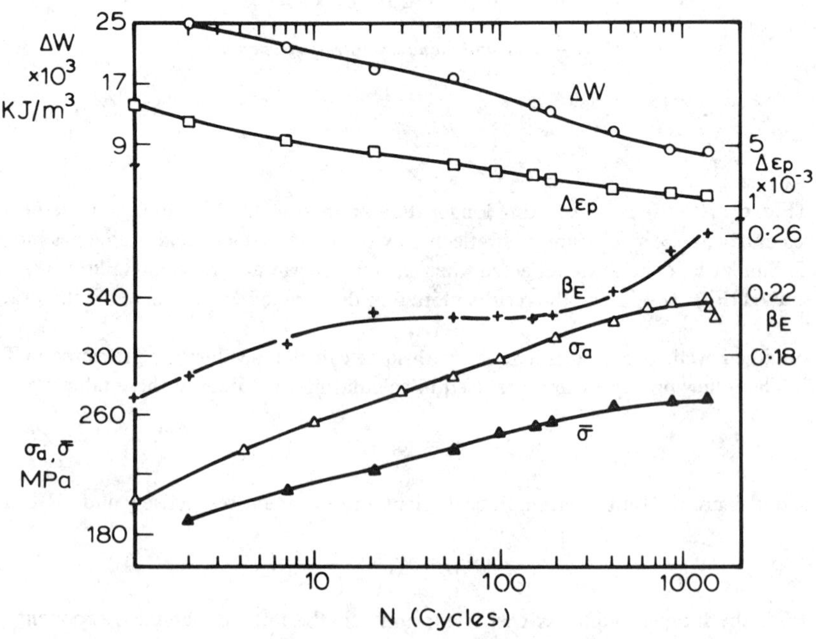

FIG. 3—*Cyclic response of Alloy 5083;* $\epsilon_T = 0.8\%$.

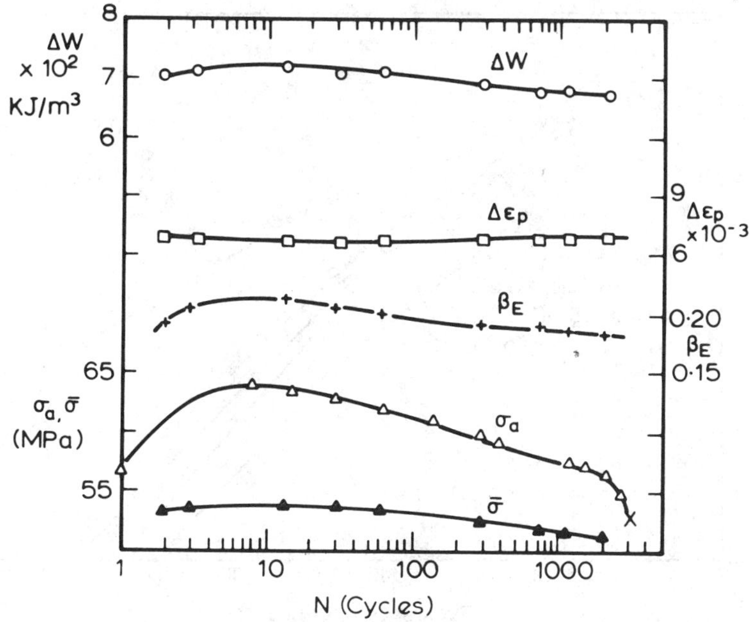

FIG. 4—*Cyclic response of Alloy 1200;* $\epsilon_T = 0.46\%$.

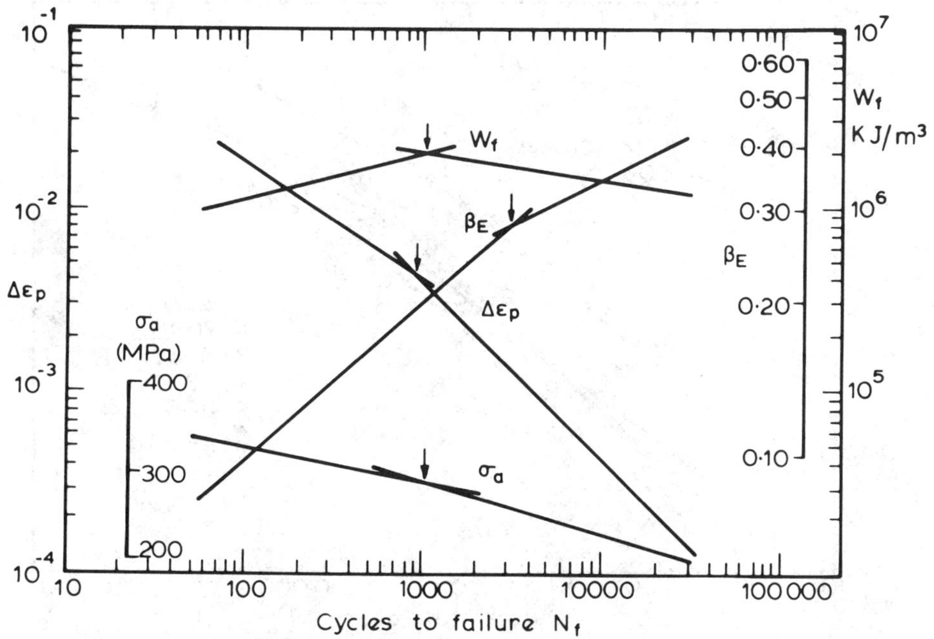

FIG. 5—*Breaks in the cycle response of Alloy 6351.*

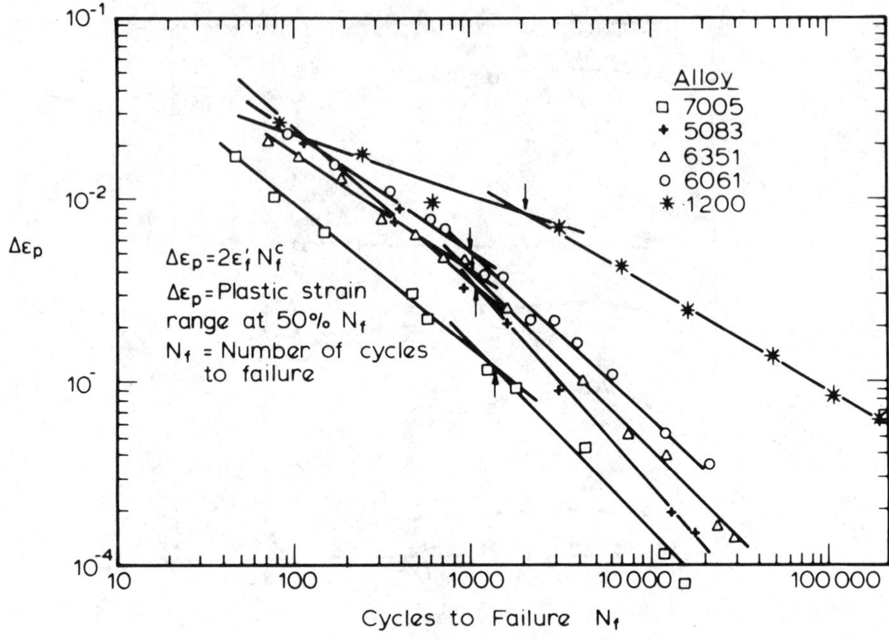

FIG. 6—*Coffin-Manson plots.*

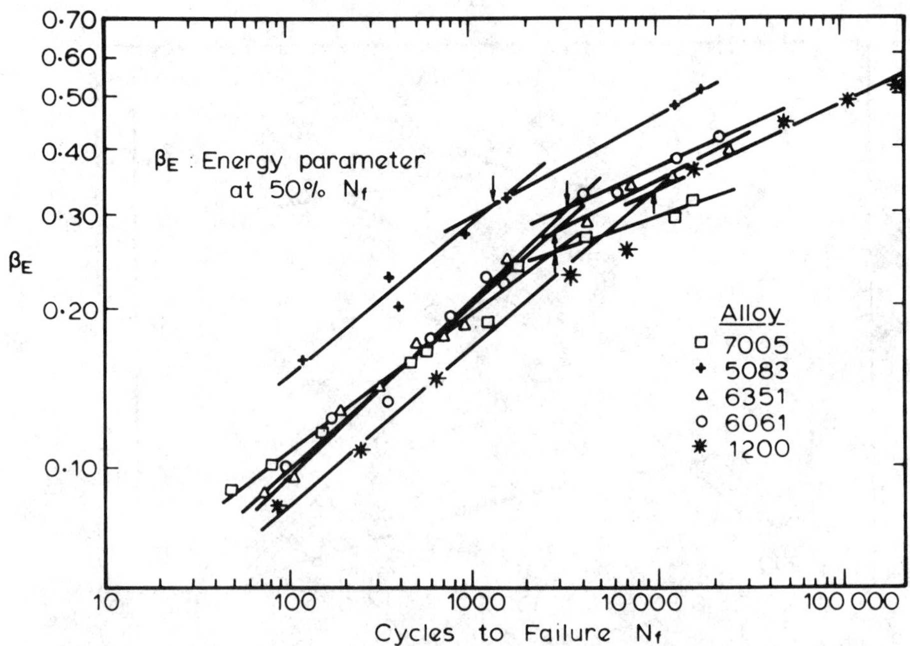

FIG. 7—*Variation of stable loop shape obtained after 50%* N_f.

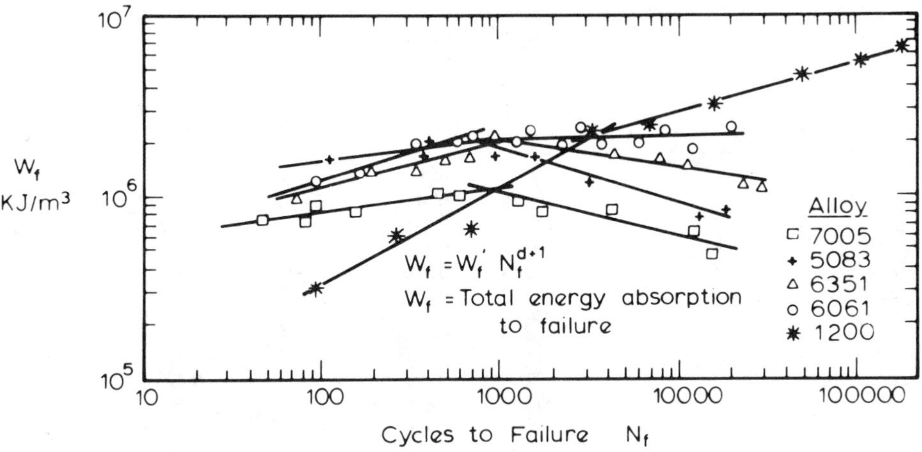

FIG. 8—*Total plastic energy versus fatigue life.*

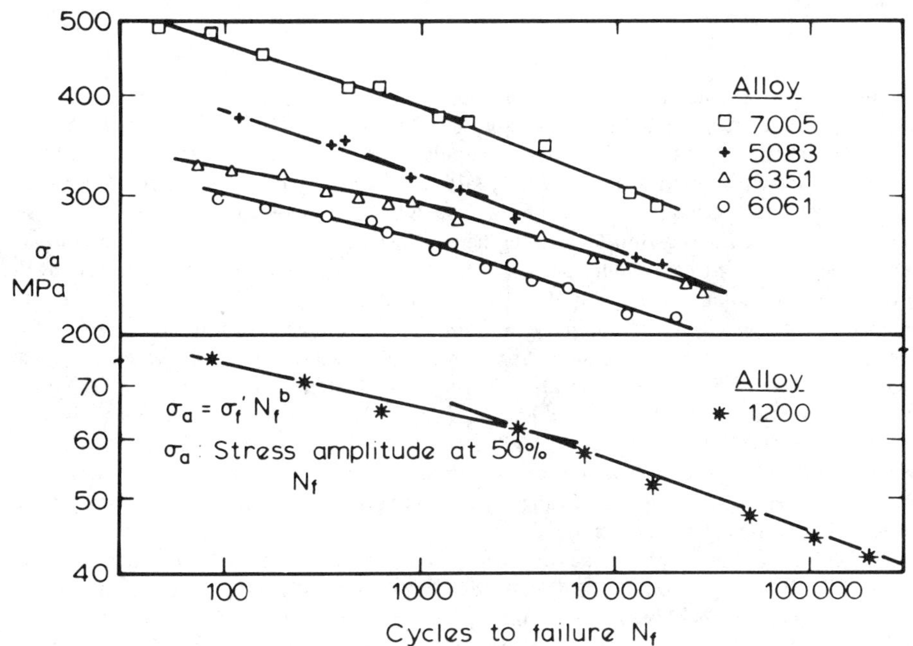

FIG. 9—*Stress amplitude versus fatigue life.*

Discussion

The present study shows that all investigated alloys exhibit cyclic hardening (Fig. 1); the demonstrated highest cyclic hardening capability of the 5083 alloy matches the highest σ_{UTS}/σ_Y ratio of 1.9. The three precipitation hardened alloys depict both an almost constant σ_{UTS}/σ_Y ratio of 1.2 and a similar extent of cyclic hardening. Although the 1200 alloys showed the least

TABLE 2—*Experimentally obtained low cycle fatigue parameters compared with various theoretical predictions.*

Present Results	Alloy 7005	5083	6351	6061	1200
n'	0.079	0.087	0.072	0.089	0.152
$-d$	0.852	0.962	0.748	0.702	0.544
$-c$	0.802	0.890	0.669	0.628	0.432
$*-c = 1/(1 + 5n')$; [*3*]	0.673	0.697	0.735	0.692	0.568
$-c = 1/(1 + 2n')$; [*4*]	0.838	0.852	0.874	0.849	0.767
$-c = d/(1 + n')$ [*5*]	0.787	0.885	0.698	0.645	0.471
ϵ_f'	0.195	0.803	0.205	0.207	0.091
$*\epsilon_f' = \epsilon_f$; [*3*]	0.494	0.150	0.905	1.335	1.50
$\epsilon_f' = \epsilon_f^{0.6}/2\sqrt{2}$; [*6*]	0.232	0.113	0.333	0.420	0.45
$\epsilon_f' = 0.35\ \epsilon_f$; [*7*]	0.173	0.053	0.317	0.467	0.53
$\epsilon_f' = \epsilon_f/2\sqrt{2}$; [*8*]	0.175	0.053	0.320	0.472	0.53
$*-(d + 1)/c$	0.185	0.043	0.377	0.475	1.058
$-(1 + n' + 1/c)$	0.150	0.037	0.423	0.503	1.163

*Values calculated in accordance with indicated references, but using present results of n', d, c, and ϵ_f as applicable.

amount of cyclic hardening, it also had a σ_{UTS}/σ_Y ratio of 1.2. In general, therefore, these results are in agreement with the ideas of Manson [*24*] and compatible with the explanation of unidirectional and cyclic hardening micromechanisms proposed by Calabresse and Laird [*25*].

The significance of microstructure [*14,25,26*] is shown clearly by the observed cyclic behavior. The precipitate strengthened 6061, 6351, and 7005 alloys show initial hardening followed by softening, while the substitutional strengthened 5083 alloys hardens throughout cycling, although the rate diminishes, and in the case of the 1200 alloy the softening nearly wipes out the initial cyclic hardening under the applied testing conditions.

Considering the initial and rapid cyclic hardening in the case of the 6351 precipitation strengthened alloy, it may be explained by the increase in dislocation multiplication [*27*]. This leads to an increase in mobile dislocation density during reverse loading and causes a rapid increase in the value of the energy parameter β_E [*22*]. This process does not last long, since the dislocation-particle interactions [*25,28–30*] become more significant. At the point where β_E reaches a maximum the dislocation-particle interaction becomes dominant. The subsequent softening process is indicated by the slightly increasing $\Delta\epsilon_p$ and decreasing σ_a, and it is also reflected by the steady decrease in the β_E value.

The 5083 alloy shows a continuous cyclic hardening (Fig. 3) that is accompanied by the increase of β_E. It is suggested that the dispersoids, characteristically with a nonuniform distribution, are responsible for this behavior. During solidification of the ingot an incoherent phase, usually dominated by Mn, Mg, or Cr in these types of alloys, precipitates at high temperatures. Transmission electron microscopy showed, in the present case, the existence of dispersoids with varying sizes, ranging between 0.1 to 0.5 μm, which must have acted as dislocation sources throughout cycling, causing the continuous hardening.

The 1200 alloy shows a relatively constant plastic strain amplitude and a relatively constant average flow stress, $\bar{\sigma}$, during cycling, so that the softening manifests itself through decreasing peak stresses and through the decreasing value of β_E. These tendencies may be explained by the back-and-forth movement of a large number of dislocations.

In relation to the fatigue performance the discussion will be limited to the high strain portions of the various $\Delta\epsilon_p$ versus N_f curves. The obtained fatigue ductility coefficients ϵ_f' are presented in

TABLE 3—*Cyclic and monotonic parameters obtained from the present study.*

Alloy	True Fracture Ductility (ϵ_f)	$\Delta\epsilon_p/2 = \epsilon_f' N_f^c$				Monotonic Fracture Toughness, 10^5 kJ/m^3	$W_f = W_f' N_f^{d+1}$				True Fracture Stress, MPa	$\sigma_a = \sigma_f' N_f^b$			
		*ϵ_{f1}'	$-C_1$	ϵ_{f2}'	$-C_2$		*W_{f1}', 10^5 kJ/m^3	$d_1 + 1$	W_{f2}', 10^6 kJ/m^3	$d_2 + 1$		*σ_{f1}'	$-b_1$	σ_{f2}'	$-b_2$
7005	0.494	0.195	0.802	1.270	1.061	0.357	4.06	0.148	5.31	−0.235	570	673	0.079	817	0.106
5083	0.150	0.802	0.890	3.685	1.108	0.280	13.8	0.038	16.7	−0.315	404	540	0.075	611	0.094
6351	0.905	0.205	0.669	2.470	1.019	0.292	3.51	0.252	6.25	−0.158	534	408	0.050	477	0.072
6061	1.335	0.207	0.628	1.285	0.901	0.285	3.081	0.299	1.84	0.018	559	383	0.053	463	0.081
1200	>1.5	0.091	0.432	0.288	0.565	0.294	0.421	0.457	0.741	0.173	180	102	0.066	120	0.085

*Subscripts 1 and 2 denote respectively the low and high N_f portions of the various plots. The change in slope is in the 1000 and 4000 N_f cycle range.

Table 2. Calculated values of the coefficient, based on the various existing proposals, are also shown. The results indicate that the use of the monotonic fracture ductility, ϵ_f, is not leading to a good estimate of ϵ_f'. In the case of the 1200 alloy, for example, the highest ϵ_f value is matched with the lowest ϵ_f'.

A similar problem arises when the monotonic fracture toughness and the fatigue toughness, W_f', coefficients are compared (Table 3). The data indicate an order of magnitude difference between the highest and lowest values of W_f' amongst the alloys in contrast to the slight variation found in their monotonic fracture toughness values.

The cyclic hardening exponent, n', is the only material property governing the value of c in the models of Morrow [3] and Tomkins [4] based on energy and crack propagation considerations respectively. The model of Saxena and Antolovich [5] incorporates both the exponent d and n' in the prediction of c; this approach seems to give the best prediction for c (Table 2). Since this approach accounts for the variation of the fraction of total plastic work stored as damage as a function of the slip mode, c will also be a function of the slip mode in agreement with the suggestions of Laird and Feltner [32] and Wells [33].

Santner and Fine [34] have obtained a fatigue toughness exponent d of -0.909 on some Al-Cu alloys. This value is reasonably close to those obtained in this study with the exception of the 1200 alloy for which d equals -0.543.

Taking the data of Halford [35], however, where all types of metals are included, the 5083 alloy seems to give the least consistent result. According to Halford the fatigue toughness of all metals increases approximately as the one-third power of the fatigue life (i.e., the exponent $(d + 1)$ in Table 3). The results are generally close to one third except the very low value given by the 5083 alloy.

The relationship between the stress amplitude, σ_a, and the total plastic work, W_f, can be obtained from Eq 2 and 3, giving the expression

$$\sigma_a = \sigma_f'(W_f/W_f')^{b(d+1)}$$

The calculated values of the exponent, $b/(d + 1)$, are -0.62, -0.25, -0.24, and -0.16 for the 7005, 6351, 6061, and 1200 alloys respectively. Using different assumptions regarding the zone of plastic material associated with a crack, Morrow [3] and Sandor [9] have obtained $b/(d + 1)$ values of -0.25 and -0.17 respectively, and these were assumed to be constant. The results obtained here suggest that there is no general relationship between W_f and σ_a and therefore $b/(d + 1)$ is material dependent.

Finally, it is evident from Figs. 5, 6, 7, and 9 that the hysteresis loop shape is related to fatigue performance. All the alloys show that as the applied strain amplitude decreases, β_E increases; that is, the loop shape becomes more and more pointed as the endurance increases. These results are consistent with results obtained on vastly different materials [22,23,31].

Conclusions

The following conclusions are drawn from the observed low cycle fatigue behavior of the five aluminum alloys studied:

1. The precipitation hardened alloys, 7005, 6351, and 6061, show cyclic hardening followed by cyclic softening. The essentially sub-grain strengthened 1200 alloy exhibits the same tendency, but the extent of cyclic hardening and softening is considerably lower. Cyclic hardening only is obtained on the solid solution strengthened 5083 alloy.
2. The cyclic hardening generally increases with the strain amplitude and with the ratio of ultimate tensile stress to yield stress for each alloy type.
3. When evaluated in terms of stress amplitude, the 7005 and 1200 alloys have the best and

worst fatigue performance respectively. This order is, however, reversed when plastic strain amplitude is used as a criterion.

4. The 7005, 6351 and 5083 alloys show a decrease in the total energy absorption to failure beyond the break in the Coffin-Manson plot, while the 6061 and especially the 1200 alloy absorb more and more energy with increasing fatigue life.

5. The hysteresis loop shape as expressed through β_E shows a break when plotted against N_f. This break approximately coincides with the break in the Coffin-Manson plot.

6. All the alloys tested show that the hysteresis loop shape becomes more pointed (that is, β_E increases) as the endurance increases.

Acknowledgments

This research was sponsored by the Australian Aluminum Development Council. The materials were generously supplied by Alcan Australia Limited and Comalco Limited.

References

[*1*] Coffin, L. F., "A Study of the Effects of Cyclic Thermal Stress on a Ductile Metal," *Transactions of ASME,* Vol. 76, 1954, pp. 931-950.

[*2*] Manson, S. S., "Behavior of Materials under Conditions of Thermal Stress," NACA TN 2933, 1953.

[*3*] Morrow, J. D., "Cyclic Plastic Strain Energy and Fatigue of Metals," in *Internal Friction, Damping, and Cyclic Plasticity, ASTM STP 378,* American Society for Testing and Materials, Philadelphia, 1965, pp. 45-84.

[*4*] Tomkins, B., "Fatigue Crack Propagation—An Analysis," *Philosophical Magazine,* Vol. 18, 1968, pp. 1041-1066.

[*5*] Saxena, A. and Antolovich, S. D., "Low Cycle Fatigue, Fatigue Crack Propagation and Substructures in a Series of Polycrystalline Cu-Al Alloys," *Metallurgical Transactions,* Vol. 6A, 1975, pp. 1809-1820.

[*6*] Manson, S. S. and Halford, G. R., "Thermal Stress and High Stress Fatigue," Institute of Metals Monograph and Reprint Series, No. 32, 1967, pp. 154-162.

[*7*] Coffin, L. F., "Low Cycle Fatigue: A Review," *Applied Materials Research,* Vol. 1, No. 3, 1962, pp. 129-141.

[*8*] Martin, D. E., "An Energy Criterion for Low Cycle Fatigue," *Journal of Basic Engineering, Transactions of ASME, Series D,* Vol. 83, 1961, pp. 565-571.

[*9*] Sandor, B. I., "Fundamentals of Cyclic Stress and Strain Behaviour," University of Wisconsin Press, Madison, 1972.

[*10*] Quesnil, D. J. and Meshii, M., "The Response of High-Strength Low Alloy Steel to Cyclic Plastic Deformation," *Materials Science and Engineering,* Vol. 30, 1977, pp. 223-241.

[*11*] Landgraf, R. W. in *Achievement of High Fatigue Resistance in Metals and Alloys, ASTM STP 467,* American Society for Testing and Materials, Philadelphia, 1970, pp. 3-36.

[*12*] Sanders, R. E. and Starke, E. A., "The Effect of Grain Refinement on the Low Cycle Fatigue Behaviour of an Aluminum-Zinc-Magnesium-(Zirconium) Alloy," *Materials Science and Engineering,* Vol. 28, 1977, pp. 53-68.

[*13*] Lin, F. S. and Starke, E. A., "The Effect of Copper Content and Degree of Recrystallization on the Fatigue Resistance of 7xxx Type Aluminum Alloys—I: Low Cycle Corrosion Fatigue," *Materials Science and Engineering,* Vol. 39, 1979, pp. 27-41.

[*14*] Sanders, T. H., Staley, J. T., and Mauney, D. A., "Strain Control Fatigue as a Tool to Interpret Fatigue Initiation of Aluminum Alloys," Alcoa Lab., Alcoa Centre, Pa., 1975.

[*15*] Sanders, T. H. and Starke, E. A., "The Relationship of Microstructure to Monotonic and Cyclic Straining of Two Age Hardening Aluminum Alloys," *Metallurgical Transactions,* Vol. 7A, 1976, pp. 1407-1418.

[*16*] Coffin, L. F., "A Note on Low-Cycle Fatigue Laws," *Journal of Materials,* Vol. 6, 1971, pp. 388-402.

[*17*] Tomkins, B., "Fatigue Failure in High Strength Metals," *Philosophical Magazine,* Vol. 23, 1971, pp. 687-703.

[*18*] Laird, C., Langelo, V. J., Hollrah, M., and DeLaVeaux, R., "The Cyclic Stress-Strain Response of Precipitation Hardened Al-15 wt. % Ag Alloy," *Materials Science and Engineering,* Vol. 32, 1978, pp. 137-160.

[*19*] Abdel-Raouf, H., Topper, T. H., and Plumtree, A., "The Influence of Interparticle Spacing on Cyclic

Deformation and Fatigue Crack Propagation in an Aluminum - 4 Pct Copper Alloy," *Metallurgical Transactions,* Vol. 10A, 1979, pp. 449-456.

[*20*] Chung, Y. S., "The Low Cycle Fatigue of Aluminum Alloys," Ph.D. Thesis, University of Sydney, Australia, 1981.

[*21*] Staley, J. T., "Influence of Microstructure on Fatigue and Fracture of Aluminum Alloys," *Aluminum,* Vol. 55, No. 4, 1979, pp. 277-281.

[*22*] Abel, A. and Muir, H., "The Bauschinger Effect and Discontinuous Yielding," *Philosophical Magazine,* Vol. 26, 1972, pp. 489-504.

[*23*] Abel, A., "Fatigue of Copper Single Crystals at Low Constant Plastic Strain Amplitude," *Materials Science and Engineering,* Vol. 36, 1978, pp. 117-124.

[*24*] Manson, S. S., *Thermal Stress and Low Cycle Fatigue,* McGraw-Hill, New York, 1966.

[*25*] Calabresse, C. and Laird, C., "Cyclic Stress-Strain Response of Two-Phase Alloys, Parts I and II," *Materials Science and Engineering,* Vol. 13, 1974, pp. 141-174.

[*26*] Brodrick, R. F. and Spiering, G. A., "Low Cycle Fatigue of Aluminum Alloys," *Journal of Materials* Vol. 7, No. 4, 1972, pp. 515-526.

[*27*] Feltner, C. E. and Laird, C., "Cyclic Stress-Strain Response of F.C.C. Metals and Alloys—II: Dislocation Structures and Mechanisms, *Acta Metallurgica,* Vol. 15, 1967, pp. 1633-1653.

[*28*] Laird, C. in *Cyclic Stress-Strain and Plastic Deformation Aspects of Fatigue Crack Growth, ASTM STP 637,* American Society for Testing and Materials, Philadelphia, 1977, pp. 3-35.

[*29*] Abel, A. and Ham, R. K., "Cyclic Strain Behaviour of Crystals of Aluminum - 4 wt. % Copper," *Acta Metallurgica,* Vol. 14, 1966, pp. 1495-1503.

[*30*] Laird, C. and Thomas, G., "On Fatigue-Induced Reversion and Overaging in Dispersion Strengthened Alloy Systems," *International Journal of Fracture Mechanics,* Vol. 3, No. 2, 1967, pp. 81-97.

[*31*] Abel, A., "Implications of Hysteresis Behaviour with Regard to Fatigue," *Scripta Metallurgica,* Vol. 13, 1979, pp. 903-905.

[*32*] Laird, C. and Feltner, C. E., "Coffin-Manson Law in Relation to Slip Character," *Metal Society of AIME-Trans.,* Vol. 239, No. 7, 1967, pp. 1074-1083.

[*33*] Wells, C. H., "An Analysis of the Effect of Slip Character on Cyclic Deformation and Fatigue," *Acta Metallurgica,* Vol. 17, 1969, pp. 443-449.

[*34*] Santner, J. S. and Fine, M. E., "Hysteretic Plastic Work as a Failure Criterion in a Coffin-Manson Type Relation," *Scripta Metallurgica,* Vol. 11, 1977, pp. 159-162.

[*35*] Halford, G. R., "The Energy Required for Fatigue," in *ASTM Proceedings,* Vol. 66, 1966, pp. 3-16.

[*36*] Grosskreutz, J. C., "Strengthening and Fracture in Fatigue (Approaches for Achieving High Fatigue Strength)," *Metallurgical Transactions,* Vol. 3, 1972, pp. 1255-1262.

[*37*] Nageswararao, M. and Gerold, V., "Fatigue Crack Propagation in Stage I in an Aluminum-Zinc-Magnesium Alloy: General Characteristics," *Metallurgical Transactions,* Vol. 7, 1976, pp. 1847-1855.

R. W. Swindeman[1]

Cyclic Stress-Strain-Time Response of a 9Cr-1Mo-V-Nb Pressure Vessel Steel at High Temperature

REFERENCE: Swindeman, R. W., **"Cyclic Stress-Strain-Time Response of a 9Cr-1Mo-V-Nb Pressure Vessel Steel at High Temperature,"** *Low Cycle Fatigue, ASTM STP 942,* H. D. Solomon, G. R. Halford, L. R. Kaisand, and B. N. Leis, Eds., American Society for Testing and Materials, Philadelphia, 1988, pp. 107–122.

ABSTRACT: Data are presented that describe the behavior of a 9Cr-1Mo-V-Nb steel under conditions of constant and cyclic loading at temperatures from 25 to 600°C. The general features of hardening under monotonic and cyclic strains are presented. Creep data for temperatures in the range 500 to 650°C and times to 10 000 h are correlated by a creep law that is used in conjunction with hardening data to construct isochronous stress versus strain curves applicable to monotonic and cyclic loading. Simple rules for hardening and recovery are postulated and used to estimate the time dependent behavior after mechanically and thermally induced reversal strains. Data from relaxation, partially restrained creep, and two-bar creep ratchetting experiments are presented, and the estimated response is compared with the test data. It is shown that simple concepts of strain hardening and reversed strain softening are adequate for estimating the material behavior under most of the loading conditions that were examined.

KEY WORDS: elevated temperature, cyclic behavior, creep, high temperature design, relaxation, thermal cycling

Since the early 1960s the concept of strain-controlled fatigue has been used successfully in the design of pressure vessels and piping systems that operate under the rules of Section III or Section VIII, Division 2, of the ASME Boiler and Pressure Vessel Code. However, at high temperatures the exponent in the Coffin-Manson law relating cyclic life to inelastic strain range becomes dependent on strain range, frequency, temperature, and environment. An often used solution to this problem for pressure vessel and piping applications requires that a calculated creep-rupture damage component be summed linearly with the fatigue damage component based on the strain fatigue curve for rapid cycling [*1–3*]. Alternative models for time-dependent fatigue have been advanced that do not explicitly involve stress. Specific examples include frequency separation, strainrange partitioning, and damage rate models [*4–6*]. In spite of the attractiveness of these newer models, it seems probable that high temperature pressure vessels and piping components will be designed on the basis of the allowable primary stress intensities for many years to come; hence any model to accommodate high-strain transients that introduce fatigue damage must recognize the importance of stress-induced damage in the overall design. Indeed, the calculation of the time- or creep-related components of cyclic strain in the strainrange partitioning model requires some sort of constitutive equation that relates the thermal and mechanical loads to strain and strain rate. At least one vessel manufacturer uses this approach [*7*]. Thus, although inelastic analysis and stress-induced damage have not been of

[1]Metallurgist, Research Staff, Oak Ridge National Laboratory, Oak Ridge, TN 37831.

primary importance in looking at low-cycle fatigue at low temperatures, they are of utmost importance at temperatures in the creep range.

In the United States typical pressure vessel steels include C-Mn-Si steels, annealed 2.25Cr-1Mo steel, and austenitic alloys such as Type 304 stainless steel and alloy 800H. These alloys are either stable or work harden under conditions of cycling or combined creep and cycling. On the other hand, interest has developed lately in new bainitic and martensitic steels for components in the fossil, petroleum, and petrochemical industries [*8–12*]. These steels exhibit strain softening under monotonic and cyclic straining conditions. Relative to steels traditionally used at high temperature, they are strain-rate sensitive and have high temperature dependencies for the flow stress. These features raise questions regarding the selection of reference strain rates and life fractions on which to base the stress versus strain relationships needed for structural design and evaluation of creep-fatigue damage.

This paper presents mechanical properties data for a 9Cr-1Mo-V-Nb (T91) steel developed for service in the temperature range 450 to 600°C [*11*]. The purpose is to offer information that will aid in the eventual development of a high-temperature design methodology that recognizes the unique behavioral features of these high strength steels and takes full advantage of their desirable features.

Material

The material was from a 15 ton heat AOD melted by Carpenter Technology, Inc., and hot rolled to 51 mm plate by Jessop Steel Company. The chemical composition in weight percent was as follows: 0.083 C, 0.46 Mn, 0.010 P, 0.004 S, 0.41 Si, 0.09 Ni, 8.46 Cr, 1.02 Mo, 0.198 V, 0.072 Nb, 0.005 Ti, 0.03 Cu, 0.002 Al, and 0.051 N. The plate, designated as heat 30383, was normalized at 1020°C, air cooled, tempered at 760°C for 2 h, and given a simulated post-weld heat treatment at 730°C for 20 h. This treatment produced a tempered martensite microstructure with a prior austenite grain size of ASTM No. 8 and an ultimate tensile strength near 700 MPa.

Specimens were machined at three thickness locations through the plate and parallel to the primary rolling direction. No differences in the through thickness direction were observed, and all specimens were considered to represent typical material. Two types of specimen geometries were employed. Tensile, creep, and relaxation tests were performed on threaded-end bar specimens having a reduced section length of 32 mm and a diameter of 6.35 mm. Strain cycling, creep tests with strain reversals, and nonisothermal plasticity tests were performed on button end bar specimens having a reduced section length of 19 mm and a diameter of 6.35 mm. Details of the test methods are provided elsewhere [*13*].

Monotonic and Cyclic Hardening

Tracings from extensometer plots produced by tensile testing are provided in Fig. 1*a* to show the general hardening features observed during monotonic straining at temperatures in the range 25 to 600°C. Consistent with the trends for other martensitic alloys, the yield strength is a very high fraction of the ultimate strength and the strain at which the ultimate strength occurs is quite low, decreasing from 8% strain at 25°C to 1% at 600°C. At strains between the proportional limit and 1% the work hardening rate is very high, but strongly dependent on the temperature and strain rate at temperatures in excess of 450°C. Necking develops very gradually beyond the strain at which the ultimate occurs, and conversion of the engineering stress versus engineering strain data to true stress versus strain, assuming constancy of volume, would reveal a range of strain of several percent in which the flow stress is essentially constant. At the highest temperatures of interest, say, 600°C, this stable flow stress is achieved at very low strains, as implied by the data shown in Fig. 1*b*.

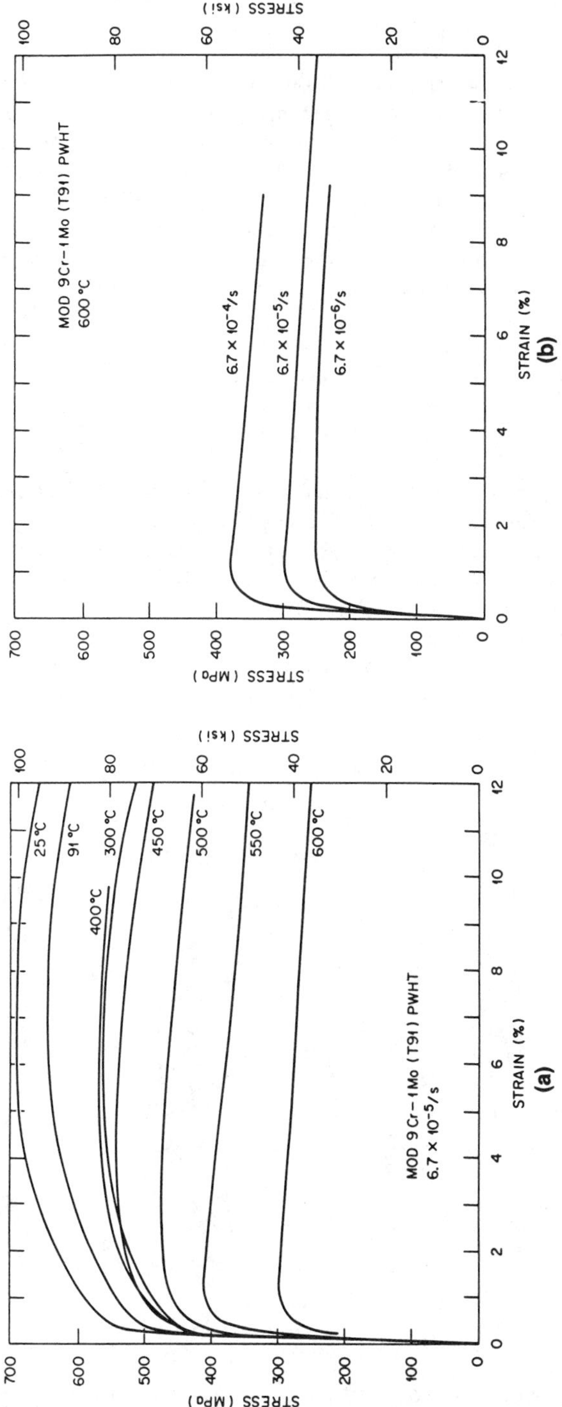

FIG. 1—*Stress versus strain tensile curves for T91. (a) Effect of temperature at 6.7 × 10^{-5}/s rate. (b) Effect of strain rate at 600°C.*

Typical stress versus strain hysteresis loops are shown in Fig. 2 for several temperatures and a strain rate of 10^{-4}/s. These loops reveal a high initial yield strength and elastic range followed by gradual cyclic softening. To expand on this point, the cyclic stress versus strain curves have been plotted in Figs. 3*a* and 3*b* to correlate the stress amplitude with strain amplitude for a constant number of cycles (1, 10, 100, and 1000) at 500 and 600°C. At the highest temperature, these curves not only reveal the cyclic softening tendency but also show that the stress amplitude decreases with increasing strain amplitude. Such curves are difficult to work with in structural analysis. Selecting a fraction of life as a basis for constructing cyclic stress versus strain curves produces a more desirable diagram (Figs. 3*c* and 3*d*). Here, the curves again indicate cyclic softening, insomuch as the curve for 5% of life falls below the first cycle curve, and the curve for half-life is even lower. A desirable feature of these curves is that they show a monotonic rise in stress amplitude with strain amplitude at strains of most interest to design.

In this paper no analytical model is developed to represent the evolution of the hysteresis loop with test parameters. However, when needed for analytical purposes, the hardening within a loop was represented by a simple power law relation between the stress and the plastic strain:

$$S = A e_p^m \tag{1}$$

where S is stress (MPa) in the tensile direction, e_p is plastic strain (%) determined from the offset between the elastic modulus line placed through the point where the loop crosses the zero stress line and the loop, and A and m are constants that vary with temperature, strain rate, and cycle number. In evaluating the constants, Bauschinger strains produced during unloading have been ignored. The equation also implies that the yield stress after stress reversal is at zero stress. Examination of the hysteresis loops show these to be rather gross approximations to the hardening curve, but the understanding of the constitutive behavior for cyclic plasticity in T91 is not sufficient to specify a more elegant model at this time. Indeed, for multiaxial, nonradial

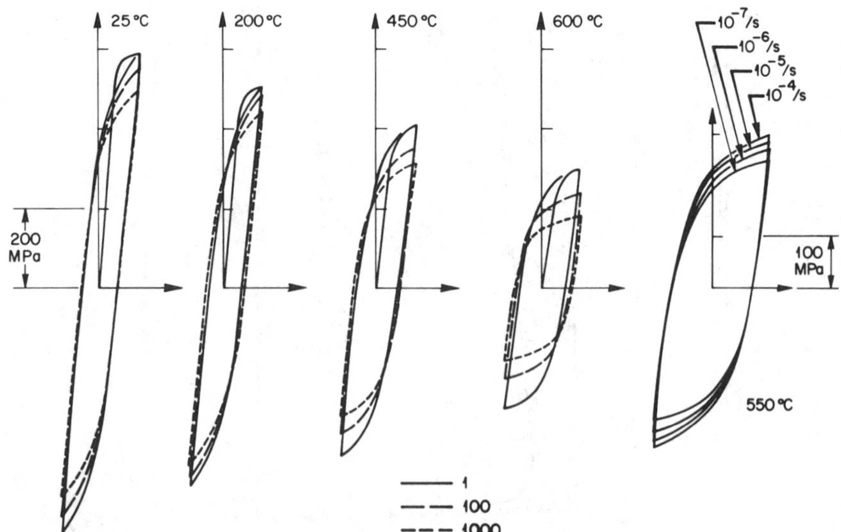

FIG. 2—*Stress versus strain hysteresis loops for 0.95% strain range, showing the effects of cycle number and temperature at 10^{-4}/s rate and strain rate at 550°C.*

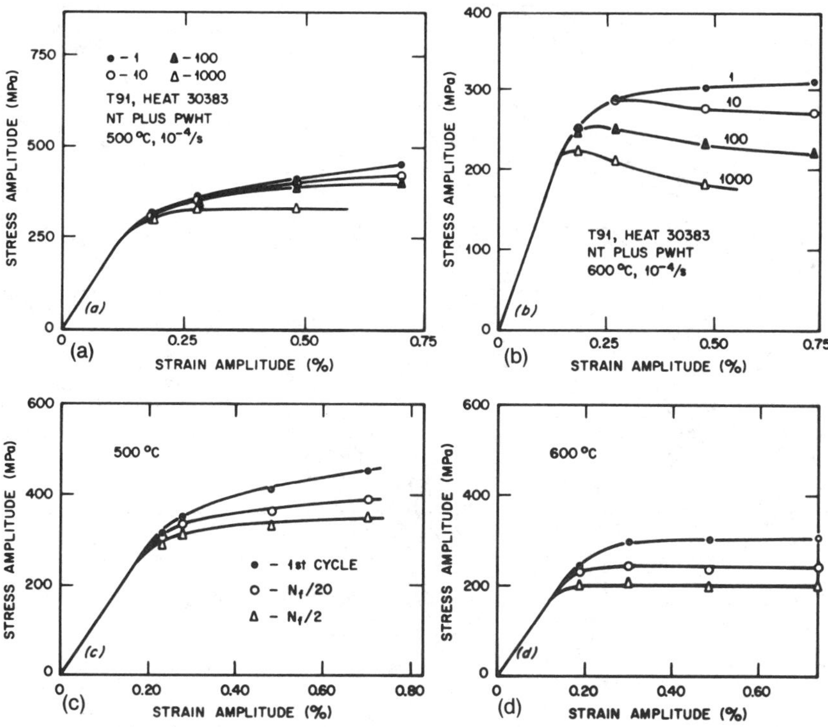

FIG. 3—*Stress amplitude versus strain amplitude for* (a) *1, 10, 100, and 1000 cycles at 500°C,* (b) *1, 10, 100, and 1000 cycles at 600°C,* (c) $N_f/20$ *and* $N_f/2$ *at 500°C, and* (d) $N_f/20$ *and* $N_f/2$ *at 600°C.*

loadings analysts today might select a bilinear or piecewise linear representation of the hysteresis loop [*14*]. Since this paper only addresses uniaxial stresses, a power law hardening model does not introduce complications. When needed, the values for A and m have been determined for the hysteresis loop just prior to the introduction of the time dependent leg of the test. Typical values are provided in the Appendix.

Creep

The behavior of T91 under high strain rate transients can be represented graphically, as indicated above, or by an analytical model. Either way, real-time data can be used. The situation is different for creep, however, since long times are involved, and some predictive capability is needed to estimate the performance of the new steel. The creep testing of T91 was largely directed toward the examination of hardening and softening in the primary and secondary stages of creep for the purpose of obtaining a creep law. Typical creep data are provided in Fig. 4, which shows log creep strain versus log time for several stresses and temperatures. A striking feature of all curves is the tendency for parabolic hardening at low strains. This feature produces very large changes in the creep rate during the primary creep stage. Although not obvious in the log-log plots in Fig. 4, the second stage of creep is established at low creep strain levels, typically below 1%, and, in this sense, behavior is consistent with the observations made earlier with respect to the low strain at which the ultimate strength is attained.

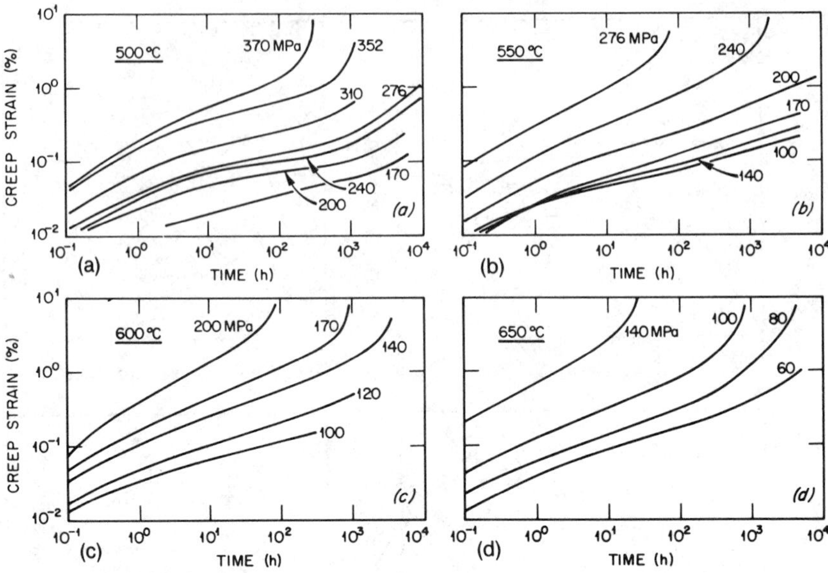

FIG. 4—*Log creep strain versus log time curves for* (a) *500°C,* (b) *550°C,* (c) *600°C, and* (d) *650°C.*

The creep data can be represented by a creep law based on simple power law hardening and secondary creep rate terms:

$$e_c = Bt^{1/3} + \dot{e}_m t \tag{2}$$

where e_c is creep strain (%), t is time (h), B is a constant, and $\dot{e}_m$ is the minimum creep rate (%/h). The constants B and $\dot{e}_m$ show the following stress and temperature dependencies:

$$B = C(V1\ S)^{n_1} \exp(V1\ S) \exp(-Q1/T) \tag{3}$$

$$\dot{e}_m = D(V2\ S)^{n_2} \exp(V2\ S) \exp(-Q2/T) \tag{4}$$

where S is stress (MPa), T is absolute temperature (K), and C, D, n_1, n_2, $V1$, $V2$, $Q1$, and $Q2$ are constants whose values are provided in the Appendix.

Isochronous Curves

For purposes of estimating structural response, isochronous stress versus strain curves show considerable promise, since these curves lend themselves to problems involving monotonic, cyclic, and combined straining. The construction and use of the curves, however, may not be very straightforward for work hardening and work softening alloys, and the selection of the curve representing the high strain rate transient requires astute judgement. In this paper the monotonic curve corresponding to a strain rate of approximately 10^{-4}/s is used as the basis for anchoring the time dependent strains. This strain rate is high relative to transients in many pressure vessel and piping applications, but is low relative to the strain rate often used in fatigue testing. Nevertheless, the cyclic stress versus strain curves for the 9Cr-1Mo-V-Nb steel will nest below the first cycle and tensile curve, as indicated by the curves shown in Fig. 3. Good reasons could be found for selecting one of the cyclic curves as anchoring lines in many applications.

Isochronous creep curves may be constructed by adding to the anchor line the creep strains corresponding to the stresses and times of interest. These strains can be read from graphs, such as provided in Fig. 4, or calculated from the model described above. The curves reflect the behavior of components deforming under essentially primary membrane stresses, and a typical set of curves for a temperature of 550°C is provided in Fig. 5*a*. In this paper, however, emphasis is placed on the relaxation of transient stresses toward the long time primary stress levels. For such problems isochronous relaxation curves can be of more value than isochronous creep curves. In constructing isochronous relaxation curves the creep law may be used in conjunction with a hardening rule to calculate the relaxed stress for the time of interest. The anchor line is used for identifying the starting stress or strain. With the assumption of a simple strain hardening rule for creep, the relaxed stress at any time can be determined by breaking the relaxation process into small increments of constant stress creep and reducing the stress by a stress increment that corresponds to the elastic equivalent of the creep strain increment. Thus

$$S = S_0 - \sum_i \delta Sr_i \tag{5a}$$

$$\delta Sr_i = \dot{e}_i \delta t_i M/100 \tag{5b}$$

where S is stress (MPa) during the relaxation process, δt is the time increment (h) at a stress S_i, S_0 is the starting stress (MPa), $\dot{e}_i$ is the creep rate (%/h) at S_i, and M is the elastic modulus (MPa). The isochronous relaxation curves constructed by this procedure at 550°C are shown in Fig. 5*b*; they may be compared with the isochronous creep curves in Fig. 5*a*. For equivalent times, the isochronous creep and relaxation curves are close together for low starting stresses and short times, but at higher starting stresses the isochronous relaxation curves fall well below the isochronous creep curves. For repetitive relaxation loading, without strain reversals, how-

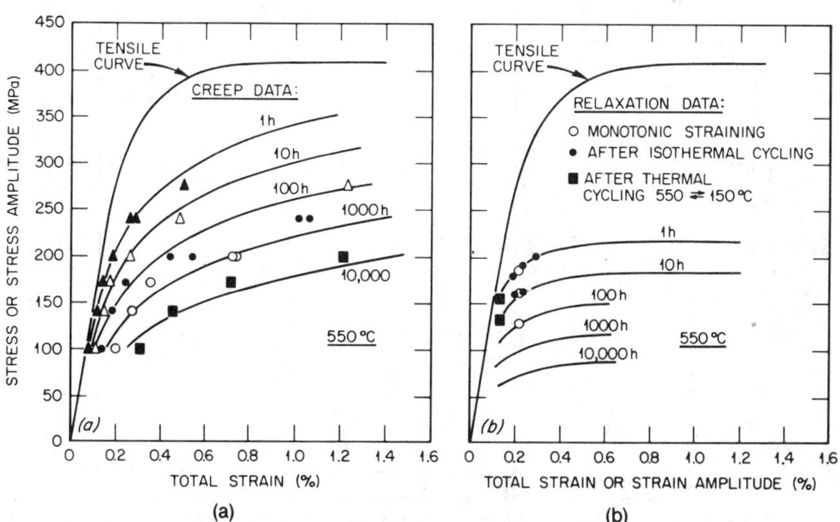

FIG. 5—*Calculated isochronous curves at 550°C:* (a) *creep curves and creep data, and* (b) *relaxation curves and relaxation data produced from monotonic, cyclic, and nonisothermal tests.*

ever, the relaxation strength should increase as the creep strain hardening diminishes the contribution of primary creep to the relaxation rate. Thus one might expect that a family of isochronous relaxation curves should exist for repetitive reloadings without reversals, and that the curves should be nested with strength increasing with the reloading number. Eventually, relaxation curves reflecting the minimum creep rate would result. A long time isochronous creep curve would represent an upper bound to the relaxation strength for monotonic strain accumulation.

As mentioned earlier, a cyclic stress versus strain curve could be selected as an anchoring line for strain reversal conditions, rather than the tensile curve. For simplicity, however, the tensile curve can still be used, provided that the correct S_0 is selected from the cyclic hardening data. In Fig. 5*b* experimental relaxation data from cyclic tests have been included and fall on the curves constructed from the creep law produced from constant load tests and the strain hardening rule. To accomplish this superimposition, the data from cyclic tests must be plotted at strains that superimpose the cyclic stress amplitude on the tensile curve. These results apply regardless of the cycle number and suggest that the reversal strain erases the creep hardening introduced in the prior relaxation period. Similar behavior has been observed in 2.25Cr-1Mo steel [*15*].

Combined Creep and Fatigue

The softening influence of reversal stresses on creep can be seen in the creep loading condition as well as the relaxation condition. A few tests were performed in which reversal stresses were periodically imposed on a creeping specimen. Results from the test series are illustrated in Fig. 6. Here, the creep stress (276 MPa) would produce rupture in a time estimated to be 10^5 h. In the first test, the stress was reversed to -276 MPa with a frequency of 12 cycles/h (3.3×10^{-3} Hz) and a ramp time of 10 s. The hysteresis loop indicated less than 10^{-5} inelastic strain during the reversal. The primary creep rate after the reversal was higher than the monotonic creep rate for the same creep strain, and eventually the hysteresis loop broadened, the creep rate accelerated, and a creep failure occurred in only 600 h. In a second test the frequency of cycling

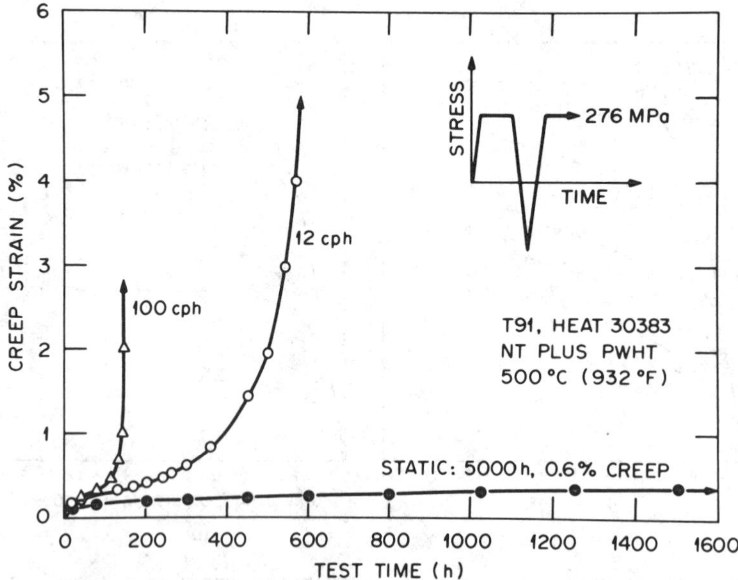

FIG. 6—*Effect of short time stress reversals on the creep at 276 MPa and 500°C.*

was increased to 100 cycles/h (2.8×10^{-2} Hz). In this test the specimen crept more rapidly, but failed in a fatigue mode in 145 h. A third test was started at 1 cycle/h (2.8×10^{-4} Hz), but was discontinued without noticeable acceleration in the creep rate after several hundred hours.

When dealing with high temperature cyclic problems involving martensitic steels, one must be aware that realistic loadings cannot always be categorized as strain cycling with interspersed relaxation or creep with occasional strain cycling. Rather, the structure may go through a transient, then gradually relax to a condition involving slow creep at the primary stress level. Thus, the isochronous relaxation must be restricted to use for relatively short times, where the relaxation strain rate is greater than the creep rate associated with the primary stresses. As a way of examining the transition from relaxation to creep after a stress reversal, several experiments were performed in which the strain rate was drastically reduced near the tensile strain amplitude, then the strain was allowed to continue beyond the tip of the hysteresis loop. An example of the specimen response is shown in Fig. 7, where the stress versus strain trajectory may be compared to the constructed 10 h isochronous creep and relaxation curves. The actual strain range for the test was near 0.6%, and the stress amplitude superimposed on the tensile loading curve near 0.2% strain. From the initial stress of 300 MPa, corresponding to a strain rate of 10^{-4}/s, the stress dropped rapidly under the imposed strain rate of 10^{-8}/s, approached the 10 h relaxation curve, and then began to increase toward the 10 h isochronous creep curve. We expect that the stress would be asymptotic to the stress to produce a minimum creep rate of 10^{-8}/s (240 MPa at 550°C).

Calculations were performed to model the stress versus strain response for large strain rate changes. It was assumed that the flow stress increases during plastic straining according to the power law described in Eq 1. Then during any small time increment the increase in flow stress,

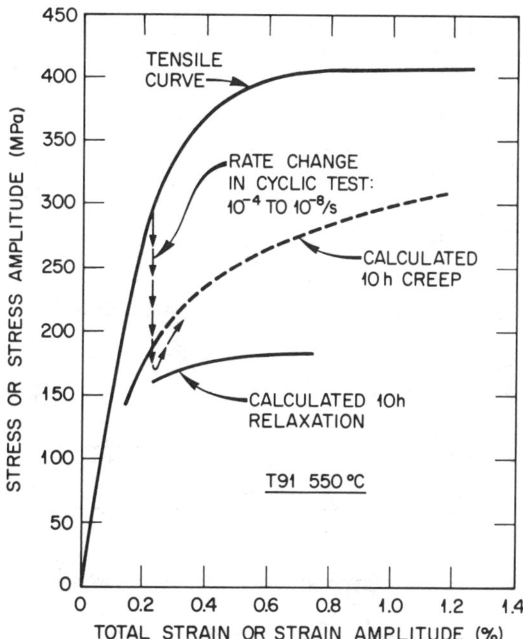

FIG. 7—*Comparison of isochronous creep and relaxation curves for 10 h to the stress versus strain response after a large strain rate decrease in a cyclic test at 550°C.*

δS_p, can be calculated from knowledge of the plastic strain at that time and the increment of plastic strain that occurs due to the imposed strain rate acting during that time. Thus

$$\delta S_p = (dS/de_p)\dot{e}_p\delta t \tag{6}$$

where (dS/de_p) can be obtained by differentiating Eq 1 and solving for the strain of interest. During the same time interval, of course, some relaxation can occur. This decrease in stress due to relaxation, ΔS_r, can be estimated by the same procedure as outlined earlier for relaxation testing. The stress at any time after the strain rate change can then be estimated by

$$S = S_0 + \sum_i \delta S_{p_i} - \sum_i \delta Sr_i \tag{7}$$

Some test data are compared to calculations from Eq 7 in Fig. 8. These calculations, although somewhat crude, capture the trend for the data and indicate that after a strain reversal initial softening is present that makes the steel behave as a monotonic test with no prior creep strain. The exhaustion of primary creep during the relaxation process, however, results in eventual hardening under strain rate controlled conditions. The curves are expected to asymptotically approach the stress amplitudes corresponding to minimum creep rates equivalent to the imposed strain rates. Again, the influence of the reversed straining is to erase the previous creep hardening.

Nonisothermal cyclic tests can be quite representative of real transient conditions, but are difficult to define, perform, and interpret. A number of different tests were performed on T91

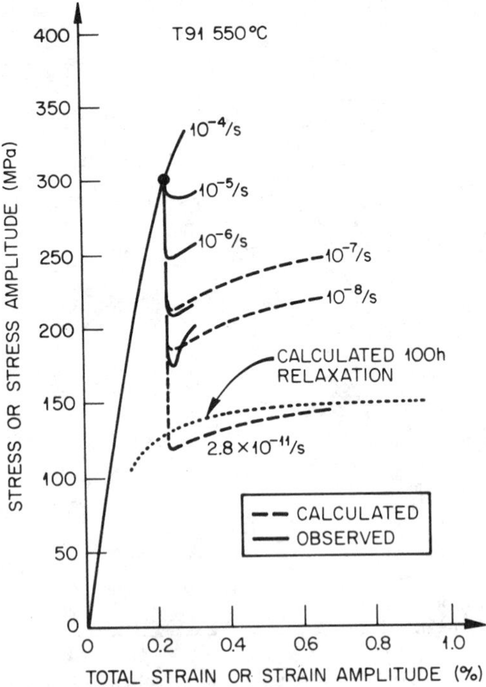

FIG. 8—*Comparison of experimental and calculated stress versus strain response to strain rate reductions in cyclic tests at 550°C.*

specimens that involved block type loading, where a number of cycles were introduced at two different temperatures, one-bar tests, where the specimen length was held constant while the temperature was cycled between two extremes, and two-bar ratchetting tests, in which two specimens were forced to have equal strains but undergo out-of-phase temperature excursions [*16,17*]. Hold times at high temperature were introduced for times up to 12 h.

Examples of hysteresis loops produced by block loadings are provided in Fig. 9. Here, the loop on the left characterizes behavior at 150°C and 0.4% strain range, while the hysteresis loop on the right characterizes the behavior after heating the same specimen to 550°C, strain cycling again at 0.4% strain range, but introducing a relaxation hold for 1 h at the compression strain amplitude. In between these two extremes is a loop produced by straining on the tensile side at 150°C and on the compression side at 550°C. The most obvious and expected feature of the loop involving both temperatures is the development of a large mean stress in tension. To a first approximation, however, the tensile peak stress matches the isothermal cycling peak stress, and the compressive peak stress matches that for the isothermal cycling at 550°C. Thus in this test isothermal data could be used to estimate behavior under isothermal cycling.

Results from a one-bar restrained thermal cycling test are provided in Fig. 10. In this experiment the specimen was held at constant length while the temperature was cycled between 550 and 150°C with a 600 s (10 min) ramp. One insert figure shows the stress versus thermal strain for the first cycle; another insert figure shows the temperature versus time pattern after 300 cycles when a 12 h hold was introduced at the 550°C limit of each cycle. The stress versus cycle pattern indicates the same trends as indicated by the data in Fig. 9. That is, a mean stress develops in tension, which occurs on cool down, and the peak stresses can be estimated from isothermal data at the mechanical strain range corresponding to the thermal strain range in the one-bar nonisothermal test. The 12 h relaxation period shifts the compressive peak toward zero stress and increases the tensile peak slightly.

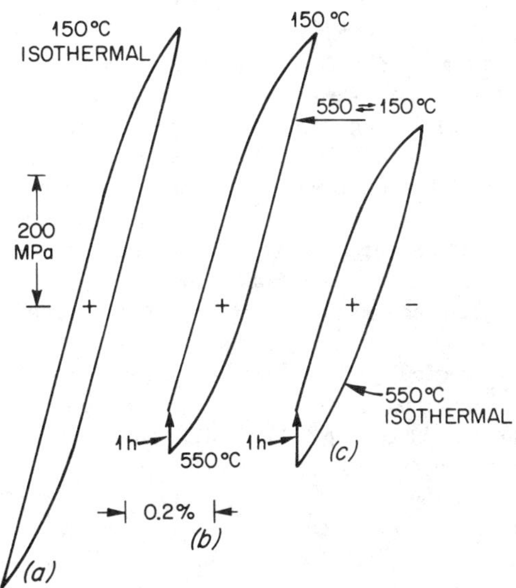

FIG. 9—*Comparison of hysteresis loops near half life and 0.42% strain range for* (a) *isothermal continuous cycling at 150°C,* (b) *tensile strain at 150°C and compressive strain at 550°C with hold time, and* (c) *isothermal cycling at 550°C with hold time.*

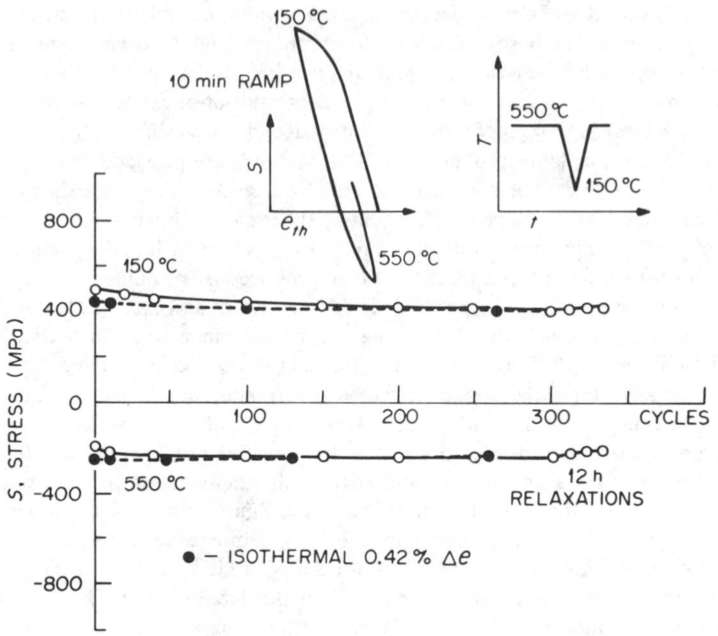

FIG. 10—*Comparison of stress amplitudes produced by isothermal cycling at 150 and 550°C with peak stresses produced in a one-bar restrained thermal cycling test between 150 and 550°C.*

The two-bar ratchetting test differs significantly from the one-bar test in the sense that the mean stress can be maintained throughout the test at a predetermined level while allowing the samples to progressively deform from cycle-to-cycle by the accumulation of plastic, relaxation, or creep strains [*16,17*]. In Figures 11 and 12 results are shown from a two-bar ratchetting test in which the temperature was cycled between 550 and 150°C, with a mean stress of 140 MPa and a hold time at 550°C of 11.5 h. Figure 11*a* shows the temperature versus time histograms for bars 1 and 2. Bar 1 cooled from 550°C to 150°C in 600 s (10 minutes). Bar 2 then cooled in a similar way. The bars were heated uniformly back to 550°C in 600 s (10 min) and held there until the repeat cycle 11.5 h later. The stress versus strain patterns for the bars during the 1st and 30th cycles are also illustrated in Figs. 11*b* and 11*c* for total strain and in Figs. 11*d* and 11*e* for mechanical strain. Here, both bars started at the mean stress of 140 MPa, but during cool down of bar 1, the thermal contraction in bar 1 increased the stress and produced a corresponding stress decrease in bar 2 to maintain equal strain and a constant mean stress. The restraint on bar 1, therefore was only partial, and, although bar 1 experienced the same thermal strain as the one-bar test specimen described earlier, the mechanical strain was less, and only a small amount of yielding occurred. The cooling of bar 2 produced essentially elastic stress changes, and, after the two bars were reheated to 550°C, the net strain was small and around 0.03%. Bar 1 developed a stress above the mean stress after the reheat, while bar 2 was below the mean stress by an equal amount. During the period of constant temperature the deformation of the two bars was a combination of relaxation and creep. About 0.03% time-dependent deformation occurred. After the first cycle little or no plastic yielding occurred, and virtually all of the ratchetting strain was produced during the creep hold period. Figure 12 compares the accumulated creep strain to a constant load creep curve for 140 MPa and to a creep curve constructed by linking segments of the first 11.5 h of primary creep from the constant load curve. This curve is

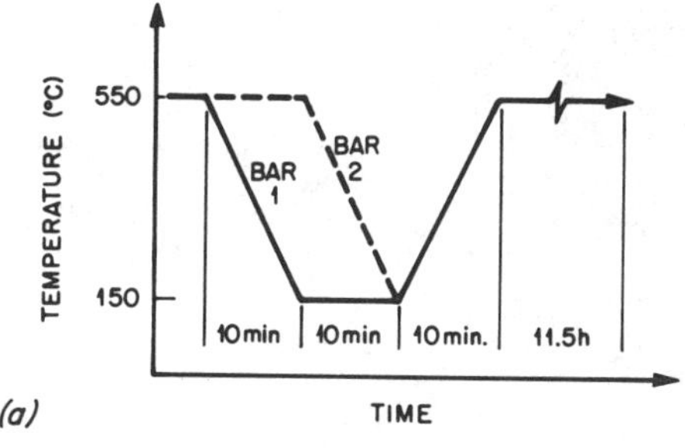

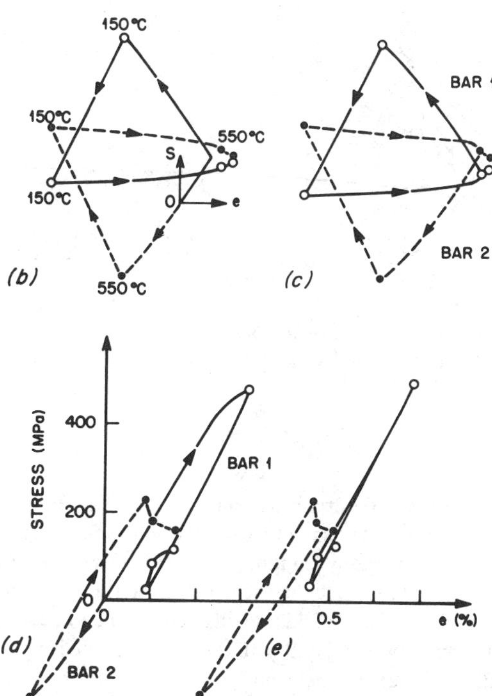

FIG. 11—*Data produced in a two-bar creep ratchetting test.* (a) *Temperature versus time histogram.* (b) *Stress versus strain for the two bars in the first cycle.* (c) *Stress versus strain for the two bars in the 30th cycle.* (d) *Stress versus mechanical strain for the two bars in the first cycle.* (e) *Stress versus strain for the two bars in the 30th cycle.*

identified as the trend if reversal strains erase creep hardening. The ratchetting creep curve falls between these two extremes, but the creep rate under these rather severe conditions is diminishing and is approaching that expected for constant load conditions.

Discussion

The T91 steel described in this paper was recently approved for construction under the rules in Section I and VIII of the ASME Boiler and Pressure Vessel Code and has attracted much interest as a boiler tube material. The very high design stress levels and good corrosion resis-

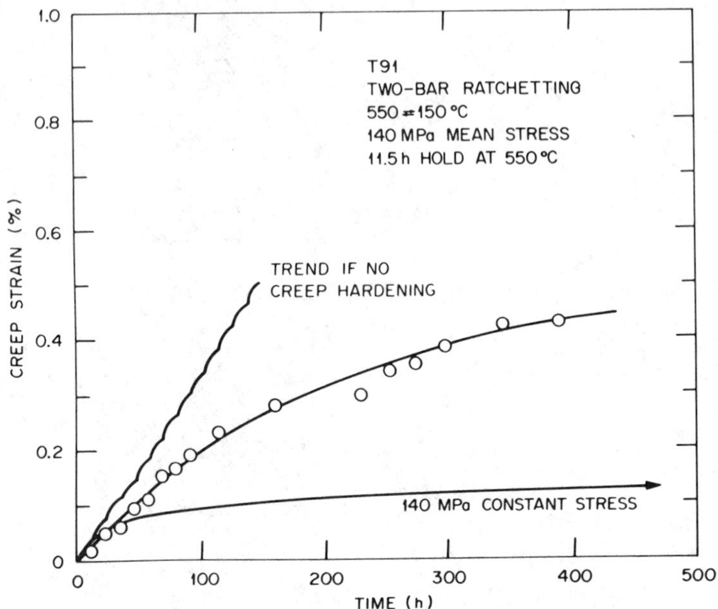

FIG. 12—*Comparison of the creep strain versus time for a two-bar ratchetting test at a mean stress of 140 MPa with creep data at the same stress and temperature but assuming either no creep hardening from cycle to cycle or full creep hardening.*

tance to steam make it an attractive replacement for both T22 and stainless steels in steam service. Although behavioral features such as deformation induced softening do not enter directly into the Code design rules, many component manufacturers recognize the importance of such information in the creep and fatigue damage assessment needed to predict residual life of serviced components. It follows that such data will eventually be needed for T91. Even for new construction, potential problems associated with the lack of service experience can be mitigated by a sound understanding of material behavior.

The high initial yield strength, low thermal expansion coefficient, and good thermal conductivity of T91 greatly reduce concerns about fatigue damage produced by thermal transients in this material. Moreover, the continuous cycling data indicate that the softening is not severe when temperatures are below 550°C and cyclic strains are less than 1%. For more severe conditions, however, engineers must carefully choose what strain rate and what fraction of life or cycle number best represents their reference design condition. With this information, and an adequate data base, the design analysis can proceed. Alternatively, more sophisticated analyses could be performed with knowledge of the load histogram in conjunction with a unified constitutive model that incorporates evolutionary state variables to accommodate hardening and softening. Much progress has been made recently in modeling cyclic softening by means of unified equations [*14,15,18*].

The data presented here indicate that the behavior of T91 is not overly complicated. A good feature is the high initial creep rate in the primary stage. This allows rapid relaxation of secondary stresses. This same feature, however, is responsible for the high strain rate dependence of the flow stress and the slow rate of shakedown under ratchetting experiments. The acceleration of the creep rate when small cyclic strains are superimposed on creep needs more study. Indications are that it is a cycle dependent phenomena and may not be significant under service conditions expected for most engineering structures [*19,20*]. It does suggest, however, that more attention should be paid to combined creep and fatigue conditions where most of the damage is to be expected from the creep loading.

Conclusions

1. After an initially high rate of hardening, T91 softens under both monotonic and cyclic strain conditions. The extent of softening increases with increasing temperature and strain rate.
2. Creep rates in T91 are initially very high but decay away as a power law. These high rates produce rapid relaxation after reversed plasticity and permit the construction of isochronous relaxation curves that are applicable to reversed loading conditions followed by periods of strain control.
3. The rapid recovery under cyclic stresses allows the use of isothermal cyclic data in the estimation of stresses developed under nonisothermal conditions.

Acknowledgments

Much of the testing was performed by B. C. Williams. The paper was reviewed by M. K. Booker and C. R. Brinkman. The research was sponsored by the U.S. Department of Energy, AR&TD Fossil Energy Materials Program, under Contract DE-AC05-840R21400 with Martin Marietta Energy Systems, Inc.

APPENDIX

The values for the constants in Eq 3 were determined by least squares fitting 0.2% primary creep data to the model and found to be as follows: $C = 2.366 \times 10^9 \%/\text{h}^{1/3}$; $n_1 = 1.26$; $V1 = 0.01316\ (\text{MPa})^{-1}$; $Q1 = 23\ 260$ K. The standard error of estimate for the fit of 20 data was 0.06 in log time. The values for the constants in Eq 4 were determined in a similar way from minimum creep rate data and found to be as follows: $D = 1.685 \times 10^{33} \%/\text{h}$; $n_2 = 6.117$; $V2 = 0.04338\ (\text{MPa})^{-1}$; $Q2 = 88\ 160$ K. The standard error of estimate for the fit of 20 data to the model was 0.23 in log rate.

TABLE 1—*Hardening, modulus, and coefficient of thermal expansion data at several temperatures.*

Temp. (°C)	Tensile Data (MPa)		Cyclic Data (MPa)[a]		Modulus (GPa)	Coef. of Thermal Expansion (μe/°C)
	A	*m*	*A*	*m*		
25	590	0.06	740	0.37	205	. . .
200	550	0.08	650	0.35	195	11.9
300	520	0.11	580	0.33	190	12.1
400	510	0.13	520	0.32	180	12.7
500	460	0.14	450	0.28	170	13.3
600	310	0.14	260	0.24	160	. . .

[a] *A* and *m* from cyclic tests were estimated at 0.4% plastic strain range. Ramp rates were approximately 10^{-4}/s.

As indicated earlier, the constants for the stress versus strain relation (Eq 1) varied significantly with the testing conditions. The calculations displayed in Fig. 8 were based on $A = 500$ MPa and $m = 0.33$. These values were obtained from hysteresis loops near 5% of life, 550°C, 0.4% plastic strain range, and 0.0001/s rate. Typical values for A and m derived from tensile and cyclic testing are provided in Table 1 along with elastic modulus and thermal expansion data.

References

[1] Campbell, R. D. in *Advances for Design for Elevated Temperature Environment,* American Society of Mechanical Engineers, 1975, pp. 45–56.

[2] Masuyama, F., Setoguchi, K., Haneda, H., and Nanjo, F. in *Residual-Life Assessment, Nondestructive Examination, and Nuclear Heat Exchanger Materials,* PVP-Vol. 98-1, American Society of Mechanical Engineers, 1985, pp. 79–90.

[3] Berman, I. and Rao, M. S. M. in *Thermal and Environmental Effects in Fatigue: Research-Design Interface,* PVP-Vol. 71, American Society of Mechanical Engineers, 1983, pp. 75–91.

[4] Coffin, Jr., L. F. in *1976 ASME-MPC Symposium on Creep-Fatigue Interaction,* MPC-3, American Society of Mechanical Engineers, 1976, pp. 349–363.

[5] Manson, S. S., Halford, G. R., and Hirsberg, M. H. in *Design for Elevated Temperature Environment,* American Society of Mechanical Engineers, 1971, pp. 12–28.

[6] Majumdar, S. and Maiya, P. S., *Journal of Engineering Materials and Technology,* Vol. 102, 1980, pp. 159–167.

[7] Lawton, C. W. in *Low-Cycle Fatigue and Life Prediction, ASTM STP 770,* American Society for Testing and Materials, Philadelphia, 1982, pp. 585–599.

[8] Ishiguro, T., Murakami, Y., Ohnishi, K., and Watanabe, J. in *Application of 2¹/₄Cr-1Mo Steel for Thick-Wall Pressure Vessels, ASTM STP 755,* American Society for Testing and Materials, Philadelphia, 1982, pp. 383–417.

[9] Wada, T. and Cox, T. B. in *Research on Chrome-Moly Steels,* MPC-21, American Society of Mechanical Engineers, 1984, pp. 77–93.

[10] Ritchie, R. O., Parker, E. R., Spencer, P. N., and Todd, J. A., *Journal of Materials for Energy Systems,* Vol. 6, 1984, pp. 151–162.

[11] Sikka, V. K., Ward, C. T., and Thomas, K. C. in *Ferritic Steels for High Temperature Applications,* American Society for Metals, 1982, pp. 65–84.

[12] Swindeman, R. W., *Journal of Testing and Evaluation,* Vol. 7, 1979, pp. 192–198.

[13] Swindeman, R. W. in *Research on Chrome-Moly Steels,* MPC-21, American Society of Mechanical Engineers, 1984, pp. 31–42.

[14] McDowell, D. L., Socie, D. F., and Lamba, H. S. in *Low-Cycle Fatigue and Life Prediction, ASTM STP 770,* American Society for Testing and Materials, Philadelphia, 1982, pp. 500–518.

[15] Pugh, C. E. and Robinson, D. N. in *Pressure Vessels and Piping: Design Technology - 1982 - A Decade of Progress,* American Society of Mechanical Engineers, 1982, pp. 171–183.

[16] Stentz, R. H. in *Cyclic Stress-Strain Behavior—Analysis, Experimentation, and Failure Prediction, ASTM STP 519,* American Society for Testing and Materials, Philadelphia, 1973, pp. 3–12.

[17] Morrow, D. L. in *Thermal and Environmental Effects in Fatigue: Design-Research Interface,* PVP-Vol. 71, American Society of Mechanical Engineers, 1983, pp. 59–73.

[18] Mróz, Z., *Journal of Engineering Materials and Technology,* Vol. 105, 1983, pp. 113–119.

[19] Jones, W. B. in *Ferritic Steels for High Temperature Applications,* American Society for Metals, 1983, pp. 221–235.

[20] Handrock, J. L. and Marriott, D. L. in *Properties of High-Strength Steels for High-Pressure Containment,* P-117-M-27, American Society of Mechanical Engineers, 1986.

E. Krempl,[1] *H. Lu,*[1] *M. Satoh,*[1] *and D. Yao*[1]

Viscoplasticity Based on Overstress Applied to Creep-Fatigue Interaction

REFERENCE: Krempl, E., Lu, H., Satoh, M., and Yao, D., **"Viscoplasticity Based on Overstress Applied to Creep-Fatigue Interaction,"** *Low Cycle Fatigue, ASTM STP 942,* H. D. Solomon, G. R. Halford, L. R. Kaisand, and B. N. Leis, Eds., American Society for Testing and Materials, Philadelphia, 1988, pp. 123–139.

ABSTRACT: The uniaxial version of the viscoplasticity theory based on overstress is applied to the deformation behavior of Type 304 stainless steel at 650°C. The effects of strain rate, waveform, and position of hold-time are computed for a cyclic steady-state as well as the ratchetting behavior under zero-to-tension load control.

The computations of deformation behavior are used in correlating and predicting the low-cycle fatigue life at 593°C using the strainrange partitioning method as well as an incremental damage accumulation law developed previously. The results of the computations exhibit the appropriate trends, but can show deviations from actual life exceeding a factor of two. Considering all the uncertainties, the results are deemed acceptable and promising.

The incremental damage accumulation law permits application to arbitrary stress or strain histories. The effect of position of hold-time on the hysteresis loop is computed and shown to exhibit the appropriate trend.

KEY WORDS: low-cycle fatigue, life prediction, stainless steel, elevated temperature, constitutive equation, mathematical modeling

Within recent years the methods of low-cycle fatigue life prediction have undergone considerable changes. In the early days of the Coffin-Manson Laws, it was only required to know the ductility of the alloy for predicting the fatigue life for a given plastic strain range. The methods were extended to elevated temperature, including hold-times with creep and relaxation, and to different waveforms such as slow-fast and fast-slow. It then became necessary to know precisely the shapes of the hysteresis loops so that relevant quantities for the various life prediction methods could be deduced. This trend started with the frequency-modified stress range and continues with the partitioned strain ranges and others. It is becoming increasingly clear that life prediction and deformation behavior are interdependent.[2] In experimentation the hysteresis loops are determined during a test, and the relevant quantities can be read directly from the test records.

In the use of the newly developed life prediction methods in the design of a component operating under severe conditions of loading and environment, deformation behavior must be obtained by calculation. In this case a realistic material model is indispensable. It should be available in multiaxial form for use in conjunction with a finite element code or equivalent so that the appropriate hysteresis loops under given loading conditions can be computed. These quantities are then used as inputs to the life prediction law. Of course, the quantities necessary to predict

[1]Department of Mechanical Engineering, Aeronautical Engineering and Mechanics, Rensselaer Polytechnic Institute, Troy, NY 12180-3590. Dr. Lu is presently with Bendix Corporation, South Bend, Indiana. Dr. Satoh is presently with Nippon Kokan KK, Tokyo, Japan.

[2]It is not the purpose of this paper to give a historical account of this development.

the life are different for different life prediction methods and the material model or constitutive equation should be capable of determining these quantities.

The life prediction methods which emanate from the original Coffin-Manson Laws [*1,2*] postulate a typical cycle for which the relevant quantities must be determined. This method has the advantage that only one cycle has to be considered in the analysis. However, the cycle must be of the type that was considered in the development of the life prediction law. If the cyclic pattern is not periodic, or periodic but different from that considered in the development of the life prediction law, uncertainties arise which are hard to resolve. The reason for these uncertainties lies in the algebraic or finite nature of these life prediction laws.

Life prediction laws which lend themselves naturally to the evaluation of the life spent under variable loading are those formulated in incremental form [*3,4*]. By virtue of their incremental nature they can be integrated for any stress or strain path and give an indication of the life used up under such paths. In the case of periodic loading only one cycle needs to be considered as in the case of algebraic laws.

Due to the path dependence of the inelastic deformation of metals, material models for the prediction of deformation must be formulated in an incremental fashion. Such an incremental formulation couples naturally with an incremental life prediction law. However, it is also possible to integrate the constitutive equation for a certain typical cycle, to plot the results in terms of stress versus strain and to determine the quantities of interest from the calculated hysteresis loops instead of from the experimental ones.

The intention of this paper is to demonstrate the use of the viscoplasticity theory based on overstress (VBO) in calculating hysteretic behavior. After an introduction of the salient features of the theory, test data obtained for AISI Type 304 stainless steel at 650°C are used to identify the material functions. With these data various hysteresis loop shapes are determined numerically by integrating the incremental equations. Command waveforms considered are hold-times at various positions around the hysteresis loop, small hysteresis loops within a large loop, and change of strain rate within a loop.

Viscoplasticity Theory Based on Overstress

Outline of the Model

This theory is of the "unified type" and does not differentiate between plastic and creep strains. Therefore

$$\dot{\epsilon} = \dot{\epsilon}^{e\ell} + \dot{\epsilon}^{in} \tag{1}$$

where a superposed dot designates differentiation with respect to time, ϵ designates the engineering strain,[3] and the superscripts "$e\ell$" and "in" denote the elastic and inelastic parts of the strain rate, respectively.

In VBO the inelastic strain rate is solely a function of the overstress defined as the difference between the applied engineering stress and the equilibrium stress[4] which is defined as the external stress obtained under very slow (mathematically infinitely slow) loading. Therefore

$$\dot{\epsilon}^{in} = \frac{\sigma - g}{Ek[\sigma - g]} \tag{2}$$

[3]The use of engineering strain permits a straightforward generalization to three dimensions. If only uniaxial versions are to be considered, then ϵ can also be interpreted as the true strain.

[4]When true strain is used, the stresses should be interpreted as true stresses.

The positive viscosity function k (square brackets following a symbol denotes "function of") has the dimension of time and controls the rate sensitivity or the viscous properties. The quantity $\sigma - g$, the overstress, is akin to the effective stress used in materials science, but has a different meaning here (see [5]). The modulus of elasticity is designated by E.

The equilibrium stress represents approximately the time-independent or plastic properties of the material and needs a growth law indicating how g changes with deformation. The growth law chosen is (see [6])

$$\dot{g} = \psi[\sigma - g]\dot{\epsilon} - \frac{(\psi[\sigma - g] - E_t)(g - E_t\epsilon)}{A}|\dot{\epsilon}^{\text{in}}| \tag{3}$$

With this growth law the equilibrium stress describes a hysteresis loop during inelastic cyclic loading. An evaluation of the properties of the model is given in Ref *6*. The function ψ has the dimension of stress and is termed the *shape (modulus) function,* since it controls the shape of the stress-strain curve at the transition from the nearly linear region to the region of predominantly inelastic deformation. It is positive and decreasing. The constant E_t designates the tangent modulus at the maximum strain of interest. The quantity A with the dimension of stress is the asymptotic value of $\{g - E_t\epsilon\}$ and will be explained shortly.

Equations 1 to 3 represent a system of nonlinear differential equations whose mathematical properties have been studied in detail [6]. Under constant but arbitrary strain rate they permit asymptotic solutions which are algebraic expressions and are useful in the identification of the material functions k and ψ and of the constant A. These expressions, to be given below, are mathematically valid as time goes to infinity but can be considered as descriptions of the deformation behavior when plastic flow is fully established. The asymptotic values, designated by $\{\ \}$, are

$$\{\dot{g}\} = \{\dot{\sigma}\} = E_t\dot{\epsilon} \tag{4}$$

and

$$\{\sigma - g\} = (E - E_t)k[\{\sigma - g\}]\dot{\epsilon} \tag{5}$$

as well as

$$\{g - E_t\epsilon\} = A \tag{6}$$

Equation 4 states that both the curves for σ and g have the same tangent modulus when plastic flow is fully established independent of the strain rate and that this modulus is equal to E_t. The model predicts no strain rate hardening. The constant A is independent of the strain rate and is a measure of how much the equilibrium stress-strain curve is above $E_t\epsilon$. It constitutes the time-independent or plastic contribution to the stress. Superposed is the viscous or rate-dependent contribution which is nonlinearly related to the strain rate as indicated in Eq 5. The value of $\{\sigma - g\}$ is strongly influenced by the function k; because of this, it is called the *viscosity function.* The above statements are reflected in the identity

$$\{\sigma\} = \{\sigma - g\} + \{g - E_t\epsilon\} + E_t\epsilon \tag{7}$$

which is also displayed in Fig. 1.

These equations do not consider recovery of state and do not model any thermal aging. If these effects are shown to be present by suitable tests, the equations can be modified to include such phenomena.

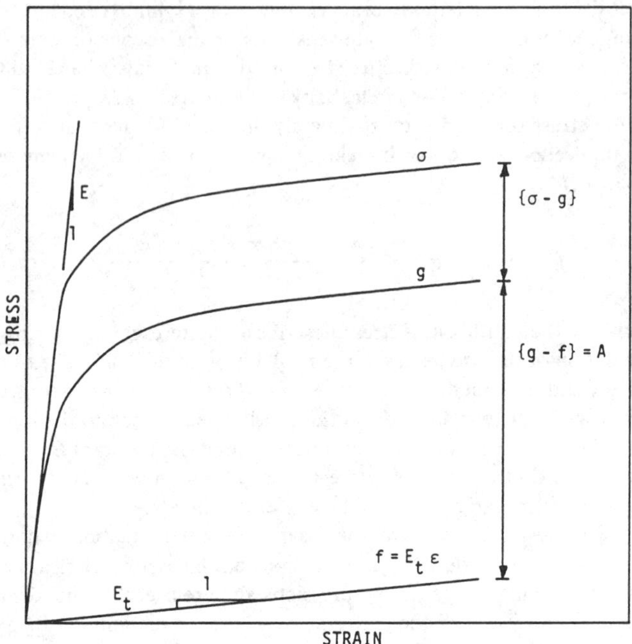

FIG. 1—*Schematic showing the slope contribution* $E_t\epsilon$, *the plastic contribution* $\{g - E_t\epsilon\}$, *and the viscous contribution* $\{\sigma - g\}$ *to the observed stress* $\{\sigma\}$; $f = E_t\epsilon$.

It is also shown in Ref *6* that these equations represent a generalized Masing hypothesis and cyclic neutral behavior. Models of cyclic hardening or softening are being developed (D. Yao, Ph.D. thesis at RPI, forthcoming). The modeling of hardening/softening in the uniaxial case at constant amplitude is straightforward; however, for multiaxial cases, especially for out-of-phase loading, difficulties are encountered due to the complexity of the observed phenomena [*7*]. Further details of the above model, including a three-dimensional isotropic formulation, are given in Refs *6* and *8*.

Identification of Material Functions for AISI Type 304 Stainless Steel at 650°C

An identification procedure for the material functions is given in [*6*] and this procedure is followed in identifying the material functions and constants of the theory from the experimental data [*9,10*] in an approximate way.[5]

The constants and material functions are listed in Table 1 and pertain to the cyclic steady-state. With these constants and functions known, the model is completely determined and can be used for the computation of the deformation behavior.

[5]In determining the material functions, recovery, although likely at this temperature, was not considered, since no data were found which would give a quantitative indication of recovery. Experiments aimed at finding the quantitative effects of recovery and aging would be necessary for a proper consideration of these effects in material modeling.

TABLE 1—*Material constants and functions.*

Material Constants:

$E = 150\ 000$ MPa; $E_t = 2\ 400$ MPa
$A = 170$ MPa (for stabilized behavior)

Viscosity Function:[a]

$$k[x] = k_1\left(1 + \frac{|x|}{k_2}\right)^{-k_3}$$

$k_1 = 314\ 200$ s; $k_2 = 60$ MPa; $k_3 = 17.59$

Shape Function:

$\psi[x] = \psi_1 + (\psi_2 - \psi_1)\exp(-\psi_3 x)$
$\psi_1 = 79\ 500$ MPa (for stabilized behavior)
$\psi_2 = 147\ 000$ MPa; $\psi_3 = 0.216$ MPa^{-1}

Equations of Growth for A and ψ_1 in Cyclic Hardening:[b]

$A = A_1 + (A_2 - A_1)\exp(-A_3 P)$
$A_1 = 170$ MPa; $A_2 = 95$ MPa; $A_3 = 20$
$\psi_1 = C_1 + (C_2 - C_1)\exp(-C_3(A - A_2))$
$C_1 = 79\ 500$ MPa; $C_2 = 60\ 000$ MPa; $C_3 = 0.08$ MPa^{-1}

[a] All x's in this table are in units of MPa.

[b] $P = \int_0^t |\dot{\epsilon}^{in}|d\tau.$

Numerical Integration of the System of Differential Equations for Various Input Histories

Equations 1 to 3 are a set of stiff differential equations which can be numerically integrated by various means. In this study the integrations were performed using the IMSL routine DGEAR and the IBM 3081D computer of RPI. A typical computation of a hysteresis loop takes less than 1 s of CPU time.

Figure 2 shows a hysteresis loop each for a strain rate of 4×10^{-3} and 4×10^{-6} s^{-1}. The influence of strain rate is clearly evident.

When the strain rates are varied between each half of the hysteresis loop to simulate the slow/fast and fast/slow waveforms, the integration yields the curves shown in Fig. 3. The near-vertical drops at the transition from fast to slow loadings and designated by A are a characteristic of the model and should also be observed in experiments. At the end of the fast/slow periods the respective curves coincide with those obtained for constant strain-rate cycling and plotted in Fig. 2.

The introduction of ten-minute relaxation times at various points of the hysteresis loops is depicted in Fig. 4. The differences in the relaxation behavior at various positions around the hysteresis loop are evident by examining the lengths of the vertical stress drops. (Since the relaxation time is constant, the stress drops are an indication of the average relaxation rate.) In regions of a small tangent modulus the relaxation is much more pronounced than in those of a high tangent modulus. Also, the direction of the relaxation is in accord with experimental observations [*18*]. The evolution of the equilibrium stress g is not shown in Fig. 4 for reasons of clarity.

In Fig. 5 the evolution of the equilibrium stress g is plotted. In this case relaxation periods of 10 min are introduced in the steep unloading portions of the hysteresis loop at points A and B. It

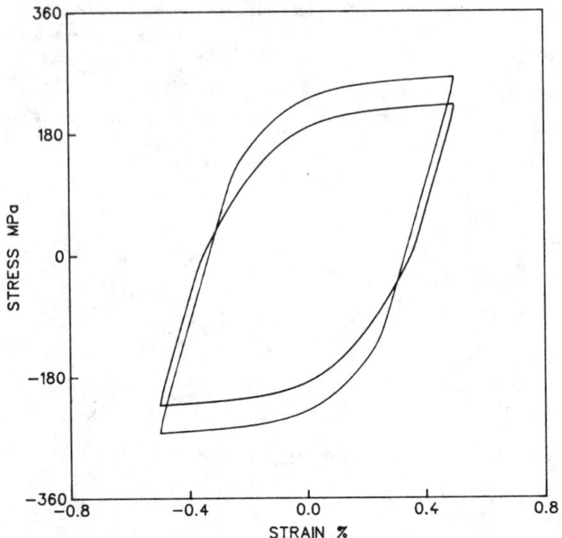

FIG. 2—*Steady-state hysteresis loops at strain rates of 4 × 10*$^{-3}$ *and 4 × 10*$^{-6}$ *s*$^{-1}$.

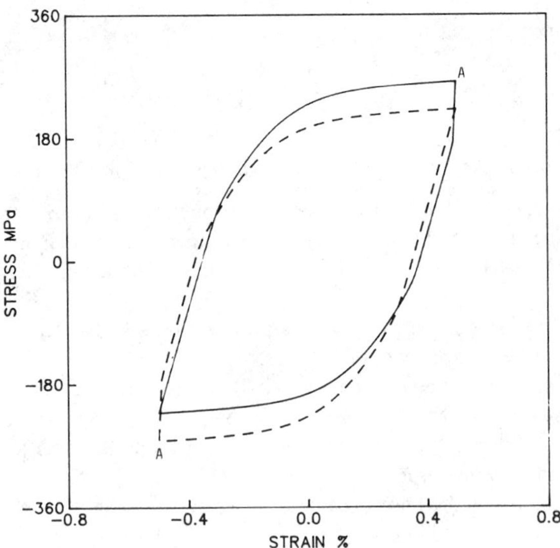

FIG. 3—*Hysteresis loops for slow/fast* (dashed line) *and fast/slow loadings. The near-vertical drops at the transitions from fast to slow loadings are marked by* A.

is seen that the stress reduces at *A* but increases at *B*, albeit by a small amount. The relaxation rate and its sign is determined by the inelastic strain rate which depends only on $\sigma - g$, and changes its sign with this quantity; see Eqs 1 and 2. Figure 5 shows that $\sigma - g$ changes sign at the unloading segments of the hysteresis loop and that there are regions where $\sigma = g$. In these regions the behavior of the model is almost elastic and no relaxation occurs. Since σ crosses the

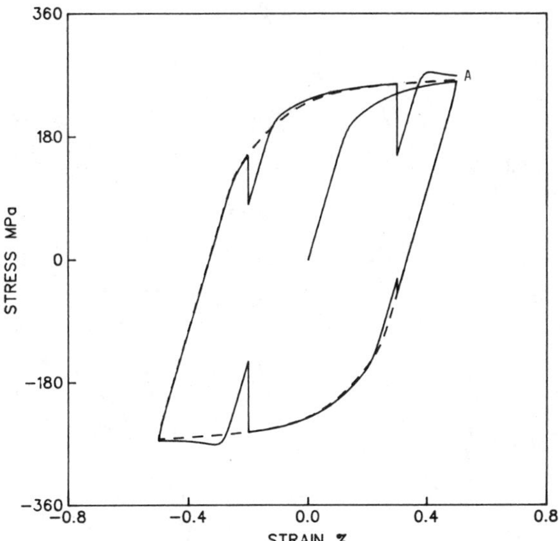

FIG. 4—*Hysteresis loop with 10-min relaxation periods at two values of strain. The relaxation is much more pronounced in regions of a small tangent modulus than in regions where the tangent modulus is close to the elastic one. The dashed line represents the hysteresis curve for a constant strain rate of* $4 \times 10^{-3}\, s^{-1}$. *The overshoots after the relaxation periods are evident. At* A *the overshoot has not yet died out at the maximum strain of the graph.*

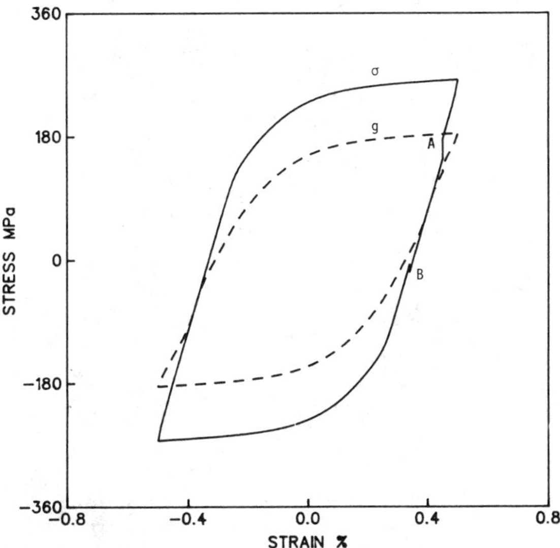

FIG. 5—*Ten-minute relaxation periods at points* A *and* B *superimposed on a straining with a constant strain rate of* $4 \times 10^{-3}\, s^{-1}$. *In addition to the stress* (full line), *the equilibrium stress* (dashed line) *is also plotted. The sign of the relaxation rate is determined by the sign of the overstress. Because of this dependence, an increase in stress magnitude is possible during relaxation.*

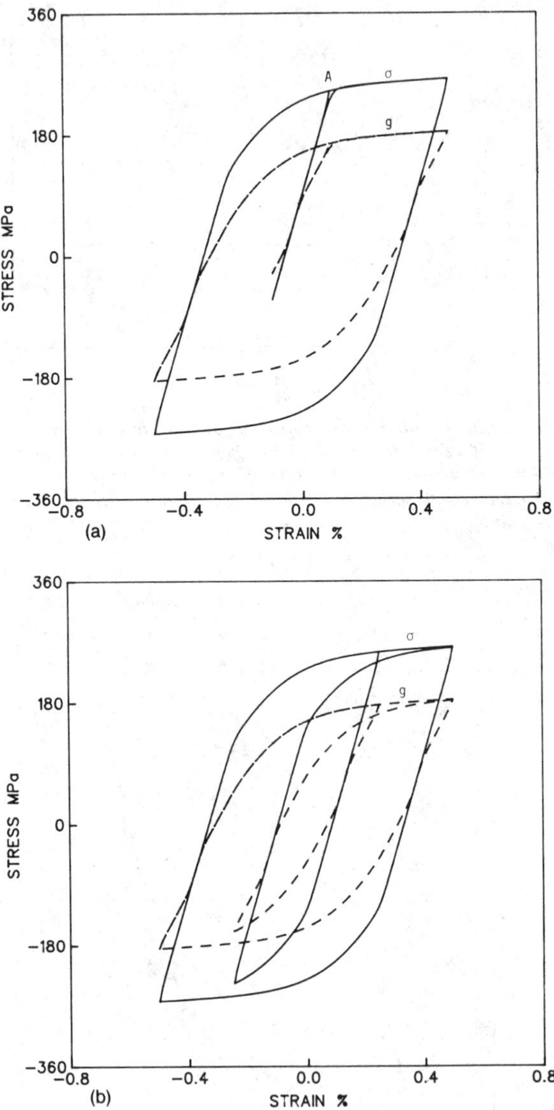

FIG. 6—*Hysteresis loops within hysteresis loops during strain-controlled loading equilibrium stress* (dashed line) *is also shown. Regions of overlap indicate almost 0.5% in Fig. 6b; in Fig. 6c, two loops with* $\Delta\epsilon = 0.4\%$ *are shown.*

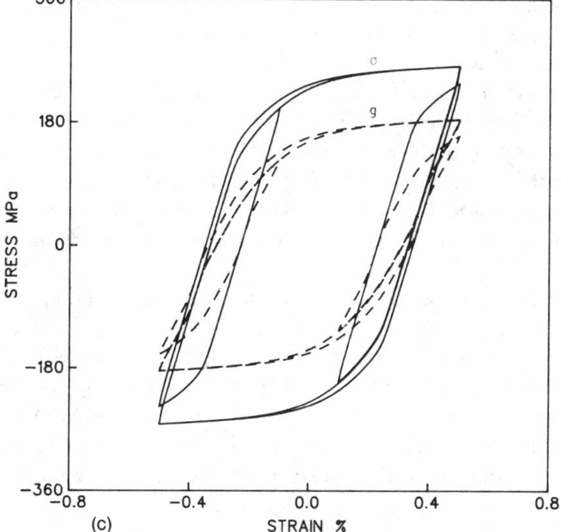

at a strain rate of 4 × 10^{-3} s^{-1}. In addition to the stress (continuous line) *the*
astic behavior. Strain range of small loops are Δϵ = 0.1% in Fig. 6a, Δϵ =

g-curve before zero is reached, relaxation can increase the stress magnitude. Such behavior is observed in experiments [*18*]. The plotting of σ and g also illustrates the differences in magnitude in the relaxation behavior around the hysteresis loop. Relaxation rate increases with the increase in the magnitude of the overstress.

Figure 6 exhibits predictions of the theory for strain-controlled loadings and unloadings, for hysteresis loops within a hysteresis loop. The almost linear loading and unloading behavior is evident in Fig. 6*a*. When the strain range of the inside loop is increased, a hysteresis loop develops (Fig. 6*b*). In Fig. 6*c* two hysteresis loops are shown which start on the unloading branches of the large hysteresis loop. In addition to the stress (continuous line) the equilibrium stress (dashed line) is also plotted. In regions of overlap of the σ- and g-curves almost linear elastic behavior is represented. It is obvious that these regions evolve with the deformation history and do not always include the zero stress level.

The transition from almost linear behavior during reloading to the plastic flow region (point *A* in Fig. 6*a*) is very sharp. When a hysteresis loop is developed the transition is gradual (Figs. 6*b* and 6*c*). The governing Eqs 1 to 3 do not contain any parameter for adjusting the transitions. They are controlled by the material functions of the theory which in turn influence the rapidity with which the asymptotic solutions (Eqs 4 to 6) are obtained. In general a sharp transition is effected by an early attainment of the asymptotic solutions. The predictions of the model shown in Fig. 6 are considered acceptable.

A direct comparison with experimental data is only possible for the stress ranges determined at various strain rates. This is done in Table 2. The biggest discrepancy is for a strain rate of $4 \times 10^{-2}\ s^{-1}$. At this strain rate the experimental data show an anomaly; the stress range is less than that obtained with a strain rate of $4 \times 10^{-4}\ s^{-1}$. The model, of course, predicts an increase in stress range with an increase in strain rate. While improvements in the matching of theoretical and experimental data are possible, the correlations shown were deemed acceptable for the present purposes.

All the examples so far were for strain control. The proposed equations can be used for stress control as well. Figure 7 shows strain-controlled loading at a strain rate of $4 \times 10^{-3}\ s^{-1}$ up to a strain of 0.5% which is reached at point *A*. At this point a "mode switch" to stress control is performed with a subsequent stress rate of 208 MPa s^{-1}. Ratchetting is evident and the ratchet strain decreases with each cycle. At point *B* the stress rate decreases by one order of magnitude and the ratchet strain increases right away. Although room temperature tests [*12*] indicate that the qualitative behavior depicted in Fig. 7 is realistic, the authors do not know of any elevated

TABLE 2—*Stress ranges observed during low-cycle fatigue tests for Type 304 stainless steel at 593°C and stress ranges calculated with the constitutive equations.*

Total Strain Range ($\Delta\epsilon$), %	Strain Rate In Tension ($\dot{\epsilon}_t$), s^{-1}	Strain Rate In Compression ($\dot{\epsilon}_c$), s^{-1}	Observed Stress Range ($\Delta\sigma$), MPa	Calculated Stress Range ($\Delta\sigma$), MPa	Difference, %
1	4×10^{-2}	←	473	695.4	36.4
1	4×10^{-3}	←	518	532.1	2.7
1	4×10^{-4}	←	496	437.6	−11.8
1	4×10^{-5}	←	460.5	426.6	−2.9
1	4×10^{-6}	←	422	409.6	−9.3
1	1×10^{-4}	1×10^{-2}	464.7	507.5	9.2
1	1×10^{-2}	1×10^{-4}	479.1	507.4	5.9
1	4×10^{-6}	4×10^{-3}	428	470.7	10.0
1	4×10^{-3}	4×10^{-6}	451.6	470.5	4.2

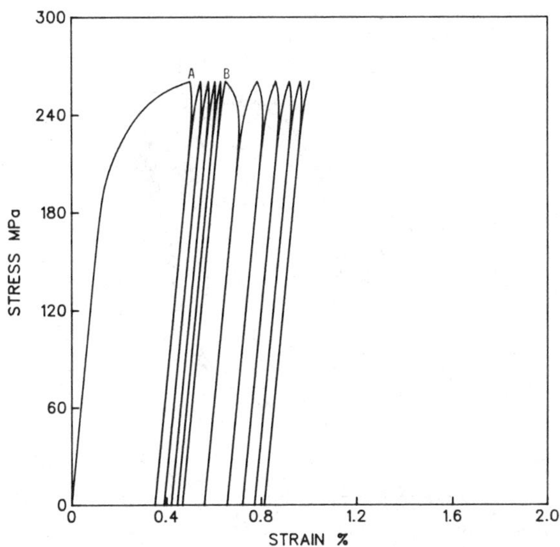

FIG. 7—*Predictions of the theory for strain-controlled loading at a strain rate of* $4 \times 10^{-3}\ s^{-1}$ *up to a strain of 0.5%, point* A. *At this point, a "mode switch" is performed with subsequent cycling at a stress rate of 208 MPa* s^{-1}. *At point* B *the stress rate is decreased by one order of magnitude. The dependence of the ratchet strain on stress rate is clearly evident.*

temperature experiments duplicating the simulation of Fig. 7. The strong dependence of the ratchetting phenomenon on the *stress rate* is obvious.

In Figs. 1 to 7 a steady cyclic condition was consistently assumed. In reality Type 304 stainless steel undergoes considerable cyclic hardening before reaching this steady-state condition. For the modeling of this phenomenon it must be known whether these changes are plastic in nature (stemming from a change in dislocation density) or whether they affect the viscous properties (change in mobile dislocation density). Room temperature experiments [*7*] indicate that the changes are plastic in nature and are therefore to be modeled by making the growth law of g, predominantly the quantity A in Eq 6, dependent on the accumulated inelastic strain. (Room temperature experiments show that the frequently used inelastic work is not an appropriate quantity for modeling hardening in the presence of time dependence [*5*].) Since no elevated temperature experiments are available, it was assumed that cyclic hardening was plastic in nature and the growth law of g was modified to accommodate cyclic hardening. The modifications are given in Table 1 and the resulting hysteresis loops during the first ten cycles are depicted in Fig. 8. It should be emphasized that all the properties delineated previously, such as the relaxation and the strain-rate behavior, are also modeled during periods of hardening. Figure 8 illustrates how the theory can be extended to model cycle-dependent changes. The simple method indicated in Table 1 is not sufficient to reproduce all the complex hardening phenomena observed with stainless steels [*7,8*].

Application to Life Prediction

Once the hysteresis loops are computed they can be treated like those obtained from experiment. They provide input data for the life prediction analysis. Two approaches will be illustrated, the use of the strainrange partitioning method and the application of the incremental damage accumulation law proposed in Ref *4*.

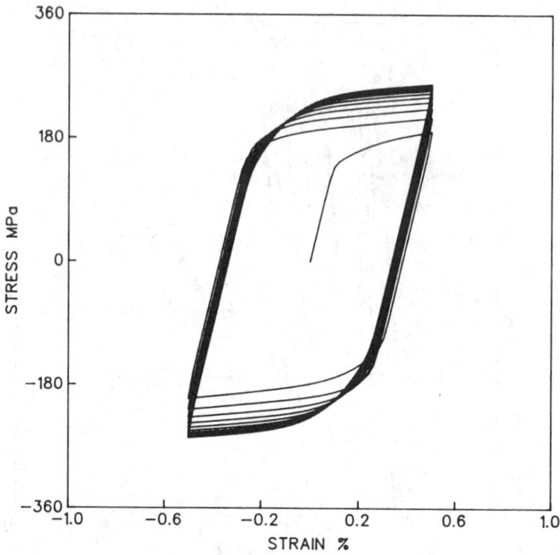

FIG. 8—*Simulation of cyclic hardening under strain control. The material data of Table 1 are used. Strain rate is* $4 \times 10^{-4}\ s^{-1}$.

Strain-Range Partitioning

To obtain the partitioned strain ranges for slow/fast loading, the method of [*13*] was employed. For the case of hold-time tests, a method of partitioning was proposed in [*14*]. These methods were used in the present analysis. With the partitioned strains known, the partitioned life relations proposed in Ref *13,* together with the linear damage rule, were employed to obtain the fatigue life. It was assumed that only pp-components evolve at a strain rate equal to or higher than $4 \times 10^{-3}\ s^{-1}$. The results are given in Table 3. The experimental results in that table were taken from Ref *10*.

It is seen that the predictions account for the life-reducing effects of hold-time and of slow/fast loading. (Use of the interaction damage rule would result in underprediction of fatigue lives [*19*].)

TABLE 3—*Predicted life using the computer hysteresis loops and strainrange partitioning and experimental results from Ref* 10.

Strain History	Strain Ranges Partitioned	Partitioned Life	Life Predicted	Experimental Results
Continuous:				
$\dot{\epsilon} = 4 \times 10^{-3}\ s^{-1}$	$\Delta\epsilon_{pp} = 0.00709$	$N_{pp} = 3600$	3600	3395
Slow-Fast:				
$\dot{\epsilon} = 4 \times 10^{-3}\ s^{-1}$	$\Delta\epsilon_{pp} = 0.00458$	$N_{pp} = 8014$	307	261
$\dot{\epsilon} = 4 \times 10^{-6}\ s^{-1}$	$\Delta\epsilon_{cp} = 0.00214$	$N_{cp} = 319$		
Tensile Hold:				
$\dot{\epsilon} = 4 \times 10^{-3}\ s^{-1}$	$\Delta\epsilon_{pp} = 0.00640$	$N_{pp} = 4343$	1696	666
$t = 900$ s	$\Delta\epsilon_{pp} = 0.00060$	$N_{cp} = 2736$		

Incremental Damage Accumulation Law

An incremental damage law was proposed in Ref *4*, which deviates from that introduced in Refs *3* and *11*, by postulating a stress dependence rather than a plastic strain dependence; details are given in Refs *4* and *15*. The proposed incremental law is

$$D_f + D_c = 1 \tag{8}$$

$$\dot{D}_f = \frac{h[\dot{\epsilon}_{\text{in}}]}{T_f}\left(\frac{|\dot{\epsilon}_{\text{in}}|}{\dot{\epsilon}_f}\right)^{n_f}\left(\frac{|\sigma|}{\sigma_f}\right)^{m_f} \tag{9}$$

$$\dot{D}_c = \frac{\text{sign}[\sigma]}{T_c}\left(\frac{|\dot{\epsilon}_{\text{in}}|}{\dot{\epsilon}_c}\right)^{n_c}\left(\frac{|\sigma|}{\sigma_f}\right)^{m_c} \tag{10}$$

In these equations, constants T_f, T_c, $\dot{\epsilon}_f$ and $\dot{\epsilon}_c$ are introduced for dimensional considerations. They are set equal to one in the appropriate units. The other constants n_f, m_f, n_c, m_c, σ_f, and σ_c must be determined from appropriate low-cycle fatigue experiments. This is done using some of the data from Refs *10, 11,* and *16*. The function $h[\ ]$ is the step function defined as

$$h[\dot{\epsilon}_{\text{in}}] = \begin{cases} 1, & \dot{\epsilon}_{\text{in}} \geq 0 \\ 0, & \dot{\epsilon}_{\text{in}} < 0 \end{cases} \tag{11}$$

and sign$[\sigma]$ is the signum function defined as

$$\text{sign}[\sigma] = \begin{cases} 1, & \sigma \geq 0 \\ -1, & \sigma < 0 \end{cases} . \tag{12}$$

When the data of Refs *10, 11,* and *16* were plotted on a cycles-to-failure time-to-failure diagram separations according to waveform were found. They are shown in Fig. 9.

This observation led to the postulate of the above damage accumulation law; details including the method used to determine the material constants are given in Ref *4*. Using the constitutive law and the damage accumulation law life-time can be calculated. Since some failure points were used to determine the constants, some of the calculated points are correlations and others are predictions [*4*]. Overall the outcome of this calculation is shown in Fig. 10 where the observed life is plotted versus the calculated life for the total strain ranges of 1% and 2% together with two lines indicating a deviation of a factor of two in life. In this calculation the definition of failure used in Refs *10, 11,* and *16* was employed. It is seen that the computations are unconservative for short lives but become conservative as life increases. It is also observed that the calculations are better for Heat 9T2796 than for the other heats. Overall, the results appear to be acceptable.

The increase in the conservative nature of the calculations with cycles is also evident when the computed results are replotted using the coordinates of Fig. 9 for a strain range of 1% (Fig. 11). The figure exhibits the same general trends as Fig. 9. At a large number of cycles the calculations are conservative, but are unconservative at short number of cycles-to-failure. The computations capture the effects of slow/fast, fast/slow, and hold-time loadings quite well. For long hold-times the calculations show a smaller number of cycles than the experiment. While improvements in the prediction are possible, they were not attempted in the present investigation. As a first attempt the results were considered acceptable.

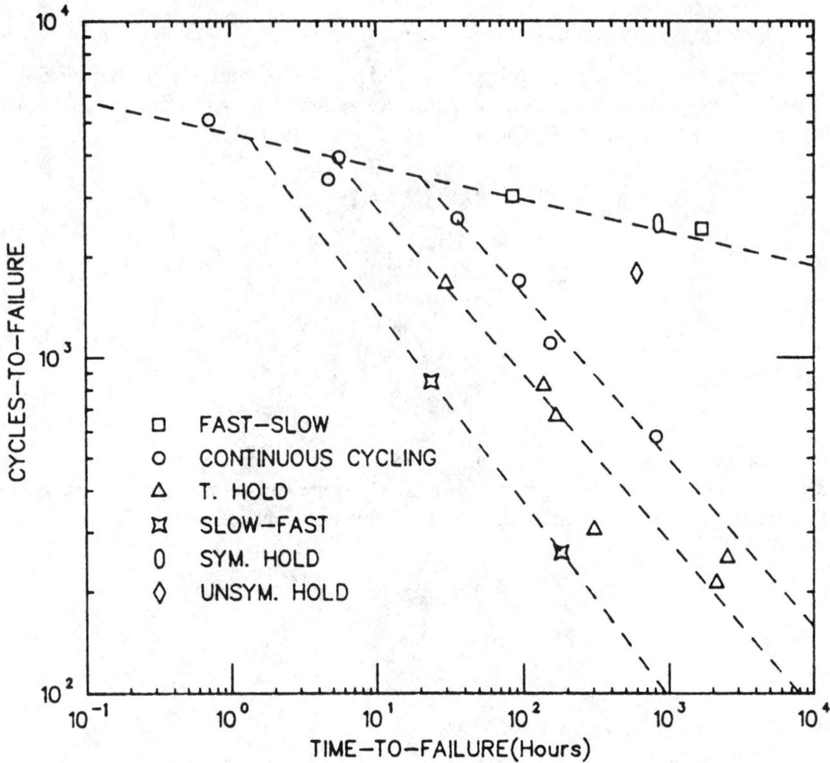

FIG. 9—*Cycles-to-failure versus time-to-failure for low-cycle fatigue tests on Type 304 stainless steel at 593°C in air at a total strain range of 1%. Data from Refs* 10, 11, *and* 16.

Due to the incremental nature of Eqs 9 and 10, the damage accumulation of any history can be computed. As an example the lifetime for 10-min relaxation periods at various positions around the hysteresis loop are computed and listed in Table 4 as a function of hold-position on the tensile-going portion of the hysteresis loop. The biggest life reduction is as the maximum strain of interest, and no life reduction is found when the hold-time is introduced at zero stress. In computing the lifetime in Table 4 a steady-state cyclic condition was assumed.

Discussion

An attempt was made to combine life prediction in low-cycle fatigue with constitutive equations for Type 304 stainless steel at elevated temperature. Since not enough material deformation data were available from the literature at 593°C where the low-cycle fatigue tests were performed [*10,11,16*], the material functions of the constitutive theory were identified at 650°C using data of Ref *9*. Although low-cycle fatigue life is generally considered to be insensitive to temperature, the deformation behavior is most likely sensitive to temperature. In the constitutive theory, effects of recovery and of aging were neglected. There are indications that these phenomena do play a role at these temperatures. A quantitative assessment of these effects, however, cannot be obtained from the available low-cycle fatigue data.

Because of these difficulties, the present paper is considered a demonstration of the feasibility of combining constitutive theory with life prediction in the context of a uniaxial state of stress.

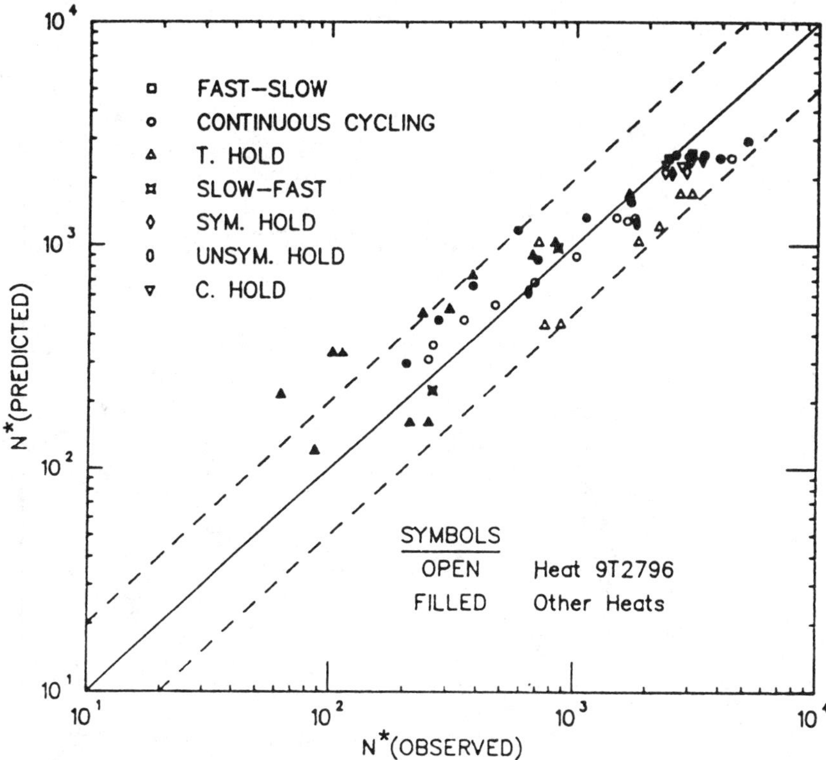

FIG. 10—*Observed fatigue life in cycles versus calculated fatigue life in cycles using the constitutive equation and the damage accumulation law for Type 304 stainless steel at 593°C. Strain range 1% and 2%. Data from Refs* 10, 11, *and* 16.

Although the calculated lives do show deviations which sometimes exceed a factor of two, the trends of the computed lives are in the proper direction. Further explorations of the proposed approaches seem to be appropriate. They should be performed on a consistent set of data, some of which are to be used to determine the constants of the constitutive theory and of the chosen life prediction law. Others must be reserved for an independent check of the predictions.

The proposed approach is not limited to the uniaxial state of stress. Indeed the biggest advantage lies in the potential use in the analysis of components at elevated temperature. A three-dimensional version of the constitutive theory is available and can be used in finite element computations (such computations have already been performed [*17*]). When combined with an appropriate life-prediction law, the lifetime of a structure as a function of loading history can be directly calculated without displaying the results of the intermediate calculation of stresses or strains. Such an approach is possible and economically feasible.

It was only possible to discuss two approaches to life prediction in this paper. Since the constitutive theory and the life prediction laws are uncoupled at the present stage, any other suitable life prediction law can be used in principle. Constitutive laws are almost always formulated in an incremental fashion when inelastic deformation is involved. Because of this fact, the first author favors the use of incremental damage accumulation laws. Incremental laws can be applied to any type of cycle (see the calculation of the effect of the position of the hold-time on life in this paper for example). It is not necessary to postulate a typical cycle; rather the actual

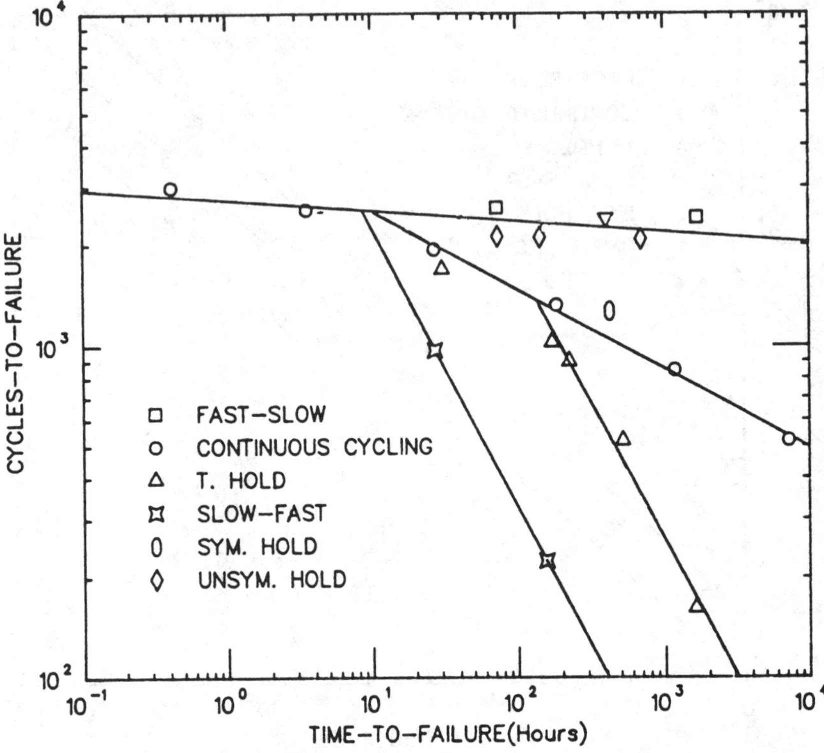

FIG. 11—*Calculated cycles-to-failure versus time-to-failure using the constitutive equation and the damage accumulation law for a strain range of 1%. Compare with the experimental results shown in Fig. 9.*

TABLE 4—*Effect of hold position on the hysteresis loop during tensile loading on lifetime.*

Hold Position Strain, %		Predicted Life, cycles
1	(ϵ_{max})	543
0.5		587
0		641
−0.5		813
−0.832	($\sigma = 0$)	885
Continuous Cycling		885

deformation history can be followed. If, however, a typical cycle exists in a periodic loading situation, this typical cycle can be used with the incremental law as well. Such a law must always be integrated, which may be a drawback in the eyes of some. However, integration can be readily performed using numerical methods. With ever-increasing computing power, such operations can easily be performed in the design office now and even more so in the future.

Acknowledgments

M. Satoh performed this work while on a leave of absence from Nippon Kokan K.K. The other authors acknowledge the support of the Solid Mechanics Program of the National Science Foundation.

References

[*1*] Coffin, L. F., Jr., "Predictive Parameters and Their Application to High Temperature, Low-Cycle Fatigue," in *Fracture 1969, Proceedings of 2nd International Conference on Fracture,* Chapman and Hall, London, 1969, pp. 643-654.

[*2*] Manson, S. S., Halford, G. R., and Hirschberg, M. H., "Creep-Fatigue Analysis by Strainrange Partitioning," NASA TM X-67838, NASA-Lewis Research Center, 1971.

[*3*] Majumdar, S. and Maiya, P. S., "A Damage Equation for Creep-Fatigue Interaction," in *Proceedings,* MPC-3, American Society for Mechanical Engineers, New York, 1976, pp. 323-326.

[*4*] Satoh, M. and Krempl, E., "An Incremental Life Prediction Law for Creep-Fatigue Interaction," *Pressure Vessels and Piping,* Vol. 60, ASME, New York, 1982, pp. 71-79.

[*5*] Krempl, E. and Kallianpur, V. V., "Some Critical Uniaxial Experiments for Viscoplasticity at Room Temperature," *Journal of the Mechanics and Physics of Solids,* Vol. 32, 1984, pp. 301-314.

[*6*] Krempl, E., McMahon, J. J., and Yao, D., "Viscoplasticity Based on Overstress with a Differential Law for the Equilibrium Stress," *Mechanics of Materials,* Vol. 5, 1986, pp. 35-48.

[*7*] Krempl, E. and Lu, H., "The Hardening and Rate-Dependent Behavior of Fully Annealed AISI Type 304 Stainless Steel under Biaxial In-Phase and Out-of-Phase Strain Cycling at Room Temperature," *Journal of Engineering Materials and Technology,* Vol. 106, 1984, pp. 376-382.

[*8*] Yao, D. and Krempl, E., "Viscoplasticity Theory Based on Overstress: The Prediction of Monotonic and Cyclic Proportional and Nonproportional Loading Paths of an Aluminum Alloy," *International Journal of Plasticity,* Vol. 1, 1985, pp. 259-274.

[*9*] Conway, J. B., "An Analysis of the Relaxation Behavior of AISI 304 and 316 Stainless Steel at Elevated Temperature," GEMP-730, USAEC, 1969.

[*10*] Majumdar, S. and Maiya, P. S., "Waveshape Effects in Elevated Temperature Low-Cycle Fatigue of Type 304 Stainless Steel," Inelastic Behavior of Pressure Vessel and Piping Components, PVP-PB-028, ASME, New York, 1978, pp. 43-54.

[*11*] Maiya, P. S. and Majumdar, S., "Elevated Temperature Low-Cycle Fatigue Behavior of Different Heats of Type 304 Stainless Steel," *Metallurgical Transactions A,* Vol. 8A, 1977, pp. 1651-1660.

[*12*] Kujawski, D., Kallianpur, V., and Krempl, E., "An Experimental Study of Uniaxial Creep, Cyclic Creep and Relaxation of AISI Type 304 Stainless Steel at Room Temperature," *Journal of the Mechanics and Physics of Solids,* Vol. 28, 1980, pp. 129-148.

[*13*] Manson, S. S., Halford, G. R., and Nachtigall, A. J., "Separation of the Strain Components for Use in Strain Range Partitioning," in *Advances in Design for Elevated Temperature Environment,* Proceedings of the ASME 2nd National Congress on Pressure Vessels and Piping, San Francisco, 1975.

[*14*] Yamaguchi, K. and Kanazawa, K., "Effect of Strain Wave Shape on High Temperature Fatigue Life of a Type 316 Steel and Application of the Strain Range Partitioning Method," *Metallurgical Transactions A,* Vol. 11A, 1980, pp. 2019-2027.

[*15*] Satoh, M., "An Incremental Life Prediction Law for Creep-Fatigue Interaction," Ph.D. thesis, Rensselaer Polytechnic Institute, Troy, N.Y., 1982.

[*16*] Majumdar, S., "Designing Against Low-Cycle Fatigue at Elevated Temperature," *Nuclear Engineering and Design,* Vol. 63, 1981, pp. 121-135.

[*17*] Little, M. M., Krempl, E., and Shih, C. F., "On the Time and Loading Rate Dependence of Crack-Tip Fields at Room Temperature—A Viscoplastic Analysis of Small Scale Yielding," in *Elastic-Plastic Fracture: Second Symposium, Volume I—Inelastic Crack Analysis, ASTM STP 803,* American Society for Testing and Materials, Philadelphia, 1983, pp. I-615 to I-636.

[*18*] Asada, Y. and Mitsuhashi, S., "Creep-Fatigue Interaction of Types 304 and 316 Stainless Steel in Air and in Vacuum," in *Proceedings,* 4th International Conference on Pressure Vessel Technology, Part II, Institution of Mechanical Engineers, London, 1980, pp. 321-327.

[*19*] Saltsman, J. F., letter to E. Krempl, 26 Nov. 1985.

Low Cycle Fatigue Damage

B. N. Leis[1]

A Nonlinear History-Dependent Damage Model for Low Cycle Fatigue

REFERENCE: Leis, B. N., **"A Nonlinear History-Dependent Damage Model for Low Cycle Fatigue,"** *Low Cycle Fatigue, ASTM STP 942,* H. D. Solomon, G. R. Halford, L. R. Kaisand, and B. N. Leis, Eds., American Society for Testing and Materials, Philadelphia, 1988, pp. 143–159.

ABSTRACT: A nonlinear damage postulate that embodies the dependence of the damage rate on cycle-dependent changes in the bulk microstructure and the surface topography is examined. The postulate is analytically formulated in terms of the deformation history dependence of the bulk behavior. This formulation is used in conjunction with baseline data in accordance with the damage postulate to predict the low cycle fatigue resistance of OFE copper. Close comparison of the predictions with experimentally observed behavior suggests that the postulate offers a viable basis for nonlinear damage analysis.

KEY WORDS: fatigue, damage, nonlinear assessment, model, history-dependent, damage postulate, similitude

Damage analysis under variable amplitude cycling requires a procedure that identifies and matches ascending and descending segments of the stress-strain history to form closed stress-strain loops similar to those developed under constant amplitude cycling. For many years the Japanese-developed rainflow cycle counting procedure [*1*] has been used to so match segments of strain history. During that time a number of authors demonstrated its utility in a variety of critical experiments (e.g., [*2*]). Subsequently, a number of numerical procedures which match segments of hysteresis to form closed loops in a manner consistent with the material's plastic flow process evolved and have tended to displace rainflow procedures in damage analysis [*3–7*]. Thus a number of experience-proven schemes are available for purposes of matching hysteresis segments to form cycles—the first step in damage analysis.

Damage analysis also requires the assessment of damage done in a given event of loading by comparing the number of times that event occurs with the resistance of the material to such damaging events. This assessment thus involves a reference data base that characterizes the material's resistance to damage. For example, in the simplest framework—uniaxial stressing—damage analysis requires a parameter that accounts for the effect of the mean stresses developed under variable amplitude cycling as compared to that of the constant amplitude reference data used in damage calculations. As with cycle counting procedures, numerous schemes are also available and continue to be developed to account for mean stress effects. The list of mean stress parameters is long, including one due to Morrow [*8*], one due to Smith et al. [*9*], and variants or alternatively developed forms similar to that of Smith et al. [*10–12*]. While successes are claimed for all mean stress parameters in the literature, the fact that such parameters continue to evolve indicates that shortcomings exist in this aspect of damage analysis. Finally, the last step in damage analyses—damage accumulation—requires a criterion that integrates damage increments over the history and indicates when failure has occurred.

[1]Research Leader, Mechanics Section, Advanced Materials Department, Batelle, Columbus Division, Columbus, OH 43201.

Regarding the latter aspects of damage analysis—damage assessment and damage accumulation—, numerous studies illustrate the controversy and uncertainties involved with damage assessment and linear damage accumulation (e.g., [*13*]). This paper focuses on these last two aspects—damage assessment and accumulation. Specifically, a nonlinear damage assessment postulate is advanced. This postulate characterizes the nonlinear dependence of the damage rate on the history through the associated history dependence of the bulk microstructure and surface residual stresses and topography. Results of experiments designed to characterize the dependence of fatigue resistance in terms of deformation history dependence are reported for OFE copper. Baseline constant amplitude strain life and cyclic deformation response data are used to predict the fatigue resistance of the material in the as-received condition and in a heat-treated state. Throughout, consideration is restricted to situations involving proportional stressing. For simplicity the behavior of small diameter uniaxial specimens is used as a vehicle for discussion.

Concepts and Assumptions in Damage Analysis

Concepts in damage analysis relevant to the present purposes include similitude, operational definitions of damage and crack nucleation, and damage parameters. These concepts have been discussed in detail previously (e.g., [*14,15*]). Only the salient aspects are included here for the sake of completeness.

Damage is herein taken as the progressive change in a material's microstructure that eventually culminates in the nucleation and growth of cracks due to the localized action of reversed microplastic strain and tensile stress. It is postulated (based on observations for OFE copper [*15*]) that changes which lead to crack nucleation also result in changes in the bulk deformation behavior and in the character of the material's surface.

The focus of this paper is on the portion of life over which bulk measures suitably reflect the damage state. For bulk measures to serve as measures of damage, it follows that the crack length chosen as "nucleation" must be small. The *operational definition of nucleation* adopted is the cycle for which the back-extrapolated asymmetric drop in the tensile load intersects the stable deformation response. Calculations of "area lost" due to cracking indicate this definition is associated with a single surface crack whose depth is on the order of 50 to 200 μm [*15*]. Nucleation based on this operational definition has been compared with results of surface studies made using single-stage gold-coated replicas interpreted at up to 5000× magnification in a scanning electron microscope (SEM) [*15*]. This comparison indicates nucleation, defined as above using bulk parameters, correlates well with the onset of growth of Mode I cracks from the slip bands which formed much earlier.

A damage parameter is introduced, because the reference data may be developed under conditions that differ from the variable amplitude event being assessed. A *damage parameter* is defined as an analytical (or empirical) relationship which establishes the link between the complex damage process in the component and the similar situation for the laboratory reference data. For the present, the only difference between constant and variable amplitude loading is mean stress. Thus a "mean stress" damage parameter, which is a special case of a more general and widely demonstrated parameter [*12,14–16*], is introduced. The parameter has the form [*12*]

$$D = s_m \Delta e^t + \Delta s \Delta e^t \tag{1}$$

The symbol s_m denotes the mean stress and Δs and Δe^t denote the stress and total strain ranges, respectively. By definition *equal values of the damage parameter at two critical locations mean equal lives to crack nucleation within statistical scatter at both locations.*

Similitude between conditions for reference and variable amplitude data is a key aspect in any damage analysis, be it for nucleation or for macrocrack growth. A general discussion of similitude, however, is beyond the scope of this paper. Interested readers should consult Refs *17* and *18*, which deal with crack growth and nucleation, respectively.

Experimental data indicate that for a given set of strain levels the fatigue resistance depends on the order of their application. It is this sequence dependence that leads to consideration of nonlinear damage analysis procedures. Figure 1*a* indicates that there are two steps in processing the input set of strain levels for purposes of damage analysis—*assessment and accumulation.* Because it is difficult to phenomenologically uncouple these steps, it is difficult to conclusively ascribe the observed nonlinearity to either step. The essence of all procedures developed to date (Fig. 1*b*) is to compare cycles applied in a variable amplitude history to the damage resistance observed under constant amplitude cycling.

If constant amplitude data are to form the basis for variable amplitude damage analyses, the damage measure must include both the control and response parameters. That is, the ordinate used as the damage measure in Fig. 1*b* must include both control and response parameters. By definition, equal values of the damage measure (ordinate in Fig. 1*b*) mean equal lives to crack nucleation and equal amounts of damage per cycle, provided that similitude is ensured. Since each cycle does an equal amount of damage for equal values of the damage parameter, it is reasonable to postulate that damage accumulation is linear. By analogy, for constant amplitude total strain control, it is reasonable to *assume* that each cycle in the history does an amount of damage which is proportional to the value of the damage parameter for that cycle. Because total strain control data are intended to serve as the basis for damage analysis for variable amplitude histories, differences between sequence-dependent development of transient action under constant and variable amplitude histories must be accounted for. The practical implication is that Δe^t may be an adequate basis for comparing constant amplitude data. But, in general, Δe^t is an inadequate basis for characterizing the driving force for damage under block sequence or variable amplitude cycling. Extending the above to deal with variable amplitude cycling requires a

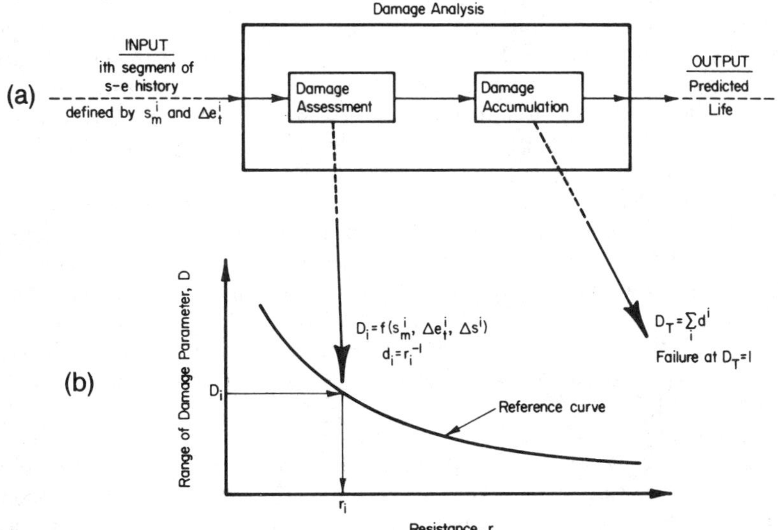

FIG. 1—*Schematic of the damage analysis procedure.* (a) *Steps in the process.* (b) *Essential ingredients.*

second *assumption*—that closed loops[2] that are developed under constant amplitude loading do the same amount of damage as their variable amplitude counterparts.

Similitude and the Inelastic Volume Fraction

Available data suggest that nucleation resistance depends on the volume of material actively accumulating fatigue damage. At very short lives (high strains) the bulk of a uniaxial sample undergoes homogeneous deformation. The more ductile the material, the more uniform the flow. As the life to nucleation increases, the inelastic action associated with the fatigue process decreases. As strain decreases, inelastic action concentrates at the surface where decreased grain to grain compatibility and the absence of restraint to slip systems with components normal to the surface enhance surface flow as compared to the interior. Flow continues to localize on the surface until at and below the so-called fatigue limit only isolated grains are inelastically activated. While slip-band decohesion may develop in these grains, constraint from surrounding elastic grains and the slip barrier effect of the grain boundary preclude the formation of life-limiting Mode I cracks. The fatigue limit thus arises due to the contained inelastic action [*19*] and the nonrotation of surface Mode II cracks to form Mode I cracks that grow into the depth of the sample.

Bulk measures of damage and observations of transient effects may be masked at smaller strains because the inelastic action becomes very localized. Thus assessment of the validity of bulk measures of damage may be confounded as the damage localizes. For example, the similitude condition would hold that the total plastic work to initiation is a constant, if plastic work was a valid damage parameter. However, as is well established [*20,21*], plastic work to initiation increases as the plastic strain per cycle decreases, based on bulk measurements. But, the real question is, Is the local total plastic work to nucleation constant in the grain(s) that initiate cracks? That is, Is similitude provided for by plastic work at the scale of grains? Certainly, as the volume fraction of inelastically strained material, V_f^i, decreases, it becomes harder to spread inelastic action beyond the grains it initially localizes in. Nearest-neighbor grains less favorably oriented for slip must therefore subsequently develop comparable damage levels before earlier developed Mode II slip cracks can propagate out of the grain(s) they initiated in. Successive grains must admit this process as the Mode II slip-band cracks join and rotate onto the plane of maximum principal stress to form Mode I cracks that propagate into the body rather than across surface grains. The use of bulk measures of damage, such as plastic work, would on this basis require an increased amount of total plastic work to initiation as V_f^i decreases, just as is observed. Apparently V_f^i tends to explain why bulk measures of damage do not provide similitude at the local level.

There also are implications for size effects related to similitude at the local level. The absolute volume of material which is favorably oriented for slip increases as the volume of material under test increases. Likewise, the probability that nearest-neighbor grains will be favorably oriented for slip increases as the absolute volume under test increases. It follows that life would decrease as the probability of nearest neighbors being oriented favorably for slip—that is, absolute volume—increases. To this extent, the above assertion of the role of V_f^i is consistent with observed fatigue resistance trends as size increases. Likewise, it is consistent with volume-dependent size effect theories (e.g., [*22*]).

One final observation is of particular interest for aluminum alloys and other materials which show very inhomogeneous slip (e.g., [*23*]). Data in the literature often show a break in the plastic strain life line, to much higher negative slopes (e.g., [*23*]). In view of the above notion, as V_f^i decreases, $\Delta\epsilon^p$ becomes smaller in grains developing cracks and thus becomes much harder to measure. Therefore the observed break in the slope of $\Delta\epsilon^p$–N_f as V_f^i decreases may be an

[2]This assumption could also be stated in terms of reversals of loading.

artifact of the bulk measurement. Consequently, the plastic (inelastic) strain life behavior in grains actively developing fatigue cracks is asserted to have a constant slope based on the low-cycle behavior for the present analysis.

Inelastic Volume Fraction and the Development of Surface Damage

The inelastic volume fraction is of particular significance in damage analysis because, as V_f^i decreases, surface features become important. Surface discontinuities tend to either focus and enhance the rate of accumulation of surface plastic strain or to develop a focused plastic zone that penetrates below the surface with an orientation very near that of a Mode I crack. In general, the surface discontinuity could be either geometric or material related, because the net effect of both is the same—focused inelastic action that develops at an increased rate.

It follows that the fatigue resistance is controlled by the bulk at large values of V_f^i and by the surface at small values of V_f^i. In between, both the bulk and the surface control. Obviously high strength materials have small V_f^i even at short lives, whereas ductile materials have large V_f^i at short lives. The bulk behavior will control fatigue behavior until the surface intervenes. This bulk surface control is shown schematically in Fig. 2*a*. The bulk versus surface control concept provides a new interpretation for the strength-ductility explanation previously invoked [*20,24*] to explain observed results [*25*], such as shown in Fig. 2*b*. More importantly, it provides the framework for a damage postulate that embodies the nonlinearities in fatigue resistance in the deformation history dependence of the bulk microstructure and surface topography.

The Damage Postulate [*14,15*]

It is *postulated* that the nonlinear nature of the material's fatigue resistance arises due to the deformation history dependence of the bulk microstructure (as manifested, for example, in hardness/ductility) and the character of the surface topography and surface residual stress field. In this respect, the nonlinear nature of the damage process involves damage assessment, but not damage accumulation. The postulate is schematically depicted in Fig. 3.

Examination of the viability of this postulate requires a data base that develops the history-dependent relationships between (1) bulk microstructure (hardness/ductility) and (2) surface residual stresses and topography, and fatigue nucleation resistance. A study has recently been completed that has begun to generate such data [*15*]. The postulate has been representative of the behavior of one material [*15,26*] on the basis of a preliminary evaluation. The purpose here is to present and examine a simple analytical framework that relates fatigue resistance to deformation behavior for one facet of the postulate—bulk control.

History-Dependent Formulation Based on Bulk Control

Experimental Aspects

The material used in this study was certified 99.9987% pure OFE copper provided as received (AR) in 1-in. diameter rods, cold worked to about 40%. The grain size was ASTM 2½ to 3, with a diamond pyramid hardness (DPH) of 101. The yield strength (YS) was 305 MPa. The AR material was modified by one of two thermal treatments or a mechanical treatment. Two anneals performed in inert conditions were used. In both cases the temperature was 270°C. The first anneal—HTA—lasted 24 min and produced a grain size of 2½ to 3 with a DPH of 103 and a YS of 278 MPa. The second anneal—HTB—lasted 50 min. It produced a grain size 2½ to 3½ with some signs of recrystallization. The DPH was 84 and the YS was 247 MPa. The mechanical treatment consisted of an incremental prestrain (IPS) from zero strain symmetrically incrementing up to the level of the amplitude for constant amplitude cycling. Observe that this

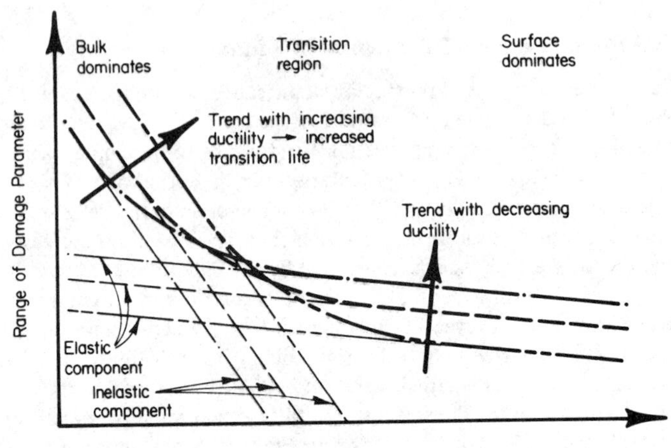

IF fatigue life were controlled only by bulk, these trends suggest ductile materials would provide greatest resistance

AND indeed at high strains all grains are active in the flow process and the surfaces of samples develop similar morphologies to an extent not strongly dependent on ductility, so that bulk response and the ability to absorb plastic deformation of ductile materials provides optimum resistance

BUT at longer lives, inelastic action is confined to a few grains favorably oriented for slip, the number of which depends strongly on the total strain. Fatigue becomes a localized process and concentrates in these grains without regard for the bulk response. Because it is surface dominated, surface discontinuities and the ease and concentration of slip at the free surface intervenes. Optimum resistance is now provided by the material least sensitive to or slowest to develop surface damage.

(a)

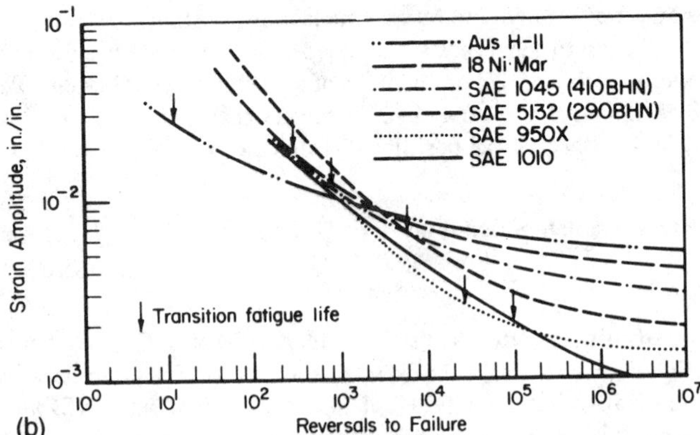

(b)

FIG. 2—*Schematic and actual indications of bulk and surface control of fatigue resistance.* (a) *Schematic of bulk and surface control of fatigue resistance.* (b) *Strain-life curves for a variety of steels: note the cross-over tendencies and that strong materials are best at long lives while ductile materials are superior at short lives (after* [25]*).*

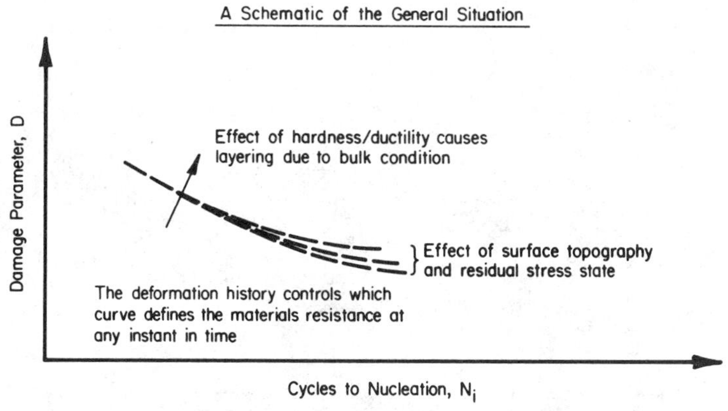

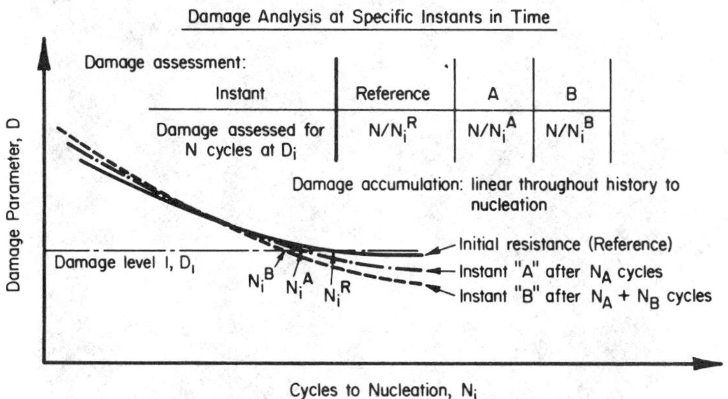

FIG. 3—*Essential features of the damage analysis postulate.*

processing did not radically alter the microstructure, but did significantly change the hardness and transient response during subsequent testing under otherwise identical conditions.

The specimen used in the study had a cylindrical gage section 12.5 mm long by 9.5 mm in diameter. As shown in Fig. 4*a* the specimen was tapered in the shoulders to reduce reflections from the ultrasonic probe buried in one grip end during the return period for indications in the gage length. Specimens failed within the gage section, and in no case was there evidence of buckling or barreling in the test section.

Testing was done in a commercial closed-loop system under total axial strain control to a sinusoidal forcing function at frequencies from 0.2 to 2 Hz. Test frequency in no case caused an increase in temperature greater than 2°C. Calibration of the test system and performance of the tests was in accordance with relevant ASTM standards. All tests were done in an argon environment to avoid intergranular initiation which occurred under ambient laboratory conditions. Testing, done under flowing argon (99.999% pure) and gettered using heated titanium chips, followed a 15-min system purge each time the chamber was opened. Figure 4*b* shows a view of the setup.

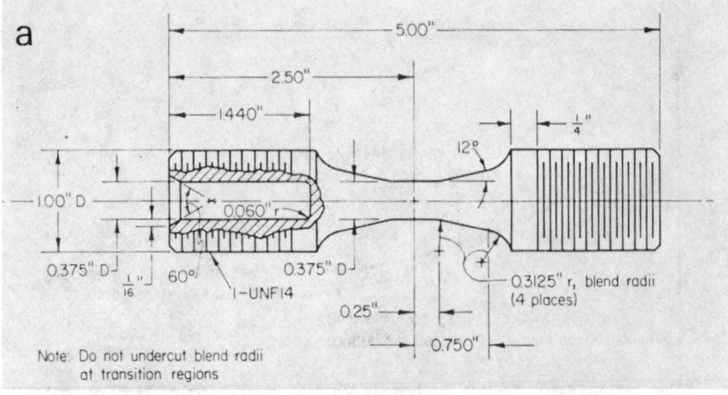

FIG. 4—*Experimental aspects.* (a) *Axial fatigue test specimen (dimensions in inches).* (b) *Overview of test cell and gripping arrangement.*

Testing was interrupted at intervals equal to one tenth of the anticipated life to make replicas of the test section. In addition to stress response and surface topography, nucleation was tracked using reflection ultrasonics. Hardness was measured on split specimens either after a given number of cycles prior to anticipated failure or after separation on tests run to failure.

The test program involved a variety of histories. Reference data were developed for all materials studied at one of three total strain ranges, denoted $\Delta\epsilon_1$, $\Delta\epsilon_2$, $\Delta\epsilon_3$, which were 1.85%, 0.66%, and 0.33%, respectively. These ranges were chosen to produce crack nucleation lives of about 500, 5000, and 50 000 cycles. The choice of strain levels was tempered by the desire that failure be dominated by bulk response in each of the materials considered. With one exception—the lowest strain and softest material—the surface morphology was found to develop similarly so that bulk behavior controlled fatigue resistance. A number of initial and subsequent prestrain histories and block cycling histories were used to develop different transient histories and therefore different bulk states [15]. The IPS history introduced earlier was particularly useful for the present purposes since it significantly changed the bulk state but did not alter the surface (at up to 5000× magnification).

Fatigue resistance and transient response data are plotted from the reference and IPS histories in Figs. 5 and 6. Use is made of a dissipated energy damage parameter as the basis to present the fatigue resistance. (Note that for large plastic strains a dissipated energy parameter is a special case of Eq 1). In adopting a dissipated energy damage parameter, the materials fatigue resistance is being correlated with dissipated energy. However, nothing fundamental is assumed which would imply that life can be related to dissipated energy through basic physics. The potential advantage of using dissipated energy as a damage parameter lies in its being nearly constant over life. This aspect is illustrated for the OFE copper in Fig. 5*a*. Note, however, that at small strains (where $\Delta e^p \rightarrow 0$ as Δe^t decreases) this constancy breaks down, at least in terms of bulk measures of Δe^p. When this breakdown occurs the damage/cycle has been integrated over the test and the resistance characterized in a manner consistent with the linear accumulation assumption. That is, the value of the damage parameter per cycle is equal to the integrated dissipated energy divided by the number of cycles to initiation.

Figure 5*b* plots dissipated energy per cycle on the ordinate and nucleation resistance on the abscissa in the manner of Fig. 1*b*. Results are shown in Fig. 5 for the as-received (AR) material and three other "bulk" conditions. One condition is the mechanically hardened AR material using the incrementally increasing prestrain (IPS). This IPS history has been applied until the subsequent CA strain level is reached [15,26]. The other conditions are HTA and HTB.

Cyclic stress-strain response developed for the three material conditions considered is shown in Fig. 6. Figure 6*a* presents the monotonic stress-strain behavior on the usual linear coordinates. Figure 6*b* presents monotonic and cyclic stress-strain response as contours of constant cumulative plastic strain, on coordinates of log plastic strain and log stress for the AR and HTB conditions. Note that the slope of log Δs-log Δe^p quickly stabilizes once cycling commences, based on data taken at the two largest strains.[3] These results suggest that the strain hardening exponent, n, depends on the cyclic deformation history through $\Sigma\Delta e^p$. Moreover, they suggest that the strain corresponding to a given plastic strain depends on the deformation history. Other data also show these trends, and have led numerous investigators to formulate models of cyclic stress strain response (e.g., [3,5]). Such models are of the form

$$\Delta e^t = \frac{\Delta s}{E} + K_1 \Delta s^{n_1} \tag{2}$$

where Δe^t, Δs, and E are as previously defined, and K_1 and n_1 are materials constants that depend on prior history through $\Sigma\Delta e^p$. The functional dependence of K_1 and n_1 follows from fitting results such as shown in Fig. 6*b* to equations of the form of Eq 2.

Model Framework

The postulate asserts that the fatigue resistance of the bulk material is controlled by the state of the microstructure herein assessed in terms of changes in hardness (ductility) due to the deformation history as macroscopically characterized by the stress at a given value of $\Sigma\Delta e^p$. In the absence of history-related differences in the surface topography it follows from the postulate

[3]Using bulk measurements of Δe^p, use of the largest strains avoids measurement errors caused by bulk elastic behavior masking inelastic action that would be encountered using results for the smallest total strain.

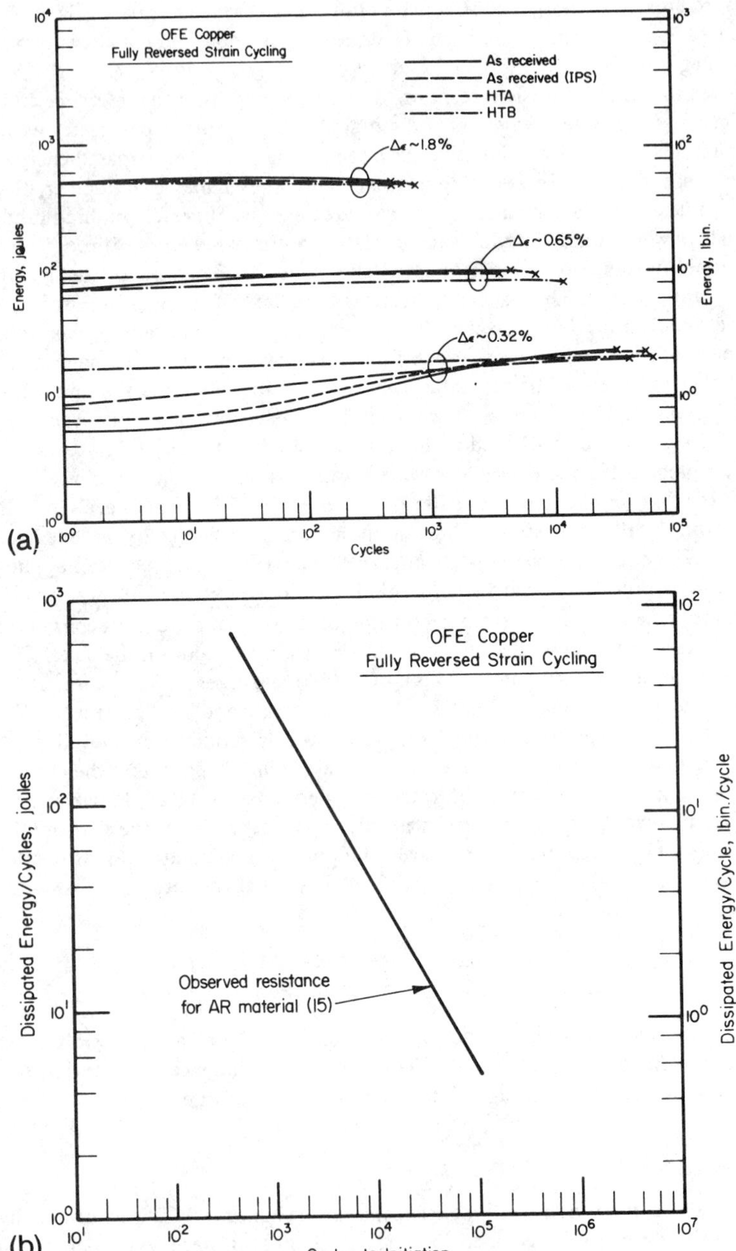

FIG. 5—*Fatigue resistance and transient flow behavior of OFE copper.* (a) *Transients evident in dissipated energy as a function of cycles.* (b) *Fatigue resistance as a function of dissipated energy.*

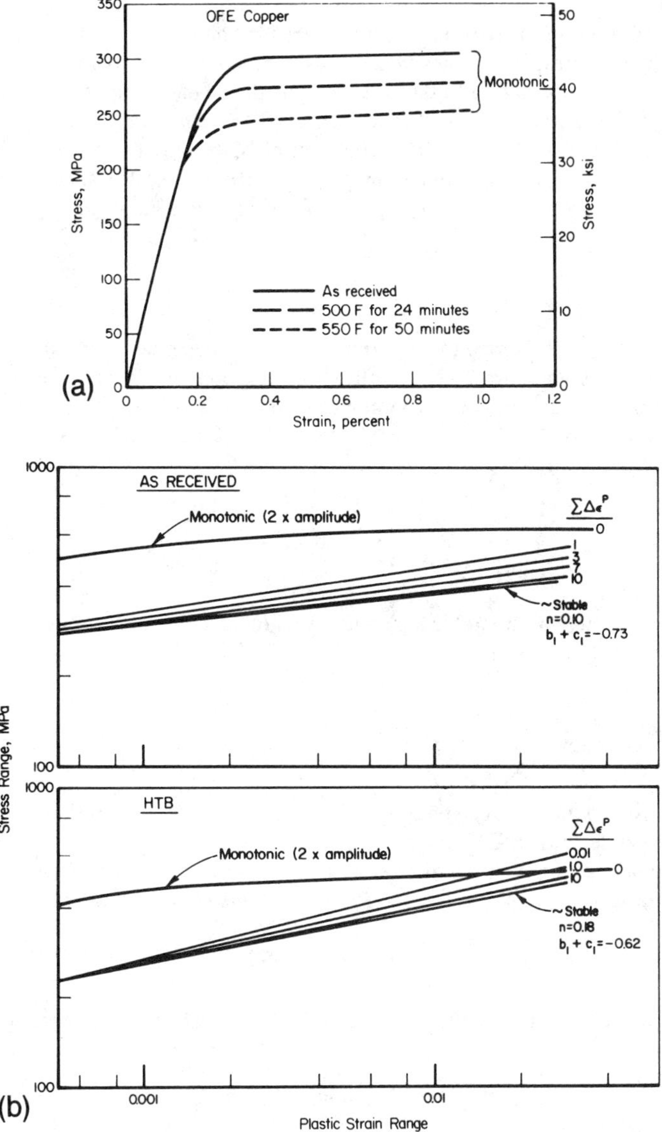

FIG. 6—*Monotonic and cyclic stress-strain behavior of OFE copper.* (a) *Stress-strain behavior.* (b) *Stress-plastic strain behavior.*

that differences in the fatigue resistance of the material can be characterized solely in terms of the current hysteresis loop shape and deformation resistance normalized to some reference condition. The best material reference condition is one for which transient effects are minimal (i.e., damage parameter control), which in light of Fig. 5*a* is reasonably achieved using dissipated energy.

As just outlined, deformation resistance and loop shape vary in a unique history dependent fashion which can be characterized as functions of $\Sigma\Delta e^p$, with respect to some reference deformation condition, say, the monotonic stress-strain curve. Differences in fatigue resistance under bulk control are due solely to innate differences in the microstructure, which are herein considered to be manifest in changes in stress as a function of the deformation history. The postulate asserts that this is due to the history dependence of the deformation response. Depending on the strain level, soft material, such as HTB, may harden under cyclic loading, whereas hard material, such as AR/IOS, may soften. Certainly these contrasting transients complicate the picture. But fortunately for present purposes, such transients can be simply tracked and accurately predicted via computer models of stress-strain behavior (e.g., [5]).

Formulation

It remains now to characterize the deformation history dependence of fatigue resistance in situations where the bulk controls (i.e., surface has not intervened). Study of prior work indicates that within the framework of an energy-based damage parameter, the slopes of the elastic and plastic strain life relationships, denoted as b and c, respectively, depend on the deformation response. Specifically, Morrow [20] has shown that

$$c = \frac{1}{1 + 5n'} \quad \text{and} \quad b = \frac{-n'}{1 + 5n'} \tag{3}$$

where n' is the exponent of the stable log cyclic stress-log plastic strain behavior. By analogy to Eq 3, it is asserted that, in general,

$$c_1 = \frac{-1}{1 + 5n_1} \quad \text{and} \quad b_1 = \frac{-n_1}{1 + 5n_1} \tag{3a}$$

where n_1 is the strain hardening exponent. Note that $n_1 = n_1\,(\Sigma\Delta e^p)$ so that, according to Eq 3a, c_1 and b_1 also depend on $\Sigma\Delta e^p$.

The damage parameter of Eq 1 can be expressed in a history-dependent form that embraces Eq 3a for the CA fully reversed situation of Fig. 5. In this case Eq 1 can have the form

$$D = \Delta s(\Delta e^e + \Delta e^p) \tag{1a}$$

Introducing the familiar strain life expressions of Morrow [20] generalized according the Eq 3a,

$$\Delta e^e = \frac{2s_f'}{E}(2N_i)^{b_1} \quad \text{and} \quad \Delta e^p = 2e_f'(2N_i)^{c_1}$$

and noting that $\Delta s = \Delta e^e E$, Eq 1a becomes

$$D = D(H) = \frac{4s_f'}{E}(2N_i)^{2b_1} + 4\,s_f'e_f'(2N_i)^{b_1+c_1} \tag{1b}$$

where s_f' and e_f' denote the fatigue strength and ductility coefficients, respectively [20]. The notation $D = D(H)$ is introduced to indicate that the magnitude of the damage parameter, denoted D, is now a function of the history for the same value of resistance, denoted N_i. In Eq 1b, $D = D(H)$, because the exponents c_1 and b_1 depend on the history through $n_1 = n_1(\Sigma e^p)$. Finally, because the focus here is lives dominated by bulk behavior, Fig. 2a implies that the contribution of the first term to $D(H)$ in Eq 1*b* is small. Consequently, consistent with the trend of Fig. 5, the ordinate—damage—and abscissa—resistance—of this figure are related by

$$D(H) = 4s_f'e_f'(2N_i)^{b_1+c_1} \tag{1c}$$

Note that Eq 1*c* predicts the value of the damage parameter associated with a given life only in terms of the deformation response of the material. That is, from Eq 3*a*, c_1 and b_1 are functions of n_1, and n_1 is a function of the deformation history only.

In conjunction with Eq 3*a*, $D(H)$ given by Eq 1*c* specifies the history dependence of the resistance. In general, s_f' and e_f' are also history dependent and can be related to the true fracture strength and ductility measured in a tension test.[4] However, it is sufficient for the present purposes to ignore this (weak) dependence. Inspection of the results presented in Fig. 6 indicates that under CA cycling the strain hardening exponent quickly stabilizes and remains nearly constant over the remainder of the test. For this reason c_1 and b_1 also quickly stabilize and remain constant thereafter. Indeed, the results of Fig. 6 suggest that the values of n_1 (and therefore also b_1 and c_1) remain nearly constant for more than 95% of the life. Given that this is the case, Eq 1*c* can be used to predict the value of the damage parameter associated with a given fatigue resistance for the AR, IPS, HTA, and HTB conditions of the material using only the deformation response of the material.

Values of the damage parameter predicted for AR and HTB conditions at constant values of fatigue resistance using the nearly stable deformation response shown in Fig. 6*b* are presented in Fig. 7 as trend lines. Comparison of the predicted trends with the observed results and the

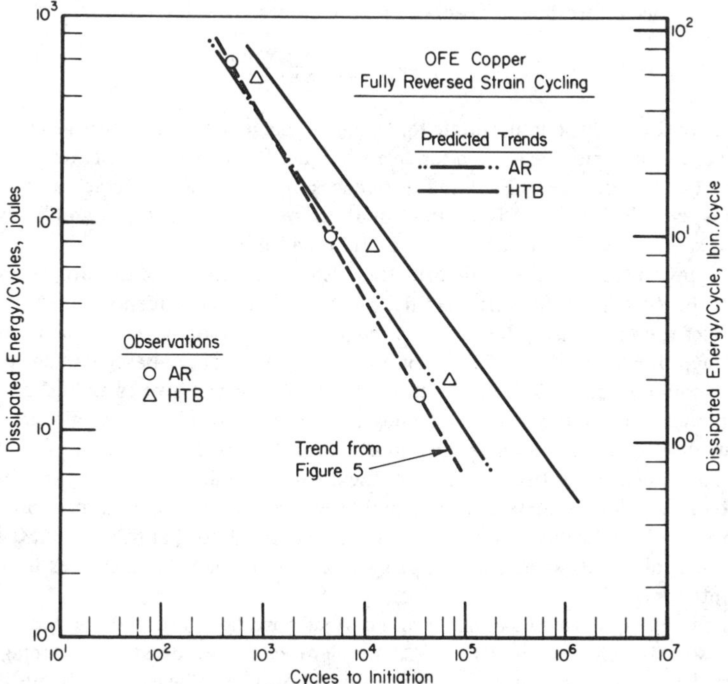

FIG. 7—*Comparison of predicted fatigue resistance with observed trends.*

[4]Details of this formulation are beyond the scope of the present paper.

observed trend line for AR material from Fig. 5 suggests that the postulate, as characterized by Eqs 1*c* and 3*c*, is consistent with observations. Moreover, as detailed in Refs *15* and *26*, this postulate is capable of independently predicting the influence of prestrains and block sequence effects in cases where the bulk controls life. Further generalization of the formulation to include the effects of surface, and to characterize when and to what extent surface control intervenes, is now in progress.

Discussion

Although the postulate has been advanced and discussed for fatigue loading, the postulate itself is rather general. For example, in the absence of surface effects one has

$$\underset{\sim}{s} = \underset{\sim}{s}(\underset{\sim}{e}) \tag{4}$$

which defines the current (history-dependent) deformation state. It has been postulated that damage per cycle, $\underset{\sim}{D}$, and the damage resistance, $\underset{\sim}{R}$, are also history dependent.

$$\begin{aligned} \underset{\sim}{D} &= \underset{\sim}{D}(\underset{\sim}{s}, \underset{\sim}{e}, \ldots) \\ \underset{\sim}{R} &= \underset{\sim}{R}(\underset{\sim}{s}, \underset{\sim}{e}, \ldots) \end{aligned} \tag{5}$$

Equations of the form of Eq 5 have been shown by data developed elsewhere [*14,15*] and the analysis presented herein to hold for a variety of conditions. In general the resistance will also have to be related to the surface topography, $\underset{\sim}{S}$, and residual stress state, $\underset{\sim}{\sigma}$, which may be step functions of $\underset{\sim}{R}$, once a threshold is reached. That is, more generally,

$$\underset{\sim}{R} = R_1(\underset{\sim}{s}, \underset{\sim}{e}, \Sigma e^p, \underset{\sim}{S}, \underset{\sim}{\sigma}, \ldots) \tag{6}$$

By analogy to the results and postulate for fatigue damage, a similar, more general postulate and similar equations could be written for creep-fatigue. But care must be taken to include the appropriate rate dependence in Eq 4 and differences in surface effects (crack wedging, oxidation area loss, etc.) in Eqs 5 and 6. Simple equations of the form of Eqs 4 and 5 have already been proposed (e.g., [*12*]) and shown to be useful on a limited basis.

The results presented indicate the deformation history dependence of the fatigue resistance of OFE copper. Furthermore, the concept of deformation history dependence was shown to qualitatively predict the relationship between the magnitude of the damage parameter and the fatigue resistance of two conditions of OFE copper, provided an energy-based damage postulate is employed. Similarly, quantitative predictions for block cycle histories published elsewhere [*15,26*] were found to accurately track the observed results. In this respect, the assertion that damage per cycle for a given level of the damage parameter is deformation history dependent appears to be physically justified. Also, use of the history-dependent damage parameter, which was consistent with the postulate provided qualitative and accurate quantitative predictions of life to form a small crack when damage was linearly accumulated. For this reason, the nonarbitrary linear accumulation assumption appears to be valid when nonlinearities in assessment were accounted for.

The phenomenology discussed as background to the postulate indicated that the volume fraction of inelastically deformed material, V_f^i, is a significant consideration in damage analysis based on bulk measurements. It is noteworthy that V_f^i also has a bearing on the utility of damage curve concepts (e.g., [*27,28*]). Recall that damage curve concepts employ a specific test geometry to develop reference curves. As with the present case, small diameter specimens are often used to develop these data. However, in contrast to the present investigation, such curves usually contain a possibly significant contribution of crack growth in the life associated with a given damage level. Also, as in the present approach, these damage curves represent resistance as a function of some bulk measure of the driving force for damage. Often, as was done in the present, strain, or plastic strain is used. In other cases, stress (an indirect measure of strain or plastic strain) is used.

Contrary to the present formulation, the significance of V_f^i is neither recognized nor accounted for in currently formulated models based on damage curves. Problems due to ignoring V_f^i become acute at small strains. In such cases the localized inelastic action leading to initiation may have a ratio of plastic to elastic strain near or in excess of unity in grains actively forming cracks. However, the bulk measure of Δe^p will depend on V_f^i, which in turn depends on the absolute volume of material under test. It follows that the bulk measure of $\Delta e^p/\Delta e^e$ for fixed Δe^t depends on V_f^i and absolute specimen size. The life to form a small crack in some localized region, therefore, also depends on the specimen size for fixed Δe^t (or Δs associated with Δe^t). Therefore, as presently formulated independent of V_f^i and related similitude requirements, damage curves will be relevant only for the specimen geometry they are based on.

Not surprisingly, data for similar geometries can be correlated using damage curve concepts. Indeed, the above argument would suggest that the breakdown in similitude due to the use of bulk measures of a localized process is not a problem when dealing with similar geometries. Similitude requirements related to V_f^i only become an issue in trying to predict (not correlate) the resistances developed for different geometries. In this respect, the true test of a damage analysis procedure is not correlation of data for similar specimens. Correlation is but a necessary condition to demonstrate the viability of a model. Rather, the viability of damage models should be demonstrated by prediction of data for other geometries (sizes and shapes). In this case, one has both a necessary and sufficient demonstration of viability.

Summary and Conclusions

The objective of this paper was to examine a damage postulate which asserted that the fatigue resistance curve of a metal is history dependent due to inelastic action. The paper focused on OFE copper because this simple model material accentuated the inelastic action central to the damage postulate. Given the novel nature of the postulate, the scope of the results reported was limited to that of a preliminary evaluation of a purely phenomenological nature. The deformation history dependence of the damage postulate was formulated in terms of a simple model containing one deformation history-dependent parameter. This parameter, which is characterized as a function of the deformation history-independent of fatigue tests, was shown to successfully predict the relationship between the magnitude of the damage parameter and the fatigue resistance for OFE copper in two initial states. The results suggest that, in the short-life regime where bulk behavior controls fatigue resistance, the sequence dependence of the damage process can be explicitly tied to the microstructural state. This tie has been affected through bulk measures of the deformation behavior, analytically consistent with the energy basis for the damage parameter employed.

The major conclusion of this study is that the sequence dependence of damage is tied, through the bulk mechanical properties, to the microstructure. History dependence in deformation response leads to history dependence of fatigue resistance. It follows then that damage mechanics and constitutive behavior are explicitly related, at least in the short-life domain controlled by the bulk.

Acknowledgments

This work has been supported in part by the Mechanics Section of the Transportation and Structures Department of Battelle's Columbus Laboratories. The facilities and the atmosphere that supports continued research by largely the result of Dr. A. T. Hopper, manager of the Mechanics Section. His support and encouragement are greatly acknowledged. NASA Lewis Research Center provided financial support for the early experimental work performed in regard to evaluating the postulate. The interest and support of Dr. R. C. Bill, program monitor for Contract NAS3-22825, are gratefully acknowledged as are the comments of Dr. G. R. Halford. Finally, the indirect support of Corporate Technical Development of Battelle Memorial Institute under Program B-1333-1430 is gratefully acknowledged.

References

[1] Matsuishi, M. and Endo, T., "Fatigue of Metals Subjected to Varying Stress," presented to the Japan Society of Mechanical Engineers, Fukouka, Japan, March 1968; see also Endo, T., Mitsunaga, K., Takahashi, K., Kobayashi, K., and Matsuishi, M., "Damage Evaluation of Metals for Random or Varying Load," presented to the 1974 Symposium on Mechanical Behavior of Materials, Kyoto, August 1974.

[2] Dowling, N. E., "Fatigue Life and Inelastic Strain Response Under Complex Histories for an Alloy Steel," *Journal of Testing and Evaluation,* Vol. 1, No. 4, July 1973.

[3] Wetzel, R. M., "A Method of Fatigue Damage Analysis," Ph.D. thesis, University of Waterloo, Ontario, Canada, 1971.

[4] Jhansale, H. R., "Inelastic Deformation and Fatigue Response of Spectrum Loaded Axial and Flexural Members," Ph.D. thesis, University of Waterloo, Ontario, Canada, March 1974.

[5] Conle, F. A., "A Computer Simulation Assisted Statistical Approach to the Problem of Random Fatigue," M.A.Sc. thesis, University of Waterloo, Ontario, Canada, March 1974.

[6] Landgraf, R. W., Richards, F. D., and La Pointe, N. R., "Fatigue Life Predictions for a Notched Member Under Complex Loading Histories, SAE AE-6, SAE, 1977, pp. 95–106.

[7] Leis, B. N., "Fatigue Analysis to Assess Crack Initiation Life for Notched Coupons and Complex Components," Ph.D. thesis, University of Waterloo, Waterloo, Ontario, Canada, Sept. 1976.

[8] Morrow, J. D. in *Fatigue Design Handbook,* SAE, Chapter 3.2, 1968.

[9] Smith, K. N., Watson, P., and Topper, T. H., "A Stress-Strain Function for the Fatigue of Metals," SMD Report 21, University of Waterloo, Ontario, Canada, Oct. 1969; see also, Smith, K. N., Watson, P., and Topper, T. H., "A Stress-Strain Function for the Fatigue of Metals," *Journal of Materials,* Vol. 5, No. 4, Dec. 1970, pp. 767–778.

[10] Haibach, E. and Lehrke, H. P., "Das Verfahren der Amplituden-Transformation," LBF Report FB-125, LBF Darmstadt, W. Germany, 1975.

[11] Ostergren, W. J., "A Damage Function and Associated Failure Equations for Predicting Hold Time and Frequency Effects in Elevated Temperature Low-Cycle Fatigue," *Journal of Testing and Evaluation,* Vol. 4, No. 5, 1976, pp. 327–339.

[12] Leis, B. N., "An Energy-Based Fatigue and Creep-Failure Damage Parameter," *Journal of Pressure Vessel Technology, Transactions of ASME,* Vol. 99, No. 4, Nov. 1977, pp. 524–533.

[13] Leve, H. L., "Cumulative Damage Theories," in *Metal Fatigue: Theory and Design,* Wiley, New York, 1969.

[14] Leis, B. N. and Forte, T. P., "Fatigue Damage Analysis Under Variable Amplitude Cycling," in *Random Fatigue Life Prediction,* ASME, PVP, Vol. 72, June 1983, pp. 89–105.

[15] Leis, B. N. and Forte, T. P., "Nonlinear Damage Analysis: Postulate and Evaluation," Final Report on Contract NAS3-22825 to NASA-Lewis Research Center, Feb. 1983.

[16] Leis, B. N. and Laflen, J. H., "An Energy Based Postulate for Damage Assessment of Cyclic Nonproportional Loadings with Fixed Principal Directions," in *Ductility and Toughness in Elevated Temperature Service,* ASME/MPC-8, 1978, pp. 371–389.

[17] Broek, D. and Leis, B. N., "Similitude and Anamolies in Crack Growth Rates," in *Materials, Experimentation, and Design in Fatigue,* Westbury House, IPC Science Press, U.K., March 1981, pp. 129–146.

[18] Leis, B. N., "Predicting Crack Initiation Fatigue Life in Structural Components," in *Methods of Predicting Fatigue Life,* ASME, 1979, pp. 57–76.

[19] Seika, M., Kitaoka, S., and Ko, H., "Change of Material Properties at the Tips of Fatigue Cracks Subjected to Macroscopic Tensile and Shear Stresses for Propagation (in Rotary Bending and Cyclic Torsion)," *Bulletin of the JSME,* Vol. 19, Oct. 1976.

[20] Morrow, J., "Cyclic Plastic Strain Energy and Fatigue of Metals," in *Internal Friction, Damping, and Cyclic Plasticity, ASTM STP 378,* American Society for Testing and Materials, Philadelphia, 1965, pp. 45–84.

[21] Halford, G. R., "The Fatigue Toughness of Metals: A Data Compilation," TAM Report 265, Department of Theoretical and Applied Mechanics, University of Illinois, Urbana, May 1964.

[22] Kuguel, R., "The Highly Stresses Volume of Material as a Fundamental Parameter in the Fatigue Strength of Metal Members," University of Illinois, TAM Report No. 169, June 1960; see also "The Relation Between the Theoretical Stress Concentration Factor, K_t, and the Fatigue Notch Factor, K_f, According to the Highly Stressed Volume Approach," University of Illinois, TAM Report No. 184, Dec. 1960.

[23] Sanders, T. H., Jr., Mauney, D. A., and Staley, J. T., "Strain Control Fatigue as a Tool to Interpret Fatigue Initiation of Aluminum Alloys," in *Fundamental Aspects of Structural Alloy Design,* Plenum Press, New York, 1975, pp. 487-519.

[24] Landgraf, R. W., "The Resistance of Metals in Cyclic Deformation," in *Achievement of High Fatigue Resistance in Metals Alloys, ASTM STP 467,* American Society for Testing and Materials, Philadelphia, 1970, pp. 3-36.

[25] Landgraf, R. W. and La Pointe, N. R., "Cyclic Stress-Strain Concepts Applied to Component Fatigue Life Prediction," *SAE Transactions,* Vol. 83, Sect. 2, 1974, pp. 1198-1207.

[26] Leis, B. N. and Forte, T. P., "Variable Amplitude Fatigue Damage Analyses: Postulate and Critical Experiments," in *Application of Fracture Mechanics to Materials and Structures,* G. C. Sih, E. Sommer, and W. Dahl, Eds., 1984.

[27] Lamaitre, J. and Plumtree, A., "Application of Damage Concepts to Predict Fatigue Failures," *ASME Journal of Engineering Materials and Technology,* Vol. 101, July 1979, pp. 284-292.

[28] Hashin, Z. and Rotem, A., "A Cumulative Damage Theory of Fatigue Failure," *Materials Science and Engineering,* Vol. 34, 1978, pp. 147-160.

Kunihiro Iida[1]

Very Low Cycle Fatigue Life Influenced by Tensile or Compressive Prestrain

REFERENCE: Iida, K., **"Very Low Cycle Fatigue Life Influenced by Tensile or Compressive Prestrain,"** *Low Cycle Fatigue, ASTM STP 942,* H. D. Solomon, G. R. Halford, L. R. Kaisand, and B. N. Leis, Eds., American Society for Testing and Materials, Philadelphia, 1988, pp. 160–172.

ABSTRACT: The effects of excessive prestrain in tension or in compression upon very low-cycle fatigue life were investigated for two mild steels and two high strength steels as an aid in the analysis of a ship failure in service, in which the bow structure was broken off due to local buckling and by low-cycle fatigue in extremely heavy sea conditions. The test series in the present investigation was: (1) uni-directional tension test after prestraining, (2) completely reversed strain cycling tests of non-prestrained material with several strain ranges, (3) tensile or compressive straining directly followed by completely reversed strain cycling, and (4) completely reversed strain cycling tests of specimens which were machined out of previously prestrained large diameter specimens.

The residual static fracture ductility decreases as a function of the amount of prestrain, showing much more loss of ductility in the case of compressive prestrain. Fairly good agreements on strain range and fatigue life were found between experimental data and estimated values calculated after Manson's and Iida's formulae. The failure life of a specimen subjected to prestrain showed a remarkable decrease from the failure life of the original material depending on three parameters: amount of prestrain, direction of prestrain, and specimen surface conditions. The worst case in the range of the present tests is that of tensile prestraining directly followed by strain cycling. In this case the low-cycle fatigue failure life of a prestrained specimen is, as an example, reduced by 93% of the failure life of the original material, when the amount of the tensile prestrain is 60% of the original static fracture ductility based on area reduction to failure.

KEY WORDS: fatigue (materials), fatigue failure life, fatigue life estimation, high strength steel, mild steel, prestrain, residual static fracture ductility, service failure, ship, strain range estimation, very low-cycle fatigue, visible crack initiation life

Nomenclature

- N_c Visible crack initiation life (number of cycles to the initiation of a surface fatigue crack 0.5 to 1 mm long) of a specimen not subjected to prestrain
- N_f Failure life (number of cycles to complete failure) of a specimen not subjected to prestrain
- N_{fp} Failure life of a specimen subjected to one cycle prestrain in advance of completely reversed strain cycling
- (*Cal.*) Calculated value after Manson's and Iida's formulas
- (*Exp.*) Experimental value
- $\Delta\epsilon$ Longitudinal, natural, and total strain range (elastic strain range plus plastic strain range) calculated by the equation

$$\Delta\epsilon = 4\epsilon_\alpha(d) + (1 - 2\nu)\sigma_R/E$$

[1]Professor, Department of Naval Architecture, University of Tokyo, 7-3-1, Hongo, Bunkyo-ku, Tokyo, Japan.

where $\epsilon_\alpha(d)$ is the diametral strain amplitude (= the controlled value in strain cycling in the present tests), ν is Poisson's ratio, σ_R is the true stress range at approximately half N_f, and E is Young's modulus

$\Delta\epsilon/2$ Longitudinal, natural and total strain amplitude (= half $\Delta\epsilon$)

ϵ_f Longitudinal, static fracture ductility of a specimen not subjected to prestrain ($\epsilon_f = \ln 100/(100 - \text{RA})$, where RA is the percent reduction in area) on the assumption of constant volume in the fully plastic condition

ϵ_{fR} Residual ductility (longitudinal, static fracture ductility of a specimen subjected to a certain amount of prestrain)

ϵ_{pre} Prestrain applied in advance of completely reversed strain cycling

m Mean value

Introduction

In connection with the possible service failure of a steel plate structure, it should be noted that the exhaustion of material ductility has not been fully considered as a key factor causing damage to a welded structure. It may occur by such excessive deformation as local plastic buckling at a geometrical discontinuity (e.g., the vicinity of fillet welds connecting a plate and a strong stiffener). In this regard, an example of service failure of a ship that suffered from serious damage by unexpected wave loads may find a possible explanation from the concept of ductility exhaustion which deleteriously affected and shortened even further the very low-cycle fatigue life of material at such discontinuities.

Recently a bulk carrier in the fully loaded condition, which was navigating with an average speed of 5.25 knots through periodical and heavy swells in the North Pacific Ocean in winter, suffered disastrous structural damage consisting of bow break off in the sea area latitude 31°N and longitude 156°11′E [*1*]. According to the evidence given by the ship's crew, the bow of the ship was struck strongly by a wave some 10 m high, ducked into the trough of a swell, and did not appear on the sea surface for a while. When the bow was observed again above the surface, it was found that the No. 1 cargo hatch had caved in and that the bow structure afore of No. 1 cargo hatch had turned upward. The stays of the foremast had broken. About 2 h later, the bow structure that had been oscillating up and down with increasing displacement by waves broke off from the ship.

On the basis of the investigation of fracture path and surface, a possible conjecture about the mechanism of the service failure may be considered as follows: (1) the material of a "hot" spot area where the fracture started had been damaged more or less by higher alternating loading due to waves than expected in design, and (2) the spot was subjected to very high and impulsive loading by waves of extraordinary height. This may have caused buckling of a structural member including the spot area. This means that the material of the spot area was subjected to excessive overstrain in tension or compression, resulting in a loss of ductility; finally, cracks started and propagated by the repetition of ever-increasing ligament stresses. It is uncertain whether the cracking occurred during the impulsive load or not. If the cracking is assumed to have been initiated by the impulsive load, the crack may have extended rapidly by the alternating stresses.

It may also be possible that cracking by very low-cycle fatigue occurred in the course of repetitions of load after the impulsive load in the sense that the ductility of the material was severely lost by excessive deformation.

The present work was initiated to systematically study the effect of prestrain in tension or in compression on the fatigue life in the very low-cycle fatigue regime for a certain range of weldable structural steels. Two possible factors were considered to have influenced the failure mechanism: microcracking on the specimen surface and loss of ductility. Two series of tests were carried out on one of the steels.

Experimental Procedure

The steels, two mild steels and two high strength steels, were received in the form of hot rolled plate and were used for the tests in the as-received condition. Table 1 shows the tensile properties, which are averaged values obtained by each of three smooth and solid cylinder specimens of 10 mm diameter and 50 mm gage length. The original plate thickness of the three steels (codes P, R, and S) was 25 mm, while the original thickness of the steel Q plate was 45 mm.

The test series in the present investigation was as follows:

1. Static tension tests were performed on hourglass-shaped, 8-mm-diameter specimens subjected to certain amounts of prestrain for the purpose of checking the residual static fracture ductility depending on prestrain (steel Q only).
2. Normal, diametral strain controlled low-cycle fatigue tests were done on hourglass-shaped, 8 to 10 mm diameter specimens in order to obtain strain range versus fatigue life curves, which provide the basic reference curves in discussing the decrease in fatigue life depending on the amount of prestrain (steels P, R, and S).
3. Diametral strain controlled fatigue tests were done using hourglass-shaped, 8 to 10 mm diameter specimens subjected to one cycle of prestrain in tension or in compression. The specimen surface was not machined and remained as it was after prestraining (steels P, Q, R, and S). Experimental parameters were chosen so as to realize the fatigue life of not-prestrained specimens of less than 100 cycles in order to consider the phenomenon of very low-cycle fatigue in the service failure of the ship. This test series is called the surface *remained* series.
4. Diametral strain controlled fatigue tests were also done using hourglass-shaped, 8 mm diameter specimens machined out from previously prestrained 15 mm diameter specimens (steel Q only). The purpose of this test series was to remove the effects of possible microcracking on the specimen surface, which is one of the superposed factors in the test series (3). Experimental parameters were similar to those in test series (3). This series is called the surface *removed* series.

All the diametral strain controlled fatigue tests were executed by completely reversed strain cycling by controlling diametral natural strain amplitude at ambient temperature. Subsequently, the given values for strain range (2× amplitude), $\Delta\epsilon$, are the longitudinal equivalent values that are calculated by the equation noted in the description of nomenclature of $\Delta\epsilon$. This equation is based on two assumptions: (1) Hook's law should exist between true stress and natural strain, and (2) the volume constant condition should hold for the plastic strain (Poisson's ratio is equal to 0.5). The cycling rate was adjusted to slower than 0.025 Hz in general and less than 0.005 Hz in the case where the failure life was expected to be shorter than 10 cycles.

TABLE 1—*Mechanical properties of steels (room temperature tensile).*

Steel		0.2% Yield Strength, MPa	Ultimate Strength, MPa	Elongation, %	Reduction of Area, %	Static Fracture Ductility
Code	Type					
P	MS[a]	242	440	41.7	68.0	1.14
Q	MS[a]	274	431	38.0	66.7	1.10
R	HT60[b]	510	597	31.0	75.0	1.39
S	HT80[c]	774	837	32.5	72.7	1.30

[a]Mild steel.
[b]600 MPa class high strength steel.
[c]800 MPa class high strength steel.

The strain cycling in test series (3) and (4) was carried out after one cycle prestraining. Attention was paid to start strain cycling with a very slow strain rate in the beginning in order to check for the occurrence of unexpected buckling. Although many tests in the case of larger amounts of prestrain were unsuccessful due to buckling, only the test results of specimens which had not shown irregular deformation in the course of both prestraining and strain cycling will be discussed later.

Figure 1 shows specimen configurations with the elastic stress concentration factor at the test section, K_t. In test series (4), smooth cylinder specimens of 15 mm in test section diameter were machined out in the rolling direction at the half-plate thickness level of the 45-mm-thick plate and used for prestraining. After prestraining, the 15-mm-diameter specimen was machined so as to form a 8-mm-diameter, hourglass-shaped specimen. The hourglass-shaped specimens, 8 to 10 mm in test section diameter, were taken in the rolling direction at half-plate thickness for the steels P, R, and S. In the comparative test series (3) and (4) for steel Q, the specimens for series (3) were machined out directly in the rolling direction at the half-plate thickness level of the 45-mm-thick plate.

Residual Static Fracture Ductility

The results of test series (1) are shown in Fig. 2. The residual ductility ϵ_{fR}, which is the static fracture ductility of a specimen subjected to prestrain, is normalized by the static fracture ductility of the specimen without prestrain ϵ_f. This is plotted against the normalized prestrain, which is the ratio of absolute value of prestrain ϵ_{pre} to ϵ_f. It is observed that data of ϵ_{fp}/ϵ_f for

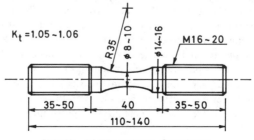

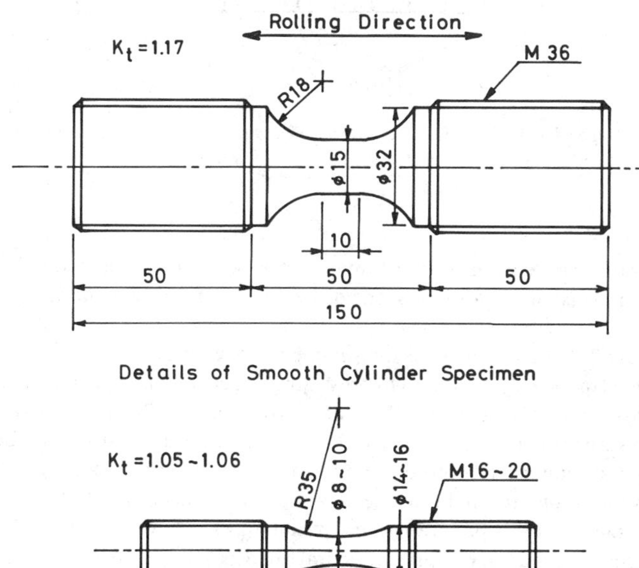

FIG. 1—*Details of specimens.*

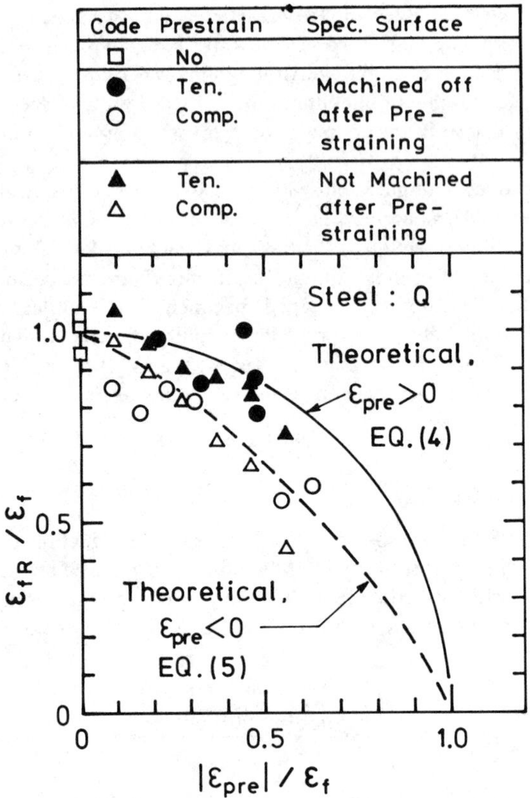

FIG. 2—*Dependency of residual static fracture ductility against prestrain (steel Q).*

specimens prestrained in tension form a family clearly separated from a family of specimens prestrained in compression, regardless of the difference in the surface condition of the tension specimens.

Circle marks exhibit the results of 8-mm-diameter specimens machined out from 15-mm-diameter specimens previously subjected to prestrain. The surfaces of the tension test specimens were smoothly machined to remove the effects of microcracks possibly initiated on the specimen surface upon the static fracture ductility. On the other hand, triangular marks show the results of 8-mm-diameter specimens tension-tested just after prestraining. As an example, the microscopic investigation of longitudinal cross sections revealed cracks 0.05 to 0.1 mm deep in the surface of a specimen that was prestrained in tension up to 55% of ϵ_f. Thus the fact that there was little difference between the residual static fracture ductilities of the surface removed series and the surface remained series suggests no considerable effects of microcracks in tensile prestrain and roughened surface irregularity in compressive prestrain upon the static fracture ductility.

The process of prestraining followed by static tension at any part of strain cycling was also considered. To do this the mechanism of reduction in static fracture ductility of a specimen subjected to prestrain was analyzed using an assumption of linear damage. If the static tension is considered as the strain cycling with $\Delta\epsilon = 2\epsilon_f$ and $n = 1/4$, where n is the number of cycles imposed, the following expression is derived:

$$(\Delta\epsilon)^k N_f = (2\epsilon_f)^k(1/4) \tag{1}$$

where k is an empirical material constant.

As shown in Fig. 3, each part of the prestrain and tension test may be regarded as strain cycling with the following parameters:

A − B: $\Delta\epsilon = \epsilon_{pre}$, $n = 1/4$ B − C: $\Delta\epsilon = \epsilon_{pre}$, $n = 3/4$
C − D: $\Delta\epsilon = 2\epsilon_{fR}$, $n = 1/4$
E − F: $\Delta\epsilon = \epsilon_{pre}$, $n = 1/4$ F − G: $\Delta\epsilon = \epsilon_{pre}$, $n = 1/2$
G − H: $\Delta\epsilon = \epsilon_{pre} + 2\epsilon_{fR}$, $n = 1/4$

For tensile prestrain:

$$(2\epsilon_{fR})^k(1/4) = (2\epsilon_f)^k(1/4) - (\epsilon_{pre})^k(1) \tag{2}$$

For compressive prestrain:

$$(\epsilon_{pre} + 2\epsilon_{fR})^k(1/4) = (2\epsilon_f)^k(1/4) - (\epsilon_{pre})^k(3/4) \tag{3}$$

The value of k was obtained as 1.926 by normal strain cycling fatigue for steel P, which is quite similar material to steel Q. Then assuming $k = 2$, Eqs 2 and 3 are deformed as follows:

For tensile prestrain:

$$\epsilon_{fR} = [(\epsilon_f)^2 - (\epsilon_{pre})^2]^{1/2} \tag{4}$$

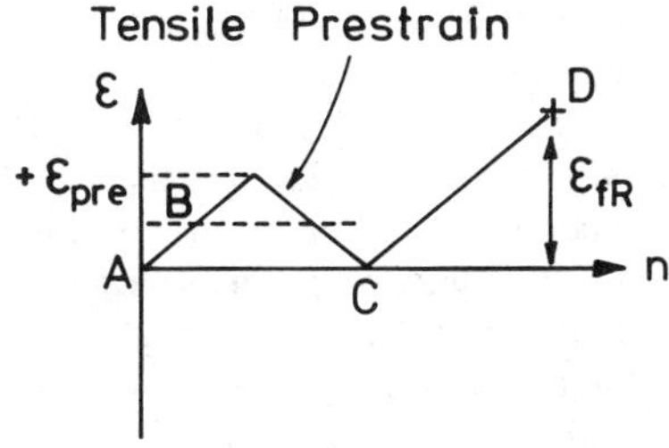

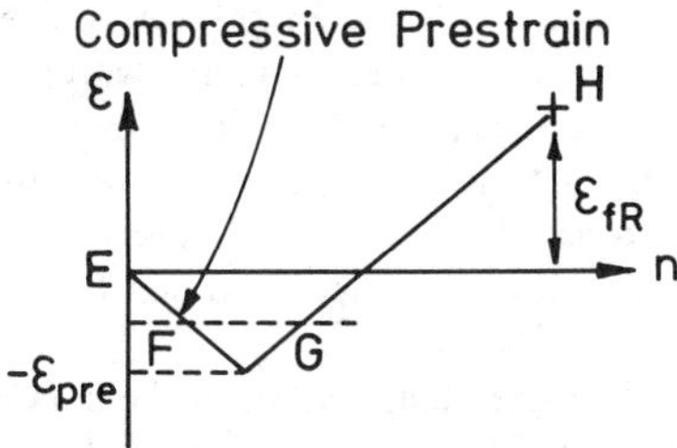

FIG. 3—*Schematic diagrams of prestraining and static tension regarded as a part of strain cycling.*

For compressive prestrain:

$$\epsilon_{fR} = -(1/2)\epsilon_{pre} + (1/2)[(2\epsilon_f)^2 - 3(\epsilon_{pre})^2]^{1/2} \tag{5}$$

The solid and dashed lines in Fig. 2 represent Eq 4 and Eq 5 respectively and show fair agreement with experimental data.

Correlations Between Experimental Results and Estimated Values after Manson's and Iida's Formulae

In test series (2), completely reversed strain cycling tests were carried out on non-prestrained 8 to 10 mm diameter hourglass specimens of steels P, R, and S. Instead of visible crack initiation life N_c, the complete failure life N_f was taken as the dominant criterion in the tests, because the results from test series (3) and (4) were evaluated on the N_f basis. On this account, N_c was estimated for some specimens by the correlation curve between N_c and N_f (discussion follows).

Analyzing the experimental results of completely reversed strain cycling of 29 materials, Manson [*2*] proposed the following expression, which is useful for estimating strain range on the basis of tensile properties:

$$\Delta\epsilon = (\epsilon_f)^{0.6}(N_f)^{-0.6} + 3.5(\sigma_u/E)(N_f)^{-0.12} \tag{6}$$

where σ_u is the ultimate tensile strength (MPa) and E is Young's modulus (MPa).

Similarly an estimation formula for the visible crack initiation life basis was proposed by Iida et al. [*3*] by analyzing accumulated data on base and weld metals of structural steels. The following formula is applicable in the range $0.7 < \epsilon_f < 1.4$ and 400 MPa $< \sigma_u <$ 1100 MPa:

$$\Delta\epsilon/2 = 0.286\epsilon_f(N_c)^{-(0.0425\epsilon_f+0.544)} + (5.26 \times 10^{-6}\sigma_u + 0.0013) \times (N_c)^{-(0.173-1.074\times10^{-4}\sigma_u)} \tag{7}$$

Figure 4 illustrates the correlation between experimental longitudinal strain range $\Delta\epsilon$(*Exp.*) and calculated longitudinal strain range $\Delta\epsilon$(*Cal.*) after Manson's formula (Eq 6). The experimental longitudinal strain range was calculated by inputting the diametral strain amplitude $\epsilon_a(d)$ and true stress range σ_R at approximately 1/2 N_f into the equation noted in the description of nomenclature of $\Delta\epsilon$. The calculated longitudinal strain range is obtained by input of tensile properties of each steel and N_f corresponding to a given experimental, longitudinal strain range into Eq 6. The mean value over total ratios of experimental strain range to calculated strain range is 1.07; all the data fall into a scatter band of which the upper and lower limits are 1.3 times and 1/1.3 fraction of the solid line. On the other hand, the estimated failure life N_f (*Cal.*) was calculated by the iterative method by inputting corresponding experimental longitudinal strain range to given N_f (*Exp.*) into Eq 6. Figure 5 plots experimental N_f against calculated N_f. In this case the agreement is not so good as when compared with Fig. 4, but all the data fall into a scatter band of 2.2 times and 1/1.7 fraction of the solid line.

In the same way, the estimated longitudinal strain amplitude after Iida's formula $\Delta\epsilon/2$(*Cal.*) was calculated by Eq 7 and correlated with the experimental longitudinal strain amplitude $\Delta\epsilon/2$(*Exp.*) (Fig. 6). All the data fall in a scatter band of 1.6 times and 1/1.4 fraction of the solid line.

Table 2 shows the results of the regression analysis of the data obtained. The third column from the left indicates the number of data analyzed, and the fourth column the mean ratio, if the regression is assumed to be the 45° line. The fifth to seventh columns show constants and coefficients of determination for the case of power curve fitting.

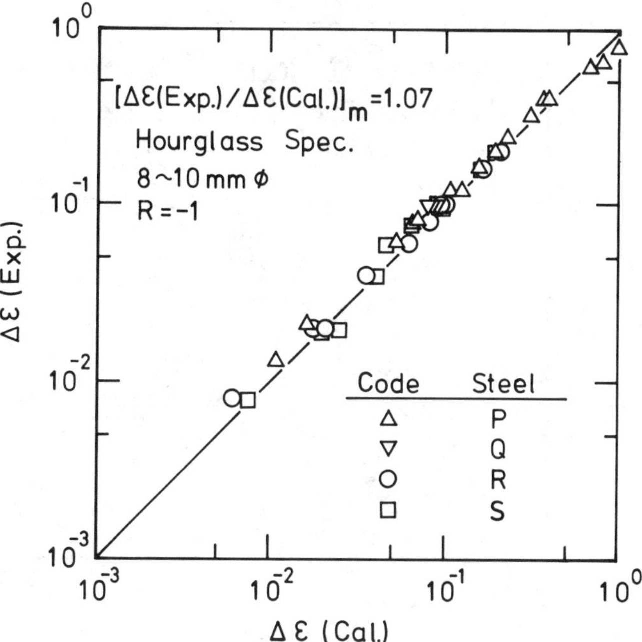

FIG. 4—*Correlation between experimental and calculated strain ranges (after Manson's formula).*

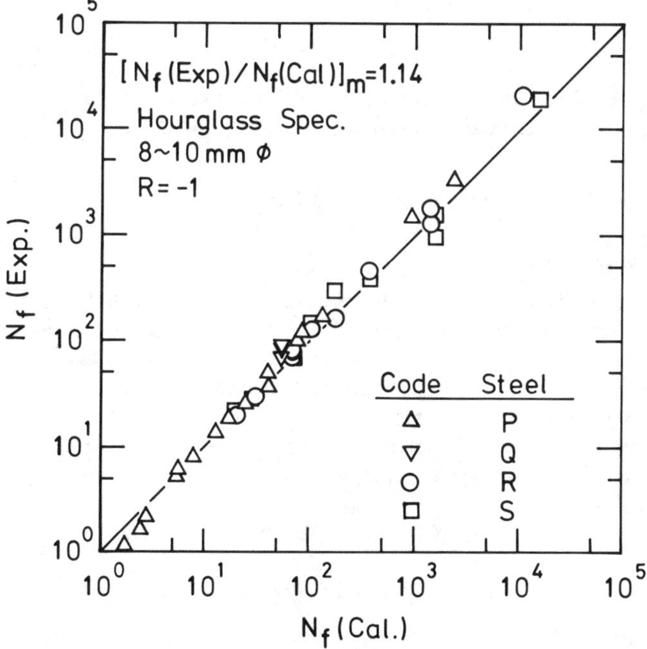

FIG. 5—*Correlation between experimental and calculated failure lives (after Manson's formula).*

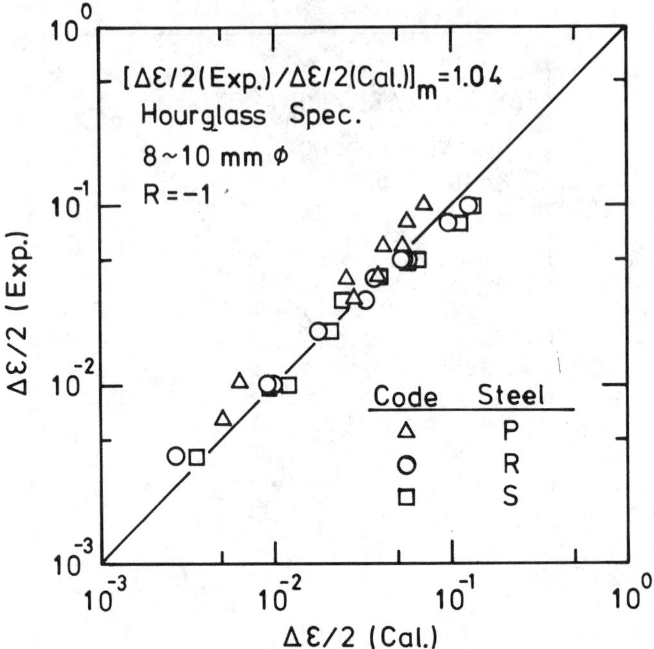

FIG. 6—*Correlation between experimental and calculated strain amplitudes (after Iida's formula).*

It may be remarked that the agreement of the estimated value with the experimental data is generally good even in the estimations of N_f or N_c, implying enough applicability of Eqs 6 and 7.

Effects of Prestrain on the Reduction of Fatigue Life

Effects of Loss of Ductility

The following analysis will be applicable for test series (4), in which the hourglass specimens taken from the previously prestrained large diameter specimens were subjected to diametral strain cycling. No microcracks by tensile prestrain or surface roughness by compressive prestrain are expected on the surface of the strain cycling specimen, and the loss of ductility due to prestraining may be the only parameter.

Assuming that the residual static ductility of the specimen just before strain cycling should be the quantity corresponding to ϵ_f in Eq 1, then

$$(\Delta\epsilon)^k N_{fp} = (2\epsilon_{fR})^k (1/4) \tag{8}$$

Substituting $k = 2$ similarly as for Eqs 4 and 5, the following expressions are derived:

For tensile prestrain:

$$(N_{fp}/N_f) = 1 - (\epsilon_{\text{pre}}/\epsilon_f)^2 \tag{9}$$

TABLE 2—*Correlation between experimental data and estimated values after Manson's and Iida's formulae.*

Category	Estimation Formula After	No. of Data	Mean Ratio	Constants and Coefficient[d]		
				a	b	R^2
N_f	Manson	41	1.14[a]	0.907	1.050	0.992
$\Delta\epsilon$	Manson	41	1.07[b]	0.933	0.949	0.992
N_c	Iida	29	0.93[c]	0.876	1.040	0.959
$\Delta\epsilon/2$	Iida	29	1.04[b]	0.751	0.912	0.950

[a]Mean of $N_f(Exp.)/N_f(Cal.)$
[b]Mean of $\Delta\epsilon(Exp.)/\Delta\epsilon(Cal.)$
[c]Mean of $N_c(Exp.)/N_c(Cal.)$
[d]Regression formula: $y = ax^b$, where combinations of y and x are $N_f(Exp.):N_f(Cal.)$, $\Delta\epsilon(Exp.):\Delta\epsilon(Cal.)$, $N_c(Exp.):N_c(Cal.)$. R^2 is the coefficient of determination.

For compressive prestrain:

$$(N_{fp}/N_f) = [(-1/2)(\epsilon_{pre}/\epsilon_f) + \{1 - (3/4)(\epsilon_{pre}/\epsilon_f)^2\}^{1/2}]^2 \quad (10)$$

Comparative plots of the results of the surface removed test series (4), of which strain cycling tests were carried out with an equivalent longitudinal strain range, $\Delta\epsilon$ of 0.1, are shown in Fig. 7. The ordinate stands for the ratio of N_{fp} to N_f. The average value of N_f is 76.7 cycles. Equations 9 and 10 are expressed by the solid and the dashed lines, respectively. The agreement between experimental and theoretical values is not always good, and it may be expressed as qualitatively fair but quantitatively poor. This result suggests that not only a loss of ductility but other factors

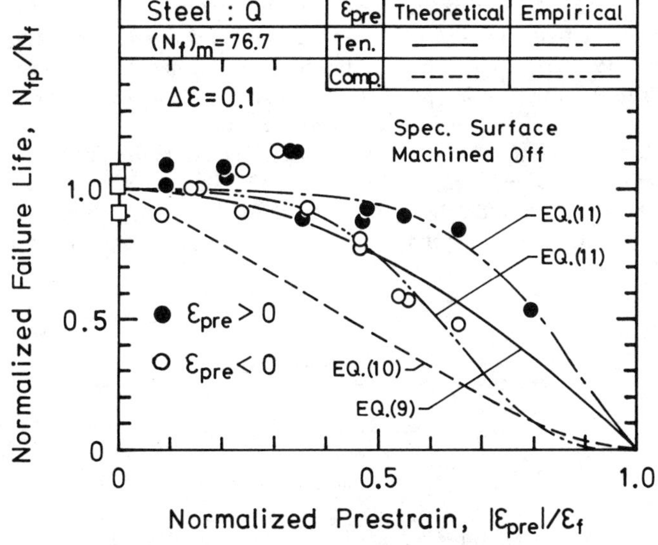

FIG. 7—*Normalized failure life against normalized prestrain (surface removed series).*

should be considered for a clear and full explanation of the deteriorating mechanism of prestraining.

The experimental results may be summarized as follows: The N_{fp}/N_f ratio is nearly equal to unity up to $\epsilon_{pre}/\epsilon_f = 0.4$ in the case of tensile prestrain and up to $|\epsilon_{pre}|/\epsilon_f = 0.35$ in the case of compressive prestrain, respectively. Here $|\epsilon_{pre}|$ means the absolute value of ϵ_{pre}. After these critical points, the N_{fp}/N_f ratio shows a gradually decreasing tendency with an increasing amount of $|\epsilon_{pre}|/\epsilon_f$. It is the general and evident tendency that a larger reduction of N_{fp}/N_f ratio is shown by compressive prestrain than by tensile prestrain. The N_{fp}/N_f at $|\epsilon_{pre}|/\epsilon_f = 0.6$ is figured as 0.87 for tensile prestrain and 0.53 for compressive prestrain, respectively. For engineering purposes, the expression of best fit curves for the experimental data was assumed to take into consideration the concept that a curve should pass 1.0 on the ordinate as well as on the abscissa:

$$N_{fp}/N_f = [1 - (|\epsilon_{pre}|/\epsilon_f)^{k_1}]^{k_2} \tag{11}$$

By the least squares method k_1 and k_2 were determined as listed in Table 3. The best fit curves are shown in Fig. 7 by the dot-dash and double dot-dash lines for tensile and compressive prestrains, respectively.

Superposed Effects of Ductility Loss and Microcracks

At least two factors, loss of ductility and microcracks on the specimen surface, should contribute to a reduction of fatigue life of strain cycling specimens in test series (3). Due to the superposed effects, Eqs 9 and 10 are not applicable to the experimental results of test series (3). For this series a summarized plot is shown in Fig. 8. The best fit curves are obtained by applying Eq 11. Constants in the expressions are listed in Table 3.

From a comparison of Fig. 7 and Fig. 8, the following remarks are derived:

1. The tensile prestrain is more deleterious than the compressive prestrain in the case of the surface remained series. The compressive prestrain is more deleterious in the case of the surface removed series. This remarkable difference may possibly result from the detrimental effects of microcracks formed by tensile prestrain.
2. The tendency of N_{fp}/N_f to decrease is considerably greater in the case of the surface remained series compared with the surface removed series. The decrease starts at about 0.2 of the $|\epsilon_{pre}|/\epsilon_f$ ratio and increases for larger prestrain ratio's. The median value of the N_{fp}/N_f ratio at $|\epsilon_{pre}|/\epsilon_f = 0.6$ is 0.07 for the tension prestrained and surface remained specimen and 0.32 for the compression prestrained and surface remained specimen.
3. As shown in Fig. 8, no significant difference in deviation from the best fit curve is observed among test results obtained by strain cycling in a certain range of the equivalent longitudinal strain range $\Delta\epsilon$ between 0.04 and 0.2, regardless of the kind of steel. This may suggest that the longitudinal strain range in strain cycling after tensile or compressive prestrain may have little

TABLE 3—*Constants in Eq 11.*

Test Series	Prestrain	k_1	k_2
Surface Removed	tension	4.695	1.459
	compression	3.868	4.211
Surface Remained	tension	3.174	12.863
	compression	2.319	3.097

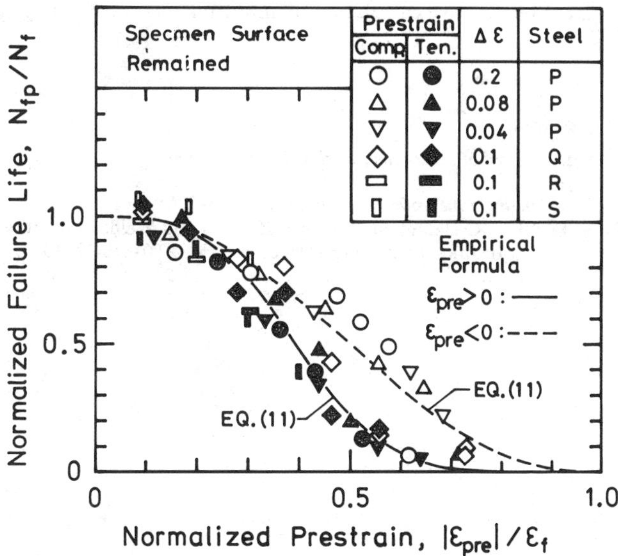

FIG. 8—*Normalized failure life against normalized prestrain (surface remained series).*

influence. In addition the best fit curves shown in Fig. 8 may be applicable to a certain range of steel, of which the mechanical properties are similar to those of the tested steels.

Conclusions

As a result of this investigation, the following conclusions were drawn:

1. One cycle prestrain in tension or in compression influences, in general, the residual static ductility and very low-cycle fatigue life.
2. The static fracture ductility retained after one cycle prestrain decreases with the increasing prestrain in the absolute value $|\epsilon_{pre}|$, showing the more deleterious effects of compressive prestrain than of tensile prestrain. The behavior can be explained by the concept of linear damage in low-cycle fatigue.
3. If the specimen surface is left as it was after prestrain in tension or in compression (surface remained test series), the low-cycle fatigue life of the prestrained specimen N_{fp} reduces remarkably as a function of $|\epsilon_{pre}|$. Tensile prestraining is more deleterious than compressive prestraining due to the preferential effects of microcracks formed on the specimen surface. For example, the ratio of N_{fp} to N_f is only 0.07 for the tensile prestrain and 0.32 for the compressive prestrain, at 0.6 of $|\epsilon_{pre}|$ to ϵ_f ratio, where N_f is the failure life of the non-prestrained specimen.
4. When the specimen surface is removed after prestraining to eliminate the effects of microcracks in the case of tensile prestrain and of roughened irregularity in the case of compressive prestrain, the N_{fp} also decreases as a function of $|\epsilon_{pre}|$, but the decreasing amount of N_{fp} is not so large compared with the surface remained test series. It is generally remarked that compressive prestrain reduces N_{fp} more than tensile prestrain. This tendency can be explained conceptually by the linear damage hypothesis and the residual fracture ductility. For example, the ratio of N_{fp}/N_f at $|\epsilon_{pre}|/\epsilon_f = 0.6$ is 0.87 for tensile prestrain and 0.53 for compressive prestrain.
5. Fairly good agreements on strain range and fatigue life are found between the experimental data and the estimated values after Manson's and Iida's formulae, even in the case where

fatigue life is estimated by iterative calculation by inputting a corresponding strain range into the fomulae.

References

[1] Yamamoto, Y. et al., *Journal of the Society of Naval Architects of Japan,* Vol. 154, 1983, pp. 516–524 (in Japanese).
[2] Manson, S. S., *Experimental Mechanics,* July 1965, pp. 193–226.
[3] Iida, K. and Fujii, E., "Low Cycle Fatigue Strength of Steels and Welds in Relation to Static Tensile Properties," International Institute of Welding, Document No. XIII-816-77, 1977.

D. L. DuQuesnay,[1] *M. A. Pompetzki,*[1] *T. H. Topper,*[1] *and M. T. Yu*[1]

Effects of Compression and Compressive Overloads on the Fatigue Behavior of a 2024-T351 Aluminum Alloy and a SAE 1045 Steel

REFERENCE: DuQuesnay, D. L., Pompetzki, M. A., Topper, T. H., and Yu, M. T., **"Effects of Compression and Compressive Overloads on the Fatigue Behavior of a 2024-T351 Aluminum Alloy and a SAE 1045 Steel,"** *Low Cycle Fatigue, ASTM STP 942,* H. D. Solomon, G. R. Halford, L. R. Kaisand, and B. N. Leis, Eds., American Society for Testing and Materials, Philadelphia, 1988, pp. 173–183.

ABSTRACT: Smooth cylindrical specimens of 2024-T351 aluminum alloy and SAE 1045 steel were tested under constant amplitude cycling to study the effect of compressive stress on fatigue life. The results show a reduction in the fatigue life at constant maximum stress as the compressive portion of the stress cycle is increased.

Tests performed to study the effect of a periodic compressive overload on the order of the yield strength show an increasing reduction in fatigue life at constant maximum stress as the frequency of application of the overloads is increased. The compressive overload cycle significantly increases the damage done by subsequent smaller cycles, including those far below the constant amplitude fatigue limit. The results indicate that the present techniques used for damage summation for variable amplitude service histories may give grossly unconservative fatigue life predictions, especially for histories containing large compressive load cycles accompanied by a relatively large number of small cycles. It is suggested that conservative life predictions for such histories can be made using the stress-life curve from constant amplitude tests with compression on the order of the yield stress accompanying every cycle.

KEY WORDS: fatigue (metals), compression, compressive overloads, variable amplitude, damage

Nomenclature

- D_o Damage due to overload cycles
- D_r Damage due to fully reversed cycles
- D_s Average damage per fully reversed cycle
- η Number of $R = -1$ cycles between overloads
- N_e Equivalent life of a $R = -1$ test (D_o removed)
- N_f Number of cycles to failure
- N_{fo} Number of overload cycles to failure
- n_o Number of applied overload cycles
- n_r Number of applied $R = -1$ cycles
- R Cyclic stress ratio (S_{min}/S_{max})

[1]M.A.Sc. Candidate, M.A.Sc. Candidate, Professor, and Ph.D Candidate, respectively, Department of Civil Engineering, University of Waterloo, Waterloo, Ontario, Canada N2L 3Gl.

S_o Compressive overload stress, MPa
S_{fat} Maximum stress at the fatigue limit, MPa
S_{max} Cyclic maximum stress, MPa
S_{min} Cyclic minimum stress, MPa

Introduction

A typical fatigue service history contains both tensile and compressive loading cycles. Recent work has shown that a compressive cycle not only contributes to damage itself but also can increase the damage caused by subsequent cycles. Several investigators, for example [*1–9*], have studied the effects of overload cycles on fatigue life and damage accumulation. In general, it was shown that fatigue life and the fatigue limit are reduced by periodic overload cycles. These effects were mostly attributed to detrimental residual stresses induced by the overload cycles. Conle and Topper [*8, 9*] showed that the omission of the effect of small strain cycles below the constant amplitude endurance limit from a variable amplitude service history could result in unconservative predictions of fatigue life. They concluded that larger strain cycles increase the damage done by the smaller cycles, and emphasized that small cycles far below the constant amplitude fatigue limit could do significant damage. It was also demonstrated [*8–11*] that a linear extension of the low-cycle portion of the constant amplitude strain-life curve, plotted on logarithmic axes, could give improved life predictions for variable amplitude histories.

Since tensile overloads are known to delay fatigue crack propagation, Topper and Au [*12*] hypothesized that the increased damage must be due to high compressive strain excursions in the fatigue service histories. Au et al. [*13*] later showed that both the compressive portion of a loading cycle and periodic compressive overloads increased crack propagation rates and decreased the threshold stress intensity in center-cracked plate specimens of 2024-T351 aluminum alloy. Recent work on the effects of compressive overloads on near-threshold crack growth [*14–16*] has shown similar effects in a 2024-T351 aluminum alloy, a SAE 1045 steel, and a mild steel (CSA G40.21), indicating that the phenomenon is general.

The purpose of the present investigation is to determine whether effects equivalent to those described above in center-cracked plate specimens can be observed in round smooth specimens, and also to determine whether the reduction in fatigue life due to compressive overloads is more severe than has previously been reported.

Materials, Equipment, and Test Techniques

The materials used in this study were a 2024-T351 aluminum alloy and a SAE 1045 steel, both in the "as received" condition. The chemical compositions of these alloys are given in Table 1.

TABLE 1—*Chemical composition (weight percent) of test material.*

2024-T351 Aluminum Alloy							
Si	Fe	Cu	Mn	Mg	Cr	Zn	Ti
0.50	0.50	4.35	0.60	1.50	0.10	0.25	0.15

SAE 1045 Steel					
C	Si	Mn	P	S	Fe
0.46	0.17	0.81	0.027	0.023	rem.

Smooth cylindrical specimens, with the dimensions given in Fig. 1, were machined from rolled plates of the materials. Specimens of each alloy were tested in monotonic tension to establish the mechanical properties given in Table 2. The specimens tested in fatigue were hand polished in the direction of the loading axis, using progressively finer grades of emery paper, to remove surface machining marks. Thin bands of M-COAT plastic coating were applied at the central gage length, in order to avoid surface scratches from the knife edges of an axial extensometer which was clipped to each specimen for strain controlled testing.

All tests were conducted in a laboratory environment at room temperature (23°C) using an axial closed-looped servo-controlled electrohydraulic testing machine. A sinusoidal loading waveform was used for all constant amplitude tests and for the periodic overload tests in the low-cycle fatigue region. In the latter case, the tests were halted and the overloads were applied manually at the appropriate intervals. In the high-cycle fatigue region, a digital process control

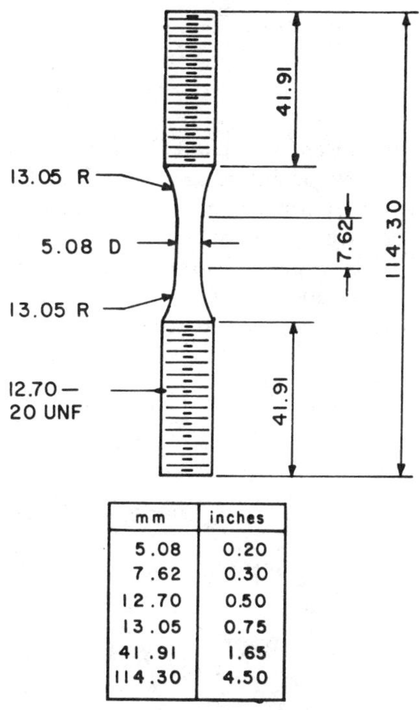

mm	inches
5.08	0.20
7.62	0.30
12.70	0.50
13.05	0.75
41.91	1.65
114.30	4.50

FIG. 1—*Smooth specimen design and dimensions.*

TABLE 2—*Mechanical properties of test materials.*

	2024-T351 Aluminum Alloy	SAE 1045 Steel
Elastic modulus (MPa)	72 400	203 500
Yield stress (MPa) (0.2% offset)	356	472
Ultimate tensile strength (MPa)	466	745
True fracture stress (MPa)	623	1046
Hardness (HB)	...	235

computer equipped with digital to analog circuits was connected to the system. The computer was programmed to output a sine-like loading waveform and to apply the periodic compressive overload histories.

Each specimen was first subjected to ten cycles at ± 1% total strain on the 7.62 mm (0.3 in.) gage length, under strain control. Such prestraining was noted [17] to reduce fatigue life and data scatter. The prestrain cycles essentially complete the crack nucleation process and short crack propagation can be considered to dominate the remaining fatigue life. The initial prestrain also allows the material to strain harden and cyclically stabilize to a degree. The prestraining cycles were terminated with zero stress in the specimen gage length as illustrated by point A in Fig. 2.

The specimens were then subjected to uniaxial stress histories under load control until fracture occurred. Cyclic frequencies between 5 and 100 Hz were used, depending upon the applied load amplitude.

Results and Discussion

General

The stress-life data in this investigation are plotted on axes with the maximum stress on the ordinate versus the number of cycles to failure on the abscissa. There are two advantages in plotting the data in this manner, rather than on the more conventional axes with the stress amplitude on the ordinate. Firstly, the constant amplitude data can be plotted along with the periodic compressive overload data on the same axes. Secondly, the results can be qualitatively compared directly with crack propagation trends reported in other investigations where only the tensile portion of the stress intensity factor was considered.

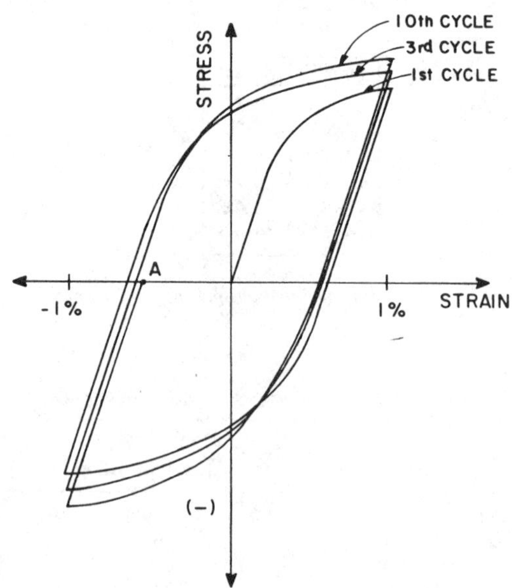

FIG. 2—*Material hysteresis behavior during initial prestrain cycles.*

Effect of Compression

Six test series were performed to study the effect of compressive stress on fatigue life during constant amplitude loading.

The aluminum alloy was subjected to the following four test series to determine the stress-life behavior:

1. Zero to tension ($R = 0$).
2. Fully reversed ($R = -1$).
3. Tests with the compressive minimum stress, S_{min}, held constant at -276 MPa (-40 ksi), while the maximum tensile stress, S_{max}, was varied from test to test.
4. As in Series 3, except that the compressive minimum stress, S_{min}, was held constant at -414 MPa (-60 ksi).

The results of the aluminum alloy test series, plotted in Fig. 3, clearly show that at a constant minimum stress the fatigue life increases as the maximum stress, and the stress range, decrease. This figure also illustrates that at a constant fatigue life, the maximum stress decreases and the stress range increases as the compressive stress is increased. The presence of compression is shown in Fig. 3 to decrease the maximum stress at the fatigue limit, S_{fat}, from 186 MPa (27 ksi) for the $R = 0$ tests to 124 MPa (18 ksi) for the $R = -1$ tests. The presence of a high compressive stress further reduces S_{fat} to 69 MPa (10 ksi) for the $S_{min} = -276$ MPa tests. For higher compressive stresses ($S_{min} = -414$ MPa) there is no evidence of an endurance limit, even at 5×10^7 cycles where the maximum stress approaches zero.

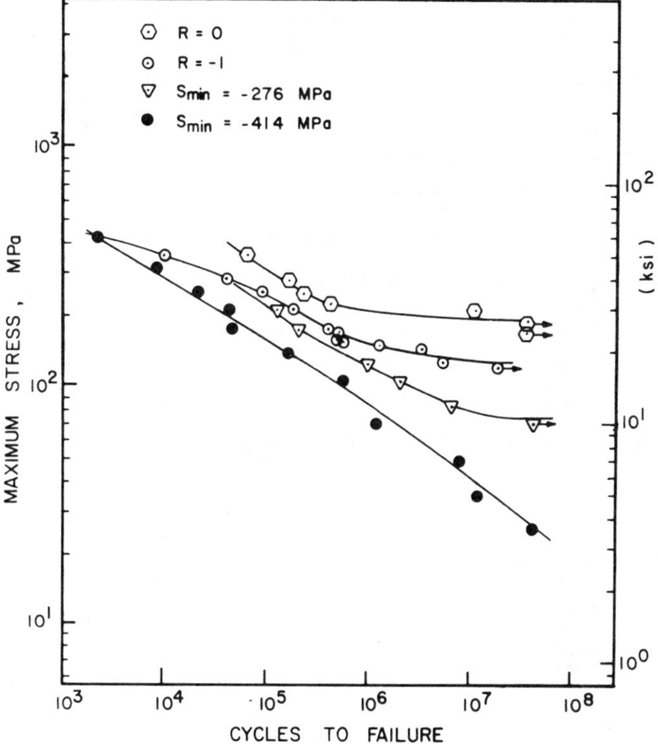

FIG. 3—*Effect of compressive stress on the constant amplitude fatigue life of 2024-T351 aluminum alloy.*

The results of Fig. 3 confirm that the compressive portion of the cycle contributes significantly to the accumulation of damage, even when the maximum stress is below the fully reversed fatigue limit.

The SAE 1045 steel was subjected to two constant amplitude test series:

1. Fully reversed ($R = -1$).
2. Compressive minimum stress, S_{min}, held constant at -552 MPa (-80 ksi), similar to Series 3 for the aluminum alloy.

The results of the steel test series, plotted in Fig. 4, show that the presence of a high compressive stress in the loading cycle reduces the maximum stress at a given fatigue life and reduces the maximum stress at the endurance limit, S_{fat}, from 296 MPa (43 ksi) for the $R = -1$ tests, to 117 MPa (17 ksi) for the $S_{min} = -552$ MPa tests.

The results of the above six test series are consistent with the observations of various authors [*13-16*] on the near-threshold crack growth behavior of center cracked plate specimens of these and other materials. They observed a decrease in the threshold stress intensity factor and an increase in crack growth rates as the compressive portion of the loading cycle was increased. It is important to note that only the tensile portion of the stress intensity factor was considered when making these observations. This behavior is similar to the reduction in the maximum stress at a given fatigue life and at the fatigue or endurance limit in smooth specimens. Furthermore, their observation that the threshold may disappear at a high compressive minimum stress in the 2024-T351 aluminum alloy is consistent with the apparent lack of an endurance limit for the $S_{min} = -414$ MPa tests on smooth specimens.

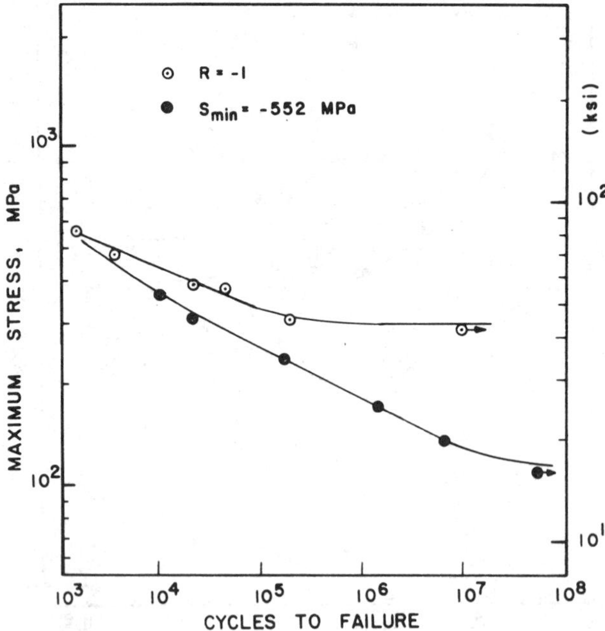

FIG. 4—*Effect of compressive stress on the constant amplitude fatigue life of SAE 1045 steel.*

Effect of Periodic Compressive Overloads

The results presented so far clearly indicate that at a constant maximum stress the fatigue life of smooth specimens decreases as the compressive portion of a constant amplitude loading cycle is increased.

A series of tests were performed to study the effect of a single compressive overload inserted periodically in a fully reversed ($R = -1$) load history. The compressive overload stress, S_o, with a magnitude equal to the stress amplitude of the final prestrain cycle ($\pm 1\%$ total strain), was -414 MPa (-60 ksi) and -552 MPa (-80 ksi) for the aluminum and steel, respectively. These compressive overload stresses are in the same order of magnitude as the monotonic yield stress for each material. In general, η fully reversed cycles with a maximum stress, S_{max}, were applied following each compressive overload. Both η and S_{max} were held constant for the duration of each test.

The aluminum alloy was tested with $\eta = \infty$, 50 000, 100, 30, 10, and 0. The data for $\eta = \infty$ and $\eta = 0$ were obtained from the tests for $R = -1$ and $S_{min} = -414$ MPa respectively, presented in Fig. 3. These define the upper and lower bounds for the periodic compressive overload tests. The steel was tested with $\eta = \infty$, 1000, 30, and 0. The upper and lower bound data defined by $\eta = \infty$ and $\eta = 0$ were taken from Fig. 4 for this material.

The results of these tests, plotted in Figs. 5 and 6, suggest that the periodic compressive overloads significantly reduce the fatigue life, and that the severity of the reduction increases with the frequency of application ($1/\eta$).

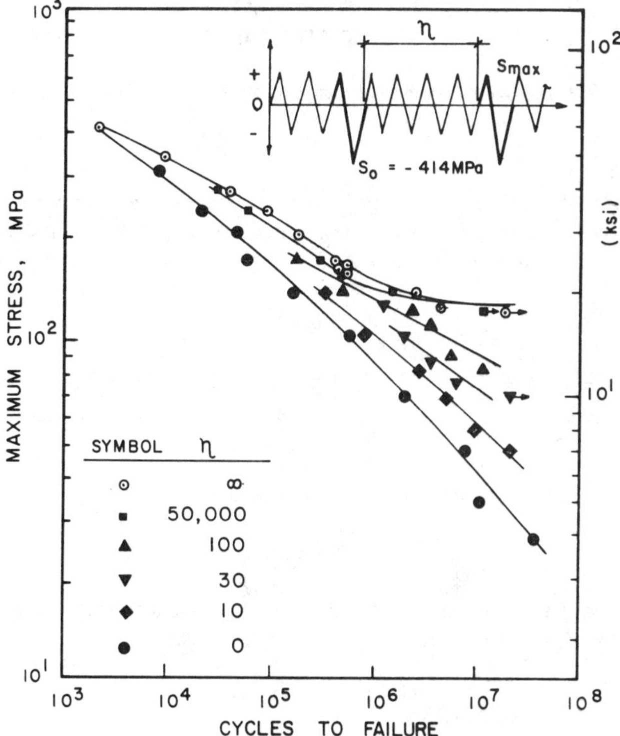

FIG. 5—*Effect of periodic compressive overloads on the fatigue life of 2024-T351 aluminum alloy.*

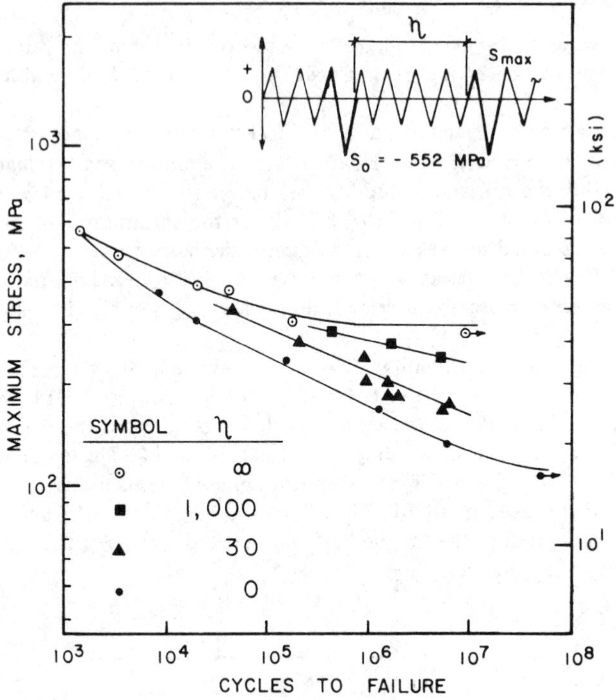

FIG. 6—*Effect of periodic compressive overloads on the fatigue life of SAE 1045 steel.*

A simple Miner's rule analysis [*18*] was performed to evaluate the effect of the overload cycles on the damage done by the fully reversed cycles. For each data point, the total damage done by the overload cycles, D_o, was determined from the $\eta = 0$ life curve:

$$D_o = n_o/N_{fo} \tag{1}$$

The remainder of the damage, D_r, was assumed to be done by the fully reversed cycles:

$$D_r = 1 - D_o \tag{2}$$

The average damage done by each fully reversed cycle, D_s, is then:

$$D_s = D_r/n_r \tag{3}$$

Alternatively, the equivalent life, N_e, of the fully reversed cycle test, with the effect of the overload damage removed, can be determined:

$$N_e = 1/D_s \tag{4}$$

The results of the damage analysis, plotted in Figs. 7 and 8, show that the equivalent life decreases, or the damage per cycle increases, with the frequency of application of the compressive overloads. It is interesting that a single application of the overload in thousands of cycles

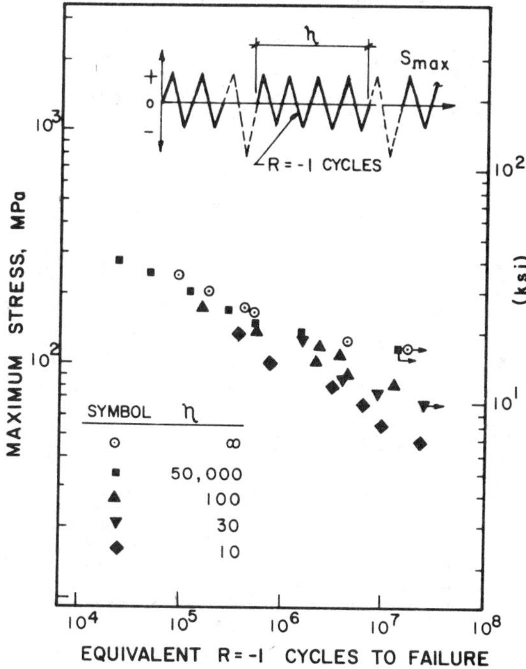

FIG. 7—*Equivalent lives of fully reversed cycles for 2024-T351 aluminum alloy.*

significantly increases the rate of damage accumulation for subsequent cycles below the constant amplitude endurance limit.

The results are again consistent with previous work [*13–16*] on center cracked plate specimens. It was noted that periodic compressive overloads increased near-threshold crack growth rates and decreased the threshold stress intensity. It was suggested that the compressive overload decreased the crack closure level so that a greater portion of the subsequent loading cycles was effective in propagating the crack. No evidence has been presented in this study to suggest the mechanism by which the compressive overload cycles increase the damage done by the subsequent cycles. However, based on the similarities observed between crack propagation and smooth specimen fatigue behavior, it is speculated that a decrease in the crack closure level following the compressive overloads may also be responsible for the increased rate of damage accumulation observed in smooth specimens.

The observations of this investigation imply that present techniques for the summation of damage in variable amplitude service histories may give grossly unconservative predictions of fatigue life. This is especially true for service histories containing a few large compressive cycles and a relatively large number of small cycles below the constant amplitude fatigue limit, since such small cycles are usually assumed not to contribute to damage. Techniques employing modified life curves [*8–11*] were shown to give improved fatigue life predictions for certain variable amplitude histories. However, these modified life curves would still overpredict fatigue lives for variable amplitude histories with large compressive overloads of the type used in this study.

Service stresses in notched or un-notched components are, from a practical point of view, not likely to exceed the order of magnitude of the monotonic yield stress of the material. The $\eta = 0$ overload curve therefore provides a stress-life relationship reflecting the maximum practical

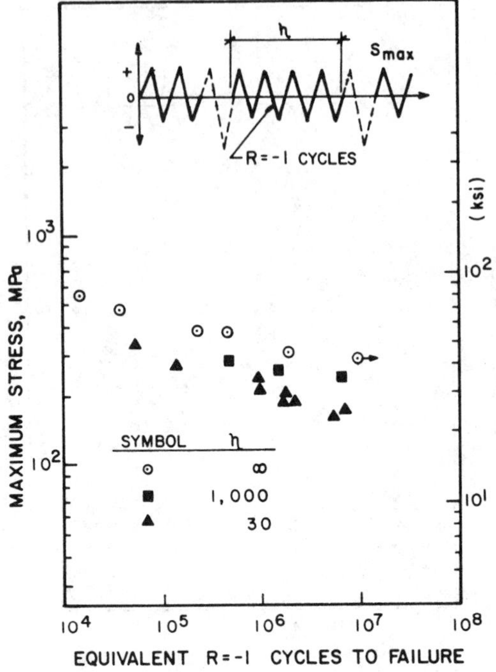

FIG. 8—*Equivalent lives of fully reversed cycles for SAE 1045 steel.*

effect of compression and of compressive overloads. Conservative life predictions for variable amplitude service histories may be made using the $\eta = 0$ overload curve.

Conclusions

1. At a constant minimum stress, the fatigue life increases as the maximum stress, and the stress range, decrease. At a constant fatigue life, the maximum stress decreases and the stress range increases as the compressive stress is increased.

2. Periodic compressive overloads, with magnitudes on the order of the monotonic yield strength, reduce the fatigue strength of smooth specimens to levels far below the constant amplitude fatigue limit.

3. The accelerated damaging effect of the compressive overloads on the subsequent smaller cycles increases with the frequency of application of the overloads. A single compressive overload applied only once in thousands of smaller cycles significantly increases the damage done by the smaller cycles.

Acknowledgments

Financial support of this research by GKN Technology Limited, The Ford Motor Company, and the Natural Sciences and Engineering Research Council (Grant A1694) is gratefully acknowledged.

References

[*1*] Dowling, N. E., "Fatigue Life and Inelastic Strain Response Under Complex Histories for an Alloy Steel," *Journal of Testing and Evaluation,* Vol. 1, No. 4, 1973, pp. 271-287.

[*2*] Brose, W. R., Dowling, N. E., and Morrow J. D., "Effect of Periodic Large Strain Cycles on the Fatigue Behavior of Steels," SAE Report 740221, 1974.

[*3*] Plumtree, A. and Martin, J. F., "Influence of Overstrains on the Fatigue Life of Low Carbon Steels," ICM Conference, Boston, 1976.

[*4*] Watson, P., Hodinott, D. S., and Norman, J. P., "Periodic Overloads and Random Fatigue Behavior," in *Cyclic Stress-Strain Behavior—Analysis, Experimentation, and Failure Prediction, ASTM STP 519,* American Society for Testing and Materials, Philadelphia, 1973, pp. 271-284.

[*5*] Dowling, N. E., "Fatigue Failure Predictions for Complicated Stress-Strain Histories," *Journal Materials,* Vol. 7, No. 1, 1972, pp. 71-87.

[*6*] Jacoby, G. H., "Fatigue Life Estimation Processes Under Conditions of Irregular Varying Loads," Technical Report AFML TR 67-215, Air Force Materials Laboratory, Aug. 1967.

[*7*] Watson, P. and Topper, T. H., "Fatigue-Damage Evaluation for Mild Steel Incorporating Mean Stress and Overload Effects," *Experimental Mechanics,* Vol. 12, No. 1, 1972, pp. 11-17.

[*8*] Conle, A. and Topper, T. H., "Overstrain Effects During Variable Amplitude Service History Testing," *International Journal of Fatigue,* Vol. 2, No. 3, 1980, pp. 130-136.

[*9*] Conle, A. and Topper, T. H., "Evaluation of Small Cycle Omission Criteria for Shortening of Fatigue Service Histories," *International Journal of Fatigue,* Vol. 1, No. 1, 1979, pp. 23-28.

[*10*] Leipholz, H. H. E., Topper, T. H., and El Menoufy, M., "Lifetime Prediction for Metallic Components Subjected to Stochastic Loading," *Computers and Structures,* Vol. 16, No. 1-4, 1983, pp. 499-507.

[*11*] El Menoufy, M., Leipholz, H. H. E., and Topper, T. H., "Fatigue Life Evaluation, Stochastic Loading and Modified Life Curves," *The Shock and Vibration Bulletin,* Bulletin 52, May 1982.

[*12*] Topper, T. H. and Au, P., "Cyclic Strain Approach to Fatigue in Metals," AGARD Lecture Series 118, Fatigue Test Methodology, The Technical University of Denmark, Denmark, 19-20 Oct. 1981.

[*13*] Au, P., Topper, T. H., and El Haddad, M., "The Effects of Compressive Overloads on the Threshold Stress Intensity for Short Cracks," AGARD Conference Proceedings No. 328, Behavior of Short Cracks in Airframe Components, Toronto, Canada, 19-24 Sept. 1982.

[*14*] Yu, M. T., Topper, T. H., and Au, P., "The Effects of Stress Ratio, Compressive Load and Underload on the Threshold Behavior of a 2024-T351 Aluminum Alloy," in *Proceedings,* Fatigue 84, C. J. Beevers, Ed., Vol. 1, 1984, pp. 179-190.

[*15*] Yu, M. T. and Topper, T. H., "The Effects of Material Strength, Stress Ratio and Compressive Overload on the Threshold Behavior of a SAE 1045 Steel," *ASTM Journal of Engineering Materials and Technology,* Vol. 107, 1985, pp. 19-25.

[*16*] Topper, T. H., Yu, M. T., and Au, P., "The Effects of Stress Cycle on the Threshold of a Low Carbon Steel," presented at the 8th Inter-American Conference on Materials Technology, San Juan, Puerto Rico, 25-29 June 1984.

[*17*] Topper, T. H. and Sandor, B. I., "Effects of Mean Stress and Prestrain on Fatigue Damage Summation," in *Effects of Environment and Complex Load History on Fatigue Life, ASTM STP 462,* American Society for Testing and Materials, Philadelphia, 1970, pp. 93-104.

[*18*] Miner, M. S., "Cumulative Damage in Fatigue," *Journal of Applied Mechanics,* Vol. 12, 1945, pp. A159-164.

M. L. Karasek,[1] *Huseyin Sehitoglu,*[1] *and D. C. Slavik*[1]

Deformation and Fatigue Damage in 1070 Steel under Thermal Loading

---

REFERENCE: Karasek, M. L., Sehitoglu, H., and Slavik, D. C., **"Deformation and Fatigue Damage in 1070 Steel under Thermal Loading,"** *Low Cycle Fatigue, ASTM STP 942,* H. D. Solomon, G. R. Halford, L. R. Kaisand, and B. N. Leis, Eds., American Society for Testing and Materials, Philadelphia, 1988, pp. 184–205.

ABSTRACT: The fatigue lives of 1070 steel (Class U wheel steel) under (1) isothermal loading, (2) thermo-mechanical constant amplitude loading, and (3) thermo-mechanical block loading have been examined. Fatigue lives for these three cases were compared based on the same mechanical strain range and maximum temperature level, with similar strain rates. Predictions of thermo-mechanical block loading cases were nonconservative based on isothermal data, while the predictions improved to within a factor of 1.5 of experimental lives when thermo-mechanical constant amplitude data were utilized.

Oxide scales readily formed at high temperatures ($>500°C$) and resulted in increased damage due to a localized high strain state and an inherent lack of ductility. Oxides of different composition and morphology form depending on the temperature and type of loading. Stratified and non-stratified oxide layers (predominantly Fe_3O_4) were identified in the experiments.

KEY WORDS: thermo-mechanical fatigue, isothermal fatigue, oxidation, block loading, life prediction, mechanical strain, oxide cracking, damage, stress-strain response

Thermo-mechanical fatigue problems are encountered in many engineering applications. Temperature and load cycling may occur in turbine blades, piping, pressure vessels, railroad wheels (due to brake shoe action), and other components which experience high temperatures in service. Often isothermal test results are used for "cyclic temperature design." Isothermal tests at high temperatures have been performed and results are available for many materials. However, conditions exist where isothermal data fail to adequately predict thermo-mechanical damage. Studies that identify thermo-mechanical loading conditions where such unfavorable life predictions occur are needed.

Since the interaction of many mechanical and environmental factors affect different materials in different ways, previous workers have used a variety of parameters to compare isothermal and thermo-mechanical fatigue. Some of the parameters used in previous studies to compare thermal fatigue lives include mechanical strain range (Δe_m) [*1–4*], plastic strain range (Δe_p) [*1,5–8*], stress amplitude ($\Delta S/2$) [*4,5,7*], combination of maximum stress and mechanical strain range ($S_{max} * \Delta e_m/2$) [*4*], and the strain range partitioning approach [*10*]. Isothermal tests performed at the maximum or mean temperature of the thermo-mechanical cycle provided life comparisons in strain-based approaches. In stress-based approaches, the TMF lives could be predicted based on isothermal tests at minimum temperature, maximum temperature, or at a temperature corresponding to maximum stress in the thermal cycle.

Microstructural damage in isothermal and thermo-mechanical fatigue has been considered in several studies [*1,2,6,8,11*]. Transgranular crack growth is dominant in most out-of-phase

[1]Graduate Assistant, Associate Professor, and Graduate Student, respectively, Department of Mechanical and Industrial Engineering, University of Illinois at Urbana-Champaign, Urbana, IL 61801.

(maximum strain at minimum temperature) thermo-mechanical fatigue tests. Intergranular cracking is dominant in many high temperature isothermal and in-phase (maximum strain at maximum temperature) thermo-mechanical fatigue cases. In general, the introduction of a hold time at the high temperature end in the out-of-phase tests increases intergranular cracking and reduces fatigue life. Mean stresses also need to be considered in damage assessment. Mean stresses develop in cases where the material strength changes with temperature. Depending on the phasing of temperature and strain, the mean stress may be tensile or compressive. This may have a significant effect on thermal fatigue behavior when crack growth dominates the life.

Oxidation is an important consideration in thermal fatigue of steels, since it can accelerate crack initiation [*9*]. While there are numerous papers on oxidation kinetics in steel [*12–20*], only a few have dealt with oxides formed during fatigue [*11,21,22*]. Decohesion or cracking of an oxide layer could affect the subsequent oxidation rate and thus affect crack initiation and crack propagation. Specimen geometry and differences in expansion coefficients for the metal and oxide can contribute to oxide layer decohesion and cracking. Carbon content may also affect oxide integrity in steels. Since carbon is insoluble in iron oxide, a layer of graphite is deposited at the metal/oxide interface during oxidation of plain carbon steels [*18,19*]. This may cause deceleration of the oxidation rate, but oxide decohesion and cracking may develop at the interface.

Railroad wheels experience thermal loading during brake shoe applications. Hot spots develop on the tread of the wheel as it passes under the brake shoe. Thermal stresses on the hot spots are higher than the surrounding cooler material. Oxide formation at the tread (hot spots) and flange regions may be extensive and influence crack nucleation. The hot spots and other critical regions experience many small thermal cycles within major thermal cycles. This could occur during repeated brake applications where the wheel material does not cool fully to room temperature after each cycle. Localized variations of temperature within the tread surface also contribute to small thermal cycles within a major cycle. This process is simulated in this study by performing thermo-mechanical block tests, where minor temperature cycles occur at the high temperature end of the major cycle. The temperatures chosen for this study are in the range 150 to 700°C and reflect those experienced by critical regions in railroad wheel service.

This study will attempt to predict (using isothermal data obtained at comparable strain rates) thermo-mechanical fatigue lives obtained under constant amplitude and block histories. Mechanical strain range at the maximum temperature of the thermal cycle will be used as a basis for comparison of isothermal and thermo-mechanical fatigue tests. The material stress-strain behavior, oxidation, and thermal cracking characteristics of thermo-mechanical block tests, thermo-mechanical constant amplitude fatigue tests (TMF), and isothermal fatigue (IF) tests will be compared.

Experimental Procedure

Material

The material examined was a 1070 steel typically used in railroad wheels (Class U wheel steel). The "U" stands for the untreated (cast and then furnace drawn) wheel condition. The chemical composition of the steel is given in Table 1. The specimens were cut from the rim of a railroad wheel in the circumferential direction. The machined specimens had a uniform gage length of 25.4 mm and a diameter of 7.62 mm. A typical pearlitic microstructure as shown in Fig. 1 was obtained. The pearlitic colony size was measured as 50 μm. The pearlite spacing was approximately 0.5 to 1.1 μm. In this study, strain rates of 0.0002 s^{-1} were considered. Material properties at higher strain rates $\dot{\epsilon} = 0.02$ and 0.002 s^{-1}) can be found in Refs *11* and *23*. The coefficient of thermal expansion is taken as constant ($= 17 \times 10^{-6}/°C$) over the 150 to 700°C temperature range of interest.

TABLE 1—*Chemical composition (weight percent) of Class U wheel steel.*

C	Mn	Si	Cn	Cr	P	S	Ni	Mo
0.64	0.8	0.2	0.08	0.06	<0.05	0.047	0.03	>0.02

Equipment

Tests were performed on a 20-kip MTS testing machine controlled with a PDP-11/05 computer. Induction heating was used to heat the specimens. In the thermo-mechanical tests, thermocouples attached to the specimen provided input to the temperature controller which received the command signal from the PDP-11/05 computer. The schematic of the test system is shown in Fig. 2. The "C/L" represents the closed loops of the system. In isothermal tests certain specimens were tested using a resistance furnace, while induction heating with temperature control was used on other specimens. The stress-strain response was similar for each heating method. Stress-strain data were recorded on floppy disk for later analysis.

After removal from the test frame, the specimens were sectioned using a low speed saw and mounted in epoxy. The compositions of oxides in the corrosion layers were identified by etching with a 1% solution of HCl in ethanol. A 2% Nital etch was used to evaluate the microstructure of the metal. A scanning electron microscope was used to obtain pictures of the specimen surface and oxidation characteristics.

Test Conditions

Experiments were conducted under isothermal and thermo-mechanical loading conditions. The isothermal tests were performed under completely reversed strain control in the temperature range 20 to 700°C at strain amplitudes ranging from 0.0015 to 0.01. The strain rates considered were 0.02, 0.002, and 0.0002 s^{-1}. The 0.0002 s^{-1} results are reported in this study.

The thermo-mechanical tests were performed under total constraint and were out-of-phase (maximum strain at minimum temperature). A total constraint experiment involves clamping the specimen at the low temperature end and cycling the temperature while the net strain is maintained at zero. All the thermal strain is converted to mechanical strain in this case. An illustration of total constraint is given in Fig. 3*a*, where Bar 1 is heated and cooled and Bar 2 has a large stiffness (A_2/A_1 large) compared with Bar 1. Other loading conditions are possible with this model (A_2/A_1 not large) and have been considered in Ref *23*.

Two sets of thermo-mechanical tests were performed: (1) constant amplitude tests with minimum temperatures of 150, 400, or 500°C and maximum temperatures ranging from 450 to 700°C, and (2) block tests of one major temperature cycle (with a minimum temperature of 150°C and maximum temperatures of 500, 600, 650, and 700°C) followed by 100 minor (sub) temperature cycles (of range 100, 150, and 200°C) at the high temperature end. Combinations of maximum and minimum temperatures were chosen to simulate thermal cycling conditions experienced in railroad wheel service. Strain rates for heating were higher than strain rates for cooling, but the average strain rate for these tests was on the order of 0.0002 s^{-1}. Heating times for thermo-mechanical tests ranged from 30 to 50 s, while cooling times ranged from 120 to 180 s. A typical example of experimental temperature-time variation is given in Figs. 3*b* and 3*c* for constant amplitude and block thermo-mechanical loading, respectively. Duration of constant amplitude thermal cycle (T_{min} = 150°C, T_{max} = 500°C) is 170 s (Fig. 3*b*), while the period of thermal block with the same major temperature range (with the subcycle temperature range equal to 100°C) was 0.76 h (Fig. 3*c*).

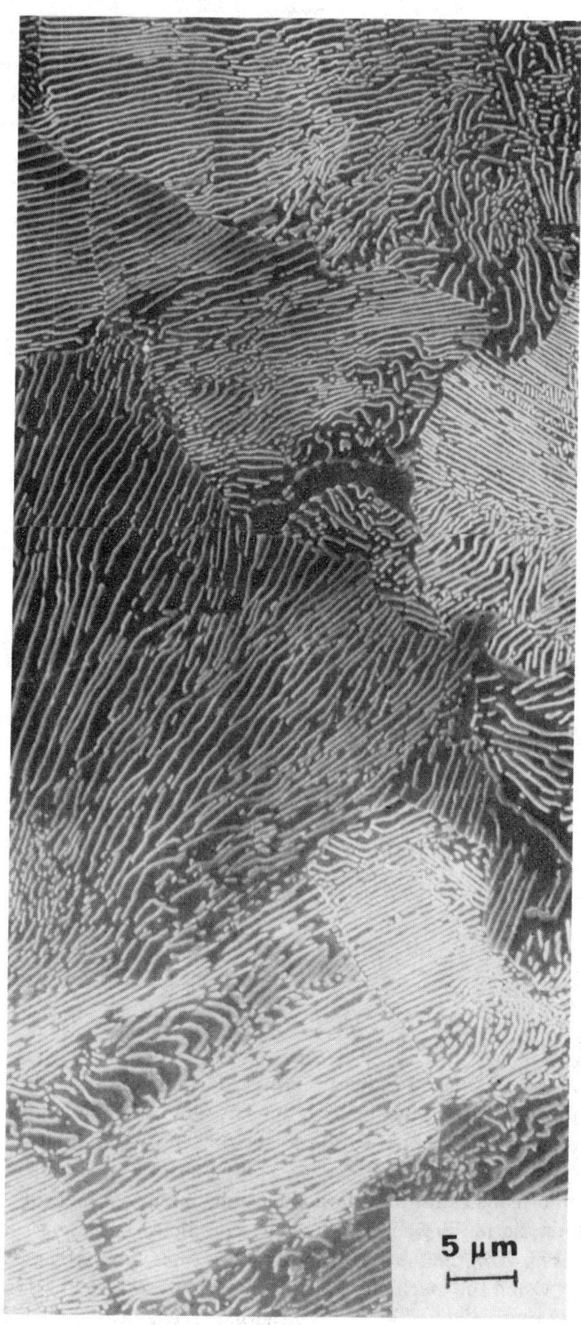

FIG. 1—*Microstructure of 1070 steel.*

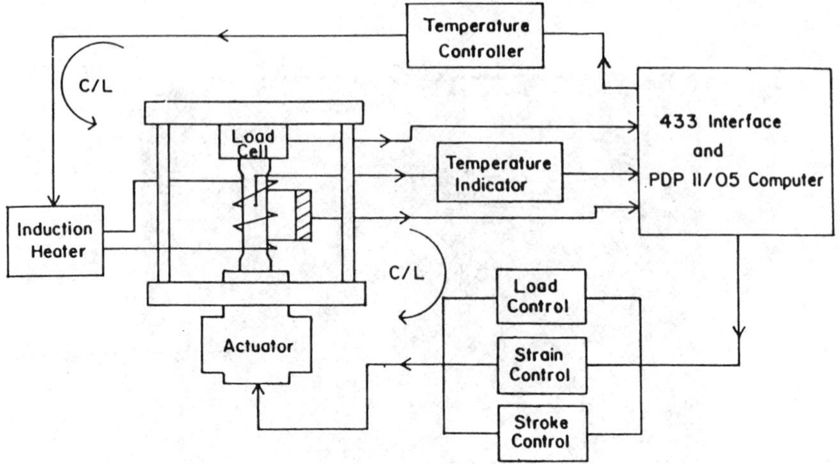

FIG. 2—*Schematic of the test system.*

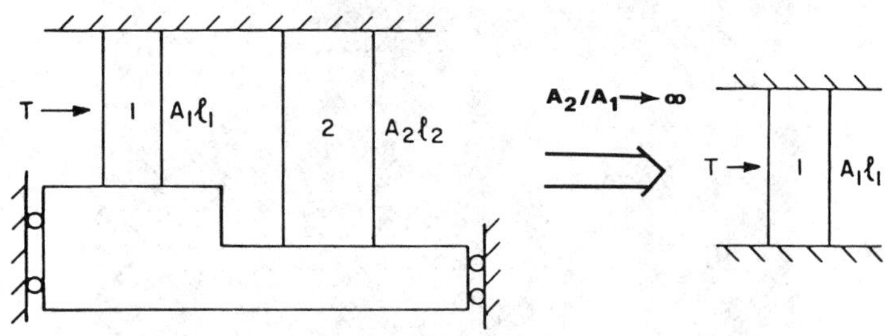

FIG. 3*a*—*Two bar model.*

Experimental Results

Fatigue Test Results and Predictions

Strain-life curves were established from isothermal constant amplitude tests. Above 400°C, a decrease in the strain rate resulted in shorter fatigue lives for all strain amplitudes considered. The strain-life curves were represented by the total strain-life equation [*24*], and the resulting coefficients and exponents were established as a function of temperature for the case $\dot{\epsilon}$ = 0.0002 s^{-1}. The isothermal strain-life equation constants are given in Table 2.

Constant amplitude thermo-mechanical lives were predicted based on isothermal lives; the results are shown in Fig. 4. In the first series of tests ("○" data points) the minimum temperature was held at 150°C and the mechanical strain range increased proportionally with increasing maximum temperature (450 to 700°C). The thermo-mechanical predicted lives were determined using isothermal data corresponding to the high temperature level of thermal cycle and the mechanical strain range of the thermal cycle. For example, a thermal-mechanical test where T_{min} = 150°C and T_{max} = 600°C experienced a mechanical strain range ($\alpha\Delta T = 17 \times 10^{-6} \times$

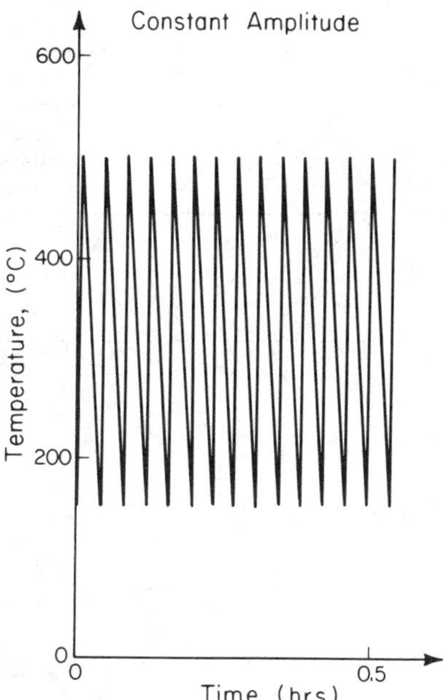

FIG. 3b—*Experimental temperature-time histories for thermo-mechanical constant amplitude loading (net strain = 0).*

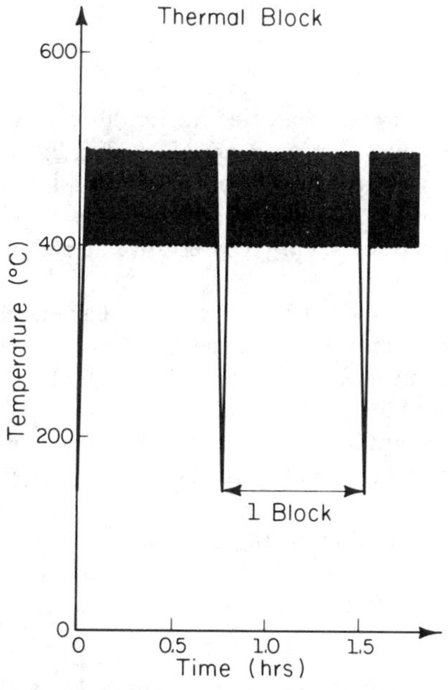

FIG. 3c—*Experimental temperature-time histories for thermo-mechanical block loading (net strain = 0).*

TABLE 2—*Strain-life equation constants for 1070 steel;* $\dot{\epsilon} = 0.0002\ s^{-1}$.

T (°C)	E (MPa)	b	c	ϵ_f'	σ_f' (MPa)
400	189 340	−0.114	−0.526	0.1426	936
600	145 200	−0.004	−0.659	0.2583	112
700	116 640	−0.062	−0.752	0.9359	98

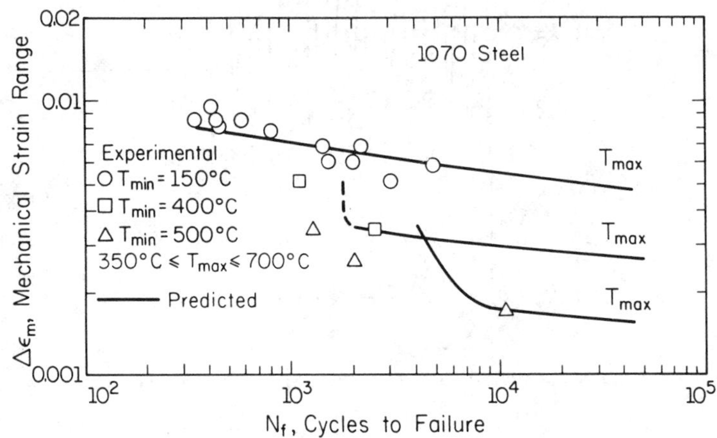

FIG. 4—*Predictions of thermo-mechanical constant amplitude lives based on isothermal data obtained at* T_{max} *of the thermal cycle.*

450) of 0.00765 and lasted 735 cycles. Predicted life for this case was established from isothermal data at 600°C corresponding to a strain range of 0.00765. Experimental and predicted lives for the case T_{min} = 150°C showed good agreement over a wide range of mechanical strain. When the mechanical strain range is small, the predicted lives are slightly nonconservative. It is noted that the predictions become slightly conservative at higher strain amplitudes (higher T_{max}).

When the minimum temperature is 400°C or above, the predicted lives are nonconservative (Fig. 4). The maximum temperatures in this case varied in the range 500 to 700°C. In Ref *11* similar results were obtained using T_{max} and T_{mean} $\dot{\epsilon} = 0.002\ s^{-1}$ isothermal data. The use of $\dot{\epsilon} = 0.0002\ s^{-1}$ data improved the predictions.

The predicted blocks-to-failure for thermo-mechanical block loading were calculated using Miner's linear damage rule as a good approximation:

$$D_t = D_{major} + D_{minor}$$

where

$D_{major} = 1/N_{f\ major}$ = damage/block due to major cycles,
$D_{minor} = 100/N_{f\ minor}$ = damage/block due to minor cycles, and
$D_t = 1/B_f$ = total damage per block where
B_f = blocks to failure.

The fatigue lives $N_{f\ major}$ and $N_{f\ minor}$ were determined based on (1) isothermal constant amplitude data (Table 2) or (2) constant amplitude thermomechanical experimental data obtained from the $T_{min} = 150$, 400, and 500°C cases indicated in Fig. 4.

Isothermal constant amplitude data result in highly nonconservative predictions of thermomechanical block test lives (Fig. 5*a*). Predictions based on constant amplitude thermo-mechanical data are within a factor of 1.5 of experimental blocks to failure (Fig. 5*b*).

Material Response

Examination of the material response provides guidelines for choice of predictive parameters and identifies inherent differences between isothermal and thermo-mechanical loading. Under strain control ($R = -1$) isothermal loading the mean stress is zero. Maximum tensile stresses (at T_{min}) in thermo-mechanical constant amplitude tests are higher than those in comparable isothermal tests (as shown in Fig. 6*a* for $T_{max} \geq 400$°C). The dashed lines in Fig. 6*a* indicate the isothermal cyclic stress-strain curve at 150°C, and the data points "○" denote the maximum stress in TMF tests. Note that the maximum stresses (upon cooling) in TMF cases are zero over a finite strain amplitude ($0 \leq \Delta\epsilon_m/2 \leq 0.0011$). When the strain amplitude is high enough to cause yielding in compression, stresses become tensile upon cooling. A comparison of stress ranges in isothermal fatigue (IF) and thermo-mechanical fatigue (TMF) is provided in Fig. 6*b*. The data points denote the TMF data. It is noted that stresses (isothermal data) at 400°C lie below that of 150°C at a strain rate of 0.0002 s^{-1}. At higher strain rates [*11*] the isothermal stress-strain curves in the range 20 to 400°C coincided.

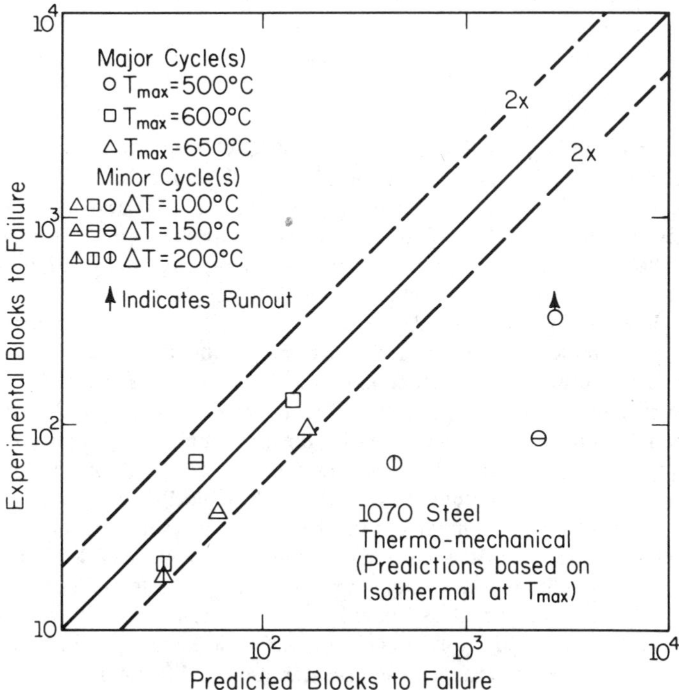

FIG. 5*a*—*Comparison of experimental lives and isothermal predictions for thermal block loading.*

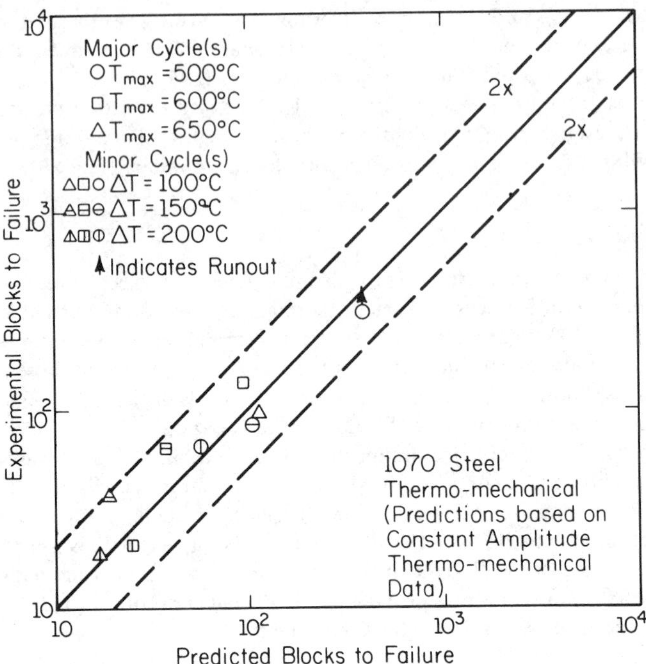

FIG. 5*b*—*Comparison of experimental lives and thermo-mechanical predictions for thermal block loading.*

Maximum stresses for $T_{min} = 400°C$ and $T_{min} = 500°C$ TMF tests are shown in Fig. 6*c* with corresponding stresses in IF cases. Comparison of stress ranges for these cases is indicated in Fig. 6*d*. The data points indicate the TMF results, and the dashed lines denote the isothermal stress-strain data at 400, 500, 600, and 700°C. Note that the strain rate in the isothermal fatigue (IF) tests is 0.0002 s^{-1}; this is lower than the strain rates given in Ref *11* (0.02 and 0.002 s^{-1}).

Stress-strain response for thermo-mechanical constant amplitude and thermo-mechanical block tests are shown in Figs. 7*a*, 7*b*, and 7*c*, respectively. Cycle and block number corresponding to these loops are indicated on the figure. The first cycle and first block are denoted by the dashed line. The hysteresis loop of the 1st minor cycle in the block is indicated in Figs. 7*b* and 7*c*. The hysteresis loops corresponding to the 100th minor cycle (not shown) exhibit a higher mean stress than that of first minor cycle. Initially, the maximum stresses (in the major temperature cycles) of the thermo-mechanical block tests are similar to the maximum stresses in thermo-mechanical constant amplitude tests. As the thermal blocks are accumulated, the maximum stresses decrease rapidly compared with the stresses measured in constant amplitude tests. The maximum stresses in the minor temperature cycles are generally higher in thermo-mechanical constant amplitude tests compared with thermal block tests. When the maximum stress is higher, predictions of blocks to failure based on constant amplitude thermo-mechanical data are generally (but not always) nonconservative (Table 3 and Fig. 5*b*). Note that the use of maximum stress and/or stress range alone has limitations in life prediction of this material. This point is discussed further in the Discussion section to this paper; see also our response to Remy's discussion appended to this paper.

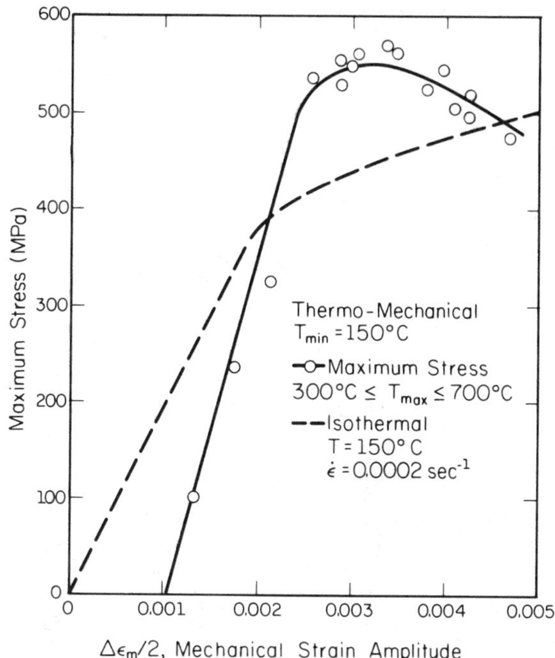

FIG. 6a—*Comparison of stresses for isothermal fatigue (150°C) and thermo-mechanical fatigue; minimum temperature = 150°C, maximum tensile stress.*

Oxide Structure

The study of oxide structure provides guidelines on the mechanisms of crack initiation and crack propagation in the temperatures of interest. Oxide thicknesses measured on failed samples were in the range 0.05 to 1.0 mm, depending on the temperature and test time. The corresponding oxidation time varied in the range 2 to 100 h, respectively. If the oxidation time is taken to be equal to the time to failure of the specimen, oxide thicknesses measured in thermo-mechanical tests are lower than those observed in corresponding isothermal tests. However, if the oxidation time is interpreted as the fraction of time the specimen was subjected to temperatures between 600 and 700°C, oxide thicknesses in isothermal and thermo-mechanical tests match closely.

After etching the oxide layers, the types of oxide present could be evaluated. In isothermal tests, oxide layers larger than 10 μm have been observed only above 400°C ($t_{ox} \geq 2$ h). The oxide layers for all isothermal tests are stratified and contain large proportions of Fe_3O_4. A stratified oxide layer observed in the $T = 600$°C, $\Delta\epsilon = 0.007$ case is shown in Fig. 8. Buckled sections of oxide are present at the oxide/air interface with occasional spalling of these sections. The majority of the oxide layer is made up of stratifications with some accompanying porosity. Numerous short, tightly closed oxide cracks which do not extend completely through the oxide layer are observed. Stratified oxide layers exhibit wide, semi-circular oxide intrusions.

Thermo-mechanical constant amplitude tests exhibit stratified oxide layers only in the case of high mean temperatures ($T_{mean} > 400$°C), with mostly Fe_3O_4 present (Fig. 9*a*). When the mean temperature is lower ($T_{mean} \leq 400$°C), the oxide layer is nonstratified (FeO and Fe_3O_4) (Fig.

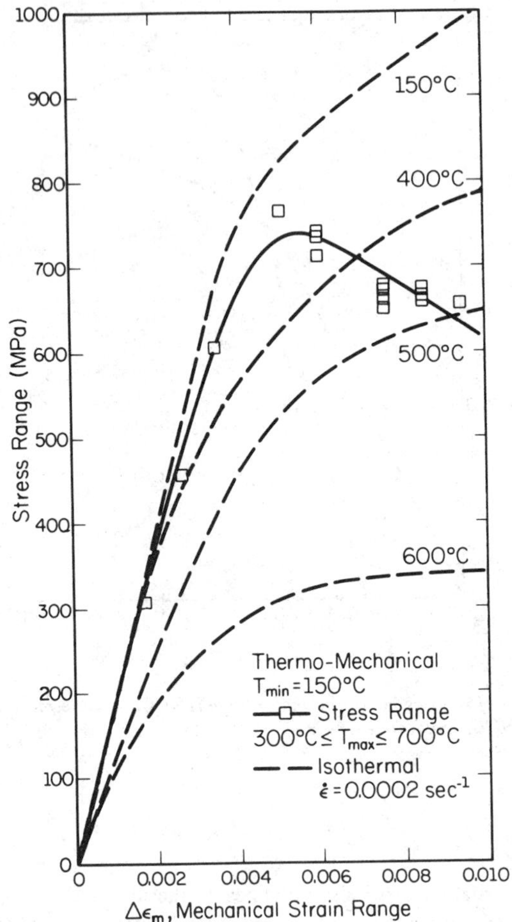

FIG. 6b—*Minimum temperature = 150°C, stress range.*

9*b*). These oxide layers exhibit small fraction of porosities, but no stratifications are observed. Oxide intrusions are long and thin as opposed to the wide intrusions observed in stratified oxide layers. Oxide cracks are open and extend completely through the oxide layer to the metal. Etching of the oxide and metal did not identify any preferential oxide intrusion growth along grain boundaries.

In thermo-mechanical block tests, oxide layers are non-stratified when the magnitude of the minor temperature cycle is 100°C, with very few oxide cracks. When the magnitude of the minor temperature cycles is greater than 100°C, fragmented oxide layers are observed.

Experiments were performed under zero applied load conditions to provide insight into oxide behavior under growth stresses. Nonstratified oxyde layers with no evidence of major oxide cracks or oxide intrusions were observed (Figs. 10*a* and 10*b*). Oxide layers in these tests are lower in depth than oxide layers in the corresponding tests where stresses are present due to applied strain or imposed constraint (400 μm versus 50 μm in isothermal case and 60 μm versus 20 μm in thermal cycling case). This indicates that oxide failure (in the form of buckling or cracking) does lead to accelerated oxide growth rate.

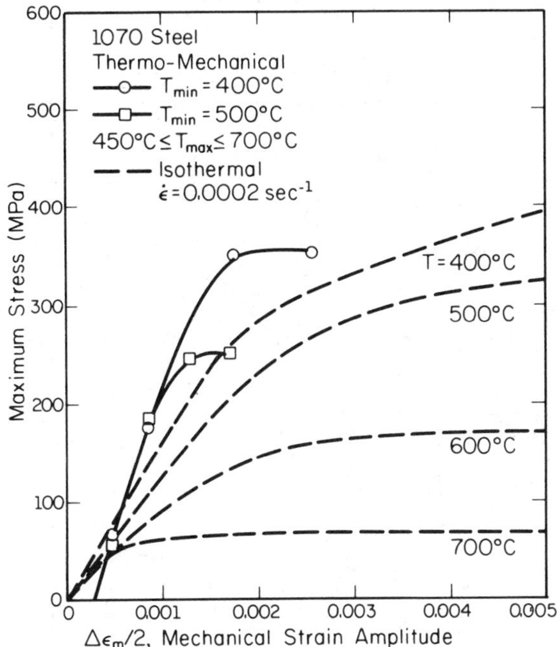

FIG. 6c—*Minimum temperature = 400°C and 500°C, maximum tensile stress.*

Discussion

The isothermal predictions of thermo-mechanical tests with T_{min} = 150°C were slightly nonconservative (Fig. 4) at lower strain amplitudes. The maximum stresses in thermo-mechanical and isothermal tests differed for these cases (Fig. 6*a*). However, the predictions became conservative for high strains (high T_{max}) where maximum stresses coincided. The predictions of T_{min} = 500°C TMF data have been nonconservative, particularly for the cases T_{max} = 650 and 700°C. This occurred despite the fact that stress ranges between TMF and the isothermal cycling at 500°C matched closely (Fig. 6*d*).

Nonconservative predictions based on isothermal data have been reported in other studies [*1,2,4,25*]. In Ref *11* the predictions of thermo-mechanical data were based on $\dot{\epsilon} = 0.002$ s^{-1} (isothermal) tests. The use of $\dot{\epsilon} = 0.0002$ s^{-1} in this study (at T_{max}) matched thermo-mechanical strain rates and improved the predictions.

It is worth noting that the use of maximum stress or stress range parameters poses limitations in prediction of TMF lives for the steel considered in this study. The predictions become highly conservative (×100 or more) if the maximum stress or stress range corresponding to the maximum temperature level of the TMF cycle (450 to 700°C isothermal data) is utilized. The results indicate that the temperature level where stress-based life predictions provide satisfactory predictions is not known *a priori*. This holds for all the $T_{min} \rightarrow T_{max}$ cases considered in this study.

The results suggest that the material's constitutive response provides guidelines on the choice of predictive parameter(s). The decrease in stress with increasing mechanical strain range has been identified in the 0.7% carbon steel studied here and in low alloy steels. This behavior is attributed to thermal recovery and other microstructural changes (such as spheroidization) ob-

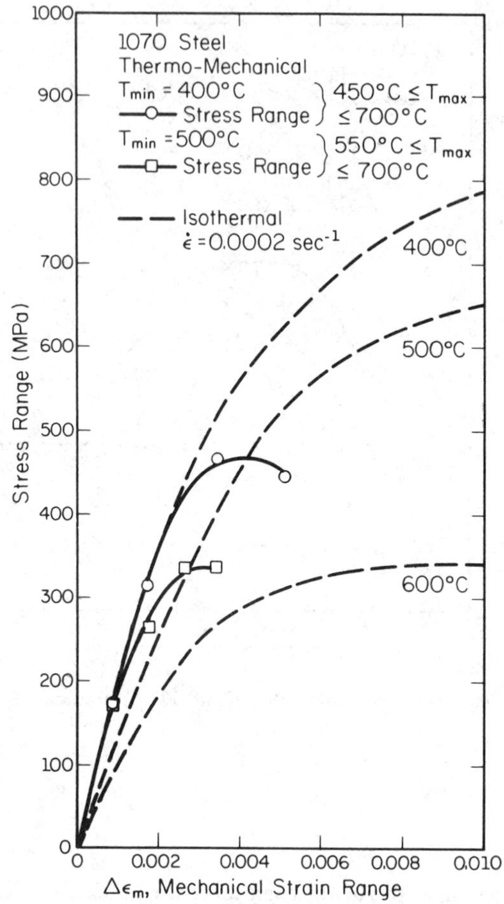

FIG. 6*d*—*Minimum temperature = 400°C and 500°C, stress range.*

served in this material. In general, Ni-based superalloys may not exhibit this behavior at the temperatures of interest. Therefore it may be possible to use stress-based approaches in those classes of alloys. Caution should be exercised in generalizing predictive parameters to all classes of alloys.

Constant amplitude thermo-mechanical data satisfactorily predicted thermo-mechanical block loading results. The minor temperature cycles (sub-cycles) of a block test (occurring at T_{max} of the major temperature cycle) are analogous to a compressive strain-hold; however, cyclic damage is induced due to the subcycles. This damage may exceed that due to the major cycle, depending on the number of subcycles. It was reported in Ref *25* that a compressive strain-hold on stainless steel essentially causes no damage when crack growth is transgranular and may result in the recovery of existing damage. Conservative predictions of thermal block tests shown in Fig. 5*b* may be attributed to a similar recovery effect. However, non-conservative predictions were observed in several cases which do not substantiate the general validity of the recovery effect.

It is noted that in previous studies non-linear damage rules have been implemented to account for major-minor cycle interactions. The results shown in Fig. 5*b* exhibit both conservative

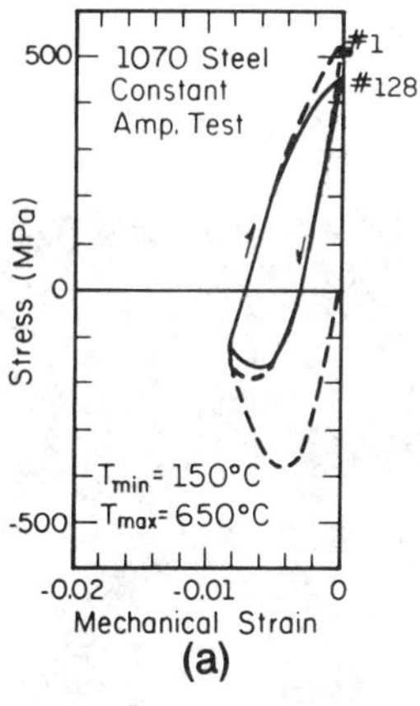

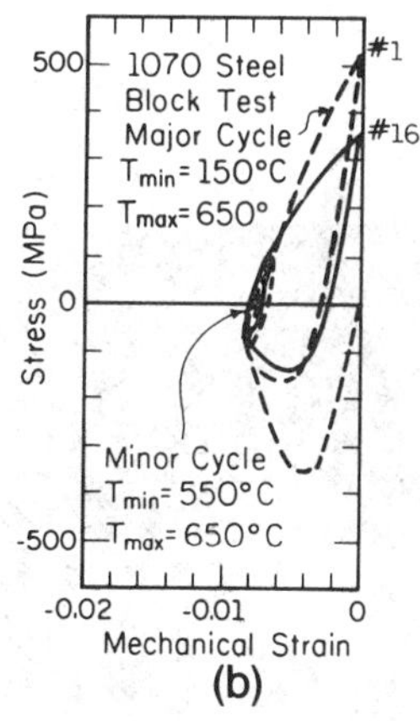

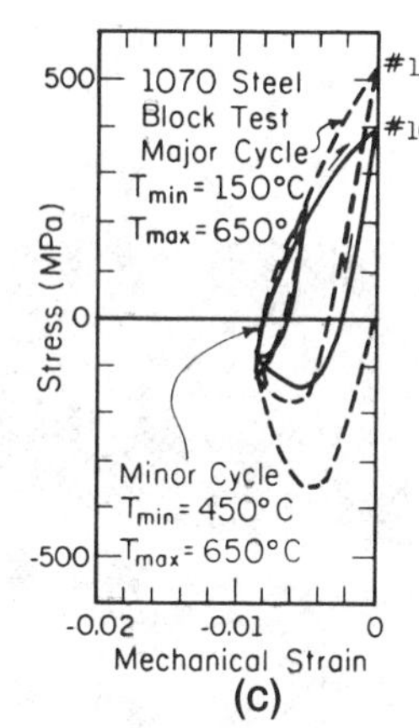

FIG. 7—*Typical stress-mechanical-strain behavior for thermo-mechanical loading.* (a) *Constant amplitude loading (*T_{min} = *150°C,* T_{max} = *650°C).* (b) *Thermal block loading (*T_{min} = *150°C,* T_{max} = *650°C,* ΔT_{sub} = *100°C).* (c) *Thermal block loading (*T_{min} = *150°C,* T_{max} = *650°C,* ΔT_{sub} = *200°C).*

TABLE 3—*Comparison of maximum stress for thermo-mechanical loading.*

Symbol in Fig. 5*b*	Minor Cycle T_{min} (°C)	Minor Cycle T_{max} (°C)	Minor Cycle: Constant Amplitude σ_{max} (MPa)	Minor Cycle: Thermal Block σ_{max} (MPa)	Experimental B_f	Predicted B_f
□	500	600	188	160	144	113 (conservative)
◫	400	600	358	330	21	24 (nonconservative)
Δ	550	650	131	151	92	104 (nonconservative)
▲	500	650	245	212	38	19 (conservative)

and non-conservative predictions, which confirms the validity of simple linear damage rule in these cases.

Crack initiation is accelerated by oxide failure. Strains due to internal constraints and applied strains influence oxide integrity and hence promote crack initiation. For cylindrical steel specimens, oxide growth results in compressive hoop strains in the oxide layer [*15*] as oxide tends to remain attached to the receding metal (as the metal diameter decreases). Differences in thermal expansion coefficients for oxide and metal would result in metal/oxide mismatch hoop strains. Under complete constraint of the assembly, axial strains in the oxide develop that are proportional to its coefficient of thermal expansion. It has been reported recently that between 470 and 610°C the thermal expansion coefficient of Fe_3O_4 is almost twice as large as that of steel [*21*].

Approximate axial strains and mismatch strains in the hoop direction in a representative constant amplitude thermo-mechanical test are shown in Fig. 11. Isothermal tests at 600°C with the same mechanical strain range as the thermo-mechanical tests are indicated in Fig. 11*a*. While the metal is clamped at the minimum temperature, the oxide is clamped at or near the maximum temperature (in the axial direction), since most oxide growth occurs above 500°C.

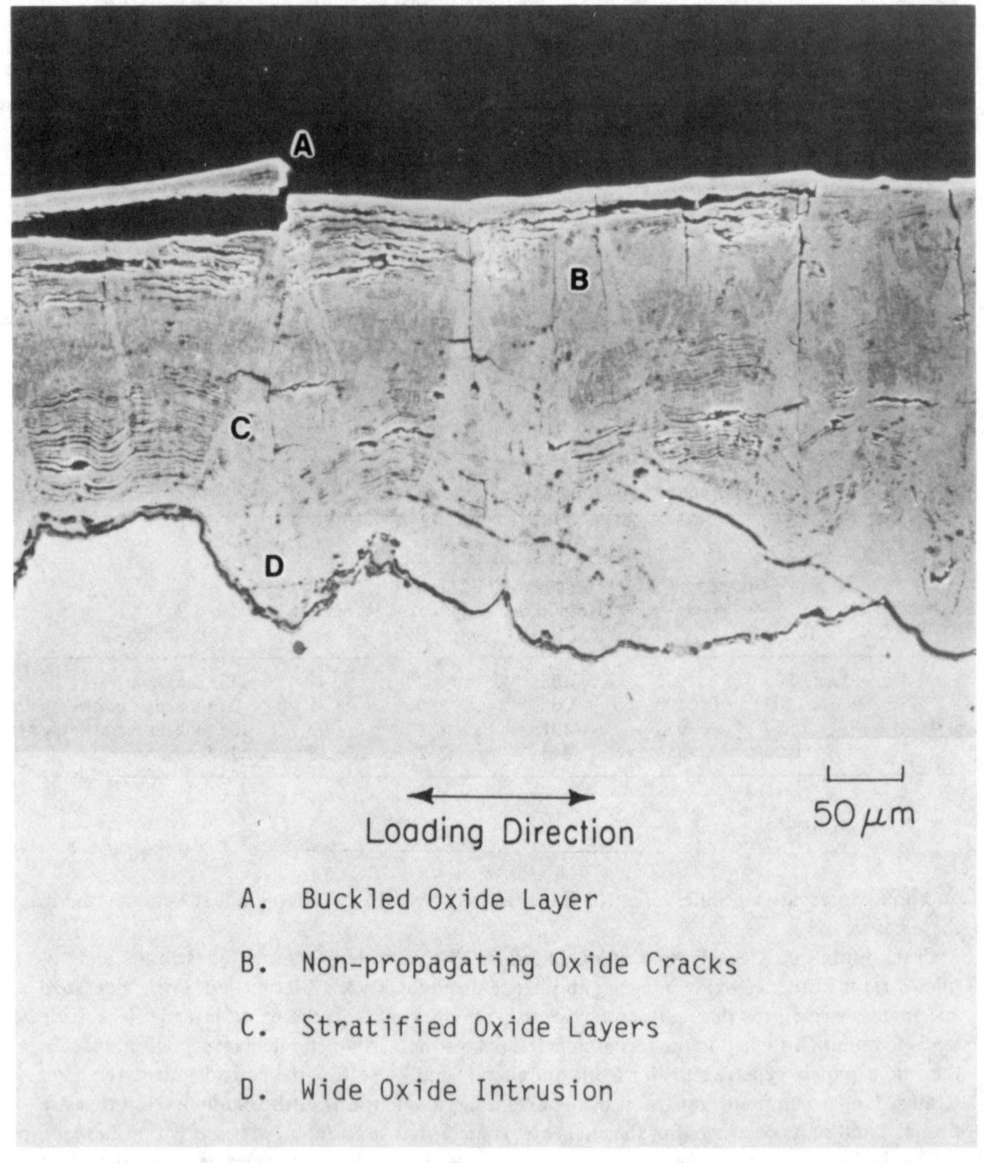

FIG. 8—*Typical stratified oxide layers resulting from isothermal constant amplitude loading,* T = *600°C,* $\Delta\epsilon = 0.007$*, test time = 19 h.* $\dot{\epsilon} = 0.0002\ s^{-1}$.

(a)

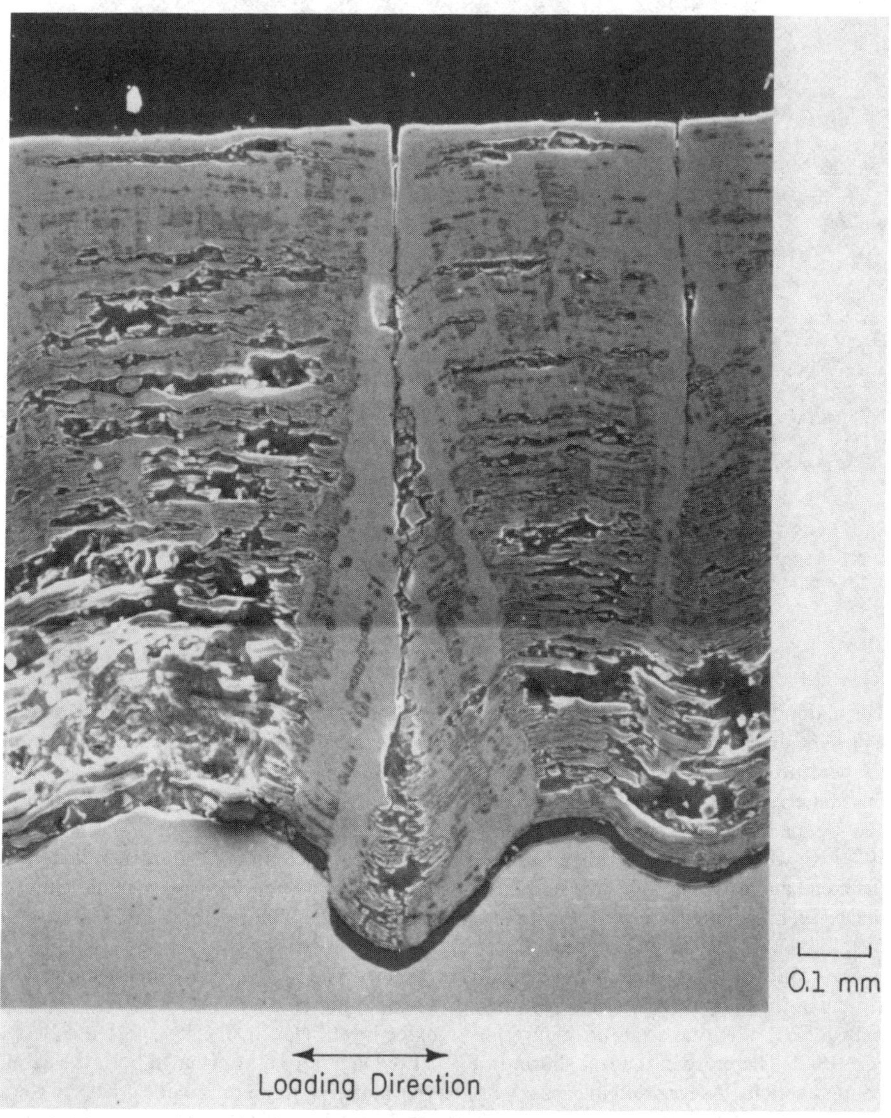

FIG. 9—*Typical oxide layers resulting from thermo-mechanical constant amplitude loading.* (a) *Stratified oxide layer,* T_{max} = *600°C,* T_{min} = *500°C, test time = 98 h.* (b) *Non-stratified oxide layer,* T_{max} = *600°C,* T_{min} = *150°C, test time = 8.2 h.*

(b)

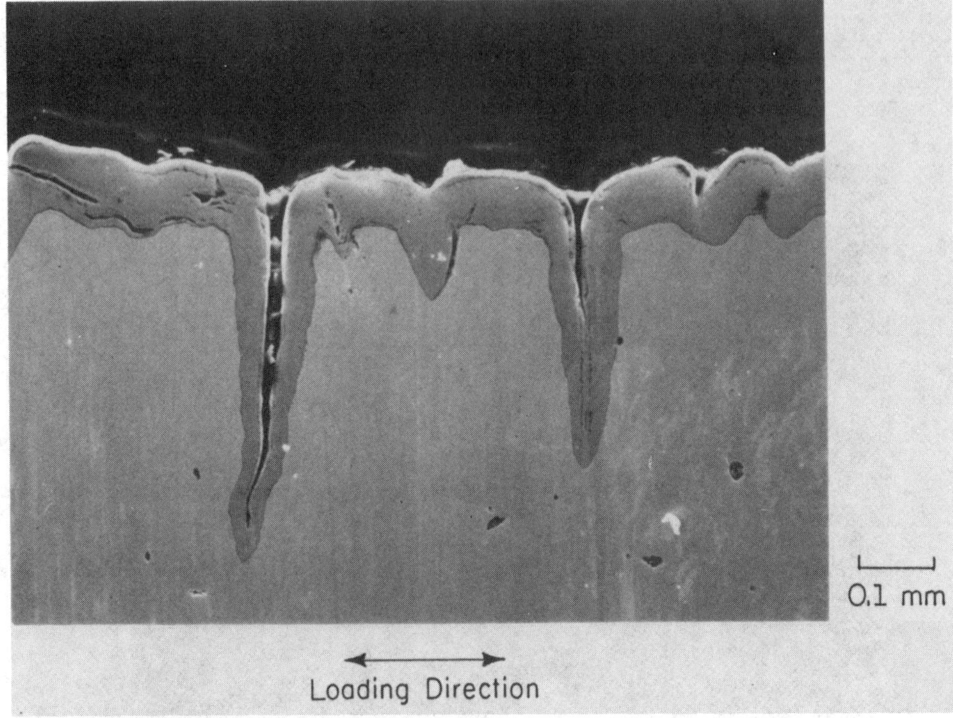

FIG. 9—*continued.*

The clamping temperature is denoted by the symbol T_c. In Figs. 11*a* and 11*b*, the solid lines represent strains in an oxide clamped at 600°C, while the dashed lines correspond to a clamping temperature of 500°C. The actual strains are probably between these values. The value of the thermal expansion coefficient of the metal in this study was 1.7×10^{-5}/°C. Between 470 and 610°C, the thermal expansion coefficient of the oxide was taken to be approximately 3.4×10^{-5}/°C, while outside this range it is approximately equal to that of the steel. It is noted that the metal and oxide strains indicated in Figs. 11*a* and 11*b* do not represent strain conditions at the oxide/metal interface or at the outer oxide surface but are indicative of gross strains on the metal and oxide layer, respectively.

The oxide/metal mismatch hoop strains are shown in Fig. 11*c*. This figure indicates that strains in the hoop direction are not small and should not be neglected.

In Ref *17*, the total strain to failure for an oxide layer below 600°C is reported as less than 3×10^{-4}. Therefore in the test shown in Fig. 11 oxide failure may occur in both the axial and hoop directions. As reported in earlier work on this material, numerous oxide cracks in the axial and hoop directions were observed in thermo-mechanical tests [*11*].

Conclusions

The following conclusions hold for the 1070 steel examined in this study:

1. Based on the mechanical strain range, isothermal data adequately predicted thermo-mechanical constant amplitude fatigue lives when the minimum temperature was 150°C. When

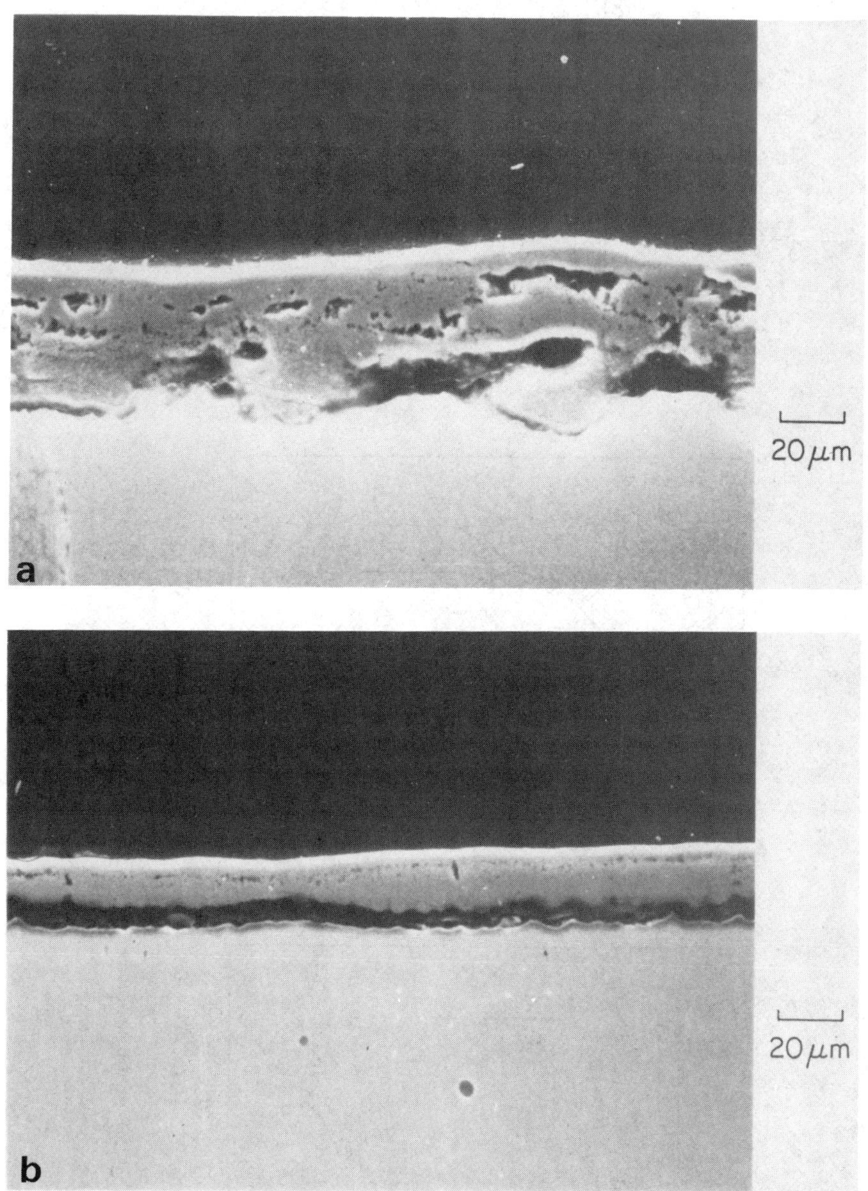

FIG. 10—*Non-stratified oxide layers observed in zero applied load tests.* (a) *Isothermal zero applied load test,* T = *600°C, test time = 16 h.* (b) *Thermal cycling zero applied load test,* T_{min} = *150°C,* T_{max} = *600°C, test time = 30 h.*

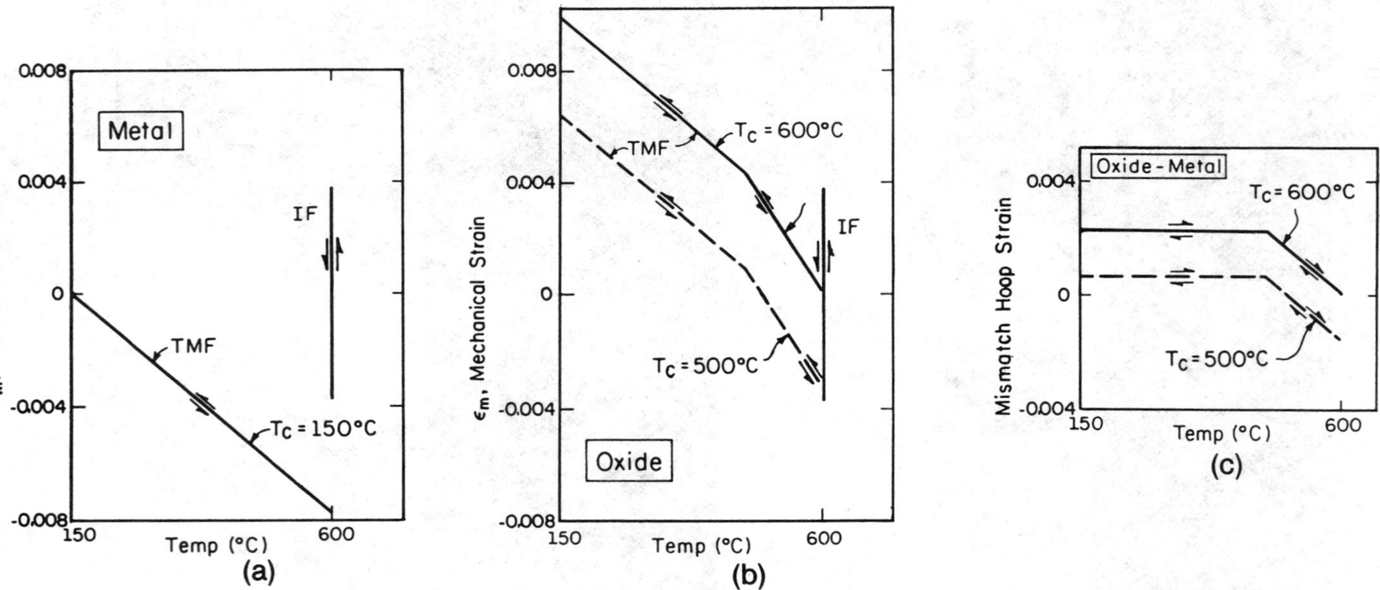

FIG. 11—(a) *Axial strains in metal in TMF and IF.* (b) *Axial strains in oxide in TMF and IF.* (c) *Mismatch hoop strains in TMF.*

the minimum temperature exceeded 400°C, isothermal predictions (at T_{max}) were nonconservative in several cases.

2. Fatigue lives of thermo-mechanical block tests were satisfactorily predicted by constant amplitude thermo-mechanical data, while isothermal predictions (at T_{max}) were highly nonconservative. Therefore isothermal data alone may not accurately predict variable amplitude thermal damage.

3. Oxide structure is dependent on the temperature and strain history. Compressive oxide strains develop during oxide film growth. The growth stresses are not sufficiently high to cause cracking of oxide layer in this material. Reversed strain in the oxide in thermal loading and during isothermal cycling are superimposed on the growth strains, resulting in buckling and cracking of the oxide. The cracking of oxides cause an increase in oxidation and crack growth rates.

Future Directions

1. Thermo-mechanical loading cases at high mean temperatures need to be investigated. Conditions where isothermal data fail to predict material damage for thermal loading cases need to be identified.

2. The contribution of oxide failure to crack nucleation needs to be identified for cases of long oxidation times and small strains.

Acknowledgments

This research was funded by the Association of American Railroads, Technical Center, Chicago, Illinois. The cooperation of Dr. Daniel Stone and Mr. Michael Fec of AAR Technical Center is gratefully acknowledged. Mr. Richard Neu assisted in some of the thermo-mechanical fatigue experiments.

References

[*1*] Kuwabara, W. and Nitta, A., "Effect of Strain Hold-Time of High Temperature on Thermal Fatigue Behavior of Type 304 Stainless Steel," in *Proceedings*, 1976 ASME - MPC Symposium on Creep-Fatigue Interaction, R. M. Curran, Ed., Dec. 1976, pp. 161–177.

[*2*] Kawamoto, M., Tanaka, T., and Nakajima, H., "Effect of Several Factors on Thermal Fatigue," *Journal of Materials*, Vol. 1, No. 4, 1966, pp. 719–758.

[*3*] Lundberg, L. and Sandstrom, R., "Application of Low Cycle Fatigue Data to Thermal Fatigue Cracking," *Scandinavian Journal of Metallurgy*, Vol. 11, No. 2, 1982.

[*4*] Jaske, C., "Thermal-Mechanical, Low-Cycle Fatigue of AISI 1010 Steel," in *Thermal Fatigue of Materials and Components, ASTM STP 612*, D. Spera and D. Mowbray, Eds., American Society for Testing and Materials, Philadelphia, 1976, pp. 170–198.

[*5*] Udoguchi, T. and Wada, T., "Thermal Effect on Low-Cycle Fatigue Strength of Steels," *Thermal Stresses and Thermal Fatigue*, D. J. Littler, Ed., Butterworths, London, 1971, pp. 109–123.

[*6*] Kuwabara, K. and Nitta, A., "Thermal-Mechanical Low-Cycle Fatigue under Creep-Fatigue Interactions on Type 304 Stainless Steels," *Fatigue of Engineering Materials and Structures*, Vol. 2, 1979, pp. 293–304.

[*7*] Leis, B., Hopper, A., Ghadiali, N., Jaske, C., and Hulbert, G., "An Approach to Life Prediction of Domestic Gas Furnace Clam Shell Type Heat Exchangers," *Thermal Stresses in Severe Environments*, Hasselman and Heller, Eds., Plenum Press, New York, 1980, pp. 207–228.

[*8*] Taira, S., Fujino, M., and Ohtani, R., "Collaborative Study on Thermal Fatigue Properties of High Temperature Alloys in Japan," *Fatigue of Engineering Materials and Structure*, Vol. 1, 1979, pp. 495–508.

[*9*] Baron, H. G., "Thermal Shock and Thermal Fatigue," *Thermal Stress*, Benham and Hoyle, Eds., Pitman, New York, 1964, pp. 182–206.

[*10*] Halford, G. R. and Manson, S. S., "Life Predictions of Thermal-Mechanical Fatigue Using Strain Range Partitioning," in *Thermal Fatigue of Materials and Components, ASTM STP 612*, D. Spera

and D. Mowbray, Eds., American Society for Testing and Materials, Philadelphia, 1976, pp. 239-254.

[11] Sehitoglu, H. and Karasek, M., "Observations of Material Behavior under Isothermal and Thermo-Mechanical Loading," *Journal of Engineering Materials and Technology, Transactions of ASME,* Vol. 108, 1986, pp. 192-198.

[12] Caplan, D. and Cohen, M., "Effect of Cold Work on the Oxidation of Iron from 400-650°C," *Corrosion Science,* Vol. 6, 1966, p. 321.

[13] Mackenzie, J. and Birchenall, C., "Plastic Flow of Iron Oxides and the Oxidation of Iron," *Corrosion,* Vol. 13, No. 12, 1957, p. 783.

[14] Price, W. R., "On the Effect of Cold Work on the Oxidation of Iron from 400° to 650°C," *Corrosion Science,* Vol. 7, 1967, p. 473.

[15] Manning, M. E., "Geometrical Effects on Oxide Scale Integrity," *Corrosion Science,* Vol. 21, No. 4, 1981, p. 301.

[16] Bruce, D. and Hancock, P., "Influence of Specimen Geometry on the Growth and Mechanical Stability of Surface Oxides Formed on Iron and Steel in the Temperature Range 570°-800°C," *Journal of the Iron and Steel Institute,* Nov. 1970, p. 1021.

[17] Bruce, D. and Hancock, P., "Mechanical Properties and Adhesion of Surface Oxide Films on Iron and Nickel Measured During Growth," *Journal of the Institute of Metals,* Vol. 94, 1969, p. 148.

[18] Caplan, D., Sproule, G., Hussey, R., and Graham, M., "Oxidation of Fe-C Alloys at 500°C," *Oxidation of Metals,* Vol. 12, No. 1, 1978, p. 67.

[19] Caplan, D., Sproule, G., Hussey, R., and Graham, M., "Oxidation of Fe-C Alloys at 700°C," *Oxidation of Metals,* Vol. 13, No. 3, 1979, p. 255.

[20] Malik, A. and Whittle, D., "Oxidation of Fe-C Alloys in the Temperature Range 600-850°C," *Oxidation of Metals,* Vol. 16, Nos. 5/6, 1981, p. 339.

[21] Skelton, R., "Environmental Crack Growth in 0.5Cr-Mo-V Steel During Isothermal High Strain Fatigue and Temperature Cycling," *Materials Science and Engineering,* Vol. 35, No. 2, 1978, pp. 287-298.

[22] Skelton, R. and Bucklow, J., "Cyclic Oxidation and Crack Growth During High Strain Fatigue of Low Alloy Steel," *Metal Science,* Feb. 1978, p. 64.

[23] Sehitoglu, H., "Constraint Effect in Thermo-Mechanical Fatigue," *Journal of Engineering Materials and Technology, Transactions of ASME,* Vol. 107, 1985, pp. 221-226.

[24] Morrow, JoDean, "Cyclic Plastic Strain Energy and Fatigue of Metals," in *Internal Friction, Damping, and Cyclic Plasticity, ASTM STP 378,* American Society for Testing and Materials, Philadelphia, 1965, pp. 45-87.

[25] Berling, J. T. and Conway, J. B., "Effect of Hold Time on the Low-Cycle Fatigue Resistance of 304 Stainless Steel at 1200°F," in *Proceedings,* First International Conference on Pressure Vessel Technology, ASME, Vol. 2, 1969.

DISCUSSION

L. Rémy[1] *(written discussion)*—You conclude from your work that thermo-mechanical fatigue (TMF) damage was greater than isothermal fatigue damage, even when using isothermal data at the maximum temperature in the cycle. This conclusion, however, is based on a total strain range criterion. The stress range developed during thermomechanical loading is often much higher than that observed under isothermal loading for a given mechanical strain range. Mean stresses are also more important in thermal-mechanical loading than in strain-controlled isothermal loading. Thus a stress-based criterion using stress range at least, or accounting for mean stresses, should be able to give a reliable prediction of thermal-mechanical fatigue life from isothermal fatigue data [*1,2*]. Would you comment on this point?

[1]Centre des Matériaux, Ecole des Mines de Paris, BP. 87, 91003 Evry Cedex, France.

TABLE 4—*Thermo-mechanical constant amplitude fatigue test results for* $\dot{\epsilon} = 0.0002\ s^{-1}$.

Temperature Range (°C)	Mechanical Strain Range ($\Delta\epsilon_m$)	σ_{max} (MPa)	$\Delta\sigma$ (MPa)	N_f (cycles)
150–450	0.0051	538	771	3048
150–500	0.0060	550	745	1524
150–600	0.0076	526	681	735
150–650	0.0085	517	668	436
150–650	0.0085	522	674	573
150–700	0.0094	477	659	410

M. L. Karasek et al. (authors' closure)—We appreciate your comments. Representative results from thermo-mechanical constant amplitude tests are listed in Table 4. It is noted that the TMF lives decrease with increasing $\Delta\epsilon_M$ as expected. However, the TMF lives decrease with decreasing σ_{max} and $\Delta\sigma$. This is in contrast to isothermal fatigue results where the fatigue lives increase with concurrent decreases in σ_{max} and $\Delta\sigma$. Therefore the use of stress-based parameters can not correlate TMF results for this alloy in a consistent manner.

Discussion References

[1] Rezai-Aria, F., Rémy, L., Herman, C., and Dambrine, B. in *Proceedings,* Mechanical Behaviour of Materials-IV, J. Carlsson and N. G. Ohlson, Eds., Pergamon Press, Oxford, 1984, Vol. I, pp. 247–253.
[2] Rémy, L., Rezai-Aria, F., Danzer, R., and Hoffelner, W., "Evaluation of Life Prediction Methods in High Temperature Fatigue," this publication, pp. 1115–1132.

Crack Propagation

R. P. Skelton[1]

Application of Small Specimen Crack Growth Data to Engineering Components at High Temperature: A Review

REFERENCE: Skelton, R. P., "**Application of Small Specimen Crack Growth Data to Engineering Components at High Temperature: A Review,**" *Low Cycle Fatigue, ASTM STP 942,* H. D. Solomon, G. R. Halford, L. R. Kaisand, and B. N. Leis, Eds., American Society for Testing and Materials, Philadelphia, 1988, pp. 209-235.

ABSTRACT: This review considers the standard low-cycle fatigue (LCF) specimen, the area of the component it represents, the empirical crack growth relations measured on that specimen, and their ranges of validity. It also presents some recent observations on the effects of crack shape, strain rate, weld material, notches, and aging upon cyclic propagation rates at high temperature.

In applying LCF growth results to estimate crack behavior in service, size effects must be considered. From crack growth tests at constant plastic strain on LCF specimens of differing width, it is shown in both cases that the cracks accelerated, then decelerated after a critical depth which occurred sooner in the smaller specimens. Accompanying variations in the shape of the hysteresis loop and crack closure effects (all of which have implications for service loading) are also reported.

Beyond a certain depth in the specimen, crack growth is no longer describable by the bulk plastic strain, so other parameters (cyclic J integral and equivalent stress intensity) are compared and their ranges of validity discussed.

The survey next considers a constant-load test where, despite equal tension and compression ranges, ratchetting of the specimen occurred, leading to quite different conditions for crack closure. The crack growth data nevertheless agree with LCF data when the correct stress range is taken into account.

Finally, a high temperature lifetime assessment route is briefly described, where the user is permitted to enter a permissible defect path which utilizes many of the experimental short crack data mentioned in this review.

KEY WORDS: high temperature alloys, thermal cycling, low-cycle fatigue, empirical cyclic crack growth relation, accelerating and decelerating growth, linear elastic fracture mechanics, equivalent stress intensity, ΔJ parameter, strain rate (frequency) effect, integrated endurance, aging effect, austenitic weld material, notch effect, crack shape, crack monitoring, shallow and deep cracks, laboratory specimen size, component size, plastic zone depth, hysteresis loop, stress drop with crack penetration, crack closure, crack opening, strain- or load-controlled test, ratchetting, total and plastic strain range, tension dwell, negative R ratio, assessment procedure

The study of cyclic crack growth behavior during the last 30 years has occurred alongside that of the more traditional approach of determining the endurance of smooth fatigue specimens. Crack growth experiments have arisen from a new philosophy which states that structures can perform their design function even though they may contain defects, provided only that the

[1]Research Officer, Technology Planning & Research Division, Central Electricity Generating Board, Central Electricity Research Laboratories, Leatherhead, Surrey KT22 7SE, England.

cyclic propagation behavior is well characterized. Appropriate experimentation has usually involved large cracks in specimens whose shape has little to do with the theme of this symposium.

The traditional low-cycle fatigue (LCF) strategy assumes that significant crack growth cannot be allowed during the lifetime of the component. The critical area (where cracking during repeated service cycles is expected to initiate) is imagined to be modelled by a smooth specimen which is then taken through stress reversals until complete separation is achieved. If a more stringent definition of service initiation is required, an earlier failure criterion is sought in the specimen itself.

A closer argument runs as follows [1,2]. In Fig. 1, *ABC* is a strain concentrating feature in the component which, as a result of temperature or mechanical strain cycling, produces a cyclic plastic zone of depth X. The effects of strain gradient in the zone are minimized by arranging that the specimen width or diameter $W \ll X$ (Fig. 1), because in the laboratory the crack grows under push-pull (uniform strain) conditions. If $X < W$, two options are available: (1) the specimen diameter may be reduced or (2) the test may be terminated at a specified load drop, indicating crack initiation in the specimen [3]. In this way service initiation in both "thin" components (*e.g.*, nuclear fuel element cladding) and "thick" components (*e.g.*, the region ahead of a heat relief groove in a turbine rotor) may be modelled [4].

From the foregoing arguments it appears that early growth rates in the component may also be reproduced by monitoring those occurring in the specimen. However, it is clear from Fig. 1 that when the crack is at a depth of, say, $3/4W$, the specimen will indicate a notable increase in compliance (inverse stiffness), whereas a very much smaller increase would be observed in a thick component for the same crack depth. Corresponding propagation rates therefore are not necessarily equivalent.

The present work incorporates developments since two previous reviews of LCF growth rates at high temperature [5,6], and concentrates on the application of these developments to the situation in service. On the one hand, the LCF growth relations may be regarded as simply supplementing the information recorded during a traditional fatigue test. For example, growth

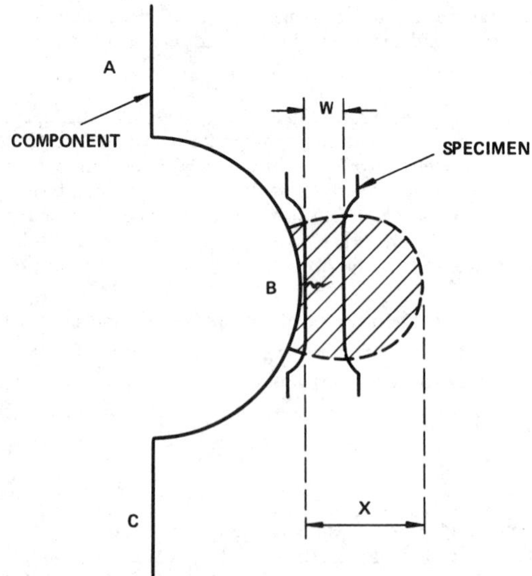

FIG. 1—*Laboratory specimen modelling fatigue crack growth in a component.*

rates may be integrated, the resulting cycles being compared with smooth specimen endurance N_f [7]. Again, by modelling crack tip events using cyclic stress-strain data, the cyclic hardening exponent may be related [8] to the slope α in the Coffin relation

$$N_f^{\alpha}\epsilon_p = \text{constant} \tag{1}$$

where ϵ_p is the plastic strain range. On the other hand, the data may be made to join the parallel studies of crack growth in differently shaped specimens by choosing alternative parameters such as ΔJ or equivalent stress intensity to describe propagation rates. Our purpose is to show that by careful attention to crack depth with respect to specimen and component size, the laboratory growth results may often be applied directly to the plant item.

Form of Crack Growth Law

Empirical Relations

In a typical LCF growth test, a smooth specimen of circular or rectangular cross section is lightly notched and crack progress monitored by one of several means (see next section). Testing is usually done between limits of total strain range ϵ_t, but a constant plastic strain range can equally well be employed. From such studies an empirical growth law is found [5] to be

$$\frac{da}{dN} = Ba^Q \tag{2}$$

where da/dN is the cyclic growth rate, B is a constant which depends on ϵ_t or ϵ_p, a is crack depth, and Q is also a constant.

A typical laboratory test might start with a chordal notch of depth 0.1 mm in a cylindrical specimen of about 10 mm diameter ($a/W = 0.01$) and finish when a significant load drop or cusp in the hysteresis loop had occurred [3]. At this point a/W is about 0.25. By plotting values of log B against log ϵ_p it may be shown [5] that a more general form of Eq 2 is

$$\frac{da}{dN} = C\epsilon_p^n a^Q \tag{3}$$

Values of C, n, and Q for several materials at high temperature are summarized in Table 1 [5,6,9–12] for growth expressed in mm/cycle.

For assessing the behavior of short cracks in a component the *total* strain range is more useful. The variation of B in Eq 2 (for growth expressed in mm/cycle at unit crack depth) with total strain is shown in Fig. 2 for several ferritic steels [5,11,12]. An upper bound relation is given by

$$B = 2.61 \times 10^4 \epsilon_t^{2.85} \tag{4}$$

Equation 4 is valid for tensile dwells up to 1/2 h, and it is perhaps important to note that cracking may be either intergranular (1/2Cr-Mo-V steel) or transgranular (2 1/4Cr-1Mo and 9Cr-1Mo steels).

Although fundamental theory predicts $Q = 1$ [8] (exponential growth law), Table 1 shows that variations can occur; these have been discussed previously [5]. Further, at very small crack depths in the specimen ($a/W < 0.01$), Eqs 2 and 3 break down and growth rates appear to be independent of crack depth for a given strain level. Figure 3 shows the effect in an austenitic

TABLE 1—*Values of constants in Eq 3.*

Material	Temperature, °C	C	n	Q	Comments	Ref
20Cr-25Ni-Nb	750	25.4	1.52	1.0	...	5
Type 316 Steel	625	13.5	1.35	1.0	...	5
High N 316 Steel	600	10.3	1.6	1.0	...	9
Aged 316 Steel	625	25	1.7	1.9	1/2 h dwell, vacuum	10
Hastelloy X	760	10.7	1.28	1.0	...	5
Alloy 800	500–760	4.6	1.37	1.0	...	6
A 286	593	0.44	0.7	1.0	...	5
9Cr-1Mo (annealed)	550	0.2	1.0	1.9	vacuum	11
9Cr-1Mo (N&T)	550	0.1	1.06	1.3	vacuum	this work
2¼Cr-1Mo (annealed)	525	0.12	0.73	0.74	...	12
2¼Cr-1Mo (N&T)	525	16.1	1.45	0.86	...	12
½Cr-Mo-V	550	1.0	1.0	1.0	...	5
½Cr-Mo-V	550	4.2	1.1	1.5	1/2 h dwell	5

steel [*13*] and a superalloy [*14*]. The validity of the equations beyond $a/W = 0.25$ in the specimen is discussed later.

Shape Effects

Crack growth rates during LCF may be determined by one or more of the following methods:

1. Potential drop [*3*].
2. Striation counting [*15*].
3. Beach marking, which may be deliberate or otherwise. The method includes heat tinting [*3*].
4. Compliance changes (e.g., by tensile load drop [*3*]).
5. Surface observation. Telescopic methods may be used for rectangular specimens [*16*]; replication techniques, which infer crack depth, are used for cylindrical specimens [*14,17*].

Methods 1 and 4 are only able to monitor the area of a growing crack [*3*] so that peak depth at any time must be verified by subsequent examination of the fracture surface. In the well-developed stage the aspect ratio a/ℓ (where ℓ is the semi-length in the surface) of a thumbnail crack in a cylindrical specimen is typically about 0.72 for several ferritic and austenitic steels in the range 300 to 600°C [*17*].

In principle one may express the growth rate equations (2 and 3) in terms of crack area, as has recently been proposed for alternating bending of a shaft [*18*]. Surface growth rates alone (Method 5) do not give sufficient information. For example, Usami et al. [*19*] found that a surface Stage I crack <0.1 mm in 304 steel at 550°C grew faster than the subsequent Stage II crack. This is in contrast with the work of Gangloff [*20*], who observed that a crack grew deeper in the same alloy before a change in surface length took place. Similarly, it has been shown that the growth rate of a planar or circumferential crack in 316 steel at 400°C and 625°C is some four times that of a discrete semicircular crack [*21,22*]. Such differences are easily detectable by incremental area methods.

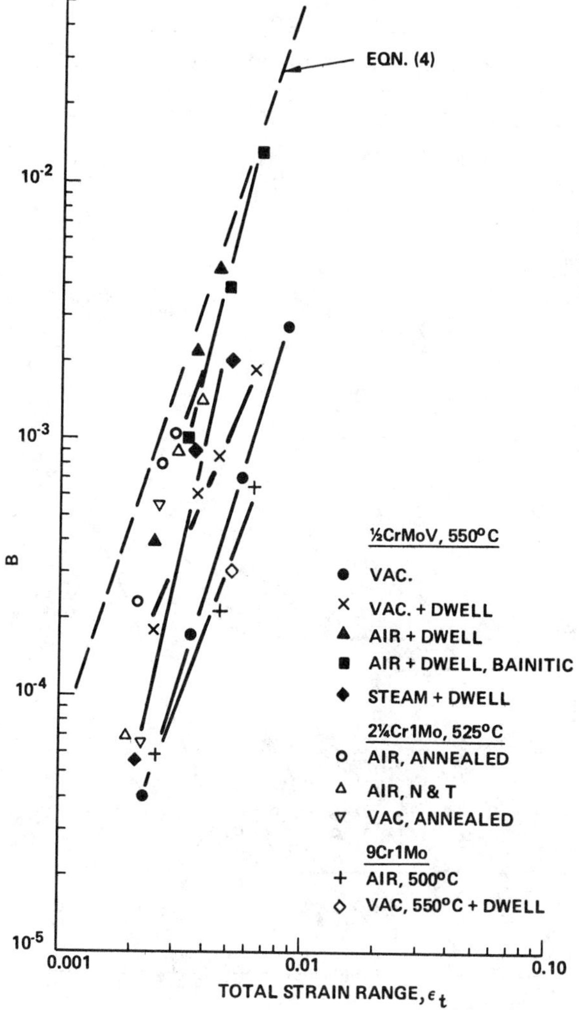

FIG. 2—*Variation of growth rate with total strain range (Eq 2). Corresponding values of Q are given in Table 1.*

Some Recent Developments

Strain Rate Effect

It is well known in high temperature LCF that for most materials the total endurance N_f decreases as the strain rate $\dot{\epsilon}_t$ is reduced, and is accompanied by intergranular cracking. Part of this effect is due to the crack growth phase; in Fig. 4 the variation of B (Eq 2) is plotted against frequency for typical data [*15,19,23,24*]. For continuous cycling (triangular waveform) the relation between total strain rate and frequency ν is given by

$$\dot{\epsilon}_t = 2\nu\epsilon_t \quad (5)$$

so in Fig. 4 the corresponding strain rate is also given.

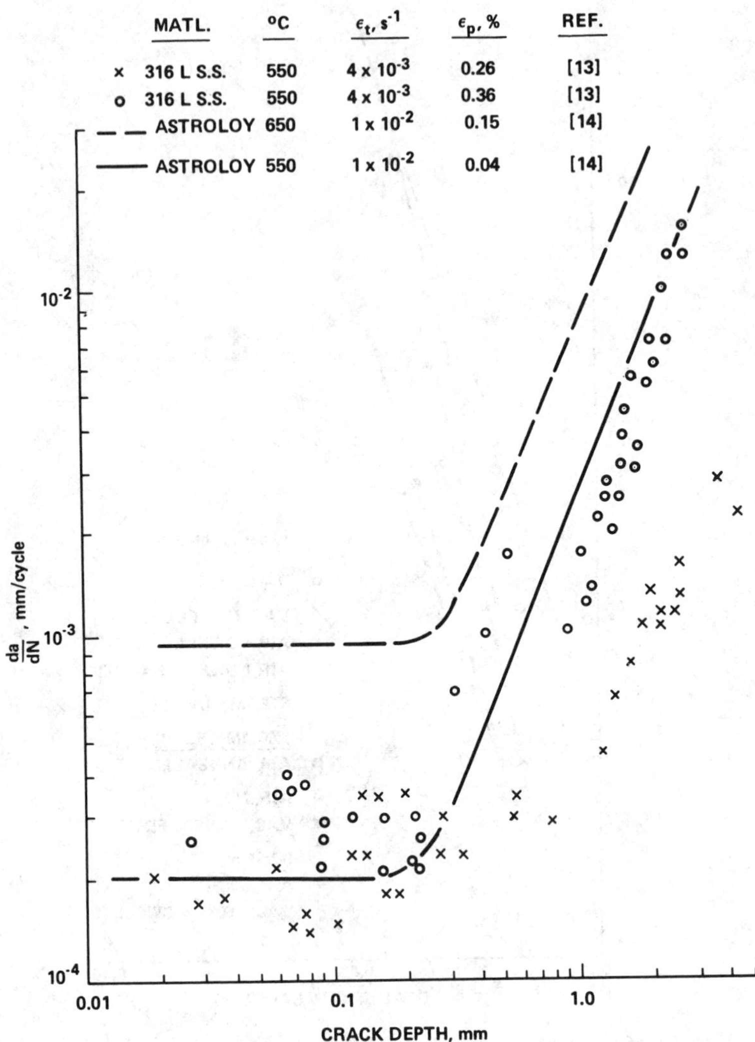

FIG. 3—*Below a critical depth, the crack growth rate is constant.*

Similar effects have been observed in unbalanced cycling. In "slow-fast" tests (tension-going $\dot{\epsilon}_t = 10^{-6}$ s^{-1}, compression-going $\dot{\epsilon}_t = 10^{-3}$ s^{-1}) on 304 steel at 550°C Usami et al. [*19*] recorded about a fourfold increase in crack growth/cycle compared with when the higher strain rate was maintained throughout.

Weld Metal

Some recent LCF growth rates measured in the author's laboratory on 316 weld metal in several combinations of orientation and heat treatment are shown in Fig. 5 as crack growth/cycle versus peak crack depth. Cracks were thumbnail-shaped in cylindrical specimens of 12.7 mm diameter, and tests were conducted at 550°C, with a 1/2 h tension dwell, under vacuum. In

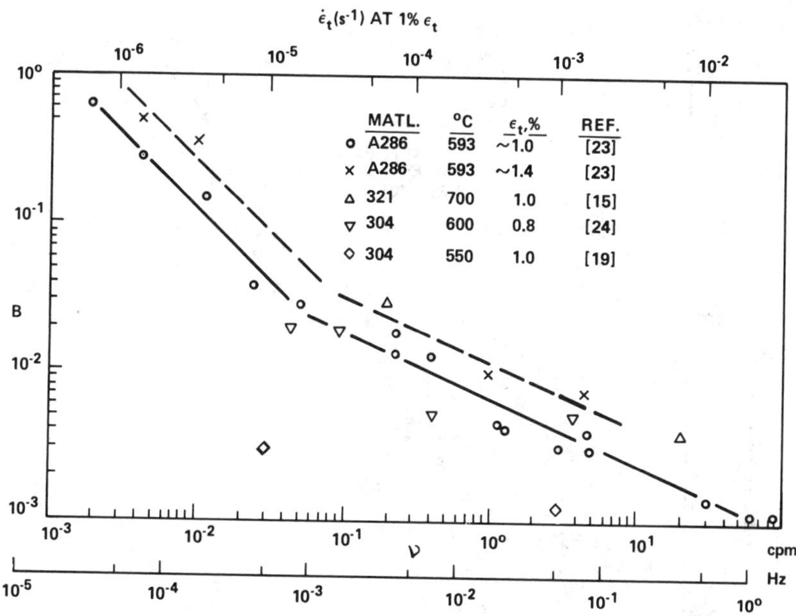

FIG. 4—*Variation of crack growth rate with frequency (or strain rate).*

Fig. 5 comparison is made with similar data on parent material [*10*], both in the solution-treated (ST, 1050°C for 1 h) and post-weld-heat-treated (PWHT, i.e. ST + 800°C for 10 h) conditions.

The only other LCF weld data reported seem to be that of Usami et al. [*19*] on Type 304 steel. Their results are compared in Fig. 6 with the present work and also with parent material data on 304. Taking the results as a whole, there does not appear to be much effect of dwell upon crack growth in the weld metal.

Notch Effect

If a stress concentrator is machined into a LCF specimen, the enhanced cyclic strain at the notch root will itself be embedded in an elastic-plastic matrix. Accordingly, early growth rates will be affected. The results of Usami et al. [*19*], again for 304 steel at 550°C, are replotted (in the form of a B versus ϵ_t diagram) in Fig. 7. They clearly show that for a given total strain range measured remotely, the local crack growth rate (in this case plotted for crack tips 0.15 mm from the notch root) increased as the elastic stress concentration factor K_t increased. As cracks grew out of the notch influence, however, propagation rates decreased towards those characteristic of a plain specimen [*19*].

Aging Effect

Yamaguchi and Kanazawa found [*15*] that aging a 321 steel at 750°C for 24 h reduced subsequent growth rates in LCF by about a factor of three compared with ST material tested at the same strain range. The testing temperature was 700°C, the strain rate was 6.7×10^{-5} s^{-1} and cracking was intergranular. When $\dot{\epsilon}_t$ was raised to 6.7×10^{-3} s^{-1}, cracking was transgranular and the rate was independent of these two heat treatments.

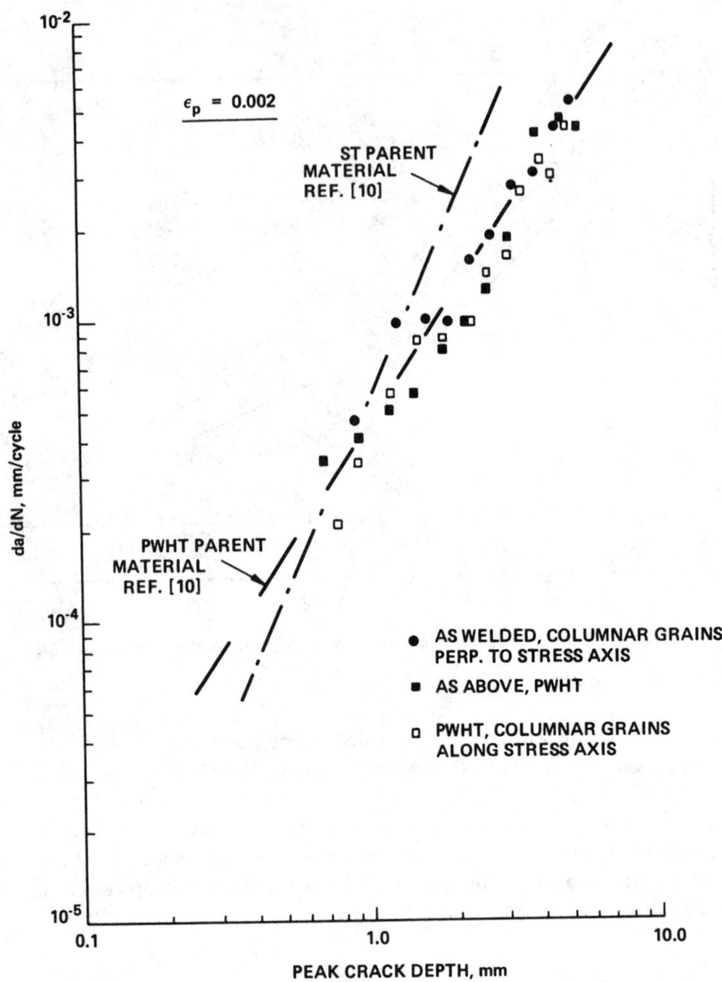

FIG. 5—*LCF crack growth in 316 weld metal, 550°C, 1/2 h tension dwell, determined under vacuum.*

Similar effects were found in 316 steel tested with dwell at 625°C except that prior aging promoted transgranular cracking [*10*]. When aging occurred during the actual test ("cyclic aging") the fine precipitation induced during creep-fatigue did not inhibit grain boundary sliding and cracking remained intergranular [*10*].

Integration of Growth Law

Separation of Initiation and Endurance Cycles

Equation 2 may be expressed in its integrated form

$$N_c = \frac{1}{(Q-1)B}(a_o^{1-Q} - a_f^{1-Q}) \tag{6a}$$

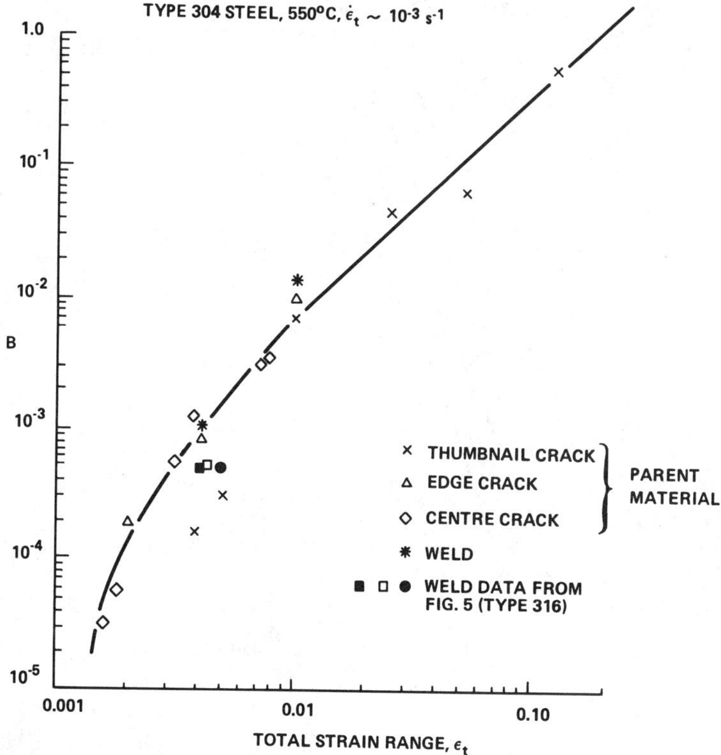

FIG. 6—*Data of Usami et al.* [19] *(for which* Q = *1 in Eq 2) compared with present work.*

if $Q \neq 1$ or

$$N_c = \frac{1}{B} \ln\left(\frac{a_f}{a_o}\right) \tag{6b}$$

if $Q = 1$, where N_c is the number of cycles involved in propagation, a_o is the initial depth, and a_f is some chosen final depth. Theoretical expressions may alternatively be used; these have been reviewed recently [*22*]. All expressions have been used to confirm total cycles to failure N_f (i.e., to place a lower bound on experimental endurance data if an initial defect is present). It is sometimes found that N_c is overestimated at low strain rates [*14*].

The value usually assumed for a_f is $3/4W$. It has been argued [*22*] that N_c calculated from Eq 6*b* is insensitive to the precise value of a_f. This reasoning suggests that most of the lifetime is occupied in the growth of short cracks and that later compliance changes have little influence. It is, however, pertinent to enquire whether the accelerating growth properties predicted by Eq 2 are valid across the whole specimen section when tests are carried out under strain control.

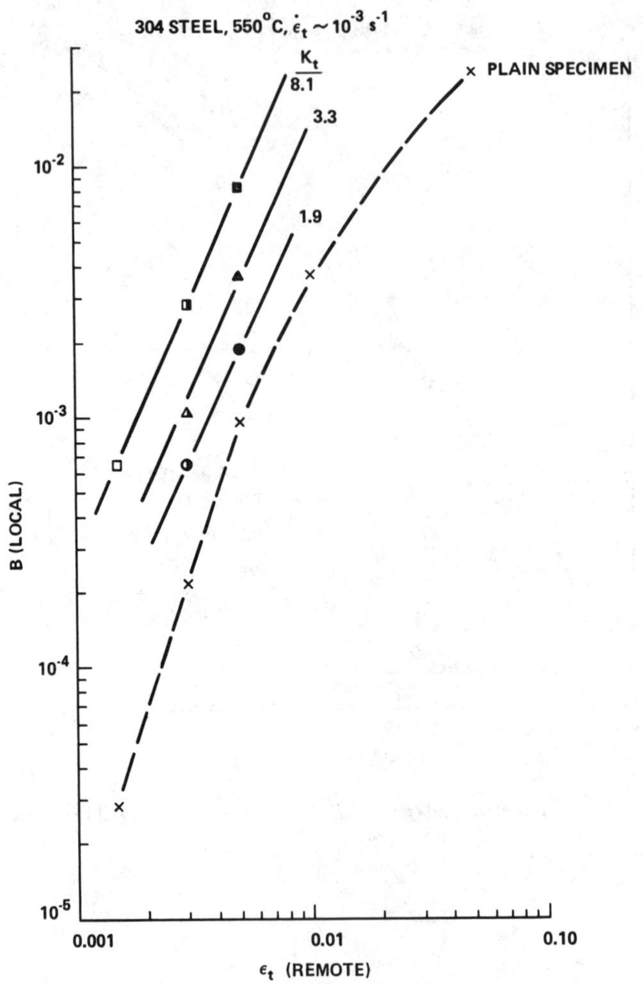

FIG. 7—*Replotted data from Ref* 19. *In these examples,* Q = *1 in Eq 2.*

Extension of Law to Deeper Cracks

There are two aspects to LCF crack growth for deep cracks:

1. *Propagation Over the Final Half of a Specimen*—The theory of crack advance by irreversible shear decohesion at the tip [*8*] predicts $da/dN = \epsilon_p D$, where D is the width of the plastic zone. As the crack grows, this zone eventually meets the back face and the rate is then given by [*25*]

$$\frac{da}{dN} = \epsilon_p(W - a) \tag{7}$$

A decreasing advance rate is then expected.

2. *Validity of the Law for Large Specimens*—When Eq 2 is expressed in the form

$$\frac{da}{dN} = B'\left(\frac{a}{W}\right)^Q \tag{8}$$

where $B' = BW^Q$, it is implied that an accelerating growth period should be obtained over a greater depth range in larger specimens. Hence in predicting the LCF growth rate in a thick component over a distance greater than W, the specimen width (Eq 2) may be extrapolated to equivalent values of normalized crack depth in the service component (i.e., within the plastic zone X of Fig. 1). Unfortunately, the only LCF data known to the author are on large (35 mm diameter) specimens in ambient temperature total endurance tests [*26,27*].

Experimental Crack Growth Across Whole Section

In order to verify Eqs 2, 7, and 8, LCF tests were performed in air at 550°C on a ½Cr-Mo-V turbine casing steel and a 1Cr-Mo-V rotor steel specimen. The former was of 8 by 25 mm rectangular cross section (19 mm gage length), while the latter was of 12.7 mm diameter and gage length. Previous work discussed tensile load drop in relation to cracked area [*3*]; this short investigation examines compression behavior and the shape of the hysteresis loop more closely.

Method

Specimens were tested in an Instron reversed stress machine at a strain rate of $\sim 10^{-4}$ s^{-1} corresponding to a frequency of 8.3×10^{-3} Hz in continuous cycling. Heating was by RF induction and crack growth began from a 0.15 mm deep notch (chordal for the cylindrical specimen and inserted in the 8 mm face for the other). Crack growth rates were monitored by d-c potential drop; in addition, the fracture surface was marked by admitting argon or evacuating the environmental chamber at certain steps (Fig. 8). The plastic strain range was kept constant at 0.002 and specimens taken to complete separation.

Crack Growth Results

Figure 9 plots the observed growth rate across the entire section for both steels. It is seen that da/dN does indeed decrease in the later stages, but that agreement with Eq 7 is not good. This probably arises because the observed plastic strain ceases to be a true measure of plastic strain in the ligament, as will be discussed in a later section. Also, a departure from an exponential growth law takes place at about 3 mm peak depth ($a/W \sim 0.25$) in the cylindrical specimen and at 6 mm depth (again, $a/W \sim 0.25$) in the rectangular shape. Since $Q \sim 1$ in the early stages (Fig. 9), the deviation occurs at a critical value of a/W in agreement with Eq 8.

Similar results would be expected under total strain control, except that the plastic strain range decreases slightly with increasing crack penetration [*3*].

Observations on Stress Range

Stress-Depth Variation—The variation of the gross peak tensile and compressive stress with normalized crack depth for both steels is shown in Fig. 10. All tensile effects are due to an increase in compliance of the remaining ligament and confirm the $(1 - a/W)$ variation of stress for the rigidly gripped push-pull situation (i.e., that the net section stress remains constant, independent of crack shape [*3*]).

By comparison, the compressive peak was observed to fall relatively slowly (Fig. 10), owing to the effect of crack closure in each cycle restoring the compliance [*28*].

FIG. 8—*Fracture surface of 12.7 mm diameter 1Cr-Mo-V specimen marked by intermittent evacuation during high temperature air test. Note that thumbnail crack becomes straight-fronted towards end of test. Notch is at bottom.*

Shape of Hysteresis Loop—A typical "hysteresis" loop, well beyond the first "cusp" formation in compression [*28*] is shown in Fig. 11. Apart from an obvious change in compliance upon unloading from tension (crack fully open) and compression (crack fully closed), other distinct features may be observed:

1. Crack closure stress (σ_c) occurs further into compression than the subsequent opening stress σ_o. This is reasonable because the plastic strain undergone by the ligament in the previous half-cycle must first be reversed before the crack can close.
2. Opening at σ_o occurs below zero stress.
3. As the crack continues to close along its length, the apparent modulus *AB* is less than the subsequent value *BC* upon compressive unloading.

This last effect gives rise to a plastic strain ϵ_{PC} associated with closing and opening the crack; that is,

$$\epsilon_{PC} = \frac{\sigma_o - \sigma_c}{E} \tag{9}$$

where *E* is Young's modulus (1.5×10^5 MPa). Experimentally it was observed that ϵ_{PC} decreased as the crack grew. Results are given in Fig. 12 along with some data from a test on 316 steel at 600°C in total strain control (0.7%) at 0.3% nominal plastic strain range.[2]

Crack Closure Effects—By detailed study of hysteresis loops such as that of Fig. 11 it was possible to plot the following parameters against normalized crack depth:

[2]Hales, R., private communication, 1985.

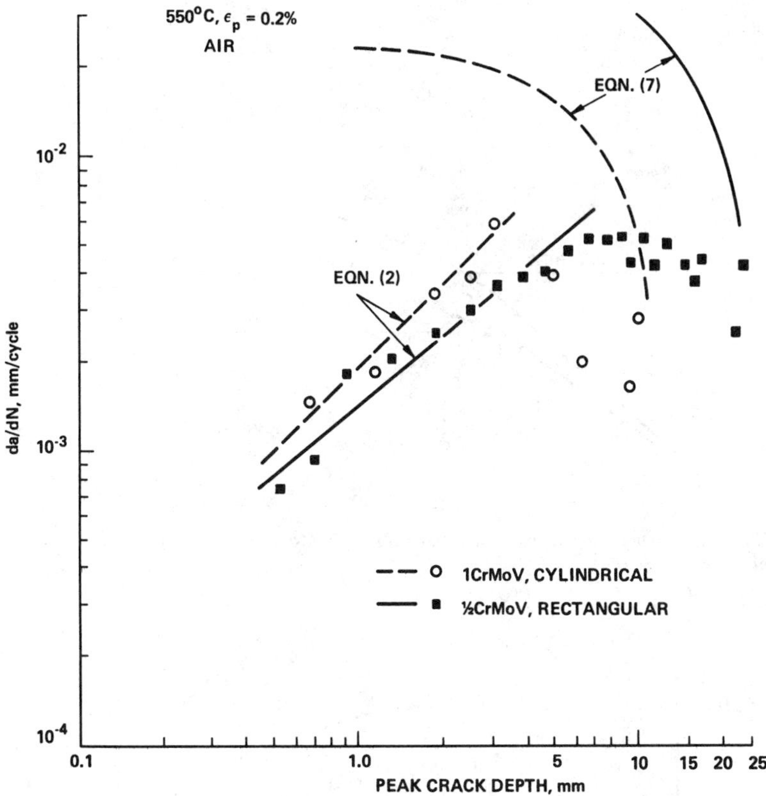

FIG. 9—*Decrease in crack growth rate occurs at certain depth, depending on specimen width. Vacuum data omitted for clarity.*

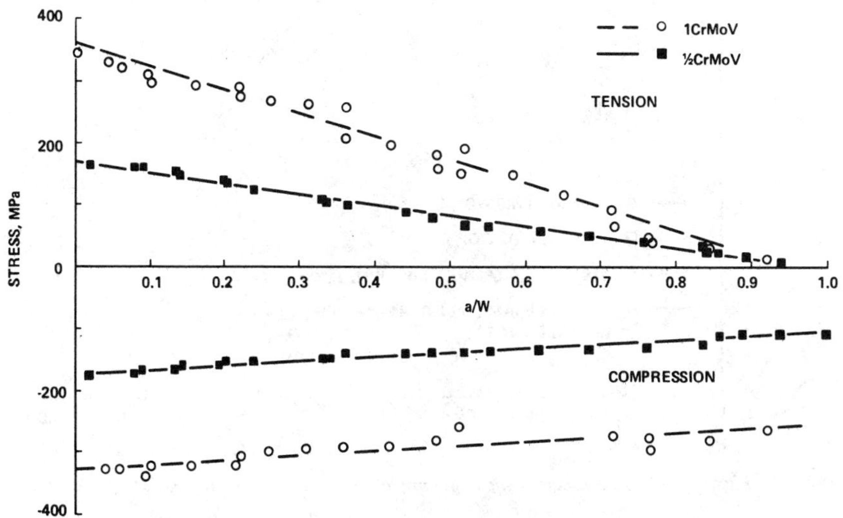

FIG. 10—*Variation of stress with crack depth during tests of Fig. 9.*

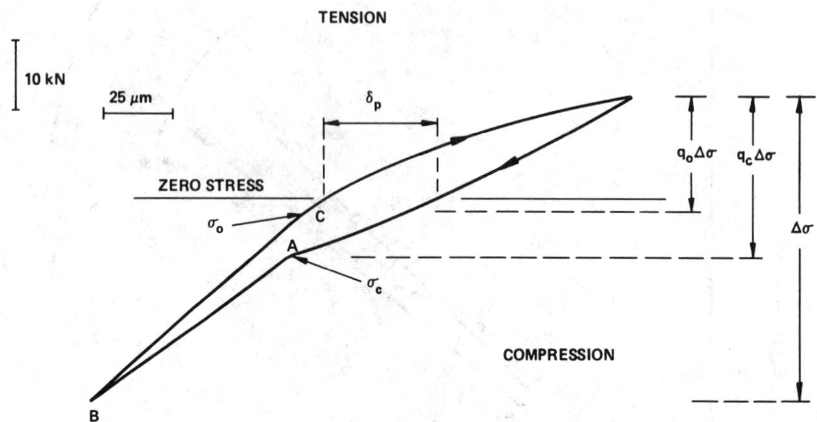

FIG. 11—*Typical load-displacement plot taken for 1/2Cr-Mo-V steel (rectangular specimen, crack depth 13.8 mm).*

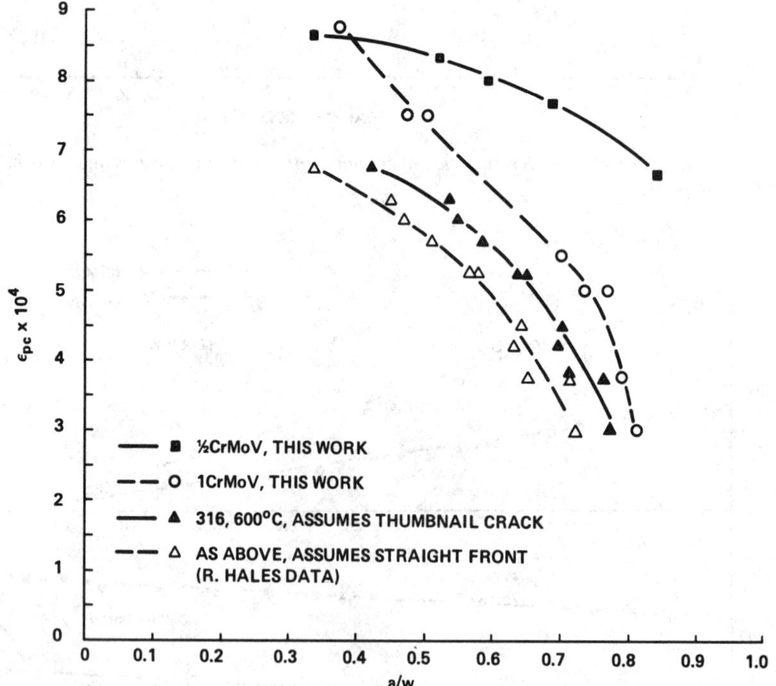

FIG. 12—*Plastic strain ϵ_{pc} associated with opening/closing (Eq 9), showing variation with depth.*

1. q_o, where $q_o\Delta\sigma$ is fractional stress range for which the crack is open, $\Delta\sigma$ being the instantaneous stress range (e.g., Fig. 10).
2. q_c, where $q_c\Delta\sigma$ is the stress range required to close the crack.

The results are given in Fig. 13 together with those of Hales. For each material, q_c gradually reduced as crack depth increased and was always greater than q_o. It is felt that the latter term is more significant in converting results to an equivalent stress intensity (see next section), and its variation for 1/2Cr-Mo-V steel for example is given by

$$q_o = 0.78\left(1 - 0.87\frac{a}{W}\right) \tag{10}$$

For very short cracks, Eq 10 thus predicts an effective opening range $\sim 3/4\,\Delta\sigma$.

In crack propagation studies, the term R is defined as minimum stress/maximum stress, and Fig. 10 shows that, although tests started out at $R = -1$ (symmetrical load range), subsequent values of R increased negatively with crack depth. The variation of $-1/R$ with a/W is depicted in Fig. 14.

It is thus helpful to plot R directly against q_o, and this is done for all results in Fig. 15. For all reasonable values of R expected in practice (i.e., $R > -4$):

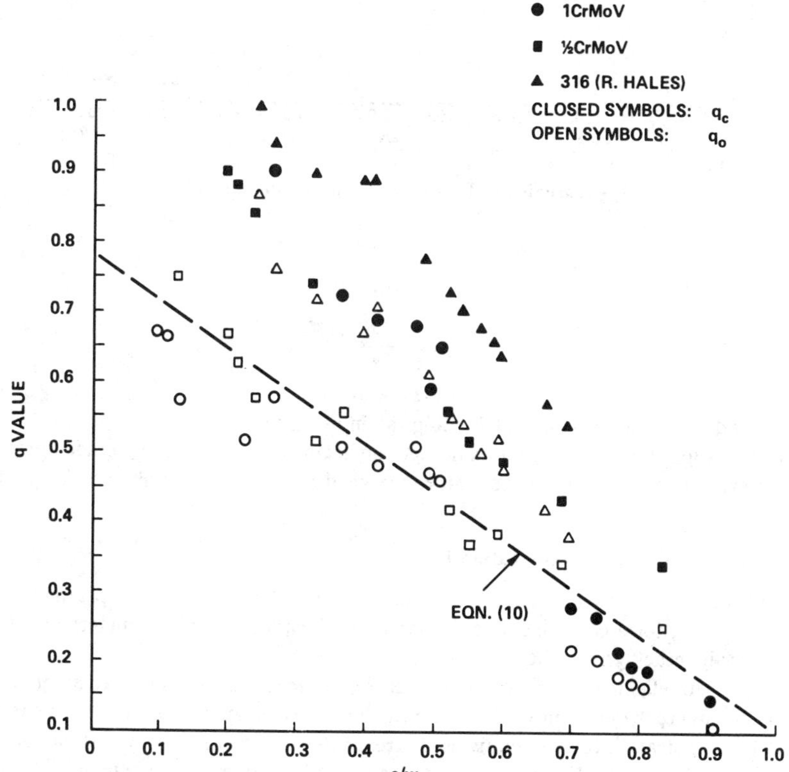

FIG. 13—*Variation of crack opening and closing with depth in specimen.*

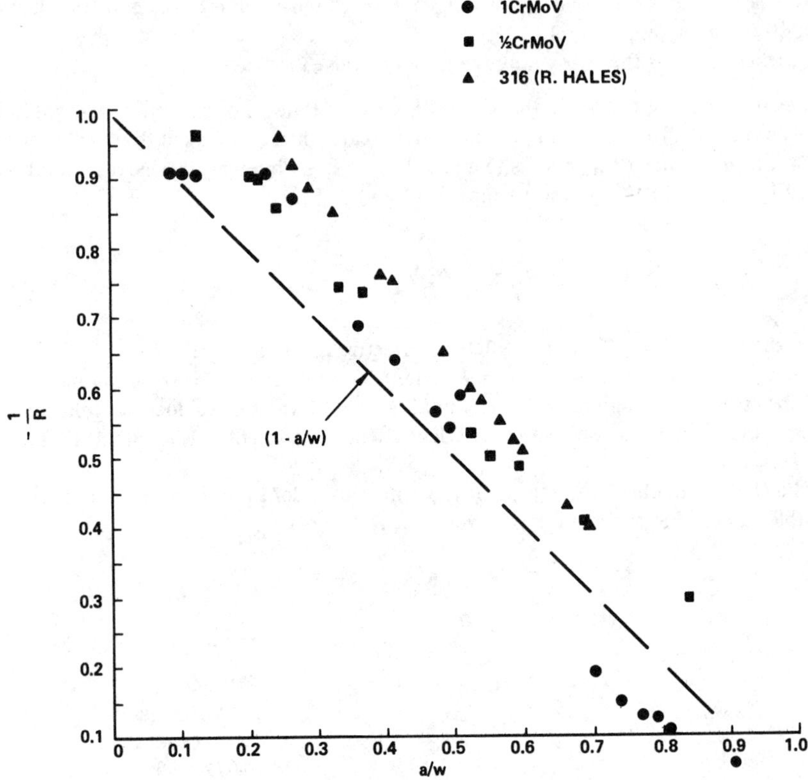

FIG. 14—*Variation of* R *ratio with crack depth in specimen.*

$$q_o = -\frac{F}{R} \tag{11}$$

an upper bound value for the constant F being taken as unity.

These results suggest that when, for example, constrained thermal fatigue cracks propagate into structures, only part of the service load range need be taken to describe crack opening.

Conversion to Stress-Intensity Based Relations

The results of Fig. 9 show that Eq 2 is neither valid for $a/W > 0.25$ in a specimen nor, by application of Eq 8, for a component at the same depth ratio. Other parameters are thus required to describe crack growth in this region.

The advantage of elastic stress intensity is that it combines the effects of a remotely applied stress with current crack depth into a single parameter. The development of alternative parameters for LCF (i.e., strain intensity, equivalent stress intensity, or the J-integral) has been discussed previously [*5,6,29*]. Although shown to apply well in laboratory specimens, one disadvantage is that the calculation of ΔJ, for example, in a component is not obvious and requires finite element analysis (e.g., for a stress gradient in thermal loading [*30*]).

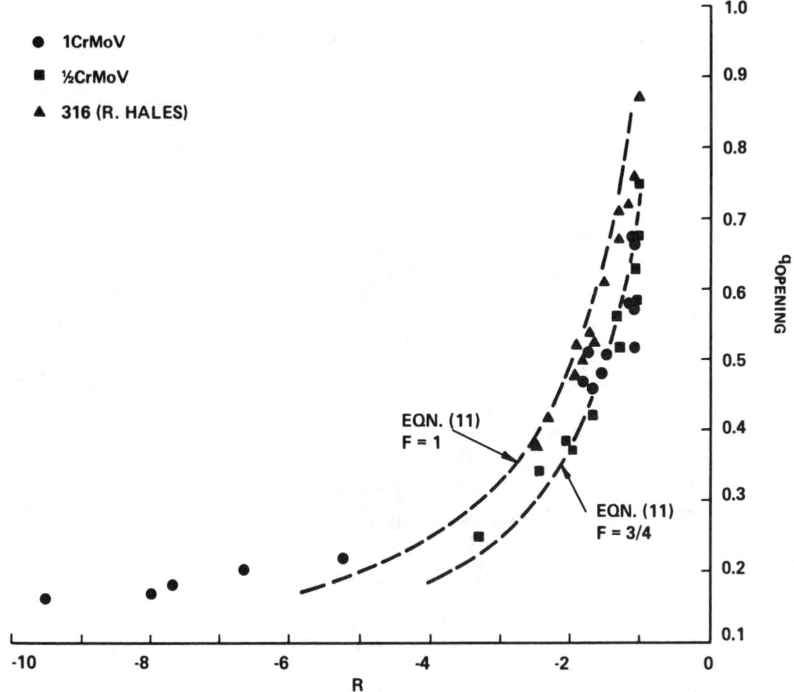

FIG. 15—*Variation of* R *with crack opening ratio,* q_o, *for push-pull conditions.*

*Short (*a/W $\leq$ *0.2) Cracks in Specimen*

The equivalent stress intensity ΔK_{eq} is given [*31*] by

$$\Delta K_{eq} = (E\epsilon_p + q_o\Delta\sigma)Y\sqrt{\pi a} \tag{12}$$

where $\Delta\sigma$ is the stress range, and Y is a compliance function which may be taken as unity for a rigidly-gripped single-edge notch specimen or $2/\pi$ for a thumbnail crack. It has been demonstrated [*29*] that the method is in good agreement with the more familiar formula of Shih and Hutchinson [*32*] for estimating ΔJ values for short cracks (see next section).

Taking values of $q_o\Delta\sigma$ from the appropriate hysteresis loops, the crack growth rates for 316 weld metal in Fig. 5 were converted to ΔK_{eq} using Eq 12 and are plotted against da/dN in Fig. 16. They are to be compared with some data obtained on the same weld metal in load control (discussed later) in the linear elastic fracture mechanics (LEFM) regime. Figure 16 also includes LEFM data for the rotor steel. The results may be expressed as

$$\frac{da}{dN} = C'\Delta K_{eq}^n \tag{13}$$

where $n \sim 2.1$ and C' is a constant.

Some workers prefer to plot their results in terms of ΔJ where

$$\Delta K_{eq} = (\Delta JE)^{1/2} \tag{14}$$

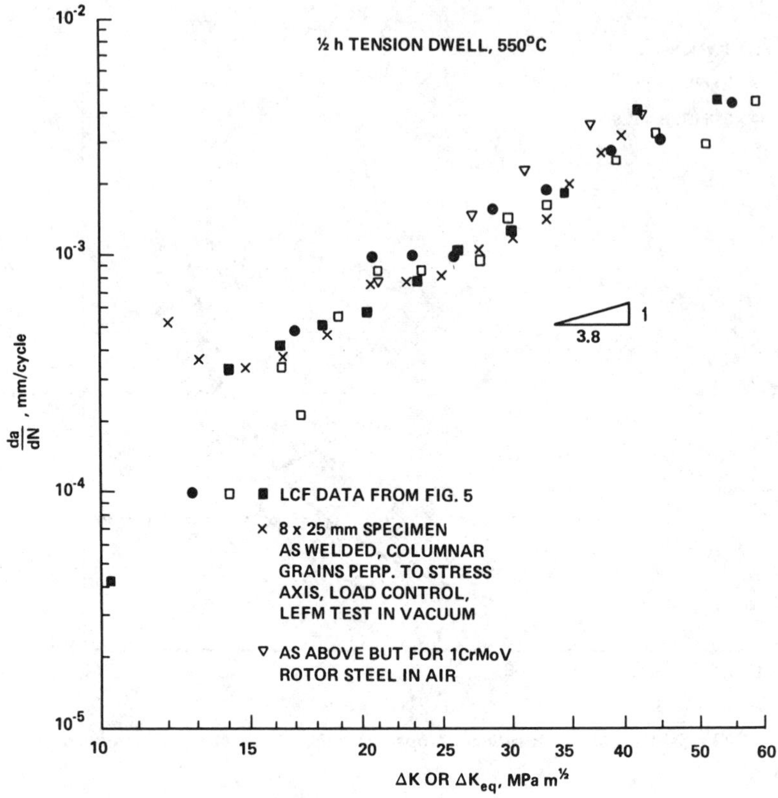

FIG. 16—*Crack growth (expressed as equivalent stress intensity range) for 316 weld metal. Note expanded horizontal scale.*

Data from several sources [*17,24,33,34*] are plotted in Fig. 17. A general growth law is thus alternatively expressed as

$$\frac{da}{dN} = C''\Delta J^m \tag{15}$$

where C'' is a constant and $m = n/2 \sim 1.4$. It should be noted, however, that investigators are not consistent in their derivation of ΔJ. Reger et al. [*33,34*] assume that the semi-stress range is effective, while Okazaki et al. [*17,24*] take the "whole range" ΔJ. Further, when creep-fatigue damage occurs during a dwell, the strain rate effect persists, so that the data separate into time-dependent and cycle-dependent bands [*35*].

Deep (a/W $\leq$ 0.8) Cracks in Specimen

Beyond an a/W value of 0.25 considerable compliance changes occur, and because the "plastic strain" range ϵ_p is not spread evenly over the ligament it is more appropriate to work in terms of the reversed plastic displacement δ_p (Fig. 11). However, it is possible to propagate the crack under nominally elastic conditions ($\delta_p = 0$) in this region.

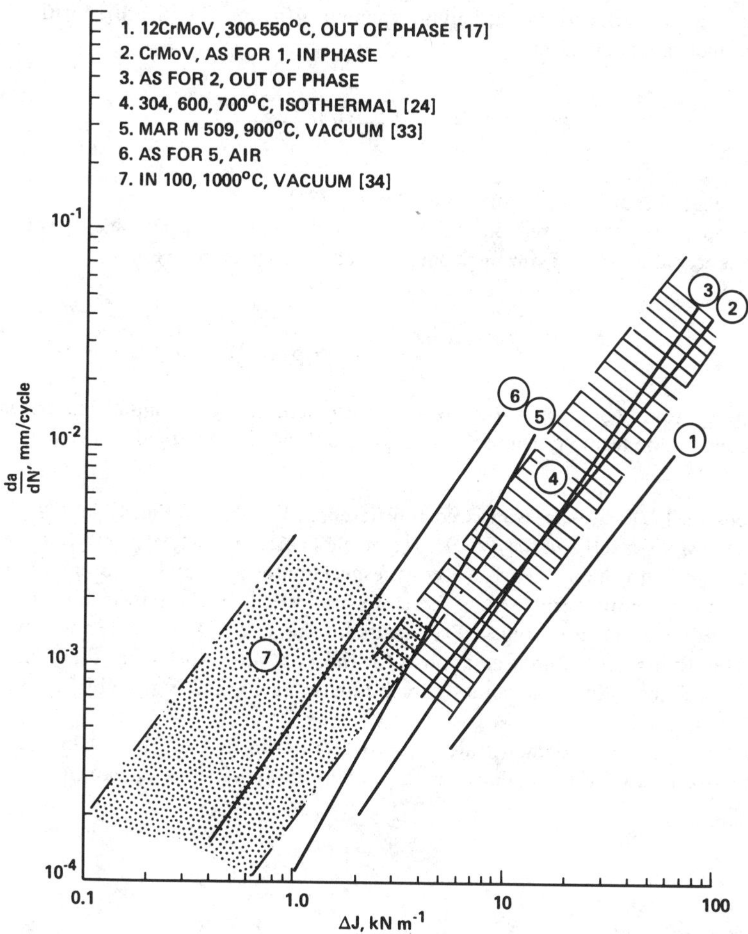

FIG. 17—*Growth rates expressed in terms of* ΔJ *for several high temperature alloys.*

Three relations are available for calculating ΔJ, or if preferred, ΔK_{eq} via Eq 14:

1. Equation 12 may alternatively be expressed [*29*] as

$$\Delta K_{eq} = q_o \Delta\sigma \sqrt{\pi a}\, Y(1 + x) \tag{16}$$

where $x = f/q_o$, f being the ratio of plastic to elastic strain range (i.e., $\epsilon_p E/\Delta\sigma$). It may be assumed that values of Y calculated elastically in the range $a/W > 0.25$ for straight-fronted [*36*] and thumbnail [*37*] cracks also apply in the LCF regime. This is reasonable since Okazaki and Koizumi have shown [*17*] that LCF crack fronts at high temperature adopt shapes expected from purely elastic considerations.

2. The relation of Shih and Hutchinson [*32*] similarly reduces to

$$\Delta K_{eq} = q_o \Delta\sigma \sqrt{\pi a} \left[g(n) + \frac{2xh(n)}{1 + \beta} \right]^{1/2} \tag{17}$$

[29], where β is the cyclic strain hardening exponent, $g(n)$ may be identified with Y, and the plastic compliance term is given by

$$h(n) = \frac{1+\beta}{2\pi} g_1\left(\frac{a}{W}, \beta\right)\left(\frac{W}{W-a}\right)^{1/\beta}$$

the values being obtainable from the original paper [32].

3. For deep cracks, the expression of Dowling [38] for center cracked plates may be modified (by the omission of a factor 2) for our edge notched specimens to give

$$\Delta K_{eq} = q_o \Delta\sigma\sqrt{\pi a}\left[Y^2 + \frac{x\ell_o W}{(1+\beta)\pi a(W-a)}\right]^{1/2} \quad (18)$$

where for purposes of comparison the cyclic displacement δ_p in the original expression [38] has been replaced by the term $x\ell_o$ where ℓ_o, the gage length of our rectangular specimen, is 19 mm and $W = 25$ mm.

Equations 16, 17, and 18 are compared at different a/W values in Fig. 18 for $\beta = 0.1$ (typical of forged 1Cr-Mo-V steel) and for $\beta = 0.25$ (typical of cast 1/2Cr-Mo-V steel). It is seen that Eq 17 is at variance with the other two methods, especially for low values of β and high values of plastic strain (i.e., x increasing). It is therefore proposed to use Eq 16 in the early stages of growth and Eq 18 in the later stages. The results for the data of Fig. 9 are shown in Fig. 19. It may be noted that Eq 18 cannot be applied for the 1Cr-Mo-V specimen (thumbnail crack in cylindrical specimen). An upper bound growth rate (in mm/cycle) is given by Eq 13 with $C' = 3.7 \times 10^{-6}$, $n = 2.0$.

The data of Fig. 19 thus demonstrate that growth rates in LCF ($a/W \leq 0.25$) may be reconciled with growth rates in the region $a/W \geq 0.25$ in the specimen by means of the parameter

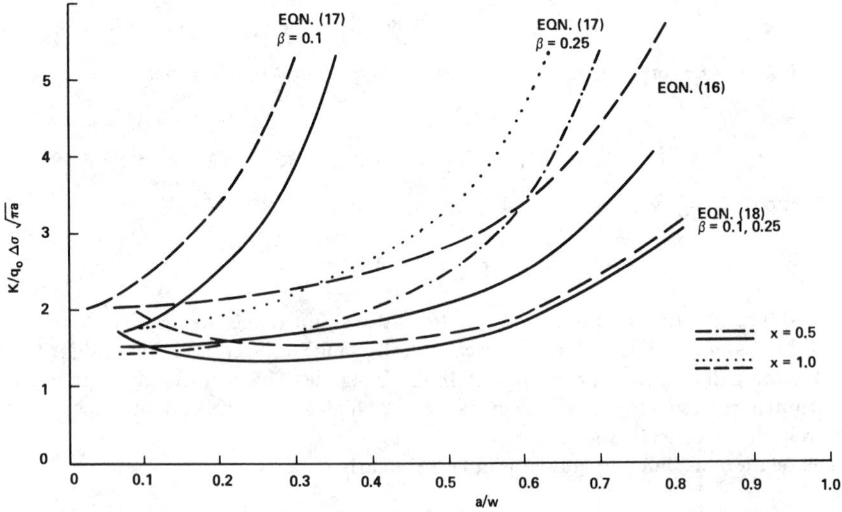

FIG. 18—*Comparison of Eqs 16, 17, and 18, assuming two values of cyclic strain hardening exponent and straight crack front.*

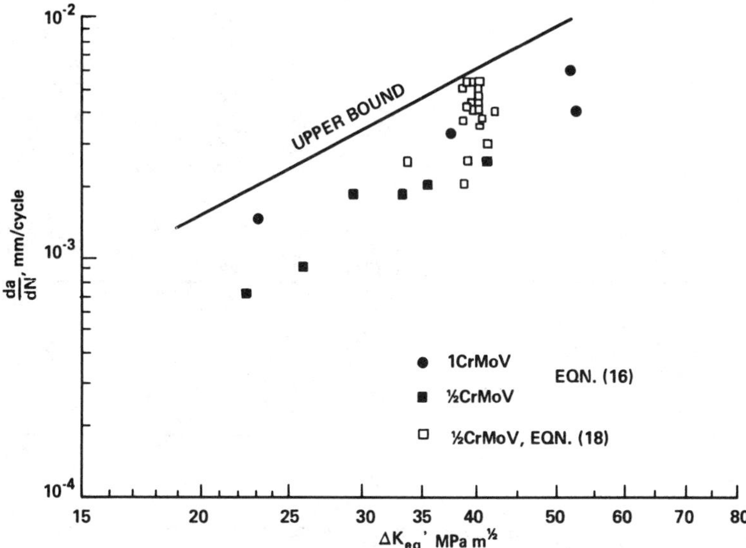

FIG. 19—*Data of Fig. 9 expressed alternatively.*

ΔK_{eq} (or ΔJ). To apply this type of data in service it would thus be necessary to calculate ΔK_{eq} or ΔJ for the component; in the meantime, a simpler route may be preferable (see Discussion section).

Relation to Load Control Tests

Thirty years of crack propagation testing at $R \geq 0$ on other specimen types using pin loading has favored load control and corresponding displacements have not often been monitored. Following the requirement to correlate short crack with long crack growth rates, push-pull tests on large center-cracked specimens have been devised by Dowling [*38*]. This author preferred to control to a "sloping load versus deflection" line (i.e., somewhere between load and strain control so that limited ratchetting (incremental plasticity) was permitted).

Load-controlled tests are easier to perform, but it follows that the crack (which causes load drop during strain control; see Fig. 10) must cause a gradual extension (ratchetting) in order to maintain these load limits. The degree of ratchetting, δ, in load-controlled tests will depend on the imposed R value, the cyclic stress-strain properties, and the cyclic displacement term δ_p, which is now not quite reversed in each cycle.

Experimental Observations

Figure 20 illustrates the deformation behavior for an austenitic and two ferritic steels under load control at $R = -1$ with a 1/2 h tension dwell in some high temperature tests at 550°C at the author's laboratory. Both ratchetting displacement δ and current displacement range Δ are expressed as a fraction of the initial displacement range Δ_i and plotted against normalized crack depth.

Figure 21 illustrates an extreme case of ratchetting which occurred suddenly, with little warning either from the displacement or accompanying potential drop crack monitoring records. It is clear that by this stage ($\delta/\Delta_i > 1$) the crack can never close and $q_o = 1$. Thus, in contrast to

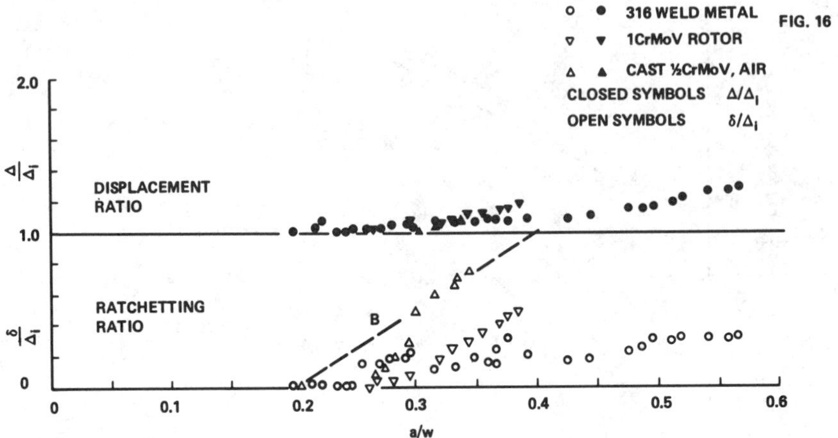

FIG. 20—*Ratchetting and increased displacement range occurring in load control tests (±90 MPa for weld and cast steel, ±140 MPa for rotor steel).*

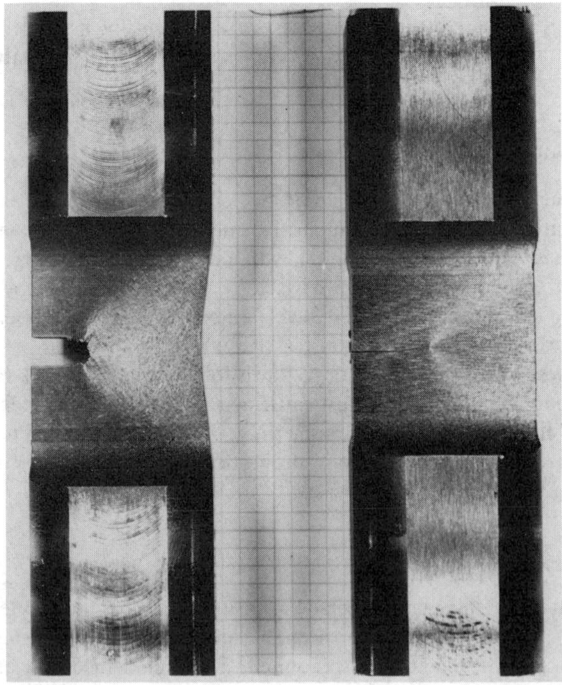

FIG. 21—*Gross ratchetting in solution treated 316 steel of 19 mm original gage length, compare notch opening displacement with specimen at right, which has not ratchetted. Ratchetting displacement has exceeded applied displacement range. Load controlled test in tension-compression at 625°C.*

the strain-controlled test, where q_o decreases with a/W (see, for example, Eq 10 and Fig. 13), in a load-controlled test q_o increases with a/W. From the ratchetting information given in Fig. 20 (e.g., the upper bound line B for cast 1/2Cr-Mo-V steel and hysteresis loops where available), it is possible to arrive at the relation

$$q_o = 0.7 + 1.5\left(\frac{a}{W} - 0.2\right) \quad (19)$$

for $0.2 < a/W < 0.4$, similar expressions applying for the other steels.

Comparison with Strain-Controlled Data

These considerations of q_o have been used in the comparison with strain-controlled data for weld material already presented in Fig. 16. For a given stress intensity, growth rates per cycle are similar in the two types of test. In previous work on solution-treated 316 steel at 625°C it was stated [*10*] that poor correlation existed between LCF and LEFM growth rates. This arose because the latter data were plotted, as is customary, using just the tension part of the load range. A reassessment in terms of q_o gives a new correlation shown in Fig. 22. In the presence of limited ratchetting, the load-control results lie within the scatterband of the data.

The 1/2 h tension dwell introduced in many of these tests produced transgranular cracking in the 1Cr-Mo-V rotor steel, but intergranular cracking in the 1/2Cr-Mo-V and 316 steels and correlation in LCF and LEFM growth rates was achieved for both types of crack path. It is not certain whether this agreement would be maintained for prolonged dwells. However, it is clear from Fig. 21 that, unless the service situation dictates otherwise, strain-controlled crack growth tests are preferable to load-controlled tests.

Discussion

Choice of Test

This review has discussed fatigue crack growth rates measured under strain control in a laboratory push-pull specimen which is small compared with the thickness of the component. Load control tests at high temperature are to be avoided because of undesired ratchetting, but it may be noted that if the yield stress of 316 steel at 625°C is assumed to be 160 MPa, the critical depth of 7.2 mm observed in Fig. 21 can be predicted from an analysis by Haigh and Richards [*39*] of notional stress intensities at general yield.

Bending tests may be used alternatively in the LCF regime to determine crack growth rates. This method was chosen by Ermi et al. [*40*] on Alloy 800 in the range 650 to 750°C, and their computed growth rates compare well with other data determined conventionally [*6*].

Sometimes the specimen is actually *thicker* than the component, an example being 6.5 mm diameter push-pull specimens representing the 0.5 mm wall of fuel element cladding [*4,25*]. In this case, the experiment should terminate at 0.5 mm crack depth (i.e., $a/W = 0.08$), which occurs during the accelerating period of growth. In an actual simulation of component behavior using 0.5 mm thick bend specimens [*41,42*] a 20% load drop was observed at 0.15 mm crack depth (i.e., $a/W = 0.3$). Thicker specimens thus give a lower bound endurance in this example because the decelerating period of growth cannot occur within the specified distance.

It may be noted that, whatever specimen thickness is chosen, the first visible load drop occurs when the crack has consumed about 1.5% of the cross-sectional area [*3*].

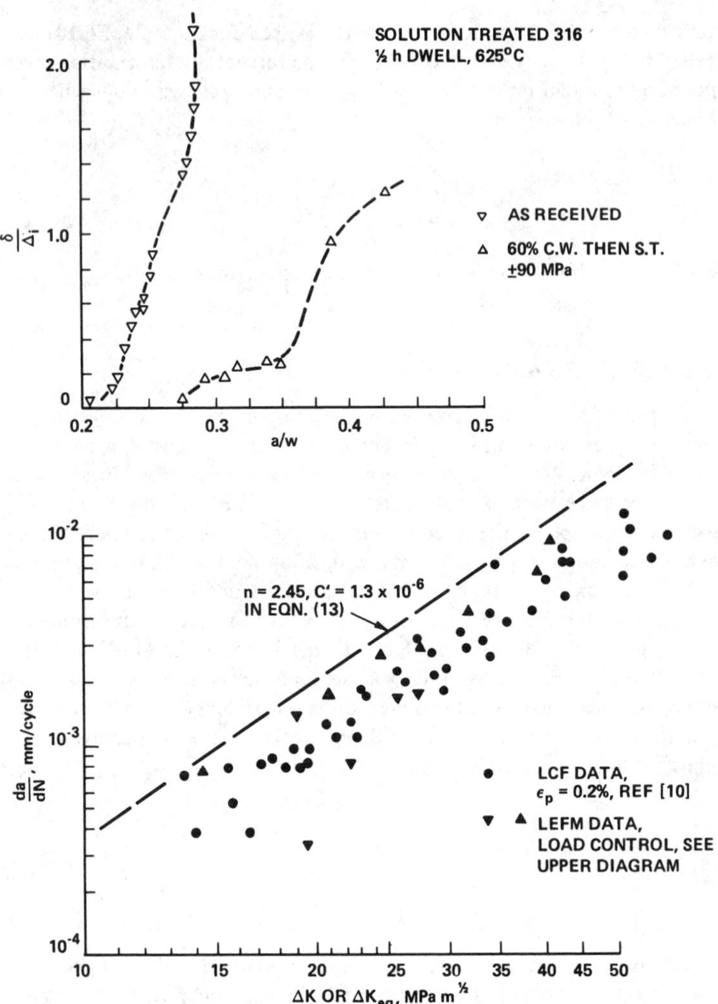

FIG. 22—*Crack growth rates in LEFM regime in presence of ratchetting* (upper diagram) *may be correlated with LCF data* (lower diagram).

Assessment Route

There is at present no comprehensive procedure for dealing with the growth of short cracks in components undergoing cyclic loading conditions at elevated temperature. Code Cases such as Division I of ASME III (N-47) [*43*] are concerned only with determining the number of creep-fatigue cycles that will avoid initiation at the location of interest. There is much incentive to extend the life of existing plants, and this can be made possible if the growth of cracks discovered in service is well characterized for future operating cycles.

It is now felt that there are enough high temperature crack growth data for coordinating into an Assessment Procedure. The Remaining Life path of this procedure enables the lifetime of a cracked component to be determined (for an assumed initial defect size and an acceptable final size) by integration of equations such as Eq 2 or Eq 13. The Tolerable Defect path expresses the integrated growth alternatively (viz., What will be the defect size after further specified service cycles, when the initial defect size is known?) (The same question may be asked in reverse: What *initial* defect size can be tolerated given the postulated service cycles that will grow it to an acceptable depth? This method is frequently used at the design stage [*44*].)

The route may be entered immediately for a known defect size or, alternatively, it may be entered after the traditional "initiation" cycles have been exhausted and which have been calculated by other means. Time may be preferable to cycles in a life assessment, since it can be associated with a repeatable group of service loadings of different magnitudes (start up/shut down etc.). If a strain gradient occurs beneath the component surface, a better estimate of initial growth rate is possible by replacing Eq 2, which is an upper bound, with Eq 3 and allowing ϵ_p at the surface to decrease to zero at the edge of the plastic zone X (Fig. 1).

With a knowledge of representative stress distributions in the component and also of surface strain conditions, lifetime may be estimated by integration of the crack growth laws. When the crack in the component leaves the LCF and enters the LEFM regime, the arguments of this review suggest that analysis be performed in terms of ΔJ (e.g., as was done for the specimens (Eq 15)). If calculation of this parameter cannot be performed easily for the component, the contributions to growth inside and outside the plastic zone X of Fig. 1 must be assessed separately, at the same time demonstrating that there is a smooth transition in cyclic crack growth rates across the edge of the zone.

Conclusions

Cyclic growth relations for short fatigue cracks (and their integrated forms) have up to now been applied solely to the laboratory specimen (e.g., in verifying the stages leading to failure). There is much merit in expressing LCF crack growth rates in terms of parameters such as ΔK_{eq} or ΔJ, which are independent of crack depth, since they can be linked with growth rates measured in the LEFM regime. However, where calculation of these parameters proves difficult in service, short crack behavior may be predicted via Eq 2:

$$\frac{da}{dN} = Ba^Q$$

for the same relative crack depth in the component.

Acknowledgments

The author is indebted to Dr. R. Hales for free use of his hysteresis loop measurements. The work was carried out at the Central Electricity Research Laboratories, and the paper is published by permission of the Central Electricity Generating Board.

References

[*1*] Coffin, L. F. in *Fatigue at Elevated Temperatures, ASTM STP 520,* American Society for Testing and Materials, Philadelphia, 1973, pp. 5–34.

[*2*] Coffin, L. F., *Proceedings of the Institution of Mechanical Engineers,* Vol. 188, September, 1974, pp. 109–127.

[*3*] Raynor, D. and Skelton, R. P. in *Techniques for High Temperature Fatigue Testing,* Elsevier Applied Science Publishers, London, 1985, pp. 143–166.

[*4*] Skelton, R. P., *Transactions of Indian Institute of Metals,* Vol. 35, 1982, pp. 519–535.

[*5*] Skelton, R. P. in *Low Cycle Fatigue and Life Prediction, ASTM STP 770,* American Society for Testing and Materials, Philadelphia, 1982, pp. 337–381.

[*6*] Skelton, R. P. in *Fatigue at High Temperature,* Applied Science Publishers, London, 1983, pp. 1–62.

[7] Wareing, J., *Metallurgical Transactions,* Vol. 6A, 1975, pp. 1367-1377.
[8] Tomkins, B., *Philosophical Magazine,* Vol. 18, 1968, pp. 1041-1066.
[9] Nilsson, J. O., *Fatigue of Engineering Materials and Structures,* Vol. 7, 1984, pp. 55-64.
[10] Skelton, R. P. in *Mechanical Behaviour and Nuclear Applications of Stainless Steel at Elevated Temperatures,* Book 280, The Metals Society, London, 1982, pp. 129-135.
[11] Skelton, R. P. in *Time and Load Dependent Degradation of Pressure Boundary Materials,* IAEA Meeting, Innsbruck, Report IWG-RRPC-79/2, 1979, pp. 73-87.
[12] Skelton, R. P. and Challenger, K. D., *Materials Science and Engineering,* Vol. 65, 1984, pp. 271-281.
[13] Pineau, A. in *Fatigue at High Temperature,* Applied Science Publishers, London, 1983, pp. 305-364.
[14] Bressers, J., De Cat, R., and Fenske, E., "Crack Initiation and Growth in High Temperature Low Cycle Fatigue," Commission of European Communities, Luxemburg, Report EUR 8808, 1984.
[15] Yamaguchi, K. and Kanazawa, K., *Transactions of the National Research Institute for Metals,* Japan, Vol. 25, 1983, pp. 143-147.
[16] Solomon, H. D. and Coffin, L. F. in *Fatigue at Elevated Temperatures, ASTM STP 520,* American Society for Testing and Materials, Philadelphia, 1973, pp. 112-121.
[17] Okazaki, M. and Koizumi, T., *Metallurgical Transactions,* Vol. 14A, 1983, pp. 1641-1648.
[18] Tingshi, Z. and Yuanhan, W., International Conference, *Fatigue and Fatigue Thresholds,* C. J. Beevers, Ed., Engineering Materials Advisory Services Ltd., Cradley Heath, U.K., 1984, pp. 625-636.
[19] Usami, S., Fukuda, Y., and Shida, S., Paper 83-PVP-97, American Society of Mechanical Engineers, New York, 1983, pp. 1-12.
[20] Gangloff, R. P., *Fatigue of Engineering Materials and Structures,* Vol. 4, 1981, pp. 15-33.
[21] Wareing, J. and Vaughan, H. G., *Metal Science,* Vol. 13, 1979, pp. 1-8.
[22] Wareing, J. in *Fatigue at High Temperature,* Applied Science Publishers, London, 1983, pp. 135-185.
[23] Solomon, H. D., *Metallurgical Transactions,* Vol. 4, 1973, pp. 341-347.
[24] Okazaki, M., Hattori, I., Shiraiwa, F., and Koizumi, T., *Metallurgical Transactions,* Vol. 14A, 1983, pp. 1649-1659.
[25] Wareing, J., Tomkins, B., and Sumner, G. in *Fatigue at Elevated Temperatures, ASTM STP 520,* American Society for Testing and Materials, Philadelphia, 1973, pp. 123-138.
[26] Rie, K. T. and Lachmann, E., *Zeitschrift für Werkstofftech.,* Vol. 13, 1982, pp. 244-253.
[27] Lachmann, E. and Rie, K. T., International Conference, *Fatigue and Fatigue Thresholds,* C. J. Beevers, Ed., Engineering Materials Advisory Services Ltd., Cradley Heath, U.K., 1984, pp. 1281-1290.
[28] Rao, K. B. S., Valsan, M., Sandhya, R., Ray, S. K., Mannan, S. L., and Rodriguez, P., *International Journal of Fatigue,* Vol. 7, 1985, pp. 141-147.
[29] Starkey, M. S. and Skelton, R. P., *Fatigue of Engineering Materials and Structures,* Vol. 5, 1982, pp. 329-341.
[30] Muscati, A., *Materials, Fracture and Fatigue, 4th International Conference on Pressure Vessel Technology,* Institution of Mechanical Engineers, London, Vol. 1, 1981, pp. 169-173.
[31] Haigh, J. R. and Skelton, R. P., *Materials Science and Engineering,* Vol. 36, 1978, pp. 133-137.
[32] Shih, C. F. and Hutchinson, J. W., *Transactions ASME Series H: Journal of Engineering Materials Technology,* Vol. 98, 1976, pp. 289-295.
[33] Reger, M. and Rémy, L., *Fracture and the Role of Microstructure, Proceedings of the 4th European Conference on Fracture,* Engineering Materials Advisory Services Ltd., Cradley Heath, U.K., Vol. 2, 1982, pp. 531-538.
[34] Reger, M., Soniak, F., and Rémy, L., International Conference, *Fatigue and Fatigue Thresholds,* C. J. Beevers, Ed., Engineering Materials Advisory Services Ltd., Cradley Heath, U.K., 1984, pp. 797-806.
[35] Ohtani, R., *Engineering Aspects of Creep,* Institution of Mechanical Engineers, London, Vol. 2, 1980, pp. 17-22.
[36] Rooke, D. P. and Cartwright, D. J., *Stress Intensity Factors,* Her Majesty's Stationery Office, London, 1976.
[37] Athanassiadis, A., Boissenot, J. M., Brevet, P., Francois, D., and Raharinaivo, A., *International Journal of Fracture,* Vol. 17, 1981, pp. 553-566.
[38] Dowling, N. E. in *Cracks and Fracture, ASTM STP 601,* American Society for Testing and Materials, Philadelphia, 1976, pp. 19-32.
[39] Haigh, J. R. and Richards, C. E., International Conference, *Creep and Fatigue in Elevated Temperature Applications,* Institution of Mechanical Engineers, London, 1974, pp. 159.1-159.9.
[40] Ermi, A. M., Nahm, H., and Moteff, J., *Materials Science and Engineering,* Vol. 30, 1977, pp. 41-48.
[41] Broomfield, G. H. and Hutchins, E. E. C., *Metals and Materials,* Vol. 6, 1972, pp. 150-154.
[42] Broomfield, G. H., Gravenor, J., Moffat, J., and Hutchins, E. E. C., Proceedings Conference, *Irradiation Embrittlement and Creep in Fuel Cladding and Core Components,* British Nuclear Energy Society, London, pp. 131-142.

[43] ASME Boiler and Pressure Vessel Code, Case N47-17, Class 1 Components in Elevated Temperature Service, Section III Division 1, American Society of Mechanical Engineers, New York, 1979, pp. 90–182.
[44] Holdsworth, S. in *High Temperature Alloys for Gas Turbines*, Applied Science Publishers, London, 1978, pp. 549–572.

DISCUSSION

J. Polák[1] *(written discussion)*—Are your cracks short enough that you cannot use the fracture mechanics parameters ΔK or ΔJ to correlate their growth rates?

R. P. Skelton (author's closure)—For high-strength materials, fracture mechanics parameters can indeed be used in the depth range discussed in this review (0.2 to 3 mm in the specimen itself). But for low-strength materials where reversed plasticity is required in each cycle to cause the cracks to grow, one has a choice. Either the parameters are converted to an equivalent stress intensity, ΔJ or whatever, or the crack growth rates may be applied directly to the component (within a specified depth) for the corresponding value of total or plastic strain.

[1]Institute of Physical Metallurgy, Czechoslovak Academy of Sciences, Brno, Czechoslovakia.

A. Toshimitsu Yokobori,[1] *Takeo Yokobori,*[2] *and Takashi Kuriyama*[2]

Life of Crack Initiation, Propagation, and Final Fracture under High-Temperature Creep, Fatigue, and Creep-Fatigue Multiplication Conditions

REFERENCE: Yokobori, A. T., Yokobori, T., and Kuriyama, T., **"Life of Crack Initiation, Propagation, and Final Fracture under High-Temperature Creep, Fatigue, and Creep-Fatigue Multiplication Conditions,"** *Low Cycle Fatigue, ASTM STP 942,* H. D. Solomon, G. R. Halford, L. R. Kaisand, and B. N. Leis, Eds., American Society for Testing and Materials, Philadelphia, 1988, pp. 236–256.

ABSTRACT: Studies were carried out on the crack initiation and propagation laws under high-temperature creep, fatigue, and creep-fatigue multiplication conditions for notched specimens of SUS 304 stainless steel.

For crack initiation life, it was found that relative notch opening displacement (RNOD) can be used. Furthermore, the remaining part of crack initiation life can be predicted by RNOD criterion; that is, by recording RNOD with respect to the suitable time within the service term and extrapolating the data with a straight line until the line intersects the critical value of RNOD.

The crack propagation life was expressed by a thermally activated type equation in explicit terms of independent parameters such as applied stress and temperature. The equation was approximately expressed in terms of the time-temperature equivalent parameter $T(\ell n\, t_p + C)$, where T = absolute temperature, t_p = crack propagation life, and C = a nearly constant numerical factor. The experimental data are well expressed by this parameter.

KEY WORDS: stainless steel, cumulative damage law, thermally activated process, crack propagation life equation, relative notch opening displacement, crack initiation life equation, evaluation, remaining life

There have been many studies of the fatigue crack propagation law. What is its purpose? The assessment of fatigue life? If this is so, then in order to obtain the fatigue life law, it may become necessary to obtain the law of the fatigue crack initiation life, t_i, and that of the fatigue crack propagation life, t_p, separately. It is especially essential when t_i is not negligible as compared with t_p. However, there have been very few studies [*1*] of this problem. Thus we have conducted studies on the crack initiation and propagation laws under high-temperature creep, fatigue, and creep-fatigue multiplication conditions for notched specimens of SUS 304 stainless steel.

For crack initiation life, it was found that relative notch opening displacement (RNOD) can be used. Furthermore, the remaining part of the crack initiation life can be predicted by RNOD criterion; that is, by recording RNOD with respect to a suitable time within the service term and extrapolating the data with a straight line until the line intersects the critical value of RNOD.

The crack propagation life was expressed by a thermally activated type equation in explicit

[1]Associate Professor, Department of Mechanical Engineering II, Tohoku University, Aoba, Sendai, Japan.

[2]Professor Emeritus and Graduate, respectively, Department of Mechanical Engineering II, Tohoku University, Aoba, Sendai, Japan.

terms of independent parameters such as applied stress and temperature. The equation was approximately expressed in terms of the time-temperature equivalent parameter

$$T(\ell n\, t_p + C) \quad (1)$$

where

T = absolute temperature,
t_p = crack propagation life, and
C = a nearly constant numerical factor.

The experimental data are well expressed by this parameter.

Materials, Experimental Apparatus, and Procedure

The material used was 304 stainless steel that was solution treated at 1050°C. The chemical composition and mechanical properties at room temperature are given in Tables 1 and 2, respectively. Double edge notched (DEN) specimens and center notched specimens (CN) were used (Figs. 1*a* and 1*b*). A high-temperature creep-fatigue testing machine designed by our laboratory was used. Tests were carried out in a vacuum of 1.3 mPa. Crack detection and the measurement of crack length were monitored continuously during the tests (without stopping the tests) by using a high-temperature microscope (100× magnification) through a peeping window. Specimen temperatures were continuously monitored with a thermocouple spot-welded in the vicinity of the notch of the specimen surface. The load waves were controlled as shown in Figs. 2*a*, 2*b*, and 2*c* for fatigue, creep, and creep-fatigue interaction tests, respectively. Thus, by changing the hold time t_H, the loading wave (Fig. 2*c*), a combination of the fatigue loading wave (Fig. 2*a*) and the creep loading wave (Fig. 2*b*), can be obtained. The frequency of the load wave (Fig. 2*a*) is 40 cpm. The minimum stress, σ_{min}, was zero for all tests, and the gross section stress, σ_g, corresponding to the maximum value of tensile stress, σ_{max}, ranged from 177 to 205 MPa. Tests were carried out at 600, 650, and 700°C; the temperatures were kept constant within ±1°C. In the present article the crack initiation instant was taken as the time when the crack from the notch root with a length of 5 μm was observed, the averaged length over both sides being used.

TABLE 1—*Chemical composition (weight percent).*

C	Si	Mn	P	S	Cr	Ni
0.094	0.56	1.28	0.027	0.013	18.0	8.81

TABLE 2—*Mechanical properties (room temperature).*

Yield Stress, kg/mm^2	Ultimate Tensile Strength, kg/mm^2	Elongation, %
33.6	66.3	47

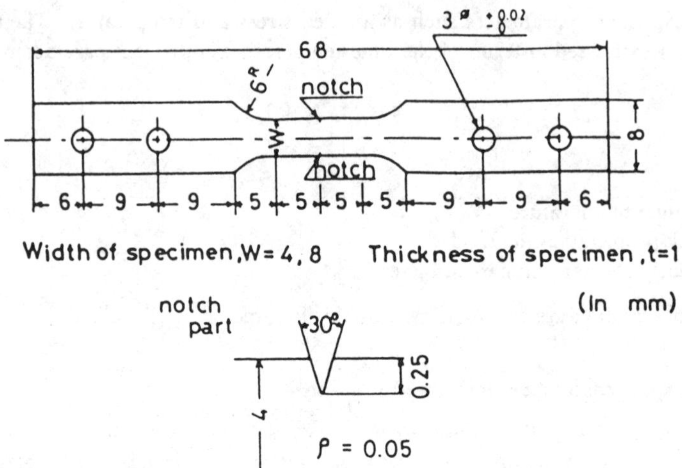

FIG. 1*a*—*Double edge notched (DEN) specimen.*

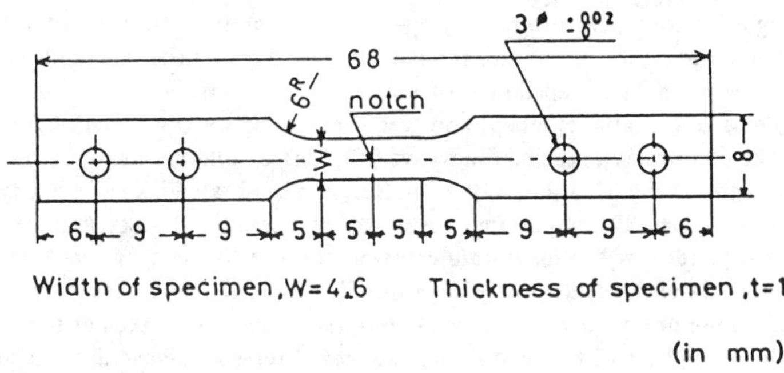

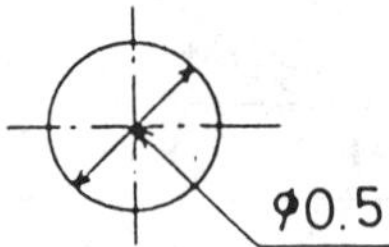

FIG. 1*b*—*Center notched (CN) specimen.*

Experimental Results

Crack Initiation Life

Figures 3*a* and 3*b* show the ratio of the crack initiation life, t_i, in terms of time, to the total fracture life, t_f, in terms of time, plotted against applied gross stress, σ_g, and temperature under the conditions concerned. It can be seen that t_i/t_f ranges from 30 to 50%; thus t_i cannot be neglected as compared with t_f and affects t_f characteristics.

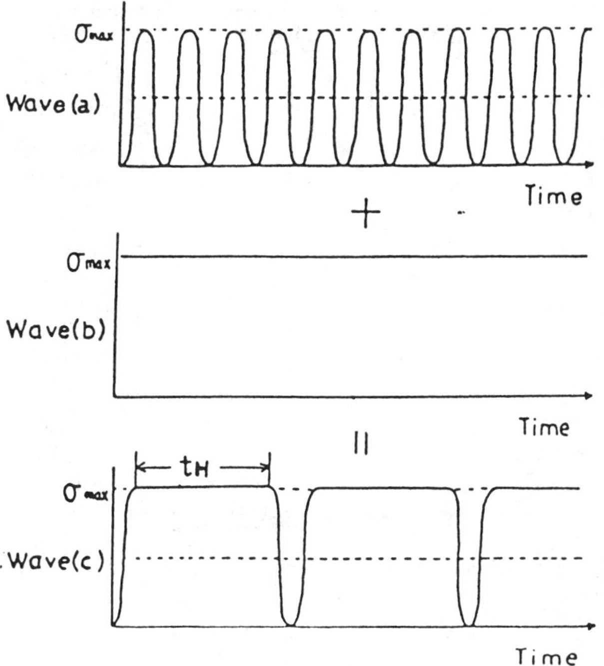

FIG. 2—*Stress wave;* t_H = *hold time.* (a) *Fatigue test.* (b) *Creep test.* (c) *Creep-fatigue interaction test.*

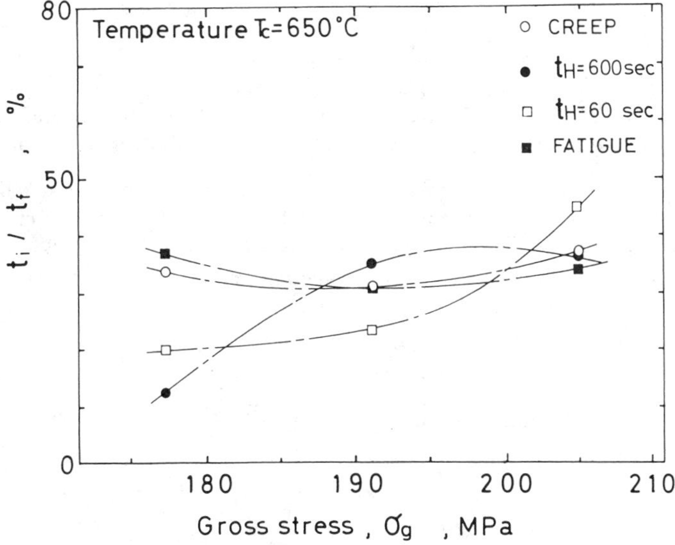

FIG. 3a—*Ratio of crack initiation life,* t_i, *in terms of time, to total life,* t_f, *in terms of time, versus gross stress,* σ_g.

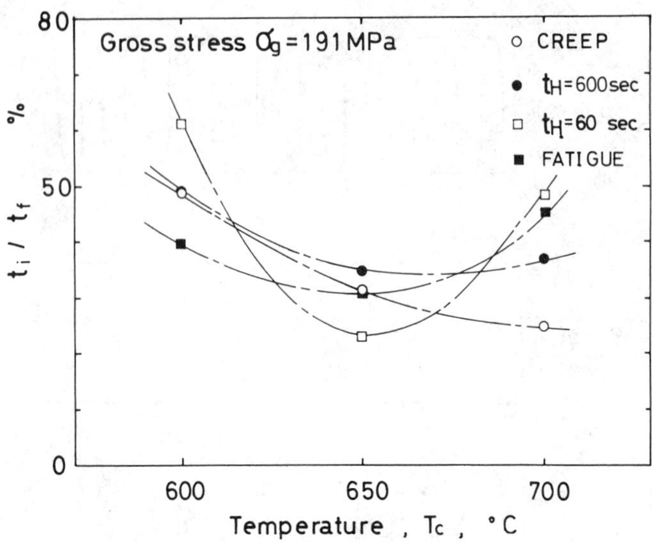

FIG. 3b—*Ratio of crack initiation life,* t_i, *in terms of time, to total life,* t_f, *in terms of time, versus temperature.*

Law of Crack Initiation Life

For criterion of crack initiation at high temperatures in notched specimens, a critical value of relative notch opening displacement was proposed [*2*]. The relative notch opening displacement (RNOD) is defined as (Fig. 4)

$$\text{RNOD} = \Delta\phi/\phi_0 = (\phi - \phi_0)/\phi_0 \tag{2}$$

where

ϕ = notch opening displacement,
ϕ_0 = initial value of notch opening (i.e., the value before applying the load), and
$\Delta\phi = \phi - \phi_0$ = increments of notch opening value ϕ from the instant of load application.

The crack initiation instant is taken as the time when the crack from the notch root with a length of 5 μm is observed. The data show that RNOD, $\Delta\phi_i/\phi_0$ at the instant of crack initiation, takes a nearly constant value $(\text{RNOD})_c$ that falls within the range 0.25 ~ 0.38 in common with high-temperature creep, fatigue, and creep-fatigue multiplication conditions (Fig. 5) [*2*].

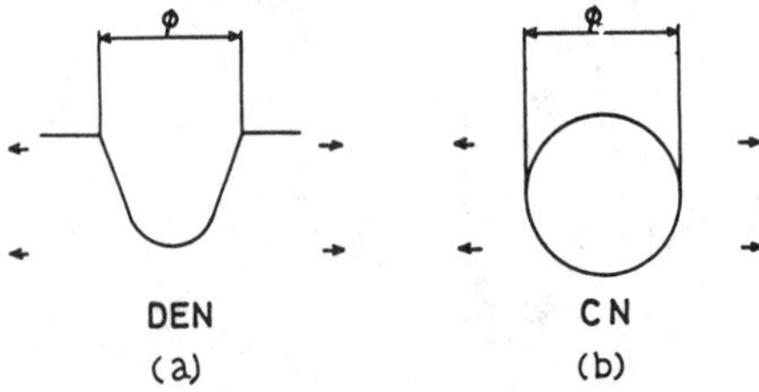

FIG. 4—*Notched opening displacement,* ϕ, *of* (a) *DEN specimen and* (b) *CN (circular hole) specimen.*

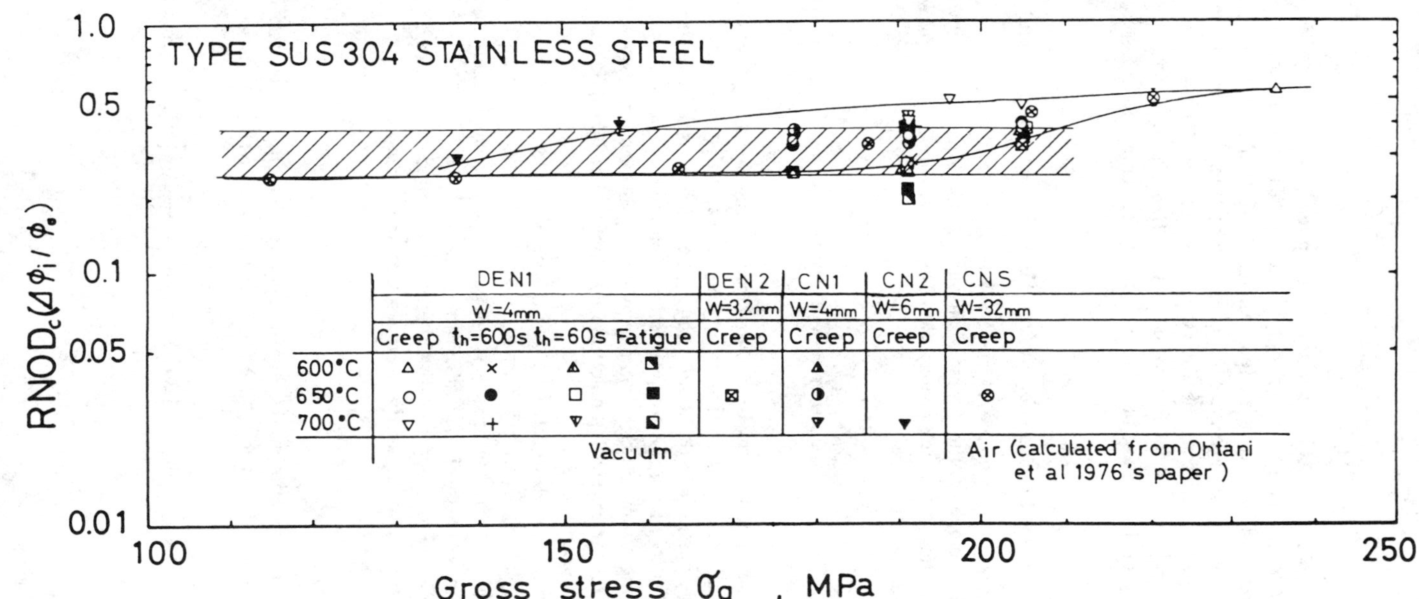

FIG. 5—*Relative notch opening displacement (RNOD)$_c$ at crack initiation versus gross stress, σ_g, for various conditions under creep, fatigue, and creep-fatigue interaction* [2], *where* W = *specimen width and CNS = slit-type center notch.*

Let us show that the (RNOD)$_c$ criterion can be used as the law of crack initiation at high temperatures. It was found experimentally [3] that RNOD increases linearly with time (Fig. 6):

$$\Delta\phi/\phi_0 = a^*t + b^* \tag{3}$$

where a^* and b^* are constants. It is important that, in spite of the variability of the coefficients a^* and b^* among the individual specimens, one and the same linear relation of Eq 3 holds for each specimen. Therefore, if we measure a few data of RNOD for the specimen concerned with respect to a suitable time within the service term beforehand, then by extrapolating the straight line connecting the data until the line intersects with the critical value line of RNOD, 0.25 ~ 0.38, as shown in Fig. 7, we can find the crack initiation life, t_i. Furthermore, the RNOD criterion can be used for assessment of the remaining life to crack initiation by using a very simple procedure.

Temperature Dependence of Crack Initiation Life, Crack Propagation Life, and Fracture Life

The inverse of the crack initiation life, t_i, the crack propagation life, t_p, and the total life, t_f, each in time, were plotted against the inverse of absolute temperature, T, as shown in Figs. 8*a*, 8*b*, and 8*c*. Figure 8*b* shows the crack propagation life, t_p, is in good agreement with the equation of a thermally activated process:

$$\ln\frac{1}{t_p} = C - \frac{Q}{RT} \tag{4}$$

where Q denotes an activation energy term. Furthermore, Fig. 8*b* shows that the slope of the straight line decreases; that is, activation energy decreases as the condition changes from a creep-dominant range to a fatigue-dominant range. The characteristics may correspond to the change of activation energy from the creep to fatigue condition in the crack growth (Table 3) [4,5], that is, the change from an activation energy for self-diffusion to dislocation movement [4,5]. Also note that for the case with a hold time, t_H, longer than 600 s, da/dt is expressed by the same formula proposed [4,5].

On the other hand, as can be seen from Fig. 8*a*, for fatigue-dominant cases log t_i does not show a straight line relationship with $1/T$ and the tangent increases with an increase of $1/T$. This shows that the crack initiation process is not expressed as a single rate-determining process, but suggests it may consist of successive rate processes [6].

The total fracture life, t_f, is in accord with a thermally activated process expression (Fig. 8*c*). This comes from its corresponding to the averaged feature of t_i and t_p.

Law of Crack Propagation Life

The crack growth rate equation under high-temperature creep, fatigue, and creep-fatigue multiplication conditions in the steady-state region has been derived [4,5] as

$$\frac{da}{dt} = B_t\sigma_g^m(\alpha\sqrt{a}\,\sigma_g)^n \exp\left[-\left\{\Delta f_1 - \Delta f_2 \ln\left(\frac{\alpha\sqrt{a}\,\sigma_g}{G\sqrt{b}}\right)\right\}\Big/RT\right] \quad \text{mm/h} \tag{5}$$

where

a_0 = notch length,
a^* = actual crack length,
$a = a_0 + a^*$,

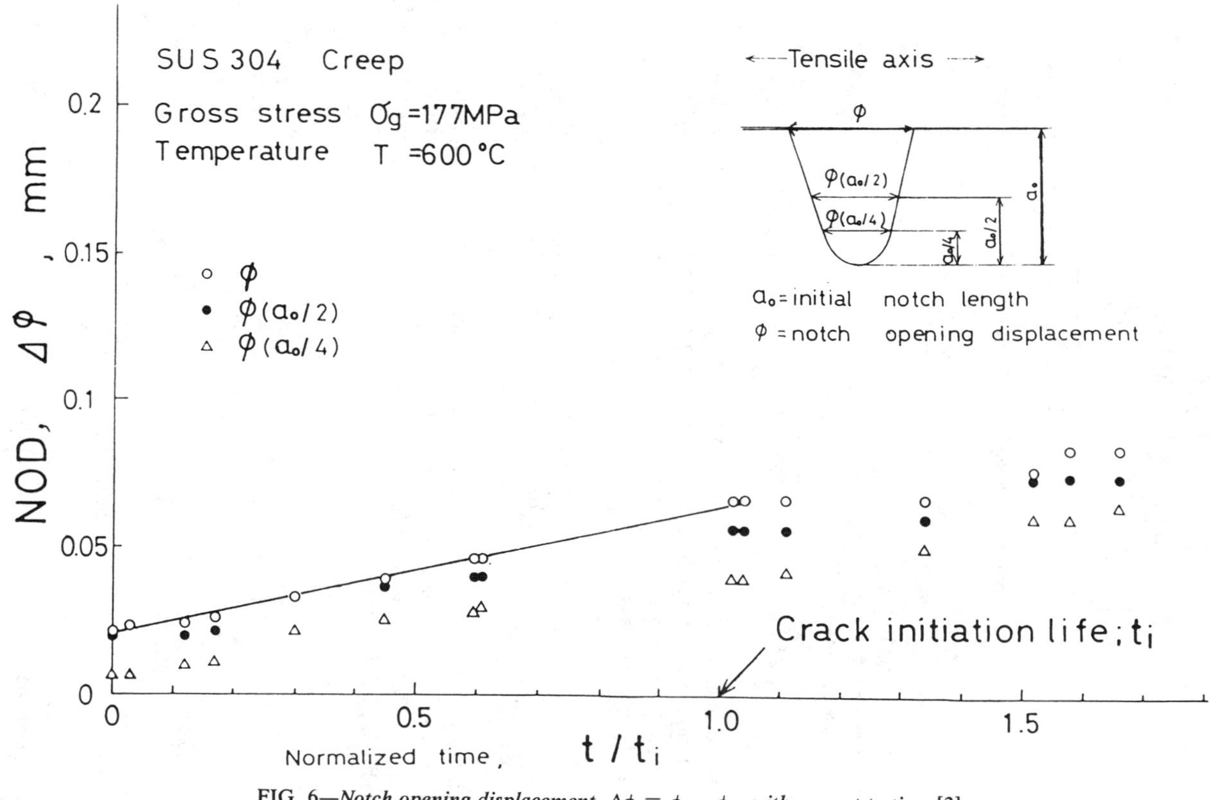

FIG. 6—*Notch opening displacement,* $\Delta\phi = \phi - \phi_0$, *with respect to time* [3].

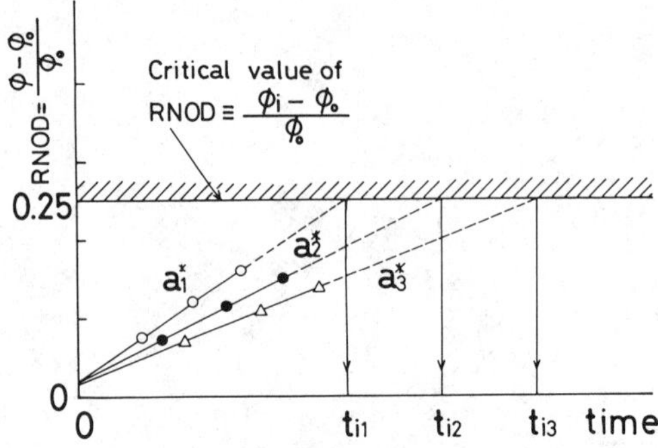

FIG. 7—*Schematic of method used to evaluate crack initiation life.*

G = modulus of rigidity,
b = Burgers vector,
R = gas constant, and

$$\alpha = 1.98 + 0.36(a/W) - 2.12(a/W)^2 + 3.42(a/W)^3 \tag{6}$$

where W is specimen width, and B_t, m, n, Δf_1, and Δf_2 are functions of t_H (Table 3).

These equations were derived [4–6] experimentally. The procedure for derivation is described in the literature [4–6]. Therefore no critical assumptions were employed.

Rewriting Eq 5, we obtain

$$\frac{da}{dt} = B_t(G\sqrt{b})^n e^{-\frac{\Delta f_1}{RT}} \sigma_g^m \left(\frac{\alpha\sqrt{a}\,\sigma_g}{G\sqrt{b}}\right)^{n+\frac{\Delta f_2}{RT}} \tag{7}$$

where a_i is the crack length at the instant of initiation of propagation, and a_f is the crack length at the final stage of propagation; then $a_i/a_f < 1$. Also, as can be seen from Table 3, $1/2(n + \Delta f_2/RT) > 1$. Therefore $(a_i/a_f)^{1/2(n+\Delta f_2/RT)} \ll 1$. Integrating Eq 3 and using the above condition, we obtain

$$T\left[\ln t_p + \ln\left\{\frac{1}{2}\left(n + \frac{\Delta f_2}{RT}\right) - 1\right\} + \ln\left(\frac{B_t\sigma_g^m(\alpha\sqrt{a_i}\sigma_g)^n}{a_i}\right)\right] = \frac{\Delta f_1 - \Delta f_2 \ln\left(\frac{\alpha\sqrt{a_i}\sigma_g}{G\sqrt{b}}\right)}{R} \tag{8}$$

Equation 8 reduces to

$$T[\ln t_p + C] = \frac{Q(\sigma_g, a_i)}{R} \tag{9}$$

where C denotes the sum of the second and third terms in the bracket on the left-hand side of Eq 4.

If C is not affected so much by σ_g, then the parameter $T(\ln t_p + C)$ becomes just like the Larson-Miller parameter [7] for fracture life of a smooth specimen. Equation 9 may be applied to each case for hold time t_H; the only difference is in the value of C and Q. For creep crack

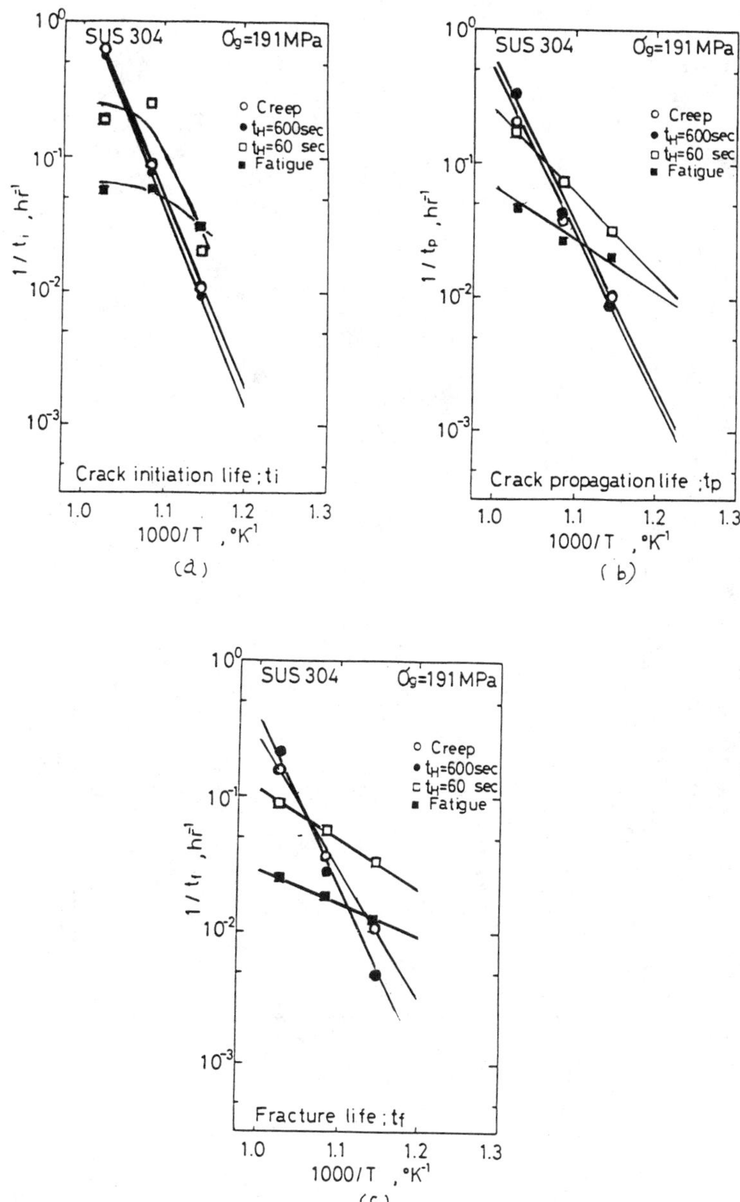

FIG. 8—*Temperature dependence of each life (DEN specimen).* (a) *Crack initiation life* (t_i). (b) *Crack propagation life* (t_p). (c) *Total life* (t_f).

propagation life, the present data and those in the literature [*8*] are shown in Fig. 9*b* versus the parameter $T(\log t_p + C)$, taking the value of C as 20. Figure 9*b* shows that t_p can be evaluated by using the parameter $T(\log t_p + 20)$, provided σ_g and T are known.

For the crack initiation life, t_i, Fig. 9*a* shows that our data under creep conditions are also in good accord with the $T(\log t_i + C)$ expression as inferred from Fig. 8*a*. The data in the litera-

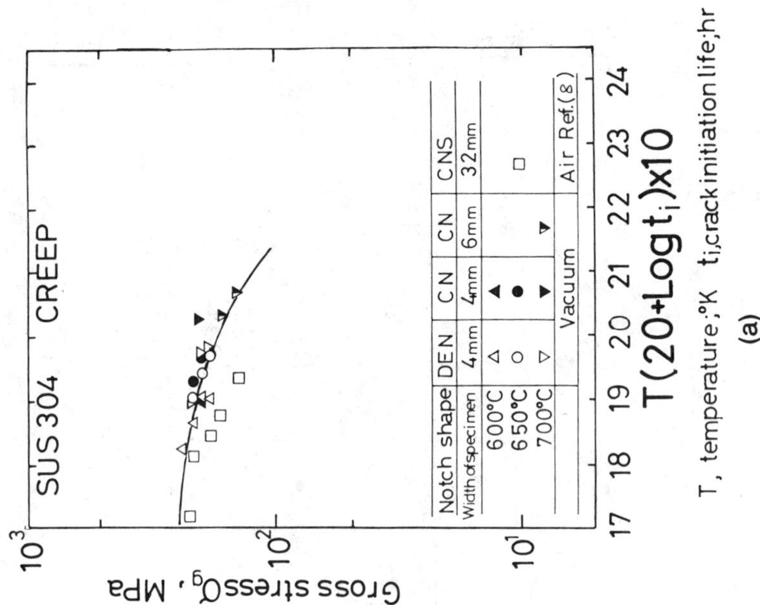
SUS 304 CREEP
Gross stress σ_g, MPa
10^3
10^2
10^1
17 18 19 20 21 22 23 24
$T(20+\log t_i)\times 10$
Notch shape | DEN | CN | CN | CNS
Width of specimen | 4mm | 4mm | 6mm | 32mm
600°C
650°C
700°C
Vacuum
Air Ref.(8)
T, temperature; °K t_i, crack initiation life; hr

(a)

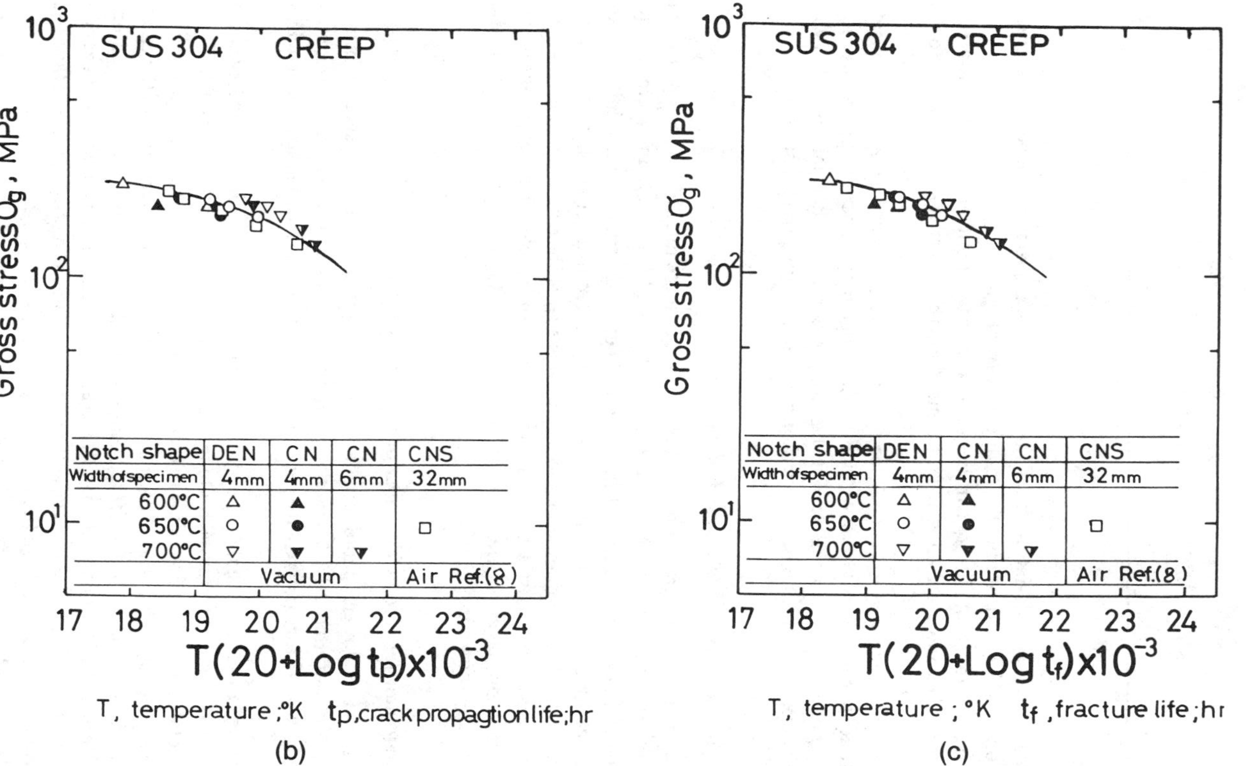

FIG. 9—*Representation by parameter* T(ln t + C). (a) *Initiation life* (t_i). (b) *Propagation life* (t_p). (c) *Fracture life* (t_f).

TABLE 3—*Values of activation energy* Δf_1 *and other constants in Eq 5.*

	B_t	m	n	Δf_1 (kJ/mol)	Δf_2 (kJ/mol)
Creep ($t_H \rightarrow \infty$)	6.52×10^2	4.14	$\simeq 0$	359	72.5
$t_H = 600$ s	3.09×10^3	3.79	$\simeq 0$	356	71.2
$t_H = 60$ s	1.90×10^{-7}	4.12	5.41	215	36.1
Fatigue ($t_H \rightarrow 0$)	6.23×10^{-3}	$\simeq 0$	4.62	71.4	7.67

ture [8] show some systematic deviation from the master curve, probably because the crack initiation definition and the measurements are different from ours.

As shown in Fig. 9*c* the total life t_f can be approximately predicted by the parameter $T(\log t_f + C)$, provided σ_g and T are known. This comes from the fact that t_f corresponds to the averaged value of $t_p + t_i$.

The range in which the $T(\log t_p + C)$ or $T(\log t_f + C)$ expression can be applied is considered wider than that shown in Figs. 9*b* or 9*c;* that is, the range may be better applied over the range $T(\log t + C)$ of 21×10^{-3}, because over this range the effect of thermal activation will become larger in connection with higher temperature and longer time (larger stress).

Life in Terms of Time as Affected by Stress Holding Time

The crack initiation life, t_i, the crack propagation life, t_p, and the total life, t_f, are plotted in terms of time versus stress holding time, t_H, as the ratio to each life t_{ic}, t_{pc}, and t_{fc} for creep, respectively, as shown in Figs. 10*a,* 10*b* and 10*c;* t_{ic} is defined as the time to crack initiation in creep measured since stress application and $t_{pc} = t_{fc} - t_{ic}$. From these figures, it can be seen that each life ratio in the fatigue-dominant condition, where t_H is short, increases with an increase of temperature and applied stress. This trend is more remarkable for the crack initiation life than for the crack propagation life. Figures 10*a,* 10*b* and 10*c* show the same systematic trends as schematically shown in Fig. 10*d*.

Cumulative Damage Law for Life in Time and in Repeated Cycles

For the prediction of life for creep and fatigue at high temperatures there are several formulae; linear [*9-11*] and non-linear [*12*] cumulative damage law and strainrange partitioning creep-fatigue analysis [*13*] for unnotched specimens and the prediction method for notched specimens [*14*].

The relation between the fractional life in time and the fractional life in repeated cycles is shown in Figs. 11*a,* 11*b,* and 11*c* for each life in a double logarithmic plot. The values t_i, t_p, and t_f denote the total sum $\Sigma\, t_H$ of hold time to crack initiation, during crack propagation, and to final fracture, respectively. The ratio of each life in time was taken as coordinate to the crack initiation life, t_{ic}, the crack propagation life, t_{pc}, and the total life, t_{fc}, under creep condition, respectively. The terms N_i, N_p, and N_f denote the crack initiation life, the crack propagation life, and the total life in terms of repeated cycles, respectively. The ratio of each life was taken as abscissa to the crack initiation life, N_{if}, the crack propagation life, N_{pf}, and the total life, N_{ff}, for fatigue only in terms of repeated cycles, respectively; N_{if} is defined as the repeated cycles when crack length becomes 5 μm in fatigue, N_{ff} is the cycles to fracture in fatigue, and $N_{pf} = N_{ff} - N_{if}$.

From Figs. 11*a,* 11*b,* and 11*c* it can be seen that in the crack initiation life the scatter is remarkable and that the data deviate considerably from the linear cumulative damage law,

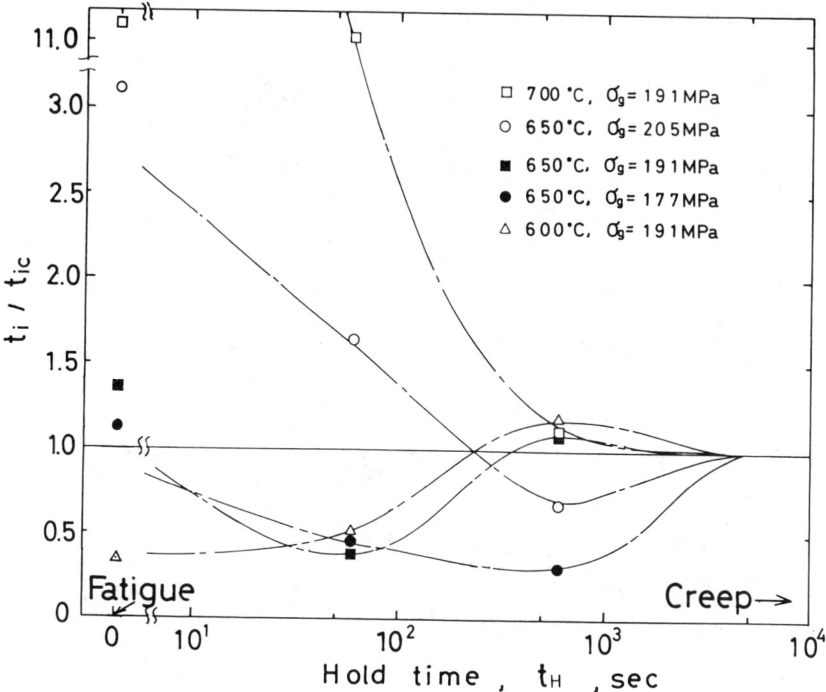

FIG. 10*a*—*Crack initiation life,* t_i, *versus hold time,* t_H *(DEN specimen);* t_{ic} *is under creep condition.*

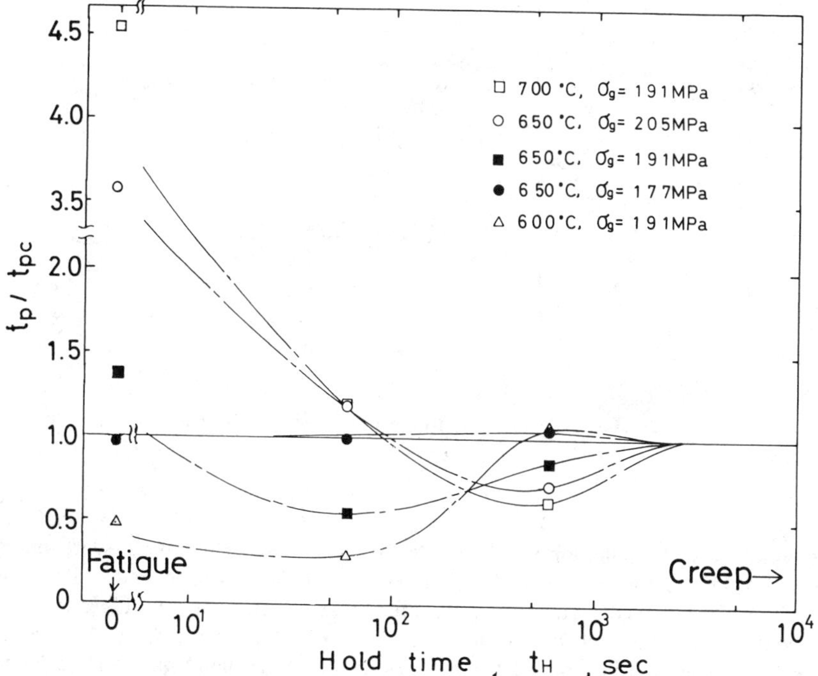

FIG. 10*b*—*Crack propagation life,* t_p, *versus hold time,* t_H *(DEN specimen);* t_{pc} *is under creep condition.*

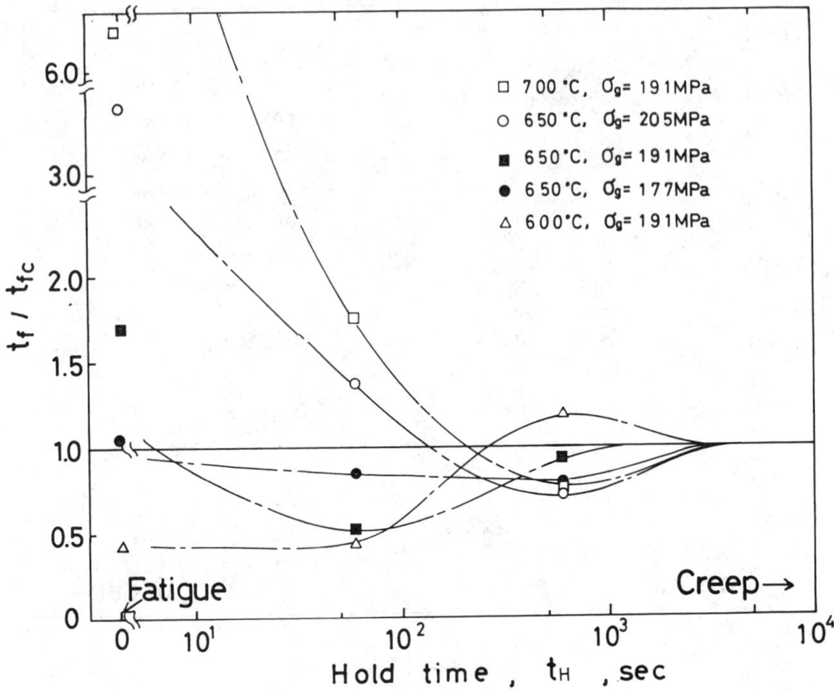

FIG. 10*c*—*Total life,* t_f*, versus hold time,* t_H *(DEN specimen);* t_{fc} *is under creep condition.*

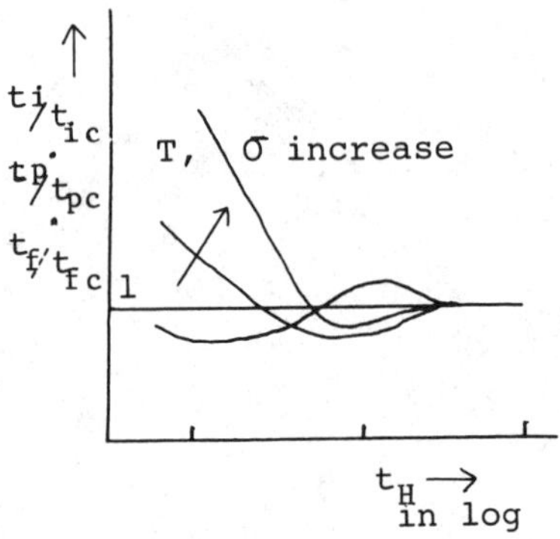

FIG. 10*d*—*Schematic illustration of life versus hold time.*

whereas in the crack propagation life the scatter is not so great. This may suggest that crack initiation life is more sensitive to creep-fatigue interaction than crack propagation life. Thus the total life shows the averaged behavior. It can be seen that Figs. 11*a*, 11*b*, and 11*c* show the same systematic trends as schematically shown in Fig. 11*d*. When t_H decreases, the damage due to the increase of cycles per unit time may increase and, on the other hand, the damage due to the creep component may decrease. The occurrence of the minimum in the damage diagram in

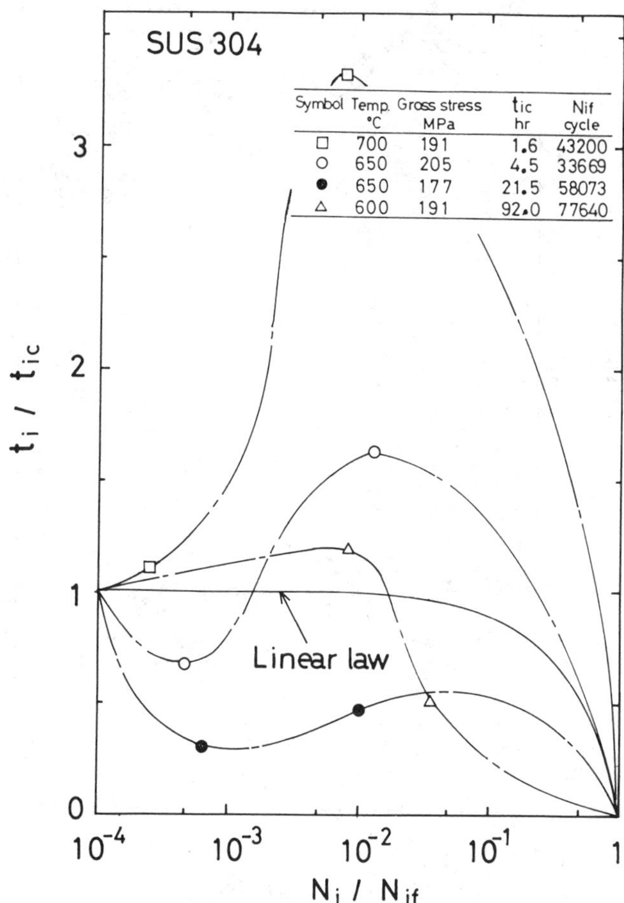

FIG. 11*a*—*Cumulative damage double logarithmic plot for life in time and in repeated cycles (DEN specimen): crack initiation life,* t_i *and* N_i.

Figs. 11*a*, 11*b*, and 11*c* may be in connection with such an interaction of the two effects. Similar characteristics as shown in Fig. 11*d* were observed for the case under combined stress corrosion cracking and corrosion fatigue conditions [*15*], and the subject may be a future study. These damage diagrams have been phenomenologically obtained in terms of a stochastic model [*16*]. In Fig. 12 the damage plot for the crack initiation life is shown linearly; the trend is similar to that shown in Fig. 11*a*.

Discussion

For the crack initiation life in creep-dominant cases, the process can be proved as expressed in terms of a single process thermally activated (Fig. 8*a*). Therefore for these cases we can derive the crack initiation life equation of the thermal activation type. However, in the present paper, we are concerned with the crack initiation life under creep, fatigue, and creep-fatigue multiplication conditions at high temperatures, and we showed that the RNOD criterion is very useful and effective.

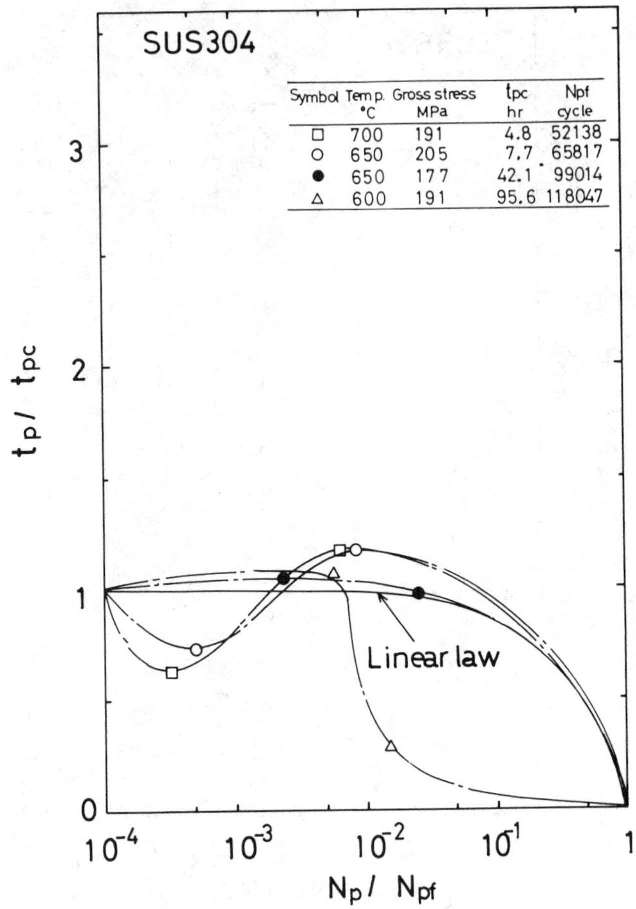

FIG. 11*b*—*Cumulative damage double logarithmic plot for life in time and in repeated cycles (DEN specimen): crack propagation life,* t_p *and* N_p.

Previously we experimentally derived [*4,5*] the equation of the crack propagation rate under high-temperature creep, fatigue, and creep-fatigue multiplication conditions and interpreted it in terms of a rate process. However, it is not *a priori* evident that the crack propagation life obtained by integrating this equation becomes a simple equation of a thermally activated process. The life equation was given in terms of such explicit parameters as applied stress and absolute temperature T; finally we approximately obtained the equivalent equation of t_p and T. On the other hand, Larson-Miller's parameter [*7*] is an experimental one; it is not derived from a thermally activated model and concerns only "smoothed specimens" and only "rupture life" in "creep". Therefore it is not *a priori* evident that this parameter should be valid in "crack propagation life" in "cracked or notched specimens" and, furthermore, "under fatigue, creep, and fatigue-creep multiplication conditions." Thus there is no inevitable reason why our equation or even parameter should be identical to the Larson-Miller parameter.

We also noted the effect of t_H on each life and cumulative damage law. However, the law for the crack initiation life and propagation life proposed above are not affected by clarifying the detailed mechanisms concerning these additional results observed.

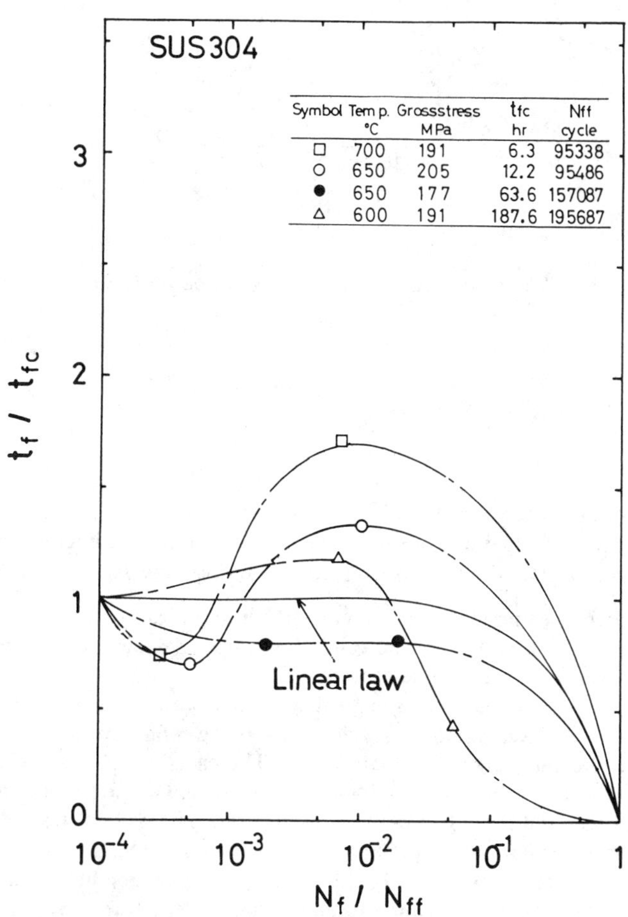

FIG. 11c—*Cumulative damage double logarithmic plot for life in time and in repeated cycles (DEN specimen): total life,* t_f *and* N_f.

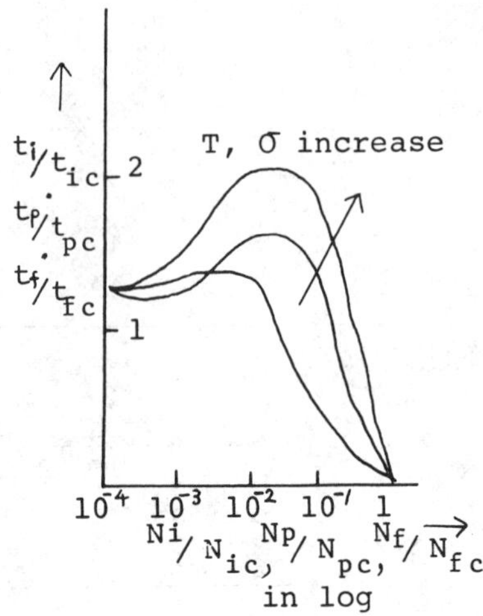

FIG. 11*d*—*Schematic illustration of life in time versus life in repeated number.*

Conclusions

From the experimental and analytical study on the life law of crack initiation and propagation, respectively, high temperature creep, fatigue, and creep-fatigue interaction conditions for notched specimens of SUS 304 stainless steel, the following conclusions were drawn:

1. Relative notch opening displacement (RNOD) criterion can be used as the law of crack initiation. Furthermore RNOD criterion can be used for assessment of the remaining life of crack initiation by using a very simple procedure.

2. The crack propagation life, t_p, is in good agreement with an equation derived by a thermally activated process. Activation energy decreases as the condition changes from the creep-dominant range into the fatigue-dominant range. The characteristics are similar to that of da/dt. On the other hand, the crack initiation life t_i obeys well a rate process equation only for creep-dominant cases and is not expressed in a single rate process for fatigue-dominant cases.

3. An equation for crack propagation life, t_p, has been derived. It was shown that applied gross stress, σ_g, absolute temperature, T, and the crack propagation life, t_p, are related to the parameter $T(\ell n\, t_p + C)$ which is similar to the Larson-Miller parameter.

4. The total life, t_f, can be evaluated by the parameter $T(\log t_f + C)$.

5. Crack initiation life, t_i, in terms of time occupies about 30 to 50% of total life, t_f, in terms of time, and thus cannot be neglected as compared with t_f; it also affects t_f characteristics.

6. The ratio of crack initiation life, t_i, crack propagation life, t_p, and total life, t_f, to each for creep in the fatigue dominant condition increases with an increase of temperature. This trend is more remarkable for the crack initiation life, t_i, than for the crack propagation life.

7. The relation between the fractional life in terms of time and that in terms of repeated cycles is plotted as predicting the cumulative damage law. In this plot, for crack initiation life the scatter is remarkable and the data deviates considerably from the linear cumulative damage law, whereas for crack propagation life the scatter is not so great.

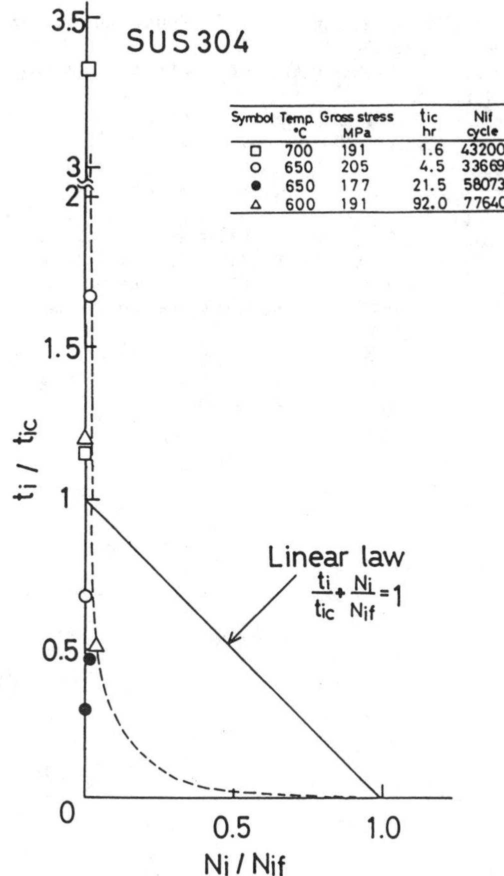

FIG. 12—*Cumulative damage linear plot for life in time and in repeated cycles for crack initiation (DEN specimen).*

References

[1] Yokobori, T., Kawasaki, T., and Horiguchi, M. in *Proceedings,* 3rd National Congress Fracture, Law Tartry, Czechoslovakia, 1976.

[2] Yokobori, A. T., Jr., Yokobori, T., Kuriyama, T., and Kako, T., "Advances in Fracture Research," in *Proceedings,* ICF 6, Vol. 3, Pergamon Press, Oxford, 1984, p. 2181.

[3] Yokobori, A. T., Jr., Kuriyama, T., Yokobori, T., and Kako, T., *Transactions of Japan Society of Mechanical Engineers A,* Vol. 52, No. 477, 1986, p. 1221 (in Japanese).

[4] Yokobori, A. T., Jr., Yokobori, T., Tomizawa, H., and Sakata, H., *Journal of Engineering Materials and Technology, Transactions of ASME,* Vol. 105, 1983, p. 13.

[5] Yokobori, T. and Yokobori, A. T., Jr., "Advances in Fracture Research," in *Proceedings,* ICF 6, Vol. 1, Pergamon Press, Oxford, 1984, p. 273.

[6] Yokobori, T., Yokobori, A. T., Jr., and Sakata, H. in *Proceedings,* ICF International Symposium on Fracture Mechanics, Beijing, 1983, p. 1025.

[7] Larson, F. R. and Miller, J., *Transactions of ASME,* Vol. 74, 1952, p. 765.

[8] Ohtani, R., Nitta, A., Nakamura, S., and Okuno, M., *Journal of the Society of Materials Science, Japan,* Vol. 25, 1976, p. 256 (in Japanese).

[9] Taira, S. in *Proceedings,* IUTAM Coll. on Creep in Structures, Springer, Berlin, 1962, p. 96.
[10] ASME Boiler and Pressure Vessel Code Case N-47.
[11] Lemaitre, J. and Plumtree, A., *Journal of Engineering Materials and Technology, Transactions of ASME,* Vol. 101, 1979, p. 285.
[12] Bui-Quoc, T. and Biron, A., *Nuclear Engineering and Design,* Vol. 71, 1982, p. 89.
[13] Manson, S. S. in *Fatigue at Elevated Temperatures, ASTM STP 520,* American Society for Testing and Materials, Philadelphia, 1973, p. 744.
[14] Sakane, M. and Ohnami, M., *Journal of Engineering Materials and Technology, Transactions of ASME,* Vol. 105, 1983, p. 75.
[15] Yokobori, A. T., Jr., Yokobori, T., Kosumi, T., Chiba, N., and Takasu, N., *Transactions of Japan Society of Mechanical Engineers A,* Vol. 51, No. 465, 1985, p. 1457 (in Japanese).
[16] Yokobori, T., Ichikawa, M., and Yokobori, A. T., Jr., *Journal of Japan Society for Structure and Fracture,* Vol. 9, No. 1, 1974, p. 13 (in Japanese); *Engineering Fracture Mechanics,* Vol. 7, 1975, p. 441.

K. Hatanaka[1] *and T. Fujimitsu*[1]

Growth of Small Cracks and an Evaluation of Low Cycle Fatigue Life

REFERENCE: Hatanaka, K. and Fujimitsu, T., "**Growth of Small Cracks and an Evaluation of Low Cycle Fatigue Life,**" *Low Cycle Fatigue, ASTM STP 942,* H. D. Solomon, G. R. Halford, L. R. Kaisand, and B. N. Leis, Eds., American Society for Testing and Materials, Philadelphia, 1988, pp. 257–280.

ABSTRACT: Smooth and through-thickness cracked specimens of JIS S35C medium carbon and SNCM 439 alloy steels, heat-treated in several conditions, were low cycle fatigued. The growth rates of small surface and through-thickness large cracks, da/dN, were analyzed in terms of the cyclic J-integral range, ΔJ, and the strain intensity factor range, $\Delta K\epsilon$.

The crack growth rates plotted against ΔJ and $\Delta K\epsilon$ are expressed by straight lines on a log-log diagram. The fairly large scatter inherent in a small crack problem hides some difference in the relations of da/dN versus ΔJ and the da/dN versus $\Delta K\epsilon$ plots between different materials. These relations in a small surface crack are slightly different from those in the through-thickness large crack.

The equations derived from combining these two relations and the cyclic stress-strain curve express well the fatigue life curves, of which slopes decrease in the lower plastic strain range in materials having a high strength and a low ductility, as well as the usual Coffin-Manson plot.

KEY WORDS: metallic materials, low cycle fatigue, crack growth, small surface crack, through-thickness large crack, cyclic J-integral, strain intensity factor, cyclic stress-strain curve, Coffin-Manson law, life prediction

Elastic-plastic fracture mechanics approaches have been vigorously attempted for evaluating a crack growth rate in the low-cycle fatigue range of various materials [*1–15*]. Consequently, the cyclic J integral, ΔJ has been proved to be a useful tool for evaluating the crack growth problem under cyclic deformation conditions accompanying a large-scale plastic strain [*2–15*].

Since a significant portion of the fatigue life is occupied by the growth of small cracks, specifically in the high-strain low-cycle region fatigue [*2,16*], some quantitative work was done to determine the small crack growth in low-cycle fatigue. The above cyclic J integral approach has been successfully extended from a large crack to a short one [*2,4–11*], and a strain intensity factor range, $\Delta K\epsilon$ has also been effectively applied to the short crack problem [*8,9,17,18*]. The appropriateness of these approaches should be studied in materials having a wide range of strength and ductility levels.

The validity of the Coffin-Manson law [*19,20*] has been extensively tested for many kinds of metals in low-cycle fatigue. Test results, however, show that the linear relation established between the logarithmic $\Delta\epsilon_p$ and the logarithmic N_f breaks at a certain plastic strain range, depending upon the kind of material, where $\Delta\epsilon_p$ is the plastic strain range and N_f is the number of cycles to failure. That is, the slope of the log $\Delta\epsilon_p$ versus log N_f plot decreases in the lower plastic strain range in some alloy steels [*19,21–25*], aluminum alloys [*22,25,26*], copper alloys [*27*], and titanium alloys [*19,28–30*]. Such a material dependence of the log $\Delta\epsilon_p$ versus log N_f relation may

[1]Professor and Senior Research Associate, respectively, Department of Mechanical Engineering, Yamaguchi University, Ube, 755, Japan.

be successfully analyzed in terms of the growth behavior of small cracks. Moreover, the physical meaning of the Coffin-Manson law will also be discussed through this approach.

Thus, the present work aims to examine the mechanical parameters controlling crack growth in low-cycle fatigue and to derive the log $\Delta\epsilon_p$ versus log N_f curve in terms of the small crack growth.

Materials and Experimental Procedure

The materials used were JIS S35C medium carbon and SNCM 439 alloy steels with the chemical compositions shown in Table 1. Tables 2 and 3 show the heat treatments performed for both steels and their static mechanical properties attained after these heat treatments. The materials have a wide range of strength and ductility levels. The grain sizes of ferrite in the normalized S35C steel and of prior austenite in the quenched and tempered SNCM 439 alloy steel were about 18 and 10 μm, respectively.

TABLE 1—*Chemical composition (weight percent) of materials used in this study.*

Material	C	Si	Mn	P	S	Cu	Cr	Ni	Mo
S35C	0.37	0.24	0.77	0.019	0.023	0.01	0.04	0.02	...
SNCM 439	0.38	0.25	0.78	0.019	0.014	0.04	0.80	1.77	0.24

TABLE 2—*Details of heat treatments.*

Material	Normalizing & Quenching	Tempering
S35C	Heated at 865°C for 30 min in air → air cooled	
SNCM 439(A)	Heated at 855°C for 30 min in air → air cooled and then Heated at 845°C for 30 min in air → oil quenched	(A) Heated at 700°C for 1 h in air → water quenched
(B)		(B) Heated at 635°C for 1 h in air → water quenched
(C)		(C) Heated at 550°C for 1 h in air → water quenched
(D)		(D) Heated at 400°C for 1 h in air → water quenched

TABLE 3—*Mechanical properties of materials.*

Material	Yield Stress, MPa	Ultimate Strength, MPa	Elongation, %	Reduction of Area, %	Fracture Ductility	Hardness Hv (300 g)
S35C	360[a]	614	33	57	0.85	181
SNCM 439(A)	675[b]	830	24	65	1.06	278
(B)	755[b]	877	23	67	1.11	315
(C)	965[b]	1066	20	60	0.91	370
(D)	1277[b]	1390	14	51	0.72	428

[a]Lower yield stress.
[b]0.2% proof stress.

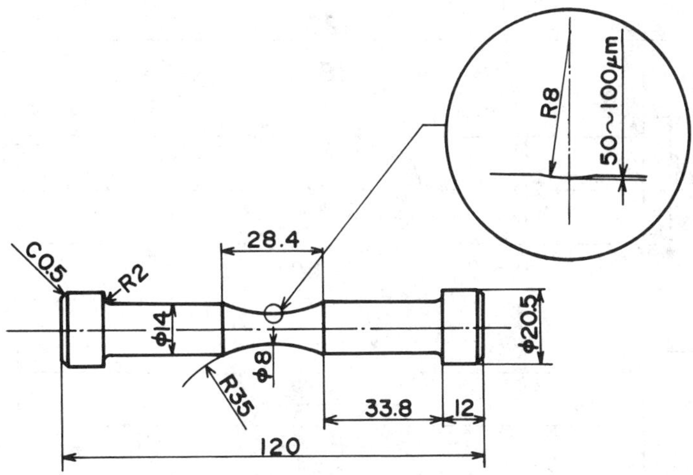

FIG. 1—*Schema of smooth specimen (dimensions in millimetres).*

Figure 1 shows the shape and size of the smooth test specimen. The shallow notch shown in the magnification was given so as to localize a crack initiation within a small area. The specimens were polished with emery paper and then electropolished before the test.

Through-thickness crack specimens are shown in Figs. 2*a* and 2*b*. In the single-edged crack specimen, a pre-fatigue crack of about 0.5 mm in size was introduced from the root of a V-shaped notch in load control mode. The V-shaped notch was then removed by machining. The center crack specimen was prepared by introducing the pre-fatigue crack of about $2a \simeq$ 1.5 mm from the small hole in the same way as the single-edged crack specimen.

The low-cycle push-pull fatigue tests were performed at a cycle frequency of 0.2 Hz at room temperature under axial and diametral strain-controlled conditions with a so-called strain ratio of

$$R_\epsilon = \epsilon_{max}/\epsilon_{min} = -1.0$$

The closed loop-type testing system was used. The axial is the test for through-thickness crack specimen and the diametral for the smooth specimen, respectively. The diametral strain was transformed into the axial one using the equation of

$$\epsilon_L^T = \frac{1}{0.5}(\epsilon_D^P) + \frac{1}{\nu}(\epsilon_D^E)$$

where ϵ_L^T, ϵ_D^P and ϵ_D^E are axial total, diametral plastic, and diametral elastic strains, and ν is a Poisson's ratio, which is determined as 0.303 in the present work. The crack sizes were measured with a travelling microscope in the through-thickness crack specimen and by examining microscopically the plastic replicas taken from the shallow notch area in the smooth specimen at proper intervals during fatigue tests.

The crack shape of the smooth specimen was assumed to be a semi-ellipse with ℓ and $2a$ for major and minor axes, respectively. The minor to major axis ratio, $2a/\ell$, was experimentally determined to be 0.76 [9]. The crack depth, a, was estimated by multiplying the measured surface crack length by one half of the above ratio.

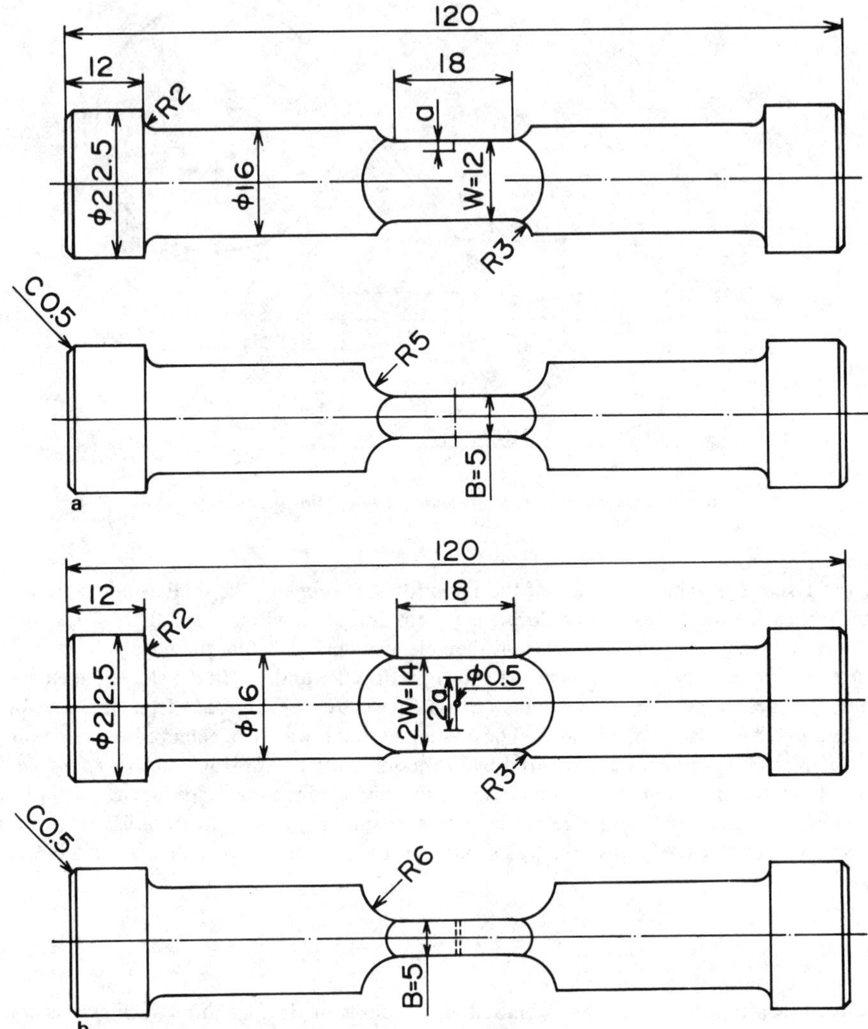

FIG. 2—*Schema of through-thickness crack specimens (dimensions in millimetres).* (a) *Single-edged crack specimen.* (b) *Center crack specimen.*

Analyses of Strain Intensity Factor Range and Cyclic *J*-Integral Range

The strain intensity factor range, ΔK_ϵ, and the cyclic J-integral range, ΔJ, were estimated by the following equations

$$\Delta K_\epsilon = \frac{M_K}{\phi(\lambda)} \Delta\epsilon_t \sqrt{\pi a} \tag{1}$$

$$\Delta J = \Delta J_e + \Delta J_p = 2\pi a \left\{\frac{M_K}{\phi(\lambda)}\right\}^2 \{\Delta W_e + f(n')\Delta W_p\} \tag{2}$$

where M_K is the numerical correction factor, $\phi(\lambda)$ a complete elliptical integral of the second kind involving $\lambda = a/\ell/2$ and was written as $\int_0^{\pi/2} (\cos^2 \xi + \lambda^2 \sin^2 \xi)^{1/2} d\xi$, where a is the crack depth; $\Delta\epsilon_t$ the total strain range; ΔJ_e and ΔJ_p the elastic and plastic components of ΔJ; ΔW_e and ΔW_p elastic and plastic components of the remote strain energy density [2]; and $f(n')$ is the function of the strain hardening exponent of the cyclic stress-strain curve, n', given as $(n' + 1)\{\pi n' + 3.85(1 - n')/\sqrt{n'}\}/2\pi$ [31]. The surface crack was assumed to be open even at the compressive tip of the stress-strain hysteresis loop [2,4,9].

The cyclic J-integral range, ΔJ, was estimated using the convenient method of Rice et al. [32] for through-thickness center crack of $a/w \geq 0.4$. Meanwhile, the strain intensity factors, ΔK_ϵ, were evaluated from the equations of $\Delta K_\epsilon = \Delta\epsilon_t\sqrt{\pi a}$ and $\Delta K_\epsilon = 1.12\Delta\epsilon_t\sqrt{\pi a}$ for the through-thickness center and the single-edged cracks of $a/w \leq 0.3$, respectively, where $\Delta\epsilon_t$ is the total strain range corresponding to the period when the crack tip is open during one strain cycle. The other detailed procedures for assessing ΔJ and ΔK_ϵ in through-thickness crack specimens are described in Ref *15*.

Test Results and Analyses of Fatigue Crack Growth Rate

Initiation and Growth Behaviors of Fatigue Cracks

In the normalized S35C medium carbon steel, slip produced intensely inside a ferrite grain causes the fatigue crack to nucleate at or near the grain boundary [33]. On the other hand, in the quenched and then tempered SNCM 439 alloy steel, the fatigue crack commonly originates at an inclusion accompanying very few slip lines, as shown in Fig. 3 [34]. Such a difference in the crack initiation behavior between both materials makes it easier to detect fatigue microcracks in the latter than in the former.

Figures 4*a* and 4*b* show the fatigue crack growth curves that were measured on the specimen surface in the normalized S35C and the quenched and tempered SNCM 439(C) steels. The crack growth rate was defined as the slope of the straight line connecting two neighboring

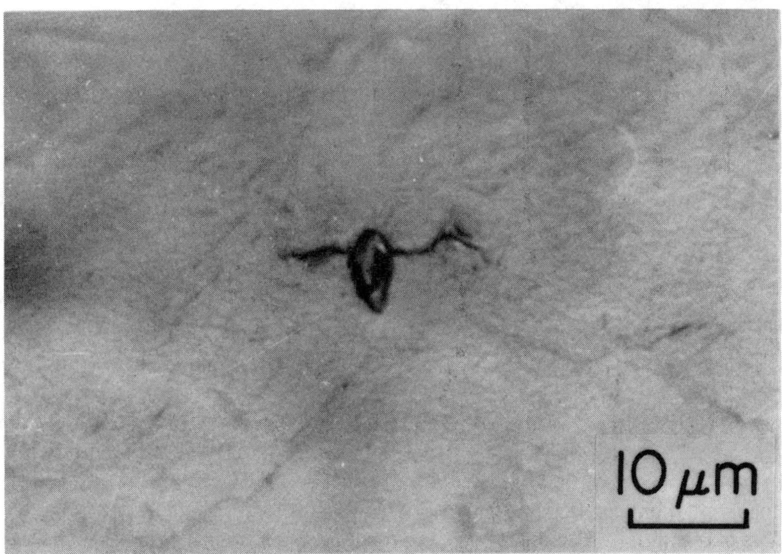

FIG. 3—*Fatigue microcrack formed in SNCM 439(C) steel.*

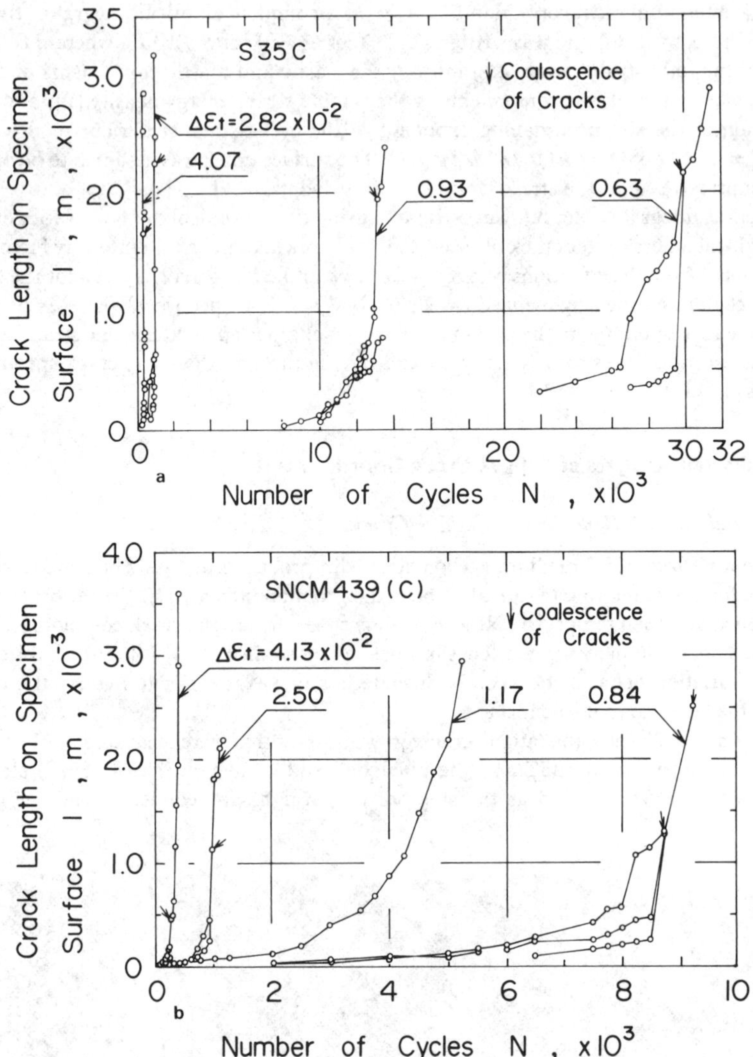

FIG. 4—*Surface crack growth curves.* (a) *Normalized S35C steel.* (b) *Quenched and tempered SNCM 439(C) steel.*

points. The crack growth rate in the direction of depth was obtained by multiplying the above slope by the aspect ratio of $a/\ell = 0.38$ [*9*]. The small arrow denotes the coalescence of the cracks.

Figure 5 shows the log $\Delta\epsilon_{ps}$ versus log N_i and log $\Delta\epsilon_{ps}$ versus log N_f relations of all the materials used, where $\Delta\epsilon_{ps}$ is the stable plastic strain range, and N_i and N_f are the numbers of strain cycles required for crack initiation and failure, respectively. The solid lines represent the log $\Delta\epsilon_{ps}$ − log N_f relations for the smooth specimens with no shallow notch [*24,35*]. The experimental data of the present specimens nearly fall on these lines, showing that the present shallow notch scarcely influences the low-cycle fatigue life. Hereupon, N_i was defined as the number of strain cycles at which the fatigue crack of $\ell = 20 \sim 30$ μm was first detected on the plastic

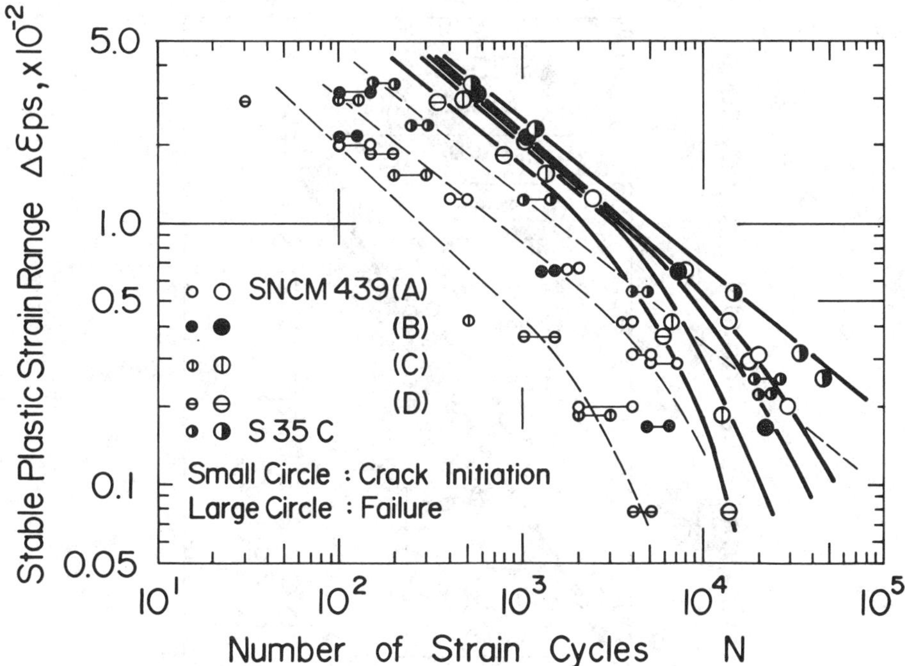

FIG. 5—*Plastic strain range versus life data for initiation of small cracks and for failure. Solid lines denote failure life data of fully smooth specimen* [24,35].

replica intermittently taken from the specimen surface. Since the experimental data yield a fairly large scatter, the log $\Delta\epsilon_{ps}$ versus log N_i relations in SNCM 439(A) and (B) steels seem to belong to a common data group and those in SNCM 439(C) and (D) steels also appear to constitute another one. Consequently, the experimental data belonging to each group were connected by a dashed line. The dashed line showing the log $\Delta\epsilon_{ps}$ versus log N_i relation in S35C steel is clearly separated from those in the alloy steel. From Fig. 5 it is known that the crack initiation to failure life ratio, N_i/N_f is about 0.15 ~ 0.20, independent of the kind of material.

The slope of the log $\Delta\epsilon_{ps}$ versus log N_i relation decreases in almost the same way as the slope of the log $\Delta\epsilon_{ps}$ versus log N_f in the lower plastic strain range for SNCM 439 steel. This means that the crack initiation process partly contributes to the break of the log $\Delta\epsilon_{ps}$ versus log N_f curve in this material. On the other hand, the crack initiation and the failure lines extend almost straight, supporting the validity of the Coffin-Manson law in the S35C steel.

Analyses of Crack Growth Rate

Figures 6*a*, 6*b*, and 6*c* show the relation between the small crack growth rate, da/dN, and the cyclic J integral range, ΔJ, in the S35C and the SNCM 439(A) and (C) steels on a log-log diagram, respectively. A linear regression analysis by the least squares method was done for the experimental data, and consequently the thick lines were determined. The band constituted by the two thin straight lines gives the range from one half to double the crack growth rate described by the thick line. Figure 7 shows the da/dN versus ΔJ relation for a through-thickness center crack of $a/w \geq 0.4$ in all the materials used. The best-fit line obtained by the least squares regression analysis of these data is expressed as the bold line in Fig. 6.

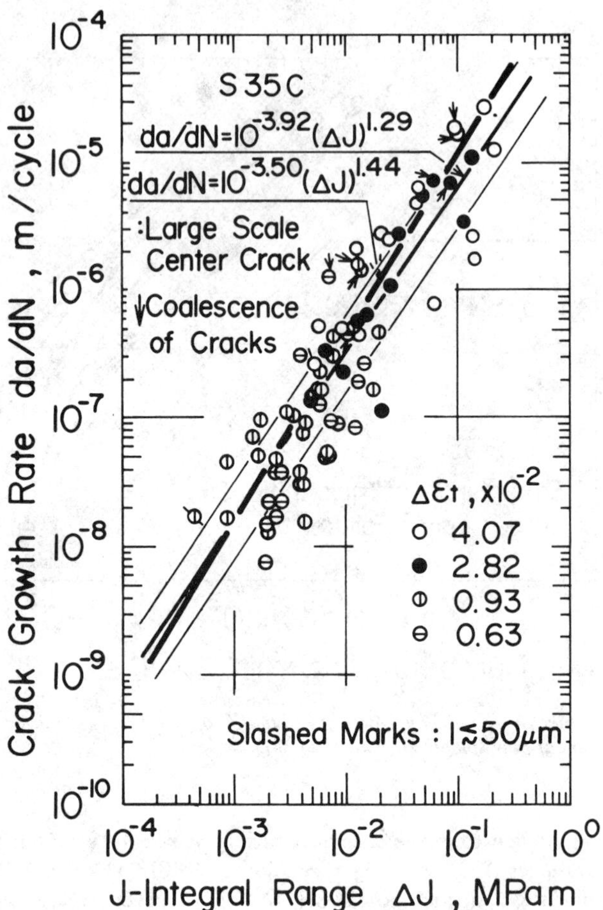

FIG. 6a—*Growth rate of small surface crack versus* Δ*J plot in smooth specimen: Normalized S35C steel.*

The slope of the log da/dN versus log ΔJ plot is smaller in a small surface crack than in a through-thickness crack. As a result, in a regime of higher crack growth, the log da/dN versus log ΔJ relation in the former is below that in the latter. Such a difference in the crack growth behavior is most evident in the SNCM 439(C) steel and becomes smaller in the SNCM 439(A) and the S35C steels in that order. Adding the log da/dN versus log ΔJ relations in the SNCM 439(B) and (D) steels to these results, it is concluded that the difference in the crack growth behavior between the surface and through-thickness cracks is more evident in materials of the higher strength and the lower ductility.

The log da/dN versus log ΔJ relations for small surface cracks in all the materials used are shown in Fig. 8. The thick, thin, and bold lines denote the same relations as they do in Fig. 6. The material dependency of the log da/dN versus log ΔJ relation is almost concealed for the fairly large scatter in the experimental data, which seems to be inherent in a small crack problem. As a result we obtain the following common equation on the da/dN versus ΔJ relation for the small surface crack in steels, independent of the kind of materials, provided that the amount of scatter is disregarded:

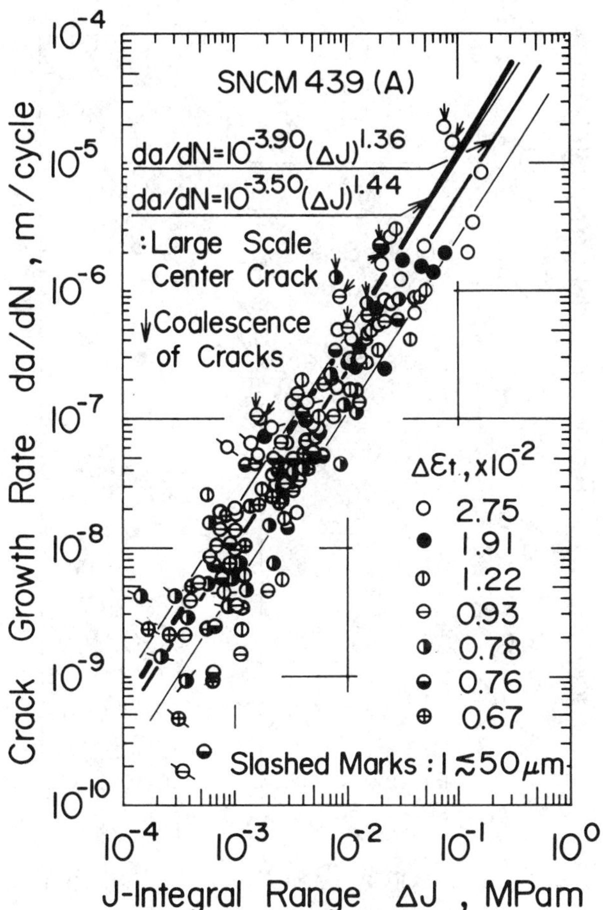

FIG. 6*b*—*Quenched and tempered SNCM 439(A) steel.*

$$da/dN = 10^{-4.15}(\Delta J)^{1.25} \quad (3)$$

This equation is a little different from the equation

$$da/dN = 10^{-3.50}(\Delta J)^{1.44} \quad (4)$$

derived for a through-thickness crack in all the materials used.

Figures 9*a*, 9*b*, and 9*c* represent the log da/dN versus log ΔK_ϵ relations in the normalized S35C, the quenched and tempered SNCM 439(A) and (C) steels. All the experimental data for the surface crack growth rate are plotted against ΔK_ϵ in Fig. 10, where the data on the SNCM 439(B) and (D) steels are added to those in Fig. 9. Again, thick, thin, and bold lines denote the same relations in Fig. 10 that they do in Figs. 6 and 8. Figure 11 shows the da/dN versus ΔK_ϵ relations for the through-thickness center and the single-edged cracks of $a/w \leq 0.3$ in all the materials used. The least squares fit line obtained from this figure is reproduced in Figs. 9 and 10, being expressed by a bold line. The bold line in the da/dN versus ΔJ plot is obtained for a

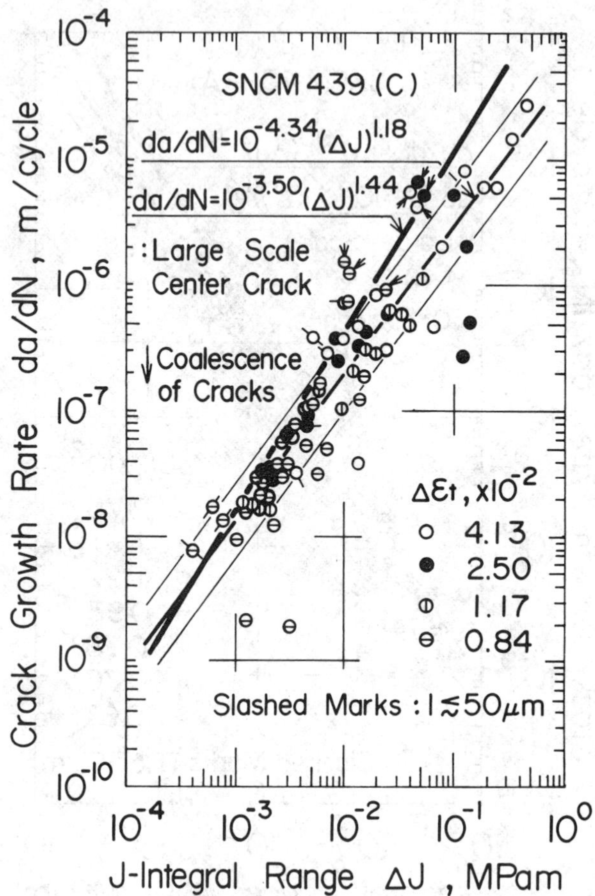

FIG. 6c—*Quenched and tempered SNCM 439(C) steel.*

through-thickness center crack of $a/w \geq 0.4$. Thus, the bold lines given in the da/dN versus ΔJ and the da/dN versus ΔK_ϵ plots are different from each other in the crack size concerned.

Some material dependence exists in the da/dN versus ΔK_ϵ relations shown in Fig. 9. We, however, hardly find a material dependence of the da/dN versus ΔK_ϵ relation in Fig. 10, since some scatter of the data shows a significant effect on the da/dN versus ΔK_ϵ relation among the materials studied. Furthermore, the experimental results on the surface crack growth rate are expressed fairly well by the bold line, which is commonly determined from the crack growth data in through-thickness center and single-edged crack specimens in all the materials used. The equations

$$da/dN = 10^{1.39}(\Delta K_\epsilon)^{2.24} \tag{5}$$

and

$$da/dN = 10^{3.25}(\Delta K_\epsilon)^{2.73} \tag{6}$$

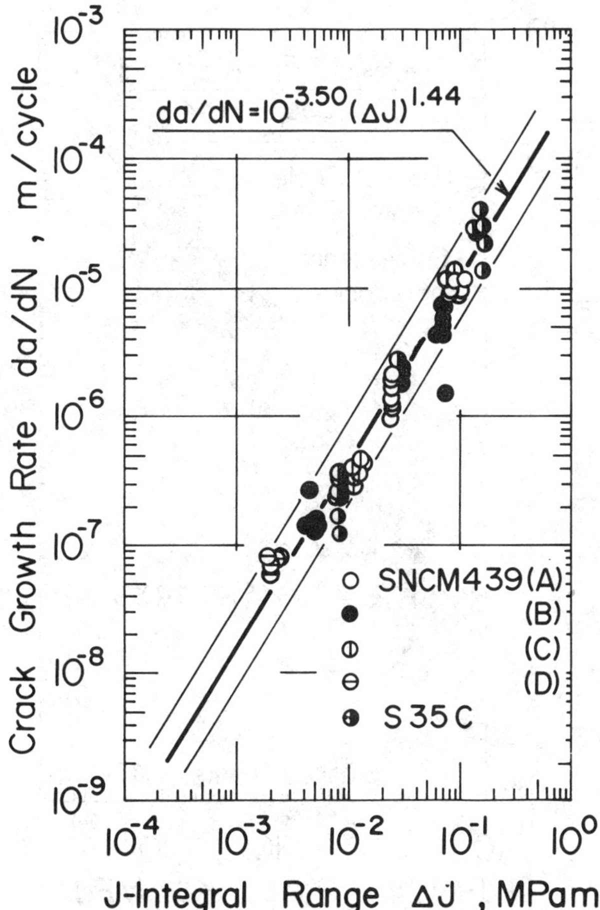

FIG. 7—*Growth rate versus* ΔJ *plot in through-thickness center crack of* a/w ≥ *0.4 in all the materials used.*

hold for the da/dN versus ΔK_ϵ relations for small surface and through-thickness cracks in all the materials used.

Derivation of the Fatigue Life Curves on the Basis of Crack Growth Process

A cyclic stress-strain curve is usually expressed by the following equation

$$\sigma_s = K'(\epsilon_{ps})^{n'} \tag{7}$$

where σ_s and ϵ_{ps} are the stable stress and plastic strain amplitudes, and K' and n' are the cyclic strength coefficient and the cyclic strain hardening exponent, respectively. Denoting the stable stress and plastic strain range as $\Delta\sigma_s$ and $\Delta\epsilon_{ps}$, Eq 7 is rewritten as

$$\Delta\sigma_s = 2K'(\epsilon_{ps}/2)^{n'} \tag{8}$$

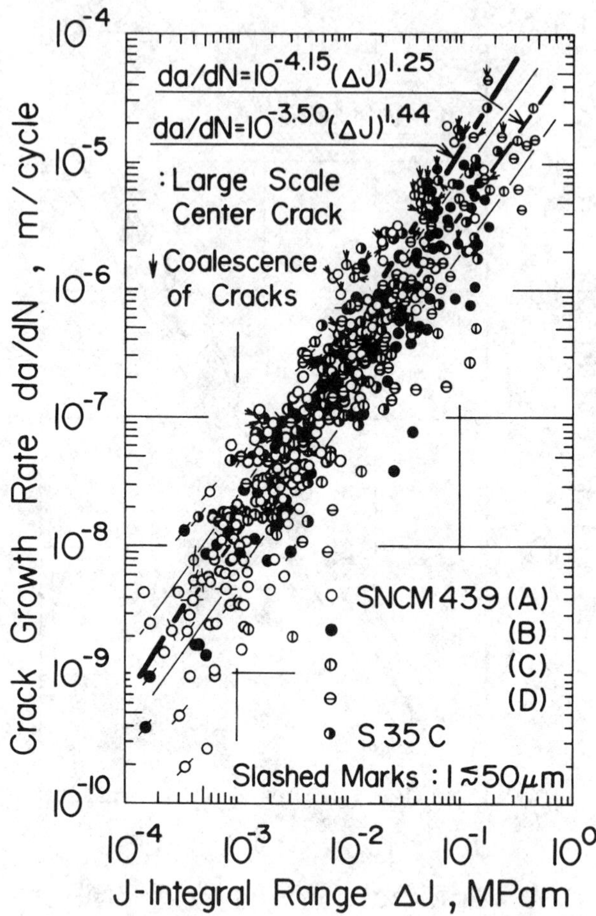

FIG. 8—*Growth rate of small surface crack versus ΔJ plot in all the materials used.*

As stated before, crack growth rate da/dN is expressed as function of ΔJ and ΔK_ϵ by the following equations

$$da/dN = C_1(\Delta J)^\beta \tag{9}$$

and

$$da/dN = C_2(\Delta K_\epsilon)^\gamma \tag{10}$$

where C_1, C_2, β, and γ are material-dependent constants.

Two types of low-cycle fatigue life curves are derived from combining Eqs 8 and 9, and Eqs 8 and 10.

By using Eq 8, ΔW_e and ΔW_p in Eq 2 are transformed as follows:

$$\Delta W_e = \frac{(\Delta\sigma_s)^2}{2E} = 2^{1-2n'} \frac{K'^2}{E} (\Delta\epsilon_{ps})^{2n'} \tag{11}$$

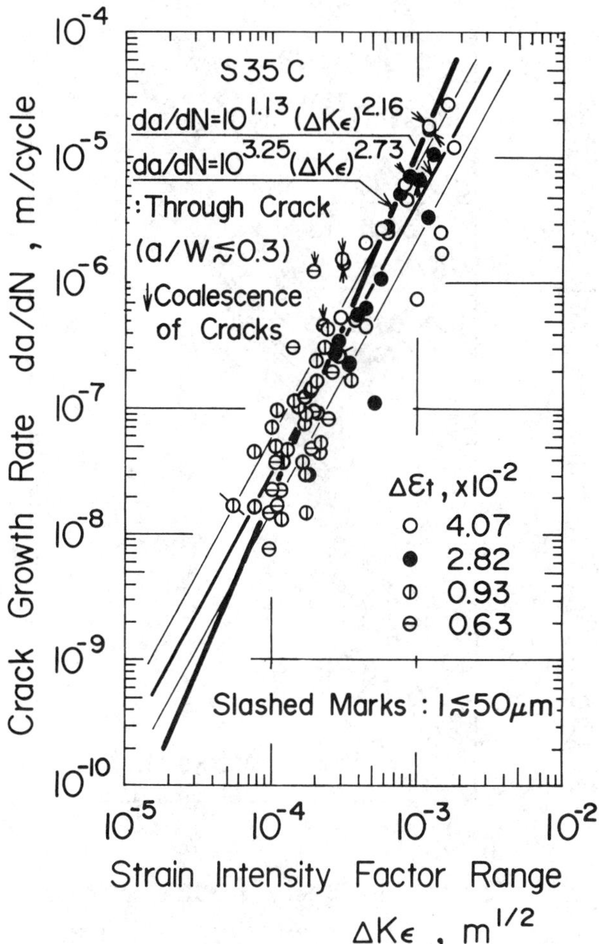

FIG. 9a—*Growth rate of small surface crack versus* ΔK_ϵ *plot in smooth specimen: Normalized S35C steel.*

and

$$W_p = \int_0^{\Delta\epsilon_{ps}} (\Delta\sigma_s)d(\Delta\epsilon_{ps}) = 2^{1-n'} \frac{K'}{1+n'} (\Delta\epsilon_{ps})^{1+n'} \tag{12}$$

Substituting Eqs 11 and 12 into Eq 2, the following equation is obtained:

$$\Delta J = 2\pi a \left\{\frac{M_K}{\phi(\lambda)}\right\}^2 \left\{2^{1-2n'} \frac{K'^2}{E} (\Delta\epsilon_{ps})^{2n'} + f(n')2^{1-n'} \frac{K'}{1+n'} (\Delta\epsilon_{ps})^{1+n'}\right\} \tag{13}$$

Since $\Delta\epsilon_t$ is expressed by the equation

$$\Delta\epsilon_t = \Delta\epsilon_e + \Delta\epsilon_{ps} = 2K'(\Delta\epsilon_{ps/2})^{n'}/E + \Delta\epsilon_{ps} \tag{14}$$

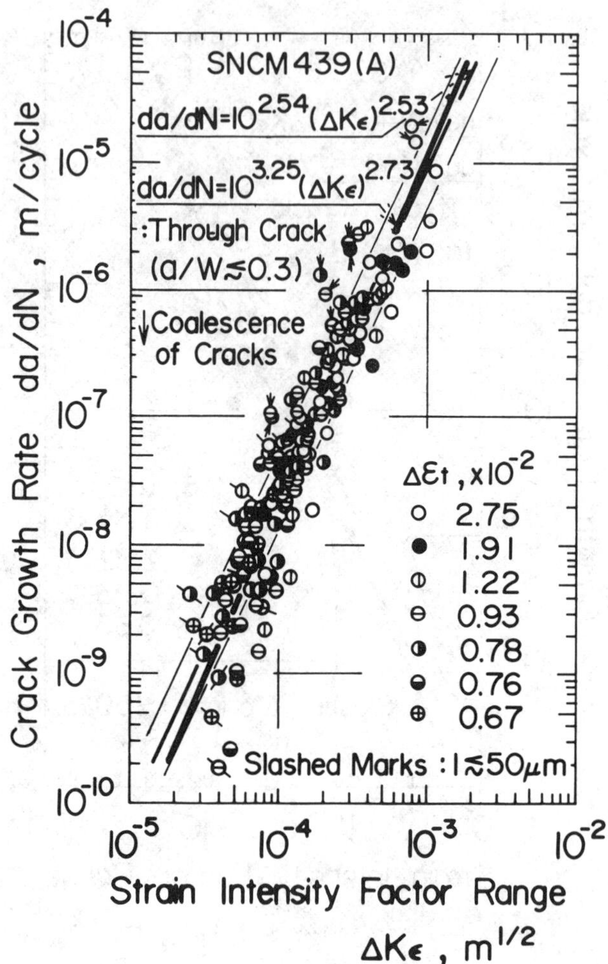

FIG. 9b—*Quenched and tempered SNCM 439(A) steel.*

as a function of $\Delta\epsilon_{ps}$, Eq 1 is rewritten as

$$\Delta K_\epsilon = \frac{M_K}{\phi(\lambda)}\left\{2^{1-n'}\frac{K'}{E}(\Delta\epsilon_{ps})^{n'} + \Delta\epsilon_{ps}\right\}\sqrt{\pi a} \tag{15}$$

We can derive the equations

$$\frac{da}{dN} = C_1\left[2\pi a\left\{\frac{M_K}{\phi(\lambda)}\right\}^2\left\{2^{1-2n'}\frac{K'^2}{E}(\Delta\epsilon_{ps})^{2n'} + f(n')2^{1-n'}\frac{K'}{1+n'}(\Delta\epsilon_{ps})^{1+n'}\right\}\right]^\beta \tag{16}$$

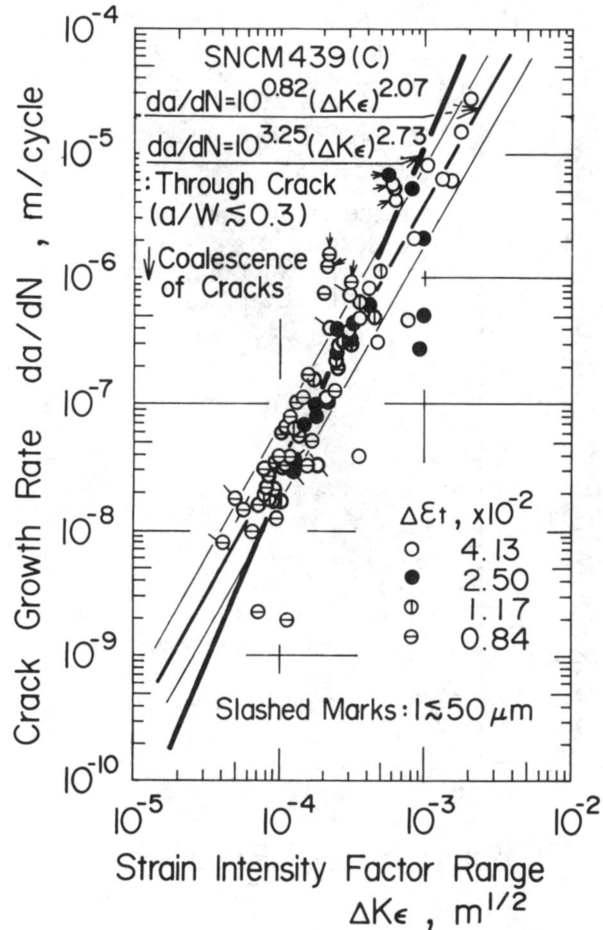

FIG. 9c—*Quenched and tempered SNCM 439(C) steel.*

and

$$\frac{da}{dN} = C_2\left[\frac{M_K}{\phi(\lambda)}\left\{2^{1-n'}\frac{K'}{E}(\Delta\epsilon_{ps})^{n'} + \Delta\epsilon_{ps}\right\}\sqrt{\pi a}\right]^{\gamma} \tag{17}$$

from combining Eqs 9 and 13, and Eqs 10 and 15, respectively. Integrating Eq 16 leads to the equation

$$\int_{N_i}^{N_f} dN = \frac{1}{C_1}\left[2\pi\left\{\frac{M_K}{\phi(\lambda)}\right\}^2\left\{2^{1-2n'}\frac{K'^2}{E}(\Delta\epsilon_{ps})^{2n'} + f(n')2^{1-n'}\frac{K'}{1+n'}(\Delta\epsilon_{ps})^{1+n'}\right\}\right]^{-\beta}\int_{a_0}^{a_f}\frac{da}{a^{\beta}} \tag{18}$$

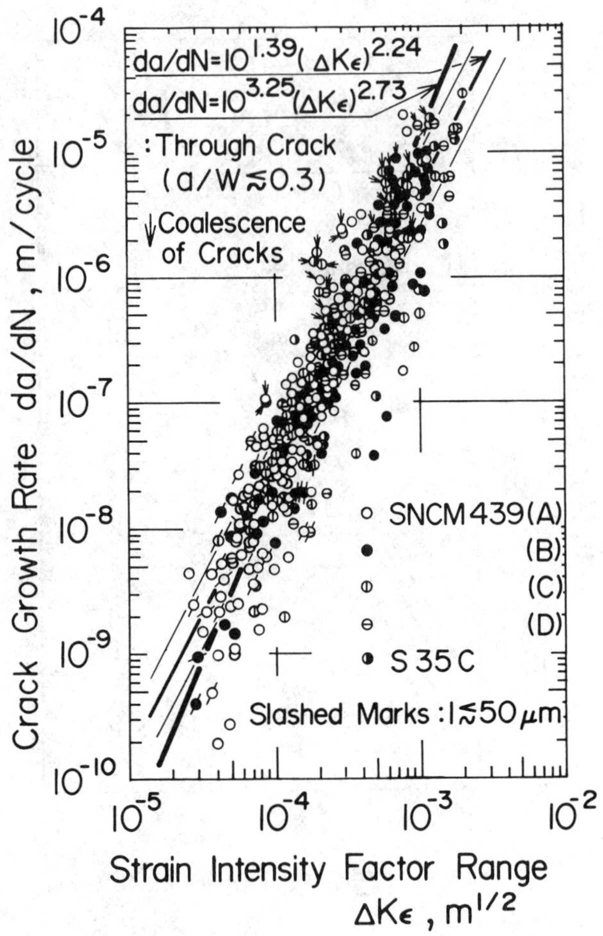

FIG. 10—*Growth rate of small surface crack versus* ΔK_ϵ *plot in all the materials used.*

where N_i is the number of strain cycles required before the crack depth grows into a_0, and a_f is the crack depth just before a final fracture occurs. Since in this instance N_i is negligibly small compared with N_f, it is thus considered that $N_f - N_i$ is nearly equal to N_f. Consequently the equation

$$N_f \cong \frac{a_f^{1-\beta} - a_0^{1-\beta}}{C_1(1-\beta)} \left[2\pi \left\{ \frac{M_K}{\phi(\lambda)} \right\}^2 \right]^{-\beta} \left\{ 2^{1-2n'} \frac{K'^2}{E} (\Delta\epsilon_{ps})^{2n'} + f(n') 2^{1-n'} \frac{K'}{1+n'} (\Delta\epsilon_{ps})^{1+n'} \right\}^{-\beta} \tag{19}$$

is obtained from Eq 18. In the same way, Eq 17 leads to the equation

$$N_f \cong \frac{a_f^{1-\gamma/2} - a_0^{1-\gamma/2}}{C_2(1-\gamma/2)} \left\{ \frac{M_K}{\phi(\lambda)} \right\}^{-\gamma} (\sqrt{\pi})^{-\gamma} \left\{ 2^{1-n'} \frac{K'}{E} (\Delta\epsilon_{ps})^{n'} + \Delta\epsilon_{ps} \right\}^{-\gamma} \tag{20}$$

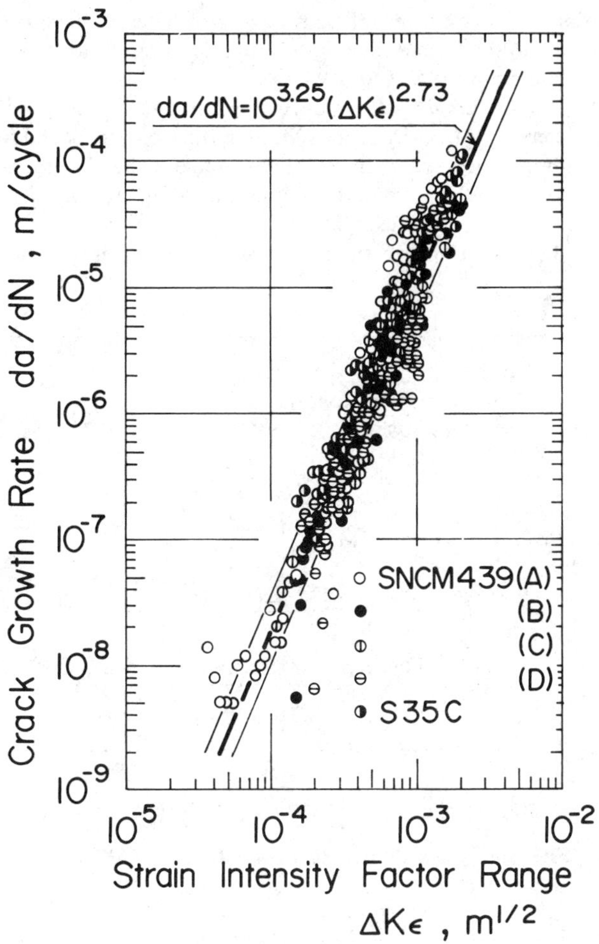

FIG. 11—*Growth rate versus* ΔK_ϵ *plot in through-thickness single-edged and center cracks of* a/w $\leq$ *0.3 in all the materials used.*

Equations 19 and 20 give the low-cycle fatigue life curves evaluated on the basis of the crack growth process.

For the surface crack having the semi-elliptical shape of $2a/\ell = 0.76$, the value of $M_K/\phi(\lambda)$ was determined to be 0.755, with reference to the study of Raju and Newman [*36*], and this value was kept constant in the whole crack growth process in the present analyses.

The value of a_f was set at 10% of the specimen diameter, $a_f = 8 \times 10^{-3} \times 0.1 = 8 \times 10^{-4}$ m [*5*]. Since we defined N_i as the number of strain cycles required for the surface crack length ℓ_0 to be 20 ~ 30 μm as stated before, a_0 was fixed upon as $0.38\ell_0 = 0.38 \times (20 \sim 30)\ \mu\text{m} \cong 8 \times 10^{-6}$ m in principle. The fatigue life is considered more sensitive to a_0 than a_f. Accordingly, it doesn't seem reasonable that a_0 remains constant independent of the material; while in steels of a high strength and a low ductility very small faults like an inclusion usually become the origin of crack initiation, an intense slip deformation originates a fatigue crack at a site like a slip band impinging on a grain boundary in steels of high ductility and low strength. The difference in the crack initiation behavior between these two types of material suggests that a_0 should be estimated to be smaller in the former than in the latter. As a result, an analysis for $a_0 = 5 \times 10^{-6}$ m was also attempted in the SNCM 439 steels.

The constants C_1 and β in the da/dN versus ΔJ relation, and C_2 and γ in the da/dN versus ΔK_ϵ relation, were determined from those for a through-thickness crack and for a small surface crack; for the through-thickness crack, the constants were obtained from Eqs 4 and 6, and for the small crack, these were obtained separately from the da/dN versus ΔJ and da/dN versus ΔK_ϵ relations in the respective materials used.

Figures 12*a*, 12*b*, 12*c*, and 12*d* show the cyclic stress-strain curves of the annealed S10C [9],

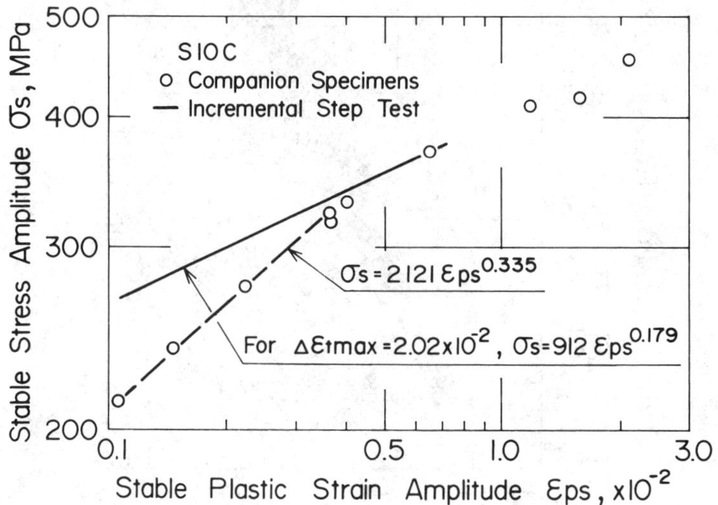

FIG. 12*a*—*Cyclic stress-strain curves: Annealed S10C steel.*

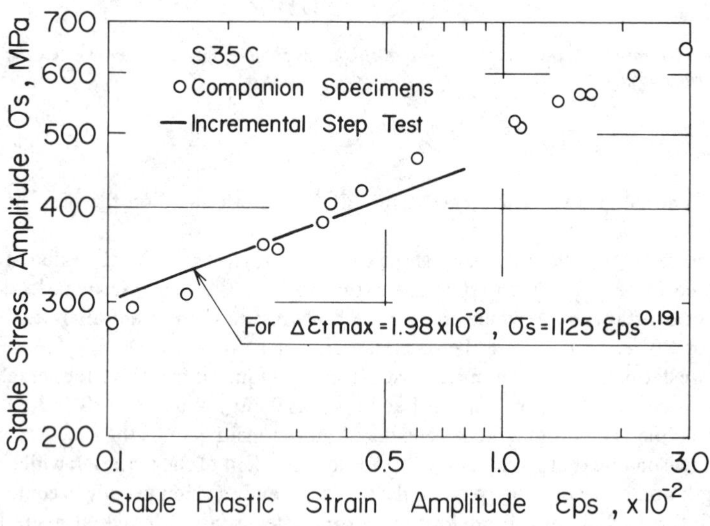

FIG. 12*b*—*Normalized S35C steel.*

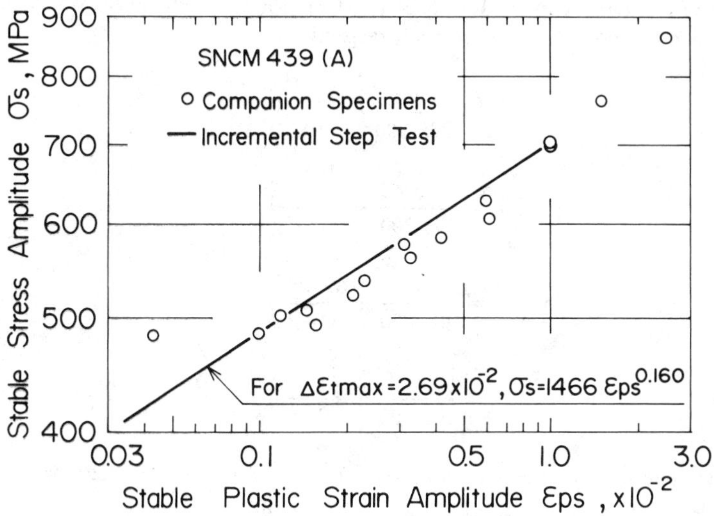

FIG. 12*c*—*Quenched and tempered SNCM 439(A) steel.*

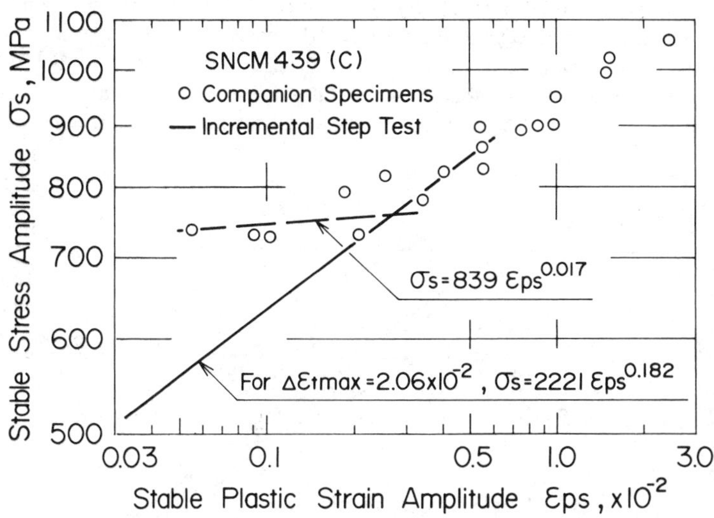

FIG. 12*d*—*Quenched and tempered SNCM 439(C) steel.*

the normalized S35C, and the quenched and tempered SNCM 439(A) and (C) steels obtained from several companion specimens and in an incremental step test. The cyclic stress-strain curves constituted through the companion specimens are nearly consistent with those determined by the incremental step tests for the normalized S35C and the SNCM 439(A) steels. On the other hand, the annealed S10C [*9*] and the SNCM 439 (C) steels exhibit marked difference in the cyclic stress-strain behavior in the lower plastic strain, depending upon the testing condition. For the two materials in the latter case, therefore, the cyclic stress-strain curves expressed by Eq 7 were determined in the two ways shown in Figs. 12*a* and 12*d*. The Ramberg-Osgood type cyclic stress-strain curves of S10C, S35C, SNCM 439(A) and (C) steels, which are obtained in the incremental step test, are given next, respectively.

$$\epsilon = \frac{\sigma}{205\ 300} + \left(\frac{\sigma}{912}\right)^{1/0.179}$$

$$\epsilon = \frac{\sigma}{199\ 700} + \left(\frac{\sigma}{1125}\right)^{1/0.191}$$

$$\epsilon = \frac{\sigma}{205\ 900} + \left(\frac{\sigma}{1466}\right)^{1/0.160} \qquad (21)$$

$$\epsilon = \frac{\sigma}{202\ 100} + \left(\frac{\sigma}{2221}\right)^{1/0.182}$$

Thus, we can determine all the material constants needed in Eqs 19 and 20 for evaluating the low-cycle fatigue life.

Figures 13*a*, 13*b*, 13*c*, and 13*d* show the predicted low-cycle fatigue lives along with the experimental data, which are plotted against the plastic strain range $\Delta\epsilon_{ps}$, in S10C, S35C, SNCM 439(A), and (C) steels. The thick and thin lines mark the log $\Delta\epsilon_{ps}$ versus log N_f relations evaluated from Eqs 19 and 20, respectively. The test results are denoted by open circles. The solid lines show the fatigue life curves estimated on the basis of the universal da/dN versus ΔJ and da/dN versus ΔK_ϵ relations of $da/dN = 10^{-3.50}(\Delta J)^{1.44}$ and $da/dN = 10^{3.25}(\Delta K_\epsilon)^{2.73}$ on through-thickness cracks. On the other hand, the fatigue life curves estimated using the da/dN versus ΔJ and da/dN versus ΔK_ϵ relations on the small surface cracks, which are determined

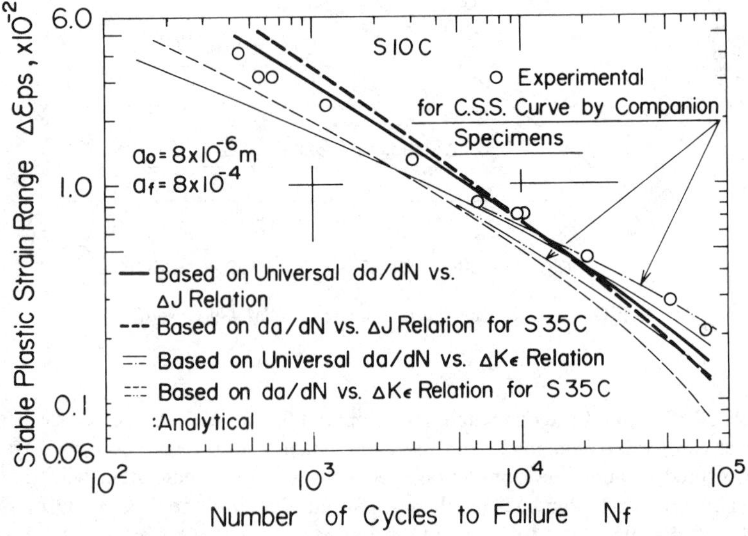

FIG. 13*a*—*Fatigue life curves evaluated by combining* da/dN *versus* ΔJ *and* da/dN *versus* ΔK_ϵ *relations, and cyclic stress-strain curve. Dash-dotted and dash-double dotted lines are derived using cyclic stress-strain curve by companion specimens, and others using one by incremental step test. Annealed S10C steel.*

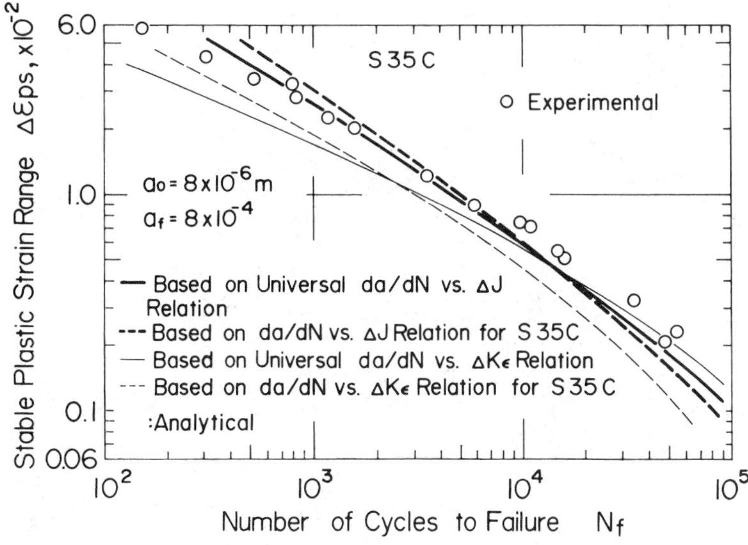

FIG. 13*b*—*Normalized S35C steel.*

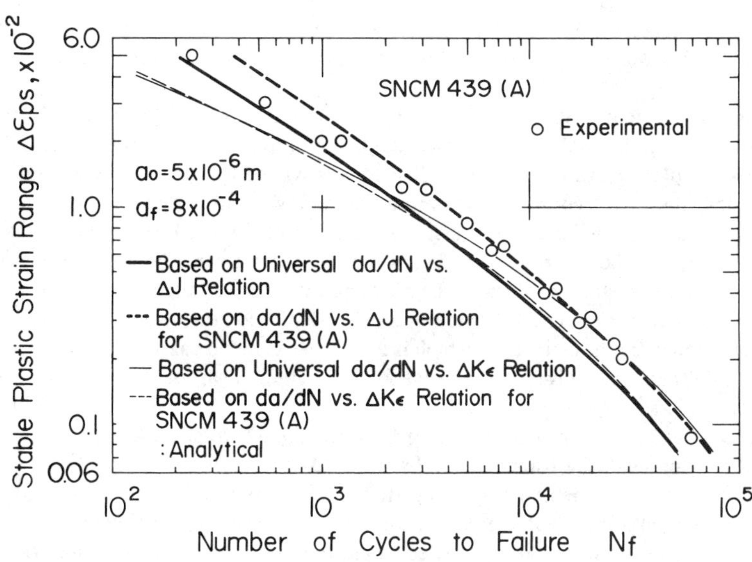

FIG. 13*c*—*Quenched and tempered SNCM 439(A) steel.*

separately from the respective materials used, are given by the dashed lines. In the present study, the crack growth of the annealed JIS S10C low carbon steel is not examined. Our previous study, however, showed that the growth behavior of the small surface crack in this material is almost the same as that in the normalized S35C steel under a rotating bending fatigue condition [*9*]. This is easily understood as the microstructure is very similar for both materials.

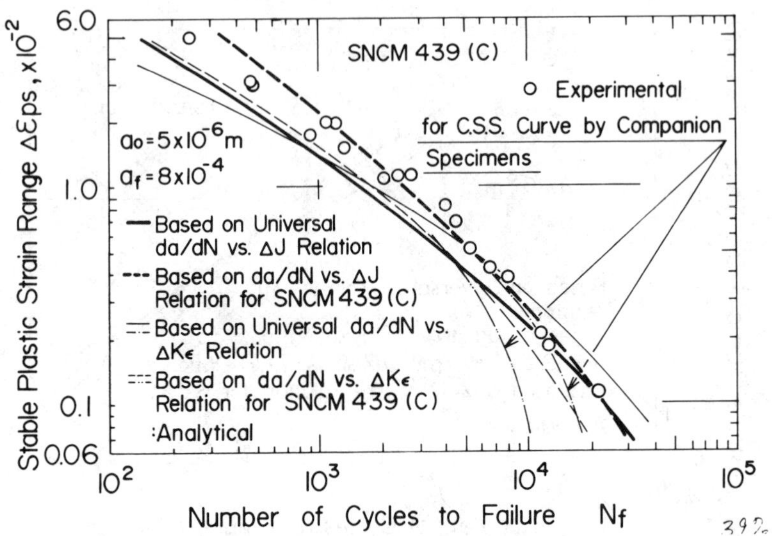

FIG. 13*d*—*Quenched and tempered SNCM 439(C) steel.*

Accordingly the da/dN versus ΔJ and the da/dN versus ΔK_ϵ relations of the S35C steel was used for evaluating the low-cycle fatigue life of the S10C steel based on the small surface crack growth.

While the experimental data on the log $\Delta\epsilon_{ps}$ versus log N_f plot extends straight to the life range of about 10^5 cycles in the S10C and the S35C steels, the slopes in this plot decrease gradually in the lower plastic strain range in the SNCM 439(A) and (C) steels. The estimated fatigue life curves describe such trends of the experimental data well. In general, the fatigue life curve estimated on the basis of the da/dN versus ΔJ relation of the small surface crack is in fairly good agreement with the test results in the lower plastic strain range, but this predicts slightly longer fatigue life than the test data in the higher plastic strain range. The da/dN versus ΔJ relation on the small surface crack is somewhat dependent upon the kind of materials, as shown in Fig. 6. Nevertheless, the estimation which is based on the universal da/dN versus ΔJ relation of through-thickness crack gives a relatively good prediction of the low-cycle fatigue life in each material used.

In general, the fatigue life curve resulting from the da/dN versus ΔJ relation has a steeper slope than the one resulting from the da/dN versus ΔK_ϵ relation, which is also in fairly good agreement with the experimental data. As stated before, a relatively large difference exists between the cyclic stress-strain curve composed of several companion specimens and the one obtained from a single specimen which is tested by an incremental step method, specifically in the lower plastic strain range; the former stress level is below the latter in the S10C steel, and the reverse situation occurs in the SNCM 439(C) steel. In these materials the cyclic stress-strain curves bent so as to fit the data for the former were also used for the present analysis. These cyclic stress-strain curves are shown in Figs. 12*a* and 12*d*. The log $\Delta\epsilon_{ps}$ versus log N_f curves derived from this type of cyclic stress strain curve and the da/dN versus ΔK_ϵ relation are represented by the dash-dotted and the dash-double dotted lines in Figs. 13*a* and 13*d*. The two lines agree with the test results in the lower plastic strain range in the S10C and the SNCM 439(C) steels. Especially, the bend behavior which appears in the log $\Delta\epsilon_{ps}$ versus log N_f relation of the SNCM 439(C) steel is described well by this estimation (Fig. 13*d*).

Conclusions

Samples of steels, with tensile strength in the range of 614 to 1390 MPa and fracture ductility in the range of 0.72 to 1.11, were low-cycle fatigued using a closed-loop-type testing system under a strain-controlled condition. The growth rates of the small and large fatigue cracks were analyzed in terms of the cyclic J integral, ΔJ, and the strain intensity factor range, ΔK_ϵ.

The logarithmic growth rate of a large through-thickness center crack, log da/dN, is successfully plotted by the straight line against the logarithmic ΔJ. This relation is scarcely influenced by the kind of materials. The logarithmic growth rate of the small surface crack is also expressed by the universal straight line against the logarithmic ΔJ, in all the materials tested. As a result, the equations of $da/dN = 10^{-3.50}(\Delta J)^{1.44}$ and $da/dN = 10^{-4.15}(\Delta J)^{1.25}$ are empirically obtained for the through-thickness center and the small surface cracks, respectively.

The strain intensity factor range ΔK_ϵ is also a good parameter for evaluating the growth rates of the small surface crack and a through-thickness center crack of $a/w \leq 0.3$. These log da/dN versus log ΔK_ϵ relations are represented by the empirical equations of $da/dN = 10^{1.39}(\Delta K_\epsilon)^{2.24}$ and $da/dN = 10^{3.25}(\Delta K_\epsilon)^{2.73}$ for all the materials used. The low-cycle fatigue lives were evaluated by combining the da/dN versus ΔJ and the da/dN versus ΔK_ϵ relations, and the cyclic stress-strain curves. The estimated fatigue life curves describe well the test results such that a break appears on the log $\Delta\epsilon_{ps}$ versus log N_f plot in the quenched and tempered alloy steel as well as on the usual Coffin-Manson plot.

References

[1] Boettner, R. C., Laird, C., and McEvily, A. J., Jr., "Crack Nucleation and Growth in High Strain-Low Cycle Fatigue," *Transactions of the Metallurgical Society of AIME*, Vol. 233, Feb. 1965, pp. 379–387.

[2] Dowling, N. E., "Crack Growth during Low Cyclic Fatigue of Smooth Axial Specimens," in *Cyclic Stress-Strain and Plastic Deformation Aspects of Fatigue Crack Growth, ASTM STP 637*, American Society for Testing and Materials, Philadelphia, 1977, pp. 97–121.

[3] Dowling, N. E., "Geometry Effects and the J-Integral Approach to Elastic-Plastic Fatigue Crack Growth," in *Cracks and Fracture, 9th Conference, ASTM STP 601*, American Society for Testing and Materials, Philadelphia, 1976, pp. 19–32.

[4] Yamada, T., Hoshide, T., Fujimura, S., and Manabe, M., "Growth in Low Cycle Fatigue of Smooth Specimen of Medium Carbon Steel," *Transactions of the Japan Society of Mechanical Engineers*, Vol. 49, No. 440, A, 1983, pp. 441–448.

[5] Mowbray, D. F., "Derivation of a Low-Cycle Fatigue Relationship Employing the J-Integral Approach to Crack Growth," in *Cracks and Fracture, 9th Conference, ASTM STP 601*, American Society for Testing and Materials, Philadelphia, 1976, pp. 33–46.

[6] Solomon, H. D., "Low Cycle Fatigue Crack Propagation in 1018 Steel," *Journal of Materials*, Vol. 7, No. 3, 1972, pp. 299–306.

[7] Huang, J. S. and Pelloux, R. M., "Low Cycle Fatigue Crack Propagation in Hastelloy-X at 25 and 760°C," *Metallurgical Transactions A*, Vol. 11A, June 1980, pp. 899–904.

[8] Imai, Y. and Matake, T., "Surface Crack Growth during Push-Pull Fatigue of Smooth Specimens," *Journal of the Society of Materials Science*, Japan, Vol. 32, No. 361, 1983, pp. 1157–1161.

[9] Hatanaka, K., Fujimitsu, T., and Watanabe, H., "Growth Behaviors of Small Surface Cracks in Low Carbon Steel Fatigued Under Rotating Bending," *Transactions of the Japan Society of Mechanical Engineers*, Vol. 50, No. 452 A, 1984, pp. 737–742.

[10] El Haddad, M. H., Smith, K. N., and Topper, T. H., "Fatigue Crack Propagation of Short Cracks," *Journal of Engineering Materials and Technology, Transactions of the ASME*, Vol. 101, Jan. 1979, pp. 42–46.

[11] Hoshide, T., Fujimura, S., and Yamada, T., "Initiation and Subsequent Growth of Surface-Cracks in Low-Cycle Fatigue of Smooth Specimen," *Transactions of the Japan Society of Mechanical Engineers*, Vol. 50, No. 451, A, 1984, pp. 320–328.

[12] Dowling, N. E. and Begley, J. A., "Fatigue Crack Growth During Gross Plasticity and the J-Integral," in *Mechanics of Crack Growth, ASTM STP 590*, American Society for Testing and Materials, Philadelphia, 1976, pp. 82–103.

[13] Taira, S., Tanaka, K., and Hoshide, T., "Evaluation of J-Integral Range and Its Relation to Fatigue Crack Growth Rate," in *Proceedings*, Twenty-Second Japan Congress on Materials Research, 1979, pp. 123–129.

[14] Hoshide, T., Yamada, A., and Tanaka, K., "Evaluation of J-Integral for Center-Cracked Plate Under Gross and General Yieldings and Its Application to Fatigue Crack Growth," *Journal of the Society of Materials Science*, Japan, Vol. 31, No. 347, 1982, pp. 810-816.

[15] Hatanaka, K., Fujimitsu, T., and Watanabe, H., "Growth of Low-Cycle Fatigue Crack in Steels for Machine Structural Use," *Transactions of the Japan Society of Mechanical Engineers*, Vol. 52, No. 475, A, 1986, pp. 579-586.

[16] Murakami, Y., Harada, S., Tani-ishi, H., Fukushima, Y., and Endo, T., "Correlations Among Propagation Law of Small Cracks, Manson-Coffin Law and Miner Rule Under Low-Cycle Fatigue," *Transactions of the Japan Society of Mechanical Engineers*, Vol. 49, No. 447, A, 1983, pp. 1411-1418.

[17] Usami, S., Kimoto, H., Enomoto, K., and Shida, S., "Low Cycle Fatigue Crack Propagation of a Notch Under Pulsating Load," *Fatigue of Engineering Materials and Structures*, Vol. 2, No. 2, 1979, pp. 155-164.

[18] Kitagawa, H., Takahashi, S., Suh, C. M., and Miyashita, S., "Quantitative Analysis of Fatigue Process-Microcracks and Slip Lines Under Cyclic Loading," in *Fatigue Mechanism, ASTM STP 675*, American Society for Testing and Materials, Philadelphia, 1979, pp. 420-449.

[19] Coffin, L. F., "A Note on Low Cycle Fatigue Laws," *Journal of Materials*, Vol. 6, No. 2, 1971, pp. 388-402.

[20] Manson, S. S., "Behavior of Materials Under Conditions of Thermal Stress," Technical Note 2933, National Advisory Committee on Aeronautics, 1954.

[21] Tomkins, B., "Fatigue Failure in High Strength Metals," *Philosophical Magazine*, Vol. 23, No. 183, 1971, pp. 687-703.

[22] Endo, T. and Morrow, JoDean, "Cyclic Stress-Strain and Fatigue Behavior of Representative Aircraft Metals," *Journal of Materials*, Vol. 4, No. 1, 1969, pp. 159-175.

[23] Thielen, P. N., Fine, M. E., and Fournelle, R. A., "Cyclic Stress Strain Relations and Strain-Controlled Fatigue of 4140 Steel," *Acta Metallurgica*, Vol. 24, No. 1, 1976, pp. 1-10.

[24] Hatanaka, K. and Fujimitsu, T., "Some Considerations on Cyclic Stress-Strain Relation and Low Cycle Fatigue Life of Metallic Materials," *Transactions of the Japan Society of Mechanical Engineers*, Vol. 50, No. 451, A, 1984, pp. 291-299.

[25] Martin, D. E., "Plastic Strain Fatigue in Air and Vacuum," *Transactions of the ASME, Series D*, Vol. 87, No. 4, 1965, pp. 850-856.

[26] Sanders, T. H., Jr. and Starke, E. A., Jr., "The Relationship of Microstructure to Monotonic and Cyclic Straining of Two Age Hardening Aluminum Alloys," *Metallurgical Transactions*, Vol. 7A, Sept. 1976, pp. 1407-1418.

[27] Lukas, P. and Klesnil, M., "Cyclic Stress-Strain Response and Fatigue Life of Metals in Low Amplitude Region," *Materials Science and Engineering*, Vol. 11, 1973, pp. 345-356.

[28] Steele, R. K. and McEvily, A. J., "The High-Cycle Fatigue Behavior of Ti-6Al-4V Alloy," *Engineering Fracture Mechanics*, Vol. 8, No. 1, 1976, pp. 31-37.

[29] Mahajan, Y. and Margolin, H., "Low Cycle Fatigue Behavior of Ti-6Al-2Sn-4Zr-6Mo: Part II. Cyclic Deformation Behavior and Low Cycle Fatigue," *Metallurgical Transactions*, Vol. 13A, Feb. 1982, pp. 269-274.

[30] Saleh, Y. and Margolin, H., "Low Cycle Fatigue Behavior of Ti-Mn Alloys: Fatigue Life," *Metallurgical Transactions*, Vol. 13A, July 1982, pp. 1275-1281.

[31] Shih, C. F. and Hutchinson, J. W., "Fully Plastic Solutions and Large Scale Yielding Estimates for Plane Stress Crack Problems," *Transactions of the ASME, Series H, Journal of Engineering Materials and Technology*, Vol. 98, No. 4, 1976, pp. 289-295.

[32] Rice, J. R., Paris, P. C., and Merkle, J. G., "Some Further Results of J-Integral Analysis and Estimates," in *Progress in Flow Growth and Fracture Toughness Testing, ASTM STP 536*, American Society for Testing and Materials, Philadelphia, 1973, pp. 231-245.

[33] Yokobori, T., Nanbu, M., and Takeuchi, N., "Observations of Initiation and Propagation of Fatigue Crack by Plastic-Replication Method," *Reports of the Research Institute for Strength and Fracture of Materials*, Tohoku University, Japan, Vol. 5, No. 1, 1969, pp. 1-17.

[34] Yokobori, T., Kuribayashi, H., Kawagishi, M., and Takeuchi, N., "Studies on the Initiation and the Propagation of Fatigue Crack in Tempered-Martensitic High Strength Steel by Plastic-Replication Method and Scanning Electron Microscope," *Reports of the Research Institute for Strength and Fracture of Materials*, Tohoku University, Japan, Vol. 7, No. 1, 1971, pp. 1-23.

[35] Hatanaka, K., Fujimitsu, T., and Shigemura, T., "Cyclic Stress-Strain Response and Low Cycle Fatigue Life of JIS SNCM 439 Alloy Steel Heat-Treated in Several Conditions," Preprints of 61st Annual Meeting of Japan Society of Mechanical Engineers, No. 830-10, Oct. 1983, pp. 231-233.

[36] Raju, I. S. and Newman, J. C., Jr., "Stress Intensity Factors for a Wide Range of Semi-Elliptical Surface Cracks in Finite-Thickness Plates," *Engineering Fracture Mechanics*, Vol. 11, No. 4, 1979, pp. 817-829.

K. M. Nikbin[1] and G. A. Webster[1]

Prediction of Crack Growth under Creep-Fatigue Loading Conditions

REFERENCE: Nikbin, K. M. and Webster, G. A., **"Prediction of Crack Growth under Creep-Fatigue Loading Conditions,"** *Low Cycle Fatigue, ASTM STP 942,* H. D. Solomon, G. R. Halford, L. R. Kaisand, and B. N. Leis, Eds., American Society for Testing and Materials, Philadelphia, 1988, pp. 281-292.

ABSTRACT: Crack growth data are presented on a brittle and a ductile low alloy steel and a nickel base superalloy which have been subjected to combined static and cyclic loading at elevated temperatures. The cyclic tests have been conducted between fixed load and fixed displacement ranges. Distinct regions of fatigue and creep dominated cracking have been identified. It is shown that crack growth eventually ceases because of load relaxation in the constant displacement amplitude tests. Fracture of a component made of the ductile steel is unlikely to occur under this type of low-cycle fatigue loading, but significant cracking is possible with the brittle steel. Where cyclic crack growth is creep controlled, it is shown that crack propagation rates can be predicted from static creep data using the creep fracture mechanics parameter C^*.

KEY WORDS: creep-fatigue crack growth, fracture mechanics, C^* parameter, low-cycle fatigue

Considerable economic savings are possible if the useful lives of engineering components can be prolonged and the period between maintenance inspections extended. If conservative assessment procedures were adopted initially, a component may still be serviceable after its design life has been reached. To determine the remaining usefulness of a component requires an estimate of the damage it has suffered and of its remaining life. This paper is concerned with establishing procedures for estimating crack extension in components subjected to combined static and cyclic loading at elevated temperatures.

The need for determining tolerable defect sizes is becoming more important as the sensitivity of non-destructive evaluation equipment improves. Smaller and smaller cracks are being detected and the question of whether a cracked component can be returned to service, or must be replaced, is being encountered more frequently in, for example, the electric power generation, chemical process, and aircraft industries.

Most of the relevant components in these industries are subjected to non-steady operating conditions which can give rise to various combinations of environmental, creep, fatigue, and thermal fatigue crack growth [*1-9*]. The dominating mode of failure in a particular circumstance will depend upon material composition, heat-treatment, cyclic to mean load ratio, frequency, temperature, and operating environment. This paper is concerned chiefly with situations where creep and fatigue mechanisms control.

Typically at room temperature under cyclic loading conditions crack growth is cycle dependent and governed by fatigue processes. In this circumstance crack growth/cycle (da/dN), where a is crack length and N the number of cycles, can be expressed in terms of stress intensity factor range ΔK by the Paris Law [*10*]

$$da/dN \propto \Delta K^m \tag{1}$$

[1]Department of Mechanical Engineering, Imperial College, London SW7, England.

where m is a material dependent constant which usually has a value between 2 and 4. The mode of fracture is normally transgranular.

In contrast under steady loading at sufficiently high temperatures ($T/T_m \geq 0.5$, where T is temperature and T_m the melting temperature in degrees absolute) creep processes dominate and intergranular fractures are usually observed. For this situation it has been found [*11–13*] that crack growth rate $\dot{a}$ can be characterized most satisfactorily in terms of the creep fracture mechanics parameter C^* by a relation of the form

$$\dot{a} \propto C^{*\phi} \tag{2}$$

where ϕ is a number slightly less than one and the proportionality factor is governed chiefly by the material creep ductility and the constraint local to the crack tip. An increase in crack growth rate is obtained with increase in degree of constraint and with decrease in ductility [*12–14*].

During combined static and cyclic loading at these elevated temperatures creep and fatigue dominated crack growth have both been observed [*6–9*]. Generally the rate of cracking is sensitive to minimum-to-maximum load ratio R and frequency f. Examples of data obtained on a hot isostatically pressed nickel base superalloy designated *AP1* at $R = 0.7$ and 0.1 are shown in Figs. 1 and 2, respectively. Additional results are included in Fig. 3 for a low alloy steel at $R = 0.8$. Similar features have been demonstrated by other alloys [*4,9*].

Two regions can be identified in Figs. 1 to 3. At high frequencies the crack growth/cycle is approximately insensitive to frequency whereas at the lower frequencies da/dN increases with decrease in frequency. This change from frequency independent crack growth to frequency sensitive cracking is often associated with a change from a transgranular to an intergranular mode of fracture. The frequency dependence can be understood by considering the relationship between da/dN and crack propagation rate $\dot{a}$; that is,

$$da/dN = \frac{1}{3600f}\,\dot{a} \tag{3}$$

where da/dN is measured in mm/cycle, f in Hz, and $\dot{a}$ in mm/h. A horizontal line in Figs. 2 and 3 therefore identifies a region where cycle-dependent processes control and da/dN is constant. A slope of -1 corresponds to when time-dependent processes dominate and $\dot{a}$ is constant.

These results indicate that for the temperatures considered, crack growth in both alloys is governed by fatigue processes at high frequencies and creep processes at low frequencies for each R. Further evidence that creep mechanisms control at the low frequencies is provided by Fig. 4 for the nickel base superalloy. It is apparent that cyclic crack propagation rate can be correlated with the creep fracture mechanics parameter C^* and Eq 2. In addition, the cyclic cracking rates can be predicted from static creep crack growth data. Similar features have been demonstrated for the low alloy steel [*6*].

The sharp transition in Figs. 2 and 3 from cycle to time dependent cracking suggests little creep-fatigue interaction for the conditions examined. It appears that the two damaging processes develop independently and cause different fracture modes so that in the transition region linear superposition can be used to predict cumulative crack growth rates as shown by the solid lines in the figures.

Further data have been obtained in this investigation, mainly in the low frequency region. In previous studies cycling was only carried out between fixed load limits. In this study cycling has been performed, in addition, between fixed displacement amplitudes so as to represent more closely the conditions often experienced in low-cycle fatigue and thermal fatigue situations.

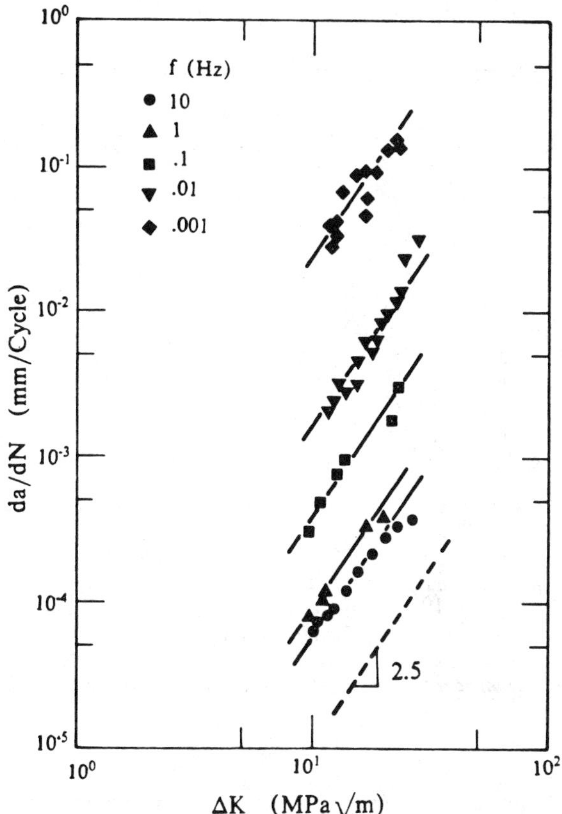

FIG. 1—*Dependence of crack growth/cycle at various frequencies on* ΔK *for Alloy AP1 at* R = *0.7 and 700°C* [8].

Experiments

Two low alloy steels have been investigated; the ½Cr-½Mo-¼V steel (½ CMV) considered earlier and a 2¼Cr-1Mo steel (2¼ CM). The ½ CMV steel was heat-treated to represent brittle stress relieved heat-affected zone material and the 2¼ CM steel ductile parent plate [*13*]. They have been tested at 565°C and 538°C, respectively. At these temperatures the ½ CMV steel had a uniaxial creep ductility in the range 0.1 to 2% and the 2¼ CM steel a creep failure strain of 45%.

Standard compact tension (CT) test-pieces of thickness $B = 25$ mm and width $W = 50$ mm were machined from each material. All samples were provided with side-grooves to give a net thickness across the grooves of $B_n = 18$ mm. A fatigue crack was introduced into each specimen at room temperature prior to testing at elevated temperature.

Crack growth experiments were conducted at various combinations of static and cyclic loading on each material. Crack extension was measured by a combination of optical and electrical potential methods to an estimated accuracy of ±0.1 mm [*15*]. Load point displacement was recorded continuously throughout each test with the aid of an electrical transducer and an extensometer attached to the loading shackles. This displacement was measured as it is used later to determine C^*.

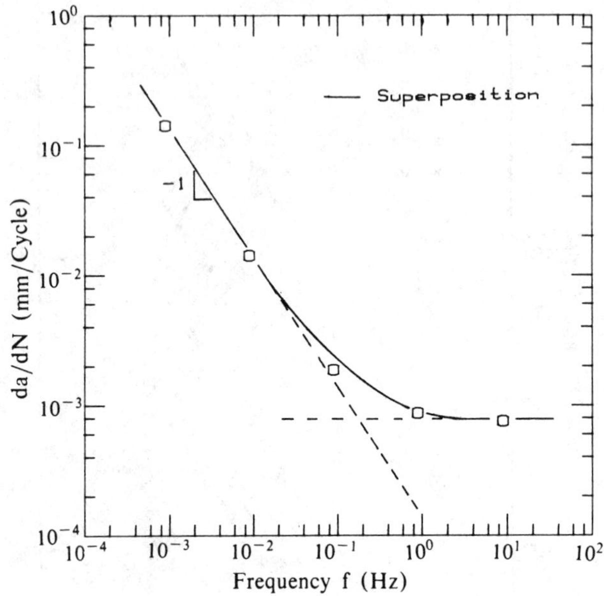

FIG. 2—*Frequency dependence of crack growth/cycle at* ΔK = *10 MPa* $m^{1/2}$ *and* R = *0.1 for Alloy AP1 at 700°C* [8].

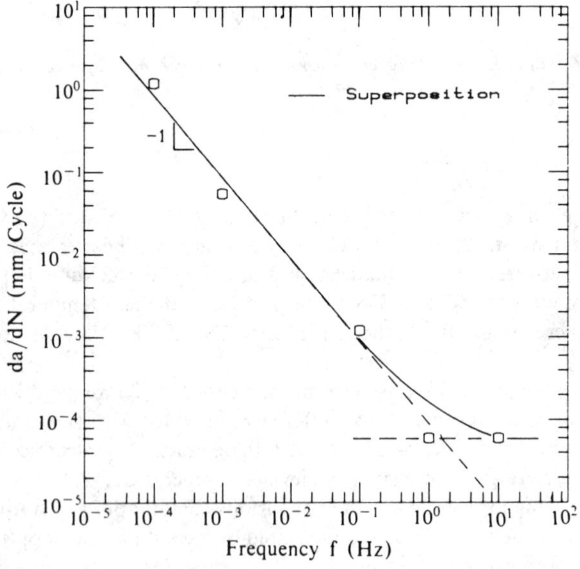

FIG. 3—*Frequency dependence of crack growth/cycle for a* $^1/_2$*Cr-*$^1/_2$*Mo-*$^1/_4$*V steel at* ΔK = *10 MPa* $m^{1/2}$, R = *0.8, and 565°C* [6].

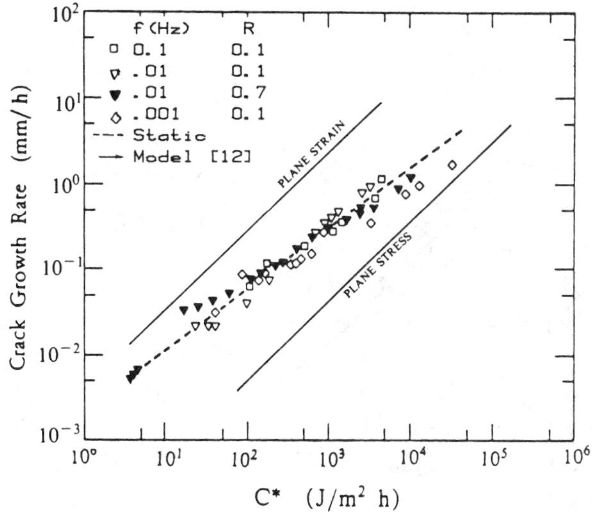

FIG. 4—*Comparison of crack growth rate versus* C* *for AP1 subjected to static and cyclic loading at 700°C* [7].

The static load tests were performed in dead-load creep machines. The cyclic tests were carried out under microprocessor control in servohydraulic testing machines or in dead-load creep machines suitably modified to allow slow load cycling. A sinusoidal waveshape was employed for frequencies of 0.1 Hz and greater and a square wave for lower frequencies. Some of the experiments in the servohydraulic machines were conducted under load control and others under displacement control using the extensometer signal output. For these latter tests an initial R of 0.8 was employed to correspond with that employed in the constant load range experiments. The test conditions were chosen to give crack propagation rates in the range 10^{-4} to 10 mm/h.

Results

The cyclic crack growth data obtained at constant load amplitude on the 2¼ CM steel are presented in Fig. 5. Although insufficient data have been collected to identify trends with certainty, similar characteristics are exhibited to those shown by the nickel base superalloy and the ½ CMV steel. There is a tendency for crack growth/cycle to be again time dependent at the low frequencies.

In the constant load range cyclic tests, the crack growth and displacement rates both accelerated with crack advance and time. An example of the results obtained in a test performed at $R = 0.8$ on the 2¼ CM steel at a frequency where time dependent processes dominate is shown in Fig. 6. In this ductile steel substantial creep deformation accompanies cracking. For the brittle ½ CMV steel much less deformation was recorded.

In the cyclic tests performed at constant displacement amplitude, relaxation of the load took place with crack extension. The behavior exhibited by the 2¼ CM steel is shown in Fig. 7 and by the ½ CMV steel in Fig. 8. The data are for tests carried out in the low frequency regions. Both materials show a rapid initial decay in load, and a corresponding decrease in crack growth rate, as elastic recovery is exchanged for creep deformation. Eventually the load drops to such a low value that creep deformation and cracking eventually cease. It was found that if the initial

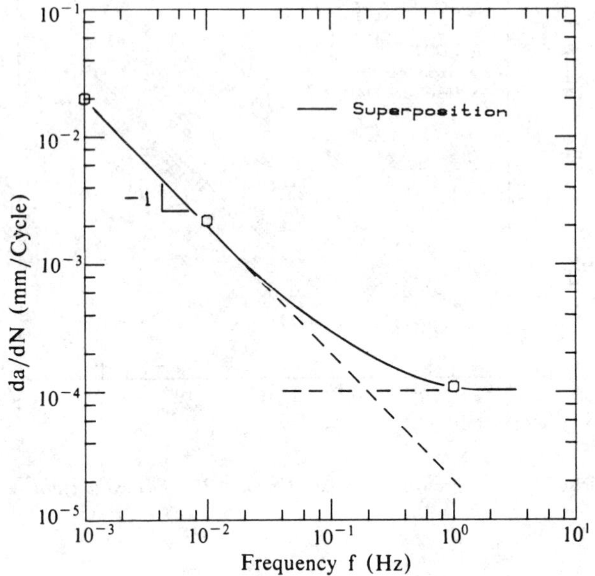

FIG. 5—*Frequency dependence of crack growth/cycle for 2¼ CM steel at* $\Delta K = 10$ *MPa* $m^{1/2}$, R = *0.8, and 538°C.*

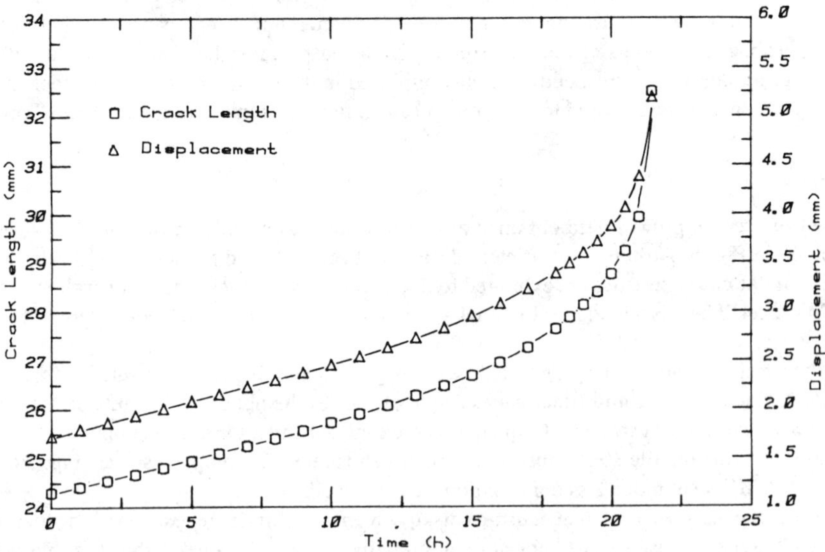

FIG. 6—*Crack growth and displacement versus time for 2¼ CM steel for constant load amplitude cycling at* R = *0.8 and 538°C.*

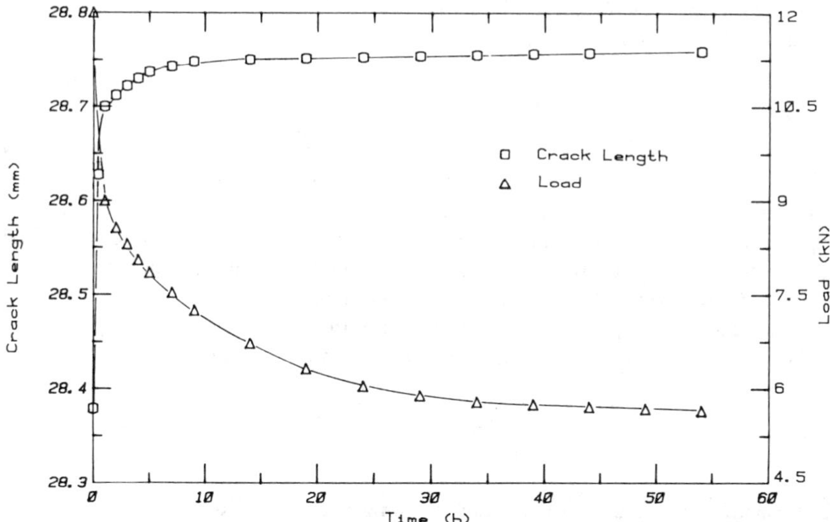

FIG. 7—*Crack growth and load relaxation for 2¹/4 CM steel for constant displacement amplitude cycling at* f = *0.001 Hz.*

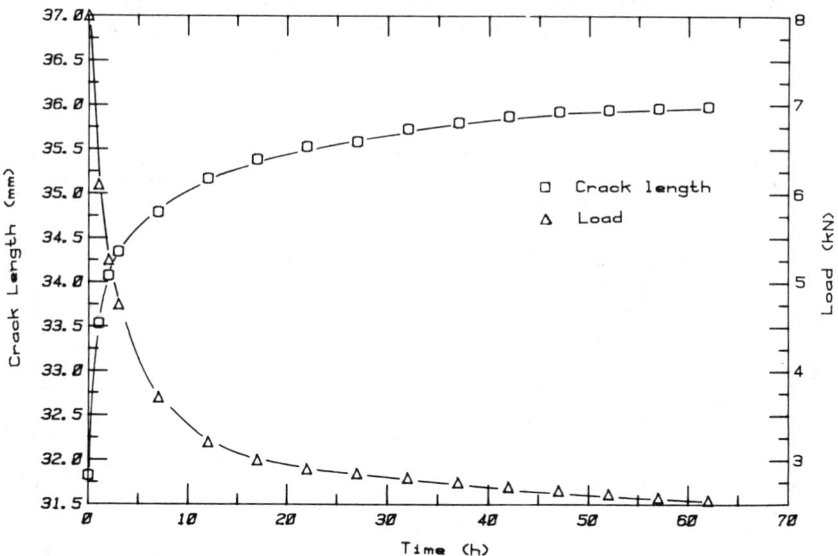

FIG. 8—*Crack growth and load relaxation for ¹/2 CMV steel during constant displacement amplitude cycling at* f = *0.001 Hz.*

load is regenerated and cycling is continued cracking can be reactivated and the whole process repeated.

A comparison of Figs. 7 and 8 indicates that the ductile 2¹/4 CM steel exhibits correspondingly less cracking and load relaxation than the brittle ¹/2 CMV steel. Only about 0.4 mm of crack extension was observed in the 2¹/4 CM steel before cracking ceased whereas about 5 mm was measured in the 1/2 CMV steel. Similar trends were demonstrated in all the constant displacement amplitude tests.

Discussion

Interpretation of Data

The experiments that have been carried out to represent low-cycle fatigue and thermal fatigue loading conditions were all performed at frequencies where it was demonstrated that time-dependent processes dominate. It would be expected, therefore, that crack growth rate in these tests should correlate with C^*. The data are shown in Figs. 9 and 10. The different symbols at the same frequency correspond to data obtained on different test-pieces.

For the static load tests C^* was obtained from the expression [*16*]

$$C^* = \frac{P\dot{\Delta}}{B_n W} \cdot F \tag{4}$$

where P is load, $\dot{\Delta}$ is the creep displacement rate at the specimen loading points, and F is a nondimensional factor which depends on specimen geometry, a/W, and the creep stress index n in the Norton creep law. For the cyclic constant load amplitude data P was taken to be the maximum load of the cycle as most creep damage is anticipated to occur there for $n \gg 1$. For the constant displacement amplitude tests a modified method of calculating C^* is required because of the load relaxation which takes place. During cycling between fixed displacements creep deformation is exchanged for elastic recovery. The instantaneous creep displacement rate at any

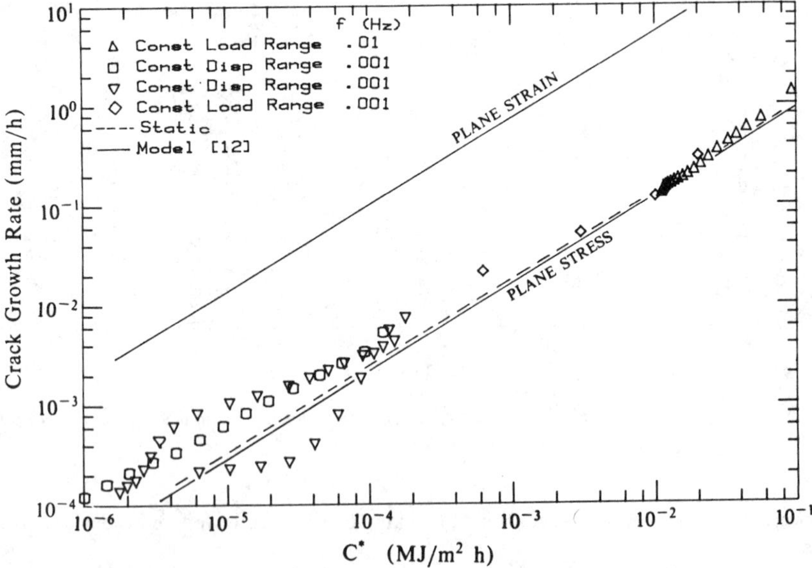

FIG. 9—*Crack growth rate versus* C* *for 2¹/₄ CM steel for low frequency constant load range and constant displacement amplitude tests.*

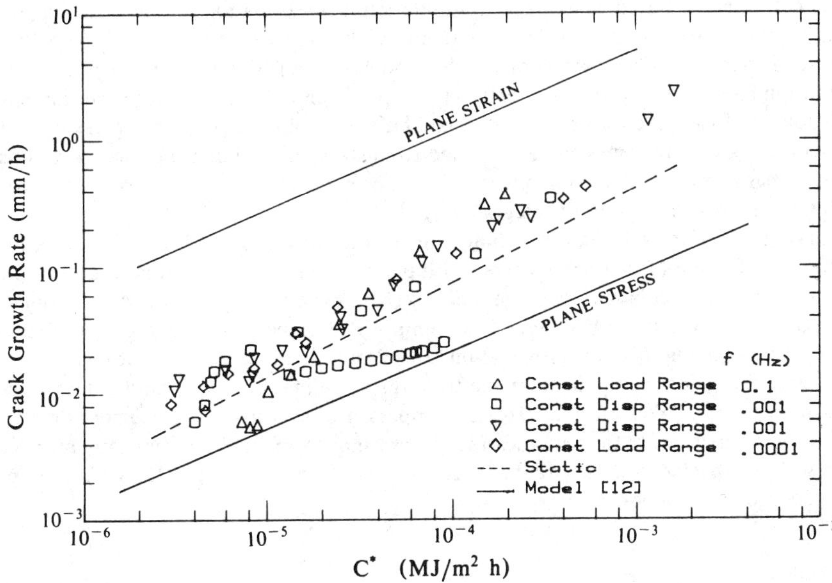

FIG. 10—*Crack growth rate versus* C* *for* 1/2 *CMV steel for low frequency constant load range and constant displacement amplitude tests.*

given crack length can therefore be obtained in terms of the elastic compliance C of a specimen by using the expression

$$\dot{\Delta} = -C(dP/dt) \tag{5}$$

Hence substituting in Eq 4 gives

$$C^* = -\frac{PC}{B_n W} \cdot F \cdot \frac{dP}{dt} \tag{6}$$

Equation 6 was used to calculate C^* in the constant amplitude displacement tests from experimentally measured values of dP/dt.

The results shown in Figs. 9 and 10 demonstrate that in the region where the cyclic crack growth is sensitive to frequency both the constant load range and displacement amplitude cyclic data correlate well with steady load creep crack propagation rates for the brittle and the ductile steels.

Application to Practical Components

The results of this investigation can be used to provide a procedure for predicting crack growth in engineering components subjected to combined static and cyclic loading at elevated temperatures. Where fatigue processes dominate, linear elastic fracture mechanics concepts and the Paris law can be used to estimate crack extension and maximum tolerable defect sizes. When creep processes govern the creep fracture mechanics parameter C^* is more relevant for determining cyclic crack propagation rates. In this region these can be estimated from a model of creep crack growth based on damage accumulation in a process zone at the crack tip [*12–14*] or from an approximate formula [*13*] which requires a knowledge of the material creep ductility and state of stress only. The predictions of the model for plane stress and plane strain loading are shown in Figs. 4, 9, and 10. They bound the experimental data. It is suggested, for components subjected to low-cycle fatigue and thermal fatigue loading where creep processes control, that cyclic crack growth rates can be predicted from static creep data using the creep fracture mechanics parameter C^*. For components, C^* can be calculated by a similar procedure to that used for estimating the J-contour integral [*16,17*].

Accurate predictions will require a knowledge of the state of stress local to a crack tip, as cracking rates 50 times faster are expected with plane strain than plane stress situations at the same value of C^*. Ductile materials are more likely to experience plane stress conditions and brittle materials plane strain loading. The assumption of plane strain loading should result in conservative predictions of crack propagation rates. It is unlikely, because of the load relaxation which takes place, that components made from ductile materials will exhibit extensive crack growth when cycling takes place at elevated temperatures at a fixed displacement amplitude, unless a mechanism exists for periodically regenerating the load. Proportionately more crack growth will be expected in these circumstances with brittle materials, particularly, when plane strain conditions are imposed.

Conclusions

Procedures for characterizing crack growth in components subjected to combined static and cyclic loading at elevated temperatures were proposed. Separate regions of fatigue and creep dominated cracking were identified in a brittle and a ductile low alloy steel and a nickel base superalloy.

At high frequencies where fatigue processes control, it was found that crack growth/cycle can be described in terms of stress intensity factor range ΔK by the Paris law. At low frequencies where time dependent processes govern crack growth rate can be characterized by the creep fracture mechanics parameter C^*. In the low frequency region the same crack propagation rates are obtained in cyclic tests carried out at fixed load range and fixed displacement amplitude as are measured in steady load creep tests performed at the same value of C^* thereby allowing cyclic crack growth rates to be predicted from static data.

During cycling between fixed displacement amplitudes representative of those likely to be experienced in some low-cycle fatigue applications, load relaxation takes place as creep deformation is exchanged for elastic recovery causing crack growth to eventually cease. Significantly more crack growth was observed in the brittle than in the ductile steel. It is unlikely that a component made out of the ductile steel will fracture under this type of loading unless the load can be regenerated periodically.

Acknowledgments

The authors would like to thank Mr. V. Dimopulos for assistance with analysis of the results and some of the testing.

References

[*1*] Solomon, H. D. and Coffin, L. F. Jr., in *Fatigue at Elevated Temperatures, ASTM STP 520,* American Society for Testing and Materials, Philadelphia, 1973, pp. 112–122.

[*2*] Sidey, D. and Coffin, L. F. Jr., in *Fatigue Mechanisms, ASTM STP 675,* American Society for Testing and Materials, Philadelphia, 1979, pp. 528–568.

[*3*] Floreen, S. and Kane, R. H., *Fatigue of Engineering Materials and Structures,* Pergamon Press, Oxford, Vol. 2, 1980, pp. 401–412.

[4] Saxena, A. and Bassani, J. L. in *Proceedings,* 113th AIME Annual Meeting, Los Angeles, J. M. Wells and J. D. Landes, Eds., 1984, pp. 357-383.
[5] Shahinian, P. and Sadananda, K. in *Proceedings,* Fifth International Symposium on Superalloys, The Metallurgical Society of AIME, 1984, pp. 741-750.
[6] Smith, D. J. and Webster, G. A. in *Mechanical Behavior of Materials - IV,* J. Carlson and N. G. Ohlson, Eds., Pergamon Press, Oxford, 1984, pp. 315-321.
[7] Nikbin, K. M. and Webster, G. A. in *Proceedings,* Second International Conference on Creep and Fracture of Engineering Materials and Structures, Pineridge Press, Swansea, 1984, pp. 1091-1103.
[8] Winstone, M. R., Nikbin, K. M., and Webster, G. A., *Journal of Materials Science,* Vol. 20, 1985, pp. 2471-2476.
[9] Pelloux, R. M. and Huang, J. S., "Creep-Fatigue Environment Interactions," in *Proceedings,* TMS-AIME, Milwaukee, Wisc., R. M. Pelloux and N. S. Stoloff, Eds., 1979, pp. 151-164.
[10] Paris, P. and Erdogan, F., *Journal of Basic Engineering, Transactions of ASME (D),* Vol. 85, 1963, pp. 528-534.
[11] Harper, M. P. and Ellison, E. G., *Journal of Strain Analysis,* Vol. 12, 1977, pp. 167-179.
[12] Nikbin, K. M., Smith, D. J., and Webster, G. A., *Proceedings of the Royal Society,* London, Vol. A396, 1984, pp. 183-197.
[13] Nikbin, K. M., Smith, D. J., and Webster, G. A., "Advances in Life Prediction Methods at Elevated Temperatures," in *Conference Proceedings,* ASME, D. A. Woodford and J. R. Whitehead, Eds., 1983, pp. 249-258.
[14] Nikbin, K. M. and Webster, G. A., "Micro and Macro Mechanisms of Crack Growth," in *Conference Proceedings,* Metallurgical Society, AIME, K. Sadananda, B. Rath, and D. J. Michael, Eds., 1981, pp. 137-147.
[15] Webster, G. A. in *Measurement of High Temperature Mechanical Properties of Materials,* M. S. Loveday, M. F. Day, and B. F. Dyson, Eds., HMSO, London, 1982, pp. 255-277.
[16] Smith, D. J. and Webster, G. A. in *Elastic-Plastic Fracture: Second Symposium, Vol. I—Inelastic Crack Analysis, ASTM STP 803,* C. F. Shih and J. P. Gudas, Eds., American Society for Testing and Materials, Philadelphia, 1983, pp. I-654-I-674.
[17] Kumar, V., German, M. D., and Shih, C. F., "Estimation Technique for the Prediction of Elastic-Plastic Fracture of Structural Components of Nuclear Systems," Report SRD-80-094, EPRI, June 1981.

DISCUSSION

J. Bressers[1] *(written discussion)*—The sharp increase in crack growth rate at 700°C in PM Astroloy with decreasing testing frequency is attributed to creep and predicted on the basis of a creep-fatigue interaction model. C^* is used to describe the results. Intergranular failure is advanced as a microstructural argument for creep-controlled behavior.

Bressers and Roth (*Proceedings,* ASME International Conference on Advances in Life Prediction Methods, 1982, pp. 85-92) analyzed the various time-dependent processes which control the cyclic life of PM Astroloy up to 730°C and determined the boundary of operation of these processes. Their work is based on vacuum versus air tests and on a detailed analysis of the deformation and failure modes. Figure 7 in the referenced paper summarizes the results. It appears that, for testing conditions comparable to yours, the strong decrease in life (or increase in crack growth rate) is due to a fatigue-oxidation interaction process. At 700°C creep deformation processes will not operate unless the deformation rate is lowered below $\dot{\epsilon}_t \cong 1 \cdot 10^{-6}\ s^{-1}$.

K. M. Nikbin and G. A. Webster (authors' closure)—Experiments were not performed in a vacuum; therefore the role of oxidation could not be distinguished in the present investigation. It is accepted that any governing process that is time dependent will give an inverse relationship between da/dN and frequency (Figs. 2, 3, and 5). The reason for attributing the time-depen-

[1]Joint Research Centre, C.E.C., Petten, The Netherlands.

dent process to creep is because of the correlation achieved between the cycling and static crack growth data when plotted against C^* (Figs. 4, 9, and 10). Since both types of test were conducted in air, the role of the environment cannot be excluded. However, the crack growth rates obtained are consistent with a model of cracking which is based on the exhaustion of the available creep ductility in a process zone at the crack tip (see Ref *12* of our paper). If oxidation reduces creep ductility, it will also be expected to give enhanced crack growth rates.

M. Nazmy[2] *(written discussion)*—What is the justification for using the energy rate integral concept C^* in correlating the creep crack growth data? Have you calculated the transition time that is required to go from the K-controlled to the C^*-controlled process?

K. M. Nikbin and G. A. Webster (authors' closure)—Transition times for the materials and conditions examined were typically <1 h, whereas test times were in the range of 100 to 1000 h. The main justification for using C^* was the improved correlations achieved and because the trends obtained are consistent with a model (see Ref *12* of our paper) of the crack growth process which is derived from uniaxial creep properties.

H. W. Liu[3] *(written discussion)*—McMeeking and Parks (1979), Shih and German (1981), and Zhuang and Liu (1985) have found that the crack tip fields at the same J-value may vary widely. The crack tip fields are strongly affected by specimen geometry and loading level. In plane-strain general yielding, the crack tip field of a center-cracked panel in tension differs considerably from that of a compact tension specimen or a three-point-bend specimen. Do you find that C^* can correlate well the data from center-cracked panels in tension with those from compact tension specimens or three-point-bend specimens?

K. M. Nikbin and G. A. Webster (authors' closure)—The same general creep crack growth behavior has been observed on a wide range of specimen geometries when crack propagation rate is plotted against C^*. It has been shown, however, that crack growth rate can be 50 times faster under plane strain conditions than when plane stress prevails at the same value of C^*. Whether plane strain or plane stress conditions exist will depend on specimen geometry and material creep ductility. Criteria have not yet been developed for specifying the degree of constraint to be expected in a given creep crack growth circumstance.

[2]Scientist, Brown Boveri Research Center, Baden-Dattwil, Switzerland.
[3]Department of Mechanical and Aerospace Engineering, Syracuse University, Syracuse, NY 13210.

J. Gayda,[1] *T. P. Gabb,*[1] *and R. V. Miner*[1]

Fatigue Crack Propagation of Nickel-Base Superalloys at 650°C

REFERENCE: Gayda, J., Gabb, T. P., and Miner, R. V., "**Fatigue Crack Propagation of Nickel-Base Superalloys at 650°C,**" *Low Cycle Fatigue, ASTM STP 942,* H. D. Solomon, G. R. Halford, L. R. Kaisand, and B. N. Leis, Eds., American Society for Testing and Materials, Philadelphia, 1988, pp. 293–309.

ABSTRACT: The 650°C fatigue crack propagation behavior of two nickel-base superalloys, René 95 and Waspaloy, were studied with particular emphasis placed on understanding the roles of creep, environment, and two key grain boundary alloying additions, boron and zirconium. Comparison of air and vacuum data showed the air environment to be detrimental over a wide range of frequencies for both alloys. In-depth analysis of René 95 showed that at lower frequencies, such as 0.02 Hz, failure in air occurred by intergranular, environmentally assisted creep crack growth, while at higher frequencies, up to 5.0 Hz, environmental interactions were still evident but creep effects were minimized. The effect of boron and zirconium in Waspaloy was found to be important where environmental and/or creep interactions were present. In those instances, removal of boron and zirconium dramatically increased crack growth. It is therefore plausible that effective dilution of these elements may explain, in part, a previously observed trend in which crack growth rates increased with decreasing grain size.

KEY WORDS: nickel-base superalloys, fatigue crack propagation, creep, environment

Fatigue is usually the life-limiting factor in modern aircraft turbine disks. Good resistance to fatigue crack propagation is important as well as resistance to crack initiation, since a crack once present must not propagate to failure between inspection intervals. Previous studies of several nickel-base, γ' strengthened disk alloys have shown that the advanced, fine grain, high strength alloys possess relatively poor fatigue crack propagation resistance at turbine disk rim temperatures, particularly when the fatigue cycle includes tensile dwell periods [*1,2*]. Since these studies were conducted on commercial alloys rather than a specially designed series of alloys, it was not possible to conclude whether the poor fatigue crack propagation resistance of the advanced alloys was due to their generally finer grain size, higher strength, or possibly lower environmental resistance. More recent studies, in which grain size and strength were systematically varied for single superalloy compositions, have shown that fine grain size, but not high strength in itself, leads to accelerated crack growth at 650°C [*3,4*]. Further, it was shown that the high fatigue crack propagation rates of the fine grain alloys were associated with an intergranular mode of failure even at relatively rapid test frequencies, and that this mode did not occur in the absence of air. In dwell tests, failure was intergranular in both large and fine grain alloys, however, larger grain size still reduced the rate of crack propagation.

To gain a more detailed understanding of the rapid crack growth behavior of the fine grain alloys it is necessary to quantitatively gage the relative importance of creep, fatigue, and environmental factors in the failure process. This question will be addressed in the first part of the present paper by comparing crack growth behavior of René 95, a fine grain, high strength nickel-base superalloy, as a function of test frequency, dwell conditions, and environment.

[1]NASA-Lewis Research Center, Cleveland, OH 44135.

To address why finer grain superalloys exhibit more rapid fatigue crack propagation, one mechanism relating grain size and grain boundary chemistry to fatigue crack propagation resistance will be assessed. The mechanism to be examined, which is essentially that proposed by Bain and Pelloux for creep crack propagation [*5*], is that since decreasing grain size increases grain boundary volume it effectively decreases the concentration of B and Zr at grain boundaries. Both elements are thought to enhance grain boundaries properties in superalloys and their effective dilution may therefore be responsible for the rapid intergranular fatigue crack propagation observed in fine grain superalloys. To test this hypothesis crack propagation tests were run on relatively fine grain Waspaloy with normal and reduced levels of B and Zr. The results of these tests and their relation to the fine grain nickel-base superalloys will be discussed in the second part of the present paper.

Materials and Procedures

Creep-Fatigue-Environment Study

Crack growth tests designed to study the effects of creep, fatigue, and environmental interactions were conducted on a single nickel-base superalloy of current interest to the aerospace community, René 95. This alloy is typical of the current generation of fine grain, high strength nickel-base superalloys used as disk materials in the hot section of aircraft turbine engines. It is strengthened by the ordered intermetallic γ' phase and is produced by powder metallurgy technology.

Argon atomized powder having the composition listed in Table 1 was used in the present study. This powder was hot isostatically pressed (HIP) at 1120°C and 105 MPa for 3 h producing essentially 100% density. The product was subsequently solution treated at 1120°C for 1 h, air cooled, and aged at 760°C for 8 h. The resulting microstructure has an average grain size of 8 microns and contains about 50 vol% of γ' precipitates about 0.1 μm in diameter plus a small amount of the MC carbide. Tensile properties for the René 95 studied are presented in Table 1 and are typical of HIP René 95.

All crack growth data were generated using the compact tension specimen shown in Fig. 1 which was modified to maximize material usage. Both cyclic (R-ratio of 0.05) and static crack growth tests were run at 650°C under load control using a closed-loop servohydraulic test system equipped with an electric resistance furnace in a chamber which may be evacuated by a diffusion pump. Vacuum tests were run at a pressure between 1×10^{-5} and 1×10^{-6} torr. All other tests were run in laboratory air.

TABLE 1—*Composition and 650°C tensile properties of René 95.*

Element	Weight Percent	Tensile Properties	
Cr	13.0	0.2% Yield Strength:	1070 MPa
Co	8.2	Ultimate Tensile Strength:	1410 MPa
Mo	3.3	Elongation:	13%
W	4.3		
Nb	3.6		
Ti	2.0		
Al	3.7		
C	0.047		
B	0.020		
Zr	0.025		
Ni	bal.		

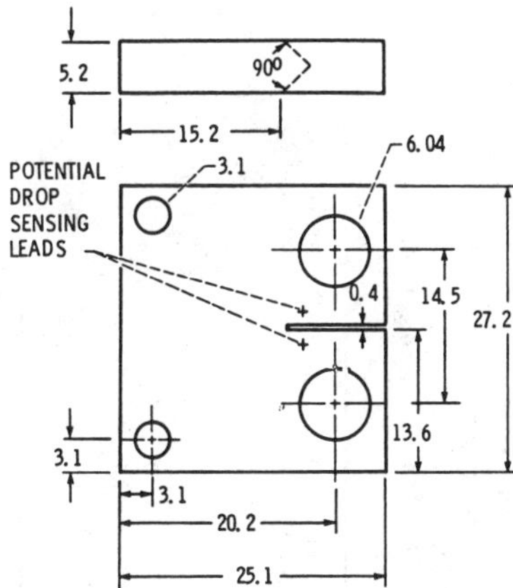

FIG. 1—*Design of the modified compact tension specimen (all dimensions in millimetres).*

All compact tension specimens were precracked at room temperature in laboratory air at 20 Hz. A symmetric, triangular waveform was used for the cyclic crack growth tests, which were run at frequencies of 0.02, 0.33, and 5 Hz. Tests with a 120 s dwell period at maximum load, but otherwise the same as the 0.33 Hz tests, were run to accentuate creep interactions. For the cyclic crack propagation tests, crack growth rates per cycle, da/dn, were monitored with a d-c potential drop system described elsewhere [*3*] and were correlated with the stress intensity range, ΔK. For the static crack propagation tests, crack growth rates per time, da/dt, were also monitored with a d-c potential drop system and were correlated with the stress intensity factor, K. The K calibration curve for the modified specimen geometry was calculated using a boundary collocation scheme developed by Newman [*6*]. At least two crack growth tests were run for every condition investigated herein.

Grain Boundary Chemistry Study

The effect of grain boundary chemistry on crack propagation rates was studied in a second nickel-base superalloy, Waspaloy. The use of HIP René 95 in this experiment was prohibited by the cost and time required to produce nonstandard compositions by powder metallurgy. However, the alloy chosen, Waspaloy, could be produced with the available facilities by casting ingots of compositions desired and extruding them to obtain a fine grain microstructure.

Waspaloy is an older generation nickel-base superalloy with lower strength than René 95 primarily due to its lower γ' content. Nevertheless, work by Lawless [*7*] on Waspaloy and Gayda, Miner, and Gabb [*4*] on several nickel-base superalloys has shown that finer grain sizes tend to promote rapid, intergranular crack growth. Thus it was felt that the behavior of Waspaloy is representative of this class of superalloys and that Waspaloy was a suitable vehicle for studying the effects of grain boundary chemistry on crack growth behavior.

After several casting iterations, ingots with "normal" and low levels of B and Zr were obtained having otherwise nearly identical compositions (Table 2). To refine the as-cast structure, these ingots were extruded at 1060°C using a 6:1 reduction ratio. The extrusions were then solutioned at 1010°C for 2 h, forced air cooled, and then given a two-step aging treatment of 845°C for 4 h and 760°C for 16 h. This heat treatment is similar to that used for commercial disk applications except the solution temperature was lowered slightly to prevent grain growth.

TABLE 2—*Composition (weight percent) and 650°C tensile properties of the two Waspaloy extrusions.*

Element	Low B+Zr Alloy	"Normal" Alloy	Material	0.2% Yield Strength, MPa	Ultimate Tensile Strength, MPa	Elongation, %
Cr	20.8	20.7	low B+Zr	810	1080	12
Co	12.8	12.8	"normal" alloy	770	1090	28
Mo	4.58	4.48				
Ti	3.53	3.55				
Al	1.81	1.79				
C	0.045	0.043				
B	0.001	0.025				
Zr	0.001	0.053				
Ni	bal.	bal.				

The resulting microstructures for both compositions are shown in Fig. 2. The grain sizes are both about 40 μm. Both alloys were found to contain approximately 25 vol% γ' and a small amount of the MC and $M_{23}C_6$ carbides. The "normal" composition also contains a small amount of the M_3B_2 boride.

Crack growth tests were run on modified compact tension specimens previously described. Test temperature and procedures were identical to that already described for René 95. Cyclic crack propagation tests were run in air and vacuum at a frequency of 0.33 Hz with and without a 120 second dwell at maximum load. The specimens were oriented so as to produce fracture perpendicular to the extrusion direction. The axes of tensile and creep specimens were also oriented to fracture in this plane.

Results and Discussion

Creep-Fatigue-Environment Study

Fatigue crack growth data for the fine grain nickel-base superalloy René 95 are plotted in Figs. 3 and 4 for tests in air and vacuum, respectively. As seen in Fig. 3, crack growth rates in

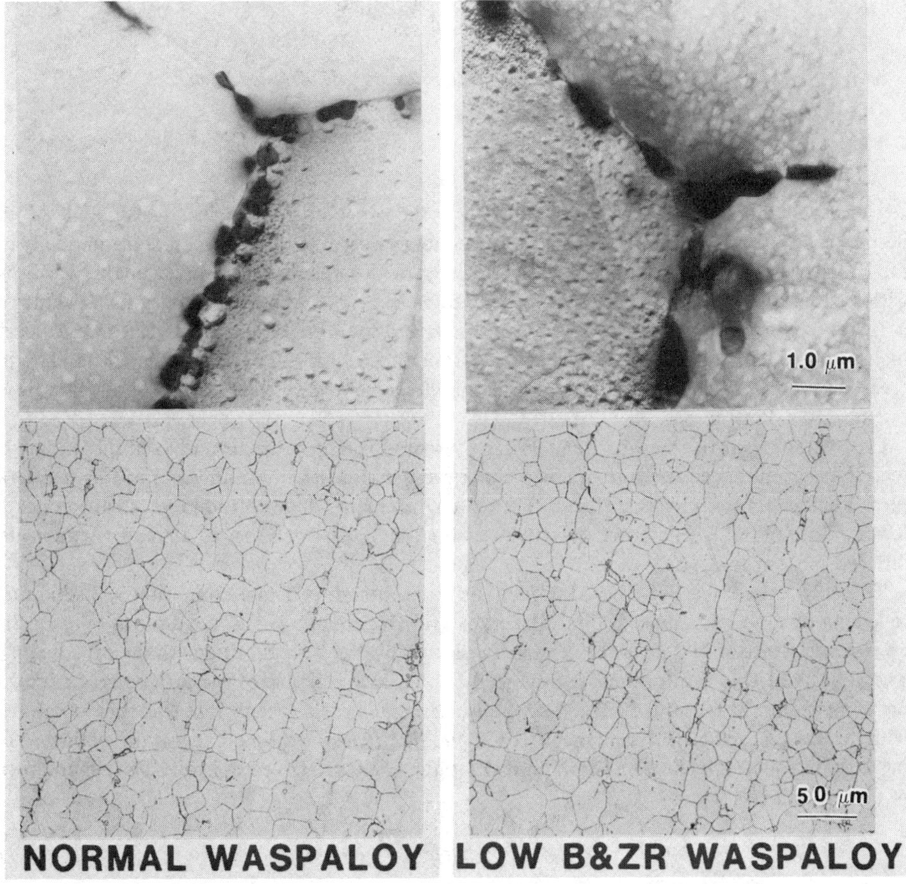

FIG. 2—*Microstructures of the two Waspaloy compositions.*

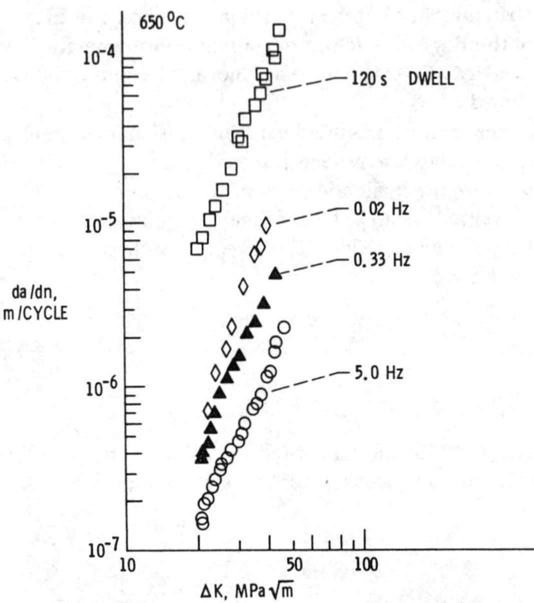

FIG. 3—*Fatigue crack growth rates of René 95 tested in air.*

air varied by about one order of magnitude over the range of frequencies investigated. As expected, crack growth rates increased as test frequency decreases and increased dramatically for the 120-s dwell test. In vacuum (Fig. 3) the crack growth rates showed a much smaller variation with frequency and the 120 s dwell test is seen to be less damaging, especially at lower values of ΔK. Comparing tests of the same frequency in air and vacuum, crack growth rates were clearly higher in air. This effect is seen to be most pronounced in the lower frequency tests. The increase of the crack growth rate in air was about a factor of two for the 5 Hz tests, but fifty times for the 120 s dwell tests.

Creep crack growth rates for René 95 are presented in Fig. 5 for air and vacuum. It may be seen that air enhanced the rate of creep crack growth almost a thousand fold at this temperature. Further, crack growth rates in the two environments showed the greatest difference at lower values of K. Cyclic crack growth rates for the 120 s dwell test in air and vacuum also differed most at low ΔK.

To analyze the effects of creep, fatigue, and environment in a more quantitative fashion the following scheme was adopted. First, the crack growth data at 5 Hz in vacuum was taken as the approximate rate of fatigue crack growth unaffected by creep or the air environment, $(da/dn)_f$. The validity of this assumption is supported by the near collapse of the fatigue crack growth data in vacuum at 5 and 0.33 Hz. This "pure" fatigue crack growth rate was then combined with the creep crack growth rate measured in air, da/dt, using a time integration method similar to that proposed by Saxena [*8*] to predict fatigue crack growth rates in air. The general form of this equation is

$$da/dn = (da/dn)_f + \int (da/dt)dt \tag{1}$$

This expression is then evaluated at a given value of ΔK.

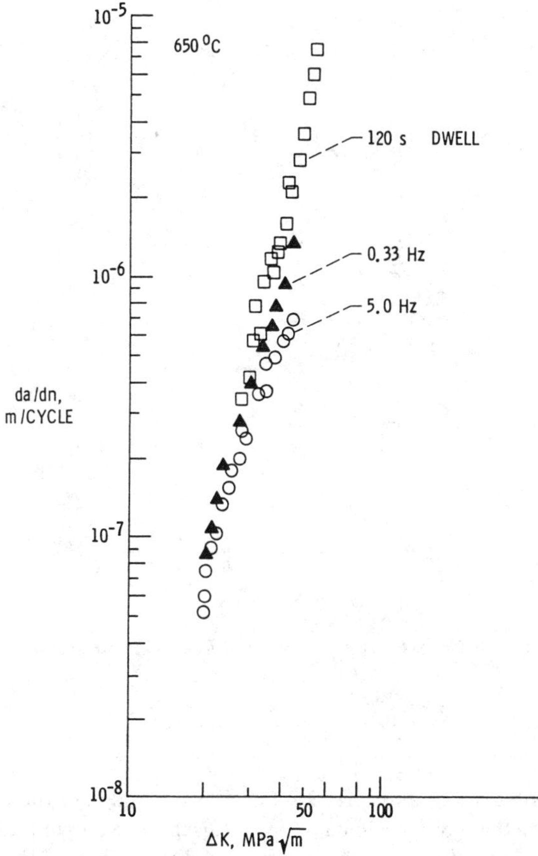

FIG. 4—*Fatigue crack growth rates of René 95 tested in vacuum.*

The first term in Eq 1 was approximated by the Paris expression

$$(da/dn)_f = B\Delta K^m$$

$$= 1.07 \times 10^{-11} \Delta K^{2.95} \tag{2}$$

The integral in Eq 1 was evaluated over one cycle from $t_1 = 0$ to $t_2 = (1/\nu) + t_{hold}$, where ν is the frequency of the test and t_{hold} is the duration of any dwell.

Before the integral can be evaluated an expression for da/dt must be formulated. In this case the following expression was adopted:

$$da/dt = AK^n$$

$$= 1.37 \times 10^{-16} K^{6.2} \tag{3}$$

This is merely a convenient form which accurately represents the creep crack growth data in air.

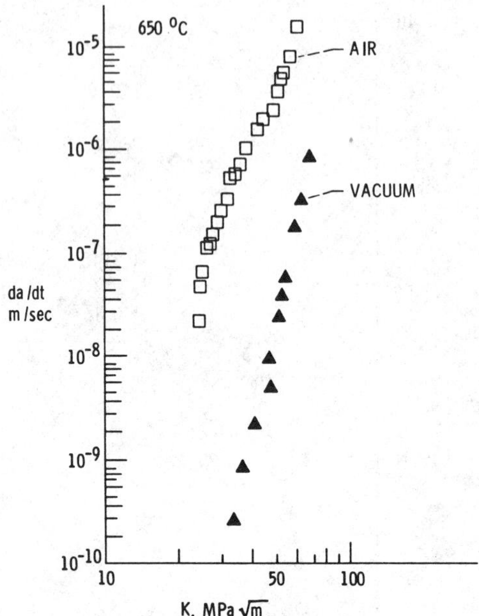

FIG. 5—*Creep crack growth rates of René 95 in air and vacuum.*

To perform the integration, an expression for K as a function of time within the cycle was then written based on the waveform in question, in this instance, a symmetric, triangular waveform with an optional dwell period at maximum load. The form of this expression and the mechanics of the integration are described in the Appendix. The final relation is

$$da/dn = B\Delta K^m + A\Delta K^n[Z/(\nu(n + 1)) + t_{hold}/(1 - R)^n] \quad (4)$$

where

$$Z = (1 - R^{n+1})/(1 - R)^{n+1}$$

Evaluating the above expression at various values of ΔK for the 5, 0.33, and 0.02 Hz tests as well as the 120 s dwell test yielded the plots shown in Fig. 6. Also shown for comparative purposes is the 5 Hz vacuum data used to approximate the "pure" fatigue crack growth rates.

Before proceeding, two points should be made about the time integration scheme used herein. First, the effect of creep crack growth is overestimated at low ΔK, since the expression for the creep crack growth rate in air, Eq 3, does not reflect the sharp decrease in da/dt below $K = 30$ MPa $m^{1/2}$ shown in Fig. 5. The effect is, however, relatively small on predicted fatigue crack growth rates at or above $\Delta K = 30$ MPa $m^{1/2}$, since the majority of the creep component in a fatigue cycle is incurred near K_{max} with the waveforms used in this study. Second, the time integration scheme as applied herein simply adds the creep component as measured in air to the "pure" fatigue component. The effects of environment are ignored except as they effect creep crack growth. Any interactive effects between the air environment and fatigue loading are not considered.

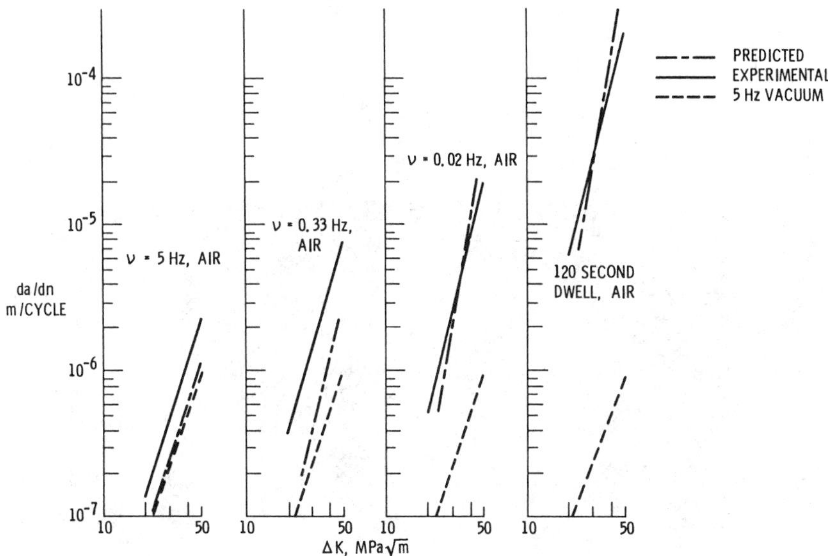

FIG. 6—*Comparison of predicted and observed crack growth rates of René 95 at 650°C.*

Examination of the predicted fatigue crack growth rates in air (Fig. 6) shows that simple addition of the air creep crack growth rates to the "pure" fatigue crack growth rate does not adequately explain the deleterious effect of the air environment at 5 and 0.33 Hz. At these higher frequencies the predicted crack growth rates are only slightly greater than the "pure" fatigue crack growth rates, especially at low ΔK. This suggests that there is an additional environmental effect in these fatigue tests which is not present in the creep crack growth tests run in air. Figure 7 shows that the 0.33 Hz tests exhibited an intergranular mode of failure, which indicates that some form of environmental and/or creep damage is occurring; however, the mode of failure was predominantly transgranular in the 5 Hz air tests, with only a limited amount of intergranular failure evident.

While simple addition of the air creep crack growth rate to the "pure" fatigue crack growth rate underpredicts fatigue crack growth for the higher frequency test it does much better for the cycles with a larger creep component, the 0.02 Hz and 120-s dwell tests (Fig. 6). In these tests it would appear that any acceleration in crack growth due to additional environmental interactions seen at higher frequencies is largely masked by the high crack growth rates produced by the large amount of creep crack growth in air. For example, the crack growth rate in air for the 120 s dwell test is over two orders of magnitude greater than the "pure" fatigue crack growth rate due to the large amount of creep crack growth per cycle.

The effect of air on the high crack growth rates of the 0.02 Hz and 120 s dwell tests in air is by no means minimal. This was demonstrated by the extremely low crack growth rates for the 120 s dwell test in vacuum relative to that in air. Furthermore, the fracture modes of the air and vacuum 120 s dwell tests were clearly different (Fig. 7), that in air being decidedly more intergranular.

On the basis of these observations it appears that several failure mechanisms are operative in air run over the range of frequencies studied. At relatively high frequencies, represented by the tests at 5 Hz, it is probable that the rate of crack growth outpaces the rate of oxygen penetration along grain boundaries, as evidenced by the transgranular fracture mode. Here the mild effect

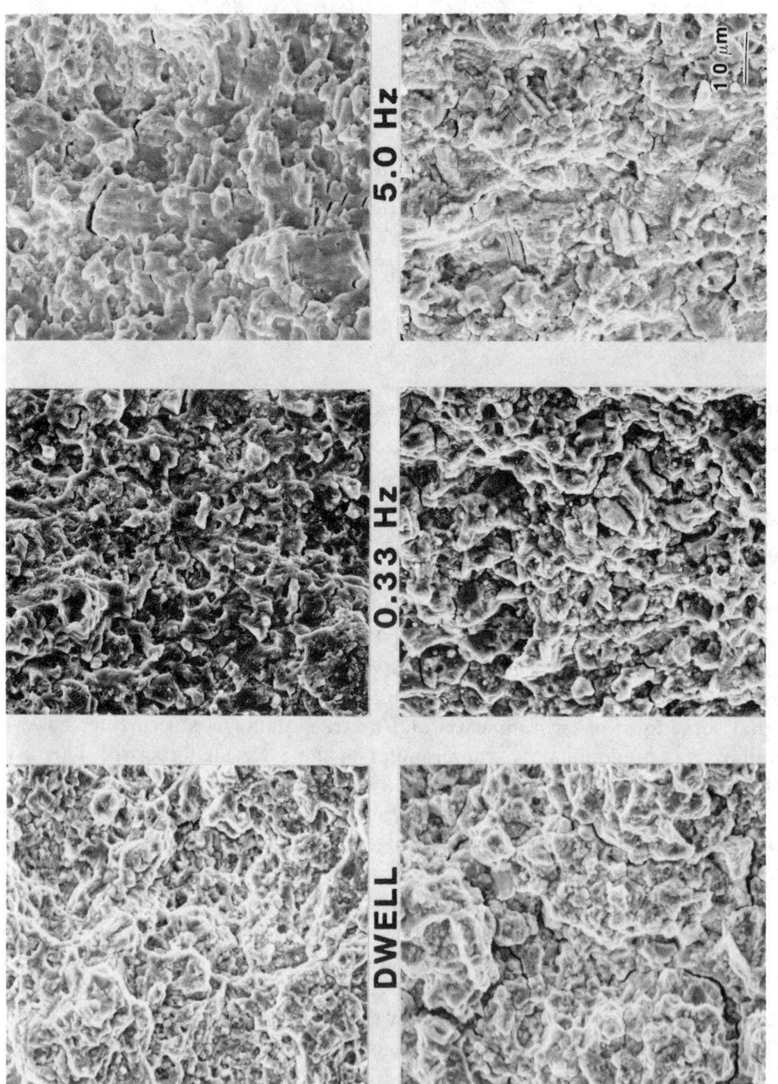

FIG. 7—*Fracture modes observed in air* (bottom) *and vacuum* (top) *for René 95 crack growth tests at 650°C.*

of air is probably related to environmental effects at or very near the crack tip. These could include penetration and embrittlement by oxygen ahead of the crack tip at slip bands or oxidation of newly formed crack surfaces that would inhibit "rewelding".

At intermediate frequencies, represented by the tests at 0.33 Hz, the rate of oxygen penetration along grain boundaries is probably significant compared with the rate of crack propagation, as indicated by the more rapid, intergranular failure mode in air. Here the crack is propagating along grain boundaries apparently affected by air. It is possible that the grain boundaries have been embrittled by the actual depletion of oxide forming elements as proposed by Gell and Leverant [*9*], or, more likely at this test temperature, just by the presence of oxygen diffused into the grain boundaries ahead of the crack tip.

At still lower frequencies, represented by the 0.02 Hz and the 120 s dwell tests, the rate of oxygen penetration along grain boundaries must again be significant compared to the rate of crack propagation. However, there appears to be sufficient time in these cycles that the mechanism of crack growth resembles that occurring during static creep in an air environment as evidenced by the success in predicting crack growth rates at lower frequencies. Under these conditions oxygen may weaken or embrittle grain boundaries as previously described, or it may also increase creep rates, and therefore crack growth rates by altering cavitation kinetics as proposed by McLean [*10*].

Regardless of the mechanisms involved, it is apparent that for René 95 fatigue crack growth rates for cycles with a substantial creep component are controlled by intergranular, environmentally assisted creep crack growth, while at higher frequencies fatigue crack growth rates are enhanced by the presence of the air environment with minimal creep interactions.

Grain Boundary Chemistry Study

Before examining the effects of B and Zr on crack growth in Waspaloy, a short discussion of tensile and creep properties is in order. As seen in Table 2, the 650°C yield and ultimate tensile strengths were virtually unaffected by lowering B and Zr concentrations. The tensile elongation of the low B and Zr alloy was diminished but still acceptable at 12%, compared with 28% for the "normal" Waspaloy. The strength levels of both alloys were comparable to commercial Waspaloy as is the ductility of the "normal" alloy. The creep properties of the low B and Zr alloy, both rupture life and ductility, were also degraded as anticipated (Fig. 8).

The crack growth behavior in air and vacuum at 0.33 Hz for both alloys is shown in Fig. 9. Lowering the B and Zr level led to a threefold increase in crack growth rates in air. There was also a corresponding transition in fracture appearance to a more intergranular mode with lowered B and Zr levels (Fig. 10). In vacuum the crack growth rates of both alloys were nearly equivalent and less than that of either alloy in air. In addition, both alloys had a transgranular fracture appearance in vacuum (Fig. 10).

The crack growth behavior in air and vacuum for the 120 s dwell test for both alloys are plotted in Fig. 11. In air the low B and Zr alloy had a crack growth rate more than one order of magnitude greater than the "normal" alloy. For both alloys the crack growth rate was greater than that observed at 0.33 Hz, although the effect in the low B and Zr alloy was much greater. For both alloys the mode of crack growth for the 120 s dwell test in air was predominantly intergranular. As in the 0.33 Hz tests, the crack growth rates of both alloys were suppressed in vacuum, but the crack growth rate of the alloy with lower B and Zr levels remained quite high, well above that obtained in air for the alloy with higher levels of B and Zr. Fracture in the low B and Zr alloy was predominantly intergranular for the vacuum dwell tests; however, that of the "normal" alloy was mixed (Fig. 10).

These results clearly show that removing B and Zr adversely effects the crack growth behavior of Waspaloy when environmental and creep components are present alone or in combination.

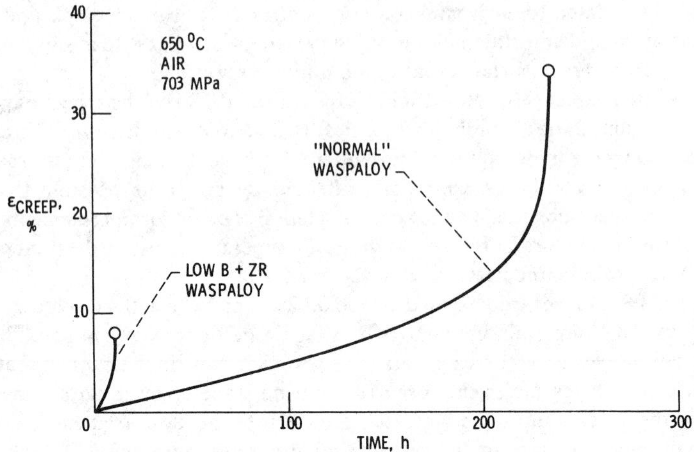

FIG. 8—*Creep curves of the two Waspaloy compositions.*

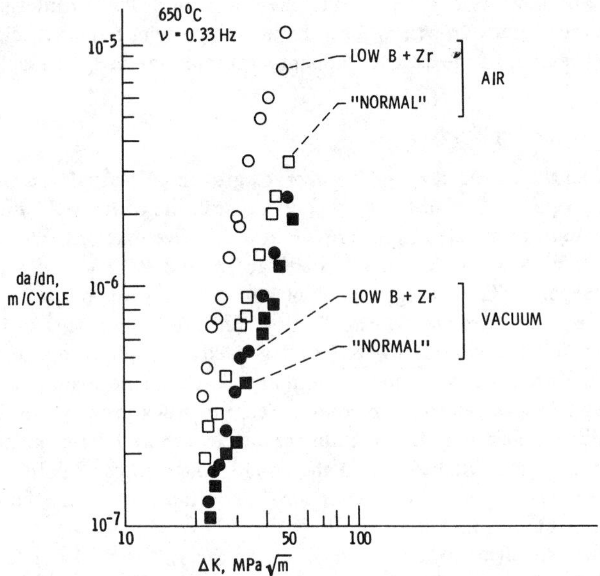

FIG. 9—*Fatigue crack growth rates of the two Waspaloy compositions.*

However, these two alloy additions have little effect on the "pure" fatigue crack growth process where failure is predominantly transgranular.

Comparison of Alloy Behaviors

The effects of air and tensile dwell on the crack growth behavior of René 95 and the two Waspaloy compositions are compared in Fig. 12. The crack growth rates at $\Delta K = 30$ MPa m$^{1/2}$ for the 0.33 Hz (non-dwell) and 120 s dwell tests in both vacuum and air are presented. The 0.33

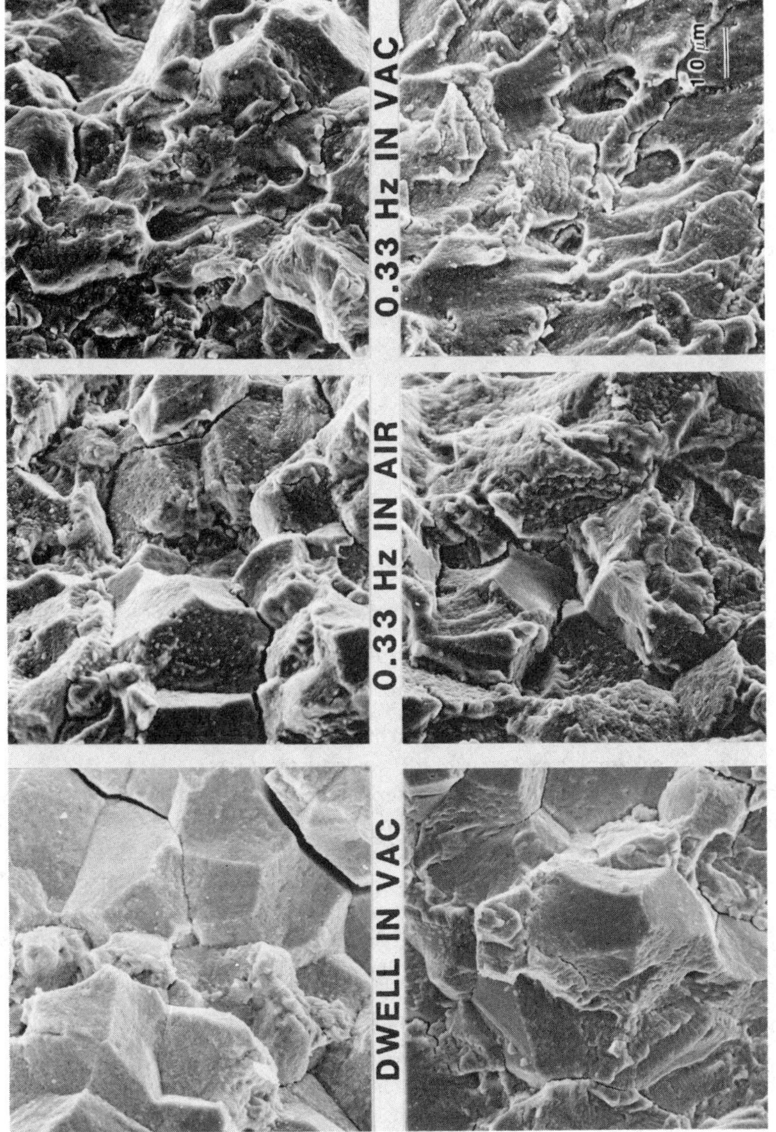

FIG. 10—*Fracture modes observed in crack growth tests run at 650°C on normal* (bottom) *and low B&Zr* (top) *Waspaloy compositions.*

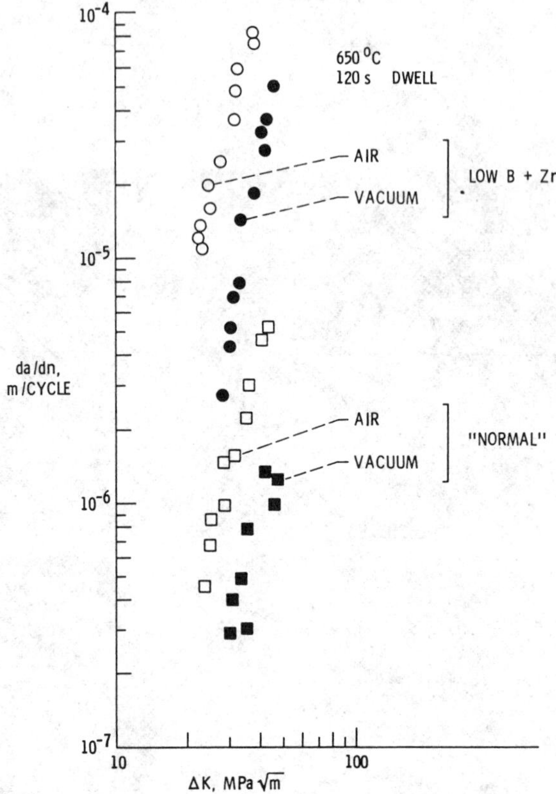

FIG. 11—*Fatigue crack growth rates in the 120 s dwell tests run on the two Waspaloy compositions.*

Hz tests are the highest frequency tests conducted on all three alloys in vacuum and are taken as the "pure" fatigue crack growth behavior in this comparison.

It may be seen that the three alloys exhibit differing effects of air and dwell individually or in combination. However, the "pure" fatigue behavior of the three alloys is very similar and, as previously stated, removing B and Zr has little effect on crack growth behavior in this regime.

In vacuum, both René 95 and "normal" Waspaloy showed essentially no effect of the 120 s tensile dwell. In the absence of air, creep loads have minimal effect on the crack growth behavior of these two alloys. However, Waspaloy with low B and Zr exhibited a more than tenfold increase in crack growth rates due to the application of the 120 s dwell in vacuum, reflecting its low creep strength and ductility.

In air, all three alloys showed an increase in crack growth rates for both test cycles. However, the increase in crack growth rates was larger for René 95 and the Waspaloy with low B and Zr than for "normal" Waspaloy, particularly for the 120 s dwell test. Low B and Zr levels were clearly detrimental to Waspaloy in both test cycles. This is unlike the findings of Floreen and Davidson [*11*] on the Ni-Fe-base superalloy they studied, NIMONIC PE16. It may be that PE16 is not as environmentally sensitive [*12*] as the general class of Ni-base superalloys with higher fractions of γ'.

As the behavior of Waspaloy with low B and Zr somewhat mimics that of the fine-grain René 95, support is given to a mechanism relating effective grain boundary composition to crack growth resistance such as that proposed by Bain and Pelloux [*5*]. The effect of decreasing grain

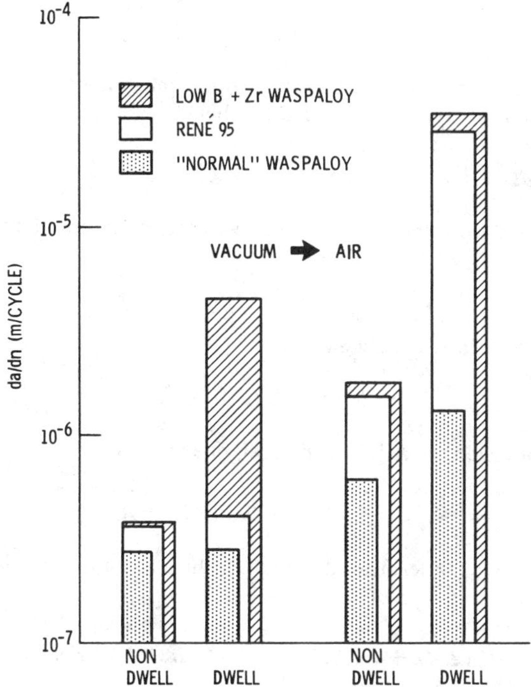

FIG. 12—*Comparison of environmental and creep effects on fatigue crack growth behavior of René 95 and the two Waspaloy compositions.*

size in increasing fatigue and creep-fatigue crack growth rates of superalloys in air [*2,3*] may result, at least partly, from an effective dilution of the B and Zr concentration at grain boundaries.

Still, the effect of environment on René 95 is even more severe than that on Waspaloy with low B and Zr under creep-fatigue conditions. The high crack growth rates for Waspaloy with low B and Zr in the 120 s dwell test in air appears to be as much due to basically poor creep behavior as to environmental sensitivity. For René 95 the 120 s dwell had no effect except in tests conducted in air. Thus, other factors, such as generally lower Cr concentration, probably also contribute to the enhanced crack growth rates of the advanced, fine-grain superalloys like René 95 when creep, fatigue, and environment interactions are prevalent.

Conclusions

The results of this and previous studies show that rapid intergranular crack propagation observed in high strength, fine grain Ni-base superalloys, such as René 95, are a result of a strong environmental interaction. For low frequencies or cycles incorporating a tensile dwell, the damage mechanism is similar to that found in creep crack growth in air. Yet even for high frequency tests where creep interactions are minimal, the damaging effect of air is quite large. It has been shown previously that such environmental sensitivity is accentuated by fine grain microstructures. The present results on the effects of B and Zr in Waspaloy lends some support to the concept which attributes diminished crack growth resistance of fine grain nickel-base superalloys, in part, to effective dilution of these elements at grain boundaries.

APPENDIX

To evaluate the creep component of crack growth, the integral is broken into three parts, the ramp up, the hold, and the ramp down, as shown below:

$$\int (da/da)dt = \int_0^{1/2v} AK^n dt + AK_{max}t_{hold} + \int_{1/2v}^{1/v} AK^n dt^n$$

Since the first and the third term are numerically equivalent they may be combined as follows:

$$=2\int_0^{1/2v} AK^n dt + AK_{max}t_{hold}$$

K is a linear function of time, t, as shown below:

$$K = 2v\Delta Kt + K_{min}$$

Substituting this expression for K in the integral, one obtains:

$$=2\int_0^{1/2v} A(2v\Delta Kt + K_{min})^n dt + AK_{max}t_{hold}$$

Expressing K_{min} and K_{max} in terms of ΔK and R, the load ratio, the following expression is obtained:

$$=2\int_0^{1/2v} A(2v\Delta Kt + (R\Delta K)/(1 - R))^n dt + (A\Delta K^n t_{hold})/(1 - R)^n$$

The results of the integration are shown below:

$$=((A\Delta K^n)/(vn + n))[(2vt + R/(1 - \mathrm{R}))^{n+1}]^{t=1/2v} + (A\Delta K^n t_{hold})/(1 - R)^{n^{t=0}}$$

Note that ΔK is constant for a given cycle and has been removed from the intergrand as ΔK^n. Evaluating the limits yields the following expression:

$$=[(A\Delta K^n)/(vn + n)][(1 + Q)^{n+1} - Q^{n+1}] + (A\Delta K^n t_{hold})/(1 - R)^n$$

where

$$Q = R/(1 - R)$$

This can be simplified by combining like terms to yield the second term of Eq 4:

$$=A\Delta K^n[Z/(v(n + 1)) + t_{hold}/(1 - R)^n]$$

where

$$Z = (1 - R^{n+1})/(1 - R)^{n+1}$$

References

[1] Cowles, B. A., Sims, D. L., Warren, J. R., and Miner, R. V., *Journal of Engineering Materials Technology*, Vol. 102, No. 4, Oct. 1980, pp. 356–363.

[2] Cowles, B. A. and Warren, J. R., "Evaluation of the Cyclic Behavior of Aircraft Turbine Disk Alloys," NASA CR-165123, National Aeronautics and Space Administration, Washington, D.C., Aug. 1980.

[3] Gayda, J. and Miner, R. V., *Metallurgical Transactions A*, Vol. 14A, No. 11, Nov. 1983, pp. 2301–2308.
[4] Gayda, J., Miner, R. V., and Gabb, T. P. in *Superalloys 1984*, M. Gell, Ed., The Metallurgical Society of the AIME, Warrendale, Pa., 1984, pp. 731–740.
[5] Bain, K. R. and Pelloux, R. M. in *Superalloys 1984*, M. Gell, Ed., The Metallurgical Society of AIME, Warrendale, Pa., 1984, pp. 387–396.
[6] Newman, J. C. in *Fracture Analysis, ASTM STP 560*, American Society for Testing and Materials, Philadelphia, 1974, pp. 105–121.
[7] Lawless, B. H., "Correlation Between Cyclic Load Response and Fatigue Crack Propagation Mechanisms in the Ni-Base Superalloy Waspaloy," Master's thesis, University of Cincinnati, Cincinnati, Ohio, 1980.
[8] Saxena, A., *Fatigue of Engineering Materials and Structures*, Vol. 3, No. 3, 1980, pp. 247–255.
[9] Gell, M. and Leverant, G. R. in *Fatigue at Elevated Temperatures, ASTM STP 520*, A. E. Garden, A. J. McEvily, and C. H. Wells, Eds., American Society for Testing and Materials, Philadelphia, 1973, pp. 37–67.
[10] McLean, D., *Metals Forum*, Vol. 4, No. 1-2, 1981, pp. 44–47.
[11] Floreen, S. and Davidson, J. M., *Metallurgical Transactions A*, Vol. 14A, No. 5, May 1983, pp. 895–901.
[12] Sadananda, K. and Shahinian, P., *Metals Technology*, Vol. 9, 1982, pp. 18–25.

DISCUSSION

N. Marchand[1] *(written discussion)*—To study the effect of boron and zirconium in René 95 you used Waspaloy. Knowing that the amounts of nickel and γ' are quite different between these two alloys, is it fair to use the results on Waspaloy to assess the response of René 95?

J. Gayda et al. (authors' closure)—The degrading effect of fine grain microstructures on crack growth rates at 650°C has been demonstrated in Waspaloy, René 95, and several other γ'-strengthened nickel-base superalloys. The present results on boron and zirconium show that dilution of these elements does degrade crack growth resistance and therefore lends support to a concept put forth by Bain and Pelloux which links dilution of these two elements to crack growth problems in fine grain microstructures. The magnitude of this effect will, as you state, depend on alloy composition, but I feel one would observe similar behavior in other γ' strengthened nickel-base superalloys, including René 95.

G. A. Webster[2] *(written discussion)*—Elimination of boron and zirconium from the grain boundaries of your alloy appears to reduce significantly its uniaxial creep ductility. Ductility exhaustion crack growth models predict propagation rates which are inversely proportional to failure strain. Is it possible that the faster cracking rates observed in the boron- and zirconium-free alloy can be attributed directly to its reduced creep ductility? It may be that the reduced ductility is a consequence of enhanced environmental grain boundary attack.

J. Gayda et al. (authors' closure)—The diminished creep ductility and strength of the Waspaloy with reduced levels of boron and zirconium have, as stated in the text, contributed to increased crack growth rates, but there is also a strong environmental effect as comparison of air and vacuum data show. The degraded creep properties may, as you suggest, be related, at least in part, to enhanced environmental grain boundary attack.

[1]Massachusetts Institute of Technology, Cambridge, MA 02139.
[2]Department of Mechanical Engineering, Imperial College, London.

High-Temperature Low Cycle Fatigue

K.-T. Rie,[1] R.-M. Schmidt,[1] B. Ilschner,[2] and S. W. Nam[3]

A Model for Predicting Low Cycle Fatigue Life under Creep-Fatigue Interaction

REFERENCE: Rie, K.-T., Schmidt, R.-M., Ilschner, B., and Nam, S. W., **"A Model for Predicting Low Cycle Fatigue Life under Creep-Fatigue Interaction,"** *Low Cycle Fatigue, ASTM STP 942,* H. D. Solomon, G. R. Halford, L. R. Kaisand, and B. N. Leis, Eds., American Society for Testing and Materials, Philadelphia, 1988, pp. 313–328.

ABSTRACT: An attempt is made to predict fatigue life under high-temperature low-cycle fatigue conditions with superimposed hold times. It is well recognized that a reduction in life in this region is caused by the nucleation and growth of cavities along grain boundaries. The cavities can be formed by vacancy clustering, dislocation pileup, and grain boundary sliding. The subsequent growth leads to grain boundary cracking and intergranular fracture.

The possibility of unstable crack advance is given if the crack tip opening displacement becomes equal to a critical cavity configuration. To describe this configuration we postulate a power function for cavity growth.

Experimental data and the predicted values show excellent agreement, indicating the important role of creep cavitation for creep-fatigue life prediction.

KEY WORDS: high-temperature low-cycle fatigue, creep-fatigue interaction, intergranular fracture, creep damage, grain boundary cavity nucleation and growth, crack growth, crack-tip opening displacement

Failure by low-cycle fatigue is of great concern in the design of structures used at elevated temperatures (e.g., gas turbines, coal gasification facilities, and nuclear reactors). A number of engineering materials suffer a reduction in life when used at elevated temperatures in low-cycle fatigue regime. It is well recognized that this reduction is caused by the nucleation and growth of cavities along grain boundaries as well as oxidation of freshly exposed fracture surfaces. It has been shown that a change of crack path from predominantly transgranular to intergranular can occur at high temperatures and low strain rates [*1,2*].

The considerable interest in creep-fatigue life prediction during the last decade has resulted in a large number of different attempts to develop prediction methods. It seems that interest has tended more or less to focus on three basic types of prediction: damage summation methods [*3*], frequency-modified fatigue life methods [*4*], and strain range partitioning methods [*5*]. Several other models [*6, 7*] based on a mechanistic aspect or failure mechanism have also been proposed in the past.

Under complex loading circumstances and, in particular, when there is a significant involvement of creep fracture in the failure process, some of the proposed life prediction methods can be in error by an order of magnitude in endurance [*8*]. It will always be difficult to obtain good

[1]Institut für Schweisstechnik und Werkstofftechnologie, TU Braunschweig, Langer Kamp 8, D-3300 Braunschweig, West Germany.

[2]Ecole Polytechnique Federale de Lausanne, Department des Matériaux, Laboratoire de Metallurgie Mechanique, CH-1007 Lausanne, Switzerland.

[3]Korea Advanced Institute of Science and Technology, Department of Materials Science, Seoul, Korea.

life prediction without detailed attention to the failure mechanism and to the prevailing fracture mode, without extensive use of a mechanistic approach.

In this paper we attempt to develop a relationship between loading history, temperature, and the number of cycles to failure by considering creep-fatigue interaction.

Basic Concept of the Model

It has been known that cavities can be formed by vacancy clustering, dislocation pileup, and grain boundary sliding under high temperature tensile loading [*9*]. The subsequent growth of these cavities leads to grain boundary cracking and intergranular fracture (Fig. 1). In a static case, failure is established by assuming that catastrophic fracture occurs when the cavities coalescence. However, in a high-temperature low-cycle fatigue regime (HT-LCF) with superimposed hold times or with very slow cyclic straining in tension phase, failure occurs when cavities reach a critical geometric configuration. The basic reason for this creep-fatigue effect can best be understood by detailed consideration of the stress-strain pattern involved in the tensile hold period (Fig. 2).

During the tensile hold period of the cycle, stress relaxation occurs at an infinite variety of strain rates and produces an additional amount of tensile strain. This additional strain, though $<0.1\%$ per cycle, has a drastic effect on life [*1*].

Because of the essentially different deformation modes of creep and fatigue during the cycle, creep and fatigue damage must be considered separately. Fatigue damage is a surface phenomenon, while creep damage is essentially an internal failure process which occurs by the initiation and growth of grain boundary cracks or cavities. The detailed mechanisms by which creep failure takes place during static or cyclic loading are complex and depend on many factors including test temperature, strain rate, and the relative strength of grain matrix to grain boundaries, which is frequently related to complex precipitate reactions. Usually creep failure results from two basic processes known as wedge type or triple point cracks and "r"-type or round cavities [*10*]. Both types of fracture occur by the nucleation of grain boundary cavities. Wedge cracking

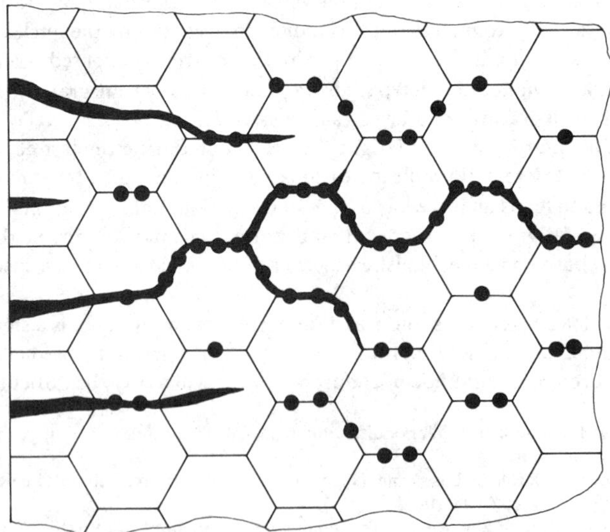

FIG. 1—*Intergranular crack path by creep-fatigue interaction.*

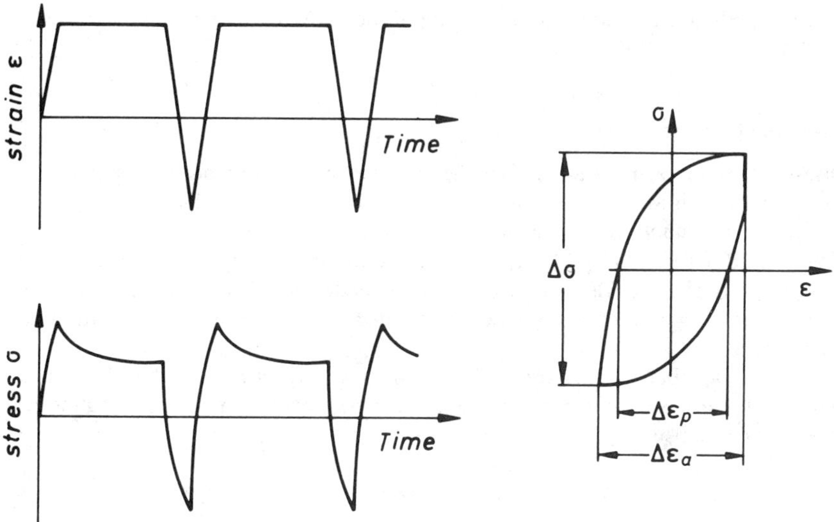

FIG. 2—*Strain-time and stress-time waveforms during high-temperature testing.*

is usually favored by high strain rate deformation while r-type cavitation occurs at lower strain rates and higher temperatures [*1*].

We confine our consideration to round cavities (r-type). The crack tip opening displacement may be seen as the upper bound to crack growth [*11*] and the relation between crack propagation rate da/dN and displacement δ can be written as

$$\frac{da}{dN} = \frac{\delta}{2} \tag{1}$$

Unstable crack advance occurs if the crack tip opening displacement becomes approximately equal to the spacing of the nucleated intergranular cavities [*12*]. The plastic strain required for the cracking of the ligament between the cavities is provided by the crack tip. For this reason the condition for unstable crack advance is thought to be represented by

$$\frac{\delta}{2} = \lambda - 2r \tag{2}$$

where λ is cavity spacing and r is radius [*12,13*]. This relation is based on the assumption that precipitations are responsible for cavity-nucleation and each particle is associated with a cavity [*14*]. It has subsequently been found that this is not always the case [*12,14*] and therefore the equation is modified to

$$\frac{\delta}{2} = \alpha(\lambda - 2r) \tag{3}$$

where α is a positive integer and a measure of the ease of nucleation of cavities [*11,12*].

The most likely reason for the large increase in crack growth rate is the attainment of the cavity configuration at the crack tip described in Eq 3. Instead of a continuous crack advance, this cavity configuration causes an unzipping of the cavitated material by the crack displacement field.

Nucleation of Cavities and Cavity Spacing

Investigations on the nucleation of cavities in the early stage of monotonic creep indicate that the number of cavities nucleated by creep is approximately proportional to the applied strain [*15–19*]. The void nucleation is not strongly sensitive to the rate of deformation.

Following Ratcliffe et al [*15*], the competing sintering processes under creep condition in tension are ineffective. Voids are created as a result of deformation at a size sufficiently great for subsequent vacancy condensation under the applied stress. The time to reach this critical size must be short.

By considering that cavities are nucleated progressively with strain and by a mechanism which is not strongly sensitive to strain rate, a quantitive approach in interpreting the cavitation process was developed [*20*]:

$$n = p \cdot \epsilon \tag{4}$$

where n is the number of cavities per unit area of the grain boundary, p is the nucleation factor for monotonic creep per unit grain boundary area, and ϵ is the creep strain.

Under repeated loading there will be an additional dependence on the number of cycles. In analogy to the Manson-Coffin relationship, we postulate a cycle-dependent cavity nucleation under cyclic creep and low-cycle fatigue condition with superimposed hold time. Assuming that only the plastic strain imposed is responsible for cavity nucleation and disregarding stress dependency, the number of cavities is given by

$$n = pN^{\kappa}\Delta\epsilon_p \tag{5}$$

where $\Delta\epsilon_p$ is the plastic strain range, N is the number of cycles, p is the cavity nucleation factor, and κ is the cyclic cavity nucleation exponent. Cavity nucleation under creep condition is favored on grain boundaries perpendicular to the load axis, and the cavity spacing is given by

$$\lambda = n^{-1/2} \tag{6}$$

so that from Eqs 5 and 6

$$\lambda = (p \cdot N^{\kappa} \cdot \Delta\epsilon_p)^{-1/2} \tag{7}$$

Cavity Growth

Nucleation of cavities is governed by a deformation of the matrix, and the cavity growth is controlled by diffusion. The max cavitation occurs on boundaries oriented normal to the tensile axis. The first detailed analysis of intergranular creep cavity growth was presented by Hull and Rimmer [*21*]. For their model they assumed a regular square array of equalized spherical cavities lying in a plane defined by the grain boundary. The cavity rate equation from Hull and Rimmer's analysis is

$$\frac{dr}{dt} = \frac{2\pi\Omega\delta_{gb} \cdot D_{gb}\left(\sigma - P - \dfrac{2\gamma}{r}\right)}{kT\lambda r} \tag{8}$$

where

D_{gb} = grain boundary diffusion coefficient,
δ_{gb} = grain boundary width for enhanced diffusion,
λ = cavity spacing,
r = cavity radius,
Ω = atomic volume,
k = Boltzmann's constant,
T = absolute temperature,
γ = surface energy,
σ = tensile stress, and
P = hydrostatic pressure.

The Hull-Rimmer model has recently been modified by neglecting surface energy and hydrostatic pressure. The rate of diffusion growth of a cavity of radius r with time t is given in a first approximation [20] as

$$\frac{dr}{dt} = \frac{2\pi\Omega\delta_{gb} \cdot D_{gb}\sigma}{kT\lambda r} \tag{9}$$

Crack-Tip Opening Displacement Due to Fatigue

The crack-tip opening displacement of the fatigue crack (δ_f) is given approximately [22] by

$$\delta_f = 2\pi \cdot a(\epsilon_p - 0.25\,\epsilon_{ys}) \tag{10}$$

where ϵ_{ys} is the yield strain.

For high strain, Eq 10 becomes approximately

$$\delta_f \approx 2\pi \cdot a \cdot \Delta\epsilon_p \tag{11}$$

The crack length a can be well described by the crack growth law developed by Manson [23]

$$\frac{da}{dN} = G \cdot \Delta\epsilon_p^{\gamma} \cdot a \tag{12}$$

where G and γ are constant.

Equation 12 may be integrated and the crack length will be given by

$$a = a_0 \cdot \exp[G \cdot \Delta\epsilon_p^{\gamma} \cdot N] \tag{13}$$

where a_0 is the initial crack length.

From Eqs 11 and 13 the crack-tip opening displacement δ is given by

$$\delta_f = a_0 \cdot \exp[G \cdot \Delta\epsilon_p^{\gamma} \cdot N] \cdot 2\pi \cdot \Delta\epsilon_p \tag{14}$$

Crack-Tip Opening Displacement Due to Creep-Fatigue

In creep-fatigue interaction there may be an additional term to δ_f due to creep:

$$\delta = \delta_f + \delta_c \tag{15}$$

The crack-tip opening displacement may be represented in analogy to the total strain by an elastic term $\Delta\delta_{el}$ plus a contribution due to plastic deformation $\Delta\delta_p$ and by thermally activated, time-dependent processes δ_c [*24*]:

$$\delta = \Delta\delta_{el} + \Delta\delta_p + \delta_c \tag{16}$$

Following Okazaki et al [*25*] the elastic term can be described by

$$\Delta\delta_{el} = 4\frac{\Delta\sigma}{E}a = 4\,\Delta\epsilon_{el}\cdot a \tag{17}$$

Using finite element calculations, Shih and Hutchinson [*26*] developed the time-independent plastic term

$$\Delta\delta_p = \Delta\epsilon_p\, a(4n' + 3.85\,(1 - n')/\sqrt{n'}) \tag{18}$$

In Eq 18 the material is assumed to obey the power-hardening equation

$$\Delta\sigma = K'\,(\Delta\epsilon_p)^{n'} \tag{19}$$

The crack-tip opening displacement due to creep δ_c is defined in analogy to the development of Taira et al [*27*] and Okazaki et al [*25*]. The creep crack opening displacement is obtained by replacing the strain field by the strain rate field. The displacement rate $\dot{\delta}$ will then be

$$\dot{\delta} = \dot{\epsilon}\cdot a(4c + 3.85\,(1 - c)/\sqrt{c}) \tag{20}$$

where the material obeys the simple power law equation

$$\dot{\epsilon} = C\cdot\sigma^{1/c} \tag{21}$$

δ_c can be obtained by the integration of Eq 20:

$$\delta_c = \int_{t_{t1}}^{t_{t2}} \dot{\delta}_c\, dt \tag{22}$$

In Eq 22, t_{t2} is the tensile time per cycle and t_{t1} is the time in cycle beyond which creep damage by cavity growth becomes dominant. From the relaxation curve, time t_{t1} can be determined as the time beyond which the deformation rate is less than 10^{-4}/s [*28*]. The influence of both creep and fatigue can be estimated by the summation of $\Delta\delta_{el}$, $\Delta\delta_p$ and δ_c.

From Eqs 16, 17, 18, and 22 the total crack-tip opening displacement is

$$\delta = 4\frac{\Delta\sigma}{E}\cdot a + \Delta\epsilon_p\cdot a\cdot(4n' + 3.85\,(1 - n')/\sqrt{n'}) + \int_{t_{t1}}^{t_{t2}} C\cdot\sigma^{1/c}\,dt\cdot a\cdot(4c + 3.85\,(1 - c)/\sqrt{c}) \tag{23}$$

The integral term in Eq 23 represents the creep strain. Equation 23 can be simplified to

$$\delta = a(K_1 \Delta\epsilon_{el} + K_2 \Delta\epsilon_p + K_3 \epsilon_c) \tag{24}$$

where

$K_1 = 4$,
$K_2 = 4n' + 3.85\,(1 - n')/\sqrt{n'}$, and
$K_3 = 4c + 3.85\,(1 - c)/\sqrt{c}$.

Prediction of Lifetime

Fatigue life in high-temperature low-cycle fatigue regime can be estimated using the criterion of the critical cavity configuration expressed in Eq 3.

The cavity spacing λ can be estimated from Eq 7 and the crack-tip opening displacement δ from Eq 24. Following Eq 9, the cavity radius r at N cycles is given by

$$r = \left[\frac{4\pi\Omega\delta_{gb}D_{gb}}{kT\lambda}\int_{t_{t1}}^{t_{t2}}\sigma(t)dt \cdot N\right]^{1/2} \tag{26}$$

If the critical number of cycles is reached which is defined as the number of cycles leading to a remarkable decrease of strength [29], Eq 3 is satisfied and catastrophic intergranular damage will occur. Equation 3 can be solved iteratively by applying the described parameters. By neglecting the creep induced displacement at the crack tip, the fatigue life can be estimated from

$$\pi \cdot \Delta\epsilon_p \cdot a_0 \cdot \exp[G \cdot \Delta\epsilon_p^{\gamma} \cdot N] = \alpha\left[(p \cdot N^{\kappa} \cdot \Delta\epsilon_p)^{-1/2} - 2\left(\frac{4\pi\Omega\delta_{gb} \cdot D_{gb}}{kT(pN^{\kappa} \cdot \Delta\epsilon_p)^{-1/2}}\int_{t_{t1}}^{t_{t2}}\sigma(t)dt \cdot N\right)^{1/2}\right] \tag{27}$$

If creep crack-tip opening displacement is taken into consideration, N can be obtained from

$$1/2[K_1\Delta\epsilon_{el} + K_2\Delta\epsilon_p + K_3\epsilon_c]a = \alpha\left[(p \cdot N^{\kappa} \cdot \Delta\epsilon_p)^{-1/2} - 2\left(\frac{4\pi\Omega\delta_{gb}D_{gb}}{kT(p \cdot N^{\kappa}\Delta\epsilon_p)^{-1/2}}\int_{t_{t1}}^{t_{t2}}\sigma(t)dt \cdot N\right)^{1/2}\right] \tag{28}$$

Experimental Procedure

The introduced method for life prediction was confirmed on results of Ermi and Moteff [30], Majumdar and Maiya [31], and on our own results. The tests of Ermi et al [30] and Majumdar et al [31] were carried out on specimens fabricated from AISI 304 stainless steel. The chemical composition is given in Table 1.

In their tests, they used hour-glass shaped specimens, 6.35 mm in diameter, which were solution annealed for 30 min at 1092°C and aged for 1000 h at 593°C.

The test conditions included strain rates of $4 \cdot 10^{-3}$ and $4 \cdot 10^{-5}$/s, total strain ranges from 0.5 to 2.0%, and temperatures of 593 and 650°C. Details of their testing procedures can be taken from Ref *30*. Our own tests were carried out on cylindrical specimens, fabricated from AISI 304 L stainless steel. The chemical composition is given in Table 1.

TABLE 1—*Chemical analysis of AISI 304 and AISI 304 L stainless steel.*

Materials	Chemical Composition, wt%						
	C	Si	Mn	P	S	Cr	Ni
304	0.047	0.47	1.22	0.029	0.012	18.5	9.58
304L	0.027	0.37	1.48	0.026	0.003	18.04	10.13

The specimens were machined from 22 mm thick plate keeping their axes parallel to the rolling direction of the plate and electrolytically polished.

The specimen size is illustrated in Fig. 3. Before testing, the specimens were solution annealed for 90 min at 1050°C. The tests were carried out on a servohydraulic push-pull machine ($R_\epsilon = -1$) and the strain rate was $4.17 \cdot 10^{-3}$/s with superimposed hold times. For the high-temperature tests, a multi-zone radiant heater was used and the test temperature was 530, 600, or 650°C.

The total strain was measured using a strain gage. The plastic strain is determined from the width of the hysteresis loop at tension zero.

Results and Discussion

The foregoing model for life prediction in the LCF regime at high temperature is checked by the experimental results obtained independently by other investigators for 304 stainless steel [*30,31*] and by our results on 304 L stainless steel. The fatigue life curves of other authors [*30,31*] are shown in Fig. 4 in double-logarithmic plotting. The results are obtained from the tests conducted at 593 and 650°C with tension hold times between 1 and 600 min. It can be seen from Fig. 4 that an increase in hold time and test temperature resulted in a remarkable reduction of fatigue life. The results obtained by cyclic creep tests are also included in Fig. 4 for comparison.

Our experimental results are plotted in Fig. 5. As found by others [*30,31*], an increase in tensile hold time and test temperature leads to a drastic reduction of fatigue life. Figure 6 shows the typical intergranular fracture observed after testing with superimposed tensile hold time.

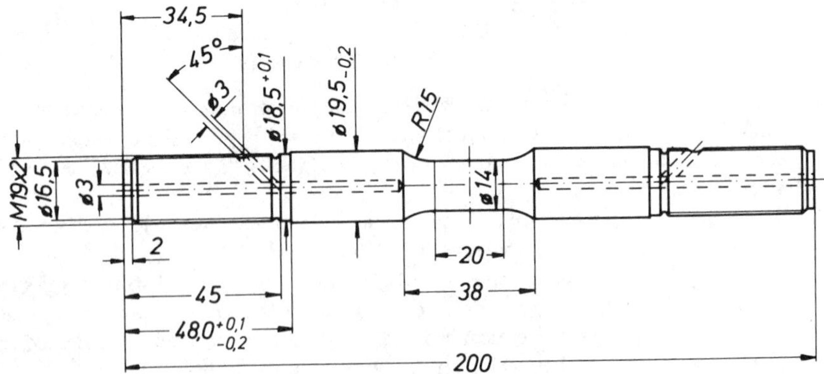

FIG. 3—*Geometry of specimen (dimensions in millimetres).*

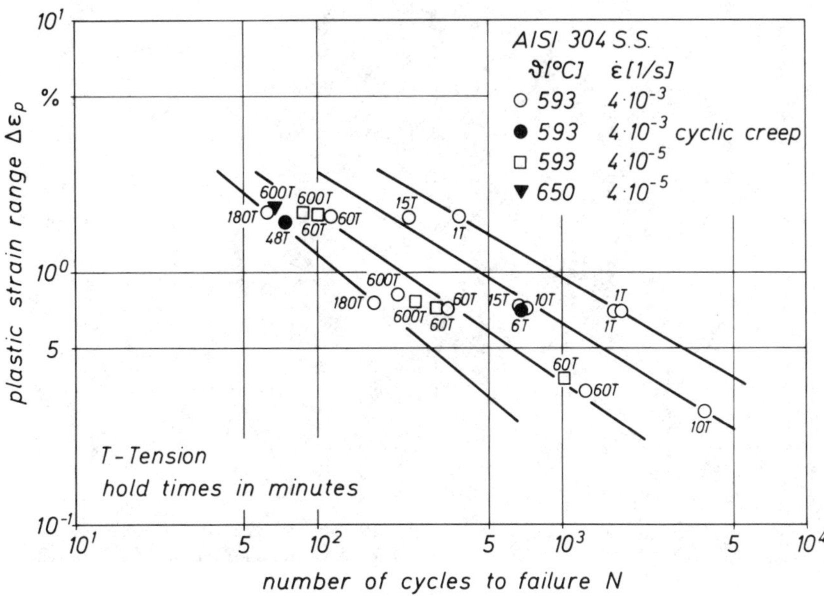

FIG. 4—*Plastic strain range versus number of cycles to failure* [30,31].

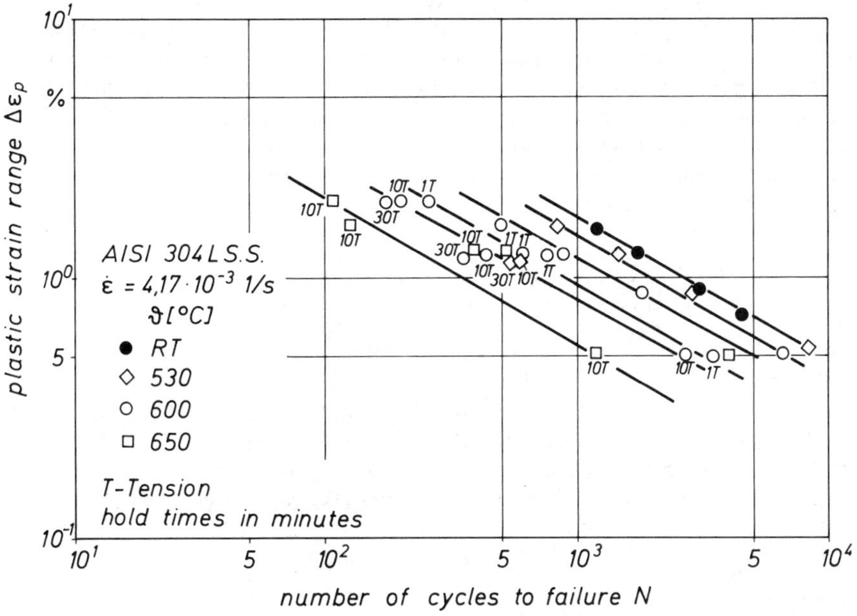

FIG. 5—*Plastic strain range versus critical number of cycles.*

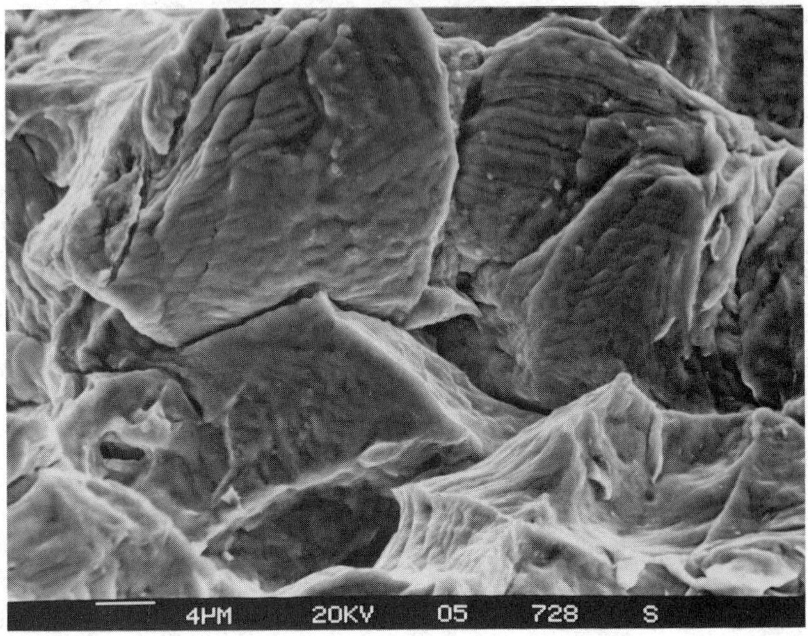

FIG. 6—*Intergranular fracture due to creep damage, 304 L, 650°C,* $\Delta\epsilon_p = 1.61\%$, $t_H =$ *10 min.*

By applying Eq 27, the results of others [*30,31*] and our experimental results are analyzed. The predicted and experimental fatigue lives are compared (Figs. 7 and 8).

Though the agreements between experimentally observed and predicted lives are satisfactory, the tendency is obvious that overestimation of fatigue life may occur by disregarding the creep induced crack tip opening displacement. By using Eq 20, which includes the creep crack opening displacement, the same experimental data are analyzed and plotted in Figs. 9 and 10. Assuming that catastrophic fracture (unstable crack growth) occurs if the crack length reaches about 10 to 12% of the specimen thickness, fatigue lives are estimated using constants

$$\alpha = 15$$
$$\kappa = 2/3$$
$$p = 5 \cdot 10^{10}\ 1/\text{m}^2\ (304),\ 2 \cdot 10^{10}\ 1/\text{m}^2\ (304\ \text{L})$$
$$\delta_{gb} \cdot D_{gb} = 2 \cdot 10^{-13} \exp(-167\ \text{kJ}/k \cdot T)\ \text{m}^3/\text{s}$$
$$\Omega = 1.18 \cdot 10^{-29}\ \text{m}^3$$

which are the values taken from the literature [*11,15–19,32–35*]. The power-hardening exponent for both materials was $n' = 0.3$ and $c \cong 0.15$. The agreement between experimental and predicted fatigue lives is again excellent. The merit of the prediction method using Eq 28 is evident if high temperature and long tensile hold periods are involved.

We recognize that an important feature in the successful application of the model is the use of appropriate values of the constant κ and the parameter p.

Conclusion

The postulated model permits the prediction of lifetime under high-temperature low-cycle fatigue conditions with superimposed creep damage. Cavity nucleation and subsequent growth

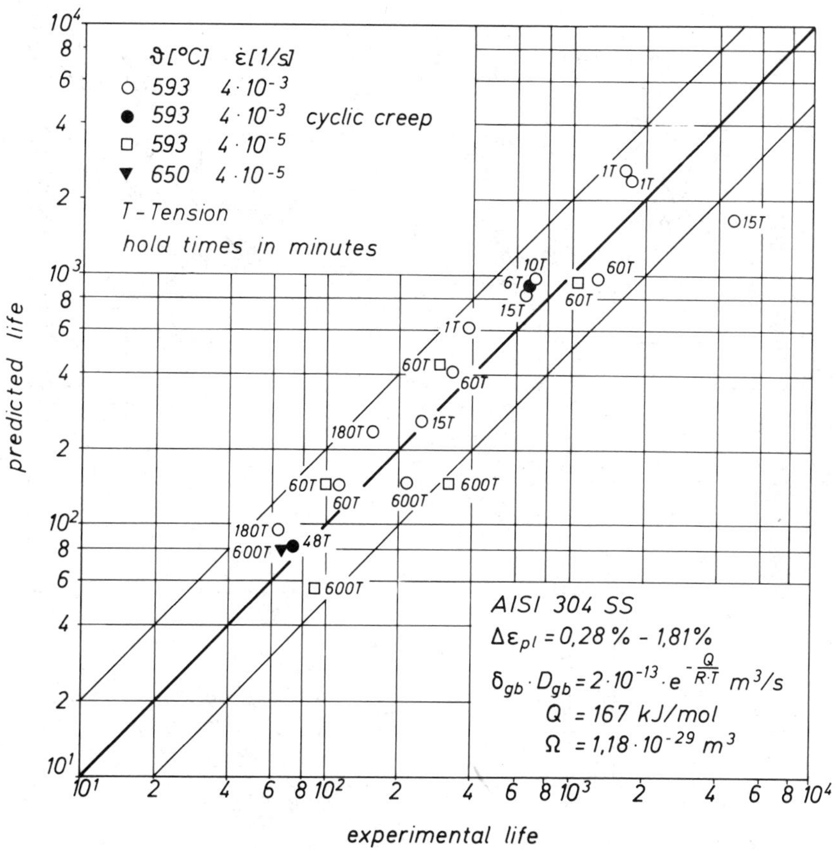

FIG. 7—*Comparison of experimental life* [30,31] *and predicted life following Eq 27.*

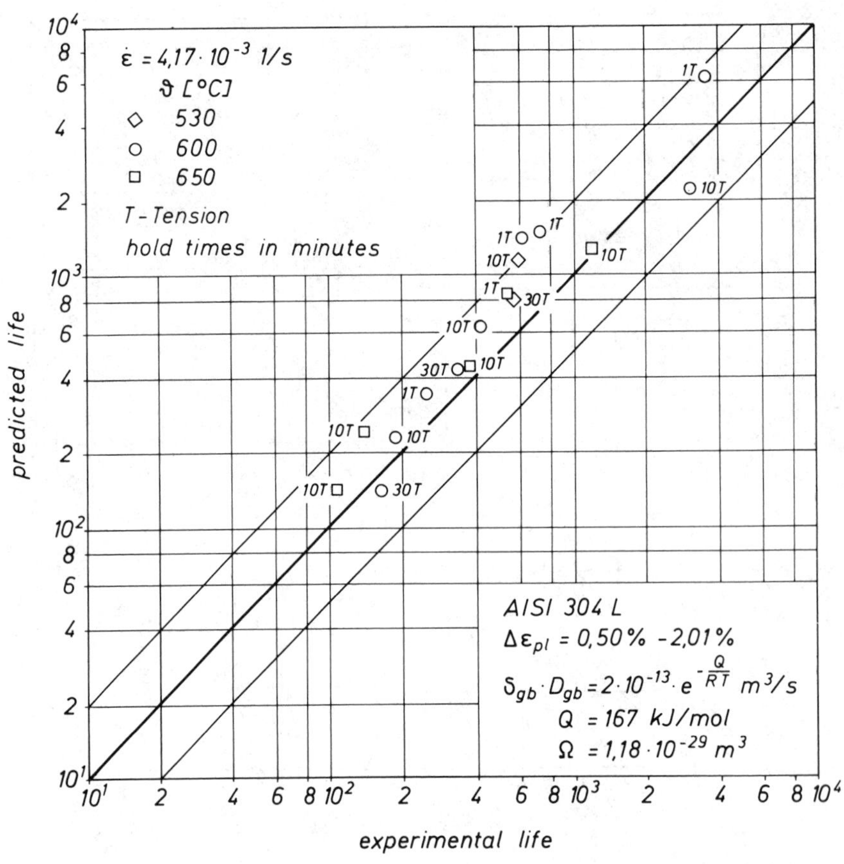

FIG. 8—*Comparison of experimental life and predicted life following Eq 27.*

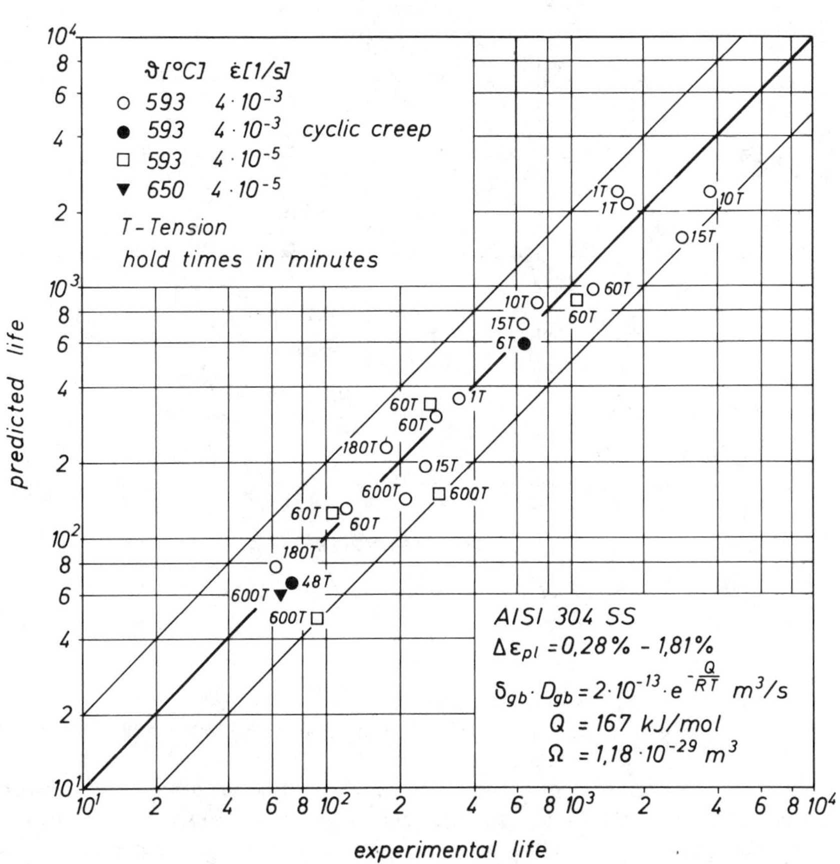

FIG. 9—*Comparison of experimental life* [30,31] *and predicted life following Eq 28.*

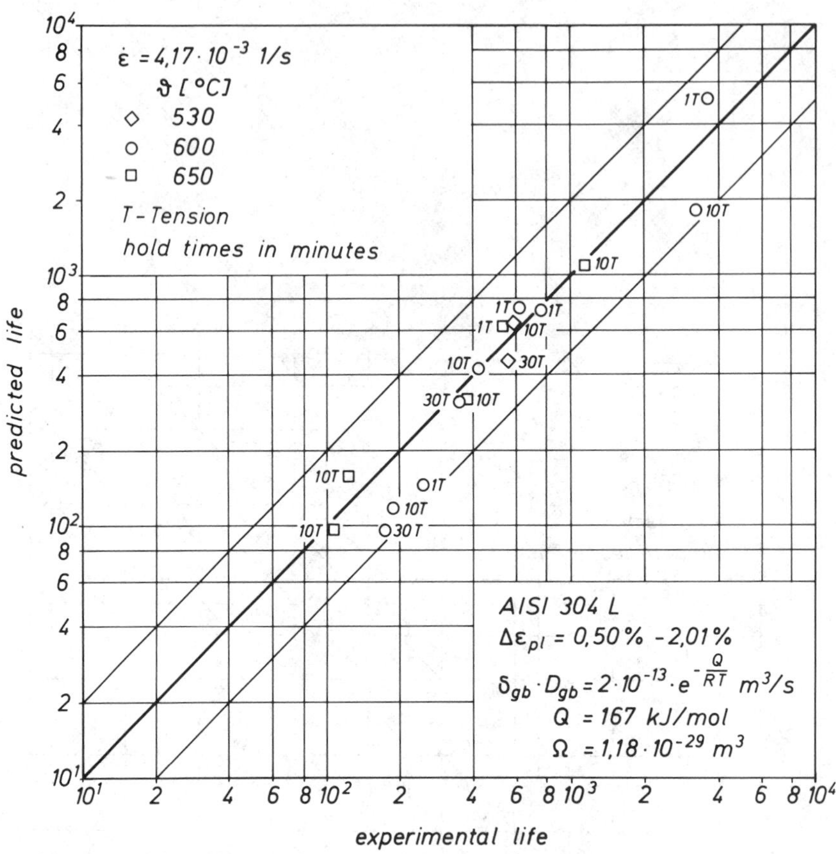

FIG. 10—*Comparison of experimental life and predicted life following Eq 28.*

are considered, and the unzipping of the cavitated grain-boundary is taken as criterion for catastrophic failure.

For cyclic cavity nucleation a power law is assumed. The crack tip opening displacement due to creep has been considered for the estimation of crack growth. It should be emphasized that better understanding of the annealing mechanism for creep damage and more information on creep crack tip opening displacement under creep-fatigue conditions are necessary for the improvement of the predictive method. The agreement between the predicted life and the observed fatigue life, however, may strongly support the suggested role of creep cavitation during high-temperature low-cycle fatigue.

References

[1] Wareing, J., "Mechanisms of High Temperature Fatigue and Creep-Fatigue Failure in Engineering Materials," in *Fatigue at High Temperatures*, R. P. Skelton, Ed., Applied Science Publishers, London and New York, 1983, pp. 97–134.

[2] Livesey, V. B. and Wareing, J., "Influence of Slow Strain Rate Tensile Deformation on Creep-Fatigue Endurance of 20Cr-25Ni-Nb Stainless Steel at 593°C," *Metal Science*, Vol. 17, 1983, pp. 297–303.

[3] Taira, S., *Creep in Structures,* N. J. Hoff, Ed., Academic Press, New York, 1962, p. 96.
[4] Coffin, L. F., "Fatigue at High Temperature-Prediction and Interpretation," in *Proceedings,* Institution of Mechanical Engineers, Vol. 188, 1974, pp. 109–127.
[5] Manson, S. S., Halford, G. R., and Hirschberg, M. H. in *Proceedings,* Symposium on Design for Elevated Temperature Environment, ASME, New York, 1971, p. 12.
[6] Majumdar, S. and Maiya, P. S., "Creep-Fatigue Interactions in an Austenitic Stainless Steel," *Canadian Metallurgical Quarterly,* Vol. 18, 1979, pp. 57–64.
[7] Miller, D. A., Hamm, C. D., and Philipps, J. L., "A Mechanistic Approach to the Prediction of Creep-Dominated Failure During Simultaneous Creep-Fatigue," *Materials Science and Engineering,* Vol. 53, 1982, pp. 233–244.
[8] Esztergar, E. P. and Ellis, J. R., "Cumulative Damage Concepts in Creep-Fatigue Life Prediction," in *Proceedings,* International Conference on Thermal Stresses and Thermal Fatigue, D. J. Littler, Ed., Butterworths, London, 1971.
[9] Evans, H. E., *Mechanisms of Creep Fracture,* Elsiver Applied Science Publishers, New York, 1984.
[10] Gandhi, C., and Ashby, M. F., "Fracture-Mechanism Maps for Materials Which Cleave: F.C.C., B.C.C. and H.C.P. Metals and Ceramics," *Acta Metallurgica,* Vol. 27, 1979, pp. 1565–1602.
[11] Lloyd, G. J., "High Temperature Fatigue and Creep Fatigue Crack Propagations Mechanics, Mechanisms and Observed Behavior in Structural Materials," in *Fatigue at High Temperatures,* R. P. Skelton, Ed., Applied Science Publishers, London and New York, 1983, pp. 187–258.
[12] Lloyd, G. J., and Wareing, J., "Stable and Unstable Fatigue Crack Propagation During High Temperature Creep-Fatigue in Austenitic Steels: the Role of Precipitation," *Journal of Engineering Materials and Technology,* Vol. 101, 1979, pp. 275–282.
[13] Tomkins, B., "The Development of Fatigue Crack Propagation Models for Engineering Applications at Elevated Temperatures," *Journal of Engineering Materials and Technology,* Vol. 97 H, 1975, pp. 289–297.
[14] Wood, D. S., Wynn, J., Baldwin, A. B., and Oriordan, P., "Some Creep/Fatigue Properties of Type 316 Steel at 625°C," *Fatigue of Engineering Materials and Structures,* Vol. 3, 1980, pp. 39–57.
[15] Ratcliffe, R. T. and Greenwood, G. W., "Formation of Voids in Creep and Competition from Sintering Processes," *Nature,* Vol. 215, 1967 p. 50.
[16] Greenwood, G. W., "Cavity Nucleation in the Early Stages of Creep," *Philosophical Magazine,* Vol. 19, 1969, pp. 423–426.
[17] Price, C. E., "On the Creep Behavior of Silver-I: The Oxygen Free Metal," *Acta Metallurgica,* Vol. 14, 1966, pp. 1781–1786.
[18] Dyson, B. F. and McLean, D., "A New Method of Predicting Creep Life," *Metal Science Journal,* Vol. 6, 1972, pp. 220–223.
[19] Cane, B. J. and Greenwood, G. W., "The Nucleation and Growth of Cavities in Iron During Deformation at Elevated Temperatures," *Metal Science,* Vol. 9, 1975, pp. 55–60.
[20] Miller, D. A., Mohamed, F. A., and Langdon, T. G., "An Analysis of Cavitation Failure Incorporating Cavity Nucleation with Strain," *Materials Science and Engineering,* Vol. 40, 1979, pp. 159–166.
[21] Hull, D. and Rimmer, D. E., "The Growth of Grain-Boundary Voids Under Stress," *Philosophical Magazine,* Vol. 4, 1959, pp. 673–687.
[22] Rolfe, S. T. and Barsom, J. M., *Fracture and Fatigue Control,* Prentice-Hall, Englewood Cliffs, N.J., 1977.
[23] Manson, S. S., "Interfaces Between Fatigue, Creep and Fracture," *International Journal of Fracture Mechanics,* Vol. 2, March 1966, pp. 327–369.
[24] Ilschner, B. "Plastic Deformation of Crystalline Materials at High Temperature," in *Mechanical and Thermal Behavior of Metallic Materials,* Soc. Italiana di Fisica, Bologna, 1982, pp. 188–216.
[25] Okazaki, M., Hattori, I., and Koizumi, T., "Effect of Strain Wave Shape on Low-Cycle Fatigue Crack Propagation of SUS 304 Stainless Steel at Elevated Temperatures," *Metallurgical Transactions A,* Vol. 14A, 1983, pp. 1649–1659.
[26] Shih, C. F. and Hutchinson, J. W., "Fully Plastic Solution and Large Scale Yielding, Estimates for Plane Stress Crack Problems," *Journal of Engineering Materials and Technology,* Vol. 98, 1976, pp. 289–295.
[27] Taira, S., Ohtani, R., and Kitamura, T., "Application of J-Integral to High-Temperature Crack Propagation: Part I—Creep Crack Propagation," *Journal of Engineering Materials and Technology,* Vol. 101, 1979, pp. 154–161.
[28] Majumdar, S. and Maiya, P. S., "A Mechanistic Model for Time-Dependent Fatigue," *Journal of Engineering Materials and Technology,* Vol. 102, 1980, pp. 159–167.
[29] Rie, K.-T. and Stüwe, H.-P., "A Note on the Influence of Dwell Time on Low-Cycle Fatigue," *International Journal on Fracture,* Vol. 10, 1974, pp. 545–548.
[30] Ermi, A. M. and Moteff, J., "Correlation of Substructure with Time-Dependent Fatigue Properties of AISI 304 Stainless Steel, *Metallurgical Transactions A,* Vol. 13A, 1982, pp. 1577–1588.

[31] Majumdar, S. and Maiya, P. S., "A Unified and Mechanistic Approach to Creep-Fatigue Damage," ANL-76-58, Argonne National Laboratory, Jan. 1976.

[32] Min, B. K. and Raj, R., "Hold-Time Effects in High Temperature Fatigue," *Acta Metallurgica*, Vol. 26, 1978, pp. 1007–1022.

[33] Frost, H. J. and Ashby, M. F., *Deformation-Mechanism Maps*, Pergamon Press, Oxford, 1982.

[34] Nai-Yong, T. and Plumtree, A., "Nucleation of Cavities Due to Vacancy Supersaturation," *Zeitschrift für Metallkunde*, Vol. 76, 1985, pp. 46–53.

[35] Ashby, M. F., "A First Report on Deformation-Mechanism Maps," *Acta Metallurgica*, Vol. 20, 1972, pp. 887–897.

DISCUSSION

N. Marchand[1] (*written discussion*)—Is your model strictly alloy specific (304 SS)? If so, is there any way to apply the model to other classes of alloys, such as nickel-based superalloys and ferritic steels?

K.-T. Rie et al (*authors' closure*)—The proposed model may be applicable for other materials if the same damage mechanisms occur. Creep-fatigue interaction results in intergranular fracture. The basic reason for intergranular crack path is the nucleation and growth of cavities or wedge cracks. Wedge cracks occur at lower temperatures and high strain rates. Our model, however, is only applicable when intergranular fracture due to round cavities is dominant.

[1]Massachusetts Institute of Technology, Cambridge, MA 02139.

James F. Saltsman[1] *and Gary R. Halford*[1]

An Update of the Total-Strain Version of Strainrange Partitioning

REFERENCE: Saltsman, J. F. and Halford, G. R., "**An Update of the Total-Strain Version of Strainrange Partitioning,**" *Low Cycle Fatigue, ASTM STP 942,* H. D. Solomon, G. R. Halford, L. R. Kaisand, and B. N. Leis, Eds., American Society for Testing and Materials, Philadelphia, 1988, pp. 329–341.

ABSTRACT: An updated procedure has been developed for characterizing an alloy and predicting cyclic life by using the total-strain-range version of strainrange partitioning (TS-SRP). The principal feature of the update is a new procedure for determining the intercept of time-dependent elastic-strain-range-versus-cyclic-life lines. The procedure is based on an established relation between failure and the cyclic stress-strain response. The stress-strain response is characterized by empirical equations presented in this report. These equations were determined with the aid of a cyclic constitutive model. The procedures presented herein reduce the testing required to characterize an alloy. Failure testing is done only in the high-strain, short-life regime; cyclic stress-strain response is determined from tests conducted in both the high- and low-strain regimes. These tests are carried out to stability of the stress-strain hysteresis loop but not to failure. Thus both the time and costs required to characterize an alloy are greatly reduced. This procedure was evaluated and verified for two nickel-base superalloys, AF2-1DA and Inconel 718. The analyst can now predict cyclic life in the low-strain, long-life regime without having to conduct expensive failure tests in this regime or resort to questionable larger extrapolations.

KEY WORDS: fatigue (metal), creep-fatigue, life prediction, strainrange partitioning, constitutive modeling, nickel-base superalloys, creep, plasticity, strain fatigue

Nomenclature

A General constant in empirical flow equations
A' General constant in empirical flow equations
B Intercept of elastic-strain-range-versus-life relations
b Exponent on cylic life for elastic-strain-range-versus-life relations
C Intercept of inelastic-strain-range-versus-life relations
C' Intercept of equivalent inelastic line for combined creep-fatigue cycles
c Exponent on cyclic life for inelastic-strain-range-versus-life relations
F Strain fraction
K Cyclic strain-hardening coefficient
N Applied cycles or number of data points in prediction
n Cyclic strain-hardening exponent
m General exponent on time in empirical flow equations
SE Standard error of estimate
t Time, s
α Exponent on total strain range in empirical flow equations

[1] Research Engineer and Senior Research Engineer, respectively, NASA-Lewis Research Center, Cleveland, OH 44135.

Δ Range of variable
ϵ Strain
σ Stress

Subscripts

c Compression
cc Creep strain in tension, creep strain in compression
cp Creep strain in tension, plastic strain in compression
el Elastic
f Failure
ij pp, cc, pc, cp
in Inelastic
obs Observed
pc Plastic strain in tension, creep strain in compression
pp Plastic strain in tension, plastic strain in compression
pre Predicted
t Tension or total

Introduction

A total-strain-range version of strainrange partitioning (TS-SRP) has been introduced by Halford and Saltsman [*1*]. This development, following the pioneering work of Manson and Zab [*2*], extends the capabilities of SRP into the low-strain, long-life regime, where the inelastic strains are small and difficult to determine by either experimental or analytical methods. In the total-strain-range approach, an alloy is characterized in much the same manner as in the original inelastic-strain-range version. The noted exception is the greater emphasis placed on knowledge of cyclic stress-strain-time behavior. It is this issue that is addressed by the present research. Since failure testing at the lower strain ranges is time consuming and expensive, it is advantageous to find alternative ways to characterize alloys with a minimum of failure tests. Cyclic stress-strain-time characterization can be carried out in considerably less time and can provide the required information.

TS-SRP requires the determination of the elastic-strain-range-versus-life relations in addition to the inelastic-strain-range-versus-life relations. For cycles involving creep the elastic lines are influenced by hold time, wave shape, and how creep is introduced into the cycle (stress hold, strain hold, slow strain rate, etc.). Analysis shows that the elastic-strain-range-versus-life relations can be related to flow response (i.e., the cyclic stress-strain-time response). Flow response can be determined by cycling a specimen until the stress-strain hysteresis loop approaches stability. These tests are much shorter and hence less expensive than failure tests.

Flow response for any conceivable cycle could be determined by using an appropriate constitutive flow model, provided that the constants are known for a material of interest. However, reliable and fully evaluated constitutive flow models are not presently available, particularly in the low-strain regime. This difficulty could be overcome if suitable empirical equations characterizing flow behavior were available. This concept has been successfully used by Brinkman et al [*3*] for modeling the behavior of 2.25Cr-1Mo steel. Our initial efforts [*1*] showed that a considerable amount of flow data would be required to determine the relations between flow and failure behavior and thus the desired empirical relations. The limited amount of data generally available from failure testing was inadequate, and additional flow testing was no longer feasible after the original testing was completed. In searching for a solution to this dilemma of seemingly insufficient information, we propose using a combination of empiricism and constitutive theory.

Walker's functional theory [4] is available, and the computer program [5] for this theory has been recently made more efficient and user friendly. Although the Walker theory has been used herein, numerous other constitutive models developed over the past few years could be incorporated, such as the Robinson model [6]. The Walker model was used to identify general *trends* in flow behavior, and we were able to develop empirical equations that describe these trends. These empirical flow equations make it possible to predict cyclic life in the low-strain, long-life regime at elevated temperature by using TS-SRP. To fully characterize the cyclic flow and failure of an alloy, failure testing is done in the high-strain, short-life regime according to the SRP guidelines of Hirschberg and Halford [7]. Flow behavior is documented along with failure behavior. In addition, cyclic flow testing is needed in the low-strain, long-life regime. This procedure greatly reduces the time and cost of characterizing the creep-fatigue behavior of alloys over a broad range of lives.

This report summarizes the updated procedure for characterizing an alloy and predicting cyclic life by using TS-SRP. Complete details are given in Ref *8*. The principal feature of this update is the method for determining the intercept of the time-dependent elastic-strain-range-versus-cyclic-life line. The method is based on an established relation between failure and the cyclic stress-strain response of an alloy. The stress-strain response is characterized by the empirical equations presented in this report.

Analysis

In the original total-strain-range version of SRP [*1*], the time-dependent elastic line intercept (elastic strain range at $N_f = 1$) for cycles involving creep is obtained from an empirical equation with constants determined exclusively from failure data.

New procedures have been developed to determine the elastic line intercept for creep-fatigue cycles from flow and failure data. Analysis shows that flow behavior can be related to failure behavior in the following manner:

Failure behavior:

$$\Delta\epsilon_{el} = B(N_f)^b \tag{1}$$

$$\Delta\epsilon_{in} = C'(N_f)^c \tag{2}$$

where

$$C' = [\Sigma F_{ij}(C_{ij})^{1/c}]^c \tag{3}$$

Flow behavior:

$$\Delta\epsilon_{el} = K_{ij}(\Delta\epsilon_{in})^n \tag{4}$$

Our experience to date [*1*] suggests that the inelastic and elastic failure lines for creep-fatigue cycles can be assumed to be parallel to the corresponding failure lines for PP cycles (Fig. 1). It then follows that n in Eq (4) is constant ($n = b/c$) for all wave shapes. However, K_{ij} may well depend on how creep is introduced into the cycle (stress hold, strain hold, etc.). Setting Eq 1 equal to Eq 4 and eliminating N_f by using Eq 2, we obtain the desired equation relating flow and failure characteristics:

$$B = K_{ij}(C')^n \tag{5}$$

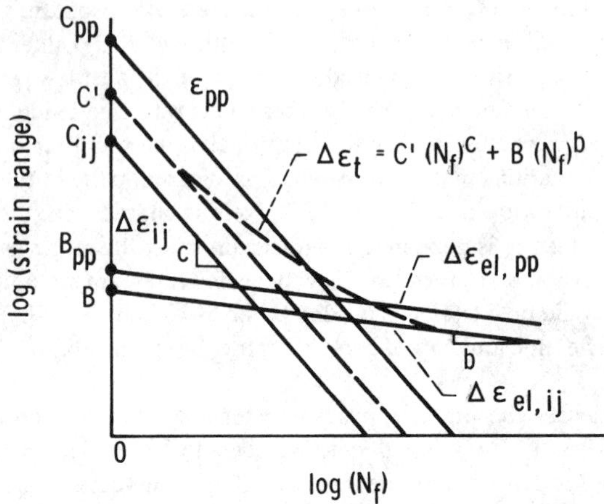

FIG. 1—*Relation between total-strain-range and life for a creep-fatigue cycle. Inelastic line intercept* (C′) *determined from Eq 3; elastic line intercept* (B) *determined from Eq 5.*

In this equation the inelastic line intercepts C_{pp}, C_{cc}, C_{pc}, and C_{cp} and the exponent c are considered to be failure terms. The strain fraction F_{ij}, the cyclic strain-hardening coefficient K_{ij}, and the strain-hardening exponent n are considered to be flow terms. Thus the elastic line intercept B can be determined for a creep cycle from a combination of flow and failure data. The cyclic strain-hardening coefficient and strain fractions will, in general, depend on waveform.

We are now in a position to establish a total-strain-range-versus-life relation and thus to predict life by using the TS-SRP approach. The total strain range is the sum of the elastic and inelastic strain ranges:

$$\Delta\epsilon_t = \Delta\epsilon_{el} + \Delta\epsilon_{in} \tag{6}$$

From Eqs 1 and 2 we obtain

$$\Delta\epsilon_t = B(N_f)^b + C'(N_f)^c \tag{7}$$

A schematic plot of Eq 7 is shown in Fig. 1. Note that the solution of this equation gives the cyclic life for a theoretical zero-mean-stress condition. The final step in a life prediction is to adjust the computed life to account for any mean stress effects that may be present. In this paper the method of mean stress correction proposed by Halford and Nachtigall [*9*] is used.

There are three basic variations of the general method for predicting life by using the TS-SRP approach, depending on what type of information is, or can be made, available.

Variant 1

1. Determine the SRP inelastic-strain-range-versus-life relations and the PP elastic-strain-range-versus-life relation from failure tests. As an alternative the ductility-normalized SRP (DN-SRP) life relations proposed by Halford and Saltsman [*10*] could be used, but they do require plastic and creep ductility information at the temperature and failure times of interest.
2. Calculate the cyclic strain-hardening coefficient (K_{ij}) and the strain fractions (F_{ij}) by using an appropriate constitutive flow model for which the material constants are known. If the

DN-SRP life relations are used, the cyclic strain-hardening exponent n must be determined from tests and the slope of the PP elastic line calculated ($b = nc$).

3. The elastic line intercept B can now be calculated by using Eq 5 and the preceding information.

4. Determine the total-strain-range-versus-life curve for the case in question (Fig. 1). Enter the curve at the appropriate total strain range and determine cyclic life for the theoretical zero-mean-stress condition. In this paper we have used the inversion method of Manson and Muralidharan [*11*] to solve Eq 7. This life is then adjusted to account for mean stress effects in accordance with Ref *9*.

Variant 2

1. Determine SRP inelastic-strain-range-versus-life relations and the PP elastic-strain-range-versus-life relation from failure tests.
2. Determine the elastic line intercept B by using the empirical equation of Halford and Saltsman [*1*]. The constants in this equation are determined from failure data. Failure tests should be performed at the lower strain ranges to reduce extrapolation errors.
3. Measure strain ranges (elastic and inelastic) and stresses from failure tests and extrapolate to lower strain ranges by using empirical equations.
4. Determine cyclic life by using Step 4 of variant 1.

Variant 3

1. Same as Step 1 of Variant 1.
2. Conduct flow tests for creep-fatigue cycles of interest and obtain from these data necessary empirical correlations describing the flow behavior. If failure data are lacking and the DN-SRP life relations are used, tests must be done to determine the strain-hardening exponent n for PP cycles. The slope of the PP elastic line b can then be calculated ($b = nc$).
3. Calculate the elastic line intercept B by using Eq 5. The strain-hardening coefficient (K_{ij}) and the strain fractions (F_{ij}) are determined from the correlations obtained in Step 2.
4. Determine cyclic life by using Step 4 of Variant 1. Although Variant 1 is the most general procedure, it is not a viable option at this time because reliable constitutive flow models in the low-strain regime are not available. Variant 2, used in our original TS-SRP paper [*1*], relies on failure data for the determination of the required equation constants. Variant 3, the subject of the present paper, represents a middle ground between Variants 1 and 2. Inelastic-strain-range-versus-life relations based on failure are necessary for reliable life predictions, but preliminary life predictions could be made by using the DN-SRP life relations and information obtained from flow tests.

Analysis Using the Walker Model

The Walker constitutive model was used to generate the cyclic stress-strain or flow information needed to identify *trends* in the required SRP flow behavior. From this information we were able to obtain the required correlations, which are based on a simple power-law equation

$$y = A(t)^m \tag{8}$$

where y is the dependent variable representing several different variables as discussed shortly, and t is the hold time per cycle. In this paper t is the *total* hold time per cycle. Thus for CC cycles t is the sum of the hold times in both tension and compression. When dealing with CC cycles in the present paper, we have treated only the case in which the cycles are balanced (i.e., PC or CP strain-range components are negligibly small). Generally the coefficient A is a function of total strain range (Fig. 2). It was found that the intercept A could be correlated with total strain range by another power law:

$$A = A'(\Delta\epsilon_t)^{\alpha} \tag{9}$$

Thus

$$y/(\Delta\epsilon_t)^{\alpha} = A'(t)^{m} \tag{10}$$

We can now normalize the dependent variable y and collapse the family of lines shown in Fig. 2 to a single line. The values of A', α, and m depend on the type of correlation and the mechanical properties of the alloy. Alloys with positive strain-rate-hardening characteristics soften under creep loading, and m will be negative. If an alloy hardens (i.e., negative strain rate hardening) under creep loading, m will be positive.

Equation 10 was used to obtain the empirical flow correlations used herein. The first two are used to determine the coefficients B and C' in Eq 7. The remainder are used to determine the mean stress for use in the mean stress correction procedure [*9*]. The complete details for determining these correlations are given in Ref *8*.

K_{ij} versus hold time
F_{ij} versus hold time } To determine B and C'

$\Delta\epsilon_{el}$ versus hold time
$\Delta\sigma$ versus hold time
σ_c versus hold time
σ_t versus hold time } To determine mean stress

Note that each of these relations could be derived directly from a reliable, fully evaluated constitutive model were it available. Although the exact form of the relations would surely differ from that selected here, the trends would be similar.

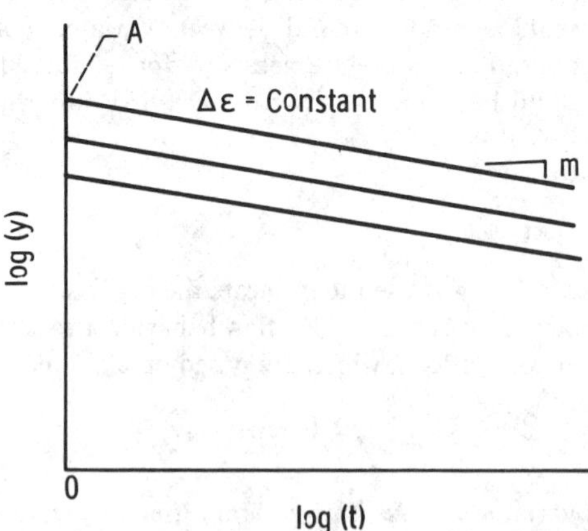

FIG. 2—*Power-law relation used to correlate flow data. Lines are parallel, and intercept* A *at* t = *1 s is a function of total strain range.*

For a specific alloy these correlations depend on temperature, creep time, wave shape, and the manner in which creep is introduced into the cycle (stress hold, strain hold, etc.). Only two specific ways of introducing creep into a cycle have had to be considered in this paper since these are the only types of waveforms for which data are available: stress hold and strain hold.

Evaluation of Method

The TS-SRP approach (Variant 3) as presented herein has been applied to two nickel-base alloys, AF2-1DA and Inconel 718, by using literature data sources. These results feature fully reversed cycles with zero mean strain. Since separate flow tests were not conducted when the data on these two alloys were generated, we were unable to use fully the procedures of Variant 3. Instead failure data were used to determine the empirical flow correlations.

Predictions made by the TS-SRP approach were compared with predictions made by the inelastic-strain-range SRP approach. In the inelastic-strain-range approach strains, strain fractions, and mean stress are measured directly from test results. In the TS-SRP approach these values, along with the elastic line intercept, are calculated by using the various correlations.

AF2-1DA at 760°C

Three data sources [*9,12,13*] were available for this alloy, and each used a different heat of the alloy. All the results reported in Ref *9* were generated from stress-hold tests; those reported in Ref *12* involved both stress-hold and strain-hold tests; and those reported in Ref *13* involved strain-hold tests and low-rate-strain cycle (LRSC) tests. We have assumed that the fatigue characteristics of the LRSC waveform are similar to the balanced-cyclic-creep-cycle (BCCR) waveform. Both waveforms feature PP and CC strain components.

The first step in characterizing this alloy is to determine the required SRP inelastic-strain-range-versus-life relations and the elastic-strain-range-versus-life relation for PP cycling. These life relations are generally insensitive to wave shape. The next step is to determine the empirical flow correlations for the stress-hold and strain-hold wave shapes. These flow correlations were determined by using the appropriate data from Refs *9* and *12,* respectively, and are reported in Ref *8*.

We now have all the equations necessary to predict the lives of stress-hold and strain-hold cycles for AF2-1DA at 760°C by using the TS-SRP approach. The predictive ability of this approach will be evaluated by comparing results with predictions made by using the inelastic-strain-range SRP approach.

If we "predict" the data used to determine the flow correlations, we will obtain a measure of both the validity of our assumptions and our ability to correlate the data. For the stress-hold data in Ref *9* the TS-SRP approach predicted 92% of the data within a factor of 2 on life and 100% within a factor of 3, and the standard error of estimate (SE) was 0.173. By comparison, the inelastic-strain-range SRP approach predicted 100% of the data within factors of 2 and the SE was 0.135. A review of these results [*8,* Fig. 21] shows that they are very nearly identical.

The "predictions" of the data in Ref *12* used to determine the strain-hold flow correlations are quite similar to the TS-SRP and inelastic-strain-range SRP approaches [*8,* Fig. 28] except for two PC points. For these two points the calculated elastic strain range was greater than the total strain range. Obviously this cannot be correct and is due to uncertainties in the correlation. To avoid this dilemma, the calculated elastic strain range was restricted not to exceed the total strain range.

Life predictions for the stress-hold tests reported in Ref *12* and for the strain-hold and LRSC tests reported in Ref *13* are shown in Fig. 3. These data were not used in any of the preceding correlations and hence serve to verify the predictive capabilities of the TS-SRP method. The

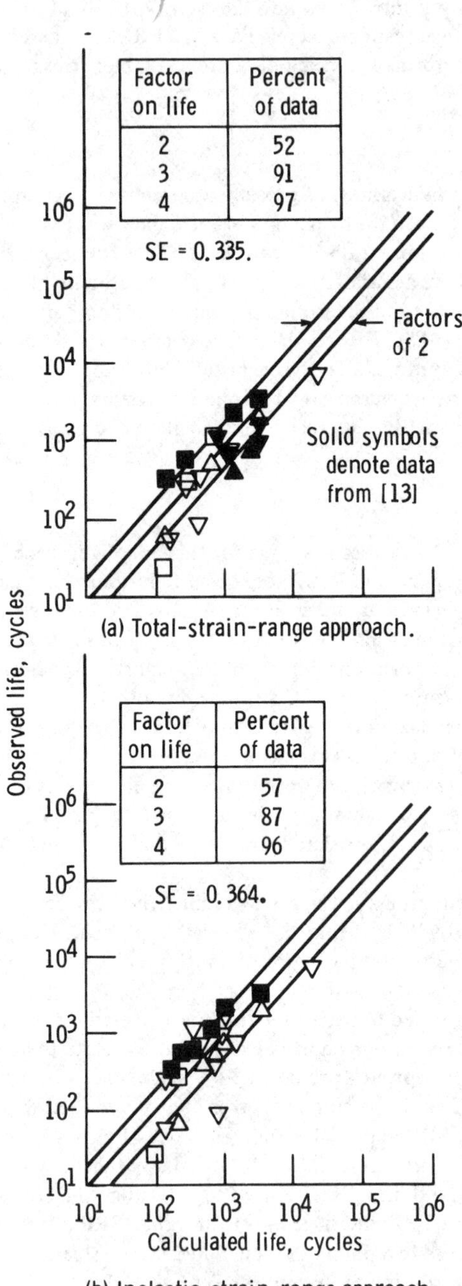

FIG. 3—*Life predictions of stress-hold data* [12] *and strain-hold and low-rate strain cycle data* [13]. *Data not used to establish correlations; AF2-1DA at 760°C.*

degree of fit is summarized in the table inset in the figure. A review of the results shows that the TS-SRP approach predicts the data in Ref *12* even better than the inelastic strain-range SRP approach. The reverse is true for the LRSC data shown by the solid symbols. The assumption that LRSC data can be predicted by using stress-hold correlations may not be accurate enough. Note that the data in Refs *12* and *13* were obtained from two heats of the alloy and can be expected to have different fatigue characteristics. Overall the predictions made with the TS-SRP approach are slightly better than those made with the inelastic-strain-range SRP approach as measured by the SE.

Inconel 718 at 650°C

Two data sources [*12,14*] are available for this alloy, and both feature the strain-hold waveform. The data reported in Ref *12* were obtained by using specimens from bar stock, and the data in Ref *14* were obtained by using three heats (Heats 1, 2, and 6) of alloy in plate form. Information on the heat treatment of Heats 1, 2, and 6 is given in Refs *14* and *15*. The different processing histories can be expected to produce different mechanical properties.

Heat 6 of Ref *14*, chosen as the reference heat because of its completeness, was used to establish the life relations and flow correlations required for TS-SRP predictions. Life "predictions" of these data made with the TS-SRP approach were slightly better than those made with the inelastic-strain-range approach. The TS-SRP approach predicted 91% of the data within a factor of 2 and the SE was 0.206. The inelastic-strain-range approach predicted 86% of the data within a factor of 2 and the SE was 0.229.

Life predictions of the data in Ref *12* and in Heats 1 and 2 of Ref *14* are shown in Fig. 4. The predictions based on the TS-SRP approach were nearly identical to those based on the inelastic-strain-range approach as measured by the SE and are summarized by the tables inset in the figure. These predictions include six LRSC tests (indicated by the solid symbols). A review of the results [*8*, Figs. 39 and 40] shows that the strain-hold data were predicted about equally well by both the TS-SRP and inelastic-strain-range approaches. The main difference was in the life prediction of the LRSC tests; the TS-SRP overpredicted, the inelastic-strain-range approach underpredicted, the results. Previously, for alloy AF2-1DA, we assumed that this type of cycle could be predicted by using stress-hold flow correlations. Here we are assuming that strain-hold correlations can be used. This was done because more appropriate correlations simply were not available. Our experience indicates that correlations for one type of waveform (stress hold, say) may be used to predict the lives of strain-hold cycles provided that the cycle types are the same (PC, CP, or CC). The accuracy of the predictions will suffer accordingly, but waveform effects should be less at the lower strain ranges.

The results of these predictions for the two alloys are encouraging and strongly suggest that the method of determining the elastic line intercept B is sound. However, a more direct verification of Eq 5 is highly desirable. The equation's derivation is based on the assumption that the inelastic- and elastic-strain-range failure lines for PC, CP, and CC cycles are parallel to the corresponding lines for PP cycles. This assumption implies that the strain-hardening exponent n in Eq 4 is constant for all cycles.

The validity of these assumptions and the ability of Eq 5 to calculate the elastic line intercept were evaluated by comparing these "calculated" values against "observed" values determined from test data. The observed values of B were determined from the reported values of elastic strain range and the observed cyclic life:

$$B = \Delta\epsilon_{el}/(N_f)^b \qquad (11)$$

The results of these calculations are quite good [*8*, Fig. 41] and indicate that Eq 5 is valid.

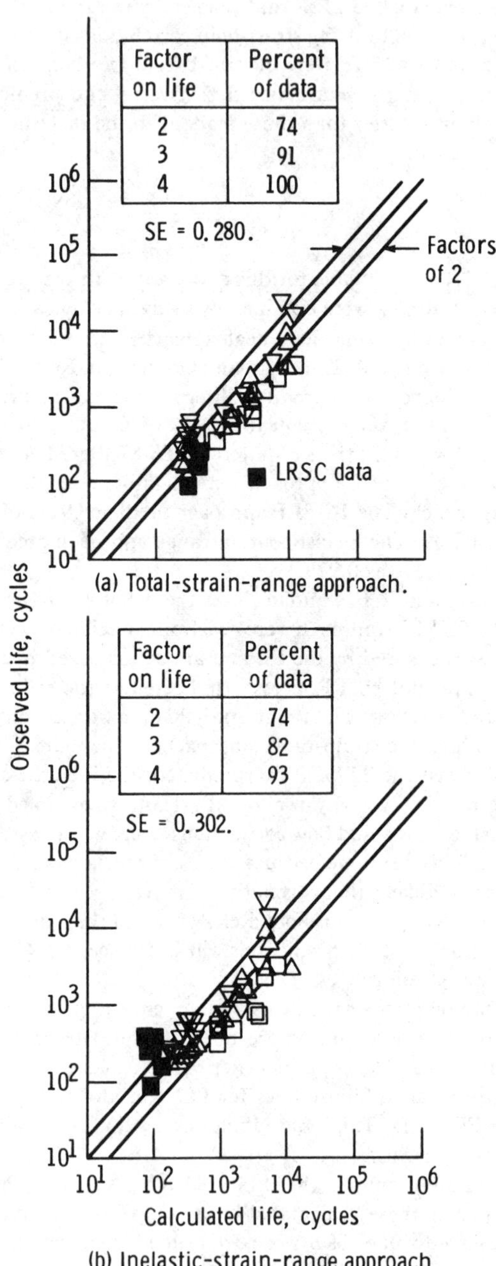

FIG. 4—*Life predictions of strain-hold data* [12] *and strain-hold and low-rate strain cycle data* [14]. *Data not used to establish correlations; Inconel 718 at 650°C.*

Concluding Remarks

An updated total-strain-range version of strainrange partitioning (TS-SRP) has been developed that makes it easier to characterize an alloy in the low-strain regime. This was accomplished by developing a set of empirical equations that characterize the constitutive behavior of an alloy. The constants in these equations can be determined from flow test results only. Data from failure tests are not required. For a given alloy these equations are a function of temperature, wave shape (PC, CP, CC), and the method of introducing creep into a cycle (stress hold, strain hold, etc.).

To fully characterize an alloy, failure tests are conducted in the high-strain-range regime, where test times and costs are reasonable. The data from these tests are used to determine the SRP inelastic-strain-range-versus-life relations and the elastic-strain-range-versus-life relation for the pure fatigue case of PP cycling. Flow testing is then done in the low-strain regime, where failure testing would be prohibitively expensive. These data are used to determine the constants in the empirical constitutive flow equations.

The analyst now has sufficient information to predict the life of a cycle of interest. In our original total-strain-range SRP paper the elastic line intercept was calculated by using an empirical equation with constants determined from failure data alone. The updated version presented herein can be expected to reduce extrapolation errors in determining the elastic line intercept, because the flow tests can be conducted at lower strain ranges than the failure tests.

References

[1] Halford, G. R. and Saltsman, J. F. in *International Conference on Advances in Life Prediction Methods,* D. A. Woodford and J. R. Whitehead, Eds., American Society of Mechanical Engineers, New York, 1983, pp. 17–26.

[2] Manson, S. S. and Zab, R. in *Proceedings of the Conference on Environmental Degradation of Engineering Materials,* M. R. Louthan, Jr., and R. P. McNitt, Eds., Virginia Polytechnic Institute and State University, Blacksburg, 1977, pp. 757–770.

[3] Brinkman, C. R., Strizak, J. P., and Booker, M. K. in *Characterization of Low Cycle High Temperature Fatigue by the Strainrange Partitioning Method,* AGARD CP-243, AGARD, Paris, France, 1978, pp. 15-1 to 15-18.

[4] Walker, K. P., "Research and Development Program for Nonlinear Structural Modeling with Advanced Time-Temperature Dependent Constitutive Relationships," PWA-5700-50, United Technologies Research Center, East Hartford, Conn., Nov. 1981 (NASA CR-165533).

[5] Chang, T. Y. and Thompson, R. L., "A Computer Program for Predicting Nonlinear Uniaxial Material Responses Using Viscoplastic Models," NASA TM-83675, NASA-Lewis Research Center, Cleveland, July 1984.

[6] Robinson, D. N. and Swindeman, R. W., "Unified Creep-Plasticity Constitutive Equations for 2¼Cr-1Mo Steel at Elevated Temperature," ORNL/TM 8444, Oak Ridge National Laboratory, Oak Ridge, Tenn., Oct. 1982.

[7] Hirschberg, M. H. and Halford, G. R., "Use of Strainrange Partitioning to Predict High-Temperature Low-Cycle Fatigue Life," NASA TN D-8072, NASA-Lewis Research Center, Cleveland, Jan. 1976.

[8] Saltsman, J. F. and Halford, G. R., "An Update of the Total Strain Version of SRP," NASA TP-2499, NASA-Lewis Research Center, Cleveland, Sept. 1985.

[9] Halford, G. R. and Nachtigall, A. J., *Journal of Aircraft,* Vol. 17, No. 8, 1980, pp. 598–604.

[10] Halford, G. R., Saltsman, J. F., and Hirschberg, M. H. in *Proceedings of the Conference on Environmental Degradation of Engineering Materials,* M. R. Louthan, Jr., and R. P. McNitt, Eds., Virginia Polytechnic Institute and State University, Blacksburg, 1977, pp. 599–612.

[11] Manson, S. S. and Muralidharan, U., "A Single-Expression Formula for Inverting Strain-Life and Stress-Strain Relationships," Case Western Reserve University, Cleveland, May 1981 (NASA CR-165347).

[12] Thaker, A. B. and Cowles, B. A., "Low Strain, Long Life Creep Fatigue of AF2-1DA and INCO 718," Pratt & Whitney Aircraft, West Palm Beach, Fla., April 1983 (NASA CR-167989).

[13] Hyzak, J. M., "The Effects of Defects on the Fatigue Initiation Process in Two P/M Superalloys,"

AFWAL-TR-80-4063, Air Force Wright Aeronautical Labs., Wright-Patterson AFB, Ohio, Sept. 1980.

[*14*] Korth, G. E. and Smolik, G. R., "Physical and Mechanical Test Data of Alloy 718," TREE-1254, Idaho National Engineering Lab., Idaho Falls, Id., March 1978.

[*15*] Brinkman, C. R. and Korth, G. E., *Journal of Testing and Evaluation*, Vol. 2, No. 4, 1974, pp. 249-259.

DISCUSSION

S. D. Antolovich[1] (*written discussion*)—Your model appears to be based on damage by plasticity and/or creep mechanisms, yet many materials, including IN 718 which you partially used to verify your model, are well-known to be very sensitive to damage by environmental degradation. Furthermore, it would seem that in the long-life, low-total-strain regime for which your model is designed, environmental effects would become dominant.

To what extent will the actual mechanism(s) of failure affect your formulation? This point may be quite important in applying your model, since short-term experimental results are used to predict long-term component life, and mechanism changes could, in some cases, lead to nonconservative estimates while in others lead to overly conservative estimates.

J. F. Saltsman and R. G. Halford (*authors' closure*)—You raise a pertinent issue that was not addressed directly in the oral presentation. As you aptly point out, environmental degradation of fatigue life may rival and even exceed creep degradation, particularly at long exposure times. However, it must also be pointed out that environmental effects are not necessarily always detrimental to fatigue. For example, results from the Naval Research Laboratories of several years ago indicate regimes where oxidation can enhance life compared with fatigue life in vacuum [*1*].

In the current formulation of the SRP method, all the life degradation encountered in the baseline tests used in evaluating the SRP constants is attributed implicitly to a single time-dependent phenomenon. We have for simplicity called this *creep*, although it is obvious that the degradation is indeed a mixture of at least two major deleterious mechanisms—creep and oxidation. To our knowledge, quantitative separation of the contributions by each has never been accomplished (see, for example, Ref *2*), although there has been some discussion written about doing so [*3*]. If we have been incapable of separating the contributions in the life regime of laboratory tests, we will likely not be able to separate them for long extrapolated times either. Critical experiments are required. Until such time as a clear separation of the variables is attained, we are forced to calibrate models using available quantitative data acquired in shorter term experiments. Extrapolation is then made to longer times. In so doing, it is tacitly assumed that no additional mechanisms of damage come into play, and that the combined creep/environmental mechanisms of degradation do not alter their relative contributions. If enough uncertainty is prevalent, "factors of safety" are added by designers/manufacturers.

We should point out that since the time of the conference, two references [*4,5*] have appeared on the subject of the time dependencies of the SRP life relations. While they do not explicity identify the time dependencies with oxidation or other time-dependent environmental degradations, they do offer greater flexibility in being able to model both waveshape (creep/plasticity) and time (creep/oxidation) effects.

Obviously, further research is required before we can confidently include a quantitative assessment of the individual contributions to cyclic damage accumulation at high temperatures.

[1]Fracture and Fatigue Research Laboratory, Georgia Institute of Technology, Atlanta, GA 30332.

Possibly Professor Antolovich will be able to expand upon his oxidation penetration/maximum tensile stress criterion for fracture [*6*] to provide a workable model for engineering life prediction.

Discussion References

[*1*] Achter, M. R., Danek, G. J., Jr., and Smith, H. H., "Effect on Fatigue of Gaseous Environments under Varying Temperatures and Pressure," *Transactions of AIME,* Vol. 227, 1963, pp. 1296–1301.
[*2*] Bernstein, H. L., "A Stress-Strain-Time Model (SST) for High-Temperature, Low-Cycle Fatigue," *Methods for Predicting Material Life in Fatigue,* ASME, New York, 1979, pp. 89–100.
[*3*] Manson, S. S., Halford, G. R., and Oldrieve, R. E., "Relation of Cyclic Loading Pattern to Microstructural Fracture in Creep Fatigue," NASA TM 83473, presented at Fatigue '84 (Second International Conference on Fatigue and Fatigue Thresholds), Birmingham, England, 3–7 Sept. 1984.
[*4*] Kalluri, S. and Manson, S. S., "Time Dependency of Strainrange Partitioning Life Relationships," Case Western Reserve University, Cleveland, NASA CR174946, Aug. 1984.
[*5*] Kalluri, S., "Generalization of the Strainrange Partitioning Method for Predicting High Temperature Low Cycle Fatigue Life at Different Exposure Times," Ph.D. thesis, Case Western Reserve University, Cleveland, Aug. 1986.
[*6*] Antolovich, S. D., Liu, S., and Baur, R., "Low Cycle Fatigue Behavior of René 80 at Elevated Temperature," *Metallurgical Transactions A,* Vol. 12, No. 3, March 1981, pp. 473–481.

H. D. Solomon[1]

Low-Frequency, High-Temperature Low Cycle Fatigue of 60Sn-40Pb Solder

REFERENCE: Solomon, H. D., **"Low-Frequency, High-Temperature Low Cycle Fatigue of 60Sn-40Pb Solder,"** *Low Cycle Fatigue, ASTM STP 942,* H. D. Solomon, G. R. Halford, L. R. Kaisand, and B. N. Leis, Eds., American Society for Testing and Materials, Philadelphia, 1988, pp. 342–370.

ABSTRACT: Solder thermal fatigue is a critical problem for large surface mounted leadless chip carriers. This fatigue results from the thermal mismatch between the Al_2O_3 chip carrier and the circuit board. This study was aimed at this problem, particularly the influence of time-dependent effects.

Isothermal tests were run on 0.15 to 0.23 mm (0.006 to 0.009 in.) solder layers tested in simple shear. It has been found, as expected, that solder exhibits Coffin-Manson low-cycle fatigue (LCF) behavior. The present study details the influence of cycling frequency on the fatigue life. Results were obtained on 60Sn-40Pb tested at 35 and 150°C. In agreement with previous studies made on Pb, it is shown that, at both temperatures, the fatigue life decreases as the frequency decreases, especially below 10^{-3} Hz. The fatigue life can be described by Coffin's frequency modified LCF law; i.e.,

$$(N_f \nu^{k-1})^{\alpha} \Delta\gamma_p = \theta$$

The data were also analyzed by means of strain range partitioning (SRP). It was found that this approach does *not* describe the data. A time-modified SRP approach, i.e., the time to failure t_f is given by

$$\frac{1}{t_f} = \frac{f_{cc}}{t_{cc}} + \frac{\nu f_{pp}}{N_{pp}}$$

was found to describe the data. The primary difference between this and the standard SRP approach is that with the time-modified SRP we assume that at low frequencies the time to failure t_{cc} is a constant, not the number of cycles to failure N_{cc}. The correlation between this approach and the frequency-modified fatigue law is discussed. There is also a characterization of the different types of fatigue fracture surfaces that are observed.

KEY WORDS: fatigue, solder, strain range partitioning, creep-fatigue

The fatigue of solder used for electrical connections has recently become a problem because of the shift to surface-mounted devices [*1–18*]. Surface mounting means directly soldering the device to pads on a printed wiring board instead of connecting the device by means of leads that are soldered to the board. Surface mounting conserves valuable board space, speeds up circuit operation due to the reduction of the lead length, allows for mounting devices on both sides of a circuit board, and (with modern vapor phase soldering) can be less expensive to perform. Unfortunately, with surface mounting there is no lead to flex and absorb thermal and mechanical strains. The solder in a surface-mounted device is therefore more prone to fatigue failure and

[1]Staff Metallurgist, General Electric Corporate Research and Development Center, Schenectady, NY 12301.

these failures are being observed. Of particular importance is thermally induced low-cycle fatigue (LCF), since the thermal strains can be quite large (depending upon the exact geometry, chip carrier and circuit board materials, thermal cycles experience, etc.).

The present study aims at understanding the high-temperature, low-cycle fatigue of 60-40 solder (60Sn-40Pb), the most commonly used solder. Since 60-40 solder melts at 185°C, even room temperature is a high temperature (25°C is a homologous temperature of 0.65 T_m). When most metals are low-cycle fatigue tested, at high homologous temperatures, they experience a significant reduction in the fatigue life when the cycling frequency is reduced. Solders are no exception. Indeed, the earliest observation of this phenomenon was reported in lead [*19,20*].

The present study was aimed at quantifying the low-cycle behavior of 60-40 solder. In a significant departure from most previous studies, solder layers, as opposed to bulk specimens, were studied. This was done because solder is used as a joining material not as a bulk structural metal. The tests were also done in shear, which is the usual type of loading encountered when electronic package/circuit board solder joints experience thermal strains. Of particular interest in these studies was the correlation of the LCF behavior with several approaches used in high-temperature LCF life prediction.

Experimental Procedure

The experiments described here were performed on solder layers, tested in simple shear. A detailed description of the test techniques, including the soldering procedures, is included in Refs *21* and *22* so only the most important points will be discussed here. Two brass or copper blocks were soldered together. The solder joints ranged in thickness from 0.152 mm (0.006 in.) to 0.229 mm (0.009 in.). The exact thickness depended upon the spacer shims that were used and the exact dimensions of the test blocks that were used. By measuring the blocks before soldering and the assembly after soldering, the solder thickness could be determined to within about 0.0076 mm (0.0003 in.) or about 5%. The solder pads were nominally 2.54 mm (0.1 in.) by 12.7 mm (0.5 in.) with the shear in the 12.7 mm (0.5 in.) direction.

The soldering was done using 0.076 mm (0.003 in.) thick foil made by the Semi Alloys Company. The solder was nominally 60Sn-40Pb (analyzed to be 60.18Sn-39.53Pb with 66 ppm Zn and 12 ppm Cu). The soldering was done on a hot plate using several steps of flow and reflowing in order to produce void-free joints [*21,22*].

One of the test blocks was bolted to a grip attached to the load cell and the other to a grip attached to a movable servo-controlled piston. The assembly was aligned so that the solder joint was in the center of the system, and the up and down motion of the piston was transmitted to the load cell via shearing stresses applied through the solder layer. The exact design of the blocks, which was soldered together, was patterned after the Iosipescu shear specimen [*23*], which eliminates the singularity that develops in lap shear tests when tensile (or compressive) stresses are translated into shear stresses. A servohydraulic testing machine was utilized with dual rate function generators. This enabled a wide range of possible waveforms to be applied. Ramp loading and unloading were employed in all the fatigue tests described here, with equal loading and unloading rates.

Because thin layers are being tested, the displacement resolution must be quite good so that reasonable strains could be utilized. Using a clip gage attached to the test blocks, it was possible to resolve 6.35×10^{-5} mm (2.5×10^{-6} in.) of displacement. A ± 907 kg (± 2000 lb) load cell was employed, which enabled a load resolution of 0.09 kg (0.2 lbs). With an average thickness of 0.19 mm (0.0075 in.) and an area of 32 mm^2 (0.05 in.^{2}) and the sensitivity ranges utilized in most tests, a strain of 6.67×10^{-4} and a stress of 55×10^{-3} MN/m^2 (8 psi) was resolved.

The test setup was enclosed within a box with a circulating fan that extracted the gas from the box and forced it through a duct which contained a heater. Liquid nitrogen could also be fed into this duct. By controlling the power fed into the duct and feeding in liquid nitrogen when below-ambient temperatures were required, any temperature between −50 and +150°C could be achieved. At above ambient temperatures, where no liquid N_2 was employed, the temperature could be held to within ±0.2°C for a few hours. Over a period of days the temperature variation was ±0.5°C. At low temperatures, the temperature variation was about ±2°C. Dry N_2 (dried with a molecular sieve) was fed into the box during all the tests. This was done primarily to prevent icing during the low-temperature tests. A flow rate of about 20 standard cubic feet per hour (SCFH) was used to provide a flow of N_2 out of the openings in the box that were required for the pull rods. This system does not completely preclude air infiltration into the box so the tests *cannot* be considered to have been run in an inert gas. The specimens were equilibrated about 15 to 18 h at the test temperature prior to testing.

Plastic strain limits were employed for the fatigue tests [*22*]. The plastic strain was computed by an analog plastic strain computer, which subtracted out the elastic displacement portion of the total displacement signal. The cycling frequency was altered by changing the ramp loading/unloading rates, which were total displacement controlled. The same system was employed to measure the creep rate. In the creep tests, the specimen was loaded (under load control at a fixed loading rate) to a predetermined level and the load was then maintained via the servo system. The load was maintained to within ±0.045 kg (±0.1 lb) or about 13.8×10^{-3} MN/m^2 (2 psi), during each test, which lasted from a few seconds to more than 12 h (depending upon the load and resulting creep rate).

Specimens were metallographically prepared using a final polish on a Jarret wheel with a 200 g load and using a ½ μm diamond paste. Mechanical polishing without etching revealed the Pb-rich and Sn-rich areas (the Pb-rich areas are dark). Additional discrimination was achieved using a 2 s swab with an etch consisting of 1 part nitric acid, 1 part acetic acid, and 8 parts glycerol, or etching with hydrol (HCl in ethyl alcohol).

Results and Discussion

Low-Cycle Fatigue

Figures 1 and 2 show the low-cycle fatigue life (N_f) as a function of plastic strain range ($\Delta\gamma_p$) for specimens cycled at 35 and 150°C. The life is defined as the number of cycles to reduce the hysteresis load range to one half its maximum value (which is reached on the initial cycle or at most within three cycles) [*21,22*]. Specimens behave in accordance with the Coffin-Manson low-cycle fatigue law [*24–29*]; i.e.,

$$N_f^{\alpha} \Delta\gamma_p = \theta \qquad (1)$$

where α and θ are constants. For $N_f = 1/4$, i.e., a tensile test where $\Delta\gamma_p = \Delta\gamma_F'$ (the tensile ductility) and $\theta = \Delta\gamma_F' (1/4)^{\alpha}$.

At 35°C, $\alpha = 0.52$, which is in the expected range of 0.5 to 0.6. At 150°C, α is only 0.37. All of these tests were run with a cycling frequency of about 0.3 Hz. When the cycling frequency is reduced, the fatigue life is also generally reduced. This is illustrated in Figs. 3, 4, and 5. As can be seen, this frequency effect is present at both 35 and 150°C. (The single X point in each of these figures are the N_f values taken from the curves of Fig. 1 or 2 for the appropriate temperature and strain; the other points were taken from individual experiments run at different cycling frequencies.)

These results can be described by the frequency-modified Coffin-Manson law [*30–33*]; i.e.,

$$(N_f \nu^{K-1})^{\alpha} \Delta\gamma_p = \theta \qquad (2)$$

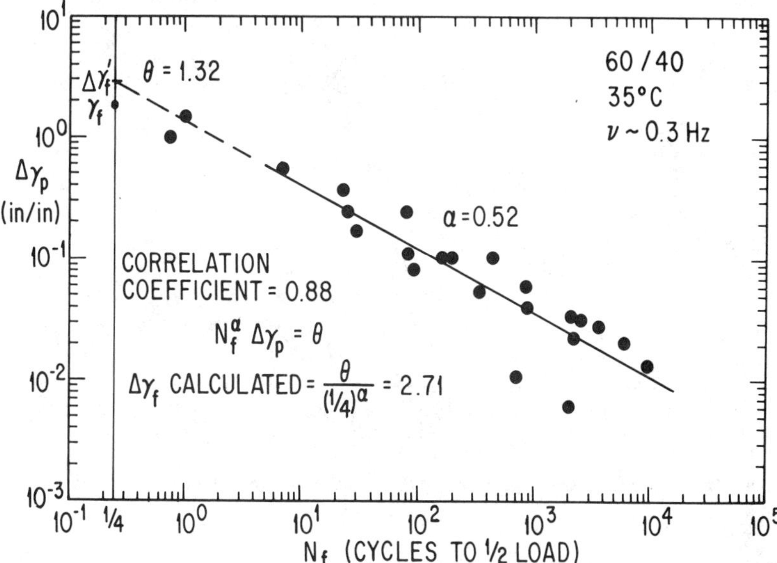

FIG. 1—*Strain life behavior for specimens tested at 35°C and at ~0.3 Hz.*

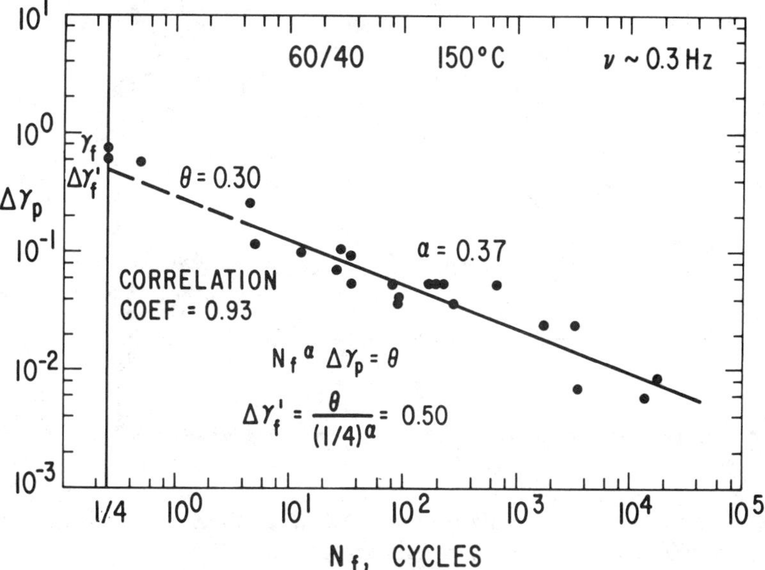

FIG. 2—*Strain life behavior for specimens tested at 150°C and at ~0.3 Hz.*

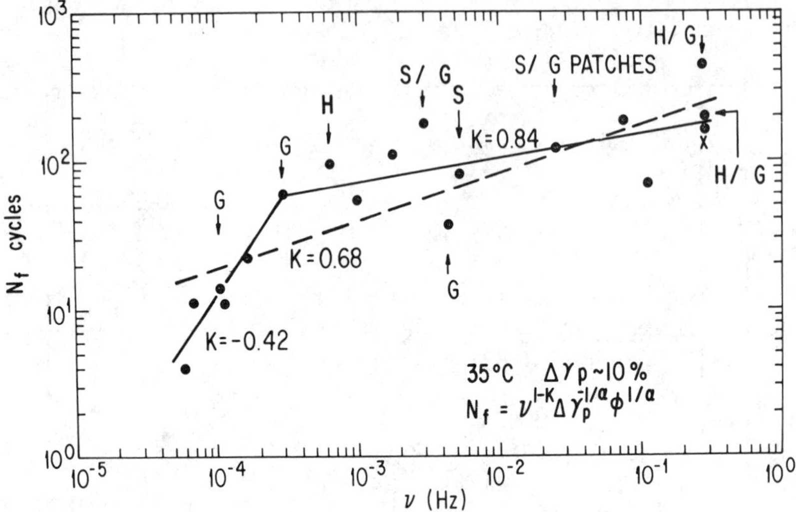

FIG. 3—*Fatigue life* (N_f) *versus cycle frequency* (ν) *for specimens tested at 35°C with* $\Delta\gamma_p \sim 10\%$.

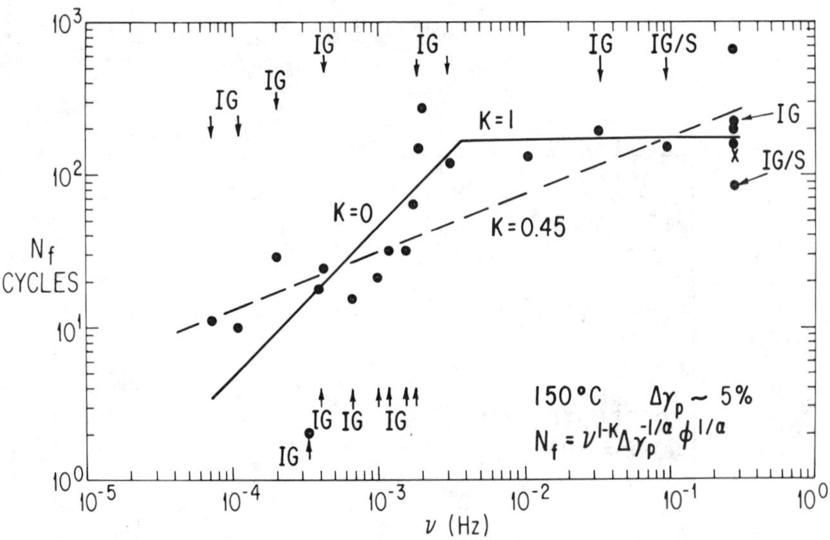

FIG. 4—*Fatigue life* (N_f) *versus cycle frequency* (ν) *for specimens tested at 150°C with* $\Delta\gamma_p \sim 5\%$.

A value of $K = 1$ denotes no frequency effect. For $K < 1$, there is a reduction in life with reductions in the cycling frequency (the reverse is true for $K > 1$). When K is between 0 and 1, the number of cycles to failure is decreased as the cycling frequency is decreased (or the cycling period is increased) but the time to failure is increased [*33*]. For $K = 0$, the time to failure is constant [*33*] (i.e., halving the cycling frequency (or doubling the cycling period) halves the number of cycles to failure so that the time to failure is constant). Negative values of K denote a

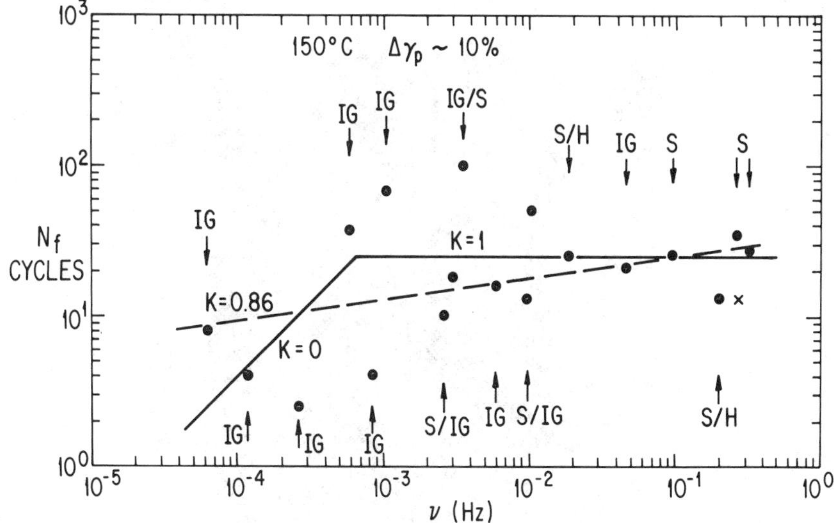

FIG. 5—*Fatigue life* (N_f) *versus cycle frequency* (ν) *for specimens tested at 150°C with* $\Delta\gamma_p \sim 10\%$.

still stronger frequency effect (i.e., halving the cycling frequency more than halves the number of cycles to failure so that the time to failure is reduced).

The data taken at 35°C have been correlated in two ways. All the data points have been fit to a single curve (the dashed curve) for which $K = 0.68$. The data suggest, however, that a better approach would be to consider two regimes (the solid curves); i.e., for frequencies above 3×10^{-4} Hz, $K = 0.84$ and below 3×10^{-4} Hz, $K = -0.42$.

Figures 4 and 5 show the data taken at +150°C (for $\Delta\gamma_p \sim 0.05$ and 0.10, respectively) as correlated with a single K value for each strain range or with the data broken into regimes with different K values. At high frequencies there is no statistically significant variation in life with cycling frequency (i.e., $K = 1$). At low frequencies, the data can be correlated with $K = 0$.

Figures 3 to 5 also contain information about the morphology of the fracture surface. No fatigue striations were observed in any specimen; instead, four general types of fracture surfaces were observed. These are illustrated in Figs. 6 to 9. Figure 6 illustrates a shear fracture (denoted *S* in Figs. 3 through 5). Shear markings are evident, although they do not generally cover the entire fracture surface (i.e., the shear fractures are found in conjunction with other morphologies). Figure 7 illustrates ductile hole growth (denoted *H* in Fig. 3). The hole bottoms are relatively smooth. Figure 8 illustrates the type of granulated fractures that are denoted as *G* in Fig. 3. At low magnifications they look like the hole growth fractures. The difference is that at higher magnifications one can observe what appear to be granulated facets at the hole bottoms. The fractures are thus characterized by the presence of granulated facets and ductile ridges. These granulated regions are finer and less angular then those observed in specimens tested at 150°C; hence the term *granulated* has been used instead of *intergranular,* which is being used for the larger more angular facets observed in the specimens tested at 150°C.

Figure 9 illustrates the intergranular type of fracture observed in specimens cycled at 150°C. At low magnifications, the fractures appear characteristic of ductile hole growth. At higher magnifications, it can be seen that there are intergranular facets at the bottom of the holes. These fractures are distinguished from the granulated fractures by the size of the facets (about 10 μm in the case of the intergranular (IG) facets versus 2 μm for the granulated fractures) and

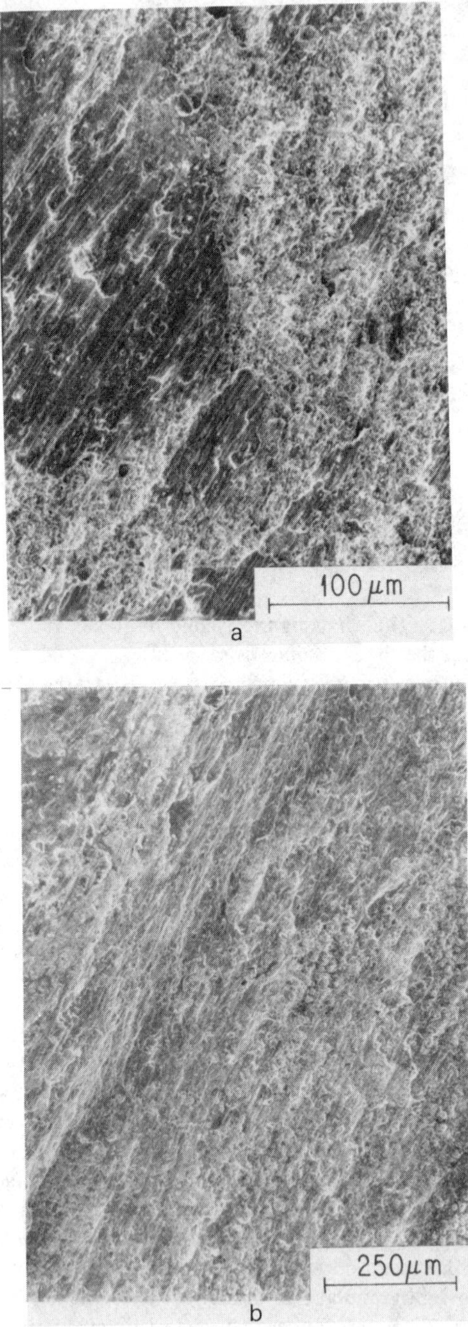

FIG. 6—*Examples of shear-type fractures observed in specimens tested at 35°C.*

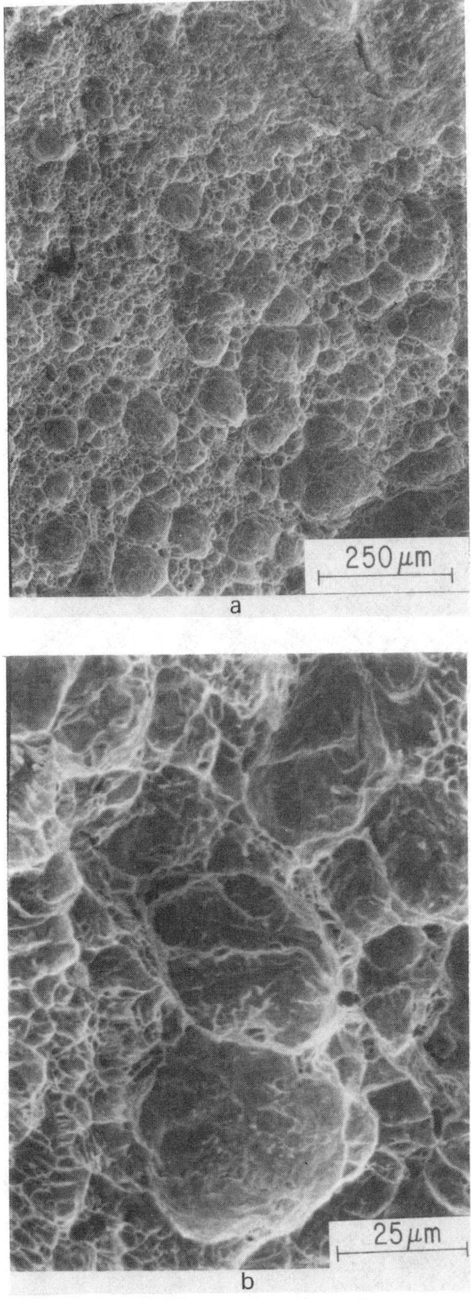

FIG. 7—*Examples of ductile-type fractures observed in specimens tested at 35°C.*

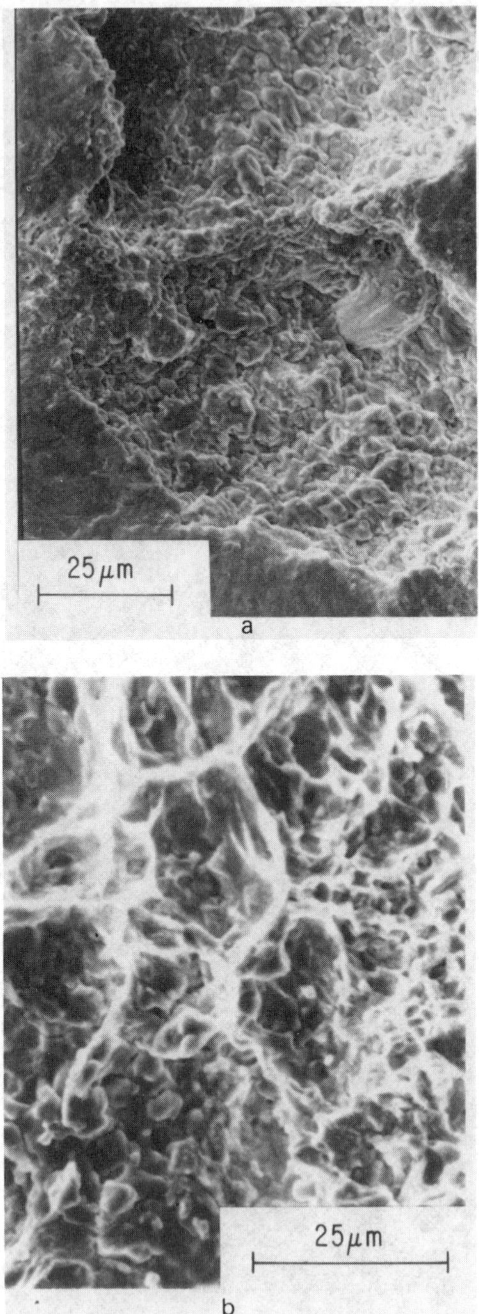

FIG. 8—*Examples of granulated-type fractures observed in specimens tested at 35°C.*

the more intergranular appearance of the intergranular fractures (they are more three-dimensional, and boundaries and triple points are more often seen). The ductile ridges which give these fracture their ductile appearance at low magnifications are about 25 to 100 μm apart. EDAX measurements were made to determine if the IG regions (and granulated facets) were predominantly Sn or Pb. Both Sn- and Pb-rich IG facets and ridges were observed.

Figures 10 and 11 show the microstructure of various specimens. Figures 10*a* and 10*b* show the structure of an untested specimen. Figures 10*c* to 10*e* show the structure of specimens cycled at 35°C at 0.28, 10^{-4}, and 10^{-3} Hz. Figures 10*e* and 10*f* show cracks that go through both Pb- and Sn-rich regions (the Pb-rich regions are dark). Figures 11*a* to 11*f* show the structure of specimens cycled at 150°C (at 0.1, 1×10^{-3}, and 1×10^{-4} Hz).

The specimens tested at both temperatures exhibit a similar cast structure with large Pb-rich dendrites (the Pb-rich phase is the dark phase) and smaller Pb-rich precipitates in an Sn matrix. The presence of Pb precipitates at grain boundaries and the orientation of precipitates within Sn grains points to a primary Sn grain diameter of 25 to 100 μm. A Hydrol etch (HCl in ethyl alcohol) used in conjunction with Nomarski interference contrast microscopy bring out some subgrains (Fig. 10*c*). The 25 to 100 μm grain diameter coincides with the ridge spacing observed on the fracture surfaces, and the fact that the EDAX analysis shows both Sn and Pb ridges is explained by the presence of Pb precipitates in the Sn grain boundaries. Heating to 150°C, even for several days, does not appear to cause any recrystallization or even any appreciable increase in the Sn grain size. There does, however, appear to have been some coarsening of the Pb-rich precipitates in the specimens tested at 150°C versus those tested at 35°C, but little or no difference between specimens tested for various lengths of time at 150°C (all of which were soaked 15 to 18 h at 150°C prior to cycling).

The somewhat finer Pb-rich precipitates and their closer spacing in the specimens tested at 35°C correlates in a general way with the finer granulated facets observed on the fracture surface of these specimens compared with the coarser Pb-rich regions and larger facets observed in the specimens tested at 150°C. In general, however, the granulated structure observed on the fracture surface of specimens tested at 35°C does not appear quite so intergranular; hence these are not being referred to as intergranular. Many of these facets do appear to be intergranular; others, however, look more like oxide particles which have initiated ductile hole growth. In contrast, the intergranular facets observed on the fracture surface of specimens tested at 150°C are clearly intergranular.

While there are slight differences in the microstructure between specimens tested at 35°C and those tested at 150°C, there are no significant differences between specimens tested at different frequencies (at the same temperature). Hence, the Sn/Pb microstructure *does not explain the observed frequency dependence of the fatigue life.*

The question of microstructural changes with fatigue cycling is an important one for solders. Zakraysek et al [*34*] have observed marked microstructural changes with fatigue cycling of eutectic solders. The lack of any significant changes in this study can be attributed to the relatively coarse initial structure, which is much more stable than the fine equiaxed structure studied by Zakraysek.

Another feature of the structure that is subject to change with time is the intermetallic (Cu-Sn [*22*]), which forms between the substrate and solder layer. Figures 12*a* to 12*f* show this intermetallic and crack surface in specimens tested at 35 and 150°C. Several points should be noted. The fatigue crack *does not* propagate through the middle of the solder layer at any of the cycling temperatures or cycling frequencies. It always propagates nearer to one or the other of the substrates that were soldered together (many specimens have cracks near both substrates). Typically the crack is about 5 to 50 μm from a substrate (the center of the solder layer is about 100 μm from the substrate). The intermetallic layer is somewhat irregular in the specimens tested at 35°C with many lenticular precipitates about 1 to 5 μm long (Fig. 12). In contrast, in the speci-

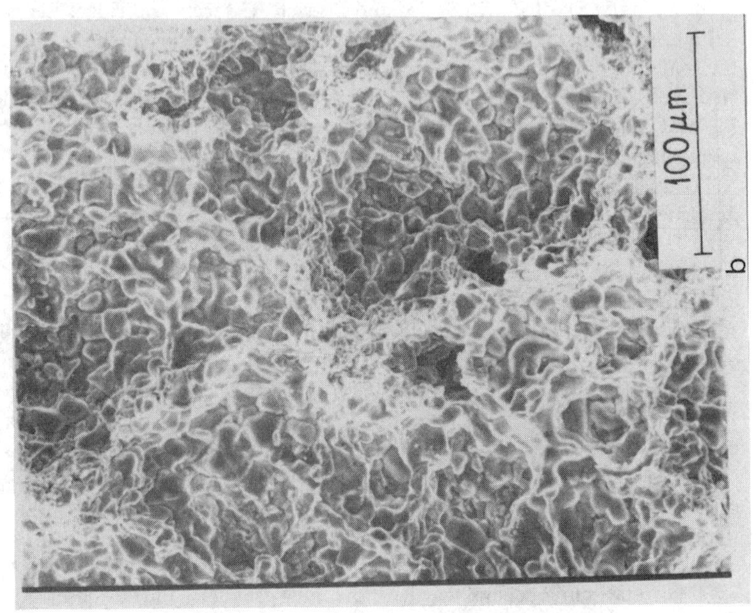
100 μm
b

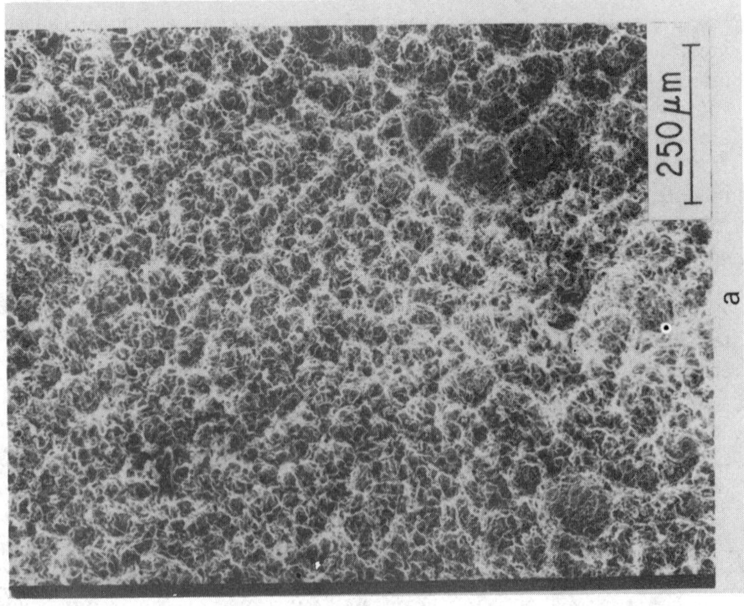
250 μm
a

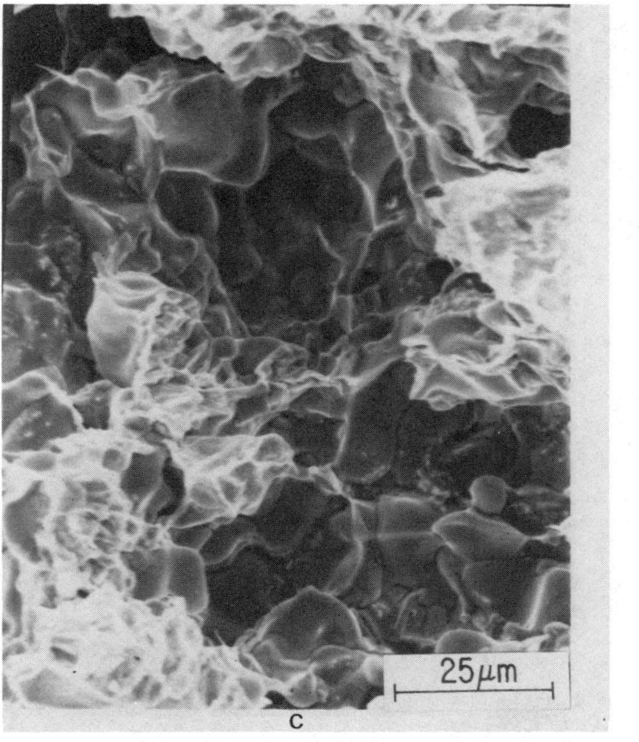

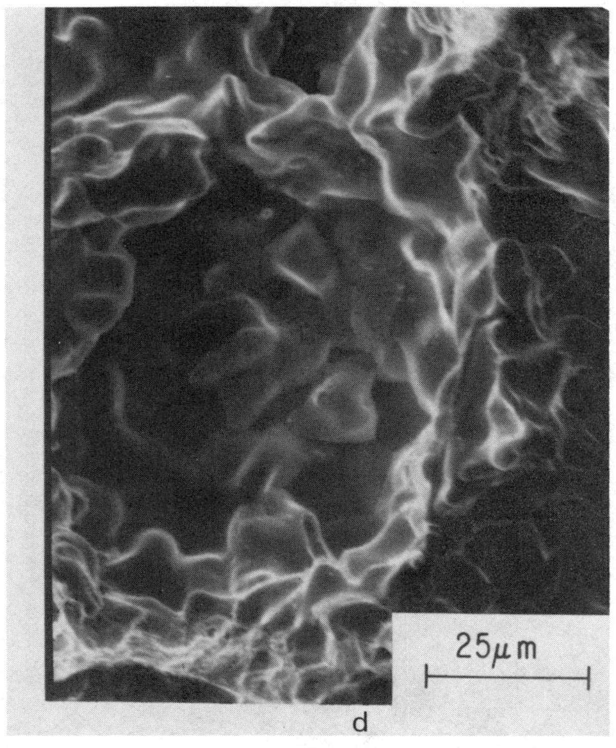

FIG. 9—*Examples of intergranular fractures observed in specimens tested at 150°C.*

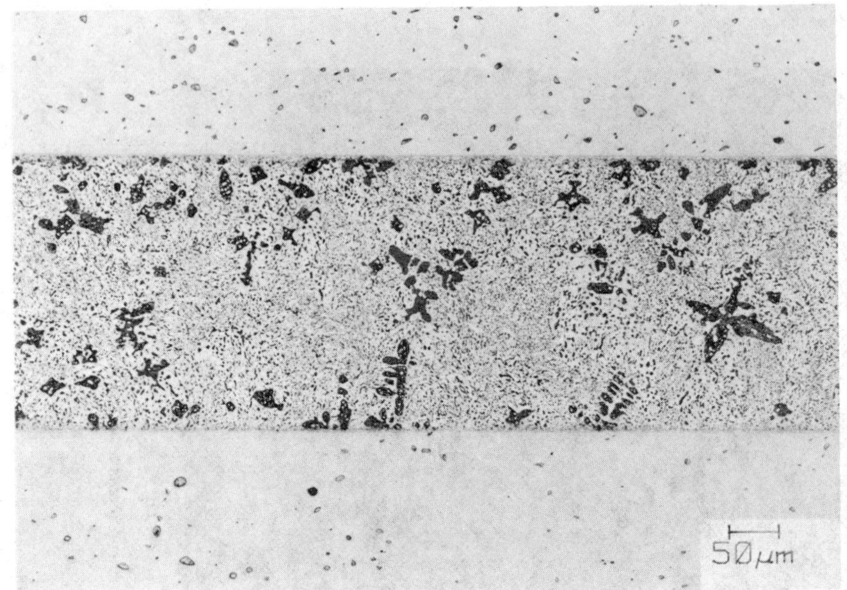

FIG. 10*a*—*Microstructure of an untested specimen.*

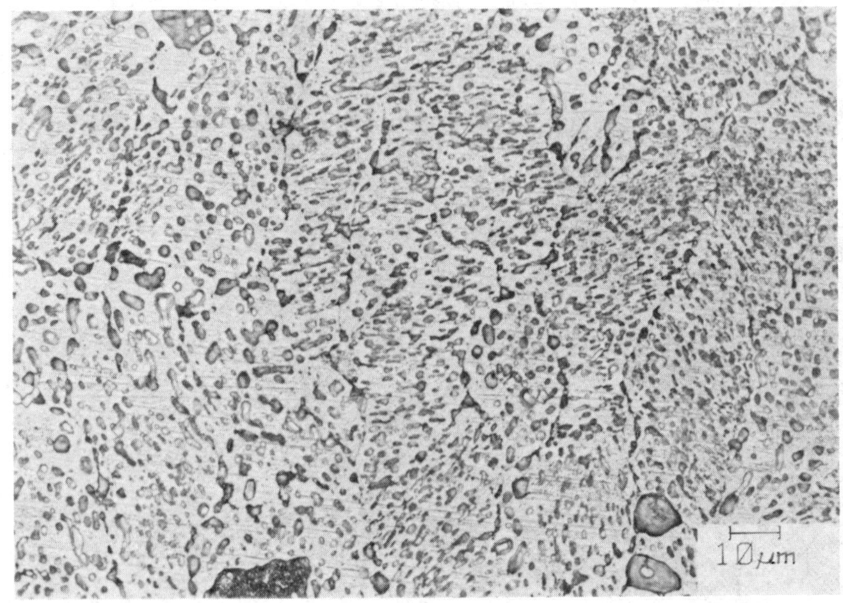

FIG. 10*b*—*Microstructure of an untested specimen (higher magnification).*

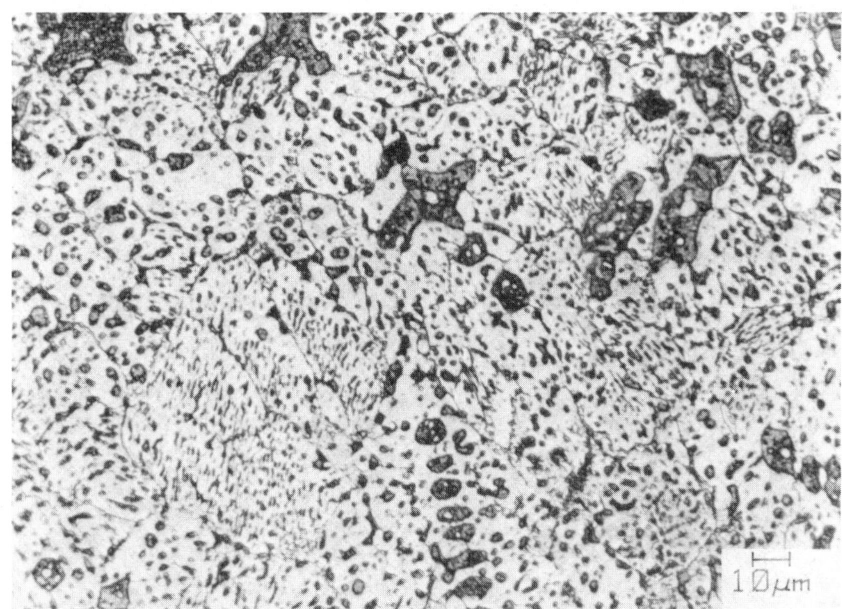

FIG. 10*c*—*Microstructure of a specimen cycled at 0.28 Hz at 35°C. Hydrol etch used to bring out subboundaries.*

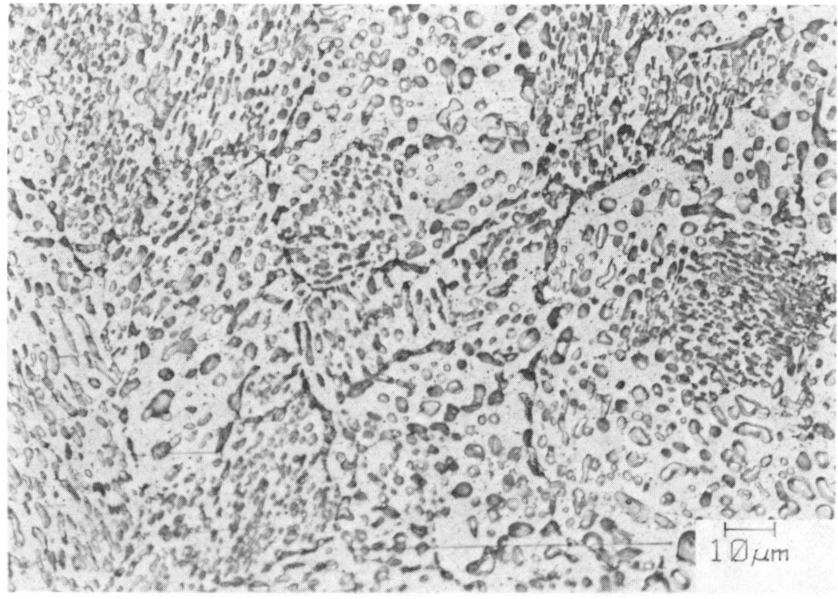

FIG. 10*d*—*Microstructure of a specimen cycled at 1×10^{-4} Hz at 35°C, glycerol etched to delineate boundaries.*

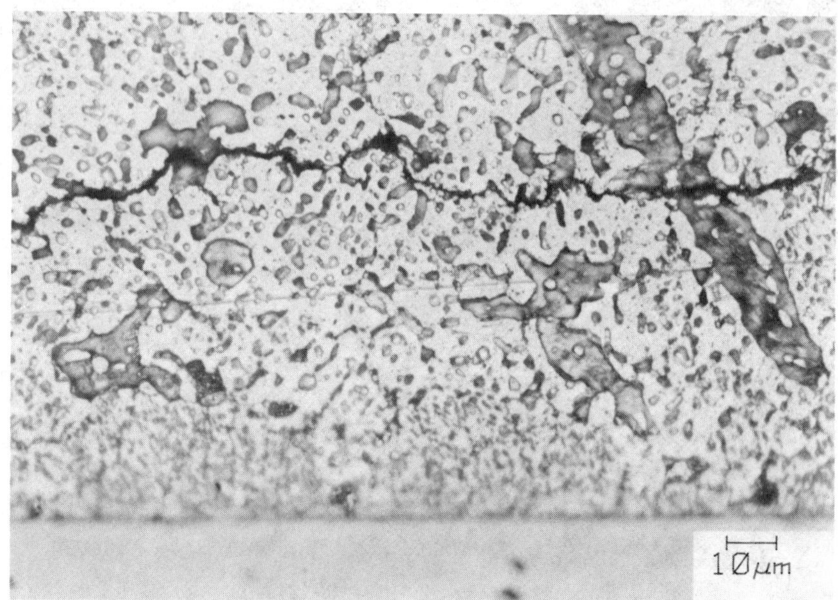

FIG. 10e—*Microstructure of a specimen tested at 35°C at 1 × 10^{-3} Hz, showing a crack.*

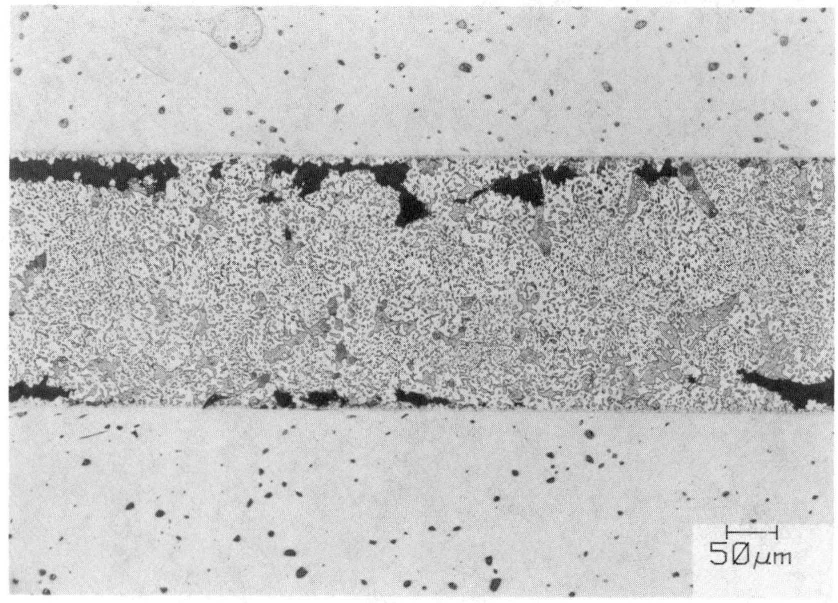

FIG. 10f—*Microstructure of a specimen tested at 35°C at 1 × 10^{-4} Hz, showing cracks.*

FIG. 11*a*—*Microstructure of a specimen tested at 0.1 Hz at 150°C.*

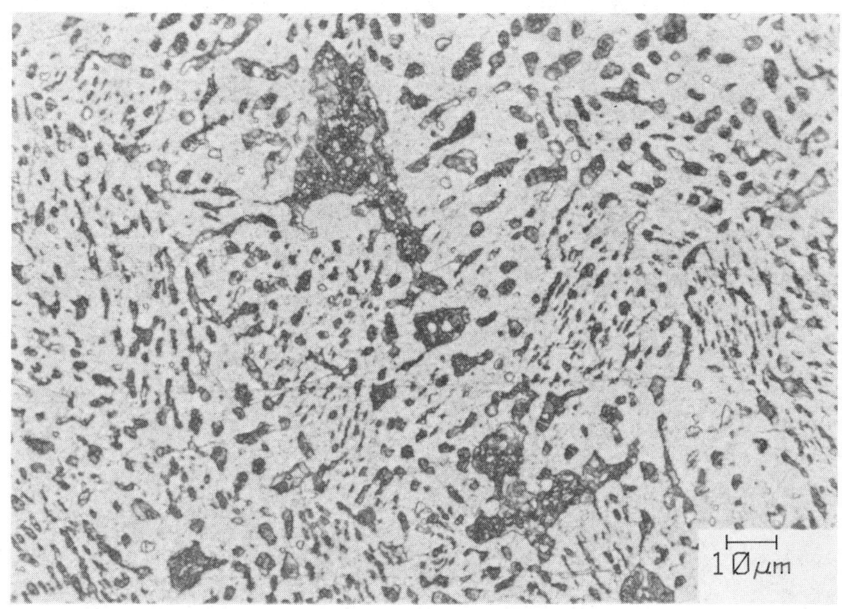

FIG. 11*b*—*Closeup of Fig. 11a.*

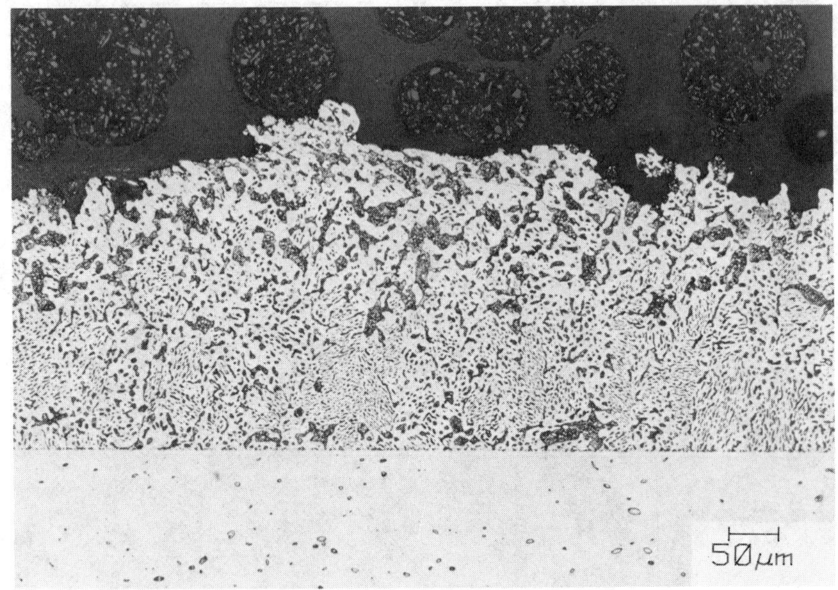

FIG. 11c—*Microstructure of a specimen tested at 1 × 10^{-3} Hz at 150°C.*

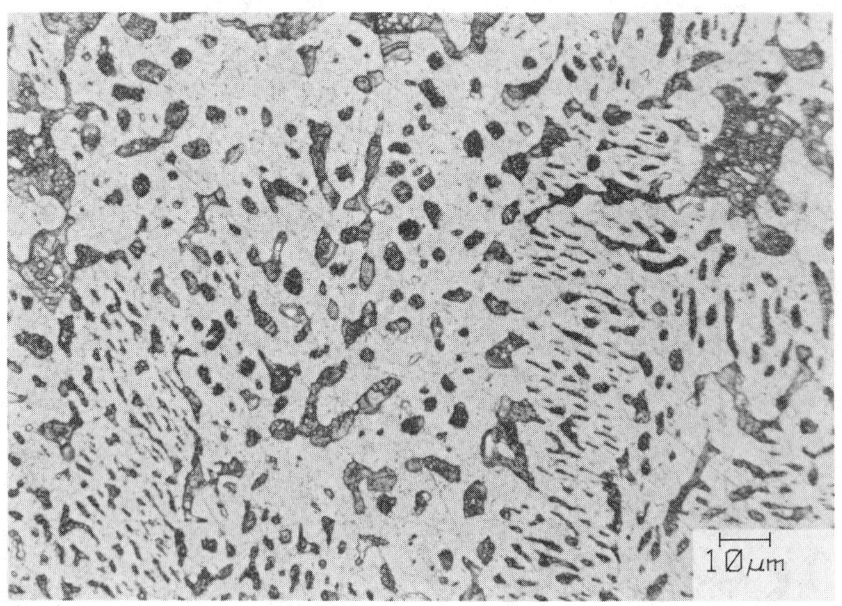

FIG. 11d—*Closeup of Fig. 11c.*

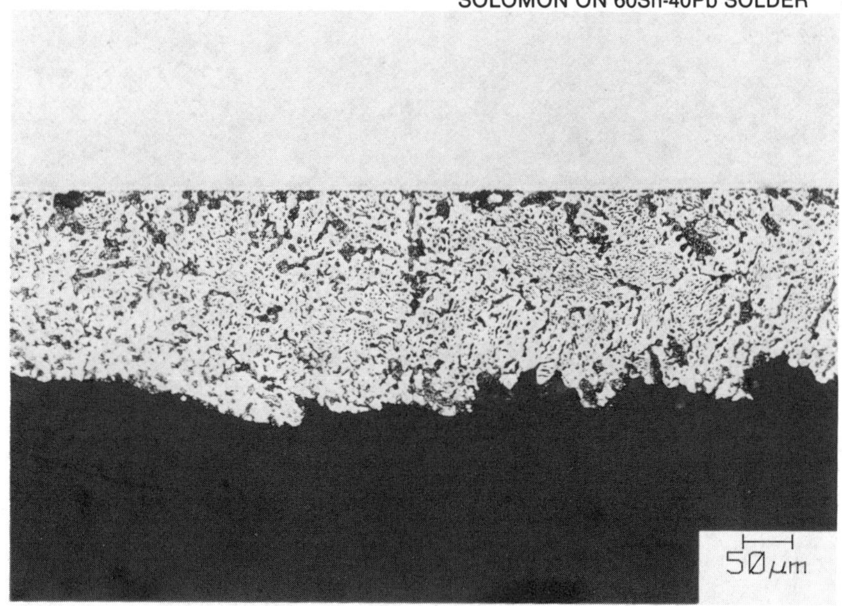

FIG. 11e—*Microstructure of a specimen tested at 1 × 10^{-4} Hz at 150°C.*

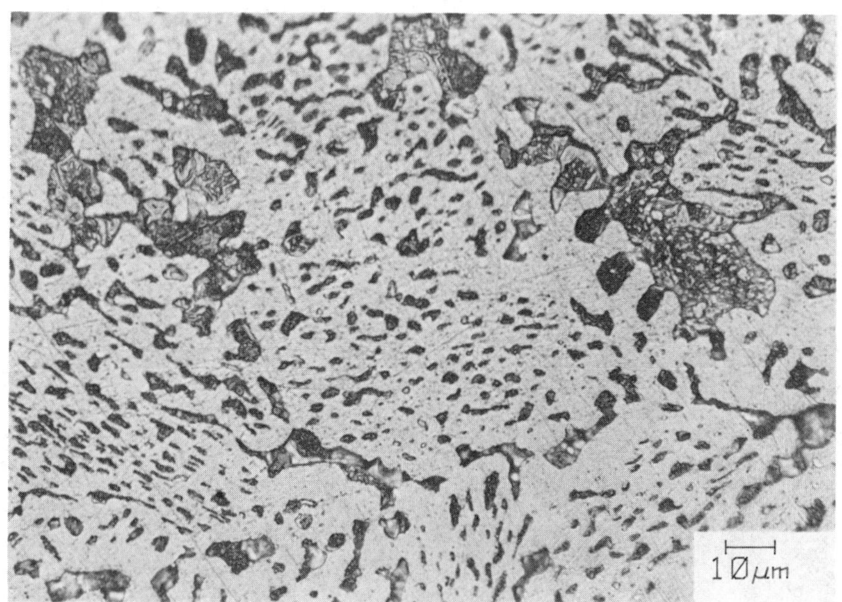

FIG. 11*f*—*Closeup of Fig. 11e.*

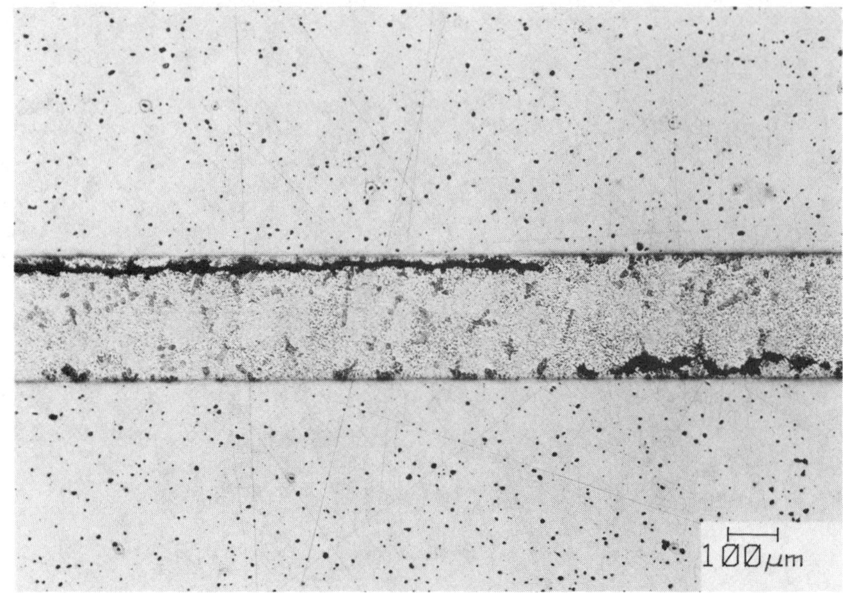

FIG. 12a—*Cracks in a specimen tested at 35°C at 1 × 10^{-4} Hz.*

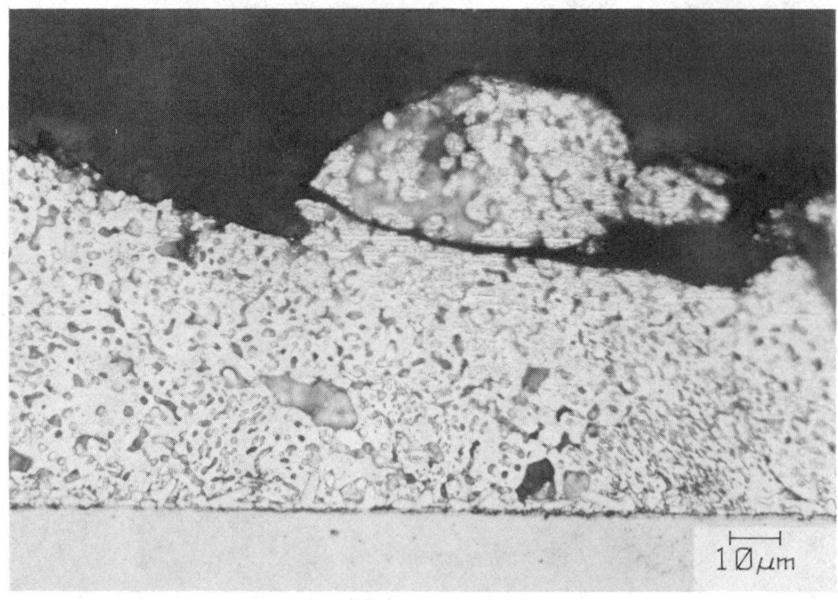

FIG. 12b—*Closeup of a specimen tested at 35°C at 0.28 Hz showing the intermetallic and the crack.*

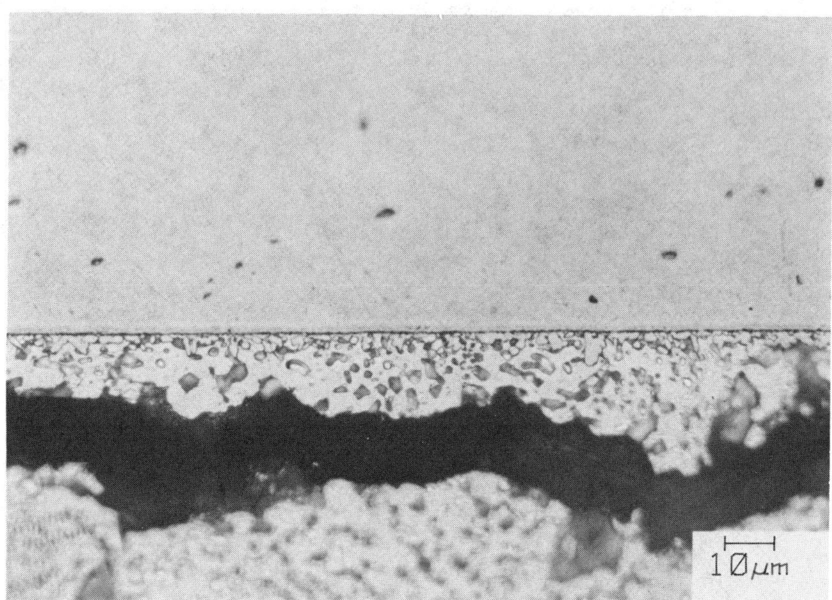

FIG. 12*c*—*Closeup of a specimen tested at 35°C at 1 × 10*$^{-4}$, *showing the intermetallic and the crack.*

FIG. 12*d*—*Closeup of a specimen tested at 150°C at 0.28 Hz, showing the intermetallic, and the crack.*

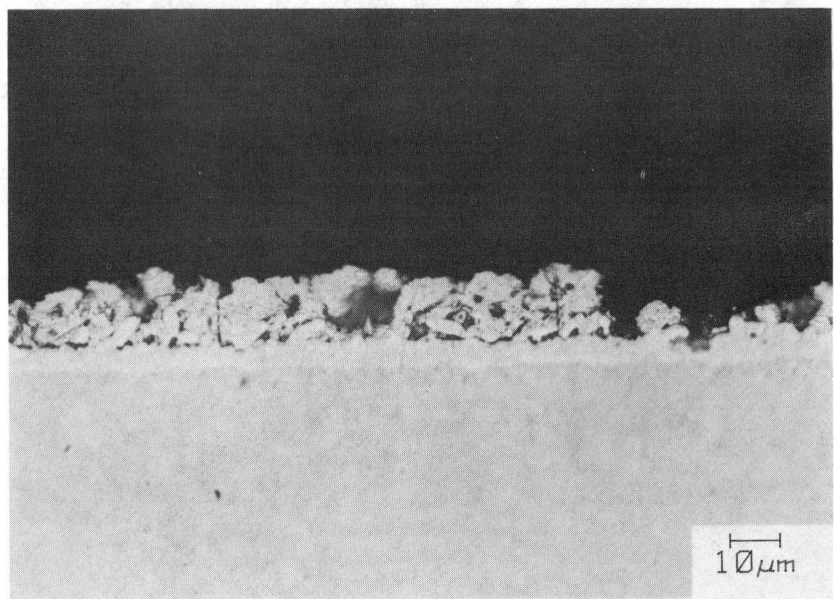

FIG. 12*e*—*Closeup of a specimen tested at 150°C at 0.1 Hz, showing the intermetallic and the crack.*

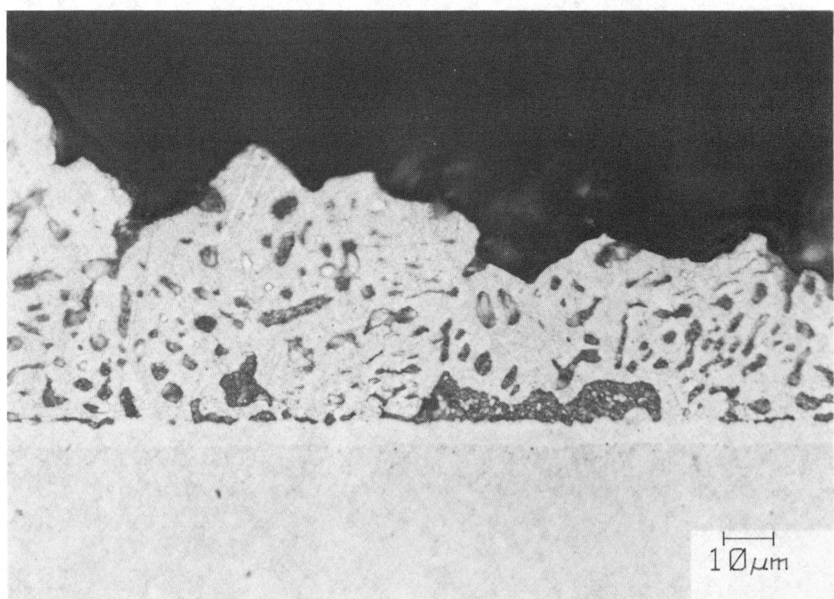

FIG. 12*f*—*Closeup of a specimen tested at 150°C at 1 $\times$ 10^{-4} Hz, showing the intermetallic and the crack.*

mens tested at 150°C it is a uniform layer 2 to 6 μm thick (depending upon the specimen but *not* the time at 150°C). It is to be expected that the longer the time at temperature the thicker the intermetallic layer [*35*]. It should be remembered, however, that all the specimens tested at 150°C were soaked 15 to 18 h prior to testing. Therefore, even though the time to failure ranged from 0.1 to about 50 h, the total time the temperature ranged from about 15 to 65 h; this is not a large enough variation in time to produce a large enough variation in the intermetallic thickness to be detected in these tests.

Time at 150°C prior to testing does not seem to influence the fatigue life (at least within the limited range studied). For instance, a specimen exposed to 150°C for 65 h prior to testing at $\Delta\gamma_p$ ~5% and $\nu = 0.265$ Hz had a life of 220 cycles, which was within the range (but on the high side) observed with specimens soaked the typical 15 to 18 h. Another specimen soaked 45 h at 150°C and then tested at $\Delta\gamma_p$ ~10% and $\nu = 0.325$ had a life of 28 cycles, also in the expected range (although on the high side). Time (as long as 65 h) at 150°C has not reduced the fatigue life; if anything, it has improved it (but not significantly). This, however, is not to say that soaking for hundreds of hours longer might not produce a great effect.

The low melting point of solders and the presence of intermetallic layers make solders prone to time-dependent (i.e., cycle frequency dependent) changes in the microstructure and intermetallic layer. The coarse structure developed as the result of relatively slow cooling (~3°C/s) following the creation of the solder joint and the 15 to 18 Hz soak at 150°C prior to testing at 150°C appear to have prevented the time-dependent microstructural changes or soak time effects that could explain the observed dependence of fatigue life with cycling frequency.

SRP, Time Modified SRP, and Frequency Modified LCF

The oral presentation of this work included creep and strain rate data and an analysis relating these data to strain range partitioning (SRP) [*36–38*]. Space limitations, however, prevent publishing these data here. This part of the work will be published elsewhere [*39*] and is also available in a GE report [*40*]. Only a few conclusions of this lengthy analysis will be mentioned here.

The creep curves (steady-state creep rate versus stress) were used to calculate the amount of creep strain developed during the LCF straining (i.e., the percent creep in each hysteresis loop). To do this each analyzed hysteresis loop was digitized (i.e., digital values of the stress and corresponding strain were determined) and divided into intervals. The mean stress of that interval was used to calculate the creep strain rate for each interval. The time duration of each interval was calculated from the cycle frequency and from this the creep strain developed in each interval was calculated from the creep strain rate. The mechanical plastic strain developed in each interval was determined directly from the input digitized hysteresis loop. The total creep strain and mechanical plastic strain was then summed for all the intervals and the percent creep (creep strain divided by the plastic strain times 100) thus determined. The results of these calculations are shown in Fig. 13, which shows the percent creep as a function of the temperature, plastic strain, and cycling frequency.

According to the SRP model the number of cycles to failure can be expressed in terms of the fraction creep (fcc) of the fatigue cycle, the fraction of non-creep strain ($f_{pp} = 1 - f_{cc}$), the fatigue life when there is no creep strain (N_{pp}), and the number of cycles to failure (N_{cc}), when all the cyclic strains are creep strains; i.e., with an interactive damage law [*36–38*].

$$\frac{1}{N_f} = \frac{f_{cc}}{N_{cc}} + \frac{f_{pp}}{N_{pp}} \tag{3}$$

Figure 14 shows the fatigue life data obtained at 150°C with $\Delta\gamma_p$ 5% (Fig. 4). Different choices were made for N_{pp} and N_{cc} but none were able to create a curve which would fit all the

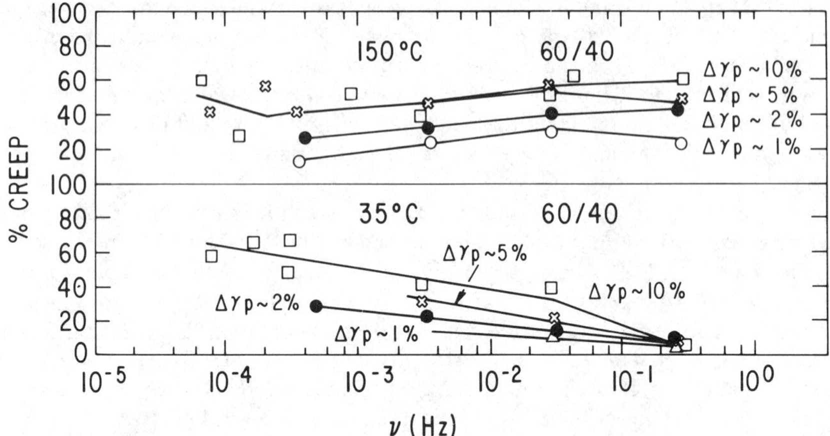

FIG. 13—*The percent creep as calculated from the creep data versus frequency.*

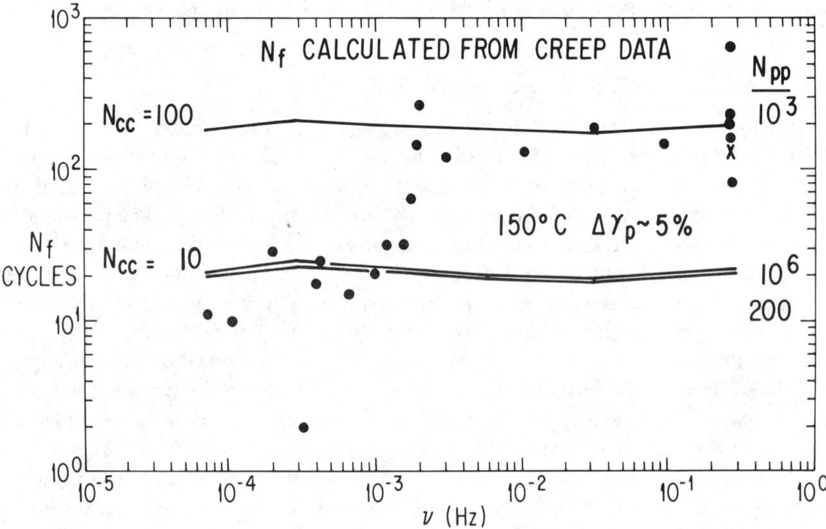

FIG. 14—N_f *versus* ν *correlated by the SRP model with* f_{cc} *calculated from creep data for 150°C and* $\Delta\gamma_p \sim 5\%$.

data. The same lack of agreement was also found for the data of Figs. 3 and 5 (see Refs *39* and *40* for the unsuccessful attempts to correlate this data with SRP model). The lack of agreement of the data with the SRP model stems from the fact that the decrease in fatigue life at low cycling frequencies does not change in a manner which is consistent with the change in the percent creep at low cycling frequencies. At 150°C the percent creep does not change appreciably as the cycling frequency is decreased, whereas there is a large reduction in life when the cycling frequency is below about 10^{-3} Hz. At 35°C there is an increase in the percent creep as the cycling frequency is reduced, but according to the SRP model this increase is not large enough to cause the reduction in the fatigue life.

The SRP model does not describe these data; however, a time-modified version of the SRP approach does. This is illustrated in Figs. 15 to 17. A critical aspect of the SRP model is that at low frequencies (where creep is assumed to dominate) it is assumed that there is a specific fatigue life N_{cc} (for pure creep on loading and unloading) which can be used to describe the data. These data and high-temperature tests on other alloys [*33*] do not agree with this assumption; rather, at low frequencies the data can be described by a constant time to failure t_{cc} [*41*]. Figures 15 to 17 show the data according to a time-modified SRP model where the time to failure $t_f = N_f \nu$ is given by

$$\frac{1}{t_f} = \frac{f_{cc}}{t_{cc}} + \frac{\nu f_{pp}}{N_{pp}} \tag{4}$$

An interactive damage law is being employed with a superposition of time-dependent LCF at low frequencies (defined by a critical time to failure t_{cc}) and a high-frequency cycle-dependent LCF (defined by a critical number of cycles N_{pp}). The ν term in Eq 4 is the frequency and is needed to maintain dimensional equality. As can be seen in Figs. 15 to 17, Eq 4 does a good job of describing the data. *The critical difference between this time-modified SRP and the conventional SRP is the switch from a constant number of cycles to failure at low frequencies to a constant time to failure.*

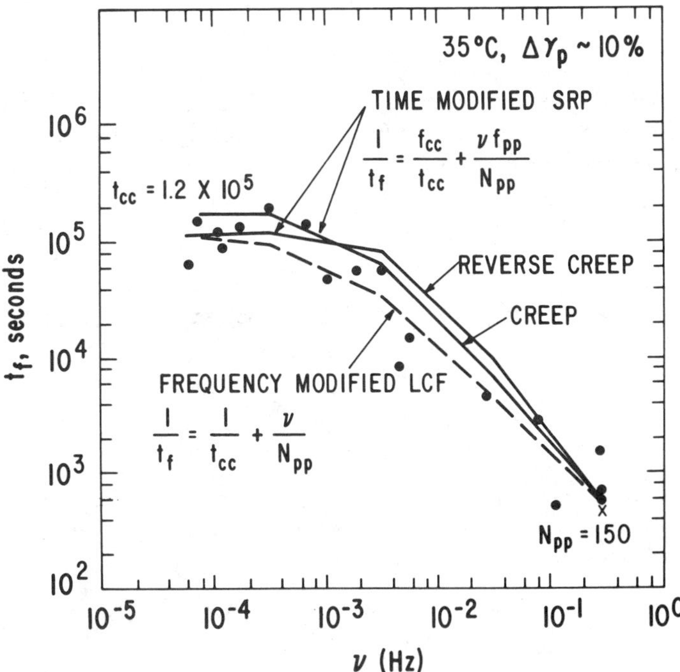

FIG. 15—*Time to failure* (t_f) *versus* ν *as correlated by time-modified SRP and frequency-modified LCF, 35°C,* $\Delta\gamma_p \sim 10\%$.

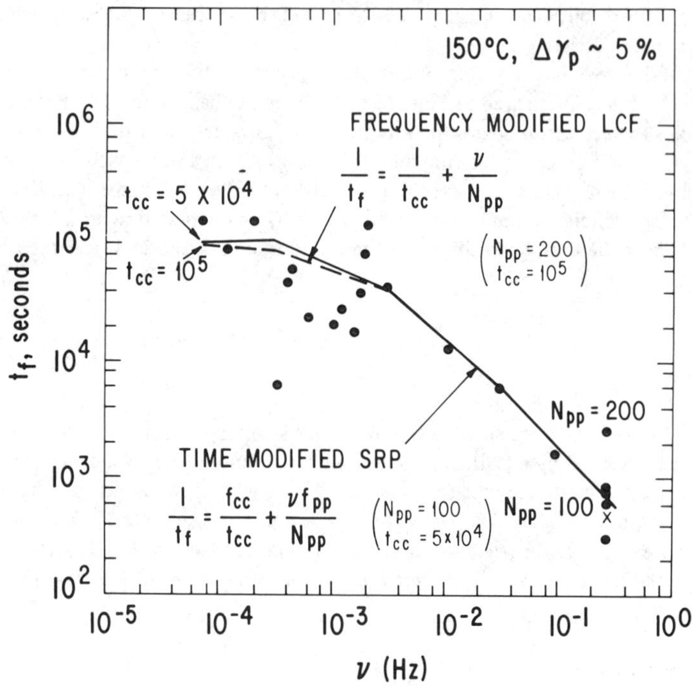

FIG. 16—*Time to failure* (t_f) *versus* ν *as correlated by time-modified SRP and frequency-modified LCF, 150°C,* $\Delta\gamma_p \sim 5\%$.

Also shown in Figs. 15 to 17 is the correlation with the frequency-modified LCF law, which is in effect the same as the time modified SRP approach except that a linear damage law is used rather than an interactive damage law; i.e.,

$$\frac{1}{t_f} = \frac{1}{t_{cc}} + \frac{\nu}{N_{pp}} \tag{5}$$

Equation 5 is a special case of the general frequency-modified LCF law [*33*]. This can be seen by writing Eq 5 in terms of the number of cycles to failure; i.e.,

$$\frac{1}{N_f} = \frac{1}{\nu t_{cc}} + \frac{1}{N_{pp}} \tag{6}$$

Equation 6 yields the $K = 1$ and $K = 0$ curves of Figs. 3 to 5. It is thus clear that the $K = 1$ and $K = 0$ curves are manifestations of the superposition of time-dependent and cycle-dependent LCF. Equation 4 can also be written in terms of N_f; i.e.,

$$\frac{1}{N_f} = \frac{f_{cc}}{\nu t_{cc}} + \frac{f_{pp}}{N_{pp}} \tag{7}$$

and will likewise correlate the N_f versus ν data of Figs. 3 to 5, yielding curves similar to the $K = 1$ and $K = 0$ curves. The foregoing illustrates a second difference between SRP and frequency-

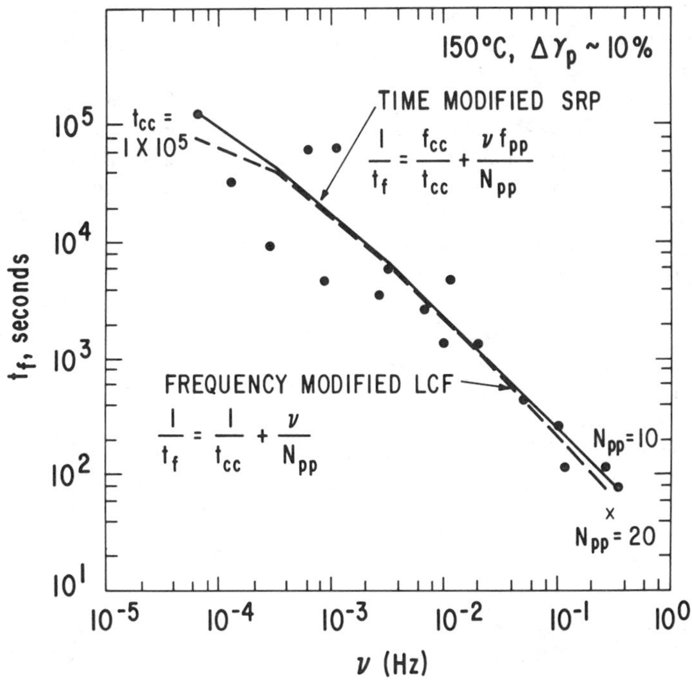

FIG. 17—*Time to failure* (t_f) *versus* ν *as correlated by time-modified SRP and frequency-modified LCF, 150°C,* $\Delta\gamma_p \sim 10\%$.

modified LCF (in addition to the N_{cc} or t_{cc} assumptions). SRP generally employs an interactive damage law which is not utilized in frequency-modified LCF. Figures 15 through 17 show that for these data at least this is not a significant difference. This is true even though at 35°C there is a significant variation in f_{cc} with ν. At high ν, where $f_{cc} \rightarrow o$, there is little influence of f_{cc} because the ν/N_{pp} term dominates and at low ν, where $f_{cc} \rightarrow 1$, the $1/t_{cc}$ term dominates and again the change in f_{cc} is relatively unimportant. At 150°C where $f_{cc} \sim 50\%$ at all frequencies there is again little influence of f_{cc}. The only influence is to change the values of N_{pp} and t_{cc} required to give the best fit to the data. *For these data a* K $= 1$, K $= 0$ *frequency-modified LCF approach is equivalent to SRP when SRP is time modified (i.e., at low frequencies it is assumed that the time to failure and not the number of cycles to failure is constant).*

The SRP model assumes that creep is the dominant contributor to time-dependent behavior. The preceding data show that for these experiments, while creep plays a significant role in the deformation, the fraction creep does *not* correlate with the fatigue behavior. The conventional SRP model does not describe the data and, while a time-modified approach does describe the data, it does so equally well in a formulation which disregards f_{cc} entirely (i.e., Eqs 5 and 6).

Creep is not the only time-dependent phenomenon that plays a role in determining the fatigue life. The SRP life prediction approach and the strain rate/creep analysis do not consider the environment that the solder experiences. The tests described here were done with an N_2 purge gas. The furnace was not completely sealed, however (there were holes in the top and bottom for the load train and input ports where the N_2 gas line, the thermocouple, and the extensometer cables were fed in). The high gas flow (20 SCFH) was meant to produce an outward N_2 flow through all these furnace openings, but such a system cannot completely preclude air entry.

Environmental influences have been shown to play an important role in solder fatigue. For instance, Neugebauer et al [*18,21*] has shown that destroying the hermiticity of some electronic packages greatly reduced their thermal fatigue life. Oxygen has been shown to greatly reduce the fatigue life of Pb [*42,43*]. Oxygen contents as small as 3×10^{-2} mm Hg reduced the fatigue life of polycrystalline Pb by about two orders of magnitude compared with tests run in vacuum. The oxygen caused intergranular fractures. This oxygen effect saturated for $P_{O_2} > 3 \times 10^{-2}$ mm Hg with tests run at this partial pressure having the same life as those run at higher pressures. An influence of cycling frequency was also noted [*43*]. Oxygen pressures as high as 3×10^{-2} mm Hg were certainly present in the tests described here, so the hypothesis that oxygen was causing the observed reduction in fatigue life is certainly quite reasonable.

Environmental influences in superalloys tested at high temperatures and low cycling frequencies have also been well documented [*32,44,45*]. Fatigue crack growth at high temperatures in superalloys is strongly frequency dependent and much of this effect is due to oxidation. Oxygen and water vapor have also shown to greatly accelerate the fatigue of aluminum and aluminum alloys; for instance, Refs *46* and *47* show a pressure dependence of water vapor similar to that of O_2 in Pb. It is quite possible that in the present solder tests, oxygen or water vapor is responsible for the reduced life at low frequencies that is not explained by SRP or creep. Only low-frequency tests run in vacuum will answer this question.

Conclusions

1. Reducing the cycling frequency, especially below about 10^{-3} Hz, greatly reduces the number of cycles to low-cycle fatigue failure. This is true of tests run at both 35 and 150°C on 60-40 solder.
2. The percent creep curves calculated from the creep data were used to predict the life via the SRP model, but this approach proved unsuccessful.
3. A time-modified SRP approach was successful in describing the fatigue life data. This approach differed from the conventional SRP approach in that at low frequencies it is assumed that the time to failure, not the cycles to failure, is constant.
4. This time-modified SRP model is shown to be equivalent to the two slope ($K = 1$, $K = 0$) frequency-modified LCF approach, the only difference being that in the former an interactive damage law is used, while in the latter a simple linear superposition approach is used. For these data at least the two approaches yield virtually the same results (i.e., the changes in the fraction creep that is considered in the interactive damage law does not significantly influence the results). This was true both in situations where there is a large change in f_{cc} with cycling frequency and in cases where f_{cc} was roughly constant.
5. The fatigue results do not correlate with the creep behavior. The standard SRP approach does *not* describe the data. While a time-modified SRP approach does describe the data, it does so equally well in a formulation that does not consider the fraction creep. This lack of a dependency with the fraction creep supports studies that ascribe the time dependency of fatigue in solders to an environmental rather than a creep effect.

Acknowledgments

The author would like to express his thanks to the GE Corporate Research and Development metallography unit. Preparing Sn-Pb alloys is an art that they have learned through patience and quite a bit of hard work. The enclosed micrographs demonstrate their considerable proficiency. The author would also like to thank Paul Emigh who helped perform these tests and helped to examine the fracture surfaces with SEM.

Lastly, but certainly not least, the author would like to acknowledge the fruitful discussions with his colleagues Drs. L. F. Coffin, Jr., F. P. Ford, and M. F. Henry. The author is particularly indebted to Dr. M. F. Henry for suggesting that he consider utilizing a time-to-failure approach to these data.

References

[1] Korb, R. W. and Ross, D. O., *IEEE Transactions,* Vol. CHMT-6, No. 3, 1983, pp. 227-231.
[2] Pankratz, J. M. and Plumlee, H. R., *IEEE Transactions,* Vol. ED-17, No. 9, 1970, pp. 793-797.
[3] Kong, S. K., Zommer, N. D., Fencht, D. L., and Heckel, R. W., *IEEE Transactions,* Vol. PHP-13, No. 3, 1977, pp. 318-321.
[4] Zommer, N. D., Feucht, D. L., and Heckel, R. W., *IEEE Transactions,* Vol. ED-23, No. 8, 1976, pp. 843-850.
[5] Levine, E. and Ordonez, J., *IEEE Transactions,* Vol. PHP-13, No. 3, 1977, pp. 318-321.
[6] Rathore, H. S., Yih, R. C., and Edenfeld, A. R., *Journal of Testing and Evaluation,* Vol. 1, No. 2, 1973, pp. 170-178.
[7] Nagesh, V. K., *IEEE Transactions,* Vol. CH-1727, No. 7, 1982, pp. 6-15.
[8] Norris, K. C. and Landzberg, A. H., *IBM Journal of Research and Development,* Vol. 13, 1969, pp. 266-271.
[9] Taylor, J. R. and Pedder, D. J., *International Journal of Hybrid Microelectronics,* Vol. 5, No. 2, 1982, pp. 209-214.
[10] Bangs, E. R. and Beal, R. E., *Welding Journal,* Vol. 10, 1975, pp. 377S-383S.
[11] Long, G. A., Fehder, B. J., and Williams, W. D., *IEEE Transactions,* Vol. ED-17, No. 9, 1970, pp. 787-793.
[12] Wild, R. N., *Welding Journal,* Vol. 51, 1972, pp. 521S-526S.
[13] Engelmaier, W., *IEEE Transactions,* Vol. CHMT-6, No. 3, 1983, pp. 232-237.
[14] Engelmaier, W., *Circuit World,* Vol. 12, No. 3, 1985, pp. 61-72.
[15] Jarboe, D. M., "Thermal Fatigue Evaluation of Solder Alloys," NTIS-BDX-613-2341, Feb. 1980.
[16] Shine, M. C., Fox, L. R., and Sofia, L. W. in *Proceedings,* 4th Annual International Electronics Packaging Conference, Oct. 1984, pp. 181-188.
[17] Ross, D. O. in *Proceedings,* 4th Annual International Electronics Packaging Conference, Oct. 1984, pp. 181-188.
[18] Burgess, J. F., Carlson, R. O., Glascock, H. H., Neugebauer, C. A., and Webster, H. F., *IEEE Transactions,* Vol. CHMT-7, No. 4, Dec. 1984, pp. 405-410.
[19] Gohn, G. R. and Ellis, W. C., *Proceedings of ASTM,* Vol. 51, 1951, pp. 721-744.
[20] Eckel, J. F., *Proceedings of ASTM,* Vol. 51, 1951, pp. 745-760.
[21] Neugebauer, C. A., Webster, H. F., Solomon, H. D., and Carlson, R. O., "The Measurement of Fatigue Suppression in Electronic Solder Joints," presented at the 8th Annual Soldering Technology Seminar, 22-23 Feb. 1984, Naval Weapons Center, China Lake, Calif.
[22] Solomon, H. D., "Low Cycle Fatigue of 60/40 Solder-Plastic Strain Limited vs. Displacement Limited Testing," in *Electronic Packaging: Materials and Processes,* J. A. Sartell, Ed., American Society for Metals, 1985, pp. 29-48.
[23] Iosipescu, N., *Journal of Materials,* Vol. 2, No. 3, 1967, pp. 537-566.
[24] Coffin, L. F., Jr. *Transactions of ASME,* Vol. 76, 1954, pp. 931-950.
[25] Manson, S. S., "Behavior of Materials under Conditions of Thermal Stress," in *Proceedings,* Heat Transfer Symposium, University of Michigan, 27-28 June 1952, University of Michigan Press, Lansing, 1952; also NACA TN2933, July 1953.
[26] Coffin, L. F., Jr., and Tavernelli, J. F., *Transactions of AIME,* Vol. 215, 1959, pp. 794-807.
[27] Tavernelli, J. F. and Coffin, L. F., Jr., *Journal of Basic Engineering, Transactions of ASME,* Vol. 84D, 1962, pp. 533-537.
[28] Manson, S. S., discussion to Ref *6,* pp. 537-541.
[29] Manson, S. S., *Mechanical Design,* Vol. 32, No. 14, 1960, pp. 139-144.
[30] Coffin, L. F., Jr., in *Fracture 1969,* Chapman and Hall, London, 1969, p. 643.
[31] Coffin, L. F., Jr., *Metallurgical Transactions,* Vol. 2, 1971, p. 3105.
[32] Coffin, L. F., Jr., in *Fatigue at Elevated Temperatures, ASTM STP 520,* American Society for Testing and Materials, Philadelphia, 1973, pp. 5-34.
[33] Solomon, H. D., *Metallurgical Transactions,* Vol. 4, 1973, pp. 341-347.
[34] Zakraysek, L., Kelsey, D. E., and DeVore, J. A., "Microstructure of Solders," presented at 9th Annual Soldering Technology Seminar, 19-20 Feb. 1985, Naval Weapons Center, China Lake, Calif.
[35] Zakraysek, L., *Welding Journal,* 1971, pp. 5225-5275.
[36] Manson, S. S. in *Fatigue at Elevated Temperatures, ASTM STP 520,* American Society for Testing and Materials, Philadelphia, 1973, pp. 744-782.

[37] Halford, G. R., Hirschberg, M. H., and Manson, S. S. in *Fatigue at Elevated Temperatures, ASTM STP 520,* American Society for Testing and Materials, Philadelphia, 1973, pp. 658-669.
[38] Halford, G. R., Hirschberg, M. H., and Manson, S. S. in *Fatigue at Elevated Temperatures, ASTM STP 520,* American Society for Testing and Materials, Philadelphia, 1973, pp. 239-254.
[39] Solomon, H. D., *Brazing and Soldering,* No. 11, Autumn 1986, pp. 68-75.
[40] Solomon, H. D., "Low-Frequency, High-Temperature, Low-Cycle Fatigue of 60 Sn/40 Pb Solder," GE Report 85CRD238, Dec. 1985.
[41] Henry, M. F., private Communication.
[42] Snowden, K. U., *Acta Metallurgica,* Vol. 12, 1964, pp. 295-303.
[43] Snowden, K. U., *Philosophical Magazine,* Vol. 10, 1964, pp. 435-440.
[44] Solomon, H. D. and Coffin, L. F. in *Fatigue at Elevated Temperatures, ASTM STP 520,* American Society for Testing and Materials, Philadelphia, 1973, pp. 112-122.
[45] Gabrielli; F. and Pelloux, R. M., *Metallurgical Transactions,* Vol. 13A, 1982, 1083-1090.
[46] Bradshaw, F. J. and Wheeler, C., *Applied Materials Research,* Vol. 5, 1966, pp. 112-120.
[47] Bradshaw, F. J. and Wheeler, C., *International Journal of Fracture Mechanics,* Vol. 5, 1969, pp. 255-268.

DISCUSSION

S. D. Antolovich[1] (*written discussion*)—Some caution must be exercised vis-à-vis your comment that it may be possible to ignore interactions when environmental effects dominate. There are certainly systems and temperature regimes where creep deformation, cyclic deformation, and environmental effects may each influence one or both of the other processes. For example, oxygen penetration may prevent re-welding of voids formed by creep and plastic deformation (with its associated higher dislocation density) may accelerate oxygen penetration into the bulk by either sweeping or by providing short-circuit diffusion paths.

H. D. Solomon (*author's closure*)—I did not mean to imply that one can ignore creep/environmental interactions. My intent was only to show that these LCF results are not well correlated by the creep data and that here environmental interactions may predominate.

[1]Fracture and Fatigue Research Laboratory, Georgia Institute of Technology, Atlanta, GA 30332.

R. V. Miner,[1] *J. Gayda,*[1] *and M. G. Hebsur*[2]

Creep-Fatigue Behavior of Ni-Co-Cr-Al-Y Coated PWA 1480 Superalloy Single Crystals

REFERENCE: Miner, R. V., Gayda, J., and Hebsur, M. G., **"Creep-Fatigue Behavior of Ni-Co-Cr-Al-Y Coated PWA 1480 Superalloy Single Crystals,"** *Low Cycle Fatigue, ASTM STP 942,* H. D. Solomon, G. R. Halford, L. R. Kaisand, and B. N. Leis, Eds., American Society for Testing and Materials, Philadelphia, 1988, pp. 371–384.

ABSTRACT: Single-crystal specimens of a Ni-base superalloy, PWA 1480, with a low pressure plasma sprayed Ni-Co-Cr-Al-Y coating, were tested in various 0.1 Hz fatigue and creep-fatigue cycles at 1015 and 1050°C. Creep-fatigue tests of the cp, pc, and cc types were conducted with various constant total strain ranges employing creep dwells at various constant stresses. Considerable cyclic softening occurred as was evidenced particularly by rapidly increasing creep rates in the creep-fatigue tests. The cycle time in the creep-fatigue tests typically decreased by more than 80% at $0.5N_f$.

Though cyclic life did correlate with $\Delta\epsilon_{in}$, a better correlation existed with $\Delta\sigma$ for both the fatigue and creep-fatigue tests, and poor correlations were observed with either σ_{max} or the average cycle time. A model containing both $\Delta\sigma$ and $\Delta\epsilon_{in}$, $N_f = \alpha\Delta\epsilon_{in}^{\beta}\Delta\sigma^{\gamma}$, with best fit values of α for each cycle type but the same values of β and γ, was found to provide good correlations. Lifelines were not greatly different among the cycle types, differing by only a factor of about three. The cp cycle life line was lowest for both test temperatures; however, among the other three cycle types there was no consistent ranking. For all test types, failure occurred predominately by multiple internal cracking originating at porosity. The strong correlation of life with $\Delta\sigma$ may reflect a significant crack growth period in the life of the specimens. The lack of improvement in the models when average cycle time was considered reflects both that there is no large effect of strain rate on the damage mechanisms in the single-crystal material nor any environmental effect due to the internal cracking mode of failure.

KEY WORDS: low-cycle fatigue, creep-fatigue, superalloys, single crystals, fatigue life prediction

Nickel-base superalloy single crystals with protective metallic coatings are employed as blades and vanes in aircraft gas turbine engines. Thermomechanical fatigue is a major life limiting factor in these components. The present study of high temperature fatigue and creep-fatigue behavior is part of a program to identify the basic features of the effects of temperature, creep, fatigue, and environment on the behavior of a single crystal superalloy, a bulk coating alloy, and the coated alloy system. The goal is to test the feasibility of a life prediction model for coated single crystal specimens in laboratory thermomechanical test cycles based on the individual behaviors of the superalloy and coating, and mechanical and thermal analyses of the coated specimen. The framework of such a model might be extended to predict the life of actual components.

To provide the best opportunity for success, it was decided to model a specific superalloy-coating system, and to select one which has had considerable commercial production experi-

[1]Materials Engineer, NASA-Lewis Research Center, Cleveland, OH 44135.

[2]National Research Council Research Associate, NASA-Lewis Research Center, Cleveland, OH 44135.

ence: the Ni-base superalloy, PWA 1480, and the Ni-Co-Cr-Al-Y coating, PWA 276, inventions of the Pratt and Whitney Aircraft Company of the United Technologies Corporation. This coating has the additional benefit of being low pressure plasma sprayed. It may be deposited in thicknesses great enough for the preparation of mechanical test specimens.

Isothermal behavior was studied first, since it was desired to base any thermomechanical fatigue model on more simple isothermal data if possible. Also, it was not clear what to expect of the creep-fatigue behavior of a superalloy without grain boundaries. For instance, the common creep damage mechanisms of grain boundary cavitation or sliding were certain not to occur.

A series of fatigue and creep-fatigue tests of the types commonly designated as pp, cp, pc, and cc were conducted. The letters "p" and "c" refer to rapid "plastic" and slow "creep" type deformation, and the first and second letters to the tensile and compressive portions of the cycle, respectively. These tests were conducted at various constant total strain ranges. The creep-fatigue cycles employed constant stress dwells at the maximum and/or minimum load.

After the first series of tests it was discovered that they had been conducted at 1015°C, rather than at 1050°C as intended. Also, it appeared that the dwell stress level effected life in the creep-fatigue cycles, but the result was somewhat confounded since the controlled dwell stresses and total strain ranges were varied commensurately in order to minimize test times. In a second series of tests, conducted at 1050°C, total strain range and dwell stress level were varied as independently as practicable. It will be seen that both series of tests indicate an important effect of stress on the cyclic life of the coated superalloy single crystals.

Discussion will center on the constitutive behavior, life behavior, and failure modes for the individual cycle types. It will be shown that creep does not greatly influence the fatigue life of this material, or at least it is difficult to introduce creep damage using stress dwells. Probably because of this, a single life relation provides a fairly good correlation for all the cycle types investigated, though it is unconventional in being based on $\Delta\epsilon_{in}$ and $\Delta\sigma$, and thus more sophisticated damage interaction models have not been investigated. Since space is not available herein, a thorough set of data from the 44 tests conducted has been published in a companion NASA paper [*1*] so that those wishing to corroborate generalizations made or try their own analysis may do so.

Materials and Procedures

Materials

The single crystal superalloy studied herein, PWA 1480, was developed by Pratt and Whitney Aircraft for application as aircraft gas turbine blades and vanes [*2*]. The alloy has the following nominal composition: 10 Cr, 5 Al, 1.5 Ti, 12 Ta, 4 W, 5 Co, and the balance Ni, in weight percent. This alloy contains about 65 vol% of the γ' phase, but essentially no carbides or borides, since C, B, and Zr are not necessary for grain boundary strengthening in single crystals. The single crystal specimens were grown as round bars about 21 mm in diameter and 140 mm long.

The bars were solution treated for 4 h at 1290°C. This was done before machining since it could lead to recrystallization after machining. Bars having their ⟨001⟩ within less than 7° of the axis were selected for the LCF specimens. After machining, the LCF specimens were coated with a Ni-Co-Cr-Al-Y alloy by low pressure plasma spraying. The coating is designated PWA 276 and is also an invention of Pratt and Whitney Aircraft. The coating composition was 20 Co, 17 Cr, 12.4 Al, 0.5 Y, and the balance Ni, in weight percent. The coating thickness was about 0.12 mm. After coating, the specimens were given a diffusion treatment of 1080°C for 4 h and then aged at 870°C for 32 h.

Interdendritic porosity is known to be a problem in single-crystal superalloys, and those studied herein were no exception. The crystals tested contained an average of about 0.3 vol% of porosity. The pore diameter averaged about 7 μm with a standard deviation of about 6 μm. The yield strengths of these crystals averaged about 10% lower from room temperature to 1050°C than those of the crystals studied by Shah and Duhl [*3*]. A larger fraction of large γ' particles in the crystals studied herein indicated that the solution treatment temperature was actually lower than expected.

Test Procedures

The testing facility used in this investigation has been described elsewhere [*4*] except for the computer control facility used for the later tests, and thus will be described only briefly herein. The 0.1 Hz fatigue and creep-fatigue tests were conducted using an hourglass type specimen and diametral strain control. The nominal diameters of the single crystal specimens before and after coating were 6.35 and 6.60 mm, respectively. Heating was produced by the passage of alternating current directly through the specimen. A thermocouple for temperature control was attached to the specimen about 5 mm from the minimum section. The control temperature was adjusted to give the desired temperature at the minimum diameter as measured by an optical pyrometer.

The diametral extensometer was placed across the specimens in a direction as close as possible to one of the two ⟨001⟩ always nearly in the diametral plane. Still, exact placement of the extensometer was not critical. As stated previously, all specimens had their axes within 7° of an ⟨001⟩, and in cubic crystals Poisson's ratio for elastic deformations is constant for any direction in the plane normal to an ⟨001⟩. Further, at the test temperatures employed, inelastic deformation in this material also appeared to be homogeneous and isotropic.

It was intended to conduct tests only at 1050°C; however, when a laboratory renovation provided new computer control, some additional tests were conducted. It was found during these additional tests that the optical pyrometer had been out of calibration during the previous testing. What had been intended as 1050°C tests where actually conducted at about 1015°C. Both because there was some uncertainty about the temperature of the first tests, and because it was suspected that life in the creep-fatigue cycles was influenced by the dwell stress level, a set of computer-controlled tests were conducted at 1050°C. In these tests, total strain range and dwell stress level were varied as independently as practicable. It will be seen, however, that life behavior in the lower temperature tests shows the same dependencies on strain and stress as in the 1050°C tests even though the two "independent" variables are more highly correlated.

All tests were controlled at constant total diametral strain range. In the creep-fatigue cycles, constant stress dwells were employed. In the first tests, those at 1015°C, the dwell stress limits were controlled using an electromechanical programmer. In the 1050°C tests, a Data General S/20 computer was employed to control the dwell stress limits.

The frequency of the pp tests at both temperatures was about 0.1 Hz. The creep-fatigue tests employed about the same ramp rates as the pp tests; however, in the early tests a roughly constant displacement rate was used, and in the computer-controlled tests a constant loading rate was used.

During the tests, hysteresis loops of axial load versus diametral displacement were recorded periodically. Also, continuous time records of axial load and diametral displacement were maintained. The strains reported herein are calculated axial strains. The axial elastic strains were calculated from the measured stress range and the modulus for the ⟨001⟩ crystal direction, 7.58×10^4 MPa. The axial inelastic strain was assumed to be twice the diametral inelastic strain based on constancy of volume.

TABLE 1—*Low-cycle fatigue data for coated PWA 1480 single crystals. Values of $\Delta\epsilon_{in}$, $\Delta\sigma$, and σ_{max} are those at half-life.*

Cycle Type	N_f	$\Delta\epsilon_{tot}$, %	$\Delta\epsilon_{in}$, %	$\Delta\sigma$, MPa	σ_{max}, MPa	t_{avg}, min
			1050°C			
pp	110	1.93	0.92	766	388	0.17
	580	1.30	0.47	631	331	0.17
	900	1.27	0.52	571	292	0.17
	950	1.05	0.32	552	272	0.17
	1000	1.18	0.47	537	274	0.17
	2900	0.85	0.30	416	210	0.17
cp	85	2.06	1.20	655	255	1.14
	160	1.52	0.82	528	197	1.28
	300	1.55	0.81	559	163	0.35
	440	1.30	0.69	465	206	0.31
	810	1.27	0.60	505	201	0.50
	1150	1.10	0.52	438	201	0.40
pc	195	1.67	0.85	625	376	0.60
	610	1.65	0.81	639	379	0.50
	930	1.54	0.76	592	287	0.63
	1250	1.56	0.91	490	266	0.60
	1410	1.49	0.90	445	225	0.40
	1650	1.26	0.71	419	225	0.40
	1700	1.11	0.52	445	247	0.24
cc	370	1.33	0.70	474	210	0.78
	790	1.41	0.87	412	206	0.64
	990	1.21	0.68	403	206	0.49
	1100	1.32	0.74	439	257	1.45
	2110	0.83	0.30	400	199	0.30
			1015°C			
pp	174	1.81	0.62	904	457	0.14
	192	2.05	0.81	941	478	0.16
	2021	1.06	0.38	521	275	0.15
	2314	0.85	0.15	529	276	0.15
	3900	0.74	0.16	438	227	0.12
	7870	0.50	0.05	344	166	0.15
cp	15	3.39	1.95	1090	429	15.3
	78	2.22	1.16	802	270	4.8
	146	1.30	0.40	680	206	4.6
	218	1.43	0.58	644	208	2.4
	610	0.92	0.30	531	345	15.0
pc	29	2.46	1.37	826	569	22.1
	54	2.01	0.88	856	394	14.7
	340	1.46	0.62	603	396	3.92
	730	1.15	0.34	613	395	1.23
	2101	0.91	0.30	468	276	1.26
cc	192	1.66	0.98	514	257	7.38
	724	1.38	0.72	499	250	0.45
	1826	0.88	0.30	436	218	0.59

Results

Some results of the 1050 and 1015°C fatigue and creep-fatigue test results are shown in Table 1. Shown are the cycle type; cyclic life (N_f); the values at half-life of the total axial strain range ($\Delta\epsilon_{tot}$), inelastic axial strain range ($\Delta\epsilon_{in}$), stress range ($\Delta\sigma$), and maximum stress (σ_{max}); and the average cycle time (t_{avg}). Other data that are discussed below, such as that for the first cycle, may be found elsewhere [*1*].

Constitutive Behavior

Cyclic stress-strain behavior at half-life, 0.5 N_f, for the pp tests is shown in Fig. 1 for both the 1015 and 1050°C tests. A cyclic strain hardening exponent of 0.4 fits the data at both temperatures reasonably well. At these high temperatures the effect of temperature on $\Delta\sigma$ is substantial. The estimated 35°C temperature difference produces about a 150 MPa difference in $\Delta\sigma$ at $\Delta\epsilon_{in}$ equal to 0.005.

The constitutive behavior of PWA 1480 at 1015 and 1050°C for all the stress dwell creep-fatigue cycle types studied is characterized by extreme cyclic softening. As a basis for discussion, stress-strain, strain-time, and stress-time plots corresponding to the first cycle, 0.5 N_f, and near failure for a typical cp test at 1050°C are presented in Fig. 2. Shown below the three loops are various other measures of the stress-strain-time behavior. This cp test also represents many of the features of the pc and cc tests.

The increase in $\Delta\epsilon_{in}$ with cycling, which occurred for all cycle types, may be seen for the cp tests represented in Fig. 2. The increase in $\Delta\epsilon_{in}$ on a percentage basis measured at 0.5 N_f was roughly independent of $\Delta\epsilon_{in}$ and test temperature. At 0.5 N_f, $\Delta\epsilon_{in}$ had increased an average of ~25% for the pp tests, ~15% for the cp and pc tests, but generally less than 5% for the cc tests.

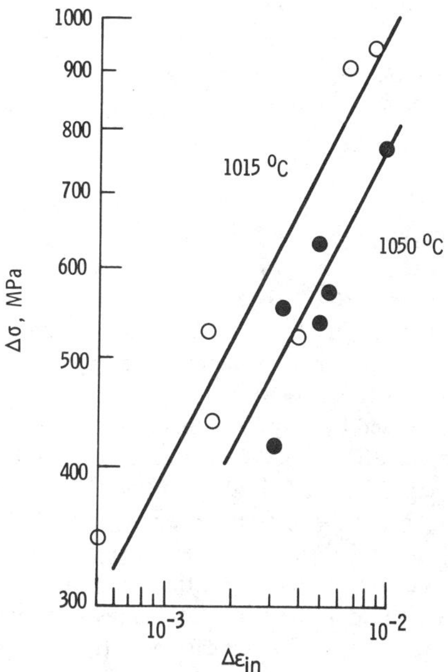

FIG. 1—*Cyclic stress-strain behavior of PWA 1480 single crystals in pp cycles at half-life.*

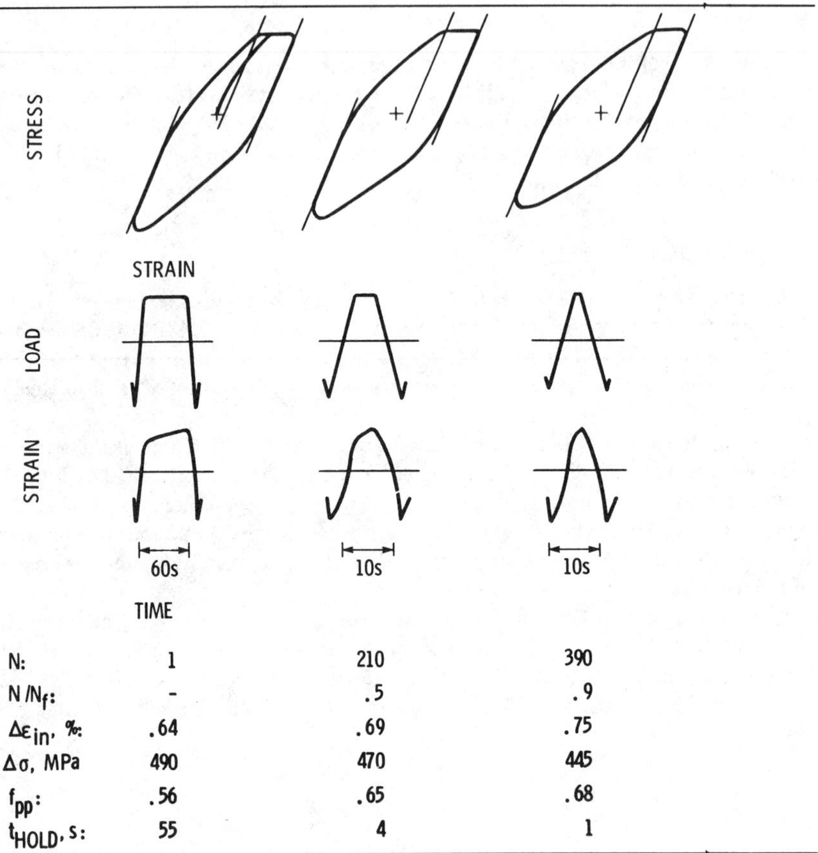

FIG. 2—*Stress-strain-time behavior of a PWA 1480 single crystal in cp tests illustrating typical cyclic softening behavior.*

For the 1050°C tests, the decrease in $\Delta\sigma$ at 0.5 N_f averaged ~10% for the pp and cp tests, but was slightly higher for the pc tests, ~15%. In the cc tests $\Delta\sigma$ was, of course, maintained constant. For the cp and pc tests at 1015°C, the decrease in $\Delta\sigma$ appeared to be smaller than for those tests at 1050°C, but there is considerable scatter in the data.

The most dramatic change with cycling in the creep-fatigue tests was the increase in creep rates. This is reflected in the cycle times shown in Fig. 2. At 0.5 N_f, the cycle time has decreased by 77%, though $\Delta\epsilon_{cp}$ has decreased by only 25%. In many tests there was no reduction in the amount of creep strain per cycle. Still, by 0.5 N_f the reduction in cycle time was typically ~80% in the cp and pc tests, and more than 95% in the cc tests. For the 1050°C tests, by 0.5 N_f the cycle times were approaching that of the pp tests. However, at 1015°C creep rates were substantially slower than at 1050°C, and average cycle times for some of the cp and pc tests with high $\Delta\epsilon_{in}$ or low creep stress were roughly 100× longer than that for a pp test.

Though the creep strain per cycle in the creep-fatigue tests typically did not decrease much during cycling on an absolute basis, it did decrease as a fraction of the total inelastic strain range. It may be seen in Fig. 2 that the fraction of pp strain, f_{pp}, increases with cycling at the expense of f_{cp}. This occurred to a greater degree in the 1050°C tests than in the 1015°C tests.

Rounding of the ends of the hysteresis loops where the direction of straining is reversed such as seen at the bottom of the hysteresis loops in Fig. 2, also occurred in the pp and pc tests at 1050°C. This effect is due to the rapid creep rates which allows continued straining while the stress is being reduced. This effect was much less evident in the 1015°C tests.

Another interesting observation in the creep-fatigue tests was that for the same absolute stress level, creep rates were higher in compression than in tension. This effect was observed in comparisons of cp and pc tests, but was most readily seen in the cc tests. Creep rates in compression were about 1.5 to 2× higher than those in tension.

Life Behavior

For both the fatigue and stress dwell creep-fatigue tests conducted in this study, the life of PWA 1480 was found to correlate well with a model including both $\Delta\epsilon_{in}$ and $\Delta\sigma$. This model provides considerably better correlations than those based on $\Delta\epsilon_{in}$ alone, $\Delta\epsilon_{in}$ and σ_{max}, or $\Delta\epsilon_{in}$ and t_{avg}.

Table 2 shows a summary of the results of regression analyses of the data for all test types at either test temperature using various models. In each case, log N_f has been regressed against the logarithims of the "independent" variables; that is, power law relationships between the variables have been assumed. These various equations test the basic dependencies of life on frequency, strain rate, or σ_{max} used in various models including Strain-Range-Partitioning [*5*], Frequency-Separation [*6*], Frequency-Modified [*7*], and Damage-Rate [*8*].

The 1050°C results will be examined first since they are clearer. As previously indicated, care was taken in the design of this series of tests to reduce as much as possible the correlation between the two "independent" variables in the creep-fatigue tests, total strain range, and dwell stress. Tests employed various total strain ranges, but only two dwell stress levels (±). For this data set, even with the pp tests included, $\Delta\sigma$ and σ_{max} are only 25 and 16% (R-values) correlated with $\Delta\epsilon_{in}$, respectively.

It may be seen in Table 2 that the model $N_f = \alpha\Delta\sigma^{\gamma}$ provides a better fit than the single variable models containing $\Delta\epsilon$, σ_{max}, or t_{avg}, or any of the two variable models not including $\Delta\sigma$. Neither the addition of σ_{max} or t_{avg} to $\Delta\epsilon_{in}$ in the model provides substantial improvement. Models containing the mean stress likewise provided little improvement. The model $N_f = \alpha\Delta\epsilon_{in}^{\beta}\Delta\sigma^{\gamma}$ provides the best fit of all. Note also that in the models containing σ_{max} or t_{avg} together with $\Delta\epsilon_{in}$, the absolute values of the *T*-ratios for their coefficients are much less than three, the usually accepted value for statistical significance. A visual example of the significance of $\Delta\sigma$ in the model is shown in Fig. 3. Plots of $N_{f,obs}$ and $N_{f,pred}$ are shown for the two models $N_f = \alpha\Delta\epsilon_{in}^{\beta}$ and $N_f = \alpha\Delta\epsilon_{in}^{\beta}\Delta\sigma^{\gamma}$.

Also, the model $N_f = \alpha\Delta\epsilon_{in}{}^{\beta}\Delta\sigma^{\gamma}$ provides the best fit for individual analyses of each cycle type. The best fit coefficients β and γ are not the same for each cycle type, as might be expected. Still, equations using the average values of β and γ, -1.26 and -3.08, provide good fits for the individual cycle types. The values of α in these equations are shown in Table 3.

It is well demonstrated by the cc tests that $\Delta\sigma$ is more significant than σ_{max} in determining life. Three tests were conducted with σ_{max} and σ_{min} of about +200/−200 MPa, or a $\Delta\sigma$ of about 400 MPa. An additional 1050°C cc test, the first listed in Table 1, had about the same σ_{max}, 210 MPa, but a larger $\Delta\sigma$, 474 MPa. The life of this test was reduced to about one third of that expected for a +200/−200 MPa test with the same $\Delta\epsilon_{in}$. In the fourth test listed in Table 1 it was intended to increase σ_{max} but keep $\Delta\sigma$ the same as for the first tests. Actually, σ_{max} was increased about 30% to 257 MPa, but $\Delta\sigma$ was also increased about 10%, and still life increased relative to the +200/−200 MPa tests. For these cc tests alone, R^2 for the model $N_f = \alpha\Delta\epsilon_{in}^{\beta}\sigma_{max}^{\gamma}$ is 40%, only slightly better than the value of 37% for $N_f = \alpha\Delta\epsilon_{in}^{\beta}$, and considerably less than the value of 76% for $N_f = \alpha\Delta\epsilon_{in}^{\beta}\Delta\sigma^{\gamma}$.

For the tests at 1015°C, the model containing $\Delta\sigma$ alone provides no better correlation than that containing $\Delta\epsilon_{in}$; however, this could be explained by the high degree of correlation between

TABLE 2—*Coefficients for models relating log N_f to the log of various single variables and combinations of log $\Delta\epsilon_{in}$ with the log of various other variables, T-ratios of the coefficients, and the standard deviations and R^2 values for the regressions.*

Temp., °C	Constant		log $\Delta\epsilon_{in}$		log $\Delta\sigma$		log σ_{max}		log t_{avg}			
	α	T	Coef.	T	Coef.	T	Coef.	T	Coef.	T	s	R^2, %
1050	−0.792	−0.8	−1.659	−3.7	...	...	...	...	...	...	0.32	37
	12.772	6.1	...	...	−3.683	−4.8	...	...	...	...	0.29	50
	5.602	2.8	...	...	...	...	−1.159	−1.4	...	...	0.39	8
	3.037	22.8	...	...	...	...	...	...	−0.466	−1.7	0.38	12
	8.396	4.2	−1.258	−3.8	−3.077	−4.9	...	...	...	...	0.23	70
	1.218	0.6	−1.578	−3.5	...	...	−0.765	−1.1	...	...	0.32	40
	−1.297	−1.1	−1.927	−3.1	...	...	...	...	0.199	0.6	0.33	38
1015	−1.505	−3.1	−1.762	−8.5	...	...	...	...	...	...	0.34	81
	16.427	9.8	...	...	−4.958	−8.2	...	...	...	...	0.33	80
	10.287	4.5	...	...	...	...	−3.091	−3.4	...	...	0.60	40
	2.689	21.7	...	...	...	...	...	...	−0.654	−4.3	0.53	52
	7.781	3.0	−1.003	−3.8	−2.696	−3.6	...	...	...	...	0.26	90
	0.424	0.2	−1.607	−6.1	...	...	−0.63	−1.0	...	...	0.34	82
	−0.706	−1.4	−1.433	−6.2	...	...	...	...	−0.251	−2.3	0.30	86

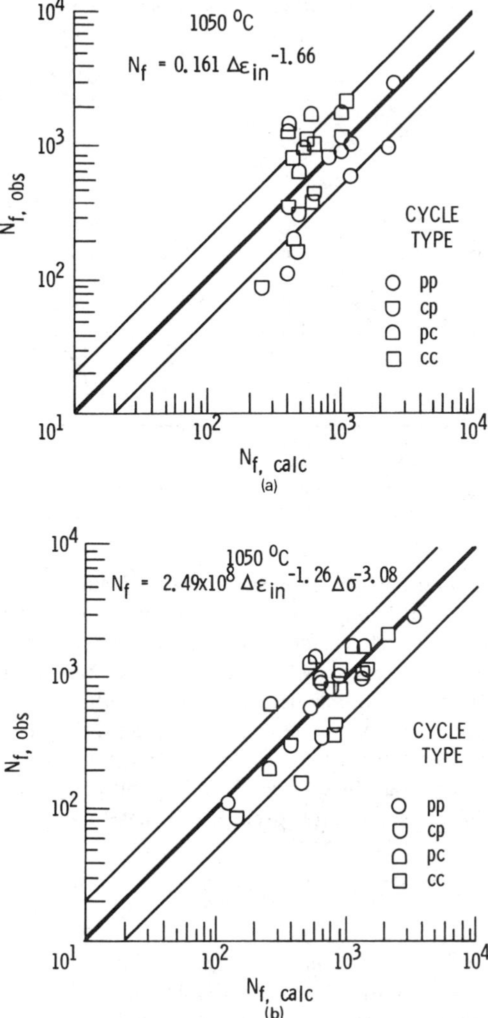

FIG. 3—*Observed versus calculated cyclic life of coated PWA 1480 single crystals for various cycle types at 1050°C.* (a) *For life model containing only* $\Delta\epsilon_{in}$. (b) *For life model containing both* $\Delta\epsilon_{in}$ *and* $\Delta\sigma$.

$\Delta\sigma$ and $\Delta\epsilon_{in}$ in these tests, R of 80%. As indicated previously, the dwell stress levels were increased more or less commensurately with $\Delta\epsilon_{in}$ among the creep-fatigue tests at 1015°C in order to reduce test times. Since $\Delta\epsilon_{in}$ is strongly correlated with $\Delta\sigma$, life correlates equally well with either variable. However, as for the tests at 1050°C, σ_{max} does not provide a good correlation, nor does t_{avg}, and the model $N_f = \alpha\Delta\epsilon_{in}^{\beta}\Delta\sigma^{\gamma}$ provides the best correlation. Figure 4 shows a comparison of the predictions of the models $N_f = \alpha\Delta\epsilon^{\beta}$ and $N_f = \alpha\Delta\epsilon^{\beta}\Delta\sigma^{\gamma}$. The best fit values of β and γ for the latter model are -1.00 and -2.70. Values of α which provide the best fit for each cycle type are shown in Table 3.

It may be seen that correlations using all the models are better for the 1015°C data than for the 1050°C data. This is largely because the 1015°C data cover a greater range of the "independent" variables.

TABLE 3—*Best-fit values of the constant α in the life models for PWA 1480 single crystals at 1050 and 1015°C.*

Cycle Type	α	95% Confidence Limits on $\alpha \times 10^{-8}$
	1050°C: $N_f = \alpha \Delta\epsilon_{in}^{-1.26} \Delta\sigma^{-3.08}$	
pp	2.25×10^{-8}	1.8 to 2.9
cp	1.41	1.0 to 2.0
pc	3.66	2.4 to 5.5
cc	1.86	1.2 to 2.9
	1015°C: $N_f = \alpha \Delta\epsilon_{in}^{-1.00} \Delta\sigma^{-2.70}$	
pp	0.489×10^{-8}	0.24 to 0.98
cp	0.236	0.16 to 0.35
pc	0.318	0.17 to 0.60
cc	0.360	0.11 to 1.18

Failure Mode

Internal crack initiation at pores was the predominant failure mode in these tests. For the creep-fatigue tests, in 80% of the specimens cracking initiated at many internal pores and linked up before the final overload, as may be seen on the fracture face shown in Fig. 5. Others appeared to have a dominant crack which initiated near the surface, possibly at a pore, but generally the fracture faces were heavily oxidized and difficult to interpret. Fracture surfaces with this appearance were more common for the pp tests. Still, the majority of pp tests failed at multiple internal pores. It is clear that in these isothermal creep-fatigue tests few, if any, failures initiated in the coating, or at the surface of the superalloy because of a defect in the coating. Of course, the failure mode of bare specimens at these high test temperatures might be quite different.

Discussion

Except in that it permits inelastic strain, creep does not have a great effect on the cyclic life of the coated single crystal superalloy, PWA 1480. On any basis of comparison, and particularly when compared on the basis of $N_f = \alpha \Delta\epsilon^{\beta} \Delta\sigma^{\gamma}$ as in Table 3, lives for the creep-fatigue cycles are not greatly, if at all, worse than those for the pp tests. Though life may be lower for the cp cycle than for the others, it is only about 30% lower than the average for the other cycles. Lives for the other cycles may all be the same.

In fact, there appears to be no time-dependent process having a great effect on life. This is shown by the lack of any substantial improvement when t_{avg} is included in the life models. Neither creep nor the environmental degradation has affected the coated single-crystal superalloy. The mechanisms of creep degradation in polycrystalline alloys such as grain boundary cavitation or sliding obviously cannot occur, and the environment cannot affect the internal crack propagation mode of failure.

The successful life model containing $\Delta\epsilon_{in}$ and $\Delta\sigma$ is unusual, and may be peculiar to the coated single crystal superalloy system studied. Crack initiation in high temperature creep-fatigue, at least for polycrystalline materials, is usually found to be determined by $\Delta\epsilon_{in}$ and some measure of time dependent damage processes such as creep cavitation at grain boundaries or oxidation attack. For instance, frequency or strain rate is introduced into the model [*6–8*]. Or, in the Strain-Range-Partitioning Model [*5*], more or less damage may be assigned to the inelastic strain pro-

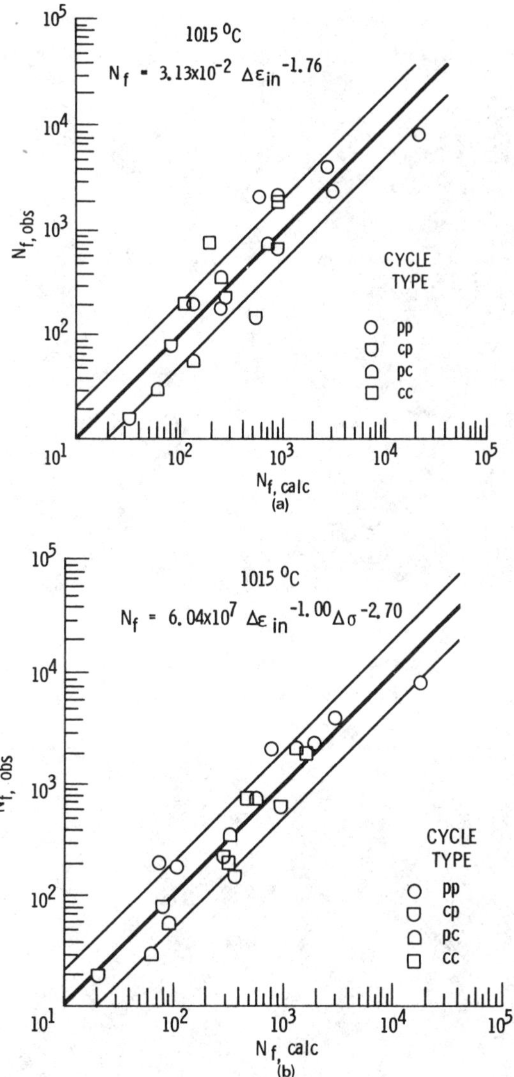

FIG. 4—*Observed versus calculated cyclic life of coated PWA 1480 single crystals for various cycle types at 1015°C.* (a) *For life model containing only* $\Delta\epsilon_{in}$. (b) *For life model containing both* $\Delta\epsilon_{in}$ *and* $\Delta\sigma$.

duced by, say, a cycle with creep in tension than that produced by the rapid straining in a "pure" fatigue cycle.

The internal crack initiation observed and the importance of $\Delta\sigma$ in the life model may reflect that crack propagation is a significant portion of life in these tests. Since the cracks are protected from the atmosphere, it might be expected the crack growth rates are relatively low, and while crack initiation is thought to be primarily driven by $\Delta\epsilon_{in}$, $\Delta\sigma$ can be tied to crack propagation rates.

Elber [9] introduced the concept that a crack may not close when the stress intensity, K, is zero, but at some positive K. The effective ΔK driving crack growth, ΔK_{eff}, is taken as

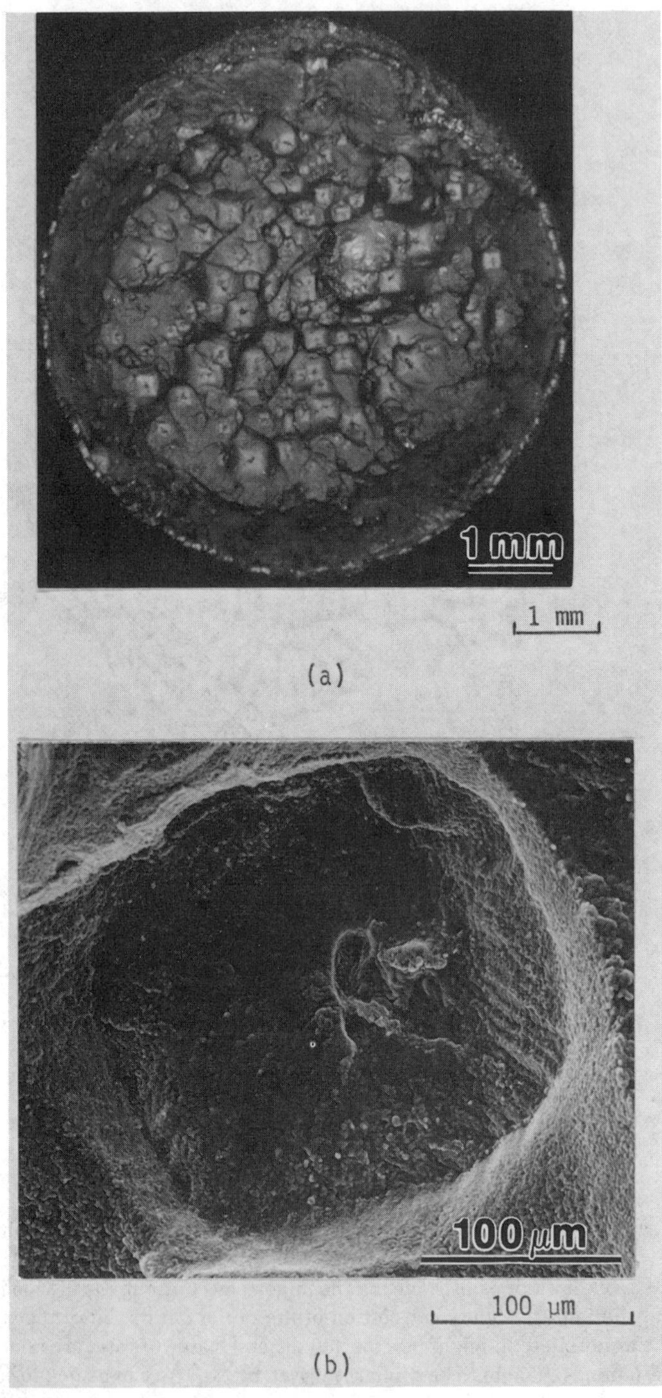

FIG. 5—*Multiple internal crack initiation in coated PWA 1480 single-crystal low-cycle fatigue test specimens.* (a) *Optical photomicrograph of fracture face.* (b) *SEM photomicrograph of a single crack initiation site.*

$K_{max} - K_{cl}$, where K_{cl} is the K at which closure occurs. It has been shown that compressive overloads increase subsequent fatigue crack growth rates and that the increase is greater with both increasing peak stress and frequency of the compressive overloads [*10–12*]. Such effects have been explained on the basis of reduced K_{cl}, which increases ΔK_{eff}. It has been proposed that compression may flatten roughness of the crack faces and thus reduce K_{cl}. Reversed plasticity may also reduce the size of the plastic zone ahead of a crack, again reducing the K_{cl}. Thus, for reversed tension-compression tests such as conducted herein, ΔK_{eff} is expected to be more directly related to $\Delta\sigma$ than to σ_{max} through the effect of compressive loading in reducing K_{cl}.

These concepts offer a basis for understanding the present results, still some caution is necessary. The studies cited on the effects of compressive overloads, and the models to explain them, apply strictly to linear elastic conditions and "long" cracks. In the present tests, general yielding occurs. Also, if crack growth does indeed constitute a major portion of life, then it would appear that for much of that time the growing cracks might be described as "short". Still speaking of the linear elastic situation, closure of the crack faces at $K > 0$ occurs behind the crack tip in its plastic wake. Since "short" cracks would not have great length of wake behind them, K_{cl} should be near zero and thus ΔK_{eff} would not be expected to be greatly affected by additional reduction of K_{cl} [*13*]. Still, the results on this particular material clearly show that cyclic life is reduced by increasing the magnitude of the compressive stress in the cycle, and closure effects on crack growth rates may well offer the explanation.

Results and Conclusions

Fatigue tests at 1.0 Hz and cp, pc, and cc type creep-fatigue tests have been conducted on Ni-Co-Cr-Al-Y coated specimens of a single-crystal superalloy, PWA 1480, at 1050°C and about 1015°C. The following results and conclusions were obtained:

1. Considerable cyclic softening occurred for all test cycles, evidenced particularly by rapidly increasing creep rates in the creep-fatigue tests.
2. Lives for the pp, cp, pc, and cc cycles were not greatly different; however, those for the cp cycle did appear to be lowest at both test temperatures.
3. A life model, $N_f = \alpha\Delta\epsilon_{in}^{\beta}\Delta\sigma^{\gamma}$, was found to provide good correlation for all cycle types, better than models based on $\Delta\epsilon_{in}$ alone, or $\Delta\epsilon_{in}$ with either σ_{max} or t_{avg}.
4. For all test types, failure occurred predominately by multiple internal cracking originating at porosity.
5. The strong correlation of life with $\Delta\sigma$ may reflect a significant crack growth period in the life of the specimens.
6. The lack of improvement in the models when average cycle time was considered appears to reflect both that there is no large effect of strain rate on the damage mechanisms in the single-crystal material nor any environmental effect due to the internal cracking mode of failure.

References

[*1*] Miner, R. V., Gayda, J., and Hebsur, M. G., NASA TM-87110, National Aeronautics and Space Administration, Washington, D.C., 1985.

[*2*] Gell, M., Duhl, D. N., and Giamei, A. F. in *Superalloys 1980,* American Society for Metals, Metals Park, Ohio, 1980, pp. 205–214.

[*3*] Shah, D. M. and Duhl, D. N. in *Superalloys 1984,* AIME, Warrendale, Pa., 1984, pp. 105–114.

[*4*] Hirshberg, M. H., *Manual on Low-Cycle Fatigue Testing, ASTM STP 465,* American Society for Testing and Materials, Philadelphia, 1969, pp. 67–86.

[*5*] Manson, S. S., Halford, G. R., and Hirshberg, M. H. in *Proceedings,* Symposium on Design for Elevated Temperature Environment, ASME, New York, 1971, pp. 12–28.

[*6*] Coffin, L. F., Jr., in *Proceedings,* 1976 ASME-MPC Symposium on Creep-Fatigue Interaction, MPC3, ASME, New York, pp. 349–364.

[7] Ostergren, W. J. in *Proceedings*, 1976 ASME-MPC Symposium on Creep-Fatigue Interaction, MPC3, ASME, New York, pp. 179–202.

[8] Majumdar, S. and Maiya, P. S. in *Proceedings, 1976 ASME-MPC Symposium on Creep-Fatigue Interaction*, MPC3, ASME, New York, pp. 323–336.

[9] Elber, W. in *Damage Tolerance in Aircraft Structures, ASTM STP 486*, American Society for Testing and Materials, Philadelphia, 1971, pp. 230–242.

[10] Au, P., Topper, T. H., and El Haddad, M. L. in *Behavior of Short Cracks in Airframe Components*, AGARD Conference Proceedings No. 328, AGARD, France, 1983, p. 11.1.

[11] Yu, M. T., Topper, T. H., and Au, P. in *Fatigue '84*, D. J. Beevers, Ed., Engineering Materials Advisory Services, Warley, U.K., Vol. 1, 1984, p. 79.

[12] Zaiken, E., *Fatigue Behavior of Long and Short Cracks in Wrought and Powder Aluminum Alloys*, Report No. UCB/RP/85/A1032, University of California, Berkeley, May 1985, pp. 35–53.

[13] Suresh, S. and Ritchie, R. O., *International Metals Reviews*, Vol. 29, No. 6, 1984, pp. 445–476.

DISCUSSION

L. A. Carol[1] *(written discussion)*—In your presentation, you remarked that there were no environmental effects (e.g., oxidation) associated with your fatigue and creep-fatigue characterizations of PWA 1480 superalloy single crystals because the Ni-Co-Cr-Al-Y coating protected the superalloy at the high test temperatures. One point that was not addressed, however, was whether coating/superalloy interactions played a role in the fatigue and creep-fatigue characterizations of PWA 1480 superalloy. Specifically, was there significant coating/superalloy interdiffusion (as noted, for example, by microstructural changes such as β-phase recession [*1–3*]), and what, if any, interactions could this interdiffusion have on your fatigue and creep-fatigue data? In addition, significant porosity can result at the coating/superalloy interface from the initial coating anneal and from Kirkendall porosity developed during interdiffusion at the high test temperatures [*1*]. Was porosity present at the coating/superalloy interface and was crack initiation ever noted at the coating/superalloy interface because of this porosity? If so, did this porosity have any effect on your experimental results?

R. V. Miner et al. (authors' closure)—Significant microstructural changes were not noted at the superalloy-coating interface after these tests, the longest of which experienced less than 200 h at 1050°C. Further, crack initiation at internal pores indicated that they controlled life, not the coating or superalloy-coating interface. However, in longer life tests it might be expected that degradation of the coating and/or interface would become life limiting. Further, failures originating in the coating do occur in thermo-mechanical fatigue tests simulating locations in gas turbine engine parts which experience tension at low temperatures during cooling. The coating possesses much less ductility at lower temperatures.

Discussion References

[1] Nesbitt, J. A., "Overlay Coating Degradation by Simultaneous Oxidation and Coating/Substrate Interdiffusion," NASA Technical Memorandum 83738, Washington, D.C., Aug. 1984.

[2] Pilsner, B. H., "The Effects of Substrate Composition (MAR-M247) on Coating Oxidation and Coating/Substrate Interdiffusion," M. S. thesis, Michigan Technological University, Houghton, Mich., 1984.

[3] Levine, S. R., *Metallurgical Transactions*, Vol. 9A, 1978, p. 1237.

[1]AC Spark Plug Division, General Motors Corporation, Flint, MI 48556.

M. Y. Nazmy[1]

High-Temperature Low Cycle Fatigue Behavior and Lifetime Prediction of a Nickel-Base ODS Alloy

REFERENCE: Nazmy, M. Y., "**High-Temperature Low Cycle Fatigue Behavior and Lifetime Prediction of a Nickel-Base ODS Alloy,**" *Low Cycle Fatigue, ASTM STP 942,* H. D. Solomon, G. R. Halford, L. R. Kaisand, and B. N. Leis, Eds., American Society for Testing and Materials, Philadelphia, 1988, pp. 385–398.

ABSTRACT: The high-temperature low-cycle fatigue (HTLCF) behavior of mechanically alloyed oxide dispersion strengthened MA 754 nickel-base alloy was studied under strain-controlled condition at 850°C. Specimens were cut from the longitudinal and the long transverse directions of the bar stock. Triangular strain cycles as well as strain cycles with hold times were employed in this study. In general, longitudinal specimens exhibited the longest lives, long transverse specimens the shortest lives. The cycles with tensile hold times were the most damaging to the HTLCF lives of specimens of both orientations. The difference between the HTLCF response of the alloy in the longitudinal and the long transverse orientations is interpreted in terms of factors such as modes of crack initiation and propagation, anisotropy in crack growth rates, and the directionality in creep properties. Three models for HTLCF lifetime were evaluated for their ability to correlate the fatigue data on MA 754 in both orientations. These models are the Strain Range Partitioning Model (SRP), the Frequency-Separation Model (FS), and the Frequency-Modified Damage Function (FMDF). The correlations of the data, having the smallest standard deviations of 0.22 and 0.14 for the longitudinal and the long transverse orientations, respectively, were by the FS model. The FMDF and the SRP models had standard deviations, for the longitudinal orientations, of 0.25 and 0.33, respectively.

KEY WORDS: MA 754, high-temperature low-cycle fatigue, texture, nickel-base oxide dispersion strengthened alloy, life prediction

The development of gas turbines requires the use of new materials with improved high temperature capabilities. Such materials are needed for blade and vane application as well as for other components. Thermal fatigue, and hence high-temperature low-cycle fatigue (HTLCF), is one of the important criteria that should be considered when selecting an alloy for a fluctuating temperature application. Directionally solidified and single crystals of nickel-base superalloys were introduced to achieve improved creep-rupture and thermal fatigue resistance in the solidification direction by developing a preferred crystallographic texture and grain boundary orientation, [*1,2*]. The marked improvement in thermal fatigue resistance of directionally solidified and single-crystal nickel-base superalloys can be attributed to (1) the lower modulus associated with the ⟨100⟩ primary dendrite growth direction, which gives a reduced thermal stress and plastic strain for a given total strain range, and (2) the absence of grain boundaries perpendicular to the principal stress axis [*3*]. The mechanically alloyed oxide dispersion strengthened (ODS) class of alloys has shown attractive high temperature properties (e.g., the nickel-base ODS MA 754 alloy is used in advanced gas turbine engines [*4*]. This class of alloys exhibits, in addition to highly elongated grain structures, a variety of strong textures [*5*]. The ODS MA 754

[1]Scientist, Brown Boveri Research Center, CH-5405 Baden/Dättwil, Switzerland.

alloy exhibits a strong [100] (110) cube-on-edge texture along the working direction of the bar [6]. Although the longitudinal, or working, direction is the prime load-bearing direction in turbine airfoils, the actual stress states are, in general, quite complex as a result of thermal stresses and geometric constraints. This complexity inevitably leads to the generation of stress in directions other than the longitudinal direction. For satisfactory airfoil performance, the material capability in these off-axis orientations must be determined.

The aim of the present study is to investigate the orientation dependence of the HTLCF behavior of the ODS alloy MA 754 and to determine the correlative ability of three different lifetime prediction models as applied to this alloy.

Experimental Procedure

The material used in this investigation is Inconel alloy MA 754. The specific composition (in weight percent) is: balance Ni, 20.32 Cr, 0.31 Al, 0.4 Ti, 0.98 Fe, 0.05 C, 0.001 S, 0.57 Y_2O_3. Thermomechanical processing results in elongated grains. The average long grain dimension parallel to the rolling direction, as determined from quantitative metallography, is 3.5 mm. The determined average grain aspect ratio is 10.2. The Y_2O_3 dispersoids are uniformly distributed throughout the matrix and range in size from 5 to 100 nm. The alloy exhibits a strong texture, with a ⟨100⟩ crystallographic direction parallel to the rolling direction [6]. A full account on the tensile mechanical properties of MA 754 is given in Ref *5*.

The HTLCF specimens, with the dimensions shown in Fig. 1, were machined from the longitudinal direction (i.e., the rolling direction) and the long transverse direction of the MA 754 bar stock. Final machining was carried out by low stress grinding to remove surface defects due to machining.

The HTLCF behavior was investigated under fully reversed axial strain control conditions using a servohydraulic testing machine. Testing was done in air at 850°C. Four types of strain wave shapes were used to examine the effect of strain wave shape on the fatigue life: triangular wave with a strain rate of 10^{-2} s^{-1}, and truncated waves with hold times in tension or in compression or in both tension and compression. The hold times varied from 30 to 300 s, and the ramp rate was always 10^{-2} s^{-1}.

Table 1 gives the results on HTLCF tests. The strain and stress ranges were determined at half-life of the specimens.

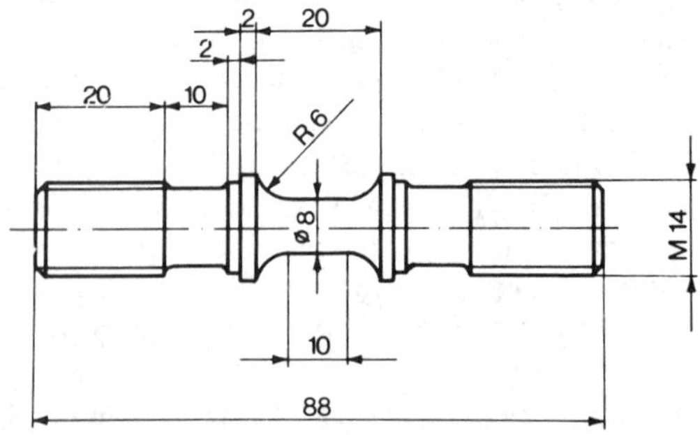

FIG. 1—*HTLCF specimen (dimensions in millimetres).*

TABLE 1—*Typical test results on HTLCF.*

Specimen Orientation	Hold Time, s		Total Strain Range ($\Delta\epsilon_t$), %	$\Delta\epsilon_{in}$, %	$\Delta\epsilon_{pp}$, %	$\Delta\epsilon_{pc}$, %	$\Delta\epsilon_{cp}$, %	$\Delta\epsilon_{cc}$, %	Stress Range ($\Delta\sigma$), MPa	Tensile Stress (σ_t), MPa	N_f
	Tension	Compression									
Longitudinal	...	...	0.640	0.245	0.245	...	...	...	586	294	1119
Longitudinal	...	...	0.776	0.323	0.323	...	...	...	588	294	969
Longitudinal	...	...	0.482	0.086	0.086	...	...	...	536	261	6277
Longitudinal	...	...	0.857	0.410	0.410	...	...	...	624	312	783
Longitudinal	...	100	0.795	0.394	0.257	0.137	...	...	628	324	1316
Longitudinal	...	50	0.743	0.418	0.269	0.149	...	...	620	314	1281
Longitudinal	...	300	0.970	0.615	0.451	0.164	...	...	628	316	777
Longitudinal	50	...	0.578	0.226	0.112	...	0.114	...	551	251	926
Longitudinal	300	...	0.602	0.315	0.196	...	0.119	...	596	278	428
Longitudinal	300	...	0.751	0.442	0.289	...	0.153	...	614	294	244
Longitudinal	30	30	1.306	1.002	0.869	...	...	0.133	618	310	239
Longitudinal	30	30	0.840	0.553	0.432	...	...	0.121	582	292	750
Long transverse	...	...	0.700	0.408	0.408	...	...	...	602	294	474
Long transverse	...	...	0.865	0.567	0.567	...	...	...	580	282	314
Long transverse	...	...	0.528	0.233	0.233	...	...	...	594	290	974
Long transverse	...	...	0.948	0.65	0.65	...	...	...	561	263	216
Long transverse	...	50	0.614	0.322	0.239	0.083	...	...	553	288	589
Long transverse	...	300	0.553	0.310	0.239	0.071	...	...	519	278	811
Long transverse	...	300	0.948	0.656	0.579	0.077	...	...	544	273	143
Long transverse	50	...	0.625	0.379	0.298	...	0.091	...	557	288	63
Long transverse	300	...	0.584	0.406	0.329	...	0.077	...	614	290	23
Long transverse	300	...	0.410	0.215	0.143	...	0.072	...	577	273	91

Results and Discussion

HTLCF Behavior

The variation of the fatigue life of MA 754 at 850°C with orientation and different types of strain waves is shown in Fig. 2. In general, the fatigue lives of the specimens of the longitudinal orientation are longer than the fatigue lives of the specimens of the long transverse orientation, for the same type of strain wave. This behavior for MA 754 is in agreement with the reported results on the influence of orientation on the HTLCF of directionally solidified René 150 [*7*]. In all the HTLCF tests on MA 754, the fracture started by surface crack initiation. This is similar to the reported observation on the high-cycle fatigue behavior of the same alloy [*8*].

The specimens of the longitudinal orientation, tested under triangular strain waves, fractured by surface transgranular crack initiation and propagation (Fig. 3). The specimens of the long transverse orientation fractured by transgranular crack initiation and mixed crack propagation. Figure 4 shows the mixed mode of crack propagation on the fracture surface of a long transverse specimen tested under a triangular strain wave condition. The fatigue crack growth in stress-controlled testing of MA 754 at 850°C exhibited anisotropic behavior [*8*]. Increased growth rates have been measured in the longitudinal orientation specimens with the lower value for Young's modulus of 94 × 10^3 MPa as compared to the growth rates determined in the specimens of long transverse orientation with the higher value for Young's modulus of 137 × 10^3 MPa [*8*]. Gemma et al., in their work on crack growth in directionally solidified MAR-M 200 alloy, presented an argument based on the effective stress intensity factor and the crack opening displacement that under strain-controlled testing, in contrast to stress-controlled testing, an increase in Young's modulus would lead to an increase in the crack growth rates [*9*]. Hence the observed difference in HTLCF lives between the longitudinal and the long transverse orientations is attributed to the difference in Young's moduli.

The HTLCF of the specimens of the longitudinal orientation under truncated strain waves with compression hold times and with equal hold times in both tension and compression showed

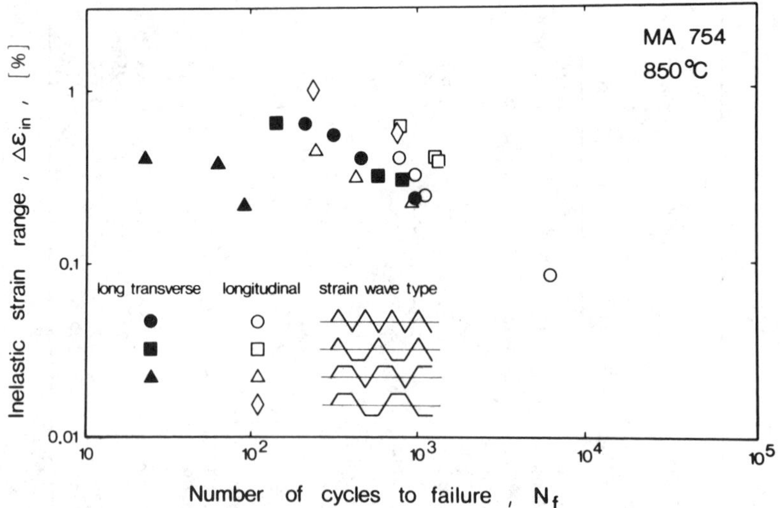

FIG. 2—*Inelastic strain range versus number of cycles to failure for the longitudinal and long transverse orientations.*

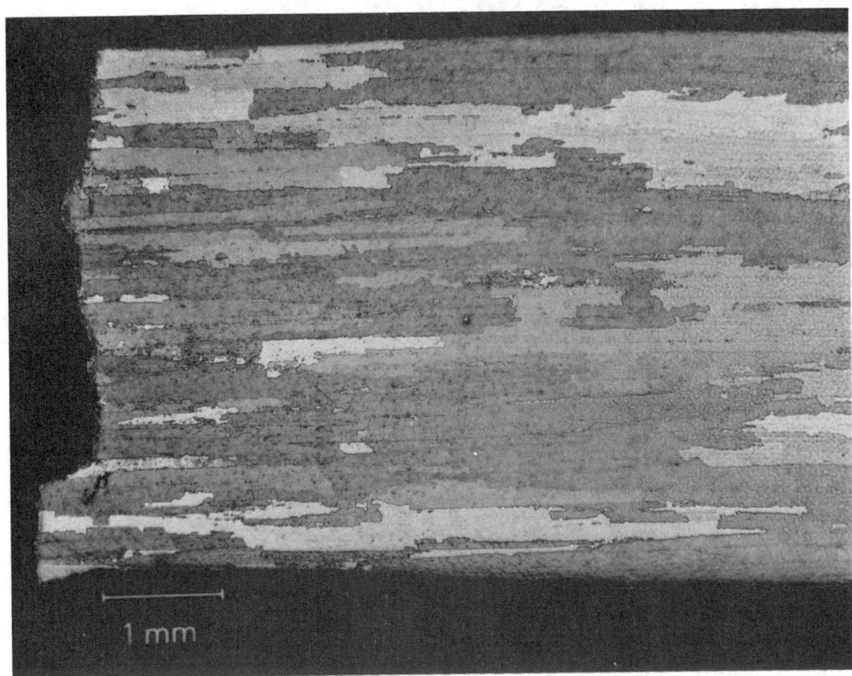

FIG. 3—*Optical photomicrograph of HTLCF specimen of the longitudinal orientation tested under triangular strain wave.*

improved cyclic life as compared to the tests conducted under triangular strain waves. The specimens tested under strain waves with compression hold were fractured by transgranular crack initiation and propagation and exhibited multiple cracking (Fig. 5). The specimens tested under strain waves with equal hold times in tension and compression fractured by transgranular initiation and mixed crack propagation. Gell and Leverant reported that HTLCF tests on Type 304 stainless steel with compressive hold time or equal tensile and compression hold times exhibited transgranular cracking and long lives [*10*]. Hence the longer lives exhibited in the HTLCF tests, with compressive dwells in the half-cycle, of the longitudinal orientation specimens are attributed to the healing effect of the compressive creep.

The HTLCF tests with compressive dwells, of the long transverse orientation, exhibited variation in endurances, depending on the strain range, as compared to the tests with triangular wave shape for the same orientation. This trend can be attributed to the change in the fracture mode from transgranular initiation and mixed propagation, at the lower strain range, to intergranular crack initiation and mixed propagation at the high strain range.

The strain cycles with tensile hold times were the most damaging cycles for both orientations (Fig. 2). The specimens of the longitudinal orientation exhibited grain boundary damage and fractured by transgranular initiation and intergranular crack propagation (Figs. 6*a* and 6*b*, respectively).

In the case of long transverse orientation, testing under tensile hold cycles exhibited reduced lives and the fracture mode was intergranular crack initiation and propagation (Fig. 7).

In general, intergranular crack initiation and propagation occur at faster rates than transgranular cracking [*10*]. Therefore the fatigue lives of the specimens tested with cycles with tensile hold times were shorter than those tested with the triangular or compressive dwells cycles.

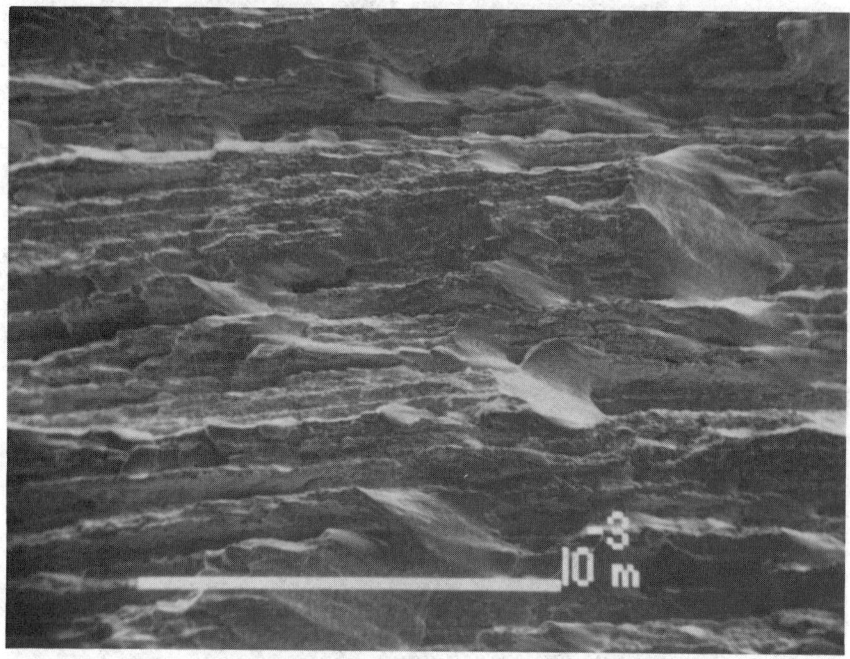

FIG. 4—*Scanning electron micrograph of fracture surface of a specimen of the long transverse orientation tested under triangular strain wave.*

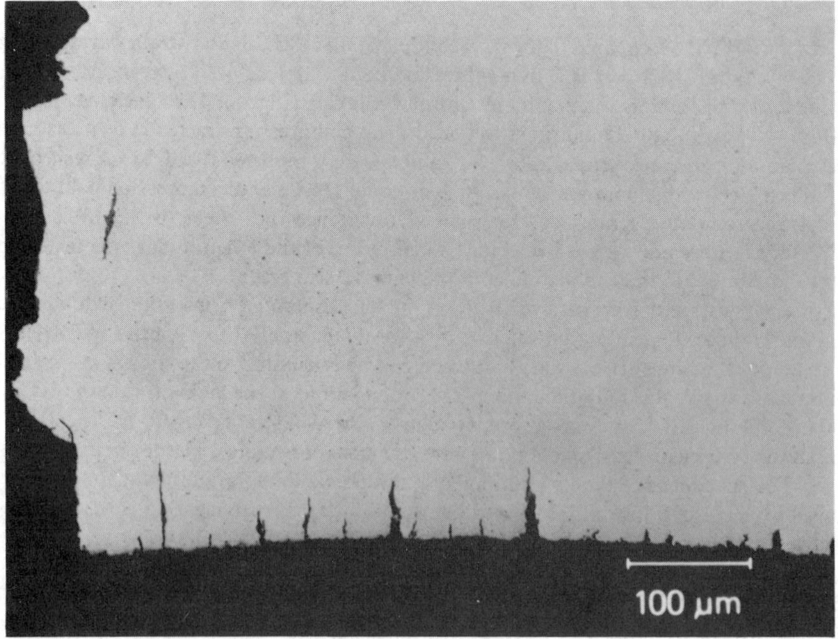

FIG. 5—*Optical photomicrograph of HTLCF specimen of the longitudinal orientation tested under strain wave with compression hold time.*

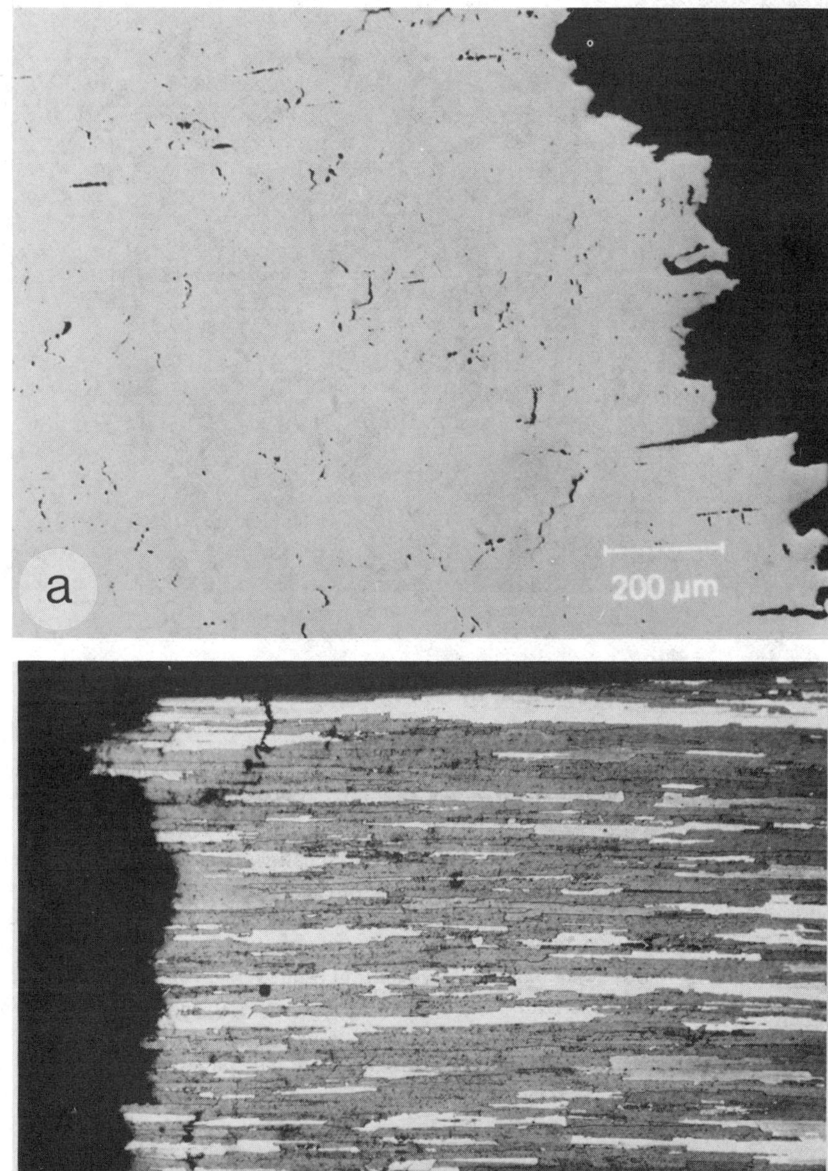

FIG. 6—*Optical photomicrographs of HTLCF specimen of the longitudinal orientation tested under strain wave with tensile hold time, showing* (a) *grain boundary damage and* (b) *mode of crack initiation and propagation.*

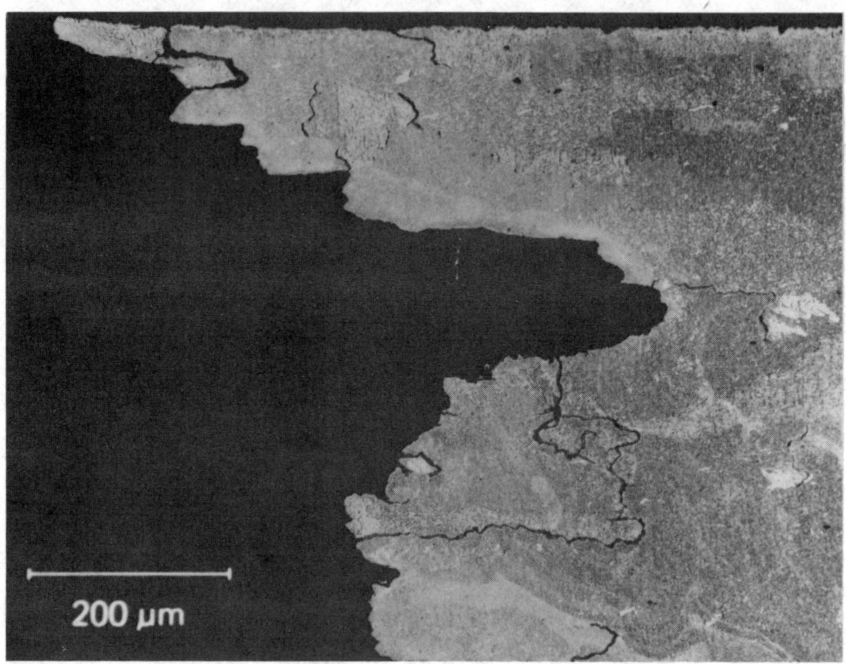

FIG. 7—*Optical photomicrograph of HTLCF specimen of long transverse orientation tested under strain cycle with tensile hold time.*

The creep rupture of MA 754 in the long transverse orientation is inferior to the creep rupture in the longitudinal direction [*11*]. Hence the shorter HTLCF lives exhibited by the long transverse orientation as compared to the longitudinal orientation, under tensile hold cycles, can be attributed to the inferior creep properties in the long transverse orientation of MA 754.

Conventionally cast IN 738 is currently used in vane applications in industrial gas turbines. In advanced military aircraft engines, the high pressure turbine vanes are made of MA 754 [*4*]. Hence it is of interest to compare the HTLCF properties of both alloys. The data on HTLCF of IN 738 have been taken from Ref *12*. MA 754 showed improved resistance to HTLCF as compared to IN 738. Figure 8 shows a plot of inelastic strain range versus cycles to failure for MA 754 of the longitudinal, long transverse orientations, and IN 738. In this figure, MA 754 assumes longer endurances than IN 738. This behavior is in agreement with the reported improved resistance to thermal fatigue exhibited by MA 754 as compared to conventionally cast superalloys [*13*].

Lifetime Prediction Models

The ability of the models to correlate the data was determined using the standard deviation, S, defined as

$$S = \left(\frac{\sum_{n} (\log N_f - \log N_p)^2}{n - 1} \right)^{1/2} \tag{1}$$

where N_f is the observed life, N_p is the predicted life, and n is the number of data points.

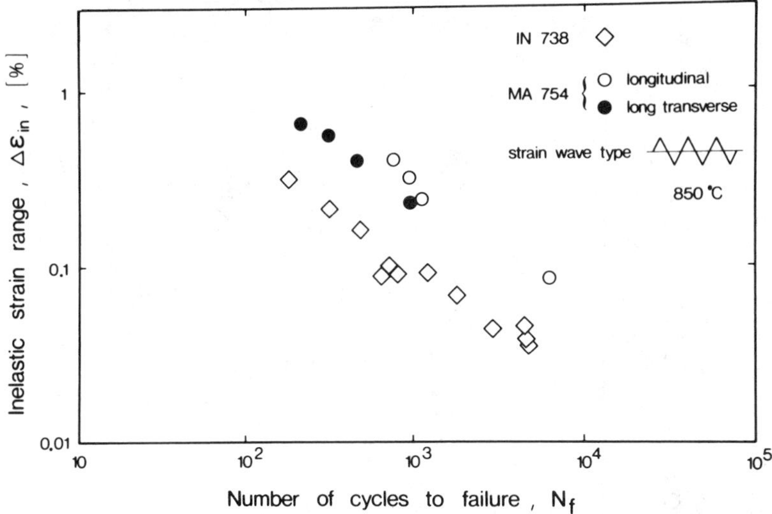

FIG. 8—*Inelastic strain range versus cycles to failure for the longitudinal and the long transverse orientations of MA 754 and cast IN 738.*

Frequency-Separation (FS) Model

The FS model postulates that the basic parameters necessary to predict the HTLCF life are the inelastic strain range $\Delta\epsilon_{in}$, the tension-going frequency ν_t, and the loop-time unbalance ν_c/ν_t, where ν_c is the compression-going frequency [*14,15*]. (The unit of ν_t and ν_c is s^{-1}.) These three parameters are combined in a power-law relationship to predict the cycles to failure:

$$N_f = C\Delta\epsilon_{in}^{\beta}\nu_t^m\left(\frac{\nu_c}{\nu_t}\right)^k \tag{2}$$

where C, β, m, and k are constants.

The values of the constants in Eq 2 that yielded the best fit to the data of the longitudinal and long transverse orientations were $C = 197.7$, $\beta = -1.37$, $m = -0.02$, and $k = -0.172$, and $C = 113.8$, $\beta = -1.65$, $m = 0.57$, and $k = 0.04$, respectively. The calculated standard deviations for the data of the longitudinal and the long transverse orientations were 0.22 and 0.14, respectively. Figure 9 shows a plot of observed life versus predicted life for the data of the longitudinal orientation.

Frequency-Modified Damage Function (FMDF)

The FMDF is based upon the premise that HTLCF is controlled by the tensile hysteretic energy absorbed by the specimen [*16,17*]. This energy is approximated by the damage function: the product of the inelastic strain range, $\Delta\epsilon_{in}$, and the maximum tensile stress, σ_t.

The life is predicted by using a power-law relationship between the damage function and the life. When time-dependent mechanisms are present, the frequency term that takes into account the time dependency is

$$N_f = C(\Delta\epsilon_{in} \cdot \sigma_t)^{\beta}\nu^m \tag{3}$$

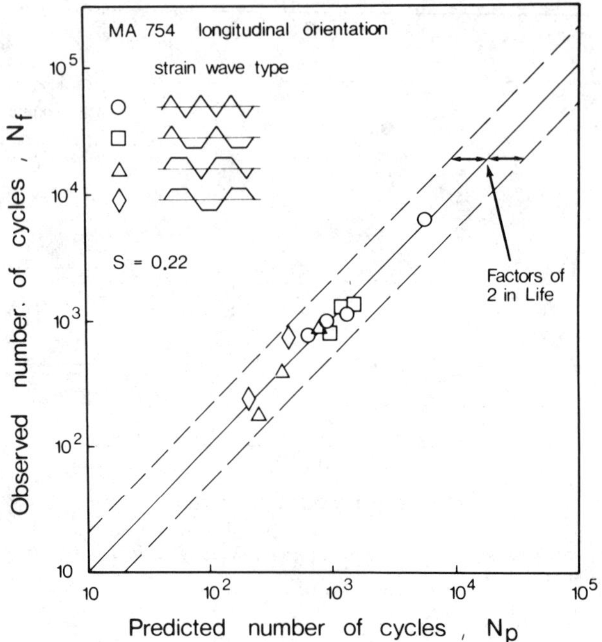

FIG. 9—*Observed versus predicted numbers of cycles to failure for HTLCF data in the longitudinal orientation using the FS model.*

where C, β, and m are constants and ν is the frequency. (The units of the variables are MPa for σ_t and s^{-1} for ν.)

The frequency term in Eq 3 is dependent upon the sensitivity of the material to different strain cycles. It was found that MA 754 is strain wave sensitive and the frequency term is defined as

$$\nu = \frac{1}{\tau_0 + \tau_t - \tau_c} \tag{4}$$

for $\tau_t > \tau_c$ and

$$\nu = \frac{1}{\tau_0} \tag{5}$$

for $\tau_t \leq \tau_c$, where τ_0 is the time for continuous cycling, τ_t is the tensile hold time, and τ_c is the compressive hold time.

The values of the constants in Eq 3 that yielded the best fit to the data of the longitudinal orientation were $C = 1.5 \times 10^5$, $\beta = -1.01$, and $m = 0.26$, and for the long transverse orientation were $C = 2.14 \times 10^6$, $\beta = -1.7$, and $m = 0.58$. The calculated standard deviations for the data of the longitudinal and the long transverse orientations were 0.25 and 0.16, respectively. Figure 10 shows a plot of observed life versus predicted life for the data of the longitudinal orientation.

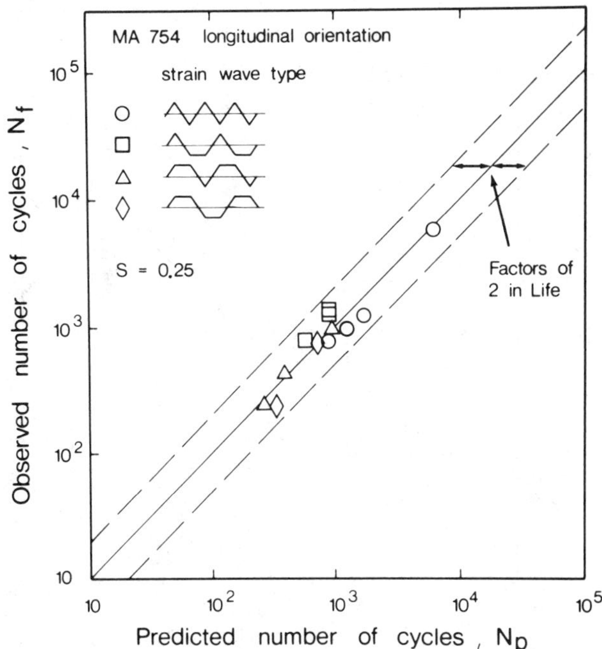

FIG. 10—*Observed versus predicted numbers of cycles to failure for HTLCF data in the longitudinal orientation using the FMDF model.*

Strain Range Partitioning (SRP) Model

The SRP model extends the Manson-Coffin Law, which is valid at room temperature, to high temperature by considering the interaction of time-dependent inelastic (or creep) strains and time-independent inelastic (or plastic) strains [*18*]. The equations of the SRP model are

$$N_{ij} = C_{ij}(\Delta\epsilon_{in})^{-\beta_{ij}} \tag{6}$$

and

$$(N_f)^{-1} = \sum_{ij} (f_{ij}/N_{ij}) \tag{7}$$

where ij represents the four strain ranges pp, cp, pc, and cc. N_{ij} is the number of cycles to failure if $\Delta\epsilon_{in}$ is all ij-strain; $f_{ij} = \Delta\epsilon_{ij}/\Delta\epsilon_{in}$, and C_{ij} and β_{ij} are constants. Equation 7 is the interaction damage rule and is used to predict the fatigue life N_f. When Eqs 6 and 7 were fit to the fatigue data of the longitudinal orientation, the following constants were determined: $C_{pp} = 12.52$, $C_{cp} = 0.8$, $C_{pc} = 2.24$, $C_{cc} = 1.48$, $\beta_{pp} = 0.55$, $\beta_{cp} = 0.34$, $\beta_{pc} = 0.35$, and $\beta_{cc} = 0.17$. The determined constants for the long transverse orientation were: $C_{pp} = 13.79$, $C_{cp} = 1.01$, $C_{pc} = 4.46$, C_{cc} = not determined, $\beta_{pp} = 0.59$, $\beta_{cp} = 0.97$, $\beta_{pc} = 0.71$, and β_{cc} = not determined. The calculated standard deviations for the data of the longitudinal and the long transverse orientations were 0.33 and 0.49, respectively. Figure 11 shows a plot of observed life versus predicted life for the data of the longitudinal orientation.

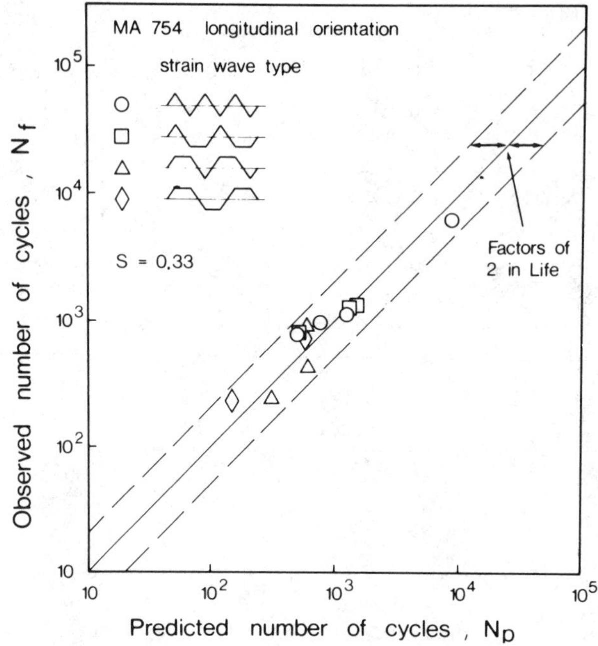

FIG. 11—*Observed versus predicted numbers of cycles to failure for HTLCF data in the longitudinal orientation using the SRP model.*

Conclusions

1. MA 754 exhibited improved HTLCF life in the longitudinal orientation as compared to the long transverse orientation.
2. Strain cycles with tensile hold times were the most damaging to the HTLCF life of MA 754 in both orientations.
3. On the basis of inelastic strain range, MA 754 showed improved HTLCF behavior as compared to cast IN 738 at 850°C and under testing with a triangular strain cycle.
4. The correlations of the data in both orientations having the smallest standard deviations were by the frequency separation model.
5. The three life prediction models correlated the data of the longitudinal orientation within the conventionally used scatter band of factor-of-two.

Acknowledgments

Helpful discussions with Drs. R. Singer and C. Wüthrich and the skillful work with experimentation of W. Meixner are greatly appreciated. This work was financed in part by the Swiss Federal Government and performed within the framework of COST-action 501.

References

[*1*] Versnyder, F. L. and Shank, M. E., *Materials Science and Engineering,* Vol. 6, 1970, pp. 213–247.
[*2*] Bizon, P. T. and Spera, D. A. in *Thermal Fatigue of Materials and Components, ASTM STP 612,* American Society for Testing and Materials, Philadelphia, 1975, pp. 106–122.

[3] Boone, D. H. and Sullivan, P. C. in *Fatigue at Elevated Temperatures, ASTM STP 520,* American Society for Testing and Materials, Philadelphia, 1973, pp. 401-414.
[4] Crawford, W., "Oxide Dispersion Strengthened Materials in Advanced Gas Turbine Engines—Inconel MA 754 Vanes," in *Proceedings,* Conference on Frontiers of High Temperature Materials II, J. Benjamin, Ed., INCO Alloy Products Co., 22-25 May 1983.
[5] Singer, R. F. and Gessinger, G. H. in *Powder Metallurgy of Superalloys,* G. H. Gessinger, Ed., Butterworths, London, 1984, Chapter 7, pp. 213-292.
[6] Nazmy, M. Y., Singer, R. F., and Török, E. in *Proceedings,* Seventh International Conference on Textures of Materials, C. M. Brakman, P. Jongenburger, and E. J. Mittemeijer, Eds., Netherlands Society for Materials Science, 17-21 Sept. 1984, pp. 275-280.
[7] Wright, P. K. and Anderson, A. F. in *Superalloys 1980,* J. K. Tien, S. T. Wlodek, H. Morrow III, M. Gell, and G. E. Maurer, Eds., American Society for Metals, Metals Park, Ohio, 1980, pp. 689-698.
[8] Nazmy, M. Y. and Singer, R. F., *Metallurgical Transactions,* Vol. 16A, 1985, pp. 1437-1444.
[9] Gemma, A. E., Langer, B. S., and Leverant, G. R. in *Thermal Fatigue of Materials and Components, ASTM STP 612,* American Society for Testing and Materials, Philadelphia, 1975, pp. 199-213.
[10] Gell, M. and Leverant, G. R. in *Fatigue at Elevated Temperatures, ASTM STP 520,* American Society for Testing and Materials, Philadelphia, 1973, pp. 37-66.
[11] Whittenberger, D. J., *Metallurgical Transactions,* Vol. 8A, 1977, pp. 1155-1163.
[12] Nazmy, M. Y., *Metallurgical Transactions,* Vol. 14A, 1983, pp. 449-461.
[13] Whittenberger, J. D. and Bizon, P. T., *International Journal of Fatigue,* Oct. 1981, pp. 173-180.
[14] Coffin, L. F., Jr., in *Symposium on Creep-Fatigue Interaction, MPC-3,* R. M. Curran, Ed., American Society of Mechanical Engineers, New York, 1976, pp. 349-364.
[15] "Time-Dependent Fatigue of Structural Alloys," ORNL-5073, Oak Ridge National Laboratory, Oak Ridge, Tenn., 1975, pp. 109-133.
[16] Ostergren, W. J., *ASTM Standardization News,* Vol. 4, No. 10, 1979, p. 327.
[17] Ostergren, W. J. in *Symposium on Creep-Fatigue Interaction, MPC-3,* R. M. Curran, Ed., American Society of Mechanical Engineers, New York, 1976, pp. 179-202.
[18] Manson, S. S. in *Fatigue at Elevated Temperatures, ASTM STP 520,* American Society for Testing and Materials, Philadelphia, 1973, pp. 744-775.

DISCUSSION

N. Marchand[1] *(written discussion)*—(1) The fatigue lives in both directions are very different. Is it safe to use this alloy for airfoil applications, since the airfoil cracking often controls the life of these components?

(2) What can be done to improve the LT properties?

(3) Is it fair to compare cast IN 738 with porosities that act as a microcrack initiation site? Is it better to compare MA 754 to HIP IN 738?

M. Y. Nazmy (author's closure)—(1) The fatigue life as well as the creep properties in the longitudinal direction are the most relevant to vane applications, since the main loading direction is the longitudinal one. In fact, this material is currently used in vane applications in advanced engines.[2]

[1]Department of Materials Science and Engineering, Massachusetts Institute of Technology, Cambridge, MA 02139.

[2]Crawford, W., "Oxide Dispersion Strengthened Materials in Advanced Gas Turbine Engines—INCONEL MA 754 Vanes," in Proceedings of Conference on Frontiers of High Temperature Materials II, J. Benjamin, Ed., 22-25 May 1983.

(2) The HTLCF and creep properties in the LT direction are mainly governed by the texture and the grain morphology of this material. Hence it is not possible to improve the LT properties without affecting the properties in the L direction.

(3) The basis for the comparison between IN 738 and MA 754 is that IN 738 is currently used in vane applications in industrial gas turbines and MA 754 is being adopted as a vane material in advanced military air craft engines. Hence it is interesting from the application side to compare the HTLCF properties of both alloys.

M. Yamauchi,[1] *T. Igari,*[1] *K. Setoguchi,*[1] *and H. Yamanouchi*[2]

Comparison of Creep-Fatigue Life Prediction by Life Fraction Rule and Strainrange Partitioning Methods

REFERENCE: Yamauchi, M., Igari, T., Setoguchi, K., and Yamanouchi, H., **"Comparison of Creep-Fatigue Life Prediction by Life Fraction Rule and Strainrange Partitioning Methods,"** *Low Cycle Fatigue, ASTM STP 942,* H. D. Solomon, G. R. Halford, L. R. Kaisand, and B. N. Leis, Eds., American Society for Testing and Materials, Philadelphia, 1988, pp. 399–413.

ABSTRACT: A comparison of two kinds of creep-fatigue life prediction methods, Life Fraction Rule and Strainrange Partitioning, was performed for the cases of creep-fatigue tests under uniaxial strain-controlled conditions at 500 to 600°C in air, using a normalized and tempered 2¹/₄Cr-1Mo steel. The main focus was on the influence of stress-strain behavior on life prediction. Two kinds of stress-strain values (i.e., experimental values and calculated values by inelastic analysis) were used for the life prediction of creep-fatigue tests. The results show that life prediction by the Life Fraction Rule was very sensitive to the stress-strain values, indicating a large discrepancy between predicted and observed lives in some cases, and that life prediction by the Strainrange Partitioning method was not so sensitive to the stress-strain values, indicating a good agreement between predicted and observed lives in all cases.

KEY WORDS: fatigue (materials), creep-fatigue, life prediction, strainrange partitioning, creep properties, plastic properties, cyclic plasticity, inelastic analysis, strain rate, hold time, low alloy steel

Many methods have been proposed for predicting the life of components under creep-fatigue interaction. The advantages and disadvantages of these various methods have been discussed using experimentally obtained stress-strain behavior from the standpoint of how they could explain quantitatively the effects of factors relating to creep-fatigue life, such as temperature, strain rate, hold time, wave shape, phase, and creep ductility. Recently, a detailed critical review of various methods has been made by Manson [*1*].

Meanwhile, service life is often predicted by inelastic analysis in the creep-fatigue design of high-temperature equipment under severe service conditions. In such cases, the accuracy of the predicted life is affected by the accuracy of inelastic analysis, including the constitutive equation, in addition to the accuracy of the life prediction method itself. Hence, in order to discuss the advantage and disadvantage of some life prediction methods, it is necessary to investigate the accuracy of life prediction when the accuracy of inelastic analysis is combined with that of the prediction method itself.

A comparison of two kinds of creep-fatigue life prediction methods, Life Fraction Rule (LFR) [*2,3*] and Strainrange Partitioning (SRP) [*4–6*], was performed for the cases of creep-fatigue tests under uniaxial strain-controlled conditions at 500 to 600°C in air, using a normalized and tempered 2¹/₄Cr-1Mo steel. This paper describes the results of comparison between observed lives and predicted lives obtained by the two aforementioned life prediction methods and by

[1]Mitsubishi Heavy Industries, Ltd., Nagasaki Research and Development Center, Nagasaki, Japan.
[2]Choryo Engineering Company, Ltd., Nagasaki, Japan.

using experimental stress-strain behavior and calculated stress-strain behavior by inelastic analysis.

Material

The material investigated was a normalized and tempered 2¹/₄Cr-1Mo steel (SA387-Gr22). Test specimens were taken from a 40-mm-thick plate in the rolling direction. Chemical composition and mechanical properties are presented elsewhere [7].

Creep-Fatigue Tests and Life Prediction Using Experimental Stress-Strain Behavior

Testing Procedure and Conditions

Table 1 summarizes the items and conditions of creep-fatigue tests performed in this study. The SRP test was performed to obtain four partitioned strainrange-cyclic life relationships and the strain-hold test was conducted to verify the life prediction methods. In the SRP test, triangular waveshapes of fast-fast (PP), fast-slow (PC), slow-fast (CP), and slow-slow (CC) types were used. In the strain-hold test, two types of test methods were used as shown schematically in Table 1; that is, Type A was perfectly strain-controlled and Type B was not perfectly strain-controlled but the slight increase of total strain occurred in the strain-hold period.

The specimens used in these tests, except for the strain-hold ones of Type B, had a parallel-sided gage length (10 and 12 mm) with a circular cross section (diameter 10 and 12 mm). The tests were performed under constant axial strain control in two servohydraulic machines, with the strain being measured by an extensometer attached to both ends of the gage length. The specimen was heated by induction, and the temperature was measured by a thermocouple spot-welded in the gage length. The methods of the strain-hold test of Type B are presented in Ref *8*.

TABLE 1—*Summary of items and conditions of creep-fatigue tests.*

TEST ITEM			WAVE SHAPE	SCHEMATIC HYSTERESIS LOOP	TEMP. deg C	STRAIN RATE (s^{-1})	HOLD-TIME
SRP TEST	PP				500	2×10^{-3}	
					550	5×10^{-3}	
					600	1×10^{-2}	
	PC		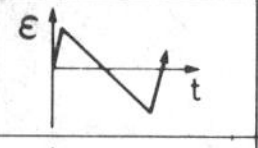	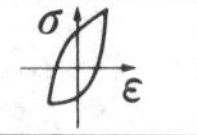	500	MIN 1×10^{-5}	
					550		
					600		
	CP		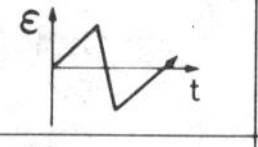	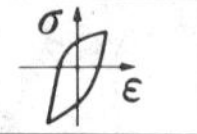	500	MIN 1×10^{-5}	
					550		
					600		
	CC		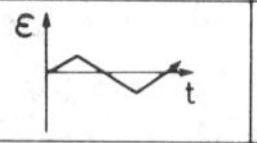	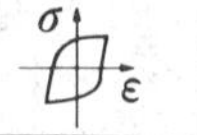	500	MIN 1×10^{-5}	
					550		
					600		
STRAIN-HOLD TEST	TYPE A	TENSILE-HOLD	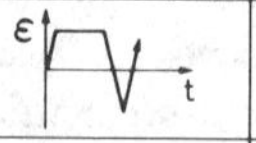	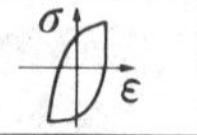	550		MAX 30 min
					600		
		COMPRESSIVE-HOLD	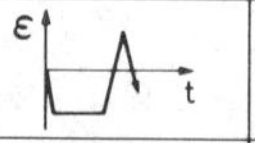	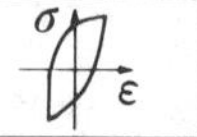	550		MAX 13 min
					600		
	TYPE B	TENSILE-HOLD	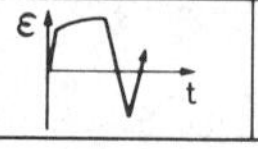	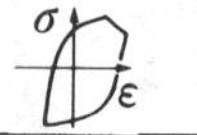	550		MAX 24 h

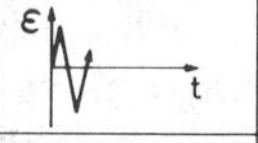

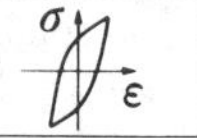

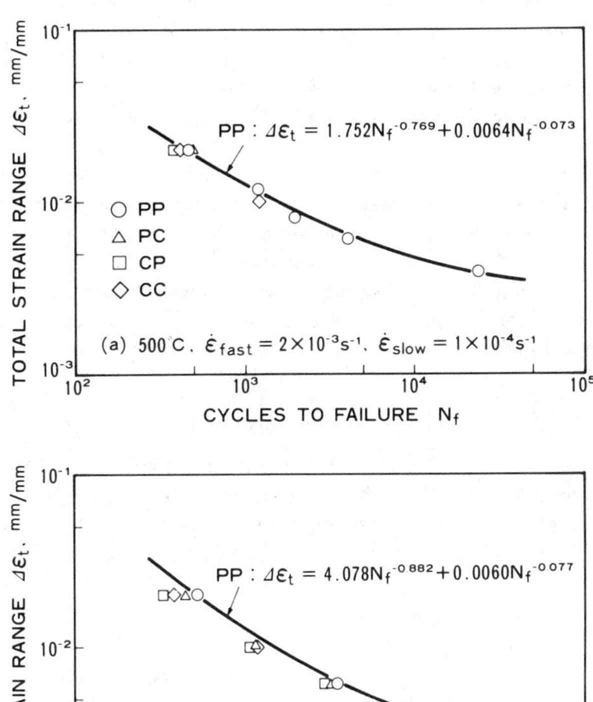

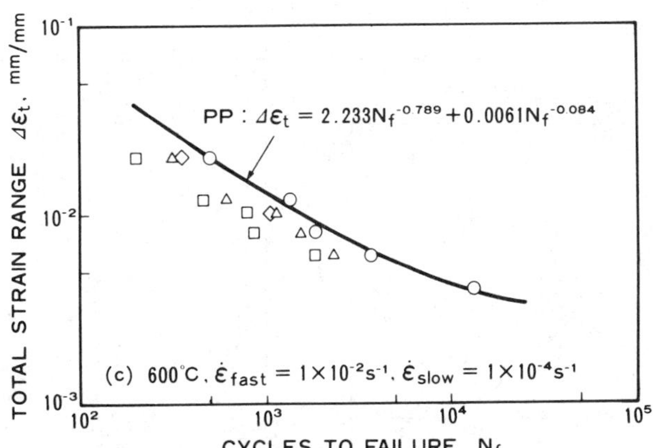

FIG. 1—*Examples of SRP test results.*

Results of Creep-Fatigue Tests

Typical examples of the SRP test results are shown in Fig. 1. The cyclic life for the PP wave shape with a strain rate of $\dot{\epsilon}_{fast}$, in which creep strain was negligible, was the longest of the four waveshapes, and the temperature dependence of the PP-type strain range-life relations was small. The cyclic life for the CP waveshape was the shortest of the four waveshapes and decreased as temperature increased under the condition that the strain rate for slow loading, $\dot{\epsilon}_{slow}$, was equal to 10^{-4} s^{-1}.

Four partitioned strain range-life relationships obtained in this study are shown in Fig. 2. Here, the half-cycle rapid load and unload method [*6*] and the Interaction Damage Rule [*5*] were used to partition strain ranges and to obtain the above relationships, respectively. There was no temperature dependence in any of the relationships (Fig. 2).

The results of the strain-hold test are shown in Fig. 3. It was observed that life reduction due to strain-hold became greater as hold time became longer. Life reduction in compressive-hold was smaller than that in tensile-hold.

Life Prediction Using Experimental Stress-Strain Behavior

Life prediction using experimental stress-strain behavior obtained at the midlife of fatigue was performed by the LFR and SRP methods. In prediction by the LFR method, the results of the PP test (Fig. 1) and creep rupture properties (Fig. 4), obtained from the same heat of material, were used for the calculation of fatigue and creep damage, respectively.

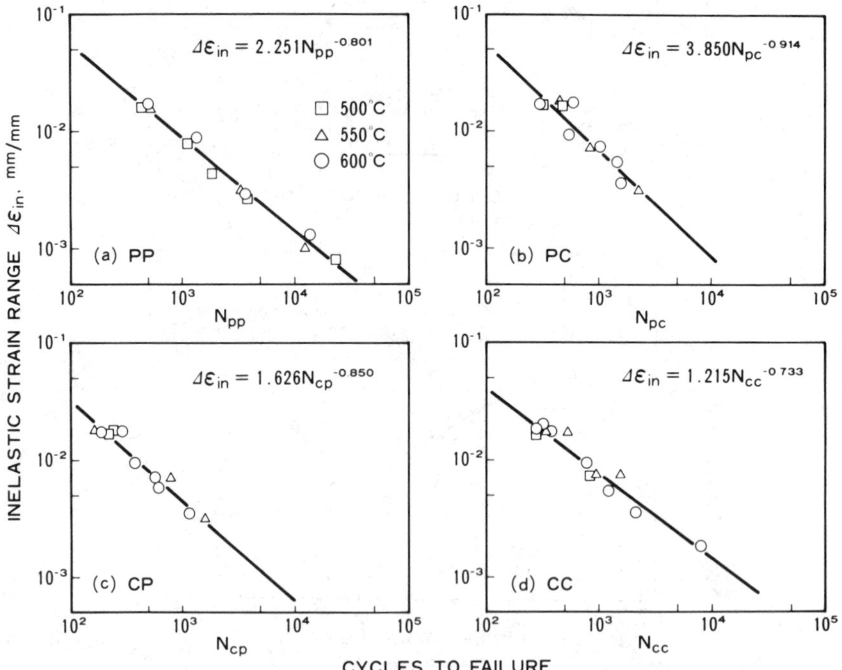

FIG. 2—*Partitioned strain range-life relationships.*

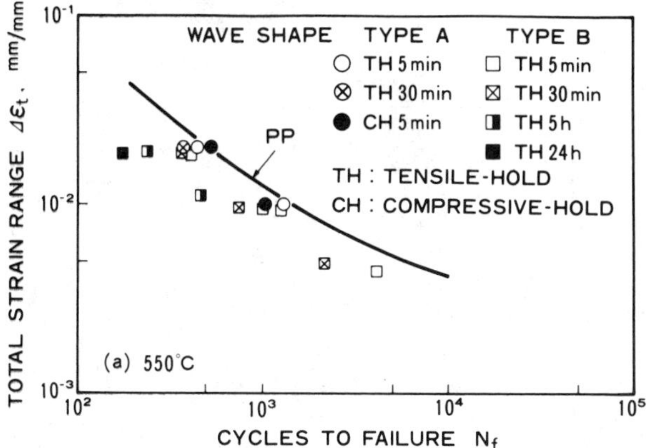

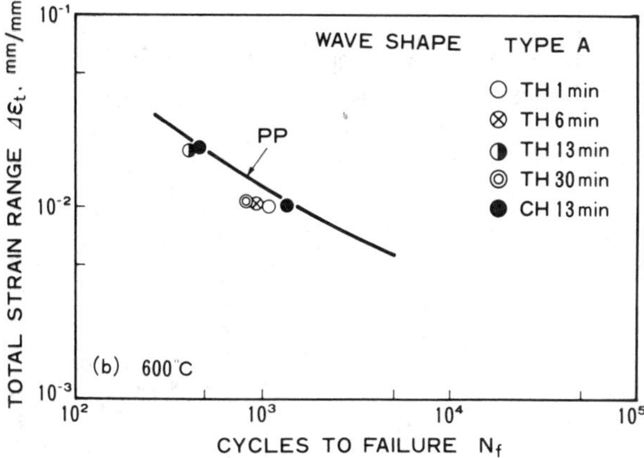

FIG. 3—*Results of strain-hold test.*

First, from the viewpoint of the applicability to the evaluation of the strain-rate effect, life prediction by the above two methods was performed for the SRP test results. In the prediction by LFR, creep damage was summed for the whole part of the hysteresis loop, which is a method often used in a practical design practice. The results are shown in Fig. 5, where predicted life is plotted against observed life. Using the LFR, predicted lives as much as 10 times lower than experimental lives were observed (Fig. 5*a*). Predicted lives tended to decrease as temperature increased and strain rate decreased. Using the SRP method, the predictions almost agreed with the experiments within factors of 1.5 (Fig. 5*b*).

Secondly, from the viewpoint of the applicability to the evaluation of hold-time effect, life prediction was made for the strain-hold test. The results are shown in Fig. 6. In the prediction by the LFR, creep damage during the strain-hold period was considered. In the prediction by the LFR, the majority of the predicted lives agreed with the observed lives within factors of 1.5, though two data points showed a somewhat greater discrepancy (Fig. 6*a*). Using the SRP method, the predicted lives agreed with the observed lives within factors of 1.5 (Fig. 6*b*).

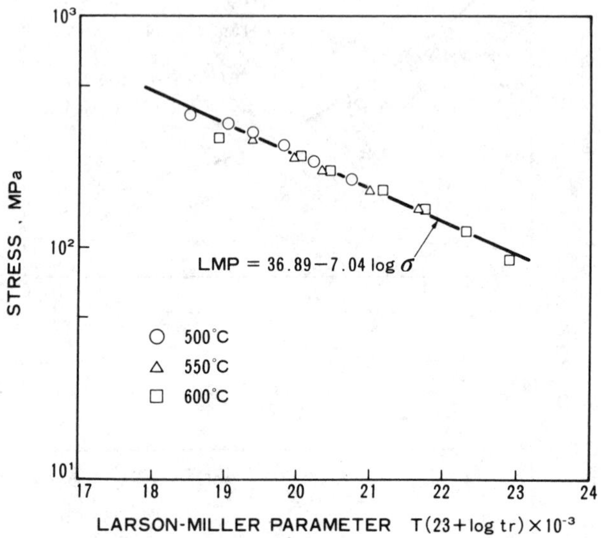

FIG. 4—*Master curve of creep rupture strength.*

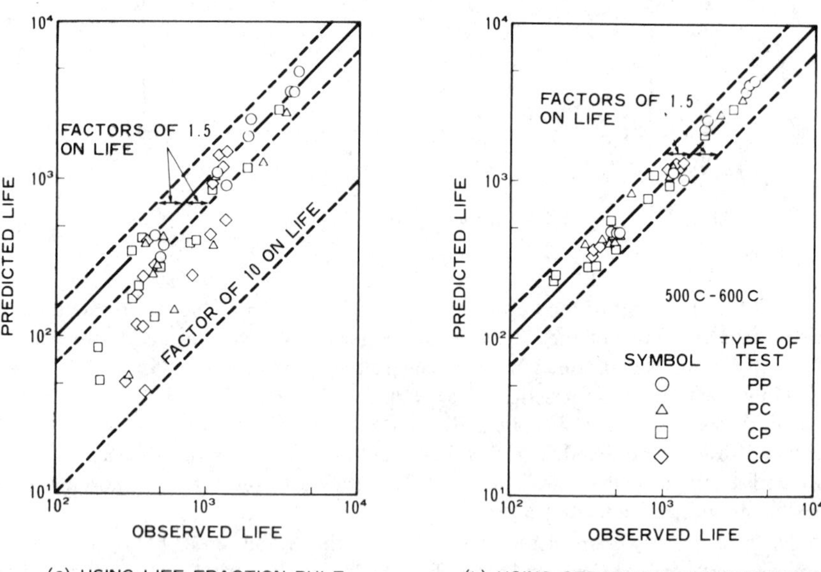

FIG. 5—*Life prediction for strain-rate effect based on experimental stress-strain behavior.*

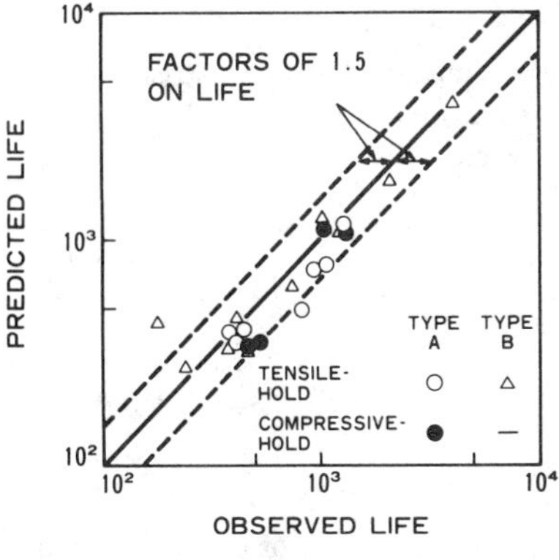

(a) USING LIFE FRACTION RULE

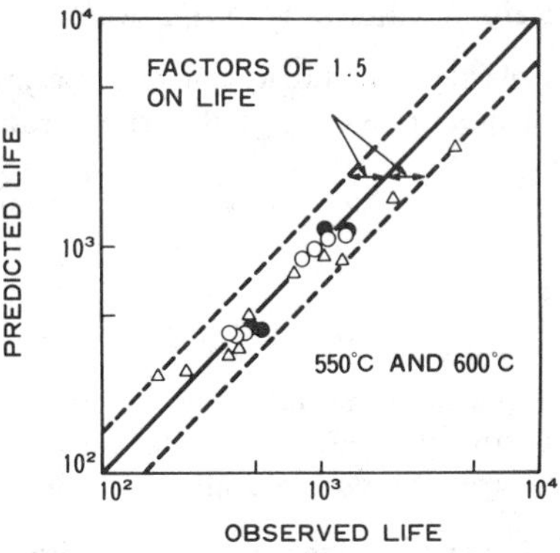

(b) USING STRAINRANGE PARTITIONING METHOD

FIG. 6—*Life prediction for hold-time effect based on experimental stress-strain behavior.*

Inelastic Analysis and Life Prediction Using Analytical Stress-Strain Behavior

Analytical Procedure

The simulation of stress-strain behavior under the test conditions described above was made by inelastic analysis. The constitutive equation used was the type separating plastic strain and creep strain as shown in the equation

$$d\epsilon_{ij} = d\epsilon_{ij}^{e} + d\epsilon_{ij}^{p} + d\epsilon_{ij}^{c} \tag{1}$$

where

$d\epsilon_{ij}$ = total strain increment,
$d\epsilon_{ij}^{e}$ = elastic strain increment,
$d\epsilon_{ij}^{p}$ = plastic strain increment, and
$d\epsilon_{ij}^{c}$ = creep strain increment.

For the plastic strain increment $d\epsilon_{ij}^{p}$, the initial yield condition of von Mises, the associated flow rule, and the kinematic hardening rule by Ziegler were used. For the plastic deformation behavior at stress reversal, the Masing model [*9,10*] was used. For the creep strain increment $d\epsilon_{ij}^{c}$, the auxiliary rule for strain hardening rule by ORNL was used. The concrete expression of the equation is presented elsewhere [*11*].

When Eq 1 is used for the inelastic analysis, the analytical method for strain-rate effect must be clarified. The strain-rate effect was expressed by making the elastic-plastic calculation and subsequent creep calculation in one load increment. First, the stress increment was predicted by the elastic-plastic calculation. Then, stress relaxation due to creep was considered, and the stress increment in this increment was finally determined. Details of the method described above and the applicability of this method are presented elsewhere [*12*].

Material Properties Used in Analysis

The plastic properties used in inelastic analysis were determined from the stress-strain behavior at a strain rate of $\dot{\epsilon}_{fast}$ and were expressed in the equation

$$\sigma = k\epsilon_p^n \tag{2}$$

where

σ = stress, MPa,
ϵ_p = plastic strain, mm/mm,
k = strength coefficient, MPa, and
n = strain-hardening exponent.

Yield stress was defined as the stress with an offset plastic strain, ϵ_p, of 0.00002 mm/mm.

The creep properties were determined from the results of creep tests by the expression

$$\epsilon_c = a\sigma^b t^c + d\sigma^e t \tag{3}$$

where

ϵ_c = creep strain, mm/mm,
σ = stress, MPa,
t = time, h, and
a, b, c, d, and e = constants.

Three sets of material properties (i.e., Case 1, Case 2, and Case 3 described below) were prepared to obtain three kinds of stress-strain behavior.

In Case 1, the plastic properties were obtained from the results of monotonic tensile tests at a strain rate of $\dot{\epsilon}_{fast}$, and the creep properties were obtained from monotonic creep tests. In Case 2,

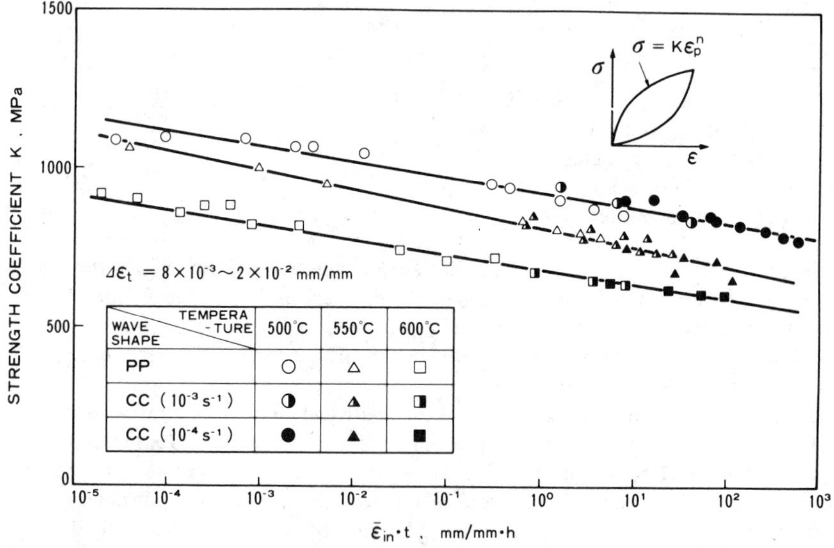

FIG. 7—*Relations of strength coefficient* K *versus* $\bar{\epsilon}_{in} \times$ t.

the plastic properties were obtained from the cyclic stress-strain curve at the midlife of PP-type fatigue, and the creep properties were the same as the ones in Case 1. In this case, the cyclic stress-strain curve was estimated from the shape of the hysteresis loop [*10*].

In Case 3, the strength coefficient k in Eq 2 was defined as a function of $\bar{\epsilon}_{in} \times t$ with the strain-hardening exponent being constant, where $\bar{\epsilon}_{in}$ is the accumulated equivalent inelastic strain and t is the time. The relations of K versus $\bar{\epsilon}_{in} \times t$ (where $K = 2^{1-n} \times k$) are shown in Fig. 7 and were obtained from PP and CC test results. Here, the values of K for the CC test results were obtained from the hysteresis loops for rapid-loading cycles introduced into CC cycling at a certain interval. The minimum creep rate was determined from the results of the creep test after PP-type plastic cycling to midlife and is shown in Fig. 8 with the results of the monotonic creep test.

Figure 7 is thought to be a master curve to predict the softened plastic property under strain cycling and thermal exposure. Figure 8 is thought to represent the softening in creep property by plastic cycling. These kinds of acceleration of softening in plastic and creep properties by both strain and thermal history are thought to be one feature of "plasticity-creep interaction".

Case 1 and Case 2 correspond to the method popularly used in design practice for high-temperature components. Case 3 was performed as one attempt to improve the prediction accuracy of stress-strain behavior. Plastic and creep constants for these three cases are listed in Tables 2 and 3, respectively.

Results of Inelastic Analysis

Typical examples of the comparison between the calculated hysteresis loop and the experimental results at midlife are shown in Fig. 9 for the strain-hold test at 600°C. In the analysis of Case 1, a large difference between the calculated and the experimental stress-strain behavior was observed and a higher stress was estimated (Fig. 9*a*). In the analysis of Case 2, in which softening by plastic cycling was considered in the plastic properties, the difference between the calculated results and the experimental ones was smaller than that in Case 1, but a fairly large

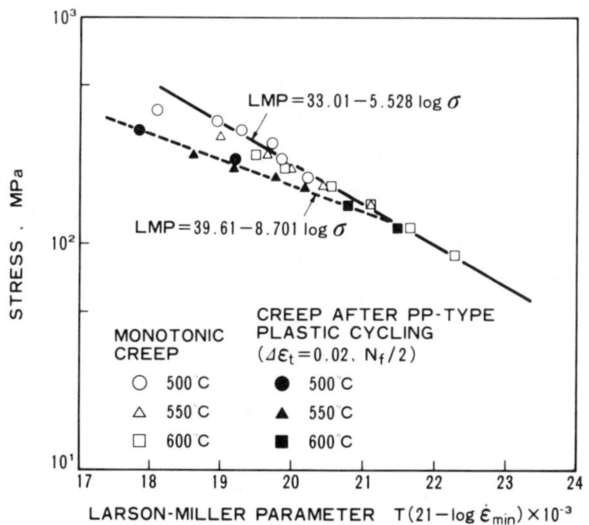

FIG. 8—*Master curve of minimum creep rate.*

TABLE 2—*Plastic properties used in inelastic analysis.*

TEMPERATURE deg C	k, MPa			n
	CASE 1	CASE 2	CASE 3[a]	CASE 1, 2, 3
500	534.5	456.7	$487.0-24.3 \log(\bar{\varepsilon}_{in} \cdot t)$	0.0635
550	520.3	419.3	$428.9-30.9 \log(\bar{\varepsilon}_{in} \cdot t)$	0.0656
600	453.1	367.7	$357.1-23.3 \log(\bar{\varepsilon}_{in} \cdot t)$	0.0575

[a] $\bar{\varepsilon}_{in}$: Accumulated Equivalent Inelastic Strain in mm/mm

t : Time in hour

TABLE 3—*Creep properties used in inelastic analysis.*

TEMP. deg C	CASE	CONSTANTS IN EQUATION 3						
		a	b	c	d		e	
					$\sigma \geq 120$ MPa	$\sigma < 120$ MPa	$\sigma \geq 120$ MPa	$\sigma < 120$ MPa
500	1				2.02×10^{-22}		7.15	
	2	8.73×10^{-16}	4.93	0.363				
	3				5.86×10^{-31}	2.02×10^{-22}	11.25	7.15
550	1				7.84×10^{-20}		6.72	
	2	2.91×10^{-14}	4.63	0.432				
	3				7.59×10^{-28}	7.84×10^{-20}	10.57	6.72
600	1				1.57×10^{-17}		6.33	
	2	6.55×10^{-13}	4.36	0.502				
	3				4.32×10^{-25}	1.57×10^{-17}	9.97	6.33

discrepancy was observed (Fig. 9*b*). In the analysis of Case 3, in which the accelerated softening due to cyclic inelastic strain and thermal exposure and the accelerated creep rate due to plastic cycling were taken into consideration, good agreement was observed between the calculated and the experimental hysteresis loop (Fig. 9*c*).

The comparison between the calculated relaxation curves and the experimental results in the strain-hold period is shown in Fig. 10 for the same testing conditions shown in Fig. 9. The calculated result in Case 3 showed a slightly better agreement with the experimental results than with the other two cases, but the calculated remaining stress was a little higher than the experimental values, even in this case. This might have come from the lack of considering accelerated behavior in transient creep and the effect of thermal exposure on creep behavior.

Life Prediction Using Analytical Stress-Strain Behavior

Life prediction using stress-strain behavior obtained by inelastic analysis was performed using the LFR and SRP methods.

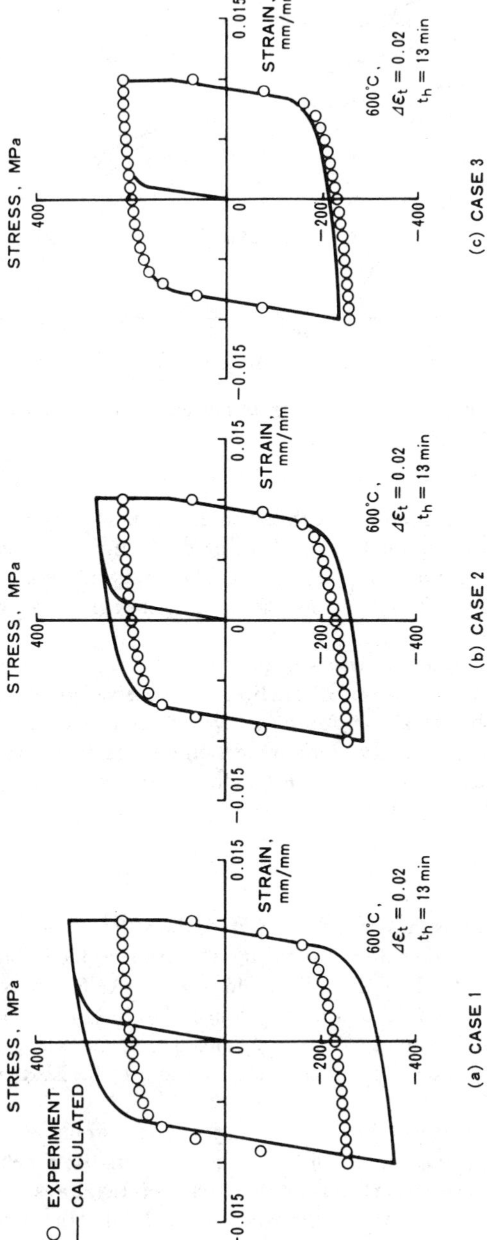

FIG. 9—*Comparison of calculated hysteresis loop with experimental result.*

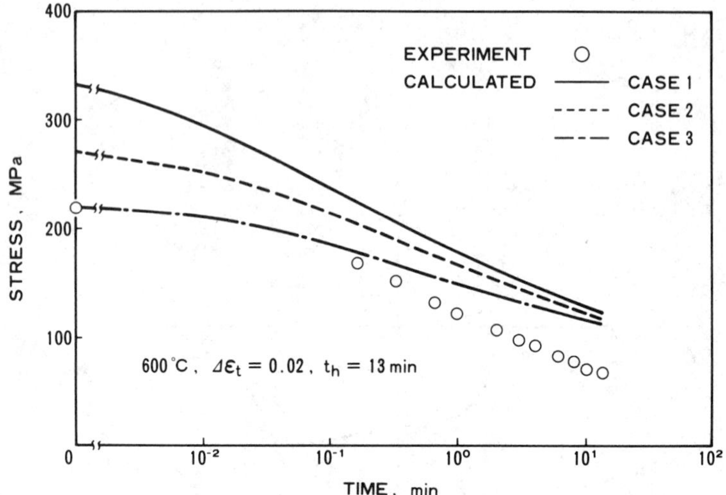

FIG. 10—*Comparison of calculated relaxation curve with experimental results.*

The comparison of the predicted lives with the observed ones for typical cases of PP and CC tests is shown in Fig. 11. Using the LFR method, predicted lives as much as 100 times lower than experimental ones were observed (Fig. 11*a*). The predicted lives tended to decrease as strain rate decreased. Using the SRP method, the predictions agreed well with the experiments within factors of 1.5 (Fig. 11*b*).

The comparison of the predicted lives with the observed ones for strain-hold test Type A is shown in Fig. 12. Using the LFR, the predicted lives were observed to be as much as 15, 7, and 3 times lower than the experimental ones for the analytical results of Case 1, Case 2, and Case 3, respectively (Fig. 12*a*). The accuracy of life prediction was greatly improved by conducting a more accurate inelastic analysis. Using the SRP method, the predictions agreed with the experiments within factors of 1.5 (Fig. 12*b*).

Discussion

When experimental stress-strain behavior was used, the SRP method gave good results in the evaluation of both the strain-rate effect and the hold-time effect; the LFR method gave similar results in the evaluation of the hold-time effect. However, the LFR gave conservative results in the evaluation of strain-rate effect, and predicted lives tended to be more conservative as the temperature increased and strain rate decreased. This discrepancy seems to be due to the reason, as pointed out by Manson [*1*], that the predicted results by the LFR included the effects of creep twice.

When life prediction is performed by using analytical stress-strain behavior, the accuracy of life prediction depends on the accuracy of the estimated stress-strain behavior. In this study, three types of analytical stress-strain behavior were prepared. In the analyses of Case 1 and Case 2, the estimated stresses were higher than the experimental results. In the analysis of Case 3, the estimated stresses almost agreed with the experimental results. When life prediction by the LFR was performed using the analytical stress-strain behavior described above, the predicted lives became more conservative as the estimated stresses became higher. This is because the calculation of creep damage in the LFR is stress-based and the damage is sensitive to stress. To im-

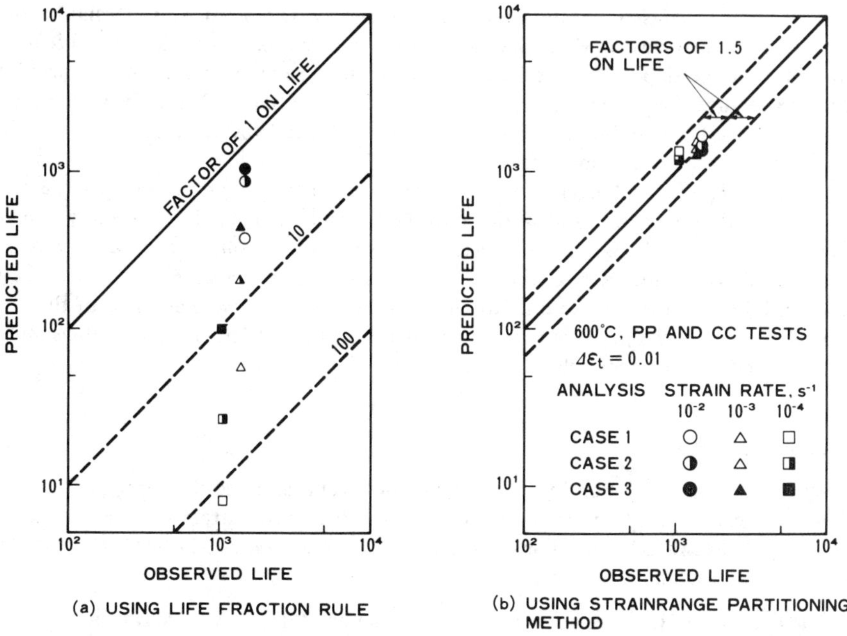

FIG. 11—*Life prediction for strain-rate effect based on analytical stress-strain behavior.*

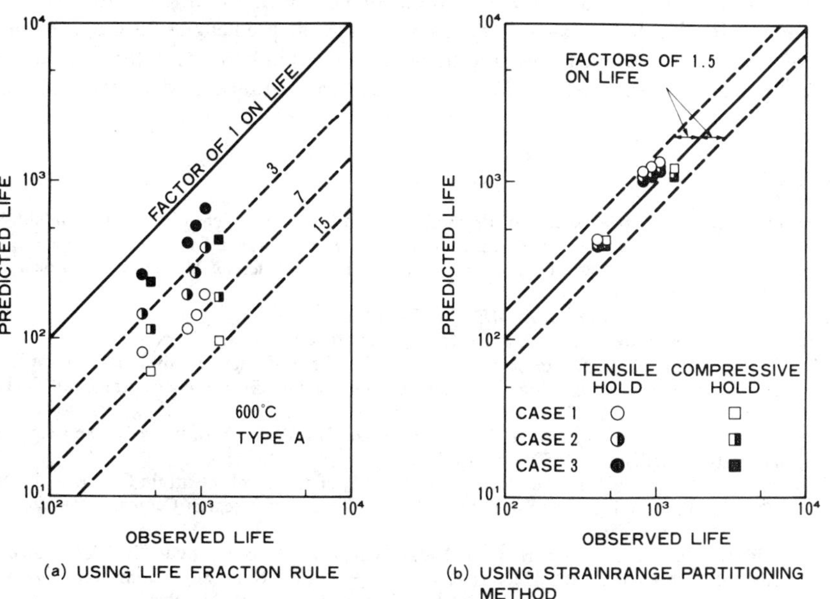

FIG. 12—*Life prediction for hold-time effect based on analytical stress-strain behavior.*

prove the accuracy of life prediction by the LFR, it is necessary to improve the accuracy of the inelastic analysis by considering the effects of softening by both strain and thermal history on the material properties. Meanwhile, the SRP method is strain-based and the strain-life relation diagram has a large slope. As a result, the predicted lives were insensitive to the variation of strain. This feature, together with the fact that the test condition in this study was strain-controlled, was thought to be the main reason why the predicted lives had a good accuracy in all cases of inelastic analysis, as mentioned above.

This study was performed in a limited condition, where a closed hysteresis loop was available in a uniformly distributed uniaxial strain-controlled condition. Thus, it is necessary to investigate other conditions. Those include: a case of stress controlled condition, a case where stress and strain are distributed, a case where a hysteresis loop is not completely closed, a case of multiaxial state of stress, and so on. However, in the case of a thermal stress of the strain-controlled type, or in the problem of notches when a strain restriction is strict, the same tendency described in this paper will be expected.

Concluding Remarks

Creep-fatigue life prediction by the Life Fraction Rule and the Strainrange Partitioning method was performed using experimental and analytical stress-strain behavior for uniaxial strain-controlled conditions of a normalized and tempered 2¹/₄Cr-1Mo steel. The results are summarized as follows:

1. The accuracy of life prediction by the Life Fraction Rule was very sensitive to the stress-strain behavior used in the prediction, and a large discrepancy was observed between the predicted lives and the observed lives in some cases. To improve the accuracy of life prediction using an inelastic analysis, it was necessary to consider the effects of softening by both strain and thermal history on material properties.
2. On the other hand, the accuracy of life prediction by the Strainrange Partitioning method was not so sensitive to the stress-strain behavior used in the prediction. Good agreement was observed between the predicted lives and the observed lives in both cases where experimental stress-strain behavior were used and where analytical stress-strain behavior were used.

References

[1] Manson, S. S., "A Critical Review of Predictive Methods for Treatment of Time-Dependent Metal Fatigue at High Temperatures," in *Pressure Vessels and Piping Design Technology—1982, A Decade of Progress,* S. Y. Zamrik and D. Dietrich, Eds., American Society of Mechanical Engineers, New York, 1982, pp. 203–225.
[2] Robinson, E. L., *Transactions of ASME,* Vol. 74, 1952.
[3] Taira, S. in *Creep in Structures,* N. J. Hoff, Ed., Academic Press, New York, 1962.
[4] Manson, S. S., Halford, G. R., and Hirschberg, M. H., "Creep-Fatigue Analysis by Strainrange Partitioning," in *Symposium on Design for Elevated Temperature Environment,* American Society of Mechanical Engineers, New York, 1971, pp. 12–28.
[5] Manson, S. S. in *Fatigue at Elevated Temperatures, ASTM STP 520,* American Society for Testing and Materials, Philadelphia, 1973, pp. 744–775.
[6] Manson, S. S., Halford, G. R., and Nachtigall, A. J., "Separation of the Strain Components for Use in Strainrange Partitioning," NASA TMX-71737, NASA-Lewis Research Center, Cleveland, Ohio, 1975.
[7] Setoguchi, K., Yamauchi, M., Igari, T., and Wakamatsu, Y., "Creep-Fatigue Life Prediction of Normalized and Tempered 2¹/₄Cr-1Mo Steel by Life Fraction Rule and Strainrange Partitioning Method," *Transactions of the Iron and Steel Institute of Japan,* Vol. 24, 1984, pp. 1063–1071.
[8] Endo, T. and Sakon, T., "Creep-Fatigue Life Prediction Using Simple High Temperature Low-Cycle Fatigue Testing Machines," *Metals Technology,* Vol. 11, 1984, pp. 489–496.
[9] Jhansale, H. R., *Journal of Engineering Materials and Technology,* Vol. 97, 1975, pp. 33–38.

[*10*] Igari, T., Setoguchi, K., and Yamauchi, M., "Study on Elastic-Plastic Deformation Analysis Using a Cyclic Stress-Strain Curve," *Journal of the Society of Materials Science*, Japan, Vol. 32, 1983, pp. 610–614.

[*11*] Setoguchi, K. and Igari, T., "Inelastic Analysis with a Generalized Plane Strain Model," *Mitsubishi Heavy Industries Technical Review*, Mitsubishi Heavy Industries, Ltd., Tokyo Japan, Vol. 14, 1977, pp. 914–921.

[*12*] Setoguchi, K., Igari, T., and Yamauchi, M., "Creep-Fatigue Life Prediction of 2¼Cr-1Mo Steel Based on Inelastic Analysis," *Journal of the Society of Materials Science*, Japan, Vol. 33, 1984, pp. 862–868.

C. Levaillant,[1] J. Grattier,[2] M. Mottot,[3] and A. Pineau[1]

Creep and Creep-Fatigue Intergranular Damage in Austenitic Stainless Steels: Discussion of the Creep-Dominated Regime

REFERENCE: Levaillant, C., Grattier, J., Mottot, M., and Pineau, A., **"Creep and Creep-Fatigue Intergranular Damage in Austenitic Stainless Steels: Discussion of the Creep-Dominated Regime,"** *Low Cycle Fatigue, ASTM STP 942,* H. D. Solomon, G. R. Halford, L. R. Kaisand, and B. N. Leis, Eds., American Society for Testing and Materials, Philadelphia, 1988, pp. 414-437.

ABSTRACT: Creep and strain controlled low-cycle fatigue tests with an imposed hold time at maximum tensile strain were carried out, mainly at 600°C, on two heats of 316L stainless steel. Very long times to failure ($\simeq 10^4$ h) corresponding to low applied strain and long dwell period (24 h) were investigated. Both fatigue and creep damage were measured by quantitative metallography.

Creep intergranular surface cracking shortens the initiation stage of fatigue cracks, while creep intergranular bulk cracking accelerates the average fatigue crack growth rate. These observations are the basis for a creep-fatigue interaction model previously proposed using intergranular damage as a life correlating parameter. The applicability of this model is tested with results of 10^4 h creep-fatigue tests showing a saturation effect. The saturation in creep-fatigue life observed for long dwell periods is explained by an improvement in creep ductility.

This intergranular damage approach is discussed in the light of the creep dominated regime concept. It is shown that pure creep failure data cannot provide reliable prediction of creep fatigue life. The need to use a model that takes into account the effect of creep intergranular damage on fatigue crack propagation is emphasized.

KEY WORDS: creep-fatigue, creep, austenitic stainless steels, intergranular cracking, damage measurements, quantitative metallography, long dwell time

Many studies have been devoted to the effect of tensile hold times on low-cycle fatigue (LCF) endurance of Type 304 and 316 stainless steels (see, e.g., [*1-4*]). As a general rule, most of these experiments were performed at strain ranges ($\simeq$0.5 to 2%) and temperatures ($\simeq$600°C) higher than those corresponding to in-service conditions. Moreover, the investigated hold-time durations, t_h, (< a few hours) were also much shorter than those anticipated in service ($\simeq 10^2$ to 10^3 h). Presently there is no generally accepted method of extrapolating laboratory short-term data to service conditions. This is a major area of concern.

In particular, it is not always known whether the fatigue life decreases continuously with increasing tensile hold times. Recent research efforts are oriented towards the investigation of possible saturation effects for very long hold times [*5,6*]. The first goal of the present study was to extend previous LCF results obtained on two heats of stainless steels with t_h < 4 h [*7*] to

[1]Centre des Matériaux - Ecole des Mines, B.P.87 - 91003 Evry Cedex, France.
[2]Centre de recherches E.D.F. Les Renardières, B.P.1, 77250 Moret Sur Loing, France.
[3]Commissariat Energie Atomique, SRMA, 91191 Gif Sur Yvette, France.

much longer hold times (24 h). A number of tests were performed by French laboratories through a cooperative research program [*8*].

The occurrence of a saturation effect in LCF life with increasing hold times is discussed in light of a model developed by two of the authors [*7,9*]. This model is essentially based upon intergranular damage as a correlating life parameter. Intergranular damage is assumed to accumulate during tensile hold times and to accelerate the fatigue crack growth rate. The approach, which has been developed in more detail elsewhere [*7,10*], involves quantitative metallographic measurements to assess the kinetics of microcrack nucleation and propagation. Both surface transgranular cracking associated with continuous fatigue and intergranular cracking observed under pure creep and creep-fatigue conditions are examined.

In the present paper, results of more extensive metallographic studies are also reported. In particular, the density of secondary cracks (i.e., the cracks present in the specimen but not responsible for the final failure) was determined. From these measurements, the crack growth rates were assessed and compared with results of striation spacing observations. The evaluation of fatigue and creep damage by using quantitative metallography is the second goal of this paper.

Finally, a discussion of the saturation effect observed in creep-fatigue is presented in light of recent creep ductility measurements on smooth and notched specimens. Qualitatively, an analogy between both types of tests can be drawn. However, a number of limitations of the creep-dominated regime concept [*6,11*], applied to long-hold-time LCF tests, are emphasized.

Experimental Procedure

Two 316L steels were examined. Their compositions and tensile properties are reported in Tables 1 and 2, respectively. VIRGO is essentially a low-carbon 316 steel, while ICL is a high creep-resistant 316 steel with a controlled nitrogen content. ICL steel is used for internals and vessels of the Superphenix nuclear reactor. The creep properties determined on smooth bars of both heats have been extensively investigated [*12,13*]. ICL steel exhibits a better creep resistance than VIRGO steel.

All specimens were cut in the longitudinal direction of 15 mm and 32 mm thick rolled plates for VIRGO steel and for ICL steel, respectively. Both steels were tested in the as-received condition—annealing treatment at 1110°C and water quench for VIRGO steel and a double annealing treatment at 1070°C and water quench for ICL steel. In both cases, the grain size was about 50 μm, although ICL steel contained a few larger grains ($\simeq$200 μm).

TABLE 1—*Chemical composition (weight percent).*

	C	S	Si	Mn	Ni	Cr	Mo	N	B	Co	Cu	P
VIRGO (316L)	0.033	0.022	0.44	1.55	13.6	16.4	2.12	0.025	0.0012	0.18	0.07	0.022
ICL	0.021	0.007	0.41	1.74	12.3	17.2	2.40	0.080	0.0032	0.21	0.15	0.030

TABLE 2—*Tensile properties at 600°C (1112°F).*

	E, 10^3 MPa	0.2% Yield Stress, MPa	UTS, MPa	Elongation, %	Reduction of Area, %
VIRGO (316L)	144	116	377	49	74
ICL	144	159	404	41	67

Low-Cycle Fatigue Tests

Strain-controlled LCF tests were carried out on both materials using cylindrical specimens with a gage length of 12 mm and a diameter of 8 mm. Most of these tests were run at 600°C. Continuous LCF experiments were performed with a triangular fully reversed waveshape signal ($\dot{\epsilon}_t = 4 \times 10^{-3}\ s^{-1}$). Creep-fatigue tests were carried out by imposing a tensile hold time at the maximum tensile cyclic strain. The details of this procedure are given elsewhere [*7,10*].

Creep Tests

Creep tests on smooth and notched bars were performed at 600°C on ICL steel. The details of these procedures are given elsewhere [*10,14,15*]. Three notched geometries were employed, with notch radii of 5 mm (Specimen FLE5), 1 mm (FLE1), and 0.1 mm (FLE0.1). An extensometer was attached to the specimen to measure the mean diametral strain at failure as

$$\bar{\epsilon}_{df} = 2\ \ell n(\phi_0/\phi_f) \qquad (1)$$

where ϕ_0 is the initial diameter of the minimum section, and ϕ_f is the minimum diameter at failure.

Metallography and Fractography

Optical microscopy observations were performed to measure the density of secondary cracks and the percentage of cracked grain boundaries both in creep specimens and in creep-fatigue specimens. The bulk intergranular damage was determined as before [*16*] by measuring the ratio between the length of damaged grain boundaries per unit of area on plane section parallel to the specimen axis and the average length of grain boundaries per unit area. Scanning electron microscopy (SEM) was used extensively to examine the fracture surfaces and to measure fatigue striation spacings when they were present.

Results

Mechanical Tests

The results of LCF tests corresponding to VIRGO and ICL steels are given in Figs. 1 and 2, respectively. The data points are plotted in terms of $\Delta\epsilon_p$ in Fig. 1 or $\Delta\epsilon_t$ in Fig. 2 to be consistent with the experimental procedures; VIRGO steel was tested under plastic strain control, while ICL steel was tested under total strain control. In Figs. 1 and 2 we have included previous results of testing on VIRGO steel [*17*] and on ICL steel [*18*]. As a general rule the second material exhibits a better resistance to the effect of tensile hold time. This reflects its better creep resistance. In Table 3 we have only reported the results for specimens that were analyzed by quantitative metallography. In this table, σ_{tmax} is the maximum tensile stress, while σ_R is the amount of stress relaxation calculated as $\sigma_{tmax} - \sigma_{tmin}$ where σ_{tmin} is the tensile stress at the end of hold time. These values are given at midlife. Three behaviors can be distinguished for results obtained for the same total strain range ($\Delta\epsilon_t \simeq 1.20\%$) and the largest time to failure:

1. For the shortest hold times the fatigue life is only slightly affected by imposed hold times. This is especially marked in the case of ICL steel, which is not sensitive to hold times less than 10 min long as shown by the comparison of 18 tests run in four different laboratories [*8,17,18,26*].

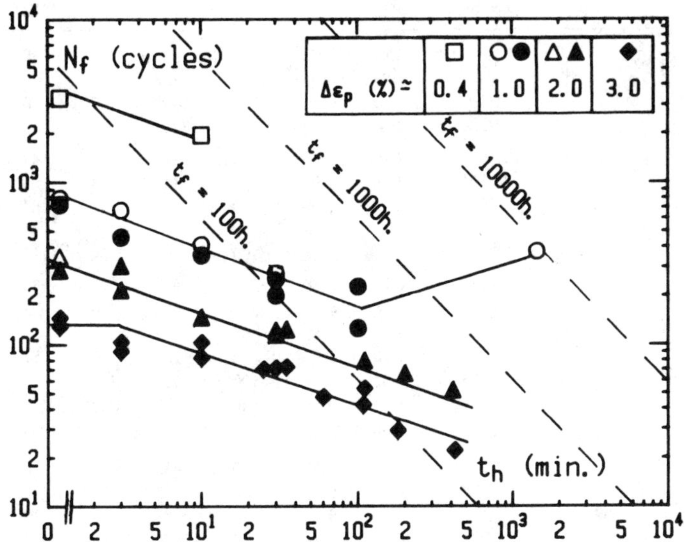

FIG. 1—*Influence of hold time on number of cycles to failure. LCF tests on VIRGO steel at 600°C. Dark symbols = data from Rezgui* [17].

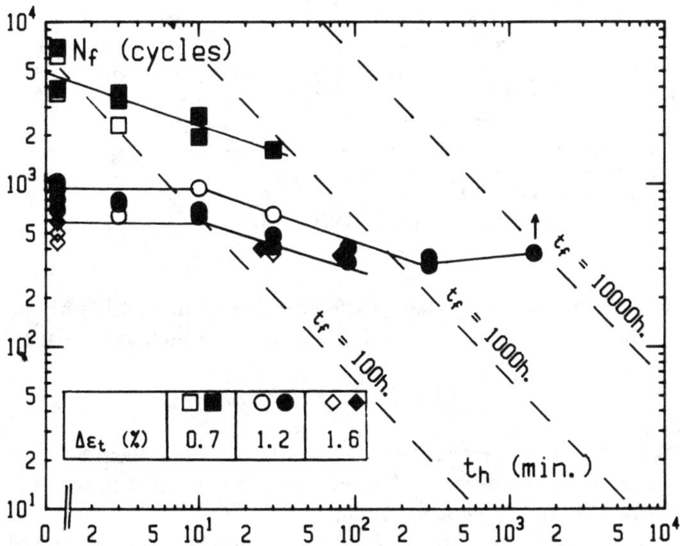

FIG. 2—*Influence of hold time on number of cycles to failure. LCF tests on ICL steel at 600°C. Dark symbols = data from Groupe de Travail Matériaux* [18].

TABLE 3—*LCF results at 550°C (1022°F) and 600°C (1112°F).*

Material	Temperature, °C	Specimen	$\Delta\epsilon_t$, %	$\Delta\epsilon_p$, %	t_h, min	N_f, cycles	σ_{tmax}, MPa*	σ_R, MPa*	D_c (10^{-4})
VIRGO	600	V54	2.40	1.96	0	348	376	0	0
		V30	1.38	0.98	0	828	306	0	0
		V19	0.90	0.60	0	1631	287	0	0
		V20	0.64	0.40	0	3288	245	0	0
		R60**	2.47	1.96	3	216	370	49	5.18
		R25**	2.46	1.96	10	146	358	69	8.08
		V55	2.45	1.96	30	122	350	84	4.60
		R28**	2.45	1.96	30	114	350	84	9.38
		R23**	2.31	1.76	200	66	324	113	24.5
		V21	1.26	0.88	3	670	276	33	1.35
		V22	1.26	0.88	3	670	276	33	1.98
		V23	1.25	0.88	30	273	264	59	5.02
		R78**	1.40	1.04	30	251	259	57	4.74
		R90**	1.51	1.14	100	125	268	73	1
		V68	1.04	0.88	1440	373	212	98	2.78
		V39	0.68	0.40	10	1941	224	28	0.239
ICL	550	SP3	1.60	1.06	0	438	404	0	0
		SP20	1.20	0.76	0	1371	327	0	0
		SP19	0.68	0.36	0	6287	236	0	0
		SP2	0.60	0.26	0	13437	246	0	0
	600	SP15	1.60	1.10	0	498	363		0
		SP6	1.20	0.72	0	704	341	0	0
		SP7	0.70	0.32	0	3624	269	0	0
		SP17	1.60	1.18	30	375	334	64	3.00
		SP8	1.20	0.78	3	634	329	33	0.31
		SP11	1.20	0.78	10	944	312	35	0.75
		SP10	1.20	0.82	30	651	294	39	1.13
		188***	1.18	0.81	90	341	310	60	1.15
		SPJ1***	1.20	0.83	300	353	311	87	2.23
		SPR5***	1.20	0.88	1440	396	274	109	...
		SP9	0.70	0.34	3	2308	271	20	0.07

*Values at mid-life.
**Specimens tested by Rezgui [*17*].
***Specimens tested by Groupe de Travail Matériaux [*18*].

2. In an intermediate domain, the fatigue life decreases continuously with hold time with almost the same slope on a $N_f - t_h$ bilogarithmic diagram for both materials:

$$N_f^{CF} = N_f^F(t_h/t_{hc})^{-0.33} \quad \text{for} \quad t_h > t_{hc} \tag{2}$$

where N_f^{CF} is the fatigue life in creep fatigue, N_f^F is the fatigue life without hold time, and t_{hc} is the critical hold time beyond which fatigue life decreases. In VIRGO steel, $t_{hc} \simeq 1$ min, while in ICL steel the critical hold time is close to 10 min.

In this domain, total times to failure, t_f, still increase with hold time as derived from Eq 2:

$$t_f \simeq N_f^{CF} \times t_h = N_f^F(t_{hc})^{0.33}(t_h)^{0.67} \tag{3}$$

Isochronous rupture curves are shown in Figs. 1 and 2 (dotted lines).

3. For large hold times (24 h) and very long tests ($\simeq 10^4$ h), there is a tendency indicating that the fatigue life increases slightly again in both materials. This statement is supported by metallographical observations, as shown later. Furthermore, this conclusion is supported by the fact that the scatterband obtained from tests run in four laboratories [*8,17,18,26*] is lower than a factor of 2. The domain of saturated or improved fatigue life is not observed for other strain range conditions. Investigated rupture times did not exceed 1000 h in those cases.

The existence of such a saturation in the detrimental effect of hold times on fatigue life seems to become more and more obvious, as confirmed by still running long duration tests at lower strain ranges. These results are notable because there are few experimental examples of this phenomenon at temperatures below or equal to 600°C on 316 type steels under annealed conditions. The same behavior was observed by Wareing [*5*] in extending the results of Brinkman et al. [*1*] at high strain levels ($\Delta\epsilon_t = 2\%$). Conversely Hales [*6*] did not report any saturation effect for up to 6000 h to rupture at low strain range ($\Delta\epsilon_t = 0.5\%$). Saturation effects are more easily observed at higher temperatures (at 625°C, see Wareing [*5*]; at 650°C, see Conway et al. [*19*], Jaske et al. [*20*], and Rezgui [*17*]). Saturation effects have also been observed on sufficiently aged material (at 566°C, see Jaske et al. [*20*]; at 593°C, see Brinkman et al. [*1*] and Wareing [*5*]; at 625°C, see Wood et al. [*4*]).

Although the three creep-fatigue domains seem to be a general rule, the extent of each domain is very sensitive to heat variations and test conditions such as temperature, strain range, and pre-aging treatment [*5*]. The influence of slight chemical composition variations similar to those encountered in our two materials also plays a dominant role.

Fractography and Metallography

Continuous Fatigue Specimens—SEM observations show transgranular fracture features with fatigue striations in specimens tested without hold time. Estimate of the fatigue crack growth rates using striation spacing measurements can be performed [*7,16*]. Figure 3 gives an example of striation spacing measurements as a function of crack length, a, defined as the distance to the crack initiation site. Figure 4 illustrates the schematic law used to fit 11 measurement results (4 on VIRGO steel at 600°C, 3 on ICL steel at 600°C, 4 on ICL steel at 550°C). In this figure an initial plateau value i_0 is taken. The existence of this plateau could be observed only at 550°C, presumably because of the oxidation effects at higher temperature. From these measurements and assuming that the striation spacing equals the macroscopic crack growth increment per cycle, the fatigue crack growth rates were written as

$$da/dN = i_0 \quad \text{for} \quad a_0 < a < a_c \tag{4}$$

$$da/dN = A(\Delta\epsilon_p/2)^{\alpha} a^{\beta} \quad \text{for} \quad a_c < a < a_f \tag{5}$$

where at 550 and 600°C: $\alpha = 1.5$, $\beta = 1.45$, and $A = 2 \times 10^{-3}$ (units: a in μm, $\Delta\epsilon_p$ in %), $i_0 = 0.27$ μm/cycle, $a_0 = 20$ μm, and $a_f = 2$ mm.

The values of α and β are in agreement with those reported in the literature on 316 stainless steels ($\alpha = 1.5$, $\beta = 1.3$ after Yamaguchi and Kanazawa [*21*]; $1.1 < \alpha < 1.5$, $\beta = 1$ after Wareing et al. [*22*]). The existence of a plateau was not reported in these studies, but was also observed by Jacquelin et al. [*23*] at room temperature. This plateau may represent a large fraction of the propagation stage, N_p, at low strain ranges. Integrating Eqs 4 and 5 and using the Manson-Coffin endurance law:

$$\Delta\epsilon_p/2 = C_p(N_f)^{-m} \tag{6}$$

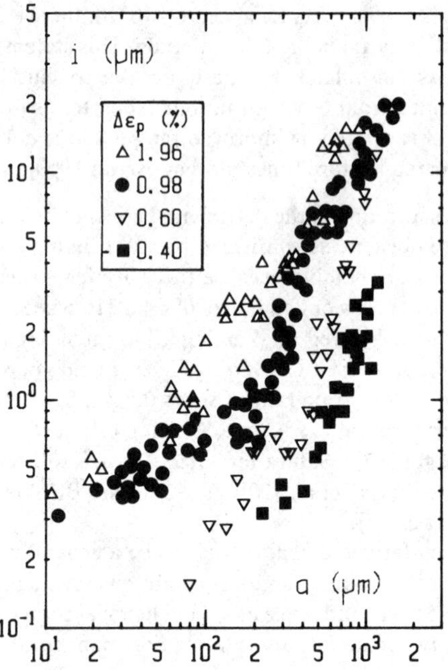

FIG. 3—*Striations spacings measurements as a function of crack length for VIRGO steel. Continuous fatigue at 600°C.*

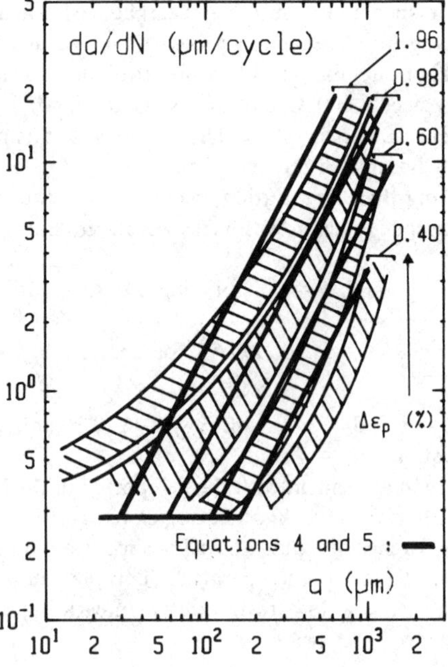

FIG. 4—*Comparison of crack propagation law (Eqs 4 and 5) with striation spacing measurements. VIRGO steel. Continuous fatigue at 600°C.*

the initiation stage, N_i, is related to the fatigue life, N_f, by

$$N_i = N_f - \frac{\beta}{\beta - 1}(i_0)^{(1-\beta/\beta)}(A)^{(-1/\beta)}(C_p)^{(-\alpha/\beta)}(N_f)^{(\alpha m/\beta)} + \frac{(a_f)^{(1-\beta)}}{(\beta - 1)A} C_p^{-\alpha} N_f^{\alpha m} + \frac{a_0}{i_0} \tag{7}$$

Numerically it is found that

$$N_i = N_f - 12N_f^{0.62} + 0.226N_f^{0.90} + 74 \tag{8}$$

As indicated in Fig. 5, Eq 8 agrees with other results obtained by striation spacing measurements [*21,23*] even at room temperature. The good agreement between room temperature and elevated temperature data may be simply explained if it is noticed that the parameters i_0, m, α, β, a_0, and a_f do not depend on temperature, while the quantity AC_p is almost constant, although the A and C_p parameters vary with temperature. The initiation criterion in Eq 7 corresponds to a crack length $a_0 = 20$ μm (i.e., about one half the mean grain size, or, which is equivalent, the mean size of grains cut by the free surface of the specimen). This value seems to agree with the physical initiation process in this material involving intense slip bands in surface grains and an intrusion-extrusion mechanism very sensitive to oxidation (cf. Levaillant et al. [*16*]).

Metallographical observations of longitudinal polished sections of continuous fatigue specimens show a distribution of secondary fatigue cracks along the specimen surface. This distribution may be characterized by histograms giving the density of cracks (i.e., the number of cracks per unit length of free surface edge) as a function of apparent crack length observed on the section, as illustrated in Fig. 6, which refers to interrupted tests on ICL steel. In these measure-

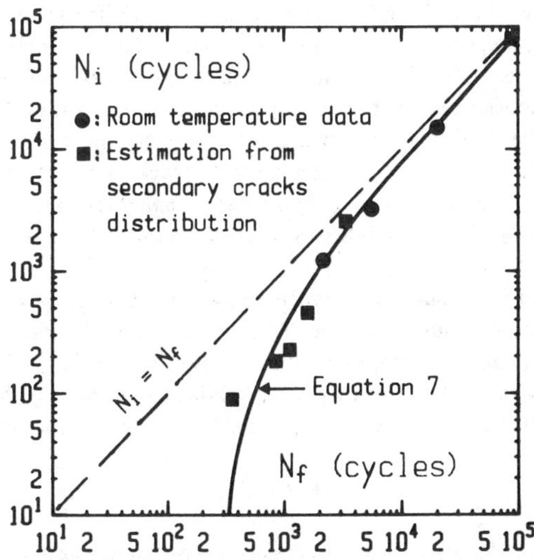

FIG. 5—*Initiation period* N_i *as a function of failure life* N_f. *Comparison of curve derived from Eq 7 (for continuous fatigue results on VIRGO and ICL steels at 550 or 600°C) with room temperature data* [23] *and estimates deduced from secondary crack distribution.*

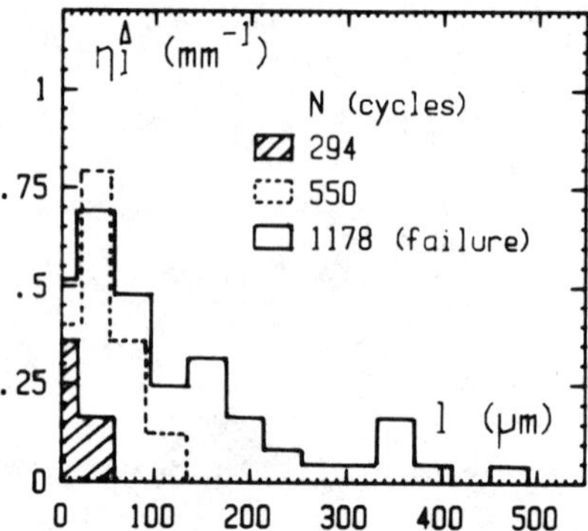

FIG. 6—*Histograms of secondary crack density at various continuous fatigue life fraction. ICL steel. Total strain range: 1.2%. Test temperature: 600°C.*

ments, the length increment was taken as $\Delta = 40$ μm. Figure 6 shows that the crack distribution extends towards large dimensions when the number of applied cycles increases. Using a general expression for crack distribution presented by Quantin and Guttmann [*24*] and several simplifying assumptions, detailed in Appendix I, estimates of the fatigue crack growth rate, da/dN, and of the number of cycles, N_i for initiation may be obtained from the partial density of cracks η^Δ. These estimates are compared with those derived from striation spacing measurements in Fig. 7 for the crack growth rate and Fig. 5 for the initiation stage. A good agreement is found between both methods, which are fully independent.

Creep Fatigue Specimens—SEM observations of fracture surfaces show that the creep-fatigue crack propagation path exhibits a more and more intergranular character as number of cycles to rupture decreases, as currently reported in the literature (cf. Hales [*6*]). The higher sensitivity of VIRGO steel to the hold time was associated with its higher tendency to intergranular cracking than ICL steel [*7*]. Longitudinal sections of creep fatigue specimens show that intergranular cracking occurs uniformly in the bulk of the material (Fig. 8). Figure 8*a* shows that intergranular decohesions appear generally as sharp wedge cracks. Round cavities appear only in the vicinity of the final fracture zone, due to large strains during final instability. At a lower magnification, Fig. 8*b* shows that intergranular cracks are more abundant ahead of the main crack, for moderate hold times. Figure 8*c* shows that intergranular damage is more homogeneously distributed in the 10 000-h creep-fatigue tests, but large cracks initiated at the free surface are still visible.

Levaillant and Pineau [*7*] proposed to use intergranular damage as a correlating parameter for creep-fatigue life. Intergranular damage, D, is measured on longitudinal polished sections of the specimens on optical micrographs at 200× magnification [*10*]. Observations are made away from the main crack in order to estimate the homogeneous damage due to creep processes, without interaction with major cracks (cf. Fig. 8*b*). Intergranular damage, D, is defined by

$$D = L_f/L_t \quad (9)$$

where L_f is the cumulative length of cracked grain boundaries per unit area of section, and L_t is the total length of grain boundaries on the same area and is a characteristic of the grain size and shape. A mean value is determined for each material. Damage measurements after various

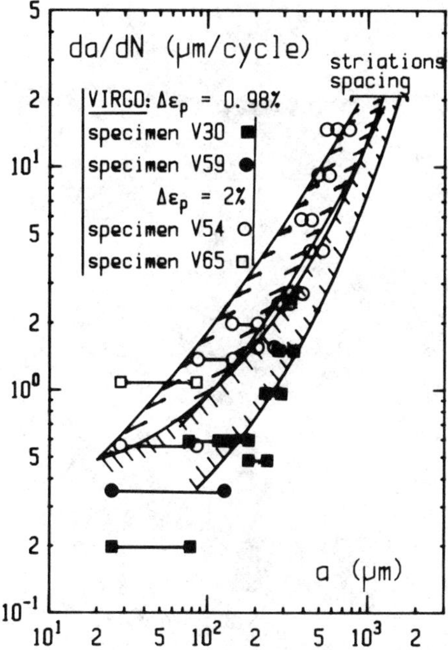

FIG. 7—*Comparison of continuous fatigue crack growth rates estimated from secondary crack density and from striation spacing measurements. Test temperature: 600°C.*

fractions of the creep fatigue life suggest that intergranular damage increases linearly throughout the life (Fig. 9):

$$D = D_c N \tag{10}$$

D_c is thus the intergranular damage per cycle, which is characteristic of the testing conditions (temperature, strain range, hold times). A similar approach was applied to the results of Hales [*6*]. This author used a slightly different definition for intergranular damage, since he determined only L_f. His results are shown in Fig. 10 where, except for low values of the parameter L_f which could correspond to a lack of sensitivity of the metallographical method, a simple linear law can also be assumed.

The measurements of intergranular damage were used in a crack growth model which takes into account creep-fatigue interactions. The details of the model are given in Appendix II. Our approach is based on two main ideas:

1. Hold times drastically reduce the initiation stage in creep fatigue tests, so that

$$N_i^{CF} \simeq 0 \tag{11}$$

This assumption is supported by observations made on interrupted tests.

2. The crack growth rate of the main crack is accelerated by intergranular bulk damage. This assumption is supported by the correlation between the intergranular damage per cycle,

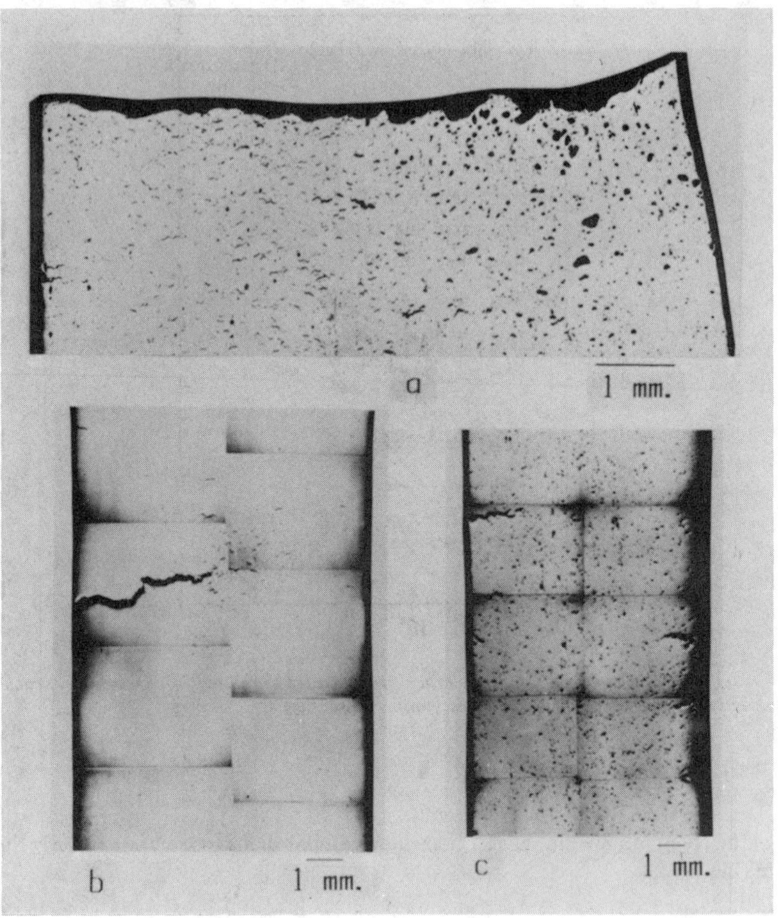

FIG. 8—*Longitudinally polished sections of creep-fatigue specimens. Optical microscopy. Test temperature: 600°C. Total strain range: 1.20%.* (a) *Detail of region located under fracture surface. Hold time: 30 min. ICL steel.* (b) *Typical aspect of a short hold time test. Hold time: 10 min. ICL steel.* (c) *Overview of specimen submitted to a very long hold time. Hold time: 24 h. VIRGO steel.*

D_c, and the normalized reduction, i_c, per cycle of the propagation stage N_p between continuous fatigue and creep fatigue

$$i_c = \frac{N_p^F - N_p^{CF}}{N_p^F \times N_p^{CF}} \quad (12)$$

Measured values of the intergranular damage per cycle D_c are reported in Table 3, while the correlation between D_c and i_c parameters is shown in Fig. 11. The correlation is fairly good, except for the shortest hold times, for which Eq 11 may be too extreme. At the opposite, results of a very long hold time test ($t_f = 24$ h) are in agreement with the general correlation expressed by

$$i_c = kD_c^p \quad (13)$$

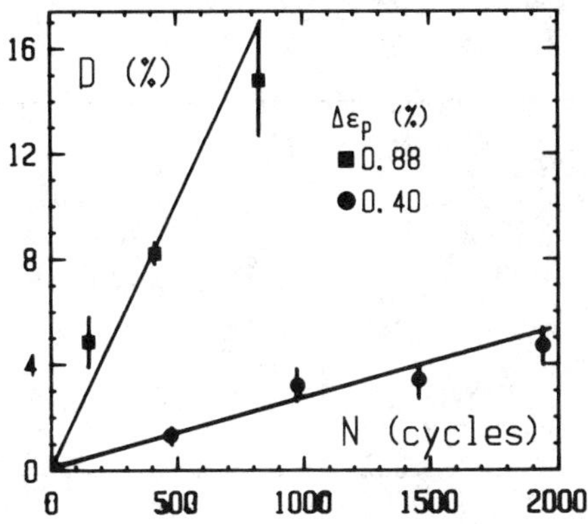

FIG. 9—*Evolution of intergranular damage with the number of creep-fatigue cycles. VIRGO steel. Hold time: 10 min.*

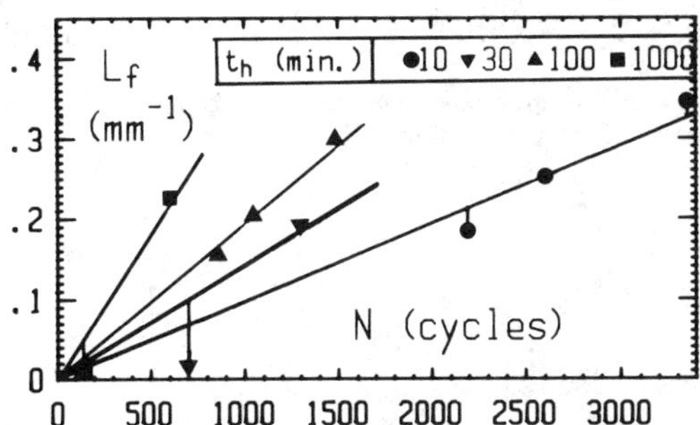

FIG. 10—*Evolution of intergranular damage with the number of creep-fatigue cycles. Results from Hales [6]. 316 stainless steel. Total strain range: 0.5%. Test temperature: 600°C.*

At least squares fit on the eleven data points (for $t_h > 3$ min) leads to $k = 2.76$ and $p = 0.923$.

The distributions of secondary cracks were also determined in specimens tested with an imposed hold time. In this case both intergranular and transgranular cracks were distinguished. Figure 12 illustrates the results obtained in the specific cases corresponding to the long hold times tests. Figure 12*a* as compared to Fig. 6 shows that intergranular damage leads to an increase in the density of secondary cracks. This result suggests that the application of hold times tends to reduce the initiation period, as already discussed. Figure 12*b* which concerns the 10^4 h rupture time test shows that cracking is fully intergranular. These measurements were used to assess the intergranular crack growth rate by using the model presented in Appendix I. The results are given in Fig. 13. We have also included in this figure the results derived from the model based on intergranular damage parameter, D_c, given in Appendix II. A good agreement between the results derived from both models is observed. The comparison with continuous fatigue shows that intergranular damage gives rise to an acceleration of about a factor of three in crack growth rate, which remains relatively moderate.

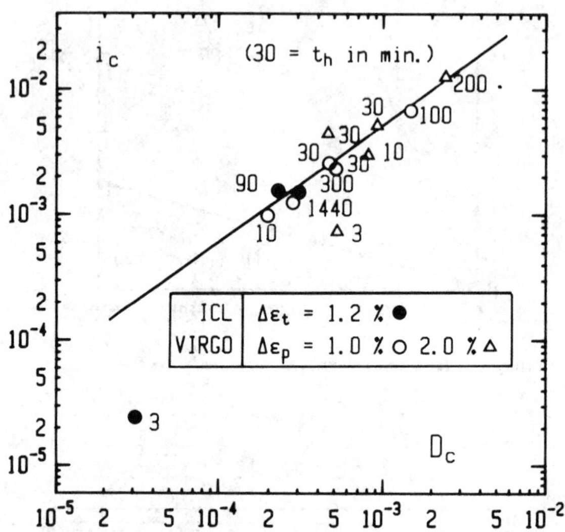

FIG. 11—*Correlation between intergranular damage per cycle* D_c *and factor of crack propagation reduction* i_c. *ICL and VIRGO steels. Test temperature: 600°C.*

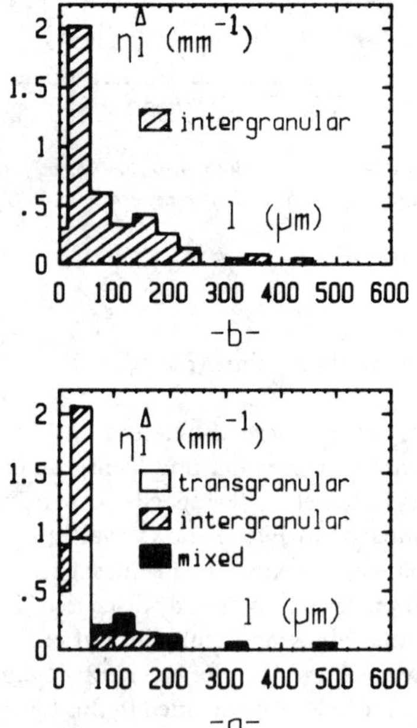

FIG. 12—*Histograms of secondary crack density. Creep fatigue specimens. Total strain range: 1.2%. Test temperature: 600°C.* (a) *VIRGO steel. Hold time: 24 h.* (b) *ICL steel. Hold time: 5 h.*

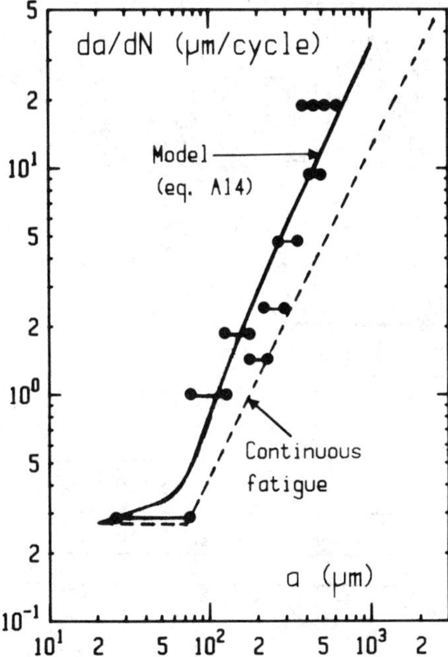

FIG. 13—*Crack propagation rates for intergranular cracking at long hold time conditions (24 h). Comparison of estimates from secondary crack distribution with the model prediction (Eq 31). Broken line indicates the continuous fatigue crack growth rate. VIRGO steel. Total strain range: 1.2%.*

Discussion

The experimental results have confirmed that there is a saturation in the detrimental effect of tensile hold times on LCF life for both investigated materials. Moreover, it has been shown that further measurements of both the density of secondary cracks and intergranular damage provide an attractive guideline for modelling creep-fatigue interactions. In this discussion we concentrate on a more global approach to the problem, often referred to as the *creep-ductility exhaustion concept* [*25*]. For this purpose we first consider the significance of stress relaxation as a creep-fatigue life correlating parameter. Then the saturation effect observed in the present study will be discussed in light of creep ductility measurements that were made on ICL steel.

Stress Relaxation as a Creep-Fatigue Life Correlating Parameter

Several life prediction approaches have been proposed in the literature to account for the effect of creep strain associated with tensile hold times on fatigue life. A recent review on this subject showed the large differences obtained between the models when extrapolations to long hold times are made [*26*]. In this respect two of the present authors proposed a simple correlation between the amount of stress relaxation, σ_R, during hold times, which is a measure of viscoplastic strain, and the relative reduction in fatigue life measured by i_c [*7*].

It was shown that this correlation provided attractive results when applied to a large number of data published in the literature. It should be emphasized, however, that when this correlation was proposed the results available were limited to short hold time duration (a few hours). This is shown in Fig. 14 where, in spite of a large scatterband, a reasonable correlation between σ_R and

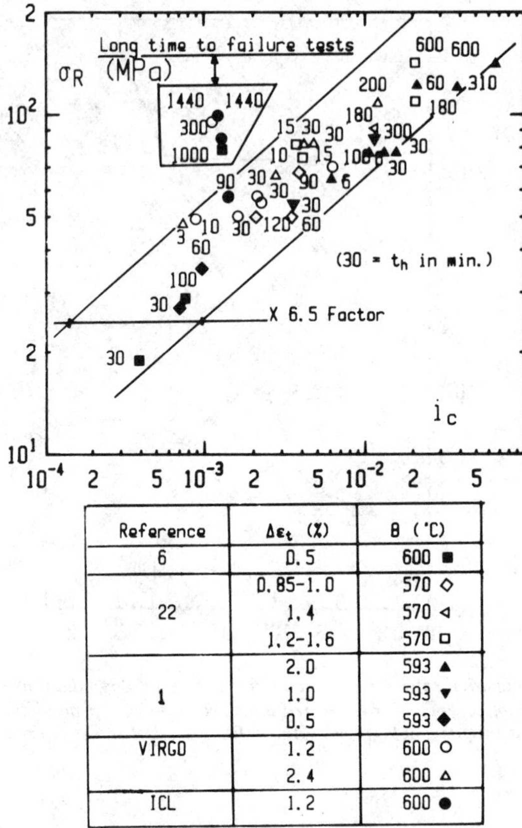

Reference	$\Delta\varepsilon_t$ (%)	θ (°C)
6	0.5	600 ■
22	0.85–1.0	570 ◇
	1.4	570 ◁
	1.2–1.6	570 □
1	2.0	593 ▲
	1.0	593 ▼
	0.5	593 ◆
VIRGO	1.2	600 ○
	2.4	600 △
ICL	1.2	600 ●

FIG. 14—*Correlation between the factor of crack propagation reduction* i_c *and the amount of stress relaxation for one cycle. Literature data on 316 stainless steels in the 570 to 600°C temperature range.*

i_c is observed when $t_h < 300$ min. In this figure we have also included the more recent results corresponding to long hold times (i.e., those published by Hales [*6*]) and those obtained in the present study on both materials. Figure 14 clearly indicates that the $\sigma_R - i_c$ correlation established previously no longer applies to the data corresponding to long hold times (24 h). Although the correlation between i_c and D_c was shown to provide reliable results over a broad t_h range (Fig. 11), a simpler method directly correlating i_c and creep strain or σ_R taking place during each cycle does not give satisfactory results. This is reflected in Fig. 15, especially for VIRGO steel, where it is observed that the amount of intergranular damage, D_c, for long hold times tends to decrease although σ_R increases. In other words, this behavior could be interpreted as if the resistance to intergranular creep damage was improved for long hold times. This raises a question related to the parallel that can be drawn between creep-fatigue resistance and creep ductility.

Parallel Between Creep-Fatigue Resistance and Creep Ductility

The creep damage cyclic component obviously increases with hold time. This general statement led several authors to introduce the creep-dominated regime concept (see, e.g., Goodall et

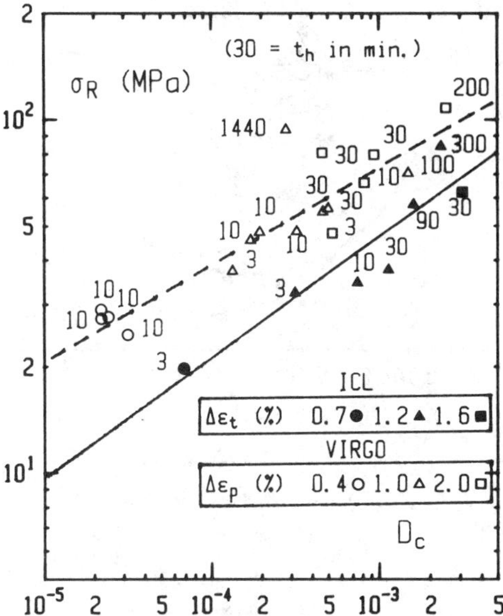

FIG. 15—*Correlation between intergranular damage per cycle* D_c *and amount of stress relaxation. ICL and VIRGO steels. Test temperature: 600°C.*

al. [*27*] and Plumbridge et al. [*11*]). In this respect it is worthwhile to consider recent results obtained on smooth specimens and notched bars of ICL steel creep tested at 600°C [*14*]. In these specimens the creep ductility exhibited a complex behavior presumably related to aging effects (Fig. 16*a*) [*10,28*]. In Fig. 16*a* it is observed that either the elongation to failure determined on smooth specimens, or the mean diametral strain measured on notched specimens, first decreases and then increases with creep life. Furthermore, the creep ductility improvement observed for sufficiently long times is clearly associated with a decrease in the maximum intergranular damage, D_{crit}, which was also measured using a metallographic procedure similar to that employed in the present study (Fig. 16*b*). These results in conjunction with those reported in Table 3 emphasize the parallel which can be drawn, at least qualitatively, between long hold times creep fatigue and creep ductility.

In order to evaluate more quantitatively the similarities between intergranular damage under creep or creep-fatigue conditions, an incremental intergranular damage law proposed elsewhere for pure creep [*14*] was used to calculate the intergranular damage, D_c, per cycle during stress relaxation in a creep fatigue test. This law can be written as

$$dD = B\sigma^q(\epsilon_f)^n d\epsilon_f \qquad (14)$$

where ϵ_f is the creep strain; that is,

$$\epsilon_f = \frac{\sigma_{tmax} - \sigma}{E} \qquad (15)$$

where σ_{tmax} is the tensile stress at the beginning of the hold time. In Eq 14 the values of the coefficients determined on ICL steel tested at 600°C are $B = 10^{-12}$, $q = 4$, and $n = -0.50$ when D and ϵ_f are expressed in true values and σ in MPa.

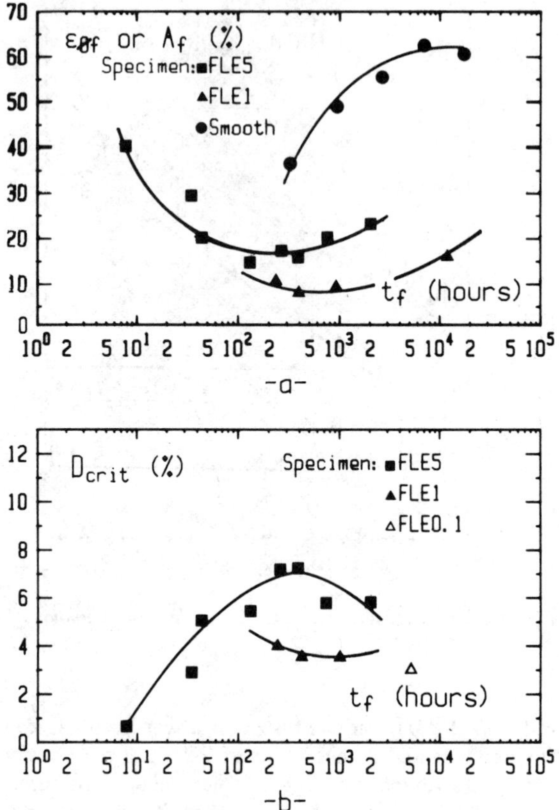

FIG. 16—*Pure creep results on smooth and notched specimens of ICL steel at 600°C.* (a) *Creep ductility versus creep life.* (b) *Intergranular creep damage versus creep life.*

Equations 14 and 15 can be integrated along the stress relaxation curve. This leads to

$$D_c = F\frac{2B}{\sqrt{E}}(\sigma_{tmax})^4\sqrt{\sigma_R} \tag{16}$$

where F is a polynomial function of (σ_R/σ_{tmax}) which varies only between 1 and 0.7 in the investigated domain [*10*]. Within a first approximation, D_c can be written as

$$D_c \simeq \frac{2B}{\sqrt{E}}(\sigma_{tmax})^4\sqrt{\sigma_R} \tag{17}$$

The comparison between measured values of D_c (Table 3) and the values calculated from Eq 17 is shown in Fig. 17. It is clearly observed that the pure creep damage law (Eq 13) does not give satisfactory results when applied to short hold times (t_h < 30 min). This is not too surprising in the sense that the corresponding creep strain rates are much larger than those associated with pure creep tests. Moreover, Eq 17 leads approximately to a one-half power dependence on σ_R, whereas the correlation between D_c and σ_R shown in Fig. 15 for short hold times tests gives

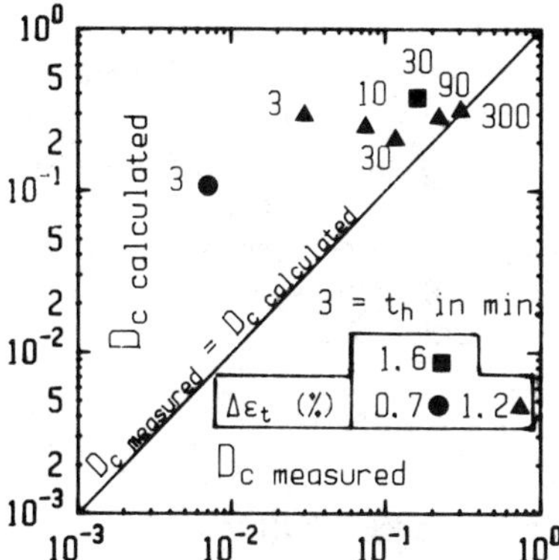

FIG. 17—*Comparison of measured and calculated (Eq 17) pure creep condition values of intergranular damage per cycle in creep fatigue. ICL steel at 600°C.*

rise to a power of about three. Conversely, a much better agreement is observed when the comparison is made for longer hold times (Fig. 17). This situation gives, therefore, some credence to the creep-dominated regime concept. However, a number of limitations must be emphasized when an attempt is made to make a very close correspondence between the results obtained from both types of loading which remain intrinsically different.

The first limitation is obviously related to the opposite limit conditions used in both cases. For creep-fatigue tests the strain is strictly limited and fully reversed. Under these conditions, failure can only occur by a crack propagation mechanism. This was observed even in the 10^4 h time-to-failure LCF test where major cracks initiated from the specimen surface were observed (Fig. 8). Conversely the elongation of creep specimens is not limited. In these conditions, failure occurs by instability of a cavitated material with no propagation of major cracks. The second limitation arises from the fact that the ductility concept remains too vague to quantify the reduction in fatigue life associated with the accumulation of creep strain. If it is accepted that the creep fatigue life is essentially controlled by crack growth, the corresponding ductility should represent the crack tip mechanical situation. The use of smooth bar data is not very relevant to the problem. Figure 16*b* shows the large modification in the results when different specimen geometries are tested.

Conclusions

1. When tensile hold times are introduced into LCF tests, progressively larger reductions in cyclic life are observed as the hold time is increased. For very long hold times (24 h) corresponding to long tests ($\simeq 10^4$ h) a saturation effect is observed.

2. In continuous fatigue a model based on quantitative measurements of secondary cracks strongly supports the results of crack growth kinetics inferred from striation spacing measurements.

3. The reduction in LCF life due to the superimposition of tensile hold times is attributed to the reduction of both crack initiation and crack propagation stages. The acceleration in crack growth is directly related to intergranular damage, D_c, per cycle taking place during hold times.

4. For very long hold times, the intergranular damage per cycle, D_c, tends to decrease although the stress relaxation taking place every cycle, σ_R, is increased. This relative improvement in intergranular damage resistance is similar to the behavior observed in pure creep tests.

5. A model based essentially on the effect of intergranular damage on fatigue crack growth rate is proposed. This model accounts for the saturation effect observed for long hold times, provided the proper value of D_c is known. The results of crack growth kinetics inferred from this model compare well with the results obtained from secondary cracks density measurements.

6. The correlation proposed previously between σ_R and the relative reduction in fatigue life, i_c, is invalid for long hold time periods. This is related to the difficulty of using concepts based essentially on creep ductility exhaustion.

Acknowledgments

This work was partly performed under SCSIN contracts. We wish to thank the "Groupe de Travail Matériaux" for many fruitful discussions with its members, especially Mr. S. Masson from E.D.F. and Mr. P. Petrequin from C.E.A. This study is also one part of the French CNRS GRECO "Grandes Déformations et Endommagement" program.

APPENDIX I

Analysis of Secondary Cracks Density in LCF Specimens

In LCF specimens, secondary cracks can be observed on a longitudinally polished section. The size distribution of these cracks may be characterized by histograms of the partial linear densities η_ℓ^Δ as function of size intervals $[\ell - \Delta/2, \ell + \Delta/2]$. The linear density is the number of cracks per unit length of free surface edge.

The interval Δ must be chosen in order to smooth the experimental scatter for low Δ values and not to give too coarse a description of the actual distribution for large Δ values. In this study, $\Delta = 40$ μm for a 200× magnification. For simplicity, ℓ is taken as an integer multiple of Δ.

Quantin [*29*], who used quantitative metallography in the study of the nucleation and growth of bulk intergranular cracking in aluminum alloys, has shown that the partial density η_ℓ^Δ can be

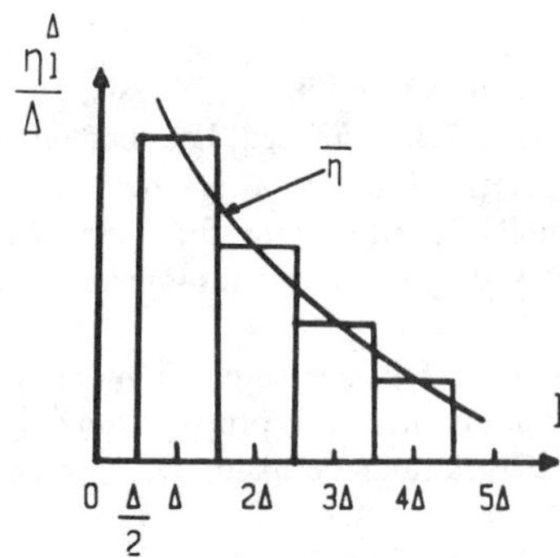

FIG. 18—*Description of secondary crack distribution by histograms or a statistical density curve.*

related to the actual density of the crack length distribution $\overline{\eta}_\ell$ (in the statistical meaning of the density) when Δ is small (cf. Fig. 18) by:

$$\overline{\eta}_\ell \simeq \frac{\eta_\ell^\Delta}{\Delta} \tag{18}$$

This equation can be derived from the mathematical definition of $\overline{\eta}_\ell$:

$$\overline{\eta}_\ell = \lim_{\Delta \to 0} \frac{\eta_\ell^\Delta}{\Delta} \tag{19}$$

For Quantin, $\Delta/2$ is the detection threshold depending on the sensitivity of the metallographical technique.

If crack length at nucleation is assumed to be smaller than $\Delta/2$, the time variations of the density $\overline{\eta}_\ell$ are governed by a classical conservation equation:

$$\frac{\partial \overline{\eta}_\ell}{\partial t} + \frac{\partial}{\partial \ell}(v \cdot \overline{\eta}_\ell) = 0 \tag{20}$$

where v is the velocity. The crack growth rate v is supposed to be the same for all cracks of the same length.

Equation 20 of conservation of the number of cracks of length belonging to the interval $[\ell - \delta\ell/2, + \delta\ell/2]$ applies only if crack coalescence does not occur. This can be easily assumed in the case of secondary cracks observed on longitudinal sections.

Quantin and Guttmann [*24*] discussed the form of the solutions of Eq 20 when the crack growth rate v depends only on the crack length l and when the density $\overline{\eta}_\ell$ is a function of separated variables l and t.

In the case of LCF testing, the variable N (i.e., the number of cycles) is preferred to the time t. The quantity v is then the crack growth per cycle. Therefore, Eq 20 becomes

$$\frac{\partial \overline{\eta}_\ell}{\partial N} + \frac{\partial}{\partial \ell}(v \cdot \overline{\eta}_\ell) = 0 \tag{21}$$

where in this case $v = dl/dN$.

For a stationary regime:

$$\frac{\partial \overline{\eta}_\ell}{\partial N} = 0 \tag{22}$$

so that:

$$v \cdot \overline{\eta}_\ell = Q \tag{23}$$

where Q is a constant that is related to the total density of cracks, and η is the total number of observable cracks per unit length of free surface edge. It can be derived from the expression

$$\frac{d\eta}{dN} = v\left(\frac{\Delta}{2}\right) \cdot \overline{\eta}_{\Delta/2} \tag{24}$$

Equation 24 expresses that the total crack density η increases only due to the newly detectable cracks flux ($\ell = \Delta/2$).

From Eqs 23 and 24, Q is the stationary crack nucleation rate.

If stationary regime is assumed, an estimation of crack velocity v can be deduced from Eqs 23 and 18:

$$v \simeq \frac{Q_\Delta}{\eta_\ell^\Delta} \tag{25}$$

In Eq 25, v is the mean crack velocity in the crack length interval $[\ell - \Delta/2, \ell + \Delta/2]$.

In the case of observations on longitudinal sections, the measured crack depth l is less than the actual crack length a. Assuming an half elliptical crack front, the mean observed crack depth $\bar{l}$ is related to the crack length a by

$$a = \lambda\bar{\ell} \tag{26}$$

where $\lambda = 4/\pi \simeq 1.274$ [*10*].

The actual crack growth rate is then estimated by

$$v_{[\lambda(\ell-\Delta/2),\lambda(\ell+\Delta/2)]} = \frac{Q\lambda\Delta}{\eta_\ell^\Delta} \tag{27}$$

Application to Continuous Fatigue Specimens

For continuous fatigue specimens, cracks are transgranular. Cracks density measurements at various life fractions [*10*] showed that the hypothesis of constant crack nucleation rate and of a stationary regime is a good approximation, so that Eq 27 may be used to estimate fatigue crack growth rate.

Since the optical detection threshold is estimated to be 10 μm at 200× magnification, the crack nucleation rate Q is related to the crack density η_0 in the interval 10 to 20 μm by

$$Q = \frac{\eta_0}{N_i} \tag{28}$$

where N_i is the number of cycles for crack initiation ($a > a_0 = 20$ μm). Equation 28 is simply derived from the assumption of a constant crack nucleation rate. It can be used to estimate the initiation period N_i from failed specimen observations. The crack nucleation rate Q is then derived from the total crack density η:

$$Q = \frac{\eta}{N_f} \tag{29}$$

Application to Creep-Fatigue Specimens

For creep-fatigue specimens, transgranular, intergranular, and mixed secondary cracks are generally observed on the same section (cf. Fig. 8). A unique crack growth rate law can no longer be assumed. In the case of very long hold times (24 h), only intergranular cracks are detected. Equations 27 and 28 may be used if a constant intergranular surface crack nucleation rate is assumed. This assumption seems reasonable because of the constant nucleation rate reported for bulk intergranular damage, characterized by the D_c parameter. Within a first approximation, consistency between surface intergranular cracks density and intergranular bulk damage, D_c, was found [*10*] through stereological considerations (i.e., the free surface of the specimen is supposed to have no particular influence on creep cracking).

Equation 28 was used to derive the results shown in Fig. 13. It must be emphasized that the crack growth rate in pure intergranular mode is not drastically higher than in continuous fatigue (less than a factor of five). Application of Eq 28 gives an initiation period of only 25 cycles, which is negligible compared with the 373 cycles to failure.

A final remark about this analysis must be made. The analysis of secondary cracks density is based on statistical considerations. This means that the number of cracks observed in a specimen must be important enough to give some statistical confidence to the results of the method [*10*].

APPENDIX II

Creep-Fatigue Interaction Model Based on Intergranular Damage

The model was extensively described in a previous publication [*7*]. Initiation and propagation periods are distinguished, with a metallographical limit crack length $a_0 = 20\ \mu\text{m}$.

For creep-fatigue, the initiation period is neglected:

$$N_\text{i}^\text{CF} \simeq 0 \tag{30}$$

The crack propagation rate in creep fatigue $(da/dN)^\text{CF}$ is accelerated by intergranular damage compared with the growth rate corresponding to continuous fatigue $(da/dN)^\text{F}$:

$$\left(\frac{da}{dN}\right)^\text{CF} = \left(\frac{da}{dN}\right)^\text{F} \times \frac{dD}{(1 - kD_\text{c}^{\text{P}-1}D)^2} \tag{31}$$

Integration of Eq 31 leads to empirical correlation:

$$i_\text{c} = \frac{N_\text{P}^\text{F} - N_\text{P}^\text{CF}}{N_\text{P}^\text{F} \cdot N_\text{P}^\text{CF}} = kD_\text{c}^\text{P} \tag{32}$$

where $N_\text{P}^\text{CF} \simeq N_\text{f}^\text{CF}$.

Integration limits a_0 and a_f are the same in continuous fatigue and in creep-fatigue. Linear cumulation of creep damage is expressed as

$$D = D_\text{c}N \tag{33}$$

References

[*1*] Brinkman, C. R., Korth, G. E., and Hobbins, R. R., *Nuclear Technology,* Vol. 16, Oct. 1972, pp. 297-307.

[*2*] Maiya, P. S. and Majumdar, S., *Metallurgical Transactions,* Vol. 8A, Nov. 1977, pp. 1651-1660.

[*3*] Wareing, J., *Metallurgical Transactions,* Vol. 6A, July 1975, pp. 1367-1377.

[*4*] Wood, D. S., Wynn, J., Baldwin, A. B., and O'Riordan, P., *Fatigue of Engineering Materials and Structures,* Vol. 3, 1980, pp. 39-57.

[*5*] Wareing, J., *Fatigue of Engineering Materials and Structures,* Vol. 4, 1981, pp. 131-145.

[*6*] Hales, R., *Fatigue of Engineering Materials and Structures,* Vol. 3, 1980, pp. 339-356.

[*7*] Levaillant, C. and Pineau, A. in *Low-Cycle Fatigue and Life Prediction, ASTM STP 770,* American Society for Testing and Materials, Philadelphia, 1982, pp. 169-193.

[*8*] Mottot, M., Petrequin, P., Amzallag, C., Rabbe, P., Grattier, J., and Masson, S. H. in *Low-Cycle Fatigue and Life Prediction, ASTM STP 770,* American Society for Testing and Materials, Philadelphia, 1982, pp. 152-168.

[*9*] Clavel, M., Levaillant, C., and Pineau, A. in *Proceedings,* Fall Meeting of the Metallurgical Society of AIME, Pelloux and Stoloff, Eds., Milwaukee, Wisc., Sept. 1979, pp. 24-45.

[*10*] Levaillant, C., Thesis, Université de Technologie de Compiègne, France, 1984.

[*11*] Plumbridge, W. J., Dean, M. S., and Miller, D. A., *Fatigue in Engineering Materials and Structures,* Vol. 5, 1982, pp. 101-114.

[*12*] Groupe de Travail Matériaux - EDF-CEA Report No. 4, 1975.

[*13*] Groupe de Travail Matériaux - EDF-CEA Report No. 13, 1982.

[*14*] Yoshida, M., Levaillant, C., and Pineau, A. in *Proceedings,* International Conference on Creep, Tokyo, 14-18 April 1986, pp. 327-332.

[15] Contesti, E., Cailletaud, C., and Levaillant, C., paper presented to 7th International SMIRT Conference, Brussels, Extended Seminar No. 5, Paris, Aug. 1985.
[16] Levaillant, C., Rezgui, B., and Pineau, A. in *Proceedings,* Miller and Smith, Eds., International Conference ICM3, Cambridge, England, Aug. 1979, pp. 163-172.
[17] Rezgui, B., Thesis, Université Paris XI, Orsay, 1982.
[18] Groupe de travail Matériaux - EDF-CEA, Report No. 18, 1985.
[19] Conway, J. B., Berling, J. T., and Stentz, R. H. in *Proceedings,* International Conference on Thermal Stresses and Thermal Fatigue, Berkeley, U.K., 1969, Butterworths, pp. 89-108.
[20] Jaske, C. E., Mindlin, H., and Perrin, J. S. in *Fatigue at Elevated Temperature, ASTM STP 520,* American Society for Testing and Materials, Philadelphia, 1973, pp. 365-376.
[21] Yamaguchi, K. and Kanazawa, K., *Metallurgical Transactions,* Vol. 11A, Dec. 1980, pp. 2019-2027.
[22] Wareing, J., Vaughan, H., and Tomkins, B. in *Proceedings,* Fall Meeting of the Metallurgical Society of AIME, Pelloux and Stoloff, Eds., Milwaukee, Wisc., Sept. 1979, pp. 129-154.
[23] Jacquelin, B., Hourlier, F., and Pineau, A., *Journal of Pressure Vessel Technology, Transactions of ASME,* Vol. 105, 1983, pp. 138-143.
[24] Quantin, B. and Guttmann, M., paper presented to 4th European Conference on Fracture, Wien, Austria, Sept. 1982.
[25] Edmunds, H. G. and White, D. J., *Journal of Mechanical Engineering Science,* Vol. 8, 1966, pp. 310-321.
[26] Cailletaud, G., Nouailhas, D., Grattier, J., Levaillant, C., Mottot, M., Tortel, J., Escaravage, C., Heliot, J., and Kang, S., *Nuclear Engineering and Design,* Vol. 83, 1984, pp. 267-278.
[27] Goodall, I. W., Hales, R., and Walters, D. J., paper presented to IUTAM Symposium on Creep of Structures, Leicester, England, 1980.
[28] Yoshida, M., Thesis, Ecole des Mines de Paris, Sept. 1985.
[29] Quantin, B., Thesis, Université Paris XI, Orsay, 1981.

DISCUSSION

N. Marchand[1] *(written discussion)*—(1) You have shown that there is a cut off hold-time period where the number of cycles to failure does not decrease with increasing hold time. This means that using short hold-time tests one could extrapolate the data to long life and obtain very conservative data. Would you use the projected short time data to predict life longer than 10 000 h (e.g., 20 years)? Would you still be confident in your predictions?

(2) This model is alloy specific. Can we use it to assess the remaining life of ferritic steels used in power plants?

C. Levaillant et al. (authors' closure)—(1) For a 1.2% total strain range, if the lowest number of cycles to failure is obtained for 300 min hold time (ICL steel), the extrapolated time to failure for 500 h hold time would be about 17 years. Is this a safe prediction? As discussed in the present paper, the saturated LCF life seems to be correlated with the increased creep ductility or with the decreased creep damage in pure creep conditions for creep-rupture times between 1000 and 50 000 h (6 years). To have more confidence in predictions based on the saturation hypothesis, it would be necessary to confirm the improved creep resistance (in terms of ductility or damage) in an experimental domain as large as possible. The predictions of the present approach may be supported by observations of interrupted specimens submitted to low strain, large hold-time cycling; estimates of the intergranular damage allow us to predict the number of cycles to failure through the $i_c - D_c$ correlation. Thus the present approach is actually a predictive methodology.

[1]Department of Materials Science and Engineering, Massachusetts Institute of Technology, Cambridge, MA 02139.

(2) The approach proposed for austenitic stainless steels in this paper has been recently applied successfully on Alloy 800 by one of the authors, J. Grattier, and A. El Garad [*1*] at EDF. For this material, no saturation of hold-time effect is observed. This behavior has been correlated with an increasing creep damage per cycle, D_c, with hold-time duration over all the experimental domain. A $D_c - \sigma_R$ correlation has been also reported for the IN 738 LC superalloy by Day and Thomas [*2*]. In ferritic steels used in power plants, at least in the situation where creep-fatigue damage is associated with bulk intergranular damage, the method could be applied.

B. Ilschner[2] *(written discussion)*—This question refers to your results on long hold times (24 h). It appears that you would have to reduce the stress level during hold time considerably in order to arrive at comparable N_r values. Under these circumstances, what is your opinion concerning the possibility of passing a borderline in the "Ashby map" of the material under examination; that is, getting into a (σ, T) range which is characterized by a different predominant mechanism of deformation (creep), either in the bulk section or in the crack tip area? Also, one might enter a different fracture mechanism area. (See also the paper by Kancfeve et al. in this conference.)

C. Levaillant et al. (authors' closure)—Ashby maps represent the different deformations or damage mechanisms in creep conditions in a (σ, T) plot. During a relaxation period, the main parameter is the viscous strain rate, which decreases drastically at the beginning of the hold time. Nevertheless, there is always a relaxation rate, although low, for very large hold-time conditions. It is clear that when the strain rate decreases, the intergranular cavitation mechanism may change from constrained diffusion cavity growth (Dyson's model [*3*]) to diffusion-controlled cavity growth (Hull and Rimmer's model [*4*]). Dyson's mechanism is consistent with Eq 14 as discussed in Ref *10* and applies to the domain of decreasing fatigue life where the damage is crack-like. Recent observations on a creep-notched specimen that failed after 17 000 h with an improved ductility revealed the existence of many small dimples on the broken grain facets. Damage would consist in this latter case of rounded cavities regularly distributed along grain boundaries. Ashby map representation is obviously an interesting way of plotting creep results, but it does not avoid the necessity of detailed and extensive metallography with high resolution for identifying the damage mechanisms.

J. Wareing[3] *(written discussion)*—The longest hold time (24 h) tensile dwell tests were conducted at strain ranges of $\simeq 1\%$. These produced a restoration in creep-fatigue endurance from those in shorter hold periods, but failure occurred due to growth of a surface crack. Would you expect the same effects and failure mechanisms with decreasing applied strain level, where surface crack initiation becomes increasingly difficult?

C. Levaillant et al. (authors' closure)—In the experimental domain, surface-initiated cracks (even intergranular cracks) grow faster than internal cracks due to the mechanically different conditions and perhaps because of some environmental influence. It is expected that this situation will also apply to low strain level conditions.

Discussion References

[*1*] Grattier, J. and El Garad, M., private communication, EDF Les Renardières.
[*2*] Day, M. F. and Thomas, G. B., *Metal Science*, Vol. 13, 1979, p. 25.
[*3*] Dyson, B., *Canadian Metallurgical Quarterly*, Vol. 18, 1979, p. 31.
[*4*] Hull, D. and Rimmer, D. E., *Philosophical Magazine*, Vol. 4, 1959, p. 673.

[2]Department des Matériaux, Ecole Polytechnique Federale de Lausanne, 1007 Lausanne, Switzerland.
[3]UKAEA, Springfields, Salwick, Preston, Lancashire, England.

J. H. Driver,[1] *C. Gorlier,*[1] *C. Belrami,*[2] *P. Violan,*[2] *and C. Amzallag*[3]

Influence of Temperature and Environment on the Fatigue Mechanisms of Single-Crystal and Polycrystal 316L

REFERENCE: Driver, J. H., Gorlier, C., Belrami, C., Violan, P., and Amzallag, C., "**Influence of Temperature and Environment on the Fatigue Mechanisms of Single-Crystal and Polycrystal 316L,**" *Low Cycle Fatigue, ASTM STP 942,* H. D. Solomon, G. R. Halford, L. R. Kaisand, and B. N. Leis, Eds., American Society for Testing and Materials, Philadelphia, 1988, pp. 438–455.

ABSTRACT: Single-crystal and polycrystal 316L specimens were fatigued at constant plastic strain amplitudes in the range 10^{-4} to 2×10^{-2} in air and in vacuum (about 3×10^{-4} to 10^{-3} Pa) at 20, 300, and 600°C. The temperature and strain dependencies of the single-crystal cyclic stress-strain response are qualitatively similar to those of the polycrystalline material.

In air, the fatigue lives of both polycrystals and single crystals decrease by about an order of magnitude at 600°C compared with the results at 20 and 300°C. The fatigue lives of polycrystal specimens were prolonged in vacuum (compared with air) by factors of ~2 or 3 at room temperature, to ~20 at elevated temperatures and low plastic strains. The crack initiation mechanism of polycrystals tends to become partially intergranular at 600°C, but this change is not thought to significantly affect fatigue life. The reduction in fatigue resistance in air of both single and polycrystals at 600°C is attributed to oxidation—enhanced Stage I crack initiation.

KEY WORDS: low-cycle fatigue, austenitic stainless steels, temperature, environment (air and vacuum), single crystals, crack initiation, oxidation

The fatigue resistance of austenitic stainless steels is known to be sensitive to temperature and environment. Thus in air low-cycle fatigue (LCF) lifetimes are reduced significantly at 600°C compared with room temperature, whereas in inert environments there appears to be little, if any, fatigue life reduction in the same temperature range. These results have been obtained on A286 steel in air and vacuum [*1,2*], Type 316 steel in air and vacuum [*3*], and Types 304 and 316 steels in air and liquid sodium [*4*]. The decrease in fatigue resistance at high temperatures in air has been attributed either to the onset of intergranular cracking or to oxidation effects (or to both). Single crystals tests could provide useful information on the controlling mechanisms, since one of the possible causes—intergranular cracking—is impossible. Monocrystal tests also enable one to study the mechanisms of slip band (Stage I) cracking in greater detail, in particular as a function of the environment.

To clarify the specific influences of temperature and environment on the fatigue mechanisms of 316 steels, low-cycle fatigue tests have been performed on both single-crystal and polycrystal 316L specimens in air and in vacuum at different temperatures (20, 300, and 600°C). Constant

[1]Department of Materials, Ecole des Mines de St-Etienne, France.
[2]Laboratory of the Mechanics and Physics of Materials, ENSMA, Poitiers, France.
[3]UNIREC Research Centre, Firminy, France.

plastic strain amplitude tests were carried out over a wide range of strains, in particular at relatively low plastic strains where the environmental effect can be particularly pronounced.

This paper describes the results concerning the cyclic stress-strain response, the strain-life characteristics, and the fatigue mechanisms of polycrystal material (in air and vacuum) and single crystals (in air), together with some preliminary work on single crystals in vacuum.

Experimental Procedure

The polycrystalline fatigue specimens were machined from a 25 mm thick plate of a nuclear-grade 316L steel, denoted 17.12 SPH. The chemical analysis and conventional mechanical properties of this steel are given in Table 1. The fatigue specimens were mechanically polished using fine SiC paper, then electropolished before testing.

Single crystals were grown by controlled solidification in horizontal furnaces using oriented seed crystals and simple laboratory-cast alloys. The presence of ferrite forming elements, such as molybdenum, which lead to α-γ dendritic solidification structures was compensated by increasing the nickel content so that reasonably sized γ single crystals could be obtained. The final composition range of the crystals after solidification was (weight percent): 16–18 Cr, 13.7–14.5 Ni, 1.6–2.8 Mo, and 0.01–0.03 C. Square section (5 by 5 mm^2) monocrystal fatigue specimens were prepared by arc welding the crystals to polycrystalline threaded heads (Fig. 1*a*). The single-crystal gage lengths were oriented close to $[\bar{1}49]$ to promote single slip on the (111) $[\bar{1}01]$ system (Schmid Factor 0.49–0.5) as shown schematically in Figs. 1*b* and 1*c*. After welding the crystals to the heads, the test specimens were solution treated for 1 h at 1050°C and finally polished, mechanically and electrolytically, before testing.

The fatigue tests in air and in vacuum were conducted in different laboratories but with identical specimen shapes, for both single and polycrystal material, and the same grip assembly and extensometry. Constant plastic strain amplitude tests were performed in symmetrical tension-compression using a triangular waveform at strain rates between 2 and 4×10^{-3} s^{-1}. The displacements were measured by LVDTs across the specimen shoulders and converted to deformation in the gage length using suitable correction factors [5].

The tests in air (Ecole des Mines, EMSE) were performed on a 40 kN Schenck servohydraulic machine with modified electronics to enable plastic strain control. A 1500 W quartz radiant heating lamp was used for the elevated temperature tests. The vacuum tests were carried out at the ENSMA laboratory using an Instron 1362 electromagnetic machine. A vacuum of between 3×10^{-4} and 10^{-3} Pa, depending upon the test temperature, was maintained in a chamber

TABLE 1—*Chemical composition and mechanical properties of 17.12 SPH steel.*

(a) Chemical Composition, wt%

C	Mn	Si	S	P	Ni	Cr	Mo	Cu	B	N_2	Co
0.022	1.69	0.31	0.002	0.023	11.9	17.45	2.25	0.11	0.009	0.069	0.19

(b) Mechanical Properties

Temperature, °C	0.2% Yield Strength, MPa	U.T.S., MPa	Elongation, %	Reduction in Area, %
20	261	583	53	80
300	156	460	37	72
600	124	396	40	68

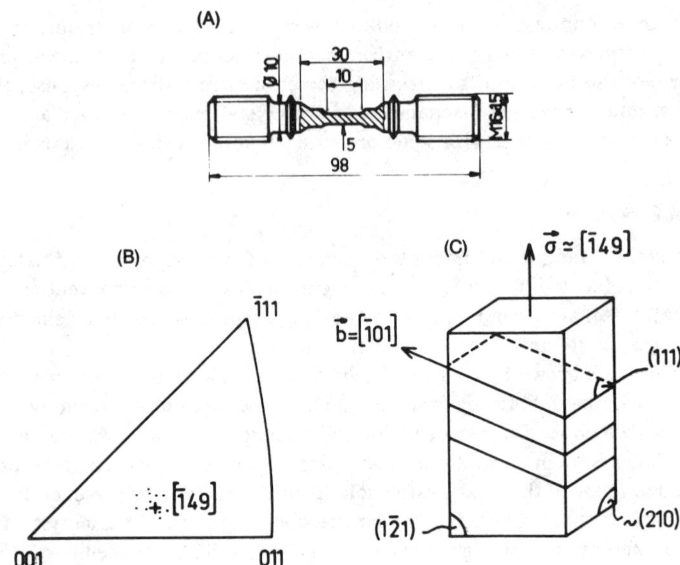

FIG. 1—*Single-crystal specimen dimensions (millimetres)* (A), *orientations* (B), *and crystal faces* (C). *The shaded area in* (A) *represents the single-crystal part.*

surrounding the entire grips, extensometry, and resistance furnace assembly. Some preliminary calibration tests were conducted at ENSMA and EMSE on polycrystalline specimens at room temperature in air to confirm that the cyclic stress-strain response and strain-life plots were the same.

Fatigue crack initiation mechanisms were characterized by optical and scanning electron microscopy (SEM). Particular attention was paid to determining the distribution of crack lengths and sites by analyzing as large a surface of the gage length as possible and by plotting histograms of the crack length distribution at different stages before final rupture. Large specimen surfaces were conveniently examined by optical microscopy of shadowed acetate paper replicas and by SEM.

Results

The results obtained on the polycrystal and single-crystal specimens will be presented separately to illustrate more clearly the respective influence of temperature and environment.

Polycrystalline Specimens

Cyclic Stress-Strain Response—The cyclic hardening, $\sigma(N)$ curves established at room temperature in air and vacuum are compared in Fig. 2 for two plastic strain amplitudes ϵ_{pa} ($= \Delta\epsilon_p/2$) of 2×10^{-3} and 5×10^{-3}. The room temperature curves exhibit a stress maximum after about 20 cycles followed by slight softening before saturation. As would be expected, the environment does not change the general form of these curves, the only difference being that in vacuum the saturation stage is prolonged leading to a significant increase in fatigue life, even at room temperature.

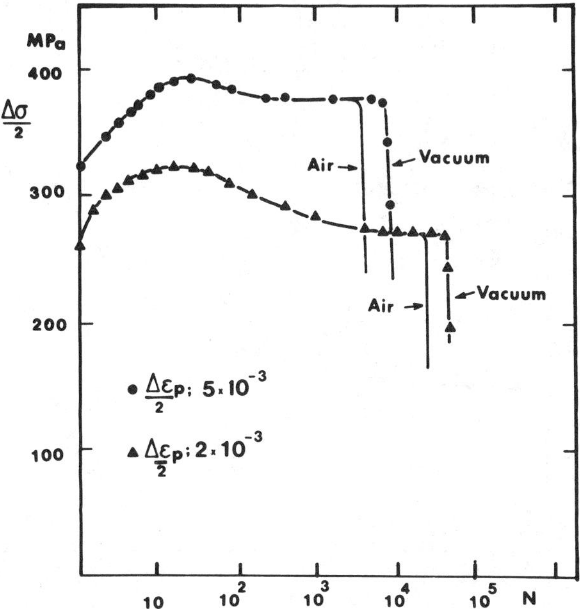

FIG. 2—*Cyclic hardening curves of 17-12 SPH polycrystal steel at room temperature in air and vacuum.*

At 300 and 600°C the saturation stresses determined in air and in vacuum at $N_f/2$ are shown in Fig. 3, where it can be seen that the environment generally has little effect on the cyclic stress-strain response at saturation. The only significant exception concerns the low plastic strain behavior at 300°C; in vacuum the saturation stress is significantly higher than the values obtained in air. As shown in the cyclic hardening plots of Fig. 4, this results from a secondary hardening phenomenon which occurs when the steel is fatigued in vacuum. Cyclic hardening rates are similar in air and vacuum up to the point at which the air-tested specimen begins to fail; the vacuum-tested specimen continues to harden during its extended fatigue life. Similar secondary hardening effects have been observed on Cu single crystals fatigued in vacuum [*6*].

Fatigue Lifetimes and Crack Initiation Mechanisms—The strain life $\epsilon_p(N_f)$ curves of the polycrystalline steel have been established in air and in vacuum at 20, 300, and 600°C (Figs. 5*a*, *b*, and *c*). These curves clearly demonstrate the significant increase of the fatigue lifetimes in vacuum (N_v) compared with those in air (N_A). As shown in Table 2 the N_v/N_A ratio is about 2 to 3 at room temperature but, at elevated temperatures varies from 3 to 5 at high plastic strains to ~20 at low plastic strains. The fatigue lives in air and vacuum at 600°C are close to the values for a 316L steel at 550°C published by Furuya et al. [*3*] for the same environments. Furthermore the fatigue lifetime ratios at 600°C in air and vacuum correspond reasonably well with the results of Smith et al. [*4*] on 316L steel at 600°C in air and liquid sodium; in the latter environment fatigue lives were 5 to 6 times those in air for plastic strain amplitudes of 2×10^{-3}. The present study thus confirms previous results on the environmental effect at moderate plastic strain amplitudes (i.e, 2 to 5×10^{-3}) and also reveals a pronounced effect at lower strains ($\epsilon_{pa} \sim 6 \times 10^{-4}$) corresponding to longer fatigue lives. The strong environmental influence for long lives at high temperatures is to be expected, since under these conditions extensive oxidation occurs in air. The increased fatigue lifetimes in vacuum at room temperature are somewhat surprising in view of the oxidation resistance usually attributed to this stainless steel.

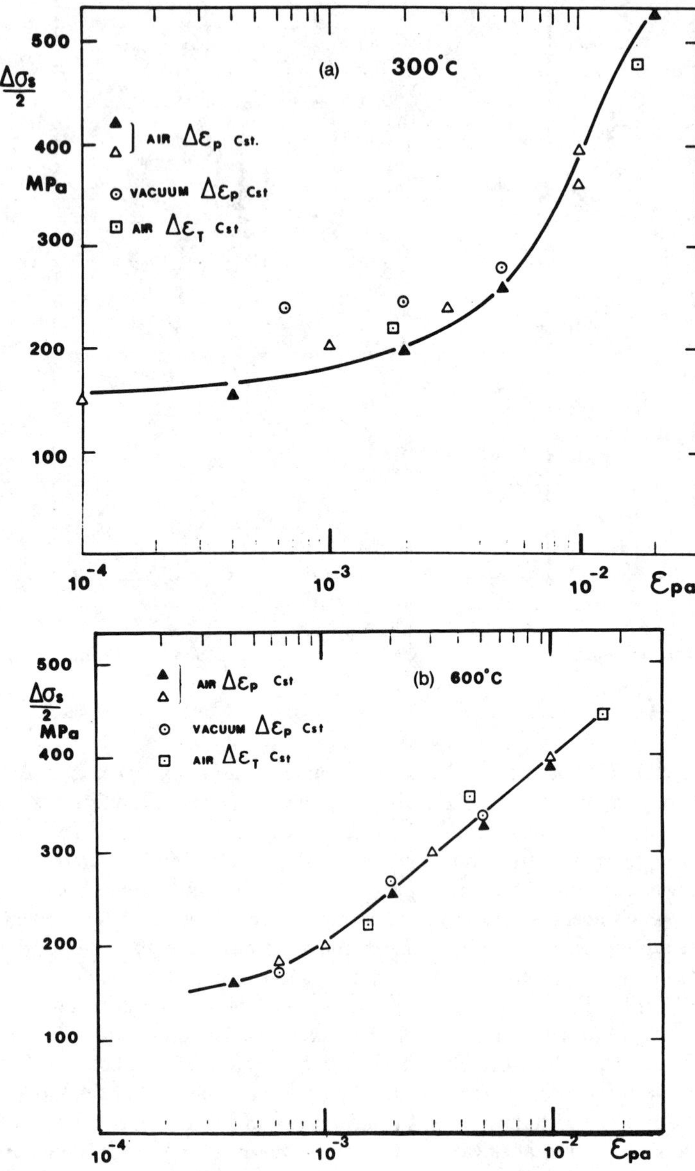

FIG. 3—*Cyclic saturation stresses of 17-12 SPH polycrystal steel in air and vacuum at* (a) *300°C and* (b) *600°C.*

Optical and SEM of the specimen gage lengths after fatigue at 20 and 300°C show that the cracks initiate predominantly along slip bands, in both air and vacuum. The microcracks initially follow intense slip bands across grain boundaries and, after crossing 2 or 3 grains, begin to propagate roughly perpendicular to the stress axis. The crack length distributions in a specimen fatigued in air at $\epsilon_{pa} = 10^{-3}$ are shown at different life fractions in Fig. 6. These histograms demonstrate that, during cycling, the secondary crack density decreases while their average

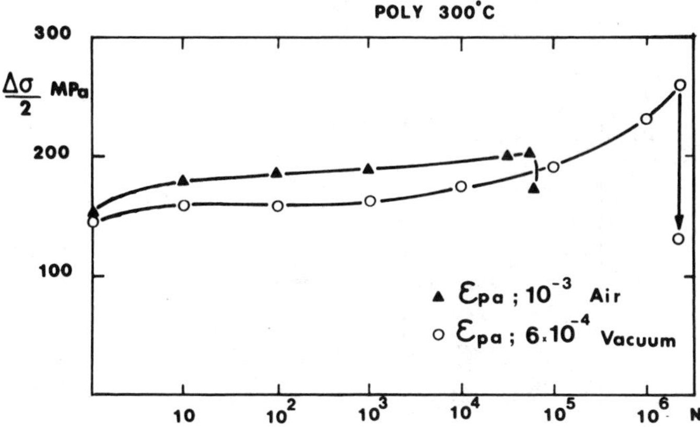

FIG. 4—*Cyclic hardening curves of 17-12 SPH at 300°C and relatively low plastic strain amplitudes, illustrating the secondary hardening effect in vacuum.*

length increases, implying that many microcracks coalesce during the period before macroscopic crack growth. Observations on other specimens cycled in air at room temperature also indicate that the secondary crack density increases with the plastic strain in agreement with previous results [*7*]. At 600°C significant differences in the crack initiation modes appear:

1. In air, crack initiation is both transgranular and intergranular, the latter mode being favored at high strains.
2. In a vacuum, however, crack initiation is always intergranular, even at the lowest strains used ($\epsilon_{pa} = 6 \times 10^{-4}$) which lead to long fatigue lives (750 000).

These observations are illustrated by the histograms of Fig. 7 which give the crack surface density as a function of crack length, and indicate for each length range the relative proportions of intergranular, transgranular, and mixed crack modes. Only a few isolated and relatively small transgranular cracks are observed, together with a certain proportion of mixed mode, predominately intergranular, cracks. The majority of the microcracks are purely intergranular and follow the grain boundaries perpendicular to the stress axis (Fig. 8).

From the point of view of the crack initiation mechanism, the behavior of this steel at 600°C is similar to that of copper at room temperature for which a previous study [*8*] has shown that crack initiation is always intergranular under vacuum.

Single Crystals

The fatigue tests on single-crystal 316L specimens were carried out with the same experimental equipment as for polycrystals. The present report describes extensive work on the behavior of single crystals in air at 20, 300, and 600°C, together with some preliminary results on single crystals in vacuum at 600°C.

Cyclic Hardening—Figure 9 presents the cyclic hardening curves, $\sigma(N)$, at 20, 300, and 600°C of the single slip oriented crystals tested in air at two plastic strain amplitudes: $\epsilon_{pa} = 3 \times 10^{-3}$ and $\epsilon_{pa} = 10^{-2}$. It should be noted that, in contrast to the cyclic hardening of annealed pure metal crystals such as Al, Cu, and Ni, there is very little hardening at low strain amplitudes and moderate temperatures ($\leq$300°C); see Fig. 9. At higher strains and particularly at elevated temperatures the crystals exhibit pronounced cyclic hardening so that the maximum stress at

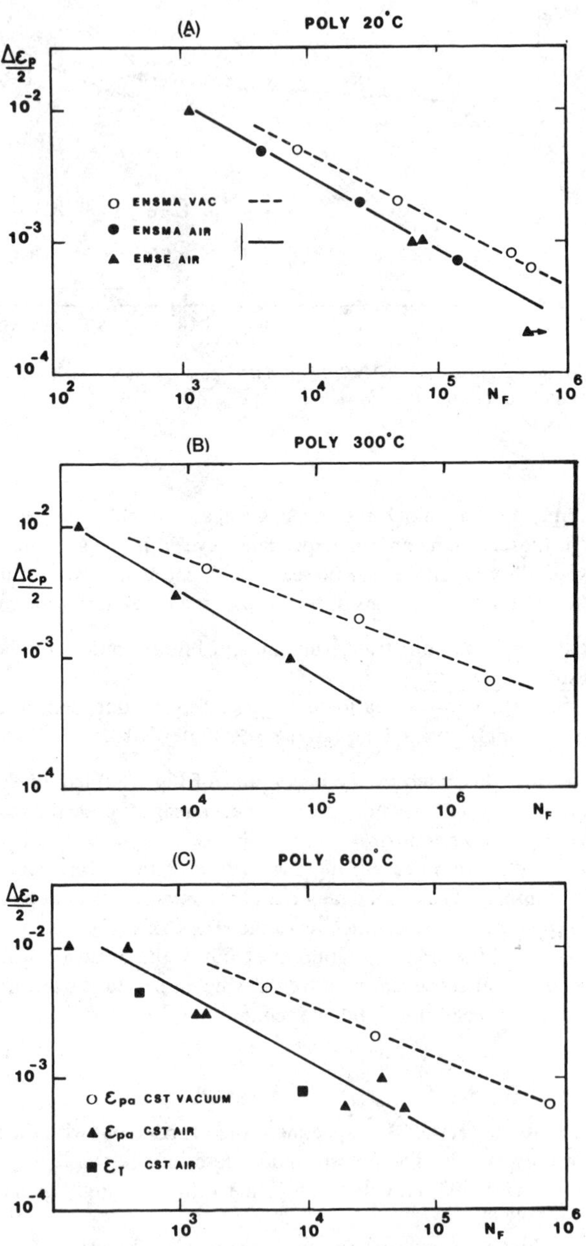

FIG. 5—*Plastic strain-life curves of 17-12 SPH steel at* (a) *20°C,* (b) *300°C, and* (c) *600°C in air and vacuum.*

TABLE 2—*Fatigue lifes of 17.12 SPH (316L) steel in air* N_A *and vacuum* N_v *at 20, 300, and 600°C.*

Temperature, °C	ϵ_{pa}	N_A*	N_v	N_v/N_A
	6.2×10^{-4}	180 000	530 000	2.9
20	2.03×10^{-3}	25 000	50 000	2.0
	5.07×10^{-3}	4 000	8 100	2.0
	6.5×10^{-4}	120 000	2 230 000	18.6
300	2.03×10^{-3}	19 000	215 000	11.3
	4.9×10^{-3}	4 000	13 200	3.3
	6.25×10^{-4}	40 000	750 000	18.7
600	2.1×10^{-3}	4 400	32 000	7.3
	5.03×10^{-3}	900	4 800	5.3

*N_A values obtained by interpolation from Figs. 5*a* to 5*c*.

600°C is often greater than that at 20°C. At 600°C and, in certain cases, at 300°C this initial rapid hardening is followed by a long period of cyclic softening for $\epsilon_{pa} \gtrsim 2 \times 10^{-3}$. The gradual softening stage, which can occupy 80 to 90% of the life, is not due to crack nucleation which generally occurs relatively late in the specimen life. Transmission electron microscopy (TEM) studies of the fatigued single crystals [*5*] show in fact that at 600°C when the softening occurs there is a significant change in the dislocation configurations towards a well-defined cell structure elongated along the $[\bar{1}01]$ slip direction.

The cyclic saturation stress-strain curves (CSSC) of the single crystals at 20, 300, and 600°C (taking "saturation" at $N_f/2$) have been previously described [*9*]. Although the absolute saturation stress levels of the single crystals are well below those of the polycrystals, there is a remarkable qualitative similarity for the strain and temperature dependencies of the cyclic stresses.

The cyclic hardening of a crystal in vacuum at 600°C and a plastic strain of 1.3×10^{-3} is shown in Fig. 10 together with the results of a test in air at $\epsilon_{pa} = 1 \times 10^{-3}$. The higher stresses in vacuum are probably due to a slightly higher plastic strain (at 600°C the σs values are very strain sensitive in the ϵ_{pa} range of 1 to 5×10^{-3}) and/or slight composition variations in the single crystals. The important feature of the $\sigma(N)$ plots, however, is the significantly extended fatigue life of the crystal in vacuum ($N_v = 95\ 000$) compared with the life in air ($N_A = 28\ 000$).

Fatigue Lifetimes and Crack Initiation—Figure 11 illustrates the strain-life plots of the $[\bar{1}49]$ single crystals in air at 20, 300, and 600°C. As for the polycrystals, between 20 and 300°C there is very little difference in fatigue life, but at 600°C the fatigue resistance is decreased by a factor of 3 to 5. The result of the test in vacuum is also indicated in Fig. 11 and shows that, as for the polycrystals, the fatigue life in vacuum at 600°C is roughly the same as the life in air at room temperature. It is clear that the temperature dependency of the single-crystal strain-life behavior is very similar to that of 316 polycrystals, although, for a given value of ϵ_{pa} the single crystals exhibit slightly longer fatigue lives, probably because of the lower stresses and shear strains involved. It is also obvious that the high temperature 600°C fall in the fatigue resistance of the single crystals in air is almost certainly related to an oxidation effect.

There are relatively few published results on the effects of environment on fatigue of single crystals at high temperatures. The present results on 316L steel crystals at different temperatures (i.e. with different degrees of oxidation) can be compared with some previous work on crystals in air and vacuum at room temperature. Duquette and Gell [*10*] observed a significant increase in the fatigue lives of Mar M200 single crystals at room temperature in vacuum, compared with air, for long-life (endurance) fatigue tests. In this γ' precipitation hardened superal-

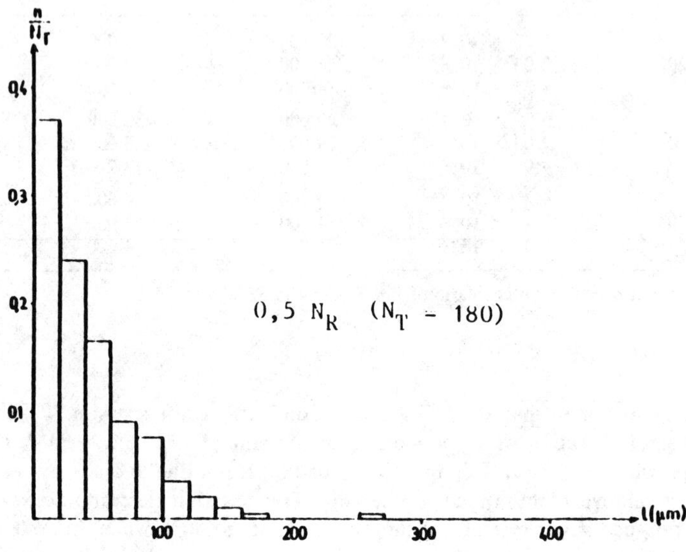

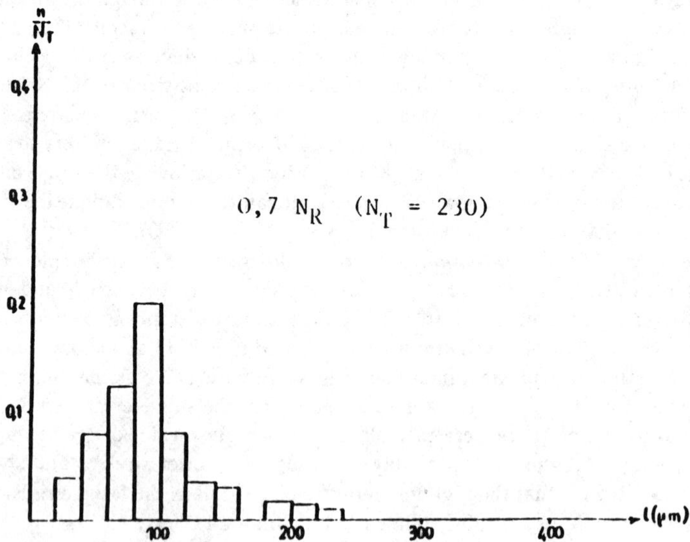

FIG. 6—*Histograms of the crack length densities at*

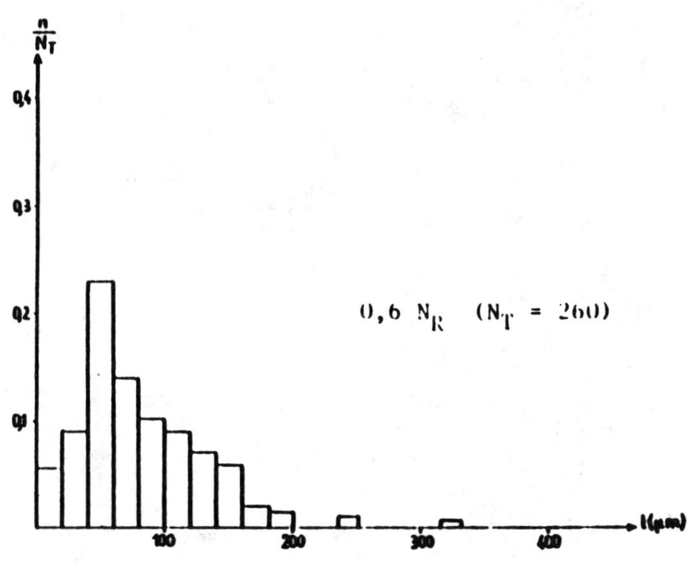

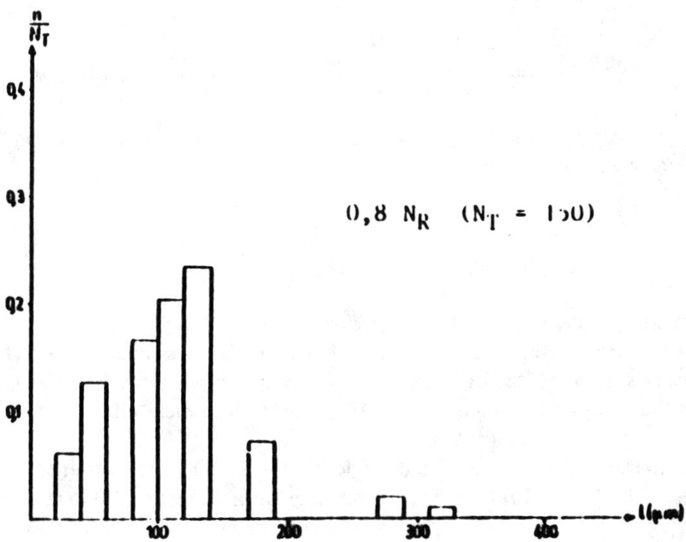

different life fractions of 17-12 SPH steel at 20°C in air.

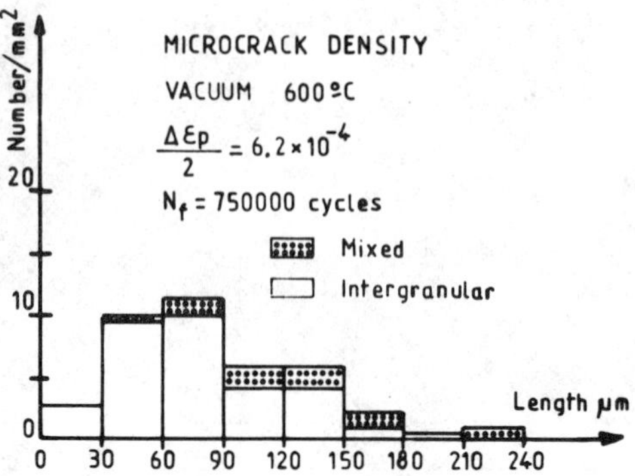

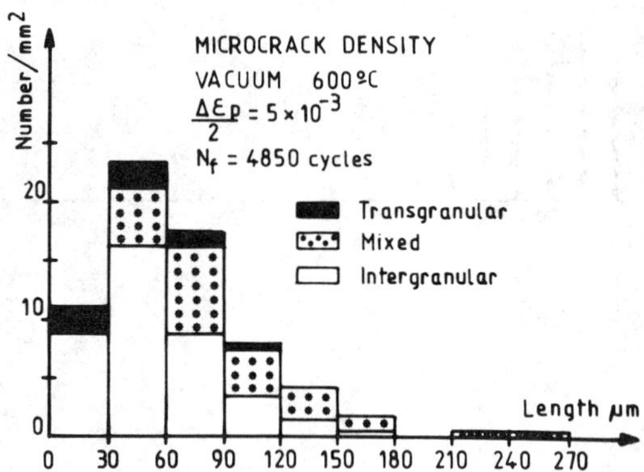

FIG. 7—*Histogram of the crack length densities in vacuum at 600°C at two strain amplitudes, showing the predominantly intergranular mode.*

loy, Stage I cracking is the dominate crack mechanism under these conditions. Similarly, more recent work by Wang and Mughrabi [*11,12*] has shown that in vacuum at 20°C the fatigue lives of Cu single crystals are increased by a factor of ~20 compared with lives in air for plastic strain amplitudes on the same order as ours ($\epsilon_{pa} \sim 10^{-3}$), an environment effect which is similar to that of stainless steels at 300 and 600°C.

The crack initiation mechanisms of the single crystals in air were examined on the "top" specimen faces, close to ~(210) at which the primary Burgers vector [$\bar{1}$01] emerges (Fig. 1*c*). Figure 12 shows typical SEM micrographs on specimens fatigued to rupture at 600°C at different strain amplitudes. Cracks always nucleate in the slip bands which, on the top face, are perpendicular to the stress axis. At high strains the oxide layer is very uneven, indicating that spalling occurs, and the density of secondary cracks is relatively high. At lower strains the crack density is much lower, so that at $\epsilon_{pa} = 10^{-3}$ only one significant crack was observed along the entire specimen length. Figure 13 gives the crack density (i.e., the number of cracks per unit

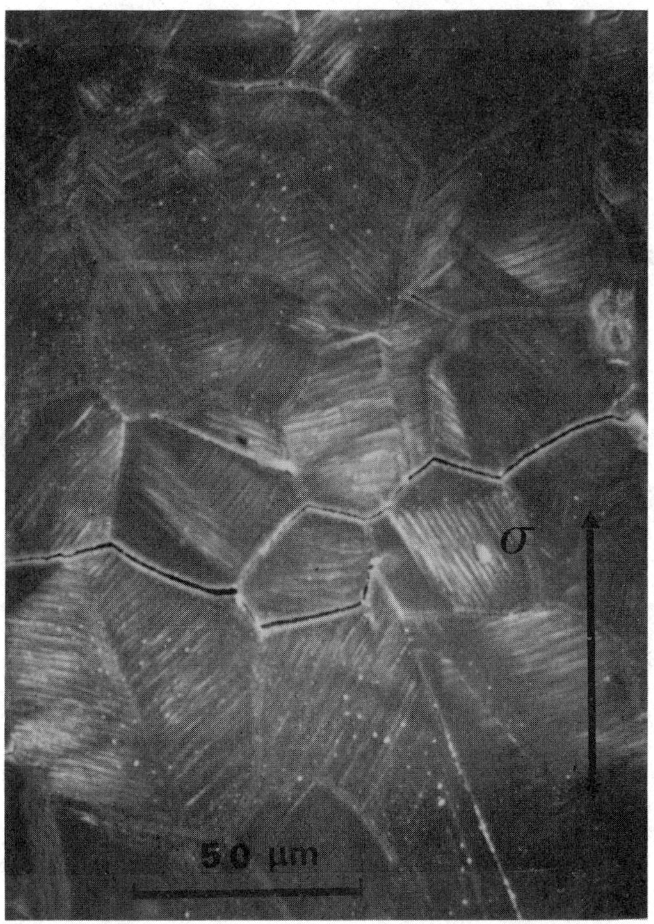

FIG. 8—*SEM micrograph of intergranular cracks in 17-12 SPH steel fatigued in vacuum at* $\epsilon_{pa} = 2 \times 10^{-3}$ *and 600°C.*

length, N/L, along the stress axis) as a function of plastic strain and indicates an approximate power-law relation: $N/L \simeq 10^8\ \epsilon_{pa}{}^3$.

An intriguing feature of the microcrack propagation process in crystals is the tendency for cracks to jump from one slip band to another at low and intermediate plastic strains. On a specimen cycled at a low plastic strain at 300°C (Fig. 14), where there are practically no secondary cracks, it is evident that the jump process does not simply involve shearing between two cracks on parallel slip planes, as in copper crystals [*13*], but is an integral part of the "Stage I" micropropagation process of an individual crack. In some cases the jump path could follow the trace of the secondary systems (for example, $(\bar{1}11)\langle 101 \rangle$) for which the Schmid factor is 0.47, although there is little external evidence of slip activity on this secondary system. A similar crack jumping process has been observed in the early stages of microcrack propagation within the grains of polycrystalline 316 steel. Tests in vacuum are currently in progress to verify this behavior in stainless steel crystals in the absence of oxidation.

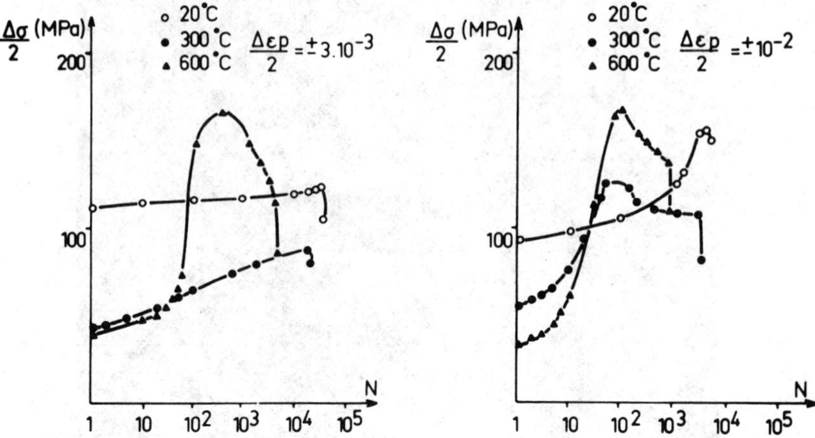

FIG. 9—*Single-crystal cyclic hardening behavior of 316 steel in air at 20, 300, and 600°C and two plastic strain amplitudes;* $\epsilon_{pa} = 3 \times 10^{-3}$ *and* 10^{-2}.

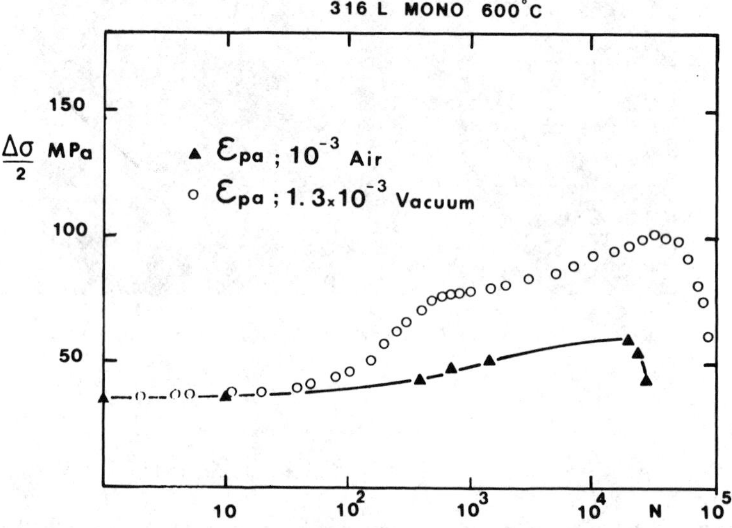

FIG. 10—*Cyclic hardening curves of 316 single crystals fatigued at 600°C and relatively low plastic strains in air and in vacuum.*

Discussion and Conclusions

This study describes on-going work aimed at understanding the mechanisms governing the fatigue resistance of austenitic stainless steels as a function of temperature and environment. The approach followed here consists of comparing the cyclic plastic fatigue behavior of polycrystals and single crystals at 20, 300, and 600°C in air and vacuum to clarify the roles of possible crack sites such as grain boundaries and slip bands. The present study shows that the re-

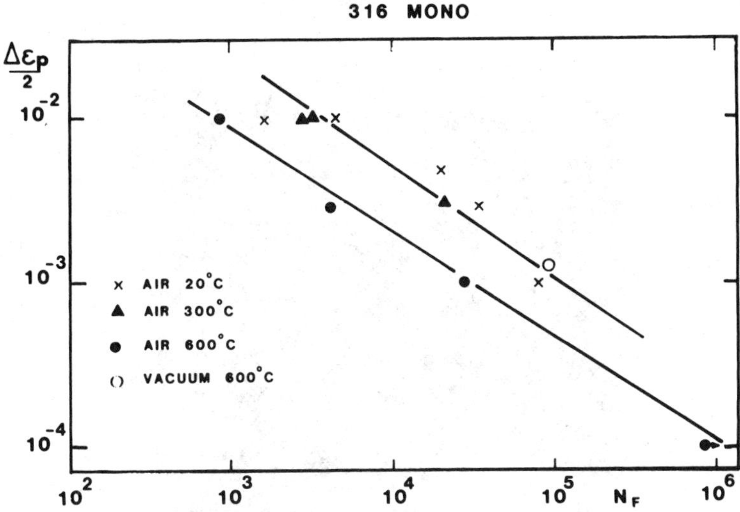

FIG. 11—*Strain-life plots of 316 single crystals at 20, 300, and 600°C in air and at 600°C in vacuum.*

spective influences of temperature and environment on the fatigue crack initiation mechanisms and strain-lifetimes of both single and polycrystal 316 steel are very similar in many respects. The significant results obtained so far can be summarized as follows:

In air:

1. Polycrystalline 316 steel exhibits a large decrease in fatigue life at 600°C when compared with results obtained at 20 and 300°C. The present results obtained using constant plastic strain tests over a wide range of ϵ_{pa} confirm previous results at constant total strain.
2. Single-crystal 316 steel specimens exhibit a similar decrease in fatigue life at 600°C. Obviously the fracture mode in single crystals is transgranular, so this decrease can reasonably be attributed to oxidation—enhanced cracking as for Cu single crystals.

In vacuum:

1. The fatigue lifetimes of polycrystalline 316L steel at 600°C are roughly the same as for those at 20 and 300°C, despite the fact that the crack initiation mode changes to intergranular at 600°C. Moreover, preliminary results on single crystals suggest that the single crystal lifetimes at 600°C are close to the values at 20 and 300°C in air.
2. The cyclic stress-strain response of polycrystalline 316L in vacuum is virtually the same as in air except when pronounced cyclic secondary hardening occurs as at 300°C.

The overall results on single-crystals and polycrystals in air and vacuum clearly demonstrate that the decrease in low-cycle fatigue lives of 316 steel at 600°C is not necessarily related to a change in crack initiation mode from transgranular to intergranular. The faster rate of crack initiation in air is almost certainly related to oxidation phenomena on the surface and close to the crack embryos. Oxidation would seem to affect Stage I crack initiation rather than Stage II propagation, since (1) in the majority of our experiments $N_f > 1000$ (i.e., most of the fatigue life is spent in crack initiation [*7*]), and (2) previous work on the environmental influence on Stage II crack propagation rates (i.e., $da/dNf(\Delta K)$ in 316 steel indicates that there is very little difference between air and vacuum at room temperature [*14*] and at elevated temperatures [*15*]. The

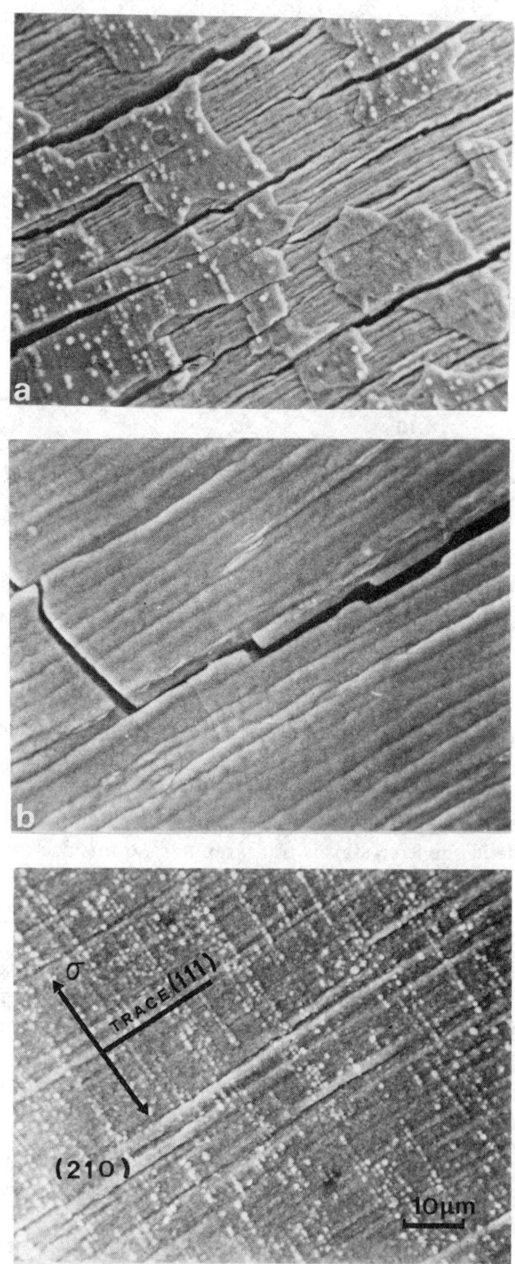

FIG. 12—*SEM micrographs of the crack initiation mode in 316 single crystals at 600°C in air at different strain amplitudes:* (a) 10^{-2}, (b) 3×10^{-3}, *and* (c) 10^{-3}. *Orientation and magnification indicated in* (c).

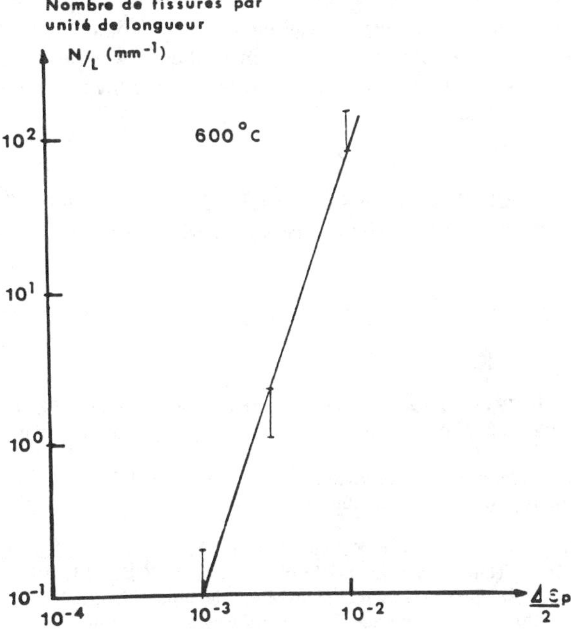

FIG. 13—*Crack density* N/L *as a function of plastic strain for 316 crystals fatigued in air at 600°C.*

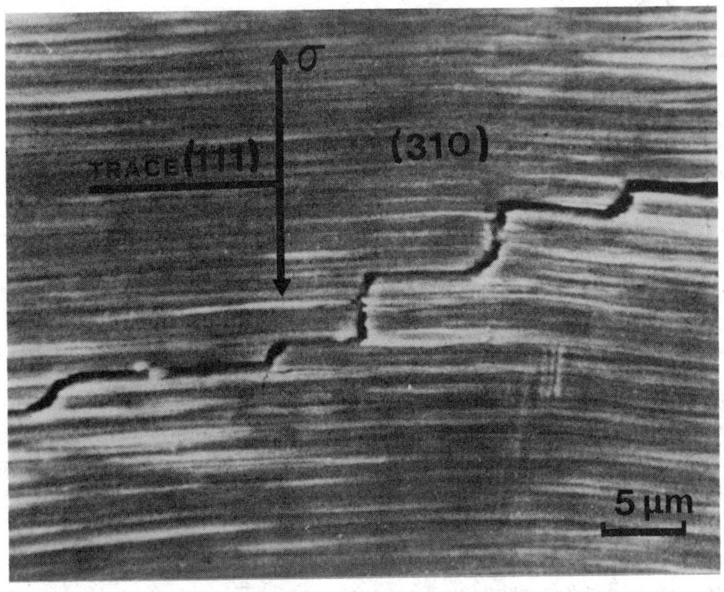

FIG. 14—*Microcrack propagation of 316 steel crystal at 300°C in air,* $\epsilon_{pa} = 3 \times 10^{-3}$, *showing crack jumping between parallel slip planes.*

influence of oxidation on Stage I crack initiation and microcrack propagation in austenitic stainless steels seems to be a key factor controlling low-cycle fatigue lives; the detailed mechanisms by which oxygen accelerates crack initiation in these steels needs to be elucidated by further fundamental work to provide a firmer basis for fatigue life time predictions.

Acknowledgments

This work has been partially supported by DGRST Grant 80-7-0526-8. The authors wish to thank Mr. P. Rieux for invaluable technical assistance with the single-crystal fatigue tests.

References

[*1*] Coffin, L. F., Jr., in *Proceedings,* International Conference on Fatigue: Chemistry, Mechanics and Microstructure, NACE-2, National Association of Corrosion Engineers, Houston, Tex., 1972, pp. 590-600.

[*2*] Coffin, L. F., Jr., *Metallurgical Transactions,* Vol. 3, 1972, pp. 1777-1778.

[*3*] Furuya, K., Nagata, N., and Watanabe, R., *Journal of Nuclear Materials,* Vol. 89, 1980, pp. 372-382.

[*4*] Smith, D. L., Zeman, G. J., Nateson, K., and Kassner, T. F. in *Proceedings,* International Conference on Liquid Metal Technology in Energy Production, Conf. 760503-PI, 1976, Paper WA-7, p. 359.

[*5*] Gorlier, C., *Docteur-Ingénieur thesis,* Ecole de Mines de Saint-Etienne, 1984.

[*6*] Wang, R. and Mughrabi, H., *Materials Science and Engineering,* Vol. 63, 1984, p. 147.

[*7*] Levaillant, C., *Doctoral thesis,* Paris, 1984.

[*8*] Mendez, J. and Violan, P. in *Basic Questions in Fatigue, Vol. II, ASTM STP 924,* forthcoming.

[*9*] Gorlier, C., Amzallag, C., Rieux, P., and Driver, J. H. in *Proceedings,* 2nd International Conference on Fatigue and Fatigue Thresholds, Fatigue '84, C. J. Beever, Ed., EMAS, 1984, Vol. 1, pp. 41-48.

[*10*] Duquette, D. J. and Gell, M., *Metallurgical Transactions*, Vol. 2, 1971, pp. 1325-1331.

[*11*] Wang, R., Mughrabi, H., McGovern, S., and Rapp, M., *Materials Science and Engineering,* Vol. 65, 1984, pp. 219-233.

[*12*] Wang, R. and Mughrabi, H., *Materials Science and Engineering,* Vol. 65, 1984, pp. 235-243.

[*13*] Basinski, Z. S. and Basinski, S. J. in *Proceedings,* Sixth International Conference on the Strength of Metals and Alloys (ICSMA6), R. C. Gifkins, Ed., Pergamon Press, Oxford, 1982, Vol. 2, p. 819.

[*14*] Amzallag, C., Rabbe, P., Bathias, C., Benoit, D., and Truchon, M. in *Fatigue Crack Growth Measurement and Data Analysis, ASTM STP 738,* American Society for Testing and Materials, Philadelphia, 1981, pp. 29-44.

[*15*] Sadanada, K. and Shahinian, P., *Metallurgical Transactions,* Vol. 11*A*, 1980, pp. 267-276.

DISCUSSION

J. Bressers[1] *(written discussion)*—Upon changing from an air to a vacuum environment at 600°C one would expect a concurrent change from an intergranular to a transgranular failure. You observed an opposite behavior. Could this possibly be explained on the bases of an increase in the degree of homogenization of the deformation in the vacuum test which, in terms of time, lasted much longer than the air test?

J. H. Driver et al. (authors' closure)—More recent work has shown that the transition to an intergranular mode of crack initiation under vacuum at 600°C is in fact related to a more homo-

[1]Joint Research Centre, J. E. C., Petten, The Netherlands.

geneous deformation *in the grains.* An SEM examination [*1*] of the specimen surfaces after fatigue in vacuum reveals that the slip lines in the grain are fine and homogeneously distributed and that the extrusions are not very pronounced, but also that severe surface relief effects are found *at the grain boundaries.* Furthermore, a current TEM study [*2*] shows that the prolonged cycling which is possible under vacuum leads to a more homogeneous dislocation structure in the grains at 600°C.

Discussion References

[*1*] Belrami, C., thèse 3ème cycle, Poitiers, France, 1986.
[*2*] Gerland, M. and Violan, P., *Materials Science and Engineering,* Vol. 64, 1986, pp. 23–33.

Glenn R. Romanoski,[1] Stephen D. Antolovich,[2] and Regis M. Pelloux[1]

A Model for Life Predictions of Nickel-Base Superalloys in High-Temperature Low Cycle Fatigue

REFERENCE: Romanoski, G. R., Antolovich, S. D., and Pelloux, R. M., **"A Model for Life Predictions of Nickel-Base Superalloys in High-Temperature Low Cycle Fatigue,"** *Low Cycle Fatigue, ASTM STP 942,* H. D. Solomon, G. R. Halford, L. R. Kaisand, and B. N. Leis, Eds., American Society for Testing and Materials, Philadelphia, 1988, pp. 456–469.

ABSTRACT: Extensive characterization of low-cycle fatigue damage mechanisms was performed on polycrystalline René 80 and IN100 tested in the temperature range from 871 to 1000°C. Low-cycle fatigue life was found to be dominated by propagation of microcracks to a critical size governed by the maximum tensile stress. A model was developed which incorporates a threshold stress for crack extension, a stress-based crack growth expression, and a failure criterion. The mathematical equivalence between this mechanistically based model and the strain-life low-cycle fatigue law was demonstrated using cyclic stress-strain relationships. The model was shown to correlate the high-temperature low-cycle fatigue data of the different nickel-base superalloys considered in this study.

KEY WORDS: nickel-base superalloys, high-temperature low-cycle fatigue, high-strain fatigue crack propagation, fatigue models, environmental effects

The low-cycle fatigue design philosophy is based on relating the fatigue life of engineering components subjected to cyclic inelastic strains to the fatigue life of test specimens subjected to the same conditions in the laboratory [*1,2*]. The number of cycles to failure of the specimen should correspond to the number of cycles to initiation of a small crack in the component. In general, the greatest fraction of cyclic life in high-temperature low-cycle fatigue (HTLCF) is dominated by the growth of microcracks. Hence, from the standpoint of design and fatigue life management, it is desirable to translate phenomenological strain-life approaches to ones which address microcrack growth. This requires experimentally determining high-strain fatigue crack growth (HSFCG) behavior under the conditions of interest and over the crack length regime of interest. Fatigue life can then be calculated by integrating between the appropriate limits.

Skelton has recently reviewed the subject of HSFCG [*3*]. Many studies have characterized crack growth rates with empirical relationships [*4–18*]. A frequently used form that appears to have broad application is

$$\frac{da}{dN} = C(\Delta\epsilon_p)^\alpha a^Q \tag{1}$$

[1]Research Assistant and Professor, respectively, Department of Materials Science and Engineering, Massachusetts Institute of Technology, Cambridge, MA 02139.

[2]Director and Professor of Metallurgy, Fracture and Fatigue Research Laboratory, Georgia Institute of Technology, Atlanta, GA 30332.

where a is the crack length, $\Delta\epsilon_p$ is the bulk plastic strain range and C, α, and Q are constants that depend on the material. The value of Q is ≥ 1 and generally thought to be related to the mechanism of crack extension. Tomkins proposed that $Q = 1$ when crack advance per cycle occurred by irreversible shear decohesion at the crack tip [*19*]. This micro-plasticity mechanism leads to ductile striation formation [*4,16,20*]. More recent studies have employed the J-integral [*21–24*] as well as the strain intensity factor [*12,25,26*] to correlate crack growth rate with test parameters. One conclusion to be drawn from a review of the published work is that there is not a universally applicable HSFCG relationship.

This paper will develop a model for life prediction of conventionally cast nickel base superalloys tested under HTLCF conditions. The model is based on a postulated crack growth expression which is in agreement with the observed crack growth mechanism.

Background

Numerous U.S. and European laboratories participated in the AGARD-Strain Range Partitioning (SRP) program to assess the applicability of the SRP method of life prediction in HTLCF. Mechanical test results were thoroughly reported in the conference proceedings [*27*]. Under the sponsorship of NASA-Lewis, a large number of these specimens were collected for characterization of cyclic damage mechanisms. Specimens were characterized by scanning electron microscopy (SEM) of fracture surfaces and gage section surfaces and metallographic sectioning. A number of additional HTLCF experiments were performed to validate certain conclusions. Conventionally cast polycrystalline René 80 and IN100 tested in the temperature range from 871 to 1000°C were studied most extensively. The model which follows is based on this investigation of the physical nature of HTLCF damage [*28*].

Some Preliminary Considerations

It is important to consider the nature of control generally employed in HTLCF testing. The material test volume is under either longitudinal or diametral displacement control, which results in the material being subjected to elastic and plastic bulk strains with total strain limits. The cyclic stress-strain response evolves throughout the experiment to reflect the mechanical conditions and the metallurgical changes that cause hardening or softening. The materials being considered for this model exhibit a stabilized cyclic stress-strain response throughout most of the experiment (Fig. 1). However, a sudden decay in the maximum load occurs when one or more macroscopic cracks appear. This is the result of crack(s) opening and accommodating part of the longitudinal displacement as well as a reduction in the effective cross-sectional area. Consequently, the test volume is no longer subjected to the same total or plastic strain range which it experienced throughout most of the experiment. At this point the test should be considered to be over since the main crack(s) will continue to grow under a reduced driving force.

The sudden drop in maximum load corresponds to the onset of rapid propagation of one or more cracks in the specimen gage section. This point is operationally defined as failure. However, the cycle number at failure cannot be determined with precision, so the convention is to take the cycle number at 10% load drop as failure. For high inelastic strain - high maximum stress tests, rapid propagation occurs for small crack sizes and specimen separation may occur with no detectable load drop.

Physical Basis of the Model

The following observations generally apply to many conventionally cast polycrystalline nickel base superalloys tested under HTLCF conditions.

(*a*) Crack initiation occurs very early in cyclic life, probably on the first cycle. This fact was

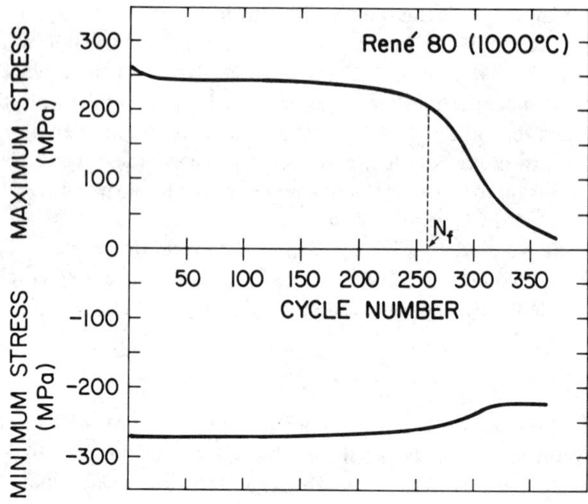

FIG. 1—*Peak tensile and compressive stresses versus cycle number for cast René 80 tested in air at 1000°C.*

established by sectioning specimens from tests stopped after a few cycles. Crack initiation generally occurs at the intersection of transverse grain boundaries and the specimen surface. Multiple initiation sites are generally observed.

(*b*) Crack propagation occurs along transverse grain boundaries but there will also be segments of transgranular propagation so that the crack plane does not deviate too strongly from a direction normal to the applied stress (Fig. 2).

(*c*) Failure is defined as the cycle number at which there was a transition to rapid propagation. This transition (failure) corresponds to a particular combination of crack size and maximum stress; the higher the maximum stress, the smaller the critical crack size. Consequently, this stress controlled failure criterion was used in the model.

The HTLCF life is essentially a microcrack propagation phenomenon for the materials and test conditions being considered. The relevant crack length regime is generally less than 2 mm.

The Model

Since failure occurs at a particular combination of crack size and maximum stress, a quasi-fracture mechanics failure criterion should apply:

$$S\sigma_{\max}(\pi a_f)^{\gamma} = K' \tag{2}$$

where S is a shape factor which depends on crack and specimen geometry, $\sigma_{\max}$ is the maximum tensile stress in the cycle, a_f is the crack depth at the onset of rapid propagation, and K' is the apparent stress intensity factor. γ will be taken as $1/2$ in the following development.

Since cyclic life is dominated by microcrack propagation, an appropriate expression for crack growth rate must be formulated. Previously published crack growth expressions, such as Eq 1, are well founded on experimental data but appear to be a function of the micromechanism of crack advance. In none of these cases was the mechanism identified to be incremental decohesion along a predominantly intergranular path. Antolovich et al. have shown that an air envi-

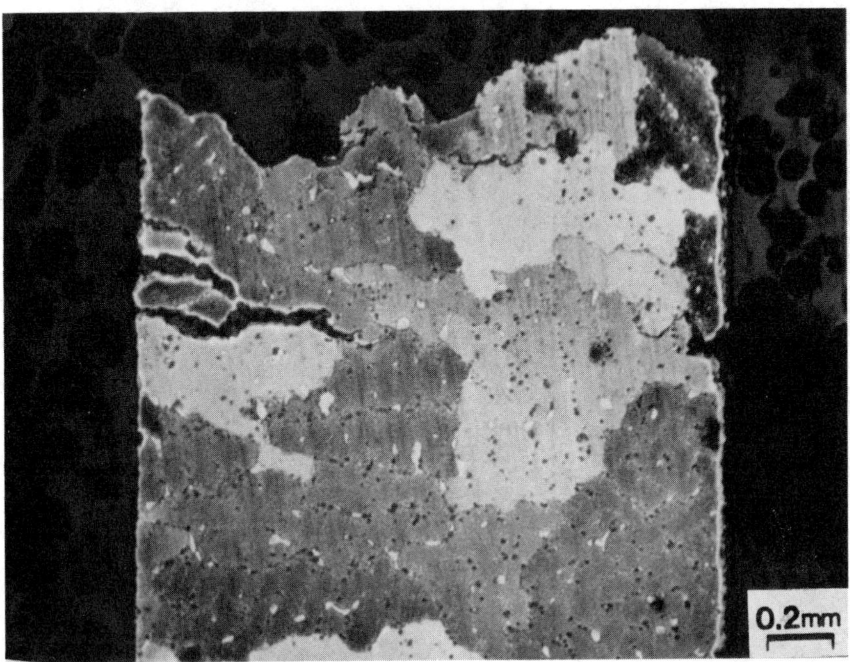

FIG. 2—*Optical micrograph showing an intergranular crack propagation path for cast IN100 tested in air at 925°C. The specimen was of tubular geometry.*

ronment has a deleterious effect on the HTLCF life of superalloys because it degrades the cohesive strength of grain boundaries [*29,30*]. This degradation of strength has been attributed to oxygen which diffuses down the boundaries [*31,32*]. Given a constant cyclic period, the environmental degradation ahead of the crack tip would be constant in each cycle and allow an equal increment of crack extension in each cycle. The crack will extend Δa in a given cycle until the cohesive strength of the grain boundary is greater than or equal to the maximum effective stress across the grain boundary. This mechanism is shown schematically in Fig. 3. Hence, the following crack growth expression may be postulated:

$$\frac{da}{dN} = A\Delta\sigma_{\text{eff}}^{m} \tag{3}$$

where A and m are constants. σ_{eff} will be taken as the effective engineering stress for ease of calculations. Calculating the true stress may be a worthwhile refinement, particularly for low $\Delta\epsilon_p$ - low $\sigma_{\max}$ tests where the presence of multiple cracks of increasing size near the end of the test obviously increases σ_{eff}.

Expressing the effective stress range in a form which incorporates a threshold stress for microcrack extension, σ_T, Eq 3 becomes

$$\frac{da}{dN} = A(\sigma_{\max} - \sigma_T)^m \tag{4}$$

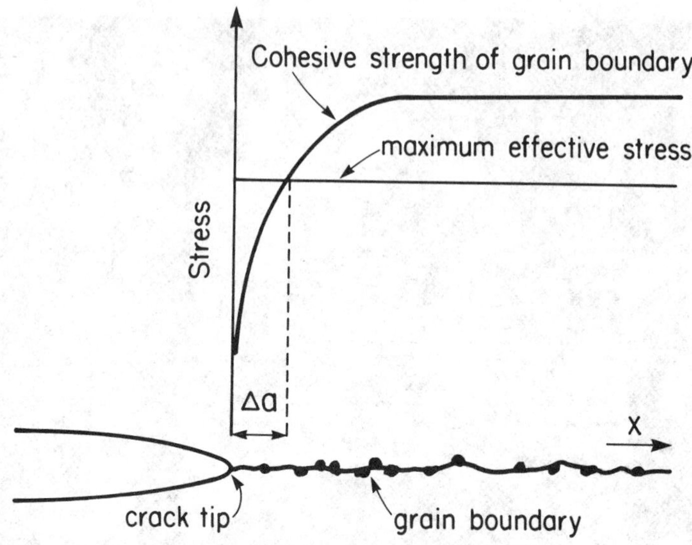

FIG. 3—*Schematic presentation of the mechanism of incremental decohesion along a favorably oriented grain boundary. The cohesive strength of the grain boundary is shown to be reduced in the near-crack tip region due to the diffusive ingression of oxygen.*

Quantification of a crack growth rate represents a balance between the mechanical driving force and the material's resistance to that driving force. A more general formulation of Eq 4 would include the cyclic period, t, to account for the time dependency of environmental degradation ahead of the crack tip. Antolovich et al. [*29*] have successfully modeled the role of environment in crack initiation in nickel-base superalloys by introducing a $\sqrt{t}$ factor to account for the diffusion of oxygen along grain boundaries. Since the present focus is on continuous cycling tests of constant frequency, time does not appear explicitly in the crack growth expression.

Equation 4 essentially states that the local mechanical driving force at the crack tip, σ_{eff}, does not increase with increasing crack length throughout most of the cyclic life. The bulk conditions dominate due to the physically small size of the cracks being considered.

The postulated crack growth rate curves are shown schematically in Fig. 4. It is apparent that the cyclic lives for high inelastic strain - high maximum stress test specimens are shorter for two reasons: first, the mechanical driving force (σ_{max}) is higher, causing a greater crack advance per cycle, and secondly, the crack length required to bring about failure is much shorter (the failure criterion). Determination of N_f and a_f can only be approximated for the low inelastic strain - low maximum stress tests.

Cyclic lives can be calculated by integrating Eq 4:

$$\int_{a_0}^{a_f} da = \int_{N_i}^{N_f} A(\sigma_{max} - \sigma_T)^m dN \quad (5)$$

where a_0 is the initial crack length and N_i is the number of cycles to the formation of a Mode I crack. a_0 will be taken as zero in the following development, since only smooth specimens with non-degraded surfaces and no known defects were studied; a_0 would be non-zero in the case of specimens pre-exposed at elevated temperature (see, for example, Ref *33*). In this case a large

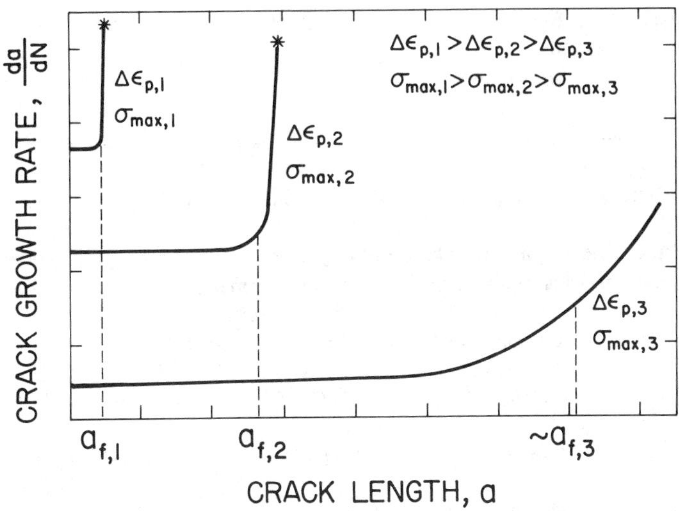

FIG. 4—*Schematic presentation of the postulated crack growth rate curves.*

increment of crack extension, $a_0 + \Delta a$, would occur in the first cycle. N_i will be taken as zero in the following development. Accounting for $a_0 \neq 0$ and $N_i \neq 0$ when appropriate would give the model greater generality. Equation 5 can now be integrated between the appropriate limits:

$$a_f = A(\sigma_{\max} - \sigma_T)^m N_f \tag{6}$$

Substituting the above expression for a_f into the failure criterion given in Eq 2:

$$S\sigma_{\max}\pi^{1/2}[A(\sigma_{\max} - \sigma_T)^m N_f]^{1/2} = K' \tag{7}$$

Combining the constants into one:

$$\sigma_{\max}[(\sigma_{\max} - \sigma_T)^m N_f]^{1/2} = K_f \tag{8}$$

At this point Eq 8 was applied to HTLCF data. The constant m was determined by trial and error. Values between 0.5 and 5 were chosen in increments of 0.1, and all the calculations subsequently described were carried out. It was found that the integer value $m = 2$ gave the best correlation for every data set considered. The physical significance of $m = 2$ will be discussed later. Equation 8 becomes

$$\sigma_{\max}(\sigma_{\max} - \sigma_T)N_f^{1/2} = K_f \tag{9}$$

Applying the model to HTLCF data is now straightforward. All that is required is $\sigma_{\max}$ and N_f for each experiment. The two constants σ_T and K_f must be determined for each data set. This is carried out by using Eq 9 and iterating on σ_T to find that value which produces the minimum variance in K_f for the data set. K_f is taken to be the average value of K_f for the data set once σ_T has been determined.

The calculations described above will be presented explicitly for two data sets: uncoated IN100 tested in air at 925°C and coated René 80 tested in air at 1000°C [*34*]. In both cases continuous cycling at a frequency of 0.5 Hz and 1.0 Hz, respectively, was the test mode. Values of σ_{max} were taken at half-life.

Expressing Eq 9 for both materials:

$$\text{IN100:} \quad \sigma_{max}(\sigma_{max} - 121.6) = 1.92 \times 10^6 N_f^{-1/2} \qquad R = 0.988 \tag{10}$$

$$\text{René 80:} \quad \sigma_{max}(\sigma_{max} - 129.6) = 1.04 \times 10^6 N_f^{-1/2} \qquad R = 0.986 \tag{11}$$

where R is the coefficient of correlation and σ_{max} is in MPa. The very high correlation is illustrated graphically in Figs. 5 and 6 by plotting the parameter $\sigma_{max}(\sigma_{max} - \sigma_T)$ against N_f. The slope is $-1/2$.

The Coffin-Manson relationships for the same data sets are:

$$\text{IN100:} \quad \Delta\epsilon_p = 4.74 N_f^{-0.555} \qquad R = 0.984 \tag{12}$$

$$\text{René 80:} \quad \Delta\epsilon_p = 4.64 N_f^{-0.478} \qquad R = 0.989 \tag{13}$$

The conventional strain-life representation shows essentially the same level of correlation. The inelastic strain range is also plotted against N_f in Figs. 5 and 6.

It is now possible to attach greater physical significance to σ_T. Consider the plot of maximum stress versus inelastic strain range shown in Fig. 7 for the René 80 data. When extrapolating this curve back to $\Delta\epsilon_p = 0$, an intercept is found at σ_T. The same observation was made for the IN100 data. This suggests that the threshold stress for microcrack propagation corresponds to a threshold stress for bulk cyclic plasticity. This may be due to the fact that crack tips were generally observed to be blunt and, consequently, the crack tip region fully loaded only when crack opening occurred to the degree made possible by plastic accommodation.

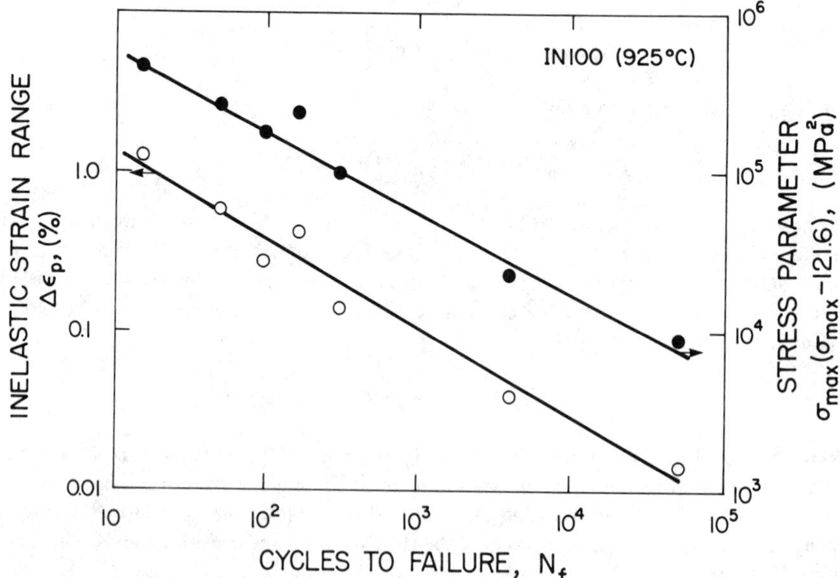

FIG. 5—*Graphical representation of Eq 9 and the Coffin-Manson equation for cast IN100 tested in air at 925°C.*

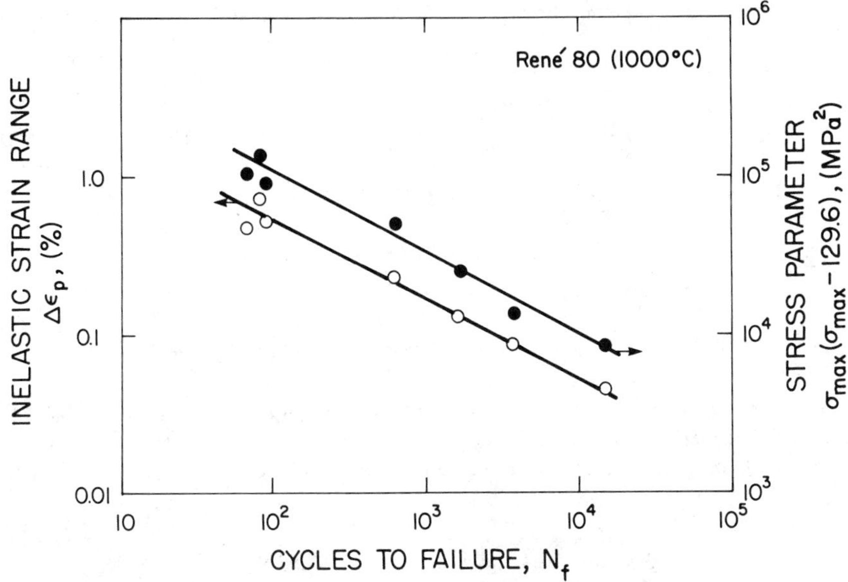

FIG. 6—*Graphical representation of Eq 9 and the Coffin-Manson equation for cast René 80 tested in air at 1000°C.*

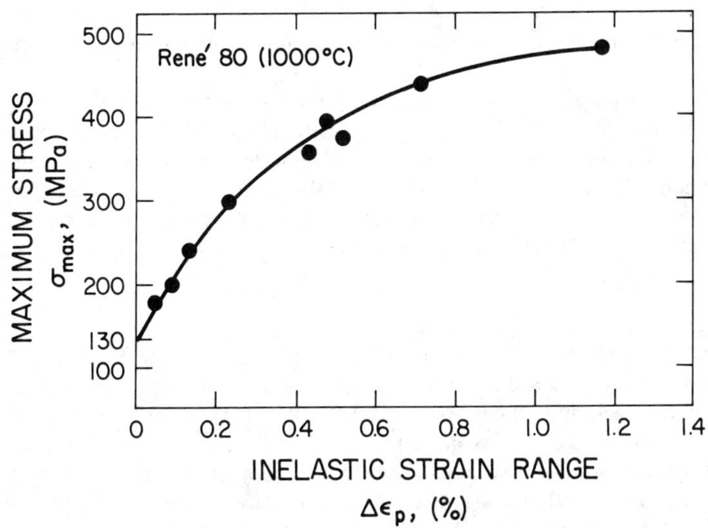

FIG. 7—*Cyclic stress-strain response for René 80 tested in air at 1000°C.*

Equations 10 and 11 are expressed in terms of maximum stress and cycles to failure. The mathematical equivalence between these equations and the Coffin-Manson equations representing the same data is not directly apparent. The relationship between the maximum stress and the inelastic strain range may be expressed in the form of the Hollomon [*35*] or Ludwik [*36*] equations:

IN100:

$$\text{Hollomon Eq:} \quad \sigma_{max} = 709.2\Delta\epsilon_p^{0.335} \quad R = 0.996$$

$$\text{Ludwik Eq (using } \sigma_T\text{):} \quad \sigma_{max} = 121.6 + 654.0\Delta\epsilon_p^{0.568} \quad R = 0.988 \tag{14}$$

René 80:

$$\text{Hollomon Eq:} \quad \sigma_{max} = 489.3\Delta\epsilon_p^{0.345} \quad R = 0.993$$

$$\text{Ludwik Eq (using } \sigma_T\text{):} \quad \sigma_{max} = 129.6 + 417.9\Delta\epsilon_p^{0.696} \quad R = 0.993 \tag{15}$$

Note the difference in the exponent of $\Delta\epsilon_p$ in the Hollomon and Ludwik equations. By making the appropriate substitutions for σ_{max} and $(\sigma_{max} - \sigma_T)$ Eqs 10 and 11 can be rewritten in strain-life format and compared directly with the Coffin-Manson equation:

IN100:

$$\text{Eq 10:} \quad \Delta\epsilon_p = 4.83N_f^{-0.554}$$
$$\text{Coffin-Manson Eq:} \quad \Delta\epsilon_p = 4.74N_f^{-0.555}$$

René 80:

$$\text{Eq 11:} \quad \Delta\epsilon_p = 4.78N_f^{-0.481}$$
$$\text{Coffin-Manson Eq:} \quad \Delta\epsilon_p = 4.64N_f^{-0.487}$$

The ability to make this conversion is the result of high correlation for all the equations employed.

The model has been applied to numerous HTLCF data sets for conventionally cast superalloys. The results are reported in Table 1. The correlation of HTLCF data with this stress based model, Eq 9, was found to be equivalent to or better than the correlation found with the Coffin-Manson equation. The conversion of Eq 9 to the strain-life format was performed in each case. The similarity to the Coffin-Manson equation was generally good, especially when there was high correlation (~0.99) for all equations employed.

Discussion

The physical significance of $m = 2$ in Eq 4 can be rationalized with the aid of Fig. 3. It implies a relationship between the cohesive strength of the grain boundary, σ_{gb}, and the distance, X, from the crack tip in the near-crack tip region. This will depend on: (*1*) the relationship between the concentration of oxygen and X, and (*2*) the relationship between this concentration of oxygen and σ_{gb}. The first relationship should follow the general solution for diffusion along a grain boundary path and has been established analytically and experimentally [*43*]. Unfortunately, the second relationship has not been established. The dependence of σ_{gb} on X may be expected to follow the approximate form depicted schematically in Fig. 3. This relationship can be expressed as

$$\sigma_{gb} = RX^{1/2} \tag{16}$$

TABLE 1—*Model test results.*

Material	Temperature (°C)	Coefficient of Correlation		$\Delta\epsilon_p = CN_f^\beta$				Reference
				C		β		
		Eq 9[a]	C-M[b]	Eq 9[c]	C-M[b]	Eq 9[c]	C-M[b]	
IN100	900	0.976	0.977	3.9	4.1	−0.50	−0.51	*37*
IN100	1000	0.994	0.998	6.0	5.4	−0.57	−0.56	*37*
MAR M002	750	0.985	0.989	19.8	15.7	−0.87	−0.84	*38*
MAR M002	850	0.978	0.987	7.5	6.6	−0.77	−0.76	*38*
MAR M002	1040	0.952	0.996	3.7	4.2	−0.52	−0.54	*38*
IN738 LC	850	0.947	0.982	4.0	4.2	−0.56	−0.58	*39*
MAR M200	1000	0.973	0.954	3.8	3.1	−0.46	−0.44	*40*
René 95 (PM)	650	0.992	0.990	102.0	84.0	−0.94	−0.92	*41*
IN100 (PM)	760	0.981	0.977	41.0	23.0	−0.86	−0.76	*42*

[a]Eq 9 determined by using σ_{max} and N_f.
[b]The Coffin-Manson relationship determined by using $\Delta\epsilon_p$ and N_f.
[c]Eq 9 converted to strain-life format using cyclic stress-strain relationships.

where R is a constant which depends on the material, environment, and cyclic period. The condition satisfied by one increment of crack extension is

$$\sigma_{eff} = \sigma_{gb} = R(\Delta a)^{1/2} \tag{17}$$

Hence

$$\frac{\Delta a}{\Delta N} = \frac{\sigma_{eff}^2}{R^2} \tag{18}$$

which is Eq 4.

It is also interesting to note that when $m = 2$ the right hand side of Eq 4 is approximately proportional to the plastic work above σ_T in a given cycle. The concept of cyclic plastic work is incorporated in the *J*-integral approach proposed by Dowling to characterize the mechanical driving force for the growth of small cracks in high strain fatigue [*22*].

In this model, crack growth rate was expressed in terms of an effective stress. This stress based approach was shown to be consistent with the observed mechanism of incremental decohesion along an environmentally degraded grain boundary path. Previous investigators of HSFCG have correlated crack growth rate with expressions which include the inelastic strain range as in Eq 1. In these studies crack propagation was observed to be transgranular and striations were usually observed on fracture surfaces. It is generally agreed that crack advance by this mechanism involves shear deformation and decohesion of the material at the crack tip along favorably oriented slip planes [*44*]. The magnitude of crack extension in each cycle is related to the magnitude of the shear strain accommodated ahead of the crack tip and in turn related to the bulk plastic strain range. Therefore, including $\Delta\epsilon_p$ in an expression for crack growth rate is appropriate. It is interesting to note that when the Ludwik equation is substituted for $(\sigma_{max} - \sigma_T)$ in Eq 4, the crack growth rate is approximately proportional to $\Delta\epsilon_p$.

The main point of departure from the more conventional approach is that crack growth rate was assumed to be independent of crack length. This was felt to be appropriate for two reasons: (*1*) the environmental degradation associated with a constant cyclic period predisposes grain boundaries to an equal increment of crack extension in each cycle, and (*2*) the near crack tip stress or strain fields are not expected to be a function of crack length for physically small cracks. It is important to point out that the crack length regime which represents most of the cyclic life is less than 1 mm. Such physically "short cracks" have been shown to exhibit anomalous crack growth behavior by many investigators. An empirical approach proposed by El Haddad et al. [*45*] to account for the observed threshold behavior of short cracks is a modification of the stress intensity parameter $K \propto \sigma\sqrt{\pi(a + \ell_0)}$. This modification results in the assumed mechanical driving force being approximately independent of crack length when $a \ll \ell_0$. When a approaches ℓ_0 there is a transition to LEFM behavior and the mechanical driving force becomes proportional to $\sqrt{a}$. This is similar to the crack growth behavior proposed by this model. The mechanical driving force is postulated to be independent of crack length with a transition to rapid propagation (defined here as failure) where the crack driving force becomes proportional to $\sqrt{a}$ (the failure criterion). Similar crack growth behavior has been observed by Feng in measuring "short crack" growth rates in PM René 95 [*46*]. The model was also successfully applied to two PM superalloys tested under HTLCF conditions (Table 1).

Finally consider some further evidence in support of a stress based approach for representing the HTLCF behavior of conventionally cast nickel base superalloys. When complex cycles which include hold times are considered, shorter cyclic lives are observed for compression holds than for tension holds with equivalent inelastic strain ranges [*27*]. This is a result of higher values of σ_{max} in compression hold cycles. In an investigation of the HTLCF behavior of conventionally cast MAR-M 200 Milligan and Bill [*40*] noted a fourfold increase in cyclic life at 1000°C compared to 927°C for all values of inelastic strain range. This was attributed to values of σ_{max} at 1000°C being about 25% lower than at 927°C, therefore reducing the dynamic crack driving force. On the basis of environmental considerations alone, one would expect shorter cyclic lives at 1000°C. These results and arguments are in essential agreement with some earlier work of Antolovich et al. [*29*] on René 80 tested at elevated temperature.

Conclusions

1. The HTLCF life of the conventionally cast nickel-base superalloys considered here is dominated by the propagation of microcracks by incremental decohesion along a grain boundary path. Failure occurs at a critical crack size governed by the maximum tensile stress.
2. The HTLCF life can be correlated with a parameter based on the maximum tensile stress. This approach is consistent with many trends observed in the data and with the HTLCF model proposed here.

Acknowledgments

This work was performed under the sponsorship of NASA-Lewis Research Center; contract NSG-3-263. We would especially like to acknowledge the many helpful discussions with Dr. M. Hirschberg of NASA during the course of the study.

References

[*1*] Coffin, L. F., *Transactions of ASME,* Vol. 76, 1954, pp. 931–950.
[*2*] Manson, S. S., *Heat Transfer Symposium,* University of Michigan Press, June 1952; also NACA TN 2933, July 1953.
[*3*] Skelton, R. P. in *Low-Cycle Fatigue and Life Prediction, ASTM STP 770,* American Society for Testing and Materials, Philadelphia, 1982, pp. 337–381.
[*4*] Price, A. T. and Elder, W. J., *Journal of the Iron and Steel Institute,* Vol. 204, 1966, pp. 594–598.
[*5*] Solomon, H. D. and Coffin, L. F. in *Fatigue at Elevated Temperatures, ASTM STP 520,* American Society for Testing and Materials, Philadelphia, 1973, pp. 112–122.

[6] Maiya, P. S., *Scripta Metallurgica,* Vol. 9, 1975, pp. 1277-1282.
[7] Solomon, H. D., *Metallurgical Transactions A,* Vol. 4, 1973, pp. 341-347.
[8] Maiya, P. S. and Busch, D. E., *Metallurgical Transactions A,* Vol. 6A, 1975, pp. 1761-1766.
[9] Wareing, J., *Metallurgical Transactions A,* Vol. 6A, 1975, pp. 1367-1377.
[10] Baudry, G. and Pineau, A., *Materials Science and Engineering,* Vol. 28, 1977, pp. 229-242.
[11] Skelton, R. P., *Materials Science and Engineering,* Vol. 19, 1975, pp. 193-200.
[12] Skelton, R. P., *Materials Science and Engineering,* Vol. 32, 1978, pp. 211-219.
[13] Skelton, R. P., *Materials Science and Engineering,* Vol. 35, 1978, pp. 287-298.
[14] Reuchet, J. and Rémy, L., *Fatigue of Engineering Materials and Structures,* Vol. 2, 1979, pp. 51-62.
[15] Skelton, R. P., *Fatigue of Engineering Materials and Structures,* Vol. 2, 1979, pp. 305-318.
[16] Yamaguchi, K. and Kanazawa, K., *Metallurgical Transactions A,* Vol. 10A, 1979, pp. 1445-1451.
[17] Chalant, G. and Rémy, L., *Acta Metallurgica,* Vol. 28, 1980, pp. 75-88.
[18] Huang, J., Pelloux, R. M., and Runkle, J., *Metallurgical Transactions A,* Vol. 11, 1980, pp. 899-904.
[19] Tomkins, B., *Philosophical Magazine,* Vol. 18, 1968, pp. 1041-1066.
[20] Yamaguchi, K., Kanazawa, K., and Yoshida, S., *Materials Science and Engineering,* Vol. 33, 1978, pp. 175-181.
[21] Dowling, N. E. and Begley, J. A. in *Mechanics of Crack Growth, ASTM STP 590,* American Society for Testing and Materials, Philadelphia, 1976, pp. 82-103.
[22] Dowling, N. E. in *Cyclic Stress-Strain and Plastic Deformation Aspects of Fatigue Crack Growth, ASTM STP 637,* American Society for Testing and Materials, Philadelphia, 1977, pp. 97-121.
[23] Hoshide, T., Yamada, T., Fujimura, S., and Hayashi, T., *Engineering Fracture Mechanics,* Vol. 21, No. 1, 1985, pp. 85-101.
[24] Leis, B. N., *Engineering Fracture Mechanics,* Vol. 21, 1985.
[25] Haigh, J. R. and Skelton, R. P., *Materials Science and Engineering,* Vol. 36, 1978, pp. 133-137.
[26] Leis, B. N. in *Fracture Mechanics: Fifteenth Symposium, ASTM STP 833,* American Society for Testing and Materials, Philadelphia, 1984, pp. 449-480.
[27] "Characterization Of Low Cycle-High Temperature Fatigue By The Strain Range Partition Method," *AGARD Conference Proceedings No. 243,* NATO, Advisory Group for Aerospace Research and Development, 1978.
[28] Romanoski, G. R., "Mechanisms of Deformation and Fracture in High-Temperature Low-Cycle Fatigue of René 80 and IN100," M.S. thesis, University of Cincinnati, 1981; also NASA Contractor Report 165498.
[29] Antolovich, S. D., Liu, S., and Baur, R., *Metallurgical Transactions A,* Vol. 12A, 1981, pp. 473-481.
[30] Antolovich, S. D., Rosa, E., and Pineau, A., *Materials Science and Engineering,* Vol. 47, 1981, pp. 47-57.
[31] Woodford, D. A., *Metallurgical Transactions A,* Vol. 12A, 1981, pp. 229-308.
[32] Woodford, D. A. and Bricknell, R. H., *Metallurgical Transactions A,* Vol. 12A, 1981, pp. 1945-1949.
[33] Antolovich, S. D., Domas, P., and Strudel, J. L., *Metallurgical Transactions A,* Vol. 10A, 1979, pp. 1859-1868.
[34] Halford, G. R. and Nachtigall, A. J., *AGARD Conference Proceedings No. 243,* NATO, Advisory Group for Aerospace Research and Development, 1978, pp. 2.1-2.14.
[35] Hollomon, J. H., *Transactions of AIME,* Vol. 162, 1945, p. 268.
[36] Ludwik, P., *Elemente der Technologischen Mechanik,* Springer, Berlin, 1909, p. 32.
[37] Chaboche, J. L., Policella, H., and Kaczmarek, H., *AGARD Conference Proceedings No. 243,* NATO, Advisory Group for Aerospace Research and Development, 1978, pp. 4.1-4.20.
[38] Atunes, V. T. A. and Hancock, P., *AGARD Conference Proceedings No. 243,* NATO, Advisory Group for Aerospace Research and Development, 1978, pp. 5.1-5.9.
[39] Day, M. F. and Thomas, G. B., *AGARD Conference Proceedings No. 243,* NATO, Advisory Group for Aerospace Research and Development, 1978, pp. 10.1-10.13.
[40] Milligan, W. W. and Bill, R. C., NASA Technical Memorandum 83769, 1984.
[41] Hyzak, J. M. and Bernstein, H. L., *AGARD Conference Proceedings No. 243,* NATO, Advisory Group for Aerospace Research and Development, 1978, pp. 11.1-11.25.
[42] Van Wanderham, M. C., Wallace, R. M., and Annis, C. G., *AGARD Conference Proceedings No. 243,* NATO, Advisory Group for Aerospace Research and Development, 1978, pp. 3.1-3.17.
[43] Martin, G. and Perraillon, B., *Grain Boundary Structure and Kinetics,* American Society for Metals, 1980, pp. 239-295.
[44] McEvily, A. J., in *Fatigue Mechanisms: Advances in Quantitative Measurement of Physical Damage, ASTM STP 811,* American Society for Testing and Materials, Philadelphia, 1983, pp. 283-312.
[45] El Haddad, M. H., Smith, K. N. and Topper, T. H., *Transactions of the American Society of Mechanical Engineers,* Vol. 101, No. 1, 1979.
[46] Feng, J., private communication.

DISCUSSION

L. Remy[1] *(written discussion)*—You assume oxidation to form the basis of your high temperature fatigue model. Oxidation is a time-dependent phenomenon; this time dependency has been explicitly described in a recent fatigue crack growth model accounting for oxidation fatigue interactions [*1*]. I do not see such a time dependency in your model. How do you take into account frequency effects in high temperature fatigue?

G. R. Romanoski et al. (authors' closure)—This time dependency is recognized with regard to Eq 4. Time does not appear explicitly in the crack growth expression, because only tests of constant cyclic period were being considered. Introduction of a $t^{1/2}$ factor to account for the time dependency of diffusion is suggested and is in agreement with the Reuchet-Remy model to which you referred [*1*] and an earlier one by Antolovich et al. [*2*]. The implication of a more general formulation which includes time is that the cohesive strength of the grain boundary would be degraded to a greater depth with increasing cyclic period. This could be represented in Fig. 3 of our paper by a series of curves for grain boundary cohesive strength which shift downward and to the right with increasing cyclic period. Sufficient data was not available to be more general at this time.

The formation of a physical oxidation product is not considered to be the necessary precursor to crack extension, but rather the environmental degradation of grain boundary cohesive strength ahead of the crack tip. Oxidation product was observed to form at the specimen surface, along crack flanks, and even at crack tips for cracks that had arrested in grain boundaries which deviated significantly from a favorable orientation. Grain boundary decohesion has been shown to precede the formation of bulk oxidation product [*3*]. Oxidation is acknowledged to play an important role in the initiation process for alloys and test conditions not giving rise to an intergranular fracture path and may play a role under any conditions where oxide wedging affects the mechanical driving force.

C. T. Sims[2] *(written discussion)*—The two superalloys you studied are similar in several ways: both are cast, they are of similar "type" chemical composition, and they both develop a grain boundary γ' film. Yet, the inelastic strain parameter versus N_f curve is much closer to the stress parameter versus N_f curve for one alloy than it is for the other. Can you attribute some physical significance in alloy behavior to this difference? Does it, for instance, relate to crack growth rate?

G. R. Romanoski et al. (authors' closure)—The greater separation of the two curves for IN100 is due to a significantly higher response stress for that alloy at any given imposed strain range. Since the scales are the same in Figs. 5 and 6 of our paper, they can be superimposed for consideration of the physical significance of these relative positions. In considering inelastic strain range versus cyclic life, the René 80 data exhibit longer cyclic lives by a factor of two to three when compared with the IN100 data. On the basis of environmental considerations alone, one would expect shorter lives for René 80 due to the higher temperature. However, the response stress for IN100 is greater than that for René 80 by approximately 50% for all values of inelastic strain range. This behavior is consistent with the proposed model. Higher values of the maximum stress result in faster crack growth rates and require a smaller critical crack size to bring about failure, hence, shorter cyclic lives.

If the stress parameter versus cyclic life is compared for these two alloys, the relative positions are inverted. The René 80 data fall to shorter cyclic lives by a factor of two to three. (This is also

[1]Centre des Matériaux, Ecole des Mines de Paris, B.P. 87, 91003 Evry Cedex, France.
[2]General Electric Co., Schnectady, NY.

true if the comparison is made on the basis of stress versus cyclic life.) When stress is recognized as the driving force for crack extension under these conditions, the René 80 data fall into a relative position consistent with an accelerated environmental degradation. A comparison of the two alloys tested at the same temperature would give a better indication of inherent resistance to HTLCF damage by this mechanism. The two governing factors would be grain boundary cohesive strength and response stress.

B. Ilschner[3] *(written discussion)*—This question relates to the depth of decohesion, Δa, introduced in one of your introductory slides. You associate this loss of cohesion to environmental interaction, probably oxygen. Is it your conclusion that this "zone of decohesion"—being a very convincing model by itself—depends on the assumption of environmental interaction? Or would you also agree to discuss a modified version stating that the decohesion zone may well be *extended* by environmental interaction *but* does exist even in vacuum due to acceleration cavitation which is produced by the local stress concentration ahead of the crack tip? See, for example, the work of Riedel. By the way, this effect may also be summarized within the context of environmental interaction, namely, as diffusion of vacancies from the outside vacuum along grain boundaries into the process zone. Finally, if a is mainly determined by oxygen diffusion, there should be a definite hold time effect, $\sqrt{D_{gb} t_h}$.

G. R. Romanoski et al. (authors' closure)—In examining specimens of the same materials tested in vacuum, we have also observed evidence of intergranular decohesion [*3*]. The mechanism of crack extension may be essentially the same, because grain boundaries are intrinsically weaker than the matrix at high temperatures. In vacuum, crack initiation was associated with a loss of material from grain boundaries at the intersection with the specimen surface, particularly for René 80 tested at 1000°C. The model was applied to these data and yielded a significantly lower correlation ($R = 0.78$). This suggests that the number of cycles to crack initiation represents a significant fraction of cyclic life and would not be accounted for by the model in its present form. Testing of these materials at high temperatures in air ensures crack initiation at grain boundaries early in cyclic life, if not on the first cycle. Dramatic evidence for the role of environment in degrading the cohesive strength of grain boundaries is manifested in the significant reductions in HTLCF life observed for specimens tested after pre-exposure at high temperature. A hold time effect is observed in these alloys.

Discussion References

[*1*] Reuchet, J. and Rémy, L., *Metallurgical Transactions A*, Vol. 14A, 1983, pp. 141–149.

[*2*] Antolovich, S. D., Liu, S., and Baur, R., *Metallurgical Transactions A*, Vol. 12A, 1981, pp. 473–481.

[*3*] Romanoski, G. R., "Mechanisms of Deformation and Fracture in High-Temperature Low-Cycle Fatigue of René 80 and IN100," M.S. thesis, University of Cincinnati, 1981; also NASA Contractor Report 165498.

[3]Department des Matériaux, Ecole Polytechnique Federale de Lausanne, 1007 Lausanne, Switzerland.

T. Bui-Quoc,[1] R. Gomuc,[1] A. Biron,[1] H. L. Nguyen,[2] and J. Masounave[2]

Elevated Temperature Fatigue-Creep Behavior of Nickel-Base Superalloy IN 625

REFERENCE: Bui-Quoc, T., Gomuc, R., Biron, A., Nguyen, H. L., and Masounave, J., **"Elevated Temperature Fatigue-Creep Behavior of Nickel-Base Superalloy IN 625,"** *Low Cycle Fatigue, ASTM STP 942*, H. D. Solomon, G. R. Halford, L. R. Kaisand, and B. N. Leis, Eds., American Society for Testing and Materials, Philadelphia, 1988, pp. 470–486.

ABSTRACT: A research program has been carried out to establish the low-cycle fatigue and creep-fatigue behaviors of Inconel 625 at elevated temperatures (650 and 815°C). The main observations were related to the effect of temperature and of hold times. Under continuous cycling, a temperature increase from 650 to 815°C caused a reduction in the fatigue life by a factor of 2 at a high strain and by a factor of 3 at a low strain.

Tension hold times had little detrimental effect on cyclic life at 650°C concerning the life reduction with respect to continuous cycling; at 815°C, on the other hand, this became more significant. Compressive hold times also had a damaging effect on fatigue life, and this was larger than that associated with tensile hold times; in particular, it was very pronounced for low strain levels at 815°C.

An analysis of data using a life prediction method previously suggested for creep-fatigue combination loadings has also been carried out. The method takes into account the damaging effect due to a compression hold time separately from that due to a tensile hold time in interspersed creep-fatigue loadings. The overall correlation between theoretical calculations and experimental results is reasonably good, for strain levels ranging from 1.2 to 0.4%.

KEY WORDS: continuous cycling, fatigue-creep, tension/compression hold time, Inconel 625

Superalloy IN 625 is a potential candidate material for use in structural components subjected to cyclic loadings at high temperatures, such as parts of jet engines and nuclear reactors [*1*]. To the authors' knowledge, little information concerning the material behavior under these loading conditions is presently available; in particular, the scarceness of experimental data under creep-fatigue combination loadings makes the use of this material in component design difficult.

The present research program has been carried out to provide the mechanical properties on the low-cycle fatigue and creep-fatigue behavior of Inconel 625 at elevated temperatures (650 and 815°C). Under continuous cycling, the data base of the material is established; the influence of hold-times introduced in each cycle (interspersed creep-fatigue) on the material life is then examined.

A life prediction procedure recently suggested [*2,3*] has been applied to the data obtained under cyclic loading, with and without hold times. The main characteristic of the approach stems from a combination of two stress- (or strain-) dependent damage concepts (for fatigue

[1]Department of Mechanical Engineering, Ecole Polytechnique, Montreal, Quebec, Canada.
[2]Industrial Materials Research Institute, National Research Council Canada, Boucherville, Quebec, Canada.

and for creep), which take into account the stress relaxation behavior during the hold period. The overall correlation between isothermal experimental results and theoretical predictions is discussed.

Experimental Procedure

Superalloy IN 625 is a nickel-chromium alloy strengthened by an addition of molybdenum. The nominal chemical composition of the material studied is given in Table 1.

Smooth cylindrical test specimens with uniform gage length of 6.35 mm in diameter conforming to ASTM E 606 recommendations were used in this experimental work. The research was a joint effort conducted in two laboratories, one at Ecole Polytechnique, the other at the Industrial Materials Research Institute.

All fatigue tests were carried out under uniaxial, strain-controlled conditions using a servohydraulic closed-loop testing machine. Strains were measured by an axial extensometer, and the load measurements were obtained by means of a load-cell in series with the specimen. Stress-strain hysteresis loops were recorded at predetermined cycles throughout the life of the specimen.

The specimen was heated by an electrical resistance furnace with three separately controlled heating zones providing a uniform temperature distribution over the specimen gage length. The test temperature was controlled within ±2°C.

The loading waveform was generated by a computer which also assumed data acquisition during the tests. All tests were performed in air under isothermal conditions, at 650 and 815°C; strain ranges with a zero mean value between 0.4 and 1.2% were considered. For continuous cycling tests, a triangular waveform with a constant strain rate (1×10^{-2} s^{-1}) was used. In interspersed creep-fatigue tests (cyclic loading with hold time in each cycle), tensile or compressive dwell periods of 1, 2, 5, 10, 15, and 30 min were imposed in each cycle at peak strain. Failure was defined as a reduction of the stabilized maximum cyclic stress by 50%.

Test Results

Continuous Cycling

A summary of the experimental results obtained under continuous cycling is given in Table 2 and the diagram of the total strain range $\Delta\epsilon$ in terms of the number of cycles at failure N_0 is shown in Fig. 1. The results show that a temperature increase from 650 to 815°C causes a reduction in fatigue life by a factor of 2 at a high strain and by a factor of 3 at a low strain.

The variation of the stress range $\Delta\sigma$ versus the number of applied cycles n, obtained at different strain ranges, is shown typically in Fig. 2. Under continuous cycling, it appears that the material cyclic characteristic depends significantly upon the temperature; in fact, for a high strain level, at 650°C, a strain hardening behavior was observed, while at 815°C a reverse trend (strain softening) was noted. In contrast, at a low strain range, the material exhibited a relatively stabilized cyclic behavior at both temperatures [*4*].

It is also noted that IN 625, when subjected to continuous cycling, developed in most cases a relatively small compressive mean stress (less than 6% of the stress range).

TABLE 1—*Chemical composition (weight percent) of Inconel 625.*

C	Mn	Fe	S	Si	Al	Ti	Mo	Cb-Ta	Cr	Ni
0.05	0.25	2.5	0.008	0.25	0.2	0.2	9.0	3.65	21.5	bal.

TABLE 2—*Continuous cycling LCF results of Inconel 625.*

Specimen No.	Strain Range, % Total ($\Delta\epsilon$)	Plastic ($\Delta\epsilon_p$)	Elastic ($\Delta\epsilon_e$)	Stress, MPa Maximum (σ_m)	Minimum (σ_{min})	Cycles to Failure (N_0)
			$T = 650°C$			
54	0.55	0.013	0.537	459	−472	29 000
55	0.78	0.130	0.650	553	−559	6 780
46	0.99	0.320	0.670	584	−598	940
29	1.19	0.400	0.790	666	−700	800
64	0.60	0.024	0.576	531	−477	16 942
			$T = 815°C$			
32	0.37	0.021	0.349	253	−239	57 342
25	0.6	0.063	0.537	387	−402	5 450
24	0.8	0.224	0.576	395	−496	1 374
34	1.0	0.360	0.640	464	−477	560
23	1.0	0.407	0.593	460	−463	421
44	1.2	0.585	0.615	466	−474	455
45	1.2	0.595	0.605	457	−465	315

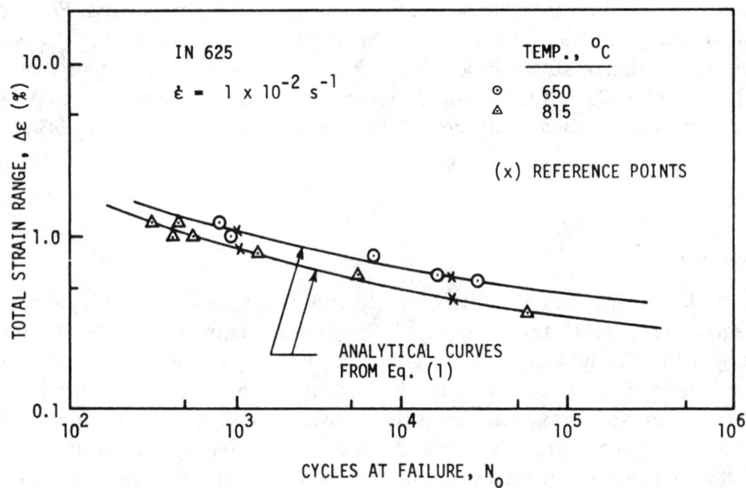

FIG. 1—*Fatigue diagram of IN 625 together with analytical curves.*

Interspersed Creep-Fatigue Tests

The results of the interspersed creep-fatigue tests at two temperatures are summarized in Table 3. On the basis of a limited number of test results, the following observations can be made:

1. IN 625 exhibited a particular stress relaxation behavior during a tension hold time (Fig. 3). The stress relaxation rate was relatively high at the beginning of the hold period and the tensile stress reached a saturation level rapidly. Furthermore, the temperature has a very strong

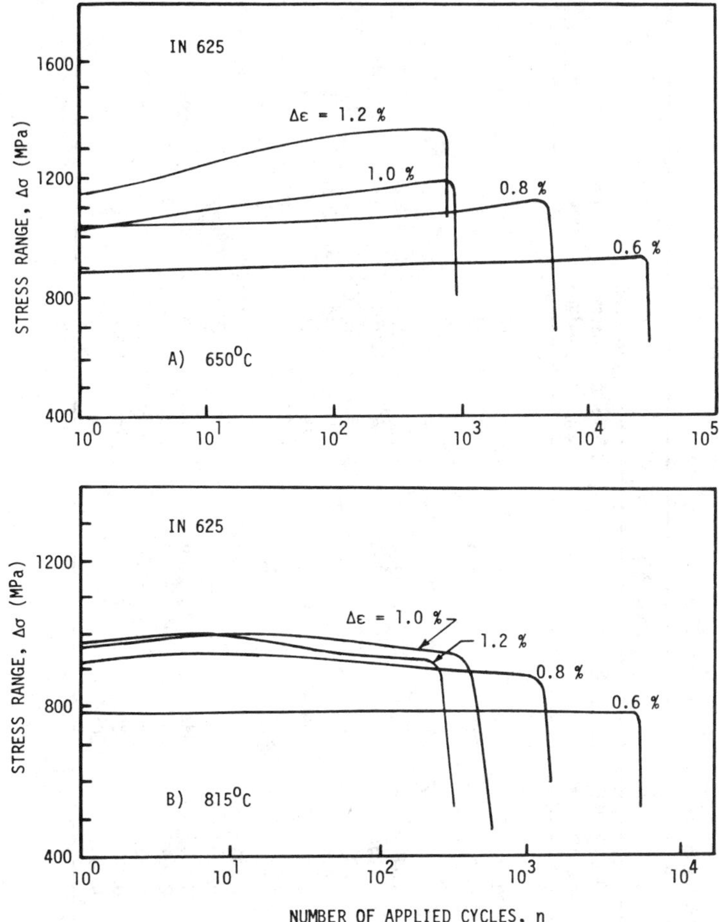

FIG. 2—*Variation of stress range with applied cycles under continuous cycling.*

effect on this relaxation rate. In fact, after an elapsed time equivalent to 10% of the total hold period, the maximum stress was relaxed by 25% at 650° and by 90% at 815°C. This particular characteristic is reflected on the material life subjected to creep-fatigue loading.

The stress relaxation behavior during compression hold times was similar to that observed during tension hold times.

2. In general, the variation of the stress range $\Delta\sigma$ with applied cycles n, in tests with a tensile hold time, follows a pattern comparable to that already shown for tests without a hold time. However, the stabilized stress range (usually considered at half-life) may depend upon the hold time periods (Fig. 4). It is also noted that the effect of the hold time period on the stabilized stress range increases with a decrease of the imposed strain level for both temperatures. The trend of variation, however, is quite different for a low strain range: with an increase of tension hold time t_d, $\Delta\sigma$ increases at 650°C, whereas it decreases at 815°C. Further investigation may be necessary in order to explain this material characteristic.

Examination of the hysteretic behavior of this material reveals that, during hold time tests, a mean stress opposite the hold direction was developed. This mean stress, already reported for

TABLE 3—*Hold time fatigue test results.*

Specimen No.	Hold Time, min[a]		Strain Range, %			Stress, MPa			Relaxed Stress, MPa	Cycles to Failure (N)	Time to Failure (t_f), h
	t_d	t_e	Total ($\Delta\epsilon$)	(Plastic) ($\Delta\epsilon_p$)	Elastic ($\Delta\epsilon_e$)	Maximum (σ_m)	Minimum (σ_{min})	Mean σ_{mean}			
						T = 650°C					
59	10	0	1.0	0.3	0.7	536	−846	−155	154	1092	183
40	30	0	1.2	0.447	0.753	599	−863	−132	208	712	356
						T = 815°C					
58	5	0	0.6	0.26	0.340	250	−456	−103	211	722	60
60	15	0	0.62	0.31	0.310	233	−405	−86	210	1041	261
51	15	0	0.82	0.428	0.392	332	−499	−83.5	297	437	109
49	15	0	1.02	0.605	0.415	371	−501	−65	332	262	66
36	1	0	1.22	0.777	0.443	469	−520	−25.5	332	220	4
30	10	0	1.21	0.633	0.577	431	−566	−67.5	346	310	52
38	10	0	1.22	0.846	0.374	388	−454	−33	308	349	58
33	15	0	1.22	0.777	0.443	441	−543	−51	357	210	53
37	15	0	1.22	0.800	0.420	432	−526	−47	356	223	56
27	30	0	1.22	0.800	0.420	441	−536	−47.5	388	170	85
75	0	5	0.39	0.086	0.304	443	−142	150.5	101	880	74
71	0	5	0.61	0.290	0.320	493	−358	67.5	303	310	26
68	0	5	0.65	0.289	0.361	489	−337	76	271	430	36
69	0	15	1.10	0.706	0.394	506	−427	39.5	384	250	63

[a] t_d: tension hold time; t_e: compression hold time.

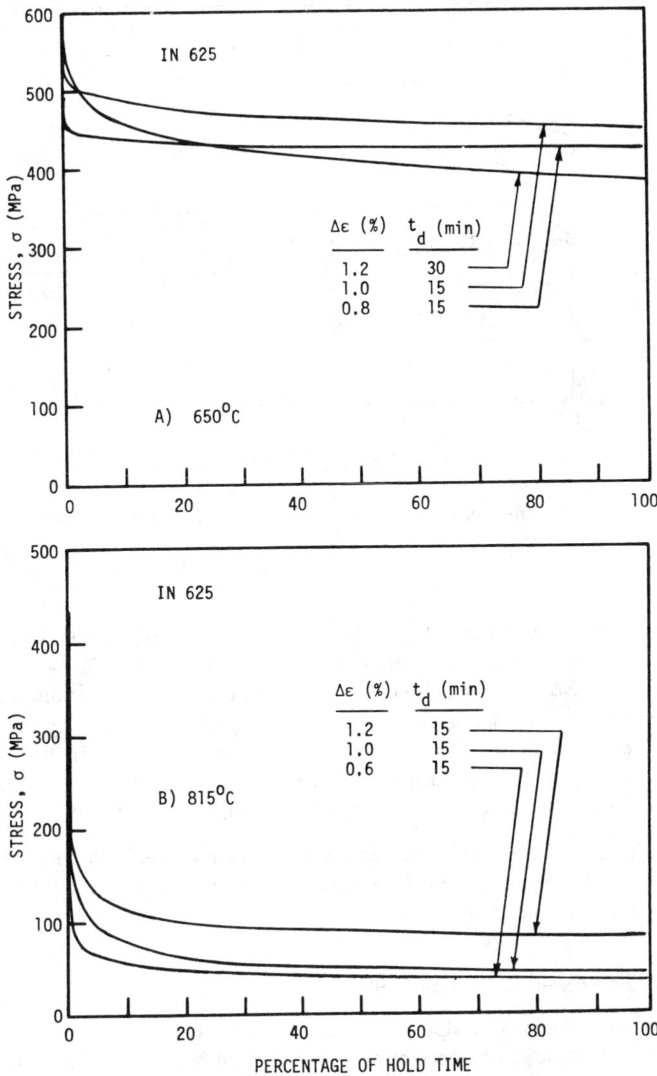

FIG. 3—*Typical stress relaxation curves during tension hold times* (t_d).

several nickel-base superalloys [*5–8*], is usually attributed to the rapid stress relaxation occurring during hold times. The cyclic stress is driven in a direction opposite the hold direction and this leads to a tensile mean stress under compressive holds and a compressive mean stress under tensile holds. As typically observed in the case of high strength and ductility materials [*7*], the magnitude of the mean stress for IN 625 is larger at lower strain levels.

3. Tension hold times (t_d) have little detrimental effect on cyclic life at 650°C with respect to continuous cycling (Fig. 5*a*). At 815°C however, the effect of tension hold times becomes more significant (Fig. 5*b*); for example, with t_d = 15 min at the latter temperature, the life is reduced by a factor of 2 at $\Delta\epsilon$ = 1% and by a factor of 5 at $\Delta\epsilon$ = 0.6%.

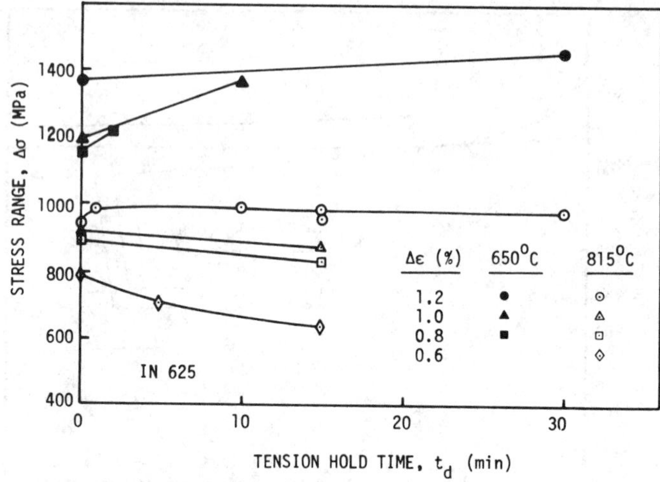

FIG. 4—*Effect of tension hold times on cyclic stress range behavior.*

Compressive hold times (t_e) also have a damaging effect on fatigue life. This effect is larger than that associated with tensile hold times and, furthermore, at 815°C is very pronounced for low strain levels (Fig. 5*b*). As an example, at this temperature and at $\Delta\epsilon = 0.6\%$, a tension hold of 5 min reduced the fatigue life significantly (in the present observation, by a factor of 7) compared with that observed under continuous cycling; this factor is approximately doubled for a 5 min hold in compression. The significant detrimental effect of compression hold times has been noted for many materials, particularly for nickel-base superalloys [*6–12*].

In the investigation of the material creep-fatigue behavior, one of the primary objectives of the study was to analyze the experimental results using a potential procedure for life prediction. This is explored in the following section.

Application of the Procedure for Life Prediction

Among many methodologies proposed in recent years for life prediction from an engineering point of view, the continuous damage approach suggested by Bui-Quoc and Biron [*2,3*] takes into consideration the order effect of loading under pure fatigue, under pure creep, under a fatigue-creep sequence, and under interspersed creep-fatigue. This method was recently applied for the analysis of the behavior of stainless steels [*13*], Alloy 800 [*14*], and 2¼ Cr-1Mo steel [*12*] under cyclic loading conditions including tension and compression hold times at temperatures up to 760°C.

The procedure is now applied to the present data under cyclic loading with and without hold time. Only the essential features of the concept are given here, since the details of the development and the physical interpretation of different parameters can be found in Refs *13* and *14*.

Fatigue Loading

In pure fatigue, the damage has been based on the reduction of the fatigue limit due to damage accumulation. The development has led to an analytical expression for describing the basic fatigue curve in a $\Delta\epsilon$-N_0 (total strain range versus number of cycles at failure) diagram. Under

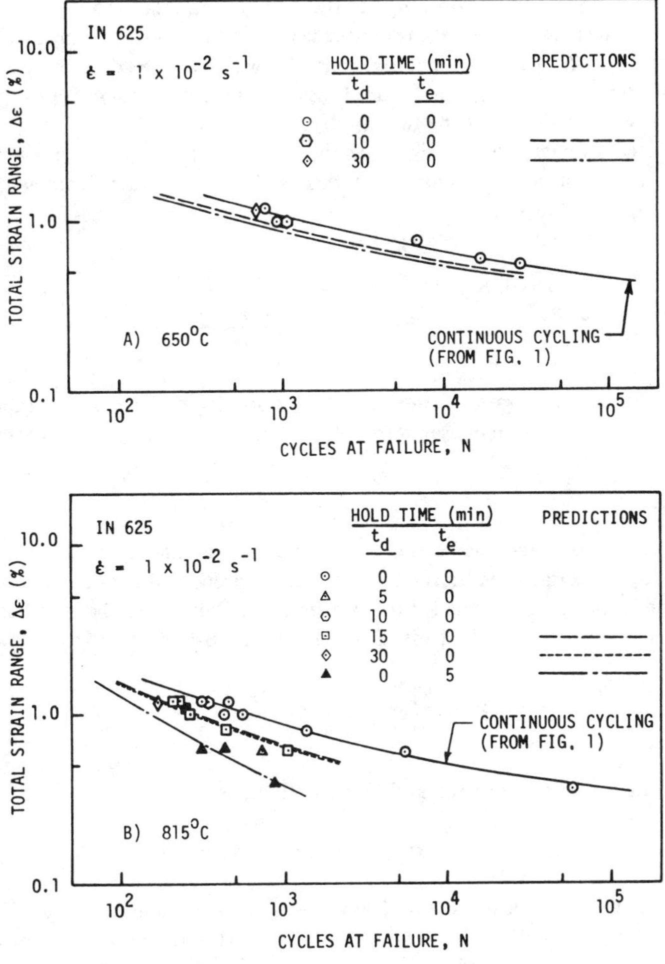

FIG. 5—*Effect of hold times on material life* (t_d *and* t_e *are the hold time periods in tension and compression, respectively*).

isothermal strain-controlled continuous cycling conditions, the analytical curve may be described by [*3*]

$$N_0 = \frac{K_f}{\gamma^a}\left[\frac{1}{\gamma - 1} - \frac{1}{\gamma - (\gamma/\gamma_u)^8}\right] \tag{1}$$

In Eq 1, the strain parameters γ and γ_u are expressed in terms of the total strain range $\Delta\epsilon$ as

$$\gamma = 1 + \ell n(\Delta\epsilon/\Delta\epsilon_0) \tag{2a}$$

$$\gamma_u = 1 + \ell n(2\epsilon_u/\Delta\epsilon_0) \tag{2b}$$

where $\Delta\epsilon_0$ is the fatigue limit strain range of the original material and ϵ_u is the true strain at fracture in a short-term tensile test at the temperature and the strain rate imposed in continuous cycling. Also in Eq 1, K_f and a are material constants which may be determined by referring to two data points in the $\Delta\epsilon$-N_0 diagram. Figure 1 also shows the behavior of Eq 1 together with the experimental results of IN 625 at the two temperatures considered.

The normalized damage due to fatigue has then been defined in terms of the instantaneous fatigue strength (instantaneous fatigue limit); this has led to the strain-dependent fatigue damage function, D_f, as follows [3]:

$$D_f(\gamma, \beta_f) = \frac{\beta_f}{\beta_f + (1 - \beta_f)\left[\dfrac{\gamma - (\gamma/\gamma_u)^8}{\gamma - 1}\right]} \tag{3}$$

where β_f is the life fraction (ratio between number of applied cycles n and number of cycles to failure N_0). An inverse relation of β_f in terms of D_f (i.e., $\beta_f = \beta_f(\gamma, D_f)$) may also be obtained.

Creep Loading

In pure creep, the damage has been characterized in a similar manner by the reduction of creep strength σ_p (material strength related to the creep process) due to damage accumulation. The development, in this case, has led to an expression which describes the basic isothermal creep-rupture curve in a σ-T(σ = stress, T = time at rupture) diagram as follows [3]:

$$T = \frac{K_c}{\theta^b}\left[\frac{1}{\theta - \theta_p} - \frac{1}{\theta^d - \theta_p}\right] \tag{4}$$

In Eq 4, the dimensionless stresses θ and θ_p are defined as

$$\theta = \sigma/\sigma_u \quad \text{and} \quad \theta_p = \sigma_p/\sigma_u \tag{5}$$

with σ_u being the ultimate tensile stress. The values of $\theta_p = 0.05$ and $d = 0.5$ have been suggested in previous investigations [15] and will also be used in the present work. Material constants K_c and b may be evaluated by considering two reference data points in a σ-T diagram. The isothermal creep-rupture data of IN 625 for two test temperatures [16] are reproduced in Fig. 6 together with the analytical curves obtained from Eq 4.

The normalized damage D_c due to creep has then been defined in terms of the instantaneous creep strength, leading to the relation [3]

$$D_c(\theta, \beta_c) = \frac{\beta_c}{\beta_c + (1 - \beta_c)\left[\dfrac{\theta - \theta_p}{\theta^d - \theta_p}\right]} \tag{6}$$

where β_c is the life fraction (ratio between exposure time t and time at rupture T). The reverse relation $\beta_c = \beta_c(\theta, D_c)$ may also be obtained.

For each type of loading, with the damage functions obtained from the basic theory, the concept takes into account the sequence effect of loading in multistep tests in a manner consistent with the trend generally observed in experiments.

The analytical results of Eqs 1 and 4 will be used to characterize the fatigue and creep damages, respectively, in combined fatigue-creep evaluation.

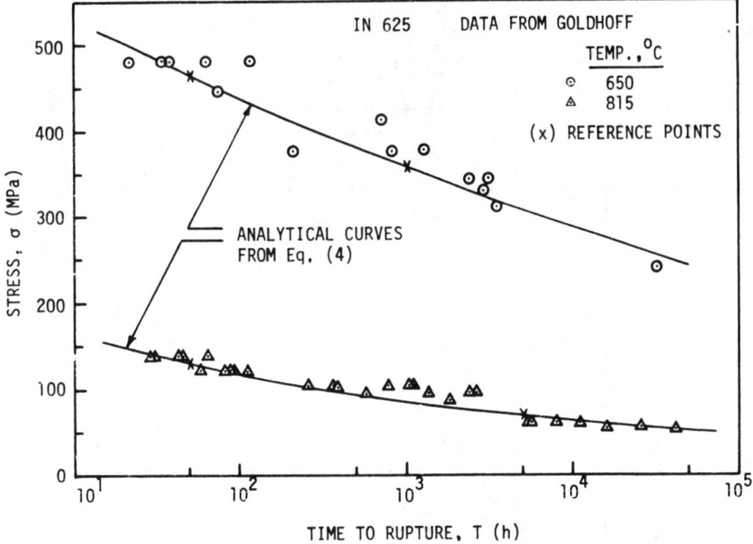

FIG. 6—*Representation of isothermal creep-rupture data by analytical curves.*

Interspersed Fatigue-Creep

The procedure for predicting the material life under combined creep-fatigue loading is formulated in a straightforward manner by combining the two damage functions. In order to take into account the strong sequence effect, such as observed on Alloy 800 and 304 SS, an interaction parameter, identified as w, has been introduced in whichever damage function is considered as the second process [*13,14*]. For example, for the creep process, the damage function with interaction D_c' is similar to D_c except that β_c is replaced by β_c^w; thus $D_c' = D_c'(\theta, \beta_c^w)$.

(a) Evaluation Procedure—Under interspersed creep-fatigue, the procedure of damage evaluation is schematically illustrated in Fig. 7. At Stage J, damage D_J is equal to that accumulated in the previous cycles. The cycle ratio β_J is given by $\beta_J = \beta_f(\gamma, D_J)$ and D_K is calculated from Eq 3; that is, $D_K = D_f[\gamma, (\beta_J + \Delta\beta_f)]$ where $\Delta\beta_f = 1/N_0$ with N_0 being obtained from Eq 1.

A similar procedure is applied for the creep portion of the cycle (Fig. 7). In particular, the equivalent cycle ratio $\Delta\beta_c$ is calculated on the basis of the relaxation stress characteristics observed during the hold period; that is, $\Delta\beta_c$ due to creep during a tensile hold time t_d and a compressive hold time t_e in a cycle is assumed to be governed by

$$\Delta\beta_c = \Delta\beta_d + (H_d + H_e)(\Delta\beta_e)^\alpha \tag{7}$$

where

$$\Delta\beta_d = \int_0^{t_d} \frac{dt}{T} \quad \text{and} \quad \Delta\beta_e = \int_0^{t_e} \frac{dt}{T} \tag{8}$$

Material constants H_d and H_e are associated with tension and compression hold times. If a particular hold time is not imposed, the associated constant takes the value of zero. Parameter α is also a material characteristic.

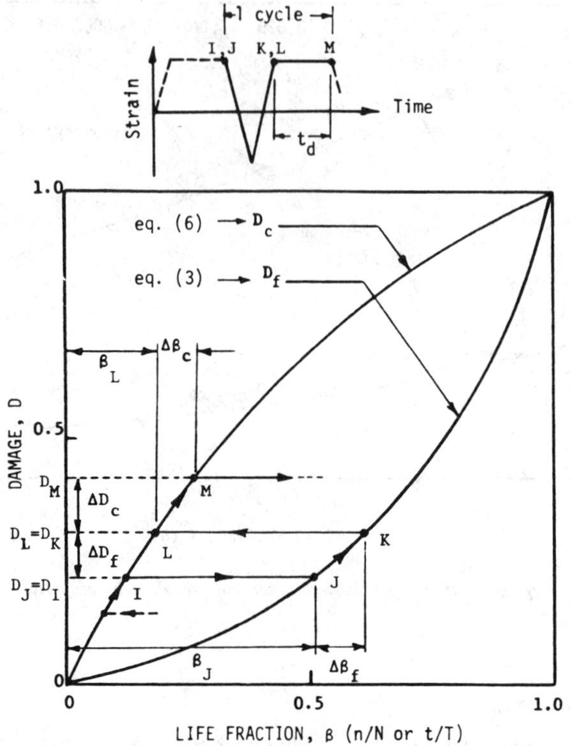

FIG. 7—*Life calculation procedure for material subjected to cyclic loading with hold times.*

Failure is assumed to occur when the accumulated damage equates unity.

With the interaction consideration, the procedure is essentially the same, except that the modified creep damage function D_c', instead of D_c, is used.

(b) Stress Relaxation Behavior—During the tensile strain hold period, the stress σ decreases from σ_m (at the beginning of the hold) to σ_d (at the end of the hold) as shown in Fig. 8*a* for a tensile hold condition. This stress characteristic may be adequately described by [*17*]

$$\ell n(\sigma_m/\sigma) = A(t^*)^p \tag{9}$$

where

$$t^* = (\Delta\epsilon/2m)^{2m}t \tag{10}$$

and m is the monotonic strain hardening exponent. A and p are material constants independent of the strain level and the hold period; these constants may be obtained from a plot of the experimental data corresponding to the limiting condition of Eq 9; that is, $\ell n(\sigma_m/\sigma_d)$ versus t_d^*, where t_d^* is calculated from Eq 10 with $t = t_d$.

As an approximation, σ_m may be taken as the cyclic stress amplitude ($\Delta\sigma/2$) obtained in continuous cycling at the same strain range and strain rate at the stabilized condition. The error on the calculated lives due to this approximation, as discussed in Ref *13,* is quite acceptable for

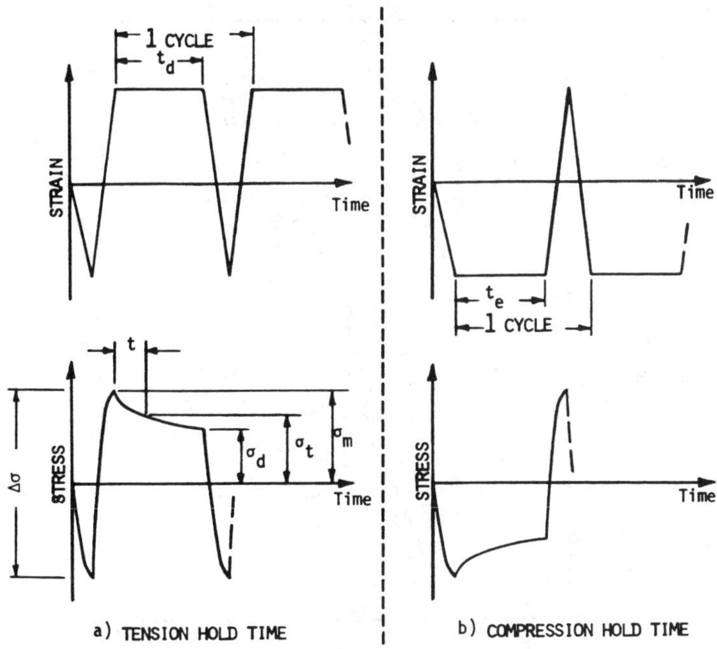

FIG. 8—*Interspersed creep-fatigue loading pattern with hold time at peak strains.*

many cases. The parametric correlation between $\Delta\sigma$ and $\Delta\epsilon$, as suggested in Ref *12,* is used again for IN 625:

$$\frac{\sigma_m}{Y} \simeq \frac{\Delta\sigma}{2Y} = k + m' \ \ell n\left(\frac{\Delta\epsilon}{2\epsilon_Y}\right) \tag{11}$$

where ϵ_Y is the ratio between the yield strength Y and the elastic modulus E, and k and m' are temperature-dependent material constants evaluated from a plot of the experimental data (Fig. 9).

(c) Life-Time Calculation with Tension Hold Times Only ($t_e = 0$)—For this case, $\Delta\beta_c = \Delta\beta_d$ and is evaluated using Eq 8 in connection with the relaxation stress characteristics. Then, the procedure outlined in Section *a* can be followed.

The interaction parameter w can be precisely evaluated only with experimental data from fatigue/creep or creep/fatigue sequences [*14*]. In the absence of this type of data, w may be determined by referring to a data point at a particular value of $\Delta\epsilon$ and a given value of t_d [*13*]. For example, for the present data, by referring to the test result at 815°C at $\Delta\epsilon = 1.0\%$ and $t_d = 15$ min, this parameter is very close to 1.0. This means that, when the presently suggested procedure is used for the analysis of the behavior of IN 625 material, the interaction effect between the two processes is negligible. In subsequent calculations, $w = 1.0$ is therefore assumed; the calculated lives of IN 625 with different values of t_d are shown in Fig. 5. The overall agreement between the calculated and observed lives is acceptable. In particular, at 815°C, the additional theoretical detrimental effect due to $t_d = 30$ min is small in comparison with the condition associated with $t_d = 15$ min. For example, at $\Delta\epsilon = 1.0\%$, a tension hold time of $t_d = 15$ min reduced the fatigue life by a factor of about 2.0 with respect to continuous cycling; with

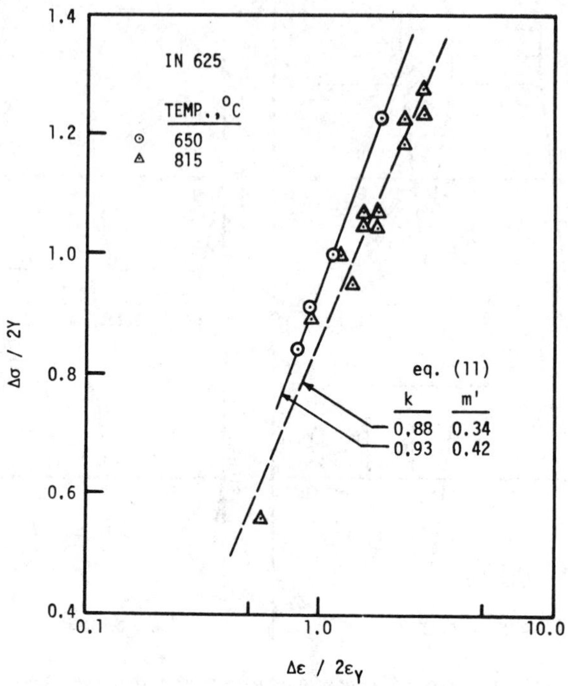

FIG. 9—*Correlation between normalized cyclic stress range* ($\Delta\sigma/2Y$) *and normalized strain range* ($\Delta\epsilon/2\epsilon_Y$).

$t_d = 30$ min, the additional reduction in life is negligible. This characteristic is obtained as a result of a rapid stress relaxation at the beginning of the hold period to a saturation stress at a very low level, as mentioned in the previous sections. Since the relaxation characteristics are taken into account in the evaluation of the creep-fatigue behavior, the analytical predictions correctly reflect the experimental data as shown in Fig. 5.

(d) Life-Time Calculation with Compression Hold Times Only ($t_d = 0$)—Under cyclic loading with a compression hold time only, IN 625 exhibits a damaging effect, and Eq 7 is reduced to $\Delta\beta_c = H_e(\Delta\beta_e)^\alpha$. With the results of two tests with two different values of $\Delta\epsilon$, constants H_e and α may be evaluated. For the present material, $H_e = 0.31$ and $\alpha = 0.65$.

The correlation between the calculated lives and the available test results are shown in Fig. 5*b* for 815°C. For a given hold period, the compression hold times are theoretically more detrimental than the tension hold times. For example, at $\Delta\epsilon = 0.6\%$, a compression hold time of 5 min reduced fatigue life 18 times with respect to continuous cycling; this factor is about 5 for a tension hold time of 15 min. This is in agreement with the experimental results.

Discussion

For the present material, a tensile hold time of larger than 5 min introduced in each cycle caused a reduction of life; this reduction was quite small at 650°C and more significant at 815°C. This may be explained by the reduction of effectiveness of the strengthening mechanisms, as usually observed in nickel-based superalloys at relatively high temperatures [*18*].

In spite of a limited number of experimental data, it may be pointed out that under creep-fatigue loading with a tensile hold time only, IN 625 exhibits a damaging effect much smaller than stainless steels and Alloy 800. For example, in the temperature range 650 to 815°C, with $\Delta\epsilon = 1\%$ and $t_d = 30$ min, the typical factor of reduction of material life is 2 for IN 625, 4 for Alloy 800, 5 for 304 stainless steel, and 10 for 316 stainless steel. On the other hand, IN 625 incurs a more damaging effect with a compression hold time than with a tensile hold time. This behavior is in contrast to that observed in the study of the three other materials; in addition, these latter materials exhibit a healing effect when subjected to a tension-compression hold time as compared to a tension hold time only. These characteristics have been taken into consideration in the development of the life prediction procedure. The overall correlation between calculated lives and the reported data is reasonably good.

As mentioned earlier, the life calculations with a hold time have been carried out using a maximum cyclic stress, σ_m, identical to that obtained under continuous cyclic loading ($\sigma_m = \Delta\sigma/2$). For some materials, such as stainless steels, experimental results have shown that σ_m decreases with an increase of the hold period and that this variation is more pronounced for a high strain than for a low strain. Nevertheless, the life predictions obtained using the actual and the approximate maximum stresses were both acceptable (with the approximate maximum stresses, the predicted lives stand on the conservative side), as already discussed in Ref *13*. In the case of IN 625, due to a rapid stress relaxation during hold times, no significant difference between stress ranges in tests with and without hold times was observed; however, due to this rapid relaxation, a mean stress opposite to the direction of the hold period occurred.

An attempt has been made to compare the predicted lives already presented in the previous sections with those obtained using the actual maximum stresses. Typical results of calculations for IN 625 are given in Table 4. For a given strain level, since the actual maximum stress is reduced due to the mean stress opposite in sign, it is expected that the predicted lives should be larger when the actual experimental maximum cyclic stresses are used. It should be noted that if the actual maximum cyclic stresses are used to compute the equivalent time ratio due to creep during the hold time, complete experimental results (i.e., for a given temperature, tests with different strain levels combined with different hold periods) must be made available. This gives a realistic material behavior, but it also involves extensive testing. The use of the maximum cyclic stress based on the stress range under continuous cycling simplifies the procedure and gives predictions smaller (on the conservative side) than those obtained with the actual maximum stress.

TABLE 4—*Comparison of predictions obtained with the approximate and the actual stresses (IN 625, 815°C,* t_d *= 15 min,* t_e *= 0).*

		Predictions (cycles)			
$\Delta\epsilon$, %	Experimental Results (cycles) (A)	Using Approximate Maximum Stress ($\sigma_m = \Delta\sigma/2$) (B)	$\frac{(B)}{(A)}$	Using Actual Maximum Stress C	$\frac{(C)}{(A)}$
1.2	210	207	0.98	236	1.12
1.0	262	325	1.24	443	1.69
0.8	437	560	1.28	917	2.10
0.6	1041	1300	1.25	3183	3.06

Conclusions

Continuous cycling and hold time tests were performed on Inconel 625 at 650 and 815°C in order to obtain the basic low-cycle fatigue properties for the material and to assess the influence of time-dependent damage. Experimental results indicated that temperature and hold time both had a significant effect on the fatigue life of IN 625.

1. Under continuous cycling, IN 625 cyclically hardened at 650°C and softened at 815°C at high strain ranges; at low strain range levels, however, it showed a relatively stable behavior throughout its life in the considered temperature range.
2. During hold times, the stress relaxed rapidly after a short period and remained relatively unchanged thereafter. The period at which the stress relaxation rate stabilized was dependent upon the temperature: at 815°C this period was about 3 min for all strain ranges investigated, whereas at 650°C it increased with the strain range and was typically 3 min at $\Delta\epsilon = 0.8\%$ and 15 min at $\Delta\epsilon = 1.2\%$.
3. At 815°C, compression hold times introduced in each cycle caused more damage than tension hold times; a drastic reduction of life due to hold times at peak compressive strain was observed at low strain ranges.
4. An analysis of data using a life prediction method previously suggested for the creep-fatigue combination loadings was carried out. In order to take into account the damaging effect due to a compression hold time separately from that due to a tensile hold time, a generalized relation has been introduced for the equivalent time ratio associated with the creep process. The overall correlation between theoretical calculations and experimental results was reasonably good, particularly for interspersed fatigue-creep loadings with tension or compression hold times at strain levels approximately ranging from 1.2 to 0.4%. A rough evaluation of the numerical values of the material constants has been made, based on limited experimental results.

References

[1] Purohit, A., Thiele, U., and O'Donnell, J. E. in *Proceedings,* 1983 Winter Meeting of the American Nuclear Society, San Francisco, Nov. 1983, pp. 1-4.
[2] Bui-Quoc, T. and Biron, A. in *Transactions,* SMIRT-6, Paris, 1981, Paper L9/1, pp. 1-7.
[3] Bui-Quoc, T. and Biron, A., *Nuclear Engineering and Design,* Vol. 70, 1982, pp. 89-102.
[4] Nguyen, H. L. and Masounave, J., *Res Mech.,* Vol. 18, 1986, pp. 201-206.
[5] Stouffer, D. C., Papernik, L., and Bernstein, H. L., "An Experimental Evaluation of the Mechanical Response Characteristics of René 95," Technical Report AFWAL-TR-80-4136, Oct. 1980.
[6] Lord, D. C. and Coffin, L. F., *Metallurgical Transactions,* Vol. 4, 1973, pp. 1647-1654.
[7] Ostergren, W. J., *Journal of Testing and Evaluation,* Vol. 4, 1976, pp. 327-339.
[8] Nazmy, M. Y., *Metallurgical Transactions,* Vol. 14*A*, 1983, pp. 449-461.
[9] Conway, J. B. and Stentz, R. H., "High Temperature Low-Cycle Fatigue Data for Three High Strength Nickel-Base Superalloys," Air Force Wright Aeronautical Laboratories, Report AFWAL-TR-80-4077, June 1980.
[10] Thakker, A. B. and Cowles, B. A., "Low Strain Long Life Creep-Fatigue of AF2-IDA and INCO 718," NASA Report CR-167989, National Aeronautics and Space Administration, Washington, D.C., April 1983.
[11] Hyzak, J. M., Hugues, D. A., and Kaae, J. L. in *Proceedings,* 35th Meeting of the Mechanical Failures Prevention Group, National Bureau of Standards, Gaithersburg, Md., 20-22 April 1982.
[12] Gomuc, R., Bui-Quoc, T., and Biron, A. in *Transactions,* SMIRT-8, Bruxelles, 1985, Paper L6/4, pp. 293-298.
[13] Gomuc, R. and Bui-Quoc, T., *Journal of Pressure Vessel Technology,* Vol. 108, 1986, pp. 280-288.
[14] Bui-Quoc, T. and Gomuc, R. in *Proceedings,* ASME International Conference on Advances in Life Prediction Methods, Albany, N.Y., 1983, pp. 105-113.
[15] Bui-Quoc, T. in *Transactions,* SMIRT-5, Berlin, 1979, Paper L5/1, pp. 1-8.
[16] Goldhoff, R. M., *Journal of Testing and Evaluation,* Vol. 2, 1974, pp. 387-424.
[17] Bui-Quoc, T. in *Fatigue des matériaux et des structures,* Maloine, Paris, 1980, pp. 497-522.
[18] Clauss, F. M., *Engineer's Guide to High-Temperature Materials,* Addison-Wesley, Reading, Mass., Chapter 6, 1969.

DISCUSSION

G. R. Halford[1] *(written discussion)*—You have shown that a compressive hold time is more damaging than a tensile hold time for IN 625. Have you examined your specimens to determine if there are observable differences in the failure mechanisms for these two types of cycles? In particular, could you observe differences in the oxidation cracking on the specimen surface?

T. Bui-Quoc et al. (authors' closure)—Attempts were made to examine the fractured surfaces of some specimens. Unfortunately, the chemical solutions used to remove the oxide layer also affected the material surface, so that no useful information could be obtained in relation to crack propagation mechanisms. The question of differences in the oxidation cracking on the specimen surface has not been addressed yet, but we plan to study a few failed specimens in the near future.

C. T. Sims[2] *(written discussion)*—Your fatigue-creep study was conducted at 650°C (1200°F) and 815°C (1500°F) for extended times. You stated that the material was not inspected metallographically, either before or after testing, for either of the two temperature variables.

According to my experience, IN 625 is a (truly) unstable material. Above temperatures on the order of 650 to 725°C (1200 to 1300°F) the alloy commences to precipitate quantities of plate-shaped phases, usually emanating from grain boundaries. The plates are principally eta-phase (NI_3Cb, Mo), but Laves and mu also occur. These phases remove strengthening elements from solution, promote initiation of cracks, and assist directly in crack propagation under load, sharply reducing creep and rupture properties. Accordingly, it is obvious that your test results at 815°C (1500°F) ("tension hold times—at 815°C—had (significant) effects on cyclic life; compressive hold times also had a damaging effect on the fatigue life—very pronounced for low strain levels at 815°C; - etc.") were directly caused by the precipitation of quantities of eta, Laves, and perhaps mu. There are private corporate documents and ASTM documents that warn against the use of IN 625 under these conditions because of degeneration of the alloy.

In short, the failure of the authors to acquire rudimentary knowledge about IN 625 and the failure of the study to conduct simple metallographic analysis of the test material have resulted in a body of data which has most questionable value as presented. If published in the present state, it becomes a most misleading document, since the paper tacitly assumes the material tested at 650°C (1200°F) was more or less the same as that tested at 816°C (1500°F). This work is an example of the errors generated when mechanists conduct studies without the guidance essential from qualified metallurgists.

I suggest the authors approach a metallurgical group competent to reframe the information obtained into its correct context. So handled, the work may become of some value; for instance, it may show the deleterious effect of plate-phase precipitation in more mechanical detail than currently in the literature. Also, of course, it gives data for IN 625 at 650°C (1200°F) which *may* be usable industrially, *assuming no plate phases formed.*

T. Bui-Quoc et al. (authors' closure)—Dr. Sims has provided metallurgical information which may help to explain the significant effect of hold times on the cyclic life of IN 625 at elevated temperatures. The additional phase formation at a high temperature would be expected to lead to a sharp reduction in creep-rupture properties of the material, and this is reflected in the relative fatigue-creep strengths at the two temperatures considered, as noted in the paper.

[1]NASA-Lewis Research Center, Cleveland, OH 44135.
[2]General Electric Company, Schenectady, NY 12345.

It should be pointed out that the main objective of the paper is to quantitatively evaluate the mechanical performance of IN 625 under creep-fatigue loading conditions at 650 and 815°C. This assessment is achieved by looking at, on the one hand, the reduction of the cyclic life at 815°C with respect to that at 650°C under continuous cycling (without hold times) and, on the other hand, the additional reduction of this life due to different hold-time periods at each temperature considered. Since the metallurgical changes are reflected by an overall modification in mechanical properties, the reduction of the creep-fatigue strength has been assessed using the fatigue (continuous cycling) and the creep-rupture properties *at the temperature considered*. With these considerations, the proposed predictive model takes into account, at least partially, the microstructural change effects.

Regarding the relevance of the results reported in the paper, there has recently been an expression of industrial interest for the fatigue-creep properties of IN 625 at the temperatures studied.

M. Morishita,[1] K. Taguchi,[2] T. Asayama,[3] A. Ishikawa,[3] and Y. Asada[3]

Application of the Overstress Concept for Creep-Fatigue Evaluation

REFERENCE: Morishita, M., Taguchi, K., Asayama, T., Ishikawa, A., and Asada, Y., "**Application of the Overstress Concept for Creep-Fatigue Evaluation,**" *Low Cycle Fatigue, ASTM STP 942,* H. D. Solomon, G. R. Halford, L. R. Kaisand, and B. N. Leis, Eds., American Society for Testing and Materials, Philadelphia, 1988, pp. 487–499.

ABSTRACT: Type 304 stainless steel was subjected to a controlled push-pull strain with a wide variety of waveforms at 650°C in 0.1 μPa vacuum. No indication of oxidation was observed in the specimens. The frequency effect on creep-fatigue behavior was diminished, but the tensile hold time and waveshape effects were still observed in this high vacuum environment. It was concluded that a reduction of fatigue life due to creep-fatigue interaction is observed when a tension going time exceeds a compression going time. A back stress was determined from the data, and an overstress was defined as the difference between an applied stress and a back stress. The overstress was related to the inelastic strain rate. It is shown that overstress can be reasonably used to describe the creep-fatigue damages for life prediction.

KEY WORDS: creep-fatigue, vacuum, Type 304 stainless steel, life prediction, overstress, back stress, time independent, time dependent, damage

The environmental effect of air is included in the usual creep-fatigue test data obtained by tests in air. It should be considered that such test data reflect the creep-fatigue-environment interaction as suggested by Coffin [*1*]. Coffin has shown in his pioneering study [*1*] that a remarkable increase is observed in the low-cycle fatigue life of metals and alloys at elevated temperatures, in a high vacuum compared with that in air, and that the frequency effect disappears in vacuum. White has also reported an increase of low-cycle fatigue life in vacuum with 0.5Mo steel at 550°C [*2*]. These studies suggest that the behaviors of creep-fatigue interaction, as well as fatigue life, are affected by the environment in which the material is tested.

The authors consider that it is necessary to obtain "pure" creep-fatigue interaction free from environmental effects in order to quantify the environmental effect. This work is necessary to improve the accuracy of creep-fatigue life prediction. Additional experimental work was done for this purpose; that is, a wide variety of creep-fatigue tests were conducted with Type 304 stainless steel at 650°C in a very high vacuum of 0.1 μPa. This provides a common base for evaluating the effect of environment on creep-fatigue interaction. The experimental data were subjected to further mechanical analysis of stress-strain response based on the overstress concept by Krempl [*3*], in which the overstress is defined by a difference between the applied stress

[1]Power Reactor and Nuclear Fuel Development Corporation, Oh-arai Engineering Center, Oh-arai-cho, Higashi-Ibaragi-gun, Ibaragi, Japan.

[2]Toshiba Corporation, Nuclear Engineering Laboratories, 4-1 Ukishima-cho, Kawasaki-ku, Kawasaki, Japan.

[3]Graduate Student, Research Fellow, and Professor, respectively, Department of Mechanical Engineering, University of Tokyo, 7-3-1 Hongo, Bunkyo-ku, Tokyo, Japan.

and the internal back stress. The evaluation procedure of the present study employs damage described with the overstress derived from the analysis.

Experimental Procedure

A brief summary of the experimental procedure and results are given here; fuller details have been reported in the authors' recent papers [*4,5*]. Hourglass specimens were machined from a Type 304 stainless steel bar product that was subjected to solution treatment of 1120°C water cool. A newly developed creep-fatigue test facility was employed that incorporated an induction heating device, a vacuum system to attain a high vacuum, and a digital computer for control and data acquisition of a closed-loop system.

Push-pull strain-controlled creep-fatigue tests have been conducted at 650°C in 0.1 μPa vacuum, where the measured diametral strain is converted to an axial strain for control. Numerical data of stress and strain are sampled at 80 to 200 points in every specified cycle and stored in the computer. The tests employed the six strain waveforms summarized in Table 1 with detail loading conditions. They are symmetric continuous, slow-fast (S-F) continuous, fast-slow (F-S) continuous, tensile hold-time, tensile and compressive hold-time, and tensile hold-time superimposed upon S-F cycling. The strain rates and hold-times were 10^{-5} to 10^{-3} s^{-1} and 60 to 600 s, respectively.

Experimental Results

Creep-Fatigue Test Results

Details of test results were reported in Refs *4* and *5*. In Fig. 1 broken lines give test results in air with corresponding test conditions. The number of cycles to failure is defined by the number of cycles at which a tensile peak stress reduces to 70% of that at saturation.

No indication of oxidation was observed on the specimens. This was checked by scanning electron microscope (SEM) and optical microscope techniques on the external surfaces and failure surfaces of the specimens. In addition, SEM observations showed that transgranular cracking was dominant in symmetric and F-S cycling, but that intergranular cracking was dominant for other waveforms. A time-dependent life reduction was observed for the latter waveforms as described later. The stress-strain responses observed in these tests were similar to those in air tests.

Trends of Test Results

The authors consider that the present test results reflect a "pure" creep-fatigue behavior of the material at this temperature, which can be assumed to be completely free from the environmental effects of air. Some interesting trends were observed. A symmetric continuous cycle gave a larger increase of fatigue life in this high vacuum than that in air [*1*]. The frequency effect diminished at a strain rate of 10^{-4} to 10^{-3} s^{-1}.

Regarding the strain waveform effect, F-S cycling showed the same fatigue life as that of symmetric continuous cycling. This trend is different from that in air [*6*]. S-F cycling showed a reduction of fatigue life compared with the former two waveforms. This life reduction became more severe with decreasing strain rate in tension.

A tensile strain hold-time reduced the fatigue life in this high vacuum when compared with that of symmetric continuous cycling. The fatigue life in this case is larger in vacuum than in air. However, the ratio of fatigue life of tensile hold-time cycling to symmetric continuous cycling with a strain rate of 10^{-3} s^{-1} was smaller in vacuum than in air. A recovery of fatigue life

TABLE 1—*Strain wave forms and experimental conditions.*

		Symmetric	Slow-Fast	Fast-Slow	Tens. Hold	Tens.-Comp. Hold	Slow-Fast & Tens. Hold
Strain Wave Form							
Strain Rate	Tension Going	1×10^{-3} to $1\times10^{-4}\,s^{-1}$	1×10^{-4} to $1\times10^{-5}\,s^{-1}$	$1\times10^{-3}\,s^{-1}$	$1\times10^{-3}\,s^{-1}$	$1\times10^{-3}\,s^{-1}$	$1\times10^{-4}\,s^{-1}$
	Compression Going	1×10^{-3} to $1\times10^{-4}\,s^{-1}$	$1\times10^{-3}\,s^{-1}$	$1\times10^{-4}\,s^{-1}$	$1\times10^{-3}\,s^{-1}$	$1\times10^{-3}\,s^{-1}$	$1\times10^{-3}\,s^{-1}$
Hold Time	Tension	0	0	0	60 to 600 s	600 s	600 s
	Compression	0	0	0	0	60 s	0

Note; Type 304 Stainless Steel, 650°C, 0.1 Micro Pascal

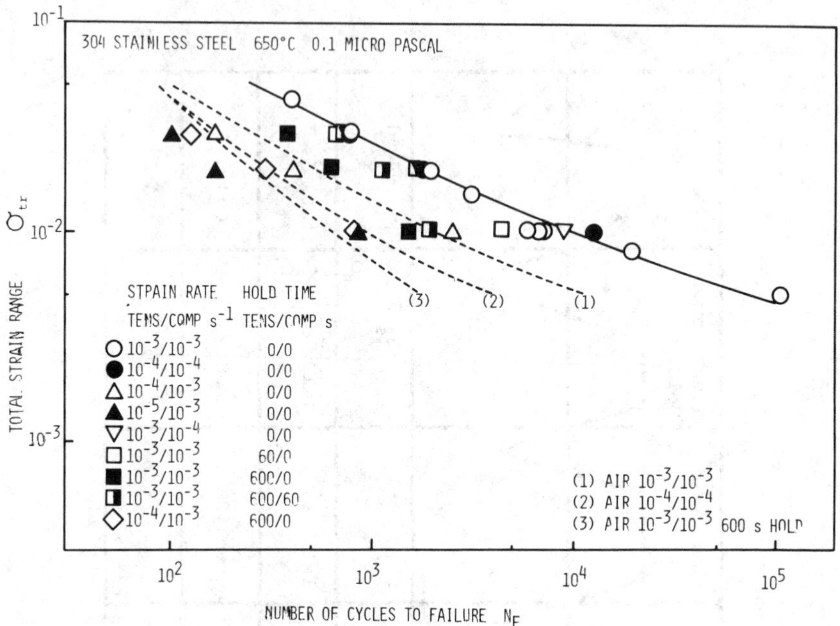

FIG. 1—*Creep-fatigue test results at 650°C in 0.1 μPa vacuum.*

occurred with 60 s hold-time in compression. A superposition of tensile hold-time resulted in a further reduction in the life of S-F cycling. The Coffin-Manson relation holds in each series of tests. However, the exponent in the equation is larger in hold-time cycling than in continuous cycling.

These observations suggest that a reduction of fatigue life occurs even in high vacuum, when the employed waveform has a longer tension going time than compression going time. This can be explained in other ways. No life reduction was observed when the operating time of a negative hydrostatic stress was longer than that of a positive one. Life reduction occurred in the opposite case. This observation corresponds to the SEM observation; that is, transgranular cracking is dominant in symmetric and F-S continuous cycling, but intergranular cracking is dominant in the other waveforms. The authors consider that this life reduction in vacuum was caused by the effect of creep.

Stress-Strain Response and Overstress

A bowing-out shape is observed in the stress-strain hysteresis curves during unloading from both tensile and compressive peaks (Fig. 2). This bowing-out shape becomes clear at high temperature and low strain rate. This shape suggests progress of an inelastic strain in the direction of the applied stress immediately after unloading. Almost all models of inelastic constitutive laws based on the unified theory interpret the phenomenon as an inelastic strain progress in the direction of the overstress at the moment, where the overstress is defined as the applied stress σ minus the internal back stress R. This overstress acts in the direction of the applied stress when unloading starts. The overstress reduces to zero as unloading develops, then changes its sign. At

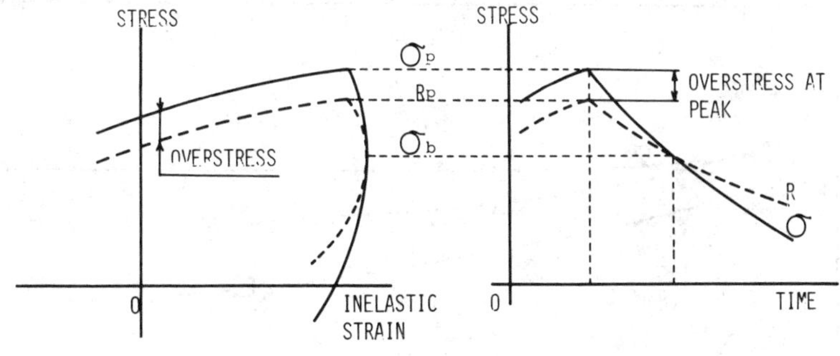

FIG. 2—*Schematic interpretation of unloading behavior.*

the change in sign of the overstress, the inelastic strain changes the direction of its development from that before unloading to the value of stress denoted by σ_b in Fig. 2, which corresponds to a value of the stress at the peak of this bowing-out shape. At this peak in inelastic strain, the overstress becomes zero; that is, the applied stress coincides with the back stress. This situation means the same procedure of determining the value of the back stress to that of the dip test [7], which determines the back stress by using a relation between a stress and inelastic strain rate. This discussion suggests that the value of the stress σ_b in Fig. 2 is the value of the back stress at the described moment, but it may be different from the value of the back stress just at unloading (that is, at the stress peaks). The values of σ_b are given in Ref *8*.

Time-Dependent Behavior of Back Stress

The value of σ_b shows a cyclic hardening behavior similar to that of the peak stress σ_p (σ_{max} and σ_{min}). The difference of these peak stresses and the corresponding σ_b is observed to be a constant as the number of cycles progresses. A similar trend is observed in cases of hold-time cycling. The stress difference is dependent upon the strain waveforms, strain range, and strain rate during unloading.

The absolute value of the stress difference between σ_p and σ_b (that is, $\sigma_{max} - \sigma_b$ or $\sigma_b - \sigma_{min}$) is plotted against the strain rate during unloading as shown in Fig. 3. This shows that the value of the stress difference decreases as the strain rate increases. But this decrease saturates at a strain rate of about 10^{-3} s^{-1}. The authors consider that Fig. 3 shows a time-dependent recovery or relaxation of the back stress during unloading. This figure suggests that the value of the overstress $\sigma - R$ at stress peaks can be obtained at an unloading strain rate of 10^{-3} s^{-1}.

Figure 4 shows a relation between the overstress at the stress peaks thus obtained and the inelastic strain rate just before unloading starts. The value of the overstress at stress peaks $\sigma_p - R_p$ takes on a different value when a strain range is different, where R_p is a value of back stress at stress peaks. A common trend is observed irrespective of strain range; that is, the overstress is almost rate-insensitive at high strain rate, but is rate-sensitive at low strain rate. A transition occurs at a strain rate of about 5×10^{-5} s^{-1}. This trend implies that the deformation is rate-independent in the high strain rate region, but is rate-dependent in a low strain rate region. The back stress is probably rate-dependent, but this is not examined at present.

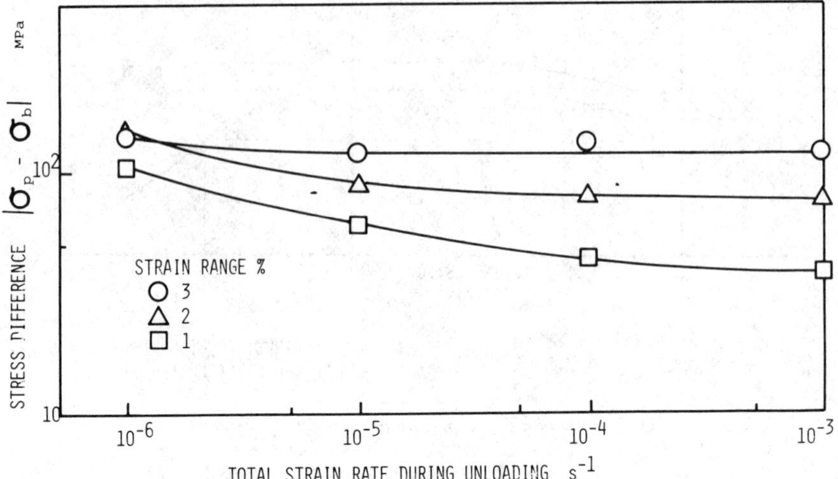

FIG. 3—*Change of σ_b according to strain rate.*

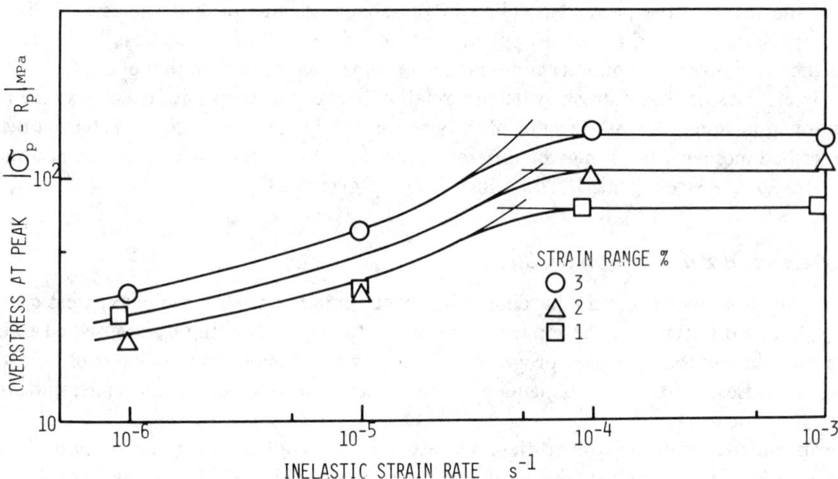

FIG. 4—*Inelastic strain-rate dependence of overstress.*

Prediction of Back Stress at Arbitrary Time of Cycling

The authors assumed that Fig. 4 represented a unique relation between the overstress and the inelastic strain rate at a denoted strain range, and that it was available to predict the value of back stress at an arbitrary time during strain cycling. Figure 5 gives an example of the back stress estimated with the aid of Fig. 4. It is obtained from a test result with a strain range of 2%, a strain rate of $10^{-5}/10^{-3}$ s^{-1} (S-F cycling), and no hold-time. The inelastic strain rate $d\epsilon_p/dt$ is evaluated by Eq 1, where ϵ_t is total strain and E is Young's modulus:

$$d\epsilon_p/dt = d\epsilon_t/dt - d(\sigma/E)/dt \quad (1)$$

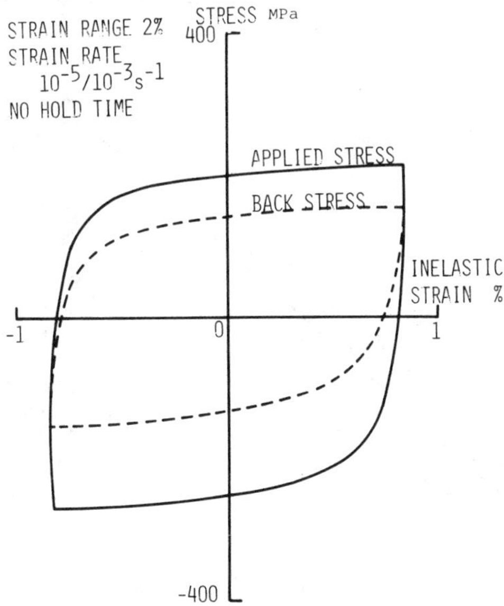

FIG. 5—*Prediction of back stress.*

Creep-Fatigue Evaluation Using Overstress

Time-Independent Damage Parameter

A damage model was developed to describe the creep-fatigue test results of the present study. The model deals with the "pure" creep-fatigue interaction in which an overstress was used to express damage components. The damage was separated into two components, time-independent and time-dependent, in accordance with the experimental results.

A time-independent damage parameter D_I was introduced to describe the time-independent failure observed with symmetric continuous and F-S cycling. D_I is defined by Eq 2 and refers to the effective inelastic strain energy:

$$D_I = \int_{\text{cycle}} (\sigma - R) d\epsilon_p \tag{2}$$

where the back stress R is obtained by the aforementioned procedure. D_I is positive because the overstress and inelastic strain rate have the same sign.

The value of D_I was computed for all specimens at around half life, and was correlated with a life fraction $1/N_{F0'}$ where N_{F0} denotes fatigue life from symmetric continuous and F-S cycling and shows the "pure" fatigue life independent of time/rate. The time-independent damage ϕ_I for one cycle was given by $1/N_{F0}$. A reasonable relation was observed between ϕ_I and D_I (Fig. 6). The following equation can be fit to the results:

$$\phi_I = C_1 D_I^{\,n_1} \tag{3}$$

where C_1 and n_1 are determined from the fitting.

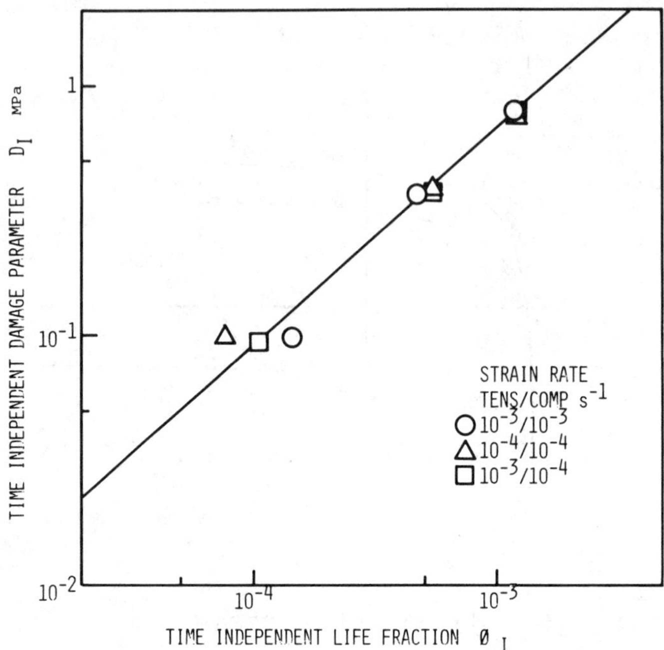

FIG. 6—*Relation between time-independent damage parameter and time-independent life fraction.*

Time-Dependent Damage Parameter

A time-dependent damage parameter, D_D, is defined by Eq 4 in order to express the time or rate dependent life reduction observed with S-F and hold-time cycling. D_D is given by a time integration of the overstress during one cycle; that is,

$$D_D = \int_{\text{cycle}} (\sigma - R)dt \tag{4}$$

where R was obtained by the aforementioned procedure. D_D can take a positive or negative value depending on the waveform. In the present tests, a positive D_D was obtained for S-F and tensile hold-time cycling, and D_D was negative for F-S cycling.

A negative D_D means an increase of fatigue life to more than that of symmetric continuous cycling, which is inconsistent with the experimental observations. The following limitation is necessary for adjusting the model to the experimental results, as proposed by Majumder et al [*9*], Ostergren et al [*10*], and Satoh et al [*11*]:

$$D_D = 0 \qquad \text{(if Eq 4 gives a negative value)} \tag{5}$$

The physical meaning of Eqs 4 and 5 is more ambiguous than Eq 2, but it seems to express the effect of a hydrostatic stress on the expansion or shrinkage of grain boundary cavities which cause grain boundary cracking as suggested by Coffin [*6*]. A positive D_D means a growth of cavities; a negative D_D means a shrinkage or vanishing of these. This idea implies that D_D must be zero when Eq 4 gives a negative value, since cavities have vanished in this case.

D_D is correlated to the rate-dependent life fraction, ϕ_D, given by $1/N_F - 1/N_{F0'}$ where N_F is the fatigue life of S-F and hold-time cycling and N_{F0} is the time-independent fatigue life aforementioned. N_{F0} can be obtained from Eq 3. In the present analysis ϕ_D was computed and plotted against D_D (Fig. 7). The following relation fits the $D_D - \phi_D$ data, although a scatter is observed:

$$\phi_D = C_2 D_D{}^{n_2} \tag{6}$$

where C_2 and n_2 are experimentally determined.

Linear Damage Rule

A failure criterion of the model was derived from the above definitions of ϕ_I and ϕ_D. The criterion is given by a linear summation of two damage components; that is,

$$N_F(C_1 D_I{}^{n_1} + C_2 D_D{}^{n_2}) = 1 \tag{7}$$

with the additional limitation that

$$D_D = 0 \qquad \text{(if computed } D_D \text{ is negative)} \tag{8}$$

where $C_1 = 1.70 \times 10^{-4}$, $C_2 = 4.56 \times 10^{-9}$, $n_1 = 1.19$, and $n_2 = 1.32$ for Type 304 stainless steel at 650°C. Here the stress and time are expressed in terms of MPa and seconds, respectively.

Equation 7 assumes that when the same hysteresis loop is repeated from the initial to final cycle it can be represented by its half-life cycle. A life prediction can be made using Eq 7 for the

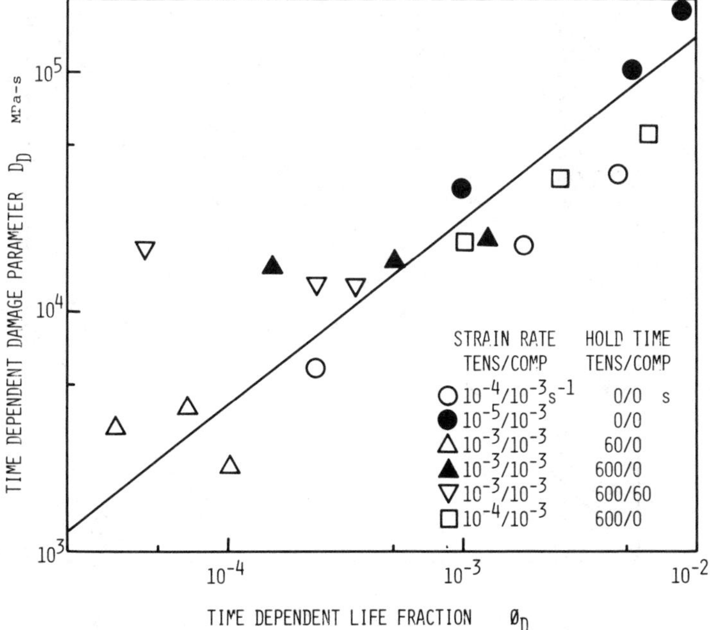

FIG. 7—*Relation between time-dependent damage parameter and time-dependent life fraction.*

present test results. Figure 8 shows that Eq 7 gives scatter of the predicted life within a factor of 2. No trend is observed in Fig. 8 with respect to the strain range, waveform, or other test conditions.

Discussion

Elastic Core

It is questioned in the present procedure whether the relation between the overstress versus inelastic strain rate given by Fig. 4 is dependent on strain range. Similar tests are being run with 2.25Cr-1Mo steel at 550°C in 0.1 μPa vacuum in the authors' laboratory. The material shows a linear region in the unloading curve (Fig. 9). The linear region has a slope equal to Young's modulus. The authors consider this as an elastic core in which the material behaves as a completely elastic body. A single relation is obtained between the overstress and the inelastic strain rate, if the overstress is defined by the difference between the peak stress and the stress corresponding to the upper end of this linear region (that is, the end nearer to the peak stress). This observation suggests that there may be an elastic core for Type 304 stainless steel, too, but it may be overlooked due to inadequate experimental accuracy. More detailed experiments and analyses are required to check this point. If this observation with 2.25Cr-1Mo steel is applicable to Type 304 stainless steel, the relation in Fig. 4 may be improved.

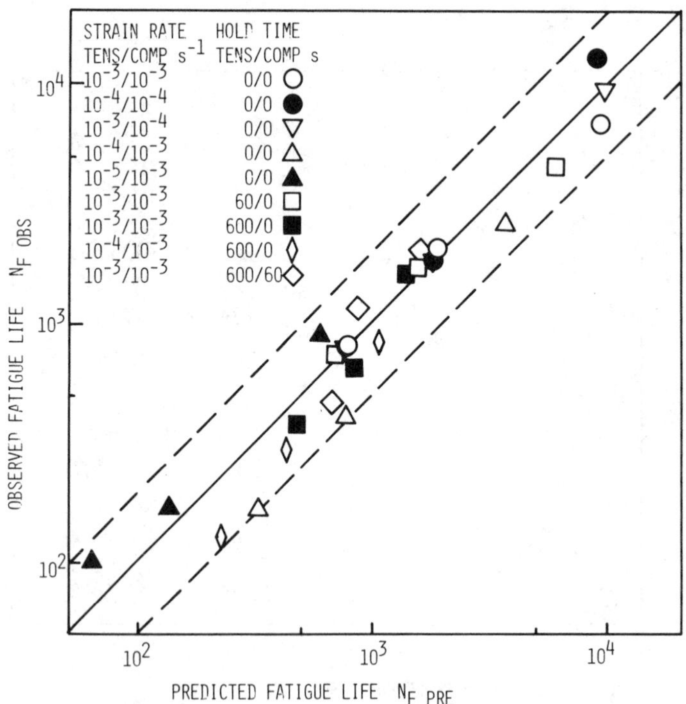

FIG. 8—*Life prediction results.*

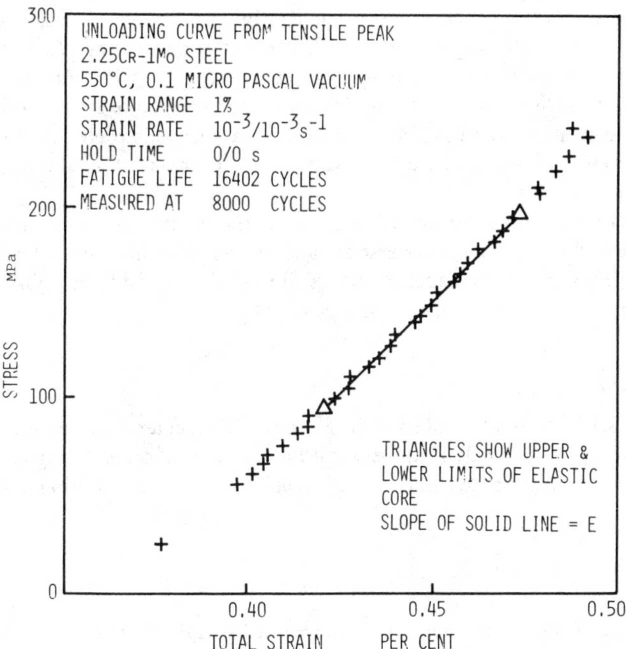

FIG. 9—*Elastic core on unloading curve.*

Damage Rate Approach

As easily anticipated, Eq 7 can be rewritten as Eq 9 to give a damage rate equation:

$$d\phi/dt = A_1(\sigma - R)^{n_1}(d\epsilon_p/dt) + A_2(\sigma - R)^{n_2} \tag{9}$$

with the failure criterion

$$\phi = 1 \tag{10}$$

where the time integration of the second term on the right-hand side of Eq 9 should be subjected to the limitation given by Eq 8. The coefficients and exponents of Eq 9 are determined experimentally. The first term on the right-hand side of the equation may correspond to the fatigue damage at high frequency and the second term may correspond to the creep damage. If these anticipations are true, the coefficients and exponents of Eq 9 can be determined from a fatigue test result at high frequency and from a creep test. The back stress R is computed with the aid of a relation such as given by Fig. 4. However, a theoretical derivation of stress and back stress is recommended for general purposes. Inelastic constitutive laws based on the unified theory seem to be promising.

Conclusions

Creep-fatigue tests have been conducted with Type 304 stainless steel in 0.1 μPa vacuum. The frequency effect is not observed but a waveshape effect and a hold-time effect still remain, reducing the fatigue life from that of symmetric continuous cycling. Fast-slow cycling shows no life reduction. Test results suggest that a life reduction due to the effect of hold-time and waveshape occurs when the time for tension going is longer than that for compression going.

An overstress was obtained by analysis of the unloading region of stress-strain curves. A relation between the overstress and the inelastic strain rate was constructed on the basis of experimental phenomenology and was used to evaluate the progression of the back stress of each specimen tested.

Damage parameters were introduced to evaluate the present creep-fatigue data. These parameters are described by using an overstress and are related with time-independent and time-dependent life fractions of the experiments. A life prediction made by a linear summation of these damage parameters gives scatter of a factor of 2.

Acknowledgments

The authors wish to thank Professor E. Krempl, Rensselaer Polytechnic Institute, for his effective comments on the analysis and evaluation of the back stress during his stay at the University of Tokyo. Mr. S. Mitsuhashi is also appreciated for his contribution in the experimental portion of the study.

References

[1] Coffin, L. F. in *Fatigue at Elevated Temperatures, ASTM STP 520,* American Society for Testing and Materials, Philadelphia, 1973, pp. 5-34.
[2] White, D. J., *Proceedings of the Institution of Mechanical Engineers,* Vol. 84, 1969-70, pp. 223-240.
[3] Krempl, E., *Journal of Engineering Materials and Technology,* Vol. 101, 1979, pp. 380-386.
[4] Morishita, M. and Asada, Y., *Nuclear Engineering and Design,* Vol. 83, 1984, pp. 367-377.
[5] Morishita, M., Taguchi, K., Satake, M., Ishikawa, A., and Asada, Y. in *Proceedings,* 5th International Conference on Pressure Vessel Technology, ASME, 1984, Part II, pp. 1109-1120.
[6] Coffin, L. F., "Creep-Fatigue-Environment Interaction," R. M. Pelloux and N. S. Stoloff, Eds., The Metallurgical Society of AIME, 1980, pp. 1-23.
[7] Ahlquist, C. N. and Nix, W. D., *Scripta Metallurgica,* Vol. 3, 1969, pp. 679-682.
[8] Morishita, M., Taguchi, K., Satake, M., Ishikawa, A. and Asada, Y., to appear in *Proceedings,* 5th International Conference on Pressure Vessel Technology, ASME, Part III.
[9] Majumdar, S. and Maiya, P. S., ASME Publication PVP-PB-028, 1978, pp. 43-54.
[10] Ostrgren, W. J. and Krempl, E., *Journal of Pressure Vessel Technology,* Vol. 101, 1979, pp. 118-124.
[11] Satoh, M. and Krempl, E., ASME Publication PVP-, Vol. 60, 1982, pp. 71-79.

DISCUSSION

N. Marchand[1] *(written discussion)*—How accurate are your back stress (σ_b) measurements? The plastic strain range is an easy quantity to measure, yet there is controversy as to how it relates to fracture under slow-fast, fast-slow, etc., cycling. Is it practical to use σ_b to correlate fatigue life? For low plastic strain ranges ($N_f \simeq$ 100 000 cycles) is your approach still valid? That is, does it have the same degree of correlation?

M. Morishita et al. (authors' closure)—Stress and strain were measured by a computer at 80 to 200 sampling points in every specified cycle. After that. a curve fitting was made to the

[1]Massachusetts Institute of Technology, Cambridge, MA 02139.

sampled values to determine the peak of the bowing-out. The accuracy of determined σ_b depends upon the strain rate during unloading. Here, it is from 1 to 10 MPa.

We consider use of the plastic strain range an acceptable method of correlating fatigue life with applied loading condition. We intend to find a new parameter to unify the difference in fatigue life when it is correlated with plastic strain range. The present approach is based on a damage theory. The present damage model can predict the fatigue behavior of the present tests. We believe this approach is applicable to low plastic strain ranges, but at present it has not been proved.

P. Agatonovic[1] *and R. Heidenreich*[2]

An Integrated Approach to Creep-Fatigue Life Prediction

REFERENCE: Agatonovic, P. and Heidenreich, R., **"An Integrated Approach to Creep-Fatigue Life Prediction,"** *Low Cycle Fatigue, ASTM STP 942,* H. D. Solomon, G. R. Halford, L. R. Kaisand, and B. N. Leis, Eds., American Society for Testing and Materials, Philadelphia, 1988, pp. 500–518.

ABSTRACT: There has been special interest recently in developing new, reliable analytical design methods for components under higher temperature conditions. However, at present, the use of material properties is still limited to arrays of single characteristics which do not interact with each other. In this work, several creep fatigue experiments on smooth specimens of IN 800 H have been carried out at 830°C. In some tests, these have also been combined with inside hysteresis loops to investigate the different effects on deformation and damage behavior which originate in a creep and fatigue environment. As a result of these tests, it has been found that the material behavior under creep-fatigue conditions can be significantly changed compared to the material behavior under simple load conditions. Therefore there is a need for life analysis methods to be expanded to include possible variations in properties; for greater accuracy, the material properties must be treated as a complex interacting system of parameters. The examination has been extended to a typical component used under high-temperature conditions. The results of the numerical analysis show that the stress-strain history in the critical area of that component is not simply strain controlled, as it is in the typical laboratory creep-fatigue interaction life test containing a tensile or compressive dwell at constant peak strain level. At high temperatures, the conditions in the component are more severe, causing the life to be reduced compared with the typical laboratory test. In this paper, these conditions are successfully simulated with the help of a generalized Neuber law: $\sigma \cdot \epsilon^p =$ constant. Based on this ratio, the engineering method for evaluating component geometry and loading conditions and their effects on material behavior can be established.

On the basis of the results of this study, it follows that a very large amount of information on material behavior and on its component dependence is needed when the material properties are time-dependent, as in the design of hot components. For that reason, the satisfactory solution to designing components for high-temperature conditions requires an integrated approach, with full consideration of different interaction effects from the various influences in the main areas of the material deformation and damage behavior, and of the component effects. The results of this study may apply to the standard methodology of life prediction in high-temperature areas, which shall be developed in the future.

KEY WORDS: creep, fatigue, life prediction, notch analysis, viscoplasticity, IN 800 H alloy

Nomenclature

- B Coefficient in creep rate equation
- E Young's modulus
- K_t Theoretical stress/strain concentration factor
- K_ϵ True strain concentration factor
- K_σ True stress concentration factor
- $N_{A S5}$ Cycles to 5% load drop

[1]MAN Technologie, GmbH, Munich, West Germany.
[2]Industrieanlagen-Betriebsgesellschaft (IABG), Ottobrunn, West Germany.

N_B Cycles to failure
T_H Hold period
U Constant in creep rate equation
d Specimen diameter
m Exponent in creep rate equation
p Exponent in generalized Neuber equation
t Time, s
α Exponent in notch equation
ϵ_{cp} CP component of inelastic strainrange
ϵ_{in} Total inelastic strainrange
ϵ_0 Tensile strain
ϵ_{pc} PC component of inelastic strainrange
ϵ_{pp} PP component of inelastic strainrange
ϵ_u Compression strain
$\Delta\epsilon_t$ Total strainrange
ϵ_c Creep strain
Ω Internal stress
σ Stress
σ_R Stress in tension
σ_0 Stress at the beginning of relaxation
σ_r Stress at the end of relaxation
σ_u Stress in compression

Introduction

The improving efficiency of power machinery and the development of future-assured power plant technologies are bringing about a trend towards increasing the working temperature of components. This innovation must be considered for novel plants (such as nuclear plants, coal gasification systems, and solar tower plants), as well as for conventional plants (steam, coal), which are all included in the MAN production program.

The design of components for conventional plants has taken place until now essentially on the basis of proven rules and standards, such as TRD (Technische Regeln für Dampfkessel), BS (British Standard), and ASME (American Society of Mechanical Engineers). These procedures are completely satisfactory, provided the temperature limits established in the rules are not exceeded. The trend towards increased temperatures noted above, however, means that this condition can no longer be fulfilled and that the previous rules and standards are unsuitable.

Therefore, new reliable analytical design methods for higher temperatures are necessary. Studies in this area considering creep-fatigue interaction behavior have made progress in recent years [*1–6*], so that some of their results could be considered within the current practical design methods [*7*]. However, very complex interaction occur under high-temperature conditions, originating from material deformation and damage behavior and from different component effects, and existing developments are still not sufficient to allow a structural design code to be established in this area.

It is generally well known that structural components operating at high temperatures are subjected to a transient thermal stress change resulting from start-up, shut down, or heat input change. These transient loadings are usually followed by long, steady-state conditions. Thus, the failure of components subjected to high-temperature loading conditions is coupled with intensive inelastic deformation. According to the majority of the damage hypotheses used for life predictions, the intensity of these deformations is the deciding factor in determining the crack initiation life. It may be assumed that, with the help of a more precise description of the material deformation behavior, the deformation history can be accurately predicted, and more reliable life prediction can be achieved [5].

The material deformations appear as time-independent plastic and time-dependent creep strain increments. These deformations occur through various influences having their origin in material structure, component geometry, and loading conditions. In isolating one of these influences under laboratory conditions, through tests on material specimens, the material scientist simplifies the execution of the tests and evaluation of the results so that the different influences can be systematically examined. In recent practice, the systematic investigations conducted to estimate the effects of creep fatigue interactions on material damage and on material deformation behavior have been carried out almost independent of each other. Moreover, the typical creep-fatigue interaction life test involves a strain-controlled test containing a tensile or compressive dwell at peak strain level. These conditions do not realistically represent the structural response of the component. Therefore, the necessary consideration of component stress-strain history, depending on its geometry and load conditions, is more or less overlooked. It is clear that the interaction effects, resulting from the influences mentioned above, are in this way not taken into account, and for practical purposes the design procedures may be inadequate.

The practical engineer will prefer a less "systematic" approach, starting with the component as a whole. Required verification testing of components under accurately controlled conditions and the development of design diagrams through the testing of original components are examples of this approach. Both procedures make the present practice of lifetime assessment less economical. They lead to demands for more efficient design procedures and increasing interest in theoretical methods capable of including these complex conditions of creep-fatigue interaction on the component for use in a stress-strain history prediction and life assessment.

More integrated approaches that account for mutual interaction of all parameters influencing the component and that lead to more relevant and competitive design methods for components under high temperature would be of great importance for industrial practice.

One possible approach to this task is presented in this paper. Several creep fatigue experiments have been carried out to investigate the different effects on deformation and damage behavior which originate in a creep and fatigue environment. The results are intended to support the improvement of current design methods and supply some basic information on creep-fatigue interaction behavior.

Test Planning and Development

Test Material and Specimen

The material investigated was the austenitic alloy Incoloy 800 H, which is used by a large spectrum of technology today. With a solution temperature of 1130°C, it was possible to adjust the average grain size up to 2 to 5 according to ASTM. The chemical composition corresponds to the German standard TRD 434.

For all tests at elevated temperature, unnotched (smooth) specimens had a test section 9.0 mm in diameter and a cylindrical length of 22 mm. The specimens were machined from a sheet with 16-mm thickness and were taken in the longitudinal (rolling) direction. After machining, the test section of each specimen was polished to produce a surface finish about 3 μm.

Test Conditions

Stress-strain histories for the tests carried out in the program are presented in Fig. 1. The first group describes conventional tests with dwell times at one side (tension or compression) of the hysteresis loop and creep displacements generated under load-controlled (cyclic creep) or strain-controlled (cyclic relaxation) conditions. The second group includes complex histories with creep or relaxation displacements at the tension side of the inside hysteresis loop. For these nonconventional test forms, standard test machine control devices and programs are not adequate, and special development is required in this area.

All tests with strain-controlled operation had a strain rate of 6 mm/m/s.

Description of the Test System

The tests were executed on a servohydraulic test machine (max load range ±100 kN, displacement range ± 125 mm). Separate control units were available for each of the three possible control signals: load, strain, and displacement. With the aid of a shift unit and an automatic compensator, these three control signals could be alternated every time during the test. The switching was shockless, with an interchanging time less than 5 ms.

The axial strain was measured with an axial high-temperature extensometer (MTS, Type 632.40 C-04) over a 21-mm gage length. Two slight punch marks on the specimen surface fixed the sensing arm targets from the extensometer, pressed on the specimen with springs, in the right position.

A furnace with an infrared radiating system was used to heat the specimen. The control measurements of axial temperature distribution along the gage section showed a temperature difference of less than 5°C.

Control of the Test Machine

In order to perform the novel tests, an adaptable control device on microprocessor basis has been developed by IABG. The device uses analog signals for command output and stress-strain measurements. The conditions of the test machine (stress or strain controlled) were switched by the digital outputs and controlled with the digital inputs. Furthermore, software was developed by IABG to control the combined stress and strain controlled tests shown in Fig. 1. With this software each control program could be generated in a short time. All test stress-strain histories in this investigation could be put together with separate program steps at the "monitor program." When the test is running, the monitor program has a control function. The necessary constant test parameters are fed into the microprocessor at the keyboard of the microprocessor.

Test Results

Cyclic Relaxation Test

The most common strain-controlled test is extended for the purpose of this investigation to a test which can be called the "incremental dwell test" (Fig. 2). During the test, the hold-time at the tension side of the hysteresis loop is step-increased between 10 s and 3 h. Through the relatively quick stabilization of stress-strain history with the chosen material the execution of the test becomes simplified and the long times, usually 50% of life, were not needed. The unchanged conditions can be reached early during test steps with a short dwell time, so that on long dwell conditions after only a few repeated cycles, which are necessary for possible changes not to be overlooked, the digital recording of the relaxation lines for further analysis can be carried out. Also, the repetition of the complete dwell series on the same specimens yield no measurable differences.

Two different strain ranges and two temperature levels are examined. The interpretation of test results is done through a regression analysis, with the evaluated value—the creep rate according to the relaxation line equation

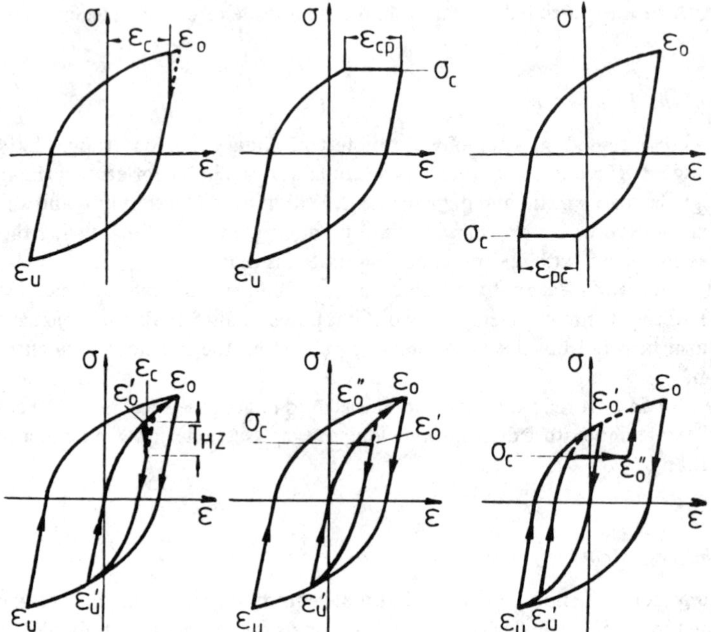

FIG. 1—*Different kinds of test hysteresis loops.*

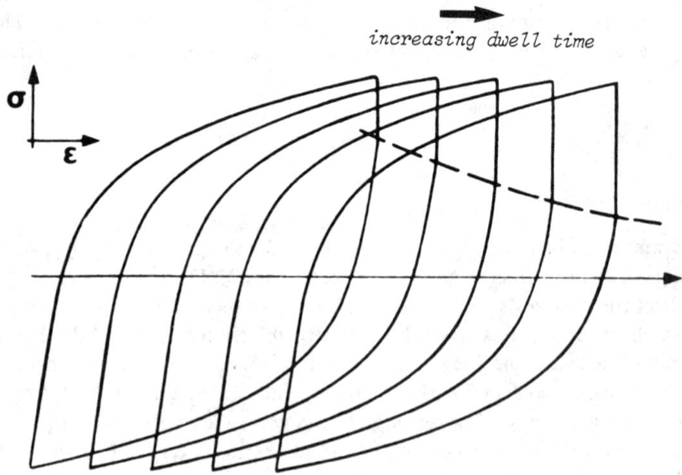

FIG. 2—*Incremental dwell test (dotted line shows the end of relaxation in case of increasing dwell time).*

$$\dot{\epsilon}_c = \frac{1}{E} \cdot \frac{d\sigma}{dt} \tag{1}$$

The results show, through the positive value of the constant (U) in the "best fit" regression line for the temporary creep rates, the trend to some "threshold" value of the stress (Fig. 3).

$$\dot{\epsilon}_c = B \cdot (\sigma - U)^m \tag{2}$$

No significant differences are seen in the results for two strain ranges. Compared with static minimal creep rates from the data of suppliers, the measured creep rates are far higher at all stress levels.

Cyclic Creep Test

Within this kind of test, the creep strain is generated at constant stress. It is well known that reversed tests with dwell at constant stress are commonly used for the evaluation of material characteristics according to the Strain-Range-Partitioning method [*8*], which we have also done in this case. So all the tests are extended through to failure of the specimen.

For the determination of cyclic creep behavior, the creep deformation lines are recorded after a different number of cycles and compared. The lines show typical creep behavior with a creep rate change similar to the static creep test (Fig. 4). But, at the same time, with ever-increasing cycle number, the time necessary to produce the equal creep strain range becomes shorter. Hence, this type of test verifies the creep rate increase caused by the cyclic loading, which is found in the preceding cyclic relaxation test.

Inside Hysteresis Test

This kind of test starts with a simple strain cycle for outer hysteresis loop without dwell. After stabilizing the stress-strain behavior the inside hysteresis loop was introduced. Figure 5*a* shows

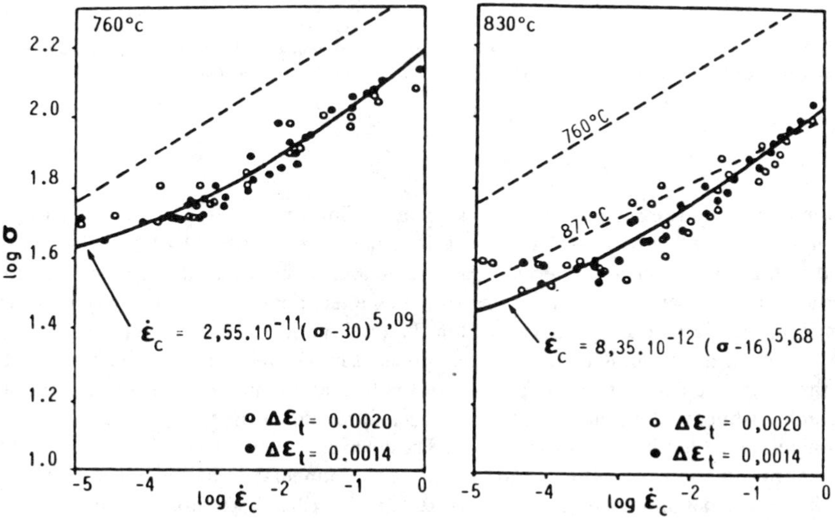

FIG. 3—*Creep rate stress from incremental dwell test (dotted line: —creep rate data from static tests).*

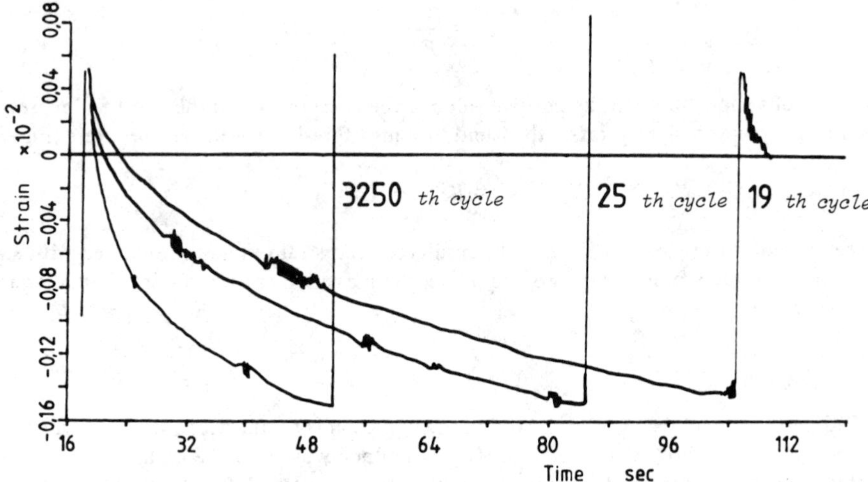

FIG. 4—*Effect of cyclic number (given in figure) on cyclic creep behavior.*

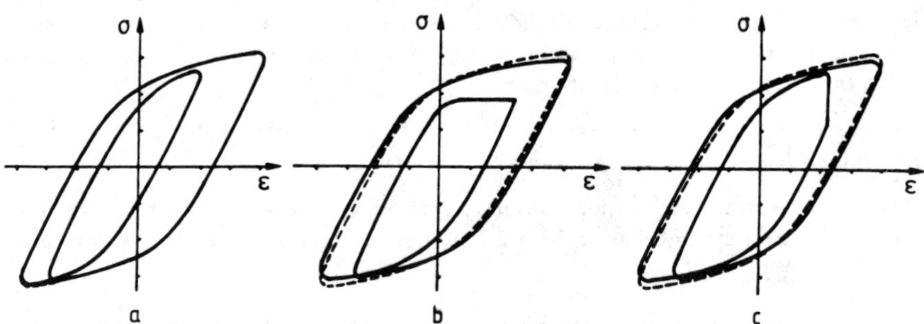

FIG. 5—*Results of test with inside hysteresis (full line) compared with the test without inside hysteresis (dotted line).* (a) *Without dwell time.* (b) *Creep dwell time.* (c) *Relaxation dwell time.*

that the material possesses full memory, since after finishing the inside hysteresis loop, the outer hysteresis is followed again. This behavior changes if the inside hysteresis loop is interrupted with creep or relaxation periods. Figures 5*b* and 5*c* show that during the dwell time a material recovery originates, so that some softening effect appears. The intensity of softening depends on the creep strain produced within the dwell periods. Under creep conditions, the creep rate changes in a similar manner, as in previous cyclic creep tests. In the case of relaxation, the creep rate depends on the position of the relaxation line. After increasing the strain range and pushing to the right the site of the relaxation line of the inside hysteresis loop, the material retains a "harder" response (Fig. 6). The creep rate relationships (Eq 2) given in Fig. 6 show that in the case of cyclic relaxation the so-called "kinematic" hardening can also be applied. The corresponding change in material state is described here through the values of the constant (U) according to Eq 2.

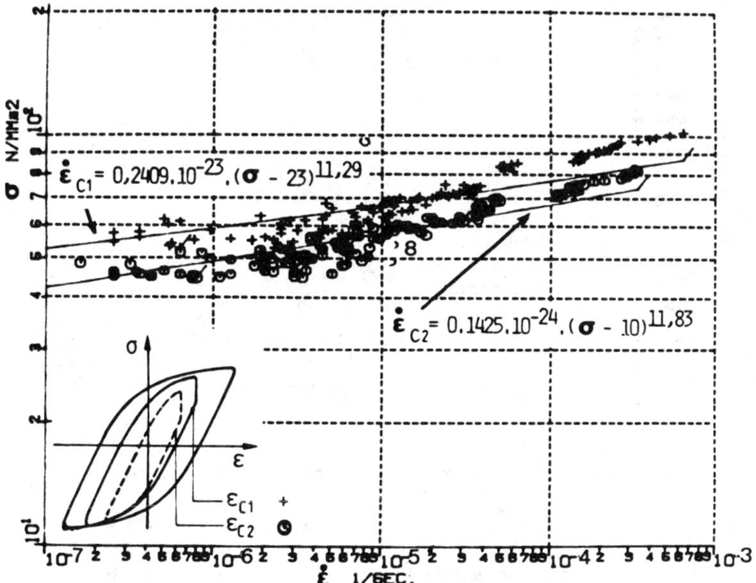

FIG. 6—*Creep-rate dependence on prestrain conditions (inside hysteresis tests).*

Evaluation of Damage Parameters

Among different life prediction methods based on a strain approach only the Strain Range Partitioning (SRP) method, developed by Manson and co-workers [*1,2*], takes the value of time-dependent inelastic strain, irrespective of how it was produced, at creep (σ = constant) or relaxation (ϵ = constant) conditions as a damage parameter. Other methods that are tied to some special test issue, as, for example, with dwell-time at ϵ = constant [*3*] or with $\dot{\epsilon}$ = constant [*5*] and take time or cycle frequency, respectively, as a damage parameter, have the disadvantage that, compared to the component histories, the test conditions do not hold. Considering that the service histories with arbitrary variations of strain with time can be examined only with the help of the SRP method, this method is chosen for further examination.

Two different testing techniques were used to obtain a *cp*-data: the basic one involving a constant creep stress and the other, less convenient, involving tensile strain hold-time relaxation cycle. The results are shown in Fig. 7. It can be seen that the slope of the *cp*-line is very well defined with the constant creep stress tests. Also, the results from strain hold tests generally lie on the same curve. There is some trend for this kind of test with short dwell (60 s) to overestimate the *cp*-values.

The two other life curves are also shown in Fig. 7. Because of limited test data (5 specimens), the slope of the *pc*-line is less well defined, but the line definitely lies between *pp*- and *cp*-data.

The duration of different tests show that the total time in a case of strain hold can be one order higher (Table 1). At the same time, the life of specimens were not influenced by these differences.

Life Analysis and Predictions

Under conditions with load repetition and strain influence due to the time-dependent deformations of values significant enough to be considered, the life of components is time- and cycle-

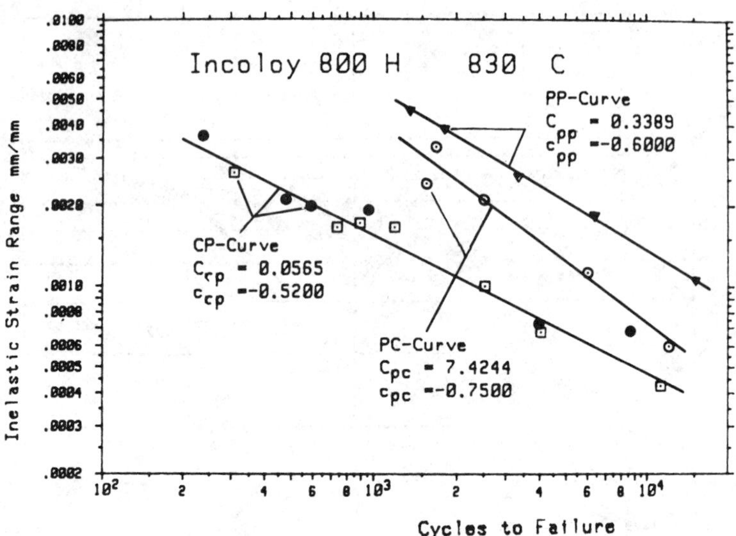

FIG. 7—*Summary of partitioned strain life relations (full symbols are from the strain-controlled test).*

dependent. But the general dependence on load sequence and hold or duration time is not so simple as is often assumed. Damage analysis for pure cycle loadings show, for example, that differences can exist between external loading and the conditions in a components-critical area resulting from these loadings. Because of this, instead of basing the life assessment on external loadings only, it is usual to consider the conditions in the critical area of the component through the so-called "local approach." In an analogous way for the proper consideration of the time effect in a life analysis one must bear in mind that entirely different results can be achieved if the particular dwell conditions vary.

Another problem in life assessment arises from the fact that the empirical (experimental) life evaluations are generally limited to some short-term values of the time and number of cycles. This limitation is invoked not only to avoid excessive experimental cost; as the duration of tests increase, the possibility of achieving the results in time for design are reduced. In these circumstances, the data relating to the component must be produced from data obtained over shorter times, sometimes with a considerable degree of extrapolation. In order to be able to successfully extrapolate these data, reliable characterization of material behavior and its analytical representation in the form of adequate constitutive laws are required. Both basic ingredients are needed for the successful introduction of the numerical simulation method as a substitute for time-consuming physical experiments. This is expected to lead to improvements compared with pure empirical extrapolation methods predominantly used up to now. Moreover, through the combination of experimental and analytical methods in a way in which these approaches support each other, a rational procedure shall be established for the transfer of material properties to the component under complex conditions found at high temperatures.

Numerical Simulation of the Strain Hold Test

To prove the possibility of the computational method to be used as a substitute for physical experiments and verify the validity of the relationship obtained from the tests, the test with the tensile strain hold-time relaxation cycle is numerically simulated first of all and compared with

TABLE 1—*Creep-fatigue test data.*

Specimen No.	d (mm)	ϵ_0 (mm/m)	ϵ_u (mm/m)	ϵ_{pp} (mm/m)	ϵ_{cp} (mm/m)	ϵ_{pc} (mm/m)	$\Delta\epsilon_t$ (mm/m)	$\Delta\epsilon_{in}$ (mm/m)	σ_0 (N/mm^2)	σ_u (N/mm^2)	σ_R (N/mm^2)	T_H (s)	N_{ASS}	N_B	T_H (h)	Test Type
2	9.02	1.90	−1.90	1.84	...	...	3.80	1.84	135	−136	...	...	6 350	7 105	...	*pp*
11	9.01	3.00	−3.00	3.86	...	...	6.00	3.86	151	−152	...	...	1 800	2 305	...	*pp*
19	9.04	3.50	−3.50	4.52	...	...	7.00	4.52	159	−161	...	...	1 350	1 804	...	*pp*
30	9.06	2.25	−2.25	2.56	...	...	4.50	2.56	136	−138	...	...	3 300	3 923	...	*pp*
39	9.02	1.40	−1.40	1.06	...	...	2.80	1.06	122	−124	...	...	15 080	16 262	...	*pp*
4	9.00	...	−2.0	0.25	2.4	...	4.25	2.65	70	−139	...	190	308	336	8	*cp*
8	9.02	...	−1.0	0.125	0.3	...	1.70	0.425	74	−110	...	4.7	11 200	11 715	15	*cp*
10	8.98	...	−1.25	0.125	0.875	...	2.4	1.00	70	−117	...	22	2 520	2 605	16	*cp*
16	9.01	...	−2.0	0.16	1.55	...	3.175	1.71	70	−133	...	70	890	948	18	*cp*
23	9.00	...	−1.0	0.075	0.6	...	2.05	0.675	70	−116	...	12	4 030	4 207	14	*cp*
35	8.99	...	−2.0	0.15	1.5	...	3.175	1.65	75	−136	...	30	1 190	1 247	10	*cp*
43	9.00	...	−2.0	0.15	1.5	...	3.05	1.65	70	−134	...	65	735	777	14	*cp*
14	9.00	2.00	...	0.30	...	2.10	4.10	2.40	149	−81.5	...	52	1 550	1 780	26	*pc*
15	9.01	1.10	...	0.20	...	0.40	2.025	0.60	121	−80.0	...	5.5	12 000	12 600	19	*pc*
24	9.00	2.05	...	0.20	...	1.90	3.80	2.10	149	−75.0	...	65	2 500	3 676	66	*pc*
32	9.00	2.50	...	0.30	...	3.00	5.50	3.30	162	−86.5	...	65	1 680	1 932	35	*pc*
37	9.00	1.25	...	0.20	...	0.925	2.55	1.125	129	−70.0	...	18.5	6 000	7 413	38	*pc*
2.8	8.99	2.50	−2.50	3.00	0.64	...	5.00	3.64	134	−166	93	60	900	900	16	*cp**
2.15	8.98	1.50	−1.50	1.40	0.52	...	3.00	1.91	120	−144	74	60	2 620	2 940	49	*cp**
2.20	8.96	0.87	−0.87	0.35	0.34	...	1.75	0.68	84	−124	47	60	14 850	15 000	250	*cp**
2.12	8.80	1.50	−1.50	1.40	0.54	...	3.00	1.98	116	−149	77	120	1 775	1 860	62	*cp**
2.17	8.91	1.00	−1.00	0.35	0.37	...	2.00	0.73	92	−117	53	120	7 100	7 500	250	*cp**
2.19	8.92	1.50	−1.50	1.50	0.58	...	3.00	2.10	125	−142	82	300	1 450	1 525	127	*cp**

*Strain hold test.

the experimental results. For this purpose, the hysteresis loops are, as Fig. 8 shows, numerically determined. This procedure differs from the usual method of life prediction where the hysteresis loops are taken directly from test. It is clear that in the latter case only the damage function would be checked. Since the stress-strain relationships cannot be considered in this way, this kind of approach has only limited relevance concerning the life assessment of components.

The cyclic stress-strain curve for the simulation is extracted from cyclic *pp*-tests presented in the preceding paragraph. Also, the evaluation from strain hold tests shows no significant differences (Fig. 9). Consequently, it was possible to assume that the material cyclic stress-strain behavior is not influenced by the hold time periods within of the range covered through the tests.

The stress at the end of strain hold is computed according to the equation

$$\sigma_r' = \sigma_0' \, [B \cdot E \cdot \sigma_0^{m-1} \cdot (m - 1) \cdot t + 1]^{1/(1-m)} \qquad (3)$$

with $\sigma' = \sigma - U$ and B, m and U from the results of cyclic relaxation tests (Fig. 3).

The agreement between numerical values and test results is very good (Fig. 10). The computed lines show how the results of this test can be extrapolated to the long dwells, which are experimentally very expensive to conduct.

Component Stress History

The material properties that are determined from smooth specimens do not remain unaffected by the condition in a critical area of the component. The standard influences, like stress gradients, multiaxiality of stress state, and support effects, originate through the form and the load combination of the component. The customary test with a strain hold cycle shown before is not representative for component history. Moreover, this test cycle represents the favorable load variant. This means that the results from constant strain tests are nonconservative compared to conditions within the component.

The stress-strain history for the critical area of the component can be determined numerically and experimentally, but at elevated temperatures under conditions of steady or slow changed

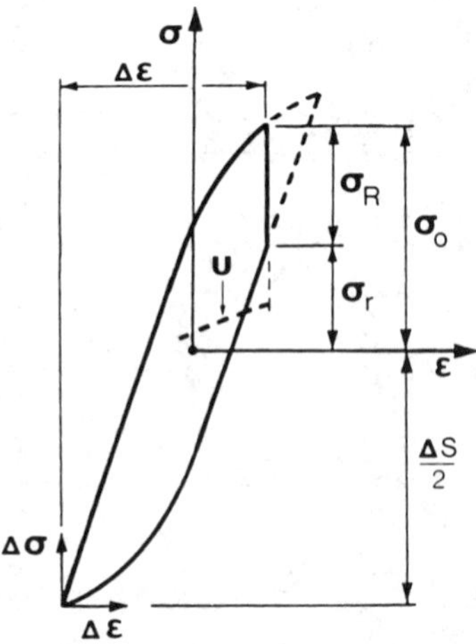

FIG. 8—*Numerical simulations of the stress hold test.*

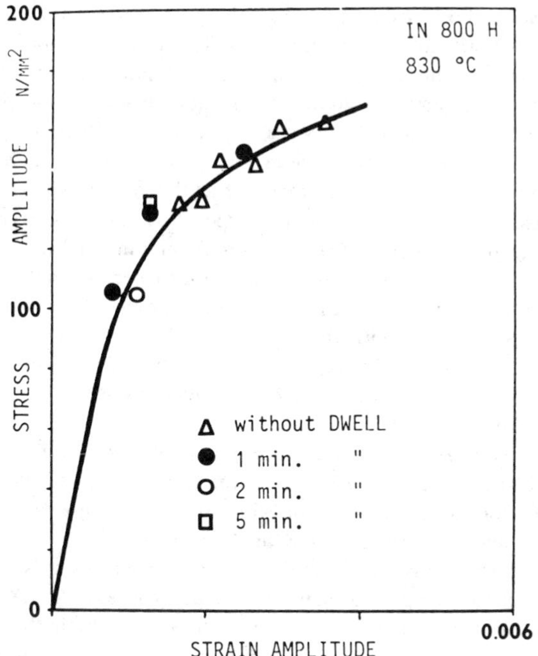

FIG. 9—*Cyclic stress-strain curve.*

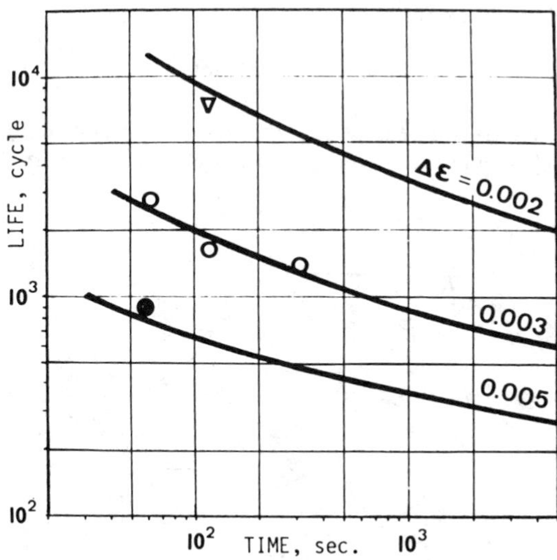

FIG. 10—*Variation of number of cycles to failure with hold time.*

loads, most measuring methods failed. On the other hand, the methods of nonlinear structure analysis are very expensive and in hitherto existing experience almost never based on sufficiently developed constitutive relations. Consequently, if the problem cannot be precisely defined and the required amount of material data for describing its behavior cannot be provided, it makes less sense to spend hours of computing time on a complex nonlinear analysis. In these circumstances, especially for preliminary design of components, an approximative procedure for predicting inelastic behavior from elastic analysis is very often used. However, in the high temperature range, because of the complexity of the general conditions, the development of realistic simplified methods and their application is more difficult. The best-known method of approximation to data is based on the so-called Neuber hyperbola, based on the assumption that the stress and strain at the critical region of the notch in developing inelastic deformations form a constant product. If one compares the stress-strain history according to this approximation with the strain hold test, it would be evident that this test cannot be related to the above method of approximation. This is explained with the aid of Fig. 11 [*10*]. Given the same time—in the strain hold test (a) as in the case of the strain history (b) where the condition $\sigma \cdot \epsilon =$ constant is to be maintained—considerably greater inelastic deformations occur which cannot be achieved at all in a test of this kind.

How the stress-strain on the component can change shall be shown here on one typical component. Considered is the heat exchanger pipe for the GAST solar collector. The computing of stress-strain history is done at the pipe section with generalized plane strain elements (MARC-28). Full nonlinear analysis is carried out for the dwell time of 6 h.

The computed stress-strain history, including warm-up conditions, is presented in Fig. 12. Because of the asymmetry of temperature distribution in the pipe section, strain is accumulated during plasticity and dwell conditions. This example shows that also in the case of pure thermal stress the stress-strain history of a component cannot be strain controlled, as one would expect. The temperature-dependent different relaxation rates through the section cause the equilibrium of the section to be changed and the strain on the "hot" side of the section to grow.

The analysis of the assigned stress-strain history show that its approximation according to the Neuber hyperbola would be extremely conservative. For this reason a correction to this method was introduced in the form of a variable exponent p:

$$\sigma \cdot \epsilon^{p} = \text{constant} \tag{4}$$

The value of the exponent p can be evaluated through a regression analysis. Figure 12 shows that the stress-strain change in this case can be very well described with Eq 4.

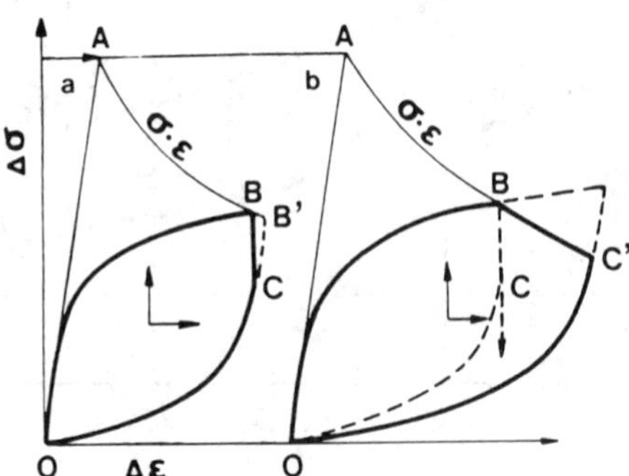

FIG. 11—*Difference between the test conditions with* (a) *strain hold and* (b) *the conditions on component.*

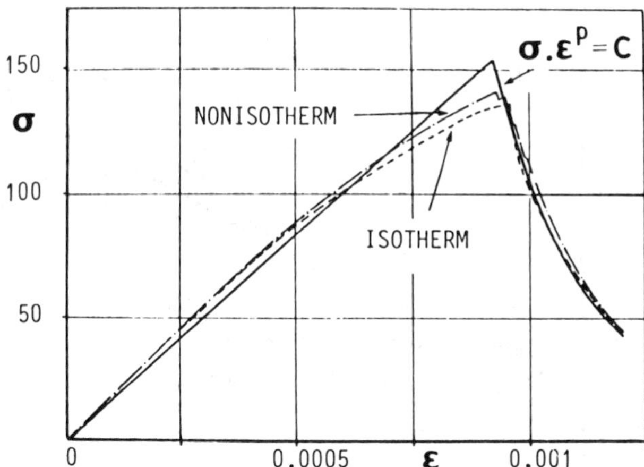

FIG. 12—*Stress-strain history at the critical area of the pipe.*

The generalized equation (Eq 4) applied to the so-called "notch" problem differs from the generalization proposed in Ref *11:*

$$K_\sigma \cdot K_\epsilon' \left[\frac{K_\epsilon}{K_\sigma}\right]^\alpha = K_t^2 \tag{5}$$

in which the exponent α related to p with

$$\alpha = \frac{p-1}{p+1} \tag{6}$$

is found [*11*] to depend on nominal stress level. Our analysis of the data of Mayer and Cruse [*12*] (used also in Ref *11*) show that the values of p in Eq 4 vary relatively slightly depending on the nominal stress. For practical purposes the mean value of p can be invoked to successfully predict the notch behavior in a large range of the nominal stress, as Fig. 13 shows. From results in Figs. 12 and 14 it can be nearly assumed that the value of p does not depend on the nominal stress level and dwell-time. For further verification of Eq 4, additional numerical calculations and strain measurements are proposed.

The exponent p has a very general meaning. Except for $p = 1$ (Neuber hyperbola), there are two special cases for $p = 0$ (stress control) and $p = \infty$ (strain control) (Fig. 14). These cases show that, through the p-variation in a given interval, very different histories can be accounted for. The values of p are in general larger than 1 (Neuber). However, agreement with the dwell test is not reached until $p = \infty$.

With the help of the same basic approach, it can be demonstrated how the life is influenced through the condition on the component. For numerical evaluation, the typical hysteresis loop is assumed with the creep strain generated in dwell, balanced with the plastic strain on the other side of the hysteresis loop. For both sides of the hysteresis loop the same value of p is adopted. The compiled life dependence on p for different strain ranges is shown in Fig. 15. These results demonstrate that the difference in the load history between the test with $\epsilon =$ constant ($p = \infty$) and the one of a typical component ($p = 1 \div 5$) can be very important.

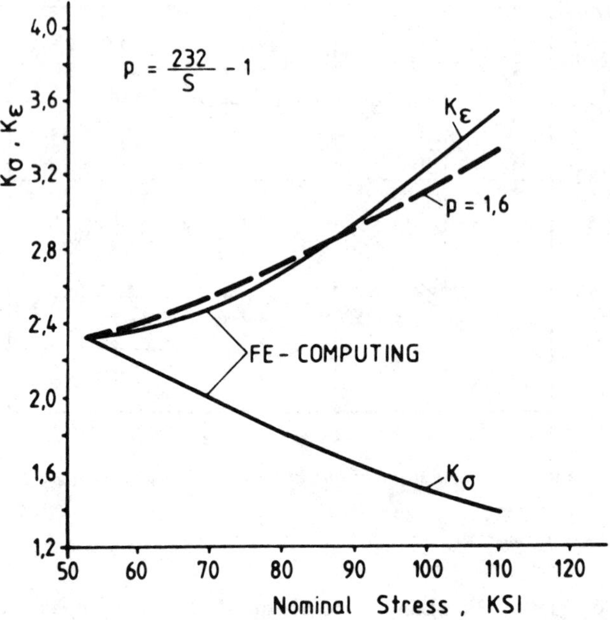

FIG. 13—*Comparison of results from finite element calculations and approximation according to "generalized notch" equation.*

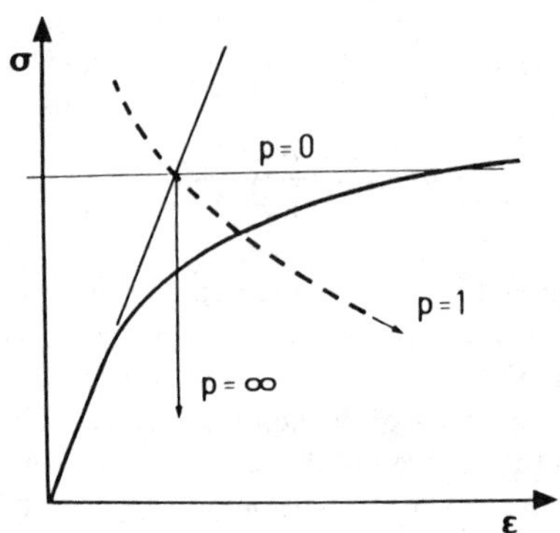

FIG. 14—*Comparison of different stress-strain behavior depending on* p*-variation.*

Stress-strain Behavior Under Operating Conditions

The results of the tests presented above show very complex interaction behavior between creep and plasticity under cyclic load conditions. As the cyclic relaxation and creep tests have shown, the cyclic hardening behavior of the material examined is followed with an increase in creep rate. On the other side the introduction of dwell periods causes the opposite effects of material recovery. But, whereas in the cyclic relaxation tests the amount of the recovery can be

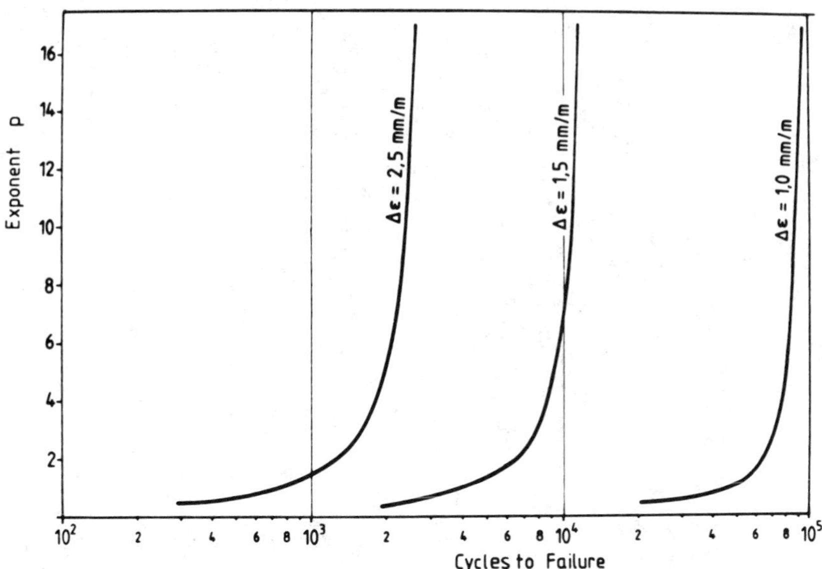

FIG. 15—*Effect of* p-*change on fatigue life.*

ignored for dwell-times within few minutes, in the tests with inside hysteresis loops a noticeable softening is achieved after only 10 to 30 s.

The existing classical constitutive theories are not capable of adequately describing this material behavior. Moreover, the separate treatment of "plastic" and "creep" deformation usual in classical mechanics is inadequate, because the mutual influence of these two types of inelastic deformation cannot be taken into account in this way.

On the basis of cyclic relaxation test results, it follows that the material behaves viscoplastically. The dependence on the material internal state of the creep rates is particularly perceptible in the relaxation test with inside hysteresis loops, where the creep rates are changed by the material kinematic hardening according to the position of the relaxation line (Fig. 6).

Figure 16 shows for the uniaxial state of stress the change of the "equilibrium" stress, Ω, under fully reversed strain-controlled conditions following the unified viscoplastic model description in Ref *13*. The shape of the $\Omega - \epsilon$ curve involves a typical creep behavior, measured also in the cyclic creep test (Fig. 4), with a decrease of creep rate (primary creep) due to overstress change, $(\sigma - \Omega)$, depending on the total strain accumulated along the A to C line. According to this simple description, which does not consider recovery effects, the equilibrium stress, Ω, stays unchanged during relaxation. The differences in the equilibrium stress under relaxation test conditions appear only when the relaxation line is moved (Fig. 6).

The described relaxation conditions in viscoplasticity explains why the calculated stress relaxation curves based on primary creep equations show significantly greater stress change than actual stress relaxation [*15*]. On the other side, as Fig. 3 shows, the creep rates in the cyclic relaxation tests are also significantly higher than the static secondary creep rates. From our

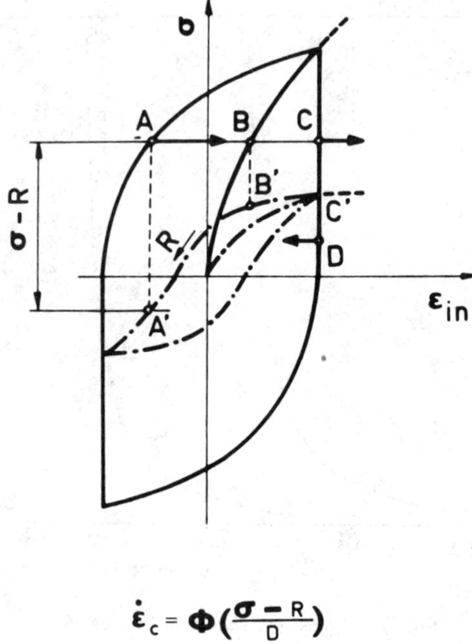

$$\dot{\varepsilon}_c = \Phi\left(\frac{\sigma - R}{D}\right)$$

FIG. 16—*Creep and relaxation rates dependence on the overstress.*

results, it appears that two effects contribute to this behavior: the nonsaturated equilibrium stress and the cyclically induced increase of creep rate as measured in the cyclic creep test (Fig. 4).

The complex stress-strain response on uniform specimens will be additionally complicated by the conditions on components. Because of generally lower strain ranges and the difference to the pure cyclic creep or relaxation conditions, the effect of primary creep (i.e., both of the unsaturated equilibrium stress value and its continual change) shall be considered. Also the results of relaxation tests could not be generalized for other relaxation conditions (i.e., for $p = \infty$), because of the possible equilibrium stress differences. With the material examined, both the creep strains are increased through the cyclic loading and the plastic strains enlarged through the creep load. It looks as if the life of a component can be significantly influenced through the described interaction behavior.

Nevertheless, for practical purposes, the possibility of using some acceptable simplifications in describing material deformation behavior must also be considered. These simplifications can be achieved if the material properties can be measured at nearly the same conditions, as have been demonstrated here through the successful numerical simulation of the strain hold test (Fig. 9). The use of representative or "mean" values, as, for example, in the case of cyclic hardening or cyclic softening of the values at mean life [*16*], can also be accepted as a useful simplification.

To allow for the simplifications to be always conservative it is important in all cases to consider the limits that must not be exceeded.

Concluding Remarks

There has been increased emphasis in recent years on managing complex, multi-faceted life analysis problems at high temperatures. However, at present, the use of material properties is still limited to the array of the single characteristics that do not interact with each other or with other influences caused by the component, loading, or material effects. Therefore, there is a need for life analysis methods to be expanded to include possible variations in material properties and, for real use, the material properties must be treated as a complex interacting system of parameters.

The results of this study show that a very large amount of information on material behavior is needed when the material properties are time-dependent, as in the design of hot components. Under these conditions the problems faced by stress engineers in choosing the correct design for a given structure to perform reliably over an established period of time and given operational conditions have grown from the simple consideration of different isolated properties. For that reason, the satisfactory solution to designing components for high-temperature conditions requires an integrated approach, with full consideration of different interaction effects from the influences to be involved in three main areas: (1) material deformation behavior, (2) component effects (geometry, size), and (3) material damage behavior.

The study presented in this paper concerning this object could be understood as a preliminary work, since several critical aspects that are required to provide a sound background for this approach are still lacking, including (1) constitutive theories with the capability of full material deformation behavior representation, (2) engineering methods for proper evaluation of material and component dependent deformation behavior, and (3) the damage hypothesis adopted to consider very different application possibilities. However, some general conclusions have been reached in this study which may apply to the standard methodology which will be developed in the future:

1. Material behaves viscoplastically. That means, the description of different features of interaction behavior between "creep" and "plasticity" can be achieved using viscoplastic theories that also make full use of different interaction, hardening, and recovery phenomena. This can be expected, even if further developments in this area are necessary.
2. For practical purposes, the material model should include some simplifications. The existing viscoplastic theories are still too complicated for these purposes and are dependent on a large number of empirical coefficients that can be reduced for particular purposes. The simplifications will provide clear limits for their applications.
3. Generalization of the Neuber hyperbola ratio as proposed and used in this study can be employed as a viable means for evaluating the effect of component geometry and loading conditions, and their effects on material behavior. Based on this ratio, an engineering method can be established.
4. With regard to damage material behavior it must be considered that only one method can be applicable to the components, which allow arbitrary variation in the strain history. More attention to this kind of selection criteria for material damage hypothesis should be given in the future.

Acknowledgments

This work has been undertaken within the GAST (Gas-Cooled Solar Tower Plant) project, which is being implemented under the contract of INTERATOM and is sponsored by the German Federal Ministry for Research and Technology (BMFT).

References

[1] Manson, S. S., Halford, G. R., and Hirschberg, M. H., "Creep-Fatigue Analysis by Strain-Range Partitioning," NASA Technical Memorandum TM X-67838, National Aeronautics and Space Administration, Washington, D.C., June 1971.

[2] Hirshberg, M. H. and Halford, G. R., "Use of Strain Range Partitioning to Predict High-Temperature Low Cycle Fatigue Life," NASA Technical Note TN D-8072, National Aeronautics and Space Administration, Washington, D.C., Jan. 1976.

[3] Jaske, C. E., Midlin, H., and Perrin, J. S., "Combined Low-Cycle Fatigue and Stress Relaxation of Alloy 800 and Type 304 Stainless Steel at Elevated Temperatures," in *Fatigue at Elevated Temperature, ASTM STP 520,* American Society for Testing and Materials, Philadelphia, 1973, pp. 365-376.

[4] "ASME Boiler and Pressure Vessel Code Cases, Case N-47 (1592)," Class 1 Components in Elevated Temperature Service, Section III, Division 1, American Society of Mechanical Engineers, New York, 1979.

[5] Coffin, L. F., "Fatigue at High Temperature," Technical Information Series Report 72CRD135, General Electric, Schenectady, N.Y., April 1972.

[6] Saltsman, J. F. and Halford, G. R., "Application of Strain Range Partitioning to the Prediction of MPC Creep-Fatigue Data for 2¼Cr-1Mo Steel," NASA Technical Memorandum TM X-73474, National Aeronautics and Space Administration, Washington, D.C., Dec. 1976.

[7] Lawton, C. W., "Use of Low-Cycle Fatigue Data for Pressure Vessel Design," in *Low-Cycle Fatigue and Life Prediction, ASTM STP 770,* American Society for Testing and Materials, Philadelphia, 1982, pp. 585-599.

[8] Agatonovic, P. in *Proceedings,* Second International Conference on Creep and Fracture of Engineering Materials and Structures, B. Wilshire and D. R. Owen, Eds., Pineridge Press, Swansea, U.K., 1984, pp. 1079-1090.

[9] Agatonovic, P. and Heidenreich, R. in "Werkstoffprüfung 1984," DVM 1985 (250), Deutscher Verband für Materialprüfung e.V., Berlin, West Germany, 1985, pp. 477-488.

[10] Agatonovic, P. and Dogigli, M. in *Proceedings,* Fourth International Conference on Numerical Methods in Thermal Problems, R. W. Lewis, Ed., Pineridge Press, Swansea, U.K., 1985.

[11] Walcher, J. and Gray, D., "Aspects of Cumulative Fatigue Damage Analysis of Cold End Rotation Structures," AIAA/SAE/ASME 15th Joint Propulsion Conference, 18-20 June 1979, Las Vegas, Nev.

[12] Cruse, T. A. and Meyer, T. G., "Structural Life Prediction and Analysis Technology, Report FR-10896, Pratt and Whitney Aircraft Group, 3 Nov. 1978.

[13] Walker, K. P., "Research and Development Program for Nonlinear Structural Modelling with Advanced Time-Temperature Dependent Constitutive Relationship," NASA Report CR-165533, National Aeronautics and Space Administration, Washington, D.C., Nov. 1981.

[14] Hart, E. W., *ASME Journal of Engineering Materials and Technology,* Vol. 98, 1976, pp. 193-202.

[15] Mottot, M., Petrequin, P., Amzallag, C., Rabbe, P., Grattier, J., and Masson, S., "Behavior in Fatigue-Relaxation of High-Creep Resistant Type 316L Stainless Steel" in *Low-Cycle Fatigue and Life Prediction, ASTM STP 770,* American Society for Testing and Materials, Philadelphia, 1982, pp. 152-168.

[16] Morrow, J. D., Wetzel, R. M., and Topper, T. H., "Laboratory Simulation of Structural Fatigue Behavior" in *Effects of Environment and Complex Load History on Fatigue Life, ASTM STP 462,* American Society for Testing and Materials, Philadelphia, 1970, pp. 74-91.

Kenji Kanazawa,[1] *Koji Yamaguchi,*[1] *and Satoshi Nishijima*[1]

Mapping of Low Cycle Fatigue Mechanisms at Elevated Temperatures for an Austenitic Stainless Steel

REFERENCE: Kanazawa, K., Yamaguchi, K., and Nishijima, S., **"Mapping of Low Cycle Fatigue Mechanisms at Elevated Temperatures for an Austenitic Stainless Steel,"** *Low Cycle Fatigue, ASTM STP 942,* H. D. Solomon, G. R. Halford, L. R. Kaisand, and B. N. Leis, Eds., American Society for Testing and Materials, Philadelphia, 1988, pp. 519–530.

ABSTRACT: Mapping of low cycle fatigue mechanisms at elevated temperatures is attempted for a better understanding of the elevated temperature, low cycle fatigue behavior of an austenitic stainless steel. Results of strain-controlled, uniaxial low cycle fatigue tests on a solution-treated Type 310 stainless steel were used in the analysis. The experiments were performed at 15 temperatures that ranged from room temperature to 800°C and under four strain rate conditions that ranged from 6.7×10^{-3} s^{-1} to 6.7×10^{-6} s^{-1}. Modes of variation, both in stress amplitude and in fatigue life, are classified and plotted against temperature-strain rate coordinates. By superposing the two maps, one can easily find the principal factors that govern the fatigue life for any given conditions of temperature and strain rate. The proposed idea of mapping is believed to have wide applicability (e.g., for selecting materials and predicting fatigue behavior under service conditions).

KEY WORDS: low cycle fatigue, stainless steel, temperature, strain rate, cyclic hardening, cyclic softening, transgranular fracture, intergranular fracture, fatigue mechanism map

Many studies have been done in recent years on the elevated temperature, low cycle fatigue properties of various engineering materials. They have revealed that fatigue life not only depends on strain range and temperature but also on factors such as frequency [*1*], strain rate [*2*], or strain hold time [*3*]. Effects of creep damage [*4*], oxidation environment [*5*], cyclic deformation behavior [*6*], structural change [*7*], and so on have been pointed out as an explanation for the time and temperature dependence of low cycle fatigue life. When low cycle fatigue tests are carried out for engineering materials, temperatures are generally limited to a narrow range that depends on the service conditions of the material. Therefore it is not yet clear which factors are dominant in explaining the low-cycle fatigue behavior for each test condition. It remains difficult, however, to understand in a comprehensive manner the elevated temperature, low cycle fatigue behaviors of metals, since the effect of these factors seems to be very complicated.

In this paper, mapping of low cycle fatigue mechanisms at elevated temperatures is attempted in order to provide a better understanding of the behavior in the case of austenitic stainless steel. The proposed idea of mapping is believed to have wide applicability (e.g., for selecting materials and predicting their fatigue behaviors under service conditions).

[1]Head of the First Laboratory, Senior Researcher, and Director, respectively, Fatigue Testing Division, National Research Institute for Metals, Tokyo, Japan.

Experimental Procedure

A commercial grade of Type 310 austenitic stainless steel, received in the form of 22-mm-diameter bar which had been solution treated at 1100°C for 1 h, was used for this study. The chemical composition of the material is given in Table 1.

The fatigue specimens had a constant diameter of 6 mm over a gage length of 15 mm. The final polishing was performed longitudinally to the surface of the gage section with 600-grade silicon carbide paper. Strain-controlled uniaxial fatigue tests were carried out in air using a servo-hydraulic fatigue testing machine equipped with a resistance furnace. The waveform of the strain cycle was triangular, the mean strain being zero. The strain rates employed were 6.7×10^{-3}, 6.7×10^{-4}, 6.7×10^{-5}, and 6.7×10^{-6} s^{-1}, and the test temperatures selected were in the range from room temperature to 800°C.

During a fatigue test in which the total strain range ($\Delta\epsilon_t$) is maintained at a constant level, the stress range changes gradually by hardening or softening with increased cycles. As a result, changes occur both in the plastic strain range ($\Delta\epsilon_p$) and the elastic strain range ($\Delta\epsilon_e$) during the test. Therefore the stress range and the plastic and elastic strain ranges were monitored. Values at about half of the fatigue life are taken as representative. The fatigue life was defined as the number of cycles to complete separation of the specimen.

After the fatigue tests, the fracture surfaces were observed in a scanning electron microscope. Dislocation structures were also examined by transmission electron microscopy in specimens taken from transverse slices cut from the gage section near the fracture surface.

Results

Temperature and Strain Rate Dependence of Deformation

Examples of the change in stress range with increasing strain cycles at a strain rate of 6.7×10^{-3} s^{-1} are shown in Fig. 1 for a total strain range of 10^{-2}. From room temperature to 300°C, the stress range increased during the initial stage and then remained at a constant value or decreased slightly (i.e., cyclic hardening behavior was observed). Cyclic hardening behavior was more pronounced at temperatures from 400 to 700°C, especially at 600°C. Figure 2 shows the effect of strain rate on the change in stress range with increasing strain cycles for a total strain range of 10^{-2}. At 450°C increasing strain rate produced a decrease in stress range, while at 700 and 800°C the reverse was true. Both effects occurred at 600°C, depending on the strain rate. Cyclic softening behavior was observed for slow strain rates at 700 and 800°C.

Temperature dependence of the cyclic 0.2% proof stress is shown in Fig. 3 for a range of strain rates together with the monotonic 0.2% proof stress obtained at a strain rate of 5×10^{-5} s^{-1}. The cyclic 0.2% proof stress was taken to be half of the stress range at a plastic strain range of 0.4×10^{-2}. This was obtained as follows. The plastic strain range and stress range were taken at about half the fatigue life from a plot such as Fig. 1. These values were then transferred to a log-log plot, and the stress range at a plastic strain range of 0.4×10^{-2} was determined by interpolation or extrapolation of this linear relationship. For a strain rate of 6.7×10^{-3} s^{-1} it had a minimum value of 200°C; this increased strongly with increasing temperature up to 600°C and then decreased drastically. As for the strain rate dependence of the cyclic 0.2% proof

TABLE 1—*Chemical composition (weight percent) of test material.*

C	Si	Mn	P	S	Ni	Cr
0.12	0.88	1.43	0.024	0.008	19.91	24.69

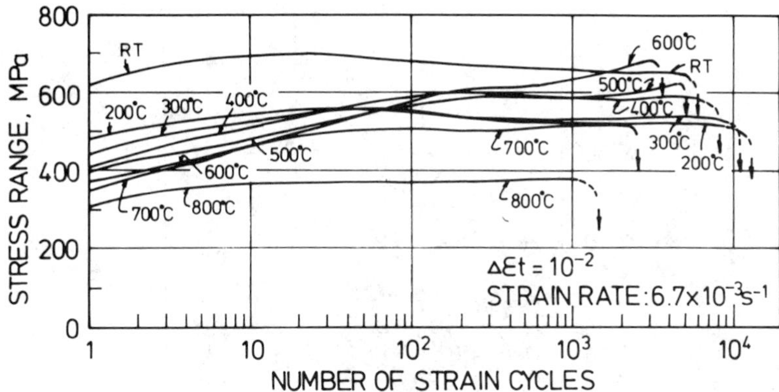

FIG. 1—*Temperature dependence of the change in stress range with increasing strain cycles.*

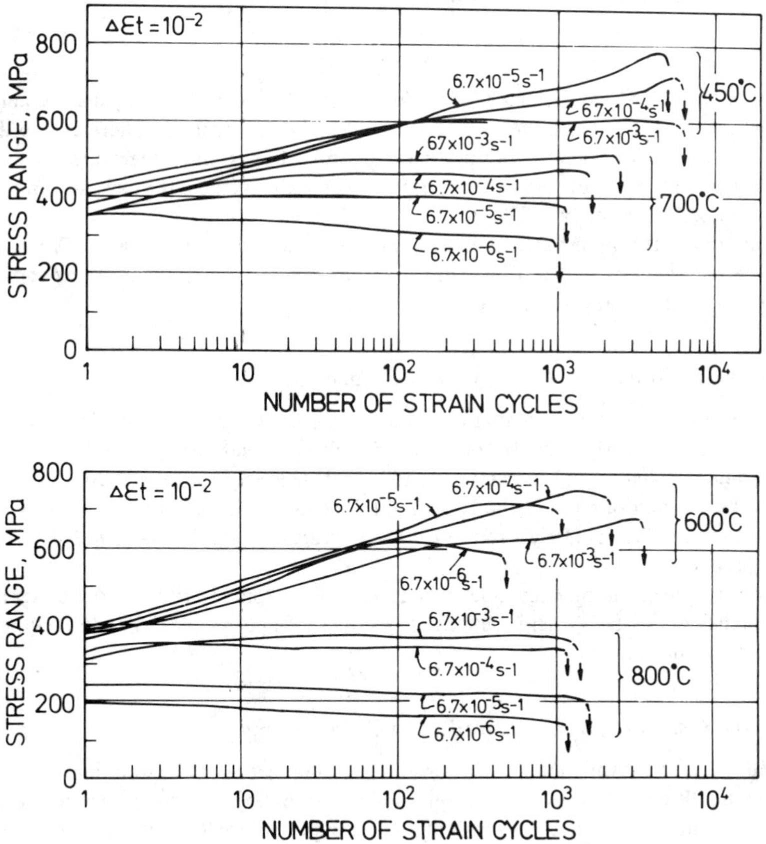

FIG. 2—*Strain rate dependence of the change in stress range with increasing strain cycles.*

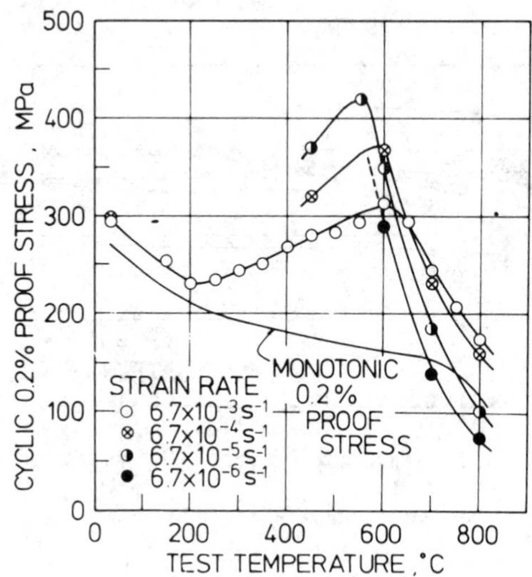

FIG. 3—*Temperature dependence of cyclic 0.2% proof stress.*

stress, it was positive where the proof stress decreased with increasing temperature and it was negative where the proof stress increased with increasing temperature. There are no data depicting the strain rate dependence of the cyclic 0.2% proof stress at temperatures below 400°C for the present material, but results from a Type 321 stainless steel [*8*] have shown the same tendency as described above.

To summarize, the temperature and strain rate dependence of the cyclic 0.2% proof stress are visualized in Fig. 4 by a three-dimensional surface where the stress axis is perpendicular to the temperature-strain rate coordinates.

Temperature and Strain Rate Dependence of Fatigue Life

Figure 5 shows the effect of temperature on fatigue life for a plastic strain range of 10^{-2} obtained by linear regression analysis of the Coffin-Manson relationship. The fatigue life has a maximum value at 200°C and decreases steadily with temperatures above 200°C, except for slow strain rate conditions where the fatigue life increases again above 650°C. The effect of strain rate on fatigue life is fairly strong in the temperature range from 400 to 600°C, but not noticeable at 800°C.

Temperature and strain rate dependence of fatigue life are summarized in Fig. 6, again using a three-dimensional surface where the axis of fatigue life is perpendicular to the temperature-strain rate coordinates.

Change in the Dislocation Substructure Due to Cyclic Straining

One important result which is needed for a more fundamental understanding of the fatigue behavior is the dislocation substructure following strain cycling at elevated temperatures [*9*].

Typical dislocation substructures after fatigue testing to failure for each test temperature are presented in Fig. 7. A cellular dislocation distribution is observed in specimens fatigued at temperatures below 200°C and at about 600°C. A planar dislocation array of high dislocation density is found at about 450°C and a subgrain structure is present above 700°C.

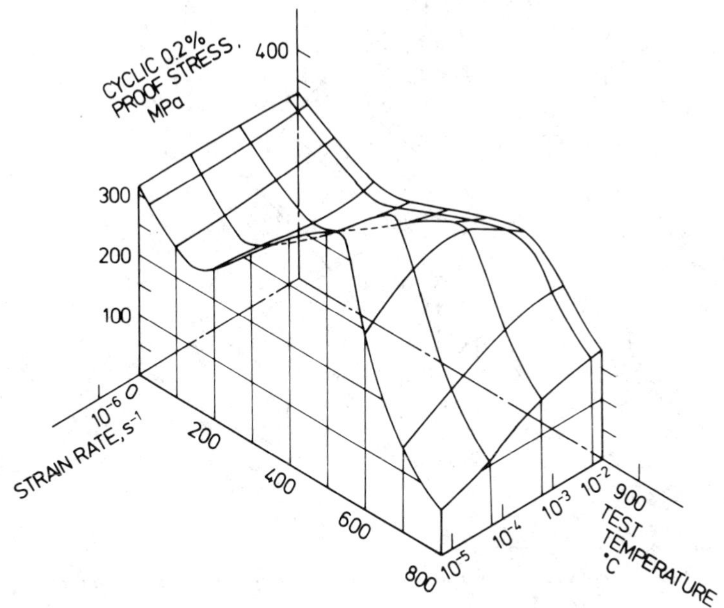

FIG. 4—*Cyclic 0.2% proof stress surface as a function of temperature and strain rate.*

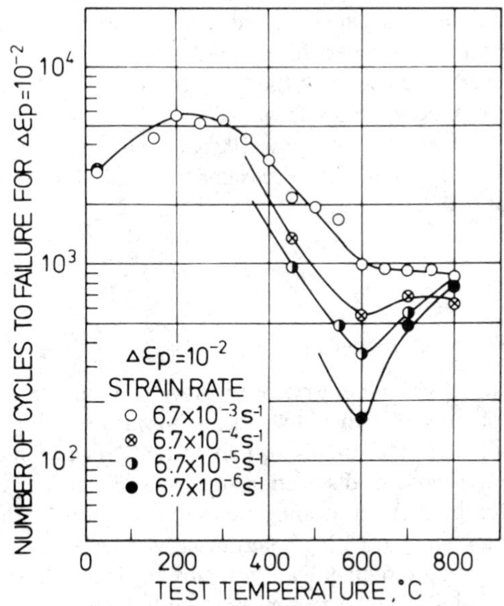

FIG. 5—*Temperature dependence of fatigue life for a plastic strain range of 10^{-2}.*

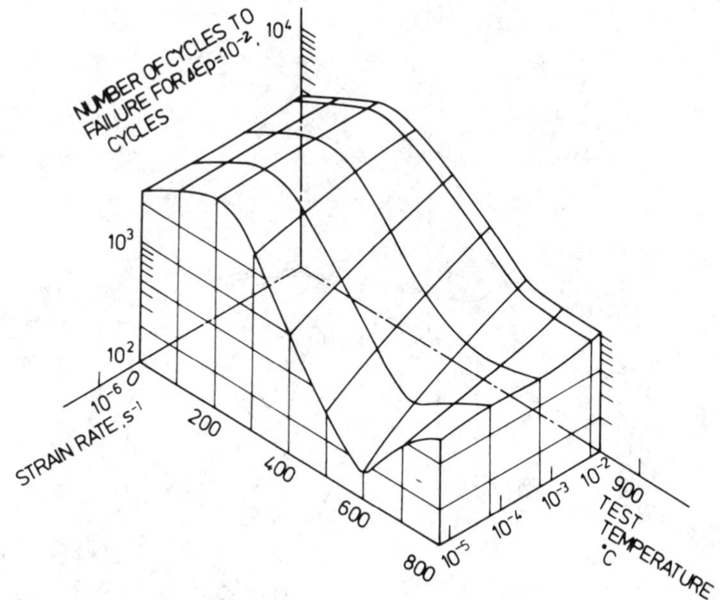

FIG. 6—*Fatigue life surface as a function of temperature and strain rate.*

Fracture Modes

Figure 8 shows examples of fracture surfaces observed by a scanning electron microscope. Obvious striations can be seen in specimens tested at temperatures less than 600°C, except for slow strain rate conditions at 600°C where intergranular fracture occurs. The creation of intergranular fracture facets seems to depend on the total strain range. When the total strain range is large, intergranular failure occurs more frequently.

For specimens fatigued at temperatures above 700°C, the fracture surfaces are covered with a thick oxidation layer and intergranular fracture is partly seen, but the fracture facets cannot be seen as clearly as those at 600°C.

Discussion

Mapping of Low-Cycle Fatigue Mechanisms

Figure 9 presents a map of cyclic deformation behavior. Solid curves indicate the contour lines for cyclic 0.2% proof stress, obtained by cutting the surface in Fig. 4 parallel to the temperature-strain rate coordinates. The map is also divided into regions indicated by dotted lines according to the dominating mode of dislocation substructure, such as cell type, planar band structure, or subgrain structure. A dot-dashed curve indicates the boundary between regions where cyclic softening and cyclic hardening behavior occurs.

Figure 10 is a map presenting fatigue life and fracture mode as a function of strain rate and test temperature. Solid lines indicate the contour lines for fatigue life at a plastic strain range of 10^{-2}. The boundary between transgranular fracture and intergranular fracture modes is shown by a dotted line. A region which exhibits extensive oxidation is also indicated by a dot-dashed line.

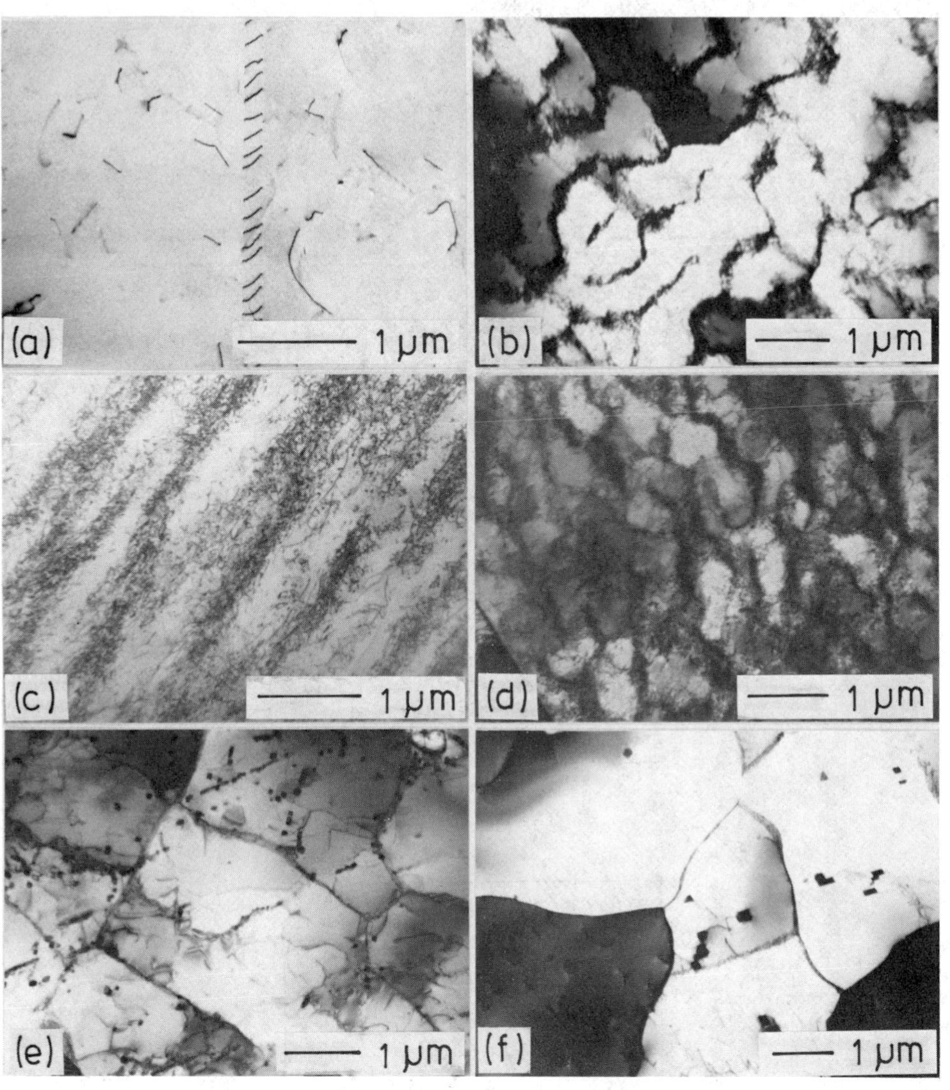

(*a*) Virgin.
(*b*) Fatigued at 200°C, $\Delta\epsilon_t = 10^{-2}$, $\dot{\epsilon} = 6.7 \times 10^{-3}\ s^{-1}$.
(*c*) Fatigued at 450°C, $\Delta\epsilon_t = 10^{-2}$, $\dot{\epsilon} = 6.7 \times 10^{-3}\ s^{-1}$.
(*d*) Fatigued at 600°C, $\Delta\epsilon_t = 10^{-2}$, $\dot{\epsilon} = 6.7 \times 10^{-3}\ s^{-1}$.
(*e*) Fatigued at 700°C, $\Delta\epsilon_t = 2 \times 10^{-2}$, $\dot{\epsilon} = 6.7 \times 10^{-5}\ s^{-1}$.
(*f*) Fatigued at 800°C, $\Delta\epsilon_t = 10^{-2}$, $\dot{\epsilon} = 6.7 \times 10^{-5}\ s^{-1}$.

FIG. 7—*Typical dislocation structures in low cycle fatigued specimens.*

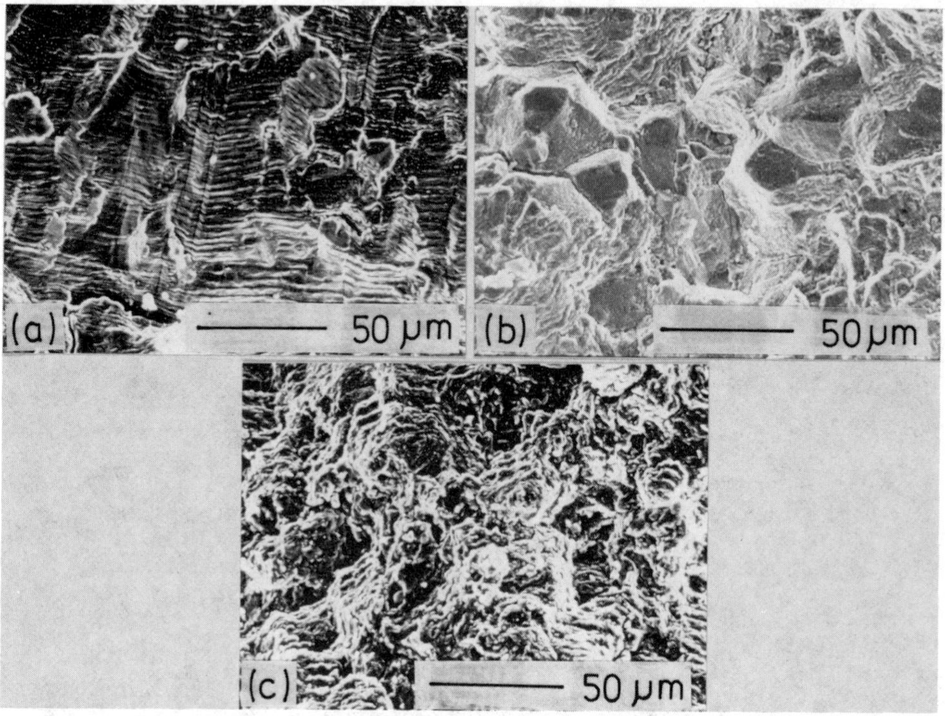

(a) Fatigued at 450°C, $\Delta\epsilon_t = 10^{-2}$, $\dot{\epsilon} = 6.7 \times 10^{-3}\ s^{-1}$.
(b) Fatigued at 600°C, $\Delta\epsilon_t = 10^{-2}$, $\dot{\epsilon} = 6.7 \times 10^{-5}\ s^{-1}$.
(c) Fatigued at 800°C, $\Delta\epsilon_t = 2 \times 10^{-2}$, $\dot{\epsilon} = 6.7 \times 10^{-5}\ s^{-1}$.

FIG. 8—*Typical fracture modes in low cycle fatigued specimens.*

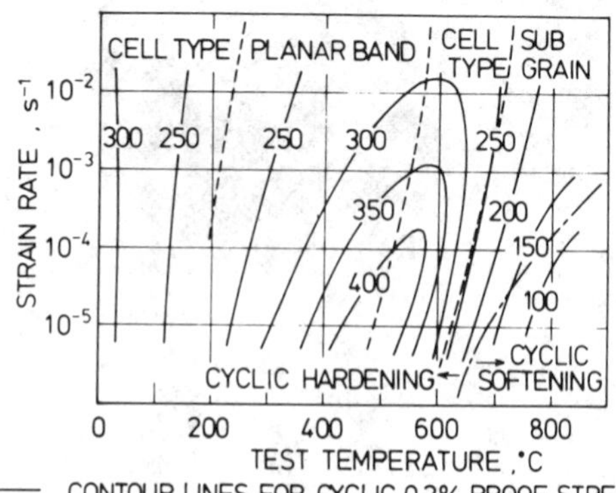

FIG. 9—*Map for cyclic deformation stress and dislocation substructures.*

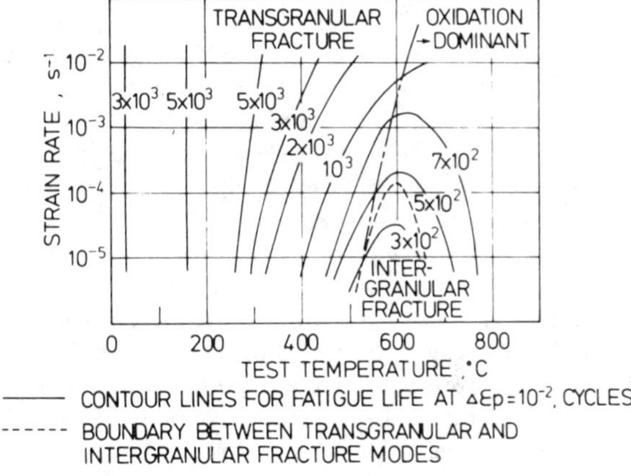

FIG. 10—*Map for fatigue life and fracture modes.*

Jenkins and Smith [*10*] observed serrated plastic flow in tensile tests for 35Cr-15Ni austenitic stainless steel at temperatures from about 200 to 650°C. They pointed out that it was caused by dynamic strain aging related to carbon-vacancy pairs at temperatures below 500°C and to chromium atoms above 500°C. An increase in the rate of dislocation generation occurs during serrated flow in austenitic stainless steels leading to increased strengthening of these alloys in this temperature range.

Remarkable cyclic hardening behavior in the present material between 200 and 650°C is also considered to be due to the dynamic strain aging. Below and above this temperature range the dislocation substructure is predominantly cellular; at intermediate temperatures it takes on a planar band structure with a high dislocation density, and this is where the negative strain rate dependence of deformation stress is observed. On the other hand, the subgrain structure observed at above 700°C indicates that a recovery process has taken place during cyclic straining. As a result, it is considered that cyclic softening behavior and a strong positive strain rate dependence of deformation stress occurs in this temperature region.

Challenger and Moteff [*11*] pointed out that the cell size (d) for Types 304 and 316 stainless steels, after elevated-temperature low-cyclic fatigue tests, was related to the saturation stress (σ) through a power function ($d \alpha \sigma^{-b}$), where the value of b depends on the test temperature. As for the present material, the following relationship has been obtained [*12*] between the normalized stress range (σ/G) and the cell or subgrain size (d), where G is the shear modulus:

$$\sigma/G_T \alpha d^{-0.5}$$

The first map (Fig. 9) expresses variations of dislocation substructure in the materials after cyclic deformation as a balance between dynamic strain aging and recovery. The second map (Fig. 10) shows the effect of temperature on fatigue life and fracture mode. Consequently, by superposing the two maps, one can easily find the principal factors which govern the fatigue life for any given conditions of temperature and strain rate. The idea of mapping is believed to have a wide application (e.g., for selecting materials and predicting fatigue behavior of the selected materials under service conditions).

Fatigue Behavior of Temperatures Below 0.5 T_m

A temperature of 0.5 T_m for Type 310 steel is 563 to 590°C, where T_m is the absolute melting temperature [*13*]. In the temperature range from room temperature to 500°C, which is lower than 0.5 T_m, transgranular fracture with striations is dominant and the fatigue life is a maximum at 200°C for a fixed value of strain range. The fatigue life decreases with decreasing strain rate at 450°C, notwithstanding the impossible effect of creep deformation. It has been pointed out [*14*] that a good correlation exists between low-cycle fatigue life and fracture ductility in tensile test; however, the ductility of the present material exhibits no maximum value around 200°C [*15*].

Comparing Fig. 10 with Fig. 9 shows that the fatigue life decreases for temperature and strain rate conditions where the stress range is increased by dynamic strain aging. In Fig. 11 the fatigue life is plotted against the stress range normalized by Young's modulus, which therefore corresponds to the elastic strain range, for a plastic strain range of 10^{-2}. An excellent correlation is obtained for temperatures below 500°C where transgranular fracture is dominant. It is recognized that fatigue life decreases with increasing stress range for a fixed plastic strain range in this temperature region. This means that the fatigue life is governed not only by plastic strain range but also by a mechanical factor related to the stress range.

Dowling and Begley [*16*] evaluated the crack propagation rate under cyclic plastic deformation by using the cyclic *J*-integral. Applying this approach to the propagation problem of a small crack, a new method was proposed to estimate the fatigue life of smooth specimens [*17*]. This suggests the importance of considering the stress or elastic strain range, in addition to the plastic strain range, because the cyclic *J*-integral contains these parameters.

Fatigue Behavior at Temperatures Above 0.5 T_m

Intergranular fracture is observed in this experiment in the region of lower fatigue life which occurs for slow strain rate conditions at 600°C, a temperature beyond 0.5 T_m. However, at temperatures higher than 600°C the intergranular fracture mode is not always evident and the reduction in fatigue life with decreasing strain rate is small.

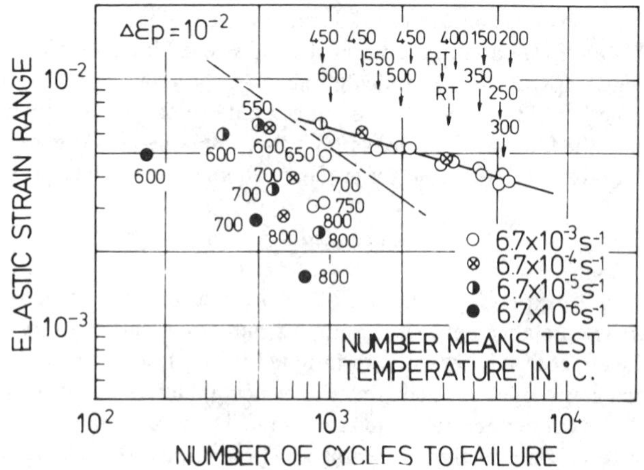

FIG. 11—*Relationship between elastic strain range and number of cycles to failure for a plastic strain range of 10^{-2}.*

Sidey and Coffin [*18*] reported that the strain rate dependence of fatigue life and the appearance of intergranular fracture above $0.5\,T_m$ in OFHC copper was attributed to the environmental effect because these phenomena were not observed in vacuum.

In general, grain boundary sliding can occur under low strain rate conditions at temperatures above $0.5\,T_m$. It has been said in the case of the creep rupture of austenitic stainless steels [*19*] that the intergranular fracture with wedge-type cracking occurs in high-stress, short-time tests and intergranular fracture with cavity formation occurs in low-stress, long-time tests. In the case of low-cycle fatigue, there would seem to be no chance for intergranular fracture to occur because of the reversibility of grain boundary sliding under cyclic loading. However, two factors can be pointed out for grain boundary fracture to occur; the ease with which grain boundary sliding can occur and oxidation in air which reduces the reversibility of the sliding. Grain boundary sliding can occur easily for slow strain rates at 600°C, because stresses applied to the grain boundary become higher as a result of matrix hardening produced by dynamic strain aging. The initiation and growth of intergranular cracks are believed to be faster than those for transgranular ones, because if that was not the case the fracture mode would be of the transgranular type.

On the other hand, at temperatures higher than 600°C it is thought that grain boundary sliding becomes difficult because stresses applied to the grain boundary decrease as a result of the recovery of the grain matrix. It therefore becomes difficult for intergranular fracture to occur for low strain rate conditions and the strain rate dependence of fatigue life is no longer distinct.

Summary

Mapping of the low-cycle fatigue mechanisms at elevated temperatures is attempted for the better understanding of the elevated temperature, low cycle fatigue behaviors. The results of strain controlled, uniaxial low cycle fatigue tests on solution-treated Type 310 stainless steel were used in the analysis. Modes of variation, both in stress amplitude and in fatigue life, are classified and plotted against temperature-strain rate coordinates. By superposing the two maps, one can easily find the principal factors with govern the fatigue life under any conditions of temperature and strain rate.

At temperatures below $0.5\,T_m$, the fracture mode was transgranular with striations and the fatigue life decreased for conditions of temperature and strain rate that produced dynamic strain aging and therefore a high stress range. At temperatures above $0.5\,T_m$, the fatigue life decreased when the fracture mode changed from transgranular to intergranular. Grain boundary sliding and the oxidation effect in air were pointed out as factors contributing to intergranular fracture.

References

[*1*] Coffin, L. F. in *Fracture 1969,* Chapman and Hall, London, 1969, pp. 643–654.

[*2*] Berling, J. T. and Slot, T. in *Fatigue at High Temperature, ASTM STP 459,* American Society for Testing and Materials, Philadelphia, 1969, pp. 3–30.

[*3*] Coles, A., Hill, G. J., Dawson, R. A. T., and Watson, S. J. in *Proceedings,* International Conference on Thermal and High Strain Fatigue, Institute of Metals, London, 1967, pp. 270–294.

[*4*] Campbell, R. D., *Journal of Engineering for Industry, Transactions of American Society of Engineers,* Vol. 93, 1971, pp. 887–892.

[*5*] Coffin, L. F. in *Corrosion Fatigue,* National Association of Corrosion Engineers, Texas, NACE-2, 1972, pp. 590–600.

[*6*] Wareing, J., Tomkins, B., and Sumner, G., in *Fatigue at Elevated Temperatures, ASTM STP 520,* American Society for Testing and Materials, Philadelphia, 1973, pp. 123–138.

[*7*] Yamaguchi, K. and Kanazawa, K., *Metallurgical Transactions,* American Society for Metals and the Metallurgical Society of AIME, Vol. 11A, 1980, pp. 1691–1699.

[8] Kanazawa, K., Iwanaga, S., Kunio, T., Iwamoto, K., and Ueda, T., *Bulletin of Japan Society of Mechanical Engineers,* Vol. 12, 1969, pp. 188-199.
[9] Feltner, C. E. and Laird, C., *Acta Metallurgica,* Vol. 15, 1967, pp. 1633-1653.
[10] Jenkins, C. F. and Smith G. V., *Transactions,* Metallurgical Society of AIME, Vol. 245, 1969, pp. 2149-2156.
[11] Challenger, K. D. and Moteff, J. in *Fatigue at Elevated Temperatures, ASTM STP 520,* American Society for Testing and Materials, Philadelphia, 1973, pp. 68-79.
[12] Yamaguchi, K. and Kanazawa, K., *Transactions of National Research Institute for Metals,* Vol. 26, 1984, pp. 210-214.
[13] American Iron and Steel Institute, Steel Products Manual: Steel and Heat Resisting Steels, 1974.
[14] Baldwin, E. E., Sokol, G. J., and Coffin, L. F. in *Proceedings,* American Society for Testing and Materials, Vol. 57, 1957, pp. 567-586.
[15] Kanazawa, K., Yamaguchi, K., Sato, M., Kobayashi, K., Suzuki, N., Shiohara, M., and Yoshida, S., *Transactions of National Research Institute for Metals,* Vol. 20, 1978, pp. 382-400.
[16] Dowling, N. E. and Begley, J. A. in *Mechanics of Crack Growth, ASTM STP 590,* American Society for Testing and Materials, Philadelphia, 1976, pp. 82-103.
[17] Yamada, T., Hoshide, T., Fujimura, S., and Manabe, M., *Journal of Japan Society of Mechanical Engineers,* Vol. 49, 1983, pp. 441-451.
[18] Sidey, D. and Coffin, L. F. in *Low-Cycle Fatigue Damage Mechanisms at High Temperature, ASTM STP 675,* American Society for Testing and Materials, Philadelphia, 1979, pp. 528-568.
[19] Shinya, N., Tanaka, H., and Yokoi, S. in *Proceedings,* 2nd International Conference on Creep and Fracture of Engineering Materials and Structures, Pineridge Press, Swansea, U.K., 1984, pp. 739-750.

DISCUSSION

J. Bressers[1] *(written discussion)*—With increasing temperature you note a change in dislocation substructure from cell type to planar bands and again to cell type. Is this related to dynamic strain-aging reactions?

K. Kanazawa et al. (authors' closure)—Yes, but we have not yet verified it for the present material. Refer to Jenkins and Smith's results (Ref *10* of our paper) for austenitic stainless steels containing high Ni and high Cr elements.

[1]Joint Research Centre, J.E.C., Petten, The Netherlands.

G. L. Chen,[1] Q. F. He,[1] L. Gao,[1] and S. H. Zhang[1]

A Maximum Stress Modified Life Equation on the Basis of a Fatigue-Creep Interaction Map

REFERENCE: G. L. Chen, Q. F. He, L. Gao, and S. H. Zhang, **"A Maximum Stress Modified Life Equation on the Basis of a Fatigue-Creep Interaction Map,"** *Low Cycle Fatigue, ASTM STP 942,* H. D. Solomon, G. R. Halford, L. R. Kaisand, and B. N. Leis, Eds., American Society for Testing and Materials, Philadelphia, 1988, pp. 531-542.

ABSTRACT: The effects of maximum stress on crack initiation and propagation, the fracture life, and the dynamic creep rate have been studied on the basis of a fatigue-creep interaction fracture map. A number of maximum stress modified equations have been deduced to describe the complete interaction behavior of a large ESR-Cast-to-Shape superalloy disk under a wide range of fatigue and creep stress combinations at a constant temperature. All the observed fracture lives fall within a range of a factor of two of fracture lives predicted by these equations.

KEY WORDS: superalloy, fatigue, creep, fatigue-creep interaction, life prediction

Several approaches to the prediction of material behavior under the cyclic condition at high temperature have been developed. One of these is the modification of a low-temperature relationship, including the early concepts of 10% rule [*1*] and the frequency-modified life equation [*2*], another is the summation of creep time and fatigue cycle fraction [*3*], a third is deduced from the concept of ductility exhaustion [*4*], and, finally, the strain-range partitioning method [*5*]. All these methods are supported by some experimental evidence and have certain limitations. Chen [*6*] reviewed the correlation between fracture life and alternating stress, mean stress under constant maximum stress condition, and suggested a description of the fatigue-creep interaction behavior of materials by using a fatigue-creep interaction curve and fracture map (Figs. 1 and 2) [*7-9*]. The term σ_i is an arbitrarily chosen parameter to make the figures on the ordinate ($\sigma_m + \sigma_i$) adaptable to different values of σ_{max}. The fatigue-creep interaction maps consist of the measured data of fracture lives and dynamic creep rates under a wide range of combinations of fatigue and creep stresses. With the different stress dependences of fracture life and dynamic creep rate, the interaction behavior of materials should be examined in three regions: interaction resulting in predominantly fatigue fracture (F region), interaction resulting in predominantly creep fracture (C region), and interaction resulting in mixed fracture (FC region). The appearances of the fracture surface also showed that a number of fracture mechanisms can occur under various combinations of fatigue and creep stresses. According to the experimental observations of crack initiation and propagation, the complete map also can be divided into five regions: (1) fatigue crack nucleation and propagation to fracture (FF region), (2) fatigue crack initiation followed by creep crack nucleation and predominantly fatigue crack propagation to fracture (FCF region), (3) creep crack initiation followed by fatigue crack nucleation and competition with creep crack, and final fatigue crack propagation to fracture (CFF

[1]Beijing University of Iron and Steel Technology, Beijing, China.

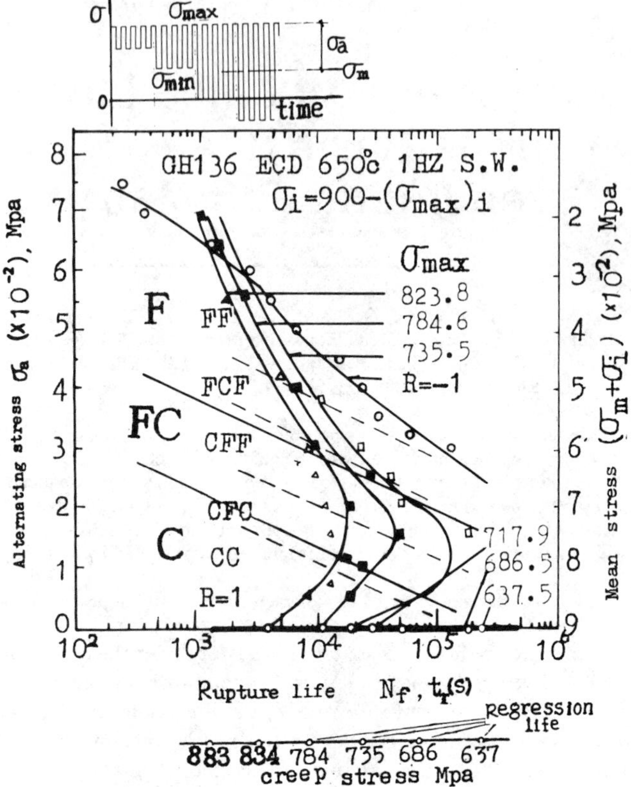

FIG. 1—*Fatigue-creep interaction map.*

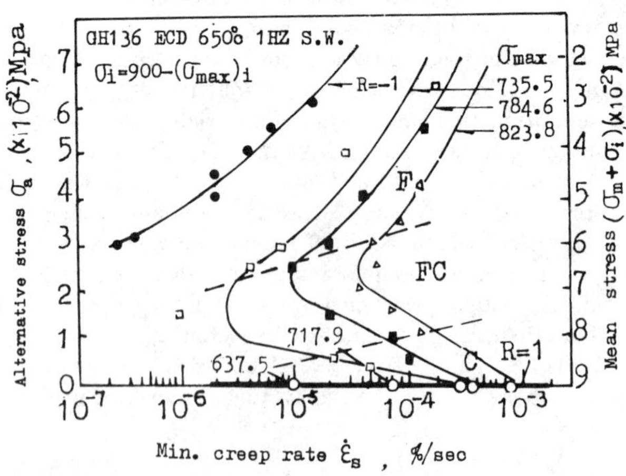

FIG. 2—*Fatigue-creep interaction strain rate map.*

region), (4) creep crack initiation followed by fatigue crack nucleation and final creep crack propagation to fracture (CFC region), and (5) creep crack initiation and propagation to fracture (CC region).

This paper is a further study of the effect of maximum stress with the intention of establishing a universal maximum stress modified life equation instead of a normal *S-N* curve.

Materials and Procedures

The material used in this investigation is a large ESR-Cast-to-Shape gas turbine superalloy disk (ECD). The chemical composition of the ECD is given in Table 1. The heat treatment scheme is 1050°C/4 h O.C. plus 700°C/32 h A.C. The fatigue-creep interaction tests are load control tests conducted at a frequency of 1 Hz. The waveform is square. The duration of maximum and minimum stresses are 0.5 s, respectively. The maximum stress (σ_{max}) is kept constant. Both the alternating stress (σ_a) and the mean stress (σ_m) can be varied simultaneously in the opposite directions by changing the minimum stress (σ_{min}); the maximum stress is equal to the sum of the alternating stress (σ_a) and mean stress (σ_m). All tests are performed at 650°C in air. The crack propagation rate is measured at 650°C using a center hole specimen.

Results and Discussion

Effect of Maximum Stress

Figures 3 and 4 show the effect of σ_{max} on fracture life and dynamic creep rate in the F and C regions. It can be seen that as σ_{max} increases, the fracture life decreases, and the dynamic creep rate increases. It is obvious that the increase of σ_{max} leads to an increase in the fraction of creep-fracture mechanism; for example, the fracture mechanism can change from FF to FCF fracture or from CFC to CC fracture. As a result of increasing σ_{max} the C region expands, the FC region moves into the region of higher alternating stress, and the F region contracts. It is also found that the increase of σ_{max} promotes the role of defects in crack initiation, which leads to an increase in the fraction of interdendritic crack due to the initiation of cracks around the interdendritic phases and other defects (Fig. 5). Under the stress condition of interaction with the predominantly fatigue fracture region, the effect of subsurface defects on crack initiation is more severe than that for the pure fatigue condition (Fig. 6*a*). Figures 6 *b, c,* and *d* show the features of crack initiation and propagation around the interdendritic phases. Both fatigue crack propagation (Fig. 6*b*) and creep void formation (Fig. 6*c*) can take place from the brittle fracture area around the interdendritic defects under suitable combined stress condition. Figure 6*d* shows that the defects can initiate creep voids and that these creep voids become the starting points for fatigue crack propagation under the stress condition of the FC region.

The effect of σ_{max} on crack propagation is also determined experimentally. It can be seen from Fig. 7 that the crack propagation rate increases with increasing maximum stress if the stress intensity factor range (ΔK) is kept constant.

Maximum Stress Modified Life Equation

On the basis of the interaction map and the effect of σ_{max}, the σ_{max} modified life equation and dynamic creep rate equation can be deduced for each of the F, C, and FC regions. The general forms of these equations are

For F region:

$$N_f = A_1 \sigma_a^{\alpha_1} \sigma_{max}^{\beta_1} \tag{1}$$

TABLE 1—*Chemical composition (weight percent) of the superalloy ECD.*

C	Si	P	S	Ni	Cr	Mo	V	Al	Ti	B	Fe
0.02	0.19	0.002	0.003	27.61	14.44	1.82	0.07	0.2	2.88	0.005–0.025	bal.

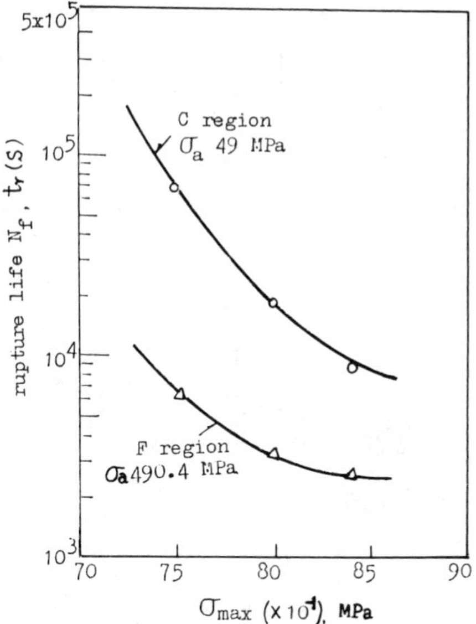

FIG. 3—*Dependence of rupture life on maximum stress.*

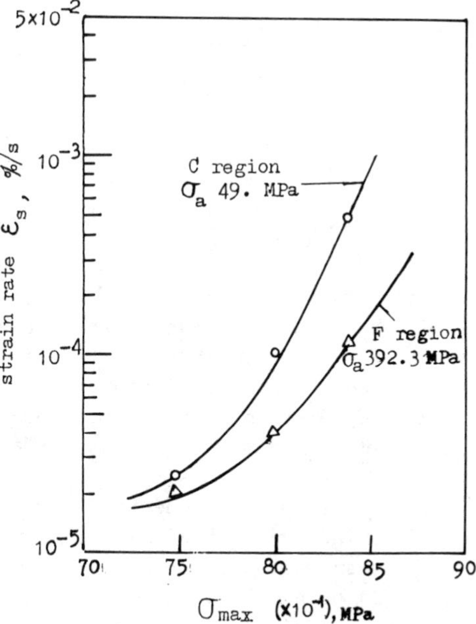

FIG. 4—*Effect of maximum stress on strain rate.*

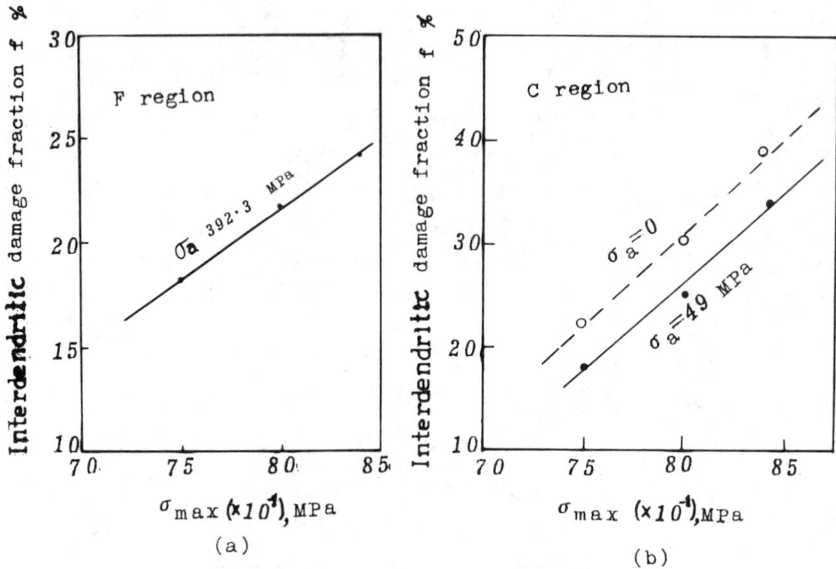

FIG. 5—*Interdendritic damage fraction as a function of maximum stress.*

$$\dot{\epsilon}_F = A_1' \sigma_a^{\alpha_1'} \sigma_{max}^{\beta_1'} \tag{2}$$

σ = MPa
$\dot{\epsilon}$ = %/s
t_r = s

For C region:

$$t_r = A_2 \sigma_m^{\alpha_2} \sigma_{max}^{\beta_2} \tag{3}$$

$$\dot{\epsilon}_c = A_2' \sigma_m^{\alpha_2'} \sigma_{max}^{\beta_2'} \tag{4}$$

The experimentally determined parameters for ECD are shown in Table 2. For the pure fatigue condition ($R = -1$), the damage also comes from the effect of alternating stress and maximum stress, but in this case the magnitude of the maximum stress is equal to that of the alternating stress. Therefore, Eq 1 will degenerate to the normal *S-N* equation. This means that the normal *S-N* equation is only a special case of the general equation of $N_f = A_1 \sigma_a^{\alpha_1} \sigma_{max}^{\beta_1}$. A similar analysis can also be applied to the test condition within the C region.

The σ_{max} modified equation is one of the ways to describe the interactions between fatigue and creep stresses. Let K equal β_i divided by α_i; if K is a minus quantity, the fracture lives will increase with increasing maximum stress. However, this is only the case with a very large alternating stress (near ultimate stress) for this large ECD. If the alternating stress is larger than 637.4 to 686.4 MPa, the fracture lives for the pure fatigue tests are less than those for constant σ_{max} tests (Fig. 1). Fractographic observation shows that the appearance of fatigue crack initiation and propagation becomes less favorable at the latter test condition, and the increase of maximum stress speeds up the shift of the fracture mechanism from fatigue fracture to the

TABLE 2—*Experimentally determined parameters for ECD.*

A_1	α_1	β_1	A_1'	α_1'	β_1'	A_2	α_2	β_2	A_2'	α_2'	β_2'
1.79×10^{17}	−3.03	−1.95	4.0×10^{-40}	2.77	16.1	8.31×10^{47}	−6.71	−8.44	1.05×10^{-54}	21.2	5.57

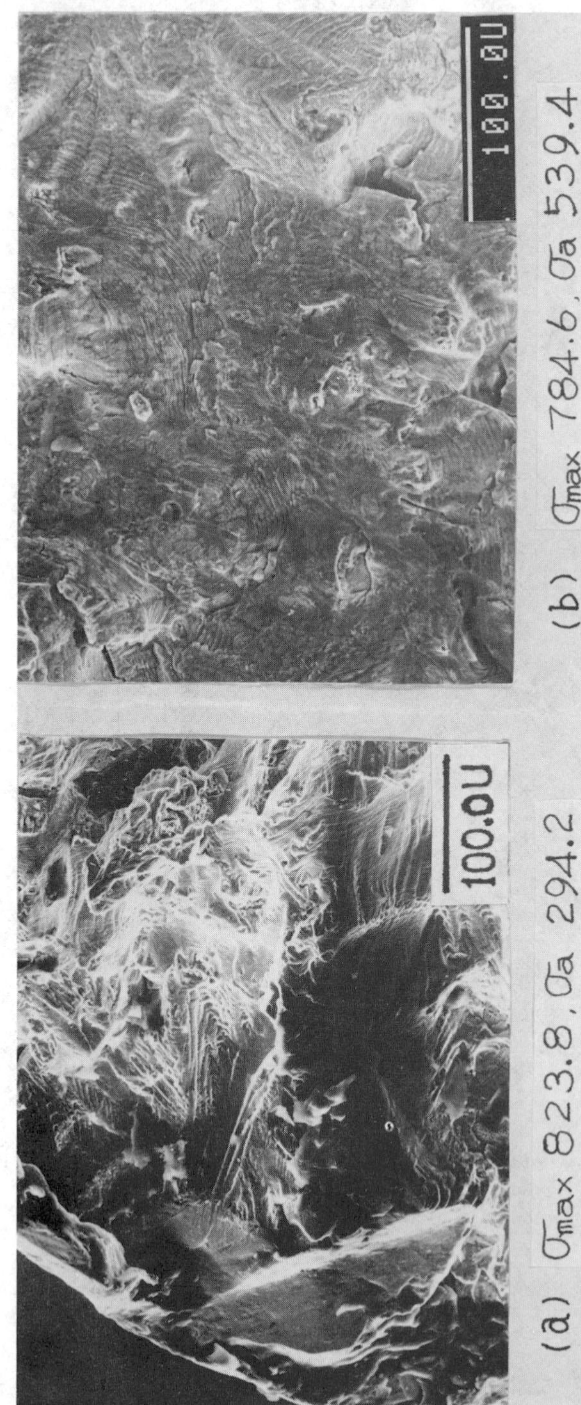
100.0U
(a) σmax 823.8, σa 294.2
100.0U
(b) σmax 784.6, σa 539.4

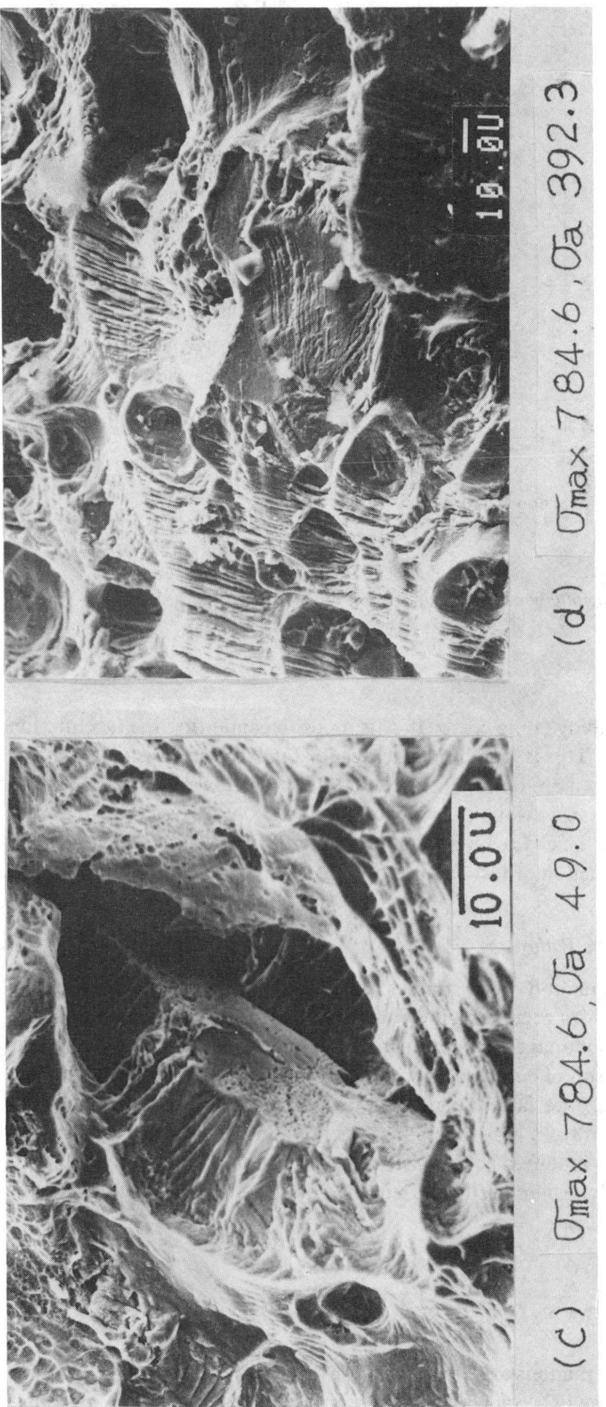

FIG. 6—*Scanning electron micrographs of fracture appearance.*

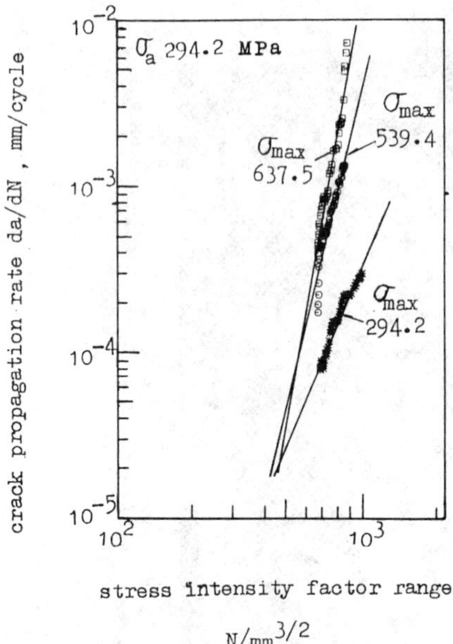

FIG. 7—*Fatigue-creep crack propagation characteristic.*

mixed fracture of tensile strain and dynamic creep. If K is a positive quantity, the fracture life will decrease with increasing σ_{max}. This is true in most cases for the large cast disk. The maximum stress modified Eqs 1 and 2 can predict fracture life successfully. However, the value of K for the C region is always larger than 1, while that for the F region is often less than 1.

An important conclusion is that the characteristics of fatigue-creep interaction are not only a function of material but also a function of test condition.

Application of Cumulative Damage Rule

Under the combination of fatigue and creep stresses, both the cumulative and competitive effect of fatigue and creep damage can occur simultaneously. As a result of the competition, the fatigue damage may occupy the dominant position in the F region, while the creep damage is dominant in the C region. As a result of cumulative damage, this predominant fatigue damage for the F region is different from pure fatigue damage or, similarly, the predominant creep damage for the C region is different from pure creep damage. The influence of interaction is to be included in the maximum stress modified equations. In the FC region both fatigue and creep damages can occur simultaneously. Therefore the cumulative damage rule should be used for predicting fracture life; that is,

$$\left(\frac{N}{N_f}\right)_F + \left(\frac{t}{t_r}\right)_C = 1 \tag{5}$$

Even though the fatigue-creep interactions are very strong for this large ECD, the linear cumulative damage rule still holds for the FC region. The reason is that N_f and t_r in Eq 5 are calculated from maximum stress modified Eqs 1 and 3, in which the influence of interactions on

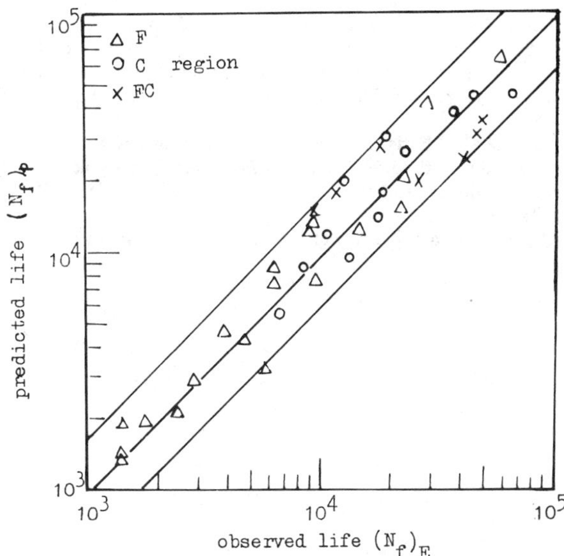

FIG. 8—*Observed rupture life versus predicted life used by the maximum stress modified life equation.*

fracture lives has already been taken into consideration. Therefore the linear cumulative rule can be used to predict fracture life successfully.

It is clearly demonstrated that all the observed lives fall within a range of a factor of two of the fracture lives predicted from Eqs 1, 3, and 5 (Fig. 8).

Conclusions

1. The fatigue-creep interaction fracture map provides an approach to study the effect of maximum stress under a wide range of combinations of fatigue and creep stresses. It is clearly demonstrated that maximum stress is an important factor influencing fracture life and dynamic creep rate.

2. Maximum stress modified equations have been deduced. All observed lives fall within a range of a factor of two of the fracture lives predicted from these equations.

3. The characteristics of fatigue-creep interaction are not only a function of material but also a function of test condition. The maximum stress modified equation is one way to express the interaction behavior of materials.

References

[1] Manson, S. S. and Halford, G. R. in *Proceedings*, International Conference on Thermal and High-Strain Fatigue, The Metals and Metallurgy Trust, London, 1967, pp. 154–170.

[2] Coffin, L. F., Jr., in *Proceedings*, Air Force Conference on Fatigue and Fracture of Aircraft Structures and Materials, AFFDL 70-144, 1970, pp. 301–311.

[3] Robinson, E. L., *Transactions of ASME*, Vol. 74, 1952, pp. 777–780.

[4] Polhemus, J. F. et al. in *Fatigue at Elevated Temperatures, ASTM STP 520*, American Society for Testing and Materials, Philadelphia, 1972, pp. 625–635.

[5] Manson, S. S. in *Fatigue at Elevated Temperatures, ASTM STP 520*, American Society for Testing and Materials, Philadelphia, 1972, pp. 744–782.

[6] Chen, G. L., *Journal of Aeronautical Materials*, Vol. 3, 1983, p. 47–52.
[7] Chen, G. L., He, Q. F., et al., "A New Fatigue-Creep Interaction Map for a Big ECD," in *Strength of Metals and Alloys,* Proceedings of 7th International Conference on the Strength of Metals and Alloys, Montreal, Canada, 12–16 Aug. 1985, H. J. McQueen, J. P. Bailon, et al., Eds., Pergamon Press, Oxford, pp. 1361–1366.
[8] Chen, G. L., He, Q. F. et al., "Fatigue-Creep Interaction Maps and Life Prediction under Combined Fatigue-Creep Stress Cycling with Dwell Time," in *Proceedings,* International Symposium on Microstructure and Mechanical Behavior of Materials, Xian, China, 21–24 Oct. 1985, Gu Haicheng and He Jiawen, Eds., pp. 393–401.
[9] He, Q. F., Ph.D. thesis, Beijing University of Iron and Steel Technology, Beijing, China, 1985.

J.-O. Nilsson[1]

Effect of Nitrogen on Creep-Fatigue Interaction in Austenitic Stainless Steels at 600°C

REFERENCE: Nilsson, J.-O., "**Effect of Nitrogen on Creep-Fatigue Interaction in Austenitic Stainless Steels at 600°C,**" *Low Cycle Fatigue, ASTM STP 942,* H. D. Solomon, G. R. Halford, L. R. Kaisand, and B. N. Leis, Eds., American Society for Testing and Materials, Philadelphia, 1988, pp. 543–557.

ABSTRACT: Three austenitic stainless steels (AISI 316, AISI 316LN, and 253MA) have been investigated at 600°C with respect to fatigue and creep-fatigue behavior. Transmission electron microscopy showed that cross-slip and climb of dislocations are inhibited in AISI 316LN and 253MA due to the presence of nitrogen in solid solution. This results in a planar slip character and a concomitant increase in fatigue strength. However, during creep-fatigue deformation, nitrogen has an adverse effect on strength. This can be explained as an enhanced interaction between creep and fatigue due to the generation of high stresses in the grain boundary regions, caused by inhibited recovery.

KEY WORDS: low-cycle fatigue, high-temperature fatigue, creep-fatigue interaction, intergranular fracture, austenitic stainless steels

The fatigue and creep-fatigue properties of Type 316 austenitic stainless steels have received much attention during the past several years, particularly in the U.K., where they have a wide usage as piping and superheater tubing in power plants and in the Commercial Fast Reactor. Attempts have been made to correlate crack propagation rates with theoretical models [*1–4*], and the relevance of life-prediction methods has been investigated [*5–7*].

During pure fatigue deformation of 316 stainless steel, the crack path is usually transgranular and striations are formed on the fracture surface. However, when creep deformation is introduced between the fatigue cycles, intergranular damage appears as cavities in grain boundaries [*8*]. Creep-fatigue interaction is envisaged as cavity formation caused by creep and cavity interlinkage during cycling [*9,10*], resulting in a crack propagation path, which is mainly intercrystalline. The degree of interaction is strongly material dependent as shown, for instance, by the creep-fatigue envelopes in the N-47 ASME Code Case.

It has been shown in previous work that nitrogen alloying in austenitic stainless steels was beneficial to the low-cycle fatigue (LCF) strength by promoting a planar slip mode [*11,12*].

Although the detailed mechanism by which nitrogen interacts with dislocations is not fully understood, it has been established that nitrogen inhibits cross-slip and climb of dislocations and therefore favors planar slip [*13*]. The purpose of the present paper is to investigate the effect of slip behavior on combined creep and fatigue. Three austenitic stainless steels were selected: AISI 316, which exhibits a wavy slip mode, and two nitrogen alloyed steels, AISI 316LN and 253MA (UNS S 30815), both of which exhibit a planar slip character.

[1]Research Metallurgist, Research & Development Centre, Mechanical Metallography, AB Sandvik Steel, S-811 81 Sandviken, Sweden.

Experimental Details

The compositions of the steels investigated are given in Table 1. It should be noted that the compositions of all three steels are within the limits specified for the corresponding commercial grades. The material was available in the form of extruded bar of 20 mm diameter. Solution heat treatment was performed according to the details shown in Table 2, in which also the resulting grain size and hardness are given.

Cylindrical specimens having 15 mm gage lengths were cyclically strained at 600°C under axial strain control in a servohydraulic testing machine. Symmetric wave shapes were used with a total strain rate of $5 \times 10^{-3}\ s^{-1}$. Creep-fatigue tests were conducted using 6-min hold times at maximum strain in tension interspersed with one fatigue cycle. Failure was defined as a 25% reduction of peak load in tension compared with the maximum load.

Creep tests were performed in the temperature range 550 to 800°C on smooth bar specimens of 30 mm gage length. The determination of creep ductility was based upon the reduction in cross-sectional area measured after the tests. Extensive creep tests were performed at 600°C in order to collect creep rupture data.

Fractography was performed using optical microscopy (OM) and scanning electron microscopy (SEM). Thin foil specimens for transmission electron microscopy (TEM) were taken from the gage sections of fatigued and creep-fatigued specimens as well as from the grip sections. Examination of a large number of foils showed that the specimen preparation technique did not introduce artefact dislocations.

Results

Creep-Fatigue Testing

Fatigue and creep-fatigue results are presented in Table 3 and Fig. 1 in terms of total strain range as a function of the number of cycles to failure. A detailed presentation of results obtained in pure fatigue has been presented elsewhere [*11*], but in order to facilitate a comparison with results from creep-fatigue tests they have been included in Fig. 1 from which it can be inferred that 253MA and AISI 316LN exhibit longer lives than AISI 316 under pure fatigue conditions.

The number of cycles to failure is reduced considerably due to the introduction of hold times. For AISI 316, the reduction is 2 to 3 times, while for 253MA the reduction can be as much as 10 times in the low-strain regime.

A notable feature is that under creep-fatigue, the relative strengths of these alloys are reversed when compared with pure fatigue. It should be pointed out that all experimental points fall to the right of the design curve suggested in the ASME Code Case N-47.

Creep Testing

Creep tests were conducted in the temperature range of 550 to 800°C. The results of ductility measurements plotted in Fig. 2 show that all three alloys exhibit a comparatively low reduction

TABLE 1—*Chemical composition (weight percent) of the experimental steels.*

Material	C	Si	Mn	P	S	Cr	Ni	Mo	Al	N	Ce
253MA	0.084	1.51	0.45	0.017	<0.003	20.85	10.97	0.03	0.014	0.21	0.02
AISI 316LN	0.025	0.66	1.90	0.010	0.010	17.60	12.70	2.74	...	0.21	...
AISI 316	0.043	0.57	1.67	0.022	0.003	16.57	12.24	2.62	0.002	0.056	...

TABLE 2—*Solution heat treatment, grain size, and hardness.*

Steel Grade	Condition	Grain Diameter, μm	Hardness (HV30)
253MA	solution heat-treated[a]	48	200
AISI 316LN	solution heat-treated[a]	43	195
AISI 316	solution heat-treated[a]	63	160

[a]Annealed for 10 min at 1100°C.

of area at 600°C. The alloys which contain nitrogen, AISI 316LN and 253MA, are less ductile than AISI 316 under creep conditions at 600°C. Long-term creep-rupture tests were performed at 600°C for all three materials. The creep-rupture curves thus obtained (Fig. 3) were used to assess the relation between stress and rupture time for each alloy.

Damage Accumulation

An attempt was made to estimate the amount of damage in each creep-fatigue test using a linear summation rule of the type

$$\sum_{i=1}^{N} \frac{t_i}{t_{ir}} + \sum_{j=1}^{M} \frac{N_j}{N_{jf}} = D \quad (1)$$

where t_{ir} refers to the time to rupture at stress level i and N_{jf} refers to the number of cycles to failure at strain level j. The first term denotes the creep damage, the second term the fatigue damage.

This attempt was made possible by comparing the number of fatigue cycles with pure fatigue tests and the sum of the hold times with the creep-rupture times obtained from Fig. 3. The creep-rupture times given in Table 3 were determined assuming that the relevant stress level was the maximum peak stress. This appears to be a reasonable assumption, since the stress decrease during the hold time is very small compared with the peak stress. Furthermore, since stress saturation is usually attained before 10% of the lifetime, more than 90% of the cyclic life is spent in the saturated stage. It was assumed in the present case that the damage evolution was linear, although it is known that damage evolution is not linear in reality [*14*]. This observation implies that the estimated damage can only be used as a comparative parameter and that values of D should not be regarded as absolute.

In Fig. 4, the damage D estimated according to the linear summation rule can be seen for all three alloys. The data points form three distinct groups, one for each alloy. Furthermore, the occurrence of failures in 253MA and 316LN at comparatively low values of the damage parameter D is interpreted as a result of a stronger creep-fatigue interaction in these alloys than in 316. The results for AISI 316 are somewhat scattered, and one point representing the lowest strain falls outside the damage envelope specified in the N-47 Code Case.

Relaxation During Hold Times

Stress relaxation was recorded in each test and each hold time period. It was found that the peak stress decrease was not only stress dependent but also dependent on material. The nitro-

TABLE 3—*Test data relevant to creep, creep-fatigue, and fatigue of all alloys.*

Alloy	Hold Time, s	Initial Stress, MPa	Maximum Stress, MPa	Total Strain, %	Stress Drop, MPa	Cycles	Testing Time, h	Creep Rupture Time, h	Fatigue Damage (D_f)	Creep Damage (D_c)	Total Damage (D)
253MA	300	224	469	3.70	29	60	6.0	13	0.24	0.46	0.70
253MA	300	211	428	2.00	17	132	13.2	20	0.14	0.65	0.79
253MA	300	195	375	1.20	10	255	25.5	45	0.08	0.57	0.65
253MA	300	187	300	0.70	6	628	62.8	150	0.05	0.42	0.47
AISI 316	300	287	354	1.30	16	241	24.1	20	0.48	1.20	1.68
AISI 316	300	290	337	1.20	13	381	38.1	40	0.63	0.95	1.58
AISI 316	300	286	315	1.00	10	504	50.4	60	0.50	0.83	1.33
AISI 316	300	279	281	0.76	7	800	80.0	200	0.34	0.40	0.74
AISI 316	300	238	237	0.58	4	1748	174.8	1000	0.35	0.17	0.52
AISI 316LN	300	237	436	2.43	7	108	10.8	20	0.16	0.54	0.70
AISI 316LN	300	228	430	2.10	6	112	11.2	22	0.12	0.51	0.63
AISI 316LN	300	213	357	1.46	6	252	25.2	150	0.12	0.17	0.29
AISI 316LN	300	203	326	1.10	4	492	49.2	300	0.13	0.17	0.30
AISI 316LN	300	190	287	0.75	2	1014	101.4	1100	0.10	0.09	0.19

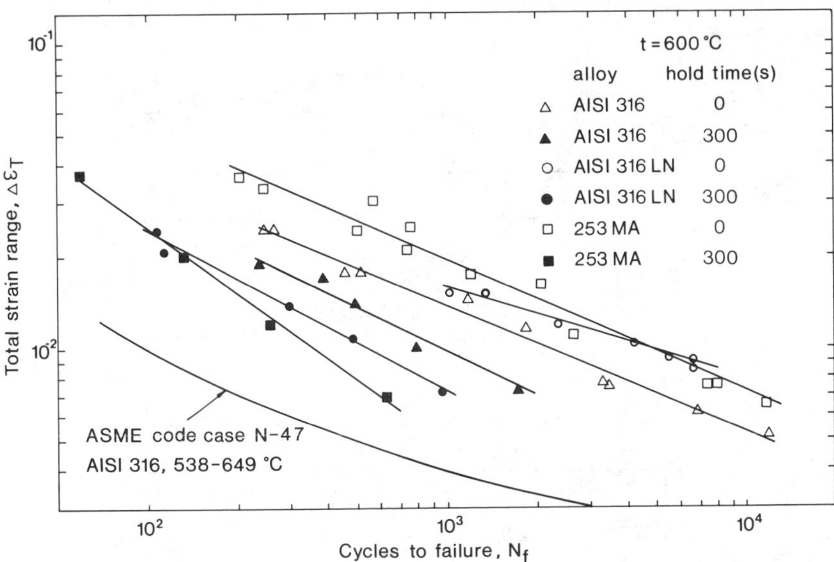

FIG. 1—*Coffin-Manson plot of fatigue and creep-fatigue results. Design curve for AISI 316 per ASME is included.*

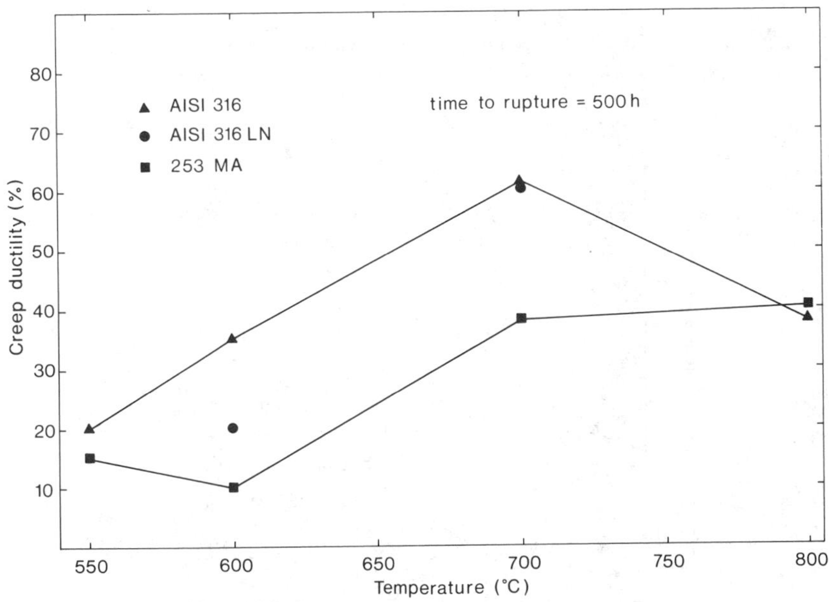

FIG. 2—*Creep ductility at temperatures in the range of 550 to 800°C.*

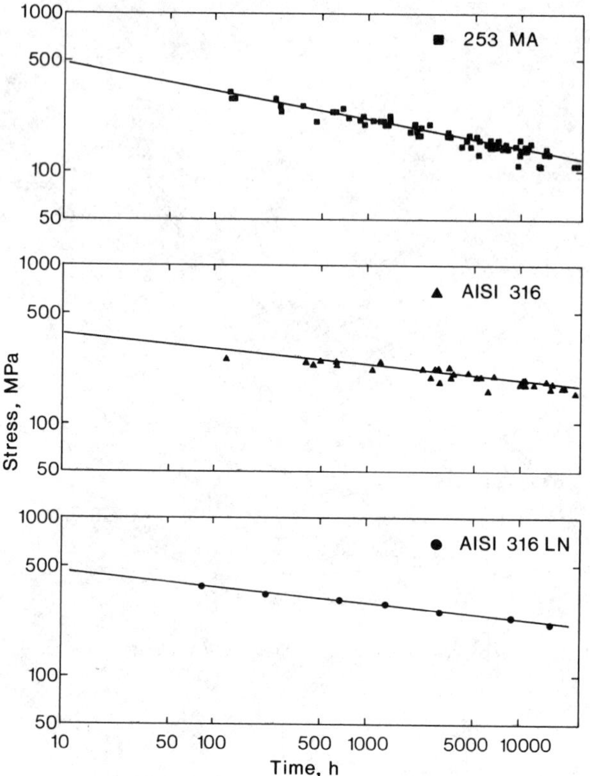

FIG. 3—*Creep rupture curves obtained at 600°C.*

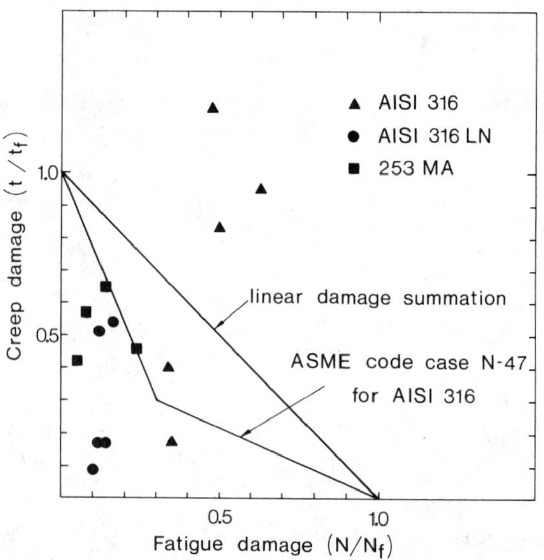

FIG. 4—*Creep-fatigue damage plot showing experimental results and damage envelope per ASME.*

gen-bearing alloys exhibited a lesser stress relaxation than 316 as shown in Fig. 5, which is a plot of stress decrease at the maximum load versus saturation stress. A notable feature is the strong effect on relaxation of nitrogen in Type 316 steels. Although the relaxation is least in 316LN it is evident that 253MA also exhibits a lower relaxation rate than 316. This effect is observed despite the fact that a higher stress level is attained in 316LN and 253MA.

Microstructure

As shown in a previous report [*11*] planar slip occurs in 253MA and 316LN during fatigue at 600°C and at moderate strain levels, whereas wavy slip occurs in 316 over a wide range of strains. This effect results in planar arrays of dislocations in 253MA and 316LN and the formation of cells in 316.

Little evidence of planar slip was found in 253MA following creep-fatigue deformation. Instead, tangles of dislocations were frequently observed, an example of which is shown in Fig. 6. Furthermore, dislocation loops could be observed at several places in regions of low dislocation density. Unlike 253MA, cells were easily formed in 316, as shown in Fig. 7. From this it can be inferred that recovery occurs more easily in 316.

Although the starting material was virtually precipitate-free, creep-fatigued material exhibited both inter- and intragranular precipitation of $M_{23}C_6$. The precipitation of grain boundary carbides was particularly pronounced in 253MA. While, in general, the intergranular precipitates were rather coarse, the intragranular precipitates were very fine. After deformation of 253MA at $\Delta\epsilon_t = 0.70\%$ for 125 h, the precipitates were typically 100 Å in diameter, as determined from dark field micrographs. Qualitatively similar observations were made in 316 after creep-fatigue deformation in terms of precipitation reactions, although the volume fraction of precipitates was greater in 253MA due to the higher carbon content. Evidence of grain boundary migration was found following creep-fatigue deformation in the form of fibrous precipitates of $M_{23}C_6$ in the grain boundaries. The example given in Fig. 8 is taken from 316 deformed at

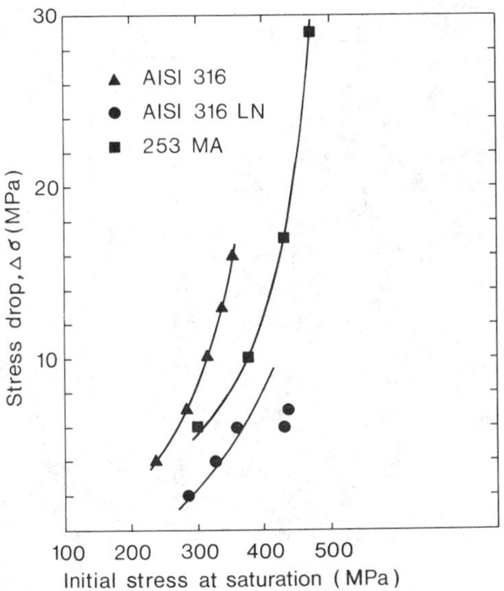

FIG. 5—*Stress relaxation during one hold time as a function of initial stress level.*

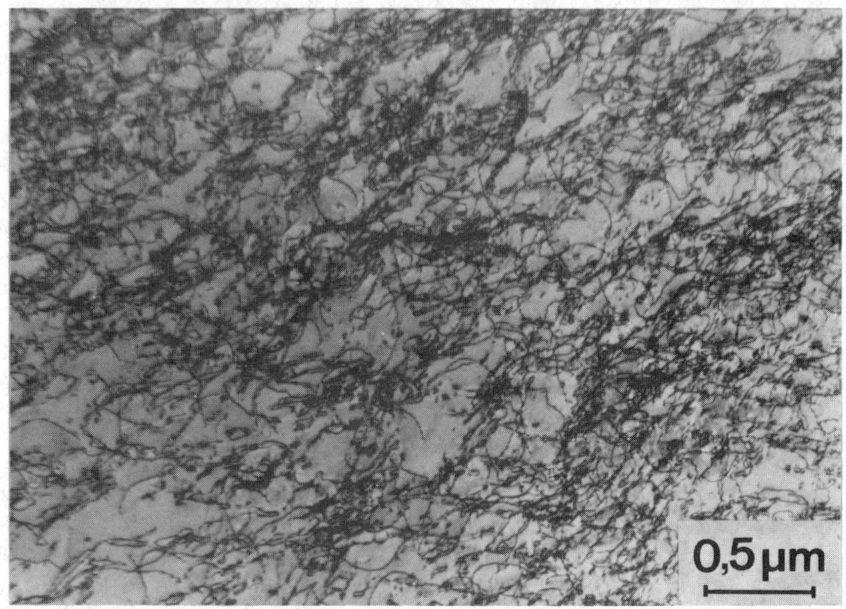

FIG. 6—*Dislocation tangles in 253MA after creep-fatigue at* $\Delta\epsilon_t = 0.70\%$.

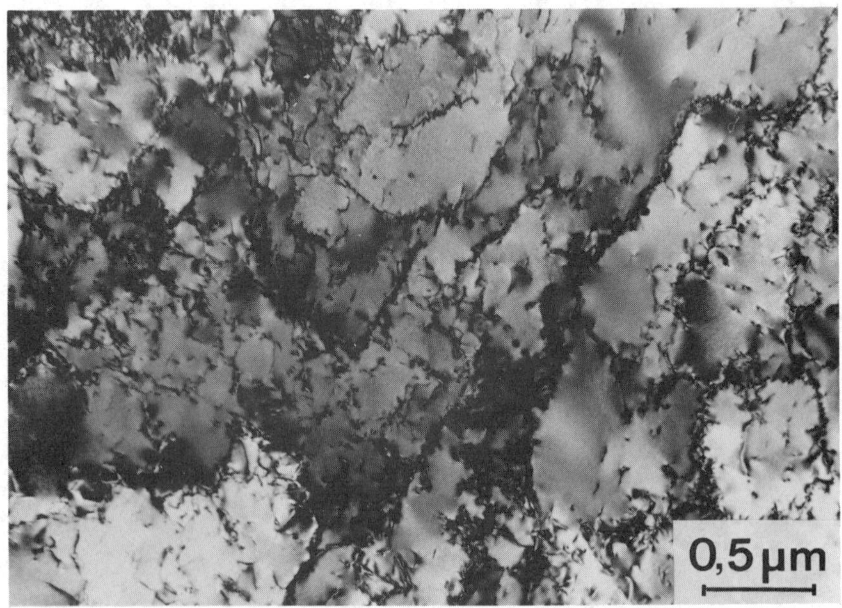

FIG. 7—*Dislocation cells in AISI 316 after creep-fatigue at* $\Delta\epsilon_t = 1.3\%$.

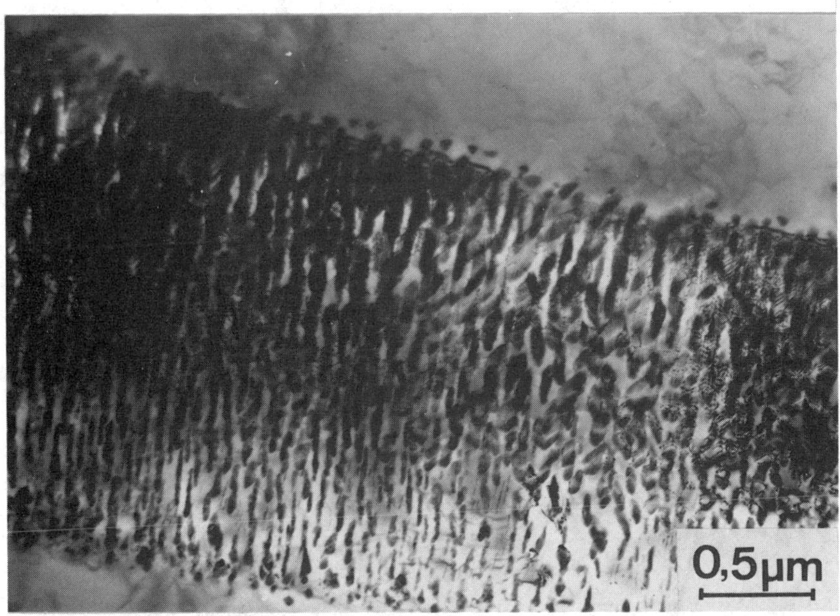

FIG. 8—*Fibrous precipitation of $M_{23}C_6$ in a high angle grain boundary in AISI 316 deformed at $\Delta\epsilon_t$ = 1.0%.*

$\Delta\epsilon_t$ = 1.0%, but similar observations were also made in 253MA. It is to be noted that no evidence of fibrous precipitation in grain boundaries was observed after pure fatigue, but only after creep and fatigue in combination.

Fractographic observations of creep-deformed specimens showed that the fracture mode in 316 differed markedly from that in 316LN and 253MA. The relatively brittle fracture behavior in 316LN and 253MA at 600°C is associated with the formation of wedge cracks, whereas the more ductile creep rupture in 316 involves formation of spherical voids. This is illustrated by the micrographs of 253MA and 316 in Figs. 9*a* and 9*b* respectively. In accordance herewith the creep fracture was found to be entirely intercrystalline in 316LN and 253MA. The crack path was essentially intercrystalline in 316, but transcrystalline cracks were observed occasionally. Furthermore, the ductile fracture behavior in 316 shows that it can accommodate larger strains.

The fracture process during creep-fatigue deformation was found to be similar to that under pure creep. Wedge-like cracks formed in 253MA not only along the main crack but also in the heavily deformed region at some distance from the main crack (Fig. 10*a*). The tendency to form wedge cracks was also pronounced in 316LN. Very few wedge cracks similar to those observed in 253MA and 316LN were found in 316. Furthermore, the crack path was occasionally transcrystalline in 316 (Fig. 10*b*), while 253MA and 316LN exhibited entirely intercrystalline cracks. Examination of a large number of fracture surfaces failed to reveal fatigue striations in any of the materials investigated.

By analogy with the creep fracture process, the creep-fatigue fracture process in 316 is much more ductile than in 253MA. While the grain boundary facets were very distorted in 316, they were easily observed in 253MA and 316LN, an example of which is shown in Fig. 11. This micrograph, taken from 253MA after creep-fatigue, also shows extensive cavitation on the grain boundary facets. The SEM micrographs taken from grain boundary regions of polished and lightly etched specimens revealed substantial intergranular cavitation in all three steels. How-

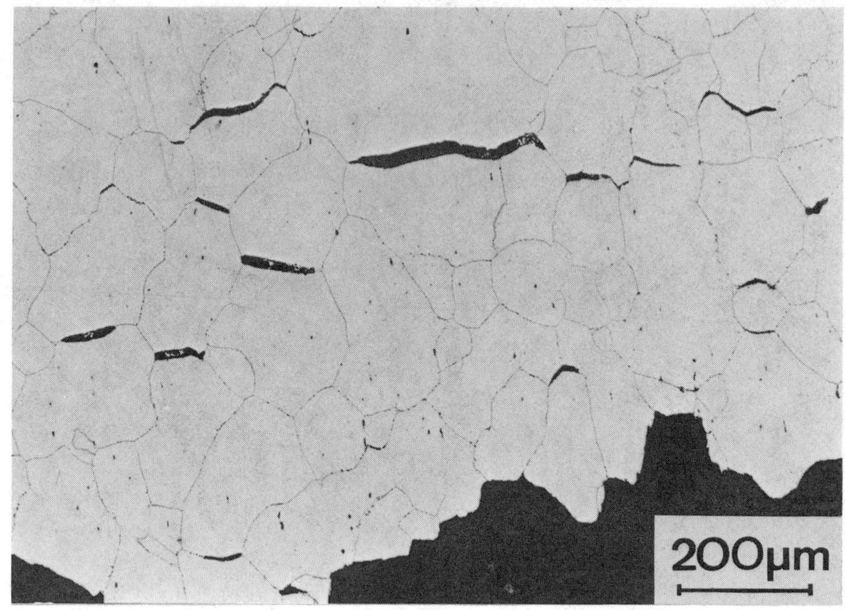

FIG. 9a—*Wedge-crack formation in 253MA after creep to fracture at 600°C and 210 MPa.*

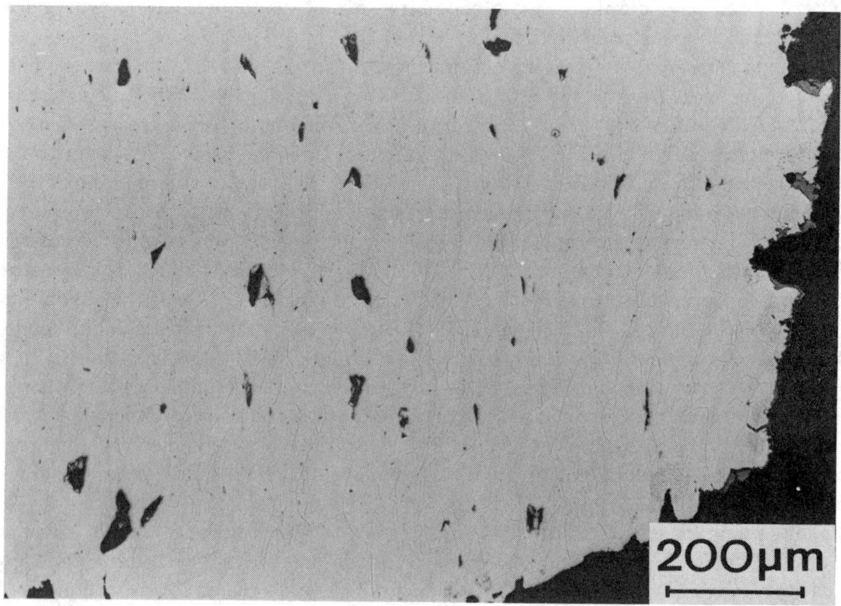

FIG. 9b—*Formation of voids near fracture surface in AISI 316 creep deformed at 600°C and 225 MPa.*

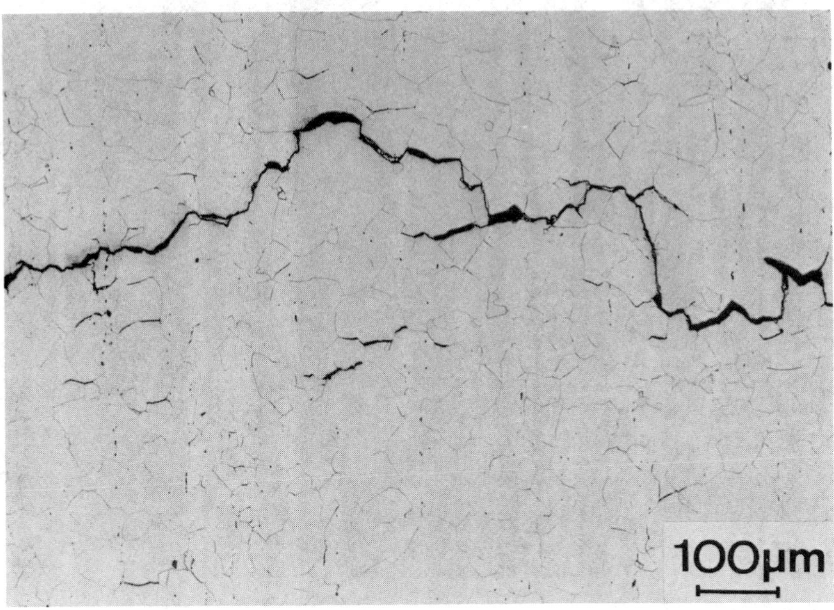

FIG. 10*a*—*Wedge-cracks in the vicinity of the intercrystalline main crack formed during creep-fatigue of 253MA of* $\Delta\epsilon_p = 0.70\%$.

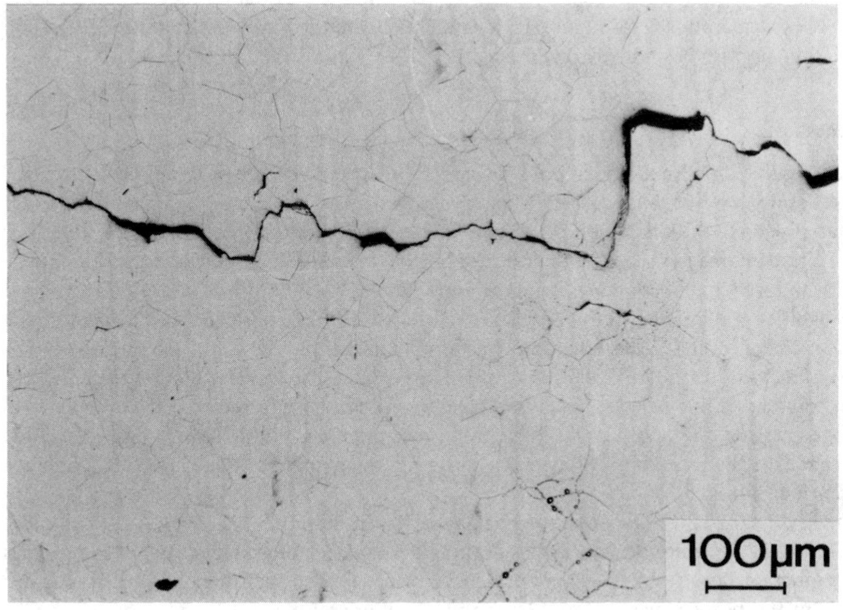

FIG. 10*b*—*Main crack path in AISI 316 after creep-fatigue at* $\Delta\epsilon_p = 0.76\%$.

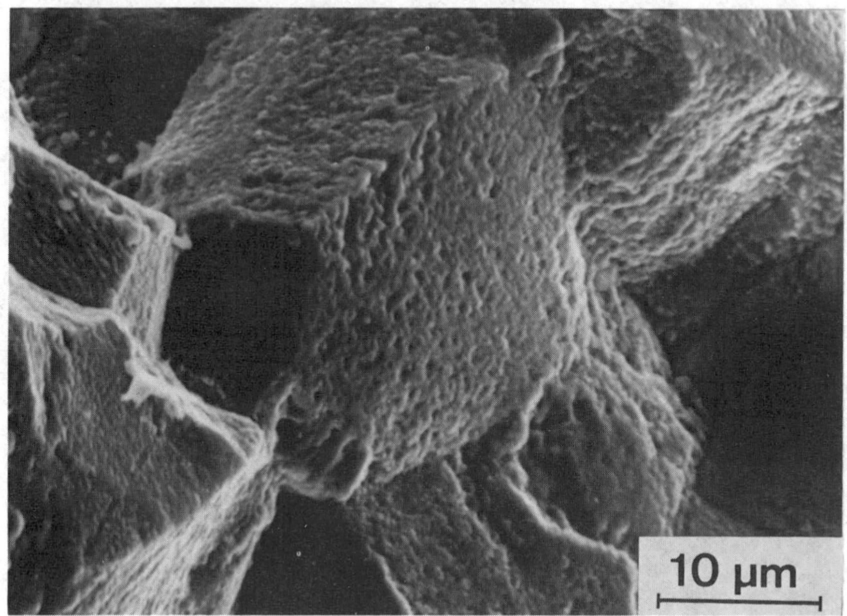

FIG. 11—*Scanning electron micrograph of fracture surface of 253MA after creep-fatigue at* $\Delta\epsilon_p$ = *0.70% showing cavities on grain boundary facets.*

ever, grain boundaries in 253MA tended to cavitate more easily than the steels of type 316. In all three alloys, cavitation was found to be located mainly to triple junctions, and evidence of cavity formation at grain boundary precipitates in 253MA was also found (Fig. 12).

Discussion

The results show that under pure fatigue 253MA and 316LN have longer lives than 316. These results have been discussed in an earlier paper [*11*] and could be explained in terms of planar slip reversibility. Beneficial effects of planar slip on fatigue strength is also found in other alloy systems as shown by the work of Saxena and Antolovich [*15*] and Laird and Feltner [*16*], who investigated copper-aluminum alloys.

A significant reduction in the number of cycles to failure is observed as a result of the introduction of hold times. The reduction is approximately three times for 316 and as much as an order of magnitude for 253MA in the low strain regime. There is also a concomitant transition from transcrystalline to intercrystalline fracture. It can be inferred that the steels containing nitrogen are not only more sensitive to hold times but also exhibit shorter lives than 316. The fact that 316LN is inferior to 316 in creep-fatigue strongly suggests that the effect is linked to the presence of nitrogen.

Creep damage is usually envisaged as cavitation in grain boundaries. As proposed by Min and Raj [*4,17*] interlinkage of cavities occurs by nucleation of wedge cracks at grain boundary triple junctions as a result of grain boundary sliding. Evidence of this mechanism was also observed in the present work (Fig. 12). Although intergranular cracks were found in all three alloys, the tendency to form wedge cracks was much more pronounced in the nitrogen-bearing steels as shown by the fractographic observations after creep and creep fatigue (Figs. 9 and 10).

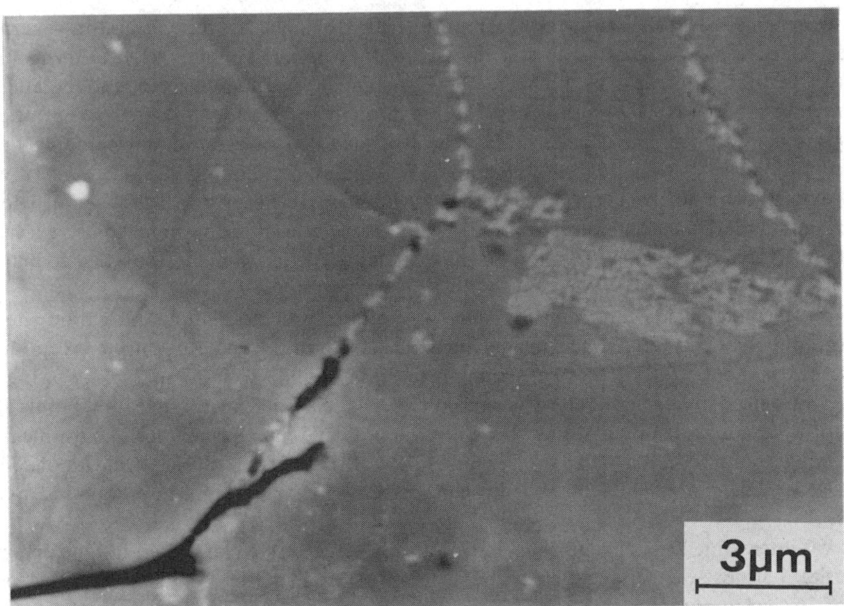

FIG. 12—*Scanning electron micrograph of grain boundary cavitation at precipitate particles and wedge-crack formation in 253MA following creep-fatigue at* $\Delta\epsilon_t = 0.70\%$.

The major difference between AISI 316 and the nitrogen-alloyed steels is the relative ease of cross-slip and climb of dislocations in the former alloy. This observation offers a reasonable explanation for the difference in fracture behavior during creep and creep-fatigue. Grain boundary sliding (GBS) at a triple junction, which is necessary for the proposed mechanism, occurs when a certain threshold stress is exceeded. Due to easier climb of dislocations in AISI 316, stress concentrations can be expected to be alleviated; therefore the stress required for sliding and wedge-crack formation is more difficult to attain.

Although 253MA exhibited some recovery during creep-fatigue, the microstructural investigations of *fatigued* material showed that dislocation cross-slip and climb were inhibited in the nitrogen-alloyed steels. This was also confirmed in material deformed in *creep-fatigue* through a lesser tendency to form intragranular dislocation cells and subgrains. Moreover, creep relaxation, which involves the motion of dislocations by climb, takes place at a slower rate in AISI 316LN and 253MA (Fig. 5).

The present results are consistent with results obtained in investigations of nickel-base alloys. When investigating a series of nickel-based alloys, Floreen observed a lower creep crack growth rate in overaged structures, an effect that was attributed to the ability of overaged particles to disperse slip [*18*]. In agreement with Floreen [*18*], Clavel and Pineau [*19*], who investigated Inconel 718, found that inhomogeneous slip promoted intergranular fatigue fracture. Dyson et al. [*20*] observed that cavitation in Nimonic 80, in which slip has a planar character, can be caused by plastic strain alone. Their results, which emphasize the role of slip character, suggest that the formation of cavities is not necessarily associated with creep deformation. From the results obtained in nickel-based alloys and the results in the present investigation it can be inferred that the slip character is very important in intergranular fracture.

The degree of interaction can be represented by the damage parameter D, which is shown for all three steel grades in Fig. 4. We can infer from this diagram that the creep-fatigue interaction is stronger in AISI 316LN and 253MA than in AISI 316, since fracture occurs, in general, at lower values of the damage parameter. These effects can now be interpreted in terms of the microstructural differences between the alloys. A planar slip character tends to enhance the interaction between creep and fatigue processes through an increase in stress concentration in the grain boundary regimes. It has, in fact, been argued that cavities can nucleate solely as an effect of the plastic strain [20] and therefore do not necessarily require GBS. Conversely, when cross-slip and climb become easier, as in the case of AISI 316, recovery processes in the vicinity of grain boundaries become more significant and stress concentrations tend to be relieved. It seems likely that this is an explanation for the weaker interaction in AISI 316, reflected in larger values of the damage parameter.

It is well established that grain boundary precipitates provide suitable nucleation sites for intergranular cavities [21]. Hales, for instance, has observed this to occur during creep-fatigue in AISI 316 at 600°C [8]. The larger amount of grain boundary $M_{23}C_6$ in 253MA owing to a higher carbon content can therefore be expected to lead to earlier cavity nucleation in this alloy. This provides an explanation of its lower creep ductility and more pronounced hold-time sensitivity when compared with AISI 316LN. Quite likely, therefore, the extreme behavior of 253MA is a result of the combined action of nitrogen, causing planar slip, and intergranular precipitates.

It was interesting to observe fibrous precipitation in creep-tested and creep-fatigue deformed material, but not after fatigue alone. This type of feature has been found by other authors in Type 316 steel after high-stress creep [22] and also after creep-fatigue [23] and is interpreted as an effect of migrating grain boundaries. The fact that this effect can be observed after both creep and creep-fatigue shows that the intergranular damage occurring in these two deformation modes is very similar, and therefore suggests that the corresponding damage mechanisms are almost identical.

Conclusions

1. Hold times in tension were found to reduce life during creep-fatigue in 253MA, AISI 316LN, and AISI 316 austenitic stainless steels.
2. The hold-time sensitivity was higher in the nitrogen-alloyed steels, 253MA and AISI 316LN, than in AISI 316. This can be explained by the difference in slip character.
3. It was found that the degree of creep-fatigue interaction could be interpreted on the basis of slip character. Planar slip tends to enhance the interaction by generation of stresses in the immediate vicinity of grain boundaries.
4. The loss of creep ductility and the high sensitivity to hold times in 253MA depend jointly upon the effect of nitrogen and grain boundary $M_{23}C_6$ precipitates.

Acknowledgments

This paper is published by permission of AB Sandvik Steel. The continuous support of R. E. Johansson and T. Andersson and the technical assistance of G. Svensk are gratefully acknowledged.

References

[1] Wareing, J., *Metallurgical Transactions A*, Vol. 6, 1975, p. 1367.
[2] Wareing, J., *Metallurgical Transactions A*, Vol. 8, 1977, p. 711.
[3] Michel, D. J. and Smith, H. H., *Acta Metallurgica*, Vol. 28, 1980, p. 339.
[4] Min, B. K. and Raj, R., *Acta Metallurgica*, Vol. 26, 1978, p. 1007.
[5] Wood, D. S., Wynn, J., Baldwin, A. B., and O'Riordan, P., *Fatigue of Engineering Materials and Structures*, Vol. 3, 1980, p. 39.

[6] Wareing, J., *Fatigue of Engineering Materials and Structures,* Vol. 4, 1981, p. 131.
[7] Plumbridge, W. J., Dean, M. S., and Miller, D. A., *Fatigue of Engineering Materials and Structures,* Vol. 5, 1982, p. 101.
[8] Hales, R., *Fatigue of Engineering Materials and Structures,* Vol. 3, 1980, p. 339.
[9] Baik, S. and Raj, R., *Metallurgical Transactions A,* Vol. 13, 1982, p. 1207.
[10] Baik, S. and Raj, R., *Metallurgical Transactions A,* Vol. 13, 1982, p. 1215.
[11] Nilsson, J.-O., *Fatigue of Engineering Materials and Structures,* Vol. 7, No. 1, 1984, p. 55.
[12] Degallaix, S., Vogt, J. B., and Foct, J., *Memoires et Etudes Scientifiques de la Revue de Metallurgie,* Vol. 9, 1982, p. 443.
[13] Nilsson, J.-O., *Scripta Metallurgica,* Vol. 17, 1983, p. 593.
[14] Plumtree, A., *Metal Science,* 1977, p. 89.
[15] Saxena, A. and Antolovich, S. D., *Metallurgical Transactions A,* Vol. 6A, 1975, p. 1801.
[16] Laird, C. and Feltner, C. E., *Transactions of the Metallurgical Society of AIME,* Vol. 239, 1967, p. 1074.
[17] Min, B. K. and Raj, R. in *Fatigue Mechanisms, ASTM STP 675,* American Society for Testing and Materials, Philadelphia, 1979, p. 569.
[18] Floreen, S., *Creep-Fatigue-Environment Interactions,* R. M. Pelloux and N. S. Stoloff, Eds., Conf. Proc. AIME, Fall Meeting, Wisc., 1979, p. 112.
[19] Clavel, M. and Pineau, A., *Metallurgical Transactions A,* Vol. 9, 1978, p. 471.
[20] Dyson, B. F., Loveday, M. S., and Rodgers, M. J., *Proceedings of the Royal Society* (London), Vol. A349, 1976, p. 245.
[21] Raj, R., *Acta Metallurgica,* Vol. 26, 1978, p. 995.
[22] Morris, D. C. and Harries, D. R., *Metal Science,* Vol. 11, 1977, p. 257.
[23] Rezgui, B., Petrequin, P., and Mottot, M. in *Proceedings,* International Conference on Fracture, Cannes, France, 1981, p. 2393.

P. K. Wright[1]

Oxidation-Fatigue Interactions in a Single-Crystal Superalloy

REFERENCE: Wright, P. K., **"Oxidation-Fatigue Interactions in a Single-Crystal Superalloy,"** *Low Cycle Fatigue, ASTM STP 942,* H. D. Solomon, G. R. Halford, L. B. Kaisand, and B. N. Leis, Eds., American Society for Testing and Materials, Philadelphia, 1988, pp. 558-575.

ABSTRACT: This research was conducted to identify, understand, and model the interactions of oxidation and low cycle fatigue (LCF) in a single crystal nickel-base superalloy, René N4. The LCF testing was conducted at 1093°C (2000°F) in three test environments: purified helium (<1 ppm oxygen), laboratory air, and enriched oxygen (80% oxygen at 2 atm pressure). Two cycle frequencies, 20 and 1/2 cpm, were investigated.

The fatigue life and behavior of René N4 at 1093°C was found to depend on both oxygen concentration and cyclic frequency. These parameters influenced the portions of fatigue life required for crack initiation and crack propagation to different extents, although both portions were decreased under conditions which were environmentally more severe. A relatively small oxygen pressure dependence of fatigue life was seen ($N_f \alpha \, PO_2^{-0.23}$). The fatigue process in René N4 showed a mixed time and cycle dependency leading to reduced fatigue life for longer cycles. Crack initiation was distinctly more time dependent than crack propagation and varied from 1/2 to 1/10 or less of the total fatigue life over the conditions investigated.

Crack initiation in René N4 occurred by fatigue assisted cracking and spalling of oxide products to produce a roughened and pitted surface. This process also accelerated the observed oxidation rate. With further cycling, these pits developed into oxide spikes and then into sharp fatigue cracks. A modification of an oxidation-fatigue model by Antolovich was satisfactory in describing this process. The crack propagation process was also influenced by oxidation. However, the precise mechanism could not be clearly defined.

KEY WORDS: low cycle fatigue, environmental effects, single crystals, high temperature fatigue, oxidation, crack growth

During normal operation, turbine blade materials in aircraft turbine engines are subjected to severe cyclic thermal stresses in a highly oxidizing environment. This combination of factors can lead to the development of fatigue cracks in turbine blades which must be controlled or reduced by proper material selection and blade design. The importance of oxidation in the high temperature fatigue process of superalloys has been recognized for some time, and in equiaxed alloys various mechanisms and models [*1-3*] have been proposed to describe the oxidation assisted grain boundary cracking process which dominates in these materials. However, the introduction of directionally solidified and single crystal superalloys for gas turbine blade applications has reduced or eliminated the importance of grain boundary failure modes. Instead, the high temperature fatigue process in these materials occurs by transgranular modes.

This research was undertaken to identify the mechanisms of fatigue failure in a single crystal superalloy at high temperature and to examine the applicability of current oxidation-fatigue models. Since cracking in blades occurs under conditions which are more oxidizing and involve longer exposure times than typically encountered in laboratory testing, the influence of oxygen pressure, cycle period, and surface condition were of particular interest.

[1]Manager, Advanced Materials Behavior, General Electric Company, Cincinnati, OH 45215.

Approach

The material selected for study was General Electric's single crystal superalloy, René N4 [*4*]. Test specimens of René N4 were fabricated from single crystal castings with the [001] growth direction within 7° of the specimen axis, and low cycle fatigue (LCF) tests were conducted at 1093°C (2000°F) using conventional servohydraulic machines equipped with RF induction heating and environmental chambers. Strain-controlled LCF testing was conducted in three environments: purified helium (< 1 ppm oxygen), laboratory air, and oxygen enriched air (80% oxygen at 2 atm pressure). At a total strain range of 0.45%, two cycle frequencies (20 and 1/2 cpm) were used (Fig. 1). While a choice of various waveshapes was possible, these cycles were felt to most faithfully simulate the waveshape imposed at critical locations in advanced turbine blades [*5*]. Some specimens (as identified in the text) were oxidized prior to test by exposing them in still air at 1093°C for 100 h. A few specimens were coated with an environmentally protective aluminide diffusion coating to determine the effect of excluding oxygen from the surface. To identify crack initiation and propagation rates and mechanisms, tests were periodically interrupted for inspection by surface replication by plastic tape at room temperature, and specimens were removed from test and destructively sectioned for metallography.

Separate crack growth tests were conducted in air and helium using the same cylindrical fatigue specimen with semi-circular surface flaws. The surface flaw geometry was selected for compatibility with the LCF testing and to permit compressive excursions in the 20 cpm, $R = -1$ cycle used. In air, an electrical potential drop method was used, and data reduction methods were as outlined in Ref *6*. In helium the cracks were monitored by periodically interrupting the test, cooling to room temperature, and measuring the cracks with an optical telescope. Crack depth was determined from surface length using observed crack shape measurements on

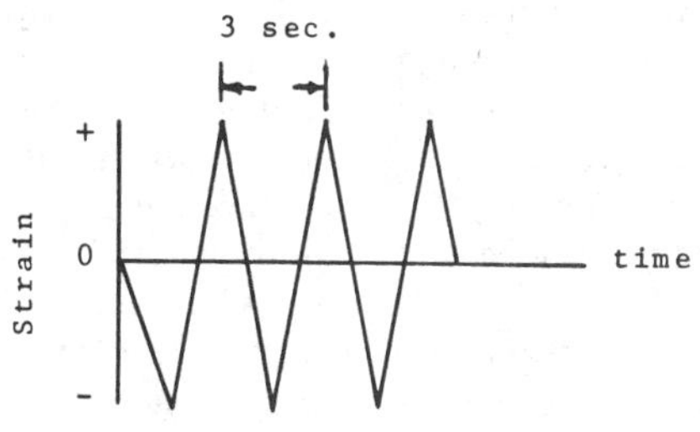

(a) 20 cpm Triangular Wave

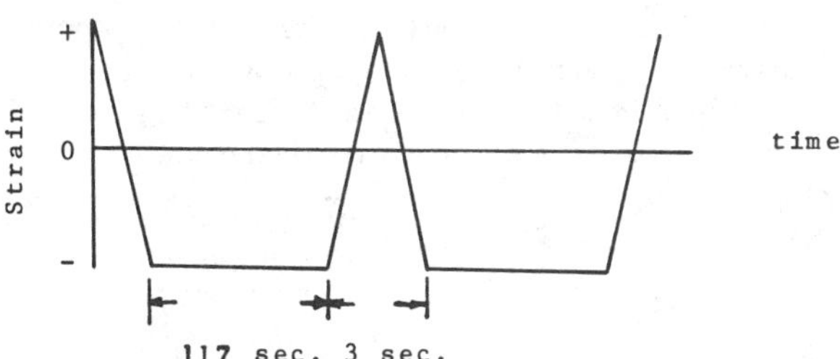

(b) 1/2 cpm Compressive Hold

FIG. 1—*LCF cycles used in this study.*

the fracture surface after testing. Data from load and crack size measurements were reduced to da/dN-ΔK data using load and crack sizes at the midpoint of the crack growth intervals and taking growth rate from a secant calculation.

Results

Air Environment

The results of specimens tested in laboratory air are shown in Table 1 and Fig. 2. Crack initiation was defined as the first surface crack detectable by replication, which was 50 μm deep by 100 μm long. Specimen failure was defined as complete separation or 50% tensile load drop, whichever occurred first.

At 20 cpm, surface cracking was first detected at about 4000 cycles (Fig. 3). This observation was obtained by surface replication (Fig. 4) and confirmed by observation of beachmarks on specimen fracture surfaces. Propagation of the initial cracks required approximately another 4000 cycles, during which period numerous other crack nuclei initiated and grew to substantial sizes.

Reducing the cycle frequency from 20 cpm to 1/2 cpm significantly reduced the fatigue life. This reduction was primarily associated with the reduction in cycles required to initiate a fatigue crack. The crack propagation period was slightly reduced in the slower cycle, but because of the particular compressive hold cycle used the tensile stresses were slightly different (Table 1) so that the difference may be due to applied stress and not to intrinsic growth rate differences.

The influence of surface condition was determined by testing uncoated, coated, and (at 20 cpm only) preoxidized specimens. As shown in Fig. 2, isolating the surface from the environment by applying an environmentally protective coating significantly improved fatigue life, while accelerating the oxidation attack by preoxidizing it reduced life. In both cases the major influence of the surface condition on life was through its effect on crack initiation. Within the scatter in experimental results the crack propagation period appeared to be unchanged. This is not surprising since once the surface layer (coating or oxide) was penetrated the substrate was identical in all cases. While the preoxidation exposure coarsened the gamma prime size from approximately 0.5 to 1.0 μm, this did not seem to have a significant effect on fatigue properties, since plastic strain and stress levels (Table 1) were the same as for unexposed material. At 1/2 cpm no significant effect of coating on fatigue life occurred, since crack initiation took place at a small fraction of life for both coated and uncoated specimens. The acceleration of crack initiation in the coated specimens at 1/2 cpm compared with 20 cpm appears to be related to the large amount of inelastic strain generated in the soft coating during the hold period. Considerable wrinkling of the coating was observed to accompany cracking at 1/2 cpm.

An instrumented crack growth test was conducted in air at 20 cpm frequency using a fully reversed loading cycle ($R = -1$). Crack growth data from this test are shown in Fig. 5. These data were used to back-calculate the crack propagation seen in the 20 cpm LCF tests, knowing the final crack depth in the LCF specimens and the applied stress range. This was accomplished by fitting a Paris Law through the data of Fig. 5, and assuming $\Delta K = 0.67\Delta\sigma\sqrt{\pi a}$ [*7*]. The calculated a versus N curves are superimposed on the actual measurements in Fig. 3 and show that good agreement is observed.

Oxygen-Enriched Environment

Testing in the oxygen-enriched environment reduced the fatigue life at 20 cpm from 8063 to 3588 cycles (Table 1 and Fig. 6). Additional specimens were tested to fractions of the observed total life to identify the contributions of crack initiation and propagation. A specimen tested to 1800 cycles (1/2 of life) had no detectable fatigue cracks, although oxidation of the surface was

TABLE 1—*Strain-controlled LCF test results.*

S/N	Environment	Cycle Frequency (cpm)	Surface Condition	Modulus		Cycle for Data	$\Delta\epsilon_p$ (%)	σ_{max}		σ_{min}		Total Cycles	Termination Mode
				10^3 MPa	(10^6 psi)			MPa	(ksi)	MPa	(ksi)		
I031-4	air	20	uncoated	75.8	(11.0)	4 000	0.04	+156	(+22.7)	−153	(−22.2)	8 063	fracture
K037-3	air	20	uncoated	76.5	(11.1)	3 000	0.04	+159	(+23.1)	−156	(−22.7)	6 000	test stopped
K037-1	air	20	uncoated	74.4	(10.8)	1 000	0.04	+156	(+22.6)	−150	(−21.8)	2 000	test stopped
J037-1	air	20	coated	70.3	(10.2)	8 800	0.05	+138	(+20.0)	−143	(−20.8)	17 636	fracture
F036-3	air	20	preoxidized	72.3	(10.5)	2 850	0.04	+147	(+21.3)	−153	(−22.2)	5 699	fracture
D033-1	air	0.5	uncoated	71.0	(10.3)	1 700	0.11	+140	(+20.3)	−101	(−14.6)	3 474	fracture
I037-3	air	0.5	uncoated	76.5	(11.1)	800	0.07	+203	(+29.4)	−115	(−16.7)	1 600	test stopped
I037-4	air	0.5	uncoated	73.7	(10.7)	200	0.03	+213	(+30.9)	−116	(−16.8)	400	test stopped
J037-2	air	0.5	coated	68.2	(9.9)	1 700	0.12	+164	(+23.8)	−106	(−15.4)	3 369	fracture
D031-1	high concentration oxygen	20	uncoated	62.7	(9.1)	1 800	0.13	+96	(+13.9)	−102	(−14.8)	3 588	fracture
D031-3	high concentration oxygen	20	uncoated	62.7	(9.1)	1 800	0.15	+92	(+13.3)	−94	(−13.6)	3 556	test stopped
D031-2	high concentration oxygen	20	uncoated	61.3	(8.9)	900	0.07	+118	(+17.2)	−123	(−17.8)	1 800	test stopped
D033-4	helium	20	uncoated	72.3	(10.5)	31 530	0.05	+163	(+23.6)	−128	(−18.6)	61 635	fracture
B035-2	helium	20	uncoated	71.0	(10.3)	23 000	0.03	+161	(+23.3)	−152	(−22.0)	30 818	test stopped
B035-1	helium	20	uncoated	69.6	(10.1)	4 100	0.03	+165	(+24.0)	−145	(−21.1)	15 408	test stopped

All tests: 1093°C (2000°F), $\Delta\epsilon_t = 0.45\%$, $R_\epsilon = -1$.

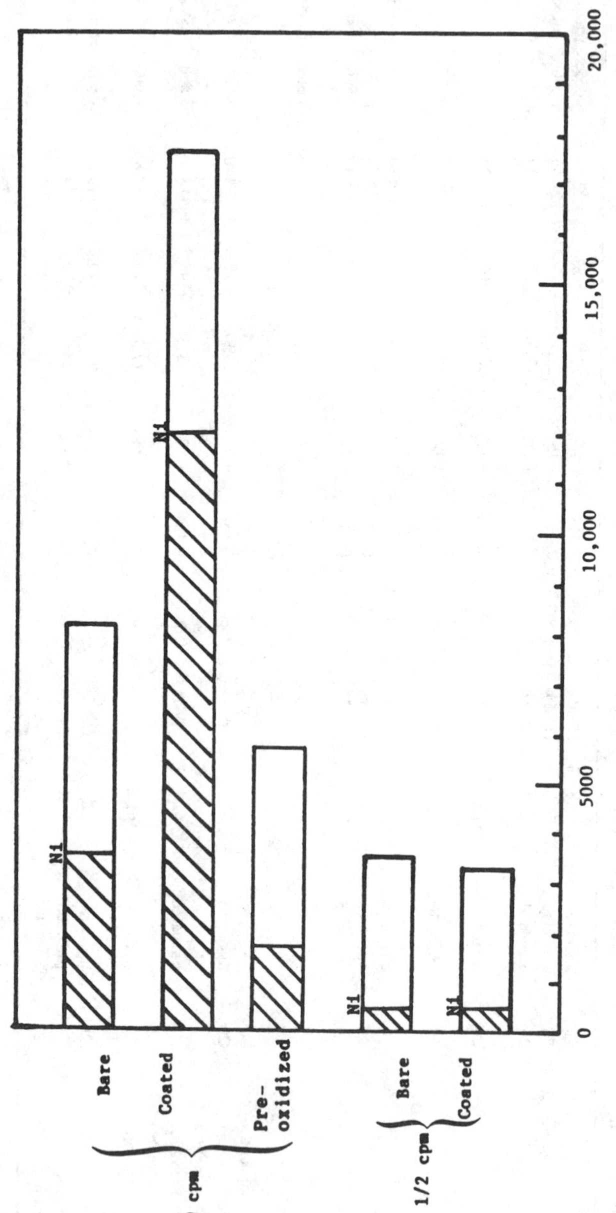

FIG. 2—*Influence of frequency and coating on life of René N4 at 1093°C (2000°F) in laboratory air* ($\Delta\epsilon_T = 0.45\%$).

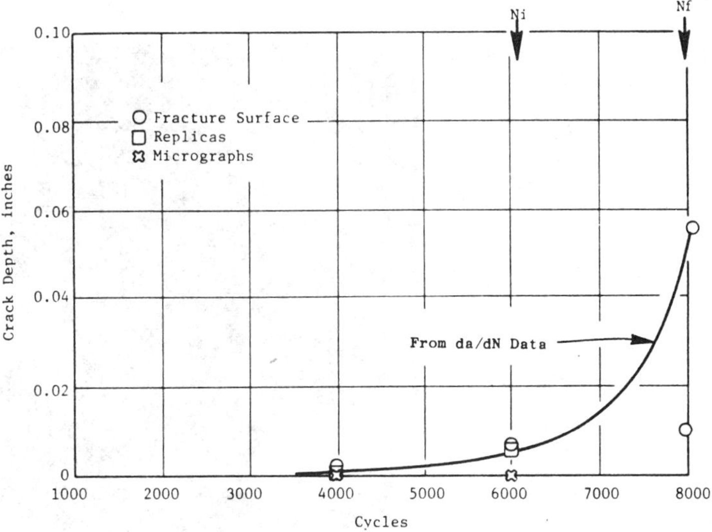

a) 20 cpm Cycling in Air

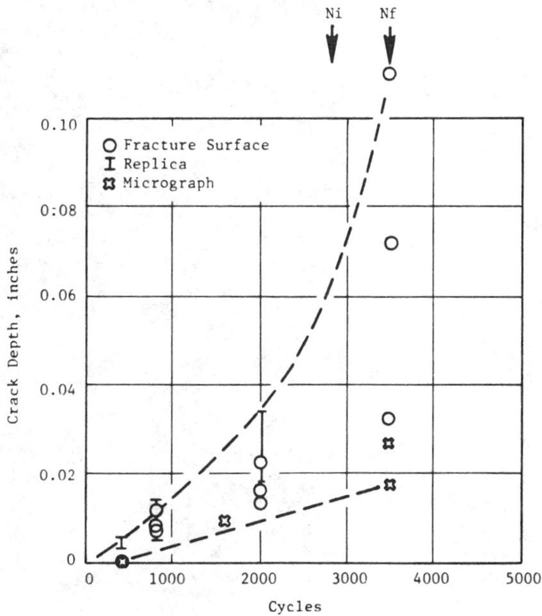

b) 1/2 cpm Cycling in Air

FIG. 3—*Crack depth versus cycles for uncoated René N4.*

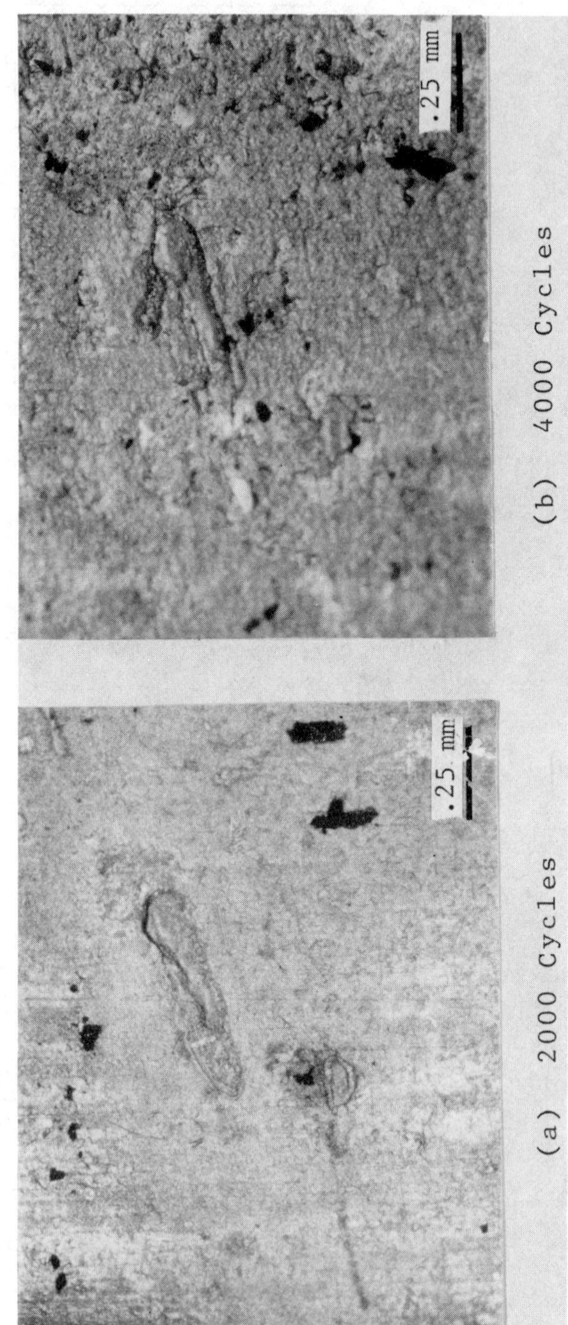

(a) 2000 Cycles

(b) 4000 Cycles

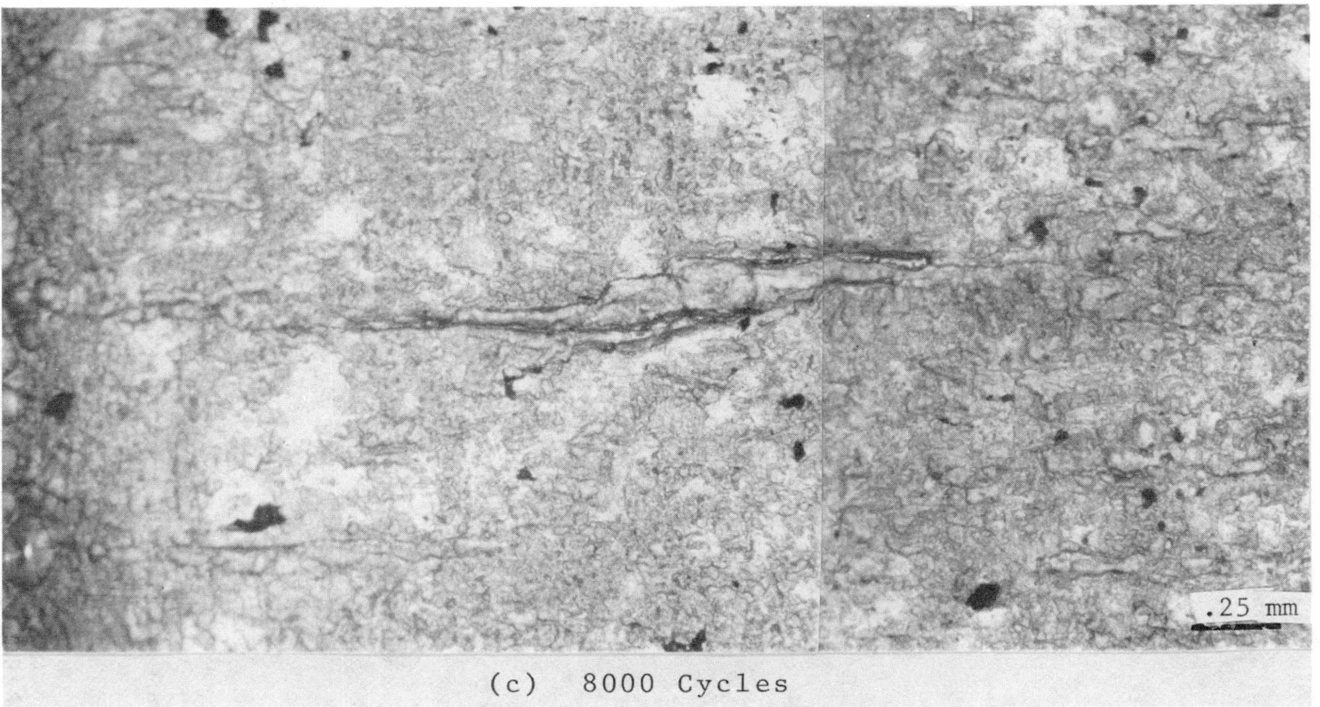

FIG. 4—*Progression of surface cracking on uncoated René N4 at 1093°C, 20 cpm (N_f = 8063).*

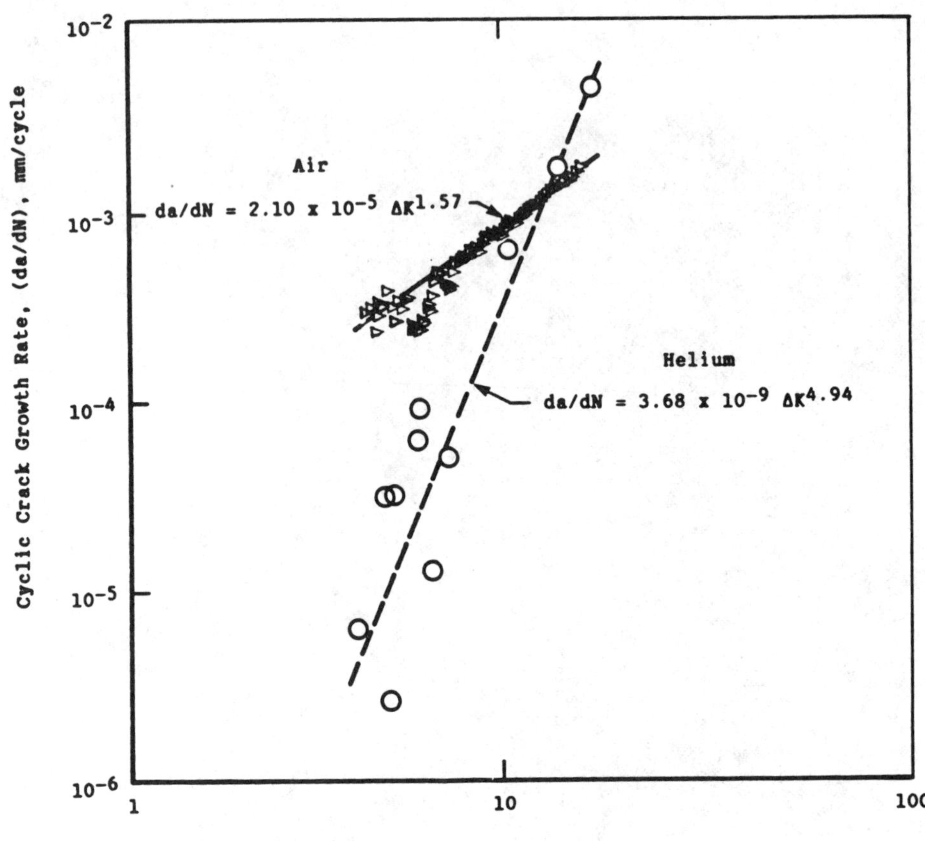

FIG. 5—*Crack growth in air and helium for René N4 at 1093°C,* R = −*1, 20 cpm.*

more severe than in air. A second specimen tested to 2400 cycles (2/3 of life) showed small crack nuclei approximately 0.125 mm long to have formed with the same appearance as in air (Fig. 4). This crack size was comparable to that defined as initiation in the air tests, so that crack initiation occurred at approximately 2400 cycles, while crack propagation consumed approximately 1200 cycles. Thus the crack propagation portion of life was slightly more affected by the enriched oxygen environment, dropping from 1/2 to 1/3 of the total life.

Tests in oxygen-enriched environment had noticeably lower moduli and higher plastic strains (Table 1) than in other environments. These characteristics suggest that the temperature in these tests was different. Comparison with other data suggests that the temperature of the oxygen-enriched environment tests was probably about 1135°C (2075°F). This error may have arisen from the different heat transfer characteristics of the oxygen-enriched environment influencing the specimen temperature. Testing of coated specimens in air environment at this temperature produces a life which is 55% of that at 1093°C (2000°F). If this same temperature effect holds for uncoated material, the anticipated life at 1093°C (2000°F) in enriched oxygen would be 6500 cycles. Thus the actual environmental effect, while still present, is significantly less.

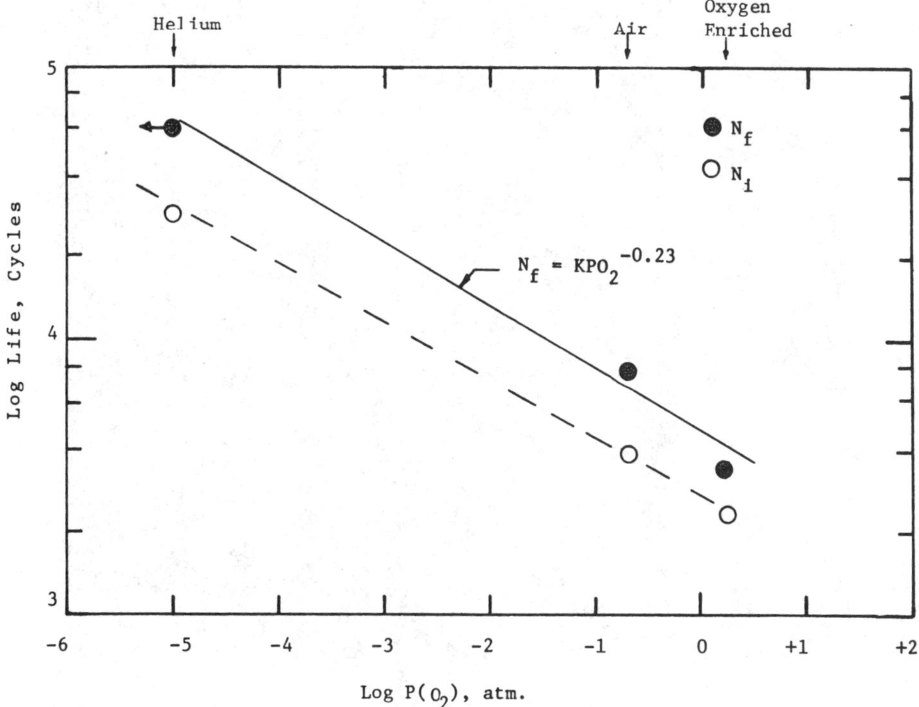

FIG. 6—*Effect of oxygen pressure on fatigue life of René N4 at 1093°C (2000°F) uncoated, 20 cpm, 0.45%.*

Helium Environment

Helium environment increased the fatigue life to 61 635 cycles (Table 1 and Fig. 6), despite the observation that some oxidation was still occurring in this environment. While oxidation rate was greatly reduced, specimens were perceptibly blackened and a tight, mostly adherent scale was formed. No cracking was detected after 15 408 cycles, but at 30 818 cycles ($N_f/2$) several small cracks up to 140 μm long and 35 μm deep were seen (Fig. 7). Unlike the tests in air, slight rumpling or plastic deformation of the surface could be observed at this point. By failure this surface cracking was profuse and typical crack sizes had grown to 0.4 to 0.8 mm long. Crack propagation data in helium are shown in Fig. 5 and show significantly lower growth rate and steeper slope in helium than in air. There appears to be a crossover in growth rate at higher ΔK, but data in helium are inadequate to confirm this.

Discussion

These test results and metallurgical observations show that oxidation had a significant effect on the high temperature, low-cycle fatigue of the single crystal superalloy René N4. The data of Fig. 6 show a consistent increase in fatigue life with decreasing oxygen level. Environmentally protective coating also increased fatigue life under some conditions (Fig. 2).

Decreasing cycle frequency or preoxidizing before testing both reduced fatigue capability. These general trends have been observed before in other studies of high-temperature fatigue. However, each of the variables of oxygen content, coating, cycle frequency, and prior oxidation

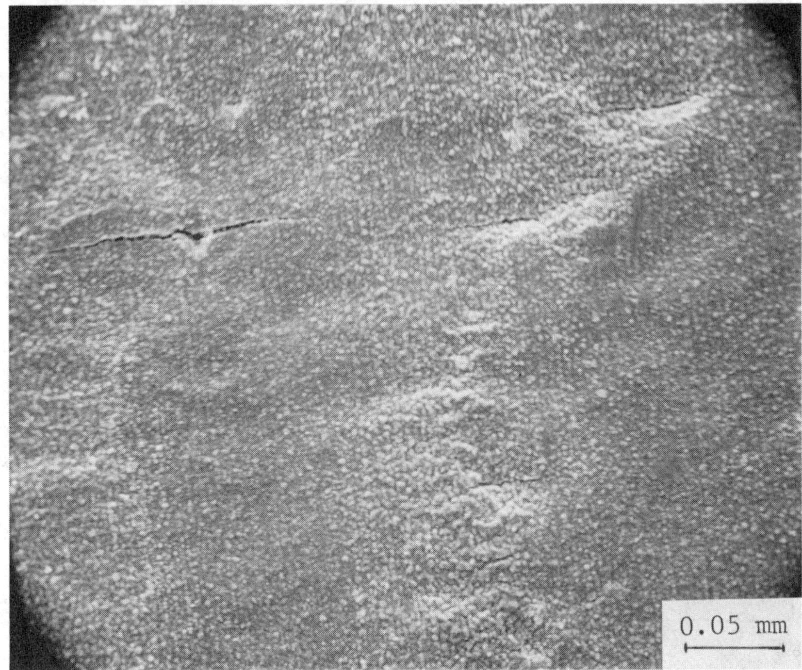

FIG. 7—*Small cracks on surface of low oxygen environment specimen after 30 818 cycles.*

influenced fatigue behavior in different ways. This discussion will examine these effects in order to try to construct a comprehensive understanding of how the oxidation/fatigue interaction processes actually take place.

Oxidation/Fatigue Interaction Mechanisms

The mechanism of crack initiation observed in oxidizing environments was a result of the combined action of fatigue cycling and oxidation. Upon heating to temperature, a protective oxide scale (typically high in Cr and Ti) rapidly formed on the surface. Cycling the specimen in fatigue caused the cracking and/or spalling of this scale and re-exposed the substrate to the environment. Further oxidation occurred at the areas where the substrate was exposed, and local differences in chemistry or microstructure (such as dendritic segregation), led to the development of preferential oxidation or pitting (Fig. 8) as the oxidation/scale rupturing process continued. At some critical depth, as seen in Fig. 8, these pits transitioned into actual fatigue cracks, as evidenced by the narrower, more elongated shape of the fatigue crack. The continuing influence of oxidation during crack growth was evident from the oxidized appearance of the fatigue crack tips.

It has been known for years [*8*] that thermal cycling accelerates oxidation by the same process of spalling of protective oxide scales. In thermal cycling the spalling is induced by thermal-expansion mismatch of the oxide and the substrate, while in isothermal LCF the mechanical cycling itself drives the spalling. If this process is occurring, it should be observable in the oxidation kinetics of the fatigue specimens. For example, accelerated oxidation of Mar-M-509 in fatigue was seen by Reuchet and Remy [*3*] and was attributed to the increased fracturing of oxide scale by cycling. Specimens undergoing combined oxidation and cycling showed considerably more appearance of oxidation than did those exposed for comparable times unstressed.

Measurements of oxide scale thicknesses and alloy depletion depths (Fig. 9) confirm this observation and suggest that the depth of oxidation as measured by alloy depletion was approximately doubled by cycling. The available measurements of scale thickness suggest less of an effect, but these measurements may be biased because thicker scales are being continually lost by spalling.

Role of Oxygen Pressure

As shown in Fig. 6, fatigue life depends on oxygen concentrations to a low power:

$$N_f = K(PO_2)^{-0.23} \tag{1}$$

A larger exponent (-0.5) is expected for an oxidation process controlled by oxygen diffusion within the substrate, or for a process controlled by oxygen molecule disassociation. However, the observed pressure dependence is very close to the ${PO_2}^{-1/6}$ dependence reported for oxidation of pure nickel [*9*]. This dependence is predicted from a lattice defect (nickel ion vacancy) model of oxygen diffusion through the nickel oxide scale and suggests that this process also controls fatigue life in oxygen environments.

The practical implication of the relatively small oxygen pressure dependence of fatigue life is that fatigue life in high-pressure engine environments should not be greatly degraded from values obtained in laboratory air. The relationship shown in Eq 1 provides a means of evaluating the influence of the increased oxygen concentration in the high-pressure turbine environment. It does, however, ignore the potentially strong effects of gas flow velocity which are known to increase oxidation rate.

Role of Cycle Period and Prior Exposure on Crack Initiation

Increased cycle period or exposure times increased the amount of oxidation observed, as expected. This increased oxidation per cycle should accelerate crack initiation (on a cyclic basis) in view of the mechanism discussed above. For the relatively slow oxidation kinetics observed here ($d = kt^{0.17}$ for oxide scale formation), the amount of oxidation increase with longer cycle period is small (about 2× for the 40× increase in cycle period from 20 to ½ cpm). On the other hand, if fatigue cycling is relatively effective in cracking and spalling oxide scale layers, the rate of oxidation will tend to be increased by more frequent cycling (shorter cycle periods). The newly spalled surface would reoxidize at the relatively rapid initial kinetics, grow to a critical thickness, and be spalled off again to repeat the process. Thus two counteracting effects are occurring: (1) the normal increase in oxidation depth with longer time per cycle and (2) a fatigue-induced increase in oxidation with more frequent cycling. The net result is a damage process intermediate between purely time dependent and purely cycle dependent.

To demonstrate this with the present results, under 20 cpm cycling 4000 cycles or 3.3 h were required for crack initiation. At ½ cpm, approximately 400 cycles or 13.7 h were needed, and with the 100-h preoxidation, 1700 cycles or 101.4 h (including preoxidation) were required. Thus crack initiation was neither a purely time-dependent (oxidation) process nor solely dependent on the cyclic (oxide fracture) rate.

This process can be modeled by an approach based on one by Antolovich [*2*] developed for oxidation-assisted, grain-boundary cracking of equiaxed, cast superalloys. In the derived form the model is given as

$$[\sigma_m + K(\Delta\epsilon_p/2)^{n'}]^8 D[t_e + (1/\nu + t_h)]N_i = C_i \tag{2}$$

0.05 mm

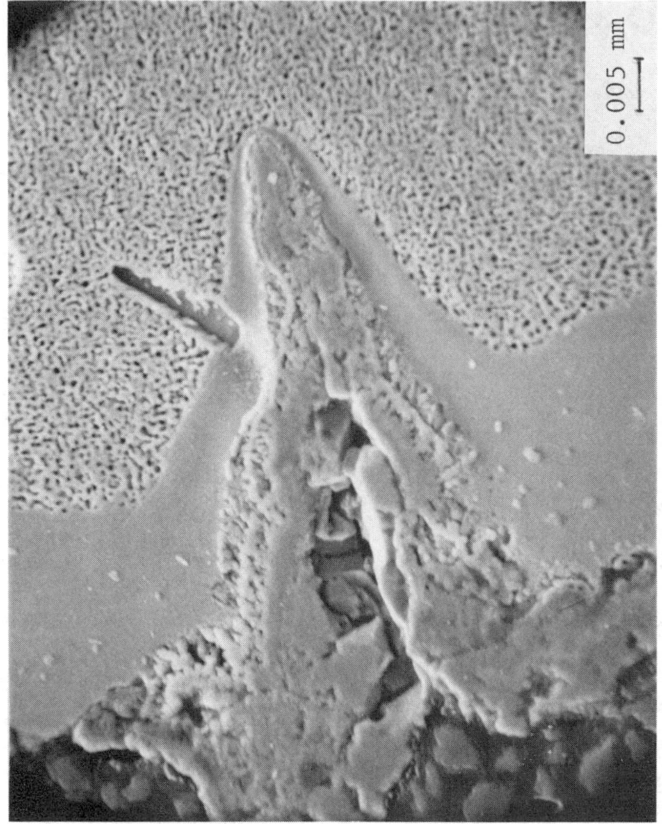

FIG. 8—Fatigue crack initiation from oxide pits in uncoated René N4.

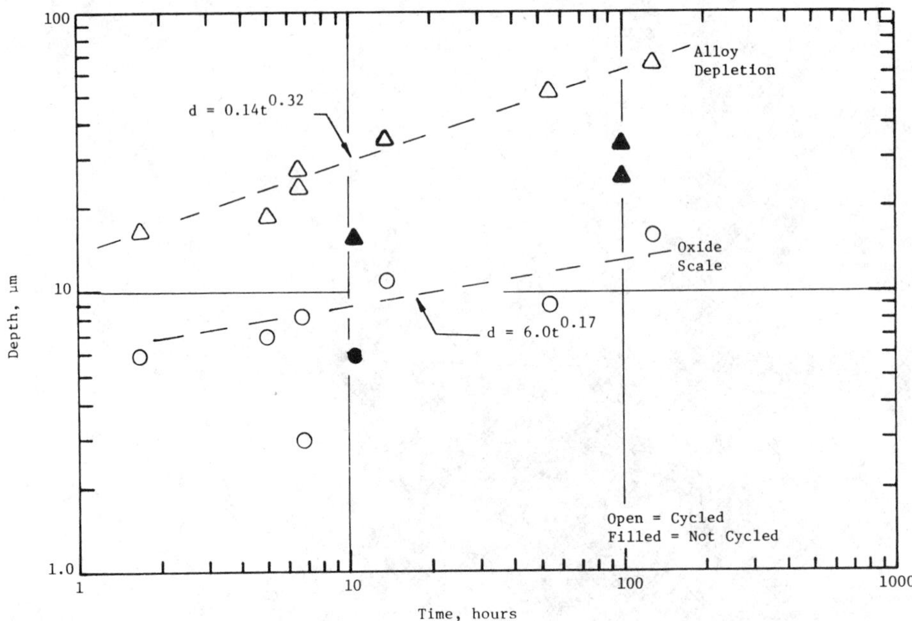

FIG. 9—*René N4 oxidation depths versus time in air environment for cycled and uncycled material at 1093°C.*

where

σ_m = mean stress,
K = cyclic stress-strain coefficient,
$\Delta\epsilon_p$ = cyclic plastic strain range,
n' = cyclic stress-strain exponent,
D = diffusion rate of oxygen = $D_0 \exp(-Q/RT)$,
t_e = prior exposure time,
ν = cycle frequency,
t_h = hold time,
N_i = cycles to crack initiation, and
C_i = constant.

When applied to the present results, this model tends to overemphasize time dependence, especially unstressed exposure (t_e), since crack initiation life is inversely proportional to cycle time. An alternative approach is to consider initiation life to occur when a certain critical depth, d_{cr}, of oxidation/fatigue damage has occurred. This depth is related to time in power-law fashion (Fig. 9):

$$d_{cr} = k_1(t_e)^{n_1} + k_2(t_{cy})^{n_2}N_i \tag{3}$$

where k_1 and n_1 are rate constants for uncycled oxidation, and k_2 and n_2 are rate constants for cycled exposure.

The form of this expression permits separation of uncycled and cycled oxidation effects as observed in this study and permits discounting of uncycled exposure degradation through the difference in rate constants k_i and n_i. This form also weighs damage more heavily for short cycle periods, since typically $1 > n > 0$. The choice of which constants, k and n, to use is less clear, however, since both oxide scale and alloy depletion may contribute to damage, and each has different kinetics. It appears physically more appropriate to use the scale-formation kinetics when considering crack initiation, since initiation is primarily a surface phenomenon. Using k_1 and n_1 values from Fig. 9 results in satisfactory accounting of the damage due to uncycled exposure, but for cycled exposure the constants $k_2 = 6.0$ μm/h$^{-0.17}$ and $n_2 = 0.17$ significantly overestimate the damage during cycling. The actual observations of crack initiation, 4000 cycles at 20 cpm, 400 cycles at 1/2 cpm, and 2300 cycles after exposure, agree much better with constants of $k_2 = 0.3$ and $n_2 = 0.6$. These constants may reflect different oxidation processes which occur at short times. These parameters imply a critical oxidation depth (Eq 3) of about 17 μm; this value is physically reasonable but somewhat smaller than the observed crack depth at initiation (50 μm).

While not specifically investigated, applied stress range should enter through two terms. First, d_{cr} is probably related to the fatigue crack propagation threshold, since it is the size at which the crack is able to grow into the substrate. Thus

$$d_{cr} = \frac{1}{\pi}\left(\frac{\Delta K_{th}}{Y\Delta\sigma}\right)^2 \tag{4}$$

where

K_{th} = crack growth threshold,
Y = geometric factor of the crack shape involved (semicircular surface flaw), and
$\Delta\sigma$ = applied stress range.

For example, if $\Delta K_{th} = 3$ MPa m$^{1/2}$, $Y = 0.7$, and $\Delta\sigma = 290$ MPa (42 ksi), as appropriate for these conditions, then $d_{cr} = 75$ μm, which is reasonably close to the observed crack initiation depth of about 50 μm.

The other contribution of applied stress may be through its effect on the cyclic oxidation parameters k_2 and n_2. In general, as stress range is decreased the cyclic oxidation kinetics should approach static kinetics as the fatigue cycling becomes less effective in cracking and spalling oxide scales.

These parameters are admittedly empirical at this stage. More careful measurements of oxidation at short times with and without cycling will be required to relate the crack initiation parameters directly to oxidation. However, in its present state a model of the form

$$N_i = \frac{d_{cr} - k_1(t_e)^{n_1}}{k_2(t_{cy})^{n_2}} \tag{5}$$

is proposed for oxidation-assisted crack initiation.

Role of Cycle Period on Crack Propagation

Crack growth behavior was significantly less influenced by cycle period than was crack initiation. This is seen by comparing the cycles for propagation in the 20 cpm and 1/2 cpm tests. The propagation lives are only slightly reduced in the 1/2 cpm cycle despite the longer time for oxidation. This observation is supported by metallographic observations of the crack tips; there is no observable difference in oxidation between 20 and 1/2 cpm. It is also supported by other measurements of crack growth rate in single-crystal superalloys [*10*], which show little or no effect of cycle frequency. On the other hand, a significant difference in crack growth rate was observed between air and helium environments.

This lack of time dependence in crack growth is difficult to resolve with a mechanism of crack growth based on oxidation alone. If the kinetics of oxidation measured for relatively long exposure times (over 1 h) in Fig. 9 can be extrapolated to the shorter times associated with a single cycle of crack growth, then the rate of oxidation (and hence oxidation-controlled crack growth) per cycle should be

$$a_{\text{oxidation}} = 4 \times (8.3 \times 10^{-4})^{0.17} = 1.2\ \mu\text{m} \tag{6}$$

for the 20 cpm cycle where 8.3×10^{-4} is the cycle period in hours. The slower oxidation kinetics of scale formation have been used here. This corresponds to a da/dN of 1.2×10^{-3} mm/cycle or 4.7×10^{-5} in./cycle. Comparing this rate with the crack growth rate obtained in air at 20 cpm (Fig. 5) shows that this oxidation-based estimate of growth rate is faster than the observed growth rate during most of the test, despite the fact that this estimate is based on the slowest rate of oxidation. This observation indicates that oxidation processes as measured on the surface of the specimens after a long time are not the same as those at the tip of an advancing crack.

Such differences in oxidation at the crack tip may be due to the restricted access of oxygen down the crack, or they may be due to the inappropriateness of extrapolating long-term oxidation rates to the short times per cycle used in these tests. However, both of these effects would still produce a time dependence in crack growth behavior. It may be that a "fully embrittled" environmental state, in the sense described by McGowan and Liu [*11*], has been reached.

The similarity of crack growth in the 20 and 0.5 cpm cycles may be due to similarities in the tensile, or opening portions, of these cycles. This is the portion of the cycle when oxygen access to the crack tip is easiest. Similarity in the crack opening time would suggest similarity in environmental contribution to crack growth; hence the two cycles used should show similar crack growth rates. However, other data at higher frequencies using a triangular wave [*10*] shows little difference from 20 cpm, suggesting that crack opening time alone is not the controlling parameter.

An alternative explanation for the lack of time dependence of crack growth is that the mechanical conditions at the crack tip are not identical in the 20 and 1/2 cpm cycles. While both cycles have 20 cpm ramp rates in the tensile or crack-opening portions of the cycles, the 1/2 cpm cycle has a long compressive-strain hold period. One effect of this hold can be observed in the bulk stress-strain response of the specimen. As observed in Table 1, a mean stress shift occurs which causes the maximum tensile stress in the 1/2 cpm cycle to be increased slightly (180 MPa or 27 ksi) compared with the 20 cpm cycle (150 MPa or 22 ksi) and the plastic strain to be increased (0.078% versus 0.057%). It is possible that such changes in mechanical conditions around the crack by time-dependent relaxation may counteract the time-dependent effects of oxidation and mask any overall cycle-frequency effects. Further study of cycle shape and frequency effects on strain-controlled crack growth are needed to clarify the role of mechanical versus environmental factors.

Summary

The fatigue life and behavior of the single-crystal superalloy René N4 at 1093°C (2000°F) depends on the environmental parameters of oxygen concentration, protective coating, and cyclic frequency. Both crack initiation and crack propagation lives are decreased under conditions which are environmentally more severe, although to different extents.

Oxygen concentration has a relatively modest effect on fatigue life ($N_f \alpha P O_2^{-0.23}$). The kinetics suggest that the controlling oxidation interaction process is one of oxygen diffusion through the protective scale. Protective coatings can significantly increase the crack initiation time. However, the protective effect of the coating is lost as soon as the coating is penetrated.

Slower frequency cycling leads to reduced fatigue life primarily due to reduction in crack initiation life. Neither crack initiation nor crack propagation processes are purely time dependent, however, but show a mixed time- and cycle-dependent behavior.

The crack initiation process in uncoated René N4 is one of fatigue-assisted cracking and spalling of oxide products to produce a roughened and pitted surface. Oxidation rates were found to increase in the presence of fatigue cycling as a result of this effect. With further cycling, these pits developed into oxide spikes and then into sharp fatigue cracks. Simultaneous oxidation and fatigue cycling were required to produce this mode. A crack initiation model was developed to express this interaction.

The crack propagation process was also influenced by oxidation; however, a simple model based on crack advance governed by bulk oxidation rate was inadequate to explain the observed behavior. It was speculated that the crack tip oxidation/fatigue process was different from the surface process characterized in this study.

Acknowledgments

This research was sponsored by the Naval Air Systems Command under Contract N0019-82-C-0269, Joe Collins, Project Engineer. This support is gratefully acknowledged.

References

[*1*] Coffin, L. F., *Metallurgical Transactions,* Vol. 3, 1972, pp. 1777-1788.

[*2*] Antolovich, S. D., Baur, R., and Liu, S. in *Superalloys 1980—Proceedings of the Fourth International Symposium on Superalloys,* J. K. Tien et al., Ed., Sept. 1980, pp. 605-614.

[*3*] Reuchet, J. and Remy, L., "Fatigue Oxidation Interaction in a Superalloy—Application to Life Prediction in High Temperature Low Cycle Fatigue," *Metallurgical Transactions,* Vol. 14A, Jan. 1983, pp. 141-149.

[*4*] Wukusick, C. S., "Directional Solidification Alloy Development," Final Report, Naval Air Systems Command, Contract N62269-78-C-0315, 25 Aug. 1980.

[*5*] McKnight, R. L., Laflen, J. H., and Spamer, G. T., "Turbine Blade Tip Durability Analysis," NASA CR-165268, Feb. 1982.

[*6*] Gangloff, R. P. in *Fatigue Crack Growth Measurement and Data Analysis, ASTM STP 738,* American Society for Testing and Materials, Philadelphia, 1981, pp. 120-138.

[*7*] Rooke, D. P. and Cartwright, D. J., *Compendium of Stress Intensity Factors,* Her Majesty's Stationery Office, London, 1976.

[*8*] Wasilewski, G. E. and Rapp, R. A., "High Temperature Oxidation," in *The Superalloys,* C. T. Sims and W. C. Hagel, Eds., Wiley, New York, 1972.

[*9*] Hauffe, K., *Oxidation of Metals,* Plenum Press, New York, 1965, p. 173.

[*10*] Jang, H. and Wright, P. K., "Fatigue and Fracture of Advanced Blade Materials," Final Report, Air Force Materials Laboratory, Contract F33615-82-C-5031, Sept. 1984.

[*11*] McGowan, J. J. and Liu, H. W. in *Proceedings,* 27th Sagamore Army Materials Research Conference on Fatigue, Environment and Temperature Effects, July 1980.

G. Engberg[1] *and L. Larsson*[2]

Elevated-Temperature Low Cycle and Thermomechanical Fatigue Properties of AISI H13 Hot-Work Tool Steel

REFERENCE: Engberg, G. and Larsson, L., **"Elevated-Temperature Low Cycle and Thermomechanical Fatigue Properties of AISI H13 Hot-Work Tool Steel,"** *Low Cycle Fatigue, ASTM STP 942,* H. D. Solomon, G. R. Halford, L. R. Kaisand, and B. N. Leis, Eds., American Society for Testing and Materials, Philadelphia, 1988, pp. 576–587.

ABSTRACT: Thermal fatigue of the martensitic hot work tool steel AISI H13 is complicated by structural degradation which in turn is influenced by plastic deformation. These properties are dependent upon plastic deformation, temperature, and time, and thus it is of vital importance to find a description which allows inter- or extrapolation to the thermal load experienced in various practical applications.

Isothermal, low cycle fatigue testing was performed mainly at 500 and 600°C and also to a limited amount at 80, 300, and 750°C. The material was tempered to three different hardness levels (43, 47, and 51 HRC) prior to testing. The main bulk of tests were performed on the 47 HRC variant. At 600°C no difference in life was observed between the different hardness variants. The cyclic stress-strain curves, evaluated at half lives, were also almost identical. Softening is believed to cause this similarity, while the initial differences will only influence the behavior during a small fraction of the life after which all material differences will have vanished. Softening is due to particle coarsening and a decrease of the initially very high dislocation density. By assuming that Ostwald ripening takes place through volume diffusion, that the Orowan mechanism for particle strengthening is valid, and that the dislocation network spacing is directly proportional to the interparticle distance, a fair functional description of the experimental data is obtained. The temperature dependence of the solubility limit for the carbides and the activation energy for self-diffusion are nicely reflected by the data. The model predicts that a low solid solubility and a high volume fraction of the carbides will give rise to a low rate of softening as will a high strain-rate and of course a low temperature. Plastic deformation is found to increase particle coarsening significantly, but the extent is independent of the magnitude of the plastic strain within the limits investigated.

Thermomechanical fatigue testing (simultaneous cycling of both strain and temperature, half a period out of phase with one another) was performed with a maximum temperature of 600°C in the cycle. The minimum temperature, T-min, was varied from 80 to 400°C. In comparison to the isothermal 600°C data, life was reduced when T-min was less than or equal to 200°C, but remained fairly unaffected when a T-min of 400°C was used. As expected, softening increased with increasing minimum temperature.

The same hardness variants as described above were also tested with the 200 to 600°C cycle. No significant differences in life were obtained, although the cyclic stress-strain curves (maximum tensile stress at half life) differed. The Ostergren method for life prediction fit our data reasonably well if the plastic work is normalized with the cube of the shear-modulus.

KEY WORDS: low cycle fatigue, thermal fatigue, stress relaxation, softening, tool steel

For a hot work tool steel, like AISI H13, thermal fatigue limits the life of the component in many practical applications. Die casting of aluminum and brass induces temperature cycling of

[1]Svenskt Stål, Domnarvet, S-78184 Borlänge, Sweden.
[2]Swedish Institute for Metals Research, Drottning Kristinas väg 48, S-114 28 Stockholm, Sweden.

the die typically between 250 to 600°C and 300 to 700°C, respectively. The corresponding cycle times are 10 to 30 s.

During thermal fatigue at the mentioned temperatures the material will soften at a rate dependent upon the specifics of the cycle. As softening will influence the life of the material it is essential to be able to understand and predict the softening in a specific application in order to optimize the material life. Investigations have shown that short cycle time will decrease the number of cycles to crack initiation [*1*] and that, as expected, the amount of softening increases with increasing cycle time.

The aim of the present work is to derive a physically based model which will be able to describe the softening process in AISI H13 and relate this to the life of the material. The experimental procedure adopted consists of isothermal low-cycle fatigue (LCF) testing and thermomechanical fatigue (TMF) testing. The latter combines temperature and strain cycling in a controlled way.

Experimental Procedure

The experimental setup and procedure has been thoroughly described in other contexts [*2*] and will not be related in detail here. All experiments were performed in *diametrical* strain control with the total strain varying linearily with time. Tubular specimens with a gage length of 34 mm, an inner radius of 2 mm, and an outer radius of 4 mm were used. In the TMF tests the temperature was half a period out of phase with the strain, which means that the maximum temperature is obtained at the maximum compressive strain. All strain values given in this paper are diametrical. Temperature control was obtained by an induction furnace with the temperature registered by spotwelded Type S thermocouples.

The material studied was of Type AISI H13 with the composition given in Table 1. After austenitization at 1020°C for 0.5 h, three different tempering procedures were employed. Double tempering (1 h at the first temperature followed by quenching and 1 h at the second temperature) was performed at 575°C + 610°C, 600°C + 590°C, and 575°C + 580°C giving hardness values of 43, 47, and 51 HRC, respectively. The normal hardness is RC 47.

The number of cycles to crack initiation has been evaluated from maximum load versus cycle curves and is taken as the point where the load starts to drop sharply as compared to the decrease due to softening (Fig. 1). The size of the crack is then on the order 0.1 to 1 mm.

The shear modulus of the material can be represented by the relation

$$G = 91\,540 - 38.7 \cdot T$$

where G is given in MPa and T is the absolute temperature.

Results and Discussion

Theory

The flow stress, σ, may be considered to consist of contributions from particle strengthening, σ_p, deformation hardening, σ_d, and from other athermal as well as thermal barriers to disloca-

TABLE 1—*Chemical composition (weight-percent) of AISI H13 steel.*

Element	C	Si	Mn	P	S	Cr	Ni	Mo	W	Co	V	Cu	N	O	Al
Amount	0.40	1.11	0.48	0.021	0.005	5.2	0.13	1.33	0.04	0.01	0.93	0.05	0.027	0.0011	0.016

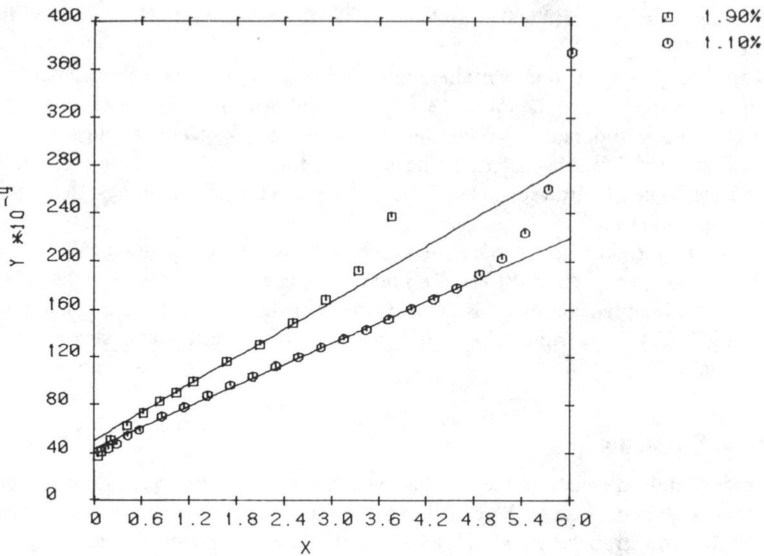

FIG. 1—$(G/\sigma)^3$ *versus* $N\Delta\epsilon/(T\dot{\epsilon})$ (Y *versus* X) *for two specimens tested at 600°C. The square symbols denote a specimen tested at a total strain amplitude of 1.90% and the circles a specimen tested at 1.10%. The solid lines describe the linear parts of the curves.*

tion slip. At elevated temperatures the contribution from thermal barriers is negligible, and as the material has a high dislocation density and a substantial particle content we may assume that only these latter contributions are of importance. The carbides precipitated are quite strong and dislocation bypassing will take place by the Orowan mechanism. We may write

$$\sigma_p = mGb\sqrt{f}/r \tag{1a}$$

where m is the Taylor factor, G the shear modulus, and f the volume fraction of particles with radius r. The high dislocation density due to the martensitic transformation will be restricted by the precipitated particles during tempering. Subsequently, deformation induced dislocations will also be held up on the particles. We may thus assume that the dislocation density, ρ, is related to the interparticle spacing, λ. If we further assume that we have a dislocation network with an interdislocation spacing equalling λ then we obtain

$$\sigma_d = m\alpha Gb\sqrt{k}/\lambda \tag{1b}$$

where $k/\lambda^2 = \rho$, b is the Burgers vector, and α is a constant related to the strength of the dislocation junctions. As λ is directly proportional to $r/\sqrt{f}$, we obtain

$$\sigma = mGb(\alpha\sqrt{k} + 1)\sqrt{f}/r \tag{1c}$$

During exposure at an elevated temperature, T, the dispersion of particles will be subject to particle growth or dissolution during which the volume fraction will change. The rate of growth for spherical particles may be approximated using a quasistationary approach to the solution of the diffusion problem [3]. This yields.

$$\frac{dr}{dt} = \frac{D}{r(1 - r/\Lambda)} \frac{x^{\alpha 0} - x^{\alpha/\beta}}{x^{\beta} - x^{\alpha/\beta}} \tag{2a}$$

where D is the diffusion coefficient and Λ is the radius of the depleted layer in the matrix, α. β denotes the particle phase and x the composition at the phase-boundary (α/β) in the matrix at a distance Λ from the particle ($\alpha 0$) and in the particle (β). We have ignored the influence of differing molar volumes in the particle and matrix phases. The solubility limit of the carbide may be written as

$$x^{\alpha/\beta} = a_1 \exp(-Q_1/RT) \tag{2b}$$

where a_1 and Q_1 are constants and R has its usual meaning. After soft impingement Λ remains constant and is directly proportional to the interparticle spacing. When the mean value of the concentration difference driving the diffusion process equals zero, subsequent growth will occur due to differences in particle size as the interface energy, γ, is finite. The solubility limit for a particle of size r as compared to an infinite particle is given by the Gibbs-Thomson equation and may be written as

$$x_r^{\alpha/\beta} = x_e^{\alpha/\beta} \exp\left(\frac{2V\gamma}{rRT}\right) \tag{2c}$$

where V is the molar volume of the particle phase. Equation 2 may now be summarized as

$$\frac{dr}{dt} = \frac{A_1}{(1 - A_2\sqrt{f})} \frac{1}{T} \exp\left(-\frac{Q_1 + Q_2}{RT}\right) \frac{1}{r^2} \tag{3}$$

where A_1 and A_2 are temperature independent. We have used a small value approximation for the exponential term in Eq 2*c* and assumed a constant particle size distribution. Consequently, we have arrived at the usual equation for Ostwald ripening except for the dependence upon the volume fraction which is normally negligible.

Isothermal LCF Results

Time, t, may be expressed in units of total strain amplitude, $\Delta\epsilon$, total strain-rate, $\dot{\epsilon}$, and the number of cycles, N:

$$t = 2N\Delta\epsilon/\dot{\epsilon} \tag{4a}$$

Equation 3 is integrated and r is substituted using Eq 1 which yields

$$\left(\frac{k_1\sqrt{f}}{r_0}\right)^3 \left(\frac{G}{\sigma}\right)^3 = 1 + B_0 \exp\left(-\frac{Q_1 + Q_2}{RT}\right) \frac{N\Delta\epsilon}{T\dot{\epsilon}} \tag{4b}$$

where

$$B_0 = 6A_1/(r_0^3(1 - A_2\sqrt{f})) \tag{4c}$$

Equation 4 thus predicts that $(G/\sigma)^3$ is linearily related to the number of cycles, the strain-amplitude, and the strain-rate.

We have analyzed our experimental data according to this relation and an example is shown in Fig. 1. A linear relation is obtained over a large part of the lifetime of the specimen. The test at 750°C displayed a somewhat different behavior, but conformed to the other results if the early part of the life was studied. The volume fraction of particles depend upon temperature in the same way as does the solubility limit (assuming equilibrium), and via the lever rule we obtain

$$f = f_0 + f_1 \exp(-Q_1/RT) \tag{4d}$$

Using LCF-data obtained at 80, 300, 500, 600, and 750°C we obtain $Q_1 = 54\,070$ J/mol. This compares favorably with literature data for the solubility of cementite in ferrite where $Q_1 = 40\,590$ J/mol has been given [4]. In our derivation of Q_1 we have assumed that f_0 is given by the low temperature values obtained at 80 and 300°C. The parameter $f \cdot (k_1/r_0)^2$ is plotted versus T in Fig. 2 and reflects a normal variation of the solubility limit in a binary system.

The same set of data was also used to evaluate B_0 and $Q_1 + Q_2$ (Fig. 3). In this case two different temperature regimes were found. At temperatures above 430°C we obtain $B_0 = 46.21 \cdot 10^6$ K/s and $Q_1 + Q_2 = 130\,120$ J/mol. Below 430°C the corresponding values are 0.0125 K/s and 1300 J/mol, respectively. The activation energy for self diffusion is 240 000 J/mol [5]. Half of this energy originates from the creation of vacancies. This part may thus be omitted if the vacancies are supplied in sufficient number by plastic deformation. The value we found for temperatures above 430°C agrees very well with this idea. At the lowest level of plastic strain-ranges tested, $\Delta\epsilon$-plastic = 0.15%, the independency of strain no longer was valid. Therefore in order to treat the very low strain-amplitudes the above treatment must be modified.

In Table 2 we have listed the results from the LCF tests treated in detail here.

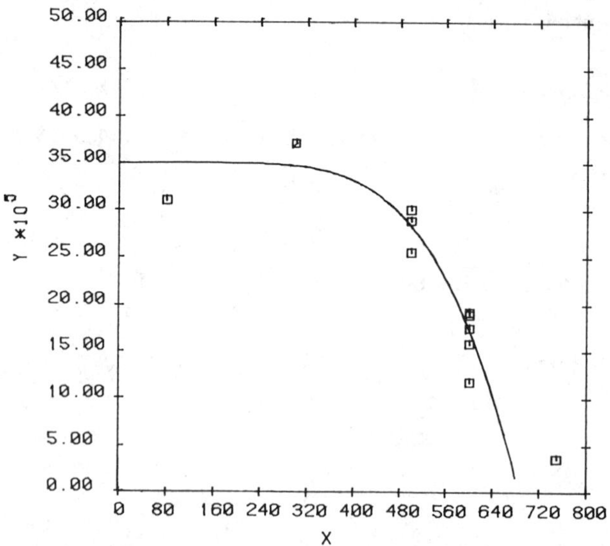

FIG. 2—$(k_1\sqrt{f}/r_0)^2$ *versus* T *(°C)* (Y *versus* X). *The solid line represents the derived analytical expression.*

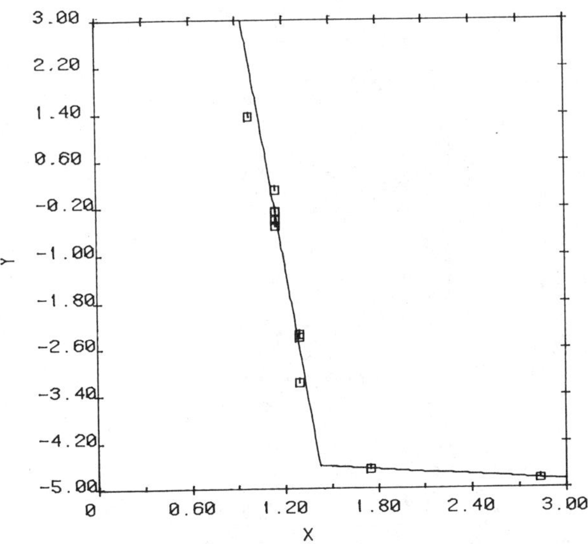

FIG. 3—*Natural logarithm of the slope obtained as in Fig. 1 versus* 10^3/T *(*K^{-1}*)* (Y *versus* X). *The straight lines represent best fit to data. The data point at 750°C has not been used in the derivation of the high temperature line.*

Two specimens with HRC 43 and 51 have also been analyzed. The high hardness specimen behaves as do the HRC 47 specimens, while annealing to HRC 43 yields quite different results, a considerably higher rate of softening. The higher annealing temperature in this case may change the carbide type distribution; that is, affect the amount and size of special carbides.

Non-Isothermal TMF Results

In our tests temperature varies linearily with time as does the total strain. We thus obtain

$$T = T_{\min} + \Delta T \frac{\dot{\epsilon}}{\Delta\epsilon} t \tag{5}$$

Integration of Eq 3, assuming equilibrium at all times, yields

$$\left(\frac{k_1\sqrt{f}}{r_0}\right)^3\left(\frac{G}{\sigma}\right)^3 = 1 + \frac{6A_1}{r_0^3}\int_{T_{\min}}^{T_{\max}} \frac{\exp\{-(Q_1 + Q_2)/RT\}}{T(1 - A_2\sqrt{f})}\, dT \frac{N\Delta\epsilon}{\Delta T\dot{\epsilon}} \tag{6}$$

This equation thus predicts a linear relationship between $(G/\sigma)^3$ and the parameter $N\Delta\epsilon/(\Delta T\dot{\epsilon})$. Figure 4 shows an example that this is indeed the case. All the parameters in Eq 6, except A_2, are known from the isothermal experiments. As seen in Fig. 4 we obtain almost similar results for the stresses at the highest and lowest temperature in the cycle. The difference may stem from changes in the particle volume fraction. At the lower temperature we expect a higher value of f. This implies a higher value of the parameter $k_1\sqrt{f}/r_0$, which is also found (see Table 3). The change of f when going from the minimum to the maximum temperature or vice

TABLE 2—*Isothermal LCF data. Values were obtained at half the number of cycles to crack initiation.*[a]

Temperature (°C)	Strain-Rate (1/s)	Strain Amplitude (%)		No. of Cycles	$(r_0/k_1\sqrt{f})^3$	B	Regression Coefficient
		Total	Plastic				
80	3.84×10^{-4}	0.96	0.50	234	18.30×10^4	14.69×10^2	0.945
300	3.85×10^{-4}	0.94	0.50	207	14.04×10^4	13.37×10^2	0.902
500	5.05×10^{-4}	1.39	1.05	113	19.25×10^4	17.35×10^3	0.973
500	4.71×10^{-4}	0.73	0.45	361	20.43×10^4	19.33×10^3	0.998
500	4.97×10^{-4}	0.41	0.15	1140	24.62×10^4	10.18×10^3	0.996
600	5.21×10^{-4}	1.90	1.70	90	50.55×10^4	38.58×10^4	0.999
600	4.40×10^{-4}	1.10	0.90	190	43.60×10^4	29.28×10^4	0.999
600	4.51×10^{-4}	0.44	0.25	490	38.62×10^4	29.92×10^4	0.999
600[b]	3.62×10^{-4}	1.03	0.88	172	78.54×10^4	47.33×10^4	0.999
600[c]	3.62×10^{-4}	1.19	1.00	138	37.81×10^4	42.08×10^4	0.997
750	8.49×10^{-4}	0.59	0.51	237	41.27×10^5	67.14×10^5	0.855

[a]Number of cycles to crack initiation:

Experimental:	234	207	113	361	1140	90	190	490	172	138
Calculated:	190	156	71	203	1196	59	117	501	131	109

[b]HRC 43.

[c]HRC 51.

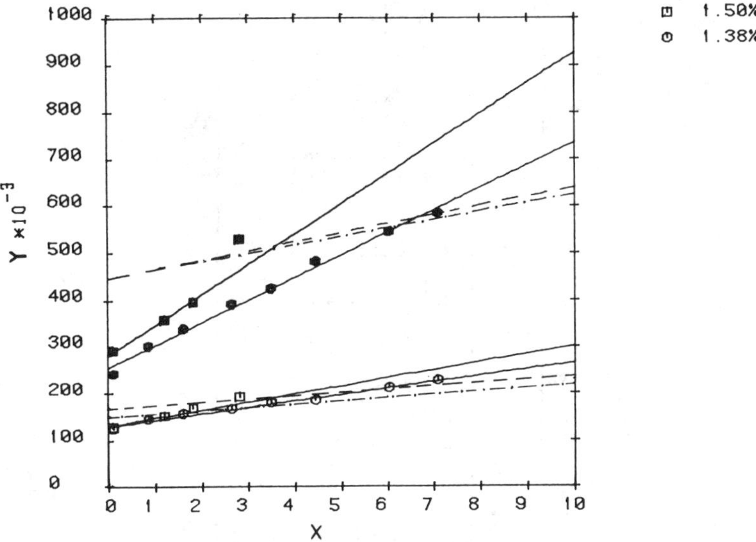

FIG. 4—$(G/\sigma)^3$ *versus* $N\Delta\epsilon/(\Delta T\dot{\epsilon})$ (Y *versus* X) *for two TMF specimens. The squares denote a specimen tested at a total strain amplitude of 1.50% between 200 and 600°C, while the corresponding numbers for the circles are 1.38% between 400 and 600°C. The open symbols originate from the minimim temperature in the cycle, while the solid points are all obtained at 600°C. The dashed and dash-dotted lines have been calculated from isothermally derived parameters. The dashed lines correspond to a minimum temperature of 200°C, while the dash-dotted lines correspond to 400°C. The uppermost lines refer to 600°C, while the lower lines refer to the minimum temperature in the cycle.*

versa can be deduced if we assume that the number of particles remain constant. As f is proportional to r^3, Eqs 2*a* and 2*b* can be integrated between the two temperatures. This yields

$$f^{2/3} = f_0^{2/3} + \frac{\Delta\epsilon}{\Delta T\dot{\epsilon}} F(T_{\max}, T_{\min}) \tag{7}$$

where F denotes the temperature integral. By plotting the parameter $(k_1\sqrt{f}/r_0)^3$ raised to a power of 4/9 versus $\Delta\epsilon/\dot{\epsilon}$ we expect a straight line and this is the case (Fig. 5). In our experiments we have only varied the minimum temperature and, since the high rate of softening was obtained above 430°C, we do not expect $F/\Delta T$ to vary much. If we ignore this strain dependence of the volume fraction and evaluate Eq 6 from LCF data only, we obtain results as shown in Fig. 4. The calculated slope is approximately half of the experimentally found slope in all cases examined which might be due to the fact that the term containing f was neglected in the temperature integral.

Lifetime Results

The number of cycles to crack initiation for the various tests performed are plotted against the total strain amplitude in Fig. 6. The results imply that the approach of Ostergren [*6*] may be appropriate. The plastic work done in the tensile part of the cycle will sum up to a constant at failure according to this model. Although the model was proposed for crack growth, we will use it for what we have designated *crack initiation*. Microscopic cracks may and probably have been

TABLE 3—*Nonisothermal TMF data. Values were obtained at half the number of cycles to crack initiation.*[a]

Temperature (°C)		Strain-Rate (1/s)	Strain Amplitude (%)		No. of Cycles	$(r_0/k_1\sqrt{f})^3$		B	
Max.	Min.		Total	Plastic		Max.	Min.	Max.	Min.
600	80	0.49×10^{-4}	0.23	0.07	1422	12.60×10^5	25.06×10^4	27.87×10^3	-48.30×10^2
600	80	1.51×10^{-4}	0.80	0.50	128	36.83×10^4	14.80×10^4	37.58×10^3	10.62×10^3
600	80	1.62×10^{-4}	1.43	1.17	40	23.72×10^4	12.36×10^4	49.76×10^3	11.05×10^3
600	200	2.56×10^{-4}	0.41	0.26	305	38.57×10^4	19.62×10^4	74.56×10^3	14.44×10^3
600	200	2.47×10^{-4}	1.00	0.65	95	31.12×10^4	13.60×10^4	44.26×10^3	16.92×10^3
600	200	2.46×10^{-4}	1.50	1.14	48	28.11×10^4	13.08×10^4	64.60×10^3	17.07×10^3
600	400	1.07×10^{-4}	0.50	0.27	467	38.32×10^4	15.79×10^4	32.53×10^3	8.27×10^3
600	400	2.76×10^{-4}	1.38	0.97	100	25.51×10^4	13.00×10^4	48.05×10^3	13.52×10^3

[a]The regression coefficients are all in the range of 0.967 to 0.999.

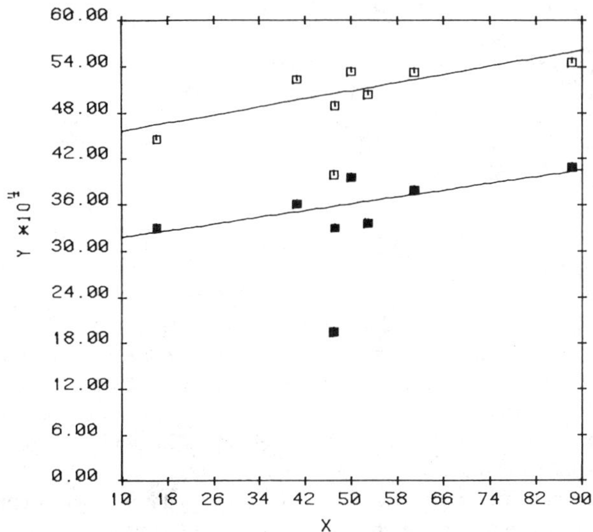

FIG. 5—$k_1\sqrt{f}/r_0$ *raised to a power of* $^4/_3$ *versus* $\Delta\epsilon/\dot{\epsilon}$ (Y *versus* X). *The open symbols are obtained at the minimum temperature and the closed symbols at 600°C. The lines have been obtained by linear regression. The specimen showing bad fit had a poor temperature control, resulting in distorted hysteresis loops.*

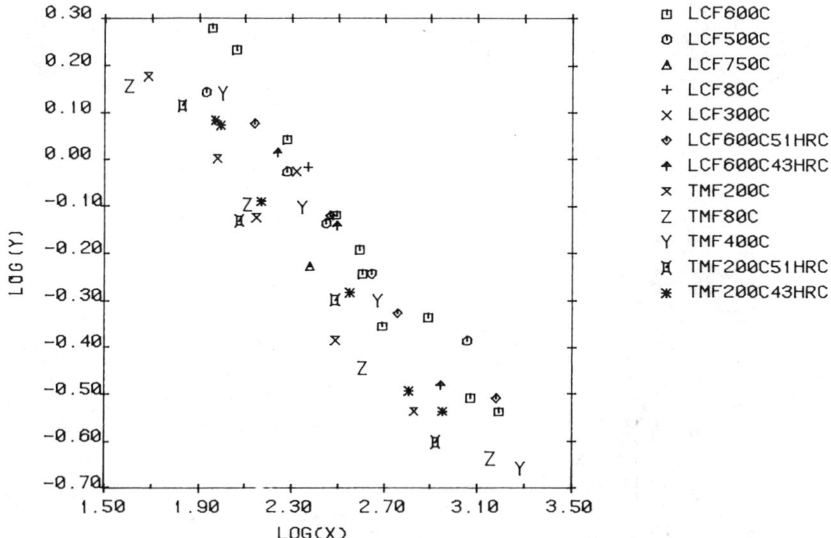

FIG. 6—*Total strain amplitude in percent* (Y) *versus the number of cycles to crack initiation* (X). *The TMF cycles all have a maximum temperature of 600°C.*

initiated much earlier than our criterion implies. As we have a good description of the variation of the maximum tensile stress with cycle number, we may easily calculate the sum of tensile plastic work for our specimens. We thus have the criterion

$$\sum_{1}^{N} (\sigma_T \Delta\epsilon_{pl}) = C \tag{8a}$$

where

$$\Delta\epsilon_{pl} = \Delta\epsilon_{tot} - \nu\left(\frac{\sigma_T}{E_T} - \frac{\sigma_C}{E_C}\right) \tag{8b}$$

where E is Young's modulus, T denotes tension and C compression, and ν is Poisson's ratio (=0.3). The sum is fairly constant for tests performed at the same temperature, but varies considerably with temperature. A normalization with the cube of the shear modulus unites the isothermally obtained data to a fair degree (Fig. 7). A least-square optimization gives $C/G^3 = 3.073 \cdot 10^{-12}$ MPa^{-2} with the sum of squares equalling 0.039 (natural logarithms used). The TMF data do not, however, conform to the isothermal data.

The failure criterion has been used together with the derived functions and parameters for the softening process to calculate the number of cycles to crack initiation for the isothermal tests performed. The calculated values were all within a factor of two from the experimentally derived numbers (Table 2).

The flow stress is proportional to the square root of the plastic strain, and the fracture criterion may be written as a cubic function of the stress equalling a constant, which lends support to

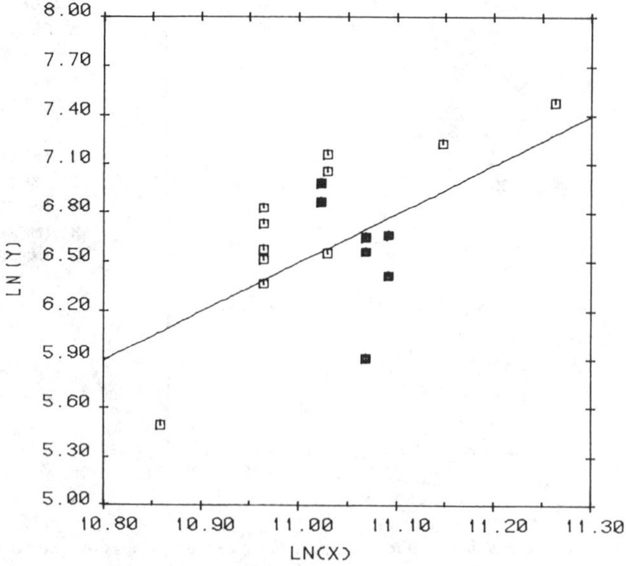

FIG. 7—*Total tensile plastic work,* $\Sigma\sigma\Delta\epsilon$ (Y), *versus shear modulus,* G (X). *The slope of the line obtained by a least squares optimization is 3. The filled symbols denote TMF tests, the open symbols LCF tests.*

the normalization that was found necessary. Equation 8 is a simplified form of Ostergren's model. Instead of σ he proposed a reduced stress, $\sigma - \sigma_0$. A least-square optimization of all our data gives $\sigma_0 = 285$ MPa and $C/G^3 = 2.43 \cdot 10^{-12}$ MPa^{-2}. The standard deviations for the two parameters are 202 MPa and $0.88 \cdot 10^{-12}$ MPa^{-2}, respectively. An improvement compared to the first approach is obtained, but the result is still not satisfactory.

Conclusions

Softening during cyclic deformation of the hot work tool steel investigated is caused by structural degradation; the rate of this degradation is well described by the theory for Ostwald ripening. The derivation of a physically motivated model for the softening allows interpolations and to some extent extrapolations of derived experimental data to practical applications.

A tensile plastic work failure criterion describes the experimental data reasonably well if normalized with the cube of the shear modulus.

The model still needs improvements concerning the softening behavior during temperature cycling. The fracture criteria also needs considerable improvement.

The results from thermomechanical testing in addition to those from ordinary isothermal testing illustrate the dangers of extrapolations from only isothermal data.

Acknowledgments

This work has been financed and guided by the Swedish Creep Committee, which is gratefully acknowledged.

References

[*1*] Malm, S., Svensson, M., and Tidlund, J., "Heat Checking in Hot Work Tool Steels," presented to Second International Colloquium on Tool Steels for Hot Working, Cercle d'Etudes des Metaux, Saint-Etienne, Dec. 1977.

[*2*] Samuelsson, A., Larsson, L., and Lundberg, L., "Thermomechanical and High Temperature Low-Cycle Fatigue of a Hot Work Tool Steel," IM-1589, Swedish Institute for Metals Research, Stockholm, May 1981.

[*3*] Engberg, G., Hillert, M., and Oden, A., *Scandinavian Journal of Metallurgy*, Vol. 4, 1975, pp. 93–96.

[*4*] *Sandvikens Handbok*, Vol. 1, No. 7, 1963, AB Sandvik Steel, Sweden, p. 29.

[*5*] Fridberg, J., Törndahl, L.-E., and Hillert M., "Diffusion in Iron," *Jernkontorets Annaler*, Vol. 153, 1969, p. 274.

[*6*] Ostergren, W. J., *Journal of Testing and Evaluation*, Vol. 4, 1976, pp. 327–339.

M. C. Shine[1] *and L. R. Fox*[1]

Fatigue of Solder Joints in Surface Mount Devices

REFERENCE: Shine, M. C. and Fox, L. R., **"Fatigue of Solder Joints in Surface Mount Devices,"** *Low Cycle Fatigue, ASTM STP 942,* H. D. Solomon, G. R. Halford, L. R. Kaisand, and B. N. Leis, Eds., American Society for Testing and Materials, Philadelphia, 1988, pp. 588–610.

ABSTRACT: Lifetime studies of a 16 I/O surface-mounted solder joint array undergoing isothermal cyclic fatigue in torsion shear under fixed plastic strain range show a strong correlation with creep fatigue and a creep-cracking mechanism. Experimental lifetime data follow an inverse dependence on matrix creep. Experimental measurement of the steady-state shear creep rate versus shear stress defines the creep characteristic that is sensitive to changes in metallurgical structure. The amounts of grain boundary and matrix creep taking place during a fatigue cycle are derived from experimental creep data combined with stress-strain hysteresis data obtained in steady-state cycling. Initially, thicker solder joints have a larger grain size than thinner solder joints, giving more matrix creep during fatigue and a faster failure rate. Fatigue increases the mean grain size of the solder joint as determined by the creep-rate-versus-stress characteristic and microstructure. Effects of grain size and joint thickness on lifetime are discussed. A maximum in the creep fatigue rate occurs at 333 K (60°C).

KEY WORDS: creep fatigue, creep cracking, matrix creep, grain boundary creep, lifetime inverse relationship with matrix creep, surface-mounted solder joint array, torsion shear, fixed plastic strain range, creep characteristic, fatigue rate

Introduction

Solder Fatigue

Until now fatigue studies of solder joints under shear have been developed by an approach in which rapid thermal cycling and/or accelerated isothermal mechanical testing regimes were extrapolated on a Coffin-Manson plot to predict the lifetime under "use" conditions of strain range and average temperature [*1–4*]. Creep and creep fatigue have been found to occur in solder joints under shear [*5*]. Wild has shown that solder joints undergo grain growth during fatigue and that large grain samples are weaker in fatigue than small grain samples [*5*]. The crack formation process was identified as being both intercrystalline and transcrystalline, originating from voids and fatigue type cracks [*5*]. Studies by Solomon on isothermal mechanical fatigue of solder joints using hysteresis loop analysis to control plastic strain range showed large cycle frequency effects on lifetime [*6*]. Fox has demonstrated accelerated fatigue rates using a square wave form of strain in the torsion shear of an array of surface-mounted solder joints [*7,8*]. Shine has demonstrated that creep is the predominant deformation mechanism under the conditions a solder joint experiences during the fatigue cycle. Shine has also shown that it is very difficult to separate, even qualitatively, matrix creep from grain boundary sliding creep or to substantiate any structural model that explains the failure mechanisms which solder or solder

[1]Principal Engineer and Consulting Engineer, respectively, Digital Equipment Corporation, Andover, MA 01810-1098.

joints experience during a typical fatigue cycle [*9,10*]. Stone has modeled solder fatigue using theoretical and empirical relationships to describe cavitation and creep phenomena: at high stresses, matrix deformation predominates; at intermediate stresses, a combination of grain boundary and matrix deformation takes place; at low stresses, matrix deformation again predominates [*11,12*]. It is not clear how the matrix creep is incorporated into the model at the very low stresses [*13,14*].

This paper describes how to calculate grain boundary and matrix creep using the experimental hysteresis loop and the creep rate versus stress characteristic. Experimental investigations were carried out on samples consisting of an array of solder joints in a surface-mounted configuration. This study shows that a correlation exists between the computed amount of creep per cycle and the isothermal mechanical fatigue rate.

Creep Fatigue/Strain Range Partitioning

For many alloys, strain-range partitioning is useful in predicting the life-times for high-temperature low-cycle fatigue conditions [*15–17*]. Rapid plastic and creep deformation are grouped in four combinations (PP, PC, CC, and CP) to characterize the Coffin-Manson relationships [*15–17*]. PP refers to a fatigue cycle in which rapid plastic tensile strain P is administered in the first half of the cycle, followed by rapid plastic compressive strain P in the second half of the cycle. PC refers to a fatigue cycle in which rapid plastic tensile strain (P) in the first half of the cycle is followed by compressive creep (C) in the second half of the cycle, and so on for CC and CP cycles. The analysis is based upon a linear superposition principal in which the damage mechanisms are additive for both types of plastic deformation in one half of the fatigue cycle. Two of the predictions of strain-range partitioning are the existence of an upper frequency limit in the lifetime-versus-frequency relationship above which frequency the lifetime is constant (PP), and a lower limit for lifetime below which value (CC, PC, or CP, depending on when the creep is introduced into the cycle) the lifetime is fixed at some minimum value. The CC condition has some associated partial healing from reverse creep, but a considerable portion of the damage produced in the first half-cycle remains, with cavitation and both intercrystalline and transcrystalline cracking [*15*]. Creep is defined in terms of grain boundary sliding aided by matrix deformation slip. The damage process has been linked to that of grain boundary-related triple point cavitation. If a lower limit of lifetime exists for solder joints, it would be valuable in defining the conditions of maximum acceleration factors. For example, because solder exhibits simultaneous grain boundary and matrix creep, the relationship of lifetime to the relative amounts of both types of creep is essential.

Creep Fatigue/General Considerations

Studies on high-temperature superalloys and ferrous and non-ferrous alloys, as well as theoretical studies, show similarities in the failure mechanisms of creep crack growth and creep fatigue [*17–33*]. Both depend upon power law creep generating grain boundary cavities that coalesce, producing intercrystalline cracks and ultimate failure. Some of the characteristics of creep crack growth and creep fatigue are:

1. An inverse fatigue crack growth rate dependence on frequency [*18–20*].
2. A simplistic linear superposition damage rule cannot be used to predict fatigue lifetimes for certain geometries and systems [*18,19,25*].
3. Symmetrical waveforms can produce reversible creep damage without any degradation in lifetime; unsymmetrical waveforms can produce reversible creep damage at high enough temperatures to activate a faster reversible creep process [*21,22*].
4. Grain boundary cavitational voids of the "*r*" type, formed at the boundary defined by two

grains, have a lenticular shape and are associated with long hold time; "*w*" type or wedge voids, formed at the intersection of three grains (triple points), are associated with grain boundary sliding which theoretically reaches a limiting value of 0.5 of total strain at low frequencies [*21*].

5. Many damage processes associated with creep fatigue can be theoretically explained by a matrix process obeying power law creep with a bulk activation energy [*26,32,33*].

6. Nucleation and growth of voids leading to cracking and failure requiring a threshold stress show decreasing strength (time to failure) with increasing grain size [*27*]. Creep rupture theory shows a maximum in strength for intermediate grain size [*28*]. Materials that are creep-brittle show decreasing strength with decreasing grain size [*26*].

7. Nucleation and growth theory for creep-generated voids predicts minimum failure time at some intermediate temperature, approximately $T/Tm = 0.5$ [*27*].

8. Both grain boundary sliding and matrix creep (power law creep) contribute to void formation; the relative contribution of matrix creep to void cavitation is expected to be large in lead [*29*].

Methodology: Strain-Range Partitioning in Solder Fatigue

Measuring Plastic Strain Components of a Fatigue Cycle

As a first approximation, strain-range partitioning will be applied to solder fatigue. This involves making accelerated lifetime tests under high creep conditions and using a linear additive rule for the actual components of the "use" fatigue cycle to predict lifetimes. In addition, since both "matrix" and "grain boundary" creep are present as well as rapid plastic strain if it exists, these types of creep must be partitioned by some additive rule. Their relative amounts are measured by an analytical technique.

The measurement technique used here is a modification of that proposed by Halford and Manson [*34*]. The experimental data input is a hysteresis tracing of the fatigue cycle and a steady-state creep rate versus stress characteristic up to the maximum stress attained in the fatigue cycle. The input data are transformed to give a profile of creep rate versus time, which in turn is integrated over the time limits of the cycle to give the total creep per cycle. We may then write a generalized expression for the total creep generated in a cycle in terms of the components of creep:

$$\underset{\text{total creep}}{\underset{\text{cycle period}}{\oint \dot{\gamma}_T dt}} = \underset{\text{grain boundary creep}}{\underset{\text{cycle period}}{\oint \dot{\gamma}_{gb} dt}} + \underset{\text{matrix creep}}{\underset{\text{cycle period}}{\oint \dot{\gamma}_M dt}} \qquad (1)$$

The total steady-state creep rate versus stress characteristic is similar for a solder joint array and solid solder of eutectic composition Sn/Pb (63/37) as will be seen in the Experimental Results section [*13,14*]. For the applicability of the experimental results, the grain boundary and matrix creep rate versus stress characteristic is needed to calculate the contribution each one makes to the creep. Grain boundary and matrix creep may each be represented by a power law expression of the form

$$\dot{\gamma} = K\tau^{1/m} \exp^{-q/kT} \qquad (2)$$

where τ is the acting shear stress, m is a creep exponent that has a value of 0.5 to 0.6 for grain boundary creep and a value of approximately 0.2 to 0.08 for matrix creep, K is a constant that has a grain-size-dependent term as well as other parameters, q is the appropriate activation energy, k is the Boltzmann constant, T is the absolute temperature, and $\dot{\gamma}$ is the steady-state shear creep rate [*13,14*].

Within the stress range of the experimental fatigue cycle, there is a mixture of grain boundary creep and matrix creep. It may be assumed that the two creep contributions are in series and additive so that the total creep rate is the sum of those two independent creep rates from the grain boundary and matrix contributions. The grain boundary creep rate at higher stress levels, where the m exponent decreases below the value of approximately 0.6, is simply obtained by extrapolation of the total creep rate curve at low stresses where the m value is approximately 0.6 (grain boundary controlled) to the higher stress levels where the m value is less than 0.6 (increasing amounts of matrix creep). The matrix contribution to creep rate is the difference between the extrapolated grain boundary creep rate curve and the total experimental creep rate curve at that stress level.

For the experimental hysteresis loop we write:

$$\tau_{\text{stress}} = f\,(\text{time}) \tag{3}$$

Using the experimental creep curves for total, grain boundary, and matrix creep (Eq 2) with the particular hysteresis curve for a given fatigue cycle (Eq 3), we compute the experimental creep profile for the total, grain boundary, and matrix creep, which may be represented for each by

$$\dot{\gamma}_{\text{steady-state creep rate}} = g\,(\text{experimental time}) \tag{4}$$

The total, grain boundary, and matrix creep contributions are obtained by numerical integration of each experimentally derived creep rate (Eq 4) over the cycle period (Eq 1).

Experimental Procedure

A torsional apparatus powered by an electric aerospace torque motor with a laser extensometer [7,8] was used to obtain isothermal fatigue hysteresis loops and creep data on the 16 I/O solder joint array (Figs. 1*a* to 1*c*). A modification for the creep excitation entailed a feedback loop to ramp up to some arbitrary displacement value, a hold period at constant torque, and a reversal to unload or to creep in the reverse direction. Ramp up time was typically 15 s, and hold times ranged from 120 to 300 s. Torque versus displacement and time, and displacement versus time, were plotted on an HP 7046 high-speed recorder.

The 7.6 by 7.6 mm 16 I/O ceramic chip carrier had solder pads approximately 0.64 by 1.0 mm. The 16 pads were arranged in a square array of four per side, at an average distance from the center of 3.43 mm, which was used to calculate the average stress on the pads due to the applied torque. These chip carriers were surface mounted to a stainless steel plate 45 by 45 mm. The stainless steel plate had electroplated copper footprints patterned to those of the chip carrier. The assembly was soldered together by solder paste of 63/37 Sn/Pb. The solder joint thickness ranged from 0.102 to 0.305 mm (0.004 to 0.012 in.). Fillets extended the shear area by approximately 50%. The assembly was securely clamped to the table with a custom clamp, which was supported by a vertical micropositioning stage for fine adjustment during setup and axial force control. The chip carrier fit snugly into a recess in the torque delivery chuck, where it was bonded with epoxy. Figure 2*a* shows the chip carrier array with filleted solder joints on a stainless steel substrate. Figure 2*b* shows the sample chuck with laser mirror surface secured to the chip carrier array.

Figure 3 shows a set of experimental creep strain versus time curves and the corresponding constant-applied torque values (for the torque axis, 1 Volt equals 1703 kPa (247 psi), for the strain axis 1 V equals 5%). Creep time curves are characterized by a transient portion (Stage I) and a steady-state portion (Stage II). The slope of the steady-state portion is the steady-state creep rate, which is constant with time. The steady-state creep rate is seen to increase with holding torque. The steady-state creep rate for Curve 1 is $2 \times 10^{-6}\ \text{s}^{-1}$, at a shear stress of 9.3

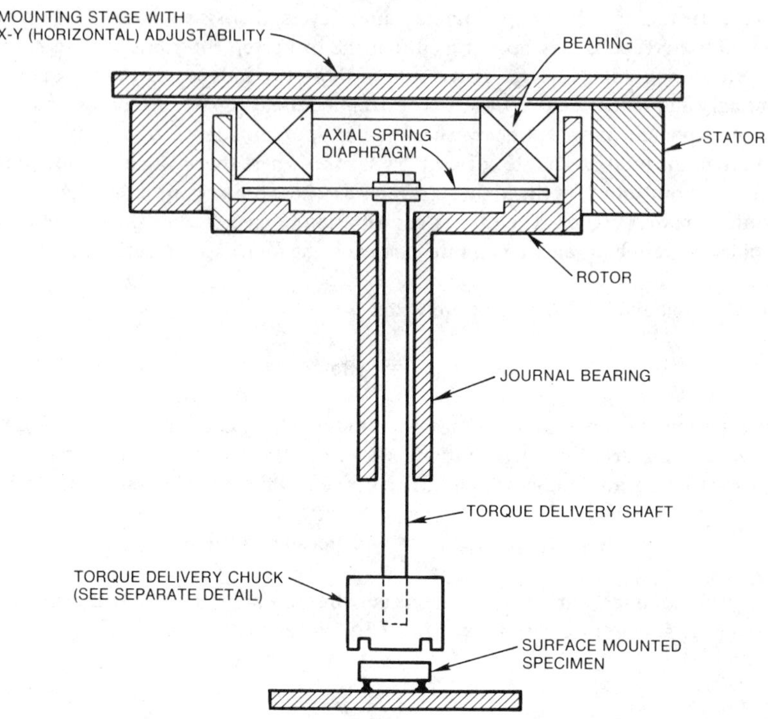

FIG. 1*a*—*Torsional apparatus for solder joint isothermal mechanical fatigue: torque apparatus physical plant.*

MPa (1350 psi); that for Curve 6 is 3×10^{-5} s^{-1}, at a shear stress of 16.2 MPa (2342 psi). Total creep strains up to 5% were generated by the torsion apparatus for the highest attained steady-state creep rate. Creep data were usually obtained at 298 K (25°C), 333 K (60°C), and 373 K (100°C) before the sample was damaged by fatigue.

Test arrays were excited in fatigue with a square wave excitation. The triangular wave and the square wave forms, with their resulting hysteresis loops, are shown in Fig. 4. For square-wave excitation, total strain with time was constant during each half period; with time elastic strain was converted to creep plastic strain during the load relaxation. It can be seen (Fig. 4) from the hysteresis loop for the square wave that the permanent set, due to creep strain, increases as the relaxation process proceeds at constant total strain. In order to perform fatigue experiments at a given plastic strain range, period, and temperature, the total displacement was adjusted until the hysteresis loop width equaled the desired plastic strain. During the course of the experiment, the width of the loop representing the total plastic strain was kept constant by manually adjusting the total displacement. The plastic strain range was varied up to 1.0% for the experiments. The maximum torque per cycle decreased during the fatigue experiment and was monitored, keeping plastic strain range constant, until the torque was reduced by 50% of its maximum value when the fatigue experiment was terminated. The number of cycles to reach this 50% value was taken as the lifetime [*6*].

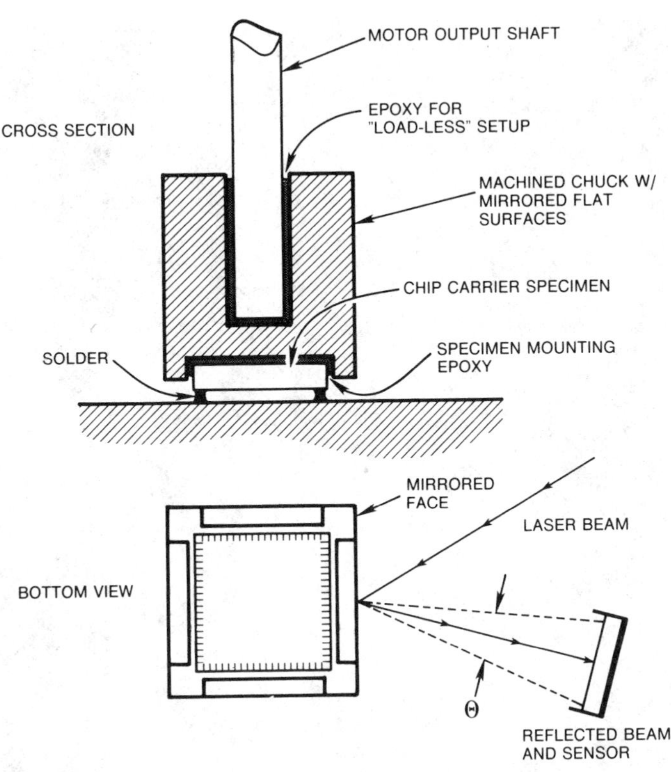

FIG. 1*b*—*Torsional apparatus for solder joint isothermal mechanical fatigue: sample check and optics.*

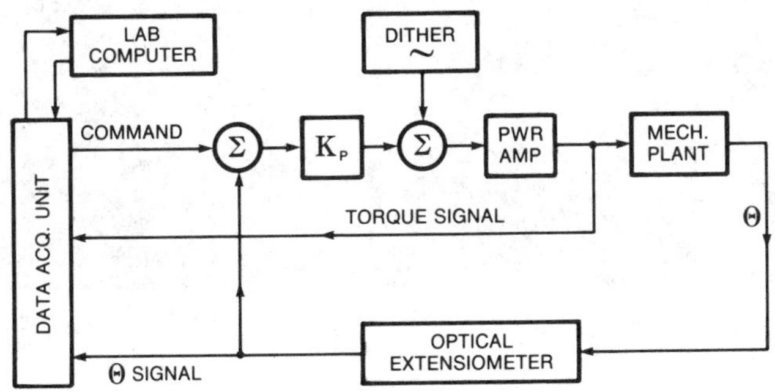

FIG. 1*c*—*Torsional apparatus for solder joint isothermal mechanical fatigue: system block diagram.*

Structural Studies

Isothermally fatigued solder joints were examined in a scanning electron microscope (SEM) *in situ* at the end of the fatigue experiment to look for void formation, crack formation, and grain size. Evidence of intergranular cracking and void formation in the grain boundaries, a product of creep fatigue, was especially sought. Two types of microvoid cavities were wedge ("*w*") voids, associated with grain boundary sliding, and "*r*" voids, which grow uniformly

FIG. 2*a*—*Chip carrier array solder joints on stainless steel substrate.*

across the grain boundaries. After SEM examination, samples were cut up and sectioned for study by light microscope to determine grain size and other structural features.

Experimental Results

Figures 5 and 6 show the experimental results of torque versus creep rate characteristics of Samples A and A3 having joint thicknesses 0.13 mm (0.005 in.) and 0.21 mm (0.0083 in.) respectively. Data are shown for samples as-received (AR) and after fatigue reduction (RED). There are two creep regions: one associated with high torques (shear stresses) and creep rates

FIG. 2*b*—*Laser mirror assembly secured to chip carrier array.*

with a shallow slope (matrix creep), and one associated with low torques (shear stresses) and creep rates with a steeper slope (grain boundary). The transition from grain boundary creep to matrix creep takes place at higher values of creep rate and lower values of stress as the temperature increases from 298 K (25°C) to 373 K (100°C). The transition creep rate is higher, the thinner the joint. The effect of fatigue is to shift the transition to lower values of creep rate for the same torque as well to shift the transition from grain boundary to matrix creep to lower creep rates. The coincidence of fatigued and unfatigued curves at 333 K (60°C) for Sample A3 is due to the instability in the fatigued state, which leads to failure shortly after the creep test. The slope of the log (stress) versus log (creep rate) relationship gave values of $m = 0.5$ to 0.7 for the grain boundary creep region and values of $m = 0.1$ to 0.30 in the region beyond the transition point where there is a mixture of both grain boundary and matrix creep. Table 1 shows the effect of temperature and sample thickness on the grain boundary to matrix creep transition and the corresponding slope m in the grain boundary creep region for Samples A, A1, A2, and A3.

To show the contribution of the grain boundary creep to the total creep at torque levels above the knee of the curve, a dotted line has been extrapolated from the below the knee, where the slope is constant at some value between 0.6 and 0.7, to stress levels where increasing amounts of matrix creep are generated, decreasing the slope above the knee as described earlier (Figs. 5 and 6). This bounding of the matrix creep and the grain boundary creep rates is believed to be accurate, because the activation energies measured for these two portions of the creep rate versus torque curves gave activation energies of 0.5 eV for the grain boundary portion and 1.0 eV for the matrix portion [*35*]. Therefore, during a fatigue cycle, the matrix creep contribution corresponding to stresses above the knee of the curve is determined by the difference between the total creep rate obtained from the actual experimental curve and the extrapolated experi-

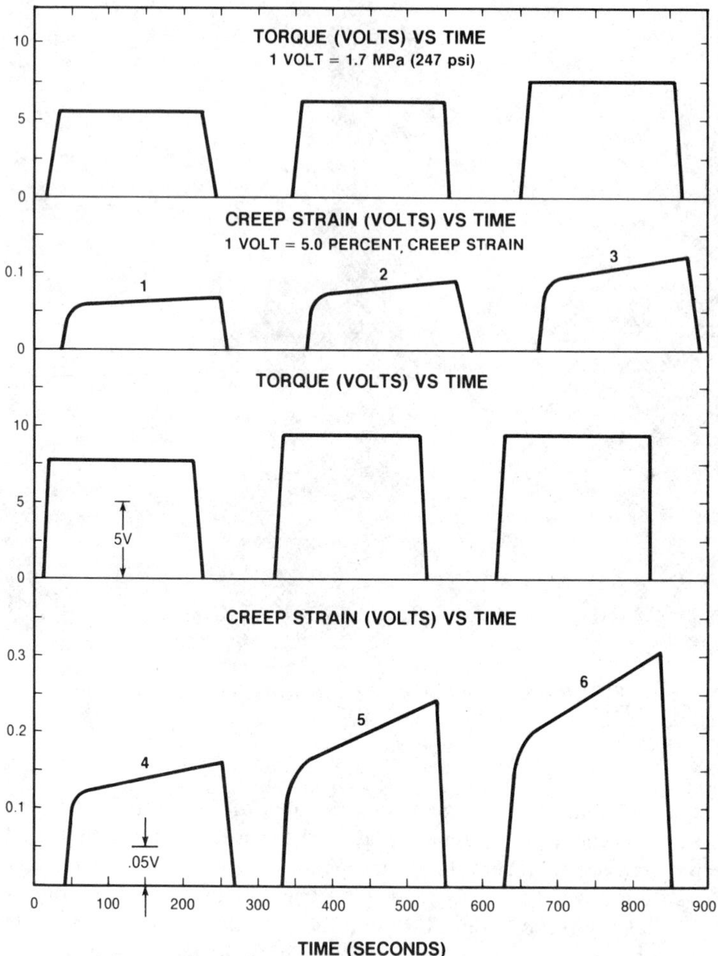

FIG. 3—*Holding torque, creep strain, versus time at 298 K (25°C).*

mental grain-boundary creep-rate curve integrated over the period of the cycle. Similarly, the grain-boundary contribution is the extrapolated value of the experimental grain-boundary creep rate integrated over the period of the cycle (Eq 1).

Figures 7 to 10 show results of the creep components for a triangular and a square wave on Samples A and A3 at 333 K (60°C) for a 600 s period at the indicated plastic strain ranges. In Figs. 7 and 9 the time variable is equivalent to the strain variable for the triangular excitation, since strain is linear with time (Fig. 4); the creep rate versus time curve for the cycle is readily obtained to calculate the creep. The relative amount of creep varies from 65 to 82%. The relative amounts of grain boundary creep and matrix creep are approximately equal for the thick Sample A3, but grain-boundary creep predominates for the thinner Sample A.

Figures 11 and 12 show the effect of temperature and period on the fraction of creep per cycle (creep strain divided by total plastic strain) for triangular-wave and square-wave forms for thick

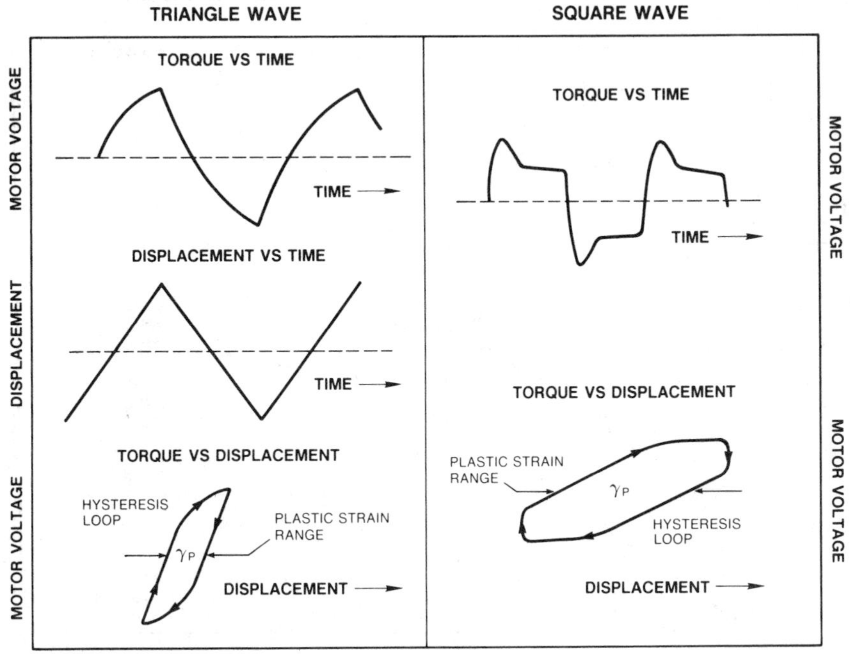

FIG. 4—*Triangular wave: torque versus time, displacement versus time, hysteresis loop. Square Wave: torque versus time, hysteresis loop.*

Sample A3 at 0.5% plastic strain range. For all periods except the very fastest, the creep per cycle has maxima at approximately 333 K (60°C). Figure 13 shows the effect of frequency on fraction creep per cycle at 333 K (60°C), at 0.5% plastic strain range, for Samples A and A3. The creep per cycle goes from 30% at 120 s period to 90% for a 1200 s period, independent of sample type.

Figure 14 shows the effect of frequency and strain range on the lifetime of solder joint Samples A, A1, A2, A3, and B0 at 333 K (60°C). The lifetimes were 1870 cycles for Sample A, 1250 cycles for Sample B0, 880 cycles for Sample A2, 530 cycles for Sample A3, and 320 cycles for Sample A1. Sample A fractured after 704 cycles, so its trend line was extrapolated to the 50% initial torque level for comparison with the other samples. All other samples followed a reasonably linear decrease in torque with cycles, so that a 50% value of torque was easily extrapolated for the lifetime. These data indicate that the lifetime decreases with both decrease in frequency and increase in strain range.

Figure 15 shows the strain life behavior of Samples A (0.35% plastic strain), B0 (0.50% plastic strain), and A1 (1.0% plastic strain) for 600 s compared with the data of Wild [*1*] and Solomon [*36*]. The Coffin-Manson slope of approximately 0.53 agrees with that of Solomon [*36*]; however, the shift to shorter times by about 1½ orders of magnitude is expected because of the much lower frequency (600 s versus 3.34 s) and the higher temperature (333 K [60°C] versus 308 K [35°C]) and also because Solomon's data were obtained with a triangular wave, while these data were obtained with a square wave. Since the work of Solomon was obtained at high plastic strain ranges (10%) compared with a maximum of approximately 1% for this work, the frequency effects are not comparable in the range of interest [*36*]. In addition, these proper-

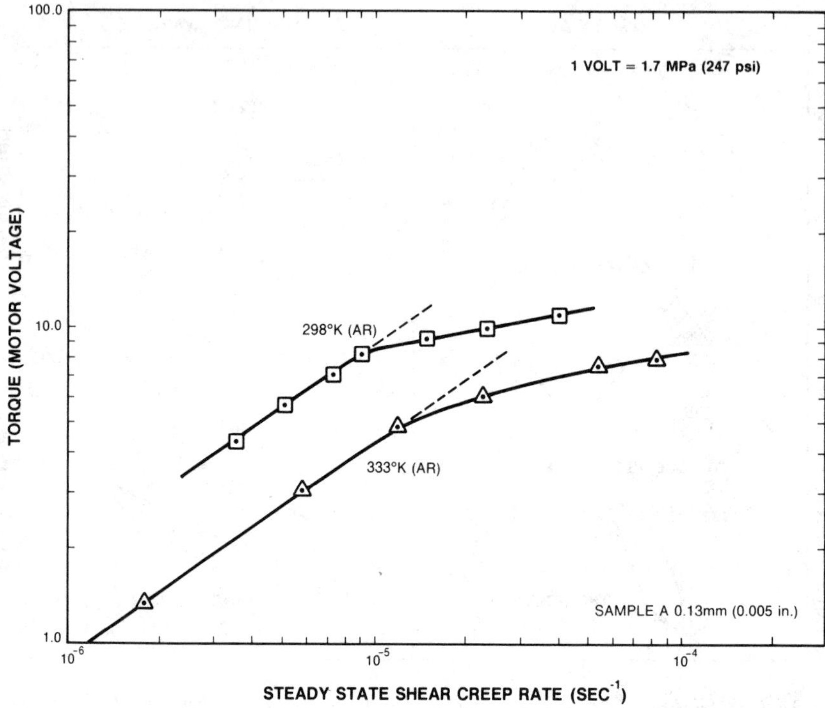

FIG. 5—*Creep characteristic (torque versus steady-state creep rate) for 0.13-mm (0.0005-in.)-thick solder joints; Sample A at 298 K (25°C) and 333 K (60°C).*

ties were obtained from a chip carrier device of 16 solder joints, a configuration different from single shear joints tested by Wild and Solomon [*1,36*].

Figure 16 correlates lifetime with total creep and grain boundary creep. The five data points were obtained from five tests run on steel substrates (Figs. 14 and 15). Frequency, strain range, and thickness of the solder joint array varied for each fatigue test. As can be seen in Fig. 16, lifetime correlated poorly with total creep and grain boundary creep. Figure 17 correlates matrix creep with lifetime for Samples A, A1, A2, A3, and B0 (steel substrates) and Samples A10, A11, A12, #1, #2, and #3 (G10 substrates) at 333 K (60°C). The matrix creep parameter, which was evaluated by the numerical integration procedure outlined in the preceding pages, correlates well with the lifetime for these varying conditions of plastic strain range (1.44/0.27) and period (120 s/1800 s). For the eleven points plotted in Fig. 17 of lifetime versus matrix creep, a monotonic relationship with a slope of -1 on a log-log plot fits well with the data. Samples A2 (120 s period), A3 (360 s period), and B0 (600 s period) had a 0.5% plastic strain range. At constant strain range, Samples A2 and A3 show a decreasing lifetime with increasing period; however, Sample B0 has a longer lifetime because the thickness of the joint is almost half that of A2 and A3, which is controlling the lifetime in spite of the longer 600 s period. For these three samples, the lifetime does not correlate with the total creep but does correlate with the matrix creep parameter (which is very sensitive to solder joint thickness). Similarly, test results at 600 s for Samples A (0.35% plastic strain), B0 (0.5% strain), and A1 (1.0% plastic strain), all within

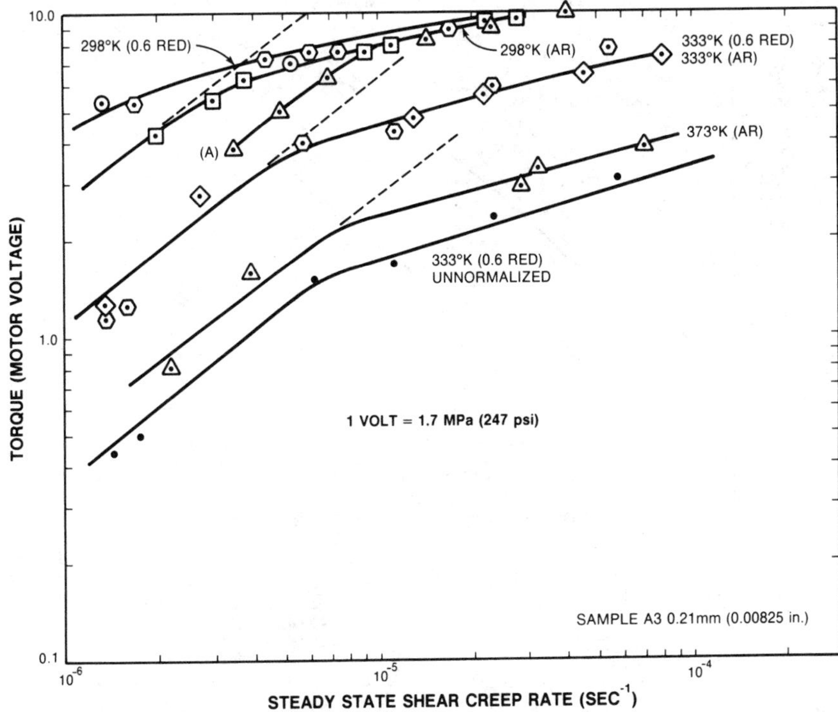

FIG. 6—*Creep characteristic (torque versus steady-state creep rate) for 0.21-mm (0.00825-in.)-thick solder joints; Sample A3 at 298 K (25°C), 333 K (60°C), and 373 K (100°C); as-received (AR) and after fatigue (RED).*

TABLE 1—*Comparison of creep characteristic transition creep rate versus sample type (thickness), temperature, and grain boundary exponent* m.

Sample	Thickness	Temperature	m	Transition Creep Rate, s^{-1}
A	0.13 mm (0.005 in.)	298 K (25°C)	0.7	1×10^{-5}
		333 K (60°C)	0.7	1.5×10^{-5}
A1	0.15 mm (0.0058 in.)	298 K (25°C)	0.5	0.8×10^{-5}
		333 K (60°C)	0.7	1.2×10^{-5}
A2	0.2 mm (0.008 in.)	298 K (25°C)	0.6	5×10^{-6}
		333 K (60°C)	0.8	1×10^{-5}
A3	0.21 mm (0.0083 in.)	298 K (25°C)	0.55	4×10^{-6}
		333 K (60°C)	0.7	6×10^{-6}
		373 K (100°C)	0.7	1×10^{-5}

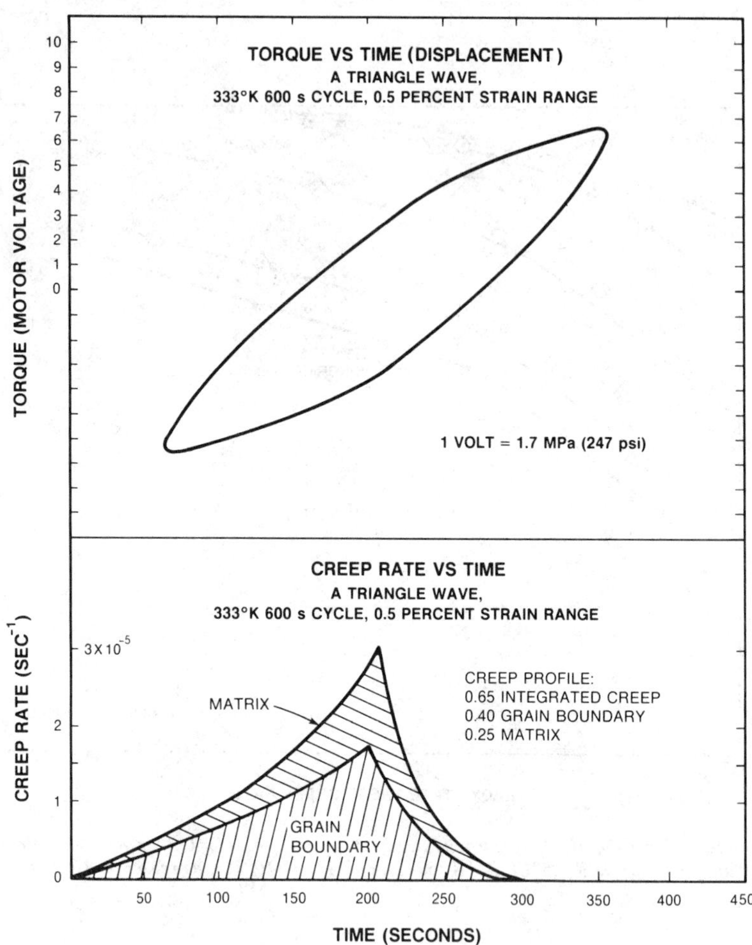

FIG. 7—*Components of creep calculated for Sample A: triangular-wave hysteresis loop, 333 K (60°C), 0.5% strain range, 600 s cycle.*

20% of each other in thickness, are comparable and show decreasing lifetimes with increasing plastic strain range as well as decreasing lifetimes with increasing matrix creep. Therefore, with comparable thicknesses at constant strain range, the lifetime decreases with increase in hold time (decrease in frequency) due to increasing amounts of matrix creep generated during the hold period. Similarly, for constant thickness and hold time (frequency), the lifetime decreases with increase in plastic strain range (Fig. 15). The same observations apply to the lifetime data obtained for the G10 substrates in terms of frequency, thickness, plastic strain range, and amount of matrix creep. These data were obtained with a square-wave excitation and therefore similar correlations with frequency and strain range may not be expected from the data of Solomon, who employed a triangular-wave ramp excitation [*36*].

The grain size was measured for a number of solder joints to determine the effects of solder joint thickness and fatigue on the mean grain size. The results correlate with the findings of the

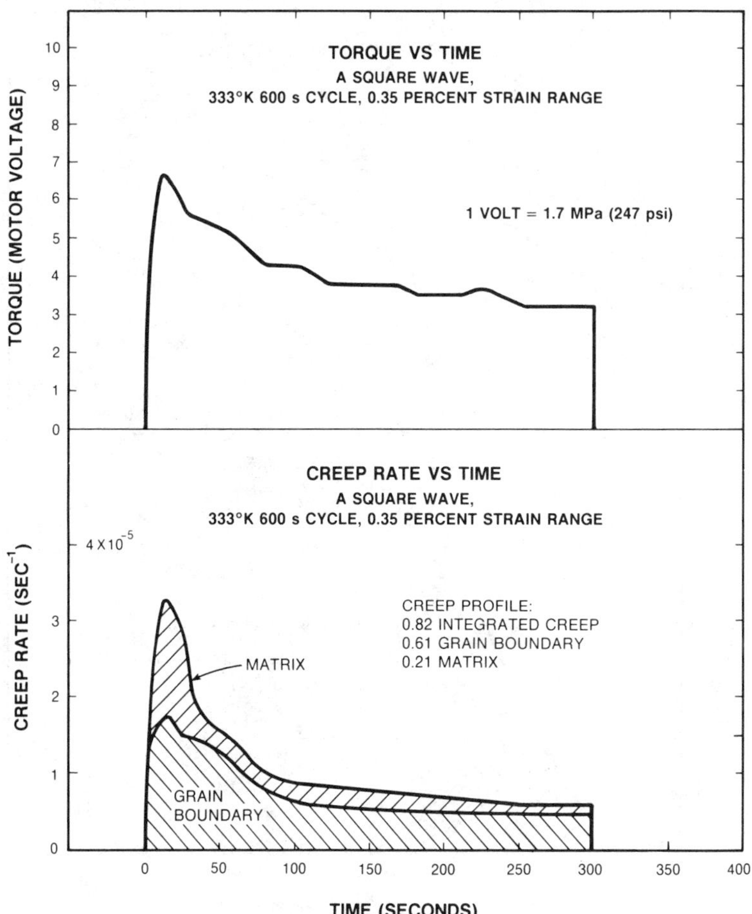

FIG. 8—*Components of creep calculated for Sample A: square-wave hysteresis loop at 333 K (60°C), 0.35% strain range, 600 s cycle.*

creep rate versus stress characteristic (i.e., thicker joints have a larger grain size and fatigue increases the grain size). For example, a 0.2-mm (0.008-in.)-thick joint had a mean grain size of 74.5 μm, while a 0.1-mm (0.004-in.)-thick joint had a mean grain size of 63.1 μm. These measurements were made in the region between the fillet and the thinnest part of the joint at the bottom. Another set of measurements on two samples at the fillet part of the joint showed a fatigued joint with an average grain size of 54 μm and an unfatigued joint with an average grain size of 45 μm. After fatigue, Sample A1 had a grain size of 12.4 μm and Sample A3 had a grain size of 27.6 μm at the thinnest part of the joint.

Figure 18 shows a SEM photo of the fractured surface of a solder joint from Sample D near the fillet and bottom part of the solder joint. This sample had a 60% reduction in strength by fatigue at an average temperature of 333 K (60°C). No unfatigued microstructure is presented for comparison; however, by the outline of the cavities, they appear to form predominantly at

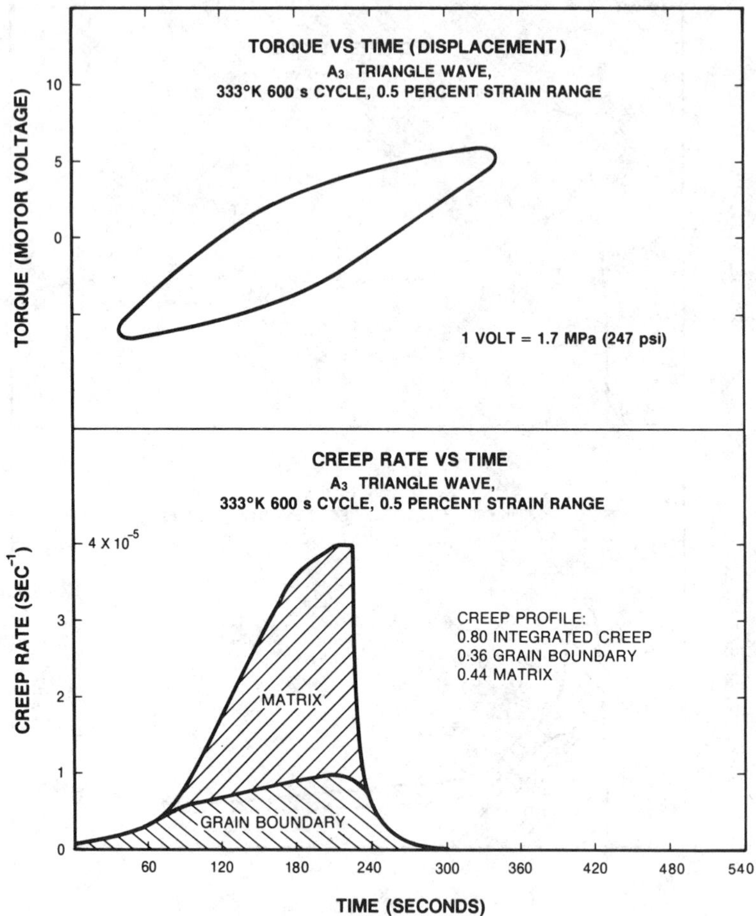

FIG. 9—*Components of creep calculated for Sample A3: triangular wave hysteresis loop at 333 K (60°C), 0.5% strain range, 600 s cycle.*

grain boundaries for this particular fatigue condition. This condition shows a similarity to the cavitational voids formed in grain boundaries when aluminum and other high-temperature alloys fail by creep fatigue [*18,26*].

Discussion

The overwhelming evidence for a creep fatigue damage mechanism for solder joints undergoing shear fatigue can be enumerated. A measurement technique has demonstrated that creep is the predominant mode of deformation occurring in the temperature range of interest, namely 298 K (25°C) to 373 K (100°C), for isothermal fatigue cycles.

Furthermore, two types of creep have been identified and measured: matrix creep, which predominates at high stresses and creep rates, and grain boundary creep, which predominates

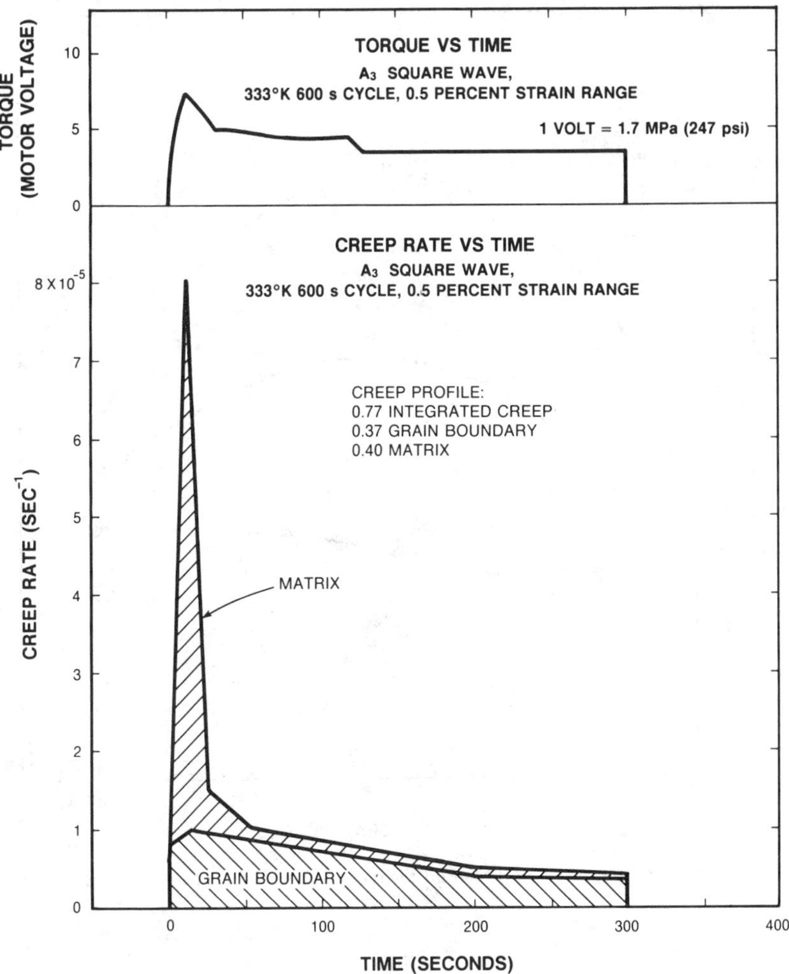

FIG. 10—*Components of creep calculated for Sample A3: square wave hysteresis loop at 333 K (60°C), 0.5% strain range, 600 s cycle.*

at much lower values of stress and creep rates. At intermediate stresses and creep rates there is a mixture of both types of creep. The relative amounts of each in a typical cycle can be calculated by the technique described above, which may be extended to a thermal cycle. Large-grain samples show more matrix creep than smaller-grain samples and consequently a higher fatigue rate for a given plastic displacement and strain range. Also, a maximum total creep is found at 333 K (60°C) which approaches 100% of the strain range for a 1200 s cycle. A maximum in fatigue rate has been found at 333 K (60°C) compared with 298 K (25°C) or 373 K (100°C), but the results are not presented here [*37*]. These two dependencies on temperature and grain size are predicted by a model for creep crack growth failure due to cavitational void growth controlled by a nucleation and growth mechanism [*27*].

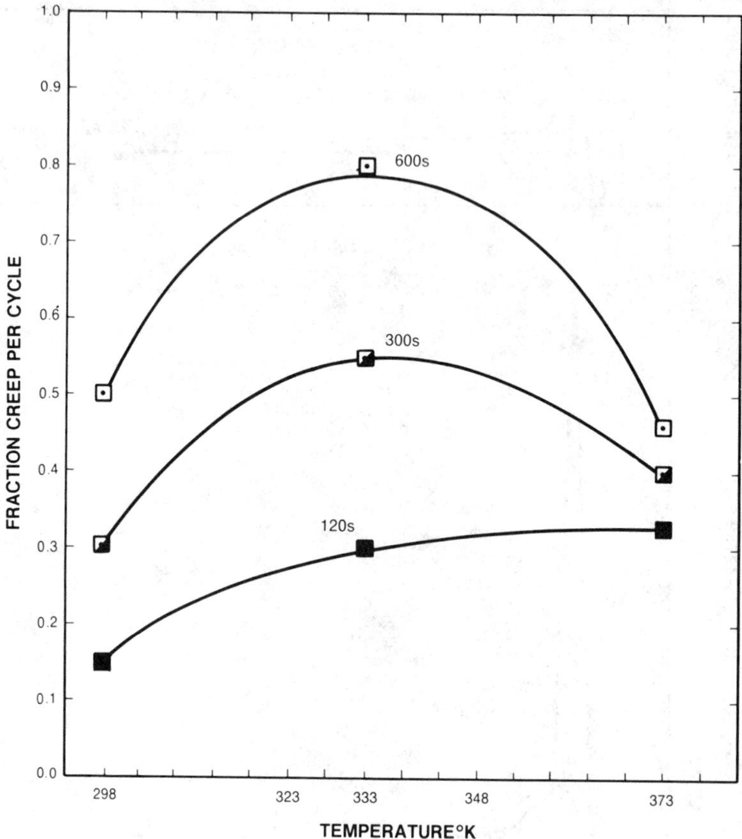

FIG. 11—*Effect of temperature and period on the creep per cycle for a triangular wave at 0.5% strain range Sample A3.*

From the lifetime data presented (Fig. 17), a linear lifetime dependence has been found for the matrix creep parameter:

$$\{Mc\} \times n_f = \text{const} \tag{5}$$

where $\{Mc\}$ is the total integrated matrix creep per cycle and n_f is the lifetime in cycles. For the eleven samples tested, the constant has a value of 125%, which may be interpreted as the creep exhaustion value or creep fatigue ductility of the samples. The experimentally obtained relationship (Eq 5) is similar to the Monkman and Grant equation found for materials which show a creep exhaustion limit due to cracking failure by creep cavitation [*28*]. Typically, the relationship is that the product of creep rate and lifetime is a constant with a value of up to 50% [*33*]. Also, a similarity exists with the strain-range partitioning relationship for the CC mode [*16*].

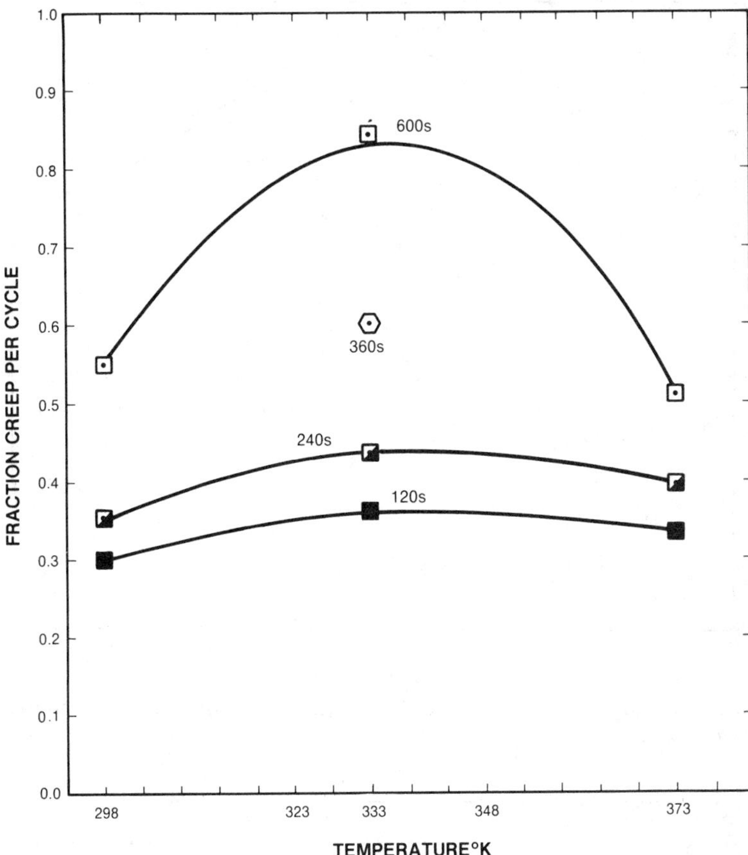

FIG. 12—*Effect of temperature and period on the creep per cycle for a square wave at 0.5% strain range Sample A3.*

Increased grain size associated with thicker joints shifts the creep-rate-versus-stress characteristic to higher values of matrix creep and lower values of grain boundary creep rates. A similar, smaller shift of the creep-rate-versus-stress characteristic after fatigue indicates an increase in grain size with fatigue. The finding of grain growth with fatigue in solder has also been reported by Wild [*5*]. Since matrix creep originates from dislocation generation within the grains, from either grain boundary or interior sources, smaller grain structures will not activate matrix creep until higher stress levels are attained than for larger-grained structures because of the inverse dependence for the flow stress on grain size [*38*]. Therefore, smaller-grained solders, which this work indicates are associated with thinner solder joints, will last longer in fatigue under the same conditions than larger grained solders, which are found to be associated with thicker solder joints, because less matrix creep is generated during the fatigue cycle.

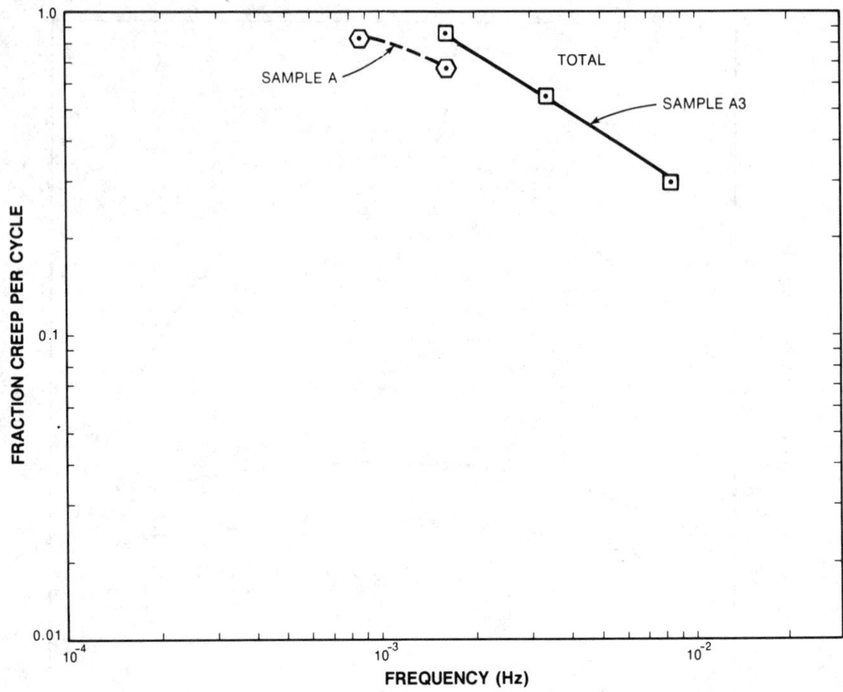

FIG. 13—*Effect of frequency on creep per cycle for a triangular wave at 333 K (60°C), for Samples A and A3 0.5% strain range.*

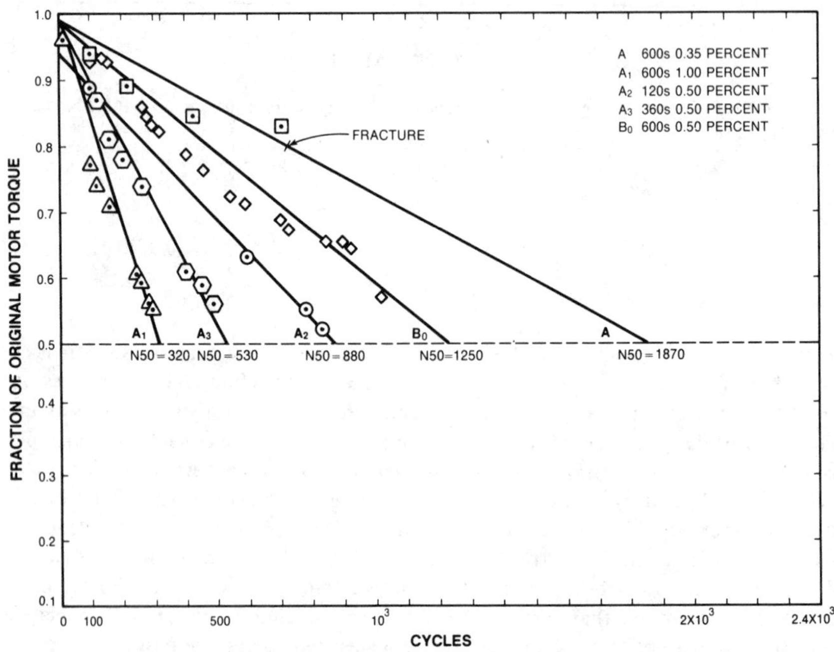

FIG. 14—*Lifetime versus frequency/strain range at 333 K (60°C), for Samples A, A1, A2, A3, and B0.*

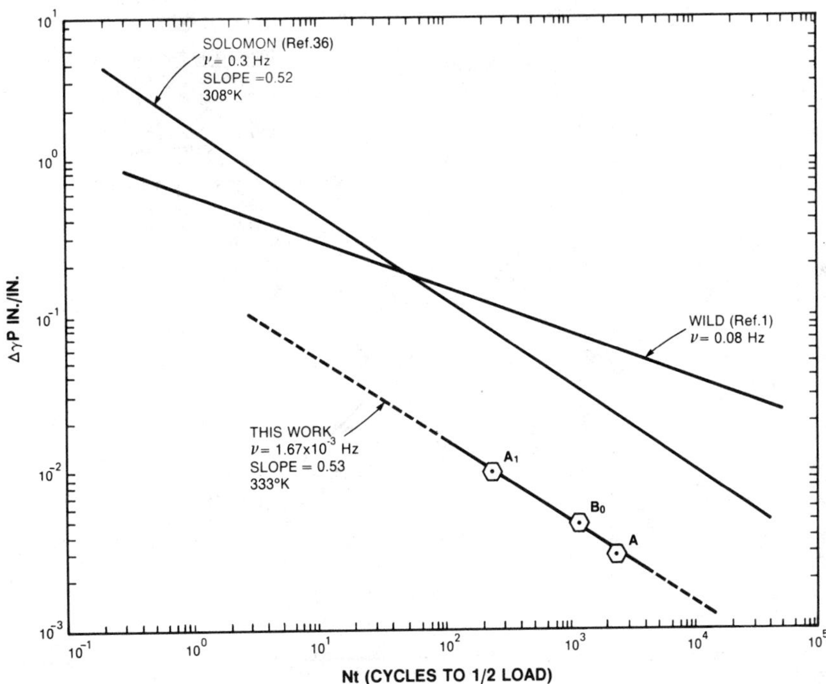

FIG. 15—*Strain life behavior with Samples A, B0, and A1 compared with data of Wild and Solomon.*

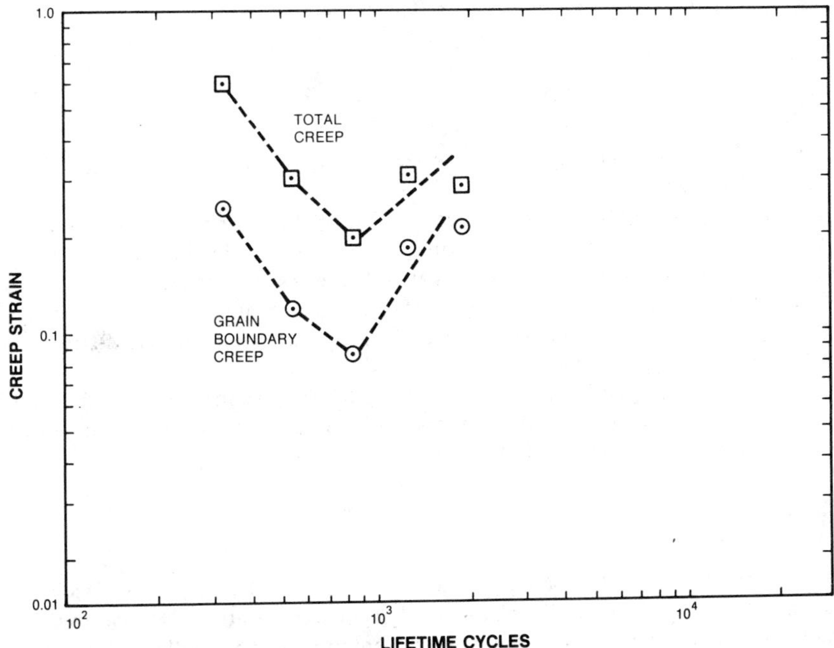

FIG. 16—*Lifetime versus total, grain boundary creep at 333 K (60°C), for Samples A, A1, A2, A3, and B0.*

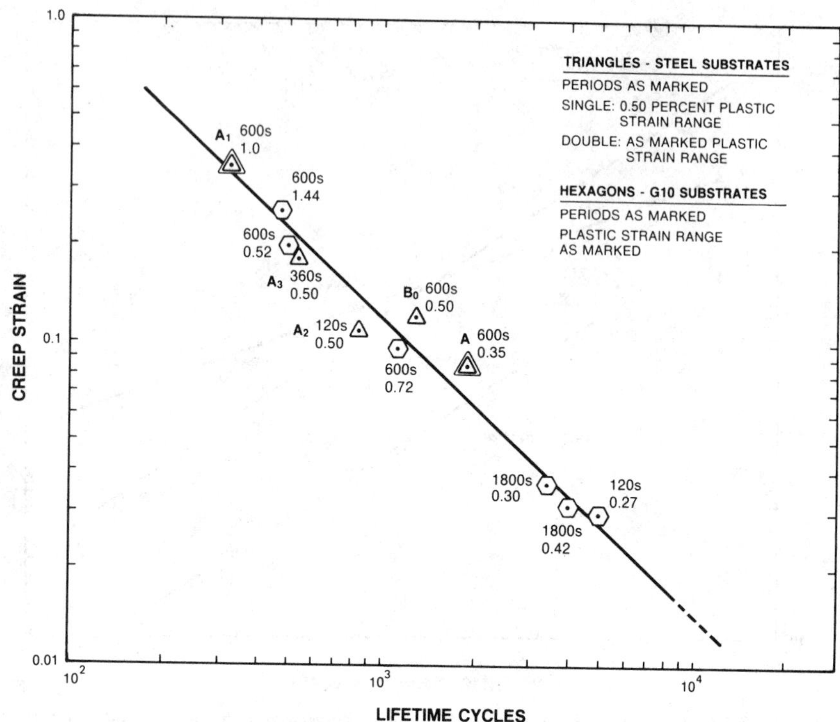

FIG. 17—*Lifetime versus matrix creep at 333 K (60°C), for Samples A, A1, A2, A3, and B0 (steel substrates); Samples A10, A11, A12, #1, #2, and #3 (G10 substrates).*

Conclusions

1. Solder joints undergoing isothermal or thermal fatigue undergo creep fatigue associated with steady-state creep.

2. The experimental creep-rate-versus-stress characteristic for the solder joint array shows two well-defined regions of creep: grain boundary creep, predominating at low stresses and creep rates, and matrix creep, predominating at high stresses and creep rates.

3. The accumulation of the two components of creep (grain boundary and matrix creep) has been calculated from the experimental hysteresis loop and the creep-rate-versus-stress characteristic for the solder joint array under a fatigue cycle.

4. The amount of creep or fraction of creep per cycle reaches a maximum at 333 K (60°C).

5. The lifetime was found to have an inverse linear relationship to the integrated matrix creep, with a constant of 125% creep ductility.

Acknowledgments

The authors wish to thank F. Boumil for experimental technical assistance, D. Waller for the program to calculate creep, D. Hallowell and R. Waple for preparation of the soldered devices, J. Sofia for assistance in the apparatus and instrumentation, and S. Laidlaw, G. Bartlett, and B. Larmouth for assistance in the metallurgical studies.

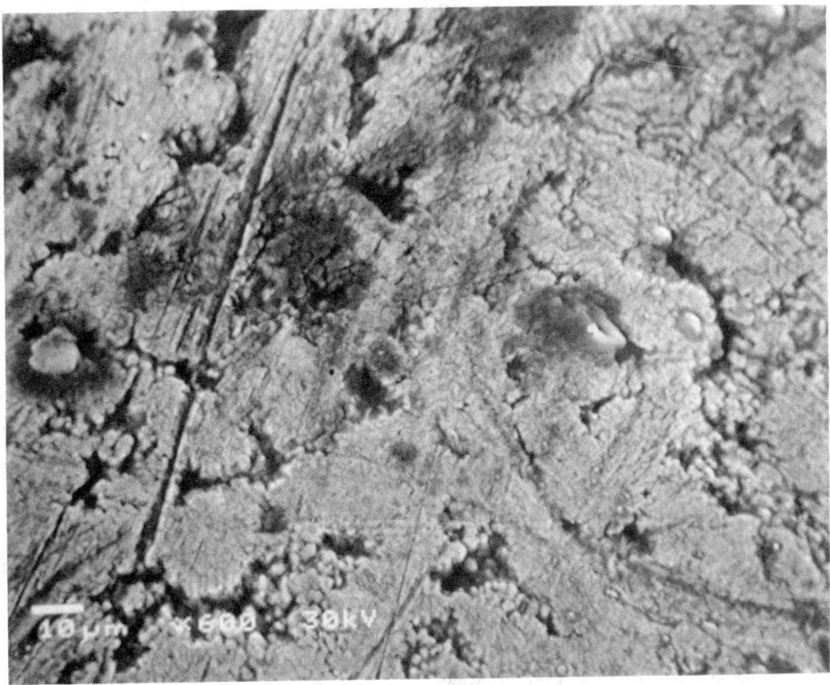

FIG. 18—*SEM of Sample D showing intercrystalline voids and cracks.*

References

[1] Wild, R. N., "Fatigue Properties of Solders," *Welding and Research,* No. II, Nov. 1972, pp. 521s-526s.

[2] Lake, J. K. and Wild, R. N., "Some Factors Affecting Leadless Chip Carrier Solder Joint Fatigue Life," in *Proceedings,* 28th National SAMPE Symposium, 12-14 April 1983.

[3] Hagge, J. K., "Predicting Fatigue Life of Leadless Chip Carriers Using Manson-Coffin Equations," *Proceedings of International Electronics Packaging Society,* 1982, pp. 199-208.

[4] Engelmaier, W., "Fatigue Life of Leadless Chip Carrier Solder Joints During Power Cycling," *IEEE Transactions on Components, Hybrids, and Manufacturing Technology,* Vol. 6, No. 3, Sept. 1983, pp. 232-237.

[5] Wild, R. N., "Some Fatigue Properties of Solders and Solder Joints," IBM Report 73z000421, RC14858, Owego, NY, 1973.

[6] Solomon, H. D., paper presented to China Lake Conference, 8th Annual Soldering Technology Seminar, Feb. 1984.

[7] Fox, L. R., Sofia, J. W., and Shine, M. C., "Investigation of Solder Fatigue Acceleration Factors," presented at CHMT Symposium, Tokyo, Oct. 1984, *IEEE Transactions on Components, Hybrids and Manufacturing Technology,* Vol. CHMT-8, No. 2, June 1985, pp. 275-286.

[8] Fox, L. R., Shine, M. C., and Sofia, J. N., "Strain Rate Loading Effects in Solder Fatigue," in *NEPCON West,* 1985, pp. 871-883.

[9] Shine, M. C., Fox, L. R., and Sofia, J. W., "A Strain Range Partitioning Procedure for Solder Fatigue," *Proceedings of the International Electronic Packaging Society,* Vol. 4, 1984, pp. 346-359.

[10] "Development of Highly Reliable Soldered Joints for Hybrid Circuit Boards," Westinghouse Report N69-25697, 1968.

[11] Stone, D. S., Homa, T. R., and Li, C.-Y., Kinetics of Cavity Growth in Solder Joints During Thermal Cycle," in *Proceedings*, Materials Research Society Meeting on Electronic Packaging, Vol. 40, E. A. Giess, Ed., Materials Research Society, Pittsburgh, 27-29 Nov. 1984, pp. 117-122.

[12] Stone, D., Hannula, S.-P., and Li, C.-Y., "The Effects of Service and Material Variables on the Fatigue Behavior of Solder Joints During the Thermal Cycle," in *Proceedings*, 35th Electronic Components Conference, Washington, D.C., 1985, pp. 46-51.

[13] Kashyap, B. P. and Murty, G. S., "Experimental Constitutive Relations for High Temperature Deformation of a Pb-Sn Eutectic Alloy," *Materials Science and Engineering*, Vol. 50, 1981, pp. 205-213.

[14] Padmanabhan, K. A. and Davies, G. J., *Superplasticity*, Springer-Verlag, Berlin, 1980, pp. 190-191.

[15] Manson, S. S., Halford, G. R., and Oldrieve, R. E., "Relation of Cyclic Loading Pattern to Microstructural Fracture in Creep Fatigue," NASA Technical Memorandum 83473, 1984.

[16] Manson, S. S., Halford, G. R., and Nachtigall, A. J., "Separation of Strain Components for Use in Strain Range Partitioning," in *Advances in Design for Elevated Temperature Environment*, ASME, New York, 1975, pp. 17-28.

[17] Halford, G. R., "High Temperature Fatigue in Metals—A Brief Review of Life Prediction Methods," *SAMPE Quarterly*, April 1983, pp. 17-25.

[18] Bensussan, P. L., Jablonski, D. A., and Pelloux, R. M., "A Study of Creep Crack Growth in 2219-T851 Aluminum Alloy Using a Computerized Testing System," *Metallurgical Transactions A*, Vol. 15a, Jan. 1984, pp. 107-120.

[19] Pelloux, R. M. and Huang, J. S. in *Creep-Fatigue Environment Interactions*, R. M. Pelloux and N. S. Stoloff, Eds., AIME, New York, 1980, pp. 151-64.

[20] Gabrielli, F. and Pelloux, R. M., "Effect of Environment on Fatigue and Crack Growth in Inconel X-750 at Elevated Temperature," *Metallurgical Transactions A*, Vol. 13a, June 1982, pp. 1083-1090.

[21] Baik, S. and Raj, R., "Wedge Type Creep Damage in Low Cycle Fatigue," *Metallurgical Transactions A*, Vol. 13a, July 1982, pp. 1207-1214.

[22] Baik, S. and Raj, R., "Mechanisms of Creep Fatigue Interaction," *Metallurgical Transactions A*, Vol. 13a, July 1982, pp. 1215-1221.

[23] Bassani, J. L. in *Creep and Fracture of Engineering Materials and Structures*, B. Wilshire and D. R. Owen, Eds., Pineridge Press, Swansea, England, 1981, pp. 329-44.

[24] Sadananda, K. and Shahinian, P., "Creep Crack Growth Behavior and Theoretical Modeling," *Metal Science*, Vol. 15, Oct. 1981, pp. 425-432.

[25] Vitek, V. and Takasugi, T., "Mechanisms of Creep Crack Growth," in *Micro and Macro Mechanics of Crack Growth*, K. Sadananda, B. B. Ruth, and D. J. Michel, Eds., AIME, New York, 1982, pp. 107-118.

[26] Huang, J. S., SC.D. thesis, Department of Materials Science and Engineering, Massachusetts Institute of Technology, Cambridge, Feb. 1981.

[27] Raj, R. and Ashby, M. F., "Intergranular Fracture at Elevated Temperature," *Acta Metallurgica*, Vol. 23, June 1975, pp. 653-666.

[28] Nix, W. D., Matlock, D. K., and Dimelf, R. J., "A Model for Creep Fracture Based on the Plastic Growth of Cavities at Tips of Grain Boundary Wedge Cracks," *Acta Metallurgica*, Vol. 25, 1977, pp. 495-503.

[29] Needleman, A. and Rice, J. R., "Plastic Creep Flow Effects in the Diffusing Cavitation of Grain Boundaries," *Acta Metallurgica*, Vol. 28, 1980, pp. 1315-1332.

[30] Dimelf, R. J. and Nix, W. D., "The Stress Dependence of Crack Growth Rate During Creep," *International Journal of Fracture*, Vol. 13, No. 103, June 1977, pp. 341-349.

[31] Tvergaard, W., "Constitutive Relations for Creep in Polycrystals with Grain Boundary Cavitation," *Acta Metallurgica*, Vol. 32, No. 11, 1984, pp. 1977-1990.

[32] Miller, D. A. and Pilkington, R., "Diffusion and Deformation Controlled Creep Crack Growth," *Metallurgical Transactions A*, Vol. 11a, Jan. 1980, pp. 177-180.

[33] Edward, G. H. and Ashby, M. F., "Intergranular Fracture During Power-Law Creep," *Acta Metallurgica*, Vol. 27, 1979, pp. 1505-1518.

[34] Halford, G. R. and Manson, S. S., "Life Prediction of Thermo-Mechanical Fatigue Using Strain-Range Partitioning," in *Thermal Fatigue of Materials and Components, ASTM STP 612*, American Society for Testing and Materials, Philadelphia, 1976, pp. 239-254.

[35] Shine, M. C., private communication, Feb. 1985.

[36] Solomon, H. D., "Low-Frequency, High-Temperature Low Cycle Fatigue of 60Sn-40Pb Solder," this publication, pp. 342-370.

[37] Shine, M. C., private communication, May 1985.

[38] Rack, H. J. and Maurin, J. K., "Mechanical Properties of Cast Tin-Lead Solder," *Journal of Testing and Evaluation*, Vol. 2, No. 5, 1974, pp. 351-353.

Manabu Tanaka[1] *and Hiroshi Iizuka*[1]

A Micromechanics Model of the Initiation of Grain-Boundary Crack in High-Temperature Fatigue

REFERENCE: Tanaka, M. and Iizuka, H., "**A Micromechanics Model of the Initiation of Grain-Boundary Crack in High-Temperature Fatigue,**" *Low Cycle Fatigue, ASTM STP 942,* H. D. Solomon, G. R. Halford, L. R. Kaisand, and B. N. Leis, Eds., American Society for Testing and Materials, Philadelphia, 1988, pp. 611-621.

ABSTRACT: A continuum mechanics model which incorporated the recovery effect by diffusion of atoms is presented to explain the initiation of grain-boundary wedge-type crack in high-temperature fatigue. The direction of deformation is periodically changed in fatigue, and the crack initiation depends on the deformation history of the grain boundary. A theoretical calculation based on the present model was made on the grain-boundary crack initiation in high-temperature fatigue where the deformation is path-dependent. The calculation results satisfactorily explained the experimental observations in polycrystalline metallic materials.

KEY WORDS: fatigue, high temperature, micromechanics model, grain boundary, wedge-type crack, crack initiation, stress concentration, grain-boundary sliding, polycrystalline metallic materials

Typical grain-boundary cracks often initiate at the grain-boundary triple junction where high-stress concentration is built up by the blocking of grain-boundary sliding. McLean [*1*] obtained the critical stress necessary for wedge-type cracking at the triple junction in high-temperature creep. The grain-boundary sliding also sets up the stress concentration that leads to cavity nucleation at second-phase particles [*2*] and at various irregularities [*3*] along the grain boundaries. The diffusion of atoms as well as the plastic accommodation results in a decrease of the stress concentration at the grain boundaries at high temperatures. Mori et al [*4*] discussed the blocking of grain-boundary sliding by second-phase particles using a continuum mechanics model in which the recovery effect by diffusion of atoms was taken into account. However, a theoretical analysis has not fully been made on the initiation of grain-corner cracks in polycrystalline metallic materials in high-temperature fatigue.

In this study, a continuum mechanics model which incorporated the recovery effect by diffusion of atoms has been developed to explain crack initiation at the grain-boundary triple junction in high-temperature fatigue. The basic equation was derived from the present model. The direction of deformation is generally periodically changed after each half-cycle in fatigue, and the crack initiation depends on the deformation history of the grain boundary. A theoretical calculation was made on the initiation of grain-boundary wedge-type crack in high-temperature fatigue. The result of the calculation was then compared with the experimental observations in polycrystalline metallic materials.

[1]Associate Professor and Research Associate, respectively, Department of Mechanical Engineering for Production, Mining College of Akita University, Akita 010, Japan.

A Model for Crack Initiation

Calculation of Elastic Strain Energy and Internal Stress of Grain Boundary

The internal stress state when grain-boundary sliding is blocked at the grain corner can be calculated by Eshelby's inclusion method [5], if a grain boundary is approximated by a flat ellipsoidal inclusion [6]. As shown in Fig. 1*a*, a grain boundary is assumed here to be a two-dimensional flat inclusion, D_0, with length $2L$ and thickness b (the magnitude of Burgers vector):

$$x_1^2/L^2 + x_2^2/(b/2)^2 = 1 \qquad (b/2L \ll 1) \tag{1}$$

We consider that plastic shear strain on the grain boundary, ϵ_{21}^* $(=\epsilon^*)$ occurs as a result of grain-boundary sliding under an external tensile stress, σ^A. If the sliding is blocked at the grain corner, the internal stress, τ_{in} with eigen strain $-\epsilon_{21}^*$ $(= -\epsilon^*)$ occurs in the domain D_0. Figure 2 shows the internal stress on the grain boundary caused by the blocking of grain-boundary sliding. The internal stress, τ_{in}, is given by [7]

$$\tau_{in} = -2A(b/2L)\mu\epsilon_{21}^* = -A(b/L)\mu\epsilon^* \tag{2}$$

where μ is the rigidity, ν is Poisson's ratio of the material, and $A = 1/(1 - \nu)$. The elastic strain energy arising from the blocking of grain-boundary sliding, E_{el}, is

$$E_{el} = A(b/L)\mu\epsilon^{*2}V \tag{3}$$

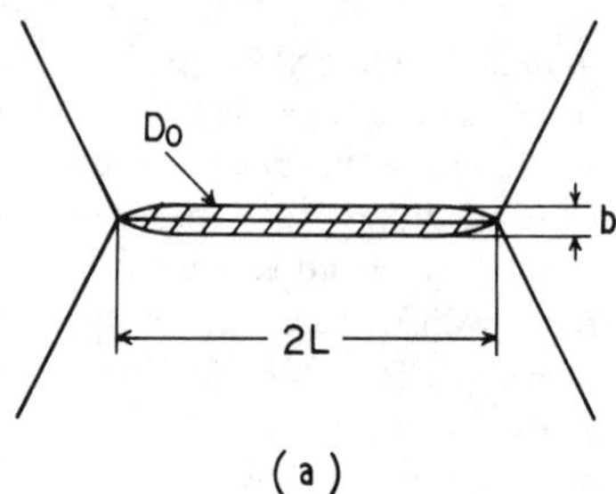

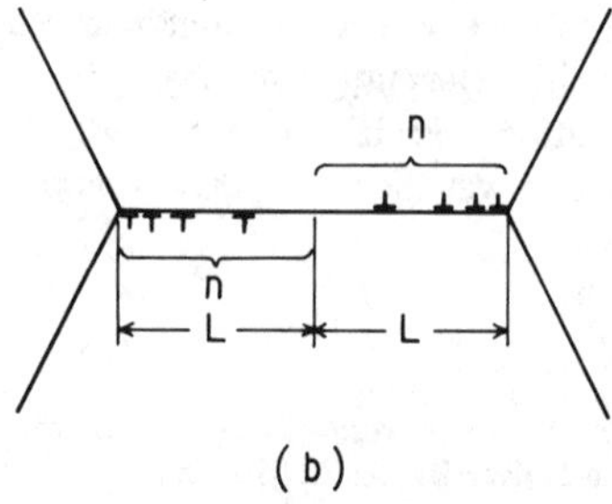

FIG. 1—*Grain boundary approximated by a two-dimensional flat inclusion,* D_0.

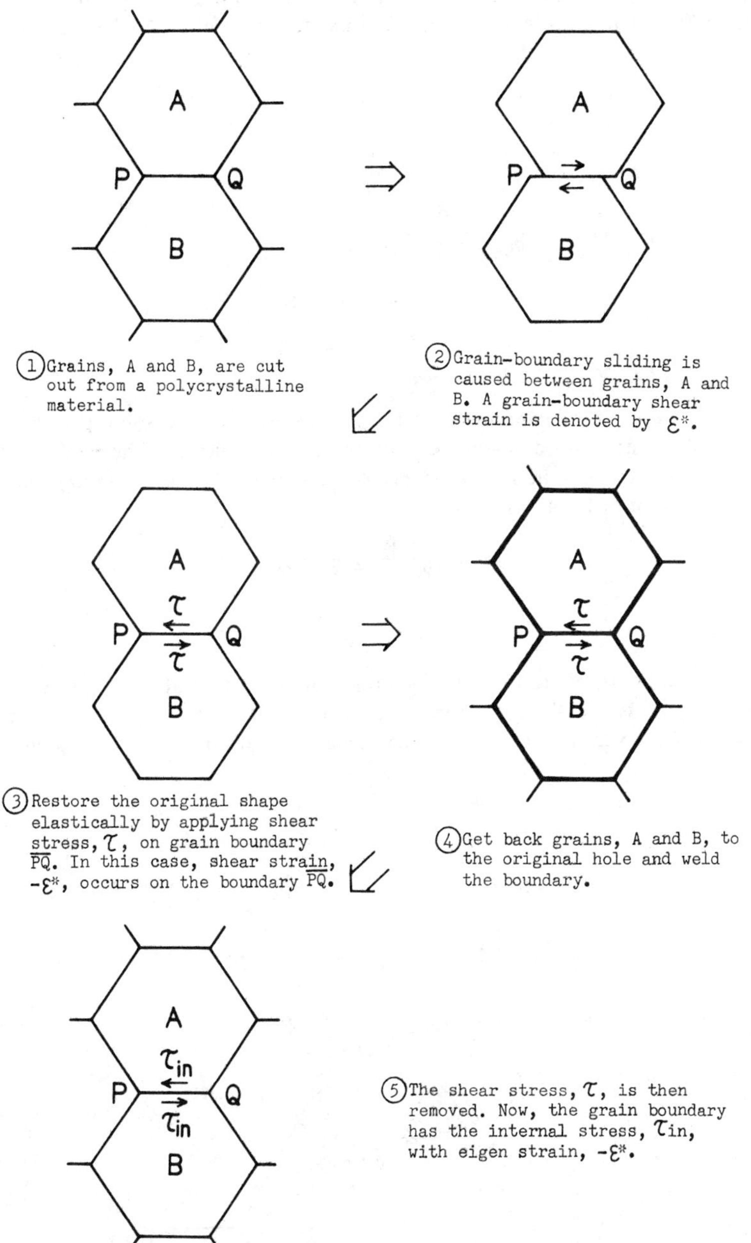

FIG. 2—*Schematic illustration of the internal stress,* τ_{in}, *on the grain boundary,* D_0, *caused by blocking of grain-boundary sliding.*

where the volume of the inclusion, V, is $\pi Lb/2$ per unit thickness of the material. The shear strain on the grain boundary, ϵ^*, can be replaced by the conjugate slip of n dislocations with the same Burger's vector, $\vec{b}$, on a half of the grain-boundary length, L (Fig. 1*b*). Since the total displacement by dislocations is nb on the grain-boundary length, $2L$, the local strain, ϵ^*, is defined by

$$\epsilon^* = \frac{\frac{nb}{2}}{b} = \frac{n}{2} \tag{4}$$

Substituting Eq 4 for Eq 3, E_{el} can be rewritten as

$$E_{el} = \pi A \mu n^2 b^2/8 \tag{5}$$

Internal Stress and Eigen Strain in Inclusion D_0

The total number of atoms contained in extra-half planes of n dislocations per unit thickness, N, is expressed by the grain diameter, D, and the atomic volume of the material, Ω. In the absence of recovery, the total number of atoms, N_0, is given by the corresponding plastic strain, ϵ, and the number of dislocations, n_0:

$$N = nbD/\Omega = 2\epsilon^* bD/\Omega \tag{6}$$

$$N_0 = 2\epsilon bD/\Omega \tag{7}$$

Figure 3 shows the deficit and excess of volume caused as a result of the blocking of grain-boundary sliding. N is also the net excess of atoms which should be transported. Emission of atoms, dN, from the side of excessive volume occurs when dislocations climb along the grain

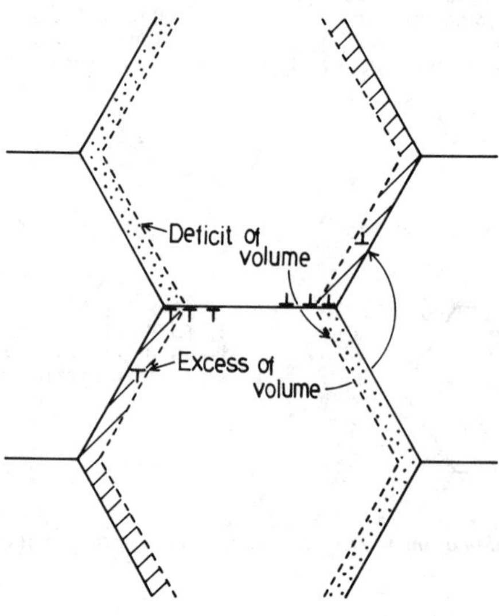

FIG. 3—*Deficit and excess of volume caused by blocking of grain-boundary sliding.*

boundary in the recovery process. Those atoms are transported by diffusion along the grain boundary or through the grain to the deficit side of volume. The difference in chemical potential between the emission and the absorption sides of atoms, μ^*, is given by [8]

$$\mu^* = \frac{dE_{el}}{dN} = \frac{dn}{dN}\cdot\frac{dE_{el}}{dn} = \frac{\pi A \mu n b \Omega}{4D} \tag{8}$$

When the recovery is controlled by the grain-boundary diffusion, the flux of atoms, J_{GB}, is expressed by the following equation, provided that the average migration distance of atoms is $4L$.

$$J_{GB} = \frac{D_{GB}}{\Omega kT}\,\text{grad}\,\mu^* = \frac{\pi A \mu \Omega D_{GB} N}{16LD^2kT} \tag{9}$$

where D_{GB} is the grain-boundary diffusion coefficient, k is Boltzmann's constant, and T is the absolute temperature. The thickness of the cross section of grain-boundary diffusion is δ ($\approx 2b$).

Since the emission of atoms occurs from a pair of the extra-half planes of dislocations, the total migration rate of atoms due to the grain-boundary diffusion, $(dN/dt)_{GB}$, is given by

$$\left(\frac{dN}{dt}\right)_{GB} \doteqdot 2\delta J_{GB} = \frac{\pi\sqrt{3}A\mu D_{GB}\Omega N\delta}{4D^3kT} \tag{10}$$

where L is approximated by $D/2\sqrt{3}$. If the climb of grain-boundary dislocation is controlled by the volume diffusion, the cross section of diffusion path is $2D$ per unit thickness and the diffusion distance is approximated to be D. Therefore, the diffusion flux, J_v, is given by

$$J_v = \frac{D_v}{\Omega kT}\,\text{grad}\,\mu^* = \frac{\pi A \mu D_v \Omega N}{4D^3kT} \tag{11}$$

where D_v is the volume diffusion coefficient. The migration rate, $(dN/dt)_v$, in this case is expressed by

$$\left(\frac{dN}{dt}\right)_v \doteqdot 2DJ_v = \frac{\pi A \mu D_v \Omega N}{2D^2kT} \tag{12}$$

From Eqs 10 and 12:

$$\left(\frac{dN}{dt}\right)_v \Big/ \left(\frac{dN}{dt}\right)_{GB} = \frac{2D_vD}{\sqrt{3}D_{GB}\delta} < 1 \tag{13}$$

if the grain-boundary diffusion controls the recovery process.

In the recovery process controlled by the grain-boundary diffusion, the net migration rate, dN/dt, is given by

$$dN/dt = dN_0/dt - (dN/dt)_{GB} \tag{14}$$

Substituting Eqs 6 and 7 into Eq 14, and using Eq 10, the following equation of primary reaction is obtained:

$$\frac{d\epsilon^*}{dt} = \frac{d\epsilon}{dt} - \frac{\pi\sqrt{3}A\mu D_{GB}\Omega\delta}{4D^3kT}\epsilon^*$$
$$= \dot{\epsilon} - C\epsilon^* \tag{15}$$

where $\dot{\epsilon} = d\epsilon/dt$ and $C = \pi\sqrt{3}A\mu D_{GB}\Omega\delta/(4D^3kT)$. If Eq 15 is solved under the initial condition of $\epsilon^* = 0$ for $t = 0$ when $\dot{\epsilon}$ is constant, then

$$\epsilon^* = (\dot{\epsilon}/C)\{1 - \exp(-Ct)\} \tag{16}$$

The internal stress, τ_{in}, is then given by

$$\tau_{in} = -A(b/L)\mu(\dot{\epsilon}/C)\{1 - \exp(-Ct)\} \tag{17}$$

If the volume diffusion controls the dislocation climb, the value of C in the above equations should be replaced by $\pi A\mu D_v\Omega/(2D^2kT)$.

Critical Strain, ϵ_c^*, *for Crack Initiation*

The stress field outside the inclusion D_0 can be easily obtained by Mura's alternative method [9]. We consider an ellipsoidal notch which has the same shape and size as the inclusion D_0. This notch can be regarded as a crack, because the inclusion D_0 is an extremely flat ellipsoid ($b \ll 2L$) and $\rho \cong 0$ (ρ is the radius of curvature at notch root). According to the calculations of Stroh [10], the stress distribution around a crack in an infinite body under a uniformly applied stress, τ_{out}, can also be obtained without difficulty by putting $\tau_{out} = -\tau_{in}$. The stress outside the flat inclusion D_0 can be obtained by superposing Stroh's solution of a crack and a uniformly applied stress field, τ_{in}, as shown in Fig. 4.

Stroh [10] also calculated the normal stress acting on the plane making an angle θ with the slip plane. This stress has a maximum when θ is 70.5°. The maximum value of the normal stress is $(2/\sqrt{3})(\sqrt{L/r})\,\tau_{out}$ at the distance r from the grain corner. For the initiation of the grain-boundary wedge-type crack, McLean [1] utilized Stroh's equation

$$\tau_{out} > \sqrt{\frac{12\gamma\mu}{\pi(1-\nu)L}} \tag{18}$$

where γ is the effective surface energy per unit area. The grain-boundary sliding in two directions (Fig. 5) is considered in this study. Wedge-type cracking can occur on the grain boundary plane $\overline{OA}$. Therefore, from Eq 2, we obtain

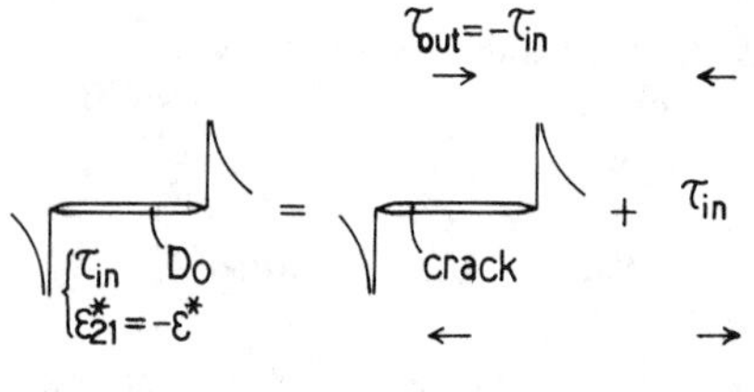

FIG. 4—*Stress field outside the inclusion (grain boundary),* D_0.

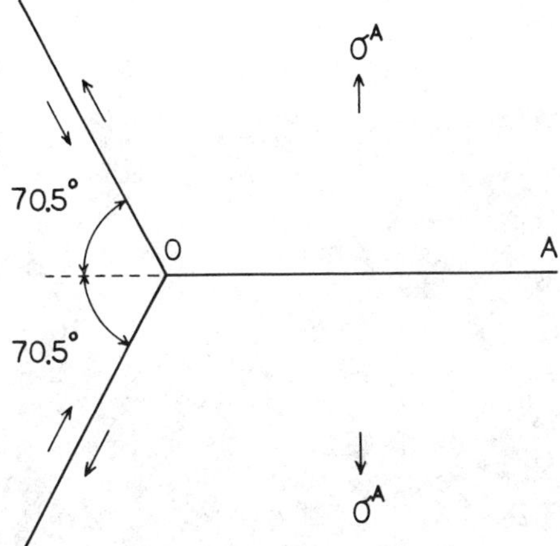

FIG. 5—*Grain-boundary sliding model considered in this study.*

$$2\tau_{out} = -2\tau_{in} = \frac{2}{1-\nu}\left(\frac{b}{L}\right)\mu\epsilon^* > \sqrt{\frac{12\gamma\mu}{\pi(1-\nu)L}} \tag{19}$$

The critical shear strain for the crack initiation is given by

$$\epsilon^* > \sqrt{\frac{3\gamma(1-\nu)L}{\pi\mu b^2}} = \epsilon_c^* \tag{20}$$

The interaction between two sliding boundaries are neglected here.

Grain-Boundary Crack Initiation in High-Temperature Fatigue

The continuum mechanics model in this study was applied to the understanding of crack initiation in high-temperature fatigue where the plastic deformation is path-dependent. The direction of deformation is periodically changed after each half-cycle in fatigue. In this case, the unidirectional accumulation of the residual shear strain on grain boundary causes the stress concentration at the grain corner due to the blocking of grain-boundary sliding, and consequently leads to the initiation of grain-boundary wedge-type crack. Figure 6 shows a typical wedge-type crack observed in high-temperature fatigue of SUS304 type steel. For simplicity, it is assumed in the present calculation that the shear strain rate on the grain boundary is constant in each half-cycle of fatigue loading, and that it is $\dot{\epsilon}_{gt}$ (>0) during the tensile-half period, $2t_t$, and $\dot{\epsilon}_{gc}$ (<0) during the compressive-half period, $2t_c$, as shown in Fig. 7. The duration of one cycle, $\overline{T}$, is equal to $2(t_t + t_c)$.

The eigen strain on the grain boundary, at ith peak strain, ϵ_i^*, is given by the following equation, if Eq 15 is solved under an initial condition of $\epsilon^* = \epsilon_0^* = 0$ for $t = 0$. Setting $\dot{\epsilon} = \dot{\epsilon}_{gt}$, one can obtain ϵ_1^* at the first peak ($t = t_t$), which is expressed as

$$\epsilon_1^* = (\dot{\epsilon}_{gt}/C)\{1 - \exp(-Ct_t)\} \tag{21a}$$

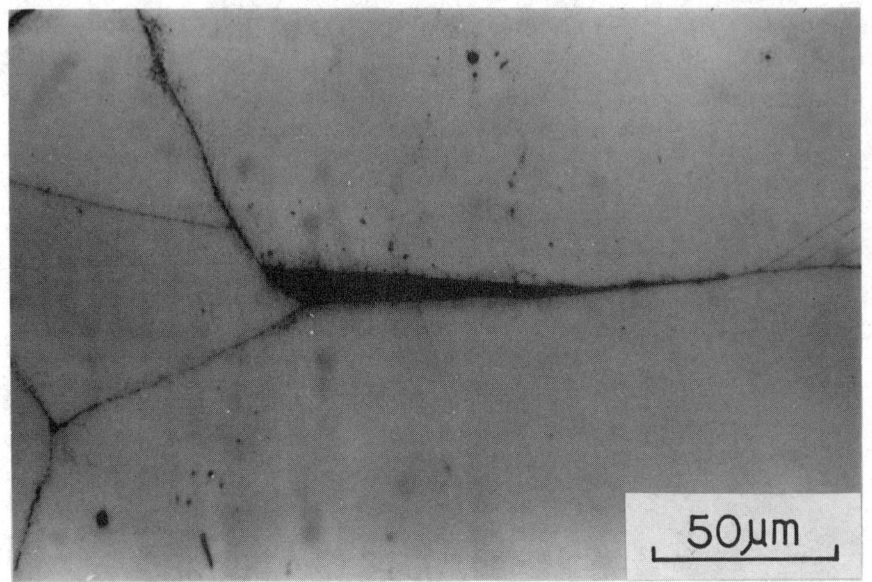

FIG. 6—*An example of wedge-type crack observed in fatigue test at 973 K of SUS304 steel (D = 400 μm) (total strain range = 1.2%; tensile strain rate = 1 × 10⁻⁵ s⁻¹; compressive strain rate = 1 × 10⁻² s⁻¹; fatigue life = 44 cycles).*

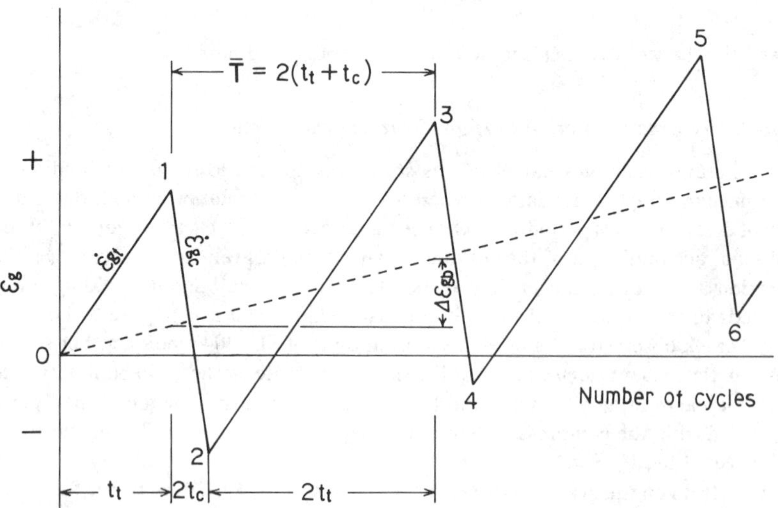

FIG. 7—*Example of change in the grain-boundary shear strain, ϵ_g, in fatigue loading.*

$\dot{\epsilon}_2^*$ at the second peak ($t = t_t + 2t_c$) is calculated by putting $\dot{\epsilon} = \dot{\epsilon}_{gc}$ and $\epsilon^* = \epsilon_1^*$ for $t = t_t$:

$$\epsilon_2^* = \frac{\dot{\epsilon}_{gc}}{C} - \left\{\left(\frac{\dot{\epsilon}_{gc} - \dot{\epsilon}_{gt}}{C}\right) \exp(Ct_t) + \frac{\dot{\epsilon}_{gt}}{C}\right\} \times \exp\{-C(t_t + 2t_c)\} \qquad (21b)$$

ϵ_3^* at the third peak ($t = 3t_t + 2t_c$, the second positive peak) is also given by

$$\epsilon_3^* = \frac{\dot{\epsilon}_{gt}}{C} - \left[\left(\frac{\dot{\epsilon}_{gt} - \dot{\epsilon}_{gc}}{C}\right) \exp\{C(t_t + 2t_c)\} + \left(\frac{\dot{\epsilon}_{gc} - \dot{\epsilon}_{gt}}{C}\right) \times \exp(Ct_t) + \frac{\dot{\epsilon}_{gt}}{C}\right] \exp\{-C(3t_t + 2t_c)\} \qquad (21c)$$

ϵ_{2m+1}^* at the $(m + 1)$th positive peak ($m \geqq 1$) is expressed as

$$\dot{\epsilon}_{2m+1}^* = \frac{\dot{\epsilon}_{gt}}{C} [1 - \exp\{-C[2m(t_t + t_c) + t_t]\}] - \frac{\left(\frac{\dot{\epsilon}_{gt} - \dot{\epsilon}_{gc}}{C}\right) \{1 - \exp[-2mC(t_t + t_c)]\} \{1 - \exp(2Ct_c)\}}{1 - \exp\{2C(t_t + t_c)\}} \qquad (22)$$

If the dynamic recovery is not significant ($mC\overline{T} \ll 1$), ϵ_{2m+1}^* is simply given by

$$\begin{aligned} \epsilon_{2m+1}^* &= 2m(\dot{\epsilon}_{gt} \cdot t_t + \dot{\epsilon}_{gc} \cdot t_c) + \dot{\epsilon}_{gt} \cdot t_t \\ &= m\Delta\epsilon_{gb} + \dot{\epsilon}_{gt} \cdot t_t \end{aligned} \qquad (23)$$

where $\Delta\epsilon_{gb} = 2(\dot{\epsilon}_{gt} \cdot t_t + \dot{\epsilon}_{gc} \cdot t_c)$ is the residual shear strain after each cycle. It was found in a stainless steel [*11*], pure copper, and zirconium [*12*] that the amount of accumulated grain-boundary sliding after each cycle, Δu_{gb} ($\simeq 2b\Delta\epsilon_{gb}$), was approximately constant in fatigue tests. If the absolute value of the first term in Eq 23 is very large compared with the second one, the number of cycles to crack initiation, N_i, is expressed as

$$N_i \cong \frac{\epsilon_c^*}{|2(\dot{\epsilon}_{gt} \cdot t_t + \dot{\epsilon}_{gc} \cdot t_c)|} = \frac{\epsilon_c^*}{|\Delta\epsilon_{gb}|} \qquad (24)$$

where $\epsilon_c^* = \sqrt{3\gamma(1 - \nu)L/(\pi\mu b^2)}$ (Eq 20). A similar relationship was found between the sliding rate and the time to fracture in high-temperature fatigue [*12*].

Discussion

Table 1 shows the number of cycles to crack initiation calculated by Eq 24, N_i, and the fatigue life, N_f, in the strain-controlled fatigue test on SUS304 type steel (D = 600 μm) at 973 K obtained by Taira et al [*11*]. It is considered in this case that the volume diffusion of atoms may control the dynamic recovery, but it is not significant, because $mC\overline{T} \ll 1$. The numerical values used in the calculation are $\mu = 5.586 \times 10^4$ MPa (973 K) [*13*], $\nu = 0.29$ (973 K) [*13*], and $2\gamma = 2\gamma_s - \gamma_{GB}$, where γ_s (1.95 J/m^2) is the surface energy and γ_{GB} (0.70 J/m^2) is the grain-boundary energy of γ-Fe [*14*]. The magnitude of Burger's vector, b, is assumed to be 2.55×10^{-10} m. The

TABLE 1—*Calculated value of number of cycles to crack initiation,* N_i, *and fatigue life,* N_f, *in the strain-controlled fatigue test on SUS304 steel (D = 600 μm) at 973 K.*[a]

Waveshape	$\bar{\dot{\epsilon}}_t/\bar{\dot{\epsilon}}_c$	N_f (cycles)	Δu_{gb} (m/cycle)	N_i (cycles)
Slow-fast	1/19	103	4.5×10^{-9}	26
Fast-slow	19	330	2.5×10^{-9}	46
Balanced	1	360	4.2×10^{-10}	280

[a] $T = 5$ min, $\bar{\dot{\epsilon}}_t$ = tensile strain rate, $\bar{\dot{\epsilon}}_c$ = compressive strain rate, Δu_{gb} = amount of grain-boundary sliding per cycle ($\simeq 2b\Delta\epsilon_{gb}$), and total strain range = $\Delta\epsilon_t = 1.0\%$.

grain-boundary crack is expected to initiate in the early stage of fatigue in unbalanced slow-fast or fast-slow wave-shape, although the strain cycling is completely reversible. A relatively small amount of grain-boundary sliding can accumulate even under a balanced and completely reversible strain cycling. The value of N_i is comparable with the actual fatigue life, N_f, in this case. Thus the calculation results in this study satisfactorily explain the experimental observations [*15*]. The amount of $\Delta\epsilon_{gb}$ usually depends on the test conditions, including waveshape, strain or stress amplitude, strain rate, temperature, and environment, as well as microstructural factors such as grain size, presence of the second-phase particles, and so on [*12,15*]. But, if $|\Delta\epsilon_{gb}|$ is proportional to the inelastic strain range in fatigue, $\Delta\epsilon_{in}$, when the grain-boundary wedge-type crack occurs, N_i is expected to be almost inversely proportional to $\Delta\epsilon_{in}$, so that $N_i \propto 1/\Delta\epsilon_{in}$, as is known from Eq 24. Similar relations were observed on the inelastic strain range against fatigue-life properties in heat-resisting alloys [*15–17*], which were obtained by strain-range partitioning method. This may suggest that the growth of the grain-boundary wedge-type crack as well as crack initiation is controlled by grain-boundary sliding in high-temperature fatigue [*11,18*].

Min and Raj [*18*] pointed out that cavitation failure is another possible fracture mechanism in high-temperature fatigue. Majumdar and Maiya [*19*] observed this type of failure in Type 304 steel at 866 K. In cavitation failure, the cavities nucleated on the grain boundaries as a result of grain-boundary sliding linked by grain deformation that is distributed rather uniformly in the entire specimen, while both crack nucleation and growth occur by the localized sliding displacement of the grain boundary in the triple junction fracture [*18*]. A fairly large amount of unidirectional sliding displacement of the grain boundary is accumulated in high-temperature low-cycle fatigue, specifically for unbalanced slow-fast and fast-slow strain cycling [*11*], and the grain-boundary wedge-type crack occurs by the concentration of stress at the grain-boundary triple junctions. The authors' model of the wedge-type cracking can be applied to the prediction of time to crack initiation in this case, if the amount of grain-boundary sliding in each fatigue cycle is properly estimated. The present model, however, is not valid under conditions where the stress concentration at the grain-boundary triple junction is too low to cause wedge-type cracking because of the small amount of grain-boundary sliding at lower temperatures, or where the stress concentration at the triple junction is greatly reduced by the diffusional recovery (or the localized plastic deformation) in materials with small grain size at very high temperatures.

Conclusions

A theoretical discussion was presented on the initiation of the grain-boundary wedge-type crack in high-temperature fatigue, using a continuum mechanics model that incorporated the recovery effect by diffusion of atoms. The periodic change in the direction of deformation occurs after each half-cycle in fatigue. Therefore, the crack initiation process depends on the deformation history of the grain boundary. The basic equation was derived from consideration of the dynamic recovery process at the grain boundary. The number of cycles to crack initiation was then calculated. The calculation results based on the present model (Eq 24) satisfactorily explain the experimental observations of high-temperature fatigue in polycrystalline metallic materials.

References

[*1*] McLean, D., *Journal of the Institute of Metals,* Vol. 85, 1956–57, p. 468.

[*2*] Gifkins, R. C., *Acta Metallurgica,* Vol. 4, 1956, p. 98.

[*3*] Harris, J. E., *Transactions of the Metallurgical Society of AIME,* Vol. 233, 1965, p. 1509.

[*4*] Mori, T., Koda, M., Monzen, R., and Mura, T., *Acta Metallurgica,* Vol. 31, 1983, p. 275.

[*5*] Eshelby, J. D., *Proceedings of Royal Society of London,* Series A, Vol. 241, 1957, p. 376.

[*6*] Sakaki, T., *Scripta Metallurgica,* Vol. 8, 1974, p. 189.

[*7*] Mori, T., *Bulletin of the Japan Institute of Metals,* Vol. 17, No. 11, 1978, p. 920.

[*8*] Tanaka, M. and Iizuka, H., *Journal of Materials Science,* Vol. 19, No. 12, 1984, p. 3976.

[*9*] Mura, T., *Micromechanics of Defects in Solids,* Martinus Nijhoff Publishers, Hague, 1982, p. 74.

[*10*] Stroh, A. N., *Proceedings of Royal Society of London,* Series A, Vol. 223, 1954, p. 404.

[*11*] Taira, S., Fujino, M., and Yoshida, M., *Journal of the Society of Materials Science,* Japan, Vol. 27, 1978, p. 447.

[*12*] Snowden, K. U., in *Cavities and Cracks in Creep and Fracture,* J. Gittus, Ed., Applied Science Publishers, London, 1981, p. 259.

[*13*] Muramatsu, A., *Journal of the Japan Society of Mechanical Engineers,* Vol. 68, 1965, p. 1623.

[*14*] Abe, H., *Recrystallization,* Kyoritsu, Tokyo, 1969, p. 162.

[*15*] Ohtani, R., *Bulletin of the Japan Institute of Metals,* Vol. 22, 1983, p. 190.

[*16*] Manson, S. S. in *Fatigue at Elevated Temperatures, ASTM STP 520,* American Society for Testing and Materials, Philadelphia, 1973, p. 744.

[*17*] Taira, S. and Ohtani, R., *High-Temperature Strength of Materials,* Ohm, Tokyo, 1980, p. 42.

[*18*] Min, B. K. and Raj, R. in *Fatigue Mechanisms, ASTM STP 675,* American Society for Testing and Materials, Philadelphia, 1979, p. 569.

[*19*] Majumdar, S. and Maiya, P. S., *Canadian Metallurgical Quarterly,* Vol. 18, No. 1, 1979, p. 57.

Thermal and Thermomechanical Low Cycle Fatigue

Gary R. Halford,[1] *Michael A. McGaw,*[1] *Robert C. Bill,*[1] *and Paolo D. Fanti*[1]

Bithermal Fatigue: A Link Between Isothermal and Thermomechanical Fatigue

REFERENCE: Halford, G. R., McGaw, M. A., Bill, R. C., and Fanti, P. D., **"Bithermal Fatigue: A Link Between Isothermal and Thermomechanical Fatigue,"** *Low Cycle Fatigue, ASTM STP 942,* H. D. Solomon, G. R. Halford, L. R. Kaisand, and B. N. Leis, Eds., American Society for Testing and Materials, Philadelphia, 1988, pp. 625-637.

ABSTRACT: Many technologically important elevated temperature service cycles are non-isothermal. Nevertheless, major design codes rely on the most severe—usually the highest—temperature of an operational cycle as being the pertinent temperature upon which to base a design. Consequently, most high-temperature fatigue data for design have been generated under isothermal conditions. There is a growing awareness of the potential inadequacy of such a simplistic approach since many thermomechanical fatigue results have been found to exhibit considerably lower fatigue lives than would be expected on the basis of isothermal results at the maximum cycle temperature. Yet, variable-temperature, low-cycle fatigue tests are difficult to conduct and to interpret. The considerable gap between isothermal and thermomechanical fatigue technology can be bridged by an approach which retains the simplicity and ease of interpretation of isothermal fatigue, but captures many of the first order effects of the greater complexities involved in thermomechanical fatigue. We have developed a procedure for conducting what has been designated as bithermal fatigue experiments. In this procedure, the tensile and compressive halves of the cycle are conducted isothermally at two significantly different temperatures. The higher temperature is chosen to be in the time-dependent creep and oxidation prone regime and the lower temperature in the regime wherein time dependencies are minimized due to lack of thermal activation.

Interestingly, bithermal fatigue tests prior to those performed for this paper have been conducted in conjunction with the evaluation of the isothermal Strainrange Partitioning characteristics of high-temperature alloys, not with thermomechanical behavior *per se.* Nevertheless, the bithermal fatigue test may well be used as an alternative to thermomechanical cycling. In this paper, we place emphasis on using the bithermal testing concept as a link between isothermal and thermomechanical testing. New bithermal fatigue data for the nickel-base superalloy B1900 + Hf are presented herein.

KEY WORDS: fatigue (metal), low-cycle fatigue, thermomechanical fatigue, creep-fatigue, Strainrange Partitioning method, testing techniques, nickel-base alloys

Major design codes [*1*] rely on maximum temperature of an operational cycle as being representative of the most severe of all possible temperatures within the cycle. Consequently, designs are frequently based upon isothermal, low-cycle fatigue data generated at the maximum temperature of the expected service cycle. The main body of high-temperature, low-cycle, creep-fatigue design data has been generated under these isothermal conditions.

There is a growing awareness of the potential inadequacy of such a simplistic approach since thermomechanical fatigue results have been found, in many cases, depending on the materials

[1]NASA-Lewis Research Center, Cleveland, OH 44135. Dr. Fanti is currently with the European Patent Office, Munich, West Germany.

and conditions, to exhibit considerably lower fatigue lives than would be expected on the basis of isothermal results at the maximum cycle temperature. A recent example was reported by Bill, et al. [2] for the nickel-base superalloy MAR M-200. For this alloy, in-phase thermomechanical cycling (495 to 1000°C) reduced the cyclic lives well below isothermal creep-fatigue lives obtained at temperatures of 650, 927, and 1000°C. Unfortunately, variable-temperature, low-cycle fatigue experiments are not easy to conduct or interpret. Furthermore, there are many more variables in a thermomechanical cycle than in an isothermal one, and hence a thermomechanical data base for design would have to be much more extensive than current data bases.

Compared to isothermal cycling, thermomechanical fatigue cycling is expected to introduce a multitude of cyclic deformation and damage mechanisms. For example, during the higher temperature portion of a thermomechanical cycle, various time-dependent creep mechanisms may be operative, along with aggressive environmental attack and metallurgical transformation.

At the lower temperature, these mechanisms are reduced in number or are excluded because of the lack of necessary thermal activation. Obviously, the alternate activation of the high and low temperature mechanisms can result in a combination of mechanisms not normally encountered in isothermal fatigue. Which ones of the host of potential mechanisms come into play depends largely upon the particular alloy at hand, the temperature range, the overall temperature level, the time rate of change of temperature, and the extent and type of inelasticity occurring at each temperature level. The situation can become exceedingly complex to interpret.

At our present state of technological development, we are forced to rely heavily on our extensive and reasonably well understood isothermal fatigue knowledge and data bases for the design of components subjected to thermal fatigue loadings. In recognition of this situation, we must seek ways for spanning the isothermal/thermal fatigue gap. An integral part of that activity requires that well controlled, documented, and thoroughly interpreted variable temperature fatigue experiments be conducted. We propose the utilization of bithermal fatigue testing procedures to accomplish this goal. The chief advantages and disadvantages of the bithermal testing philosophy are presented in the next section.

The Case for Bithermal Cycling

Figure 1 is a schematic illustration of the temperature versus strain and resultant stress versus mechanical strain relationships for the bithermal cycles employed in studies to date. Both in-phase, wherein all of the tensile inelastic strain is imposed at the highest temperature, and out-of-phase, wherein the converse is true, bithermal cycles are illustrated. In the figure, the bithermal cycles are compared to conventional thermomechanical cycles. While bithermal cycling in the past has been used to introduce time-dependent creep strains at the high temperature and time-independent plastic strains at the low temperature (i.e., a CP or PC type Strainrange Partitioning (SRP)[2] cycle), it can be used to introduce plastic strains only at both temperatures. This latter condition is extremely difficult to achieve with conventional thermomechanical cycling, since thermal cycling a specimen usually cannot be accomplished with a high enough rate to preclude the introduction of creep. In the bithermal testing technique, the specimen is held at zero stress during the temperature excursions and hence no mechanical deformation takes place. Once at the high temperatures, the deformation can be imposed at high enough rates to preclude creep. Such PP bithermal experiments should help to isolate time-independent from time-dependent damage mechanisms.

[2]The terminology of the Strainrange Partitioning method is:
CP = Creep in tension, Plasticity in compression.
PC = Plasticity in tension, Creep in compression.
CC = Creep in tension, Creep in compression.
PP = Plasticity in tension, Plasticity in compression.

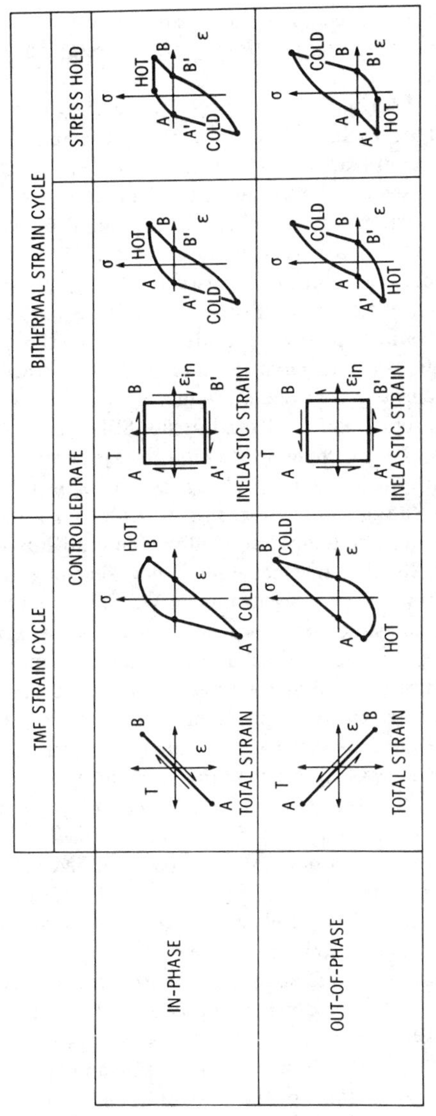

FIG. 1—*Bithermal and TMF cycles.*

Among the important effects of thermomechanical loading that are naturally encompassed by bithermal fatigue experiments are: (1) influence of alternate high and low temperatures on the cyclic stress-strain and cyclic creep response characteristics; (2) possibility of introducing both high temperature and low temperature deformation mechanisms within the same cycle but limiting the number of active mechanisms at high temperature by proper choice of the temperature level, stress level, and hold time; and (3) thermal free expansion mismatch straining (i.e., temperature range times the difference between the coefficients of thermal expansion) between surface oxides and the substrate (or coatings and substrate). Much of the relative simplicity of testing and interpretation of isothermal results, however, is retained by the bithermal testing procedures.

Many of the troublesome experimental aspects of TMF cycling can be readily overcome by bithermal testing. During TMF cycling, precisely controlled temperature and strain histories must be synchronized. If the temperature control is not accurate, the resultant mechanical hysteresis loop becomes highly distorted. Distortion of this loop hampers measurement and interpretation of the stresses and strains. This problem is minimized by use of bithermal cycling, since temperature is held constant during the application of mechanical strain. Obviously, the temperature must be accurately controlled at a constant value or the hysteresis loop will distort in this case as well. However, constant temperature control is easier to achieve than is accurate variable temperature and strain, both in magnitude and phase relationship. For in-phase or out-of-phase TMF cycles, the peak temperatures and strains must occur simultaneously and repeatedly. If this is not accurately controlled, the mechanical hysteresis loop again becomes distorted and accurate interpretation of results is difficult. Still another experimental disadvantage of TMF cycling that can be overcome by the bithermal procedure has to do with the uniformity of the temperature profile along the axial gage section of a specimen (when utilizing an axial extensometer) during heating and cooling. It is difficult to maintain a uniform temperature profile while also changing the temperature, since the gradients tend to increase as the cycling becomes more rapid. Should a thermal gradient develop along the gage length in a TMF test, the hysteresis loop again will experience distortion with subsequent difficulties in interpretation. For bithermal experiments, no mechanical strain changes are imposed during heating or cooling; hence temperature gradients imposed at zero stress are assumed to be unimportant. Finally, bithermal cycles can be run and thermal free expansion strains can be subtracted from the extensometer signal in an extremely simple manner, not requiring nearly as high an accuracy as for TMF cycling. Because the thermal free expansion strain of both the specimen and the extensometer can be considerably larger than the mechanical strain, it is common practice to first document this component of the strain signal (by thermally cycling the specimens under zero load control) and then subtracting it from the total strain signal. The difference is the mechanical strain. The differential output can then be amplified before it is recorded, thus providing a more accurate mechanical strain signal. For a TMF test, the subtraction process becomes critical and the accuracy of the mechanical strain signal depends upon the accuracy of the total signal and the apparent thermal free expansion signal. The accuracy of the thermal free expansion signal and the accuracy of its subtraction from the total signal is relatively unimportant to the bithermal cycle. If the subtraction process is 100% perfect and complete, the resultant mechanical hysteresis loops will appear as in Fig. 1. On the other hand, if the subtraction is imperfect, the only effect is to have the upper (tensile) and lower (compressive) halves of the hysteresis loop displaced horizontally by the amount of the error. This error does not affect the accuracy of the mechanical strain signal.

The major difference between TMF and bithermal fatigue has to do with what happens during the change in temperature. Both types of cycling impose the same range of temperature, and thus any thermal expansion mismatch between dissimilar constituents of the sample results in the same local range of strain. However, if it is important to the life of the sample that both the differential expansion strains and the externally applied mechanical strains be applied simultaneously, then the bithermal cycle would be an over simplification of the TMF cycle. The simplification aspects of the bithermal cycle, however, are desirable when it comes to isolating and interpreting the micromechanisms of cyclic deformation and cracking. It must be kept in mind that the short hold period at zero load to permit thermal equilibrium may allow the dislocation structure to relax to a lower energy state, thus exerting an undesirable second-order effect on the cyclic stress-strain response and possibly the cyclic life.

Interestingly enough, the bithermal testing technique was not originally proposed as an alternative to thermomechanical fatigue testing or even as a link between isothermal and thermomechanical fatigue. Instead, it was used in the evaluation of the isothermal Strainrange Partitioning (SRP) characteristics of alloys. Hence, numerous bithermal fatigue tests were performed on a variety of alloys, but comparative thermomechanical fatigue tests were seldom performed since that was not the subject under investigation at the time. The bithermal testing technique has been applied to several wrought and cast alloys. Wrought alloys included 304 [*3*] and 316 [*4,5*] austenitic stainless steels, A-286 [*3*], 2¼Cr-1Mo steel [*5*], and the tantalum alloys ASTAR 811C [*6*] and T-111 [*6*]. Cast superalloys examined were the aluminide coated cobalt-base alloy MAR-M-302 [*7*] and for the current paper, B-1900 + Hf, a nickel-base alloy.

Prior Bithermal Fatigue Results

As early as 1971 [*5*], the bithermal fatigue cycling technique was used to characterize the isothermal SRP behavior of 316 austenitic stainless steel and 2¼Cr-1Mo steel. The peak temperature in the cycle was considered to be the important temperature and results were compared with isothermal results for approximately the same temperature. The concept behind the technique was to ensure that only time-independent plastic deformation was encountered in the half cycle of interest by reducing the temperature below the creep range. No attempt was made at the time to consider the bithermal cycle as an approximation to a thermomechanical fatigue cycle. However, since these bithermal results agreed well with isothermal results [*8*] for the two alloys at the temperature and time conditions studied, we anticipated that isothermal behavior could be used to predict thermomechanical fatigue lives. This assumption was reinforced in 1976 for the 316 austenitic stainless steel [*9*].

The agreement between bithermal and thermomechanical results for this alloy is illustrated in Fig. 2. Results for both in-phase (CP) and out-of-phase (PC) cycles are shown. For the in-phase cycles, it appears that bithermal cycling produces slightly lower lives than TMF cycling. This is rationalized by noting that the bithermal cycle contain a greater proportion of the highly damaging CP strain range component than do the TMF cycles. For example, in the case of inelastic strain ranges below 0.01, the fraction of the inelastic strain range that is of the CP type is approximately 0.225 [*9*] for the two TMF tests, whereas for the two bithermal tests this fraction is 0.365 and 0.475 [*4*]. By applying the interaction damage rule of the SRP method, these two sets of results can be brought into even closer agreement. In Ref *5*, photo-micrographs were presented and showed the extensive intergranular cracking present in the in-phase bithermal experiments. Similar intergranular cracking was also documented for the in-phase TMF tests reported in Ref *9*. Since the basic failure mechanism was the same for the two types of testing, it is not surprising that the cyclic lives were comparable. Differences in lives between the in-phase and the out-of-phase cycles are explained by the more benign transgranular cracking mechanisms in the bithermal and TMF out-of-phase experiments. For this material and associated conditions, then, the damage accumulation behavior was reasonably independent of both the temperature and temperature-time path taken and, instead, depended upon the magnitude and type of deformation experienced by the material. This can be explained on the basis of the relatively temperature-independent nature of the creep and tensile ductilities [*8*].

In 1972, Sheffler and Doble [*6*] of TRW, under contract to NASA, also performed bithermal fatigue tests on the tantalum alloys T-111 and ASTAR 811C in ultrahigh vacuum. Comparisons

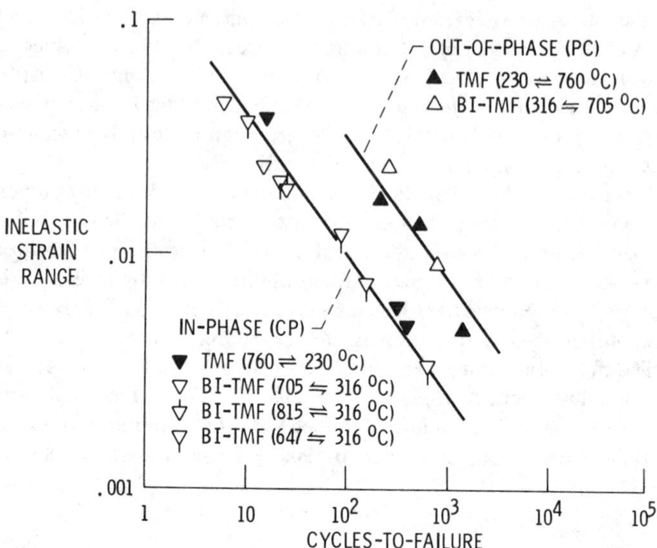

FIG. 2—*Comparison of bithermal and TMF LCF behavior of AISI Type 316 stainless steel. Bithermal results reported by Saltsman and Halford* [4]; *TMF results are NASA data generated by Halford and Manson and partially reported in Ref* 9.

of bithermal and thermomechanical fatigue tests could be made only for the ASTAR 811C; results are shown in Figs. 3*a* and *b* for in-phase (CP) and out-of-phase (PC) cycles, respectively. The bithermal cycles are more severe than the TMF cycles. Again, this would be expected based upon the greater degree of damaging creep strains present in the bithermal cycles compared to the TMF cycles. Greater creep strains occur in the bithermal tests since all the inelastic deformation is imposed at the peak temperature, whereas in the case of linear thermal ramping for TMF cycling, the inelastic deformation within a half cycle is imposed at a considerably lower average temperature.

The results discussed above are the only known examples for which both bithermal and TMF tests have been conducted for the same alloy.

Bithermal only data in ultra high vacuum have been reported for 304 austenitic stainless steel [*3*] and A-286 [*3*] as well as for aluminide coated MAR-M-302, a cast cobalt-base superalloy [*7*]. For the three alloys noted above, we have no valid basis for judging whether the bithermal cycling results are compatible with TMF or isothermal creep-fatigue behavior, since the latter type tests were not performed.

While the proposed bithermal cycles offer several advantages over conventional thermomechanical cycling, other forms of strain-temperature cycles have also been used to advantage by researchers in the field. For example, Lindholm and Davidson [*10*] have employed phased strain-temperature cycles that are similar to the bithermal cycle, the chief exception being that both tensile and compressive deformation is imposed at both the maximum and minimum temperatures in their work. This aspect can introduce features not found in thermally driven service cycles. Another form of cycling, referred to as a *dog-leg cycle,* is one in which the tensile and compressive straining is imposed at a single temperature (either maximum or minimum), and at some point in each cycle, the specimen is subjected to a thermal excursion during which no change in mechanical strain is allowed to be experienced by the specimen (although stress may

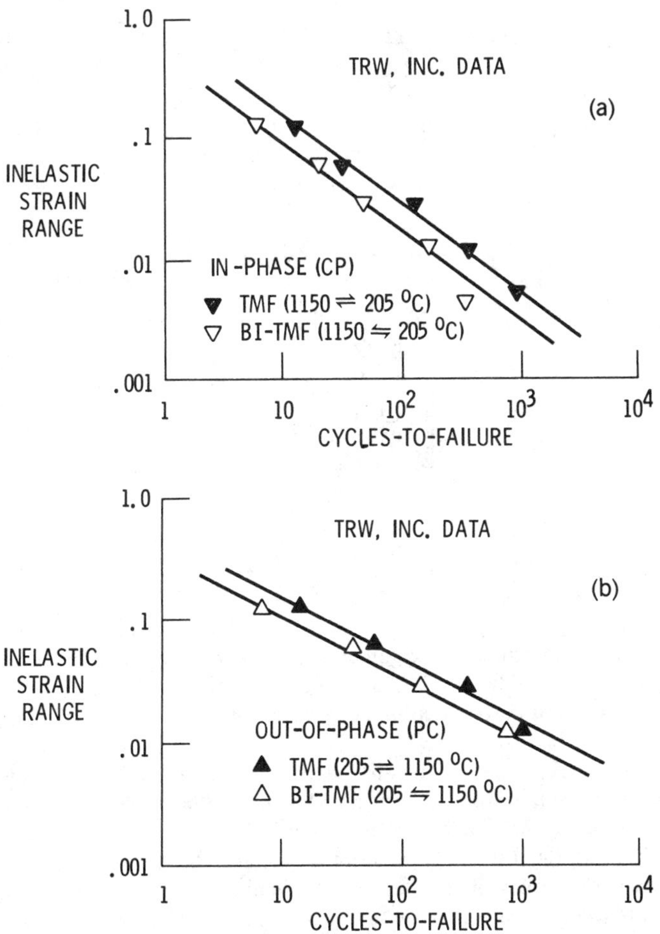

FIG. 3—*Comparison of bithermal and TMF LCF behavior of a ductile refractory alloy in ultra-high vacuum, after Sheffler and Doble* [6].

change). This form of cycling does introduce some aspects of a service cycle, but not to the same extent as for the bithermal cycle.

Details of Current Experiments

The B-1900 + Hf remelt stock for this program was procured by United Technologies, Inc. under NASA HOST Contract NAS-23288 [*11*]. Individually cast specimens were heat treated according to a schedule of: 1080 ± 14°C in air, 4 h air cool; 900 ± 14°C in air, 10 h air cool. All machining was done after heat treatment. All experiments were conducted on hourglass specimens with a geometry as shown in Fig. 4. These were tested in uniaxial push-pull using the servo-controlled hydraulic testing equipment and diametral extensometry described in [*12*]. Direct-resistance heating was used. Temperature at the specimen's minimum diameter was measured using optical and infrared pyrometry. The bulk of the current testing was accomplished

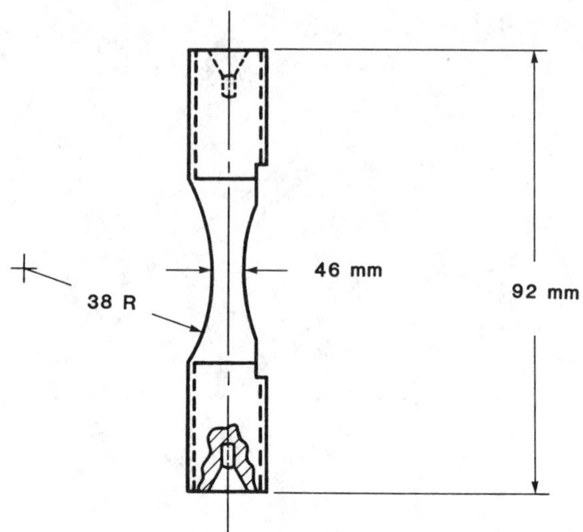

FIG. 4—*Specimen geometry.*

using computer control techniques. A commercially available digital computer, equipped with appropriate A/D, D/A, and discrete I/O devices, as well as secondary storage, was interfaced to the analog servo-hydraulic test equipment. All programming was done in (multitasked) PASCAL.

The procedures followed in the conduct of the bithermal tests consisted of computer controlling the temperature, load, and strain limit time history. Initially, the specimen was thermally cycled under zero load control, and the temperature setpoints, strain zero, and time delays adjusted. At this point, the test was begun. In the case of bithermal CP, the initial ramping of the load was positive, at a specified rate, until the tensile creep load was reached. Once the strain limit was reached, the specimen was unloaded to zero load and the temperature commanded to the lower value. A specified amount of time elapsed, allowing the specimen and extensometer to thermally stabilize, whereupon the specimen was load-ramped compressively until the compressive strain limit was reached. Again, the specimen was unloaded to zero load, the temperature commanded to the higher level, and another time period elapsed for thermal stabilization. The average time period for heating, cooling, and stabilization was approximately 1.8 min. The case of bithermal PC was identical, save for the fact that the initial loading was compressive. The initial loading direction for bithermal PP tests was always positive.

Results for B-1900 + Hf

Isothermal

The isothermal SRP results generated to date for the B1900 + Hf alloy have been limited to PP data at 483 and 871°C and PC and CP stress-hold, strain-limited data at 871°C. These results are displayed in Figs. 5*a* and *b*, respectively, as inelastic strain range versus cycles to failure (complete separation). At this stage of our investigation, we have not attempted to establish the pure SRP life relations (using the interaction damage rule and validity requirements of Ref *13*), since the data are too sparse. As more results are obtained, the life relations can be more readily defined, and attempts made to correlate and predict all the results.

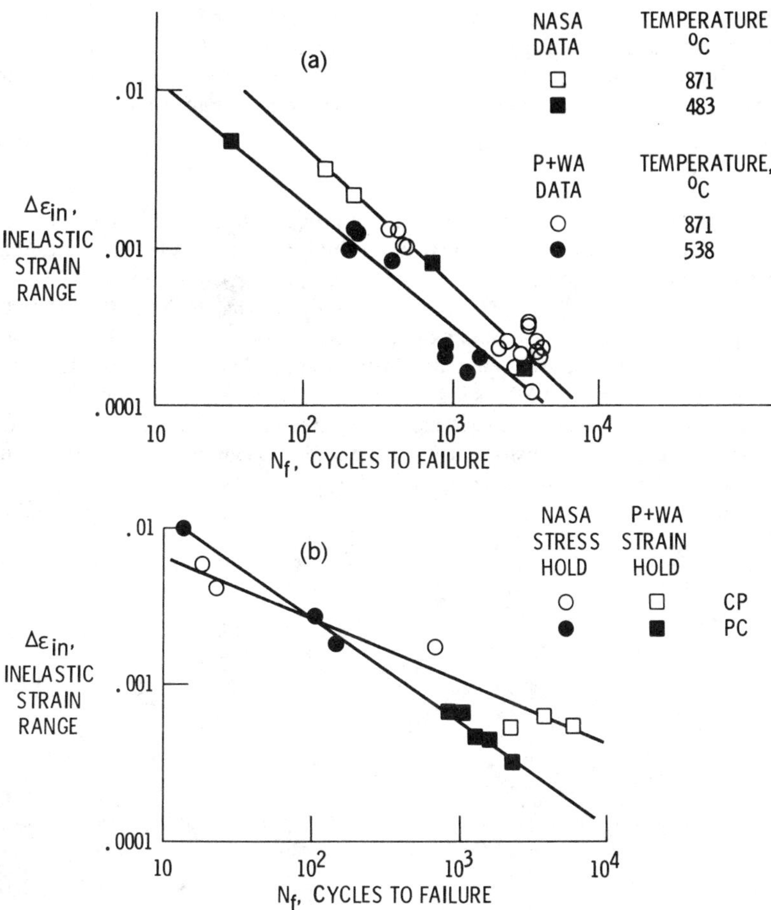

FIG. 5—*Isothermal data for B-1900 + Hf. P&WA data from Moreno et al.* [11]. (a) *PP data.* (b) *CP and PC data.*

For the time being, we can examine how the isothermal results of Moreno et al. [*11*] compare with our data. In NASA-LeRC sponsored work at Pratt & Whitney, Moreno used uniform gage length specimens with an axial extensometer. Heating was by radio frequency induction. Since the specimen design was different from what we have used, and since different strain measurement and heating techniques were employed, we are reluctant to pool the data into a single base for SRP life relation evaluation. Nevertheless, Moreno et al.'s data are included in Figs. 5*a* and *b* to provide a broader data base. PP tests were performed at 871 and 538°C, and PC and CP tests were performed at 871°C, using 60 s strain hold periods in compression and tension, respectively.

The four sets of lines drawn through the data are simply trend lines and do not represent the results of regression analyses. The first observation is that the PP results at 871°C from both investigations can be approximated by a single line; this line is above the PP line for the 483 and 538°C data. The combined 871°C PC results from both sources exhibit a slightly shallower slope than the PP results, with a cross-over at about 1000 cycles to failure. This is not a conclusive observation; however, an even shallower slope is found for the 871°C CP data from either or both studies. A cross-over of the PP and CP lines occurs at about 300 cycles to failure. As can be seen, the isothermal behavior is somewhat complex in appearance, and conclusions should not be drawn until a sounder data base has been obtained from a single source.

Bithermal

Only a few bithermal fatigue tests have been performed on the B1900 to date, and the testing is continuing. The first bithermal tests were of the PP type, with tensile plastic strain at 483°C and compressive plastic strain at 871°C (Fig. 6*a*). Surprisingly, the bithermal PP test results agree better with the higher life 871°C isothermal PP results than with the 483°C PP data. The higher temperature compressive plasticity in the bithermal PP test apparently is beneficial compared to the lower temperature compressive plasticity applied in the 483°C PP tests. For direct comparison with the bithermal PP results, bithermal PC test data were generated using 483°C in tension for plasticity and 871°C in comparison for creep (Fig. 6*b*). The results of the two sets of bithermal tests are indistinguishable from one another as can be seen from the figures. Our isothermal and bithermal results are summarized in Table 1.

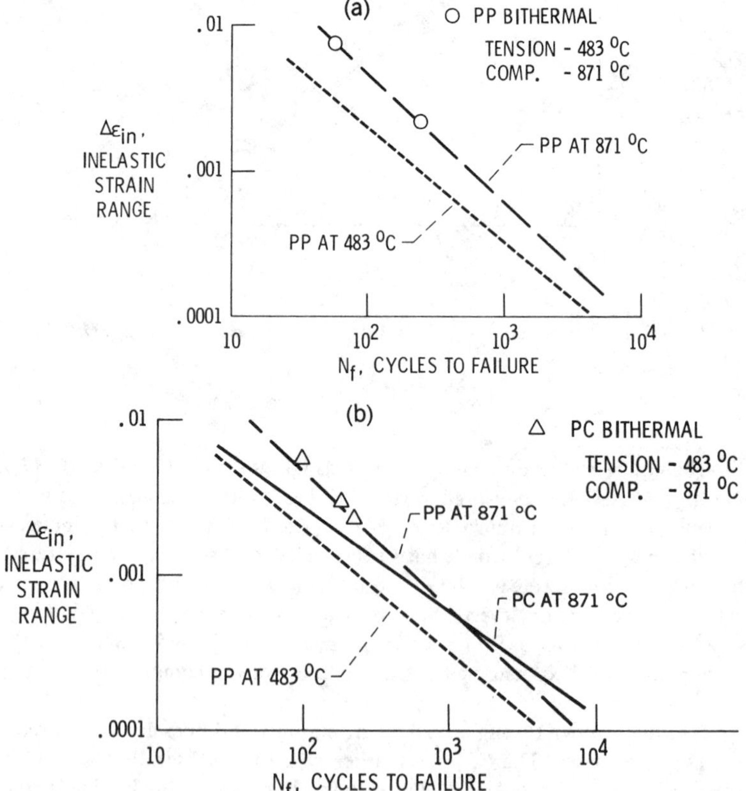

FIG. 6—*Bithermal data for B-1900 + Hf. P&WA data from Moreno et al.* [11]. (a) *PP bithermal data.* (b) *PC bithermal data.*

TABLE 1—*Isothermal and bithermal creep-fatigue results for B-1900 + Hf.*

Temperature, °C (Tension/ Compression)	Specimen No.	Total Strain Range	Inelastic Strain Range	F_{PP}	Elastic Strain Range	Stress Range, MPa	Creep Stress, MPa	Cycles to Failure
483/483	1907	0.00822	0.00017	1.00	0.00805	1527.	...	3 904
483/483	1933	0.00986	0.00081	1.00	0.00905	1717.	...	742
483/483	1935	0.01574	0.00475	1.00	0.01099	2085.	...	33
871/871	1929	0.00391	0.00014	1.00	0.00377	616.	...	17 731
871/871	1930	0.00967	0.00212	1.00	0.00755	1235.	...	214
871/871	1931	0.01081	0.00314	1.00	0.00767	1254.	...	139
871/871	1906	0.01770	0.00978	0.30	0.00792	1296.	−443.	14
871/871	1939	0.00960	0.00276	0.62	0.00684	1119.	−435.	105
871/871	1941	0.00711	0.00177	0.45	0.00534	874.	−251.	146
871/871	1903	0.01108	0.00411	0.47	0.00697	1140.	+474.	23
871/871	1913	0.01331	0.00591	0.46	0.00740	1210.	+467.	19
871/871	1924	0.00786	0.00176	0.19	0.00610	997.	+480.	697
483/871	1918	0.01964	0.00760	1.00	0.01204	2116.	...	56
483/871	1927	0.01183	0.00216	1.00	0.00967	1694.	...	234
483/788	1914	0.01247	0.00295	0.07	0.00952	1724.	−753.	179
483/871	1928	0.01060	0.00241	0.13	0.00819	1441.	−712.	220
483/871	1940	0.01321	0.00561	0.07	0.00760	1353.	−560.	96

Discussion

In this paper we adopted the philosophy that complex thermomechanical fatigue behavior can be approximated by simpler bithermal fatigue experiments. As an important criteria for the approximation to be acceptable, the same basic deformation, crack initiation, and crack propagation mechanisms must be present in both types of cycles. And, in order to directly compare the results of one type of test to the other, a damage accumulation (i.e., life prediction) theory must be invoked. We have utilized the SRP framework herein.

The principal reason for promoting bithermal fatigue testing as a viable technique is to provide a simple test procedure that captures the first order aspects of more complex TMF cycling, and does so by taking advantage of isothermal testing philosophy.

From the limited experience to date, it is apparent that bithermal fatigue testing can be used to impose more creep-fatigue damage per cycle than a comparable TMF cycle. Consequently, bithermal cycling could be used to determine lower bounds on TMF life, or at least to determine a conservative life provided creep damage is the dominant damaging mechanism. The bithermal technique is expected to be advantageous from the standpoint of interpreting the mechanisms of cyclic deformation and cracking, since only two discrete temperatures are involved. The TMF cycling includes the complete spectrum of temperature from minimum to maximum, and mechanistic interpretation could be hampered by the integrated effects of a spectrum of damage mechanisms. Our program is in its early stages of development, with considerably more work to be carried out. Micromechanistic interpretation of the results will play an important role in our further studies.

Summary and Conclusions

1. Bithermal fatigue testing can be a simple alternative to thermomechanical fatigue and can provide a conservative determination of TMF life for creep damage dominated failure nodes.
2. Bithermal fatigue results can be directly related to TMF results through the use of an appropriate damage rule.
3. Bithermal fatigue may be related to isothermal fatigue, provided no new mechanisms of deformation and cracking are introduced by the change in temperature required by the bithermal experiments. That is, bithermal fatigue may be related to isothermal fatigue provided the life behavior does not depend on the temperature-deformation-time path taken.
4. The mechanisms of cyclic deformation and cracking should be easier to interpret for bithermal cycles than for thermomechanical cycles because of the discrete nature of bithermal cycling compared with the continuous spectrum of TMF cycling.
5. Bithermal fatigue testing has been performed using two different laboratory testing techniques, one at NASA-LeRc and the other at TRW, Inc. Such testing could be conducted in almost any high-temperature fatigue laboratory with a minimum of alterations to the instrumentation. The current testing procedures utilize computer control and data acquisition and handling for greatest flexibility.

References

[1] ASME Code Case N-47.

[2] Bill, R. C., Verrilli, M. J., McGaw, M. A., and Halford, G. R., "A Preliminary Study of the Thermomechanical Fatigue of Polycrystalline MAR M-200," NASA TP-2280, AVSCOM TR 83-C-6, Feb. 1984.

[3] Sheffler, K. D., "Vacuum Thermal-Mechanical Fatigue Testing of Two Iron Base High Temperature Alloys," NASA CR-134524, 1974 (TRW Inc., ER-7697).

[4] Saltsman, J. F. and Halford, G. R., "Application of Strainrange Partitioning to the Prediction of Creep-Fatigue Lives of AISI Types 304 and 316 Stainless Steel," *Transactions of ASME,* Vol. 99, Ser. J, No. 2, May 1977, pp. 264–271.

[5] Manson, S. S., Halford, G. R., and Hirschberg, M. H., "Creep-Fatigue Analysis by Strain-Range Partitioning" in *Proceedings*, Symposium on Design for Elevated Temperature Environment, ASME, 1971, pp. 12–28.
[6] Sheffler, K. D. and Doble, G. S., "Influence of Creep Damage on the Low Cycle Thermal-Mechanical Fatigue Behavior of Two Tantalum-Base Alloys," NASA CR-121001, 1972 (TRW-ER-7592).
[7] Sheffler, K. D., "The Partitioned Strainrange Fatigue Behavior of Coated and Uncoated MAR-M-302 at 1000°C (1832°F) in Ultrahigh Vacuum," NASA CR-134626, 1974 (TRW Inc., ER-7723).
[8] Halford, G. R., Manson, S. S., and Hirschberg, M. H., "Temperature Effects on the Strainrange Partitioning Approach for Creep Fatigue Analysis" in *Fatigue at Elevated Temperatures, ASTM STP 520,* A. E. Carden, A. J. McEvily, and C. H. Wells, Eds., American Society for Testing and Materials, Philadelphia, 1972, pp. 658–667.
[9] Halford, G. R. and Manson, S. S., "Life Prediction of Thermal-Mechanical Fatigue Using Strainrange Partitioning" in *Thermal Fatigue of Materials and Components, ASTM STP 612,* D. A. Spera and D. F. Mowbray, Eds., American Society for Testing and Materials, Philadelphia, 1976, pp. 239–254.
[10] Lindholm, U. S. and Davidson, D. L., "Low-Cycle Fatigue with Combined Thermal and Strain Cycling" in *Fatigue at Elevated Temperatures, ASTM STP 520,* A. E. Carden, A. J. McEvily, and C. H. Wells, Eds., American Society for Testing and Materials, Philadelphia, 1973, pp. 473–481.
[11] Moreno, V., Nissley, D. M., and Lin, L. S., "Creep Fatigue Life Prediction for Engine Hot Section Materials (Isotropic)," NASA CR-174844, Dec. 1984 (NAS3-23288, P&WA).
[12] Hirschberg, M. H., "A Low-Cycle Fatigue Testing Facility," *Manual on Low-Cycle Fatigue Testing, ASTM STP 465,* American Society for Testing and Materials, Philadelphia, 1969, pp. 67–86.
[13] Hirschberg, M. H. and Halford, G. R., "Use of Strainrange Partitioning to Predict High-Temperature Low-Cycle Fatigue Life," NASA TN D-8072, 1976.

DISCUSSION

S. Antolovich[1] *(written discussion)*—I agree that the bithermal test is an excellent one for capturing the major effects in a TMF cycle. In particular, the test appears to have great potential for evaluating damage interaction mechanisms. The test seems to be most realistic where the difference in the two temperatures is large and where the maximum temperature is high. This is a consequence of the fact that the high temperature damage mechanism(s) (oxidation, creep) are thermally activated and their rates obey an Arrhenius-type equation. Thus the preponderant majority of high temperature damage will occur within a few degrees of the maximum temperature. When the temperature difference is reduced, this "end-point" approximation becomes less valid and the relationship of the bithermal test to the conventional TMF test may be less direct.

G. R. Halford et al. (authors' closure)—As Professor Antolovich acknowledges, both a bithermal test and a TMF test would be expected to exhibit the greatest deformation and damage interaction effects when the maximum temperature is high and the temperature range is large (and hence when the deformation and damage mechanism differences are exaggerated). This behavior is aptly demonstrated by the results shown in Figs. 3 and 4 when these results are compared with isothermal fatigue behavior at the maximum temperature of the cyclic temperature tests [*4, 6, 9*]. However, this does not mean that the two test types lose their close-knit relationship under less dramatic thermal cycling conditions. For example, if the temperature extremes were close together and the mechanisms of deformation and damage accumulation did not change, it is expected that negligible life reductions would be incurred by either a bithermal or a TMF test.

[1]Fracture and Fatigue Research Laboratory, Georgia Institute of Technology, Atlanta, GA 30332.

N. Marchand,[1] *G. L'Espérance,*[2] *and R. M. Pelloux*[3]

Thermal-Mechanical Cyclic Stress-Strain Responses of Cast B-1900+Hf

REFERENCE: Marchand, N., L'Espérance, G., and Pelloux, R. M., **"Thermal-Mechanical Cyclic Stress-Strain Responses of Cast B-1900+Hf,"** *Low Cycle Fatigue, ASTM STP 942,* H. D. Solomon, G. R. Halford, L. R. Kaisand, and B. N. Leis, Eds., American Society for Testing and Materials, Philadelphia, 1988, pp. 638–656.

ABSTRACT: A study was undertaken to develop an understanding of the fatigue response of superalloy B-1900+Hf under combined thermal and mechanical strain cycling in air. Comparative evaluations were made with existing thermal-mechanical data of B-1900+Hf and with results of a comprehensive study of the fatigue behavior of the same alloy under isothermal conditions. The thermal-mechanical fatigue (TMF) response was investigated for constant amplitude, fully reversed, mechanically strained cycling of uniaxially loaded specimens in the temperature range from 400 to 925°C. Experiments were conducted both with maximum strain in-phase with maximum temperature and out-of-phase with maximum temperature.

The TMF cycling was observed to cause more cyclic hardening than in isothermal fatigue experiments at the maximum and minimum temperatures. In terms of mean stress or plastic strain range, out-of-phase cycling was shown to be more deleterious than in-phase or isothermal cycling. However, few differences were observed in terms of the stabilized stress ranges. The asymmetric cyclic hardening/softening behavior is explained in terms of coarsening of the γ' and associated strain field. For TMF cycling, the high temperature flow stress depends on the density of the misfit dislocations, whereas the low temperature flow stress is controlled by the magnitude and sign of the applied stress. The TMF cracking modes are discussed. The results show that the fracture criterion under TMF cycling is stress based.

KEY WORDS: thermal-mechanical fatigue, superalloy, cyclic hardening, damage mechanisms, directional coarsening

Most high temperature components operate under conditions of cyclic mechanical non-isothermal loading. The design of these components must consider the effect of plastic strains that are associated with transients as well as with the steady-state operating stresses and strains that cause time-dependent failure. Figure 1 [*1*] shows schematically the stress and strain cycles to which gas turbine disk and blade materials are subjected as a result of rapid start, steady-state operation, and rapid shut-down. These stress-strain-temperature cycles produce high strain low-cycle fatigue which leads to crack initiation, slow crack growth, and failure by fast fracture. As a consequence, it is necessary to investigate the cyclic stress-strain behavior of these structural materials under cyclic temperature conditions. High temperature fatigue studies generally bypass real thermal fatigue loading partly because isothermal tests are simpler and much less expensive to perform, but also because it was felt that such tests carried out at the maximum

[1]Research Assistant, Department of Material Science and Engineering, Massachusetts Institute of Technology, Cambridge, MA 02139.

[2]NSERC Research Fellow, Department of Metallurgical Engineering, Ecole Polytechnique, Montreal, Quebec, Canada H3C-3A7.

[3]Professor, Department of Material Science and Engineering, Massachusetts Institute of Technology, Cambridge, MA 02139.

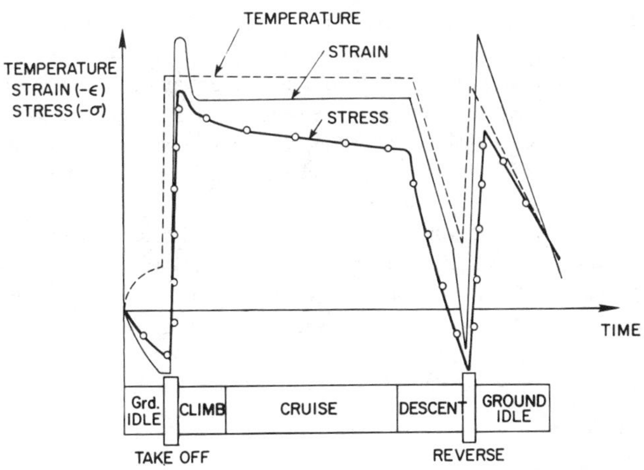

FIG. 1—*Stress-strain temperature cycle experienced by turbine components.*

service temperature would give worst-case results. However, several studies which have compared the fatigue resistance under thermal cycling conditions with that of isothermal tests have shown that, in many cases, the latter, rather than giving the worst-case situation, can seriously overestimate the fatigue life [*2–19*].

This paper describes the results of a study where the low-cycle fatigue behavior of superalloy B-1900+Hf was investigated for conditions of combined thermal and mechanical strain cycling. Experiments were limited to the two extreme conditions of in-phase and out-of-phase cycling. The objective of this study was to obtain information on thermal-mechanical fatigue (TMF) behavior of a commonly used superalloy. In turn, this information is used to assess the suitability of various parameters for correlating high temperature TMF crack growth. However, since thermal-mechanical fatigue testing is more time consuming, costly, and complicated than isothermal fatigue testing, the approach taken in this program was to conduct a limited study of TMF behavior over the temperature range of 400 to 925°C for correlation with previous isothermal work on the same material.

Review of Experimental Data

For superalloy IN-738, high strain fatigue tests, in-phase and out-of-phase TMF cycling between 400° and 950°C, all gave the same lives at a given $\Delta\epsilon_t$ level [*20*]. However, Nitta et al. [*18–19*] showed that the life of IN-738 under in-phase, out-of-phase, and isothermal testing depends on the applied strain range $\Delta\epsilon_t$ being the lowest for out-of-phase cycling at high $\Delta\epsilon_t$, but the longest at low $\Delta\epsilon_t$. Their results pointed out that thermal fatigue life of (nine) nickel-base alloys depends more upon the strength of the alloy then on the inelastic strain range. The data obtained by Lindholm and Davidson [*7*] on conventional B-1900 agreed with this conclusion. The dependence of thermal fatigue on strength may not be restricted to superalloys. Jaske [*11*] found that in order to correlate in-phase, out-of-phase, and isothermal fatigue data for a low carbon steel, a knowledge of both cyclic stress and cyclic strain was necessary. On the basis of stabilized stress range versus cycles to failure, little difference was observed in the behavior of specimens subjected to in-phase and out-of-phase cycling. However, the cyclic hardening characteristics of the alloy tested were very complex, indicating difficulty in predicting the stabilized stress range.

It thus appears that there is no hard and fast rule for relating the thermal-mechanical method to isothermal testing. There are many uncertainties in comparing total lives only; for example, the number of cycles to initiation may differ, and cracking may be intergranular in isothermal tests but transgranular in the cyclic temperature tests. Obviously, low life data will depend on initiation/propagation ratios, ductility, including sensitivity to strain rate, hardening behavior, etc. It should be emphasized that temperature cycling has a marked effect upon the σ-ϵ loop shape and deformation mode. The materials tested may or may not accommodate differing amounts of deformation before failure, depending upon its microstructural condition, which in turn is dependent upon the condition of cyclic deformation and temperature history. This points out the importance of knowing the temperature and strain history on the instantaneous material response to assess the kinetics of damage accumulation.

Base Material Description

The material for this program was taken from a special quality melt of B-1900+Hf obtained from Certified Alloy Products, Inc., Long Beach, California. The chemical analysis, heat treatment, and tensile properties are shown in Tables 1 to 3. The structure of the material was documented in both the as-cast and fully heat-treated conditions.

The grain size is about 1 to 2 mm. The replica technique was used to measure and study the gamma prime (γ') size and distribution. The fully heat-treated material showed the γ' size to be about 0.6 to 0.9 μm. The structure has an interdendritic spacing of about 100 μm, and islands of γ'-eutectic surrounded by a zone of fine γ' (0.9 μm) were observed. MC carbides near the coarse γ' islands were also observed.

Apparatus and Test Conditions

The TMF specimens had a rectangular cross section of 11.7 by 4.4 mm^2. The test section of each specimen was polished with successively finer grades of silicon-carbide paper to produce a bright finish, with finishing marks parallel to the longitudinal axis of the specimen. Specimens were degreased with trichlorethylene, followed by reagent-grade acetone before being heated to temperature.

The thermal-mechanical fatigue experiments were conducted using the same basic servocontrolled electrohydraulic test system used for TMF crack growth [*23*]. Details of the equipment, method of heating, gripping, alignment procedure, and general experimental procedures are given elsewhere [*22,23*].

TABLE 1—*Chemical composition (weight percent) of test material.*

C	Cr	Co	Mo	Al	Ti	Ta	B	Zr	Fe	W	Nb	Bi	Pb	Hf	Ni
0.09	7.72	9.91	5.97	6.07	0.99	4.21	0.016	0.04	0.17	0.04	0.08	0.1	0.1	1.19	bal.

TABLE 2—*Heat treatment of test material.*

Temperature, °C	Time, h	Quench
1080	4	air cool
900	10	air cool

TABLE 3—*Tensile properties of test material.*

Temperature, °C	$\dot{\epsilon}$, min^{-1}	$E \times 10^3$, MPa	0.2% Yield Strength, MPa	UTS, MPa	Elongation, %	Reduction of Area, %
RT	0.005	187.5	714	...	4.9	5.9
260	0.005	169.6	702	888	8.3	10.7
538	0.005	149.6	727	...	...	...
649	0.005	143.4	701	...	7.7	7.2
760	0.005	146.8	709	950	7.9	8.4
871	0.005	138.9	633	785	5.7	6.1
982	0.005	123.4	345	480	7.1	6.9

Temperature was measured with 0.2 mm diameter chromel-alumel thermocouples, which were spot-welded along the gage length. By computer control, the temperature in the gage length was maintained within ±5°C of the desired temperature for both axial and transverse directions throughout the duration of a test. For axial strain measurement, a gage length of 15 mm was selected on the middle part of the specimen and a contact-type extensometer was employed. Temperature and mechanical strain were computer-controlled by the same triangular waveform with in-phase or out-of phase cycling. The mechanical strain was obtained by subtracting the thermal strain from the total strain. The thermal strain was measured by cycling the temperature at zero load and stored into the computer. The mechanical strain ($\Delta\epsilon_{mec}$) was maintained constant at each strain level. The mechanical strain was increased by approximately 10% of its previous value after saturation of the total stress range. Saturation was defined as no change in stress range within 50 consecutive cycles. The accuracy and reproducibility of the results were checked by repeating the tests at two strain ranges only (0.25 and 0.50%). The results of these tests were in good agreement (within 10%) of those obtained by strain-incremental technique (13 strain ranges). The strain ranges used in this investigation are typical of those experienced by in-service components and therefore their corresponding plastic strain ranges are very small. The frequency was kept constant at 0.0056 Hz (1/3 cpm) for all strain ranges, which resulted in varying the strain rates between 0.002 and 0.007 s^{-1}.

Results

Cyclic Responses

Results of all the experiments conducted in this program are summarized in Tables 4 and 5, where the mechanical strain range, the number of applied cycles at each strain range, the initial and final stress range, and final plastic strain range are listed for both in-phase and out-of phase cycling. The plastic strain range was taken as the width of the hysteresis loop at zero stress. Although $\Delta\epsilon_p$ at zero stress is not the maximum plastic strain ($\Delta\epsilon_{pmax}$), achieved within each cycle (Fig. 3), $\Delta\epsilon_p$ was chosen over $\Delta\epsilon_{pmax}$ because the temperature at which $\Delta\epsilon_{pmax}$ was reached was a function of not only the applied strain range but also a function of the strain-temperature relationship. Figure 2 shows the thermal strain, the mechanical strain, the total strain, and the stress amplitude as a function of time (one cycle) for both in-phase and out-of-phase cycling. From the load-time and mechanical strain (ϵ_{mec})-time curves the hysteresis loops were obtained.

Even though strain cycling was fully reversed, the stress cycle was not symmetric about zero because the temperature was different at each extreme of the cycle. Figure 3 shows examples of

TABLE 4—*Summary of thermal-mechanical testing (in-phase).*

$\Delta\epsilon_t$, %	N	$\Delta\sigma_i$, MPa	$\Delta\sigma_f$, MPa	$\Delta\epsilon_{pf}$, %
0.2000	347	328	356	0.000
0.2515	337	426	436	0.000
0.3030	556	527	524	0.000
0.3580	388	611	620	0.000
0.3830	490	657	664	0.005
0.4075	303	706	708	0.0120
0.4330	340	745	748	0.0150
0.4525	280	777	782	0.0275
0.4825	160	831	842	0.0475
0.5650	54	862	880	0.0700

TABLE 5—*Summary of thermal-mechanical testing (out-of-phase).*

$\Delta\epsilon_t$, %	N	$\Delta\sigma_i$, MPa	$\Delta\sigma_f$, MPa	$\Delta\epsilon_{pf}$, %
0.1765	475	304	310	0.0000
0.1925	426	450	369	0.0000
0.2155	312	416	408	0.0000
0.2500	360	460	455	0.0000
0.2760	260	492	485	0.0030
0.2880	274	565	558	0.0080
0.3280	274	605	585	0.0120
0.3655	316	650	640	0.0075
0.4040	374	711	684	0.0145
0.4375	326	754	745	0.0145
0.4675	222	794	787	0.0325
0.5375	142	885	876	0.0485
0.6000	60	981	958	0.0880

the loops obtained for both in-phase and out-of-phase cycling. For out-of-phase cycling, a positive (tensile) mean stress is observed and a negative mean stress is observed for the in-phase cycling. That is, for the in-phase cycle, the magnitude of peak compressive stress was greater than the magnitude of peak tensile stress, or $|\sigma_{min}| > |\sigma_{max}|$. The opposite was true for the out-of-phase cycle where $|\sigma_{max}| > |\sigma_{min}|$. Figures 4*a* and 4*b* show the type of stress response for in-phase and out-of-phase cycling at a fixed strain range. From Fig. 4, the following conclusions can be drawn. First there is a significant change in mean stress ($\bar{\sigma}$) with N and $\Delta\epsilon_{mec}$ for out-of-phase cycling as compared to in-phase cycling. Furthermore, at equal strain range, the absolute value of $\bar{\sigma}$ is higher for out-of-phase than for in-phase cycling. Another conclusion that can be drawn is that for all $\Delta\epsilon_{mec}$, in-phase cycling shows σ_{max} to harden, whereas σ_{min} stayed almost unchanged except at high applied strain range. For out-of-phase cycling, σ_{max} hardened and σ_{min} softened. The softening of σ_{min} being more important than the hardening of σ_{max} results in a drift of $\bar{\sigma}$ to higher value of tensile stress. On the other hand, because σ_{min} stayed almost unchanged and σ_{max} hardened for in-phase cycling, the net result is a softening of the mean stress, that is $|\bar{\sigma}|$ decreases.

As stated before, the "saturation" was defined such that the change in stress range ($\Delta\sigma$), in about 50 cycles, was lower than 2%. Because the plastic strain ranges are always very small

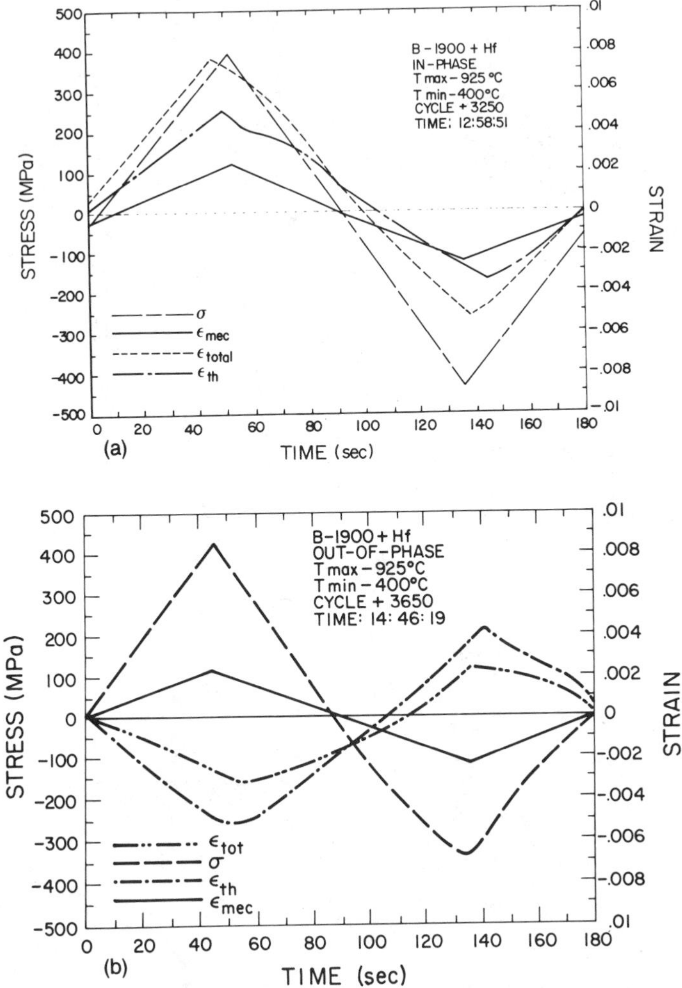

FIG. 2—*Stress, mechanical strain, thermal strain, and total strain as a function of time (one cycle).* (a) *In-phase.* (b) *Out-of-phase.*

compared with the mechanical strain ranges (i.e., $\Delta\epsilon_e \gg \Delta\epsilon_p$), the values of σ_{max}, σ_{min}, and $\overline{\sigma}$ are plotted against the strain amplitude ($\Delta\epsilon_{mec}/2$) to obtain the cyclic stress-strain (CSS) curves for in-phase and out-of-phase cycling. These curves are shown in Fig. 5 along with the isothermal CSS curves obtained on the same alloy at 871 and 538°C [*21*]. Interesting conclusions can be drawn from Fig. 5. First, one can see that the CSS curves of in-phase, out-of-phase, and isothermal testing converge at low $\Delta\epsilon_{mec}/2$ (<0.0012), but diverge as $\Delta\epsilon_{mec}/2$ increases. At higher $\Delta\epsilon_{mec}/2$ (but lower than 0.28%), the maximum stress (σ_{max} at 925°C) for in-phase cycling is higher than for isothermal fatigue at 871°C. For $\Delta\epsilon_{mec}/2 > 0.28\%$, the inverse behavior is observed (i.e., a higher hardening rate for isothermal fatigue than for σ_{max} of in-phase cycling). The hardening rate of σ_{min} of in-phase cycling ($T = 400°C$) is identical to the hardening rate measured for isothermal fatigue at 538°C. For out-of-phase cycling, σ_{min} (at $T = 925°C$)

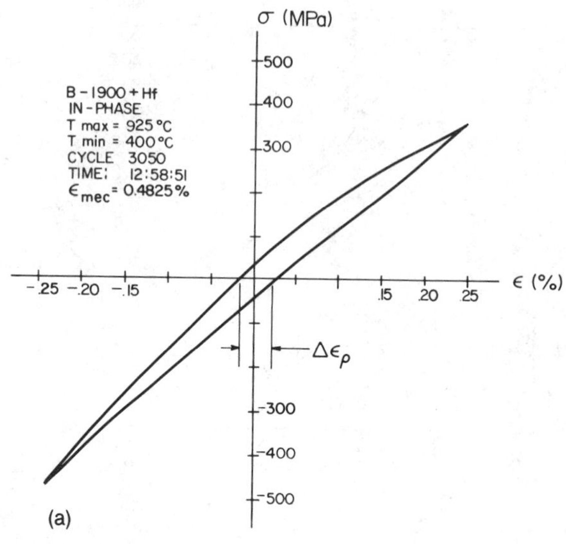

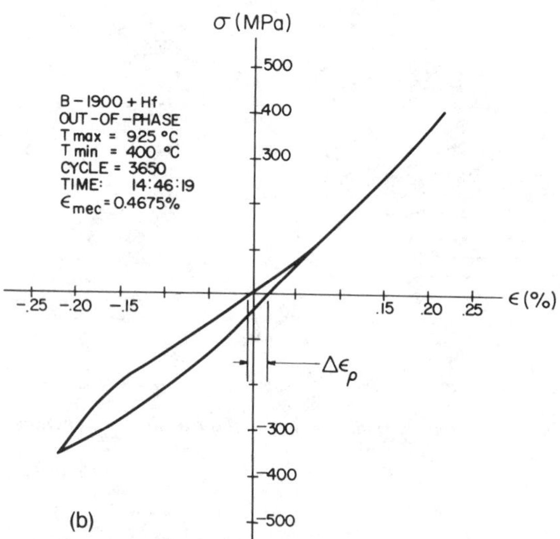

FIG. 3—*Typical hysteresis loop obtained under TMF conditions.* (a) *In-phase.* (b) *Out-of-phase.*

also shows a higher hardening rate than isothermal fatigue at 871°C (for $\Delta\epsilon_{mec}/2 < 0.25\%$) and a lower hardening rate for $\Delta\epsilon_{mec}/2 > 0.25\%$ than isothermal fatigue. However, σ_{max} ($T =$ 400°C) for out-of-phase shows a higher hardening rate than isothermal fatigue at 538°C.

The hardening behaviors of in-phase and out-of-phase cycling were compared by plotting σ_{max} of in-phase and $|\sigma_{min}|$ of out-of-phase, both measured at 925°C (Fig. 6). σ_{min} of in-phase cycling and $-\sigma_{max}$ of out-of-phase cycling are also plotted. One can see that the hardening rate at 925°C is higher for in-phase (in tension) than for out-of-phase cycling (in compression). How-

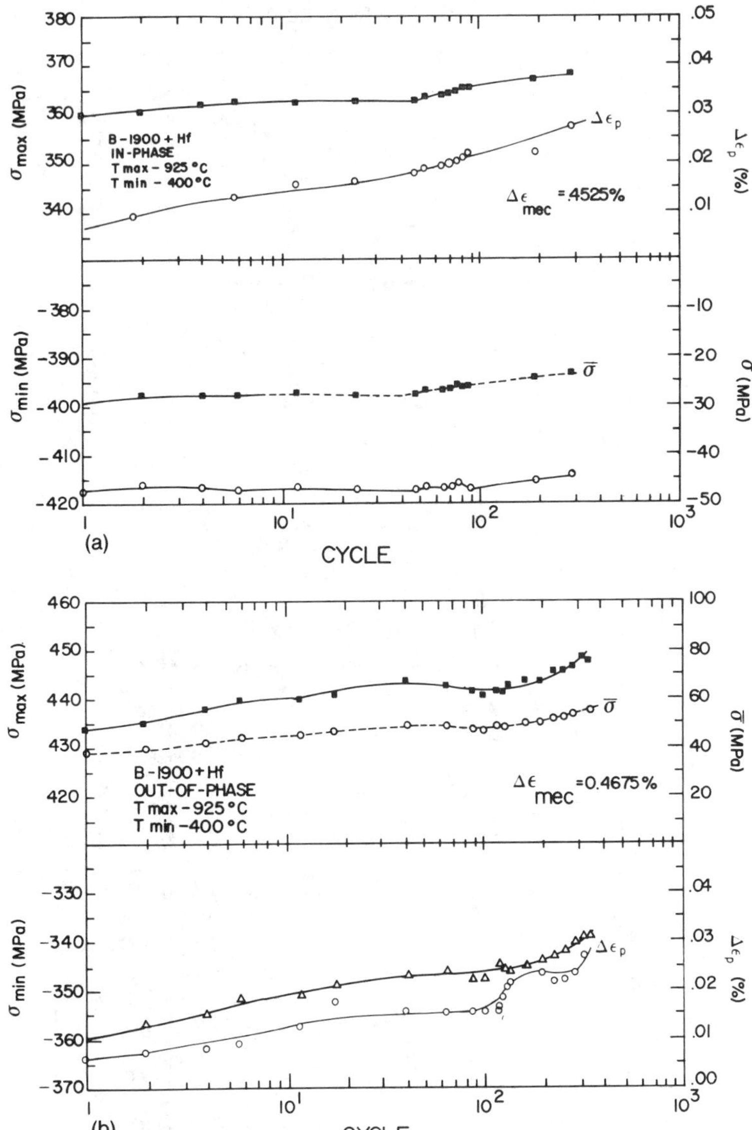

FIG. 4—*Typical cyclic hardening/softening curves obtained under in-phase* (a) *and out-of-phase* (b) *cycling.*

ever, the hardening rate at 400°C is higher for out-of-phase cycling (in tension) than for in-phase cycling (in compression) and isothermal testing.

Fractographic and TEM Observations

In order to identify the cracking process, the long transverse and longitudinal sections perpendicular to the fracture surface were mounted for metallographic observation. For out-of-

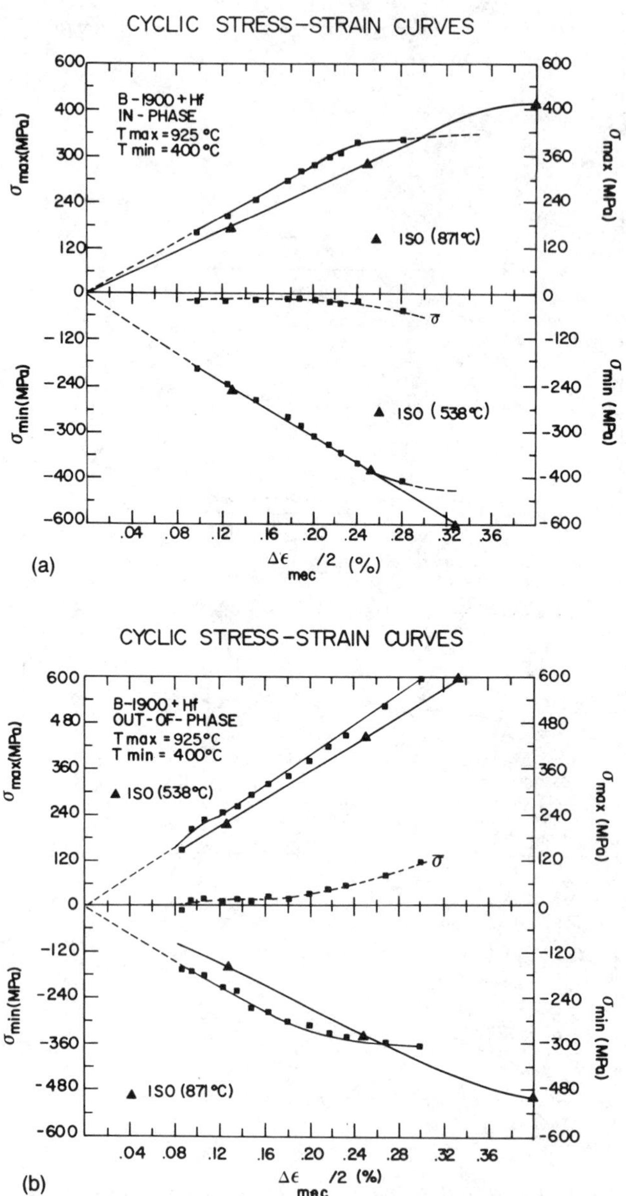

FIG. 5—*Cyclic stress-strain curves obtained under TMF and isothermal cycling.*

phase cycling, multiple cracks were observed along the gage length (Figs. 7*a* and 7*b*). The propagation path is transgranular and appears to proceed interdendritically (Fig. 7*c*). Examination of the specimens that failed under in-phase cycling reveals a varying degree of transgranular and intergranular cracking (Fig. 8) with a density of surface cracks much lower than in specimens that failed under out-of-phase cycling. The fracture path, however, appears mainly inter-

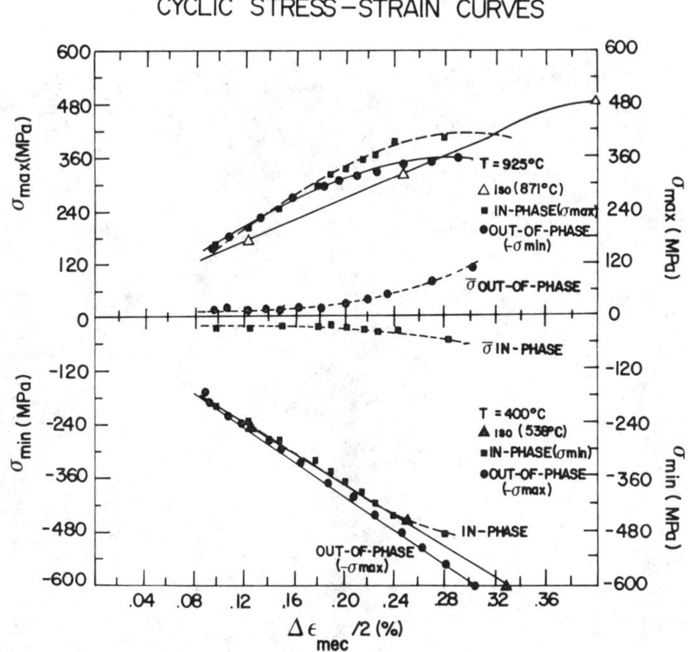

FIG. 6—*Comparison of TMF and isothermal cycling.*

granular (Fig. 8*c*). These conclusions were supported by scanning electron microscope (SEM) fractographic observations.

The TEM observations were made on the specimens fatigued to fracture. Foils were taken parallel to the loading axis and observed with a JEOL 100CX operated at 120 keV. As expected from the studies of large-grained materials, the dislocation substructure was not uniform in every grain. However, the dislocation structures presented in Figs. 9 and 10 represent an average of what was observed from grain to grain. In that sense they are taken as typical of those observed under out-of-phase and in-phase cycling. Figures 9 and 10 show the dislocation substructures obtained under in-phase and out-of-phase conditions. In all cases, coarsening of the γ' phase has taken place, being more pronounced under in-phase conditions. In some grains of the specimens cycled under in-phase conditions, directional coarsening (rafting) was observed. An extreme case of rafting is shown in Fig. 11. Little or no rafting was observed under out-of-phase cycling. A tight dislocation network encapsulating the γ' (rafted and unrafted) can be observed in Figs. 9 and 11 (in-phase cycling), whereas a looser dislocation network is observed in Fig. 10 (out-of-phase cycling), as can be seen from the dislocation spacing. These observations indicate that the dislocation density is higher under in-phase than under out-of-phase cycling. The octahedral active systems were found to be {111}[110], {111}[101], and possibly {111}[011] in both cases. Three Burgers vectors were identified using the invisibility criterion and assuming that the dislocations were screw in character. These Burgers vectors are $a/2[110]$, $a/2[101]$, and $a/2[101]$, indicating that at least three slip systems were operative. Comparison with specimens that failed under isothermal cycling shows that (on an average basis from grain to grain) the dislocation density ranked in increasing order: isothermal (T_{min} and T_{max}), out-of-phase, and in-phase cycling.

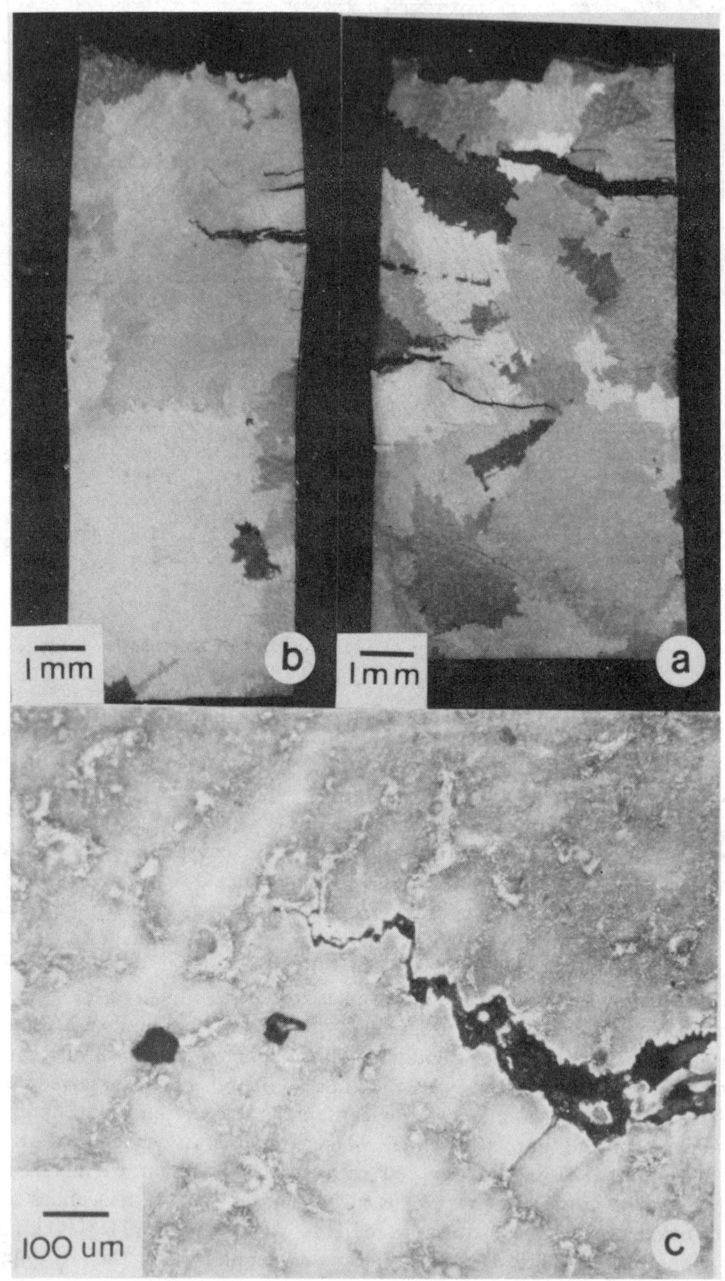

FIG. 7—*Fractographic observations of out-of-phase cycled specimen.* (a) *Longitudinal section.* (b) *Transverse section.* (c) *Interdendritic cracking.*

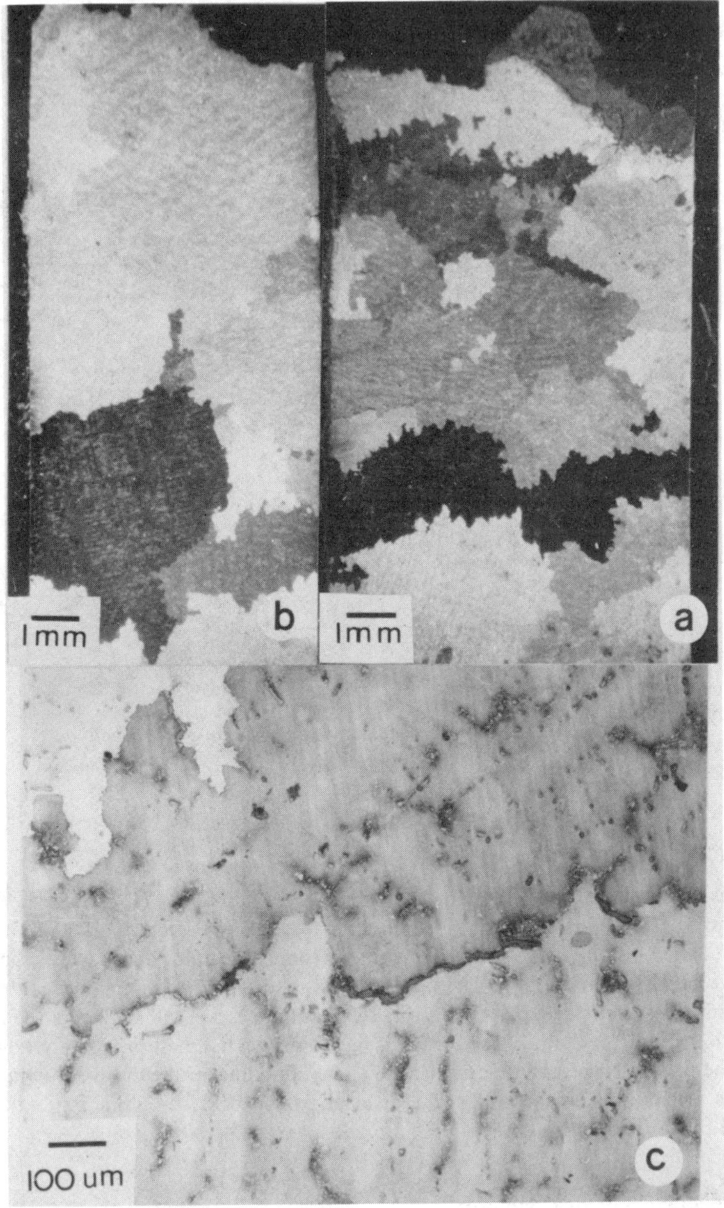

FIG. 8—*Fractographic observations of in-phase cycled specimen.* (a) *Longitudinal section.* (b) *Transverse section.* (c) *Interdendritic cracking.*

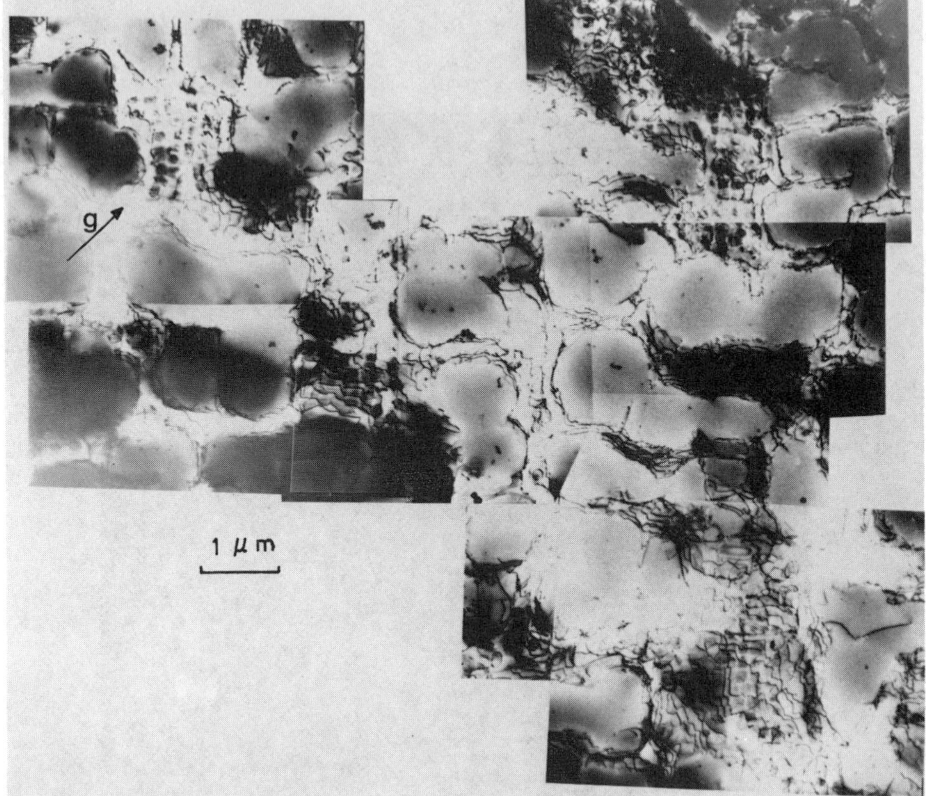

FIG. 9—*Dislocation structure observed in a sample cycled under in-phase conditions;* B = *[112];* g = *(111).*

Discussion

These results show that the two major microstructural features are (1) changes in γ' precipitate morphology (Figs. 9 to 11) and (2) introduction of a dislocation network about the γ' precipitates. It is well known [*24,25*] that coarsening of the γ' precipitates influences the mechanical behavior of nickel-base alloys. The task at hand is to separate the relative contribution of each structural feature to the cyclic behavior. In what follows, we will show that the increase in the cyclic flow stress during in-phase cycling is due to dislocation networks surrounding the γ', and that the lower flow stress under out-of-phase cycling and pronounced softening behavior (Fig. 4*b*) is a consequence not only of the dislocation networks strengthening but also of the directional strain field around the γ' precipitates.

Cyclic Hardening/Softening Behavior

It is generally accepted that coherent particles can be sheared by dislocations and, consequently, the work done in forcing the first dislocations through the particles will be important in determining the flow stress. The resistance to shear is governed by several factors:

1. The interaction of the cutting dislocation with the stress field of the precipitates.

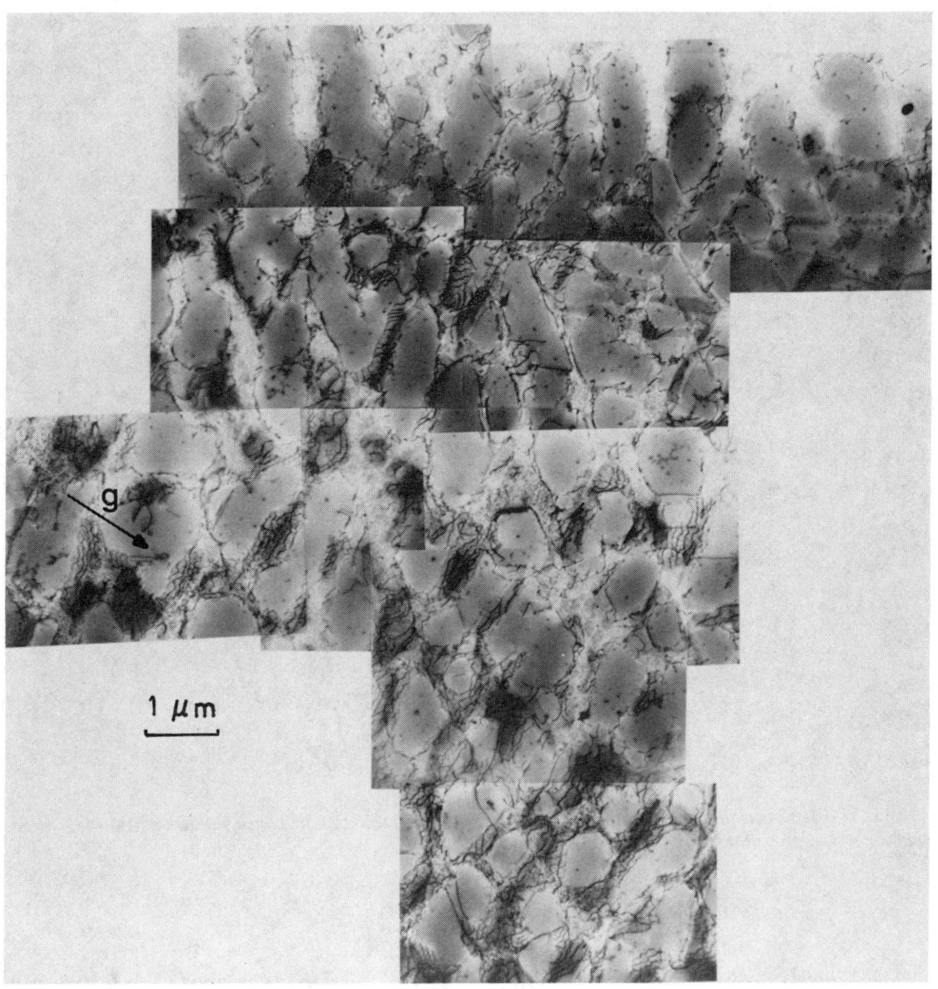

FIG. 10—*Dislocation structure observed in a sample cycled under out-of-phase conditions;* B = *[112];* g = *(111).*

2. If the lattice parameters of matrix and precipitate differ, then during shearing of the particles, misfit dislocations must be created at the precipitate-matrix interface. The magnitude of the Burgers vector of the interface dislocation will be the difference between the Burgers vector of the slip dislocation in the matrix and in the precipitate; that is, $(b_m - b_p)$.

3. If the matrix and precipitate possess different atomic volumes, a hydrostatic interaction would be expected between a moving dislocation and the precipitate.

In the case of superalloys with high volume fraction of γ' (>50%), γ'-shearing is the primary strengthening mechanism. With the mean free edge-to-edge distance in the matrix between the precipitates being smaller than the average precipitate size itself, dislocation shearing of the particle is favored over dislocation looping around the particles.

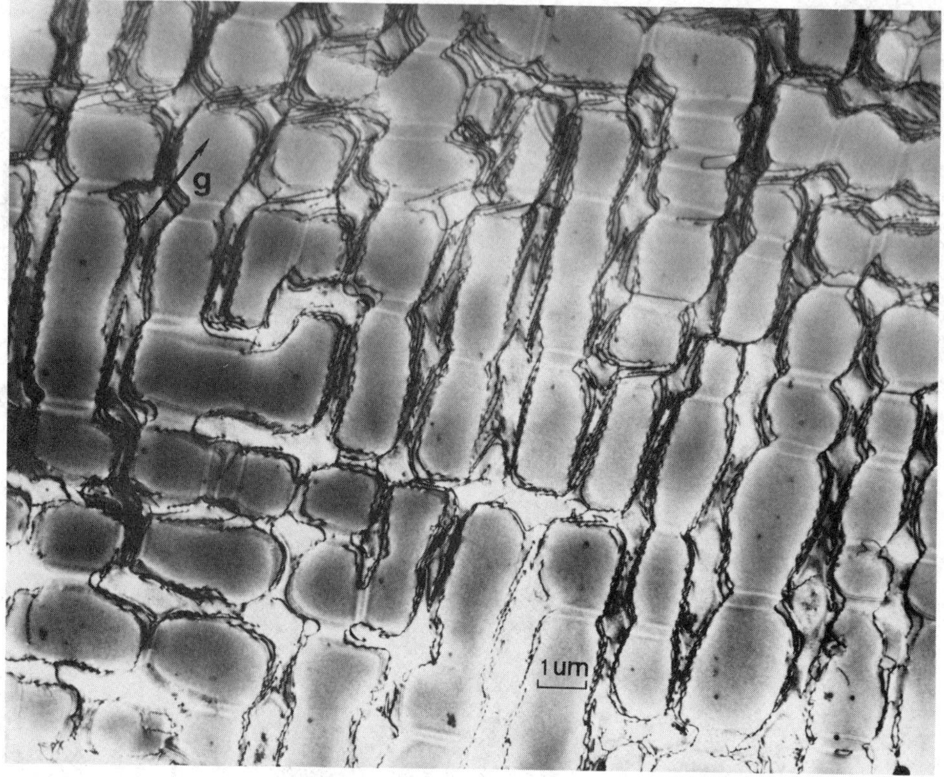

FIG. 11—*Dislocation structure observed in a sample cycled under in-phase conditions showing directional coarsening of the* γ' *phase;* B = *[112];* g = *(111).*

As previously mentioned, coarsening and rafting of γ' develop in this alloy. This feature can be attributed to the large lattice misfit (misfit ~ −0.25% [*26*]), which generates sufficient interfacial strain to produce misfit dislocations at elevated temperature. Significant deformation can occur only by dislocation penetration of the γ' phase, and it is postulated that the misfit dislocation nets at the interface retard this process. Therefore more hardening should be expected when rafting takes place, because the dislocation networks surrounding the γ' are more intense (Figs. 9 to 11). On the other hand, it has been shown by Shah and Duhl [*27*] that the flow stress decreases as γ' size increases, provided the cubic shape of the γ' is conserved. These two superimposed phenomena, with opposite effects with regard to the flow stress, determine the apparent flow stress. Figure 4 shows that the maximum stress (σ_{max}) of in-phase cycling (T_{max}) display continuous hardening with little appearance of stabilization. On the other hand, the σ_{min} curve of out-of-phase (T_{max}) displays continuous softening. This behavior has been previously observed on B-1900+Hf cycled in TMF [*28*], where continuous hardening of σ_{max} and softening of σ_{min} occurred until final fracture without evidence of saturation. This raises the question of the validity of the strain increment technique for measuring the CSS curves of low stacking-fault energy materials where continuous hardening (or softening) is observed until fracture [*28–30*] at low applied strain. This behavior is usually rationalized in terms of planar configuration of dislocations [*29,31*]. At high strain, where cross-slip takes place, saturation of the CSS curves is observed. Although B-1900+Hf is also a low stacking fault energy material, the planar configuration of dislocation alone cannot explain the observed cyclic hardening/softening behavior (Fig. 4).

At high temperature ($T > 600°C$) the flow stress depends on the APB energy and thermally activated cross slip of the glide dislocations [*27,32*]. The directionality of the internal stress field around the γ' is small because thermal activation is important. Therefore, the flow stress will depend on the factors controlling the internal stress. The density of misfit dislocations around the γ' particles, which depends on the γ' size and shape (raft), affect the flow stress because it controls the internal stress on the glide dislocations. On the other hand, when coarsening takes place, the particle spacing increases, which leads to weakening because of the increased probability of avoiding shearing by Orowan-type mechanisms [*34*]. Under in-phase cycling, coarsening of the γ' takes place along with rafting, leading to high density of misfit dislocations without significant increase in interparticle spacing (Fig. 11). The net result is an increase in the flow stress, because the internal stress increases faster than the relaxation time required by the glide dislocation to overcome the barrier created by the misfit dislocations. During out-of-phase cycling, isotropic coarsening takes place leading to a smaller increase of misfit dislocations and a more significant increase of γ' particle spacing. The net result is a decrease in flow stress.

At low temperature the directionality of the resultant stress field around the γ' particles will also contribute to the internal stress acting on the glide dislocations [*26,27,33*]. With coarsening, the hydrostatic tensile stress field around the γ' increases, and that is one of the reasons why misfit dislocations are required. It is well known that the resistance to the movement of glide dislocation (i.e., the frictional force) increases with increasing hydrostatic stress [*35*]. Therefore the superposition of an external hydrostatic stress field will increase or decrease the flow stress depending on the magnitude and sign of the applied stress. If an external tensile stress field is applied, the frictional stresses increase and so does the flow stress. This corresponds to the out-of-phase cycling case where the applied stresses are tensile at low temperature (Fig. 4*b*). When the applied stress field is compressive, the flow stress decreases or remains unchanged, depending on the magnitude of the net stress field [*33*]. During in-phase cycling, the stresses are compressive at T_{min} and cancelled with the hydrostatic tensile stress field around the γ'. The net result is that the frictional forces on the glide dislocation are lower and the flow stress does not change much as cycling proceeds and coarsening takes place (at T_{max}), as can be seen in Fig. 4*a*. It is important to recognize at this point that for the previous argument to hold, all the plastic deformation must take place in the γ-matrix, which means that there is no dislocation in the γ'-phase. This requirement was supported by the TEM observations, which show no evidence of dislocations in the γ'-phase. This is consistent with the observations made by Raguet et al. [*38*] which show for René 80 above 760°C that there is no evidence of dislocation in the γ'-particles and that all deformation occurs in the γ-matrix.

To fully understand the flow behavior with change in sign of the applied stress, we must also consider the resolved constriction stress of partial dislocations [*27,32*]. The direction of the glide force per unit length, $F_{g/L}$, will reverse upon reversing the applied stress, σ [*27*]. While this has no physical meaning in a macroscopic sense, since it only alters the direction of glide for a particular dislocation, reversing the direction of the glide forces acting on the partial dislocation, it leads to a distinctly different physical situation. The resulting force tends to constrict the partials under an applied tensile stress and extend them under a compressive stress [*27*]. Since constriction of the partials is required by the cross-slip process, the flow stress appears stronger in tension than in compression where the extended partials retard cross-slip activity. The previous argument implies that the flow behavior of B-1900+Hf is governed by octahedral slip activity. The behavior cited above occurs for single crystal oriented with their stress axes near ⟨001⟩ (where octahedral slip prevailed) and not for all orientations [*27*]. Orientation near ⟨011⟩, for instance, exhibit a tension-compression anisotropy opposite to that of ⟨001⟩ orientation [*27*]. However, the opposite behavior observed for other orientation involves deformation by cube slip. The fact that only octahedral slip activity was observed means that the flow behavior of B-1900+Hf is controlled by the parameters that affect the activity of the octahedral slip system. In that case, the flow stress can be written as

$$\sigma \alpha (1/R + 1/\lambda) \tag{1}$$

where R is the particle size and λ the mean free distance between the two constricted nodes where the cross-slip event occurs [27]. At high temperature, λ is always smaller than R and the flow stress is governed by the dislocation network and internal stress which fixed λ [27,36]. At low temperature, the flow stress depends on which of these two parameters (R, λ) is the smallest. If a tensile stress is applied, the partials tend to be constricted and the mean free distance between cross-slip event decreases. If a compressive stress is applied, the partials are pulled further apart (λ increases), and the flow stress is controlled by R, the particle size.

Cyclic Stress-Strain Behavior

The cyclic stress-strain curves (Figs. 4 and 5) show that the flow stress, as a function of the strain amplitude ($\Delta \epsilon_t/2$), is higher for in-phase than out-of-phase or isothermal cycling at the maximum temperature (T_{max}). This is consistent with the fact that dislocation density increases in the order of isothermal, out-of-phase and in-phase cycling (Figs. 9 to 11). At low temperature (T_{min}), because cross-slip is a function of the magnitude and sign of applied stress, the flow stress is higher under out-of-phase cycling than under in-phase or isothermal cycling (Fig. 6).

The issue that needs to be addressed now is the cracking mode. As shown in Figs. 7 and 8, fracture is transgranular and proceeds interdentritically in out-of-phase cycling and intergranularly under in-phase cycling conditions. As expected, fracture is controlled by the favored mode of rupture in the tensile part of the cycle. Under in-phase cycling, tension occurs at high temperature where the cohesive strength of the grain boundary is low, which obviously promotes intergranular cracking. During out-of-phase cycling, the specimen is under tensile loading at low temperature and the weakest transgranular features (carbide film, secondary dendrites, inclusion stringers, etc.) control the rupture mode. The fact that the fracture mode depends on the maximum tensile stress, rather than on a critical plastic strain range, suggests that the failure criteria for B-1900+Hf is stress-based rather than strain-dependent. In other words, the testing condition (temperature and strain relationship) leading to the maximum tensile stress will determine the number of cycles to initiation and the propagation rates.

As mentioned earlier, the major critical turbine components operate under strain-controlled conditions and, more specifically, under displacement-control. The stresses are not known *a priori*. Therefore, if the failure criterion in B-1900+Hf is stress-dependent, the relative crack growth rates for a given crack length can be determined from the CSS curves. Figure 8 shows that at low strain amplitude ($<0.25\%$) the stress range for TMF cycling is higher than for isothermal fatigue. Consequently, faster crack growth rates should be obtained for TMF cycling [38]. Under fully plastic conditions ($\Delta \epsilon_t/2 > 0.25\%$), Fig. 8 shows that the isothermal stress range is higher than the stress ranges obtained under TMF cycling and faster crack growth rates are expected under isothermal cycling (T_{max}).

Conclusion

The cyclic stress-strain behavior under TMF cycling differs from the isothermal behavior and shows more hardening, both on a high and low temperature basis. This indicates that it is difficult to predict the cyclic stress-strain behavior under realistic conditions (Fig. 1) from isothermal data. The synergistic coupling between the cyclic strains and temperatures cannot be ignored.

It was shown that the cyclic flow stress at elevated temperature (T_{max}) is primarily controlled by the density of misfit dislocations, which depends on the amount of isotropic and directional coarsening. At low temperature (T_{min}) the flow stress is controlled by the directionality of the stress field around the γ', the magnitude of which depends on the sign of the applied stress. The cracking modes observed have indicated that TMF fracture is controlled by a critical tensile stress (σ_c) rather than a critical tensile strain.

Acknowledgments

This research work was sponsored by the National Aeronautics and Space Administration under Grant NAG3-280.

References

[*1*] McKnight, R. L., Laflen, J. H., and Spaner, G. T., "Turbine Blade Tip Durability Analysis," NASA CR-165268, Feb. 1981, 112 pp.

[*2*] Udoguchi, T. and Wada, T., "Thermal Effect on Low-Cycle Fatigue Strength of Steels," in *Thermal Stresses and Thermal Fatigue,* D. J. Littler, Ed., Butterworths, London, 1971, pp. 109-123.

[*3*] Sheffler, K. D., "Vacuum Thermal-Mechanical Fatigue Behavior of Two Iron-Base Alloys," in *Thermal Fatigue of Materials and Components, ASTM STP 612,* American Society for Testing and Materials, Philadelphia, 1976, pp. 214-226.

[*4*] Taira, S., "Relationship Between Thermal Fatigue and Low-Cycle Fatigue at Elevated Temperature," in *Fatigue at Elevated Temperatures, ASTM STP 520,* American Society for Testing and Materials, Philadelphia, 1973, pp. 80-101.

[*5*] Fujino, N. and Taira, S., "Effects of Thermal Cycle on Low-Cycle Fatigue Life of Steels and Grain Boundary Sliding Characteristics," in *ICM 3,* Vol. 2, 1979, pp. 49-58.

[*6*] Stentz, R. N., Berling, J. T., and Conway, J. B., "A Comparison of Combined Temperature and Mechanical Strain Cycling Data with Isothermal Fatigue Results," in *Proceedings,* Conference on Structural Mechanics in Reactor Technology, Berlin, Vol. 6, 1971, pp. 391-411.

[*7*] Lindholm, U. S. and Davidson, D. L., "Low-Cycle Fatigue with Combined Thermal and Strain Cycling," in *Fatigue at Elevated Temperatures, ASTM STP 520,* American Society for Testing and Materials, Philadelphia, 1973, pp. 473-481.

[*8*] Kuwabara, K. and Nitta, A., "Thermal-Mechanical Low-Cycle Fatigue Under Creep-Fatigue Interaction in Type 304 Stainless Steel," in *ICM 3,* Vol. 2, 1979, pp. 69-78.

[*9*] Kuwabara, K., Nitta, A., and Kitamura, T., "Thermal-Mechanical Fatigue Life Prediction in High Temperature Component Materials for Power Plants," in *Proceedings,* Conference on Advances in Life Prediction Methods, ASME-MPC, Albany, N.Y., 1983, pp. 131-141.

[*10*] Troschenko, V. T. and Zaslotskaya, L. A., "Fatigue Strength of Superalloys Subjected to Combined Mechanical and Thermal Loading," in *ICM 3,* Vol. 2, pp. 3-12.

[*11*] Jaske, C. E., "Thermal-Mechanical Low Cycle Fatigue of AISI 1020 Steel," in *Thermal Fatigue of Materials and Components, ASTM STP 612,* American Society for Testing and Materials, Philadelphia, 1976, pp. 170-198.

[*12*] Sheinker, A. A., "Exploratory Thermal-Mechanical Fatigue Results for René 80 in Ultrahigh Vacuum," NASA CR-159444, 1978.

[*13*] Bhongbhobhat, S., "The Effect of Simultaneously Alternating Temperature and Hold Time in the Low Cycle Behavior of Steels," in *Low-Cycle Fatigue Strength and Elasto-Plastic Behavior of Materials,* K. T. Rie and E. Harbach, Eds., DVM Pub., 1979, pp. 73-82.

[*14*] Westwood, H. J. and Lee, W. K., "Creep Fatigue Crack Initiation in 1/2 Cr-Mo-V Steel," in *Creep and Fracture of Engineering Materials and Structure,* B. Wilshire and W. Owen, Eds., Pineridge Press, Swansea, U.K., 1982, pp. 517-530.

[*15*] Kloos, K. H., Granacher, J., Barth, H., and Rieth, P., "Simulation of the Service Conditions of Heat Resistant Steels by Creep-Rupture Tests Under Variable Stress or Temperature and by Strain Controlled Service-Type Fatigue Tests," in *Proceedings,* 5th International Conference on Fracture, Vol. 5, Pergamon Press, Oxford, 1982, pp. 2355-2369.

[*16*] Adams, W. R. and Stanley, P., "A Programmable Machine for Simulated Thermal Fatigue Testing," *Journal of Physics E, Scientific Instruments*, Vol. 7, 1974, pp. 669-673.

[*17*] Dawson, R. A. T., Elder, W. J., Hill, G. J., and Price, A. T., "High Strain Fatigue of Austenitic Steels," in *Thermal and High Strain Fatigue,* Metals and Metallurgy Trust, London, 1967, pp. 239-269.

[18] Kuwabara, K. and Nitta, A., "Effect of High Temperature Tensile Strain Holding in Thermal Fatigue Fracture," in *Creep-Fatigue Interactions,* R. M. Curran, Ed., ASME-MPC, 1976, pp. 161–177.
[19] Nitta, A., Kuwabara, K., and Kitamura, T., "The Characteristic of Thermal-Mechanical Fatigue Strength in Superalloys for Gas Turbines," in *Proceedings,* International Conference on Gas Turbines, Tokyo, Japan, 1983.
[20] Speidel, M. O. and Pineau, A., "Fatigue of High Temperature Alloys for Gas Turbines," in *High Temperature Alloys for Gas Turbines,* Coutsouradis, Felix, Tischmeister, Habraken, Lindblom, and Speidel, Eds., Applied Science, London, 1978, pp. 469–512.
[21] Moreno, V., "Creep-Fatigue Life Prediction for Engine Hot Section Materials," NASA CR-168228, 1983, 85 pp.
[22] Marchand, N. and Pelloux, R. M., "A Computerized Testing System for Thermal-Mechanical Fatigue Crack Growth," *Journal of Testing and Evaluation,* Vol. 14, No. 6, 1986, pp. 303–311.
[23] Marchand, N. and Pelloux, R. M., "Thermal-Mechanical Fatigue Crack Growth in Inconel X-750," in *Time-Dependent Fracture,* K. Krausz and A. S. Krausz, Eds., Martinus Nijoff, 1985, pp. 167–178.
[24] Tien, J. K. and Gamble, R. P., "Effects of Stress Coarsening on Coherent Particle Strengthening," *Metallurgical Transactions,* Vol. 3, 1972, pp. 2157–2162.
[25] Pearson, D. D., Lemkey, F. D., and Kear, B. H., "Stress Coarsening of γ' and Its Influence on Creep Properties of a Single Crystal Superalloy," in *Proceedings,* 4th International Symposium on Superalloys, J. K. Tien, Ed., American Society for Metals, Metals Park, Ohio, 1980, pp. 513–520.
[26] Lin, L. S. and Walsh, J. M., "Lattice Misfit by Convergent-Beam Electron Diffraction—Part I: Geometry Measurement," in *Proceedings,* 42nd Annual Meeting of EMSA, 1984, pp. 520–523.
[27] Shah, D. M. and Duhl, D. N., "The Effect of Orientation, Temperature, and Gamma Prime Size on the Yield Strength of a Single Crystal Nickel-Base Superalloy," in *Superalloys 1984,* M. Gell et al., Eds., AIME, Warrendale, Pa., 1984, pp. 107–116.
[28] Hill, J. and Masci, R. M., private communication, Pratt & Whitney Aircraft, 1985.
[29] Nilson, J. O., "The Influence of Nitrogen of High Temperature Low Cycle Fatigue Behavior of Austenitic Stainless Steels," *Fatigue of Engineering Materials and Structures,* Vol. 7, No. 1, 1984, pp. 55–64.
[30] Hatanaka, K. and Yamada, T., "Effect of Grain Size on Low Cycle Fatigue in Low Carbon Steel," *Bulletin of JSME,* Vol. 24, No. 196, 1981, pp. 1692–1699.
[31] Magnin, T., Driver, J., Lepinoux, J., and Kubin, L. P., "Aspects Microstructturaux de la Déformation Cyclique dans les Métaux et Alliages C. C. et C.F.C. I. Consolidation Cyclique," *Revue de Physique Appliquee,* Vol. 19, 1984, pp. 467–482.
[32] Leverant, G. R., Kears, B. H., and Oblack, J. M., "Creep of Precipitation-Hardened Nickel-Base Alloy Single Crystals at High Temperatures," *Metallurgical Transactions,* Vol. 4, 1973, pp. 355–362.
[33] Miyazaki, T., Nakamura, K., and Mori, H., "Experimental and Theoretical Investigations on Morphological Changes of γ'-Precipitates in Ni-Al Single Crystals During Uniaxial Stress-Annealing," *Journal of Materials Science,* Vol. 14, 1979, pp. 1827–1837.
[34] Jensen, R. R. and Tien, J. K., "Temperature and Strain Rate Dependence of Stress-Strain Behavior in a Nickel-Base Superalloy," *Metallurgical Transactions,* Vol. 16A, 1985, pp. 1049–1068.
[35] Honeycombe, R. W. K., *The Plastic Deformation of Metals,* 2nd ed., Edward Arnold, London, 1984.
[36] Gell, M. and Duhl, D. N., "The Development of Single Crystal Superalloy Turbine Blades," in *Proceedings,* N. J. Grant Symposium on Processing and Properties of Advanced High-Temperature Alloys, 17–18 June 1985, ASM Pub., forthcoming.
[37] Marchand, N. and Pelloux, R. M., "Thermal-Mechanical Fatigue Crack Growth in B-1900+Hf," paper presented to Conference on High Temperature Alloys for Gas Turbines, Liege, Belgium, 6–9 Oct. 1986.
[38] Raguet, M., Antolovich, S. D., and Kelley Payne, R., "Fatigue and Deformation Behavior of Directionally Solidfied René 80," *Superalloys 1984,* M. Gele et al., Eds., AIME, Warrendale, Pa., 1984, pp. 231–241.

J. L. Malpertu[1] *and L. Rémy*[1]

Thermomechanical Fatigue Behavior of a Superalloy

REFERENCE: Malpertu, J. L. and Rémy, L., "**Thermomechanical Fatigue Behavior of a Superalloy,**" *Low Cycle Fatigue, ASTM STP 942,* H. D. Solomon, G. R. Halford, L. R. Kaisand, and B. N. Leis, Eds., American Society for Testing and Materials, Philadelphia, 1988, pp. 657–671.

ABSTRACT: A thermal-mechanical fatigue test facility is described to study the stress-strain behavior and fatigue damage under anisothermal conditions of materials for gas turbine blading applications. Hollow specimens are heated by a radiation furnace; a microcomputer is used to generate simultaneous strain and temperature signals and tests are conducted under closed-loop control of axial strain. A typical mechanical strain-temperature loop has been used from 600 to 1050°C (873 to 1323 K) with peak strains at intermediate temperatures. Application to IN 100, a nickel base superalloy, is reported. Crack initiation and early crack growth are shown which are deduced from a plastic replication technique. The role of oxidation has been emphasized and the life of thermal mechanical fatigue specimens was found to be in good agreement with low-cycle fatigue results at 1000°C (1273 K).

KEY WORDS: thermal-mechanical fatigue, high temperature fatigue, nickel base superalloy, crack initiation, microcrack propagation, stress-strain response

Thermal fatigue is one of the primary life limiting factors of machinery components, due to start up and shut down operations. This is especially the case of blades in jet engines. For a long time the resistance to thermal fatigue of candidate materials has been assessed using Glenny-type tests with fluidized beds [*1,2*] and circular disks or wedge type specimens, or flame heating and wedge-type specimens (e.g., [*3*]). In these thermal fatigue (TF) tests the critical part of the specimen (edge or periphery) is submitted to thermal stress arising from transient temperature differences between various points of the test specimen, as occurs in the actual components.

More recently thermal mechanical fatigue (TMF) tests were introduced to simulate the behavior of critical parts of real components [*4,5*]. At present both kinds of tests, TF and TMF, have been used by different laboratories. Each type of test has its own advantages and drawbacks. The TF tests give an inexpensive simulation of real components, while TMF tests enable the measurement of both stress and strain. However, stress and strain in a TF specimen cannot be measured and have to be derived from a structure computation using finite element analysis as for a real component. Thus the prediction of thermal fatigue life depends upon the validity of the damage model used as much as upon the computation of the stress-strain behavior.

In TMF tests the behavior of a critical part of a component can be simulated. There is no temperature gradient across the specimen section and the stress is induced by a mechanical strain to simulate the constrained free thermal expansion of a component part (which is due to temperature gradients across the component section). The period of the TMF cycle is usually longer than that of thermal transients in real components. Thus a good choice of the mechanical

[1]Centre des Matériaux de l'Ecole des Mines de Paris, UA CNRS 866, BP. 87, 91003 EVRY Cédex, France.

strain to be used in TMF cycling relys heavily upon the temperature-stress-strain history computations of real components.

Both kinds of tests complement each other and should be used together. A detailed investigation of thermal life of wedge specimens using a burner rig has been carried out in the Centre des Matériaux in cooperation with the SNECMA Materials Department [*3,6*] and has led to our development of a TMF test facility. This test can actually simulate the behavior of a volume element under anisothermal conditions and then gives a check on both stress-strain behavior models and damage models to be used in real components.

This paper reports on these thermal-mechanical fatigue facilities and preliminary results on the thermal mechanical fatigue behavior of IN100, a cast nickel base superalloy used in jet engine turbine blades.

Material and Specimens

The alloy studied is a cast nickel based superalloy IN100. Its composition is in weight percent: 0.18C, 14.7Co, 10.3Cr, 3.15Mo, 1.01V, 4.6Ti, 5.68Al, 0.01B, 0.05Zr, 0.1Mn, 0.1Si, 0.15Fe, and balance Ni. Specimens were taken from cylindrical castings 20 mm in diameter. In this study this alloy was uncoated, but it was given a heat treatment 1150°C (1423 K) for 3 h. The average grain size is about a few millimetres for the average conditions used. Its microstructure is composed of a face-centered cubic matrix which is hardened by a high volume fraction of fine γ' precipitates, interdendritic MC carbides, and large areas of blocky γ'. Details concerning the microstructure of this alloy and its high temperature low-cycle fatigue behavior can be found elsewhere [*7*].

Because much data have been accumulated on bulk cylindrical low-cycle fatigue (LCF) specimens 8 mm in diameter, the TMF specimens were designed to be similar to the external shape of LCF specimens. Hollow specimens 15 mm in gage length were machined from bulk castings down to an external diameter of 9 mm and a thickness of 1 mm. Details of the machining procedure have been thoroughly investigated to keep hardening of the surface layers to a minimum [*8*]. Specimens are internally polished, and the external surface is polished down to 3 μm diamond paste to make easier metallographic observations.

Experimental Procedure

Principle of the TMF Test

To carry out a thermal mechanical fatigue test, a thermal and strain cycle have to be applied simultaneously to a specimen. In earlier TMF facilities the temperature and strain input were given by Data-Track systems [*4*] which limited the accuracy of both signals. The recent development of microcomputers affords more versatility and increased accuracy.

Therefore a microcomputer was used to generate two synchronous temperature and mechanical strain signals. The closed-loop control was ensured for each signal by a temperature controller and by the testing machine, respectively (Fig. 1). The imposed total strain ϵ_t is the algebraic sum of the mechanical strain ϵ_m and the thermal strain ϵ_{th}. This free thermal expansion on heating (or contraction on cooling) is memorized by the computer and the mechanical strain can be chosen at will. The test principle is shown in Fig. 1. The microcomputer imposes a temperature cycle on the specimen through the temperature controller of the furnace after digital/analogic conversion. Simultaneously the microcomputer puts a total strain cycle on the specimen through the testing machine. Both strain and temperature signals have the same period. Only the mechanical strain is "seen" by the specimen; if the thermal expansion/contraction signal is put out from the microcomputer, the specimen is thermally cycled under no stress.

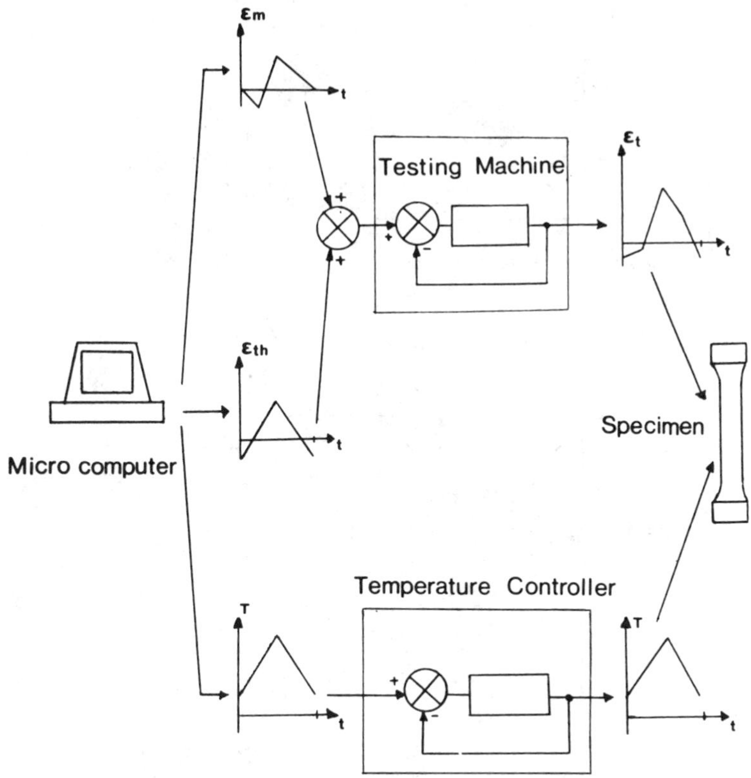

FIG. 1—*Principle of thermal-mechanical fatigue test.*

Thermal Mechanical Strain Cycle

The computation of real blades under service conditions was actually used to choose the shape of thermal and mechanical strain cycles to be applied in TMF tests. It turns out that both types of transient can be simulated by linear piecewise variations.

A basic cycle has been used with no hold time at high temperature to simulate only the transient behavior of a critical part of a blade, and was considered to be typical of pure thermal fatigue loading. A triangular waveshape temperature signal was used with maximum and minimum temperatures of 1050°C and 600°C (1323 and 873 K), respectively (Fig. 2). The ratio R_ϵ of minimum to maximum mechanical strain is −1, and the mechanical strain is zero at minimum temperature. The mechanical strain cycle has three linear parts; on heating it becomes compressive down to a peak at 900°C (1173 K) and then it increases to near zero at maximum temperature; on cooling it increases further up to a tensile peak at 700°C (973 K) and then decreases to zero at 600°C (Fig. 2). It is worth emphasizing that this cycle is more complex than simple "out of phase" or "in phase" temperature-strain cycles used by most authors [*4,5*]. This results in a temperature mechanical strain cycle with considerable hysteresis (Fig. 2), which is rather similar to published computations of Glenny-type specimens [*9*] or of wedge-type specimens [*3*] submitted to thermal fatigue.

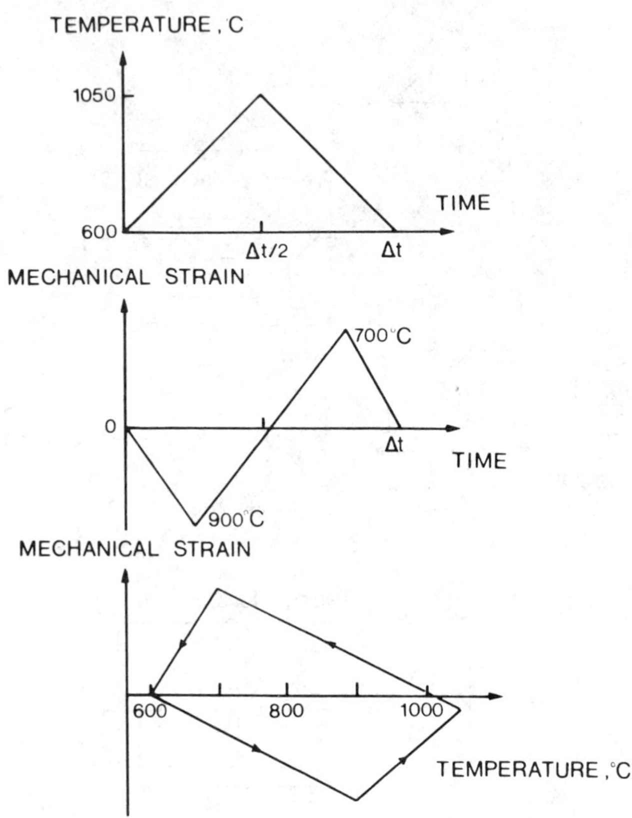

FIG. 2—*Shape of thermal-mechanical fatigue cycle.* (a) *Temperature versus time* (Δt *is the cycle period*). (b) *Mechanical total strain versus time*. (c) *Mechanical strain versus temperature.*

Experimental Procedure Facilities

Most TMF facilities previously described in the literature used induction heating [*4*]. In the present case a radiation furnace was used for TMF tests since we have gained a good deal of experience with this means of heating in high temperature LCF cycling of up to 1100°C (1373 K) [*10*]. No forced cooling was allowed, since it was necessary to have a negligible temperature difference across the specimen thickness (say, about 10 K) to ensure that the test was representative of the behavior of a volume element. Temperature was servocontrolled using a thermocouple hung on specimen gage length for the feed-back signal.

The choice of a thermal transient of constant rate from 600 to 1050°C (873 to 1323 K) imposes a cycle period ranging from a few minutes to 10 min, depending upon specimen design and computer program. In this paper early specimen design results in a period of 9.5 min, which has been used for all the tests reported. This period is, of course, much larger than real transients in blades, but this test provides a good check on the effect of the mechanical strain-temperature cycle.

The microcomputer was used to put a strain signal to the hollow specimen through either servohydraulic or screw-driven machines. The long duration of TMF tests at the rather slow displacement rate has led to the use of mainly a screw-driven machine. This screw-driven machine was completely modified to allow fatigue cycling at a variable strain rate under closed-loop control. Longitudinal strain control was ensured using a strain gage extensometer with alumina knives in contact with the specimen of improved design, developed in our group.

Testing Procedure

The TMF testing procedure was as follows. First the testing machine is put under load control and the hollow specimen is submitted to thermal cycles under zero load. The microcomputer is used only to generate temperature cycles. Within five cycles both the specimen and extensometer reached a stabilized thermal transient (dynamic thermal equilibrium) so that the fifth thermal strain cycle is digitally recorded by the micro-computer (using 710 points). Then the testing machine is put under strain control and the TFM test starts. The microcomputer generates a temperature cycle and a total strain cycle (a mechanical strain is added to the memorized thermal strain cycle) which induces a stress cycle on the specimen. This stress is continuously recorded as a function of time.

Crack growth has been monitored in all the specimens using a plastic replication technique which necessitates test interruptions at regular intervals. This technique enables cracks as small as 10 to 20 μm in length to be detected.

Surface replicas have been observed using a shadowing technique under optical microscopy and scanning electron microscopy. A number of specimens have been sectioned and broken in order to get an experimental calibration curve between surface crack length a_s and crack depth a ($a = 0.36a_s$) [*8*].

Results and Discussion

Stress-Strain Behavior

The instantaneous stress which results from TMF cycling has been continuously recorded throughout the tests. The peak stress amplitude has been found to remain constant throughout testing whatever the mechanical strain amplitude in the range investigated 0.2 to 1%. Thus stress-mechanical strain loops are stabilized after a couple of cycles; some are shown in Fig. 3. Let us consider the stabilized stress-strain loop observed for a mechanical strain range of 1%, for instance. Heating starts from 600°C (873 K) at zero mechanical strain from a residual compressive stress which has been generated during the first cycle (on the first cycle there is no stress at zero mechanical strain). On heating the mechanical strain decreases linearly and the stress becomes more compressive down to a minimum at 900°C (1173 K). Then the mechanical strain increases and so does the stress which is almost zero at maximum temperature 1050°C (1323 K). On cooling from maximum temperature the mechanical strain continues to increase and the stress becomes tensile up to a maximum at 700°C (973 K). Then the stress decreases down to a compressive value at minimum temperature when the mechanical strain decreases to zero. For this high strain amplitude, absolute values of peak stresses in compression and in tension are almost the same. The alloy is more plastic and viscoplastic at high temperatures, say, in the 900°C to 1050°C (1173 to 1323 K) range. Thus a compressive inelastic strain developed on heating, which results in a compressive residual stress at 600°C (873 K).

The curvature of the stress-strain loop, which might look anomalous with respect to usual isothermal loops, results from the variation of Young's modulus with temperature. Nevertheless, it has to be recalled that the tangent to the stress-strain curve in elastic areas is not simply equal to the elastic modulus but includes a contribution from the derivative of mechanical strain with respect to temperature.

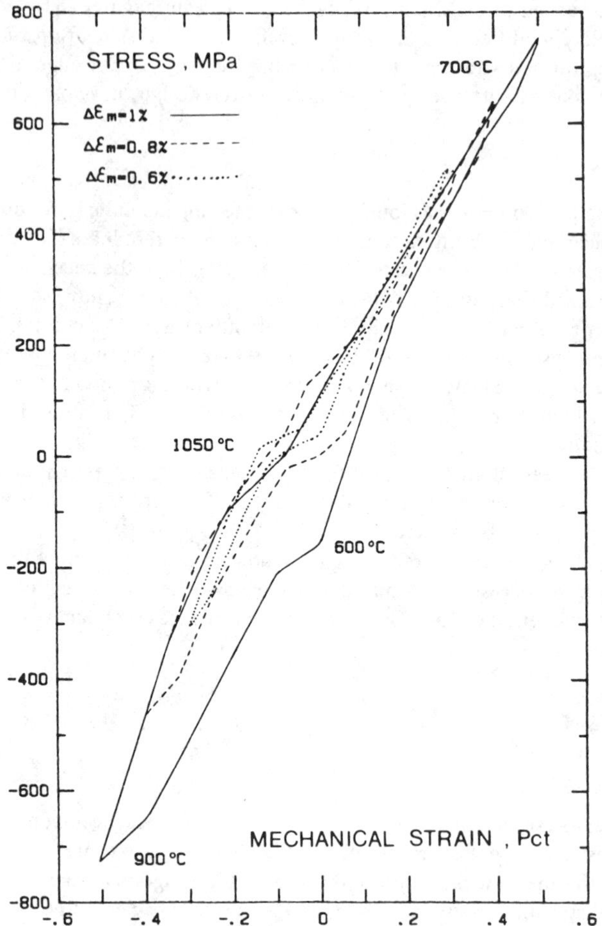

FIG. 3—*IN100: Stabilized stress-mechanical strain loops under TMF cycling from 600 to 1050°C (873 to 1123 K) for various strain amplitudes.*

Decreasing the total mechanical strain amplitude gives rise to smaller inelastic strain amplitudes and to higher mean tensile stress. The development of a mean tensile stress is caused by the shape of the mechanical strain-temperature cycle (Fig. 2*c*). The absolute value of yield stress in compression is smaller at 900°C (1173 K), where the mechanical strain is minimum compared with that of yield stress in tension at 700°C (973 K) where the mechanical strain is maximum.

Observations on the Development of Cracks

Metallographic observations have been carried out on specimens and on plastic replicas taken at various fractions of life using optical and scanning electron microscopy. Figure 4 shows the development of the main crack throughout cycling which can be observed through plastic replicas of the specimen surface. Figure 5 is an electron micrograph of the same crack in the

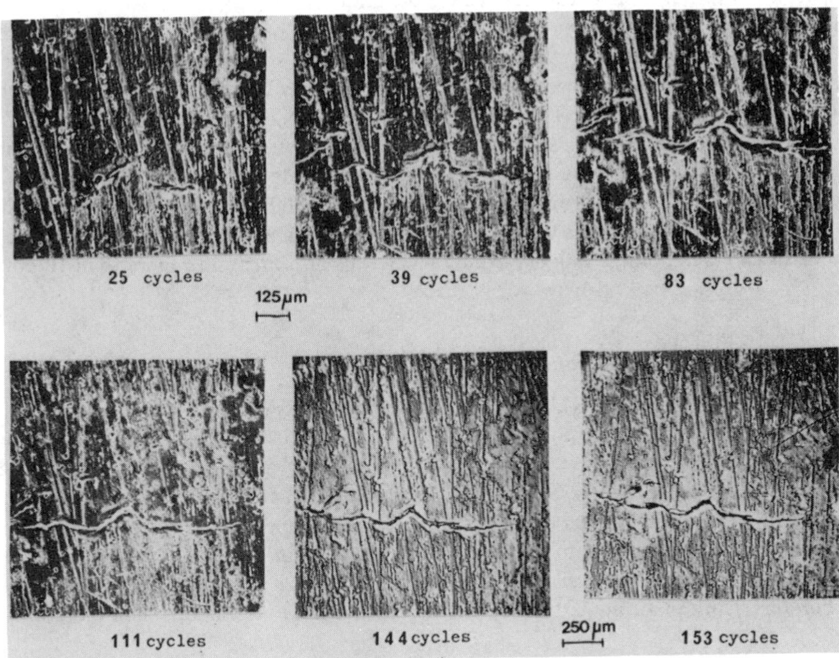

FIG. 4—*IN100: Evolution of the main crack on a TMF specimen surface for a mechanical strain range of 0.6% (scanning electron micrographs of plastic replicas at the indicated number of cycles).*

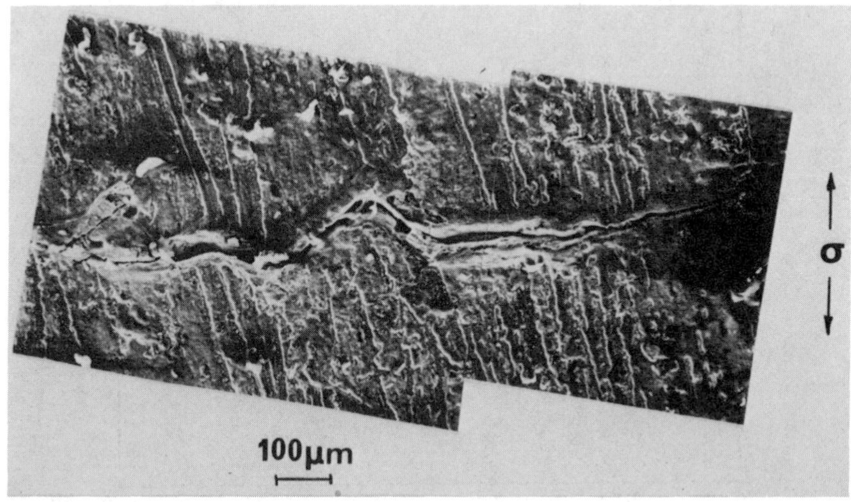

FIG. 5—*IN100: Scanning electron micrograph of the main crack as seen on specimen gage length after 160 TMF cycles at a mechanical strain range of 0.6% (this is the same crack as shown in Fig. 4).*

final stage as directly observed on the specimen surface. This indicates the good resolution that can be achieved through the plastic replication technique.

Plastic replicas have enabled us to determine the evolution of the surface length of the main crack with the number of cycles. A large number of broken specimen sections have been observed by scanning electron microscopy and a relationship has been determined between crack depth and surface crack length. The depth of the main crack was plotted as a function of the number of cycles for the various specimens studied. Some curves are shown in Fig. 6 and 7. There is a true crack initiation period where no crack can be detected on the specimen surface. This can be negligible at high strains, but can reach about 1300 cycles for a strain amplitude of 0.2%. Crack growth then occurs very rapidly, but a wide range of crack growth rates are observed at small crack lengths ranging from a few tenths of a micron to more than 10 μm per cycle.

Definition of Life to Crack Initiation

The definition of fatigue life is always a matter of controversy even for simple isothermal LCF tests. The situation is still more complex with TMF tests on hollow specimens. Complete fracture of the specimens was always avoided to enable metallographic observations to be made. The systematic use of plastic replicas makes a great number of definitions possible. Therefore a first conventional fatigue life was defined to 0.3 mm crack depth, referred to as N_i. This criterion can be used, in particular, to compare isothermal LCF and TMF test results. Data to 0.3 mm crack depth have been extensively generated on bulk LCF specimens on various alloys in our laboratory using an a-c potential drop technique [*11,10,6*]. This should allow more straight-

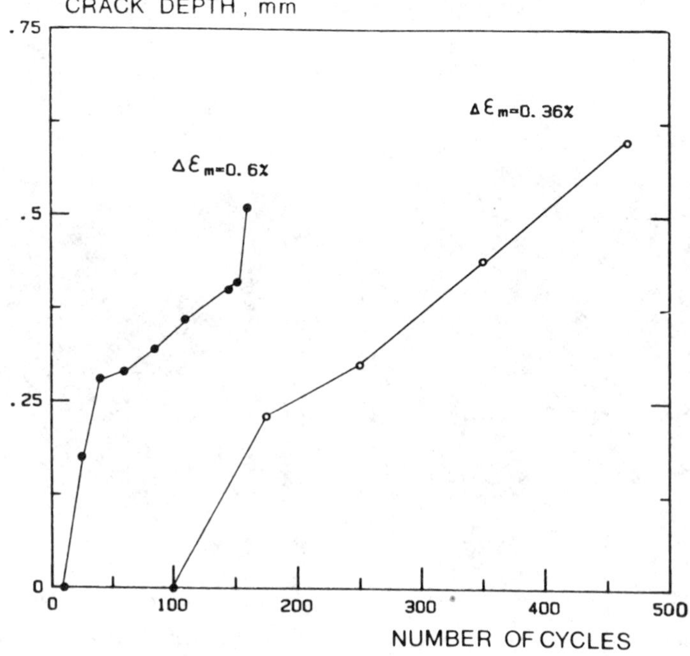

FIG. 6—*IN100: Variation of crack depth with the number of cycles for TMF strain ranges of 0.6 and 0.36%.*

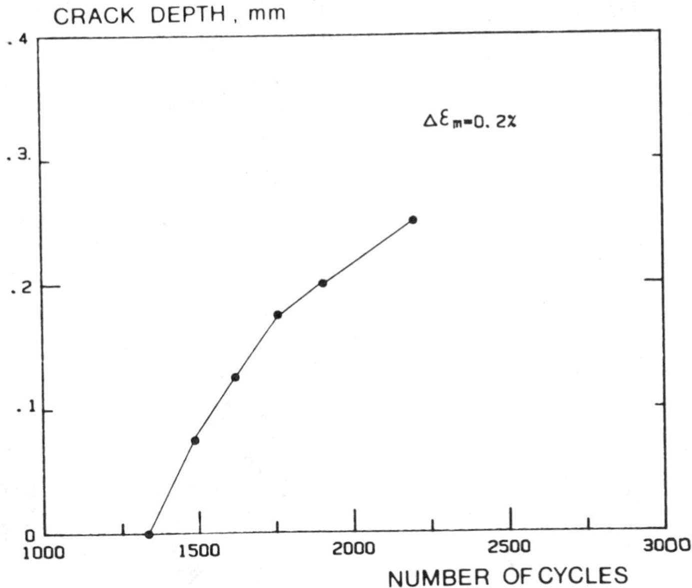

FIG. 7—*IN100: Variation of crack depth with the number of TMF cycles for a strain range of 0.2%.*

forward comparisons with isothermal LCF than using conventional total life which can be composed of a large part of crack propagation (between 1 and 2 mm fatigue crack depth).

Plastic replicas enable still better definition of crack initiation. A second crack initiation criterion was so defined at 50 μm crack depth referred to as N_a. Early crack growth takes a negligible part in life up to this crack length (Figs. 6 and 7) so that this criterion should provide a good description of true crack initiation.

Both definitions of life to crack initiation N_i and N_a have been plotted as a function of mechanical strain and stress amplitudes in Figs. 8 and 9, respectively. Life to crack initiation under TMF cycling is found to be in good agreement with Manson-Coffin and Basquin equations whatever the criterion. The difference between N_i and N_a gives a measurement of early crack growth in life. Early crack growth dominates at high strain ranges, while the crack initiation period increases from about 10 to 50% of life with decreasing strain amplitude.

Comparison with Isothermal Low-Cycle Fatigue

A number of life prediction methods have been proposed which assume that thermal fatigue or TMF damage is basically LCF damage [*11*]. Others assume thermal fatigue damage to be essentially creep damage [*12*]. A detailed investigation of the LCF behavior of IN100 has been carried out from 20 to 1000°C (293 to 1273 K) in our laboratory [*7*]. Crack initiation at high temperature has been found to occur mainly at MC carbides which are preferentially oxidized (Fig. 10). In the present work longitudinal sections of TMF specimens have been observed by scanning electron microscopy to check the locations of crack initiation which occur at an oxidized carbide. Thus crack initiation mechanisms under TMF cycling from 600 to 1050°C (873 to 1323 K) are actually the same as under LCF cycling at 1000°C (1273 K). The hypothesis that TMF damage is basically the same as LCF damage at high temperature has a physical basis for the present material.

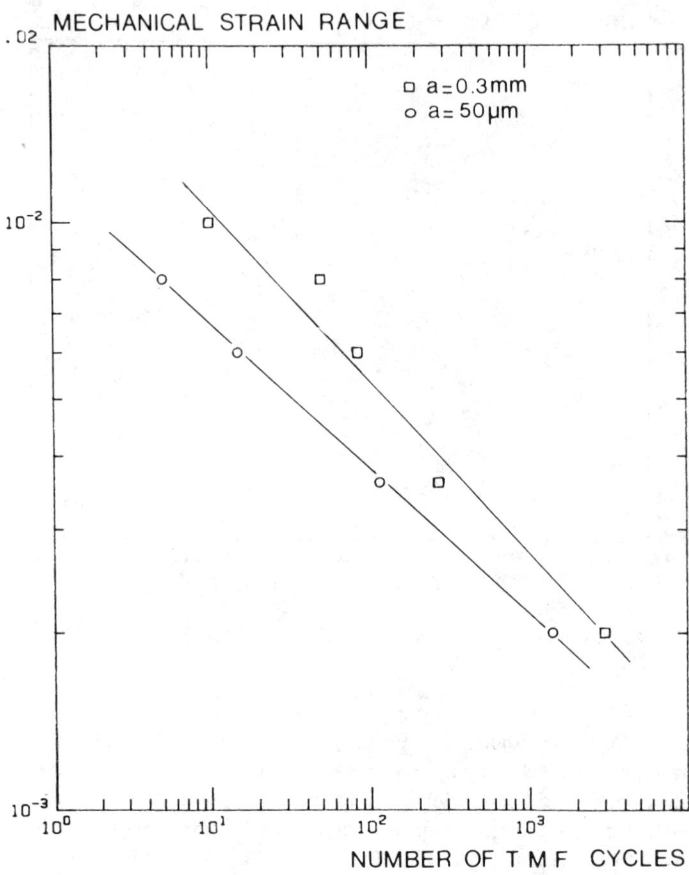

FIG. 8—*IN100: Variation of the number of TMF cycles to 50 μm and 0.3 mm crack depth with the mechanical strain range.*

A quantitative comparison has been made between TMF data to crack initiation and LCF results. The LCF data established at 1000°C (1273 K) have been used for this purpose in the absence of sufficient information at the maximum temperature of TMF cycles, 1050°C (1323 K). Life to 0.3 mm crack depth was used in this comparison, since relevant data were available for LCF tests from potential drop measurements [*13*]. Figure 11 shows TMF and LCF life as a function of mechanical strain range. A unique curve is observed for LCF results at 4×10^{-3} Hz (which is almost the frequency of TMF tests 1.7×10^{-3} Hz) and at 5×10^{-2} Hz. The TMF and LCF data are in good agreement within a factor of 2. In the present material a good estimate of life under TMF cycling can be deduced from isothermal LCF near the maximum temperature in the same frequency range using mechanical strain range, when crack initiation mechanisms are the same. This agreement is probably due to the use of a life criterion to a fairly short crack depth (0.3 mm). This agreement could nevertheless be partly coincidental, since life to 0.3 mm crack depth still encompasses a crack growth period. A better and surer agreement should be observed using a more proper crack initiation criterion—for example, to 50 μm crack depth—to compare LCF and TMF lives. Current work is under progress along these lines.

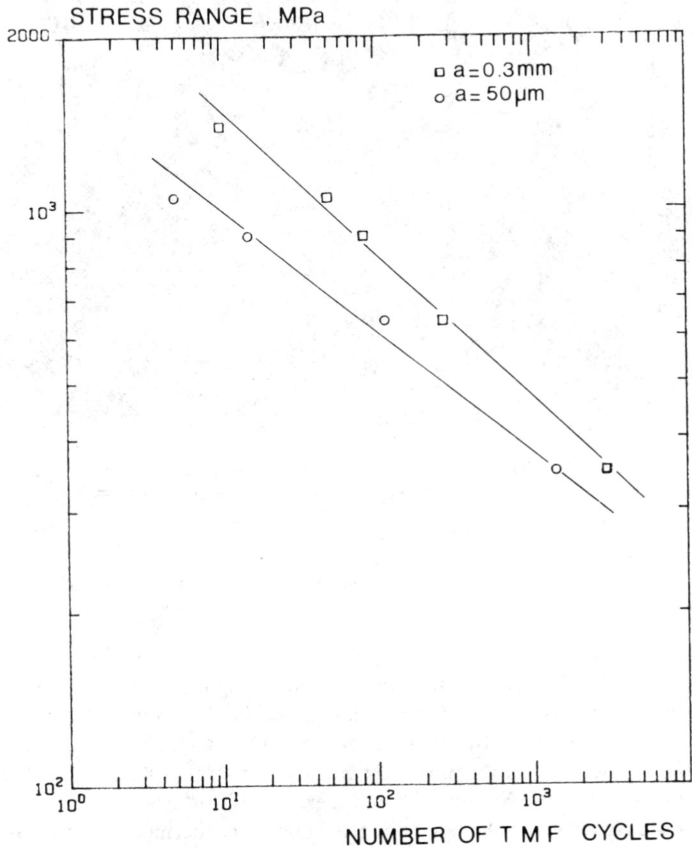

FIG. 9—*IN100: Variation of the number of TMF cycles to 50 μm and 0.3 mm crack depth with the stress range.*

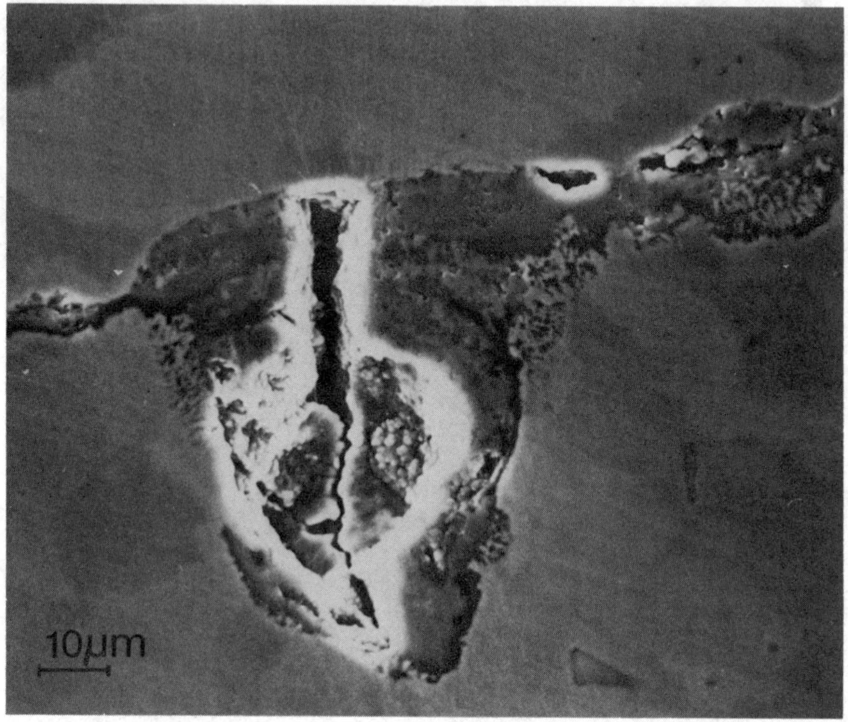

FIG. 10—*IN100: Crack initiation at preferentially oxidized interdendritic areas (scanning electron micrograph on a longitudinal TMF specimen section; mechanical strain amplitude = 0.36%).*

Conclusions

A thermal mechanical fatigue test facility has been designed which enables mechanical strain-temperature transients occurring in real components to be simulated.

This test facility complemented by a plastic replication technique has been shown to give a reliable definition of crack initiation and early crack growth data under thermal-mechanical fatigue cycling in the case of IN100 a nickel base superalloy. Life to crack initiation was found to obey a classical Manson-Coffin behavior as in conventional isothermal low-cycle fatigue tests.

Crack initiation was found to occur at preferentially oxidized MC carbides in thermal-mechanical fatigue as in low-cycle fatigue at high temperature. In addition, the life of thermal-mechanical fatigue specimens was found to be in good agreement with that of low-cycle fatigue specimens when tested near the maximum temperature in the same frequency range.

Acknowledgments

This work is part of a research project which has been defined in collaboration with SNECMA (Société Nationale d'Etude et de Construction de Moteurs d'Aviation). The authors are indebted to SNECMA engineers for their continuous interest throughout this work. Financial support of the overall project by SNECMA is gratefully acknowledged.

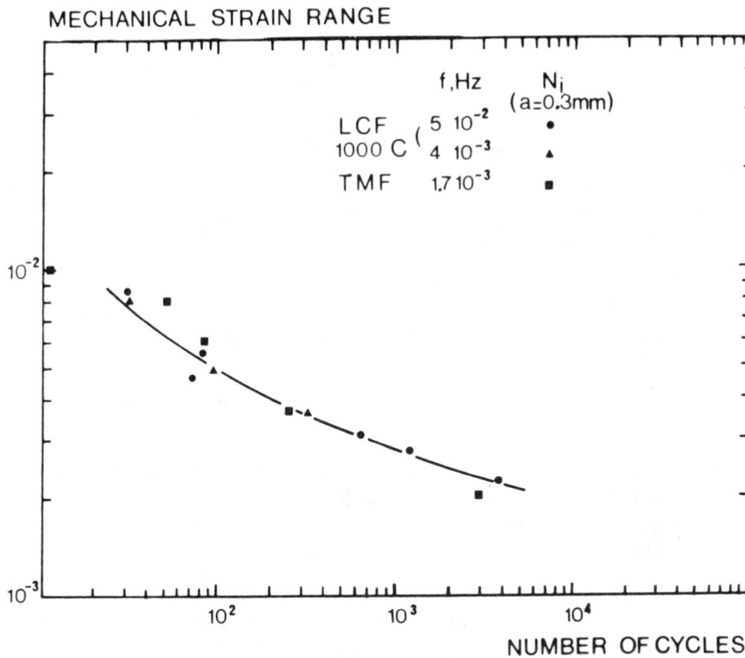

FIG. 11—*IN100: Comparison of crack initiation life versus total mechanical strain range under TMF cycling and LCF crack initiation life at 1000°C (1273 K).*

References

[1] Glenny, E., Northwood, J. E., Shaw, S. W. K., and Taylor, T. A., *Journal of the Institute of Metals,* Vol. 87, 1958–1959, pp. 294–302.

[2] Woodford, D. A. and Mowbray, D. F., *Materials Science and Engineering,* Vol. 16, 1974, pp. 5–43.

[3] Rezai-Aria, F., Rémy, L., Herman, C., and Dambrine, B. in *Mechanical Behavior of Materials - IV,* J. Carlsson and N. G. Ohlson, Eds., Pergamon Press, Oxford, Vol. 1, 1984, pp. 247–253.

[4] Hopkins, S. W. in *Thermal Fatigue of Materials and Components, ASTM STP 612,* D. A. Spera and D. F. Mowbray, Eds., American Society for Testing and Materials, Philadelphia, 1976, pp. 157–169.

[5] Rau, C. A., Jr., Gemma, A. E., and Leverant, G. R. in *Fatigue at Elevated Temperatures, ASTM STP 520,* American Society for Testing and Materials, Philadelphia, 1973, pp. 166–178.

[6] Rémy, L., Rezaï-Aria, F., François, M., Herman, C. Dambrine, B., and Honnorat, Y., "Application of Isothermal Fatigue to the Study of Thermal Fatigue," Final Report, F6, COST50-Round III, 1984.

[7] Reger, M., thesis, Ecole des Mines de Paris, 1984.

[8] Malpertu, J. L. and Rémy, L., unpublished results, Centre des Matériaux, 1983.

[9] Mowbray, D. F. and McConnelee, J. E. in *Thermal Fatigue of Materials and Components, ASTM STP 612,* D. A. Spera and D. F. Mowbray, Eds., American Society for Testing and Materials, Philadelphia, 1976, pp. 10–29.

[10] Reuchet, J. and Rémy, L., *Materials Science Engineering,* Vol. 58, 1983, pp. 19–32.

[11] Taira, S. in *Fatigue at Elevated Temperatures, ASTM STP 520,* A. E. Carden, A. J. McEvily, and C. H. Wells, Eds., American Society for Testing and Materials, Philadelphia, 1973, pp. 80–101.

[12] Halford, G. R. and Manson, S. S. in *Thermal Fatigue of Materials and Components, ASTM STP 612,* D. A. Spera and D. F. Mowbray, Eds., American Society for Testing and Materials, Philadelphia, 1976, pp. 239–254.

[13] Rémy, L., Reger, M., Reuchet, J., and Rezaï-Aria, F. in *Proceedings,* Conference on High Temperature Alloys for Gas Turbines, Liège, 4–6 Oct. 1982, R. Brunetaud, D. Coutsouradis, T. B. Gibbons, Y. Lindblom, D. B. Meadowcroft, and R. Stickler, Eds., Reidel, Dordrecht, 1982, pp. 619–632.

DISCUSSION

J. Polák[1] *(written discussion)*—What is your mechanism of crack initiation assisted by carbide oxidation? Is it a formation of a notch-like geometry or does the crack grow from the oxide into the metal?

J. L. Malpertu and L. Rémy (authors' closure)—Your question is quite difficult to answer. In this kind of alloy primary interdendritic carbides are preferentially oxidized; however, it is not only the carbide which is oxidized at a higher rate than the dendrite material but also the interdendritic area, which has a local composition different from the average. A number of observations at the outer surface of the specimens suggest that cracks nucleate inside the carbides, and this may result from a notch-like geometry effect. Very rapidly, however, the crack grows from the outer surface into the metal and this occurs mainly in interdendritic areas, often along the carbides and not inside the carbides.

N. Marchand[2] *(written discussion)*—(1) Every time you make a replica you have to stop the test, wait for the specimen to cool, make the replica, heat the specimen again, wait for the temperature to stabilize, and start the test again. Are you concerned about introducing transient phenomena in your test?

(2) Your $\Delta T(T_{max} - T_{min})$ is about 450°C. You also show no difference between isothermal life and TMF lives. This behavior has also been shown by some Japanese workers; in particular, they show that the difference in life between isothermal and TMF increases with increasing ΔT. With higher ΔT use for your tests, are you expecting to obtain the same results (i.e., a difference between N_{iso} and N_{TMF})?

J. L. Malpertu and L. Rémy (authors' closure)—(1) You raise an interesting point. We carry out only interrupted thermal-mechanical fatigue tests to take replicas, and we do not have results of tests without any interruption. However, a number of experiments have been carried out under isothermal conditions, and in most cases test interruptions do not bring any alteration in the total life of the specimens. Test interruptions certainly introduce transient strains in the test, but the hysteresis stress-strain loop is stabilized within a few cycles.

(2) We have found no life difference between isothermal and thermal mechanical fatigue. Our temperature of 450°C is fairly high already, and we do not think that a higher temperature variation will give very different results. The main problem is not the temperature difference itself but rather what life criterion is used. In our case we used replicas to achieve a life criterion to 0.3 mm crack depth, one rather sensitive to initiation. A criterion to a shorter crack depth should still be better, but relevant data were not available for isothermal fatigue. If your life criterion encompasses a large part of both initiation and propagation, however, you may expect great difficulties when you compare isothermal fatigue and thermomechanical fatigue data.

Eric Jordan[3] *(written discussion)*—Is your radiant heating based on light bulb or furnace heating? How do you keep the shadow of the extensometer off the specimen?

J. L. Malpertu and L. Rémy (authors' closure)—We are aware that most people use induction heating for thermal-mechanical fatigue tests, but we have gained much experience in our labo-

[1]Institute of Physical Metallurgy, Czechoslovak Academy of Sciences, Brno, Czechoslovakia.
[2]Massachusetts Institute of Technology, Cambridge, MA 02139.
[3]University of Connecticut, Storrs, CT 06268.

ratory on high temperature fatigue testing using radiation furnaces. Radiation using four bulbs (power: 1500 W each bulb) enables specimens to be heated rapidly up to 1100°C (1373 K). The inner cross section of the furnace is made of four ellipses. The number of bulbs used minimizes the shadow of the extensometer on the specimen, as checked by thermocouple measurements at various specimen locations.

John W. Holmes,[1] Frank A. McClintock,[2] Kevin S. O'Hara,[3] and Maureen E. Conners[2]

Thermal Fatigue Testing of Coated Monocrystalline Superalloys

REFERENCE: Holmes, J. W., McClintock, F. A., O'Hara, K. S., and Conners, M. E., **"Thermal Fatigue Testing of Coated Monocrystalline Superalloys,"** *Low Cycle Fatigue, ASTM STP 942,* H. D. Solomon, G. R. Halford, L. R. Kaisand, and B. N. Leis, Eds., American Society for Testing and Materials, Philadelphia, 1988, pp. 672–691.

ABSTRACT: Induction heating of stepped-disk specimens to study thermal fatigue of coated superalloys is well suited for studying anisotropic alloys and is easily adaptable for corrosive or inert atmosphere testing. Details of the experimental apparatus and results of a thermoelastic finite element analysis to find the stress-strain history are given.

To illustrate the technique, the effect of cyclic thermal strains on the durability of an aluminide coating applied to a typical monocrystalline nickel-base superalloy was studied in air. Results show that both the compressive strain encountered on specimen heatup and the tensile strain encountered on cooldown critically affect the aluminide coating degradation. After 6000 cycles of heating from 520 to 1080°C in 5 s, followed by 30 s cooling, 80% coating penetration by scalloping was observed. A similar treatment with 6 s cooling resulted in alumina-filled "cracks" extending into the substrate.

KEY WORDS: thermal fatigue, induction heating, stepped-disk specimen, aluminide coating, monocrystalline, nickel-base, superalloy, René N4, finite element

Coated monocrystal blades and vanes are currently used in advanced gas turbines. The improved high-temperature creep and thermal fatigue characteristics of the low modulus ⟨100⟩ oriented airfoils allow higher turbine operating temperatures and improved thermodynamic efficiency.

Currently, the leading and trailing edges of first stage blades and vanes can rise from 500 to 1100°C in 4 to 10 s during takeoff. During turbine shutdown, after landing and thrust reverse, the airfoil temperature can decrease from 1080 to 500°C in under 10 s. As a result of high metal temperatures and the thermal strains due to temperature transients and internal blade cooling, coating degradation has become a dominant life-limiting phenomenon for gas turbine blades and vanes.

The properties of a substrate are in many cases affected by the thermal cycles associated with coating application [*1–5*]. The oxidation, corrosion, and mechanical properties of the protective coating itself are strongly influenced by substrate composition due to diffusion of substrate elements into the coating during coating application or interdiffusion between the coating and

[1]Graduate Student, Department of Materials Science and Engineering, Massachusetts Institute of Technology, Cambridge, MA 02139.

[2]Professor and Undergraduate Student, respectively, Department of Mechanical Engineering, Massachusetts Institute of Technology, Cambridge, MA 02139.

[3]Metallurgist, General Electric Company, Lynn, MA 01901.

substrate during high-temperature exposure [6–10]. Therefore, when determining the mechanical behavior of a substrate or the merit of using a specific coating, the substrate and coating should be examined as a unit. This testing should take into consideration the temperature and strain history of the coated component as well as the operating environment.

Currently there are several techniques available for studying the effect of thermal fatigue on coating-substrate durability. These include gas-burner rigs [11–13] and fluidized beds [14–16]. Gas-burner rigs suffer from high costs and a complex heat-transfer analysis, which makes it difficult to correlate coating durability with the stress-strain history of test specimens. The fluidized bed technique of studying thermal fatigue has two serious drawbacks: (1) incompatibility with corrosive atmosphere testing, and (2) lack of versatility in studying arbitrary temperature histories or mission profiles due to lack of easy control over specimen heating and cooling rates. (For a given maximum and minimum bed temperature, changing the heating or cooling rate of a specimen requires a change in specimen dimensions or geometry.) Thermomechanical fatigue tests are expensive and limit heating rates to about 500°C/min.

Induction heating of stepped-disk specimens is an alternative to the use of burner rigs and fluidized beds in thermal fatigue studies. The induction heating technique [17] allows one to subject test specimens to a variety of temperature and strain histories typical of those encountered by the leading and trailing edges of gas turbine blades and vanes. The technique can also be easily modified to allow thermal fatigue testing in inert or corrosive gas atmospheres.

While the main point here is the induction heating technique, it will be demonstrated by using it to study the effect of various strain histories on the durability of an aluminide-coated monocrystalline superalloy (René N4) [18,19].

Experimental Procedure

Test Specimens

The constant thickness disk specimens used in an earlier feasibility study [17] have been changed to stepped-disk specimens (Fig. 1) to allow periphery radii typical of the trailing edge of a gas turbine airfoil and to allow modeling of the 4 to 10 s heating and 6 to 30 s cooling experienced by the edges of internally cooled gas turbine airfoils.

For the present study, specimens were machined from a single crystal rod of René N4 such that their faces were normal to the [001] crystal growth direction (Fig. 1). Due to the anisotropic elasticity and radially inhomogeneous temperatures in the substrate, the strain history along the specimen periphery is a function of angular position, varying from a maximum along ⟨100⟩ directions to a minimum along ⟨110⟩ directions. Thus, this specimen orientation, which includes four radial ⟨100⟩ and ⟨110⟩ directions, allows use of a *single* specimen to study thermal fatigue in the coating under the different strain amplitudes along the specimen periphery.

Alloy Composition, Heat-Treatment, and Coating

The chemical composition of the monocrystalline René N4 used in this study is given in Table 1. Prior to the machining operations, the 19 mm diameter crystal rod was given a solutionizing heat treatment in vacuum: 1270°C for 2 h, followed by furnace cooling (average cooling rate measured at the surface of the rod was 40°C/min to 500°C).

After machining, specimens were aluminide coated [20,21]. The coating was applied by a low-activity pack-aluminization technique (CODEP B-1). The coating cycle was: 1050°C for 4 h in argon, followed by cooling to 25°C. To obtain the correct substrate microstructure after aluminization, specimens were heated in vacuum to 1050°C, held for 15 min, then cooled to 25°C in 5 min. Specimens were next aged for 16 h at 870°C in vacuum.

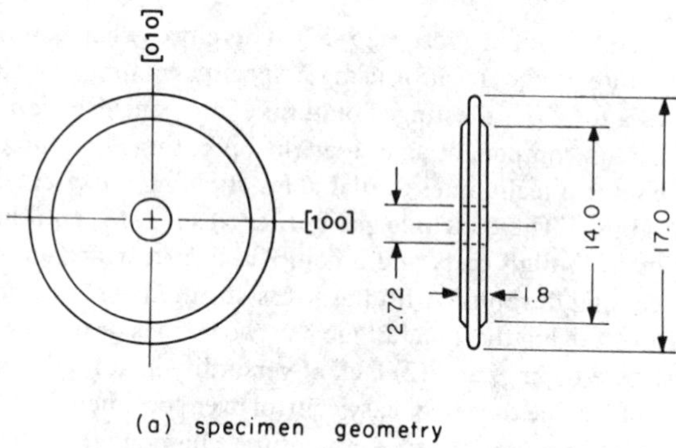

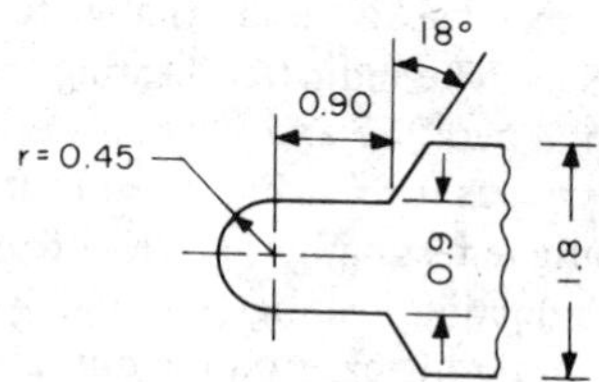

FIG. 1—*Detailed drawing and orientation of stepped-disk specimen used in coating durability studies (all dimensions in millimetres).*

TABLE 1—*Chemical composition (weight percent) of René N4.*

Al	Cb	Cr	Co	Cu	Fe	Hf	Mo	Ni	Ta	Ti	V	W
3.7	0.5	9.3	7.5	0.1	0.2	0.1	1.5	bal.	4.0	4.2	0.1	6.0

Test Apparatus

Thermal fatigue was obtained by inductively heating stepped disk specimens (Fig. 1) around their peripheries using a Lepel[4] solid-state 2.5 kw (450 kHz) induction generator coupled to a plate concentrator coil (see Ref *17* and Figs. 2 to 4). This technique relies on the skin effect obtained with high frequency induction heating of metals (at 450 kHz the skin depth[5] obtained in nickel-base superalloys is approximately 1 mm). The concentrator coil is constructed with five turns of 3.2 mm diameter copper tubing and has a base-plate thickness of 1.5 mm. The base-plate was machined with a 0.3-mm-wide radial air-gap, from the inner to outer radius. To

[4]Lepel Corporation Model T-2.5-1-KC1-B3W-T; supplied with a Research Incorporated Set-Point Programmer Model 73211 and a West Temperature Controller Model 1646A-1-9441A.

[5]The skin depth (the depth at which the strength of the magnetic field falls to 0.3679 of its surface value) is given by $\delta = \sqrt{(\rho/4\pi^2 \times 10^{-7} f\mu_r)}$, where ρ = resistivity ($\Omega \cdot$ m), f = frequency (Hz), and μ_r = relative permeability (=1 for nonmagnetic materials).

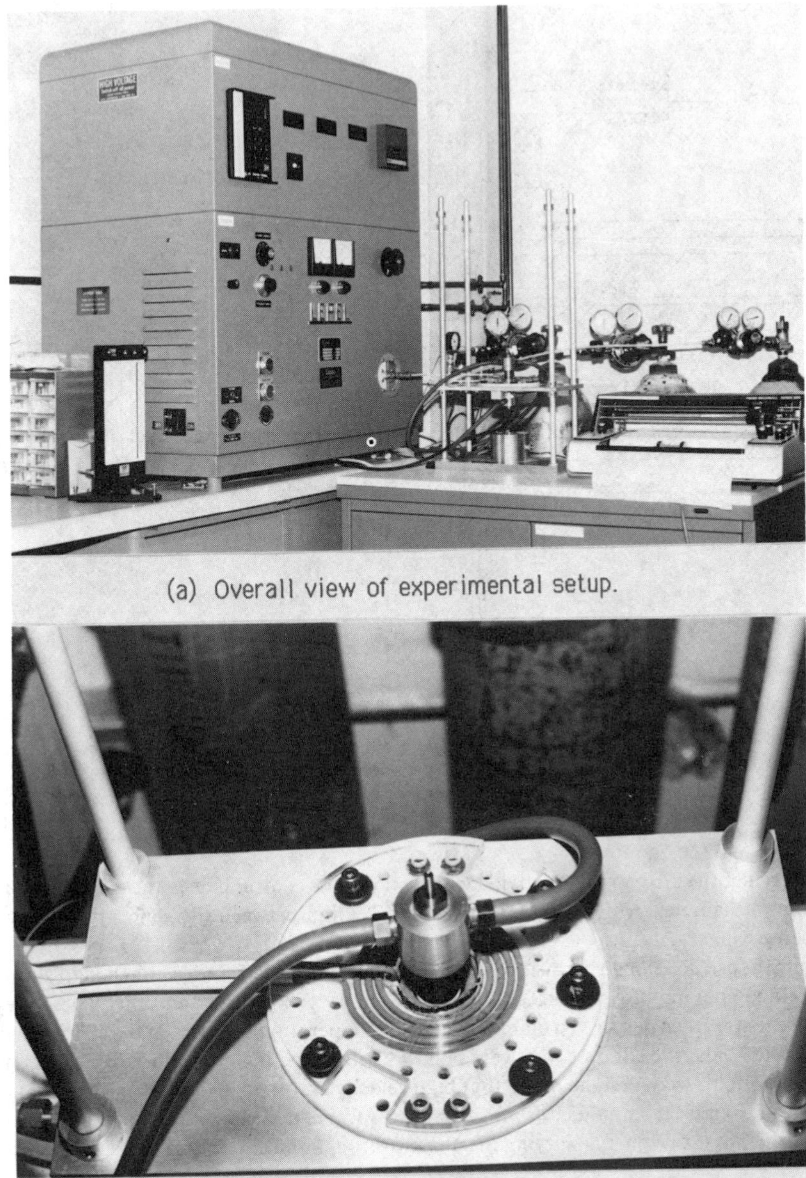

(a) Overall view of experimental setup.

(b) Top view of apparatus showing spiral-coil and cooling manifold.

FIG. 2—*Induction heating apparatus used to study thermal fatigue of coated monocrystals.* (a) *Overall view of experimental setup.* (b) *Top view of apparatus showing spiral-coil and cooling manifold.*

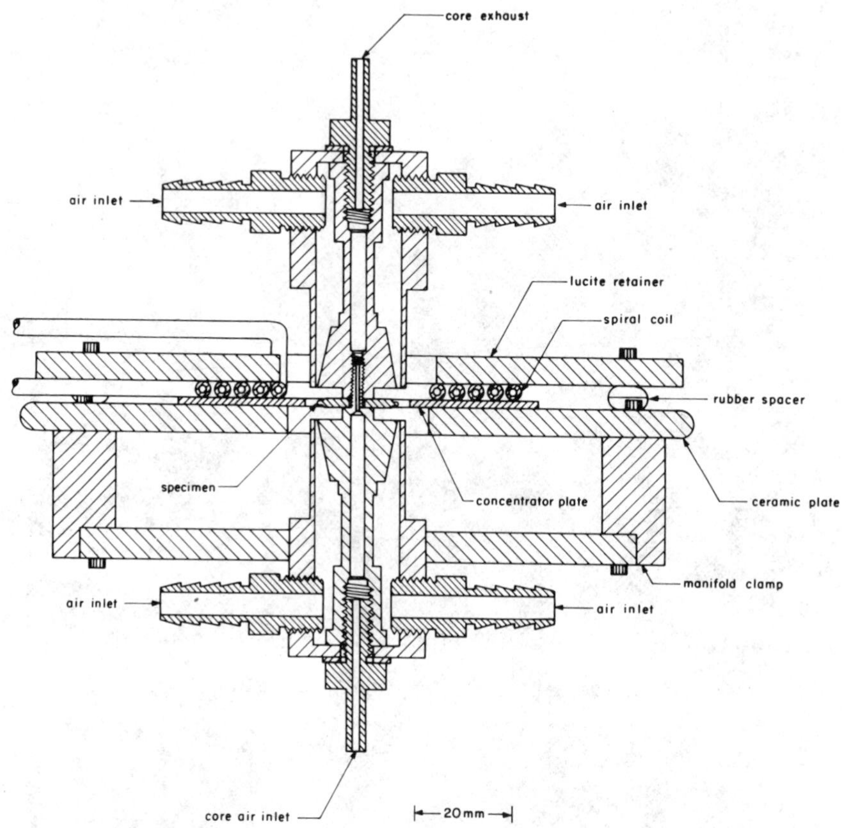

FIG. 3—*Schematic of test apparatus for studying thermal fatigue in air.*

prevent arcing, the air-gap was insulated with Teflon. The central hole in the concentrator plate is 22 mm in diameter, which gives an air-gap of 2.5 mm between the concentrator plate and specimen periphery.

Specimens are positioned in the load coil using 304 stainless-steel rods attached to aluminum[6] forced-air cooling manifolds (Figs. 3 and 4). The support rods thread together where they pass through the specimen center. To maintain a radial temperature gradient in the specimen during steady-state holds, the support rods were hollow to allow cooling air to flow through them. The cooling manifolds were designed such that high velocity air (up to 200 m/s), exiting from a 0.17-mm gap in a circular annulus 2.5 mm from the specimen surface, impinged directly on the reduced thickness periphery. Varying the airflow through the annulus allows control of the ten-

[6]Due to the close proximity of the cooling manifolds to the specimen surface and the concentrator plate, radiation from the specimen and fringing of the magnetic field cause heating of the manifolds. For the aluminum manifolds the maximum manifold temperature reached was 350°C (for a specimen temperature of 1080°C). Had the manifolds been constructed of a metal of higher resistivity (e.g., stainless steel or nickel) the magnetic-field fringing would have resulted in much higher manifold temperatures, which could result in fatigue damage of the cooling manifolds. The expected increase in manifold temperature with resistivity was shown by tests with a 304 stainless steel manifold; for a specimen temperature of 1080°C the stainless steel manifold reached a temperature of 750°C versus 350°C for the aluminum manifold.

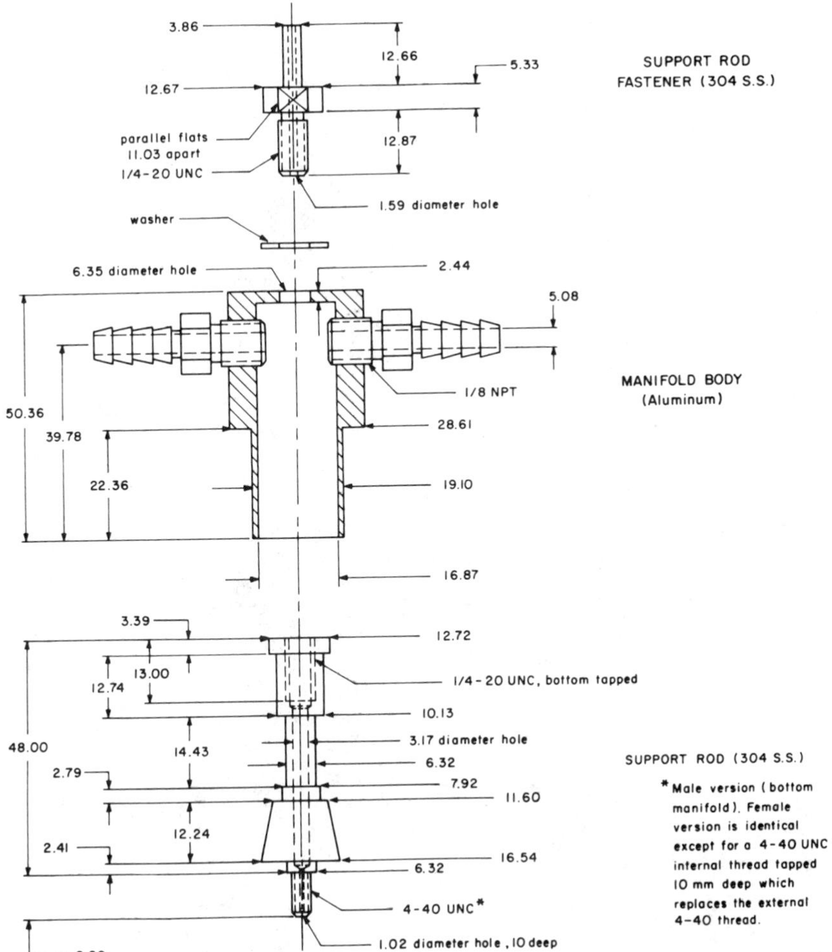

FIG. 4—*Detailed drawing of cooling manifold showing manifold body, specimen support rod, and support rod fastener (all dimensions in millimetres).*

sile strains encountered during the cool-down portion of the temperature cycle. A microprocessor-based temperature controller was used to give automatic control of the induction generator and the on/off air supply to the cooling manifolds.

The rapid temperature transients (Figs. 5*a*, *b*, *c*, *d*) and the dependence on measured temperatures rather than heat transfer calculations for the stress-strain analysis necessitated extreme care in using thermocouples for temperature measurement and control. Several procedures were used in an effort to increase the accuracy and response time of the thermocouple measurements. To increase the contact area between the junction and specimen and to position the thermocouples accurately, 0.3 mm diameter by 0.15 mm deep holes were drilled at radii of 3.5, 4.0, 5.0, 6.0, and 6.5 mm, and chromel-alumel thermocouples (0.13 mm wire diameter) having 0.3 mm diameter ball junctions were spot-welded into the positioning holes. Had the positioning holes been drilled much deeper, say, 0.25 mm, the increased surface area inter-

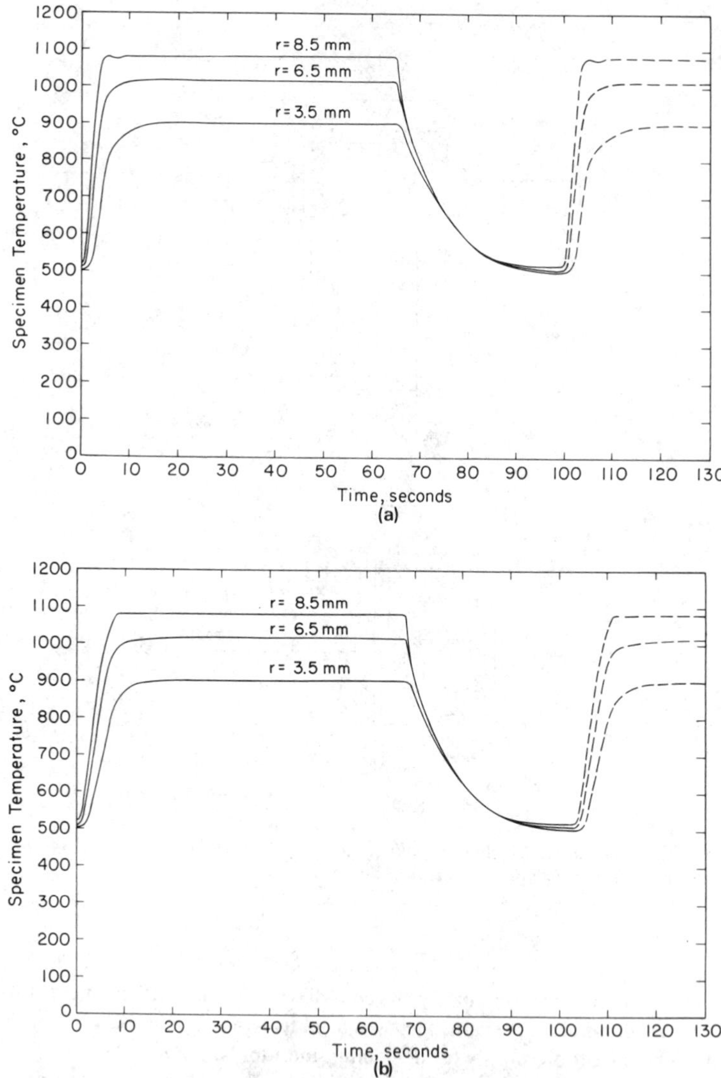

FIG. 5—*Temperature history of stepped-disk specimens used in coating durability study (air cooling.* (b) *8 s heating and 30 s cooling.* (c) *35 s heating and 30 s cooling.* (d) *8 s heating and*

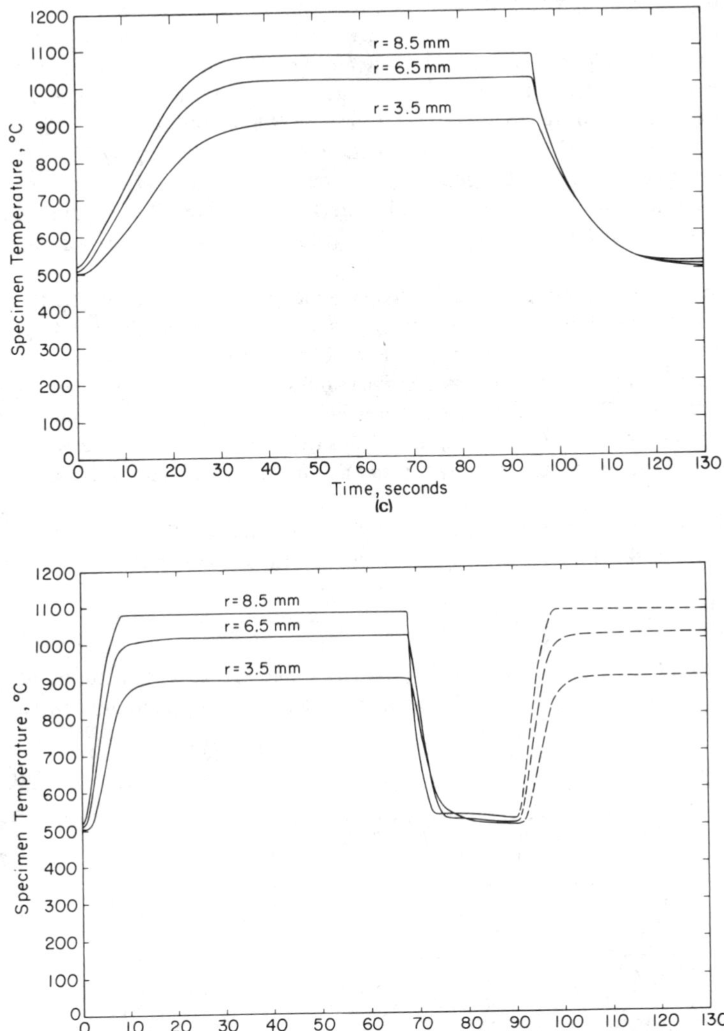

environment). Heating from 520 to 1080°C, with 60 s hold at 1080°C. (a) *5 s heating and 30 s 6 s cooling.*

cepted by the magnetic field would have resulted in localized heating. It was undesirable to locate positioning holes along the highly stressed periphery (radius = 7.15 to 8.5 mm). To optimize the response time and accuracy of temperature measurements made in this region, the thermocouple junctions were thinned prior to attachment so that the overall junction size measured 0.3 mm square by 0.1 mm thick.

To ensure that the specimens heated uniformly, six thermocouples were located 60° apart at a distance of 2 mm from the specimen periphery. The specimen was then positioned such that one of these thermocouples was adjacent to the base-plate air-gap. The lateral position of the specimen relative to the coil was adjusted[7] until the temperature difference between these six thermocouples was less than 5°C. For the temperatures of interest in this study (up to 1080°C) the 0.13 mm diameter chromel-alumel thermocouples could only be used for roughly 25 h before oxidation led to reading errors and embrittlement. Therefore, for the long-term tests, two Pt/Pt-10% Rh control thermocouples (0.20 mm wire diameter) were positioned at a radius of 6.5 mm, and a temporary chromel-alumel thermocouple was located in the periphery as an initial check.

In contrast to other coil geometries (e.g., helical coils), the magnetic field produced by the plate concentrator coil did not interfere with temperature measurements. For the 520 to 1080°C temperature range of interest in this study, the response time of the thermocouple/chart-recorder system used to record temperature history was less than 0.3 s. The thermocouple temperature measurements were within 2.1% of the temperatures indicated by an optical pyrometer (width of pyrometer filament = 0.8 mm) and within 1.1% of the temperature indicated by melting-point standards applied to the specimen by stencil and airbrush as 0.5 mm wide annular rings.

For this specimen, the control system could be tuned to allow periphery heating by 600°C in 3 s with less than a 2.5% overshoot and cooling by 600°C in 4 s with 3% overshoot.

Thermal Fatigue Tests

To determine the effect of strain history on coating oxidation and cracking, four different periphery temperature histories were used (Table 2). The strain histories corresponding to these

TABLE 2—*Periphery heating and cooling times, between 520 and 1080°C, with a 60 s hold at 1080°C.*

Heating Time, s	Cooling Time, s	Figure
5	30	5*a*
8	30	5*b*
35	30	5*c*
8	6	5*d*

[7]Due to the reduction in magnetic field intensity near the 0.3 mm wide air-gap in the concentrator-coil base-plate, uniformity of periphery heating was improved with the specimen offset approximately 0.8 mm from the geometric center of the base-plate hole, towards the *base-plate air-gap.* The variation in circumferential periphery temperature was approximately 25°C with the specimen centered in the coil versus 5°C with the specimen offset. Uniformity of periphery heating can also be improved by increasing the *specimen-to-base-plate air-gap.* Note, however, that heating efficiency decreases with increasing air-gap. Magnetic-field fringing effects are also accentuated by increasing this air-gap.

temperature histories are discussed in the next section. The rapid heating of Figs. 5*a, b* and the slow heating of Fig. 5*c* are included to show how the magnitude of the compressive periphery strains, encountered during specimen heatup, affect coating durability. The slow cooling of Fig. 5*b* and the fast cooling of Fig. 5*d* (30 s versus 6 s) are included to show the effect of tensile strains on coating durability.

Stress and Strain History of Test Specimens

The thin skin analysis used in Ref *17* was found to be inadequate for the diffuse radial temperature gradients encountered in this study, since it does not account for core deformation, which can have a large effect on strain history. It was therefore necessary to perform a thermoelastic finite element analysis.

Finite-Element Model

The elastic stress-strain history of the test specimens was determined using the finite-element program ABAQUS [*22*]. To determine whether or not it was appropriate to use generalized plane-stress elements for the stress-strain analysis of the stepped-disk specimen, an axisymmetric *z-r* isotropic mesh with the actual specimen thickness was compared to the same mesh with one tenth the thickness. The results show that, except in the step itself, the through-the-thickness stresses were negligible and the circumferential stresses agreed within 3%.

Due to symmetry, it was only necessary to model a 45° octant of the anisotropic stepped-disk specimen. The 136 element mesh used (Fig. 6) was made up of 8-node, bi-quadratic, plane-stress elements, each with nine Gaussian integration points. The element width near the hole was chosen to ensure that the stress concentration at the hole, linearly extrapolated from integration points, would be accurate to within 1%. To maintain symmetry, the displacements of all nodes along the ⟨100⟩ and ⟨110⟩ radial directions were constrained to the radial direction and the shear stresses were set to zero.

The finite element analysis was simplified by noting that the thin coating (<0.1 mm versus a specimen radius of 8.5 mm) gives negligible constraint. It can therefore be assumed the coating undergoes the same *strain* history as the substrate periphery. Prior to yielding of the coating, the *stress* history of the coating differs by the ratio of the elastic moduli of the coating and substrate. For the temperature range 25 to 1100°C the elastic modulus of the coating is approximately 40% higher than the ⟨100⟩ elastic modulus of the substrate and approximately 20% lower than the ⟨110⟩ substrate modulus. Since René N4 and CODEP B-1 coating have similar thermal expansion coefficients (Table 3) over the temperature range of interest, the effect of thermal expansion mismatch on coating stress history would be small in comparison to elastic modulus effects.

Material Properties

The stiffness constants and mean coefficient of thermal expansion of René N4, which are required as input data for the finite element program, are given in Fig. 7 and Table 3, respectively. The original data for elastic moduli and Poisson's ratio, determined by an ultrasonic technique, indicated an abrupt increase in bulk modulus for temperatures greater than 870°C. This abrupt increase in bulk modulus was due to Poisson's ratio approaching 0.5, apparently caused by inelasticity present even at the 5000 Hz at which the data were taken. To circumvent this problem, we extrapolated the bulk modulus K versus temperature, upward from 870°C, and used these extrapolated values to normalize E_{100} and G_{100}. E_{100}/K and G_{100}/K were next

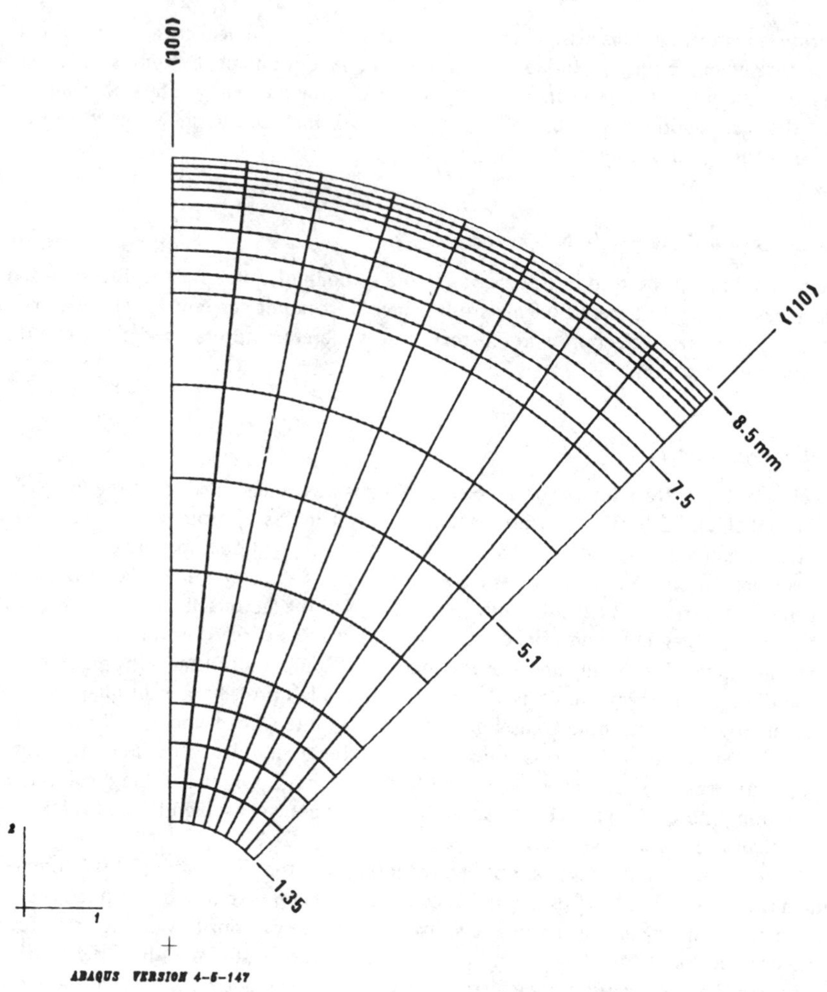

FIG. 6—*Mesh used in anisotropic thermoelastic finite-element analysis of specimen stress-strain history.*

extrapolated to obtain the corrected values for E_{100} and G_{100} as a function of temperature (for details see Ref *23*). These corrected values were then used to obtain the stiffness constants C_{mn} of Fig. 7. (In ABAQUS [*22*] these stiffness constants are denoted E_{ijkl}.) A trial run with isotropic values of the stiffness constants in the orthotropic finite element program gave results identical to the isotropic solution.

Temperature History

As described earlier, the temperature history of the test specimens was determined from thermocouple data. Because of the radial heating and thin specimen (only 1.8 mm), it was assumed that there was no temperature gradient through the thickness of the specimens.

TABLE 3—*Mean coefficient of thermal expansion versus temperature for René N4 and CODEP coating. (Values given are for a reference temperature of 25°C.)*

Temperature, °C	α_{N4}, °C^{-1} (René N4)	α_C, °C^{-1} (CODEP)
100	11.97×10^{-6}	...
200	12.06×10^{-6}	...
300	12.15×10^{-6}	...
400	12.51×10^{-6}	...
500	12.78×10^{-6}	13.82×10^{-6}
600	13.14×10^{-6}	14.04×10^{-6}
700	13.41×10^{-6}	14.31×10^{-6}
800	13.97×10^{-6}	14.47×10^{-6}
900	14.62×10^{-6}	14.78×10^{-6}
1000	15.41×10^{-6}	15.39×10^{-6}
1100	16.52×10^{-6}	16.41×10^{-6}

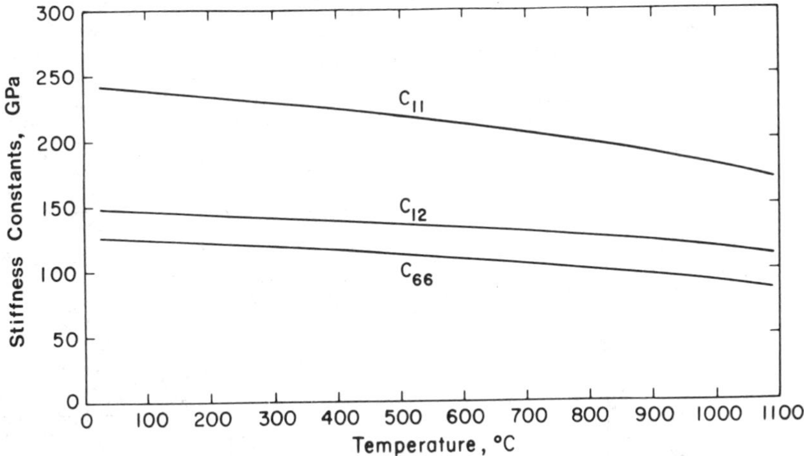

FIG. 7—*Stiffness constants for René N4 as a function of temperature.*

Results of Thermoelastic Finite Element Analysis

The circumferential periphery stress and strain histories along the ⟨100⟩ and ⟨110⟩ directions for the different heating and cooling rates of Figs. 5*a, b, c, d* are given in Figs. 8*a, b, c, d.*[8] Note that the loops are caused by the interaction of anisotropy and the changing thermal strains in the core relative to the periphery, and by the change of modulus with temperature, and not by material hysteresis. For the rapid heating rates of Figs. 5*a, b, d,* the maximum circumferential

[8]Note that these strain histories assume entirely *elastic* substrate behavior. For the moderate heating rate of Figs. 5*b* and 5*d,* data supplied by General Electric, Lynn, MA, show that the initial compressive yield strength of René N4 (0.1% offset, 1093°C) is exceeded by approximately 8.5% along ⟨100⟩ directions. These data also show that for a total strain range of 0.4% the widths of the 1093°C hysteresis loops increased from an initial value of 0.05% to a maximum of 0.15% at 1000 cycles. It is important to note that inelastic strains in the thin periphery will have relatively little effect on the circumferential strains induced by the thicker core of the specimen. A more complete stress-strain analysis currently in progress takes into account the possibility of microplastic flow and the resulting shift in mean stress.

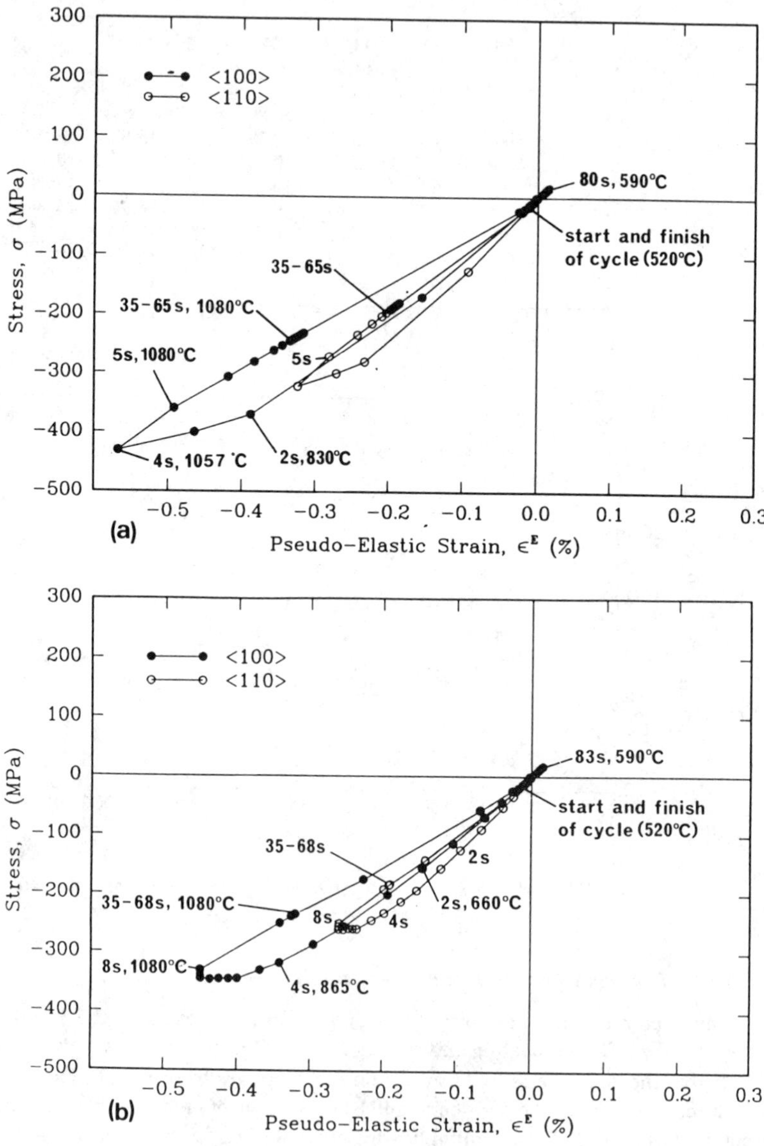

FIG. 8—*Circumferential stress-strain history for the* ⟨*100*⟩ *and* ⟨*110*⟩ *directions of the specimen 1080°C.* (a) *5 s heating and 30 s cooling.* (b) *8 s heating and 30 s cooling.* (c) *35 s heating and 30 s*

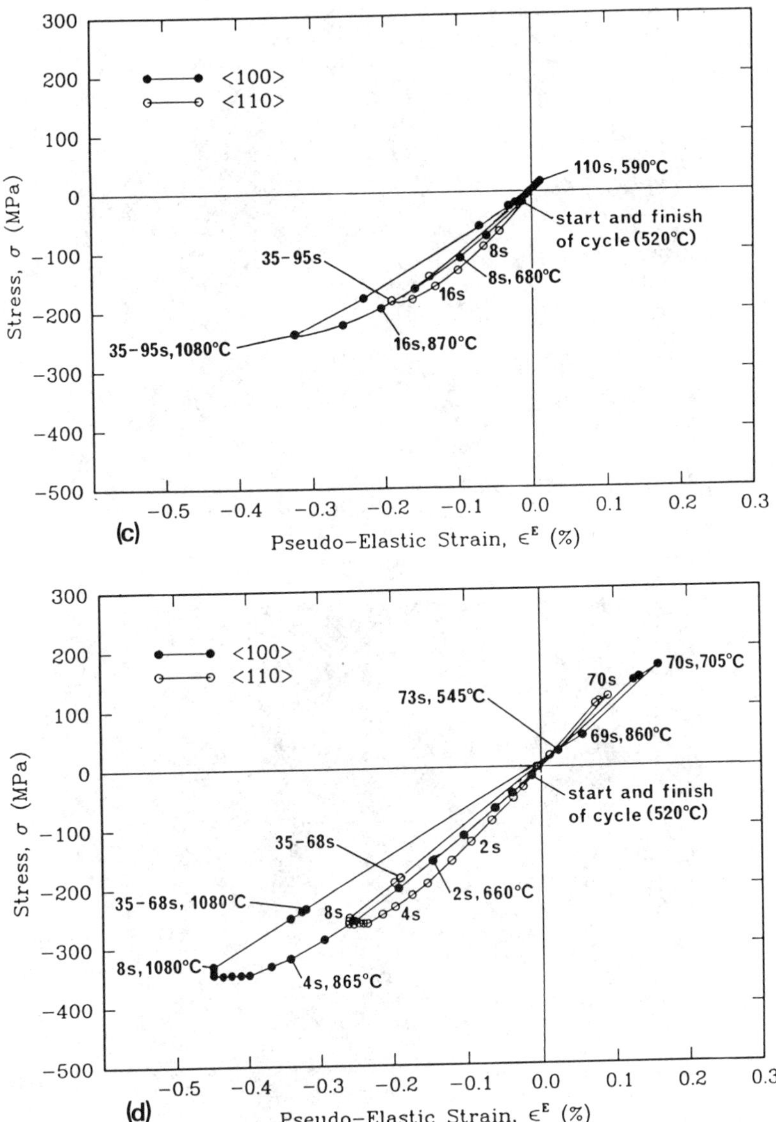

periphery. Thermoelastic finite-element analysis for heating from 520 to 1080°C, with 60 s hold at cooling. (d) *8 s heating and 6 s cooling.*

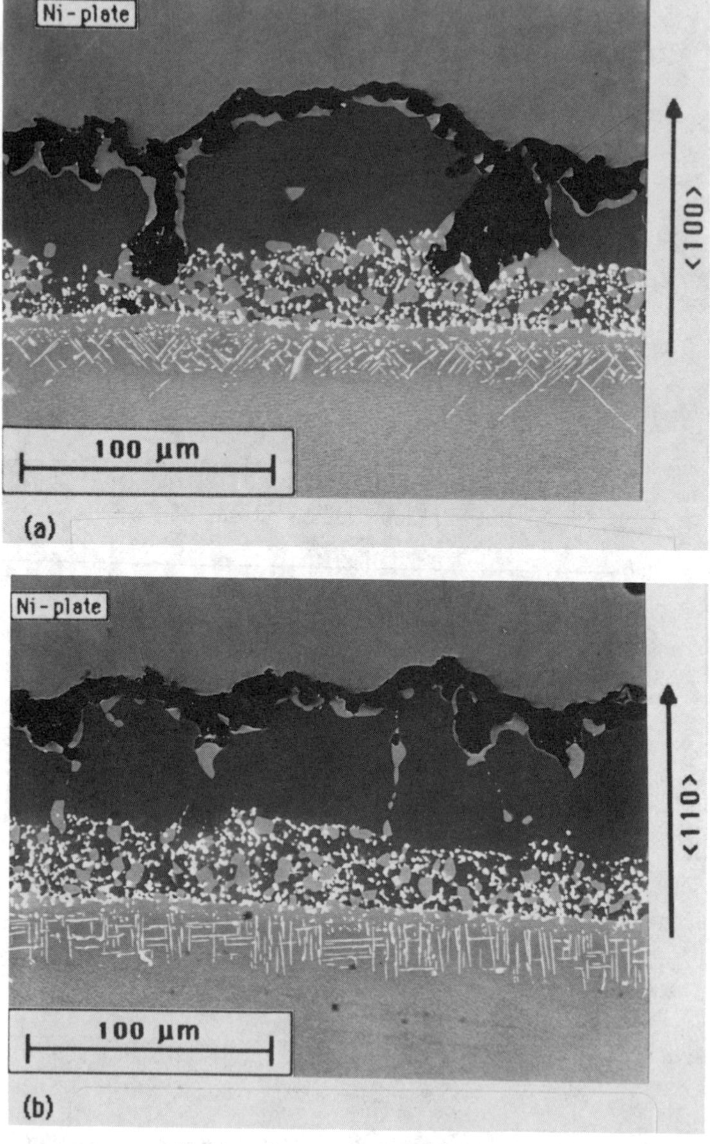

FIGS. 9*a*–9*d*—*Coating degradation along periphery after 6000 fatigue cycles of heating from 520 to* $\epsilon < 0.01\%$*).* (b) *5 s heating and 30 s cooling. ⟨110⟩ periphery direction (*$-0.33\% < \epsilon < 0.01\%$*).* (c) *30 s cooling. ⟨110⟩ periphery direction (*$-0.26\% < \epsilon < 0.01\%$*).*

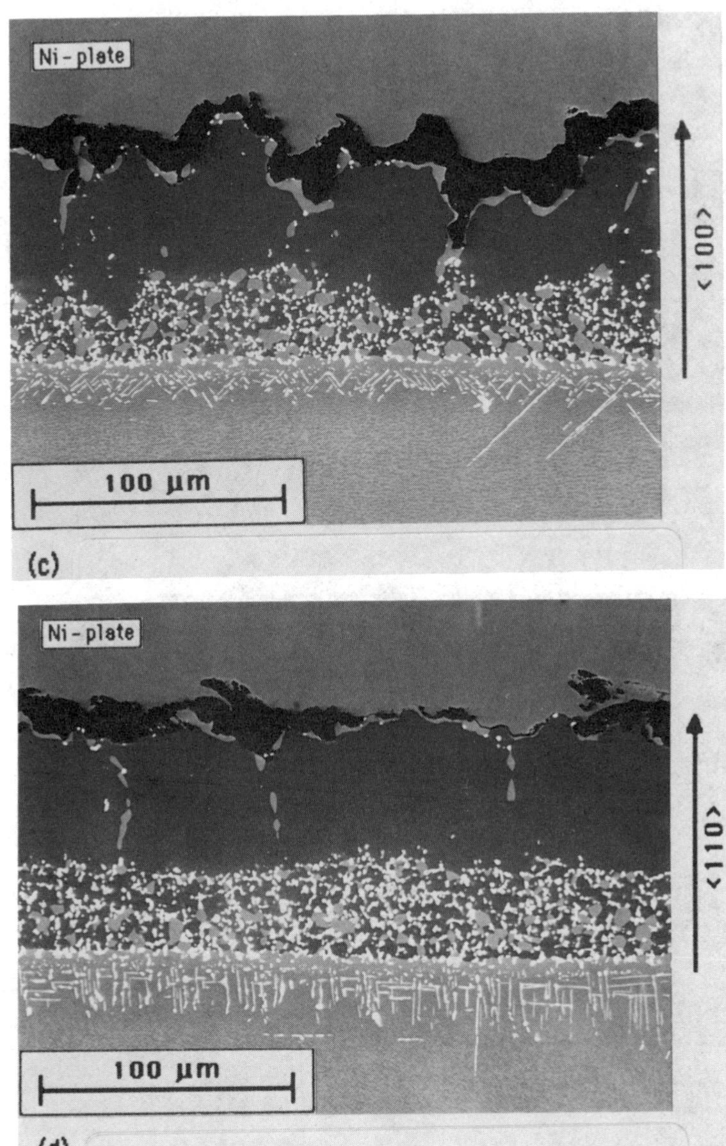

1080°C, with 60 s hold at 1080°C. (a) *5 s heating and 30 s cooling. ⟨100⟩ periphery direction (−0.56% <*
8 s heating and 30 s cooling. ⟨100⟩ periphery direction (−0.45% < ε < 0.01%). (d) *8 s heating and*

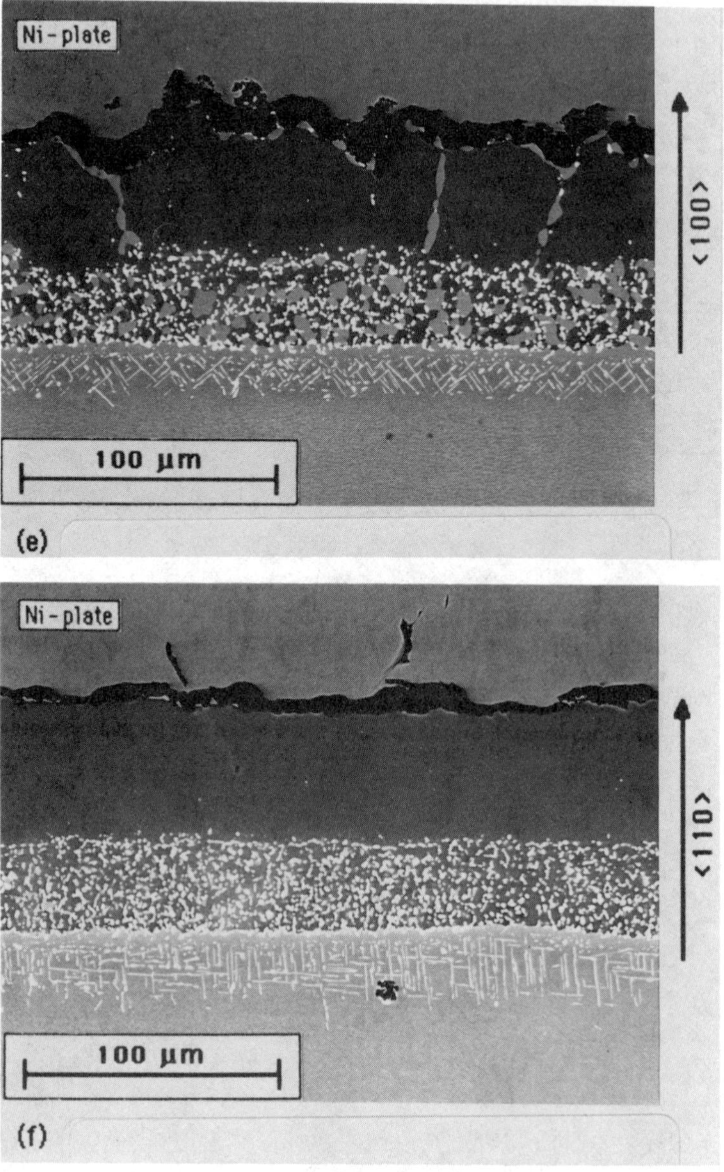

FIGS. 9e–9h—*Coating degradation along periphery after 6000 fatigue cycles of heating from 520 to* *ϵ < 0.01%).* (f) *35 s heating and 30 s cooling. ⟨110⟩ periphery direction (−0.20% < ϵ < 0.01%).* (g) *8 s* *⟨110⟩ periphery direction (−0.26% < ϵ < 0.09%).*

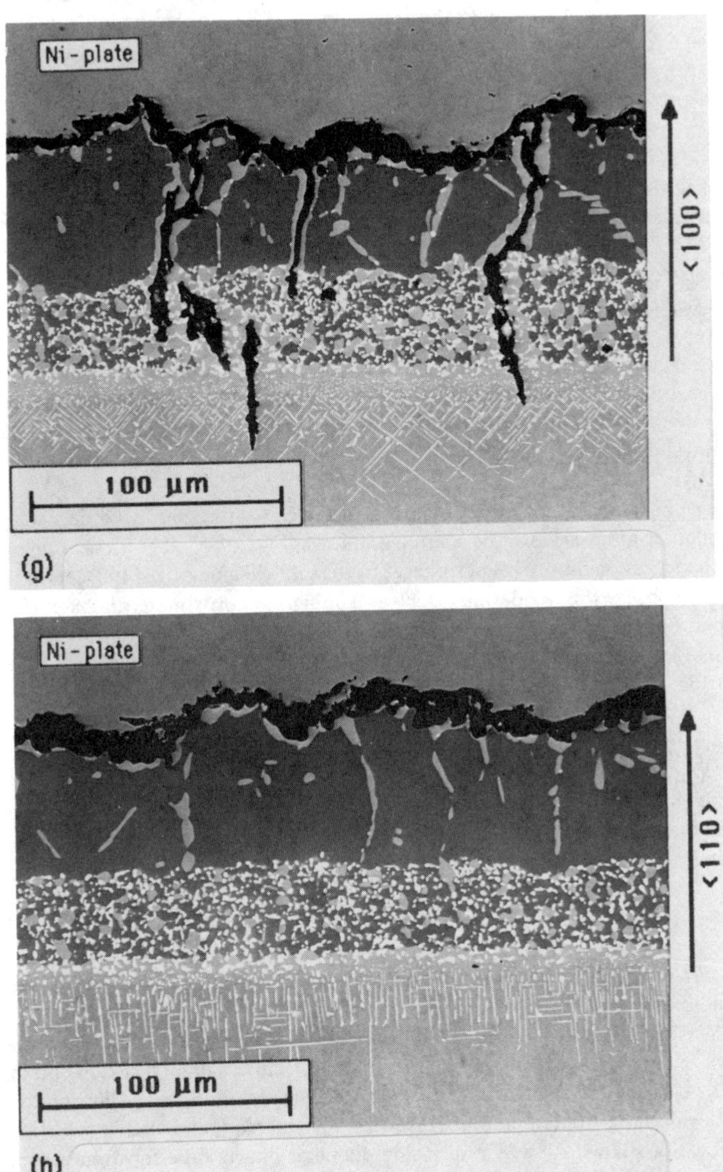

1080°C, with 60 s hold at 1080°C. (e) *35 s heating and 30 s cooling. ⟨100⟩ periphery direction (−0.32% < heating and 6 s cooling. ⟨100⟩ periphery direction (−0.45% < ε < 0.16%).* (h) *8 s heating and 6 s cooling.*

compressive strain in the periphery occurs along the ⟨100⟩ directions at 1057 to 1080°C and decreases as the specimen core increases in temperature. For the slow heating rate of Fig. 5*c*, where the transient radial temperature is minimized, the peak circumferential compressive strain occurs at the peak temperature of 1080°C. Prior to substrate yielding, the coating stress along the ⟨100⟩ directions will be 40% higher than the substrate, due to the modulus differences mentioned above. Cyclic stress-strain data for stress in the aluminide coating could not be found.

Results and Discussion

After completion of testing, specimens were nickel-plated and then ground parallel to their faces, down to the mid-plane. The specimens were next diamond polished to a final finish of 0.25 μm.

The backscattered-electron micrographs of Figs. 9*a* to 9*h* show two distinct types of coating degradation: surface scalloping and through-thickness cracking. These micrographs clearly show the greater degree of coating degradation along ⟨100⟩ periphery directions compared to the ⟨110⟩ directions. The initial coating was found to be isotropic, and a microprobe analysis along the circumference of an as-coated specimen showed no variation in coating composition with crystallographic orientation. Thus the difference in degradation can be attributed to the larger substrate strains and, hence, larger coating strains along ⟨100⟩ directions, as given by the thermoelastic analysis.

The *compressive* periphery strain encountered during specimen heating produced degradation in the form of scalloping at the scale of the coating grain size (Figs. 9*a–f*). The 5 s heating of Fig. 5*a*, which gave the largest compressive periphery strains (0.56% along the ⟨100⟩ peripheries), caused severe scalloping, with oxidation attack extending to approximately 80% of the coating thickness along ⟨100⟩ directions. The moderate (8 s) heating of Fig. 5*b* (0.45% compressive strain) resulted in ⟨100⟩ scalloping extending to 50% of the coating thickness. The slow (35 s) heatup of Fig. 5*c* (0.20% compressive strain) produced very little coating degradation.

Tensile periphery strain increments were produced by decreasing the cooling time from the 30 s of Fig. 5*b* to 6 s (Fig. 5*d*). This rapid cooling gave tensile strains of 0.16% along the ⟨100⟩ peripheries and resulted in crack-like defects that extend into the substrate (Figs. 9*g, h*). Microprobe analysis showed these defects to be alumina-filled. This "cracking" is to be contrasted to the coating scalloping (Fig. 9*c, d*) found after the 30 s cooling of Fig. 5*b*.

Conclusions

Induction heating of stepped-disk specimens provides a versatile technique for studying the thermal fatigue degradation of coated anisotropic superalloys. The apparatus can heat low thermal mass disk specimens through a temperature difference of 600°C in as little as 3 s, with less than a 2.5% overshoot in temperature. The cooling rate of the specimen can be varied to obtain rates as rapid as a 600°C drop in 4 s, with less than 3% overshoot. In addition to allowing the study of the effect of a wide variety of temperature histories, the technique can also be modified (by use of a 22 mm outside diameter by 19 mm inside diameter quartz tube for atmosphere containment) to study the effect of corrosive or inert atmospheres on coating or substrate durability.

Tests performed on aluminide-coated René N4 to illustrate the induction heating technique show that after 6000 cycles, the maximum compressive strains of 0.56% along the ⟨100⟩ periphery directions due to 5 s heating gave coating scalloping and accelerated coating oxidation to 80% of the coating thickness but did not cause coating or substrate cracking. For a compressive strain of 0.20% due to 35 s heating, no scalloping was observed.

After 6000 cycles the peak tensile strains of 0.16% encountered during rapid (6 s) cool-down resulted in alumina-filled "cracks" in the coating and substrate. The results obtained for the various stress-strain histories clearly illustrate that *both* the compressive strains encountered on specimen (or component) heatup and the tensile strains encountered on cooldown can have a profound effect on useful coating life.

Acknowledgments

It is a pleasure to acknowledge the support for this work provided by the National Science Foundation through Grant DMR 81-19295 to the Center for Materials Science and Engineering at MIT. We greatly appreciate the support of General Electric Company through the donation of the René N4 and the supplying of relevant technical information. The use of the finite element program ABAQUS made available to MIT by Hibbitt, Karlsson, and Sorensen, Inc. is gratefully acknowledged. We thank E. P. Busso for his valuable help with the finite element analysis.

References

[*1*] Schneider, H., Von Arnim, H., and Grunling, H. W., *Thin Solid Films,* Vol. 84, 1981, pp. 29-36.
[*2*] Hsu, L. and Stetson, A., *Thin Solid Films,* Vol. 73, 1980, pp. 419-428.
[*3*] Castillo, R. and Willett, K. P., *Metallurgical Transactions,* Vol. 15A, 1984, pp. 229-236.
[*4*] Felix, P. C. in *Materials and Coatings to Resist High Temperature Corrosion,* D. R. Holmes and A. Rahmel, Eds., Applied Science, London, 1978, pp. 199-212.
[*5*] Nicoll, A. R., Wahl, G., and Hildebrandt, U. W. in *Materials and Coatings to Resist High Temperature Corrosion,* D. R. Holmes and A. Rahmel, Eds., Applied Science, London, 1978, pp. 233-252.
[*6*] Smeggil, J. G. and Bornstein, N. S. in *High-Temperature Protective Coatings,* C. Singhal, Ed., The Metallurgical Society of AIME, Warrendale, Penn., 1982, pp. 61-74.
[*7*] Levine, S. R., *Metallurgical Transactions A,* Vol. 9A, 1978, pp. 1237-1250.
[*8*] Fleetwood, M. J., *Journal of the Institute of Metals,* Vol. 98, 1970, pp. 1-7.
[*9*] Boone, D. H., Crane, D. A., and Whittle, D. P., *Thin Solid Films,* Vol. 84, 1981, pp. 39-47.
[*10*] Steinmetz, P., Roques, B., Dupre, B., Duret, C., and Morbioli, R. in *High-Temperature Protective Coatings,* S. C. Singhal, Ed., The Metallurgical Society of AIME, Warrendale, Penn., 1982, pp. 135-157.
[*11*] Mom, A. J. A. in *Behaviour of High Temperature Alloys in Aggressive Environments,* The Metals Society, London, 1979, pp. 363-374.
[*12*] Hart, A. B., Laxton, J. W., Stevens, C. G., and Tidy, D. in *High Temperature Alloys for Gas Turbines,* D. Coutsouradis et al, Eds., Applied Science, London, 1978, pp. 81-107.
[*13*] Saunders, S. R. J., Hossain, M. K., and Fergusion, J. M. in *High Temperature Alloys for Gas Turbines 1982,* R. Brunetaud et al, Eds., D. Reidel Publishing Company, Dordrecht/Holland, 1982, pp. 177-206.
[*14*] Mowbray, D. F., Woodford, D. A., and Brandt, D. E. in *Fatigue at Elevated Temperatures, ASTM STP 520,* American Society for Testing and Materials, Philadelphia, 1973, pp. 416-426.
[*15*] Howes, M. A. H. in *Thermal Fatigue of Materials and Components, ASTM STP 612,* D. A. Spera and D. F. Mowbray, Eds., American Society for Testing and Materials, Philadelphia, 1976, pp. 86-105.
[*16*] Kaufman, A., "Elastic-Plastic Finite-Element Analyses of Thermally Cycled Single-Edge Wedge Specimens," NASA Technical Paper 1982, National Aeronautics and Space Administration. Washington, D.C., 1982.
[*17*] Holmes, J. W. and McClintock, F. A., *Scripta Metallurgica,* Vol. 17, 1983, pp. 1365-1370.
[*18*] Miner, R. V., Voigt, R. C., Gayda, J., and Gabb, T. P., *Metallurgical Transactions A,* Vol. 17A, 1986, pp. 491-496.
[*19*] Gabb, T. P., Gayda, J., and Miner, R. V., *Metallurgical Transactions A,* Vol. 17A, 1986, pp. 497-505.
[*20*] Pichoir, R. in *High Temperature Alloys for Gas Turbines,* D. Coutsouradis et al, Eds., Applied Science, London, 1978, pp. 191-208.
[*21*] Goward, G. W., and Boone, D. H., *Oxidation of Metals,* Vol. 3, No. 5, 1971, pp. 475-495.
[*22*] *ABAQUS Users Manual,* Hibbitt, Karlsson and Sorensen, Inc., Providence, R.I., July 1982.
[*23*] Holmes, J. W., "Thermal Fatigue Oxidation and SO_2 Corrosion of an Aluminide Coated Superalloy," Doctoral thesis, Department of Materials Science and Engineering, Massachusetts Institute of Technology, Cambridge, Mass., 1986.

T. S. Cook,[1] *K. S. Kim,*[1] *and R. L. McKnight*[1]

Thermal Mechanical Fatigue of Cast René 80

REFERENCE: Cook, T. S., Kim, K. S., and McKnight, R. L., **"Thermal Mechanical Fatigue of Cast René 80,"** *Low Cycle Fatigue, ASTM STP 942,* H. D. Solomon, G. R. Halford, L. R. Kaisand, and B. N. Leis, Eds., American Society for Testing and Materials, Philadelphia, 1988, pp. 692–708.

ABSTRACT: Hot path components of aircraft engines contain complex thermal and mechanical fields which can interact to produce thermal mechanical fatigue (TMF) failure. Development of a model of thermal mechanical fatigue for cast René 80 is underway through TMF experiments, a series of isothermal tests, and finite element analysis. The modelling effort consists of two parts, the determination of a suitable damage parameter and a technique for calculating the quantities required by the damage parameter. Classical plasticity theory was employed in a finite element code to model the TMF deformation. The predicted hysteresis loops were compared to the experimental values; the correlation was excellent for both the in-phase and out-of-phase cycles at 871 to 982°C but not as good as 760 to 871°C. The thermal cycle variation in these tests is limited, but the model's ability to correlate the TMF lives and predict the constitutive behavior of a material whose behavior changes with temperature provides optimism that this analysis may be applied to other TMF situations.

KEY WORDS: nickel-base superalloy, elevated temperature fatigue, thermomechanical fatigue, fatigue life prediction, inelastic finite element analysis, cyclic constitutive behavior

The goal of gas turbine engineers is the improvement of performance without a sacrifice in reliability. The performance goal has been pursued through the elevation of turbine temperatures; the durability has been sought through increased cooling of turbine hot path members, particularly blades. These cooled blades can contain complex thermal and mechanical fields which can interact to produce fatigue failure. The difficulty in understanding and predicting this behavior has stimulated a significant research effort in thermal mechanical fatigue (TMF). This research has not produced a clear picture of damage development during TMF. Not only are the differences in material characteristics important, but the enormous variety of temperature and stress cycles can alter the behavior of a single alloy, depending on the specific cycle the material undergoes. The resulting complexity of the experimental data has precluded the development of a general model of thermal mechanical fatigue. Given the variety of damage observed, a number of isothermal models have been adapted to specific situations, and have been reasonably successful. These models are usually applied to a certain type of cycle or material; when the same model is used in other situations, it is often much less successful. A broader class of models needs to be developed to allow extrapolation of the data base. The resulting improvements in predicted lives will enhance reliability, but, in addition, the high cost of TMF testing makes this a mandatory goal.

The objective of this investigation is to develop a model of thermal mechanical fatigue of cast René 80. This is being accomplished by carrying out TMF experiments and also a series of isothermal tests to provide data to support modelling efforts. The modelling effort consists of

[1]General Electric Company, AEBG, Cincinnati, OH 45215-6301.

two parts, the determination of a suitable damage parameter and a technique for calculating the quantities required by the damage parameter. It must be emphasized that this phase of the investigation examined only the classical plasticity model in thermomechanical applications. As will be shown, the comparison between test and analysis is excellent in some, but not all, cases. The results of this investigation will help to identify those aspects of thermomechanical analysis which need improvement.

In addition, a damage parameter has been found which correlates this set of data. The generality of this approach remains to be seen. Substantiation of the damage mechanisms through metallography and microscopy remains to be done. Nevertheless, the progress to date has been encouraging, and it is hoped that future studies will demonstrate the soundness of the approach.

Experimental Procedure

Material

The material selected for this study is conventionally cast superalloy René 80. The composition (weight percent) is: 3.0 Al, 5.0 Ti, 14.0 Cr, 3.9 W, 4.0 Mo, 9.8 Co, 0.17 C, and the balance Ni. This nickel-base material is a typical turbine blade alloy and is quite widely used. The specimens were given the following heat treatment:

1218°C (2225°F) - 2 h - He quench
1093°C (2000°F) - 4 h - He quench
1052°C (1925°F) - 4 h - furnace cool to
649°C (1200°F) - in 20 min, air cool to room temperature
843°C (1550°F) - 16 h - He quench or air cool

The microstructure of this alloy has been described by Antolovich and coworkers [*1,2*]; it consists of large (ASTM 2 to 3), irregular grains with a distribution of grain boundary carbides. The strengthening mechanism is uniformly distributed cuboidal gamma prime precipitates. These precipitates tend to coarsen with increasing temperature and are at least partially responsible for the temperature dependence of cyclic deformation. At the lowest test temperature, the deformation is largely within the matrix, leading to matrix failure when a critical dislocation density is achieved. At the higher temperatures, damage accumulates at grain boundaries, leading to the formation of oxide spikes [*1*]. This temperature-dependent behavior presents a challenge to modelling the TMF experiments and makes the current experiments, despite their restricted temperature cycle, a true test of any TMF model.

The specimens were cast to size. Following heat treatment, the specimens were machined using low stress procedures. The solid cylindrical specimens had a diameter of 6.35 mm and were 92.1 mm in overall length. Threaded ends were used for gripping.

Mechanical Testing

The objective of this program is to develop a predictive capability for thermal mechanical fatigue. The TMF tests are difficult to carry out and expensive, making it desirable to base the model on isothermal data insofar as possible. In addition to TMF tests, therefore, creep, cyclic stress-strain (CSS), and low-cycle fatigue (LCF) tests were carried out under isothermal conditions. The LCF and CSS tests were run in longitudinal strain control; total strain was controlled. The creep tests were run in the same servohydraulic testing machines as the displacement control tests. In the creep tests the load was controlled while displacements were monitored using the same extensometer as the LCF tests. Creep data in both tension and compression were acquired. Since the data were intended to help analyze the TMF experiments, the stress levels employed were considerably higher than normal test levels for design purposes.

Low-cycle fatigue data was acquired at several strain rates and at temperatures of 760, 871, and 982°C. There was only limited isothermal testing at the lower two temperatures, but appreciable testing was done at 982°C. A total of 38 LCF tests covering four strain rates and three strain ratios ($R_\epsilon = \epsilon_{min}/\epsilon_{max}$) were tested at this temperature. Triangular waveforms were used.

The René 80 TMF specimens were subjected to the three temperatures and strain waveforms shown in Fig. 1. The general testing procedure has been described by Embley and Russell [*3*] and consists of two control loops, one for strain and one for temperature. Induction heating is employed and controlled by calibrated thermocouples mounted on the specimen shoulders. The TMF cycle was limited by the solid specimens; forced air cooling would induce undesirable temperature gradients and was not employed. A strain rate of 1%/min was used in the TMF experiments; this resulted in a period of 1.6 to 2.6 min. This consideration restricted the temperature cycle to 100 to 150°C. This cycle is, of course, not representative of an entire flight cycle, but it does simulate parts of the cycle—for example, a thrust reversal [*4*].

Data Analysis

Load-displacement hysteresis loops were recorded periodically throughout the TMF and LCF tests. Load and strain data were also recorded on strip charts and used to determine cycles to crack initiation, N_i, and failure, N_f. The René 80 softened at a constant rate for much of the fatigue test; an acceleration in the softening rate was used to define crack initiation. Cycles to failure was defined as 50% load drop. The hysteresis loop nearest $N_f/2$ was used as the definition of cyclic values. Stress values and strain ranges were obtained from this loop. The measured plastic strain range was defined as the width of the loop at zero load. The calculated plastic strain was used in the data analysis.

The creep tests were run more recently and employed conventional *X-Y* recorders and strip charts, as well as computerized data acquisition. This was also the case with the CSS tests, where the specimens were cycled in strain control with open hysteresis loops. The test was halted at an estimated half life and the specimen pulled monotonically to failure. The intent was to obtain the cyclic curve over a larger strain range than the cycling approach would allow [*5*], but operational problems limited the success of this approach. Thus the CSS behavior was obtained from the last hysteresis loop before the specimen was pulled to failure.

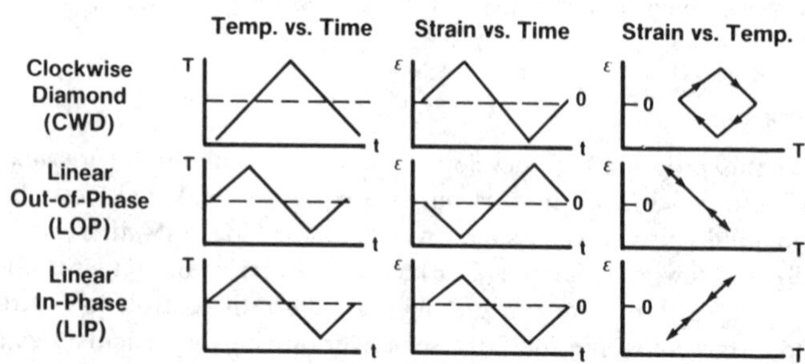

FIG. 1—*Thermal mechanical cycles used in this investigation.*

Experimental Results

The cyclic stress-strain behavior of René 80 was obtained at 760, 871, and 982°C at a strain rate of 1%/min to correspond to the TMF testing rate. In addition, the cyclic behavior at two other frequencies was obtained from the constant amplitude LCF hysteresis loops. The resulting data (Fig. 2) show a mixed picture of strain rate sensitivity. At 760 and 982°C the material has little sensitivity, while at 871°C the cyclic amplitude of René 80 is quite strongly influenced by strain rate. In addition to the strain rates plotted in Fig. 2, a few LCF tests were run at 0.2%/min at 982°C. The strain ranges employed in these tests were larger than those in Fig. 2, thus making direct comparisons difficult. The stress ranges measured in these low rate LCF tests are approximately 40% lower than the extrapolation of the 1%/min curve. This observation suggests that the low frequency allows significant stress relaxation to take place within the cycle. In general, this agrees with Antolovich et al. [*1*], who found a small effect of frequency on cyclic behavior at 982°C, but the two strain rates used were two orders of magnitude apart. The influence of frequency at 871°C is well known [*1,6*] and documented.

Figure 2 indicates that temperature plays a much greater role determining the cyclic behavior of René 80 than does the strain rate. This temperature influence is reflected in other properties (e.g., the tensile elongation which undergoes a minimum around 760°C and then increases to a maximum around 982°C). An analysis of the hysteresis loops from the CSS tests also shows this change. At 760°C the material cyclically hardens, while at 871 and 982°C René 80 softens.

The cycles to failure for the elevated temperature LCF specimens are plotted in Fig. 3 as a function of plastic strain; unlike the stress-strain behavior, the data separate according to both temperature and strain rate. It does appear that the effect of strain rate decreases as the temperature increases. Note that the life generally increases with temperature; there does not appear to be any crossover effect unless it happens at lives longer than were considered in this program. Such a crossover was found by Antolovich et al. [*7*] and reflects the temperature dependent changes in deformation and fracture mechanisms.

The effect of frequency shown in Fig. 3 suggested the use of a frequency-modified parameter to correlate the LCF lives. At the same time, Domas and Antolovich [*8*] argued that the oxide

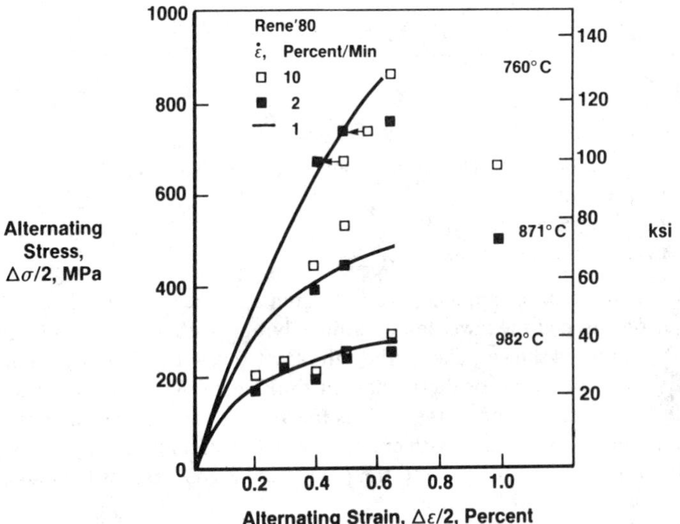

FIG. 2—*Cyclic stress-strain behavior for René 80.* $R_\varepsilon = 0$ *at the strain rates of 2 and 10%/min. and* $R_\varepsilon = -1$ *for the lowest strain rate.*

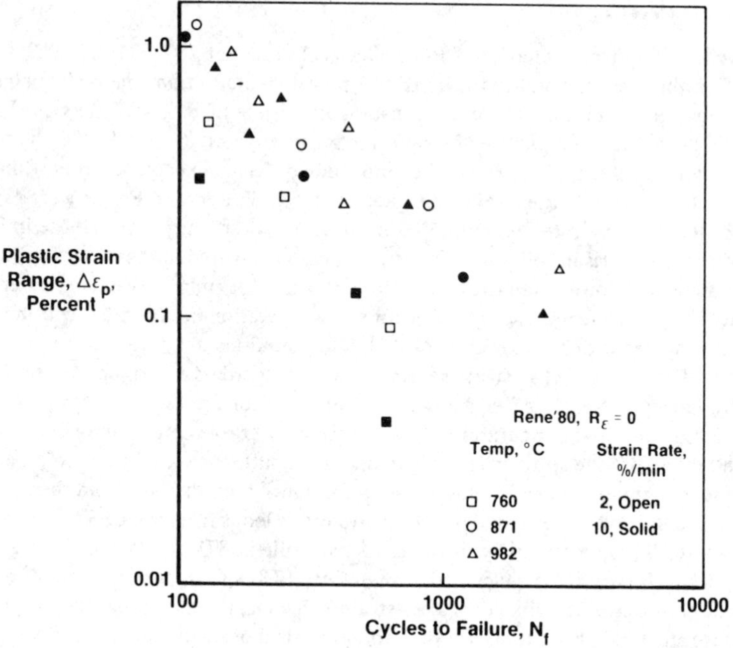

FIG. 3—*Coffin-Manson plot for LCF of René 80.*

spikes fracture at the maximum stress in the LCF cycle. In order to incorporate both concepts in predicting thermomechanical fatigue life, an approach to LCF damage incorporating these parameters was used. The approach suggested by Ostergren [*9*] can be written as

$$C = N_f^{\beta} \nu^{\beta(K-1)} \Delta\epsilon_p \sigma_T \tag{1}$$

where

C, β, K = constants,
N_f = cycles to failure (cycles to initiation, N_i, may also be used),
ν = frequency,
$\Delta\epsilon_p$ = plastic strain range, and
σ_T = maximum stress in the cycle.

Figure 4 shows the result of applying the Ostergren approach to the 982°C LCF data. The same technique was used at the lower temperatures, but because of the much larger data set at 982°C, only this figure is shown. The method fits Eq 1 to the data and determines the three constants, C, β, and K, that provide the best fit to the data. The results of this analysis are given in Table 1. The error listed in the table is the average error of the curve fit to each data point. The same approach was used with cycles to initiation and the result was not significantly altered. Figure 4 indicates that the fit to the data is quite good except for four points; these points account for the bulk of the error.

The values of the constants for the René 80 data set predict a changing effect of frequency with temperature. For the two lower temperatures, K is greater than one, so Eq 1 predicts cycles to failure decreases with increasing frequency. At 982°C, K is less than one, so N_f should in-

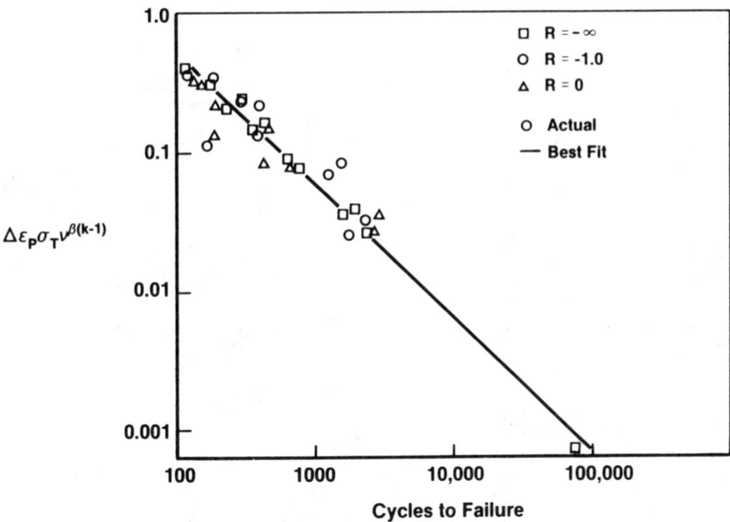

FIG. 4—*Ostergren approach to LCF of René 80 at 982°C.*

TABLE 1—*Results of fitting Ostergren method.*

Temp. (°C)	C	K	β	Error (%)
760	280.73	1.202	1.255	24.2
871	87.07	1.027	0.978	17.1
982	40.14	0.924	0.952	35.8

crease with frequency. This does not agree with the data of Ref *1,* which found N_f decreased with frequency at both 871 and 982°C. It may be that the prediction at 982°C is purely the result of the particular curve fit and does not reflect the material behavior. A sensitivity study of this point needs to be carried out.

The next step taken was to attempt to predict the TMF results using the Ostergren parameter. However, applying Eq 1 to a variable temperature situation poses two problems: first, what values of the three constants should be used, and second, what value of plastic strain should be selected? Making the simplest assumption that the maximum temperature in the cycle is related to the damage, the three constants were chosen on that basis. Similarly, the plastic strain range was calculated from the stress range in the cycle and the elastic modulus at the maximum temperature. Using this approach, the damage was calculated and the life of each TMF test predicted. The results are shown in Fig. 5 and indicate a surprisingly good correlation. All but two points lie within a factor of two of the unit correlation line; there does not appear to be any particular pattern or layering to the data.

Since the choice of maximum temperature for calculation purposes was somewhat arbitrary, a rationale for the choice was sought. Some insight into the damage process can be gained by examining the maximum stress in the TMF cycle. Figure 6 compares the maximum stress in the TMF hysteresis loop at half-life to the cyclic stress-strain curve for René 80 at three tempera-

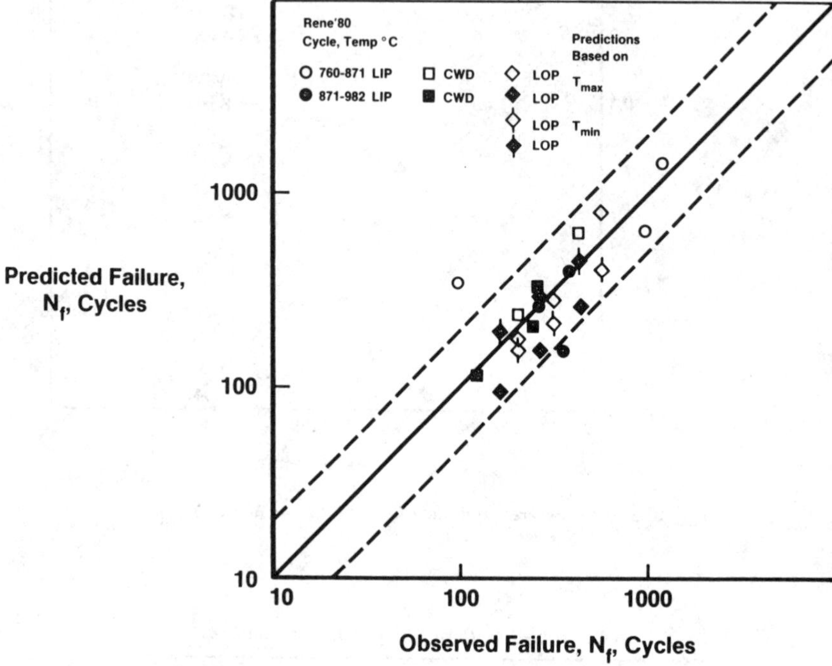

FIG. 5—*Predicted versus observed life for TMF of René 80.*

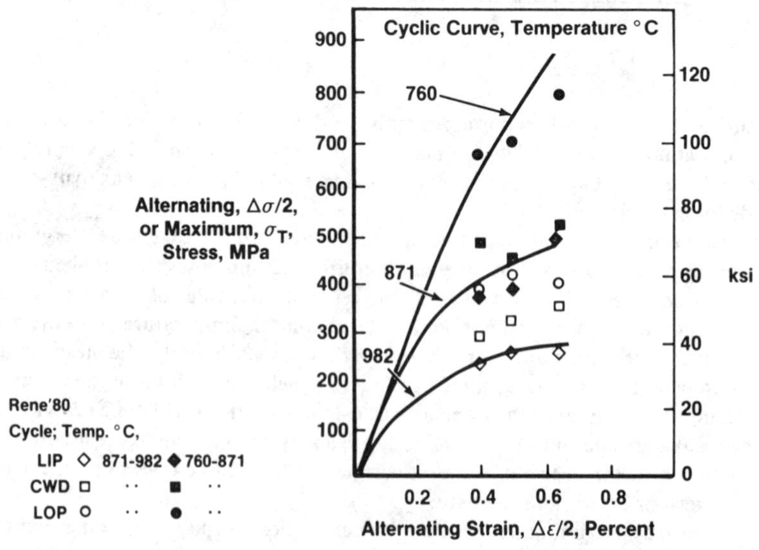

FIG. 6—*Comparison of isothermal CSS properties with maximum stress in TMF cycle.*

tures. In general, the in-phase maximum stresses agree with the CSS curve at the maximum temperature, while for the 180° out-of-phase the agreement is with the low temperature CSS curve. The third cycle type, CWD has the temperature lagging the strain by 90° and is closer to the in-phase than the out-of-phase cycle. This trend is carried over in a plot of the maximum stress versus cycles to initiation (Fig. 7). For example, the life data for the 760 to 871°C LOP cycle agrees with the 760°C LCF values. The higher temperature LIP cycle produces lives very similar to those of the 982°C LCF cycle. The data for the other cycles agree well with that of Fig. 6.

These observations suggest that the lower temperature data are more appropriate for the LOP cycle than the maximum temperature data used. Making this change, the predicted values for the 871 to 982°C cycle are significantly improved, while the values at 760 to 871°C are slightly worse. The revised predictions for the LOP cycles based on minimum temperature are included in Fig. 5. While not shown, the effect of minimum temperature on the predictions for the CWD was also checked; there was little effect on the 760 to 871°C cycle, but the predictions were somewhat poorer for the higher temperature TMF cycles. This suggests that there may be temperatures appropriate to each cycle that need to be selected rather than simply selecting the maximum temperature LCF line as suggested by Kuwabara et al. [*10*]. Clearly, the changing deformation/failure mechanisms undergone by the material during the TMF cycle must be considered in both the selection of the life prediction model and the data used.

Modeling Procedure

Accurate prediction of structural response in the high temperature environment is a prerequisite for reliable life prediction of hot section components of gas turbine engines. In many situations, this involves the use of a general purpose finite element code capable of modeling nonlinear material behavior. The constitutive models of material behavior may be either classical (i.e., separation of creep and plastic deformation) or unified (i.e., inelastic deformation as a whole is considered). While there is considerable interest in the unified theories [*11,12*] they require

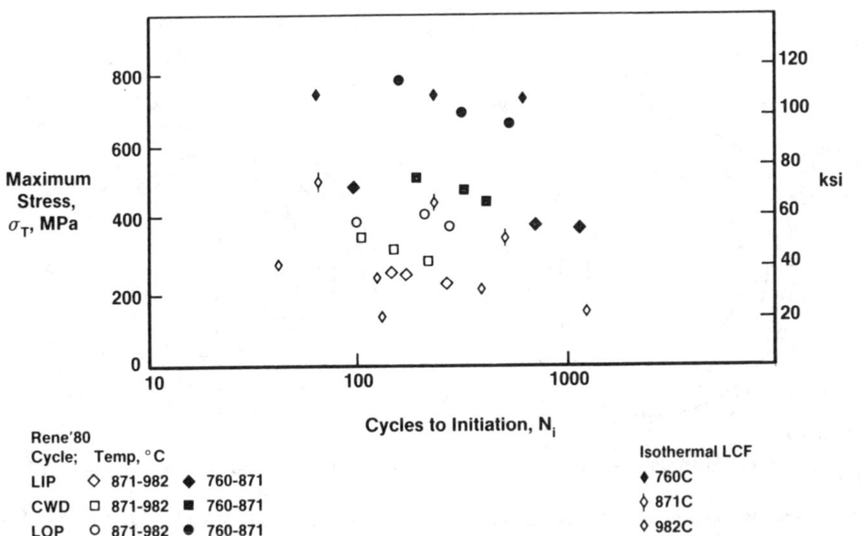

FIG. 7—*Initiation life of LCF and TMF specimens.*

further development of capabilities and experimental verification under realistic loading conditions. Their implementation in numerical analysis is not widespread. The classical theories, on the other hand, are relatively well established and available in many finite element codes. Therefore the modeling effort in this study will be made using a classical model to predict cyclic deformation rather than a unified theory. The objective is to examine the accuracy of a classical model in thermomechanical applications of René 80 within the test data generated in this program.

In the current TMF tests, the strain rate was kept constant at 1%/min. Cyclic stress-strain curves at this rate were input to the finite element code. Since the appropriate rate data were available, the analysis was elastic-plastic with temperature-dependent constitutive properties; creep was not explicitly included. Separate consideration of plasticity and creep is also underway for different types of loading, but is beyond the scope of this paper.

The finite element program used here is that developed by McKnight [*13*]. The outline of the underlying theory in the program will now be briefly reviewed.

Constitutive Theory

The classical incremental plasticity theory which utilizes the Prandtl-Reuss flow rule, the Von Mises yield criteria, and the kinematic hardening rule in the strain space is the basis of the constitutive model employed in this analysis. The Besseling's subvolume method [*14*] is used within this constitutive framework. In the subvolume method a strain-hardening material is considered as a composition of several subvolumes, each of which is an elastic-ideally plastic material. The subvolumes have identical elastic constants but different yield stresses. The yield function for the kth subvolume is given by (Fig. 8)

$$f_k = (e_{ij} - e^p_{ijk})(e_{ij} - e^p_{ijk}) - P^2_k \tag{2}$$

where e_{ij} is the strain deviator, e^p_{ijk} is the plastic component of e_{ij} for the kth subvolume, and P_k is the radius of the kth subvolume yield surface ($P_1 < P_2 < \ldots$). Notice that the kinematic

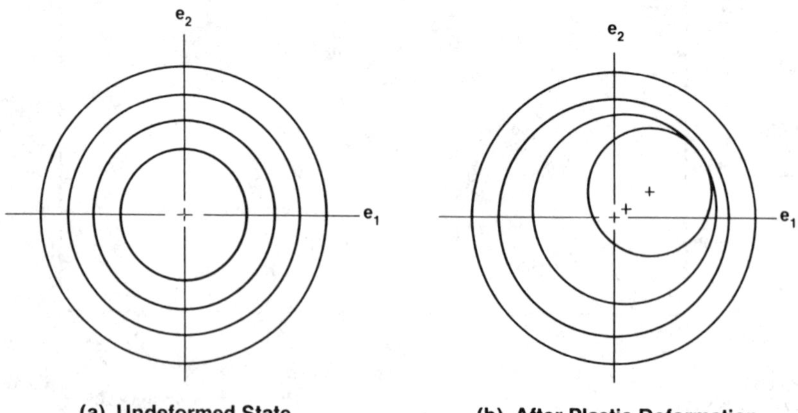

FIG. 8—*Representation of yield surfaces in Besseling's model.* (a) *Undeformed state.* (b) *After plastic deformation.*

hardening rule in the strain space is used in Eq 2. The elastic strain increment for the material is given by

$$de^e_{ij} = dS_{ij}/2G \tag{3}$$

where S_{ij} is the stress deviator and G is the modulus of rigidity. The plastic strain increment of the kth subvolume is given by

$$de^p_{ijk} = \frac{(e_{ij} - e^p_{ijk})(e_{mn} - e^p_{mnk})de_{mm}}{P_k^2} \tag{4a}$$

if

$$f_k = 0 \quad \text{and} \quad (e_{ij} - e^p_{ijk})de_{ij} > 0$$

and

$$de^p_{ijk} = 0 \tag{4b}$$

otherwise. The total plastic strain can be written as the weighted sum of the plastic strains of the subvolumes satisfying Eq 4*a:*

$$de^p_{ij} = \sum_k \psi_k de^p_{ijk} \tag{5}$$

The determination of ψ_k, and also P_k in Eq 2, can be done by considering the uniaxial stress-strain curves. For details, the reader is referred to Besseling [*14*].

In the loading process beyond the elastic limit, the first subvolume is subjected to plastic deformation and other subvolumes may yield in sequence, depending on the severity of loading; see, for example, Fig. 8. It is noted that the yield surfaces in the Besseling's model contact at a point with common tangency if the loading is proportional, but they can intersect each other in nonproportional loading.

The elastic properties and the stress-strain curves are input at several temperatures. In a time-varying temperature field (for example, the thermomechanical process under consideration) the material data, G, P_k and ψ_k, are linearly interpolated using the input data at nearest temperatures.

Finite Element Equations

The numerical scheme of the plasticity analysis used is the pseudo-force method in which the right-hand side force vector is revised in the plasticity iterations. The finite element equation of equilibrium for an element can be written as

$$K_{ij}\Delta U_j = \Delta f_i + \Delta f^p_i + \Delta f^\theta_i \tag{6}$$

where ΔU_j is the increment of displacement, and

$$K_{ij} = \int_v B_{ki}C_{kl}B_{lj}dv \tag{7a}$$

$$\Delta f_i^p = \int_v B_{ki} C_{kl} \Delta \epsilon_l^p \, dv \tag{7b}$$

$$\Delta f_i^\theta = \int_v B_{ki} C_{kl} \Delta \epsilon_l^\theta \, dv \tag{7c}$$

In these equations, B_{ij} is given in terms of the derivatives of the interpolation functions for the displacement field in an element, C_{ij} is the elastic properties matrix,

$$\{\Delta \epsilon_i^p\} = \{\Delta \epsilon_{11}^p, \Delta \epsilon_{22}^p, \Delta \epsilon_{33}^p, \Delta \gamma_{12}^p\}^T$$

$$\{\Delta \epsilon_i^\theta\} = \alpha \Delta T \{1,1,1,0\}^T$$

α is the thermal expansion coefficient, ΔT is the temperature increment, Δf_i, Δf_i^p, and Δf_i^θ are the increments of the external force, plastic force, and thermal force, respectively, and v is the volume of the element. The detailed expressions for B_{ij} and C_{ij} can be found in any finite element textbook and will not be given here. The plastic force vector is updated in every iteration. Notice that the stiffness must be recomputed if temperature changes between loading steps; however, it need not be computed repeatedly in plasticity iterations.

Model Results

The thermomechanical cycles were analyzed using the four-constant-strain element model shown in Fig. 9. The analyses were made for the 760 to 871°C, 871 to 982°C in-phase (LIP), 90° out-of-phase (CWD), and 180° out-of-phase (LOP) cycles shown in Fig. 1. The hysteresis loops for an applied mechanical strain range of 1% are presented in Figs. 10 to 15 in comparison with experimental results. The test hysteresis loops are approximately those at half life. Each hysteresis loop obtained analytically contains 36 loading steps with a constant strain increment. The results are quite satisfactory. In particular, excellent correlation between analysis and test results are seen for 871 to 982°C cases. The individual cases will be discussed in some detail below.

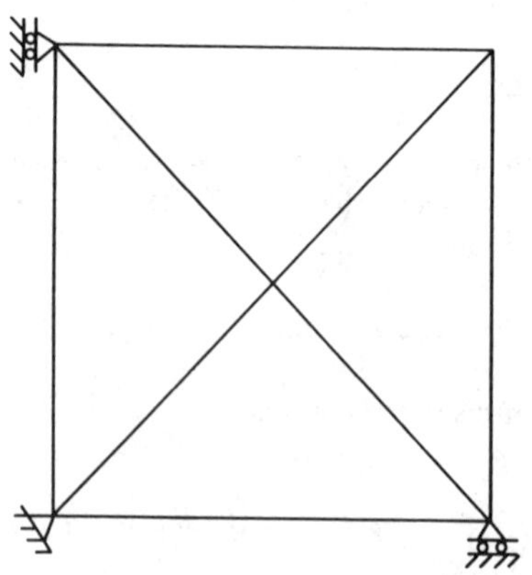

FIG. 9—*Finite element model used in TMF modeling.*

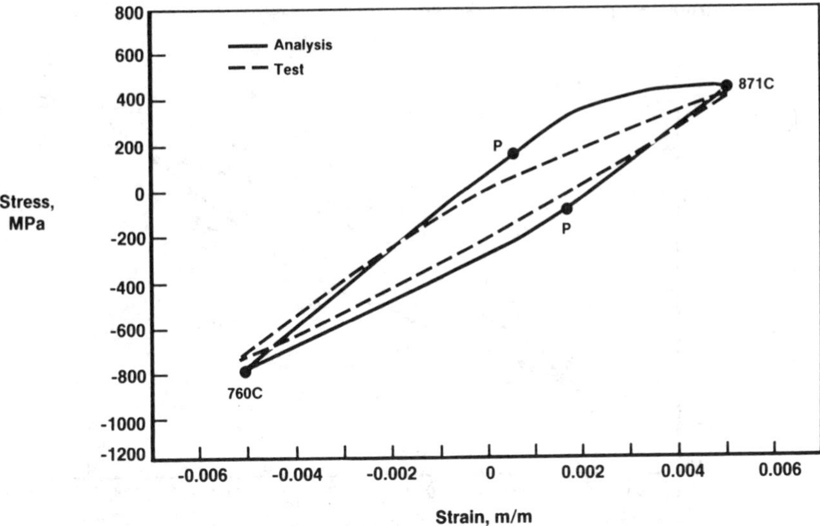

FIG. 10—*Predicted versus observed constitutive behavior for 760 to 871°C LIP TMF cycle.*

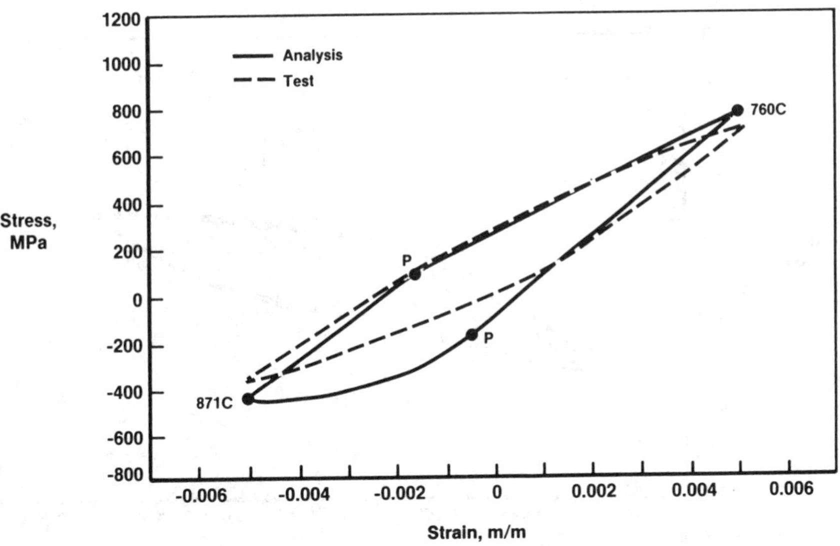

FIG. 11—*Predicted versus observed constitutive behavior for 760 to 871°C LOP TMF cycle.*

760 to 871°C, Linear In-Phase

The predicted stress amplitude for the LIP cycle is acceptable, but the computed plastic strain range is significantly lower than the test value. The deviation becomes more severe as the maximum temperature is reached. The shape of the test hysteresis loop is much thinner in the upper region than the analytical prediction. The shape of the analytical curve is in agreement with expectations. The curve becomes stiffer ($d^2\sigma/d\epsilon^2 < 0$) in the unloading process as temper-

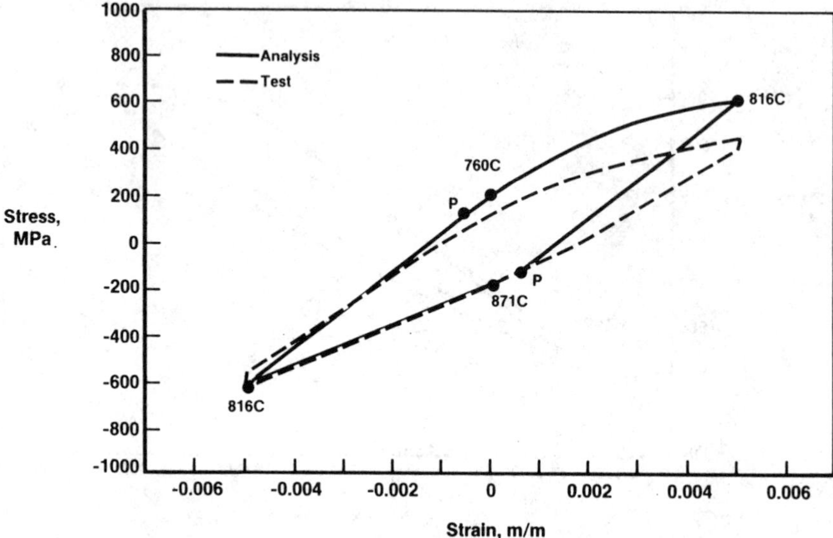

FIG. 12—*Predicted versus observed constitutive behavior for 760 to 871°C CWD TMF cycle.*

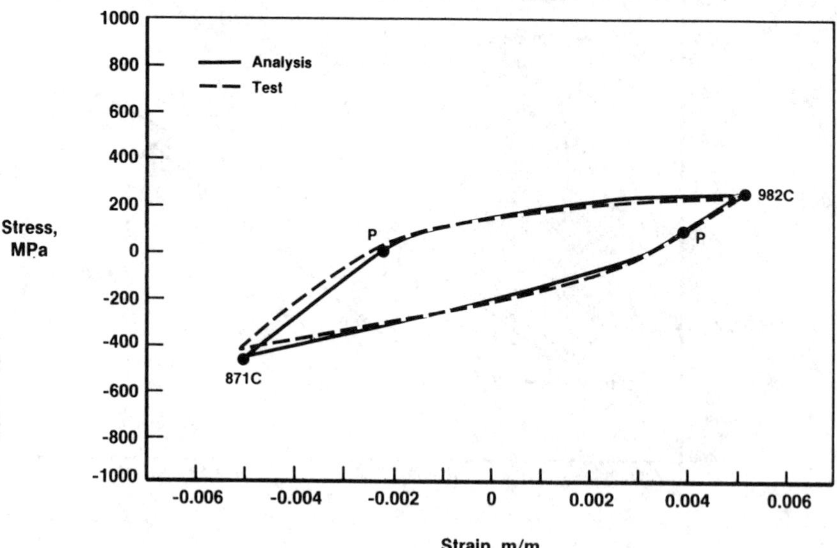

FIG. 13—*Predicted versus observed constitutive behavior for 871 to 982°C LIP TMF cycle.*

ature goes down from 871°C and the elastic modulus increases. The plasticity reverses at approximately $\epsilon = 0.17\%$ and $T = 834°C$ (denoted by P) and the slope starts to decrease ($d^2\sigma/d\epsilon^2 > 0$). The degree of strain hardening is maintained relatively high due to decreasing temperature. As the loading and temperature reverse their directions at the lower peak, the material is again subjected to elastic unloading. The material becomes less stiff ($d^2\sigma/d\epsilon^2 < 0$),

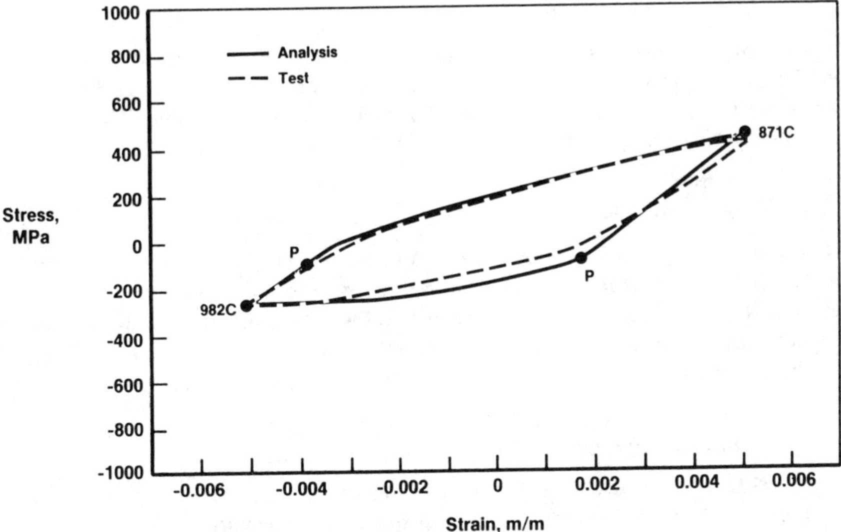

FIG. 14—*Predicted versus observed constitutive behavior for 871 to 982°C LOP TMF cycle.*

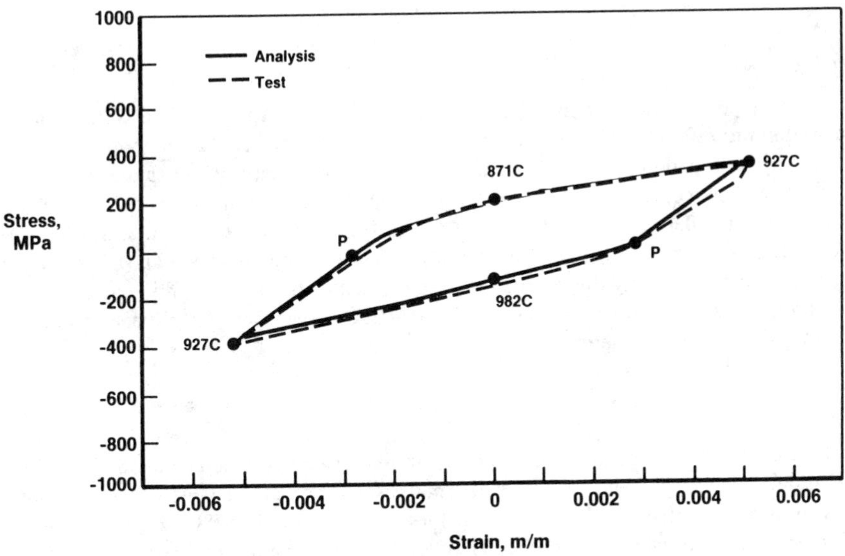

FIG. 15—*Predicted versus observed constitutive behavior for 871 to 982°C CWD TMF cycle.*

meaning that the elastic modulus decreases due to increasing temperature. Then, plastic deformation occurs at P and the curve becomes flatter as the upper peak is approached.

760 to 871°C, 180° Out-of-Phase

The LIP and LOP hysteresis loops obtained analytically are symmetric with respect to the origin. This results from the assumption of symmetric stress-strain behavior in tension and compression. The stress amplitude for the LOP cycle is again acceptable. The plastic strain range is in better agreement with test results than in the in-phase case.

760 to 871°C, 90° Out-of-Phase

The test maximum tensile stress in the CWD test is substantially lower than the computed value. The derivative $d^2\sigma/d\epsilon^2$ becomes negative, negative, positive, and negative as the point moves from 816°C at the top toward P and 816°C at the bottom and then loads along the upper branch of the hysteresis loop. An interpretation similar to the case of in-phase loading is possible for these changes.

871 to 982°C, In-Phase and Out-of-Phase

All 871 to 982°C hysteresis loops show remarkable correlation with the test results. Both the stress range and the plastic strain range are satisfactory. The general trend of the behavior of $d^2\sigma/d\epsilon^2$ is analogous to that of 760 to 871°C except that there is a region in 90° out-of-phase near 982°C on the lower loop branch where $d^2\sigma/d\epsilon^2$ becomes negative. This is caused by the change in the rate of strain hardening as the temperature increases from 951°C at P to 982°C, then decreases to 927°C at the lower peak. This phenomenon was not appreciable in the 760 to 871°C, CWD cycle.

Other Remarks

It was observed that the experimental elastic modulus was smaller than the average analytical value used in the 760 to 871°C cases. An investigation of the effects of a reduction of the modulus on the hysteresis loops was made. It was found that the plastic strain range is significantly affected. For instance, the plastic strain range in Fig. 10 can be reduced to 0.2% by changing the moduli from 170.1 to 160.3 MPa at 760°C and from 160 to 146.5 MPa at 871°C. The reduced moduli are within the range of scatter of the data base for René 80. The stress amplitude did not change significantly due to the increase of elastic strain. No significant changes have been found, however, in any variables for 871 to 982°C cases where plastic deformation is the dominant part of the total deformation.

Conclusions

In the course of this program, a series of tests were carried out to determine the high-temperature fatigue behavior of René 80. The isothermal results showed the constitutive behavior to be quite strain-rate sensitive at 871°C but much less so at 760 and 982°C. Temperature plays a much more important role in the cyclic behavior. The low-cycle fatigue results were rate sensitive at all temperatures. The Ostergren damage approach, which includes the effect of frequency explicitly, was found to correlate the LCF data at each temperature. Using the isothermal data at the temperature of maximum cyclic stress, an approach based on the Ostergren method was developed to predict the TMF lives; the correlation between observed and computed life was quite good, with the majority of the values falling within the ± 2 scatterband. An examination of the stresses in the TMF cycle and the isothermal CSS values provides a rationale for this agreement.

An attempt has been made in this paper to examine the utility of the classical plasticity theory in modeling the thermomechanical response of René 80. Although there has been a general research trend putting more emphasis on unified theories in recent years for high-temperature applications, this study indicates that the classical constitutive model is a good method for the type of thermomechanical loading considered. The correlation between experimental and analytical results was found to be excellent for the 871 to 982°C in-phase and out-of-phase loading cases. The correlation is not as good for the 760 to 871°C cases, but it might be improved by increasing the number of stress-strain curves input to better define the intermediate temperature behavior. As was discussed, the material deformation mode changes between 760 and 871°C, perhaps rendering it inadvisable to linearly interpolate between these end point values. Unfortunately, additional data were not available at this time. It is thought that further investigation including other aspects of deformation such as mean strain, mean stress, and hold time would be worthwhile to further evaluate the capability of the classical model. This requires the inclusion in the analysis of creep deformation; this has not been considered separately here.

While the thermal cycle variation in these tests is limited, the ability to correlate the TMF lives and predict the constitutive behavior provides optimism that this analysis may be applied to other TMF situations.

References

[1] Antolovich, S. D., Liu, S., and Baur, R., *Metallurgical Transactions A*, Vol. 12A, March 1981, pp. 473 to 481.

[2] Antolovich, S. D., Domas, P., and Strudel, J. L., *Metallurgical Transactions A*, Vol. 10A, Dec. 1979, pp. 1859–1868.

[3] Embley, G. T. and Russell, E. S. in *Proceedings*, First Parsons International Turbine Conference, Dublin, Institute of Mechanical Engineers, 1984, p. 157.

[4] McKnight, R. L., Laflen, J. H., and Spamer, G. T., "Turbine Blade Tip Durability Analysis," NASA Report 165268, National Aeronautics and Space Administration, Feb. 1982.

[5] Cook, T. S. in *Mechanical Testing For Deformation Model Development, ASTM STP 765*, R. W. Rohde and J. C. Swearengen, Eds., American Society for Testing and Materials, Philadelphia, 1982, pp. 269–283.

[6] Coffin, L. F., *Metallurgical Transactions*, Vol. 5, May 1974, pp. 1053–1060.

[7] Antolovich, S. D., Baur, R., and Liu, S. in *Superalloys 1980*, J. K. Tien et al., Eds., American Society for Metals, Metals Park, Ohio, 1980, pp. 605–613.

[8] Domas, P. A. and Antolovich, S. D., *Engineering Fracture Mechanics*, Vol. 21, No. 1, 1985, pp. 203–214.

[9] Ostergren, W. J., *Journal of Testing and Evaluation*, Vol. 4, No. 5, Sept. 1976, pp. 327–337.

[10] Kuwabara, K., Nitta, N., and Kitamura, T. in *Advances in Life Prediction Methods*, D. A. Woodford and J. R. Whitehead, Eds., ASME, 1983, pp. 131–141.

[11] Walker, K. P., "Research and Development Program for Nonlinear Structural Modeling with Advanced Time-Temperature Dependent Constitutive Relationships," NASA CR-165533, 1981.

[12] Proceedings of NASA Conference on Nonlinear Constitutive Relations For High Temperature Applications, NASA-CP-2271, 1983.

[13] McKnight, R. L., "Finite Element Cyclic Thermoplasticity Analysis by the Method of Subvolumes," Ph.D. dissertation, Department of Aerospace Engineering, University of Cincinnati, Ohio, 1975.

[14] Besseling, J. F., "A Theory of Plastic Flow for Anisotropic Hardening in Plastic Deformation of an Initially Isotropic Material," Report S. 410, National Aeronautical Research Institute, Amsterdam, 1953.

DISCUSSION

M. Nazmy[1] *(written discussion)*—(1) Have you tried approaches of lifetime prediction other than that of Ostergren? How was the fit?

(2) Have you applied the double damage rule, developed by Halford and Manson, to your results?

T. S. Cook et al. (authors' closure)—(1) A number of different parameters were fit to the low cycle fatigue for René 80. In addition to the Coffin-Manson and Ostergren approaches noted in the paper, the frequency-modified (FM) method of Coffin [*1*], the Leis parameter [*2*], the pseudostress ($E\Delta\epsilon$) approach, and the Basquin-Coffin-Manson [*3*] method were examined. In addition to our data, other René 80 data, including hold time results from the literature, were checked [*1,4*]. The average error of the curve fit was defined as

$$(\text{error})^2 = \frac{1}{n} \Sigma(1 - N_o/N_c)^2$$

where n is the number of data points, and N_o and N_c are the observed and correlated lives respectively. In virtually all situations examined, the approaches explicitly containing the frequency gave the smallest average error for the curve fits. The Ostergren approach was selected over the FM method for the reasons specified in the paper. However, a similar analysis could presumably be carried out using the FM method.

(2) We did not attempt to apply the double linear damage rule to the TMF analysis. As just noted, once the curve fits to the isothermal results are obtained, a double damage model could be applied. However, given the nature of the TMF cycle, a hysteretic energy analysis seems more satisfying.

Discussion References

[*1*] Coffin, L. F., *Metallurgical Transactions*, Vol. 5, May 1974, pp. 1053–1060.

[*2*] Leis, B. N., *Journal of Pressure Vessel Technology, Transactions of ASME*, Vol. 99, No. 4, Nov. 1977, pp. 524–533.

[*3*] Cook, T. S., Paper 84-PVP-27, 1984 ASME Pressure Vessel and Piping Conference, San Antonio, Tex.

[*4*] Antolovich, S. D., Liu, S., and Baur, R., *Metallurgical Transactions A*, Vol. 12A, March 1981, pp. 473–481.

[1]Brown Boveri Research Center, Baden/Dättwil, Switzerland.

Microstructural Effects

J. Wareing[1]

Influence of Material Microstructure on Low Cycle Fatigue Failure, with Particular Reference to Austenitic Steel

REFERENCE: Wareing, J., **"Influence of Material Microstructure on Low Cycle Fatigue Failure, with Particular Reference to Austenitic Steel,"** *Low Cycle Fatigue, ASTM STP 942,* H. D. Solomon, G. R. Halford, L. R. Kaisand, and B. N. Leis, Eds., American Society for Testing and Materials, Philadelphia, 1988, pp. 711–727.

ABSTRACT: A study has been made of the low-cycle fatigue behavior of austenitic stainless steels and a magnesium alloy over a wide range of temperature. The effects of microstructural variation on fatigue life of these materials which exhibit different crystal structures and deformation modes have been studied by thermo-mechanically treating material prior to testing and by inducing grain boundary creep cavitation during elevated temperature unbalanced creep-fatigue cycles. For balanced cycles microstructural changes result in moderate variations in fatigue life, despite changes in failure mode from trans- to intercrystalline. On the other hand, the introduction of discrete creep-type grain boundary cavities during elevated temperature unbalanced creep-fatigue cycles can cause an order of magnitude reduction in fatigue life. The effects observed in these two materials are rationalized in terms of current theories of fatigue failure and serve as a basis of assessing other materials.

KEY WORDS: fatigue, creep, stainless steels, magnesium alloy, grain size precipitates, crack initiation, crack propagation, transgranular, intergranular

Ambient temperature smooth specimen fatigue failure takes place by the relatively rapid nucleation and slow controlled growth of a surface crack until material separation occurs or the crack achieves a critical size for fast fracture. Thus the important factors determining fatigue endurance are the initiated crack size, rate of crack growth, and critical crack size. As the temperature is increased this mechanism still persists, but additional effects take place which may modify the rates at which the phases of failure develop. In particular, the weakening of grain boundaries can promote intergranular cracking and enhance crack initiation and growth. This tendency is controlled to a large extent by material microstructural changes resulting from the precipitation of second phase particles, which are governed by prior thermo-mechanical treatments and the exposure to elevated temperature during cycling. Such changes modify the materials general strength level and the relative strength of grain interiors to grain boundaries, the latter relationship has a major influence on the material's propensity to the formation of inter-crystalline cavitation, particularly if the cycle contains an element of creep deformation. Grain boundary cavitation can substantially increase the rate of fatigue crack growth [*1*]. Alternatively, internal cavitation can lead to structural degradation and creep-type failures [*2*]. It is therefore essential to quantify the influence of these changes on the fatigue failure process,

[1]Principal Scientific Officer, Springfields Nuclear Power Development Laboratories, United Kingdom Atomic Energy Authority (Northern Division), Springfields, Salwick, Preston PR4 ORR, Lancashire, England.

particularly if short term data are to be extrapolated in terms of time, for example, for the design of an elevated temperature plant.

Ambient temperature fatigue crack growth is a well understood, closely controlled process [3]. New crack surface is generated by shear decohesion at the crack tip, the extent of this decohesion is related only to the extent of intense crack tip deformation and the applied stress-strain field. The factors determining crack growth rate per cycle are the magnitude of the applied stress-strain field, the current crack length and a material strength parameter. This simple picture of growth is a good starting point to assess elevated temperature smooth specimen failure when the secondary processes, previously referred to, come into effect. If the basic surface crack growth process occurs, it would be expected that those factors still strongly influence behavior.

This paper considers the low-cycle fatigue failure process in austenitic stainless steels at temperatures from ambient to 750°C and how it is influenced by material property and microstructural changes introduced by prior thermo-mechanical treatments and by creep cavitation induced during cycling. Additionally, the response of a magnesium alloy will be considered at temperatures up to 170°C. The smooth specimen response of the materials will be assessed in terms of a theoretical crack propagation model, developed to quantify ambient temperature failure. Deviations from the theory can be readily seen and rationalized in terms of metallographic observation.

Mechanism of Fatigue Failure

Crack Initiation

Smooth specimen fatigue failure occurs by the nucleation and growth of a crack. In general, nucleation is a surface phenomenon and in ductile metals results from a microscopically irregular surface formed as a result of the slip activity. Cracks form early in life and are of the order of 1 to 10 μm deep. Other microstructural sites for initiation are discontinuities, such as grain boundaries and twin boundaries: the latter being particularly operative in hcp metals. For stronger more complex alloys, planar slip dominates and localized slip bands become initiation sites of the random notch peak topography generated by the shearing of a pack of cards. The interaction of slip bands with second phase particles (inclusions, precipitates) can also produce a local stress concentration which cracks the interface, producing a surface or sub-surface crack. Additionally, pores or large carbides can act as nucleation sites at either the surface or within the material. For this type of nucleation the initiated size depends on the particle or pore size.

Crack Propagation

For high strain fatigue, the model proposed by Tomkins [3] adequately describes the response of many materials. Here crack growth rate per cycle, da/dN, is given as

$$\frac{da}{dN} = \Delta\epsilon_p D = \Delta\epsilon_p \left[\sec\left(\frac{\pi\sigma}{2T}\right) - 1 \right] a \tag{1}$$

where D is the extent of intense deformation at the crack tip, $\Delta\epsilon_p$ is the applied plastic strain range, σ is the maximum tensile stress in the cycle, T is a material strength parameter approximately equal to the cyclic UTS, and a is the current crack length.

If the number of cycles to initiate a crack of length a_o is N_o and N_p is the number of cycles to propagate the crack to a final size a_f, the number of cycles to failure N_f is given as

$$N_f = N_o + N_p \tag{2}$$

However, if life is dominated by crack propagation, N_p approximates to N_f, then integration of Eq 1 yields

$$\ell n\left(\frac{a_f}{a_o}\right) = N_f \Delta\epsilon_p \left[\sec\left(\frac{\pi\sigma}{2T}\right) - 1\right] \tag{3}$$

In ductile materials a_o is approximately 10 μm and a_o approximately $2/3 \times$ specimen section [*4*].

For a material with a cyclic stress-strain response of the form $\sigma = k\Delta\epsilon_p^n$, where k is a constant and n the cyclic hardening exponent, and for $\sigma/T < 0.5$ Eq 3 reduces to

$$\Delta\epsilon_p N_f^\alpha = \left[\left(\frac{T}{k}\right)^2 \frac{8}{\pi^2}(2n + 1)\ell n\left(\frac{a_f}{a_o}\right)\right]^\alpha \tag{4}$$

where

$$\alpha = \frac{1}{2n + 1}$$

This is the Coffin-Manson equation. Note that for rapid initiation the slope of the curve is related only to the slope of the cyclic stress-strain curve: the position of the curve depends on T/k, a_f/a_o, and n. As T and k are strength parameters, variations in strength levels alone, with microstructural or temperature changes, will not radically alter the endurance for cycling at a given strain range. Regarding the ratio a_f/a_o, this appears in the logarithmic form and this ratio must alter drastically to bring about a significant positional change in the fatigue curve.

In the remaining part of the paper, fatigue data, generated on austenitic stainless steels and a magnesium alloy over a wide range of temperature, will be assessed against the model of failure described above. When theoretical integrated crack propagation fatigue life (Eqs 3 and 4) are used, a value of 10 μm for a_o (*i.e.*, typical of that observed in initiation studies of ductile materials) and a value of a_f of $2/3 \times$ the specimen diameter, as measured from failed specimens, are used unless otherwise stated.

All fatigue data were generated in air in push-pull, fully reversed, strain controlled loading using hydraulic or screw driven testing machines. Strain control was effected using diametral extensometers. For elevated temperature tests, the temperature of the specimen gage length was maintained within ±2.5°C during testing.

Contribution to Total Endurance of Crack Initiates and Propagation Phases of Failure

Figure 1*a* illustrates smooth specimen endurance data for a type 316 stainless steel at 625°C, where a break is observed in the plastic strain range - endurance relationship [*4*]. Now if the crack propagation phase of failure alone dominates life, and Eq 4 were to describe fatigue life, a similar break would be expected in the cyclic stress-strain curve: this is not observed as indicated by Fig. 2. However, also illustrated in Fig. 1*a* are measured crack propagation data, integrated between an initial crack size of 10 μm and a final size of two-thirds the smooth speci-

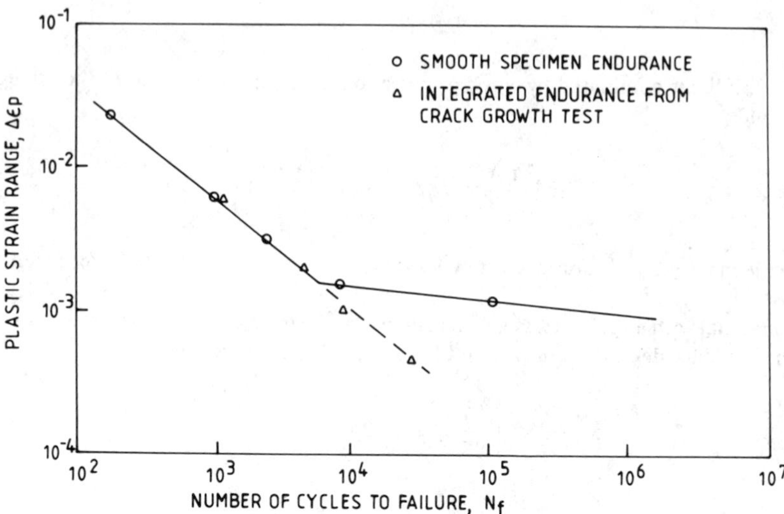

FIG. 1*a*—*Coffin-Manson plot for Type 316 stainless steel at 625°C from Wareing* [4].

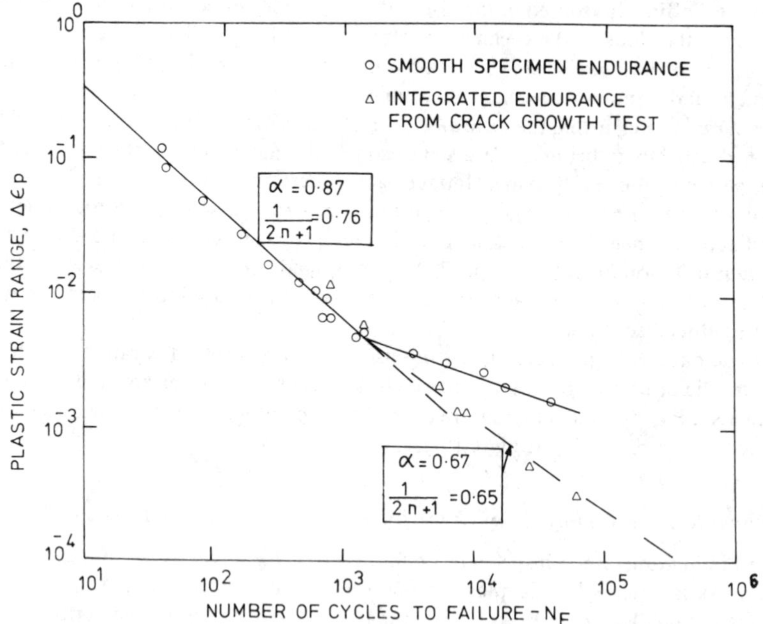

FIG. 1*b*—*Coffin-Manson plot for 20/25/Nb stainless steel at 750°C.*

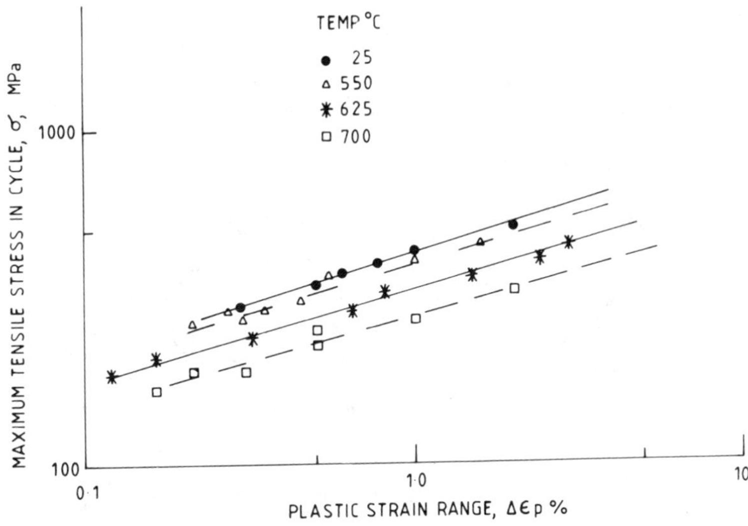

FIG. 2—*Cyclic stress-strain curves for Type 316 stainless steel at various temperatures.*

men diameter. The integrated crack growth fatigue life is in good agreement with smooth specimen endurance at high strain ranges but diverges at low ranges. Thus the crack initiation life, *i.e.*, the difference in cycles between the two curves, is very short at high strain ranges: at low strain ranges approaching the fatigue limit, crack initiation consumes an increasing fraction of life. The two-slope form of the *plastic* strain - endurance relationship is therefore a good indication of the strain ranges at which initiation makes a significant contribution to total life. Similar data on solution annealed 20/25/Nb stainless steel tested at 750°C and a balanced strain rate of 10^{-3} s^{-1} support this premise: they are illustrated in Fig. 1*b*. Note again the discrepancy between integrated endurance and smooth specimen life at low strain ranges. For this material a break occurs in the cyclic stress-strain curve (see Fig. 6), and the difference between lives cannot be wholly attributable to initiation effects. However, the change in cyclic hardening exponent n with strain range is not sufficient to explain the differences in life obtained from the two types of test. This is illustrated in Fig. 1*b* by comparison between the exponent of the Coffin-Manson equation α and the predicted exponent $(1/(2n + 1))$ of crack propagation dominated fatigue life, further indicating the influence of crack initiation at low strain ranges. An alternative explanation of this behavior is that as these tests were conducted in air, oxide blocking at small crack lengths would reduce crack growth rate and enhance fatigue endurance, particularly at low strain ranges; this would also give rise to a discontinuity in the curve. However, similar differences between integrated growth endurance and smooth specimen life have been observed in ambient temperature tests where oxidation effects are minimal [*4*]. The break in the Coffin-Manson curve does therefore appear to be related to enhanced numbers of cycles required for crack initiation at low strain ranges.

Further evidence to support this premise is indicated from work on a 2¼ Cr-1Mo ferritic steel tested at 538°C by Brinkman et al [*5*]. Figure 3 illustrates data from continuous cycling tests and from tests including a hold period at the maximum compressive strain in the cycle. As for the austenitic stainless steel a discontinuity is observed in the continuous cycling curve. This is not so, however, for the compressive hold period data. Metallographic examination indicated that oxidation of specimen surfaces took place during testing, particularly in the hold period

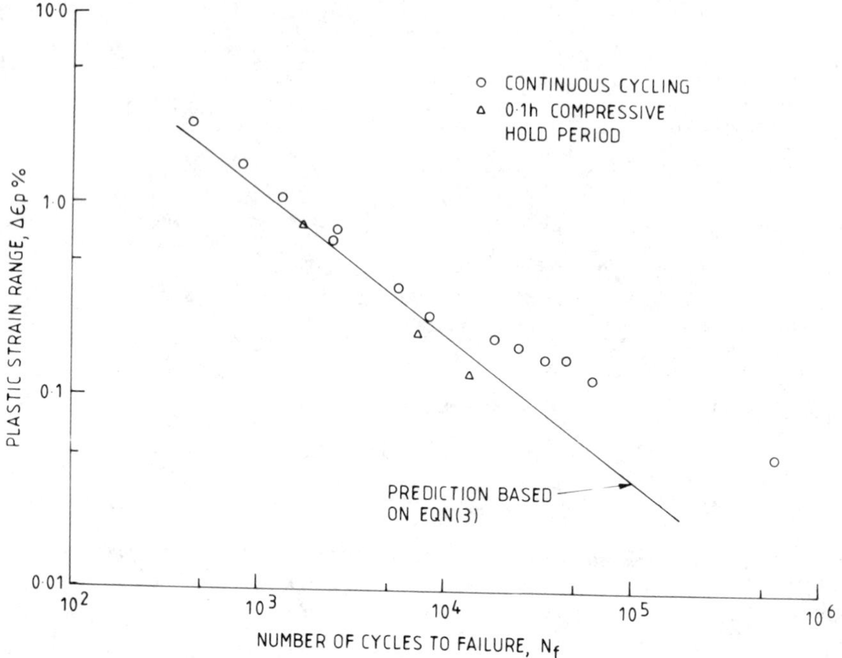

FIG. 3—*Effect of compressive hold period on fatigue behavior of a 2¼Cr-1Mo ferritic steel at 538°C, after Brinkman et al.* [5] *(1 h = 3600 s).*

[5]. The surface oxide, formed during a compressive hold period, both adheres to the specimen surface and cracks during subsequent tensile fatigue straining. Thus enhanced initiation in low strain compressive hold period tests is responsible for the reduction in life from that in continuous cycling. At high strain ranges, where crack nucleation during continuous cycling is a rapid process, data for each cycle type are similar. Also included in Fig. 3 is a prediction of fatigue life from Eq 3, which is in good agreement with all data at high strain ranges and the low strain range crack propagation dominated compressive hold period data. For tests including a tensile hold period, endurances were similar to those in continuous cycling tests and a break occurred in the Coffin-Manson plot. Here the oxide formed during the tension dwell spalled off during compressive straining; thus enhanced initiation did not occur.

The above evidence regarding the contribution of the crack initiation phase to total life cannot be unequivocally proven from the above data. However, reference to fatigue data on a magnesium alloy (see Section on Fatigue Failure in a Magnesium Alloy) reveals a strong relationship between slope changes in Coffin-Manson plots and cyclic stress strain plots at high strain ranges where crack initiation is extremely rapid. This is further indirect corroboration of the premise outlined above.

Fatigue Failure in Austenitic Stainless Steels

Effect of Test Temperature

Figure 4 illustrates the influence of test temperature in the range 25 to 700°C on the low-cycle fatigue endurance of a cast of solution annealed (1050°C for 1 h) type 316 stainless steel, having

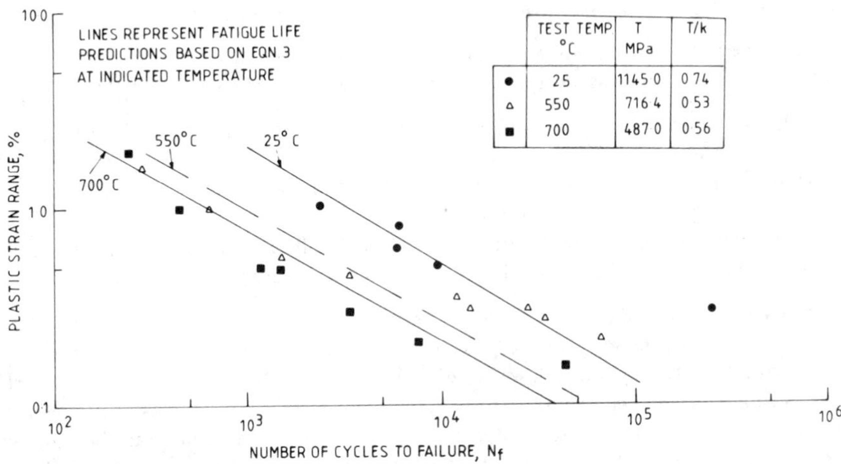

FIG. 4—*Coffin-Manson plots for Type 316 stainless steel at various temperatures.*

an average grain size of 38 μm, for tests at a constant strain rate of $1.5 \times 10^{-3}\ s^{-1}$. Discontinuities occur in the Coffin-Manson plots at low strain ranges for each considered temperature indicating the influence of crack initiation. At higher strain ranges, the data for each temperature are simply translated with respect to one another: fatigue life is reduced by a factor of three as the temperature is increased from 25 to 550°C. A further increase in temperature to 700°C caused only a slight decrease in life. In all cases failure occurred by the growth of a surface crack: crack initiation and propagation was transgranular and fracture surfaces were characterized by ductile fatigue striations. Cyclic stress-strain curves based on half-life stress levels at the three temperatures are illustrated in Fig. 2. The cyclic stress-strain curves at 25 and 550°C are very similar and are displaced from that at 700°C by a factor of about 1.5 in stress for a given plastic strain range: the slopes of all curves are virtually identical. Transmission electron microscopy of failed specimens revealed a well-defined dislocation cell structure at 700 and 550°C, and an ill-defined structure together with incidences of mechanical twinning at 25°C. This change in micro-structure reflects the increase in stacking fault energy and ease of cross-slip with increase in temperature. The similar cyclic stress-strain response at 25 and 550°C is probably a result of dynamic strain-aging, commonly observed in austenitic stainless steels in the temperature range 500 to 600°C, enhancing cyclic strength [*6*].

Figure 4 illustrates predictions of fatigue life based on Eq 3, for a_o of 10 μm and a_f of two-thirds of the smooth specimen diameter. Note the good agreement between data and theory, particularly at high strain ranges. From Eq 4 the slope, α, of the Coffin-Manson equation should be related to the cyclic hardening exponent n. Values of α, at the three temperatures, are approximately, 0.6: this compares with a predicted value $1/(2n + 1)$ of approximately 0.64. The variation in fatigue life with temperature results primarily from the relatively weak variation of T/k despite the large spread in T with temperature, namely 1145 MPa (25°C) to 487 MPa (700°C), as indicated in Fig. 4.

The ability of the crack propagation model, based primarily on material strength parameters, to predict fatigue endurance at ambient and elevated temperature is a consequence of the balanced nature of the cycles, i.e., equal tensile/compressive strain rates, this precluding the formation of creep damage at elevated temperature (see section on the effect of creep cavitation on fatigue endurance).

Effect of Grain Size on Fatigue Life of a 20/Cr/25Ni/Nb Stainless Steel

To study the influence of grain size on low-cycle fatigue behavior, specimens machined from bar stock 20/25/Nb stainless steel were given the following heat treatments:

(*a*) Solution annealed (SA) followed by a furnace cool: 1000°C for 3.6×10^3 s (1 h) – average grain size 25 μm.

(*b*) High-temperature solution treatment (HTST) followed by an oil quench: 1150°C for 3.6×10^4 s (10 min) – average grain size 150 μm.

(*c*) High-temperature solution treatment, oil quench followed by thermal aging: 1150°C for 3.6×10^4 s (10 min) plus 800°C for 1.0×10^6 s (300 h) – average grain size 150 μm.

Heat treatments (*b*) and (*c*) were chosen to produce a constant large grain size by the use of a solution treatment sufficiently high to take virtually all of the carbon in solution and permit grain growth. The rapid oil quench was intended to retain carbon in solution and impart solid solution strengthening. The aging treatment included in (*c*) was selected as one known to substantially complete the precipitation of carbon, in the form of M_6C, on grain boundaries and dislocations [*7*] thereby retaining a large grain size but reducing the material strength level. In this way the influence of strength and grain size could be studied.

Coffin-Manson plots for the three material heat treatments are illustrated in Fig. 5, for tests in air at 750°C and a strain rate of 10^{-3} s^{-1}. The fatigue life of SA material exceeds that of the HTST material by about a factor of four, for a given strain range, while that of aged material is only marginally reduced from that of SA material. There is therefore no unequivocal effect of grain size on fatigue life. All failures took place by the growth of a surface crack. Scanning electron microscopy of failed specimens revealed that the SA material failed in a predominantly transgranular manner, HTST material produced totally intergranular failures while aged material failures were predominantly intergranular. Fatigue striations were observed in all cases, even where failure was intercrystalline. There does not therefore appear to be a correlation between failure mode and fatigue life.

Figure 6 includes cyclic stress-strain curves for all material conditions, based on half-life stress levels: initial rapid hardening was followed in all cases by a saturation in stress level. The

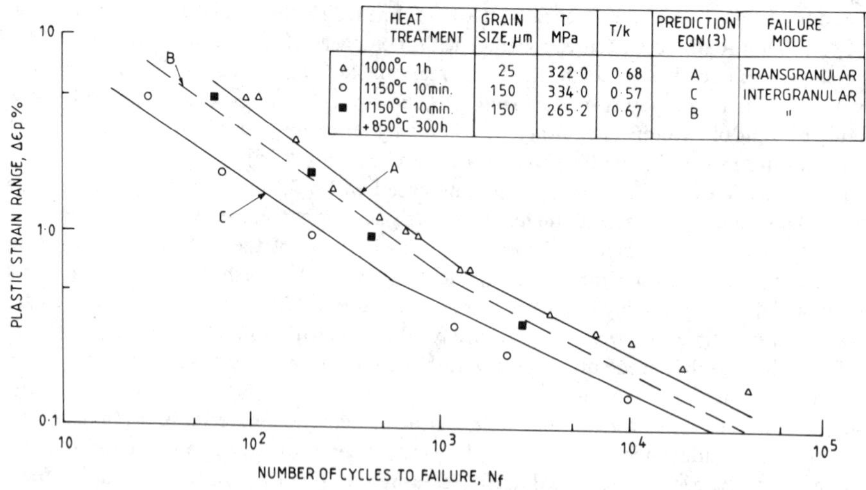

FIG. 5—*Effect of grain size on fatigue life of a 20/25/Nb stainless steel at 750°C (1 h = 3600 s, 1 min = 60 s).*

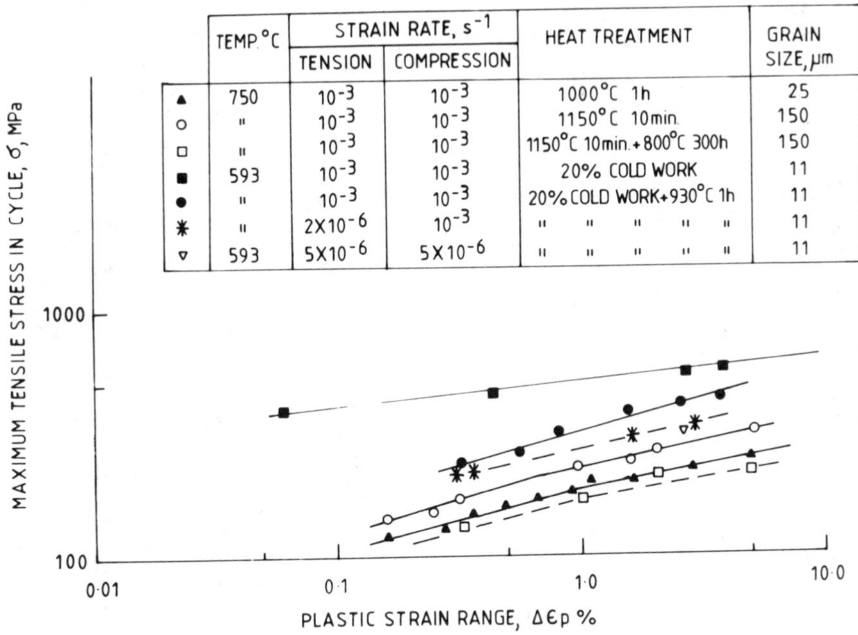

FIG. 6—*Cyclic stress-strain curves for 20/25/Nb stainless steel at various temperatures and strain rates (1 h = 3600 s, 1 min = 60 s).*

HTST material is strongest, reflecting solid solution strengthening, while the SA and the aged material are respectively weaker. It will be noted that a break occurs in the cyclic stress-strain curve for the SA material in line with findings at other temperatures for this steel [*8*]. A break is also suggested for the other two heat treatments. The reason for this break was not found: transmission electron microscopy revealed dislocation sub-cell formation at all strain levels. Figure 5 indicates that a break exists in the Coffin-Manson plot at the same strain level as that in the stress-strain curve. Also included in Fig. 5 are fatigue life predictions based on Eq 3, where good agreement is obtained for all heat treatments. It is worth noting that at this elevated temperature, 750°C, a significant deviation between predicted integrated growth rate life and smooth specimen endurance is not observed at low plastic strain ranges suggesting that crack initiation is accelerated, perhaps by oxidation effects. The variation in fatigue life with heat treatment can be explained solely in terms of material strength changes T/k with heat treatment, Fig. 5. Grain size per se and mode of crack propagation (trans or intercrystalline) do not significantly alter fatigue life.

Effect of Cold Work on Fatigue Life

Figure 7 compares the low-cycle fatigue behavior of a 20/25/Nb stainless steel in the 20% cold worked condition with that of annealed material (20% cold work plus 930°C for 1 h) for tests at a strain rate of 10^{-3} s^{-1} [9]. The average grain size of both materials was 11 μm. Note that the lives, at the highest strain ranges for the two conditions, are very similar. With decreasing strain range, the life of annealed material exceeds that of cold worked material by an increasing amount, *i.e.*, the exponent α of the Coffin-Manson equation is higher for the cold worked material. Failure in both cases occurred by the transgranular propagation of a surface

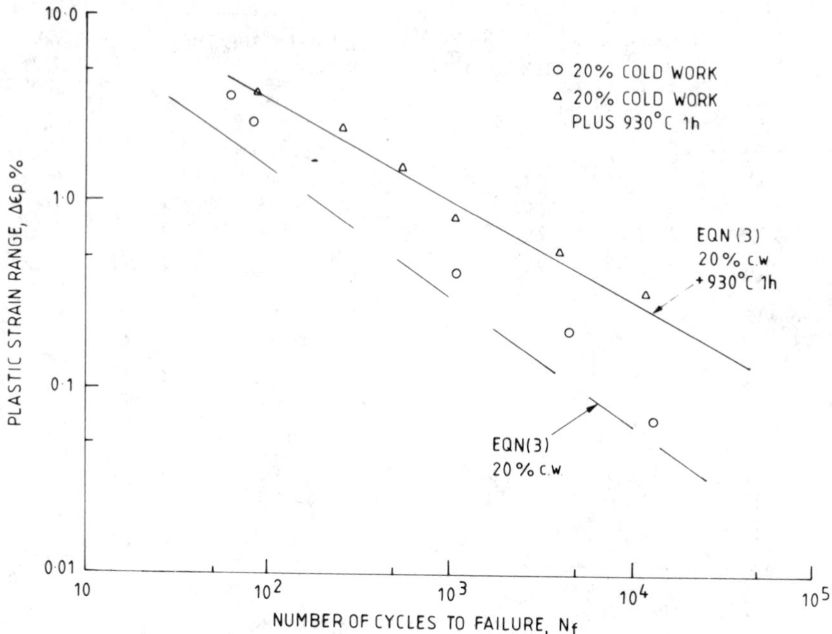

FIG. 7—*Effect of 20% cold work on fatigue life of a 20/25/Nb stainless steel at 593°C.*

initiated transgranular crack, and striations were observed on fracture surfaces. Figure 6 illustrates cyclic stress-strain data for the two material conditions, based on half-life stress levels: cyclic hardening occurred in annealed material, while softening was observed for cold worked material. Note the higher strength levels and reduced hardening exponent of the cold worked material compared to that of annealed material. In Fig. 7, predictions of fatigue life based on Eq 3 are compared with data and show fairly good agreement. This confirms, once again, that the difference in fatigue life for the two material conditions results primarily from differences in material strength levels.

Fatigue Failure in a Magnesium Alloy

Tests have also been conducted on a magnesium 0.8% aluminum alloy that has a close packed hexagonal structure, melts at 650°C, and exhibits brittle characteristics at ambient temperature but has a ductility that increases with increasing temperature. Results are included for comparison with austenitic stainless steels. At room temperature, the alloy deforms by (0001) ⟨1120⟩ slip and $\{10\bar{1}2\}$ twinning: at elevated temperature non-basal slip can occur and cross-slip takes place. The material was given a heat treatment of 400°C for 3600 s (1 h), which resulted in an average grain size of approximately 450 μm.

Figure 8 illustrates Coffin-Manson plots for tests in air at 25, 135, and 170°C, at a strain rate of approximately 1.5×10^{-3} s^{-1}. Unlike the austenitic stainless steel, fatigue life increased with increasing temperature. Furthermore, the data at 25°C exhibit three separate slopes, that at 135°C a single slope, while a two slope form represents the data at 170°C. Reference to Fig. 9 reveals that distinct slope changes occurred in the cyclic stress-strain curves at the same strain ranges as those of the endurance plots. At ambient temperature, deformation was controlled by mechanical twinning and the shape changes in the cyclic stress strain curve arose due to tensile

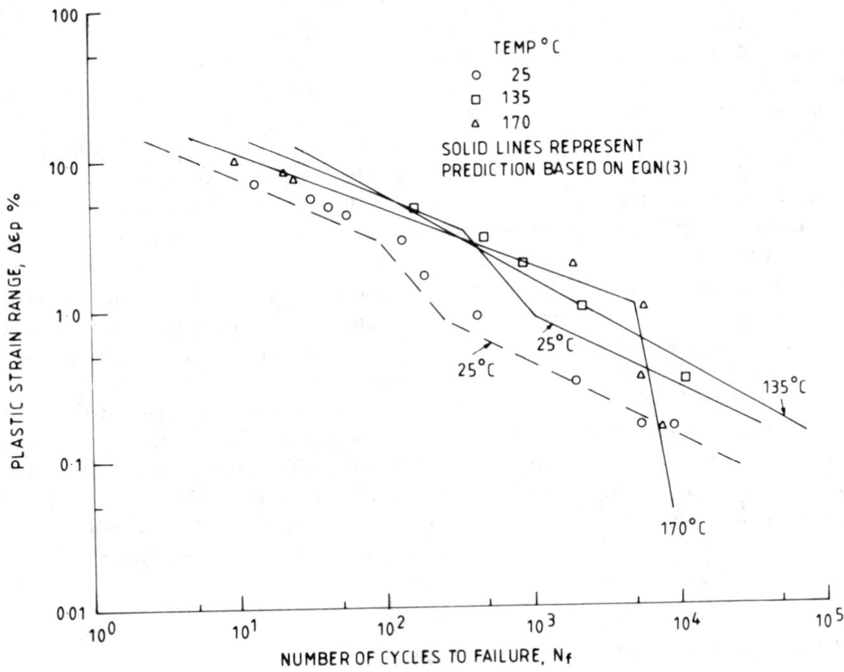

FIG. 8—*Effect of temperature on fatigue life of a magnesium-aluminum alloy.*

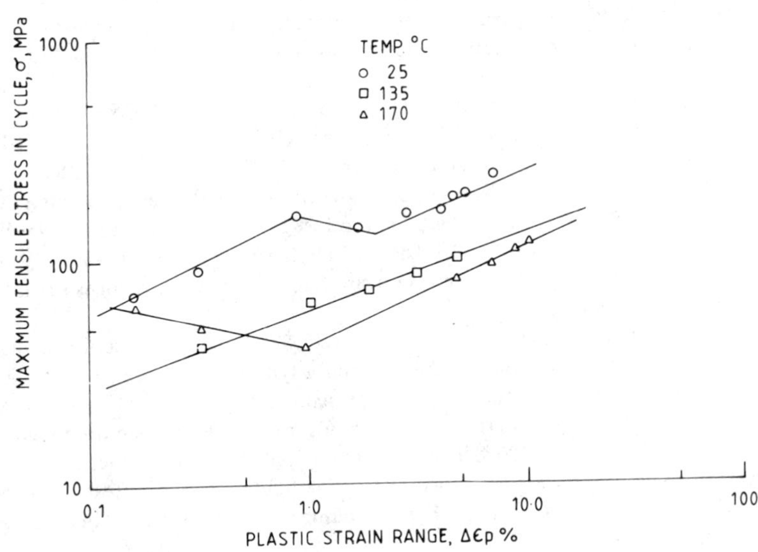

FIG. 9—*Cyclic stress-strain curves for a magnesium-aluminum alloy at various temperatures.*

and compressive twinning and untwinning during cycling [10]. At 135°C slip dominates, and a single slope is observed as in the cyclic stress-strain curve of type 316 stainless steel (Fig. 2). A similar effect would be expected at 170°C, but this was not the case. However, metallographic examination of failed specimens revealed that, for plastic strain ranges between 1 and 0.1%, grain growth had occurred during testing presumably as a result of a critical strain annealing process. This did not occur for strain ranges greater than 1%.

Fatigue life predictions based on Eq 3 (a_o of 10 μm and a_f of 2/3 × specimen diameter) are included in Fig. 8. These show slope changes in good agreement with all data. Regarding the positional fit, this is good at 135 and 170°C. At 25°C, however, prediction overestimates life by about a factor of four. Examination of failed specimens indicated that failure at all temperatures was by the growth of a surface crack. At 25°C, alone however, areas of cleavage were observed on fracture surfaces. It is therefore possible that the final crack nucleated as a cleavage crack across a single grain. This would increase a_o considerably, and with a slightly reduced a_f to account for fast fracture, $\ell n\ a_f/a_o$ is decreased by a factor of four to five. The dashed line, Fig. 8, represents prediction based on this correction.

This material therefore exhibits a significantly different response in terms of plastic strain endurance relationships to that of the austenitic stainless steels. The drastic slope changes observed in the Coffin-Manson plots at 170 and 25°C reflect slope changes in cyclic stress-strain response resulting from grain growth and mechanical twinning, respectively. At 170 and 135°C fatigue life is ultimately governed by material strength property through its influence on fatigue crack propagation, while at 25°C strength level and enhanced crack initiation account for the relatively poor fatigue response at this temperature.

Effect of Creep Cavitation on Fatigue Endurance

In previous sections attention has focused on balanced cycling conditions, *i.e.*, ones in which the tensile and compressive going strain rates are equal. Here fatigue crack growth rate was shown to be governed by crack tip continuum mechanics effects and for reasons outlined above, fatigue life is weakly dependent on temperature and the induced metallurgical conditions considered. However, for tests at temperatures within the creep regime, fatigue life was shown to be strongly dependent on the precise shape of the applied cycle [11,12]. This is illustrated in Figs. 10 and 11.

Figure 10 illustrates data for annealed 20/25/Nb stainless steel at 593°C for equal strain rate (fast-fast and slow-slow) and unbalanced (slow-fast) cycles [9]. The majority of the fast-fast data was reproduced from Fig. 7, *i.e.*, that at a strain rate of 10^{-3} s^{-1}, but a single test at a strain rate of 5 × 10^{-6} s^{-1} is included for comparison. The slow-fast data were generated at a compressive strain rate of 10^{-3} s^{-1}, as for the balanced fast-fast cycle tests, but the tensile strain rate was varied from 2 × 10^{-6} to 10^{-7} s^{-1}. The life in balanced cycle tests is virtually identical, but the addition of an unbalanced slow tensile strain rate into the cycle reduces the fatigue life by about an order of magnitude.

In Fig. 11, test data are shown for continuous cycling tests and for tests involving a hold period at the maximum tensile strain of the cycle on a type 316 stainless steel at 570°C: the strain rate of the fatigue portion of the cycle was approximately 3 × 10^{-4} s^{-1} [13]. Here fatigue life decreases with increasing hold period and this results in an order of magnitude reduction in life for hold periods in excess of 5400 s. The hold period cycle is also unbalanced in that the low strain rate tensile deformation is introduced during the hold period, by stress relaxation processes, at variable strain rates similar to those encountered in creep tests. This is reversed by rapid compressive straining.

The large reductions in life shown above occur as a result of the introduction of grain boundary creep damage during low strain rate tensile deformation, which modifies the failure process. This is best illustrated by reference to Fig. 12, a section from a partially failed 20/25/Nb stain-

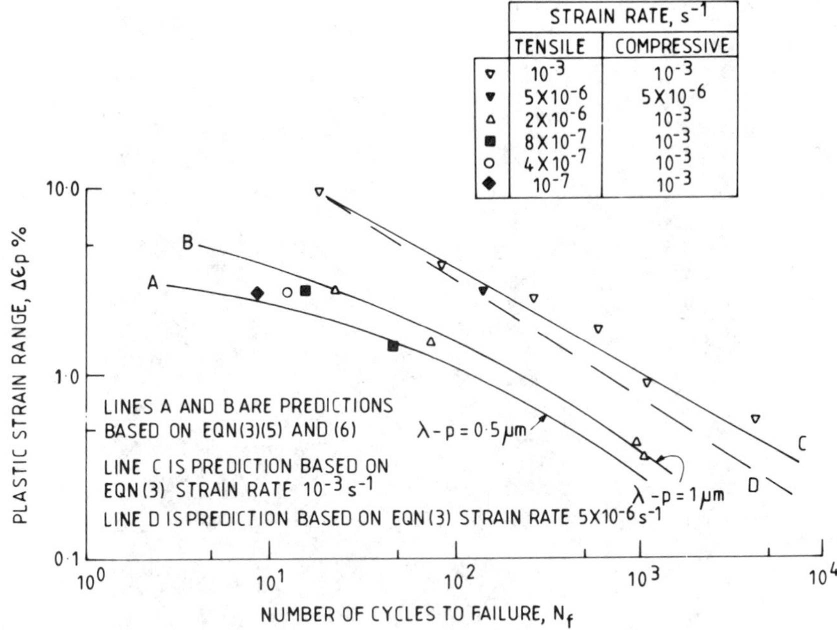

FIG. 10—*Effect of cycle shape on fatigue behavior of a 20/25/Nb stainless steel at 593°C.*

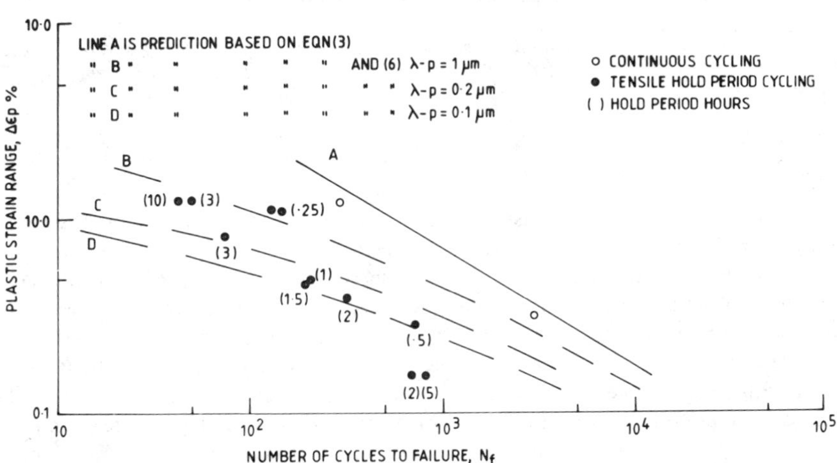

FIG. 11—*Effect of tensile hold period on fatigue life of a Type 316 stainless steel at 570°C (1 h = 3600 s).*

less steel specimen subjected to slow-fast cycling conditions at 593°C [*9*]. Note that in addition to the surface nucleated cracks, there exists abundant intergranular creep cavitation.

Tomkins and Wareing [*14,15*] have extended the model of fatigue crack growth, outlined previously, to include the influence of cavitation on fatigue crack growth rate. During slow-fast or tensile hold period cycling, surface crack initiation and growth take place together with internal cavity nucleation and extension. However, unlike simple fatigue, propagation occurs in es-

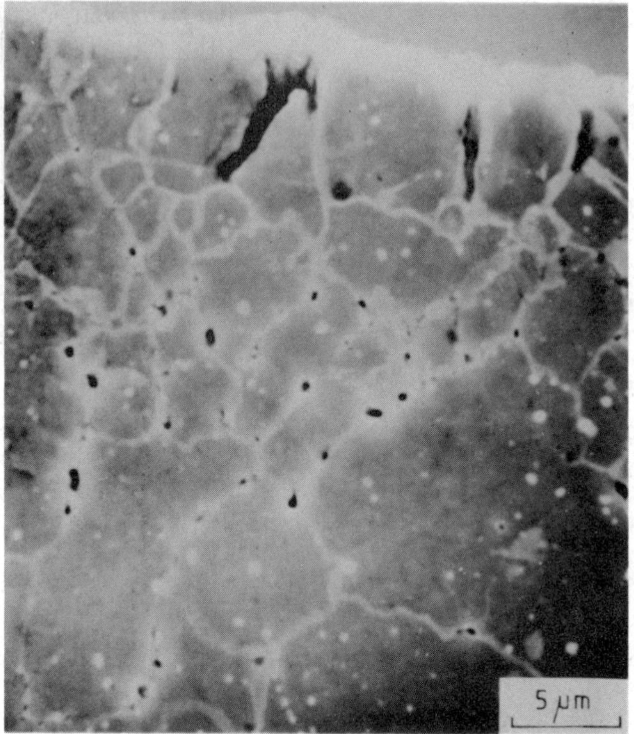

FIG. 12—*Intergranular creep cavitation and surface cracking resulting from slow-fast test on 20/25/Nb stainless steel at 593°C. Specimen cycled to 75% of predicted life, from Livesey and Wareing* [9].

sentially three distinct phases. In Phase I, the crack path is normal to the axis of tensile straining and the growth rate in tension hold period tests, as evidenced by fatigue striations, is independent of hold period and is similar to that in continuous cycling tests [*15*]. Phase II, which occupies a major portion of the fracture surface, follows a more irregular intergranular path in which grain boundaries on fracture surfaces are extensively cavitated. Finally Phase III, unstable fracture, occurs in a single cycle. Because of the large reductions in the fatigue life associated with creep-fatigue cycles compared to continuous cycling tests, it follows that Phase II growth occurs very rapidly compared to that of Phase I. It is postulated that the reason for the large increase in growth rate is the attainment of cavity linkage at the crack tip. This triggers the unzipping of the cavitated material by the crack tip displacement field: the crack tip condition is given by

$$\frac{\delta}{2} = \lambda - p \tag{5}$$

where δ is the crack tip opening displacement and p and λ are the average cavity size and spacings, respectively. Because the shear displacement along the crack tip plastic zone does not decrease rapidly, it follows that crack advance can be of the order of 0.4× the plastic zone size on the attainment of the crack tip opening condition. An approximate solution for δ in a fully plastic field is given by [16]

$$\frac{\delta}{2} = \ell n \sec\left(\frac{\pi\sigma}{2T}\right)\left[\frac{2T}{E} + \frac{2\Delta\epsilon_p}{1+n}\right]a \tag{6}$$

where E is Young's modulus. Because Phase II growth is so rapid, the life of a specimen failing by this mechanism is therefore the cycles for initiation and Phase I growth, *i.e.*, the number of cycles to attain the critical crack size a_f' commensurate with the prevailing cavity spacing. For rapid crack initiation this is given by integration of Eq 3 between an initial crack size a_o (10 μm) and a final size a_f' given by Eq 6.

Turning to the data for 20/25/Nb stainless steel, Fig. 10 includes predictions for fatigue failure based on Eq 3 and for creep-fatigue failure based on Eqs 3, 5, and 6 [9]. The $10^{-3}/10^{-3}$ s^{-1} data are reproduced from Fig. 7 and as noted good agreement exists between data and prediction. For the $5 \times 10^{-6}/5 \times 10^{-6}$ s^{-1} (slow-slow) test condition only one datum point is available. The prediction for this strain rate therefore utilized the cyclic stress-strain data generated for the $2 \times 10^{-6}/10^{-3}$ s^{-1} slow-fast cycle, which is shown in Fig. 6. Note that a reduction in tensile strain rate from 10^{-3} to 2×10^{-6} has caused considerable material softening. Data for other slow-fast cycles showed similar softening but have been omitted for clarity. Thus prediction for slow-slow cycling (a_o = 10 μm and a_f = 2/3 × specimen diameter) is that which would be obtained for slow-fast cycling conditions if bulk grain boundary cavitation did not occur. Note the good agreement between experiment and prediction. Observations of failed specimens from equal strain rate tests indicated that failure had occurred by the growth of a surface crack. No bulk cavitation was observed, and thus any cavities formed during slow strain rate tensile straining are sintered during compressive straining at the same strain rate.

The slow-fast data are bounded by predictions based on cavity spacings of 1 and 0.5 μm. Observations of failed specimens revealed extensive grain boundary cavitation whose spacings were within this range. Thus sintering of cavities does not take place when the compressive strain rate is rapid. The reduced life with decreasing tensile strain rate and a constant compressive strain rate implies that enhanced cavity growth takes place at the reduced tensile strain rate: the condition for Phase II growth is therefore achieved at a shorter crack length a_f' (Eq 6), and life is reduced.

Prediction, (Eq 3), is compared with continuous cycling data and tensile hold period data for type 316 stainless steel (Eqs 3, 5, and 6) and cavity spacings of 0.1, 0.2, and 1.0 μm in Fig. 11. Continuous cycling data are in good agreement with prediction. For the hold period tests, the data at the highest strains are in line with prediction for a 1 μm spacing. With decreasing strain range the data are more in line with prediction based on spacings of 0.2 and 0.1 μm, respectively, until at the lowest strain range (0.15%) prediction overestimates data. Examination of fracture surfaces of specimens tested at strain ranges in excess of 0.15% revealed predominantly intergranular failures containing fatigue striations close to specimen surfaces, thus confirming the mode of creep-fatigue failure outlined above. At the lowest strain range, failure was totally intergranular and no striations were evident. Sections through failed specimens revealed extensive grain boundary cavitation associated with carbide particles. Cavity spacings varied from grain boundary to grain boundary but were in the range 0.1 to 0.5 μm, which coincided with the grain boundary carbide spacings: these were identified as $M_{23}C_6$ [15].

An explanation for this hold period behavior is as follows. During cycling, fatigue crack nucleation takes place and low rate creep strain accumulates from stress relaxation. Creep cavity nucleation is a continuous process and the number of cavities increase with strain. The onset of Phase II growth is related to crack tip opening displacement, which is primarily related to plastic strain range and crack length, Eq 6, and to cavity spacing, Eq 5. At high plastic strain ranges the crack tip opening condition for Phase II growth is satisfied as soon as the first cavities are nucleated. With decreasing strain range the crack length necessary to achieve a given crack opening increases, and hence a larger number of cycles are required, thereby resulting in greater creep strain accumulation. Thus a larger fraction of grain boundary carbides nucleate cavities, and the spacing is reduced. For this reason data at lower strain ranges are in agreement with predictions based on a decreasing cavity spacing. At the lowest considered strain ranges, prediction based on a cavity spacing of 0.1 μm (carbide spacing) overestimates life: failure here is totally intergranular. It has been suggested that these low strain range failures occur by a creep type fracture mechanism and surface fatigue cracking plays no part in the failure. Failure occurs when the accumulated relaxation strain exhausts the material uniaxial creep ductility [*13*]. This is consistent with the low fatigue strain range employed, where fatigue crack nucleation consumes an increasing fraction of life, Fig. 1.

Thus, for cycles involving creep deformation, fatigue life can be drastically reduced by the presence of grain boundary creep cavities. The spacing of these cavities dictates the onset of rapid crack growth and effectively reduces the final crack size to failure. The extent of life reduction therefore reflects cavity spacing which itself is linked to material creep rupture ductility. Small reductions in creep-fatigue life are therefore associated with materials exhibiting high creep rupture ductilities and vice versa [*1*]. This is well demonstrated by the work of Brinkman et al. [*17,18*] on type 316 stainless steel, in which fatigue and tensile hold period tests were conducted on material in the solution treated condition and after thermal aging. For continuous cycling tests, little difference was noted in lives of aged and unaged material, because strength changes with aging were small. On the other hand, the life reduction factor for tension hold period cycling tests on aged material was significantly smaller than that for solution annealed material. This is consistent with the observation that pre-aging significantly increases uniaxial creep rupture ductility compared to that of solution annealed material [*19*].

General Comments

An assessment was made of the fatigue behavior of austenitic stainless steel and a magnesium alloy over a wide range of temperatures and how this behavior was affected by changes in material microstructure. It was shown that under balanced continuous cycling conditions, failure occurs by the growth of a surface crack. During low-cycle fatigue, crack initiation is a rapid process and failure is governed to a large extent by the rate of fatigue crack propagation. For this type of failure the induced microstructural changes considered have affected crack growth rate by at most a factor of four, despite changes in fracture path from transgranular to intercrystalline. The use of a theory of crack propagation has shown that these observed variations in crack growth rate are directly related to variations in material strength levels with microstructural change. This is reflected in fatigue life variations.

The most significant effect on fatigue endurance can occur when a component of tensile creep deformation is introduced into the fatigue cycle. For this type of cycle, internal grain boundary creep cavitation damage accompanies the normal surface fatigue crack nucleation and growth process. Under these creep-fatigue failure conditions, fatigue life can be reduced by over an order of magnitude from that observed in the absence of creep cavitation. This occurs because the presence of creep cavitation modifies the normal fatigue crack growth process and causes a rapid increase in crack growth rate when the fatigue crack tip opening displacement is of the order of the cavity spacing. This effectively reduces the final fatigue crack size at failure and hence the fatigue endurance. Because the creep failure process is sensitive to microstructural variation, through its effect on cavity nucleation rate and cavity distribution, creep-fatigue life is extremely sensitive to material microstructural changes.

Acknowledgments

I wish to acknowledge the valuable discussions with colleagues, particularly Dr. V. B. Livesey, Mr. I. Bretherton, and Mr. G. B. Heys who also carried out some of the experimental work.

References

[*1*] Wareing, J., *Fatigue of Engineering Materials and Structures*, Vol. 4, 1981, p. 131.
[*2*] Hales, R., *Fatigue of Engineering Materials and Structures*, Vol. 3, 1981, p. 399.
[*3*] Tomkins, B., *Philosophical Magazine*, Vol. 18, 1968, p. 1041.
[*4*] Wareing, J. in *Fatigue at High Temperature*, R. P. Skelton, Ed., Applied Science, London, 1983, Chapter 4, pp. 135–187.
[*5*] Brinkman, C. R., Strizak, J. P., Booker, M. K., and Jaske, C. E., *Journal of Nuclear Materials*, Vol. 62, 1976, p. 181.
[*6*] Pineau, A. in *Fatigue at High Temperature*, R. P. Skelton, Ed., Applied Science, London, 1983, Chapter 7, pp. 305–365.
[*7*] Dewey, M. A. P., Sumner, G., and Brammar, I. S., *Journal of the Iron and Steel Institute*, Vol. 203, 1965, p. 938.
[*8*] Wareing, J., Tomkins, B., and Sumner, G. in *Fatigue at Elevated Temperatures, ASTM STP 520*, American Society for Testing and Materials, Philadelphia, 1972, pp. 123–137.
[*9*] Livesey, V. B. and Wareing, J., *Metal Science*, Vol. 17, 1983, p. 297.
[*10*] Azam, N., Didout, G., Leveque, J. P., Sanchez, P., and Weiss, M., 7th Colloque de Metallurgie, 1963, p. 79.
[*11*] Dawson, R. A. T., Elder, W. J., Hill, G. J., and Price, A. T., *Proceedings*, International Conference on Thermal and High Strain Fatigue, Metals and Iron and Steel Institute, 1967, p. 239.
[*12*] Majumdar, S. and Maiya, P., *Canadian Metallurgical Quarterly*, Vol. 18, 1979, p. 57.
[*13*] Wareing, J., Tomkins, B., and Bretherton, I. in *Proceedings*, TMS—AIME Symposium Fall Meeting, R. Raj, Ed., Flow and Fracture at Elevated Temperatures, 1983, p. 251.
[*14*] Tomkins, B. and Wareing, J., *Metal Science*, Vol. 11, 1977, p. 414.
[*15*] Wareing, J., Vaughan, H. G., and Tomkins, B., Creep-fatigue Environment Interactions, *Proceedings*, TMS-AIME Symposium Fall Meeting, R. M. Pelloux and N. Stollof, Eds., 1980, p. 129.
[*16*] Tomkins, B., *Journal of Engineering Materials and Technology*, Vol. 97, 1975, p. 289.
[*17*] Brinkman, C. R., Korth, G. E., and Hobbins, R. R., *Nuclear Technology*, Vol. 16, 1972, p. 299.
[*18*] Brinkman, C. R. and Korth, G. E., *Metallurgical Transactions*, Vol. 5, 1974, p. 972.
[*19*] Sikka, V. K., Brinkman, C. R., and McCoy, H. E., *Proceedings*, Symposium on Structural Materials for Service at Elevated Temperatures in Nuclear Power Generation, ASME Winter Meeting, Houston, 1975, p. 316.

D. R. Stahl,[1] *M. Mirdamadi,*[2] *S. Y. Zamrik,*[2] *and S. D. Antolovich*[1]

Effect of Temperature, Microstructure, and Stress State on the Low Cycle Fatigue Behavior of Waspaloy

REFERENCE: Stahl, D. R., Mirdamadi, M., Zamrik, S. Y., and Antolovich, S. D., **"Effect of Temperature, Microstructure, and Stress State on the Low Cycle Fatigue Behavior of Waspaloy,"** *Low Cycle Fatigue, ASTM STP 942,* H. D. Solomon, G. R. Halford, L. R. Kaisand, and B. N. Leis, Eds., American Society for Testing and Materials, Philadelphia, 1988, pp. 728–750.

ABSTRACT: The low-cycle fatigue (LCF) behavior of Waspaloy was studied in uniaxial and torsional loading for two heat treatments at 24 and 649°C. It was shown that for both heat treatments deformation and failure mechanisms were independent of stress state at 24°C. Deformation occurred by precipitate shearing and the formation of intense shear bands for the coarse-grain/small-precipitate (CG-SP) condition and by precipitate looping and loosely defined shear bands for the fine-grain/large-precipitate (FG-LP) condition. Failure in both microstructures was associated with the formation of shear cracks.

At 649°C deformation and failure mechanisms for the FG-LP condition were independent of stress state, and the mechanisms were similar to those observed at 24°C. For the CG-SP condition, the situation was different. Failure occurred on principal planes when tested in torsion and on shear planes when tested in uniaxial tension. The mechanism transition is interpreted in terms of deformation mode and microstructural instability.

In general, there was a more pronounced decrease in life with increasing temperature for uniaxial specimens than for torsional specimens. This result is interpreted in terms of an environmental interaction that is accelerated by a comparatively significant dilatational strain component.

KEY WORDS: Waspaloy, superalloys, low cycle fatigue, elevated temperature, torsion, microstructure, environment

Many nickel-base superalloys have been developed for use in critical components of gas turbine engines. These components may be subjected to simple or complex loading systems such as uniaxial, torsional, or combined loading. Mechanical and microstructural behavior under these operating conditions are important factors in design analysis. Experience has shown that these materials have relatively short low cycle fatigue (LCF) lives and that the life depends on state of stress, temperature, and material properties, among other factors.

Waspaloy was chosen for investigating the mechanical/microstructural behavior in the high-temperature low-cycle fatigue regime under uniaxial and torsional strain cycling because of its wide use and because it can be heat treated to produce a variety of microstructures. It is strengthened by a low volume fraction of coherent γ' precipitates having a small mismatch with the γ matrix. Two different microstructures that promote different deformation modes were produced by heat treatment in order to study damage mechanisms that might be microstructurally related.

The results of the investigation are interpreted in terms of a mechanics/materials interaction.

[1]Georgia Institute of Technology, Atlanta, GA 30332.
[2]The Pennsylvania State University, University Park, PA 16802.

Experimental Procedure

Materials, Specimen Fabrication, and Heat Treatment

Specimens were machined tangentially from a forged pancake into solid uniaxial specimens with a 6.35 mm diameter gage section and tubular torsional specimens with a 12.70 mm outer diameter and 1.53 mm wall thickness in the gage section. Heat treatments shown in Table 1 were performed to produce two different microstructures: (1) coarse grain, small γ' (CG-SP), and (2) fine grain, large γ' (FG-LP). Final machining and electropolishing was then done with details provided elsewhere [*1,2*]. The Waspaloy was of standard composition with the following analysis (in weight percent): Ni-56.57, Cr-19.32, Co-14.41, Mo-3.89, Ti-3.83, Al-1.35, Fe-0.75, Zr-0.05, C-0.04, Si-0.04, V-0.04, B-0.03, Mn-0.03, P-0.02, S-0.02, Cu-0.01.

Test Matrix and Procedures

The uniaxial test matrix consisted of fully reversed, continuously cycled LCF tests to failure with a constant plastic strain range of 0.30% at temperatures of 24 and 649°C. A similar torsional test matrix was followed with an equivalent von Mises constant plastic shear strain range of 0.52%. Tests were run on conventional uniaxial and tension-torsion servohydraulic machines with zero axial load maintained for the torsional tests. High-temperature tests were performed with careful temperature control using RF induction heating. Details of the testing are provided elsewhere [*1,2*].

In this study, initiation was defined from the load-time records to be at the point of intersection between a line tangent to the steady-state tensile load peaks and a line tangent to the tensile load peaks produced by the final load drop at failure. Life to failure was defined as the approximate cycle showing a 10% tensile load decrease from the half-life tensile amplitude.

Metallurgical Evaluation Techniques

Standard techniques were followed for the optical microscopy, scanning electron microscopy (SEM), and transmission electron microscopy (TEM). Cellulose acetate replicas were taken after testing, and stereopairs were extensively used to determine the fracture surface characteristics. Other details are provided elsewhere [*1,2*].

Results

Initial Microstructures

The CG-SP condition had a grain size approximately ASTM #3 (125 μm) and ASTM #9 (15 μm) for the FG-LP condition. The grain structure was found to be essentially isotropic. The FG-LP condition contained a duplex spherical γ' morphology with most precipitates approximately 800 Å and a small volume fraction approximately 2600 Å in diameter (Fig. 1). The CG-SP condition had spherical γ' precipitates approximately 50 to 80 Å in diameter. Discrete carbides were present on some of the grain boundaries in the FG-LP condition (Fig. 2*a*). The CG-SP condition contained smaller carbides along some grain boundaries (Fig. 2*b*).

Uniaxial LCF Tests of CG-SP Microstructure

LCF Test Results—Test results at 24 and 649°C are shown in Table 2. High-temperature tests showed a decrease in life by about a factor of three compared with room temperature tests. There was a short duration of initial hardening followed by gradual softening over the life. This

TABLE 1—*Waspaloy heat treatments.*

Heat Treatment	Grain Size	γ' Size
Fine-Grain/Large-Precipitate (FG-LP)		
Heat Treatment A [*3,13*]		
1010°C (1850°F)/2 h, oil quench 875°C (1607°F)/24 h, oil quench	fine (ASTM #9, 15 μm)	large duplex (400–1200, 2200–3000 Å)
Large-Grain/Small-Precipitate (CG-SP)		
Heat Treatment F [*3,13*]		
1100°C (2012°F)/2 h, oil quench 730°C (1346°F)/6 h, oil quench	coarse (ASTM #3, 125 μm)	small (50–80 Å)

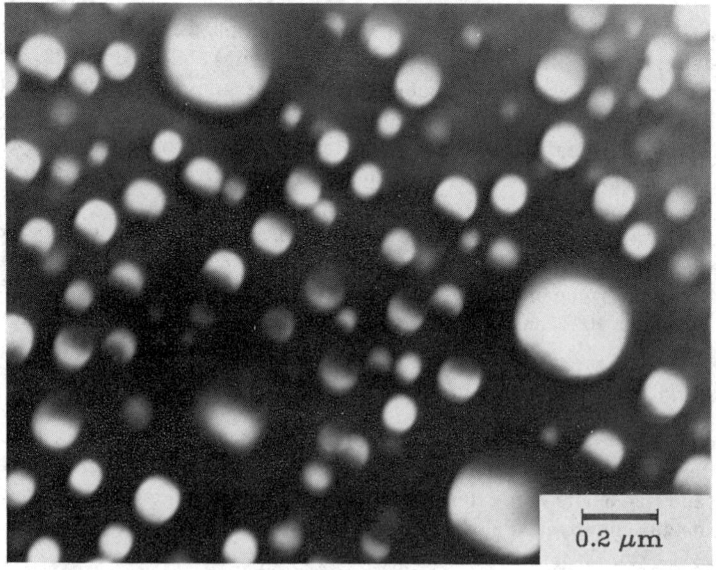

FIG. 1—*TEM micrograph of γ' morphology in FG-LP condition. The CG-SP condition had a similar morphology, except the γ' was smaller, more closely spaced, and more uniform in size.*

behavior was observed at both temperature levels, strongly indicative that precipitate shearing was taking place during cycling. This behavior has previously been documented elsewhere [*3*].

Crack Formation—Replicas of the gage surface showed slip inhomogeneously distributed with numerous planar slip bands intersecting the surface at both temperatures. Crack initiation occurred upon the slip planes at both temperatures (Fig. 3).

The SEM studies of the fracture surfaces showed that, in general, the mode of cracking was transgranular at both temperatures, and there were large facets following crystallographic planes. The similarity in fracture mode between the two temperatures, along with the reduction in life at high temperature, would seem to imply very strong thermally activated environmental interactions.

Deformation Substructures—Well-defined planar slip bands were present at both test temperatures (Fig. 4). Paired dislocations can be seen in the slip bands and are indicative of precipitate shearing.

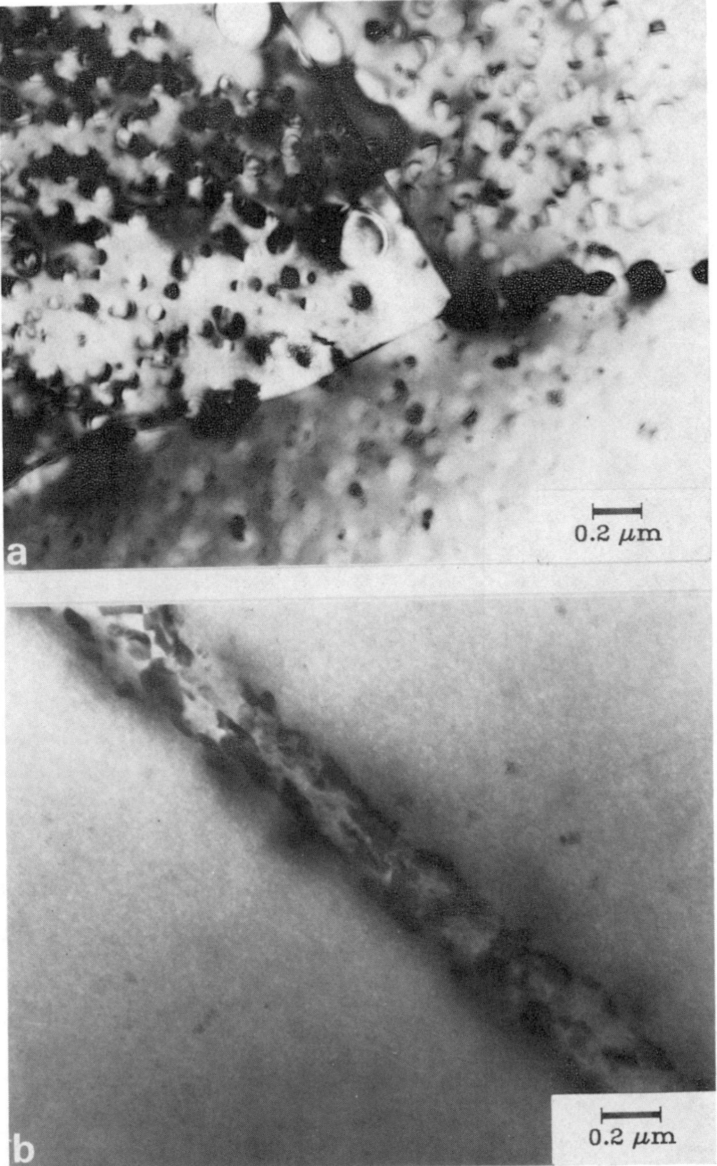

FIG. 2—*TEM micrographs of grain boundaries. Both conditions exhibited grain boundaries with and without carbides.* (a) *FG-LP. Note large discrete grain boundary carbides.* (b) *CG-SP. Note smaller, more continuous grain boundary carbides.*

Uniaxial LCF Tests for FG-LP Microstructure

LCF Test Results—Test results at 24 and 649°C are shown in Table 2. Again, the results showed a marked decrease in life at 649°C by approximately a factor of six compared with tests at 24°C. The stress range tended to be stable through the duration of the test.

Crack Formation—Examination of the gage surface using replica techniques revealed that

TABLE 2—*Waspaloy uniaxial LCF data (all values at half-life).*

Micro-structure	Specimen No.	Temperature, °C	$\Delta\epsilon_t$, %	$\Delta\epsilon_e$, %	$\Delta\epsilon_p$, %	N_i	N_f	σ_t, MPa	σ_c, MPa	$\dot{\epsilon}$, %/min	t_f, min
CG-SP	UF 7	24	0.80	0.50	0.30	12 175	14 000	535.9	554.7	38	585
CG-SP	UF 8	24	0.80	0.50	0.30	10 900	12 075	545.3	560.8	48	355
CG-SP	UF 13	649	0.70	0.40	0.30	3 825	5 125	354.5	357.6	47	155
CG-SP	UF 14	649	0.71	0.41	0.30	3 475	4 225	366.7	372.9	41	145
FG-LP	UA 4	24	1.08	0.78	0.30	9 125	9 600	767.7	783.6	54	385
FG-LP	UA 5	24	1.14	0.84	0.30	7 175	8 625	834.2	865.3	53	375
FG-LP	UA 1	649	1.06	0.76	0.30	1 450	1 750	666.5	678.7	50	70
FG-LP	UA 2	649	0.97	0.67	0.30	1 700	1 800	617.6	624.0	49	80

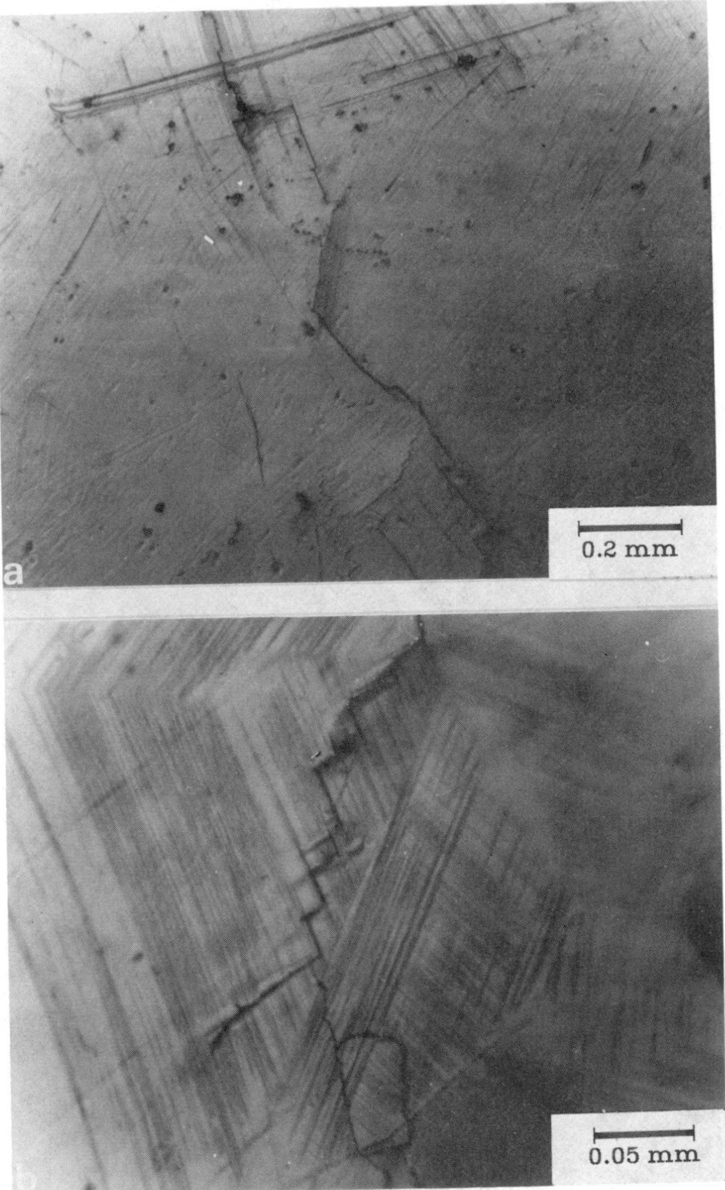

FIG. 3—*Replicas of gage section in uniaxial CG-SP condition. Note inhomogeneous deformation with cracking along slip bands. Specimen axis was in the horizontal direction.* (a) *24°C (Specimen UF8).* (b) *649°C (Specimen UF13).*

the slip was more homogeneous and that the slip bands were less well-defined than in the CG-SP microstructure. These features are shown in Fig. 5.

Crack initiation occurred both at slip lines and at grain boundaries, with grain boundary initiation more predominant at 649°C. Crack extension was transgranular at both temperatures. The fracture surface was less crystallographic than in the CG-SP uniaxial condition. Al-

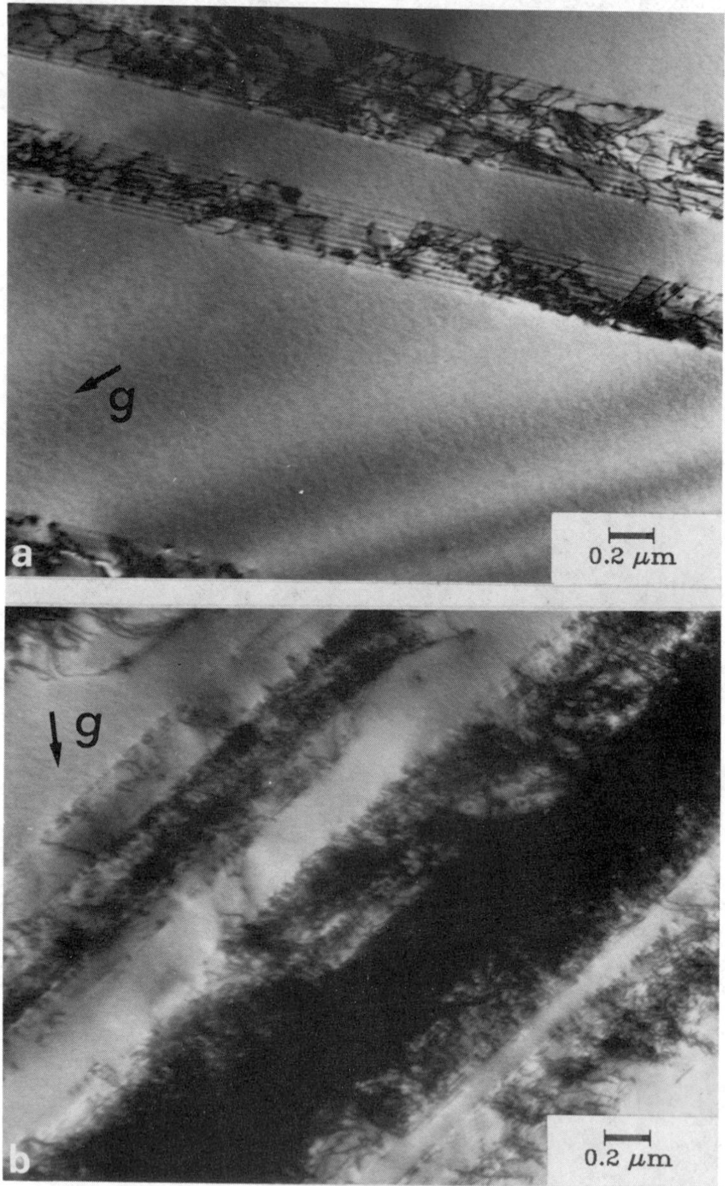

FIG. 4—*TEM micrographs of uniaxial CG-SP condition. Note well-defined planar slip-band formation. The dislocation pairs are indicative of shearing.* (a) *24°C (Specimen UF8),* g = *⟨200⟩.* (b) *649°C (Specimen UF13),* g = *⟨200⟩.*

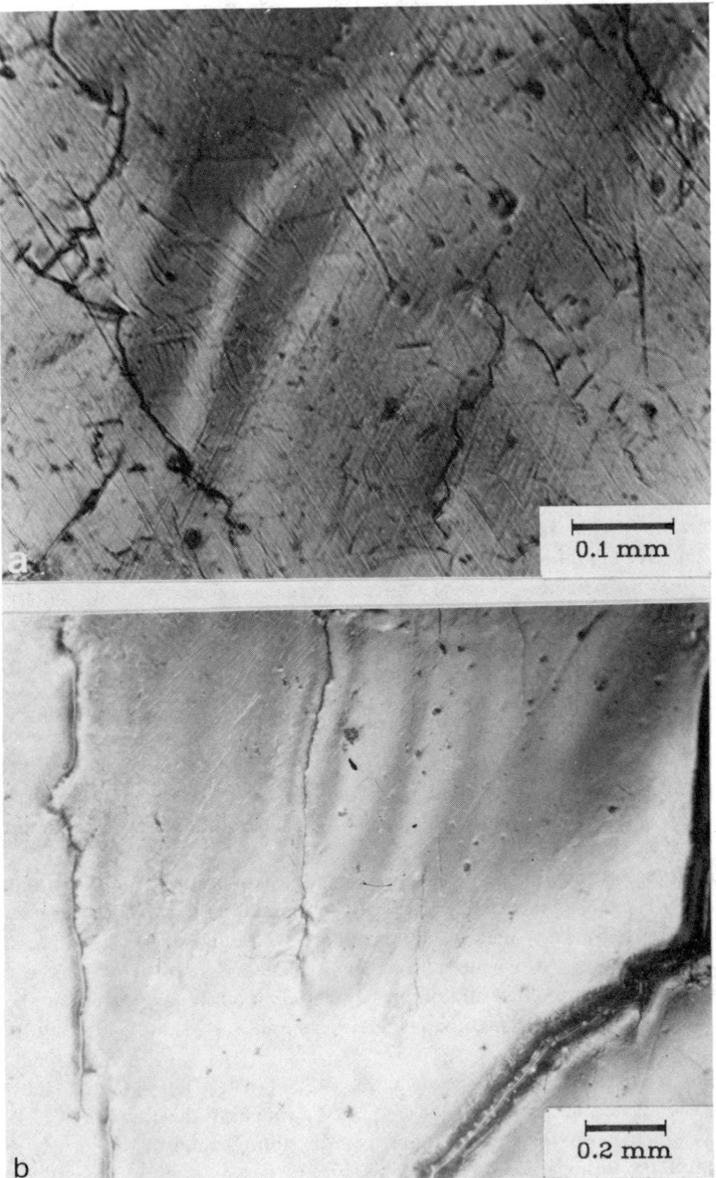

FIG. 5—*Replicas of gage surface in FG-LP uniaxial condition. Slip is more homogeneous than in CG-FP condition, though there was more linking of slip-band cracks at 24°C. Specimen axis was in the horizontal direction.* (a) *24°C (Specimen UA4).* (b) *649°C (Specimen UA1).*

though grain boundary initiation was somewhat more pronounced at 649°C, the similarity in fracture mode along with the reduction in life at high temperature would again indicate strong thermally activated environmental interactions.

Deformation Substructures—Loosely defined slip bands were present at both temperatures. Bowing and looping of the dislocations around the γ' particles were also observed. Representative TEM micrographs are shown in Fig. 6.

Torsional LCF Tests of CG-SP Microstructure

LCF Test Results—Test results at 24 and 649°C are listed in Table 3. The results showed only slightly longer lives at 24°C. As in the uniaxial tests, there was a short duration of initial hardening followed by gradual softening over the life at both temperatures indicative of precipitate shearing.

Crack Formation—At 24°C, crack initiation occurred on maximum shear planes (at 0 and 90° to the specimen axis) as shown in Figs. 7*a* and 8*a*. This was similar to the uniaxial case. The fracture surface was generally transgranular.

At 649°C, replicas of the gage surface showed cracks initiating mainly on planes of maximum shear (0 and 90° to specimen axis) and principal planes (±45° to specimen axis) as shown in Fig. 7*b*. The crack paths on a macroscopic scale followed principal planes as shown in Fig. 8*b*. This behavior in the CG-SP torsion tests at 649°C deviates from that shown by all the other testing conditions and will be discussed later. There appears to be a somewhat greater proportion of intergranular fracture than was observed for the case of torsional loading at 24°C and uniaxial loading at both 24 and 649°C for the CG-SP microstructure.

Deformation Substructures—Planar slip bands were present at both test temperatures (Fig. 9). The slip bands appeared to be somewhat less defined than in the uniaxial CG-SP specimens. Dislocation pairs were again present, which is indicative of precipitate shearing.

Torsional LCF Tests of FG-LP Microstructure

LCF Test Results—Test results at 24 and 649°C are listed in Table 3. The results showed a decrease in lives at 649°C by approximately a factor of two compared with room temperature tests. The stress range was relatively stable throughout the test indicative of a looping deformation mode.

Crack Formation—Crack initiation at both temperatures occurred on the maximum shear planes (Fig. 10). The 649°C test specimen displayed more numerous, shorter cracks compared with the longer, more widely spaced cracks shown at 24°C. Following initiation and growth on the maximum shear planes, it is believed that due to local stress redistributions resulting from the cracking, the cracks branched off and propagated on the principal planes as shown in Fig. 11 for both temperatures. The fracture surface was similar at 24 and 649°C and was mainly transgranular.

Deformation Substructures—As shown in Fig. 12, both test temperatures showed similar substructures comprised of dislocation looping and bowing as in the uniaxial FG-LP condition. The dislocation debris was more homogeneously distributed than in the loosely defined slip bands present in the uniaxial FG-LP condition (Fig. 6).

Discussion

General Discussion

LCF Life Comparisons—LCF studies have traditionally found torsional lives to exceed uniaxial lives at room temperature by approximately a factor of two using a von Mises effective strain

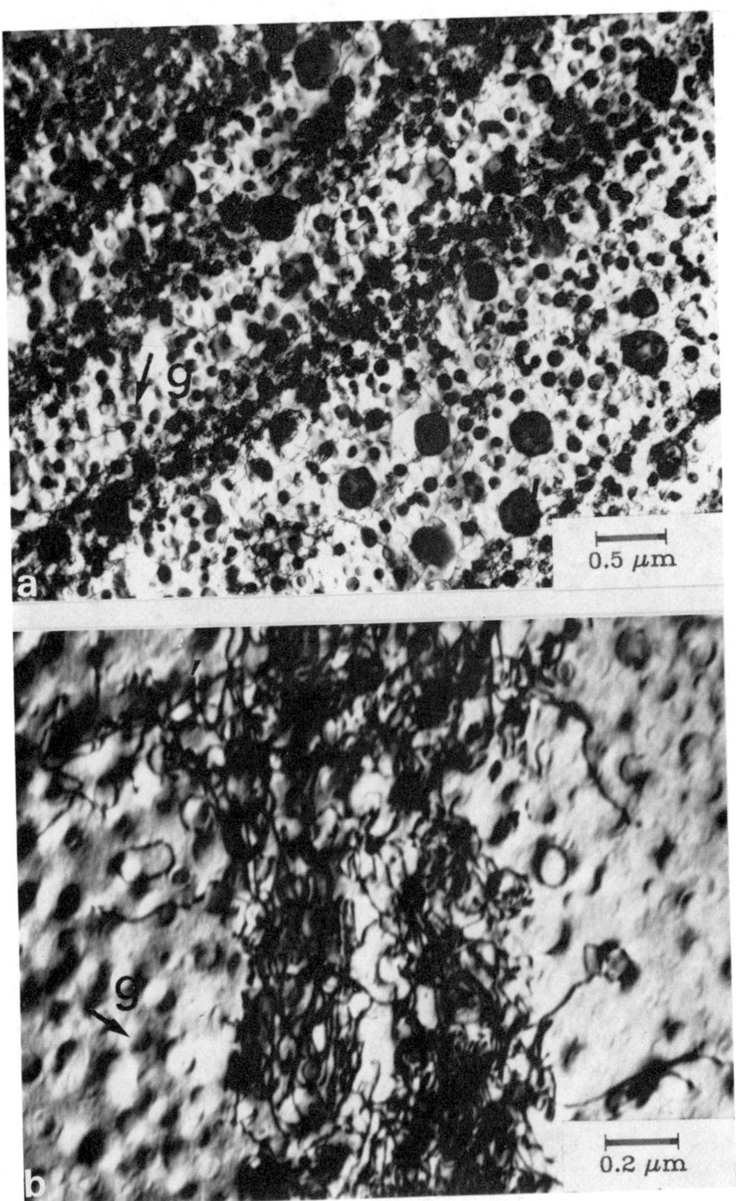

FIG. 6—*TEM micrographs of uniaxial FG-LP condition. Loosely defined slip bands were present with much more dislocation looping and bowing than in the CG-SP condition.* (a) *24°C (Specimen UA4),* g = ⟨*200*⟩. (b) *649°C (Specimen UA2),* g = ⟨*200*⟩.

TABLE 3—*Waspaloy torsional LCF data (all values at half-life).*

Microstructure	Specimen No.	Temperature, °C	$\Delta\gamma_t$, %	$\Delta\gamma_e$, %	$\Delta\gamma_p$, %	N_i	N_f	$\Delta\tau$, MPa	$\dot{\gamma}$, %/min	t_f, min
CG-SP	BF 10	24	2.03	1.14	0.89	...	4983	788.1	16	1246
CG-SP	BF 12	24	1.53	1.01	0.52	4200	5980	779.1	31	598
CG-SP	BF 27	649	1.58	0.93	0.65	3876	4528	556.4	32	453
CG-SP	BF 4	649	1.34	0.82	0.52	...	5200	468.8	13	620
CG-SP	BF 6	649	1.93	1.01	0.92	2100	2500	603.3	19	310
FG-LP	BA 15	24	1.60	1.30	0.30	8260	8331	946.0	32	833
FG-LP	BA 16	24	1.96	1.47	0.49	3875	3965	964.6	39	396
FG-LP	BA 17	24	1.66	1.10	0.56	4200	4529	1038.4	33	452
FG-LP	BA 29	649	1.94	1.38	0.56	1940	2104	774.3	39	210

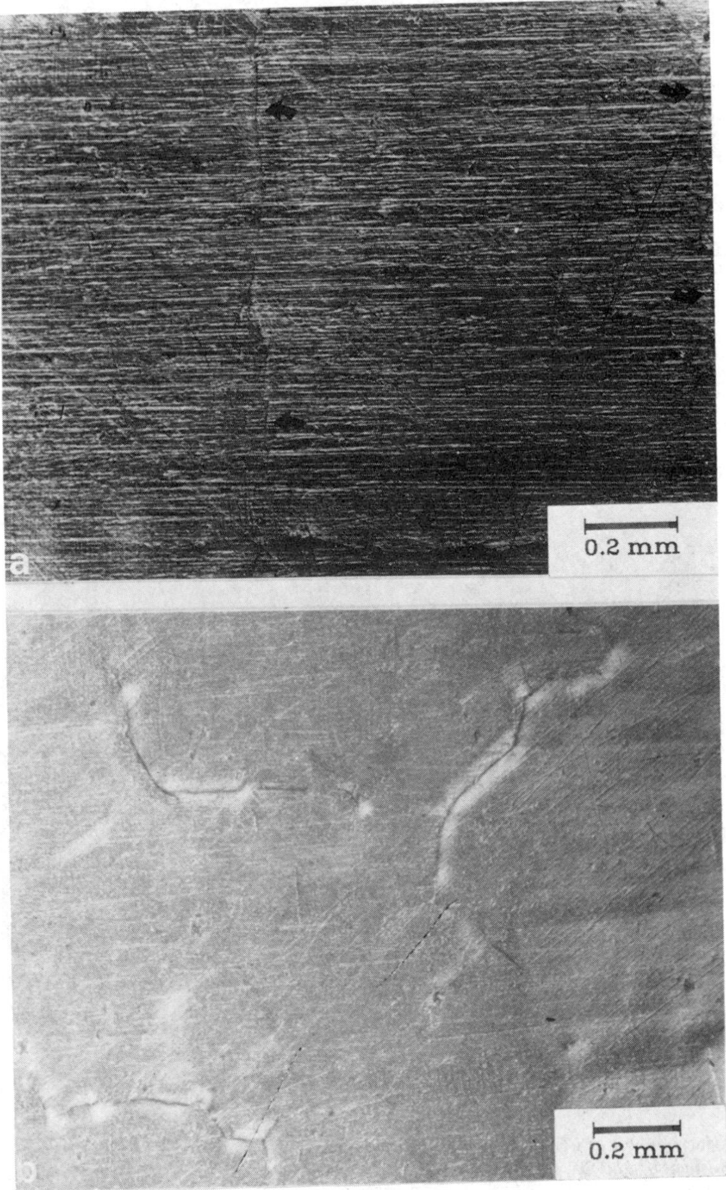

FIG. 7—*Replicas of gage surface of CG-SP torsional specimens. The specimen axis is in the horizontal direction. Some of the cracking is shown by arrows.* (a) *24°C (Specimen BF10). Cracking is only on maximum shear planes 0° and 90° to specimen axis.* (b) *649°C (Specimen BF27). Cracking occurs on a combination of maximum shear and principal planes, usually giving an "average" crack direction on principal planes.*

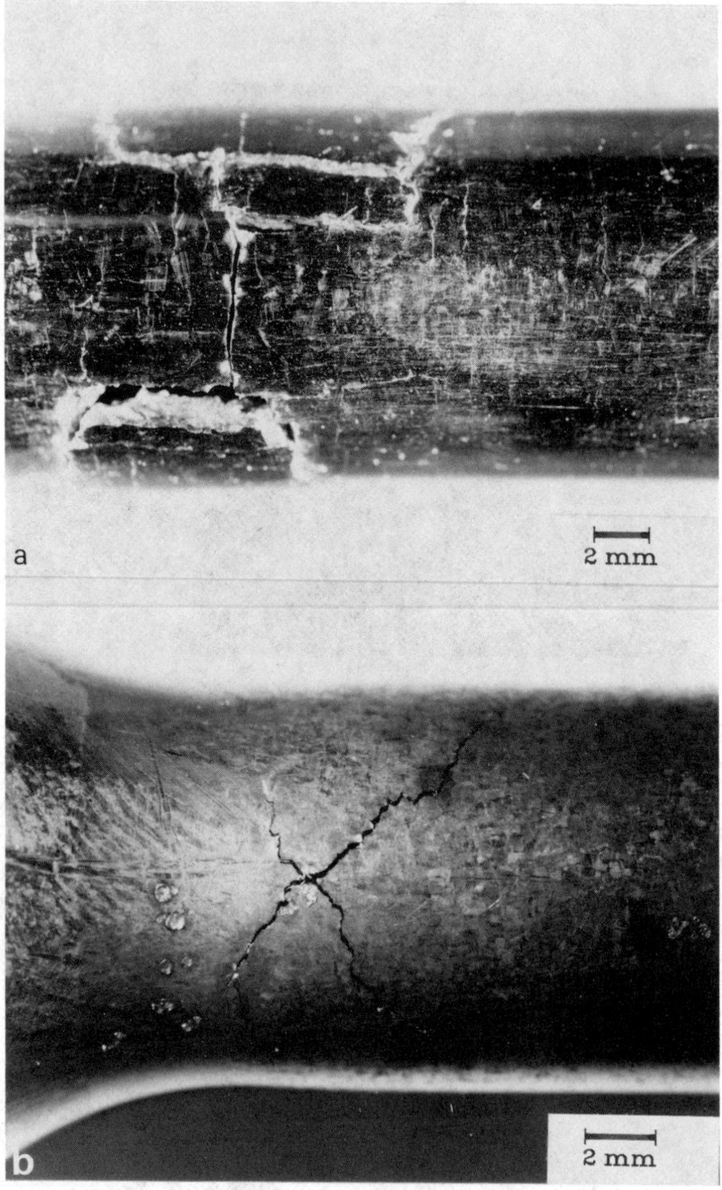

FIG. 8—*Macrophotos of CG-SP torsional specimens.* (a) *24°C (Specimen BF10). Cracking is on planes of maximum shear 0° and 90° to specimen axis.* (b) *649°C (Specimen BF27). Primary cracking at ±45° to specimen axis on principal planes.*

comparison criteria [*4–6*]. Results of this study show torsional lives at 24°C approximately one half those of the uniaxial lives at room temperature. This can be explained, at least in part, by a specimen size effect in the torsional tests. Miller and Chandler [*7*] found a significant decrease in the torsional fatigue life with a decrease in relative wall thickness. Wall thickness in the torsional specimens was 1.53 mm with a ratio of inside diameter to outside diameter of 0.76. At

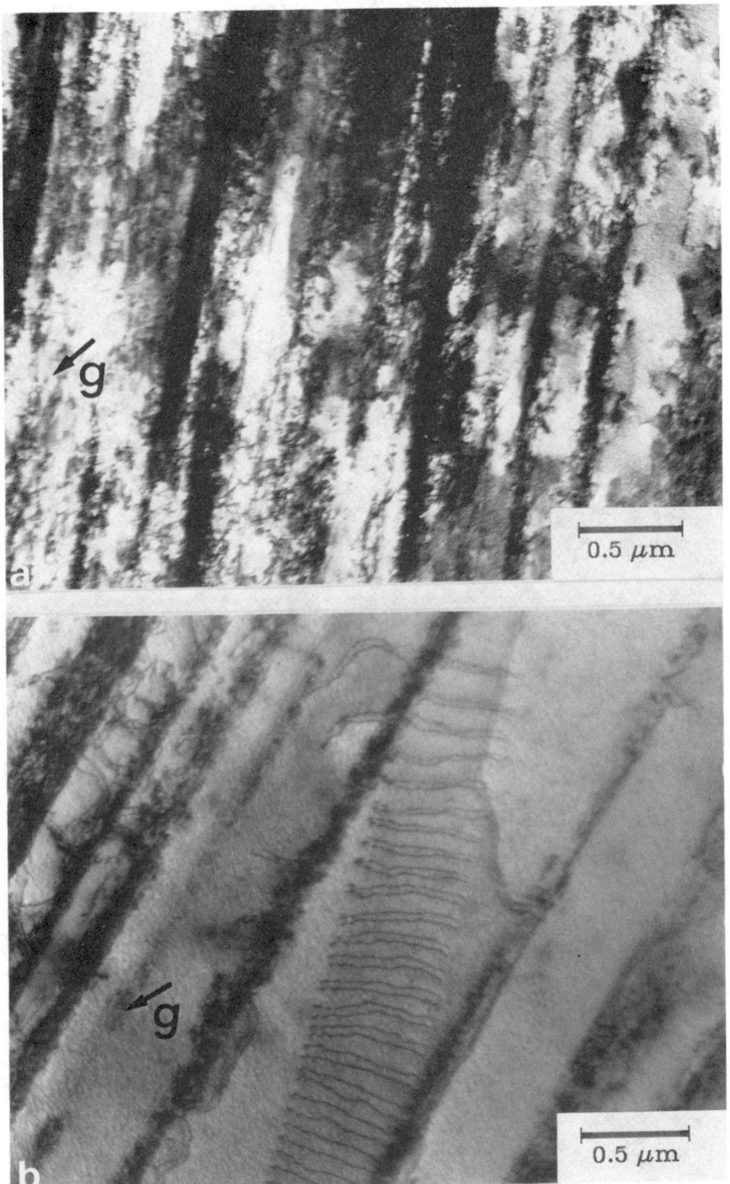

FIG. 9—*TEM micrographs of torsional CG-SP condition showing planar slip bands. The slip bands appear less well-defined than in the uniaxial CG-SP condition. Note dislocation pairs.* (a) *24°C (Specimen BF12),* g = ⟨*200*⟩. (b) *649°C (Specimen BF27),* g = ⟨*200*⟩.

649°C, fatigue lives were approximately equal for both microstructures. The relative effect of temperature on uniaxial and torsional tests will be discussed later.

Crack Formation Comparisons—Crack initiation occurred as a result of shear on slip bands or at grain boundaries in all test conditions except CG-SP torsional at 649°C, where there was cracking on both maximum shear planes and principal planes. This case will be discussed later.

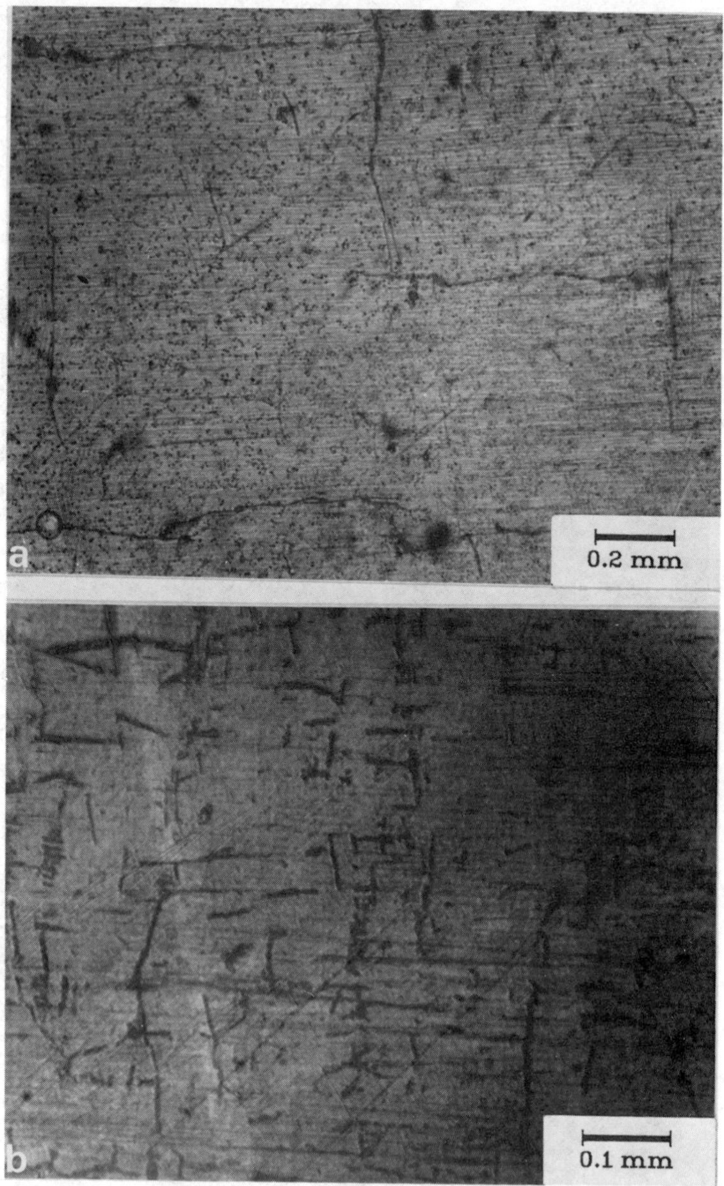

FIG. 10—*Replicas of torsional FG-LP specimens. All secondary cracking was on maximum shear planes. Specimen axis was in the horizontal direction.* (a) *24°C (Specimen BA16).* (b) *649°C (Specimen BA29).*

In the CG-SP uniaxial 24 and 649°C tests and the CG-SP torsional tests at 24°C, cracking always occurred on slip bands. This would suggest that when a grain was oriented to have {111} planes parallel to the maximum shear planes, crack formation was favored. Since Waspaloy is basically isotropic, there will always be some grains favorably oriented for early crack formation. It is also clear that cracks can form on planes oriented as much as 15° off the maximum shear planes. On such planes the shear stress is reduced by about 13% from the maximum if

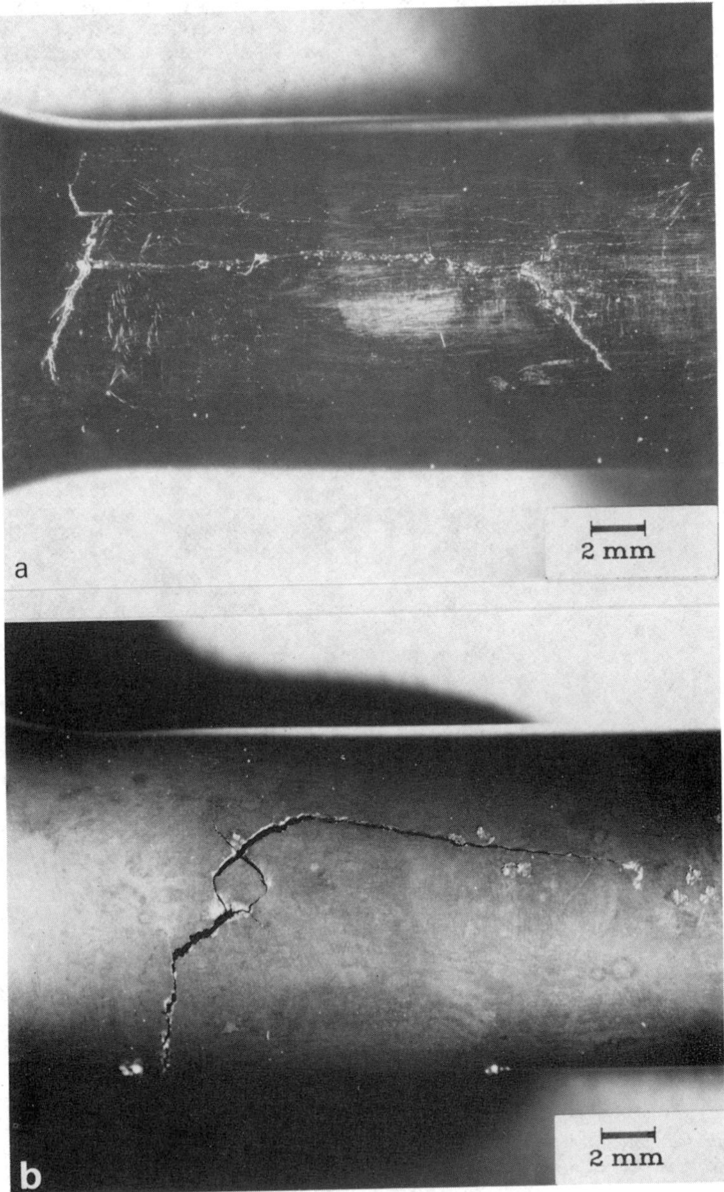

FIG. 11—*Macrophotos showing primary cracks in FG-LP torsional specimens. Cracks are believed to form on maximum shear planes and then link or grow on principal planes.* (a) *24°C (Specimen BA16).* (b) *649°C (Specimen BA29).*

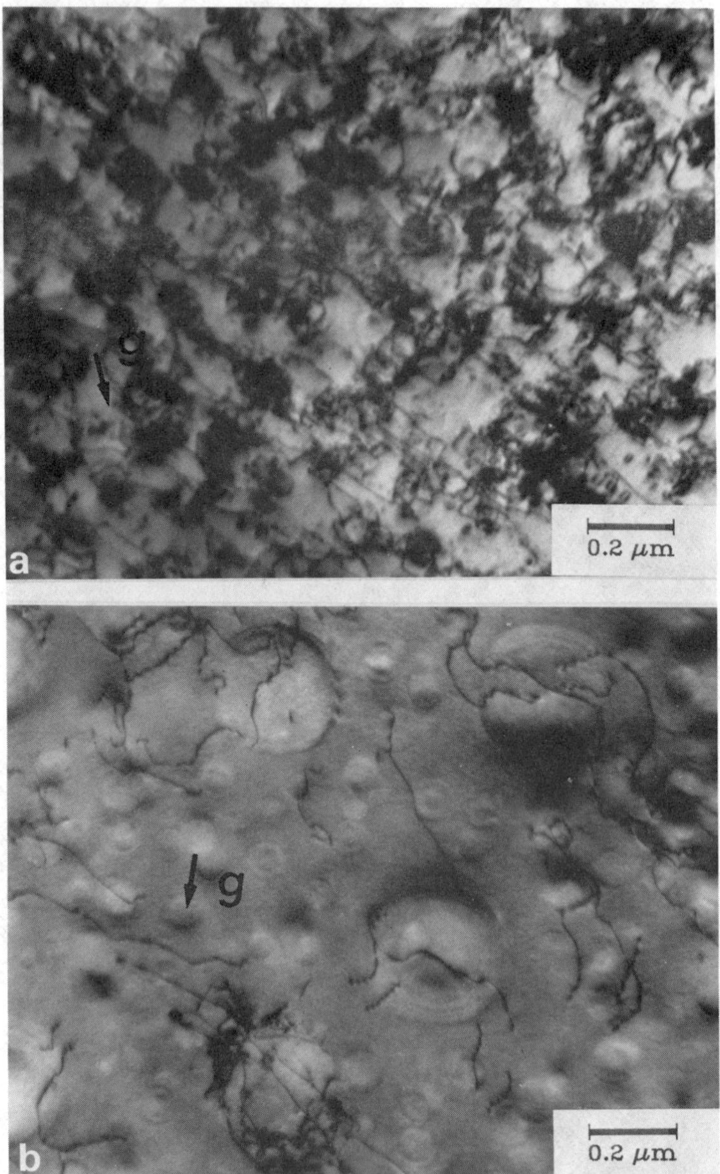

FIG. 12—*TEM micrographs of torsional FG-LP condition. The dislocation substructures are homogeneously distributed, and there is much bowing and looping.* (a) *24°C (Specimen BA17),* g = ⟨*200*⟩. (b) *649°C (Specimen BA29),* g = ⟨*200*⟩.

grain boundary effects are ignored. This is still apparently sufficient to cause dislocation motion and damage accumulation (though at a much lower rate than on the maximum shear planes). Another interpretation is that grain boundary effects cause high local stresses which arise from accommodation requirements and differences in elastic moduli. Such stresses have been shown to be important in uniaxial tensile deformation [*8–10*] and have also been shown to play a major role in crack formation during fatigue [*11,12*]. They should also be significant in torsional loading. In the FG-LP uniaxial and torsional conditions, crack initiation also occurred at grain boundaries, especially at 649°C.

Crack propagation was mainly transgranular in all cases except CG-SP torsional at 649°C, where there was evidence of a somewhat greater amount of intergranular fracture.

Deformation Substructure Comparisons—Deformation substructures were generally independent of temperature for all combinations of loading and microstructure. CG-SP uniaxial and torsional conditions showed well-defined planar slip bands with dislocation pairs indicative of precipitate shearing. FG-LP uniaxial and torsional conditions showed dislocation bowing and looping around the γ' precipitates. There was a slight trend for both microstructures to exhibit more homogeneous deformation substructures in torsional tests compared with uniaxial tests. The slip bands in the CG-SP torsional conditions are somewhat less well-defined than in the uniaxial conditions, and very loosely defined slip bands were present in the FG-LP uniaxial condition while the dislocation substructure was completely homogeneous in the FG-LP torsional conditions. In previous studies [*3,13*], the homogeneous substructure in the FG-LP condition was shown to be associated with a longer life to initiation.

Principal Plane Cracking in the CG-SP Torsional Condition at 649°C

As previously stated, only in the CG-SP torsional condition at 649°C was cracking seen on the principal planes (except as a result of stress redistributions in the FG-LP conditions after large cracks had formed).

Thus an interesting difference in failure modes was observed. At room temperature, failure was by shear-type Modes II and III cracking along maximum shear planes; however, at elevated temperature, failure (at least macroscopically) was by tensile-type Mode I cracking along principal planes. The failure mode at elevated temperature is thus different from that observed at room temperature for the CG-SP microstructure and also different from that observed for torsional FG-LP tests at 649°C (to be discussed in the next section).

A similar transition in the mode of cracking has been observed to be a function of strain amplitude in torsion and combined tension-torsion [*14*], showing a transition from principal plane cracking to maximum shear plane cracking with increasing strain amplitude. This behavior was found in 1045 steel and 304 and 316 stainless steels tested at room temperature. The transition in cracking mode observed in the CG-SP Waspaloy at 649°C was not found to be a function of strain amplitude. Principal plane cracking was also observed in the CG-SP condition tested at 649°C at a very high strain amplitude (Specimen BF6, $\Delta\gamma_p = 0.92\%$) in a regime in which the transition to shear cracking would be expected if such a mode were applicable to this material at this temperature. Waspaloy is relatively similar to IN 718, which showed no transition from maximum shear plane cracking over a wide range of strain amplitudes in torsion and combined tension-torsion at room temperature [*14,15*]. These observations would suggest that the transition seen in the CG-SP Waspaloy at 649°C was a result of a different type of mechanism, probably associated with temperature-activated environmental effects.

A tentative model may be advanced that incorporates state-of-stress effects and previously documented [*3*] microstructural damage mechanisms. As the temperature increases, there is a tendency for carbides to precipitate on grain boundaries and in slip bands (Fig. 13) during LCF strain cycling. Thus two competing damage mechanisms can be envisioned:

1. Damage by ordinary slip processes within a grain, perhaps aided by environmental factors.
2. Damage on the grain boundaries due to precipitation of carbides and accelerated by the environment. For this mechanism, slip bands embrittled by carbide precipitation would also be expected to behave like grain boundaries.

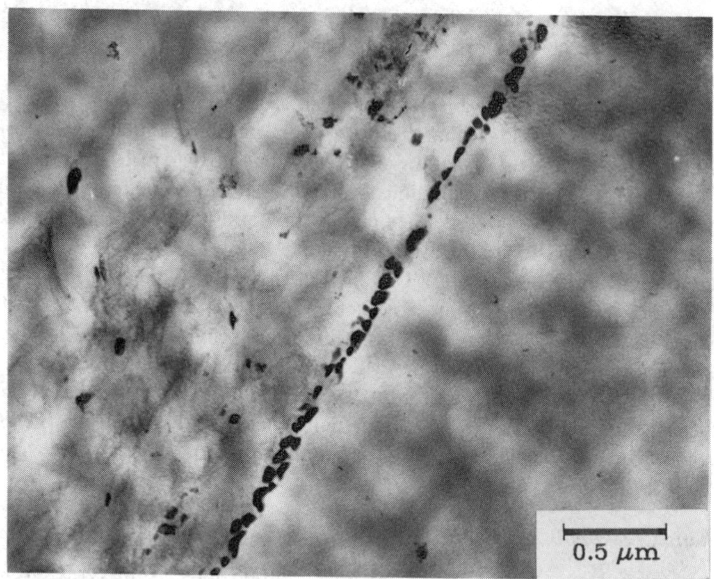

FIG. 13—*TEM micrograph showing carbide formation on grain boundaries and slip bands in CG-SP condition (700°C, Ref 3).*

It is known that brittle zones in a structure are very sensitive to normal stresses [*16*]. At yield, the maximum normal stresses in a uniaxial test are higher than the maximum normal stresses at yield in a torsional test. Thus the uniaxial test would ordinarily be expected to activate normal stress fracture mechanisms more easily for equivalent degrees of shear, as is the case in this study. However, the normal stress fracture mechanisms in torsional loading may actually be favored because of the very rough fracture surface produced by the intense slip in the CG-SP material. It has been shown that fracture surface roughness can indeed influence shear crack propagation [*17,18*]. Numerous small cracks form on the surface of this material when tested in torsion. This result has been seen elsewhere [*15,19*] for IN 718, which also deforms by formation of intense slip bands. These cracks form on the maximum shear planes at 0 and 90° to the specimen axis. They run for a short distance until they intersect another short crack propagating on a maximum shear plane, at which point they are arrested. For this reason, one would expect to see a rather high density of small shear cracks on the surface of materials that deform by intense shear, since the blocking mechanism would necessitate nucleation of more slip bands and shear cracks.

At room temperature, environmental effects should not be too important and the structure should be stable. The expected fracture surface would then consist primarily of shear facets, with some of the 0°-90° intersections giving rise to regions in which the "average" crack surface is inclined at ±45° to the specimen axis. (The crack in such regions would be opened by the principal stresses acting normal to the crack plane.) Such fracture morphologies were in fact observed at 24°C.

At high temperatures, carbides precipitate on slip bands and grain boundaries [*3*] for the CG-SP specimens. Carbides, in addition to being brittle, can also be oxidized [*16*], leading to further embrittlement. As mentioned previously, embrittled regions will be sensitive to normal stresses and strains, and the first regions to fail by this mechanism should be those embrittled boundaries and slip bands that lie normal to the principal stresses. We thus have a situation in which a different fracture mechanism, sensitive to normal stresses, is also operative. This mechanism becomes more predominant with increasing temperatures and is furthermore unaffected by the slip-band blocking mechanism described previously. The net result is that normal stress mechanisms become more important with increased temperature and the fracture surface takes on a Mode I orientation. The situation is illustrated schematically in Fig. 14.

This mechanism would not be expected for the FG-LP material, since the carbides are well distributed and deformation does not occur by the formation of inhomogeneously distributed bands of intense shear. This difference underlines the specificity of damage mechanisms. Depending on the microstructure, temperature, and state of stress, Waspaloy can exhibit different failure mechanisms. These results demonstrate that great care must be exercised when predicting failure under multiaxial loads from uniaxial test results.

Effect of Temperature on Uniaxial and Torsional Fatigue Lives

Another interesting and important observation is that the LCF lives of the uniaxial specimens decreased much more rapidly with increasing temperature than did the torsional specimens although the deformation mechanisms were the same, as seen in the TEM micrographs. This is graphically shown in Fig. 15. One possible explanation of this behavior relates to the different states of dilatation for the two stress states and the effect of dilatation on the susceptibility to degradation by environmental effects. In the case of uniaxial loading, the dilatational component is relatively high and there is a corresponding reduction in the activation energy for diffusion of oxygen along grain boundaries and slip bands. This means that oxygen, which is well known to have an embrittling effect on Ni-base superalloys [*20–24*], enters the material more rapidly and has a higher equilibrium concentration. Thus the effect of oxygen becomes more

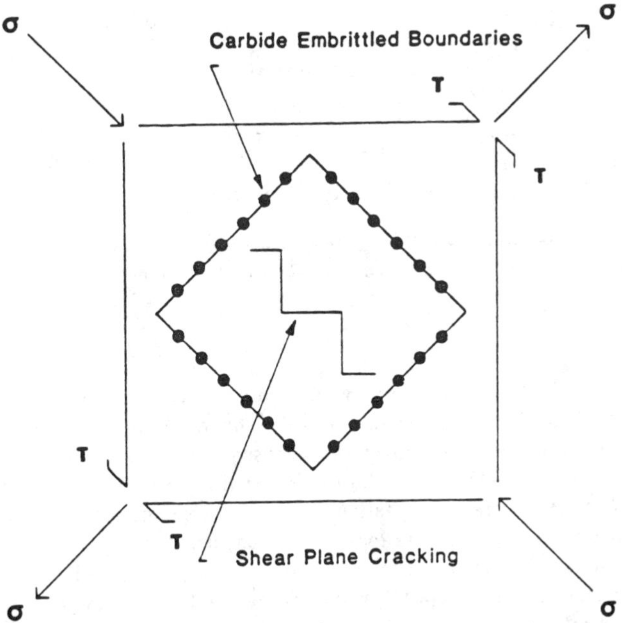

FIG. 14—*Schematic showing embrittled boundaries suitably oriented for cracking due to the normal stress (σ). The major crack has formed by link-up, and propagation is difficult because of its roughness.*

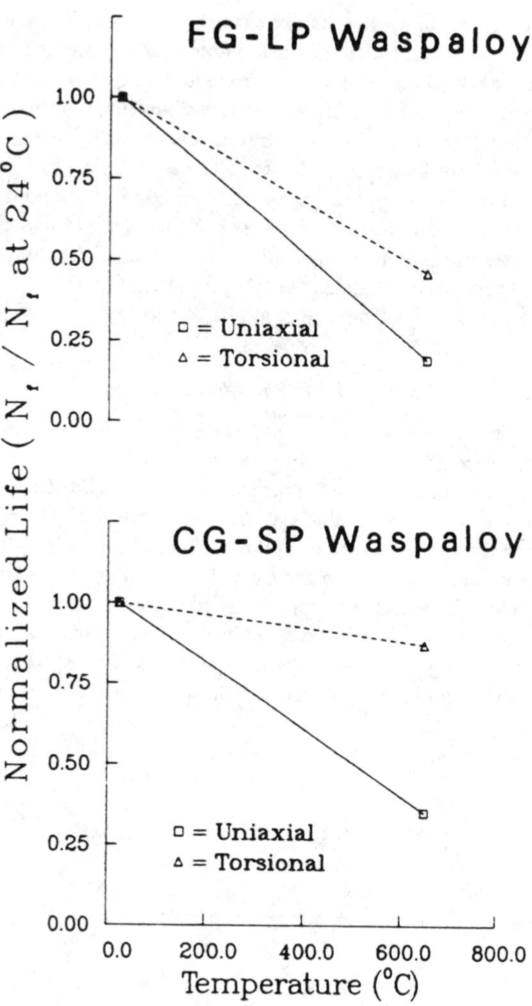

FIG. 15—*Graphic comparison of normalized lives for each microstructure and stress state. Only tests at 0.30%* $\Delta\epsilon_p$ *equivalent strain were averaged and plotted for each condition.*

pronounced with increasing temperature for this stress state. On the other hand, for pure torsional loading the dilatation is ideally zero and the enhanced effects of an increased oxygen penetration rate and increased concentration are absent for this stress state. This means that the embrittling effects are not significant for the torsional stress state and there is a correspondingly reduced effect of temperature on fatigue life. This explanation is consistent with various triaxiality parameters that are used to force agreement between uniaxial and multiaxial test results. This explanation is also in agreement with published results on high strength steels [*25*], where it has been shown that susceptibility to environmental effects is significantly reduced for torsional modes of loading.

There is also a greater effect of temperature on the FG-LP microstructure specimens than on the CG-SP microstructure specimens in both stress states. This is also shown in Fig. 15. This effect is a result of the greater environmental effect on grain boundary crack initiation due to grain boundary oxidation and embrittlement of grain boundary carbides in the FG-LP microstructure than on the slip plane crack initiation in the CG-SP microstructure.

Verification of these hypotheses will depend on detailed Auger spectroscopy to characterize oxygen concentrations as a function of stress state and microstructure for equivalent test conditions as well as tests performed in which the dilatational component is systematically varied.

Summary and Conclusions

1. Deformation occurred by Orowan looping for the fine grain microstructure in both uniaxial and torsional loading at 24 and 649°C. Loosely defined slip bands were evident in the uniaxial specimens, but were absent in the torsional specimens.
2. Deformation occurred by precipitate shearing and formation of intense slip bands for the coarse grain microstructure in both uniaxial and torsional loading at 24 and 649°C.
3. In the fine grain material, cracks initiated primarily transgranularly at 24°C for both stress states. Initiation at 649°C was mainly intergranular for both stress states.
4. Cracks initiated transgranularly for the coarse grain material at both temperatures for both uniaxial and torsional loading. In torsion at 649°C, there was also crack formation on grain boundaries that were oriented parallel to principal planes, indicating a normal stress mechanism rather than a shear stress mechanism.
5. In the coarse grain material in torsion at 649°C, the fatal cracks were macroscopically parallel to principal planes due to the effect of (1) fracture surface roughness, (2) lack of a normal stress component to open shear cracks, (3) intersecting shear cracks limiting shear crack propagation, and (4) principal plane grain boundary cracking.
6. The low cycle fatigue life decreased more rapidly with increasing temperature (independent of microstructure) for the uniaxial stress state. This result is attributed to a thermally-activated environmental interaction, which is favored by the higher dilatational field of the uniaxial stress state compared with that of the torsional stress state.
7. There was a greater decrease in life at elevated temperature (for both stress states) for the fine grain material due to the greater environmental effect on the grain boundaries in the fine grain microstructure than on the slip bands in the coarse grain microstructure.
8. These results demonstrate that low cycle fatigue damage depends on the material, heat treatment, temperature, environmental interactions, and state of stress. They also underline some intrinsic problems in making life predictions for different stress states from a single data set without consideration of these influences.

Acknowledgments

The authors wish to acknowledge the support received from NASA-Lewis Research Center, Cleveland, Ohio, under a joint grant NAG3-264 to the Pennsylvania State University and Georgia Institute of Technology. The reviews and technical discussions with Dr. R. Bill of NASA were greatly appreciated. The authors also wish to thank Dr. Ted Nicholas of Air Force Materials Lab, Dayton, Ohio, for donating materials and the General Electric Company, Evendale, Ohio, for assisting in specimen fabrication.

References

[1] Stahl, D. R., "The Effect of Temperature, Microstructure and Stress State on the Low Cycle Fatigue Behavior of Waspaloy," M.S. thesis, Georgia Institute of Technology, Atlanta, Dec. 1985.
[2] Antolovich, S. D. and Zamrik, S. Y., "Fatigue Damage Under a Complex Loading System," NASA Technical Progress Report, NASA-CR-174787, Oct. 1984.

[3] Lerch, B. A., Jayaraman, N., and Antolovich, S. D., *Materials Science Engineering*, Vol. 66, 1984, pp. 151-166.
[4] Blass, J. J. and Zamrik, S. Y. in *Proceedings*, ASME-MPC Symposium on Creep-Fatigue Interactions, American Society of Mechanical Engineers, 1976, pp. 129-159.
[5] Taira, S., Inoue, T., and Yoshida, T. in *Proceedings*, Twelfth Japan Congress on Materials Research, 1969, pp. 50-55.
[6] Yokobori, T., Yamanouchi, H., and Yamamoto, S., *International Journal of Fracture Mechanics*, Vol. 1, 1965, pp. 3-13.
[7] Miller, K. J. and Chandler, D. C., *Proceedings of the Institution of Mechanical Engineers*, Vol. 184, 1969-70, pp. 433-448.
[8] Hirth, J. P., *Metallurgical Transactions*, Vol. 3A, 1972, pp. 3047-3067.
[9] Hook, R. E. and Hirth, J. P., *Acta Metallurgica*, Vol. 15, 1967, pp. 535-551.
[10] Lee, T. D. and Margolin, H., *Metallurgical Transactions*, Vol. 8A, 1977, pp. 157-167.
[11] Swearingen, J. C. and Taggart, R., *Acta Metallurgica*, Vol. 19, 1971, pp. 543-559.
[12] Wang, Z. and Margolin, H., *Metallurgical Transactions*, Vol. 16A, 1985, pp. 873-880.
[13] Lawless, B. H., "Correlation Between Cyclic Load Response and Fatigue Crack Propagation in the Ni-Base Superalloy Waspaloy," M.S. thesis, University of Cincinnati, Ohio, Aug. 1980.
[14] Bannatine, J. A. and Socie, D. F., "Observations of Cracking Behavior in Tension and Torsion Low Cycle Fatigue," this publication, pp. 899-921.
[15] Socie, D. F., Waill, L. A., and Dittmer, D. F. in *Multiaxial Fatigue, ASTM STP 853*, K. J. Miller and M. W. Brown, Eds., American Society for Testing and Materials, Philadelphia, 1985, pp. 463-481.
[16] McMahon, C. J. and Coffin, L. F., Jr., *Metallurgical Transactions*, Vol. 1A, 1970, pp. 3443-3450.
[17] Tschegg, E. K., Ritchie, R. O., and McClintock, F. A., *International Journal of Fatigue*, Vol. 6, 1983, pp. 29-35.
[18] Nayeb-Hashemi, H., McClintock, F. A., and Ritchie, R. O., *Metallurgical Transactions*, Vol. 13A, 1982, pp. 2197-2204.
[19] Beer, T. A., "Crack Shapes During Biaxial Fatigue," UILU-ENG 84-3606, Report No. 106, University of Illinois at Urbana-Champaign, May 1984.
[20] Antolovich, S. D. and Jayaraman, N. in *Fatigue: Environment and Temperature Effects*, J. J. Burke and V. Weiss, Eds., Plenum Press, New York, 1983, pp. 119-144.
[21] Antolovich, S. D. and Jayaraman, N., *High Temperature Technology*, Vol. 2, 1984, pp. 3-13.
[22] Raquet, M., Antolovich, S. D., and Payne, R. K. in *Superalloys 1984*, Gell, Kortovich, Bricknell, Kent, Radavich, Eds., American Institute of Mechanical Engineers, Warrendale, Pa., 1984, pp. 233-246.
[23] Antolovich, S. D., Domas, P., and Strudel, J. L., *Metallurgical Transactions*, Vol. 10A, 1979, pp. 1859-1868.
[24] Antolovich, S. D., Liu, S., and Baur, R., *Metallurgical Transactions*, Vol. 12A, 1981, pp. 473-481.
[25] St. John, C. and Gerberich, W. W., *Metallurgical Transactions*, Vol. 4A, 1973, pp. 589-594.

T. Kunio,[1] M. Shimizu,[1] N. Ohtani,[1] and T. Abe[2]

Microstructural Aspects of Crack Initiation and Propagation in Extremely Low Cycle Fatigue

REFERENCE: Kunio, T., Shimizu, M., Ohtani, N., and Abe, T., **"Microstructural Aspects of Crack Initiation and Propagation in Extremely Low Cycle Fatigue,"** *Low Cycle Fatigue, ASTM STP 942,* H. D. Solomon, G. R. Halford, L. R. Kaisand, and B. N. Leis, Eds., American Society for Testing and Materials, Philadelphia, 1988, pp. 751-764.

ABSTRACT: A study has been made of the crack initiation and propagation in a low cycle fatigue of annealed carbon steel including extremely short fatigue life ($N_f < 100$), with an emphasis on the establishment of the relation between the microfracture behavior and the damage accumulation process depending on the plastic strain range, $\Delta\epsilon_p$, as well as the mean strain, ϵ_m.

It was found that at the small $\Delta\epsilon_p$ the surface damage due to the initiation and propagation of surface microcracks is predominant, while at the very large $\Delta\epsilon_p$ giving fatigue life less than $N_f =$ 10, the internal cracking originated from the fracture of a pearlite becomes a primary source of the damage which results in the reduction of the residual ductility, ϵ_{FR}. The transition from the surface damage to the internal one takes place when $\Delta\epsilon_p$ becomes so large that the pearlite cracking may start inside the material at each level of ϵ_m

Good correlation was obtained between the reduction of residual ductility and the cracked pearlite ratio newly defined as a parameter for the evaluation of the internal microfracture behavior.

KEY WORDS: low cycle fatigue, microfracture process, damage accumulation, surface crack, internal crack, residual ductility, mean strain, pearlite cracking

Several laws of the damage accumulation have been proposed for the prediction of the low cycle fatigue behavior of metals [*1-8*]. In those hypotheses, it has been supposed that a fatigue fracture takes place when a fatigue damage associated with the repetition of a plastic strain or with the plastic work done during strain cycling accumulates to a certain critical amount. However, the relation between the assumed processes of damage accumulation in those hypotheses and the actual fatigue behavior under strain cycling has not been clarified yet.

Kikukawa et al. has pointed out that two basic mechanisms can be considered for the actual damage accumulation in low cycle fatigue of the steel and that the actual fatigue life is determined by the competition between them [*9*]. One is the mechanism associated with the irreversible cyclic slip, by which microcracks occur at the surface layer of the material; the other is the mechanism associated with the exhaustion of the ductility of a material, by which internal crack occurs.

In relation to the evaluation of the surface damage, many works have recently been published on the estimation of low cycle fatigue life based on the results of the observation of the behavior of small surface cracks [*10-13*]. But only limited knowledge has been obtained about the internal fatigue damage which causes the internal crack.

[1]Department of Mechanical Engineering, Keio University, Yokohama, Japan.
[2]Technical Research Center, Nippon Kokan K. K., Kawasaki, Japan.

The aims of the present study are (1) to make clear the essential feature of the internal damage at very high plastic strain range $\Delta\epsilon_p$ through the metallographic observation of the internal fracture behavior; (2) to evaluate the development of such a damage through the measurement of the residual ductility ϵ_{FR} during fatigue; and (3) to study the role of the microstructure in the transition from the surface damage accumulation to internal one with increase of $\Delta\epsilon_p$.

Experimental Procedure

A plain carbon steel having a carbon content of 0.20% was employed for the present study, the chemical composition of which is given in Table 1. The ferrite-pearlite microstructure with an ferrite-grain diameter of about 90 μm was obtained by the full annealing at 1200°C for 3 h. Figure 1 shows the shape and dimensions of a specimen. A vacuum annealing was given to the specimen at 600°C for 30 min after polishing the specimen surface with an alumina powder having a particle size of 0.16 μm in diameter prior to the fatigue test. The mechanical properties of the material are shown in Table 1.

Strain controlled low cycle fatigue tests were carried out using the servocontrolled test machine by detecting the change of a minimum diameter of a specimen. Both the plastic strain range $\Delta\epsilon_p$ and the mean strain ϵ_m were widely changed from 0.0065 to 0.45 ($\Delta\epsilon_p$) and from 0 to 0.50 (ϵ_m), respectively, where the maximum ϵ_m was as large as 70% of the initial fracture ductility of the material.

Efforts were made for the establishment of the relationship between the change in a residual ductility ϵ_{FR} and the microfracture processes. The microfracture behavior was observed by an optical microscope at a magnification of 400× both on the surface and the longitudinal section of the specimen; also the measurement of ϵ_{FR} was made on the specimens which had been given strain cycling to the various stages of fatigue.

The ϵ_{FR} value was calculated by the equation

$$\epsilon_{FR} = 2\,\ell n(D_m/D_f) \tag{1}$$

where D_m and D_f are the diameters of the specimen before and after the static fracture test for the partially fatigued specimen.

Results and Discussion

Dependence of Fatigue Life on Mean Strain at Very High Plastic Strain Range

Figure 2 shows the results for the relationship between the plastic strain range $\Delta\epsilon_p$ and the number of strain cycles to failure N_f at various levels of mean strain ϵ_m. The results can be well represented by the following equation in which the value of 0.72 on right-hand side is in good agreement with the fracture ductility of the virgin material:

$$\Delta\epsilon_p N_f^{0.55} = 0.72 - \epsilon_m \tag{2}$$

TABLE 1—*Chemical composition and mechanical properties.*

Chemical Composition, wt%					Mechanical Properties			
C	Si	Mn	P	S	Y.S., MPa	T.S., MPa	El., %	R.A., %
0.20	0.24	0.47	0.016	0.017	210	433	30.7	54.8

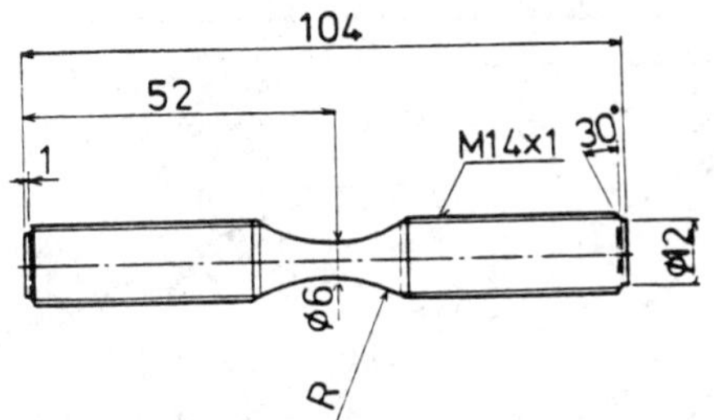

FIG. 1—*Shape and dimensions of specimen.*

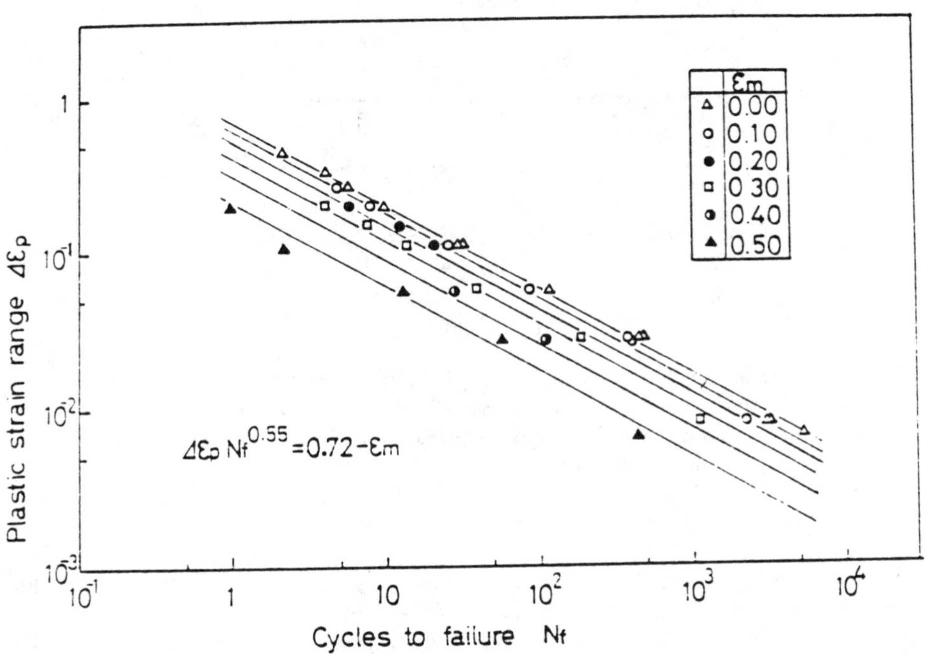

FIG. 2—*Fatigue life curves under various mean strain.*

This equation shows that even in the very short fatigue life range, the mean strain ϵ_m gives the same effect as that of the reduction in effective fracture ductility by amount of ϵ_m to the fatigue life of the steel, as proposed by Sachs and Weiss [2].

Change in the Residual Ductility During Fatigue ($\epsilon_m = 0$) and the Microfracture Processes

A measurement of ϵ_{FR} at the various stages of fatigue showed that the changes of ϵ_{FR} with strain cycling depend strongly on $\Delta\epsilon_p$. Figure 3 shows such a result, where the behavior of ϵ_{FR} can be classified as of three types:

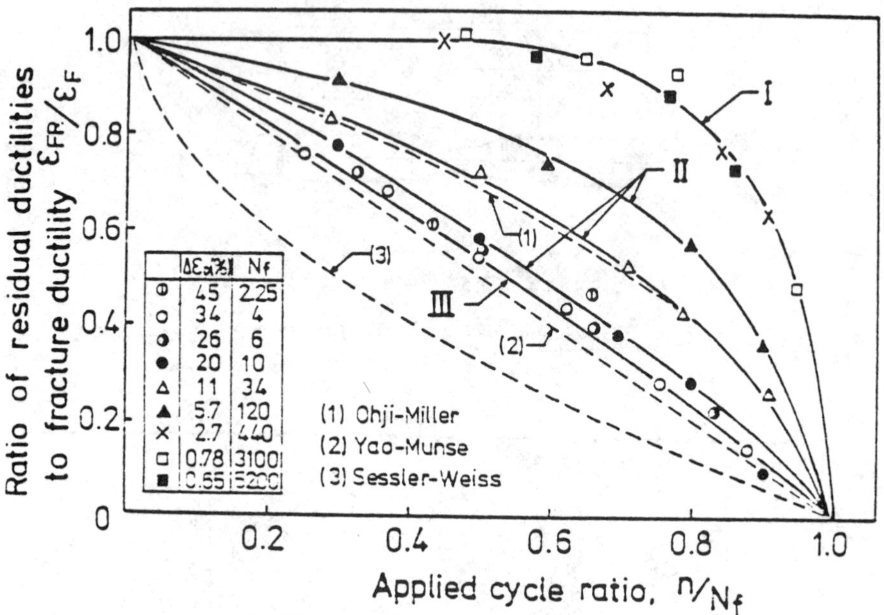

FIG. 3—*Change of residual ductility with strain cycling at various plastic strain ranges.*

1. *Type I*—At the small level of $\Delta\epsilon_p(\Delta\epsilon_p < 0.027)$, no drastic change of ϵ_{FR} occurs, and about 80% of initial fracture ductility is preserved until final stage of fatigue life. No dependence of ϵ_{FR} upon $\Delta\epsilon_p$ appears for this type of behavior, and the results can be characterized by the single curve (solid line) with Mark I.
2. *Type II*—In the intermediate strain range (about $0.057 < \Delta\epsilon_p < 0.20$), a gradual decrease of ϵ_{FR} with strain cycling occurs depending on the magnitude of $\Delta\epsilon_p$. Three solid curves with Mark II show this behavior.
3. *Type III*—At the large level of $\Delta\epsilon_p(\Delta\epsilon_p > 0.26)$ which results in an extremely short life ($N_f < 10$), the decrease in ϵ_{FR} appears at the early stage of fatigue, and ϵ_{FR} continues to decrease linearly with strain cycling before final fracture. The dependence of ϵ_{FR} on $\Delta\epsilon_p$ is not observed in this behavior, and the results can be well represented by a single solid line with Mark III.

The variation of ϵ_{FR} with strain cycling can be predicted using some of the hypotheses which have been developed to describe the cumulative effect of plastic strain cycles on the low cycle fatigue behavior of metals.

First, based on the Yao-Munse's hypothesis [*3*] in which the cyclic tensile change in plastic strain ϵ_t has been assumed to be effective for the damage accumulation and also the damage per each cycle is given by $(\epsilon_t/\epsilon_f)^{1/\alpha}$ where α is an exponent in the Manson-Coffin equation $\Delta\epsilon_p N_f^\alpha = C$, the following equation can be derived for the variation of ϵ_{FR} with strain cycling [*15*]:

$$\epsilon_{FR}/\epsilon_F = 1 - n/N_f \tag{3}$$

Next, in Ohji-Miller's hypothesis [*4*], which describes the damage per each cycle as $(\Delta\epsilon_p)^{1/\alpha}$, it has been postulated that both tensile and compressive changes in plastic strain have the same contribution to the damage accumulation. From this, the following prediction can be made for the behavior of ϵ_{FR} under strain cycling [15]:

$$\epsilon_{FR}/\epsilon_F = (1 - n/N_f)^\alpha \quad (4)$$

In the above two hypotheses the linear accumulation of damage with strain cycles has been assumed, but in Sessler-Weiss's law [6] the non-linear one has been postulated and the damage after n cycles of $\Delta\epsilon_p$ has been given by $\Delta\epsilon_p n^\alpha$. This hypothesis gives the prediction of ϵ_{FR} as in the following equation [15]:

$$\epsilon_{FR}/\epsilon_F = 1 - (n/N_f)^\alpha \quad (5)$$

Results of these predictions of ϵ_{FR} have been shown by the dashed lines in Fig. 3. It should be noted that the Yao and Munse law [3] gives a good estimation for Type III behavior of ϵ_{FR}, while no single law gives the estimation of the change in ϵ_{FR} during fatigue over a wide range of $\Delta\epsilon_p$. This could be because the process whereby damage accumulates in fatigue is extremely complex and the mechanisms of damage accumulation may be different between high and low levels of $\Delta\epsilon_p$ as pointed out by Kikukawa et al. [9]. To make clear the actual process of damage accumulation, the direct observation of microfracture behavior was performed over a wide range of $\Delta\epsilon_p$ values. As a result, it was confirmed that there are three different types of fracture processes depending on the level of $\Delta\epsilon_p$ as described below:

1. *Type A*—Crack initiation and its propagation in the surface microstructure (Fig. 4). At the small level of $\Delta\epsilon_p$, the crack behavior of this type leads to the final fracture. Figure 4*b* shows a typical feature of such a crack observed on the longitudinal section of the specimen which was subjected to strain cycling of $n = 4800 (n/N_f = 0.92)$ at $\Delta\epsilon_p = 0.0065$.
2. *Type B*—Surface crack initiation followed by the depth direction growth, which is assisted by the cracking of a pearlite. Typical feature of such a crack is shown in Fig. 5. Fracture behavior of this type appears at the intermediate level of $\Delta\epsilon_p$ between the low $\Delta\epsilon_p$ for Type A and the high $\Delta\epsilon_p$ for Type C.
3. *Type C*—Development of internal crack resulted from the coalescence of the micro-voids which originated from the shear cracking of a pearlite. Figure 6 shows a sequence of the events in the fracture of this type at $\Delta\epsilon_p = 0.34$.

The condition for the occurrence of each type of fracture behavior was then investigated and compared with that of the three types of ϵ_{FR} variation (I, II, III) with strain cycling mentioned earlier. As a result, it was clarified that the fracture behaviors of Type A, B, and C appear at the respective levels of $\Delta\epsilon_p$ giving the three types of ϵ_{FR} variation with n/N_f shown in Fig. 3.

Further study of the microfracture behavior in relation to the residual ductility ϵ_{FR} revealed that the ϵ_{FR} is not affected so much only by the development of a surface crack of Type A before it grows to a certain large length (Fig. 7). This implies that the small surface crack has little contribution to a loss of ductility of the material and that the residual ductility ϵ_{FR} is not appropriate as a measure for the evaluation of the damage accumulation resulting from the surface crack development. Another approach would be required to evaluate such a surface damage accumulation.

From these results, the following description could be made for the nature of the damage accumulation in the present steel. At small $\Delta\epsilon_p$ as employed in the ordinary low cycle fatigue test, the surface damage due to the initiation and propagation of surface microcracks is predominant, while at the very large $\Delta\epsilon_p$ giving fatigue life which is less than $N_f = 10$, the internal cracking originated from the fracture of a pearlite becomes a primary source of the damage, which results in the loss of the ductility. The transition from the surface damage to the internal

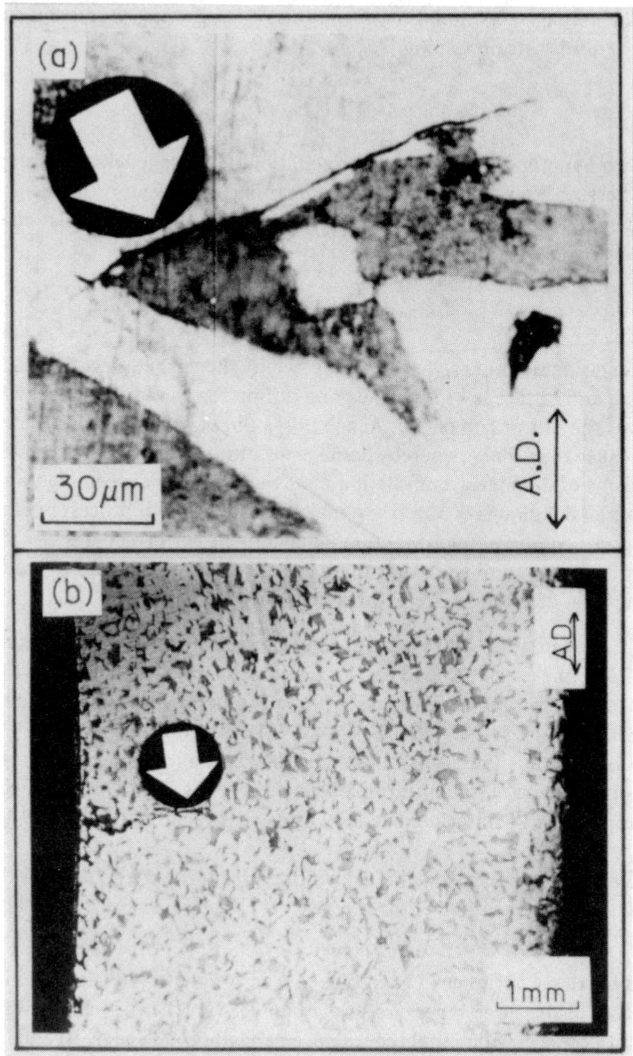

FIG. 4—*Microscopic and macroscopic observations of surface crack for Type A fracture process* ($\Delta\epsilon_p$ = *0.0065,* N_f = *5200*). (a) *Crack initiation at the ferrite-pearlite boundary* (n/N_f = *0.19*). (b) *Crack propagation from surface to the depth direction perpendicular to tensile axis* (n/N_f = *0.92*).

one takes place when the $\Delta\epsilon_p$ becomes so large that the pearlite cracking may start inside the material.

Effect of Mean Strain on the Residual Ductility and the Microfracture Behavior

Since it has been known that the fracture of a pearlite occurs more easily at a higher strain level in a static tensile test [*14*], it could be inferred that the shift of a microfracture mode from Type A to B or Type B to C appears at the smaller $\Delta\epsilon_p$ with larger ϵ_m. This inference is confirmed by observing the fracture mode map shown in Fig. 8, which was obtained through the

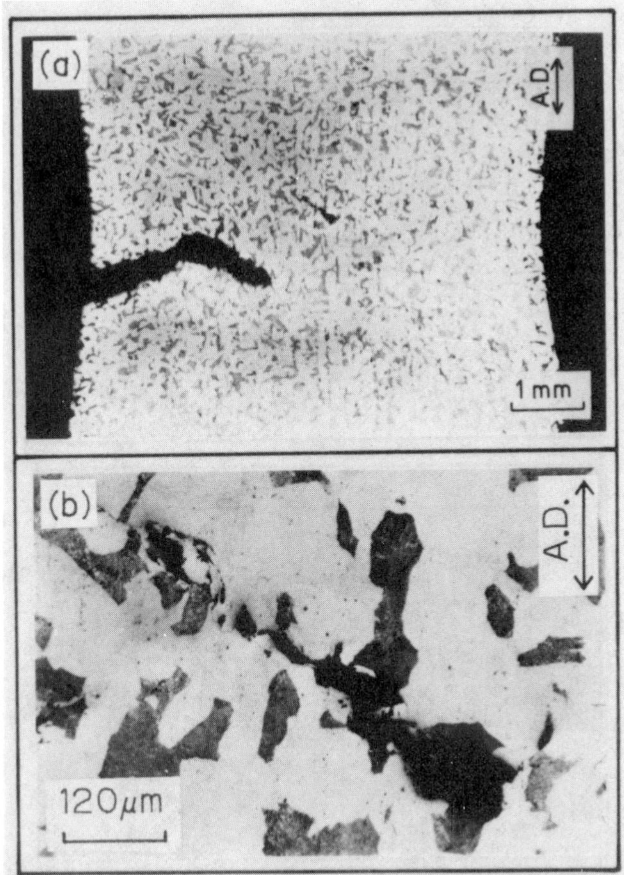

FIG. 5—*Microscopic and macroscopic observations of surface and internal crack for Type B fracture process* ($\Delta\epsilon_p$ = *0.11*, N_f = *34*, n/N_f = *0.94*). (a) *Coexistence of macrosurface crack and internal cracks.* (b) *Coalescence of internal microcracks originated from the pearlite cracking.*

careful observation of the microfracture processes under strain cycling with various levels of mean strain ϵ_m.

Figure 9 shows a typical example of such an observation where the change in fracture mode from Type B to C can be observed at a constant $\Delta\epsilon_p$ as ϵ_m increases.

For making clear the feature of an internal fatigue damage at a very high level of $\Delta\epsilon_p$, the quantitative study of the pearlite cracking was then conducted through the introduction of a parameter "cracked pearlite ratio," which is defined as the ratio of the number of cracked pearlite to the total number of pearlite in the microstructure at the region shown in Fig. 10.

Figure 11 shows the results. It is evident that (1) the cracked pearlite ratio n_{cp}/n_p increases with the applied cycle ratio n/N_f; (2) such an increase of n_{cp}/n_p is remarkable at the early stage of fatigue process; and (3) the relation between n_{cp}/n_p and n/N_f becomes independent of the level of $\Delta\epsilon_p$ when the $\Delta\epsilon_p$ is so high that the fracture behavior of Type C occurs (see the data with solid marks).

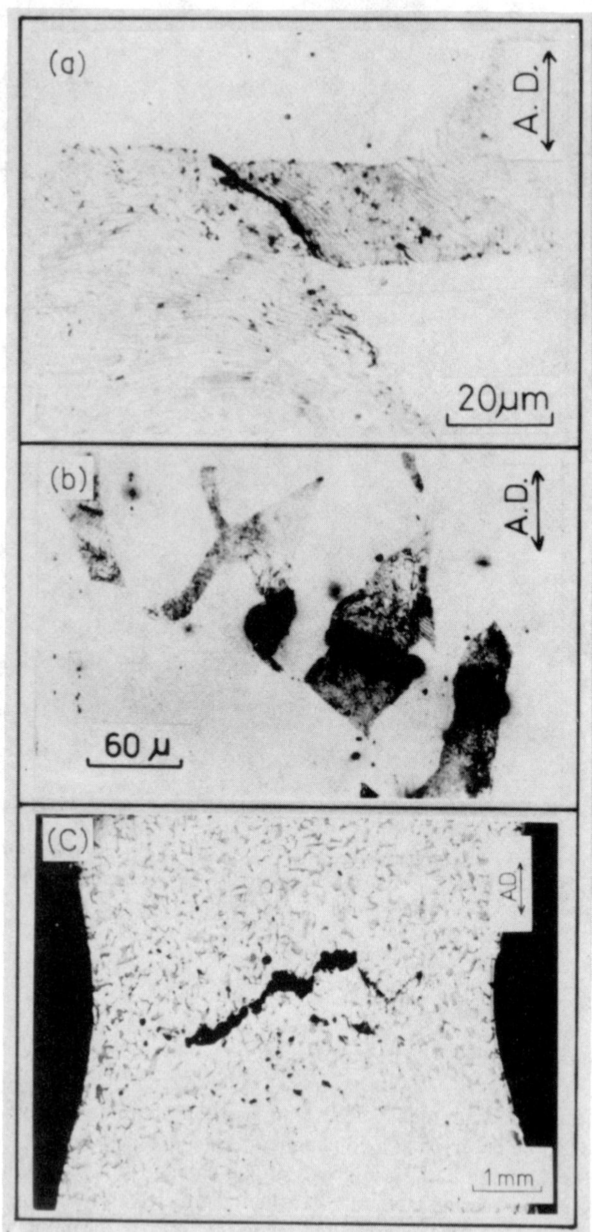

FIG. 6—*Microscopic and macroscopic observations of internal cracks for Type C fracture process* ($\Delta\epsilon_p$ = *0.34*, N_f = *4*). (a) *Shear cracking of pearlite due to compression in the first loading* (n/N_f = *0.19*). (b) *Void formation due to strain reversals* (n/N_f = *0.75*). (c) *Development of internal crack by void coalescence* (n/N_f = *0.81*).

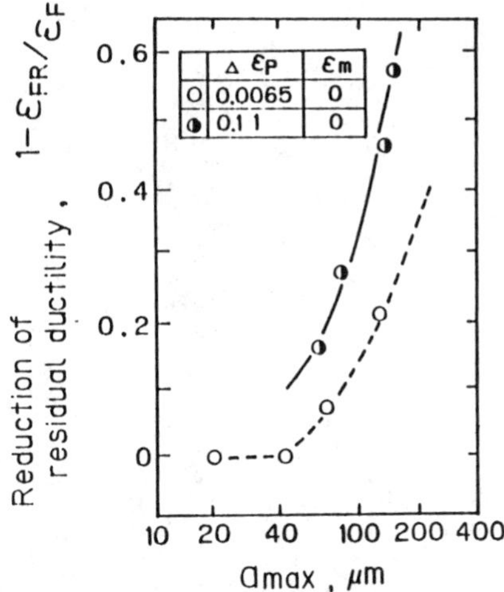

FIG. 7—*Change of residual ductility with increase in surface crack length.*

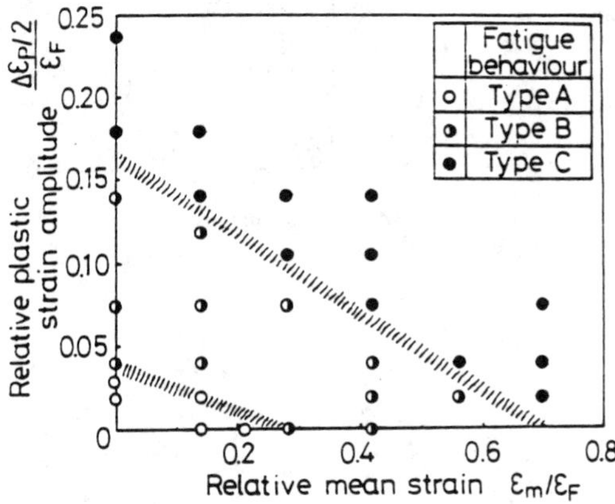

FIG. 8—*Microfracture mode map in low cycle fatigue.*

Such a feature of pearlite cracking at a high level of $\Delta\epsilon_p$ can be seen more clearly in Fig. 12, where the results of the n_{cp}/n_p for various $\Delta\epsilon_p$ have been well represented by a single curve for each ϵ_m, and at the final stage of fatigue life ($n/N_f = 0.95$), the cracked pearlite ratio n_{cp}/n_p reaches a certain critical value ($=0.40$) irrespective of the ϵ_m. This implies that there may be a close relationship between ϵ_{FR} and n_{cp}/n_p at such a high level of $\Delta\epsilon_p$.

To investigate this point, the measurement of ϵ_{FR} under strain cycling with various mean strains was then performed. The results are shown in Fig. 13. As ϵ_m becomes higher, the ductility decreases more remarkably at the early stage of fatigue, but the slower change in ϵ_{FR} comes out at the later stage of fatigue.

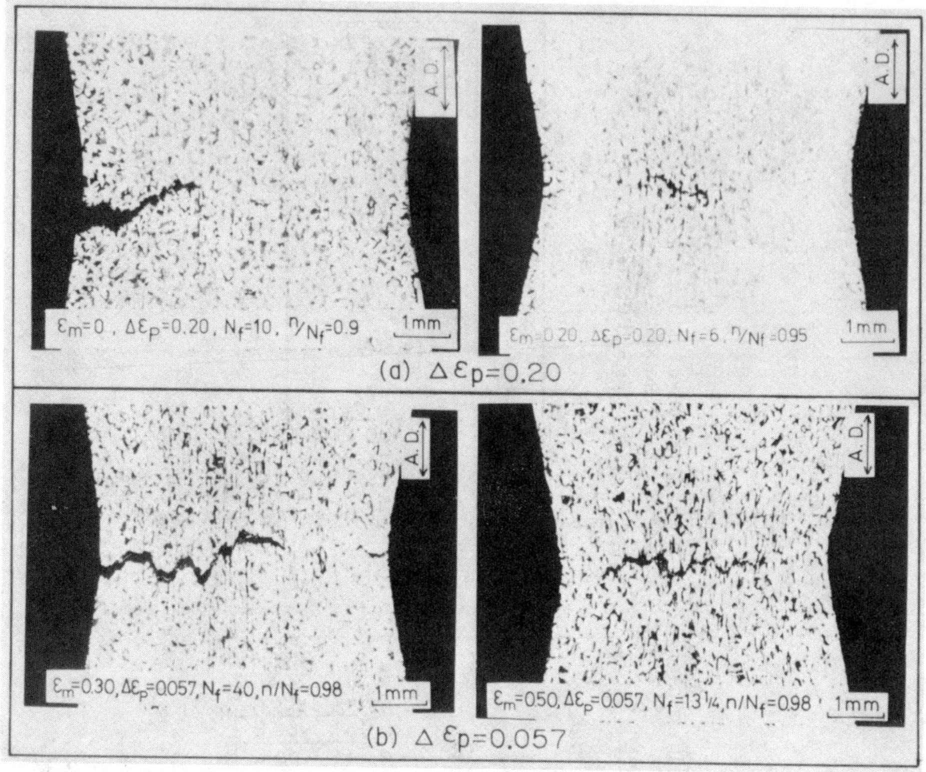

FIG. 9—*Variation of fracture mode with the change of mean strain.* (a) ϵ_m *changed from 0 to 0.20 under constant* $\Delta\epsilon_p$ *of 0.20.* (b) ϵ_m *changed from 0.30 to 0.50 under constant* $\Delta\epsilon_p$ *of 0.057.*

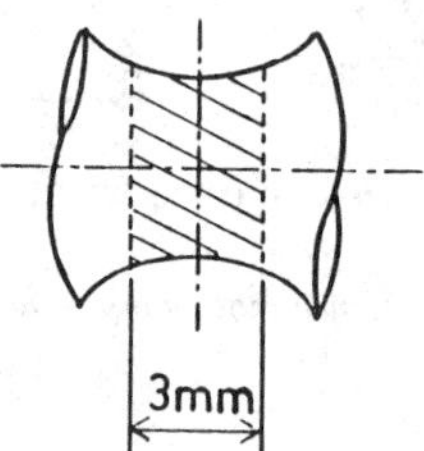

FIG. 10—*Region for the measurement of cracked pearlite ratio.*

In addition, the relation between ϵ_{FR} and n/N_f becomes independent of the $\Delta\epsilon_p$ irrespective of the level of ϵ_m when $\Delta\epsilon_p$ or ϵ_m is high enough to produce the fatigue behavior of Type C. These behaviors of ϵ_{FR} are quite similar to those of the n_{cp}/n_p.

Finally, from this result together with that in Fig. 12, the relation between the reduction of residual ductility and the cracked pearlite ratio n_{cp}/n_p was examined. The result is given in Fig. 14, which shows that the reduction of ϵ_{FR} in the extremely low cycle fatigue of the annealed carbon steel is governed by the parameter, cracked pearlite ratio.

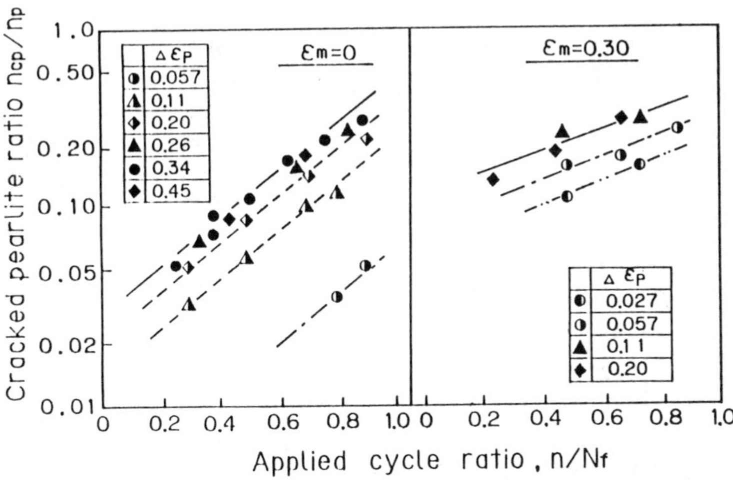

FIG. 11—*Relation between the number of cracked pearlite and the cycle ratio.* (a) $\epsilon_m = 0$. (b) $\epsilon_m = 0.30$.

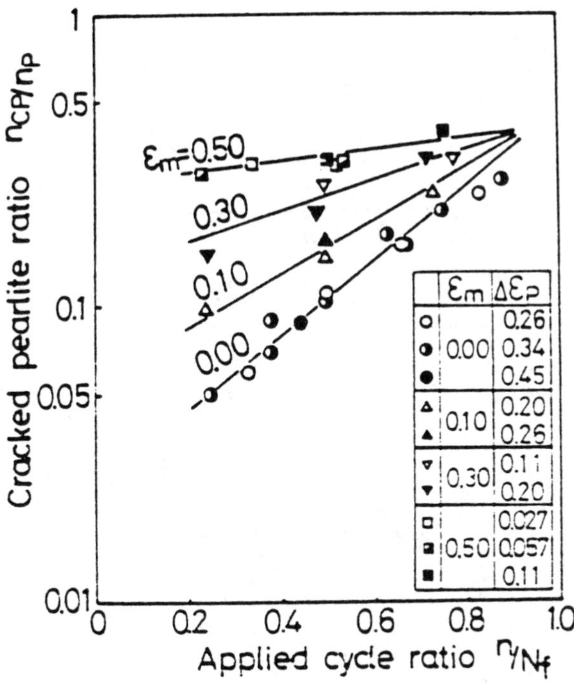

FIG. 12—*Variation of cracked pearlite ratio with applied cycle ratio for Type C fracture process.*

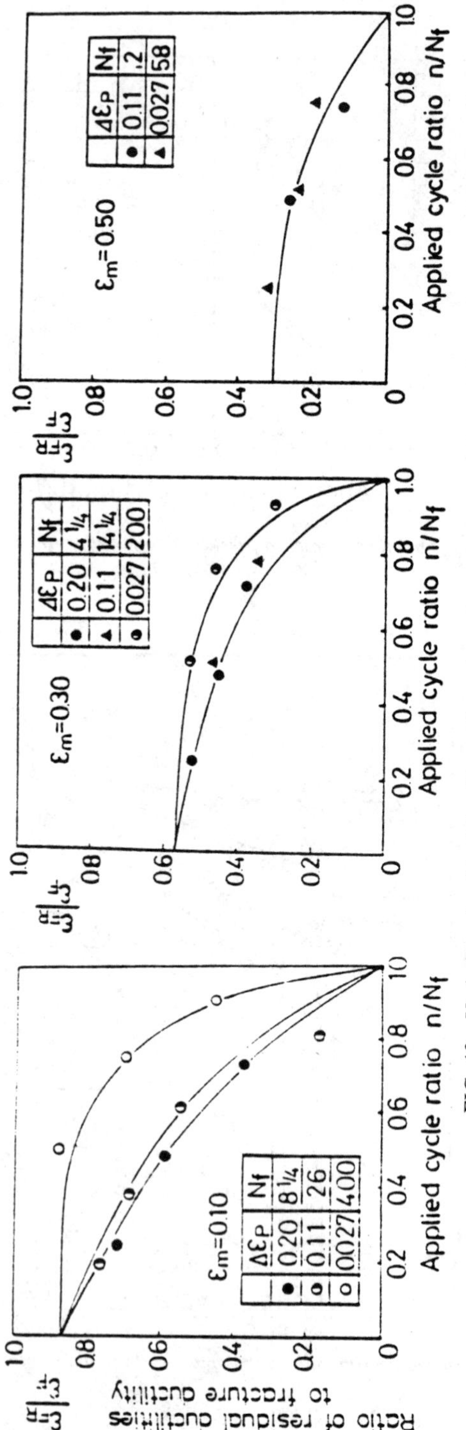

FIG. 13—*Variation of residual ductility with strain cycling under various mean strains.*

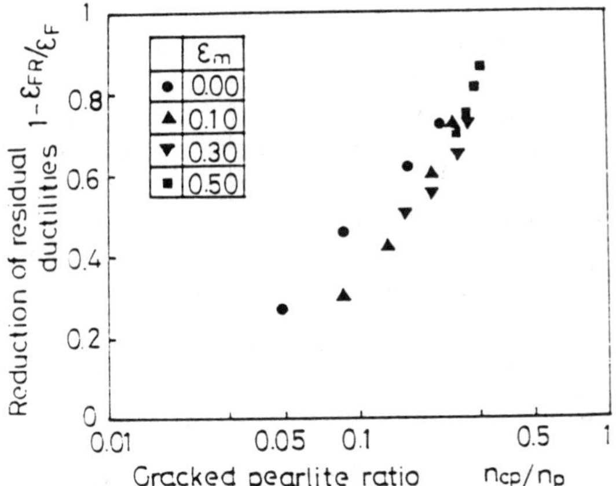

FIG. 14—*Relation between reduction of residual ductility and cracked pearlite ratio for Type C fracture process.*

From this and the results mentioned so far, it could be concluded that the loss of ductility with strain cycling at high $\Delta\epsilon_p$ and ϵ_m is closely related to the behavior of pearlite cracking inside the material, which is a primary cause for the damage responsible for the final fracture of an annealed carbon steel due to the extremely low cycle fatigue.

Conclusions

The following conclusions can be made as the results of the study on the microstructural aspects of crack initiation and propagation in the low cycle fatigue of annealed carbon steel including very short fatigue life ($N_f < 10$).

1. There are three different types of fracture processes depending on plastic strain range $\Delta\epsilon_p$ and mean strain ϵ_m in the range of 0.0065 to 0.45 for $\Delta\epsilon_p$ and 0 to 0.50 for ϵ_m:

(a) *Type A*—Crack initiation and its propagation in the surface microstructure.
(b) *Type B*—Surface crack initiation followed by the depth direction growth which is assisted by the cracking of a pearlite.
(c) *Type C*—Internal cracking caused by the void formation originated from the shear cracking of pearlite.

The shift of microfracture processes from Type A to B and Type B to C occurs as $\Delta\epsilon_p$ and ϵ_m increase.

2. The change of residual ductility with strain cycling depend strongly on $\Delta\epsilon_p$ at each level of ϵ_m. For the fully reversed condition ($\epsilon_m = 0$), these behaviors are classified into three types:

(a) *Type I*—For small $\Delta\epsilon_p$, no drastic change of ϵ_{FR} occurs and about 80% of initial fracture ductility is preserved before final stage of the fatigue life.
(b) *Type II*—In the case of intermediate $\Delta\epsilon_p$, the gradual decrease of ϵ_{FR} with strain cycling appears depending on the level of $\Delta\epsilon_p$.
(c) *Type III*—For large $\Delta\epsilon_p$ resulting in extremely short life ($N_f < 10$), ϵ_{FR} continues to decrease linearly with strain cycles before final fracture.

These three types of ϵ_{FR} variation with strain cycling appear at the respective levels of $\Delta\epsilon_p$ giving the fracture processes of Type A, B, and C.

3. Good correlation is obtained between the reduction of ϵ_{FR} and the cracked pearlite ratio newly defined as a parameter for the evaluation of the internal microfracture behavior.

4. On the basis of the results obtained, the following conclusion can be drawn for the nature of the damage accumulation in the low cycle fatigue of the present steel. At the small $\Delta\epsilon_p$, the surface damage due to the initiation and propagation of surface microcracks is predominant, while at the very large $\Delta\epsilon_p$ giving fatigue life less than $N_f = 10$, the internal cracking originated from the fracture of a pearlite becomes a primary source of the damage which results in the reduction of ϵ_{FR}. The transition from the surface damage to the internal one takes place when the $\Delta\epsilon_p$ becomes so high that the pearlite cracking may start inside the material.

References

[1] Martin, D. E., *Transactions, American Society of Mechanical Engineers, Series D,* Vol. 83, 1961, pp. 565–571.

[2] Sachs, G. and Weiss, V., *Zeitschrift für Metallkunde,* Vol. 53, 1962, pp. 37–47.

[3] Yao, J. T. P. and Munse, W. H. in *Fatigue of Aircraft Structures, ASTM STP 338,* American Society for Testing and Materials, Philadelphia, 1963, pp. 5–24.

[4] Ohji, K., Miller, W. R., and Martin, J., *Transactions, American Society of Mechanical Engineers, Series D,* Vol. 88, 1966, pp. 801–810.

[5] Manson, S. S., *Machine Design,* Vol. 32, No. 16, 1960, pp. 129–135; Vol. 32, No. 17, 1960, pp. 160–166.

[6] Sessler, J. G. and Weiss, V., *Transactions, American Society of Mechanical Engineers, Series D,* Vol. 85, 1963, pp. 539–547.

[7] Kikukawa, M., Ohji, K., Sumiyoshi, A., and Asai, M., *Journal of the Japan Society of Mechanical Engineers,* Vol. 70, 1967, pp. 1495–1509.

[8] Obataya, Y. and Shiratori, H., *Transaction of the Japan Society of Mechanical Engineers,* Vol. 36, 1970, pp. 1452–1462.

[9] Kikukawa, M., Ohji, K., Ohkubo, H., Yokoi, T., and Morikawa, T., *Transactions of the Japan Society of Mechanical Engineers,* Vol. 38, 1972, pp. 8–15.

[10] Murakami, Y., Harada, S., Tani-ishi, H., Fukushima, Y., and Endo, T., *Transactions of the Japan Society of Mechanical Engineers, Series A,* Vol. 49, 1983, pp. 1411–1419.

[11] Yamada, T., Hoshide, T., Fujimura, S., and Manabe, M., *Transactions of the Japan Society of Mechanical Engineers, Series A,* Vol. 49, 1983, pp. 441–451.

[12] Hatanaka, K., Fujimitsu, T., and Watanabe, H., *Transactions of the Japan Society of Mechanical Engineers, Series A,* Vol. 51, 1985, pp. 790–798.

[13] Harada, S., Murakami, Y., Fukushima, Y., Ishimatsu, Y., and Endo, T., *Transactions of the Japan Society of Mechanical Engineers, Series A,* Vol. 51, 1985, pp. 1215–1223.

[14] Kobayashi, Y., Takashima, Y., Tsuzuki, S., Shimizu, M., and Kunio, T., *Transactions of the Japan Society of Mechanical Engineers,* Vol. 40, 1974, pp. 2117–2126.

[15] Ohji, K., *Journal of the Japan Society of Mechanical Engineers,* Vol. 70, 1967, pp. 36–47.

Klaus Detert[1] *and Rainer Scheffel*[1]

Low Cycle Fatigue of Al-Mg-Si Alloys

REFERENCE: Detert, K. and Scheffel, R., "**Low Cycle Fatigue of Al-Mg-Si Alloys,**" *Low Cycle Fatigue, ASTM STP 942,* H. D. Solomon, G. R. Halford, L. R. Kaisand, and B. N. Leis, Eds., American Society for Testing and Materials, Philadelphia, 1988, pp. 765–775.

ABSTRACT: The low cycle fatigue behavior of several age-hardened aluminum alloys with 1% Mg and 1% Si as alloying elements and varying Mn additions of up to 1% was studied. The alloy with the smallest grain size showed the highest strength as well as the highest low cycle fatigue strength. The number of cycles to failure represented by its dependency on strain range did not show any significant differences. However, one could derive from these results that one should try to reduce the number and size of inclusion particles in order to improve fatigue properties. In such a case one can expect that a small grain size of 20 μm diameter, Mn additions of 0.2 to 0.5%, and a polished surface with a roughness below 0.5 μm would be beneficial for low cycle fatigue properties.

KEY WORDS: fatigue, low-cycle fatigue, precipitation-hardened alloys, aluminum alloys with up to 1% Si and up to 1% Mg, age-hardened Al-Mg-Si alloys, Mn addition

Age-hardened Al alloys with Mg and Si as alloying elements of Type 6061-T6 or 6151-T6 are commonly used as a material with medium strength for light-weight construction, in particular for ground transport systems when good corrosion resistance is required. Fatigue properties are important in regard to service conditions when alternating stresses are imposed. This is regularly the case of components in such constructions. Material degradation by cyclic stress is called *low-cycle fatigue* when subjected to stress amplitudes which will lead to failure within a number of cycles below 5×10^4.

In order to evaluate the low-cycle fatigue properties and to define the optimum composition and structure of these alloys, different Al alloys with Mg and Si as alloying elements have been investigated. Published results of low cycle fatigue behavior of those alloys are very limited and contradictory [*1–6*].

Procedure

Several age-hardened Al-Mg-Si alloys with various additions of Mn, grain sizes, and heat treatments were studied in low-cycle fatigue tests in accordance with ASTM Recommended Practice for Constant-Amplitude Low-Cycle Fatigue Testing (E 606). The chemical compositions of the alloys investigated are listed in Table 1. The grain structure was made visible by anodic oxidation [*7*] of a polished surface. Figures 1*a* to 1*c* give examples of the grain structure. Low-cycle fatigue was applied by an automatic controlled tensile testing machine in a push-pull cycle with constant strain amplitude on uniaxial loaded specimens with a uniform cross section of 10 mm in diameter on a 30 mm length. The form and dimension of the specimens are shown in Fig. 2. In a few cases, extra care was taken to obtain a smooth surface by polishing. The strain amplitude was measured by a mechanical extensometer of 25 mm gage length. The strain

[1]Full Professor of Materials Technology and Assistant Professor in the Institute of Materials Technology, University of Siegen, Germany.

TABLE 1—*Chemical composition, heat treatment, grain size, and surface finish of alloys investigated.*

Alloy No.	Heat	Chemical Composition (wt%)								Heat Treatment	Hardness HB	Grain Diameter, μm	Surface Finish (R_z), μm
		Si	Mg	Mn	Fe	Ti	Cu	Zn	Cr				
41.1	2141	0.92	0.91	0.01	0.25	0.011	0.01	0.02	0.01	1.0 h-540°C, solution annealed 2.0 h-160°C, aged	101	280	1.02
41.2	2141	0.92	0.91	0.01	0.25	0.011	0.01	0.02	0.01	1.0 h-540°C, solution annealed 2.0 h-160°C, aged	95	280	0.55
42	2142	0.98	0.91	0.21	0.26	0.011	0.01	0.02	0.01	1.0 h-540°C, solution annealed 2.5 h-160°C, aged	95	306	2.08
44.1	2144	0.95	0.95	0.44	0.29	0.011	0.01	0.02	0.01	1.0 h-540°C, solution annealed 2.5 h-160°C, aged	98	260	3.50
44.2	2144	0.95	0.95	0.44	0.29	0.011	0.01	0.02	0.01	1.0 h-540°C, solution annealed 2.0 h-160°C, aged	95	260	0.28
43	2588	1.05	1.02	0.70	0.21	0.007	0.01	0.01	...	1.0 h-540°C, solution annealed 2.0 h-160°C, aged	100	175	3.30
45	2145	0.88	0.89	0.97	0.30	0.011	0.01	0.03	0.01	1.0 h-540°C, solution annealed 6.0 h-160°C, aged	96	210	2.45

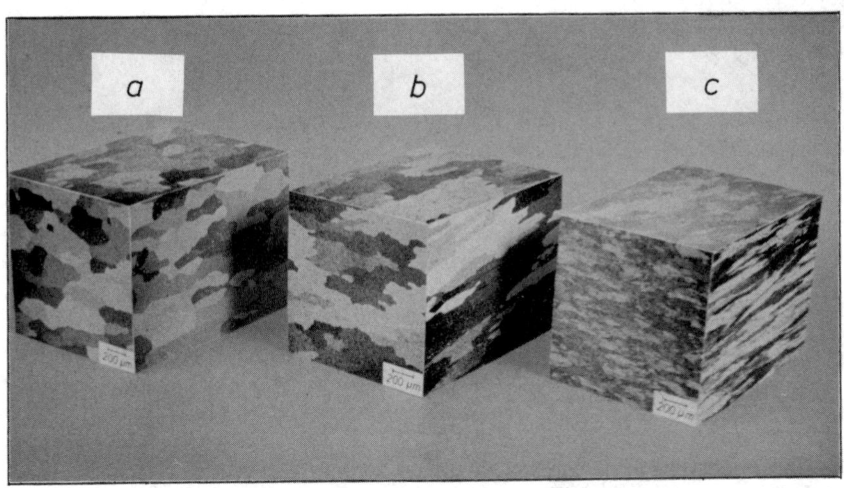

FIG. 1—*Grain structure of Al-Mg-Si alloys (scale mark indicates 200 µm).* (a) *Alloy 41—0.01 wt% Mn.* (b) *Alloy 42—0.2 wt% Mn.* (c) *Alloy 43—0.7 wt. % Mn.*

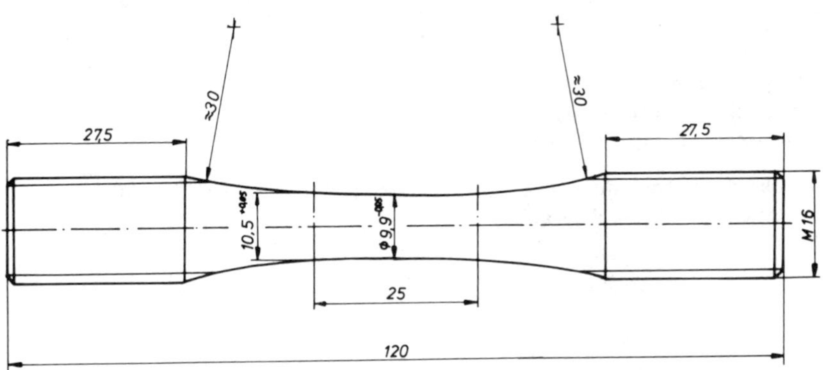

FIG. 2—*Specimen dimensions (in millimetres).*

amplitude was kept constant for each specimen. It varied, however, from ±0.25 to ±1.0% for different samples with strain ratio $\epsilon_{min}/\epsilon_{max} = -1$. The frequency was between 0.01 and 0.03 cycles/s, depending on the strain amplitude, in order to keep the plastic strain rate at 0.02%/s. The number of cycles to failure were determined by a drop of the peak tensile stress to 80% of its maximum value.

This drop of the peak tensile stress is associated with a crack of a few millimetres depth. This number of cycles to failure, N_f, is defined as the number of cycles until a macroscopic crack can be detected by either visual observation or dye penetration technique in the larger components. Closed-loop process control and data recording were accomplished automatically by a microprocessor and microcomputer of the PC type. The data recorded for each cycle of a stress-strain hysteresis diagram are shown in Fig. 3. The number of cycles to failures, N_f, as defined above were listed dependent on total strain range (Fig. 4) to be evaluated by the formula

$$\Delta\epsilon_{tot} = (\sigma_f'/E)\cdot N_f^b + \epsilon_f' N_f^c \qquad (1)$$

where

$\Delta\sigma$ = total stress range,
$\Delta\epsilon_{tot}$ = total strain range,
N_f = number of cycles to failure,
E = modulus of elasticity,
σ_f' = fatigue strength coefficient,
ϵ_f' = fatigue ductility coefficient,
b = fatigue strength exponent, and
c = fatigue ductility exponent.

The stress amplitude of the hysteresis loop at the number of cycles $\approx 1/2\ N_f$ was associated with the stabilized cyclic stress strain loop in dependence of the total strain amplitude to be evaluated by the formula

$$\Delta\sigma/2 = \sigma_a = k'(\Delta\epsilon_{pl}/2)^{n'} \qquad (2)$$

where

σ = stress amplitude,
$\Delta\epsilon_{pl}$ = plastic strain range,
k' = cyclic strength coefficient, and
n' = cyclic strain hardening exponent.

The cyclic stress-strain diagram can be described with the same mathematical expression as for the tensile stress-strain diagram:

$$\sigma = k(\epsilon_{pl})^n \qquad (3)$$

where

σ = true tensile stress,
ϵ_{pl} = true plastic strain,
k = strength coefficient, and
n = strain hardening exponent.

Tensile tests were performed on tensile specimens with a 10 mm diameter and a 50 mm gage length. The tensile stress-strain diagram should correlate to the peak tensile stress and strain during the first positive half cycle of alternating stresses.

Results

The results confirm that Eqs 1 to 3 describe the experimental data satisfactorily within the experimental scatter. A typical example of tensile and cyclic stress strain diagram is given in Figs. 5*a* and 5*b*. Table 2 lists the data obtained from tensile tests as mean values from three single specimens tested for each alloy. Table 3 lists the data derived from low-cycle fatigue tests as mean values of five specimens tested at each given total strain range for each alloy. In this table a so-called cyclic yield strength, $R'_{p0.2}$, derived from the cyclic stress-strain curve, is also

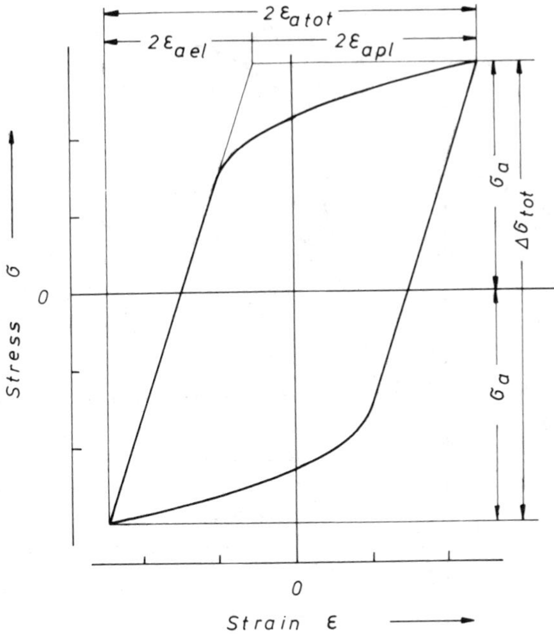

FIG. 3—*Stress-strain hysteresis diagram.*

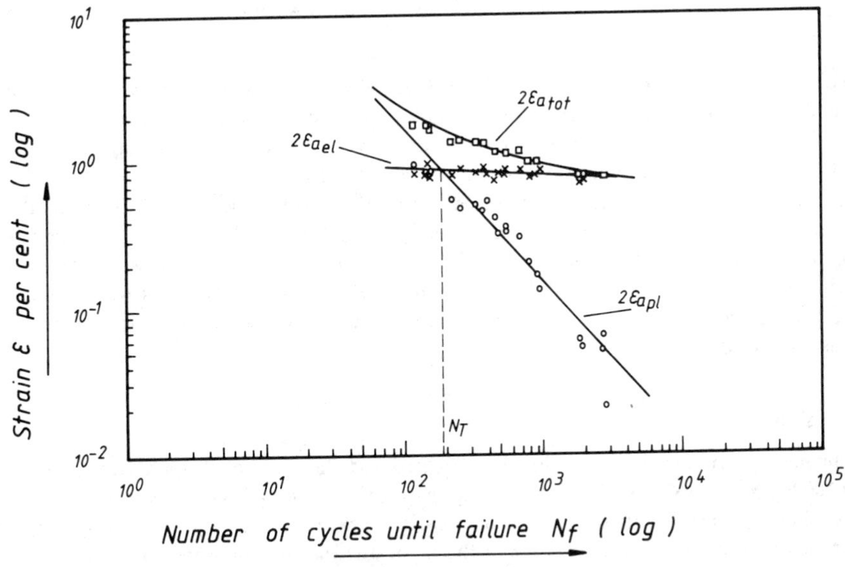

FIG. 4—*Fatigue-life relationship of Alloy 41.2.*

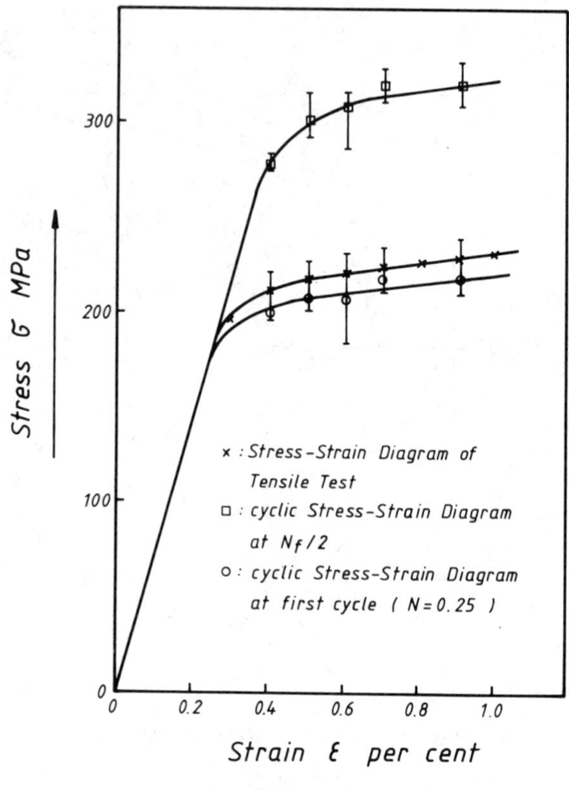

FIG. 5a—*Stress-strain diagram of Alloy 41.2.*

listed. A comparison of $R'_{p0.2}$ with yield strength $R_{p0.2}$ demonstrates the work hardening by cyclic deformation in those alloys. Also, a number of cycles, the so-called transition life, N_T (Fig. 4), is listed in Table 3. When the number of cycles to failure is less than N_T then cyclic plastic strain dominates. Above N_T elastic strain dominates.

Examination by optical microscopy revealed that the first microcracks of 20 μm depth could be observed after about 5% of the number of cycles to failure (cf. Fig. 6 at a somewhat later stage). Those microcracks appeared to originate at sites where inclusions were present at the specimen surface. It was observed that inclusions of about 5 μm in diameter were present in all the alloys studied. In alloys with a Mn content above 0.5% a somewhat larger population of inclusions was observed.

Discussion

The heat treatment of all the alloys investigated was chosen to obtain a hardness between 95 and 100 HBN. Alloy 45 required a longer aging time than other alloys with less Mn content. The alloys showed considerable differences of yield strength, tensile strength, and ductility despite their hardness similarities. Alloy 43 had the highest strength values and lowest ductility. One may associate the higher strength of Alloy 43 with the lower grain size of this alloy. It can, however, depend on the different manufacturing technique of this alloy which resulted also in

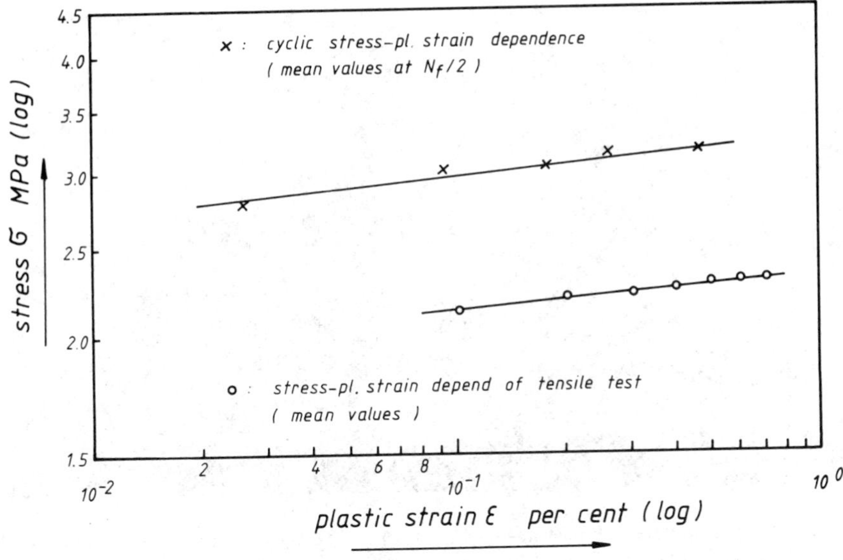

FIG. 5*b*—*Stress-plastic strain dependence of Alloy 41.2 (log scale).*

TABLE 2—*Data derived from tensile tests.*

Alloy No.	$R_{p0.2}$, MPa	R_m, MPa	A, %	Z, %	k, MPa	n	σ_f, MPa	ϵ_f
41.1	221	327	25.3	43.2	315.4	0.0551	520	0.560
41.2	219	305	27.0	42.0	280.0	0.040	469	0.541
42	198	291	26.0	40.0	298.5	0.0655	420	0.511
44.1	207	281	23.0	33.5	267.1	0.041	375	0.410
44.2	207	281	23.0	33.5	267.1	0.0412	375	0.410
43	310	405	13.0	27.0	398.2	0.0401	510	0.400
45	198	291	20.0	40.0	267.1	0.0475	444	0.511

NOTE: σ_f = true fracture stress.
ϵ_f = local fracture strain = ln (1/1-Z).

TABLE 3—*List of data derived from fatigue tests.*

Alloy No.	$R'_{p0.2}$, MPa	N_T	k' MPa	n'	σ_f'	ϵ_f'	b	c
41.1	328	138	526.9	0.0771	1046.4	1.194	−0.0914	−0.978
41.2	311	172	426.0	0.051	870.0	1.555	−0.0559	−1.0048
42	314	258	403.5	0.0406	737.2	2.370	−0.0296	−1.014
44.1	300	302	374.1	0.0348	708.6	1.230	−0.02515	−0.881
44.2	288	325	354.7	0.0344	681.7	1.866	−0.0284	−0.952
43	390	53	492.1	0.0371	1072.3	3.880	−0.0603	−1.4641
45	308	288	366.4	0.0298	662.9	1.377	−0.02604	−0.91274

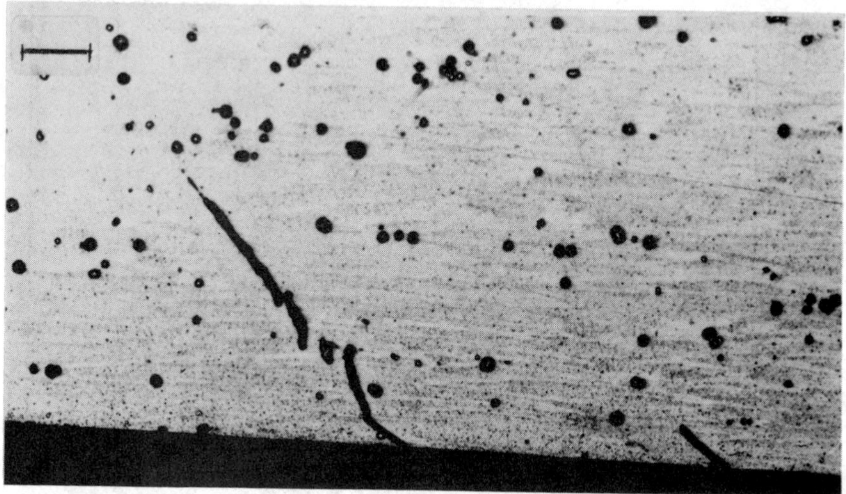

FIG. 6—*Cross section through specimen of Alloy 45 after 20 000 cycles at strain range of 0.5% (N_f = 115 000) (scale mark indicates 50 μm).*

the smaller grain size. The higher strength of this alloy leads to a higher strength coefficient k and k'. The strain hardening exponent appeared to be similarly low within $n = 0.04$ to 0.0655. This rather low hardening exponent depends on the evaluation of the stress-strain tensile diagram in the strain range up to 1% strain. One can show that a higher tensile strain of up to 20% can better be approximated by a form

$$\sigma = \sigma_0 + A \cdot \varphi^n$$

where σ = true stress and φ = true plastic strain, with a strain hardening exponent n on the order of 0.5 to 0.6 as shown elsewhere [*9*].

The low-cycle fatigue behavior showed the same difference regarding strength in the cyclic tensile diagram with the highest cyclic strength coefficients k' and σ_f'. Comparing $R_{p0.2}$ with $R'_{p0.2}$ showed that considerable work hardening was obtained by cyclic deformation in the age-hardened alloys.

It can be questioned whether the differences of the data obtained by evaluation of the fatigue tests are experimental scatter or material-dependent deviations. The low value of N_T for Alloy 43 appears to be material dependent. From this comparison of the fatigue data one can determine that the number of cycles to failure, associated with the stress amplitude, will differ for the materials investigated. Since Alloy 43 also has the highest fatigue strength coefficient, σ_f', a higher number of cycles to failure for a given stress amplitude can be expected. The results show, however, that the number of cycles to failure for all specimens tested fall within the same scatterband as shown in Fig. 7 when represented as dependent on strain amplitude. This scatterband is essentially similar to that of former experimental results [*1*]. A disagreement with the results of Edwards and Martin [*2*] should be mentioned. A particularly poor low-cycle fatigue resistance of Alloy 41 was not found, although in this alloy the absence of Mn dispersoids of 0.1 μm was confirmed by electron microscopic examination. It was confirmed, however, that coarse slip occurs during plastic strain [*8*] in this alloy without the Mn addition. In other alloys the distance of glide band is much smaller, indicating that a 0.2% Mn addition is sufficient to provide fine slip during plastic strain.

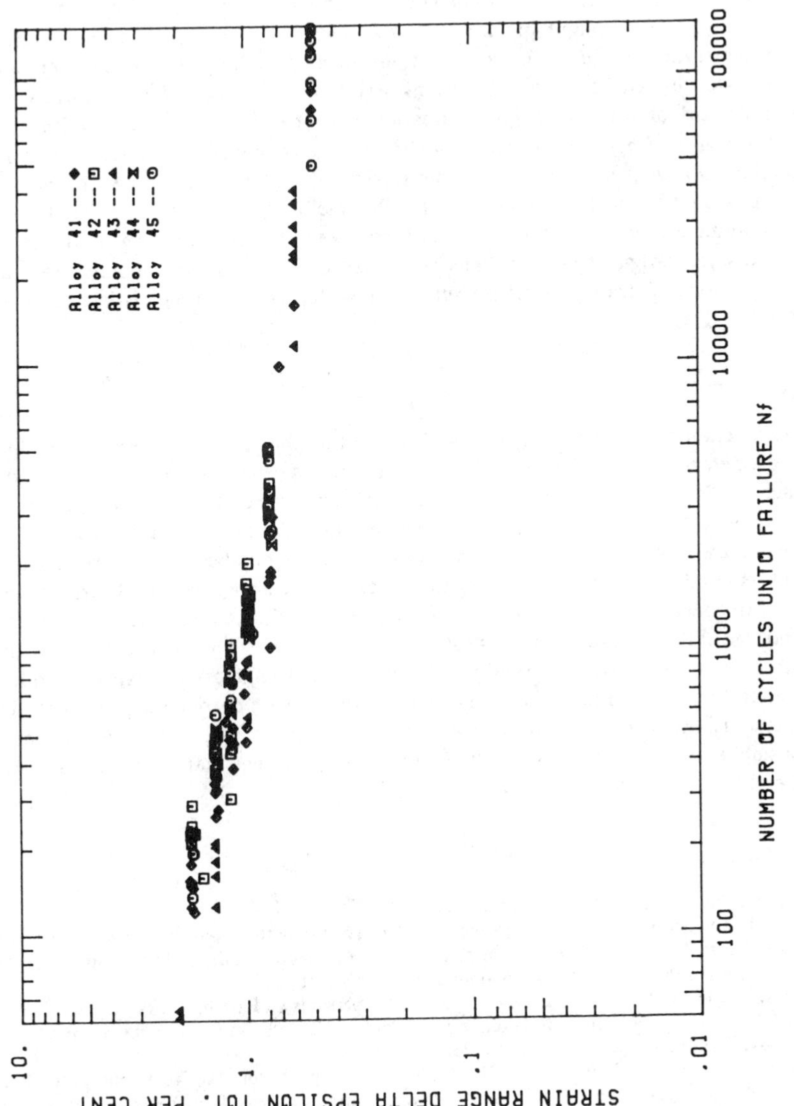

FIG. 7—*Number of cycles to failure of Al-Mg-Si alloys at different strain ranges.*

Also, the results of Ruch and Gerold [*6*] could not be confirmed, namely, that a small grain size is beneficial for the low-cycle fatigue behavior of Al-Mg-Si alloys because of the longer incubation period needed to form small fatigue cracks. The results of Alloy 43 with the smallest grain size showed that the number of cycles to failure, N_f, of this alloy was even at the lower side of the scatterband. There appears to be an approximate relation of 2 times k' equal σ_f' which may depend on the fact that k' applies to the stress-strain relation using the strain amplitude ϵ_a rather than the strain range $\Delta\sigma = 2\sigma_a$. The latter is defined by Eq 1 according to the original work of Manson [*10,11*], though Appendix X1 of ASTM E 606 has modified this equation by introducing the strain amplitudes instead of strain range. There appears to be no relation between the terms of the tensile test and the terms derived from low-cycle fatigue. As one would expect from the work of Manson [*11*] the evaluation of low-cycle fatigue data according to Eq 1 should lead to values of $b = -0.12$ and $c = -0.6$ for the exponents of fatigue strength and fatigue ductility, which is not the case for the reported investigation. Neither the method of universal slopes [*11,12*] nor the four-point correlation method [*10,11*] can therefore be used to calculate the number of cycles to failure which correlates to the experimental results. Such an estimate of low-cycle fatigue would yield an endurance limit from the data of tensile tests about 10 times longer for the strain range between 0.8 and 2%. The difference is less pronounced below 0.8%.

Conclusions

It can be concluded from these experimental results that the observed presence of inclusions of 5 μm in diameter obscures all other influences regarded as beneficial, such as fine slip due to the presence of Mn dispersoids, small grain size, and better smoothness of the specimen surface. The presence of those fairly large inclusions may also be regarded as the reason that the first microcracks were clearly observed as early as at 5% of the number of cycles to failure. Very often one observed that a substantial part of the cycles to failure are necessary to create the first microcracks. In conclusion, one can conclude from these experiments that one has to reduce the number and size of the inclusion particles substantially to improve low-cycle fatigue properties. Then a Mn content of 0.2 to 0.5% and a reduced surface roughness are expected to have a beneficial effect. Since slip lines are visible across the whole grain despite the presence of inclusions, dispersoids, and precipitation particles, one should also strive to manufacture these alloys with a smaller grain size to improve the fatigue properties. The manufacture of such alloys is in progress, but they have not been tested yet.

References

[*1*] Wellinger, K. and Sautter, S., *Aluminum,* Vol. 47, 1971, p. 741.

[*2*] Edwards, L. and Martin, J. W., "The Influence of Dispersoids on the Low-Cycle Fatigue Properties of Al-Mg-Si Alloys," in *Proceedings,* International Conference on Strength of Metals and Alloys, No 6. Melbourne, Australia, Aug. 1982, Pergamon Press, Oxford, p. 873.

[*3*] Edwards, L. and Martin, J. W., "The Influence of Dispersoids on Fatigue Crack Propagation in Al-Mg-Si Alloys," in *Proceedings,* 5th International Conference on Fracture, Cannes, France, March/April 1981, Pergamon Press, Oxford, p. 323.

[*4*] Titchener, A. L. and Ponniah, C. D., "The Effect of Thermomechanical Treatment on the Fatigue Behavior of an Al-Mg-Si-Mn Alloy," in *Proceedings,* 3rd International Conference on the Strength of Metals and Alloys, Cambridge, England, Aug. 1973, Pergamon Press, Oxford, p. 432.

[*5*] Bomas, H. and Mayr, P., "Einfluβ der Wärmebehandlung auf die Schwingfestigkeitseigenschaften der Legierung AlMgSiO.7," *Zeitschrift für Werkstofftechnik* (VCH Verlagsges. Weinheim), Vol. 16, 1985, p. 88.

[*6*] Ruch, W. and Gerold, V., "Einfluβ des Gefüges auf die Ermüdungseigenschaften von AlMgSi1 Legierungen," *Zeitschrift für Metallkunde,* Vol. 76, 1985, p. 338.

[*7*] Barker, L. J., *Transactions of ASM,* Vol 42, 1950, p. 347.

[*8*] Detert, K., Scheffel, R., and Stünkel, R., "The Influence of Grain Size and Dispersion of Small

Particles on Crack Initiation and Growth During Fatigue in Age Hardened AlMgSi Alloys," in *Proceedings,* 7th International Conference Strength of Metals and Alloys, Montreal, Canada, Aug. 1985, Pergamon Press, Oxford, p. 1219.

[9] Scheffel, R., Stünkel, R., and Detert, K., "Einsatz von Personalcomputern zur Automatischen Meβdatenerfassung und Auswertung von werkstofftechnischen Versuchen," Berichte der Tagung Werkstoffprüfung Bad Nauheim, Germany, Dez. 1984 (DVM Berlin).

[10] Duggan, T. V. and Byrne, J., *Fatigue as a Design Criterion*, Macmillan, London, 1977, p. 57 (Eq 3.3).

[11] Manson, S. S., "Fatigue: A Complex Subject—Some Simple Approximation," *Experimental Mechanics,* Vol. 5, No. 7, 1965, p. 193.

[12] Coffin, L. F., *Transactions of ASME,* Vol. 76, 1954, p. 931.

S. M. Pickard,[1] *F. Guiu*,[1] *and A. P. Blackie*[1]

Low Cycle Fatigue of Strain Aging Ferrous Alloys

REFERENCE: Pickard, S. M., Guiu, F., and Blackie, A. P., **"Low Cycle Fatigue of Strain Aging Ferrous Alloys,"** *Low Cycle Fatigue, ASTM STP 942,* H. D. Solomon, G. R. Halford, L. R. Kaisand, and B. N. Leis, Eds., American Society for Testing and Materials, Philadelphia, 1988, pp. 776–797.

ABSTRACT: The fatigue behavior of two high purity Fe alloys, one with 75 ppm (by mass) of both N and C and the other with 0.1% (by mass) C, has been investigated at 5 Hz. These alloys were heat treated to retain different solute concentrations, and it was found that the effect of increasing solute concentration was to increase the fatigue life and to suppress the "knee" in the *S-N* curve. It is shown that the flat appearance of the *S-N* curve is due to dynamic strain aging of the alloys. The Coffin-Manson analysis of the data shows that dynamic strain aging hardly affects the fatigue ductility of the material but produces an increase in yield stress sufficient to increase the fatigue life.

Fatigue predeformation and static aging at 60°C for 16 h produces a precipitation of metastable nitrides, or carbides, both at matrix sites and dislocations and an increase in the yield stress of the alloys. On retesting at the same stress amplitude, the plastic strain is found to have been considerably reduced and the fatigue life greatly increased, with a return of a "knee" in the *S-N* curve. The Coffin-Manson analysis reveals a great loss in fatigue ductility, but it is found that the cumulative strain to failure remains nearly constant at each stress level. This suggests that the influence of the increased strength by strain aging on fatigue life is more important than the reduction in ductility. The dislocation substructures developed at different amplitudes both before and after aging are correlated with the formation of slip bands and the nucleation of fatigue cracks.

KEY WORDS: fatigue properties, ferrous alloys, interstitial solutes, strain aging, dynamic strain aging, fatigue crack nucleation, dislocation substructures, increase in fatigue life

The phenomenon of strain aging in dilute Fe alloys containing C and N in solution has been known for many years [*1*]. It produces an increase in strength with an accompanying reduction in ductility and for this reason may be an undesirable effect in those cases where loss of ductility and fracture toughness is to be avoided. So far as fatigue properties are concerned, the strengthening achieved by strain aging could be more beneficial because strength may be more important than ductility under small strain amplitude testing conditions. In fact, the strain aging which occurs during stress-controlled fatigue is believed to be responsible for such effects as the increase in fatigue life after "resting periods" during the fatigue test [*2–4*], the "coaxing" of strain aging metals [*5,6*], and the occurrence of a fatigue limit in ferrous alloys [*7–9*]. Strain aging that takes place during fatigue, or "dynamic" strain aging, has been studied by several workers [*4,10,11*]. They observed that the microstructure which evolves during long duration fatigue testing is an ineffective means of strengthening at higher stress levels, because it is unstable under high amplitude dislocation motion.

The possibility of strengthening and increasing fatigue life in dilute interstitial alloys of iron is further investigated in this paper, where not only the effects of "dynamic" aging are considered but also a "fatigue predeformation" and static aging is used to develop more rapidly and effec-

[1]Department of Materials, Queen Mary College, London, England.

tively a strengthened microstructure. The use of fatigue deformation prior to aging has the advantage of providing a very large cumulative prestrain, hundreds of times in excess to that which would be possible in unidirectional deformation, and thus producing a more developed dislocation substructure, with higher dislocation densities, and increased strain age strengthening [*12*]. The two alloys chosen for this investigation have the composition given in Table 1 and were chosen because they offer different fatigue strain aging capabilities which have been previously determined and studied in low frequency strain amplitude controlled cyclic deformation [*12,13*]. It was shown in those studies that a large increase in cyclic peak stress can be achieved by a cyclic predeformation and low temperature (60°C) aging treatment due to the formation of a dispersion of metastable nitride or carbide particles, both at matrix sites and at dislocations. The strongly aged and precipitation pinned dislocation substructure was unstable under repeated cyclic loading at high strain amplitudes. The possibility of it being stable, however, under constant stress amplitude fatigue testing is investigated in this paper.

Experimental Procedure

The two alloys used in this investigation will be referred hereafter as the N-alloy and the C-alloy to distinguish the main interstitial element present. Fatigue test specimens of a dumbbell shape with a central gage of 3 mm diameter and 5 mm length were subjected to the following initial heat treatments: The N-alloy was heated in a vacuum of 10^{-6} torr at 700°C for 3 h and cooled to room temperature at a rate of about 100°C/min; this produced an average grain size of 95 μm. The C-alloy was heated for 1 h at 950°C and either cooled to room temperature at a rate of about 100°C/min (the furnace cooled C-alloy) or quenched in ice brine (the quenched C-alloy). Both these treatments produced an average grain size of 30 μm in the C-alloy.

Both transmission electron microscopy (TEM) observations and consideration of existing references [*14,15*] indicate that most of the 75 ppm (by mass) N content in the N-alloy remained in solution, whilst in the C-alloy about 150 ppm and less than 5 ppm C (by mass) are in solution in the quenched and furnace cooled alloy, respectively.

Prior to mechanical testing all the specimens were chemically or electrolytically polished to produce a uniform mirror-like surface finish. In order to avoid aging at room temperature, all the specimens were either tested shortly after the initial heat treatment or stored at −10°C before testing.

Constant stress amplitude push-pull fatigue tests were carried out in a Mayes servohydraulic machine at a frequency of 5 Hz with zero mean stress and sinusoidal wave form. The load-elongation hysteresis loops were recorded by means of a digital storage oscilloscope, so that the evolution of plastic strain amplitude throughout the tests could be followed.

The aging treatment of the prefatigued specimens was carried out at 60°C in a silicon oil filled capsule submerged in a water bath.

Surface slip lines were examined both by light and scanning electron microscopy (SEM) without special surface preparation. However, some sections of the deformed samples were examined after etching them in 3% nital to reveal surface and subsurface dislocation etch patterns.

TABLE 1—*Composition (ppm by mass) of the iron alloys.*

Element	C	N	O	S	Mn
N-alloy	75	75	25	2	3
C-alloy	1000	10	150	50	. . .

Small disks (<2 mm in diameter), with electron transparent thin areas, were prepared from the deformed specimens by a two-stage technique involving pre-thinning and final perforation. They were examined by TEM in a Jeol 100CX electron microscope operating at 100 kV.

Results

Fatigue Properties of Unaged N-alloy

The *S-N* curve for the as heat-treated (unaged) N-alloy is shown in Fig. 1. It can be noticed that the "fatigue limit" is approached gradually and that there is no sharp "knee" in this curve. As shown later, the absence of this knee is a characteristic of iron alloys which exhibit strong dynamic strain aging.

The evolution of the plastic strain amplitude, ϵ_p, measured by the half-width of the hysteresis curve is shown in Fig. 2. The relatively large values of ϵ_p measured on the first cycle of the test, at all the stress levels, indicate that the elastic limit of this alloy is exceeded at all the amplitudes of the test. This "soft" behavior of the material is clearly revealed in the shape of the tensile stress-strain curve; this does not display a sharp yield point but is rounded with a deviation from linear behavior between 85 and 90 MPa which can be taken as its elastic limit.

At the highest amplitudes of the test there is a rapid cyclic hardening rate which continues up to specimen failure. At the intermediate levels a near-saturation level is reached, but at the lower amplitudes the cyclic hardening persists throughout the test and at the fatigue limit it reduces the plastic strain amplitude to below 10^{-4} before the test is eventually stopped. This effect is believed to be due to dynamic strain aging [*11*] as the microscopical observations described below confirm.

An attempt can be made to represent the fatigue data in terms of the Coffin-Manson relationship

$$\epsilon_p = \epsilon_f(2N_f)^n$$

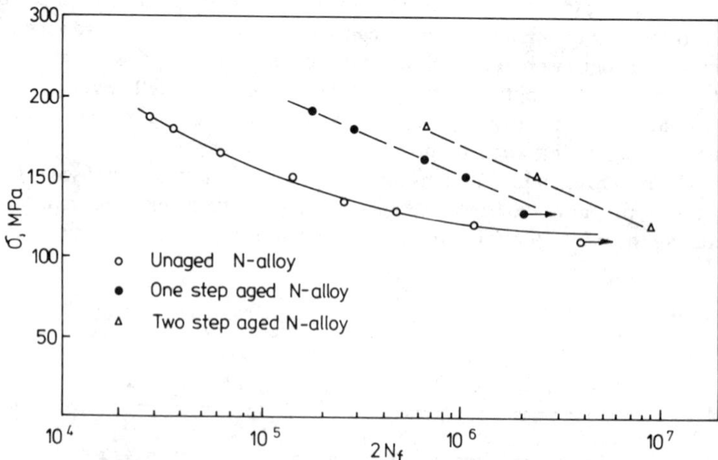

FIG. 1—S-N *curves for the N-alloy in the unaged condition, after one-step fatigue aging and after a two-step fatigue aging.*

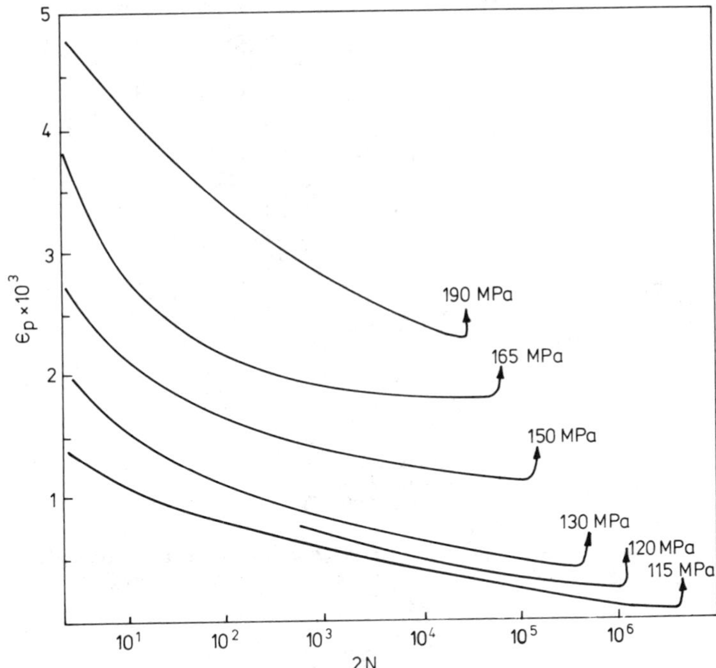

FIG. 2—*Variation of the plastic strain amplitude with number of cycles (half-width of the hysteresis loop) during the fatigue tests of the unaged N-alloy.*

where ϵ_f and n are ductility coefficient and exponent, respectively, and $2N_f$ is the number of stress reversal cycles. The representative values of ϵ_p plotted in Fig. 3 are the average values during the tests which lie within 10% of the actual values for more than 95% of the fatigue life. Within the limits of experimental accuracy the log ϵ_p versus log $2N_f$ plot defines a unique relationship with constant values of ϵ_f and n throughout the amplitude range investigated (Table 2). The constancy of ϵ_f implies that the dynamic aging has not reduced significantly the ductility of the alloy, whilst it has produced a relatively large reduction in plastic strain amplitude (from 10^{-3} to 10^{-4}) which must have lengthened the fatigue life. Thus the effect of dynamic aging on the position of the fatigue limit, as first suggested by Wilson [*9*], is clearly illustrated. The increase in fatigue life is achieved by a moderate strengthening and small absolute reduction in plastic strain amplitude. The limited and precarious effect of dynamic aging on strength is demonstrated by the fact that this is lost (as detected by an increase in ϵ_p to the level of the unaged alloy) when the stress is increased from 120 to 150 MPa.

The surface observations of specimens fatigued to failure at the high amplitudes reveal a large density of diffuse slip lines associated with heavily deformed grains of puckered appearance. These surfaces features are associated with the formation of dislocation cell structures in the bulk of the material (Fig. 4). As the amplitude is reduced, the slip lines become coarser and sharper, and at the lowest amplitudes of the test they are distinctly sharp and straight as previously observed in quenched and fatigued mild steel [*4,16*]. These straight and sharp slip lines, indicative of the predominance of single glide inside each grain, are associated with the formation of "cloud" and "vein" dislocation substructures with free channels as regions in which slip is concentrated, thus leading to the formation of the straight and coarse slip markings (Fig. 5).

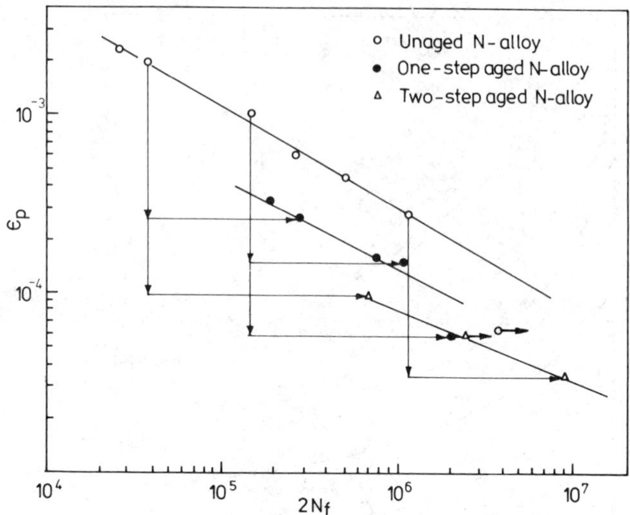

FIG. 3—*Coffin-Manson plot of fatigue data for the unaged, one-step fatigue aged, and two-step fatigued aged N-alloy. Arrows indicate the change in* ϵ_p *and* N_f *produced by the fatigue aging.*

TABLE 2—*Values of fatigue ductility coefficients,* ϵ_f, *fatigue ductility exponent,* n, *and cumulative strain to failure at different stress amplitudes for the various alloys after different treatments.*

	ϵ_f	n	σ, MPa	ϵ_c
Unaged N-alloy	1.009	−0.585	190	29.9
			180	41.4
			165	58.0
			150	78.8
One-step aged N-alloy	0.240	−0.542	190	30.5
			180	38.5
			165	52.7
			150	71.7
Two-step aged N-alloy	0.027	−0.426		
F.C. C-alloy	2.363	−0.678		
Q. C-alloy	6.551	−0.821		

At the high amplitudes, specimen failure occurs by the formation of grain boundary cracks due to incompatible grain deformation [*17*]. At intermediate and low amplitudes the incompatible grain deformation is less severe and, although grain boundary cracks are still observed, the fracture path tends to follow the coarse slip bands.

Specimens examined by TEM, immediately following testing, showed that at stress levels close to the fatigue limit dynamic strain aging had proceeded to the stage where small disk-shaped precipitates ($\sim$15 nm in diameter) of α'' $Fe_{16}N_2$ had formed in the matrix and at the edge of dislocation clouds, although the dislocation free channels remained precipitate free (Fig. 6). Slip activity inside these channels is responsible for this effect [*15*]; this can be con-

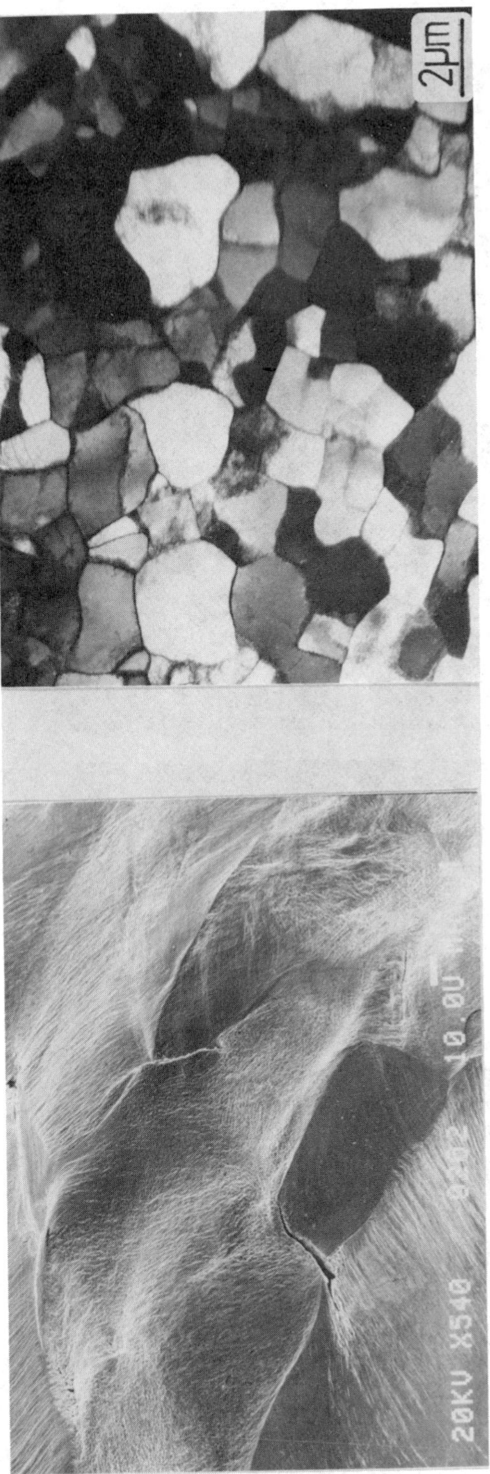

FIG. 4—*Diffuse slip associated with a dislocation cell structure in the unaged N-alloy fatigued at 190 MPa.*

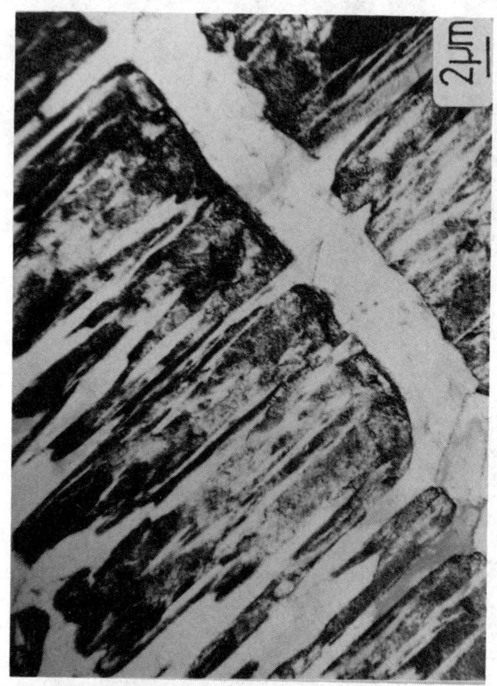

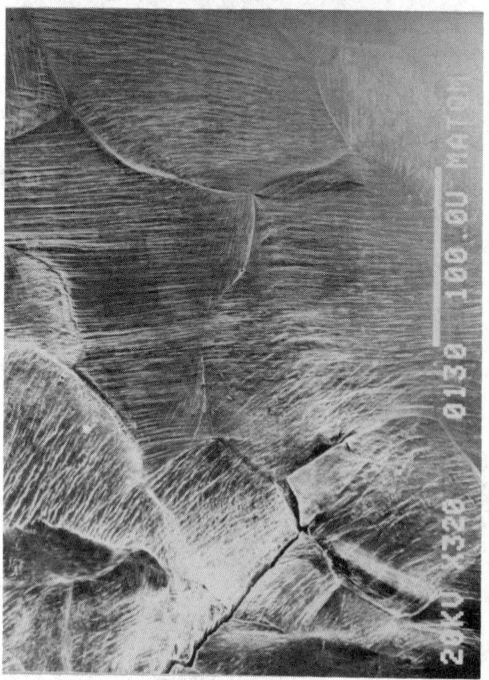

FIG. 5—Straight and sharp slip bands associated with channels in a "vein" matrix structure. Unaged N-alloy fatigued between 130 and 115 MPa.

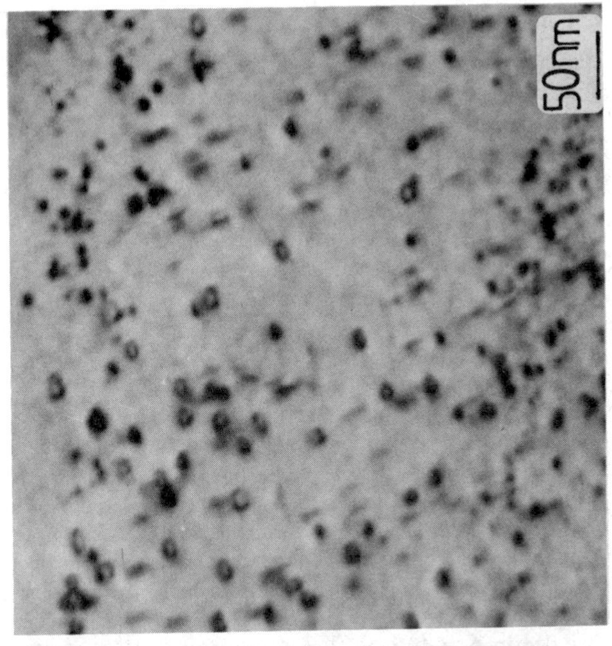

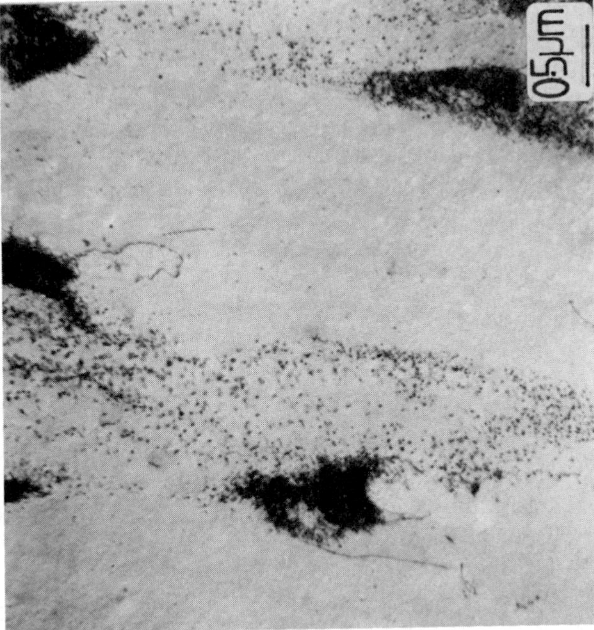

FIG. 6—*Precipitates of the metastable α" $Fe_{16}N_2$ formed by dynamic aging during fatigue testing at the lowest stress amplitudes.*

firmed by the fact that upon static aging of the fatigued alloy precipitation occurs inside the channels.

Fatigue Properties of Fatigued and Aged N-alloy

Specimens of the furnace cooled N-alloy were fatigued for 500 cycles at various stress amplitudes and then aged for 16 h at 60°C. The samples were retested immediately or subjected to a second fatigue and aging sequence. The influence of this fatigue aging treatment is displayed by the *S-N* curves of Fig. 1. A large increase in fatigue life has been achieved, especially at the high strain amplitudes and after a two-step fatigue aging treatment. This is less effective at the lower strain amplitudes, suggesting that the fatigue limit of the alloy is relatively unchanged.

Fatigue aging has reduced drastically the plastic strain amplitude at each stress level as a result of the increase in yield stress (Fig. 3). At the stress $\sigma = 180$ MPa, ϵ_p is reduced by a factor of ten with a single fatigue aging step. The evolution of the plastic strain amplitude during testing of the fatigue aged alloy is shown in Fig. 7. Since the plastic strain amplitude remains relatively unchanged on retesting, it can be concluded that the age strengthening is effective up to specimen failure.

The fatigue data are represented in terms of the Coffin-Manson relationship in Fig. 3, where the reduction in ϵ_p and increase in N_f brought about by the fatigue aging treatment is shown by arrows. One can see from these plots and Table 2 that the fatigue ductility coefficient has been reduced by a factor of 4.2 with a single-step treatment. A two-step treatment reduced ϵ_f even further. This can be taken as evidence for the embrittling effect of strain aging. The TEM observations of the fatigue aged specimens showed that the dislocation substructures developed during the first 500 cycles of testing (ranging from cells at high stresses to isolated entanglements at low stresses) were heavily decorated with α'' $Fe_{16}N_2$ precipitates. In addition, the matrix contains a particle density $\delta = 10^{13}$ cm^{-3} of α'' precipitates of about 140 nm diameter, which is the same at all the predeformation levels (Fig. 8). This means that differences in the degree of fatigue age strengthening as measured by the reduction in ϵ_p is due to differences in the "locked" dislocation substructure. The effect of repeated fatigue aging is that of increasing the volume fraction and size of the precipitates by about 30%.

Examination of the failed specimen surfaces showed that sharp and coarse slip bands are characteristic of the fatigue-aged alloy (Fig. 9). This is evidence of concentrated slip activity which contrasts with the more homogeneous deformation of the unaged alloy. The higher plas-

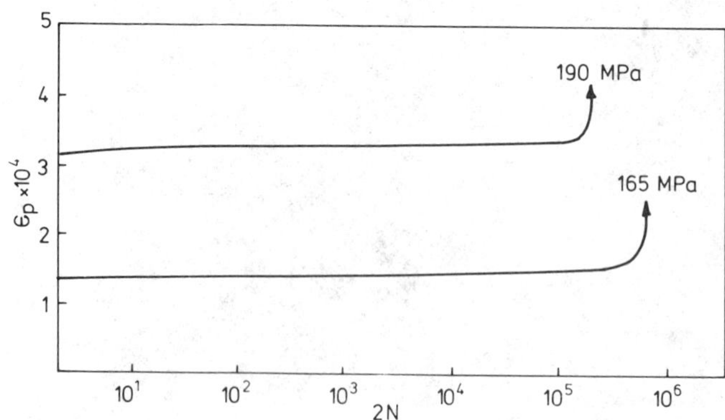

FIG. 7—*Variation of the plastic strain amplitude with number of cycles on retesting the single-step fatigue aged N-alloy at two stress amplitudes.*

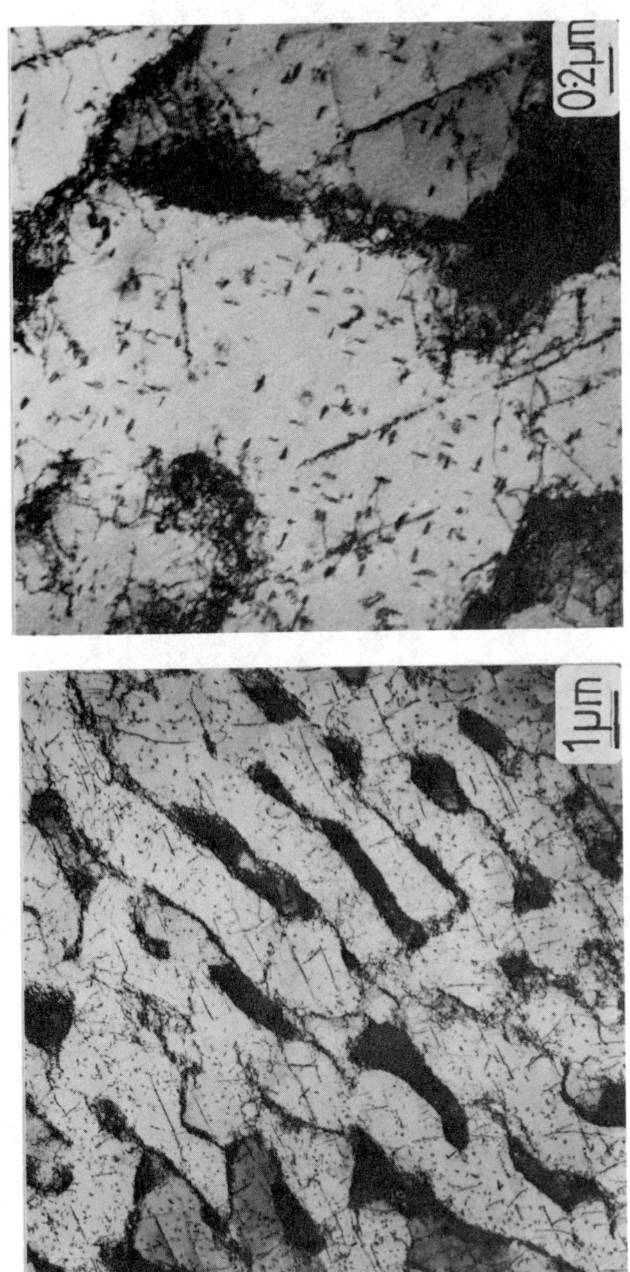

FIG. 8—Precipitates of α″, $Fe_{16}N_2$, formed after 500 cycles fatigue at 165 MPa and aging for 16 h at 60°C.

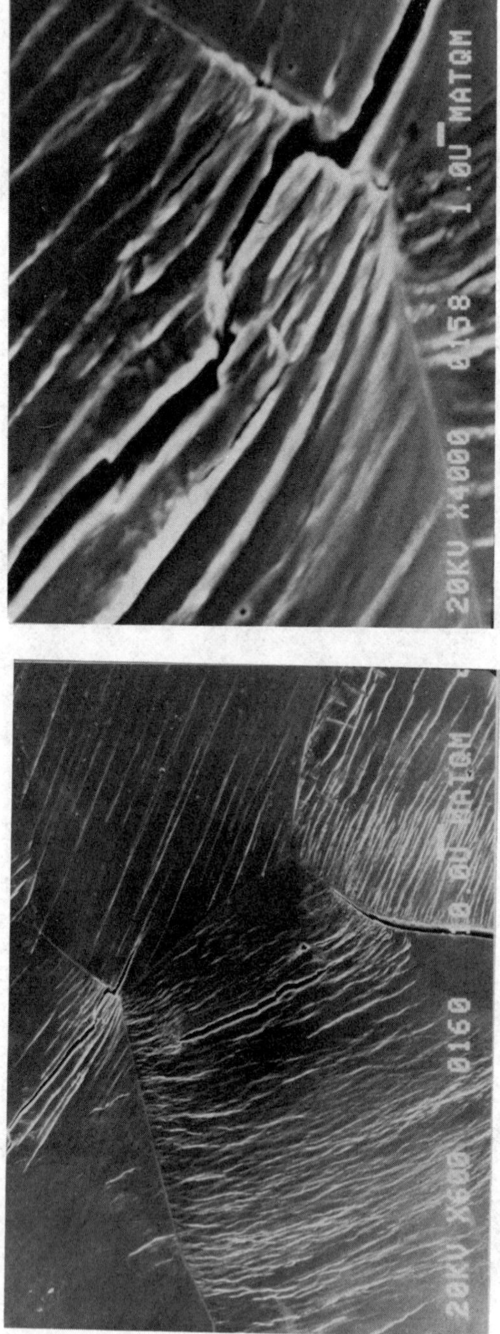

FIG. 9—*Slip bands and crack nucleation sites in a fatigue aged N-alloy fatigued to failure at 165 MPa.*

tic strain amplitude during the first 500 cycles of predeformation at the higher amplitudes had produced some surface deformation in the form of fine slip traces. The coarse slip bands which appear on retesting after aging follow the same path as those slip traces, suggesting that they occur by the re-establishment of localized slip activity in the favorably orientated slip planes. This is confirmed by TEM observations of longitudinal sections of failed specimens which show that localized precipitate shearing occurred within channels which correspond to surface slip bands.

Fatigue deformation of the fatigue aged alloy is rather inhomogeneous, with the extent of deformation varying from grain to grain, the percentage of deformed grains increasing with stress amplitude, and there being no evidence of grain shape changes. Fatigue failure of the aged specimens was found to take place by both grain boundary and slip band cracking (Fig. 9).

Fatigue Behavior of Unaged C-alloy

The *S-N* curves of both the furnace cooled and quenched C-alloys are shown in Fig. 10. The influence of the high concentration of dissolved C in the quenched alloy is that of increasing the fatigue life at all stress levels as a result of the solution strengthening effect of C+N and not simply of dynamic aging which would only affect the high-cycle fatigue life. The *S-N* curve of the quenched alloy is much flatter than that of the furnace cooled, which displays a more distinct "knee" near the fatigue limit, as in many ferrous alloys [*18,19*]. The flat appearance of the *S-N* curve of the quenched alloy is believed to be due to the dynamic strain aging which occurs at the intermediate and low stress levels. This is also supported by the evolution of the plastic strain amplitude in both alloys (Figs. 11 and 12). Both alloys showed delayed yielding over all the ranges of stress amplitude investigated, so that the plastic strain amplitude builds up by gradual "plastification" of the grains after an initial "incubation" period, as observed in other normalized plain carbon steels and explained by Veith [*20*] in terms of gradual fatigue assisted diffusion of solute along dislocation lines. This behavior shows that all the stress amplitudes used in the test are below the monotonic yield point of the materials.

In the furnace cooled C-alloy a saturation regime is established immediately after yielding, because there is negligible dynamic strain age hardening in this low interstitial C-alloy. However, in the quenched C-alloy cyclic hardening continues throughout the test after yielding, with

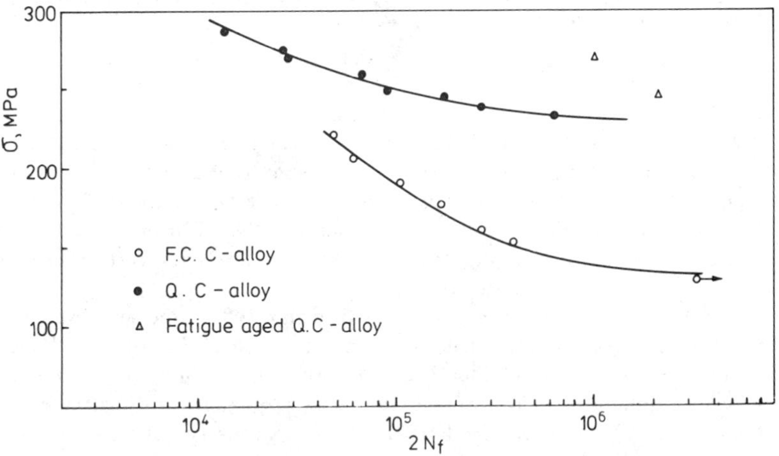

FIG. 10—S-N *curves for the furnace cooled (FC), quenched, and quenched fatigue aged C-alloy.*

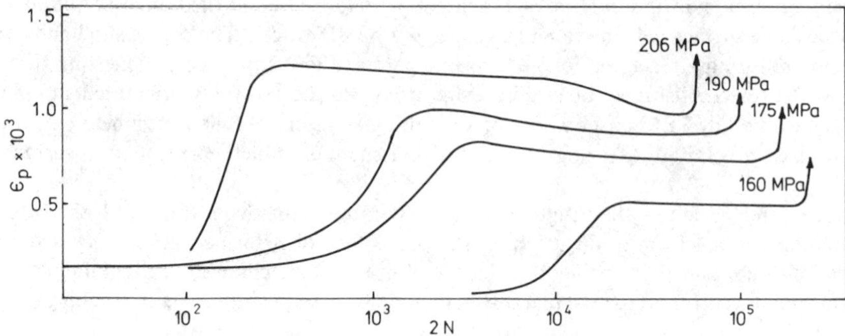

FIG. 11—*Variation of the plastic strain amplitude with number of cycles during fatigue of the furnace cooled C-alloy at different stress amplitudes.*

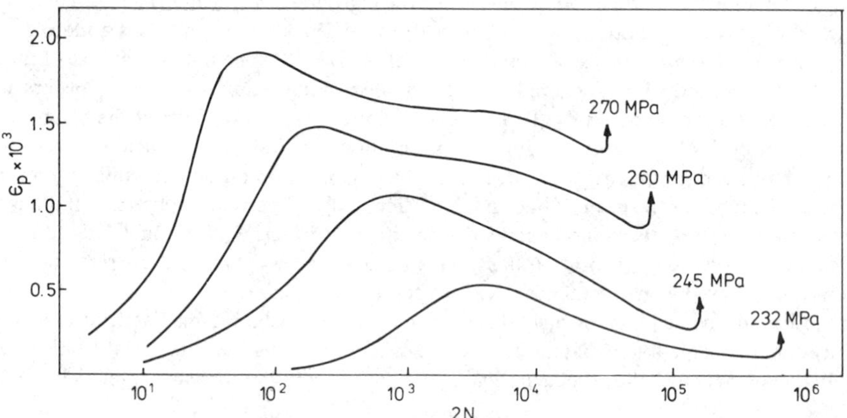

FIG. 12—*Variation of the plastic strain amplitude with number of cycles during fatigue of the quenched C-alloy at different stress amplitudes.*

the greatest reduction in plastic strain being observed at the intermediate stress levels. At the lower stress levels, approaching the fatigue limit, the plastic strain is reduced to less than 10^{-4} due to dynamic strain aging which in this case is more effective than in the N-alloy. The occurrence of dynamic strain aging, which is also confirmed by the TEM observations described below, reduces the plastic strain amplitude to non-damaging levels and is the cause of the fatigue limit in the interstitially strengthened alloy.

The average values of ϵ_p at each stress amplitude lie within 10% of the actual values for over more than 95% of the fatigue life; they have been plotted against $2N_f$ in full logarithmic coordinates in Fig. 13. The data can be fitted to a Coffin-Manson relationship showing that at equivalent values of N_f the cumulative plastic strain to failure, ϵ_p, is greater in the furnace cooled alloy, except at the high amplitudes where the data for both alloys converge. It seems that the divergence at low stress amplitudes can be attributed to dynamic aging which reduces the ductility.

Surface observations on the failed specimens show that there are clear differences in the slip

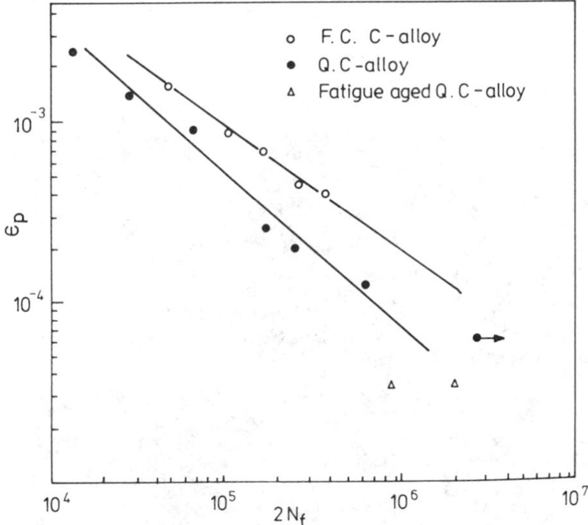

FIG. 13—*Coffin-Manson plot of the fatigue data for both the furnace cooled and quenched C-alloy. Two data points for the quenched fatigue aged alloy are also shown.*

traces of the furnace cooled and the quenched C-alloy which arise from the evolution of different dislocation substructures. Following fatigue at all amplitudes, the furnace cooled alloy develops diffuse slip lines covering all the grains, with widespread surface rumpling and incompatible grain deformation at the high stress amplitudes, which produce grain boundary cracks. The associated dislocation substructure is that of cells with an average diameter of 2.5 μm.

The slip lines in the quenched C-alloy fatigued at high stress amplitudes have also a somewhat diffuse appearance, but when dynamic aging occurs, at intermediate and low amplitudes, slip activity tends to be concentrated in coarse slip bands similar to those generally observed in fatigued mild steel [*4*] and which become the sites for crack nucleation (Fig. 14). By examining etched sections taken at various depths from the specimen surface it can be seen that the diffuse slip markings are associated with the formation of a fine dislocation (sometimes elongated) cell structure in the bulk and that the coarse slip bands correlate with dislocation free channels that have formed within a dense vein structure (Fig. 15).

A notable difference between the furnace cooled and the quenched C-alloy is that in the latter a much higher dislocation density is stored in the sample at all amplitudes, and that a dense dislocation cloud or vein structure develops in the quenched alloy when dynamic fatigue aging occurs. This dislocation vein substructure persists down to the stress amplitude approaching the fatigue limit. Specimens which were fatigued to failure in this stress range were examined by TEM immediately after testing; it was established that for fatigue lifes greater than 10^6 cycles (3 days) the dynamic aging had proceeded to the extent of formation of small precipitates of ϵ carbide.

Fatigue Behavior of Fatigue Aged C-alloy

Specimens of the quenched C-alloy were fatigued at stress amplitudes of 270 and 245 MPa until they had fully yielded as measured by the attainment of maximum plastic strain amplitude. This relatively high cycle predeformation is necessary because plastic deformation is re-

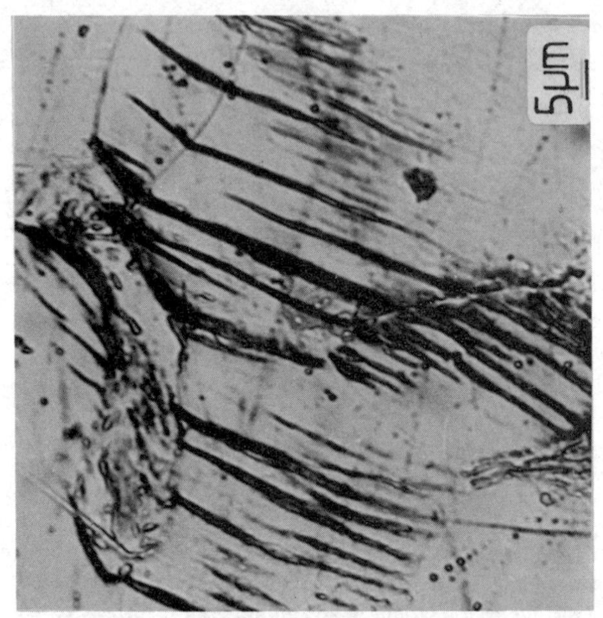

FIG. 14—*Coarse slip bands formed in quenched C-alloy that has undergone dynamic strain aging in fatigue at 245 MPa.*

quired for strain aging to develop. The specimens were subsequently aged for 16 h at 60°C and retested at the same stress levels. The number of cycles to failure are plotted in Fig. 10 showing that the fatigue life has been increased by a factor of 33 and 12 at the higher and lower amplitudes, respectively. This represents a more effective increase in N_f than achieved in the N-alloy, and microstructural observations suggest that this is due to the precipitation of a larger volume fraction of fine metastable ϵ carbide particles. The trend is again that of greater fatigue aging effectiveness at the higher amplitudes.

On retesting the fatigue-aged alloy, the plastic strain amplitude is found to have decreased to less than 3×10^{-5}, which is the detection limit of the measuring device, without any measurable increase during the fatigue life. These reductions in ϵ_p are also far greater than those measured in the fatigue-aged N-alloy. The two data points entered in the Coffin-Manson plot of Fig. 13 are for reference only and, if anything, they reveal a reduction in the ductility of the alloy.

Similarly to the N-alloy, the fatigue-aging treatment produces disk-shaped precipitates of metastable ϵ carbide (of approximately 35 nm diameter) both within the matrix and at dislocation lines (Fig. 16). The density of precipitate particles is about 1.3×10^{-15} cm^{-3}, and this seems to be independent of frequency and strain amplitude [*13*]. The TEM examination of the fatigue-aged alloy tested to failure at 245 MPa shows that most of the grains contain dislocation clusters similar to those observed at equivalent levels of plastic strain in the N-alloy. Although the fatigue-aging has restricted very effectively deformation inside the grains, where there is no evidence of fresh slip activity, localized redissolution of precipitate can be seen close to the grain boundaries where the deformation is concentrated.

The initial fatigue predeformation produced faint and fine slip traces far less visible than in the N-alloy. After retesting the fatigue-aged alloy to failure, no signs of widespread slip activity was detected, the limited deformation observed being confined to the grain boundaries and adjoining the pearlite regions (Fig. 17). This localized deformation takes the form of diffuse slip and coarse slip bands extending short distances from these boundaries. The grain interiors appear undeformed, and cracks can be seen along the slip bands at the ferrite-pearlite interface and along grain boundaries.

Discussion and Conclusions

The effect of dynamic strain aging upon room temperature fatigue behavior of interstitial iron alloys has been clearly illustrated by the experiments described above. It has been possible to follow the dynamic aging sequence by the gradual and continuous reduction of plastic strain amplitude during the test which reaches non-damaging levels at stresses approaching the fatigue limit. At these low stress levels the aging can proceed to the stage of formation of metastable nitride, or carbide, particles at both matrix and dislocations. The restricted mobility of dislocations promotes the formation of clouds or veins with high dislocation densities and dislocation free channels where the slip activity is concentrated and which produce the typical coarse slip bands observed often in fatigued mild steel. Dynamic strain aging is impossible during fatigue at high stress or strain amplitudes because high amplitude dislocation motion prevents the formation of fine particle dispersions and the fatigue life is too short for effective aging to develop. The prolongation of the fatigue life at the intermediate and low stress amplitudes due to the dynamic aging is responsible for the flat appearance of the *S-N* curve.

By means of fatigue predeformation and low temperature static aging it has been possible to produce both a matrix and a dislocation substructure, strengthened by the precipitation of fine metastable nitrides or carbides, which is stable against further fatigue deformation at all stress levels. This fatigue aging treatment can produce a very large increase in fatigue life, even at the high stress amplitudes, so that it has the effect of reintroducing a knee in the *S-N* curve. In fact, the relative increase in fatigue life by fatigue aging is less effective at low stress amplitudes

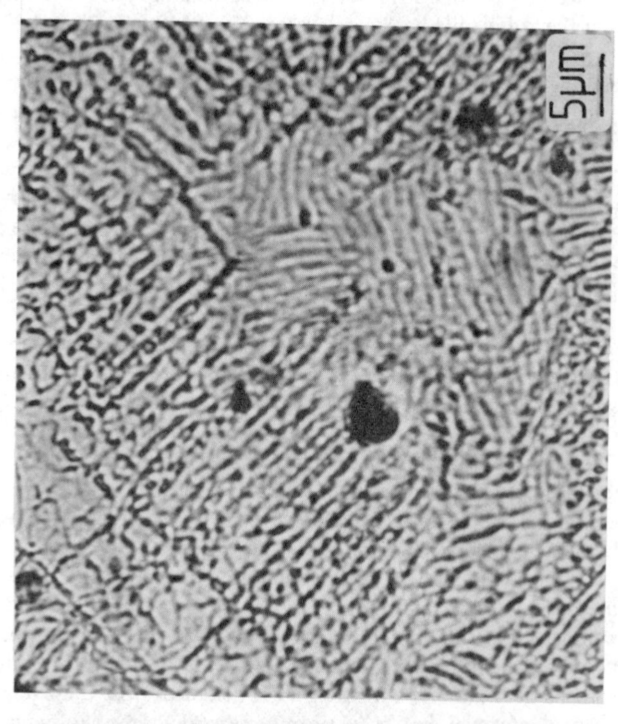
5μm

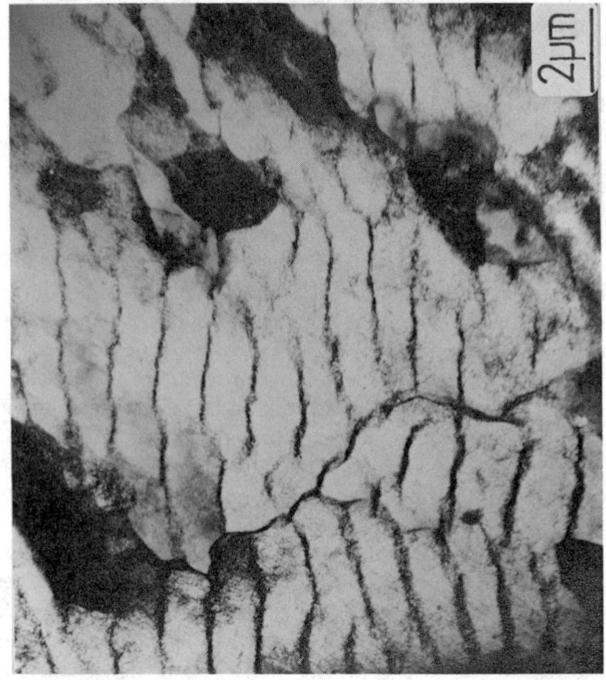
2μm

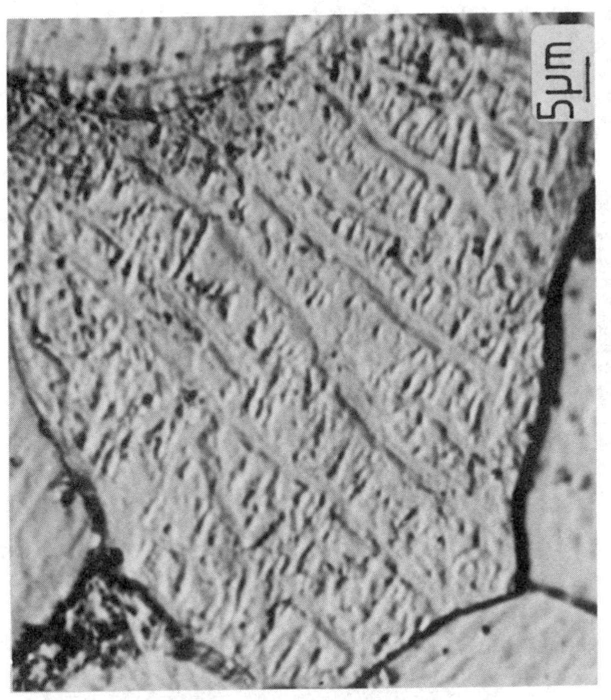

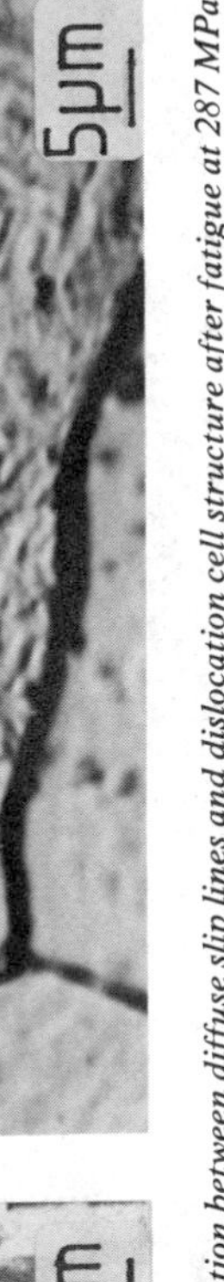
FIG. 15—*Etched sub-surface and TEM micrographs.* (top) *Correlation between diffuse slip lines and dislocation cell structure after fatigue at 287 MPa.* (bottom) *Correlation between coarse slip bands and dislocation free channels after fatigue at 245 MPa. Quenched C-alloy.*

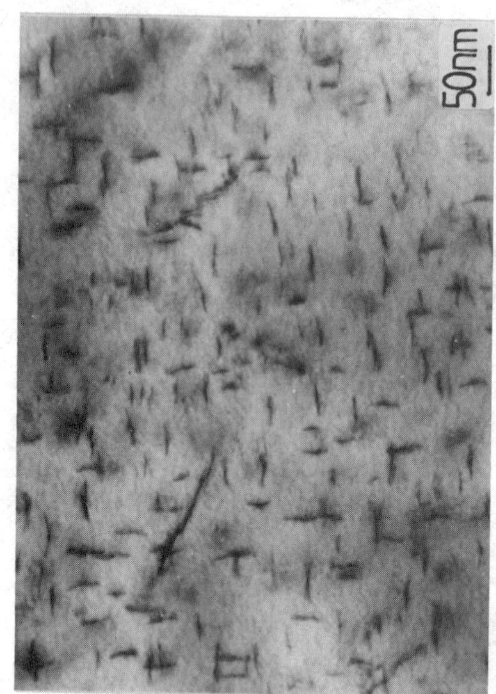

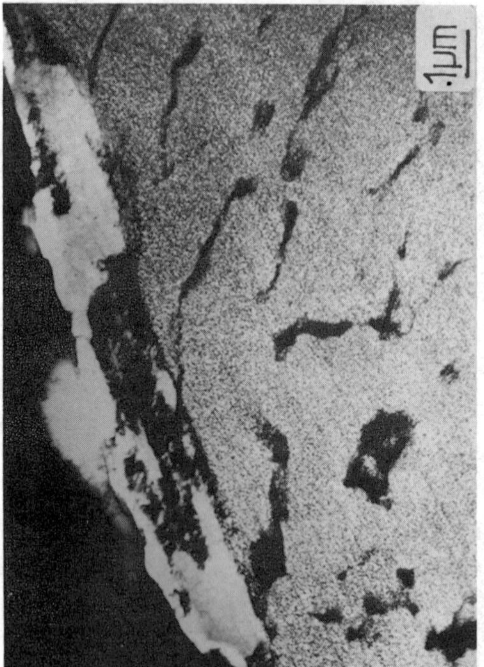

FIG. 16—*Fine precipitates of the metastable ε carbide formed on fatigue aging the quenched C-alloy.*

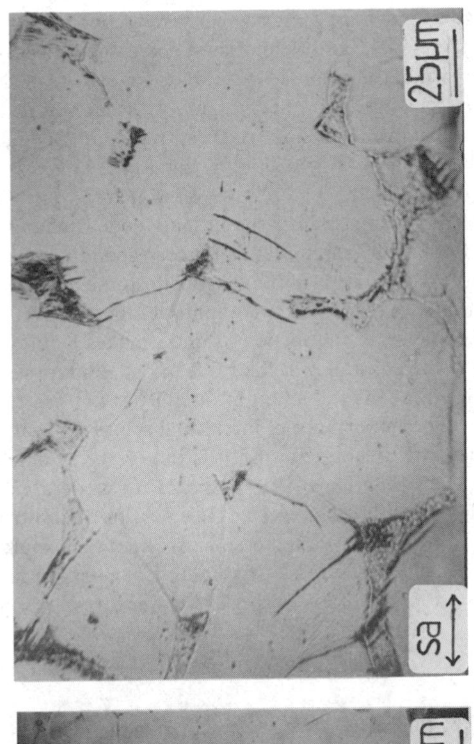

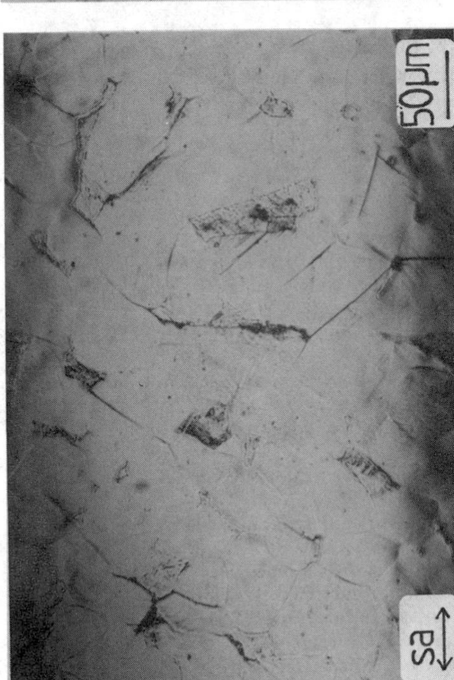

FIG. 17—*Deformation confined at the grain boundary regions after fatigue failure at 270 MPa (left) and 245 MPa (right) in the quenched and fatigue-aged C-alloy.*

because near the fatigue limit the reduction in plastic strain amplitude due to dynamic aging is as important as the reduction achieved by a fatigue aging treatment.

It is obvious that the fatigue aging has produced a large reduction in plastic strain amplitude, when retesting at the same stress level, and a significant decrease in the fatigue ductility coefficient of the alloys. However, when the cumulative strain to failure, ϵ_c, of the unaged N-alloy is compared with that of the fatigue aged alloy at the same stress amplitude (Table 2) it is found that it has remained nearly unchanged in spite of the large reduction in ductility. It seems clear, therefore, that the beneficial effect of fatigue aging is due to the increase in strength, or yield stress, and the ensuing reduction in strain amplitude, ϵ_p, which far outweighs the detrimental ductility loss. As a result of this decrease in ϵ_p the fatigue mode is changed from one of high strain amplitude to one of low strain amplitude with a corresponding change in the modes of crack nucleation. Fatigue cracks form early in the fatigue life at the high strain amplitudes, very often at grain boundaries, whilst at the low strain amplitudes most of the fatigue life is spent in nucleating the cracks. This change in the mode of failure makes it difficult for any physical significance to be attached to the constancy in the value of ϵ_c upon fatigue aging which should simply be taken as indication that the reduction in ductility is not greatly detrimental to the overall fatigue performance. The importance of interstitial solute concentration on the fatigue aging capability has also been demonstrated, and it is important to note that in this respect nitrogen is a more effective age strengthener than carbon after moderate cooling rates. This is undoubtedly due to the higher solubility of N in α-Fe. It is possible that nitrogen in solution may be the element responsible for the observed increase in fatigue life in some ferrous alloys after "resting periods" [2,4]. The beneficial effect of retaining free nitrogen in steels when fatigue properties are considered is a topic perhaps worthy of further investigation.

Acknowledgments

This work was supported by the SERC and the AERE, Harwell. The authors are indebted to Dr. B. C. Edwards, Harwell, for his continued interest, assistance, and monitoring of the research project.

References

[1] Baird, J. D., *Metallurgical Reviews*, Vol. 16, 1971, p. 1.
[2] Daeves, K., Gerold, E., and Schultz, E. H., *Stahl und Eisen*, Vol. 60, 1940, p. 100.
[3] Kuzmanovic, B. O. and Williams, N., *Engineering Fracture Mechanics*, Vol. 4, 1972, p. 687.
[4] Wilson, D. V., *Metal Science*, Vol. 11, 1977, p. 321.
[5] Sinclair, G. M. in *Proceedings of ASTM*, American Society for Testing and Materials, Philadelphia, Vol. 52, 1952, p. 743.
[6] Levy, J. C. and Kanitkar, S. L., *Journal of the Iron and Steel Institute*, Vol. 195, 1961, p. 296.
[7] Oates, G. and Wilson, D. V., *Acta Metallurgica*, Vol. 12, 1964, p. 21.
[8] Klesnil, M., Holtzman, M., Lukas, P., and Rhys, P., *Journal of the Iron and Steel Institute*, Vol. 203, 1965, p. 47.
[9] Wilson, D. V., *Philosophical Magazine*, Vol. 22, 1970, p. 643.
[10] Levy, J. C., *Metallurgia*, Vol. 56, 1957, p. 71.
[11] Mintz, B. and Wilson, D. V., *Acta Metallurgica*, Vol. 13, 1965, p. 947.
[12] Pickard, S. M., Blackie, A., Guiu, F., and Edwards, B. C. in *Proceedings*, 4th RISØ International Symposium on Metallurgy and Materials Science, Denmark, 1983, p. 485.
[13] Pickard, S. M. and Guiu, F., to be published.
[14] Wilson, D. V. and Mintz, B., *Acta Metallurgica*, Vol. 20, 1972, p. 985.
[15] Thomas, W. R. and Leak, R. G., *Proceedings of the Physical Society*, Vol. B68, 1955, p. 1001.
[16] Wilson, D. V. and Tromans, J. K., *Acta Metallurgica*, Vol. 18, 1970, p. 1197.
[17] Guiu, F., Dulniak, R., and Edwards, B. C., *Fatigue of Engineering Materials and Structures*, Vol. 5, 1982, p. 311.

[*18*] Thomson, N. and Wandsworth, N. J., *Advances in Physics*, Vol. 7, 1970, p. 117.
[*19*] Yoshikawa, A. and Sugeno, T., *Transactions of the Metallurgical Society of AIME*, Vol. 233, 1965, p. 1314.
[*20*] Veith, H. in *Proceedings*, 4th RISØ International Symposium on Metallurgy and Materials Science, Denmark, 1983, p. 563.

Suzanne Degallaix,[1] *Gérard Degallaix,*[1,2] *and Jacques Foct*[1]

Influence of Nitrogen Solutes and Precipitates on Low Cycle Fatigue of 316L Stainless Steels

REFERENCE: Degallaix, S., Degallaix, G., and Foct, J., **"Influence of Nitrogen Solutes and Precipitates on Low Cycle Fatigue of 316L Stainless Steels,"** *Low Cycle Fatigue, ASTM STP 942,* H. D. Solomon, G. R. Halford, L. R. Kaisand, and B. N. Leis, Eds., American Society for Testing and Materials, Philadelphia, 1988, pp. 798–811.

ABSTRACT: Improvement of monotonic and cyclic mechanical properties of 316 steels by nitrogen is studied for different nitrogen contents between 0.03 and 0.25 wt% at 20 and 600°C. At 20°C, yield and ultimate tensile stress increase almost linearly with nitrogen content to the detriment of ductility. Low-cycle fatigue (LCF) life also increases linearly with nitrogen solutes until 0.12 wt% nitrogen where a saturation of this effect occurs. This improvement is attributed to a more homogeneous distribution and to a better reversibility of plastic strain.

The influence of aging treatment at 600°C is studied on tensile and LCF properties at 20 and 600°C. An LCF life decrease with aging is more noticeable when nitrogen content is high. Furthermore, the higher the strain range, the more marked is the decrease in life at room temperature; this tendency is inverted at 600°C.

At 20°C, stress evolution during the hardening-softening stage is virtually independent of aging time but strongly dependent on nitrogen content. For high strain ranges at 600°C and high nitrogen content, the stress increases continuously until rupture, due to significant intergranular precipitation and embrittlement.

KEY WORDS: austenitic stainless steels, nitrogen steels, low-cycle fatigue, aging, high temperature

It is well known that nitrogen in solid solution can considerably improve the low mechanical characteristics of austenitic stainless steels (yield strength, ultimate tensile strength) at room temperature or high temperature [*1–3*]. Nitrogen addition, unlike carbon which induces intergranular corrosion, has no detrimental effect on the high resistance to corrosion of AISI 316L steels (0.02 to 0.03 wt% C). Moreover nitrogen, which is a strong γ-austenite stabilizer, delays the precipitation of chromium carbides and intermetallic compounds [*4–6*].

Our work aims to show: (1) the influence of interstitial nitrogen content on the low-cycle fatigue (LCF) behavior at 20°C of 316L and 316LN steels (the investigated composition range extends from 0.03 to 0.25 wt% N); and (2) the influence of a previous aging treatment at 600°C on LCF behavior at 20 and 600°C of two 316L steels containing 0.08 and 0.25 wt% N, respectively.

[1]Laboratoire de Métallurgie - Physique, C6, Université de Lille, 59655 Villeneuve d'Ascq Cedex, France.
[2]Institut Industriel du Nord, B.P. 48, 59651 Villeneuve d'Ascq Cedex, France.

Experimental Methods

The compositions and grain sizes of the steels are given in Table 1. Before machining, the four steels were solution-treated for 1 h at 1100°C and then water-quenched in order to have both carbon and nitrogen in solid solution. Some specimens of Steels B and D were also aged at 600°C for 2000 and 10 000 h.

Tension and LCF tests were carried out on a servohydraulic testing machine equipped with a resistance furnace. Room-temperature specimens were cylindrical and button-headed. High-temperature specimens were cylindrical, round-ridged, and threaded (Fig. 1). An axial strain-gage extensometer is used at room temperature and a LVDT extensometer at high temperature.

Tension tests were carried out at a total strain rate of $\dot{\epsilon}_t = 4 \cdot 10^{-3}\ s^{-1}$. The LCF tests were performed under total axial strain control with a triangular fully reversed waveshape ($\dot{\epsilon}_t$ =

TABLE 1—*Material data.*

Steel	Weight Percent of						Grain Size, μm
	N	C	Mn	Ni	Cr	Mo	
A	0.029	0.020	1.66	11.50	16.90	2.02	45
B	0.080	0.026	1.54	11.54	17.10	2.19	135
C	0.120	0.014	1.72	13.13	18.84	3.35	110
D	0.250	0.024	1.59	12.98	17.01	2.62	100

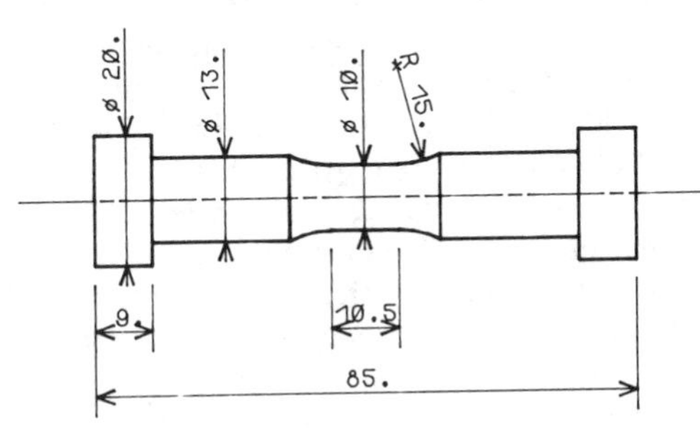

a – Room temperature

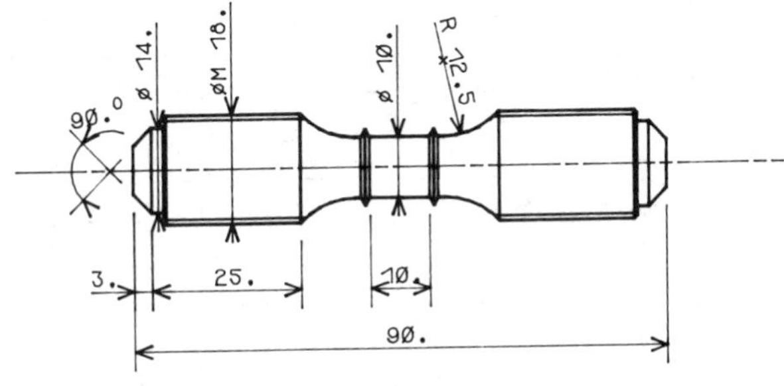

b – High temperature

FIG. 1—*Specimen configurations (dimensions in millimetres).*

$4 \cdot 10^{-3}$ s^{-1}). Total strain ranges ($\Delta\epsilon_t$) of $6 \cdot 10^{-3}$, 10^{-2}, 1.6 10^{-2}, and 2.5 10^{-2} were chosen at 20°C, and $6 \cdot 10^{-3}$, 10^{-2}, and 1.6 10^{-2} at 600°C (two tests for each level).

The tests carried out on the four unaged (UA) steels and on Steels B and D aged for 2000 h (A) and 10 000 h (LA) at 600°C are described in Table 2. More details of experimental results of the tension and LCF tests at room temperature on the unaged steels have been reported elsewhere [*7,8*]. The tensile and fatigue fracture surfaces were examined by a scanning electron microscope (SEM) for Steels B and D in the three structural states.

Experimental Results

Mechanical Behavior

Tensile Properties—The variation of tensile characteristics with interstitial nitrogen content is given in Fig. 2. The 0.2% yield stress (YS) and the ultimate tensile stress (UTS) show a quasi-linear increase with the interstitial content. This is to the detriment of ductility and is consistent with a hardening by solid interstitial solution.

Figure 3 shows the variation of YS, UTS, and elongation of Steels B and D with aging time at 20°C and 600°C, respectively. Aging treatment increases yield stress and has little effect on ultimate tensile stress, but lowers ductility. Age hardening is due to intragranular precipitation of nitrides-carbides-carbonitrides [*9*].

Room-Temperature LCF—We have shown [*7*] that, at room temperature, cyclic stress-strain behavior of the as-quenched 316L-316LN steels is slightly altered by increasing nitrogen content (the stabilized stress amplitude σ_a varies from 410 MPa for Steel A to 435 MPa for Steel D at the plastic strain amplitude of $\Delta\epsilon_p/2 = 0.75\%$), unlike the monotonic stress-strain behavior (stress σ varies from 310 MPa for Steel A to 410 MPa for Steel D at the plastic strain of $\epsilon_p = 0.75\%$). This is explained by the different hardening-softening behaviors of low (<0.10 wt% N) and high nitrogen steels (>0.10 wt% N). At the end of the first half cycle the stress reached increases in the same way as yield stress with increasing nitrogen content. During the following cycles and before stress stabilization, at low nitrogen content, the alloy shows considerable hardening; at higher content, a strong initial hardening (lasting about 20 cycles) followed by a softening is observed. The maximum duration of this hardening-softening stage is 20% of fatigue life. The high nitrogen steel softening is typical of a destabilization of a hardening process due to cyclic strain. This is attributed to dislocation-interstitial interaction mechanisms considering the sensitivity to temperature, strain rate, and interstitial content. The stabilized stress levels thus obtained vary slightly according to nitrogen content.

Such cyclic hardening or hardening and softening behaviors at room temperature are only slightly altered by previous aging at 600°C (Steels B and D); only a small reduction in intensity

TABLE 2—*Test program (A, B, C, D = index of the steels).*

Tests	Test Temperature, °C	States: Water-Quenched Unaged (UA)	Water-Quenched Aged (A)	Water-Quenched Long-Aged (LA)
Tension	20	A, B, C, D	B, D	B, D
	600	B, D	B, D	B, D
Low-cycle fatigue	20	A, B, C, D	B, D	B, D
	600	D only at $\Delta\epsilon_t = 1.6\%$	B, D	B, D

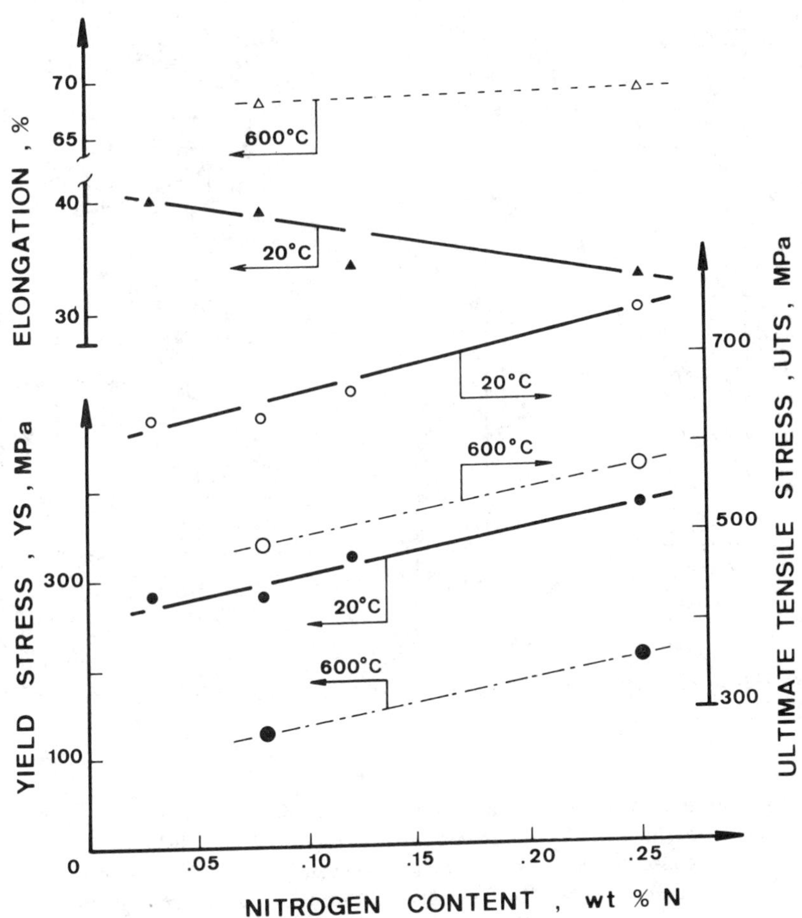

FIG. 2—*Tensile characteristics at 20 and 600°C versus interstitial nitrogen content.*

is observed. The shapes of the cyclic stress-strain curves shown in Fig. 4 are a consequence of these phenomena. The stress level being stabilized before 20% of the fatigue life, the saturation cyclic hysteresis loop is chosen at $0.2N_f$ for tests at room temperature.

The fatigue life of 316L steels at room temperature shows a continuous increase with the nitrogen content, which is more marked when the strain range is high (Fig. 5), according to the relation [*8*]

$$N_f(X) = N_f(0.03) + 3 \times 10^4 \Delta\epsilon_p^{-1.9}(X - 0.03)$$

where X is the weight percent nitrogen content and

$$N_f(0.03) = 1835\Delta\epsilon_p^{-2.39} \qquad (\Delta\epsilon_p \text{ in } \%)$$

is the Manson-Coffin law of Steel A. This relationship is valid between 0.03 and 0.12 wt% N; saturation appears above 0.12 wt% N.

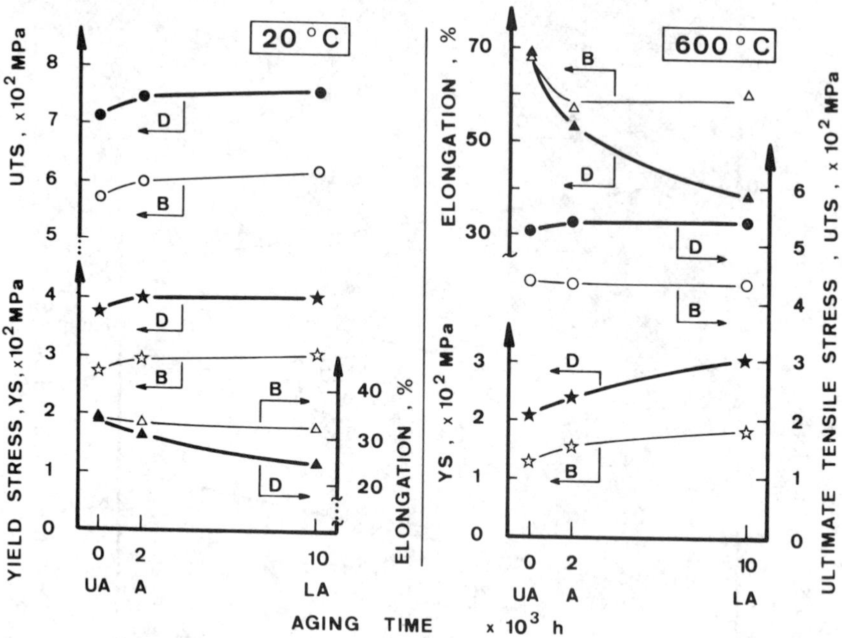

FIG. 3—*Tensile characteristics at 20 and 600°C versus time aging (B and D: index of the steels).*

The fatigue strength increase according to the nitrogen content is attributed to a strain process becoming more planar as the nitrogen content rises [*10,11*]. The dislocation planar arrays are probably the consequence of the stacking fault energy (SFE) decrease due to nitrogen [*12,13*]. This decrease is confirmed by the appearance of deformation twins and intrinsic stacking faults in fatigued unaged Steel D but not in Steel B [*10*]. Aging for 2000 h at 600°C causes no change in fatigue life. On the other hand, an aging treatment of 10 000 h at 600°C tends to reduce the fatigue strength of these steels. This reduction is marked at high strain range; at $\Delta\epsilon_t = 2.5\%$ the reduction factor of the fatigue life is 1.75 for Steel D and 1.45 for Steel B. At low strain range ($\Delta\epsilon_t = 0.6\%$) no further reduction in fatigue life due to this aging treatment is noted (Fig. 6).

High-Temperature LCF—At 600°C, the stress evolution during the fatigue test is strongly dependent on the nitrogen content, especially at high strain range. For Steel B, the stress evolution is only slightly affected by the aging time (Fig. 7). This evolution is characterized by a considerable hardening which increases with the strain range and decreases with the aging time. For Steel D, the stress evolution is greatly affected by the aging time at high strain range ($\Delta\epsilon_t = 1.6\%$) (Fig. 7); for the UA specimens and for one of the 2000 h aged specimens, the strong hardening is followed by a stress stabilization before the appearance of a main crack and fracture. But for the second test of a 2000 h aged specimen and for the two 10 000 h aged specimens, the hardening is continuous and final tensile failure occurs without the formation of a main crack becoming evident by a fall in stress. Stress stabilization does not appear. The reference hysteresis loop at 600°C is then taken at $0.5N_f$. Cyclic and monotonic stress-strain curves at 600°C are given in Fig. 4 for Steels B and D.

The fatigue lives obtained for Steels B and D at 600°C are represented by the curves $\Delta\epsilon_t - N_f$

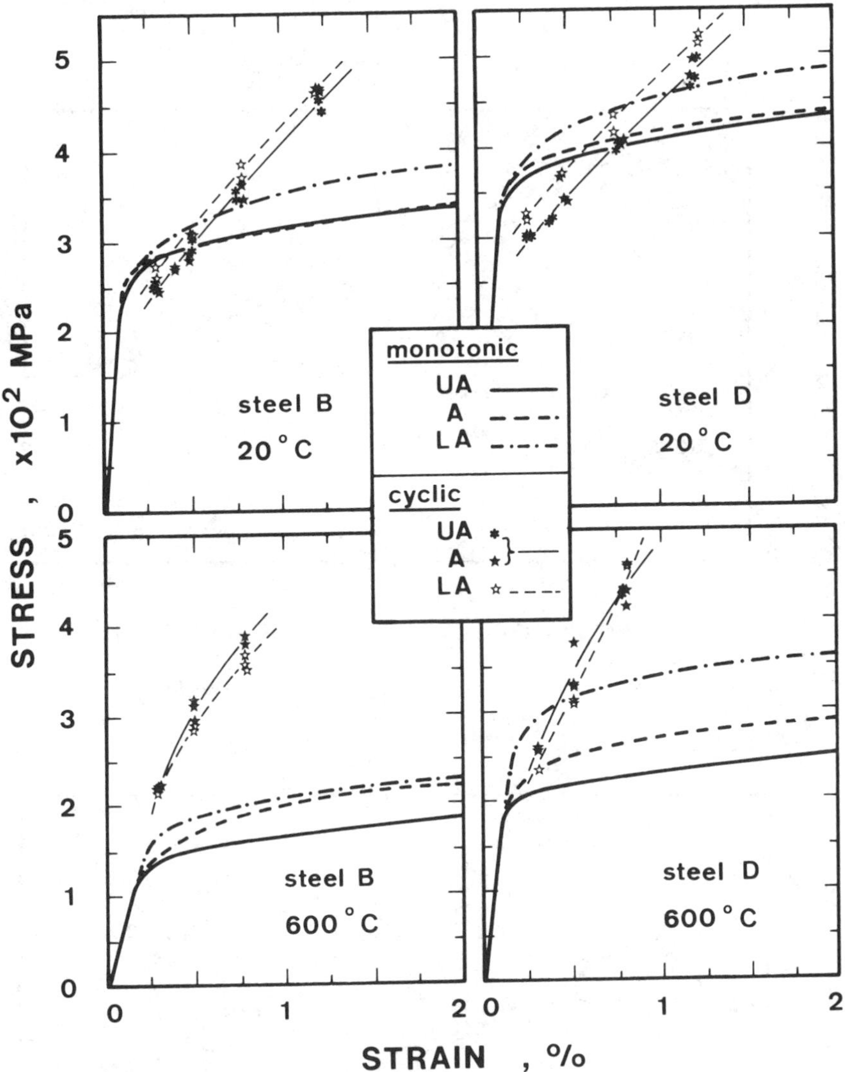

FIG. 4—*Monotonic and cyclic stress-strain curves at 20 and 600°C for steels B and D. At 20°C the reference cycle corresponds to 0.2*N_f*, at 600°C to 0.5*N_f.

in Fig. 6. It can be seen that for both steels LCF resistance is reduced by aging at 600°C. Furthermore, for the same aging time this resistance at high strain range ($\Delta\epsilon_t \geqq 0.8\%$) decreases with nitrogen content but increases at low strain range ($\Delta\epsilon_t < 0.8\%$).

It should be noted that the fatigue life of Steels B and D at 20 and 600°C are not comparable, since the two types of specimens have highly different geometries. In particular, the existence of ridges on the high temperature specimens leads to crack initiation at the base of the ridges. However, this drawback, a consequence of the available extensometers, does not hinder the comparison of fatigued steels at the same temperature.

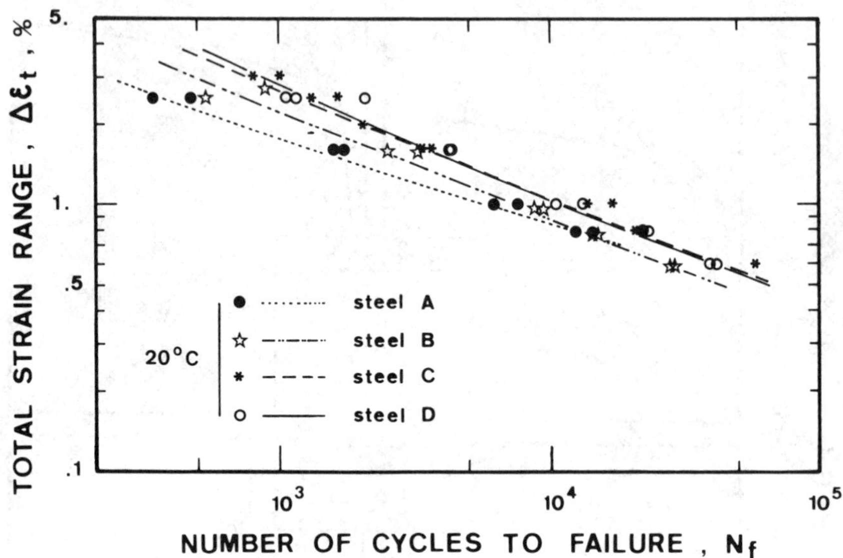

FIG. 5—*Fatigue life curves at 20°C for Steels A, B, C and D.*

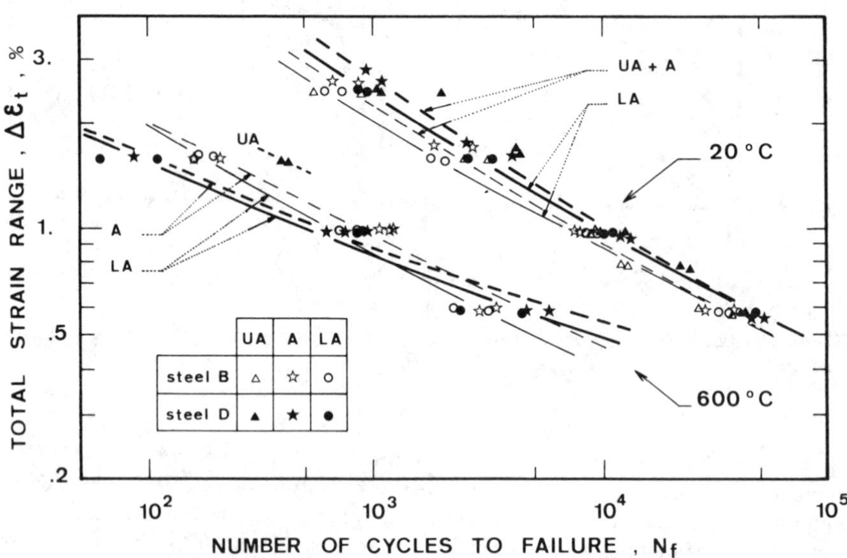

FIG. 6—*Effect of aging treatment on fatigue life curves at 20 and 600°C for Steels B and D.*

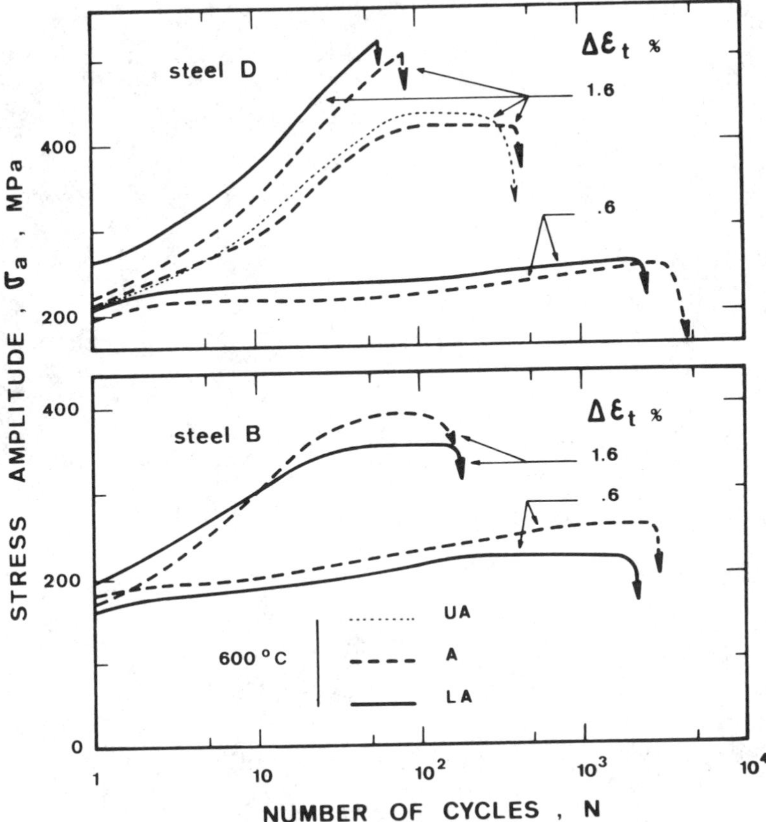

FIG. 7—*Effect of aging treatment on cyclic hardening-softening curves at 600°C for Steels B and D.*

Fractographic Analysis

Figure 8 shows the fracture surfaces of unaged and aged specimens that failed in tension and LCF. This analysis reveals an important intergranular precipitation induced by the aging treatment, leading to intergranular embrittlement of the aged steels. The intergranular embrittlement becomes all the more significant with nitrogen content increase and testing temperature decrease. This applies in tension and LCF (Fig. 9). At room temperature, as well as at 600°C, the tensile fracture surfaces of unaged steels are purely transgranular and ductile with dimples. Intergranular rupture appears after aging and covers a part of the surface, the area of which increases with aging time and decreases with rising temperature (Fig. 9*a*). In LCF, fracture surfaces of both UA and aged steels show a transgranular fracture with striations, followed, at high strain range ($\Delta\epsilon_t \geqq 1.6\%$), by a final pure tension failure. Final tensile fracture surface is ductile with dimples for the quenched steels, while ductile intergranular rupture appears after aging treatment (Fig. 9*b*).

Discussion

The YS and UTS mechanical properties of 316L steels are improved by nitrogen addition in solid interstitial solution and by previous aging at 600°C which is responsible for the precipita-

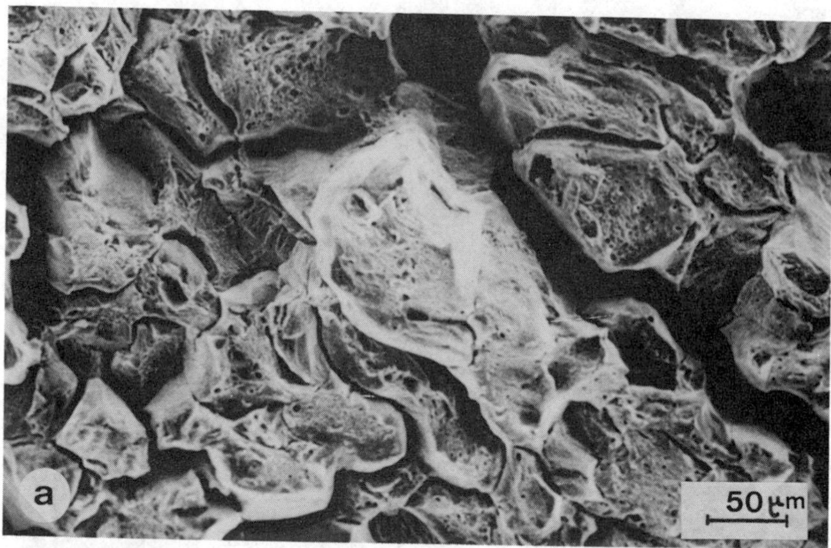

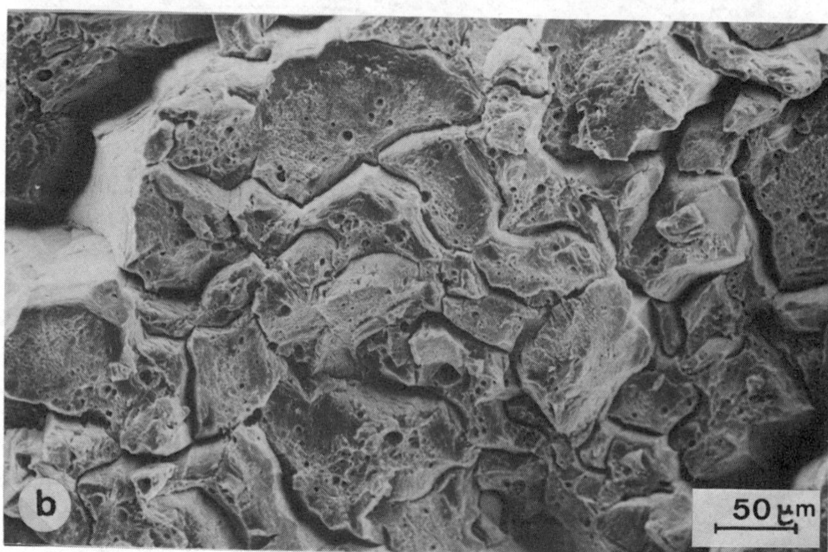

FIG. 8—*Scanning electron micrographs; fracture surfaces of tension specimens and final tensile fracture surfaces of fatigue specimens.* (a) *Steel D, State A, tension, 20°C, intergranular fracture.* (b) *Steel B, State LA, tension, 20°C, intergranular fracture.*

tion of carbides, nitrides, and intermetallic compounds (σ, χ, and η) [5]. The interest in this improvement is unfortunately reduced by a loss of ductility. While the loss of ductility caused by the nitrogen solid solution treatment is largely compensated for by the improvement in YS and UTS, the important loss in ductility caused by previous aging at 600°C is not compensated for by the increase in YS and UTS. The SEM analysis shows such a drop in ductility to be due to intergranular embrittlement of aged steels.

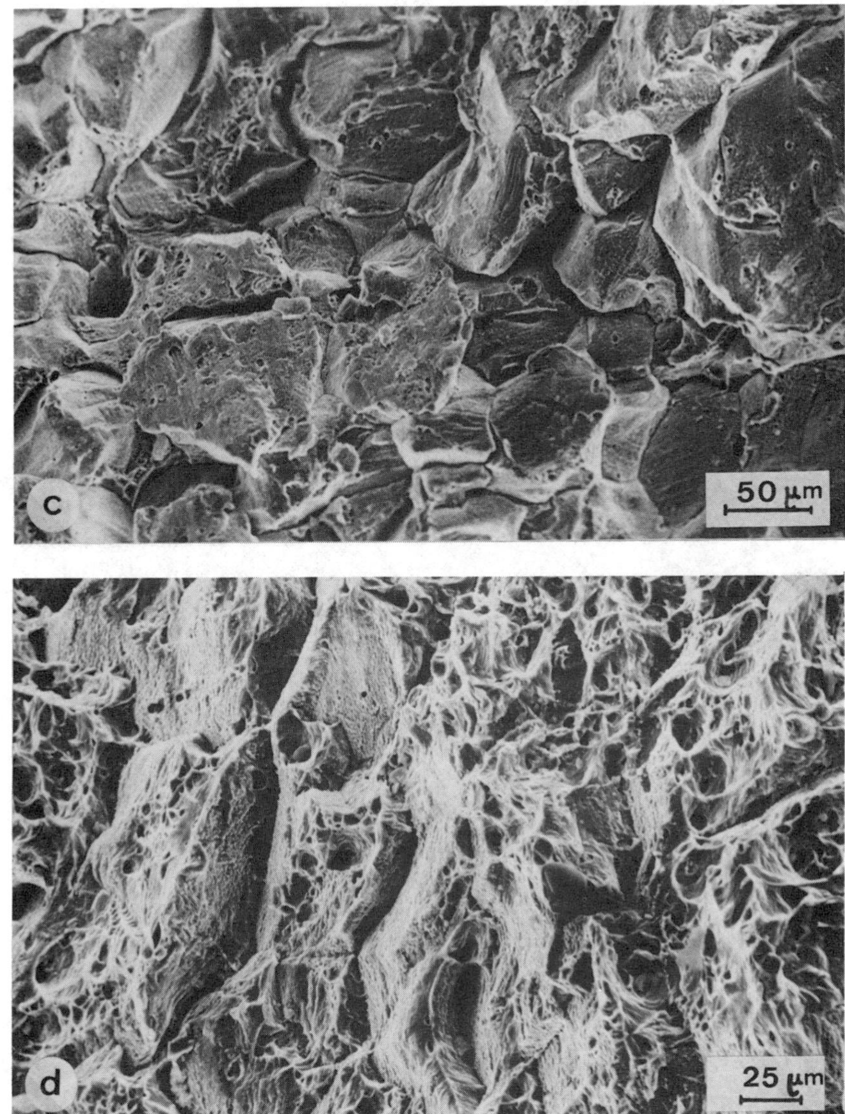

FIG. 8 (cont.)—(c) *Steel D, State A, LCF, 20°C, $\Delta\epsilon_t = 2.5\%$, intergranular fracture.* (d) *Steel B, State LA, LCF, 600°C, $\Delta\epsilon_t = 1.6\%$, mixed rupture (essentially ductile transgranular).*

The LCF resistance of 316L steels at room temperature is clearly improved by the addition of interstitial nitrogen which ensures a more equal distribution of the strain according to two mechanisms:

1. Interstitial-dislocation interactions remain significant, although they are of lesser importance in a face-centered cubic matrix than in a body-centered cubic matrix.
2. A decrease in the stacking fault energy (SFE) consistent with the stability of ϵ Fe $-$ N phases. The reduction of the SFE due to nitrogen is essentially shown by the planar nature of

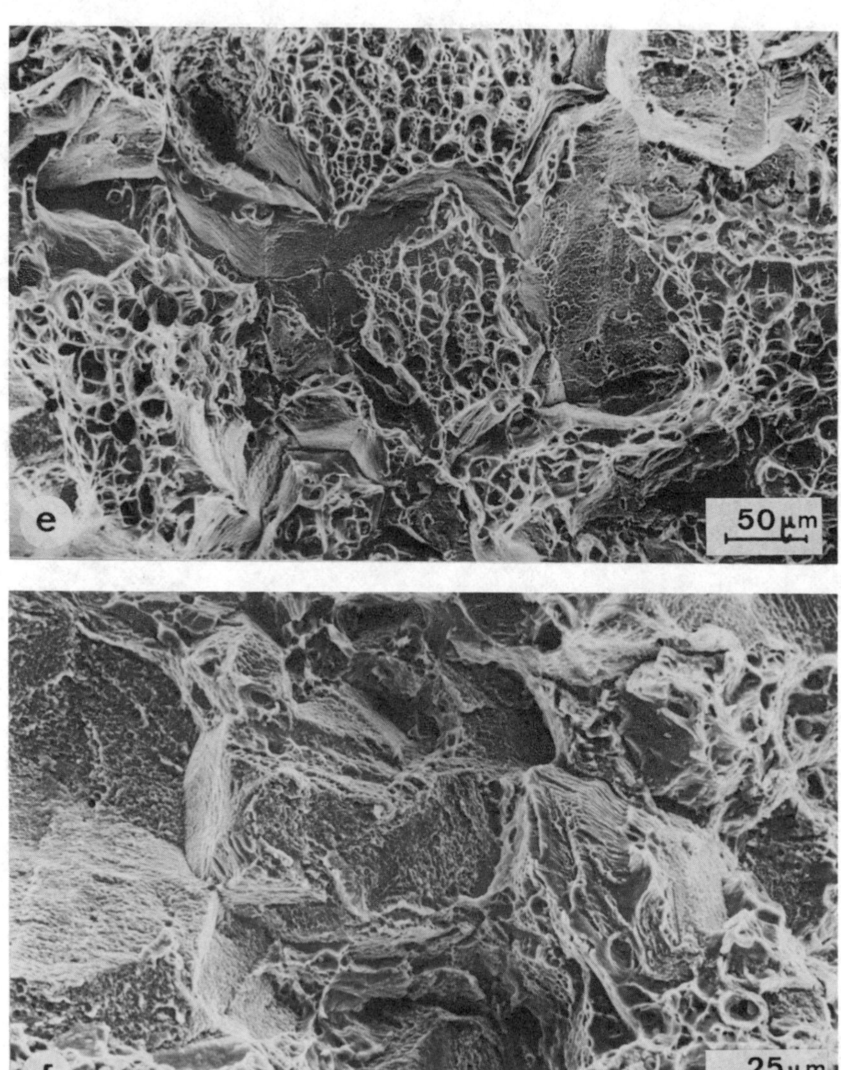

FIG. 8 (cont.)—(e) *Steel D, State A, LCF, 600°C,* $\Delta\epsilon_t$ = *1.6%, mixed rupture (intergranular-ductile transgranular).* (f) *Steel D, State LA, LCF, 600°C,* $\Delta\epsilon_t$ = *1.6%, mixed rupture (essentially intergranular).*

the slip process as seen by transmission electron microscopy in the 0.25 wt% N. This explanation is consistent with Nilsson's results [*14*].

The saturation in fatigue life improvement with nitrogen content can be explained by a nonlinearity of nitrogen effect on the SFE or a nonlinearity of SFE on fatigue life. Below the cross-slip threshold, planar arrays of dislocations are not changed by further lowering of SFE. No additional improvement can be expected.

Previous aging at 600°C involves a reduction in the fatigue life of Steels B and D. At room temperature this loss of resistance only appears beyond an aging for 2000 h. However, this

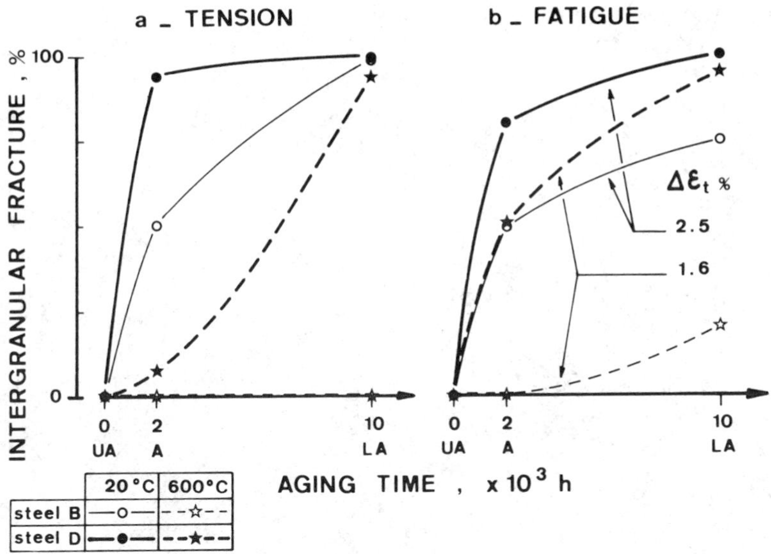

FIG. 9—*Estimation of percentage of intergranular fracture in final tensile fracture surfaces.*

reduction in fatigue life is moderate even after 10 000 h aging. The improvement in fatigue resistance at 20°C due to nitrogen remains noticeable in aged steels. All this is consistent with slow precipitation at 600°C; even after 10 000 h aging, little intragranular precipitation is observed, the nitrogen content in solid solution remains high for Steel D, and the strain process is slightly altered. The similarity between the stress evolution curves at 20°C, before and after aging, and the low reduction in the intensity of transient hardenings and softenings, confirm that there is incomplete nitrogen precipitation in both steels.

We attribute the loss in fatigue resistance after aging, both at 20 and 600°C, to intergranular precipitation (Fig. 10) and to the more rapid appearance of monotonic fracture at the end of the test by the lowering of the fracture toughness K_C. Indeed, a difference is only observed between the final tensile fracture surfaces of the fatigued specimens. It is well known that intergranular fracture proceeds by cavity nucleation by grain boundary sliding (particularly on the precipitates) and at triple points, followed by a growth of these cavities as a result of the applied stress. The presence of precipitates at the grain boundaries, a reduction of the grain boundary cohesion, and the presence of a high stress favor intergranular rupture. Similarly, the lower ductility of the matrix at room temperature makes boundary sliding less easy and favors intergranular rupture. At 600°C, the stress evolution is characterized by a considerable hardening, which is all the more intense as nitrogen content is high. The duration of the high temperature tests is very short and does not permit any precipitation, even with the influence of strain. Thus, the high mobility of interstitial atoms at 600°C, confirmed by a strong Portevin-Le Chatelier effect, seen in tension as well as LCF, ensures a high dislocation pinning by Cottrell atmospheres which doubtless explains the considerable hardening of Steel D. In high strain fatigue, the final tensile rupture occurs all the more rapidly in cases of intense boundary precipitation and high strain level. This explains the decreased fatigue resistance at 600°C of Steel D compared to Steel B at high strain level. Nevertheless, at low strain level, where intergranular rupture is nonexistent, the gain in fatigue life due to the presence of nitrogen still remains noticeable even in the 10 000 h aged state.

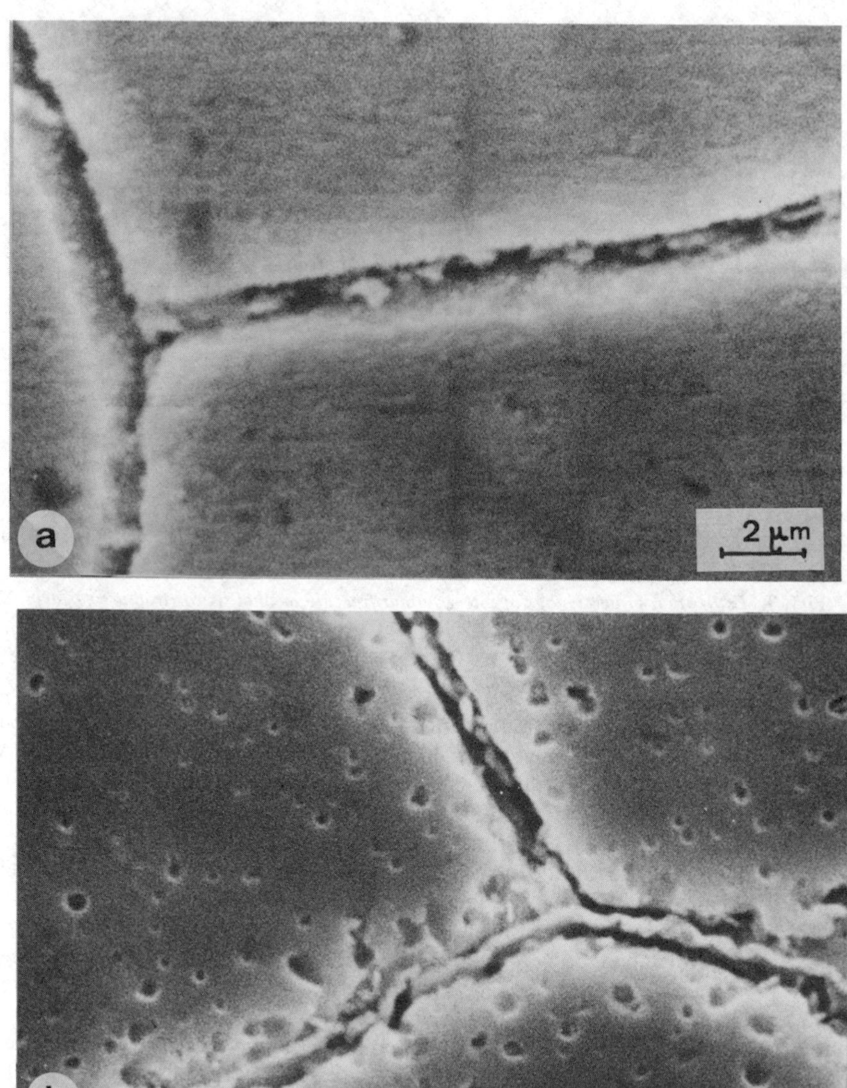

FIG. 10—*Intergranular precipitation in 10 000 h aged Steels B and D.* (a) *Steel B, etched in $FeCl_3$.* (b) *Steel D, etched in $CH_3COOH + HClO_4$.*

Conclusions

We have shown that the presence of nitrogen in AISI 316L-316LN austenitic stainless steels considerably changes static and dynamic properties in three ways:

1. In nitrogen solid solution the interstitial-dislocation interactions cause considerable monotonic hardening and ensure a homogeneous distribution of cyclic plastic deformation.

2. After a partial nitrogen precipitation in the matrix and, more particularly, at the grain boundaries, a matrix hardening and, especially, an intergranular embrittlement are observed which lower the critical stress intensity factor. Thus there is a strong reduction in ductility as well as in fatigue life.

3. Finally, the SFE lowering due to nitrogen favors dislocation planar sliding by delaying the appearance of cross-slip until a higher strain level or higher temperature is reached. Planar nature of plastic deformation ensures an improved reversibility of strain and delays crack initiation.

Acknowledgments

This research was sponsored partly by CREUSOT-LOIRE (Centre de Recherches d'Unieux). The authors express their appreciations to Mr. P. Rabbe for his support during this program.

References

[*1*] Bourrat, J., Demestre, J., Mercier, A., and Rémy, G., *Revue de Métallurgie,* Vol. 64, No. 12, Dec. 1967, pp. 1095-1114.

[*2*] Mercier, M. A., *Aciers Spéciaux,* Vol. 10, 1969, pp. 13-20.

[*3*] Norström, L. A., *Metal Science,* June 1977, pp. 208-212.

[*4*] Weiss, B. and Stickler, R., *Metallurgical Transactions,* Vol. 3, No. 4, April 1972, pp. 851-866.

[*5*] Tůma, H., Landa, V., and Löbl, K., *Mémoires Scientifiques de la Revue de Métallurgie,* Vol. 78, No. 5, May 1981, pp. 255-259.

[*6*] Mallick, "Phase Identification in Nitrogen Stainless Steels," M.Sc. thesis, Newcastle-upon-Tyne University, England, 1979.

[*7*] Degallaix, S., Vogt, J. B., and Foct, J., *Mémoires Scientifiques de la Revue de Métallurgie,* Vol. 80, No. 11, Nov. 1983, pp. 619-633.

[*8*] Vogt, J. B., Degallaix, S., and Foct, J., *International Journal of Fatigue,* Vol. 6, No. 4, Oct. 1984, pp. 211-215.

[*9*] Degallaix, S., Taillard, R., and Foct, J., in *Fatigue at High Temperature,* Paris, International Spring Meeting, Société Française de Métallurgie, June 1986, pp. 53-63.

[*10*] Degallaix, S., Taillard, R., and Foct, J. in *Fatigue 84,* Birmingham, Beevers, C. J., Ed., International Editorial Panel, Vol. 1, Sept. 1984, pp. 49-59.

[*11*] Nilsson, J. O., *Scripta Metallurgica,* Vol. 17, 1983, pp. 593-596.

[*12*] Schramm, R. E. and Reed, R. P., *Metallurgical Transactions,* Vol. 6A, No. 7, July 1975, pp. 1345-1351.

[*13*] Stoltz, R. E. and Vander Sande, J. B., *Metallurgical Transactions,* Vol. 11A, No. 6, June 1980, pp. 1033-1037.

[*14*] Nilsson, J. O., *Fatigue of Engineering Materials and Structures,* Vol. 7, No. 1, 1984, pp. 55-64.

T. Magnin,[1] J. M. Lardon,[1] and L. Coudreuse[1]

A New Approach to Low Cycle Fatigue Behavior of a Duplex Stainless Steel Based on the Deformation Mechanisms of the Individual Phases

REFERENCE: Magnin, T., Lardon, J. M., and Coudreuse, L., **"A New Approach to Low Cycle Fatigue Behavior of a Duplex Stainless Steel Based on the Deformation Mechanisms of the Individual Phases,"** *Low Cycle Fatigue, ASTM STP 942,* H. D. Solomon, G. R. Halford, L. R. Kaisand, and B. N. Leis, Eds., American Society for Testing and Materials, Philadelphia, 1988, pp. 812-823.

ABSTRACT: This paper concerns the low-cycle fatigue (LCF) behavior of a duplex (austenitic-ferritic) stainless steel with a 50% ferrite content. The study focuses on the damage mechanisms leading to crack initiation in the duplex alloy cycled at imposed plastic strain amplitude. Furthermore, the behavior of two other alloys of compositions close to those of the α and γ phases is simultaneously examined and correlated to the duplex alloy cyclic properties. The LCF properties of the duplex alloy approach those of the ferrite phase at high strain amplitudes and those of the austenite phase at low strain amplitudes. Many observations confirm this result: the Coffin-Manson curves, the hardening/softening curves, the cyclic stress-strain curves, and the crack initiation sites indicate a change in the duplex behavior near a plastic strain amplitude of 10^{-3}. The proposed analysis of the LCF behavior of the duplex alloy is based on the cyclic plastic deformation mechanisms of the α phase (which exhibits twinning and pencil glide) and of the γ phase (with planar slip). Finally, the influence of a 3.5% NaCl solution on the LCF properties of the duplex alloy clearly underlines the occurrence in this alloy of two different types of behavior according to the applied plastic strain amplitude.

KEY WORDS: low-cycle fatigue, dislocation behavior, fatigue mechanisms, crack initiation, corrosion-fatigue, duplex stainless steel

Duplex austenitic-ferritic stainless steels were developed for their good resistance to intergranular corrosion and chloride stress corrosion cracking and also for their high room-temperature strength. These properties are related to complex electrochemical and mechanical coupling effects between the two phases [*1,2*] and are closely dependent upon the ferrite content. In practical applications, duplex stainless steels are often exposed to both electrochemical and mechanical damages, the latter usually being fatigue. The purpose of this paper is to study the low-cycle fatigue (LCF) behavior of a duplex alloy with a 50% ferrite content, paying particular attention to the damage mechanisms leading to crack initiation in air and a 3.5% NaCl solution.

As for corrosion properties, the LCF behavior of the duplex stainless steels must be related to the cyclic properties of the individual phases. Plastic deformation of the fcc austenite phase is accommodated by planar configurations of dislocations in contrast to the bcc ferrite phase for which screw dislocations control the deformation [*3*]. This leads to quite different cyclic properties in each phase. The aim of this study is to propose an approach of the LCF behavior of a

[1]Département Matériaux, Ecole des Mines, 158, cours Fauriel, 42023 Saint-Etienne, France.

duplex alloy based on the deformation mechanisms and the fatigue properties of the austenitic and ferritic phases.

Experimental Procedure

The LCF tests were performed under plastic strain control ($10^{-4} < \Delta\epsilon_p/2 < 10^{-2}$) at constant strain rate, $\dot{\epsilon}$, on smooth specimens (5 mm diameter, 10 mm gage length) symmetrically cycled in tension-compression on a servohydraulic machine. To quantitatively determine the crack initiation, the number of cycles, N_i, corresponding to a rapid 1% decrease of the cyclic peak stress in tension, σ, has been selected. The corresponding crack length is about 500 μm. It has been shown elsewhere that such cracks are able to propagate, which is not always the case of shorter cracks [*4,5*].

The chemical composition of the materials examined in this study and their respective heat treatments are given on Table 1 (for more details concerning the influence of heat treatments, see Ref *6*).

The compositions of the ferritic and austenitic alloys are close to those of the α and γ phases of the 50% α duplex alloy. In the duplex alloy elongated ferritic bands are observed in the austenitic matrix. Finally, the grain sizes are 50 μm for the austenitic alloy and 200 μm for the ferritic alloy.

Some ferritic and duplex specimens were also aged at 400°C for 2 h, within the theoretical ($\alpha_1 + \alpha_2$) region of the Fe-Cr diagram [*7*]. This usual treatment induces a large hardening of the α phase.

Finally, some LCF tests have been performed in an aerated 3.5% NaCl solution under potentiostatic control. Sensitive testing equipment, which is described in detail elsewhere [*8*], is used to record, at imposed electrochemical potential, the cyclic evolution of the current transients. The perturbations of the metal-solution interface during cycling can be quantified [*8*] using the parameter J_t which is the peak value of the current density J corresponding to the tensile part of each cycle ($J = I/1.5$ cm^2, where I is the current and 1.5 cm^2 the specimen area exposed to the corrosive solution).

LCF Properties of the Duplex Alloy

Mechanical Results

The main mechanical result is observed on the Coffin-Manson curves where $N_i = f(\Delta\epsilon_p/2)$ is described in Fig. 1 for the three alloys. The curve of the duplex alloy clearly exhibits a change of slope in the vicinity of $\Delta\epsilon_p/2 = 10^{-3}$. For $\Delta\epsilon_p/2 > 10^{-3}$, the slope $m = -0.7$ and for $\Delta\epsilon_p/2 < 10^{-3}$ the slope $m' = -0.4$. It can be noticed that:

1. At high plastic strain amplitudes ($\Delta\epsilon_p/2 > 10^{-3}$) the behavior of the duplex alloy seems to approach that of the ferritic alloy.

TABLE 1—*Chemical compositions (weight percent) and heat treatment (WQ = water-quenched) of alloys.*

Alloy	Cr	Ni	Mo	Mn	Cu	Si	Co	C	N	Heat Treatment
Duplex (50% α)	21.7	7.0	2.5	1.7	0.08	0.39	0.06	0.020	0.07	WQ 1150°C
Ferritic	25.5	5.0	1.2	...	...	0.2	...	0.002	0.006	WQ 1250°C
Austenitic	18.1	14.4	2.3	...	...	0.07	...	0.02	0.005	WQ 1050°C

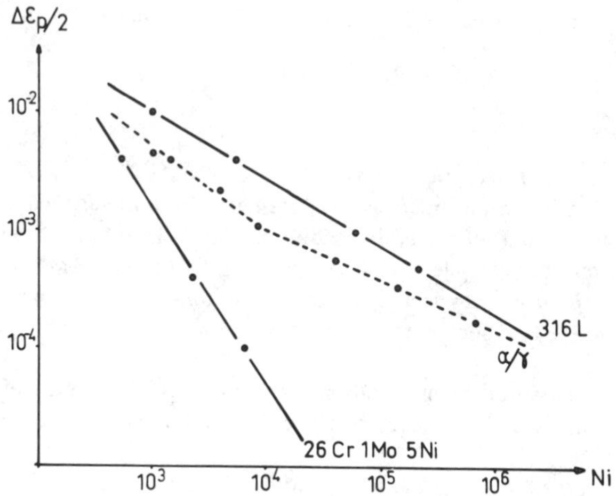

FIG. 1—*Coffin-Manson curves of the duplex* (α/γ)*, ferritic (26Cr-1Mo-5Ni), and austenitic (316L) alloys at* $\dot{\epsilon} = 2 \cdot 10^{-3}\ s^{-1}$ *in air (each point represents an average of three to five tests).*

2. At low plastic strain amplitudes ($\Delta\epsilon_p/2 < 10^{-3}$), the behavior of the duplex alloy approaches the "austenite behavior." To our knowledge, such a result has never been published. It is confirmed by many other observations. Thus the hardening/softening curves of the three alloys clearly show that:

(*a*) At low strain amplitudes, the duplex alloy exhibits rapid hardening during the first cycles, followed by slow softening, as it is observed for the austenitic alloy (Fig. 2).

(*b*) The softening, which can be quantified by the parameter $\delta = (\sigma_{max} - \sigma_s/\sigma_{max})$ (Fig. 3), is significant for the duplex and the austenitic alloys for $\Delta\epsilon_p/2 < 10^{-3}$ but becomes negligible for the duplex alloy at higher strain amplitudes, as for the ferritic Fe-26Cr-1Mo-5Ni alloy.

Finally, the critical value $\Delta\epsilon_p/2 = 10^{-3}$ is also pointed out by the cyclic stress-strain curve of the duplex stainless steel (Fig. 4).

All the mechanical results suggest a change in the LCF mechanisms of the duplex alloy in the vicinity of $\Delta\epsilon_p/2 = 10^{-3}$: an "austenitic behavior" at $\Delta\epsilon_p/2 < 10^{-3}$ and a "ferritic behavior" at $\Delta\epsilon_p/2 > 10^{-3}$.

Crack Initiation Sites

Observations by scanning electron microscopy of the crack initiation sites underline the conclusions of the mechanical results. At low plastic strain amplitudes for which the fatigue resistance of the duplex alloy is good, cracks only nucleate in the austenitic phase (Fig. 5), but at high strain amplitudes the first cracks nucleate principally in the α phase (Fig. 6). Figure 6 also indicates that the cracks in the α phase of the duplex alloy are often intergranular at α/α grain boundaries [*1*] and sometimes transgranular [*2*].

Thus, to analyze the LCF mechanisms of the duplex stainless steel, it seems particularly important to determine the cyclic deformation mechanisms of the individual phases.

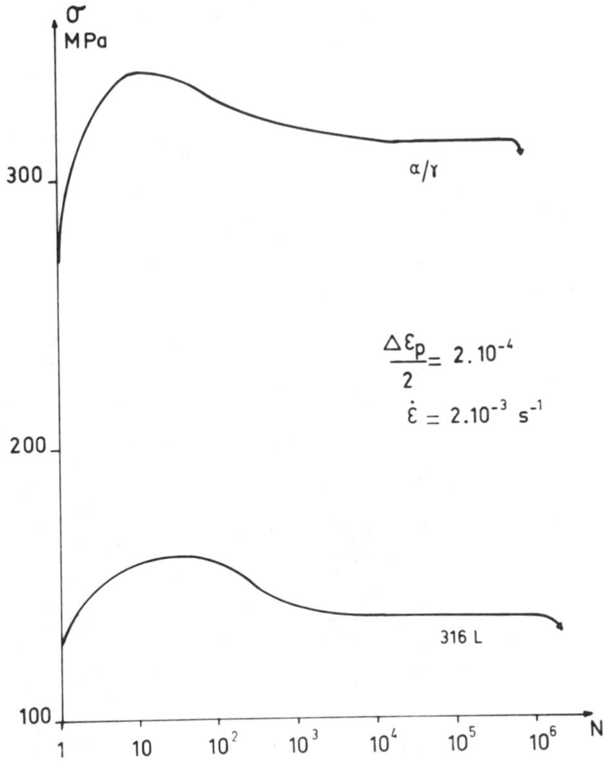

FIG. 2—*Hardening/softening curves of the duplex* (α/γ) *and the austenitic (316L) stainless steels at* $\Delta\epsilon_p/2 = 2\cdot10^{-4}$ *and* $\dot{\epsilon} = 2\cdot10^{-3}\ s^{-1}$.

Deformation Mechanisms of the Individual Phases

Cyclic Deformation of the Ferritic Fe-26Cr-1Mo-5Ni Alloy

It has been recently shown [*9,10*] that twinning occurs in the ferritic Fe-26Cr-1Mo-5Ni alloy deformed in tension at 300 K. This original result has been studied in detail as a function of temperature, strain rate, heat treatment, and alloying elements [*9*]. Twinning was attributed to the introduction of nickel in the bcc Fe-Cr matrix at 300 K (primary effect), which is associated with large friction stresses [*10*]. Deformation twins can then nucleate and grow from dissociated screw dislocations as for bcc crystals in the low temperature regime. An example of the observed {112} ⟨111⟩ twins is presented on Fig. 7. Moreover, it has been shown [*9*] that an aging at 400°C promotes twinning (secondary effect).

When a cyclic strain is applied, twinning occurs during the first cycles and then rapidly decreases as indicated on Fig. 8. This has been analyzed in terms of slip/twinning interactions in the Fe-Cr matrix [*10*]. The cumulative plastic deformation by twinning $\Sigma_i \epsilon_{pt}^i$ which can be approximatively determined from hysteresis curves (Fig. 8) is constant whatever the applied cyclic plastic deformation (i.e., $\Sigma_i \epsilon_{pt}^i \simeq 3 \cdot 10^{-3}$ at room temperature). In the saturation regime, only pencil glide occurs. This deformation mode is related to the screw dislocations which accommo-

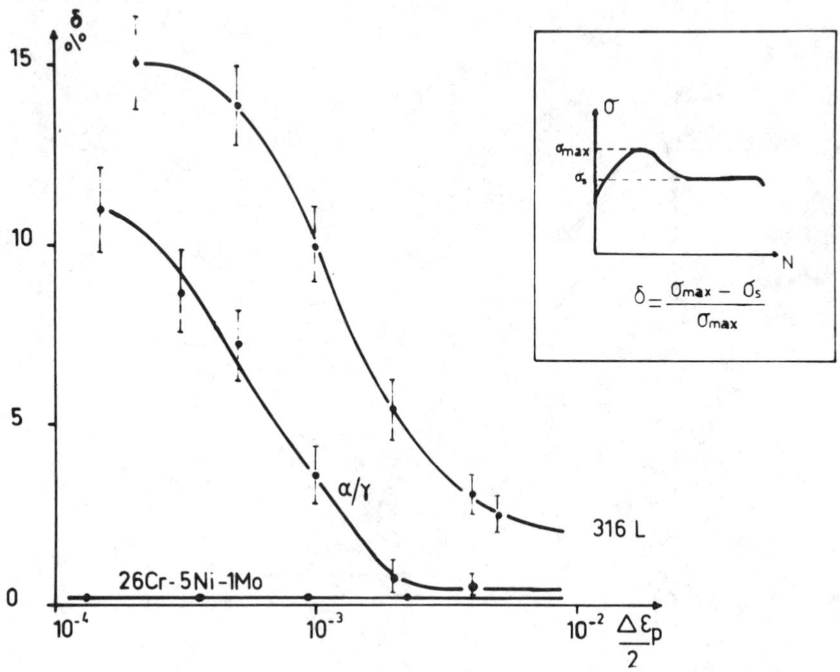

FIG. 3—*Influence of $\Delta\epsilon_p/2$ on the amount of softening (δ) in the duplex, ferritic, and austenitic stainless steels at $\dot{\epsilon} = 2 \cdot 10^{-3}\ s^{-1}$.*

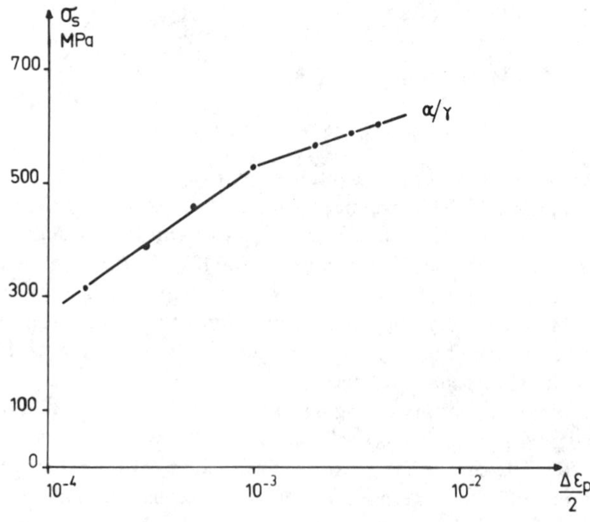

FIG. 4—*Cyclic stress-strain curve of the duplex alloy at $\dot{\epsilon} = 2 \cdot 10^{-3}\ s^{-1}$.*

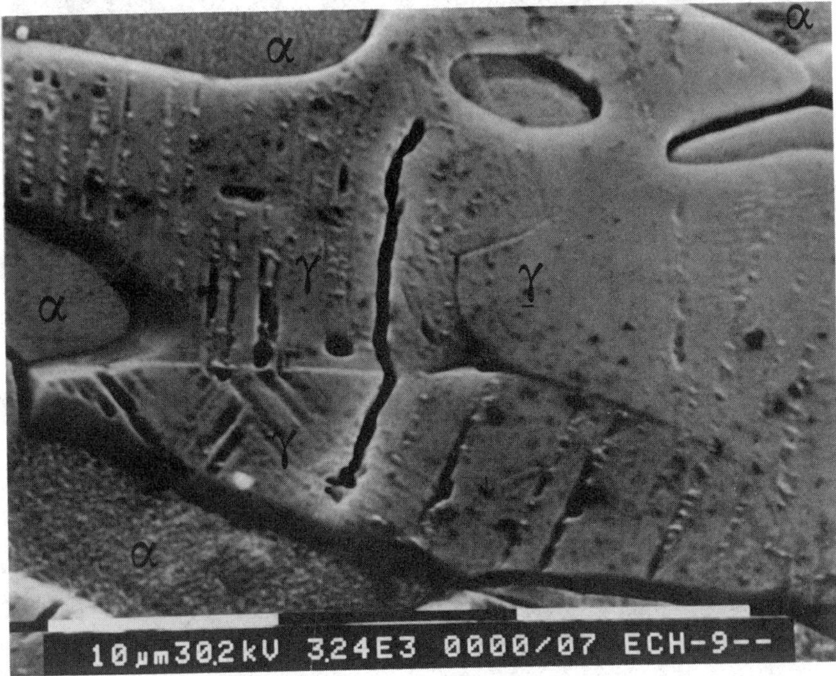

FIG. 5—*Crack initiation in the austenitic phase of the duplex stainless steel at* $\Delta\epsilon_p/2 = 3 \cdot 10^{-4}$ *and* $\dot{\epsilon} = 2 \cdot 10^{-3}\ s^{-1}$.

date the plastic strain (Fig. 9) and which exhibit an asymmetrical behavior between tension and compression [3]. This phenomenon promotes intergranular cracking because of the high strain irreversibility in each grain [3]. Figure 9 describes the saturation regime of the ferritic alloy at $\Delta\epsilon_p/2 = 4 \cdot 10^{-3}$. Screw dislocations move in channels between well oriented walls which are formed of bundles of edge multipoles. Thus the presence of twins and the occurrence of pencil glide explain the early crack initiation in the ferritic alloy (Fig. 1).

Cyclic Deformation of the Austenitic 316L Alloy

Planar slip occurs in fcc alloys with low stacking fault energy such as the 316L stainless steel cycled at room temperature [*11*]. During the first cycles, planar areas of dislocations are observed, essentially confined to {111} slip lines. It has been shown that the nitrogen which is present in the γ phase of the duplex alloy in contrast to the 316L alloy favors planar slip [*12,13*], which increases the fatigue resistance [*12*]. The planar aspect of the deformation subsists up to crack initiation which occurs in sharp slip bands [*11*].

Consequences on the LCF Mechanisms in the Duplex Alloy

In the low plastic strain range, the softer austenitic phase accommodates the deformation. Moreover, because of the presence of nickel in the α phase, screw dislocations are very dissociated in this phase which favors an "austenitic behavior" of the duplex stainless steel. Because of the low degree of slip irreversibility in the γ phase, delayed transgranular crack initiation will

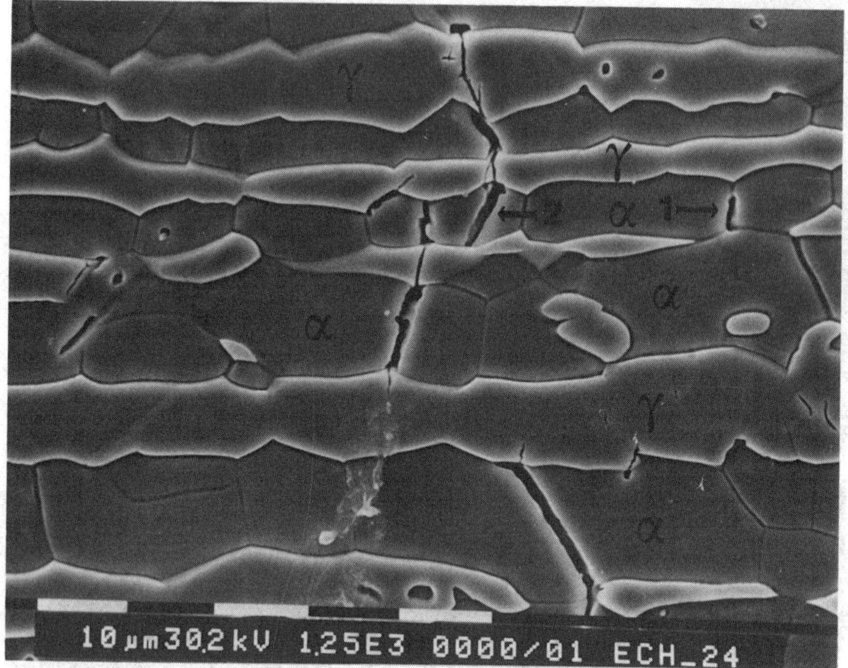

FIG. 6—*Crack initiation in the ferritic phase of the duplex stainless steel at* $\Delta\epsilon_p/2 = 4\cdot10^{-3}$ *and* $\dot{\epsilon} = 2\cdot10^{-3}\ s^{-1}$.

occur in this phase. This explains the good fatigue resistance of the duplex alloy at $\Delta\epsilon_p/2 < 10^{-3}$ as shown by the Coffin-Manson curve (Fig. 1).

At high strain amplitudes, both the austenitic and ferritic phases accommodate the plastic deformation. Screw dislocations in the ferrite phase promote twinning and pencil glide which induce early crack nucleation in the α phase of the duplex alloy. This explains the LCF properties of the duplex at $\Delta\epsilon_p/2 > 10^{-3}$ and the change of slope of the Coffin-Manson curve (Fig. 1). Finally, it has been noticed that at high strain amplitudes crack initiation often occurs at α/α boundaries. This confirms the role of pencil glide which induces intergranular cracking in the ferritic Fe-26Cr-1Mo-5Ni alloy.

It now seems particularly interesting to see if this analysis is relevant to the study of LCF properties of the duplex alloy in a corrosive solution.

LCF Behavior of the Duplex Alloy in a 3.5% NaCl Solution

Fatigue tests were performed in a 3.5% NaCl solution maintained at pH 2 by addition of HCl, at imposed electrochemical potential E, in two different conditions: (1) at $E = -120$ mV/SCE (SCE = saturated calomel electrode), which corresponds to the passive region and to a cathodic protection of the austenitic phase by the ferritic phase [*2*]; and (2) at $E = -500$ mV/SCE, which favors hydrogen embrittlement of the ferritic phase.

Tests were conducted for two different values of $\Delta\epsilon_p/2$ corresponding respectively to a "ferritic" and to an "austenitic" behavior of the duplex alloy. The experimental results are summarized in Table 2. It can be observed that:

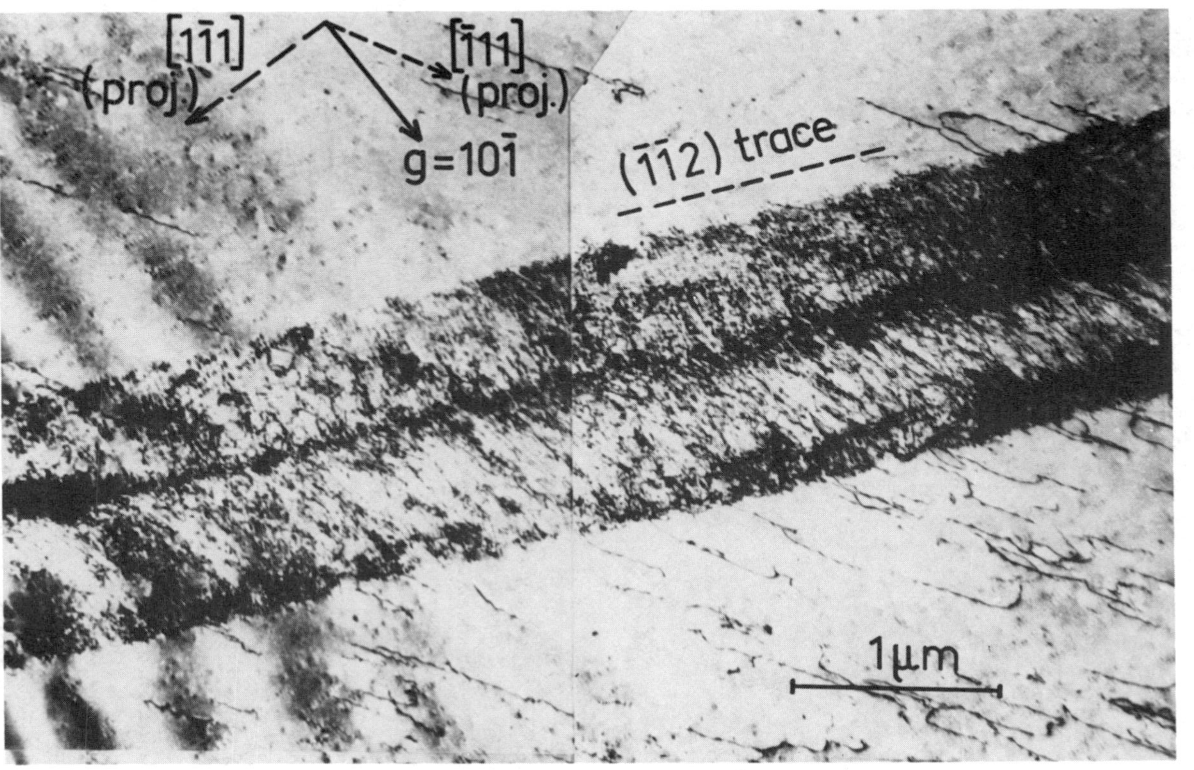

FIG. 7—*TEM observation of* {112} ⟨111⟩ *twins in the ferritic Fe-26Cr-1Mo-5Ni alloy deformed 1% in tension at 300 K.*

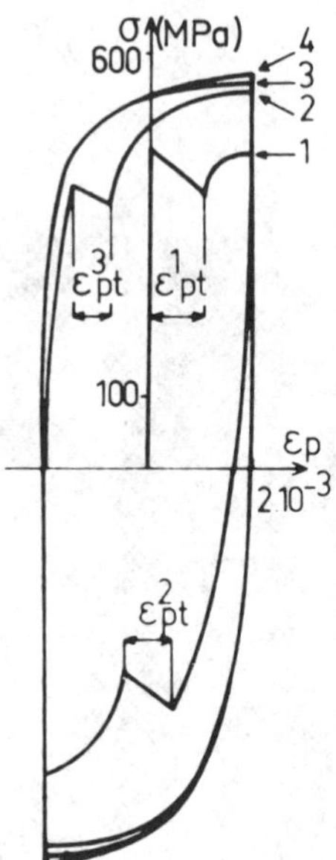

FIG. 8—*Stress-strain response of the ferritic Fe-26Cr-1Mo-5Ni alloy during the first cycles ($\Delta\epsilon_p/2$ = $2 \cdot 10^{-3}$, $\dot{\epsilon}$ = $2 \cdot 10^{-3}\ s^{-1}$).*

1. At $E = -120$ mV/SCE, the reduction of the fatigue life in the corrosive solution is observed at $\Delta\epsilon_p/2 = 4 \cdot 10^{-3}$, but is not significant at $\Delta\epsilon_p/2 = 10^{-3}$ for water-quenched specimens. Since austenite is cathodically protected by ferrite, the dissolution effects can only occur in ferrite. At low strain amplitudes, only the austenite is depassivated by the cyclic strain and no corrosion-fatigue can occur. At high strain amplitudes, depassivation takes place in both the ferrite and austenite phases. Thus the dissolution effects are localized in the ferrite phase, which induces a reduction of the fatigue life in this case. To confirm this explanation, tests have been conducted at high strain amplitudes on specimens aged at 400°C for 2 h. In such conditions, the ferrite is harder and twinning is favored ($\Sigma_i \epsilon^i_{pt} = 2.5 \cdot 10^{-2}$). Table 2 clearly shows that the reduction of the fatigue life in the corrosive solution is very pronounced for aged specimens ($N_c/N_{air} = 0.48$). If we now look at the cyclic evolution of the current density J_T (Fig. 10), it can be observed that (*a*) during the first cycles, J_T is very similar for specimens with and without the aging treatment; and (*b*) when the saturation regime is reached, J_T increases much earlier for the aged specimens. This corresponds to an earlier and stronger localization of the dissolution effects in the aged specimens, which can be related to the larger amount of twinning in these specimens. The obtained results clearly underline the role of ferrite at high strain amplitudes.

2. At $E = -500$ mV/SCE, the reduction of the fatigue life is much more pronounced at high than at low strain amplitudes (Table 2). This effect can be related to the observed hydrogen embrittlement of the ferritic Fe-26Cr-1Mo-5Ni alloy in such conditions. In the duplex alloy when ferrite is deformed, a local hydrogen embrittlement promoted by the plastic strain can occur. At low strain amplitudes, however, this effect is considerably reduced.

The LCF tests in the corrosive solution clearly show that (1) the duplex alloy exhibits a good corrosion-fatigue resistance at low strain amplitudes, when ferrite is not strained; and (2) the

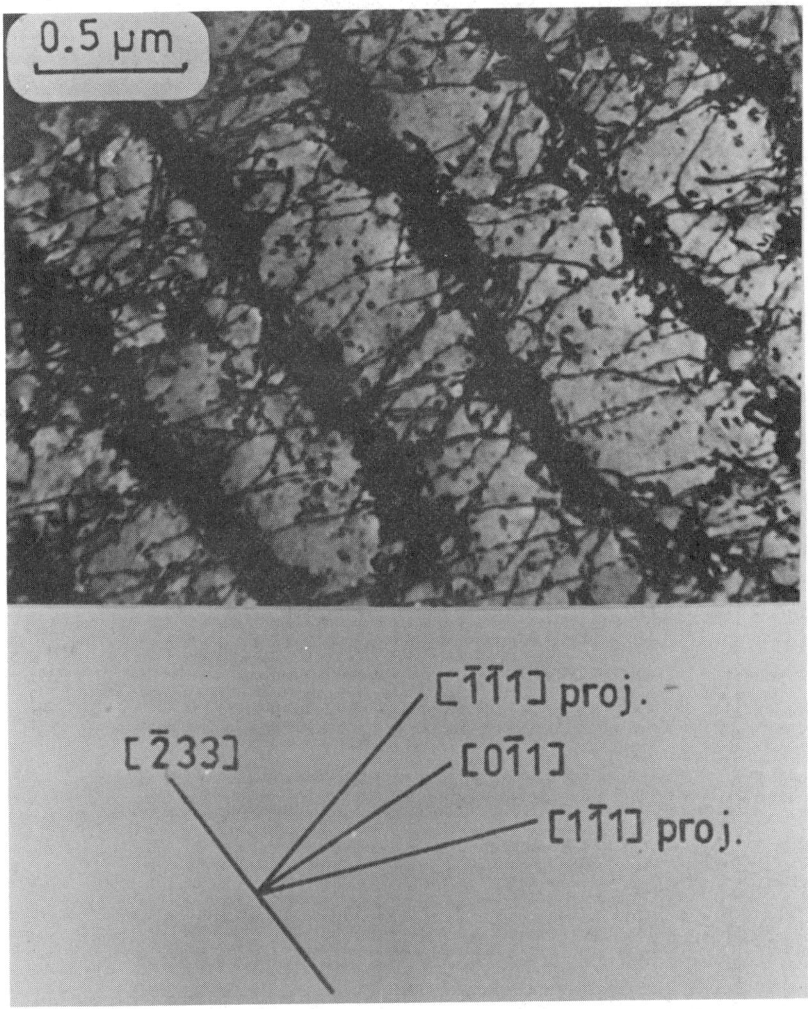

FIG. 9—*Dislocation structures in the ferritic Fe-26Cr-1Mo-5Ni alloy cycled at saturation* ($\Delta\epsilon_p/2 = 4 \cdot 10^{-3}$, $\dot{\epsilon} = 2 \cdot 10^{-3}\ s^{-1}$, *50 cycles).*

analysis of the fatigue properties of the duplex alloy in terms of "austenitic behavior" at low strain amplitudes and "ferritic behavior" at high strain amplitudes is quite relevant.

Conclusions

This study has demonstrated the existence of two distinct types of mechanical behavior in a duplex stainless steel with 50% ferrite content subjected to low-cycle fatigue tests at room temperature:

1. At low strain amplitudes ($\Delta\epsilon_p/2 < 10^{-3}$), the cyclic deformation of the austenitic phase controls the LCF properties of the duplex alloy. Because of the large reversibility of the cyclic

TABLE 2—*Influence of the plastic strain amplitude on the reduction of fatigue life of the duplex alloy at $\dot{\epsilon} = 10^{-2}\ s^{-1}$ in a 3.5% NaCl solution at pH 2 (N_c = number of cycles in the corrosive solution; N_{air} = number of cycles in air; WQ = water-quenched).*

$\Delta\epsilon_p/2$	Heat Treatment	E	N_c/N_{air}
$4 \cdot 10^{-3}$	WQ	−120 mV/SCE	0.68
		−500 mV/SCE	0.3
	WQ + 2 h at 400°C	−120 mV/SCE	0.48
		−500 mV/SCE	0.15
10^{-3}	WQ	−120 mV/SCE	1
		−500 mV/SCE	0.75

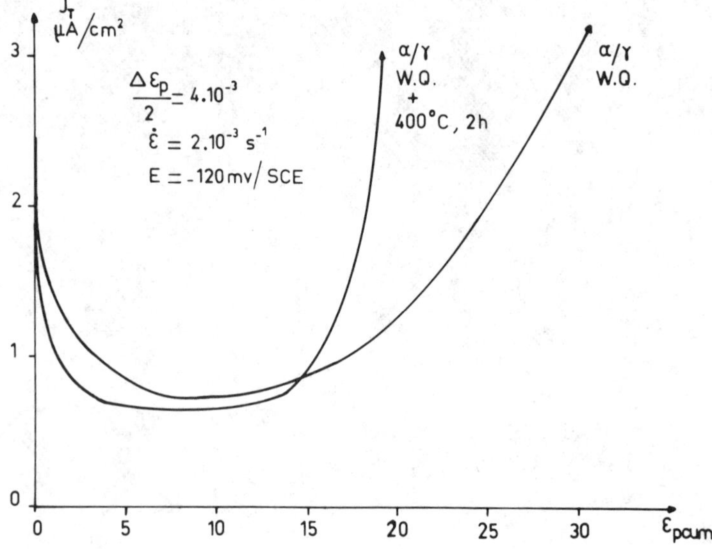

FIG. 10—*Influence of aging on the evolution of the peak current density in tension J_T as a function of the cumulative plastic strain ($\epsilon_{pcum} = 2\Delta\epsilon_p \cdot N$) in duplex stainless steel specimens (3.5% NaCl solution at pH 2, E = −120 mV/SCE, $\Delta\epsilon_p/2 = 4 \cdot 10^{-3}$ and $\dot{\epsilon} = 2 \cdot 10^{-3}\ s^{-1}$).*

strain in the γ phase, good fatigue resistance of the duplex alloy is observed. Crack nucleation is delayed and takes place in the γ phase. Moreover, the corrosion-fatigue properties of the duplex alloy in a 3.5% NaCl solution are very interesting because of the cathodic protection of the austenitic phase by the ferritic phase.

2. At high strain amplitudes ($\Delta\epsilon_p/2 > 10^{-3}$), the cyclic deformation of the ferritic phase controls the LCF properties of the duplex alloy. Because of twinning and pencil glide in the α phase, an early crack initiation takes place in the ferritic phase. The fatigue life is then reduced, and a marked change in the slope of the Coffin-Manson curve is observed. In a 3.5% NaCl solution, the localization of the dissolution effects and of the hydrogen embrittlement in the ferrite phase is favored, which induces a strong reduction of the LCF properties.

References

[1] Magnin, T., Le Coze, J., and Desestret, A. in *Duplex Stainless Steels,* R. A. Lula, Ed., American Society for Metals, Metals Park, Ohio, 1983, p. 535.

[2] Desestret, A. and Oltra, R., *Corrosion Science,* Vol. 20, 1980, p. 799.

[3] Magnin, T., Driver, J., Lepinoux, J., and Kubin, L. P., *Revue de Physique Appliquee,* Vol. 19, 1984, p. 467.

[4] Suresh, S. and Ritchie, R. O., *International Metals Review,* Vol. 29, No. 6, 1984, p. 445.

[5] Magnin, T., Coudreuse, L., and Lardon, J. M., *Scripta Metallurgica,* Vol. 19, 1985, p. 1487.

[6] Solomon, H. D. in *Duplex Stainless Steels,* R. A. Lula, Ed., American Society for Metals, Metals Park, Ohio, 1983, p. 41.

[7] Fisher, R. M., Dulis, G. J., and Carrol, K. G., *Transactions of AIME,* Vol. 197, 1953, p. 690.

[8] Magnin, T. and Coudreuse, L., *Materials Science and Engineering,* Vol. 72, No. 2, 1985, p. 125.

[9] Magnin, T. and Moret, F., *Scripta Metallurgica,* Vol. 16, 1982, p. 1225.

[10] Magnin, T., Coudreuse, L., and Fourdeux, A., *Materials Science and Engineering,* Vol. 63, 1984, p. L5.

[11] Gorlier, C., thesis, Saint-Etienne, France, 1984.

[12] Nilsson, J. O., *Scripta Metallurgica,* Vol. 17, 1983, p. 593.

[13] Degallaix, S., Taillard, R., and Foct, J. in *Proceedings,* 2nd International Conference Fatigue and Fatigue Thresholds, Vol. 1, Sept. 1984, p. 49.

D. L. Anton[1] *and L. H. Favrow*[1]

Effect of HIP on Elevated-Temperature Low Cycle Fatigue Properties of an Equiaxed Cast Superalloy

REFERENCE: Anton, D. L. and Favrow, L. H., **"Effect of HIP on Elevated-Temperature Low Cycle Fatigue Properties of an Equiaxed Cast Superalloy,"** *Low Cycle Fatigue, ASTM STP 942,* H. D. Solomon, G. R. Halford, L. R. Kaisand, and B. N. Leis, Eds., American Society for Testing and Materials, Philadelphia, 1988, pp. 824–837.

ABSTRACT: A typical high strength equiaxed cast superalloy, B-1900, was tested in the fully heat-treated condition, preceded by an optional hot isostatic pressing (HIP) cycle. Fully reversed constant total strain amplitude testing was carried out at temperatures ranging from 760 to 982°C in air. The HIP cycle successfully eliminated all non-surface connected porosity, and specimens, machined from the center of test bars, were totally pore free. Changes in grain boundary carbide morphology and γ' size also ensued from the HIP treatment. Fatigue crack initiation sites were identified in scanning electron microscopy (SEM) fractographic analysis along with modes of crack propagation.

An enhancement of fatigue life by a factor of 6 to 8 was realized in the HIP alloy at temperatures of 871°C and below. Above this temperature no effect of HIP was observed. Cracks of HIP specimens initiated at grain boundary-surface intersections. Fatigue crack initiation was enhanced by local oxidation of surface-connected carbides, with fatigue crack propagation being predominantly intergranular. The non-HIP specimens typically contained surface cracks which propagated through near surface clusters of microporosity which in turn were formed adjacent to the interdendritic carbide networks. The HIP treating of equiaxed cast superalloys for turbine blade applications completely closed interdendritic and intergranular microporosity and caused grain growth to occur. This led to enhanced fatigue life at low temperatures where oxidation of grain boundary carbides was minimized. Fatigue life enhancement due to HIP was separated into microstructural modifications effecting yield stress and grain boundary structure.

KEY WORDS: superalloy mechanical properties, low-cycle fatigue, microstructure, porosity, castings

Low-cycle fatigue crack initiation sites in smooth sections can be divided into roughly three types, having their origins in (1) surface slip, (2) internal stress concentrations, or, in the case of an aggressive environment, (3) localized surface attack. In many cases, when either the test temperature or strain amplitude is varied, the mode of crack initiation changes. Ni-base superalloys display all three of these crack initiation modes, but at elevated temperatures in an oxidizing environment, the competing mechanisms are internal stress concentrations versus localized surface attack.

Studies conducted on the elevated temperature fatigue response of high strength cast superalloys have revealed that fatigue crack initiation usually occurs at grain boundary-surface intersections [*1,2*] and transgranular surface sites [*3,4*]. Additionally, it was shown in Ref *4* that the strain rate and strain wave profile affected whether initiation was intergranular or transgranular. All of these studies employed fractography and the technique of random longitudinal sec-

[1]United Technologies Research Center, East Hartford, CT 06108.

tioning of failed specimens to observe small cracks which did not lead to failure. This latter type of analysis, while rendering an excellent view of "embryonic" cracks, ofttimes fails to reveal the predominant fatigue crack initiation site. Microporosity, for instance, was not identified as potential fatigue crack initiation sites in any of these studies, although its presence is unquestionable in superalloy castings.

A number of more recent articles on the effect of hot isostatic pressing (HIP) on low-cycle fatigue (LCF) response identified microporosity originating from shrinkage during casting as the primary fatigue crack initiation sites [*5,6*]. However, these studies, which utilized various fatigue cycles and alloys, came to differing conclusions regarding the effect of HIP treating. Burke et al. [*5*] found little effect of HIP on LCF life at 650°C with substantial enhancement at 850°C. In this study, factors resulting from HIP other than pore removal, such as elemental homogenization and carbide redistribution, were suspected to influence fatigue behavior. Furthermore, this study was complicated by using a constant cyclic frequency at differing strain amplitudes, resulting in tests conducted at varying strain rates which was shown to profoundly affect initiation sites [*4*]. This is a serious consideration in testing Ni-base superalloys, since a pronounced strain rate effect on yield stress at 760°C and above was first observed by Leverant et al. [*7*]. In high-cycle fatigue testing, Schneider et al. [*6*] concluded that the elimination of porosity was the only important factor in fatigue life enhancement.

In the directional solidification of these alloys into columnar grain or single crystal form, where thermal gradients and therefore porosity are strictly controlled, microporosity has been cited as the initiator of fatigue cracks [*8–10*]. Furthermore, proponents of oxide dispersion strengthened (ODS) Ni-base alloys regularly identify the absence of microporosity in these alloys with longer fatigue lives than cast alloys [*11*].

As briefly described above, the literature regarding microporosity in low-cycle fatigue life of superalloys is clouded. The purpose of this study is to determine the effect of HIP treating a cast superalloy and to determine by what role HIP modifies the microstructure and fatigue properties.

Experimental Procedure

Investment cast bars 1.6 cm in diameter and 15 cm long of Alloy B-1900, having the nominal composition (weight percent) of 8.0 Cr, 10.0 Co, 6.0 Mo, 4.0 Ta, 6.0 Al, 1.0 Ti, 0.1 C, 0.015 B, 0.10 Zr, and balance Ni, were fabricated. These bars were given either a standard high temperature heat treatment consisting of solutioning at 1232°C/15 min/AC followed by two-step aging at 1079°C/4 h/AC + 900°C/10 h/AC where AC denotes air cooling or the above heat treatment preceded by a HIP cycle consisting of 1232°C/103 MPa/2 h. These heat treatments shall be designated SHT for the standard heat treatment and HHT for the HIP heat treatment. Subsequent to heat treatment, LCF specimens having a gage section 0.64 cm in diameter and 2.54 cm in length were machined using low residual stress grinding techniques.

Constant total strain amplitude LCF tests were conducted at 760, 871, and 982°C (1400, 1600, and 1800°F) in air using a servohydraulic test system. A fully reversed ramp wave was used keeping the strain rate constant at 0.005 s^{-1}. Strain was monitored using a high temperature extensometer and conical point extension rods [*12*]. Tests were run at total strain amplitudes, $\Delta\epsilon_T/2$, of 0.2, 0.4, 0.6, and 0.8% at all three temperatures and at 0.1 and 0.15% at 982°C. Plastic strain measurements were taken at Cycle 50 or as soon as stabilized hysteresis loops were obtained.

Optical and scanning electron microscopy (SEM) metallography were conducted on the fracture surfaces to assess fatigue crack initiation sites and crack propagation paths. Microscopy was also conducted on the base microstructures to determine any microstructural differences attributed to the HIP cycle.

Results

Metallography

It is of primary interest to understand the microstructural differences brought about by the HIP treatment. Optical microscopy shows that the SHT alloy does indeed contain a substantial degree of porosity estimated at approximately 1 to 2 vol% (Fig. 1*a*). The pores are distributed throughout the alloy but exist primarily in the vicinity of grain boundaries. These pores are irregular in shape as would be expected from their formation in the interdendritic spaces of the last solidifying liquid. For this reason, pores are sometimes grouped together where liquid feeding was particularly poor (Fig. 1*b*). These pores range in size from 20 μm to 0.15 mm in their longest dimension.

Another feature of the SHT microstructure is the grain boundary carbide morphology. These carbides, while not continuous as may be inferred from Figs. 1*a* and *b*, are elongated in the grain boundary as shown in Fig. 1*c*. Higher magnification SEM micrographs of the SHT grain boundaries are given in Fig. 2. The large Ti- and Ta-rich, elongated primary carbides reside in the grain boundary shown in Fig. 2*a* with fine Mo-rich carbides interspersed between them (see arrows in Fig. 2*b*).

The γ' size was very uniform throughout the SHT specimens. They ranged in size from 0.1 to 0.4 μm as measured on cube edge.

As expected in the HHT alloy, no pores were present. Grain growth has clearly taken place with the prior grain boundaries delineated by the primary carbides as in Fig. 3*a*. Close examination of this micrograph shows that no grain boundary is present where expected. Instead, the grain boundaries took on the appearance of that given in Fig. 3*b*. Here the boundary is free of all primary carbides and has precipitated the much finer Mo-rich carbide (arrowed in Fig. 3*b*). The grain size did change, but not drastically, with a typical grain diameter of 3 to 5 mm. Also of interest was the precipitation of an acicular topologically close pack (TCP) phase within the grains.

The γ' size and distribution were significantly altered in the HHT alloy. Figure 4 shows a drastic change in the γ' size within a small distance. Here the precipitates range in size from 0.3 to 1.7 μm on the cube edge.

Monotonic Data

Monotonic compression data was taken from the first cycle of tests run at $\Delta\epsilon_T/2 = 0.8\%$ at all three temperatures and for both HHT and SHT specimens. These data are given in Table 1, where σ_y is defined as the 0.2% offset yield stress, σ_p is the proportional limit yield stress, and E is Young's modulus. A significant difference in yield properties is observed between the two heat treatments. At the lower two temperatures the SHT material is stronger in both σ_y and σ_p than the HHT alloy. This is reversed at 982°C where the HHT alloy was stronger. The trend of the data shows that increasing temperature favors the strengthening increment of the HHT alloy as $\sigma_{(HHT)}$-$\sigma_{(SHT)}/\sigma_{(SHT)}$ in percent gives -8, -4, and $+7\%$ for σ_y and -17, -10, and $+19\%$ for σ_p in order of ascending temperature. Young's moduli for specimens of both heat treatments were essentially identical, with the HHT alloy being 2 to 7% stiffer.

Low-Cycle Fatigue

The low-cycle fatigue data are given in Table 2 and in Fig. 5. No significant cyclic hardening or softening was observed with stable hysteresis loops obtained within the first few cycles. It is readily apparent from the total strain analysis of Fig. 5*a* that the HHT material (solid symbols) possesses greater LCF life at 760 and 871°C, while at 982°C the SHT and HHT data are compa-

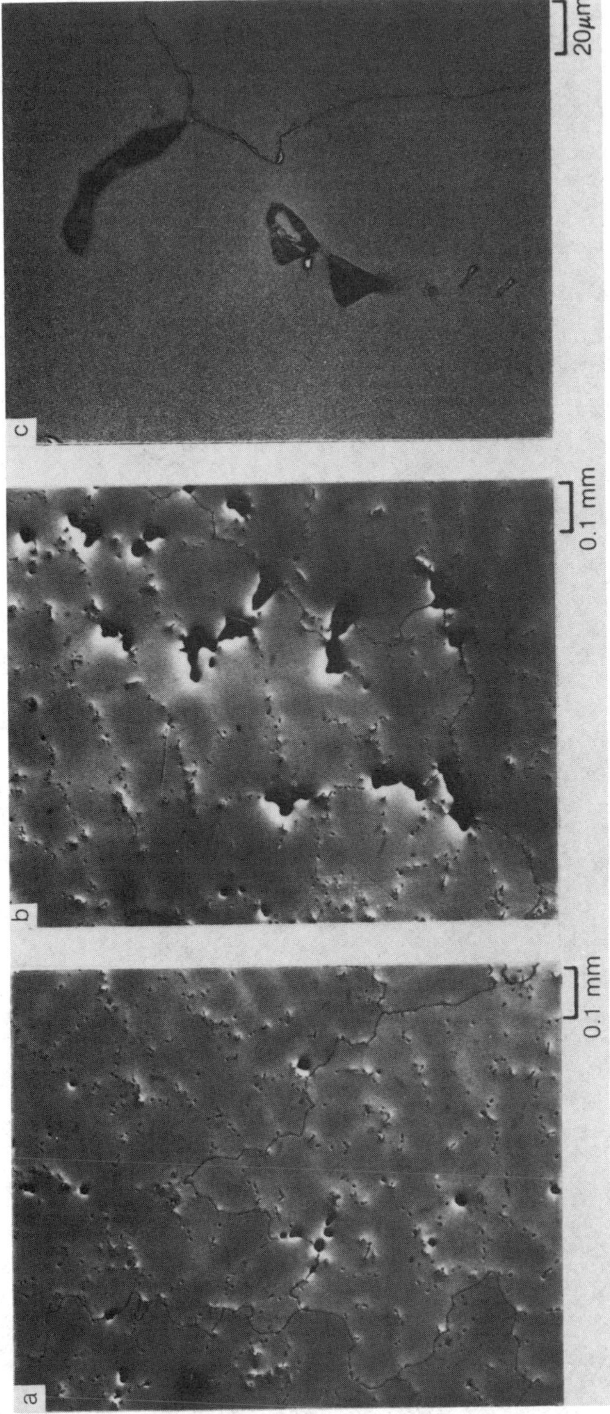

FIG. 1—*Representative microstructures of B-1900 after the standard heat treatment, SHT.*

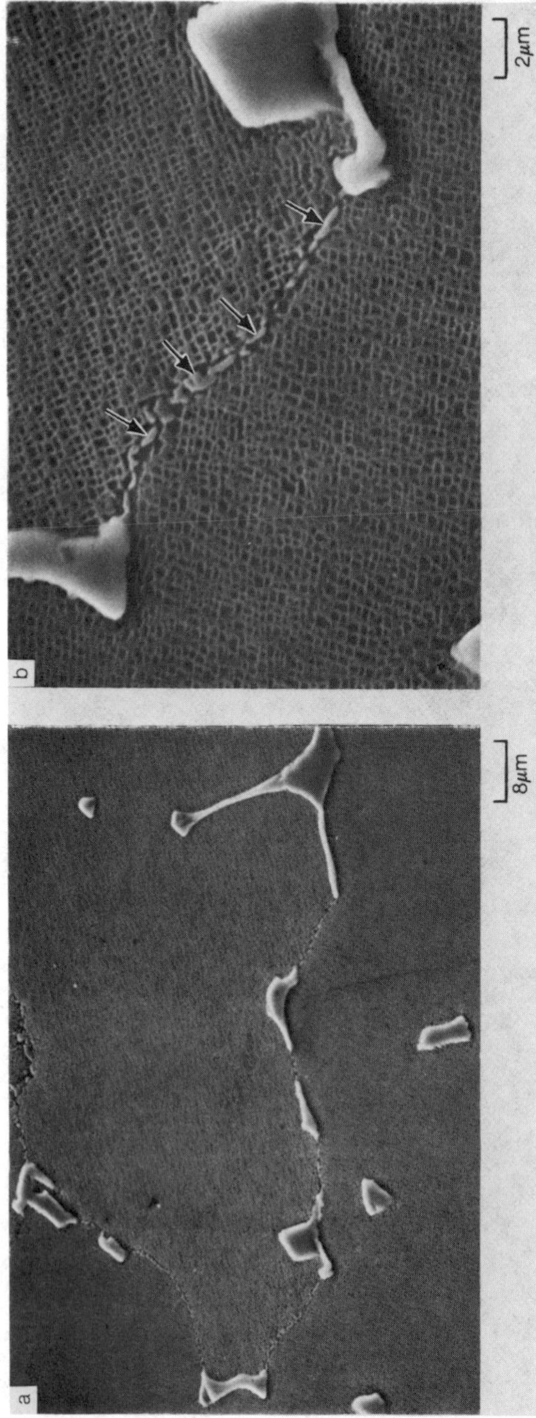

FIG. 2—*Grain boundary structure of SHT alloy showing* (a) *large primary carbides and* (b) *fine carbide dispersion.*

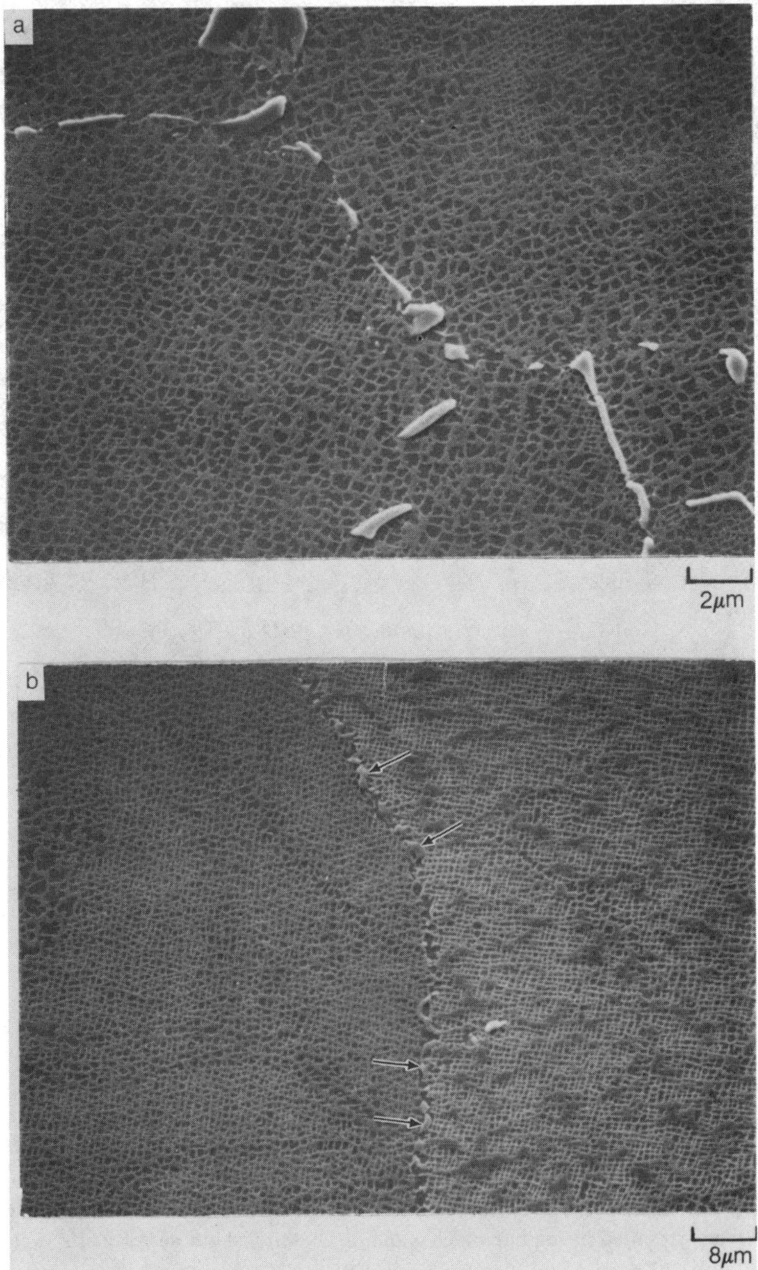

FIG. 3—*Grain boundary structure of HHT B-1900.* (a) *Prior grain boundary showing carbide decoration.* (b) *Grain boundary with fine carbide dispersion.*

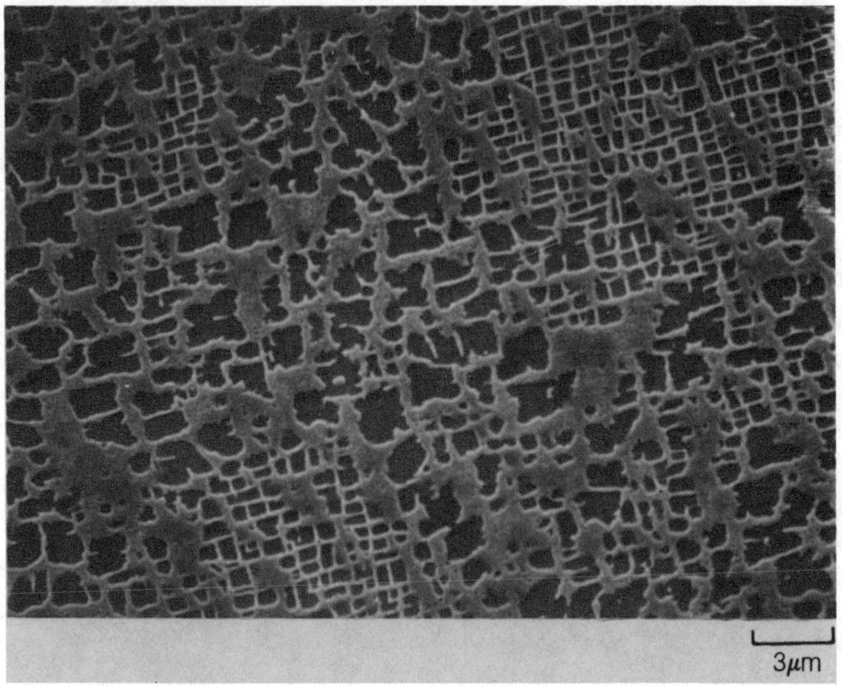

FIG. 4—*Bi-modal γ′ distribution in HHT B-1900 alloy.*

TABLE 1—*Monotonic data taken from first compression loading to 0.8% strain.*

	Temp. (°C)	σ_y (MPa)	σ_p (MPa)	E (GPa)
SHT	760	855	688	157
	871	818	611	151
	982	536	359	130
HHT	760	787	569	168
	871	783	547	158
	982	572	428	133

rable. At a strain amplitude of 0.8% the increase in fatigue life is roughly a factor of 6 to 8. At lower strain amplitudes, such as 0.4%, this advantage diminished for the 760°C tests while it was maintained at a factor of 6 for the 871°C test. Tests conducted on HHT specimens at the lowest strain amplitudes ($\leq$0.2%) ran substantially longer than their SHT counterparts. Two of the HHT tests resulted in run out which could not be carried out past 3×10^5 cycles due to time constraints.

Also apparent from Fig. 5*a* is the change in fatigue life with temperature and strain range for the SHT specimens. At high-strain amplitudes the fatigue life increased with increasing temperature. As the strain amplitude was decreased this difference in fatigue life became negligible as at $\Delta\epsilon_T/2 = 0.4\%$. Further decreases in strain amplitude revealed that lower temperature tests

TABLE 2—*Low-cycle fatigue data for SHT and HHT B-1900.*

		SHT		HHT	
Temp. (°C)	$\Delta\epsilon_T/2$ (%)	$\Delta\epsilon_p/2$ (%)	N_f	$\Delta\epsilon_p/2$ (%)	N_f
760	0.2	0.0000	225 000	0.0000	328 648$^+$
	0.2	0.0029	60 602	...	...
	0.4	...	512	0.0031	632
	0.6	0.0780	32	0.0980	85
	0.8	0.2210	4	0.1730	33
871	0.2	0.0005	33 823	0.0013	298 992
	0.4	0.0058	347	0.0095	1 160
	0.6	0.1210	42	0.0760	253
	0.6	...	...	0.1110	232
	0.8	0.2090	12	0.2370	57
982	0.1	...	...	0.0012	1 222 980$^+$
	0.15	0.0010	56 923	0.0015	71 095
	0.2	0.0038	11 438	0.0042	7 018
	0.4	0.0924	444	0.0790	317
	0.6	0.2550	97	0.2050	65
	0.8	0.3460	17	0.3810	81
	0.8	...	...	0.3470	72

$^+$Indicates run-out, no failure.

resulted in enhanced fatigue lives. The HHT data showed the same trend with the exception of the 982°C test data, which degraded faster (reflecting the importance of HIP at 982°C).

A plastic strain amplitude analysis is given in Fig. 5*b*. The best straight line was drawn by eye for the few test results here. Generally, the data for the two heat treatments at a given temperature fell on the same slope with the HHT alloy having a benefit in cycle life of five times at 871°C. At 982°C the data were coincident for the HHT and SHT material, while at 760°C run out tests and scatter at the low strain amplitudes of both heat treatments made curve fitting unreliable, but a definite benefit of a factor of 4 was achieved at plastic strain amplitudes of 0.08% and greater. The degree of scatter in these data is attributable to the difficulty in measuring such low plastic strain amplitudes.

Fractography

The SEM fractography revealed grain boundaries to be the predominant early crack growth path in all specimens tested (in agreement with the findings of Refs *1–4*), except the SHT specimens tested at $\Delta\epsilon_T/2 = 0.2\%$, where internal porosity was unambiguously observed as the fatigue failure origin. Although it was obvious that early fatigue crack growth occurred along grain boundaries transverse to the stress axis, the exact initiation site was unclear. These observations were obscured by partial oxidation of the origin due to its early exposure to the atmosphere. It can be expected that initiation in the grain boundaries stemmed from either large cracked carbides, preferential oxidation of the grain boundaries or in the case of the SHT specimens, grain boundary porosity.

An extensive discussion of the grain boundary cracking phenomenon will be foregone here because of its thorough coverage in the previously cited literature. Instances where specific fatigue crack initiation sites could be identified will be discussed along with differences between the SHT and HHT alloys. The 760 and 871°C tested fracture surfaces and initiation sites were similar and will be discussed simultaneously.

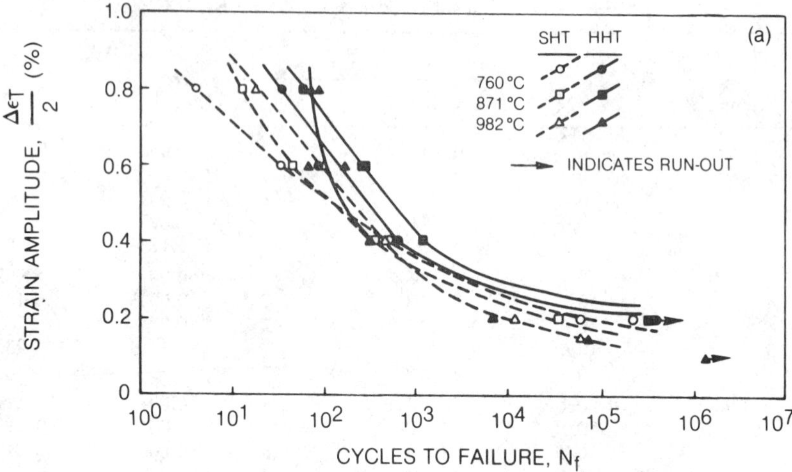

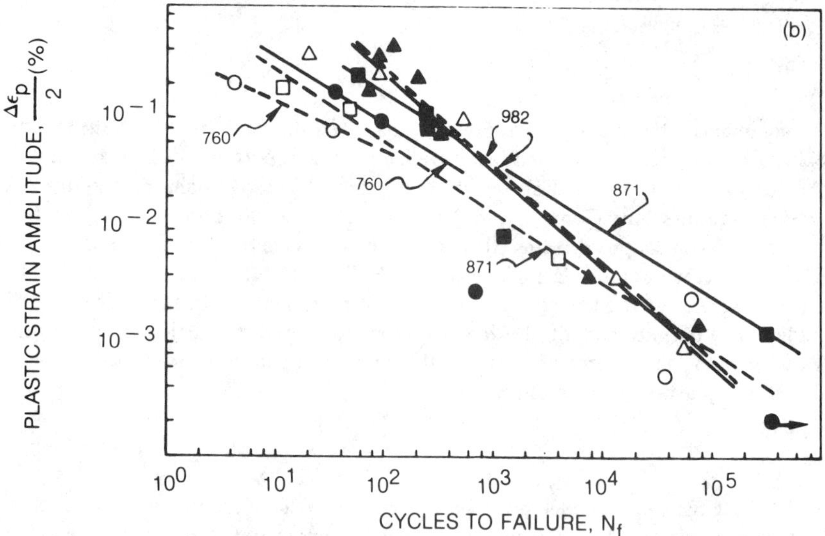

FIG. 5—S/N *curves showing advantage of HHT* (solid symbols) *over SHT* (open symbols). (a) *Total strain amplitude.* (b) *Plastic strain amplitude.*

Typical of the SHT fractures was the presence of many pores encountered along the crack path. Figure 6 shows a low magnification (Fig. 6*a*) of the fracture surface revealing crack features characteristic of initial growth along a transverse grain boundary with mixed intergranular and transgranular crack growth perpendicular to the stress axis. The region indicated by the arrow in Fig. 6*a* is shown enlarged in Fig. 6*b* where a number of pores are readily identified (arrowed). This region is believed to be one of the initiation sites, with the probability of multiple initiation followed by crack coalescence being high.

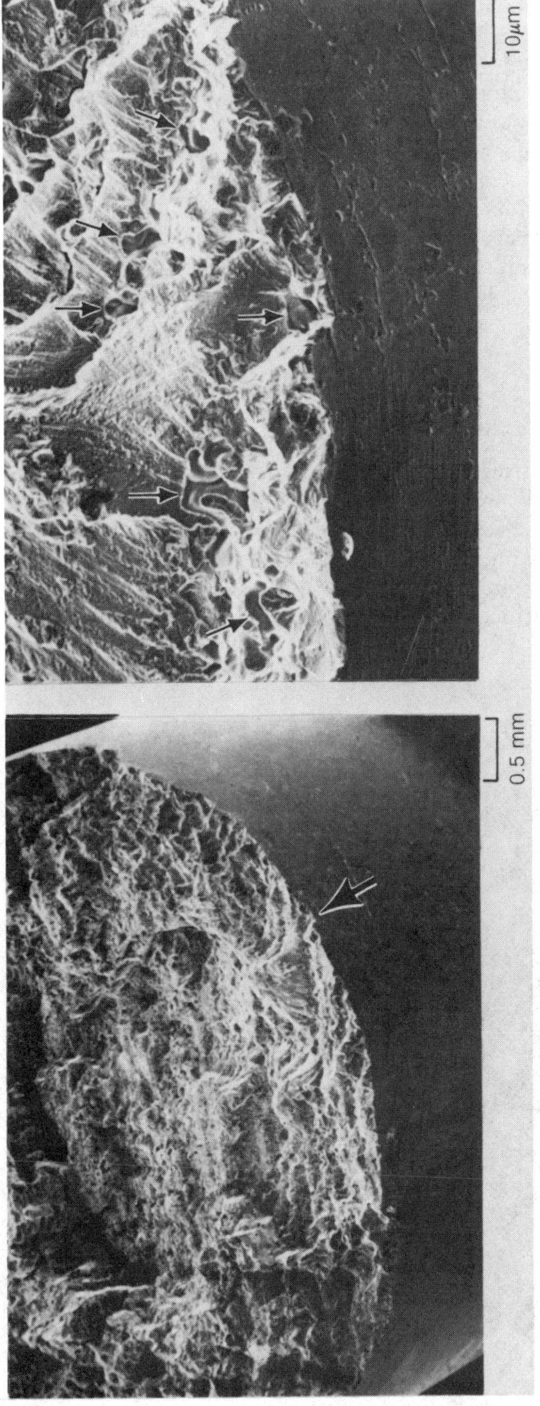

FIG. 6—*Fractograph of SHT specimen tested at* $\Delta\epsilon_T/2 = 0.4\%$ *and* T = 781°C. (a) *Fatigue crack growth region with initiation site at arrow.* (b) *Near-surface initiation site showing numerous pores.*

One of a number of probable initiation sites found in the HHT specimen tested at 760°C and $(\Delta\epsilon_T/2) = 0.4\%$ is given in Fig. 7. A number of sites were found in this specimen and others tested at 760 and 871°C of the HHT alloy. From these crack initiation sites the crack grew on a plane of maximum shear ($\sim 45°$ to the stress axis) in a small thumbnail region before propagating perpendicular to the stress axis. From the orientation of this Stage I crack growth plane it is expected that initiation was transgranular or, possibly, at a grain boundary-surface intersection with growth strictly through one grain.

Testing at 982°C resulted in fracture surfaces which were much more oxidized than those previously shown but with clear evidence of total transgranular crack propagation. Initiation sites were not easily distinguishable and when they could be identified as in Fig. 8, no obvious reason for their initiation (i.e., grain boundary, large carbides, pores, etc.) could be determined. A possible cause may be localized corrosion such as the oxidation "spikes" identified in Ref *3*. Such localized oxidation of large surface carbides may have obliterated the initiating carbide.

Discussion

The hot isostatic pressing treatment given to the B-1900 alloy of this study did more than eliminate the shrinkage porosity for which it was intended. The γ' size and distribution were altered as were the grain size and grain boundary carbide morphology. Grain growth resulted from the HIP cycle. This phenomenon was also observed in Ref *6* and attributed to recrystallization, where the driving force for recrystallization was attributed to the large localized strain required to close the microporosity. We believe this to be simply grain growth, since HIP closing

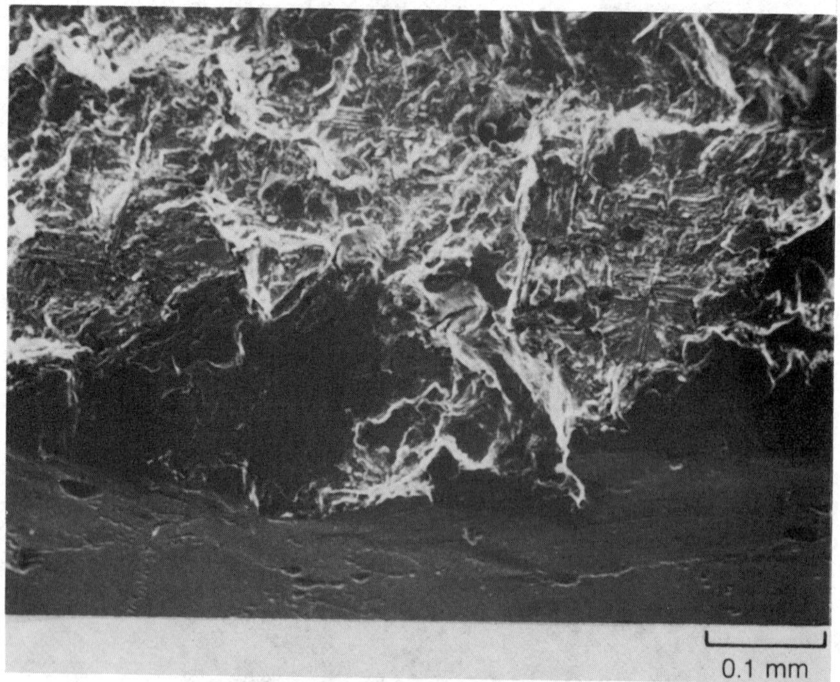

FIG. 7—*Fatigue crack initiation site of HHT specimen tested at* $\Delta\epsilon_T/2 = 0.4\%$ *at 760°C.*

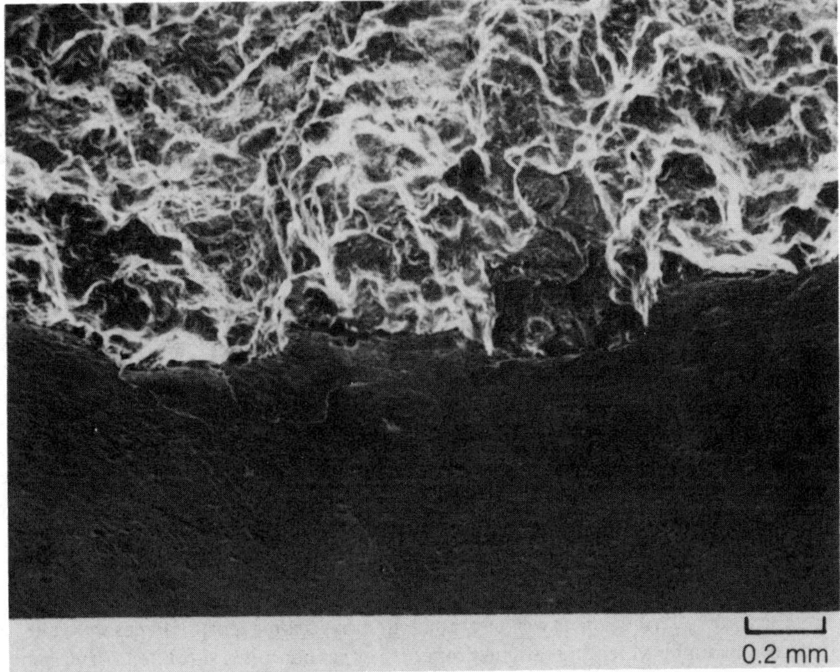

FIG. 8—*Fatigue crack initiation site of SHT specimen tested at* $\Delta\epsilon_T/2 = 0.6\%$ *at 982°C.*

of porosity in single crystal superalloys has never caused recrystallization to occur. Subsequent low temperature ages re-precipitated the Mo-rich fine M_6C carbides at the new grain boundaries. Determining the influence of these two effects on fatigue life is not straightforward. A possible route of investigation would be to monotonically strain the as-cast bars. A high temperature solution will trigger recrystallization and grain growth while leaving the pores essentially intact. Subsequent fatigue testing would determine if the distribution of grain boundary carbides plays a significant role in the enhancement of fatigue life.

Another by-product of HIP is the homogenization of chemical gradients built up during dendritic solidification [*13,14*]. Typically this entails the diffusion of Mo, Ta, and Al from the interdendritic regions into the dendrite cores. It was noted that the SHT alloy contained a uniform γ' distribution. Since this is the point of highest chemical segregation, chemistry differences cannot account for the non-uniformity of γ' size in the HHT alloy. The slow furnace cool of the HIP cycle allowed for rapid γ' coarsening just below the solvus temperature. The following 15 min age at 1232°C was not long enough to totally solution the γ'. These differences were manifest in the yield data. The finer γ' dispersion of the SHT alloy resulted in higher yield stresses at 871°C and below and lower strain amplitudes. It is expected that the slow furnace cool played a more significant role in determining the γ' size than did the leveling of chemical gradients.

Since HIP is affecting the microstructure beyond simply closing porosity, it is necessary to separate those changes associated with yield strength (i.e., γ' size and distribution as well as chemical homogeneity) and those affecting crack initiation and growth, such as grain size and grain boundary carbide distribution. Under constant total strain amplitude testing $\Delta\epsilon_p$ (HHT) > $\Delta\epsilon_p$ (SHT) at 760 and 871°C due to the HHT alloy's lower yield stress. Thus the yield stress variation should cause a variation in life given by the Coffin-Manson Law, $\Delta\epsilon_p = AN_f^{-a}$. By plotting $\Delta\epsilon_p$ versus N_f one will then negate the microstructural effects influencing yield stress and be left with a comparison of microstructure affecting crack initiation and growth.

Figure 5*a* clearly shows that at 760 and 871°C the HIP alloy displayed a greater fatigue life than the standard heat treated alloy. This advantage was on the order of an 8 times improvement in life at all the total strain ranges tested. When the γ' size differences and their effect on yield stress are taken into account through plotting cycles to failure versus the plastic strain amplitude (Fig. 5*b*), the advantage of HIP was reduced to a factor of 4. These results are directly comparable since no change in cyclic yield properties was noted and the moduli are essentially identical. This advantage is due in part to both the loss of porosity after HIP and grain growth which left the grain boundaries free of primary carbides. The new grain boundaries contained fewer M_6C carbides than in the SHT condition (compare Figs. 2*b* and 3*b*), yielding more ductile grain boundaries which were less susceptible to crack growth and not influenced by preferential oxidation of primary carbides. It has thus been demonstrated that HIP enhances LCF life via the yield strength mechanisms of γ' refinement and chemical homogenization and by grain boundary modifications of closed porosity and carbide redistribution. The former was shown to increase life by a factor of 2, while grain boundary modifications were responsible for a factor of 4 benefit. HIP resulted in a total factor of 8 enhancement in LCF life.

The fractographic evidence given here supports the claim that closing porosity extends the low-cycle fatigue life. Because the HHT initiation sites were not clearly delineated and the γ' precipitate size and distribution as well as grain boundary carbide morphologies were altered by HIP, the mechanism by which the fatigue properties are enhanced is shrouded. Arguments can be put forth for a slip dispersion mechanism by large γ' particles in which either the variation in γ' size from area to area or simply larger precipitates results in a lower degree of planar slip and small stress concentrations at brittle grain boundaries due to dislocation pile-ups. Another plausible explanation would be a decrease in grain boundary carbides yielding more ductile boundaries or those less susceptible to oxidation attack. It has been pointed out previously that all these factors come into play.

Conclusions

1. HIP enhances total strain-controlled LCF life by a factor of 6 to 8 at 871°C and below at all strain amplitudes tested and extended fatigue life at 982°C and the highest strain amplitude, 0.8%, but had no influence at lower strain amplitudes.
2. HIP treating of B-1900 alters the microstructure besides other than nearly closing shrinkage porosity. Grain boundary carbide morphologies change as well as γ' size and distribution during this high temperature process.
3. Fatigue crack initiation in the standard heat treated alloy was grain boundary related, with microporosity found to reside in the vicinity of grain boundaries. Since crack growth, especially in its early stages, is through transverse grain boundaries, many pores could be identified in the vicinity of fatigue crack origins.
4. Fatigue crack initiation and propagation in the HIP material was primarily grain boundary related. We infer that the HIP treatment sufficiently alters the grain boundaries through recrystallization and grain growth to enhance fatigue life by a factor of 4.
5. Low-cycle fatigue crack growth at 982°C was completely transgranular, while crack initiation sites were obliterated by oxidation. Since HIP's major impact on LCF life is modification of the grain boundary structure, the lack of intergranular cracking leads to minimal cyclic life enhancement by HIP at 982°C.

Acknowledgments

The authors would like to acknowledge the assistance of Mr. H. Becker of the Pratt and Whitney experimental foundry for preparation of the cast bars and Mr. C. Bankroft for assistance in servohydraulic testing.

References

[*1*] Leverant, G. R. and Gell, M., *Transactions of the Metallurgical Society of AIME,* Vol. 245, 1969, p. 1167.
[*2*] Antolovich, S. D., Domas, P., and Strudel, J. L., *Metallurgical Transactions,* Vol. 10A, 1979, p. 1859.
[*3*] Antolovich, S. D., Liu, S., and Baur, R., *Metallurgical Transactions,* Vol. 12A, 1981, p. 473.
[*4*] Nazmy, M. Y., *Metallurgical Transactions,* Vol. 14A, 1983, p. 449.
[*5*] Burke, M. A., Beck, C. G., Jr., and Crombie, E. A. in *Superalloys 1984,* M. Gell, C. S. Kortovich, R. H. Bricknell, W. B. Kent, and F. Radavich, Eds., TMS-AIME, Warrendale, Pa., 1984, p. 63.
[*6*] Schneider, K., Gnirb, G., and McColvin, G. in *High Temperature Alloys for Gas Turbines—1982,* R. Brunetand, D. Coutsouradis, T. B. Gibbons, Y. Lindblom, D. B. Meadowcraft, and R. Stickler, Eds., D. Reidel Publishing Co., London, 1982, p. 319.
[*7*] Leverant, G. R., Gell, M., and Hopkins, S. W., *Materials Science and Engineering,* Vol. 8, 1971, p. 125.
[*8*] Gell, M. and Leverant, G. R., *Acta Metallurgica,* Vol. 16, 1968, p. 553.
[*9*] Leverant, G. R. and Gell, M., *Metallurgical Transactions,* Vol. 6A, 1975, p. 367.
[*10*] Anton, D. L., *Acta Metallurgica,* Vol. 32, 1984, p. 1669.
[*11*] Hoffelner, W. and Singer, R. F., *Metallurgical Transactions,* Vol. 16A, 1985, p. 393.
[*12*] Anton, D. L., *Scripta Metallurgica,* Vol. 16, 1982, p. 479.
[*13*] Kabalbal, R., Montoya-Cruz, J. J., and Kattamis, T. Z., *Metallurgical Transactions,* Vol. 11A, 1980, p. 1547.
[*14*] Montoya-Cruz, J. J., Kabalbal, R., Kattamis, T. Z., and Giamei, A. F., *Metallurgical Transactions,* Vol. 13A, 1982, p. 1153.

S. Q. Zhang,[1] *C. H. Tao,*[1] *and M. G. Yan*[1]

Effect of Microstructures on Low Cycle Fatigue Behavior in a TC6 (Ti-6Al-2.5Mo-2Cr-0.5Fe-0.3Si) Titanium Alloy

REFERENCE: Zhang, S. Q., Tao, C. H., and Yan, M. G., **"Effect of Microstructures on Low Cycle Fatigue Behavior in a TC6 (Ti-6Al-2.5Mo-2Cr-0.5Fe-0.3Si) Titanium Alloy,"** *Low Cycle Fatigue, ASTM STP 942,* H. D. Solomon, G. R. Halford, L. R. Kaisand, and B. N. Leis, Eds., American Society for Testing and Materials, Philadelphia, 1988, pp. 838–852.

ABSTRACT: The results of an investigation on the low cycle fatigue (LCF) behavior, at room temperature and 400°C for four conventional microstructures (Widmannstatten, basket-weave, equiaxed, and duplex) in a TC6 titanium alloy are presented. The fatigue crack nucleation and propagation in fatigue-tested specimens have been observed by scanning electron microscopy (SEM). The duplex microstructure is associated with the longest LCF life at room temperature and 400°C, while the Widmannstatten microstructure has the shortest. The crack initiation sites and propagation paths were examined and discussed. The cracks primarily initiated along slip bands on the specimen surface for all four microstructures. In addition, many voids appeared along slip bands for the equiaxed microstructure. By linking-up these voids, the formation of microcracks is realized. The propagation of interior cracks in specimens with Widmannstatten structure proceeded by cross-cutting $W\alpha$ platelets by way of a plastic blunting mechanism, whereas for the equiaxed microstructure interior cracks grew by the linking-up of voids by way of a renucleation mechanism.

KEY WORDS: microstructure, low-cycle fatigue, titanium alloy, grain boundary, slip bands, voids, crack initiation, crack propagation, cyclic softening, renucleation mechanism

TC6 alloy (Ti-6Al-2.5Mo-2Cr-0.5Fe-0.3Si) has been used in aircraft engines as compressor disks operating at 400°C. Although several papers concerning the relationship between microstructures and low cycle fatigue properties in titanium alloys can be found in the literature [*1–12*], there is still a lack of information about the correlation of strain-controlled low cycle fatigue properties with the microstructures in alloys with a chemical composition analogous to TC6 alloy; however, some data of stress-controlled low cycle fatigue (LCF) related to VT3-1 alloy have been published [*13, 14*]. This paper presents the results of an investigation on strain-controlled LCF behavior at room temperature and 400°C for four conventional microstructures (Widmannstatten, basket-weave, equiaxed, and duplex) in a TC6 titanium alloy; the initiation and propagation of fatigue cracks have been studied as well.

Experimental Methods

The alloy was received in the form of bars 40 mm in diameter. The chemical composition of the TC6 alloy is listed in Table 1. The diameter of the bars was further reduced to 20 mm by rolling, then the bars were heat treated in order to obtain the required microstructures. The

[1]Institute of Aeronautical Materials, Beijing, China.

TABLE 1—*Chemical composition (weight percent) of TC6 alloy.*

Al	Cr	Mo	Si	Fe	C	N	H	O	Ti
6.08	2.01	2.66	0.24	0.32	0.018	0.018	0.003	0.09	bal.

rolling temperatures and heat treatment procedures are shown in Table 2. Heat treatments were carried out in a muffle furnace. The β transus temperature of the alloy is about 965°C.

The geometry of the fatigue specimens is shown in Fig. 1. The fatigue tests were performed on a MTS servohydraulic testing machine under strain control with $R = -1$. A triangular strain-time waveform was used. The cyclic frequency was 0.5 Hz. The cyclic number at which the load applied to the specimen dropped to 30% of its starting value was referred to as *fatigue life*, N_f. Part of the specimens were mechanically polished prior to testing for the purpose of observing slip bands and fatigue cracks on the specimen surface after LCF testing. Slip bands and cracks were observed in a JSM-35 scanning electron microscope.

Experimental Results

Metallography

Four photomicrographs of the TC6 alloy are shown in Fig. 2. In specimens with Widmannstatten microstructure, thick grain boundary α (GBα) phase was observed. The prior β grains consisted of zones called *colonies,* in which the $W\alpha$ platelets are parallel to each other along a common orientation. Grain boundaries in the basket-weave microstructure were broken up, the short and thick α platelets appeared as a weave, and there were also zones of equiaxed α phase

TABLE 2—*Regimes of rolling and heat treatment.*

Microstructure	Rolling Temperature, °C	Heat Treatment
Widmannstatten	940	1000°C, 1 h, FC→800°C, FC[a]→650°C, 2 h, AC
Basket-weave	990	920°C, 1 h, FC→800°C, FC[a]→650°C, 2 h, AC
Equiaxed	940	870°C, 1 h, FC→800°C, FC[a]→650°C, 2 h, AC
Duplex	940	920°C, 2 h, AC→550°C, 6 h, AC

[a]With furnace door opened.

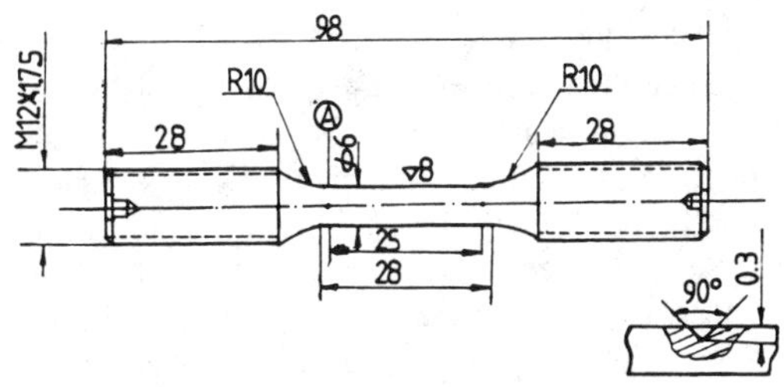

FIG. 1—*Geometry of fatigue specimen.*

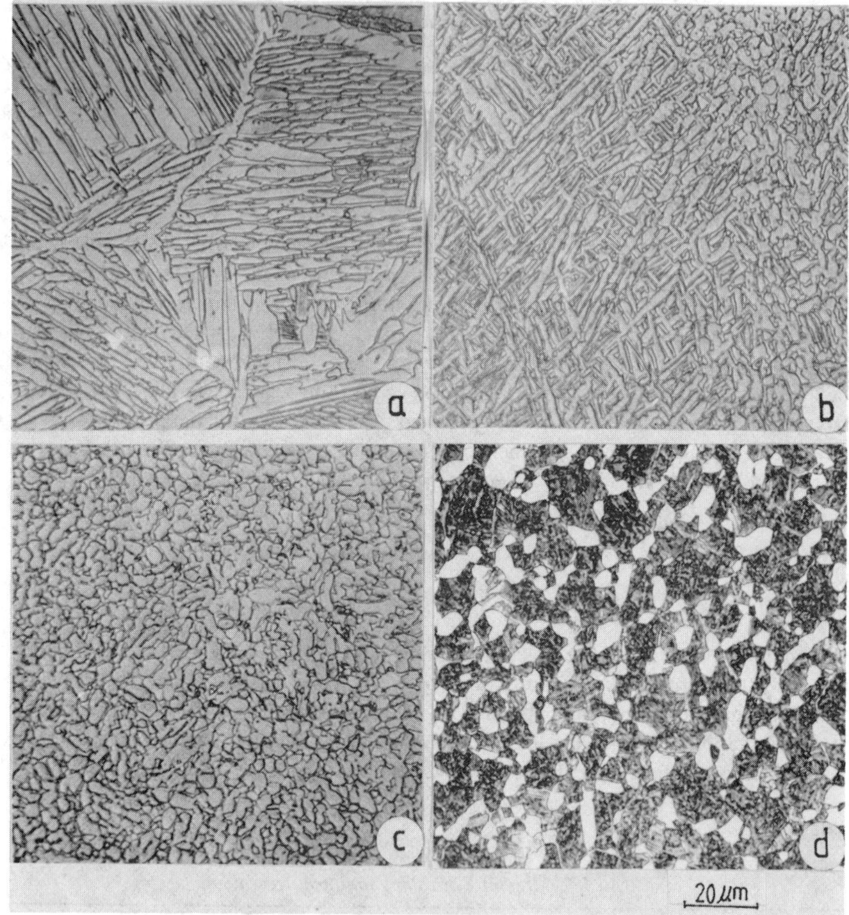

FIG. 2—*Microstructures of TC6 alloy.* (a) *Widmannstatten.* (b) *Basket-weave.* (c) *Equiaxed.* (d) *Duplex.*

in this microstructure. The equiaxed structure contained fine particles of α and β phases, and the duplex structure consisted of equiaxed primary α and transformed β regions.

Tensile Properties

Results of the tensile test for the four microstructures are given in Table 3. In general, the Widmannstatten structure had the lowest room temperature tensile ductility and the equiaxed structure had the highest. The duplex structure possessed the combination of the highest strength with adequate ductility of the four microstructures. Although at 400°C the Widmannstatten structure had good tensile strength, its ductility was significantly lower than that of the other structures. Figure 3 shows the tensile stress-strain curves of the four microstructures. It is worth noting that for the duplex and Widmannstatten structures, the strain hardening occurred when the stress exceeded the yield stress, whereas for the equiaxed and basket-weave structures, the upper and low yield stress points appeared at the stress-strain curves.

TABLE 3—*Tensile properties of the four microstructures.*

Microstructure	Test Temperature, °C	$\sigma_{0.2}$, MPa	σ_u, MPa	Elongation, %	Reduction of Area, %	E, MPa
Widmannstatten	18	992	1047	12.1	13.9	1.27×10^5
	400	. . .	837	10.6	25.1	. . .
Basket-weave	18	1005	1046	19.1	45.6	1.19×10^5
	400	. . .	809	19.1	55.7	. . .
Equiaxed	18	1058	1082	19.7	51.2	1.15×10^5
	400	. . .	820	21.2	57.6	. . .
Duplex	18	1168	1187	16.9	52.5	1.20×10^5
	400	. . .	858	16.8	66.5	. . .

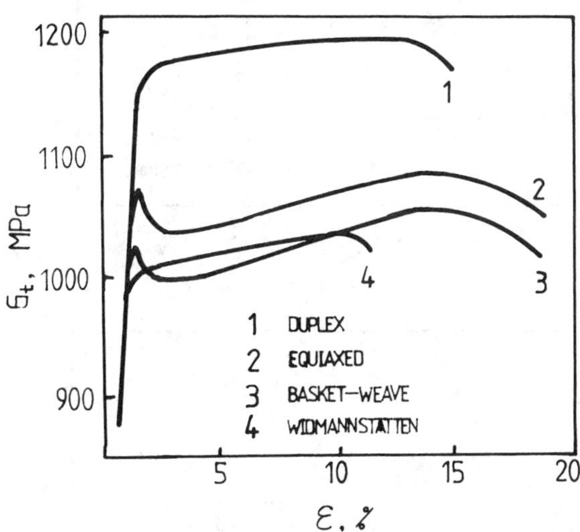

FIG. 3—*Tensile stress-strain curves tested at room temperature for four microstructures.*

LCF Properties

The relationships between total strain $\Delta\epsilon_T$ and fatigue life N_f at room temperature and 400°C are illustrated in Fig. 4. The $\Delta\epsilon_T - N_f$ curves for the four microstructures are straight lines in logarithmic coordinates. By using the least-squares method, the expression of $\Delta\epsilon_T - N_f$ curves for the four structures can be expressed as

$$\Delta\epsilon_T \cdot N_f^{\alpha} = c$$

where c and α are constants related to materials and testing conditions. The values of c and α are listed in Table 4.

The results of LCF tests (Fig. 4) have shown that at room temperature and 400°C the duplex structure is associated with the highest fatigue life, while the Widmannstatten microstructure is

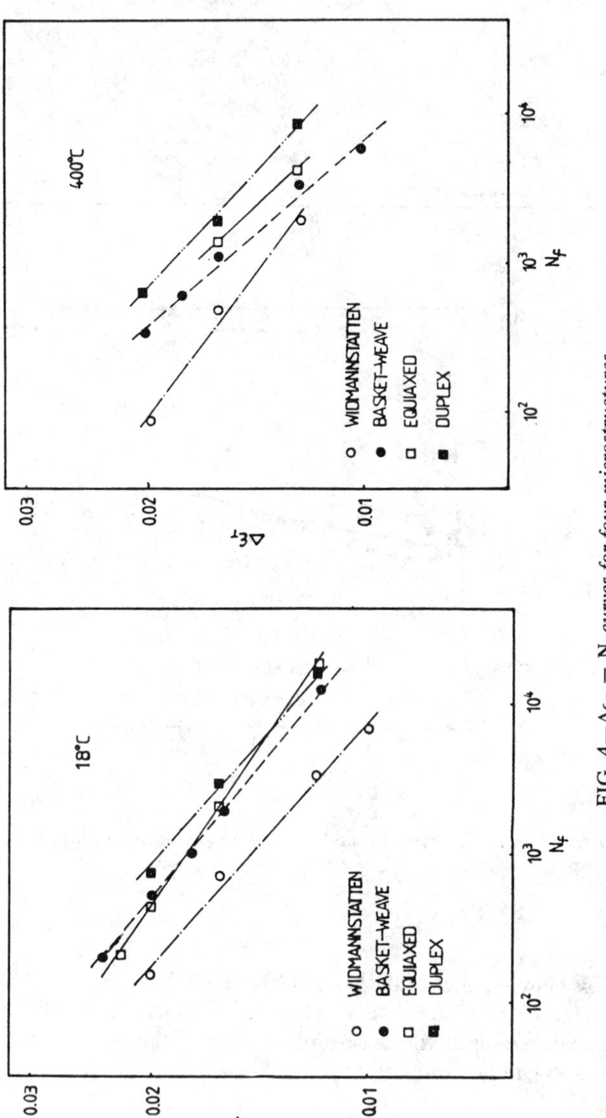

FIG. 4—$\Delta\epsilon_T$ − N_f *curves for four microstructures.*

TABLE 4—*Values of* c *and* α *for the four microstructures.*

Microstructure	Test Temperature, °C	c	α
Widmannstatten	18	0.05547	−0.19178
	400	0.03933	−0.14757
Basket-weave	18	0.05971	−0.17205
	400	0.09752	−0.25578
Equiaxed	18	0.04807	−0.14006
Duplex	18	0.06147	−0.16535

associated with the lowest. Similar results for other alloys were reported by Eylon [*1,9*], Hoffmann [*2*], and Funkenbusch [*3*], among others. As for the other two microstructures, basket-weave exhibits a fatigue strength greater than that of equiaxed at a large strain ($\Delta\epsilon_T > 0.02$); however, the fatigue life of the latter is higher than that of the former at a small strain ($\Delta\epsilon_T <$ 0.02).

Cyclic softening was encountered during LCF testing at room temperature for all these microstructures. The degree of softening was increased with increasing strains. It is well known that the cyclic characteristics are associated with the tensile properties of materials [*15*]. Softening will take place in materials in which the $\sigma_u/\sigma_{0.2}$ ratio is less than 1.2. Therefore the softening of specimens of TC6 alloy with the aforementioned microstructures can be predicted due to their $\sigma_u/\sigma_{0.2} < 1.2$. Figure 5 illustrates the $\sigma_t - N$ curves for the four microstructures at room temperature and $\Delta\epsilon_T = 0.02$. The $\sigma_t - N$ curve can be divided into three stages (initiation, slow propagation, rapid propagation).

Initiation Stage—This stage includes the number of cycles from the beginning to the first abrupt point on the curve. This stage is approximately one tenth of the total fatigue life and softening was remarkable in the materials. It is assumed that microcracks emerged at the speci-

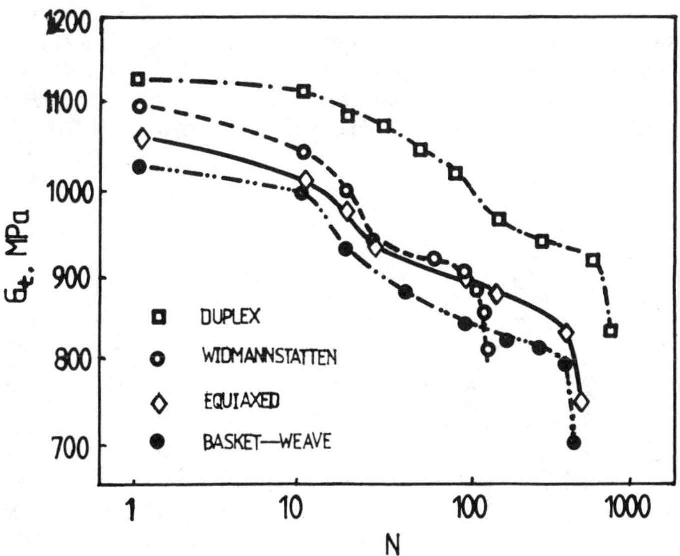

FIG. 5—$\sigma_t - N$ *curves of four microstructures at room temperature and* $\Delta\epsilon_T = 0.02$.

men surface. Judging from the curves of Fig. 5, it appears that at $\Delta\epsilon_T = 0.02$ the initiation life, N_i, was about 20 cycles for the Widmannstatten, equiaxed, and basket-weave microstructures, but about 80 cycles for the duplex one. This shows the superiority of the latter structure over the other three. These results are consistent with the data shown by Sattar et al. [*16*]. These authors indicated that the number of cycles to crack initiation in $\alpha + \beta$ forged Ti-6Al-4V alloy was two to four times higher than that in β forged materials.

Slow Propagation Stage—In this stage the stress does not change remarkably. Fatigue life is mainly dependent on this stage (fatigue propagation life N_p). The fatigue crack initiation and propagation lives at large strains are listed in Table 5.

Rapid Propagation Stage—In this stage the stress dropped rapidly and propagated unstably. This stage occupied only a small part of the number of cycles, which can be neglected.

Observation of Initiation and Growth of Fatigue Cracks

The microstructures, slip bands, and cracks on the specimen surfaces were observed by scanning electron microscopy (SEM). The vertical direction of SEM photographs is parallel to the specimen axis.

Figure 6 shows the slip bands and cracks on the surface of a specimen of Widmannstatten structure tested at 400°C and $\Delta\epsilon_T = 0.02$. The cracks run along the direction that makes a 45° angle with the main stress axis. The cracks are branched and a number of microcracks appeared in the vicinity of the main crack (Fig. 6*a*). Slip bands were distributed closely on the specimen surface. Extrusions and intrusions were produced along slip bands, and microcracks initiated along the intrusions of slip bands during cyclic loading (Fig. 6*b*). Once the microcracks formed, they propagated along the slip bands. When the slip direction was consistent with the direction of long axis of the $W\alpha$ platelets, cracks proceeded preferentially along the interface of $W\alpha$ platelets. These platelets could be cut transversely by slip bands, if the slip direction were perpendicular to the length of the platelets. At large strains, wavy slip bands can be observed. In the case of small strain ($\Delta\epsilon_T = 0.012$) the slip bands appear to be straight.

At room temperature and $\Delta\epsilon_T = 0.016$, the microcracks on the surface of specimens with Widmannstatten structure branched and developed along the direction that makes a 45° angle with the main stress axis. In comparison with the test results at 400°C, the slip bands on the specimen surface appeared to be sparser and occurred locally. Fatigue cracks were formed and propagated along slip bands straightforwardly regardless of the orientation of $W\alpha$ platelets (Fig. 7). In addition, the wavy slip was rarely observed. The slip bands became sparser (e.g., $\Delta\epsilon_T = 0.016$) with a decreasing amount of deformation. Propagation of fatigue cracks along $GB\alpha$ has been observed as well.

TABLE 5—Ni *and* N_p *values for the four microstructures.*

Microstructure	$\Delta\epsilon_T$	N_i	N_p	N_f
Widmannstatten	0.0205	20	136	156
	0.0162	80	664	744
Basket-weave	0.0205	20	486	506
	0.0161	120	1098	1218
Equiaxed	0.0205	20	504	524
	0.0160	200	1903	2103
Duplex	0.0200	80	648	728

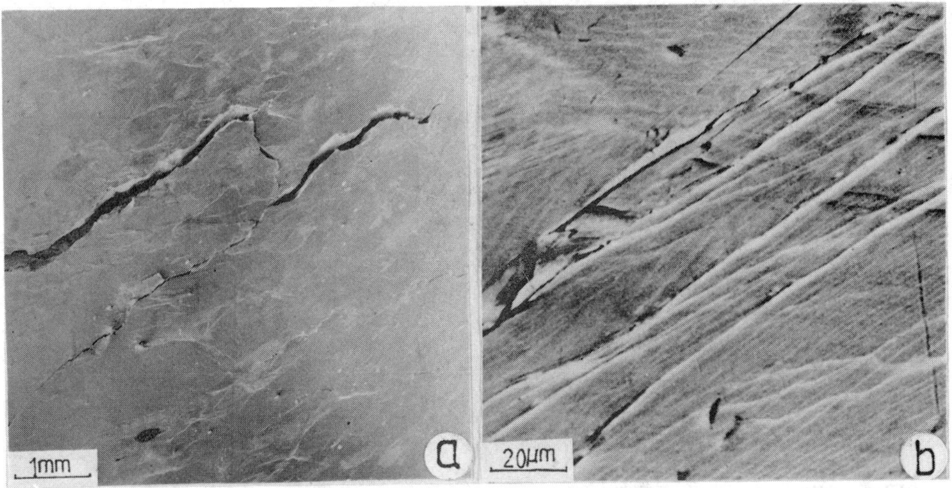

FIG. 6—*Slip bands and microcracks on specimen surface with Widmannstatten microstructure tested at 400°C,* $\Delta\epsilon_T = 0.02$, *and* N = *84.*

FIG. 7—*Slip bands and microcracks on specimen surface with Widmannstatten microstructure tested at room temperature,* $\Delta\epsilon_T = 0.016$, *and* N = *459.*

The slip in basket-weave microstructure was not so intensively exhibited as in the Widmannstatten microstructure under the same loading conditions. At 400°C the slip bands were sparse and distributed along the direction of maximum shear stress. Fatigue microcracks initiated at the sites of intrusions of slip bands (Fig. 8*a*). The tendency to crack along the interfaces of α/β was very obvious, in particular as the interfaces were consistent with the maximum shear stress direction (Fig. 8*b*). At room temperature the slip bands were relatively thin, long, and dense, and distributed along the direction of maximum shear stress.

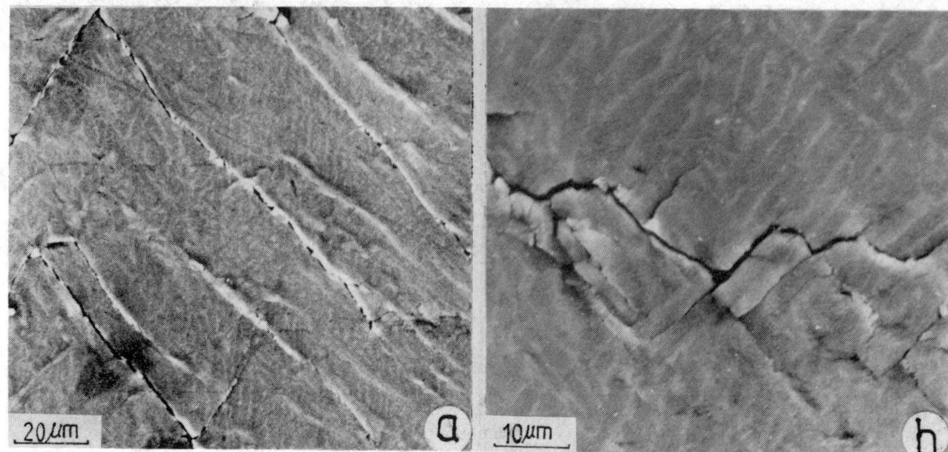

FIG. 8—*Microcracks on specimen surface with basket-weave microstructure tested at 400°C,* $\Delta\epsilon_T$ = *0.016, and* N = *819.*

After LCF testing at 400°C, slip bands on the surface of the specimen with equiaxed microstructure formed at a 45° angle with the main stress axis. It is worthwhile to point out that a large number of voids were present along slip bands on the surface of the specimen with a large strain ($\Delta\epsilon_T = 0.016$) (Fig. 9). But at small strains (e.g., $\Delta\epsilon_T = 0.012$) no voids could be found along slip bands, and the slip bands became thinner.

For the duplex microstructure, the macrocrack on the specimen surface propagated approximately perpendicular to the direction of main stress axis. The slip bands on the surface of the specimen which suffered large deformation ($\Delta\epsilon_T = 0.20$) were thin and dense, and along the slip bands thin cracks formed. A significant feature of the duplex structure is the presence of many voids on the specimen surface when tested at large strains (Fig. 10).

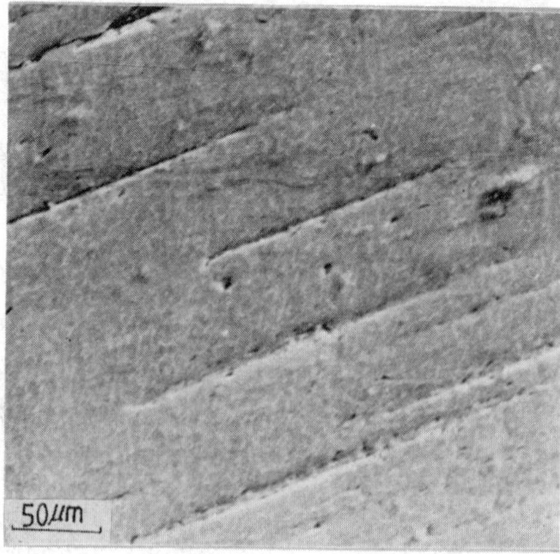

FIG. 9—*Slip bands on specimen surface with equiaxed microstructure tested at 400°C,* $\Delta\epsilon_T$ = *0.016, and* N = *1714; voids along slip bands.*

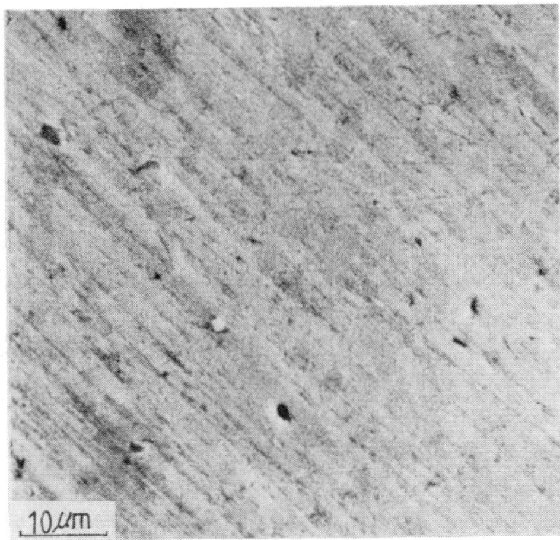

FIG. 10—*Slip bands on specimen surface with duplex microstructure tested at room temperature,* $\Delta\epsilon_T = 0.020$, *and* N = *728; voids on specimen surface.*

For the observation of crack propagation, fatigue-tested specimens were sectioned along the longitudinal axis of the specimen and perpendicularly to the trace of the surface crack. The section surfaces were mechanically ground and polished, and then examined by SEM or an optical microscope.

Figure 11 shows a photograph of fatigue cracks in a specimen with Widmannstatten structure tested at room temperature and $\Delta\epsilon_T = 0.02$. The main crack propagated tortuously along the

FIG. 11—*Optical photomicrograph of fatigue cracks in specimen with Widmannstatten microstructure tested at room temperature,* $\Delta\epsilon_T = 0.02$, *and* N = *156.*

direction of maximum shear stress, cross-cutting the $W\alpha$ platelets of colonies and running along the grain boundary α phase. Secondary cracks in the vicinity of the main crack and in front of it can be observed to be along the grain boundary α phase and the boundary of colonies. This phenomenon was observed for tests at 400°C as well. The morphology of a fatigue crack tip in the Widmannstatten structure is shown in Fig. 12. The crack tip appears to be deformed during crack propagation by the mechanism of plastic blunting [*17*].

For the basket-weave microstructure the fatigue cracks propagated predominantly along the direction of macroscopic shear stress, mainly crossing over α/β interfaces and sometimes propagating along α/β interfaces. The cracks propagated intermittently and some microcracks that had not yet linked-up with the main crack were limited in local regions (Fig. 13).

For equiaxed and duplex microstructures the fatigue crack was perpendicular to the main stress axis. Many voids were present in the vicinity of the crack tip in the equiaxed microstructure. Figure 14 shows the cluster of voids in front of the fatigue crack in the equiaxed structure. Under the action of cyclic stresses microcracks nucleated in front of the main crack; these microcracks will finally link-up with the main crack with an increasing number of cycles. For the duplex microstructure the crack propagates ahead, cutting through the primary α phase grains and transformed β regions (Fig. 15).

Discussion

Initiation of LCF Cracks

In the experimental conditions of this study, the fatigue cracks frequently formed along surface slip bands at the intrusions; thus it is reasonable to consider that crack nucleation is associated with slip path length. For the equiaxed microstructure, because of its fine grain size and indefinite orientation relationship between α and β, slip is obstructed by phase boundaries. Therefore the slip distance is short and microcracks initiate preferentially at the phase interfaces where the stress concentration exists [*4,8*]. Besides, the deformation of the polycrystal leads to strain accommodation among the grains and requires more slip systems to operate simultaneously. However, there is only a single slip system in the equiaxed microstructure, so at

FIG. 12—*Fatigue crack tip in specimen with Widmannstatten microstructure tested at 400°C,* $\Delta\epsilon_T$ = *0.016, and* N = *560.*

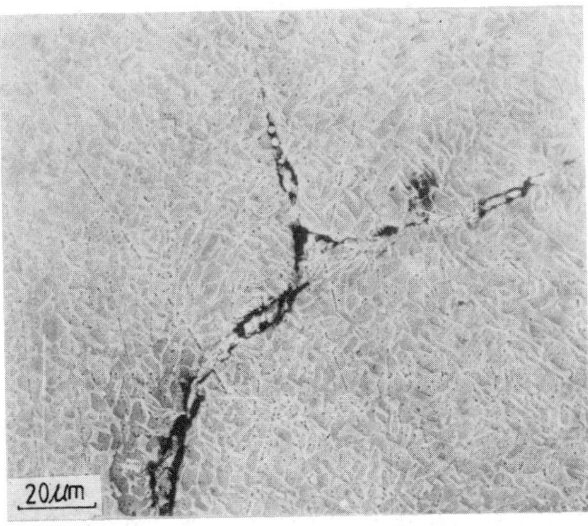

FIG. 13—*Fatigue crack tip in specimen with basket-weave microstructure tested at room temperature,* $\Delta\epsilon_T = 0.024$, *and* N = *212.*

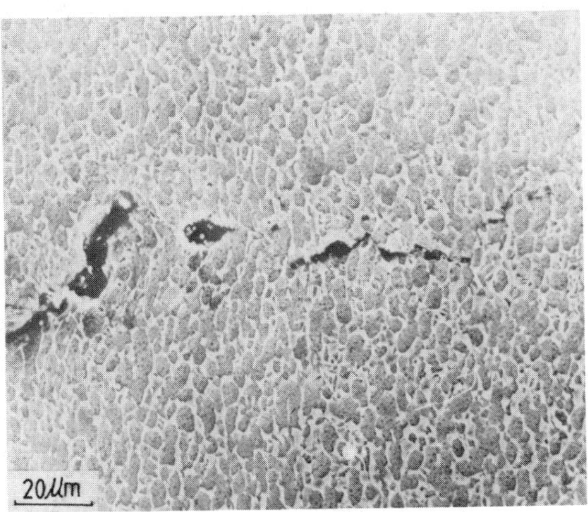

FIG. 14—*Voids in front of the fatigue crack in specimen with equiaxed microstructure tested at room temperature,* $\Delta\epsilon_T = 0.02$, *and* N = *524.*

the equiaxed α and β phase interfaces along slip bands, cracking took place due to stress concentration. The LCF properties of equiaxed primary α structures are superior to those of the coarse colony structures because of the blockage of slip at α/β interfaces. Within this phase, plastic deformation can be more readily accommodated and some dispersal of the slip process takes place [*3*]. As a result, crack initiation occurs at a larger number of cycles than for β-processed materials [*7,12*].

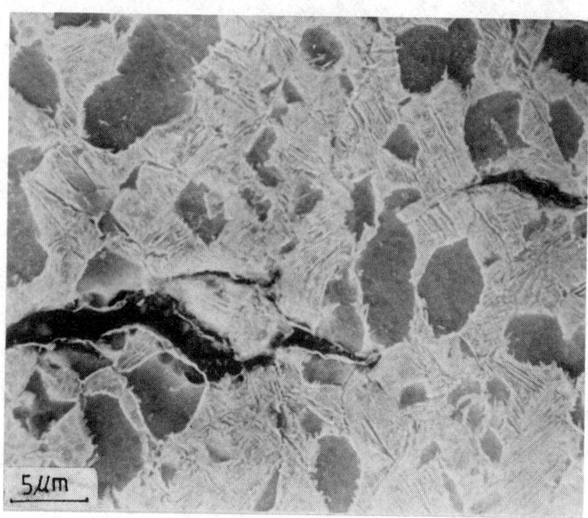

FIG. 15—*Void formation in front of the fatigue crack in specimen with duplex microstructure,* $\Delta\epsilon_T$ = *0.016, and* N = *2063.*

In the case of the Widmannstatten microstructure, the crystallographic orientation relationship between α and β in a colony is $\{0001\}_\alpha//\{110\}_\beta$, $\langle 11\bar{2}0\rangle_\alpha//\langle 111\rangle_\beta$ and $\{10\bar{1}0\}_\alpha//\{112\}_\beta$, $\langle 11\bar{2}0\rangle_\alpha//\langle 111\rangle_\beta$. Consequently, the slip system of α is parallel to that of β, and as a result the slip passed readily through the colonies of $W\alpha$ platelets [*5*]. Especially when the slip direction is consistent with the length direction of $W\alpha$ platelets, the α/β interfaces will serve as the preferential paths of crack propagation. Therefore the slip in a Widmannstatten microstructure has a longer path, and the crack initiation life in Widmannstatten structure is shorter than that in other microstructures [*6, 7*].

It was indicated that, except for the duplex microstructure, there is no great difference among the crack initiation lives of the other three microstructures at room temperature and large strains (Table 5). It is assumed that because of the complete yield of the specimens (i.e., the stress applied to the specimens exceeded the yield stress), under LCF testing at large strains, slip happened in more grains, and the greater driving force made the slip propagate across the grain boundaries. As a result, it seems that the effect of grain size and morphology on path length is reduced. Therefore it appears that the initiation of LCF cracks is not susceptible to microstructures in the present test condition.

Propagation Path of LCF Cracks

The propagation behavior of microcracks is dependent on the degree of deformation. At large strains, the surface fatigue cracks propagated basically along the slip direction; hence the crack propagation path appeared not to be influenced by the microstructure. For example, for the equiaxed microstructure the cracking of α/β interfaces took place along slip bands.

A great number of secondary cracks and the zigzag of crack propagation paths are the notable features of Widmannstatten structure due to the multiple slip. It is due to the higher strength and the susceptibility to stress concentration that the propagation path of the duplex structure is relatively straight. For the equiaxed and basket-weave microstructures, the propagation paths behaved in an intermediate manner between the Widmannstatten and duplex structures.

In regard to the propagation of interior cracks, the specimens microstructure has a significant effect on the propagation paths. The propagation of cracks in specimens with Widmannstatten structure proceeded by cross-cutting $W\alpha$ platelets and grain boundary α by way of a plastic blunting mechanism (Fig. 12), whereas in the equiaxed microstructure the crack extended by linking-up the voids ahead of the crack tip (Fig. 14).

The mechanism of void formation can be described as follows. When the specimen suffered from cyclic loading, the dislocations glide on slip planes. The jogs are formed as they meet and interact with other dislocations. Thus the dislocations with jogs produce voids when they move. The dislocations can be stopped at obstructions such as grain boundaries and phase interfaces. As the stress sign changes reversely, the dislocations leave the obstructions and move reversely on the slip planes and stop again when they meet another grain boundary. In this way, the cyclic stress changes alternatively from negative to positive, and the jog dislocations move repeatedly towards or off the obstructions; thus new voids are produced constantly. With an increasing number of cycles, the voids gradually grew and linked-up to form microcracks, and then the microcracks propagated. Hence it is considered that void formation is the result of the obstruction of grain boundaries to dislocations.

LCF Lives of Four Microstructures

It is generally recognized that LCF life is mainly dependent on the plasticity of materials. At room temperature the Widmannstatten structure has the lowest ductility and lower yield strength, and at 400°C its strength is approximately equal to other microstructures, but its ductility is lower than that of other microstructures. Consequently, it is clear that the LCF life of Widmannstatten microstructure is the shortest. The equiaxed and duplex microstructures have higher strength and good ductility, and therefore better LCF properties. As for the basketweave structure, because its ductility value is in the range of the ductility values of Widmannstatten and equiaxed microstructures, it has the intermediate LCF value of the Widmannstatten and equiaxed structures.

Conclusions

1. The duplex microstructure is associated with the highest LCF life at room temperature and 400°C, while the Widmannstatten microstructure is the lowest.
2. In LCF testing, the fatigue cracks initiated basically along slip bands on the specimen surface for all four microstructures. In addition, voids appeared along slip bands for the equiaxed microstructure. The microcracks are formed by linking-up these voids.
3. The propagation of interior cracks in specimens with Widmannstatten structure proceeded by cross-cutting $W\alpha$ platelets and grain boundary α phase by way of a plastic blunting mechanism, whereas for the equiaxed structure interior cracks grew by the linking-up of voids.
4. Cyclic softening was observed under LCF testing for all four microstructures. The amount of softening increased with increasing cycles.

References

[1] Eylon, D., Rosenblum, M. E., and Fujishiro, S. in *Titanium '80,* Proceedings of 4th International Conference on Titanium, Vol. 3, 1980, p. 1845.

[2] Hoffmann, C., Eylon, D., and McEvily, A. J. in *Low-Cycle Fatigue and Life Prediction, ASTM STP 770,* American Society for Testing and Materials, Philadelphia, 1982, p. 5.

[3] Funkenbusch, A. W. and Coffin, L. F., *Metallurgical Transactions,* Vol. 9A, 1978, p. 1159.

[4] Yashwant, Mahajan, and Margolin, H., *Metallurgical Transactions,* Vol. 13A, 1982, p. 257.

[5] Margolin, H., Williams, J. C., Chesnutt, J. C., and Luetjering, G. in *Titanium '80,* Proceedings of 4th International Conference on Titanium, Vol. 1, 1980, p. 169.

[6] Yashwant, Mahajan, and Margolin, H., *Metallurgical Transactions*, Vol. 13A, 1982, p. 269.
[7] Sprague, R. A., Ruckle, D. L., and Smith, M. P. in *Titanium Science and Technology*, Proceedings of 2nd International Conference on Titanium, Vol. 3, 1973, p. 2069.
[8] Salen, Y. and Margolin, H., *Metallurgical Transactions*, Vol. 14A, 1983, p. 1481.
[9] Eylon, D., Bartel, T. L., and Rosenblum, M. E., *Metallurgical Transactions*, Vol. 11A, 1980, p. 1361.
[10] Eylon, D., Froes, F. H., Heggle, D. G., Blenkinsop, P. A., and Gardiner, R. W., *Metallurgical Transactions*, Vol. 14A, 1983, p. 2497.
[11] Krafft, J. M., *Fatigue of Engineering Materials and Structures*, Vol. 4, No. 2, 1981, p. 111.
[12] Eylon, D. and Hall, J. A., *Metallurgical Transactions*, Vol. 8A, 1977, p. 981.
[13] *Metallografija titanovyh splavov*, C. G. Glazunov and V. A. Kolachov, Eds., Moscow, « Metallurgija », 1980.
[14] Solonina, O. P. and Glazunov, S. G., *Zharoprochnye titanovye splavy*, Moscow, « Metallurgija », 1976.
[15] Hertzberg, R. W., *Deformation and Fracture Mechanics of Engineering Materials*, Wiley, New York, 1976.
[16] Sattar, S. A., Kellogg, D. H., Oberle, K. J., and Green, G. N., ASM Technical Report No. D-8-24, American Society for Metals, Metals Park, Ohio, 1968, p. 4.
[17] Laird, C. in *Fatigue Crack Propagation, ASTM STP 415*, American Society for Testing and Materials, Philadelphia, 1967, p. 131.

William G. Fricke, Jr. [1]

Quantification of Slip Lines in Fatigue

REFERENCE: Fricke, W. G., Jr., **"Quantification of Slip Lines in Fatigue,"** *Low Cycle Fatigue, ASTM STP 942,* H. D. Solomon, G. R. Halford, L. R. Kaisand, and B. N. Leis, Eds., American Society for Testing and Materials, Philadelphia, 1988, pp. 853–858.

ABSTRACT: Although fatigue, even low-cycle fatigue, can occur at stresses below the engineering yield strength of a material, plasticity is involved because slip lines may appear on the surface of a test material. These slip lines increase in number in the early stages of fatigue and are indeed the precursors of ultimate failure. In a cyclic stress test, the initial rapid increase in number of slip lines produces slip saturation, after which other phenomena such as crack initiation become predominant. New slip lines are added at a decreasing rate.

Quantitative measurements have been made of slip lines in fatigue specimens and compared with the number in a tensile test of the same material. In fatigue, the equation describing the rate of addition of new slip lines requires that the stress be described in terms of how much the test stress exceeded a quantity, characteristic of the material, which can be thought of as a "yield strength" in fatigue. This quantity corresponded closely to the measured endurance limit of the material.

A comparison is made with similar measurements in a tensile test.

KEY WORDS: fatigue, fatigue limit, low-cycle fatigue, quantification, slip, tensile aspects

Low-cycle fatigue covers that interesting region between true fatigue and tensile deformation. As we increase the stress imposed in each load application we need fewer applications to cause failure, and the system resembles more and more delayed overload fracture.

Looking in the other direction, as we decrease the cyclic stress and increase the life of a specimen we do not completely eliminate the resemblance to tensile deformation in the fatigue mechanism. It is clear that fatigue on a microscopic scale involves plastic deformation in much the same way that overload failure is preceded by plastic deformation. The same slip systems are used and rules such as maximum resolved shear stress laws operate. Despite the fact that the cycle stresses are below the engineering yield stress, yield—that is, slip—does occur in fatigue.

A microscope and suitable sample preparation are needed to actually see the slip produced by fatigue, and the evidence is not resolvable in all alloys. However, if a suitable alloy is watched under a microscope during a fatigue test, a number of progressive stages are noted. Figure 1 [*1*] shows what happens in an Al-0.97Mg alloy sheet. Within a few cycles slip lines begin appearing on the surface. Shortly after, some of them become cracks: "persistent slip bands" in modern parlance. The number of slip bands increases rather rapidly during these early cycles until a point is reached at about 10% of the life when "slip saturation" occurs and new slip bands are added at a decreased rate. Fatigue cracks which join and pass the confines of individual grains are a consequence of saturation of slip. All these stages are entered at fewer number of cycles as the test stress increases (Fig. 1).

The accumulation of slip with subsequent cracking can be thought of as a type of cyclic work-hardening. This is more evident when the amount of slip was measured quantitatively during tests.

[1]Aluminum Company of America, Alcoa Laboratories, Alcoa Center, PA 15069.

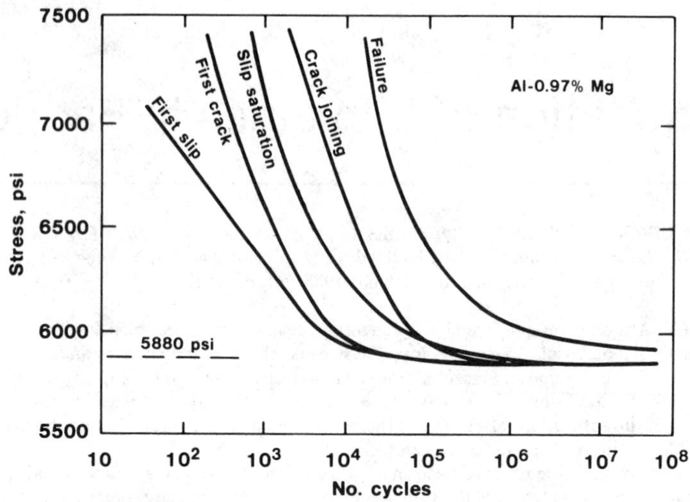

FIG. 1—*Curves of progressive change in fatigue tests. Below 5880 psi no changes occur.*

Method

A random line of known length was projected onto the microscope image of the specimen surface. The number of intersections of fatigue slip bands with that random line was counted. The stereographic quantity L_A, total length of slip bands per unit area, which is a measure of the amount of fatigue deformation, was given by

$$L_A = \frac{\pi}{2} N_l \tag{1}$$

where N_l is the number of intersections per unit length of test line. The method is described more fully in Refs *1* and *2* along with micrographs of fatigue slip lines.

Figure 2, also from Ref *1*, shows quantitatively how the amount of slip increased during fatigue tests of the Al-0.97Mg alloy. The break in the two straight line portions for a single test is defined as the slip saturation point. By comparison, in pure aluminum the second portion is almost horizontal.

Since the data fall on straight lines on the log-log plot (base 10) of Fig. 2, each portion can be represented by an equation of the type

$$L_A = cN^m \tag{2}$$

where m and c represent the slope and intercept of the straight line and N is the number of cycles. The values are obviously properties of the material. Values for the Al-0.097Mg alloy are given in Table 1.

The fatigue deformation coefficients varied with the test cyclic stress. In Ref *1* it was stated that no simple relation with stress was found. Such a relation is now known to exist.

Results

Instead of trying to find a relation between a deformation coefficient and test stress, one should find a relation with stress in excess of a critical amount. Figure 3 shows with the filled

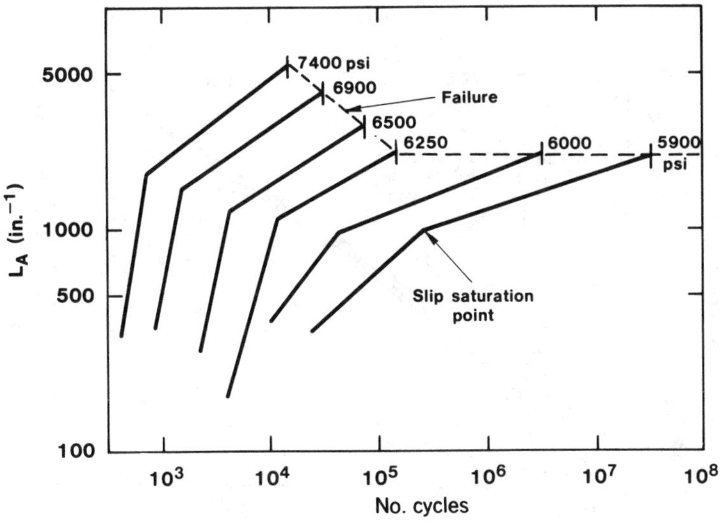

FIG. 2—*Increase in amount of slip during fatigue tests. Dashline connects failure points.*

TABLE 1—*Fatigue deformation coefficients for Al-0.97Mg.*

Stress, psi[a]	m_1	m_2	$\log_{10} c_1$	$\log_{10} c_2$
5900	0.44	0.14	0.63	2.24
6000	0.63	0.21	0.06	2.00
6250	1.69	0.27	−3.83	2.04
6500	2.22	0.30	−4.99	2.00
6900	2.40	0.34	−4.43	2.16
7400	2.88[b]	0.37	−4.94[b]	2.20

[a]Multiply psi by 6895 to obtain pascals.
[b]Estimated value.

circles a Cartesian linear plot (left and bottom scales) of the coefficient m from Table 1 against cyclic stress. The line is curved. If, on the other hand, the coefficient is plotted on a logarithmic scale (top of figure) and the stress in excess of certain levels is also plotted on a logarithmic scale (right), a series of curves results. The curvature of the log-log plot is very sensitive to the exact value subtracted from the cyclic stress. Only when a constant 40.5 MPa (5880 psi) was subtracted from the applied stress was there a straight line on the log-log plot for the alloy.

This value of 40.5 MPa (5880 psi) has special significance. It is the stress in Fig. 1 at which the curves of progressive change become horizontal. In particular, it is the stress above which slip lines will in fact appear on the surface during the test and below which no slip will appear even at very large numbers of cycles. It can be thought of as a yield point in fatigue, insofar as a

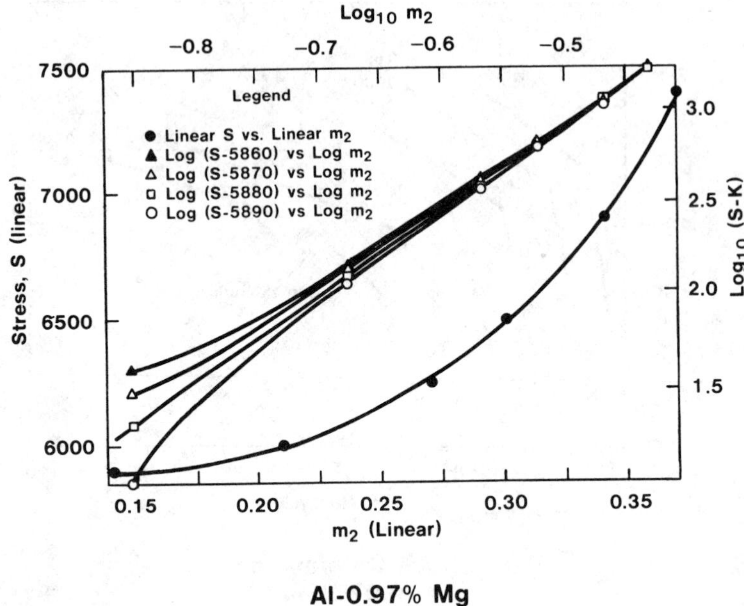

FIG. 3—*Change in value of deformation coefficient* m_2 *with stress. Various scales are used.*

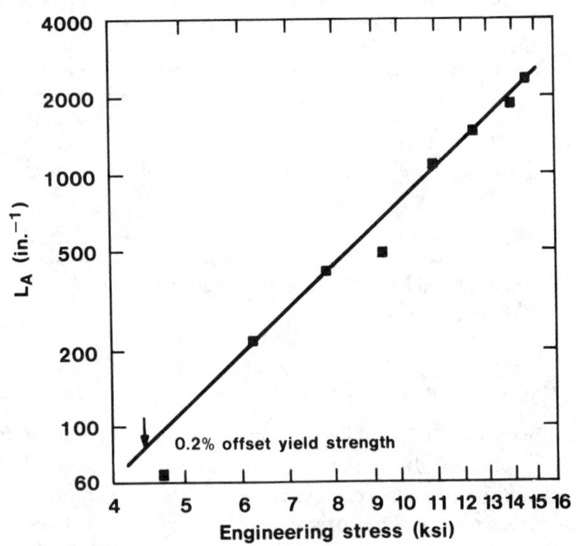

FIG. 4—*Increase in amount of slip during a tensile test.*

yield point is the dividing line between elastic and plastic behavior. If slip is a precursor of cracks and subsequent failure, this critical stress is the stress which must be exceeded for failure to occur. The failure curve in Fig. 1 is becoming asymptotic to this stress of 40.5 MPa (5880 psi) and is, perhaps, the fatigue limit at infinite cycles.

We can carry the analogy further by measuring the amount of slip in a tensile test of the same material. Just as a straight line resulted when amount of slip was plotted against number of cycles on a log-log plot (Fig. 2), a straight line resulted when amount of slip was plotted against stress in a tensile test on a log-log scale (Fig. 4). This implies that the number of cycles is analogous to an increasing internal stress during fatigue.

If one now makes the assumption that the amount of slip in fatigue is equivalent to the amount of slip in tension—and this can only be a crude approximation—one can estimate how this internal stress was built up. For a particular test one could replace the L_A scale in Fig. 2 by using Fig. 4 to give the stress producing that amount of slip in a tensile test. Figure 5 resulted.

The change between Fig. 2 and Fig. 5 is not dramatic (other than the change in ordinate) with one exception. An interpretation can now be put on those tests in the 43.1 to 40.7 MPa (6250 to 5900 psi) range, which the dashed line in Fig. 2 shows had approximately the same amount of cyclic slip at failure. The tensile test at failure also had about this amount of slip.

The curves of Fig. 5 suggest, therefore, that for cyclic stresses around the knee of the *S-N* curve deformation occurs, raising the internal stress to a point where the material can no longer

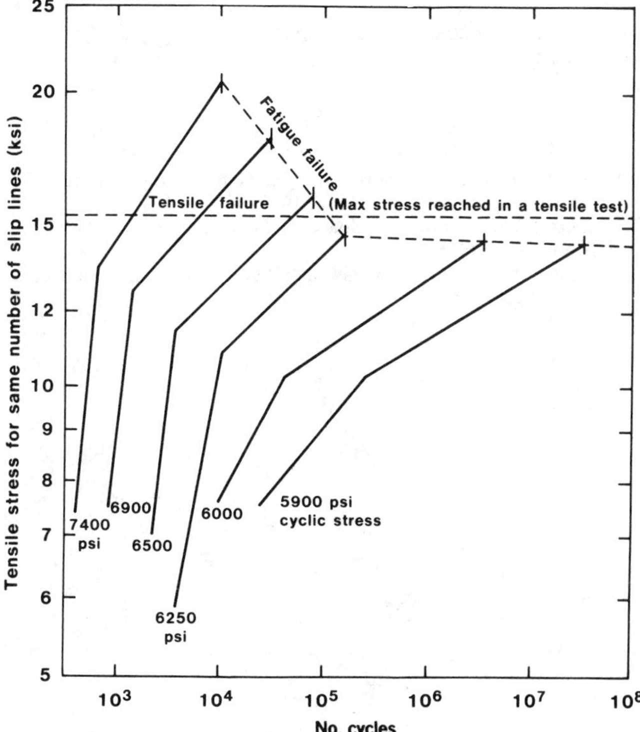

FIG. 5—*Increase in amount of slip during fatigue test compared with the stress in a tensile specimen to give that amount of slip.*

withstand additional deformation. At higher stresses, more slip is present than is encountered in a tensile test. The "stresses" in Fig. 5, above those needed for tensile failure, could only be estimated by extrapolating Fig. 4. The greater number of slip bands in fatigue might be due to a difference in failure mechanism, fatigue specimens failing by crack propagation rather than local necking.

Discussion

The analogy should not be carried too far. Although the curves of Fig. 2 are showing how the bulk material was deforming, ultimate fatigue failure was due to cracks leaving the confines of their parent grains at times somewhat after the slip saturation point. In fact, it might be the change in emphasis from making new slip bands to propagating cracks instead which keeps the number of slip lines in high-cycle fatigue failures at somewhat less than in tensile failure. (The failure points are below the horizontal dashed line in Fig. 5.)

It is useful, however, to think of the fatigue limit (the endurance limit) as resembling a yield point in fatigue. This is illustrated by the following phenomena:

1. No fatigue-induced slip is produced at stresses much below the fatigue limit (Fig. 1).
2. Cracks, because they must be preceded by plastic deformation, are not initiated by stresses below this limit, although cracks will propagate at lower stresses if they are already present [*3*].
3. The law governing the addition of new slip lines involves the portion of the stress in excess of the fatigue limit (Fig. 3).
4. The estimated internal stress at failure for specimens tested just above the fatigue limit is equivalent to that of tensile failure.

References

[*1*] Hunter, M. S. and Fricke, W. G., Jr., "Effect of Alloy Content on the Metallographic Changes Accompanying Fatigue," *Proceedings of ASTM,* Vol. 55, 1955, pp. 942–953.

[*2*] DeHoff, R. T., "Length of Lineal Features in Space," NBS Special Publication 431, National Bureau of Standards, Washington, D.C., 1976, p. 455.

[*3*] Hunter, M. S. and Fricke, W. G., Jr., "Fatigue Crack Propagation in Aluminum Alloys," *Proceedings of ASTM,* Vol. 56, 1956, pp. 1038–1050.

Multiaxial and Variable Amplitude Loading

G. E. Leese[1]

Engineering Significance of Recent Multiaxial Research

REFERENCE: Leese, G. E., "**Engineering Significance of Recent Multiaxial Research,**" *Low Cycle Fatigue, ASTM STP 942,* H. D. Solomon, G. R. Halford, L. R. Kaisand, and B. N. Leis, Eds., American Society for Testing and Materials, Philadelphia, 1988, pp. 861-873.

ABSTRACT: Recent advances in multiaxial fatigue testing capabilities have provided the means with which to evaluate multiaxial life prediction techniques. As a result of the recent surge in testing, several key concepts have emerged as common threads in the results of contemporary multiaxial fatigue investigators. These concepts are discussed within the framework of three major areas: (1) effective stress-strain, (2) plastic work and energy, and (3) critical plane approaches. While general agreement within the technical community has not been reached on many fine points of analytical techniques and predictive expressions, their sensitivity with respect to multiaxial spectrum characteristics, physical interpretation, and ease of application has been described. It is suggested that surveying the present accomplishments will provide valuable insight for steering the future directions of multiaxial low cycle fatigue research.

KEY WORDS: fatigue (materials), multiaxial fatigue, biaxial fatigue, low cycle fatigue, high cycle fatigue, life prediction

The quest for a description of cyclic life as a function of multiaxial stress states imposed upon metal structural components dates back to the 1800s, as is well documented in several excellent interpretative reviews [*1,2*]. Research has continued through the years to accommodate the ever-increasing need for improved life prediction methods, thus enabling more exacting design. Only recently have advances in multiaxial fatigue testing capabilities provided the means with which to compare the proposed life prediction theories with experimental results. Surveying these comparisons with respect to their apparent engineering significance may provide valuable insight for steering the future directions of multiaxial low cycle fatigue research.

While experimentalists and theoreticians disagree on many fine points of analytical techniques and predictive expressions, there are some commonly accepted concepts that may be summarized and supported with the independent results of many contemporary researchers investigating the field of multiaxial fatigue. The following is not a literature survey, but rather highlights several key concepts that have emerged, and addresses their impact on the current engineering needs.

Methods of representing multiaxial fatigue life response will be discussed within three general categories: (1) those in which the three-dimensional stress state is reduced to an effective or equivalent one-dimensional scalar value which is, in turn, related to life; (2) plastic work and energy methods; and (3) critical plane approaches dealing directly with the multiaxial aspects of the failure anticipated. In lieu of presenting analytical details of individual approaches (these are available in the open literature as referenced), a qualitative perspective of their roles in the overall picture of multiaxial fatigue life technology is offered.

[1]MTS Systems Corporation, Minneapolis, MN 55424.

Discussion

Effective/Equivalent Concepts

An approach that has remained popular throughout the years is the reduction of the three-dimensional stress state to an "effective" or "equivalent" scalar stress parameter. The resulting information is thus "formatted" in a manner to allow a direct comparison to be made with uniaxial test results, which may also be expressed in the same effective/equivalent terms. A power law effective stress-life relationship may be presented in much the same way as axial stress-life, for the purposes of correlating the data with life, or even for purposes of using axial data as a baseline response. This similitude with axial fatigue life methodology is a point of great engineering practicality. In essence, it suggests that the existing data base of uniaxial fatigue data could be extrapolated and/or applied to multiaxial situations.

Early acceptance of this concept was gained in application to high cycle fatigue life regimes. The effective or equivalent parameters were frequently extensions of yield criteria, such as the Tresca or von Mises (also called octahedral shear and distortion energy) criteria. Of contemporary researchers, Sines's [*3,4*] work is exemplary of refinement in this type of approach. His criterion establishes permissible amplitudes of alternating principal stresses, σ_1, σ_2, and σ_3, corrected for the orthogonal static stresses, S_1, S_2, and S_3, according to a linear relationship where

$$(1/6[(\sigma_1 - \sigma_2)^2 + (\sigma_2 - \sigma_3)^2 + (\sigma_3 - \sigma_1)^2])^{1/2} \leq A - \alpha(S_1 + S_2 + S_3) \tag{1}$$

The material's constants A and α are functions of the cyclic life level. Figure 1 illustrates this criterion in the case of biaxial fatigue for a given life level, showing that the ellipse representing permissible alternating stresses decreases in size as the sum of the static orthogonal stresses, ΣS,

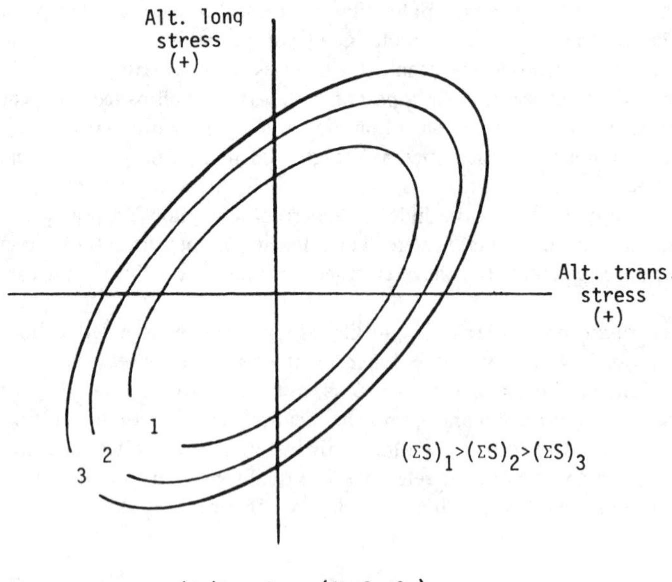

FIG. 1—*Constant life lines for a given cyclic biaxial stress state with varying levels of static mean stresses.* A and α *are materials constants* [3,4].

increases. He has recently extended his approach from high cycle to low cycle life regimes of fatigue [*4*].

The recognized value of exploiting existing (or relatively easily obtainable) axial fatigue data for multiaxial circumstances, along with the early success in high cycle fatigue applications, contributed to the survival of the effective/equivalent approaches as multiaxial fatigue life prediction methods were extended to low cycle regimes. Proceeding within the framework of local strain-life concepts, effective/equivalent strain parameters have been investigated. These, in turn, are frequently integrated into Coffin-Manson type power law strain-life expressions, as are strain parameters in axial fatigue. Figure 2 shows a summary of test results published by Taira et al. [*5*] comparing the equivalent strain range, $\Delta\epsilon_{eq}$, for a low-carbon steel in terms of a von Mises criterion in the form

$$\Delta\epsilon_{eq} = \sqrt{\Delta\epsilon^2 + 1/3\Delta\gamma^2} \tag{2}$$

and a Tresca criterion in the form

$$\Delta\epsilon_{eq} = \sqrt{\Delta\epsilon^2 + 1/4\Delta\gamma^2} \tag{3}$$

where $\Delta\epsilon$ and $\Delta\gamma$ are the axial and torsional strain ranges, respectively.

Much multiaxial data has been generated over the years by many different investigators solely for the purpose of determining the "best" equivalent/effective criteria. Overall, it appears that octahedral shear strain criteria are most frequently used for multiaxial fatigue life correlation in the literature, but conflicting reports of the "best" criterion may be pointed out repeatedly. Only a few examples will be cited here for illustrative purposes. Dowling [*6*] found that the stress-strain and strain-life curves for torsion could be adequately estimated from uniaxial data using the octahedral shear stresses and strains for a Ni-Cr-Mo-V steel. Leese and Morrow [*7*] established that the slopes of the elastic and plastic strain-life relationships were essentially the same in shear-strain-controlled torsion and in axial-strain-controlled tension for a 1045 steel. In that case, the choice of the effective criterion for estimating torsional strengths and ductility made very little difference in the total strain-life curves. However, the Tresca criterion more nearly fit the experimental data for representation of cyclic stress-strain response. The maximum shear criterion (Tresca) is incorporated into the ASME pressure vessel code for the estimation of fatigue life in biaxial conditions. Havard, Williams, and Topper [*8*] studied biaxial fatigue data from different sources in the open literature and found that the ASME code is less conservative than anticipated. At the time, an approach based on a power law relationship between life and von Mises' (octahedral) equivalent elastic and plastic strain, modified for a hydrostatic stress effect was suggested. Mowbray [*9*] also proposed a modification to the distortion energy elastic and plastic strain-life criterion to include the effects of the hydrostatic stress component and biaxiality on fatigue strength and ductility.

Modifications to materials constants, additional terms to account for mean and/or hydrostatic stresses, and considerations for ductile versus brittle materials' responses have consumed large quantities of investigators' time and energy to make their particular permutation of effective stress/strain versus life relationships fit experimental data more carefully. Many met with considerable success for the particular materials and stress states being investigated. Undoubtedly, more of this work will continue and more useful refinements will be made. However, at this time, some generalizations may be offered regarding the merits of the approaches:

1. In high cycle fatigue, effective/equivalent stress parameters are useful for life correlation and for comparison with life data obtained in uniaxial tests.
2. In low cycle fatigue, the extrapolation of materials response from one simple stress state to another can be quite useful. Given the scatter inherent in fatigue data, the representation of

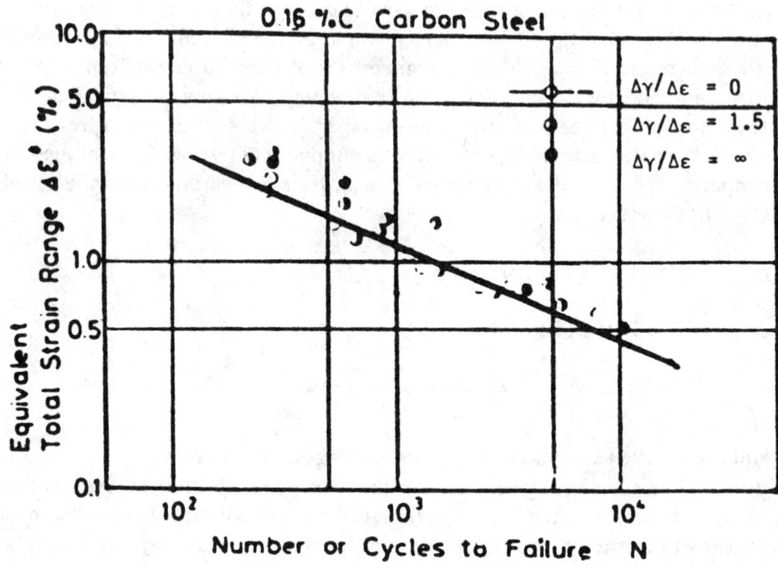

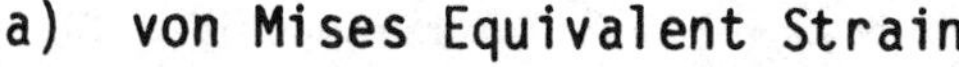
a) von Mises Equivalent Strain

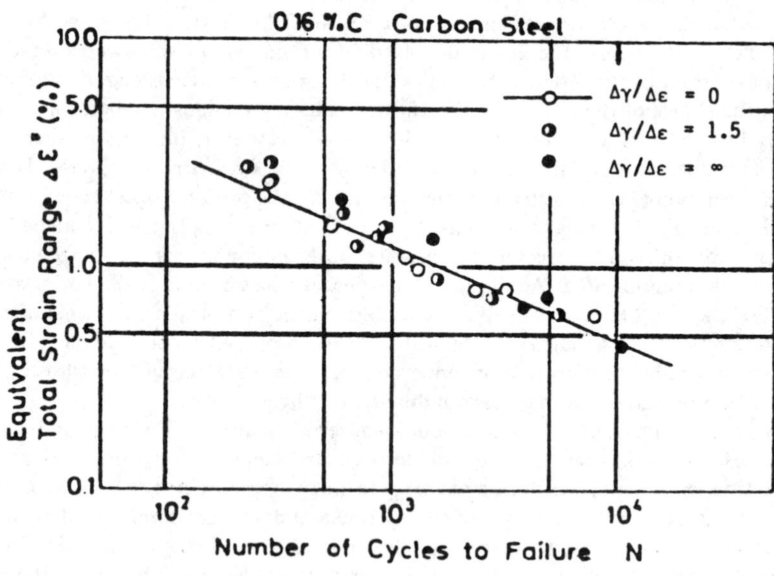

b) Tresca Equivalent Strain

FIG. 2—*Effective strain versus life data using* (a) *von Mises criterion and* (b) *Tresca maximum shear* [5].

multiaxial strain-life behavior via conversion of uniaxial materials properties through effective/equivalent criteria may be adequate for many engineering applications.

3. There may not be a "best" effective/equivalent criteria. The differences between such criteria with respect to strain-life correlation are often insignificant once incorporated into design methodology. This may not be the case in modeling cyclic stress-strain response.

Despite the great economic benefits to be derived from utilizing existing uniaxial data, as well as the technical convenience of these approaches, there are recognized limitations inherent in their analytical formulation and physical interpretation:

1. The ambiguity in defining crack initiation and failure, which has been a point of controversy for years in uniaxial testing, is greatly accentuated in the multiaxial case. Failure in axial strain-controlled tests is frequently defined as a predetermined percentage drop in load required to enforce the given strain range. In multiaxial testing, many surface cracks may be present throughout most of the full load carrying lifetime of a smooth, unnotched laboratory specimen [*7,10*]. Hence, it is difficult, and may be misleading to apply a conventional failure criterion to all stress states.

2. There is no obvious way to extend these approaches to more complex, non-proportional loading environments. It has been shown that in-phase multiaxial loading is often less severe than out-of-phase loading. There is documentation from many independent efforts of cases where stresses with rotating principal axes in a loading spectrum are more severe than comparable stationary principal axis conditions, depending on the actual multiaxial stress state and materials under consideration [*2,11,12,13,14*].

3. The relationships established between effective/equivalent stress/strain and life lack physical interpretation with respect to observed multiaxial failures, particularly in the low cycle regime. Here, multiple cracks are observed to initiate and grow on discrete planes, as a function of the multiaxial stress state. Effective/equivalent criteria treat all planes similarly.

Plastic Work, Energy Concepts

The concept that a material's resistance to fatigue can be quantified by the amount of work/energy required to cause failure is the basis of several recent efforts in multiaxial fatigue. Garud [*14*] recently proposed a new "hardening rule" which he incorporated into material constitutive relations to determine plastic work per cycle in multiaxial stress states. His correlation with fatigue life for in-phase and out-of-phase loading for a 1% Cr-Mo-V steel is shown in Fig. 3. His analytical technique requires only properties obtained from uniaxial tests, and could be altered to include the effects of known material anisotropy. Of course, one might choose to correlate multiaxial life with plastic work, substituting a constitutive model deemed more appropriate.

Jordan [*15*] recently evaluated plastic work, along with several other criteria included in this discussion, as multiaxial life criteria for elevated temperature, combined tension-torsion data generated on Hastelloy X. He found that a modified plastic work theory was the most successful for correlating lives in the environment studied.

Despite the experimentally verified success of these investigators (as well as others), energy approaches are not universally viewed as the ultimate course to follow. A great deal of attention was given to plastic work and energy expressions in quantifying fatigue life in the early years of axial LCF research. Feltner, Morrow, and Halford [*16-18*] are just a few who laid excellent foundations for the work that followed in this field. However, many of these same investigators have chosen not to pursue this direction for the characterization of axial fatigue life. Reasons, ranging from the very practical to the technically limiting, are suggested below:

1. Plastic work approaches are difficult to integrate into design and testing procedures involving component analysis.

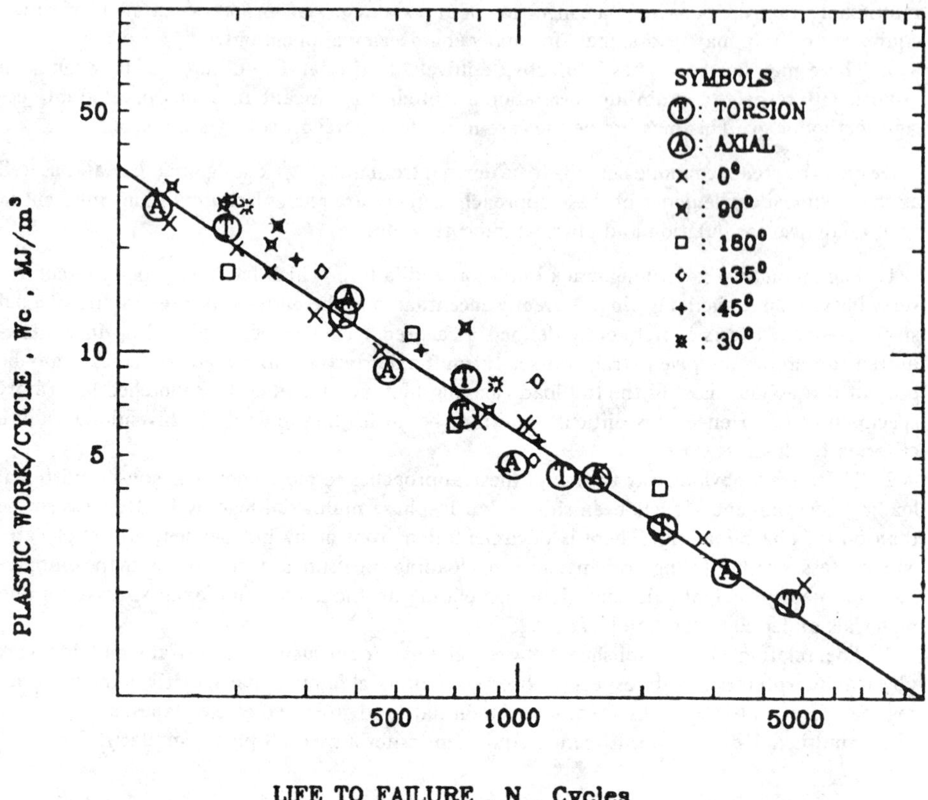

FIG. 3—*Plastic work per cycle versus life for tension, torsion and combined tension/torsion in-phase and out-of-phase cycling* [14].

2. Small values of plastic strain, and, therefore, small quantities of plastic work, characteristic of long life situations are difficult to quantify with much confidence.
3. The plastic work required to cause failure is not a material property or constant at all life levels. Therefore, even in the simplest loading configuration, one would need to characterize the relationship between plastic work and life, as well as the plastic work per cycle in that given spectrum.

These points are of major significance from the engineering perspective. It is these same points that are likely to pace any widespread acceptance of plastic work approaches for predicting multiaxial fatigue life.

Critical Plane Concepts

Commenting on the shear stress criterion for multiaxial fatigue failure, Little [*19*] points out that when the ratio of shear to normal stress is $\sqrt{3}$, and there is a phase difference of 90° for sinusoidal waveforms, the alternating amplitude of the distortion energy is zero. Jordan [*15*] reiterates the observation that in 90° phasing tests, the maximum shear strain amplitude remains constant, with only the orientation of the principal planes changing, adding that finite life in this situation can only be explained by a critical plane or plastic work approach. Certainly, many finite failures in this nonproportional loading condition have been documented.

The critical plane concepts demand that the failure plane be identified and the load spectra with respect to that plane be analyzed, as contrasted with the effective stress/strain approaches which (at least from a qualitative viewpoint) "average" the mechanical response across multiple planes. Among those who have explored this philosophy, there is some disagreement on the choice of that plane and the critical parameters describing its response.

Grubisic and Simburger [*13*] define an "interference plane" as that with the most severe combination of equivalent shear stress amplitude and mean normal stress, and proceed to evaluate the conditions on this plane for life predictions. Different criteria, though similar in form, are required for ductile, semi-ductile, and brittle materials. They have apparently found considerable success in handling multiaxial in-phase loading in steel, and have recently extended their concepts to cover out-of-phase loading. Strain-controlled response is not addressed in their approach.

Using alternating principal strains, ϵ_1, ϵ_2 and ϵ_3, Brown and Miller [*20*] define their critical plane (called the Γ-plane) based on two parameters, the maximum shear strain governing initiation, $(\epsilon_1 - \epsilon_3)/2$ and the normal strain component on that plane which assists propagation $(\epsilon_1 + \epsilon_3)/2$. It is within the identification of the latter that two different critical planes are introduced, depending on the loading situation. In their Case A, fatigue cracks would initiate at a free surface on the plane of maximum shear strain, defined by the difference between the principal strain extremes. The intermediate principal strain acts normal to the surface, allowing Mode II crack growth along the surface and Mode III into the depth. While their Case B still acknowledges crack initiation at the free surface on the plane of maximum shear strain, the intermediate principal strain is parallel to the surface, causing cracks to be driven inward and away from the surface.

A considerable amount of life data, including some from severely nonproportional strain-time spectra, has been shown to correlate well with the Γ-plane criterion by several different investigators [*10,15,20*]. The major drawback of this approach from a practical viewpoint is that it would require large amounts of multiaxial data to be generated to serve as baseline information for life prediction purposes, as can be ascertained from Fig. 4, showing only one constant life contour.

In contrast to this dual plane consideration, Lohr and Ellison [*21*] suggest that while cracks will initiate and grow along a free surface in shear, it is their propagation into the stressed body oriented at 45° to the surface that is most detrimental to its load carrying capability. Hence, two parameters are necessary in their criterion; the maximum shear strain into the specimen and the strain normal to that plane. Although the Lohr and Ellison criterion has been found useful for correlation with total biaxial fatigue lives in other investigations, actual cracking observations do not exclusively support their premise. Initiation *and* growth of cracks on maximum shear strain planes in the case of torsional loading for several different materials have been reported recently in the open literature [*7,10,15*].

In essence, the qualitative difference between the criterion of Brown and Miller and that of Lohr and Ellison is that the maximum shear plane of the latter criterion, on which cracks are driven through the thickness of a specimen, is not always the plane of maximum absolute shear strain. This is most easily shown in a Mohr's circle representation of a three-dimensional strain state as shown in Fig. 5.

More recently, Socie et al. [*22,23*] have modified the Brown and Miller criterion based on cracking and failure behavior in biaxial strain controlled fatigue tests in Inconel 718. Their modification accounts for the mean stress normal to the through-thickness critical plane. Hence, their multiaxial fatigue life criterion requires the definition of three parameters; the maximum shear strain amplitude, $\hat{\tau}_p$, the strain amplitude normal to that plane, $\hat{\epsilon}_{np}$, and the mean stress normal to that plane, $\hat{\sigma}_{np}$. Figure 6 shows their parameter correlated with axial and

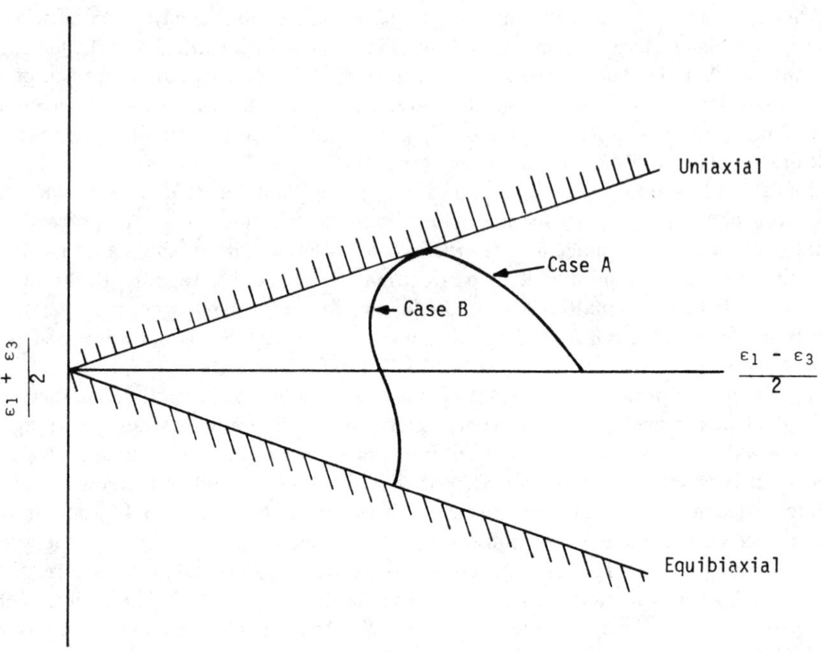

FIG. 4—*Constant life line representing Brown and Miller's critical planes* [20].

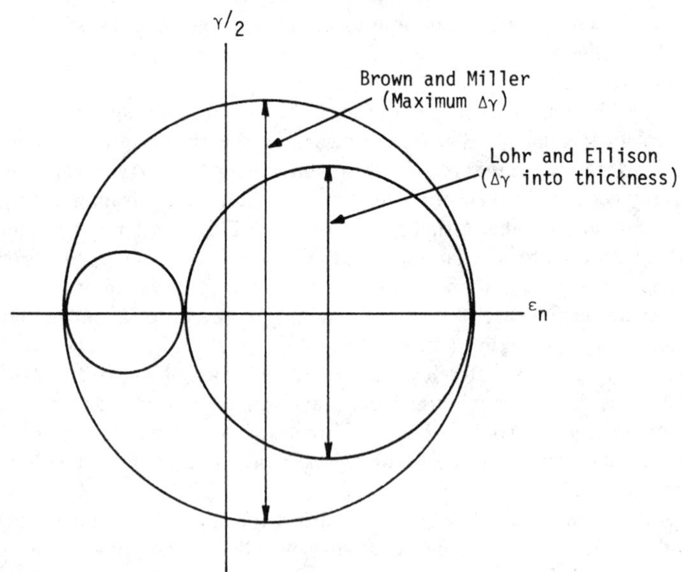

FIG. 5—*Difference in critical planes of Brown and Miller and of Lohr and Ellison for tension/torsion loading on Mohr's circles of strain.*

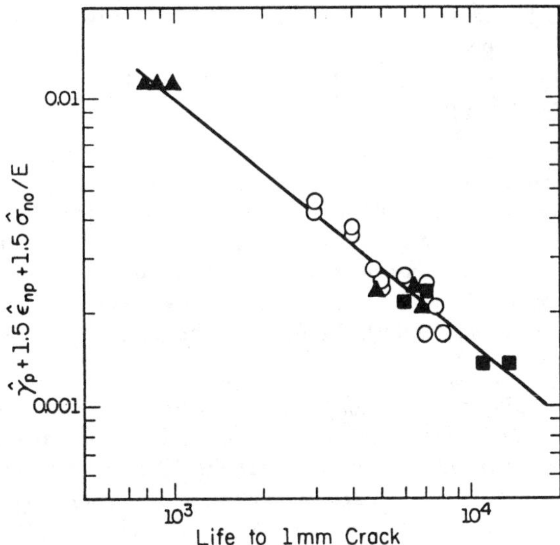

FIG. 6—*Combined tension/torsion data from Inconel 718 represented by the plastic shear and normal strains as well as the mean stress across critical plane* [22].

torsional test data. The baseline materials properties necessary for using this criterion to predict life are the strain-life properties generated from shear strain controlled torsion tests.

Jordan [*15*] found that the discrete plane containing life-limiting cracks may be a function of temperature. His research showed that the planes on which cracks formed for Hastelloy X in the same biaxial stress state changed from maximum shear stress planes at room temperature to maximum normal stress planes at elevated temperature.

Despite the controversies over the definition of the critical plane(s), and the parameters necessary to describe them, there are several commonalities of engineering significance:

1. Multiaxial fatigue failure is a phenomenon of crack initiation and growth on discrete planes as a function of stress state.
2. The fundamental reason for pursuing a critical plane approach is to establish a criterion based on, and correlating to, the actual physical cracking response observable in controlled multiaxial fatigue tests.
3. The critical plane must be established as a precursor to life prediction, demanding that one consider what the controlling parameters are in a given stress state.

Future Directions

Progress has been made in recent years in multiaxial fatigue life technology, from the perspective of verifying existing concepts, placing some boundaries on those concepts, and pushing beyond those boundaries to model complex multiaxial loading spectra life behavior. For instance, it is clear that materials properties and responses may often be extrapolated from one simple stress state to another. It is also clear that this is not the case in complex, nonproportional loading situations. Cyclic hardening/softening may be a function of phasing, making prediction of stress response more difficult. Multiple crack systems may form, from which those of significance to the load-carrying ability of the material must be distinguished from others.

Those identified as significant may be on a discrete "critical" plane that is a function of stress state, straining path, temperature, and other as yet undefined parameters.

With these points as a baseline, there are several "gaps" in our acquired or established multiaxial fatigue technology that become obvious:

1. While the new capabilities in multiaxial testing have opened many doors, it is also true that experimental methods vary greatly between the different investigators and laboratories. Establishing common frameworks within which to perform multiaxial fatigue tests and report results (as was done in axial fatigue testing through several technical organizations) would encourage more widespread utilization of actual data published.

2. More extensive identification of multiaxial histories of concern in today's applications would help direct future research efforts. Very few have been presented in the open literature for analysis.

In Fig. 7, Cook and Laflen [*24*] show the complex stress state that would evolve in a gas turbine engine disk bore during a simulated mission, and suggest that vanes and combustor liners would likely experience equally complex profiles. Their information was gleaned from structural analysis. The gas turbine engine community is continuing to identify the multiaxial stress states pertinent to their operating conditions. The ground vehicle industry has also begun to address this question [*25*].

Overall, there has been very little published documentation of measured or calculated multiaxial spectra from actual components. This kind of information is crucial to the planning of research programs. For example, laying multi-leg strain gages on components of specific engineering interest to measure details of multiaxial load spectra would be quite helpful, not only in placing boundaries on the usefulness of existing life correlation and predictive techniques, but also in narrowing the scope of future research efforts. The additional information would not in any way distract from addressing fundamental questions about multiaxial fatigue response.

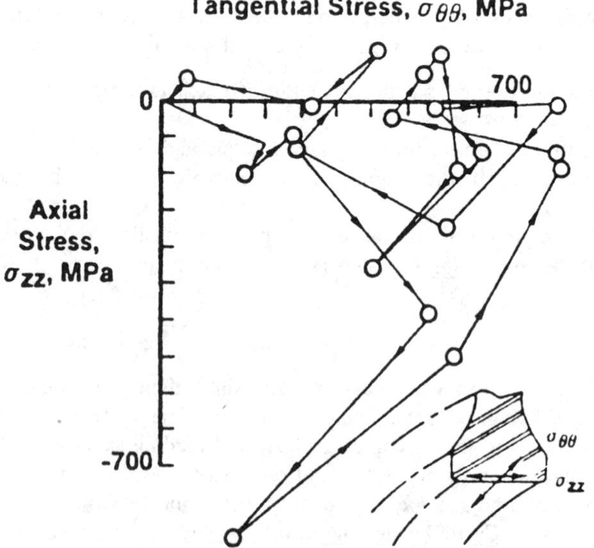

FIG. 7—*Biaxial stress state generated in a gas turbine engine disk bore* [24].

Implementation of this technology would certainly not be limited to the component testing arena. There will be many "spin-offs" from the advances made or anticipated in multiaxial fatigue research. For example, engineering design relies heavily on analytical procedures such as finite element analysis to streamline the design and testing procedures. The results of these analyses may be the input to fatigue life predictions made prior to prototype development. Redefining the critical parameters necessary for fatigue life predictability will certainly affect a large base of existing software, not to mention the actual analytical procedures. The designs that are complex enough to require finite element modeling may be the very applications that require a fatigue life predictive approach using other than the classical criteria.

Multiple crack systems early in the full load-carrying life of components will have to be acknowledged within the framework of enforced engineering philosophies, such as damage tolerance.

Another example of a "spin-off" in a related field would be the treatment of information from nondestructive inspection describing a cracked body. Detecting the mere existence of a crack of given dimensions is a small piece of the puzzle to be solved.

Conclusions

1. There is currently no universal parameter for correlation of cyclic multiaxial stress/strain with fatigue life.
2. The existing data base of axial strain controlled, fully reversed constant amplitude data may be helpful, but not sufficient, for describing baseline properties for multiaxial life prediction.
3. Effective or equivalent criteria are often useful tools for extrapolating materials' responses from one stress state to another and for life correlation in high cycle multiaxial regimes.
4. The phasing of multiaxial loading has been shown to greatly affect low cycle fatigue life in a number of experimental investigations in controlled laboratory testing. Although effective/equivalent criteria are insensitive to nonproportional aspects of complex loading environments, the effects of phasing on life may be reflected in plastic work and/or critical plane approaches.
5. Plastic work criteria have been found to be useful for multiaxial fatigue life correlation, but are difficult to apply in practice.
6. Critical plane approaches lend themselves to direct physical comparison and interpretation with cracking phenomena observed in actual tests.
7. More extensive characterization of the multiaxial loading spectra of actual components of industrial significance would be useful in directing the application-oriented research efforts.

Summary

The technology base for multiaxial cyclic testing has been adequately established to allow addressing fundamental aspects of multiaxial fatigue, rather than minute questions of curve fitting. To that extent, the significance of recent multiaxial fatigue research results, with respect to life prediction methodology, has been discussed.

References

[1] Krempl, E., *The Influence of State of Stress on Low-Cycle Fatigue of Structural Materials: A Literature Survey and Interpretive Report, ASTM STP 549,* American Society for Testing and Materials, Philadelphia, 1974.

[2] Garud, Y. S., "Multiaxial Fatigue: A Survey of the State of the Art," *Journal of Testing and Evaluation,* Vol. 9, No. 3, 1981, pp. 165–178.

[3] Sines, G., "Failure of Materials Under Combined Repeated Stresses With Superimposed Static Stresses," National Advisory Committee for Aeronautics Technology, Note 3495, Nov. 1955.

[4] Sines, G., "Fatigue Criteria Under Combined Stresses or Strains," *Transactions of ASME,* Vol. 103, April 1981, pp. 82–90.

[5] Taira, S., Inoue, T., and Yoshida, T., "Low Cycle Fatigue Under Multiaxial Stresses" (in the Case of Combined Cyclic Tension-Compression and Cyclic Torsion at Room Temperature), *Proceedings,* The Twelfth Japan Congress on Materials Research-Metallic Materials, March 1969, pp. 50–55.

[6] Dowling, N. E., "Torsional Fatigue Life of Power Plant Equipment Rotating Shafts," DOE/RA/29353-1, Westinghouse R&D Center, Pittsburgh, Pa., Sept. 1982.

[7] Leese, G. E. and Morrow, JoDean, "Low Cycle Fatigue Properties of a 1045 Steel in Torsion" in *Multiaxial Fatigue, ASTM STP 853,* American Society for Testing and Materials, Philadelphia, 1985, pp. 482–495.

[8] Havard, D. G., Williams, D. P., and Topper, T. H., "Biaxial Fatigue of Mild Steel: Data Synthesis and Interpretation," *Ontario Hydro Research Quarterly,* Second Quarter, 1975, pp. 11–18.

[9] Mowbray, D. F., "A Hydrostatic Stress-Sensitive Relationship for Fatigue Under Biaxial Stress Conditions," *Journal of Testing and Evaluation,* Vol. 8, No. 1, 1980, pp. 3–8.

[10] Socie, D. F., Waill, L. A., and Dittmer, D. F., "Biaxial Fatigue of Inconel 718 Including Mean Stress Effects" in *Multiaxial Fatigue, ASTM STP 853,* American Society for Testing and Materials, Philadelphia, 1985, pp. 463–478.

[11] Kanazawa, K., Miller, K. J., and Brown, M. W., "Low-Cycle Fatigue Under Out-of-Phase Loading Conditions," *Journal of Engineering Materials and Technology, Transactions of ASME,* Vol. 99, July 1977, pp. 222–228.

[12] Fatemi, A., "Fatigue and Deformation Under Proportional and Nonproportional Biaxial Loading," Ph.D. thesis, University of Iowa, Ames, Aug. 1985.

[13] Grubisic, V. and Simburger, A., "Fatigue Under Combined Out of Phase Multiaxial Stresses" in *Fatigue Testing and Design,* Vol. 2, R. G. Bathgate, Ed., Society of Environmental Engineers Fatigue Group, London, pp. 27.1–27.7.

[14] Garud, Y. S., "A New Approach to the Evaluation of Fatigue Under Multiaxial Loadings," in *Proceedings,* Symposium on Methods for Predicting Materials Life in Fatigue, W. J. Ostergren and J. R. Whitehead, Eds., ASME, New York, 1979, pp. 113–125.

[15] Jordan, E. H., "Elevated Temperature Biaxial Fatigue," NASA CR-175009, National Aeronautics and Space Administration, Lewis Research Center, Cleveland, Ohio, Oct. 1985.

[16] Feltner, C. E. and Morrow, JoDean, "Microplastic Strain Hysteresis Energy as a Criterion for Fatigue Fracture," *Journal of Basic Engineering, Transactions of ASME,* Vol. 83D, March 1961, pp. 15–22.

[17] Morrow, JoDean, "Cyclic Plastic Strain Energy and Fatigue of Metals," in *Internal Friction, Damping and Cyclic Plasticity, ASTM STP 378,* American Society for Testing and Materials, Philadelphia, 1965, pp. 45–87.

[18] Halford, G. R., "The Energy Required for Fatigue," *Journal of Materials,* Vol. 1, No. 1, March 1966, pp. 3–18.

[19] Little, R. E., "A Note on the Shear Stress Criterion for Fatigue Failure Under Combined Stress," *The Aeronautical Quarterly,* Feb. 1969, pp. 57–60.

[20] Brown, M. W. and Miller, K. J., "A Theory for Fatigue Failure Under Multiaxial Stress-Strain Conditions," *Proceedings of the Institution of Mechanical Engineers,* Vol. 187, 1973, pp. 745–755.

[21] Lohr, R. D. and Ellison, E. G., "A Simple Theory for Low Cycle Multiaxial Fatigue," *Fatigue of Engineering Materials and Structures,* Vol. 3, 1980, pp. 1–17.

[22] Socie, D. F. and Shield, T. W., "Mean Stress Effects in Biaxial Fatigue of Inconel 718," *Journal of Engineering Materials and Technology,* Vol. 106, July 1984, pp. 227–232.

[23] Socie, D. F., "Multiaxial Cyclic Deformation and Fatigue," NASA CR-179483, National Aeronautics and Space Administration, Lewis Research Center, Cleveland, Ohio, 1986.

[24] Cook, T. S. and Laflen, J. H., "Considerations for Damage Analysis of Gas Turbine Hot Section Components," The American Society of Mechanical Engineers, Report 84-PVP-77, 1984.

[25] Fash, J. W., Conle, F. A., and Minter, G. L., "Analysis of Irregular Loading Histories for the SAE Biaxial Fatigue Program," to be published by the Society for Automotive Engineers, Warrendale, Pa.

DISCUSSION

Edwin Haibach[1] *(written discussion)*—How may the presented models account for directional properties (anisotropy) of materials? Which model was best?

G. E. Leese (author's closure)—The author acknowledges the fact that the perspectives offered in this paper are based primarily on results of research treating the subject material as isotropic. However, the effects of material anisotropy may be accounted for within the framework of the three general approaches discussed. Most notably is the effort directed towards three-dimensional constitutive modeling, which in turn provides input for predicting finite life. In some cases, the goal may be to adequately represent inherent anisotropic response for input into an effective stress/strain type of parameter. In other cases, description of anisotropic yield surface response, ultimately to quantify the plastic strain as well as the stress, may be required for correlation with life in any of the approaches discussed. Detailed descriptions of the modeling of materials, including anisotropy, are available in the open literature [e.g., *1-3*], with some specifically addressing multiaxial fatigue life [*4,5*].

Certainly, as extremely anisotropic materials (e.g., composites and single crystals) become more prominent in the engineering world, including the effects of anisotropy in durability and/or design analyses will become increasingly important. This is especially so for multiaxial environments.

Discussion References

[*1*] Robinson, D. N., "Constitutive Relationships for Anisotropic High Temperature Alloys," *Nuclear Engineering and Design Journal,* Vol. 83, 1984, pp. 389-396.

[*2*] McDowell, D. L., Socie, D. F., and Lamba, H. S., "Multiaxial Nonproportional Cyclic Deformation," in *Low-Cycle Fatigue and Life Prediction, ASTM STP 770,* American Society for Testing and Materials, Philadelphia, 1983, pp. 500-518.

[*3*] Swanson, G. A. et al., "Life Prediction and Constitutive Models for Engine Hot Section Anisotropic Materials," NASA CR-174952, NASA-Lewis Research Center, Cleveland, Ohio.

[*4*] Harvey, S. J., Toor, A. P., Adkin, P. A., "The Use of Anisotropic Yield Surfaces in Cyclic Plasticity," in *Multiaxial Fatigue, ASTM STP 853,* American Society for Testing and Materials, Philadelphia, 1985, pp. 49-63.

[*5*] Found, M. S., "A Review of the Multiaxial Fatigue Testing of Fiber Reinforced Plastics," in *Multiaxial Fatigue, ASTM STP 853,* American Society for Testing and Materials, Philadelphia, 1985, pp. 381-395.

[1]WBK-S, Bochum, FRG.

J. W. Fash,[1] N. J. Hurd,[2] C. T. Hua,[3] and D. F. Socie[4]

Damage Development During Multiaxial Fatigue of Unnotched and Notched Specimens

REFERENCE: Fash, J. W., Hurd, N. J., Hua, C. T., and Socie, D. F., **"Damage Development During Multiaxial Fatigue of Unnotched and Notched Specimens,"** *Low Cycle Fatigue, ASTM STP 942,* H. D. Solomon, G. R. Halford, L. R. Kaisand, and B. N. Leis, Eds., American Society for Testing and Materials, Philadelphia, 1988, pp. 874-898.

ABSTRACT: Life prediction methodologies based on low-cycle fatigue concepts are well established for the evaluation of component fatigue behavior. A critical assumption made in applying these concepts is that the fatigue damage process in the critical region of the component is similar to that which occurs in the smooth specimen used to characterize the material fatigue behavior. This approach has been verified experimentally for uniaxial loading conditions but methods for applying these concepts to multiaxial fatigue problems are not yet established.

Multiaxial fatigue theories that relate to the physical damage processes have shown the most promise for reliable design criteria. Two test programs are reported and discussed to further the understanding of the physical damage processes. First, thin-wall tube specimens have been tested in combined tension-torsion strain controlled loading. Second, notched shaft specimens, designed to represent a typical component, have been tested in combined moment control torsion-bending. For both test series, the development and growth of fatigue damage from initiation to failure is reported.

The thin-wall tube geometry is considered the smooth specimen for multi-axial fatigue. Tests on this geometry establish the expected damage behavior for the normalized SAE 1045 steel used in this investigation. Damage initiates on planes of maximum shear strain. At long lives a single crack develops and eventually grows perpendicular to the maximum principal strain. At short lives extensive damage develops on shear planes and failure occurs rapidly by a crack linking process. For equivalent local strain states similar behavior is expected for the notched shaft. However, initiation and early damage development in the notched shaft occurs in the circumferential direction of the notch, rather than on maximum shear planes. At long lives crack growth resulting in failure occurs on planes perpendicular to the maximum principal strain. At short lives extensive damage occurs in the notch and failure results by crack linking perpendicular to the primary bending stress.

Specimen geometry and stress gradients have a significant influence on damage development during multiaxial loading of components. The damage processes in the notched shaft are not represented by the thin-wall tube tests. Increased understanding of the detailed micromechanisms of fatigue represents an important direction for future improvements in life prediction methods.

KEY WORDS: Biaxial fatigue, multiaxial fatigue, notch effects in fatigue, stress gradients, crack initiation, crack growth, damage, Stage I-Stage II cracking

[1]Research Engineer, General Electric Company, Aircraft Engine Business Group, Cincinnati, OH 45215.

[2]Research Scientist, G.K.N. Technology Limited, Wolverhampton, England.

[3]Engineering Science Engineer, Garrett Turbine Engine Company, Phoenix, AZ 85001; formerly Graduate Research Assistant, Department of Mechanical and Industrial Engineering, University of Illinois at Urbana-Champaign, Urbana, IL 61801.

[4]Professor, Department of Mechanical and Industrial Engineering, University of Illinois at Urbana-Champaign, Urbana, IL 61801.

Low-cycle fatigue (LCF) concepts are well established for the evaluation of engineering components and structures. Component life estimates are made by relating the local stress-strain in the critical area of the component to smooth specimen behavior (Fig. 1*a*). Equal fatigue lives are predicted when the local stress-strain histories are equivalent. Recently a major area of research has been the extension of these concepts to situations that experience multiaxial stress-strain conditions [*1*]. The majority of work has been performed to determine the parametric form of a damage parameter to relate multiaxial stress-strain conditions to uniaxial baseline data. It is attractive to develop multiaxial design methods which implement existing LCF data bases. The thin-wall tube specimen is often employed in such studies and can be considered the smooth specimen for multiaxial fatigue since a large volume of material is subjected to a uniform stress-strain state.

Classical LCF analysis applies the assumption of similitude shown in Fig. 1*a* for component analysis. Extension of these concepts to multiaxial fatigue evaluation suggests the similitude shown in Fig. 1*b*. Two arguments can be made for applying smooth uniaxial LCF data to the analysis of components subjected to complex loading conditions. First, through an appropriate multiaxial fatigue theory the smooth uniaxial data and damage behavior should represent the thin-wall tube behavior. Secondly, similitude is expected when the multiaxial strain state in the critical area of the component and in the thin-wall tube are equivalent (Fig. 1*b*). These arguments can be made both in terms of fatigue life and in the development of fatigue damage.

In a previous study [*2*], five multiaxial fatigue theories were evaluated on their ability to predict the lives of tests performed on thin-wall tube and notched shaft specimens. Smooth specimen uniaxial fatigue data were used in the analysis. Good agreement was obtained between

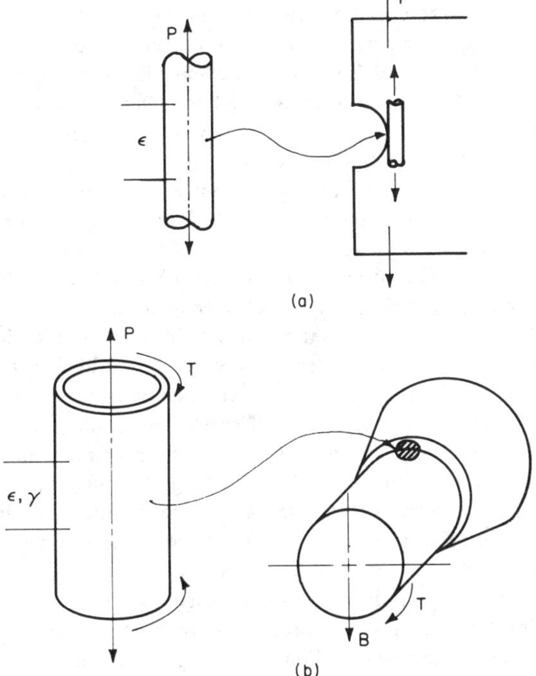

FIG. 1—*Similitude assumptions for smooth and notched specimens.* (a) *Uniaxial loading.* (b) *Multiaxial loading.*

experimental results and theoretical life predictions for the thin-wall tube tests but considerably less agreement was obtained for the notched shaft tests.

This paper investigates the development of fatigue damage in light of the two similitude arguments described above. Crack initiation and early growth have been observed for various loading conditions of the thin-wall tube and notched shaft specimens. The relationship between the local stress-strain state and crack development are identified from surface observations to illuminate the stages of damage development. Thin-wall tube behavior represents the smooth specimen behavior for multiaxial fatigue. The influence of strain state and amplitude on crack initiation and early growth are presented. Similitude arguments suggest that for similar local strain states the damage in the notched shaft should be equivalent to that in the thin-wall tube. Similarities and differences in the damage process have been observed between the two geometries. These differences are described and discussed in relation to the strain state and local conditions.

Background

Fatigue life analyses are currently separated into crack initiation and crack propagation methods. A general mechanism for crack initiation [*3,4*] is based on coarse slip processes. Coarse slip occurs on favorably oriented crystallographic planes within single grains of a material. Cyclic loading causes reversed slip and the formation of persistent slip bands. Material is displaced in local regions forming intrusions and extrusions and ultimately decohesion occurs, forming a microcrack. The applied shear stress and resulting shear strains are the dominant parameters for this initiation process.

Forsyth [*5*] has identified two distinct stages of the crack growth process. Stage I is crystallographic growth primarily related to slip processes similar to the initiation phase and is related to the applied shear stress. In fracture mechanics terminology Stage I growth is primarily by Mode II-Mode III loading. Stage II is a period of continuum crack growth in a plane perpendicular to the maximum principal stress. Models that describe Stage II growth are based on slip processes at the crack tip [*6*]. Stage II growth is often studied and characterized in Mode I fracture mechanics terminology.

Low-cycle fatigue methods are employed to predict the crack initiation life of a component. Definition of crack initiation is rather ambiguous but is usually taken as the formation of an "engineering size" crack (usually between 0.1 mm and 5.0 mm in surface length). This definition often includes the initiation process, some or all of Stage I growth, and in some cases, a portion of Stage II growth. The portions of microgrowth that are included in a life analysis depend on the development of damage in smooth specimen baseline data and the definition of failure employed. Development of cracks to the size defined as initiation has been documented in a few studies [*7–9*] but is not well understood. Strain-based LCF methods account for this microcrack development without actually quantifying the behavior.

Previous research suggests that the stages of crack development are dependent on strain state, strain amplitude, environment and vary with material. Brown and Miller [*10*] have reviewed several studies that investigated the Stage I-Stage II transition in growth behavior, as well as presenting further results of their own. A fracture mechanics argument was developed that suggested the transition from Stage I to Stage II propagation is dependent on the applied stress state and is not influenced by crack length or stress level.

Hua and Socie [*11*] have shown differences in damage development in the high cycle fatigue (HCF) and LCF regime for several strain states. Single dominant crack systems developed in the HCF regime, and damage could be based on crack length. In LCF conditions multiple cracks developed and interacted to cause failure. Microcrack growth was interpreted to be Stage I for all test conditions. For this condition, crack density was used as a measure of damage.

Very little work has been reported on the influence of specimen geometry on damage development in multiaxial fatigue. Similitude arguments (Fig. 1) require that the crack behavior in the notched component be similar to that observed in the thin-wall tube. The effect of stress-strain gradients and geometric constraint currently are not explicitly included in LCF life analysis.

Experimental Program

Material

All tests in this investigation were carried out on a normalized SAE-1045 steel. The material was provided as 63 mm diameter hot rolled bars as part of the Society of Automotive Engineers (SAE) Fatigue Design and Evaluation (FDE) Committees round-robin biaxial fatigue test program [*12*]. Mechanical and metallurgical properties are given in Table 1. Manganese sulfide inclusions up to approximately 100 μm in length were present in the longitudinal direction. It has been shown [*2*] that these result in anisotropy in terms of uniaxial fatigue life.

Thin-Wall Tube Program

The thin-wall tube geometry given in Ref *2* (25.0 mm inner diameter, 2.54 mm wall thickness) has been tested as the smooth specimen for multiaxial fatigue. Tension, torsion, and combined tension torsion tests have been performed in strain control [*2*] for strain ratios of λ (shear strain amplitude/axial strain amplitude) of 0.0, 0.5, 1.0, 2.0, and ∞. Test levels were chosen to give constant effective strains of $\bar{\epsilon}$ = 1.0, 0.43, 0.22, and 0.15%. In addition, two tests were performed at $\bar{\epsilon} = 0.13\%$, $\lambda = \infty$ [*14*]. All tests were completely reversed in-phase loading. Failure was defined as a 10% drop in axial load (torsional load for $\lambda = \infty$).

Specimens were polished to eliminate surface scratches that might be interpreted as cracks. Final polishing was done with 0.5 μm alumina. Fatigue damage development has been monitored using standard acetyl cellulose film replicating techniques. Surface replicas were taken at

TABLE 1—*Static tensile properties.*

BHN	= 153
E	= 205 000 MPa
%Ra	= 51
σ_y	= 380 MPa (lower)
σ_{uts}	= 621 MPa
σ_f	= 985 MPa
ϵ_f	= 0.71
n	= 0.23
K	= 1185 MPa

Microstructure:[a]
Grain size: 6–7 (per ASTM E 112)
(approximately 35–40 μm)

Inclusion (per ASTM E 45):

Type	Rating
A & C Thin	2–3
B Thin	1

[a]Courtesy of Deere and Company.

approximately every 10% of the expected life. Observation of crack development from the replicas was accomplished by transmission optical microscopy.

A complete list of test results can be found in Refs *2* and *13*. Axial strains are controlled to the reported values. In torsion, there is a small strain gradient between the inside and outside surfaces (18%) of the specimen. Reported torsional strains are the average or mid-thickness values. Surface strains are calculated by extrapolating these values to the surface. Maximum shear strains, principal strains, and associated directions are calculated using simple Mohr's circle concepts. An effective Poisson's ratio given in Ref *13* is used in the calculation.

Notched Shaft Program

The notched shaft geometry [*2*] was provided by the SAE FDE committee for testing in conjunction with the round-robin test program. Tests were performed in load control to provide constant amplitude bending and torsion moments at the tangency point of the 40 mm gage section and the 5.0 mm fillet radius. Tests have been performed in bending, torsion, and combined loading conditions identified as XR and ZR. These three groups represent proportions of torsional moment to bending moment as derived from the nominal stresses and stress concentration factors employing a distortion energy criterion. The XR group represents equal components of bending and torsion, and ZR twice as much torsion as bending. The test conditions reported are given in Table 2 along with local strain values. Load amplitudes have been chosen for each test condition to provide test lives of approximately 10^4, 10^5, and 10^6 cycles to failure. All tests were performed under completely reversed, in-phase loading. Crack initiation was detected using an ultrasonic surface wave transducer [*13*].

The final surface finish on the shafts was a low stress surface grinding operation. Some specimens were polished using techniques similar to those used on the thin-wall tube specimens to make the observations of crack development easier. Polished and unpolished shafts behaved in a similar manner. Observations of the development of fatigue cracks in the notch region have been made using a replicating procedure developed for this specimen. A silicon based precision impression material was used to obtain surface replicas that were then mounted on a conducting base and sputtered with a thin layer of 50% platinum/50% gold. Examination of the replicas was performed by scanning electron microscopy.

Stress-strain analysis for the notched shaft was performed using an elastic-plastic finite element model (FEM) [*2,14*]. Stabilized notch root strains were determined by implementing the uniaxial cyclic stress-strain curve in the plasticity analysis. Maximum shear strains, principal strains, and associated directions are determined by standard tensor operations. A more detailed description of the FEM and the results are given in Ref *13*.

TABLE 2—*Notched shaft loading conditions.*

Test ID	Bending Moment (M_b), N · m	Torsion Moment (M_t), N · m	Principal Strains in Notch ($\mu\epsilon$)		
			ϵ_1	ϵ_2	ϵ_3
BR2	1730	0	2559	−366	−1273
XR2	1220	1710	2660	−940	−1076
TR1	0	2000	1397	0	−1397
XR3	1850	2550	6437	−2633	−2830
ZR3	1150	2700	4511	−1266	−2659

Results of Crack Observations

The development of surface damage is reported in relation to the applied stress-strain state (direction) and the number and density of cracks that develop. Schematic representations of the local stress-strain state on the crack figures indicate the direction that the maximum shear strain planes and the maximum principal stress plane intersect the surface. Since all loading is completely reversed and in phase, the stress and strain tensors are colinear.

Thin-Wall Tube

Damage development in the thin-wall tube specimen has been shown to differ in the HCF and LCF regimes [*11*]. For different strain states tested in HCF ($\bar{\epsilon} = 0.22\%$), a single damage nucleus was observed. This site grew into a single dominant crack at failure. In LCF tests ($\bar{\epsilon} = 1.0\%$), many microcracks were observed early in life and new damage nuclei were found throughout the life. Microcracks that formed early in life did not grow appreciably in surface length; rather, they became coarser and better defined with increasing cycles. Final failure was a result of rapid linking of these multiple damage sites with a crack of a few millimetres forming in very few cycles once the linking process started. Further observations of a wider range of test conditions has resulted in a more complete understanding of the influence of strain state and strain level on the development of fatigue damage.

Damage development in terms of the HCF and LCF processes described above has been found to vary with strain state (Fig. 2). Strain state is described by the ratio of maximum-to-minimum principal strain values ($\xi = \epsilon_1/\epsilon_3$) so that torsion can be included. Test conditions that fall to the right of the shaded region in Fig. 2 experience the HCF damage process described above. Tests to the left of the shaded region show the multiple cracking of LCF. Test conditions falling within the shaded region show mixed behavior.

Examples of typical HCF damage development are shown in Figs. 3 and 4. Initiation and early growth occur on planes of maximum shear strain range. Approximately half of the life required to form a 1.0 mm crack is spent in Stage I growth of a crack less than 100 μm in length. A portion of the Stage I growth is shown at 40 000 cycles in Fig. 3. The crack then changes direction to grow to failure in Stage II perpendicular to the maximum principal stress (80 000 cycles in Fig. 3). Some deviation from this plane is observed with the crack occasionally extending on maximum shear planes. Figure 4 is a test condition that experiences mixed HCF and LCF characteristics as evidenced by the multiple cracks present. A few damage nuclei are ob-

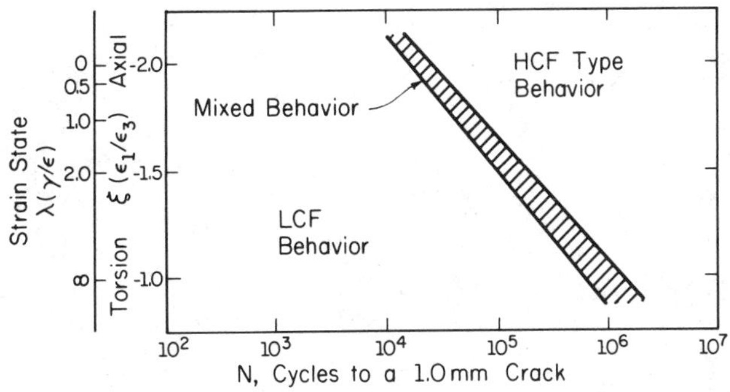

FIG. 2—*Schematic representation of the thin-wall tube damage state as a function of strain state and strain amplitude.*

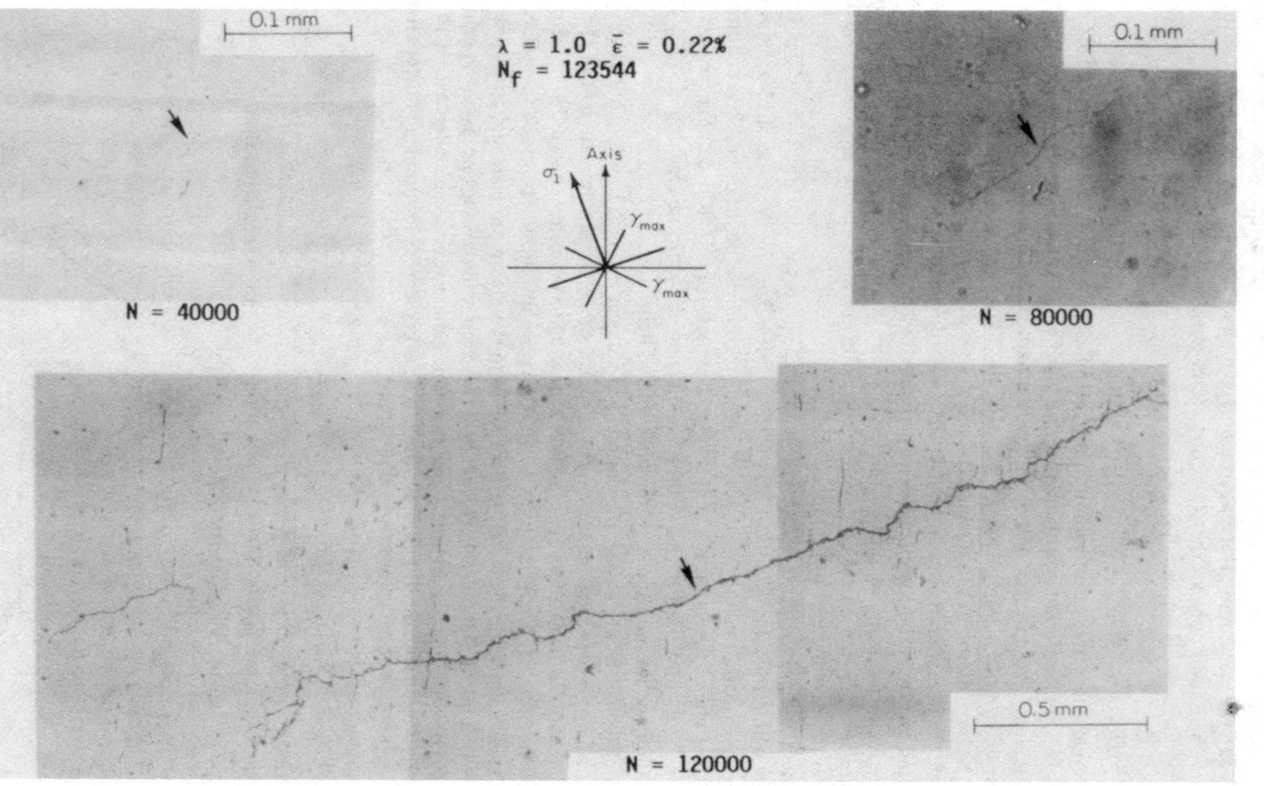

FIG. 3—*Thin-wall tube crack development for* $\lambda = 1.0$, $\bar{\epsilon} = 0.22\%$.

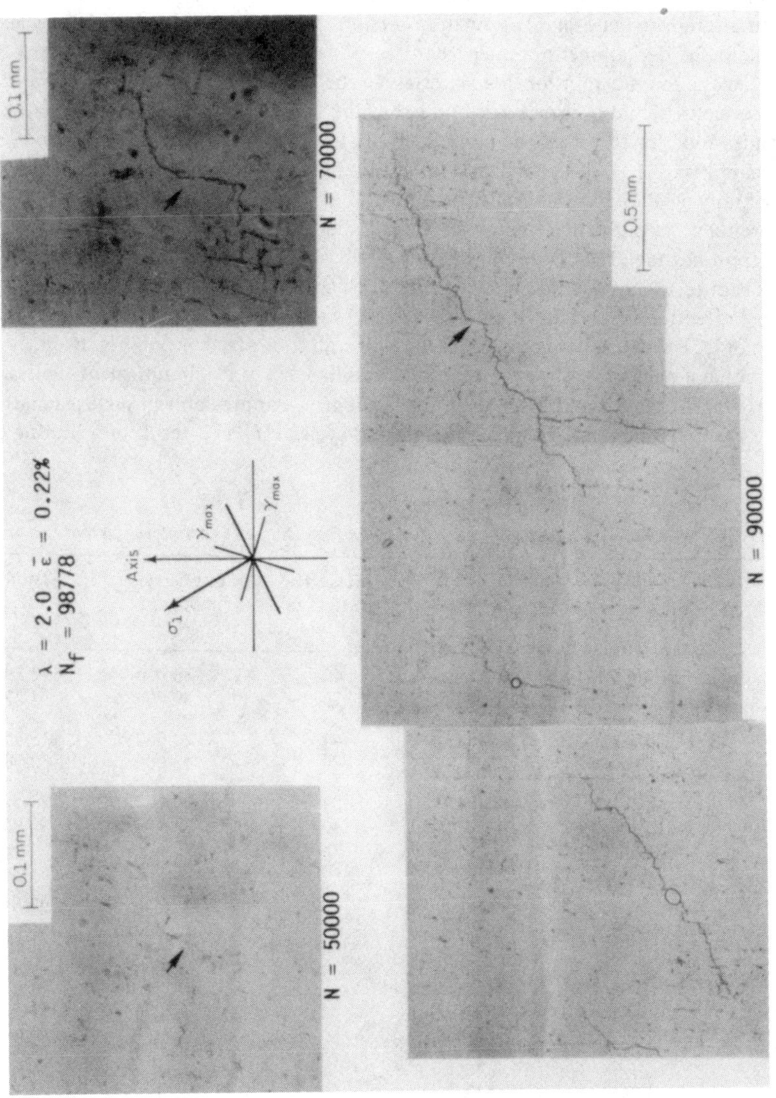

FIG. 4—*Thin-wall tube crack development for* $\lambda = 2.0$, $\bar{\epsilon} = 0.22\%$.

served early in life. Individual sites grow independently until other developing cracks are encountered. Crack linking plays a part in the final failure process late in life. The Stage I crack length and the life when the crack changed direction to Stage II growth are given in Table 3.

Damage development in torsion has been reported [*11*] to occur only on shear planes in this material. Results of the combined loading conditions described above showing the Stage I to Stage II transition in growth and the results of other researchers [*15*] suggest that Stage II crack growth in torsion may be expected at very long lives. Two tests were performed at $\lambda = \infty$, $\bar{\epsilon} = 0.13\%$ in an attempt to get Stage II growth in torsion. In the first test, 7.5×10^6 cycles were applied without the development of a clearly discernible crack or cracks. This effective strain level was below the fatigue limit for this material. In the second torsion test at this amplitude a large single cycle torsional overload was applied at the beginning of the test. It is well known that overloads reduce or eliminate the fatigue limit in mild steels [*16*]. After 3.1×10^6 cycles at a strain level of $\bar{\epsilon} = 0.13\%$, the crack system shown in Fig. 5 had developed. Crack development was initially on a longitudinal shear plane. After approximately 2.0 mm of Stage I growth the crack branched at both ends to grow in Stage II perpendicular to the maximum principal stress. Preferential Stage II growth on one of the 45° planes was a result of a small torsional mean stress that resulted from the initial overload. Near failure the main Stage II cracks once again change direction to grow as Stage I shear cracks (Stage II to Stage I transition).

Examples of LCF damage development in the thin-wall tube are shown in Fig. 6. Early in the life, many damage nuclei are observed on both complementary maximum shear planes. Very near failure (Fig. 6) these have coarsened but have not grown appreciably in surface length. The final failure crack develops in a very few number of cycles [*11*] as a result of a linking of the

TABLE 3—*Crack sizes and life at the transition from Stage I crack development.*

Effective Strain ($\bar{\epsilon}$), %	Strain State (λ)	Specimen ID	Stage I Crack Size (a_s), μm	Life Spent in Stage I (N_s)	Total Life (N_f)
0.22	0.0	4511	40	38 000	142 541
0.22	0.0	4529	25	30 000	94 525
0.22	0.5	4516	44	20 000	115 462
0.22	0.5	4528	70	40 000	80 000
0.22	1.0	4514	84	60 000	123 544
0.22	1.0	4550	40	30 000	90 000
0.15	0.5	4519	25	<100 000	611 780
0.15	1.0	4517	32	200 000	595 613
	1.0	4554	30	5 000	393 633
0.43	0.0	4545	40	7 000	7 839
0.43	0.5	4523	64	11 000	11 777
0.43	1.0	4515	64	11 000	11 611
1.0	0.0	4527	64	1 100	1 137
1.0	0.0	4553	64	1 100	1 107
1.0	0.5	4524	50	1 200	1 258
1.0	1.0	4533	40	1 200	1 229
1.0	1.0	4525	50	1 600	1 616
1.0	2.0	4526	32	1 700	1 758

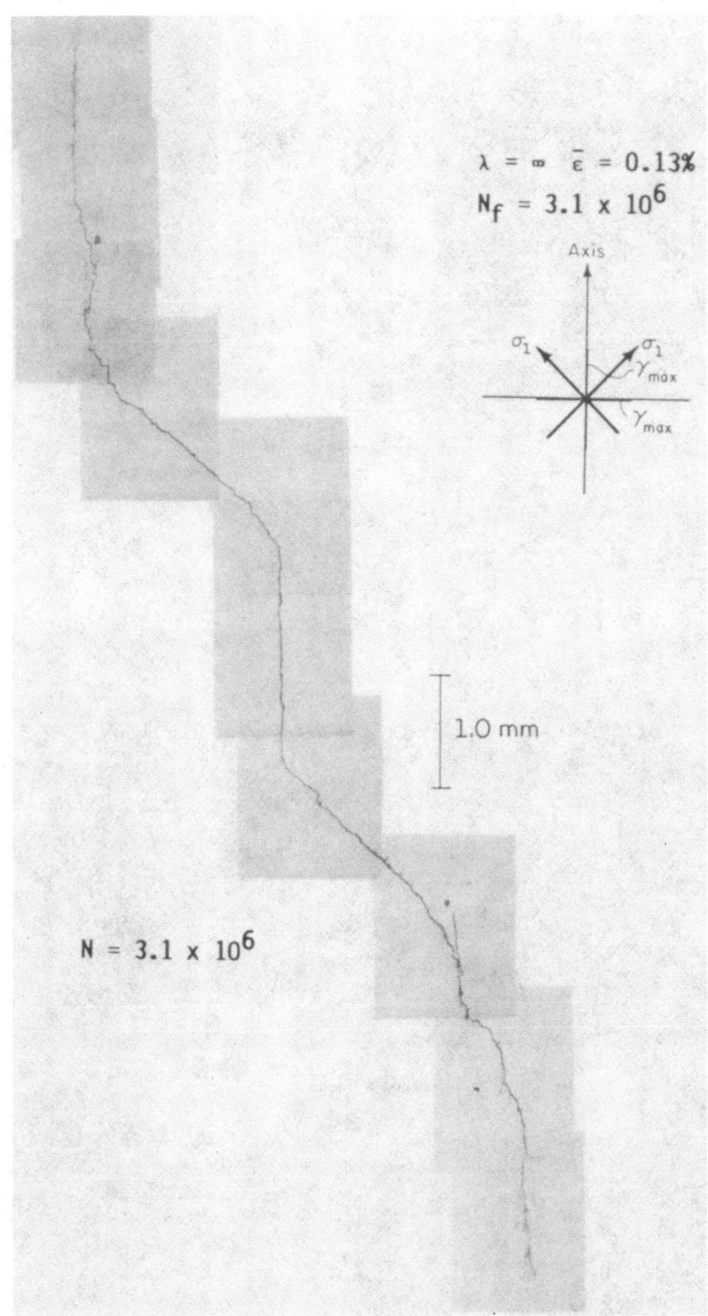

FIG. 5—*Thin-wall tube crack development for* $\lambda = \infty$, $\bar{\epsilon} = 0.13\%$.

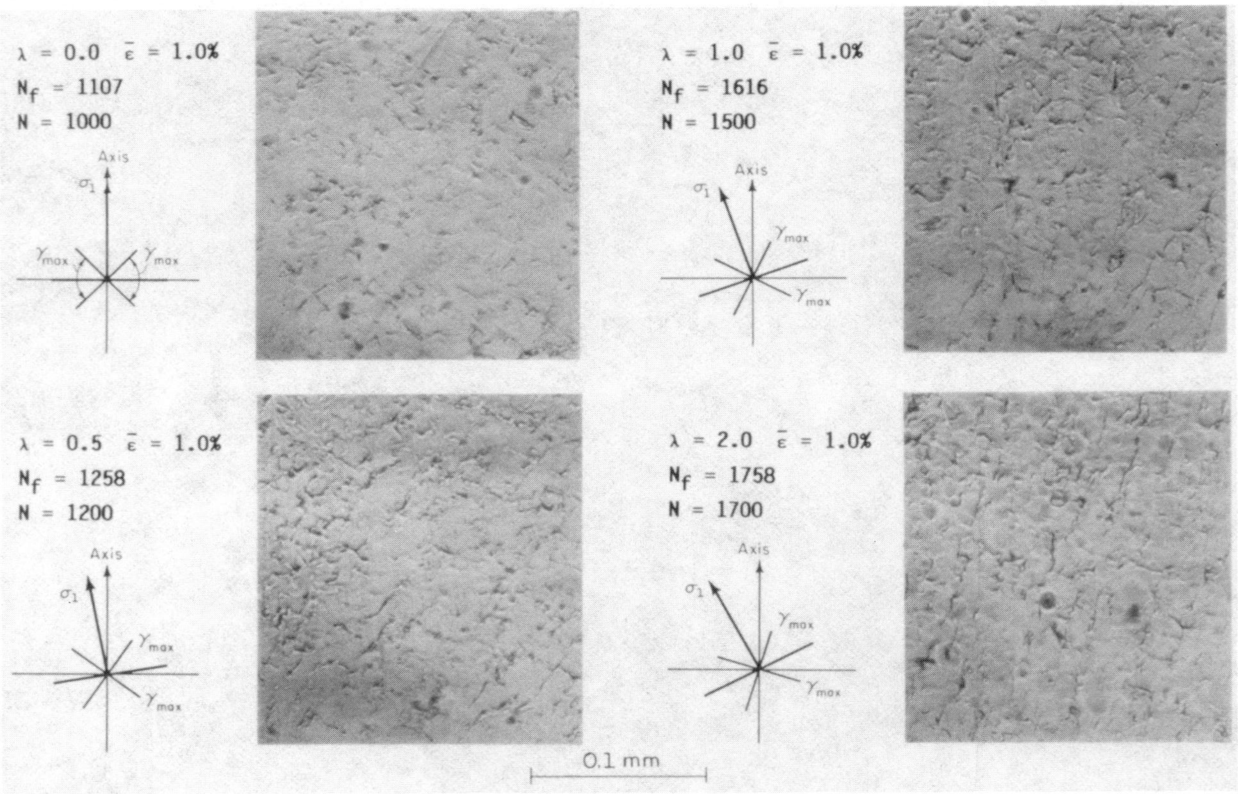

FIG. 6—*Thin-wall tube damage development at short lives,* $\bar{\epsilon} = 1.0\%$ *for* $\lambda = 0$, $\lambda = 0.5$, $\lambda = 1.0$, $\lambda = 2.0$.

many damage sites. The overall macrocrack direction is in general perpendicular to the maximum principal stress but changes direction often as the crack seeks the weakest, most highly damaged path through the material. Torsional cracking for lives less than approximately 10^6 cycles to failure develops in a similar manner with the final linking process occurring entirely along longitudinal shear planes. The manganese sulfide inclusions are important in the torsional crack development. Stage I crack length prior to the rapid failure process and the lives when linking started are given in Table 3 for tests performed at $\bar{\epsilon} = 1.0\%$.

Notched Shaft

Damage development in the notched shaft has been observed for several loading conditions. Observations at intermediate lives ($N_f \sim 10^5$ cycles) and at short lives ($N_f \sim 10^4$ cycles) are presented below to show the characteristics of crack development in these tests.

Intermediate to Long Life Load Cases

Typical intermediate life behavior for a bending test at an amplitude 1730 Nm is shown in Fig. 7. Early crack development intersects the notch surface in the circumferential direction. The maximum shear strain also intersects the notch surface in this direction since the maximum shear strain is Type B in the terminology of Brown and Miller [*17*]. Three cracks developed along the grinding marks and eventually linked to grow to final failure as one dominant crack. Macrocrack growth is in a plane perpendicular to the maximum principal stress, σ_1.

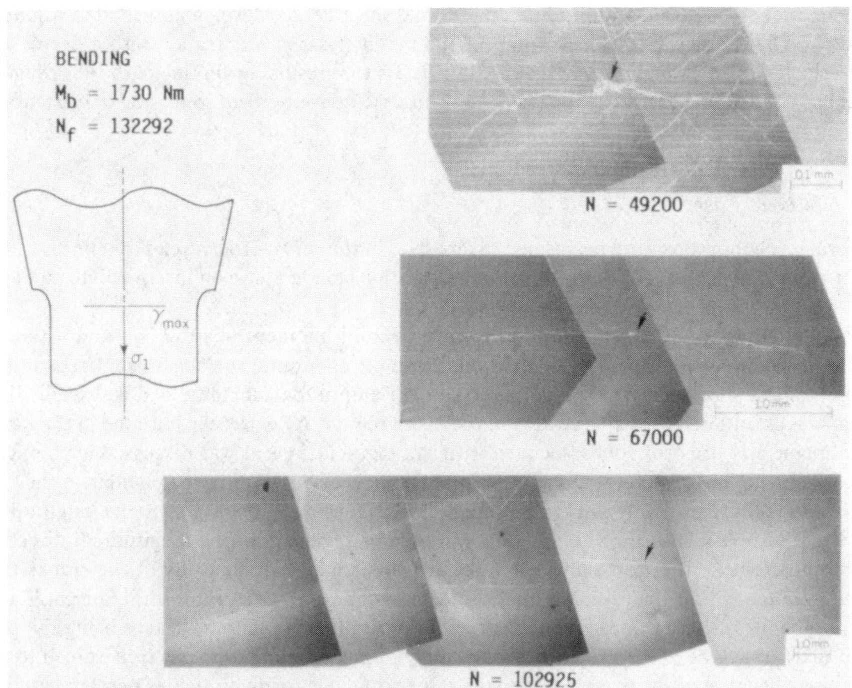

FIG. 7—*Crack development for notched shaft in bending.*

A mixed mode loading case (XR2 condition) was tested at a bending amplitude of 1220 Nm and a torsion amplitude of 1710 Nm. This specimen was also polished in the notch to facilitate replica observations. Early crack development occurred in two ways. Figure 8 indicates typical crack initiation in the remnants of the circumferential grinding marks and then growth for a period (with a small amount of linking) in the circumferential direction. These cracks then changed direction (Fig. 8) to grow out of the notch on the plane perpendicular to the maximum principal stress, σ_1. The extent of propagation within the machining marks was 1.13 mm and 0.4 mm for two of the main cracks that developed in this way. This implies that there is no obvious critical length when the mode of propagation changes. Once Stage II growth started the cracks in the center of the multiple crack system did not develop as quickly and most of the growth occurred at the tips of crack at the extreme ends of the crack systems.

Initiation also occurred by the formation of a longitudinal shear crack (Fig. 9). The two independent shear cracks shown at 20 000 cycles (first observed on a replica taken at 5000 cycles) are aligned in the longitudinal direction and are thought to be associated with inclusions. A small amount of extension occurred in the longitudinal direction together with growth along a machining mark from the larger of the two. After less than 100 μm growth in the machining mark this crack system changed direction to grow in Stage II, similar to the other cracks observed in this test.

Crack growth versus life plots [*21*] have shown that although these cracks develop by different mechanisms the subsequent damage development is very similar with similar crack lengths being reached at similar numbers of cycles. Final failure occurs by shearing of the material between the growing Stage II cracks and results in small wedge-shaped fragments as well as the two main fracture surfaces.

Crack development for long life tests in torsion show different behavior than the bending or mixed loading conditions. The entire circumference of the specimen in the notch area experiences the same shear strain amplitude. Longitudinal shear cracks develop at several locations in the notch. These Stage I cracks grow out of the notch stress concentration into a decreasing stress field and branch to grow in Stage II (Fig. 10). Failure results by linking of the 45° Stage II cracks leaving a star pattern fracture surface. This crack development in torsion only occurs at very long lives.

Short Life Load Cases

Several test conditions were performed to give lives in the range of 10^4 cycles to failure. Crack development in these tests differed from the observations made above on intermediate and long life tests.

A high amplitude (ZR3) test condition with a bending moment of 1150 Nm and a torsion amplitude of 2700 Nm was performed on a shaft with the as-ground surface finish. Crack initiation on this test showed typical behavior of the high amplitude short life load cases (Fig. 11). Several cracks initiated over a large area in the notch region. A few cracks initiated in the longitudinal plane and are probably associated with sulfide inclusions as was discussed for the XR2 load case above. These were observed first because their orientation perpendicular to the surface finish makes them easily discernible. A crack that formed at about 30° to the longitudinal direction is also visible in Fig. 11. This is composed of segments in the longitudinal direction and circumferential direction that have linked at this angle. The majority of the cracks that formed initiated within the grinding marks and were not easily discernible until later in life. A large number of microcracks developed early in life with linking of these cracks being the process of macrocrack development. Macrocrack development and growth was constrained to the notch plane because of the extensive damage rather than occurring on planes perpendicular to the maximum principal stress as found in the long life tests.

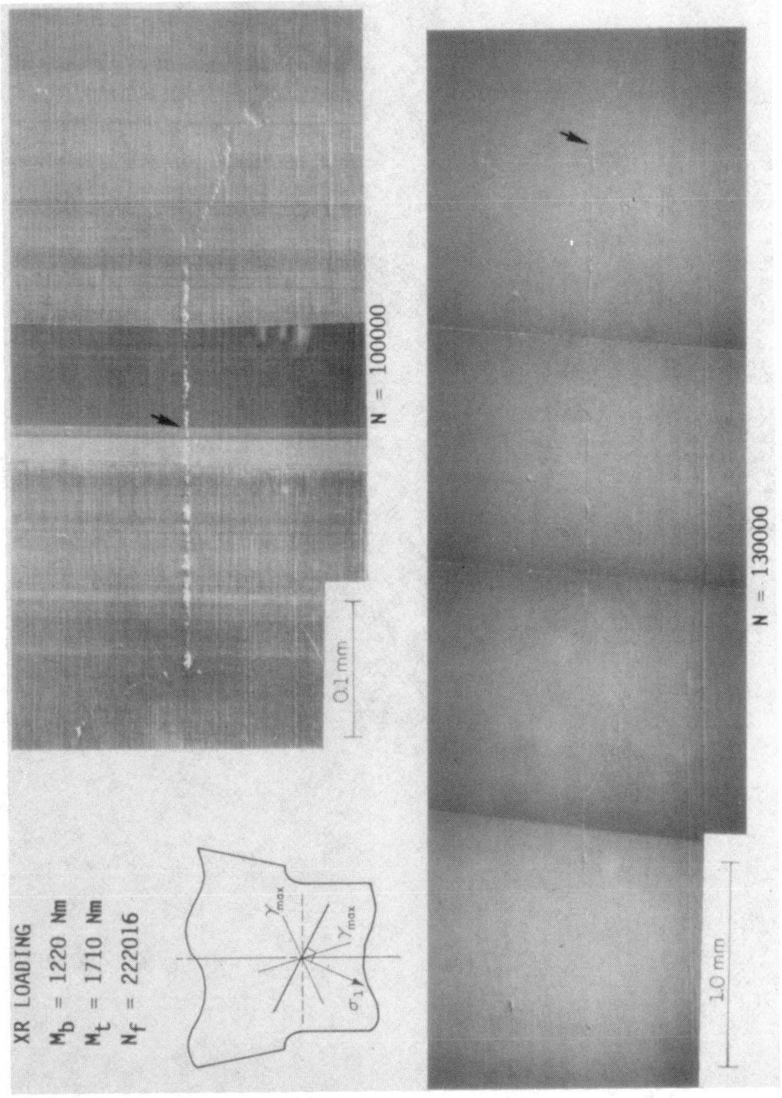

FIG. 8—Crack development in notched shaft; combined (XR2) loading condition.

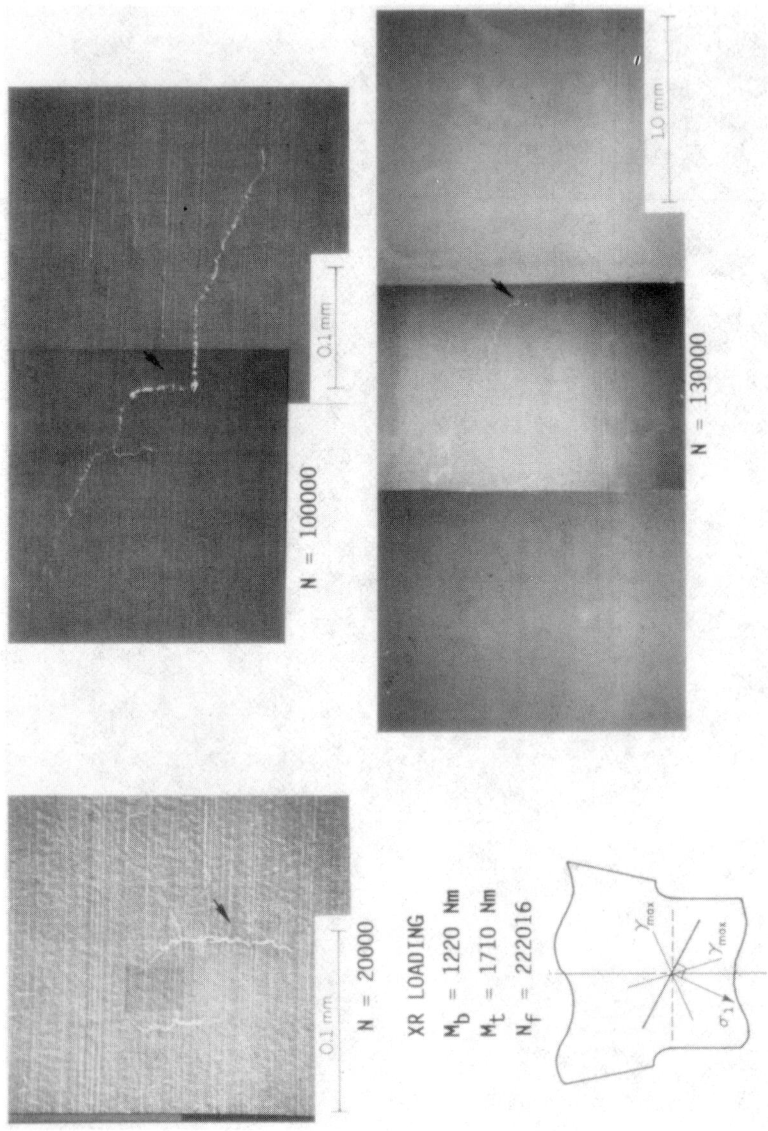

FIG. 9—*Crack development from inclusions in notched shaft; combined (XR2) loading condition.*

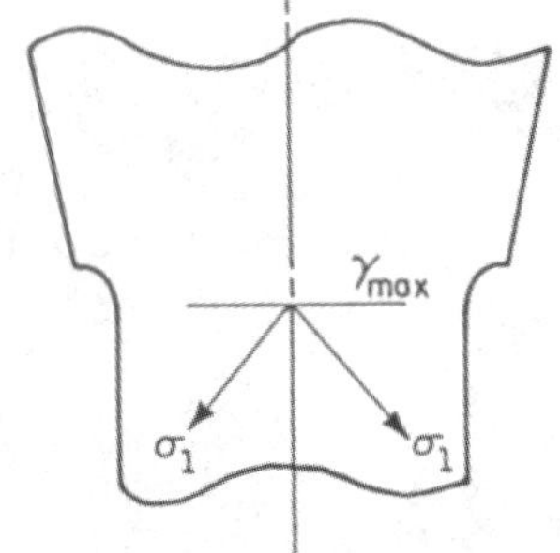

TORSION

M_t = 2000 Nm

$N_f = 2.13 \times 10^6$

FIG. 10—*Torsional cracking behavior of the notched shaft at long lives.*

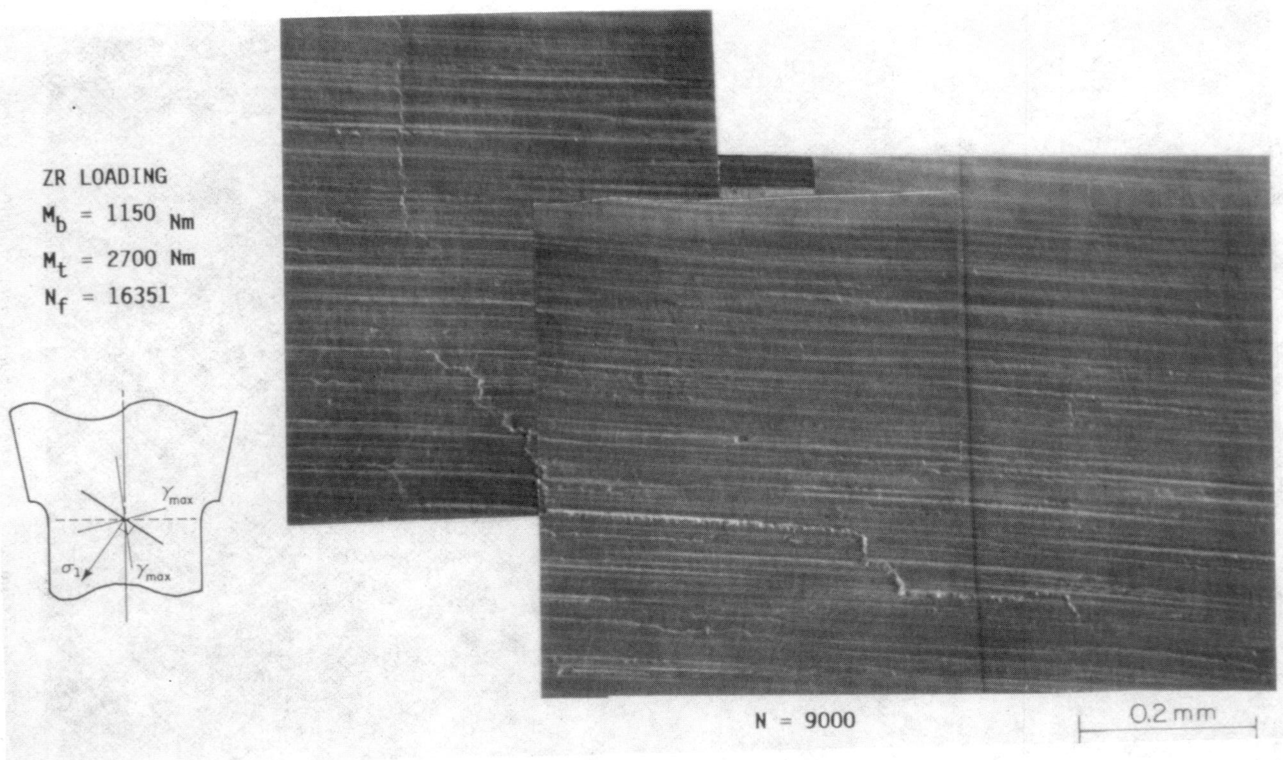

FIG. 11—*Crack development in notched shaft; combined (ZR3) loading condition.*

Another high load amplitude (XR3) with a bending amplitude of 1850 Nm and a torsional amplitude of 2550 Nm was tested in the polished condition (Fig. 12). Although the surface grinding marks were almost entirely removed by the polishing operation, microcrack initiation occurred primarily in the circumferential direction. Several cracks shown in Fig. 12 developed on parallel planes in the notch over the small area shown in this figure. Growth on longitudinal planes occurred, linking these microcracks and forming a single macrocrack of several millimeters in the notch plane. The final failure is approximately perpendicular to the specimen axis rather than occurring in the Stage II growth direction.

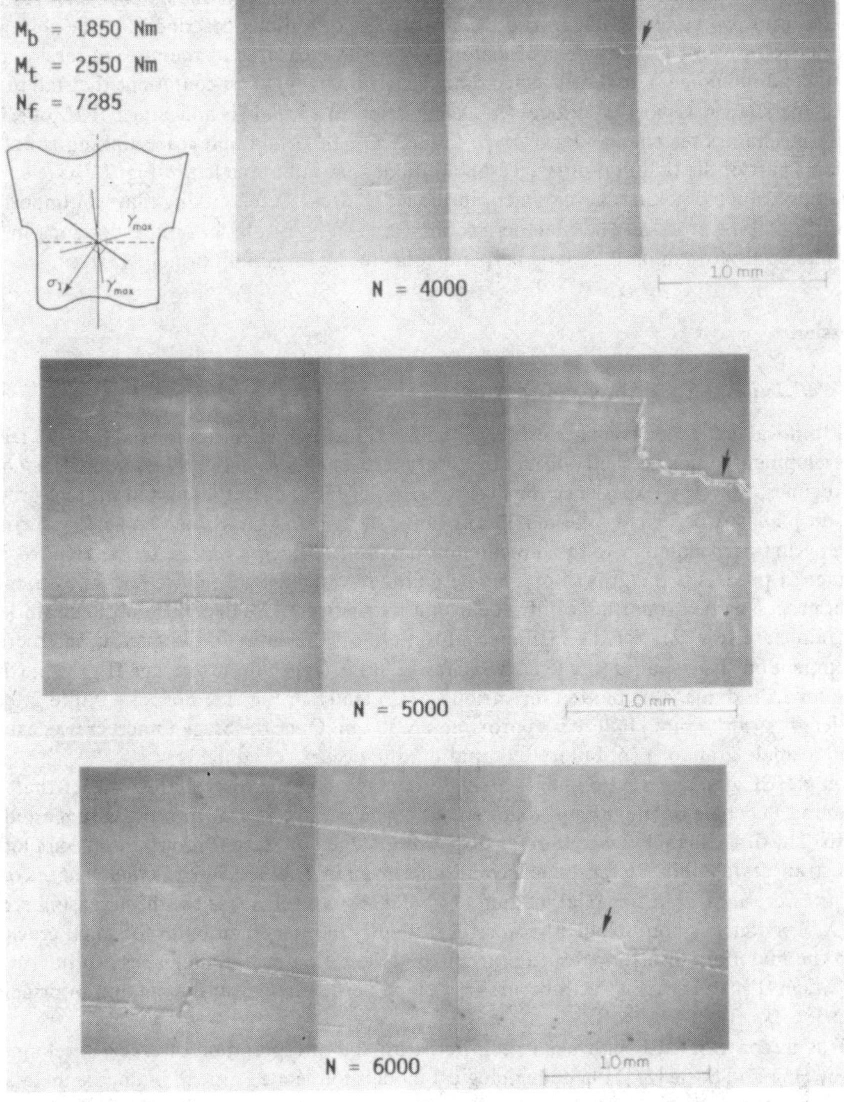

FIG. 12—*Crack development in notched shaft; combined (XR3) loading condition.*

Damage development in torsion again differs from the other load cases. Multiple cracking occurs on longitudinal shear planes. The microcrack development was most dense in the concentrated stress field of the notch but also occurred to a lesser degree over the entire gage section. Failure of the high amplitude torsion tests was a result of linking of the many shear cracks in the longitudinal direction. Several cracks can be present over the entire length of the gage section while still maintaining the load carrying capacity (although the compliance of the specimen increases).

Macroscopic observation of the failure surfaces shows the difference in the final failure process between short and intermediate life tests (Fig. 13). Although the load ratio and the angles that the maximum shear planes and the maximum principal stress plane make with the specimen axis are the same for the two test conditions the failure process is completely different.

Based on these and other observations [*13*] damage development in the notched shaft can be separated into two cases similar to the LCF and HCF conditions described for the thin-wall tube. Figure 14 shows the influence of loading condition measured as the ratio of torsion moment to bending moment on the expected damage development. Test conditions that fall to the right of the shaded region experience crack initiation in the notch and Stage II (Mode I in fracture mechanics terminology) growth to failure. For this condition crack linking is not an important part of the failure process. Test conditions that fall to the left in Fig. 14 have extensive damage (microcracking) development in the notch area. Crack linking plays an important role in the failure process. The macrocrack that eventually results in failure develops in the notch plane rather than on the plane perpendicular to the maximum principal stress.

Discussion

Thin-Wall Tube

The thin-wall tube specimen is considered the smooth specimen for multiaxial fatigue. Damage development in this geometry occurs by different mechanisms in the HCF and LCF regions. Both regimes, however, experience the classic stages of fatigue. Initiation and Stage I growth occur on planes experiencing the maximum range of cyclic shear strain. Stage II growth to failure occurs perpendicular to the direction of maximum principal stress. In the HCF regime the Stage II process is the growth of a single dominant crack. LCF failure occurs by a crack linking process. The proportion of life spent in a particular phase depends on the strain level and strain state (Fig. 2). For the LCF mechanism almost the entire life is spent in microcrack development on shear planes (Stage I). The Stage I crack length prior to Stage II growth (final failure for LCF damage) is plotted for various strain states in Fig. 15. For $\lambda = 0.0$ to 2.0 the crack length prior to rapid failure is approximately 50 μm. Once the Stage I microcracks extend beyond a single grain or two, failure by rapid linking occurs very quickly.

During HCF damage development, a single damage nucleus (or very few nuclei that did not interact) became dominant and followed the Stage I-Stage II growth process suggested by Forsyth. The Stage I crack size at the transition from Stage I to Stage II growth has been plotted versus strain state in Fig. 16 for the test conditions given in Table 3. Stage I crack lengths of 25 to 30 μm were measured for axial loading ($\bar{\epsilon} = 0.22\%$) and increased with increasing strain state, λ. Torsion was not included because of difficulty in interpretation of torsional cracking due to the material anisotropy resulting from the orientation of the manganese sulfide inclusions. Stage II macrocrack growth occurs in a plane perpendicular to the maximum principal stress.

This is in contrast, although not in conflict, with the interpretation of crack development given by Hua and Socie [*11*]. Phenomenological models for Mode I growth in ductile metals [*6*] suggest that the crack tip growth process is a shear phenomena not unlike the Stage I growth

FIG. 13—*Macroscopic growth behavior for combined (XR2 and XR3) loading conditions of the notched shaft.*

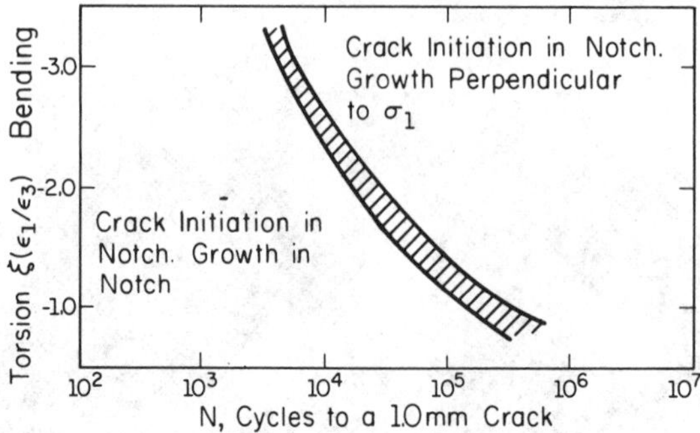

FIG. 14—*Schematic representation of damage state as a function of the ratio of applied moments and life regime for the notched shaft.*

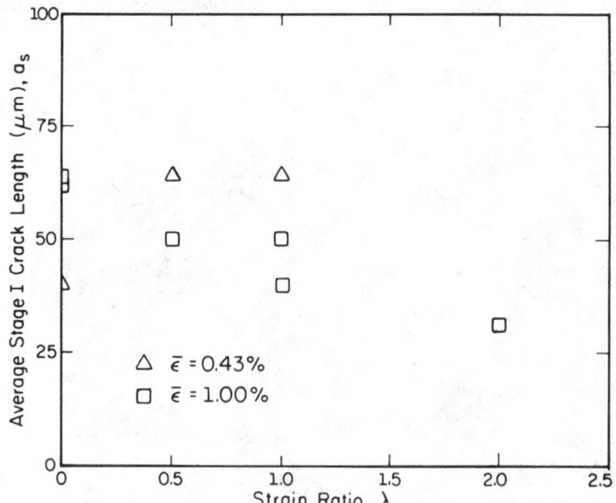

FIG. 15—*Crack length versus strain state for LCF-type damage in the thin-wall tube.*

process on a microscale. Comparison of the development of HCF damage at $\bar{\epsilon} = 0.22\%$ with damage curve approaches [*11*] has shown that there are two distinct phases except in torsion. Crack development was interpreted to be all Stage I growth and the two distinct phases were identified as initiation and propagation. Interpretation of crack development to be Stage I followed by Stage II growth (Figs. 2 to 4) suggests a slightly different interpretation of the two phase damage curves for HCF. The transition from Stage I to Stage II growth (Table 3) corresponds very closely both in terms of crack size and life fraction to the transition point of the two phase damage model reported in [*11*]. This represents a change in the growth mechanism from shear processes to Mode I opening processes. From the damage curve analysis the rate of damage development is much slower for the initiation/Stage I portion of life than for Stage II growth. With this interpretation it is logical that the LCF loading conditions and the torsion test at $\bar{\epsilon} = 0.22\%$ are satisfactorily described by a single damage curve [*11*] since Stage I growth occupied nearly the entire life.

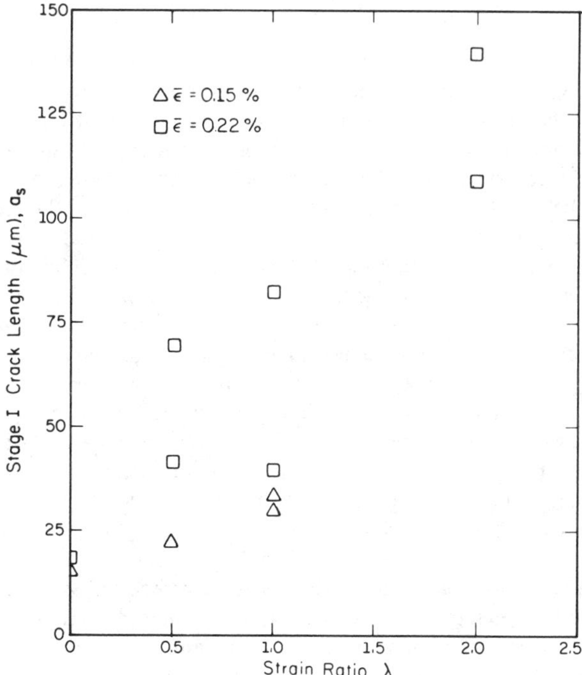

FIG. 16—*Crack length versus strain ratio for HCF-type damage in the thin-wall tube.*

Figure 16 indicates that the extent of Stage I development increases with increasing strain ratio. This must be a result of the micromechanisms influencing the crack behavior. A promising multiaxial fatigue model suggested by Brown and Miller [*17*] identifies the maximum shear strain as the primary influence on crack development but also includes the normal strain to the plane of maximum shear as having a secondary influence. Quite possibly it is this normal strain or opening (Mode I) strain that most influences the transition from Stage I to Stage II crack behavior. Shear crack lengths (Fig. 16) increase with strain ratio which corresponds to a decrease in the normal strain. However, this also corresponds to a slight increase in shear strain since the test conditions were based on equal effective strain ranges. A fracture mechanics analysis [*10*] developed to predict regions of stable Stage I and Stage II crack growth suggests that the crack behavior is dependent only on strain ratio and being a continuum approach is inadequate to describe microstructural factors that may influence this behavior.

Damage development in LCF was found to be almost entirely Stage I growth. Once the cracks change to a Stage II growth direction, crack linking resulted in very rapid failure. Damage curves for LCF showed a single phase [*11*] and were similar for different strain ratios tested. Again, torsional behavior differed being all Stage I crack growth because of the influence of the longitudinal inclusions. Fracture mechanics concepts do not adequately describe the LCF damage process.

Similitude between the uniaxial loading condition and other strain states appears to exist for LCF damage. For HCF damage the Stage I-Stage II transition occurs at different Stage I crack sizes (Fig. 16) for different strain states. Consequently similitude arguments in terms of damage development do not strictly hold between the smooth axial specimen and the thin wall tube for multiaxial loading.

Notched Shaft

Damage development in the notched shaft varies with amplitude and loading conditions (Fig. 14) in a manner similar to that observed in the thin-wall tube. Crack initiation has been observed to occur by two processes. The most frequent and apparently most significant is the initiation and development of cracks in the circumferential direction in the notch. Additionally, crack initiation was observed along the longitudinal axis apparently aligned with manganese sulfide inclusions lying in this direction. Both processes can influence the growth to failure for intermediate and long life tests but the longitudinal cracking has little or no effect in the short life tests (except torsion). In test condition XR2 cracking was observed in both directions (Figs. 8 and 9) with similar contributions to the failure process. Many more cracks initiate in the short life tests (Fig. 12) than at lower load amplitudes. Initiation does not occur on planes of maximum shear strain as predicted by continuum analysis and as observed in the thin-wall tube test.

Growth to failure is dependent on the loading amplitude. For intermediate to long life tests, the cracks (one or a few) that initiate in the circumferential direction change to grow on planes perpendicular to the maximum principal stress. When more than one crack is present they may all change direction, but growth occurs primarily at the two extreme ends of the crack system. This maximizes the Mode I (Stage II) opening component of loading similar to the behavior observed in the HCF damage in the thin-wall tube. For the short life conditions crack linking in the notch plane plays an important role in the final failure. Extensive initiation damage occurs both in the longitudinal and circumferential directions. Crack linking constrains the failure crack to develop in the notch plane perpendicular to the longitudinal axis rather than by Stage II growth. This is similar to the failure process in LCF damage reported for the thin-wall tube.

The lack of similitude in the initiation process may be related to a number of factors. The surface finish present in the notched shaft has had a marked effect. Machining marks have been found to influence the orientation of crack initiation and growth for torsional loading of steels [*18*]. Specimens polished in the circumferential direction developed cracks and failed in the circumferential direction while those polished longitudinally failed by longitudinal shear cracks. Shear strain values on the notch plane do not differ greatly from the maximum shear strain (approximately 20% for the XR2 test) consequently, cracking in the grinding marks is not surprising. However, circumferential cracking was also observed in tests that were polished so that nearly all of the surface texture from grinding was removed. This did not change the direction of crack development but did result in slightly fewer cracks.

Another factor influencing the damage development are stress gradients. The loading configuration and specimen geometry result not only in multiaxial stresses and strains in the notch, but also multidimensional stress-strain gradients. Figure 17 indicates the three dimensional gradients determined from an elastic bending analysis [*13*] and is representative of the combined loading cases. The lower left hand schematic represents the radial gradient into the specimen that is analogous to the stress gradient often considered in uniaxial loading of notched plates. The schematic at the upper right is the gradient of stress around the circumference of the notch. More significant to crack development is the gradient along the length of the shaft shown at the upper left. The maximum stress occurs in the notch slightly up the radius [*20*] but decreases rapidly in either direction longitudinally. This gradient tends to constrain the early crack development to the circumferential direction since a crack growing at an angle to the notch is trying to grow into a rapidly decreasing stress field. In tests on an unnotched shaft [*13*] that did not have this longitudinal gradient (it did have the radial and circumferential gradients) crack initiation was on maximum shear planes for both polished and as ground specimens. This indicates a significant influence of the stress gradients on the micromechanics of crack development.

Material anisotropy due to inclusions aligned in the longitudinal direction also influences the

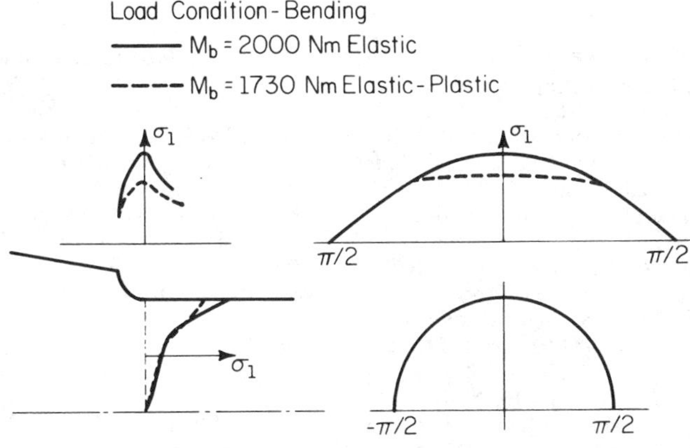

FIG. 17—*Principal stress gradients in the notched shaft for bending.*

crack behavior. Initiation was observed in the longitudinal direction in several places for short life tests but only in a few places for intermediate and long life tests. These do not appear to have a significant influence on the life to failure or crack development. Pascoe and deVilliers [*20*] showed a factor of two to three difference in life when the maximum shear planes were aligned with the inclusions during torsion tests of a medium carbon steel. Similar differences in life have been shown [*2*] for the material reported here in smooth specimens taken from the longitudinal and transverse directions of the hot rolled bar stock.

Overall these tests have shown that surface finish effects, material anisotropy, and notch effects are very important in multiaxial fatigue. The quantitative significance of effects such as surface finish and anisotropy are liable to be greater for higher strength materials given the trends observed for uniaxial fatigue. Testing and prediction methods require development in order to account for observed differences between smooth specimen and component behavior such as were found here. Such work remains an important future direction for LCF studies.

Conclusions

Crack development during multiaxial loading of the thin-wall tube specimen follows the classic stages of fatigue development, shear initiation, and Stage I and Stage II growth.

The exact process of this development differs in the HCF and LCF regimes and is also strain state dependent.

Crack initiation and early growth in the notched shaft is constrained to the circumferential direction in the notch; hence similitude does not exist between the thin-wall tube and the notched shaft.

At intermediate and long lives the failure process in both specimens is by Stage II crack growth; hence similitude exists for this portion of life.

At short lives multiple crack initiation occurs and crack linking is important in the failure process for both specimens.

Further understanding of the question of similitude is need for the development of improved life methods that relate to phenomenological damage development.

References

[1] *Multiaxial Fatigue, ASTM STP 853,* K. J. Miller and M. W. Brown, Eds., American Society for Testing and Materials, Philadelphia, 1985.
[2] Fash, J. W., Socie, D. F., and McDowell, D. L., "Fatigue Life Estimates for a Single Notched Component Under Biaxial Loading," in *Multiaxial Fatigue, ASTM STP 853,* K. J. Miller and M. W. Brown, Eds., American Society for Testing and Materials, Philadelphia, 1985, p. 497.
[3] Ewing, J. A. and Humfrey, J. C. W., "The Fracture of Metals Under Repeated Alternations of Stress," *Philosophical Transactions of the Royal Society,* Vol. 200, 1903, pp. 241-253.
[4] Wood, W. A., "Recent Observations on Fatigue Fracture in Metals," in *Basic Mechanisms of Fatigue, ASTM STP 237,* American Society for Testing and Materials, Philadelphia, 1958, pp. 110-121.
[5] Forsyth, P. J. E., "Two Stage Process of Fatigue Crack Growth," in *Proceedings,* Crack Propagation Symposium, Cranfield, England, 1961, p. 76.
[6] Neumann, P., "On the Mechanics of Crack Advance in Ductile Materials," in *Proceedings,* 3rd ICF Conference, 1973, Vol. III, p. 233.
[7] Fash, J. W., "Fatigue Crack Initiation and Growth in Gray Cast Iron," FCP Report No. 35, College of Engineering, University of Illinois at Urbana-Champaign, Urbana, Ill., Oct. 1980.
[8] Dowling, N. E., "Crack Growth During Low-Cycle Fatigue of Smooth Axial Specimens," in *Cyclic Stress-Strain and Plastic Deformation Aspects of Fatigue Crack Growth, ASTM STP 637,* American Society for Testing and Materials, Philadelphia, 1977, pp. 97-121.
[9] Starkey, M. S. and Irving, P. E., "Prediction of Fatigue Life of Smooth Specimens of SG Irons Using a Fracture Mechanics Approach," in *Low-Cycle Fatigue and Life Prediction, ASTM STP 770,* American Society for Testing and Materials, Philadelphia, 1982, pp. 382-398.
[10] Brown, M. W. and Miller, K. J., "Initiation and Growth of Cracks in Biaxial Fatigue," *Fatigue of Engineering Materials and Structures,* Vol. 1, 1979, p. 231.
[11] Hua, C. T. and Socie, D. F., "Fatigue Damage in 1045 Steel Under Constant Amplitude Biaxial Loading," *Fatigue of Engineering Materials and Structures,* Vol. 7, No. 3, 1984, p. 165.
[12] Downing, S. D. and Galliart, D. R., "A Fatigue Test System for a Notched Shaft in Combined Bending and Torsion," in *Multiaxial Fatigue, ASTM STP 853,* K. J. Miller and M. W. Brown, Eds., American Society for Testing and Materials, Philadelphia, 1985, pp. 24-32.
[13] Fash, J. W., "An Evaluation of Damage Development During Multiaxial Fatigue of Smooth and Notched Specimens," Ph. D. thesis, Department of Mechanical Engineering, University of Illinois, Urbana, Aug. 1985.
[14] Abaques Finite Element Code, Hibbitt, Karlson, and Sorenson, Inc., Providence, R.I., 1980.
[15] Nishihara, Yo, and Kawamoto, M., "The Strength of Metals Under Combined Alternating Bending and Torsion," Memoirs of the College of Engineering, Kyota Imperial University, Vol. 10, No. 6, 1941, p. 117.
[16] Brose, W. R., Dowling, N. E., and Morrow, J., "Effect of Periodic Large Strain Cycles on the Fatigue Behavior of Steels," Paper No. 740221, Society of Automotive Engineers, 1974.
[17] Brown, M. W. and Miller, K. J., "A Theory for Fatigue Failure Under Multiaxial Stress-Strain Conditions," *Proceedings of the Institute of Mechanical Engineers,* Vol. 187, 1973, p. 745.
[18] Miller, K. J. and Chandler, D. C., "High Strain Torsion Fatigue of Solid and Tubular Specimens," *Proceedings of the Institute of Mechanical Engineers,* Vol. 184, 1969-70, pp. 433-448. (Discussion by D. J. Hatter.)
[19] Tipton, S. M., "Fatigue Behavior Under Multiaxial Loading in the Presence of a Notch: Methodologies for the Prediction of Life to Crack Initiation and Life Spent in Crack Propagation," Ph.D. thesis, Stanford University, Calif., 1984.
[20] Pascoe, K. J. and deVilliers, J. W. R., "Low Cycle Fatigue of Steels Under Biaxial Straining," *Journal of Strain Analysis,* Vol. 2, No. 2, 1967, p. 117.
[21] Hurd, N. J., "Micromechanisms and Multiaxial Fatigue Life Prediction for the SAE Notched Shaft," GKN Technology Report 1728, May 1985.

J. A. Bannantine[1] *and D. F. Socie*[1]

Observations of Cracking Behavior in Tension and Torsion Low Cycle Fatigue

REFERENCE: Bannantine, J. A. and Socie, D. F., **"Observations of Cracking Behavior in Tension and Torsion Low Cycle Fatigue,"** *Low Cycle Fatigue, ASTM STP 942,* H. D. Solomon, G. R. Halford, L. R. Kaisand, and B. N. Leis, Eds., American Society for Testing and Materials, Philadelphia, 1988, pp. 899–921.

ABSTRACT: A series of tension and torsion low-cycle fatigue tests were conducted on several engineering materials. Detailed observations of nucleation and early crack growth were made. Cracking behavior is shown to depend upon loading mode (tension or torsion), strain amplitude, and material type.

Multiaxial models developed for shear sensitive materials did not correlate multiaxial test results of a tension sensitive material. Differences in cracking behavior for different materials and loading conditions need to be considered in successful life predictions for components subjected to multiaxial fatigue.

KEY WORDS: fatigue (materials), multiaxial fatigue, torsion, cracking behavior, life prediction, stainless steels, Inconel, gray cast iron, SAE 1045 steel

The majority of fatigue research has been conducted under uniaxial loading conditions. To a lesser extent, work has been done in the area of torsional fatigue. In many applications, engineering components are subjected to complicated states of stress and strain. Successful multiaxial fatigue life predictions which are based upon physical observations require an understanding of the cracking behavior of materials. This paper seeks to aid in the development of that understanding.

Cracking observations reported in the literature are reviewed. Results of tension and torsion strain-controlled low cycle fatigue tests conducted on thin wall tubular specimens at room temperature are presented. Observations of nucleation and early crack growth are made for several engineering materials. Cracking behavior is observed to depend on loading mode (tension or torsion), strain amplitude, and material type. The impact on multiaxial theories is discussed.

Background

Extensive research in uniaxial smooth specimen fatigue has shown that fatigue crack nucleation usually occurs at stress concentrations on or immediately below the surface of the material. These stress concentrations can result from surface roughness (due either to manufacturing or to intrusions and extrusions caused by cyclic slip), grain boundaries, or inclusions. Nucleation mechanisms vary for different materials.

Typically, ductile materials form slip bands. In 1902, Ewing and Humfrey [*1*] reported observations of slip bands in fatigue tests of Swedish iron. They initially observed formation of a few slip bands. As the test progressed, they observed an increased number, with the earlier observed

[1]Department of Mechanical and Industrial Engineering, University of Illinois at Urbana-Champaign, Urbana, IL 61801.

bands becoming broader and more distinct until cracks formed in the bands. These slip bands or planes, commonly referred to as persistent slip bands (PSBs), vary from the surrounding matrix by their dislocation structure [*2–5*]. Intrusions and extrusions associated with the PSBs are produced on the surface of the specimen. These tend to produce a notch effect, and it is believed that this stress concentration causes cracking to occur in the PSBs.

Although this mechanism is the most general and widely observed, other modes of crack initiation exist. By varying alloying constituents or test conditions (i.e., very high strains [*6*]), fatigue cracks may be initiated at grain boundaries rather than from PSBs. In addition, PSBs are not observed in such materials as pure body-centered cubic (bcc) metals at low temperatures [*4*]. Inclusions may also influence fatigue crack initiation [*7,8*]. Fine and Ritchie [*2*] reported that in aluminum alloys and high-strength low alloy (HSLA) steels, inclusions aid in slip band fatigue crack initiation, while cracked inclusions or debonding between inclusions and matrix were not observed.

Microstructure has a significant effect on crack initiation. Cold work, stacking fault energy, grain size, and alloying have all been shown to exhibit an influence. Crack initiation mechanisms have been widely discussed and reviewed [*2–5, 7, 9*].

The formation of PSBs first occurs, for most materials, in those grains whose slip planes are most closely aligned with the plane of maximum shear stress. Slip on these planes occurs as the result of the shear stress applied to the component. Gough [*10*] reported that crack nucleation is dependent upon the shear stress acting on the slip plane rather than upon the normal stress.

Forsyth designated crack initiation and growth on shear planes as Stage I growth [*11*]. He reported that Stage I crack growth, which may dominate a significant portion of the fatigue life, continues until reversal of dislocation movement is prevented. The crack may then turn and propagate in a Stage II growth direction, a plane normal to the maximum tensile stress. This transition often occurs at grain boundaries [*12,13*]. Forsyth further reported that the criterion for Stage II growth is the value of the maximum tensile stress acting on the specimen in the region near the crack tip. When the ratio of shear stress to tensile stress reaches a critical value, the transition from Stage I to Stage II occurs.

As early as the 1920s, Gough [*13*] reported that components subjected to torsional fatigue fail in two general modes. In one case, the crack propagates in the axial or circumferential direction of a shaft subjected to cyclic torsional loading. In the second mode, the crack propagates 45° to the axis of the shaft.

Materials that form slip bands in uniaxial fatigue also produce slip bands when tested in torsion. These bands occur on the planes of maximum shear, which are the axial or circumferential directions for a tube or shaft. Peterson [*14*] observed that microcracks develop from these PSBs which then grow in an axial, transverse (circumferential), or stepwise 45° direction. He reported, as did others, that the harder materials tend to fail in the 45° or diagonal direction (plane normal to maximum tensile stress), while ductile materials branch only at stress values not far above the endurance limit. Alternatively, Gough [*20*] believed it to be erroneous to make any such general statement. He believed 45° or spiral fractures were probably due to material inclusions or defects. Frost, March, and Pook [*7*] also noted the tendency for harder materials to crack on tensile planes. They state that obstacles to crack propagation, such as intermetallics, flaws, and inclusions, are much more likely to dominate in harder, more complex alloys. Cast iron is an example of a material with inherent flaws in the form of graphite flakes. Torsion tests show 45° or tensile cracking at all strain levels.

One explanation [*15,16*] for the observed propensity of high hardness or brittle materials to fail in the tensile mode or 45° mode is related to the increase of shear strength and decrease in tensile strength with increasing hardness. The 45° failure occurs when the tensile stress rises to the cohesive strength before the shear stress reaches the shear strength. This explanation is analogous to that used to account for differences in the monotonic torsion fracture modes of ductile and brittle materials.

Recently, increased research has been conducted in the area of torsional fatigue [*17–21*]. Hurd and Urving [*17*] observed a general dependency of cracking behavior on microstructure and strain amplitude for EN16 tempered to three strength levels. They noted that Mode III or shear cracks were most stable in the lowest strength steel, while the initiation of a 45° crack depended on the fatigue life. At high stress intensities, or shorter lives, the shear mode was favored. They also reported that induction hardening may improve fatigue lives of components subjected to torsional loading. The Mode I or 45° cracking is suppressed due to the compressive residual stresses developed during hardening.

Tschegg, Ritchie, and McClintock [*19*] observed that cracking behavior was dependent on microstructure and applied stress. Circular notched shafts made of AISI 4340 steel were subjected to torsional fatigue. Cracks were reported to initiate in a "macroscopically flat" manner in all specimens tested. At lower stress intensities and larger crack lengths, fracture surfaces developed into a local hill-and-valley morphology which they termed a "factory roof" fracture. This fracture mode was associated with Mode I, 45° branch cracks. Tschegg [*20*] postulated several reasons for this transition from Mode III, "macroscopically flat" fracture, to Mode I, "factory roof." These include mutual support of inclined surfaces, fretting fatigue, and mutual support of debris. In AISI 4340 steel, Tschegg believed that the change of fracture mode was influenced by inclusions which act as initiation sites for the branch cracks in the plastic zone of the main crack.

These same authors observed a decrease in crack growth rate at constant values of stress intensity. This was explained by an increase in roughness-induced crack closure. These results are in agreement with the work of Hult [*22*], in 1958, who stated that the rate of growth will decrease due to the friction between crack surfaces. This friction is believed to reduce the stresses at the crack tip and consequently reduce growth rate.

In 1976, Parsons and Pascoe [*12*] reported observations of crack initiation and growth under biaxial stress for AISI 304 and QT35 using cruciform specimens. They observed two stages of cracking. In the primary stage, the more prominent surface cracks developed until displacements of the crack faces could be observed. The secondary stage then occurred with the linking up of primary cracks.

In the pure shear condition, which is the strain state for torsion, Parsons and Pascoe reported that the primary and secondary cracks in QT35 were of shear type. At high strains and later stages of propagation, they noted a transition to the tensile mode by linking of primary shear cracks. The AISI 304, at very high strains (above 2.5%), developed a large number of shear cracks which linked to form a tensile secondary crack. At strains between 1.13 and 2.5%, shear type primary and secondary cracks developed. Below strains of 1.13%, a transition from shear to tensile mode was observed. After this transition, four crack legs which branched off the main shear crack continued to grow until failure. No secondary stage cracking was observed. Parsons and Pascoe believed shear cracking was favored at high strains due to a greater effect of crack tip plasticity and the large number of slip band cracks in neighboring grains.

Recently, an increased number of multiaxial theories have been developed, including those found in Refs *23* to *27*. Brown and Miller [*23*] proposed a theory which they reported was based upon physical crack observations. The primary damage parameter in this model for both Stage I and Stage II crack growth was the maximum shear strain. The tensile strain across the maximum shear strain plane was believed to have a lesser but important effect. Lohr and Ellison [*24*] later modified the theory based upon the premise that the shear strain on the plane growing into the thickness of the component was the most damaging. Jacquelin, Hourlier, and Pineau [*25*] developed a multiaxial fatigue theory for Stage I crack initiation life based on results of stainless steel tests. They stated the belief that Stage I and Stage II behavior corresponds to two different physical processes and that it is important to distinguish between the two.

Although the importance of cracking behavior has been recognized, none of the multiaxial theories reviewed account for differences in this behavior. Instead, all material types and loading modes have been merged. Successful theories must correlate parameters with observed physical damage. Differences in crack behavior obviously will affect the importance of various parameters.

The purpose of this research was to study the effect of loading mode, strain amplitude, and material type on crack behavior and to evaluate the impact of the differences on multiaxial fatigue life predictions.

Experimental Procedure

Strain-controlled tension and torsion tests were conducted on thin-walled tubular specimens using internal extensometry and an automated servohydraulic, axial-torsional test system. The gage section and internal diameter of all specimens measured 25 mm, while the wall thickness differed for each material. The monotonic tensile properties for four of the materials studied are given in Table 1.

Stainless steel AISI 304 and AISI 316 tubular specimens were machined from hot-rolled bar stock to a wall thickness of 3.8 mm. The AISI 304 specimens were cut from two bars taken from the same heat of material. Chemistry and inclusion size remained essentially constant between the bars. Manganese sulfide stringers and spherical oxide inclusions were present in the AISI 304, the size and distribution of which were similar between bars. The AISI 316 contained fewer inclusions, although the size was approximately the same as those in the AISI 304. The first bar of AISI 304, designated 304(A), had roughly a 20-μm grain size while the second bar of the 304, designated 304(B), and the bar of AISI 316 both had a grain size of roughly 25 μm. The smaller grain AISI 304 material, 304(A), achieved cyclically stable stresses up to 20 percent higher than the AISI 304(B) material. Small tensile specimens were cut from the grip sections of tested AISI 304(A) and 304(B) tubular specimens. Monotonic tensile tests were conducted to verify baseline properties. Virtually no difference between the monotonic tensile properties of the two AISI 304 materials was observed. Both the AISI 304 and 316 underwent a stress-induced phase transformation to martensite, although the amount of transformation was greater in the AISI 304.

The surfaces of failed AISI 304 specimens tested in torsion were lightly polished and etched to reveal crack interaction with microstructure. Several solid, 6.4-mm outside diameter, AISI 304 specimens were tested under axial strain-controlled conditions. The fracture surface of these specimens was examined using scanning electron microscopy (SEM).

Comparison of AISI 304 and 316 results were made to other materials recently tested with similar specimen geometries and loading conditions. These materials, Inconel 718 [*27*], SAE 1045 [*28*], and gray cast iron [*29*], are briefly described below. A detailed description of each may be found in the respective references.

Inconel 718 specimens were cut from a forged ring purchased to Aerospace Material Specification AMS 5663. The wall thickness was 2 mm and the grain size ranged from 20 to 100 μm. The SAE 1045, with a specimen wall thickness of 2.5 mm, had a grain size of 35 to 40 μm and contained stringers of magnesium sulfide inclusions. The wall thickness of the gray cast iron specimens was 3 mm.

TABLE 1—*Mechanical properties.*

	0.2% Offset Yield Strength	Fracture Strength	Fracture Strain	Reduction in Area (%)
AISI 304	325 MPa	650 MPa	1.61	80%
AISI 316	555 MPa	725 MPa	1.73	82%
SAE 1045	380 MPa	985 MPa	0.71	50%
Inconel 718	1160 MPa	1850 MPa	0.33	28%

All specimen surfaces were polished to a 0.5-μm finish in order to reduce surface roughness and enable cracks to be detected clearly. Crack nucleation and growth was monitored using acetate replicas taken at 10% intervals of expected life. Final failure was defined as a 10% drop in the stable load. Replicas were examined under an optical microscope using transmitted light.

Test Results and Observations

Three modes of cracking were observed in torsion. These are shown schematically in Fig. 1. In Region I, cracks initiated and remained on planes of maximum shear. Cracks in Region II initiated on shear planes, but branched to grow on tensile planes or Stage II planes by the linking of previously initiated shear cracks. Region III crack behavior was characterized by an initial shear crack growth followed quickly by branching and crack propagation on Stage II planes. No major linking of Stage II cracks occurred in Region III.

In tension, final failure occurred on planes normal to the maximum tensile stress. In terms of the three regions described above, only Regions II and III crack behavior were observed in the tensile loading case.

Observations of the four material types are presented separately below. General discussion and comparison of cracking behavior between the two material types follows.

Stainless Steel AISI 304 and 316

Detailed observations of crack growth in torsion were made at four fully reversed shear strain amplitudes, 1.7, 0.8, 0.6, and 0.35%, in the AISI 304 and 316 stainless steels. Cracking at the

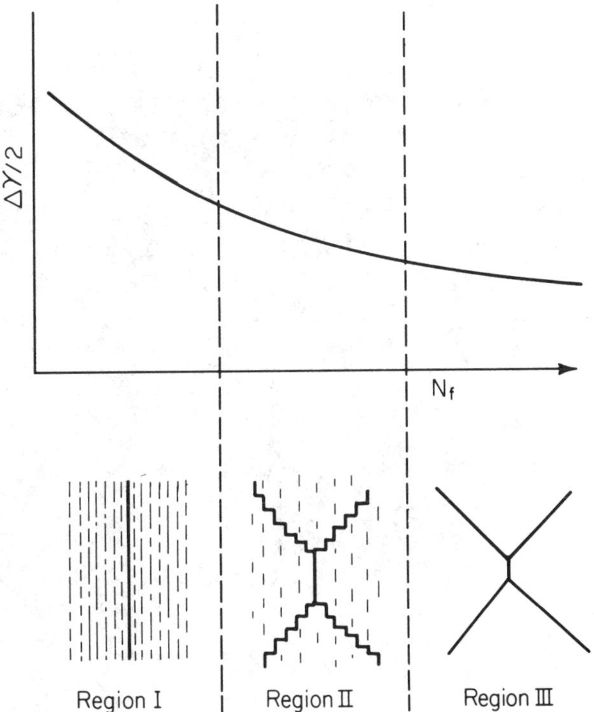

FIG. 1—*Three regions of cracking behavior observed in torsion.*

largest strain amplitude was characterized by Region I crack behavior. Cracks initiated in slip bands and at grain boundaries. Once initiated, the cracks become more distinct, but showed no significant increase in length. At failure, a large density of small, coarse cracks dominated the surface of the specimen (Fig. 2). A small amount of branching onto tensile planes (Stage II planes) was observed. However, the failure crack grew on the Stage I plane by a slow linking of previously initiated shear cracks.

Cracking behavior at 0.8% strain was characteristic of Region II. Cracks again initiated in slip bands and at grain boundaries. Bifurcation of Stage I cracks generally occurred at obstructions such as grain boundaries or triple points. Growth on Stage II planes occurred by a linking of Stage I cracks such that the macroscopic crack was approximately 45° to the maximum shear planes (Fig. 3). Crack density was much smaller at 0.6% strain than that at 1.7 or 0.8%. Cracks again branched at obstructions. Final failure occurred in a manner similar to the 0.8% test (i.e., Region II crack behavior).

Region III crack behavior was observed at the lowest strains. The fraction of life spent growing the crack on shear planes was reduced, as was the crack density. A small number of cracks initiated on shear planes but quickly branched to Stage II planes. Growth on these planes occurred by the propagation of the main crack rather than by a linking process (Fig. 4).

General cracking behavior and final failure was the same for both AISI 304 and 316. The difference in the cyclic stresses between the two bars of AISI 304 affected the initiation and early growth of the Stage I cracks rather than overall crack characteristics. The material with the lower strength, AISI 304(B), exhibited shear or Stage I cracks which were longer and often initiated at inclusion sites. The shear cracks in the higher strength stainless, AISI 304(A), were less than 0.1 mm for Regions II and III. In the AISI 304(B) material, the shear cracks developed up to 2 mm for these same regions. In general, the fraction of life spent in shear growth increased with increasing strain amplitude (Fig. 5).

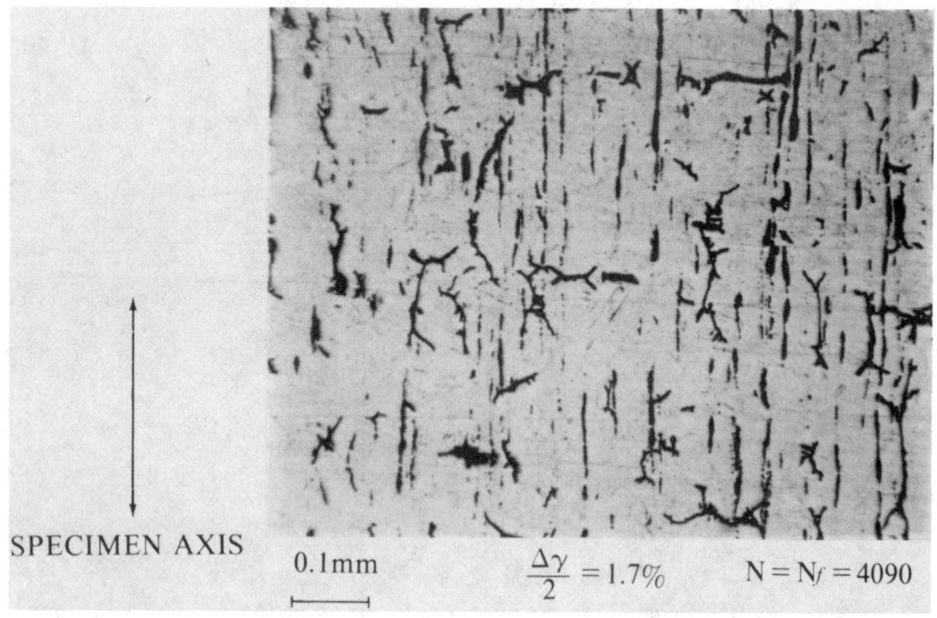

FIG. 2—*AISI 304 stainless steel specimen surface at failure. Microcracks and stringer inclusions visible.*

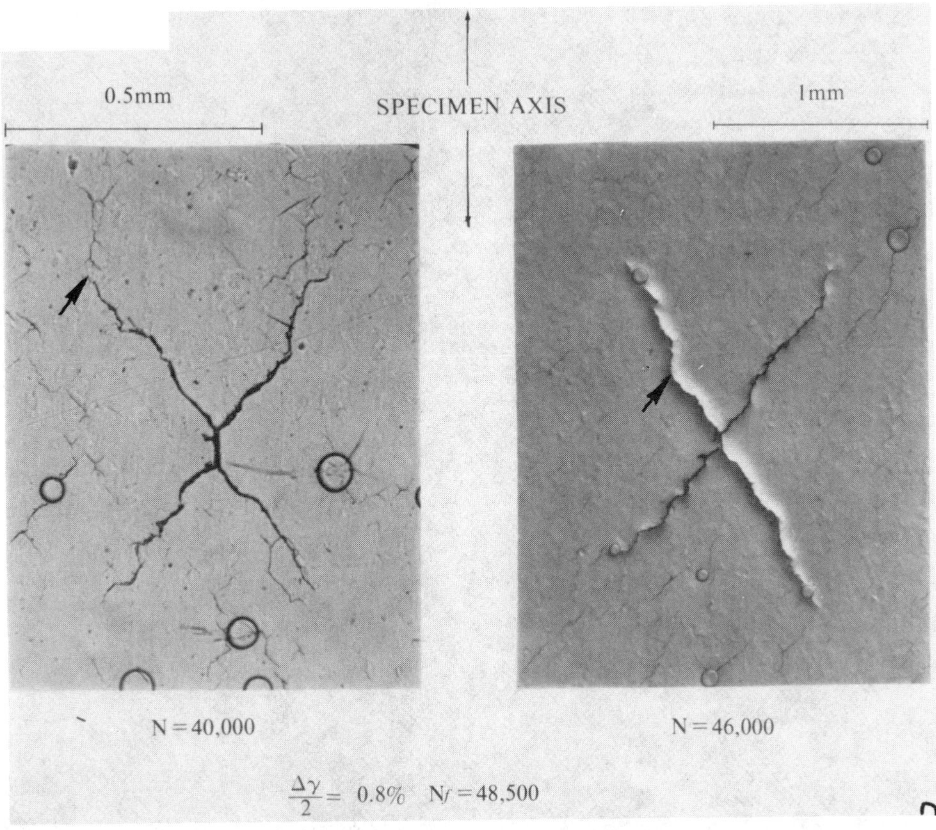

FIG. 3—*AISI 304 stainless steel region II behavior. Growth on tensile planes occurs by linking. (Arrows indicate same location in specimen.)*

Some scatter in this trend of increased portions of life spent in shear growth with increased strain amplitude was due to the variation in inclusion size. Arrows in Fig. 5 indicate specimens which were particularly susceptible to inclusions. For example, in an AISI 316 specimen tested at 0.6%, a severe stringer reduced the fraction of life spent in shear crack growth. The stringer caused a 4 mm shear crack to quickly develop before the crack propagated in the Stage II direction.

Surface replicas and SEM examination of fracture surfaces of the AISI 304, solid 6.4 mm diameter specimens tested in tension showed no perceptible evidence of Stage I growth. The fracture surfaces appeared to be almost entirely dominated by Stage II growth. It has been reported that at low strain amplitudes up to 90% of life may be taken up in initiation and Stage I growth, while at high strain amplitudes a similar fraction may be spent in Stage II crack growth [*30*].

Inconel 718

Unlike the stainless steel which displayed mixed behavior, results of the Inconel 718 tests showed that cracks initiated and remained on maximum shear planes, Region I behavior, at all values of shear strain. Torsion tests were conducted on the Inconel 718 at shear strain levels of

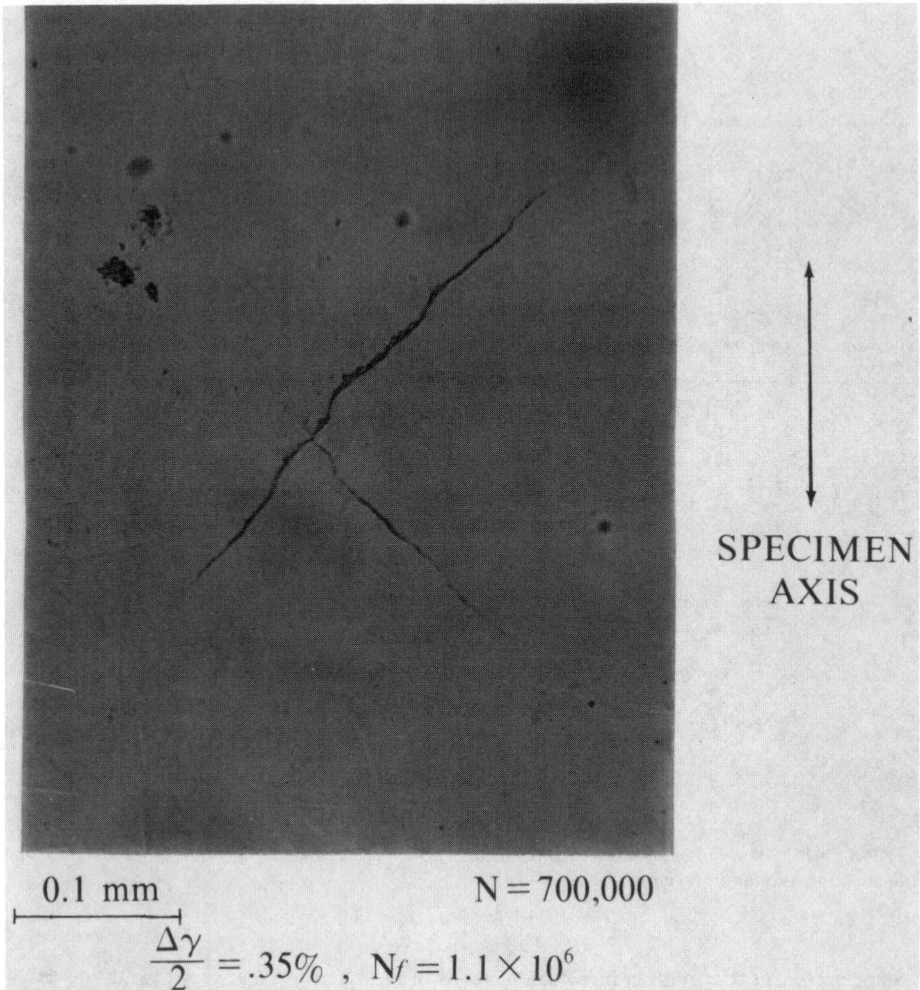

FIG. 4—*AISI 304 region III behavior. Growth on tensile planes occurs by crack propagation rather than linking.*

1.7, 0.8, 0.43, and 0.38%. Even at the lowest strain amplitude, in which the stress-strain was essentially all elastic, cracks initiated and remained on shear planes throughout the life. Figure 6 shows the failure crack developed in an Inconel specimen tested at 0.8% shear strain. As seen at failure, the crack never deviated from the plane of maximum shear.

Even under tensile loading, cracks remained on shear planes for the majority of fatigue life. Axial tests were conducted at two strain amplitudes, 1.0 and 0.5%. Final failure in all tension tests was in a macroscopic tensile direction comprised of large portions of microscopic shear growth. This is evident in Fig. 6 which also shows the failure crack in a specimen tested in tension at 0.5% strain. Large amounts of shear growth are observed at failure. Growth on Stage II planes occurred only late in life.

Damage accumulation in Inconel appears to be shear dominated. This is attributed to localized shear deformation bands developed during cyclic loading. Reversed movement of disloca-

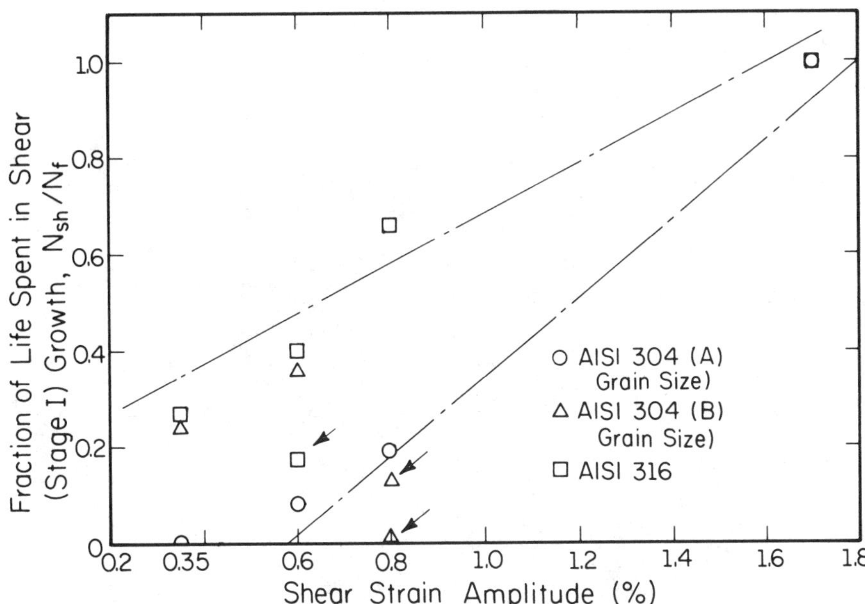

FIG. 5—*Fraction of life spent in Stage I growth as a function of shear strain amplitude.*

tions progressively shears precipitates in these bands. Crack propagation then occurs along the bands with extensive shear crack growth exhibited throughout the fatigue life [*31,32*].

SAE 1045

Two types of cracking systems have been observed in SAE 1045 [*28*]. A large density of microcracks was reported at high amplitudes, with the final failure occurring by a very rapid linking of these cracks (Fig. 7). This type of damage has been termed the R system [*33*]. Alternatively, the S system, [*33*], which dominated crack behavior at low strain amplitudes, exhibited one dominant crack which grew until failure (Fig. 8).

In torsion, at high amplitudes, the R system crack behavior was characteristic of Region I. The failure was similar to that observed in the stainless steels at high amplitudes except that the linking of microcracks and final failure in 1045 occurred over a very few cycles, while the growth of the Region I failure crack in stainless steels occurred progressively throughout the life. At lower amplitudes, progressive growth of a single crack (S system) occurred by a linking process on the shear plane. Both R and S type crack systems resulted in Region I behavior.

Region II behavior was only observed at long lives. At the lowest strain amplitude, 0.26%, the crack branched and growth occurred on the tensile plane by a linking of previously initiated shear cracks. After a period of tensile growth, the crack linked with a large shear crack which had been developing simultaneously. Final failure occurred by a mixture of Regions I and II behavior (Fig. 9).

In tension, failure occurred in both the R and S systems on Stage II planes. Microcracks initiated on shear planes at high amplitudes, 1.0 percent strain, in a manner representative of the R crack system. A very rapid linking of these microcracks occurred immediately prior to failure such that the failure crack was on tensile (Stage II) planes. At low amplitudes, 0.2%, cracks initiated on shear planes but progressive growth occurred on Stage II planes (Fig. 8).

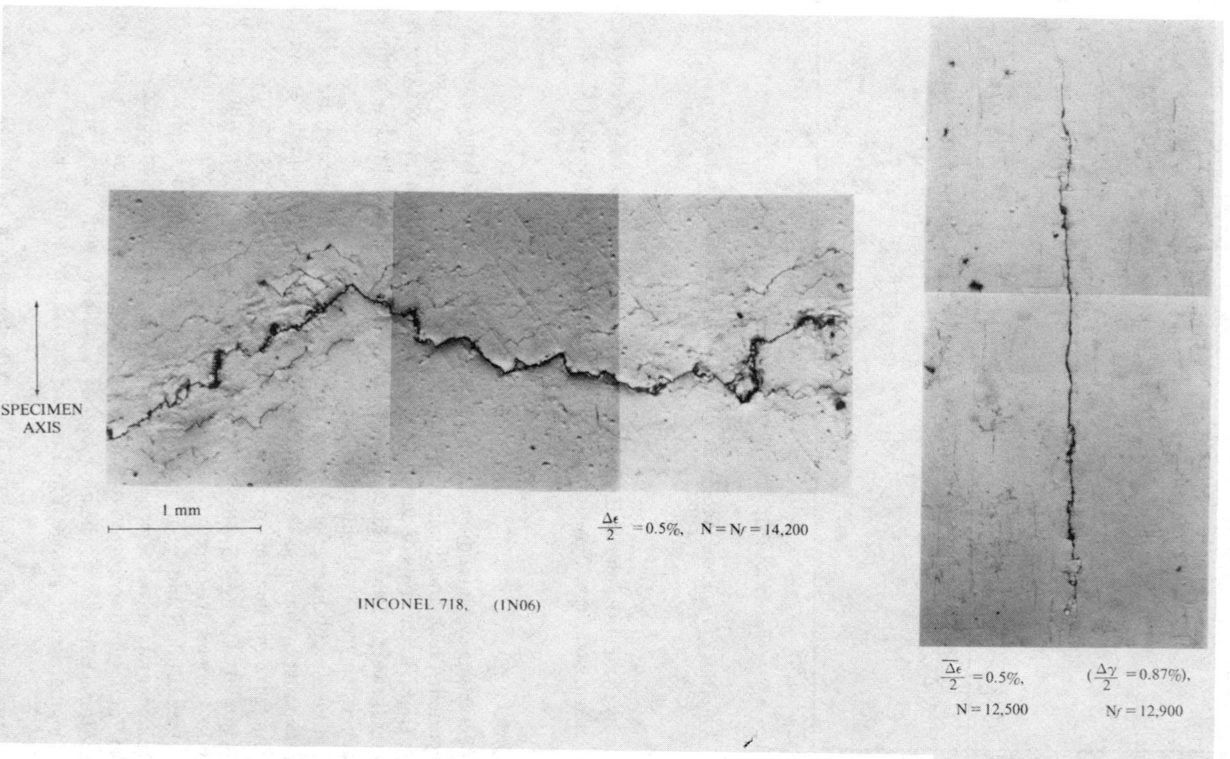

FIG. 6—*Cracking behavior of Inconel 718 in tension* (left) *and torsion* (right).

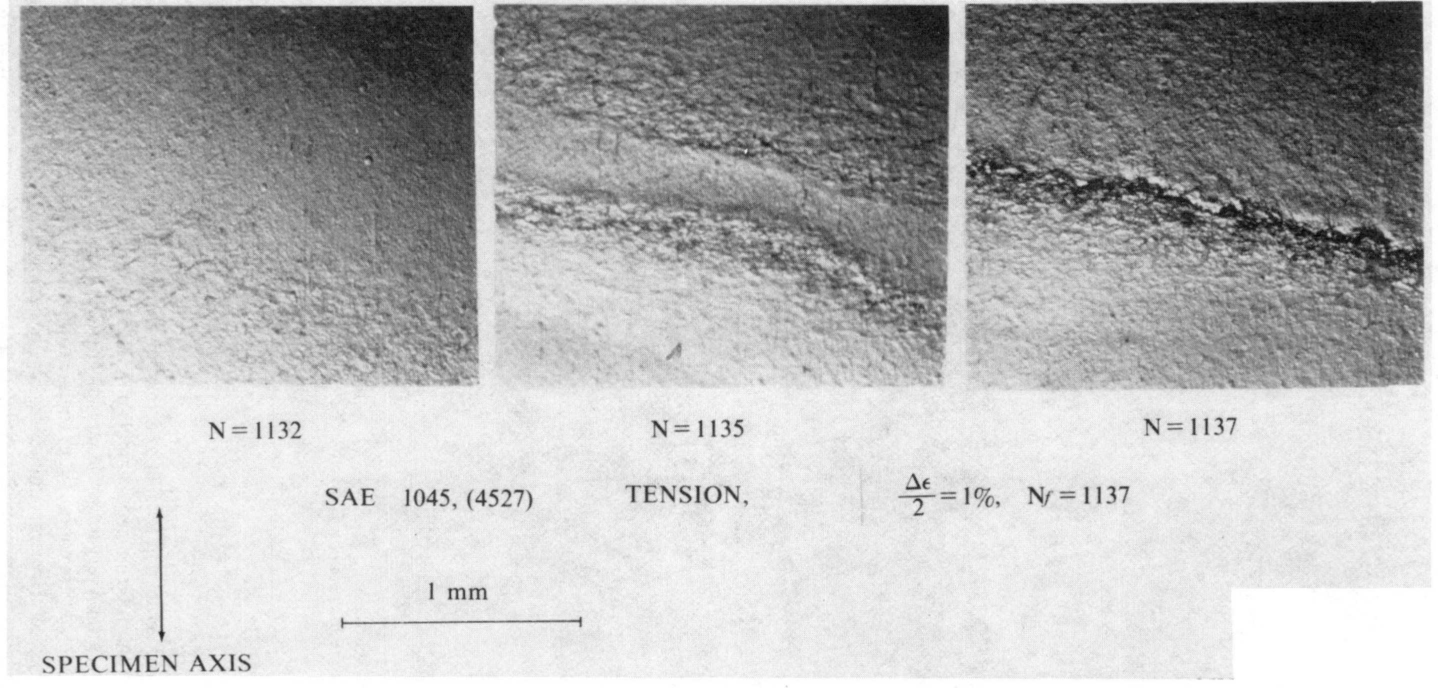

FIG. 7—*SAE 1045, tension, Type R crack system. Failure occurs by linking of microcrack late in life.*

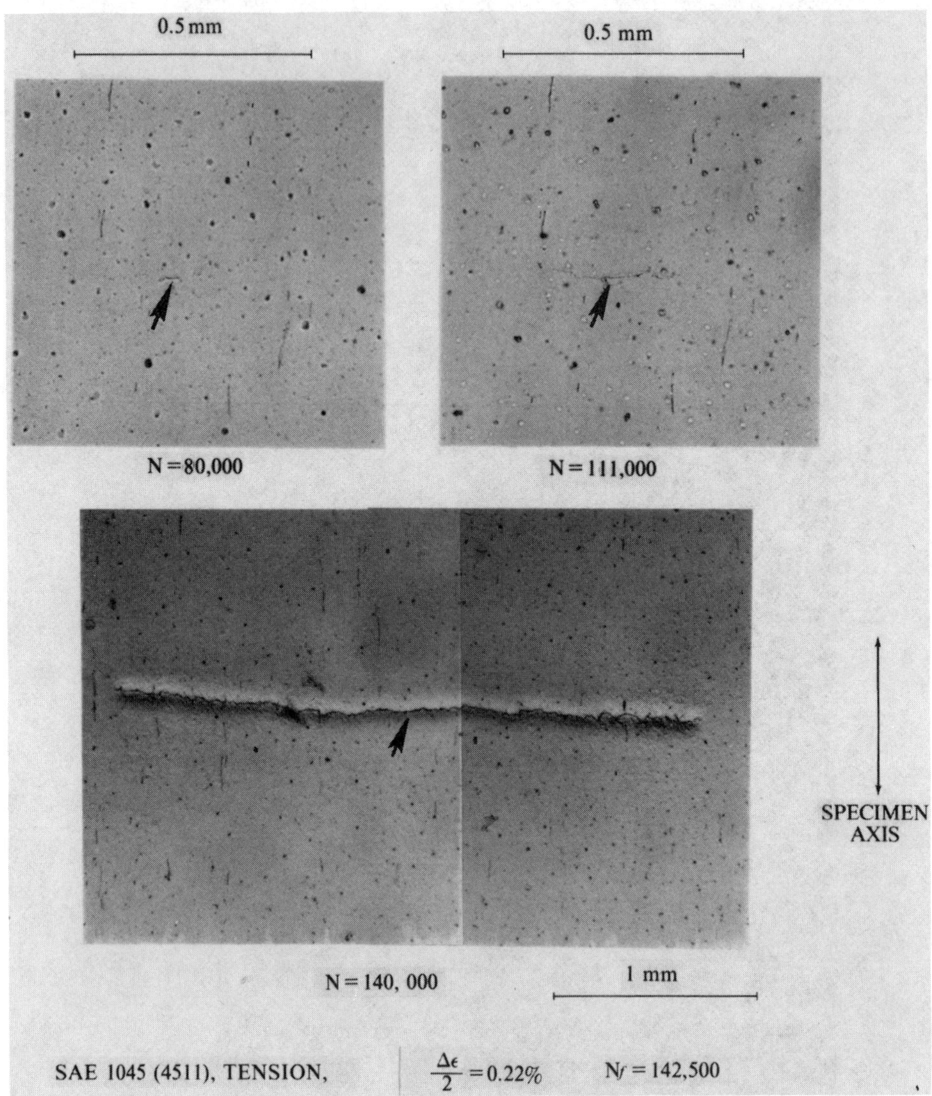

FIG. 8—*SAE 1045, tension, Type S crack system. One dominant crack grows until failure. (Arrows indicate same location in specimen.)*

Gray Cast Iron

In gray cast iron, cracks in tension and torsion propagated on Stage II planes at all strain levels [*29*]. Graphite flakes acted as small microcracks and linked to form growth on tensile planes. In torsion, after a very few cycles at high shear strain amplitudes (0.6%), graphite flakes became very distinct. These continued to coarsen similar to the behavior of the Region I microcracks in stainless steel. However, unlike Region I behavior in stainless steel where microcracks linked on shear planes, in gray cast iron Stage II cracks developed from the ends of the graphite

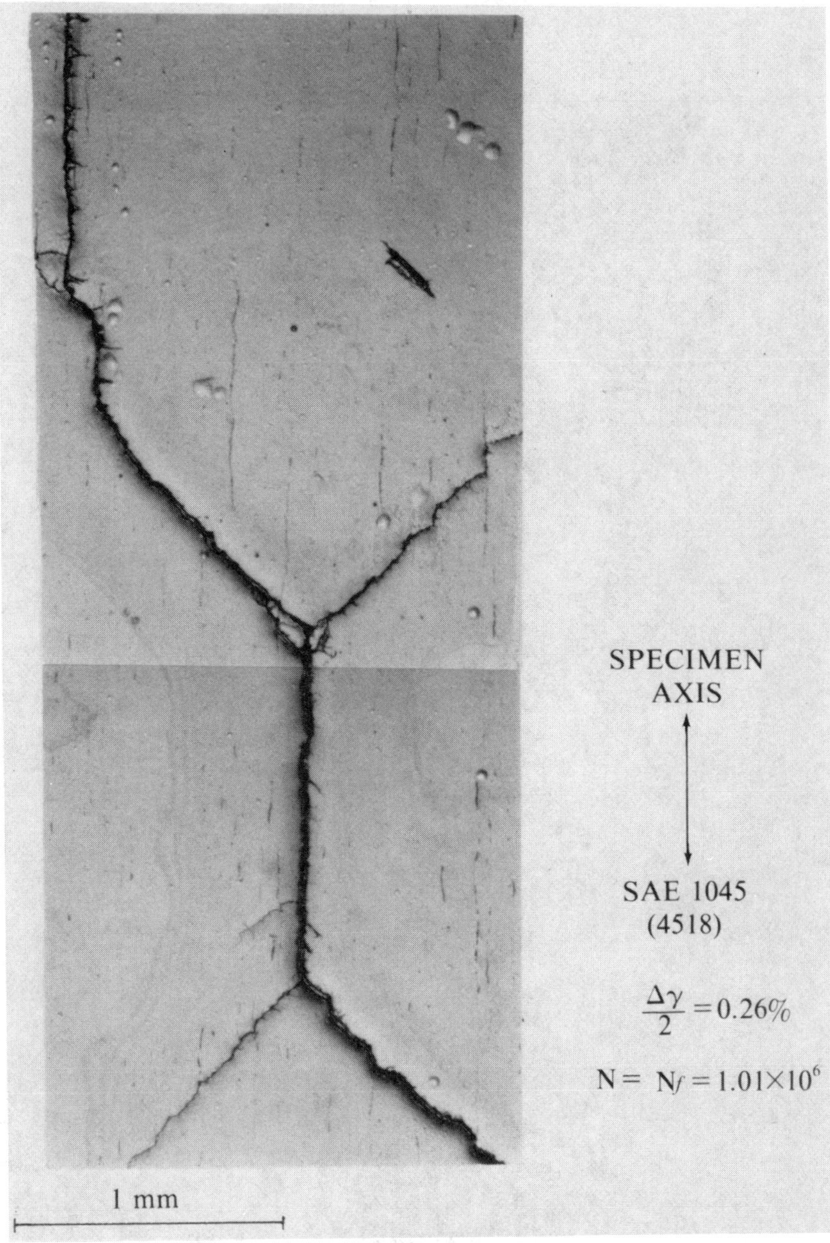

FIG. 9—*In SAE 1045, region II behavior is observed at long lives in torsion.*

flakes. These then linked up with other graphite flakes on tensile planes (Fig. 10). At low amplitudes, 0.2% shear strain, failure occurred by the same crack mechanism, although the number of cycles spent in the development and linking of cracks from the graphite flakes increased. Components tested in tension or torsion, at all amplitudes, exhibited crack initiation from graphite flakes and subsequent growth on tensile planes with no significant shear growth.

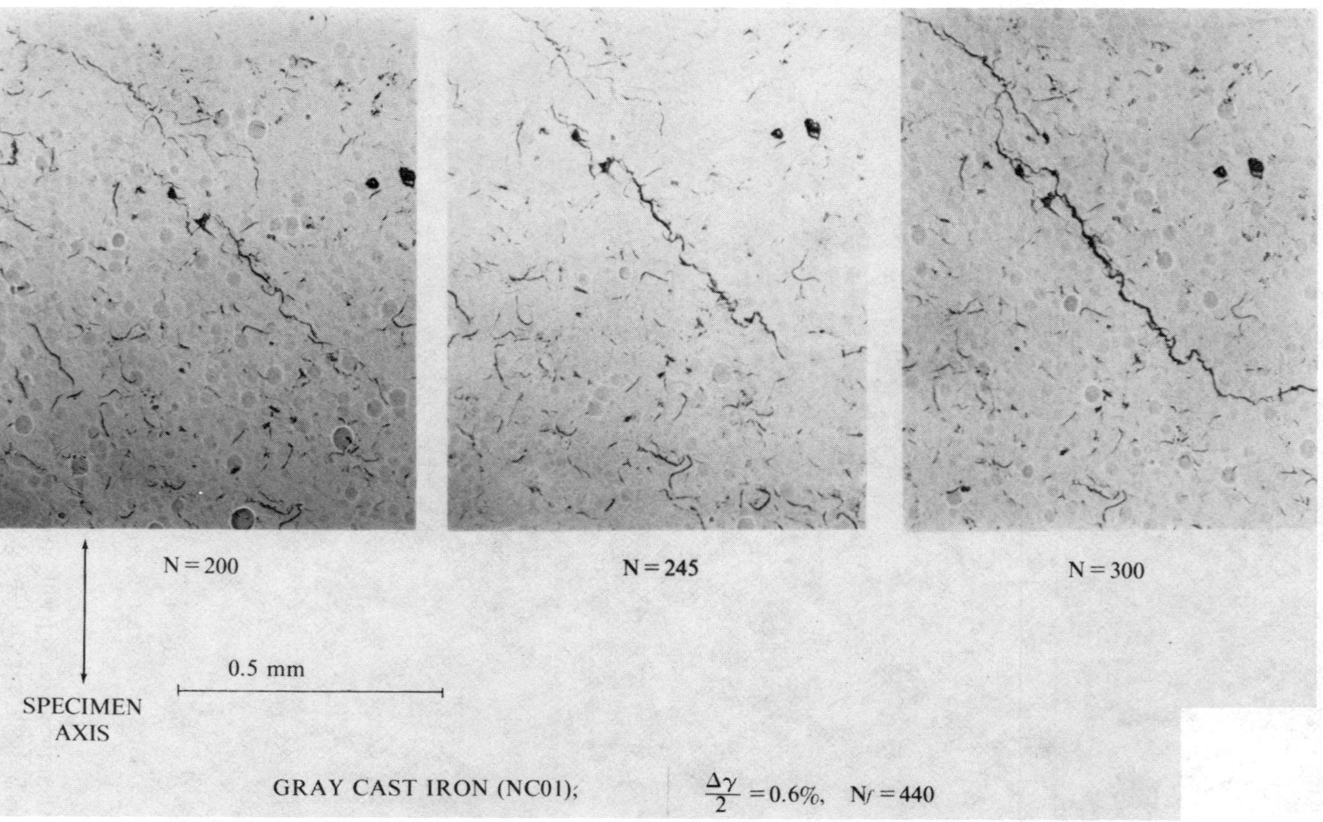

FIG. 10—*In gray cast iron, crack growth occurs by linking of graphite flakes on tensile planes.*

Discussion

Two types of cracking are generally observed in components subjected to low-cycle fatigue—shear cracking associated with Stage I behavior and tensile cracking associated with Stage II. Tension and torsion loading represent extreme cases for the development of the two crack types. Crack observations made in tests conducted in tension and torsion show that cracking behavior is dependent upon loading mode, strain amplitude, and material type.

A major effect of loading mode on cracking direction is attributed to the normal strain on the plane of maximum shear (Fig. 11). As discussed previously, in tension, cracks generally initiate on shear planes but turn and propagate on planes perpendicular to the maximum tensile stress or strain. Growth on tensile (Stage II) planes occurs primarily as a result of this normal strain across the maximum shear plane. In torsion, no normal strain exists on the maximum shear plane. This enables the shear (Stage I) crack to grow to a much greater length than the shear crack in tension.

Final failure in tension for all materials was characterized by macroscopic growth in Stage II planes. The fraction of fatigue life spent on Stage I planes varied with material and strain amplitude. Cracks in cast iron grew on Stage II planes for virtually the entire life. Shear growth was eliminated as cracks originated at graphite flakes, which acted as microcracks, and grew on tensile planes. Stainless steels, tested in tension, also exhibited little or no Stage I growth. Cracks in SAE 1045, tested at high tensile strain amplitudes (R crack system), initiated on shear planes. These cracks coarsened until linking on Stage II planes occurred immediately prior to failure. At lower amplitudes (S crack system), cracks initiated on shear planes but quickly turned to Stage II planes. Lastly, Inconel crack behavior was characterized by a dominance of shear growth. Stage II growth occurred only late in life. Differences in crack behavior for each of the material types are summarized in Fig. 12. (In this figure, tension and torsion results for each material are compared at the same von Mises effective strain.)

Forsyth [*11*] suggested that the transition from Stage I to Stage II behavior occurs when the ratio of shear to tensile stress reaches a critical value. As discussed previously and shown in Fig. 11, for torsional loading, no normal or tensile strain exists globally on the plane of maximum shear. Nevertheless, tensile cracks developed in stainless steels tested in torsion. In these materials, shear cracks were often observed to branch at grain boundaries or other obstructions. These obstructions may have caused a local tensile strain to develop on the maximum shear plane at the crack tip such that the crack branched from a shear plane to a tensile plane. In an AISI 304 specimen with a particularly severe spherical inclusion, tested in torsion at 0.8% shear strain, a tensile crack initiated at a large spherical inclusion after a very short fraction of the

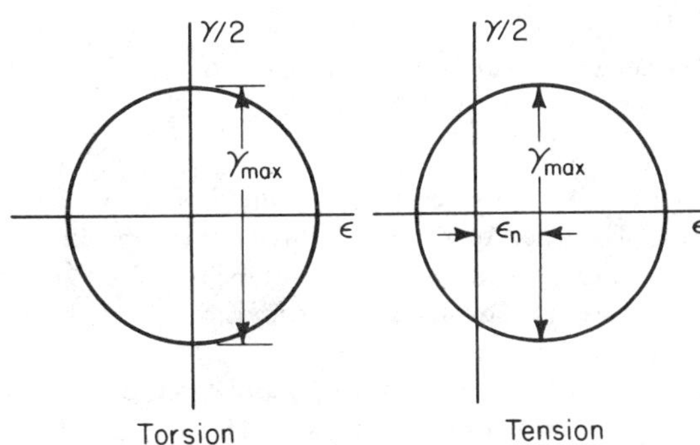

FIG. 11—*Torsion and tension states.*

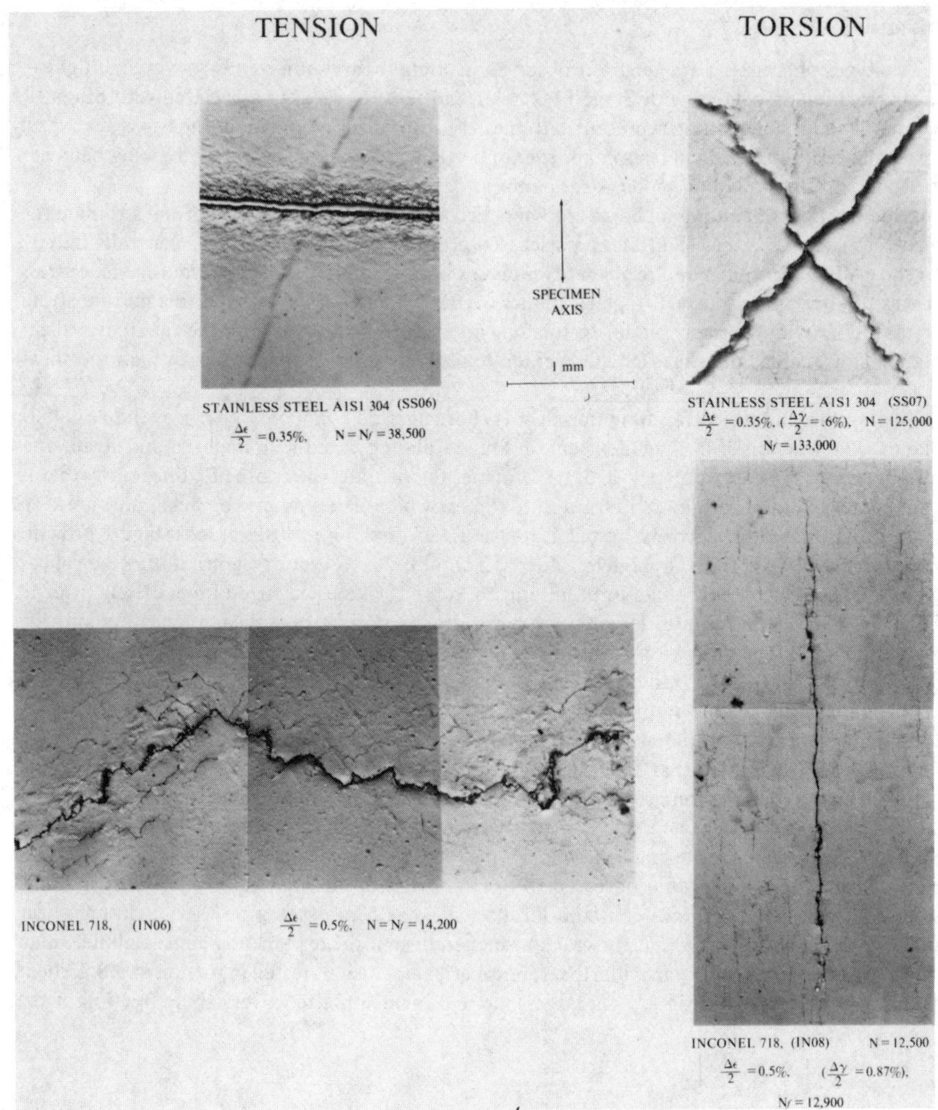

FIG. 12*a*—*Tension and torsion behavior of AISI 304 at* $\Delta\bar{\epsilon}/2 = 0.35\%$ and Inconel 718 at $\Delta\bar{\epsilon}/2 = 0.5\%$.

life. It is probable that a tensile stress developed locally around the inclusion which caused growth to shift to the Stage II plane. An alternate source of local tensile stresses was presented by Tschegg [*20*]. He reported that the mutual support of inclined faces, or the rubbing between crack faces, may cause local tensile stresses to occur and, therefore, may explain the shift to tensile growth in materials tested in torsion.

The dependency of cracking behavior on strain amplitude is clearly evident from surface replicas of specimens tested in torsion. In torsion, the planes of maximum shear strain intersect the surface of the specimen at 0° and 90°. Planes of maximum tensile strain intersect the surface at 45°. Consequently, surface observations in torsion allowed the strain amplitude dependency of

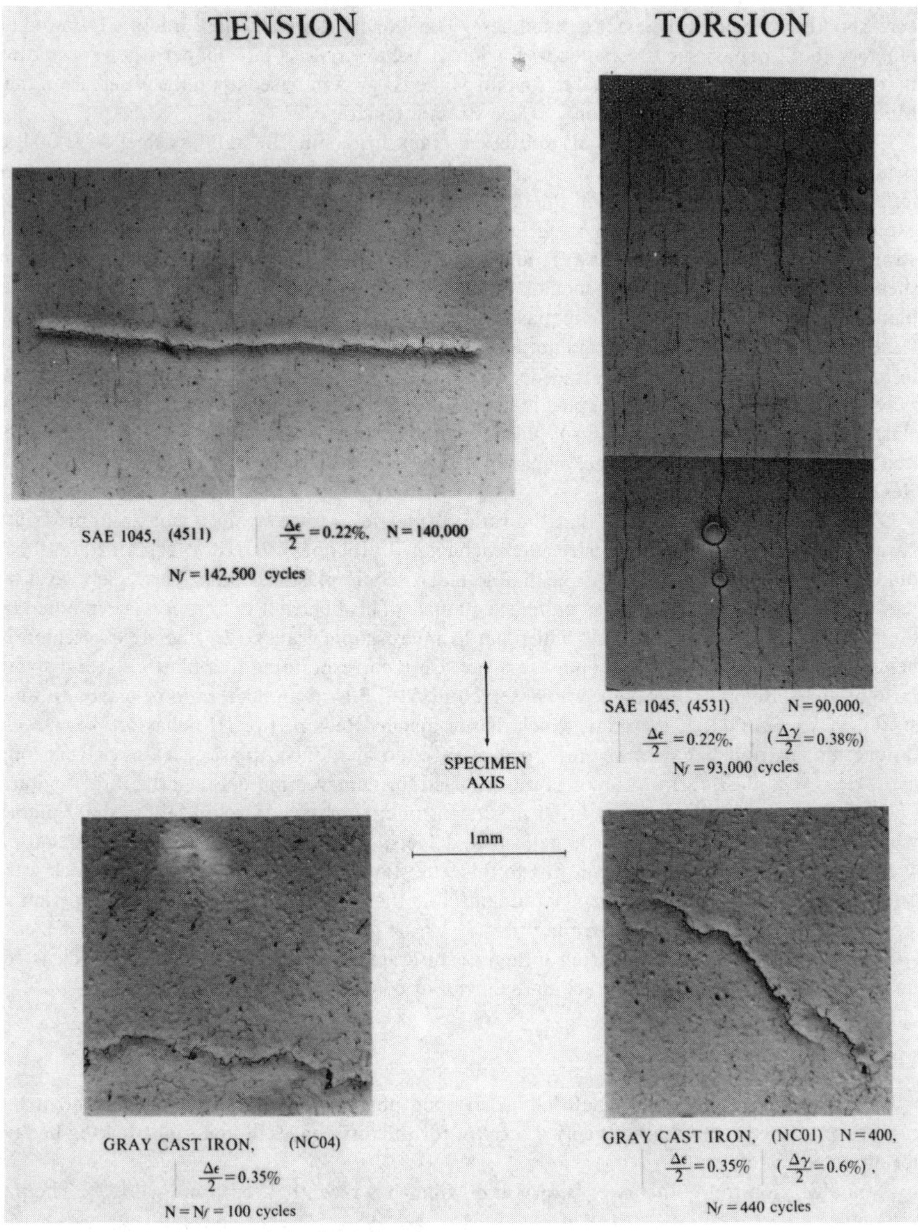

FIG. 12*b*—*Tension and torsion behavior of SAE 1045 at $\Delta\bar{\epsilon}/2$ = 0.22% and gray cast iron at $\Delta\bar{\epsilon}/2$ = 0.35%.*

Stage I and Stage II behavior to be closely monitored. From observations of stainless steel, a correlation between strain amplitude and cracking behavior was seen. In general, the fraction of life spent in shear growth decreased with a corresponding decrease in plastic strain.

Similar observations of the effect of plasticity were made by Tschegg, Ritchie, and McClintock [*19*]. The "macroscopically flat" fracture which they observed at higher stress intensities corresponds to Region I behavior defined here. The "macroscopically flat" fracture followed by "factory roof" behavior corresponds to Region II behavior. In a later paper, they speculated that extensive plasticity was needed to sustain Mode III growth based upon the observation that Mode I or tensile growth occurs only at low stress intensities.

Parsons and Pascoe [*12*], who also observed crack branching in stainless steel AISI 304 at strains below 1.3%, postulated that shear cracking was favored at high strains due to a greater influence of crack tip plasticity and the large number of slip bands in neighboring grains. This explanation seems reasonable for materials which fail in shear modes at high values of plastic strain but branch to tensile planes at reduced strain amplitudes. Other factors must account for shear crack growth in materials, such as Inconel 718, in which crack growth remains on shear planes even at very low macroscopic plastic strains.

In addition to the effects of strain amplitude and loading mode, cracking behavior is largely influenced by material type. This is evident from the torsion tests results in which the materials were compared at similar ratios of plastic strain to total strain. At approximately equal values of elastic and plastic strain (50% ratio of plastic to total strain), AISI 304 and gray cast iron exhibited Region III behavior while crack behavior in SAE 1045 and Inconel 718 was characteristic of Region I (Fig. 13).

Observations further indicate that the ratio of plastic to total strain is not an appropriate parameter with which to characterize crack behavior. In Inconel 718, cracks remained on shear planes, Region I behavior, even at negligible macroscopic plastic strains. Conversely, even at very short lives, the cast iron failed on tensile planes. (This behavior in cast iron is explained by the presence of the graphite flakes which act as microscopic cracks). In SAE 1045, Region II behavior was observed only at the lowest shear strain corresponding to a plastic to total strain ratio of 30%. Region II behavior was observed in AISI 304 at an 80% ratio of plastic to total strain. It is possible that materials which do not display Region II or III behavior in low-cycle fatigue may exhibit behavior of these regions if tested at reduced plastic strains or very long lives. However, plastic strain values cannot be used for demarcation between the three regions.

It has been previously suggested that in torsion, ductile materials would fail on shear planes while brittle materials or those with inclusions, defects, or other obstacles to crack propagation would fail on tensile planes. In contrast to this, the stainless steels and SAE 1045, which were both ductile and contained inclusions, did not fail in the same manner when tested in torsion at similar ratios of plastic to total strain.

Material type plays a dominating influence on cracking behavior. Further research is required to explain variations of cracking behavior at constant values of plastic strain.

Life Predictions

Multiaxial fatigue theories should be based upon physically observed damage. None of the recently proposed strain-based theories account for the differences in material cracking behavior observed in this study.

A modification to the theory of Brown and Miller has recently been proposed [*27*]. The parameter

$$\hat{\gamma} + \hat{\epsilon}_n + \frac{\sigma_{no}}{E} \tag{1}$$

in which $\hat{\gamma}$ is the maximum shear strain amplitude, $\hat{\epsilon}_n$ is the strain amplitude normal to the plane of maximum shear, and σ_{no} is the mean stress normal to this same plane, accurately correlates the multiaxial test results of Inconel 718 (Fig. 14). As discussed previously and shown in Fig. 6, damage accumulation in Inconel 718 was shear dominated. Even in tension, large

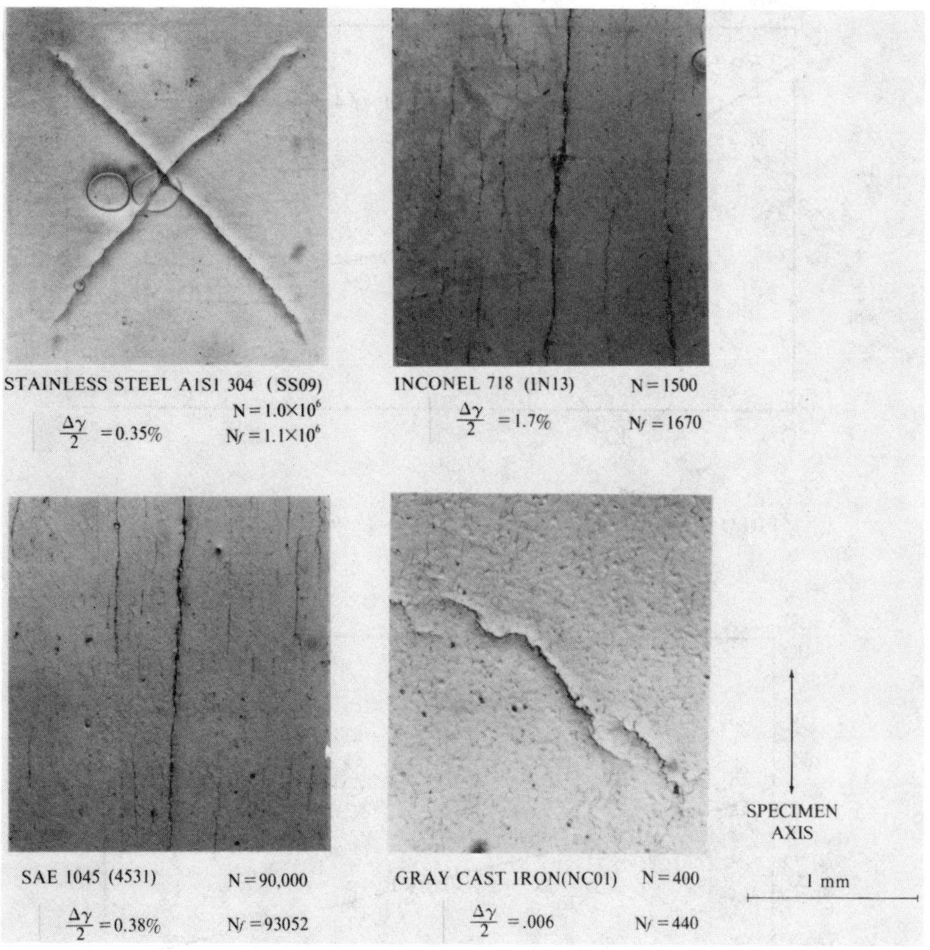

FIG. 13—*Cracking behavior in torsion of four materials at 50% ratio of plastic/total strain.*

portions of the fatigue life are spent in shear. It is to be expected that a shear-based parameter would correlate results of a shear-dominated material. For torsion tests, the parameter reduces to the shear strain amplitude. The solid line in Fig. 14 is the torsional strain life curve developed from torsion tests. The test results and model are described in detail in Ref *34*.

The same parameter was used to correlate results of multiaxial fatigue tests conducted using AISI 304 [*35*], a material which fails in a tensile mode. The predictions made using this parameter vary significantly from test results and are often nonconservative by an order of magnitude (Fig. 15). It is not surprising that the shear-based parameter did not correlate test results for a material dominated by tensile cracking.

Another model must be used to correlate the AISI 304 results. Life predictions for the tensile-crack-dominated material displayed improved correlation when analyzed with a combined principal stress-strain parameter

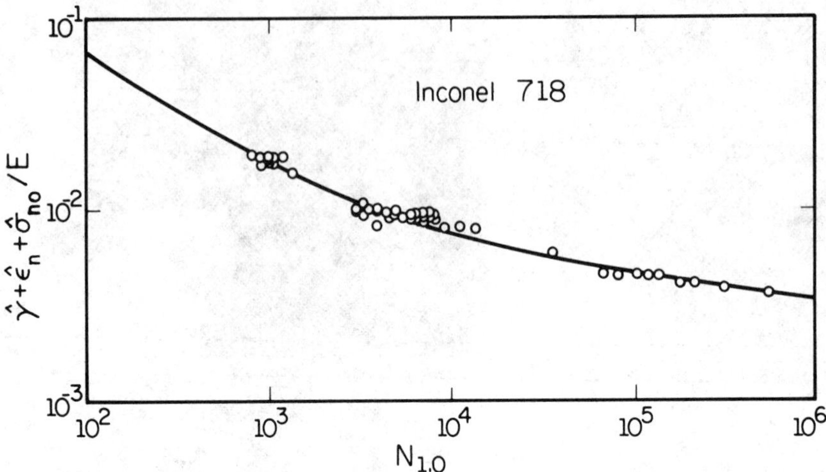

FIG. 14—*Correlation of Inconel 718 results with shear strain parameter.*

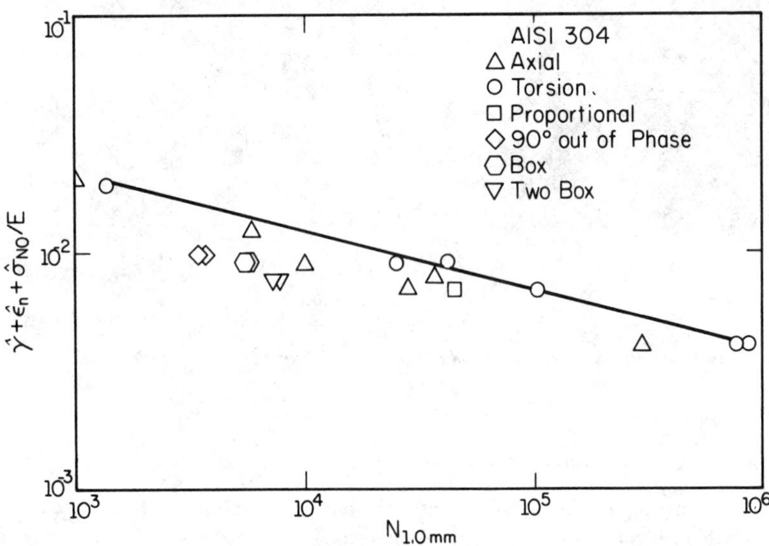

FIG. 15—*Correlation of AISI 304 test results with shear strain parameter.*

$$\frac{\Delta\epsilon_1}{2}\,\sigma_1^{max} \tag{2}$$

developed by Smith et al. [*36*] for uniaxial tests. Failure occurred on tensile planes for all axial and multiaxial tests. In Fig. 16, the solid line was developed from axial test results. Failure on shear planes occurred only in torsion tests conducted at high strain amplitudes. It is not unexpected that torsion tests at these amplitudes displayed the poorest correlation. The torsion tests

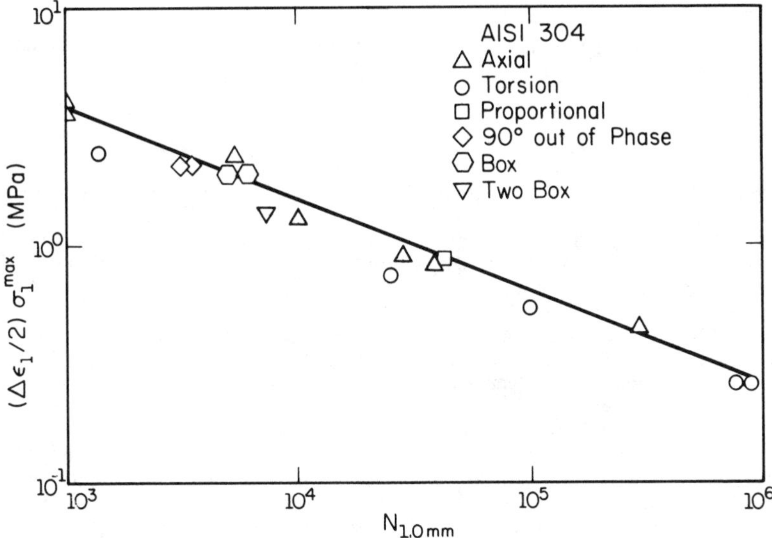

FIG. 16—*Correlation of AISI 304 test results with maximum principal stress parameter.*

at low strain amplitudes, where the failure crack developed on tensile planes, exhibited good correlation with the predictions.

Recognition of cracking characteristics is equally important when employing fracture mechanics analyses to predict fatigue life. Growth rates in Mode I have been shown to differ from Mode II and III rates [*17*]. Accurate life predictions will require an understanding of cracking behavior before correct stress intensity factors can be chosen for use in growth rate models.

Thus, it appears that successful life predictions, using either fracture mechanics concepts or strain-based multiaxial theories, will require careful considerations of cracking behavior of the material in question.

Conclusions

1. Fatigue cracking behavior is dependent upon loading mode, strain amplitude, and material type.
2. Three regions of crack behavior are observed in stainless steel AISI 304 and 316 tested in torsion. At decreasing plastic strains, there is a larger propensity for tensile growth. Many materials may exhibit these three regions if tested at sufficiently long lives in torsion.
3. Cracking behavior of different material types varies at a constant ratio of plastic to total strain. Plastic strain alone cannot be used to predict crack behavior for different material types.
4. A multiaxial fatigue model developed from test results of a material that fails in shear mode does not correlate multiaxial test results of a tensile-crack-dominated material. A maximum principal stress-strain parameter was successfully used to correlate results of the material which fails in a tensile mode.
5. Differences in cracking behavior for different materials and loading conditions will need to be considered in successful life predictions for components subjected to multiaxial fatigue.

Acknowledgments

The work was conducted at the University of Illinois and supported by the Department of Energy Grant DE-AC02-76ER01198.

References

[*1*] Ewing, J. A. and Humfrey, J. C., "Fracture of Metals Under Repeated Alternations of Stress," *Philosophical Transactions of the Royal Society of London,* Vol. 200, Series A, 1903, pp. 241-253.
[*2*] Fine, M. E. and Ritchie, R. O., "Fatigue-Crack Initiation and Near-Threshold Crack Growth," *Fatigue and Microstructure,* ASM Materials Science Seminar, St. Louis, Mo., 1978, pp. 245-278.
[*3*] Laird, C., "Mechanisms and Theories of Fatigue," *Fatigue and Microstructure,* ASM Materials Science Seminar, St. Louis, MO, 1978, pp. 149-203.
[*4*] Mughrabi, H., "Dislocations in Fatigue," presented at the 50th Anniversary Meeting on Dislocations and Properties of Real Materials, Royal Society, London, 11-12 Dec. 1984; to be published in the conference proceedings by the Metal Society, London.
[*5*] Laird, C., "The Application of Dislocation Concepts in Fatigue," in *Dislocations in Solids,* F. R. N. Nabarro, Ed., North-Holland, Amsterdam, 1983.
[*6*] Boettner, R. C., Laird, C., and McEvily, A. J., "Crack Nucleation and Growth in High Strain-Low Cycle Fatigue," *Transactions of Metallurgical Society of AIME,* Vol. 233, 1965, pp. 379-387.
[*7*] Frost, N. E., Marsh, K. J., and Pook, L. P., *Metal Fatigue,* Oxford University Press, Oxford, 1974.
[*8*] Tipnis, V. A. and Cook, N. H., "The Influence of Stress-State and Inclusion Content on Ductile Fracture with Rotation," *Journal of Basic Engineering, Transactions of American Society of Mechanical Engineers, Series D,* Sept. 1967, pp. 533-540.
[*9*] Stanzl, S. E., "Fatigue Crack Growth (Physical Principles, Models and Experimental Results)," *Res Mechanica,* Vol. 5, 1982, pp. 241-292.
[*10*] Gough, H. J., "Crystalline Structure in Relation to Failure of Metals—Especially by Fatigue," Edgar Marburg Lecture, *Proceedings of ASTM,* Vol. 33, Part II, 1933, pp. 3-114.
[*11*] Forsyth, P. J. E., "A Two Stage Process of Fatigue Crack Growth," *Proceedings,* Symposium on Crack Propagation, Cranfield, England, 1961, pp. 76-94.
[*12*] Parsons, M. W. and Pascoe, K. J., "Observations of Surface Deformation, Crack Initiation and Crack Growth in Low-Cycle Fatigue Under Biaxial Stress," *Materials Science and Engineering,* Vol. 22, 1976, pp. 31-50.
[*13*] Gough, H. J., *The Fatigue of Metals,* Van Nostrand, New York, 1926.
[*14*] Peterson, R. E., "Fatigue Cracks and Fracture Surfaces—Mechanics of Development and Visual Appearance," in *Metal Fatigue,* G. Sines and J. L. Waisman, Eds., McGraw-Hill, New York, 1959, pp. 68-86.
[*15*] Morrow, J., Halford, G. R., and Millan, J. F., "Optimum Hardness for Maximum Fatigue Strength of Steel," in *Proceedings,* First International Conference on Fracture, Sendai, Japan, Vol. 3, 1965, pp. 1611-1635.
[*16*] Wulpi, D. J., *How Components Fail,* American Society for Metals, Metals Park, Ohio, 1966, pp. 43-47.
[*17*] Hurd, N. J. and Irving, P. E., "Factors Influencing Propagation of Mode III Fatigue Cracks Under Torsional Loading," in *Design of Fatigue and Fracture Resistant Structures, ASTM STP 761,* American Society for Testing and Materials, Philadelphia, 1982, pp. 212-233.
[*18*] Hurd, N. J. and Irving, P. E., "Smooth Specimen Fatigue Lives and Microcrack Growth Modes in Torsion," in *Multiaxial Fatigue, ASTM STP 853,* K. J. Miller and M. W. Brown, Eds., American Society for Testing and Materials, Philadelphia, 1985, pp. 267-284.
[*19*] Tschegg, E. K., Ritchie, R. O., and McClintock, F. A., "On the Influence of Rubbing Fracture Surfaces on Fatigue Crack Propagation in Mode III," *International Journal of Fatigue,* Vol. 5, No. 1, 1983, pp. 29-35.
[*20*] Tschegg, E. K., "Mode III and Mode I Fatigue Crack Propagation Behavior Under Torsional Loading," *Journal of Materials Science,* Vol. 18, 1983, pp. 1604-1614.
[*21*] McClintock, F. A., Nayeb-Hashetii, H., Ritchie, R. O., Wu, E., and Wood, W. T., "Assessing Expired Fatigue Life in Large Turbine Shafts," Oak Ridge National Laboratory, ORNL/SUB/83-9062/1984.
[*22*] Hult, J., "Experimental Studies on Fatigue Crack Propagation in Torsion," *Transactions of the Royal Institute of Technology,* Stockholm, Vol. 119, 1958.
[*23*] Brown, M. W. and Miller, K. J., "A Theory for Fatigue Failure Under Multiaxial Stress-Strain Conditions," *Proceedings of the Institution of Mechanical Engineers,* Vol. 187, 1973, pp. 745-755.

[24] Lohr, R. D. and Ellison, E. G., "A Simple Theory for Low Cycle Multiaxial Fatigue," *Fatigue of Engineering Materials and Structures,* Vol. 3, 1980, pp. 1-17.

[25] Jacquelin, B., Hourlier, F., and Pineau, A., "Crack Initiation Under Low-Cycle Multiaxial Fatigue in Type 316L Stainless Steel," *Journal of Pressure Vessel Technology, Transactions of ASME, Series J,* Vol. 105, May 1983, pp. 138-143.

[26] Kandil, F. A., Brown, M. W., and Miller, K. J., "Biaxial Low-Cycle Fatigue Fracture of 316 Stainless Steel at Elevated Temperatures," Book 280, The Metals Society, London, 1982, pp. 203-210.

[27] Socie, D. F., Waill, L. A., and Ditimer, D. F., "Biaxial Fatigue of Inconel 718 Including Mean Stress Effects," *Multiaxial Fatigue, ASTM STP 853,* K. J. Miller and M. W. Brown, Eds., American Society for Testing and Materials, 1985, pp. 463-481.

[28] Hua, C. T. and Socie, D. F., "Fatigue Damage in 1045 Steel Under Variable Amplitude Biaxial Loading," *Fatigue of Engineering Materials and Structures,* Vol. 8, No. 2, 1985, pp. 101-114.

[29] Weinacht, D. J., "Torsional Fatigue of Gray Cast Iron," M. S. Thesis, Department of Mechanical Engineering, University of Illinois, Champaign-Urbana, 1985.

[30] Plumbridge, W. J., "Review: Fatigue-Crack Propagation in Metallic and Polymeric Materials," *Journal of Materials Science,* Vol. 7, 1972, pp. 939-962.

[31] Worthem, D. W., Altstetter, C. J., Robertson, I. M., and Socie, D. F., "Cyclic Deformation and Damage in Inconel 718," to be published.

[32] Fournier, D. and Pineau, A., "Low Cycle Fatigue Behavior of Inconel 718 at 298 K and 823 K," *Metallurgical Transactions A,* Vol. 8A, 1977, pp. 1095-1105.

[33] Marco, S. M. and Starkey, W. L., "A Concept of Fatigue Damage," *Transactions of American Society of Mechanical Engineers,* Vol. 76, No. 4, 1954, pp. 627-632.

[34] Koch, J. L., "Proportional and Non-Proportional Biaxial Fatigue of Inconel 718," M. S. Thesis, Department of Mechanical Engineering, University of Illinois, Champaign-Urbana, 1985.

[35] Socie, D. F., "Multiaxial Fatigue Damage Models," *Journal of Engineering Materials and Technology,* in press.

[36] Smith, R. N., Watson, P., and Topper, T. H., "A Stress-Strain Function for the Fatigue of Metals," *Journal of Materials,* Vol. 5, No. 4, Dec. 1970, pp. 767-778.

J. Polák,[1] A. Vašek,[1] and M. Klesnil[1]

Effect of Elevated Temperatures on the Low Cycle Fatigue of 2.25Cr-1Mo Steel—Part II: Variable Amplitude Straining

REFERENCE: Polák, J., Vašek, A., and Klesnil, M., **"Effect of Elevated Temperatures on the Low Cycle Fatigue of 2.25Cr-1Mo Steel—Part II: Variable Amplitude Straining,"** *Low Cycle Fatigue, ASTM STP 942,* H. D. Solomon, G. R. Halford, L. R. Kaisand, and B. N. Leis, Eds., American Society for Testing and Materials, Philadelphia, 1988, pp. 922–937.

ABSTRACT: A 2.25Cr-1Mo steel was subjected to variable amplitude block straining at test temperatures of 22, 350, 450, and 550°C. A loading block having 500 strain peaks with an exponential probability density distribution and a constant diametral strain rate was generated. The loading block was analyzed by the rain flow procedure, closed hysteresis loops were identified, and peaks forming a cycle with stress amplitude smaller than half of the fatigue limit were deleted. The strain block characterized by the maximum strain amplitude was applied to a set of cylindrical specimens in a computer-controlled electrohydraulic testing system.

Stress amplitudes of all closed hysteresis loops versus the number of blocks (i.e., the variable amplitude cyclic hardening-softening curves) were plotted. Similar hardening-softening behavior as in constant amplitude straining was observed. A plot of stress amplitudes versus plastic strain amplitudes in a block at the half-life (i.e., the service stress-strain curve) was obtained. The parameters of the service stress-strain curves were evaluated in dependence on the maximum strain amplitude and on the temperature. They were compared with the basic cyclic stress-strain curves.

The fatigue life in variable amplitude block loading (i.e., the number of blocks to failure) was plotted versus maximum strain amplitude. The experimental results were compared with the predictions using a linear cumulative rule combined with a rain flow counting procedure and the Manson-Coffin fatigue life curve. The actual to predicted life ratio was 0.4. The observed discrepancy was discussed in terms of the mechanism of fatigue damage under spectrum loading.

KEY WORDS: low cycle fatigue, variable amplitude loading, service stress-strain curves, fatigue life prediction, 2.25Cr-1Mo steel

Knowledge of data on material behavior in variable amplitude loading is essential for life prediction of components subjected to service conditions. For an elevated temperature environment, the effect of temperature also has to be accounted for. The understanding or even the evaluation of the variable amplitude and the elevated temperature influence on the fatigue resistance of many engineering materials poses a difficult task and not many authors have approached that problem despite the practical needs in this area [*1*]. Up to now most of the effort has been limited to room-temperature variable amplitude behavior of materials important in the aircraft and automotive industries [*2,3*].

Considerable progress in understanding variable amplitude loading has been achieved by analyzing the stress-strain response of materials subjected to random block straining [*4–6*]. Various methods of counting cycles, such as rain flow counting or range-pair counting [*4,5*],

[1]Institute of Physical Metallurgy, Czechoslovak Academy of Sciences, 616 62 Brno, Czechoslovakia.

have been introduced. The rain flow analysis of a block of variable strain amplitudes is often suggested to obtain pairs of peaks that form full cycles. Using this procedure, Polák and Klesnil [7] measured the cyclic hardening-softening behavior of low carbon steel during the fatigue life and also obtained the cyclic stress-strain curves (CSSCs) relevant to variable amplitude loading.

In this work we adopt a similar experimental procedure. At temperatures up to 500°C, we study the cyclic stress-strain response and determine the fatigue life in variable amplitude loading of a low alloy steel. Measured fatigue lives are compared with lives predicted by damage summation using the linear damage rule and the low-cycle fatigue data taken from the constant strain amplitude testing.

Experimental Procedure

The material and specimens were the same as used in the constant amplitude testing reported in Part I of this paper [8]. The cylindrical specimens (10 mm diameter along a 5 mm gage length) were heated in air and cycled in an MTS computer controlled electrohydraulic machine. The diametral strain was controlled using an extensometer. All tests as well as multiple data acquisition were run under an application program written in Basic.

Considerable attention was paid to the generation of the strain block that was repeatedly applied to the specimen. The peaks were generated using a random number generator without any ordering. The probability of the peak having the magnitude x was

$$p(x) = \frac{1}{2\bar{x}} \exp\left(-\frac{|x|}{\bar{x}}\right) \tag{1}$$

where $\bar{x}$ is the average value of the positive peaks. Therefore a distribution of 500 peaks with exponential probability density, symmetric relative to zero, resulted. From this source block, actual blocks characterized by the maximum diametral strain amplitude, ϵ_{adm}, were derived. In order not to feed the testing machine with too many small cycles of only elastic excursions which did not cause any fatigue damage, the critical cut-off strain amplitude, ϵ_{adc}, was chosen. It corresponds approximately to half of the fatigue limit at the highest temperature. After scaling the block to ϵ_{adm}, all amplitudes identified by the rain flow procedure smaller than ϵ_{adc} were omitted. The number of peaks left in a block after this procedure using $\epsilon_{adc} = 1.5 \times 10^{-4}$ is shown in Table 1 for different maximum diametral strain amplitudes. For the highest ϵ_{adm} the reduction is only 14%, for the lowest ϵ_{adm} up to 72%.

The strain in between the peaks changed linearly with time, the diametral strain rate being $\dot{\epsilon}_d = 5 \times 10^{-3}$ s^{-1}, which is the same as in constant amplitude tests [8]. During each experiment the block of peaks with maximum diametral strain amplitude, ϵ_{adm}, was repeatedly applied to the specimen, the stress and strain peaks were continuously monitored and, for selected block numbers, stored on disk. The data were stored in sequence close to the geometric series (i.e., 1, 2, 4, 7, 10, 20, 40, 70, 100, 200, . . .).

As the rain flow analysis of the block was known at run time, the asymmetry factor

$$P = \frac{\sigma_m}{\sigma_a}$$

could be continuously evaluated, where σ_m is the mean stress and σ_a is the stress amplitude of the largest loop in a block. The onset of a macroscopic crack resulted in the characteristic change of the hysteresis loop shape and in the decrease of P. The specimen was considered as broken when the asymmetry factor reached the value $P \leq -0.2$ in two subsequent blocks. This corresponds to the macroscopic crack the surface length of which was greater than the diameter

TABLE 1—*Dependence of the number of peaks in a modified block on the maximum diametral strain amplitude* ϵ_{adm}.

ϵ_{adm}	0.009	0.006	0.004	0.0025	0.0016	0.0010	0.0008
n	432	402	376	316	248	176	144

of the specimen as found by subsequent inspection of the specimen. In most cases, however, sudden fracture took place before this criterion became effective.

The stress and strain peaks stored on the disk and information obtained from the rain flow analysis were used for calculating the stress and strain amplitudes of all closed hysteresis loops during the fatigue life. The stress amplitudes plotted versus the number of blocks represent the variable amplitude fatigue hardening-softening curves [7]. At each temperature the effective modulus for diametral strain, E_{ef}, was measured from the slope of the stress versus diametral strain readings in the elastic region before the start of the actual test.

The longitudinal plastic strain amplitudes were calculated using the relation

$$\epsilon_{ap} = 2\left(\epsilon_{ad} - \frac{\sigma_a}{E_{ef}}\right) \tag{2}$$

The stress amplitudes could be plotted versus plastic strain amplitudes for any block. This plot for a block number close to half of the number of blocks to fracture represents the service CSSC for the particular maximum diametral strain amplitude and temperature.

Experimental Results

Both cyclic plasticity and low-cycle fatigue life in variable amplitude loading were studied using the repeated block loading at 22, 350, 450, and 550°C.

Figure 1 shows the variable amplitude fatigue hardening-softening curves of specimens cycled at room temperature with two maximum diametral strain amplitudes, ϵ_{adm}. For lower ϵ_{adm} the initial hardening is followed by long term softening. For higher ϵ_{adm} the hardening was finished during the first block and only slow long-term softening was observed.

Figure 2 shows the variable amplitude fatigue hardening-softening curves of specimens cycled at 550°C. The initial hardening can still be detected for the lowest maximum strain amplitude only. The softening behavior was more pronounced than that at room temperature or 350 and 450°C. The long-term softening was present for all amplitudes except the highest (Fig. 2*b*). At temperatures of 350 and 450°C the fatigue hardening-softening curves have a similar character and are not shown here.

Figure 3 plots the experimental points of the service CSSCs obtained at four temperatures for two different maximum diametral strain amplitudes. The straight lines drawn through each set of experimental points were obtained by least squares fitting the dependence

$$\sigma_a = K_s \epsilon_{ap}^{n_s} \tag{3}$$

The weights of individual points depending on the error of stress and strain measurement (described in more detail in the Appendix) were accounted for. The points with small ϵ_{ap} and having high scatter then affect the position of the fitted line and parameters K_s and n_s only in a proper way.

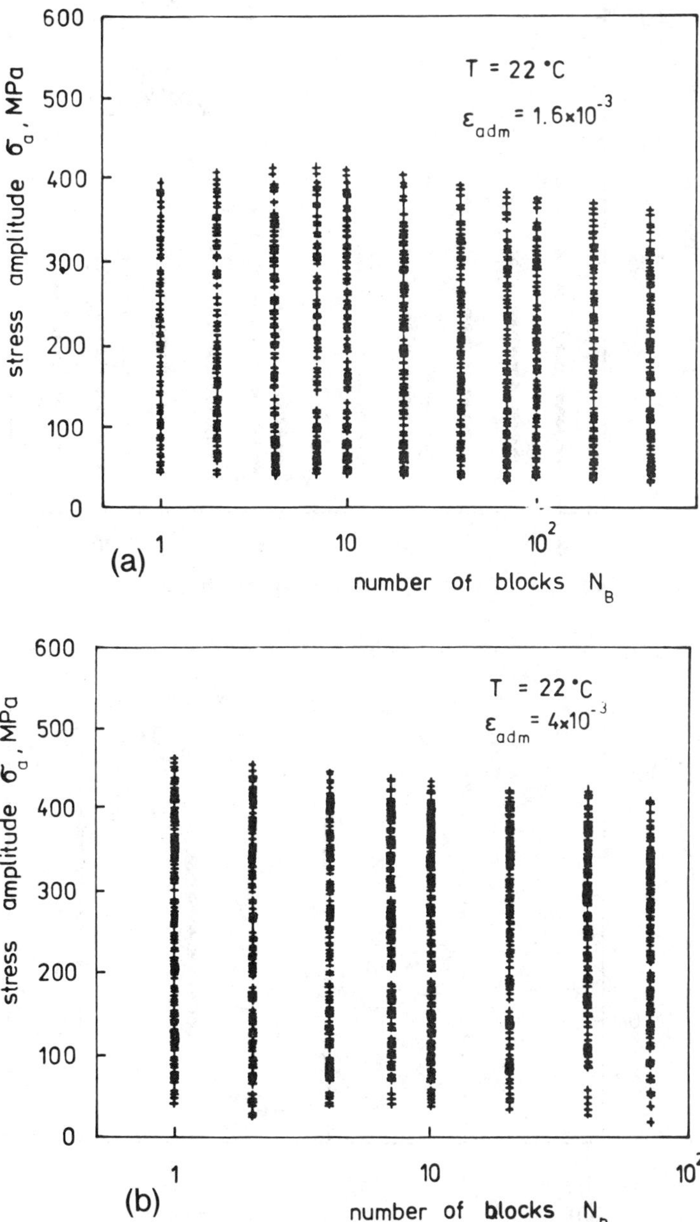

FIG. 1—*Variable amplitude hardening-softening curves at room temperature for two maximum diametral strain amplitudes.*

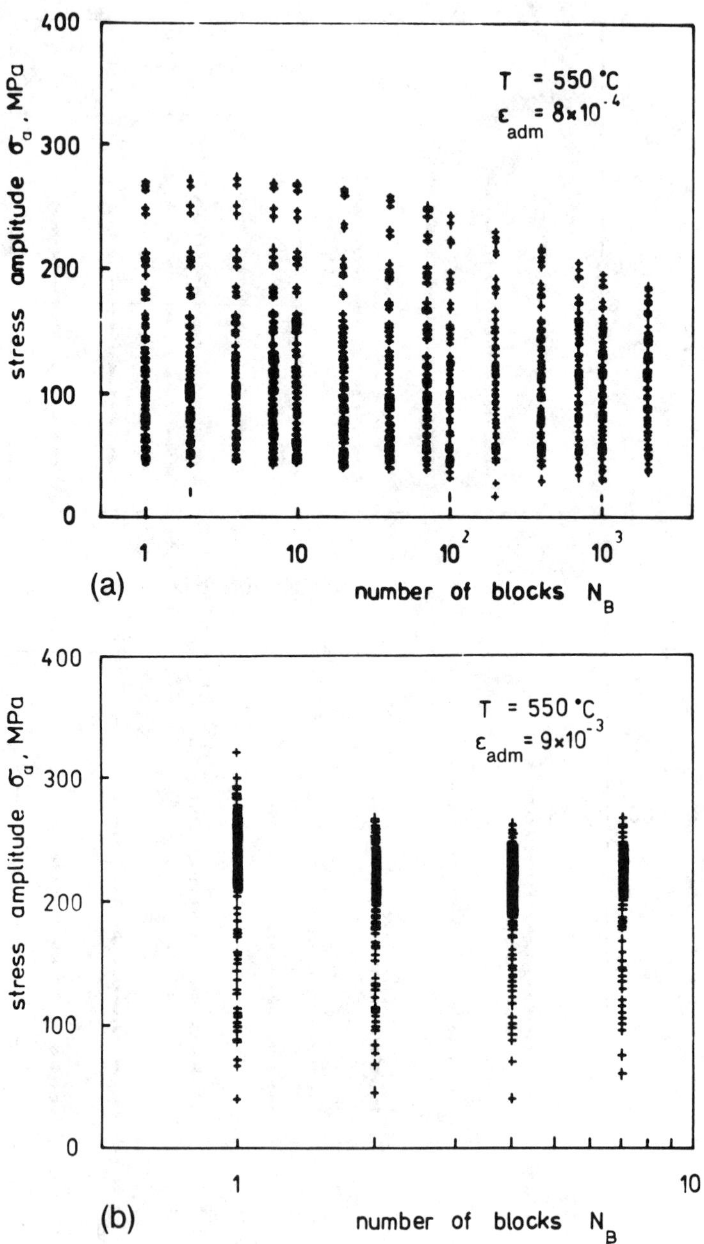

FIG. 2—*Variable amplitude hardening-softening curves at temperature 550°C for two maximum diametral strain amplitudes.*

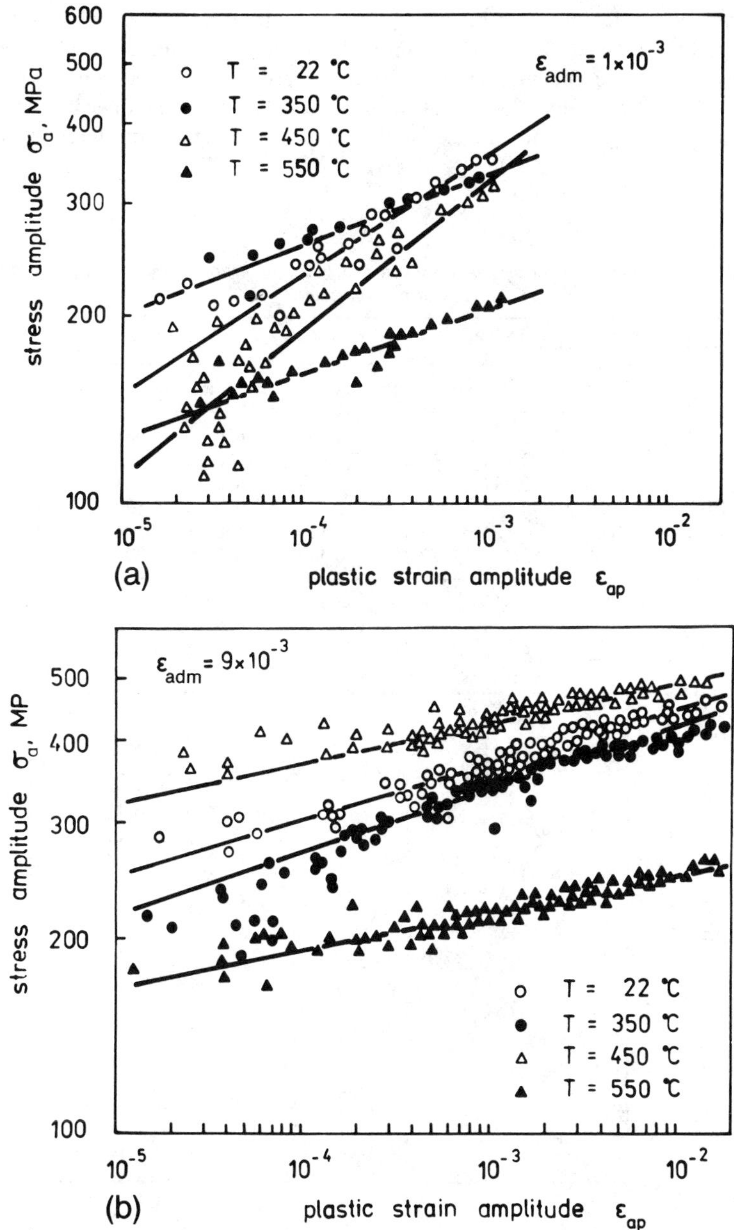

FIG. 3—*Service CSSCs at various temperatures for two maximum diametral strain amplitudes.*

The coefficients K_s and exponents n_s of the individual service CSSCs for different maximum diametral strain amplitudes and temperatures are summarized in Table 2. The standard deviations of the parameters were also calculated, but were not included in Table 2. It is enough to note that they increase considerably with decreasing ϵ_{adm}.

The fatigue life in loading with blocks of variable amplitudes can be characterized by the dependence of the number of blocks to failure, N_{Bf}, on the maximum diametral strain amplitude in a block. The experimental points for all temperatures are plotted in Fig. 4. The fatigue life is reduced when the temperature of testing increases. This drop of fatigue life with temperature is in qualitative agreement with the tendency found in constant amplitude testing [7].

TABLE 2—*Parameters of the service CSSCs for various temperatures and maximum diametral strain amplitudes.*

	22°C		350°C		450°C		550°C	
ϵ_{adm}	K_s, MPa	n_s	K_s, MPa	n_s	K_s, MPa	n_s	K_s, MPa	n_s
9.0×10^{-3}	654	0.083	640	0.095	650	0.061	325	0.057
6.0×10^{-3}	750	0.114	600	0.074	700	0.072	358	0.079
4.0×10^{-3}	650	0.087	640	0.088	640	0.104	369	0.077
2.5×10^{-3}	930	0.148	970	0.118	650	0.106	370	0.082
1.6×10^{-3}	800	0.129	810	0.129	1160	0.188	360	0.074
1.0×10^{-3}	1300	0.190	700	0.110	1600	0.230	410	0.101

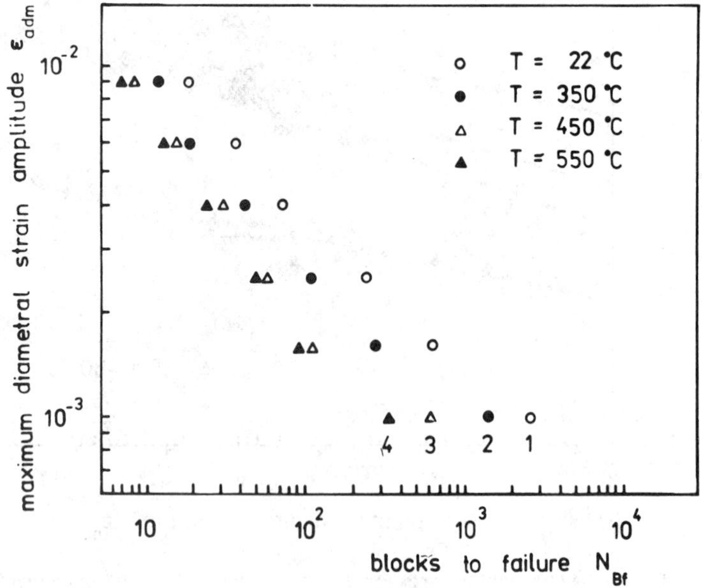

FIG. 4—*Experimental fatigue life versus maximum diametral strain amplitude.*

Discussion

Cyclic Plastic Stress-Strain Response

The rain flow analysis makes it possible to show the cyclic plastic stress-strain response in variable amplitude loading in a representation allowing direct comparison with that in constant amplitude loading. The hardening-softening behavior at all temperatures investigated is quite similar to the constant amplitude behavior. The initial hardening was observed for small maximum diametral strain amplitudes only. This is apparently due to the fact that the rate of hardening is quite intensive and the hardening can be completed even within the first block. The long-term softening for temperatures up to 450°C, however, is slower, especially for blocks with small maximum strain amplitudes. At 550°C the softening is more pronounced than in constant amplitude loading. The time factor plays an important role. Due to the higher average time to run a closed hysteresis loop in variable amplitude loading tests compared with constant amplitude tests, recovery contributes to the softening.

In order to compare the service CSSCs with the basic CSSC, the fitted power laws are shown in Fig. 5 for all temperatures. At 22°C all points corresponding to maximum strain amplitude of each service CSSC (denoted by full circles) are very close to the basic CSSC. The exponents differ slightly, but except the one characterizing the service CSSC with lowest ϵ_{adm}, all of them are smaller than the exponent n_b of the basic CSSC. The considerable scatter in the positions of the service CSSCs is caused by the inhomogeneity of the original batch of the material. This inhomogeneity is due to different local thermal treatment of a large plate and results in a slightly different percentage of bainite. It can be checked by hardness measurement on the individual specimens.

In variable amplitude straining some specimens exhibited higher stress amplitudes in the beginning of the straining. In spite of considerable softening the service CSSC of such a specimen was shifted to higher stress amplitudes. An example is the service CSSC for $\epsilon_{adm} = 2.5 \times 10^{-3}$ and temperature $T = 350$°C. In spite of this scatter due to macroscopic material inhomogeneity, a reasonable analytical approximation of the set of service CSSC can be found supposing that all service CSSCs have a single exponent n_s and points corresponding to maximum strain amplitudes lie on the basic CSSC [7]. If K_b and n_b are the coefficient and exponent of the basic CSSC, the analytical expression for a set of service CSSCs is

$$\sigma_a = K_b \epsilon_{apm}^{n_b - n_s} \epsilon_{ap}^{n_s} \tag{4}$$

The exponent n_s of service CSSCs was calculated as the average value of the exponents found for three highest maximum strain amplitudes, because they are obtained with the highest precision. They are shown in Table 3 simultaneously with exponents n_b for all four temperatures.

At 550°C the service CSSC lies systematically under the basic stress-strain curve. This is the result of more intensive softening in variable amplitude loading at high temperature. The small loops represent the interruptions of the large loops, and the average strain rate corresponding to the maximum cycle is considerably reduced, which leads to the decrease of the maximum stress amplitude and service CSSC. Instead of a set of the service CSSCs represented by Eq 4, a single service CSSC at 550°C with exponent n_s shown in Table 3 and coefficient $K_s = 353$ MPa characterizes the variable stress-strain response at this temperature.

Fatigue Life

In order to check the fatigue life prediction methods it is necessary to compare the predicted and experimental life obtained under strictly defined conditions. As the fatigue life, both in

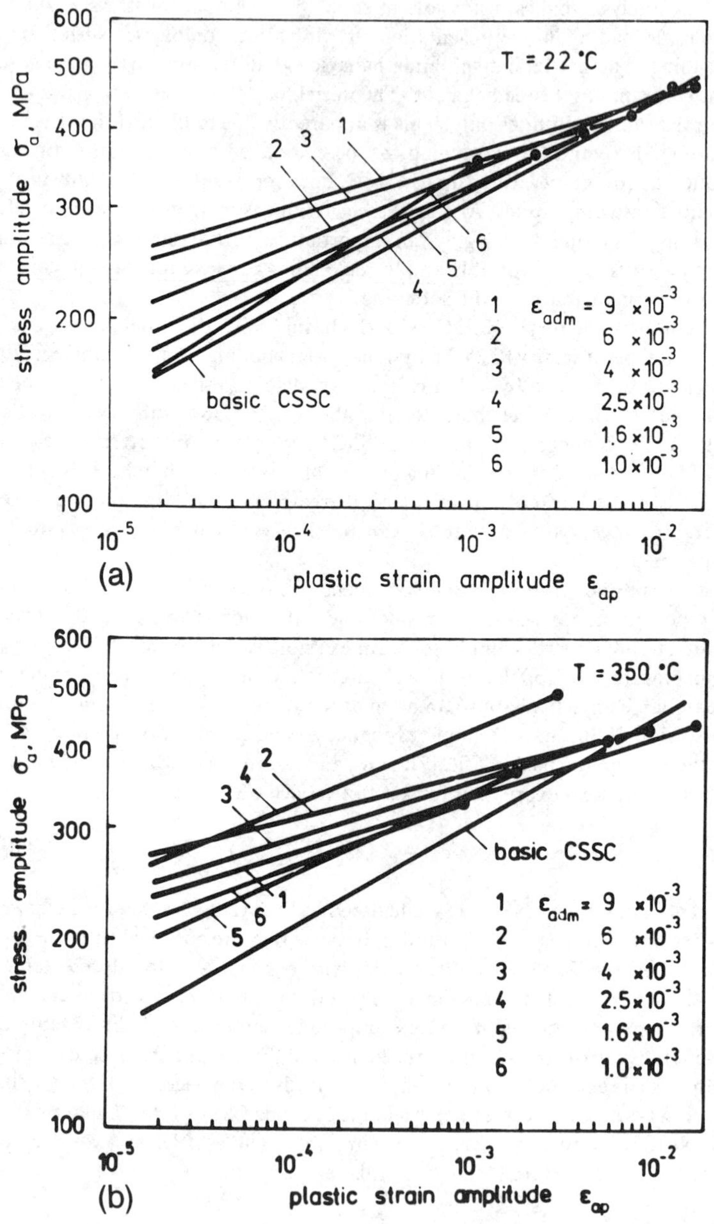

FIG. 5—*Plot of the basic and service*

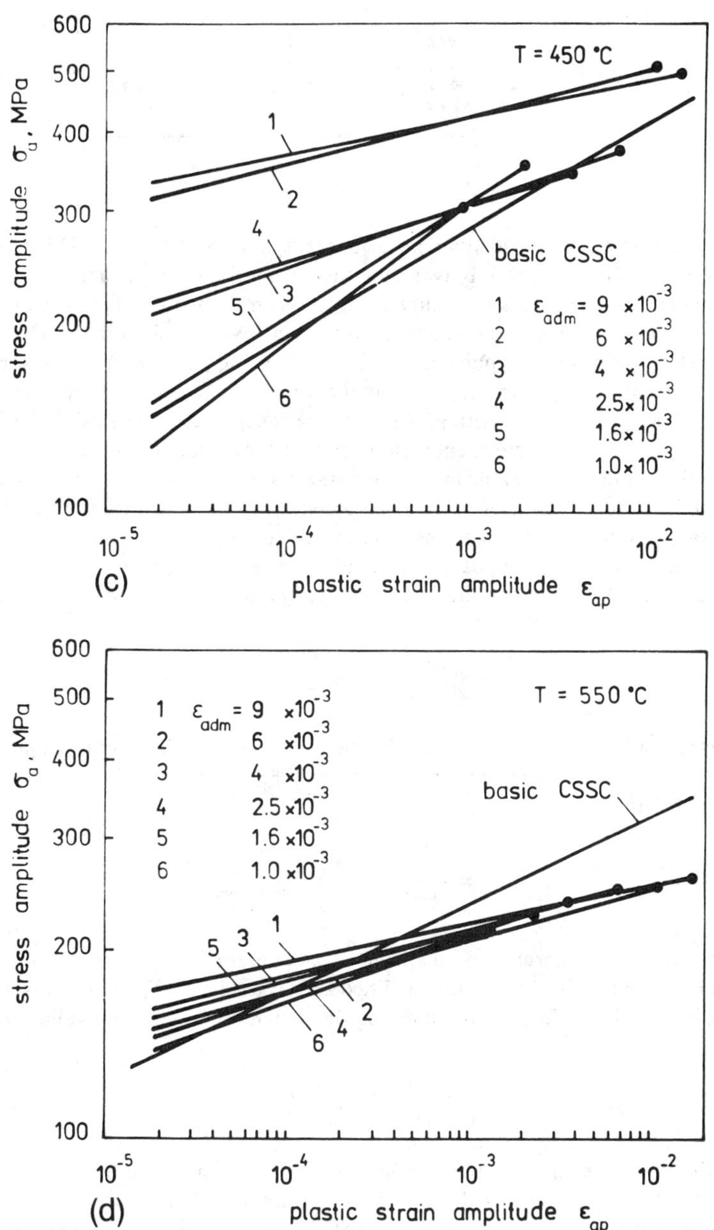

CSSCs for temperatures investigated.

TABLE 3—*Parameters of the basic CSSC and average exponents of the service CSSCs for various temperatures.*

T, °C	K_b, MPa	n_b	n_s
22	922	0.158	0.097
350	946	0.165	0.086
450	912	0.170	0.079
550	613	0.138	0.071

variable and constant amplitude loading, at various temperatures for 2.25Cr-1Mo steel and the knowledge of the loading spectrum is available, this comparison can be made.

Fatigue life prediction methods for complicated stress-strain histories have been reviewed by Dowling [*5*]. The most promising method is based on rain flow analysis of the spectrum yielding full cycles to which damage is assigned and summed. Dowling [*5,6*] and Polák and Klesnil [*7*] showed that for room temperature straining of an aluminum alloy, a low alloy steel, and a low carbon steel this method together with a linear summation of the damage using the Palmgren-Miner rule give predictions in agreement with the fatigue life found experimentally. Therefore a similar prediction method just taking into account the diametral strain control was adopted. At first, the actual measured block of peaks is analyzed using the rain flow procedure, and the diametral strain amplitudes of full cycles, ϵ_{adi}, are calculated. Then the longitudinal plastic strain amplitude is calculated supposing the validity of the basic CSSC or service CSSC. If the basic CSSC is adopted, the ϵ_{api} is evaluated numerically from the relation

$$\epsilon_{adi} = \frac{1}{2}\epsilon_{api} + \frac{K_b}{E_{ef}}\epsilon_{api}^{n_b} \tag{5}$$

In case the stress-strain relation is specified by the set of service CSSCs approximated analytically by Eq 4, which is applicable to the temperatures 22, 350, and 450°C only, the ϵ_{api} is evaluated using the relation

$$\epsilon_{adi} = \frac{1}{2}\epsilon_{api} + \frac{K_b}{E_{ef}}\epsilon_{apmi}^{n_b - n_s}\epsilon_{api}^{n_s} \tag{6}$$

where ϵ_{apmi} is calculated according to Eq 5 from the maximum diametral strain amplitude in a block ϵ_{adm}. For a temperature of 550°C a modified Eq 5 is used with K_s and n_s replacing K_b and n_b. The predicted fatigue life given by summing the damage caused by individual cycles is

$$N_{Bf} = \left[\sum_{i=1}^{n} 2\left(\frac{\epsilon_{api}}{\epsilon_f'}\right)^{-1/c}\right]^{-1} \tag{7}$$

where n is the total number of amplitudes in a block and ϵ_f' and c are the parameters of the Manson-Coffin law. Their numerical values were found for all temperatures investigated [*8*]. The prediction based on the use of the basic CSSC is shown in Fig. 6*a* and that on the use of service CSSC in Fig. 6*b*. Both predictions are very close to each other. The reason is that, due to the high effective modulus, E_{ef}, of the diametral strain, the elastic amplitudes are small; moreover, the service CSSCs do not differ significantly from the basic CSSC except at 550°C.

A comparison of Fig. 6 with Fig. 4 shows that predicted fatigue life at all temperatures is

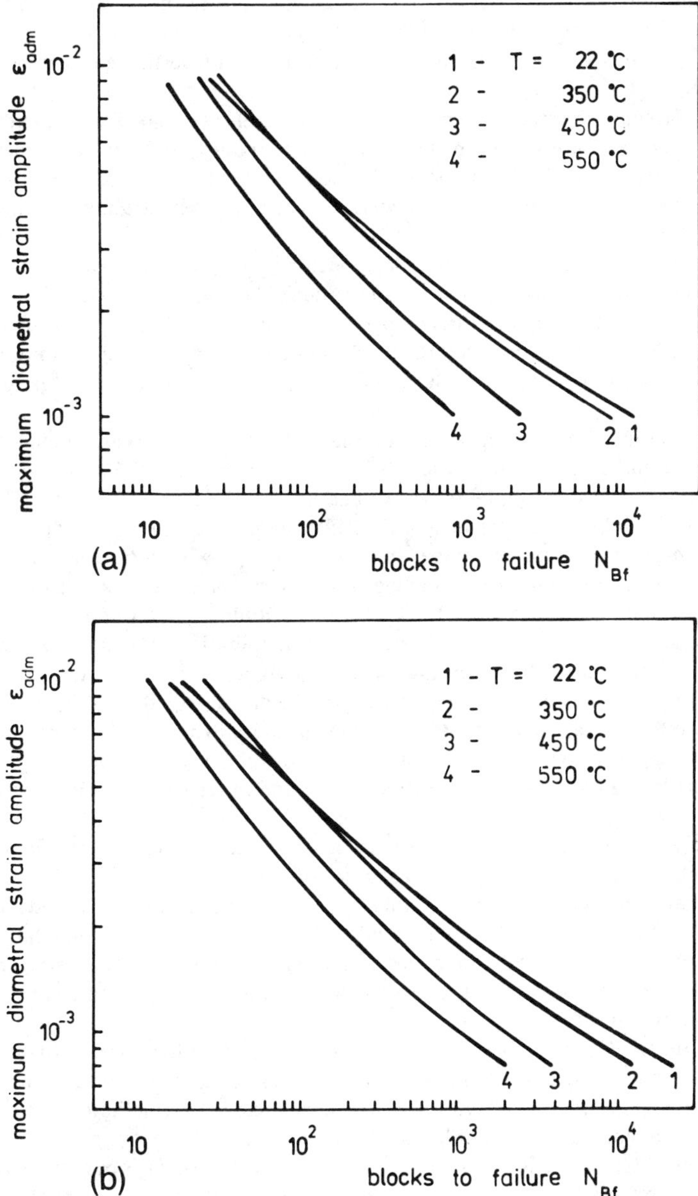

FIG. 6—*Fatigue life versus maximum diametral strain amplitude calculated adopting the validity of* (a) *basic CSSC and* (b) *service CSSC.*

higher than actual fatigue life (i.e., the prediction is nonconservative). In order to reach agreement with the experiments it could be necessary to multiply the predicted fatigue life by a factor of 0.4. Using this factor, the ratio of the actual fatigue life to the predicted life is in the interval 0.5 to 2 for all temperatures investigated, which corresponds to the usual scatter of low-cycle testing. From the practical point of view it can be used for the prediction in the case of this particular type of block of variable amplitudes and the steel investigated; however, it has no general validity.

Taking into account that the same prediction method yields very good agreement with actual fatigue life in the case of low carbon steel [*7*], aluminum alloy, and low alloy steel [*5,6*], two reasons for the discrepancy in the predicted and the actual life can be considered: the difference in loading spectrum and the difference in material behavior in cyclic straining leading to nonlinear accumulation of fatigue damage.

The adopted block of peaks, in spite of its modelling purpose, represents a reasonable long block with exponential probability density distribution. The number of cycles with small amplitudes in a block is high. Therefore, in comparison to the previously used short block [*7*] (only 100 peaks), the long block approaches the block consisting of many small amplitudes and one large amplitude. This so-called "periodic overstressing" is known to reduce the fatigue life considerably [*6*].

The reduced fatigue life in comparison with that predicted using linear summation could be explained if the micromechanism of the fatigue accumulation in cyclic plastic straining were known. At this stage at least a qualitative explanation can be offered. The study of the micromechanism of the fatigue process in this steel [*9*] showed that three important stages were present; the microcrack initiation stage, the short crack growth stage, and the macrocrack growth stage. In constant amplitude loading, the majority of cycles is spent in the stages of crack initiation and short crack growth. The high amplitudes are very effective in microcrack initiation. Once a crack is initiated, however, low amplitudes contribute effectively to its growth, starting from the short crack growth stage. The steel investigated here cyclically softens, contrary to low carbon steel [*7*]. Therefore the damaging effect of high amplitudes is higher in the crack initiation and short crack growth stages, while the damaging effect of low amplitudes is higher in the crack propagation stage in comparison to low carbon steel. This behavior leads to the reduction of the actual fatigue life in regard to the linear damage rule prediction.

Several more complicated rules have been proposed to take into account the nonlinear damage accumulation. The double linear damage rule [*10*] separates the whole fatigue process into the crack initiation and the crack propagation stage with damage summed up independently in both stages. Another concept elaborated recently [*11*] uses the damage function that is a nonlinear function of the relative number of cycles and depends on the stress or plastic strain amplitude. The damage curves were obtained experimentally under different loading modes [*12*], but they were not used for life prediction.

Recently Kliman [*13*] introduced another simple concept describing fatigue damage in variable amplitude loading by postulating the total hysteresis energy to fracture to be equal to the hysteresis energy in the so-called "equivalent harmonic cycle." This method yields better agreement with experiments in the case of "periodic overstressing."

As the aforementioned methods either have no sound theoretical basis or require a number of parameters that are difficult to obtain experimentally, none of them has been applied. For practical applications the linear damage rule with the adjusting coefficient for a particular material, loading spectrum, and temperature is a satisfactory temporary solution. In order to obtain a better quantitative description of the fatigue process, and thus more precise fatigue life prediction, further study of the fatigue mechanism both in constant and in variable amplitude loading is necessary.

Conclusions

Investigation of the fatigue behavior of 2.25Cr-1Mo steel in variable amplitude loading at temperatures from room temperature up to 550°C allows us to draw several conclusions:

1. Cyclic plastic response of the material cycled with the blocks of variable amplitudes at different temperatures can be represented using the rain-flow analysis of the block. The fatigue hardening-softening curves in variable amplitude loading as well as the service CSSCs can be found.
2. Fatigue hardening-softening curves in variable amplitude loading have a similar character as in constant amplitude loading; initial hardening is followed by long-term softening for all temperatures and maximum strain amplitudes in a block. Exponents of the service CSSCs are smaller than the exponents of the basic CSSCs and decrease with increasing temperature.
3. The life prediction in variable amplitude loading based on the constant amplitude life curves and linear damage rule is nonconservative. The quantitative description of the fatigue damage leading to precise life prediction can be obtained by studying the mechanism of the fatigue process.

APPENDIX

Weights of the Points of the CSSC to which the Power Law is Fitted

We are trying to fit a power law

$$\sigma_a = K \epsilon_{ap}^{n} \tag{8}$$

to a series of experimental points, the coordinates of which are σ_a, ϵ_a, both measured with equal relative error

$$z = \frac{\delta \epsilon_a}{\epsilon_a} = \frac{\delta \sigma_a}{\sigma_a} \tag{9}$$

First ϵ_{ap} is calculated using

$$\epsilon_{ap} = \epsilon_a - \frac{\sigma_a}{E} \tag{10}$$

If $F(x, y)$ is the function of x and y that were determined from experiment as $\bar{x} \pm \delta x$ and $\bar{y} \pm \delta y$, the mean deviation of F is

$$(\delta F)^2 = \left(\frac{\partial F}{\partial x}\right)^2 \delta^2 x + \left(\frac{\partial F}{\partial y}\right)^2 \delta^2 y \tag{11}$$

and the weight of the particular measurement is

$$w = \frac{1}{(\delta F)^2} \tag{12}$$

The function minimized in the least-squares procedure is

$$F = \log \sigma_a - \log K - n \log \epsilon_{ap} \tag{13}$$

Therefore the weight of the ith measurement is

$$w_i = \frac{1}{\left(\frac{\delta\sigma_a}{\sigma_a}\right)^2 + n^2\left(\frac{\delta\epsilon_{ap}}{\epsilon_{ap}}\right)^2} \tag{14}$$

As ϵ_{ap} is calculated from Eq 10 it is

$$(\delta\epsilon_{ap})^2 = \left(\frac{\delta\epsilon_a}{\epsilon_a}\right)^2 \epsilon_a^2 + \frac{\sigma_a^2}{E^2}\left(\frac{\delta\sigma_a}{\sigma_a}\right)^2 \tag{15}$$

Inserting Eq 9 into Eq 15 and Eq 15 into Eq 14 we obtain

$$w_i = \left[z^2 \left(1 + n^2 \frac{\epsilon_a^2 + \frac{\sigma_a^2}{E^2}}{\left(\epsilon_a - \frac{\sigma_a}{E}\right)^2} \right) \right]^{-1} \tag{16}$$

To calculate the weights of the individual points the value of the parameter n must thus be known. The iteration procedure is therefore used, putting in zero approximation all weights equal to 1. Usually in the first or second approximation K and n are obtained with satisfactory precision.

References

[1] Jaske, C. E. in *Low-Cycle Fatigue and Life Prediction, ASTM STP 770,* American Society for Testing and Materials, Philadelphia, 1982, pp. 600–611.
[2] Landgraf, R. W. in *Cyclic Stress-Strain Behavior—Analysis, Experimentation, and Failure Prediction, ASTM STP 519,* American Society for Testing and Materials, Philadelphia, 1973, pp. 213–228.
[3] Novack, J., Hauschmann, D., Foth, J., Lütjering, G., and Jacoby, G. in *Low Cycle Fatigue and Life Prediction, ASTM STP 770,* American Society for Testing and Materials, Philadelphia, 1982, pp. 269–295.
[4] Matsuishi, M. and Endo, T., "Fatigue of Metals Subjected to Varying Stress," paper presented to Japan Society of Mechanical Engineers, Fukuoka, Japan, 1968.
[5] Dowling, N. E., *Journal of Materials,* Vol. 7, 1972, pp. 71–87.
[6] Dowling, N. E., *Journal of Testing and Evaluation,* Vol. 1, 1973, pp. 271–287.
[7] Polák, J. and Klesnil, M., *Fatigue of Engineering Materials and Structures,* Vol. 1, 1979, pp. 123–133.
[8] Polák, J., Helešic, J., and Klesnil, M., "Effect of Elevated Temperatures on the Low Cycle Fatigue of 2.25Cr-1Mo Steel—Part I: Constant Amplitude Straining," this publication, pp. 43–57.
[9] Polák, J., Lukáš, P., Klesnil, M., Obrtlík, K., Helešic, J., Kunz, L., and Knésl, Z., Report No. 578/697, Institute of Physical Metallurgy, Brno, Czechoslovakia, 1984.
[10] Manson, S. S. and Halford, G. R., *International Journal of Fracture,* Vol. 17, 1981, pp. 169–192.
[11] Chaboche, J. L., in *Low Cycle Fatigue and Life Prediction, ASTM STP 770,* American Society for Testing and Materials, Philadelphia, 1982, pp. 81–104.
[12] Hua, C. and Socie, D. F., *Fatigue of Engineering Materials and Structures,* Vol. 7, 1984, pp. 165–179.
[13] Kliman, V., *Materials Science and Engineering,* Vol. 68, 1984, pp. 1–10.

DISCUSSION

G. R. Halford[1] *(written discussion)*—The block loading produced numerous small hysteresis loops within the major loops. Tensile and compressive mean stress were present. How were the mean stress effects on life accounted for in assessing fatigue damage?

J. Polák, A. Vašek, and M. Klesnil (authors' closure)—Since the spectrum of the strain peaks was symmetrical to zero, the mean stresses were not high; moreover, the number of smaller loops with tensile and compressive mean stresses was approximately equal. Even if some damage law taking into account the mean stress effects were adopted, the effect on the fatigue life prediction would be, in this case, negligible for all histories investigated. Therefore the mean stresses were not considered in life prediction.

M. Nazmy[2] *(written discussion)*—(1) Do you think that the cyclic stress-strain diagrams depend on the block size as well as on the maximum strain amplitude in the block?

(2) Have you tried to apply the double damage rule, developed by Halford and Manson, to your results?

J. Polák, A. Vašek, and M. Klesnil (authors' closure)—(1) The cyclic stress-strain diagrams, or, as we prefer to call them, the "service stress-strain curves," depend considerably on the maximum strain amplitude in the block. The dependence on the block size has not been investigated, but it can be expected that it will be weak for reasonable block sizes.

(2) Not in the form presented by Halford and Manson. The modified version of the double linear damage rule is being developed, and the predictions will be compared with room temperature data on several materials. The results will be published elsewhere.

[1]NASA-Lewis Research Center, Cleveland, OH 44135.

[2]Brown Boveri Research Center, Baden-Dättwil, Switzerland.

J.-Y. Guedou[1] *and J.-M. Rongvaux*[1]

Effect of Superimposed Stresses at High Frequency on Low Cycle Fatigue

REFERENCE: Guedou, J.-Y. and Rongvaux, J.-M., **"Effect of Superimposed Stresses at High Frequency on Low Cycle Fatigue,"** *Low Cycle Fatigue, ASTM STP 942,* H. D. Solomon, G. R. Halford, L. R. Kaisand, and B. N. Leis, Eds., American Society for Testing and Materials, Philadelphia, 1988, pp. 938-960.

ABSTRACT: The effects of superimposed stresses at high frequency on low cycle fatigue behavior were studied on two aeronautical alloys (Ti 6-4 and Inconel 718) in terms of crack initiation, crack propagation, and failure. It is shown that, if the minor cycles have limited amplitudes, not more than 10 to 15% of low cycle fatigue stresses, the life is monitored by the maximum stress of the cycles. This is correlated with the fact that, in this case, the endurance limit of the material is not exceeded. The same criterion is suitable for crack initiation of Inconel 718, but not for Ti 6-4 which is, in all cases, much more sensitive to mixed fatigue than Inconel 718.

The crack propagation regime is ruled by minor cycles only beyond an observed transition ΔK value. For both materials, linear cumulation rules are inadequate for predicting the number of cycles to failure and for calculating the crack growth rates in mixed fatigue.

KEY WORDS: low cycle fatigue, high cycle fatigue, mixed fatigue, crack initiation, crack propagation, titanium base alloy, nickel base alloy, life prediction

During the operation of a large number of industrial machines, low amplitude and high frequency vibration is frequently combined with high level, low frequency stresses. This applies, in particular, to all rotating machines, either for ground-based or aircraft applications. The operating conditions of turbine engine disks include various phases (take-off, cruise, landing) schematized by a trapezoidal cycle. The variation of centrifugal and thermal loads corresponding to the various flight phases is the main cause of low cycle fatigue (LCF) damage in the critical areas of parts. Every motorist wishes to eliminate all vibratory phenomena. However, during the flight cycle, low amplitude, high frequency vibration stresses due, in particular, to blade/vane interferences, may combine with the main LCF cycle. Therefore the cumulated effects of LCF and high frequency fatigue must be considered to the extent where these effects can generate unscheduled failures in service; this will be referred to as *mixed fatigue.*

In order to better understand these phenomena and to derive criteria applicable to parts stressing, a study was made on mixed fatigue. This study covered two forged alloys for turbo-engine disks: the Ti 6-4 (TA6V) titanium base alloy used to manufacture fan and compressor disks, and the Inconel 718 nickel base superalloy (NC19FeNb) used for turbine disks operating up to 650°C.

[1]SNECMA, Materials and Processes Laboratory, B.P. 81 - 91003 Evry CEDEX, France.

Experimental Procedure

Material and Specimens

The two alloys investigated were heat treated in conditions similar to the heat treatment of actual engine parts:

- *Ti 6-4*—TRc condition 955°C/700°C providing an α/β structure with homogenous size α nodules.
- *Inconel 718*—Solution treated and aged (955°C/720°C $\longrightarrow$ 620°C) to obtain an ASTM 6 grain size.

Specimens were taken from 60-mm diameter bars.

Test Program and Notations

Tests were done at 20°C (Ti 6-4) and 550°C (Inconel 718) using servo-controlled hydraulic test machines. The three test types were plain LCF tests (trapezoidal cycles), pure high-frequency vibration fatigue tests, and mixed fatigue tests.

Three types of specimens were used (Fig. 1):

1. Smooth specimens (Fig. 1*a*) for fatigue failure tests.
2. $Kt = 2$ notched specimens (Fig. 1*b*) for fatigue tests to crack initiation. The incipient crack starts from the notch, and specimens were instrumented to detect crack initiation by the electrical potential method ($a = 0.2$ mm crack length $\Leftrightarrow \Delta v/V_0 = 15 \cdot 10^{-4}$).
3. ASTM CT20B10 specimens (Fig. 1*c*) were used to determine crack growth rates by electrical resistimetric measurements at periodic times. Two or three specimens were used for drawing one curve in specific conditions.

The test parameters are summarized in Table 1. The minor cycle amplitude levels were representative of those which might be observed in some cases in engines (i.e., $\pm 8\%$ of the LCF component). However, tests with higher amplitudes (11 and 25%) were also performed in order to provide a better understanding of the mixed fatigue effects.

Test Analysis

The fatigue cycles are characterized by the ratio $R = \sigma_{min}/\sigma_{max}$. In the mixed fatigue condition, the ratio R_{mix} is defined by analogy with pure high frequency fatigue (Fig. 2):

$$R_{mix} = \frac{\sigma_{min} \text{ at high level}}{\sigma_{max}}$$

Thus, for the minor cycle amplitudes considered, the following load ratios are obtained:

$$\begin{array}{rll} \Delta\sigma \text{ HCF} = & 8\% \quad \sigma_{max} \text{ LCF} \longrightarrow & R_{mix} = 0.85 \\ & 11\% & 0.80 \\ & 25\% & 0.60 \end{array}$$

In addition to conventional curves (Wöhler), the test results were presented similarly to Haig diagrams: $\Delta\sigma/2$ HCF versus the component σ_{max} LCF. Crack propagation test data were plotted as da/dN versus ΔK curves where ΔK refers to the maximum applied load amplitude (i.e., including a term due to low cycle fatigue and another term correlated to vibration cycles).

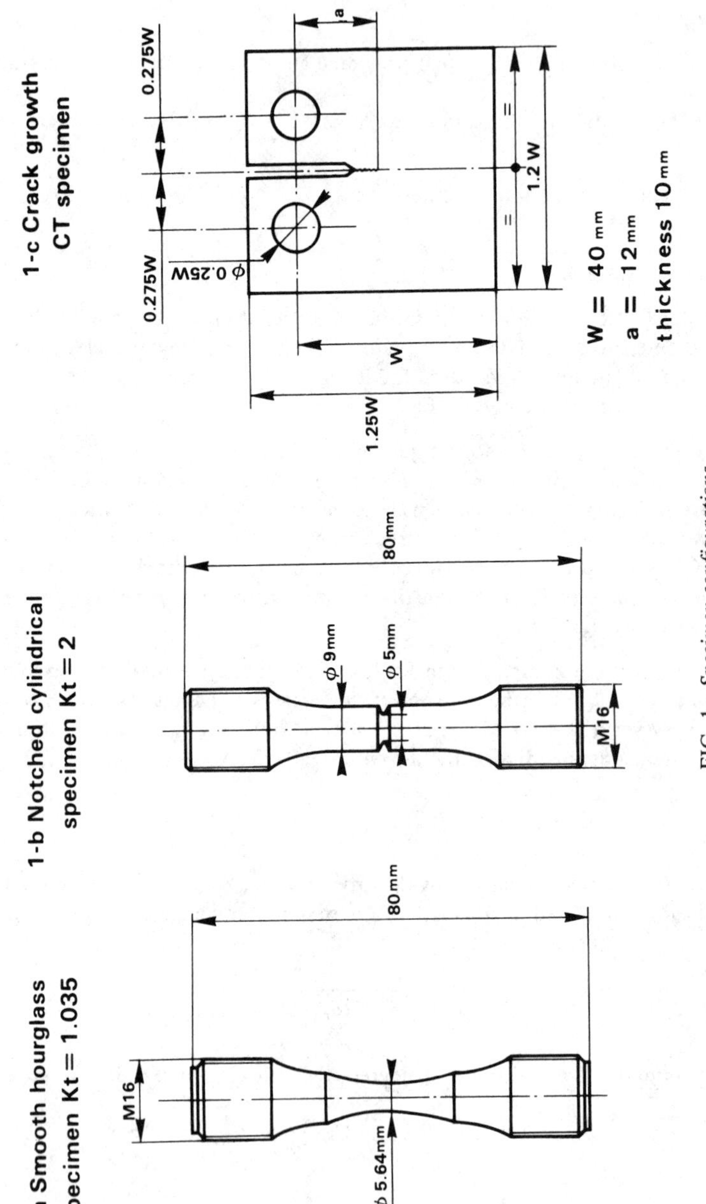

FIG. 1—*Specimen configurations.*

TABLE 1—*Test schedule.*

WAVE FORM	SMOOTH HOURGLASS SPEC. $K_t = 1.035$	NOTCHED CYLINDRICAL SPECIMEN $K_t = 2$	CRACK GROWTH CT SPEC.
LCF $R_\sigma = \frac{\sigma_{mini}}{\sigma_{maxi\,LCF}} = 0.05$	Trap. wave 10 - [90] - 10 10 - [180] - 10	Trap. wave 10-[90]-10	Trap. wave 10 - [10] - 10 10 - [90] - 10
HCF $R_\sigma = \frac{\sigma_{mini\,HCF}}{\sigma_{maxi}}$	Sine wave ν = 100 Hz R_σ: 0.8; 0.6	—	Sine wave ν = 20 Hz - R_σ = 0,85
MIXED FATIGUE $R_{LCF} = \frac{\sigma_{mini}}{\sigma_{maxi\,LCF}} = 0.05$ $R_{mix.} = \frac{\sigma_{mini\,HCF}}{\sigma_{maxi}}$	LCF: 10-[90]-10; Trap. wave 10-[180]-10 HCF: R mix. / ν(Hz): 0.85 / 20; 0.8 / 30; 0.6 / 30; 0.85 / 20	LCF: Trap. wave 10-[90]-10 HCF: Sine wave ν = 20 Hz Rmix.=0.85	LCF: 10-[10]-10; Trap. wave 10-[90]-10 HCF: Sine wave ν = 20 Hz, R mix.=0.85

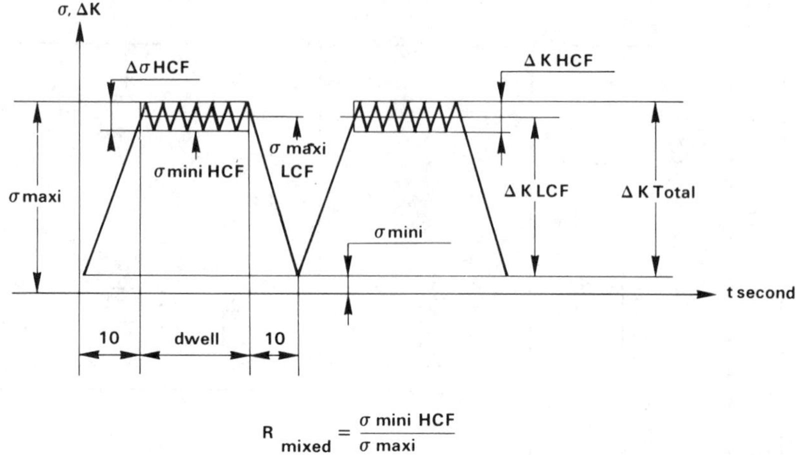

FIG. 2—*Definition of mixed fatigue.*

Tests Results and Discussion

Number of Cycles to Failure of a Smooth Specimen

Ti 6-4 Alloy—Plain LCF and mixed fatigue test data with $R_{mix} = 0.85$ are shown in Table 2. When the number of minor cycles applied during an LCF period is increased by a factor of 2 (from 1800 to 3600), the number of cycles to failure is reduced by a factor of 3. However, this tendency, derived from a limited number of tests, must be moderated since it falls within the data scatter band which, practically, results in a single Wöhler curve (Fig. 3).

Plotting the constant life curves in a Haig diagram for the various minor cycle amplitudes investigated shows (Fig. 4) that these curves follow 45° slope straight lines for low vibration amplitudes (i.e., less than 10% of the maximum LCF stress). The equation of a straight line of this type is

$$\frac{\Delta\sigma}{2}\text{ HCF} + \sigma_{max}\text{ LCF} = \sigma_{max}\text{ cycled} = \text{constant}$$

This isolife is then directly correlated with the σ_{max} parameter. This means that, in the minor cycle amplitude range to which engine components are submitted, *the criterion governing the number of cycles to failure is the maximum stress applied during the cycle.* In any case, this rule remains applicable only for a few vibration cycles applied during LCF period, since increasing the number of minor cycles apparently reduces the mixed fatigue life.

If the endurance limit at 10^7 cycles for various levels of mean stress is plotted on the same Haig diagram, the above criterion involving the maximum stress can be observed to remain satisfactory as long as *the amplitude of small vibration cycles is too low to reach the endurance limit.* As soon as the endurance limit at 10^7 cycles is exceeded, the effect of minor cycles becomes more detrimental to the total life; the maximum permissible stress drops by about 25% when a limited number (few thousand) of small vibration cycles of an amplitude equal to 25% of the LCF component is combined with the latter.

Scanning electron microscope examination of the rupture surfaces show that interstriation spacings are 5 to 6 times higher in the case of mixed fatigue than in the case of plain LCF (Fig.

TABLE 2—*Results of fatigue testing on smooth hourglass specimens Ti 6-4 at 20°C.*

Maximum cycled stress (MPa)	PLAIN LCF		MIXED FATIGUE $\Delta\sigma vib. = \pm 8\% \; \sigma maxi$ LCF	
	10 - 90 - 10 Nf	10 - 180 - 10 Nf	10 - 90 - 10 Nf	10 - 180 - 10 Nf
1145	static rupture	—	—	—
1075	—	—	153	—
1050	266	—	940	327
1025	1150	—	2899	844
1000	3436	1340	6268	2676
975	8011		1531	
950		15495		
	90 s	180 s	1800 HCF	3600 HCF

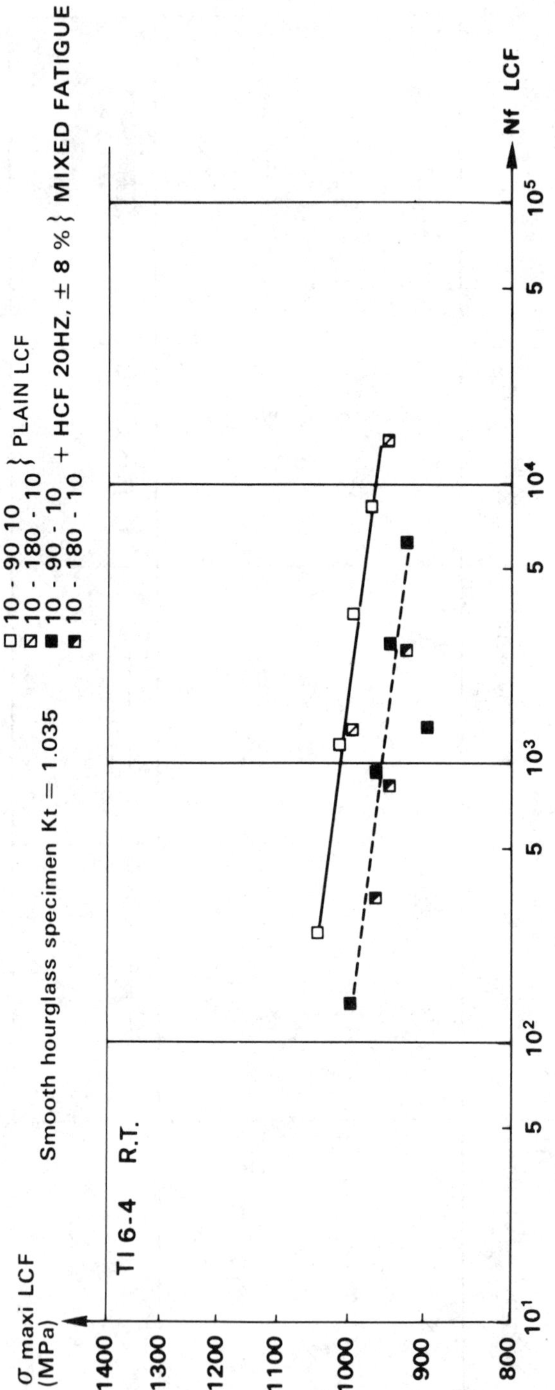

FIG. 3—*Fatigue diagram for Ti 6-4 smooth specimens.*

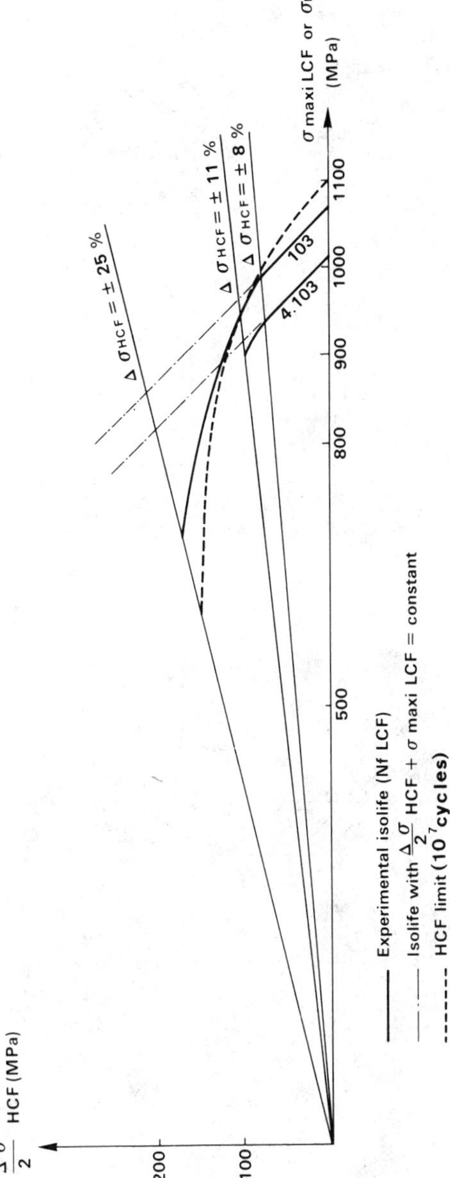

FIG. 4—Haig diagram for mixed fatigue on Ti 6-4 smooth specimens.

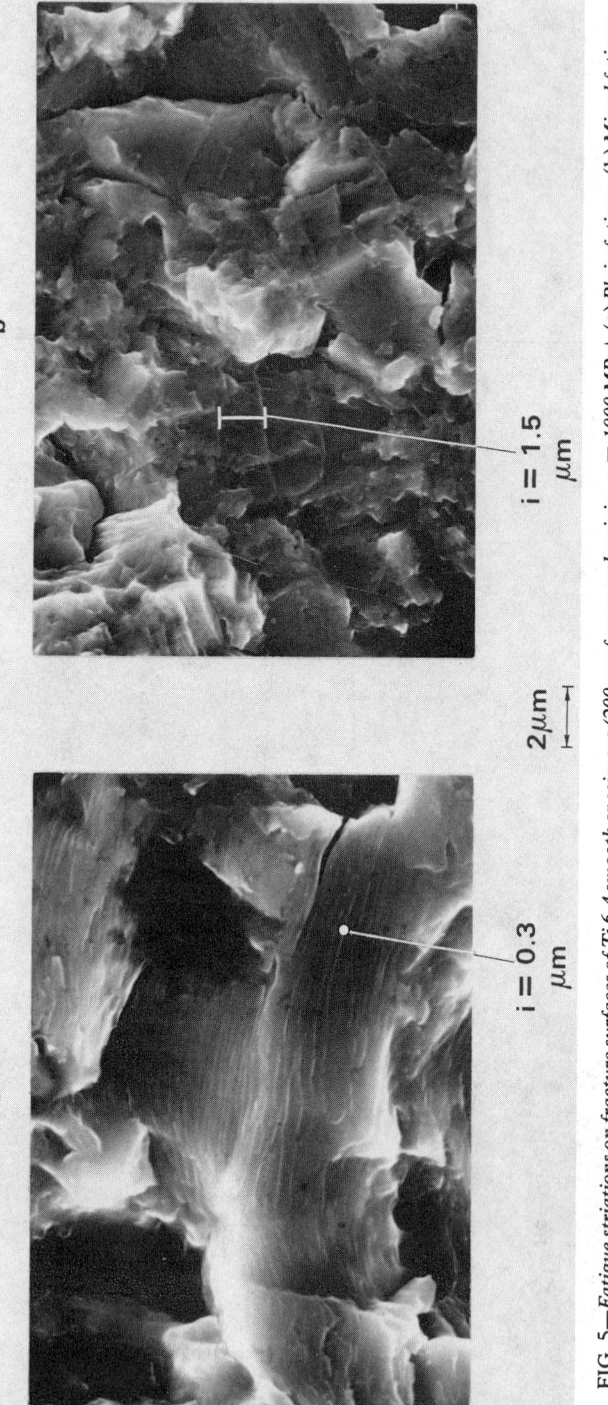

FIG. 5—*Fatigue striations on fracture surfaces of Ti 6-4 smooth specimens (200 μm from crack origin, σ = 1000 MPa).* (a) *Plain fatigue.* (b) *Mixed fatigue.*

5) for the same maximum stress being constant. This demonstrates the effect of minor cycles on crack propagation. This effect will be investigated in more detail subsequently.

Inconel 718 Alloy—With this alloy, a wide scatter of the cycle number to failure can be observed for very small stress differences under both plain LCF and mixed fatigue conditions (Fig. 6). However, the same tendency observed for Ti 6-4 about the number of minor cycles applied for the LCF period can be noticed; when this number is increased from 1800 to 3600, the total life is reduced by a factor of 2 to 3.

The isolife curves plotted on the Haig diagram (Fig. 7) are not very different from the 45° straight lines, even for high vibration amplitudes up to 25% of the LCF component. Plotting the 10^7 cycles endurance curves shows that the vibration stress levels of small cycles are consistently below the endurance limit. Therefore, as in the case of Ti 6-4 with, however, higher amplitudes of minor cycles, the life is apparently governed by the *maximum stress* applied during the combined cycle. This is true for vibration amplitudes less than 10% of the LCF component, which is what is observed in an engine; for an amplitude of up to 25% of the LCF component, the reduction in maximum stress is only 5%.

The examination of the fracture surfaces shows mixed fatigue interstriations that are more perceptible than with Ti 6-4 and significantly higher (by a factor of about 8) than those observed in the case of plain LCF (Fig. 8); this reflects the marked effect of minor cycles on crack propagation.

Discussion—A criterion that can be used for life calculation can be derived from combined fatigue test data:

> *When low amplitude cycles (≤10% of maximum LCF stress) are combined with LCF cycles, the number of cycles to failure is controlled by the maximum stress value applied during the cycle.*

When the minor cycle amplitudes are higher (up to 25% of the LCF component), the maximum allowable stress for a given life is slightly lowered for the Inconel 718 and significantly higher for the Ti 6-4.

The criterion involving the maximum stress remains applicable as long as the amplitude of minor cycles is not allowed to exceed the endurance limit for 10^7 cycles at the same R ratio. In all instances, the contribution of minor cycles to the propagation of fatigue cracks was demonstrated, but that does not mean that minor cycles play no role in crack initiation.

The test data analysis covered cases where the number of vibration cycles per LCF period remained low, on the order of a few thousand, which is fairly far from actual engine operating conditions. A tendency for the number of cycles to failure to decrease when the high-frequency, low-amplitude cycle number increased was observed (Table 2). This appears to be quite consistent with the effect of small cycles on crack propagation.

The "maximum stress criterion" derived from this analysis was also true for low carbon steels submitted to rotary bending stresses [*1–3*] with minor cycle amplitudes up to 30% of the LCF component. On the other hand, a considerable drop in maximum allowable LCF stress was recorded for high strength steels [*4,5*]. Considering crack initiation apart from crack propagation allows us to apply a damaging stress criterion to low carbon steels S15C and S25C with satisfactory results [*6*] therefore, it is interesting to consider the effect of mixed fatigue on the two life phases: crack initiation and propagation to failure.

Crack Initiation on Notched Specimens

Tests were done for the load ratio $R_{mix} = 0.85$ only, corresponding to minor cycles of an amplitude equal to 8% of the LCF component. Crack initiation is defined as a 50 μm length crack, using calibrated potential drop measurements.

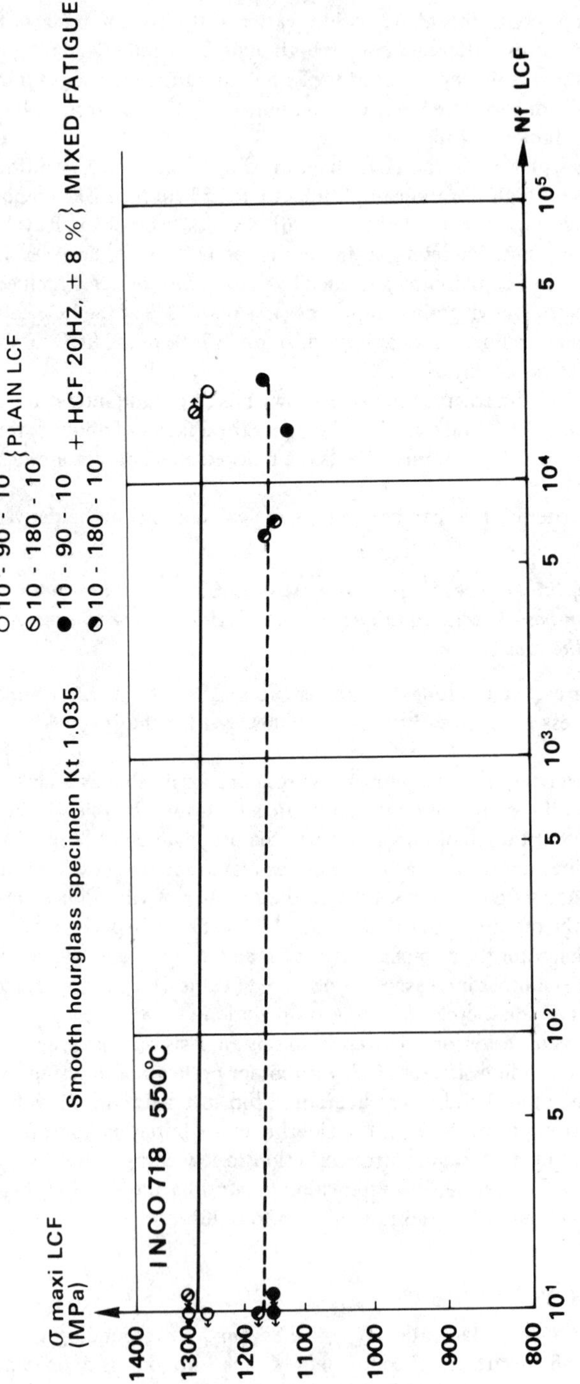

FIG. 6—*Fatigue diagram for Inconel 718 smooth specimens.*

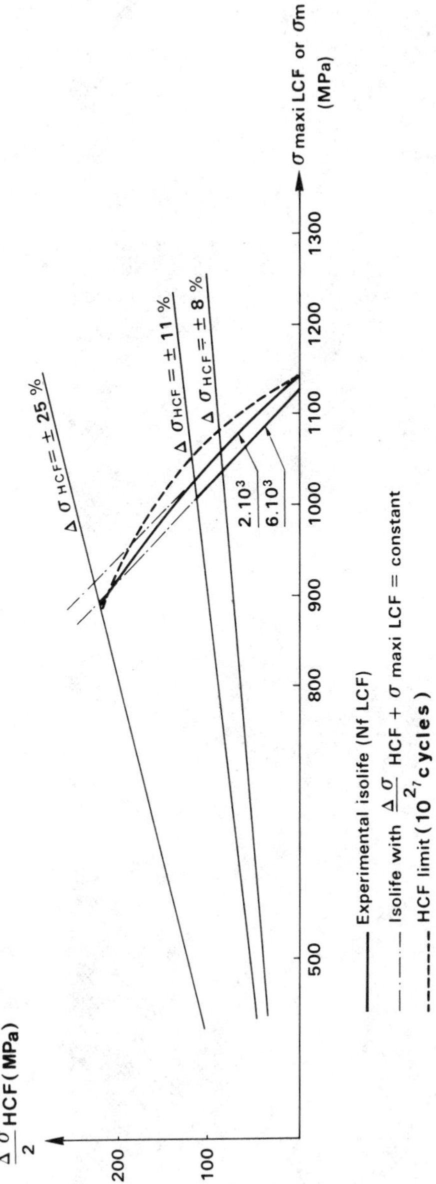

FIG. 7—Haig diagram for mixed fatigue on Inconel 718 smooth specimens.

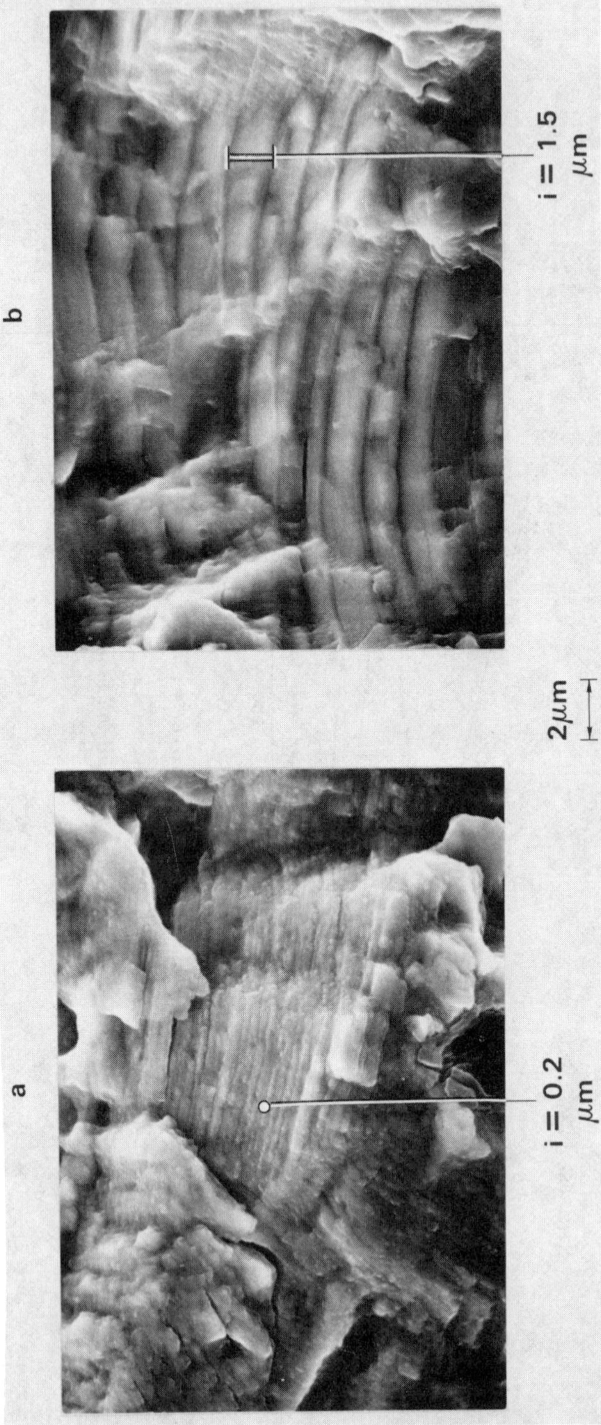

FIG. 8—*Fatigue striations on fracture surfaces on Inconel 718 smooth specimens (200 μm from crack origin, σ = 1250 MPa).* (a) *Plain fatigue.* (b) *Mixed fatigue.*

Ti 6-4 Alloy—The combination of low amplitude minor cycles with high stresses entails a significant reduction of the maximum allowable stress for a given life (Fig. 9); the drop reaches about 20%. In terms of life, the number of cycles to crack initiation on notched specimens is reduced by a factor of 2 to 3 under mixed fatigue at a given maximum stress.

When plotting the number of cycles to failure of the same specimens on the same diagram, it can be seen that the crack initiation period represents 85% of the total life under mixed fatigue conditions in contrast to only 70% in the plain LCF condition. Again, this reflects a quite marked accelerating effect of minor cycles on the crack propagation period.

The criterion of maximum stress reached during the combined cycle, although satisfactorily accounting for life to failure on the smooth specimen, is inadequate to predict the number of cycles to failure or to crack initiation of the notched specimens, if the nominal stress is considered in contrast to smooth specimens.

Inconel 718 Alloy—In contrast to the Ti 6-4 alloy, the effect of low amplitude minor cycles on crack initiation and failure of Inconel 718 notched specimens remains very limited (Fig. 10). With this alloy, the maximum stress criterion is approximately confirmed for life to crack initiation. Here again, the crack initiation portion of the overall life is much longer in the mixed fatigue condition (90%) than in the plain LCF condition (65%), which indicates a faster crack growth in the first case. This is also observed on the rupture face of failed specimens (Fig. 11). A record interstriation of spacing shows a value four times higher in mixed fatigue than under plain fatigue, which indicates the predominant effect of minor cycles on crack growth.

Discussion—The effect of minor cycles combined with LCF cycles on crack initiation is much larger for the Ti 6-4 alloy than for Inconel 718. For the latter alloy a criterion involving the maximum stress remains applicable. For the Ti 6-4 specimens, the effects of minor cycles are significantly enhanced when a stress gradient is present. This alloy is quite sensitive to notches, which may explain this difference in behavior between the two materials. Similar findings were made with the S45C steel [*3*] for which the superposition of minor cycles was as damaging as the stress intensity factor was high. The applied nominal stress is no longer the only parameter to be considered, since the stress status is triaxial. In addition, the local strain is quite large and minor cycles are more damaging if applied at a high level of mean strain. The Inconel 718 is not very sensitive to damaging effects caused by mean strain.

Crack Propagation

Crack propagation tests were conducted at a mixed fatigue load ratio of 0.85. The LCF segment represented 92% of the total ΔK and the high frequency vibration segment $\pm 8\%$.

Ti 6-4 Alloy—Basic plain LCF curves were established for two different durations of maximum load application (i.e., 10 and 90 s). Apparently, this time has no effect on the crack growth rate of the Ti 6-4 alloy at 20°C. On the other hand, the combination of low amplitude minor cycles accelerates the crack growth rate. This rate increases as the number of minor cycles applied per LCF period is increased. In the steady region of the Paris curve, da/dN is multiplied by 3 for 200 vibration cycles and by 20 for 1800 cycles during the steady maximum load segment (Fig. 12). Therefore, when the number of minor cycles is varied in a ratio of 1 to 9, the growth rate varies by a ratio of about 7 to 1 for $\Delta K \geq 20$ MPa $\sqrt{\text{m}}$. This means that with this stress intensity factor range, cracking is governed by the minor cycles. Below $\Delta K = 20$ MPa $\sqrt{\text{m}}$, a convergence of mixed and plain fatigue cracking curves was observed [*8*], which could not be confirmed by our tests since curves were established for ΔK values exceeding 20 MPa $\sqrt{\text{m}}$. In the work described in Ref *7*, a transitional value of ΔK was demonstrated below which cracks propagate under LCF condition and beyond which cracking is governed by minor cycles; this is consistent with our findings. This transitional value is reached when the threshold of the minor cycle is reached. For a ratio $R = 0.85$ (Fig. 13) we have determined that this threshold is below 3 MPa $\sqrt{\text{m}}$. Applying the formula of Hopkins et al. [*8*] would give a threshold ΔK of 2.5

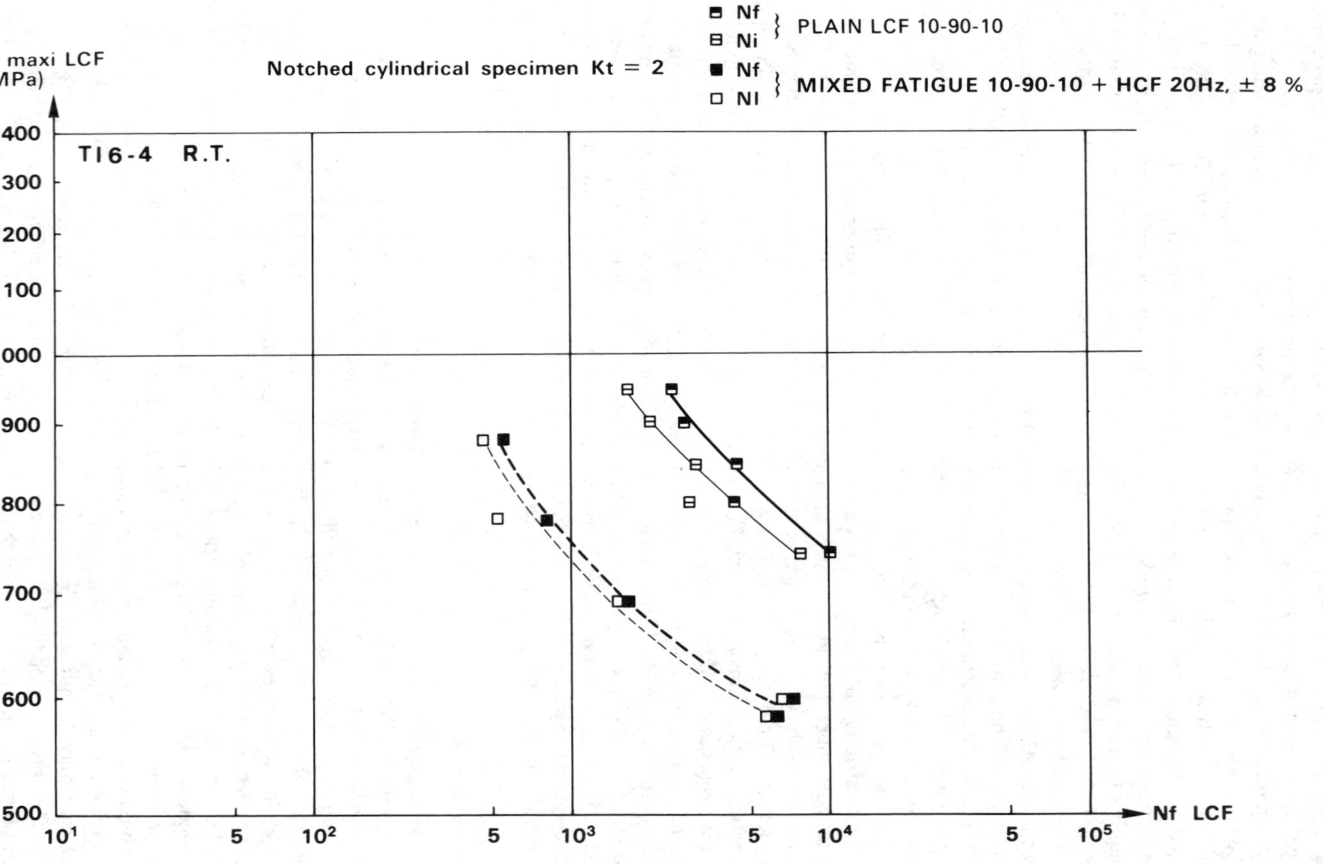

FIG. 9—*Fatigue diagram for Ti 6-4 notched specimens.*

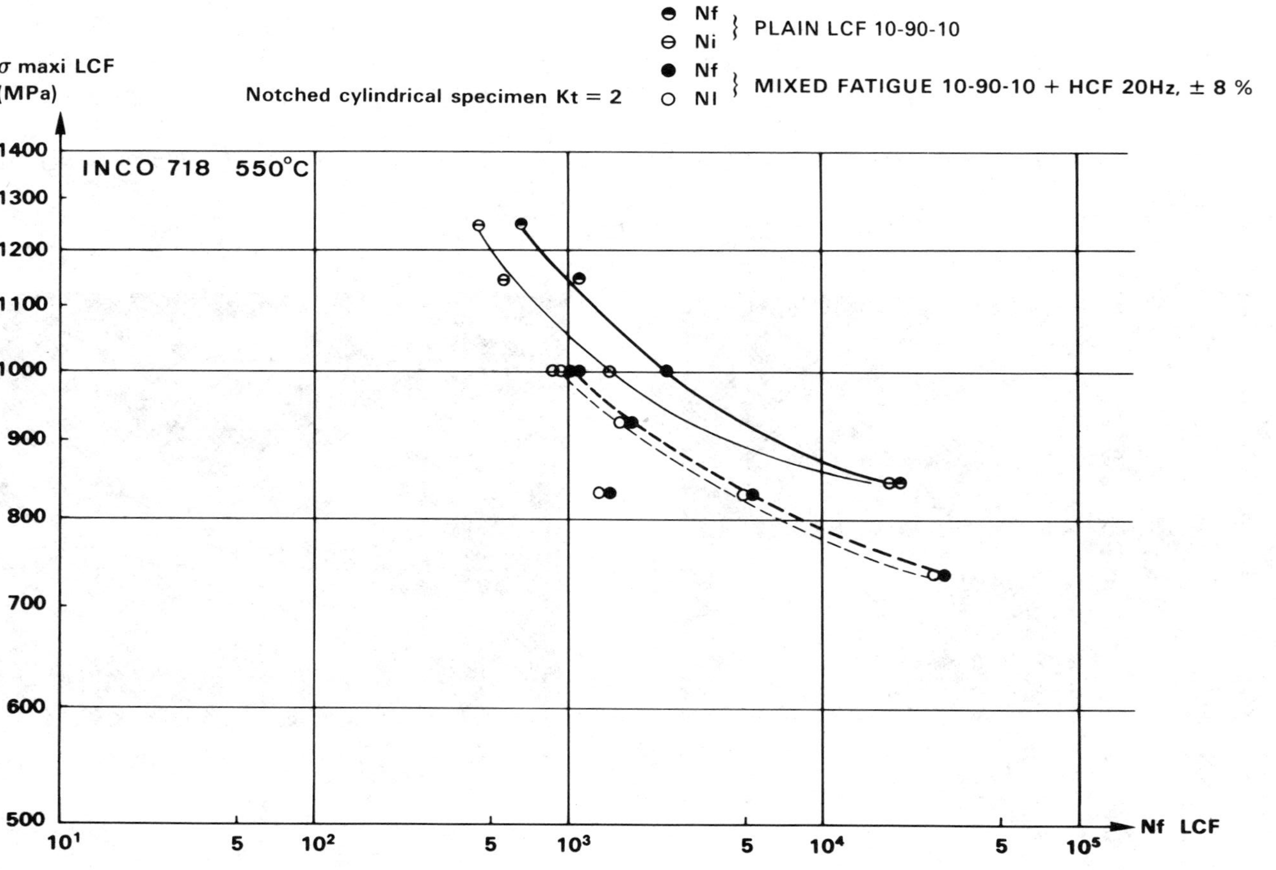

FIG. 10—*Fatigue diagram for Inconel 718 notched specimens.*

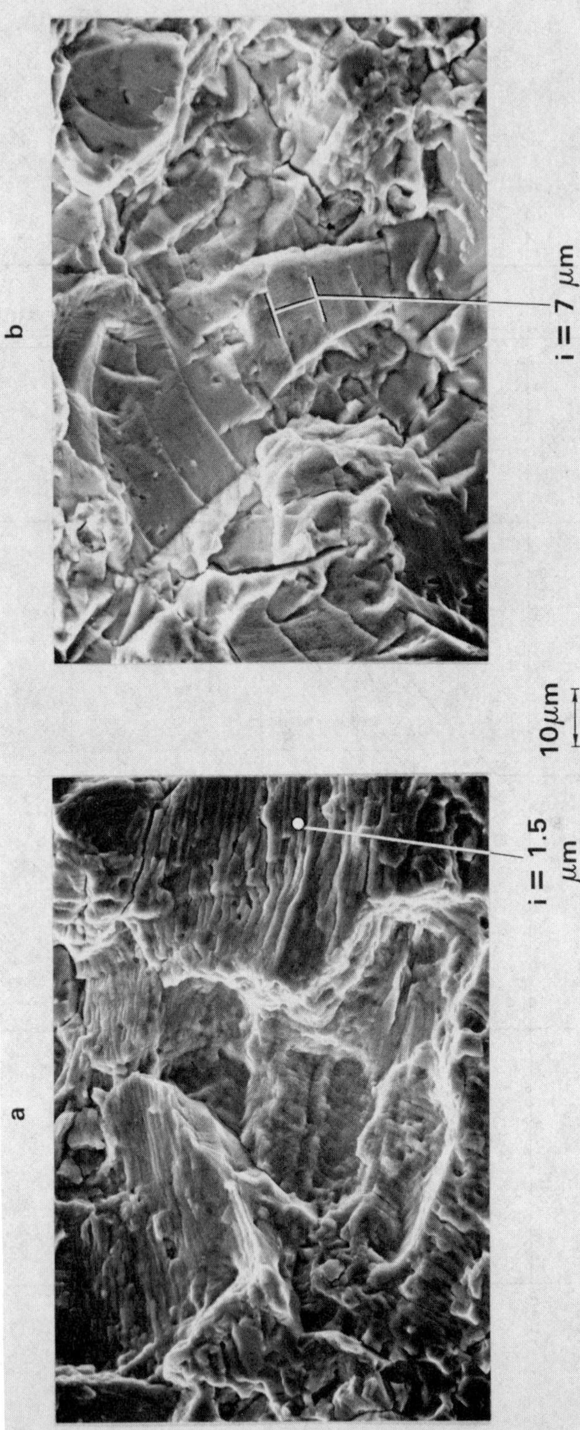

FIG. 11—*Fatigue striations on fracture surfaces of Inconel 718 notched specimens (200 μm from crack origin, σ = 900 MPa).* (a) *Plain fatigue.* (b) *Mixed fatigue.*

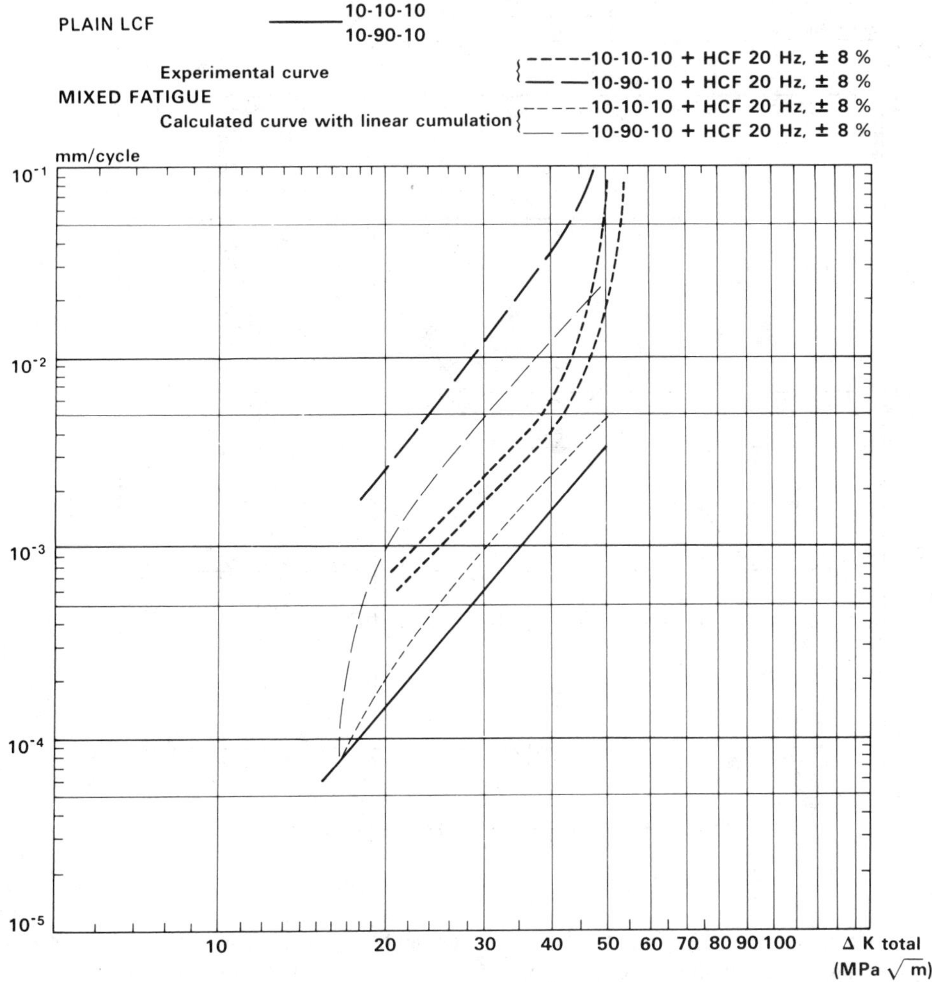

FIG. 12—*Crack propagation diagram for plain and mixed fatigue on Ti 6-4.*

MPa $\sqrt{m}$. This allows the calculation of the transitional ΔK. In our test conditions, ΔK HCF due to minor cycles is equal to 0.15 ΔK total, which yields ΔK_t transitional = 16 MPa $\sqrt{m}$ for ΔK threshold = 2.5 MPa $\sqrt{m}$.

A cumulative linear rule was applied for mixed fatigue: $da/dN = da/dN$ LCF + $P\ da/dN$ HCF where P is the number of minor cycles applied for the LCF period. With this rule, theoretical cracking curves could be plotted with the transitional ΔK_t value of 16 MPa $\sqrt{m}$. Growth rates obtained (Fig. 12) are approximately twice less than those measured experimentally. In contrast to the study described in Ref 7 the application of a cumulated linear growth criterion is not adequate to describe the crack growth in Ti 6-4 under mixed fatigue condition. Such a rule cannot be applied to component stressing, as it is too optimistic.

Inconel 718 Alloy—The effect of a hold time at maximum load on Inconel 718 cracking at 550°C is quite marked (Fig. 14). The growth rate is multiplied by 2 when the hold time is varied

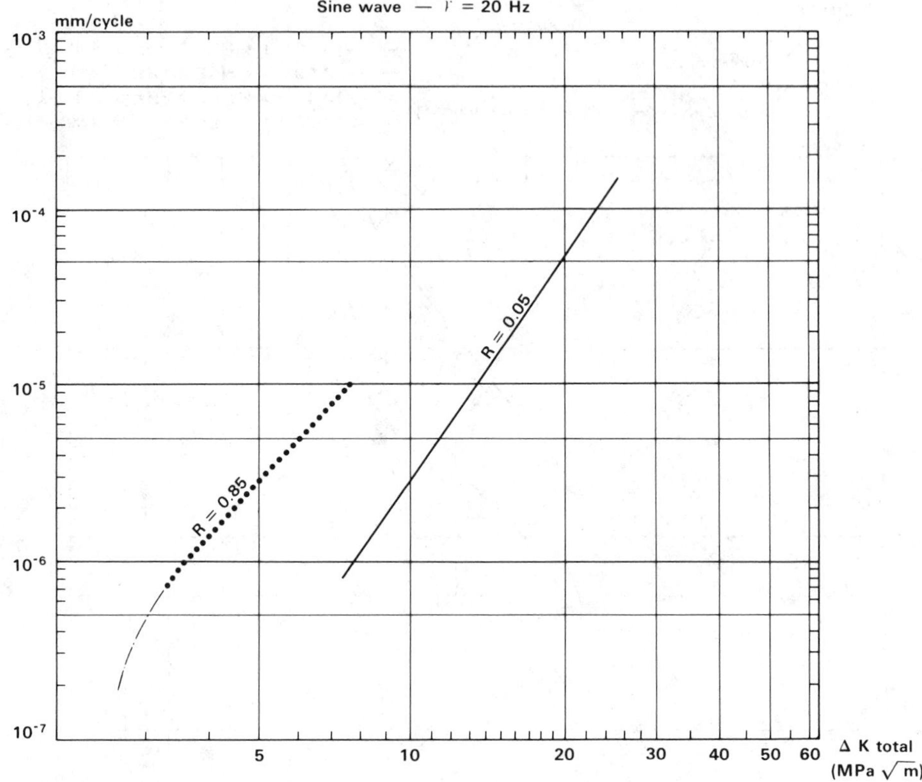

FIG. 13—*Crack growth rate on Ti 6-4 at elevated* R *ratio.*

from 10 to 90 s. On the other hand, in mixed fatigue condition, the growth rate is not influenced by superimposing a small number of minor cycles (e.g., 200) per LCF period. When 1800 cycles are applied during a LCF period, the crack growth rate is increased by a factor of about 5.

A convergence of crack propagation curves da/dN versus ΔK is observed below a ΔK value of 22 MPa $\sqrt{m}$ as already observed [*7*] with Ti 6-4. Assuming that the concept of a transitional ΔK applies here, as in the case of Ti 6-4, the threshold for the minor cycles growth should be at 3.4 MPa $\sqrt{m}$. This is significantly lower than the value of 5 MPa $\sqrt{m}$ we measured (Fig. 15) for $R = 0.85$.

The linear growth rate model applied above $\Delta K = 22$ MPa $\sqrt{m}$ is applicable for a small number of low amplitude cycles (Fig. 14), but underestimates the crack propagation rates by a factor 2 to 4 for a higher number of minor cycles (1800).

Discussion—Minor cycles have a predominant effect on crack growth, since they are the only factor governing the cracking regime beyond a given value of the stress intensity factor. Our findings for Ti 6-4 alloy are consistent with those of other experiments which highlighted a transitional ΔK, in turn related to the growth threshold of minor cycles. This concept should be correlated to the endurance limit of the material under minor cycles: at ΔK values below ΔK threshold of minor cycles, which is equivalent to a stress below the endurance limit, the cracking regime is the LCF regime. This phenomenon is again observed if a failure criterion which involves the maximum stress only is considered.

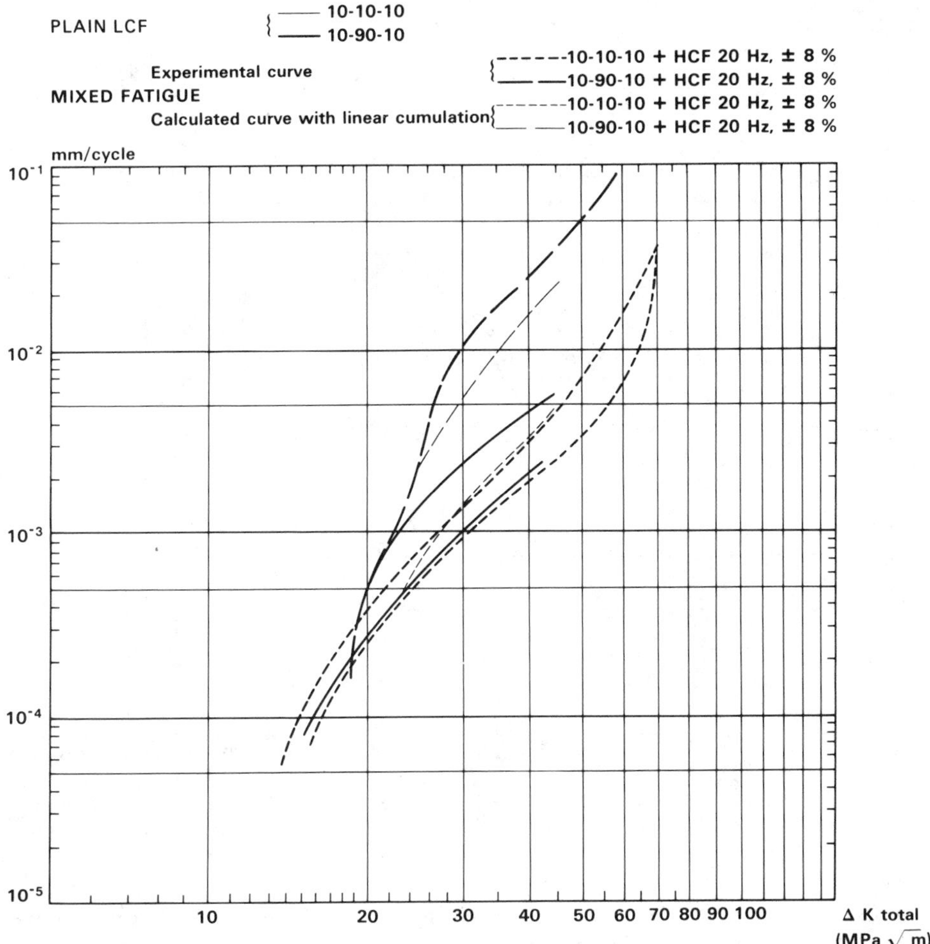

FIG. 14—*Crack propagation diagram for plain and mixed fatigue on Inconel 718.*

When ΔK is above ΔK threshold, which is equivalent to a stress above the endurance limit, the effect of minor cycles is more severe in terms of cycles to failure and the minor cycles are the only factor governing the cracking regime.

For the Inconel 718 alloy, the mixed fatigue mechanisms are less clearly defined. The effect of minor cycles is much less than for the Ti 6-4 alloy, both in terms of initiation and growth. The transitional ΔK value could not be clearly established and correlated to the ΔK threshold of minor cycles. However, beyond a certain ΔK value, the cracking regime is apparently governed by the minor cycles only. With this alloy, the environment effect, which is observed through the effect of the maximum stress hold time, should also be considered since it interferes with fatigue damage.

The rule of cumulated linear crack growth rates is not satisfactory for both materials, in particular for the Ti 6-4 alloy where the results are not consistent with those of another study [7]. This may be caused by the different test conditions in both studies, in particular due to frequency effects.

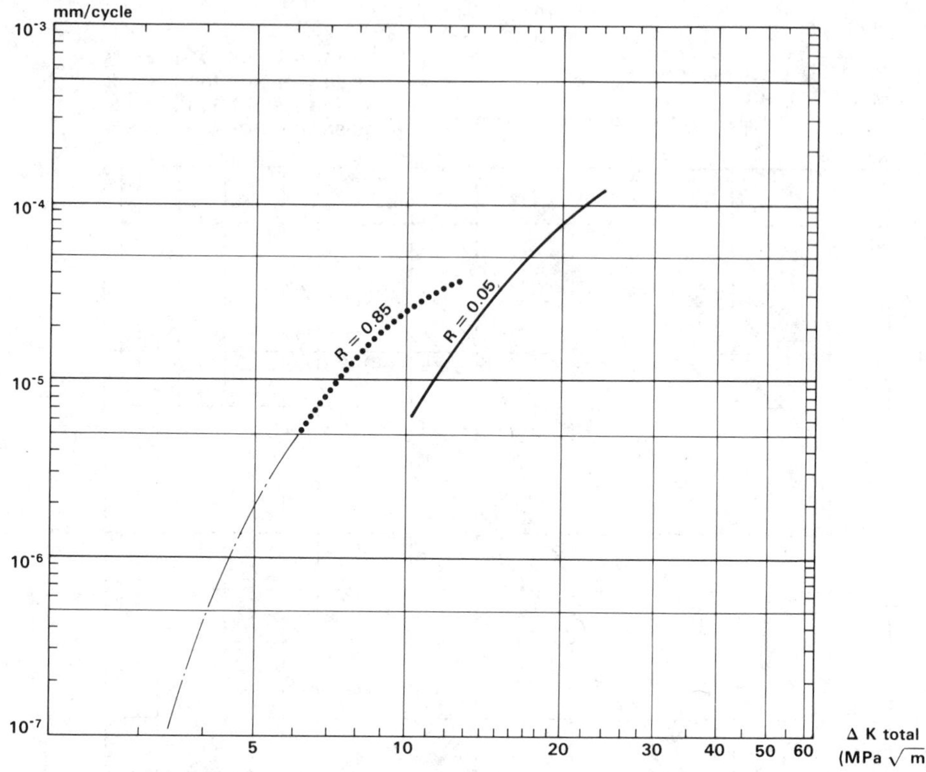

FIG. 15—*Crack growth rate on Inconel 718 at elevated* R *ratio.*

In fact, the threshold value for $R = 0.85$ has to be extrapolated in order to calculate the theoretical curves for mixed fatigue using a linear cumulative model. This model enables us to draw easily the Paris curves (straight lines), which reveal a factor of about 2 between experimental and calculated curves for both alloys. To connect those straight Paris lines with the plain LCF curves at low ΔK (i.e., when HCF has no effect on crack propagation), obviously it is necessary to extrapolate the HCF crack propagation curves to low ΔK values. At all events, those portions of extrapolated curves have no influence on the conclusion of inadequacy of linear crack growth rate cumulation, since they only concern the lower parts of the calculated curves (ΔK total less than 20 MPa $\sqrt{\text{m}}$ for Ti 6-4 and 28 MPa $\sqrt{\text{m}}$ for Inconel 718), where HCF superimposed to LCF has little or no effect on crack propagation.

Prediction Models for Mixed Fatigue Conditions

The design engineer may wish to be able to calculate the life of a component submitted to LCF stresses superimposed to high frequency, low amplitude stresses; damage models may be available for this purpose. Several models have been developed to quantify fatigue damage. Some of these are based on strains [*9*] and can therefore be applied to only materials of a sufficient ductility, such as 18-8 stainless steel. The simplest model is the linear cumulation rule of Palmgren-Miner [*10*]. However, it cannot be applied to the calculation of life to failure of Ti 6-4 and Inconel 718 smooth specimens (Fig. 16). This is due, in particular, to the fact that minor

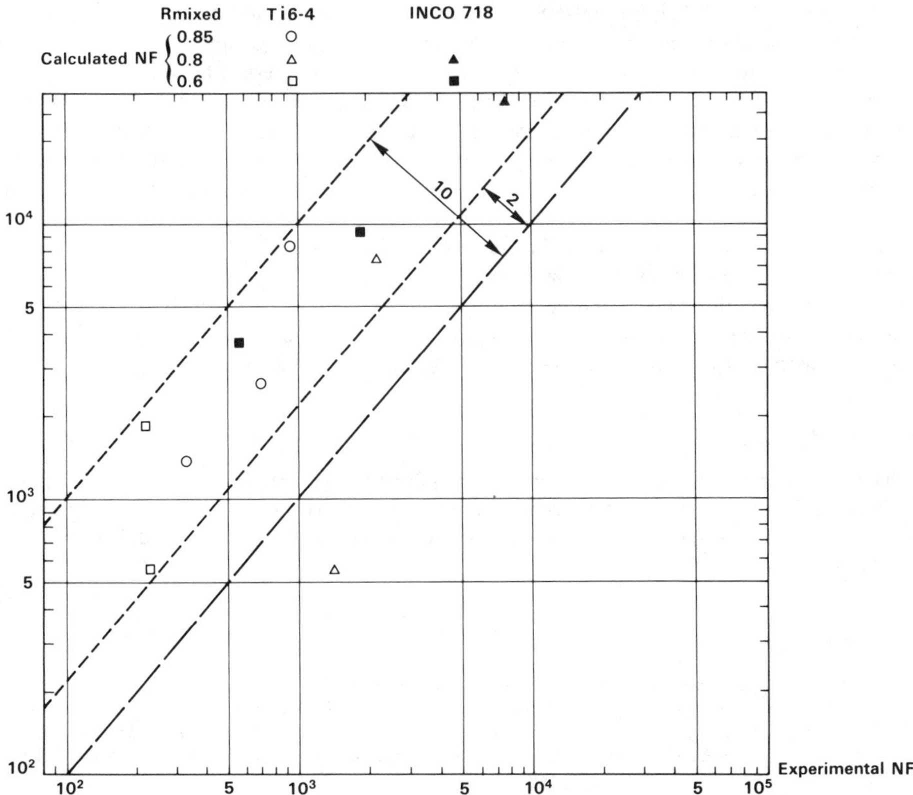

FIG. 16—*Comparison between experimental and calculated number of cycles for failure on smooth specimens.*

vibration cycles produce damage beyond their endurance limit only. Therefore the model does not account for interaction of minor cycles, which would mean that life would be determined by the LCF stress level. This is not confirmed by experience. With the model developed by ONERA [*11*], based on a local approach of damage under stress, the nonlinear cumulation of elementary damage accounts for each type of stress and it is possible to consider minor cycles, even below the endurance limit [*12*]. This model was applied to the cast IN 100 alloy submitted to mixed fatigue at high temperature and it gave data consistent with test results [*13*]. However, the complete identification of the model requires specific tests which were not performed in this study. This work will be undertaken with the Inconel 718 alloy first, and it will be possible to apply these concepts to this type of stress.

Conclusion

The present study has shown that the combination of high frequency minor cycles with high stress level LCF cycles has a significant influence on the life of Ti 6-4 alloy at room temperature and on Inconel 718 alloy at 550°C. However, the number of cycles to failure can be determined by considering only the maximum stress reached during the mixed cycle as long as the amplitude of minor cycles remains below 10% of the LCF component. The effect of minor cycles, much larger for Ti 6-4 than Inconel 718, is observed both on crack initiation and crack growth. In the latter case, there exists a transitional ΔK which is correlated to minor cycle threshold and the endurance limit. Beyond this value, the cracking regime is governed by minor cycles, making such minor cycles all the more damaging. In all cases, models implying a linear damage cumulation are not satisfactory in regards to failure and cracking. The application of more sophisticated rules, such as those developed by ONERA, should allow for better predictions.

The criterion involving the maximum stress to determine the life is valid only for a small number of low amplitude cycles per LCF period (i.e., a few thousand only). This differs from actual components during operation. Therefore calculation methods should be implemented in order to understand these mixed fatigue phenomena applied to concrete cases. The results of this analysis shows that understanding the various effects is not always sufficient and that more characterizations of materials for turbine engine disks under a mixed fatigue spectrum are needed, together with investigations of the metallurgical mechanisms involved.

Acknowledgments

The authors wish to thank the "Direction des Recherches, Etudes et Techniques" of the "Ministère de la Défense" who financed this study and authorized its publication, and the Materials Laboratory of the "Centre d'Essais Aéronautiques de Toulouse" who did most of the mechanical tests.

References

[*1*] Tanaka, T. and Denoh, S., "Effect of Superimposed Stress of High Frequency on Fatigue Strength," *Bulletin Japan Society of Mechanical Engineering,* Vol. 11, No. 43, 1968, p. 77.

[*2*] Tanaka, T. and Denoh, S., "Effect of Superimposed Stress of High Frequency on Fatigue Strength of Annealed Carbon Steel," *Bulletin Japan Society Mechanical Engineering,* Vol. 12, No. 54, 1969, p. 1309.

[*3*] Tanaka, T. and Denoh, S., "Notch Dependency of the Effect of Superimposed Stress of High Frequency of Fatigue Strength," *Bulletin Japan Society Mechanical Engineering,* Vol. 14, No. 77, 1979, p. 1139.

[*4*] Zaitsev, G. Z. and Faradzhov, R. M., "The Fatigue Limit of Steels Under Loads of Differing Frequencies," *Metallovedenie I Obrabotka Metallov,* Vol. 2, 1970, p. 44.

[*5*] Zaitsev, G. Z. and Yatsenko, V. K., "Effect of Stress Concentrations on Fatigue Strength of Steel During Two Frequencies Loadings," *Zavodskaia Laboratoriia,* Vol. 11, No. 43, 1977, p. 1398.

[*6*] Yamada, T., Kitagawa, S., and Ichihara, Y., "Fatigue Life of Metals Under Varying Mean Stress," *Journal of Society of Materials Science Japan,* Vol. 17, 1968, p. 117.

[*7*] Powell, B. E., Duggan, T. V., and Jeal, R., "The Influence of Minor Cycles on Low-Cycle Fatigue Crack Propagation," *International Journal of Fatigue,* 1982, p. 4.

[*8*] Hopkins, J. M., Rau, C. A., Leverant, C. R., and Yeun, A., "Effect of Various Programmed Overloads on the Threshold for High Frequency Fatigue Crack Growth," in *Fatigue Crack Growth under Spectrum Loads, ASTM STP 595,* American Society for Testing and Materials, Philadelphia, 1976, p. 125.

[*9*] Makhukov, N., Romanov, A., and Gadenin, M., "High Temperature LCF Resistance Under Superimposed Stresses at Two Frequencies," *Fatigue of Engineering Materials and Structures,* Vol. 1, Pergamon Press, London, 1979, p. 281.

[*10*] Miner, M. A., "Cumulative Damage in Fatigue," *Journal of Applied Mechanics, Transactions of ASME* Vol. 67, 1945, p. 159.

[*11*] Chaboche, J. L., "Life Time Predictions and Cumulative Damage Under High Temperature Conditions" in *Low-Cycle Fatigue and Life Prediction, ASTM STP 770,* American Society for Testing and Materials, Philadelphia, 1982, p. 81.

[*12*] Baptiste, D., Cailletaud, G., Chaudronneret, M., Policella, H., and Savalle, S., "Durée de Vie des Aubes et des Disques de Turbine," Report ONERA 40/1765 RY 085 R, May 1982.

[*13*] Savalle, S. and Cailletaud, G., "Microamorçage, Micropropagation, et Endommagement," *Rech. Aérosp.,* No. 6, 1982, p. 395.

J. Ouyang,[1] Z. Wang,[1] D. Song,[1] and M. Yan[1]

Influence of High Frequency Vibrations on the Low Cycle Fatigue Behavior of a Superalloy at Elevated Temperature

---

REFERENCE: Ouyang, J., Wang, Z., Song, D., and Yan, M., **"Influence of High Frequency Vibrations on the Low Cycle Fatigue Behavior of a Superalloy at Elevated Temperature,"** *Low Cycle Fatigue, ASTM STP 942,* H. D. Solomon, G. R. Halford, L. R. Kaisand, and B. N. Leis, Eds., American Society for Testing and Materials, Philadelphia, 1988, pp. 961–971.

ABSTRACT: The influence of relatively high-frequency, low-amplitude vibrations on the low-cycle fatigue (LCF) behavior and its fractographic feature for a GH36 Fe-Ni alloy at 600°C is studied. It is found that the influence is rather complicated and significant. The nature and extent of the influence depended upon the ratio of the amplitude of high-cycle to low-cycle stress range, Q. As Q approaches Q_c (the critical value of Q), there is no appreciable influence on the total LCF life; as $Q < Q_c$, it will prolong the lifetime slightly; and as $Q > Q_c$, the fatigue life will decrease progressively. The variation of fatigue life was ascribed to the occurrence of some considerable interactions between creep, high-cycle fatigue (HCF), LCF, and environmental effect under different values of Q at elevated temperature, and the change of damage mechanism of the material. The total LCF life, especially crack propagation life, will be markedly reduced by the high-cycle vibrations when Q is much greater than Q_c. Meanwhile, the mixed failure of creep, HCF, LCF, and environmental attack will gradually change into combined cycle fatigue failure with increasing Q. The fractographic analysis is in agreement with these results.

KEY WORDS: superalloy, elevated temperature, high cycle fatigue, low cycle fatigue, combined cycle fatigue, interaction, creep, crack propagation, mechanism, fractography

It is known that most rotating machines, such as gas turbines, compressors, power plant equipment, and some chemical processing plant and pressure restraining systems, usually operate under conditions where high-frequency, small-amplitude vibrations are superimposed on low-frequency, high-stress amplitude cycles (i.e., combined cycle fatigue [CCF]). The application and investigation have shown that the high-cycle vibration fatigue (HCF) has a considerable effect on the behavior of low-cycle fatigue (LCF). Thus, in recent years, this subject has been studied with increasing intensity and certain methods for evaluating the effect on fatigue life have been proposed [*1–8*]. However, a study of the process and mechanism of the influence of HCF on LCF life at high temperature, which involves creep and environmental effects, have not yet been found. This paper is intended to explore the process and mechanism of this influence for a heat-resistant alloy at elevated temperature.

Experimental Procedure

The material used is a forged Fe-Ni based alloy, GH36. The chemical composition and mechanical properties of the alloy are listed in Tables 1 and 2. The specimens were heated at 1140°C for 80 min water quenching, and then at 670°C for 14 h and 780°C for 12 h air cooling.

[1]Institute of Aeronautical Materials, Beijing, China.

TABLE 1—*Chemical composition (weight percent) of test material.*

C	Mn	Si	Fe	Ni	Cr	Ti	Mo	Nb	V
0.37	8.5	0.6	bal.	8.3	12.5	0.10	1.24	0.35	1.35

TABLE 2—*Mechanical properties of test material at 600°C.*

E, MPa	σ_{ys}, MPa	σ_u, MPa	Elongation, %	Reduction of Area, %
136 000	522	661	16.1	45.6

The geometries of the specimens with single or double edge notches (SEN, DEN) are shown in Fig. 1. The tests are performed on a electromagnetic machine for DEN specimens and a servo-hydraulic combined fatigue machine for SEN specimens at 600°C. The specimens are heated by an electric heater and induction heating, respectively. The crack length is monitored continuously using an optical microscope with a magnification of 100 times, of which the resolution is about 0.01 mm.

The basic trapezoidal cycle waveform used in the tests is shown in Fig. 2. The frequency of the low-cycle load is 3 cpm and the dwell period is 14 s. The stress ratio is $R = 0.1$. The high-cycle vibrations are at a frequency of 30 Hz. The stress range ratio is $Q = {}^1\!/_2\Delta\sigma_2/\Delta\sigma_1$. The vibrations are plane bending in the DEN specimen tests and push-pull in the SEN specimen tests.

The cracks and fracture surfaces are observed and analyzed by means of stereo and optical microscopes and scanning and transmission electron microscopes in terms of micrography and fractography. The oxide on the fracture surfaces was taken off before observation.

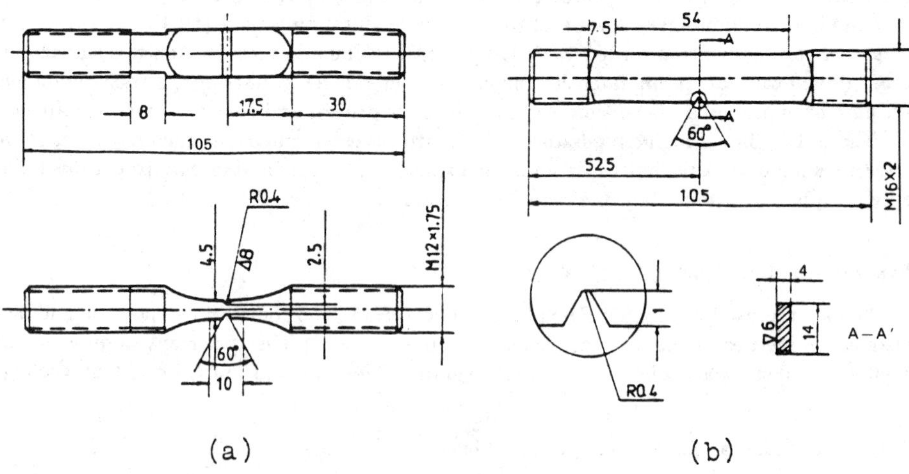

FIG. 1—*Geometry of specimens.* (a) *DEN.* (b) *SEN.*

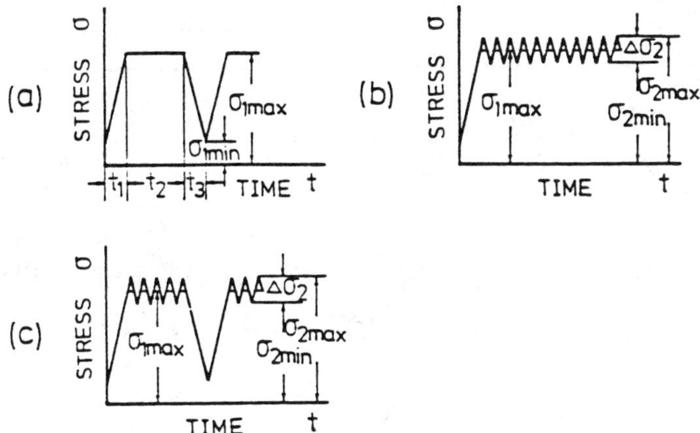

FIG. 2—*Load spectra.* (a) *Stress-hold LCF.* (b) *HCF.* (c) *CCF* ($Q = {}^1\!/_2\Delta\sigma_2/\Delta\sigma_1$).

Results

Fatigue Behavior

Figure 3 shows the total life versus LCF stress, σ_{1max}, curves for low-cycle fatigue, high-cycle vibration fatigue, and a combination of low- and high-cycle fatigue with different values of Q for the DEN specimens. The LCF life decreases with increasing stress and is evidently affected by the stress amplitude ratio, Q. The CCF life may decrease or increase at different values of Q under a certain maximum LCF stress. Furthermore, it is reduced with increasing the maximum low-cycle stress under a certain value of Q. For instance, as $Q = 0.07$, both total CCF and LCF lives are equivalent if the maximum LCF stress equals 630 MPa; when the LCF stress is less or greater than that, the CCF life appears to be higher or lower than the LCF life, respectively. On the other hand, as the maximum LCF stress equals 600 MPa, the CCF life is a little longer than the LCF life when $Q = 0.07$; it is much shorter than that when $Q = 0.13$ and markedly reduced when $Q = 0.18$. While there are relatively little data in Fig. 3 to support drawing separate $Q = 0$ and $Q = 0.07$ curves, there is crack propagation data which support this approach. These data will be presented and discussed later (at the end of the Discussion section).

In addition, the total HCF life is more than 6 times as long as the corresponding total LCF life ($Q = 0$) and more than 10 times as long as the total CCF life for $Q = 0.13$; the total HCF life is 3 times longer than the corresponding total LCF life ($Q = 0$) and 38 times longer than the total CCF life for $Q = 0.18$. It is quite clear that the reverse impact of LCF on HCF is very severe and that the greater the Q value the greater the influence. Here, for comparison, the total HCF lives are converted into equivalent LCF and CCF cycles according to the equation

$$N = T_2/t_1$$

where t_1 denotes the period of a cycle of LCF and CCF and T_2 infers the total lifetime of HCF.

Figure 4 illustrates the performance of LCF ($\sigma_{1max} = 350$ MPa) and CCF ($Q = 0.11$) for SEN specimens. The number of cycles for crack propagation (from 0.5 to 2.0 mm long) decreases

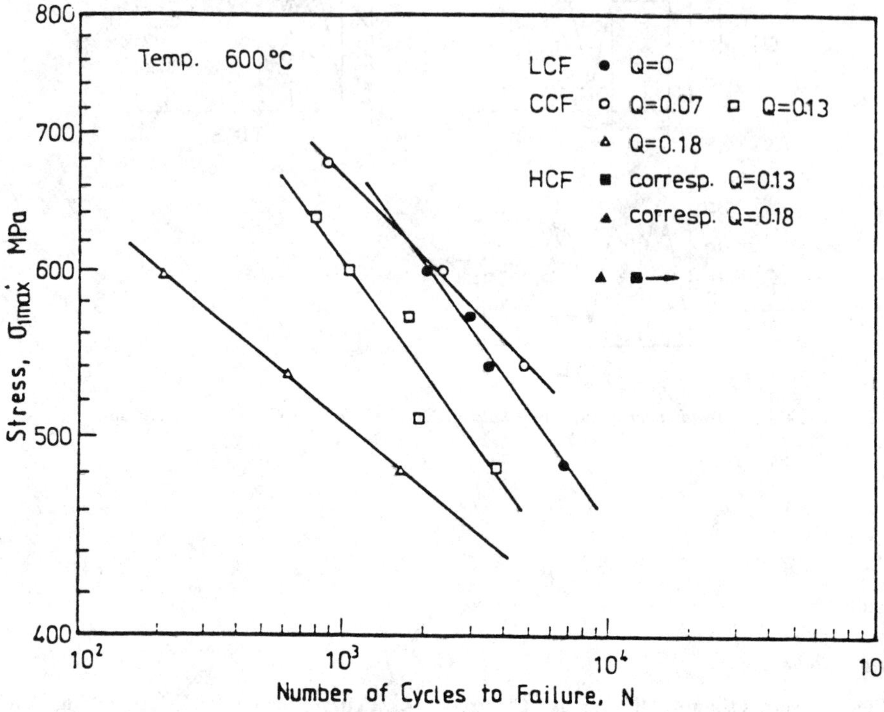

FIG. 3—*Fatigue life versus stress for stress-hold LCF, HCF, and CCF for DEN specimens.*

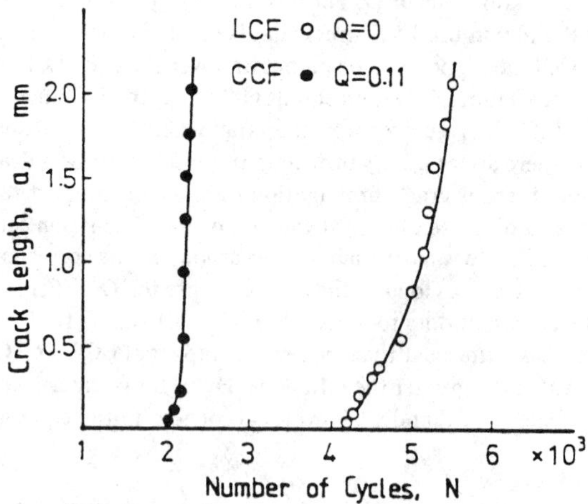

FIG. 4—*The* a-N *curves of stress-hold LCF (σ_{1max} = 350 MPa,* Q = 0*) and CCF (σ_{1max} = 350 MPa,* Q = *0.11) for SEN specimens at 600°C.*

sharply from 16% of the total cycles for LCF to 3% for CCF, although the number of cycles for crack initiation (a = 0.5 mm) and total cycles for CCF reduce only roughly to 40% of those for LCF, respectively. The crack length, a, is measured from the root of the notch to the crack tip on the SEN specimen.

Fatigue Crack Paths

The macro features of LCF and CCF cracks are shown in Fig. 5. The extent and area of the plastic deformation of the material on both sides of the LCF crack (Fig. 5*a*) are much greater than those of the CCF crack (Fig. 5*b*), and the extent of the crack opening of the former is much greater than that of the latter. Figure 6 provides the longitudinal profile of LCF (*a*) and CCF (*b*) specimens. The tortuous path of the LCF crack shows that the crack advances along the grain boundaries, while the CCF crack runs across the grains. Also, there exist a few intergranular crack branches associated with the LCF crack and an individual branch even grows simultaneously for a rather long distance. In addition, the oxidation of the LCF crack tip is much heavier.

FIG. 5—*Surface morphologies of* (a) *stress-hold LCF* (σ_{1max} = *350 MPa, Q = 0) and* (b) *CCF* (σ_{1max} = *350 MPa, Q = 0.11) cracks for SEN specimens.*

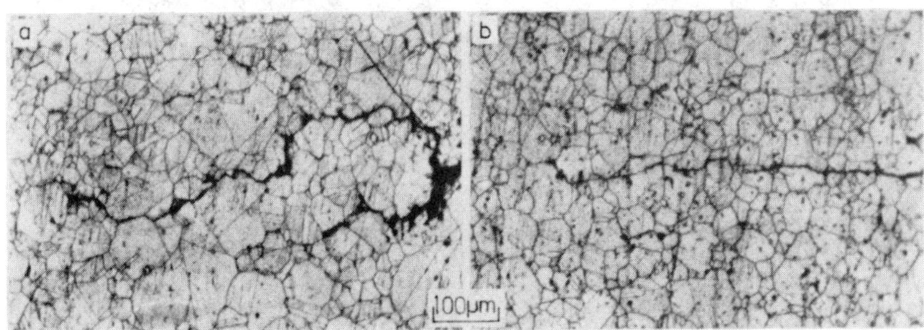

FIG. 6—*Longitudinal profile of* (a) *stress-hold LCF* (σ_{1max} = *350 MPa, Q = 0) and* (b) *CCF* (σ_{1max} = *350 MPa, Q = 0.11) cracks. Growth direction is from right to left.*

Fracture Surface Morphologies

The macroscopic appearances of LCF and CCF fracture surfaces are illustrated in Fig. 7. The LCF fracture surface is much rougher and the heavy brown surface indicates that oxidation is rather strong, while the CCF fracture surface is fairly even and the light yellow surface implies that it is less oxidized. On the other hand, the fracture portion of the LCF induced by the plane stress condition increased as the crack advanced, so that the crack front progressively appears to be tongue-like and the specimen becomes to be necking. The CCF crack front becomes less curved by degrees as the ratio Q increases, and the fracture has almost simply occurred under the plane strain state.

The micro-morphology of the LCF fracture surface illustrates that crack initiation and propagation are in a mixed type of intergranularity and transgranularity, with the intergranularity being dominant (Fig. 8*a*). Figure 8*b* is the aspect of the HCF fracture surface. Crack initiation is also a mixture of intergranularity and transgranularity, with the former dominating. It is interesting to note that some cracks originate at the subsurface grain boundaries, then extend across the grains in all directions, and that many subcracks initiated at the grain boundaries at some distance ahead of the main crack in the same way. The main crack advances through connection with them.

The features of CCF fracture surfaces with different values of Q are shown in Fig. 9. It is obvious that, as Q rises, the modes of crack initiation and propagation change from the mixed type into simple transgranularity. If the value of Q is high enough (e.g., greater than 0.13), the striations may be observed clearly even at the initial region. As Q increases further, the space between striations enhances steeply and the simple LCF striations will become combined ones, step by step, in which the wide-spaced, coarse ones correspond to the LCF stress and the narrow-spaced, fine ones between them correspond to the HCF stress.

Discussion

Characteristics of Stress-Hold LCF, HCF, and CCF at Elevated Temperature

Figure 10 shows the schemes of resolving of the stress-hold LCF (*a*), HCF (*b*), and CCF (*c*) spectra. The stress-hold LCF spectrum may be resolved into two parts: static creep load and

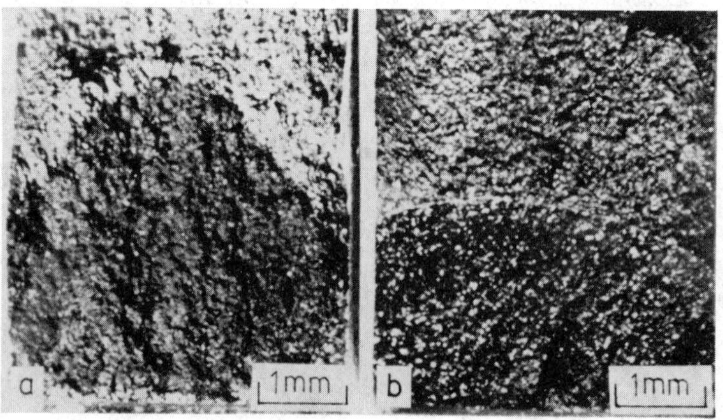

FIG. 7—*Macrofractographs of* (a) *stress-hold LCF* (σ_{1max} = *350 MPa,* Q = *0) and* (b) *CCF* (σ_{1max} = *350 MPa,* Q = *0.11).*

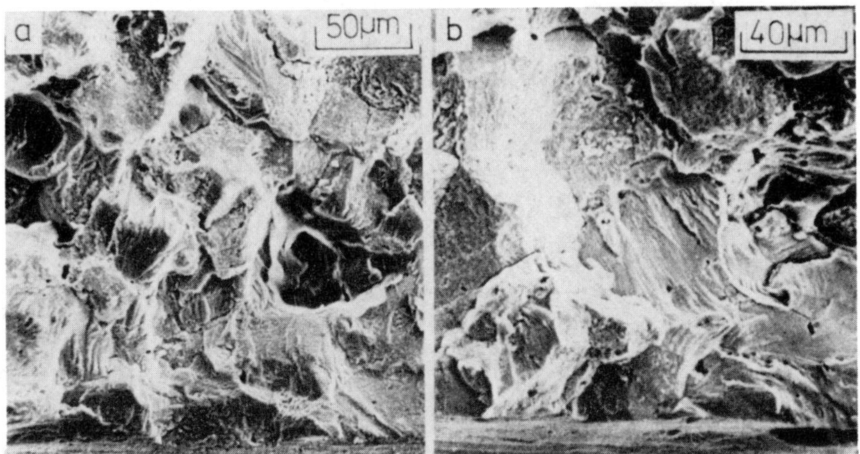

FIG. 8—*Fractographs of the initial region of* (a) *stress-hold LCF* (σ_{1max} = *600 MPa,* Q = *0) and* (b) *HCF* (σ_{1max} = *600 MPa, corresponding to Q = 0.18).*

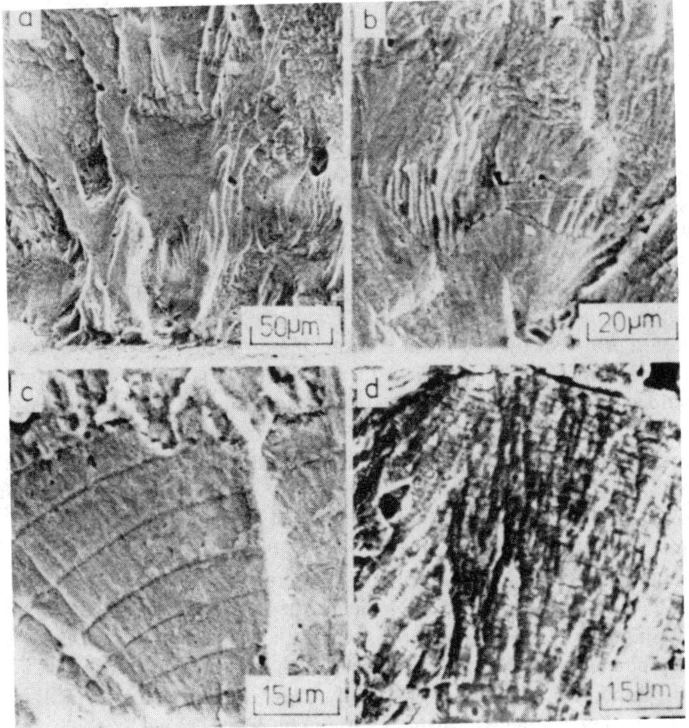

(*a*) $Q = 0.07$ ($\Delta K_{total} = 20$ MPa $\sqrt{m}$).
(*b*) $Q = 0.13$ ($\Delta K_{total} = 18$ MPa $\sqrt{m}$).
(*c*) $Q = 0.18$ ($\Delta K_{total} = 12$ MPa $\sqrt{m}$).
(*d*) $Q = 0.18$ ($\Delta K_{total} = 28$ MPa $\sqrt{m}$).

FIG. 9—*Fractographs for different values of* Q.

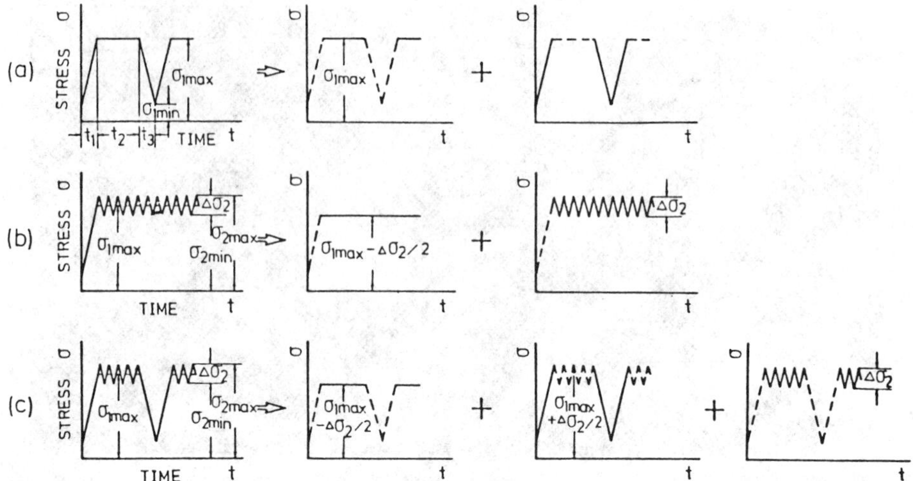

FIG. 10—*Schematic diagrams of resolving of the stress-hold LCF spectrum* (a), *HCF load* (b), *and combined spectra of LCF and HCF* (c).

LCF load. Both act alternatively. Consequently, creep deformation could not be fully developed and the damage would not be evident because the period of creep is intermitted by the LCF load. Furthermore, the reverted deformation or recovery will occur and the damage will be restored to a certain extent as the LCF load fell. As a result, the stress-hold LCF life may sometimes be slightly longer than the corresponding creep life [*9–12*], although creep will be accelerated slightly by compressive residual stresses ahead of the crack after unloading. On the other hand, the held stress would cause the LCF crack tip to deform sufficiently, hence blunting and opening more extensively. Thus the crack growth rate will increase slightly. This may be the main reason that the life of stress hold LCF is shorter than that of the continuous LCF [*13*].

The role of high temperature, apart from inducing creep damage, is to speed up crack initiation and extension primarily by intergranular cracking through oxidation, because oxygen may play a dual role; it might diffuse into a metal and harden the structure or it might, in some subtle way, reduce the equicohesive temperature by attacking the grain boundaries, which are the easiest diffusion paths [*14,15*]. The latter mechanism is dominant for stress-hold LCF.

From the aspects of fracture surface, which is a mixture of transgranularity and intergranularity with the latter dominating, it is clear that the damage of stress-hold LCF is the combined results of creep, LCF, oxidation, and the interactions between them.

The HCF load at high temperature may also be divided into static creep and HCF loads. As compared with the resolved loads of stress-hold LCF, the creep load is lower by half the amplitude of the HCF load (Figs. 10*a* and 10*b*); hence the extent of creep damage should be less. However, its acting period is continuous throughout the whole fatigue process; thus it may enhance the creep damage. On the other hand, although the amplitude of the HCF load is much smaller and the damage produced per cycle will be less, the frequency of HCF loading is very high, so the accumulated fatigue damage is rather considerable; again, the cyclic stress-strain rate of the material ahead of the crack tip is also quite high. The HCF load, therefore, apart from decreasing the creep damage to some degree by reducing the amplitude of the static creep load, will increase the fatigue damage and cause material hardening and embrittlement. As a result of the combined actions, the specimen failed in a mixed original type of transgranularity and intergranularity, with the latter dominating, and in a mixed propagation type, primarily with transgranular fracture (Figs. 8 and 9). This indicates that crack initiation is mainly time dependent and that crack propagation is primarily cycle dependent. The problem is why most original cracks initiate at grain boundaries in the subsurface and most microcracks nucleate again at grain boundaries at some distance ahead of the main crack and, then, both of them propagate transgranularly. It may be ascribed to the fact that only a few weaker grain boundaries, which are favorable in direction and suffer from higher stresses, would preferentially be damaged and cracked mainly under a long period of creep stress. As soon as the crack initiated it would propagate mainly by HCF stress. Thus both the creep and fatigue deformation are less.

In the case of high frequency loading the degree of oxidation of the material both at the specimen surface and at the crack tip is lighter, and its influence on crack initiation and propagation are also much less. This may be another reason that some cracks do not originate on the surface.

As shown in Fig. 10*c* the combined spectrum of LCF and HCF at high temperature may also be resolved into three parts. The mode of creep loading is discontinuous and identical to that in LCF, but its magnitude decreases by $\Delta\sigma_2/2$ and equals that in HCF. Hence the creep damage is less than that in single LCF or HCF. Secondly, the mode of LCF loading remains the same with that in single LCF, but the magnitude increases by $\Delta\sigma_2/2$; thus it will promote LCF damage. Thirdly, the mode of HCF loading changes from continuity into discontinuity and the total damage will be slightly reduced correspondingly. However, the damage of HCF is obviously enhanced by the off-load period of the raised LCF load, which is superimposed by HCF stress itself. Besides, both the LCF and HCF load will markedly suppress the creep and oxidation damage and promote cyclic hardening and embrittlement. Thus the nature of CCF fracture shifts from partially to fully transgranular as the Q value increases.

According to the above analysis, it is evident that not only the damage of creep, HCF, LCF, and environmental attack has to be separately considered, but it is probably more important that the interactions between them in the combined behavior of LCF and HCF at high temperature with creep and oxidation effects also have to be taken into account.

Effect of the Amplitude of HCF Stress on LCF Behavior

It is obvious that the stress amplitude of HCF has a complex influence upon the stress-hold LCF behavior. The greater the stress amplitude, the more serious the influence. For a certain value of Q at a fixed LCF maximum stress level, the increase of damage induced by HCF is approximately compensated for by the decrease of damage in creep and oxidation; thus there is no appreciable effect on the total life. This ratio Q is defined as a critical value, Q_c. As it decreases from Q_c the fatigue damage is markedly reduced, but the creep and oxidation damage increases slightly; thus the total effect is beneficial. As Q decreases to a certain value, the longest CCF life will be obtained. As Q continues to decrease, the fatigue damage will gradually be reduced and the damage of creep and oxidation will progressively be enhanced, then CCF life will return to that of simple LCF as Q decreases to zero. This regularity has been confirmed in our laboratory [*13*] in a study of crack propagation of CCF (Fig. 11) in which $Q = m/(1 - R)(1 - m)$ and $m = 0.06$ corresponds to $Q = 0.07$.

As the value of Q rises over Q_c, the fatigue damage increases significantly and the damage of creep and oxidation becomes less; the CCF life becomes more cycle dependent and will be obviously reduced. As Q is higher than a certain value where the creep and oxidation damage is approximately eliminated, the fatigue damage will be increased very fast; thus the CCF life appears to be fully cycle dependent and will be decreased sharply. As a result, creep-fatigue-oxidation failure will progressively change into pure fatigue—the combination of LCF and HCF.

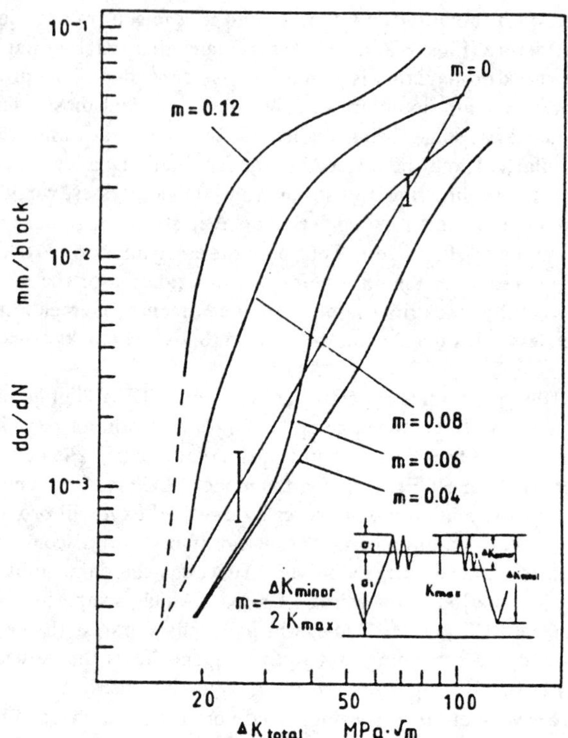

FIG. 11—*Comparison of LCF and CCF crack propagation behavior for different values of* m *at 600°C (after Ref* 13*).*

Conclusions

1. The effect of HCF on the behavior of stress-hold LCF at 600°C is complicated and significant. The nature and amplitude of the effect depend upon the ratio of the stress range of HCF to that of LCF, *Q*.

2. This influence mainly results from the fact that there exist obvious and complex interactions between creep, HCF, LCF, and oxidation at high temperature, which make the process and mechanism of material damage change from creep-fatigue-oxidation into nearly pure fatigue—combined HCF and LCF.

3. The roles of HCF in CCF behavior are that it can restrain the damage of creep and oxidation, promote damage of LCF, and cause material hardening and embrittlement, as well as induce damage by itself. Meantime, it was found that LCF stress has a very strong, reverse impact on HCF behavior.

4. The superimposed, higher amplitude of HCF stress is rather harmful at high temperatures for a component, because it may considerably reduce the total LCF life, especially the crack propagation life.

Acknowledgments

We wish to thank B. Tu, J. Yang, S. Zheng, K. Niu, and our other colleagues in the Fatigue Testing and Fractography groups for their effective help in experimental work.

References

[*1*] Duggan, T. V., "Fatigue Integrity Assessment," *International Journal of Fatigue*, Vol. 3, 1981, pp. 61–70.

[*2*] Powell, B. E., Duggan, T. V., and Jeal, R., "The Influence of Minor Cycle on Low-Cycle Fatigue Crack Propagation," *International Journal of Fatigue*, Vol. 4, 1982, pp. 4–14.

[*3*] Makhutov, N., Romanov, A., and Gadenin, M., "High Temperature Low-Cycle Fatigue Resistance Under Superimposed Stresses at Two Frequencies," *Fatigue of Engineering Materials and Structures*, Vol. 1, 1979, pp. 281–285.

[*4*] Takasugi, S., Horikawa, T., and Nakamura, H., "Results of Fatigue Tests and Prediction of Fatigue Life Under Superposed Stress Wave and Combined Superposed Stress Wave (Case of SCM 435 and 2 1/4 Cr-1Mo Steel, Room Temperature and 500°C)," *Journal of the Society of Materials Science*, Japan, Vol. 32, 1983, p. 471.

[*5*] Troshohenko, V. I. and Zaslotskaya, L. A., "Fatigue Strength of Superalloys Subjected to Combined Mechanical and Thermal Loading," in *Proceedings*, Third International Conference on Mechanical Behavior of Materials, Cambridge, England, Vol. 2, 1979, pp. 3–12.

[*6*) Coles, A., "Material Considerations for Gas Turbine Engines," in *Proceedings*, Third International Conference on Mechanical Behavior of Materials, Cambridge, England, Vol. 1, 1979, pp. 3–11.

[*7*] Serensen, S., Makhutov, N., and Romanov, A., "Crack Propagation at High Temperature Low-Cycle Loading," in *Proceedings*, Fourth International Conference on Fracture, Vol. 2B, 1977, pp. 785–790.

[*8*] Padran, J. P. and Pinean, A., "The Effect of Microstructure and Environment on the Crack Growth Behavior of Inconel 718 Alloy at 650°C under Fatigue Creep and Combined Loading," *Materials Science and Engineering*, Vol. 56, 1982, pp. 143–156.

[*9*] Atanmo, P. N. and McEvily, A. J., "On Creep Fatigue Interaction During Crack Growth," AD 753 688, Oct. 1972.

[*10*] Koterazawa, R., "A Static-to-Dynamic Transition in Creep of Metallic Materials under Cyclic Stress Conditions as Caused by an Interaction of Creep and Creep Recovery," in *Proceedings*, International Conference on Mechanical Behavior of Materials, Kyoto, Japan, Vol. 3, 1972, p. 135.

[*11*] Koterazawa, R. and Shimohata, T., "A Study on Creep Under Combined Creep-Fatigue Conditions with Special Reference to Static-to-Dynamic Transition," in *Proceedings*, International Conference on Creep and Fatigue in Elevated Temperature Applications, Philadelphia, Sept. 1973, and Sheffield, England, April 1974, C214/73. pp. 214.1–214.10.

[*12*] Zhou, H., "Interaction of Creep and Fatigue and Fracture Analysis of Some Metallic Materials," Technical Report 80-007, Institute of Aeronautical Materials, Beijing, 1980.

[*13*] Niu, K., Tu, B., and Yan, M., "A Study of Low- and High-Cycle Combined Fatigue Crack Propagation for a Superalloy GH36," *Acta Aeronautica et Astronautica Sinica*, Vol. 7, No. 3, 1986, pp. 273–280.

[*14*] Forsyth, P. J. E., "The Physical Basis of Metal Fatigue," Blackie & Son, London, 1969, pp. 121–126.

[*15*] Sidey, D. and Coffin, L. F., Jr., "Low-Cycle Fatigue Damage Mechanisms at High Temperature," in *Fatigue Mechanisms, ASTM STP 675*, American Society for Testing and Materials, Philadelphia, 1979, pp. 528–568.

Tokihiko Mori,[1] *Hiroshi Toyoda,*[1] *and Sadao Ohta*[1]

Fatigue Life and Short Crack Propagation of Turbine Rotor Steels under Variable Strain Loading

REFERENCE: Mori, T., Toyoda, H., and Ohta, S., **"Fatigue Life and Short Crack Propagation of Turbine Rotor Steels under Variable Strain Loading,"** *Low Cycle Fatigue, ASTM STP 942,* H. D. Solomon, G. R. Halford, L. R. Kaisand, and B. N. Leis, Eds., American Society for Testing and Materials, Philadelphia, 1988, pp. 972–983.

ABSTRACT: Fatigue behavior of steam turbine rotor steels with and without defects has been investigated at 550°C with emphasis on the effects of service loadings. It is shown that the life of defect-free materials is reduced remarkably by a factor of 10 to 100 due to the accumulation of the plastic strain range, which becomes conspicuous under a programmed fatigue test which simulates service loadings. A linear damage cumulative law in terms of the plastic strain range to estimate the fatigue life is presented. It is also shown that in order to evaluate the effect of defects, the conventional low-cycle fatigue test is not appropriate but a variable strain test is necessary. The fatigue life of the material with a defect is discussed in light of crack initiation and growth periods.

KEY WORDS: fatigue (materials), high temperature fatigue, service loadings, turbine rotor steels, 1Cr-Mo-V steels, 12Cr steels, stress-strain diagrams, crack propagation, short-crack, defects, inclusions, life prediction

Recent changes in industry and lifestyle have caused a greater difference in the amount of electricity consumed in the day and late at night. To satisfy peak demand during the day time, some combustion-type generators have been put into use. This type of generator does not only have to start up and stop every day but also has to start up quickly, so that the materials exposed to high temperature are required to have even better thermal fatigue properties.

Figure 1 shows an estimated total strain of such a turbine rotor at the time of startup [*1*]. In this example, a large circumferential crack was detected at 1/60 of the predicted life. The prediction was made by using the conventional linear damage cumulative law on creep and fatigue damages, which is recommended by ASME Code Case N-47.

Although the strain appearing on this turbine rotor varies (Fig. 1), few studies have been made on the influence of variable strains [*2*], while constant strain low-cycle fatigue at high temperature, thermal fatigue, and fatigue creep interaction have been the main subjects of earlier studies. It is well known that a remarkable reduction in fatigue life can take place under variable strains. These strain patterns and examples of life reduction (Fig. 1) highlight the need for studies on the influence of variable strains on turbine rotors.

This study was intended to elucidate the predominant mechanisms of life reduction under variable strains on turbine rotor steels and to propose a method for life prediction based on the results of the study.

[1]Technical Development Group, Kobe Steel, Ltd., Kobe, Japan.

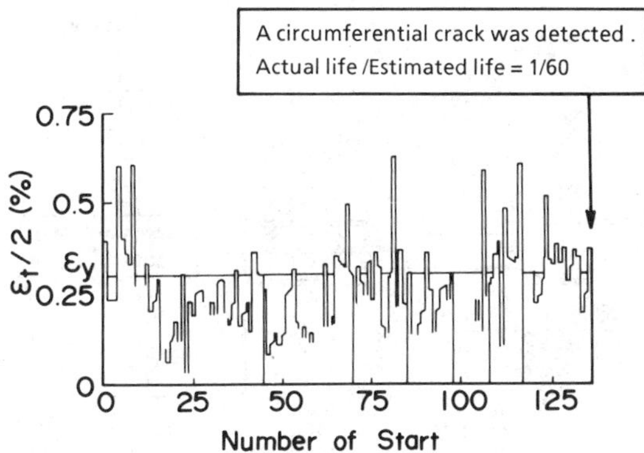

FIG. 1—*Estimated total strains of a turbine rotor at start-up.*

Experimental Procedure

The materials for this investigation were 1Cr-1Mo-1/4V steels (ASTM A470, Class 8) and a 12Cr-Mo-V-Nb turbine rotor steel. The chemical compositions are shown in Table 1. One of the 1-Cr steels, I, was made to contain large inclusions (ϕ300 ~ 500 μm) in a specimen.

A round bar specimen was used for the 1Cr-steel and 12Cr-steel tests; a center-notched specimen was used for the latter only (Fig. 2). A micro-drill hole was made on the 12Cr steel round bar specimens as an artificial defect.

The specimen was heated by means of an induction furnace, and strain control push-and-pull fatigue tests were carried out at 550°C. Crack length was measured either with a traveling microscope or using a compliance method.

Multiple two-step strain (MTSS) tests were employed as well as constant strain range low-cycle fatigue tests (Fig. 3). The large strains in the MTSS tests are referred to as the *primary strains* and the small ones as the *secondary strains.* One block is defined as consisting of one primary strain and many secondary strains. The strain rate of the triangular primary strain is 0.1%/s, and the frequency of the sinusoidal secondary strain is 1 Hz. The number of the secondary strains in one block, n_2, were 20, 300, and 600. As for the secondary strains, a strain range that would give a life of more than 10^7 cycles if tested at only that strain range was employed. Symbols concerning stress and strain are given in Fig. 4.

TABLE 1—*Chemical composition and heat treatment of test materials.*[a]

	C	Si	Mn	P	S	Ni	Cr	Mo	V	Nb
1Cr-S	0.30	0.24	0.72	0.002	0.005	0.40	1.08	1.15	0.23	...
1Cr-I	0.28	0.32	0.77	0.007	0.003	0.39	1.13	1.14	0.25	...
12Cr	0.20	0.27	0.73	0.010	0.003	0.52	10.30	0.91	0.20	0.039

[a] 1Cr: 970°C at 1 h, A.C.; 710°C at 4 h, F.C.
12Cr: 1080°C at 1 h, A.C.; 680°C at 4 h, F.C.

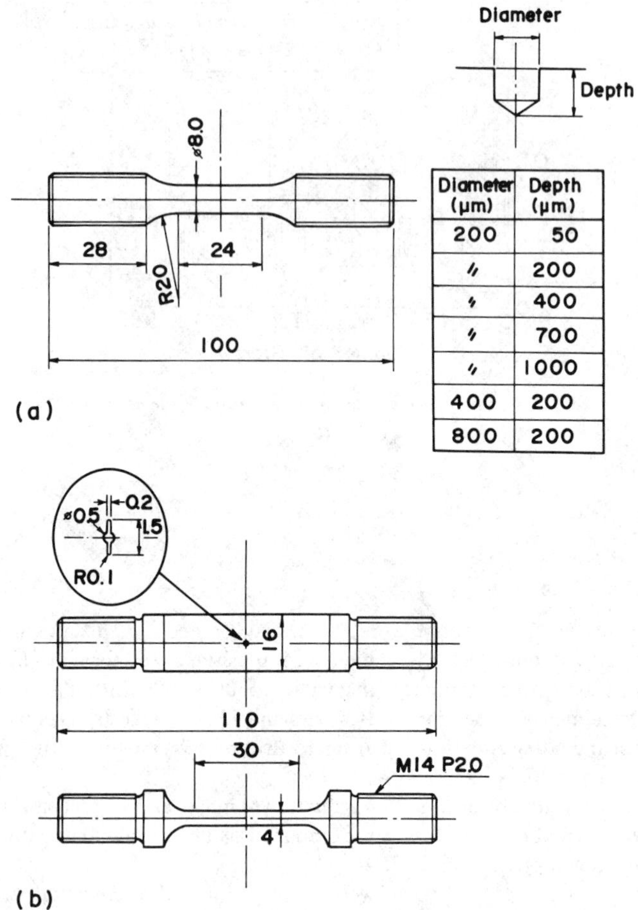

Diameter (μm)	Depth (μm)
200	50
〃	200
〃	400
〃	700
〃	1000
400	200
800	200

FIG. 2—*Test specimens.*

Results

Effects of Variable Strain Ranges (Defect-Free Material)

The fatigue life for a 1Cr turbine rotor steel plotted against the total strain range is shown in Fig. 5. For the MTSS tests, the relation between the primary strain range ($\Delta\epsilon_{1t}$) and the number of strain cycles to failure (n_{1f}), is plotted in the figure. Under variable strain conditions, a remarkable reduction in life (by a factor of 10 to 100) took place compared with the results of the constant strain range tests.

The same data are replotted in a relationship between the plastic strain range and an equivalent number of cycles to failure (N^*_{1f}) in Fig. 6. The equivalent number of cycles to failure is defined as

$$N^*_{1f} = \{1 + (\Delta\epsilon_{2p}/\Delta\epsilon_{1p})^{1/\beta}(n_2/n_1)\}n_{1f} \qquad (1)$$

where n_1, n_2, $\Delta\epsilon_{1p}$, and $\Delta\epsilon_{2p}$ are defined as in Figs. 3 and 4.

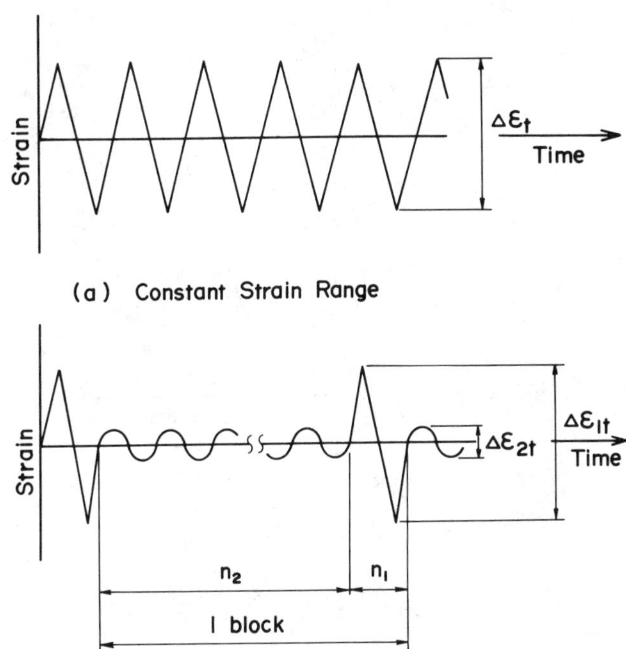

FIG. 3—*Programmed strain patterns.* (a) *Constant strain range.* (a) *Multiple two-step strain.*

With an assumption that fatigue damage based on $\Delta\epsilon_p$ can be linearly accumulated [*3*], Eq 1 was derived from the strain-life fatigue law:

$$\Delta\epsilon_p N_{1f}^{\beta} = C \tag{2}$$

The derivation is given in the Appendix.

The results of the MTSS tests agreed fairly well with those of the constant strain range tests in Fig. 6.

Effects of Variable Strain Ranges (Materials with Defects)

1Cr-1Mo-1/4V Steel with Inclusions (Denoted as I)—A fractographic study revealed that cracks tended to initiate from the large inclusions near the specimen surface. Under variable strain conditions, the life tended to be shorter than that predicted by the above method (Fig. 7). The degree of the reduction, however, is limited to one half of the prediction. As seen by comparing the results of Figs. 6 and 7, little reduction in life due to the inclusions was observed in the constant strain test.

12Cr-Mo-V-Nb Steel with a Micro-Drill Hole—The equivalent number of cycles to failure of the 12Cr steel specimen with a micro-drill hole is plotted against the plastic strain range in Fig. 8 together with the results of the defect-free specimen (shown as a line). Again, for this material the specimen with a defect shows little reduction in life under constant strain, whereas a larger reduction (1/2 ~ 1/5) took place under variable strain conditions. The results are replotted in

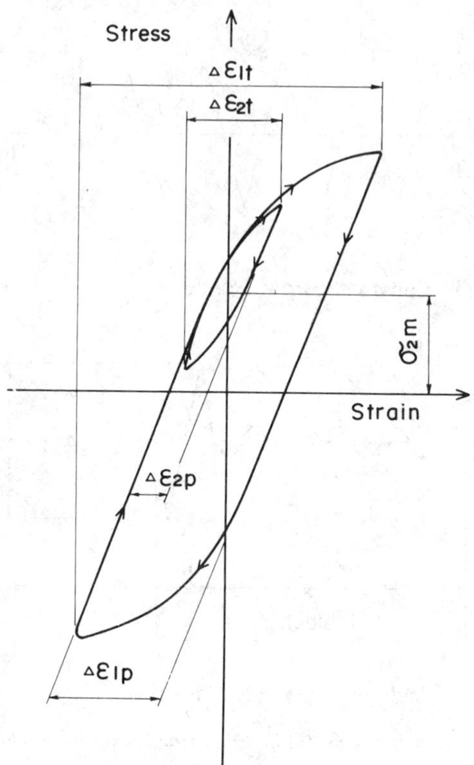

FIG. 4—*Schematic illustration of stress-strain curve.*

Fig. 9 as a relation between the fatigue life normalized by the life of the defect-free specimen tested under the same strain condition and the defect size (depth of defect[2]). With a micro-hole of about 100 μm or more in depth, the life was reduced to 30 to 40% of the defect-free specimen, whereas no reduction appeared under constant strain with the same notch size.

The crack growth rates of the round bar specimens and of the center crack plate (CCP) specimens are plotted against the cyclic J-integral (ΔJ) in Fig. 10. The cyclic J-integral was estimated experimentally in the same way as by Dowling [*4*]. All crack growth rates agreed on this cyclic J diagram regardless of crack size (100 μm to 10 mm), crack shape (through-thickness and part-through crack), and strain patterns. Plotted against the linear elastic stress intensity factor range (ΔK) the crack growth rate of CCP and that of a round bar specimen under variable strain coincided, whereas a round bar specimen under constant strain conditions exhibited a 5 to 10 times higher crack growth rate.

Discussion

Mechanisms of Life Reduction under Variable Strains

Defect-Free Material—Fatigue life was remarkably reduced under the variable strain condition. Furthermore, an increased plastic strain range in the secondary strains was measured after

[2]A similar trend appeared when the diameter of the defect was used.

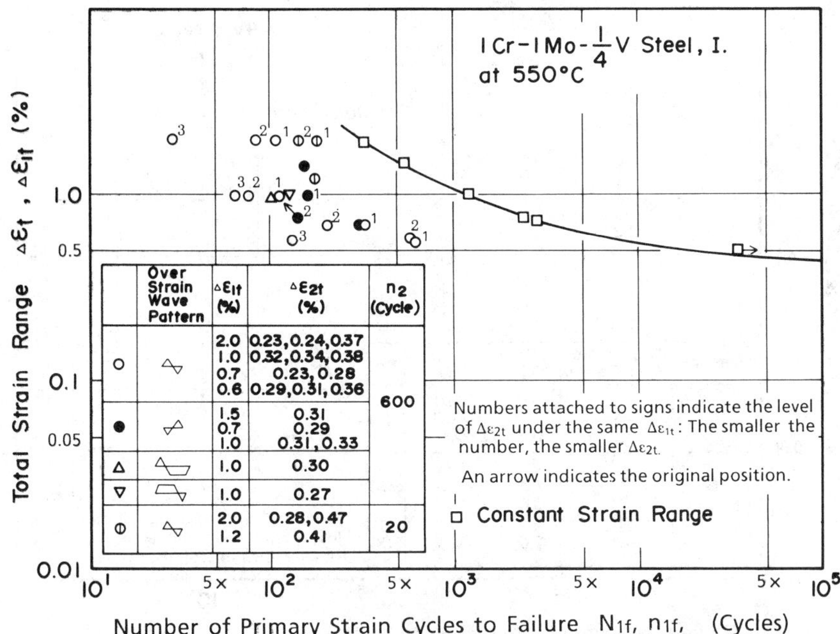

FIG. 5—*Comparison of numbers of cycles to failure of 1Cr-Mo-V steel-S under constant strains and under variable strains as a function of total strain range* ($\Delta\epsilon_t$, $\Delta\epsilon_{1t}$).

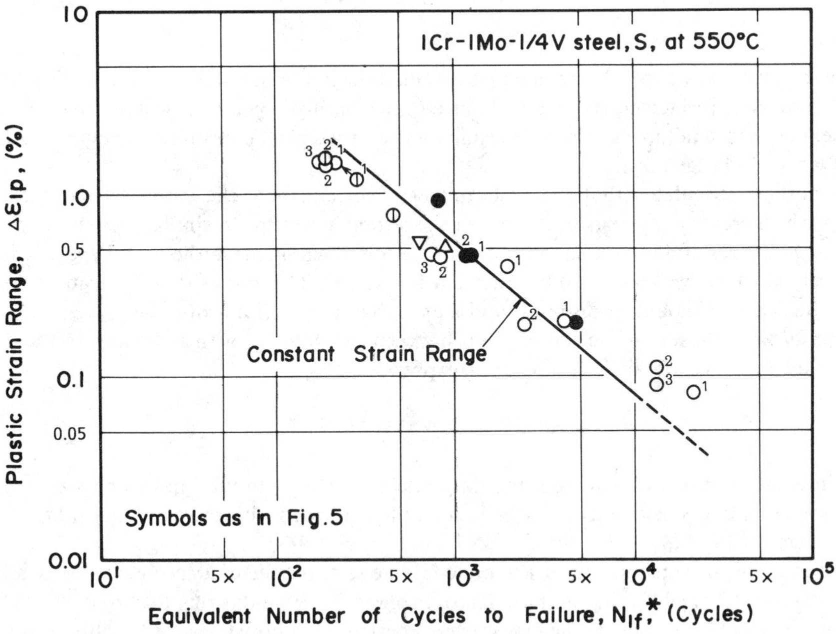

FIG. 6—*Comparison of equivalent number of cycles to failure* (N^*_{if}) *of 1Cr-Mo-V steel-S under constant strains and under variable strains as a function of plastic strain range* ($\Delta\epsilon_{1p}$).

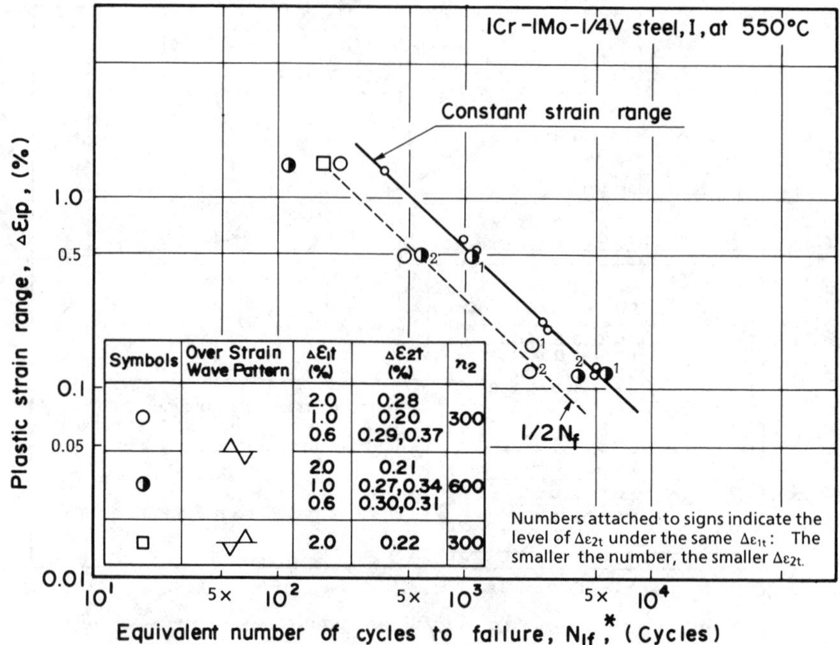

FIG. 7—*Comparison of equivalent number of cycles to failure* (N_{1f}^*) *of 1Cr-Mo-V steel-I under constant strains and under variable strains as a function of plastic strain range* ($\Delta\epsilon_{1p}$).

a primary strain was applied, indicating the reduction in life due to the accumulation of the plastic strains of the secondary strain. Figure 6 proves this hypothesis, where the equivalent number of cycles to failure estimated by using the measured plastic strain ranges agrees with the life of low-cycle fatigue test.

The authors speculated that the Bauschinger effect and/or cyclic strain softening should result in the increase of the plastic element of the secondary strains. A similar increase in plastic strains and the resulting reduction in life were also observed by the authors on Type 304 stainless steel, which shows conspicuous strain hardening [5]. This observation indicates that the Bauschinger effect should be of more significance than cyclic strain softening.

Material with Defects—The relationship between the crack growth rate (da/dN) and the cyclic J-integral (ΔJ) obtained in Fig. 10 is expressed as

$$da/dN = 1.3 \times 10^{-1} \Delta J^{1.18} \tag{3}$$

By integrating this equation from the initial defect size (a_1) to the final crack size (a_2) the crack growth life was calculated. The estimated life coincided with the actual total life within the accuracy of Eq 3 [5].

In the case of the constant strain low-cycle fatigue test, the life of a specimen with a micronotch was about 70 to 90% that of the defect-free ones (Fig. 9), indicating that 70 to 90% of the total life of defect-free material is a crack growth period. In contrast, under variable strain, the life of a specimen with a defect was about 30 to 40% of that of a defect-free one, so that the

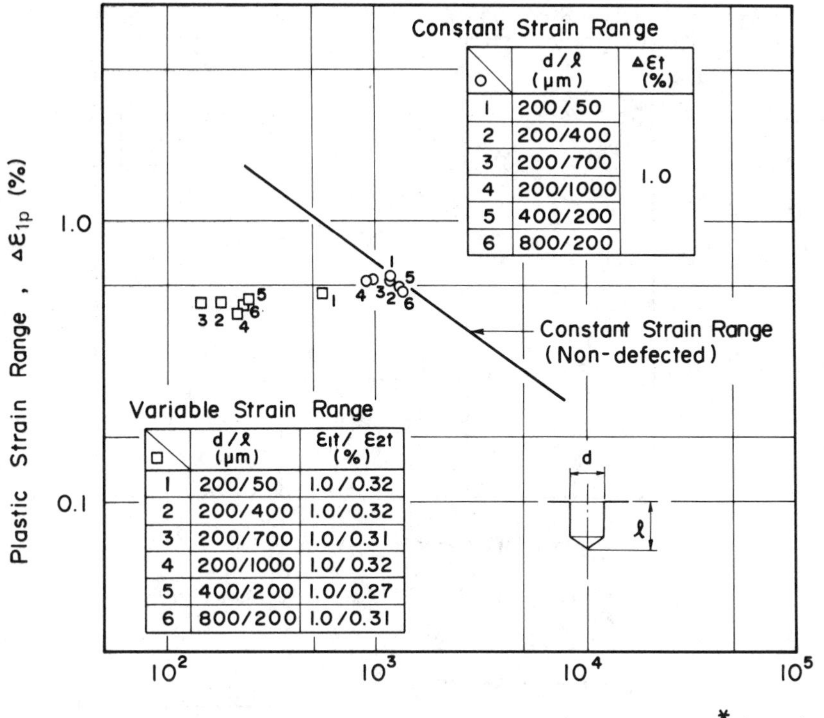

FIG. 8—*Comparison of equivalent number of cycles to failure* (N_{1f}^*) *of 12Cr steel under constant strains and under variable strains as a function of plastic strain range* ($\Delta\epsilon_{1p}$).

remaining 60 to 70% of the life should be regarded as crack initiation and micro-crack ($a < a_1$) growth periods.

Under variable straining, fatigue damage was predominantly made by secondary strains.[3] Since the secondary strains were small, most of the life was consumed by crack initiation and micro-crack growth, as in the case of high-cycle fatigue, since no strain concentrators or defects existed. The authors speculated that, under these conditions, fatigue damage was induced by macroscopic plastic strain. Thus the damage rule based on macroscopic plastic strains (Eq 1) proved to be effective in predicting the life (Fig. 6). For materials with defects, however, the crack initiation and micro-crack growth periods were reduced and a period of crack growth, which was controlled by the linear elastic stress intensity factor, namely the total strain range rather than plastic strain, dominated the total life. The damage rule based on plastic strain was therefore invalid for this case (Figs. 7 and 8).

Despite the fact that most of the life was dominated by crack growth, the strain-life fatigue law is valid under the constant strain range. The authors speculate that this is because the crack growth is accompanied by large plastic deformation, where the same formula as the strain-life fatigue law can be derived from the $da/dN - \Delta J$ relation [*6*].

[3]Beach marks on the fracture surface revealed cracks propagated predominantly by small strains.

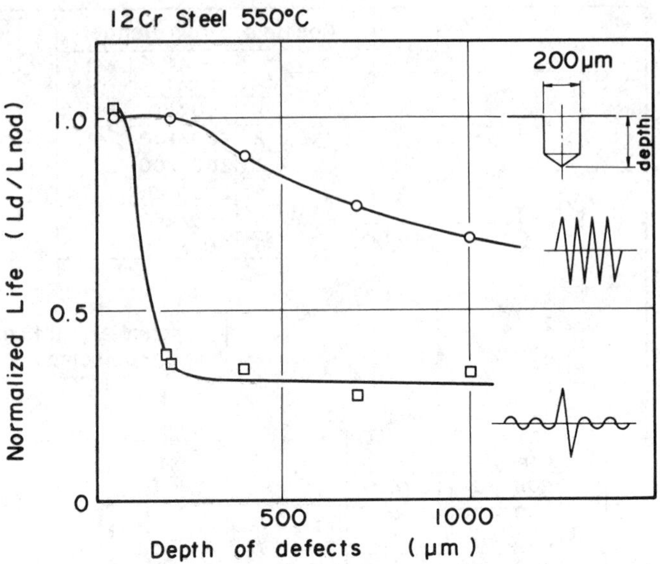

FIG. 9—*Effect of a defect size on the fatigue life of 12Cr steel under constant strains and under variable strains.*

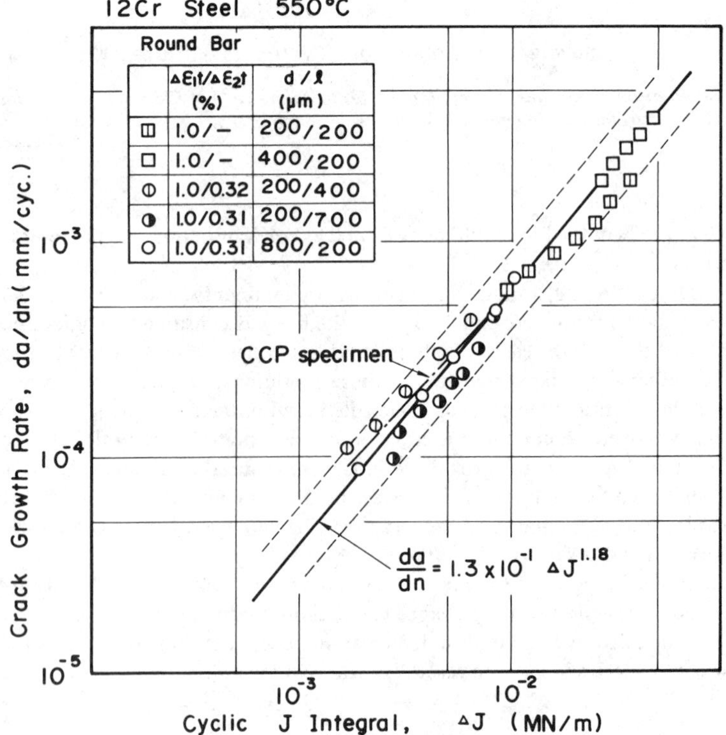

FIG. 10—*Comparison of crack growth rates of a center cracked panel and of a round bar specimen in 12Cr steel as a function of the cyclic J-integral.*

A New Life Prediction under Variable Strains

A fatigue damage monitoring system for real turbine rotors can be developed based on this result. A schematic illustration of the system is given in Fig. 11, where the plastic strain range ($\Delta\epsilon_{2p}$) increased by strain variation can be estimated by

$$\Delta\epsilon_{2p} = A\Delta\epsilon_{2t} + f(\Delta\epsilon_{1t}, n_2) \quad (4)$$

where A is a constant and $f(\Delta\epsilon_{1t}, n_2)$ is an increasing function in terms of $\Delta\epsilon_{1t}$ and a decreasing one in terms of n_2. This relationship was derived experimentally; an example obtained in this test is given in Fig. 12.

In the case of severe defects or strain concentrators being unavoidable, fracture mechanics is effective in predicting the life as presented.

Future Work

This study suggests the need for research concerning:

1. Interaction between variable straining and creep.
2. Fatigue behavior under real operation (random straining) at elevated temperatures.
3. Establishment of life prediction method under variable straining.

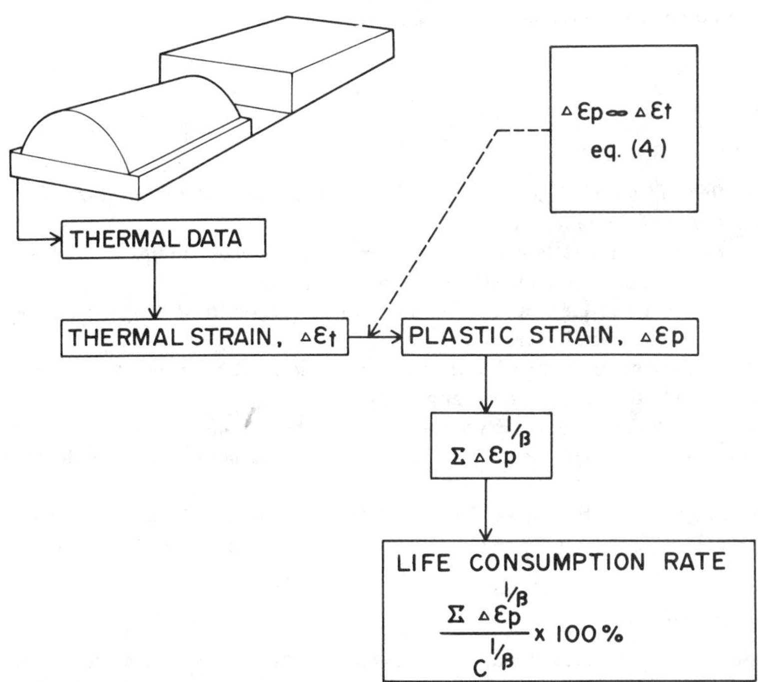

FIG. 11—*Schematic illustration of fatigue damage monitoring system for a turbine rotor.*

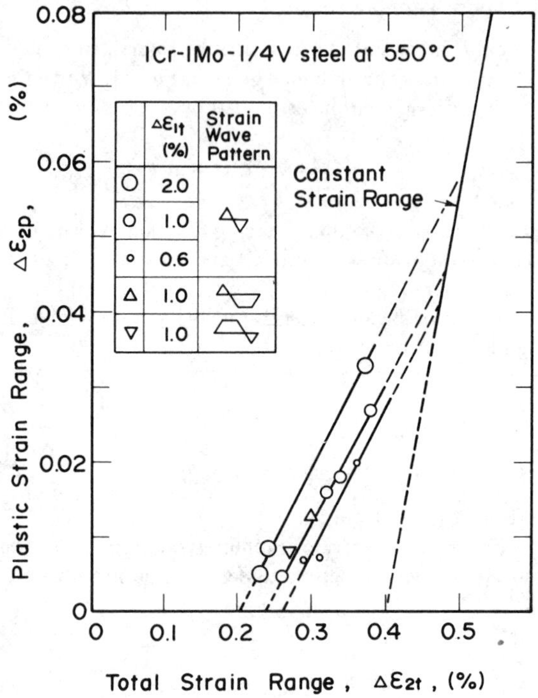

FIG. 12—*Relationship between plastic strain range* ($\Delta\epsilon_{2p}$) *of 1Cr-Mo-V steel-S and total strain range* ($\Delta\epsilon_{2t}$) *under variable straining.*

Conclusions

1. The effects of service loadings on the fatigue behavior of steam turbine rotor steels with and without defects were investigated at 550°C.
2. Under a programmed fatigue test that simulates service loadings the life of defect-free materials was remarkably reduced (by a factor of 10 to 100).
3. The reduction was due to an accumulation of the plastic strain range, which became conspicuous as a result of strain variation.
4. A linear damage cumulative law in terms of plastic strain range was presented that proved to be applicable to the estimation of fatigue life.
5. The defects had little effect in reducing the life when a constant strain range was applied. Under a variable strain, however, life reduction was even larger than that of defect-free material.
6. The fatigue life of the material with a defect was successfully estimated by short crack growth behavior expressed in terms of the cyclic *J*-integral, regardless of the straining patterns.

Acknowledgments

We wish to thank Dr. M. Kikukawa, Professor Emeritus, Osaka University, for his valuable discussions.

APPENDIX

Derivation of a Linear Damage Cumulative Law in Terms of Plastic Strain

The strain-life fatigue law (Eq 2) is reduced to the form

$$(\Delta\epsilon_p/C)^{1/\beta}N_f = 1 \tag{5}$$

Assuming that this equation is applicable to general cases where the plastic strain range varies, Eq 5 is expressed as

$$\sum_{i=1}^{N_f} (\Delta\epsilon_{pi}/C)^{1/\beta} = 1 \tag{6}$$

where $\Delta\epsilon_{pi}$ is the plastic strain range of the ith strain cycle.

In the case of this test, plastic strain ranges can be reduced to only $\Delta\epsilon_{p1}$ and $\Delta\epsilon_{p2}$. Thus Eq 6 gives

$$(\Delta\epsilon_{1p}/C)^{1/\beta}\, n_{1f} + (\Delta\epsilon_{2p}/C)^{1/\beta}\, n_{2f} = 1 \tag{7}$$

where n_{1f} and n_{2f} are the numbers of the primary and the secondary strain cycles to failure, respectively.

Defining N^*_{1f} as in Eq 1, Eq 7 is reduced to

$$\Delta\epsilon_{1p}N^{*\beta}_{1f} = C \tag{8}$$

References

[1] Kuwahara, K. and Nitta, A., "Estimation of Thermal Fatigue Damage of Steam Turbine Rotor," CRIEPI Report E277001, July 1977.

[2] Araki, R., Kadoya, Y., Yamamoto, S., Endo, T., and Fujii, H., "Low-Cycle Fatigue of High Pressure Steam Turbine Rotor," Mitsubishi Juko Giho, Vol. 10, No. 4, 1973, pp. 465–474.

[3] Kikukawa, M., Ohji, K., Kamata, T., and Johno, M., "The Cumulative Fatigue Damage under Axial Stationary Random Loading," *Transactions of the Japan Society of Mechanical Engineers,* Vol. 35, No. 278, 1969, pp. 2020–2031.

[4] Dowling, N. E., *Cyclic Stress-Strain and Plastic Deformation Aspects of Fatigue Crack Growth, ASTM STP 637,* American Society for Testing and Materials, Philadelphia, 1977, pp. 97–121.

[5] Toyoda, H. and Mori, T., "High Temperature Fatigue Properties of a Turbine Rotor Steel under Variable Strains," Kobe Steel Central Research Laboratory Report 3835, 1984.

[6] Yamada, T., Hoshide, T., Fujimura, S., and Manabe, M., "Investigation of the Fatigue Life Prediction Based on Analysis of Surface-Crack Growth in Low-Cycle Fatigue of Smooth Specimen of Medium Carbon Steel," *Transactions of the Japan Society of Mechanical Engineers,* Vol. 49, No. 440, 1983, pp. 441–451.

Notches

H. Nowack,[1] W. Ott,[1] J. Foth,[2] H. Peeken,[3] and T. Seeger[4]

Some Contributions to the Further Development of Low Cycle Fatigue Life Analysis Concepts for Notched Components under Variable Amplitude Loading

REFERENCE: Nowack, H., Ott, W., Foth, J., Peeken, H., and Seeger, T., **"Some Contributions to the Further Development of Low Cycle Fatigue Life Analysis Concepts for Notched Components under Variable Amplitude Loading,"** *Low Cycle Fatigue, ASTM STP 942,* H. D. Solomon, G. R. Halford, L. R. Kaisand, and B. N. Leis, Eds., American Society for Testing and Materials, Philadelphia, 1988, pp. 987–1006.

ABSTRACT: In order to minimize the expenses of the development of engineering components and to reduce the risk of their operation, reliable fatigue life prediction methods have to be available in the design stage. Presently two groups of methods can be distinguished which explicitly account for the local stresses/strains in the fatigue critical areas: the so-called notch analysis concepts or "local approaches," and new developments which are based on finite element methods (FEM).

In the first part of the paper typical areas in application of the notch analysis methods and some aspects of their further development are described. The significance of the short crack stage is briefly characterized from a physical and an engineering point of view.

In the second part the present state in the development of a new three-dimensional elastic-plastic FEM-based fatigue analysis concept, FEMFAT, is described. FEMFAT uses a material model according to Mróz's proposal, and the fatigue damage is calculated on the basis of the plastic work as performed at the fatigue critical locations of components. A calculation scheme for the damage increments under arbitrary out-of-phase loading based on uniaxial data is proposed. Two examples of the application of FEMFAT are given.

KEY WORDS: fatigue life prediction, constant and variable amplitude loading, notch analysis concepts, multiaxial stress states, FEM-based fatigue analysis

The fatigue process at the notch (surface) of specimens or components which are loaded in the limited life range usually covers four essential stages [*1*]. These are shown as an example for a notched Al 2024-T3 specimen in Fig. 1 [*2,3*]. In the first stage (I) small cracks are initiated that reach a size of 10 to 50 μm. Generally one of the three microscopical mechanisms is observed, which are indicated in the lower part of Fig. 1. The initiation process is significantly influenced by the specific thermomechanical treatment, which is applied to the material or to the test piece [*4*]. However, it has to be considered that such treatments which improve life

[1]Head and Scientist, respectively, DFVLR, Institute of Materials Research, Fatigue Department, D-5000 Köln 90, FRG.

[2]Project Engineer, IABG, Fatigue and Fracture Mechanics Division, D-8012 Ottobrunn, FRG.

[3]Professor, Technical University of Aachen, Institute for Machine Construction, D-5100 Aachen, FRG.

[4]Professor, Technical University of Darmstadt, Institute of Steel Constructions and Material Mechanics, D-6100 Darmstadt, FRG.

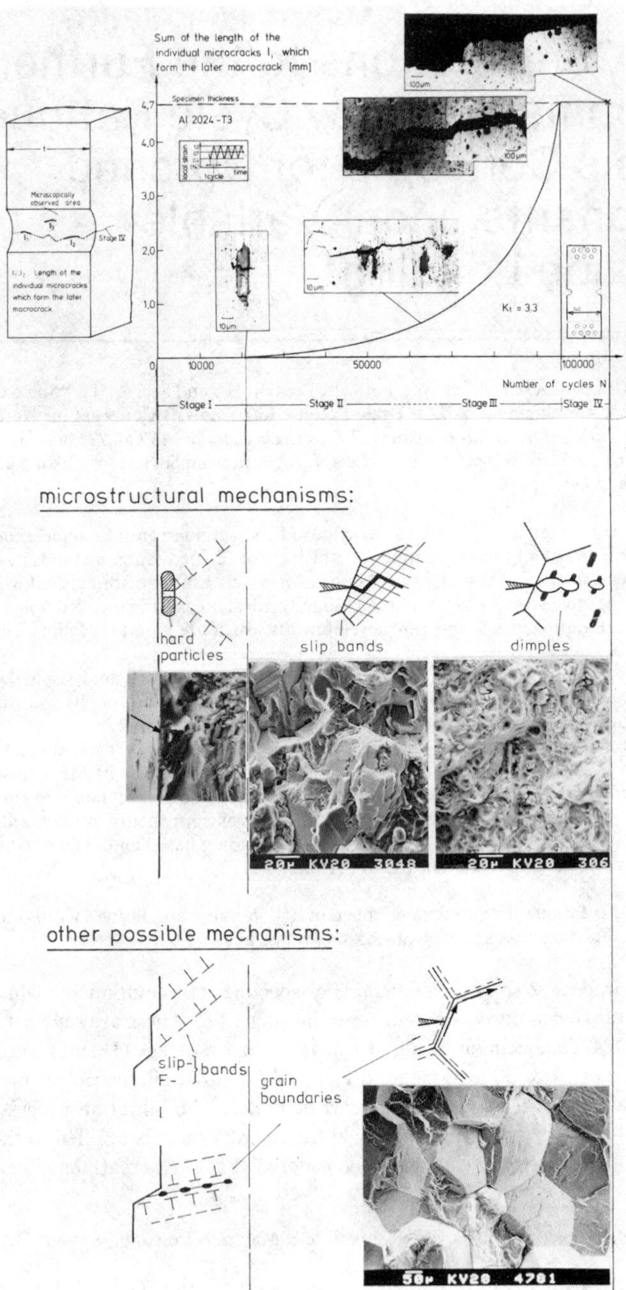

FIG. 1—*Individual stages of the fatigue process at the notch of Al 2024-T3 specimens and typical microscopical mechanisms.*

behavior in Stage I have very often negative effects on the subsequent stages. In the limited life range usually more than one small crack starts to grow. During Stage II the small cracks grow independently of each other; however, after they have reached a certain size, interactions with other small cracks occur. This terminates Stage II. In Stage III some of the small cracks coalesce and the path of the later macrocrack begins to form. In Stage IV a macrocrack is present.

The microstructural phenomena in Stages I to III are fairly well known at the magnification level used in Fig. 1. However, there still exists a considerable lack in knowledge about the fatigue processes which occur at a magnification level an order of magnitude higher.

The macroscopic continuum mechanics condition is generally seen as the most important influence on the fatigue behavior. Advanced analytical analysis methods for the detailed evaluation of the stresses and strains are under development now. (The method being developed at DFVLR will be described later.) These usually do not account explicitly for the individual stages of the fatigue process as described before, but take the "technical crack initiation stage" (Stages I–III(IV)) as a basis.

The presently available knowledge of the microstructural processes and the developments in mechanics initiated a further direction of analyzing the fatigue process: the derivation and use of micromechanical models. In this context such analyses as given in Refs *5* and *6* are of importance. In a recent publication the effect of hard inclusions in a ductile matrix has been modelled [*7*].

Under variable amplitude loading histories, load sequence induced nonlinearities in the damage accumulation occur. As has been shown in Ref *8,* these nonlinearities are observed in all stages of the fatigue process (compare Figs. 2 and 3 for Stages I to III).

Besides those factors already mentioned, the fatigue process is further influenced by environmental conditions such as temperature, corrosion, irradiation, wear, and abrasion. These will not be further considered in this paper.

In contrast to a detailed physical treatment of the fatigue processes, which tries to consider all possible influences in an interdisciplinary manner, stand the interests, demands, and possibilities of engineers. They prefer as simple as possible fatigue models. In order to compensate the

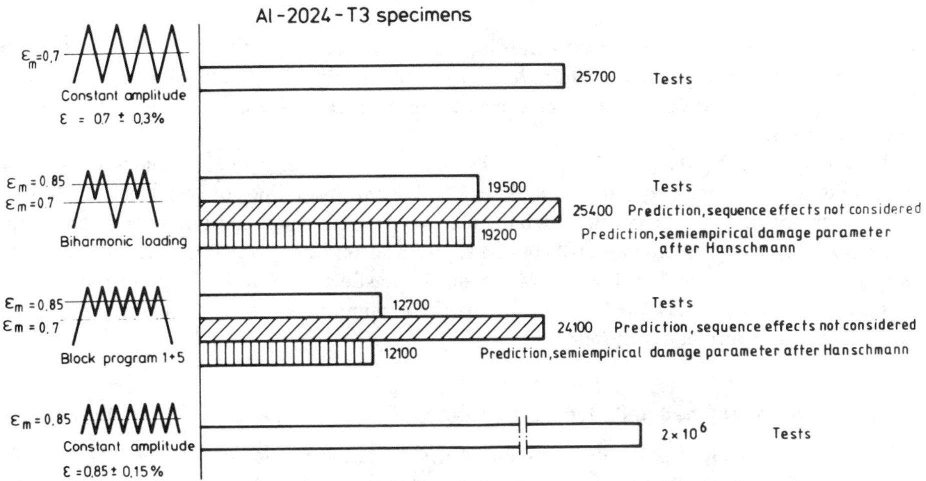

FIG. 2—*Load sequence effects in the precrack stage.*

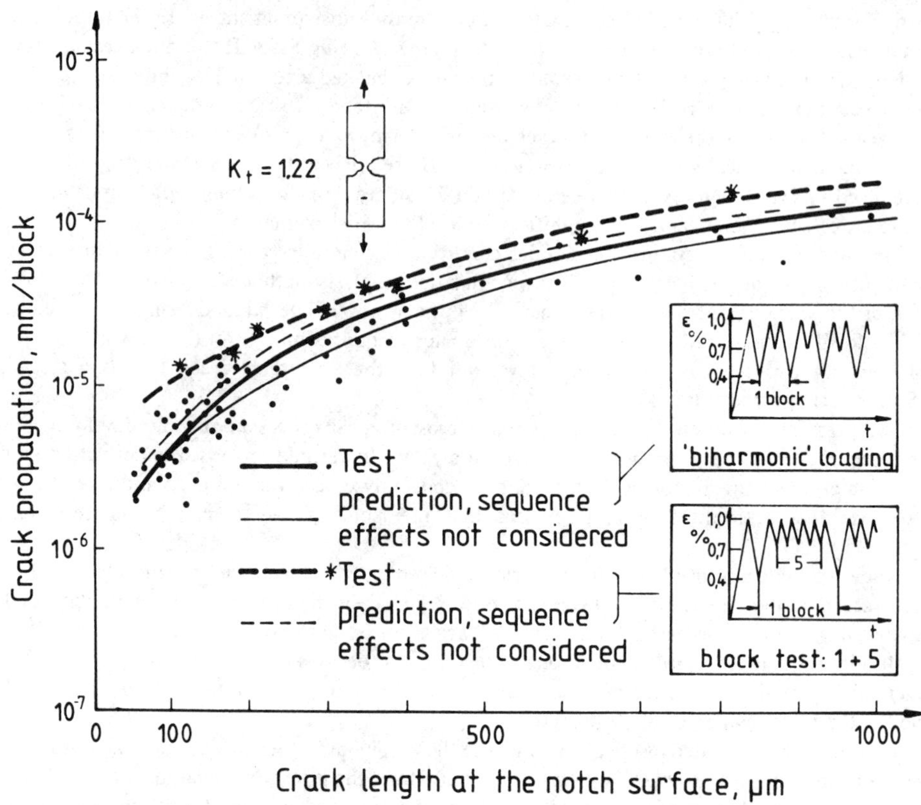

FIG. 3—*Load sequence effects in the small crack stage.*

consequences of oversimplifications regarding the consideration of the physical causes of the fatigue process, special design philosophies have been established (damage tolerance design, fail safe, etc.). In those cases where exact predictions became necessary, expensive experiments became inevitable.

A most significant pulse, which changed the situation for the engineers, is due to the general availability of computers for fatigue analyses. Today two basic lines of computerized fatigue analyses can be distinguished, one which uses small or medium size computers and applies the so-called "local" or "notch analysis" concepts, and a second line, which uses large-scale computers to perform detailed evaluations of the elastic-plastic stress/strain fields (including multiaxiality) at all fatigue critical locations of structural components and which tries to consider micromechanical influences as well.

Fatigue Analysis on Small and Medium Size Computers Using Notch Analysis Concepts

Notch analysis concepts, the development of which started in the late 1960s [*9–11*], usually incorporate three individual modules, one for the generation of the basic material stress/strain properties, a second for the consideration of notches, and a third where the fatigue damage is evaluated. One striking advantage of these concepts is that there is an enormous number of basic data for the application of these analyses already available in a standardized format [*12*].

Additionally, for some common engineering materials multiple individual data sets were already generated. This allows for the application of suitable statistical methods to derive a highly efficient data base.

The first two modules of the notch analyses are often seen as one macro-module, where the stresses and strains at the fatigue critical locations are determined. This macro-module has reached a comparatively high state in development, which shall be briefly shown in the following for typical engineering situations:

Geometrical Notches (Nominal Stresses and Notch Stresses Uniaxial)—For this simple case alternative concepts in connection with the basic monotonic or cyclic material data and with the Masing criterion can be used, as they are shown in Fig. 4, provided that the stress concentration factor of the component is known [*13–15*]. The net section stresses of the specimens or components can also be bending or other. The notch factor has adequately to be defined. In some cases a fatigue notch factor or a variable notch factor [*16*] is used.

Sometimes transient effects in the materials have to be considered as well. However, considerable reductions of the expenses of the local stress/strain calculations are achieved if they can be neglected. In such cases the correlation between the nominal stress and the local strains at the notch needs to be determined only once [*17*].

Notch root stresses and strains
(local approach)

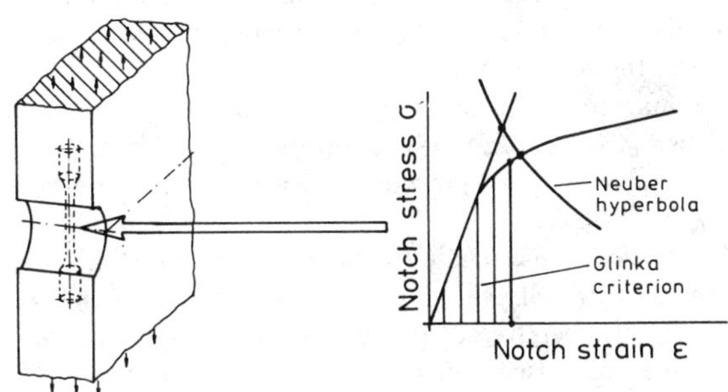

Neuber-hyperbola:

$$\sigma\varepsilon = K_t^2 S e \quad \text{(specimen net section elastic)}$$

$$\sigma\varepsilon = \frac{S^2 K_t^2}{E} \cdot \frac{e^* E}{S^*} \quad \text{(specimen net section elastic - plastic)}$$

Work criterion after Glinka:

$$W_\sigma = K_t^2 W_S \; ; \quad W_\sigma = \int_0^\varepsilon \sigma \, d\varepsilon \; ; \quad W_S = \int_0^e S \, de$$

with:

- S : nominal stress
- e : nominal strain
- K_t: stress concentration factor
- W : work

FIG. 4—*Analytical expressions for the consideration of notches in notch analysis concepts.*

The material memory behavior is always important and has to be considered.

Geometrical Notches (Stresses and Strains Multiaxial but Proportional)—In order to account for proportional multiaxial stresses and strains at notched configurations it has been suggested to use the equivalent or "effective" stresses and strains (for example, after the von Mises hypothesis) together with the uniaxial material stress/strain curve [*18*]. A corresponding effective notch factor has to be defined.

Due to the free surface of the notch, some of the components of the local stress/strain tensors become zero. The relations between the other components remain fixed, because the stresses and strains are assumed to remain proportional during the loading of the test piece. For the explicit evaluation of the components of the stress/strain tensors in the elastic-plastic range suitable hypotheses have to be used [*19-21*].

Geometrical Notches (Residual Stresses at the Notch Root)—For the consideration of residual stresses at notches several approaches have been proposed. Either a fictitious nominal stress (which is assumed to be the cause of the residual stresses) is introduced into the equations after Fig. 4 [*22*] or the nominal stress term in Neuber's equation is modified in a certain manner (compare, for example, [*23*]).

Although the proposals exhibit deficiencies from a physical viewpoint, they can be seen as acceptable first-order approaches. The third approach is limited to the case of a thin surface layer in which the residual stresses act [*24,46*]. The thin surface layer is approximately under strain control by the bulk material; the local layer strain can therefore be calculated by simply adding the residual strain to the local surface layer strain due to the load.

Geometrical Notches (Thermomechanical Treatment and/or Metallurgical Changes of the Notch Root Material)—Due to thermomechanical treatments or, for example, due to diffusion processes, oxidation, etc., the properties of the material at the notch root may change significantly. Although the complete notch problem has not been treated, some promising developments have been published [*24*]. Starting with the postulate of compatibility of deformations of the outer zone and of the core zone of probes, some averaging of the material properties is performed.

Geometrical Notches (Filled)—Especially for high performance lightweight constructions, fasteners are sometimes used which exhibit some interference with the fastener holes. Such cases have already successfully been treated in the literature [*25*]. Interactions between the fasteners and the hole surfaces may also occur and lead to fretting corrosion. For such situations proposals also exist in the literature (compare [*26*]).

Welds—The notch effect of welds is the result of the combined action of the geometrical notch effect, metallurgical notches, residual stresses, and defects. In spite of the complexity of this problem notch analysis methods were often successfully applied [*27*]. Spot weld connections could be treated by local approaches successfully as well, provided that representative notch factors could be derived.

The list of examples of applications of the notch analysis concepts could be extended much further. In various fields the developments are still going on.

If the results of the application of the local approaches are considered more generally, it can be stated that the accuracy, which is commonly reached in predicting the local stresses and strains at the fatigue critical locations of structural components, is usually quite satisfactory.

Fatigue Life Evaluation

For the determination of the crack initiation life behavior based on the local stresses and strains "damage parameters" or damage indicators are defined, which correlate to the crack initiation life when testing unnotched specimens under constant amplitude loading. The conventional notch analyses utilize damage parameters, which are derived from the stress/strain hysteresis behavior, and are based on total or plastic work, plastic strain range, elastic and plastic strain range, etc. The functional relationship between these parameters from the hysteresis loops and the crack initiation life in constant amplitude tests are graphically or analytically presented.

Mean stresses/strains may have a significant effect on the crack initiation life. They are considered by additional terms in the functional relationships (compare, for example, [*28*]).

If crack initiation life predictions are performed for variable amplitude loading, the rainflow counting method is commonly applied to identify closing stress/strain hysteresis loops in the course of the loading history. The resulting damage parameters from the loops are then calculated. The damage per loop is taken as the inverse of the cycle number to crack initiation from the plot of the cycle numbers to crack initiation from the basic constant amplitude tests, which are presented in the form of the applied damage parameter. Usually the damage increments per the consecutive loops are linearly summed up. Failure is assumed to occur at a damage sum equal to one.

However, from experience it is known that by applying the described procedure, the actual fatigue lives are quite often overestimated. Up to now a fully satisfactory explanation for this behavior has not existed. But it is known that there are some important influences which are not adequately accounted for, such as the presence of stress gradient effects, size effects, and, under variable amplitude loading, load sequence effects. In order to overcome these limitations, several concepts have been proposed and included into advanced computer programs. These are:

1. Performance of relative damage calculations, similar to the "Relative Miner Rule" [*29*], where the sum of the amount of damage increments at final failure is adjusted on the basis of the experience from actual tests with the same or a very similar loading history and with similar test pieces.
2. Application of one of numerous proposals to modify the basic constant amplitude crack initiation life curves, which are similar to the modifications that are often formally applied to conventional *S/N* curves.
3. Utilization of constant amplitude data determined with prestrained specimens, whereby the amount of prestrain is taken as high as the maximum strain under the variable amplitude loading.
4. Utilization of nonlinear damage evaluation concepts such as the double linear damage rule [*30*].
5. Introduction of correction terms derived from a comparison of the results of the application of different cycle counting procedures (for example, level crossing counting and range pair counting) [*31*].
6. Application of semiempirical damage calculation procedures. An example of such a procedure [*32*] is shown in Fig. 5. The damage per hysteresis loop which closes in the course of the loading history is first calculated in the conventional manner. In order to account for the influence of load sequence effects, an additional amount of damage is added. Long and short term sequence effects are considered.
7. Use of basic damage parameter versus life curves from tests with variable amplitude loading histories (standard random tests) instead of data from constant amplitude tests.

Development of the fatigue damage evaluation procedures as one essential module of the notch analysis concepts is still going on.

$$\boxed{D_{tot}(i) = D_{lin}(i) + ZS(i)}$$

$ZS(i) = 0$ for $N(i) \leq N(i-1)$ and $N(i) \leq K \cdot N(max)$

$$ZS(i) = FMAT \left[\frac{\ln \frac{N(i)}{N(i-1)}}{\sqrt{N(i) \quad N(i-1)}} + \frac{\ln \frac{N(i)}{N(max)}}{\sqrt{K \; N(i) \quad N(max)}} \right]$$

additional damage (ZS(i))

(i) = i-th cycle

$D_{tot}(i)$ = Total damage per i-th cycle

$D_{lin}(i)$ = Damage per cycle (i) corresponding to 1/N(i) from basic damage parameter vs. life curve from constant amplitude tests

$N(i)$ = Crack initiation life in constant amplitude tests with ε_a (strain amplitude) and ε_m (mean strain) of the i-th cycle

$N(i-1)$ = Crack initiation life in constant amplitude tests with ε_a and ε_m of cycle (i-1)

$N(max)$ = Crack initiation life in constant amplitude tests with ε_a and ε_m of the last high peak load

K = Number of cycles after the last high peak load

$FMAT$ = Empirical material dependent constant from some basic tests with biharmonic loading

FIG. 5—*Semiempirical damage parameter* [32] *for the consideration of sequence effects.*

Short Crack Stage

As mentioned before, the technical crack initiation stage includes three individual stages with different physical mechanisms (also compare Fig. 1). Presently a large amount of research work is done on the short crack stages (II–III), because it is hoped that a better physical understanding of the damage processes in these stages leads to improvement of life prediction. In order to fit into a concept, where the short crack stages II and III are explicitly considered, the range in application of the notch analysis methods as described in the previous sections could be limited to the precrack stage I (this is normally not done). This has been done as an example for the data shown in Fig. 2; it can be seen that the predictions coincide well with the results if the damage parameter concept from Ref *32* (compare Fig. 5) is used.

A first satisfactory prediction for the short crack stages (II–III) could be obtained, if "equivalent" stress intensity factors for the short cracks were derived from the cyclic J-integral and if these "equivalent" stress intensity factors were introduced into a crack propagation equation of Elber's type [*33*]. In these tests the loading histories started with a very high load at the beginning, which was followed by smaller cycles predominantly in the elastic range. The results clearly indicated that crack closure is important.

In another investigation a straightforward cyclic J-integral approach was applied together with a rough correction to account for crack closure [*34*]. This approach also led to acceptable results.

The given results and various other investigation results reported in the literature clearly indicate that it is important from a physical point of view to consider all individual stages within the technical crack initiation stage separately.

From a technical point of view it has been very difficult to explicitly consider the short crack stage, because simple and reliable prediction methods have not been available. Except in a very limited number of cases, where the design must consider the short crack stage, as for example in the case of turbine disks [*35*], engineers presently prefer to follow a concept proposed in Ref *36*. It is suggested that so long as the notches remain blunt ($K_t < 3.0$), crack initiation life predictions should be made according to the notch analysis concepts for the whole *technical* crack initiation stage, because it is the important portion of the total fatigue life. The following macrocrack stage is separately considered by starting with an initial crack length of the notch depth and by integrating the crack increments; this is done on the basis of one of the advanced equations for the macrocrack behavior under variable amplitude loading (compare [*37*]). For sharp notches with stress concentration factors $K_t > 3$, the significance of the technical crack initiation stage decreases, and predictions of the macrocrack propagation behavior become more important.

FEM-Based Fatigue Analysis for Large Scale Computers (FEMFAT)

Stress/Strain Analysis

Stress/strain analyses based on FEM give a complete description of the stress and strain distributions at structural components so that critical areas (hot spots) can be identified. From such analyses stress concentration factors can be derived, which may then be used, for example, for the application of the notch analysis methods as described before. Or the FEM results can be used for an evaluation of the prediction capability of simple stress/strain evaluation approaches. A most important field in the application of FE methods, however, is where the stress/strain situation becomes so complicated that simple evaluation methods are no longer available.

The FE analyses may be simply elastic, elastic-plastic, or two- or three-dimensional. Professional FEM systems (ASCA, NASTRAN, PERMAS, etc.) are available; however, such systems are usually too expensive to operate when applied to fatigue problems, where the loading and the local stress/strain situations can vary from half cycle to half cycle. That is why at DFVLR a special development, FEMFAT, has been started in cooperation with the Technical University of Aachen [*38*]. The basic feature of FEMFAT shall be briefly characterized:

1. For the FE representation of structural parts several typical elements are available. Commonly three-dimensional isoparametric elements are used; however, from an economical standpoint sometimes shell elements are advantageous. Suitable connection elements are also available (compare Fig. 6).

2. Under elastic-plastic stress/strain conditions the elastic part of the structure can be represented by an (elastic) substructure. The stiffness of the elastic substructure remains unchanged

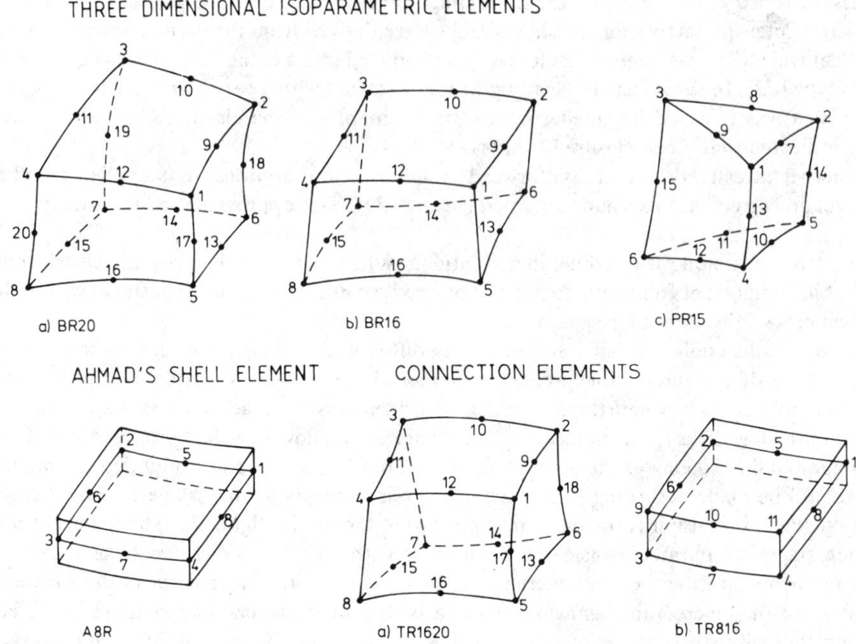

FIG. 6—*Elements used in the three-dimensional elastic-plastic FEM-based fatigue analysis method FEMFAT.*

during the incremental-iterative solution of the elastic-plastic problems. FEMFAT uses this substructure technique (which condensates the number of degrees of freedom) in connection with the frontal equation solution.

3. An important part of elastic-plastic FE analyses is the selection of a suitable material model having three essential parts: yield function, flow rule, and hardening rule. For FEMFAT the material model after Mróz [*39*] in connection with the von Mises yield criterion and the flow rule after Drucker [*40*] is used, and the tangent stiffness radial return concept [*41*] is applied. The working principle and the application of the Mróz model are briefly outlined in Appendix I. Also given in Appendix I is Garud's solution [*42*].

4. FEMFAT calculations are performed on a CRAY 1S computer with an 8 MB storage capacity; however, less powerful computers with a similar storage capacity may be used as well.

Two examples of the practical application of FEMFAT—once to a notched box beam and once to an automotive axle, both under bending conditions—are given in Figs. 7 and 8. For some specific points measured strains and FEMFAT calculations are compared. It can be seen that the data are in good agreement.

Fatigue Analysis

Depending on the loading and local stress/strain situation of structural components fatigue analyses can lead to considerable difficulties. So long as the stresses and strains remain predominantly elastic and uniaxial or multiaxial, conventional fatigue life calculation methods may be applied [*43,44*]. When the stresses and strains become elastic-plastic but still remain propor-

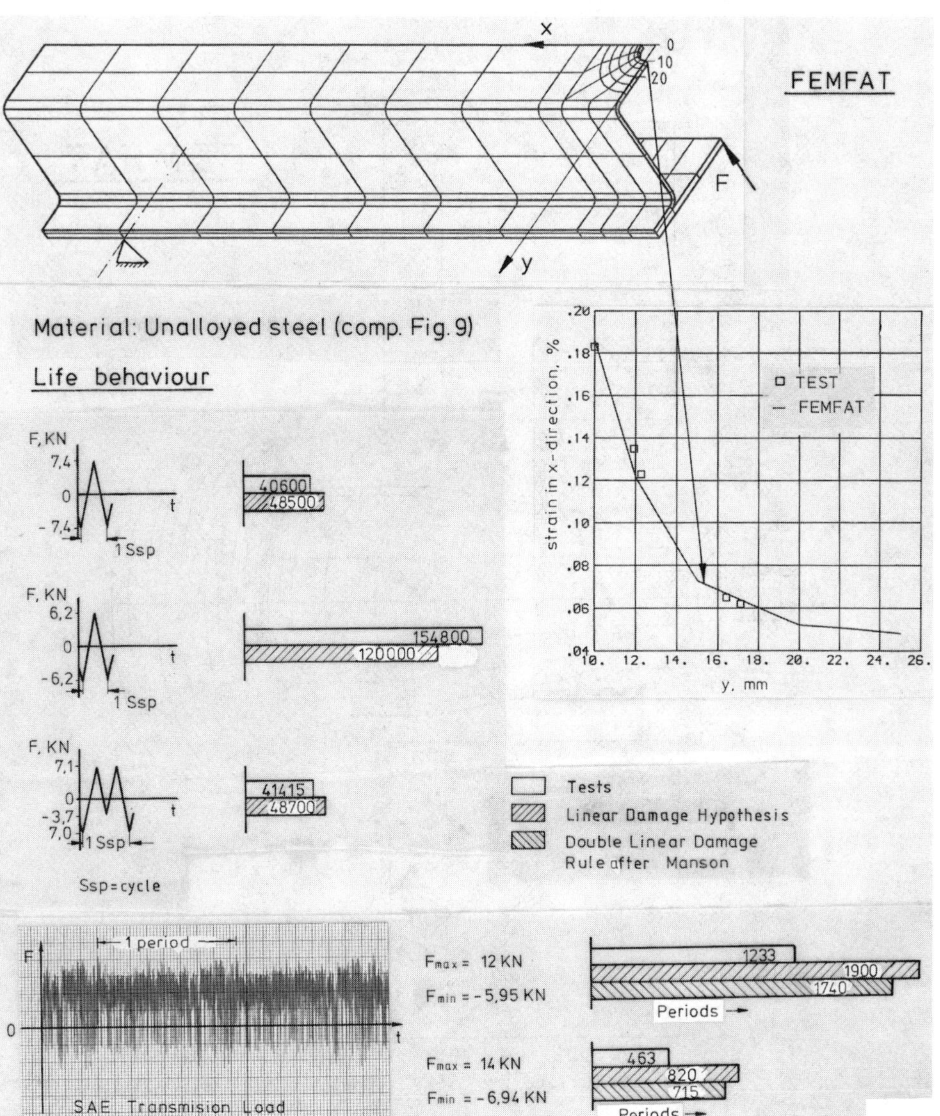

FIG. 7—*Application of FEMFAT to a box beam.*

tional, concepts on the basis of the equivalent stresses and strains are often used. However, if the loading and the local stresses and strains become non-proportional, suitable prediction methods are still in an early stage of development. The essential problem is to define suitable damage parameters and damage events if closed loops no longer exist in the stress/strain space which can be correlated to uniaxial data.

The present version of FEMFAT takes the plastic work performed during a loading history as a basis for the evaluation of the damage. However, it is unnecessary that closed hysteresis loops exist, because the individual amounts of plastic work are determined for every transition of the

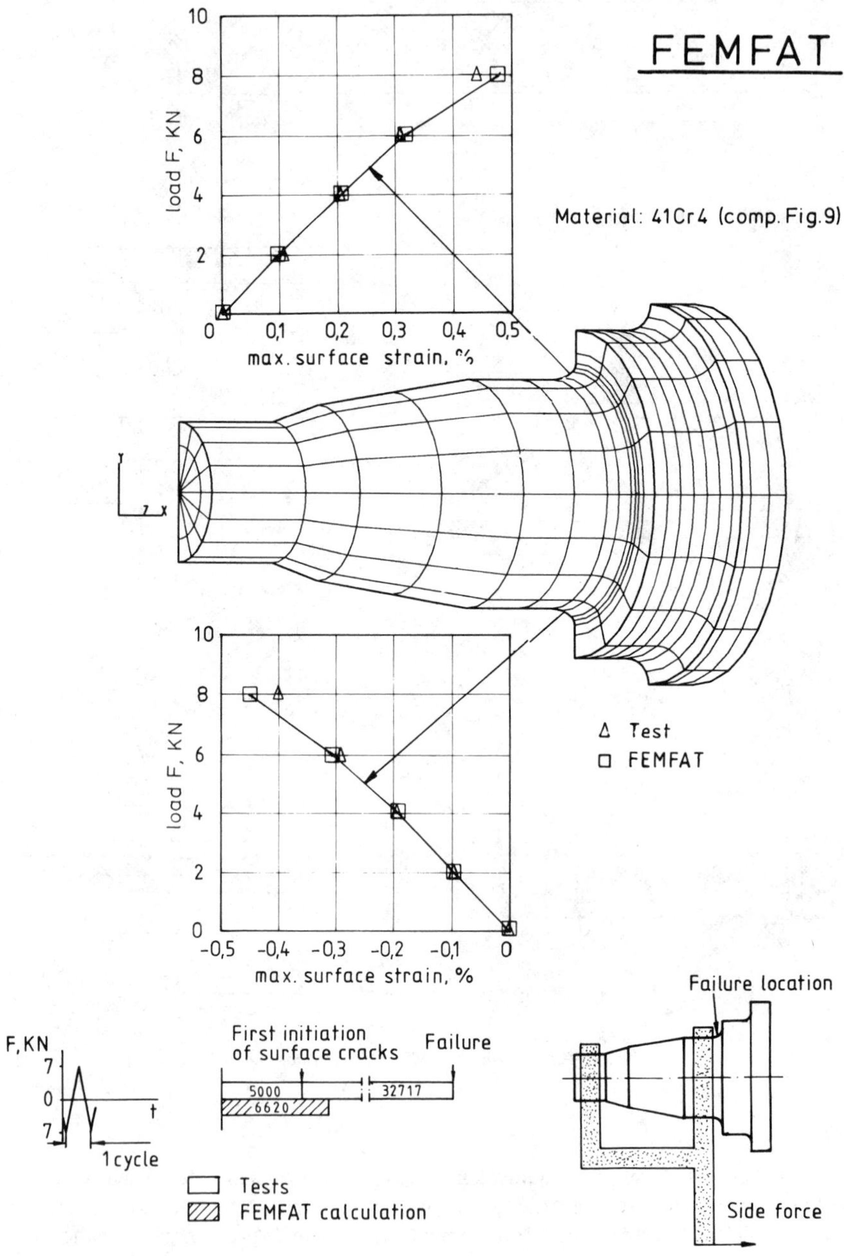

FIG. 8—*Application of FEMFAT to an automotive axle.*

stresses through certain Mróz segments with a constant hardening parameter (compare Appendix I). For the evaluation of the damage caused by the amounts of plastic work within the Mróz segments, the damage caused by one unit work per Mróz segment must be known; this is determined on the basis of uniaxial plastic work versus cycle number to crack initiation data. This special procedure is described in Appendix II in more detail.

Under arbitrary loading histories the damage caused by the transitions of the local stresses through all complete Mróz segments or through parts of them is on-line determined and continuously summed up. Thus at every instant during a loading history the damage state is known.

For an evaluation of the described damage accumulation hypothesis, various predictions and comparisons have been performed:

1. Uniaxially loaded specimens were considered, where the crack initiation life was determined, once using the FEMFAT procedure and once by the conventional notch analysis method. For the notch analysis the complete hysteresis loops, which were formed during the loading history, were taken in connection with uniaxial plastic work versus crack initiation life curves for the determination of the fatigue damage. Under constant amplitude loading conditions and for the SAE loading history [*45*], only very small differences between both types of analyses were observed, which are due to the finite width of the Mróz segments and the number of stress increments applied within one Mróz segment.
2. Comparisons were made for biaxial loading conditions, where complete hysteresis loops as under uniaxial loading could be determined. Under such circumstances differences in the life predictions between FEMFAT and conventional methods were below 20%.

FEMFAT was also applied to predict the fatigue life of the test pieces in Figs. 7 and 8. The properties of the materials are given in Fig. 9. For constant amplitude loading conditions the results show that the predictions and the actual test results do not deviate more than is known from the application of the notch analysis methods under uniaxial loading conditions.

Under variable loading mean stress effects and load sequence effects become additionally important. In the next step of the development of FEMFAT both influences will be considered.

Conclusions

The present paper considered two basic lines in the development of computerized fatigue life prediction methods: the notch analysis concepts, which need small or medium size computers and allow for the consideration of various specific design problems, and FEM-based methods (such as the FEMFAT program), which need large computers. Both lines exhibit advantages but also certain limitations, and further development is needed to overcome existing limits.

In the future an increasing amount of computer software originally developed for large computer systems will be run on smaller systems due to the steadily increasing hardware capacity. The specialty of large computers will then be the consideration of complicated structures with complicated loading and local stress/strain conditions, and the ability to apply and to further develop complicated multiaxial material models and fatigue life evaluation schemes.

Acknowledgments

The authors express their thanks to R. Marissen, P. Jouvin, Professor D. Socie (University of Illinois), and K. H. Trautmann for very fruitful discussions and to Mrs. W. Schmitz and Mrs. C. Kotauschek for their help in preparing the manuscript.

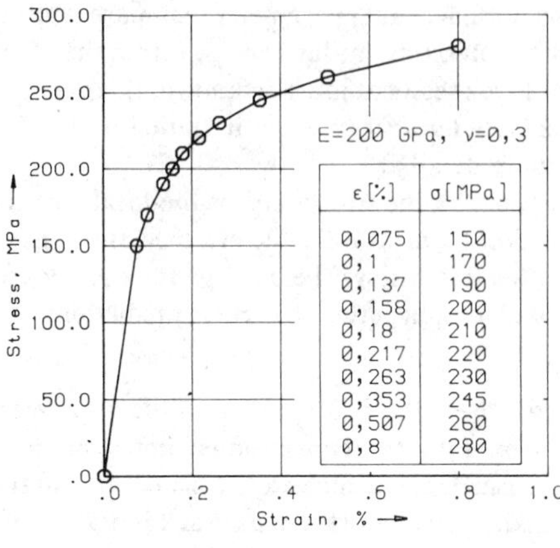

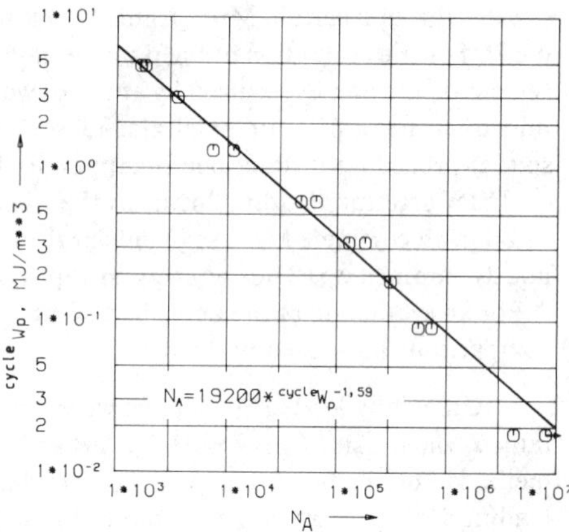

a) Unalloyed steel (0,03% C;0,02% Si;0,39% Mn;0,006% P;0,011% S)

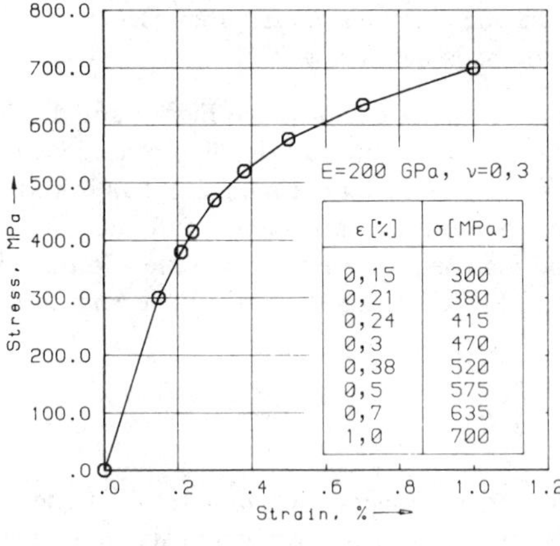

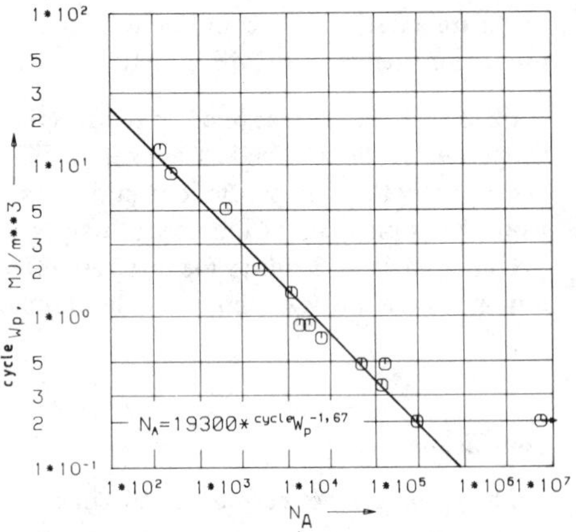

b) 41Cr4

FIG. 9—*Material properties.*

APPENDIX I

Determination of the Elastic-Plastic Multiaxial Stress/Strain Behavior by FEMFAT

In the following paragraphs the working principal and the application of the Mróz model [*39*] within the FEMFAT program are briefly described:

1. In the Mróz model hypersurfaces $f^{(l)}$ are introduced that form the boundaries of segments with a constant hardening parameter $H'^{(l)}$ (Mróz segments) and that are circles (Mróz circles) if the von Mises yield criterion is used and if they are represented in the π plane for the general

three-dimensional case. For the simpler two-dimensional tension-torsion case, the hypersurfaces after Mróz form circles in the $\sigma_x - \sqrt{3} \cdot \tau_{xy}$ plane (Fig. 10).

2. The hypersurfaces after Mróz (Mróz circles) can be represented by the expression

$$f^{(l)} = \sqrt{3/2(\sigma'_{ij} - \alpha'^{(l)}_{ij}) \cdot (\sigma'_{ij} - \alpha'^{(l)}_{ij})} = k^{(l)}$$

where

$\alpha'^{(l)}_{ij}$ = variable center coordinate of each hypersurface, and
$k^{(l)}$ = measure of the extension of each hypersurface according to the selected ranges with a constant hardening parameter for the material under consideration.

3. The shift of hypersurfaces through Mróz segments due to an applied stress occurs by the application of a series of stress increments $d\sigma_{ij}$. The shift of the hypersurface due to the application of a stress increment is shown in Fig. 11, where ${}^{r-1}\sigma_{ij}$ defines the stress point on the hypersurface ${}^{r-1}f^{(l)}$ before the stress increment $d\sigma_{ij}$ is applied.

4. The new position of the Mróz hypersurface ${}^{r}f^{(l)}$ is determined in a stepwise manner:

(*a*) One point of the new hypersurface is determined on the basis of the equation

$$d\sigma_{ij} = D^{e}_{ijkl}(d\epsilon_{kl} - d\epsilon^{P}_{kl})$$

where

D^{e}_{ijkl} = stiffness matrix for the elastic part of the stress/strain behavior of the material within a given Mróz segment,
$d\epsilon_{ij}$ = total strain increment as calculated from the finite element program (external input into the calculation after the Mróz model), and
$d\epsilon^{P}_{ij}$ = component of the plastic strain perpendicular to the hypersurface ${}^{r-1}f^{(l)}$, and is determined after the flow rule after Drucker and evaluated by the expression

$$d\epsilon^{P}_{ij} = \frac{1}{A}\left(d\sigma_{kl}\frac{\partial\, {}^{r-1}f^{(l)}}{\partial \sigma_{kl}}\right)\frac{\partial\, {}^{r-1}f^{(l)}}{\partial \sigma_{ij}} \qquad (A \sim H'^{(l)})$$

(*b*) The shift of the previous hypersurface ${}^{r-1}f^{(l)}$ through the new stress point occurs by a translation along the connecting line of the origination points of the normals once on the hypersurface ${}^{r-1}f^{(l)}$ in the previous stress point and once on the outer hypersurface $f^{(l+1)}$.

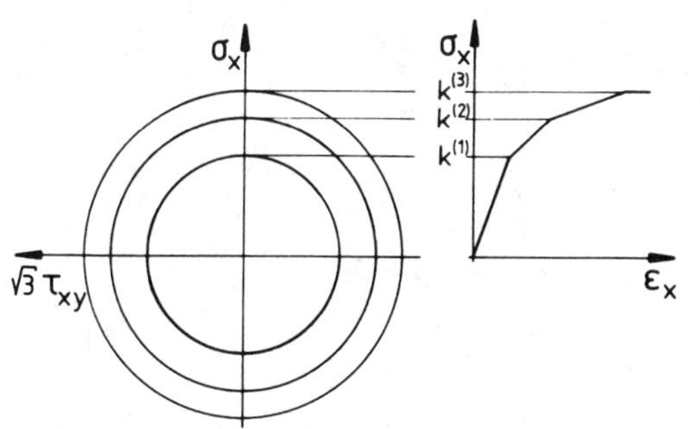

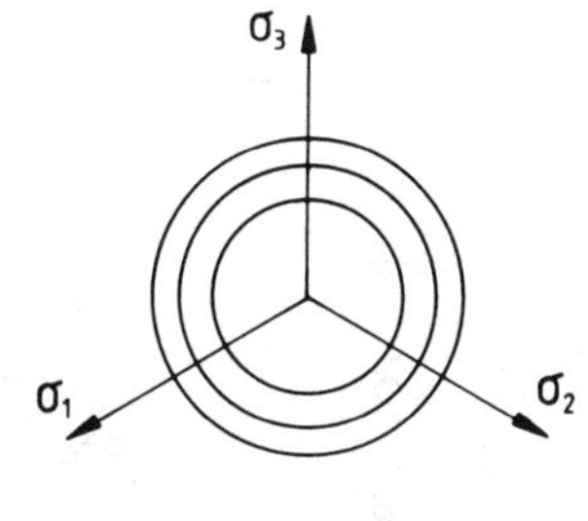

FIG. 10—*Multi(yield) surface model after Mróz* [39].

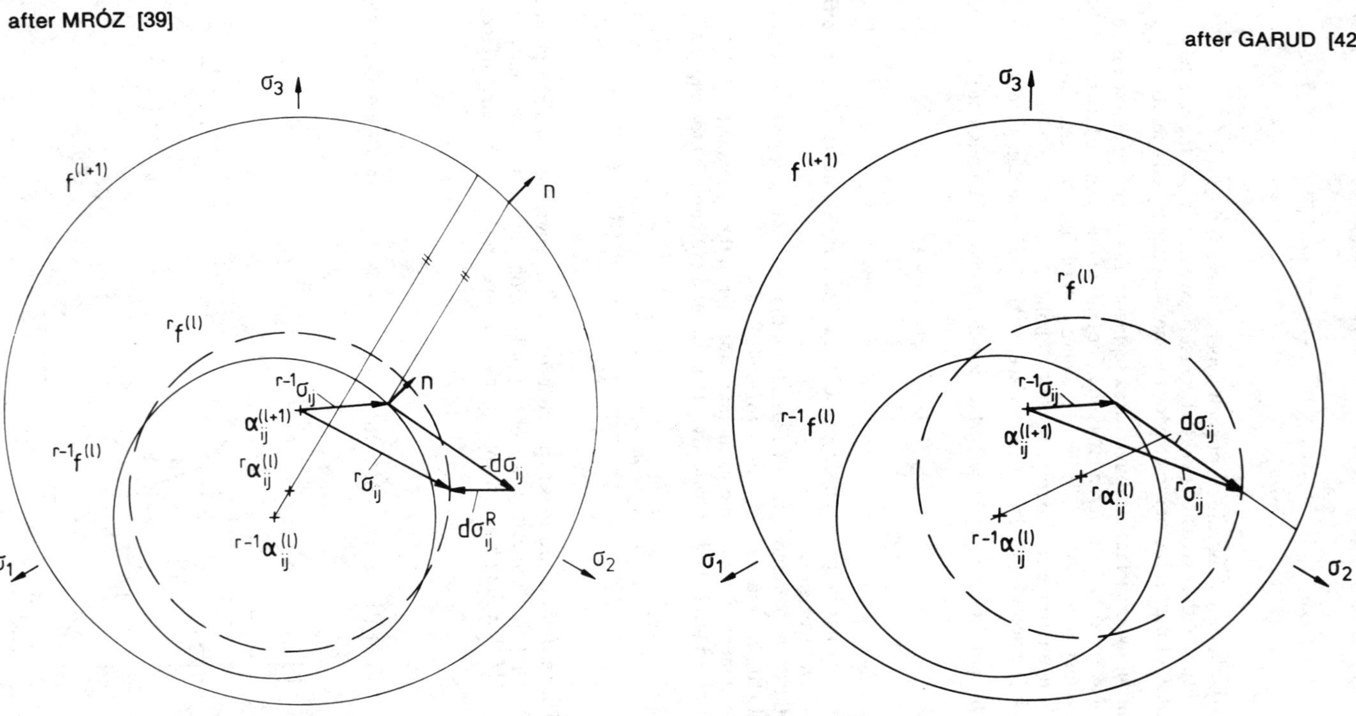

FIG. 11—*Translation of the hypersurfaces after Mróz in FEMFAT and after Garud* [42].

(*c*) Generally the new hypersurface ${}^{r}f^{(l)}$ does not pass through the endpoint of $d\sigma_{ij}$. In order to get the new stress point back on ${}^{r}f^{(l)}$, a correction has to be applied. In FEMFAT the tangent stiffness radial return method is used, and the deviation (due to the finite length of $d\sigma_{ij}$) is compensated by $d\sigma_{ij}^{R}$ (Fig. 11).

In conclusion, the stress at the time step r, which defines the stress point on the new hypersurface ${}^{r}f^{(l)}$ for the application of the next stress increment $d\sigma_{ij}$, is determined by

$$ {}^{r}\sigma_{ij} = {}^{r-1}\sigma_{ij} + d\sigma_{ij} + d\sigma_{ij}^{R} $$

Garud [*42*] applied the Mróz model in a different manner, as indicated in the right-hand side of Fig. 11.

APPENDIX II

Evaluation of the Damage per Unit Plastic Work per a Given Mróz Segment from Uniaxial Constant Amplitude Data

As described in Appendix I, the Mróz model comprises hypersurfaces (Mróz circles) and ranges with a constant hardening parameter between the hypersurfaces. Along the Mróz segments plastic work is performed which causes damage. In the following it is shown how the damage per unit work in Mróz segments is determined from uniaxial data. In FEMFAT the damage calculation is based on these specific amounts of damage.

Damage Occurring within the First Elastic-Plastic Mróz Segment Only

The plastic work per Mróz segment $l = 1$ is equal to the hatched area in Fig. 12*a*, or, more specifically, equal to the sum of the areas above and below the $\sigma = 0$ axis. For the evaluation of the damage for the hysteresis loop with the Mróz segment $l = 1$, a conventional plastic work versus cycle number to crack initiation diagram from constant amplitude loading with unnotched specimens is used (Fig. 12*b*). The curve in Fig. 12*b* is entered at W_{p1} and N_1 is read out. The amount of damage due to the plastic work per one complete hysteresis loop is taken as $D_1 = 1/N_1$. In order to obtain the damage per 1 unit work in Mróz segment $l = 1$, d_1, D_1 is divided by the total area of the loop $W_{p(1,1)}$:

$$ d_1 = \frac{D_1}{W_{p(1,1)}} $$

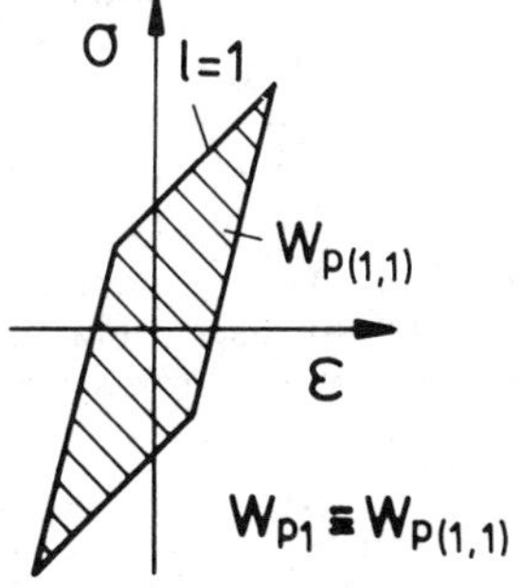

FIG. 12*a*—*Plastic work in hysteresis loops with* l = *1 Mróz segment.*

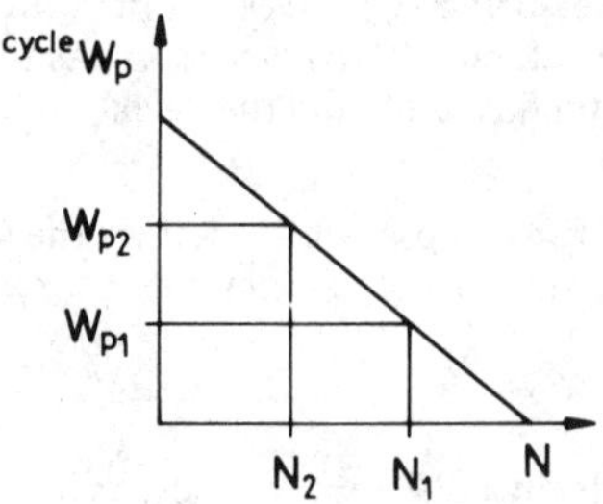

FIG. 12*b*—*Basic (uniaxial) plastic work versus cycle number to crack initiation relationship.*

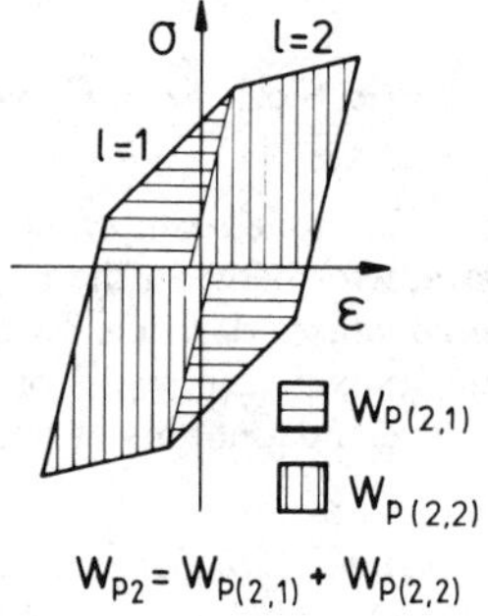

FIG. 12*c*—*Plastic work in hysteresis loops with two Mróz segments,* l = *1 and* l = *2.*

Damage in Mróz Segments l = *1 and* l = *2*

For the evaluation of the damage in Mróz segment $l = 2$ corresponding to the vertically hatched area $W_{p(2,2)}$ in Fig. 12*c* the damage as caused by the complete hysteresis loop W_{p2} has first to be determined from Fig. 12*b*: $D_2 = 1/N_2$. From this total damage the damage due to the plastic work under segment $l = 1$ (horizontally hatched area) has to be subtracted. (It must be noted that $W_{p(2,1)}$, the area below Mróz segment $l = 1$, is now smaller than $W_{p(1,1)}$ in Fig. 12*a*!) The damage per one unit of plastic work due to segment $l = 2$, d_2, is

$$d_2 = \frac{1/N_2 - W_{p(2,1)} \cdot d_1}{W_{p(2,2)}}$$

The described procedure is further continued for hysteresis loops with $l = 3, 4, \ldots, n$ Mróz segments in order to derive the amounts of the damage per unit plastic work $d_3, d_4, \ldots, d_n$. In general, the damage per unit plastic work within a Mróz segment l, d_l, can be written as:

For l = *1:*

$$d_1 = \frac{1/N_1}{W_{p(1,1)}}$$

For $l > 1$:

$$d_l = \frac{1/N_l - \sum_{k=1}^{l-1} W_{p(l,k)} \cdot d_k}{W_{p(l,l)}}$$

Under arbitrary variable amplitude loading histories with multiaxial stress/strain conditions the plastic work, which is performed during the transition of one Mróz segment after the other, is determined. The plastic work per individual Mróz segment is then multiplied by the specific amount of damage as it was derived above. The damage increments as derived in this manner are then linearly summed up.

References

[1] Nowack, H. in *Proceedings,* 2nd International Conference on Fatigue and Fatigue Thresholds, Birmingham, U.K., 3-7 Sept. 1984, pp. 961-986.
[2] Foth, J., "Einfluβ von einstufigen und nicht-einstufigen Schwingbelastungen auf Riβbildung und Mikroriβausbreitung in Al 2024-T3," DFVLR Report DFVLR-FB 84-29, DFVLR, Cologne, FRG, 1984.
[3] Foth, J., Marissen, R., Nowack, H., and Lütjering, G. in *Proceedings,* European Conference on Fracture (ECF)-5, Lisbon, Portugal, 1984.
[4] Starke, J. and Lütjering, G. in *Fatigue and Microstructure,* American Society for Metals, Metals Park, Ohio, pp. 205-244.
[5] Suresh, S. and Ritchie, R. O., "A Geometric Model for Fatigue Crack Closure Induced by Fracture Surface Roughness," *Metallurgical Transactions,* Vol. 13A, 1982, pp. 937-940.
[6] Suresh, S., "Micromechanisms of Fatigue Crack Growth Retardation Following Overloads," *Engineering Fracture Mechanics,* Vol. 18, No. 3, 1983, pp. 577-593.
[7] Amstutz, H., "Ellipsenförmige Einschlüsse in elastischer und elastisch-plastischer Matrix mit ebenem Spannungszustand," Ph.D. thesis, Technical University Darmstadt, 1983.
[8] Foth, J., Marissen, R., Nowack, H., and Lütjering, G. in *Proceedings,* 14th Congress of the International Council of the Aeronautical Sciences, Toulouse, France, 9-14 Sept. 1984.
[9] Crews, J. H. and Hardrath, H. F., *Experimental Mechanics,* Vol. 23, 1966, pp. 313-320.
[10] Wetzel, R. M., "Smooth Specimen Simulation of Fatigue Behavior of Notches," T&A.M. Report 295, Aeronautical Structures Laboratory, University of Illinois, May 1967.
[11] Dowling, N. E. in *Proceedings,* SAE Fatigue Conference P-109, Dearborn, Mich., 14-16 April 1982, pp. 161-174.
[12] Conle, A. and Landgraf, R. W., "Material Property Data Bank," Internal Report, Engineering and Research Staff, Ford Motor Comp., Dearborn, Mich., 1980.
[13] Neuber, H., "Notch Stress Theory," Report AFML-TR-65-225, Wright-Patterson Air Force Base, Ohio, 1965.
[14] Seeger, T. and Heuler, P., *Journal of Testing and Evaluation,* Vol. 3, No. 4, 1980, pp. 199-204.
[15] Molski, K. and Glinka, G., *Materials Science and Engineering,* Vol. 50, 1981, pp. 93-100.
[16] Conle, A. and Nowack, H., *Experimental Mechanics,* Vol. 17, No. 2, 1977, pp. 57-63.
[17] Bergmann, J. W., "Zur Betriebsfestigkeitsbemessung gekerbter Bauteile auf der Grundlage der örtlichen Beanspruchungen," Report 37, Technical University Darmstadt, Inst. Stahlbau u. Werkstoffmech., FRG, 1983.
[18] Leese, G. E., "Engineering Significance of Recent Multiaxial Fatigue Research," this publication, pp. 861-873.
[19] Hoffmann, M. and Seeger, T., "A Generalized Method for Estimating Multiaxial Elastic-Plastic Notch Stresses and Strains—Part 1," *ASME Journal of Engineering Materials and Technology,* Vol. 107, No. 4, 1985, pp. 250-254.
[20] Hoffmann, M. and Seeger, T., "A Generalized Method for Estimating Multiaxial Elastic-Plastic Notch Stresses and Strains—Part 2," *ASME Journal of Engineering Materials and Technology,* Vol. 107, No. 4, 1985, pp. 255-260.
[21] Hoffmann, M. and Seeger, T., "Estimating Multiaxial Elastic-Plastic Notch Stresses and Strains in Combined Loading," Report FB-2, Technical University, Darmstadt, FRG, 1986.
[22] Lawrence, F. V., Jr., Burk, J. D., and Yung, J. Y. in *Residual Stress Effects in Fatigue, ASTM STP 776,* American Society for Testing and Materials, Philadelphia, 1981, p. 33.
[23] Reemsnyder, H. S., *Materials, Experiments and Design in Fatigue,* IPC Sci. and Techn. Press Ltd., Guildford, England, 1981, p. 273.

[24] Seeger, T. and Heuler, P. in *Proceedings,* DGM-Symposium Ermüdungsverhalten metallischer Werkstoffe, Bad Nauheim, FRG, 8-9 Nov. 1984, pp. 213-235.

[25] Weisgerber, D., Keerl, P. N., Hanschmann, D., Trautmann, K.-H., and Nowack, H., "Investigation of Fatigue Life Prediction for Light Alloy Joints," ICAF Doc. No. 1216, National Aerospace Laboratory (NLR), The Netherlands, pp. 2.3/1-2.3/26.

[26] Leis, B. N., *Journal of Testing and Evaluation,* Vol. 5, No. 4, July 1977, pp. 309-319.

[27] Nelson, D. V., "Fatigue Considerations in Welded Structure," *Transactions,* Society of Automotive Engineers, Warrendale, Pa., 1982, pp. 197-211.

[28] Smith, K. N., Watson, P., and Topper, T. H., *Journal of Materials,* Vol. 5, No. 4, 1970, pp. 767-778.

[29] Schütz, W., "The Fatigue Life under Three Different Load Spectra-Tests and Calculations," AGARD-Rep. No. AGARD CP 118, Symposium on Random Load Fatigue, Lyngby, Denmark, 13 April 1972.

[30] Manson, S. S. and Halford, G. R., "Practical Implementation of the Double Linear Damage Rule and Damage Curve Approach for Treating Cumulative Fatigue Damage," NASA Technical Memorandum 81517, NASA-Lewis, Cleveland, Ohio, April 1980.

[31] Buxbaum, O. in *Proceedings,* 4th LBF-Kolloquium, Fraunhofer-Institut für Betriebsfestigkeit, 31 Jan.-1 Feb., Darmstadt, FRG, 1984, pp. 5-18.

[32] Hanschmann, D., "Ein Beitrag zur rechnergestützten Anrißlebensdauervorhersage schwingbeanspruchter Kraftfahrzeugbauteile aus Aluminiumwerkstoffen," DFVLR Report DFVLR-FB 81-10, DFVLR, Cologne, FRG, 1981.

[33] Foth, J., Marissen, R., Trautmann, K.-H., and Nowack, H., "Short Crack Phenomena in Fatigue Loaded High Strength Aluminum Alloys and Some Analytical Tools for Their Prediction," in *The Behavior of Short Fatigue Cracks,* EGF Pub. 1, K. J. Miller and E. R. de los Rios, Eds., 1986, Mechanical Engineering Publications, London, pp. 353-368.

[34] Neumann, P., Vehoff, H., and Heitmann, H.-H., "Untersuchungen zur Betriebsfestigkeit von Stahl: Reihenfolgeeinflüsse während der Rißinitiationsphase," Forschungsbericht T 84-040, Max-Planck-Institut für Eisenforschung GmbH, Düsseldorf, FRG, March 1984.

[35] Wanhill, R. J. H. in *Proceedings,* 2nd International Conference on Fatigue and Fatigue Thresholds, University of Birmingham, U.K., 3-7 Sept. 1984, pp. 1671-1682.

[36] Socie, D. F., Dowling, N. E., and Kurath, P. in *Fracture Mechanics: Fifteenth Symposium, ASTM STP 833,* American Society for Testing and Materials, Philadelphia, 1984, pp. 284-299.

[37] Nowack, H. in *Proceedings,* International Conference on Application of Fracture Mechanics to Materials and Structures, Freiburg, FRG, 20-24 June 1983, pp. 57-102.

[38] Ott, W. and Nowack, H. in *Proceedings,* VDI-Jahrestagung '85: Werkstoff-Bauteil-Schaden, München, FRG, 7-8 March 1985, pp. 91-96.

[39] Mróz, Z., *Acta Mechanica,* Vol. 7, 1969, pp. 199-212.

[40] Drucker, D. C., *Quarterly of Applied Mathematics,* Vol. 7, No. 4, 1950.

[41] Marcal, P. V., *International Journal of Mechanical Sciences,* Vol. 7, 1965, pp. 229-238.

[42] Garud, Y. S. *Transactions of ASME,* Vol. 103, 1981, pp. 118-125.

[43] Troost, A., El-Magd, E., and Mielke, S., *Material und Technik,* Vol. 9, 1981, pp. 63-71.

[44] Grubisic, V. and Simbürger, A. in *Proceedings,* International Conference on Fatigue Testing and Design, City University, London, 5-9 April 1976, pp. 27.1-27.28.

[45] Landgraf, R. W., Richards, F. D., and La Pointe, N. R., "Fatigue Life Predictions for a Notched Member under Complex Load Histories," Report 750040, Society of Automotive Engineers, Warrendale, Pa., 1975.

[46] Vormwald, M. and Seeger, T., "Crack Initiation Life Estimations for Notched Specimens with Residual Stresses Based on Local Strains," in *Proceedings,* International Conference on Residual Stresses, Garmisch-Partenkirchen, FRG, 1986, to be published.

Yi-Sheng Wu[1]

The Improved Neuber's Rule and Low Cycle Fatigue Life Estimation

REFERENCE: Wu, Y.-S., **"The Improved Neuber's Rule and Low Cycle Fatigue Life Estimation,"** *Low Cycle Fatigue, ASTM STP 942,* H. D. Solomon, G. R. Halford, L. R. Kaisand, and B. N. Leis, Eds., American Society for Testing and Materials, Philadelphia, 1988, pp. 1007–1021.

ABSTRACT: An elastic-plastic finite element (FE) analysis was performed on the Society for Automotive Engineers (SAE) keyhole specimen in order to evaluate the prediction capability of Neuber's rule. Twenty different material properties were considered in the FE analyses. It was found that the accuracy of Neuber's rule is dependent on the material properties. In order to improve the prediction capability a new parameter concerning the material properties was introduced into Neuber's rule. The further improved Neuber's rule exhibits the format, $K_t = K_\sigma^m K_\epsilon^{(1-m)}$. The relation between m (exponent) and the material properties was studied. For the RQC-100 SAE keyhole specimen, stress-strain analyses and low cycle fatigue (LCF) life estimations were performed by both the conventional Neuber's rule and the extended rule.

KEY WORDS: fatigue, low cycle, life estimation, finite element, stress concentration, notch

Elastic-Plastic Finite Element Analysis of Notch Specimen

Neuber's rule [*1*], $K_t^2 = K_\sigma K_\epsilon$, is widely used for local stress-strain analysis and low cycle fatigue (LCF) life estimation [*2*]. However, it was found that Neuber's rule can lead to a large error. In Ref *3* Neuber's rule was used for notch stress-strain analyses under reverse cyclic loading. It was found that the notch strain can be overestimated if the conventional Neuber's rule was adopted. In Ref *4* Neuber's rule was used for stress-strain analyses of Man-Ten and RQC-100 SAE notch specimens under both plane stress and plane strain conditions. The same results were obtained as in Ref *3.* In the present paper, 20 different material properties were selected together with a FE analysis. The specimen was SAE keyhole compact tension specimen (CCT specimen). The FE analyses were performed on both CCT thin plate specimens (plane stress) and CCT thick plate specimens (plane strain) in order to check Neuber's rule in both conditions. The deformation theory of plasticity and eight-node quadrilateral element was applied. The FE nets for the CCT specimen are shown in Fig. 1 (the half of the CCT specimen). The width of the near notch element was 2.5 mm. The distance of the Gaussian stress point from the notch edge was 0.28 mm. To save compution time, the multiple-substructure was used in the computer program. Then degrees of freedom in the elastic area were condensed without loss in accuracy. The material cyclic stress-strain curve used was

$$\epsilon_a = \sigma_a/E + (\sigma_a/k')^{1/n'} \tag{1}$$

where E is the elastic modulus, k' is the cyclic strength coefficient, and n' is the cyclic hardening exponent. The parameters E, k', and n' for the 20 materials considered are shown in Table 1.

[1]Scientific Work, Institute of Mechanics, Chinese Academy of Sciences, Beijing, China.

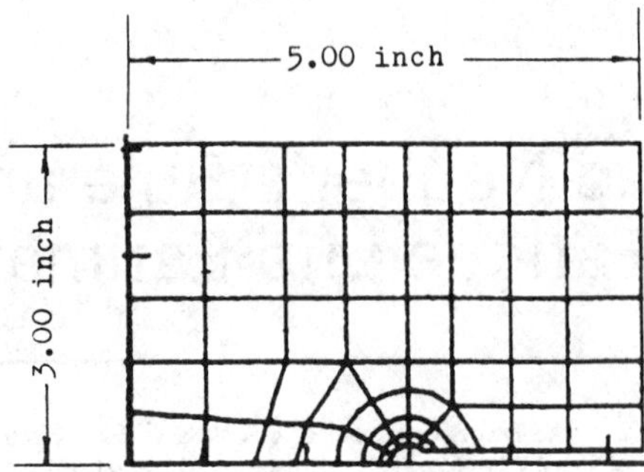

FIG. 1—*FE nets of CCT specimen.*

TABLE 1—*Material parameters* E, k′, n′.

Materials	*E* (ksi)	*k′* (ksi)	*n′*	Reference
Gainex	29 200	114	0.11	*9*
10B62	28 000	309	0.16	*9*
1005-1009	30 000	71	0.12	*9*
1015	30 000	137	0.22	*9*
1020	29 500	112	0.18	*9*
1045	29 000	195	0.18	*9*
1541-F	29 900	255	0.17	*9*
1541-F	29 900	235	0.16	*9*
5160	28 000	335	0.15	*9*
9262-2	28 000	197	0.12	*9*
9262-3	29 000	292	0.089	*9*
950x	29 500	134	0.114	*9*
980x	28 200	181	0.134	*9*
2024-T351	10 600	95	0.065	*9*
Man-Ten	29 500	160	0.19	*4*
RQC-100	29 500	199	0.15	*5*
RQC-100	29 500	167	0.10	*4*
30CrMnSiNi2a	26 700	288	0.11	*12*
30CrMnSiNia	28 700	216	0.10	*11*
LY12-CZ	10 300	77	0.03	*11*

Neuber's Rule and the Proposed Extension of Neuber's Rule

The conventional Neuber's rule is

$$K_t^2 = K_\sigma K_\epsilon \tag{2}$$

with

$$K_\sigma = \sigma_a / S_a \tag{3}$$

and

$$K_\epsilon = \epsilon_a / e_a \tag{4}$$

where K_t is the theoretical stress concentration factor, K_σ is the notch stress concentration factor, K_ϵ is the notch strain concentration factor, σ_a and ϵ_a are the notch local stress and strain range and, S_a and e_a are the nominal stress and strain range. Equation 2 can be rewritten as

$$C1C2 = 1 \tag{5}$$

with

$$C1 = K_\sigma/K_t \tag{6}$$

and

$$C2 = K_\epsilon/K_t \tag{7}$$

In the elastic range, $K_\sigma = K_t$ and $C1 = 1$. In the plastic range $K_\sigma < K_t$, $C1 < 1$. $C1$ decreases with increasing load. In the elastic range $K_\epsilon = K_t$ and $C2 = 1$. In the plastic range $K_\epsilon > K_t$ and $C2 > 1$. $C2$ increases with increasing load. According to the conventional Neuber's rule, the product of $C1$ and $C2$, denoted as $C4$, should be equal to unity. The value of $C1$ and $C2$ can be obtained from FE analyses. By substituting Eqs 3 and 4 into Eqs 6 and 7, then

$$C1 = K_\sigma/K_t = \sigma_a/K_tS_a = \sigma_a/\sigma_a^* \tag{8}$$

$$C2 = K_\epsilon/K_t = \epsilon_a/K_te_a = \epsilon_a/\epsilon_a^* \tag{9}$$

where σ_a^* and ϵ_a^* are false elastic stress and strain range, σ_a and ϵ_a can be obtained from elastic-plastic analyses, and σ_a^* and ϵ_a^* can be obtained from elastic analyses. As an example, for the 9262-3 CCT specimen the FE results are shown in Tables 2 and 3 and Figs. 2 and 3. In Tables 2 and 3 σ_{1a}, ϵ_{1a} and σ'_{1a}, ϵ'_{1a} are notch first principal stress-strain range under plane stress and plane strain conditions, respectively. We can see that the products of $C1$ and $C2$, $C4$, vary from unity. The suggested improved Neuber's rule is

$$C1^m C2^{(1-m)} = 1 \tag{10}$$

TABLE 2—*FE results of 9262-3 CCT specimen in plane stress condition.*

S_a (ksi)	ϵ_a	σ_a (ksi)	$C1$	$C2$	$C3$[a]	$C4$[a]
1.59	0.00016	4.69	1.000	1.000	1.000	1.000
17.52	0.00175	51.70	1.000	1.000	1.000	1.000
35.04	0.00351	103.17	0.998	1.000	0.999	0.999
52.57	0.00538	148.00	0.954	1.023	0.993	0.988
70.09	0.00766	172.15	0.882	1.091	0.973	0.953
87.61	0.01054	186.23	0.720	1.201	0.968	0.930
105.13	0.01399	196.16	0.632	1.329	0.971	0.917
122.66	0.01803	203.87	0.563	1.468	0.980	0.909
140.18	0.02270	210.28	0.508	1.618	0.993	0.907
157.70	0.02817	215.89	0.464	1.784	1.011	0.910
175.22	0.03446	220.89	0.427	1.964	1.032	0.916
					$\bar{C}3 = 0.993$	$\bar{C}4 = 0.948$

[a] $\bar{C}3$ and $\bar{C}4$ are average of $C3$ and $C4$, respectively.

TABLE 3—*FE results of 9262-3 CCT specimen in plane strain condition.*

S_a (ksi)	ϵ_a	σ_a (ksi)	$C1$	$C2$	$C3^a$	$C4^a$
1.59	0.00014	4.76	1.000	1.000	1.000	1.000
17.52	0.00156	52.47	1.000	1.000	1.000	1.000
35.04	0.00312	104.91	1.000	1.000	1.000	1.000
52.57	0.00474	156.48	0.994	1.010	1.003	1.002
70.09	0.00650	196.00	0.934	1.041	0.993	0.986
87.61	0.00860	223.28	0.851	1.102	0.984	0.968
105.13	0.01115	243.09	0.772	1.190	0.985	0.958
122.66	0.01408	257.49	0.701	1.288	0.988	0.950
140.18	0.01726	270.05	0.643	1.381	0.990	0.942
157.70	0.02066	282.88	0.599	1.470	0.994	0.938
175.22	0.02435	295.82	0.564	1.559	1.000	0.937
192.75	0.02830	308.68	0.535	1.647	1.008	0.938
210.27	0.03250	321.19	0.510	1.734	1.017	0.940
					$\bar{C}3 = 0.997$	$\bar{C}4 = 0.966$

[a] $\bar{C}3$ and $\bar{C}4$ are average of $C3$ and $C4$, respectively.

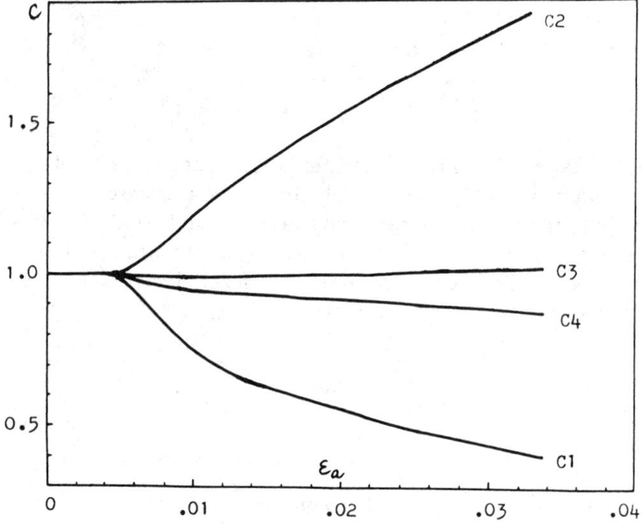

FIG. 2—*Relation between* C1, C2, C3, C4 *and notch strain in the 9262-3 CCT plane stress specimen.*

where m is a parameter concerned with material properties. If $m = 0.5$, Eq 10 reduces to Eq 5. Equation 10 can be rewritten as

$$\text{Log } C1 = -(1 - m)/m \text{ Log } C2 \tag{11}$$

The m value can be determined by use of the least squares method. The linear regression of m is shown in Figs. 4 and 5. For the 9262-3 CCT specimen under plane stress condition, m is equal to 0.42 and the linear correlation factor r is equal to 0.9920. Under plane strain condition, m is equal to 0.44 and the linear correlation factor r is equal to 0.9970. Here $C1$ and $C2$ were computed based on notch principal stress-strain. The products of $C1^m$ and $C2^{(1-m)}$, denoted as $C3$,

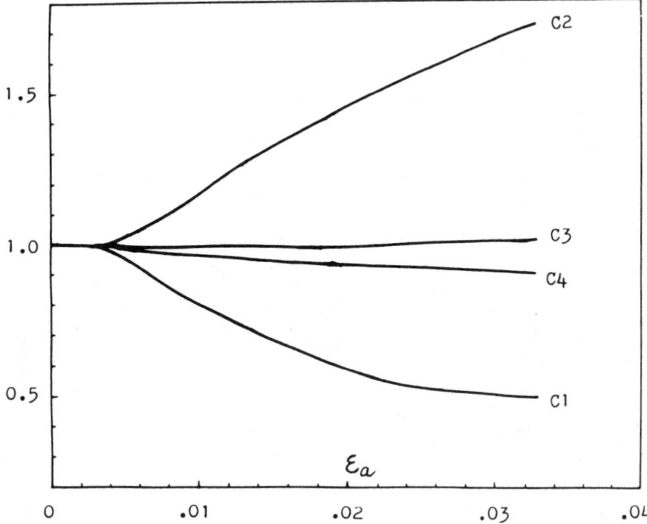

FIG. 3—*Relation between* C1, C2, C3, C4 *and notch strain in 9262-3 CCT plane strain specimen.*

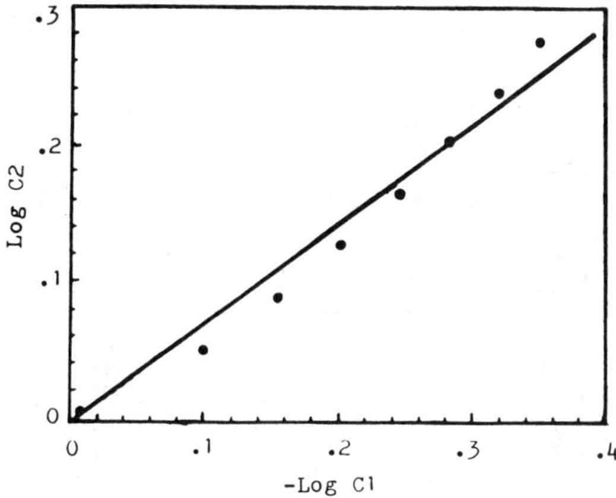

FIG. 4—*Regression of* m *in 9262–3 CCT plane stress specimen* m = *0.42,* r = *0.9924).*

are also shown in Tables 2 and 3. The values of m for 20 material CCT specimens are shown in Table 4. It can be found that m varies for different materials. By use of a linear regression, the relation between m, n', and k'/E was obtained as follows:

For the CCT plane stress condition and the notch principal stress-strain:

$$m = 0.48 + 0.31n' - 8.56k'/E \qquad (12)$$

where the linear correlation factor $r = 0.9407$.

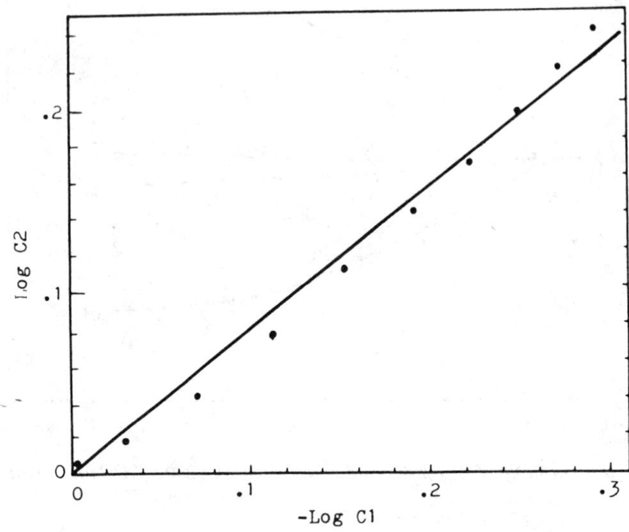

FIG. 5—*Regression of* m *in 9262-3 CCT plane strain specimen* m = *0.44*, r = *0.9970).*

TABLE 4—C3, C4 *and* m *values of CCT specimens of 20 materials (linear regression of* m *was based on notch principal stress-strain).*

	CCT Thin Plate (Plane Stress)			CCT Thick Plate (Plane Strain)		
Material	*C3*	*C4*	*m*	*C3*	*C4*	*m*
Gainex	0.982	0.965	0.49	0.989	0.984	0.50
10B62	0.993	0.951	0.44	0.996	0.968	0.45
1005-1009	0.975	0.984	0.51	0.979	1.017	0.53
1015	0.979	0.997	0.51	0.980	1.036	0.55
1020	0.977	0.994	0.51	0.981	1.014	0.53
1045	0.983	0.961	0.48	0.989	0.977	0.49
1541-F	0.990	0.954	0.46	0.994	0.972	0.47
1541-F	0.989	0.954	0.46	0.993	0.972	0.47
5160	0.993	0.950	0.43	0.997	0.967	0.45
9262-2	0.989	0.949	0.45	0.994	0.966	0.46
9262-3	0.993	0.948	0.42	0.997	0.966	0.44
950x	0.984	0.959	0.48	0.990	0.977	0.49
980x	0.988	0.952	0.46	0.993	0.969	0.47
2024-T351	0.996	0.957	0.43	0.999	0.974	0.44
Man-Ten	0.980	0.975	0.50	0.985	0.996	0.51
RQC-100	0.988	0.958	0.47	0.993	0.972	0.47
RQC-100	0.988	0.950	0.46	0.993	0.967	0.47
30CrMnSiNi2a	0.993	0.949	0.42	0.997	0.967	0.44
30CrMnSiNia	0.991	0.946	0.44	0.996	0.964	0.45
LY12-CZ	0.995	0.960	0.43	0.999	0.974	0.44
Average	0.987	0.96	0.46	0.992	0.980	0.47

For the CCT plane strain condition and the notch principal stress-strain:

$$m = 0.49 + 0.36n' - 9.17k'/E \tag{13}$$

where the linear correlation factor $r = 0.9126$.

For the CCT plane stress condition the relations between m, n', and k'/E are shown in Fig. 6. We can see that m increases with increasing n' and decreases with increasing k'/E.

In local stress-strain analysis and LCF life estimation, effective stress-strain is very useful. Fatigue is initiated by grain slip, so effective stress-strain corresponding to shear deformation is responsible for fatigue. This point will be discussed later. According to the von Mises yield criterion, the effective stress range is defined as

$$\sigma_{ma} = 1/\sqrt{2}\cdot((\sigma_{1a} - \sigma_{2a})^2 + (\sigma_{2a} - \sigma_{3a})^2 + (\sigma_{3a} - \sigma_{1a})^2)^{1/2} \tag{14}$$

In the plane stress condition the effective stress range σ_{ma} is approximately equal to the first principal stress range σ_{1a}, because the notch stress condition is approximately the uniaxial stress condition. But in the plane strain condition, the first principal stress increases under the effect of biaxiality, so there are significant differences between the principal stress range and the effective stress range, as shown in Fig. 7.

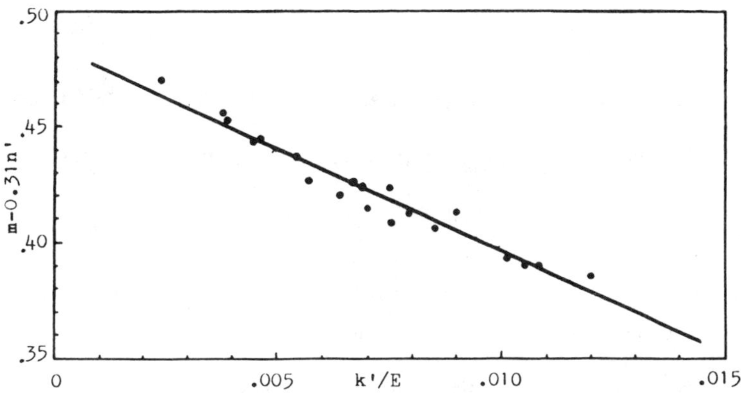

FIG. 6*a*—*Relation between* m *and* k'/E.

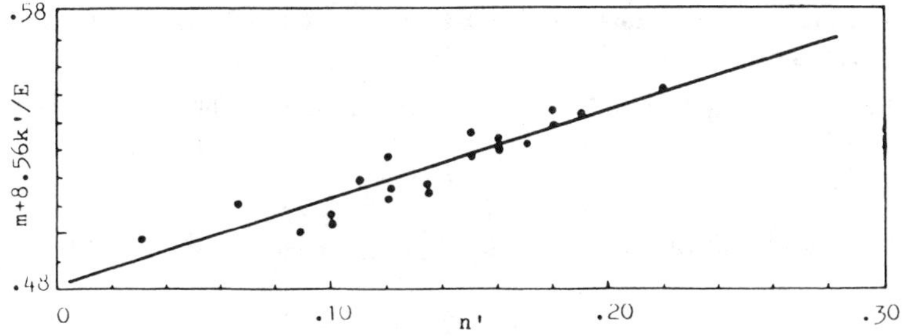

FIG. 6*b*—*Relation between* m *and* n'.

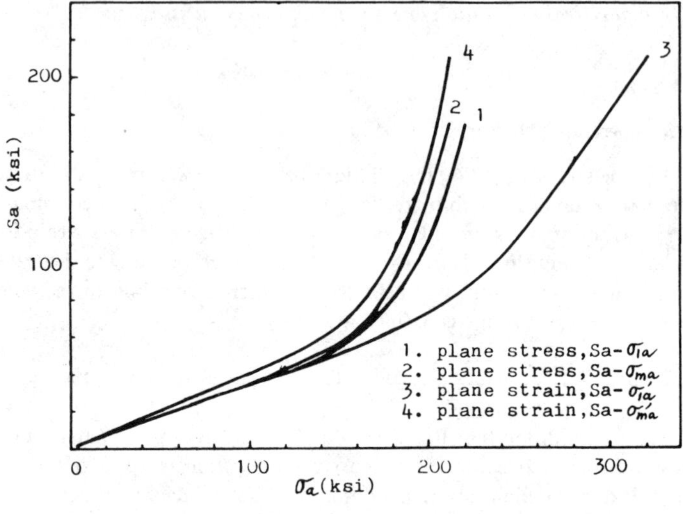

σ_{1a} = notch first principle stress range under plane stress.
σ_{ma} = notch effective stress range under plane stress.
σ'_{1a} = notch first principal stress range under plane strain.
σ'_{ma} = notch effective stress range under plane strain.

FIG. 7—*Variation of notch stress in 9262-3 CCT plane stress and plane strain condition.*

For notch effective stress-strain range, the values of m for 20 material CCT specimens are given in Table 5.

For the CCT plane stress condition and the notch effective stress-strain range:

$$m = 0.48 + 0.31n' - 8.60k'/E \tag{15}$$

$$r = 0.9421$$

For the CCT plane strain condition and the notch effective stress-strain range:

$$m = 0.42 + 0.25n' - 3.92k'/E \tag{16}$$

$$r = 0.9259$$

Local Stress-Strain Analysis of Notch Specimen Using the Conventional Neuber's Rule and the Improved Neuber's Rule

By substituting Eqs 3 and 4 into Eq 2, the following equation is obtained:

$$\sigma_a \epsilon_a = K_t^2 S_a e_a \tag{17}$$

Under nearly elastic deformation of the specimen net section $e_a = S_a/E$ and Eq 17 becomes

$$E\sigma_a \epsilon_a = K_t^2 S_a^2 \tag{18}$$

TABLE 5—C3, C4 *and* m *of CCT specimens of 20 materials (linear regression of* m *is based on notched effective stress-strain).*

Material	CCT Thin Plate (Plane Stress)			CCT Thick Plate (Plane Strain)		
	C3	*C4*	*m*	*C3*	*C4*	*m*
Gainex	0.981	0.961	0.48	0.991	0.915	0.44
10B62	0.993	0.948	0.44	0.994	0.936	0.42
1005-1009	0.974	0.981	0.50	0.985	0.912	0.45
1015	0.978	0.994	0.51	0.985	0.936	0.47
1020	0.976	0.991	0.51	0.986	0.925	0.46
1045	0.983	0.958	0.48	0.990	0.924	0.44
1541-F	0.989	0.951	0.46	0.993	0.931	0.43
1541-F	0.988	0.951	0.46	0.993	0.929	0.43
5160	0.993	0.947	0.43	0.994	0.937	0.42
9262-2	0.988	0.946	0.45	0.993	0.925	0.42
9262-3	0.992	0.945	0.42	0.994	0.937	0.41
950x	0.983	0.956	0.48	0.991	0.916	0.43
980x	0.987	0.949	0.46	0.992	0.923	0.43
2024-T351	0.995	0.954	0.43	0.995	0.947	0.41
Man-Ten	0.980	0.971	0.49	0.988	0.926	0.45
RQC-100	0.987	0.954	0.47	0.992	0.926	0.43
RQC-100	0.988	0.947	0.46	0.993	0.919	0.42
30CrMnSiNi2a	0.993	0.946	0.42	0.994	0.938	0.41
30CrMnSiNia	0.990	0.943	0.44	0.993	0.928	0.41
LY12-CZ	0.994	0.957	0.43	0.994	0.945	0.41
Average	0.987	0.958	0.46	0.992	0.929	0.43

Using Eq 18 and Eq 1, the local stress-strain range can be computed (Fig. 8). The analogous formulation of the improved Neuber's rule is

$$\sigma_a^m (E\epsilon_a)^{1-m} = K_t S_a \tag{19}$$

Notch stress-strain can be computed by the improved Neuber's rule as easily as by Neuber's rule (Fig. 8). As an example, if the nominal stress range S_a is equal to 551.6 MPa (80 ksi), the notch strain range is 0.0174 and 0.0157 by Neuber's rule and the improved Neuber's rule, respectively. The notch strain computed by Neuber's rule is larger than the strain computed by the improved Neuber's rule by about 10%.

In the plane strain condition the revised cyclic stress-strain curve should be used instead of the uniaxial cyclic stress-strain curve in order to consider the effect of biaxiality. According to Ref *4*, the revised cyclic stress-strain curve is

$$\epsilon'_{1a} = \frac{\sigma'_{1a}(1-\mu_e^2)}{E} + \left(\frac{\sigma'_{1a}}{k'}\right)^{1/n'} \tag{20}$$

where σ'_{1a} and ϵ'_{1a} are the notch first principal stress-strain range in plane strain and μ_e is Poisson's ratio; the value of k' and n' are different from the value of k' and n' under a uniaxial condition that can be determined using the method presented by Ref *4*. Notch stress-strain analyses were performed using Eqs 19 and 20.

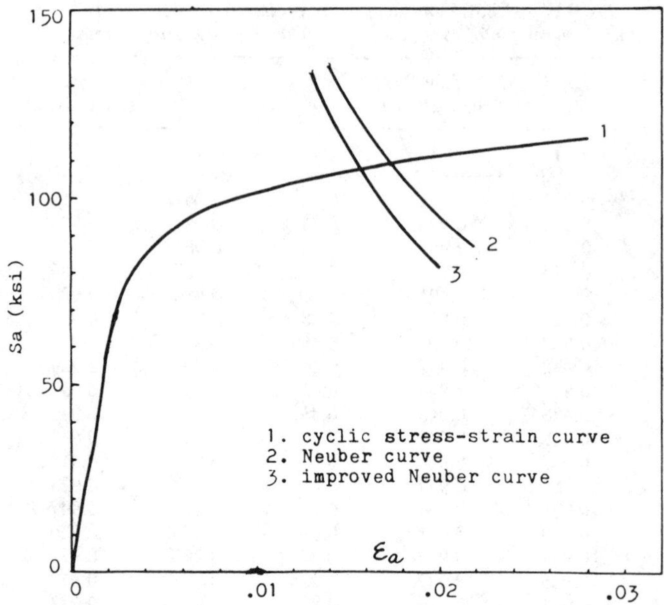

FIG. 8—*Notch stress-strain analyses using Neuber's rule and improved Neuber's rule in RQC-100 CCT plane stress condition (*E = *203 396.6 MPa (29 500 ksi),* k′ = *1151.46 MPa (167 ksi),* n′ = *0.10,* K_t = *2.95,* m = *0.46,* S_a = *551.58 MPa (80 ksi)).*

Various load-strain curves computed by FE, Neuber's rule, and the improved Neuber's rule are shown in Figs. 9, 10, and 11. The parameters used are shown in Table 6. For different stress conditions the values of m were computed using Eqs 12, 13, 15, and 16. The values for K_t were obtained from elastic analysis. An important point is that the theoretical stress concentration factor, K_t, is different for different stress conditions and different notch stress ranges (principal stress or effective stress range). From Figs. 9 to 11 some results can be obtained as follows:

1. The FE results in this paper are exactly consistent with the FE results in Ref *4*.
2. In the plane stress condition the improved Neuber's rule has given better results than Neuber's rule.
3. In the plane strain condition the benefit of the improved Neuber's rule is not obvious. The reason is that the revised cyclic stress-strain curve is approximately equal to the notch principal stress-strain curve only when n' is low, such as for LY12-CZ. Generally, the notch principal stress-strain curve under plane strain is different from the revised stress-strain curve. Therefore, theoretical results cannot be obtained using Neuber's rule and the improved Neuber's rule with the revised stress-strain curve.
4. In the plane strain condition the notch effective strain predicted by the improved Neuber's rule is close to the FE results.

LCF Life Estimation

As an example, the fatigue crack initiation life of a RQC-100 CCT specimen in Ref *4* was estimated using the local strain method [*4–7*]. The load histories are for vehicles transmission loads [*5*]. There are 1708 reversals in a block. Reference *4* points out that it appears that the behavior of an SAE CCT specimen is intermediate between plane stress and plane strain, so

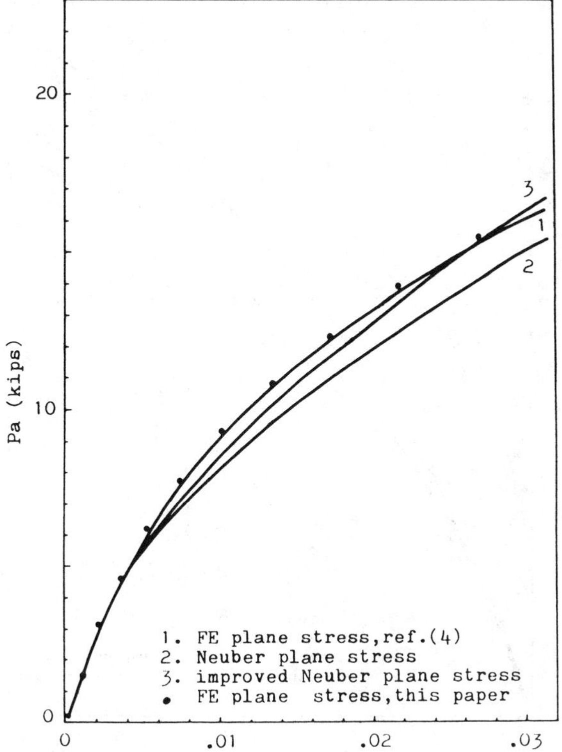

FIG. 9—*Cyclic load-notch principal strain range curve in RQC-100 plane stress condition.*

notch stress-strain were analyzed in both the plane stress and plane strain condition. The stress cycles under complex loading were distinguished using Wetzel's method introduced in Ref *6*. The damage per cycle was computed based on the notch strain range and material strain-life curve [*8*]. The effects of mean stress were considered.

$$\epsilon_a = \frac{\sigma_f' - \sigma_o}{E}(2N_f)^b + \epsilon_f'(2N_f)^c \tag{21}$$

where σ_f' is fatigue strength factor, b is fatigue strength exponent, ϵ_f' is fatigue ductility factor, c is fatigue ductility exponent, σ_o is mean stress, and N_f is number of cycles to failure. The parameters of RQC-100 are shown in Table 7. The N_f values corresponding to ϵ_a can be determined from Eq 21 using an iteration method.

The damage per cycle is

$$D_i = 1/N_f \tag{22}$$

The damage per block is

$$D_b = \Sigma 1/N_f \tag{23}$$

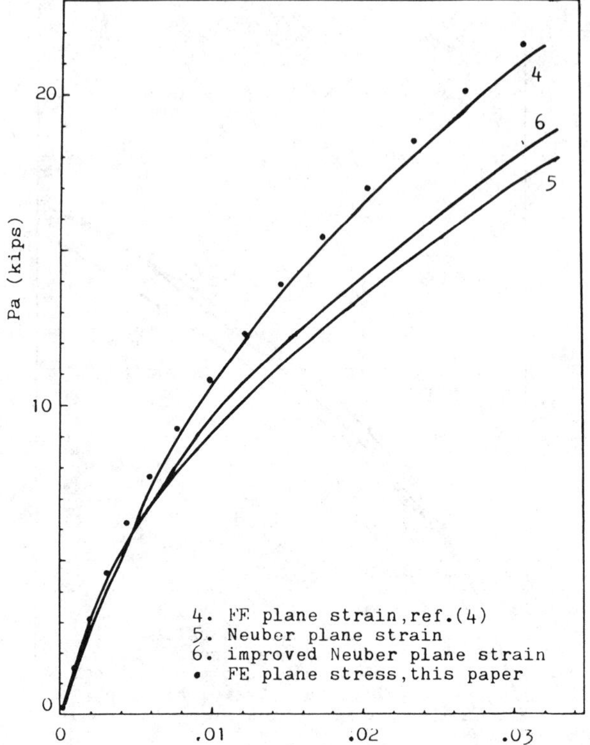

FIG. 10—*Cyclic load-notch principal strain range curve in RQC-100 plane strain condition.*

TABLE 6—*Parameters using notch stress-strain analyses of RQC-100 CCT specimen.*

Term	E (ksi)	k'	n'	K_t	m	Eq of m	Eq of Stress-Strain
Neuber plane stress	29 500	167	0.10	2.95	0.5	...	1
Neuber plane strain	29 500	200	0.105	2.99	0.5	...	20
Improved Neuber plane stress	29 500	167	0.10	2.95	0.46	12	1
Improved Neuber plane strain	29 500	200	0.105	2.99	0.476	13	20
Improved Neuber plane stress (effective stress-strain)	29 500	167	0.10	2.88	0.46	15	1[a]
Improved Neuber plane strain (effective stress-strain)	29 500	167	0.10	2.50	0.42	16	1[a]

[a]The effective stress-strain curve is the same as the uniaxial stress-strain curve.

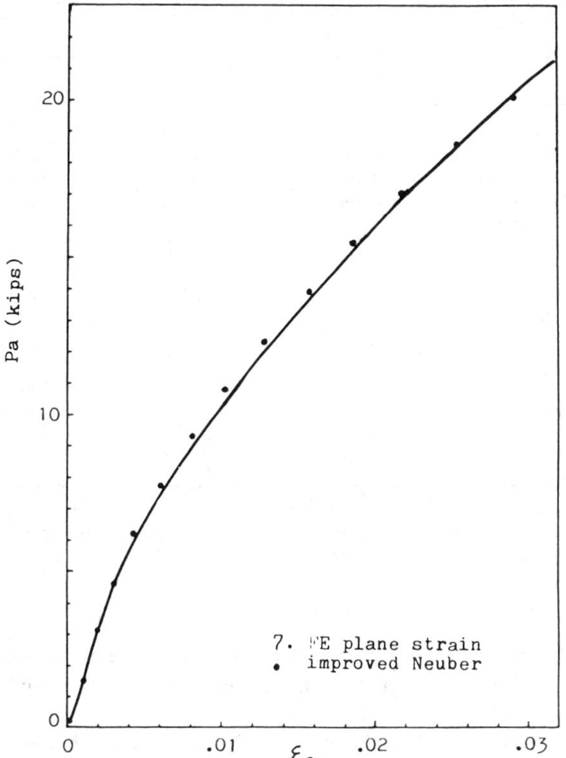

FIG. 11—*Cyclic load-notch effective strain range curve in RQC-100 plane strain condition.*

TABLE 7—*RQC-100 strain fatigue properties.*

σ_f' (ksi)	b	ϵ_f'	c
141	−0.094	1.0	−0.75

The number of blocks to failure (BTF) is

$$\text{BTF} = 1/D_b \tag{24}$$

The local stress-strain analyses and life estimation were computed using the computer program in Ref *10*. The predicted results and test results are shown in Table 8.

We can see that there are no significant differences among the predicted results using the different methods. However, the results based on notch effective stress-strain and the improved Neuber's rule are closer to test results in the LCF range.

Conclusions

1. The m in the improved Neuber's rule is concerned with the material properties. The m increases with increasing n' and decreases with increasing k'/E.

TABLE 8—*Test and predicted results of fatigue crack initiation life of RQC-100 specimen (in block).*[a]

P_{max}	16 (kips)	8 (kips)	3.5 (kips)
1. FE plane stress	8.23	130.10	42 509.51
2. FE plane strain	10.57	137.96	30 029.93
3. Neuber plane stress	6.24	130.42	67 774.31
4. Neuber plane strain	8.14	163.83	56 365.93
5. Improved Neuber plane stress	6.25	133.14	69 206.56
6. Improved Neuber plane strain	8.28	164.01	56 836.78
7. Improved Neuber plane stress	6.81	152.60	87 143.37
8. Improved Neuber plane strain	11.60	375.84	380 013.56
9. Test results	29.90	269.00	<57 090.00
	23.50	460.00	>88 020.00
	22.20	374.00	

[a]1-6 Fatigue crack initiation lifes were estimated based on notch first principal stress-strain range. 7-8 Fatigue crack initiation lifes were estimated based on notch effective stress-strain range.

2. For 20 materials the averages for m are 0.46 and 0.47 in the plane stress and plane strain condition, respectively. This means that on an average the notch strain will be overestimated by Neuber's rule ($m = 0.5$) by about 10%.

3. Notch strain analyses based on the improved Neuber's rule is better than Neuber's rule, especially in plane stress analyses and in effective stress-strain analyses under plane strain.

4. In the LCF range, fatigue life estimations based on the notch effective strain are better than the others.

References

[*1*] Neuber, H., "Theory of Stress Concentration for Shear Strained Prismatical Bodies with Arbitrary Non-Linear Stress Strain Law," *Journal of Applied Mechanics,* 1961, pp. 544–550.

[*2*] Topper, T. H., Wetzel, R. M., and Morrow, J. D., "Neuber's Rule Applied to Fatigue of Notched Specimens," *Journal of Materials,* 1969, pp. 200–209.

[*3*] Mowbray, D. F. and McConnelee, J. E., "Application of Finite Element Stress Analysis and Stress-Strain Properties in Determining Notch Fatigue Specimen Deformation and Life" in *Cyclic Stress-Strain Behavior—Analysis, Experimentation, and Failure Prediction, ASTM STP 519,* American Society for Testing and Materials, Philadelphia, pp. 151–169.

[*4*] Dowling, N. E., Brose, W. R., and Wison, W. K., "Notch Member Fatigue Life Predictions by the Local Strain Approach," *Advances in Engineering,* Vol. 6, pp. 55–84.

[*5*] Tucker, L., "The SAE Cumulative Fatigue Damage Test Program," *Advances in Engineering,* Vol. 6, pp. 1–54.

[*6*] Wetzel, R. M., Ed., "Fatigue Under Complex Loading: Analysis and Experiments," *Advances in Engineering,* Vol. 6, 1977.

[*7*] Socie, D. F., "Fatigue Life Estimation Techniques," Technical Report 145, University of Illinois, Urbana.

[*8*] Morrow, J. D., "Cyclic Plastic Strain Energy and Fatigue of Metals," in *Internal Friction, Dampling, and Cyclic Plasticity, ASTM STP 378,* American Society for Testing and Materials, Philadelphia, 1965, pp. 45–87.

[*9*] "Technical Report on Fatigue Properties," SAE J 1099, Feb. 1975.

[*10*] Wu, Y. S., "A Local Strain Fatigue Analysis Method and Computer Program," *Acta Aeronautica et Astronautica Sinica,* Vol. 3, No. 1, 1982, Beijing, China; also see N83-11040.
[*11*] Zhao, M. P., Wu, Y. S., Sun, S. T., and Chou, Z. Y., "Estimation of Fatigue Crack Initiation Life by Local Strain Method," First National Fatigue Congress in China, Huan-Shan, 7-12 Sept. 1982.
[*12*] Wu, Y. S., Jiang, X. C., Xu, S. Y., and Li, H., "Application of the Local Strain Method to Fatigue Life Estimation of Wing Main Beam Specimen," First National Fatigue Congress in China, Huan-Shan, 7-12 Sept. 1982.

G. Glinka[1]

Relations Between the Strain Energy Density Distribution and Elastic-Plastic Stress-Strain Fields Near Cracks and Notches and Fatigue Life Calculation

REFERENCE: Glinka, G., "**Relations Between the Strain Energy Density Distribution and Elastic-Plastic Stress-Strain Fields Near Cracks and Notches and Fatigue Life Calculation,**" *Low Cycle Fatigue, ASTM STP 942,* H. D. Solomon, G. R. Halford, L. R. Kaisand, and B. N. Leis, Eds., American Society for Testing and Materials, Philadelphia, 1988, pp. 1022–1047.

ABSTRACT: Relations between the strain energy density distribution and the elastic-plastic strain-stress fields near notches have been analyzed. It was found that the strain energy density distribution in the plastic zone ahead of the crack or notch tip can be satisfactorily estimated on the basis of the theoretical elastic stress field. The strain energy density can be translated into the elastic-plastic strain and stress in the plastic zone if the nonlinear material stress-strain curve is known. Good agreement with finite element calculations has been achieved for cracks under plane strain conditions. It has also been shown that the equivalent strain energy approach can be useful for the notch-tip elastic-plastic strain calculation under both monotonic and cyclic loading.

Very good agreement between the experimentally measured and calculated notch-tip strains has been obtained up to general yielding. The elastic-plastic notch-tip strain histories calculated on the basis of the equivalent strain energy density concept were used for fatigue life calculation of key-hole specimens tested under SAE transmission loading history. Significant improvement of fatigue life prediction has been achieved in comparison to the calculations based on Neuber's rule.

Possible applications of the equivalent strain energy density concept for fatigue crack growth prediction are also discussed.

KEY WORDS: cracks, notches, elastic-plastic strain-stress analysis, strain energy density, fatigue life prediction

Nomenclature

a	Half of crack length
Δa	Fatigue crack increment
B	Thickness
b	Fatigue strength exponent
C	Parameter of Paris's equation
C_p	Strain energy density correction factor accounting for plastic yielding
c	Fatigue ductility exponent
D_{ij}	Deviatoric stress tensor component
da/dN	Fatigue crack growth rate
E	Modulus of elasticity

[1]Lecturer, Department of Mechanical Engineering, University College London, Torrington Place, London WC1E 7JE, England.

FEM	Finite element method
H	Plastic modulus for linear strain hardening materials
I_n	Parameter of HRR model
K	Stress intensity factor
ΔK	Range of stress intensity factor
K_t	Stress concentration factor
K_w	Strain energy density concentration factor
k	Strength coefficient
L	Width of a flat specimen or diameter of a cylindrical bar
l	Width (diameter) of the net section
$N = 1/n$	Strain hardening coefficient
$n = 1/N$	Strain hardening coefficient
N_f	Number of cycles to failure (fatigue life)
P	Force (load)
P_{max}	Highest peak load in the loading history
P_i	Current load peak
r	Radial distance from the origin of polar coordinates
r_p	First estimation of the plastic zone size based on elastic stress field and Hencky-Mises-Huber plastic yielding criterion
Δr_p	Increment of the plastic zone size due to local stress redistribution
S	Remote stress for cracks or average net section stress for notches
U	Parameter dependent of fatigue crack tip closure, stress state, etc.
W	Strain energy density
W_e	Elastic strain energy density
W_p	Plastic strain energy density
ΔW_p	Plastic strain energy density range
x	Distance from the notch tip
α	Parameter of Ramberg-Osgood relation
δ_{ij}	Kronecker's delta, $\delta_{ij} = 1$ for $i = j$ and $\delta_{ij} = 0$ for $i \neq j$
ϵ	Strain
ϵ_{ij}	Strain tensor
$\epsilon_x, \epsilon_y, \epsilon_z$	Strain components
$\epsilon_x^p, \epsilon_y^p, \epsilon_z^p$	Plastic strain components
$\epsilon_x^e, \epsilon_y^e, \epsilon_z^e$	Elastic strain components
ϵ_i	Equivalent strain
ϵ_i^p	Equivalent plastic strain
ϵ_f'	Fatigue ductility coefficient
$\Delta\epsilon$	Strain range
$\tilde{\epsilon}_y(\theta, n)$	Parameter of HRR model
$\pi = 3.14$	Constant
ν	Poisson's coefficient
ρ	Notch tip radius
σ	Stress
σ_i	Equivalent von Mises stress
σ_{ij}	Stress tensor
$\sigma_x, \sigma_y, \sigma_z, \tau_{xy}$	Stress components
σ_{YS}	Yield strength
$\sigma_y(r_p)$	Value of stress component σ_y at the elastic plastic boundary
σ_f'	Fatigue strength coefficient
σ_m	Mean stress
θ	Angular coordinate

Introduction

It has been generally accepted in the field of mechanics of solids that singular asymptotic solutions give good stress and strain estimations near cracks with the tip radius $\rho = 0$. These calculations are usually based on the stress intensity factor K [1]. Notches are understood to have tip radii $\rho > 0$, and the stress concentration factor K_t is frequently used for the notch-tip stress calculation. Despite different mathematical descriptions, the stress-strain fields near notches and cracks exhibit similar features. Very often, the same strain energy or strain-stress criteria [2,3] are employed in fracture and fatigue analysis of cracks and notches.

Therefore it is important to know the strain energy density distribution ahead of a crack or a notch tip. However, high stress concentrations near cracks and notches may cause localized plastic yielding which requires an elastic-plastic stress-strain analysis. In spite of significant progress in this field [4-6] such analyses are difficult and rigorous calculations seldom possible. Therefore approximate methods are mostly used in engineering practice. The equivalent strain energy density method, discussed below, can be also used for the solution of such problems.

Strain Energy Distribution Ahead of a Crack Tip

It was shown by Hutchinson [7] that for the case of a through crack (Fig. 1) in an infinite plate the strain energy density in the crack tip plastic zone was always dependent on the parameter $(1/r)$ regardless of the material stress-strain behavior. It was also shown that for material characterized by the bi-linear stress-strain curve (Fig. 2) the strain energy density W in the plastic zone was the same as that calculated on the basis of the perfectly linear elastic stress-strain field. It was concluded that similar equivalence of the strain energy density in the crack tip plastic

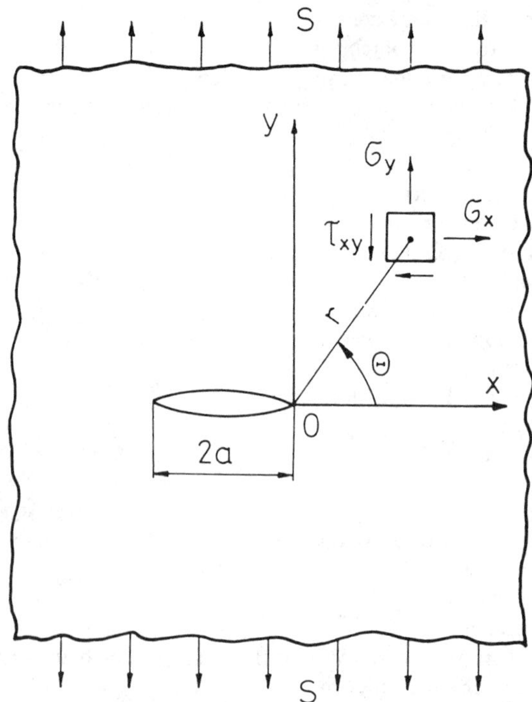

FIG. 1—*Notation for a crack in an infinite plate.*

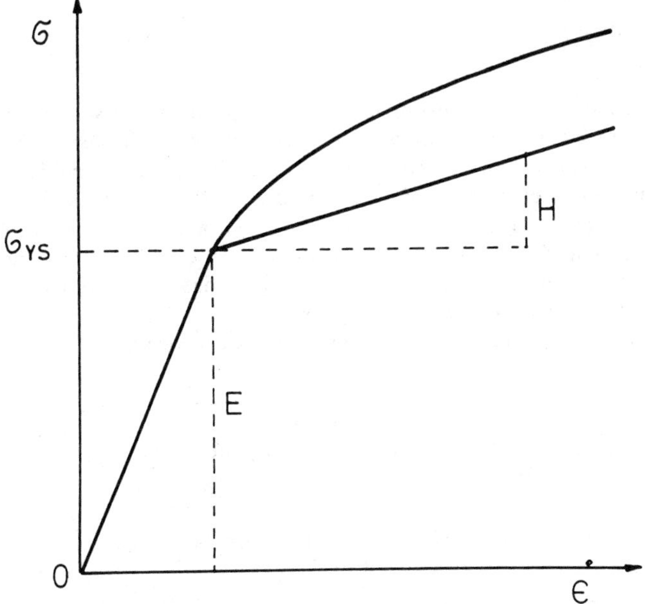

FIG. 2—*Bilinear and power law hardening stress-strain curves.*

zone might have occurred in the power law hardening material, assuming that the plastic zone was sufficiently small in comparison to the surrounding elastic field.

Elastic Strain Energy Distribution Near Cracks

The strain energy density ahead of a crack tip in a perfectly elastic body can be calculated from the well-known [*8*] asymptotic stress field solution:

$$\sigma_x = \frac{K}{\sqrt{2\pi r}} \cos\frac{\theta}{2}\left[1 - \sin\frac{\theta}{2}\sin\frac{3\theta}{2}\right]$$

$$\sigma_y = \frac{K}{\sqrt{2\pi r}} \cos\frac{\theta}{2}\left[1 + \sin\frac{\theta}{2}\sin\frac{3\theta}{2}\right]$$

$$\tau_{xy} = \frac{K}{\sqrt{2\pi r}} \sin\frac{\theta}{2}\cos\frac{3\theta}{2} \tag{1}$$

$$\sigma_z = 0 \qquad \text{for plane stress}$$

$$\sigma_z = \nu(\sigma_x + \sigma_y) \qquad \text{for plane strain}$$

Thus the energy density in the crack plane ($\theta = 0$) is

$$W = \int \sigma_{ij} d\epsilon_{ij} = \frac{K^2}{2\pi r}\left(\frac{1-\nu}{E}\right) \qquad \text{for plane stress} \tag{2}$$

$$W = \int \sigma_{ij} d\epsilon_{ij} = \frac{K^2}{2\pi r}\left(\frac{1 - \nu - 2\nu^2}{E}\right) \quad \text{for plane strain} \tag{3}$$

However, the plastic yielding occurring at the crack tip causes stress redistribution and an increase of plastic zone size [*8*] in comparison to that estimated on the basis of the elastic stress field. According to Irwin's [*9*] plastic zone correction factor (Fig. 3), the additional plastic zone increment is equivalent to the increase of the elastic stress field near a crack tip by factor of $\sqrt{2}$. It is worth noting that this correction factor was derived for perfectly elastic-plastic body under anti-plane shear loading. Therefore it can be lower than $\sqrt{2}$ for power law hardening materials. Nevertheless, it was shown by Hilton and Hutchinson [*10*] that this correction was satisfactory for a wide range of power law hardening materials.

The increase of the near-tip local stresses by factor of $\sqrt{2}$ increases the strain energy density by factor of 2:

$$W = \frac{K^2}{\pi r}\left(\frac{1 - \nu}{E}\right) \quad \text{for plane stress} \tag{4}$$

$$W = \frac{K^2}{\pi r}\left(\frac{1 - \nu - 2\nu^2}{E}\right) \quad \text{for plane strain} \tag{5}$$

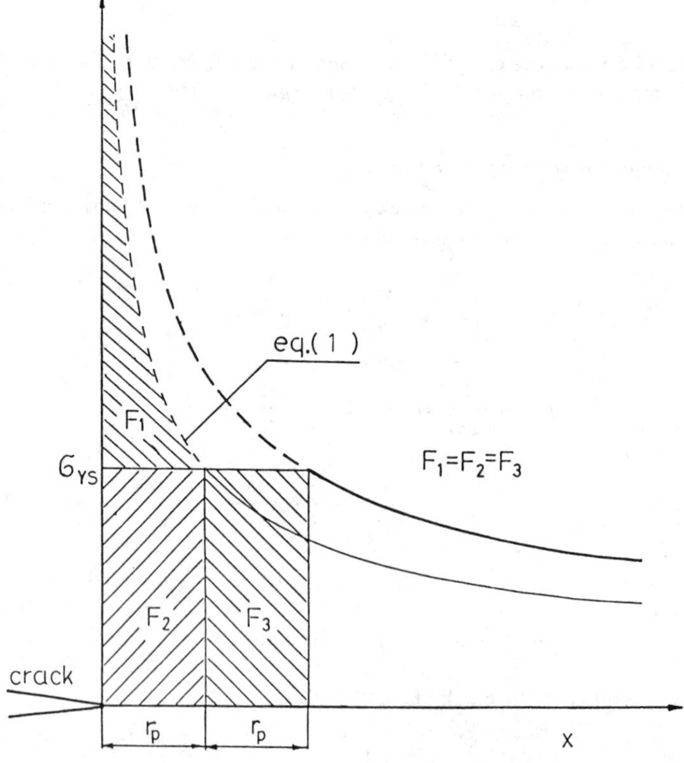

FIG. 3—*Idealized crack-tip stress redistribution and plastic zone correction.*

Thus, according to Hutchinson's [7] findings, the energy density in the plastic zone ahead of a crack tip under plane stress in a linearly strain hardening materials can be calculated on the basis of the perfect elasticity solution (see Eq 4). In order to verify this hypothesis for power law hardening materials and different stress states the strain energy density calculated from Eq 5 was compared with plane-strain elastic-plastic finite element calculations.

Elastic-Plastic Strain Energy Distribution Near Cracks

Finite element results obtained by Sih et al. [11-13] were taken for comparison. The strain energy density was calculated for two different crack geometries in a power law hardening material of the Ramberg-Osgood type:

$$\begin{cases} \epsilon = \dfrac{\sigma}{E} & \text{for} \quad \sigma \leq \sigma_{YS} \qquad (6) \\[2ex] \epsilon = \dfrac{\sigma}{E} + \dfrac{\alpha\sigma_{YS}}{E}\left(\dfrac{\sigma}{\sigma_{YS}}\right)^n & \text{for} \quad \sigma > \sigma_{YS} \qquad (7) \end{cases}$$

The material properties were as follows: $E = 2.0684 \times 10^5$ MPa, $\sigma_{YS} = 5.17 \times 10^2$ MPa, $\nu = 0.3$, $\alpha = 0.02$, and $n = 5$. Calculations were conducted for a penny-shape crack in a cylindrical bar under tension (Fig. 4*a*) and for a through crack in a finite width plate (Fig. 4*b*). The strain energy density W was calculated for the nominal stress $S = 413.7$ MPa in the case of the cylindrical bar and for stresses $S = 103.4$ MPa and $S = 426.6$ MPa in the case of the flat plate. Plane strain and axisymmetric stress state were assumed for the flat and cylindrical specimens, respectively. The corresponding stress intensity factors were calculated [14] according to

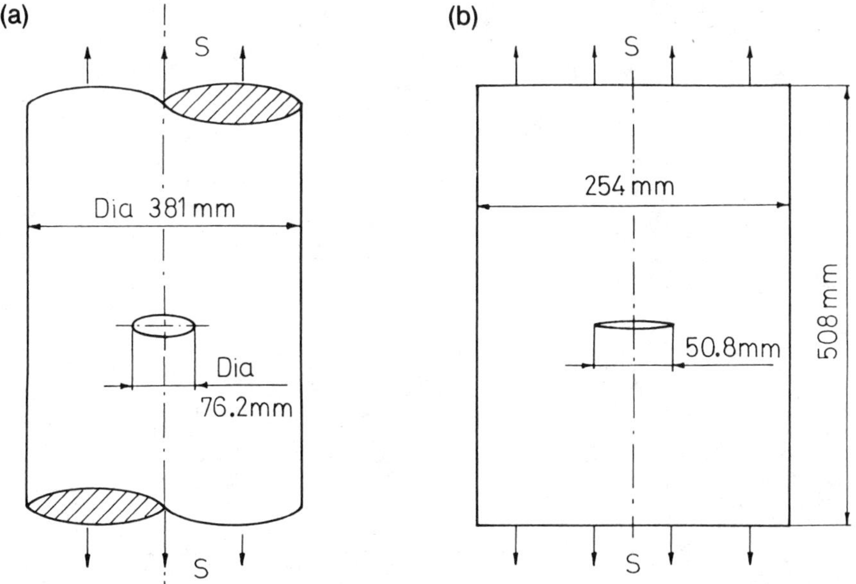

FIG. 4—*Geometry and dimensions of analyzed crack configurations.* (a) *Penny-shape crack in a cylindrical bar.* (b) *Through-crack in a finite width plate.*

$$K = S_{\text{net}}\sqrt{\pi a}\, F\left(\frac{2a}{L}\right) \qquad \text{for cylindrical bar} \tag{8}$$

where $F(2a/L) = 0.62$. For $2a/L = 0.2$,

$$K = S\sqrt{\pi a}\sqrt{\sec(\pi a/L)} \qquad \text{for flat plate} \tag{9}$$

The distributions of the strain energy density calculated by Sih et al. [*11–13*] and the strain energy density calculated from Eq 5 are shown in Figs. 5 and 6.

It is apparent that the strain energy density calculated on the basis of linear elasticity (Eq 5) was very close to that calculated for the power law hardening material by the finite element method.

The largest differences between these two methods ($\sim$20%) occurred remotely from the crack tip close to the elastic-plastic boundary. These were mostly due to the finite element strain energy density calculation procedure [*11*] based on the integration of equivalent stress σ_i and the equivalent strain ϵ_i which excludes the dilational energy and underestimates the total strain energy density in the region of relatively high elastic strains. The differences might have also resulted from the approximate nature of the plastic zone correction factor $\sqrt{2}$ and the accuracy of the finite element method. Nevertheless, good agreement between these two types of calculation was found over a large part of the crack-tip plastic zone. Therefore it was concluded that the strain energy density near a crack tip is controlled by the elastic field surrounding the crack tip and can be satisfactorily estimated on the basis of Eq 5.

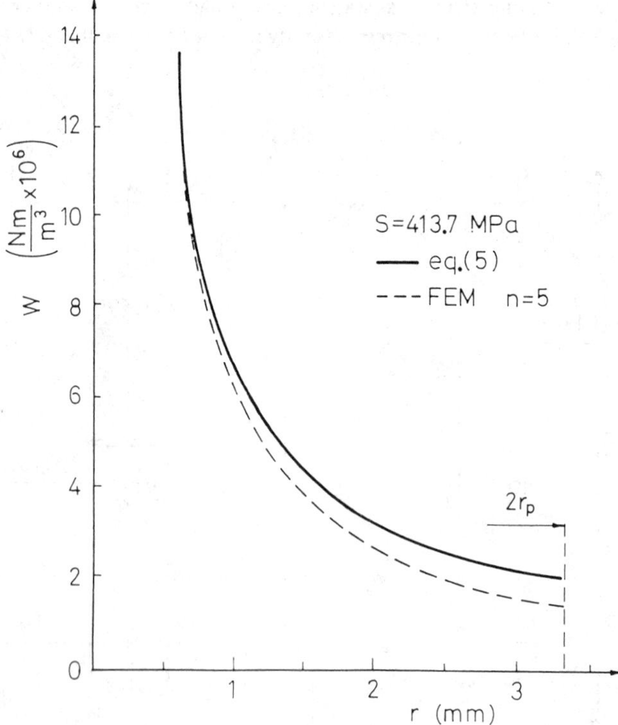

FIG. 5—*Strain energy density ahead of the penny shape crack in the cylindrical bar; elastic-plastic finite element calculations* [11] *versus linear elastic analysis (Eq 5).*

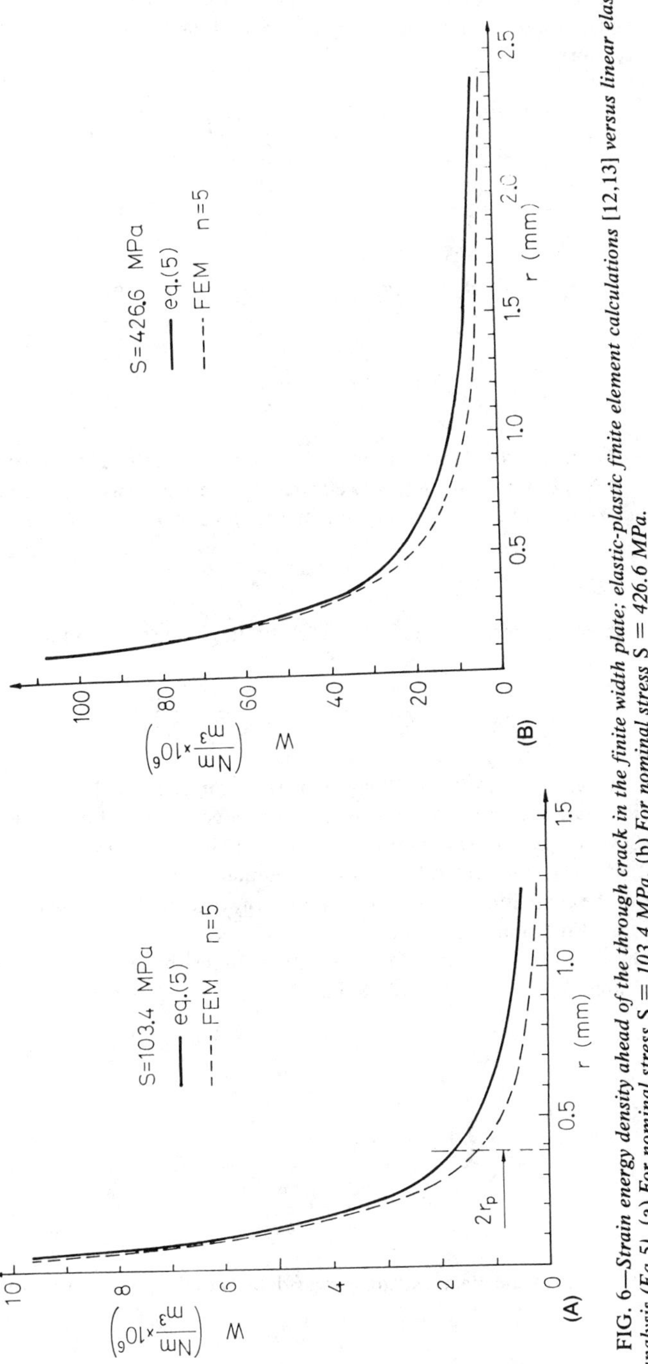

FIG. 6—*Strain energy density ahead of the through crack in the finite width plate; elastic-plastic finite element calculations* [12,13] *versus linear elastic analysis (Eq 5).* (a) *For nominal stress* S = 103.4 *MPa.* (b) *For nominal stress* S = 426.6 *MPa.*

Application of the Equivalent Strain Energy Density Hypothesis for Cracks in Plane Strain

Calculation of the strain energy density requires integration of the stress-strain relations. The constitutive relations for a power law hardening material of the Ramberg-Osgood type are usually given [7] in the form

$$\epsilon_{ij} = \frac{1+\nu}{E}\sigma_{ij} - \frac{\nu}{E}\delta_{ij}\sigma_{kk} + \frac{3}{2}\frac{\alpha}{E}\left(\frac{\sigma_i}{\sigma_{YS}}\right)^{n-1}\cdot D_{ij} \tag{10}$$

where

$$\sigma_i^2 = \frac{3}{2}D_{ij}\cdot D_{ij} = \frac{1}{2}[(\sigma_x-\sigma_y)^2 + (\sigma_y-\sigma_z)^2 + (\sigma_z-\sigma_x)^2 + 6(\tau_{xy}^2 + \tau_{yz}^2 + \tau_{zx}^2)] \tag{11}$$

$$D_{ij} = \sigma_{ij} - \frac{1}{3}\sigma_{kk} \tag{12}$$

Unfortunately, direct analytical integration of Eq 10 is not possible. Therefore the total strain energy density W was split into two parts: the elastic strain energy density W_e due to the elastic deformations and the plastic energy density W_p due to the plastic deformations. Hence

$$W = W_e + W_p \tag{13}$$

The elastic strain energy density ahead of a crack tip in plane strain can be calculated from

$$W_e = \int \sigma_{ij}d\epsilon_{ij} = \frac{\sigma_i^2}{E}\left(\frac{1-\nu-2\nu^2}{1-4\nu+4\nu^2}\right) \tag{14}$$

Equation 14 is accurate up to the yielding point (i.e., for $\sigma_i \leqq \sigma_{YS}$). However, analysis of available numerical data indicates that Eq 14 also gives a good estimation of the elastic strain energy density above the plastic yielding limit. Equation 14 is less accurate in the region of high plastic strains due to the different ratio of stress components (σ_x/σ_y), but it does not cause significant error in the estimation of the total strain energy density W because the plastic energy term W_p dominates (Eq 13) in this region.

On the other hand, the plastic strain energy density under the assumption of the "proportional plastic straining" [*15*] can be calculated as

$$W_p = \int_0^{\epsilon_i^p} \sigma_i d\epsilon_i^p \tag{15}$$

where

$$\epsilon_i^p = \frac{\alpha\sigma_{YS}}{E}\left(\frac{\sigma_i}{\sigma_{YS}}\right)^n \tag{16}$$

Integration of Eq 15 gives the plastic strain energy density in the terms of the equivalent stress σ_i:

$$W_p = \frac{\alpha\sigma_{YS}}{E}\left(\frac{n}{n+1}\right)\sigma_i\left(\frac{\sigma_i}{\sigma_{YS}}\right)^n \tag{17}$$

Finally, the total strain energy density can be written in the form

$$W = W_e + W_p = \frac{\sigma_i^2}{E}\left(\frac{1-\nu-2\nu^2}{1-4\nu+4\nu^2}\right) + \frac{\alpha\sigma_{YS}}{E}\left(\frac{n}{n+1}\right)\sigma_i\left(\frac{\sigma_i}{\sigma_{YS}}\right)^n \tag{18}$$

If the equivalent strain energy density hypothesis is valid, Eqs 5 and 18 make it possible to relate the stress intensity factor K and the equivalent stress σ_i in the crack tip plastic zone:

$$\frac{K^2}{\pi r}\left(\frac{1-\nu-2\nu^2}{E}\right) = \frac{\alpha\sigma_{YS}}{E}\left(\frac{n}{n+1}\right)\sigma_i\left(\frac{\sigma_i}{\sigma_{YS}}\right)^n + \frac{\sigma_i^2}{E}\left(\frac{1-\nu-2\nu^2}{1-4\nu+4\nu^2}\right) \tag{19}$$

In the case of plastic materials with high strain hardening exponent the equivalent stress σ_i in the plastic zone is close to the material yield strength σ_{YS}. Therefore it can be assumed that the elastic strain energy density W_e is almost constant during plastic yielding. This makes it possible to solve easily Eq 19 in respect to the equivalent stress σ_i:

$$\sigma_i = \left[\frac{(n+1)E}{n\alpha}\right]^{1/(n+1)} \cdot \sigma_{YS}{}^{(n-1)/(n+1)}\left[\frac{K^2}{\pi r}\left(\frac{1-\nu-2\nu^2}{E}\right) - W_e\right]^{1/(n+1)} \tag{20}$$

where

$$W_e = \frac{\sigma_{YS}^2}{E}\left(\frac{1-\nu-2\nu^2}{1-4\nu+4\nu^2}\right) \tag{20a}$$

Subsequently, the equivalent plastic strain ϵ_i^p can be calculated from Eq 16:

$$\epsilon_i^p = \left(\frac{n+1}{n}\right)^{n/(n+1)}\left(\frac{\alpha}{E}\right)^{1/(n+1)} \sigma_{YS}{}^{(1-n)/(1+n)}\left[\frac{K^2}{\pi r}\left(\frac{1-\nu-2\nu^2}{E}\right) - W_e\right]^{n/(n+1)} \tag{21}$$

The equivalent plastic strain ϵ_i^p can be rewritten in the form

$$\epsilon_i^p = \frac{\sqrt{2}}{3}\sqrt{(\epsilon_x^p - \epsilon_y^p)^2 + (\epsilon_y^p - \epsilon_z^p)^2 + (\epsilon_z^p - \epsilon_x^p)^2} \tag{22}$$

For plane strain conditions, the following relations between plastic strain components occur:

$$\epsilon_z^p = 0 \quad \text{and} \quad -\epsilon_x^p = \epsilon_y^p \tag{23}$$

Thus the equivalent plastic strain under plane strain conditions can be written as

$$\epsilon_i^p = \frac{2}{\sqrt{3}}\,\epsilon_y \tag{24}$$

Equations 21 and 24 make it possible to calculate the plastic strain components for a given stress intensity factor K at distance r from the crack tip:

$$-\epsilon_x^p = \epsilon_y^p = \frac{\sqrt{3}}{2}\left(\frac{n+1}{n}\right)^{n/(n+1)}\left(\frac{\alpha}{E}\right)^{1/(n+1)} \times \sigma_{\text{YS}}{}^{(1-n)/(1+n)}\left[\frac{K^2}{\pi r}\left(\frac{1-\nu-2\nu^2}{E}\right)-W_e\right]^{n/(n+1)} \tag{25}$$

The total strain component ϵ_y can be calculated as the sum of the elastic ϵ_y^e and plastic ϵ_y^p strain:

$$\epsilon_y = \epsilon_y^e + \epsilon_y^p \tag{26}$$

Assuming that under stress higher than the yield limit ($\sigma_i > \sigma_{\text{YS}}$) material deformation occurs mostly due to plastic yielding, the elastic strain ϵ_y^e at the yield point can be considered as the estimation of the elastic strain term:

$$\epsilon_y^e = \frac{\sigma_i}{E}\left(\frac{1-\nu-2\nu^2}{1-4\nu-4\nu^2}\right) = \frac{\sigma_{\text{YS}}}{E}\left(\frac{1-\nu-2\nu^2}{1-4\nu+4\nu^2}\right) \tag{27}$$

Thus an expression for the approximate value of the total strain exponent ϵ_y can be written in the form

$$\epsilon_y = \frac{\sqrt{3}}{2}\left(\frac{n+1}{n}\right)^{n/(n+1)}\left(\frac{\alpha}{E}\right)^{1/(n+1)} \sigma_{\text{YS}}{}^{(1-n)/(1+n)}\left[\frac{K^2}{\pi r}\left(\frac{1-\nu-2\nu^2}{E}\right)-W_e\right]^{n/(n+1)} + \frac{\sigma_{\text{YS}}}{E}\left(\frac{1-\nu-2\nu^2}{1-4\nu+4\nu^2}\right) \tag{28}$$

It should be noted that the elastic terms in Eq 28 are not accurate in the presence of high plastic strains, but they do not have much influence on the total strain estimation because they are relatively small. However, they give good estimation of the elastic strain contribution near the elastic-plastic boundary.

Strains calculated from Eq 28 were verified against elastic-plastic strains calculated by Walker [*16*] using an elastic-plastic finite element method. The results given in Ref *16* were obtained for a wedge opening specimen (WOL) made from power law hardening material Zircaloy-4 characterized by the stress-strain relation

$$\begin{aligned} \epsilon &= \frac{\sigma}{E} && \text{for } \sigma \leq \sigma_{\text{YS}} \\ \epsilon &= \frac{\sigma}{E} + \left(\frac{\sigma}{k}\right)^{1/N} && \text{for } \sigma > \sigma_{\text{YS}} \end{aligned} \tag{29}$$

The material data were as follows: $\sigma_{\text{YS}} = 298$ MPa, $E = 96\,561.2$ MPa, $k = 742.8$ MPa, and $N = 0.147$. The stress intensity factor, for the given load and geometric configuration, was $K = 23.1$ MPa$\sqrt{\text{m}}$. The asymptotic solution (Eq 30) for blunt deep notches, given by Creager and Paris [*17*], was used for the strain energy density calculation because the finite element calculations were carried out for a blunt crack with the tip radius $\rho = 2.54 \times 10^{-6}$ m:

$$\sigma_x = \frac{K}{\sqrt{2\pi r}} \cos\frac{\theta}{2}\left[1 - \sin\frac{\theta}{2}\sin\frac{3\theta}{2}\right] - \frac{K}{\sqrt{2\pi r}}\frac{\rho}{2r}\cos\frac{3\theta}{2}$$

$$\sigma_y = \frac{K}{\sqrt{2\pi r}} \cos\frac{\theta}{2}\left[1 + \sin\frac{\theta}{2}\sin\frac{3\theta}{2}\right] + \frac{K}{\sqrt{2\pi r}}\frac{\rho}{2r}\cos\frac{3\theta}{2}$$

$$\tau_{xy} = \frac{K}{\sqrt{2\pi r}} \sin\frac{\theta}{2}\cos\frac{\theta}{2}\cos\frac{3\theta}{2} - \frac{K}{\sqrt{2\pi r}}\frac{\rho}{2r}\sin\frac{3\theta}{2} \qquad (30)$$

$$\sigma_z = 0 \qquad \text{for plane stress}$$

$$\sigma_z = \nu(\sigma_x + \sigma_y) \qquad \text{for plane strain}$$

It should be noted that these equations (Eq 30) have been derived for the coordinate system shown in Fig. 7 with the origin shifted from the notch tip by a distance $\rho/2$. Comparisons of the

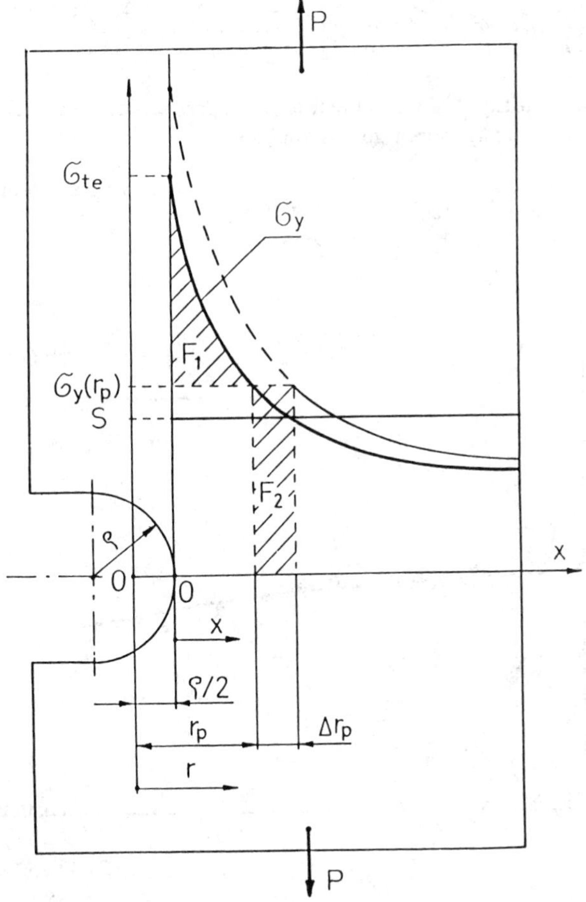

FIG. 7—*Notation for Creager and Paris* [17] *stress field near blunt notches.*

equivalent stress σ_i and the strain component ϵ_y calculated by Walker [*16*] with those calculated from Eqs 20 and 28, respectively, are shown in Figs. 8 and 9.

Very good agreement between the finite element and the energy based results has been achieved in the region of high plastic strains, close to the crack tip. The largest differences in estimation of equivalent stress σ_i and strain ϵ_y occurred near the elastic-plastic boundary. The maximum difference in equivalent stress σ_i was about 20%. Smaller differences occurred for stresses σ_i calculated from Eq 19, but it did not affect the stress and strain estimations in the highly strained area near the crack tip.

The limited data shown below indicate that good estimation of plastic strains ahead of a crack tip can be achieved on the basis of the equivalent strain energy density hypothesis. However, an additional assumption is necessary for calculating the stress components, but this is beyond the scope of this paper.

Consequences of the Energy Density Equivalence Near a Crack Tip

In the presence of high plastic strains all elastic terms in Eq 28 are relatively small and can be neglected:

$$\epsilon_y \cong \epsilon_y^p = \frac{\sqrt{3}}{2E}(\alpha)^{1/(n+1)}\sigma_{YS}{}^{(1-n)/(1+n)}\left[\frac{(1-\nu-2\nu^2)(n+1)}{n\cdot\pi}\right]^{n/(n+1)}\left(\frac{K^2}{r}\right)^{n/(n+1)} \tag{31}$$

It is worth noting that Eq 31 is very similar to the expression derived by Hutchinson [*7*] and Rice and Rosengren [*5*] using more rigorous analysis:

$$\epsilon_y = \frac{\tilde{\epsilon}_y(\theta, n)}{E}(\alpha)^{1/(n+1)}\sigma_{YS}{}^{(1-n)/(1+n)}\left[\frac{1}{I_n}\right]^{n/(n+1)}\left(\frac{K^2}{r}\right)^{n/(n+1)} \tag{32}$$

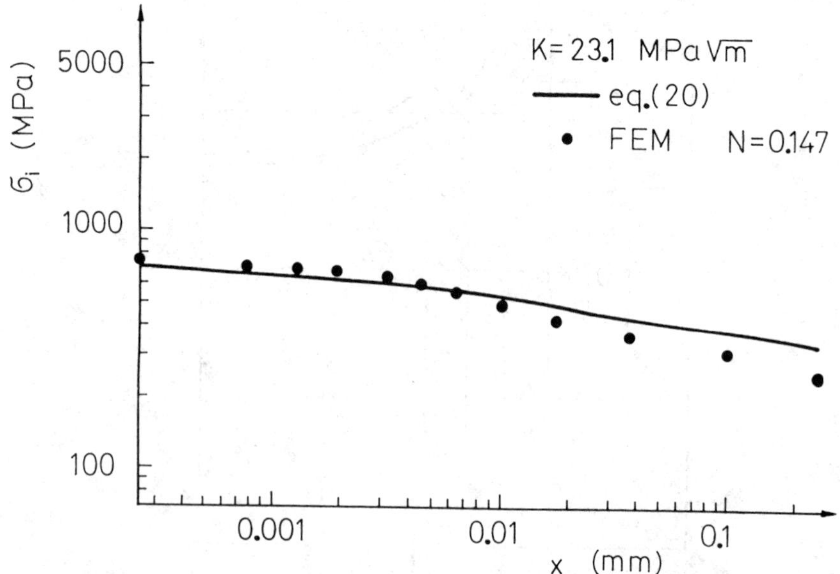

FIG. 8—*Equivalent stress* σ_i *in the plastic zone ahead of the crack tip in the WOL specimen calculated by the finite element* [16] *and the equivalent strain energy density method (Eq 20).*

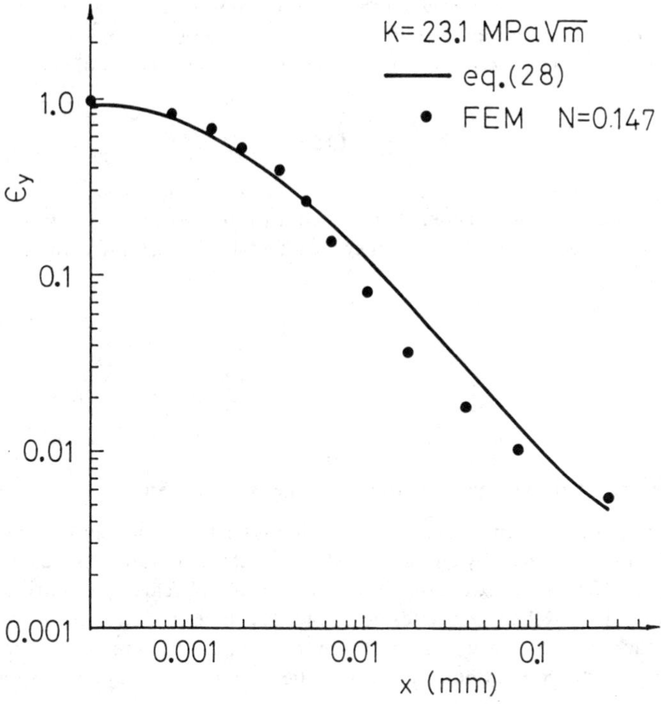

FIG. 9—*Strain* ϵ_y *in the plastic zone ahead of the crack tip in the WOL specimen calculated by the finite element* [16] *and the equivalent strain energy density method (Eq 28).*

The equivalence of the strain energy density at the crack-tip plastic zone and the strain energy density obtained from the elastic stress-strain analysis indicate that the energy density in the plastic zone depends mainly on the elastic properties of the surrounding material. As a result the plastic strain energy density in the region of the crack tip is controlled by the modulus of elasticity E, Poisson's coefficient ν, and the stress intensity factor K:

$$W_p \cong \frac{K^2}{\pi r}\left(\frac{1-\nu-2\nu^2}{E}\right) \quad \text{for plane strain} \tag{33}$$

$$W_p \cong \frac{K^2}{\pi r}\left(\frac{1-\nu}{E}\right) \quad \text{for plane stress} \tag{34}$$

Plastic strain energy density near the tip of a fatigue crack should also be controlled by the same parameters:

$$\Delta W_p \cong U\frac{\Delta K^2}{\pi r E} \tag{35}$$

where U = factor depending on the stress state, closure, etc.

This leads to the conclusion that the parameter $\Delta K/E$ is the controlling factor in fatigue crack growth as the fatigue process depends mainly on the accumulated plastic work. This very strong dependence of fatigue crack growth rate on the parameter $\Delta K/E$ was pointed out several years ago by Paris [*18*]. Equation 35 indicates also that prospective models of fatigue crack growth based on plastic energy criterion should lead to a power relation of the type

$$\frac{da}{dN} = C(\Delta K)^2 \tag{36}$$

Some of the models discussed in Refs *3* and *19* confirm this conclusion. On the other hand, fatigue crack growth models based on strain criteria may lead [*20,21*] to fatigue crack growth according to

$$\frac{da}{dN} = (\Delta K)^m \tag{37}$$

where $m = f(n)$.

Strain Energy Density Distribution Near Blunt Notches in Plane Stress

It was shown by Schijve [*22*] that the elastic stress fields near blunt notches are similar to each other and they can be satisfactorily characterized by the notch tip radius ρ and the stress concentration factor K_t. The set of equations (Eqs 30) derived by Creager and Paris [*17*] for deep sharp notches indicate that the stress fields near notches can be characterized by two dominant terms similar to the stress fields near cracks. These two conclusions were implemented in Ref *23* for derivation of generalized relations for the stress field near notches under tension and bending.

Analysis of these equations (Eq 30) has led to the conclusion [*23*] that the stresses in the symmetry plane ($\theta = 0$) ahead of the notch tip can be characterized by the notch tip radius ρ and the stress concentration factor K_t:

$$\begin{aligned}
\sigma_x &= \frac{K_t S}{2\sqrt{2}}\left[\left(\frac{\rho}{r}\right)^{1/2} - \frac{1}{2}\left(\frac{\rho}{r}\right)^{3/2}\right] + \cdots \\
\sigma_y &= \frac{K_t S}{2\sqrt{2}}\left[\left(\frac{\rho}{r}\right)^{1/2} + \frac{1}{2}\left(\frac{\rho}{r}\right)^{3/2}\right] + \cdots \\
\sigma_z &= 0 \qquad \text{for plane stress} \\
\sigma_z &= \nu(\sigma_x + \sigma_y) \qquad \text{for plane strain}
\end{aligned} \tag{38}$$

It was shown in Ref *23* that Eqs 38 give good near-tip stress estimation for a wide variety of notches under axial loading. Modified relations (Eqs 38) for notches in bending have also been discussed in Ref *23*.

Elastic Strain Energy Density Distribution Near Blunt Notches in Plane Stress

Equations 38 make it possible to calculate the strain energy density W near a notch tip. For notches under plane stress axial loading it can be written in the form

$$W = \int \sigma_{ij} d\epsilon_{ij} = \frac{1}{8}\frac{(K_t S)^2}{E}\left(\frac{\rho}{r}\right)\left[(1-\nu) + \frac{1}{4}\left(\frac{\rho}{r}\right)^2(1+\nu)\right] \tag{39}$$

Stress concentration near a notch tip may cause localized plastic yielding. This subsequently leads to a local stress redistribution (Fig. 7) in a manner similar to that discussed for cracks. Therefore a plastic zone correction factor C_p has been introduced [*23*] to account for the stress redistribution and plastic zone increment. The correction factor C_p given below was derived for a perfectly elastic-plastic body under plane stress:

$$C_p = \frac{2 - \frac{1}{2}\left(\frac{\rho}{r_p}\right) + \frac{1}{4}\left(\frac{\rho}{r_p}\right)^2}{1 + \frac{1}{2}\left(\frac{\rho}{r_p}\right)} \tag{40}$$

where

$$\sigma_{YS} = \frac{K_t S}{2\sqrt{2}}\left[\frac{\rho}{r_p} + \frac{3}{4}\left(\frac{\rho}{r_p}\right)^3\right]^{1/2} \tag{41}$$

Parameter r_p in Eq 40 is the first approximation of the plastic zone size based on the elastic plane stress field (Eq 38) and Hencky-Mises-Huber plastic yielding criterion (Eq 41).

For very sharp notches ($\rho \to 0$) the plastic zone correction factor C_p tends to $C_p \cong 2$ as Irwin proposed for cracks.

Thus, the strain energy density for small scale plastic yielding at the notch tip in plane stress can be given in the form

$$W = \frac{C_p}{8}\frac{(K_t S)^2}{E}\left(\frac{\rho}{r}\right)\left[(1 - \nu) + \frac{1}{4}\left(\frac{\rho}{r}\right)^2(1 + \nu)\right] \tag{42}$$

The energy density calculated from Eq 42 was compared with the strain energy density calculated for strain hardening materials.

Elastic-Plastic Strain Energy Distribution Near Blunt Notches in Plane Stress

The elastic-plastic strain energy density was calculated using a finite element package PAFEC 75, detailed in Ref *24*. The elastic-plastic plane stress solutions were obtained using incremental loading, Hencky-Mises-Huber yielding criterion and Prandtl-Reuss flow rule. Increments of corresponding stresses and strains were used for strain energy density calculations. These calculations were carried out for a plate with two edge notches shown in Fig. 10. Complementary elastic finite element analysis was also conducted out. The elastic stress distribution ahead of the notch tip, corresponding to the nominal (average) stress in the net section S = 303.03 MPa, are shown in Fig. 11. The stress concentration factor was $K_t = 2.552$. The stresses calculated from Eqs 38, also shown in Fig. 11, were in very good agreement with the finite element results over a distance exceeding the notch tip radius ρ.

Three different stress-strain curves, shown in Fig. 12, were used for the finite element elastic-plastic analysis. The modulus of elasticity $E = 2.06 \times 10^5$ MPa and the yield strength $\sigma_{YS} = 4.173 \times 10^2$ MPa were the same for all strain-stress curves. Corresponding strain energy density distributions, calculated for the same nominal stress $S = 303.03$ MPa, are shown in Fig. 13. It is apparent that the strain hardening properties as the plastic modulus H or the strain hardening exponent n do not have a significant effect on the strain energy density distribution. Differences between all curves, shown in Fig. 13, were less than 20% over the entire plastic zone and they were close to the numerical errors which could be introduced during the finite element

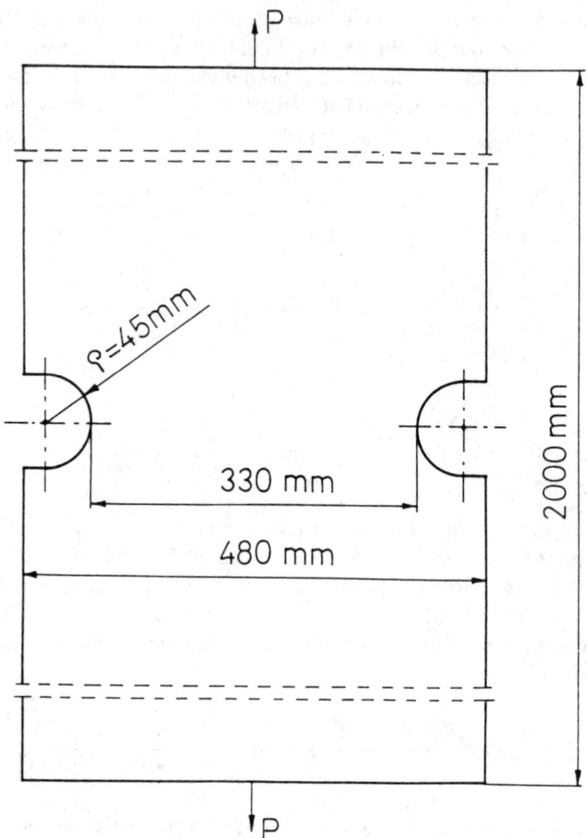

FIG. 10—*Geometry and dimensions of the specimen with two symmetrical edge notches;* B = *4 mm.*

analysis. The accuracy of the calculations depended on the required accuracy of the equivalent stress σ_i ($\sim$0.4%), which in the case of the low modulus of plasticity H, could cause errors in plastic strains and the strain energy density.

The strain energy density calculated from Eq 42 is also shown in Fig. 13. The small differences between these curves indicate that the energy density in the notch tip plastic zone can also be satisfactorily estimated on the basis of the elastic stress-strain fields.

Application of the Equivalent Strain Energy Density Hypothesis for Notches in Plane Stress

The notch-tip stress state in plane stress conditions is reduced to the uniaxial one, $\sigma = \sigma_y$. Thus it is possible to integrate both the linear elastic and power law hardening stress-strain relations. Subsequently, Eq 42 takes the form of Eq 43 for $r = \rho/2$ at the notch tip (Fig. 7):

$$W = \frac{C_p}{2} \frac{(K_t S)^2}{E} \tag{43}$$

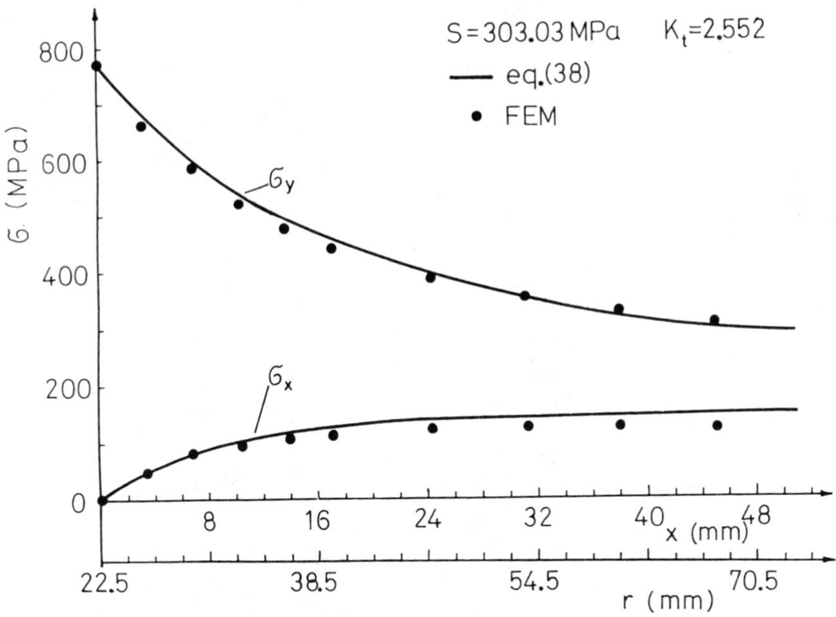

FIG. 11—*Elastic stress distribution ahead of the edge notch under nominal stress* S = *303.03 MPa.*

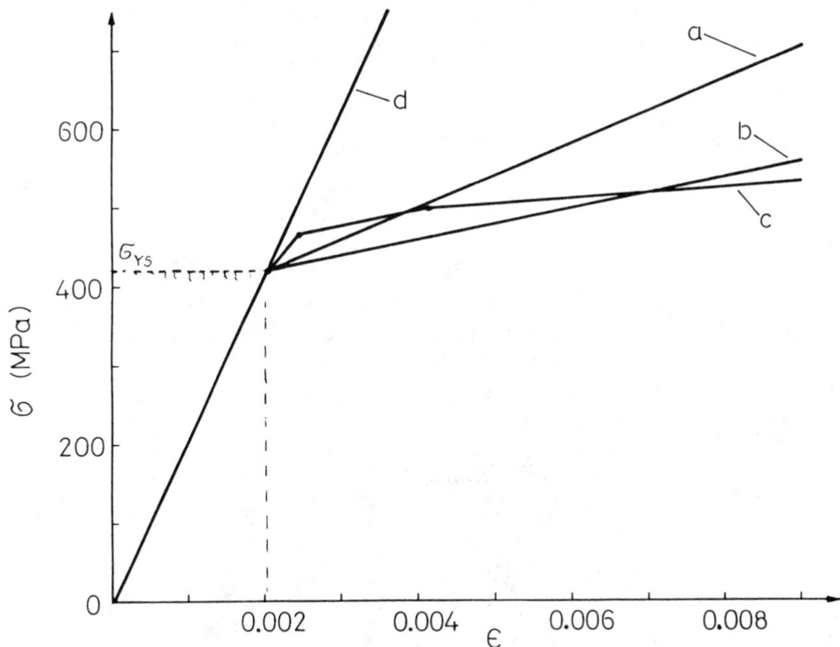

FIG. 12—*Stress-strain curves used for the finite element calculations.* (a) *Bilinear stress-strain curve,* H = *4.12* × *10*[4] *MPa.* (b) *Bilinear stress-strain curve,* H = *2.06* × *10*[4] *MPa.* (c) *Piecewise stress-strain curve.* (d) *Linear elastic stress-strain curve.*

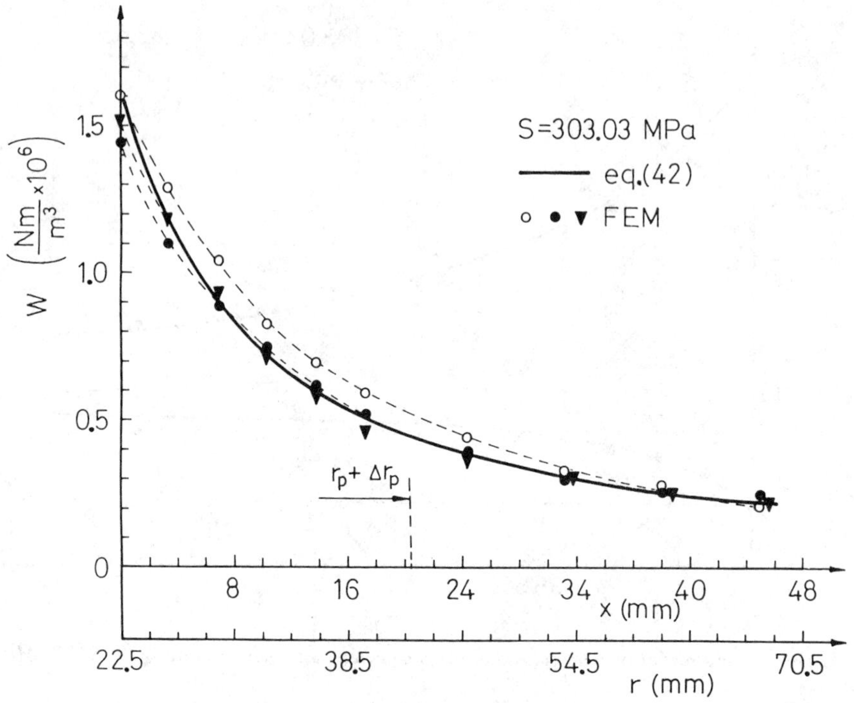

FIG. 13—*Strain energy density ahead of the edge notch calculated for:* (●) *bilinear stress-strain curve with* H = *4.12* × *10*[4] *MPa;* (○) *bilinear stress-strain curve with* H = *2.06* × *10*[4] *MPa;* (▼) *piecewise stress-strain curve;* (—) *linear elastic stress-strain curve (Eq 42).*

The strain energy density calculated for the Ramberg-Osgood relation (Eq 7) is given by

$$W = \int_0^{\epsilon_y} \sigma_y d\epsilon_y = \frac{\sigma_y^2}{2E} + \left(\frac{n}{n+1}\right)\frac{\alpha\sigma_{\text{YS}}}{E}\sigma_y\left(\frac{\sigma_y}{\sigma_{\text{YS}}}\right)^n \tag{44}$$

Since the energies given by Eqs 43 and 44 are equal, according to the equivalent energy density hypothesis, it is possible to calculate the notch tip stress σ_y for any given nominal stress S. Correspondingly elastic plastic strain ϵ_y can be calculated from Eq 7. Thus two equations have to be solved in order to calculate the stress component σ_y and the corresponding elastic-plastic strain component ϵ_y at the notch tip:

$$\begin{cases} \dfrac{\sigma_y^2}{2E} + \left(\dfrac{n}{n+1}\right)\left(\dfrac{\alpha\sigma_{\text{YS}}}{E}\right)\sigma_y\left(\dfrac{\sigma_y}{\sigma_{\text{YS}}}\right)^n = \dfrac{C_p}{2}\dfrac{(K_tS)^2}{E} \\ \epsilon_y = \dfrac{\sigma_y}{E} + \dfrac{\alpha\sigma_{\text{YS}}}{E}\left(\dfrac{\sigma_y}{\sigma_{\text{YS}}}\right)^n \end{cases} \tag{45}$$

Unfortunately, a closed solution to the set of equations (Eq 45) is not available. Therefore, the numerical procedure described in Ref *25*, has been used. The elastic-plastic strains ϵ_y, calculated from Eqs 45, were verified against notch tip strains experimentally measured in different specimens shown in Fig. 14. Two specimens with a semicircular single notch made of aluminum alloy Al HS30 and one specimen made of low alloy steel were used in the experiments. The aluminum specimens were tested under excentric loadings inducing tensile and bending stresses while the steel specimens was tested under axial loading. The notch tip strains were measured using strain gages 0.3 mm long. The results, corresponding to several nominal stress levels S, are presented in the form of plots K_tS versus ϵ_y in Fig. 14. The material stress-strain curves are also included in Fig. 14. It is interesting that good agreement has been achieved almost up to the general yielding of the net section. This indicates that the equivalent strain energy density concept can be practically used within the entire nominal stress range $0 \leq S \leq \sigma_{YS}$.

For materials that exhibit Massings type behavior, the relations in Eqs 45 can be used to calculate strains under cyclic loading [*23,26*] as it was proposed for Neuber's rule [*27,28*]. The method can therefore be used for fatigue life prediction near notches.

Fatigue test results obtained from key-hole CT specimens [*29*] manufactured from SAE 0030 cast steel and tested under the SAE transmission (SAE T/H) loading history [*30*] were used (Fig. 15) for comparisons. Two types of specimens tested, under two different load levels, were chosen for analysis. Two specimens with the stress concentration factor $K_t = 4.01$ were tested under the transmission loading history with the highest peak load $P_{max} = 22.24$ kN and three specimens with $K_t = 3.12$ were tested under a similar load history but with the peak load $P_{max} = 17.8$ kN. All other load levels in the history were proportional to the peak load P_{max}. The stress concentration factors were determined [*31*] by the finite element method. The average experimental fatigue lives to crack initiation, defined as a crack of length $\Delta a = 0.25$ mm, where $N_f = 14\,535$ cycles and $N_f = 107\,730$ cycles for the specimens with the high and low stress concentration, respectively.

Miner's linear damage cumulation rule and the rainflow cycle counting procedure were used in fatigue life calculations. Fatigue damage was calculated on the basis of the modified Manson-Coffin strain-life relation (Eq 46). The material fatigue properties were as follows: $\sigma_f' = 653$ MPa, $b = -0.082$, $\epsilon_f' = 0.28$, and $c = -0.51$, where

$$\frac{\Delta\epsilon}{2} = \frac{\sigma_f' - \sigma_m}{E}(2N_f)^b + \epsilon_f'(2N_f)^c \tag{46}$$

Notch tip strain ranges were calculated simultaneously with the rainflow counting procedure using several elements of the method described in references [*23,32*]. The material stress-strain curve was given in the form of Eq 29 with the values of $E = 2.07 \times 10^5$ MPa, $\sigma_{YS} = 317$ MPa, $k = 708$ MPa, and $N = 0.13$.

The calculated and experimental fatigue lives are compared in Fig. 16. It is apparent that the energy density based calculations gave better estimation of the fatigue lives than those based on Neuber's rule. The good agreement was achieved for both short and long fatigue lives. It is also worth noting that the energy based calculations gave good notch-tip strain and fatigue life estimation on the basis of the theoretical elastic stress concentration factor K_t, contrary to Neuber's [*28*] rule where the experimentally determined fatigue notch factor K_f is often used.

Consequences of the Strain Energy Density Equivalence Near Blunt Notches

In the case of small plastic zones or total absence of plastic yielding ($C_p \cong 1$) the strain energy density at the notch tip can be calculated as

$$W = K_t^2 W_s \qquad \text{for plane stress} \tag{47}$$

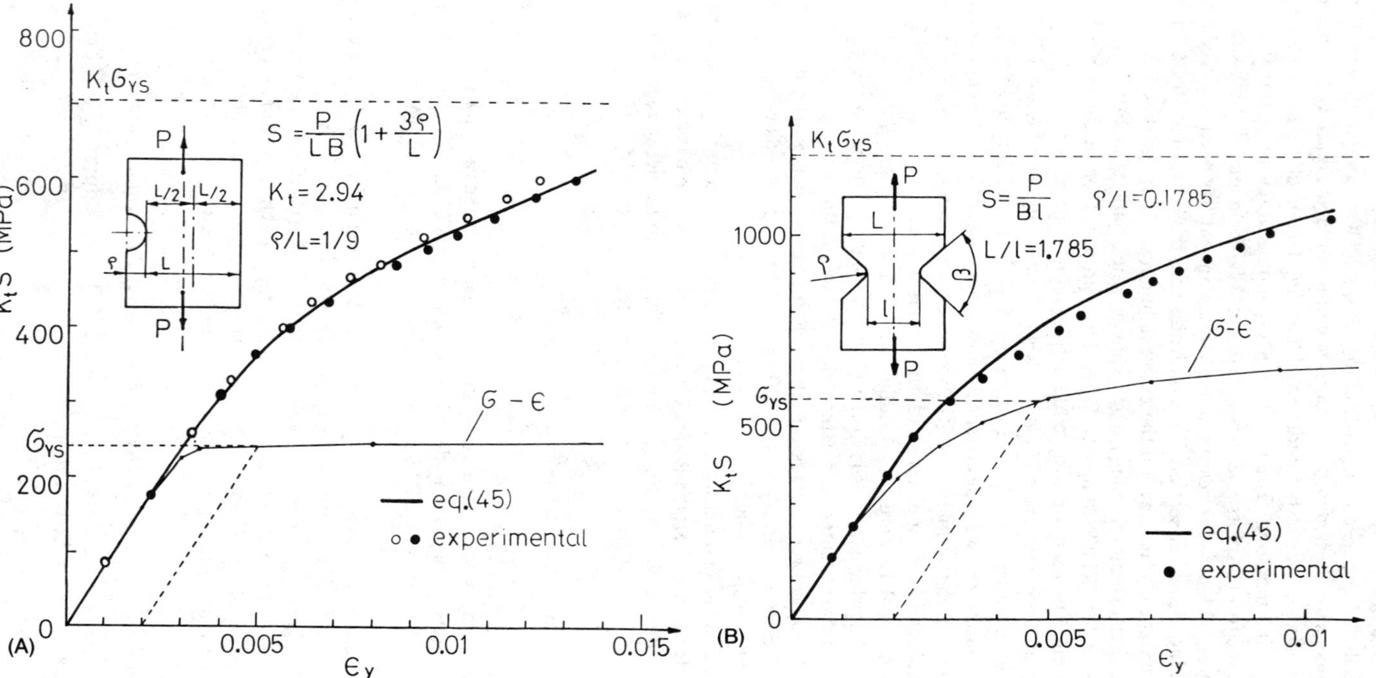

FIG. 14—*Calculated and experimentally measured notch tip strains.* (a) *Aluminum specimen,* $B = 4$ *mm,* $\rho = 5$ *mm,* $K_t = 2.94$. (b) *Steel specimen,* $B = 5$ *mm,* $\rho = 5$ *mm,* $\beta = 120°C$, $K_t = 2.33$.

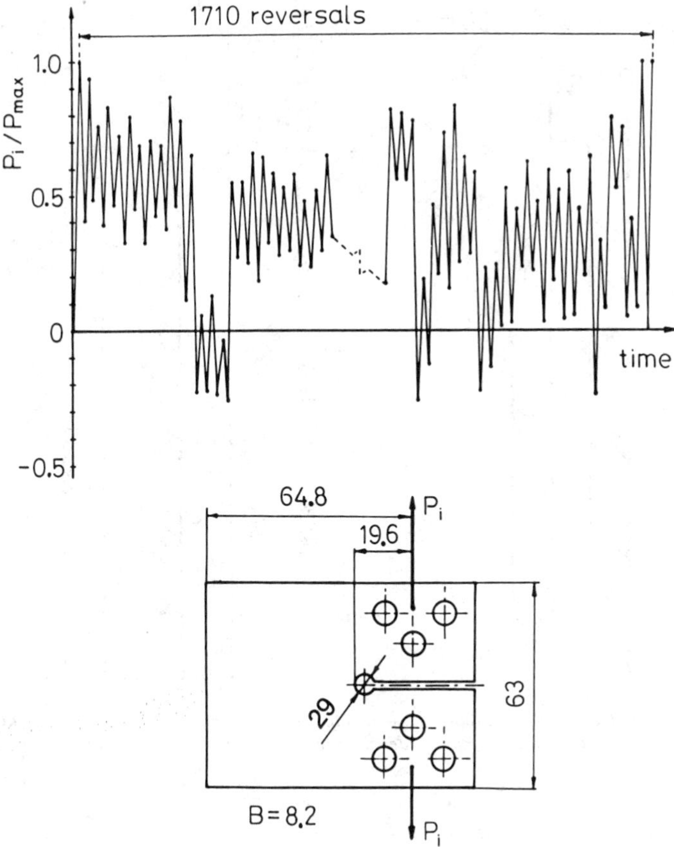

FIG. 15—*Non-dimensional SAE transmission loading history and geometry of the key-hole specimen; all dimensions in millimetres.*

$$W = K_t^2 W_s(1 - \nu^2) \qquad \text{for plane strain} \tag{48}$$

where

$$W_s = S^2/2E \tag{49}$$

It is obvious that the strain energy density at the notch tip depends not only on the nominal stress S and the stress concentration factor K_t but also on the stress state. It is apparent that the stress concentration factor K_t is not able to account for the stress multiaxiality in the notch tip. Therefore it could be useful to complement the stress concentration factor K_t by the strain energy density concentration factor K_w or the equivalent stress concentration factor K_i as it was proposed in Refs *33* and *34:*

$$K_t = \frac{\sigma_y}{S} \tag{50}$$

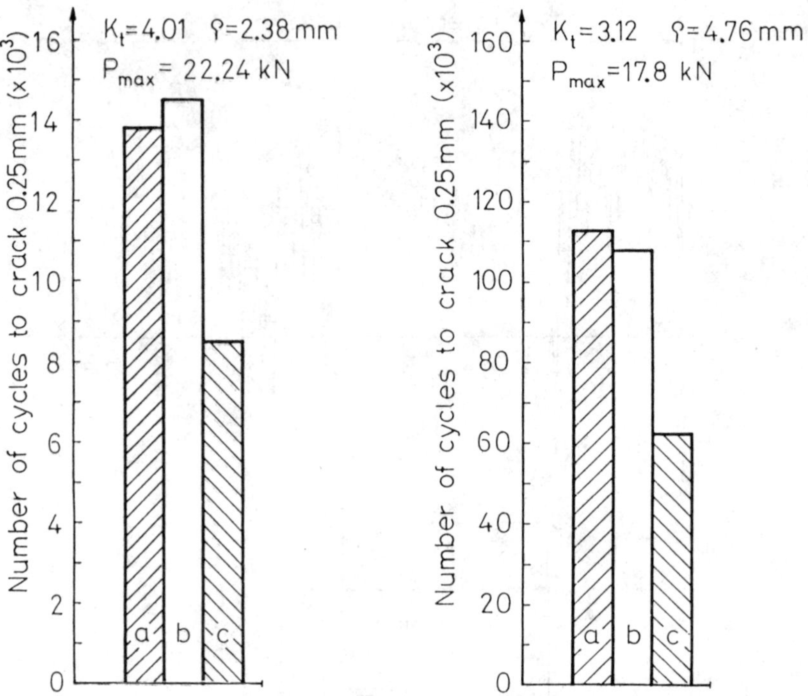

FIG. 16—*Calculated and experimental fatigue lives to crack length* $\Delta a = 0.25$ *mm in the key-hole specimens made of SAE 0030 cast steel.* (a) *Energy density method.* (b) *Experimental results.* (c) *Neuber's rule.*

$$K_w = \left(\frac{W}{W_s}\right)^{1/2} = \left(\frac{\sigma_y^2 - 2\nu\sigma_y\sigma_z + \sigma_z^2}{S^2}\right)^{1/2} \tag{51}$$

$$K_i = \frac{\sigma_i}{S} = \left(\frac{\sigma_y^2 - \sigma_y\sigma_z + \sigma_z^2}{S^2}\right)^{1/2} \tag{52}$$

These factors make it possible to determine all stress and strain components at the notch tip in the elastic regime. By using the equivalent strain energy density concept and one more assumption, similar to that suggested in Ref *33,* all stress and strain components can also be calculated for the case of localized plastic yielding. This makes it potentially possible to apply the method to fatigue life prediction taking into account a multiaxiality of stress state in the notch tip [*34,35*].

However, one must not forget that the analysis presented in this paper was based on the assumption of proportional loading in effect. It means that the deviatoric stresses in the notch tip must be proportional during loading. It also means that the principal stresses are fixed and do not rotate. This is not true in general, but the deviations from the above assumption are not large in the case of relatively blunt notches [*34,35*]. The latest results published by Glinka [*23*] and Hoffman [*34*] indicate that the assumption of proportional loading holds also under cyclic loading providing that the external loads are proportional. For arbitrary non-proportional loading further developments are necessary where the equivalent strain energy density hypothesis should be used together with the incremental theory of plasticity. This is especially true in the case of cracks and, therefore, the good agreement between the finite element calculations based on incremental theory of plasticity and the energy density results shown in Fig. 9 was somewhat of a surprise. However, it should be noted that the analysis was conducted for monotonic loading only.

Conclusions

It has been shown that the strain energy density in a plastic zone ahead of a crack or a notch tip is controlled by the surrounding elastic strain-stress field. In the case of small scale plastic yielding the strain energy distribution in the plastic zone is very close to that calculated for a perfectly elastic stress field. It is, therefore, possible to calculate the strain energy density at any point of the plastic zone without solving the complicated problem of the elastic-plastic stress-strain field. Subsequently, the strain energy density can be used for calculation of the elastic-plastic strains in the plastic zone. Good results were obtained for both sharp cracks and blunt notches. The method of notch tip strains calculations, based on the equivalent energy density concept, can be easily applied to fatigue notch analysis. Comparisons of calculated and experimental fatigue lives have shown that significant improvements in fatigue life prediction can be achieved by using the equivalent strain energy density concept.

However, it should be noted that the concept is valid for small scale yielding. Therefore, it requires that the plastic zone should be small in comparison to the surrounding elastic field. Nevertheless, available experimental data indicate that the hypothesis of the equivalent strain energy density holds almost up to general yielding (i.e., for nominal net stress $S \lesssim \sigma_{YS}$). The method is also restricted in general to cases where proportional loading is in effect. Fortunately, it was found that this assumption is satisfied in a wide variety of engineering notched elements.

The following conclusions can be drawn from the results presented above:

1. Elastic notch-tip stresses can be satisfactorily represented by approximate formulae consisting of the stress concentration factor K_t and notch-tip radius ρ.
2. The strain energy density in the notch-tip plastic zone can be determined from the elastic stress-strain field.
3. Elastoplastic strains and stresses ahead of a crack tip or in the notch root can be calculated with a sufficient accuracy on the basis of the equivalent strain energy density equation and the constitutive Ramberg-Osgood relations.
4. The equivalent strain energy density hypothesis makes it possible to calculate inelastic notch-tip strains and stresses solely on the basis of the stress-strain curve and the stress concentration factor.
5. The equivalent strain energy concept was superior to the Neuber based notched tip analysis, especially regarding the determination of the strain.
6. It is expected that this method can also be applied under some restrictions for the notch-tip stress-strain simulation under cyclic loading. The method can be included into fatigue life prediction computer programs which explicitly consider the notch-tip strains and stresses.

References

[1] Sih, G. C., "A special theory of crack propagation," in *Methods of Analysis and Solutions of Crack Problems*, G. C. Sih, Ed., Noordhoff International Publishers, The Netherlands, 1973.

[2] Theocaris, P. S., Kardomates, G., and Adrianpoulos, N. P., "Experimental study of the T-criterion in ductile fractures," *Engineering Fracture Mechanics*, Vol. 17, No. 5, 1982, pp. 439–447.

[3] Weertman, J., "Fatigue crack propagation theories," in *Fatigue and Microstructure*, Papers of the 1978 ASM Seminar, ASM, Metals Park, Ohio, 1979, pp. 279–306.

[4] Hutchinson, J. W., "Fundamentals of the Phenomenological Theory of Nonlinear Fracture Mechanics," *Journal of Applied Mechanics,* Vol. 50, No. 4, 1983, pp. 1042-1051.
[5] Rice, J. R. and Rosengren, G. F., "Plane-Strain Deformation Near a Crack Tip in a Power Law Hardening Material," *Journal of Mechanics and Physics of Solids,* Vol. 16, No. 1, 1968, pp. 1-30.
[6] Achenbach, J. D. and Dunayevsky, V., "Crack Growth Under Plane Stress Conditions in an Elastic Perfectly-Plastic Material," *Journal of Mechanics and Physics of Solids,* Vol. 32, No. 2, 1984, pp. 89-100.
[7] Hutchinson, J. W., "Singular Behavior at the End of a Tensile Crack in a Hardening Material," *Journal of Mechanics and Physics of Solids,* Vol. 16, No. 1, 1968, pp. 13-31.
[8] Irwin, G. R., "Linear Fracture Mechanics, Fracture Transition and Fracture Control," *Engineering Fracture Mechanics,* Vol. 1, No. 2, 1968, pp. 241-257.
[9] Knott, J. F., *Fundamentals of Fracture Mechanics,* Butterworths, London, 1973.
[10] Hilton, P. D. and Hutchinson, J. W., "Plastic Intensity Factors for Cracked Plates," *Engineering Fracture Mechanics,* Vol. 3, No. 2, 1971, pp. 435-451.
[11] Sih, G. C. and Chao, C. K., "Size Effect of Cylinderical Specimens with Fatigue Cracks," *Theoretical and Applied Fracture Mechanics,* Vol. 1, No. 3, 1984, pp. 239-247.
[12] Sih, G. C. and Moyer, E. T., "Path Dependent Nature of Fatigue Crack Growth," *Engineering Fracture Mechanics,* Vol. 17, No. 3, 1983, pp. 269-280.
[13] Sih, G. C. and Madenci, E., "Crack Growth Resistance Characterised by the Strain Energy Density Function," *Engineering Fracture Mechanics,* Vol. 18, No. 6, 1983, pp. 1139-1171.
[14] Tada, H., Paris, P. C., and Irwin, G. R., *The Stress Analysis of Cracks Handbook,* Del Research Corp., Hellertown, Pa., 1973.
[15] Hill, R., *The Mathematical Theory of Plasticity,* Oxford University Press, Oxford, 1971.
[16] Walker, T. J., "A Quantitative Strain-and-Stress Criterion for Failure in the Vicinity of Sharp Cracks," *Nuclear Technology,* Vol. 23, No. 8, 1974, pp. 189-203.
[17] Creager, M. and Paris, P. C., "Elastic Field Equations for Blunt Cracks with Reference to Stress Corrosion Cracking," *International Journal of Fracture Mechanics,* Vol. 3, No. 2, 1967, pp. 247-252.
[18] Paris, P. C., "Twenty Years of Reflection on Questions Involving Fatigue Crack Growth—Part I: Historical Observations and Perspective," in *Fatigue Thresholds, Fundamentals and Engineering Applications,* Vol. 1, J. Bäcklund, A. F. Bolm, C. J. Beevers, Eds., Engineering Materials Advisory Services Ltd., U.K., 1982.
[19] Irving, P. E. and McCartney, L. N. M., "Prediction of Fatigue Crack Growth Rates: Theory, Mechanisms and Experimental Results," *Metal Science,* Vol. 11, 1977, pp. 351-361.
[20] Kujawski, D. and Ellyin, F., "A Fatigue Crack Propagation Model," *Engineering Fracture Mechanics,* Vol. 20, No. 5/6, 1984, pp. 695-704.
[21] Glinka, G., "Notch Stress Strain Analysis Approach to Fatigue Crack Growth," *Engineering Fracture Mechanics,* Vol. 21, No. 2, 1985, pp. 245-264.
[22] Schijve, J., "Stress Gradients Around Notches," *Fatigue of Engineering Materials and Structures,* Vol. 5, No. 2, 1982, pp. 325-332.
[23] Glinka, G., "Calculation of Inelastic Notch-Tip Strain-Stress Histories Under Cyclic Loading," *Engineering Fracture Mechanics,* Vol. 22, No. 5, 1985, pp. 839-854.
[24] Hensell, R. D., Ed., PAFEC 75 Manuals, PAFEC Limited, Nottingham, 1984.
[25] Glinka, G., "Energy Density Approach to Calculation of Inelastic Strain-Stress near Notches and Cracks," *Engineering Fracture Mechanics,* Vol. 22, No. 3, 1985, pp. 485-508.
[26] Molski, K. and Glinka, G., "A Method of Elastic-Plastic Stress and Strain Calculation of a Notch Root," *Materials Science and Engineering,* Vol. 50, No. 2, 1981, pp. 93-100.
[27] Topper, T. H., Wetzel, R. M., and Morrow, J., "Neuber's Rule Applied to Fatigue Notched Specimens," *Journal of Materials,* Vol. 4, No. 1, 1969, pp. 200-209.
[28] Neuber, H., "Theory of Stress Concentration for Shear-Strained Prismatical Bodies with Arbitrary Nonlinear Stress-Strain Law," *Journal of Applied Mechanics,* Vol. 28, No. 4, 1961, pp. 544-550.
[29] Glinka, G. and Stephens, R. I., "Total Fatigue Life Calculations in Notched SAE 0030 Cast Steel Under Variable Loading Spectra," in *Fracture Mechanics: Fourteenth Symposium—Volume I: Theory and Analysis, ASTM STP 791,* J. C. Lewis and G. Sines, Eds., American Society for Testing and Materials, Philadelphia, 1983, pp. I-427-I-445.
[30] Wetzel, R. M., Ed., *Fatigue Under Complex Loading: Analyses and Experiments,* Society of Automotive Engineers, Warrendale, Pa., 1977.
[31] Stephens, R. I., private communications.
[32] Glinka, G., "Fatigue Crack Propagation and Initiation," *Scientific Reports of The Warsaw Technical University, Mechanics,* No. 75, Warsaw, 1981 (in Polish).
[33] Hoffman, M. and Seeger, T., "A Generalized Method for Estimating Elastic-Plastic Notch Stresses and Strains," *Journal of Engineering Materials and Technology,* ASME, Vol. 107, No. 4, 1985, pp. 250-260.

[*34*] Hoffman, M., "*Ein Näherungsverfahren zur Ermittlung mehrachsig elastisch-plastischer Kerbbeanspruchungen,*" Ph.D. dissertation, Technische Hochschule Darmstadt, West Germany, 1986 (in German).

[*35*] Glinka, G., Ott, W., and Nowack, H., "Elastoplastic Plane Strain Analysis of Stresses and Strains at the Notch Root," *Journal of Engineering Materials and Technology,* ASME (to be published).

Yukitaka Murakami[1]

Correlation Between Strain Singularity at Crack Tip under Overall Plastic Deformation and the Exponent of the Coffin-Manson Law

REFERENCE: Murakami, Y., **"Correlation Between Strain Singularity at Crack Tip under Overall Plastic Deformation and the Exponent of the Coffin-Manson Law,"** *Low Cycle Fatigue, ASTM STP 942,* H. D. Solomon, G. R. Halford, L. R. Kaisand, and B. N. Leis, Eds., American Society for Testing and Materials, Philadelphia, 1988, pp. 1048–1065.

ABSTRACT: This paper is concerned with the interpretation of the Coffin-Manson law $\Delta\epsilon_p \cdot N_f^\alpha = C$ [*1*] from the standpoint of the behavior of small cracks and also with the explanation of the uniqueness of the exponent α from the strain singularity at crack tip under overall plastic deformation.

Firstly, the characteristic behavior of small cracks in low-cycle fatigue range is discussed; then it is demonstrated that the Coffin-Manson law is virtually identical to the growth law of a small crack. This suggests that we must pay attention to the behavior of a small crack in order to solve low-cycle fatigue problems. Secondly, the singularity of strain field at crack tip is analyzed by a finite element method. It is well known that, denoting the distance from the crack tip by r, the elastic stress strain field near crack tip has the singularity of $r^{-0.5}$ and under the elastic-plastic condition the stress and strain has HRR singularity. However, under overall plastic deformation (i.e., under low-cycle fatigue range) the singularity in strain distribution near crack tip deviates from HRR singularity to $r^{-(0.5 \sim 0.7)}$ with increasing plastic deformation regardless of hardening exponents of materials.

It is concluded from the viewpoint of the propagation of a small crack that this exponent (0.5 ~ 0.7) in singular strain distribution is closely related to the fact that the exponent α in $\Delta\epsilon_p \cdot N_f^\alpha = C$ ranges from 0.5 to 0.7 for various materials.

KEY WORDS: low-cycle fatigue, the Coffin-Manson law, small cracks, crack propagation, crack tip, strain singularity, finite element analysis, fracture ductility

In 1963, at a conference held in Sagamore, Pennsylvania, Coffin [*2*] described in the discussion of the paper presented by Manson and Hirschberg that the uniqueness of the exponent α in the low-cycle fatigue law $\Delta\epsilon_p \cdot N_f^\alpha = C$ for so many metals is strongly suggestive that some simple mechanism is operating to cause failure, and this is of extreme fundamental importance. Although since that time many studies have been done to find the answer to this question, the problem remains unsolved. Early in the history of the study of low-cycle fatigue, this question was expected to be solved soon, but today the difficulty and the importance of this question are well recognized.

This paper is concerned with the interpretation of the Coffin-Manson law from the standpoint of small crack behavior, and with the explanation of the uniqueness of the exponent from the strain singularity at crack tip under overall plastic deformation.

[1]Professor, Department of Mechanics and Strength of Solids, Faculty of Engineering, Kyushu University, Higashi-ku, Fukuoka, 812 Japan.

The author and his co-workers [3,4] have shown by successive observation of the small crack growth process that the Coffin-Manson law is virtually identical to the growth law of a small crack. Their results suggest that we must pay attention to the behavior of a small crack in order to solve low-cycle fatigue problems.

In the present paper, some typical behaviors of a small crack [3,4] are briefly reviewed. The singularity of strain field at crack tip is analyzed by the finite-element method (FEM), and the results are compared with the growth mechanism of a small crack.

It is well known that denoting the distance from the crack tip by r the elastic stress and strain field near crack tip has the singularity of $r^{-0.5}$ and under the elastic-plastic condition the stress and strain singularity has HRR singularity [5,6]. This is the reason that the J integral [7] is frequently used in elastic-plastic fracture mechanics as well as stress intensity factor K in linear fracture mechanics.

We can approximate the true stress/logarithmic strain curve (σ-ϵ) in a tensile test by the Ramberg-Osgood law [8]:

$$\epsilon = \frac{\sigma}{E} + \beta\frac{\sigma_Y}{E}\left(\frac{\sigma}{\sigma_Y}\right)^n \tag{1}$$

where

σ_Y, β = material constants,
E = Young's modulus, and
n = hardening exponent.

The asymptotic solution for stress and strain near crack tip by deformation theory in small-scale yielding is expressed by the equations (so-called HRR singularity) [4,5]

$$\sigma_{ij} = \sigma_Y\left(\frac{EJ}{\beta\sigma_Y^2 I_n r}\right)^{1/n+1} \tilde{\sigma}_{ij}(\theta, n) \tag{2}$$

$$\epsilon_{ij} = \frac{\beta\sigma_Y}{E}\left(\frac{EJ}{\beta\sigma_Y^2 I_n r}\right)^{n/n+1} \tilde{\epsilon}_{ij}(\theta, n) \tag{3}$$

where (r, θ) is the polar coordinate defined at crack tip, I_n is a constant obtained from integral, $\tilde{\sigma}_{ij}$ and $\tilde{\epsilon}_{ij}$ are the functions of θ and n, and J is the J integral which prescribes the intensity of HRR singular stress and strain field.

In recent studies on fracture, J integral is frequently applied not only to the problems of small-scale yielding but also to some problems of large-scale yielding [9–13]. However, it is also pointed out that the zone of HRR singularity decreases with increasing the plastic deformation and accordingly the correlation between crack opening displacement (COD) and J integral is likely to be lost [13].

Although for the J integral many analytical and experimental studies have been done, the question of whether the stress-strain singularity near the crack tip really obeys HRR singularity has never been discussed in detail in overall plastic deformation. This question is of importance in the range of low-cycle fatigue, because a whole specimen is subject to overall plastic deformation. Therefore, before the mechanical parameters or the equations based on linear elasticity or theory of small-scale yielding are formally applied, the stress and strain field near crack tip must be investigated in detail. This kind of investigation will reveal the controlling factors that influence the propagation of a small crack under overall plastic deformation or high stress amplitude.

It is well known that in low-cycle fatigue the Coffin-Manson law given by Eq 4 holds for various materials:

$$\Delta\epsilon_p \cdot N_f^{\alpha} = C \qquad (\alpha, C = \text{material constants}) \tag{4}$$

where $\Delta\epsilon_p$ is plastic strain range and N_f is the cycle to failure. The exponents α ranges from 0.5 to 0.7 for many metallic materials. According to Coffin [*14*], C is approximately dependent on the ductility of a material and is determined by

$$C = \frac{1}{2}\,\epsilon_F; \qquad \epsilon_F = \ell n[1/(1-\psi)] \tag{5}$$

where ϵ_F is the fracture ductility and ψ is the reduction of area in tensile test. Since Coffin had $\alpha \cong 1/2$ in his experiments regardless of materials, he guessed that a simple mechanism must predominate in the process of low-cycle fatigue. However, Manson [*15*] described that the best approximation for many materials is $\alpha = 0.6$. In any case, the exponents α ranges only from 0.5 to 0.7 for most materials.

Considering this fact and also considering that the Coffin-Manson law is virtually equivalent to small crack growth law [*3,4*], we may expect a simple mechanism to predominate in the process of the propagation of a small crack.

Process of Small Crack Propagation in Low-Cycle Fatigue

Push-pull low-cycle fatigue tests were carried out on a closed-loop electrohydraulic servocontrolled testing machine. Figures 1 to 4 show the growth behavior of a fatal crack in low-cycle fatigue of 0.46% carbon steel and 70-30 brass. These data were obtained by the successive observation of fatigue process using the plastic replica method [*3,4*]. Crack initiation and growth behavior were studied by investigating the early-stage replicas of the main crack which led a specimen to final fracture. The important conclusions derived from these figures are as follows:

1. a crack as long as grain size already exists in the early stage of the total life ($N < 0.02\,N_f$), and more than 90% of the total life is consumed when the fatal crack grows about 1 mm in length.

2. From the growth behavior in Figs. 1 to 4, the crack growth rate da/dN may be formulated by

$$\frac{da}{dN} = f(\Delta\epsilon_p) \cdot a \tag{6}$$

For 0.46% carbon steel $f(\Delta\epsilon_p) = 2.58\,\Delta\epsilon_p^{1.74}$ and for 70-30 brass $f(\Delta\epsilon_p) = 5.83\,\Delta\epsilon_p^{2.0}$. Then, in general we may write

$$\frac{da}{dN} = C_1 \Delta\epsilon_p^{\gamma} \cdot a \tag{7}$$

The expression in Eq 7 was indicated by Manson [*16*], and essentially the same conclusion was also obtained by Weiss [*17*], Tomkins [*18*], and Solomon [*19*]. However, the smallest size of a crack discussed in the studies of Weiss [*17*] and Solomon [*19*] were larger than 200 μm. As seen from Figs. 1 to 4, when a fatal crack has become 200 μm, 50% of the total life is already

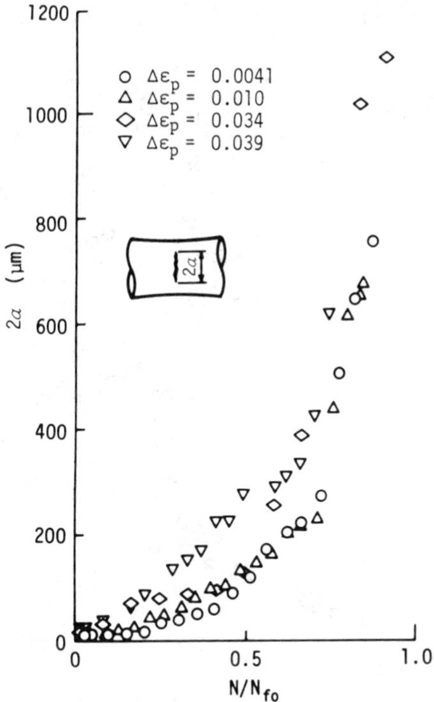

FIG. 1—*Crack propagation curve of a plain specimen (0.46% carbon steel)* [3]; N_{f_0} = *526, 704, 8053, and 35 079 for* $\Delta\epsilon_p$ = *0.039, 0.034, 0.010, and 0.0041, respectively; specimen diameter = 8 mm.*

consumed. The detailed experimental investigations on the behavior of a small crack from the very early stage of life to the final stage in a plain specimen were first done by the authors [3].

Integrating Eq 7, we have

$$\Delta\epsilon_p^\gamma(N_f - N_0) = C_2 \log \frac{a_f}{a_0} \tag{8}$$

where a_0 is the half length of a fatal crack when we detect the existence of the crack for the first time ($N = N_0$), and a_f is the half length of the fatal crack at final fracture. We may assume $N_0 \ll N_f$ and accordingly $N \cong 0$, $2a_0 \cong$ grain size, and $2a_f \cong$ specimen diameter [3,4]; see Table 1.

Consequently, we have

$$\Delta\epsilon_p^\gamma \cdot N_f = C_3 \quad \text{or} \quad \Delta\epsilon_p \cdot N_f^\alpha = C_4 \tag{9}$$

This is the derivation of the Coffin-Manson law from the experimental data. We can recognize that the exponent γ in Eq 7 is responsible for the exponent of the Coffin-Manson law. Previous studies suggest that the value of the exponent γ in Eq 7 varies little; for example, $\gamma = 1.74$ for 0.46% carbon steel [3], $\gamma = 2.0$ for 70-30 brass [4], and $\gamma = 1.86$ for low carbon steel [19]. Considering many other data, we have $\alpha = 1/\gamma = 0.5 \sim 0.7$ regardless of materials. The reason that α ranges only from 0.5 to 0.7 will be discussed in the next section.

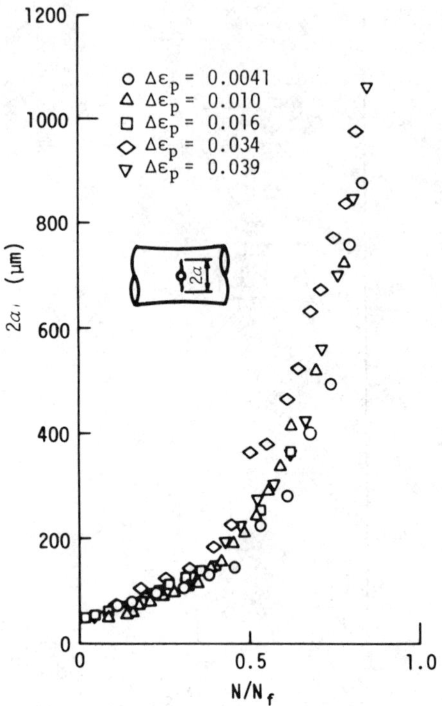

FIG. 2—*Crack propagation curve of a specimen containing a hole with a 40 μm diameter (0.46% carbon steel)* [3]; N_f = *419, 556, 2249, 5711, and 26 197 for* $\Delta\epsilon_p = 0.039, 0.034, 0.016, 0.010,$ *and 0.0041, respectively; specimen diameter = 8 mm.*

TABLE 1—*Final crack length* (ℓ_f).[a]

$\Delta\epsilon_p$	ℓ_f
0.46% Carbon Steel[b]	
0.0041	12.5
0.01	11.1
0.016	9.8
0.034	7.4
0.039	7.5
70-30 Brass[b]	
0.0067	6.88
0.01	6.54
0.02	6.38
0.05	6.26

[a]It may be understood that although ℓ_f is not exactly equal to specimen diameter D, the difference between ℓ_f and D in this table does not result in substantial changes in the descriptions from Eqs 8 and 9, because the crack growth rate near $2a = D$ is very high and accordingly the error caused by the difference between ℓ_f and D becomes very small. It should be also noted that the crack shape is approximately semicircular.

[b]Specimen diameter = 8 mm.

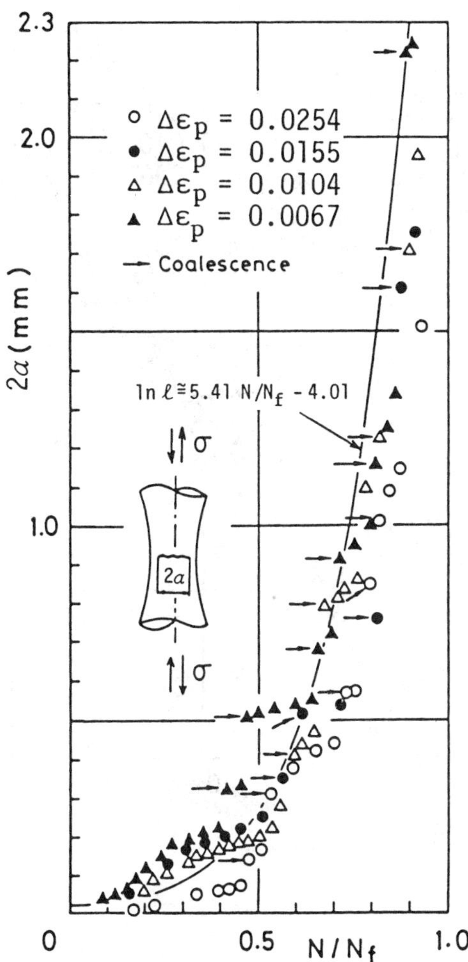

FIG. 3—*Crack propagation curve of a plain specimen (70-30 brass)* [4]; $\ell = 2a$; N_f = *1762, 3925, 8852, and 17 590 for* $\Delta\epsilon_p$ = *0.0254, 0.0155, 0.0104, and 0.0067, respectively; specimen diameter* = *8 mm.*

Singularity of Strain Field Near Crack Tip in Overall Plastic Deformation

From the previous discussion, we may expect that the narrow range of α must be attributed to the unique characteristics of strain field near crack tip regardless of materials, because the process of low-cycle fatigue is virtually equivalent to that of the propagation of a small crack. In order to reveal the singularity of the strain field near crack tip in overall plastic deformation as illustrated in Fig. 5*c*, a plate containing a small crack was analyzed by FEM.

Figure 6 shows the shape and dimension of a specimen for FEM analysis. In order to treat a small crack problem, the ratio of the crack length $2a$ and plate width $2W$ was kept $a/W = 0.1$. The mesh pattern in the vicinity of the crack tip is shown in Fig. 7. The length of the smallest side of a triangular element near crack tip is $a/400$. The application of FEM results to a very short crack comparable to the grain size should, of course, be understood with respect to the average behavior of the material. The division patterns of other portions are shown in Fig. 8.

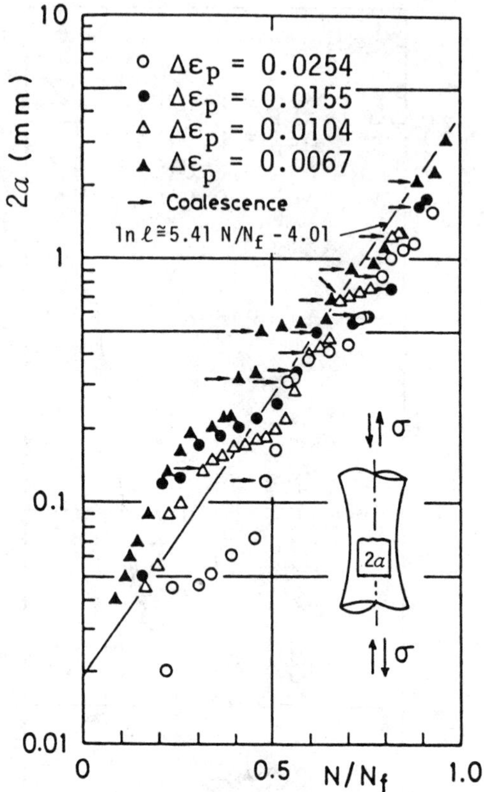

FIG. 4—*Relationship between logarithm of crack length* (2a) *and nondimensional numbers of cycles (same data as in Fig. 3);* $\ell = 2a$.

The numerical procedure is as follows. First, the elastic calculation is carried out to find the critical load in which the crack tip element starts to yield. Next, the definite small increment of loading is given. For instance, the increase of 1.5 MPa in nominal stress was given for the material with hardening exponent $n = 5$. The elements which satisfy von Mises's yield criterion within the accuracy of 2% are regarded as yielded elements, and the elastic-plastic stiffness matrix is applied to these elements. The material constants in Eq 1 are assumed to be $\sigma_y = 100$ MPa, $E = 10^4$ MPa, and $\beta = 0.02$.

Figures 9*a* to 9*e* show the variation of the distribution of the strain near crack tip with increasing the remote strain ϵ_0. The variation of the strain distribution from Figs. 9*a* to 9*e* indicates that the moderate slope of the distribution in log-log scale in the elastic distribution (*a*) tends to change to steep slope in the small-scale yielding (*b*) and (*c*), and finally the slope tends to become moderate with increasing overall plastic deformation (*e*). This tendency was also confirmed for the cases of $n = 6$, 7, 9, and 13. The slope in log-log scale in Figs. 9*a* to 9*c* corresponds to the singularity in strain distribution. If we denote this slope by α^* and calculate α^* from the strains at points *A* and *B* in Fig. 7, we have

$$\alpha_1^* = -\frac{\log(\epsilon_{yA}/\epsilon_{yB})}{\log(r_A/r_B)} = \frac{\log(\epsilon_{yA}/\epsilon_{yB})}{\log 2} \tag{10}$$

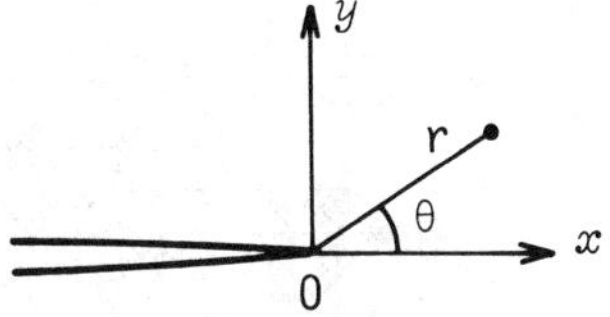

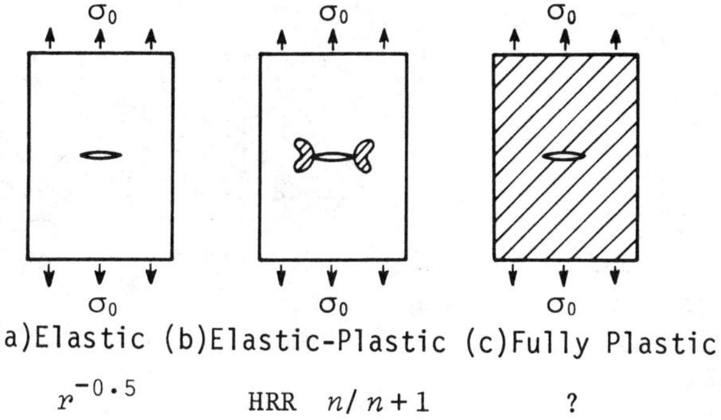

FIG. 5—*Strain singularity near crack tip.*

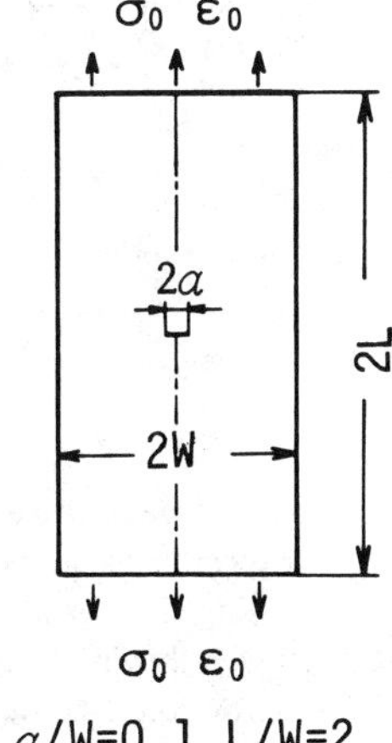

FIG. 6—*Dimension of the specimen for the analysis.*

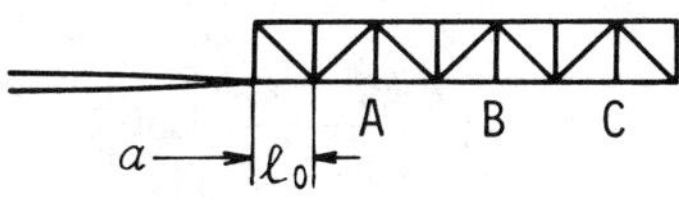

FIG. 7—*Detail of division pattern near crack tip.*

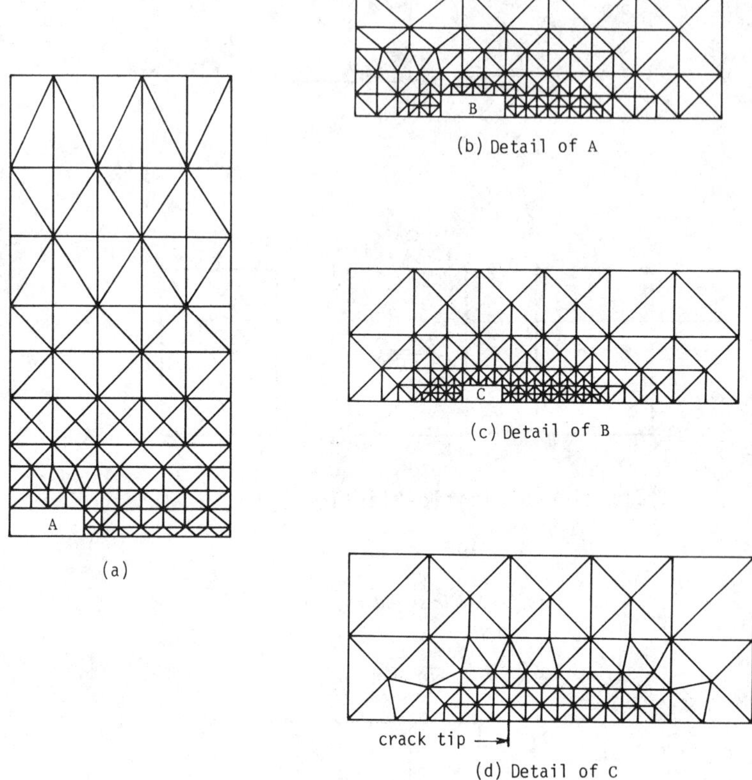

FIG. 8—*Division pattern for finite element analysis.*

where the negative value of the slope is assumed to be positive. Since the strain ϵ_y in the vicinity of the crack tip includes not only the singular term but also the normal term, the accurate value of the singularity will be determined by excluding the normal term from ϵ_y, particularly in the case of the FEM. In the elastic calculation, it was shown, on the basis of the principle of superposition, that the accurate stress intensity factor could be obtained by excluding the uniform remote stress from the stress at crack tip element [*20*]. Although in elastic-plastic problems the principle of superposition is not applicable, the use of the same procedure may result in the improvement of the determination of the strain singularity in comparison with Eq 10. Therefore, if we denote the value of singularity determined in this way by α_2^*, we have

$$\alpha_2^* = \log\left(\frac{\epsilon_{yA} - \epsilon_0}{\epsilon_{yB} - \epsilon_0}\right)\Big/\log 2 \tag{11}$$

where ϵ_0 is the remote strain (see Fig. 6).

Table 2 shows the variations of α_1^* and α_2^* with progress of the plastic deformation for hardening exponent $n = 5$. From the reason described above and Table 2, α_2^* expresses both elastic

TABLE 2—*Variation of singularity α^* with increasing plastic deformation (n = 5).*[a]

σ_0/σ_Y	ϵ_0	α_1^*	α_2^*
0.0401 (elastic)	4.00×10^{-5}	0.43	0.51
0.340	3.40×10^{-4}	0.71	0.79
0.640	6.40×10^{-4}	0.74	0.80
0.940	9.39×10^{-4}	0.75	0.79
1.24	1.28×10^{-3}	0.75	0.79
1.84	2.23×10^{-3}	0.76	0.78
2.74	5.77×10^{-3}	0.70	0.71
3.34	1.16×10^{-2}	0.59	0.60
3.64	1.64×10^{-2}	0.53	0.54
3.94	2.30×10^{-2}	0.47	0.48
4.23	3.13×10^{-2}	0.42	0.43

[a] σ_0, ϵ_0 = remote stress and strain, respectively.

and HRR singularity better than α_1^*. Accordingly, we adopt α_2^* as α^* in the following discussion.

Figures 10 to 13 show the variation of α^* with the progress of the plastic deformation for $n =$ 5, 7, 9, and 13. In these figures it should be noted that although the strain distribution near crack tip in small-scale yielding condition shows HRR singularity of $n/(n + 1)$, the singularity becomes $\alpha^* = 0.5 \sim 0.7$ for $\epsilon_0 > 10^{-2}$ in overall plastic deformation, regardless of the value of hardening exponent n. This means that the strain distribution ϵ_y near crack tip in large plastic deformation may be expressed as

$$\epsilon_y = f(n, \epsilon_0) \cdot \left(\frac{a}{r}\right)^{\alpha^*} \tag{12}$$

where $\alpha^* = 0.5 \sim 0.7$ and $\epsilon_0 = 10^{-2} \sim 3 \times 10^{-2}$.

From the investigation of the calculated results in Figs. 10 to 13, it was confirmed that $f(n, \epsilon_0)$ could approximate a linear function of ϵ_0. Accordingly, we have

$$\epsilon_y = C_0 \epsilon_0 \left(\frac{a}{r}\right)^{\alpha^*} \tag{13}$$

Equation 13 does not necessarily mean the exact linear relationship between ϵ_y and ϵ_0, because as seen in Figs. 10 to 13 α^* is a weak function of n and ϵ_0. The average values of α^* for various values of n and for $\epsilon_0 = 10^{-2} \sim 3 \times 10^{-2}$ are evaluated as in Table 3.

From the value of α^* in Table 3, the value of C_0 in Eq 13 can be plotted as in Fig. 14, where the values of ϵ_y at $r/a = 0.05$ were adopted for the calculation. As an approximate formulation for calculated points in Fig. 14, we have

$$C_0 = \frac{23}{n} \tag{14}$$

Relationship Between the Strain Singularity and the Coffin-Manson Law

The low-cycle fatigue life is almost entirely dominated by the process of small crack propagation [3]. From this fact, it was concluded that the Coffin-Manson law (Eq 4) is virtually equiva-

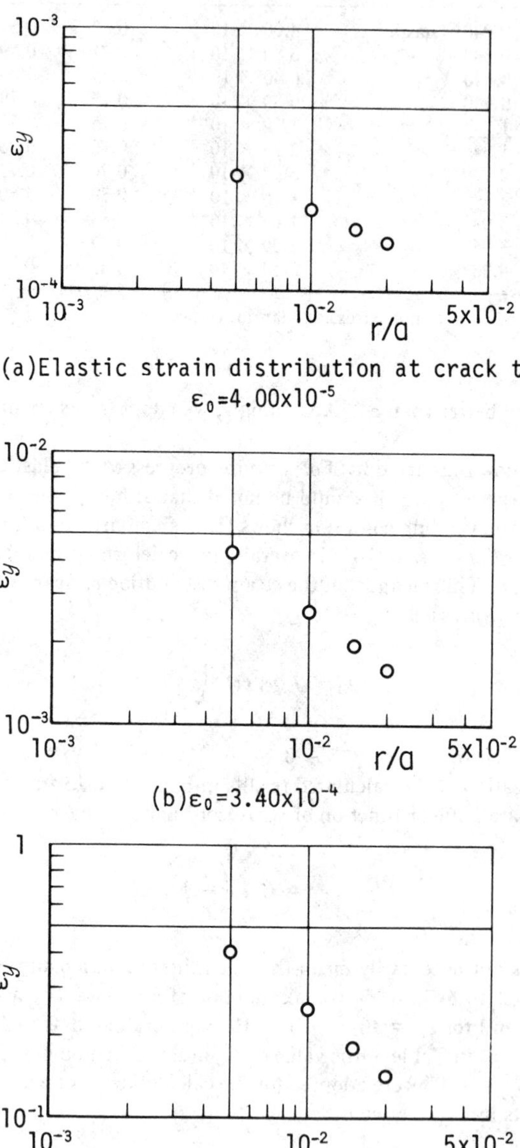

FIG. 9—*Variation of strain distribution near crack tip with increasing plastic deformation* (n = 5).

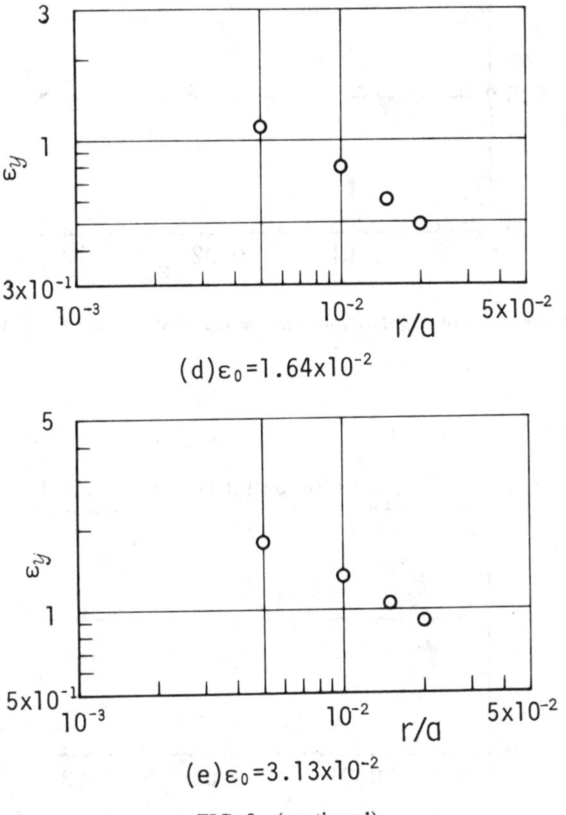

FIG. 9—(continued).

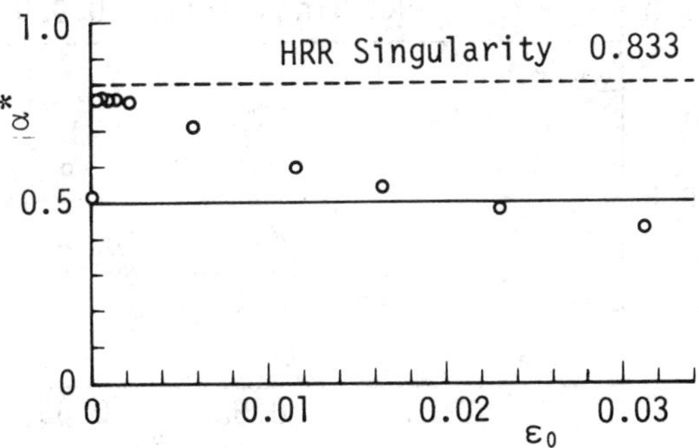

FIG. 10—*Variation of singularity with increasing plastic deformation* (n = 5).

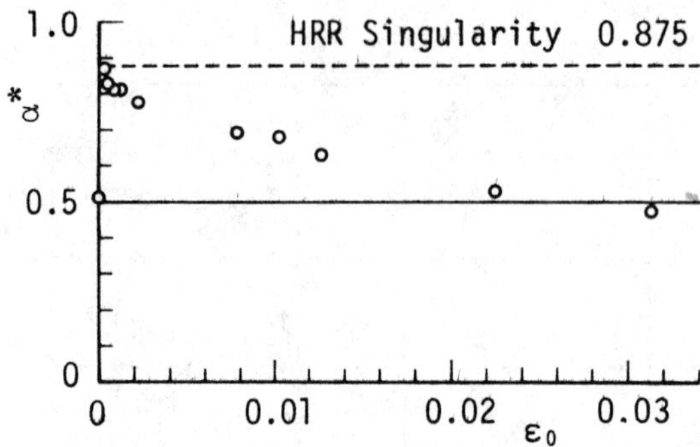

FIG. 11—*Variation of singularity with increasing plastic deformation (*n *= 7).*

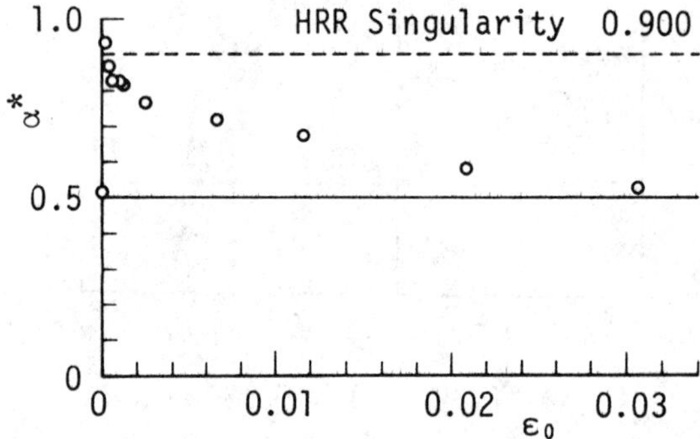

FIG. 12—*Variation of singularity with increasing plastic deformation (*n *= 9).*

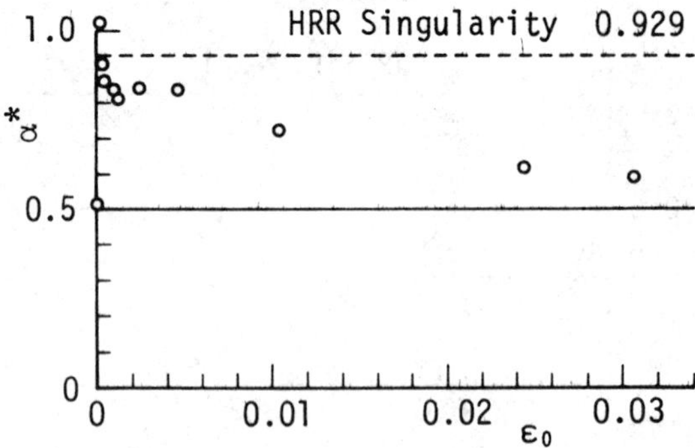

FIG. 13—*Variation of singularity with increasing plastic deformation (*n *= 13).*

TABLE 3—*Average values of singularity* α^* *for* $\epsilon_0 = 10^{-2} \sim 3 \times 10^{-2}$ *and for different hardening exponent* n.

n	α^*
5	0.50
6	0.52
7	0.56
9	0.59
13	0.64

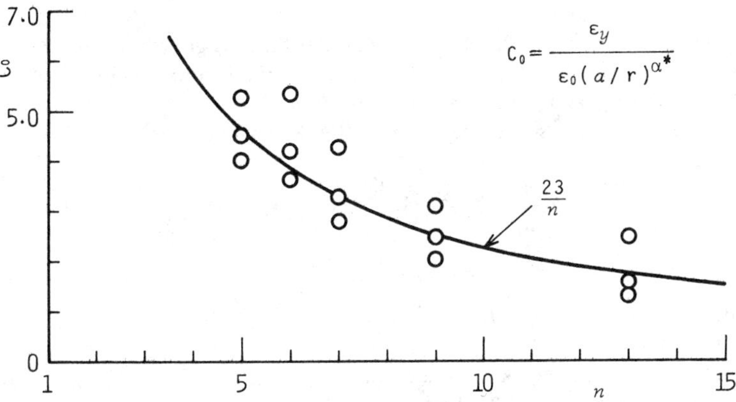

FIG. 14—*Dependence of the constant* C_0 *of Eq 13 on hardening exponent* n.

lent to the propagation law of a small crack. However, the question as to why the exponent α in Eq 4 ranges from 0.5 to 0.7 has remained unsolved.

If we seek a simple mechanism which may be operating to cause failure, as predicted by Coffin [2], we naturally should pay attention to Eq 13. Assuming a small crack of $2a$ in length in a specimen which is subject to the strain amplitude ϵ_0 and neglecting the crack closure,[2] the strain cycled near crack tip is given by Eq 13. Denoting the strain range by $\Delta\epsilon$, we can write

$$\Delta\epsilon_y = C_0 \Delta\epsilon_t \left(\frac{a}{r}\right)^{\alpha^*} \tag{15}$$

where $\Delta\epsilon_t = 2\epsilon_0$.

Although α^* is not constant and ranges from 0.5 to 0.7, it may be a reasonable assumption that the crack propagates by the distance r_c where $\Delta\epsilon_y$ exceeds a critical value ϵ_{crit} (Fig. 15). Therefore from Eq 15 it follows that

$$\frac{da}{dN} = r_c = \left(\frac{C_0 \Delta\epsilon_t}{\epsilon_{crit}}\right)^{1/\alpha^*} \cdot a \tag{16}$$

[2]In the low-cycle fatigue range, a small crack is almost completely open during the whole loading cycle [3,21].

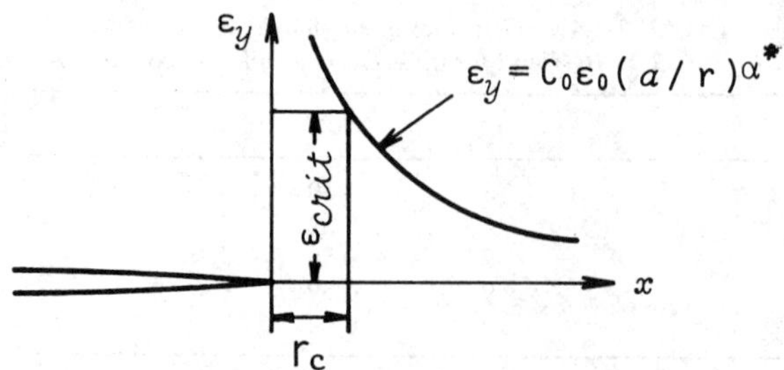

FIG. 15—*Strain distribution near crack tip under overall plastic deformation and fracture criterion.*

where N is the number of cycles. Equation 16 indicates that the growth rate of a small crack is proportional to $\Delta\epsilon_t^{1/\alpha^*} \cdot a$. Recalling $C_0 = 23/n$, the material factors which influence the crack propagation rate are ϵ_{crit} and the hardening exponent n.

Integrating Eq 16 by the same procedure as in Ref *3,* we obtain

$$\log \frac{a_f}{a_0} = \left(\frac{C_0}{\epsilon_{crit}}\right)^{1/\alpha^*} \cdot \Delta\epsilon_t^{1/\alpha^*} \cdot (N_f - N_0) \tag{17}$$

where

a_0 = half crack length at $N = N_0$,
a_f = half crack length at $N = N_f$,
N_0 = number of cycles at crack initiation, and
N_f = number of cycles at final fracture.

Considering that most of N_f is consumed for the process of crack propagation and then assuming $N_0 \cong 0$, we can rewrite Eq 14 as

$$\Delta\epsilon_t \cdot N_f^{\alpha^*} = \frac{\epsilon_{crit}}{C_0}\left(\log \frac{a_f}{a_0}\right)^{\alpha^*} \tag{18}$$

If $\Delta\epsilon_e$ is negligible in comparison with $\Delta\epsilon_p$, we may write $\Delta\epsilon_t \cong \Delta\epsilon_p$ and Eq 18 becomes

$$\Delta\epsilon_p \cdot N_f^{\alpha^*} = C^* \tag{19}$$

Equation 19 is identical to Eq 4, and the exponent α in Eq 4 is the same as the exponent α^* in Eq 13, which indicates the singularity of strain distribution near crack tip under overall plastic deformation. The foregoing discussion explains the reason that the exponent α in the Coffin-Manson law $\Delta\epsilon_p \cdot N_f^{\alpha} = C$ ranges from 0.5 to 0.7 for many materials.

The constant C in the right-hand side of Eq 4 is written from Eq 18 and Eq 19 as

$$C = \left(\frac{\epsilon_{crit}}{C_0}\right)\left(\log \frac{a_f}{a_0}\right)^{\alpha^*} = \frac{n}{23}\left(\log \frac{a_f}{a_0}\right)^{\alpha^*} \cdot \epsilon_{crit} \tag{20}$$

Equation 20 shows several factors which influence the value of C. Considering $n = 5 \sim 7$ for common metallic materials and estimating $2a$ = the order of grain size and $2a_f$ = the diameter of a specimen (~10 mm) [*3,4*], we obtain

$$C \cong (0.5 \sim 1.2)\epsilon_{crit} \tag{21}$$

Of many factors, C is the most dependent on ϵ_{crit}. If C can be approximated by Eq 5, as proposed by Coffin, it follows from the comparison between Eq 5 and Eq 21 that ϵ_{crit} must be on the order of the fracture ductility ϵ_F. This prediction may be examined from the experimental results of the crack growth rate. Thus, from Eq 16, we can write

$$\epsilon_{crit} = C_0 \Delta\epsilon_t^{\alpha^*} \Big/ \left(\frac{da}{dN}\right)^{\alpha^*} \tag{22}$$

Although the value of C_0 in the three-dimensional problems may be somewhat different from those of two-dimensional problems, we adopt Eq 14 for the estimation of C_0.

For 70-30 brass [*4*]:

$$\frac{da}{dN} = 5.83\ \Delta\epsilon_p^{2.0} \cdot a \tag{23}$$

and

$$n \cong 5.56$$

For annealed 0.46% carbon steel [*3*]:

$$\frac{da}{dN} = 2.58\ \Delta\epsilon_p^{1.74} \cdot a \tag{24}$$

and

$$n \cong 3.83$$

Therefore, regarding $\Delta\epsilon_t \cong \Delta\epsilon_p$ and $\alpha^* = \alpha$, the values of ϵ_{crit} can be approximately evaluated as in Table 4. The value of ϵ_{crit} of 70-30 brass is on the same order as ϵ_F. However, it is not the case for 0.46% carbon steel. This difference may be due to the fact that 70-30 brass is a single-phase material whereas 0.46% carbon steel has a two-phase matrix. This point needs more investigation on various materials in the future.

The experimental results and the related discussions in the present study are concerned with not only low-cycle fatigue but also with the clarification of material factors in ductile fracture and the problem of fracture criterion. The future studies in the line of present study will contribute to solve these nonlinear fracture problems.

Conclusions

The view [*3,4*] that the Coffin-Manson law is substantially equivalent to the growth of a small crack was extended to explain why the exponent α in the law ($\Delta\epsilon_p \cdot N^\alpha = C$) ranges only from 0.5 to 0.7 for many metallic materials.

TABLE 4—*Tensile fracture ductility ϵ_F and crack tip fracture strain ϵ_{crit}.*

	ϵ_{crit}	ϵ_F
70-30 brass	1.7	1.57
0.46% C steel	3.5	0.67

The conclusions drawn from the present study are:

1. According to the elastic-plastic finite element analysis on the singular strain field near crack tip of a small crack, the HRR singularity is lost with increasing plastic deformation. After the strain ϵ_0 at distance exceeds $\epsilon_0 \cong 10^{-2}$, the strain singularity α^* in the expression $(a/r)^{\alpha^*}$ becomes $\alpha^* = 0.5 \sim 0.7$ regardless of materials.

2. If the assumption that the crack growth rate da/dN is equal to the distance where the strain exceeds the critical strain ϵ_{crit} is adopted, we can obtain the relationship

$$\frac{da}{dN} = \left(\frac{C_0}{\epsilon_{crit}}\right)^{1/\alpha^*} \Delta\epsilon_t \cdot a$$

where

$\Delta\epsilon_t$ = total strain range at distance,
$C_0 = 23/n$, and
n = hardening exponent.

This relationship indicates that the material parameters which influence the crack growth rate are n and ϵ_{crit}.

3. Considering the experimental fact that the process of a small crack propagation is predominant in total life of low-cycle fatigue and also considering Conclusion 2, it can be concluded that the exponent α in the Coffin-Manson law is equivalent to the singularity α^* in the strain distribution near crack tip. Consequently, we have $\alpha \cong 0.5 \sim 0.7$ for a wide range of materials.

References

[1] Coffin, L. F., Jr., *Transactions of ASME,* Vol. 76, 1954, p. 931; Coffin, L. F., Jr., *Transactions of ASME,* Vol. 78, 1956, pp. 527-532; Manson, S. S., NACA Technical Note 2933, National Advisory Committee for Aeronautics, Washington, D.C., 1953; NACA Technical Report 1170, National Advisory Committee for Aeronautics, Washington, D.C., 1954; Manson, S. S. in *Fatigue—An Interdisciplinary Approach,* Syracuse University Press, N.Y., 1964, pp. 133-178.

[2] Coffin, L. F., Jr., "Discussion on Fatigue Behavior in Strain Cycling in the Low- and Intermediate-Cycle Range," in *Fatigue—An Interdisciplinary Approach,* S. S. Manson and M. H. Hirshberg, Eds., Syracuse University Press, N.Y., 1964, pp. 133-178.

[3] Murakami, Y., Harada, S., Endo, T., Tani-Ishi, H., and Fukushima, Y., *Engineering Fracture Mechanics,* Vol. 18, No. 5, 1983, pp. 909-924.

[4] Murakami, Y., Makabe, C., and Nisitani, H., *Transactions of JSME (A),* Vol. 50, No. 459, 1984, pp. 1828-1837.

[5] Hutchinson, J. W., *Journal of the Mechanics and Physics of Solids,* Vol. 16, 1968, pp. 13-31.

[6] Rice, J. R. and Rosengren, G. F., *Journal of the Mechanics and Physics of Solids,* Vol. 16, 1968, pp. 1-12.

[7] Rice, J. R., *Journal of Applied Mechanics, Transactions of ASME,* Vol. 35, 1968, pp. 379-386.

[8] Ramberg, W. and Osgood, W. R., NACA Technical Note 902, National Advisory Committee for Aeronautics, Washington, D.C., 1943.

[9] Berger, C. et al. in *Elastic-Plastic Fracture, ASTM STP 668,* 1979, American Society for Testing and Materials, Philadelphia, pp. 378-405; *Fracture Mechanics—Vols. I and II, ASTM STP 791,* American Society for Testing and Materials, Philadelphia, 1983.

[10] Begley, J. A. and Landes, J. D. in *Fracture Toughness, ASTM STP 514,* American Society for Testing and Materials, Philadelphia, 1972, pp. 1-23.

[11] Landes, J. D. and Begley, J. A. in *Fracture Toughness, ASTM STP 514,* American Society for Testing and Materials, Philadelphia, 1972, pp. 24-39.

[12] Dowling, N. E. in *Cyclic Stress-Strain and Plastic Deformation Aspects of Fatigue Crack Growth, ASTM STP 637,* American Society for Testing and Materials, Philadelphia, 1977, pp. 97-121.

[13] Sih, C. F., *Journal of the Mechanics and Physics of Solids,* Vol. 29-4, 1981, pp. 305-326.

[14] Coffin, L. F., Jr., *Transactions of ASME,* Vol. 78, 1956, pp. 527-532.

[*15*] Manson, S. S., *Thermal Stress and Low-Cycle Fatigue,* McGraw-Hill, New York, 1966, p. 159.
[*16*] Manson, S. S. in *Proceedings,* 1st International Conference on Fracture, Sendai, Japan, 1965, pp. 1387-1431.
[*17*] Weiss, V. in *Fatigue—An Interdisciplinary Approach,* Syracuse University Press, N.Y., 1964, pp. 179-186.
[*18*] Tomkins, B., *Philosophical Magazine,* Vol. 23, 1971, pp. 687-703.
[*19*] Solomon, H. D., *Journal of Materials,* Vol. 7, No. 3, 1972, pp. 299-306.
[*20*] Murakami, Y., *Engineering Fracture Mechanics,* Vol. 8, 1976, pp. 643-655.
[*21*] Yamada, T., Hoshide, T., Fujimura, S., and Manabe, M., *Transactions of JSME(A),* Vol. 49, No. 440, 1983, pp. 441-451.

Masateru Ohnami,[1] *Yasuhide Asada,*[2] *Masao Sakane,*[1]
Masaki Kitagawa,[3] *and Toshio Sakon*[4]

Notch Effect on Low Cycle Fatigue in Creep-Fatigue at High Temperatures: Experiment and Finite Element Method Analysis

REFERENCE: Ohnami, M., Asada, Y., Sakane, M., Kitagawa, M., and Sakon, T., **"Notch Effect on Low Cycle Fatigue in Creep-Fatigue at High Temperatures: Experiment and Finite Element Method Analysis,"** *Low Cycle Fatigue, ASTM STP 942,* H. D. Solomon, G. R. Halford, L. R. Kaisand, and B. N. Leis, Eds., American Society for Testing and Materials, Philadelphia, 1988, pp. 1066–1095.

ABSTRACT: This paper describes the notch effect on low cycle fatigue in creep-fatigue conditions by experiments and finite element method (FEM) analysis. FEM analyses for cyclic and monotonic loadings were made, and the analytical results were compared with the measured strain at the notch root using a high temperature strain gage. The calculated strain by FEM agreed well with the measured strain if we used the proper FEM model and proper constitutive equation. In order to examine the accuracy of life prediction by FEM analysis, low cycle fatigue tests were also carried out using SUS 304 stainless steel for notched plate and notched round bar specimens. Life prediction by FEM analysis had sufficient accuracy; the applicability of other conventional life prediction methods was also discussed.

KEY WORDS: low-cycle fatigue, notch effect, elevated temperature, finite element method, creep fatigue interaction

Geometrical discontinuities are inevitable in practical applications operating at elevated temperatures, so that we should have a proper design method for evaluating the local stress/strain at notched parts. Continuum mechanics, unfortunately, gives almost no information about the stress/strain at the notched part for relatively complex geometry and especially in creep-fatigue conditions. However, recent progress in computer technology and the finite element method (FEM) strongly assist the designers in evaluating the stress/strain concentrations of the notched members used in creep-fatigue conditions. We can evaluate the local stress/strain by FEM even in creep-fatigue conditions.

Extensive studies [*1–4*] have been carried out on the notch effect in creep-fatigue conditions, and the local strain concept appears to be successful for the low cycle case. However, integrated studies, those combining the evaluation of local stress/strain with life prediction in creep-fatigue conditions, are scarce. Therefore the Notch Effects Test Subcommittee of the Iron and Steel Institute of Japan planned to collaborate on studies of the notch effect in creep-fatigue conditions. The main objects of the subcommittee are:

[1]Professor and Assistant Professor, respectively, Ritsumeikan University, Kyoto, Japan.
[2]Professor, The University of Tokyo, Tokyo, Japan.
[3]Chief Researcher, Research Institute, Ishikawajima-Harima Heavy Industries Co., Ltd., Tokyo, Japan.
[4]Senior Research Engineer, Technical Institute, Mitsubishi Heavy Industries, Ltd., Takasago, Japan.

1. To determine how accurately we can estimate the local stress/strain at the notched part in creep-fatigue conditions using FEM analysis. Is there any possibility of fatal error in the calculated stress or strain? We should compare the calculated strain with the strain actually measured in order to obtain a guarantee that the calculated strain is accurate enough for the design.
2. To clarify what the relation is between the calculated local stress/strain and the crack initiation or failure life.

In order to achieve these objects, the subcommittee planned a three-year project from 1981 to 1984 to collaborate on the following goals:

1. To make a cyclic FEM analysis of the notched specimen for a strain wave having hold-times in five cycles and to compare the calculated strain with measured strain using a high temperature strain gage.
2. To check the replaceability of cyclic FEM analysis, which needs much computing time, by monotonic analysis.
3. To clarify the relationship between the calculated strain and the actual crack initiation/failure fatigue lives in low cycle fatigue. At the same time to check the applicability of conventional life prediction methods to low cycle fatigue data of the notched specimens. Results of the study are presented herein.

Elastic-Plastic-Creep FEM Analysis under Cyclic Loading

In this section, we discuss an inelastic cyclic stress/strain response at the notch root based on FEM analysis for a trapezoidal strain wave up to five cycles. Three computer programs examined how the cyclic hardening rule affects the inelastic stress/strain concentration factors. The results of FEM analysis were compared with the measured local strain using high temperature strain gage. The FEM analysis for monotonic loading was also carried out to check the replaceability of cyclic analysis by monotonic analysis, since the cyclic analysis is more expensive than the monotonic analysis.

Cyclic FEM Analysis and Measurement of Strain

Table 1 lists the three hardening rules used for the inelastic stress/strain analysis together with the material constants and FEM programs: kinematic hardening, MARC-type combined hardening, and ORNL kinematic hardening. The material used in the present study is SUS 304 austenitic stainless steel, which is well known as a typical cyclic strain hardening material; the ORNL kinematic hardening rule takes account of this cyclic strain hardening behavior. Figure 1 shows the shape and dimensions of the specimen analyzed, where the shaded part is modelled to the FEM mesh. The specimen for the cyclic stress/strain analysis is a double side grooved plate of 6 mm thickness having an elastic stress concentration factor, K_t, of 1.5, which is the same as used in the experiments measuring the strain at the notch root by the high temperature strain gage. Figure 2 shows the FEM meshes used in the analysis, where Fig. 2*a* describes the kinematic hardening rule, Fig. 2*b* the MARC-type combined hardening rule, and Fig. 2*c* the ORNL kinematic hardening rule. The number of elements and nodes are given in the figure caption. Three computer programs, HINAPS (made by Hitachi) and MARC-J2, and TEPICC-4 (made by Babcock-Hitachi), were used in the calculation of the kinematic hardening rule, the MARC-type combined hardening rule, and the ORNL kinematic hardening rule, respectively. The model lies on the *Y-Z* plane, where the axial direction of the specimen agrees with the *Z* axis. The strain wave used in the analysis is a fully reversed triangular wave of the push-pull mode (strain ratio $R = -1$) with a 30-min strain hold-time at tension peak in each cycle.

Figure 3*a* shows bilinear constitutive relations for the kinematic hardening rule and the ORNL kinematic hardening rule, while Fig. 3*b* shows a multilinear constitutive relation for the

TABLE 1—*Constitutive equations and FEM programs for cyclic analysis.*

Program		HINAPS	MARC-J2	TEPICC-4
Elastic – plastic characteristic	Hardening rule	Kinematic hardening 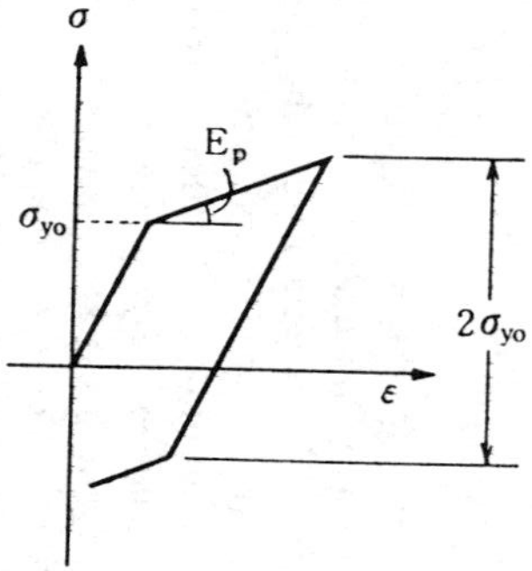	MARC type combined hardening 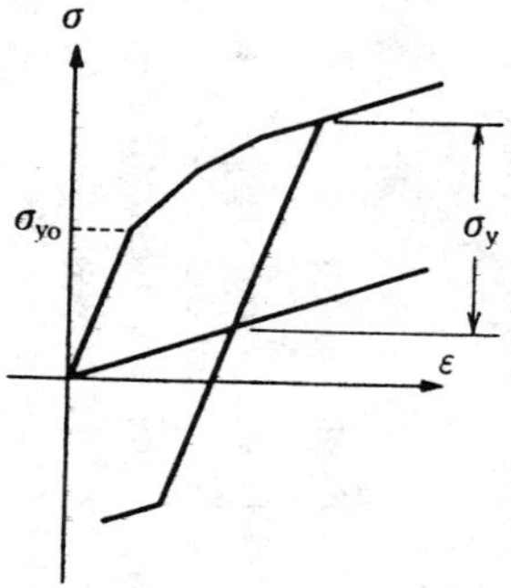	ORNL kinematic hardening 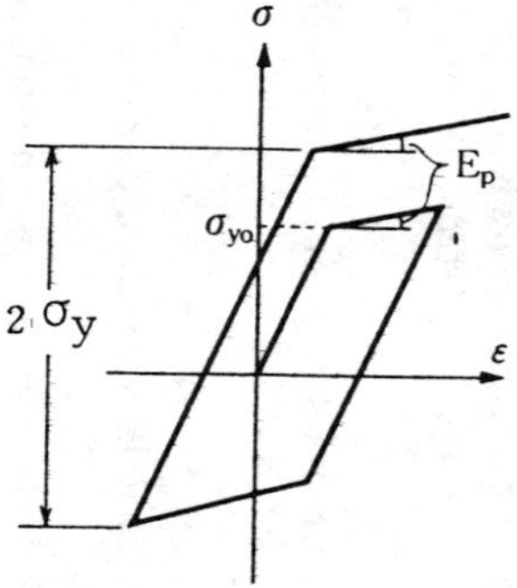
	σ_{yo} (MPa)	132.4	Multilinear approximation	132.4
	σ_y (MPa)	132.4		$\sigma_y =$ 230.5
	E (MPa)	1.499×10^5	1.499×10^5	1.499×10^5
	ν	0.3	0.31	0.31
	E_p (MPa)	2530		2530
Creep equation		ε_c (1/h) $= 2.036 \times 10^{-14}\ \sigma^{4.674}$		(σ : MPa)

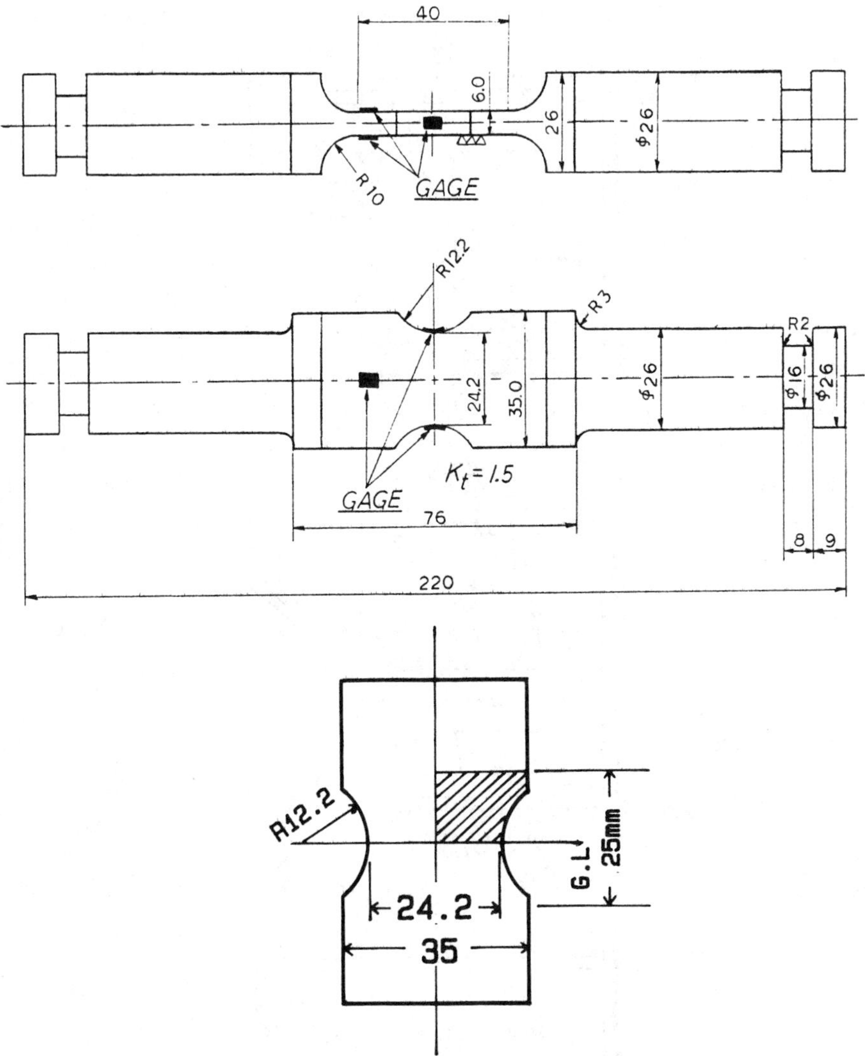

FIG. 1—*Shape and dimensions (millimetres) of test specimen used to measure strain. Shaded area is modelled to the FEM mesh.*

MARC type combined hardening rule. In the ORNL kinematic hardening rule, two yield stresses and one strain hardening modulus are defined in order to take account of the cyclic strain hardening behavior of the material. Though there exists a slight difference between the bilinear constitutive relation and the experimental results around yield stress in Fig. 3*a*, the overall fitting of the bilinear constitutive relation is good. The multilinear constitutive relation shown in Fig. 3*b* very closely approximates the actual material behavior.

Figure 4 compares the measured strain with FEM analysis. The ordinate is the nominal stress, σ_n, applied to the specimen and the abscissa is the nominal strain, ϵ_{GL}, along the 25 mm gage length of the extensometer. Tension and compression peak stresses in the FEM analysis agree well with those in experiments other than the ORNL kinematic hardening rule, but the

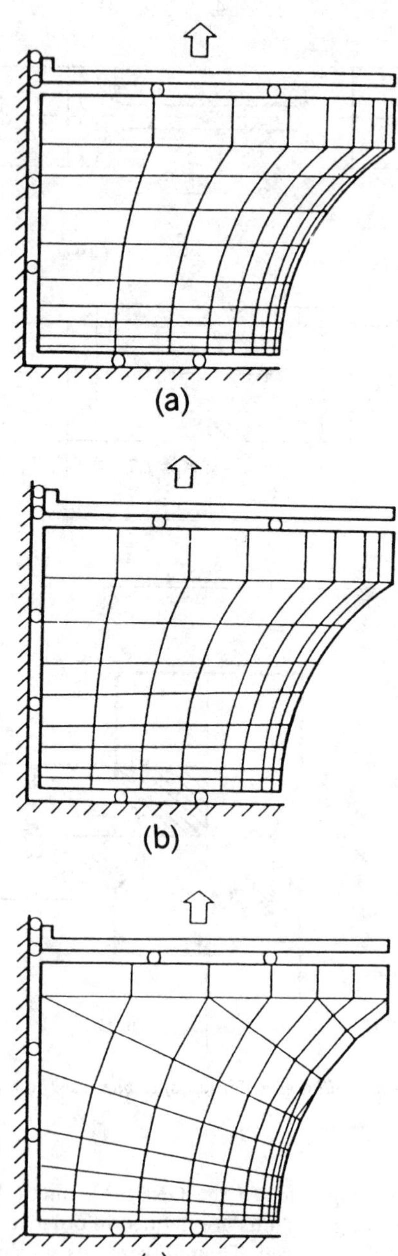

FIG. 2—*Finite element mesh.* (a) *8-node plane stress element with 4 integration points for kinematic hardening. Number of elements and nodes are 80 and 277, respectively.* (b) *8-node plane stress element with 9 integration points for MARC-type combined hardening. Number of elements and nodes are 72 and 251, respectively.* (c) *4-node plane stress element with 4 integration points for ORNL kinematic hardening. Number of elements and nodes are 77 and 92, respectively.*

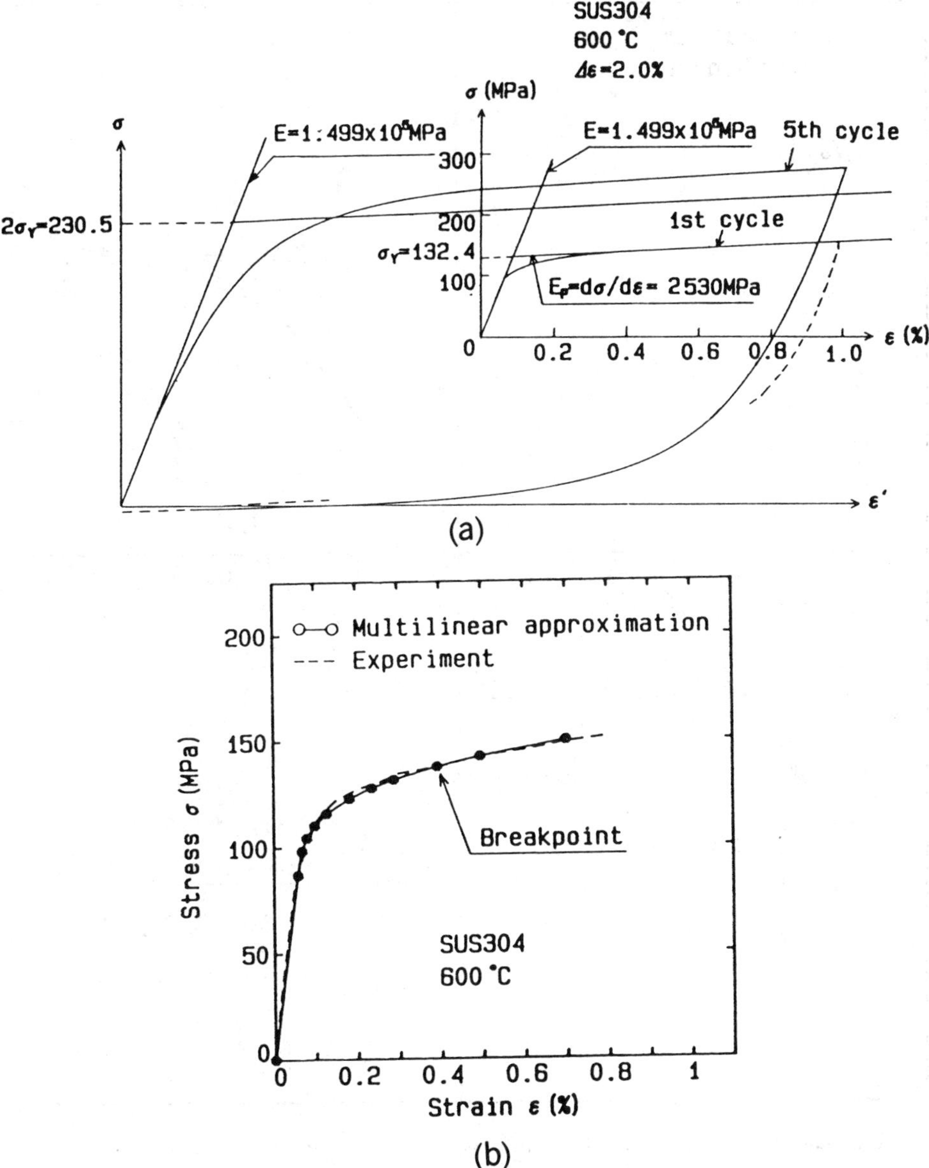

FIG. 3—*Constitutive relation used for cyclic FEM analysis.* (a) *Bilinear approximation for kinematic and ORNL kinematic hardening.* (b) *Multilinear approximation for MARC-type combined hardening.*

shape of the hysteresis loops in the analysis is somewhat different from the experimental results. The FEM analysis estimates a larger stress amplitude around zero strain amplitude in comparison with the experiments. This difference may be attributed to FEM mesh modelling. In this regard, a detailed discussion will be presented later, including the recalculated results using the improved FEM mesh. The ORNL kinematic hardening rule cannot simulate the actual strain hardening behavior of the material.

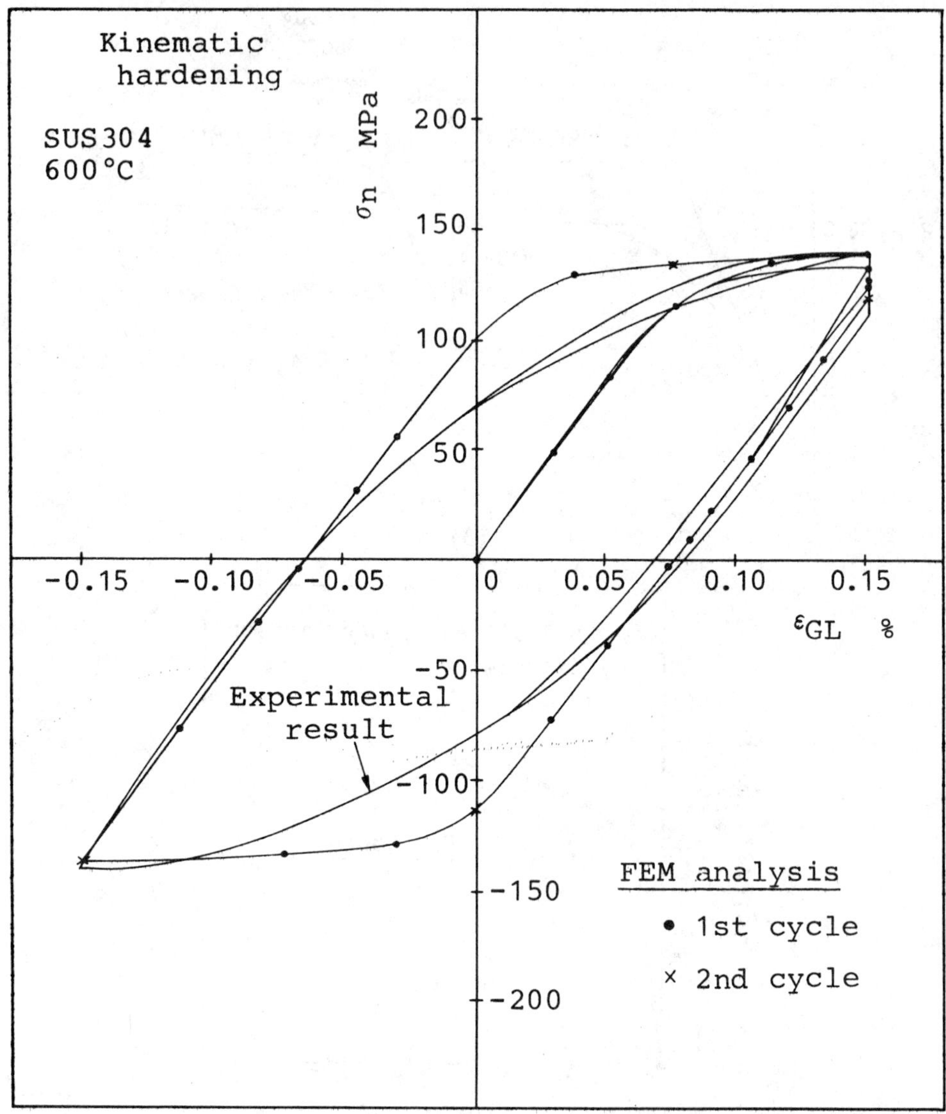

FIG. 4a—*Experimental and FEM analytical results for σ_n–ϵ_{GL} correlation: kinematic hardening.*

Figure 5 compares the measured hysteresis loops at the notch root with those of FEM analysis. Special high temperature resistant strain gages were attached to the bottom of the notch grooves. We employed two gage methods to compensate for the thermal strain and change in gage factor due to the temperature variation. The strain gage specifications are: gage length, 3.18 mm; gage width, 1.57 mm; material, P_t–W alloy; resistance, 120 Ω; and gage factor, 4.0. The strain gage was attached to the specimen by the Rokid Process (i.e., by the flame spraying of aluminum oxide). Strain controlled tests at ±0.15% along the gage length were carried out at 873 K in air by using an electro-hydraulic servo-machine with an infrared furnace.

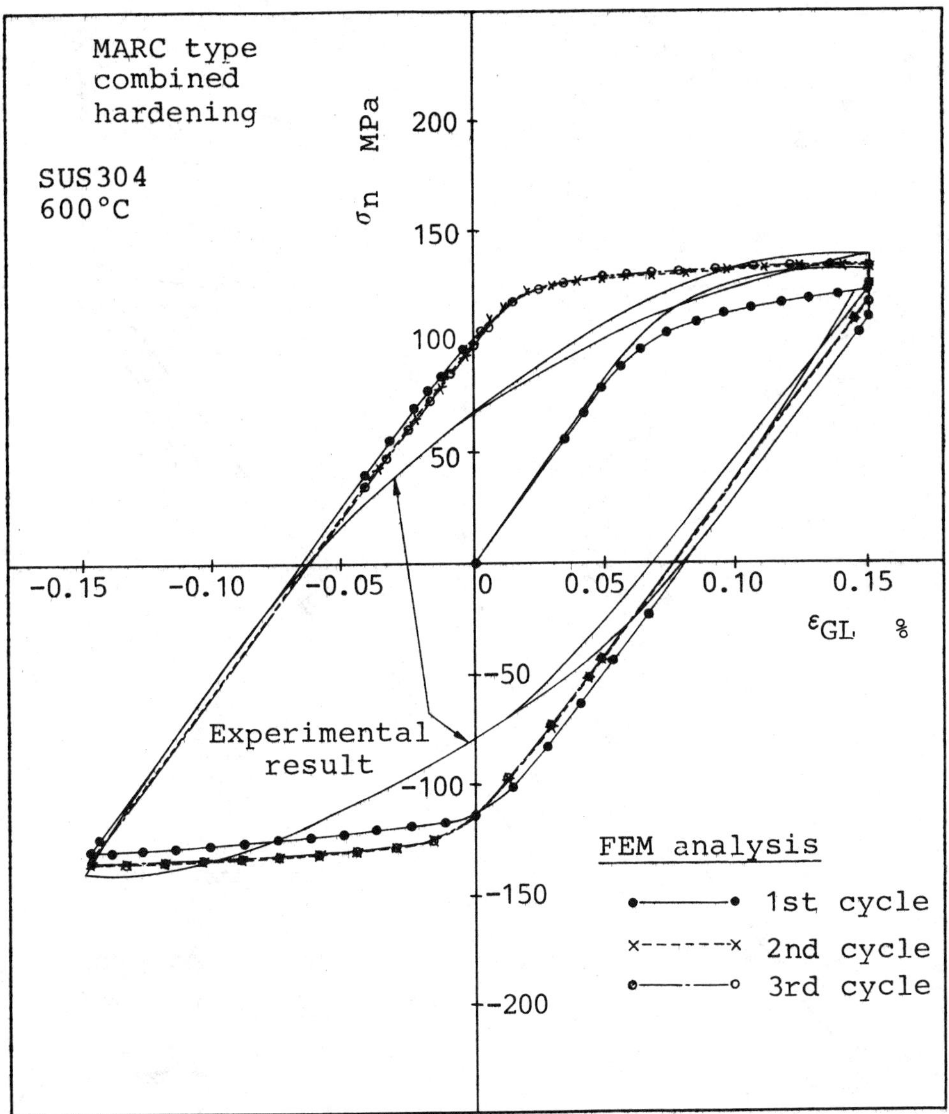

FIG. 4*b*—*Experimental and FEM analytical results for* σ_n-ϵ_{GL} *correlation: MARC-type combined hardening.*

Regarding the comparison between the analytical results, kinematic strain hardening and MARC-type combined hardening exhibit almost a closed hysteresis loop with a slight movement to the compression side. This movement is caused by the strain hold-time because in the analysis with no hold-time, completely closed hysteresis loops were found. Stress relaxation during the 30-min hold-time is small; a slight increase in the axial strain during the hold-time yields, whereas the strain along the gage length keeps a constant value. For the case of the ORNL kinematic hardening rule, on the other hand, the hysteresis loop exhibits a different behavior. At the first cycle a smaller cyclic peak stress is found, but at the second cycle significantly larger

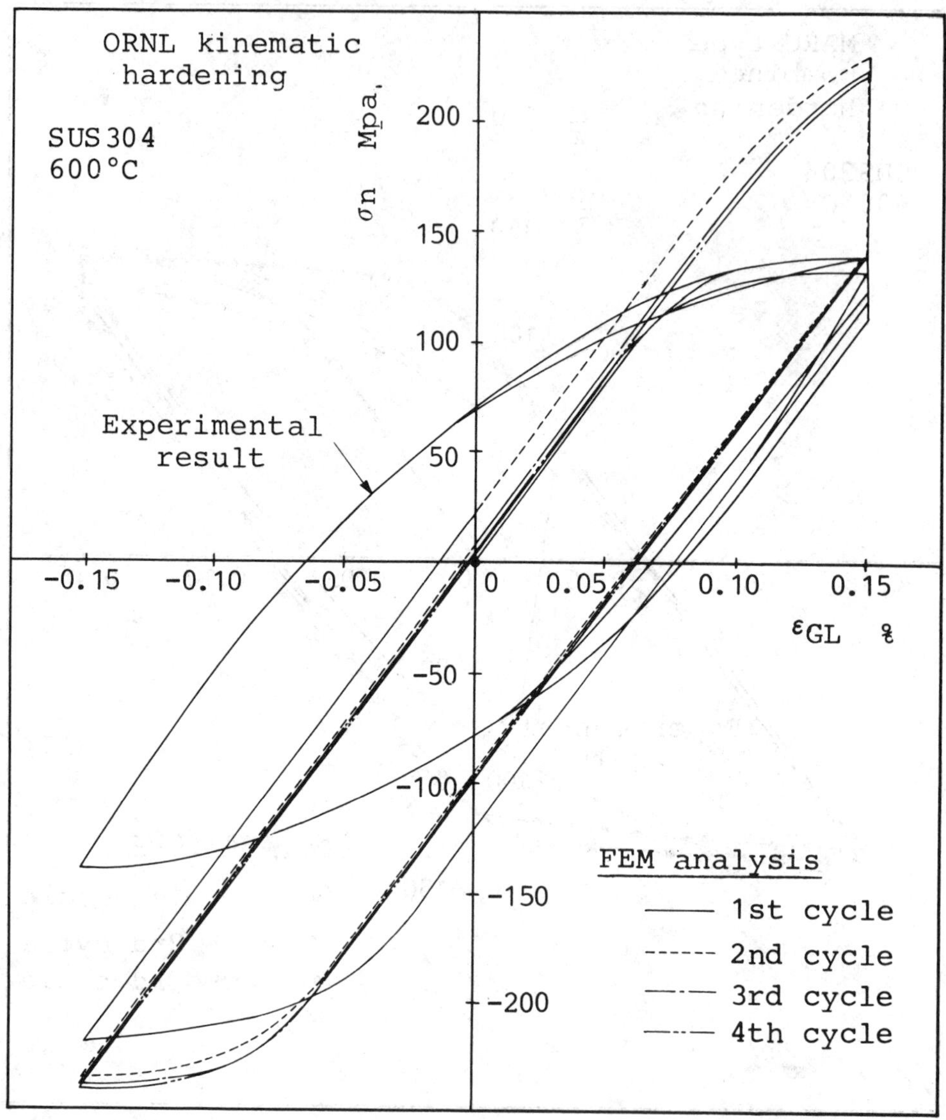

FIG. 4c—*Experimental and FEM analytical results for σ_n–ϵ_{GL} correlation: ORNL kinematic hardening.*

cyclic stress is found. After the second cycle there is no large difference in peak stress, but the hysteresis loop moves to the compression side with the proceeding cycle. Stress relaxation in the ORNL kinematic hardening rule during the hold-time is larger compared with those in the other two hardening rules.

Measured peak nominal stress agrees well with the FEM analysis except the ORNL hardening rule, but FEM analysis overestimates the total strain range about 1.6 times. The main cause of this disagreement may be attributed to improper modelling of FEM mesh. As shown in Fig. 1, only the notched part whose height is 12.5 mm was modelled to FEM mesh so that the FEM model does not simulate precisely the actual deformation behavior. The FEM analysis evaluates

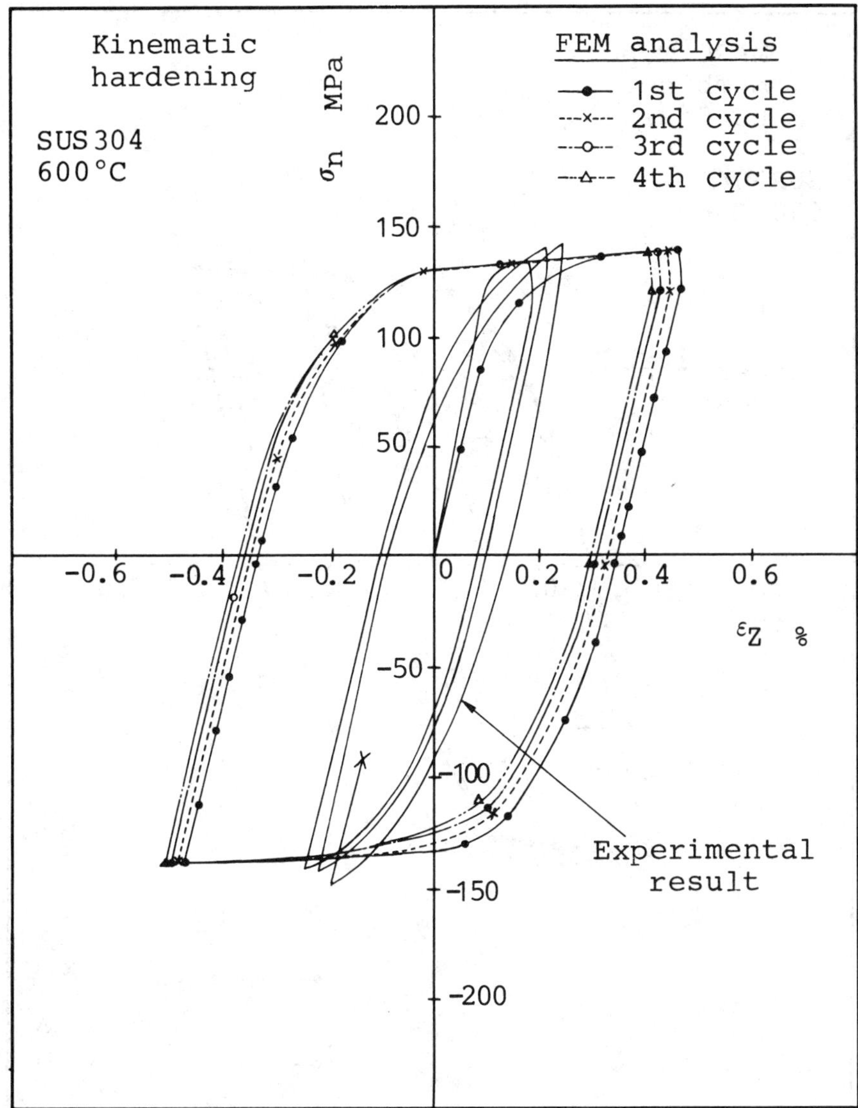

FIG. 5*a—Experimental and FEM analytical results for* σ_n-ϵ_z *correlation: kinematic hardening.*

the excessive notch opening so that the larger inelastic strain concentration occurs. In order to check how this deformation behavior affected the strain concentration factor, additional FEM analyses were made for the improved model whose height was 30 mm (i.e., 17.5 mm in height was added to the original model shown in Fig. 1, in order to improve the constant axial deformation near the notched part).

The results of the recalculation are shown in Fig. 6, where Fig. 6*a* is the relation between nominal stress and nominal strain and Fig. 6*b* is the relation between nominal stress and axial strain at the notch root. As is seen from Fig. 6*a,* peak strains of the two models agree with each other but the shape of the stress-strain curve is different. The FEM analysis in Fig. 4 estimates

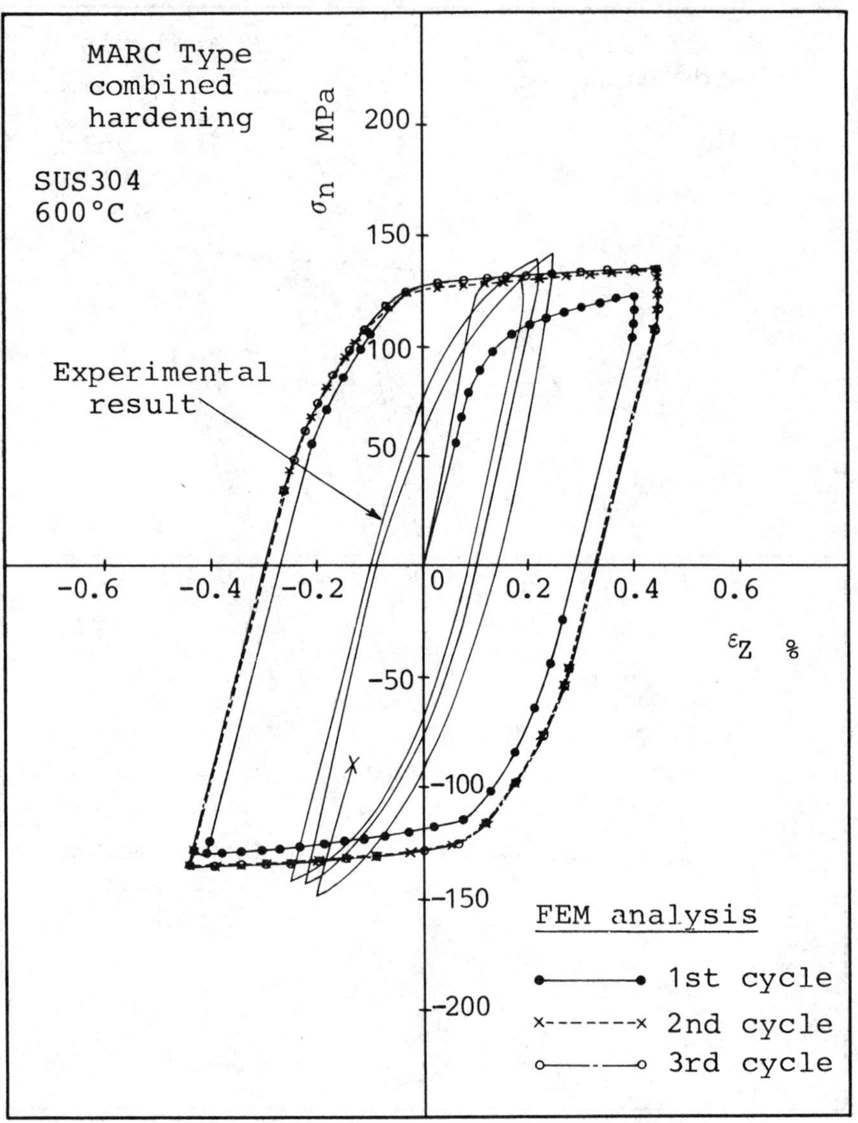

FIG. 5*b*—*Experimental and FEM analytical results for* σ_n–ϵ_z *correlation: MARC-type combined hardening.*

the larger stress around the zero strain, but the analysis of the improved model estimates the smaller stress amplitude (Fig. 6*a*). Thus the shape of the hysteresis loop of the FEM analysis more closely expresses the actual material response if we use the improved FEM mesh.

Regarding the relation between nominal stress and axial strain at the notch root as in Fig. 6*b*, the improved mesh yields smaller strain amplitude and slightly larger stress at the peak strain in comparison with the original mesh. Figure 5*b* indicates that the peak tensile strain in experiments is 0.25%, while the FEM analysis estimates that the tensile peak strain is 0.4% in original mesh. Figure 6*b*, on the other hand, shows that the tensile peak strain of the improved mesh

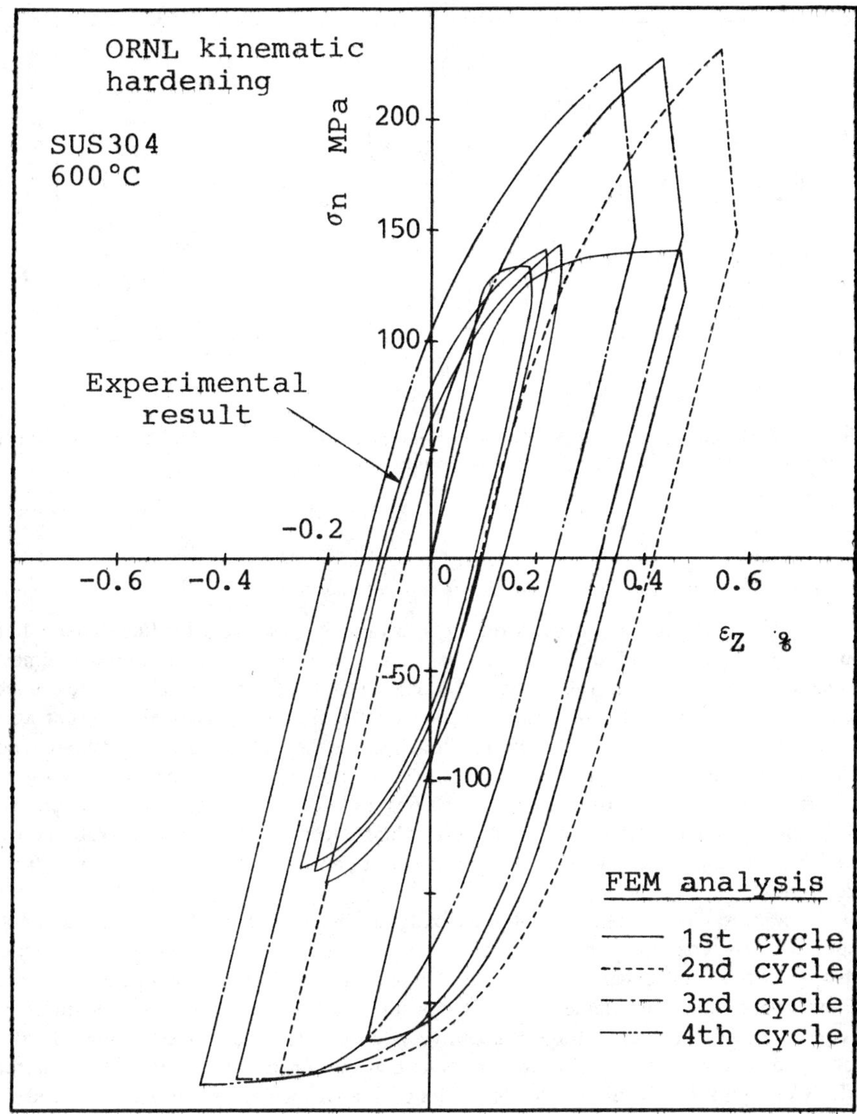

FIG. 5*c*—*Experimental and FEM analytical results for* σ_n–ϵ_z *correlation: ORNL kinematic hardening.*

is 0.30%, which is a comparatively good estimate considering that the strain measurement is carried in a difficult condition. However, there still exists a 20% error between the FEM analysis and the experimental results. This disagreement is partly the result of the strain distribution in the gage length of the strain gage. The gage length of the strain gage is 3.18 mm, and the strain gage measures the average strain along the gage length of 1.69 mm. According to the FEM analysis of the strain distribution in gage length, an approximately 10% smaller strain occurs at the point 1.59 mm from the notch root. Therefore, in conclusion, we can estimate the local strain at the notch root in cyclic elastic-plastic-creep condition if we use the proper FEM mesh and a constitutive relation.

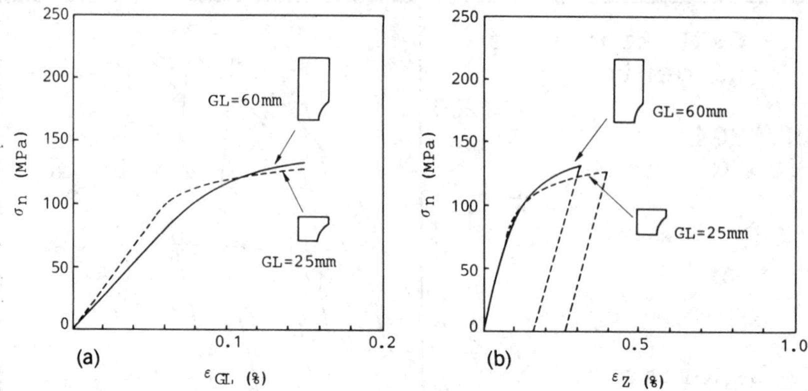

FIG. 6—*Comparison of FEM analysis between original and improved meshes.* (a) *Nominal-stress/nominal-strain relation.* (b) *Nominal-stress/local-axial-strain relation.*

Stress/Strain Concentration Factors under Monotonic Loading

Earlier we discussed inelastic stress/strain concentrations under cyclic loading that simulates the operating condition and found that the axial stress-strain hysteresis loop exhibited almost a closed loop, which means that the analysis of a quarter cycle is sufficient to evaluate the local stress in a complete cycle. In the present section we discuss the stress/strain concentrations under monotonic loading instead of cyclic loading, because the cyclic analysis needs much more calculation time. Note that the replacement of cyclic analysis by monotonic analysis is not generally permitted except for instances where there is a closed hysteresis loop and small creep effect. In the case reported here a closed hysteresis loop occurs and the stress relaxation during the hold time is small so that we can estimate the strain amplitude of a complete cycle from a quarter cycle.

The elastic stress concentration factor of the plate specimen is 2.0 and 2.2 for the round notch bar specimen. In order to examine how the stress state (i.e., plane stress and plane strain) affects the stress/strain concentration, the analysis for plate and round notched bar specimens were carried out. The finite element mesh is shown in Fig. 7, and Table 2 lists the number of nodes and elements together with the constitutive relations. The cyclic stress-strain relation in the saturated stage of cyclic strain hardening is used as the constitutive relation. The programs are ADINA for plate specimens and MARC-J2 for round bar specimens. The stress wave used in monotonic analysis is static tension up to nominal stresses of 165 MPa and 225 MPa followed by a stress hold-time of 60 min. This stress wave simulates the experimental condition mentioned later.

Figures 8 and 9 show the variations of inelastic stress/strain concentration factors with an increase in nominal stress, where the two different inelastic stress/strain concentration factors are defined. The definitions of them are

$$\left.\begin{aligned} K_\sigma &= \frac{\sigma_z}{\sigma_n} \qquad & K_\epsilon &= \frac{\epsilon_z}{\epsilon_n} \\ K_\sigma^{**} &= \frac{\sigma^*}{\sigma_n^*} \qquad & K_\epsilon^{**} &= \frac{\epsilon^*}{\epsilon_n^*} \end{aligned}\right\} \tag{1}$$

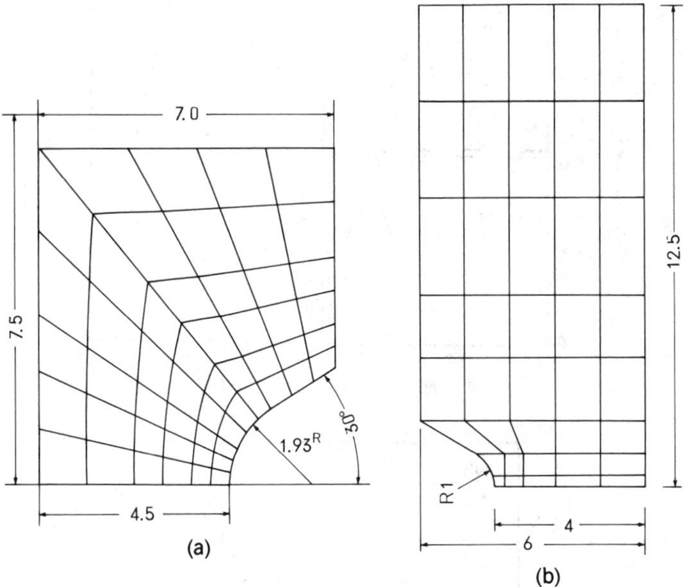

FIG. 7—*Finite element mesh for monotonic analysis.* (a) *Plate specimen (K_t = 2.0).* (b) *Round notched bar specimen (K_t = 2.2).*

TABLE 2—*Finite element mesh and constitutive equation for monotonic analysis.*

Element type	8 node plane stress	8 node ring
Total number of nodal points	193	147
Total number of elements	54	40
Computer code	ADINA	MARC - J2
Young's modulus E (MPa)	149800	149800
Yield stress σ_Y (MPa)	206.8	98.0
Stress-strain curve	$\varepsilon = \frac{\sigma}{E} + \left(\frac{\sigma - \sigma_Y}{22290}\right)$	$\varepsilon = \frac{\sigma}{E} + \left(\frac{\sigma - \sigma_Y}{3148}\right)^{\frac{1}{0.497}}$
Creep equation	$\varepsilon_c = 2.042 \times 10^{-14}\ \sigma^{4.674}$	

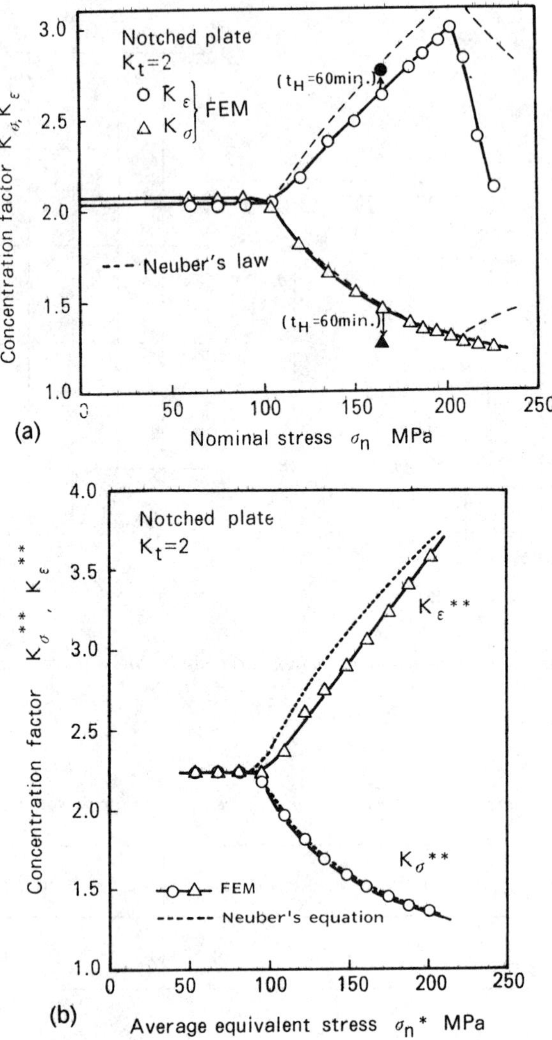

FIG. 8—*Relation between stress/strain concentration factors and nominal stress.* (a) *Axial stress/strain concentration factors.* (b) *Equivalent stress/strain concentration factors based on von Mises' criteria (notched plate* K_t *= 2.0).*

where

σ_n and ϵ_n = nominal axial stress and strain,
σ_z and ϵ_z = axial stress and strain,
σ^* and ϵ^* = equivalent stress and strain based on von Mises' criteria,
σ_n^* = average von Mises equivalent stress in notch section, and
ϵ_n^* = equivalent strain that corresponds with σ^* on cyclic stress-strain plotting based on von Mises' criteria.

The dashed lines in Figs. 8 and 9 are the stress/strain concentration factors by Neuber's rule.

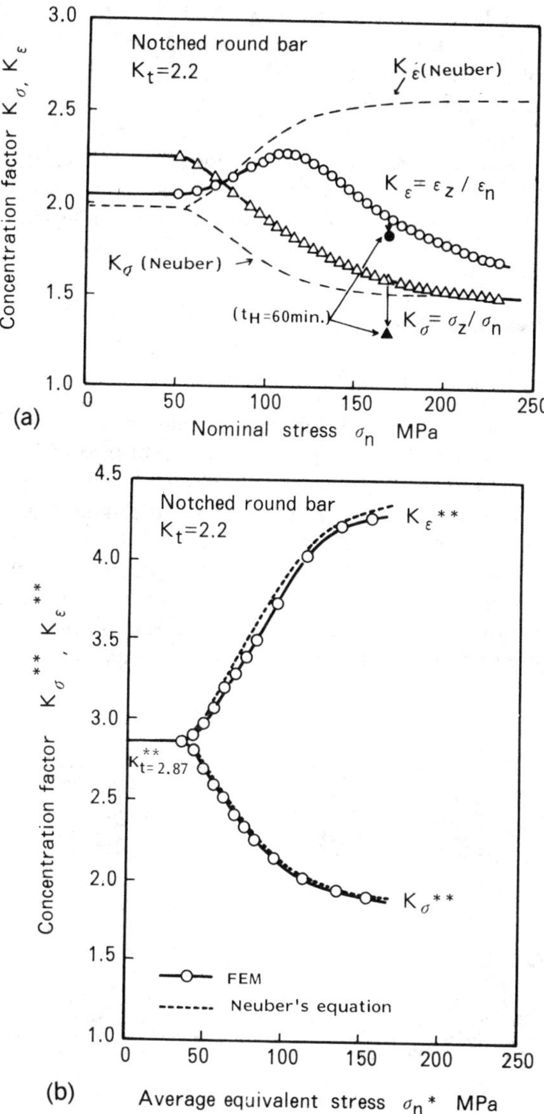

FIG. 9—*Relation between stress/strain concentration factors and nominal stress.* (a) *Axial stress/strain concentration factors.* (b) *Equivalent stress/strain concentration factors based on von Mises' criteria (notched round bar* K_t *= 2.2).*

Figures 8*a* and 9*a* show that the axial stress/strain concentration factors of the plate specimen agree well with those of Neuber's rule, while Neuber's rule overestimates the stress/strain concentrations for the round bar specimens. The difference between the Neuber's rule and FEM analysis is attributed to the stress multiaxiality at the notch root, since the stress state there for the plate specimen is uniaxial but that for the round bar specimen is biaxial. In order to improve the gap between the Neuber's rule and FEM analysis, the definition of stress/strain concentration factors is modified as K_σ^{**} and K_ε^{**}, which takes account of the stress multiaxiality into the concentration factors. The results are shown in Fig. 9*b*, where the gap disappears, so if we take into account the stress multiaxiality, Neuber's rule, which was developed in longitudinal shear for an infinite prismatical bar with hyperbolic grooves, agrees well with the stress/strain concentration factors based on von Mises' criteria.

Regarding the comparison of stress/strain concentration factors between Neuber's rule and FEM analysis, Sakon and Endo [12] reported a gap between Neuber's rule and FEM analysis in some cases even though the definition of K_σ^{**} and K_ϵ^{**} was used. Their report suggests that agreement between FEM results and Neuber's rule is delicately affected by the shape of the notch and the constitutive relation.

Life Prediction of Notched Specimen

Life Prediction in No Hold-Time Tests

The material used in the study is SUS 304 austenitic stainless steel solution treated at 1327 K. This is the standard material of the Iron and Steel Institute of Japan for calibrating creep test machines, and reports on this material have been published by the Institute [15,16].

Figure 10 shows total strain range $\Delta\epsilon$- N_f, N_c relation of the unnotched specimens in strain-controlled experiments at a strain rate of 10^{-3}/s, where N_f is the number of cycles to failure, defined as the cycle at which the tensile stress amplitude decreases to three quarters of the maximum value, and N_c is the crack initiation cycle defined as the cycle of 0.1 mm crack length on crack extension plotting. The crack extension curves were made by reading the striation spacing on the fracture surface. The failure curve is approximated by Eq 2, and the ratio of

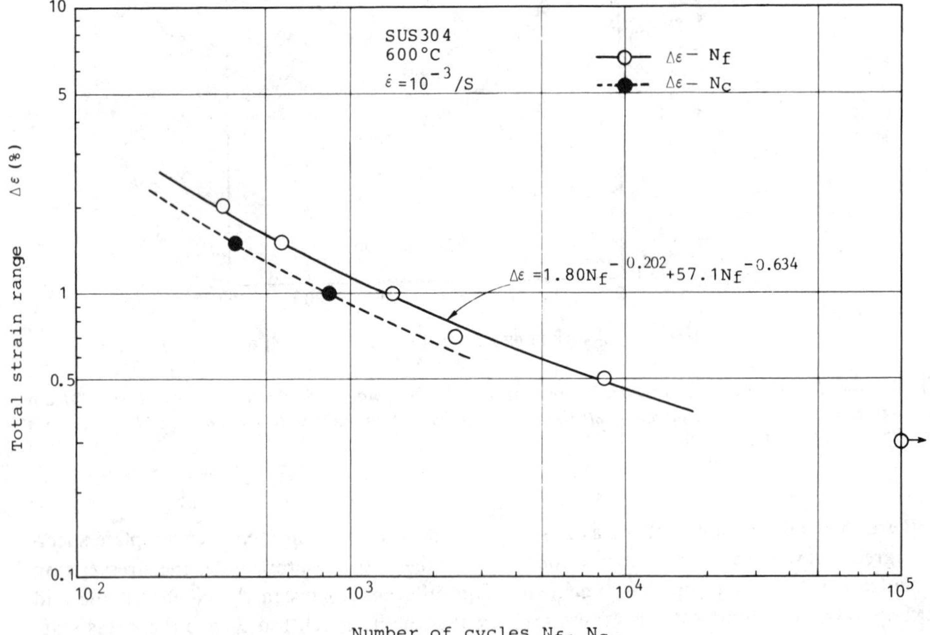

FIG. 10—*Experimental relation between total strain range and number of cycles to failure and crack initiation for unnotched specimen.*

crack initiation to failure life is about 0.5 to 0.6. The data in Fig. 10 are used for the life prediction mentioned later.

$$\Delta\epsilon = 0.018N_f^{-0.202} + 0.571N_f^{-0.673} \tag{2}$$

Nominal stress controlled tests were carried out for an unnotched and three notched specimens with elastic stress concentration factors of 2.0, 4.0, and 6.0. Stress waves are a triangular wave with stress hold-times of 10, 30, and 60 min at the tension peak in each cycle. The frequency of the triangular wave is 0.05 Hz. In order to check the notch effect in plane stress and plane strain conditions, tests using the notched plate and notched round bar specimens were made. The notched plate specimen is 14 mm wide and 4 mm thick with 60° V-shaped double-side grooves of 2.5 mm depth, while the notched round bar specimen has a 12 mm diameter with 60° V-shaped circumferential grooves of 2 mm depth. For both specimens only the notch root radius is changed to obtain the three elastic stress concentration factors. The notch root radii are 1.93 mm ($K_t = 2.0$), 0.35 mm ($K_t = 4.0$), and 0.14 mm ($K_t = 6.0$) for the plate specimens and 1.0 mm ($K_t = 2.2$), 0.22 mm ($K_t = 4.0$), and 0.09 mm ($K_t = 6.0$) for the round bar specimens.

Tables 3 and 4 summarize the test results of the plate and round bar specimens, respectively, where N_c is determined by the direct measurement of the crack initiation with a microscope for plate specimens, while N_c of the round bar specimens is defined as the cycle where the total strain range takes the minimum value on total strain range-number of cycles plotting. Note that we prechecked the relation between the variation of total nominal strain range and crack initiation cycle using the d-c potential drop method. The result obtained was that the crack initiation cycle, defined as the cycle of 0.1 mm crack length, and the cycle at which the total strain range takes the minimum value agree well with each other. Type 304 stainless steel is a typical strain hardening material, and in stress-controlled tests total strain range is reduced according to strain hardening. However, when the crack is initiated at the notch root the material turns to softening material so that the minimum value of the total strain indicates the crack initiation cycle. This correspondence was also confirmed in another report [*17*].

Figures 11*a* and 11*b* show the relationship between nominal stress range and failure cycle in no hold-time tests for plate and round bar specimens, respectively. The slope of the data of the notched specimen is steeper in comparison with the unnotched specimen for both the plate and round bar specimens. The specimen with the larger elastic stress concentration factor exhibits a smaller failure life, but the failure cycle of $K_t = 4.0$ does not differ much from that of $K_t = 6.0$.

Figure 12 shows the plotting of failure cycle as a function of K_t in no hold-time and hold-time tests. In the no hold-time test, the specimen with larger K_t exhibits a smaller fatigue life, but in the hold-time test some notch strengthening occurs. Notch strengthening is the larger fatigue life of the notched specimen compared with the unnotched specimen.

Five prediction methods were applied to fatigue data with no hold-time tests: the FEM method, Neuber's rule [*18*], frequency-elastic stress concentration factor modified fatigue life equation (ν-K_t modified equation) [*11*], Stowell-Hardrath-Ohman's rule [*19*], and Koe's method [*20*]. All the life prediction methods are based on the local strain theory. The equations used to calculate the local stress/strain are

Neuber's rule:

$$K_t^2 = K_\sigma K_\epsilon \tag{3}$$

ν-K_t equation:

$$N_f = N_{fo} K_t^m \nu^\ell \tag{4}$$

TABLE 3—*Summary of experimental results on plate specimens.*

Elastic stress concentration factor Kt	Nominal stress range $\Delta\sigma$ (MPa)	Hold-time t_H (min)	Number of cycles to failure Nf	Number of cycles to crack initiation Nc	Nominal total strain range $\Delta\varepsilon$ (%)	Nominal plastic strain range $\Delta\varepsilon p$ (%)	Nominal creep strain during hold-time εc (%)
1	420	0	3692		0.290	0.110	
	400	0	6184		0.473	0.188	
	330	0	89573		0.250	0.025	
2	450	0	1500	800	0.380	0.150	—
	400	0	2180	1030	0.350	0.100	—
	330	0	5360	3320	0.250	0.050	—
	330	10	1860	1080	0.220	0.045	0.010
	330	60	627	270	0.230	0.040	0.005
4	450	0	604	309	0.360	0.120	—
	400	0	1000	506	0.320*1	0.110*1	—
	330	0	2367	1245	0.230*2	0.030*2	—
	330	10	1030	200	0.553	0.097	0.020
	330	60	296	61	0.627	0.136	0.030
6	450	0	494	—	0.330	0.115	
	400	0	1208	135	0.286	0.088	
	400	0	1306	—	0.230	0.058	
	330	0	2136	—	0.180	0.030	

*1 at N = 300 *2 at N = 828

TABLE 4—*Summary of experimental results on round bar specimens.*

Elastic stress concentration factor Kt	Nominal stress range $\Delta\sigma$ (MPa)	Hold-time t_H (min)	Number of cycles to failure Nf	Number of cycles to crack initiation Nc	Nominal total strain range $\Delta\varepsilon$ (%)	Nominal plastic strain range $\Delta\varepsilon p$ (%)	Nominal creep strain during hold-time εc (%)
1	450	0	6918		0.557	0.264	—
	400	0	47370		0.352	0.068	—
	330	0	>252700		0.219 *1	0.011 *1	—
	330	10	> 1700		0.283 *2	0.052 *2	0.002 *2
	330	30	922		0.319	0.077	0.017
	330	60	347		0.271	0.094	0.021
2.2	450	0	1931	430	0.192	0.030	—
	400	0	2917	870	0.150	0.015	—
	330	0	10663	—	0.124 *3	0.007 *3	—
	330	10	3200	1000	0.124	0.006	4.31×10^{-4}
	330	30	2137	510	0.128	0.006	5.55×10^{-5}
	330	60	1622	480	0.123	0.005	7.71×10^{-5}
4	450	0	1052	200	0.203	0.035	—
	400	0	2047	550	0.165	0.018	—
	330	0	3643	900	0.125	0.006	—
	330	10	1670	480	0.136	0.007	0.002
	330	30	1021	360	0.138	0.009	0.002
	330	60	810	210	0.137	0.007	0.004
6	450	0	1048	250	0.222	0.045	
	400	0	1702	400	0.173	0.021	
	330	0	3392	800	0.149	0.010	

*1 at N = 126350   *2 at N = 850   *3 at N = 7845

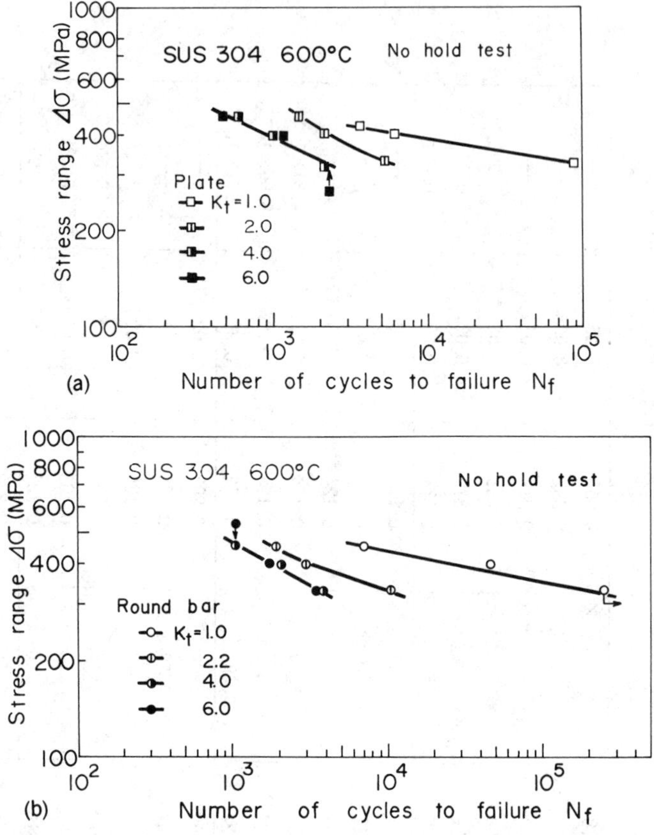

FIG. 11—*Experimental relation between stress range and number of cycles to failure in no hold-time test.* (a) *Plate specimen.* (b) *Round bar specimen.*

Stowell-Hardrath-Ohman's (SHO) equation:

$$K_\sigma = \frac{K_\epsilon}{K_\epsilon - K_t + 1} \tag{5}$$

Koe's equation:

$$\frac{K_t - 1}{K_\sigma - 1} - 1 = 2.4\left(\frac{K_\epsilon - 1}{K_t - 1} - 1\right) \tag{6}$$

where

K_σ = inelastic stress concentration factor,
K_ϵ = inelastic strain concentration factor,
K_t = elastic stress concentration factor,
N_f = number of cycles to failure of the notched specimen,
N_{fo} = reference number of cycles to failure of the unnotched specimen, and
ν = frequency of the test.

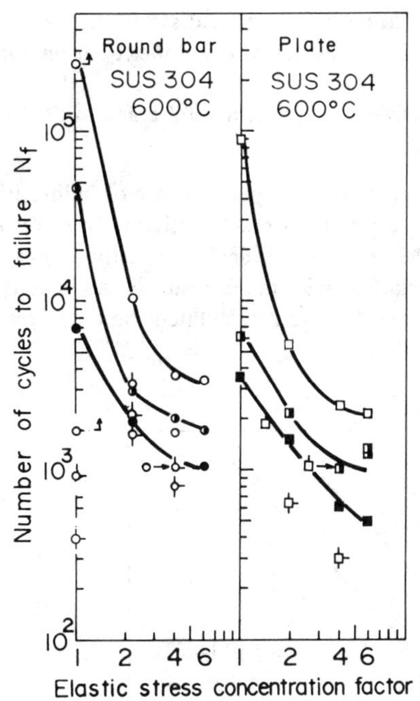

$\Delta\sigma$ (MPa)	t_H (min)	Round bar	Plate
330	0	○	□
400	0	◑	◨
450	0	●	■
330	10	○-	□-
330	30	○-'	□-'
330	60	○-+	□-+

FIG. 12—*Plotting of number of cycles to failure as a function of elastic stress concentration factor.*

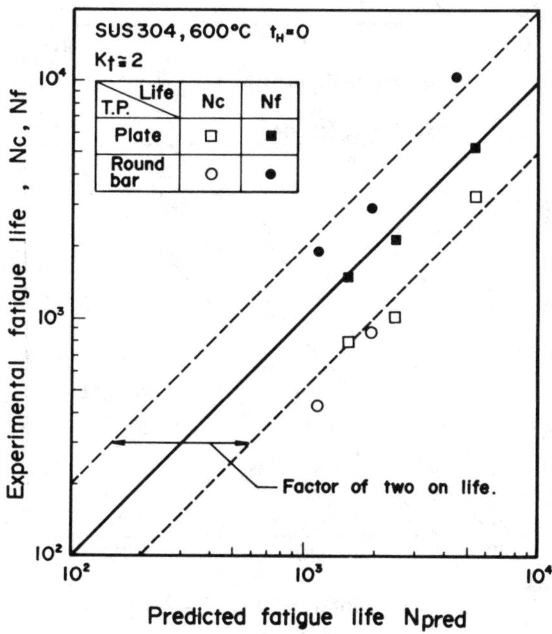

FIG. 13—*Comparison between predicted and experimental fatigue lives by FEM method.*

Note that only the ν-K_t equation does not appear to be based on local strain theory, but the equation implicitly includes the concept of local strain in the derivation of the equation. On this point refer to Ref *11*.

Figures 13 to 16 compare the predicted fatigue lives with experimental crack initiation and failure lives. The main points the figures indicate are:

1. The prediction method based on FEM analysis gives a proper estimate of failure life for $K_t = 2.0$, while it gives a slightly unconservative estimate of crack initiation life. Since the bilinear stress-strain relation was used in the FEM analysis, the deviation of the stress-strain relation used in the analysis from the actual material behavior might result in the unconservative estimate of the crack initiation life. To examine this, we recalculated the local strain by

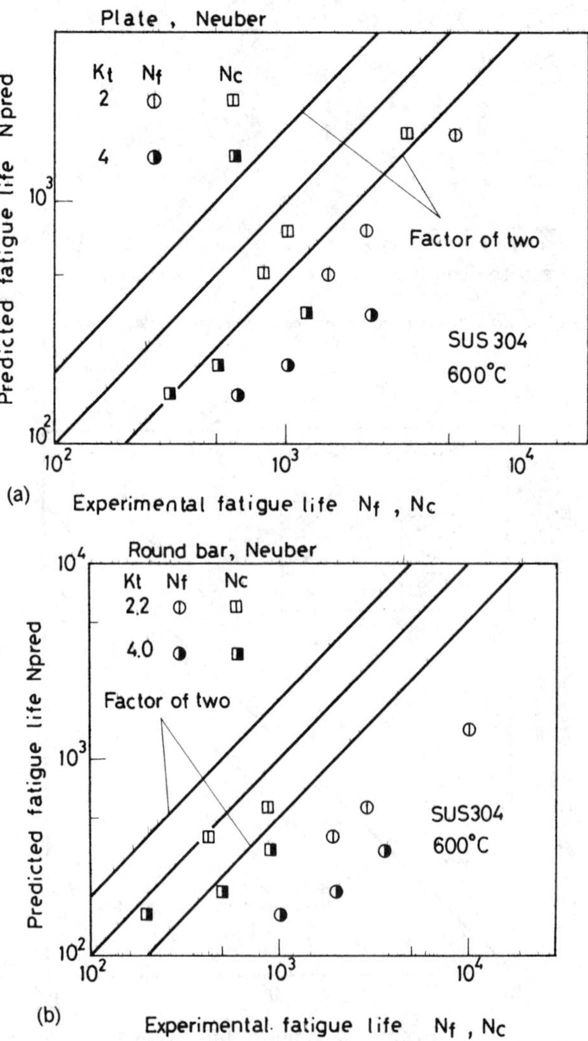

FIG. 14—*Comparison between predicted and experimental fatigue lives by Neuber's rule.* (a) *Plate specimen.* (b) *Round bar specimen.*

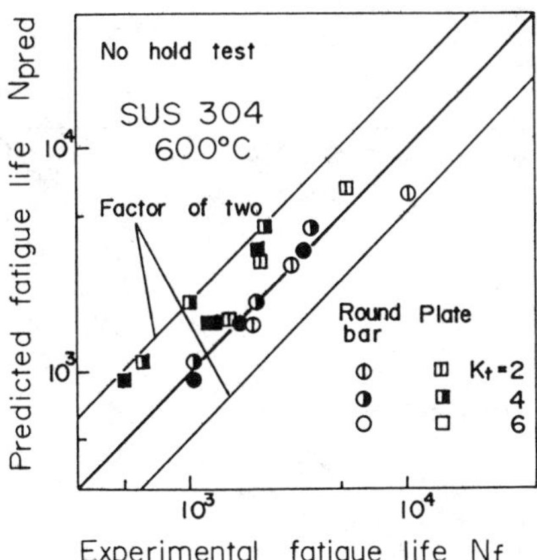

FIG. 15—*Comparison between predicted and experimental fatigue lives by* ν-K_t *modified fatigue life equation.*

using a multilinear stress-strain relation for the plate specimen with a $K_t = 2.0$. The result of life prediction obtained is $N_{pred} = 1/3 N_f$ and $N_{pred} = 1/2 N_c$ so that the accurate constitutive relation is essential to predicting the failure or crack initiation life of the notched specimen.

2. In the case of round bar specimens, Neuber's rule gives a one quarter to one tenth conservative estimate for failure lives. As K_t increases the estimate becomes more conservative. Neuber's rule predicts properly the crack initiation life for $K_t = 2.2$, but it gives a slightly conservative estimate for $K_t = 4.0$. As for the life prediction of plate specimens, Neuber's rule gives the same life prediction as for round bar specimens so that the predicted results of N_c and N_f for the plate specimens do not differ much from the round bar specimens, with a slight decrease in experimental failure lives.

3. The ν-K_t modified fatigue life equation can predict the failure lives of round bar and plate specimens properly. The equation properly follows the reduction in failure lives with an increase in K_t and nominal stress.

4. The SHO and Koe methods properly predict the crack initiation life of the round bar specimens. The experimental crack initiation life is within the scatter band of a factor of three. For plate specimens the arrangement of the data is better than that for the round bar specimens. The data are within the scatter band of a factor of two. However, for the prediction of failure lives, both the SHO and Koe methods give conservative estimates. The failure data are spread over a scatter band between the lines of $N_{pred} = N_f$ and $N_{pred} = 1/10 N_f$.

Life Prediction in Hold-Time Tests

In this section three life prediction methods are applied to the data in hold-time tests: the FEM method, the method following the ASME N-47 Code Case [*21*] and the ν-K_t modified fatigue life equation. Figures 17 to 19 compare the predicted and experimental failure lives.

Figure 17 shows the results of life prediction using FEM analysis. The prediction procedure is as follows: fatigue damage D_f is evaluated by applying the calculated strain to $\Delta\epsilon$-N_f curve of the unnotched specimen, creep damage is calculated from the equation $D_c = \int_0^{tH} dt/tr$ from the stress relaxation curve, and then predicted life is obtained from the D_c-D_f diagram in the ASME N-47 Code Case. Predicted life clearly agrees well with crack initiation life so that the FEM method has a good accuracy for predicting the crack initiation life. However, the FEM method

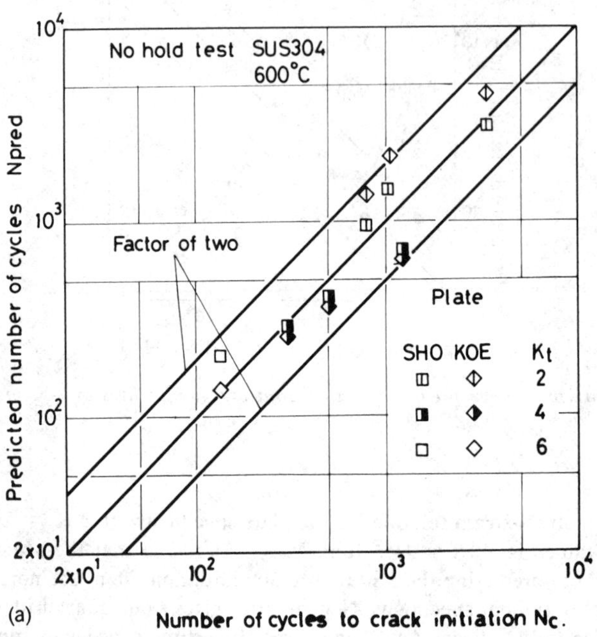

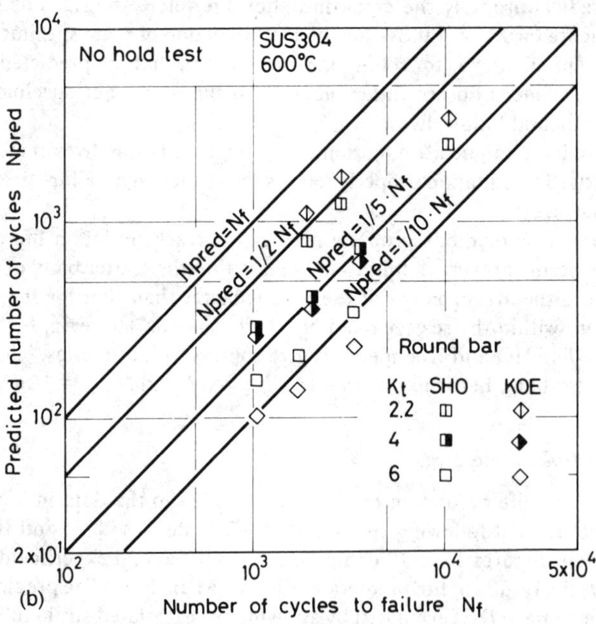

FIG. 16—*Comparison between predicted and experimental fatigue specimen.* (b) *Failure lives of plate specimen.* (c) *Crack initiation*

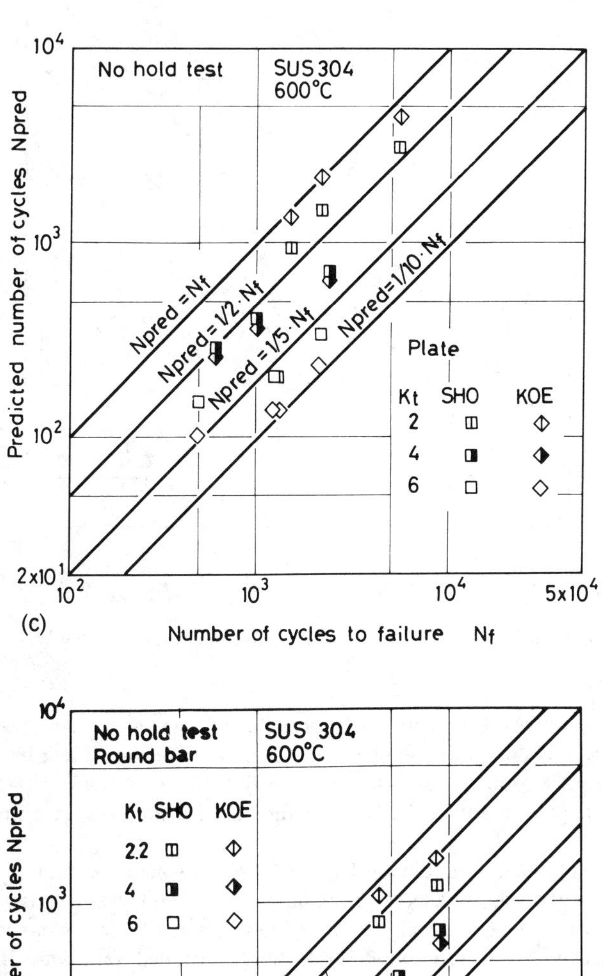

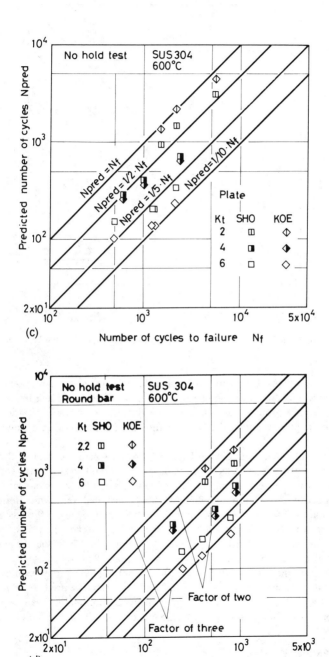

lives by SHO and Koe methods. (a) *Crack initiation lives of plate lives of round bar specimen.* (d) *Failure lives of round bar specimen.*

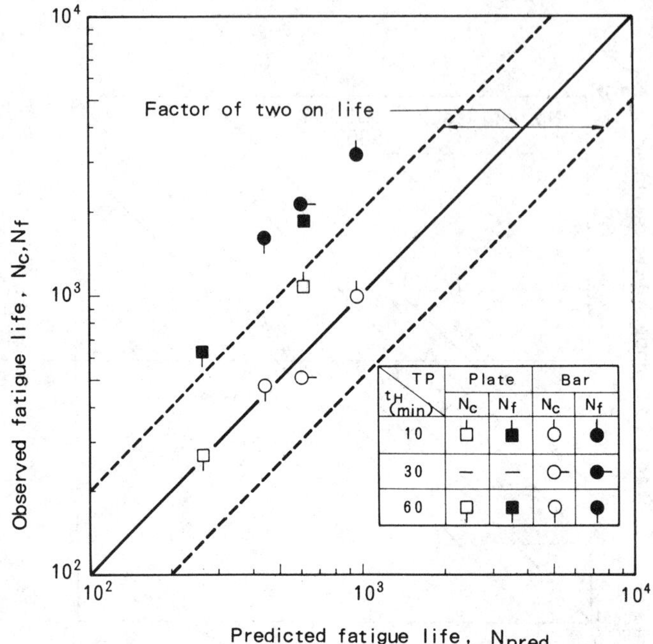

FIG. 17—*Comparison between predicted and experimental fatigue lives by FEM method.*

gives a slightly conservative estimate for the failure life. Sakane and Ohnami [*22*] reported that the stress concentrations due to the notches had their main effect on crack initiation and had almost no effect on crack propagation behavior. As N_f includes both crack initiation and propagation periods in the present study, the FEM method gives a conservative estimate for failure life.

Regarding the results of life prediction by ASME N-47, where the data of the unnotched specimen in the present study are used as basic data, the method gives a conservative estimate especially for $K_t = 6.0$. In calculating the predicted fatigue life, stress is specified as load-controlled stress because the tests are made under load-controlled conditions. Besides this estimate, we evaluated the life prediction of the notched specimen by using the basic data of the unnotched specimen in N-47. The results of the evaluation are that the prediction is far too conservative and has an unnecessarily large safety factor margin. The value of the safety factor margin is 22 on average, and the value becomes larger than 22 in some cases.

Life prediction using the ν-K_t modified fatigue life equation gives a proper estimate for failure life, where the material constants are $m = -0.59$ and $\ell = 0.31$. The predicted data are almost within a scatter band of a factor of two, so that this prediction method can be valid for the prediction of not only the life in the no hold-time test but also life in the hold-time test.

Recommended Future Work

At the end of this paper we briefly point out the questions left for the future relating to the notch effect in creep-fatigue conditions:

1. Sometimes notch strengthening occurs in the tests with a longer hold-time (Tables 3 and 4). All conventional life prediction methods always estimate less predicted life for the notched

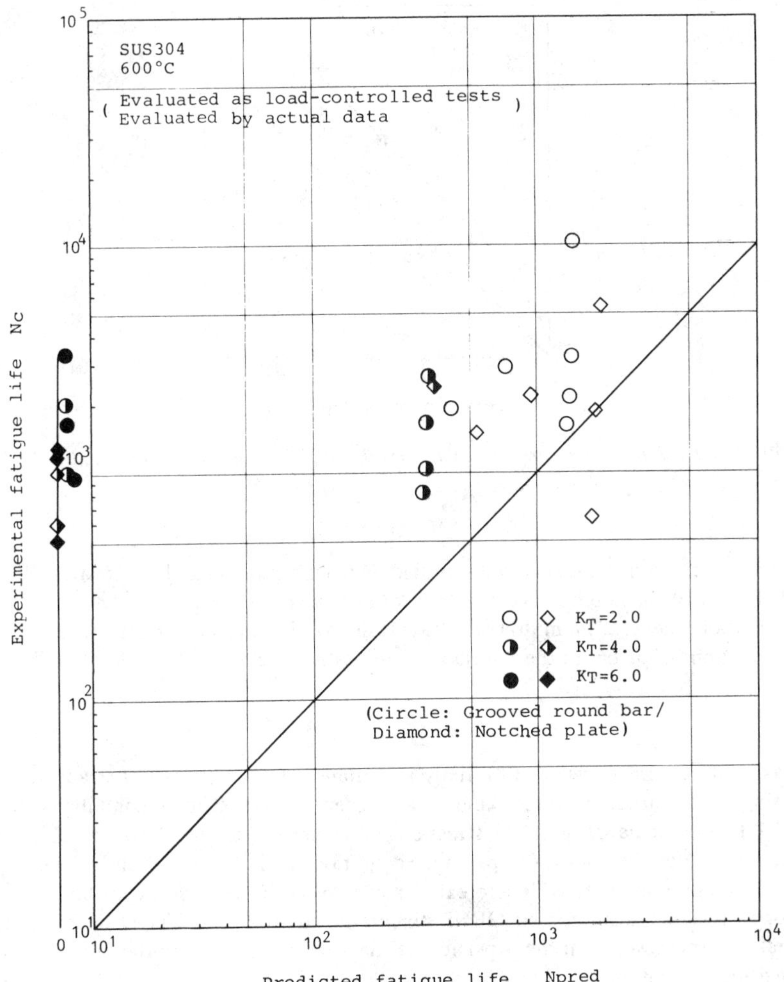

FIG. 18—*Effect of* K_t *on accuracy of life estimation by N-47 method using actual material data.*

specimens than for the unnotched specimens. How do we treat the notch strengthening in creep-fatigue conditions? What is the cause of the notch strengthening? We should clarify the cause of the notch strengthening in connection with the crack initiation/propagation behaviors.

2. This paper used a trapezoidal stress wave in the test using notched specimens, but there exists a possibility of smaller fatigue life in other stress waves. We should study further the relation between the wave shape and notch effect.

3. Accurate formulation of a cyclic constitutive relation is the key to precise estimates of the stress/strain concentrations of the notched members. An error in the constitutive equation leads to a fatal error in the calculated results.

4. Development of a simple and accurate estimation of the local stress/strain of the notched members is urgent. The FEM analysis for elastic-plastic-creep loading needs a lot of computing time for large structures.

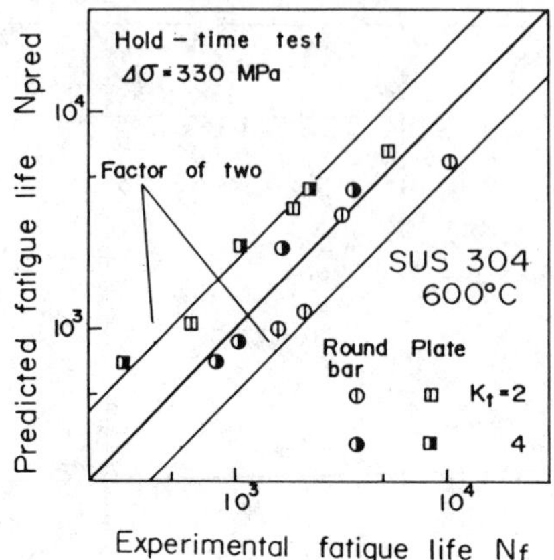

FIG. 19—*Comparison between predicted and experimental fatigue lives by* ν-K_t *modified fatigue life equation.*

5. The notch effect has not been fully studied by crack initiation and propagation behaviors. A detailed study of the notch effect from crack behavior is necessary.

6. Notch root is generally a multiaxial stress state. We have to study further the crack initiation/propagation behaviors in the multiaxial stress/strain state.

Conclusions

1. Elastic-plastic-creep cyclic FEM analysis is made for the trapezoidal strain wave in five cycles using three hardening rules: kinematic hardening, MARC-type combined hardening, and ORNL kinematic hardening. The stress-strain response at the notch root exhibits almost a closed hysteresis loop even in the creep-fatigue condition, so that we can replace the costly cyclic analysis by monotonic analysis. There exists almost no difference between the cyclic stress/strain concentrations except for the ORNL kinematic hardening rule. The calculated strain by FEM analysis agrees well with the experimental measurement if we use the proper FEM mesh and constitutive equation.

2. The inelastic stress and strain concentration factors agree well with those from Neuber's rule if we consider the concentration factors on von Mises' equivalent stress/strain basis.

3. Five life prediction methods (i.e., FEM method, Neuber's rule, ν-K_t modified fatigue life equation, SHO method, and Koe's method) are applied to the test results in a no hold-time test. There is no prediction method that gives an unconservative estimate. Among them the FEM method and the ν-K_t method have comparatively good prediction accuracy.

4. Three life prediction methods (i.e., FEM method, ASME N-47 method, and ν-K_t modified fatigue life equation) are applied to the hold-time data. The ν-K_t modified fatigue life equation and the FEM analysis have good accuracy, while ASME N-47 has an excessive safety factor margin.

Acknowledgments

This work was made possible by the collaboration of the Notch Effects Test Subcommittee of the Iron and Steel Institute of Japan. The authors wish to thank the members of the subcommittee for participating in the experiments and FEM analysis.

References

[1] Wundt, B. M. in *Effect of Notches on Low Cycle Fatigue, ASTM STP 490,* American Society for Testing and Materials, Philadelphia, 1972.

[2] Krempl, E. in *The Influence of State of Stress on Low-Cycle Fatigue on Structural Materials: A Literature Survey and Interpretive Report, ASTM STP 549,* American Society for Testing and Materials, Philadelphia, 1974.

[3] Mowbray, D. F. and McConnelee, J. E., *Cyclic Stress-Strain Behavior—Analysis, Experimentation, and Failure Prediction, ASTM STP 519,* American Society for Testing and Materials, Philadelphia, 1971, pp. 151-169.

[4] Coffin, L. F. Jr., in *Fatigue at Elevated Temperatures, ASTM STP 520,* American Society for Testing and Materials, Philadelphia, 1973, pp. 5-34.

[5] Ohnami, M. and Sakane, M., *Journal of the Society of Materials Science,* Japan, Vol. 24, No. 261, 1975, pp. 545-550.

[6] Gamble, R. M. and Paris, P. C., *International Conference on Creep Fatigue in Elevated Temperature Applications,* Vol. 1, 1975, p. 182.1.

[7] Ohnami, M. in *Proceedings,* 1976 Elevated Temperature Design Symposium, Mexico City, ASME, 1976, pp. 1-6.

[8] Maiya, P. S., *Metal Science and Engineering,* Vol. 38, 1978, pp. 289-294.

[9] Sakane, M. and Ohnami, M., *Journal of the Society of Materials Science,* Japan, Vol. 29, No. 320, 1980, pp. 458-464.

[10] Ohnami, M. and Sakane, M., *Journal of the Society of Materials Science,* Japan, Vol. 30, No. 336, 1981, pp. 915-921.

[11] Sakane, M. and Ohnami, M., *ASME Journal of Engineering Materials Technology,* Vol. 105, No. 1, 1983, pp. 75-80.

[12] Sakon, T. and Endo, T., Preprint of *Japan Society of Materials Science,* 1984, pp. 110-112.

[13] *Transactions of Japan Society of Mechanical Engineers,* Vol. 51A, No. 471, 1985, pp. 2425-2432.

[14] Johnson, R. and Lawton, C. W., ASME MPC, Vol. 25, 1984, pp. 1-58.

[15] Report of the High Temperature Creep and Fatigue Subcommittee, The Iron and Steel Institute of Japan, 1981.

[16] Report of the High Temperature Tensile Subcommittee, The Iron and Steel Institute of Japan, 1981.

[17] Sakane, M. and Ohnami, M. in *Proceedings,* 23rd Symposium on the Strength of Materials at Elevated Temperatures, The Japan Society of Materials Science, 1985, pp. 137-141.

[18] Neuber, H., *ASME Journal of Applied Mechanics,* Vol. 28, 1961, pp. 544-550.

[19] Stowell, E. Z., NASA TN 2073, 1950.

[20] Koe, S., Nakamura, H., and Tsunenari, T., *Journal of the Society of Materials Science,* Japan, Vol. 27, No. 300, 1978, pp. 847-852.

[21] Cases of ASME Boiler and Pressure Vessel Code, Case N-47, Class I, Section III, Division I, 1978, American Society of Mechanical Engineers, New York.

[22] Sakane, M. and Ohnami, M., *ASME Journal of Engineering Materials Technology,* Vol. 108, No. 4, 1986, pp. 279-284.

S. N. Singhal[1] *and R. A. Tait*[2]

Evaluation of Low Cycle Creep-Fatigue Degradation for Notched Components

REFERENCE: Singhal, S. N. and Tait, R. A., **"Evaluation of Low Cycle Creep-Fatigue Degradation for Notched Components,"** *Low Cycle Fatigue, ASTM STP 942,* H. D. Solomon, G. R. Halford, L. R. Kaisand, and B. N. Leis, Eds., American Society for Testing and Materials, Philadelphia, 1988, pp. 1096–1111.

ABSTRACT: Four approaches have been examined to predict the endurance of circumferentially notched, uniaxially loaded test pieces subjected to fully reversed strain-controlled cyclic loading at elevated temperatures. Results are compared with uniaxial data derived from smooth test pieces. The strain-time waveforms include both trapezoidal and triangular shapes to distinguish between fatigue and fatigue-creep damage. The four materials investigated are HP-45 (1144 K), Incoloy 800H (1144 K), Incoloy 800 (922 K), and 1¼Cr-½Mo (811 K). The analysis has reflected the various degrees of cyclic hardening and creep deformation exhibited by these materials. The predictive approaches include Neuber notch strain analysis and finite element analyses to identify the maximum plastic strain range, the equivalent plastic strain range, and the total plastic work per cycle.

The study provides a simple, pragmatic approach to predicting whether cyclic plant operation is lifetime limiting. The primary conclusion is that the total plastic work-per-cycle approach is the preferred method for evaluation of fatigue-limited plant lifetimes.

KEY WORDS: creep, cyclic loads, endurance, fatigue-creep tests, high temperature fatigue tests, notch tests, plastic deformation, thermal fatigue

Nomenclature

A	Material constant (kN/m^2)
a	Material constant (m)
D	Far field diameter of test piece (m)
d	Diameter of test piece at notched section (m)
E	Young's modulus (N/m^2)
K_ϵ	Strain concentration factor
K_f	Fatigue notch factor
K_t, K_σ	Stress concentration factor
k	Cyclic hardening slope (N/m^3)
N_f	Number of cycles to failure
r	Notch radius (m)
t_h	Strain hold duration (s)
W_c	Plastic work per cycle (N/m^2)
α	Thermal expansion coefficient (m/m/K)
β	Material constant
$\Delta\sigma$	Total stress range (N/m^2)

[1]Senior Research Engineer, AMOCO Production Company, Research Center, Tulsa, OK 74102.
[2]Shell Development Company, Houston, TX 77251.

σ_{pk}	Maximum tensile stress (N/m^2)
σ_y	Cyclic yield stress (N/m^2)
$\sigma_x, \sigma_y, \sigma_z, \tau_{xy}, \tau_{yz}, \tau_{zx}$	Stress components (N/m^2)
ΔS	Far field (nominal) stress range (N/m^2)
Δe	Far field (nominal) strain range (m/m)
$\Delta \epsilon$	Total strain range (m/m)
$\Delta \epsilon_p$	Plastic strain range (m/m)
$\Delta \epsilon_{eq}$	Equivalent strain range (m/m)
$\dot{\epsilon}_p$	Plastic strain rate (s^{-1})
$\dot{\epsilon}_t$	Tension-going strain rate (s^{-1})
$\epsilon_x, \epsilon_y, \epsilon_z, \gamma_{xy}, \gamma_{yz}, \gamma_{zx}$	Strain components (m/m)
ρ	Density (kg/m^3)
ν	Poisson's ratio

The utilization of an endurance approach in the problem of predicting plant lifetimes for the case of multiaxial thermomechanical loading usually relies upon the use of materials property data determined under conditions of simple uniaxial loading. The work described here is an attempt to evaluate four simple criteria for the prediction of the life-limiting damage induced during multiaxial cyclic loading of metals at elevated temperatures in terms of uniaxial endurance data. The samples chosen are HP-45, Incoloy 800H, Incoloy 800, and 1¼Cr-½Mo. The study is in no way all encompassing, but instead endeavors to provide a simple, pragmatic approach to the question of whether or not cyclic plant operation is lifetime limiting. In this regard, attention to conservatism is given priority over analytical precision. This course of action is felt to be prudent in the face of all the uncertainties associated with materials data collection, component stress analysis, and the actual plant operating scenario.

Outline of Approach

Basic Philosophy

The overall approach used here was to take the data derived from the testing of smooth samples and attempt to predict the endurance of notched samples in terms of the number of cycles to failure, N_f, by using criteria which are to be described later. The use of a circumferentially notched test geometry to derive a state of multiaxial loading represents a compromise between efficiency in use of laboratory facilities and a reduced flexibility in allowable stress state conditions. That the state of loading in the region of the notch is multiaxial is undeniable. However, the test procedure does not allow the investigation of non-proportional loading, mean stress effects, and moving principal axes. Nevertheless, by using the MARC finite element computer program to analyze, and so predict, the behavior of notched test pieces, and then comparing the results with experimental data, it is possible to test the applicability of the various approaches and to select the best calculation scheme to predict the failure conditions of the elevated temperature components.

Uniaxial Low Cycle Fatigue Parameters

It is pertinent at this stage to review the parameters of uniaxial low cycle fatigue testing which relate to the shape of the stress-strain hysteresis loop that results from fully reversed cyclic loading under strain control. Figure 1 illustrates the shapes of the hysteresis loops produced in response to the three strain waveforms that were employed in this study. The first shown in Fig. 1*a*

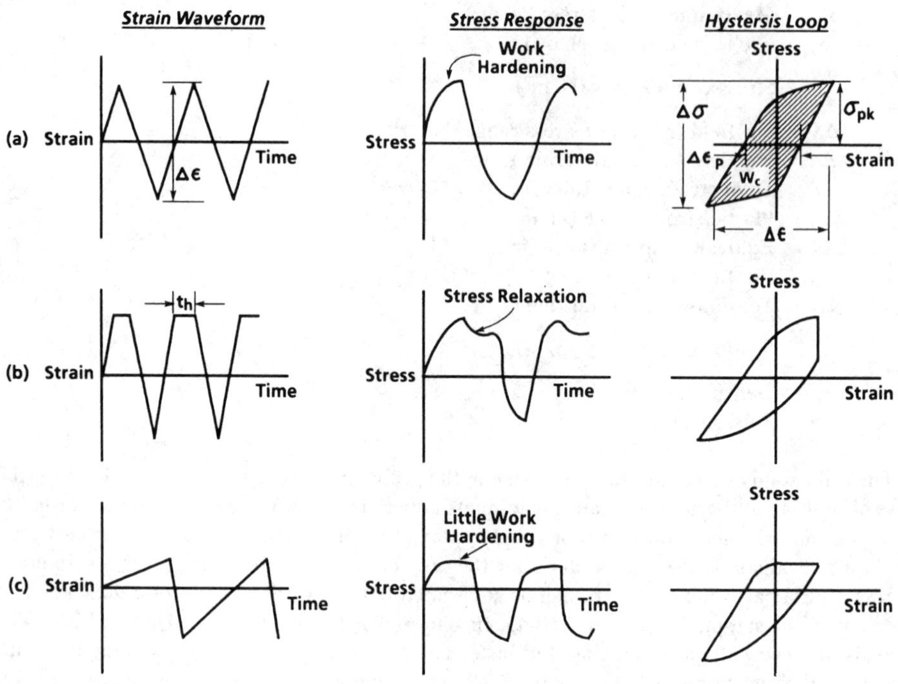

FIG. 1—*Strain waveforms and associated stress responses used in notch study.*

is the simple triangular waveform in which the tension-going strain rate is made equal to the compression-going strain rate. For the present testing program, this rate was generally in excess of $5 \times 10^{-4}\ s^{-1}$ to ensure that the plastic deformation processes were entirely glide related. The graphical stress response shown in Fig. 1*a* is meant to indicate a transition from initial elastic behavior to work hardening plastic behavior, typical of a dislocation glide mechanism. This work hardening behavior causes the stress response to become nonlinear with respect to the linear strain input waveform. As a result, the stress-strain behavior forms a symmetrical hysteresis loop as shown. The major parameters illustrated are $\Delta\epsilon$ (the total strain range), $\Delta\sigma$ (the total stress range), $\Delta\epsilon_p$ (the plastic strain range), σ_{pk} (the maximum tensile stress), and W_c (the plastic work per cycle).

The trapezoidal strain waveform shown in Fig. 1*b* is intended to introduce some degree of creep-type deformation into the cycle to simulate a creep-fatigue interaction. This requirement is achieved as a result of the very slow plastic strain rates which occur during the tensile strain hold of duration t_h. During this time stress relaxation occurs as the recoverable elastic strain is converted to plastic strain.

The plastic strain rates developed during stress relaxation are typically in the range $10^{-10}\ s^{-1} < \dot{\epsilon}_p < 10^{-6}\ s^{-1}$ for hold times $t_h \geq 900$ s. These relatively slow deformation rates are essential for the activation of creep-related deformation and damage processes, such as may be experienced during prolonged plant operation. The reversal strain rates are equivalent to those used in the simple triangular strain waveform described previously.

The saw-tooth waveform illustrated in Fig. 1*c*, represents an alternative method of achieving a creep-fatigue interaction by using an imposed tension-going strain rate $\dot{\epsilon}_t \leq 10^{-6}\ s^{-1}$. The flat tension-going response shows less evidence of work hardening behavior, as does the shape of the corresponding hysteresis loop. The reversal or compression-going strain rate is generally made equivalent to that used in the simple triangular strain waveform ($\dot{\epsilon}_t \geq 5 \times 10^{-4}\ s^{-1}$) to ensure the activation of glide processes and the reappearance of work hardening in the compression-going part of the stress-response and the hysteresis loop.

The cyclic hardening of metals is conventionally expressed as the locus of the value of the maximum tensile stress, σ_{pk}, as a function of the peak tensile strain. For the waveforms employed here, the loading was fully reversed about zero mean strain so that the peak tensile strain corresponds to $\Delta\epsilon/2$, and the peak tensile stress corresponds to $\Delta\sigma/2$. In this study, a simple linear hardening rule was adopted in which the data relating the stress amplitude, $(\Delta\sigma/2)$, to the strain amplitude, $(\Delta\epsilon/2)$, were fitted to an equation of the form

$$\Delta\sigma/2 = \sigma_y + k(\Delta\epsilon/2) \tag{1}$$

where

σ_y = cyclic yield stress, and
k = cyclic hardening slope.

Each of the four materials was tested using smooth specimens over a range of total strain amplitudes using the three waveforms described in this section. In addition, several circumferentially notched specimens were tested using the aforementioned strain waveforms. The details of the testing conditions and the results of each test are presented later in this report.

Analysis Methods

Lack of space precludes detailed presentation of the current state of knowledge of multiaxial cyclic loading in this report. Several excellent reviews are to be found in the open literature [*1-4*].

Neuber Notch Strain Range Analysis

The question to be resolved by notch-strain analysis is, Given the nominal elastic stress range (ΔS) or strain range (Δe), what are the local stress range ($\Delta\sigma$) and the local strain range ($\Delta\epsilon$) at the notch root? The approach adopted here is to use Neuber's rule [*5*]. Application of this rule requires the solution of two simultaneous equations:

$$\Delta\epsilon\,\Delta\sigma = K_t^2\,\Delta e\,\Delta S \tag{2}$$

$$\Delta\sigma/2 = \sigma_y + k(\Delta\epsilon/2) \tag{1}$$

In $\Delta\sigma$-$\Delta\epsilon$ space, Eq 2 describes a hyperbola and Eq 1 describes the cyclic stress-strain curve. The intersection of these two curves defines the values of notch root stress ($\Delta\sigma$) and notch root strain ($\Delta\epsilon$) corresponding to the far field stress (ΔS) and strain (Δe) at the condition of peak cyclic loading. In order to make allowance for the elastic contribution to the notch root strain ($\Delta\epsilon$), which is not accounted for in Eq 1, the plastic strain range at the notch root ($\Delta\epsilon_p$) is evaluated according to the equation

$$\Delta\epsilon_p = \Delta\epsilon - \frac{\Delta\sigma}{E} \tag{3}$$

This procedure is illustrated graphically in Fig. 2, which incorporates an elastic unloading line to show $\Delta\epsilon_p/2$. This most conservative procedure is the one adopted in this study to predict the plastic strain range and, hence, the lifetime for each notch test.

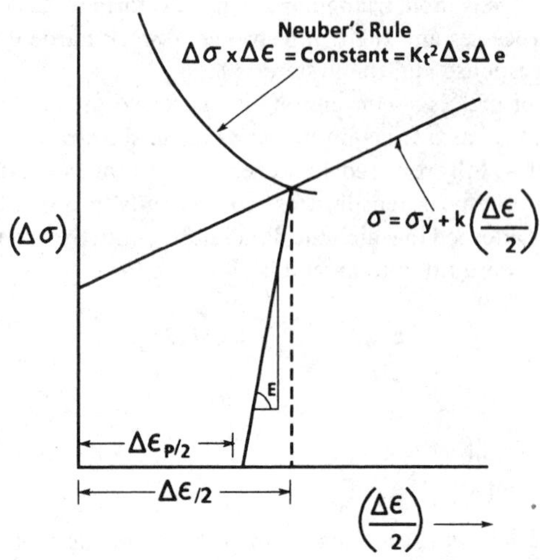

FIG. 2—*Illustration of the application of Neuber's rule to determine the notch root plastic strain range* ($\Delta\epsilon_p$).

Finite Element Analysis

A primary concern of this study is to evaluate the use of the MARC finite element computer program in situations involving cyclic loading which should lead to high strain creep-fatigue interaction. For this purpose, the axisymmetric eight-node distorted quadrilateral element (Element 28 in the MARC program) was used. The mesh for one of the test piece geometries is illustrated in Fig. 3. Several runs were made for the first loading period to determine the optimum number of steps. The results converged by using 25 steps for each loading or unloading period and 8 steps for the creep hold time. The materials data required as input for the program included elastic, cyclic work hardening, and creep properties at the temperature of interest in each particular material. The loading is achieved by prescribing a displacement waveform at locations corresponding to the 0.0127 m gage length used in the testing. The output gives values of stress and strain at various locations throughout the structure for the various steps of each cycle. The stresses and strains are computed in four directions: axial, circumferential, radial, and shear. Further computation allows construction of the stress-strain hysteresis loops at the region of interest. The maximum values of stress and strain were found close to the notch root. Figure 4 shows typical hysteresis loops for the various components of stress and strain generated close to the notch root. It is immediately apparent that the axial loading component dominates.

The maximum plastic strain range corresponds to the width of the hysteresis loop at the position of the horizontal strain axis. This value is used to compute the expected lifetime of the test piece of this geometry by inspecting the smooth $\Delta\epsilon_p$-N_f data for this particular material to find the value of N_f corresponding to this maximum plastic strain range. Essentially, this approach ignores the details of loading multiaxiality, but is regarded as providing a rapid assessment of component life expectancy. Its reliability is examined in more detail in the section dealing with results.

The MARC finite element computer program contains a provision for the calculation of the von Mises equivalent plastic strain throughout the loading cycle. This calculation is similar to one provided by the ASME Boiler and Pressure Vessel Code [*8*] for the evaluation of component fatigue damage. Since the magnitude of the equivalent plastic strain is computed for the current loading condition, the value calculated at the tension peak of the second cycle corresponds to

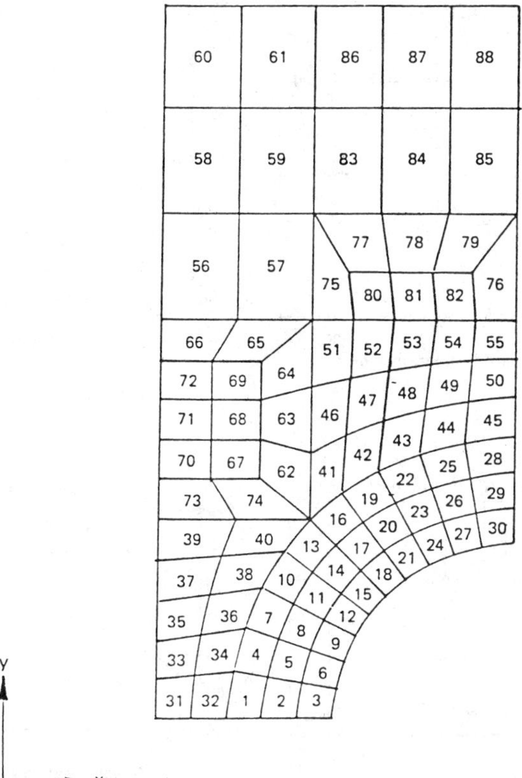

FIG. 3—*Finite element mesh for Type C specimen.*

the equivalent plastic strain range. Figure 5 shows contours of equivalent plastic strain calculated at this point in a cycle. The maximum value is found close to the notch root, and is the one used to evaluate the expected lifetime of the test piece by direct reference to the appropriate uniaxial $\Delta\epsilon_p$-N_f data.

The idea of relating hysteresis energy and fatigue is not new and many investigators [*9–12*] have dealt with the subject. Most recently, Garud [*13,15*] has related the fatigue lifetime (cycles to crack initiation) to the total plastic work per cycle (W_c) defined as

$$W_c = (\sigma_x \cdot d\epsilon_x^p + \sigma_y \cdot d\epsilon_t^p + \sigma_z \cdot d\epsilon_z^p + \tau_{xy} \cdot d\gamma_{xy}^p + \tau_{yz} \cdot d\gamma_{yz}^p + \tau_{zx} \cdot d\gamma_{zx}^p) \text{ cycle} \quad (4)$$

where

$\epsilon_x, \epsilon_y, \epsilon_z, \gamma_{xy}, \gamma_{yz}, \gamma_{zx}$ = strain components, and
$\sigma_x, \sigma_y, \sigma_z, \tau_{xy}, \tau_{yz}, \tau_{zx}$ = stress components.

The procedure is based upon the "incremental plasticity theory" and has been found to be in good agreement with available results [*14*] on fatigue under complex loading. Uniaxial data have been found [*15*] to relate the plastic work per cycle, W_c, to the lifetime, N_f, by a simple power law of the form

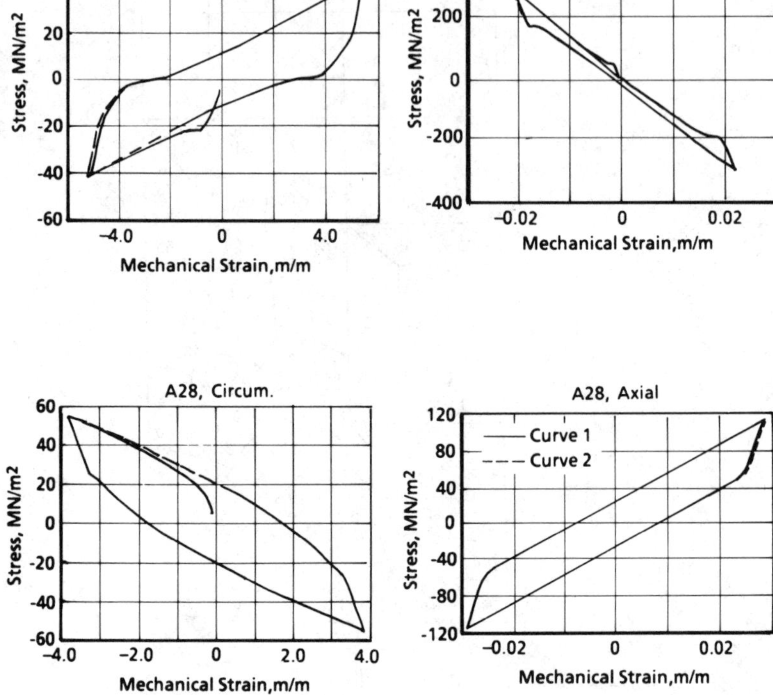

FIG. 4—*Hysteresis loops constructed using the MARC finite element program for Test A28.*

$$W_c = AN_f^{-\beta} \tag{5}$$

here A and β are constants determined by curve fitting uniaxial data for a particular material.

The MARC finite element computer program provides for a calculation of W_c for each of the directional components by evaluating separately the areas of each of the computed hysteresis loops for the axial, radial, circumferential, and shear cases. The approach adopted here was simply to add all these values together, so that

$$W_c = \Sigma W_{c_i} \tag{6}$$

The computed total plastic work for the critical region of the component (in this case the notch root area) is then used to estimate the lifetime by inspection of materials data derived for the appropriate material under appropriate conditions and described by Eq 5.

Experimental Work

The materials chosen for this study were HP-45, Incoloy 800H, Incoloy 800, and 1¼Cr-½Mo. Smooth specimens having a parallel gage section 0.03175 m long and 0.006125 m in diameter were used to generate the uniaxial data. Notched specimens were manufactured by

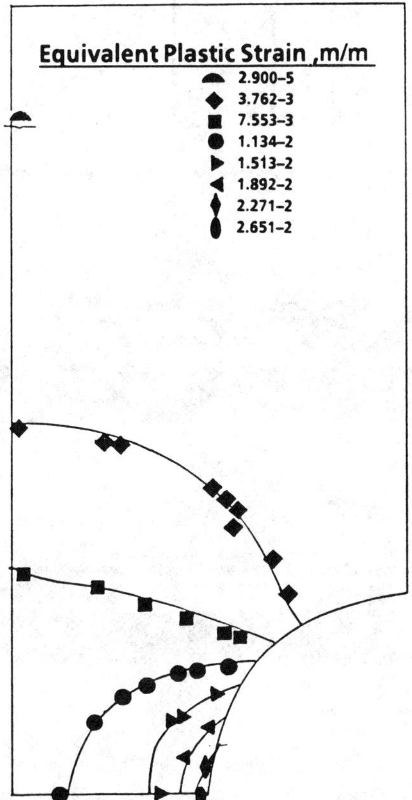

FIG. 5—*Contours of equivalent plastic strain calculated for Test C151.*

introducing the circumferential notches described in Table 1 into smooth specimens using a grinding technique specified to produce a low level of residual stress in the notch region. Figure 6 shows typical examples of the notched specimens. The specimens were attached into the test frame very carefully so as to minimize the possibility of introducing notch root deformation during the set-up procedure. Mechanical testing was carried out using an Instron Model 1310 electro-servohydraulic universal testing machine. Fully reversed cyclic loading was performed in closed-loop strain control mode using an MTS quartz rod extensometer arrangement to provide the feedback signal of specimen longitudinal strain. Heating of the HP-45, Incoloy 800H, and Incoloy 800 specimens was achieved using a LEPEL® radio frequency (RF) heater with a coil arrangement designed to keep temperature variation within ±3°C of the set point level along

TABLE 1—*Nominal notch specimen geometries.*

Notch Type	C	B	A
D (m)	0.00635	0.00635	0.00635
d (m)	0.00315	0.00315	0.00508
r (m)	0.0016	0.00046	0.00046
K_t [6,7]	1.60	2.28	2.50

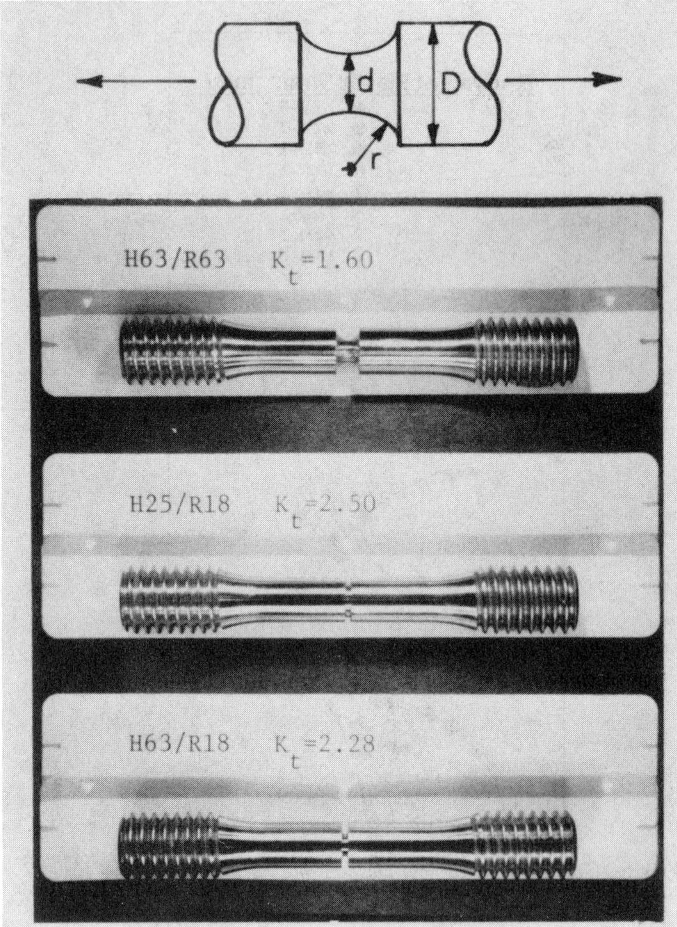

FIG. 6—*Notched specimens used in this study.*

the 0.0127 m extensometer gage length. The 1¼Cr-½Mo specimens were heated using an MTS three-zone radiant furnace, because of difficulties in achieving the required flat temperature profile using the RF arrangement. Temperatures were monitored using three chromel-alumel thermocouples maintained in good contact with the specimen surface.

Testing of smooth specimens was conducted at a variety of total strain ranges in the range $10^{-2} > \Delta\epsilon > 10^{-3}$. Testing of notched specimens was conducted in strain control, which was achieved using the signal from the quartz rod extensometer spanning the notch with an effective gage length of 0.0127 m. Both notched and smooth specimens were loaded until visible crack-like damage had been initiated. Lifetime, N_f, is here defined as the number of cycles to produce a 10% reduction in maximum tensile stress, σ_{pk}, or a 10% reduction in measured Young's modulus, whichever is less. Metallographic examination of failed specimens was conducted to ensure that the notched specimens exhibited equivalent primary failure mechanisms to those found in the smooth specimens loaded using identical strain waveforms.

Results and Discussion

The baseline uniaxial materials property data for each of the four materials tested are presented in Table 2.

The data relating lifetime (N_f) to inelastic strain range ($\Delta\epsilon_p$) and plastic work per cycle (W_c) for the smooth specimens are presented in Figs. 7 to 10. Curve fittings of the W_c-N_f (Eq 5) and $\Delta\epsilon_p$-N_f (Manson-Coffin relation) are also given in Figs. 7 to 10.

TABLE 2—*Materials property data used in finite element analysis.*

Material		Incoloy 800H	Incoloy 800	HP-45	1¹/4Cr-¹/2Mo
Temperature, °F		1144	922	1144	811
Young's modulus, GN/m²		132.4	151.7	144.8	164.1
Cyclic					
hardening constants	σ_y	62.6	188.3	92.1	159.3
$[\sigma = k(\Delta\epsilon/2) + \sigma_y]$	k	10.1	21.1	9.2	32.1
Power law					
creep constants	C	7.5×10^{-12}	10^{-18}	10^{-20}	1.2×10^{-14}
$[\dot{\epsilon} = C\sigma^b\ (s^{-1})]$	b	5.32	8.8	13.7	4.05
Density, kg/m³		7958	7944.2	7861.1	7833.4
Poisson's ratio		0.394	0.377	0.32	0.3182
Thermal expansion coefficient, m/m/K		18.36×10^{-6}	17.28×10^{-6}	21.42×10^{-6}	13.14×10^{-6}

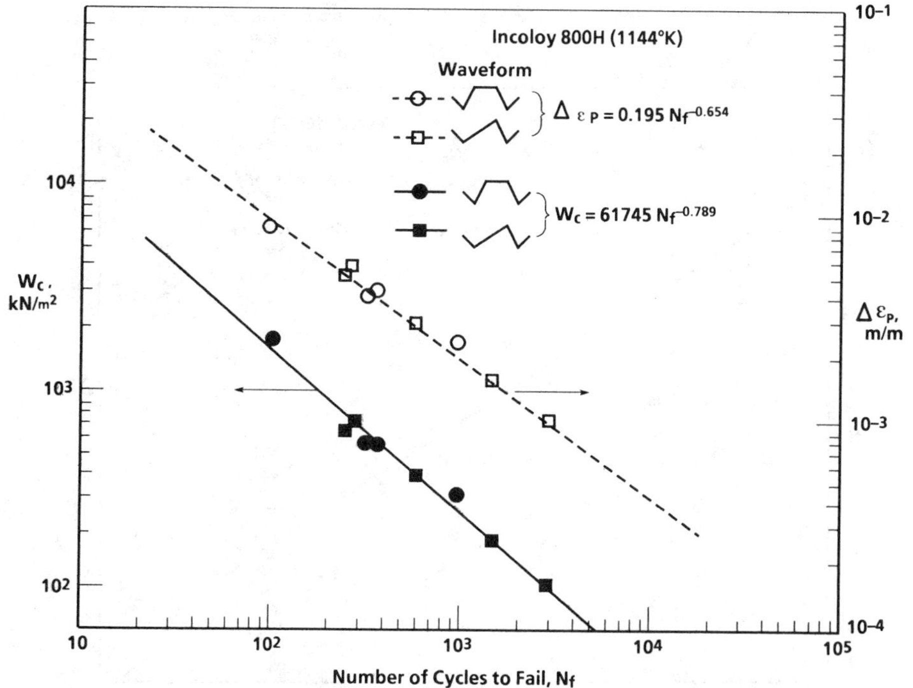

FIG. 7—$\Delta\epsilon_p$-N_f *and* W_c-N_f *results for Incoloy 800H.*

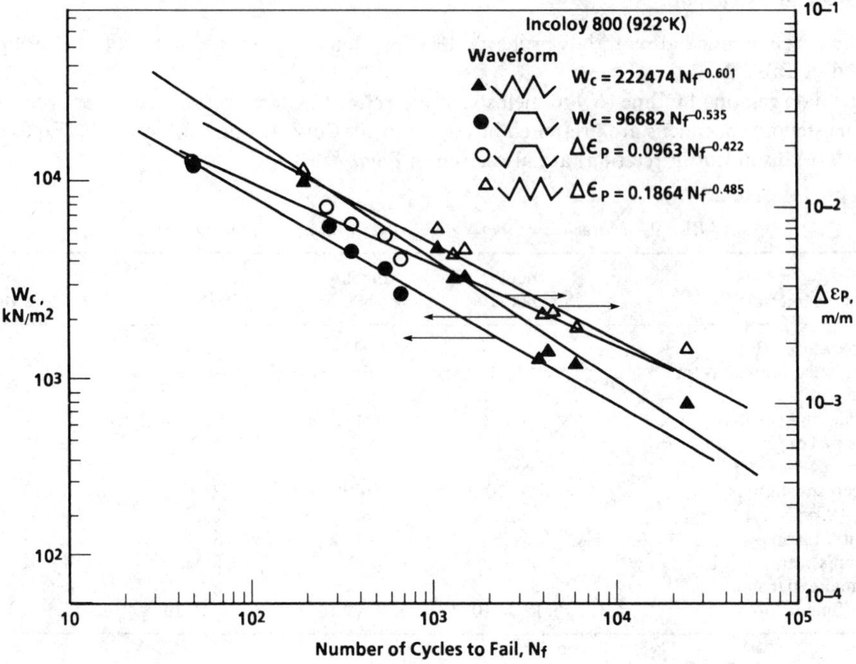

FIG. 8—$\Delta\epsilon_p$-N_f *and* W_c-N_f *results for Incoloy 800.*

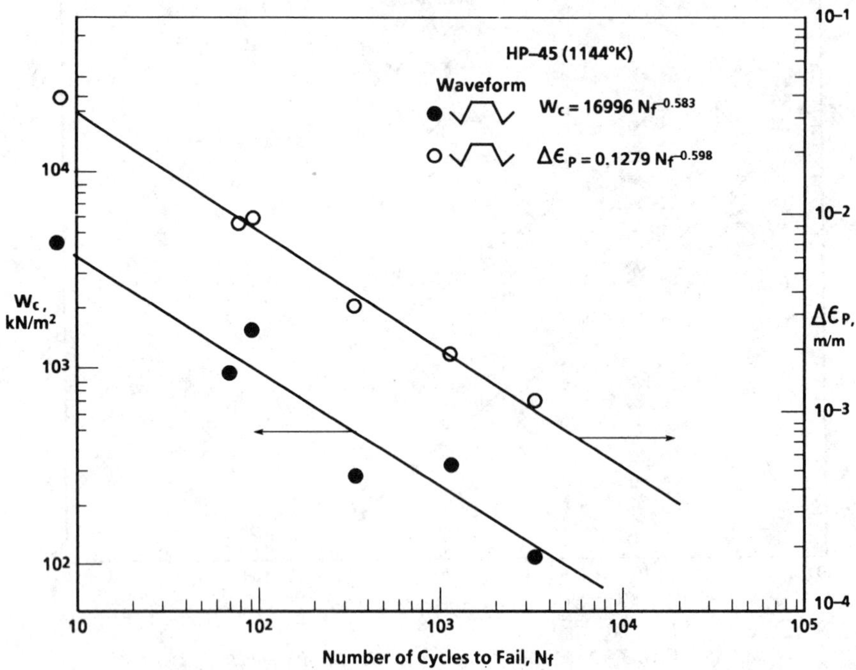

FIG. 9—$\Delta\epsilon_p$-N_f *and* W_c-N_f *results for HP-45.*

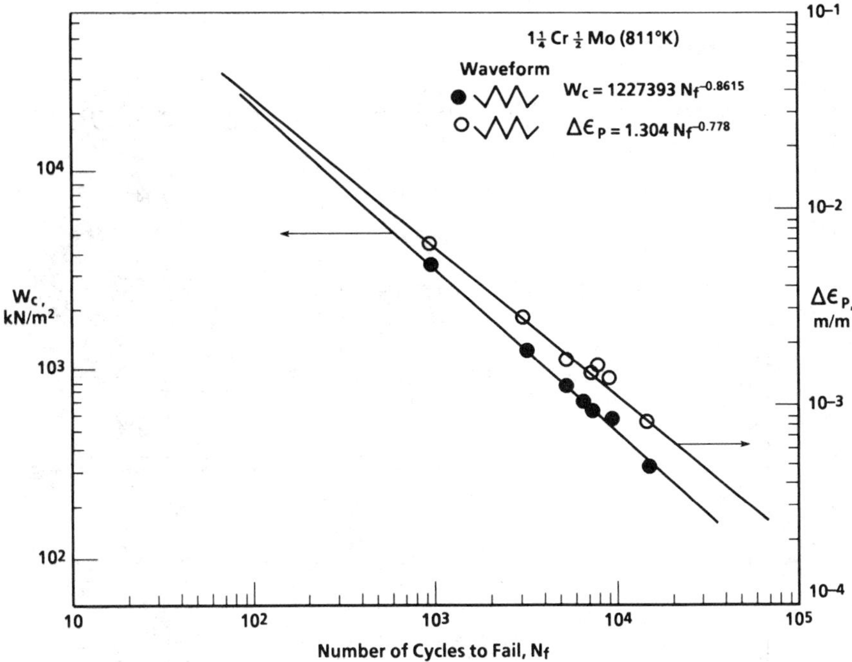

FIG. 10—$\Delta\epsilon_p$-N_f *and* W_c-N_f *results for 1¼Cr-½Mo.*

Lifetime data for the notched specimen testing program are presented in Table 3 along with corresponding lifetimes for each of the tests predicted using each of the four criteria described previously. These data are further illustrated graphically in Figs. 11 to 14, which plot N_f (predicted) versus N_f (observed) for these four criteria.

Examination of Figs. 11 to 14 reveals several points which are used to guide the decision regarding the choice of an appropriate criterion for the evaluation of component lifetime.

The use of the simple maximum plastic strain range is completely unreliable, since the predicted lifetimes for a particular material appear to be unrelated to those observed, so that it is impossible to say whether such a prediction is conservative or nonconservative.

The use of a Neuber notch strain analysis has produced predictions which generally lie within $\pm$ a factor of three on observed lifetimes for the range of materials and test waveshapes employed. The individual data sets for each material examined using the Neuber analysis were equally scattered within the $\pm$ factor of three on lifetime.

The predictions derived using the equivalent plastic strain range criterion generally lay within the $\pm$ factor of three upon lifetime for the range of materials and test waveshapes employed. The scatter on the predictions for individual data sets for each material using the equivalent plastic strain range approach was observed to be much less than the $\pm$ factor of three a lifetime. In this way, predictions for a particular material were tightly distributed along the 45° trend lines.

The lifetime predictions made using the total plastic work per cycle approach were without exception conservative. The predictions for the HP-45 material were extremely conservative, underestimating the observed lifetimes by a factor of ten or fifteen. Such a predictive discrepancy most likely arises from unreliable W_c-N_f data for the HP-45 material. Individual data sets were generally tightly distributed about the trend lines, with scatter well below $\pm$ factor of three on lifetime.

TABLE 3—*Notch test results.*

Material	Test Temperature (K)	K_t	Waveform	Far-Field Strain ($\times 10^{-3}$) ($\Delta\epsilon$)	N_f	Prediction: Neuber N_f	Prediction: $\Delta\epsilon_p^{max}$ N_f	Prediction: $\Delta\epsilon$ eq N_f	Prediction: W_c N_f	Analysis Code No.
Incoloy 800H	1144	1.6	t_h = 3600 s	4.99	30	59	87	111	14	B151
		1.6		3.00	47	81	85	68	22	B131
		1.6		1.15	203	210	78	308	72	B111
		2.5		5.01	68	41	115	80	26	A151
		2.5		3.09	84	68	141	91	28	A131
		2.28		3.11	27	25	120	38	12	C131
		2.28		5.07	20	21	109	35	9	C151
		2.28		0.97	125	123	75	151	31	C111
		2.5		1.73	133	127	205	322	76	A111
HP-45	1144	2.28	t_h = 3600 s	1.96	41	32	26	21	2	C321
		2.28		5.03	10	12	30	8	1	C351
		2.28		9.89	3	6	35	2	1	C3101
		2.5		1.96	189	101	65	65	9	A321
		2.5		4.99	64	32	65	23	3	A351
		2.5		9.98	29	15	68	10	1	A3101
		1.6		2.12	83	129	19	48	6	B321
		1.6		5.00	22	50	26	13	2	B351
		1.6		9.99	8	29	28	3	1	B3101
Incoloy 800	922	1.6		5.01	155	194	116	72	39	B25
		1.6		3.94	229	235	112	111	54	B24
		1.6		7.96	62	104	121	34	22	B28
		2.28		3.94	91	47	124	48	22	C24
		2.28		4.99	79	30	130	36	18	C25
		2.28		7.98	23	14	146	13	11	C28
		2.5		3.89	347	154	183	192	70	A24
		2.5		4.97	286	78	180	135	52	A25
		2.5		7.90	95	46	180	64	28	A28
		1.6	t_h = 900 s	8.00	42	51	52	12	7	B282
		2.28		7.98	14	7	64	6	3	C282
		2.5		7.98	69	27	139	25	9	A282
1¹/₄Cr-¹/₂Mo	811	1.6		4.47	222	401	442	225	129	B45
		2.28		5.04	100	110	474	153	81	C45
		2.5		5.01	345	208	503	326	155	A45

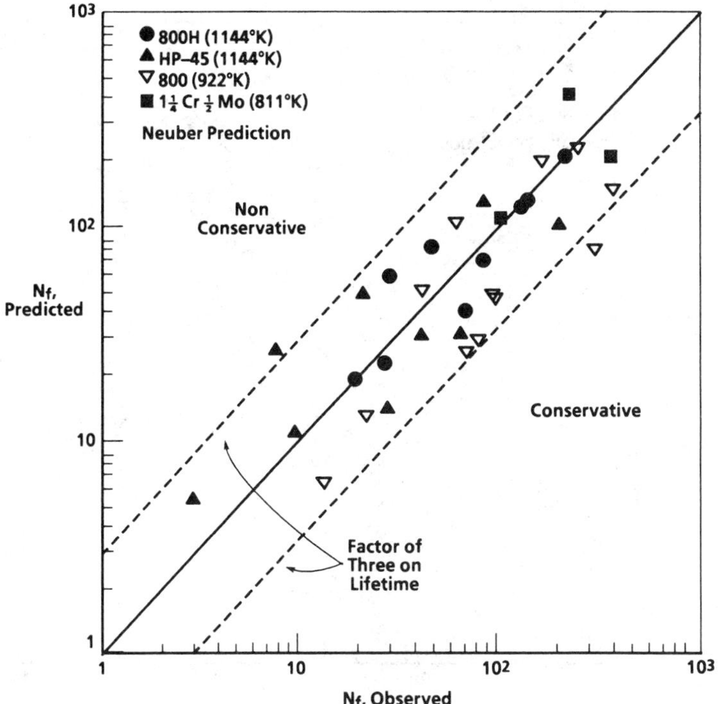

FIG. 11—*Observed lifetime versus lifetime predicted using Neuber notch strain analysis.*

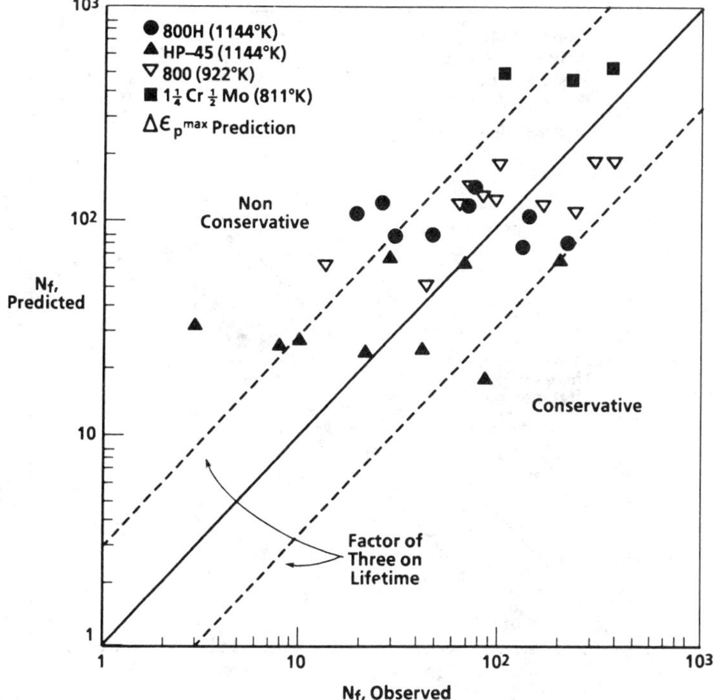

FIG. 12—*Observed lifetime versus lifetime predicted using the maximum plastic strain range* ($\Delta\epsilon_p^{max}$).

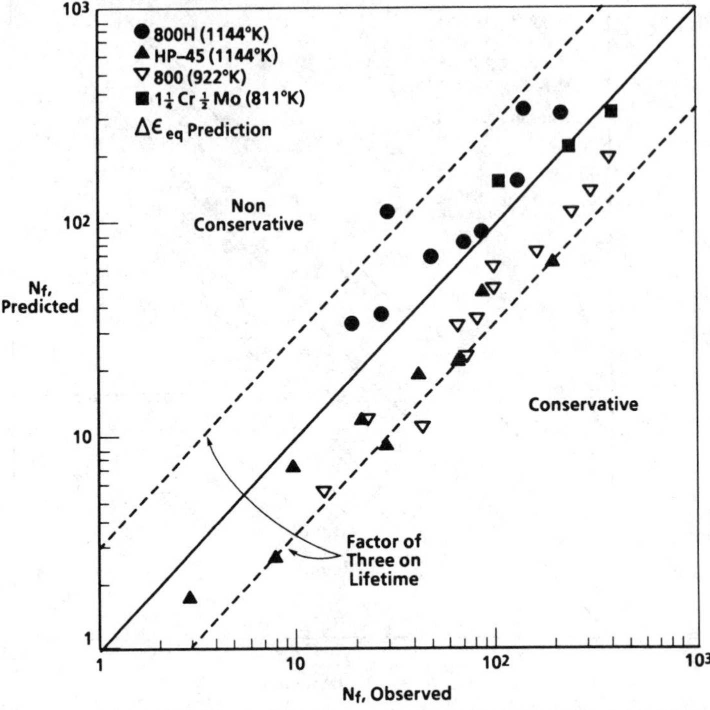

FIG. 13—*Observed lifetime versus lifetime predicted using the equivalent plastic strain range* ($\Delta\epsilon_p^{eq}$).

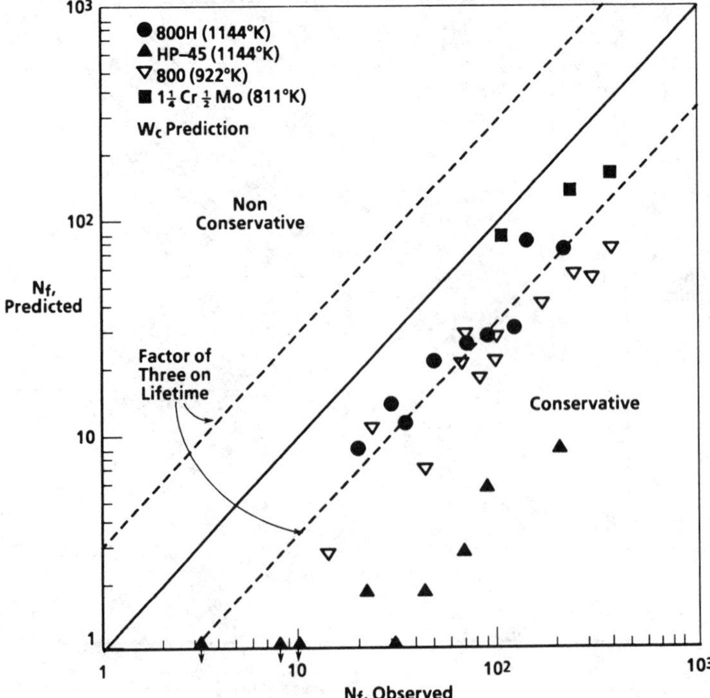

FIG. 14—*Observed lifetime versus lifetime predicted using the total plastic work per cycle* (W_c).

Conclusions

The evaluation of component lifetimes using an approach based upon the calculation of total plastic work per cycle appears to provide a consistently conservative prediction. The use of the equivalent plastic strain range determined by the MARC finite element computer program for the evaluation of component lifetime is able to provide a predictive capability well within acceptable limits of scatter and reliability. It appears to be less conservative than the life predicted using the work-per-cycle calculation. The Neuber notch strain analysis provides a reasonable first-pass prediction of component lifetime for a variety creep-fatigue interaction conditions. The use of a simple maximum plastic strain range to estimate component lifetime is completely unreliable.

The total plastic work-per-cycle calculation is recommended for use in the determination of component low-cycle fatigue lifetimes.

References

[*1*] Brown, M. W. and Miller, K. J., "A Theory for Fatigue Failure Under Multiaxial Stress-Strain Conditions," *Proceedings of the Institution of Mechanical Engineers,* Vol. 187, No. 65, 1973, pp. 745–755.

[*2*] Krempl, E., *The Influence of State of Stress on Low-Cycle Fatigue of Structural Materials: A Literature Survey and Interpretive Report, ASTM STP 549,* American Society for Testing and Materials, Philadelphia, 1974.

[*3*] Garud, Y. S., "Multiaxial Fatigue: A Survey of the State of the Art," *Journal of Testing and Evaluation,* Vol. 9, No. 3, 1981, pp. 165–178.

[*4*] Brown, M. W. and Miller, K. J., "Two Decades of Progress in the Assessment of Multiaxial Low-Cycle Fatigue Life," in *Low-Cycle Fatigue and Life Prediction, ASTM STP 770,* American Society for Testing and Materials, Philadelphia, 1982, pp. 482–499.

[*5*] Neuber, M., "Theory of Stress Concentration for Shear-Strained Prismatical Bodies with Arbitrary Nonlinear Stress-Strain Laws," *Journal of Applied Mechanics, Transactions of ASME,* Vol. 28, 1961, p. 544.

[*6*] Peterson, R. E., "Analytical Approach to Stress Concentration Effect in Aircraft Materials," U.S. Air Force-WADC Symposium on Fatigue of Metals, Technical Report 59-507, Dayton, Ohio, 1959, p. 273.

[*7*] Peterson, R. E., *Stress Concentration Factors,* Wiley, New York, 1974.

[*8*] ASME Boiler and Pressure Vessel Code Case N-47-17 (1592-17), Sect. T-1413, 1979, p. 168.

[*9*] Feltner, C. E. and Morrow, J. D., "Microplastic Strain Hysteresis Energy as a Criterion Fatigue Fracture," *Journal of Basic Engineering, Transactions of ASME,* Vol. 83D, 1961, pp. 15–22.

[*10*] Stowell, E. Z., "A Study of the Energy Criterion for Fatigue," *Nuclear Engineering and Design,* Vol. 3, 1966, pp. 32–40.

[*11*] Halford, G. R., "The Energy Required for Fatigue," *Journal of Materials,* Vol. 1, No. 1, 1966, pp. 3–18.

[*12*] Damali, A. and Esin, A., "Micro Plastic Strain Energy Criterion Applied to Reversed Biaxial Fatigue," in *Fracture 1977—Advances in Research on the Strength and Fracture of Materials,* Vol. 213, D. M. R. Taplin, Pergamon Press, New York, 1977, pp. 1201–1206.

[*13*] Garud, Y. S., "Prediction of Stress-Strain Response Under General Multiaxial Loading," in *Mechanical Testing for Deformation Model Development, ASTM STP 765,* R. W. Rohde and J. C. Swearengen, Eds., American Society for Testing and Materials, Philadelphia, 1982, pp. 223–238.

[*14*] Kanazawa, K., Miller, K. J., and Brown, M. W., "Low-Cycle Fatigue Under Out-of-Phase Loading Conditions," *Journal of Materials Engineering Technology, Transactions of ASME,* Vol. 99, 1977, p. 22.

[*15*] Garud, Y. S., "A New Approach to the Evaluation of Fatigue Under Multiaxial Loadings," *Journal of Engineering Materials and Technology, Transactions of ASME,* Vol. 103, No. 4, pp. 118–125.

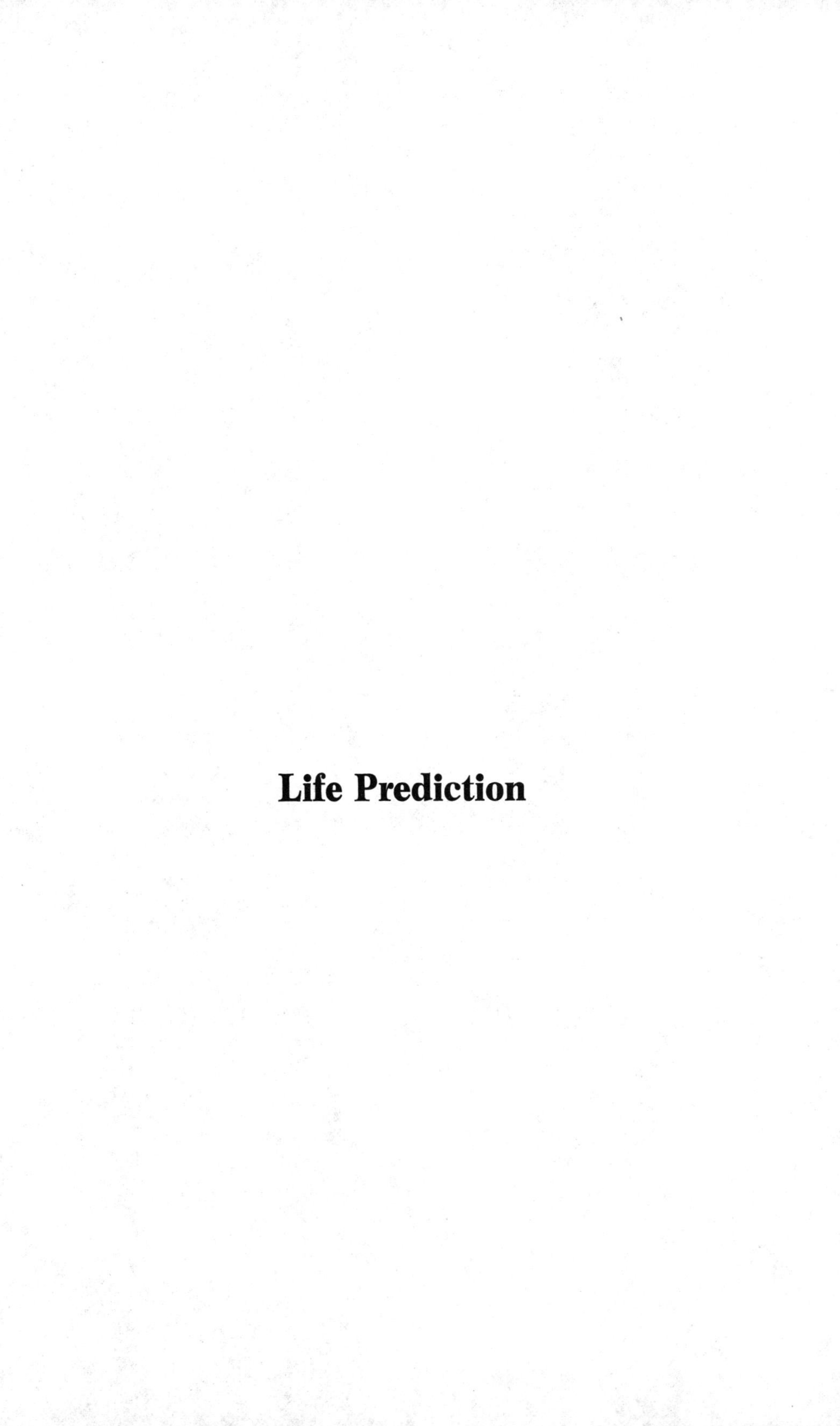

Life Prediction

L. Rémy,[1] F. Rezai-Aria,[1] R. Danzer,[2] and W. Hoffelner[3]

Evaluation of Life Prediction Methods in High Temperature Fatigue

REFERENCE: Rémy, L., Rezai-Aria, F., Danzer, R., and Hoffelner, W., "**Evaluation of Life Prediction Methods in High Temperature Fatigue,**" *Low Cycle Fatigue, ASTM STP 942*, H. D. Solomon, G. R. Halford, L. R. Kaisand, and B. N. Leis, Eds., American Society for Testing and Materials, Philadelphia, 1988, pp. 1115–1132.

ABSTRACT: A collaborative study supported by the European Community COST 50 program has been conducted on life prediction methods under high-temperature low-cycle fatigue and thermal fatigue cycling in the case of MAR-M509, a cast cobalt-base superalloy. Low-cycle fatigue tests were carried out by the Ecole des Mines de Paris at 900°C in air under longitudinal strain control and included sawtooth triangular shape cycles at three strain rates (frequency range 10^{-3} to 20 Hz), tensile strain hold cycles, and compressive strain hold cycles. Life to initiation data deduced from a-c potential drop measurements were used to evaluate three types of predictive models: strainrange partitioning, studied by BBC-Baden; creep damage models, studied by the University of Leoben; and an oxidation-fatigue interaction model, studied by the Ecole des Mines de Paris. All the methods have been found to correlate experimental data within a factor of three. However, they range from simple correlating methods with adjustable parameters to be fitted to fully predictive methods with no fitting parameter. Thermal fatigue data on wedge-type specimens (i.e., components instead of laboratory specimens) have been used to check the prediction capability of the various methods and have shown the superiority of oxidation-fatigue interaction models for this material.

KEY WORDS: life prediction, high-temperature low-cycle fatigue, thermal fatigue, cobalt-base superalloy, crack initiation, oxidation

The assessment of life to crack initiation of engineering components which are operating at high temperatures has been the object of considerable effort. Various prediction methods have been proposed to account for low-cycle fatigue damage at high temperature. Both metallurgists and engineers are aware of the necessity of taking into account the interaction between fatigue, creep, and environment. Mechanical engineers have proposed predictive methods which claim to account for creep effects, such as Manson's strainrange partitioning method [*1*], or oxidation effects, such as Coffin's frequency modified rule [*2*]. In spite of the value of each method for design, none of these methods can be expected from a metallurgical standpoint to be universal and to apply to every kind of material. The validity of a particular prediction method must be assessed for a particular material and for an application in service. Therefore the purpose of the present study was to evaluate different life prediction methods using a single set of experimental data generated under isothermal low-cycle fatigue (LCF) or under thermal fatigue (TF).

This work was completed as a collaborative project supported by the European Community COST 50 Round III program on materials for gas turbines. The basic work was devoted to the study of the thermal fatigue behavior of MAR-M509, a cast cobalt superalloy used in guide

[1]Centre de Matériaux de l'Ecole des Mines de Paris, UA CNRS 866, BP. 87, 91003 EVRY Cédex, France.
[2]Montanuniversität Leoben, Institut für Metallkunde und Werkstoffprüfung Institut für Physik, A-8700 Leoben, Austria.
[3]BBC Brown Boveri, CH5401 Baden, Switzerland.

vanes, and involved the Ecole des Mines and Société Nationale d'Etude et de Construction de Moteurs d'Aviation (SNECMA) [*3*]. The international cooperation with BBC Baden and the University of Leoben was concerned with the comparison of life prediction methods for this material. Therefore the Ecole des Mines has carried out a detailed investigation of the low-cycle fatigue behavior of MAR-M509 at 900°C (1173 K) involving low- and high-frequency sawtooth tests and tensile and compressive strain-hold tests.

Most damage rules describe the life to crack initiation. In most instances, however, their predictions are compared with life to fracture of low-cycle fatigue specimens, which often corresponds to the presence of 1 or 2 mm deep cracks in the specimens. So the application of predictive methods can be somewhat obscured by a comparison with life to fracture, which often encompasses a large part of crack propagation at high temperatures. Therefore an a-c potential drop technique was applied to all the tests. This technique enables an operational life to initiation, N_i, to be defined which has been shown to correspond to a maximum crack size about 0.3 mm in depth [*4*].

These initiation data at 900°C (1173 K) were used to evaluate different life prediction methods. The strainrange partitioning method [*1*] was used by BBC Baden. The University of Leoben applied two methods based on the prediction of fatigue life from creep rupture data, Spera's accumulated creep damage [*5*], and Franklin's cyclic creep damage [*6*] methods. The Ecole des Mines de Paris used prediction methods based on a description of oxidation fatigue interactions [*7*]. The latter methods have a physical basis in the present alloy, since the influence of oxidation on crack initiation and early growth has been clearly shown [*8*]. In particular, primary interdendritic MC carbides (tantalum rich) are preferentially oxidized and are responsible for early crack initiation in high-temperature fatigue under continuous sawtooth cycling [*7*].

Life to fracture and life to crack initiation data under low-cycle fatigue at 900°C in air are presented for MAR-M509. These data on laboratory specimens are then analyzed using the various methods mentioned before, and the results of every method are discussed. Low-cycle fatigue tests under sawtooth cycling show a fatigue life almost independent of temperature for a given inelastic strain range in the temperature range 900 to 1100°C (1173 to 1373 K, [*8*]). Thus the strainrange partitioning relationship estimated at 900°C should apply to higher temperatures. Creep and oxidation data were available up to 1100°C (1373 K, [*3*]). Thus the life to crack initiation of thermal fatigue wedge specimens cycled up to 1100°C (1373 K) was used to evaluate the prediction capabilities of the different methods in the case of a component.

Experimental Procedure

The composition of the master heat used in this study was (in weight percent) 0.59C-11Ni-23.2Cr-6.55W-3.31Ta-0.30Zr-0.22Ti-0.17Fe-0.008B-0.005P-0.003Si-bal.Co. Specimens were in the form of cast cylinders 20 mm in diameter or of special castings in which two wedge-type specimens for thermal fatigue or two cylindrical LCF specimens could be machined. A single casting microstructure was used for TF and LCF specimens. All specimens were heat-treated at 1230°C (1503 K) for 6 h. The corresponding microstructure has been shown by Reuchet and Rémy [*8*], and the grain size is 0.8 mm.

Low-cycle fatigue tests were carried out at the Ecole des Mines at 900°C (1173 K). The LCF specimens were cylindrical specimens 8 mm in diameter and 12 mm in gage length. The tests were carried out under total axial strain control. Specimens were heated by a radiation furnace. Longitudinal strain was measured using an extensometer with alumina knives in contact with the specimen and a linear variable-differential transformer. Most tests were made on a screw-driven Instron tensile machine that had been modified to conduct push-pull axial loading. Only high-frequency tests (above 1 Hz) were carried out on an electrohydraulic machine under closed-loop control.

Low-frequency tests included: (1) sawtooth continuous cycling tests at a nearly constant total strain rate of 7×10^{-4} s^{-1} (intermediate strain rate, cycle period ranging from 10 to 40 s) or of 4×10^{-5} s^{-1} (low strain rate, cycle period ranging from 100 to 900 s); (2) tests with a 2 min tension hold time, where the maximum total strain was held constant; and (3) tests with a 2 min compression hold time, where the minimum total strain was held constant.

High-frequency tests at 20 Hz were also carried out on a servohydraulic machine in order to get pp-type loops, according to Manson's terminology. In that case testing was conducted under load control. In these experiments the strain was not directly measured on the specimen gage length but was deduced from a calibration curve using a transducer applied to the whole loading system. A microcomputer was used for data monitoring, but the use of a calibration gives more uncertainty on the hysteresis stress-strain loops.

An a-c potential drop technique was used to monitor the cracking process. Stabilized current input was applied to both sides of the specimen which had been insulated from the loading frame. The potential drop was measured across specimen gage length and was continuously recorded throughout the test. Initiation was arbitrarily defined as corresponding to a relative variation of 0.1% of the potential drop. This definition has been shown to correspond to a maximum crack size about 0.3 mm in depth [*4*] using plastic replicas and a sectioning technique on various specimens. The accuracy of such values was usually about 15%, but it was poorer in some instances, especially for the high-frequency tests.

In all instances tests were interrupted before rupture occurred, and the fatigue life was conventionally defined as corresponding to the onset of tensile load drop or to a relative variation of the potential drop about 1%. Fatigue life corresponds to a maximum crack depth in the range of 1 to 2 mm.

A standard thermal transient was used for TF tests, and the geometry of wedge-type specimens was varied to decrease the maximum temperature difference between the thin edge and the bulk of the specimen from specimen types A to B to C [*3,9*]. A and B were standard wedge-type specimens with an edge radius of 0.25 and 1 mm, respectively; C had an edge radius of 1 mm but a smaller cross section to reduce thermal inertia. These variations in specimen geometry were designed to decrease the transient strain range and to increase fatigue life. Experiments were carried out on the thermal fatigue rig of SNECMA. Maximum and minimum temperatures were 1100 and 200°C (1373 and 473 K) respectively for a cycle period of 80 s. Tests were periodically interrupted to monitor the cracking process. Computation of the transient temperature at every point of the TF specimens was achieved using a two-dimensional finite element analysis. The stress-strain analysis was made using the viscoplastic constitutive equation proposed by Chaboche [*10*] to account for the Bauschinger effect. Further details concerning experimental details, computation procedure, and results can be found elsewhere [*3,9*].

Results and Discussion

High-Temperature Low-Cycle Fatigue Behavior

Conventional fatigue life results at 900°C in air are shown in Fig. 1 as a function of the inelastic strain range, $\Delta\epsilon_{in}$, for the various frequencies and waveshapes studied. The number of cycles to fracture is nearly independent of test frequency for continuous sawtooth cycles carried at medium and low strain rates ($\dot{\epsilon}_t \simeq 7 \times 10^{-4}$ s^{-1} and 4×10^{-5} s^{-1}, respectively). The fatigue life of strain-hold (tensile or compressive) tests is almost the same as for sawtooth cycles carried out at similar frequencies (i.e., corresponding to $\dot{\epsilon}_t = 7 \times 10^{-4}$ s^{-1}). The influence of cycle shape on fatigue life is therefore negligible. All the low-frequency tests (sawtooth and strain-hold cycles) show a bilinear total life-inelastic strain range curve in a log-log plot. The slope changes from a value of about -0.6, for strain ranges higher than 3×10^{-3}, to a value of about -1 for the lower strain ranges.

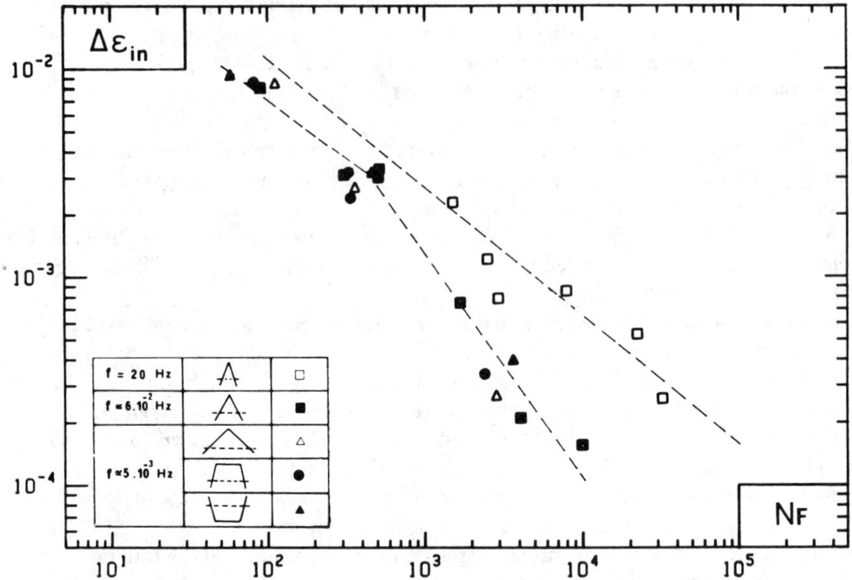

FIG. 1—*Variation of the number of cycles to failure* (N_f) *with the inelastic strain range* ($\Delta\epsilon_{in}$) *at 900°C (1173 K).*

High-frequency tests exhibit a very different life behavior. The exponent of the Manson-Coffin equation was about -0.6 for these tests, while it was about -1 for low-frequency tests for inelastic strain ranges smaller than 3×10^{-3}. It is worth emphasizing that the $\Delta\epsilon_{in}$-N_f curve extrapolates at higher strain ranges to a life rather close to that observed for low-frequency tests. For a plastic strain range about 0.2% the life at 20 Hz is about two times that at low frequency. For a plastic strain range of about 0.03%, the high-frequency life is ten times longer than at low frequency. However, fatigue life in air at 20 Hz, though much higher than at low frequency, is somewhat shorter than that observed under vacuum (under vacuum, no frequency dependence of fatigue life was observed in the range of 5×10^{-2} to 1 Hz, [*3*]).

The life to initiation of a fatigue crack 0.3 mm deep is shown in Fig. 2 as a function of the inelastic strain range $\Delta\epsilon_{in}$. Typically, the initiation period thus defined ranges from 10 to 20% of total life at high strain amplitudes to about 50% of total life at low strain ranges.

It is worth noting that the cycle frequency and cycle shape dependency of life to crack initiation are negligible for all low-frequency tests (sawtooth and strain hold tests). The curve of life to crack initiation versus the inelastic strain range show a bilinear shape for the low-frequency tests. But low-frequency data for a strain range higher than 3.10^{-3} and high-frequency data at a lower strain ranges seem to fit a single straight line. This observation suggests that crack initiation is only cycle dependent at high strain ranges. Below this critical strain range of 3×10^{-3}, differences in life to crack initiation between low- and high-frequency tests increase with decreasing strain ranges.

The number of cycles to crack initiation is shown in Fig. 3 as a function of the stress range $\Delta\sigma$. Results show more scatter than as a function of the inelastic strain range. Hold-time tests exhibit crack initiation periods about half that of sawtooth cycles ($\dot{\epsilon}_t = 7 \times 10^{-4}$ s) at intermediate strain rate. The difference between these tests and tests at high frequency is still larger than for an inelastic strain criterion.

Metallographic observations were carried out on specimen gage length and on longitudinal

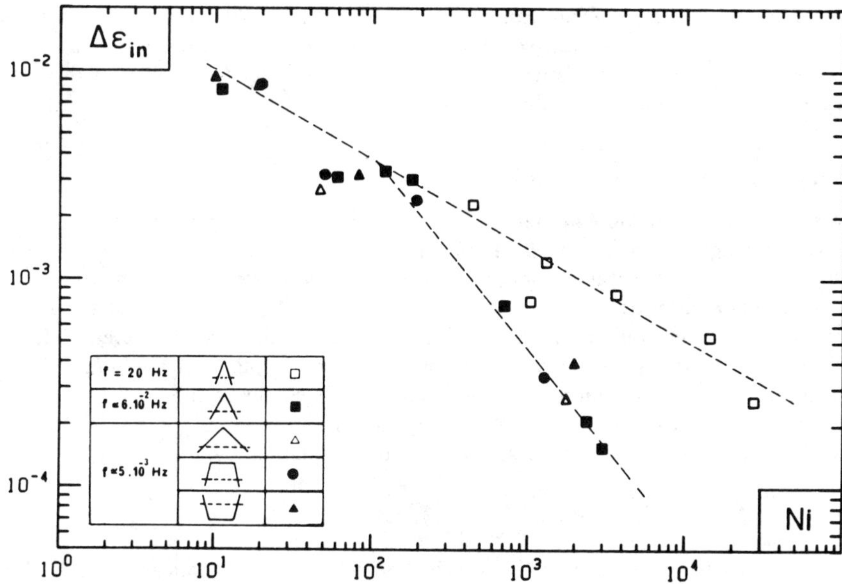

FIG. 2—*Variation of the number of cycles to initiation* (N_i) *with the inelastic strain range* ($\Delta\epsilon_{in}$) *at 900°C (1173 K).*

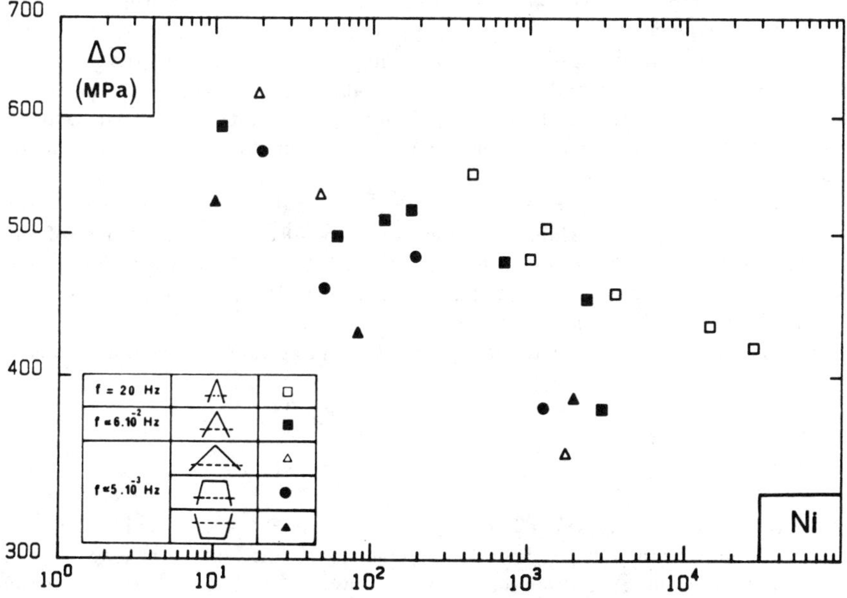

FIG. 3—*Variation of the number of cycles to initiation* (N_i) *with the peak stress range* ($\Delta\sigma$) *at 900°C (1173 K).*

sections of fatigue specimens using scanning electron microscopy [*3*]. These observations have shown the major role of preferential oxidation of primary carbides on crack initiation and early crack growth for the various waveshapes. These results are in good agreement with previous observations at an intermediate strain rate [*8*].

Strainrange Partitioning Analysis

In the strainrange partitioning method [*1*], the inelastic strain range is considered to be the relevant damage parameter, provided time-independent plastic strain is distinguished from time-dependent creep strain and damage in tension is assumed to be different from that in compression. Four basic strain-range components are so considered: $\Delta\epsilon_{pp}$ (plastic strain in tension reversed by plastic strain in compression), $\Delta\epsilon_{pc}$ (plastic strain in tension reversed by creep strain in compression), $\Delta\epsilon_{cp}$ (creep strain in tension reversed by plastic strain in compression) and $\Delta\epsilon_{cc}$ (creep strain in tension reversed by creep strain in compression).

Each pure basic type of strain range $\Delta\epsilon_{ij}$, where *i* and *j* are either "p" or "c," is related to a fatigue life N_{ij} according to the classical Manson-Coffin equation

$$N_{ij} = A_{ij}\Delta\epsilon_{ij}\alpha_{ij} \tag{1}$$

The strainrange partitioning method has been applied to the present data assuming the linear damage rule

$$1/N = \sum_{i,j} (1/N_{ij}) \tag{2}$$

to hold. Previous work at BBC on IN738 has shown this rule to give results as good as those of the interaction damage rule [*11*].

High-frequency tests (at 20 Hz) were assumed to be pure pp cycles according to Manson's procedure. Low-frequency sawtooth cycles were partitioned assuming the pp strain component to obey the same stress-strain relationship as in the high-frequency test, and the cc component was deduced by subtraction from the total inelastic strain range. The same procedure was used for the ramp in hold-time tests, and the cp or pc component corresponds to the inelastic strain variation during the tension or compression hold (that is the amount of stress relaxation divided by Young's modulus).

In usual procedures, the various N_{ij}-$\Delta\epsilon_{ij}$ relationships are determined step-by-step starting from the pure $\Delta\epsilon_{pp}$-N_{pp} relationship. The fatigue group of BBC Baden has developed a least-square fit computer program in order to get simultaneously the eight coefficients involved in the SRP analysis [*11*]. Many authors have actually reported calculation difficulties in the step-by-step identification of SRP coefficients due to fatigue data scatter.

The least square fit of experimental data gives the following values of the coefficients:

$A_{pp} = 4.7$ $\quad A_{pc} = 10$ $\quad A_{cp} = 10$ $\quad A_{cc} = 14.6$

$\alpha_{pp} = -2.48$ $\quad \alpha_{pc} = -3.97$ $\quad \alpha_{cp} = -4.13$ $\quad \alpha_{cc} = -1.12$

The component strain-range life relationships are shown in Fig. 4. Fatigue life of cp loops is found somewhat longer than that of pc loops, but much longer than that of pp loops. It is worth noting that the shorter life is observed for cc loops for number of cycles to initiation of practical interest. The cc component was also found to be the most damaging in a cast nickel-base superalloy IN738 tested at 850°C (1123 K) [*12*]. In the latter case, the nature of the strain on the

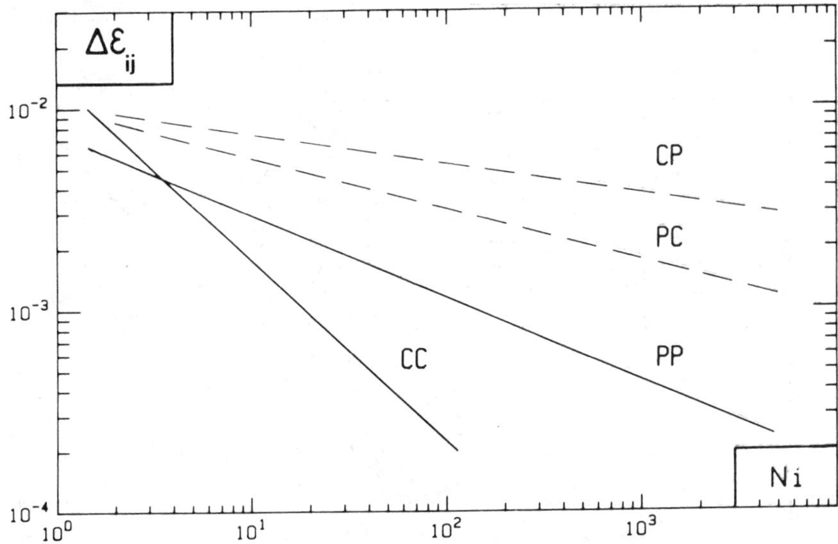

FIG. 4—*SRP analysis: number of cycles to initiation* (N_i) *as a function of component strain range* ($\Delta\epsilon_{ii}$) *at 900°C (1173 K).*

tensile stroke was predominant: $\Delta\epsilon_{cc}$ and $\Delta\epsilon_{cp}$ gave similar lives, while $\Delta\epsilon_{pp}$ and $\Delta\epsilon_{pc}$ gave (similar) much longer lives.

The life to crack initiation predicted from Eq 1 and Eq 2 is compared with the experimental life in Fig. 5. A good agreement between experimental and calculated values is observed within a factor of three. It must be emphasized, however, that this is a simple correlation of data and not a real prediction. The eight coefficients have been deduced from the whole set of data, and the prediction capability of the strainrange partitioning method should have to be tested over new cycle shapes.

Creep Damage

Two different methods based on creep damage have been used. The simplest one, which has been used for instance by Spera for thermal fatigue [*5*], is to estimate creep damage by the equation of Robinson and Taira [*13,14*]:

$$N_{f,S_p} = \left[\int_0^{\Delta t} \frac{1}{t_f(\sigma, T)}\, dt\right]^{-1} \tag{3}$$

where t_f is the time to rupture in creep experiments and Δt is the fatigue cycle period. In the following discussion, this method will be referred to as *Spera's accumulated creep damage method*. Danzer has shown, on nickel-base superalloys, that this method yields reasonable predictions, provided the strain rate in the fatigue experiment is not much higher than that in creep [*15*]. He has introduced the ratio r:

$$r = \Delta\epsilon_{in}/\Delta\epsilon_s \tag{4}$$

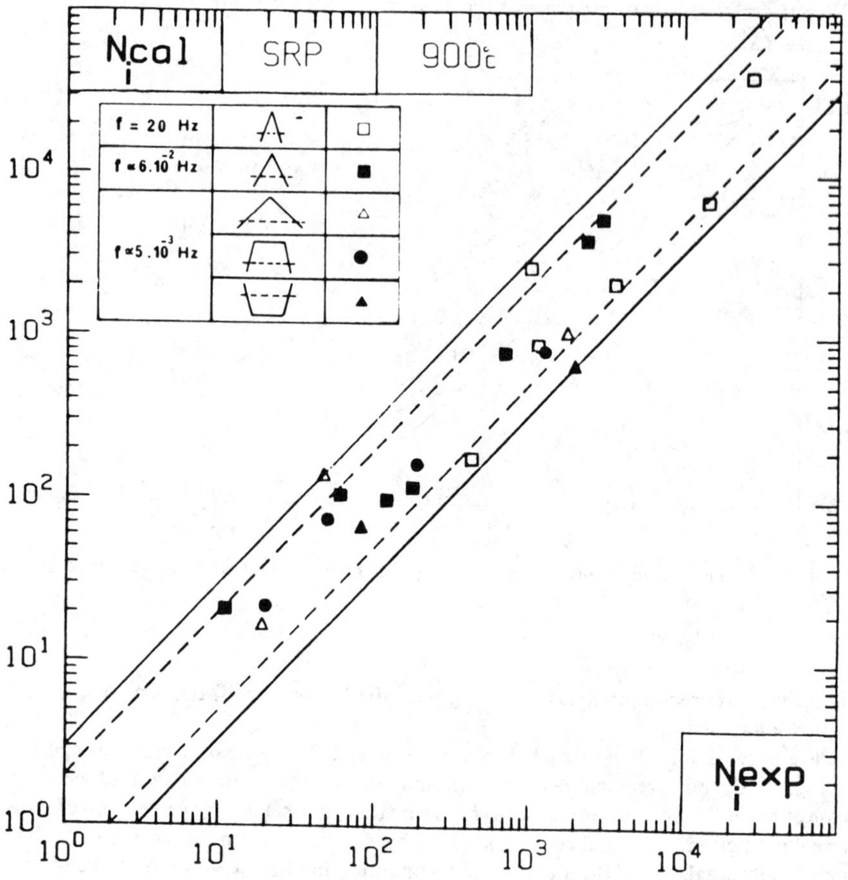

FIG. 5—*SRP analysis: comparison between calculated and experimental life to crack initiation at 900°C (1173 K).*

where

$$\Delta\epsilon_s = \int_0^{\Delta t} \dot{\epsilon}_s[\sigma(t),\ T(t)] \cdot dt$$

where $\dot{\epsilon}_s$ is the secondary creep rate. The applicability of Spera's method has been shown to be restricted to r values smaller than 10.

A second method based on creep damage has been proposed by Franklin to account for the differences between the instantaneous plastic strain rate in fatigue, $\dot{\epsilon}_p$, and the secondary creep rate, $\dot{\epsilon}_s$ [*6*]. He has introduced the concept of forced creep and derived the following equation for cyclic life:

$$N_{f.Fr} = \left[\int_0^{\Delta t} \frac{1}{t_f(\sigma,\ T)} \times \left(\frac{\dot{\epsilon}_p(\sigma,\ T)}{\dot{\epsilon}_s(\sigma,\ T)}\right)^{\nu} dt\right]^{-1} \quad (5)$$

where $\dot{\epsilon}_p$ is the instantaneous plastic strain rate and where ν is a constant. This equation will be referred to in this discussion as *Franklin's cyclic creep damage equation.* The correcting factor of the Robinson-Taira relationship so introduced reflects the experimental observation that the damage accumulation rate is higher in fatigue than in creep when the plastic strain rate is higher than the creep rate.

Stress rupture tests carried out by SNECMA in the temperature range of 800 to 1100°C (1073 to 1373 K) have been analyzed according to Spera's method and have been fitted by the equation

$$t_f(\sigma, T) = A\sigma^{-p} \exp(Q/RT) \tag{6}$$

where A, p, and Q/R are constants.

Application of Franklin's method and estimation of the r parameter require the secondary creep rate to be known. As the relevant information for all the stresses and temperatures was not available, the secondary creep rate was deduced from time to 1% strain as $\dot{\epsilon}_s \simeq 0.01/3\ t_{1\%}$. The creep rate so estimated was found to obey the Monkman-Grant rule:

$$t_f \times \dot{\epsilon}_s \simeq 0.1 \tag{7}$$

The parameter ν in Franklin's equation has been deduced from a fitting procedure to the high frequency experiments and it was found that $\nu = 0.5$.

Negative stresses were assumed not to be damaging in all the experiments. For all triangular waveshape cycles the calculation was straightforward. For compression hold-time cycles computations were carried out using the tension peak stress and the period in tension (a mean stress is induced by the compression hold). For tension hold-time cycles, the damage during the transient tension period and the damage during the relaxation period were calculated separately. In all cases, the damage accumulated during the transient period was higher than that during the strain hold.

Predicted lives according to Spera's and Franklin's methods are compared with experimental results in Figs. 6 and 7, respectively. It is worth emphasizing that Spera's method does not apply to all the experiments, which explains why some values are overestimated. This method can be applied to relatively high strain ranges (with high peak stresses), and then the prediction is relatively accurate.

The standard deviation s of the predicted values can be defined as [*11*]

$$s^2 = \sum_{i=1}^{n} (\log N_{\text{cal}}(\Delta\epsilon_i) - \log N_{\text{exp}}(\Delta\epsilon_i))^2/(n-1) \tag{8}$$

where N_{cal} and N_{exp} are the computed and experimental life for a strain range $\Delta\epsilon_i$, and n is the number of experimental values. For all the cases where the parameter r is smaller than 10, this standard deviation is $s = 0.24$. All intermediate and low strain rate sawtooth cycles and high strain range cycles with a strain-hold compressive or tensile are in this case. The r parameter is higher than 10 for strain-hold tests with inelastic strain ranges smaller or equal to 0.2% and for all high-frequency tests.

All experimental lives can be predicted by Franklin's method, within a factor of three with a standard deviation (s) of 0.36. However, in that case only high-frequency data are interpolated and all other experimental conditions are actual predictions from creep rupture and high-frequency tests. Spera's method is not so widely applicable, but it yields true predictions with no fitting parameter. It is worth emphasizing that these creep damage methods involving at most one fitting parameter yield a correlation almost as good as strainrange partitioning ($s = 0.26$).

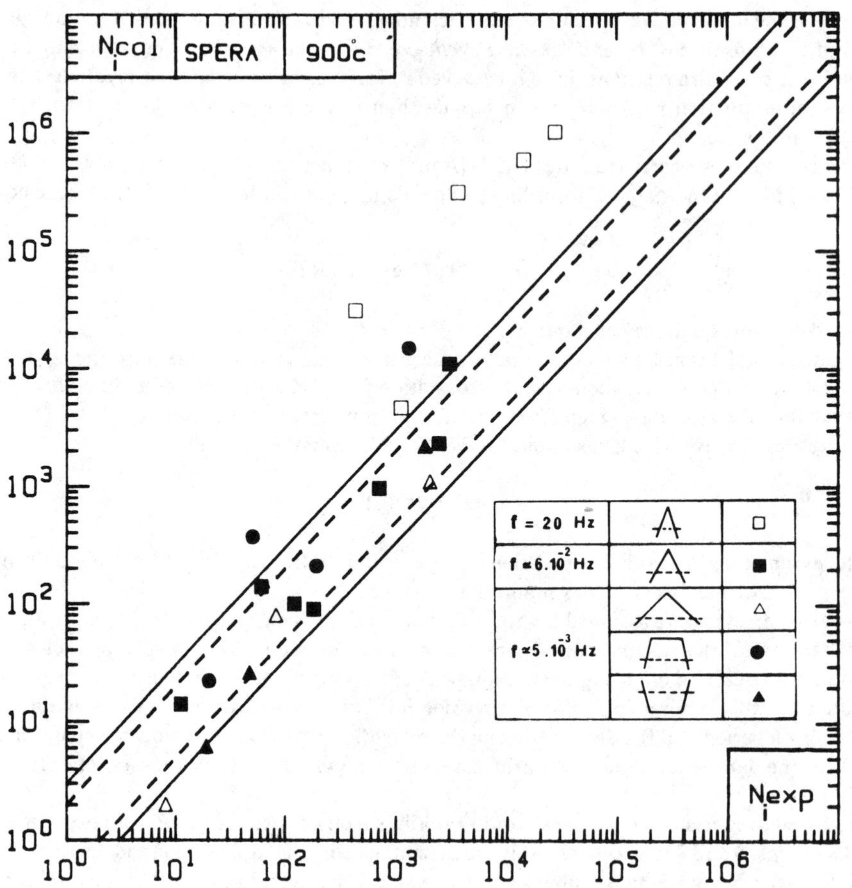

FIG. 6—*Spera's creep damage model: comparison between calculated and experimental life to crack initiation at 900°C (1173 K).*

But the latter method is only an interpolation of the whole fatigue data set and requires eight fitting parameters.

Oxidation-Fatigue Interaction Crack Growth Model

A study of the interaction between fatigue and oxidation was carried out by Reuchet and Rémy [*7*] on the present alloy at 900°C in air. The influence of fatigue cycling on oxidation of this alloy was studied by quantitative metallography on polished specimens exposed to air in a furnace and on strain-cycled low-cycle fatigue specimens. The oxidation kinetics was determined by thickness measurements for matrix oxidation and by oxidized depth measurements for the preferential oxidation of MC carbides. Matrix oxidation was found to obey a $t^{1/2}$ relationship and preferential oxidation along carbides a $t^{1/4}$ relationship. In both cases the oxidation kinetics was found to be enhanced by cycling, and the oxidation constant for the matrix increased linearly with inelastic strain amplitude [*7*]. The oxidation constant for carbides was found to be unaffected below a threshold inelastic strain; above, it is proportional to the inelastic strain amplitude [*16*].

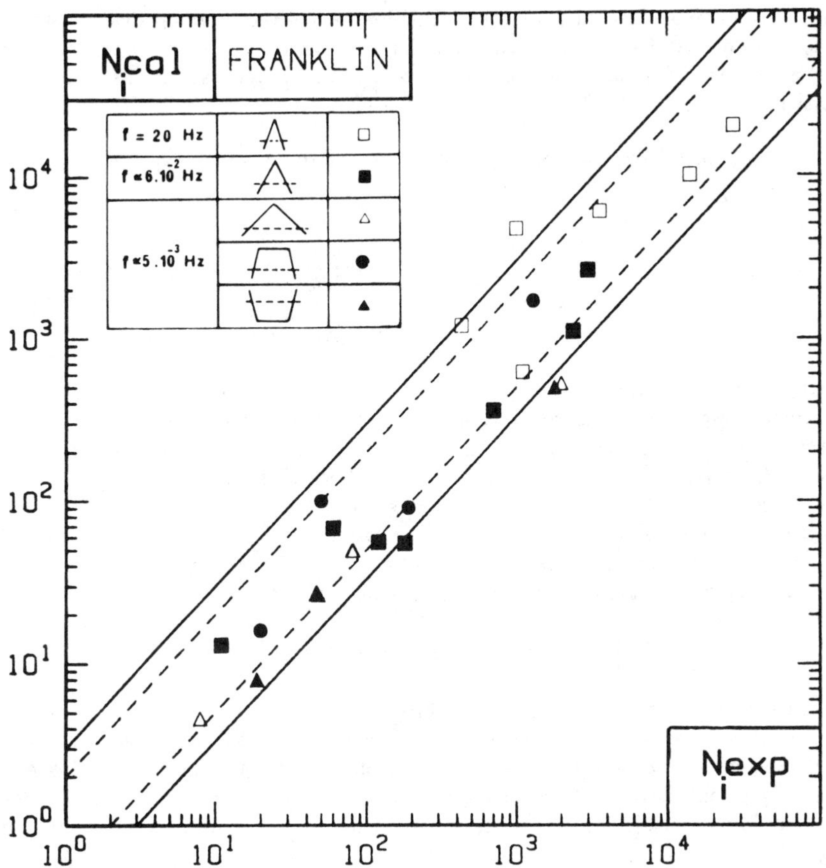

FIG. 7—*Franklin's cyclic creep damage model: comparison between calculated and experimental life to crack initiation at 900°C (1173 K).*

From these observations, a damage equation has been proposed which describes fatigue damage as a crack growth process. The elementary crack advance is assumed to result from both an advance due to crack opening under a cyclic stress, which is typical of fatigue damage, and from a supplementary crack advance due to oxidation at the crack tip. Thus the crack growth rate was derived as

$$da/dN = (da/dN)_{\text{fat}} + (da/dN)_{\text{ox}} \tag{9}$$

An approximate expression was given using Tomkin's model [*17*] for the fatigue contribution to crack advance:

$$da/dN = B \times a \tag{10}$$

where

$$B = 0.51\ \Delta\epsilon_{\text{in}}(1/\cos(\pi\sigma/2T) - 1)$$

where σ is the peak tensile stress and T the tensile fracture stress.

The contribution of crack advance due to oxidation was derived assuming that fracture of the oxide spike formed at the crack tip occurs at each tensile stroke. The oxidation rate at the crack tip was approximated to that observed at the outer surface of cycled specimens. The following expression was used for $(da/dN)_{ox}$:

$$(da/dN)_{ox} = (1 - f_c)\alpha_M(1 + K_M\Delta\epsilon_{in}) \times \Delta t^{1/2} + f_c\alpha_c g(\Delta\epsilon_{in})\Delta t^{1/4} \quad (11)$$

where

$$g(\Delta\epsilon_{in}) = \begin{cases} \Delta\epsilon_{in}/\Delta\epsilon_o & \text{if} \quad \Delta\epsilon_{in} > \Delta\epsilon_o \\ 1 & \text{if} \quad \Delta\epsilon_{in} \leq \Delta\epsilon_o \end{cases}$$

These two terms describe the oxidation contribution to crack growth rate due to the matrix and carbides, respectively. The values α_M and α_c are the oxidation constants of the matrix and of the carbides under zero stress; f_c is the effective volume fraction of carbides on the crack path (about 0.12); K_M and $\Delta\epsilon_o$ are two constants; and Δt is the cycle period.

The oxidation contribution to crack growth rate is constant for a given cycle shape and frequency. In the presence of oxidation, the crack growth rate is controlled by oxidation at short crack lengths and is therefore different from zero for $a = 0$. Therefore Eq 9 can be integrated for a crack length variation from 0 to a, and the fatigue life is given as

$$N_f = (1/B) \times \ell n[B \times a/(da/dN)_{ox} + 1] \quad (12)$$

where the constants are given by Eqs 10 and 11.

This equation involves five constants under isothermal conditions for the oxidation contribution, all of which are deduced from metallographic observations and the fracture tensile stress for the fatigue contribution, which is known from simple tensile tests. This equation does not involve any fit parameter, and it yields real predictions that are completely independent from low-cycle fatigue data.

Predictions of life to crack initiation (i.e., to 0.3 mm crack depth) have been made and are compared with experimental data in Fig. 8. Most predicted values are within a factor of three in agreement with experimental data. Predicted life to initiation is somewhat pessimistic for high-frequency results above 10^4 cycles. There are two possible reasons for this behavior: at low strains and high frequency, a significant initiation period might exist before microcracks start to grow; at low strains, the influence of oxidation might be overestimated by the present model.

This crack growth model can be applied to any crack length, which was not the case for creep-damage-based methods or strainrange partitioning. This model was used to compute total life that corresponds to a crack depth of about 2 mm. Predicted total life is compared in Fig. 9 with experimental results. Predicted and experimental values are in good agreement within a factor of two. The standard deviation s as defined by Eq 8 is 0.18 instead of 0.38 for initiation data. The larger standard deviation of initiation data could be partly due to the larger scatter of experimental data.

Application to Thermal Fatigue Life Predictions

Thermal fatigue lives have been defined to 1 mm crack length to allow comparison with conventional isothermal fatigue life. Life to crack initiation has also been defined to 0.3 mm crack depth as in isothermal tests. The experimental life to 0.3 mm and to 1 mm crack depth is reported in Fig. 10 as a function of total strain range computed at the thin edge of wedge specimens [9] and compared with the scatterband of initiation and rupture lives of low-frequency

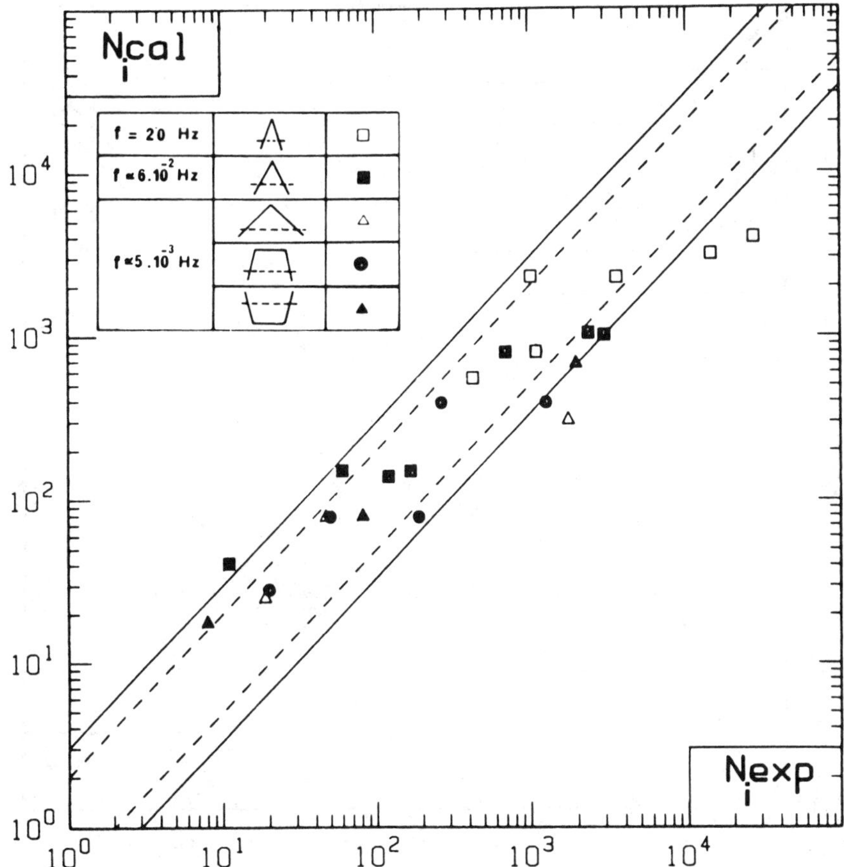

FIG. 8—*Fatigue oxidation damage model: comparison between calculated and experimental life to crack initiation at 900°C (1173 K).*

isothermal tests at 900°C. Life under thermal fatigue cycling is always much lower for a given total strain range than isothermal fatigue life at 900°C (1173 K). The same discrepancy is observed using fatigue data at 1100°C (1373 K) [*9*]. Therefore all criteria based on total strain amplitude are unable to predict the behavior under thermal fatigue. Use of the inelastic strain range does not show any improvement in the comparison between low-cycle and thermal fatigue data. Indeed, maximum and minimum stresses in thermal fatigue occurred at medium temperatures (say, 600 to 800°C [873 to 1073 K]) so that for a given mechanical total strain range, the inelastic strain range in a thermal fatigue cycle is smaller than that in an isothermal cycle at 900°C (since the yield strength is lower than at medium temperatures). The inelastic strain range is therefore rather small in thermal fatigue and even negligible for the specimen with the highest life studied [*3*]. The strainrange partitioning has not been studied, since it cannot be used for all thermal fatigue experiments. Only creep damage methods and oxidation damage methods have been applied to thermal fatigue data.

The temperature distribution and stresses and strains have been computed. The stress and inelastic strain rate variation as a function of time have been used by Leoben University for creep damage calculations. It must be said that, on heating, the thin edge of the TF specimen is

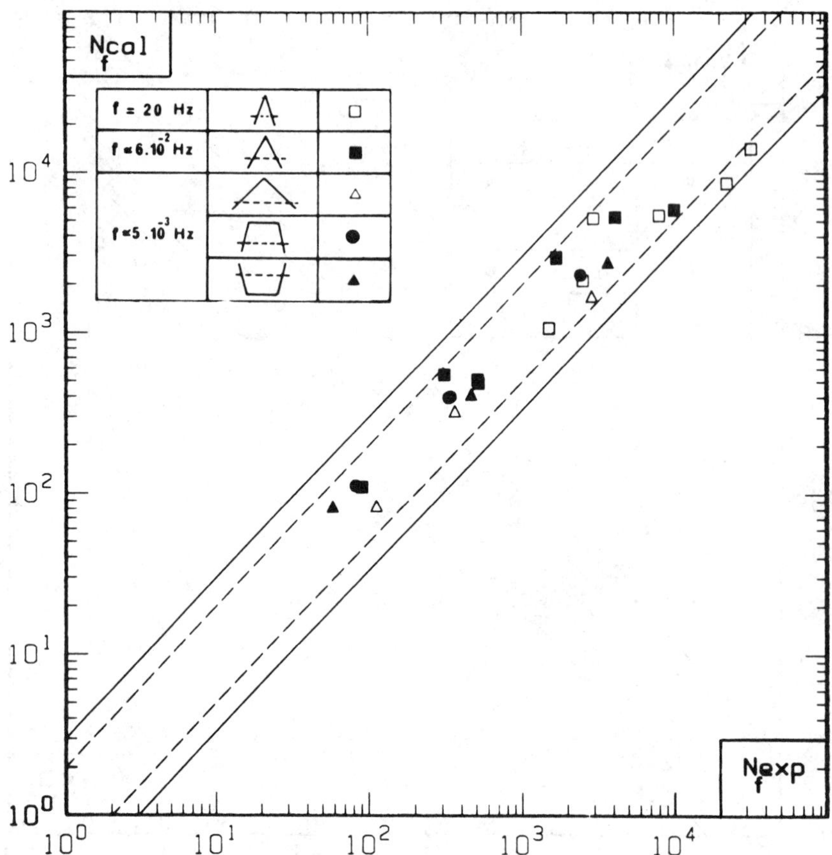

FIG. 9—*Fatigue oxidation damage model: comparison between calculated and experimental life to fracture at 900°C (1173 K).*

submitted to a compressive stress which relaxes at maximum temperatures but remains compressive. On cooling, the stress changes drastically from compressive to tensile to reach a maximum at intermediate temperatures, say, 500 to 700°C (i.e., below the creep range). Negative stresses have been assumed to be not damaging in isothermal fatigue. In the case of thermal fatigue the combination of high temperature and important tensile stress only occurs for a very short time (about 1 s) after the cooling has started. The calculation of the integrals in Eqs 3 and 5 has been performed numerically, but the results are very sensitive to the computations of transient stress and temperatures.

Predicted lives using either Spera's or Franklin's method are compared with experimental data in Fig. 11. The prediction is not so good as for isothermal data. Both methods give comparable predictions which overestimate actual thermal fatigue life to crack initiation by a factor of 3 to 5.

The differential equation proposed for oxidation fatigue damage has been applied to thermal fatigue results. In that case, only the tensile fracture stress has to be known and oxidation constants have to be measured from metallographic observations.

The oxidation kinetics under zero stress has been studied at various temperatures between 700 and 1100°C (973 and 1373 K). The contribution of oxidation to crack growth rate has been

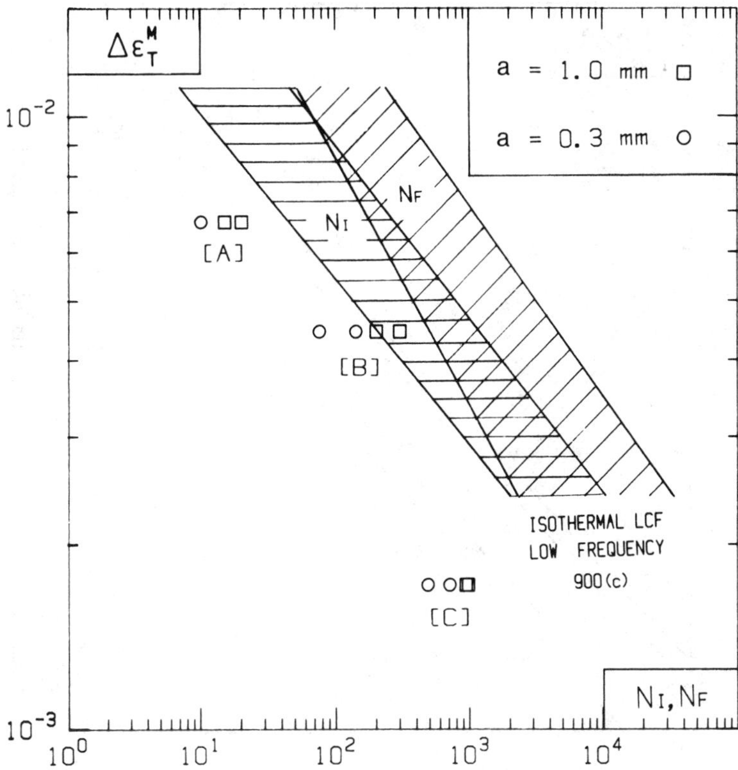

FIG. 10—*Variation of the number of cycles to various crack depth in thermal fatigue with the total mechanical strain range. Comparison with isothermal data at 900°C (1173 K).*

calculated by integration of the static oxidation equation over the entire temperature-time cycle. This gives equivalent average oxidation constants $\bar{\alpha}_M$ and $\bar{\alpha}_c$ to be used instead of isothermal α_M and α_c in equations. The interaction between fatigue and oxidation was accounted for by using the same values of K_M and $\Delta\epsilon_o$ as at 900°C, the maximum tensile stress, and the inelastic strain range of the stress-strain loop. The fatigue contribution was deduced from Tomkins's model applied to the whole stress-strain loop at the temperature of the peak tensile stress. It is worth noting that when the inelastic strain range becomes negligible, Eq 12 can be developed using an asymptotic expression of the logarithm so that it reduces to $N_f = a/(da/dN)_{ox}$.

The lifetime to 0.3 mm and to 1 mm crack depth has been so computed and is compared with actual life in Fig. 12. For the most severe test condition, the crack initiation period is largely overestimated, but this life range (less than 50 cycles) is of very little practical interest. On the other hand, for the other test conditions the life to crack initiation is correctly predicted within a factor of two. Therefore such a physically based model gives reliable predictions in the life range of real interest.

Conclusions

Life to crack initiation in high-temperature low-cycle fatigue of MAR-M509 at 900°C have been used to evaluate a number of life prediction models: Manson's strainrange partitioning,

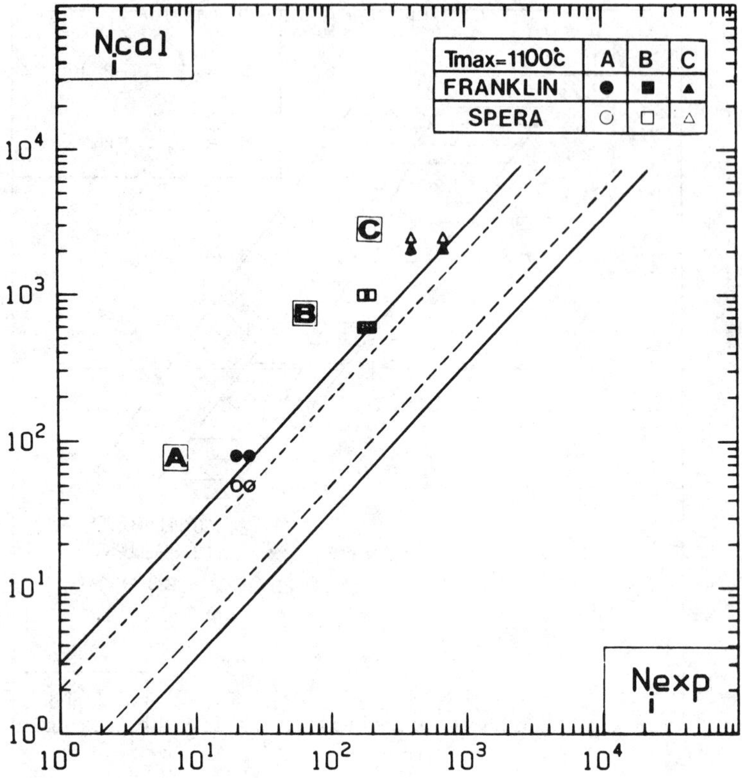

FIG. 11—*Creep damage models: comparison between experimental lives to crack initiation in thermal fatigue and predictions of Franklin's and Spera's models.*

Spera's accumulated creep damage model, Franklin's cyclic creep damage model, and an oxidation-fatigue interaction crack growth model proposed by Reuchet and Rémy.

All these models have been shown to give a reasonable correlation of these LCF data, except for some high-strain-rate tests in the case of Spera's method. However, this correlation involves eight fitting parameters for strainrange partitioning, one fit parameter for Franklin's method, which is fitted to high-frequency LCF data, and no fit parameter for Spera's method and the oxidation fatigue model.

The prediction capability of creep damage and oxidation damage models has been tested against thermal fatigue life data of wedge specimens using stress-strain computations. Oxidation models gave more reliable predictions than creep models.

Acknowledgments

The authors are indebted to the European Community COST 50 program organization and to their respective Ministries of Industry and/or Research for financial support of this work.

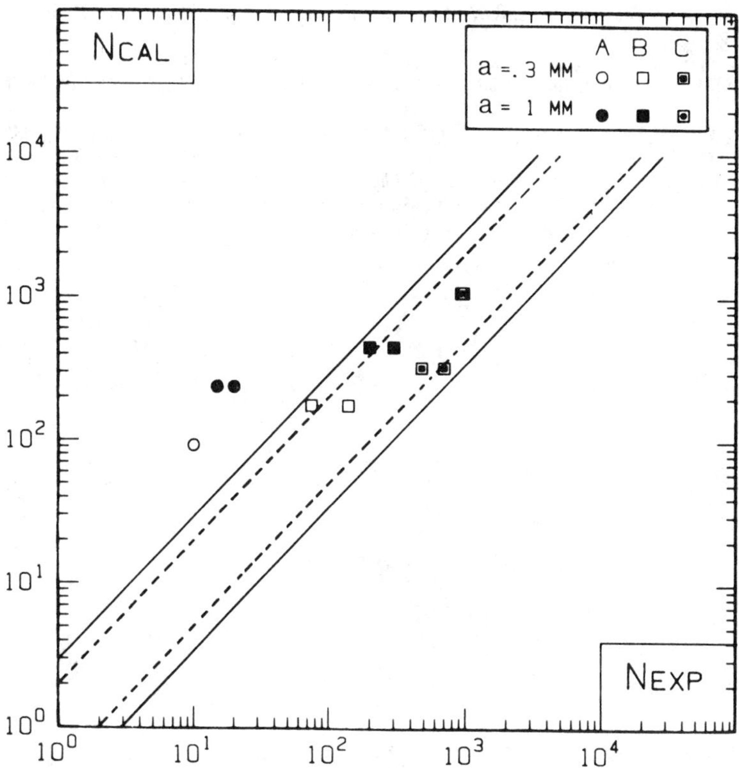

FIG. 12—*Fatigue oxidation damage model: comparison between experimental lives to 0.3 and 1 mm crack depth in thermal fatigue and predicted values.*

References

[1] Manson, S. S. in *Fatigue at Elevated Temperature, ASTM STP 520*, American Society for Testing and Materials, Philadelphia, 1983, pp. 744–782.

[2] Coffin, L. F. in *Fatigue at Elevated Temperatures, ASTM STP 520*, American Society for Testing and Materials, Philadelphia, 1973, pp. 5–34.

[3] Rémy, L., Rezai-Aria, F., François, M., Herman, C., Dambrine, B., and Honnorat, Y., "Application of Isothermal Fatigue to the Study of Thermal Fatigue," Final Report F6, COST 50-Round III, 1984.

[4] Rémy, L., Réger, M., Reuchet, J. and Rezai-Aria, F. in *Proceedings*, Conference on High Temperature Alloys for Gas Turbines, Liège, 4–6 October 1982, R. Brunetaud, D. Coutsouradis, T. B. Gibbons, Y. Lindblom, D. B. Meadowcroft, and R. Stickler, Eds., Reidel, Dordrecht, 1982, pp. 619–632.

[5] Spera, D. A. in *Fatigue at Elevated Temperature, ASTM STP 520*, American Society for Testing and Materials, Philadelphia, 1973, pp. 648–657.

[6] Franklin, C. J. in *High Temperature Alloys for Gas Turbines*, D. Coutsouradis, P. Felix, H. Fischmeister, L. Habraken, Y. Lindblom, and M. O. Speidel, Eds., Applied Science, London, 1978, pp. 513–547.

[7] Reuchet, J. and Rémy, L., *Metallurgical Transactions A*, Vol. 14A, 1983, pp. 141–149.

[8] Reuchet, J. and Rémy, L., *Materials Science and Engineering*, Vol. 58, 1983, pp. 19–32 and 33–42.

[9] Rezai-Aria, F., Rémy, L., Herman, C., and Dambrine, B. in *Mechanical Behaviour of Materials, Part IV*, Vol. 1, J. Carlson and N. G. Ohlson, Eds., Pergamon Press, Oxford, 1984, pp. 247–253.

[10] Chaboche, J. L., *Bulletin de l'Academie Polonaise des Sciences*, Vol. 25, 1977, pp. 33–42.

[*11*] Hoffelner, W., Melton, K. N., and Wüthrich, C., *Fatigue of Engineering Materials and Structures*, Vol. 6, 1983, pp. 77-87.
[*12*] Nazmy, M. Y., *Metallurgical Transactions A*, Vol. 14A, 1983, pp. 449-461.
[*13*] Robinson, E. L., *Transactions of the American Society of Mechanical Engineers*, Vol. 74, 1952, pp. 777-780.
[*14*] Taira, S. in *Creep in Structures*, N. J. Hoff, Ed., Springer Verlag, Berlin, 1962, pp. 96-119.
[*15*] Danzer, R., *Res. Mechanica*, Vol. 12, 1984, pp. 259-273; Vol. 13, 1985, pp. 63-76.
[*16*] Réger, M., François, M., and Rémy, L. in *Physical Chemistry of the Solid State: Application to Metals and Their Compounds*, P. Lacombe, Ed., Elsevier, Amsterdam, 1984, pp. 413-420.
[*17*] Tomkins, B., *Philosophical Magazine*, Vol. 18, 1968, pp. 1041-1066.

Z. Duan,[1] *J. He,*[1] *Y. Ning,*[1] *and Z. Dong*[1]

Strain Energy Partitioning Approach and Its Application to Low Cycle Fatigue Life Prediction for Some Heat-Resistant Alloys

REFERENCE: Duan, Z., He, J., Ning, Y., and Dong, Z., **"Strain Energy Partitioning Approach and Its Application to Low Cycle Fatigue Life Prediction for Some Heat-Resistant Alloys,"** *Low Cycle Fatigue, ASTM STP 942,* H. D. Solomon, G. R. Halford, L. R. Kaisand, and B. N. Leis, Eds., American Society for Testing and Materials, Philadelphia, 1988, pp. 1133–1143.

ABSTRACT: An improved life prediction approach of time-dependent fatigue, strain energy partitioning (SEP), was verified with three alloys of different strengths. SEP synthesizes the energy concept of the Ostergren model and the thought of inelastic strain partition of the SRP model, and, thus, the partitioned strain energy component (U_{ij}) is the single independent variable in SEP. The results show that the life predictability of SEP to all three alloys appears to be the most promising in comparison with those of strainrange partitioning (SRP) and frequency separation (FS) methods. It seems that the higher the strength level of the tested material, the better the predictability of SEP over that of SRP and FS.

KEY WORDS: strain energy partitioning (SEP), strainrange partitioning (SRP), frequency separation (FS), Ostergren Model, time-dependent fatigue

It is well known that high-temperature low-cycle fatigue has been considered as a limiting factor of the life of turbine engine disk since the 1960s. The testing technology and life prediction methods of time-dependent fatigue of materials recently tend to simulate their service conditions. The cycles of start-run-stop make the disk material suffer two different damages: fatigue from cyclic strain of start and stop and creep damage from the action of peak sustained loading. The interaction of creep and fatigue, sometimes with environmental attack, is called *time-dependent fatigue,* the component life of which has been proved to be much lower than that of pure fatigue or pure creep. For this reason, the study of time-dependent fatigue life prediction methods for high-temperature alloys, especially those for turbine disks, has currently become an important subject in the field of aeroengine research.

Since the early 1970s, a number of time-dependent fatigue life prediction methods have been proposed, such as strainrange partitioning (SRP) [*1*], frequency separation (FS) [*2*], damage function method (Ostergren model) [*3*], damage rate model (DRM) [*4*], and others. Although each of these approaches has its limitations, the first two, SRP and FS, are the most popular; the former has been verified on about 20 materials in different laboratories in many countries. In order to make SRP appropriate to high-strength and low-ductility materials, an improved method, strain energy partitioning (SEP), was developed by He and his co-workers [*5*]. The present paper focuses on verifying the prediction of LCF lives by SEP in comparison with those by SRP and FS for several turbine engine materials used for casing (1Cr-18Ni-9Ti) and disks (GH36, GH33A).

[1]Institute of Aeronautical Materials, Beijing, China.

Materials and Experimental Procedure

The chemical compositions, heat treatments, and mechanical properties of the materials used in the present investigation are shown in Tables 1, 2, and 3, respectively.

The time-dependent fatigue tests were conducted in our laboratory using a 50 kN MTS servo-controlled electrohydraulic testing system with a PDP 11/04 computer. Specimens 6.35 mm in diameter and 25 mm in gage length were used in this investigation. The axial strain of specimens was measured by an extensometer which was connected with two quartz bars through the furnace wall, attached closely to both ends of specimen gage length by two springs. The tests were conducted in air; the test temperatures were 600°C for 1Cr-18Ni-9Ti, 650°C for GH36, and 700°C for GH33A.

The test procedure recommended in Refs *5* and *6* was adopted. Baseline tests were performed according to the following parameters of the loading waveforms: pp-20 cpm for 1Cr-18Ni-9Ti and pp-30 cpm for all other alloys, cp-1 min/0, pc-0/1 min, and cc-1 min/1 min. The predic-

TABLE 1—*Chemical composition (weight percent).*

Materials	Al	Si	C	Ti	Cr	Fe	Cu	Mn	Nb	Ni
1Cr-18Ni-9Ti	...	1.50	0.12	...	18.00	Balance	...	...	...	9.20
GH36	...	0.61	0.37	...	12.00	Balance	...	8.22	0.46	8.06
GH33A bar	0.89	0.12	0.04	2.85	20.90	...	0.07	0.10	...	bal.
GH33A disk, arc-furnace	0.93	...	0.04	2.88	21.27	...	...	...	1.42	bal.
GH33A disk, duo-vacuo	0.97	0.35	0.05	2.84	20.90	...	0.01	0.13	1.45	bal.

TABLE 2—*Heat treatment of alloys.*

Materials	Heat Treatment
1Cr-18Ni-9Ti	quenched at 1100°C, water cooled
GH36	quenched at 1140°C for 80 min, water cooled; aged at 670°C for 12 h + 790°C for 14 h, air cooled
GH33A	1080°C for 8 h, air cooled; 750°C for 16 h, air cooled

TABLE 3—*Mechanical properties at room temperature.*

Materials	0.2% Yield, MPa	UTS, MPa	Reduction of Area, %
1Cr-18Ni-9Ti	274	608	63
GH36	807	1039	40
GH33A bar	770	1191	26
GH33A disk, arc-furnace	895	1216	34
GH33A disk, duo-vacuo	774	1196	21

tion relationships were established from these test results. The verification tests were conducted according to more complex and different waveforms (Fig. 1) in order to verify the predictability of the above life prediction relationships. In the verification tests the Fast/Slow (F/S) cycles consist of unsymmetrically continuous cycles (i.e., the tension-going frequency is 30 cpm and the compression-going frequency is 1 cpm). The term, S/F, represents reversed cycles of F/S. In the test of cp, the creep strain was imposed by controlling the tensile strain at 90% peak value and holding it for 1 min before the peak strain, then reaching compressive peak strain rapidly (cp, 90%, 1/0); for the test of cc, both compressive and tensile strain were held for 1 min at 90% peak value (cc, 90%, 1/1). For the pc-type test, the waveform was the reverse of cp (pc, 90%, 0/1).

Strain Energy Partitioning (SEP)

Strain energy partitioning is the synthesis of SRP and the Ostergren model. It has been considered in SEP that neither the partitioned strainrange nor the whole inelastic strain energy (so-

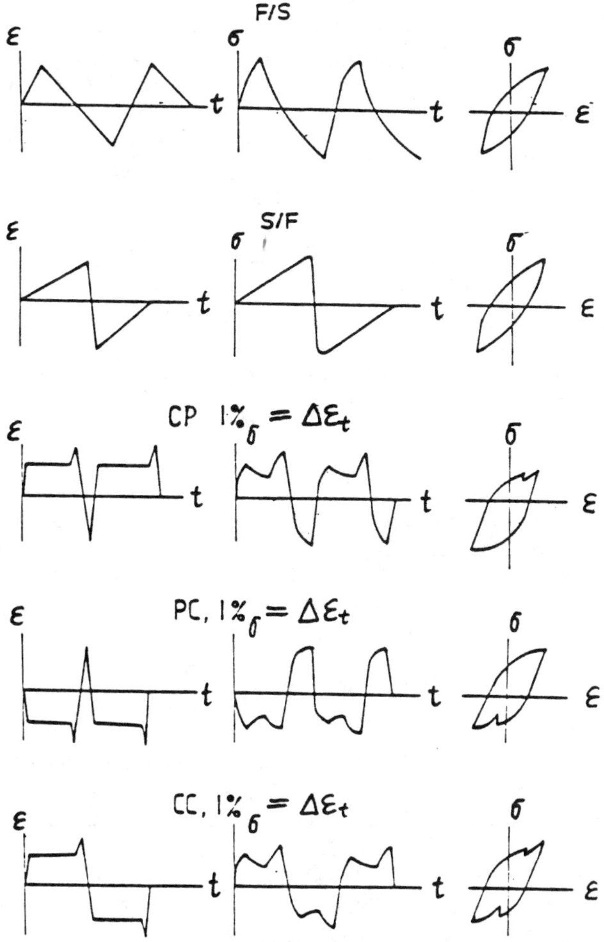

FIG. 1—*Waveforms of verification test.*

called damage function in the Ostergren model) could be the single variable in time-dependent fatigue life prediction methods. The right variable is the partitioned inelastic strain energy (U_{ij}) involving a plastic strain energy component and a creep strain energy component. The different strain energy components result in different damages and different lives in time-dependent fatigue and, thus, lead to different relationships in SEP.

The relationship in Eq 1 has been established in the baseline pp test cycle:

$$N_{pp} = A_1(\sigma_t \cdot \Delta\epsilon_{pp})^{B_1} \tag{1}$$

Similarly, three relationships for pc, cp, and cc are

$$N_{pc} = A_2(\sigma_t \cdot \Delta\epsilon_{pc})^{B_2} \tag{2}$$

$$N_{cp} = A_3(\sigma_t \cdot \Delta\epsilon_{cp})^{B_3} \tag{3}$$

$$N_{cc} = A_4(\sigma_t \cdot \Delta\epsilon_{cc})^{B_4} \tag{4}$$

where σ_t are tensile peak stresses, Ai and Bi are material constants. N_{pc}, N_{cp} and N_{cc} may be derived with a linear damage rule (LDR) or an interaction damage rule (IDR). The former was adopted in the present investigation, and the following relationships were obtained:

$$N_{pc} = (1/N_f - 1/N_{pp})^{-1} \tag{5}$$

$$N_{cp} = (1/N_f - 1/N_{pp})^{-1} \tag{6}$$

$$N_{cc} = (1/N_f - 1/N_{pp} - 1/N_{pc}(\text{or } 1/N_{cp}))^{-1} \tag{7}$$

$$N_{pre} = (1/N_{pp} + 1/N_{pc}(\text{or } 1/N_{cp}) + 1/N_{cc})^{-1} \tag{8}$$

where N_f is the observed life for pc, cp, and cc cycles, respectively, and N_{pre} is the predicted life. N_{pp} is derived from Eq 1.

Results and Discussion

In order to compare the predictability of SEP with that of SRP and FS, the life prediction equations of SRP and FS models are given as

$$N_{ij} = A_i(\Delta\epsilon_{ij})^{B_i} \tag{9}$$

$$N_f = A(\Delta\epsilon_{in})^B V_t^M \left(\frac{V_c}{V_t}\right)^K \tag{10}$$

where $\Delta\epsilon_{in}$ is total inelastic strain range; V_t and V_c are tensile and compressive-going frequencies, respectively; and A, B, M, and K are material constants.

The constants of the life predictive equations of the above three models are given in Tables 4 and 5.

The constants were determined using a least squares curve fit technique, and the relationships were obtained at the same time. The dependence of the strain energy component of the SEP model, U_{ij}, on life, N_{ij}, for different alloys is shown in Figs. 2 to 4.

The predictive ability of a model was generally assessed using the following criteria [*6*]: (1)

TABLE 4—*Coefficients and exponents of SEP and SRP life prediction equations.*

Materials	Constants	Cycle Type of Baseline Test							
		pp		pc		cp		cc	
		SEP	SRP	SEP	SRP	SEP	SRP	SEP	SRP
1Cr-18Ni-9Ti bar at 600°C	*A*	87.940	0.148	27.066	0.515	6.858	0.226	30.414	1×10^{-3}
	B	−1.485	−1.708	−1.046	−1.077	−1.227	−1.147	−0.977	−1.873
GH36 bar at 650°C	*A*	129.446	2.133	125.896	0.144	73.156	0.883	105.022	3.835
	B	−0.974	−1.032	−0.830	−1.202	−0.714	−0.924	−0.692	−0.773
GH33A bar at 700°C	*A*	84.693	0.031	10.943	2×10^{-3}	0.806	8×10^{-6}	7.123	1×10^{-4}
	B	−1.454	−1.771	−1.296	−1.634	−1.715	−2.273	−1.569	−2.170
GH33A disk, arc-furnace, at 700°C	*A*	33.321	0.012	2.047	3×10^{-4}	4.232	8×10^{-4}	2.762	1×10^{-5}
	B	−1.465	−1.730	−1.581	−1.813	−1.329	−1.704	−2.012	−2.490
GH33A disk, duo-vacuo, at 700°C	*A*	168.749	0.951	1.106	1×10^{-4}	1.170	6×10^{-5}	2.480	4×10^{-5}
	B	−1.081	−1.194	−1.690	−1.881	−1.674	−2.028	−1.811	−2.249

TABLE 5—*Coefficients and exponents of FS life prediction equations.*

Materials	Temperature, °C	*A*	*B*	*M*	*K*
1Cr-18Ni-9Ti	600	0.107	−1.519	0.226	0.094
GH36	650	7.146	−0.758	0.206	0.057
GH33A bar	700	0.027	−1.654	0.150	0.030
GH33A disk, arc-furnace	700	0.001	−1.675	0.654	0.391
GH33A disk, duo-vacuo	700	0.003	−1.600	0.652	0.328

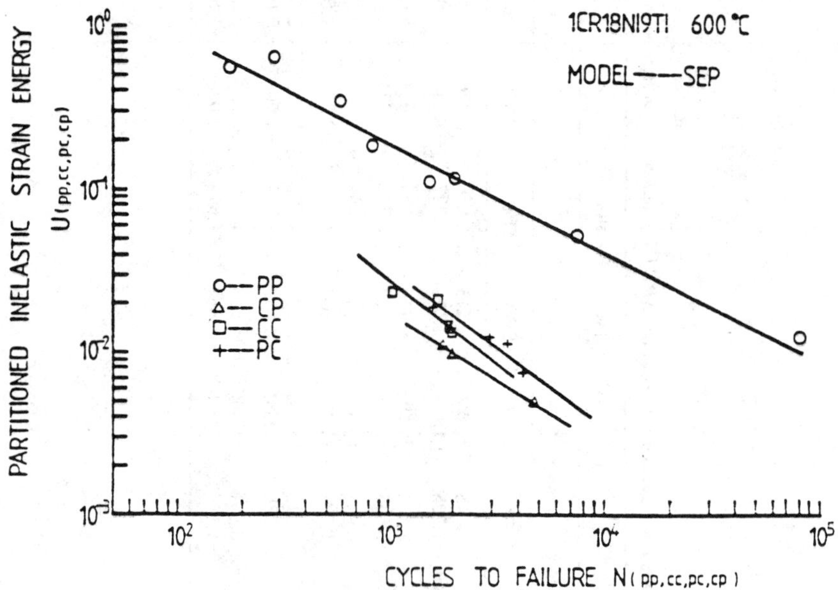

FIG. 2—*SEP life relationships for a 1Cr-18Ni-9Ti bar at 600°C.*

correlation of the baseline data, and (2) prediction of the vertification data. The ability for a model to correlate the baseline data was determined in two different ways: (1) the ratios of the predicted versus observed lives, and (2) the standard deviations between the predicted and observed lives. Generally, the ratios are limited within a factor of two. Obviously, the smaller the ratio and the standard deviation of an alloy, the better the correlation and the predictability of a model.

In Figs. 5 and 6, it can be seen that SEP method successfully correlated the time-dependent fatigue lives of these alloys (i.e., the ratios between observed and predicted lives are not greater than two). However, some ratios obtained with the SRP method are above a factor of two (Figs. 7 and 8).

Table 3 shows that, for the three alloys tested, the strength increases and the ductility decreases in the order 1Cr-18Ni-9Ti, GH36, and GH33A. The maximum scatter bands and total standard deviations for these alloys obtained with the SEP, SRP, Ostergren, and FS models are given in Table 6. For 1Cr-18Ni-9Ti stainless steel with low strength and high ductility, there is little difference in scatter band standard deviation between SEP and SRP; results are nearly the

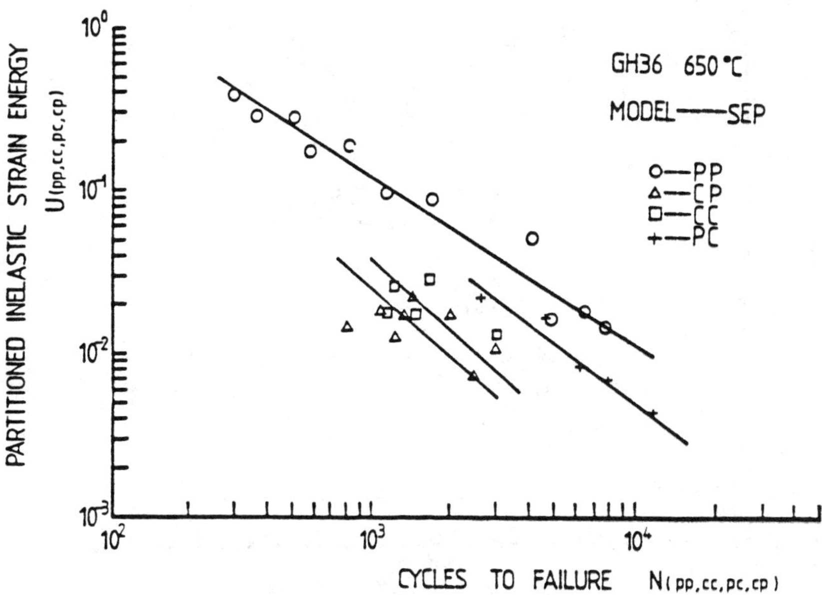

FIG. 3—*SEP life relationships for a GH36 bar at 650°C.*

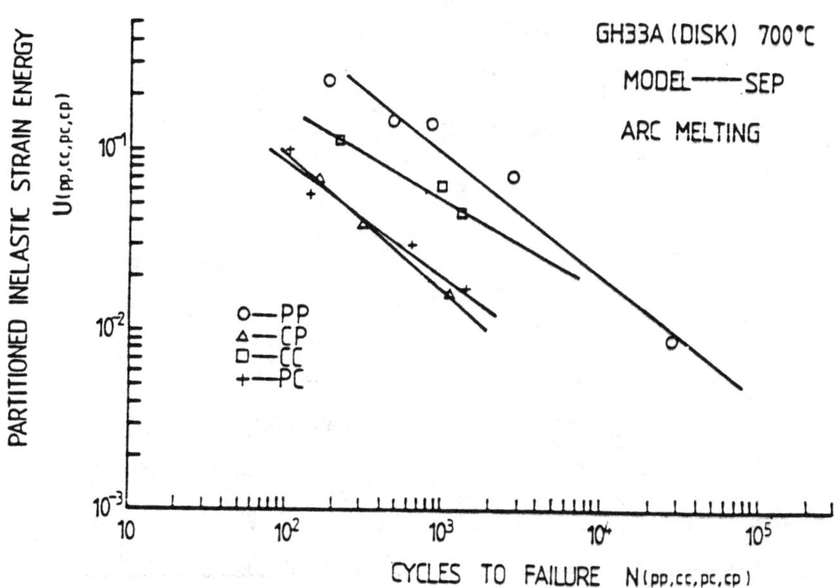

FIG. 4—*SEP life relationships for a GH33A forging disk by arc furnace melting, at 700°C.*

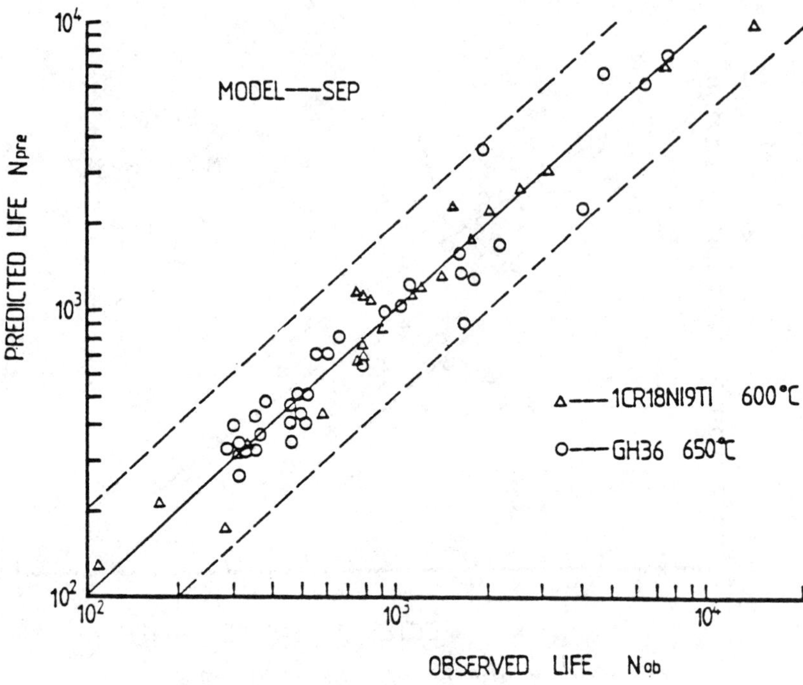

FIG. 5—*Observed versus predicted lives for 1Cr-18Ni-9Ti and GH36 by the SEP model.*

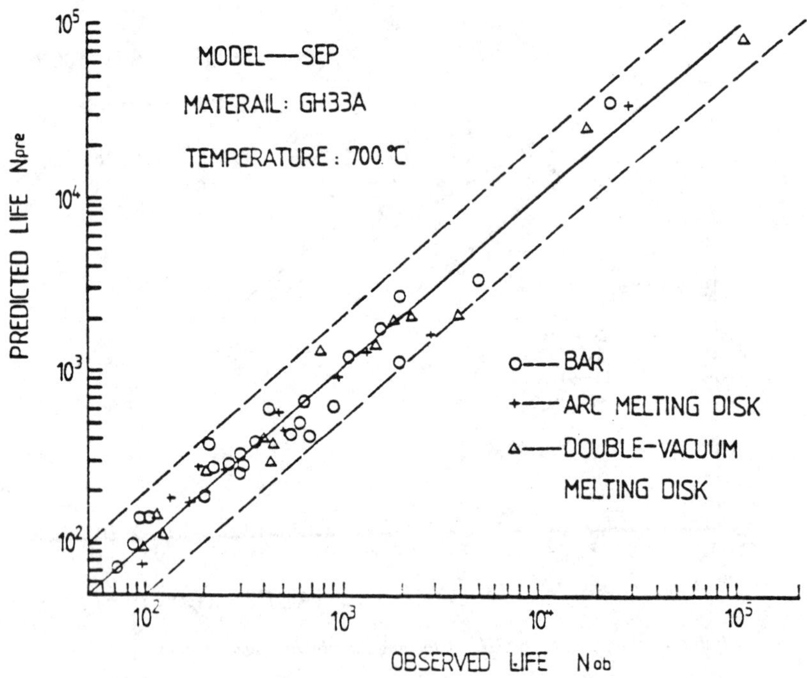

FIG. 6—*Observed versus predicted lives for GH33A by the SEP model.*

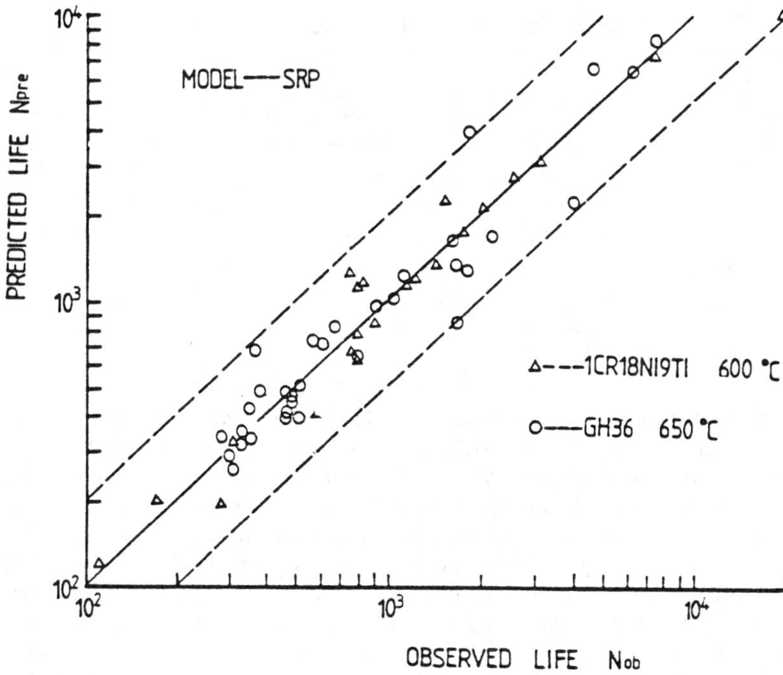

FIG. 7—*Observed versus predicted lives for 1Cr-18Ni-9Ti and GH36 by the SRP model.*

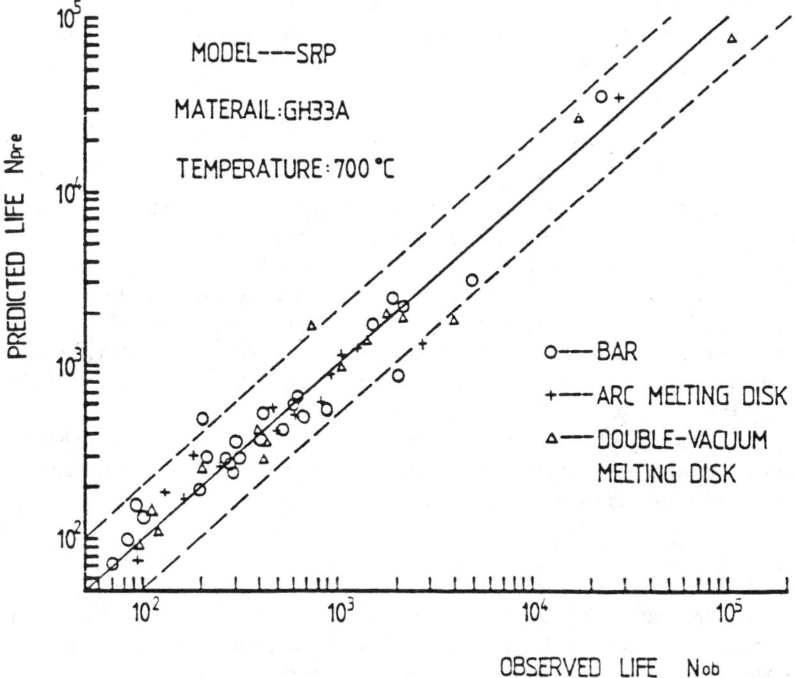

FIG. 8—*Observed versus predicted lives for GH33A by the SRP model.*

TABLE 6—*Life predictabilities of SEP, SRP, and FS.*

Materials	Temperature, °C	Scatter Band				Standard Deviation		
		SEP	SRP	Os.M.	FS	SEP	SRP	FS
1Cr-18Ni-9Ti	600	1.60	1.70	2.15	2.05	0.088	0.083	0.150
GH36	650	2.00	2.08	...	2.39	0.113	0.116	0.182
GH33A bar	700	1.95	2.38	2.44	2.90	0.121	0.141	0.241
GH33A disk, arc-furnace	700	1.70	2.04	1.75	1.88	0.107	0.127	0.142
GH33A disk, duo-vacuo	700	1.97	2.12	2.25	2.40	0.132	0.151	0.214

same for GH36 with intermediate strength and ductility. However, for higher-strength and lower-ductility nickel-base superalloy GH33A, all the scatter bands obtained by SEP are within a factor of two, while some data obtained by SRP and the Ostergren model are outside this range. The corresponding values predicted by FS exhibit the greatest scattering among these four models. With increase in strength and decrease in ductility, the advantage of SEP over SRP, the Ostergren model, and FS becomes more obvious. The reason that SEP is rather promising for higher-strength and lower-ductility alloys is not yet very clear. SEP is based upon the premise that under high strain cyclic loading the work of the external forces partially transforms into heat energy, dissipates, and partially is stored within the specimen in the form of strain energy which in turn is measured from the area of stress-strain hysteresis loop. The time-dependent fatigue life is mainly spent in growing cracks to a critical size. Only the tensile hysteresis energy, which is the summation of partitioned plastic strain energy and creep strain energy, contributes to small crack growth and, thus, to time-dependent fatigue damage. Therefore the partitioned tensile inelastic energy (U_{ij}), which is the product of maximum tensile stress and responded strain (plastic or creep), is used as the single independent variable of SEP.

Conclusions

1. The strain energy partitioning (SEP) approach is a synthesis and improvement of the SRP and Ostergren models. The partitioned tensile strain energy (U_{ij}) is considered as a single independent variable for crack initiation life of time-dependent fatigue in SEP.
2. In comparison with the SRP, Ostergren, and FS models, the SEP method appears to be best for predicting fatigue life. Only with SEP were the scatter bands of all materials used in this investigation totally within a factor of two and with smaller standard deviations.
3. The lowest fatigue lives of 1Cr-18Ni-9Ti stainless steel at 600°C and GH36 iron-base alloy at 650°C are cp-type cycles; the lowest fatigue life of GH33A nickel-base superalloy at 700°C is either a cp- or pc-type cycle.
4. The stress variable plays a more important role in predicting the life of higher-strength and lower-ductility materials than lower-strength and higher-ductility ones. Therefore SEP has a wider range of applicability than SRP or FS.

Acknowledgments

The authors are grateful for the helpful discussion given by Professor Yan Minggao and Dr. Han Ya Fang. Thanks are also due Yan Shimei, Feng Chuan, and Zhang Chen for their assistance in performing the tests and preparing this manuscript.

References

[*1*] Manson, S. S., "Creep Fatigue by Strainrange Partitioning," NASA TMX-67838, 1971.

[*2*] Coffin, L. F., "The Concept of Frequency Separation in Life Prediction for Time-Dependent Fatigue," 1976 ASME-MPC-3, Dec. 1976, pp. 349-364.

[*3*] Ostergren, W. J., "Correlation of Hold-Time Effects in Elevated Temperature, Low Cycle Fatigue Using a Frequency Modified Damage Function," in Ref *2*, pp. 179-202.

[*4*] Majumdar, S. and Maiya, P. S., "A Mechanistic Model for Time-Dependent Fatigue," *Journal of Engineering Materials and Technology, Transactions of ASME,* Vol. 102, No. 1, Jan. 1980, pp. 159-167.

[*5*] He, J., Duan, Z., Ning, Y., and Zhao, D., "Strain Energy Partitioning and Its Application to GH33A Nickel-Base Superalloy and 1Cr-18Ni-9Ti Stainless Steel," in *Proceedings,* ASME International Conference on Advances in Life Prediction Methods, New York, April 1983, pp. 27-32.

[*6*] Bernstein, H. L., "An Evaluation of Four Current Models to Predict the Creep Fatigue Interaction in René 95," AFML TR-79-4075, 1979.

S. G. Lee[1] *and R. I. Stephens*[2]

Fatigue Life Estimations for Cast Steels Using Low Cycle Fatigue Behavior

REFERENCE: Lee, S. G. and Stephens, R. I., "**Fatigue Life Estimations for Cast Steels Using Low Cycle Fatigue Behavior,**" *Low Cycle Fatigue, ASTM STP 942,* H. D. Solomon, G. R. Halford, L. R. Kaisand, and B. N. Leis, Eds., American Society for Testing and Materials, Philadelphia, 1988, pp. 1144–1162.

ABSTRACT: Fatigue notch factors for compact keyhole specimens of five common cast steels were estimated using an elastic stress distribution around the notches and the fatigue tests of simple specimens. The values were used with modified Neuber's rule to calculate the stress and strain at an equivalent characteristic length under variable loadings. Another method, which uses the load-strain relation obtained from elastic-plastic finite element method (FEM) analysis, was also used to calculate the strain at the equivalent characteristic length. Both methods were used to calculate the fatigue damage at the keyhole notch using the Palmgren-Miner linear damage rule and low cycle fatigue properties obtained from smooth specimens. The results showed that both methods could be used for the crack initiation calculation within a reasonable accuracy.

KEY WORDS: fatigue, life calculations, notch strain analysis, Neuber's rule, notches, spectrum loading, elastic/plastic, finite element methods, characteristic length, cumulative damage, cast steels

Nomenclature

A	Neuber's material constant
B	Thickness
b	Fatigue strength exponent
C	Stress or strain gradient constant
c	Fatigue ductility exponent
D	Damage
D_B	Damage per block
E	Modulus of elasticity
e	Nominal strain
K_t	Elastic stress concentration factor
K_E	Linear constant of the load-strain curve
K_σ	Stress or equivalent stress concentration factor
K_ϵ	Strain or equivalent strain concentration factor
K'	Cyclic strength coefficient
K_p'	Cyclic strength coefficient of the load-strain curve
K_t'	Elastic stress concentration factor for the simple notch
L	Distance from loading line to the notch tip
N_f, N_i	Cycles to failure, cycles to failure at the ith amplitude

[1]Senior Researcher, DaeJeon Machine Depot, DaeJeon, Korea.
[2]Professor, Mechanical Engineering Department, The University of Iowa, Iowa City, IA 52242.

n_i	Cycles at the ith amplitude
n'	Cyclic strain hardening exponent
n'_p	Cyclic strain hardening exponent for the load-strain curve
P	Load
P_{max}	Maximum load
q	Fatigue notch sensitivity
r	Notch root radius
S_f	Fatigue limit of a smooth specimen
S_Y	Monotonic yield stress
S_N, S'_N	Nominal stress, nominal stress at the simple notch tip
S'_y	Cyclic yield stress
W	Distance from loading line to the edge of the keyhole specimen
x	Coordinate along the notch axis originating at the notch tip
y	Coordinate normal to x-axis
α	Characteristic length
α_{eq}	Equivalent characteristic length
$\Delta\epsilon$	Total strain range
$\Delta\epsilon_i$	Increment of strain at the ith element of the cyclic stress-strain curve
$\Delta\sigma_y$	Average stress in the characteristic length
$\Delta\epsilon_e, \Delta\epsilon_p$	Elastic and plastic strain range
Δe	Nominal strain range
ΔS	Nominal stress range
$\Delta\sigma_i$	Increment of stress at the ith element of the cyclic stress-strain curve
ϵ_a	Strain amplitude
ϵ_{eq}	Equivalent strain
ϵ_i	Strain at the ith element of the cyclic stress-strain curve
ϵ'_f	Fatigue ductility coefficient
σ'_f	Fatigue strength coefficient
σ_a	Stress amplitude
σ_i	Stress at the ith element of the cyclic stress-strain curve
$\sigma_1, \sigma_2, \sigma_3$	Principal stresses
σ_m	Mean stress
σ_{eq}	Equivalent stress
σ_{max}	Maximum stress
δ	Ratio of stress at α to maximum stress at the notch tip
δ'	Ratio of stress at α to σ_{max} of the simple notch
ν	Poisson's ratio

Introduction

The fatigue behavior of cast steels has received much less attention than that of wrought steels. This may be because cast steels have a greater number of defects or discontinuities and low fatigue notch sensitivities [*1,2*], which create greater difficulties in fatigue life calculations. Mitchell [*3*] has reviewed the effect of imperfections in cast steels under constant amplitude fatigue conditions, and Barnby and Holder [*4*] have used the stress intensity factor range and the notch radius to describe the crack initiation behavior of notched cast steels. However, it is difficult to inspect the defects around notches and to analyze these effects under variable amplitude loadings. Thus a macroscopic approach similar to the techniques used for wrought steels is needed. Recently, Stephens et al. [*5*] have shown that low-cycle fatigue behavior of five common cast steels (SAE 0030, SAE 0050, C-Mn, Mn-Mo, and AISI 8630) are similar to those of wrought steels. Glinka and Stephens [6] then tried to apply the notch strain analysis method used for wrought steels to the SAE-0030 cast steel.

The objective of this study is to find proper methods of calculating stress and/or strain which act effectively at a notch regardless of the notch sensitivity and to apply them to fatigue life calculations at notches for cast steels. So far the methods used to calculate the stress and strain at a notch, Hardrath and Ohman equation [7] or Neuber's rule [8], have often been used along with the cyclic stress-strain relation. These two methods have limited rational reasoning and fail to show the effects of notch geometry, thickness, state of stress, and plastic deformation. To take account of these effects, an elastic finite element method (FEM) analysis is used for better analysis. When plastic deformation becomes significant, an elastic-plastic FEM analysis is used for the stress and strain calculations. In this study fatigue notch factors were obtained using an elastic stress analysis, and these were used to calculate stress-strain under variable amplitude loadings with the modified Neuber's rule. A load-strain relation was also obtained from elastic-plastic FEM analysis. These two models were then used along with low-cycle fatigue properties to calculate fatigue crack initiation lives of compact keyhole specimens using the aforementioned five cast steels under variable amplitude loadings. Fatigue crack initiation was defined as surface cracks of a length equal to an effective characteristic length which was between 0.25 and 0.7 mm. The calculations and experiments involved both room temperature and a typical low climatic temperature of −45°C.

Fatigue Notch Factor K_f and Characteristic Length α

Fatigue at a notch is controlled by the elastic stress concentration factor K_t and the stress gradient at the notch tip. The fatigue notch factor K_f is defined as

$$K_f = \frac{S_{f(\text{unnotched})}}{S_{f(\text{notched})}} \tag{1}$$

Neuber [9] proposed that the average strain within a small region at the notch tip controls fatigue at the notch and he developed the formula

$$K_f = 1 + \frac{K_t - 1}{1 + \sqrt{\dfrac{A}{r}}} \tag{2}$$

where A is a material constant. But Peterson [10] observed that the material in a ligament should be subject to a stress greater than the fatigue limit of a smooth specimen, S_f, for possible crack initiation, and he developed the formula

$$K_f = 1 + \frac{K_t - 1}{1 + \dfrac{\alpha}{r}} \tag{3}$$

For wrought steels, the characteristic length, α, obtained from test results, has agreed well with Eq 3 and the empirical relation [11]

$$\alpha = \left(\frac{2068}{S_u}\right)^{1.8} \times 0.025 \text{ mm} \tag{4}$$

where S_u is the ultimate strength in MPa units and d is in mm. For cast steels, Eqs 3 and 4 overestimate K_f and therefore are not applicable. There are also various other formulae for the fatigue notch factors. Schijve [*12*] tried to relate them with K_t, α, and r, but he could not draw a closed form equation for K_f.

The elastic stress gradient at a circular notch in a plate can be written as

$$\left(\frac{d\sigma_y}{dx}\right)_{x=0} = -C\frac{\sigma_{max}}{r} = -C\frac{K_t S_N}{r} \tag{5}$$

where the nominal stress S_N is the stress on the net area along the notch axis. As the characteristic length gets bigger, the average stress, $\Delta\sigma_y$, in it can be expressed simply as

$$\frac{\Delta\sigma_y}{\alpha} = -\sigma_{max}(1 - \delta)/\alpha \tag{6}$$

If σ_{max} and $\Delta\sigma_y/\alpha$ are the same for two different notches, the fatigue damage is assumed to be the same. Thus the stress similarity for the notches should satisfy the relations

$$\sigma_{max} = K_t'S_N' = K_t S_N \tag{7}$$

$$\frac{\Delta\sigma_y}{\alpha} = -K_t'S_N'\frac{(1-\delta')}{\alpha} = -K_t S_N\frac{(1-\delta)}{\alpha} \tag{8}$$

where the primed symbols are for a simple notch. If the elastic stress distributions of a simple notch and a complex notch were chosen to satisfy both relations, the nominal stresses for both notches at the same fatigue damage could be obtained.

Elastic Stress Distributions of the Notches

A compact keyhole specimen (Fig. 1) was used for crack initiation life tests of five cast steels under variable loadings. To estimate K_f values of the keyhole specimen, a simple notched strip specimen (Fig. 2) was designed to satisfy the stress and gradient similarities assuming α to be 0.5 mm. The elastic stress distribution for the keyhole notch obtained from several different FEM analyses is shown in Fig. 3. Socie's FEM model [*13*], which can handle elastic-plastic analysis, was used for both elastic and elastic-plastic analysis. The resulting elastic stress concentration factors for the keyholes were 4.0 and 4.6 for H/w equal to 0.6 and 0.49, respectively. The stress gradients for the two H/w ratio specimens were almost the same. The stress distribution of the notched strip specimen was calculated with Howland's theoretical formula [*14*] (Fig. 4). The elastic stress concentration factor of the notched strip specimen was 2.37. For both specimens, plane stress conditions were assumed. As the distance from the notch tip increases, σ_x increases, and the multiaxial stress state should be considered. Here the von Mises equivalent stress for yielding was used as

$$\sigma_{eq} = \frac{1}{\sqrt{2}}[(\sigma_1 - \sigma_2)^2 + (\sigma_2 - \sigma_3)^2 + (\sigma_3 - \sigma_1)^2]^{1/2} \tag{9}$$

The σ_{eq} curves for both specimens are shown in Figs. 3 and 4 along with the σ_y curves. The σ_{eq} curve may be used instead of σ_y for materials which have large characteristic lengths.

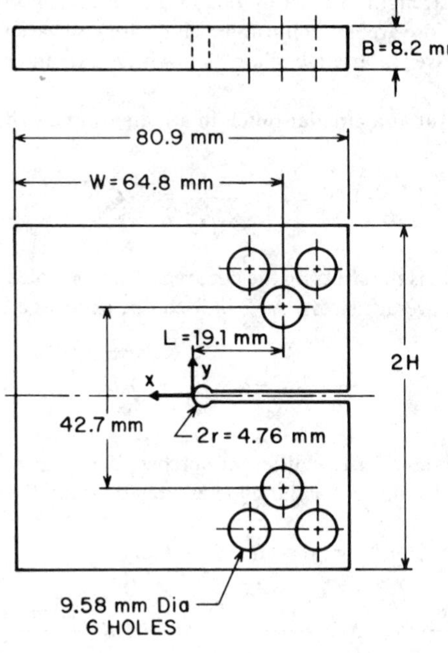

FIG. 1—*Keyhole notch specimen.*

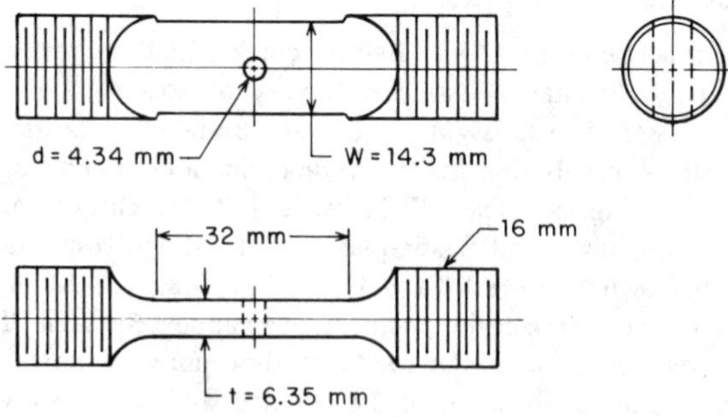

FIG. 2—*Notched strip specimen.*

Cast Steel Properties

The chemical compositions and mechanical properties of the five cast steels are given in Tables 1 and 2, respectively. The yield stresses range from 303 to 988 MPa. The cyclic stress-strain curves were expressed as

$$\epsilon_a = \frac{\sigma_a}{E} + \left(\frac{\sigma_a}{K'}\right)^{1/n'} \tag{10}$$

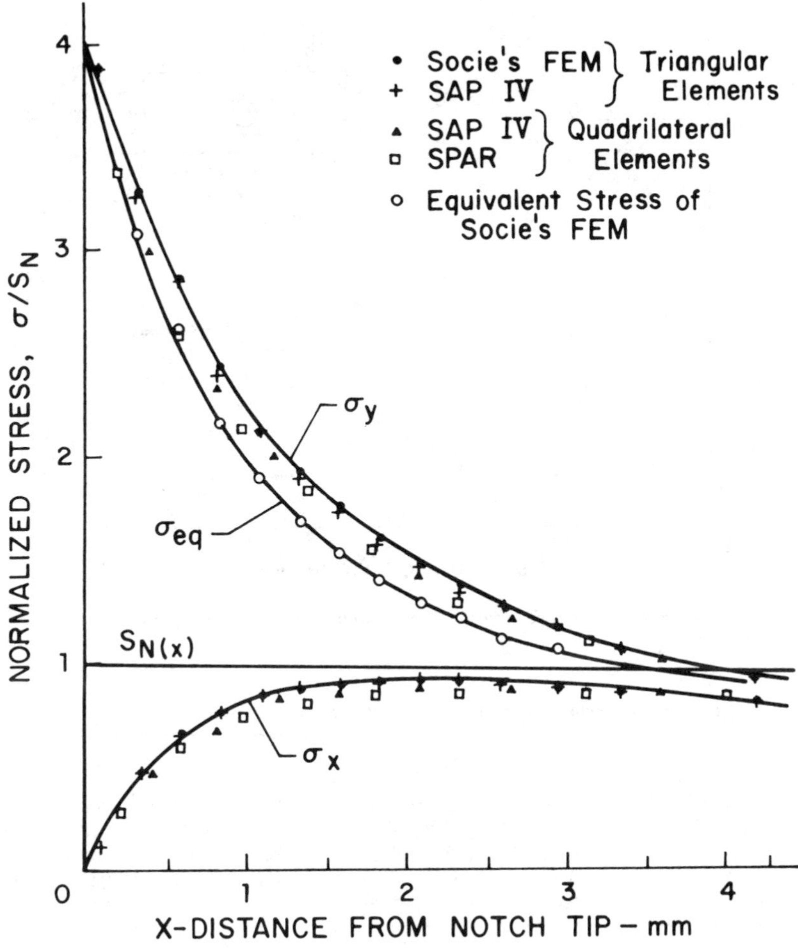

FIG. 3—*Elastic stress distribution at the keyhole notch for* y = *0.*

and the low cycle fatigue test data related well with

$$\frac{\Delta\epsilon}{2} = \frac{\sigma_f'}{E}(2N_f)^b + \epsilon_f'(2N_f)^c \tag{11}$$

The cyclic and low cycle fatigue properties are given in Table 3. In general, the five cast steels showed low cycle fatigue behavior similar to that of the wrought steels. These properties are used in the fatigue life calculations.

Fatigue Test of Notched Strip Specimen

High-cycle fatigue tests of the notched strip specimen were carried out for the five cast steels to obtain the fully-reversed fatigue limits and to use these limits to obtain the characteristic lengths. The notch was left in the as-drilled condition without additional reaming or polishing

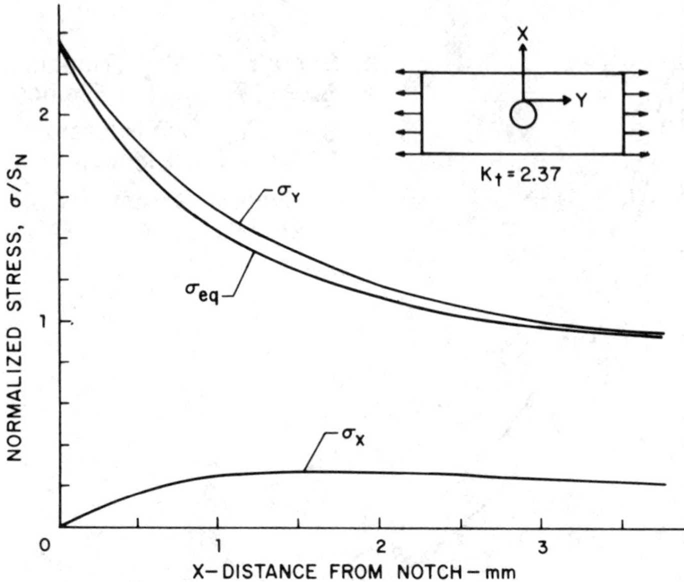

FIG. 4—*Elastic stress distribution of the notched strip specimen for* y = *0.*

TABLE 1—*Chemical composition (weight percent) of five cast steels.*

	C	Mn	Si	S	P	Cr	Ni	Mo	Al
0030	0.24	0.71	0.44	0.026	0.015	0.10	0.10	0.08	0.06
0050A	0.49	0.93	0.61	0.023	0.024	0.11	0.08	0.04	0.08
C-Mn	0.23	1.25	0.39	0.028	0.036	0.10	0.09	0.04	0.02
Mn-Mo	0.34	1.32	0.40	0.035	0.024	0.11	0.11	0.22	0.06
8630	0.30	0.84	0.53	0.022	0.021	0.51	0.61	0.17	0.08

TABLE 2—*Monotonic tensile properties of five cast steels at room temperature.*

Material	E, GPA	S_u, MPa	0.2% S_y, MPa	σ_f, MPa	% RA	ϵ_f
0030	207	496	303	705	46	0.62
0050A	209	787	415	866	19	0.21
C-Mn	209	583	402	703	26	0.34
Mn-Mo	211	702	542	749	31	0.37
8630	207	1145	988	1265	29	0.35

to better simulate the actual component field situation used with the keyhole specimens. Residual stress measurements were not attempted with either the notched strip or the keyhole specimen. The machining procedures, however, were essentially the same for both specimens. A sinusoidal fully-reversed load was applied at 30 to 40 Hz using a closed-loop electrohydraulic test system. The load amplitudes were chosen to obtain the life to crack initiation at the notch tip at about 10^7 reversals. The crack initiation length here is α or α_{eq} which is obtained with a

TABLE 3—*Cyclic and fatigue properties of five cast steels at room temperature.*

Material	S_y', MPa	K', MPa	n'	σ_f', MPa	ϵ_f'	b	c
0030	317	710	0.130	655	0.28	−0.083	−0.552
0050A	379	1151	0.174	1338	0.30	−0.127	−0.569
C-Mn	331	800	0.136	869	0.15	−0.101	−0.514
Mn-Mo	386	1041	0.153	1117	0.78	−0.101	−0.729
8630	662	2271	0.185	1940	0.42	−0.121	−0.693

method to be explained later. The crack lengths at both sides of the notch were checked and the average was used. The fatigue limit was taken as the stress at 10^7 reversals to crack initiation. Six or seven specimens were used to obtain the fatigue limit of each cast steel. The fatigue limits of the notched strip specimens and the unnotched specimens for the cast steels are given in Table 4.

Characteristic and Equivalent Characteristic Length for the Cast Steels

The fatigue limit, S_f, for the notched strip specimen implies that there will be no crack initiation with a stress less than S_f. We may assume that the stress at a certain distance from the notch tip should be less than the fatigue limit of a smooth specimen in order to have no crack initiation within 10^7 reversals. The distance can be defined as a material constant, characteristic length α. If the equivalent stress is used, it is called the equivalent characteristic length, α_{eq}. Using the stress distribution curve of the notched strip specimen at the fatigue limit, the distance from the notch tip where σ_y or σ_{eq} is equal to the fatigue limit of smooth specimens is α or α_{eq}, respectively. Since the stress at α or α_{eq} controls fatigue at a notch, the fatigue notch factor would be the same as the normalized σ_y or σ_{eq} at α or α_{eq}, respectively. The values of α, α_{eq}, and K_f for the notched strip specimen obtained using this method are given in Table 5. Notch sensitivities were calculated with the equation

$$q = \frac{K_f - 1}{K_t - 1} \tag{12}$$

and they are also given in Table 5. The values of α or α_{eq} for the five cast steels are much larger than those for wrought steels calculated using Eq 4; they range from 0.25 to 0.84 mm.

The material constant A of the Neuber Eq 2 can be obtained by checking the distance at which the average stress at the fatigue limit is equal to the fatigue limit of the smooth specimen.

TABLE 4—*Fully-reversed fatigue limits of notched and unnotched axial specimens of five cast steels at room temperature.*

Material	S_u, MPa	S_f (unnotched), MPa	S_f/S_u	S_f (notched), MPa	Notch Sensitivity (q)
0030	496	200	0.40	124	0.45
0050A	787	238	0.30	128	0.63
C-Mn	583	249	0.43	131	0.66
Mn-Mo	702	232	0.33	127	0.60
8630	932	292	0.26	147	0.74

TABLE 5—*Characteristic lengths and fatigue notch factors of notched strip specimens for five cast steels at room temperature.*

Material	K_f	q	α_{eq}, mm	α, mm
0030	1.62	0.45	0.69	0.84
0050A	1.86	0.63	0.41	0.48
C-Mn	1.91	0.66	0.38	0.43
Mn-Mo	1.82	0.60	0.46	0.56
8630	2.02	0.74	0.25	0.33

The values of A ranged from 0.7 to 2.0 mm (Table 6). In general, the stress at A is much less than the fatigue limit S_f, while the notch tip stress is much larger than S_f. Thus it seems unreasonable to assume the average stress in such a large length controls the damage at notches for notch insensitive materials.

Fatigue Notch Factor for the Keyhole Specimen

Assuming α or α_{eq} is a material constant, the fatigue notch factor, K_f, for an arbitrary notch can be estimated from the elastic stress distribution curve. It can be obtained by checking the normalized σ_y or σ_{eq} at α or α_{eq}, respectively. The K_f values for the compact keyhole specimen were obtained by this method and are given in Table 6. The K_f values calculated from Eqs 2 and 3 are compared in the table. Both these α and K_f values were greater than those of the authors' method by 15 to 30%. The K_f values based upon Neuber's concept show values very close to those obtained by the authors. The validity of this method was ascertained by checking the stress at the notch tip as to whether it is greater than the yield stress. The maximum stresses for the specimens were found to be less than the cyclic yield stresses for all five cast steels and thus the method was valid. This method to obtain K_f was also verified by Shin [*15*] using the SAE keyhole specimen and a simple notched strip for 6061-T6 aluminum alloy.

Modified Neuber's Model for Fatigue Damage Calculations

The fatigue damage at a notch under variable loading is dependent upon the effective stress-strain loops of the hysteresis curves. The curves will show sequence effects of high and low loads and will make closed loops of stress and strain. To count pairs of loads which make a closed loop, the rainflow counting method [*16*] was used. The stress and strain at each reversal were

TABLE 6—*Fatigue notch factors of the keyhole specimen from various methods.*

			Neuber's Concept			Peterson's Concept	
Material	α_{eq}, mm[a]	K_f	q	A, mm	K_f	α, mm	K_f
0030	0.69	2.4	0.46	2.0	2.38	0.84	3.22
0050A	0.41	2.86	0.62	1.2	2.85	0.48	3.5
C-Mn	0.38	2.9	0.63	1.0	2.96	0.43	3.54
Mn-Mo	0.46	2.74	0.58	1.2	2.8	0.56	3.43
8630	0.25	3.22	0.74	0.7	3.2	0.33	3.63

[a] From Table 5.

calculated during the cycle counting in order to keep the memory effects. To calculate the stress and strain at a notch, Neuber's rule [*8*] was used as

$$\sigma \cdot \epsilon = K_t^2 Se = (K_\sigma S)(K_\epsilon e) \tag{13}$$

For the effective stress and strain, Eq 13 was modified by replacing K_t with K_f and stress and strain with their ranges as

$$\Delta\sigma \cdot \Delta\epsilon = K_f^2 \Delta S \cdot \Delta e = (K_\sigma \Delta S)(K_\epsilon \Delta e) \tag{14}$$

This is called the *modified Neuber's rule* [*17*]. The stress and strain ranges in Eq 14 can be solved with the outer loop of the cyclic stress-strain curve, which can be written as

$$\frac{\Delta\epsilon}{2} = \frac{\Delta\epsilon_e}{2} + \frac{\Delta\epsilon_p}{2} = \frac{\Delta\sigma}{2E} + \left(\frac{\Delta\sigma}{2K'}\right)^{1/n'} \tag{15}$$

Under variable loadings, the stress and strain response was found to fit better with the incremental step cyclic curve than with the companion specimen stress-strain curve for steels [*18*]. Thus damage calculations here were done with the incremental step cyclic curves.

For the first reversal Eqs 10 and 14 were used, while Eqs 14 and 15 were used for the following reversals. Sometimes it is better to use the actual curve than the simulated Eqs 10 and 15; thus the actual cyclic curve was elementized into multiple line elements. To calculate the stress and strain for the first reversal, the element should be chosen to satisfy the relation

$$\sigma_{i-1} \cdot \epsilon_{i-1} \leqq K_f^2 \cdot \Delta S \cdot \Delta e = \Delta\sigma \cdot \Delta\epsilon < \sigma_i \cdot \epsilon_i \tag{16}$$

Once the element i is chosen, the stress and strain can be solved from the linear relation of the ith element as

$$\sigma = \frac{\Delta\sigma_i}{\Delta\epsilon_i}(\Delta\epsilon - \epsilon_{i-1}) + \sigma_{i-1} \tag{17}$$

and Eq 14. For the following reversals, the stress and strain ranges should satisfy the relation

$$2\sigma_{i-1} \cdot 2\epsilon_{i-1} \leqq K_f^2 \cdot \Delta S \cdot \Delta e = \Delta\sigma \cdot \Delta\epsilon < 2\sigma_i \cdot 2\epsilon_i \tag{18}$$

Then, the stress and strain can be solved from

$$\Delta\sigma = \frac{\Delta\sigma_i}{\Delta\epsilon_i}(\Delta\epsilon - 2\epsilon_{i-1}) + 2\sigma_{i-1} \tag{19}$$

combined with Eq 14. Once the range of strain and mean stress of each closed loop are determined, the plastic and elastic strain range can be calculated using Eq 15.

Fatigue Damage Calculations Using the Linear Damage Rule

Once the strain range of a closed loop is obtained, the damage of the loop can be calculated using Eq 11 by an iterative calculation method. But a simple closed-form equation could be obtained using the elastic and plastic strain range versus the cycles to failure, as Landgraf's expression [*19*]

$$D = \frac{1}{N_f} = 2\left(\frac{\sigma_f'}{\epsilon_f' E} \frac{\Delta\epsilon_p}{\Delta\epsilon_e}\right)^{1/(b-c)} \tag{20}$$

The mean stress correction was done with

$$D = \frac{1}{N_f} = 2\left(\frac{\sigma_f'}{\epsilon_f' E} \frac{\Delta\epsilon_p}{\Delta\epsilon_e} \frac{\sigma_f'}{\sigma_f' - \sigma_m}\right)^{1/(b-c)} \tag{21}$$

Assuming the damage of various closed loops can be added linearly as in the Palmgren-Miner rule, the damage at α or α_{eq} can be written as

$$D_B = \sum_i \frac{n_i}{N_i} \tag{22}$$

The number of blocks to form a crack of length equal to α or α_{eq} is the inverse of D_B. The FORTRAN listing for the fatigue damage calculation during the rainflow counting is given in the Appendix of Ref *20*.

Load-Strain Model for Fatigue Damage Calculations

Dowling et al. [*21*] indicated that the plot of load versus strain at a notch tip under a variable loading shows closed loops and memory effects in a similar way as stress-strain hysteresis loops, assuming the material at the notch is cyclically stabilized. The plot of load versus strain at the characteristic length will also show a similar behavior with that at the notch tip and is used here to also calculate fatigue damage.

To predict the strain at α or α_{eq}, an elastic-plastic FEM analysis was used. Socie's [*13*] FEM model was selected for the elastic-plastic analysis which uses the kinematic hardening model to describe strain hardening and the von Mises yield condition for new yielding loads. As the plastic deformation increases at a notch, σ_y shows an increase in a small distance from the notch due to the multiaxial stress state. Thus the stress is not proper to use in the calculation of fatigue damage along the notch axis. Sine's equivalent strain [*22*]

$$\epsilon_{eq} = \frac{1}{\sqrt{2}(1-\nu)}[(\epsilon_1 - \epsilon_2)^2 + (\epsilon_2 - \epsilon_3)^2 + (\epsilon_3 - \epsilon_1)^2]^{1/2} \tag{23}$$

showed a smooth decrease as Santhanam and Bate's [*23*] analytical form of strain distribution

$$\epsilon_{eq}(x) = \epsilon_{eq}(\alpha_{eq})e^{-Cn'x/r} \tag{24}$$

The constant C ranged from 16 to 18 at the vicinity of α_{eq} for the keyhole specimen of the cast steels.

The relation between load and equivalent strain for the first and the next reversal was similar to the stress-strain curves of the same reversals. The load-strain for the first reversal could be written as

$$\Delta\epsilon = \Delta\epsilon_e + \Delta\epsilon_p = \frac{\Delta P}{K_E} + \left(\frac{\Delta P}{K_p'}\right)^{1/n_p'} \tag{25}$$

For the next reversals:

$$\frac{\Delta\epsilon}{2} = \frac{\Delta\epsilon_e}{2} + \frac{\Delta\epsilon_p}{2} = \frac{\Delta P}{2K_E} + \left(\frac{\Delta P}{2K_p'}\right)^{1/n_p'} \tag{26}$$

A typical example of P versus ϵ is shown in Fig. 5. The equivalent strains at the keyhole notch for the first loading are shown in Fig. 6.

Once the shape of the load-strain hysteresis is known, the strain ranges of the closed loops can be calculated in a similar manner as the stress-strain calculation. An example of the load-plastic strain calculated with the FEM analysis is shown in Fig. 7. The strain ranges of the closed loops can be divided into elastic and plastic components using Eq 26 and the fatigue damage can be calculated using Eq 20 without mean stress correction. The mean stresses can be calculated using Eq 15, but the values tend to deviate excessively due to cumulative errors. The damage with and without mean stress corrections showed only small differences according to the modified Neuber's model calculations. Thus the mean stress correction was not done for the load-strain model. The FORTRAN listing for the fatigue damage calculation is given in the Appendix of Ref *20*.

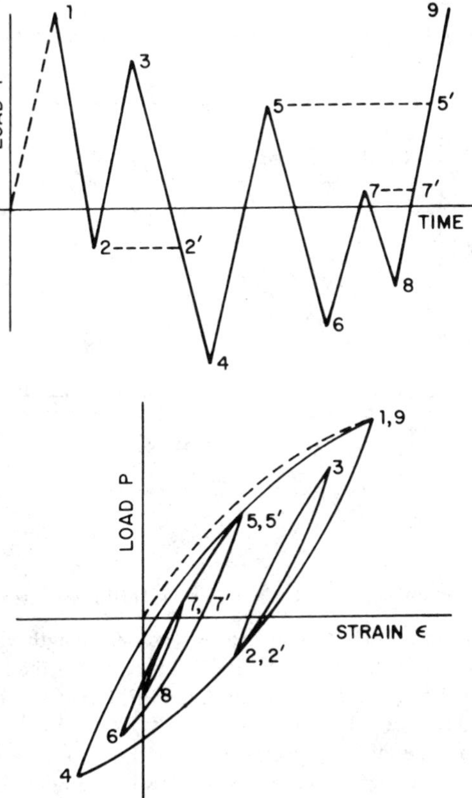

FIG. 5—*Load-strain hysteresis for C-Mn cast steel at the equivalent characteristic length.*

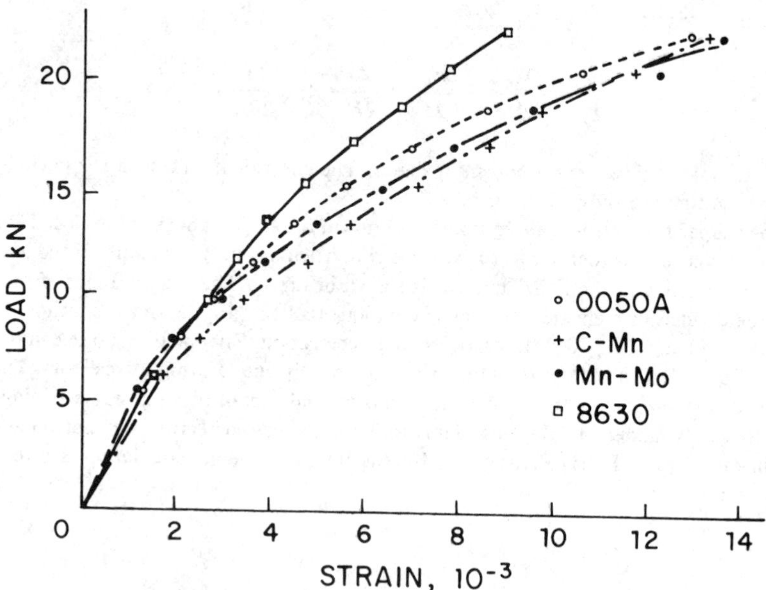

FIG. 6—*Load versus equivalent strain at the equivalent characteristic length for the cast steel keyhole specimens for the first loading.*

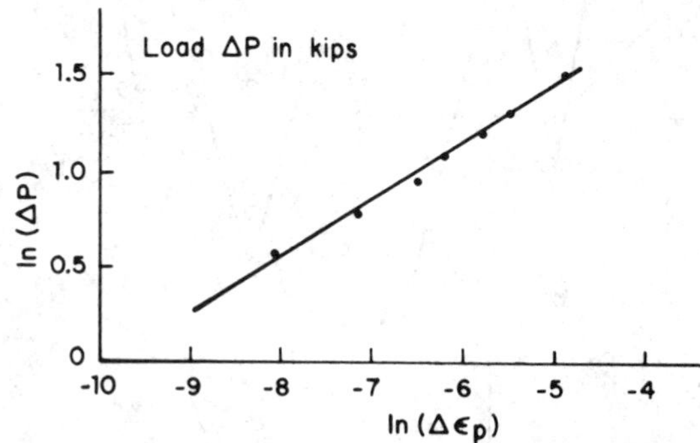

FIG. 7—*Load versus plastic strain relation for C-Mn cast steel.*

Comparison of the Experimental and Calculated Crack Initiation Lives

The load versus strain curves calculated at the characteristic length with various methods are shown in Fig. 8. For a given load, the modified Neuber's rule has the largest strains while the linear rule has the least. The equivalent strain from the FEM analysis under plane stress shows less strain than that of the modified Neuber's rule.

The two variable amplitude loading spectra used for the tests were the SAE Transmission History (T/H) [*24*] and the modified Transmission History (mod-T/H). The mod-T/H was formed from the T/H by eliminating all the compressive loadings (Fig. 9). The maximum loads of the spectra applied to the keyhole specimens were 15.6, 17.8, and 22.2 kN. The nominal stresses at the notch tip for these loads were 276, 317, and 393 MPa, respectively. The loading

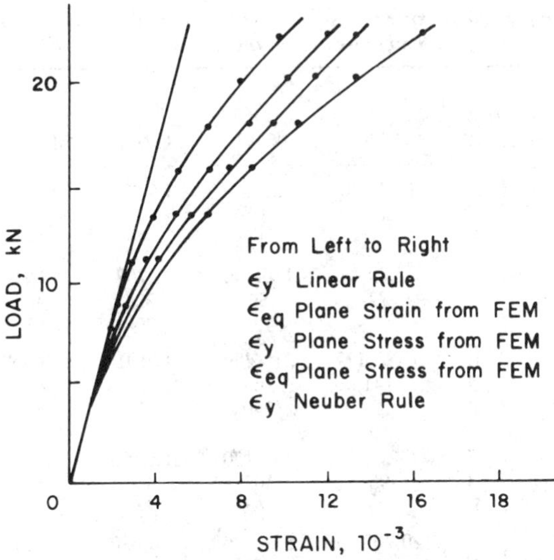

FIG. 8—*Comparison of load versus strain at the equivalent characteristic length for C-Mn cast steel.*

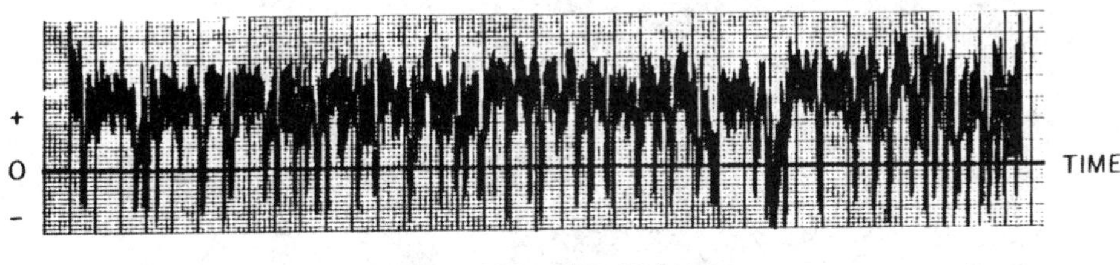

a. T/H 1,710 REVERSALS

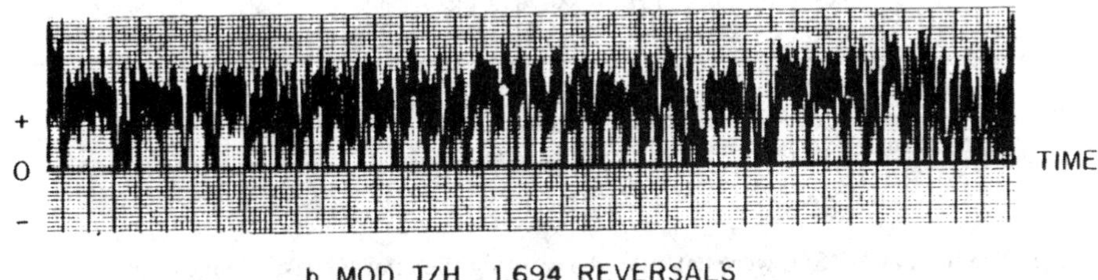

b. MOD T/H 1,694 REVERSALS

FIG. 9—*Variable amplitude load spectra for the transmission and modified transmission history.*

spectra were applied using the closed-loop electrohydraulic test system with a programmable calculator. It took about 55 s to complete one block. Surface crack initiation was monitored with a telemicroscope.

Two or three test data at each load level and spectrum were available, and they were compared with calculated data in Table 7 for both the modified Neuber's model and the load-strain

TABLE 7—*Comparison of experimental and calculated crack initiation blocks using the modified Neuber's rule and the load-strain method.*

Spectrum and Temperature	Peak Nominal Stress, MPa		0030	0050A	C-Mn	Mn-Mo	8630
T/H R.T.	393	E[a]	18,21,40	23,57	35,33	38,45	116,125
		N[b]	18	19	10	19	34
		P[c]	...	43	8	17	86
	317	E	87,70,43	110,190	93,107	112,211	250,258
		N	72	80	32	46	88
		P	...	127	29	45	230
	276	E	55,95,127	216,250	90,101	249,359	495,696
		N	141	163	70	83	165
		P	...	249	65	89	388
Mod-T/H R.T.	393	E	...	130,130	105,108	175,183	420,490
		N	95	103	46	51	79
		P	...	165	44	57	233
	317	E	120,200,270	538,330	420,430	580,519	615,1310
		N	328	375	194	157	223
		P	...	516	162	163	501
T/H −45°C	393	E	...	70,80	36,45	22,100	210,217
		N	30	17	11	26	82
		P	...	14	8	30	127
	317	E	108,75	126,55	98,139	263,310	390,696
		N	102	45	43	73	246
		P	...	41	37	94	353
	276	E	...	215,267	185,198	370,460	760
		N	234	86	110	146	503
		P	...	80	103	201	585
Mod-T/H −45°C	393	E	...	240,201	228,308	308,320	616,620
		N	149	54	68	94	272
		P	...	53	72	133	418
	317	E	...	630,735	664,1350	661,1346	2262
		N	737	172	406	373	987
		P	...	184	372	435	1072

[a]E = Experimental data.
[b]N = Calculated values using modified Neuber's rule.
[c]P = Calculated values using load-strain method.

model. The lives at −45°C were calculated assuming the characteristic lengths are the same as those at room temperature (R.T.); however, the low temperature properties were used in the low temperature life calculations. The load-strain model was not applicable to the 0030 cast steel due to excessive plastic deformation.

Modified Neuber's Model

The room temperature crack initiation lives calculated using the modified Neuber's model are compared with test data in Fig. 10. The calculated lives are within factors of 7 to −3 of the test lives. The majority of the data are in the conservative region within a factor of 4. This can be

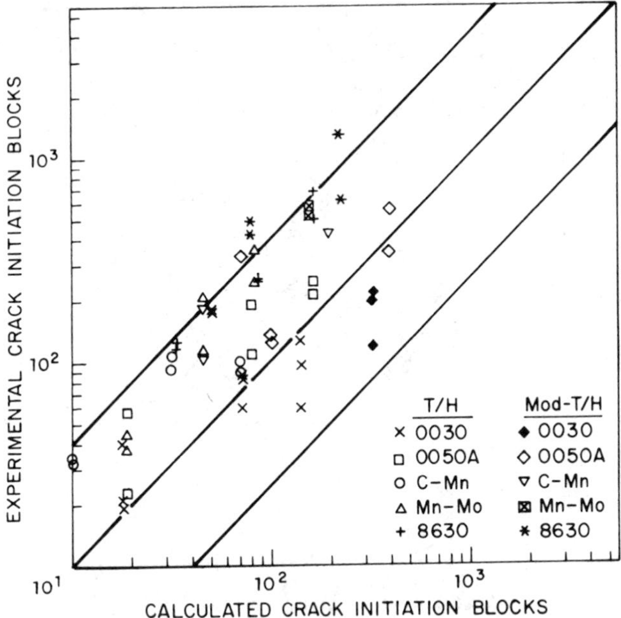

FIG. 10—*Experimental and calculated crack initiation lives using the modified Neuber's model at room temperature.*

explained in that the model overestimates the stress and strain if the plastic deformation is large. It is known that the fatigue notch factor tends to decrease as the strain range increases, and the assumption of constant K_f will cause greater stress and strain estimations. Most of the crack initiation lives at −45°C were within a factor of 4 in the conservative region. This indicates that the life calculation at a low temperature can be done with the same equivalent characteristic length at room temperature.

Load-Strain Model

The plot of the test and calculated lives at room temperature is given in Fig. 11. Most of the data are within factors of 4 and −2, the majority are in the conservative region. Less scatter of the data was obtained compared with the previous model. The Mn-Mo and 8630 cast steel showed substantial improvements. Assuming the equivalent characteristic length remains the same, the crack initiation lives at −45°C were calculated and compared with test data in Table 7. They showed similar results with the previous model.

The damage distribution along the notch axis could be calculated using the strain distribution in the form of Eq 24. The damage showed a linear decrease on a semi-log coordinate (Fig. 12). This implies that the damage in a small distance from the characteristic length may not be neglected in the calculation of crack growth.

Conclusions

1. The concept that stress and strain at the equivalent characteristic length controls the fatigue damage at a notch was established. An elastic FEM analysis can be used for the estimation

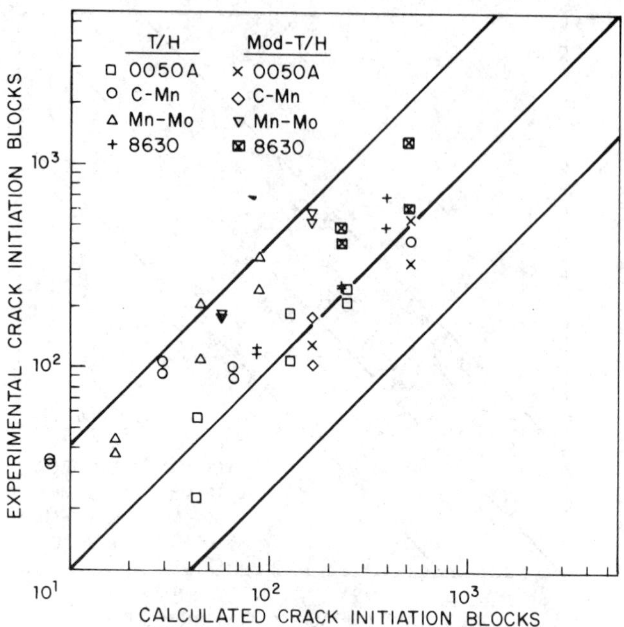

FIG. 11—*Experimental and calculated crack initiation lives using the load-strain model at room temperature.*

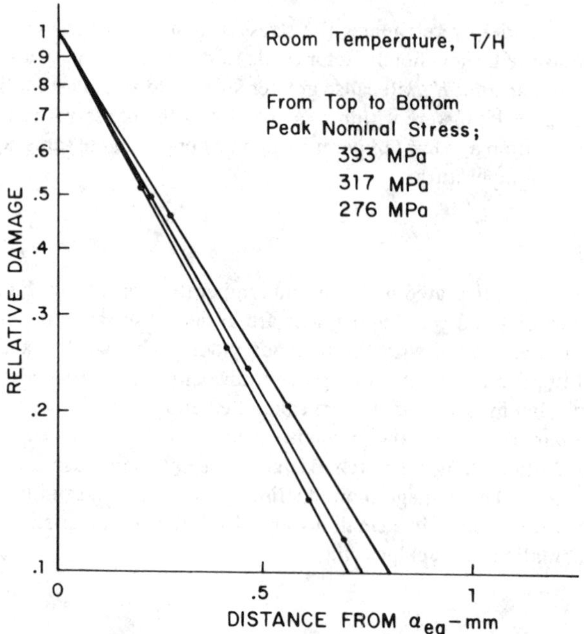

FIG. 12—*Distribution of relative damage beyond the equivalent characteristic length for the keyhole specimen of C-Mn cast steel.*

of the fatigue notch factor for a complicated notch using the results of a simple notch fatigue test.

2. The modified Neuber's model, which uses the estimated fatigue notch factor from the new method, proved to be applicable for the fatigue damage analysis of low notch sensitive metals such as the five common cast steels.

3. An elastic-plastic FEM analysis can be used for the strain estimation along the notch axis. The load-strain model, which uses the load versus strain curve to calculate the strain at the equivalent characteristic length, gave better fatigue life calculations for these low notch sensitive cast steels than the modified Neuber's model.

4. Fatigue damage at $-45°C$ could be calculated with both models within a factor of 4, assuming the equivalent characteristic length remains the same as that at room temperature.

5. The five common cast steels showed large equivalent characteristic lengths ranging from 0.25 to 0.84 mm and low notch sensitivities ranging from 0.46 to 0.74.

References

[1] Evans, E. B., Ebert, L. J., and Briggs, C. W., "Fatigue Properties of Cast and Wrought Steels," in *Proceedings*, American Society for Testing and Materials, Vol. 56, 1956, p. 979.

[2] Wallace, J. F., Vishevsky, C., and Briggs, C. W., "A Study of the Notch Effect and of Specimen Design and Loading on the Fatigue Properties of Cast Steel," *Journal of Basic Engineering*, Vol. 90, March 1968, p. 51.

[3] Mitchell, R. M., "Review of the Mechanical Properties of Cast Steels with Emphasis on Fatigue Behavior and Influence of Microdiscontinuities," *Journal of Engineering Materials and Technology*, Oct. 1977, p. 329.

[4] Barnby, J. T. and Holder, R., "Fatigue from Notches in Cast Steels," *Metal Science*, Jan. 1977, p. 5.

[5] Stephens, R. I., Chung, J. H., Fatemi, A., Lee, H. W., Lee, S. G., Vaca-Oleas, C., and Wang, C. M., "Constant and Variable Amplitude Fatigue Behavior of Five Cast Steels at Room Temperature and −45°C," *Journal of Engineering Materials and Technology*, Vol. 106, Jan. 1984, p. 25.

[6] Glinka, G. and Stephens, R. I., "Total Fatigue Life Calculations in Notched SAE 0030 Cast Steel under Variable Loading Spectra," in *Fracture Mechanics: Fourteenth Symposium. Vol. I: Theory and Analysis, ASTM STP 791*, American Society for Testing and Materials, Philadelphia, 1983, p. I-427.

[7] Hardrath, H. F. and Ohman, L., "A Study of Elastic Plastic Stress Concentration Factors due to Notches and Fillets in Flat Plates," NACA TR 1117, 1953.

[8] Neuber, H., "Theory of Stress Concentration for Shear Strained Plastic Bodies with Arbitrary Nonlinear Stress Strain Laws," *Journal of Applied Mechanics*, Vol. 28, Dec. 1961, p. 544.

[9] Neuber, H., Kerbspannungslehre, Springer, Berlin, 1958, *Translation Theory of Notch Stress*, U.S. Office of Technical Service, 1961.

[10] Peterson, R. E., "Analytical Approach to Stress Concentration Effect in Fatigue of Aircraft Structures," WADC TR 59-507, Aug. 1959, p. 273.

[11] Graham, J. A., Ed., *Fatigue Design Handbook*, Society of Automotive Engineers, 1968.

[12] Schijve, J., "Stress Gradient Around Notches," *Fatigue of Engineering Materials and Structures*, Vol. 3, No. 4, 1980, p. 325.

[13] Socie, D. F., "Estimating Fatigue Crack Initiation and Propagation Lives under Variable Loading History," Ph.D. thesis, University of Illinois, Urbana, 1977.

[14] Howland, R. C. J., "On the Stress in the Neighborhood of a Circular Hole in a Strip under Tension," *Philosophical Transactions of the Royal Society of London*, Vol. A, 1929–1930, p. 229.

[15] Shin, Y. J., "Evaluation of Fatigue Notch Factor for a Compact Keyhole Specimen," Master's thesis, Choongnam National University, Korea, Aug. 1984.

[16] Matsuishi, M. and Endo, T., "Fatigue of Metals Subject to Varying Stress," paper presented to Japan Society of Mechanical Engineers, Fukuoka, Japan, Mar. 1968.

[17] Topper, T. H., Wetzel, R. M., and Morrow, J., "Neuber's Rule Applied to Fatigue Notched Specimen," *Journal of Materials*, Vol. 4, March 1969, p. 200.

[18] Koibuchi, K. and Kitani, S., "The Role of Cyclic Stress-Strain Behavior on Fatigue Damage under Varying Load," in *Cyclic Stress-Strain and Plastic Deformation Aspects of Fatigue Crack Growth, ASTM STP 519*, American Society for Testing and Materials, Philadelphia, 1973, p. 213.

[19] Landgraf, R. W., Richards, F. D., and RaPointe, N. R., "Fatigue Life Prediction for a Notched Member under Complex Load Histories," in *Fatigue under Complex Loading*, Society for Automotive Engineers, Vol. 6, 1975, p. 95.

[*20*] Lee, S. G., "Fatigue Crack Initiation and Propagation Life of Cast Steels under Variable Loading Histories," Ph.D. thesis, University of Iowa, Iowa City, Dec. 1982.
[*21*] Dowling, N. E., Brose, W. E., and Wilson, W. K., "Notched Member Fatigue Life Prediction by the Local Strain Approach," in *Fatigue under Complex Loading,* Society for Automotive Engineers, Vol. 6, 1975, p. 55.
[*22*] Sines, G., "Failure of Material under Combined Repeated Stress with Superimposed Static Stress," NACA TN 4395, 1955.
[*23*] Bates, R. C. and Santhanam, A. T., "Notch Tip Stress Distribution in Strain Hardening Materials," *International Journal of Fracture,* Vol. 19, 1978, p. 501.
[*24*] Tucker, L. and Bussa, S., "The SAE Cumulative Fatigue Damage Test Program," in *Fatigue under Complex Loading,* Society for Automotive Engineers, Vol. 6, 1975, p. 1.

R. Ohtani,[1] *T. Kitamura,*[1] *A. Nitta,*[2] *and K. Kuwabara*[2]

High-Temperature Low Cycle Fatigue Crack Propagation and Life Laws of Smooth Specimens Derived from the Crack Propagation Laws

REFERENCE: Ohtani, R., Kitamura, T., Nitta, A., and Kuwabara, K., **"High-Temperature Low Cycle Fatigue Crack Propagation and Life Laws of Smooth Specimens Derived from the Crack Propagation Laws,"** *Low Cycle Fatigue, ASTM STP 942,* H. D. Solomon, G. R. Halford, L. R. Kaisand, and B. N. Leis, Eds., American Society for Testing and Materials, Philadelphia, 1988, pp. 1163–1180.

ABSTRACT: The objectives of this work are (1) to make clear the behavior of high-temperature low-cycle fatigue (LCF) crack propagation, (2) to verify the applicability of the *J*-integral to the fracture mechanics equations of crack propagation rates, (3) to derive two types of LCF life laws of smooth specimens based on the two types of crack propagation equations (i.e., cycle-dependent and time-dependent), (4) to show a resemblance to or a difference from the Manson-Coffin equation and the strain-range partitioning equations, and (5) to characterize the cycle-dependent and time-dependent fatigue laws. The effectiveness of the proposed failure-life equations was examined using experimental data on isothermal and thermal fatigue of smooth specimens.

KEY WORDS: fatigue crack propagation rate, *J*-integral, fatigue life equation, strain energy parameter, isothermal fatigue, thermal fatigue

The authors have been conducting crack propagation tests under high-temperature low-cycle fatigue (LCF) conditions using center cracked plate or tubular specimens of many kinds of steels and alloys for about ten years. This paper summarizes the test results and their evaluation based on a nonlinear fracture mechanics concept; emphasis is placed on the derivation of LCF life laws of smooth specimens from the crack propagation laws. For cycle-dependent crack propagation, the fatigue *J*-integral range, ΔJ_f, is suitable for a fracture mechanics parameter, and for time-dependent propagation, the creep J-integral, J^*, or the creep J-integral range, ΔJ_c, is an available parameter. The two types of the crack propagation rate equations characterize our approach to high-temperature LCF. The integral of the equations with respect to crack length and number of cycles leads to two types of fatigue failure lives, N_f, as a function of the strain energy parameter, $\Delta \tilde{W}_f$ or $\Delta \tilde{W}_c$, which have similar meanings to ΔJ_f and ΔJ_c. In this study, the characterization of the two types of failure laws of $\Delta \tilde{W}_f$ versus N_f and $\Delta \tilde{W}_c$ versus N_f was given, and the validity of the laws was discussed on the basis of experimental data on isothermal fatigue for four types of strain waveforms and data on thermal fatigue for two types of temperature-strain waveforms.

[1]Professor and Instructor, respectively, Department of Engineering Science, Kyoto University, Kyoto, Japan.

[2]Senior Research Engineer and Manager, respectively, Nuclear Energy Department, Central Research Institute of Electric Power Industry, Komae, Japan.

Crack Propagation Behavior

Crack propagation behavior of high-temperature LCF can, in general, be classified into cycle-dependent (F-type) and time-dependent (C-type) ones. Characteristics of these two types of crack propagation (Table 1) include frequency (or strain rate) dependence, stress (or strain) waveform dependence, temperature dependence, fracture morphology, crack propagation mechanisms, fracture mechanics laws, evaluating methods of the fracture mechanics parameters, creep-fatigue interaction, transition from one type to the other, and limit of application of the fracture mechanics laws. These characteristics yield the validity and applicability of fracture mechanics laws of Eqs 1 and 2 in Table 1 (Item 6), which are based on the fatigue J-integral range (or cyclic J-integral) for F-type and the creep J-integral (or C^*-parameter) for C-type. The fatigue J-integral range represents the elastic-plastic strain-range field in the vicinity of the crack tip and is well known as a controlling parameter of $d\ell/dN$ in room-temperature LCF. The creep J-integral represents the creep strain rate field in the vicinity of the crack tip and is available to the static creep crack propagation. Details of the characterization can be found in the references given in Table 1.

The experimental relations between crack propagation rate, $d\ell/dN$, and fatigue J-integral range, ΔJ_f, for F-type, and creep J-integral range, ΔJ_c, for C-type, are shown in Figs. 1 and 2 for a 1Cr-1Mo-1/4V cast steel and a 304 stainless steel. Here, ΔJ_c is the integral of the creep J-integral, J^*, with respect to tension period of a cycle as shown by Eq 3 in Table 1 (Item 6). The values of ΔJ_f and ΔJ_c were evaluated by the simple graphic method by making use of a load versus crack center opening displacement (CCOD) loop [*3,5*].

Figures 3 and 4 show the summation of experimental data on several kinds of steels and alloys obtained by the authors. Figure 5 shows the creep crack propagation rates under constant load as a function of creep J-integral for the same materials.

Fatigue Life Laws of Smooth Specimens Derived from Crack Propagation Laws

The authors' investigation of the microcrack initiation and propagation in a smooth specimen under creep and LCF conditions at high temperature has obtained the following experimental results: [*32–35*]:

1. The microcracks initiate on the surface of the smooth specimens at the very early stage of strain cycle. The continuous initiation and propagation of microcracks bring on the distributions of crack length and crack propagation rate. The fatigue failure is brought on by the propagation of a main crack.
2. The coalescence of microcracks sometimes occurs, but the failure life is scarcely affected since it gives rise to the last stage of the fatigue process.
3. The propagation rates of surface microcracks longer than a few grains in diameter are quite close to that estimated by the extrapolation of the through-macrocrack propagation laws. Therefore the range in which macrocrack propagation laws are applicable is about 70% of the fatigue life of the smooth specimen. Figure 6 shows a schematic curve of crack propagation in a smooth specimen in the application of the fracture mechanics laws of macrocrack propagation.

Based on these findings, the fatigue life laws of smooth specimens are derived from the macrocrack propagation laws in this section.

Cycle-Dependent Fatigue (F-type)

An elastic-plastic body with a relatively short crack compared with specimen size is considered. The fatigue J-integral range, ΔJ_f, for the crack can be evaluated by Eq 4 in Table 1 (Item 7). Here, $\Delta\epsilon_e$ and $\Delta\epsilon_p$ are elastic and plastic strain range, respectively, $\Delta\sigma$ is stress range, E is

TABLE 1—*Characteristics of cycle-dependent (F-type) and time dependent (C-type) fatigue crack propagation at high temperature.*

Items	Cycle-Dependent (F-type) Fatigue Crack Propagation	Time-Dependent (C-type) Fatigue Crack Propagation	Ref
1. Frequency dependence (strain-rate dependence)	High frequency High strain rate in tension-going period irrespective of strain rate in compression	Low frequency Low strain rate in tension-going period irrespective of strain rate in compression	*1–6*
2. Stress or strain waveform dependence	pp-type waveform: fast-fast or no hold pc-type waveform: fast-slow or compression hold	cc-type waveform: slow-slow or tension and compression hold cp-type waveform: slow-fast or tension hold	*7–9*
3. Temperature dependence	Little Effect of temperature dependence of ductility, elastic modulus, and/or yield stress	Strong Activation energy for crack propagation ≅ activation energy for creep	*8, 10, 11*
4. Fracture morphology	Transgranular Fatigue striation	Intergranular In some steels, transgranular fracture with dimples In some cases (in Region I), mixed fracture	*1–3, 7, 8, 12–14*
5. Mechanism of crack propagation	Crack opening and crack-tip blunting Crack closure and crack-tip resharpening	Creep strain accumulation Stable propagation of ductile crack	*15–18*
6. Fracture mechanics law	$d\ell/dN = C_f(\Delta J_f)^{m_f}$ (1) ΔJ_f: fatigue J-integral range (cyclic J-integral) C_f, m_f: material constants Eq 1 includes $d\ell/dN$-ΔK relation for high cycle fatigue Usually $1 < m_f < 2$	$d\ell/dt = C_c(J^*)^{m_c}$ (2) J^*: creep J-integral (modified J-integral or C^*-parameter) C_c, m_c: material constants $d\ell/dN = C_c\Delta J_c \quad (m_c = 1)$ (3) $\Delta J_c = \int_0^{t_t} J^*\,dt$: creep J-integral range Eq 2 nearly coincides with that of static (monotonic tensile) creep Usually $m_c \cong 1$	*3, 5–10, 13, 17–19*

TABLE 1—*Continued.*

Items	Cycle-Dependent (F-type) Fatigue Crack Propagation	Time-Dependent (C-type) Fatigue Crack Propagation	Ref
7. Evaluation of *J*-integral for relatively short cracks	$\Delta J = \Delta J_{elastic} + \Delta J_{plastic} = 2\pi M_J \left\{ \frac{\Delta\sigma\Delta\epsilon_e}{2} + \frac{N+1}{2\pi} f(N) \frac{\Delta\sigma\Delta\epsilon_p}{N+1} \right\} \ell$ (4) $M_J = M_k^2/Q$ = (boundary correction factor)/(flaw shape correction factor for K_I) (5) $\Delta\epsilon_p/2 = A\,(\Delta\sigma/2)^N$ (6) $f(N) = 3.85\sqrt{N}\,(1 - 1/N) + \pi/N$ (7)	$J^* = J^*_{transient} + J^*_{steady\ creep} = 2\pi M_J \left\{ \frac{1+2\beta n}{(n+1)t} \frac{\sigma\epsilon_e}{2} + \frac{n+1}{2\pi} f(n) \frac{\sigma\dot{\epsilon}_c}{n+1} \right\} \ell$ (8) $\dot{\epsilon}_c = B\sigma^n$ (9) $\sigma = \sigma_{max}\,(t/t_0)^{\beta}$: stress waveform (10) $\Delta J^* = 2\pi M_J \left\{ \frac{\sigma_{max}}{\Delta\sigma} \frac{\sigma_{max}\Delta\epsilon_e}{2} + \frac{n+2}{n+3} \frac{n+1}{2\pi} f(n) \frac{\sigma_{max}\Delta\epsilon_c}{n+1} \right\} \ell$ for $m_c = 1$ (11) $J^*_{transient}$: creep *J*-integral in the stress redistribution period in the vicinity of crack tip (*J*-integral for small-scale creep)	*20–29*
8. Creep-fatigue interaction	Values of A and N of Eq 6 change with creep. Interaction with creep deformation The interaction causes the increase in $\Delta J_{plastic}$ and thus $d\ell/dN$ Little or no interaction between creep fracture and fatigue fracture	Values of B and n in Eq 9 change with changing stress. Interaction with plastic deformation The interaction causes the increase in J^* or ΔJ_c and thus $d\ell/dt$ or $d\ell/dN$	*5, 25–27, 30, 31*
9. Transition from F-type to C-type	$\Delta J_f > (C_c \Delta J_c / C_f)^{1/m_f}$ (12)	$\Delta J_c > C_f \Delta J_f^{m_f} / C_c$ (13)	*5–7, 13, 32*

10. Limit of crack length in which J-integral is applicable	2 to 4 grain diameter length (100 to 200 μm in 304 stainless steel of mean grain diameter of 50 μm)	3 to 5 grain boundary length (100 to 200 μm in 304 stainless steel of mean grain diameter of 40 μm) Similar to creep	*33–40*
11. Range of fatigue life in which Eqs 1 and 2 or 3 are applicable	Not yet ascertained, but supposed to be similar to C-type	Number of cycles from 0.7 N_f to N_f (about 70% of fatigue life) Number of cycles to crack initiation, N_c, is less than 0.1 N_f	*36, 37*
12. Fatigue life derived from Eqs 1 and 2 or 3	$(\Delta\tilde{W}_f)^{m_f} N_f = D_f$ (14) $\Delta\tilde{W}_f = \frac{\Delta\sigma\Delta\epsilon_e}{2} + \frac{N+1}{2\pi} f(N) \frac{\Delta\sigma\Delta\epsilon_p}{N+1}$ (15) $D_f = \begin{cases} \frac{\ell n\,(\ell_f/\ell_0)}{C_f^2 M_j} \quad (m_f = 1) \\ \frac{\ell_0^{1-mf} - \ell_f^{1-mf}}{C_f(mf-1)(2M_J)^{m_f}} \quad (m_f \neq 1) \end{cases}$ (16) ℓ_0: initial crack length at $N = 0$ ℓ_f: final crack length at $N = N_f$ $\Delta\tilde{W}_f$: strain energy parameter	$(\Delta\tilde{W}_c)^{m_c} N_f = D_c t_0^{1-m_c}$ (17) $\Delta\tilde{W}_c = \begin{cases} \left\{\int_{-1}^{1} \left[\frac{1}{2^{2/3}} \frac{\sigma_{\max}}{\Delta\sigma} \frac{\sigma_{\max}\Delta\epsilon_e}{2} + \frac{n+2}{2^{(n+1)/2+5/3}} (1+x)^{(n+1)/2} \frac{n+1}{2\pi} f(n) \frac{\sigma_{\max}\Delta\epsilon_c}{n+1}\right]^{3/2} dx\right\}^{2/3} \quad (m_c \neq 1) \\ \frac{\sigma_{\max}}{\Delta\sigma} \frac{\sigma_{\max}\Delta\epsilon_e}{2} + \frac{n+2}{n+3} \frac{n+1}{2\pi} f(n) \times \frac{\sigma_{\max}\Delta\epsilon_c}{n+1} \quad (m_c = 1, \beta = 0.5) \end{cases}$ (18) $D_c = \begin{cases} \frac{\ell_0^{1-m_c} - \ell_f^{1-m_c}}{C_c(m_c-1)(2\pi M_j)^{mc}} \quad (m_c \neq 1) \\ \frac{\ell n\,(\ell_f/\ell_0)}{C_c^2 \pi M_J} \quad (m_c = 1) \end{cases}$ (19) t_0: see Eq 10 $\Delta\tilde{W}_c$: strain energy parameter	*29, 41, 42*

Cr-Mo-V cast steel (823 K)

Test No.	Stress waveform	Stress (MPa) σ_{max}	Stress (MPa) σ_{min}	Loading time (s) t_U	t_{TH}	t_D	t_{CH}	Symbol
PP1	PP	245	−245	ν = 1 Hz				○
PP2		198	−196					□
PC1	PC	284	−206	1	0	1	600	△
CP1	CP	245	−294	1	600	1	0	▼
CC1	CC	245	−245	1	600	1	600	●
CC2		245	−245	1	60	1	60	◑
CC3		245	−245	1	10	1	10	⊕
CC4		245	−245	1	2	1	2	⦶
CC5		196	−196	1	3600	1	3600	◆
CC6		196	−196	1	600	1	600	■
CC7		196	−196	1	60	1	60	◧
CC8		196	−196	1	10	1	10	⊞
CC9		196	−196	1	1	1	1	◫
S1	S	216	—	Static creep				+

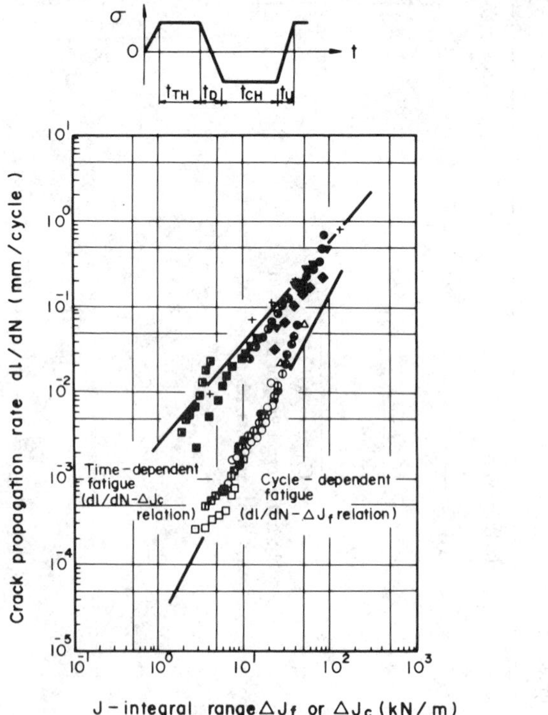

FIG. 1—*Crack propagation rate in a Cr-Mo-V cast steel of F-type as a function of fatigue J-integral range, ΔJ_f, and that of C-type as a function of creep J-integral range, ΔJ_c. It can be known from the dependence on the stress waveforms that the compression-going deformation in every cycle hardly affects the crack propagation rates.*

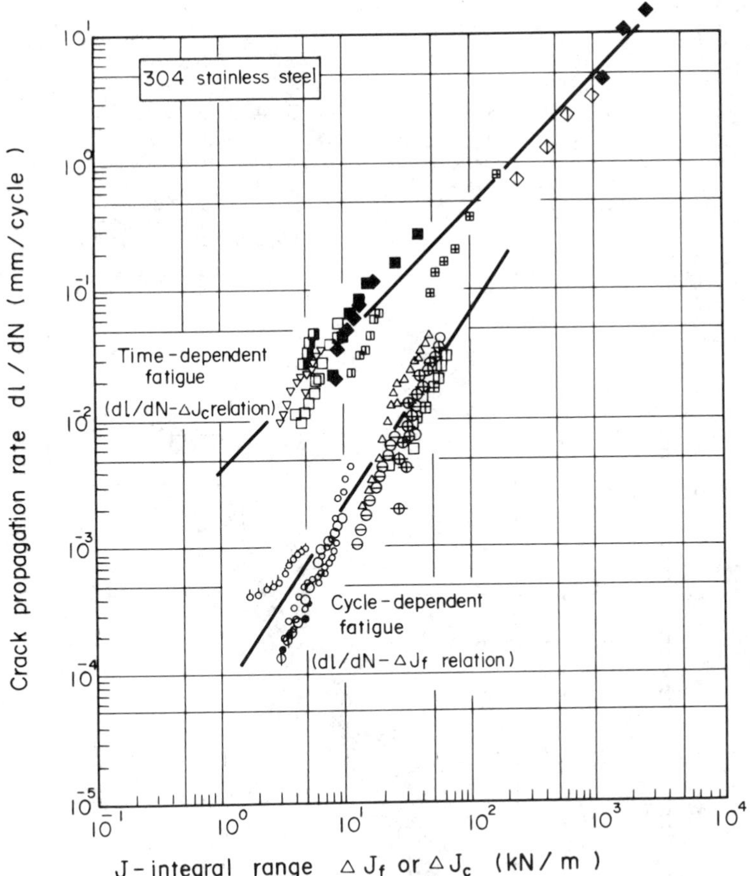

FIG. 2—*Crack propagation rates in a 304 stainless steel at temperatures from 773 to 973 K as functions of fatigue* J-*integral range and creep* J-*integral range.* $d\ell/dN$ *and* ΔJ_f *for F-type depend little on temperature.* $d\ell/dN$ *and* ΔJ_c *for C-type change to the same extent with changing temperatures. The dependence of* $d\ell/dN$-ΔJ *curves on temperature, therefore, is not explicitly evident.*

Young's modulus, and A and N are material constants. Assuming that the fatigue life of smooth specimens is equal to the crack propagation life from the initial crack length, ℓ_0, to the final crack length, ℓ_f, Eqs 14 to 16 in Table 1 (Item 12) are derived by substituting Eq 4 into Eq 1 and integrating the equation.

Time-Dependent Fatigue (C-type)

A relatively short crack is embedded in a power law creep body, obeying Eq 9 in Table 1 (Item 7), where $\dot{\epsilon}_c$ is creep strain rate, σ is stress, and B and n are material constants. The creep J-integral, J^*, under the stress waveform given in Eq 10, where σ_{max} is maximum stress, t is time, t_o is tension time, and β is constant, can be evaluated by Eq 8 in Table 1 (Item 7). Here, $J^*_{transient}$ is J^* for the stress redistribution period in the vicinity of crack tip (i.e., J^* for the transient condition from small-scale creep to large-scale creep) and $J^*_{steady\ creep}$ is J^* for the large-scale creep or the steady-state creep condition. The value of $J^*_{transient}$ calculated by the first term

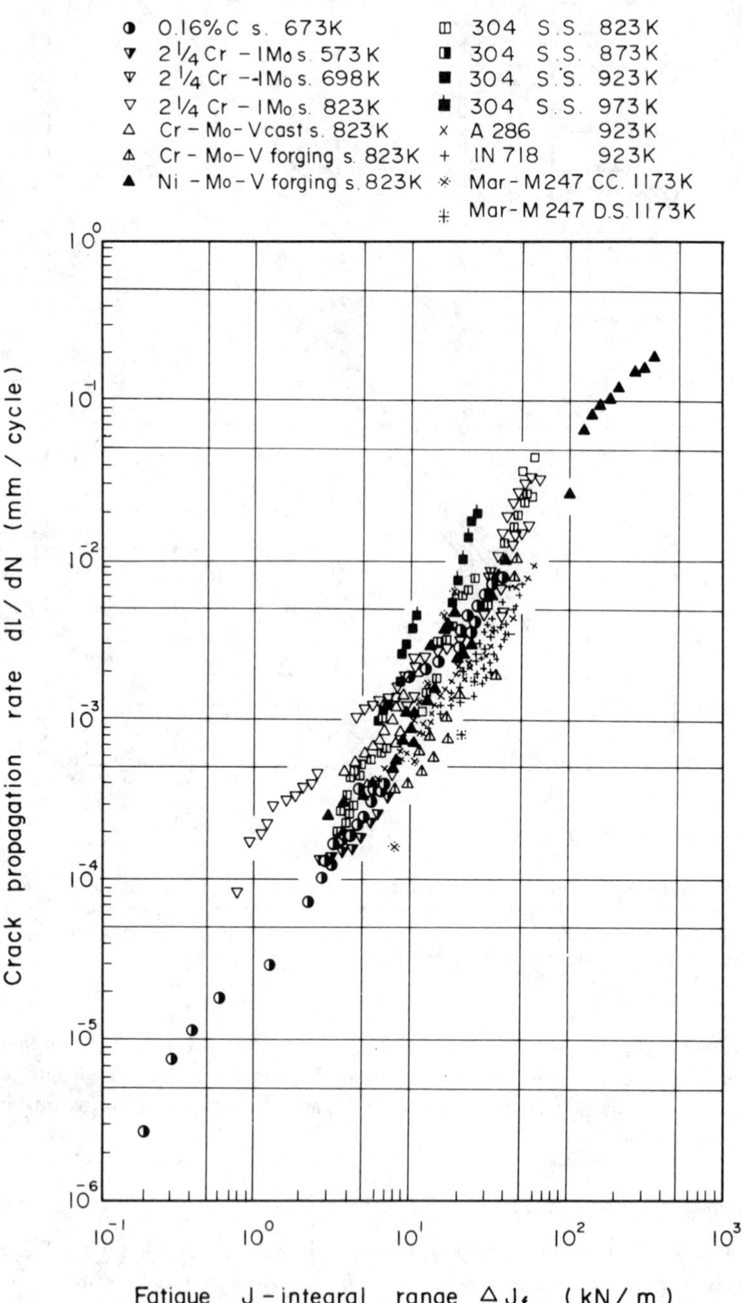

FIG. 3—*Crack propagation rate versus fatigue J-integral range for F-type fatigue on several kinds of steels and alloys. There is little difference in the propagation curves between the materials.*

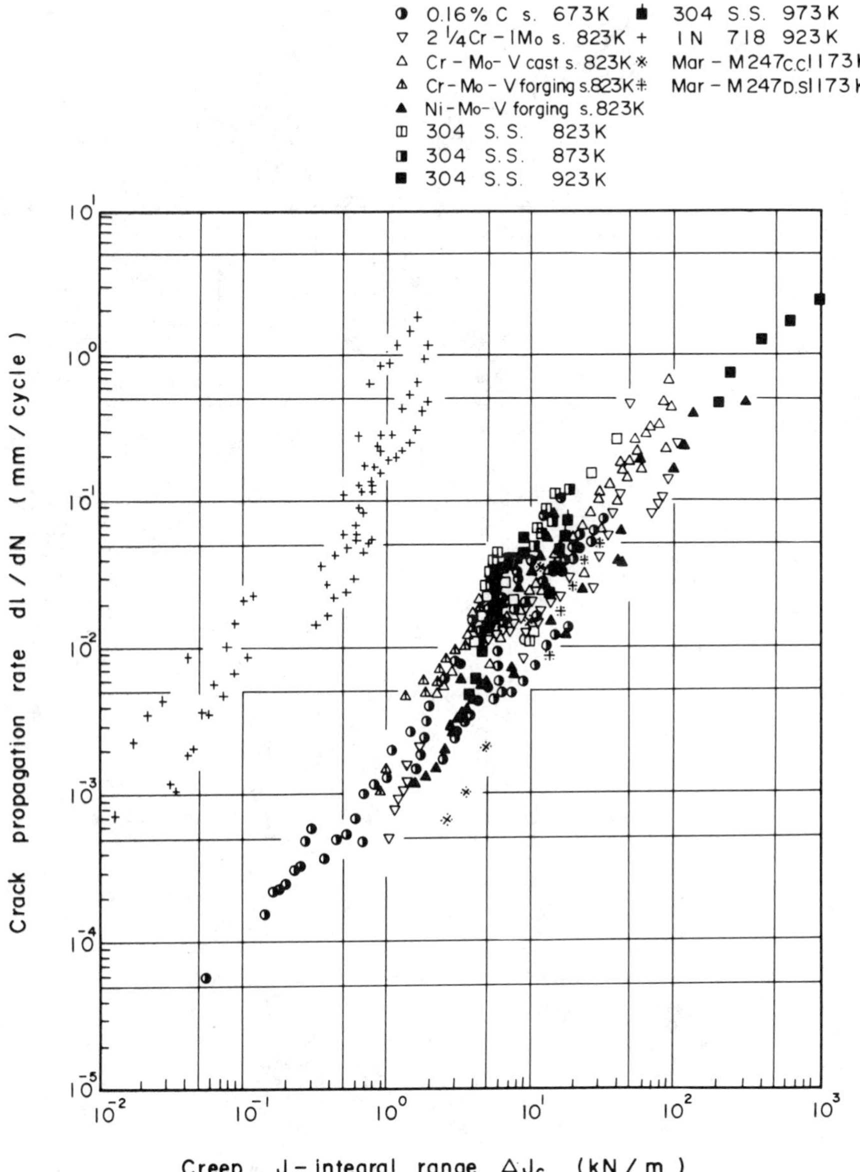

FIG. 4—*Crack propagation rate versus creep J-integral range for C-type fatigue. The difference in propagation rates between the materials is not great except for IN 718, but a detailed examination shows that the propagation rate becomes high in order of a low carbon steel, low alloy steels, stainless steels, and superalloys.*

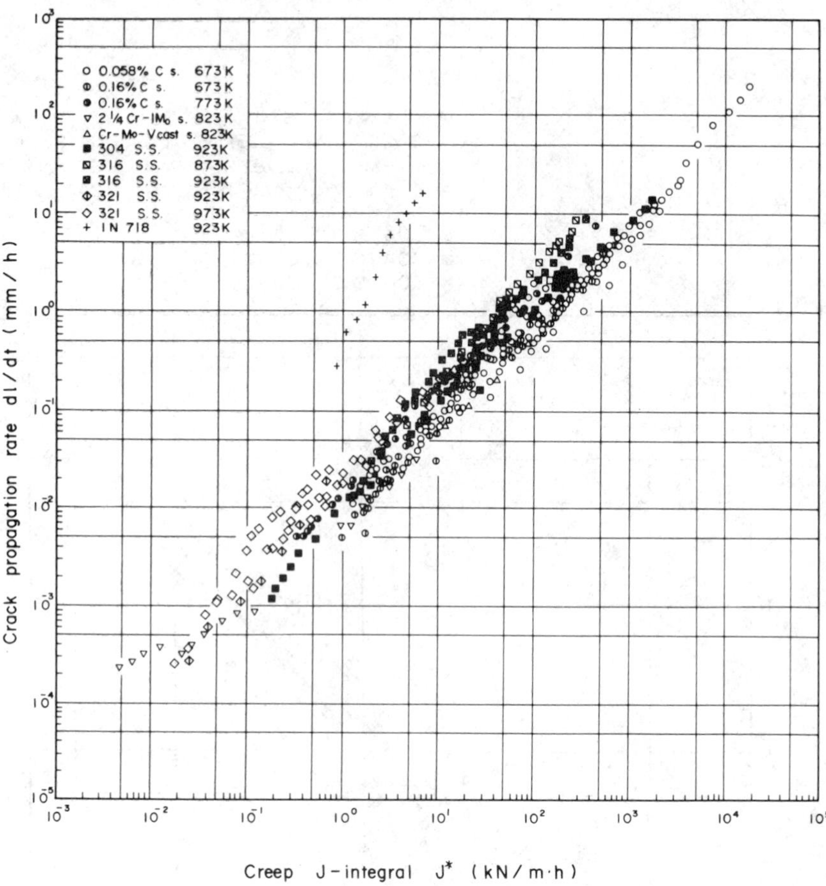

FIG. 5—*Crack propagation rate,* dℓ/dt, *versus creep* J*-integral,* J*, *for constant load creep. The curves are quite close to those of C-type fatigue (Fig. 4).*

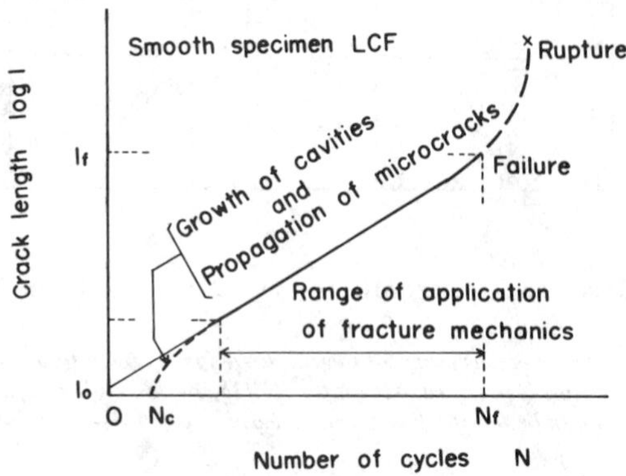

FIG. 6—*Schematic diagram showing the crack propagation behavior of a smooth specimen.*

of Eq 8 is much larger than $J^*_{\text{steady creep}}$ obtained by the second term in the transient stage and *vice versa* in the steady stage. Therefore J^* values are approximately given by their sum.

The following equation is given for an arbitrary ith cycle by substituting Eq 8 into Eq 2 and integrating during a tension period (from 0 to t_0):

$$\int_{\ell_{i-1}}^{\ell_i} \frac{d\ell}{C_c\,(2\pi M_J \ell)^{m_c}} = \int_0^{t_0} \left\{ \frac{1 + 2\beta n}{(n+1)t} \frac{\sigma \epsilon_e}{2} + \frac{n+1}{2\pi} f(n) \frac{\sigma \dot{\epsilon}_c}{n+1} \right\} dt$$

where ℓ_i is the final crack length in the ith cycle. Equations 17 to 19 can be derived by summing the above equation from the first cycle to the failure cycle. Here, to be exact, $\Delta\tilde{W}_c$ depends on m_c and β. Numerical analysis, however, reveals that the effect of m_c and β is practically negligible in the ranges $1 \leq m_c \leq 2$ and $0.3 \leq \beta \leq 1$, respectively. Therefore Eq 18 is given by assuming $m_c = 1$, $\beta = 0.5$ or $m_c = 1.5$, $\beta = 0.5$. The value of the integral in Eq 18 can be calculated by the Legendre-Gauss method.

Comparison of the Derived Life Laws with the Existing Laws

The laws expressed by Eqs 14 to 16 and 17 to 19 have a similar form to the Manson-Coffin LCF relationship [*43,44*] and the strain-range partitioning (SRP) relationship [*45,46*]. Especially when the plastic strain range, $\Delta\epsilon_p$, or the creep strain range, $\Delta\epsilon_c$, has a large part of the total strain range, there should be a close correspondence between the existing and the derived laws. However, Eqs 14 to 16 and 17 to 19 imply that the life is controlled not only by the plastic or creep strain range but also by the elastic strain range, the stress range, and the maximum tensile stress, or a combination thereof; the controlling variables depend only on the tension-going deformation in every cycle. Moreover, it should be noticed that the constants in Eq 14 and 17 can be predicted by crack propagation laws.

Some life laws on LCF have been derived by a similar procedure on the basis of other types of crack propagation equations [*47–49*]. It is stressed in our investigation, however, that the two kinds of failure life laws are logical because they are led by the reasonable fracture mechanics laws of crack propagation.

It is also noteworthy that Eqs 14 to 16 and 17 to 19 are applicable for thermal fatigue because Eqs 1 and 2 are almost independent of temperature (Fig. 2).

Verification of the Life Laws by Experimental Results of Smooth Specimens

Figure 7 shows the relationship between inelastic strain range, $\Delta\epsilon_{in}$, and the number of cycles to failure, N_f, of smooth specimens of a 2¼Cr-1Mo steel under isothermal fatigue for four types of triangular strain waveforms (fast-fast or pp-type, fast-slow or pc-type, slow-slow or cc-type, and slow-fast or cp-type) and thermal fatigue for two types of temperature-strain phases (in-phase and out-of-phase). In isothermal fatigue, the pp-type shows the longest life and the other types yield shorter lives. The results in both types of thermal fatigue fall in the band of the isothermal fatigue data. Straight-line relations between $\Delta\epsilon_{in}$ and N_f can be found in each type of strain waveform of isothermal fatigue as well as temperature-strain waveforms of thermal fatigue. According to the SRP method, the straight-line relations are revealed more definitely for the four types of strain waveforms of isothermal fatigue. The results of out-of-phase thermal fatigue tend to coincide with those of the pc-type of isothermal fatigue, the results of in-phase thermal fatigue with the cp-type. The difference between the four kinds of partitioned failure relations is not so distinct in this material.

Figure 8 shows the relationship between $\Delta\tilde{W}_f$ or $\Delta\tilde{W}_c$ and N_f. It is seen that the fatigue lives are separated into only two groups of F-type fatigue (pp- and pc-type isothermal and out-of-

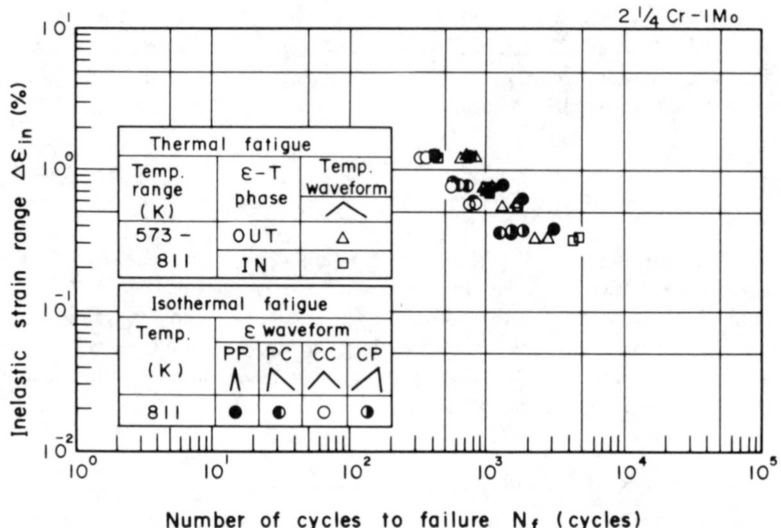

FIG. 7—*Inelastic strain range versus number of cycles to failure of smooth specimens in isothermal and thermal fatigue of 2¼Cr-1Mo steel.*

phase thermal fatigue) and C-type fatigue (cc- and cp-type isothermal and in-phase thermal fatigue); the life of the former type is longer than that of the latter. The strain energy parameter approach was also examined for many kinds of steels and alloys such as 1¼Cr-1Mo, 1Cr-1Mo-¼V cast and forging, and 2½Ni-½Mo-¹⁄₁₀V forging steels, 304 and 321 stainless steels, and IN 718, IN 738, IN 939, René 80, FSX414, 430, and Mar M247 alloys. Good results were obtained in isothermal and thermal fatigue. They showed a similar pattern of $\Delta\tilde{W}_f$ or $\Delta\tilde{W}_c$-N_f relation. The test results of 321 stainless steel and Cr-Mo-V steel are shown in Figs. 9 and 10.

The experimental results of Mar M247 are shown in Figs. 11 and 12. The strain and temperature-strain waveforms are the same as those of the 2¼Cr-1Mo steel. It is found from Fig. 11 that the life is independent of the strain waveforms in isothermal fatigue, although the transgranular fracture was observed in F-type (pc- and pp-types) and the intergranular fracture in C-type (cc- and cp-types). Consequently, the SRP relations of this alloy are equal to the $\Delta\epsilon_{in}$-N_f relation. Figure 11 also shows that the life of thermal fatigue is much shorter than that of isothermal fatigue at the temperature equal to the maximum temperature of thermal fatigue for equal inelastic strain range. It signifies that the life prediction of thermal fatigue based on the inelastic strain range or the partitioned strain range is not satisfactory for this material. Since Mar M247 has high strength and low ductility, the contribution of the elastic strain and the stress to the fatigue life is not negligible.

On the other hand, the failure lives of two types of fatigue well correlated with each strain energy parameter in both isothermal and thermal fatigue (Fig. 12). The fracture morphology proved to be different between the F-type and the C-type. This shows that the strain energy parameter approach is effective in particular for high-strength low-ductility alloys.

Since the constants D_f, m_f, D_c and m_c in Eqs 10 and 17 can be determined using the constants of the crack propagation equations, the $\Delta\tilde{W}_f$-N_f and $\Delta\tilde{W}_c$-N_f relations can be predicted by assuming fixed magnitude of the crack length, ℓ_0 and ℓ_f, and a fixed shape of the crack for a set of experiments. The lines in Figs. 8 and 10 give the predicted failure lives for $\ell_0 = 0.1$ mm as

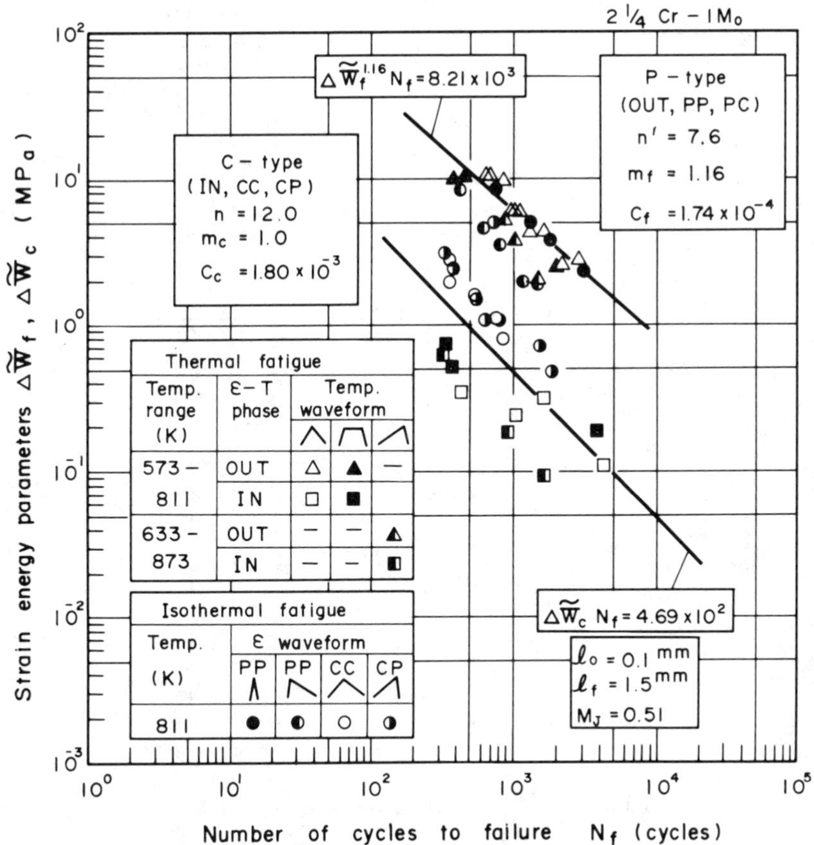

FIG. 8—*Strain energy parameters versus number of cycles to failure of smooth specimens in isothermal and thermal fatigue of 2¹/₄Cr-1Mo steel.*

the initial crack length at $N = 0$ and $\ell_f = 1.5$ mm as the final length at $N = N_f$ (Fig. 6) and for $M_J = 0.51$ which is the value for a semi-circular surface crack. The predicted lines agree fairly well with the experiments.

Conclusions

1. The crack propagation behavior was classified into cycle-dependent (F-type) and time-dependent (C-type), and the different fracture mechanics equations were obtained in each type for many steels and alloys.

For F-type:

$$d\ell/dN = C_f \Delta J_f^{m_f}$$

For C-type:

$$d\ell/dt = C_c J^{*m_c} \quad \text{or} \quad d\ell/dN = C_c \Delta J_c (m_c = 1)$$

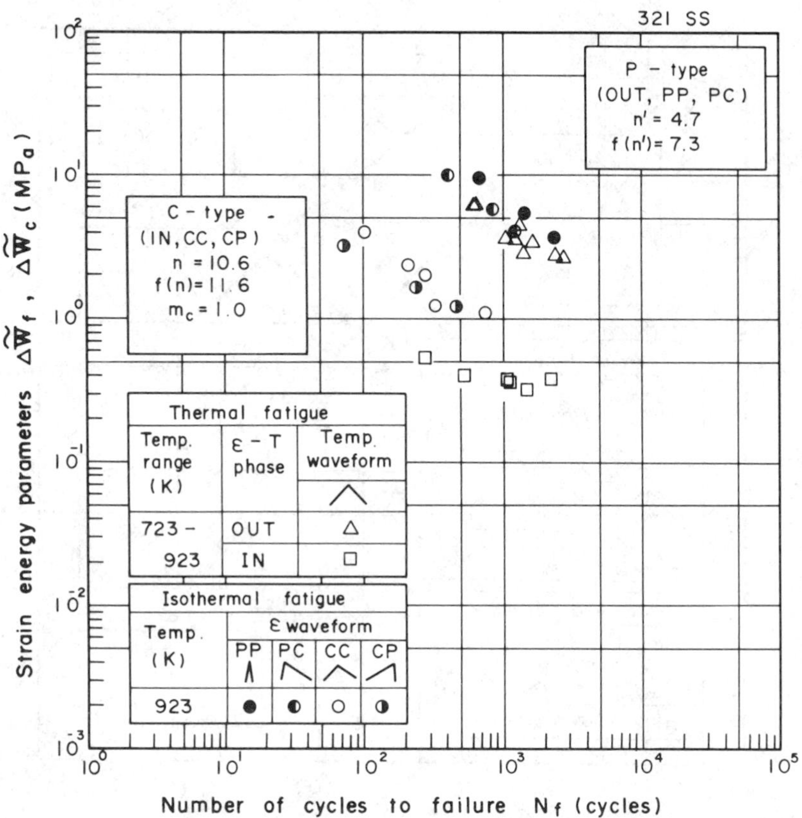

FIG. 9—*Strain energy parameters versus number of cycles to failure of smooth specimens in isothermal and thermal fatigue of 321 stainless steel.*

Here, ΔJ_f and J^* represented the elastic plastic strain range field and the creep strain rate field in the vicinity of the crack tip, respectively.

2. The life laws of smooth specimens were derived from the above crack propagation equations and given as:

For F-type:

$$\Delta\tilde{W}_f^{m_f} N_f = D_f$$

For C-type:

$$\Delta\tilde{W}_c^{m_c} N_f = D_c t_o^{1-m_c}$$

3. The validity of the life laws based on the strain energy parameters, $\Delta\tilde{W}_f$ and $\Delta\tilde{W}_c$, were confirmed by the experimental results of isothermal and thermal fatigue for many kinds of steels and alloys.

4. The other method which directly gives the relation between $\Delta\tilde{W}_f$ and N_f or $\Delta\tilde{W}_c$ and N_f by using the constants, m_f, D_f, m_c, D_c and t_0 in the crack propagation equations, proved to be successful.

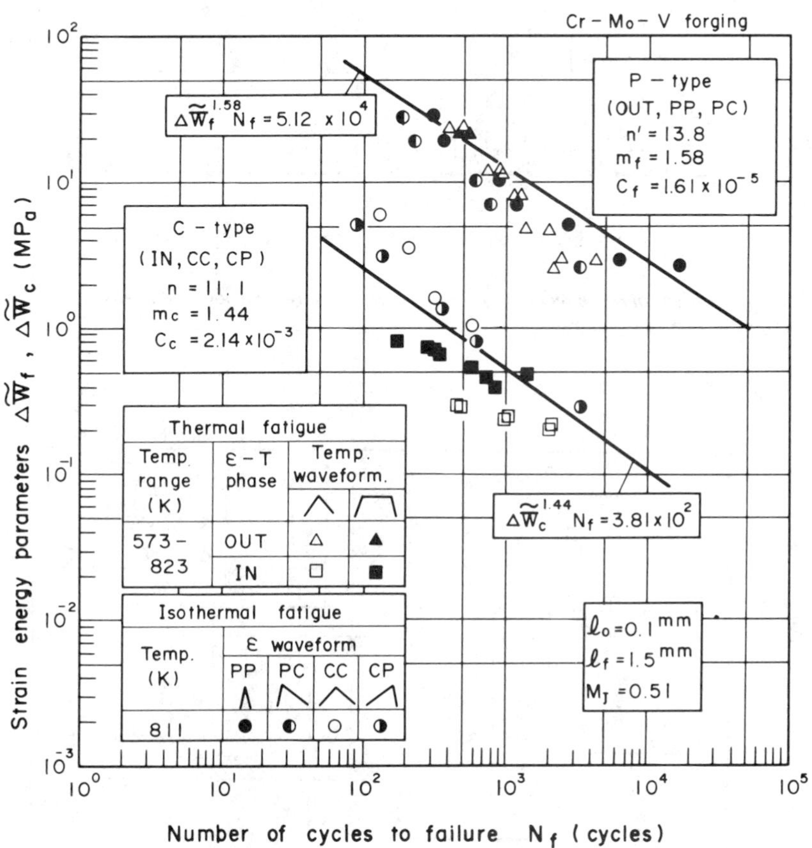

FIG. 10—*Strain energy parameters versus number of cycles to failure of smooth specimens in isothermal and thermal fatigue of Cr-Mo-V forging steel.*

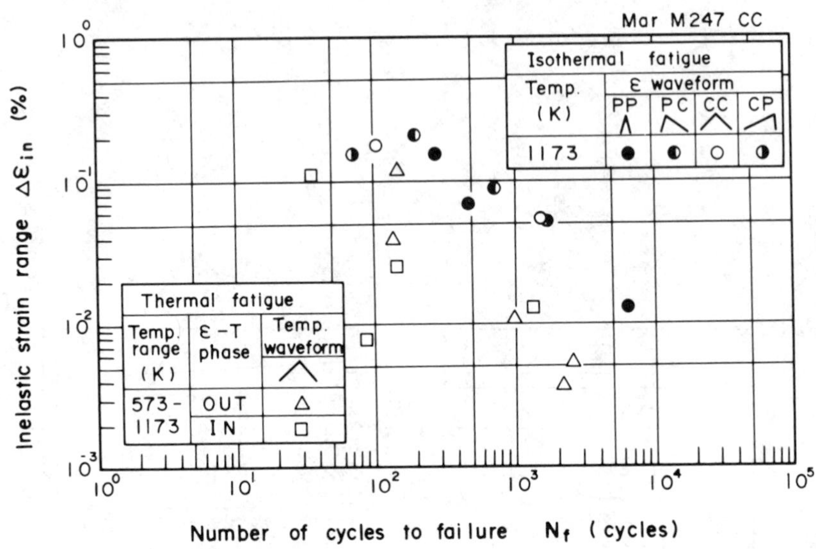

FIG. 11—*Inelastic strain range versus number of cycles to failure of smooth specimens in isothermal and thermal fatigue of Mar M247 alloy.*

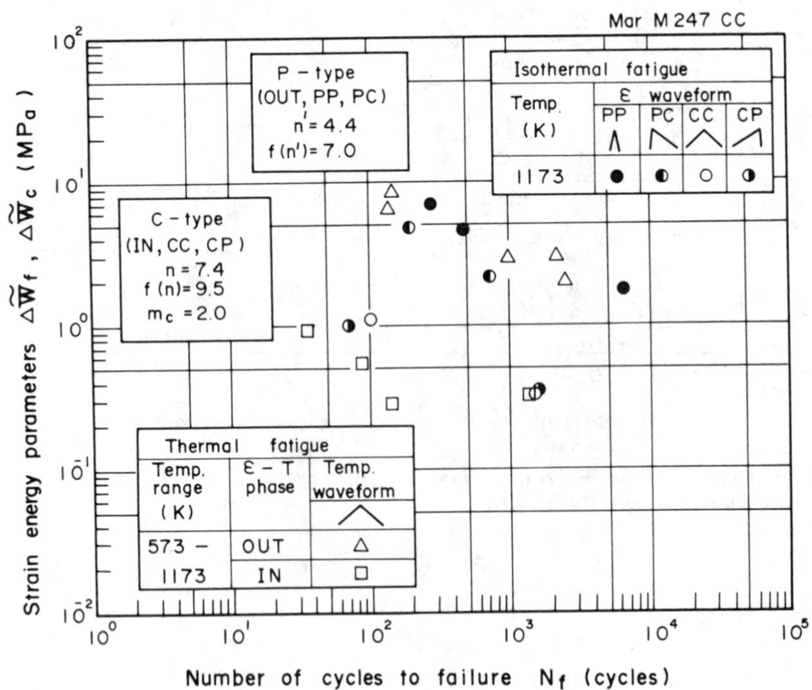

FIG. 12—*Strain energy parameters versus number of cycles to failure of smooth specimens in isothermal and thermal fatigue of Mar M247 alloy.*

Acknowledgments

This work was supported in part by a Grant in Aid of the scientific research work of the Ministry of Education, Japan. The remaining part was supported by the funds of the Science and Technology Agency, Japan.

References

[*1*] Ohmura, T., Pelloux, R. M., and Grant, M. J., *Engineering Fracture Mechanics,* Vol. 5, 1973, pp. 909-922.

[*2*] Solomon, H. D. and Coffin, L. F., Jr., in *Fatigue at Elevated Temperatures, ASTM STP 520,* American Society for Testing and Materials, Philadelphia, 1973, pp. 112-122.

[*3*] Taira, S., Ohtani, R., and Komatsu, T., *Journal of Engineering Materials and Technology, Transactions of ASME,* Vol. 101, 1979, pp. 162-167.

[*4*] Shiga, M., Sasaki, R., Hataya, F., and Kuriyama, M., *Journal of Society of Materials Science,* Vol. 28, 1979, pp. 407-413 (in Japanese).

[*5*] Ohtani, R., Kitamura, T., and Yamada, K. in *Proceedings,* U.S.-Japan Joint Seminar, Hayama, Japan, 1979, L. W. Liu, T. Kunio, V. Weiss, and H. Okamura, Eds., Martinus Nijhoff, Amsterdam, 1981, pp. 263-270.

[*6*] Ohji, K., Kubo, S., Yamakawa, N., and Kuri, T., *Transactions of JSME, Series A,* Vol. 50, 1984, pp. 1218-1227 (in Japanese).

[*7*] Ohtani, R., Yamada, K., Kashiwagi, T., and Matsubara, H., *Transactions of JSME, Series A,* Vol. 48, 1982, pp. 1378-1390 (in Japanese).

[*8*] Kuwabara, K., Nitta, A., and Kitamura, T., *Journal of Society of Materials Science,* Vol. 33, 1984, pp. 338-344 (in Japanese).

[*9*] Okazaki, M., Hattori, I., Shiraiwa, F., and Koizumi, T., *Metallurgical Transactions A,* Vol. 14-A, 1983, pp. 1649-1659.

[*10*] Kitamura, T., Nitta, A., Ogata, T., and Kuwabara, K., Technical Report of Central Research Institute of Electric Power Industry, No. 283063, 1984 (in Japanese).

[*11*] Yokobori, T., Sakata, H., and Yokobori, T., *Transactions of JSME, Series A,* Vol. 46, 1980, pp. 1062-1073 (in Japanese).

[*12*] Koterazawa, R., Nishikawa, Y., and Mori, T., *Journal of Society of Materials Science,* Vol. 27, 1978, pp. 92-98 (in Japanese).

[*13*] Ohji, K., Ogura, K., Kubo, S., Saito, H., and Fukumoto, M., *Journal of Society of Materials Science,* Vol. 33, 1984, pp. 145-151 (in Japanese).

[*14*] Horiguchi, M. and Kawasaki, T., *Journal of Society of Materials Science,* Vol. 31, 1982, pp. 277-282 (in Japanese).

[*15*] Laird, C. and Smith, G. S., *Philosophical Magazine, Series 8,* Vol. 7, 1962, pp. 847-857.

[*16*] Tanaka, K., Hoshide, T., and Sakai, N., *Engineering Fracture Mechanics,* Vol. 19, 1984, pp. 805-825.

[*17*] Taira, S., Ohtani, R., Yonekura, T., Osada, M., and Kitamura, T., *Transactions of JSME, Series A,* Vol. 46, 1980, pp. 861-869 (in Japanese).

[*18*] Ohtani, R. in *Creep in Structures,* 3rd IUTAM Symposium, Leicester, U.K., A. R. S. Ponter and D. R. Hayhurst, Eds., Springer, Berlin, 1981, pp. 542-563.

[*19*] Neate, G. J. in *Advances in Life Prediction Methods,* ASME Conference, Albany, N.Y., D. A. Woodford and J. R. Whitehead, Eds., 1983, pp. 123-129.

[*20*] Shih, C. F. and Hutchinson, J. W., *Journal of Engineering Materials and Technology, Transactions of ASME,* Vol. 98, 1976, pp. 289-295.

[*21*] Dowling, N. E. in *Crack and Fracture, ASTM STP 601,* American Society for Testing and Materials, Philadelphia, 1976, pp. 19-32.

[*22*] Dowling, N. E. in *Cyclic Stress-Strain and Plastic Deformation Aspects of Fatigue Crack Growth, ASTM STP 637,* American Society for Testing and Materials, Philadelphia, 1977, pp. 97-121.

[*23*] Harper, M. P. and Ellison, E. G., *Journal of Strain Analysis,* Vol. 12, 1977, pp. 167-179.

[*24*] Ellison, E. G. and Harper, M. P., *Journal of Strain Analysis,* Vol. 13, 1978, pp. 35-51.

[*25*] Ohji, K., Ogura, K., and Kubo, S., *Journal of Society of Materials Science,* Vol. 29, 1980, pp. 456-471 (in Japanese).

[*26*] Riedel, H. and Rice, J. R. in *Fracture Mechanics (Twelfth Conference), ASTM STP 700,* American Society for Testing and Materials, Philadelphia, 1980, pp. 112-130.

[*27*] Fujino, M., Ishikawa, F., and Ohtani, R., *Journal of Society of Materials Science,* Vol. 29, 1980, pp. 113-119 (in Japanese).

[28] Kubo, S. in *Creep in Structures,* 3rd IUTAM Symposium, Leicester, U.K., A. R. S. Ponter and D. R. Hayhurst, Eds., Springer, Berlin, 1981, pp. 606–609.
[29] Ohtani, R. and Kitamura, T., *Journal of Society of Materials Science,* Vol. 34, 1985, pp. 843–849 (in Japanese).
[30] Inoue, T., *Journal of Society of Materials Science,* Vol. 32, 1983, pp. 594–604 (in Japanese).
[31] Kitamura, T., Ohtani, R., and Sugihara, H. in *Proceedings,* 34th Annual Symposium of the Society of Materials Science, Japan, 1985, pp. 49–51 (in Japanese).
[32] Ohji, K., Kubo, S., Tokuhiro, K., and Iwai, S. in *Proceedings,* 34th Annual Symposium of the Society of Materials Science, Japan, 1985, pp. 58–60 (in Japanese).
[33] Ohtani, R., Okuno, M., and Shimizu, R., *Journal of Society of Materials Science,* Vol. 31, 1982, pp. 505–509 (in Japanese).
[34] Ohtani, R. and Nakayama, S., *Journal of Society of Materials Science,* Vol. 32, 1983, pp. 635–639 (in Japanese).
[35] Ohtani, R., Nakayama, S., and Taira, T., *Journal of Society of Materials Science,* Vol. 33, 1984, pp. 590–595 (in Japanese).
[36] Ohtani, R., Kinami, T., Sakamoto, H., and Yano, A. in *Proceedings,* 34th Annual Symposium of the Society of Materials Science, Japan, 1985, pp. 61–63 (in Japanese).
[37] Usami, S., Fukuda, Y., and Shida, S., *Journal of Society of Materials Science,* Vol. 33, 1984, pp. 685–691 (in Japanese).
[38] Chen, I-W. and Argon, A. S., *ACTA Metallurgica,* Vol. 29, 1981, pp. 1759–1768.
[39] Ohtani, R., *Bulletin of JIM,* Vol. 22, 1983, pp. 190–196 (in Japanese).
[40] Okazaki, M., Hattori, I., Ikeda, T., and Koizumi, T. in *Proceedings,* 22nd Symposium of High Temperature Strength, Society of Materials Science, Japan, 1984, pp. 137–141 (in Japanese).
[41] Mowbray, D. F. in *Cracks and Fracture, ASTM STP 601,* American Society for Testing and Materials, Philadelphia, 1976, pp. 33–46.
[42] Yamada, T., Hoshide, T., Fujimura, S., and Manabe, M., *Transactions of JSME, Series A,* Vol. 49, 1983, pp. 441–451 (in Japanese).
[43] Manson, S. S. in *Proceedings,* Heat Transfer Symposium, University of Michigan, Ann Arbor, Engineering Research Institute, 1953, p. 9.
[44] Coffin, L. F., Jr., *Transactions of ASME,* Vol. 76, 1954, pp. 931–950.
[45] Halford, G. R., Hirshberg, M. H., and Manson, S. S. in *Fatigue at Elevated Temperatures, ASTM STP 520,* American Society for Testing and Materials, Philadelphia, 1973, pp. 685–669.
[46] Manson, S. S. in *Fatigue at Elevated Temperatures, ASTM STP 520,* American Society for Testing and Materials, Philadelphia, 1973, pp. 744–782.
[47] Tomkins, B. and Wareing, J., *Metal Science,* Vol. 11, 1977, pp. 414–424.
[48] Majumder, S. and Maiya, P. S., *Journal of Engineering Materials and Technology, Transaction of ASME,* Vol. 102, 1980, pp. 159–167.
[49] Romanoski, G. R., Antolovich, S. D., and Pelloux, R. M., "A Model for Life Predictions of Nickel-Base Superalloys in High-Temperature Low Cycle Fatigue," this publication, pp. 456–468.

Shoji Harada,[1] *Yukitaka Murakami,*[2] *Yoshihiro Fukushima,*[1] *and Tatsuo Endo*[1]

Reconsideration of Macroscopic Low Cycle Fatigue Laws Through Observation of Microscopic Fatigue Process on a Medium Carbon Steel

REFERENCE: Harada, S., Murakami, Y., Fukushima, Y., and Endo, T., **"Reconsideration of Macroscopic Low Cycle Fatigue Laws Through Observation of Microscopic Fatigue Process on a Medium Carbon Steel,"** *Low Cycle Fatigue, ASTM STP 942,* H. D. Solomon, G. R. Halford, L. R. Kaisand, and B. N. Leis, Eds., American Society for Testing and Materials, Philadelphia, 1988, pp. 1181–1198.

ABSTRACT: In order to correlate macroscopic low-cycle fatigue laws such as the Manson-Coffin relation and Miner's rule with the actual microscopic fatigue process, cyclic strain-controlled low-cycle fatigue tests were conducted on a medium carbon steel under constant-, two-step-, and random-strain amplitude conditions. Through continual observation of the microscopic fatigue process of plain specimens and holed specimens with a very small hole (40 μm diameter), it was found that the total fatigue life is mostly occupied by the microcrack propagation life. Both the Manson-Coffin relation and Miner's rule are derived using the microcrack propagation law. Then, background for the applicability of Miner's rule is discussed.

The results of the random loading fatigue tests showed that microcrack propagation life can be predicted on the basis of the microcrack growth law in conjunction with the rain flow method for counting fatigue damage. Finally, the microcrack growth law-aided approach was applied to interpret anisotropy in fatigue strength of a thick forged steel plate.

KEY WORDS: carbon steel, microcrack initiation and propagation, microcrack propagation law, Manson-Coffin relation, Miner's rule, fatigue damage, two-step loading, block-random and random loading, rain flow method, anisotropy in fatigue strength

Extensive studies have been performed on low-cycle fatigue problems since Manson [*1*] and Coffin [*2*] independently formulated an empirical fatigue law, simply referred to as the *Manson-Coffin relation.* Validity of this law has been experimentally verified on various kinds of materials [*3*]. Parallel to the experimental analyses, several attempts have also been made to understand the Manson-Coffin relation from a theoretical point of view [*4–9*]. In those theoretical analyses, the assumption is mostly made that the entire fatigue process can be divided into a crack initiation and a propagation stage. Those theories, however, have not necessarily been supported by experiments.

Prediction of fatigue life of structural members subjected to varying strain amplitude is of great importance in engineering practice. It is usually done using a method of counting fatigue damage [*5,10–13*] evaluated in conjunction with the Palmgren-Miner cumulative damage law

[1]Associate Professor, Research Assistant, and Professor, respectively, Department of Mechanical Engineering, Kyushu Institute of Technology, Tobata, Kitakyushu, 804 Japan.

[2]Professor, Department of Mechanics and Strength of Solids, Faculty of Engineering, Kyushu University, Higashi-ku, Fukuoka, 812 Japan.

[*14,15*] (simply designated as *Miner's rule* later). In such a case, the concept of "fatigue damage" is, in general, implicitly regarded as a certain parameter representing reduction or decrease in residual fatigue strength of a material. Owing to ambiguity of the fatigue damage concept, confusion arises in defining crack initiation life. For example, definition of initiation life of an engineering-size crack is usually taken as the number of stress or strain reversals until initiation of a crack 1 mm long. Obviously such a definition is ambiguous, because the crack initiation life varies with the size of a specimen.

For a better understanding of fatigue damage, Kikukawa et al. [*16*], Nisitani and Morita [*17*], and Yamada et al. [*18*] showed through microscopic observation of the fatigue process on a specimen that the physical meaning of fatigue damage actually corresponds to microcrack initiation and propagation at the specimen surface in the course of the fatigue process. Therefore, previous studies [*5,10–13*] related to fatigue life prediction under random loading should be reexamined from the standpoint of whether prior history or pre-fatigue damage by stress or strain reversal affects the process of subsequent crack initiation and propagation.

Apart from the discussion stated above, increasing interest is currently focused on problems of short or small cracks [*19–22*]. In particular, elasto-plastic fracture mechanics-aided approaches are applied to evaluate fatigue life on the basis of a small crack growth law [*18,23*]. It seems at present fundamentally possible to predict fatigue life of structural members subjected to various kinds of loading with an accuracy that is of practical use.

It should be clarified to what extent the macroscopic fatigue laws such as the Manson-Coffin relation and Miner's rule actually reflect and involve the microscopic fatigue process. In this regard, fundamental questions arise: (1) How can the Manson-Coffin relation be correlated with the growth law of a small crack? (2) What does fatigue damage physically mean in conjunction with the actual fatigue process? The present authors have already studied these problems to some extent [*24*]. The main objective of the present study is to throw light on the questions stated above.

Cyclic strain-controlled low-cycle fatigue tests were performed on a medium carbon steel so as to clarify first on a microscale what happens in the actual fatigue process and then correlate the results with macroscopic and mechanics-oriented fatigue laws.

In order to demonstrate actual applications of the experimental finding obtained through the fundamental tests, low-cycle fatigue tests were carried out under random loading. Furthermore, fatigue tests were also conducted on a thick forged steel plate where the growth of microcracks originated at the elongated manganese-sulfide inclusions strongly control the total fatigue process.

Material and Test Procedure

The material used during the main course of this investigation is a medium carbon steel round bar (S45C; 0.46%C, diameter 22 mm). The material was first annealed at 844°C for 1 h and then turned. The chemical composition and mechanical properties of this material are tabulated in Table 1 and Table 2, respectively. The mean grain size of ferrite and pearlite of the material are about 19 μm and 17 μm, respectively. The hour-glass shaped specimens (minimum diameter 8 mm in the cross section; stress concentration factor 1.027 [*25*]) shown in Fig. 1*a* were used in the fatigue tests. The specimen surface was finished with a grit paper (Grade No. 06) and then about 30 μm per diameter was removed by electropolishing. For comparison, holed specimens which have a very small drilled hole (diameter 40 μm and depth 40 μm) on the periphery in the central portion of plain specimens (Fig. 1*b*) were also prepared.

Cyclic total strain range-controlled, push-pull, low-cycle fatigue tests were conducted in a closed-loop electro-hydraulic servo-controlled fatigue testing machine. Cyclic strain amplitude applied to each specimen was controlled by detecting the change of the minimum diameter with

TABLE 1—*Chemical composition (weight percent) of the material tested.*

C	Si	Mn	P	S	Cu	Ni	Cr
0.46	0.27	0.74	0.015	0.02	0.02	0.01	0.10

TABLE 2—*Mechanical properties of the material tested.*

σ_S	σ_B	σ_T	ψ
304	580.5	966.5	49.4

σ_S: Yield stress, MPa
σ_B: Ultimate tensile strength, MPa
σ_T: True rupture stress, MPa
ψ: Reduction of area, %

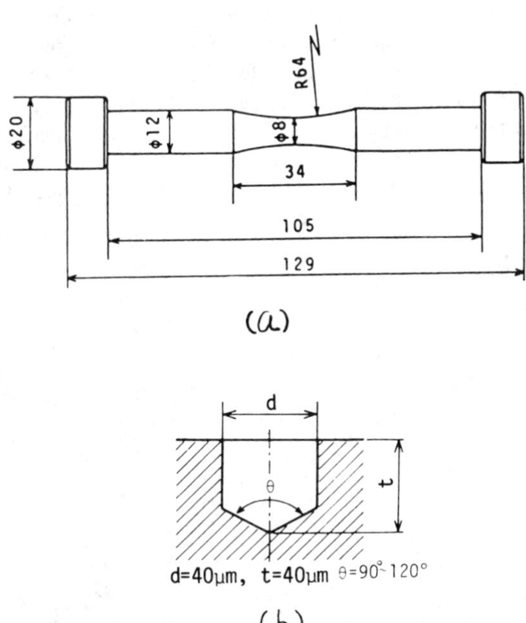

FIG. 1—(a) *Geometry of specimens.* (b) *Shape of a drilled hole.*

two pairs of linear differential transformers. The speed of the strain reversal was varied from 0.06 to 0.3Hz, depending on the case.

Continual observation of the microcrack initiation and propagation behavior was performed by a plastic replication technique on each specimen. For all the plain and holed specimens, the total crack length of the main crack, which includes the length of a zigzag-like crack and the diameter of the initial hole, was measured by projecting the crack length in a direction perpendicular to the loading axis.

Experimental Results and Discussion

Fatigue tests were conducted on the plain and holed specimens, mainly at three levels of the cyclic plastic strain range $\Delta\epsilon_p$ (i.e., $\Delta\epsilon_p = 0.65$, 1.56, and 2.73%), where $\Delta\epsilon_p$ denotes the average bulk strain in the cross section of both specimens. Figure 2 shows the log-log relation between the cyclic plastic strain range ($\Delta\epsilon_p$) and the number of strain cycles to failure (N_{fo} and N_f) for a plain and a holed specimen, respectively. It is found that both the $\Delta\epsilon_p$-N_{fo} and $\Delta\epsilon_p$-N_f relations of the plain and holed specimens show a log-log linear relation of the Manson-Coffin type. They can be given as

$$\text{Plain specimen: } N_{fo} \cdot \Delta\epsilon_p^{1.81} = 1.05 \tag{1}$$

$$\text{Holed specimen: } N_f \cdot \Delta\epsilon_p^{1.85} = 0.883 \tag{2}$$

The fact that the holed specimens show almost the same failure lives as those of the plain specimens means that the initial hole (40 μm in diameter), as a defect or a flaw, has only a slight effect on fatigue life. A broken line depicted in the figure indicates an initiation life of a small crack about 20 μm long. This crack size corresponds to the size of the ferrite grain of the material tested. From this experimental evidence that a very small crack of the order of the ferrite grain size initiates at a very early stage of the fatigue process, it is judged that the total fatigue life is mostly dominated by the crack propagation life. In other words, the Manson-Coffin relation virtually represents a growth law of a small crack.

Using the plastic replication technique, the fatigue process of each specimen was continually observed at the specimen surface on a microscale. Figure 3 shows an example of the continual observation of initiation and propagation of a leading crack for the plain specimen at $\Delta\epsilon_p =$ 1.53%. In the plain specimens, the leading crack mainly initiated in the ferrite grain or at a

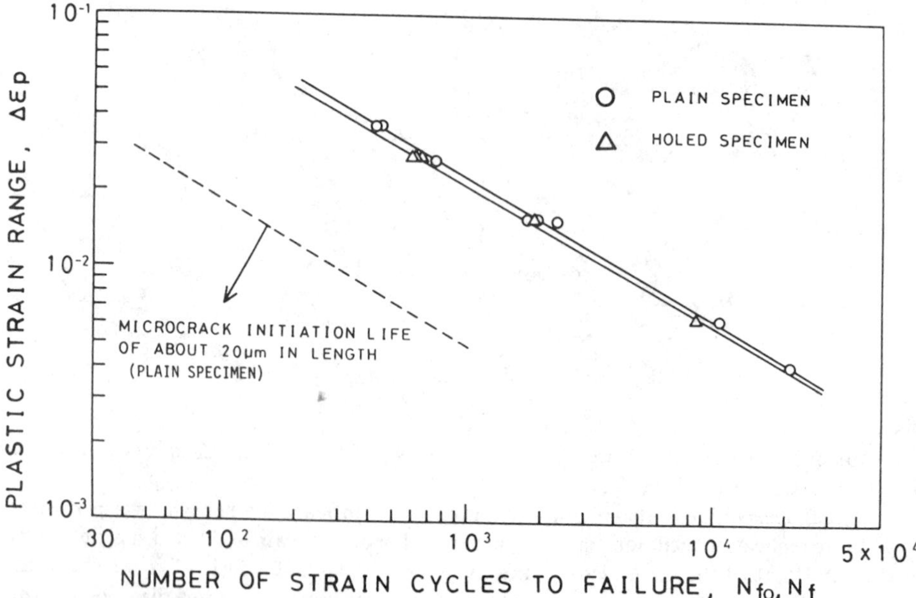

FIG. 2—*Relation between the cyclic plastic strain range* ($\Delta\epsilon_p$) *and the number of strain cycles to failure* (N_{fo} and N_f) *for the plain and holed specimens, respectively.*

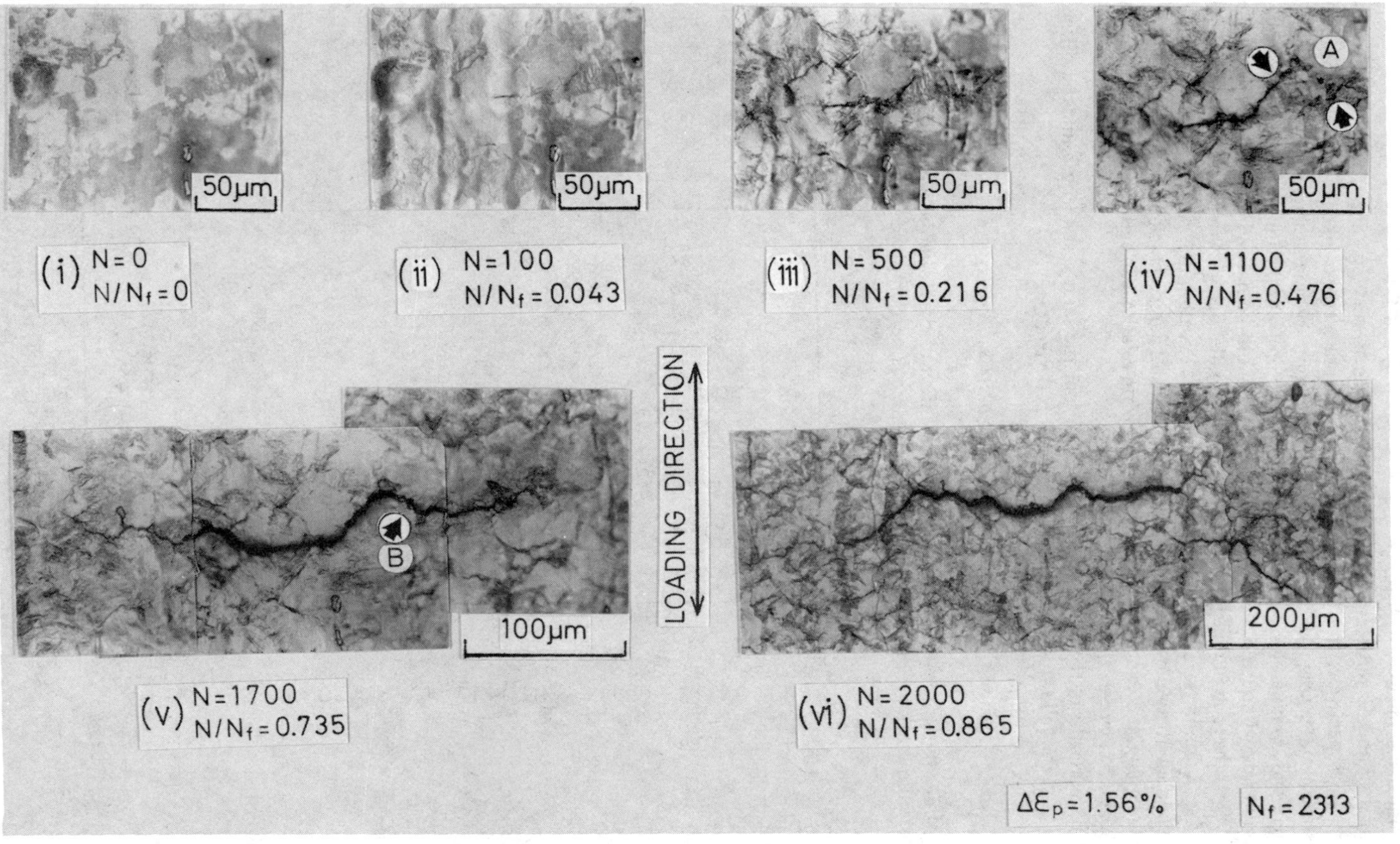

FIG. 3—*Example of the continual observation of initiation and propagation of a leading crack for a plain specimen at* $\Delta\epsilon_p = 1.53\%$.

position near the ferrite-pearlite grain boundary, while it originated on the periphery of the hole in the holed specimens.

Figures 4*a* and 4*b* represent crack extension curves at each level of $\Delta\epsilon_p$ for plain and holed specimens, respectively. In the plain specimens, the main crack linked with rather large sub-cracks at the later stage of crack growth. On the contrary, the fatigue process of the holed specimen was dominated by a leading crack emanating from the initial hole and propagated predominantly without linking with subcracks. It should be noticed in the figures that more than 90% of the total fatigue life is occupied by the small crack growth process up to a length of about 1 to 2 mm in the specimens used in this study. Recalling the possibility of a one-to-one correspondence between the Manson-Coffin relation and a microcrack growth law as postulated in the foregoing, it is expected that the Manson-Coffin relation can actually be derived using the crack extension curves shown in Figs. 4*a* and 4*b*.

From the results of the crack growth rate $d\ell/dN$ for the same crack length under a different level of $\Delta\epsilon_p$, the following experimental equation is derived:

$$\text{Holed specimen: } d\ell/dN \propto \Delta\epsilon_p^{1.58} \cdot f(\ell) \tag{3}$$

where $f(\ell)$ denotes a function of ℓ. To avoid the effect of the initial hole on subsequent crack growth, valid crack growth data for the holed specimens were employed after a main crack grew

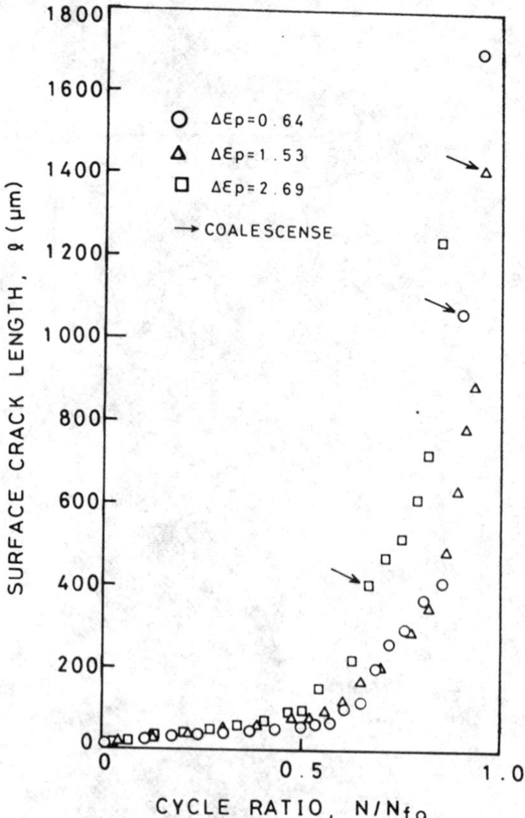

FIG. 4*a*—*Crack extension curves for plain specimens.*

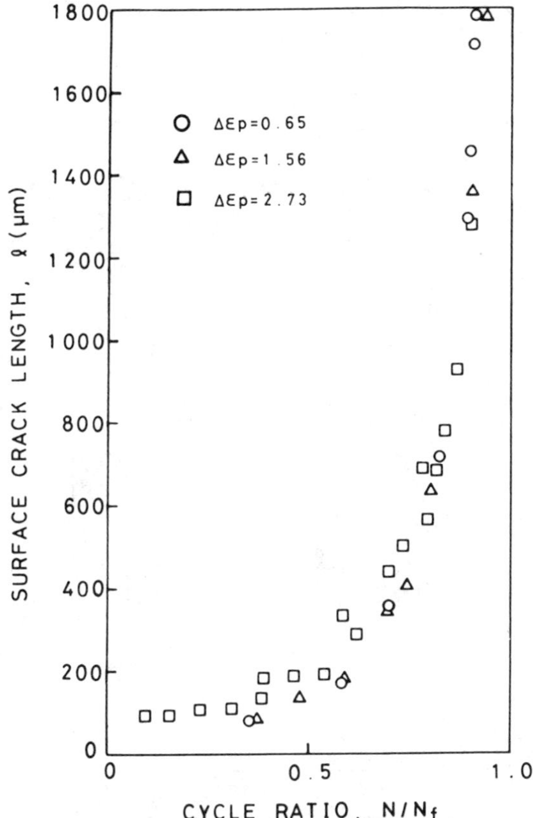

FIG. 4*b*—*Crack extension curves for holed specimens.*

up to 100 μm in length. Furthermore, to obtain the $d\ell/dN$-ℓ relation, the quantity $(d\ell/dN)/\Delta\epsilon_p^{1.58}$ was plotted against ℓ on the log-log diagram (Fig. 5). In interpreting the data, the following factors should be considered as discussed in the previous paper [*24*]: (1) those data generally scatter because of the nature of the microcrack (i.e., influence of the microstructure upon the crack growth rate at the early stage as observed in the results of the plain specimens) and variation of the aspect ratio or crack shape, (2) the total fatigue life is dominated by a period of microcrack propagation until the microcrack grows to 500 to 600 μm (Fig. 4), and (3) one of the necessary conditions for Miner's rule to hold is that the microcrack propagation law shows a linear type of the relation $d\ell/dN = C\ell$.

From the data shown in Fig. 5, the microcrack propagation law can be expressed by $(d\ell/dN)/\Delta\epsilon_p^{1.58} = C \cdot \ell^m$, where C and m are material constants. A best fit to all the data gives $m \sim 1.4$. However, considering the fact that the value of $m \sim 1.4$ represents the overall growth behavior of the microcrack where the effects of the factors aforementioned are reflected, a linear type of microcrack propagation law ($m = 1$), which has already been proposed by other researchers [*26–30*], is simply assumed. Then, the microcrack propagation law governing the main crack growth stage can be established as follows:

$$(d\ell/dN)/\Delta\epsilon_p^{1.58} = 2.4 \cdot \ell \tag{4}$$

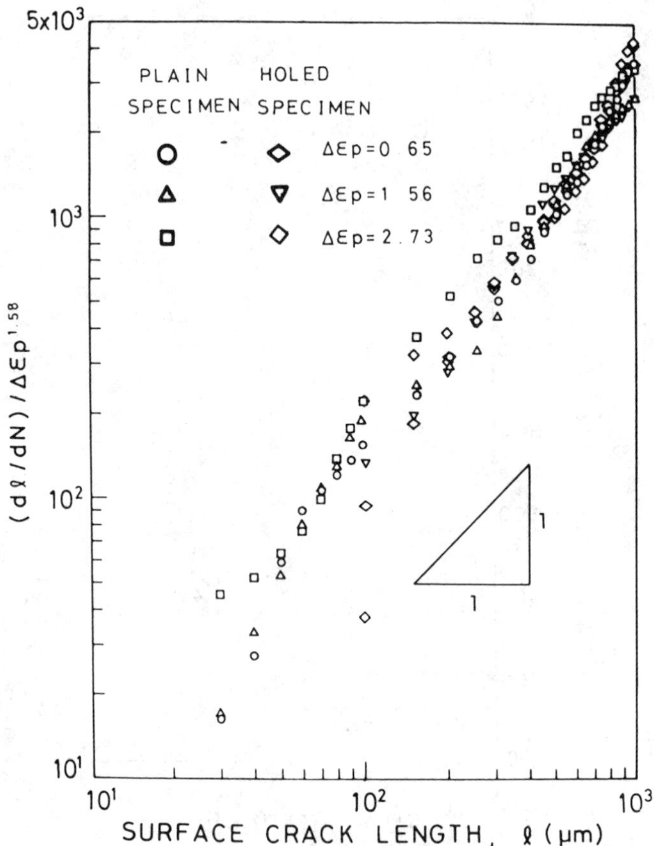

FIG. 5—*Relation between the quantity* $(d\ell/dN)/\Delta\epsilon_p^{1.58}$ *and the crack length* ℓ *of the plain and holed specimens.*

Equation 4 overestimates the crack growth rate at the early stage and underestimates it at the final stage. After integration, Eq 4 reduces to

$$\ell n(\ell_f/\ell_o) = 2.4 \cdot \Delta\epsilon_p^{1.58}\,(N_f - N_o) \tag{5}$$

where ℓ_o denotes a crack length at the microcrack initiation life N_o and ℓ_f denotes a crack length at specimen failure N_f. Taking the foregoing discussion (Fig. 4) into account, the following conditions are substituted into Eq 5: $\ell_o = 40\ \mu$m (diameter of the initial hole) at microcrack initiation of $N = N_o$, $N_o \cong 0$ ($\because N_o \ll N_f$), and $\ell_f = 8$ mm (measured value) at failure of $N = N_f$. Then, finally the following equation of the Manson-Coffin type is obtained:

$$N_f \cdot \Delta\epsilon^{1.58} = 2.21 \tag{6}$$

Figure 6 shows a comparison of the fatigue lives of the plain specimens determined experimentally with those predicted by using an equation similarly derived for the plain specimens as was done for the holed specimens. Evidently the predicted lives are in good agreement with the experimental ones.

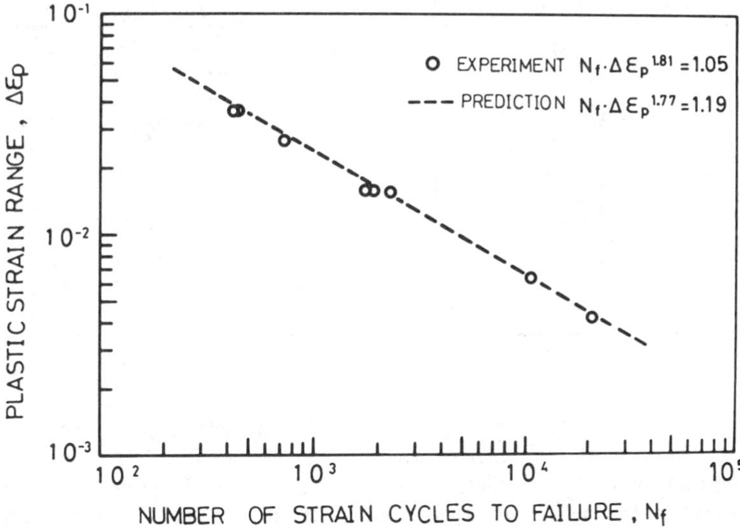

FIG. 6—*Comparison of the fatigue lives of the plain specimens predicted on the basis of the microcrack propagation law with those determined experimentally.*

The discussion aforementioned is expected to be applicable to other materials (e.g., a low carbon steel [*17*] and a 70-30 brass [*31*]). Indeed, Murakami et al. [*32*] recently demonstrated that the present method can be applied to treat a 70-30 brass of which the fatigue process is different from that of the material used in this study.

In considering the effect of the main crack linking with subcracks on subsequent crack growth rate, it is generally thought that the coalescence accelerates main crack growth, as is apparently observed in Fig. 4*b*. However, the coalescence of the main crack with subcracks had only a slight effect on the subsequent crack growth rate in this study. In this respect, a simple model is taken to have a general view for understanding the coalescence effect. Suppose a large (main crack) and small (subcrack) semi-circular surface crack; even if these cracks were joined, the internal shape of the main crack front hardly changes. This is the reason that the coalescence only slightly affects the overall growth rate of the main crack. In fact, this nature of the main crack growth rate at joining has been verified experimentally [*24,32*].

Derivation of Miner's Rule on the Basis of the Microcrack Propagation Law

Two-step loading fatigue tests were conducted to examine Miner's rule in conjunction with the microcrack propagation law. The results obtained are listed in Table 3. The combination of cyclic plastic strain range $\Delta\epsilon_{p1}$ and $\Delta\epsilon_{p2}$ of first- and second-step strain cycling and fatigue damage N_1/N_{f1} (N_1 = number of first-step strain cycles, N_{f1}: failure life at $\Delta\epsilon_{p1}$) were varied as tabulated in the table. According to the previous report by Kikukawa et al. [*5*], Miner's rule is judged generally to be satisfied if the value of the linear cumulative damage ($N_1/N_{f1} + N_2/N_{f2}$) is between 0.6 and 1.5. Comparing the present results with this criterion, it is recognized that Miner's rule holds within a high degree of accuracy in this study.

In what follows, Miner's rule is discussed on the basis of the microcrack propagation law. Figure 7 shows the $(d\ell/dN)/\Delta\epsilon_p^{1.58}$-$\ell$ relation after switching of the cyclic strain level, which is derived in the same manner as in the case of the constant strain amplitude. A solid line shown in the figure implies the $(d\ell/dN)/\Delta\epsilon_p^{1.58}$-$\ell$ relation under the constant strain amplitude. It seems

TABLE 3—*Results of fatigue tests under two-step strain-amplitude condition.*

	$\Delta\epsilon_{p1}$ (%)	$\Delta\epsilon_{p2}$ (%)	N_1/N_{f1}	N_2/N_{f2}	$\Sigma N_1/N_{f1}$
Plain specimens	0.65	1.54	0.50	0.66	1.16
	1.53	0.64	0.50	0.59	1.09
	2.73	0.65	0.49	0.59	1.08
Holed specimens	0.66	2.74	0.50	0.51	1.01
	2.74	1.57	0.50	0.47	0.97
	1.59	0.66	0.50	0.34	0.84
	2.74	0.66	0.50	0.34	0.84
	2.75	1.56	0.69	0.34	1.03
	2.74	1.57	0.31	0.64	0.95

that the crack growth rate after switching is decelerated at the early stage and accelerated at the final stage, as compared with the crack growth rate under constant strain amplitude. However, remembering that the crack propagation law represented by Eq 4 is applied to estimate the crack growth rate of the main growth stage and not the early and final stage, it is judged, as a whole, that the crack growth rate after switching can be characterized by the crack growth rate under constant strain amplitude, irrespective of the combinations of the strain level $\Delta\epsilon_{p1}$, $\Delta\epsilon_{p2}$, and regardless of the value of N_1/N_{f1}. In addition, it is suggested that the crack growth just after switching may be affected by crack closure behavior to some extent.

Taking into account the above discussion together with Eq 4 and Table 3, the following equations are obtained for the two-step loading tests:

$$\begin{aligned} \ell n(\ell_f/\ell_0) &= 2.4 \cdot \Delta\epsilon_{p1}^{1.58} \cdot N_{f1} \\ \ell n(\ell_f/\ell_0) &= 2.4 \cdot \Delta\epsilon_{p2}^{1.58} \cdot N_{f2} \\ \ell n(\ell_1/\ell_0) &= 2.4 \cdot \Delta\epsilon_{p1}^{1.58} \cdot N_1 \\ \ell n(\ell_f/\ell_1) &= 2.4 \cdot \Delta\epsilon_{p2}^{1.58} \cdot N_2 \end{aligned} \tag{7}$$

where ℓ_1 denotes a crack length at the switching of the strain level. We obtain

$$\Sigma \frac{N_i}{N_{fi}} = \frac{N_1}{N_{f1}} + \frac{N_2}{N_{f2}} = \frac{\ell n(\ell_1/\ell_0)}{\ell n(\ell_f/\ell_0)} + \frac{\ell n(\ell_f/\ell_1)}{\ell n(\ell_f/\ell_0)} = 1 \tag{8}$$

Miner's rule is derived on the basis of the microcrack propagation law.

It seems that the derivation of Miner's rule is carried out automatically as stated above, only by applying the microcrack propagation law. However, it should be noted under what conditions the microcrack propagation law, under varying strain amplitudes such as in Eq 4, holds. Those conditions are essential in examining the background for applicability of Miner's rule. In this respect, the present authors have already reported [*24*] that the required conditions to satisfy Miner's rule are as follows: (1) prior fatigue history due to reversals of the first-step strain amplitude has no effect on subsequent microcrack growth after switching of the strain level, and (2) the microcrack propagation law is of the linear type (i.e., expressed by $d\ell/dN = C\ell$). The reason for condition (1) is as follows: (*a*) macroscopically a unique cyclic hysteresis loop is rapidly reached within a few early strain cycles after switching of the strain level, (*b*) microscopically a unique dislocation structure appears irrespective of prior history, and (*c*) experimental

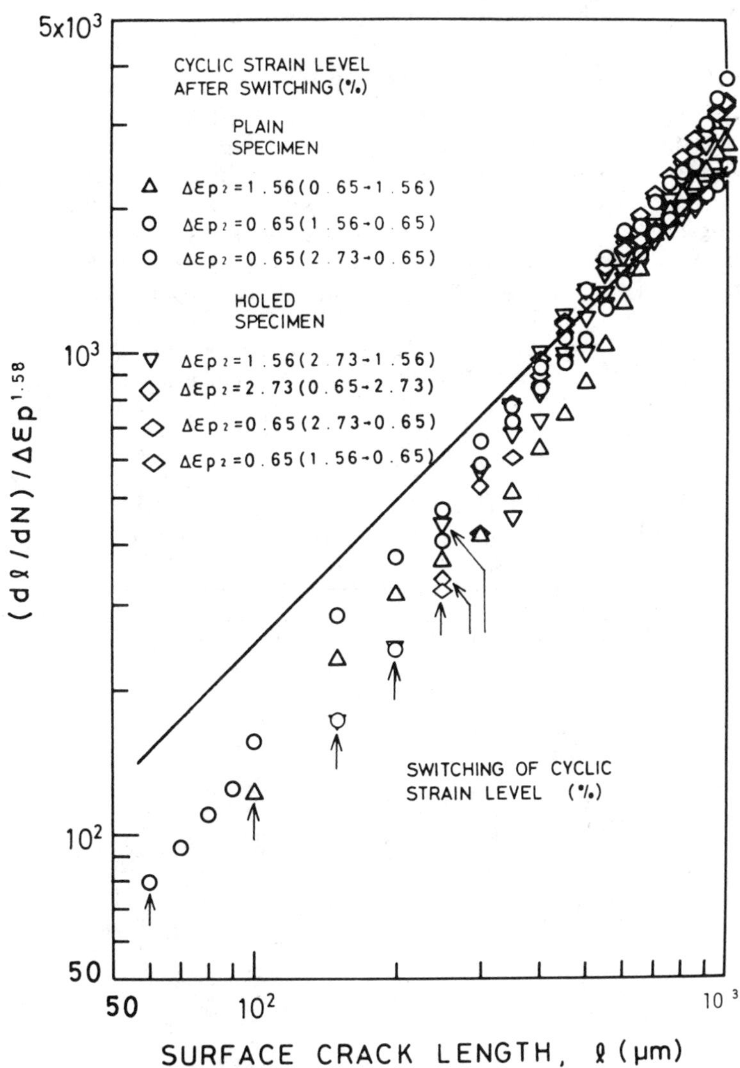

FIG. 7—*The* $(d\ell/dN)/\Delta\epsilon_p^{1.58}$–$\ell$ *relation after switching of the cyclic strain level under two-step strain cycling for plain and holed specimens.*

evidence shown by Kikukawa et al. [*16*], Nisitani and Morita [*17*], and Yamada et al. [*18*] that removal of the fatigue-damaged surface layer prolongs fatigue life, also supports condition (*1*).

Regarding condition (*2*), although the microcrack propagation law, $d\ell/dN = C\ell^m$, can also be obtained by dimensional analysis [*27,29*], such dimensional analysis is judged invalid unless condition (*1*) holds. Furthermore, if we take a crack propagation law of the form $d\ell/dN = C\ell^m$ ($m \neq 1$), we develop discussion in the same manner as that done for $m \neq 1$. However, the possibility of $m \neq 1$ is denied for a wide range of ℓ, because $m \neq 1$ means the influence of prior fatigue history.

Application of Microcrack Propagation Law-Aided Approach

I—Random Fatigue and Interpretation of Rain Flow Method

In order to actually demonstrate the validity of the microcrack propagation law-aided approach, fatigue tests were conducted under varying strain-amplitude conditions. The material used and other general test procedures are the same as those under the constant and two-step strain cycling. Figures 8*a*, 8*b*, and 8*c* show waveforms of the cyclic strain applied to the specimens: (*a*) trigonometric wave (fundamental wave), (*b*) block-random wave, and (*c*) random wave. A unit of the block-random wave was generated using a table of random numbers, and the random wave was generated with the aid of a microcomputer program. For simplicity, to ignore the effect of mean strain on microcrack growth, both the block-random and random waves were established to generate around the level of zero mean strain.

Microcrack growth characteristics under constant strain amplitude were examined on plain and holed specimens to gather fundamental data for counting fatigue damage under random

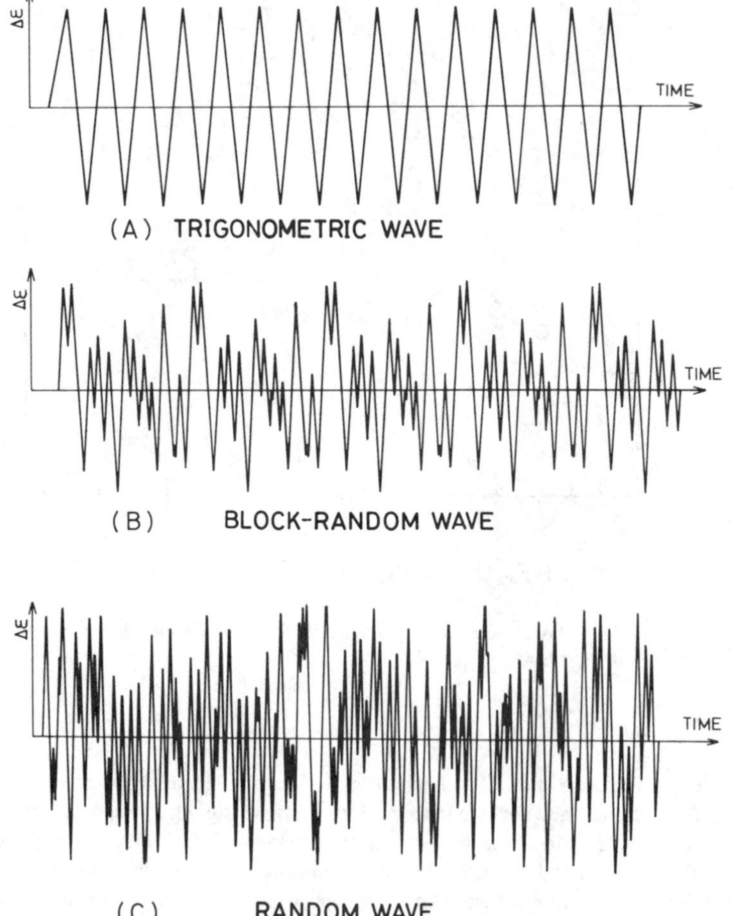

FIG. 8—*Waveforms for random loading.* (a) *Trigonometric wave (fundamental wave).* (b) *Block-random wave.* (c) *Random wave.*

loading. It was found that the crack propagation life of both plain and holed specimens satisfies the Manson-Coffin-type relation, provided the crack propagation life is defined as the number of strain reversals until a microcrack grows up to a prescribed length.

For random loading, the rain flow method proposed by Endo et al. [*11,12*] was applied to count fatigue damage. This method being originally developed to predict final failure life, the definition of fatigue damage D was changed for prediction of microcrack propagation life at an arbitrary stage of crack growth. That is, in predicting microcrack propagation life, $D = 1$ means that a microcrack with an initial crack length ℓ_o grows up to a prescribed crack length ℓ^*.

Figure 9 shows crack extension curves of the holed specimens under different levels of constant strain amplitude, plotted on the semi-log diagram. The abscissa N^*/N_f^* denotes a cycle ratio calculated by assuming that crack growth stage at $\ell = 120$ μm tentatively means the initial stage. Crack extension curves of the holed specimens under block-random and random loading plotted on the semi-log diagram are indicated in Fig. 10, where the abscissa denotes fatigue damage D evaluated by applying the rain flow method. From these figures, it is noticed that both the results coincide, in spite of the difference in measurement of the abscissa. The one-to-one correspondence between fatigue damage D and cycle ratio N^*/N_f^* means: (*a*) the rain flow method is appropriate in counting fatigue damage, (*b*) fatigue damage can be correlated with the microcrack propagation law, and (*c*) fatigue damage D is probably a suitable parameter in predicting microcrack propagation life under random loading.

In order to examine the validity of the above discussion, fatigue tests for predicting microcrack propagation life were conducted under block-random and random strain cycling. Maximum cyclic plastic strain ranges $\Delta\epsilon_{p\max}$ were established as $\Delta\epsilon_{p\max} = 0.6\%$, 1.5%, and 4.6% for the block-random strain cycling and 1.5% for the random-strain cycling. Actual fatigue damage applied to each specimen was counted in terms of the rain flow method. The results are shown in Table 4. In the holed specimens, the measured values of D range from 0.98 to 1.20. It

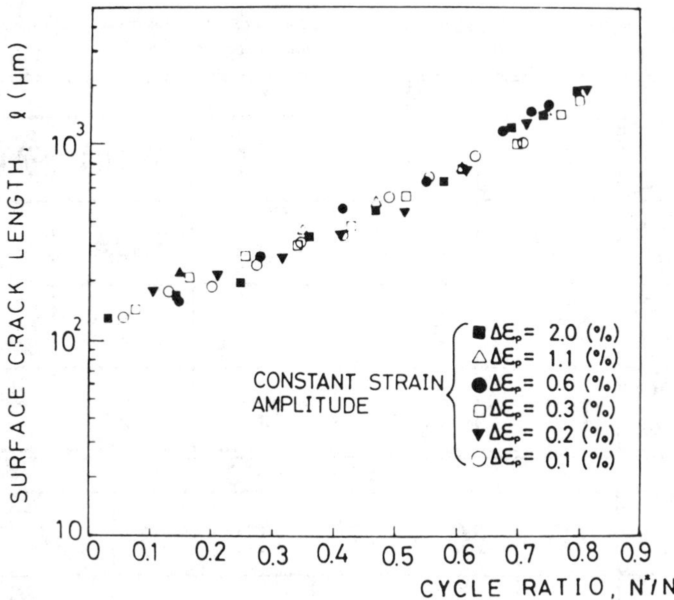

FIG. 9—*Relation between the crack length ℓ and the cycle ratio* N^*/N_f^* *plotted on the semi-log diagram where the number of strain cycles were counted after the crack grew to 120 μm long.*

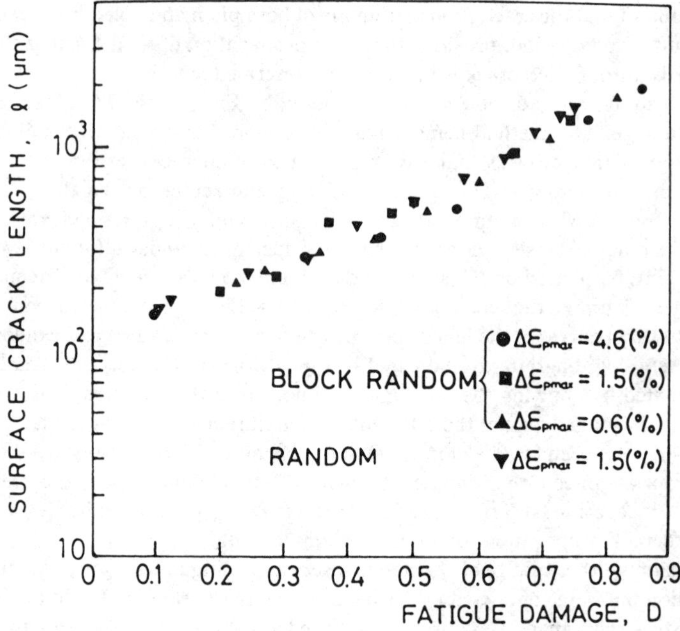

FIG. 10—*Relation between the crack length ℓ and the fatigue damage D plotted on the semi-log diagram under block-random strain cycling.*

TABLE 4—*Comparison of microcrack propagation lives experimentally determined with those predicted for block-random and random strain cycling conditions.*

	HOLED SPECIMEN				
WAVE FORM AND $\Delta\varepsilon_{pmax}$ \ CRACK LENGTH (μm)	300	600	1000	1500	FAILURE
BLOCK RANDOM $\Delta\varepsilon_{pmax}$ = 4.6 (%)	1.10	1.08	1.05	1.04	1.20
BLOCK RANDOM $\Delta\varepsilon_{pmax}$ = 1.5 (%)	1.08	0.98	1.03	1.00	1.15
BLOCK RANDOM $\Delta\varepsilon_{pmax}$ = 0.6 (%)	1.20	1.10	1.07	1.05	1.14

	PLAIN SPECIMEN						
WAVE FORM AND $\Delta\varepsilon_{pmax}$ \ CRACK LENGTH (μm)	20	30	52	75	141	234	565
RANDOM $\Delta\varepsilon_{pmax}$ = 1.5 (%)	2.00	0.97	0.90	0.96	0.86	0.86	0.82

means that microcrack propagation life can be predicted with a high degree of accuracy by applying the rain flow method. The same discussion holds for the plain specimen, except for prediction of a very small crack 20 μm long. This exception may be caused by crack growth retardation due to the effect of microstructure.

From the results and discussion aforementioned, it should be noticed that the required conditions for applicability of Miner's rule described previously for constant and two-step strain cycling are also satisfied in the case of block-random or random strain cycling.

II—Anisotropy in Fatigue Strength of a Forged Steel Plate

In general, rolled or forged steel plates show a sharp drop in fracture ductility in their thickness direction. This is mainly caused by elongated manganese-sulfide inclusions responsible for lamellar tearing. Ohji and Harada [*33*] have reported that a thick rolled steel plate for boiler and pressure vessel use shows a difference in its low-cycle fatigue life between the transverse direction *T* and thickness direction *S*, and that this difference can be evaluated by substituting the difference in fracture ductility in the *T* and *S* directions into a Manson-Coffin-type fatigue law.

To examine this kind of anisotropy in fatigue strength in conjunction with the microcrack propagation law, fatigue tests were conducted on a 200-mm-thick forged steel plate. This material has a chemical composition (weight percent) of 0.21C, 0.29Si, 1.44Mn, 0.006P, 0.013S, 0.65Ni, 0.03Cr, and other constituents. The mechanical properties are as follows: ultimate tensile strength (MPa): 664(*T*), 653(*S*); yield stress (MPa): 476(*T*), 469(*S*); and fracture ductility (%): 78.1(*T*), 49.0(*S*) (*T*: transverse direction, *S*: thickness direction). Microcrack initiation and propagation behavior was continually observed during the course of the fatigue tests. Fatigue failure lives in the *T* and *S* directions, indeed, show differences (Fig. 11). The predicted fatigue lives based on the microcrack propagation law are also indicated with broken lines in

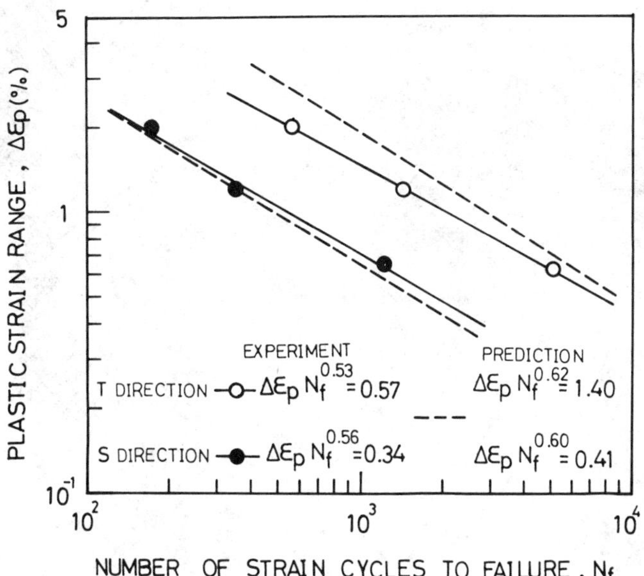

FIG. 11—*Comparison of the fatigue lives determined experimentally with those predicted on the basis of the microcrack propagation law in the* T *and* S *directions of a thick forged steel plate.*

Fig. 11. The predicted life in each direction is in good agreement with the measured life within the accuracy of engineering use.

Figures 12*a* and 12*b* indicate an example of a leading crack initiated at a manganese-sulfide inclusion observed at the specimen surface and fractured plane. The specimens with orientation *S* showed this type of fracture pattern, while the leading crack did not necessarily originate at the manganese-sulfide inclusions in the specimens with orientation *T*. The correlation between

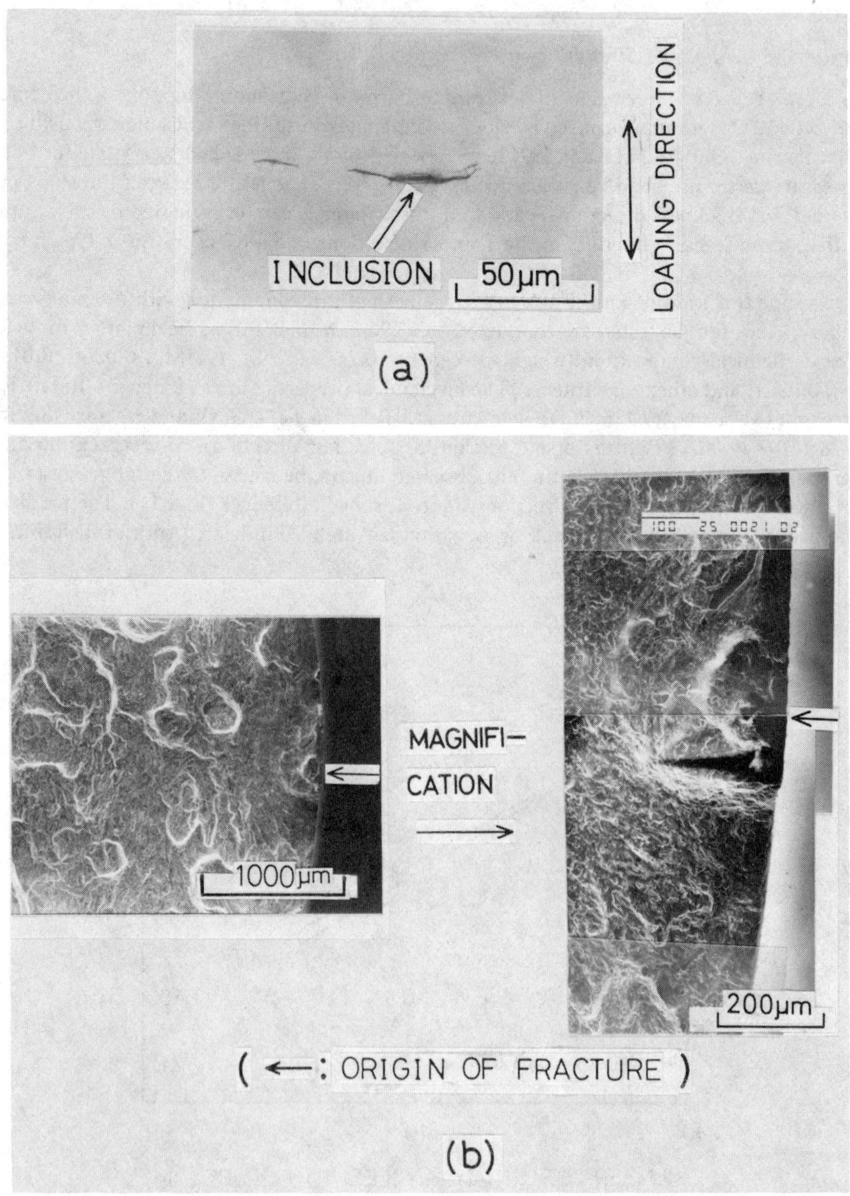

FIG. 12—*Example of a leading crack originated at a manganese-sulphide inclusion.* (a) *Specimen surface.* (b) *Fractured plane.*

anisotropy in fracture ductility and the microcrack propagation law will be examined in further study.

Finally, it should be noted that although the effect of crack closure plays an important role in estimating crack growth rate, that effect was ignored throughout this study, because the crack tip is assumed to be kept open during the full process of load cycling, as was experimentally verified in the previous study [*24*]. Further study is needed for a detailed discussion.

Recommended Future Research in Low-Cycle Fatigue

The microcrack propagation law-aided simple approach developed and substantiated by experiments here is believed to be applied widely in assessing low-cycle fatigue life. To widen the applicability without loosing simplicity the present method may be supplemented by the fracture mechanics-aided approach as required. Although some experiments [*17,31,32*] support the validity of the present results, it is better to examine it on a variety of materials, in particular, on the cyclic softening or hardening type of materials. Moreover, further studies are needed to take a loading condition into account. In this respect, how to evaluate the effect of a mean stress or strain in the microcrack propagation law is an urgent matter.

Conclusions

Fatigue tests were conducted under constant, two-step, and random strain cycling conditions. Through observation of the microscopic fatigue process, fundamental low-cycle fatigue laws such as the Manson-Coffin relation and Miner's rule were reconsidered on the basis of the microcrack propagation law-aided approach. The results and discussions obtained so far are summarized as follows:

1. The low-cycle fatigue life of an annealed medium carbon steel was mostly occupied by the propagation process of a crack from a grain-size microcrack to a crack about 1 mm long.
2. The Manson-Coffin relation is derived on the basis of the microcrack growth law. From the results of the two-step loading fatigue, Miner's rule was also found to be derived on the basis of the microcrack propagation law.
3. The required conditions for applicability of Miner's rule have shown that prior history or pre-fatigue damage hardly has an effect on the subsequent microcrack growth rate and that the microcrack propagation law takes a form of the linear type $d\ell/dN = C\ell$.
4. Crack propagation life of holed or plain specimens under block-random or random loading conditions can be predicted by applying the microcrack growth law under constant strain cycling in combination with the rain flow method for counting fatigue damage.
5. The microcrack propagation law-aided approach is also demonstrated to be applicable in evaluating anisotropy in the fatigue strength of a forged steel plate.

Acknowledgments

The authors wish to thank Messrs. Hikofumi Tani-Ishi, Yutaka Ishimatsu, and Kazuyuki Takeshima, formerly graduate students in the Department of Mechanical Engineering, Kyushu Institute of Technology (KIT), for their assistance in carrying out part of the experiments. They also wish to acknowledge the assistance of Mr. Toshimitsu Sakata, presently a graduate student in the same department, KIT, in preparing the manuscript as well as in conducting a part of the experiments.

References

[1] Manson, S. S., NACA TN 2933, 1953.

[2] Coffin, L. F., Jr., *Transactions of the American Society of Mechanical Engineers*, Vol. 76, 1954, pp. 931-950.

[3] Manson, S. S., *Thermal Stress and Low-Cycle Fatigue*, McGraw-Hill, New York, 1966.

[4] Morrow, J. in *Internal Friction, Damping, and Cyclic Plasticity, ASTM STP 378*, American Society for Testing and Materials, Philadelphia, 1965, pp. 45-87.

[5] Kikukawa, M., Ohji, K., Kamata, T., and Johno, M., *Journal of the Japan Society of Mechanical Engineers*, Vol. 70, 1967, pp. 1495-1509.

[6] Tanaka, K. and Mura, T., *Transactions of the American Society of Mechanical Engineers*, Vol. 48, 1981, pp. 97-103.

[7] Manson, S. S., Freche, J. C., and Ensign, C. R. in *Fatigue Crack Propagation, ASTM STP 415*, American Society for Testing and Materials, Philadelphia, 1967, pp. 384-412.

[8] Weiss, V. in *Fatigue—An Interdisciplinary Approach*, Syracuse University Press, New York, 1964.

[9] Irwin, G. R. and Paris, P. C. in *Fracture*, Vol. 3, Academic Press, New York, 1971.

[10] Dowling, N. E., *Journal of Materials*, Vol. 7, 1972, pp. 71-87.

[11] Anzai, H. and Endo, T., *International Journal of Fatigue*, Vol. 1, 1979, pp. 49-57.

[12] Endo, T. and Anzai, H., *Journal of the Society of Materials Science Japan*, Vol. 30, 1981, pp. 89-93.

[13] Okamura, H. and Sakai, N., *Preprint of the Japan Society of Mechanical Engineers*, No. 790-8, 1979, pp. 89-94.

[14] Palmgren, A., *Zeitschrift des Verein Deutscher Ingenieure*, Vol. 68, 1924, pp. 339-341.

[15] Miner, M. A., *Journal of Applied Mechanics*, Vol. 12, 1945, A159-A164.

[16] Kikukawa, M., Ohji, K., Okubo, N., Yokoi, T., and Morikawa, T., *Transactions of the Japan Society of Mechanical Engineers*, Vol. 38, 1972, pp. 8-15.

[17] Nisitani, H. and Morita, N., *Transactions of the Japan Society of Mechanical Engineers*, Vol. 39, 1973, pp. 1711-1719.

[18] Yamada, T., Hoshide, T., Fujimura, and Manabe, M., *Transactions of the Japan Society of Mechanical Engineers*, Vol. 49, 1983, pp. 441-451.

[19] Miller, K. J., *Fatigue of Engineering Materials and Structures*, Vol. 5, 1982, pp. 223-232.

[20] Taylor, D. and Knott, J. F., *Fatigue of Engineering Materials and Structures*, Vol. 4, 1981, pp. 147-155.

[21] Rios, E. R., Tang, Z., and Miller, K. J., *Fatigue of Engineering Materials and Structures*, Vol. 7, 1984, pp. 97-108.

[22] Nisitani, H. and Goto, M., *Transactions of the Japan Society of Mechanical Engineers*, Vol. 51, 1985, pp. 332-341.

[23] Hatanaka, K., Fujimitsu, T., and Watanabe, H., *Transactions of the Japan Society of Mechanical Engineers*, Vol. 51, 1985, pp. 790-797.

[24] Murakami, Y., Harada, S., Endo, T., Tani-Ishi, H., and Fukushima, Y., *Engineering Fracture Mechanics*, Vol. 18, 1983, pp. 909-924.

[25] Yurakami, Y., Noda, N., and Nisitani, H., *Transactions of the Japan Society of Mechanical Engineers*, Vol. 47, 1981, pp. 1194-1205.

[26] Frost, N. E. and Dugdale, D. S., *Journal of the Mechanics and Physics of Solids*, Vol. 6, 1958, pp. 92-110.

[27] Liu, W. H., *Transactions of the American Society of Mechanical Engineers*, Series D, Vol. 83, 1961, pp. 23-31.

[28] Manson, S. S. in *Proceedings*, International Conference on Fracture, Sendai, Japan, 1965, pp. 1387-1431.

[29] Yang, C. T., *Transactions of the American Society of Mechanical Engineers*, Series D, Vol. 89, 1967, pp. 487-495.

[30] Tomkins, B., *Philosophical Magazine*, Vol. 18, 1968, pp. 1041-1066.

[31] Murakami, Y. and Yano, T., Undergraduate thesis, Faculty of Engineering, Kyushu University, 1966.

[32] Murakami, Y., Makabe, C., and Nisitani, H., *Transactions of the Japan Society of Mechanical Engineers*, Vol. 50, 1984, pp. 1828-1837.

[33] Ohji, K. and Harada, S., *Bulletin of the Japan Society of Mechanical Engineers*, Vol. 16, 1973, pp. 787-796.

Y. Oshida[1] and H. W. Liu[1]

Grain Boundary Oxidation and an Analysis of the Effects of Oxidation on Fatigue Crack Nucleation Life

REFERENCE: Oshida, Y. and Liu, H. W., "**Grain Boundary Oxidation and an Analysis of the Effects of Oxidation on Fatigue Crack Nucleation Life,**" *Low Cycle Fatigue, ASTM STP 942,* H. D. Solomon, G. R. Halford, L. R. Kaisand, and B. N. Leis, Eds., American Society for Testing and Materials, Philadelphia, 1988, pp. 1199–1217.

ABSTRACT: A study on the effects of pre-oxidation on subsequent fatigue life was conducted. The surface oxidation and grain boundary oxidation of a nickel-base superalloy (TAZ-8A) were studied at 600 to 1000°C for 10 to 1000 h in air. The surface oxides were identified and the kinetics of surface oxidation was discussed.

The grain boundary oxide penetration and morphology were studied. The pancake-type grain boundary oxide penetrates deeper and its size is larger, therefore, it is more detrimental to fatigue life than the cone-type grain boundary oxide. The oxide penetration depth, a_m, is related to oxidation temperature, T, and exposure time, t, by an empirical relation of the Arrhenius type. The effects of T and t on the statistical variation of a_m were analyzed according to the Weibull distribution function. Once the oxide is cracked, it serves as a fatigue crack nucleus. The statistical variation of the remaining fatigue life, after the formation of an oxide crack of a critical length, could be related directly to the statistical variation of grain boundary oxide penetration depth.

KEY WORDS: oxidation, grain boundary oxidation, oxidation kinetics, oxide morphology, oxide crack, fatigue life, elevated temperatures, Weibull distribution function

Fatigue life is often shortened at elevated temperatures. The shortened fatigue life is usually attributed to oxidation and/or creep. Creep deformation creates grain boundary voids, which is said to weaken the material. In conjunction with the applied cyclic fatigue load, the grain boundary voids may accelerate the initiation and the propagation of fatigue cracks, and thereby shorten fatigue life. The subject of creep-fatigue interaction has been studied extensively [*1–5*].

In a vacuum of 10^{-8} torr, the effects of cyclic frequency and test temperature on fatigue lives of A286 at elevated temperatures are minimal in comparison with those in air [*6, 7*]. Therefore oxidation is more detrimental, at least for some materials, than creep.

This paper deals with surface and grain boundary oxidation and the effects of the grain boundary oxidation on fatigue life.

Gibb's free energies of metal oxide formation are negative. No metal or alloy is stable when exposed to an oxidizing environment. Hence a high oxidation resistance of an alloy implies a low oxidation rate. A low oxidation rate is generally due to the slow diffusion rate through the oxide film.

Because grain boundary is a path of rapid diffusion, grain boundary oxidation rate is higher and grain boundary oxide penetration is deeper than the surface oxidation. When an oxide reaches a critical size at an applied stress level, the oxide will fracture [*8*]. The oxide crack may

[1]Research Assistant Professor and Professor of Mechanical Engineering, respectively, Syracuse University, Syracuse, NY 13244.

serve as a nucleus of a fatigue crack, and will grow by the subsequent cyclic fatigue load. Therefore, an oxide crack may shorten fatigue life, and grain boundary oxide penetration depth can serve as a "measure" of high-temperature fatigue damage.

McMahon and Coffin [*9*] have shown ruptured oxide film to act as a fatigue crack nucleus. Shida et al. [*10*] and Stott et al. [*11*] have studied grain boundary oxidation. In this paper, the surface and grain boundary oxidations of a nickel-base superalloy (TAZ-8A) were investigated. The surface oxides were identified by an electron-probe X-ray energy dispersion spectrometry analysis, oxide weight gain and weight loss analysis, and reflection electron diffraction method. The three-dimensional grain boundary oxide morphology was studied. The grain boundary oxidation kinetics were analyzed, and the statistical distribution of the grain boundary oxide penetration depth was studied. The effects of oxide penetration depth and oxide crack on subsequent fatigue life are discussed.

Surface Oxidation

A nickel-base superalloy (TAZ-8A) was subjected to high temperature oxidation in air under the stress-free condition. The chemical composition of TAZ-8A is listed in Table 1. The oxidation temperatures were 600, 700, 800, 900, and 1000°C, controlled within ±8°C. The oxidation time varied from 10 to 1000 h.

The average weight gain per unit surface area, ΔW_{+O} (mg/cm^2), due to both surface and grain boundary oxidations was obtained for each disk coupon, 15 mm in diameter and 5 mm in thickness. At a given temperature, a linear relationship between the square of ΔW_{+O} and oxidation exposure time, t, was found (Fig. 1). The slopes of the lines in the figure are known as the parabolic rate constant, K_p. The data in the figure indicate that the oxidation of TAZ-8A obeys the parabolic rate law as many other superalloys and refractory materials do. The parabolic oxidation rate implies that the oxidation is controlled by diffusion.

Figure 2 shows the parabolic rate constant versus inverse absolute temperature of TAZ-8A and several other superalloys. The data of TAZ-8A seem to follow two line segments similar to Mar-M200. Additional studies are needed to verify whether these two line segments indicate two different oxidation kinetics in the high and low temperature regimes.

The measured weight gain during oxidation, ΔW_{+O}, reflects the amount of oxygen that reacted with the substrate to form an oxide. On the other hand, the weight loss of the substrate due to the oxidation can be measured by descaling. For rapid descaling, the oxidized sample was immersed into an acid aqueous solution (50% HCl, 49% H_2O, 1% H_2O_2) at 70°C [*12*] for 3 min and subsequently in a solution of 30% HNO_3, 10% H_2O_2, 10% H_3PO_4, and 50% CH_3COOH at 90°C [*12*] for an additional 3 min. The descaling process was followed by an examination with an optical microscope. If the amount of oxide removal was not enough, this process was repeated. The final descaling process is accomplished at a much slower rate by cathodic descaling to avoid the removal of the substrate; oxidized sample at cathode; lead plate at anode; 5% H_2SO_4 electrolyte; and 1 mA/cm^2 current density [*13*].

The measured weight loss, ΔW_{-M}, reflects the amount of metal loss due to oxidation. The weight gain to weight loss ratio, $\Delta W_{+O}/\Delta W_{-M}$, should be equal to the atomic weight ratio of oxide if the oxide is stoichiometric. The measured ratios are 0.32, 0.44, and 0.49 for oxides

TABLE 1—*Chemical composition (wt%).*

	Ni	Cr	Mo	W	Ta	Cb	Al	C	B
TAZ-8A	68	6	4	4	8	2.5	6	0.12	.004

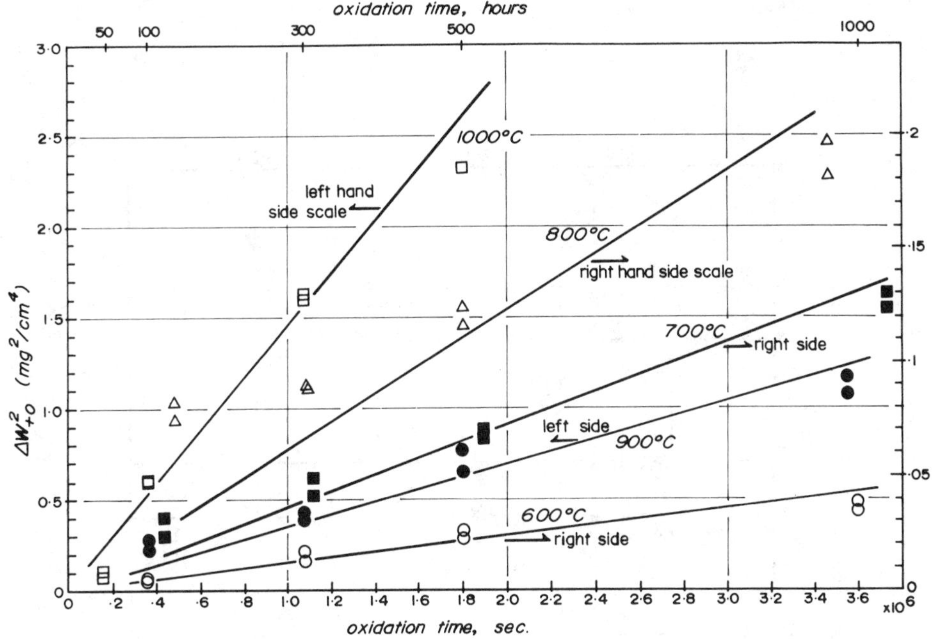

FIG. 1—*Relationship between the square of weight gain and oxidation exposure time.*

formed at 600, 800, and 1000°C, respectively. The atomic weight ratio for Cr_2O_3 is 0.462. Similar ratios of other oxides are 0.273 (NiO), 0.889 (Al_2O_3), and 0.393 ($NiCr_2O_4$).

The reflection electron diffraction pattern identified that the outermost layer of the surface oxide formed at 800°C for 960 h consists mainly of Cr_2O_3 (corundum type of sesquioxide) with a small amount of spinel-type oxide such as $NiO \cdot Cr_2O_3$ or $NiO \cdot (Cr,Al)_2O_3$. Some of chromium ions in Cr_2O_3 might be substituted by aluminum ions to form $(Cr,Al)_2O_3$.

Figure 3*a* shows the picture of a cross section, under an optical microscope, of a test coupon oxidized at 1000°C for 500 h. Figure 3*b* is the cross section under a scanning electron microscope. The microstructures of the substrate shows the hardening precipitates (γ'-phase or Ni_3Al) in γ-phase matrix with relatively large carbides. After the sample was oxidized, the decomposition of γ'-phase and the chromium depletion were observed in a substrate layer next to the substrate-oxide interface. By increasing the oxidation temperature and prolonging oxidation time, these microstructural changes due to high-temperature oxidation became remarkable (e.g., the thickness of 100 μm of γ'-phase decomposition zone was observed after 1000°C × 960 h oxidation).

Figure 4 shows the line profiles for NiK_α, CrK_α, and AlK_α constructed from the electron-probe spot analyses by an x-ray energy dispersion spectrometer. The spot analyses were made at five locations on each one of the three test coupons oxidized at 600, 800, and 1000°C. The five locations are (1) in the substrate far away from the substrate-oxide interface; (2) and (3) next to the substrate-oxide interface, one in the substrate and the other in the oxide; (4) at the mid-section of the surface oxide layer; and (5) at the outermost layer of the surface oxide. The oxide layer at 600°C is thin; the spots (4) and (5) merged together.

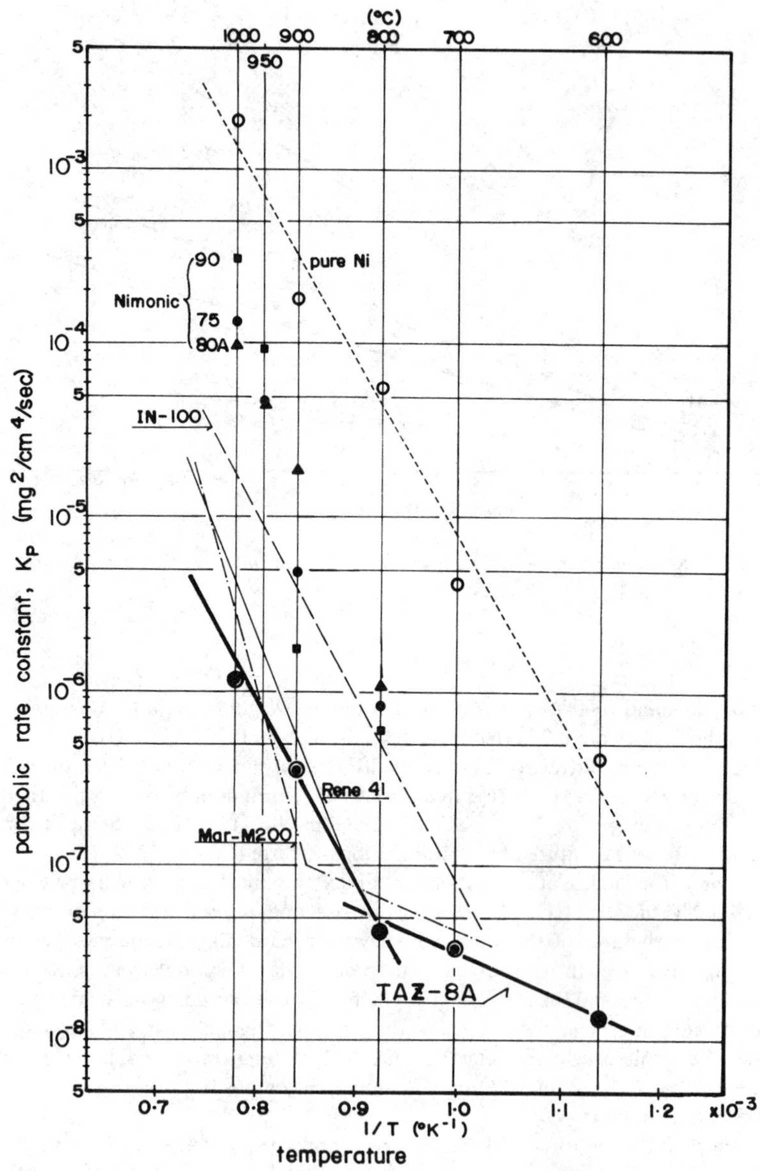

FIG. 2—*Parabolic rate constant versus inverse absolute temperature.*

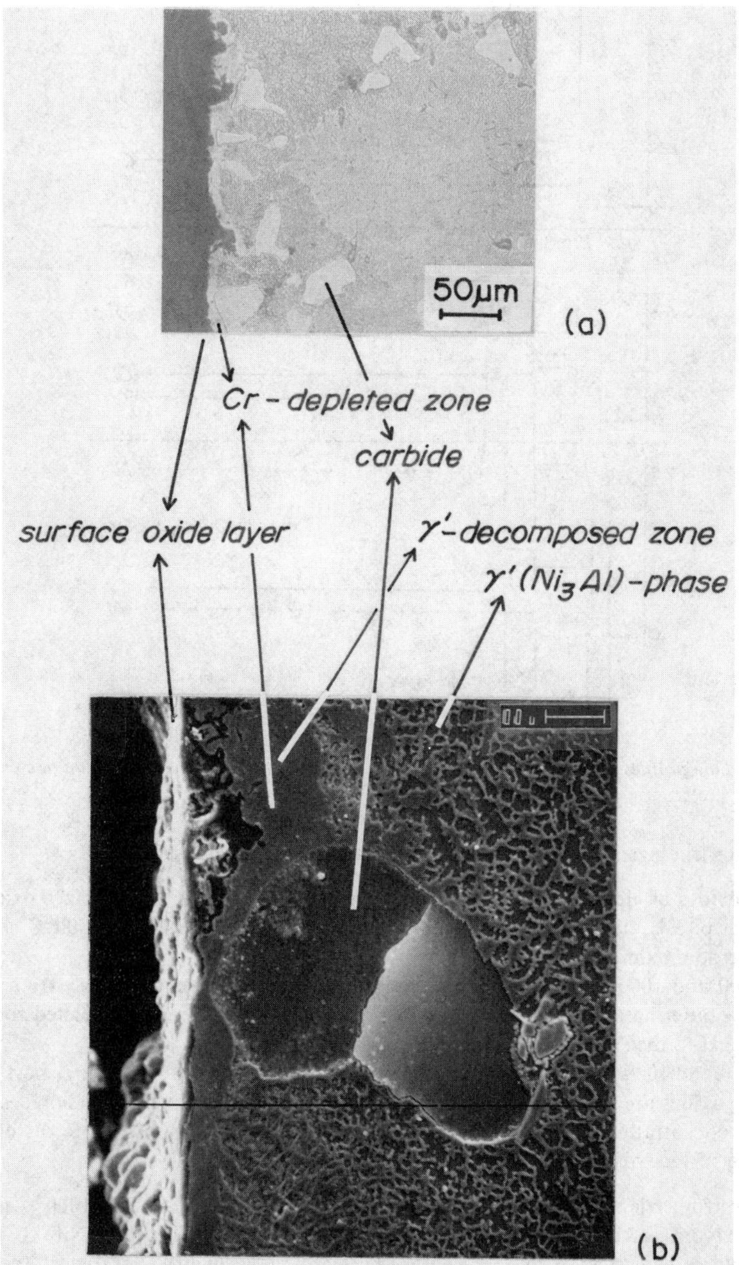

FIG. 3—*Micrographs of cross-section of a test coupon oxidized at 1000°C for 500 h.*

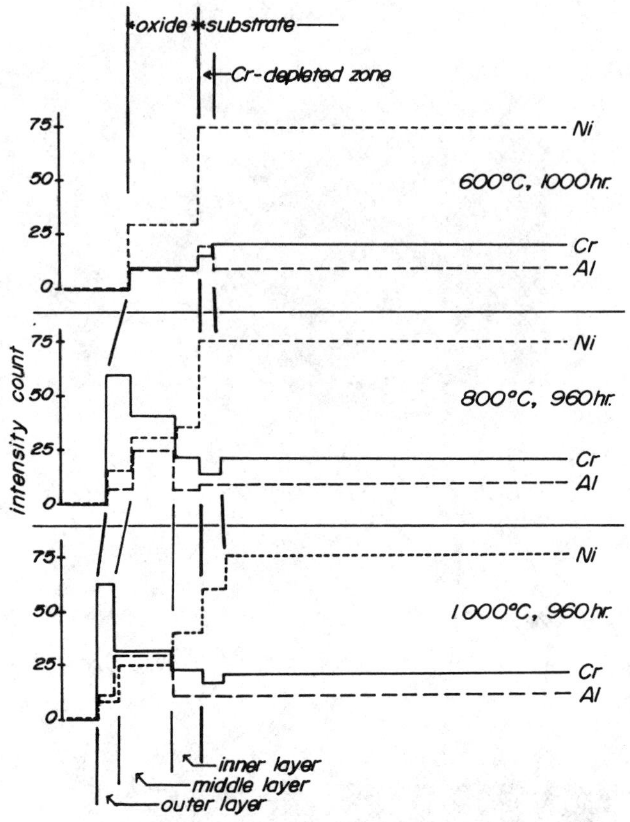

FIG. 4—*Line profiles of Ni, Cr, and Al on coupons oxidized at three different oxidation conditions.*

The major findings from the spectrometric analysis are as follows:

1. The oxides of aluminum, nickel, and chromium are present in the surface oxide layer. However, at 600°C, nickel oxide is the dominant component, at 800 and 1000°C chromium oxide is the dominant one.
2. At 800 and 1000°C, the chromium concentration increases continuously from the substrate to the outermost layer of the surface oxide with a minor chromium depleted zone in the substrate next to the substrate-oxide interface.
3. At 800 and 1000°C, the concentrations of the chromium oxide at the outermost layers of the surface oxides are much higher than that of either the aluminum or nickel oxides.
4. The concentration of nickel decreases continuously from the substrate to the outermost layer of the surface oxide.

This spectrometric analysis indicates that the oxides formed at 800 and 1000°C in air are mainly Cr_2O_3 or $(Cr,Al)_2O_3$ with a much lower amount of $NiO \cdot Cr_2O_3$ and/or $NiO \cdot (Cr,Al)_2O_3$. This finding is consistent with the measured weight-gain/weight-loss ratio, if the relative amounts of the various oxides are taken into consideration. For example, at 1000°C, the relative portions of the chromium, nickel, and aluminum oxides are 2:1:1. This proportion will give a weight gain to weight loss ratio of 0.52, which is very close to the measured ratio of 0.49. The activation energy (52.6 kcal/mol) in Fig. 2 in the high-temperature regime is very close to the activation energies of the diffusion of chromium in either Cr_2O_3 (55.3 kcal/mol) or $NiO \cdot Cr_2O_3$ (56.3 kcal/mol) [*14*].

At 600°C, the spectrometric analysis indicates that nickel oxide is the dominant component, and the ratio $(\Delta W_{+O}/\Delta W_{-M})$ is 0.32, which is between the values of NiO (0.273) and $NiO \cdot Cr_2O_3$ (0.393). Therefore, in the low-temperature regime, it is speculated that the oxidation kinetic is controlled by the diffusion of nickel atoms through the oxide layer.

A Statistical Study on Grain Boundary Oxidation Penetration

Figure 5 shows a picture of an oxidized surface. The grain boundary oxide penetrated much deeper than the surface oxide. A cracked grain boundary oxide can serve as a nucleus, which will grow by the subsequent fatigue cycles, and will shorten the fatigue life. Fatigue crack growth rate is related to ΔK, and ΔK is related to the size and shape of the microcrack. Therefore the shape and the depth of oxide penetration at grain boundaries are important factors for fatigue life prediction.

The three-dimensional shapes and sizes of the grain boundary oxides were traced optically. First a microhardness indentation was placed close to the oxide to be studied, and the size and shape of the oxide on the surface were traced. After the initial measurements, the surface was mechanically polished. The thickness of the surface layer polished off can be monitored by the change in the size of the hardness indentation. After polishing, the size and shape of the oxide were traced again. This process was repeated to trace the three-dimensional morphology of the grain boundary oxides. As shown in Figs. 6 and 7, two types of grain boundary oxides have been observed; the cone type and the pancake type. The pancake type was larger, seemed to penetrate deeper, and due to its shape and size, would give a higher value of stress intensity factor and would reduce fatigue life more than the cone type.

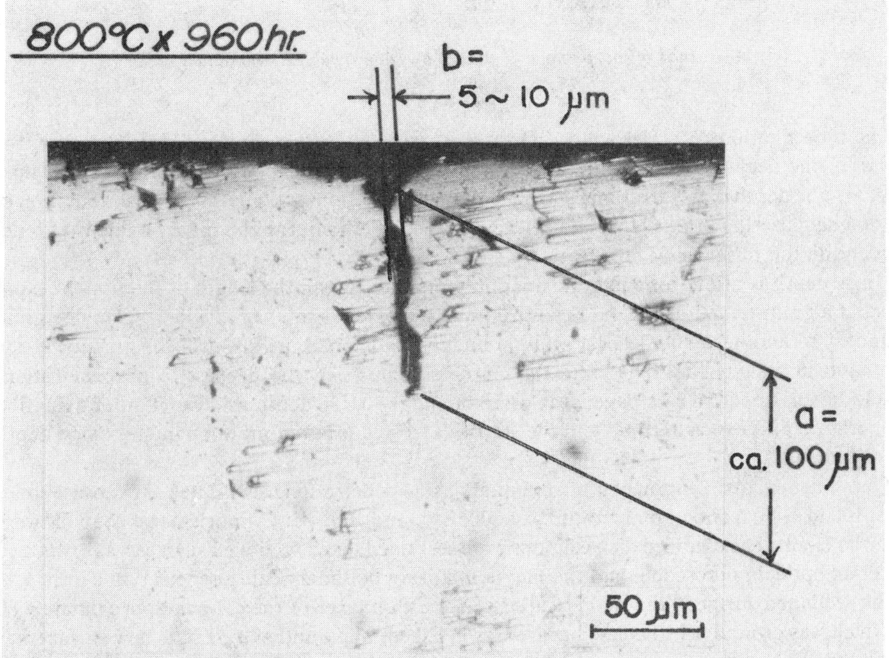

FIG. 5—*Cross-sectional micrograph showing a deep penetration by a grain boundary oxidation.*

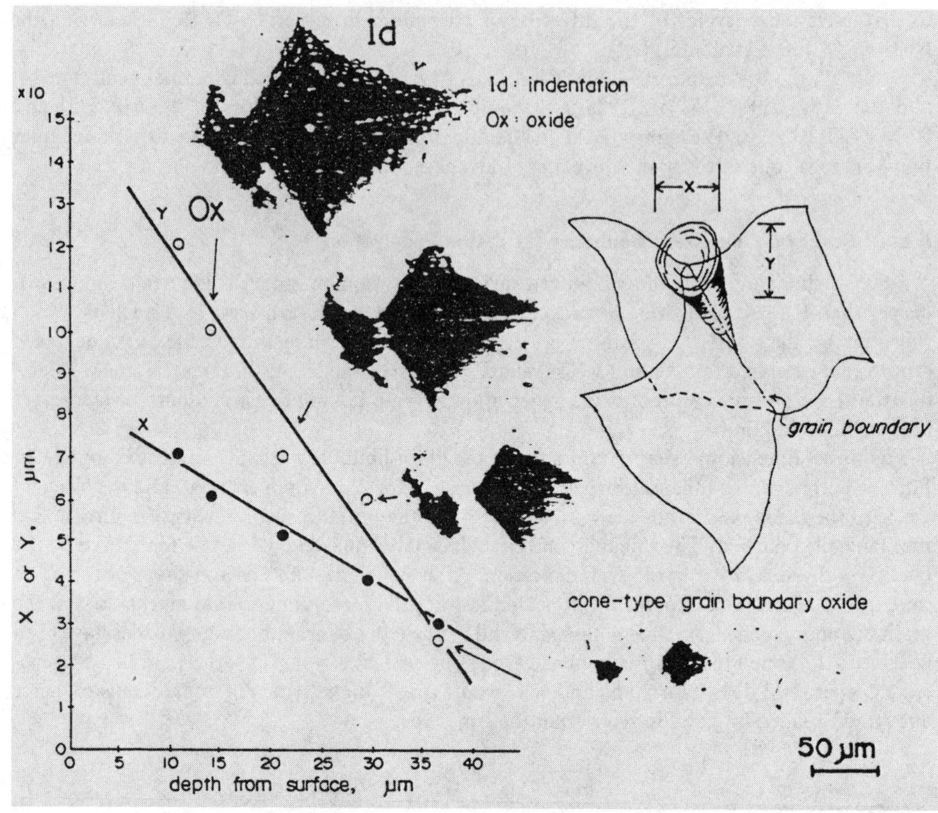

FIG. 6—*Results of morphological studies on cone-type grain boundary oxide.*

The aspect ratio a/b varies widely. On the average, a/b is close to one [*15*]. However, the ratios for the deeper oxides are much larger than one, often in the range of 10 to 20 or more (Fig. 5). The depth of penetration by grain boundary diffusion depends on the misorientation of the two neighboring grains [*16,17*]. Conceivably, the depth of grain boundary oxide is also dependent on the misorientation.

Oxide depth is also a function of oxidation temperature and the length of the time of exposure. At a given temperature and exposure time, the oxide depth varied widely from one grain boundary to another. The deepest grain boundary oxide crack will control the amount of the reduction in fatigue life. Therefore the extreme value (i.e., the deepest oxide penetration) should be analyzed. In this paper, the grain boundary oxide depth as a function of oxidation temperature and exposure time and the statistical distribution of grain boundary oxide depth were studied.

The statistical distribution of grain boundary oxide penetration was studied at exposure times from 100 to 1000 h and at temperatures of 600, 800, and 1000°C. Cylindrical test coupons were oxidized in air. The oxidized disk coupons were sectioned, each sectioned surface was examined under an optical microscope, and the *maximum* grain boundary oxide penetration depth, a_{mi}, of the sectioned surface "i" was measured. Then a thin layer of the coupon approximately 80 µm thick was ground off, the new surface was polished, and another a_{mi} of the new surface was measured. This process was repeated many times (often 12 times) for each test coupon to collect enough data for the statistical analysis. The grain size of the nickel alloy is between 50 to 80 µm.

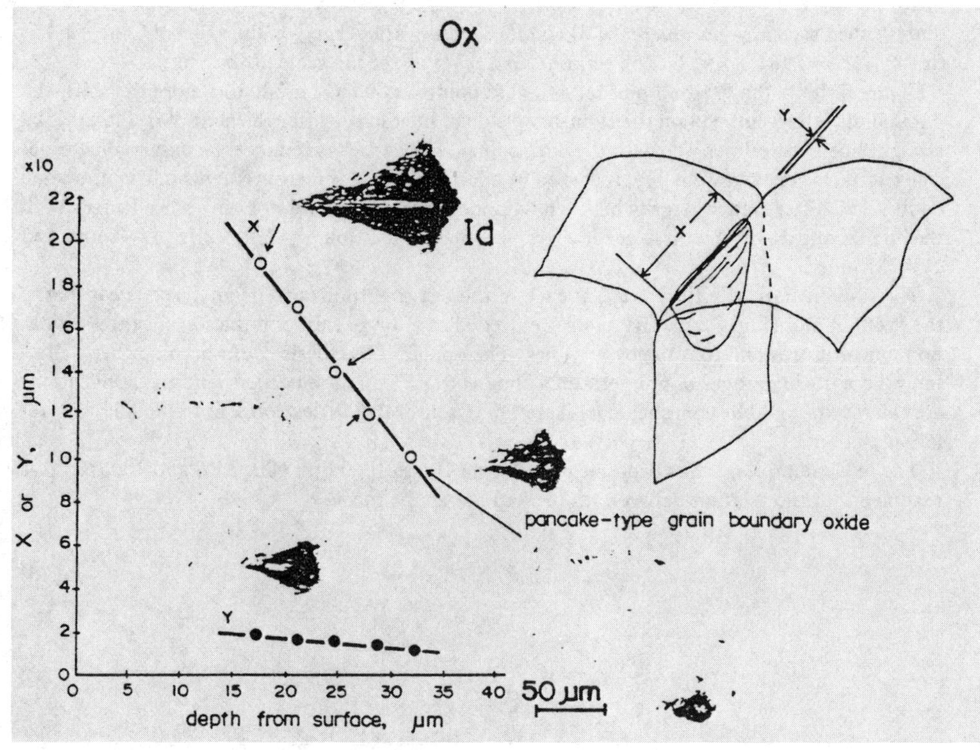

FIG. 7—*Results of morphological studies on pancake-type grain boundary oxide.*

After a removal of 80 μm, an entirely new set of grain boundaries was exposed. Therefore each sectioned surface can be considered as an independent sample. It samples an exposed area of $\pi \delta d$, where δ is coupon diameter and d is grain size.

Let N be the total number of sectioned surfaces examined. The measured a_{mi}'s are ranked from the shallowest penetration depth to the deepest penetration. The term M_i is the ranking number for a_{mi}. If N is large, the probability of finding an oxide depth less than a_{mi} on a sectioned surface is $P(a_{mi}) = M_i/(1 + N)$. The probability of having a depth equal to or deeper than a_{mi} is $[1 - P(a_{mi})]$.

Weibull assumes the distribution function

$$[1 - P(a_{mi})] = \exp\left[-\left(\frac{a_{mi} - a_u}{a_0}\right)^b\right] = \exp\left[-\frac{(a_{mi} - a_u)^b}{\eta_a}\right] \tag{1}$$

and

$$\ell n\, \ell n \frac{1}{[(1 - P(a_{mi})]} = b[\ell n(a_{mi} - a_u) - \ell n\, a_0] \tag{2}$$

In a Weibull plot of $\ell n\, \ell n\{1/[1 - P(a_{mi})]\}$ versus $\ell n(a_{mi} - a_u)$, Eq 2 gives a straight line. The term a_u is the horizontal shift for each of the data points, so that all the points will be on a straight line in the plot. It is known as the *location parameter.* The term b is the slope of the line and is called the *slope parameter* or *Weibull modulus.* The term a_0 is the value of $(a_{mi} - a_u)$ at $\ell n\, \ell n\{1/[1 - P(a_{mi})]\} = 0$. The terms a_0 and $\eta_a = a_0^b$ set the scale of the plot.

Figure 8 shows the Weibull plot for 480 data points at 800°C for 500 h. Before the shift, the data points follow curve A on the right-hand side of the figure. After the shift of a_u (40 μm) the straight line B correlates well with the data points. The vertical spread of the data points, especially at the lower end of the line, is caused by a large number of measurements having the same depth. The difference in depths between neighboring points could not be resolved easily with the microscope used. The three parameters were observed as follows: $b = 2.0$, $a_u = 40$ μm, and $a_0 = 31$ μm.

The statistical distribution of a_{mi} at each of the test conditions can be analyzed according to the Weibull plot (Fig. 8). Twelve plots are needed to analyze the combinations of three oxidation temperatures and four exposure times. The amount of analysis could become overwhelmingly large if wide ranges of temperatures and exposure times are needed, and the applications of such results could be complicated. Therefore it is desirable to develop a simplified data correlation method.

The oxide depth, a_{mi}, has been measured from 100 to 1000 h at 600, 800, and 1000°C. It is assumed that the relation between a_{mi}, t, and T has the form

$$a_{mi} = \alpha_i t^n \exp\left(-\frac{Q}{RT}\right) \tag{3}$$

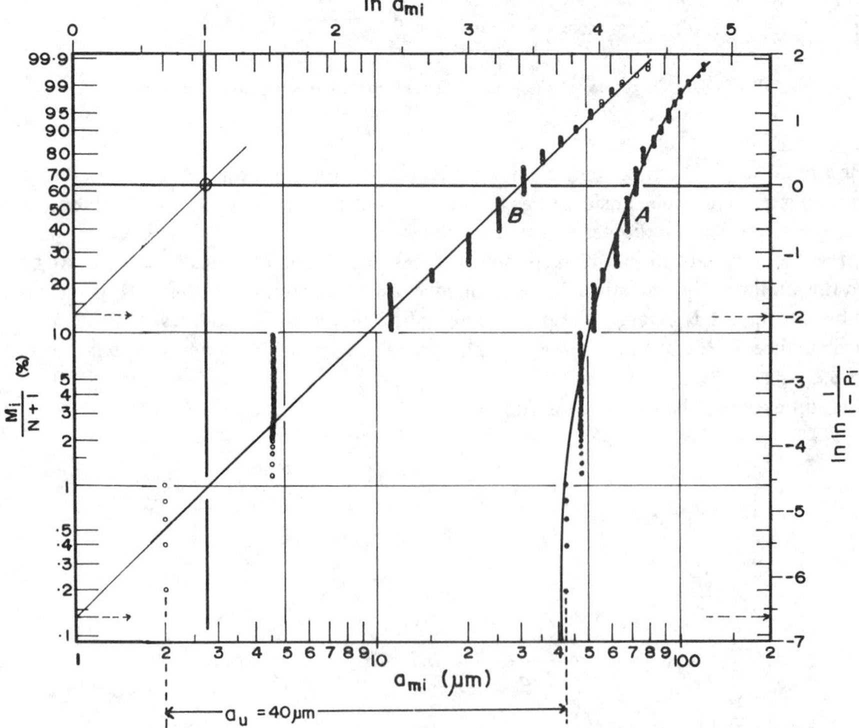

FIG. 8—*Weibull plot for 480 data points obtained from oxidized coupon at 800°C for 500 h.*

The regression analysis of the data gives the empirical relation

$$a_{mi}(\text{cm}) = 1.34 \times 10^{-3}\, t^{0.25} \exp(-4.26/RT) \tag{4}$$

where t is in seconds, the activation energy is in kcal/mol, and T is in Kelvin. The coefficient of correlation is 0.96. The plot of $\ell n(a_{mi}/t^{0.25})$ versus $(1/T)$ is shown in Fig. 9. The slope of the line is $(4.26/R)$.

Grain boundary diffusion can be considered as a channeled one-dimensional flow, with a constant concentration, C_0, at the interface between the surface oxide layer and the substrate (Fig. 10). The oxygen concentration in the grain boundary is [*18*]

$$C(x, t) = C_0\left[1 - \text{erf}\left(\frac{x}{2\sqrt{D_{gb}t}}\right)\right] \tag{5a}$$

where "erf" is the error function and D_{gb} is the grain boundary diffusion coefficient. The oxygen in the grain boundary will form bulk oxide if the concentration reaches a certain critical value, C_c. According to Eq 5*a*:

$$C_c = C_0\left[1 - \text{erf}\left(\frac{x_c}{2\sqrt{D_{gb}t}}\right)\right] \tag{5b}$$

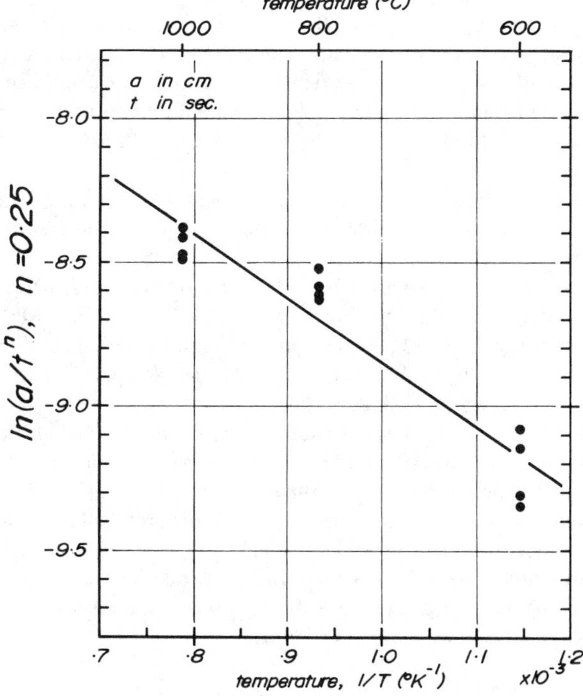

FIG. 9—*Relationship between* $\ell n(a_{mi}/t^{0.25})$ *versus* (1/T).

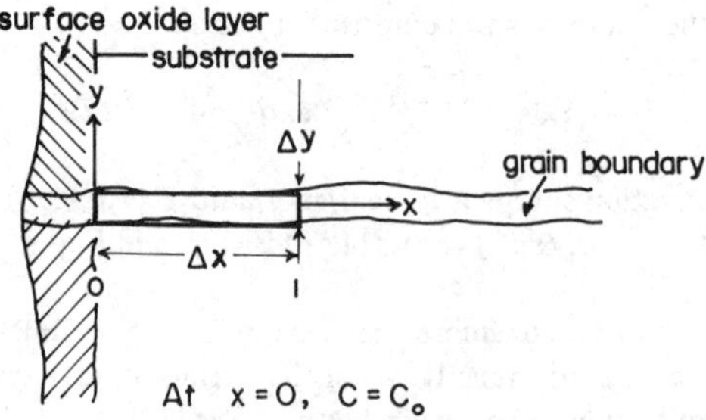

At $x = 0$, $C = C_0$

The two dimensional fluxes for the area $\Delta x\,\Delta y$:
Δy is finite, Δx is infinitesimal

$D_{gb} >> D_b$

Flux in $\quad -\left.\frac{\partial C}{\partial x}\right|_0 D_{gb}\Delta y$

Flux out $\quad \left\{-\left.\frac{\partial C}{\partial x}\right|_1 D_{gb}\Delta y - 2\frac{\partial C}{\partial y} D_b \Delta x\right\} \approx -\left.\frac{\partial C}{\partial x}\right|_1 D_{gb}\Delta y$

Bulk diffusion is negligible near $x = 0$.

FIG. 10—*A channeled one-dimensional flow model for grain boundary diffusion.*

Here x_c is the depth of the oxide penetration, where $C = C_c$. The quantity $x_c/2\sqrt{D_{gb}t}$ is a constant, and x_c must be proportional to $\sqrt{D_{gb}t}$. This is different from Eq 4.

The orientation of the grain boundary oxide is determined by the orientation of the substrate grain into which the oxide is growing. The orientations of the two halves of the grain boundary oxide growing into two "neighboring" substrate grains are different, and an oxide grain boundary exists between these two "halves." The diffusion of oxygen will pass through the oxide grain boundary.

Equations 5*a* and 5*b* are true only for a one-dimensional diffusion flow. The grain boundary oxidation is further complicated by the bulk diffusion of oxygen outward from the grain boundary oxide into the substrate.

Figure 10 shows an element of the grain boundary, $\Delta x \times \Delta y$. The grain boundary diffusion coefficient, D_{gb}, is much larger than the bulk diffusion coefficient, D_b, often several orders of magnitude larger. The large aspect ratio, a/b, in Fig. 5 indicates that the flux along the grain boundary is much larger than the flux into the substrate. Whipple [*19*] has analyzed grain boundary diffusion as affected by bulk diffusion.

The dependence of a_{mi} on t is closely related to the oxide morphology. For example, the dependence on t for the cone-type oxide will be different from the pancake type. The detailed oxidation process, the oxide structures, and the effect of bulk diffusion will affect the dependence of a_{mi} on both t and T. In addition, the chemical process of oxidation and its effect on the gradient of chemical potential have yet to be studied.

In other words, the functional relationship between the depth of oxide penetration, a_m, and the quantity, Dt, depends on the detailed diffusion process. However, a simple dimensional analysis indicates that a_m can be written in the form

$$\frac{a_m}{c} = F\left(\frac{Dt}{c}\right) \tag{6a}$$

where c is the jumping distance of the diffusing species, D is the diffusion coefficient, and t is time. Assume a power relation

$$a_m = \beta c \left(\frac{D_0 t}{c} \right)^n \exp\left(-\frac{n \Delta H}{RT} \right) \tag{6b}$$

or

$$a_m = \alpha t^n \exp\left(-\frac{Q}{RT} \right) \tag{6c}$$

where D_0 is the diffusion coefficient at $T = 0$; α, β, and n are constants; and $\alpha = \beta D_0^n C^{1-n}$. If the penetration depth is dependent on both grain boundary and bulk diffusions, the quantity (a_m/c) could be a function of both $(D_{gb}t/c)$ and $(D_b t/c)$.

Equation 4 is in the form of Eq 6c. For the moment, Eq 4 can be treated as an empirical relation, which correlates the data reasonably well. However, some scatter is shown in Fig. 9. The scatter band is ± 0.2. Perhaps the functional relation of Eq 6a is more complicated than the power relation assumed.

It is well known that grain boundary diffusion is a function of the relative orientations of the two neighboring grains. Couling and Smoluchowski [*16*] measured grain boundary diffusion penetration of radioactive silver in copper as a function of the angle of (100) tilt boundaries. Their results are shown in Fig. 11. Turnbul and Hoffman [*17*] have shown that the activation energy of grain boundary diffusion is a function of the angle of misorientation, θ. Therefore the activation energy Q in Eq 3 is not a constant but dependent on θ. Since we are measuring the deepest oxide penetration, perhaps Q is close to the minimum value. However, a statistical variation of Q from one a_{mi}-value of a cross section to another is expected.

Equation 4 is an empirical equation that correlates the measured oxide penetration. The deviation from the empirical relation of each measurement can be lumped into the term

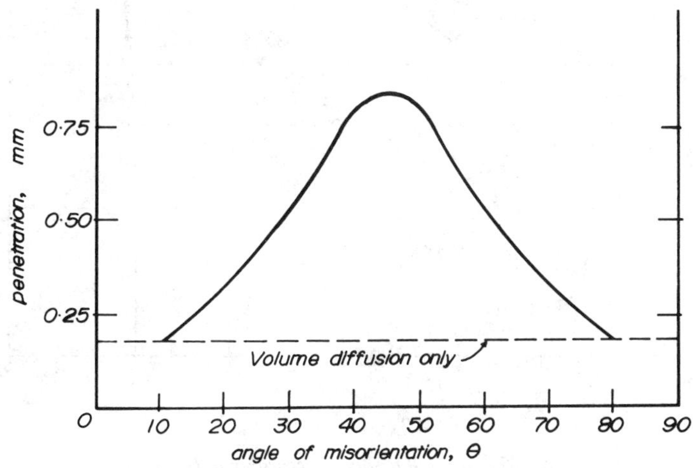

FIG. 11—*Penetration of silver along (100) tilt boundaries in copper* [16].

$$\alpha_i = a_{mi} t^{-n} \exp(Q/RT) \tag{7a}$$

At any temperature T and exposure time t, with each of the measured a_{mi}, the value of the corresponding α_i can be calculated with the empirically determined constants n and Q.

The data thus obtained fit the Weibull distribution function very well:

$$[1 - P_c(\alpha_i)] = \exp\{-[(\alpha_i - \alpha_u)/\alpha_0]^b\} = \exp[-(\alpha_i - \alpha_u)^b/\eta_\alpha] \tag{7b}$$

where $P_c(\alpha_i)$ is the probability of finding an α-value less than α_i on the cross-sectional surface of a test coupon. The plot of all the 144 values of α_i of 12 data sets, with 12 data points each, is shown in Fig. 12. The values of b, α_u, and α_0 are 1.85, 0.53×10^{-3}, and 0.51×10^{-3} respectively. These numerical values are for a_{mi} in micrometres and t in seconds in Eq 7*a*. It is interesting to note that the b-value of 1.85 for α_i for the 144 data points is close to the average b-value of 1.8 of the 12 data sets for a_{mi}. The value a_{mi} is linearly proportional to α_i, and the statistical distribution of α_i is directly related to the statistical distribution of a_{mi}.

One may consider the maximum value of α along the periphery of a sectioned surface i as the α_i of an area of $\pi\delta d$ of the test coupon, where δ is the coupon diameter and d is the grain size. Another sectioned surface at a distance one grain size away can be considered a section containing an entirely different set of grain boundaries, and it is an independent sample.

The probability of finding an oxide depth a_m is directly related to the size of the exposed area. Let $P(\alpha_i)$ be the probability of finding the maximum α-value less than α_i on a unit surface area

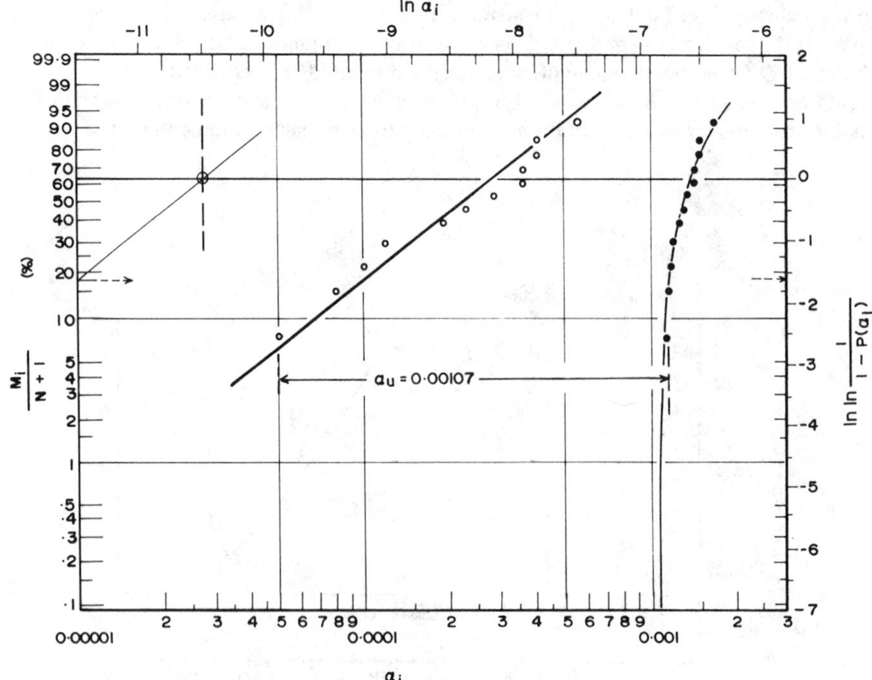

FIG. 12—*Weibull plot of* α_i *according to Eq 7.*

of a test coupon. The value $P(\alpha_i)$ is related to the probability of a sectioned coupon surface, $P_c(\alpha_i)$:

$$[1 - P_c(\alpha_i)] = [1 - P(\alpha_i)]^{\pi \delta d} \tag{8a}$$

$P_S(\alpha_i)$ is the probability of finding an α-value less than α_i on a surface area, S:

$$[1 - P_S(\alpha_i)] = [1 - P(\alpha_i)]^S = [1 - P_c(\alpha_i)]^{S/\pi \delta d} \tag{8b}$$

and

$$\ell n[1 - P_S(\alpha_i)] = (S/\pi \delta d)\, \ell n[1 - P_c(\alpha_i)] \tag{9}$$

We define R as

$$R = -(S/\pi \delta d)\, \ell n[1 - P_c(\alpha_i)] \tag{10}$$

Following Eq 1, we have

$$\begin{aligned} [1 - P_c(\alpha_i)] &= \exp[-(\alpha_i - \alpha_u)/\alpha_0]^b \\ R &= (S/\pi \delta d)[(\alpha_i - \alpha_u)/\alpha_0]^b \end{aligned} \tag{11}$$

$$P_S(\alpha_i) = 1 - \exp(-R) \tag{12}$$

The value of $P_S(\alpha)$ thus obtained can be taken as the value of $P_S(a)$ for the oxide depth $a = \alpha t^n \exp(-Q/RT)$. The $P_S(a)$ thus determined is not exactly the same as the probability of finding an oxide depth less than a among a set of measurements determined at a given oxidation temperature and exposure time. It incorporates the additional uncertainties of T and t. A more detailed statistical study is warranted.

Effects of Pre-Oxidation on Subsequent Fatigue Life

Dowling and Begley [*20*] and Dowling [*21*] have shown that, for general-yielding cyclic-loading, fatigue crack growth rate correlates very well with ΔJ. In the intermediate ΔJ region, fatigue crack growth rate is related to ΔJ by a power relation

$$\frac{da}{dN} = A \Delta J^m \tag{13}$$

For a small crack in a Ramberg-Osgood elastic-plastic solid in general yielding [*22,23*],

$$\frac{J}{\sigma_Y \epsilon_Y a} = \eta_1 \frac{W_D}{\sigma_Y \epsilon_Y} + \eta_2 \tag{14}$$

where η_1 and η_2 are functions of strain hardening exponent n. The value of η_2 is usually much smaller and negligible. W_D is the deformation work density defined as

$$W_D = \int_0^\epsilon \sigma \, d\epsilon \tag{15}$$

where σ and ϵ are the applied stress and strain away from the cracked region.

Zheng and Liu [24] have shown that Eqs 14 and 15 are also applicable to piecewise power hardening materials for small single-edge cracks in both plane stress and plane strain cases.[2]

For cyclic loading, Eq 14 is modified to the form

$$\Delta J = \eta \, \Delta W_D a \tag{16}$$

Substituting Eq 16 into Eq 13 and integrating,

$$(\Delta W_D)^m (N_f - N_i) = \frac{1}{A\eta^m(1 - m)} (a_f^{(1-m)} - a_i^{(1-m)}) \tag{17}$$

The remaining fatigue life after a crack reaches the "initial" size a_i is $(N_f - N_i)$. The value a_f is more or less a constant. For low-cycle fatigue, the applied strain range is rather high, the crack initiation period is short, and N_i is negligible.

Zheng and Liu [24] have analyzed the fatigue lives of precracked cylindrical aluminum specimens measured by Chen, Zheng, and Tsai [25]. Both the smooth and precracked cylindrical specimens are shown in Fig. 13. The specimens were precracked by a jeweler's saw [25].

The measured fatigue reversals, $2N_f$, are shown in Fig. 14 as a function of deformation work density, ΔW_D. The fatigue lives were measured in the low-cycle fatigue regime. The straight lines for the smooth and the precracked specimens are parallel in the log-log plot. Thus the ratio between the fatigue lives of the precracked specimens, N_{fi}, and the fatigue lives of the smooth specimens, N_{f0}, is a constant:

$$\frac{N_{fi}}{N_{f0}} = \text{constant} \tag{18}$$

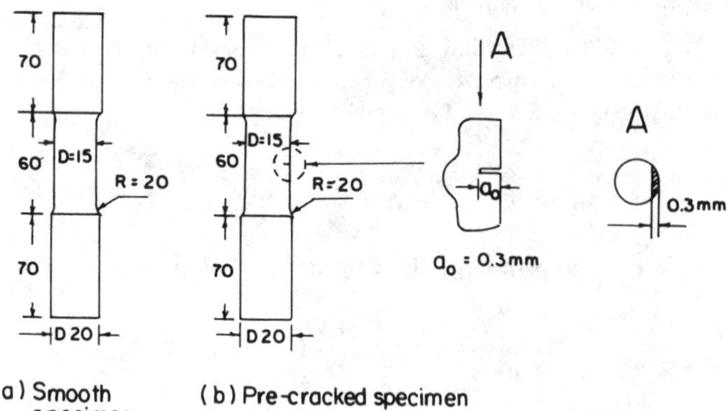

FIG. 13—*Smooth and precracked low-cycle fatigue specimen configurations.*

[2]For piecewise power hardening materials, $\sigma = E\epsilon$ for $\sigma \leq \sigma_Y$ and $(\sigma/\sigma_Y) = (\epsilon/\epsilon_Y)^n$ for $\sigma > \sigma_Y$.

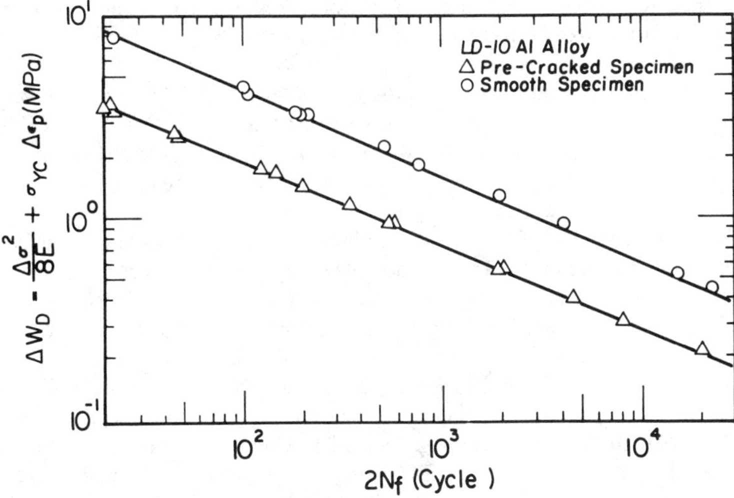

FIG. 14—*Correlation of* $\Delta W_D = (\Delta\sigma^2/8E + \sigma_{YC}\Delta\epsilon_p)$ *versus* $(2N_f)$ *for LD-10 aluminum alloy* [25].

With a_0 as a fictitious initial crack size for the "smooth" specimens, Eq 17 indicates that the ratio N_{fi}/N_{f0} is a function of the precrack size a_i:

$$\frac{N_{fi}}{N_{f0}} = f\left(\frac{a_0}{a_i}\right) \tag{19}$$

For a given precrack size, the function $f(a_0/a_i)$ is a constant. It is independent of the applied stress and strain ranges. The analysis (i.e., Eq 19 agrees with the experimental observation given by Eq 18. The detailed functional relationship of Eq 19 can be determined empirically.

Oxide is brittle. If the applied stress is high enough, a grain boundary oxide may crack. A grain boundary oxide crack, a_m, can be considered a precrack. Once the functional relation of Eq 19 is determined, the fatigue life of the pre-oxidized specimens can be predicted.

If the relation given in Eq 19 is deterministic, the probability of having a fatigue life of N_{fm} of a specimen with an oxide crack a_{mi} is also the probability of having an oxide crack size a_m on a specimen surface. Perhaps grain boundary oxidation and grain boundary oxide cracking contribute significantly to the statistical scatter of fatigue lives of engineering components at high temperatures.

Summary and Conclusions

The effects of pre-oxidation on subsequent fatigue life were studied. Nickel-base superalloy (TAZ-8A) was subjected to high-temperature oxidation (at 600 to 1000°C for 10 to 1000 h) in air. Surface oxidation kinetics and chemical and structural studies on surface oxides were carried out. In the high-temperature regime (i.e., 800 to 1000°C), outward diffusion of chromium ions through the surface oxide layer appears to control the oxidation kinetics. In the low-temperature regime (i.e., 600 to 800°C), nickel ion diffusion through the oxide layer seems to govern the parabolic oxidation rate.

Through the three-dimensional morphological observations, it is found that two different types of grain boundary oxides exist: cone type and pancake type. The pancake-type oxide is more damaging because of its size and shape.

The grain boundary oxide penetration depth, a_m, is related to oxidation temperature, T, and exposure time, t (i.e., $a_m = \alpha t^n \exp(-Q/RT)$) with $n = 0.25$ and $Q = 4.26$ kcal/mol.

A sufficiently large grain boundary oxide will be cracked by an applied stress. The largest and deepest of the oxide cracks is the most damaging to fatigue life. The effects of precracking on the subsequent fatigue life have been analyzed.

The statistical analysis using the Weibull distribution function for the grain boundary oxide penetration depth gives the probability of finding an oxide crack of a certain size on a surface area exposed to an oxidation temperature, T, for an exposure time, t. The statistical distribution of the size of the oxide crack is related to the statistical scatter of the remaining fatigue life.

Grain boundary oxidation kinetics, grain boundary oxide morphology, the statistical distribution of grain boundary oxide penetration depth, and the effects of precracking on the subsequent fatigue life are important elements in the development of a quantitative and mechanistic model for the prediction of fatigue life as affected by oxidation. Additional research on grain boundary oxide fracture strength, a detailed study on the precrack size and the reduction of fatigue life, and a mechanistic study on the effects of oxidation on fatigue crack growth are needed in the development of such a model.

Acknowledgments

The financial support by the NASA-Lewis Research Center, Grant NAG3-348, is gratefully acknowledged. The authors wish to thank Paul Liu for his elaborate preparation and performance for metallographic investigations and Mrs. Cindy Porter for her assistance in manuscript preparation.

References

[1] *Manual on Low-Cycle Fatigue Testing, ASTM STP 465,* American Society for Testing and Materials, Philadelphia, 1969.
[2] *Fatigue at Elevated Temperatures, ASTM STP 520,* American Society for Testing and Materials, Philadelphia, 1972.
[3] *Proceedings,* International Conference on Creep and Fatigue in Elevated Temperature Applications, 1973.
[4] ASME-MPC-3 Symposium on Creep-Fatigue Interaction, 1976.
[5] Pelloux, R. M. and Stoloff, N. S., Eds., *Creep-Fatigue-Environment Interactions,* American Institute of Mechanical Engineers, 1980.
[6] Coffin, L. F., Jr., in *Proceedings,* International Conference on Fatigue: Chemistry, Mechanics and Microstructure, 1972, pp. 590–680.
[7] Coffin, L. F., Jr., *Metallurgical Transactions,* Vol. 3, 1972, p. 1778.
[8] Antolovich, S. D., Domas, P., and Strudel, J. L., *Metallurgical Transactions,* Vol. 10A, 1979, pp. 1859–1868.
[9] McMahon, C. J. and Coffin, L. F., Jr., *Metallurgical Transactions,* Vol. 1, 1970, pp. 3443–3450.
[10] Shida, Y., Wood, G. C., Stott, F. H., Whittle, D. P., and Bastow, B. D., *Corrosion Science,* Vol. 21, No. 8, 1981, pp. 581–597.
[11] Stott, F. H., Wood, G. C., Shida, Y., Whittle, D. P., and Bastow, B. D., *Corrosion Science,* Vol. 21, No. 8, 1981, pp. 599–624.
[12] Hales, R. and Hill, A. C., *Corrosion Science,* Vol. 12, 1985, pp. 843–853.
[13] Rundell, G. R., *Metal Progress,* Vol. 127, No. 4, 1985, pp. 39–54.
[14] Kofstad, P., *Nonstoichiometry, Diffusion, and Electrical Conductivity in Binary Metal Oxides,* Wiley-Interscience, New York, 1972, pp. 101–362.
[15] Oshida, Y. and Liu, H. W., "Oxidation and Low-Cycle Fatigue Life Prediction," NASA Conference Publication No. 2339 on Turbine Engine Hot Section Technology, October 1984, pp. 321–329.
[16] Couling, L. and Smoluchowski, R., *Journal of Applied Physics,* Vol. 25, 1954, p. 1538.

[*17*] Turnbull, D. and Hoffman, R., *Acta Metallurgica,* Vol. 2, 1954, p. 419.
[*18*] Shewmon, P. G., *Diffusion in Solids,* McGraw-Hill, New York, 1963.
[*19*] Whipple, R. T. P., *Philosophical Magazine,* Vol. 45, 1954, p. 1225.
[*20*] Dowling, N. E. and Begley, J. A., "Fatigue Crack Growth During Gross Plasticity and the J-Integral," in *Mechanics of Crack Growth, ASTM STP 590,* American Society for Testing and Materials, Philadelphia, 1976, pp. 80–103.
[*21*] Dowling, N. E., "Crack Growth During Low-Cycle Fatigue of Smooth Axial Specimens," in *Cyclic Stress-Strain and Plastic Deformation Aspects of Fatigue Crack Growth, ASTM STP 637,* American Society for Testing and Materials, Philadelphia, 1977, pp. 97–121.
[*22*] Shih, C. F. and Hutchinson, J. W., "Fully Plastic Solutions and Large Scale Yielding J Estimates for Plane Stress Crack Problems," *Journal of Engineering Materials and Technology, Transactions ASME,* Vol. 98, 1976, pp. 289–295.
[*23*] Shih, C. F., "J-Integral Estimates for Strain Hardening Materials in Antiplane Shear Using Fully Plastic Solution," in *Mechanics of Crack Growth ASTM STP 590,* American Society for Testing and Materials, Philadelphia, 1976, pp. 3–26.
[*24*] Zheng, M. and Liu, H. W., "Crack Tip Field and Fatigue Crack Growth in General Yielding and Low-Cycle Fatigue," NASA Contract Report 174686, 1984.
[*25*] Chen, X. X., Zheng, M. Z., and Tsai, C. K., "Study of Strain Fatigue in LD-10 Alloy," *Journal of Acta Mechanica Sinica* (China), Vol. 5, 1981.

Albrecht Conle,[1] *Thomas R. Oxland,*[2] *and Timothy H. Topper*[2]

Computer-Based Prediction of Cyclic Deformation and Fatigue Behavior

REFERENCE: Conle, A., Oxland, T. R., and Topper, T. H., **"Computer-Based Prediction of Cyclic Deformation and Fatigue Behavior,"** *Low Cycle Fatigue, ASTM STP 942,* H. D. Solomon, G. R. Halford, L. R. Kaisand, and B. N. Leis, Eds., American Society for Testing and Materials, Philadelphia, 1988, pp. 1218–1236.

ABSTRACT: The paper briefly reviews the history of rheological spring-slider based, stress-strain models, and the early computerized versions used to simulate cyclic deformation and predict fatigue life. A set of rules is introduced which allows cycle by cycle simulation of the local stress-strain behavior at the root of a notched component subjected to any variable amplitude load history; predicts the crack initiation life; and, with two modifications to the notch analysis and the fatigue damage summation routines, models the local stress-strain behavior at the crack tip and resulting crack propagation rates. Supporting software that has resulted in practical applications for automotive design is discussed.

KEY WORDS: fatigue (materials), metals, stress, strain, models, crack initiation, crack propagation, variable amplitude, computer simulation

Material deformation models have been the focus of research since the earliest observations of elastic and plastic material stress-strain behavior. The realization that fatigue failures were due to reversed loading led to an interest in experiments of reversed stress-strain behavior, and Bauschinger [*1*] and others observed that the material stress-strain curves were dependent upon prior history. Studies such as those by Smith and Wedgwood [*2*] included accurate measurements of stress-strain locus movements during complex loading histories, and attempted to correlate hysteresis loop shape with Wöhler type fatigue data. One of the earliest attempts to predict this loop shape behavior was based on a rheologic model of springs and slider elements. Jenkin [*3*] showed that his model successfully predicted many of the cyclic deformation effects. Masing [*4*] modified a theory proposed by Heyn [*5*] to relate the initial deformation curve to the subsequent "cyclic" stress-strain loop shapes and showed that the effect was not due to any hardening or softening of the material, but was a consequence of residual stresses built up by the deformation history. Heyn, Jenkin, and Masing all based their deformation models on analogs for the rheological model types shown in Fig. 1. It would appear that the only reason the models did not achieve practical usage for the simulation of hysteresis behavior was the tedium of the necessary arithmetic.

The use of spring-slider models was suggested at various times between the 1920s and the 1960s [*6,7*], and Stulen [*8*] reviewed the subject and demonstrated that a modified rheologic representation could account for a number of other mechanical properties beyond simple stress-strain hysteresis behavior. Ivlev [*9*] showed that the model could be extended to multiaxial deformation, and Iwan [*10,11*] showed that the model could be adapted to simulate the movement of multiaxial yield surfaces in the same manner as the theory proposed by Mroz [*12*]. Further

[1]Ford Scientific Research Laboratories, Dearborn, MI 48121-2053.
[2]Department of Civil Engineering, University of Waterloo, Waterloo, Ontario, Canada N2L 3G1.

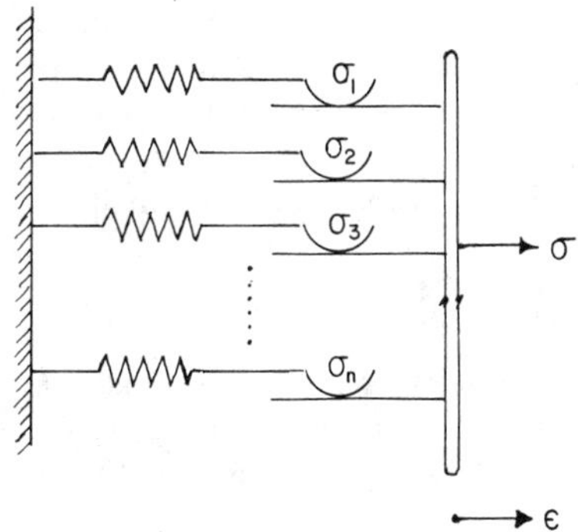

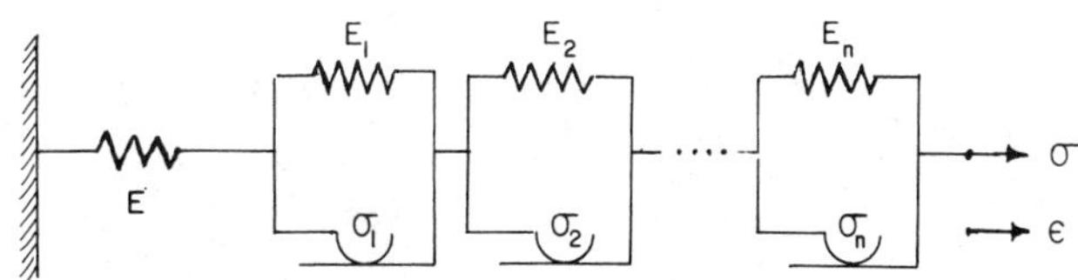

FIG. 1—*Two forms of rheological spring-slider models suitable for cyclic deformation simulation.*

development of the multiaxial model versions is underway [*13,14*], but application to fatigue awaits the development of a reliable multiaxial fatigue damage assessment procedure.

Developments in axial specimen testing, during the late 1960s, allowed improved test control and monitoring of hysteresis behavior, and Martin et al. [*15*] simulated axial cyclic deformation using a spring-slider model and applied it to the prediction of the fatigue life of notched samples. The flexibility of computer programs allowed them to model not only the stress-strain path for any arbitrary strain (or stress) history, but also cyclic hardening/softening phenomena, and cyclic mean stress relaxation and cyclic creep. An analysis based on that of Neuber [*16*] was superimposed [*17*] onto the stress-strain model in order to predict the local stress-strain behavior at a notch root of a component subjected to an arbitrary nominal stress history, and the simulated strain ranges of the loops were used to calculate fatigue damage.

While these developments were taking place in deformation modelling, studies of optimum variable amplitude history cycle counting methods were also in progress [*18*]. These efforts eventually culminated in the "Rainflow" cycle counting scheme, which related the cycles to closed stress-strain hysteresis loop observations [*19*], but at this time the two techniques, stress-strain modelling and cycle counting, were still not well integrated. Wetzel [*20*] increased the processing efficiency of the Martin et al. model by replacing the spring-slider force/deformation tracking system with "stress-strain segments" (one segment for each spring/slider) and a set of segment manipulation rules. He also overcame the cycle counting problem by counting fatigue damage on the basis of the segment usage, rather than by applying a counting method subsequent to simulation. Jhansale [*21*] refined a set of rules proposed by Jennings [*22*], which duplicated the spring-slider analog for loop shapes, and did not require the incremental stress-strain curve approaches used by Martin and Wetzel. The memory rules were simplified, however, to retain recognition of the stress-strain boundary created by only the largest previously occurring hysteresis loop.

The present study presents a set of rules for an axial deformation model which encompasses many of the features of the previous works, and in addition is able to extend the material memory accounting procedures to track memory effects in the nominal load versus local strain coordinates of components with stress raisers. The same memory rules are shown to provide a framework for monitoring the stress-strain behavior in the region at the tip of a propagating crack.

Material Cyclic Deformation Phenomena

The computerized models of references [*15,20,21,23*] are all able to simulate the various features of cyclic deformation behavior exhibited by an axially loaded smooth specimen (Fig. 2). The two features vital for good simulation in the majority of fatigue analyses are:

1. Hysteresis loop shape is the shape of a closed path of the stress-strain locus during inelastic straining. This shape is described by a nonlinear function or digital representation, the origin of which is at one point of stress-strain reversal and which passes through the next point of reversal.

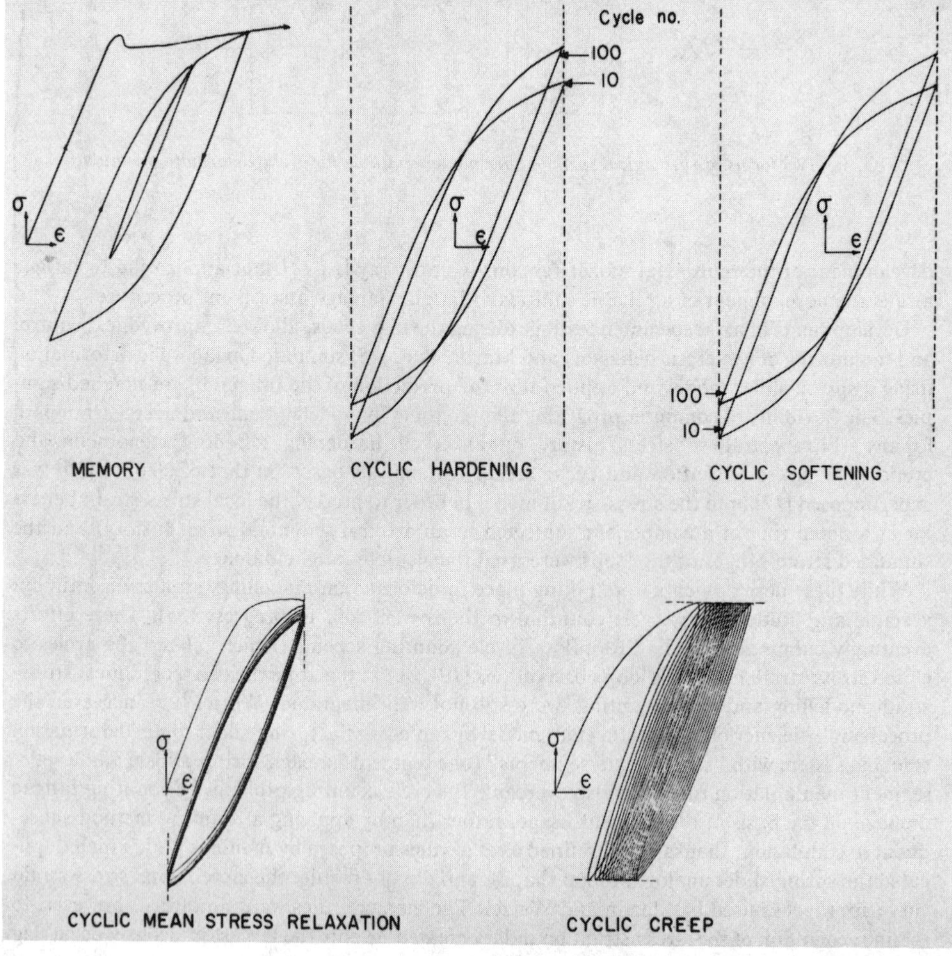

FIG. 2—*Cyclic deformation phenomena.*

2. Memory during cyclic straining refers to the tendency of a material, upon returning to the strain at which a previous stress-strain path was interrupted, to follow a continuation of the previous stress-strain path rather than an extension to the current path.

Unless specific fatigue environments require simulation of cyclic hardening/softening, or cyclic mean stress relaxation or cyclic creep, it is common practice to omit these transient effects. In most cases, the loss of simulation accuracy is small [*24*]. For cases where large overall mean stresses and small local stress-strain amplitudes characterize the fatigue conditions, the cyclic relaxation of mean stress can become important [*25*]. Due to a general lack of basic cyclic relaxation data, it is difficult to incorporate the effect in practice, and it is often more advantageous to "bracket" the life predictions by analyzing once with worst case mean stress assumptions, and once with best case assumptions.

Material properties for cyclic hardening/softening behavior are also not well documented in the literature. Present fatigue orientated Material Data Banks, in addition to monotonic data, contain only the stabilized cyclic stress-strain curves, and the stabilized strain life curves. To our knowledge none of the data banks presently contain descriptions of how a particular material gradually hardens or softens as cyclic deformation accumulates. Fatigue calculations, consequently, usually assume stabilized hardening behavior for both initial and subsequent cycle simulations and damage counting.

Simulation of these transitional properties can also cause severe problems when a notch analysis is superimposed onto the stress-strain simulation: both the notched sample and the local simulation must experience material memory events at the same instant in time. If the local simulation is hardening or softening between the interruption and later completion of a half-cycle, it becomes very difficult to compute the correct nominal and local coordinates of the memory event.

Modelling Material Memory Behavior

In a smooth specimen, as in more complex components, the material's monotonic stress-strain response is different from the cyclic stress-strain behavior, not just in terms of the effects of cyclic hardening and softening, but primarily in the magnitude of the stress-strain path. During the initial half-cycle (e.g., points 0 to 1 of Fig. 3) the shape of the stress-strain path is similar to the shape of a hysteresis loop path (e.g., points 2 to 3) but reduced by a factor of two when the origin lies at the start of the half-cycle. This assumption, termed *Masing's hypothesis*

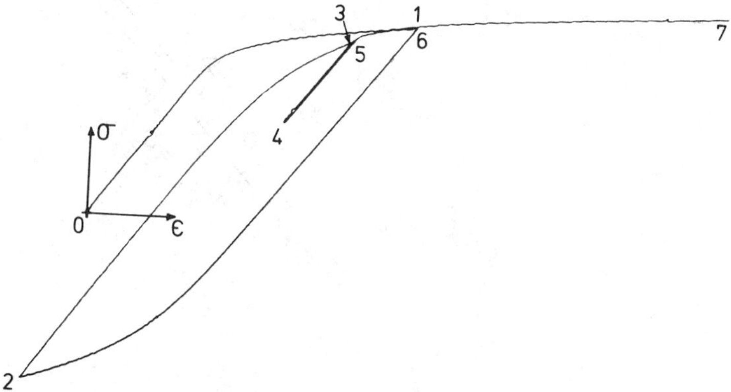

FIG. 3—*Smooth specimen stress-strain sequence showing material memory effects.*

[4], has been shown to agree well with test observations and consequently is used throughout this work.

Material memory events can be subdivided into two types. One is memory of a previously interrupted monotonic (often termed *skeleton*) curve, and the other is memory of a previously interrupted hysteresis loop curve. Although the first occurs relatively few times in a random loading history, the distinction is necessary for an accurate description of material behavior and it is therefore incorporated in the present model.

The effect of transient material phenomena, such as cyclic hardening and so forth, on material memory behavior is presently not well understood, and the complexity of the topic does not allow in-depth discussion here. Because such transient material behavior is usually confined to a small fraction of the fatigue life, it can normally be ignored by the simulation model. This is achieved by using only the stabilized hysteresis curve shape for deformation path prediction.

On the basis of the above assumptions, further definition of memory features is possible. In Fig. 3 one can observe that the stress-strain locus can return to the skeleton curve (path 6–7). Another aspect of skeleton curve memory is shown in Fig. 4. Initial deformation occurs along the skeleton curve from 0–1. After the reversal of deformation direction at point 1, the locus will first follow the hysteresis loop curve (double the monotonic curve) along the path 1–2. At point 2 (the mirror image of point 1, in compression) the path of the locus will again adopt the monotonic shape. From these observations one can hypothesize some general rules:

1. In any deformation history the maximum point of excursion on the monotonic or skeleton curve will represent an absolute limit or bound which will exist on both sides of the strain or stress axis, irrespective of whether the actual reversal occurred on the tension or compression side.
2. All deformation between these absolute bounds will utilize the cyclic hysteresis loop shape.
3. Whenever deformation exceeds these absolute bounds, the locus will again follow the shape of the monotonic curve and, after another reversal, will set a new absolute bound.

In Fig. 5*a*, for example, the reversal at point 1 sets the bounds at points 1 and 2. The limits are shown in terms of strains. As deformation proceeds to point 3 (Fig. 5*b*) and another reversal occurs, new bounds will be set.

The problem of the second type of memory event, that of a return to an interrupted hysteresis

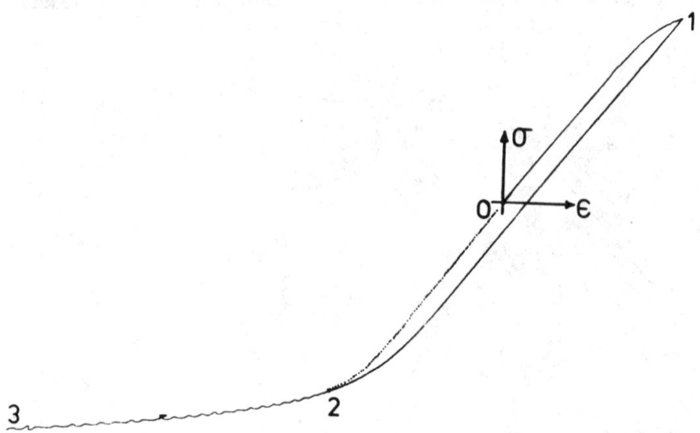

FIG. 4—*Material memory of the compressive monotonic curve after prior plastic deformation in the tensile direction.*

loop, can be described in similar terms. If a hysteresis loop half-cycle is interrupted (point 3, Fig. 3) by a smaller full cycle (3-4-5), the stress-strain locus will resume the previous half-cycle at the point of interruption (point 5). Each reversal point thus sets up a boundary which when intersected at some later time will signal a memory event. This concept of boundaries set by reversal of deformation direction can be combined into a uniform procedure for stress-strain modelling.

Modelling Using Push-Down List Counting

A "push-down list", in computer terminology, refers to a list wherein values are repeatedly being inserted and withdrawn and where the last value inserted is always the first value taken out. More explicitly, in terms of strain as a history variable, the treatment [*27*] in a computer storage would be as follows.

Two one-dimensional arrays are required, one for the storage of the tensile limits, called "TLIM", and one for compressive limits, "CLIM". Rules for the use of the arrays during stabilized cyclic straining are:

1. If there are no entries in the arrays, deformation is taking place on the monotonic curve.
2. If a reversal occurs on the monotonic curve at a strain magnitude larger than any previous reversal strain (i.e., both arrays are empty), the positive value of the limit is stored as the first entry in the TLIM array, and the negative value of the limit is stored as the first entry in the CLIM array.
3. If a reversal occurs between these absolute bounds, the value of the strain is used as the

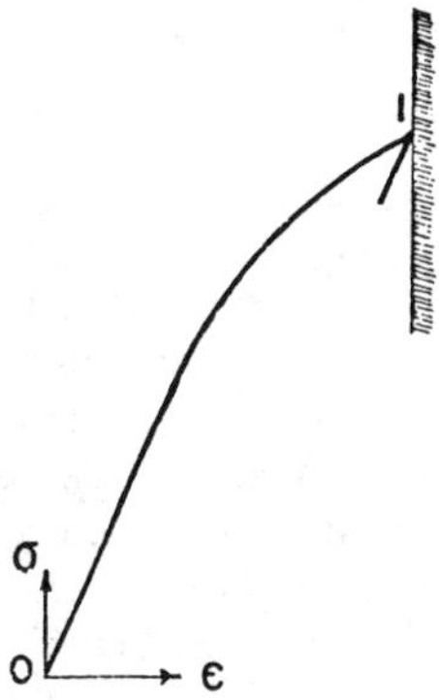

FIG. 5*a*—*Monotonic boundary points generated by a reversal of deformation direction during deformation on the monotonic curve.*

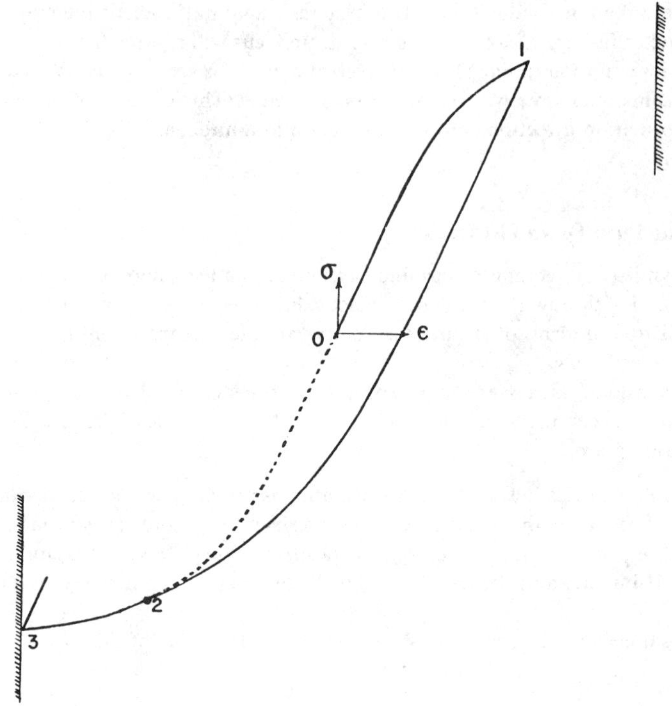

FIG. 5*b*—*Generation of new monotonic bounds due to deformation outside of the old bounds shown in Fig.* 5a.

next entry in the respective limit array, (i.e., reversals occurring during straining in the compressive direction are stored in CLIM and tensile reversals are stored in TLIM).

4*a*. As the stress-strain locus moves along any half-cycle, in the tensile direction, for example, the most recent entry in TLIM must be checked to see if its value corresponds to the current value of the strain. (If deformation is in the compressive direction, the most recent entry in CLIM must be checked.) Should the locus exceed the limit in the array, the most recent limits of both CLIM and TLIM must be removed from the arrays. The exact conditions of the removal are subject to rules 4*b*, 4*c*, and 4*d* below.

4*b*. If the limit exceeded is not the only one remaining in that array, a hysteresis loop has been closed. The most recent entries in TLIM and CLIM define the limits of that hysteresis loop. These entries must be removed from the arrays (the loop is forgotten).

4*c*. If the limit exceeded is also the only remaining entry in that array and in the opposite array two entries remain, a single hysteresis loop has been closed and the monotonic curve is incurred again. All entries are eliminated from both arrays.

4*d*. If the limit exceeded is also the only remaining entry in that array and in the opposite array only one entry remains, no hysteresis loop is being closed but the monotonic curve is incurred again. All values must be eliminated from the arrays. (In effect an unmatched half-cycle has been "forgotten".)

Rule application can be demonstrated by use of the following sequence of strain values, which is assumed to be applied to an unnotched sample initially at zero stress and strain:

Point No.	1	2	3	4	5
Strain	+0.8%	−0.22%	+0.6%	0.0	+1.0%

During the first half-cycle (0.0 to +0.8%) the arrays TLIM and CLIM would be empty and the locus would move along the monotonic curve (Rule 1). At reversal 1, Rule 2 would apply and the first entry would be made in each of CLIM and TLIM (Fig. 6*a*: $CLIM_1 = -0.8$, $TLIM_1 = 0.8$). At reversal 2 the value −0.22% would be entered into $CLIM_2$ (Fig. 6*b*) as per Rule 3. Reversals at point 3 and 4, having similar effects, are also managed by Rule 3; the results are depicted in Figs. 6*c* and 6*d*.

Assuming then, that after the straining to point 5 the locus continues to move in the tensile direction, then, by Rule 4*a*, the tensile limits are being checked for exceedance. The first limit to be incurred (Rule 4*a*) on this upward path is the +0.6% strain stored in $TLIM_2$ (point 3, Fig. 6*d*). At 0.6% strain the inner hysteresis loop between points 3 and 4 is closed and forgotten; $CLIM_3 = 0.0$ and $TLIM_2 = 0.6$ are eliminated as per rule 4*b* and the locus continues on the extension of the previous path 2–3. If the deformation direction remains tensile, the next limit is incurred at 0.8% strain. Rule 4*c* would apply; a loop is closed and forgotten ($CLIM_2 = -0.22$, $TLIM_1 = 0.8$, and $CLIM_1 = -0.8$ are eliminated), and the path of the locus is redirected onto the monotonic curve (Fig. 7*b*). A subsequent reversal would begin the procedure anew.

Analysis of Notch Effects

In order to enable engineers to analyze components other than those that can be modelled by an axially loaded sample, it became necessary to add some form of notch analysis to the cyclic deformation model described above. The Neuber analysis [*16,17*] that is commonly used employs a total deformation concept (i.e., the notch analysis computes from end point to end point of a half-cycle and cannot be applied in an incremental fashion [*26*]). The problems outlined in Ref *26* can be overcome, as suggested in Ref *27*, by incorporating the notch analysis within the same set of rules that controls the material deformation simulation (i.e., the notch analysis must be aware of prior deformation induced memory events in the same way that the stress-strain simulation must account for the events). Socie [*28*] used essentially two Wetzel-type spring-slider simulations: one for nominal load and local strain, another in parallel for local strain and local stress, to achieve the correct result. The Wetzel-type of model, however, is faced with a compromise between accuracy of reversal points and computational speed. The push-down list type deformation rules do not require this compromise.

The concept that allows integration of component notch analysis and an axial specimen deformation model [*29*] is: *When the nominal load experiences a reversal in the loading direction, both local stress and local strain will also reverse direction.* Thus the push-down list counting rules that apply to the axial deformation model also automatically provide memory for the notch analysis, or vice versa. In practice [*30,31*] the lists used to store reversal points in strain are expanded as shown in Fig. 8. The counting rules developed above are usually applied only to the nominal load variable, rather than to the axial strains. When a nominal load is entered or removed from the stack, all the corresponding local stress, local strain, and fatigue damage/half-cycle points are also entered or removed at the same time. Because both local and nominal coordinates experience memory events at the same instant in the history, many of the previously observed interaction problems between the notch analysis and the material simulation are eliminated. An added benefit is that computational requirements are significantly reduced.

Crack Propagation Simulation

Studies of constant amplitude crack growth [*32,33*] have demonstrated an improved ability to predict the behavior of short cracks propagating in the stress concentration region of notches by

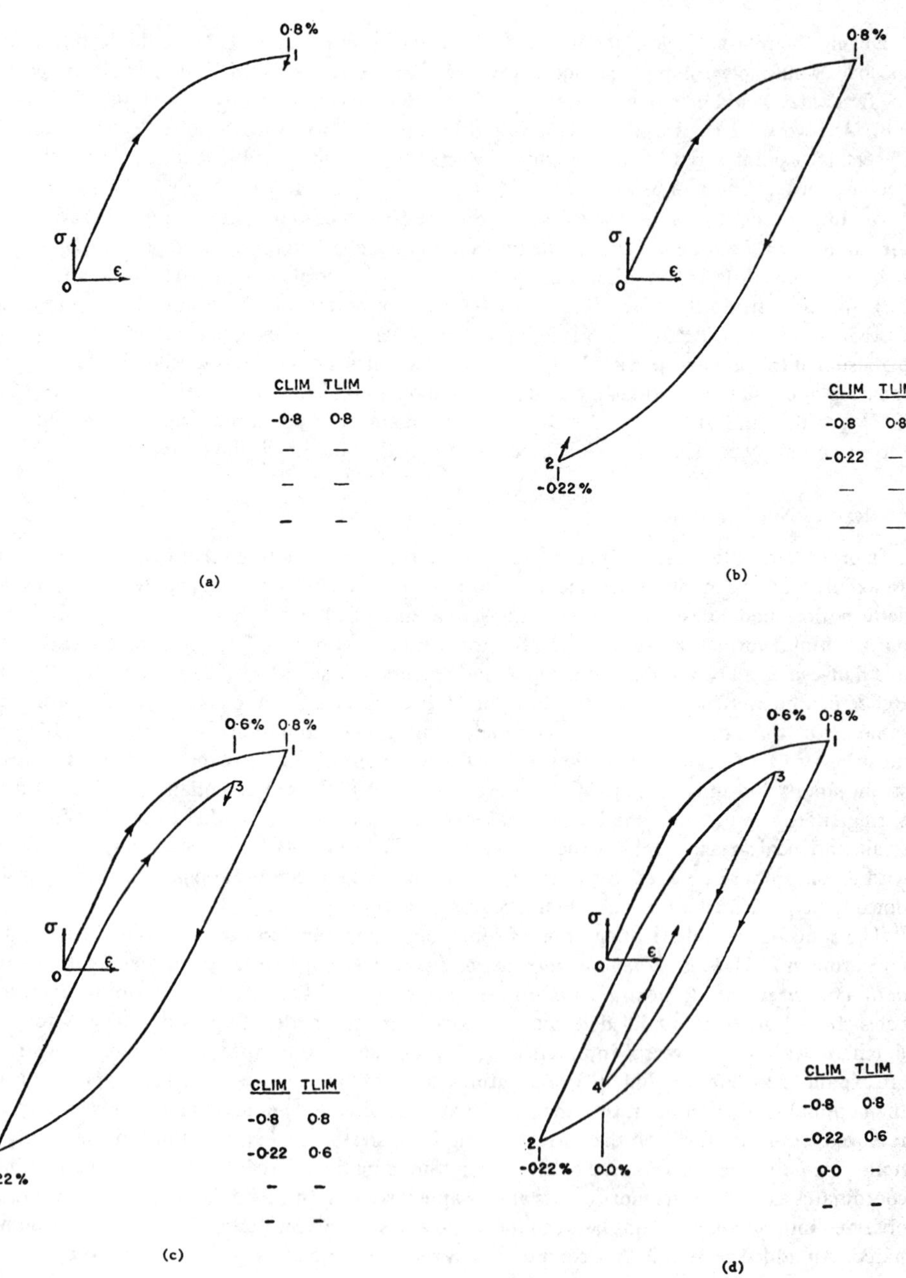

FIG. 6—*Strain reversal stacking in the "push-down" lists during the deformation sequence.*

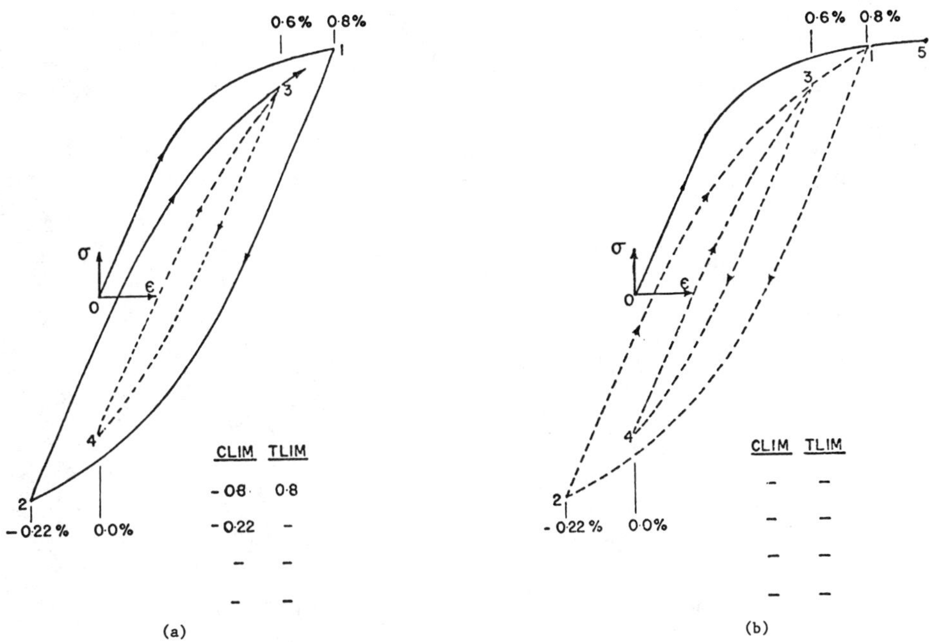

FIG. 7—*Effect of memory events on the listing procedure.*

substitution of a strain-based stress intensity factor for the usually applied stress intensity factor. The success of this approach and that of the cycle by cycle initiation model suggested that a combination of the two concepts might result in a crack propagation model that could predict the effects of variable amplitude load histories on crack propagation [*30*]. As shown in Fig. 9, all the aforementioned notch component analysis model features are maintained. The only changes are that the nominal load versus local strain transformation is replaced by a nominal stress-strain intensity transformation and that the damage summation is replaced by crack growth increment per cycle calculations. The nominal load term of the notch analysis model becomes $F_w K'S$: a product of finite width correction, the concentration factor for a crack in a notch, and the nominal (gross) stress. Strains in the vicinity of the crack tip are determined by applying the Neuber relationship

$$K' \cdot F_w \cdot \Delta S = (\Delta\sigma \cdot \Delta\epsilon \cdot E)^{1/2} \quad (1)$$

To solve this relationship for crack tip strains, constant amplitude smooth specimen data are used to plot $(\Delta\sigma \cdot \Delta\epsilon \cdot E)^{1/2}$ versus $\Delta\epsilon$ [*32*]. In the computer model the transformation from $F_w \cdot K' \cdot \Delta S$ to $\Delta\epsilon$ is made by interpolation between stored values of the curve in exactly the same way as the nominal load versus local strain curve is interpolated in the crack initiation model (Fig. 9). The local stress and strain variables are computed by applying the push-down list counting rules in the normal manner. When a half-cycle is completed, the strain-based stress intensity factor of Ref *32* is computed as

$$\Delta K = E \cdot \Delta\epsilon\sqrt{\pi(\ell + \ell_0)} \cdot F_s \cdot F_p \quad (2)$$

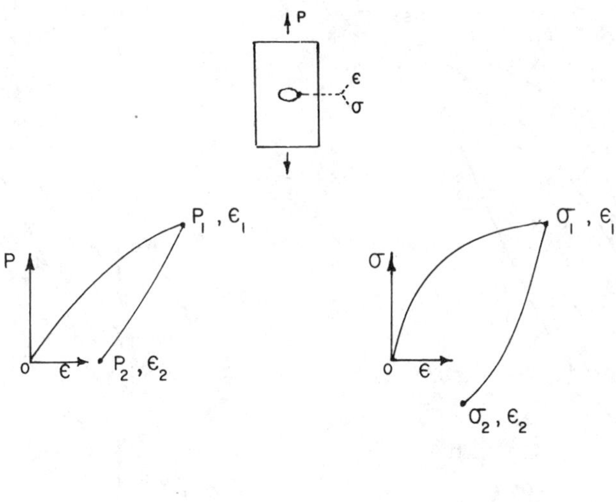

STACKS

COMPRESSION				TENSION			
DAMAGE	STRESS	STRAIN	LOAD	LOAD	STRAIN	STRESS	DAMAGE
D_{01}	$-\sigma_1$	$-\epsilon_1$	$-P_1$	P_1	ϵ_1	σ_1	D_{01}
D_{12}	σ_2	ϵ_2	P_2	–	–	–	–
–	–	–	–	–	–	–	–
–	–	–	–	–	–	–	–

FIG. 8—*Combined stacking procedure for both nominal and local variables.*

where E is Young's modulus, ℓ is the crack length, ℓ_0 is a material constant, F_s is a mean stress correction factor, and F_p is a plastic zone size correlation factor for elastic conditions around the crack tip.

As depicted in the lower section of Fig. 9 the changes to compute crack propagation damage per cycle required minimal alteration of the modelling program. The relationship between $d\ell/dn$ and ΔK was taken from the constant amplitude crack propagation data. Crack growth was calculated by fitting the crack growth equation suggested by McEvily [*34*] to the present material data:

$$d\ell/dn = \left(\frac{K_c}{K_c - \Delta K/2}\right) \cdot A \cdot (\Delta K - \Delta K_{\text{th}})^m \tag{3}$$

where K_c is the critical stress intensity factor, A and m are material constants, and ΔK_{th} is the threshold stress intensity.

Recent work [*35,36*] has shown that the large compressive cycles of a variable amplitude load

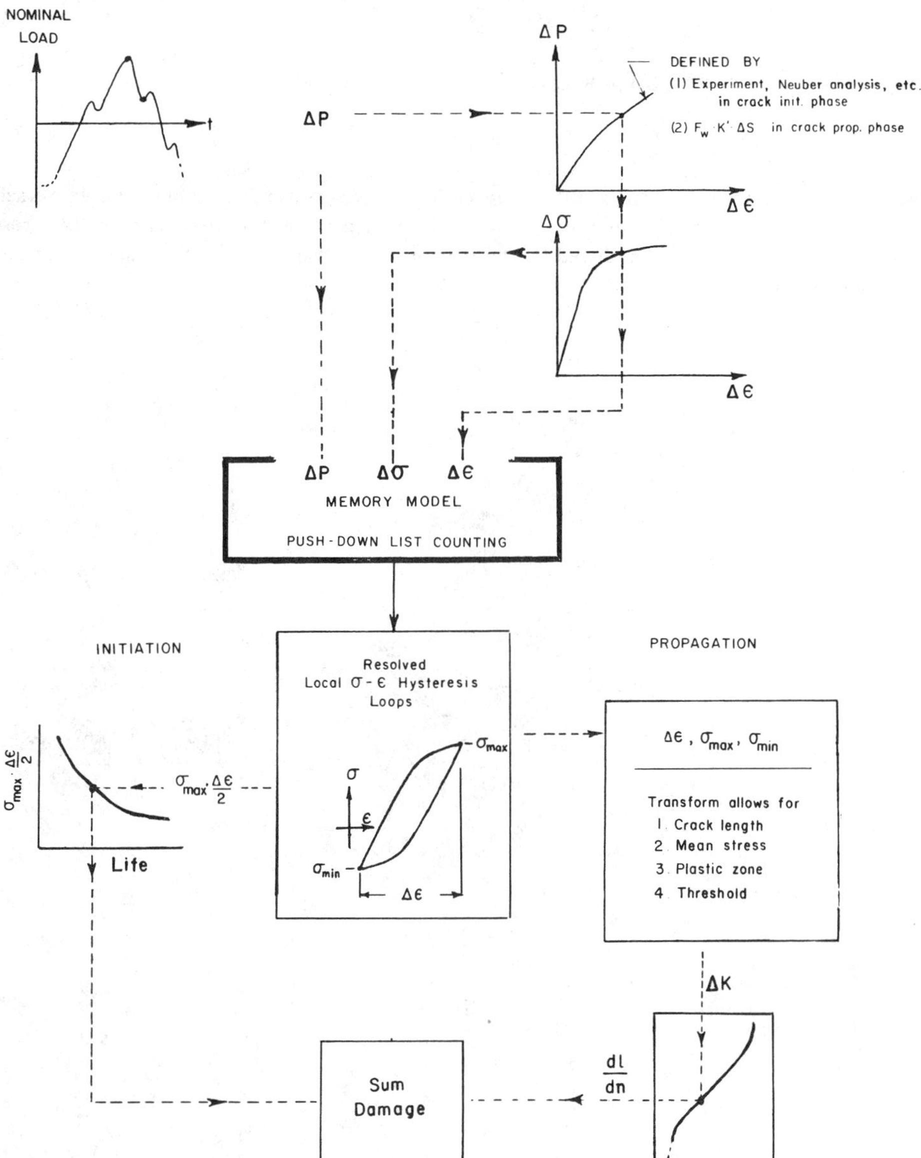

FIG. 9—*Flow and transformation of nominal and local variables and the methods of damage computation contrasted for the crack initiation and the crack propagation phases in the computer simulation.*

history reduce crack closure stresses and result in a significant lowering of threshold stress intensities. Work is in progress to define the degree and duration of the lowering of crack opening stresses due to "underload." As a simple approximation, the crack growth curve of Eq 3 was modified by setting the threshold stress intensity equal to zero. In addition, a mean stress correction of the same form as used in the crack initiation computations of the smooth specimen simulation

$$F_s = \sqrt{\sigma_{max}/(1/2(\sigma_{max} - \sigma_{min}))} \tag{4}$$

was applied to short cracks. For long cracks a similar correction

$$F_s = \sqrt{\sigma_{max}/(\sigma_{max} - \sigma_x)} \tag{5}$$

was applied to the tensile portion of the stress cycle. The constants taken from the fully reversed crack growth data were also modified to use only the tensile portion (one-half) of the stress cycle. The strain range, as predicted by the memory rules, was modified for crack increment calculations by

$$\left.\begin{array}{l} \Delta\epsilon \rightarrow \Delta\epsilon - |\sigma_{min}/E| \\ \sigma_x = 0 \end{array}\right\} \text{if} \quad \sigma_{min} < 0 \tag{6}$$

$$\left.\begin{array}{l} \Delta\epsilon = \Delta\epsilon \\ \sigma_x = \sigma_{min} \end{array}\right\} \text{if} \quad \sigma_{min} \geq 0 \tag{7}$$

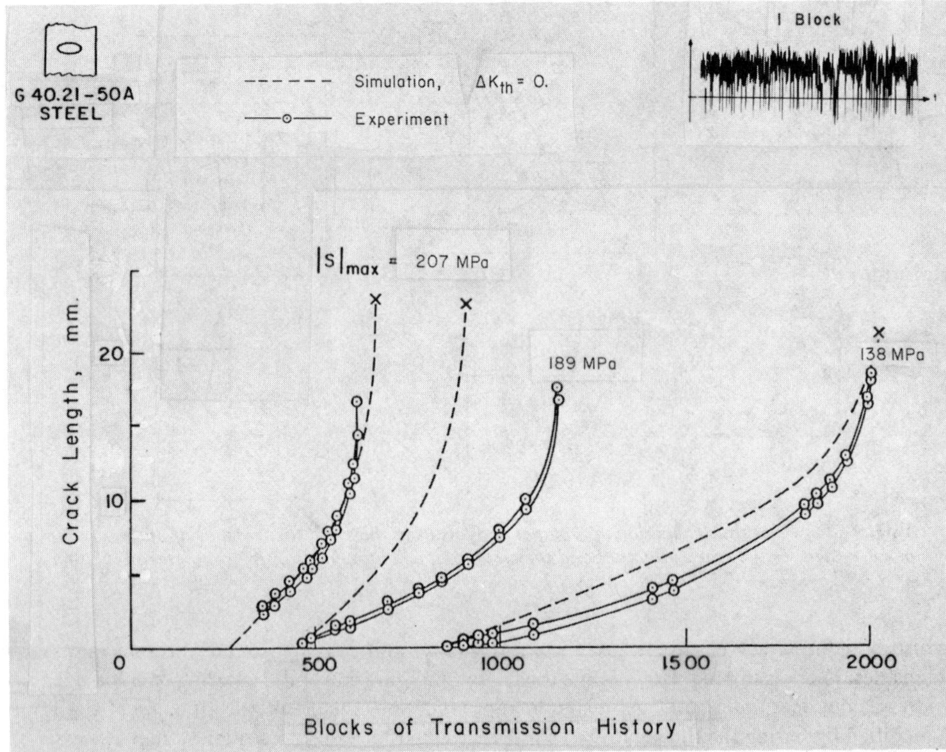

FIG. 10—*Observed and predicted crack propagation behavior in a central notched* ($K_t = 4.6$) *sample subjected to repeated blocks of SAE transmission load history.*

This has the effect of using the strain above a stress of zero to compute the strain intensity for inelastic strain cycles. For elastic cycles the method is equivalent to the assumption of using only the tension part of the stress cycle.

Variable amplitude testing and simulations were done using the SAE transmission history documented in Ref *37*. Figure 10 compares actual with predicted crack propagation behavior. The comparisons start at the observed initiation point of the test samples and show good agreement between simulation and experiment.

Applications in Automotive Design

Relatively few cycle by cycle stress-strain simulation programs are presently being used by component designers in any of the durability oriented engineering industries. This is because of a number of reasons:

- Simulation programs are more difficult to build than simpler damage summation routines.
- Service load or strain data are not usually provided in the form of peak/valley sequences. Most designers are supplied with some type of summary histogram, and the necessary computer facilities for acquisition of sequences usually do not exist.
- Supporting software that enables usage of the routines by non-experts is often lacking or expensive to acquire.

Efforts at Ford Motor Company have resulted in a set of software routines [*31*] that have overcome the first and the last of the above problems. Work on the second problem is underway. At present the software programs greatly simplify the problem definition and result presentation aspects for a wide variety of fatigue problem types, including the stress-strain and fatigue simulation programs described in this paper. The stress-strain and fatigue simulation accepts four different types of notch or geometric descriptions and several forms of load history sequences. An interactive format allows a user to specify the component load and graphically display the resulting local stress-strain behavior and the fatigue damage per cycle (Fig. 11). For notch geometry effects defined by finite element or other elastic analyses, an optional load history superposition may be performed for multiple load input points prior to the history simulation run. Very long histories use a packed format for efficient storage of peaks and valleys. Displays of the histories and damage per cycle, as in Fig. 12, are optional and, because the push-down list scheme is similar to rainflow counting, a summary of the history is collected in matrix format and then displayed as in Fig. 13. In this figure the graph at the lower right depicts range-mean-occurrence sets sorted according to largest range first, and then plotted on a cumulative cycle basis. This format is very similar to the older level crossing count plots and is usually preferred by designers over matrix printouts. Figure 13 also correlates the damage per level at the top left of the figure and the predicted "block" (history repetition) life versus history test level in the graph at the right.

The crack propagation version of the model has not yet been made available for usage by durability designers. Two essential ancillary features are lacking; namely, a data bank of crack propagation materials parameters and a data bank of stress intensity equations for differing component geometries.

Concluding Remarks

In this paper we have presented the historical development of stress-strain modelling and its usage for fatigue life prediction. A set of deformation rules were shown to govern not only the stress-strain behavior, but also the nominal force/local strain transformations of notched components and the conditions of deformation at the tip of a propagating crack. The uniformity of

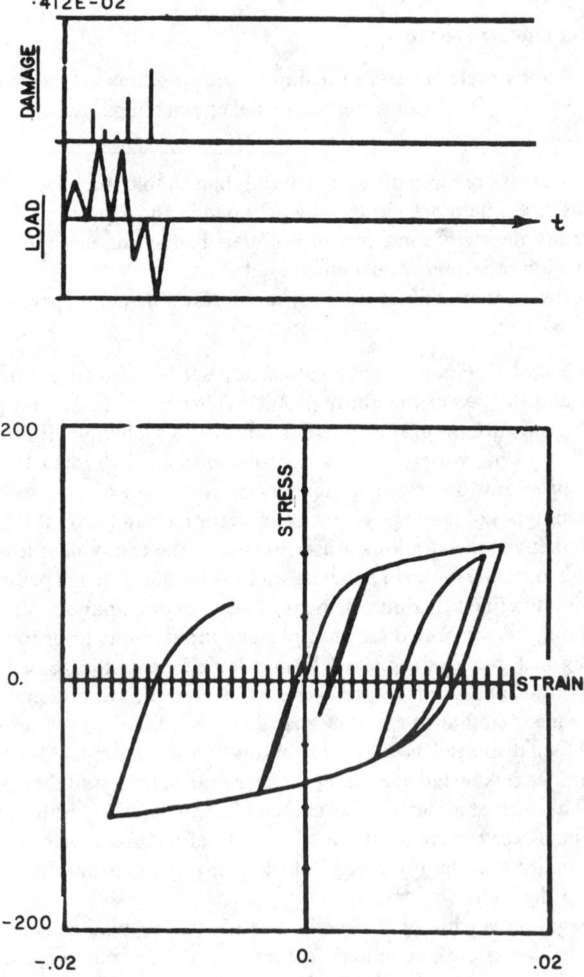

FIG. 11—*Interactive load history specification with plots of local stress-strain behavior and fatigue damage per half-cycle.*

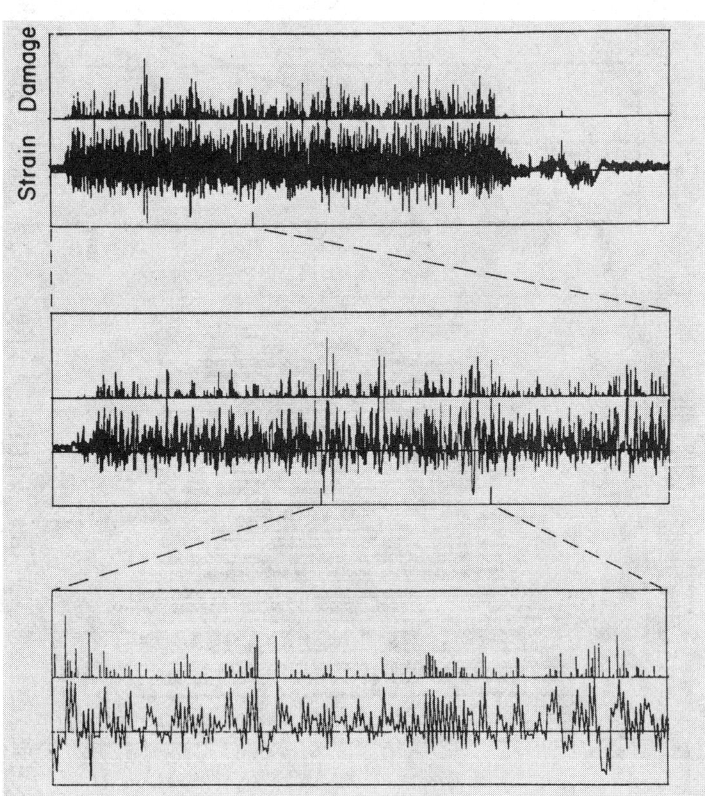

FIG. 12—*Plots of nominal load history and simulation predicted damage per half-cycle.*

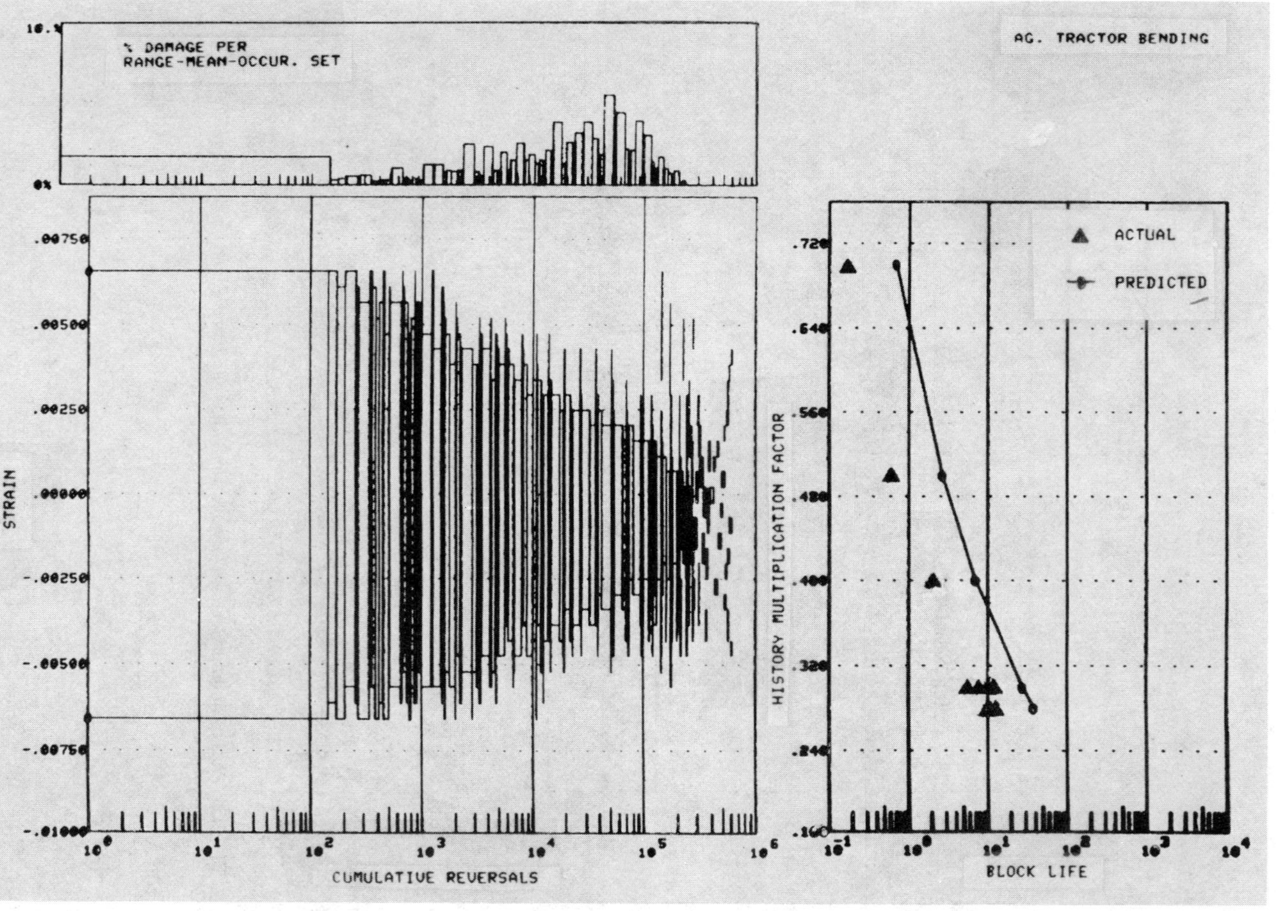

FIG. 13—*Plots of "push-down" list generated range-mean-occurrence matrix with damage per set display and effect of history magnification.*

the simulation concepts has simplified their incorporation in a user-orientated fatigue analysis program for automotive component durability design.

References

[1] Bauschinger, J., "Uber die Veraenderung der Elasticitaetsgrenze und der Festigkeit des Eisens und Stahls durch Strecken und Quetschen, durch Erwaermen und Abkuehlen und durch oftmal wiederholte Beanspruchung," *Mitteilungen Mechanishes-technisches Laboratorium,* Muenchen, Vol. 13, No. 8, 1886, pp. 1-115.

[2] Smith, J. H. and Wedgwood, G. A., "Stress-Strain Loops for Steel in the Cyclic State," *Journal of the Iron and Steel Institute,* Vol. 91, No. 1, 1915, pp. 365-397.

[3] Jenkin, C. F., "Fatigue in Metals," *The Engineer,* Vol. 134, No. 3493, Dec. 1922, pp. 612-614.

[4] Masing, G., "Eigenspannungen und Verfestigung beim Messing," in *Proceedings* 2nd International Congress of Applied Mechanics, Zurich, 1926.

[5] Heyn, E., "Eine Theorie der Verfestigung von metallischen Stoffen infolge Kaltreckens," *Festschrift der Kaiser Wilhelm Gesellschaft,* 1921, pp. 121-131.

[6] Duwez, P., "On the Plasticity of Crystals," *Physical Review,* Vol. 47, March 1935, pp. 494-501.

[7] Whiteman, I. R., "A Mathematical Model Depicting the Stress-Strain Diagram and the Hysteresis Loop," *Journal of Applied Mechanics, ASME Transactions,* March 1959, pp. 95-100.

[8] Stulen, F. B., "A Model for the Mechanical Properties of Metals," *ASTM Materials Research and Standards,* Vol. 2, No. 2, Feb. 1962, pp. 102-108.

[9] Ivlev, D. D., "The Theory of Complex Media," *Soviet Physics - Doklady,* Vol. 8, No. 1, July 1963, pp. 28-30.

[10] Iwan, W. D., "A Distributed-Element Model for Hysteresis and Its Steady State Dynamic Response," *Journal of Applied Mechanics, ASME Transactions,* Dec. 1966, pp. 893-900.

[11] Iwan, W. D., "On a Class of Models for the Yielding Behavior of Continuous and Composite Systems," *Journal of Applied Mechanics, ASME Transactions,* Sept. 1967, pp. 612-617.

[12] Mroz, Z., "On the Description of Anisotropic Work-Hardening," *Journal of the Mechanics and Physics of Solids,* Vol. 15, 1967, pp. 163-175.

[13] Chu, C.-C., "A Three-Dimensional Model of Anisotropic Hardening in Metals and Its Application to the Analysis of Sheet Metal Formability," *Journal of the Mechanics and Physics of Solids,* Vol. 32, 1984, pp. 197-212.

[14] McDowell, D. L., "A Two Surface Model for Transient Nonproportional Cyclic Plasticity," *Journal of Applied Mechanics, ASME Transactions,* Vol. 52, June 1985, p. 298.

[15] Martin, J. F., Topper, T. H., and Sinclair, G. M., "Computer Based Simulation of Cyclic Stress-Strain Behavior," Department of Theoretical and Applied Mechanics, Rep. No. 326, University of Illinois, July 1969. Also published as "Computer Based Simulation of Stress-Strain Behavior with Applications to Fatigue," *ASTM Materials Research and Standards,* Vol. 11, No. 2, Feb. 1971.

[16] Neuber, H., "Theory of Stress Concentration for Shear Strained Prismatical Bodies with Arbitrary Non-Linear Stress Strain Law," *Journal of Applied Mechanics,* Dec. 1961, pp. 544-550.

[17] Topper, T. H., Wetzel, R. M., and Morrow, J., "Neuber's Rule Applied to Fatigue of Notched Specimens," *ASTM Journal of Materials,* Vol. 4, No. 1, March 1969, pp. 200-209.

[18] Schijve, J., "The Analysis of Random Load-Time Histories with Relation to Fatigue Tests and Life Calculations," in *Proceedings,* 2nd ICAF-AGARD Symposium, Paris, 1961.

[19] Matsuishi, M. and Endo, T., "Fatigue of Metals Subjected to Varying Stress," presented at Japan Society of Mechanical Engineering, Fukuoka Japan, March 1968.

[20] Wetzel, R. M., "A Method of Fatigue Damage Analysis," Ph.D. thesis, University of Waterloo, Canada, 1971.

[21] Jhansale, H. R., "Inelastic Deformation and Fatigue Resistance of Spectrum Loaded Strain Controlled Axial and Structural Members," Ph.D. thesis, University of Waterloo, Canada, 1971. Also published as: Jhansale, H. R. and Topper, T. H., "Cyclic Deformation and Fatigue Behavior of Axial and Flexural Members—A Method of Simulation and Correlation," in *Proceedings,* 1st International Conference on Structural Mechanics in Reactor Technology, Berlin, Sept. 1972, Part L, pp. 433-455.

[22] Jennings, P. C., "Earthquake Response of a Yielding Structure," *ASCE Journal of the Engineering Mechanics,* Vol. EM4, Aug. 1965, pp. 41-68.

[23] Hanschmann, D., Ph.D. thesis, Technishe Hochschule Aachen, W. Germany, 1981.

[24] Conle, A. and Topper, T. H., "Sensitivity of Fatigue Life Predictions to Approximations in the Representation of Metal Cyclic Deformation Response in a Computer-Based Fatigue Analysis Model," in *Proceedings,* 2nd International Conference on Structural Mechanics in Reactor Technology, Berlin, Sept. 1973.

[25] Landgraf, R. W. and Francis, R. C., "Material and Processing Effects on Fatigue Performance of Leaf Springs," *SAE Transactions*, Vol. 88, Sect. 2, 1979, p. 1711.

[26] Conle, A. and Nowack, H., "Verification of a Neuber-Based Notch Analysis by the Companion Specimen Method," *Experimental Mechanics*, Vol. 17, No. 2, Feb. 1977, pp. 57-63.

[27] Conle, F. A., "A Computer Simulation Assisted Statistical Approach to the Problem of Random Fatigue," M.Sc. thesis, University of Waterloo, Canada, 1974.

[28] Socie, D. F., "Fatigue-Life Prediction Using Local Stress-Strain Concepts," *Experimental Mechanics*, Vol. 17, No. 2, Feb. 1977, pp. 50-56.

[29] Williams, D. P. et al., "Structural Cyclic Deformation Response Modelling," in *Proceedings*, ASCE & EMD Specialty Conference on Mechanics in Engineering, May 1976, pp. 291-311. Also published as: Williams, D. P. and Topper, T. H., "A Generalized Model of Structural Reversed Plasticity," *Experimental Mechanics*, April 1981, pp. 145-154.

[30] Conle, F. A., "An Examination of Variable Amplitude Histories in Fatigue," Ph.D. thesis, University of Waterloo, Canada, 1979.

[31] Conle, A. and Landgraf, R. W., "A Fatigue Analysis Program for Ground Vehicle Components," in *Digital Techniques in Fatigue, Conference Proceedings*, Society of Environmental Engineers, City University, London, March 1983.

[32] El-Haddad, M. H., Smith, K. N., and Topper, T. H., "A Strain Based Intensity Factor Solution for Short Fatigue Cracks Initiating from Notches," in *Fracture Mechanics, ASTM STP 677*, C. W. Smith, Ed., American Society for Testing and Materials, Philadelphia, 1979, pp. 274-289.

[33] El-Haddad, M. H., Smith, K. N., and Topper, T. H., "Fatigue Crack Propagation of Short Cracks," *ASME Journal of Engineering Materials and Technology*, Vol. 101, No. 1, Jan. 1979.

[34] McEvily, A. J., "Current Aspects of Fatigue," *Metal Science*, Aug./Sept. 1977, pp. 274-284.

[35] Conle, A. and Topper, T. H., "Overstrain Effects During Variable Amplitude Service History Testing," *International Journal of Fatigue*, July 1980, pp. 130-136.

[36] Au, P., Topper, T. H., and El-Haddad, M. L., "The Effects of Compressive Overload on the Threshold Intensity for Short Cracks," AGARD Structures and Materials Panel Specialist Meeting Proc. No. 328, *Behavior of Short Cracks in Airframe Components*, Toronto, Sept. 1982, pp. 11.1-11.7.

[37] *Fatigue under Complex Loading*, R. M. Wetzel, Ed., SAE Advances in Engineering Report AE-6, 1977.

V. Bicego,[1] C. Fossati,[1] and S. Ragazzoni[2]

Low Cycle Fatigue Characterization of a HP-IP Steam Turbine Rotor

REFERENCE: Bicego, V., Fossati, C., and Ragazzoni, S., **"Low Cycle Fatigue Characterization of a HP-IP Steam Turbine Rotor,"** *Low Cycle Fatigue, ASTM STP 942,* H. D. Solomon, G. R. Halford, L. R. Kaisand, and B. N. Leis, Eds., American Society for Testing and Materials, Philadelphia, 1988, pp. 1237–1260.

ABSTRACT: Results of an extensive low cycle fatigue characterization of HP and IP stages of a 1Cr-Mo-V steam turbine rotor are presented. Uniaxial strain controlled fatigue tests were carried out at 480 and 540°C, with hold times up to 24 h and strain rates down to 3×10^{-6} s^{-1}. Moderate life reductions due to long hold periods were found only in tests with strain ranges less than 0.8%, though microstructural observations revealed evidence of creep damage morphologies in a few tests with 24 h hold times carried out at high strain ranges. Life reductions caused by the lowest strain rate were more significant. Generally, inferior fatigue resistance was provided by the rotor bore material, particularly in the more heavily segregated zones of the intermediate pressure stage.

Application of the linear damage summation (LDS) rule to hold time data was not satisfactory. Better results were generally obtained by the strainrange partitioning (SRP) method, with a few unconservative life predictions mainly due to inclusions that dominated early failures. A life correlation was attempted by an energy-based criterion suitable to provide rationalization of creep and fatigue damage in terms of thermodynamic concepts.

KEY WORDS: fatigue (materials), high temperature fatigue, creep, microstructure, damage, fracture, life prediction

The current trend in electric network operation is towards an increase of cycling in thermal power plant units, including those of the largest sizes. The required increased operational flexibility may, however, be mainly limited by the steam turbine; frequent thermal transients may enhance the occurrence of thermal fatigue crack initiation. Though the present steelmaking and manufacturing processes generally result in high quality rotor forgings with improved toughness and ductility, thus greatly reducing the potential for catastrophic failures, creep and low-cycle fatigue phenomena have still to be considered of great concern for safe and reliable operation.

Results are presented here of an extensive low-cycle fatigue characterization of HP and IP stages of a 1Cr-Mo-V 320 MW steam unit rotor. This activity, carried out by CISE laboratories, is in the framework of an extensive ENEL (Italian Electricity Board) project aimed at a complete mechanical characterization of the component. Tensile, creep, fracture mechanics, and low-cycle fatigue tests, as well as extensive microstructural examinations, have been carried out at different sampling positions in the forging. The low-cycle fatigue activity, here of main concern, was in particular intended to clarify the role of microstructural inhomogeneities and the influence of varying test parameters (strain rate, hold time, temperature) on material behavior. Moreover, an analysis of damage assessment criteria was expected to verify the direct applica-

[1]CISE, Via Reggio Emilia, 39 - 20090 Segrate, Milan, Italy.
[2]ENEL/CRTN, Via Rubattino, 54 - 20134, Milan, Italy.

bility of currently available life prediction methods for components presently in service in power units.

Material

A schematic drawing of the rotor is shown in Fig. 1. The material is a 1Cr-Mo-V steel of ASTM A470 Cl8 type, with the chemical composition shown in Table 1. In the as-received condition, it underwent the conventional post-forging operations typical of this class of components: austenitizing at 970°C + air cool + tempering at 660°C + furnace cool. Specimens were sampled with a longitudinal orientation from the Rim region of the High Pressure stage (hereafter called HPR material) of the rotor, and with a tangential orientation from the Bore regions of the High Pressure and Intermediate Pressure stages (HPB and IPB specimens, respectively); these orientations were selected on the basis of the stress-strain patterns developed during in service thermal transients. Tensile and impact properties are summarized in Tables 2 and 3; creep resistance curves are shown in Fig. 2.

A complete microstructural characterization, including quantitative analysis of secondary phases, was carried out in order to investigate the evolution of microstructure during cyclic loading, as well as its role on LCF resistance. Large scale inhomogeneities were observed throughout the rotor; this finding, which is typical of this class of components, is generally attributed to small local differences in chemical compositions, leading to different morphology evolution during standard heat treatments [*1*]. The HP stage revealed (Figs. 3*a* to 3*c*) a bainitic structure together with more evoluted bainitic regions parallel to the rotor axis, rich in carbides and characterized by ferritic grains. Different grain boundary carbides were found in the two types of microstructures: mainly M_3C in the bainitic zones and $M_{23}C_6$ and M_7C_3 (with bigger dimensions than M_3C) in the temper evoluted bainite. The IPB material showed (Figs. 3*d* to 3*f*) only the temper evoluted bainitic structure; in addition, high amounts of heavily segregated zones were found, characterized by:

1. Highly evoluted bainite, with a great number of recrystallization nuclei.
2. Smaller grain size than all other zones, with large $M_{23}C_6$ carbide precipitation.

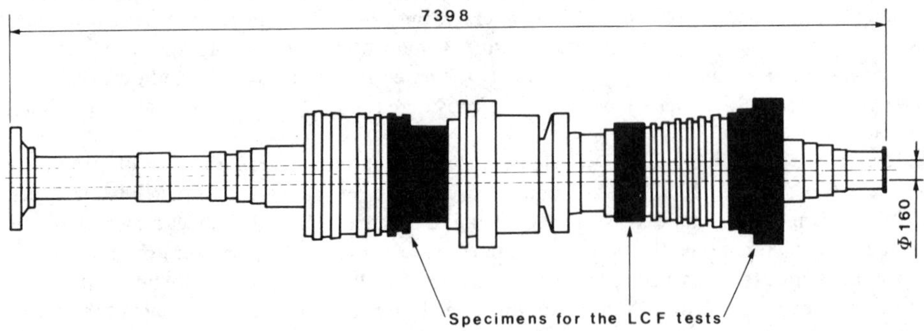

FIG. 1—*Schematic drawing of the 320 MW unit rotor (dimensions in millimetres).*

TABLE 1—*Chemical composition (weight percent) of 1Cr-Mo-V steel.*

C	Si	Mn	P	S	Cr	Mo	Ni	V	Cu	As	Sn
0.33	0.22	0.77	0.009	0.006	1.25	1.18	0.06	0.27	0.05	0.01	0.005

TABLE 2—*Summary of tensile properties.*

Temperature, °C	Rotor Zone[a]	Orientation	0.2% Yield Strength, MPa	Ultimate Tensile Strength, MPa	Elongation, %	Reduction of Area, %
RT	HPR	axial	660	806	17	62
	HPB	transverse	643	789	13	34
	IPB	transverse	618	800	12	34
500	HPR	axial	506	563	22	79
	HPB	transverse	504	553	19	69
	IPB	transverse	472	564	12	45
600	HPR	axial	411	426	34	91
	HPB	transverse	402	419	27	82
	IPB	transverse	356	446	19	85

[a] HPR = High Pressure Rim; HPB = High Pressure Bore; IPB = Intermediate Pressure Bore.

TABLE 3—*Summary of Charpy impact properties.*

Rotor Zone[a]	FATT, °C[b]	Upper Shelf Energy, J
HPR	96	107
HPB	95	105
IPB	95	78

[a] Orientation: transverse.
[b] FATT = Fracture Appearance Transition Temperature.

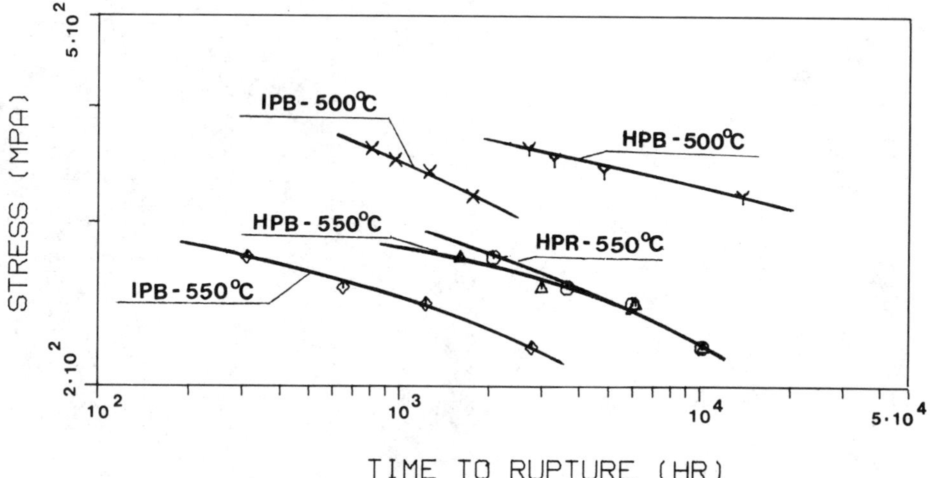

FIG. 2—*Creep resistance curves of material from different sampling positions on the rotor.*

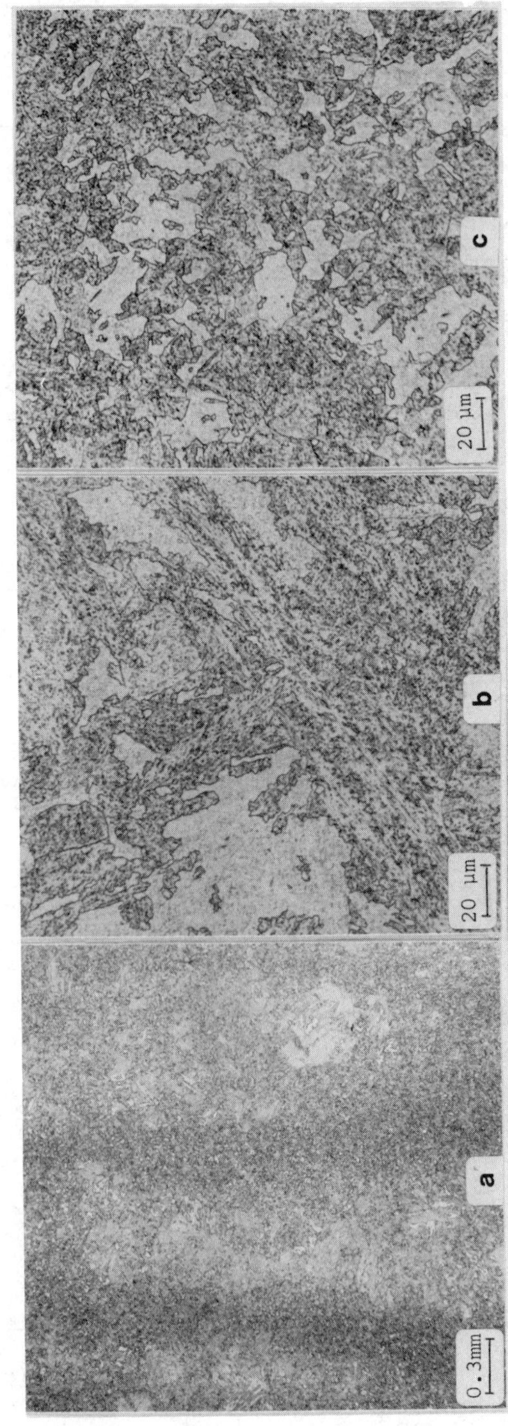
a
0.3mm
b
20 µm
c
20 µm

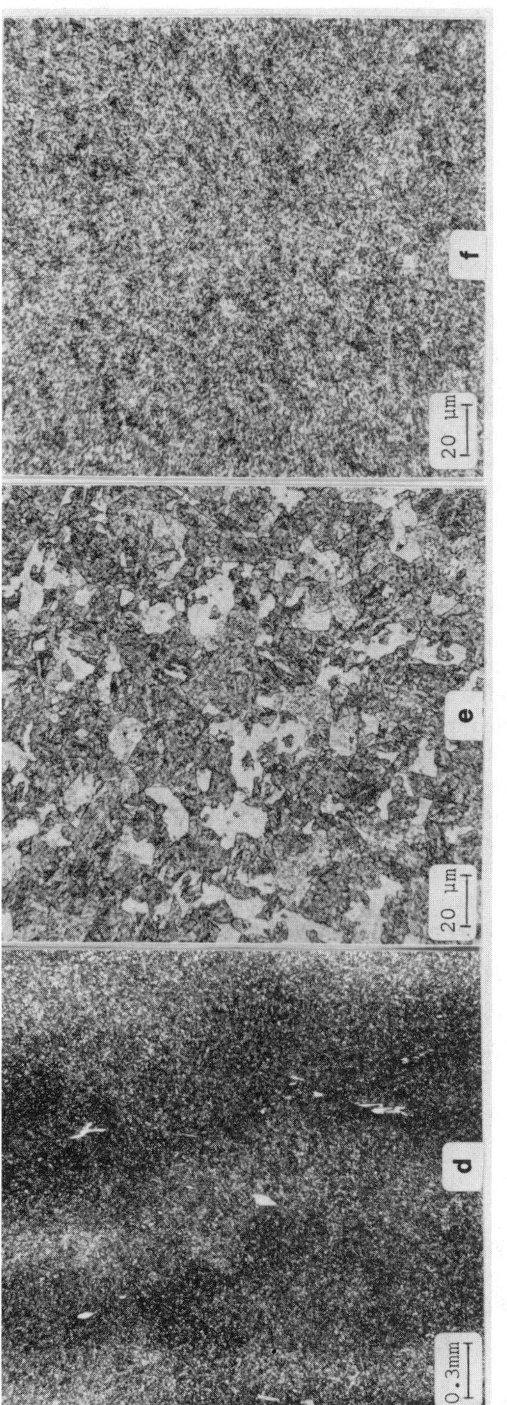

FIG. 3—*Different types of microstructure of HPR and IPB rotor regions, 2% Nital etched.* (a) *HPR macrograph.* (b) *HPR bainitic zone.* (c) *HPR temper evoluted zone.* (d) *IPB macrograph.* (e) *IPB temper evoluted zone.* (f) *IPB highly evoluted zone.*

3. Large non-metallic inclusions (mainly MnS) elongated along the rotor axis. (These were also present in the HPB region, see Fig. 4.)
4. High hardness Ti and V-rich intermetallic secondary phases.

The radial distribution of these highly segregated zones in the IP stage is shown in Fig. 5; the percent volume content ranges from 2% (rim) to 10% (bore).

An overview of quantitative metallographic information is given in Table 4.

It is noteworthy to observe that, in spite of all these microstructural inhomogeneities, conventional mechanical properties were within the specified limits for this class of components.

Experimental Procedure

The test facility consisted of six servohydraulic MTS units, with horizontal load frames of ±100 kN load capability. Specimen heating was accomplished by single zone electric resistance furnaces; the thermal gradient was within ±1°C over a region of 30 mm, and thermal stability during long-term tests was within ±2°C. Button-head hourglass gage section specimens, 9 mm minimum diameter and 72 mm hourglass radius, were used; the axial strain was derived from measurement of the transverse deformation in accordance with ASTM E 606. Undesired bending strains induced by rigid clamping of the specimen heads were within 0.01%. A computer-controlled data acquisition system connected to the different test units assured continuous on-line monitoring of all relevant test parameters.

Fully reversed strain-controlled fatigue tests involved both continuous cycling at constant strain rate ($\dot{\epsilon}$) and hold times (HT) at the peak tensile strain. Values of total strain ranges $\Delta\epsilon_t$ were in the interval 0.5 to 4%. Intervals of $\dot{\epsilon}$ (3×10^{-3} to 3×10^{-6} s^{-1}) and of HT (0 to 24 h) included values typically used in short-term laboratory tests as well as values closer to the realistic operating conditions of plant components. Tests were run in air at 480 and 540°C; these values were intended to bound the expected maximum temperatures of critical zones of the rotor (540°C is the inlet steam nominal temperature). Table 5 summarizes the various test parameters covered in this experiment and the number of tests performed.

The overall number of tests performed was 196. Test duration ranged from a minimum of a few minutes up to a maximum of 9000 h, with more than 50 000 h spent in the whole experiment. The experiment is still in progress, with the particular aim of running tests with even longer durations. A few tests (particularly tests with long cycle periods) were interrupted at various intervals to allow investigation of damage evolution. For specimens brought to rupture, the number of cycles to failure, N_f, was defined from the observed 50% peak tensile stress drop with respect to the mid-life value. Data reduction for the interpolation of $\Delta\epsilon_t - N_f$ curves made use of the usual Manson-Coffin type regression analysis. Most significant specimens were analyzed both via optical and electron (scanning and 200 kV transmission) microscopes in order to clarify the role of microstructure on fatigue resistance. Moreover, fracture mechanisms and the origin of cracks were investigated, and a quantitative evaluation of microcavitation was performed.

Results

Figure 6 shows the range of cyclic lives obtained in the various tests performed. Data are discriminated only in terms of test temperature. The effects of sampling position, strain rate, and hold time are analyzed in the plots of Figs. 7 and 8. Data of continuous cycling tests with $\dot{\epsilon} = 3 \times 10^{-3}$ s^{-1} and of tests with HT = 1/6 h were not considered, as they did not appreciably differ from similar results with $\dot{\epsilon} = 3 \times 10^{-4}$ s^{-1} and HT = 0 h. These last data were therefore assumed representative of time-independent fatigue and plotted in the form of Manson-Coffin reference curves in Fig. 8.

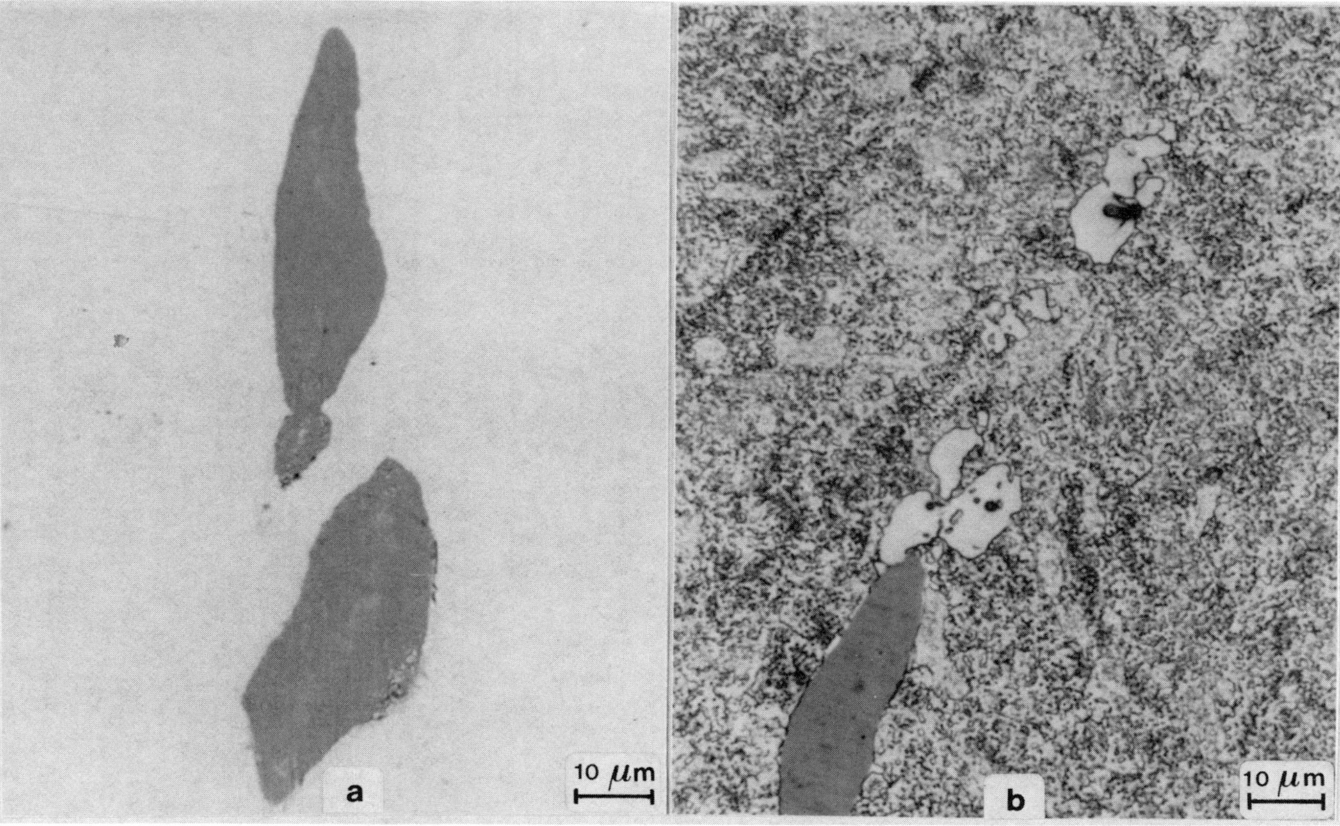

FIG. 4—*Different types of inclusions at rotor bore.* (a) *HPB material (polished).* (b) *IPB material (2% Nital etched).*

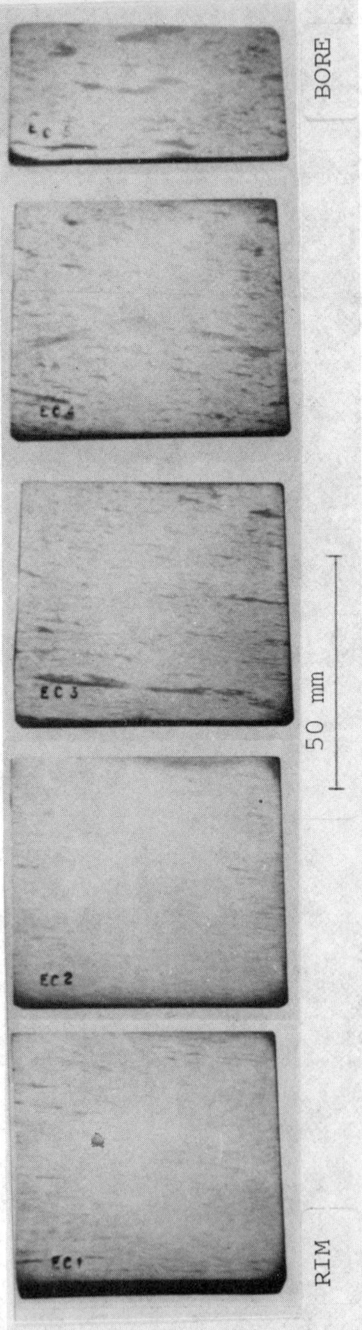

FIG. 5—*Distribution of heavily segregated zones along the rotor radius in IP stage.*

TABLE 4—*Summary of the microstructural parameters of the different regions of the rotor.*

Material			HP stage					IP stage			
			Rim		Bore			Bore			
								Normal TEB			Heavily TEB
Structure[a]			B ~ 1/3	TEB ~ 2/3	B ~ 1/3	TEB ~ 2/3	F little	B very little	TEB ~1	F very little	TEB ~1
Austenitic grain diameter		No ASTM	< 2	9	< 2	9	~10.5	-	9	-	10
		Øm (μm)	> 179	15.9	> 179	15.9	~10	-	15.9	1	12.2
Main grain boundary carbide particles		Types	M_3C	$M_{23}C_6$ M_7C_3 M_3C	M_3C	$M_{23}C_6$ M_7C_3 M_3C	-	-	$M_{23}C_6$ M_7C_3 M_3C	-	mostly $M_{23}C_6$
	Stereology	N_A (cm^{-2})	13.9 10^8	7.3 10^8	-	-	-	-	7.2 10^8	-	1.18 10^8
		Diameter Øm (μm)	0.21	0.28	-	-	-	-	0.28	-	0.49
		Interparticle spacing (μm)	0.16	0.22	-	-	-	-	0.21	-	0.61
Inclusions	LR section	N_A (cm^{-2})	1.57 10^4		0.97 10^4			1.62 10^4			1.13 10^4
		Diameter $Ø_{mL}$ (μm)	1.54		1.69			1.35			4.86
		Surface fraction	6.10 10^{-4}		6.16 10^{-4}			4.76 10^{-4}			42.0 10^{-4}
	CR section	N_A (cm^{-2})	1.18 10^4		1.18 10^4			1.09 10^4			0.68 10^4
		Diameter $Ø_{mC}$ (μm)	1.47		1.55			1.58			2.57
		Surface fraction	5.63 10^{-4}		6.11 10^{-4}			4.69 10^{-4}			6.70 10^{-4}
		N_v (cm^{-3})	9.9 10^7		5.0 10^7			8.96 10^7			4.07 10^7
Vickers Hardness HV10			270		262			250			262

[a] B = Bainite; TEB = Temper Evoluted Bainite; F = Ferritic grains

TABLE 5—*Summary of low cycle fatigue tests.*

Test	Conditions	HPR Tests at 480°C	HPR Tests at 540°C	HPB Tests at 480°C	HPB Tests at 540°C	IPB Tests at 480°C	IPB Tests at 540°C
HT = 0 h	$\dot{\epsilon} = 3 \cdot 10^{-3}$/s	7	11	...	1	...	1
	$\dot{\epsilon} = 3 \cdot 10^{-4}$/s	7	11	7	11	7	11
	$\dot{\epsilon} = 3 \cdot 10^{-5}$/s	7	11	...	6	4	6
	$\dot{\epsilon} = 3 \cdot 10^{-6}$/s	6	8	1	3	3	3
$\dot{\epsilon} = 3 \cdot 10^{-4}$/s	HT = 1/6 h	6	6	...	...	...	...
	HT = 1 h	...	15[a]	3	4	4	5
	HT = 3 h	1	4	...	...	4	4
	HT = 24 h	...	4	...	...	2	2

[a]Ten of these tests with compressive HT; tensile HT in all other cases.

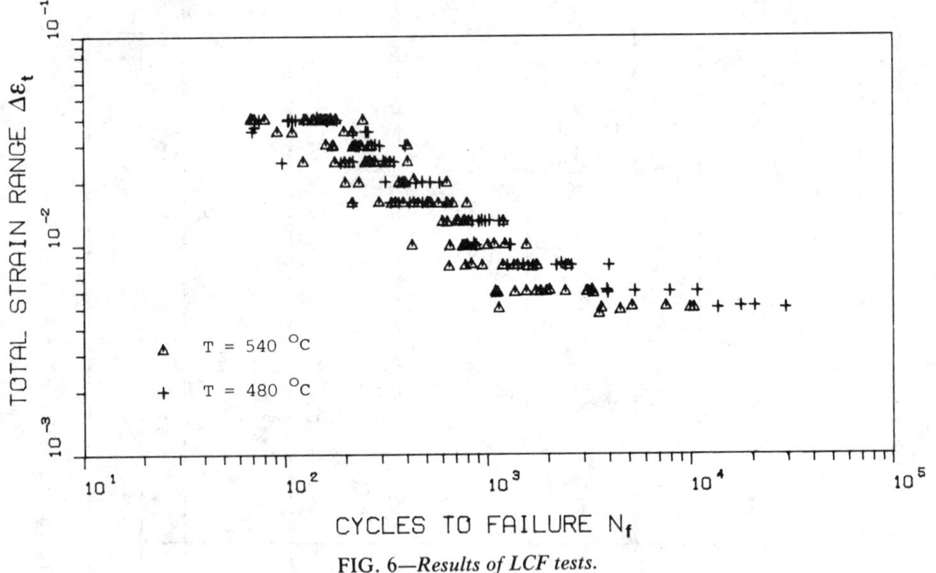

FIG. 6—*Results of LCF tests.*

Material Effect

At both temperatures, fatigue lives of specimens from the bore region of HP and, even more significantly, the IP rotor stages show higher intrinsic scatter than the HPR data (Fig. 7). For the data in Fig. 7*a* the residual variability *S* of logarithmic lives around Manson-Coffin fitting curves was evaluated

$$S = \frac{1}{n} \sum_{i} (\log N_{fi} - \log N_i)^2$$

where

N_{fi} = actual life at $\Delta\epsilon_{ti}$,
N_i = corresponding value on Manson-Coffin curve, and
n = number of test data.

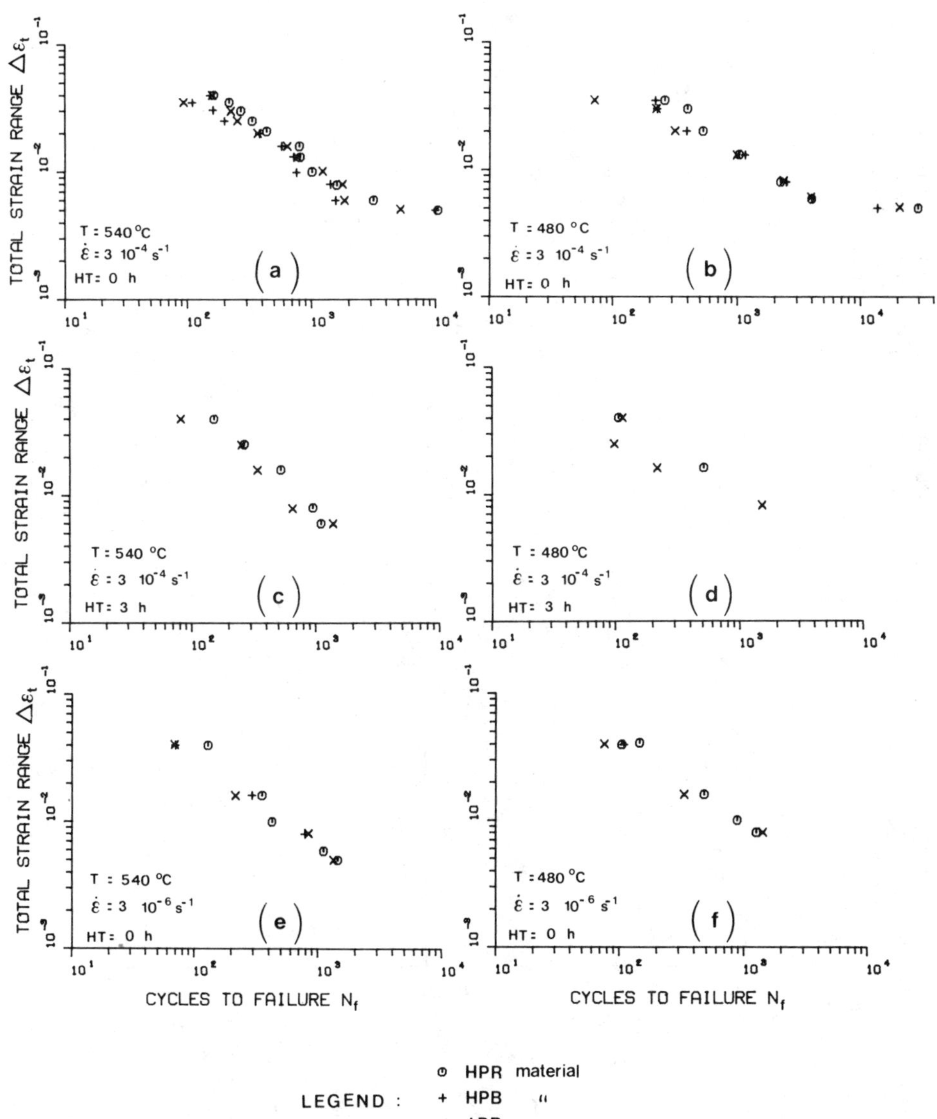

FIG. 7—*Comparisons of fatigue endurances for the different sampling regions under identical test conditions.*

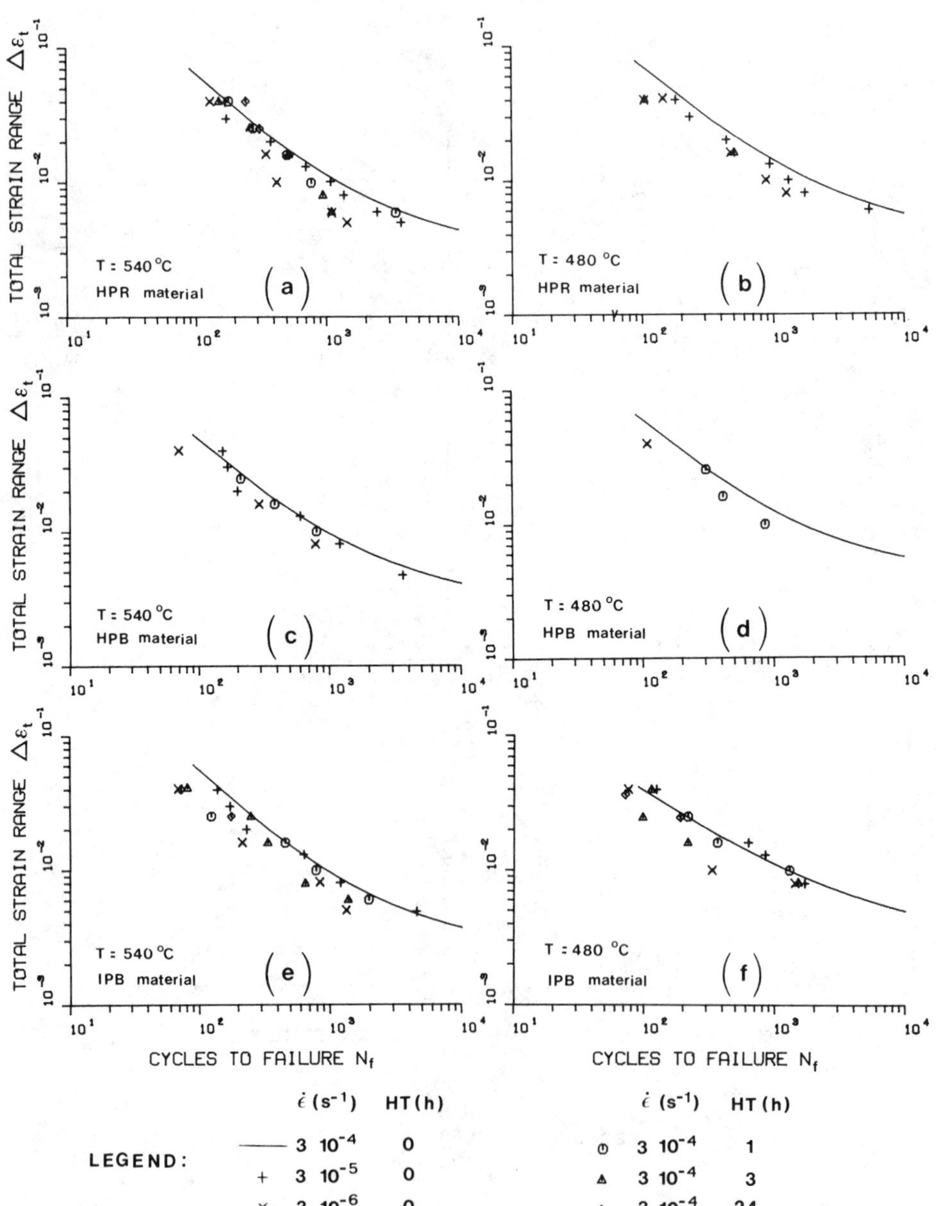

FIG. 8—*Effects of strain rate and of hold time on fatigue endurances of specimens sampled from the different regions of the rotor.*

S was found to be 0.043, 0.063, and 0.092 for HPR, HPB, and IPB specimens, respectively. The results of additional tests carried out on a few tangential HPR specimens (not presented here) demonstrated that this difference could not be merely attributed to specimen orientation.

Time Effect

Only slight life reductions are observed in HPR tests with $\dot{\epsilon} = 3 \times 10^{-5}\ s^{-1}$ (Figs. 8*a* and 8*b*), whereas a clearer effect is recognizable in tests with $\dot{\epsilon} = 3 \times 10^{-6}\ s^{-1}$, particularly at 540°C (with a maximum reduction factor of 7, Fig. 8*a*). The same effect can be noted even in tests on HPB and IPB specimens, but with less general evidence (in relation to the higher intrinsic scatter of these data). For identical time durations of the fatigue cycle, the effect of hold periods is less significant than the effect related to variations in $\dot{\epsilon}$. In some instances, short fatigue lives were obtained in tests on HPB and IPB specimens with long HT; however, life shortening was often not greater than the natural dispersions of tests with $\dot{\epsilon} = 3 \times 10^{-4}\ s^{-1}$ and HT = 0 h. In any case, time-dependent effects seemed to increase progressively at low $\Delta\epsilon_t$ values for HPR specimens at 540°C; again, such an observation can be pointed out with less confidence for the more scattered HPB and IPB data.

With reference to microstructure evolution, the patterns of hardness data along the specimens axis (Fig. 9) show a consistent material softening and some evidence of the influence of time parameters. Cavitation phenomena were also observed, dependent on test conditions and microstructural features of the different rotor zones. Cavities were essentially connected with the amount of creep plastic flow. The effect was evident in specimens which had spent a long time under stress and temperature. It was quite negligible in tests with HT = 0 h even at the lowest $\dot{\epsilon}$, and cavitation was found in all specimens tested with HT = 24 h (Fig. 10).

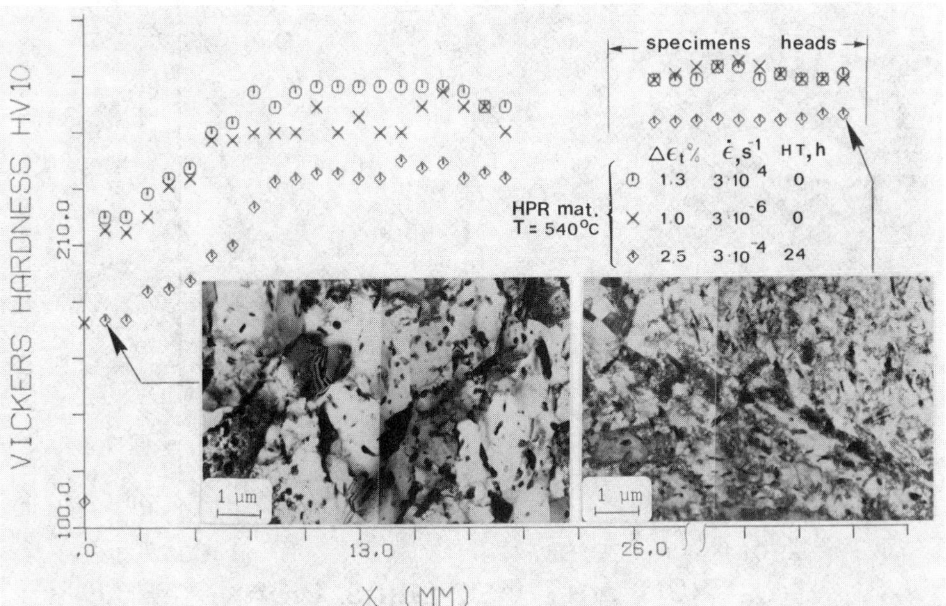

FIG. 9—*Hardness measured on three hourglass specimens as a function of distance* x *along the specimen axis from the main crack.*

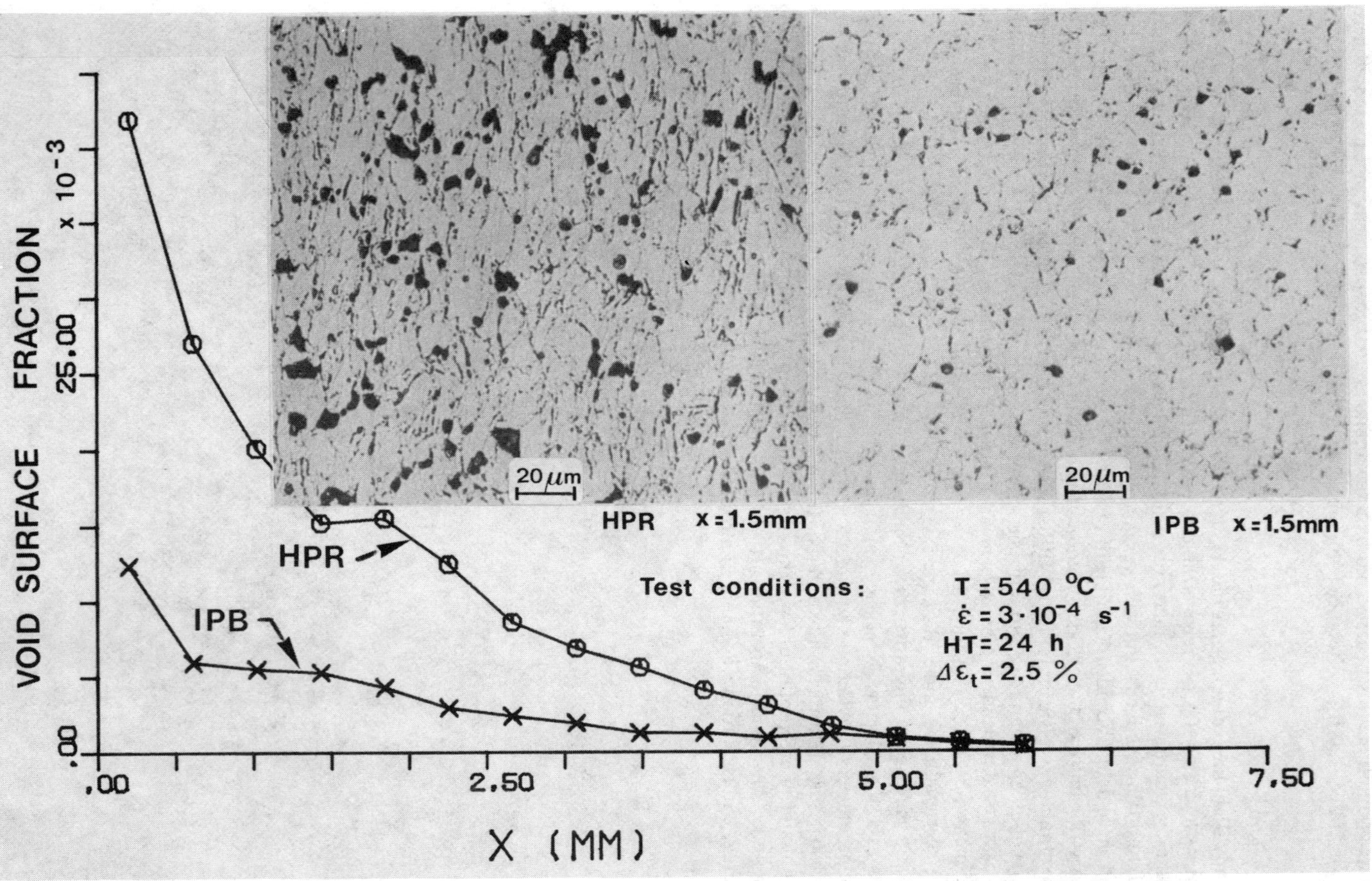

FIG. 10—*Cavitation data (void area per unit area) as a function of distance* x *along the specimen axis from the main crack.*

Discussion

Results are hereafter discussed in terms of the microstructure of the different zones of the rotor, which affects damage mechanisms and fracture events. The large amount of fatigue endurance data gained from the experiments provided the opportunity to check the applicability of current methods of damage evaluation and life prediction to the inhomogeneous material of a real rotor with a standard heat treatment history. The results of the application of three of these methods are discussed here: the linear damage summation rule, which is a traditional engineering approach to damage analysis, and two more advanced methods, the strainrange partitioning approach and an energy-based criterion, which relate material fracture to the exhaustion of ductility and to irrecoverable internal energy content, respectively.

Role of Microstructure

Microstructure was found to significantly affect, together with the testing conditions, the fatigue lives of the tests performed. Segregated zones, particularly in the IPB region, caused a spread and reduction in the N_f values by heavily affecting the fracture mechanisms. Figure 11 shows the fracture surfaces of two specimens tested under identical conditions, sampled from HPR and IPB regions. Both of them show the striation patterns typical of high load fatigue; however, while for the HPR specimen many initiation points on the external surface are evident,

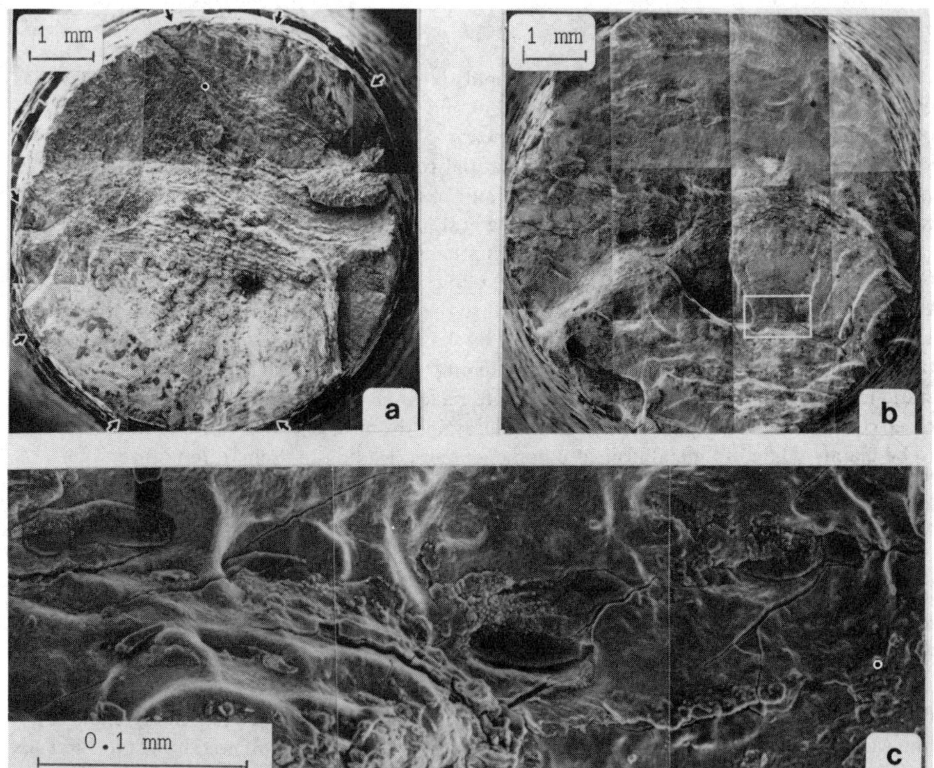

FIG. 11—*Fractographs of specimens tested at 540°C under identical conditions:* $\Delta\epsilon_t = 1.6\%$, $\dot{\epsilon} = 3 \times 10^{-6}\ s^{-1}$, *HT* = *0 h.* (a) *HPR material (starting cracks indicated by arrows).* (b) *IPB material.* (c) *Magnification of the starting crack region in* (b).

on the IPB specimen the crack started from an internal inclusion cluster and quickly propagated through a fracture plane containing other big inclusions. This latter mechanism was common to high $\Delta\epsilon_t$ tests on IPB samples whenever large inclusion clusters were present near the minimum cross sections.

The observed decrease of hardness seems mainly associated with the extent of fatigue-induced material recovery (see the micrographs in Fig. 9), this being enhanced by a long period of the fatigue cycle. Cavitation was found only in specimens which suffered a certain amount of creep damage. No evidence of significant cavitation was found in specimens with HT < 3 h, even for the lowest deformation rate which caused the maximum life reduction effect; on the contrary, cavitation was quite evident in all specimens tested with a 24 h hold. This finding is consistent with the analysis of Priest and Ellison [*2*], who derived for a similar material (with a comparable grain size, at least for the temper evoluted bainite [TEB] zones, see Table 4) the deformation mechanisms map shown in Fig. 12. The superimposed line of plastic strain rate values determined from the relaxation data of a 24 h HT test shows that diffusion assisted cavitation phenomena are activated during a large portion of the hold period. Figure 13 shows typical examples of microvoids observed in bainitic HPR and segregated IPB regions. In the former case intergranular cracks obtained by coalescence of grain boundary microvoids led to large cavitation (Fig. 10). In IPB material the linkage of microcracks that originated at large inclusions highly accelerated the fracture process and reduced the overall hold time necessary for void formation and growth.

Linear Damage Summation (LDS) Rule

The LDS rule was applied to tests with hold times. The creep component of damage was evaluated by integrating over the hold period of mid-life cycles the time fractions $1/t_R(\sigma)$, $t_R(\sigma)$ being the monotonic time-to-rupture corresponding to the instantaneous stress σ. In order to exclude from creep damage evaluation the initial rapid strain accumulation during the hold period, as also proposed by other authors [*3*], creep was assumed to occur only after the strain rate had relaxed below a critical value. On the basis of Fig. 12, 10^{-6} s^{-1} was taken as representative of the transition between matrix deformation (pure fatigue) and grain boundary sliding (creep) mechanisms. Results mostly lie on the unconservative side (Fig. 14). In all cases the calculated creep damage fractions were small, so that the consistent life reductions observed in long HT tests are not correctly accounted for by the method (Table 6). It should be considered that, for materials which undergo cyclic softening, evaluation of the creep damage term via static creep resistance data applied at mid-life relaxation behavior underestimates the actual creep contribution [*4*]. Apart from this consideration, the most unconservative predictions seem to be mainly related to metallurgical inhomogeneities, leading to premature failures.

Strain Range Partitioning (SRP) Method

According to this method, a life prediction is obtained from a fractional analysis of the material's cyclic stress-strain hysteresis loop on the basis of four equations of damage determined from specific tests [*5*]. Here, four sets of strain-controlled fatigue tests were carried out on HPR specimens. The adopted test conditions (strain hold) and the good creep characteristics of the alloy caused some problems of scatter in the determination of the creep baselines according to the conventional step-by-step regression analysis. Therefore an improved method of data reduction was applied, consisting of a preliminary smoothing of life data, followed by a simultaneous minimization of a unique error term. This latter improvement, already adopted by other authors [*6*], is actually only a minor reformulation, suitable for computer implementations, of the traditional algorithm [*7*].

The four basic SRP relationships for HPR material are shown in Fig. 15, together with the

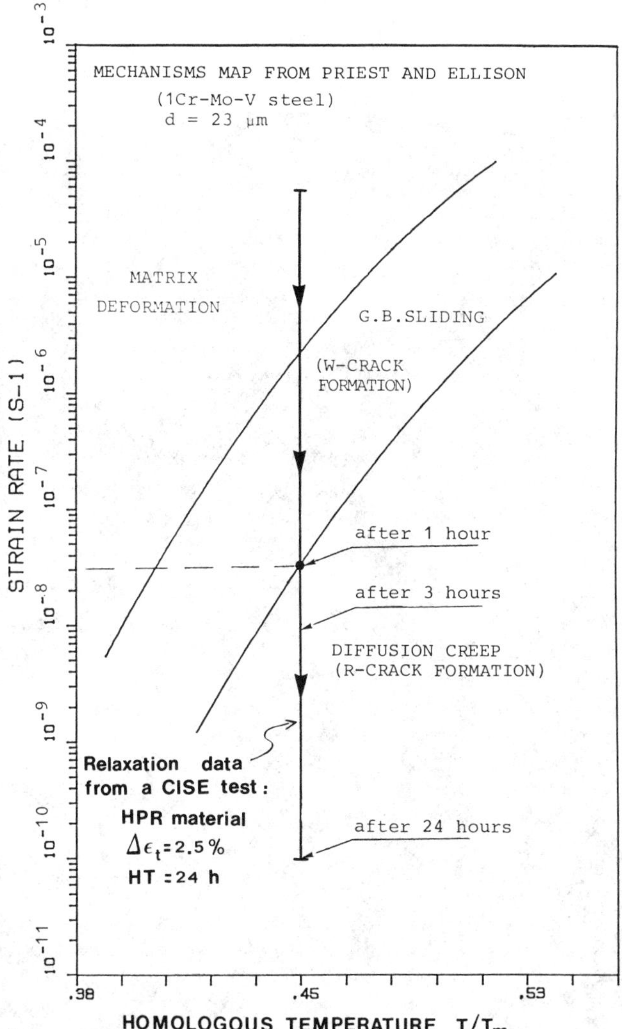

FIG. 12—*Map of dominant damage mechanisms* [2] *with superimposed stress relaxation curve from a test on a HPR specimen.*

predicted lives calculated for the tests with various strain rates [*8*] and HT on specimens from all the different sampled positions. It can be seen that the life shortening effects found in tests on HPR specimens with $\dot{\epsilon} = 3 \times 10^{-6}\ s^{-1}$ tend to be overestimated by the SRP method, leading to conservative predictions. So far as HT effects are concerned, the moderate life reductions found in HPR tests with $\Delta\epsilon_t > 0.8\%$ are correctly predicted, whereas unconservative results are obtained at lower strain ranges, (Table 6). Life predictability seems systematically better than with the LDS rule. The life data of specimens from HPB and IPB rotor regions, predicted on the basis of the HPR damage relationships, cannot be conclusively discussed. In most cases the absence of creep effects is fully confirmed, whereas in a few other cases the occurrence of unconservative predictions seems to be related more to inclusion-dominated early failures than to inappropriate evaluation of creep damage.

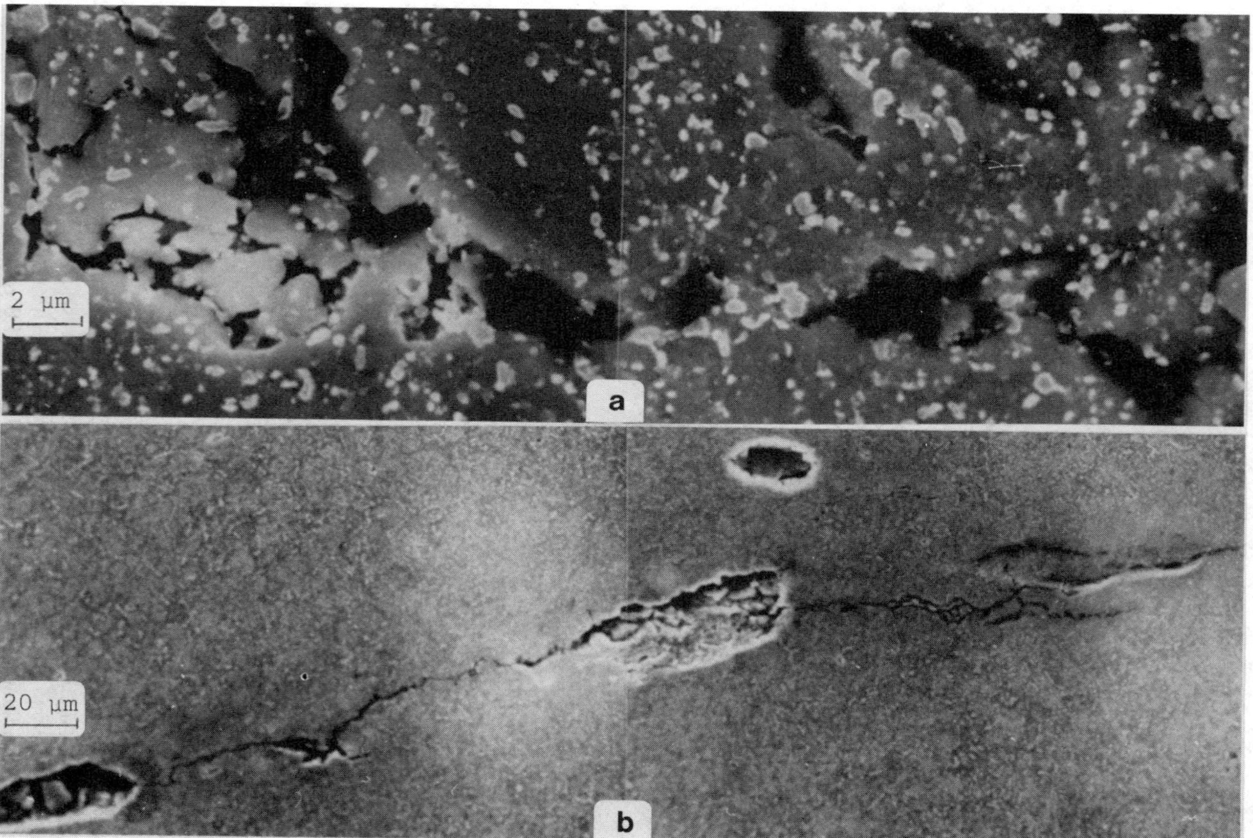

FIG. 13—*Interaction between cavitation and crack growth in two specimens tested at 540°C under identical conditions:* $\Delta\epsilon_t = 2.5\%$, $\dot{\epsilon} = 3 \times 10^{-4}\ s^{-1}$, *HT* = 24 *h.* (a) *HPR material.* (b) *IPB material.*

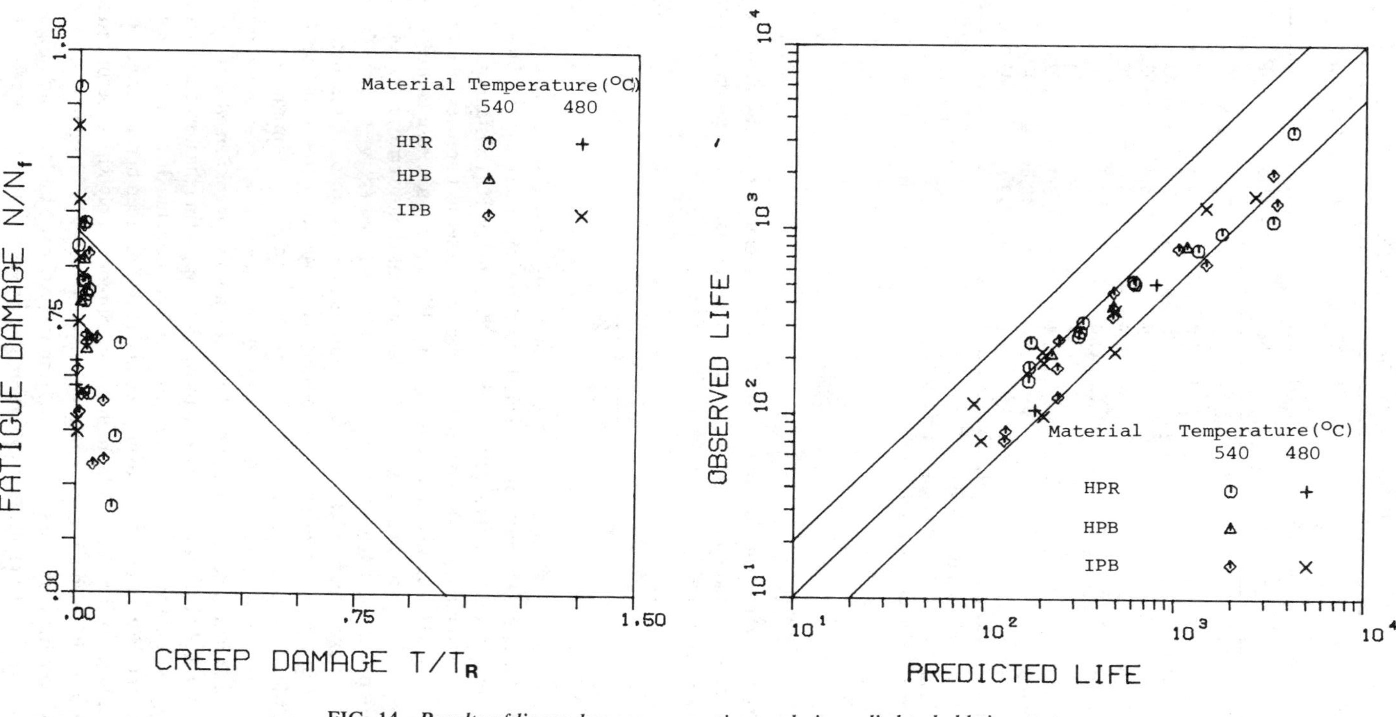

FIG. 14—*Results of linear damage summation analysis applied to hold-time tests.*

TABLE 6—*Comparison between actual life reductions and predictions obtained via linear damage summation (LDS) and strain range partitioning (SRP) method for HT tests (HPR material, 540°C).*[a]

$\Delta\epsilon_t$, %	HT, h	$(N_{fo} - N_f)/N_{fo}$, %	$(N_{fo} - N_{pred})/N_{fo}$, %	
			LDS	SRP
4.01	1	−3.0	1.9	7.2
2.41	1	14.0	1.5	12.5
1.59	1	19.5	2.5	22.1
1.00	1	44.9	5.8	32.3
0.59	1	30.9	14.4	38.4
4.00	3	13.8	2.3	8.1
2.53	3	17.7	2.8	13.5
1.60	3	16.4	3.4	22.1
0.80	3	56.6	19.6	34.3
0.60	3	76.1	29.1	47.9

[a] N_f = actual fatigue life in HT test.
N_{pred} = predicted fatigue life in HT test.
N_{fo} = fatigue life for HT = 0.

Energy-Based Criterion

Following a line of investigation directed towards the development of energy-based phenomenological criteria of fatigue damage [*9–14*], in a previous paper the authors presented some preliminary results of life predictability through a damage rule of the type [*15*]

$$\frac{1}{N_{pred}} = \frac{w}{W_t} \tag{1}$$

where w and W_t are the strain energy densities of a fatigue cycle and up to complete failure, respectively. This approach relies on the hypothesis that there is a constant value of w throughout a fatigue test. In addition, its application is convenient if W_t is a constant quantity, so that it can be easily determined from a short time test (or a test set to improve the estimate, $W_t = 1/n\ \Sigma_i\ (w \times N_f)_i$. The former point was confirmed by all the test results presented here, in agreement with other published data [*16*]. The latter point is more critical. Use of Eq 1 reflects the assumption that the different contributions to the detailed energy balance during fatigue straining (heat fluxes, mechanical energy, and internal energy stored in the material) are constant fractions of the externally supplied mechanical energy for all test conditions. Under this assumption the fundamental hypothesis of a fixed critical amount of internal energy necessary for material failure can be equivalently expressed in terms of fixed mechanical work. Experimentally, strain range independence of total work to failure was confirmed in the present investigation (within ±50% maximum scatter bands), thus supporting the use of Eq 1 for life prediction. The degree of correlation shown in Fig. 16 between the actual lives and values from Eq 1 is quite significant. When time-dependent phenomena due to creep become effective the W_t values obtained in short time (zero hold, high $\dot{\epsilon}$) tests cannot comply with experimental results. A slight decrease of W_t with respect to continuous cycling data was found in tests for which hold periods produced life reductions. This effect is more systematic for low $\dot{\epsilon}$ data. Figure 17 shows the W_t values (average from test series at different $\dot{\epsilon}$) versus deformation rate for the three rotor zones. In addition, the total works to failure in creep tests as a function of minimum creep rates

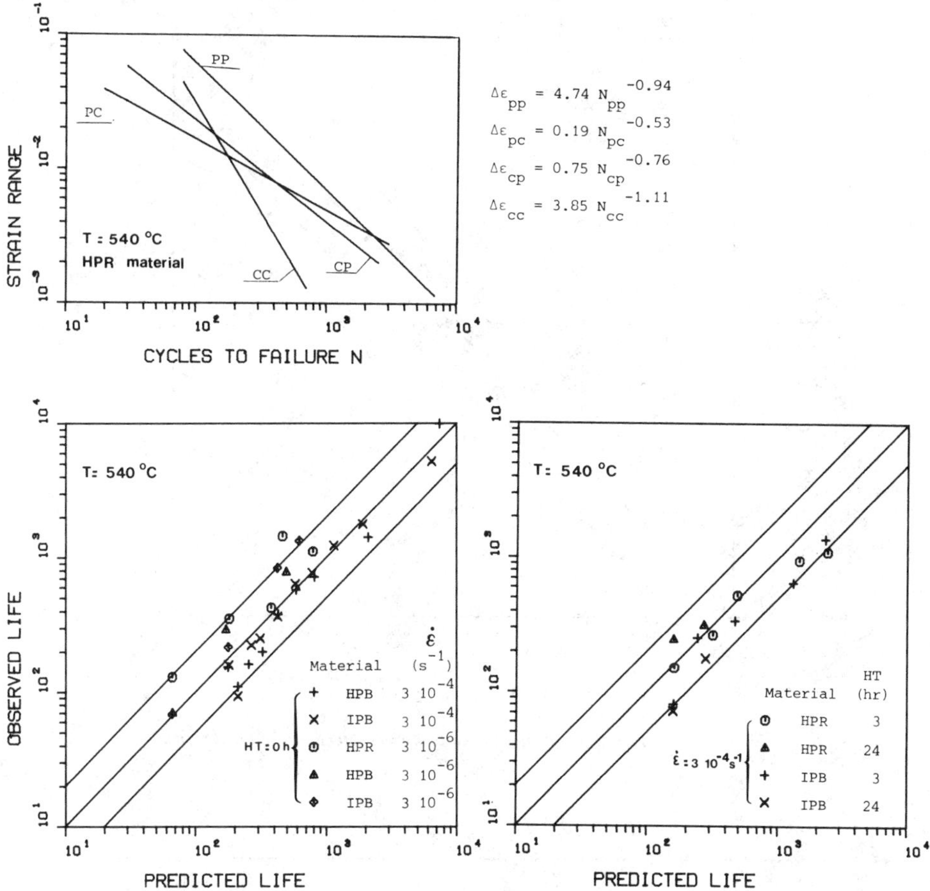

FIG. 15—*Application of SRP analysis: basic curves and life predictions.*

are plotted. A well defined trend is evident. These results seem dictated by quite general principles of irreversible thermodynamics. If creep flow has to be seen as a stable steady-state deformation process, obeying a principle of minimum entropy production rate [*17*], a decrease of failure energy W_t in slow straining tests reflects the reduction of heat dissipative losses as far as creep conditions of flow are approached. This change in the mode of damage must be reflected in Eq 1 if creep effects have to be properly taken into account in life predictions. The most obvious reformulation of Eq 1 could assume the familiar form of a linear damage summation:

$$\frac{1}{N_{\text{pred}}} = \frac{w_c}{W_c} + \frac{w_f}{W_f} \tag{2}$$

where c and f label the limiting modes of plastic flow (creep and time-independent fatigue, respectively). Therefore an energy partition of the hysteresis loops would be suggested by this criterion. Further work is needed to set up appropriate energy constants and energy partitioning criterion to verify the life predictability of Eq 2 to cyclic failures affected by consistent creep damage.

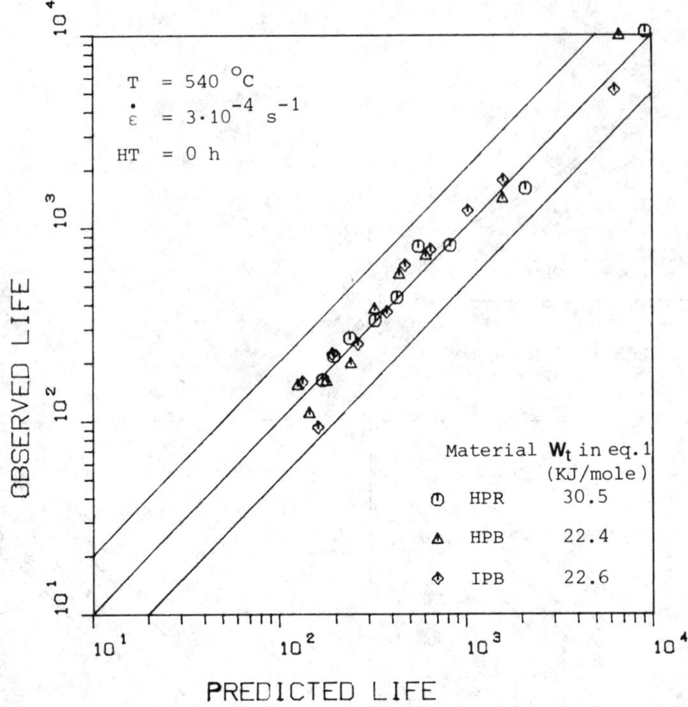

FIG. 16—*Correlation of fatigue life data by the energy-based criterion.*

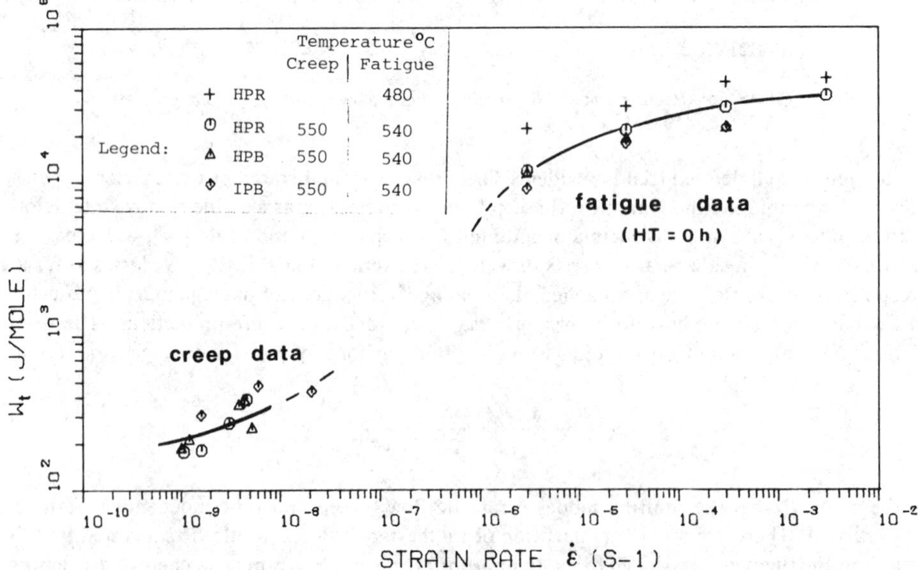

FIG. 17—*Results of total strain energy density to failure in pure creep and in LCF tests (LCF energies are average values from several tests with identical $\dot{\epsilon}$).*

Conclusions

Microstructural inhomogeneities were found to exert a marked influence on the fatigue resistance of specimens sampled from different zones of a 1Cr-Mo-V steam turbine rotor. Highly scattered results, with consistent life reductions, were obtained for the bore regions, particularly in the IP stage, which is the most heavily segregated zone. Tensile strain hold periods had only moderate life reduction effects, whereas more consistent and systematic life reductions were found in continuous cycling tests with the lowest strain rates. Despite the only moderate effects of even long hold times, the best evidence of cavitation damage was observed in the longest tests at 540°C with a 24 h hold time. This finding was consistent with the supposed dominant damage mechanism suggested by the analysis of plastic strain rate values during stress relaxation. It seems possible to conclude that the amount of creep-induced cavitation during tests with hold periods up to 24 h is still insufficient to significantly affect fatigue damage mechanisms and to shorten the lives; this could happen at longer hold times. On the other hand, the life reductions found in slow strain rate tests, where no creep cavities were detected, could not be rationalized; environmental and creep-influenced microcrack growth could be of concern. In all cases time effects had a minor role in determining lives of IP material, since failures were mainly governed by the poor fatigue ductility characteristics of the microstructure in this zone.

Application of linear damage summation and strainrange partitioning approaches to life prediction confirmed only poor creep contributions to total damage whenever no consistent life shortening occurred. In the few cases where definite reductions were observed, better data rationalization was provided by the SRP approach. Good correlation of fatigue lives was obtained by an energy-based criterion, which suggests a physically significant rationalization of rate dependent effects in terms of thermodynamic concepts.

Present experimentation confirms the need for creep-fatigue tests with quite long durations to obtain an appropriate physical understanding of damage mechanisms and conclusive verification of life prediction methods applicable to the time scale of plant components under realistic operating conditions. Moreover, wide-scale microstructural inhomogeneity of a large component should be carefully considered, both in characterization and in life prediction analysis.

Acknowledgments

The authors wish to thank A. Barbieri for precious contributions in testing activity, and C. A. Be, A. Benvenuti, L. Borghi, N. Ricci, and G. Chevallard for technical contributions in microstructural analyses and helpful discussions. The authors are also grateful to ENEL (Italian Electricity Board) for financial support of this research and for permission to publish previously unpublished data, and to TERNI Company for their cooperation with the project.

References

[1] McCann, J., Boardman, J. W., and Norton, J. F., *Journal of the Iron and Steel Institute,* Jan. 1973, pp. 66–74.

[2] Priest, R. H. and Ellison, E. G., *Materials Science and Engineering,* Vol. 49, 1981, pp. 7–17.

[3] Batte, A. D., Murphy, M. C., and Stringer, M. B., *Metals Technology,* Vol. 5, Vol. 12, 1978, pp. 405–413.

[4] Nitta, A. and Kuwabara, K., "Effect of Creep-Fatigue Interaction on Life Expenditure of a Turbine Rotor," CRIEPI Report E277002, July 1977.

[5] Manson, S. S., Halford, G. R., and Hirschberg, M. H., NASA Tech. Memo. X-67838, 1971.

[6] Hoffelner, W., Melton, K. N., and Wuthrich, C., *Fatigue of Engineering Materials and Structures,* Vol. 6, No. 1, 1983, pp. 77–87.

[7] Bicego, V. in *Proceedings,* International Conference on Fatigue, Sheffield, 3–7 Sept. 1984, Vol. 3, pp. 1257–1268.

[8] Manson, S. S., Halford, G. R., and Nachtigall, A. J., NASA Tech. Memo. X-71737, 1975.

[*9*] Fong, J. T., *Journal of Pressure Vessel Technology, Transactions of ASME,* Vol. 97, Aug. 1975, pp. 214–222.
[*10*] Lindholm, U. S. and Davidson, D. L. in *Fatigue at Elevated Temperatures, ASTM STP 520,* American Society for Testing and Materials, Philadelphia, 1973, pp. 473–481.
[*11*] Martin, D. E., *Journal of Basic Engineering, Transactions of ASME,* Dec. 1961, pp. 565–571.
[*12*] Stowell, E. Z., *Nuclear Engineering and Design 9,* North-Holland, 1969, pp. 239–257.
[*13*] Radakrishnan, V. M., *Experimental Mechanics,* Aug. 1978, pp. 292–296.
[*14*] Czoboly, E., Havas, I., and Ginsztler, J. in *Proceedings,* International Conference on Fracture, Lisbon, 16–20 Sept. 1984, pp. 481-491.
[*15*] Bicego, V., Fossati, C., and Ragazzoni, S. in *Proceedings,* P.V.P. Division Conference of ASME, Orlando, Fla., 27 Jun.-2 Jul. 1982, Vol. 60, pp. 65–70.
[*16*] Ellyin, F. and Kujawski, D., *Journal of Pressure Vessel Technology, Transactions of ASME,* Vol. 106, Nov. 1984, pp. 342–347.
[*17*] Prigogine, I. and Nicolis, G., *Self-Organization in Nonequilibrium Systems,* Wiley, New York, 1977.

Author Index

Subject Index

D

H

J

K

L

M

Q

R

S

T

U

V

W

Y

Z